GUIDED EXPLORATIONS

Mechanism of Serine Proteases

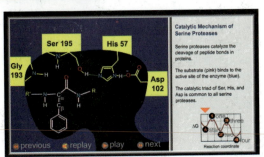

Catalytic Mechanism of Serine Proteases

Serine proteases catalyze the cleavage of peptide bonds in proteins.

The substrate (pink) binds to the active site of the enzyme (blue).

The catalytic triad of Ser, His, and Asp is common to all serine proteases.

▸ **Learning Objectives**

▸ **Script**

Guided Explorations.
These 30 self-contained presentations enhance key topics from the text with extensive animated computer graphics, which in many cases are narrated.

Interactive Exercises.
60 molecular structures from the text have been chosen for presentation in a browser-independent format that allows students to manipulate the structures and answer questions based on that experience.

INTERACTIVE EXERCISES

Ribosome

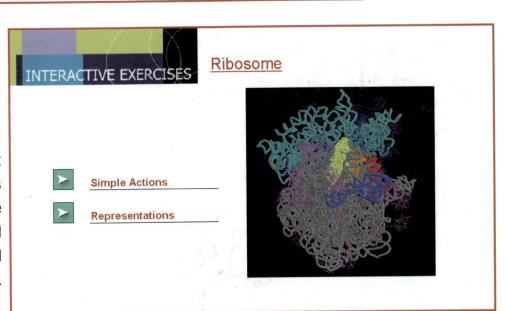

▸ **Simple Actions**

▸ **Representations**

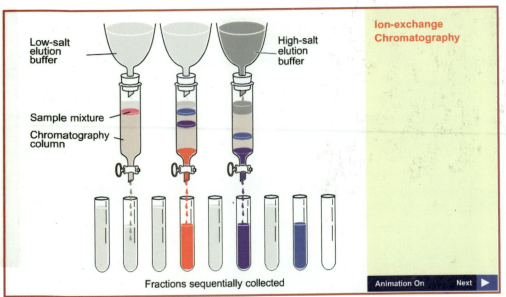

Low-salt elution buffer

High-salt elution buffer

Sample mixture

Chromatography column

Ion-exchange Chromatography

Fractions sequentially collected

Animation On   Next ▶

Animated Figures.
76 figures from the text, illustrating various concepts, techniques, and processes, are presented as brief animations.

eGrade Plus

www.wiley.com/college/voet

Based on the Activities You Do Every Day

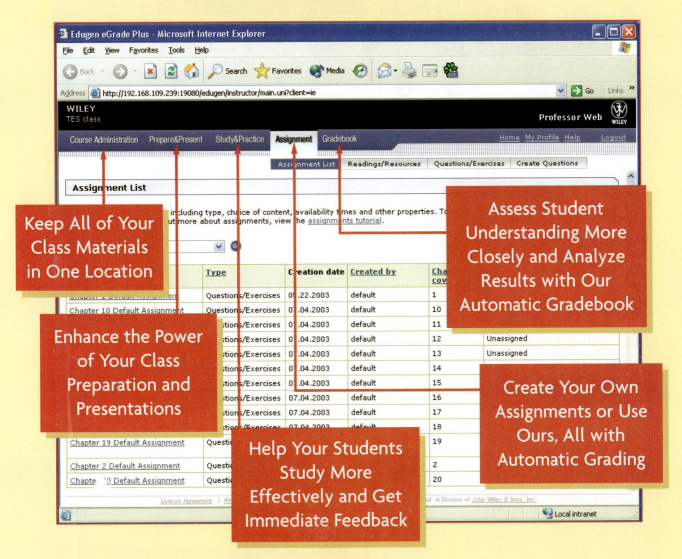

Keep All of Your Class Materials in One Location

Enhance the Power of Your Class Preparation and Presentations

Help Your Students Study More Effectively and Get Immediate Feedback

Assess Student Understanding More Closely and Analyze Results with Our Automatic Gradebook

Create Your Own Assignments or Use Ours, All with Automatic Grading

All the content and tools you need, all in one location, in an easy-to-use browser format.

Choose the resources you need, or rely on the arrangement supplied by us.

Now, many of Wiley's textbooks are available with eGrade Plus, a powerful online tool that provides a completely integrated suite of teaching and learning resources in one easy-to-use website. eGrade Plus integrates Wiley's world-renowned content with media, including a multimedia version of the text, PowerPoint slides, and more. Upon adoption of eGrade Plus, you can begin to customize your course with the resources shown here.

See for yourself!

Go to www.wiley.com/college/egradeplus for an online demonstration of this powerful new software.

Students,
eGrade Plus Allows You to:

Study More Effectively

Get Immediate Feedback When You Practice on Your Own

Our website links directly to **electronic book content,** so that you can review the text while you study and complete homework online. Additional resources include **self-assessment quizzing** with detailed feedback, **Guided Explorations, Interactive Exercises, Animated Figures,** and **Kinemages** to better understand and review key topics.

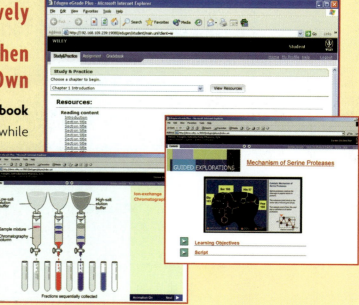

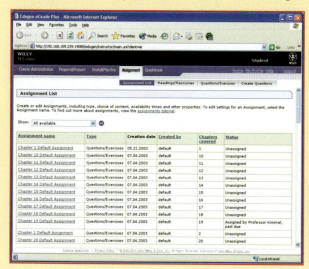

Complete Assignments / Get Help with Problem Solving

An **"Assignment"** area keeps all your assigned work in one location, making it easy for you to stay on task. In addition, many homework problems contain a **link** to the relevant section of the **electronic book,** providing you with a text explanation to help you conquer problem-solving obstacles as they arise.

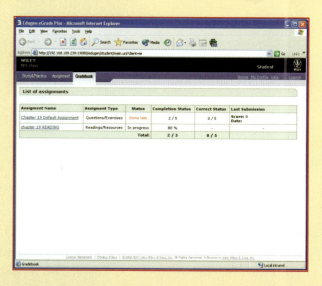

Keep Track of How You're Doing

A **Personal Gradebook** allows you to view your results from past assignments at any time.

FUNDAMENTALS OF BIOCHEMISTRY

LIFE AT THE MOLECULAR LEVEL

2ND EDITION

DONALD VOET
University of Pennsylvania

JUDITH G. VOET
Swarthmore College

CHARLOTTE W. PRATT
Seattle, Washington

WILEY

JOHN WILEY & SONS, INC.

Senior Acquistions Editor:	Patrick Fitzgerald
Senior Developmental Editor:	Ellen Ford
Marketing Manager:	Amanda Wygal
Cover/Text Designer:	Madelyn Lesure
Production Editor:	Sandra Dumas
Media Editor:	Thomas Kulesa
Photo Editor:	Hilary Newman
Photo Researcher:	Elyse Rieder
Illustration Editor:	Sigmund Malinowski
Production Management Services:	Suzanne Ingrao/Ingrao Associates

Cover Photos:

(Left photo): Geoffrey Gove/The Image Bank/Getty Images.

(Right photos, top to bottom): DNA polymerase courtesy of Gabriel Waksman, Washington University School of Medicine; GlnRS·tRNA^{Gln} based on an X-ray structure by Thomas Steitz, Yale University; RNA polymerase II based on an X-ray structure by Roger Kornberg, Stanford University School of Medicine; large ribosomal subunit courtesy of Harry Noller, University of California at Santa Cruz; DNA © Irving Geis. Image from the Irving Geis Collection/Howard Hughes Medical Institute; rights owned by HHMI. Reproduction by permission only.

This book was typeset in 10/12 Times Ten with Trade Gothic at GTS Companies and printed and bound by Von Hoffmann Corporation. The cover was printed by Lehigh Press.

The paper in this book was manufactured by a mill whose forest management programs include sustained yield harvesting of its timberlands. Sustained yield harvesting principles ensure that the number of trees cut each year does not exceed the amount of new growth.

This book is printed on acid-free paper. ∞

Voet, Donald; Voet, Judith G.; Pratt, Charlotte W.
Fundamentals of Biochemistry, Second Edition

ISBN 0-471-21495-7

Printed in the United States of America.

10 9 8 7 6 5 4 3 2

PREFACE

In writing the second edition of *Fundamentals of Biochemistry*, we have endeavored to preserve the clarity and scope of the first edition while updating its contents to reflect what is most exciting in modern biochemistry. As instructors, we recognize the importance of providing students with a textbook that is carefully organized, clearly written, and generously illustrated. Although it is not intended to be encyclopedic, *Fundamentals of Biochemistry 2nd edition* presents a broad survey of biochemical topics, using a rigorous chemical approach to explain the structures of biological molecules, the metabolic activities of cells, and the principles of molecular biology. We have taken care to include descriptions of major analytical techniques and, wherever possible, to correlate biochemical knowledge with human health and disease.

What's New in the Second Edition

The text of the first edition has been substantially rewritten in order to incorporate newly discovered information on molecular structure and function and newly developed techniques for probing the organization and regulation of biological systems. Many of these updates reflect advances in the field of bioinformatics, which will undoubtedly continue to shape the future of biochemistry. *Fundamentals of Biochemistry 2nd edition* describes the major databases for proteins and nucleic acids, along with expanded discussions of how such information is obtained by mass spectrometry, X-ray crystallography, and genome sequencing. The use of structure and sequence information is highlighted in sections describing the process of drug design and applications of DNA microarray technology. *Fundamentals of Biochemistry 2nd edition* also pays homage to the growing body of information concerning the mechanisms of signal transduction and the role of RNA in regulating gene expression by phenomena such as RNA interference. In the spirit of making information easily available to students, a comprehensive set of computer-based exercises, which include animations and interactive molecular graphics, are available online in a browser-independent format.

In addition to changes designed to cover recent discoveries in virtually all areas of biochemistry, the second edition of *Fundamentals of Biochemistry* introduces a number of features that are intended to promote student understanding of this rapidly advancing field. Thus, the second edition includes more problems at the end of each chapter, additional sample calculations, a plethora of new molecular graphics representing the latest findings in structural biology, a selection of biographical sketches to illuminate the process of scientific discovery, and a set of exercises that use the databases and tools of bioinformatics. These exercises are available online as well as in the textbook. Our goal throughout has been to provide students with a solid understanding of biochemistry as well as to foster a sense of wonder for the chemistry of life.

Organization

The 28 chapters of *Fundamentals of Biochemistry 2nd edition* cover all areas of biochemistry. Two introductory chapters discuss the origin of life, evolution, the properties of water, and acid–base chemistry. Thermodynamics is introduced in Chapter 1 since it is needed to understand the hydrophobic effect (Chapter 2), protein structure (Chapter 6), membrane transport (Chapter 10), and metabolism (Chapters 13–22). Early coverage of nucleotides and nucleic acids (Chapter 3) reflects the role these molecules play in protein evolution and in metabolism. Reviews of the principles of molecular biology and of genetic engineering provide the background for understanding some of the experimental approaches used to study protein structure and enzyme function. Chapter 3 may also be covered in conjunction with Chapter 23, which takes up the finer points of nucleic acid structure.

A sequence of three chapters (4 through 6) explores amino acid chemistry, methods for analyzing protein structure and sequence, protein secondary, tertiary, and quaternary structure, and protein folding and stability. Chapter 7 focuses on various aspects of protein function using myoglobin and hemoglobin as subjects. This focus continues in Chapter 28. Chapter 8 (Carbohydrates) and Chapter 9 (Lipids and Biological Membranes) round out the coverage of the basic molecules of life. Recent advances in understanding membrane transport phenomena are showcased in Chapter 10.

Two chapters take up enzyme chemistry. The discussion of enzyme mechanisms (Chapter 11) precedes the treatment of enzyme kinetics (Chapter 12) because it is easier to show students how enzymes work before describing kinetic parameters and how they are altered by the presence of inhibitors or by regulatory mechanisms.

Metabolism is the topic of ten chapters, beginning with an introductory chapter (Chapter 13) that provides an overview of metabolic pathways, the thermodynamics of "high-energy" compounds, and redox chemistry. The central metabolic pathways are presented in detail (e.g., glycolysis, glycogen metabolism, and the citric acid cycle in Chapters 14, 15, and 16) so that students can appreciate how individual enzymes catalyze reactions and work in concert to perform complicated biochemical tasks. These chapters are followed by Chapter 17 (Electron Transport and Oxidative Phosphorylation) and Chapter 18 (Photosynthesis) to complete a sequence that emphasizes energy-producing pathways. Not all pathways are covered in full detail, particularly those related to lipids (Chapter 19) and amino acids (Chapter 20). Instead, key enzymatic reactions are highlighted for their interesting chemistry or regulatory importance. This section of the book also includes a chapter on the integration of metabolism (Chapter 21) to illustrate organ specialization and metabolic regulation in mammals. Mechanisms of signal transduction are included here. Chapter 22 covers nucleotide metabolism as a prelude to the study of processes involving nucleic acids.

Five chapters describe the biochemistry of nucleic acids, beginning with Chapter 23, which describes the structure of DNA and its interactions with proteins. Chapters 24, 25, and 26 cover the processes of replication, transcription, and translation, incorporating considerable new information on the structures and mechanisms of the RNA and protein molecules that carry out these processes. Chapter 27 deals with a variety of mechanisms for regulating gene expression, discussing the histone code and the roles of transcription factors and their relevance to cancer and development. A new final chapter on protein function (Chapter 28) covers cytoskeletal and motor proteins to complement earlier discussions of structure-function relationships, and

describes antibody structure and generation to further illustrate the connection between genes and proteins.

Pedagogical Features

We have built several features into the text to guide students and help them study. The material in each chapter is divided into **numbered sections** to make it easy for readers to locate particular subjects and to discern the thematic links among them. The names of biochemical processes, compounds, enzymes, and diseases are highlighted in boldface at their first appearance. The most important terms are included in a list of **key terms** at the end of each chapter. Definitions for these and other terms are provided in a **glossary** for easy reference. **Key sentences** emphasizing experimental conclusions or biochemical principles are italicized.

Key figures and tables focusing on structure, function, and metabolism are identified so that they can be carefully studied. **Overview figures** in several chapters help students follow complicated metabolic processes. Wherever molecular graphics are shown, the figure legend includes the relevant **PDBid,** which allows a student or instructor to obtain the molecular structure file so that it can be visualized and explored using an appropriate program. Figure legends that include a mouse icon (🐭) prompt the reader to consult the book's website for animations and other computer-based exercises, many of which are interactive.

Optional enrichment material is placed in **boxes** so that the main text contains fewer digressions. **Biochemistry in Health and Disease** boxes include some of the lengthier clinical correlations; **Perspectives in Biochemistry** boxes offer additional information or food for thought; and **Pathways of Discovery** boxes illuminate some of the great stories and key players in the history of biochemistry.

Each chapter concludes with a **summary,** a set of **study exercises** to prompt students to identify the major themes of the chapter and to check their mastery of the information, and a set of **problems** that demand the application of knowledge rather than the mere recollection of facts. **Sample calculations** are placed at appropriate locations within chapters. **Detailed solutions** to all problems are provided in an appendix. Finally, a few **references,** which are predominantly review articles, are listed to provide students with additional information. Likewise, the URLs of **relevant websites** are provided for resources that students may wish to explore online.

Online Supplements

The book's website (http://www.wiley.com/college/voet) contains an extensive set of resources for enhancing student understanding of biochemistry. These are keyed to the text and in most cases are specifically called out with a red mouse icon (🐭).

The text is also available with eGrade Plus, a powerful online tool that provides instructors and students with an integrated suite of teaching and learning resources in one easy to use website. eGrade Plus includes the entire book on line and is organized around the essential activities instructors and students perform in class.

Bioinformatics Exercises. Because modern biochemistry is not simply a collection of facts but a way of gathering information, students should be exposed to the science of bioinformatics.

Accordingly, *Fundamentals of Biochemistry 2nd edition* provides a set of exercises covering the contents and uses of databases related to nucleic acids, protein sequences, protein structures, enzyme inhibition, and other topics. These exercises, which were prepared by Paul Craig of the Rochester Institute of Technology, use real data sets, pose specific questions, and prompt students to obtain information from online databases and to access the software tools for analyzing such data. The text of the bioinformatics exercises is printed in an appendix, but we strongly encourage website access to allow users to directly interact with the necessary online resources.

Additional web assets to enrich the learning of biochemistry are detailed in an appendix and are described below:

Guided Explorations. These 30 self-contained presentations enhance key topics from the text with extensive animated computer graphics.

Interactive Exercises. 60 molecular structures from the text have been chosen for presentation in a manner that allows students to manipulate the structures in three dimensions and answer questions based on that experience.

Animated Figures. 76 figures from the text, illustrating various concepts, techniques, and processes, are presented as brief animations.

Kinemages. A set of 21 exercises comprising 54 three-dimensional images of selected proteins and nucleic acids can be manipulated by users as suggested by an accompanying script.

Online Quizzes. Students can take a computer-graded quiz to test their knowledge of each chapter.

Other Supplements

- A CD containing the Guided Explorations, Interactive Exercises, Animated Figures, and Kinemages, available to instructors on request by contacting your local Wiley sales representative.

- A Student Companion to *Fundamentals of Biochemistry 2nd edition* by Akif Uzman, Joseph Eichberg, William Widger, Kathleen Cornely, Donald Voet, Judith Voet, and Charlotte Pratt. It contains a summary of each chapter, a review of essential concepts, answers to the end of chapter study exercises in the text, and additional problems with answers.

- An Instructor's CD containing all the figures in the textbook in both JPEG and PowerPoint formats. The figures have been optimized for classroom projection with bold leader lines and large labels. Figures are broken down into consecutive steps, where appropriate, to simplify complex processes.

- Take Note! Student workbook provides black and white images from the text so students can take notes directly on them while the instructor is presenting the color images in the classroom.

- A test bank, new to this edition, contains a variety of question types. Question types include multiple choice, matching, fill in the blank and short answer. Each question is keyed to the relevant section in the text and is rated by difficulty level. The test bank can be found on the instructor's CD.

ACKNOWLEDGMENTS

This textbook is the result of the dedicated effort of many individuals, several of whom deserve special mention:

Laura Ierardi cleverly combined text, figures, and tables in designing each of the textbook's pages. Suzanne Ingrao, our Production Coordinator, skillfully managed the production of the textbook. Ellen Ford, our Developmental Editor, coordinated both the art and writing programs and kept our noses to the grindstone. Madelyn Lesure designed the book's typography and cover. Patrick Fitzgerald, our Acquisitions Editor, skillfully organized and managed the entire project. Hilary Newman and Elyse Rieder acquired many of the photographs in the textbook and kept track of all of them. Connie Parks, our copy editor, put the final polish on the manuscript and eliminated large numbers of grammatical and typographical errors. Sandra Dumas was our in-house Production Editor at Wiley. Edward Starr coordinated the illustration program. Special thanks to Geraldine Osnato, Project Editor, who coordinated and managed an exceptional supplements package, and to Tom Kulesa, Media Editor, who substantially improved and developed the media resources, website, and e-Grade Plus program.

The atomic coordinates of many of the proteins and nucleic acids that we have drawn for use in this textbook were obtained from the Research Collaboratory for Structural Bioinformatics Protein Data Bank. We created these drawings using the molecular graphics programs RIBBONS by Mike Carson; GRASP by Anthony Nicholls, Kim Sharp, and Barry Honig; and INSIGHT II from BIOSYM Technologies. Many of the drawings generously contributed by others were made using either these programs or MIDAS by Thomas Ferrin, Conrad Huang, Laurie Jarvis, and Robert Langridge; MOLSCRIPT by Per Kraulis; and O by Alwyn Jones.

The interactive computer graphics diagrams that are presented on the website that accompanies this textbook are either Jmol images or Kinemages. Jmol is a free, open source molecule viewer for students, educators, and researchers in chemistry and biochemistry. It is cross-platform, running on Windows, Mac OS X, and Linux/Unix systems. The Jmol application is a standalone Java application that runs on the desktop. Jmol, which is based on the program RasMol by Roger Sayle, was developed and generously made publically available. Kinemages are displayed by the program KiNG, which was written and generously provided by David C. Richardson who also wrote and provided the program PREKIN, which DV and JGV used to help generate the Kinemages. KiNG (Kinemage, Next Generation) is an interactive system for three-dimensional vector graphics.

We wish especially to thank those colleagues who reviewed this edition of the text:

Fazal Ahmad, *University of Miami School of Medicine*
Ruma Banerjee, *University of Nebraska*
Donald Beitz, *Iowa State University*
Glenn Cunningham, *University of Central Florida*
Joseph Eichberg, *University of Houston*
Thomas Goyne, *Valparaiso University*
J. Norman Hansen, *University of Maryland*
Edward D. Harris, *Texas A&M University*
Martin Horowitz, *New York Medical College*
Frans Huijing, *University of Miami School of Medicine*
Barrie Kitto, *University of Texas at Austin*
Lisa C. Kroutil, *University of Wisconsin, River Falls*
Robert C. MacDonald, *Northwestern University*
Douglas McAbee, *California State University, Long Beach*
Stephen Meredith, *University of Chicago*
Laura Mitchell, *St. Joseph's University*
Angelika Niema, *Keck Graduate Institute*
Robert Renthal, *University of Texas, San Antonio*
Gale Rhodes, *University of Southern Maine*
Thomas L. Selby, *University of Central Florida*
Ann E. Shinnar, *Barnard College/Columbia University*
Jessup M. Shivley, *Clemson University*
Daniel Smith, *University of Akron*
Gerald Stubbs, *Vanderbilt University*
Michael Sypes, *Pennsylvania State University*
John Turchi, *Wright State University*
Linette M. Watkins, *Southwest Texas State University*
Ryland E. Young, *Texas A&M University*

Reviewers of the first edition:

Marjorie A. Bates
University of California at Los Angeles

Charles E. Bowen
California Polytechnic University

Caroline Breitenberger
The Ohio State University

Scott Champney
East Tennessee State University

Kathleen Cornely
Providence College

Bonnie Diehl
The Johns Hopkins University

Jacquelyn Fetrow
University of Albany

Jeffrey A. Frick
Illinois Wesleyan University

Michael E. Friedman
Auburn University

Arno L. Greenleaf
Duke University

Michael D. Griswold
Washington State University

James Hageman
New Mexico State University

Lowell P. Hager
University of Illinois at Urbana-Champaign

LaRhee Henderson
Drake University

Diane W. Husic
East Stroudsburg University

Larry L. Jackson
Montana State University

Jason D. Kahn
University of Maryland at College Park

Barrie Kitto
University of Texas

Anita S. Klein
University of New Hampshire

Paul C. Kline
Middle Tennessee State University

W. E. Kurtin
Trinity University

Robley J. Light
Florida State University

Robert D. Lynch
University of Massachusetts-Lowell

Dave Mascotti
John Carroll University

Gary E. Means
The Ohio State University

Laura Mitchell
Saint Joseph's University

Tim Osborne
University of California at Irvine

Graham Parslow
University of Melbourne

Allen T. Phillips
Pennsylvania State University

Leigh Plesniak
University of San Diego

Stephan Quirk
Georgia Institute of Technology

Raghu Sarma
State University of New York at Stony Brook

Bryan Spangelo
University of Nevada at Las Vegas

Gary Spedding
Butler University

Pam Stacks
San Jose State University

Scott Taylor
University of Toronto

David C. Teller
University of Washington

Steven B. Vik
Southern Methodist University

Jubran M. Wakim
Middle Tennessee State University

Joseph T. Warden
Rensselaer Polytechnic Institute

William Widger
University of Houston

Bruce Wightman
Muhlenberg College

Kenneth O. Willeford
Mississippi State University

Robert P. Wilson
Mississippi State University

Adele Wolfson
Wellesley College

Cathy Yang
Rowan University

Leon Yengoyan
San Jose State University

Ryland F. Young
Texas A&M University

BRIEF CONTENTS

CONTENTS

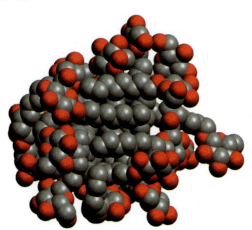

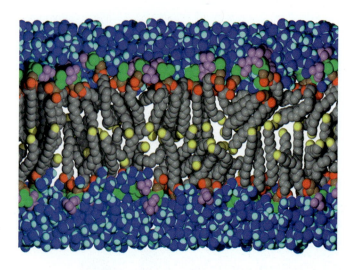

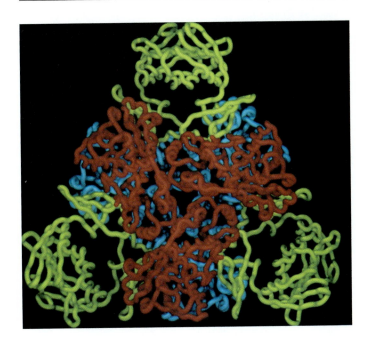

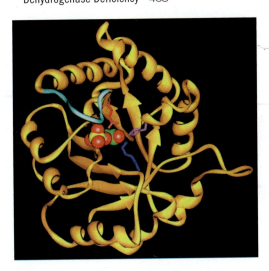

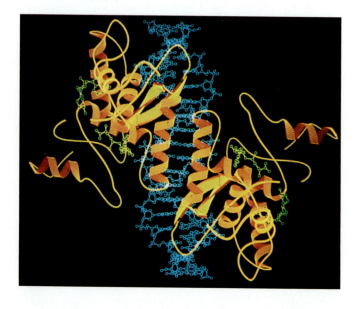

APPENDIX

FUNDAMENTALS OF

BIOCHEMISTRY

LIFE AT THE MOLECULAR LEVEL

2ND
EDITION

Early earth, a tiny speck in the galaxy, contained simple inorganic molecules that gave rise to the first biological macromolecules. These, in turn, gained the ability to self-organize and self-replicate, eventually forming cellular life forms. [Lynette Cook/Photo Researchers.]

Introduction to the Chemistry of Life

Biochemistry is, literally, the study of the chemistry of life. Although it overlaps other disciplines, including cell biology, genetics, immunology, microbiology, pharmacology, and physiology, biochemistry is largely concerned with a limited number of issues:

1. What are the chemical and three-dimensional structures of biological molecules?
2. How do biological molecules interact with each other?
3. How does the cell synthesize and degrade biological molecules?
4. How is energy conserved and used by the cell?
5. What are the mechanisms for organizing biological molecules and coordinating their activities?
6. How is genetic information stored, transmitted, and expressed?

Biochemistry, like other modern sciences, relies on sophisticated instruments to dissect the architecture and operation of systems that are inaccessible to the human senses. In addition to the chemist's tools for separating, quantifying, and otherwise analyzing biological materials, biochemists take advantage of the uniquely biological aspects of their subject by examining the evolutionary histories of organisms, metabolic systems, and individual molecules. In addition to its obvious implications for human health, biochemistry reveals the workings of the natural world, allowing us to understand and appreciate the unique and mysterious condition that we call life.

 1 The Origin of Life

Certain biochemical features are common to all organisms; for example, the way hereditary information is encoded and expressed, and the way biological molecules are built and broken down for energy. The underlying genetic and biochemical unity of modern organisms suggests they are descended from a single ancestor. Although it is impossible to describe exactly how life first arose, paleontological and laboratory studies have provided some insights about the origin of life.

A The Prebiotic World

Living matter consists of a relatively small number of elements (Table 1-1). For example, C, H, O, N, P, Ca, and S account for ~97% of the dry weight of the human body (humans and most other organisms are ~70% water). Living organisms may also contain trace amounts of many other elements, including B, F, Al, Si, V, Cr, Mn, Fe, Co, Ni, Cu, Zn, As, Se, Br, Mo, Cd, I, and W, although not every organism makes use of each of these substances.

The earliest known fossil evidence of life is ~3.5 billion years old (Fig. 1-1). The preceding **prebiotic era,** which began with the formation of the earth ~4.6 billion years ago, left no direct record, but scientists can experimentally duplicate the sorts of chemical reactions that might have given rise to living organisms during that billion-year period.

The atmosphere of the early earth probably consisted of small, simple compounds such as H_2O, N_2, CO_2, and smaller amounts of CH_4 and NH_3. In the 1930s, Alexander Oparin and J. B. S. Haldane independently suggested that ultraviolet radiation from the sun or lightning discharges caused the molecules of the primordial atmosphere to react to form simple **organic** (carbon-containing) compounds. This process was replicated in 1953 by

Table 1-1 Most Abundant Elements in the Human Body[a]

Element	Dry Weight (%)
C	61.7
N	11.0
O	9.3
H	5.7
Ca	5.0
P	3.3
K	1.3
S	1.0
Cl	0.7
Na	0.7
Mg	0.3

[a]Calculated from Frieden, E., *Sci. Am.* 227(1), 54–55 (1972).

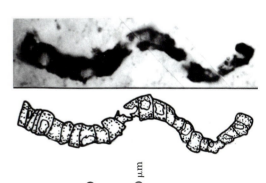

Figure 1-1 Microfossil of filamentous bacterial cells. This fossil (shown with an interpretive drawing) is from ~3.4-billion-year-old rock from Western Australia. [Courtesy of J. William Schopf, UCLA.]

Figure 1-2 A hydrothermal vent. Such ocean-floor formations are known as "black smokers" because the metal sulfides dissolved in the superheated water they emit precipitate on encountering the much cooler ocean water. [© J. Edmond. Courtesy of Woods Hole Oceanographic Institution.]

Stanley Miller and Harold Urey, who subjected a mixture of H_2O, CH_4, NH_3, and H_2 to an electric discharge for about a week. The resulting solution contained water-soluble organic compounds, including several amino acids (which are components of proteins) and other biochemically significant compounds.

The assumptions behind the Miller–Urey experiment, principally the composition of the gas used as a starting material, have been challenged by some scientists who have suggested that the first biological molecules were generated in a quite different way: in the dark and under water. Hydrothermal vents in the ocean floor, which emit solutions of metal sulfides at temperatures as high as 400°C (Fig. 1-2), may have provided conditions suitable for the formation of amino acids and other small organic molecules from simple compounds present in seawater.

Whatever their actual origin, the early organic molecules became the precursors of an enormous variety of biological molecules. These can be classified in various ways, depending on their composition and chemical reactivity. At this point, it is helpful to review the **functional groups** and **linkages** (bonding arrangements) that ultimately determine the biological activity of these molecules (Table 1-2).

B Chemical Evolution

During a period of **chemical evolution,** simple organic molecules condensed to form more complex molecules or combined end-to-end as **polymers** of repeating units. In a **condensation reaction,** the elements of water are lost. The rate of condensation of simple compounds to form a stable polymer must therefore be greater than the rate of **hydrolysis** (splitting by adding the elements of water; Fig. 1-3). In the prebiotic environment, minerals such as clays may have catalyzed polymerization reactions and sequestered the reaction products from water. The size and composition of prebiotic macromolecules would have been limited by the availability of small molecular starting materials, the efficiency with which they could be joined, and their resistance to degradation.

Obviously, *combining different functional groups into a single large molecule increases the chemical versatility of that molecule,* allowing it to perform chemical feats beyond the reach of simpler molecules. (This principle of emergent properties can be expressed as "the whole is greater than the sum of its parts.") Separate macromolecules with complementary arrangements of functional groups can associate with each other (Fig. 1-4), giving rise to more complex molecular assemblies with an even greater range of functional possibilities.

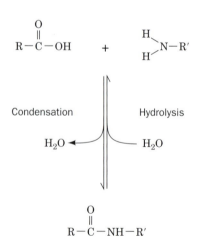

Figure 1-3 Reaction of a carboxylic acid with an amine. The elements of water are released during condensation. In the reverse process—hydrolysis—water is added to cleave the amide bond. In living systems, condensation reactions are not freely reversible.

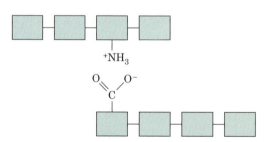

Figure 1-4 Association of complementary molecules. The positively charged amino group interacts electrostatically with the negatively charged carboxylate group.

Specific pairing between complementary functional groups permits one member of a pair to determine the identity and orientation of the other member. *Such **complementarity** makes it possible for a macromolecule to **replicate**, or copy itself, by directing the assembly of a new molecule from*

Table 1-2 Key to Structure. Common Functional Groups and Linkages in Biochemistry

Compound Name	Structure[a]	Functional Group or Linkage
Amine[b]	RNH_2 or $R\overset{+}{N}H_3$ R_2NH or $R_2\overset{+}{N}H_2$ R_3N or $R_3\overset{+}{N}H$	$-N\big<$ or $-\overset{+}{\underset{\mid}{N}}-$ (amino group)
Alcohol	ROH	—OH (hydroxyl group)
Thiol	RSH	—SH (sulfhydryl group)
Ether	ROR	—O— (ether linkage)
Aldehyde	$R-\overset{O}{\overset{\|}{C}}-H$	$-\overset{O}{\overset{\|}{C}}-$ (carbonyl group)
Ketone	$R-\overset{O}{\overset{\|}{C}}-R$	$-\overset{O}{\overset{\|}{C}}-$ (carbonyl group)
Carboxylic acid[b]	$R-\overset{O}{\overset{\|}{C}}-OH$ or $R-\overset{O}{\overset{\|}{C}}-O^-$	$-\overset{O}{\overset{\|}{C}}-OH$ (carboxyl group) or $-\overset{O}{\overset{\|}{C}}-O^-$ (carboxylate group)
Ester	$R-\overset{O}{\overset{\|}{C}}-OR$	$-\overset{O}{\overset{\|}{C}}-O-$ (ester linkage) $R-\overset{O}{\overset{\|}{C}}-$ (acyl group)[c]
Thioester	$R-\overset{O}{\overset{\|}{C}}-SR$	$-\overset{O}{\overset{\|}{C}}-S-$ (thioester linkage) $R-\overset{O}{\overset{\|}{C}}-$ (acyl group)[c]
Amide	$R-\overset{O}{\overset{\|}{C}}-NH_2$ $R-\overset{O}{\overset{\|}{C}}-NHR$ $R-\overset{O}{\overset{\|}{C}}-NR_2$	$-\overset{O}{\overset{\|}{C}}-N\big<$ (amido group) $R-\overset{O}{\overset{\|}{C}}-$ (acyl group)[c]
Imine (Schiff base)[b]	$R{=}NH$ or $R{=}\overset{+}{N}H_2$ $R{=}NR$ or $R{=}\overset{+}{N}HR$	$\big>C{=}N-$ or $\big>C{=}\overset{+}{N}\big<$ (imino group)
Disulfide	R—S—S—R	—S—S— (disulfide linkage)
Phosphate ester[b]	$R-O-\overset{O}{\underset{OH}{\overset{\|}{P}}}-O^-$	$-\overset{O}{\underset{OH}{\overset{\|}{P}}}-O^-$ (phosphoryl group)
Diphosphate ester[b]	$R-O-\overset{O}{\underset{O^-}{\overset{\|}{P}}}-O-\overset{O}{\underset{OH}{\overset{\|}{P}}}-O^-$	$-\overset{O}{\underset{O^-}{\overset{\|}{P}}}-O-\overset{O}{\underset{OH}{\overset{\|}{P}}}-O^-$ (phosphoanhydride group)
Phosphate diester[b]	$R-O-\overset{O}{\underset{O^-}{\overset{\|}{P}}}-O-R$	$-O-\overset{O}{\underset{O^-}{\overset{\|}{P}}}-O-$ (phosphodiester linkage)

[a]R represents any carbon-containing group. In a molecule with more than one R group, the groups may be the same or different.

[b]Under physiological conditions, these groups are ionized and hence bear a positive or negative charge.

[c]If attached to an atom other than carbon.

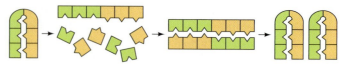

Figure 1-5 Replication through complementarity.   In this simple case, a polymer serves as a template for the assembly of a complementary molecule, which, because of intramolecular complementarity, is an exact copy of the original.

smaller complementary units. Replication of a simple polymer with intramolecular complementarity is illustrated in Fig. 1-5. A similar phenomenon is central to the function of DNA, where the sequence of bases on one strand (e.g., A-C-G-T) absolutely specifies the sequence of bases on the strand to which it is paired (T-G-C-A). When DNA replicates, the two strands separate and direct the synthesis of complementary daughter strands. Complementarity is also the basis for transcribing DNA into RNA and for translating RNA into protein.

A critical moment in chemical evolution was the transition from systems of randomly generated molecules to systems in which molecules were organized and specifically replicated. Once macromolecules gained the ability to self-perpetuate, the primordial environment would have become enriched in molecules that were best able to survive and multiply. The first replicating systems were no doubt somewhat sloppy, with progeny molecules imperfectly complementary to their parents. Over time, **natural selection** would have favored molecules that made more accurate copies of themselves.

2 Cellular Architecture

The types of systems described so far would have had to compete with all the other components of the primordial earth for the available resources. A selective advantage would have accrued to a system that was sequestered and protected by boundaries of some sort. How these boundaries first arose, or even what they were made from, is obscure. One theory is that membranous **vesicles** (fluid-filled sacs) first attached to and then enclosed self-replicating systems. These vesicles would have become the first cells.

A The Evolution of Cells

The advantages of **compartmentation** are several. In addition to receiving some protection from adverse environmental effects, an enclosed system can maintain high local concentrations of components that would otherwise diffuse away. More concentrated substances can react more readily, leading to increased efficiency in polymerization and other types of chemical reactions.

A membrane-bounded compartment that protected its contents would gradually become quite different in composition from its surroundings. Modern cells contain high concentrations of ions, small molecules, and large molecular aggregates that are found in only traces—if at all—outside the cell. For example, the ***Escherichia coli* (*E. coli*)** cell contains millions of molecules representing some 3000 to 6000 different compounds (Fig. 1-6). A typical animal cell may contain 100,000 different types of molecules.

Early cells depended on the environment to supply building materials. As some of the essential components in the prebiotic soup became scarce,

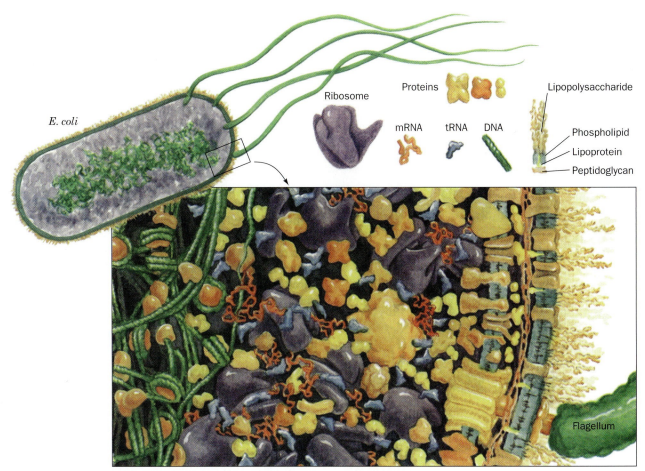

Figure 1-6 Cross section of an *E. coli* cell. The right side of the drawing shows the multilayered cell wall and membrane. The cytoplasm in the middle region of the drawing is filled with ribosomes engaged in protein synthesis. The left side of the drawing contains a dense tangle of DNA. This drawing corresponds to a millionfold magnification. Only the largest macromolecules and molecular assemblies are shown. In a living cell, the remaining space in the cytoplasm would be crowded with smaller molecules and water (the water molecules would be about the size of the period at the end of this sentence). [After a drawing by David Goodsell, UCLA.]

natural selection favored organisms that developed mechanisms for synthesizing the required compounds from simpler but more abundant **precursors.** The first metabolic reactions may have used metal or clay **catalysts** co-opted from the inorganic surroundings (a catalyst is a substance that promotes a chemical reaction without itself being changed). In fact, metal ions are still at the heart of many chemical reactions in modern cells. Some catalysts may also have arisen from polymeric molecules that had the appropriate functional groups.

In general, biosynthetic reactions require energy; hence the first cellular reactions also needed an energy source. The eventual depletion of preexisting energy-rich substances in the prebiotic environment would have stimulated the development of energy-producing metabolic pathways. For example, photosynthesis evolved relatively early to take advantage of a practically inexhaustible energy supply, the sun. However, the accumulation of O_2 generated from H_2O by photosynthesis (the modern atmosphere is 21% O_2) presented an additional challenge to organisms adapted to life in an oxygen-poor atmosphere. Metabolic refinements eventually permitted organisms not only to avoid oxidative damage but to use O_2 for oxidative metabolism, a much more efficient form of energy metabolism than anaer-

obic metabolism. Vestiges of ancient life can be seen in the anaerobic metabolism of certain modern organisms.

Early organisms that developed metabolic strategies to synthesize biological molecules, conserve and utilize energy in a controlled fashion, and replicate within a protective compartment were able to propagate in an ever-widening range of habitats. Adaptation of cells to different external conditions ultimately led to the present diversity of species. Specialization of individual cells also made it possible for groups of differentiated cells to work together in multicellular organisms.

B Prokaryotes and Eukaryotes

All modern organisms are based on the same morphological unit, the cell. There are two major classifications of cells: the **eukaryotes** (Greek: *eu,* good or true + *karyon,* kernel or nut), which have a membrane-enclosed **nucleus** encapsulating their DNA; and the **prokaryotes** (Greek: *pro,* before), which lack a nucleus. *Prokaryotes, comprising the various types of bacteria, have relatively simple structures and are almost all unicellular* (although they may form filaments or colonies of independent cells). *Eukaryotes, which are multicellular as well as unicellular, are vastly more complex than prokaryotes.* (**Viruses** are much simpler entities than cells and are not classified as living because they lack the metabolic apparatus to reproduce outside their host cells.)

Prokaryotes are the most numerous and widespread organisms on the earth. This is because their varied and often highly adaptable metabolisms suit them to an enormous variety of habitats. Prokaryotes range in size from 1 to 10 μm and have one of three basic shapes (Fig. 1-7): spheroidal (cocci), rodlike (bacilli), and helically coiled (spirilla). Except for an outer cell membrane, which in most cases is surrounded by a protective cell wall, nearly all prokaryotes lack cellular membranes. However, the prokaryotic **cytoplasm** (cell contents) is by no means a homogeneous soup. Different metabolic functions are believed to be carried out in different regions of the cytoplasm (Fig. 1-6). The best characterized prokaryote is *Escherichia coli,* a 2 μm by 1 μm rodlike bacterium that inhabits the mammalian colon.

Eukaryotic cells are generally 10 to 100 μm in diameter and thus have a thousand to a million times the volume of typical prokaryotes. It is not size, however, but a profusion of membrane-enclosed **organelles** that best characterizes eukaryotic cells (Fig. 1-8). In addition to a nucleus, eukaryotes have an **endoplasmic reticulum,** the site of synthesis of many cellular components, some of which are subsequently modified in the **Golgi apparatus.** The bulk of aerobic metabolism takes place in **mitochondria** in almost all eukaryotes, and photosynthetic cells contain **chloroplasts.** Other organelles, such as **lysosomes** and **peroxisomes,** perform specialized functions. **Vacuoles,** which are more prominent in plant cells, usually function as storage depots. The **cytosol** (the cytoplasm minus its membrane-bounded organelles) is organized by the **cytoskeleton,** an extensive array of filaments that also gives the cell its shape and the ability to move.

The various organelles that compartmentalize eukaryotic cells represent a level of complexity that is largely lacking in prokaryotic cells. Nevertheless, prokaryotes are more efficient than eukaryotes in many respects. Prokaryotes have exploited the advantages of simplicity and miniaturization. Their rapid growth rates permit them to occupy ecological niches in which there may be drastic fluctuations of the available nutrients. In contrast, the complexity of eukaryotes, which renders them larger and more slowly growing than prokaryotes, gives them the competitive advantage in

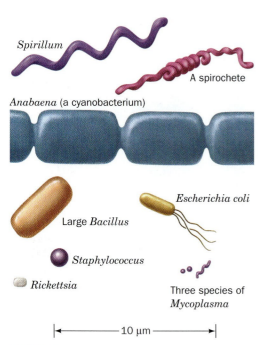

Spirillum

A spirochete

Anabaena (a cyanobacterium)

Escherichia coli

Large *Bacillus*

Staphylococcus

Rickettsia

Three species of *Mycoplasma*

|←——— 10 μm ———→|

Figure 1-7 Scale drawings of some prokaryotic cells.

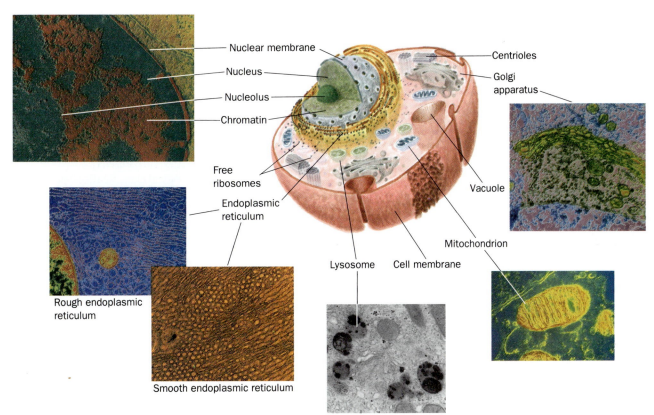

Figure 1-8 Diagram of a typical animal cell accompanied by electron micrographs of its organelles.   Membrane-bounded organelles include the nucleus, endoplasmic reticulum, lysosome, peroxisome (not pictured), mitochondrion, vacuole, and Golgi apparatus. The nucleus contains chromatin (a complex of DNA and protein) and the nucleolus (the site of ribosome synthesis). The rough endoplasmic reticulum is studded with ribosomes; the smooth endoplasmic reticulum is not. A pair of centrioles help organize cytoskeletal elements. A typical plant cell differs mainly by the presence of an outer cell wall and chloroplasts in the cytosol. [Nucleus: Tektoff-RM, CNRI/Photo Researchers; rough endoplasmic reticulum and Golgi apparatus: Secchi-Lecaque/Roussel-UCLAF/CNRI/Photo Researchers; smooth endoplasmic reticulum: David M. Phillips/Visuals Unlimited; mitochondrion: CNRI/Photo Researchers; lysosome: Biophoto Associates/Photo Researchers.]

stable environments with limited resources. It is therefore erroneous to consider prokaryotes as evolutionarily primitive compared to eukaryotes. Both types of organisms are well adapted to their respective lifestyles.

3 Organismal Evolution

Tracing the origins of different species, that is, defining their probable evolutionary history, is valuable because *the biological purpose of a particular biochemical adaptation is often best appreciated by examining how it evolved.* Such evolutionary information is often as useful as information about structure and chemistry for understanding how life operates at the molecular level.

A Taxonomy and Phylogeny

The practice of lumping all prokaryotes in a single category based on what they lack—a nucleus—obscures their metabolic diversity and evolutionary history. Conversely, the remarkable morphological diversity of eukaryotic

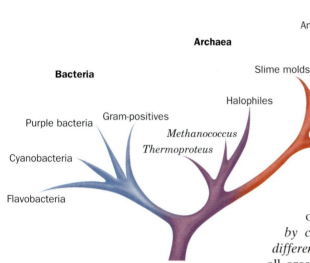

Figure 1-9 **Phylogenetic tree showing three domains of organisms.** The branches indicate the pattern of divergence from a common ancestor. The archaea are prokaryotes, like bacteria, but share some features with eukaryotes. [After Wheelis, M.L., Kandler, O., and Woese, C.R., *Proc. Natl. Acad. Sci.* **89**, 2931 (1992).]

organisms (consider the anatomical differences among, say, an amoeba, an oak tree, and a human being) masks their fundamental similarity at the cellular level. Traditional taxonomic schemes (**taxonomy** is the science of biological classification), which are based on gross morphology, have proved inadequate to describe the actual relationships between organisms as revealed by their evolutionary history (**phylogeny).**

Biological classification schemes based on reproductive or developmental strategies more accurately reflect evolutionary history than those based solely on adult morphology. But *phylogenetic relationships are best deduced by comparing polymeric molecules—RNA, DNA, or protein—from different organisms.* For example, analysis of RNA led Carl Woese to group all organisms into three domains (Fig. 1-9). The **archaea** (also known as **archaebacteria**) are a group of prokaryotes that are as distantly related to other prokaryotes (the **bacteria,** sometimes called **eubacteria**) as both groups are to eukaryotes **(eukarya).** The archaea include some unusual organisms: the **methanogens** (which produce CH_4), the **halobacteria** (which thrive in concentrated brine solutions), and certain **thermophiles** (which inhabit hot springs). Evidence that the archaea are not just unusual bacteria lies in their genetic material, some of which is more closely similar to that of eukaryotes than bacteria.

The pattern of branches in Woese's diagram indicates the divergence of different types of organisms (each branch point represents a common ancestor). The three-domain scheme also shows that animals, plants, and fungi constitute only a small portion of all life-forms. Such phylogenetic trees supplement the fossil record, which provides a patchy record of life prior to about 600 million years before the present (multicellular organisms arose about 700–900 million years ago).

B   The Origins of Complexity

The last common ancestor of bacteria, archaea, and eukaryotes was no doubt a relatively complex organism, which accounts for the shared characteristics of all present-day organisms. *It is unlikely that eukaryotes are descended from a highly developed prokaryote because the differences between bacteria and eukaryotes are so profound.* The first eukaryote instead appears to have evolved from a primordial life-form that was relatively rare, according to fossil evidence. Only after it developed complex membrane-bounded organelles did it become successful enough to generate significant fossil remains.

Among the significant evolutionary developments that produced the present-day variety of bacteria, archaea, and eukaryotes is the emergence of mechanisms for sexual reproduction. The exchange of genetic material between organisms increased their adaptability to changing conditions. A related development was the appearance of multiple chromosomes (prokaryotes typically have only one chromosome), which allowed eukaryotes to efficiently store and replicate greater amounts of genetic material.

Additional clues to the origin of eukaryotic cellular complexity lie in the mitochondria and chloroplasts. Both types of organelles resemble bacteria in size and shape, and both contain their own genetic material and protein synthetic machinery. These observations led Lynn Margulis to hypothesize

that mitochondria and chloroplasts evolved from free-living aerobic bacteria that formed **symbiotic** (mutually beneficial) relationships with primordial eukaryotes. Presumably, the eukaryotic host provided nutrients and a protected environment to the prokaryotic symbiont and was repaid severalfold by the highly efficient aerobic metabolic feats of the prokaryote. This hypothesis is corroborated by the observation that certain eukaryotes that lack mitochondria or chloroplasts permanently harbor symbiotic bacteria.

At some point in evolutionary history, individual eukaryotic cells acting in a mutually beneficial manner gave rise to multicellular organisms for whom the division of labor provided a competitive advantage. Similar principles no doubt characterized the development of higher order systems, such as societies of individuals and interacting species within an ecosystem.

C   How Do Organisms Evolve?

The natural selection that guided prebiotic evolution continues to direct the evolution of organisms. Richard Dawkins has likened evolution to a blind watchmaker capable of producing intricacy by accident, although such an image fails to convey the vast expanse of time and the incremental, trial-and-error manner in which complex organisms emerge. Small **mutations** (changes in an individual's genetic material) arise at random as the result of chemical damage or inherent errors in the replication process. *A mutation that increases the chances of survival of the individual increases the likelihood that the mutation will be passed on to the next generation.* Beneficial mutations tend to spread rapidly through a population; deleterious changes tend to die along with the organisms that harbor them.

The theory of evolution by natural selection, which was first articulated by Charles Darwin in the 1860s, has been confirmed through observation and experimentation. Bacteria are particularly useful for studies of evolution, since they rapidly reproduce under laboratory conditions, with generation times as short as 20 min. Richard Lenski, for example, has documented changes in fitness (adaptation to certain conditions) over thousands of bacterial generations. The concurrence of experimental evidence, molecular information, and the fossil record highlight some important—and often misunderstood—principles of evolution:

1. *Evolution is not directed toward a particular goal.* It proceeds by random changes that may affect the ability of an organism to reproduce under the prevailing conditions. An organism that is well adapted to its environment may fare better or worse when conditions change.

2. *Evolution requires some built-in sloppiness,* which allows organisms to adapt to unexpected changes. This is one reason why genetically homogeneous populations (e.g., a corn crop) are so susceptible to a single challenge (e.g., a fungal blight). A more heterogeneous population has greater means to resist the adversity and recover.

3. *Evolution is constrained by its past.* New structures and metabolic functions emerge from preexisting elements. For example, insect wings did not erupt spontaneously but appear to have developed gradually from small heat-exchange structures.

4. *Evolution is ongoing,* although it does not proceed exclusively toward complexity. An anthropocentric view places human beings at the pinnacle of an evolutionary scheme, but a quick survey of life's diversity reveals that simpler species have not died out or stopped evolving.

4 Thermodynamics

The normal activities of living organisms—moving, growing, reproducing—demand an almost constant input of energy. Even at rest, organisms devote a considerable portion of their biochemical apparatus to the acquisition and utilization of energy. The study of energy and its effects on matter falls under the purview of **thermodynamics** (Greek: *therme,* heat + *dynamis,* power). Although living systems present some practical challenges to thermodynamic analysis, *life obeys the laws of thermodynamics.*

A The First Law of Thermodynamics: Energy Is Conserved

In thermodynamics, a **system** is defined as the part of the universe that is of interest, such as a reaction vessel or an organism; the rest of the universe is known as the **surroundings.** *The first law of thermodynamics states that energy **(U)** is conserved;* it can be neither created nor destroyed. The energy change of a system is defined as the difference between the **heat (q)** absorbed by the system from the surroundings and the **work (w)** done by the system on the surroundings.

$$\Delta U = U_{\text{final}} - U_{\text{initial}} = q - w \qquad [1\text{-}1]$$

Heat is a reflection of random molecular motion, whereas work, which is defined as force times the distance moved under its influence, is associated with organized motion. Force may assume many different forms, including the gravitational force exerted by one mass on another, the expansional force exerted by a gas, the tensional force exerted by a spring or muscle fiber, the electrical force of one charge on another, and the dissipative forces of friction and viscosity.

Most biological processes take place at constant pressure. Under such conditions, the work done by the expansion of a gas (pressure–volume work) is $P\Delta V$. Consequently, it is useful to define a new thermodynamic quantity, the **enthalpy** (Greek: *enthalpein,* to warm in), abbreviated **H:**

$$H = U + PV \qquad [1\text{-}2]$$

Then at constant pressure,

$$\Delta H = \Delta U + P\Delta V = q_P - w + P\Delta V \qquad [1\text{-}3]$$

where q_P is defined as the heat at constant pressure. Now, admitting only pressure–volume work (other types of work in chemical reactions are usually negligible),

$$\Delta H = q_P - P\Delta V + P\Delta V = q_P \qquad [1\text{-}4]$$

Moreover, the volume changes in most biochemical reactions are insignificant, so the differences between their ΔU and ΔH values are negligible. Enthalpy, like energy, heat, and work, is given units of joules (some commonly used units and biochemical constants and other conventions are given in Box 1-1).

Thermodynamics is useful for indicating the spontaneity of a process. A **spontaneous process** occurs without the input of additional energy from outside the system. (Thermodynamic spontaneity has nothing to do with how quickly a process occurs.) The first law of thermodynamics, however, cannot by itself determine whether a process is spontaneous. Consider two objects of different temperatures that are brought together. Heat spontaneously flows from the warmer object to the cooler one, never vice versa.

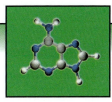

BOX 1-1

Perspectives in Biochemistry

Biochemical Conventions

Modern biochemistry generally uses Système International (SI) units, including meters (m), kilograms (kg), and seconds (s) and their derived units, for various thermodynamic and other measurements. The following table lists the commonly used biochemical units, some useful biochemical constants, and a few conversion factors.

Units

| Energy, heat, work | joule (J) | $kg \cdot m^2 \cdot s^{-2}$ or $C \cdot V$ |
| Electric potential | volt (V) | $J \cdot C^{-1}$ |

Prefixes for units

mega (M)	10^6	nano (n)	10^{-9}
kilo (k)	10^3	pico (p)	10^{-12}
milli (m)	10^{-3}	femto (f)	10^{-15}
micro (μ)	10^{-6}	atto (a)	10^{-18}

Conversions

angstrom (Å)	10^{-10} m
calorie (cal)	4.184 J
kelvin (K)	degrees Celsius (°C) + 273.15

Constants

Avogadro's number (N)	6.0221×10^{23} molecules $\cdot$ mol^{-1}
Coulomb (C)	6.241×10^{18} electron charges
Faraday ($\mathscr{F}$)	96,485 C $\cdot$ mol^{-1} or 96,485 J $\cdot$ V^{-1} $\cdot$ mol^{-1}
Gas constant (R)	8.3145 J $\cdot$ K^{-1} $\cdot$ mol^{-1}
Boltzmann constant (k_B)	1.3807×10^{-23} J $\cdot$ K^{-1} (R/N)
Planck's constant (h)	6.6261×10^{-34} J $\cdot$ s

Throughout this text, molecular masses of particles are expressed in units of **daltons (D),** which are defined as l/12th the mass of a ^{12}C atom (1000 D = 1 **kilodalton, kD**). Biochemists also use **molecular weight,** a dimensionless quantity defined as the ratio of the particle mass to l/12th the mass of a ^{12}C atom, which is symbolized M_r (for relative molecular mass).

Yet either process would be consistent with the first law of thermodynamics since the aggregate energy of the two objects does not change. Therefore, an additional criterion of spontaneity is needed.

B The Second Law of Thermodynamics: Entropy Tends to Increase

According to the second law of thermodynamics, spontaneous processes are characterized by the conversion of order to disorder. In this context, disorder is defined as the number of energetically equivalent ways, **W,** of arranging the components of a system. To make this concept concrete, consider a system consisting of two bulbs of equal volume, one of which contains molecules of an ideal gas (Fig. 1-10). When the stopcock connecting the bulbs is open, the molecules become randomly but equally distributed between the two bulbs. The equal number of gas molecules in each bulb is not the result of any law of motion; it is because the probabilities of all other distributions of the molecules are so overwhelmingly small. Thus, the probability of all the molecules in the system spontaneously rushing into the left bulb (the initial condition) is nil, even though the energy and enthalpy of this arrangement are exactly the same as those of the evenly distributed molecules.

The degree of randomness of a system is indicated by its **entropy** (Greek: *en*, in + *trope*, turning), abbreviated **S:**

$$S = k_B \ln W \qquad [1\text{-}5]$$

where k_B is the **Boltzmann constant.** The units of S are $J \cdot K^{-1}$ (absolute temperature, in units of Kelvin, is a factor because entropy varies with tem-

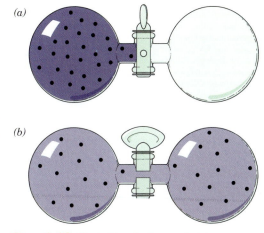

Figure 1-10 Illustration of entropy. In (*a*), a gas occupies the leftmost of two equal-sized bulbs and hence the entropy is low. When the stopcock is opened (*b*), the entropy increases as the gas molecules diffuse back and forth between the bulbs and eventually become distributed evenly, half in each bulb.

perature; e.g., a system becomes more disordered as its temperature rises). The most probable arrangement of a system is the one that maximizes W and hence S. Thus, if a spontaneous process, such as the one shown in Fig. 1-10, has overall energy and enthalpy changes (ΔU and ΔH) of zero, its entropy change (ΔS) must be greater than zero; that is, the number of equivalent ways of arranging the final state must be greater than the number of ways of arranging the initial state. Furthermore, because

$$\Delta S_{system} + \Delta S_{surroundings} = \Delta S_{universe} > 0 \qquad [1\text{-}6]$$

all processes increase the entropy—that is, the disorder—of the universe.

In chemical and biological systems, it is impractical, if not impossible, to determine the entropy of a system by counting all the equivalent arrangements of its components (W). However, there is an entirely equivalent expression for entropy that applies to the constant temperature conditions typical of biological systems: for a spontaneous process,

$$\Delta S \geq \frac{q}{T} \qquad [1\text{-}7]$$

Thus, the entropy change in a process can be experimentally determined from measurements of heat.

C Free Energy

The spontaneity of a process cannot be predicted from a knowledge of the system's entropy change alone. For example, 2 mol of H_2 and 1 mol of O_2, when sparked, react to form 2 mol of H_2O. Yet two water molecules, each of whose three atoms are constrained to stay together, are more ordered than are the three diatomic molecules from which they formed.

What, then, is the thermodynamic criterion for a spontaneous process? Equations 1-4 and 1-7 indicate that at constant temperature and pressure

$$\Delta S \geq \frac{q_P}{T} = \frac{\Delta H}{T} \qquad [1\text{-}8]$$

Thus,

$$\Delta H - T\Delta S \leq 0 \qquad [1\text{-}9]$$

This is the true criterion for spontaneity as formulated, in 1878, by J. Willard Gibbs. He defined the **Gibbs free energy** (**G,** usually called just **free energy**) as

$$G = H - TS \qquad [1\text{-}10]$$

Consequently, spontaneous processes at constant temperature and pressure have

$$\boxed{\Delta G = \Delta H - T\Delta S < 0} \qquad [1\text{-}11]$$

Such processes are said to be **exergonic** (Greek: *ergon,* work). Processes that are not spontaneous have positive ΔG values ($\Delta G > 0$) and are said to be **endergonic;** they must be driven by the input of free energy. If a process is exergonic, the reverse of that process is endergonic and vice versa. Processes at **equilibrium,** those in which the forward and reverse reactions are exactly balanced, are characterized by $\Delta G = 0$. For the most part, only changes in free energy, enthalpy, and entropy (ΔG, ΔH, and ΔS) can be measured, not their absolute values.

A process that is accompanied by an increase in enthalpy ($\Delta H > 0$), which opposes the process, can nevertheless proceed spontaneously if the

Table 1-3 Variation of Reaction Spontaneity (Sign of ΔG) with the Signs of ΔH and ΔS

ΔH	ΔS	$\Delta G = \Delta H - T\Delta S$
−	+	The reaction is both enthalpically favored (exothermic) and entropically favored. It is spontaneous (exergonic) at all temperatures.
−	−	The reaction is enthalpically favored but entropically opposed. It is spontaneous only at temperatures *below* $T = \Delta H/\Delta S$.
+	+	The reaction is enthalpically opposed (endothermic) but entropically favored. It is spontaneous only at temperatures *above* $T = \Delta H/\Delta S$.
+	−	The reaction is both enthalpically and entropically opposed. It is unspontaneous (endergonic) at all temperatures.

entropy change is sufficiently positive ($\Delta S > 0$; Table 1-3). Conversely, a process that is accompanied by a decrease in entropy ($\Delta S < 0$) can proceed if its enthalpy change is sufficiently negative ($\Delta H < 0$). It is important to emphasize that *a large negative value of ΔG does not ensure that a process such as a chemical reaction will proceed at a measurable rate. The rate depends on the detailed mechanism of the reaction, which is independent of ΔG.*

Free energy as well as energy, enthalpy, and entropy are **state functions.** In other words, their values depend only on the current state or properties of the system, not on how the system reached that state. Therefore, *thermodynamic measurements can be made by considering only the initial and final states of the system and ignoring all the stepwise changes in enthalpy and entropy that occur in between.* For example, it is impossible to directly measure the energy change for the reaction of glucose with O_2 in a living organism because of the numerous other simultaneously occurring chemical reactions. But since ΔG depends on only the initial and final states, the combustion of glucose can be analyzed in any convenient apparatus, using the same starting materials (glucose and O_2) and end products (CO_2 and H_2O) that would be obtained *in vivo*.

D Chemical Equilibria and the Standard State

The entropy (disorder) of a substance increases with its volume. For example, a collection of gas molecules, in occupying all of the volume available to it, maximizes its entropy. Similarly, dissolved molecules become uniformly distributed throughout their solution volume. Entropy is therefore a function of concentration.

If entropy varies with concentration, so must free energy. Thus, *the free energy change of a chemical reaction depends on the concentrations of both its reactants and its products.* This phenomenon has great significance because many biochemical reactions operate in either direction depending on the relative concentrations of their reactants and products.

Equilibrium Constants Are Related to ΔG. The relationship between the concentration and the free energy of a substance A is approximately

$$\overline{G}_A = \overline{G}_A^\circ = RT \ln[A] \qquad [1\text{-}12]$$

where $\overline{G}_A$ is known as the **partial molar free energy** or the **chemical potential** of A (the bar indicates the quantity per mole), $\overline{G}_A^\circ$ is the partial molar free energy of A in its **standard state,** R is the gas constant, and [A] is the molar concentration of A. Thus, for the general reaction

$$aA + bB \rightleftharpoons cC + dD$$

the free energy change is

$$\Delta G = c\overline{G}_C + d\overline{G}_D - a\overline{G}_A - b\overline{G}_B \qquad [1\text{-}13]$$

and

$$\Delta G^\circ = c\overline{G}_C^\circ + d\overline{G}_D^\circ - a\overline{G}_A^\circ - b\overline{G}_B^\circ \qquad [1\text{-}14]$$

because free energies are additive and the free energy change of a reaction is the sum of the free energies of the products less those of the reactants. Substituting these relationships into Eq. 1-12 yields

$$\Delta G = \Delta G^\circ + RT \ln\left(\frac{[C]^c[D]^d}{[A]^a[B]^b}\right) \qquad [1\text{-}15]$$

where ΔG° is the free energy change of the reaction when all of its reactants and products are in their standard states (see below). Thus, the expression for the free energy change of a reaction consists of two parts: (1) a constant term whose value depends only on the reaction taking place and (2) a variable term that depends on the concentrations of the reactants and the products, the stoichiometry of the reaction, and the temperature.

For a reaction at equilibrium, there is no *net* change because the free energy change of the forward reaction exactly balances that of the reverse reaction. Consequently, $\Delta G = 0$, so Eq. 1-15 becomes

$$\boxed{\Delta G^\circ = -RT \ln K_{eq}} \qquad [1\text{-}16]$$

where K_{eq} is the familiar **equilibrium constant** of the reaction:

$$K_{eq} = \frac{[C]_{eq}^c[D]_{eq}^d}{[A]_{eq}^a[B]_{eq}^b} = e^{-\Delta G^\circ/RT} \qquad [1\text{-}17]$$

The subscript "eq" denotes reactant and product concentrations at equilibrium (the equilibrium condition is usually clear from the context of the situation, so equilibrium concentrations are usually expressed without this subscript). *The equilibrium constant of a reaction can therefore be calculated from standard free energy data and vice versa (see Sample Calculation 1-1).*

K Depends on Temperature. The manner in which the equilibrium constant varies with temperature can be seen by substituting Eq. 1-11 into Eq. 1-16 and rearranging:

$$\ln K_{eq} = \frac{-\Delta H^\circ}{R}\left(\frac{1}{T}\right) + \frac{\Delta S^\circ}{R} \qquad [1\text{-}18]$$

where H° and S° represent enthalpy and entropy in the standard state. Equation 1-18 has the form $y = mx + b$, the equation for a straight line. A plot of $\ln K_{eq}$ versus $1/T$, known as a **van't Hoff plot,** permits the values of ΔH° and ΔS° (and hence ΔG°) to be determined from measurements of K_{eq} at two (or more) different temperatures. This method is often more practical than directly measuring ΔH and ΔS by calorimetry (which measures the heat, q_P, of a process).

SAMPLE CALCULATION 1-1

The standard free energy change for a reaction is -15 kJ·mol^{-1}. What is the equilibrium constant for the reaction?

Since ΔG° is known, Eq. 1-17 can be used to calculate K_{eq}. Assume the temperature is 25°C (298 K):

$K_{eq} = e^{-\Delta G^\circ/RT}$

$K_{eq} = e^{-(-15,000 \text{ J·mol}^{-1})/(8.314 \text{ J·K}^{-1}\cdot\text{mol}^{-1})(298 \text{ K})}$

$K_{eq} = e^{6.05}$

$K_{eq} = 426$

Standard-State Conventions in Biochemistry. In order to compare free energy changes for different reactions, it is necessary to express ΔG values relative to some standard state (likewise, we refer the elevations of geographic locations to sea level, which is arbitrarily assigned the height of zero). According to the convention used in physical chemistry, a solute is in its standard state when the temperature is 25°C, the pressure is 1 atm, and the solute has an **activity** of 1 (activity of a substance is its concentration corrected for its nonideal behavior at concentrations higher than infinite dilution).

The concentrations of reactants and products in most biochemical reactions are usually so low (on the order of millimolar or less) that their activities are closely approximated by their molar concentrations. Furthermore, because biochemical reactions occur near neutral pH, biochemists have adopted a somewhat different standard-state convention:

1. The activity of pure water is assigned a value of 1, even though its concentration is 55.5 M. This practice simplifies the free energy expressions for reactions in dilute solutions involving water as a reactant, because the [H_2O] term can then be ignored.

2. The hydrogen ion activity is assigned a value of 1 at the physiologically relevant pH of 7. Thus, the biochemical standard state is pH 7.0 (neutral pH, where [H^+] = 10^{-7} M) rather than pH 0 ([H^+] = 1 M), the physical chemical standard state, where many biological substances are unstable.

3. The standard state of a substance that can undergo an acid–base reaction is defined in terms of the total concentration of its naturally occurring ion mixture at pH 7. In contrast, the physical chemistry convention refers to a pure species whether or not it actually exists at pH 0. The advantage of the biochemistry convention is that the total concentration of a substance with multiple ionization states, such as most biological molecules, is usually easier to measure than the concentration of one of its ionic species. Since the ionic composition of an acid or base varies with pH, however, the standard free energies calculated according to the biochemical convention are valid only at pH 7.

Under the biochemistry convention, the standard free energy changes of reactions are customarily symbolized by $\Delta G^{\circ\prime}$ to distinguish them from physical chemistry standard free energy changes, ΔG°. If a reaction includes neither H_2O, H^+, nor an ionizable species, then $\Delta G^{\circ\prime} = \Delta G^\circ$.

E Life Obeys the Laws of Thermodynamics

At one time, many scientists believed that life, with its inherent complexity and order, somehow evaded the laws of thermodynamics. However, elaborate measurements on living animals are consistent with the conservation of energy predicted by the first law. Unfortunately, experimental verification of the second law is not practicable, since it requires dismantling an organism to its component molecules, which would result in its irreversible death. Consequently, it is possible to assert only that the entropy of living matter is less than that of the products to which it decays. *Life persists, however, because a system (a living organism) can be ordered at the expense of disordering its surroundings to an even greater extent.* In other words, the total entropy of the system plus its surroundings increases, as required by the second law. Living organisms achieve order by disordering (breaking

down) the nutrients they consume. Thus, the entropy content of food is as important as its energy content.

Living Organisms Are Open Systems.

Classical thermodynamics applies primarily to reversible processes in **isolated systems** (which cannot exchange matter or energy with their surroundings) or in **closed systems** (which can only exchange energy). An isolated system inevitably reaches equilibrium. For example, if its reactants are in excess, the forward reaction will proceed faster than the reverse reaction until equilibrium is attained ($\Delta G = 0$), at which point the forward and reverse reactions exactly balance each other. In contrast, **open systems,** which exchange both matter and energy with their surroundings, can reach equilibrium only after the flow of matter and energy has stopped.

Living organisms, which take up nutrients, release waste products, and generate work and heat, are open systems and therefore can never be at equilibrium. They continuously ingest high-enthalpy, low-entropy nutrients, which they convert to low-enthalpy, high-entropy waste products. The free energy released in this process powers the cellular activities that produce the high degree of organization characteristic of life. If this process is interrupted, the system ultimately reaches equilibrium, which for living things is synonymous with death. An example of energy flow in an open system is illustrated in Fig. 1-11. Through photosynthesis, plants convert radiant energy from the sun, the primary energy source for life on the earth, to the chemical energy of carbohydrates and other organic substances. The plants, or the animals that eat them, then metabolize these substances to power such functions as the synthesis of biomolecules, the maintenance of intracellular ion concentrations, and cellular movements.

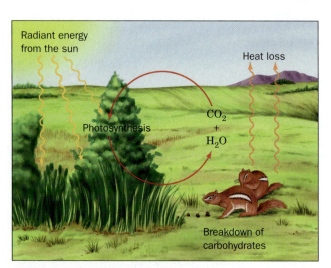

Figure 1-11 Energy flow in the biosphere.   Plants use the sun's radiant energy to synthesize carbohydrates from CO_2 and H_2O. Plants or the animals that eat them eventually metabolize the carbohydrates to release their stored free energy and thereby return CO_2 and H_2O to the environment.

Living Things Maintain a Steady State.

Even in a system that is not at equilibrium, matter and energy flow according to the laws of thermodynamics. For example, materials tend to move from areas of high concentration to areas of low concentration. This is why blood takes up O_2 in the lungs, where O_2 is abundant, and releases it to the tissues, where O_2 is scarce.

Living systems are characterized by being in a **steady state.** This means that all flows in the system are constant so that the system does not change with time. Energy flow in the biosphere (Fig. 1-11) is an example of a system in a steady state. Slight perturbations from the steady state give rise to changes in flows that restore the system to the steady state. In all living systems, energy flow is exclusively "downhill" ($\Delta G < 0$). In addition, nature is inherently dissipative, so the recovery of free energy from a biochemical process is never total and some energy is always lost to the surroundings.

Enzymes Catalyze Biochemical Reactions.

Nearly all the molecular components of an organism can potentially react with each other, and many of these reactions are thermodynamically favored (spontaneous). Yet only a subset of all possible reactions actually occur to a significant extent in a living organism. The rate of a particular reaction depends not on the free energy difference between the initial and final states but on the actual path through which the reactants are transformed to products. Living organisms take advantage of catalysts, substances that increase the rate at which the reaction approaches equilibrium without affecting the reaction's ΔG. Biological catalysts are referred to as **enzymes,** most of which are proteins.

Enzymes accelerate biochemical reactions by physically interacting with the reactants and products to provide a more favorable pathway for the transformation of one to the other. Enzymes increase the rates of reactions by increasing the likelihood that the reactants can interact productively. However, enzymes cannot promote reactions whose ΔG values are positive.

A multitude of enzymes mediate the flow of energy in every cell. As free energy is harvested, stored, or used to perform cellular work, it may be transferred to other molecules. And although it is tempting to think of free energy as something that is stored in chemical bonds, chemical energy can be transformed into heat, electrical work, or mechanical work, according to the needs of the organism and the biochemical machinery with which it has been equipped through evolution.

 ## SUMMARY

1. A model for the origin of life proposes that organisms ultimately arose from simple organic molecules that polymerized to form more complex molecules capable of replicating themselves.

2. Compartmentation gave rise to cells that developed metabolic reactions for synthesizing biological molecules and generating energy.

3. All cells are either prokaryotic or eukaryotic. Eukaryotic cells contain a variety of membrane-bounded organelles.

4. Phylogenetic evidence groups organisms into three domains: archaea, bacteria, and eukarya.

5. Natural selection determines the evolution of species.

6. The first law of thermodynamics (energy is conserved) and the second law (spontaneous processes increase the disorder of the universe) apply to biochemical processes. The spontaneity of a process is determined by its free energy change ($\Delta G = \Delta H - T\Delta S$): spontaneous reactions have $\Delta G < 0$ and nonspontaneous reactions have $\Delta G > 0$.

7. The equilibrium constant for a process is related to the standard free energy change for that process.

8. Living organisms are open systems that maintain a steady state.

 ## REFERENCES

Origin and Evolution of Life

de Duve, C., *Blueprint for a Cell. The Nature and Origin of Life,* Carolina Biological Supply Co. (1991).

Nisbet, E.G. and Sleep, N.H., The habitat and nature of early life, *Nature* **409,** 1083–1091 (2001). [Explains some of the hypotheses regarding the early earth and the origin of life, including the possibility that life originated at hydrothermal vents.]

Orgel, L.E., The origin of life—a review of facts and speculations, *Trends Biochem. Sci.* **23,** 491–495 (1998). [Reviews the most widely accepted hypotheses on the origin of life and discusses the evidence supporting them and their difficulties.]

Szathmáry, E. and Smith, J.M., The major evolutionary transitions, *Nature* **374,** 227–232 (1995). [A summary of the evolution of complexity, from the origin of prebiotic systems to the emergence of language.]

Cells

Baldauf, S.L., The deep roots of eukaryotes, *Science* **300,** 1703–1706 (2003). [Describes some recent discoveries that raise questions about the taxonomy of eukaryotes.]

Campbell, N.A., Reece, J.B., and Mitchell, L.G., *Biology* (5th ed.),

Benjamin/Cummings (1999). [This and other comprehensive general biology texts provide details about the structures of prokaryotes and eukaryotes.]

Goodsell, D.S., A look inside the living cell, *Am. Scientist* **80,** 457–465 (1992); *and* Inside a living cell, *Trends Biochem. Sci.* **16,** 203–206 (1991).

Lodish, H., Berk, A., Zipursky, S.L., Matsudaria, P., Baltimore, D., and Darnell, J., *Molecular Cell Biology* (4th ed.), Chapter 5, W.H. Freeman (2000). [This and other cell biology textbooks offer thorough reviews of cellular structure.]

Pace, N.R., A molecular view of microbial diversity and the biosphere, *Science* **276,** 734–740 (1997). [Describes the three domains of life and their general metabolic strategies.]

Thermodynamics

Tinoco, I., Jr., Sauer, K., Wang, J.C. and Puglisi, J.C., *Physical Chemistry. Principles and Applications in Biological Sciences* (4th ed.), Chapters 2–5, Prentice-Hall (2002). [Most physical chemistry texts treat thermodynamics in some detail.]

van Holde, K.E., Johnson, W.C., and Hu, P.S., *Principles of Physical Chemistry,* Chapters 2 and 3, Prentice-Hall (1998).

KEY TERMS

organic compound	endoplasmic reticulum	halobacteria	k_B
polymer	Golgi apparatus	thermophiles	G
condensation reaction	mitochondrion	symbiosis	exergonic
hydrolysis	chloroplast	mutation	endergonic
replication	lysosome	thermodynamics	equilibrium
natural selection	peroxisome	system	state function
vesicle	vacuole	surroundings	$\overline{G}_A$
compartmentation	cytosol	U	$\overline{G}^\circ_A$
catalyst	cytoskeleton	q	standard state
eukaryote	taxonomy	w	equilibrium constant
nucleus	phylogeny	H	isolated system
prokaryote	archaea	q_P	closed system
virus	bacteria	spontaneous process	open system
cytoplasm	eukarya	W	steady state
organelle	methanogens	S	enzyme

STUDY EXERCISES

1. Summarize the major stages of chemical and organismal evolution.

2. Describe the process of evolution by natural selection.

3. What kinds of organisms are found in each of the three major evolutionary domains?

4. Explain the first and second laws of thermodynamics.

5. How does the free energy change in a process depend on its enthalpy and entropy changes?

6. What is the biochemistry standard state?

7. How does life persist despite the second law of thermodynamics?

PROBLEMS

1. Identify the circled functional groups and linkages in the compound below.

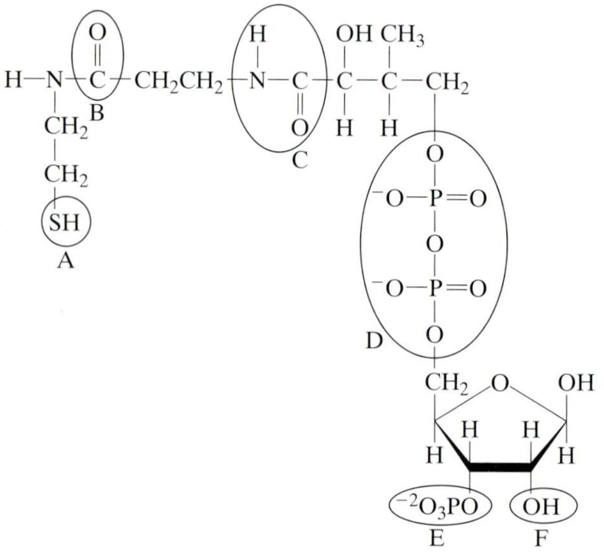

2. Why is the cell membrane not an absolute barrier between the cytoplasm and the external environment?

3. A spheroidal bacterium with a diameter of 1 μm contains two molecules of a particular protein. What is the molar concentration of the protein?

4. How many glucose molecules does the cell in Problem 3 contain when its internal glucose concentration is 1.0 mM?

5. (a) Which has greater entropy, liquid water at 0°C or ice at 0°C? (b) How does the entropy of ice at −5°C differ, if at all, from its entropy at −50°C?

6. Does entropy increase or decrease in the following processes?

(a) $N_2 + 3\,H_2 \longrightarrow 2\,NH_3$

(b) $H_2N-\overset{\overset{\textstyle O}{\|}}{C}-NH_2 \;+\; H_2O \longrightarrow CO_2 \;+\; 2\,NH_3$

Urea

(c)
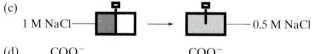
1 M NaCl ——→ 0.5 M NaCl

(d)
$$\begin{array}{ccc} COO^- & & COO^- \\ | & & | \\ HC-OH & \longrightarrow & HC-OPO_3^{2-} \\ | & & | \\ H_2C-OPO_3^{2-} & & H_2C-OH \end{array}$$

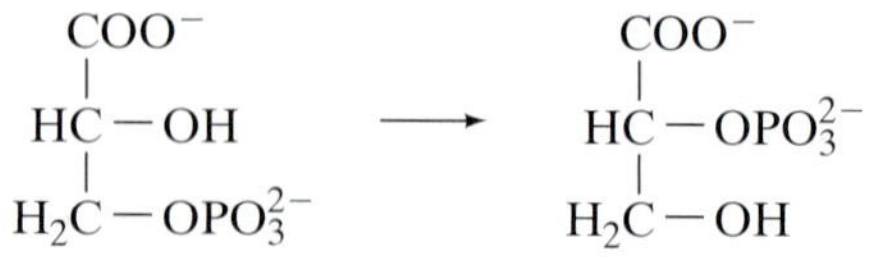

3-Phosphoglycerate   **2-Phosphoglycerate**

7. Consider a reaction with $\Delta H = 15$ kJ and $\Delta S = 50$ J·K^{-1}. Is the reaction spontaneous (a) at 10°C, (b) at 80°C?

8. Calculate the equilibrium constant for the reaction

glucose-1-phosphate + H_2O ⟶ glucose + $H_2PO_4^-$

at pH 7.0 and 25°C ($\Delta G^{\circ\prime} = -20.9$ kJ·mol^{-1}).

9. Calculate $\Delta G^{\circ\prime}$ for the reaction $A + B \rightleftharpoons C + D$ at 25°C when the equilibrium concentrations are $[A] = 10$ μM, $[B] = 15$ μM, $[C] = 3$ μM, and $[D] = 5$ μM. Is the reaction exergonic or endergonic under standard conditions?

10. $\Delta G^{\circ\prime}$ for the isomerization reaction

    glucose-1-phosphate (G1P) $\rightleftharpoons$ glucose-6-phosphate (G6P)

    is -7.1 kJ·mol^{-1}. Calculate the equilibrium ratio of [G1P] to [G6P] at 25°C.

11. For the reaction $A \longrightarrow B$ at 298 K, the change in enthalpy is -7 kJ·mol^{-1} and the change in entropy is -25 J·K^{-1}·mol^{-1}. Is the reaction spontaneous? If not, should the temperature be increased or decreased to make the reaction spontaneous?

12. For the conversion of reactant A to product B, the change in enthalpy is 7 kJ·mol^{-1} and the change is entropy is 20 J·K^{-1}·mol^{-1}. Above what temperature does the reaction become spontaneous?

13. Label the following statements true or false:

    (a) A reaction is said to be spontaneous when it can proceed in either the forward or reverse direction.

    (b) A spontaneous process always happens very quickly.

    (c) A nonspontaneous reaction will proceed spontaneously in the reverse direction.

    (d) A spontaneous process can occur with a large decrease in entropy.

Coral reefs support a variety of vertebrates and invertebrates. The water that surrounds them is critical for their existence, acting as a solvent for biochemical reactions and, to a large extent, determining the structures of the macromolecules that carry out these reactions. [Georgette Douwma/Taxi/Getty Images.]

Water

Any study of the chemistry of life must include a study of water. Biological molecules and the reactions they undergo can be best understood in the context of their aqueous environment. Not only are organisms made mostly of water (about 70% of the mass of the human body is water), they are surrounded by water on this, the "blue planet." Aside from its sheer abundance, water is central to biochemistry for the following reasons:

1. Nearly all biological molecules assume their shapes (and therefore their functions) in response to the physical and chemical properties of the surrounding water.

2. The medium for the majority of biochemical reactions is water. Reactants and products of metabolic reactions, nutrients as well as waste products, depend on water for transport within and between cells.

3. Water itself actively participates in many chemical reactions that support life. Frequently, the ionic components of water, the H^+ and OH^- ions, are the true reactants. In fact, the reactivity of many functional groups on biological molecules depends on the relative concentrations of H^+ and OH^- in the surrounding medium.

All organisms require water, from the marine creatures who spend their entire lives in an aqueous environment to terrestrial organisms who must guard their watery interiors with protective skins. Not surprisingly, living organisms can be found wherever there is liquid water—in springs as hot as 105°C and in the cracks and crevices between rocks hundreds of meters beneath the surface of the earth. Organisms that survive desiccation do so only by becoming dormant, as seeds or spores.

An examination of water from a biochemical point of view requires a look at the physical properties of water, its powers as a solvent, and its chemical behavior—that is, the nature of aqueous acids and bases.

1 Physical Properties of Water

The colorless, odorless, and tasteless nature of water belies its fundamental importance to living organisms. Despite its bland appearance to our senses, water is anything but inert. Its physical properties—unique among molecules of similar size—give it unparalleled strength as a solvent. And yet its limitations as a solvent also have important implications for the structures and functions of biological molecules.

A Structure of Water

A water molecule consists of two hydrogen atoms bonded to an oxygen atom. The O—H bond distance is 0.958 Å (1 Å = 10^{-10} m), and the angle formed by the three atoms is 104.5° (Fig. 2-1). The hydrogen atoms are not arranged linearly, because the oxygen atom's four sp^3 hybrid orbitals extend roughly toward the corners of a tetrahedron. Hydrogen atoms occupy two corners of the tetrahedron, and the nonbonding electron pairs of the oxygen atom occupy the other two corners (in a perfectly tetrahedral molecule, such as methane, CH_4, the bond angles are 109.5°).

Water Molecules Form Hydrogen Bonds. The angular geometry of the water molecule has enormous implications for living systems. Water is a **polar** molecule: the oxygen atom with its unshared electrons carries a partial negative charge (δ^-) of $-0.66e$, and the hydrogen atoms each carry a partial

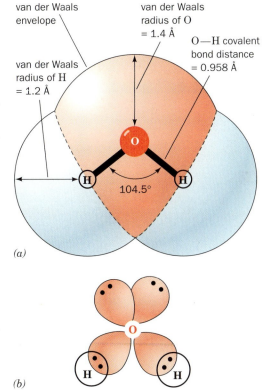

Figure 2-1 **Structure of the water molecule.** (*a*) The shaded outline represents the van der Waals envelope, the effective "surface" of the molecule. (*b*) The oxygen atom's sp^3 orbitals are arranged tetrahedrally. Two orbitals contain nonbonding electron pairs.

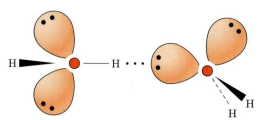

Figure 2-2 A hydrogen bond in water.   The strength of the interaction is maximal when the O—H covalent bond of one molecule points directly toward the lone-pair electron cloud of the other.

positive charge (δ^+) of $+0.33e$, where e is the charge of the electron. Electrostatic attractions between the dipoles of water molecules are crucial to the properties of water itself and to its role as a biochemical solvent. Neighboring water molecules tend to orient themselves so that the O—H bond of one water molecule (the positive end) points toward one of the electron pairs of the other water molecule (the negative end). The resulting directional intermolecular association is known as a **hydrogen bond** (Fig. 2-2).

In general, a hydrogen bond can be represented as D—H⋯A, where D—H is a weakly acidic "donor" group such as O—H, N—H, or sometimes S—H, and A is a weakly basic "acceptor" atom such as O, N, or occasionally S. Hydrogen bonds are structurally characterized by an H⋯A distance that is at least 0.5 Å shorter than the calculated **van der Waals distance** (the distance of closest approach between two nonbonded atoms). In water, for example, the O⋯H hydrogen bond distance is ~1.8 Å versus 2.6 Å for the corresponding van der Waals distance.

A single water molecule contains two hydrogen atoms that can be "donated" and two unshared electron pairs that can act as "acceptors," so each molecule can participate in a maximum of four hydrogen bonds with other water molecules. Although the energy of an individual hydrogen bond ($\sim 20\ kJ \cdot mol^{-1}$) is relatively small (e.g., the energy of an O—H covalent bond is $460\ kJ \cdot mol^{-1}$), the sheer number of hydrogen bonds in a sample of water is the key to its remarkable properties.

Ice Is a Crystal of Hydrogen-Bonded Water Molecules.   The structure of ice provides a striking example of the cumulative strength of many hydrogen bonds. X-Ray and neutron diffraction studies have established that water molecules in ice are arranged in an unusually open structure. Each water molecule is tetrahedrally surrounded by four nearest neighbors to which it is hydrogen bonded (Fig. 2-3). As a consequence of its open structure, water is one of the very few substances that expands on freezing (at 0°C, liquid water has a density of $1.00\ g \cdot mL^{-1}$, whereas ice has a density of $0.92\ g \cdot mL^{-1}$).

The expansion of water on freezing has overwhelming consequences for life on the earth. Suppose that water contracted on freezing, that is, became more dense rather than less dense. Ice would then sink to the bottoms of lakes and oceans rather than float. This ice would be insulated from the sun so that oceans, with the exception of a thin surface layer of liquid in warm weather, would be permanently frozen solid (the water at great depths, even in tropical oceans, is close to 4°C, its temperature of maximum density). Thus, the earth would be locked in a permanent ice age and life might never have arisen.

The melting of ice represents the collapse of the strictly tetrahedral orientation of hydrogen-bonded water molecules, although hydrogen bonds between water molecules persist in the liquid state. In fact, liquid water is

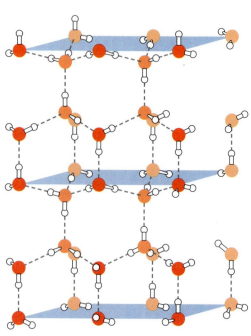

Figure 2-3 The structure of ice.   Each water molecule interacts tetrahedrally with four other water molecules. Oxygen atoms are red and hydrogen atoms are white. Hydrogen bonds are represented by dashed lines. [After Pauling, L., *The Nature of the Chemical Bond* (3rd ed.), p. 465, Cornell University Press (1960).]

only ~15% less hydrogen bonded than ice at 0°C. Indeed, the boiling point of water is 264°C higher than that of methane, a substance with nearly the same molecular mass as H_2O but which is incapable of hydrogen bonding (substances with similar intermolecular associations and equal molecular masses should have similar boiling points). This difference reflects the extraordinary internal cohesiveness of liquid water resulting from its intermolecular hydrogen bonding.

The Structure of Liquid Water Is Irregular.   Because each molecule of liquid water reorients about once every 10^{-12} s, very few experimental techniques can explore the instantaneous arrangement of these water molecules. Theoretical considerations and spectroscopic evidence suggest that molecules in liquid water are each hydrogen bonded to four nearest neighbors, as they are in ice. These hydrogen bonds are distorted, however, so the networks of linked molecules are irregular and varied. For example, three- to seven-membered rings of hydrogen-bonded molecules commonly occur in liquid water (Fig. 2-4), in contrast to the six-membered rings characteristic of ice (Fig. 2-3). Moreover, these networks continually break up and reform every 2×10^{-11} s or so. *Liquid water therefore consists of a rapidly fluctuating, three-dimensional network of hydrogen-bonded H_2O molecules.*

Hydrogen Bonds and Other Weak Interactions in Biological Molecules.   Biochemists are concerned not just with the strong covalent bonds that define chemical structure but with the weak forces that act under relatively mild physical conditions. The structures of most biological molecules are determined by the collective influence of many individually weak interactions. The weak electrostatic forces that interest biochemists include ionic interactions, hydrogen bonds, and van der Waals forces.

The strength of association of ionic groups of opposite charge depends on the chemical nature of the ions, the distance between them, and the polarity of the medium. In general, the strength of the interaction between two charged groups (i.e., the energy required to completely separate them in the medium of interest) is less than the energy of a covalent bond but greater than the energy of a hydrogen bond (Table 2-1).

The noncovalent associations between neutral molecules, collectively known as **van der Waals forces,** arise from electrostatic interactions among permanent or induced dipoles (the hydrogen bond is a special kind of dipo-

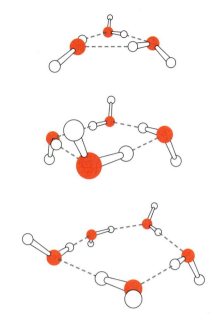

Figure 2-4 **Structure of a water trimer, tetramer, and pentamer.**   These models are based on theoretical predictions and spectroscopic data. [After Liu, K., Cruzan, J.D., and Saykally, R.J., *Science* **271**, 929 (1996).]

Table 2-1 Bond Energies in Biomolecules

Type of Bond	Example	Bond Strength ($kJ \cdot mol^{-1}$)
Covalent	O—H	460
	C—H	414
	C—C	348
Noncovalent		
Ionic interaction	$-COO^- \cdots {}^+H_3N-$	86
van der Waals forces		
Hydrogen bond	$-O-H \cdots O\diagup$	20
Dipole–dipole interaction	$\diagdown C=O \cdots \diagdown C=O$	9.3
London dispersion forces	$-\overset{\overset{\displaystyle H}{\vert}}{C}-H \cdots H-\overset{\overset{\displaystyle H}{\vert}}{C}-$	0.3

Figure 2-5 Dipole–dipole interactions. The strength of each dipole is indicated by the thickness of the accompanying arrow. (*a*) Interaction between permanent dipoles. (*b*) Dipole–induced dipole interaction. (*c*) London dispersion forces.

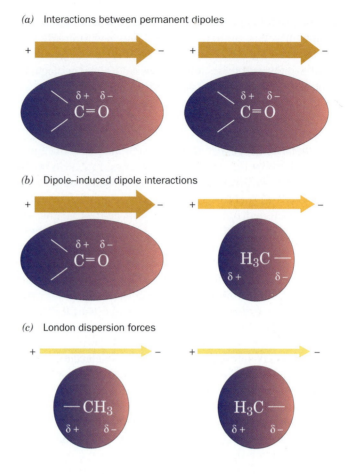

(*a*) Interactions between permanent dipoles

(*b*) Dipole–induced dipole interactions

(*c*) London dispersion forces

lar interaction). Interactions among permanent dipoles such as carbonyl groups (Fig. 2-5*a*) are much weaker than ionic interactions. A permanent dipole also induces a dipole moment in a neighboring group by electrostatically distorting its electron distribution (Fig. 2-5*b*). Such dipole–induced dipole interactions are generally much weaker than dipole–dipole interactions.

At any instant, nonpolar molecules have a small, randomly oriented dipole moment resulting from the rapid fluctuating motion of their electrons. This transient dipole moment can polarize the electrons in a neighboring group (Fig. 2-5*c*), so that the groups are attracted to each other. These so-called **London dispersion forces** are extremely weak and fall off so rapidly with distance that they are significant only for groups in close contact. They are, nevertheless, extremely important in determining the structures of biological molecules, whose interiors contain many closely packed groups.

B Water as a Solvent

Solubility depends on the ability of a solvent to interact with a solute more strongly than solute particles interact with each other. Water is said to be the "universal solvent." Although this statement cannot literally be true, water certainly dissolves more types of substances and in greater amounts than any other solvent. In particular, the polar character of water makes it an excellent solvent for polar and ionic materials, which are said to be **hydrophilic** (Greek: *hydro*, water + *philos*, loving). On the other hand, nonpolar substances are virtually insoluble in water ("oil and water don't mix") and are consequently described as **hydrophobic** (Greek: *phobos*, fear). Nonpolar sub-

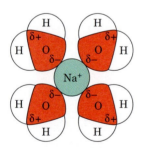

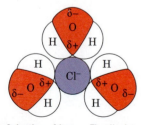

Figure 2-6 Solvation of ions. The dipoles of the surrounding water molecules are oriented according to the charge of the ion. Only one layer of solvent molecules is shown.

stances, however, are soluble in nonpolar solvents such as CCl_4 and hexane. This information is summarized by another maxim, "like dissolves like."

Why do salts such as NaCl dissolve in water? Polar solvents, such as water, weaken the attractive forces between oppositely charged ions (such as Na^+ and Cl^-) and can therefore hold the ions apart. (In nonpolar solvents, ions of opposite charge attract each other so strongly that they coalesce to form a solid salt.) An ion immersed in a polar solvent such as water attracts the oppositely charged ends of the solvent dipoles (Fig. 2-6). The ion is thereby surrounded by one or more concentric shells of oriented solvent molecules. Such ions are said to be **solvated** or, when water is the solvent, to be **hydrated.**

The water molecules in the hydration shell around an ion move more slowly than water molecules that are not involved in solvating the ion. In bulk water, the energetic cost of breaking a hydrogen bond is low because another hydrogen bond is likely to be forming at the same time. The cost is higher for the relatively ordered water molecules of the hydration shell. The energetics of solvation also play a role in chemical reactions, since a reacting group must shed its **waters of hydration** (the molecules in its hydration shell) in order to closely approach another group.

The bond dipoles of uncharged polar molecules make them soluble in aqueous solutions for the same reasons that ionic substances are water soluble. The solubilities of polar and ionic substances are enhanced when they carry functional groups, such as hydroxyl (OH), carbonyl (C=O), carboxylate (COO^-), or ammonium (NH_3^+) groups, that can form hydrogen bonds with water as illustrated in Fig. 2-7. Indeed, water-soluble biomolecules such as proteins, nucleic acids, and carbohydrates bristle with just such groups. Nonpolar substances, in contrast, lack hydrogen-bonding donor and acceptor groups.

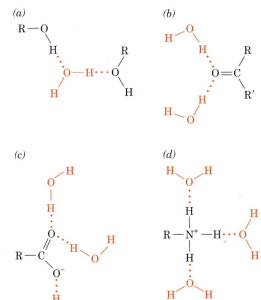

Figure 2-7 Hydrogen bonding by functional groups. Water forms hydrogen bonds with (a) hydroxyl groups, (b) keto groups, (c) carboxylate ions, and (d) ammonium ions.

C The Hydrophobic Effect

When a nonpolar substance is added to an aqueous solution, it does not dissolve but instead is excluded by the water. *The tendency of water to minimize its contacts with hydrophobic molecules is termed the* **hydrophobic effect.** Many large molecules and molecular aggregates, such as proteins, nucleic acids, and cellular membranes, assume their shapes at least partially in response to the hydrophobic effect.

Consider the thermodynamics of transferring a nonpolar molecule from an aqueous solution to a nonpolar solvent. In all cases, the free energy change is negative, which indicates that such transfers are spontaneous processes (Table 2-2). Interestingly, these transfer processes are either en-

Table 2-2 Thermodynamic Changes for Transferring Hydrocarbons from Water to Nonpolar Solvents at 25°C

Process	ΔH (kJ · mol^{-1})	$-T\Delta S$ (kJ · mol^{-1})	ΔG (kJ · mol^{-1})
CH_4 in $H_2O \rightleftharpoons CH_4$ in C_6H_6	11.7	−22.6	−10.9
CH_4 in $H_2O \rightleftharpoons CH_4$ in CCl_4	10.5	−22.6	−12.1
C_2H_6 in $H_2O \rightleftharpoons C_2H_6$ in benzene	9.2	−25.1	−15.9
C_2H_4 in $H_2O \rightleftharpoons C_2H_4$ in benzene	6.7	−18.8	−12.1
C_2H_2 in $H_2O \rightleftharpoons C_2H_2$ in benzene	0.8	−8.8	−8.0
Benzene in $H_2O \rightleftharpoons$ liquid benzene^a	0.0	−17.2	−17.2
Toluene in $H_2O \rightleftharpoons$ liquid toluene^a	0.0	−20.0	−20.0

^{a}Data measured at 18°C.
Source: Kauzmann, W., *Adv. Protein Chem.* **14,** 39 (1959).

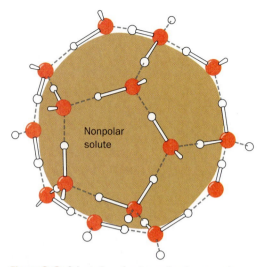

Figure 2-8 Orientation of water molecules around a nonpolar solute.   In order to maximize their number of hydrogen bonds, water molecules form a "cage" around the solute. Dashed lines represent hydrogen bonds.

dothermic (positive ΔH) or isothermic ($\Delta H = 0$); that is, it is enthalpically more or less equally favorable for nonpolar molecules to dissolve in water as in nonpolar media. In contrast, the entropy change (expressed as $-T\Delta S$) is large and negative in all cases. Clearly, the transfer of a hydrocarbon from an aqueous medium to a nonpolar medium is entropically driven (i.e., the free energy change is mostly due to an entropy change).

Entropy, or "randomness," is a measure of the order of a system (Section 1-4B). If entropy increases when a nonpolar molecule leaves an aqueous solution, entropy must decrease when the molecule enters water. This decrease in entropy when a nonpolar molecule is solvated by water is an experimental observation, not a theoretical conclusion. Yet the entropy changes are too large to reflect only the changes in the conformations of the hydrocarbons. Thus the entropy changes must arise mainly from some sort of ordering of the water itself. What is the nature of this ordering?

The extensive hydrogen-bonding network of liquid water molecules is disrupted when a nonpolar group intrudes. A nonpolar group can neither accept nor donate hydrogen bonds, so the water molecules at the surface of the cavity occupied by the nonpolar group cannot hydrogen bond to other molecules in their usual fashion. In order to maximize their hydrogen-bonding ability, these surface water molecules orient themselves to form a hydrogen-bonded network enclosing the cavity (Fig. 2-8). This orientation constitutes an ordering of the water structure since the number of ways that water molecules can form hydrogen bonds around the surface of a nonpolar group is fewer than the number of ways they can form hydrogen bonds in bulk water.

Unfortunately, the ever-fluctuating nature of liquid water's basic structure has not yet allowed a detailed description of this ordering process. One model proposes that water forms icelike hydrogen-bonded "cages" around the nonpolar groups. The water molecules of the cages are tetrahedrally hydrogen bonded to other water molecules, and the ordering of water molecules extends several layers beyond the first hydration shell of the nonpolar solute.

The unfavorable free energy of hydration of a nonpolar substance caused by its ordering of the surrounding water molecules has the net result that the nonpolar substance tends to be excluded from the aqueous phase. This is because the surface area of a cavity containing an aggregate of nonpolar molecules is less than the sum of the surface areas of the cavities that each of these molecules would individually occupy (Fig. 2-9). *The aggregation of the nonpolar groups thereby minimizes the surface area of the cavity and therefore maximizes the entropy of the entire system.* In a sense, the nonpolar groups are squeezed out of the aqueous phase.

Amphiphiles Form Micelles and Bilayers.   *Most biological molecules have both polar (or charged) and nonpolar segments and are therefore simultaneously hydrophilic and hydrophobic.* Such molecules, for example, fatty acid ions

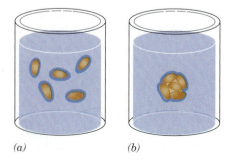

(a)   (b)

Figure 2-9 Aggregation of nonpolar molecules in water.   (a) The individual hydration of dispersed nonpolar molecules (*brown*) decreases the entropy of the system because their hydrating water molecules (*dark blue*) are not as free to form hydrogen bonds. (b) Aggregation of the nonpolar molecules increases the entropy of the system, since the number of water molecules required to hydrate the aggregated solutes is less than the number of water molecules required to hydrate the dispersed solute molecules. This increase in entropy accounts for the spontaneous aggregation of nonpolar substances in water.

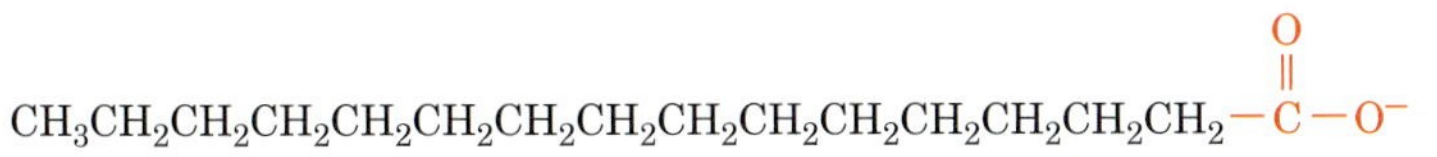

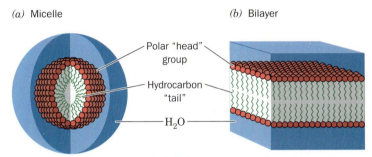

$$CH_3CH_2CH_2CH_2CH_2CH_2CH_2CH_2CH_2CH_2CH_2CH_2CH_2CH_2CH_2-\overset{\overset{\displaystyle O}{\|}}{C}-O^-$$

Palmitate ($C_{15}H_{31}COO^-$)

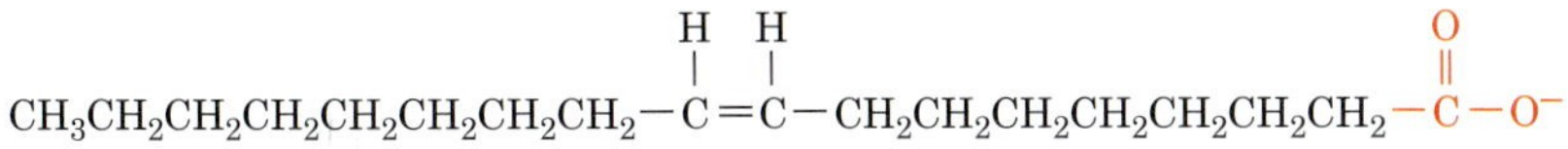

$$CH_3CH_2CH_2CH_2CH_2CH_2CH_2CH_2-\overset{\overset{\displaystyle H}{|}}{C}=\overset{\overset{\displaystyle H}{|}}{C}-CH_2CH_2CH_2CH_2CH_2CH_2CH_2-\overset{\overset{\displaystyle O}{\|}}{C}-O^-$$

Oleate ($C_{17}H_{33}COO^-$)

Figure 2-10 Fatty acid anions (soaps). Palmitate and oleate are amphiphilic compounds; each has a polar carboxylate group and a long nonpolar hydrocarbon chain.

(soaps; Fig. 2-10), are said to be **amphiphilic** or **amphipathic** (Greek: *amphi*, both; *pathos*, suffering). How do amphiphiles interact with an aqueous solvent? Water tends to hydrate the hydrophilic portion of an amphiphile, but it also tends to exclude the hydrophobic portion. Amphiphiles consequently tend to form structurally ordered aggregates. For example, **micelles** are globules of up to several thousand amphiphilic molecules arranged so that the hydrophilic groups at the globule surface can interact with the aqueous solvent while the hydrophobic groups associate at the center, away from the solvent (Fig. 2-11a). Of course, the model presented in Fig. 2-11a is an oversimplification, since it is geometrically impossible for all the hydrophobic groups to occupy the center of the micelle. Instead, the amphipathic molecules pack in a more disorganized fashion that buries most of the hydrophobic groups and leaves the polar groups exposed (Fig. 2-12).

Alternatively, amphiphilic molecules may arrange themselves to form **bilayered** sheets or vesicles in which the polar groups face the aqueous phase (Fig. 2-11b). In both micelles and bilayers, the aggregate is stabilized by the hydrophobic effect, the tendency of water to exclude hydrophobic groups.

The consequences of the hydrophobic effect are often called hydrophobic forces or hydrophobic "bonds." However, the term *bond* implies a discrete directional relationship between two entities. The hydrophobic effect acts indirectly on nonpolar groups and lacks directionality. Despite the temptation to attribute some mutual attraction to a collection of nonpolar groups excluded from water, their exclusion is largely a function of the entropy of the surrounding water molecules, not some "hydrophobic force" among them (the London dispersion forces between the nonpolar groups are relatively weak).

D Osmosis and Diffusion

The fluid inside cells and surrounding cells in multicellular organisms is full of dissolved substances ranging from small inorganic ions to huge molecular aggregates. The concentrations of these solutes affect water's **colligative properties,** the physical properties that depend on the concentration of dissolved substances rather than on their chemical features. For example, solutes depress the freezing point and elevate the boiling point of water by making it more difficult for water molecules to crystallize as ice or to escape from solution into the gas phase.

Osmotic pressure also depends on the solute concentration. When a solution is separated from pure water by a semipermeable membrane that

(a) Micelle   (b) Bilayer

Polar "head" group
Hydrocarbon "tail"
H_2O

Figure 2-11 Structures of micelles and bilayers. In aqueous solution, the polar head groups of amphipathic molecules are hydrated while the nonpolar tails aggregate by exclusion from water. (a) A micelle is a spheroidal aggregate. (b) A bilayer is an extended planar aggregate.

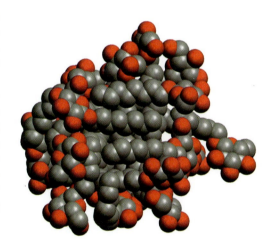

Figure 2-12 Model of a micelle. Twenty molecules of the detergent octyl glucoside (an eight-carbon chain with a sugar head group) are shown in space-filling form in this computer-generated model of a micelle. The polar O atoms of the glucoside groups are red and C atoms are gray. H atoms have been omitted for clarity. Computer simulations indicate that such micelles have rapidly fluctuating structures in which the hydrophobic octyl groups pack in an irregular manner (rather than as the sterically impossible spherically symmetric aggregate pictured in Fig. 2-11a) such that portions of the hydrophobic tails are exposed on the micelle surface at any given instant. [Courtesy of Michael Garavito and Shelagh Ferguson-Miller, Michigan State University.]

Figure 2-13 Osmotic pressure. (*a*) A water-permeable membrane separates a tube of concentrated solution from pure water. (*b*) As water moves into the solution by osmosis, the height of the solution in the tube increases. (*c*) The pressure that prevents the influx of water is the osmotic pressure (22.4 atm for a 1 M solution).

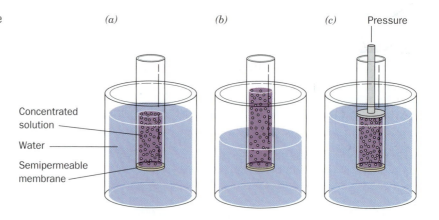

permits the passage of water molecules but not solutes, water tends to move into the solution in order to equalize its concentration on both sides of the membrane. **Osmosis** is the movement of solvent from a region of high concentration (here, pure water) to a region of relatively low concentration (water containing dissolved solute). The **osmotic pressure** of a solution is the pressure that must be applied to the solution to prevent the inward flow of water; it is proportional to the concentration of the solute (Fig. 2-13). For a 1 M solution, the osmotic pressure is 22.4 atm. Consider the implications of osmotic pressure for living cells, which are essentially semipermeable sacs of aqueous solution. In order to minimize osmotic influx of water, which would burst the relatively weak cell membrane, many animal cells are surrounded by a solution of similar osmotic pressure (so there is no net flow of water). Another strategy, used by most plants and bacteria, is to enclose the cell with a rigid cell wall that can withstand the osmotic pressure generated within.

When an aqueous solution is separated from pure water by a membrane that is permeable to both water and solutes, solutes move out of the solution even as water moves in. The molecules move randomly, or **diffuse,** until the concentration of the solute is the same on both sides of the membrane. At this point, equilibrium is established; that is, there is no further *net* flow of water or solute (although molecules continue to move in and out through the membrane).

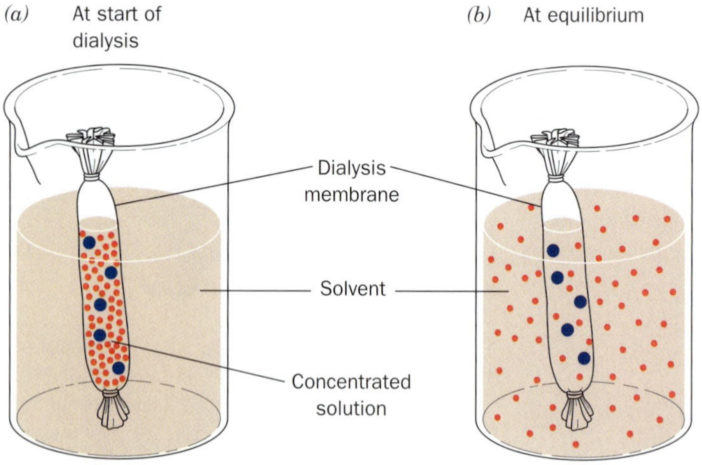

Diffusion of solutes is the basis for the laboratory technique of **dialysis.** In this process, solutes smaller than the pore size of the dialysis membrane freely exchange between the sample and the bulk solution until equilibrium is reached (Fig. 2-14). Larger substances cannot cross the membrane and remain where they are. Dialysis is particularly useful for separating large molecules, such as proteins or nucleic acids, from smaller molecules. Because small soluble particles move freely between the sample and the surrounding medium, dialysis can be repeated several times to replace the sample medium with another solution.

Figure 2-14 Dialysis. (*a*) A concentrated solution is separated from a large volume of solvent by a dialysis membrane (shown here as a tube knotted at both ends). Only small molecules can diffuse through the pores in the membrane. (*b*) At equilibrium, the concentrations of small molecules are nearly the same on either side of the membrane, whereas the macromolecules remain inside the dialysis bag.

The tendency for solutes to diffuse from an area of high concentration to an area of low concentration (i.e., down a concentration gradient) is thermodynamically favored because it is accompanied by an increase in entropy. However, diffusion, the result of the random wandering of particles, does have constraints (see Box 2-1). For example, increasing the viscosity

B O X   2 - 1

Perspectives in Biochemistry

Diffusion Rates and the Sizes of Organisms

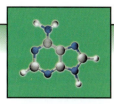

Tetrahymena

Small organisms such as the protozoan *Tetrahymena* depend on diffusion to acquire nutrients and dispose of wastes. But diffusion is efficient only when the ratio of surface area (where exchange with the surroundings takes place) to internal volume is relatively large. For any solid object, surface area is proportional to the square of the length, but volume is proportional to the cube of length.

Solid	Surface Area	Volume	Surface Area/Volume
Cube	$6l^2$	l^3	$6/l$
Cylinder	$2\pi rh + 2\pi r^2$	$\pi r^2 h$	$2/r + 2/h$
Sphere	$4\pi r^2$	$4/3\ \pi r^3$	$3/r$

Therefore, a large object has less surface area relative to its volume than a small object of the same shape. It is hardly surprising that although the sizes of organisms range over some eight orders of magnitude, the diameter of the basic unit of life—the cell—varies by only about 1000-fold. Clearly, a certain level of organization has been retained through evolutionary history.

In multicellular organisms, specialization of form and function helps overcome the surface-to-volume ratio problem. Structures where gas exchange and nutrient absorption take place, such as gills, lungs, and intestines, are characterized microscopically by tremendous surface areas. For example, the human lungs, with an air capacity of ~6 L divided among ~30 million alveoli (small sacs), have a surface area of 50–100 m². Nevertheless, the rates of diffusion of nutrients and wastes are high enough to support life only in organisms that are no larger than about 1 mm in their smallest dimension. Larger organisms require a circulatory system such as the bloodstream to actively transport substances from sites where they are produced or absorbed to sites where they are utilized or eliminated.

[Photo courtesy of Rupal Thazhath and Jacek Gaertig, University of Georgia.]

of the medium decreases the rate of diffusion. Furthermore, since diffusion is a random process, the rate of diffusion varies with the square of the distance diffused. Thus, if a particle, on average, diffuses 1 cm in 1 s, the same particle would require 100 s to diffuse 10 cm.

2 Chemical Properties of Water

Water is not just a passive component of the cell or extracellular environment. By virtue of its physical properties, water defines the solubilities of other substances. Similarly, water's chemical properties determine the behavior of other molecules in solution.

A Ionization of Water

Water is a neutral molecule with a very slight tendency to ionize. We express this ionization as

$$H_2O \rightleftharpoons H^+ + OH^-$$

There is actually no such thing as a free proton (H^+) in solution. Rather, the proton is associated with a water molecule as a **hydronium ion,** H_3O^+. The association of a proton with a cluster of water molecules also gives rise to structures with the formulas $H_5O_2^+$, $H_7O_3^+$, and so on. For simplicity, however, we often represent these ions by H^+.

The proton of a hydronium ion can jump rapidly to another water molecule and then to another (Fig. 2-15). For this reason, the mobilities of H^+ and OH^- ions in solution are much higher than for other ions, which must

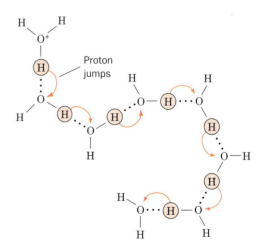

Figure 2-15 Proton jumping. Proton jumps occur more rapidly than direct molecular migration, accounting for the observed high ionic mobilities of hydronium ions (and hydroxide ions) in aqueous solutions.

move through the bulk water carrying their waters of hydration. **Proton jumping** *is also responsible for the observation that acid–base reactions are among the fastest reactions that take place in aqueous solution.*

The ionization (dissociation) of water is described by an equilibrium expression in which the concentration of the parent substance is in the denominator and the concentrations of the dissociated products are in the numerator:

$$K = \frac{[H^+][OH^-]}{[H_2O]} \qquad [2\text{-}1]$$

K is the **dissociation constant** (here and throughout the text, quantities in square brackets symbolize the molar concentrations of the indicated substances, which in many cases are only negligibly different from their activities; Section 1-4D). Because the concentration of the undissociated H_2O ($[H_2O]$) is so much larger than the concentrations of its component ions, it can be considered constant and incorporated into K to yield an expression for the ionization of water,

$$K_w = [H^+][OH^-] \qquad [2\text{-}2]$$

The value of K_w, the ionization constant of water, is 10^{-14} at 25°C.

Pure water must contain equimolar amounts of H^+ and OH^-, so $[H^+] = [OH^-] = (K_w)^{1/2} = 10^{-7}$ M. Since $[H^+]$ and $[OH^-]$ are reciprocally related by Eq. 2-2, when $[H^+]$ is greater than 10^{-7} M, $[OH^-]$ must be correspondingly less and vice versa. Solutions with $[H^+] = 10^{-7}$ M are said to be **neutral,** those with $[H^+] > 10^{-7}$ M are said to be **acidic,** and those with $[H^+] < 10^{-7}$ M are said to be **basic.** Most physiological solutions have hydrogen ion concentrations near neutrality. For example, human blood is normally slightly basic with $[H^+] = 4.0 \times 10^{-8}$ M.

The values of $[H^+]$ for most solutions are inconveniently small and thus impractical to compare. A more practical quantity, which was devised in 1909 by Søren Sørenson, is known as the **pH:**

$$pH = -\log[H^+] = \log \frac{1}{[H^+]} \qquad [2\text{-}3]$$

The higher the pH, the lower is the H^+ concentration; the lower the pH, the higher is the H^+ concentration (Fig. 2-16). The pH of pure water is 7.0,

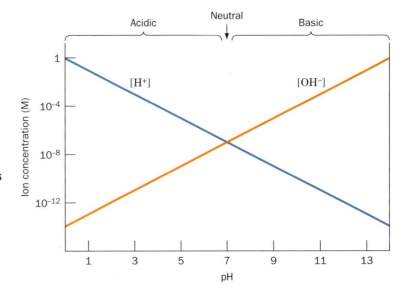

Figure 2-16 Relationship of pH and the concentrations of H$^+$ and OH$^-$ in water.   Because the product of [H$^+$] and [OH$^-$] is a constant (10^{-14}), [H$^+$] and [OH$^-$] are reciprocally related. Solutions with relatively more H$^+$ are acidic (pH < 7), solutions with relatively more OH$^-$ are basic (pH > 7), and solutions in which [H$^+$] = [OH$^-$] = 10^{-7} M are neutral (pH = 7). Note the logarithmic scale for ion concentration.

whereas acidic solutions have pH < 7.0 and basic solutions have pH > 7.0 (see Sample Calculation 2-1). Note that solutions that differ by one pH unit differ in $[H^+]$ by a factor of 10. The pH values of some common substances are given in Table 2-3.

B Acid–Base Chemistry

H^+ and OH^- ions derived from water are fundamental to the biochemical reactions we shall encounter later in this book. Biological molecules, such as proteins and nucleic acids, have numerous functional groups that act as acids or bases, for example, carboxyl and amino groups. These molecules influence the pH of the surrounding aqueous medium, and their structures and reactivities are in turn influenced by the ambient pH. An appreciation of acid–base chemistry is therefore essential for understanding the biological roles of many molecules.

An Acid Can Donate a Proton. According to a definition coined in the 1880s by Svante Arrhenius, an **acid** is a substance that can donate a proton, and a **base** is a substance that can donate a hydroxide ion. This definition is rather limited. For example, it does not account for the observation that NH_3, which lacks an OH group, exhibits basic properties. In a more general definition, which was formulated in 1923 by Johannes Brønsted and Thomas Lowry, *an acid is a substance that can donate a proton* (as in the Arrhenius definition), *and a base is a substance that can accept a proton.* Under the Brønsted–Lowry definition, an acid–base reaction can be written as

$$HA + H_2O \rightleftharpoons H_3O^+ + A^-$$

An acid (HA) reacts with a base (H_2O) to form the **conjugate base** of the acid (A^-) and the **conjugate acid** of the base (H_3O^+). Accordingly, the acetate ion (CH_3COO^-) is the conjugate base of acetic acid (CH_3COOH), and the ammonium ion (NH_4^+) is the conjugate acid of ammonia (NH_3). The acid–base reaction is frequently abbreviated

$$HA \rightleftharpoons H^+ + A^-$$

with the participation of H_2O implied. An alternative expression for a basic substance B is

$$HB^+ \rightleftharpoons H^+ + B$$

The Strength of an Acid Is Specified by Its Dissociation Constant. The equilibrium constant for an acid–base reaction is expressed as a dissociation constant with the concentrations of the "reactants" in the denominator and the concentrations of the "products" in the numerator:

$$K = \frac{[H_3O^+][A^-]}{[HA][H_2O]} \qquad [2\text{-}4]$$

In dilute solutions, the water concentration is essentially constant, 55.5 M ($1000 \text{ g} \cdot L^{-1}/18.015 \text{ g} \cdot mol^{-1} = 55.5$ M). Therefore, the term $[H_2O]$ is customarily combined with the dissociation constant, which then takes the form

$$K_a = K[H_2O] = \frac{[H^+][A^-]}{[HA]} \qquad [2\text{-}5]$$

For brevity, however, we shall henceforth omit the subscript "*a*."

SAMPLE CALCULATION 2-1

10^{-4} moles of H^+ (as HCl) are added to 1 liter of pure water. Determine the final pH of the solution.

Pure water has a pH of 7, so its $[H^+] = 10^{-7}$ M. The added H^+ has a concentration of 10^{-4} M, which overwhelms the $[H^+]$ already present.

The total $[H^+]$ is therefore 1.0×10^{-4} M, so that the pH is equal to $-\log[H^+] = -\log(1.0 \times 10^{-4}) = 4$

Table 2-3 pH Values of Some Common Substances

Substance	pH
1 M NaOH	14
Household ammonia	12
Seawater	8
Blood	7.4
Milk	7
Saliva	6.6
Tomato juice	4.4
Vinegar	3
Gastric juice	1.5
1 M HCl	0

Table 2-4 Dissociation Constants and pK Values at 25°C of Some Acids

Acid	K	pK
Oxalic acid	5.37×10^{-2}	1.27 (pK_1)
H_3PO_4	7.08×10^{-3}	2.15 (pK_1)
Formic acid	1.78×10^{-4}	3.75
Succinic acid	6.17×10^{-5}	4.21 (pK_1)
Oxalate$^-$	5.37×10^{-5}	4.27 (pK_2)
Acetic acid	1.74×10^{-5}	4.76
Succinate$^-$	2.29×10^{-6}	5.64 (pK_2)
2-(N-Morpholino)ethanesulfonic acid (MES)	8.13×10^{-7}	6.09
H_2CO_3	4.47×10^{-7}	6.35 (pK_1)^a
Piperazine-N,N'-bis(2-ethanesulfonic acid) (PIPES)	1.74×10^{-7}	6.76
$H_2PO_4^-$	1.51×10^{-7}	6.82 (pK_2)
3-(N-Morpholino)propanesulfonic acid (MOPS)	7.08×10^{-8}	7.15
N-2-Hydroxyethylpiperazine-N'-2-ethanesulfonic acid (HEPES)	3.39×10^{-8}	7.47
Tris(hydroxymethyl)aminomethane (Tris)	8.32×10^{-9}	8.08
NH_4^+	5.62×10^{-10}	9.25
Glycine (amino group)	1.66×10^{-10}	9.78
HCO_3^-	4.68×10^{-11}	10.33 (pK_2)
Piperidine	7.58×10^{-12}	11.12
HPO_4^{2-}	4.17×10^{-13}	12.38 (pK_3)

Source: Dawson, R.M.C., Elliott, D.C., Elliott, W.H., and Jones, K.M., *Data for Biochemical Research* (3rd ed.), pp. 424–425, Oxford Science Publications (1986); *and* Good, N.E., Winget, G.D., Winter, W., Connolly, T.N., Izawa, S., and Singh, R.M.M., *Biochemistry* **5,** 467 (1966).
^{a}The pK for the overall reaction $CO_2 + H_2O \rightleftharpoons H_2CO_3 \rightleftharpoons H^+ + HCO_3^-$; see Box 2-2.

The dissociation constants of some common acids are listed in Table 2-4. Because acid dissociation constants, like [H$^+$] values, are sometimes cumbersome to work with, they are transformed to **pK** values by the formula

$$pK = -\log K \qquad [2\text{-}6]$$

which is analogous to Eq. 2-3.

Acids can be classified according to their relative strengths, that is, their abilities to transfer a proton to water. The acids listed in Table 2-4 are known as **weak acids** because they are only partially ionized in aqueous solution ($K < 1$). Many of the so-called mineral acids, such as $HClO_4$, HNO_3, and HCl, are **strong acids** ($K \gg 1$). Since strong acids rapidly transfer all their protons to H_2O, *the strongest acid that can stably exist in aqueous solutions is H_3O^+*. Likewise, *there can be no stronger base in aqueous solutions than OH^-*. Virtually all the acid–base reactions that occur in biological systems involve H_3O^+ (and OH^-) and weak acids (and their conjugate bases).

The pH of a Solution Is Determined by the Relative Concentrations of Acids and Bases. The relationship between the pH of a solution and the concentrations of an acid and its conjugate base is easily derived. Equation 2-5 can be rearranged to

$$[H^+] = K\frac{[HA]}{[A^-]} \qquad [2\text{-}7]$$

Taking the negative log of each term (and letting pH $= -\log[H^+]$; Eq. 2-3) gives

$$pH = -\log K + \log \frac{[A^-]}{[HA]} \qquad [2\text{-}8]$$

SAMPLE CALCULATION 2-2

Calculate the pH of a 2 L solution containing 10 mL of 5 M acetic acid and 10 mL of 1 M sodium acetate.

First, calculate the concentrations of the acid and conjugate base, expressing all concentrations in units of moles per liter.

Acetic acid: (0.01 L)(5 M)/(2 L)
                = 0.025 M

Sodium acetate: (0.01 L)(1 M)/(2 L)
                = 0.005 M

Substitute the concentrations of the acid and conjugate base into the Henderson–Hasselbalch equation. Find the pK for acetic acid in Table 2-4.

pH = pK + log([acetate]/[acetic acid])

pH = 4.76 + log(0.005/0.025)

pH = 4.76 − 0.70

pH = 4.06

Substituting pK for $-\log K$ (Eq. 2-6) yields

$$pH = pK + \log \frac{[A^-]}{[HA]} \qquad [2\text{-}9]$$

This relationship is known as the **Henderson–Hasselbalch equation.** *When the molar concentrations of an acid (HA) and its conjugate base (A^-) are equal, log ($[A^-]/[HA]$) = log 1 = 0, and the pH of the solution is numerically equivalent to the pK of the acid.* The Henderson–Hasselbalch equation is invaluable for calculating, for example, the pH of a solution containing a known quantity of a weak acid and its conjugate base (see Sample Calculation 2-2). However, since the Henderson–Hasselbalch equation does not account for the ionization of water itself, it is not useful for calculating the pH of solutions of strong acids or bases. For example, in a 1 M solution of a strong acid, $[H^+]$ = 1 M and the pH is 0. In a 1 M solution of a strong base, $[OH^-]$ = 1 M, so $[H^+]$ = $[OH^-]/K_w$ = 1×1^{-14} M so the pH is 14.

C Buffers

Adding a 0.01-mL droplet of 1 M HCl to 1 L of pure water changes the pH of the water from 7 to 5, which represents a 100-fold increase in $[H^+]$. Such a huge change in pH would be intolerable to most biological systems, since even small changes in pH can dramatically affect the structures and functions of biological molecules. Maintaining a relatively constant pH is therefore of paramount importance for living systems. To understand how this is possible, consider the titration of a weak acid with a strong base.

Figure 2-17 shows how the pH values of solutions of acetic acid, $H_2PO_4^-$, and ammonium ion (NH_4^+) vary as OH^- is added. **Titration curves** such as these can be constructed from experimental observation or by using the Henderson–Hasselbalch equation to calculate points along the curve (see Sample Calculation 2-3). When OH^- reacts with HA, the products are A^- and water:

$$HA + OH^- \rightleftharpoons A^- + H_2O$$

Several details about the titration curves in Fig. 2-17 should be noted:

1. The curves have similar shapes but are shifted vertically along the pH axis.

2. The pH at the midpoint of each titration is numerically equivalent to the pK of its corresponding acid; at this point, $[HA] = [A^-]$.

3. The slope of each titration curve is much lower near its midpoint than near its wings. This indicates that *when $[HA] \approx [A^-]$, the pH of the solution is relatively insensitive to the addition of strong base or strong acid.* Such a solution, which is known as an acid–base **buffer,** resists pH changes because small amounts of added H^+ or OH^- react with A^- or HA, respectively, without greatly changing the value of log($[A^-]/[HA]$).

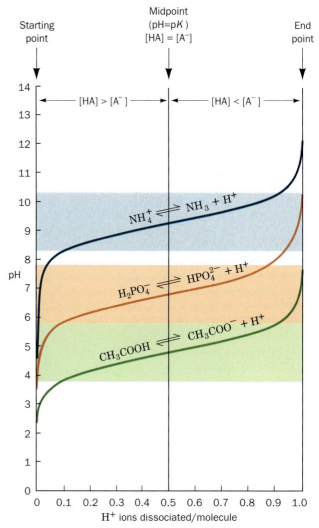

Figure 2-17 Titration curves for acetic acid, phosphate, and ammonia. At the starting point, the acid form predominates. As strong base (e.g., NaOH) is added, the acid is converted to its conjugate base. At the midpoint of the titration, where pH = pK, the concentrations of the acid and the conjugate base are equal. At the end point (equivalence point), the conjugate base predominates, and the total amount of OH^- that has been added is equivalent to the amount of acid that was present at the starting point. The shaded bands indicate the pH ranges over which the corresponding solution can function as a buffer.

🔁 **See the Animated Figures.**

SAMPLE CALCULATION 2-3

Calculate the pH of a 1 L solution containing 0.1 M formic acid and 0.1 M sodium formate before and after the addition of 1 mL of 5 M NaOH. How much would the pH change if the NaOH were added to 1 L of pure water?

According to Table 2-4, the pK for formic acid is 3.75. Since [formate] = [formic acid], the $\log([A^-]/[HA])$ term of the Henderson–Hasselbalch equation is 0 and pH = pK = 3.75. The addition of 1 mL of NaOH does not significantly change the volume of the solution, so the [NaOH] is (0.001 L)(5 M)/(1 L) = 0.005 M.

Since NaOH is a strong base, it completely dissociates, and [OH$^-$] = [NaOH] = 0.005 M. This OH$^-$ reacts with formic acid to produce formate and H$_2$O. Consequently, the concentration of formic acid decreases and the concentration of formate increases by 0.005 M.

The new formic acid concentration is 0.1 M − 0.005 M = 0.095 M, and the new formate concentration is 0.1 M + 0.005 M = 0.105 M. Substituting these values into the Henderson–Hasselbalch equation gives

$$pH = pK + \log([\text{formate}]/[\text{formic acid}])$$
$$pH = 3.75 + \log(0.105/0.095)$$
$$pH = 3.75 + 0.04$$
$$pH = 3.79$$

In the absence of the formic acid buffering system, the [H$^+$] and therefore the pH can be calculated directly from K_w. Since K_w = [H$^+$][OH$^-$] = 10^{-14},

$$[H^+] = \frac{10^{-14}}{[\text{OH}^-]} = \frac{10^{-14}}{(0.005)} = 2 \times 10^{-12} \text{ M}$$
$$pH = -\log[H^+] = -\log(2 \times 10^{-12}) = 11.7$$

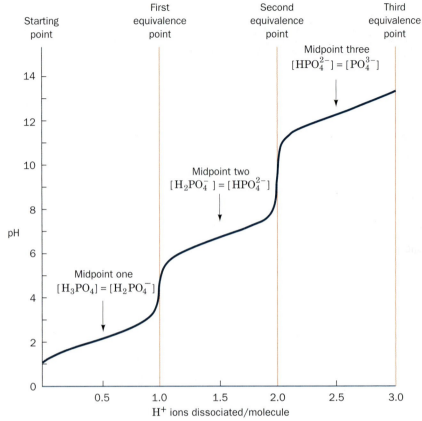

Figure 2-18 Titration of a polyprotic acid. The first and second equivalence points for titration of H$_3$PO$_4$ occur at the steepest parts of the curve. The pH at the midpoint of each stage provides the pK value of the corresponding ionization. 🔊 **See the Animated Figures.**

Substances that can lose more than one proton, or undergo more than one ionization, such as H$_3$PO$_4$ or H$_2$CO$_3$, are known as **polyprotic acids.** The titration curves of such molecules, as illustrated in Fig. 2-18 for H$_3$PO$_4$, are more complicated than the titration curves of monoprotic acids such as acetic acid. A polyprotic acid has multiple pK values, one for each ionization step. H$_3$PO$_4$, for example, has three dissociation constants because the ionic charge resulting from one proton dissociation electrostatically inhibits further proton dissociation, thereby increasing the values of the corresponding pKs. Similarly, a molecule with more than one ionizable group has a discrete pK value for each group. In a biomolecule that contains numerous ionizable groups with different pK values, the many dissociation events may yield a titration curve without any clear "plateaus."

Biological fluids, both intracellular and extracellular, are heavily buffered. For example, the pH of the blood in healthy individuals is closely controlled at pH 7.4 (see Box 2-2). The phosphate and bicarbonate ions in most biological fluids are important buffering agents because they have pKs in this range (Table 2-4). Moreover, many biological molecules, such as proteins and some lipids, as well as numerous small organic molecules, bear multiple acid–base groups that are effective buffer components in the physiological pH range.

The concept that the properties of biological molecules vary with the acidity of the solution in which they are dissolved was not fully appreciated before the twentieth century. Many early biochemical experiments

BOX 2-2

Biochemistry in Health and Disease

The Blood Buffering System

Bicarbonate is the most significant buffer compound in human blood; other buffering agents, including proteins and organic acids, are present at much lower concentrations. The buffering capacity of blood depends primarily on two equilibria: (1) between gaseous CO_2 dissolved in the blood and carbonic acid formed by the reaction

$$CO_2 + H_2O \rightleftharpoons H_2CO_3$$

and (2) between carbonic acid and bicarbonate formed by the dissociation of H^+

$$H_2CO_3 \rightleftharpoons H^+ + HCO_3^-$$

The overall pK for these two sequential reactions is 6.35. (The further dissociation of HCO_3^- to CO_3^{2-}, p$K = 10.33$, is not significant at physiological pH.)

When the pH of the blood falls due to metabolic production of H^+, the bicarbonate–carbonic acid equilibrium shifts toward more carbonic acid. At the same time, carbonic acid loses water to become CO_2, which is then expired in the lungs as gaseous CO_2. Conversely, when the blood pH rises, relatively more HCO_3^- forms. Breathing is adjusted so that increased amounts of CO_2 in the lungs can be reintroduced into the blood for conversion to carbonic acid. In this manner, a near-constant hydrogen ion concentration can be maintained. The kidneys also play a role in acid–base balance by excreting HCO_3^- and NH_4^+.

Disturbances in the blood buffer system can lead to conditions known as **acidosis,** with a pH as low as 7.1, or **alkalosis,** with a pH as high as 7.6. (Deviations of less than 0.05 pH unit from the "normal" value of 7.4 are not significant.) For example, obstructive lung diseases that prevent efficient expiration of CO_2 can cause respiratory acidosis. Hyperventilation accelerates the loss of CO_2 and causes respiratory alkalosis. Overproduction of organic acids from dietary precursors or sudden surges in lactic acid levels during exercise can lead to metabolic acidosis.

Acid–base imbalances are best alleviated by correcting the underlying physiological problem. In the short term, acidosis is commonly treated by administering $NaHCO_3$ intravenously. Alkalosis is more difficult to treat. Metabolic alkalosis sometimes responds to KCl or NaCl (the additional Cl^- helps minimize the secretion of H^+ by the kidneys), and respiratory alkalosis can be ameliorated by breathing an atmosphere enriched in CO_2.

---

were undertaken without controlling the acidity of the sample, so the results were often poorly reproducible. Nowadays, biochemical preparations are routinely buffered to simulate the properties of naturally occurring biological fluids. A number of synthetic compounds have been developed for use as buffers; some of these are included in Table 2-4. The **buffering capacity** of these weak acids (their ability to resist pH changes on addition of acid or base) is maximal when pH = pK. It is helpful to remember that a weak acid is in its useful buffer range within one pH unit of its pK (e.g., the shaded regions of Fig. 2-17). Above this range, where the ratio $[A^-]/[HA] > 10$, the pH of the solution changes rapidly with added strong base. A buffer is similarly impotent with the addition of strong acid when its pK exceeds the pH by more than one unit.

In the laboratory, the desired pH of the buffered solution determines which buffering compound is selected. Typically, the acid form of the compound and one of its soluble salts are dissolved in the (nearly equal) molar ratio necessary to provide the desired pH and, with the aid of a pH meter, the resulting solution is fine-tuned by titration with strong acid or base.

SUMMARY

1. Water is essential for all living organisms.
2. Water molecules can form hydrogen bonds with other molecules because they have two H atoms that can be donated and two unshared electron pairs that can act as acceptors.
3. Liquid water is an irregular network of water molecules that each form up to four hydrogen bonds with neighboring water molecules.
4. Hydrophilic substances such as ions and polar molecules dissolve readily in water.
5. The hydrophobic effect is the tendency of water to minimize its contacts with nonpolar substances.
6. Water molecules move from regions of high concentration to regions of low concentration by osmosis; solutes move from regions of high concentration to regions of low concentration by diffusion.
7. Water ionizes to H^+ (which represents the hydronium ion, H_3O^+) and OH^-.
8. The concentration of H^+ in solutions is expressed as a pH value; in acidic solutions pH < 7, in basic solutions pH > 7, and in neutral solutions pH = 7.
9. Acids can donate protons and bases can accept protons. The strength of an acid is expressed as its pK.
10. The Henderson–Hasselbalch equation relates the pH of a solution to the pK and concentrations of an acid and its conjugate base.
11. Buffered solutions resist changes in pH within about one pH unit of the pK of the buffering species.

 ## REFERENCES

Gerstein, M. and Levitt, M., Simulating water and the molecules of life, *Sci Am.* **279**(11), 101–105 (1998). [Describes the structure of water and how water interacts with other molecules.]

Good, N.E., Winget, G.D., Winter, W., Connolly, T.N., Izawa, S., and Singh, R.M.M., Hydrogen ion buffers for biological research, *Biochemistry* **5**, 467–477 (1966). [A classic paper on laboratory buffers.]

Halperin, M.L. and Goldstein, M.B., *Fluid, Electrolyte, and Acid–Base Physiology: A Problem-Based Approach* (3rd ed.), W.B. Saunders (1999). [Includes extensive problem sets with explanations of basic science as well as clinical effects of acid–base disorders.]

Jeffrey, G.A. and Saenger, W., *Hydrogen Bonding in Biological Structures,* Chaps. 1, 2, and 21, Springer (1994). [Reviews hydrogen bond chemistry and its importance in small molecules and macromolecules.]

Kropman, M.F. and Bakker, H.J., Dynamics of water molecules, *Science* **291**, 2118–2120 (2001). [Describes how water molecules in solvation shells move more slowly than bulk water molecules.]

Segel, I.H., *Biochemical Calculations* (2nd ed.), Chap. 1, Wiley (1976). [An intermediate level discussion of acid–base equilibria with worked-out problems.]

Tanford, C., *The Hydrophobic Effect: Formation of Micelles and Biological Membranes* (2nd ed.), Chaps. 5 and 6, Wiley–Interscience (1980). [Discusses the structures of water and micelles.]

 ## KEY TERMS

polar
hydrogen bond
van der Waals distance
van der Waals forces
London dispersion forces
hydrophilic
hydrophobic
solvation
hydration
waters of hydration
hydrophobic effect
amphiphilic
amphipathic
micelle
bilayer
colligative properties
osmosis
osmotic pressure
diffusion
dialysis
hydronium ion
proton jumping
dissociation constant
K_w
neutral solution
acidic solution
basic solution
pH
acid
base
conjugate base
conjugate acid
pK
weak acid
strong acid
Henderson–Hasselbalch equation
titration curve
buffer
polyprotic acid
acidosis
alkalosis
buffering capacity

 ## STUDY EXERCISES

1. Compare hydrogen bonding in ice and hydrogen bonding in liquid water.
2. Explain why polar substances dissolve in water while nonpolar substances do not.
3. Describe the contribution of entropy to the hydrophobic effect.
4. Why do amphiphiles form micelles in water?
5. How does osmosis differ from diffusion?
6. Compare the Arrhenius and Brønsted–Lowry definitions of acids and bases.
7. Explain why a 1 M solution of HCl has a pH of 0.

PROBLEMS

1. Identify the potential hydrogen bond donors and acceptors in the following molecules:

   (a)

   (b)

   (c) $H-\underset{\underset{NH_3^+}{|}}{C}-CH_2-OH$

2. Occasionally, a C—H group can form a hydrogen bond. Why would such a group be more likely to be a hydrogen bond donor group when the C is next to N?

3. Rank the water solubility of the following compounds:

   (a) $H_3C-CH_2-O-CH_3$

   (b) $H_3C-\underset{\underset{O}{\|}}{C}-NH_2$

   (c) $H_2N-\underset{\underset{O}{\|}}{C}-NH_2$

   (d) $H_3C-CH_2-CH_3$

   (e) $H_3C-CH_2-\underset{\underset{O}{\|}}{C}H$

4. Where would the following substances partition in water containing palmitic acid micelles? (a) $^+H_3N-CH_2-COO^-$, (b) $^+H_3N-(CH_2)_{11}-COO^-$, (c) $H_3C-(CH_2)_{11}-COO^-$.

5. Explain why water forms nearly spherical droplets on the surface of a freshly waxed car. Why doesn't water bead on a clean windshield?

6. Describe what happens when a dialysis bag containing pure water is suspended in a beaker of seawater. What would happen if the dialysis membrane were permeable to water but not solutes?

7. Compare the surface-to-volume ratios for a bacterium (length 3 μm, diameter 0.5 μm) and a fish (length 30 cm, diameter 5 cm). Assume each organism is shaped like a cylinder.

8. Draw the structures of the conjugate bases of the following acids:

   (a) $\underset{\underset{\underset{COOH}{|}}{\overset{\overset{CH}{|}}{\underset{HC}{\|}}}}{COO^-}$

   (b) $H-\underset{\underset{NH_3^+}{|}}{\overset{\overset{COOH}{|}}{C}}-H$

   (c) $H-\underset{\underset{NH_3^+}{|}}{\overset{\overset{COO^-}{|}}{C}}-H$

   (d) $H-\underset{\underset{NH_3^+}{|}}{\overset{\overset{COO^-}{|}}{C}}-CH_2-COOH$

9. Indicate the ionic species that predominates at pH 4, 8, and 11 for (a) ammonia and (b) phosphoric acid.

10. Calculate the pH of a 200 mL solution of pure water to which has been added 50 mL of 1 mM HCl.

11. Calculate the pH of a 1 L solution containing (a) 10 mL of 5 M NaOH, (b) 10 mL of 100 mM glycine and 20 mL of 5 M HCl, and (c) 10 mL of 2 M acetic acid and 5 g of sodium acetate (formula weight 82 g·mol^{-1}).

12. Calculate the standard free energy change for the dissociation of HEPES.

13. A solution is made by mixing 50 mL of 2.0 M K_2HPO_4 and 25 mL of 2.0 M KH_2PO_4. The solution is diluted to a final volume of 200 mL. What is the pH of the final solution?

14. What is the pK of the weak acid HA if a solution containing 0.1 M HA and 0.2 M A$^-$ has a pH of 6.5?

15. How many grams of sodium succinate (formula weight 140 g·mol^{-1}) and disodium succinate (formula weight 162 g·mol^{-1}) must be added to 1 L of water to produce a solution with pH 6.0 and a total solute concentration of 50 mM?

16. Estimate the volume of a solution of 5 M NaOH that must be added to adjust the pH from 4 to 9 in 100 mL of a 100 mM solution of phosphoric acid.

17. (a) Would phosphoric acid or succinic acid be a better buffer at pH 5? (b) Would ammonia or piperidine be a better buffer at pH 9? (c) Would HEPES or Tris be a better buffer at pH 7.5?

A DNA molecule consists of two strands that wind around a central axis, shown here as a glowing wire. A complete set of genetic instructions contains just four types of monomeric units. The sequence in which these monomers are linked constitutes a form of biological information that can be efficiently decoded and faithfully copied. [Illustrations, Irving Geis. Image from the Irving Geis Collection/Howard Hughes Medical Institute. Rights owned by HHMI. Reproduction by permission only.]

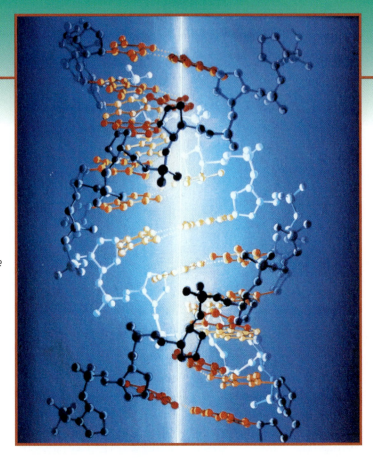

Nucleotides, Nucleic Acids, and Genetic Information

Despite obvious differences in lifestyle and macroscopic appearance, organisms exhibit striking similarity at the molecular level. The structures and metabolic activities of all cells rely on a common set of molecules that includes amino acids, carbohydrates, lipids, and nucleotides, as well as their polymeric forms. Each type of compound can be described in terms of its chemical makeup, its interactions with other molecules, and its physiological function. We begin our survey of biomolecules with a discussion of the **nucleotides** and their polymers, the **nucleic acids.**

Nucleotides are involved in nearly every facet of cellular life. Specifically, they participate in oxidation–reduction reactions, energy transfer, intracellular signaling, and biosynthetic reactions. Their polymers, the nucleic acids DNA and RNA, are the primary players in the storage and decoding of genetic information. Nucleotides and nucleic acids also perform structural and catalytic roles in cells. No other class of molecules participates in such varied functions or so many functions that are essential for life.

Evolutionists postulate that the appearance of nucleotides permitted the evolution of organisms that could harvest and store energy from their surroundings and, most importantly, could make copies of themselves. Although the chemical and biological details of early life-forms are the subject of speculation, it is incontrovertible that life as we know it is inextricably linked to the chemistry of nucleotides and nucleic acids.

In this chapter, we briefly examine the structures of nucleotides and the nucleic acids DNA and RNA. We also consider how the chemistry of these molecules allows them to carry biological information in the form of a sequence of nucleotides. This information is expressed by the transcription of a segment of DNA to yield RNA, which is then translated to form protein. Because a cell's structure and function ultimately depend on its genetic makeup, we discuss how genomic DNA sequences provide information about evolution, metabolism, and disease. Finally, we consider some of the techniques used in manipulating DNA in the laboratory. In later chapters, we examine in greater detail the participation of nucleotides and nucleic acids in metabolism and the storage and expression of genetic information.

1 Nucleotides

Nucleotides are ubiquitous molecules with considerable structural diversity. *There are eight common varieties of nucleotides, each composed of a nitrogenous base linked to a sugar to which at least one phosphate group is also attached.* The bases of nucleotides are planar, aromatic, heterocyclic molecules that are structural derivatives of either **purine** or **pyrimidine** (although they are not synthesized *in vivo* from either of these organic compounds).

Purine   **Pyrimidine**

The most common purines are **adenine (A)** and **guanine (G),** and the major pyrimidines are **cytosine (C), uracil (U),** and **thymine (T).** The purines form

Table 3-1 Names and Abbreviations of Nucleic Acid Bases, Nucleosides, and Nucleotides

Base Formula	Base (X = H)	Nucleoside (X = ribose[a])	Nucleotide[b] (X = ribose phosphate[a])
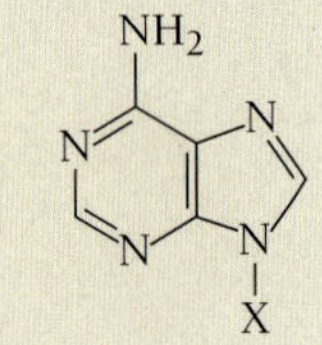	Adenine Ade A	Adenosine Ado A	Adenylic acid Adenosine monophosphate AMP
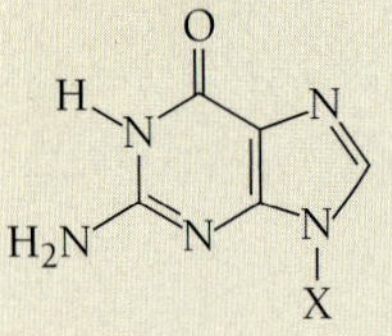	Guanine Gua G	Guanosine Guo G	Guanylic acid Guanosine monophosphate GMP
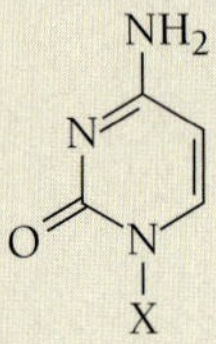	Cytosine Cyt C	Cytidine Cyd C	Cytidylic acid Cytidine monophosphate CMP
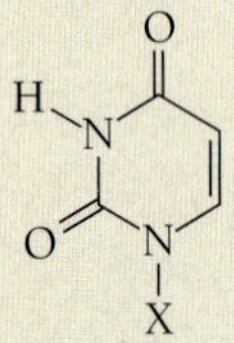	Uracil Ura U	Uridine Urd U	Uridylic acid Uridine monophosphate UMP
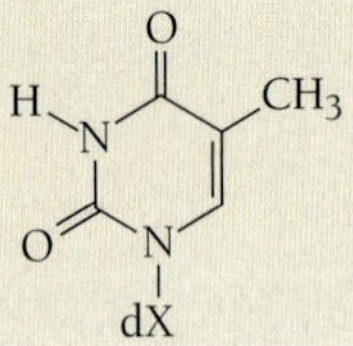	Thymine Thy T	Deoxythymidine dThd dT	Deoxythymidylic acid Deoxythymidine monophosphate dTMP

[a] The presence of a 2'-deoxyribose unit in place of ribose, as occurs in DNA, is implied by the prefixes "deoxy" or "d." For example, the deoxynucleoside of adenine is deoxyadenosine or dA. However, for thymine-containing residues, which rarely occur in RNA, the prefix is redundant and may be dropped. The presence of a ribose unit may be explicitly implied by the prefix "ribo."

[b] The position of the phosphate group in a nucleotide may be explicitly specified as in, for example, 3'-AMP and 5'-GMP.

bonds to a five-carbon sugar (a pentose) via their N9 atoms, whereas pyrimidines do so through their N1 atoms (Table 3-1).

In **ribonucleotides,** the pentose is **ribose,** while in **deoxyribonucleotides** (or just **deoxynucleotides**), the sugar is **2'-deoxyribose** (i.e., the carbon at position 2' lacks a hydroxyl group).

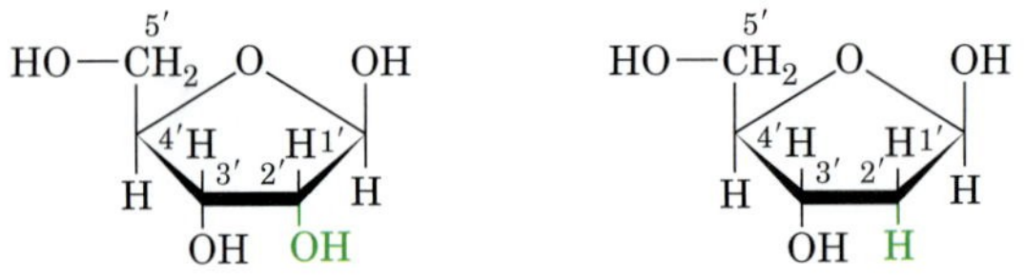

Ribose   **Deoxyribose**

Note that the "primed" numbers refer to the atoms of the pentose; "unprimed" numbers refer to the atoms of the nitrogenous base.

In a ribonucleotide or a deoxyribonucleotide, one or more phosphate groups is bonded to atom C3' or atom C5' of the pentose to form a 3'-

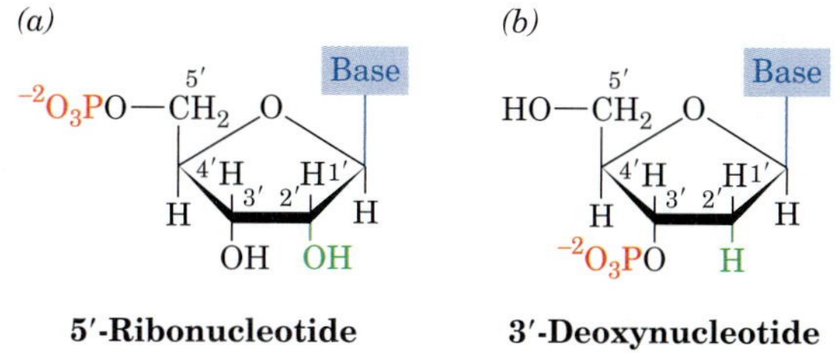

(a)   *(b)*

5'-Ribonucleotide   3'-Deoxynucleotide

Figure 3-1 Chemical structures of nucleotides. *(a)* A 5'-ribonucleotide and *(b)* a 3'-deoxynucleotide. The purine or pyrimidine base is linked to C1' of the pentose and at least one phosphate *(red)* is also attached. A nucleoside consists only of a base and a pentose.

nucleotide or a 5'-nucleotide, respectively (Fig. 3-1). When the phosphate group is absent, the compound is known as a **nucleoside.** A 5'-nucleotide can therefore be called a nucleoside-5'-phosphate. Nucleotides most commonly contain one to three phosphate groups at the C5' position and are called nucleoside monophosphates, diphosphates, and triphosphates.

The structures, names, and abbreviations of the common bases, nucleosides, and nucleotides are given in Table 3-1. Ribonucleotides are found in **RNA (ribonucleic acid),** whereas deoxynucleotides are found in **DNA (deoxyribonucleic acid).** Adenine, guanine, and cytosine are found as both ribonucleotides and deoxynucleotides (accounting for six of the eight common nucleotides), but uracil is found primarily as a ribonucleotide and thymine as a deoxynucleotide. Free nucleotides, which are anionic, are usually associated with the counterion Mg^{2+} in cells.

ATP and Nucleotide Derivatives. The bulk of the nucleotides in any cell are found in polymeric forms, as either DNA or RNA, whose primary functions are information storage and transfer. However, free nucleotides and nucleotide derivatives perform an enormous variety of metabolic functions not related to the management of genetic information.

Perhaps the best known nucleotide is **adenosine triphosphate (ATP),** a nucleotide containing adenine, ribose, and a triphosphate group. ATP is often mistakenly referred to as an energy-storage molecule, but it is more accurately termed an energy carrier or energy transfer agent. The process of photosynthesis or the breakdown of metabolic fuels such as carbohydrates and fatty acids leads to the formation of ATP from **adenosine diphosphate (ADP):**

Adenosine diphosphate (ADP)   Adenosine triphosphate (ATP)

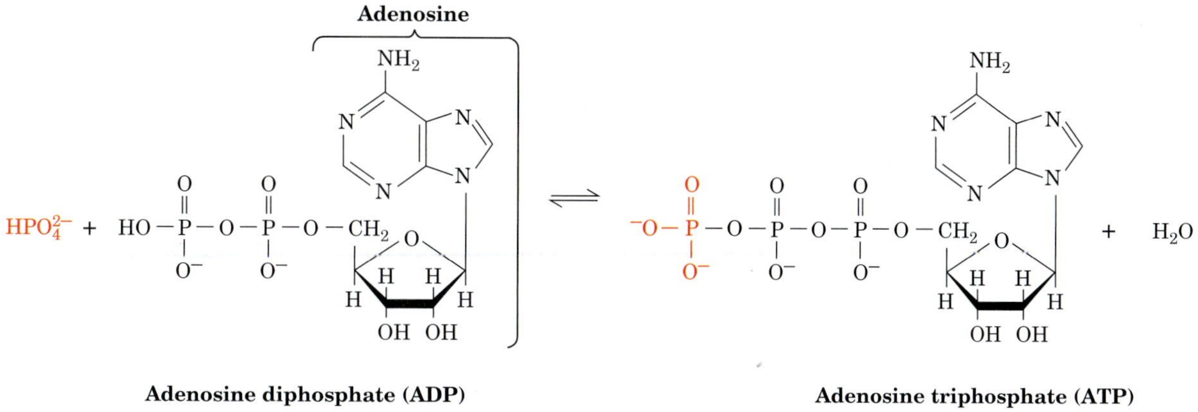

ATP diffuses throughout the cell to provide energy for other cellular work, such as biosynthetic reactions, ion transport, and cell movement. The chem-

ical potential energy of ATP is made available when it transfers one (or two) of its phosphate groups to another molecule. This process can be represented by the reverse of the preceding reaction, namely, the hydrolysis of ATP to ADP. (As we shall see in later chapters, the interconversion of ATP and ADP in the cell is not freely reversible, and free phosphate groups are seldom released directly from ATP.) The degree to which ATP participates in routine cellular activities is illustrated by calculations indicating that while the concentration of cellular ATP is relatively moderate (~5 mM), the average human recycles his or her own weight of ATP each day.

Nucleotide derivatives participate in a wide variety of metabolic processes. For example, starch synthesis in plants proceeds by repeated additions of glucose units donated by ADP–glucose (Fig. 3-2). Other nucleotide derivatives, as we shall see in later chapters, carry groups that undergo oxidation–reduction reactions. The attached group, which may be a small molecule such as glucose (Fig. 3-2) or even another nucleotide, is typically linked to the nucleotide through a mono- or diphosphate group.

2 Introduction to Nucleic Acid Structure

Nucleotides can be joined to each other to form the polymers that are familiar to us as RNA and DNA. The nucleic acids are chains of nucleotides whose phosphates bridge the 3′ and 5′ positions of neighboring ribose units (Fig. 3-3). The phosphates of these **polynucleotides** are acidic, so at physiological pH, nucleic acids are polyanions.

The linkage between individual nucleotides is known as a **phosphodiester bond,** so named because the phosphate is esterified to two ribose units. Each nucleotide that has been incorporated into the polynucleotide is known as a **nucleotide residue.** The terminal residue whose C5′ is not linked to another nucleotide is called the **5′ end,** and the terminal residue whose C3′ is not linked to another nucleotide is called the **3′ end.** By convention, the sequence of nucleotide residues in a nucleic acid is written, left to right, from the 5′ end to the 3′ end.

The properties of a polymer such as a nucleic acid may be very different from the properties of the individual units, or **monomers,** before polymerization. As the size of the polymer increases from **dimer, trimer, tetramer,** and so on through **oligomer** (Greek: *oligo,* few), physical properties such as charge and solubility may change. In addition, *a polymer of*

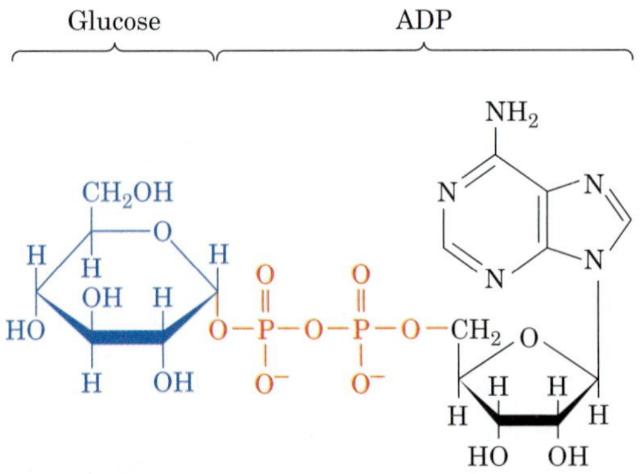

Figure 3-2 ADP–glucose.   In this nucleotide derivative, glucose (*blue*) is attached to adenosine (*black*) by a diphosphate group (*red*).

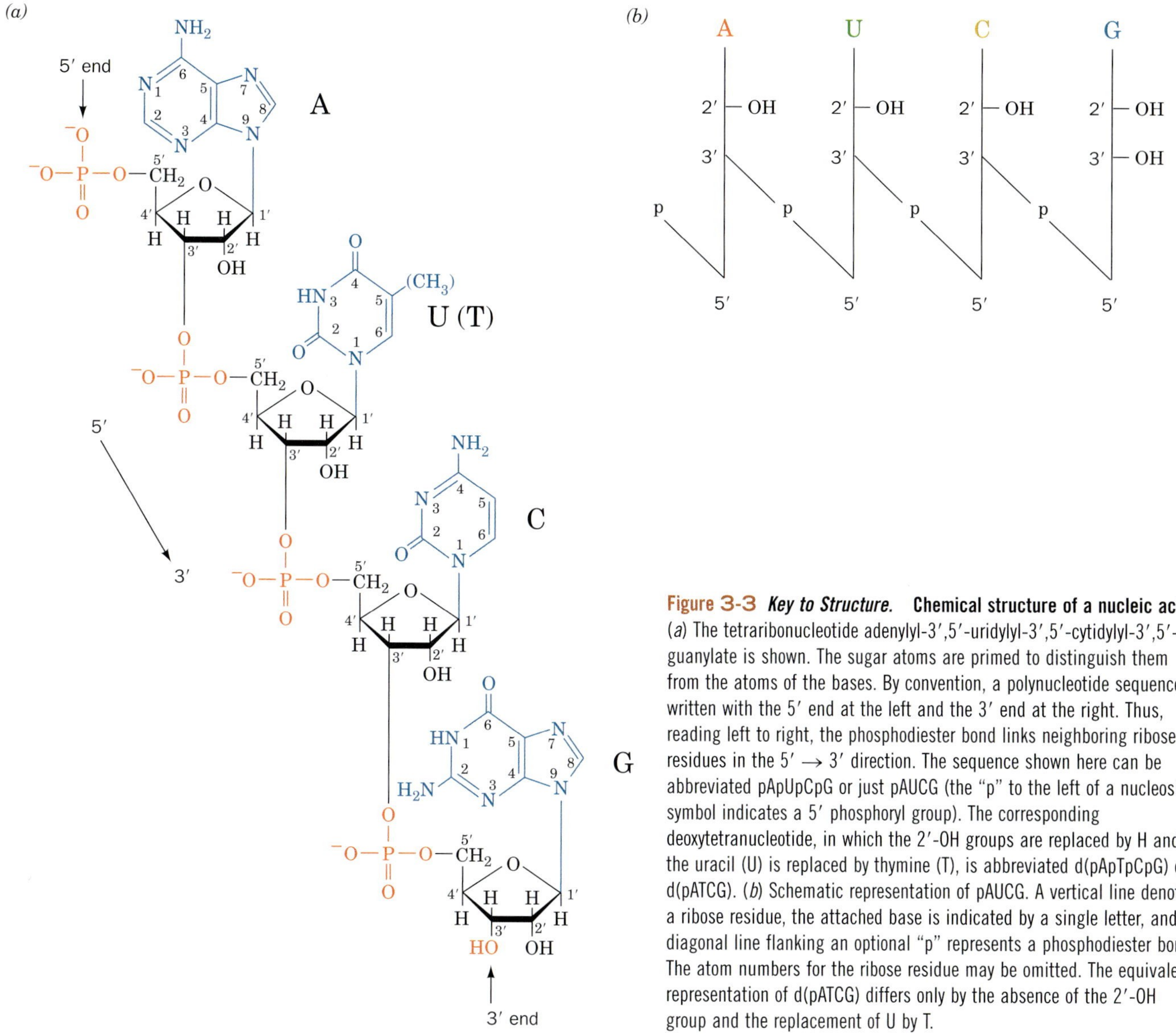

(a)

(b)

Figure 3-3 *Key to Structure.* Chemical structure of a nucleic acid. (*a*) The tetraribonucleotide adenylyl-3',5'-uridylyl-3',5'-cytidylyl-3',5'-guanylate is shown. The sugar atoms are primed to distinguish them from the atoms of the bases. By convention, a polynucleotide sequence is written with the 5' end at the left and the 3' end at the right. Thus, reading left to right, the phosphodiester bond links neighboring ribose residues in the 5' → 3' direction. The sequence shown here can be abbreviated pApUpCpG or just pAUCG (the "p" to the left of a nucleoside symbol indicates a 5' phosphoryl group). The corresponding deoxytetranucleotide, in which the 2'-OH groups are replaced by H and the uracil (U) is replaced by thymine (T), is abbreviated d(pApTpCpG) or d(pATCG). (*b*) Schematic representation of pAUCG. A vertical line denotes a ribose residue, the attached base is indicated by a single letter, and a diagonal line flanking an optional "p" represents a phosphodiester bond. The atom numbers for the ribose residue may be omitted. The equivalent representation of d(pATCG) differs only by the absence of the 2'-OH group and the replacement of U by T.

nonidentical residues has a property that its component monomers do not have—namely, it contains information in the form of its sequence of residues.

A The Base Composition of DNA

Although there appear to be no rules governing the nucleotide composition of typical RNA molecules, DNA has equal numbers of adenine and thymine residues (A = T) and equal numbers of guanine and cytosine residues (G = C). These relationships, known as **Chargaff's rules,** were discovered in the late 1940s by Erwin Chargaff, who devised the first reliable quantitative methods for the compositional analysis of DNA.

DNA's base composition varies widely among different organisms. It ranges from ~25 to 75 mol % G + C in different species of bacteria. However, it is more or less constant among related species; for example, in mammals G + C ranges from 39 to 46%. The significance of Chargaff's rules was not immediately appreciated, but we now know that the structural basis for the rules derives from DNA's double-stranded nature.

B The Double Helix

The determination of the structure of DNA by James Watson and Francis Crick in 1953 is often said to mark the birth of modern molecular biology. The **Watson–Crick structure** of DNA not only provided a model of what is arguably the central molecule of life, it also suggested the molecular mechanism of heredity. Watson and Crick's accomplishment, which is ranked as one of science's major intellectual achievements, was based in part on two pieces of evidence in addition to Chargaff's rules: the correct tautomeric forms of the bases and indications that DNA is a helical molecule.

The purine and pyrimidine bases of nucleic acids can assume different tautomeric forms (**tautomers** are easily converted isomers that differ only in hydrogen positions; Fig. 3-4). X-Ray, nuclear magnetic resonance (NMR), and spectroscopic investigations have firmly established that the nucleic acid bases are overwhelmingly in the keto tautomeric forms shown in Fig. 3-3. In 1953, however, this was not generally appreciated. Information about the dominant tautomeric forms was provided by Jerry Donohue, an office mate of Watson and Crick and an expert on the X-ray structures of small organic molecules.

Evidence that DNA is a helical molecule was provided by an X-ray diffraction photograph of a DNA fiber taken by Rosalind Franklin (Fig. 3-5). The appearance of the photograph enabled Crick, an X-ray crystallographer by training, to deduce (a) that DNA is a helical molecule and (b) that its planar aromatic bases form a stack that is parallel to the fiber axis.

The limited structural information, along with Chargaff's rules, provided but few clues to the structure of DNA; Watson and Crick's model sprang mostly from their imaginations and model-building studies. Once the Watson–Crick model had been published, however, its basic simplicity combined with its obvious biological relevance led to its rapid acceptance.

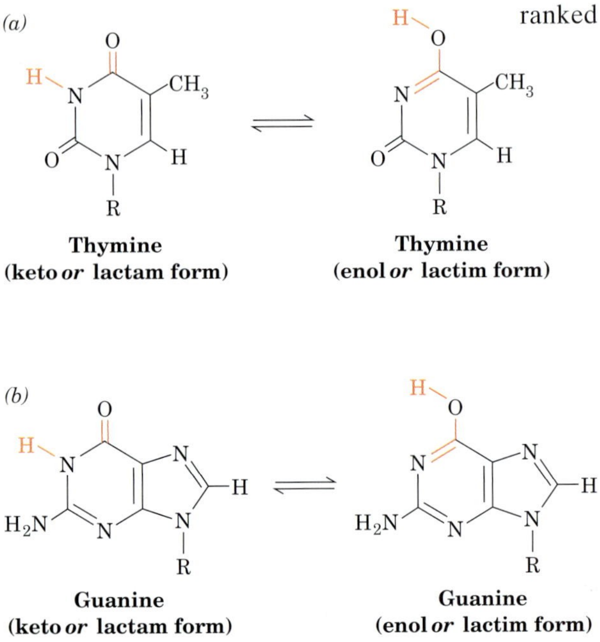

Figure 3-4 Tautomeric forms of bases. Some of the possible tautomeric forms of (*a*) thymine and (*b*) guanine are shown. Cytosine and adenine can undergo similar proton shifts.

(*a*)

Thymine
(keto *or* lactam form)

Thymine
(enol *or* lactim form)

(*b*)

Guanine
(keto *or* lactam form)

Guanine
(enol *or* lactim form)

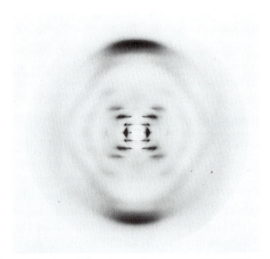

Figure 3-5 An X-ray diffraction photograph of a vertically oriented DNA fiber. This photograph, taken by Rosalind Franklin, provided key evidence for the elucidation of the Watson–Crick structure. The central X-shaped pattern indicates a helix, whereas the heavy black arcs at the top and bottom of the diffraction pattern reveal the spacing of the stacked bases (3.4 Å). [Courtesy of Maurice Wilkins, King's College, London.]

Later investigations have confirmed the general accuracy of the Watson–Crick model, although its details have been modified.

The Watson–Crick model of DNA has the following major features:

1. Two polynucleotide chains wind around a common axis to form a **double helix** (Fig. 3-6).

2. The two strands of DNA are **antiparallel** (run in opposite directions), but each forms a right-handed helix. (The difference between a right-handed and a left-handed helix is shown in Fig. 3-7.)

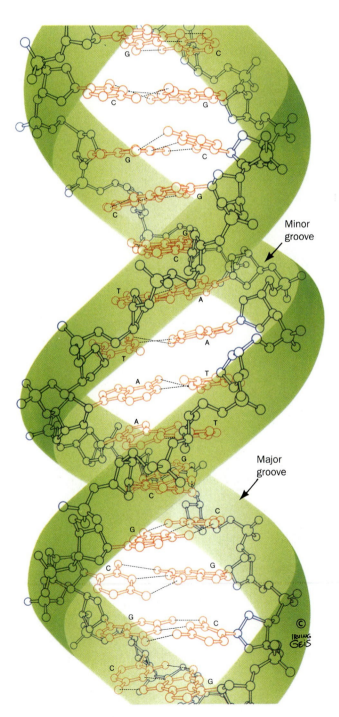

Minor groove

Major groove

Figure 3-6 **Three-dimensional structure of DNA.**   The repeating helix is based on the structure of the self-complementary dodecamer d(CGCGAATTCGCG) determined by Richard Dickerson and Horace Drew. The view in this ball-and-stick model is perpendicular to the helix axis. The sugar–phosphate backbones *(blue, with green ribbon outlines)* wind around the periphery of the molecule. The bases *(red)* form hydrogen-bonded pairs that occupy the core. H atoms have been omitted for clarity. The two strands run in opposite directions. [Illustration, Irving Geis, Image from the Irving Geis Collection/Howard Hughes Medical Institute. Reproduction by permission only.] **See the Interactive Exercises and Kinemage Exercise 2-1.**

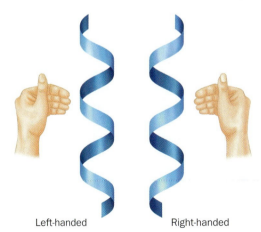

Left-handed   Right-handed

Figure 3-7 **Diagrams of left- and right-handed helices.** In each case, the fingers curl in the direction the helix turns when the thumb points in the direction the helix rises. Note that the handedness is retained when the helices are turned upside down.

3. The bases occupy the core of the helix and sugar–phosphate chains run along the periphery, thereby minimizing the repulsions between charged phosphate groups. The surface of the double helix contains two grooves of unequal width: the **major** and **minor grooves.**

4. Each base is hydrogen bonded to a base in the opposite strand to form a planar **base pair.** The Watson–Crick structure can accommodate only two types of base pairs. Each adenine residue must pair with a thymine residue and vice versa, and each guanine residue must pair with a cytosine residue and vice versa (Fig. 3-8). These hydrogen-bonding interactions, a phenomenon known as **complementary base pairing,** result in the specific association of the two chains of the double helix.

The Watson–Crick structure can accommodate any sequence of bases on one polynucleotide strand if the opposite strand has the complementary

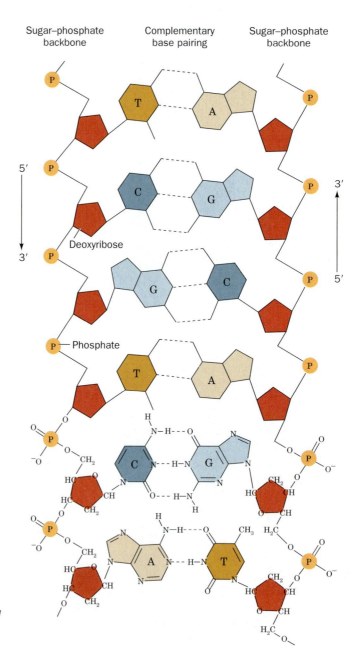

Figure 3-8 Complementary strands of DNA. Two polynucleotide chains associate by base pairing to form double-stranded DNA. A pairs with T, and G pairs with C by forming specific hydrogen bonds. ✎ **See Kinemage Exercise 2-2.**

base sequence. This immediately accounts for Chargaff's rules. More importantly, it suggests that *each DNA strand can act as a **template** for the synthesis of its complementary strand and hence that hereditary information is encoded in the sequence of bases on either strand.*

Most DNA molecules are extremely large, in keeping with their role as the repository of a cell's genetic information. Of course, an organism's **genome,** its unique DNA content, may be allocated among several **chromosomes** (Greek: *chromos,* color + *soma,* body), each of which contains a separate DNA molecule. Note that many organisms are **diploid;** that is, they contain two equivalent sets of chromosomes, one from each parent. Their content of unique **(haploid)** DNA is half their total DNA. For example, humans are diploid organisms that carry 46 chromosomes per cell; the haploid number is therefore 23.

Because of their great lengths, DNA molecules are described in terms of the number of base pairs **(bp)** or thousands of base pairs **(kilobase pairs, or kb).** Naturally occurring DNAs vary in length from ~5 kb in small DNA-containing viruses to well over 250,000 kb in the largest mammalian chromosomes. Although DNA molecules are long and relatively stiff, they are not completely rigid. We shall see later that the DNA double helix forms coils and loops when it is packaged inside the cell. Furthermore, depending on the nucleotide sequence, DNA may adopt slightly different helical conformations. Finally, in the presence of other cellular components, the DNA may bend sharply or the two strands may partially unwind. (We consider the structure of DNA in greater detail in Chapter 23.)

C Single-Stranded Nucleic Acids

Single-stranded DNA is rare, occurring mainly as the hereditary material of certain viruses. In contrast, RNA occurs primarily as single strands, which usually form compact structures rather than loose extended chains (double-stranded RNA is the hereditary material of certain viruses). An RNA strand—which is identical to a DNA strand except for the presence of 2'-OH groups and the substitution of uracil for thymine—can base-pair with a complementary strand of RNA or DNA. As expected, A pairs with U (or T in DNA), and G with C. Base pairing often occurs intramolecularly, giving rise to **stem–loop** structures (Fig. 3-9) or, when loops interact with each other, to more complex structures.

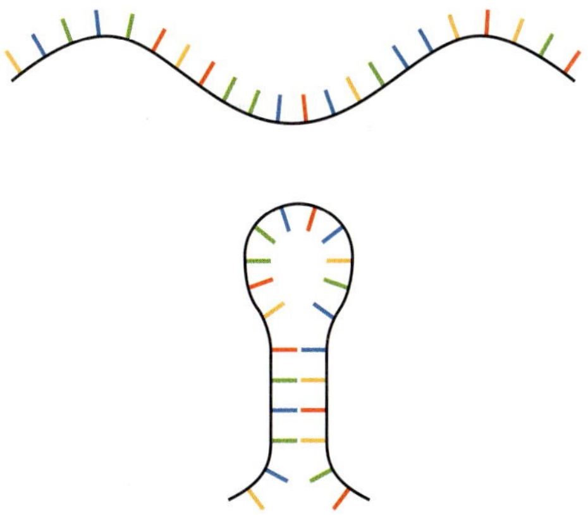

The intricate structures that can potentially be adopted by single-stranded RNA molecules provide additional evidence that RNA can do more than just store and transmit genetic information. Numerous investigations have found that certain RNA molecules can specifically bind small organic molecules and can catalyze reactions involving those molecules. These findings provide substantial support for theories that *many of the processes essential for life began through the chemical versatility of small polynucleotides* (a situation known as the RNA world). We will further explore RNA structure and function in Section 23-2E.

Figure 3-9 Formation of a stem–loop structure.   Base pairing between complementary sequences within an RNA strand allows the polynucleotide to fold back on itself.

3 Overview of Nucleic Acid Function

DNA is the carrier of genetic information in all cells and in many viruses. Yet a period of over 75 years passed from the time the laws of inheritance were discovered by Gregor Mendel until the biological role of DNA was elucidated. Even now, many details of how genetic information is expressed and transmitted to future generations are still unclear.

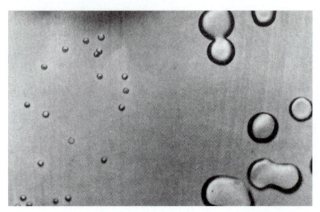

Figure 3-10   **Transformed pneumococci.**   The large colonies are virulent pneumococci that resulted from the transformation of nonpathogenic pneumococci (smaller colonies) by DNA extracted from the virulent strain. We now know that this DNA contained a gene that was defective in the nonpathogenic strain. [From Avery, O.T., MacLeod, C.M., and McCarty, M., *J. Exp. Med.* **79**, 153 (1944). Copyright © 1944 by Rockefeller University Press.]

Mendel's work with garden peas led him to postulate that an individual plant contains a pair of factors (which we now call **genes**), one inherited from each parent. But Mendel's theory of inheritance, reported in 1866, was almost universally ignored by his contemporaries, whose knowledge of anatomy and physiology provided no basis for its understanding. Eventually, genes were hypothesized to be part of chromosomes, and the pace of genetic research greatly accelerated.

A   DNA Carries Genetic Information

Until the 1940s, it was generally assumed that genes were made of protein, since proteins were the only biochemical entities that, at the time, seemed complex enough to serve as agents of inheritance. Nucleic acids, which had first been isolated in 1869 by Friedrich Miescher, were believed to have monotonously repeating nucleotide sequences and were therefore unlikely candidates for transmitting genetic information.

It took the efforts of Oswald Avery, Colin MacLeod, and Maclyn McCarty to demonstrate that DNA carries genetic information. Their experiments, completed in 1944, showed that DNA—not protein—extracted from a virulent (pathogenic) strain of the bacterium *Diplococcus pneumoniae* was the substance that **transformed** (permanently changed) a nonpathogenic strain of the organism to the virulent strain (Fig. 3-10). Avery's discovery was initially greeted with skepticism, but it influenced Erwin Chargaff, whose rules (Section 3-2A) led to subsequent models of the structure and function of DNA.

The double-stranded, or duplex, nature of DNA facilitates its **replication.** When a cell divides, each DNA strand acts as a template for the assembly of its complementary strand (Fig. 3-11). Consequently, every progeny cell contains a complete DNA molecule (or a set of DNA molecules in organisms whose genomes contain more than one chromosome). Each DNA molecule consists of one parental strand and one daughter strand. Daughter strands are synthesized by the stepwise polymerization of nucleotides that specifically pair with bases on the parental strands. The mechanism of replication, while straightforward in principle, is exceedingly complex in the cell, requiring a multitude of cellular factors to proceed with fidelity and efficiency, as we shall see in Chapter 24.

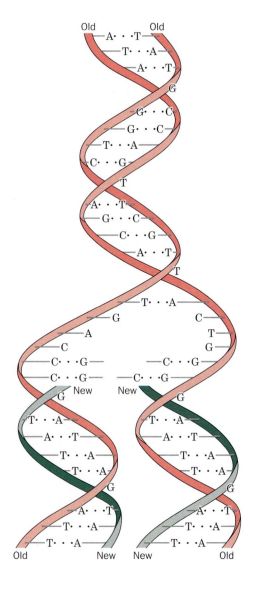

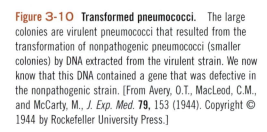

Figure 3-11   **DNA replication.**   Each strand of parental DNA (*red*) acts as a template for the synthesis of a complementary daughter strand (*green*). Thus, the resulting double-stranded molecules are identical.

B Genes Direct Protein Synthesis

The question of how sequences of nucleotides control the characteristics of organisms took some time to be answered. In experiments with the mold *Neurospora crassa* in the 1940s, George Beadle and Edward Tatum found that *there is a specific connection between genes and enzymes, the one gene–one enzyme theory.* Beadle and Tatum showed that mutant varieties of *Neurospora* that were generated by irradiation with X-rays required additional nutrients in order to grow. Presumably, the offspring of the radiation-damaged cells lacked the specific enzymes necessary to synthesize those nutrients.

The link between DNA and enzymes (nearly all of which are proteins) is RNA. *The DNA of a gene is **transcribed** to produce an RNA molecule that is complementary to the DNA. The RNA sequence is then **translated** into the corresponding sequence of amino acids to form a protein* (Fig. 3-12). These transfers of biological information are summarized in the so-called **central dogma of molecular biology** formulated by Crick in 1958 (Fig. 3-13).

Just as the daughter strands of DNA are synthesized from free deoxynucleoside triphosphates that pair with bases in the parent DNA strand, RNA strands are synthesized from free ribonucleoside triphosphates that pair with the complementary bases in one DNA strand of a gene (transcription is described in greater detail in Chapter 25). The RNA that corresponds to a protein-coding gene (called **messenger RNA,** or **mRNA**) makes its way to a **ribosome,** an organelle that is itself composed largely of RNA **(ribosomal RNA,** or **rRNA).** At the ribosome, each set of three nucleotides in the mRNA pairs with three complementary nucleotides in a small RNA molecule—a **transfer RNA,** or **tRNA** (Fig. 3-14). Attached to each tRNA molecule is its corresponding amino acid. The ribosome catalyzes the joining of amino acids, which are the monomeric units of proteins (protein synthesis is described in detail in Chapter 26). Amino acids

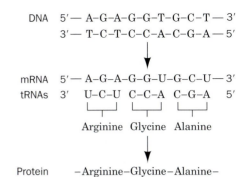

Figure 3-12 Transcription and translation. One strand of DNA directs the synthesis of messenger RNA (mRNA). The base sequence of the transcribed RNA is complementary to that of the DNA strand. The message is translated when transfer RNA (tRNA) molecules align with the mRNA by complementary base pairing between three-nucleotide segments known as **codons.** Each tRNA carries a specific amino acid. These amino acids are covalently joined to form a protein. Thus, the sequence of bases in DNA specifies the sequence of amino acids in a protein.

> **See Guided Exploration 1**
>
> **Overview of transcription and translation.**

Figure 3-13 The central dogma of molecular biology. Solid arrows indicate the types of information transfers that occur in all cells: DNA directs its own replication to produce new DNA molecules; DNA is transcribed into RNA; RNA is translated into protein. The dashed lines represent information transfers that occur only in certain organisms.

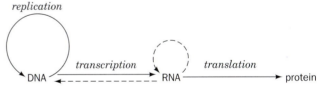

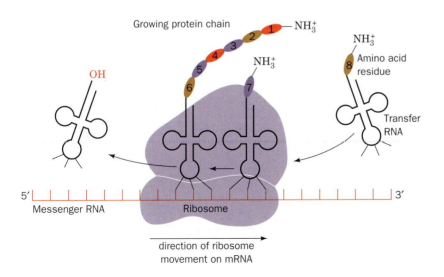

direction of ribosome movement on mRNA

Figure 3-14 Translation. tRNA molecules with their attached amino acids bind to complementary three-nucleotide sequences (codons) on mRNA. The ribosome facilitates the alignment of the tRNA and the mRNA and catalyzes the joining of amino acids to produce a protein chain. When a new amino acid is added, the preceding tRNA is ejected, and the ribosome proceeds along the mRNA.

are added to the growing protein chain according to the order in which the tRNA molecules bind to the mRNA. Since the nucleotide sequence of the mRNA in turn reflects the sequences of nucleotides in the gene, DNA directs the synthesis of proteins. It follows that alterations to the genetic material of an organism (**mutations**) may manifest themselves as proteins with altered structures and functions.

4 Nucleic Acid Sequencing

Much of our current understanding of protein structure and function rests squarely on information gleaned not from the proteins themselves, but indirectly from their genes. *The ability to determine the sequence of nucleotides in nucleic acids has made it possible to deduce the amino acid sequences of their encoded proteins and, to some extent, the structures and functions of those proteins. Nucleic acid sequencing has also revealed information about the regulation of genes.* Portions of genes that are not actually transcribed into RNA nevertheless may influence how often a gene is transcribed and translated, that is, **expressed.** Moreover, efforts to elucidate the sequences in hitherto unmapped regions of DNA have led to the discovery of new genes and new regulatory elements. *Once in hand, a nucleic acid sequence can be duplicated, modified, and expressed, making it possible to study proteins that could not otherwise be obtained in useful quantities.* In this section, we describe how nucleic acids are sequenced and what information the sequences may reveal. In the following section, we discuss the manipulation of purified nucleic acid sequences for various purposes.

The overall strategy for sequencing any polymer of nonidentical units is

1. Cleave the polymer into specific fragments that are small enough to be fully sequenced.
2. Determine the sequence of residues in each fragment.
3. Determine the order of the fragments in the original polymer by repeating the preceding steps using a degradation procedure that yields a set of fragments that overlap the cleavage points in the first step.

The first efforts to sequence RNA used nonspecific enzymes to generate relatively small fragments whose nucleotide composition was then determined by partial digestion with an enzyme that selectively removed nucleotides from one end or the other (Fig. 3-15). Sequencing RNA in this manner was tedious and time-consuming. Using such methods, it took Robert Holley seven years to determine the sequence of a 76-residue tRNA molecule.

After 1975, dramatic progress was made in nucleic acid sequencing technology. The advances were made possible by the discovery of enzymes that could cleave DNA at specific sites and by the development of rapid sequencing techniques for DNA. The advent of modern molecular cloning techniques (Section 3-5) also made it possible to produce sufficient quantities of specific DNA to be sequenced. These cloning techniques are necessary because most specific DNA sequences are normally present in a genome in only a single copy.

```
GCACUUGA
         | snake venom
         | phosphodiesterase
         ↓
GCACUUGA
GCACUUG
GCACUU
GCACU
GCAC
GCA
GC    + Mononucleotides
```

Figure 3-15 Determining the sequence of an oligonucleotide using nonspecific enzymes. The oligonucleotide is partially digested with snake venom phosphodiesterase, which breaks the phosphodiester bonds between nucleotide residues, starting at the 3′ end of the oligonucleotide. The result is a mixture of fragments of all lengths, which are then separated. Comparing the base composition of a pair of fragments that differ in length by one nucleotide establishes the identity of the 3′-terminal nucleotide in the larger fragment. Analysis of each pair of fragments reveals the sequence of the original oligonucleotide.

A Restriction Endonucleases

Many bacteria are able to resist infection by **bacteriophages** (viruses that are specific for bacteria) by virtue of a **restriction–modification system.** The bacterium modifies certain nucleotides in specific sequences of its own

DNA by adding a methyl (—CH$_3$) group in a reaction catalyzed by a **modification methylase.** A **restriction endonuclease,** which recognizes the same nucleotide sequence as does the methylase, cleaves any DNA that has not been modified on at least one of its two strands. (An **endonuclease** cleaves a nucleic acid within the polynucleotide strand; an **exonuclease** cleaves a nucleic acid by removing one of its terminal residues.) This system destroys foreign (phage) DNA containing a recognition site that has not been modified by methylation. The host DNA is always at least half methylated, because although the daughter strand is not methylated until shortly after it is synthesized, the parental strand to which it is paired is already modified (and thus protects both strands of the DNA from cleavage by the restriction enzyme).

Type II restriction endonucleases are particularly useful in the laboratory. These enzymes cleave DNA within the four- to eight-base sequence that is recognized by their corresponding modification methylase. (Type I and Type III restriction endonucleases cleave DNA at sites other than their recognition sequences.) Over 3000 Type II restriction enzymes with over 200 different recognition sequences have been characterized. Some of the more widely used restriction enzymes are listed in Table 3-2. A restriction enzyme is named by the first letter of the genus and the first two letters of the species of the bacterium that produced it, followed by its serotype or strain designation, if any, and a roman numeral if the bacterium contains more than one type of restriction enzyme. For example, *Eco*RI is produced by *E. coli* strain RY13.

Interestingly, most Type II restriction endonucleases recognize and cleave palindromic DNA sequences. A **palindrome** is a word or phrase that reads the same forward or backward. Two examples are "refer" and "Madam, I'm Adam." In a palindromic DNA segment, the sequence of nu-

Table 3-2   Recognition and Cleavage Sites of Some Restriction Enzymes

Enzyme	Recognition Sequence[a]	Microorganism
*Alu*I	AG↓CT	*Arthrobacter luteus*
*Bam*HI	G↓GATCC	*Bacillus amyloliquefaciens* H
*Bgl*I	GCCNNNNN↓NGGC	*Bacillus globigii*
*Bgl*II	A↓GATCT	*Bacillus globigii*
*Eco*RI	G↓AATTC	*Escherichia coli* RY13
*Eco*RII	↓CC(^A_T)GG	*Escherichia coli* R245
*Eco*RV	GAT↓ATC	*Escherichia coli* J62 pLG74
*Hae*II	RGCGC↓Y	*Haemophilus aegyptius*
*Hae*III	GG↓CC	*Haemophilus aegyptius*
*Hind*III	A↓AGCTT	*Haemophilus influenzae* R$_d$
*Hpa*II	C↓CGG	*Haemophilus parainfluenzae*
*Msp*I	C↓CGG	*Moraxella* species
*Pst*I	CTGCA↓G	*Providencia stuartii* 164
*Pvu*II	CAG↓CTG	*Proteus vulgaris*
*Sal*I	G↓TCGAC	*Streptomyces albus* G
*Taq*I	T↓CGA	*Thermus aquaticus*
*Xho*I	C↓TCGAG	*Xanthomonas holcicola*

[a]The recognition sequence is abbreviated so that only one strand, reading 5′ to 3′, is given. The cleavage site is represented by an arrow (↓). R, Y, and N represent a purine nucleotide, a pyrimidine nucleotide, and any nucleotide, respectively.

Source: Roberts, R.J. and Macelis, D., REBASE—the restriction enzyme database, http://rebase.neb.com.

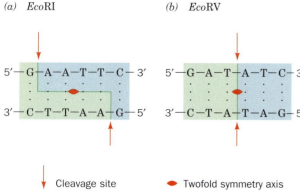

(a) *Eco*RI

(b) *Eco*RV

↓ Cleavage site   ⬤ Twofold symmetry axis

Figure 3-16 Restriction sites. The recognition sequences for Type II restriction endonucleases are palindromes, sequences with a twofold axis of symmetry. (*a*) Recognition site for *Eco*RI, which generates DNA fragments with sticky ends. (*b*) Recognition site for *Eco*RV, which generates blunt-ended fragments.

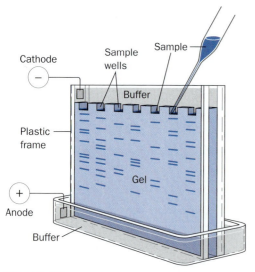

Figure 3-17 Apparatus for gel electrophoresis. Samples are applied in slots at the top of the gel and electrophoresed in parallel lanes. Negatively charged molecules such as DNA migrate through the gel matrix toward the anode in response to an applied electric field. Because smaller molecules move faster, the molecules in each lane are separated according to size. Following electrophoresis, the separated molecules may be visualized by staining, fluorescence, or a radiographic technique.

cleotides is the same in each strand, and the segment is said to have twofold symmetry (Fig. 3-16). Most restriction enzymes cleave the two strands of DNA at positions that are staggered, producing DNA fragments with complementary single-strand extensions. Restriction fragments with such **sticky ends** can associate by base-pairing with other restriction fragments generated by the same restriction enzyme. Some restriction endonucleases cleave the two strands of DNA at the symmetry axis to yield restriction fragments with fully base-paired **blunt ends.**

B Electrophoresis and Restriction Mapping

Treating a DNA molecule with a restriction endonuclease produces a series of precisely defined fragments that can be separated according to size. **Gel electrophoresis** is commonly used for the separation. In principle, a charged molecule moves in an electric field with a velocity proportional to its overall charge density, size, and shape. For molecules with a relatively homogeneous composition (such as nucleic acids), shape and charge density are constant, so the velocity depends primarily on size. Electrophoresis is carried out in a gel-like matrix, usually made from **agarose** (carbohydrate polymers that form a loose mesh) or **polyacrylamide** (a more rigid cross-linked synthetic polymer). The gel is typically held between two glass plates (Fig. 3-17). The molecules to be separated are applied to one end of the gel, and the molecules move through the pores in the matrix under the influence of the electric field. Smaller molecules move more rapidly through the gel and therefore migrate farther in a given time.

Following electrophoresis, the separated molecules may be visualized in the gel by an appropriate technique, such as addition of a stain that binds tightly to the DNA or by radioactive labeling. Depending on the dimensions of the gel and the visualization technique used, samples containing less than a nanogram of material can be separated and detected by gel electrophoresis. Several samples can be electrophoresed simultaneously. For example, the fragments obtained by digesting a DNA sample with different restriction endonucleases can be visualized side by side (Fig. 3-18). The sizes

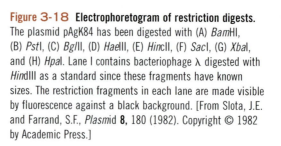

Figure 3-18 Electrophoretogram of restriction digests. The plasmid pAgK84 has been digested with (A) *Bam*HI, (B) *Pst*I, (C) *Bgl*II, (D) *Hae*III, (E) *Hinc*II, (F) *Sac*I, (G) *Xba*I, and (H) *Hpa*I. Lane I contains bacteriophage λ digested with *Hin*dIII as a standard since these fragments have known sizes. The restriction fragments in each lane are made visible by fluorescence against a black background. [From Slota, J.E. and Farrand, S.F., *Plasmid* **8**, 180 (1982). Copyright © 1982 by Academic Press.]

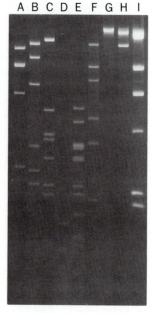

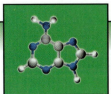

BOX 3-1

Perspectives in Biochemistry

Restriction Fragment Length Polymorphisms

Individuality in humans and other species derives from their high degree of genetic polymorphism. Homologous human chromosomes (e.g., the pairs of maternally and paternally inherited chromosomes) differ in sequence, on average, every 1250 bp. These genetic differences create or eliminate restriction sites. Restriction enzyme digests of homologous chromosomes therefore contain fragments with different lengths; that is, these DNAs exhibit **restriction fragment length polymorphisms (RFLPs).**

The two homologous chromosomal segments shown here differ in the number of restriction sites. An individual with two copies of chromosome I would yield fragments A and B in RFLP analysis; an individual with two copies of chromosome II would yield fragment C. An individual with one of each chromosome would yield fragments A, B, and C.

RFLPs are particularly valuable for diagnosing inherited diseases for which the molecular defect is unknown. If a particular RFLP is closely linked to a defective gene, detecting that RFLP in an individual indicates that there is a high probability that the individual has also inherited the defective gene. For exam-

ple, Huntington's disease, a fatal neurological disorder whose symptoms first appear around age 40, is caused by a dominant genetic defect (Box 27-1). The identification of an RFLP that is closely linked to the defective Huntington's gene has permitted the children of Huntington's disease victims (50% of whom inherit this devastating condition) to make informed decisions in ordering their lives.

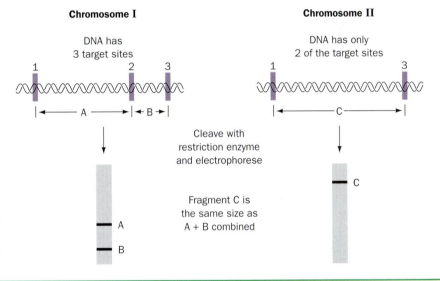

of the various fragments can be determined by comparing their electrophoretic mobilities to the mobilities of fragments of known size.

The results of gels such as the one in Fig. 3-18 can be used to construct a diagram called a **restriction map.** Consider as an example a 4-kb linear DNA molecule cleaved by *Bam*HI, *Hin*dIII, or both and subjected to gel electrophoresis (Fig. 3-19*a*). The sizes of the restriction fragments are used to deduce the positions of the restriction sites in the intact DNA and to construct the restriction map diagrammed in Fig. 3-19*b*. Restriction maps are useful laboratory tools because restriction sites are physical reference points on a DNA molecule. Restriction maps are therefore a convenient framework for locating particular base sequences or genes on a chromosome and for comparing different chromosomes (see Box 3-1).

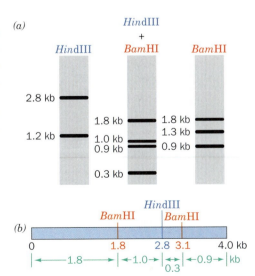

Figure 3-19 Construction of a restriction map. (*a*) The gel electrophoretic pattern of digests of a hypothetical 4-kb DNA molecule with *Hin*dIII, *Bam*HI, and their mixture. The sizes of the various fragments are indicated. (*b*) A restriction map of the DNA resulting from the information in Part *a*. The distance in kb between each restriction site corresponds to the size of the corresponding restriction fragment.

C The Chain-Terminator Method of DNA Sequencing

Here we discuss the most commonly used procedure for sequencing DNA, the **chain-terminator method,** which was devised by Frederick Sanger. The first step in this procedure is to obtain single polynucleotide strands. Complementary DNA strands can be separated by heating, which breaks the hydrogen bonds between bases. Next, polynucleotide fragments that terminate at positions corresponding to each of the four nucleotides are generated. Finally, the fragments are separated and detected.

The Chain-Terminator Method Uses DNA Polymerase.   The chain-terminator method (also called the **dideoxy method**) uses an *E. coli* enzyme to make complementary copies of the single-stranded DNA being sequenced. The enzyme is a fragment of **DNA polymerase I,** one of the enzymes that participates in replication of bacterial DNA (Section 24-2A). Using the single DNA strand as a template, DNA polymerase I assembles the four deoxynucleoside triphosphates (**dNTPs**), dATP, dCTP, dGTP, and dTTP, into a complementary polynucleotide chain that it elongates in the 5′ → 3′ direction (Fig. 3-20).

DNA polymerase I can sequentially add deoxynucleotides only to the 3′ end of a polynucleotide. Hence, replication is initiated in the presence of a short polynucleotide (a **primer**) that is complementary to the 3′ end of the template DNA and thus becomes the 5′ end of the new strand. The primer base-pairs with the template strand, and nucleotides are sequentially added to the 3′ end of the primer. If the DNA being sequenced is a restriction fragment, as it usually is, it begins and ends with a restriction site. The primer can therefore be a short DNA segment with the sequence of this restriction site.

DNA Synthesis Terminates after Specific Bases.   In the chain-terminator technique (Fig. 3-21), the DNA to be sequenced is incubated with DNA polymerase I, a suitable primer, and the four dNTP **substrates** (reactants in enzymatic reactions) for the polymerization reaction. The reaction mixture also includes a "tagged" compound, either one of the dNTPs or the primer.

See Guided Exploration 2

DNA sequence determination by the chain-terminator method.

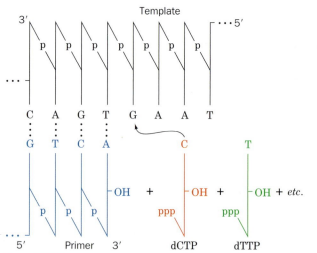

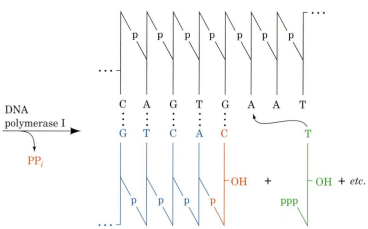

Figure 3-20 Action of DNA polymerase I.   Using a single DNA strand as a template, the enzyme elongates the primer by stepwise addition of complementary nucleotides. Incoming nucleotides pair with bases on the template strand and are joined to the growing polynucleotide strand in the 5′ → 3′ direction. The polymerase-catalyzed reaction requires a free 3′-OH group on the growing strand. **Pyrophosphate** ($P_2O_7^{4-}$; PP_i) is released with each nucleotide addition.

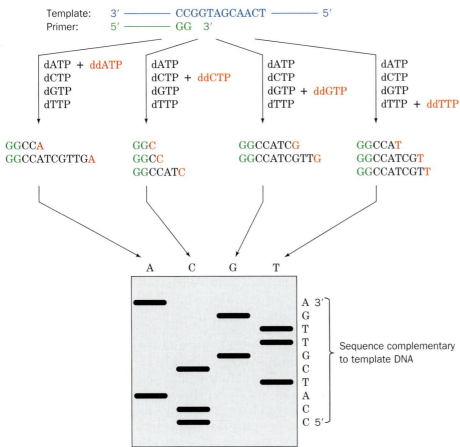

Figure 3-21 The chain-terminator (dideoxy) method of DNA sequencing. Each of the four reaction mixtures includes the single-stranded DNA to be sequenced (the template), a primer, the four deoxynucleoside triphosphates (represented as dATP, etc.), and one of the four dideoxynucleoside triphosphates (ddATP, etc.). Extension of the primer by the action of DNA polymerase generates stretches of DNA terminating with a dideoxynucleotide. Gel electrophoresis in parallel lanes of the fragments from the four reaction mixtures yields a set of polynucleotides whose 3′ terminal residues are known. The sequence obtained by "reading" from the smallest fragment to the largest (i.e., from the bottom to the top of the gel) is complementary to the sequence of the template DNA.

The tag, which may be a radioactive isotope (e.g., ^{32}P) or a fluorescent label, permits the products of the polymerase reaction to be easily detected.

The key component of the reaction mixture is a small amount of a **2′,3′-dideoxynucleoside triphosphate (ddNTP),**

2′,3′-Dideoxynucleoside triphosphate

which lacks the 3′-OH group of deoxynucleotides. *When the dideoxy analog is incorporated into the growing polynucleotide in place of the corresponding normal nucleotide, chain growth is terminated because addition of the next nucleotide requires a free 3′-OH.* By using only a small amount of the ddNTP, a series of truncated chains is generated, each of which ends

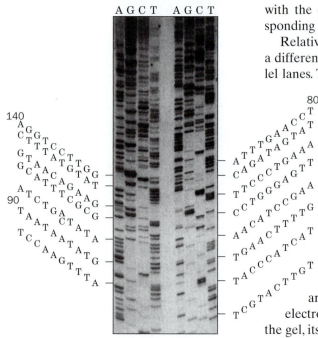

Figure 3-22 An autoradiogram of a sequencing gel. The positions of radioactive DNA fragments produced by the chain-terminator method were visualized by laying X-ray film over the gel after electrophoresis. A second loading of the gel (the four lanes at right) was made 90 min after the initial loading in order to obtain the sequences of the smaller fragments. The deduced sequence of 140 nucleotides is written along the side. [From Hindley, J., DNA sequencing, *in* Work, T.S. and Burdon, R.H. (Eds.), *Laboratory Techniques in Biochemistry and Molecular Biology,* Vol. 10, p. 82, Elsevier (1983). Used by permission.]

with the dideoxy analog at one of the positions occupied by the corresponding base.

Relatively modest sequencing tasks use four reaction mixtures, each with a different ddNTP, and the reaction products are electrophoresed in parallel lanes. The lengths of the truncated chains indicate the positions where the dideoxynucleotide was incorporated. Thus, the sequence of the replicated strand can be directly read from the gel (Fig. 3-22). The gel must have sufficient resolving power to separate fragments that differ in length by only one nucleotide. Two sets of gels, one run for a longer time than the other, can be used to obtain the sequence of up to 800 bases of DNA. Note that the sequence obtained by the chain-terminator method is complementary to the DNA strand being sequenced.

Automated Sequencing. Large-scale sequencing operations are accelerated by automation. In a variation of the chain-terminator method, the primers used in the four chain-extension reactions are each linked to a different fluorescent dye. The separately reacted mixtures are combined and subjected to gel electrophoresis in a single lane. As each fragment exits the bottom of the gel, its terminal base is identified by its characteristic fluorescence (Fig. 3-23) with an error rate of ~1%. In the most advanced systems, the sequencing gel is contained in an array of up to 96 capillary tubes rather than in a slab-shaped apparatus, sample preparation and sample loading are performed by robotic systems, and electrophoresis and data analysis are fully automated. These systems can simultaneously sequence 96 DNA samples averaging ~600 bases each with a turnaround time of ~2.5 hrs and hence can identify up to 550,000 bases per day—all with only ~15 min of human attention (vs. the ~25,000 bases per year that a skilled operator can identify using the above-described manual methods). Nevertheless, one such system would require ~30 years of uninterrupted operation to sequence the 3.2 billion-bp human genome with only two sets of overlapping fragments. However, to ensure the complete coverage of a large tract of DNA and to reduce its error rate to <0.01%, at least 10 sets of overlapping segments must be sequenced (Section 3-5B). Hence, the major sequencing centers, where most genome sequencing is carried out, each have over 100 ad-

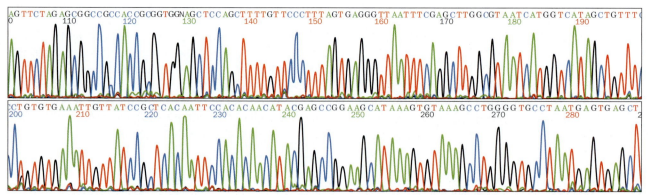

Figure 3-23 Automated DNA sequencing. In this variant of the technique, a different fluorescent dye is attached to the primer in each of the four reaction mixtures in the chain-terminator procedure. The four reaction mixtures are combined for electrophoresis. Each of the four colored curves therefore represents the electrophoretic pattern of fragments containing one of the dideoxynucleotides: Green, red, black, and blue correspond to fragments ending in ddATP, ddTTP, ddGTP, and ddCTP, respectively. The 3′-terminal base of each oligonucleotide is identified by the fluorescence of its gel band. This portion of the readout corresponds to nucleotides 100–290 of the DNA segment being sequenced. [Courtesy of Mark Adams, The Institute for Genomic Research, Rockville, Maryland.]

vanced sequencing systems. Consequently, there has been such a flood of DNA sequence data—over 43 billion nucleotides in 39 million sequences as of late 2004, and approximately doubling every two years—that only computers can keep track of them.

Nucleic acid sequencing has become so routine that directly determining a protein's amino acid sequence (Section 5-3) is generally far more time-consuming than determining the base sequence of its corresponding gene. In fact, nucleic acid sequencing is invaluable for studying genes whose products have not yet been identified. If the gene can be sequenced, the probable function of its protein product may be deduced by comparing the base sequence to those of genes whose products are already characterized (see Box 3-2).

BOX 3-2

Pathways of Discovery

Francis Collins and the Gene for Cystic Fibrosis

Francis S. Collins (1950–)

By the mid-twentieth century, the molecular basis of several human diseases was appreciated. For example, sickle-cell anemia (Section 7-4) was known to be caused by an abnormal hemoglobin protein. Studies of sickle-cell hemoglobin eventually revealed the underlying genetic defect, a mutation in a hemoglobin gene. It therefore seemed possible to trace other diseases to defective genes. But for many genetic diseases, even those with well-characterized symptoms, no defective protein had yet been identified. One such disease was cystic fibrosis, which is characterized mainly by the secretion of thick mucus that obstructs the airways and creates an ideal environment for bacterial growth. Cystic fibrosis is the most common inherited disease in individuals of northern European descent, striking about 1 in 2500 newborns and leading to death by early adulthood due to irreversible lung damage. It was believed that identifying the molecular defect in cystic fibrosis would lead to better understanding of the disease and the ability to design more effective treatments.

Enter Francis Collins, who began his career by earning a doctorate in physical chemistry but then enrolled in medical school to take part in the molecular biology revolution. As a physician-scientist, Collins developed methods for analyzing large stretches of DNA in order to home in on specific genes, including the one that, when mutated, causes cystic fibrosis. By analyzing the DNA of individuals with the disease (who had two copies of the defective gene) and of family members who were asymptomatic carriers (with one normal and one defective copy of the gene), Collins and his team localized the cystic fibrosis gene to the long arm of chromosome 7. They gradually closed in on a DNA segment that appears to be present in a number of mammalian species, which suggests that the segment contains an essential gene. The cystic fibrosis gene was finally identified in 1989. Collins had demonstrated the feasibility of identifying a genetic defect in the absence of other molecular information.

Once the cystic fibrosis gene was in hand, it was a relatively straightforward process to deduce the probable structure and function of the encoded protein, which turned out to be a membrane channel for chloride ions. When functioning normally, the protein helps regulate the ionic composition and viscosity of extracellular secretions. Discovery of the cystic fibrosis gene also made it possible to design tests to identify carriers so that they could take advantage of genetic counseling.

Throughout Collins' work on the cystic fibrosis gene and during subsequent hunts for the genes that cause neurofibromatosis and Huntington's disease, he was mindful of the ethical implications of the new science of molecular genetics. Collins has been a strong advocate for protecting the privacy of genetic information. At the same time, he recognizes the potential therapeutic use of such information. In his tenure as director of the human genome project, he was committed to making the results freely and immediately accessible, as a service to researchers and the individuals who might benefit from new therapies based on molecular genetics.

Riordan, J.R., Rommens, J.M., Kerem, B.-S., Alon, N., Rozmahel, R., Grzelczak, Z., Zielensky, J., Lok, S., Plavsic, N., Chou, J.-L., Drumm, M.L., Iannuzzi, M.C., Collins, F.S., and Tsui, L.-C., Identification of the cystic fibrosis gene: Cloning and characterization of complementary DNA, *Science* **245,** 1066–1073 (1989).

D Genome Sequencing

The advent of large-scale sequencing techniques brought to fruition the dream of sequencing entire genomes. However, the major technical hurdle in sequencing all the DNA in an organism's genome is not the DNA sequencing itself but, rather, assembling the tens of thousands to tens of millions of sequenced segments (depending on the size of the genome) into contiguous blocks and assigning them to their correct chromosomal positions. To do so required the development of automated sequencing protocols and mathematically sophisticated computer algorithms.

The first complete genome sequence to be determined, that of the bacterium *Haemophilus influenzae,* was reported in 1995 by Craig Venter. By the end of 2004, the complete genome sequences of over 200 prokaryotes had been reported (with many more being determined) as well as those of 28 eukaryotes, including *Saccharomyces cerevisiae* (baker's yeast), *Caenorhabditis elegans* (a nematode worm), *Drosophila melanogaster* (a fruit fly), *Arabidopsis thaliana* (a flowering plant), and humans (Table 3-3).

The determination of the ~3.2 billion-nucleotide human genome sequence was a gargantuan undertaking involving hundreds of scientists working in two groups, one led by Venter and the other by Francis Collins (Box 3-2), Eric Lander, and John Sulston. After over a decade of intense effort, the "rough draft" of the human genome sequence was reported in early 2001 and the "finished" sequence was reported in mid-2003. This stunning achievement promises to revolutionize the way both biochemistry and medicine are viewed and practiced, although it is likely to require many

Table 3-3 Some Sequenced Genomes

Organism	Genome Size (kb)	Number of Chromosomes
Mycoplasma genitalium (human parasite)	580	1
Rickettsia prowazekii (putative relative of mitochondria)	1,112	1
Methanococcus jannaschii (thermophilic methanogen)	1,665	1
Haemophilus influenzae (human pathogen)	1,830	1
Synechocystis sp. (cyanobacterium)	3,573	1
Escherichia coli (human symbiont)	4,639	1
Saccharomyces cerevisiae (baker's yeast)	11,700	16
Plasmodium falciparum (protozoan that causes malaria)	30,000	14
Caenorhabditis elegans (nematode)	97,000	6
Arabidopsis thaliana (dicotyledonous plant)	117,000	5
Drosophila melanogaster (fruit fly)	137,000	4
Danio rerio (zebrafish)	1,700,000	25
Homo sapiens	3,200,000	23

years of further effort before its full significance is understood. Nevertheless, numerous important conclusions can already be drawn, including:

1. About half the human genome consists of repeating sequences of various types.

2. Only ~28% of the genome is transcribed to RNA.

3. Only 1.1% to 1.4% of the genome (~5% of the transcribed RNA) encodes protein.

4. The human genome appears to contain only ~30,000 protein-encoding genes [also known as **open reading frames (ORFs)**] rather than the 50,000 to 140,000 ORFs that had previously been predicted based mainly on extrapolations. This compares with the ~6000 ORFs in yeast, ~13,000 in *Drosophila,* ~18,000 in *C. elegans,* and ~26,000 in *Arabadopsis* (although note that these numbers will almost certainly change as our presently imperfect ability to recognize ORFs improves).

5. Only a small fraction of human proteins are unique to vertebrates; most occur in other if not all life-forms.

6. Two randomly selected human genomes differ, on average, by only 1 nucleotide per 1250, that is, any two people are likely to be >99.9% genetically identical.

The obviously greater complexity of humans (vertebrates) relative to "lower" (nonvertebrate) forms of life is unlikely to be due to the not much larger numbers of ORFs that vertebrates encode. Rather, it appears that vertebrate proteins themselves are more complex than those of nonvertebrates, that is, vertebrate proteins tend to have more domains (modules) than invertebrate proteins, and that these modules are more often selectively expressed through **differential gene splicing** (a process in which a given gene transcript can be processed in multiple ways so as to yield different proteins when translated; Section 25-3A). Thus, many vertebrate genes encode several different although similar proteins.

E Sequences, Mutation, and Evolution

One of the richest rewards of nucleic acid sequencing technology is the information it provides about the mechanisms of evolution. The chemical and physical properties of DNA, such as its regular three-dimensional shape and the elegant process of replication, may leave the impression that genetic information is relatively static. In fact, *DNA is a dynamic molecule, subject to changes that alter genetic information.* For example, the mispairing of bases during DNA replication introduces errors known as **point mutations** in the daughter strands. Mutations also result from DNA damage by chemicals or radiation. More extensive alterations in genetic information are caused by faulty **recombination** (exchange of DNA between chromosomes) and the **transposition** of genes within or between chromosomes and, in some cases, from one organism to another. All these alterations to DNA provide the raw material for natural selection. When a mutated gene is transcribed and the messenger RNA is subsequently translated, the resulting protein may have properties that confer some advantage to the individual. As a beneficial change is passed from generation to generation, it becomes part of the standard genetic makeup of the species. Of course, many changes occur as a species evolves, not all of them simple and not all of them gradual.

Phylogenetic relationships can be revealed by comparing the sequences of similar genes in different organisms. The number of nucleotide differ-

Figure 3-24 Maize and teosinte. Despite the large differences in phenotype—maize (*bottom*) has hundreds of easily chewed kernels whereas teosinte (*top*) has only a few hard, inedible kernels—the plants differ in only a few genes. The ancestor of maize is believed to be a mutant form of teosinte in which the kernels were more exposed. [John Doebley/Visuals Unlimited.]

ences between the corresponding genes in two species is roughly indicative of the degree to which the species have diverged through evolution. The regrouping of prokaryotes into archaea and bacteria (Section 1-3A) according to rRNA sequences present in all organisms illustrates the impact of sequence analysis.

Nucleic acid sequencing also reveals that species differing in **phenotype** (physical characteristics) are nonetheless remarkably similar at the molecular level. For example, humans and chimpanzees share 98–99% of their DNA. Studies of corn (maize) and its putative ancestor, teosinte, suggest that the plants differ in only a handful of genes governing kernel development (teosinte kernels are encased by an inedible shell; Fig. 3-24).

Small mutations in DNA are apparently responsible for relatively large evolutionary leaps. This is perhaps not so surprising when the nature of genetic information is considered. A mutation in a gene segment that does not encode protein might interfere with the binding of cellular factors that influence the timing of transcription. A mutation in a gene encoding an RNA might interfere with the binding of factors that affect the efficiency of translation. Even a minor rearrangement of genes could disrupt an entire developmental process, resulting in the appearance of a novel species. Notwithstanding the high probability that most sudden changes would lead to diminished individual fitness or the inability to reproduce, the capacity for sudden changes in genetic information is consistent with the fossil record. Ironically, the discontinuities in the fossil record that are probably caused in part by sudden genetic changes once fueled the adversaries of Charles Darwin's theory of evolution by natural selection.

5   Manipulating DNA

Along with nucleic acid sequencing, techniques for manipulating DNA *in vitro* and *in vivo* (in the test tube and in living systems) have produced dramatic advances in biochemistry, cell biology, and genetics. In many cases, this **recombinant DNA technology** has made it possible to purify specific DNA sequences and to prepare them in quantities sufficient for study. Consider the problem of isolating a unique 1000-bp length of chromosomal DNA from *E. coli*. A 10-L culture of cells grown at a density of $\sim 10^{10}$ cells·mL^{-1} contains only $\sim$0.1 mg of the desired DNA, which would be all

but impossible to separate from the rest of the DNA using classical separation techniques (Sections 5-2 and 23-3). *Recombinant DNA technology, also called* **molecular cloning** *or* **genetic engineering,** *makes it possible to isolate, amplify, and modify specific DNA sequences.*

A   Cloning Techniques

The following approach is used to obtain and amplify a segment of DNA:

1. A fragment of DNA of the appropriate size is generated by a restriction enzyme, by PCR (Section 3-5C), or by chemical synthesis.
2. The fragment is incorporated into another DNA molecule known as a **vector,** which contains the sequences necessary to direct DNA replication.
3. The vector—with the DNA of interest—is introduced into cells, where it is replicated.
4. Cells containing the desired DNA are identified, or **selected.**

Cloning refers to the production of multiple identical organisms derived from a single ancestor. The term **clone** refers to the collection of cells that contain the vector carrying the DNA of interest or to the DNA itself. In a suitable host organism, such as *E. coli* or yeast, large amounts of the inserted DNA can be produced.

Cloned DNA can be purified and sequenced (Section 3-4). Alternatively, if a cloned gene is flanked by the properly positioned regulatory sequences for RNA and protein synthesis, the host may also produce large quantities of the RNA and protein specified by that gene. Thus, cloning provides materials (nucleic acids and proteins) for other studies and also provides a means for studying gene expression under controlled conditions.

Cloning Vectors.   A variety of small, autonomously replicating DNA molecules are used as cloning vectors. **Plasmids** are circular DNA molecules of 1 to 200 kb found in bacteria or yeast cells. Plasmids can be considered molecular parasites, but in many instances they benefit their host by providing functions, such as resistance to antibiotics, that the host lacks.

Some types of plasmids are present in one or a few copies per cell and replicate only when the bacterial chromosome replicates. However, the plasmids used for cloning are typically present in hundreds of copies per cell and can be induced to replicate until the cell contains two or three thousand copies (representing about half of the cell's total DNA). The plasmids that have been constructed for laboratory use are relatively small, replicate easily, carry genes specifying resistance to one or more antibiotics, and contain a number of conveniently located restriction endonuclease sites into which foreign DNA can be inserted. Plasmid vectors can be used to clone DNA segments of no more than ~10 kb. The *E. coli* plasmid designated **pUC18** (Fig. 3-25) is a representative cloning vector ("pUC" stands for plasmid-Universal Cloning).

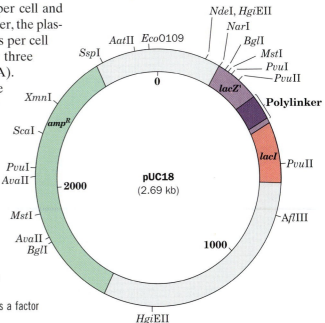

Figure 3-25   The plasmid pUC18.   As shown in this diagram, the circular plasmid contains multiple restriction sites, including a **polylinker** sequence that contains 13 restriction sites that are not present elsewhere on the plasmid. The three genes expressed by the plasmid are *amp*^R, which confers resistance to the antibiotic **ampicillin;** *lacZ*, which encodes the enzyme **β-galactosidase;** and *lacI*, which encodes a factor that controls transcription of *lacZ* (as described in Section 27-2A).

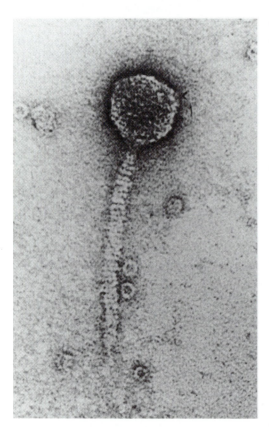

Figure 3-26 Bacteriophage λ.   During phage infection, DNA contained in the "head" of the phage particle enters the bacterial cell, where it is replicated ~100 times and packaged to form progeny phage. [Electron micrograph courtesy of A.F. Howatson. From Lewin, B., *Gene Expression,* Vol. 3, Fig. 5.23, Wiley (1977).]

Bacteriophage λ (Fig. 3-26) is an alternative cloning vector that can accommodate DNA inserts up to 16 kb. The central third of the 48.5-kb phage genome is not required for infection and can therefore be replaced by foreign DNAs of similar size. The resulting **recombinant,** or **chimera** (named after the mythological monster with a lion's head, goat's body, and serpent's tail), is packaged into phage particles that can then be introduced into the host cells. One advantage of using phage vectors is that the recombinant DNA is produced in large amounts in easily purified form. **Baculoviruses,** which infect insect cells, are similarly used for cloning in cultures of insect cells.

Much larger DNA segments—up to several hundred kilobase pairs—can be cloned in large vectors known as **bacterial artificial chromosomes (BACs)** or **yeast artificial chromosomes (YACs).** YACs are linear DNA molecules that contain all the chromosomal structures required for normal replication and segregation during yeast cell division. BACs, which replicate in *E. coli*, are derived from circular plasmids that normally replicate long regions of DNA and are maintained at the level of approximately one copy per cell (properties similar to those of actual chromosomes).

Ligation.   A DNA segment to be cloned is often obtained through the action of restriction endonucleases. Most restriction enzymes cleave DNA to yield sticky ends (Section 3-4A). Therefore, as Janet Mertz and Ron Davis first demonstrated in 1972, *a restriction fragment can be inserted into a cut made in a cloning vector by the same restriction enzyme* (Fig. 3-27). The complementary ends of the two DNAs form base pairs **(anneal)** and the sugar–phosphate backbones are covalently **ligated,** or spliced together, through the action of an enzyme named **DNA ligase.** (A ligase produced by a bacteriophage can also join blunt-ended restriction fragments.) A great advantage of using a restriction enzyme to construct a recombinant DNA

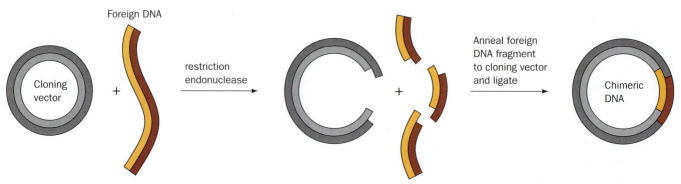

Figure 3-27 Construction of a recombinant DNA molecule.   The cloning vector and the foreign DNA are cut by the same restriction endonuclease. The sticky ends of the vector and the foreign DNA fragments anneal and are covalently joined by DNA ligase. The result is a chimeric DNA containing a portion of the foreign DNA inserted into the vector. 🔗 **See the Animated Figures.**

molecule is that the DNA insert can later be precisely excised from the cloned vector by cleaving it with the same restriction enzyme.

Transformation and Selection.   The expression of a chimeric plasmid in a bacterial host was first demonstrated in 1973 by Herbert Boyer and Stanley Cohen. A host bacterium can take up a plasmid when the two are mixed together, but the vector becomes permanently established in its bacterial host (transformation) with an efficiency of only ~0.1%. However, a single transformed cell can multiply without limit, producing large quantities of recombinant DNA. Bacterial cells are typically plated on a semisolid growth medium at a low enough density that discrete colonies, each arising from a single cell, are visible.

It is essential to select only those host organisms that have been transformed and that contain a properly constructed vector. In the case of plasmid transformation, selection can be accomplished through the use of antibiotics and/or chromogenic (color-producing) substances. For example, the *lacZ* gene in the pUC18 plasmid (see Fig. 3-25) encodes the enzyme β-galactosidase, which cleaves the colorless compound **X-gal** to a blue product:

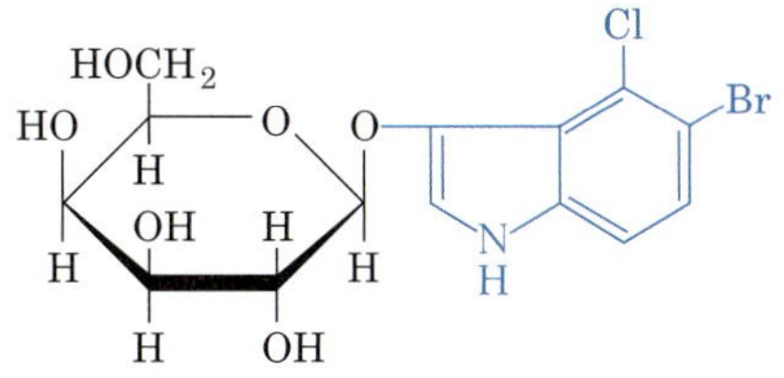

5-Bromo-4-chloro-3-indolyl-β-D-galactoside (X-gal)
(*colorless*)

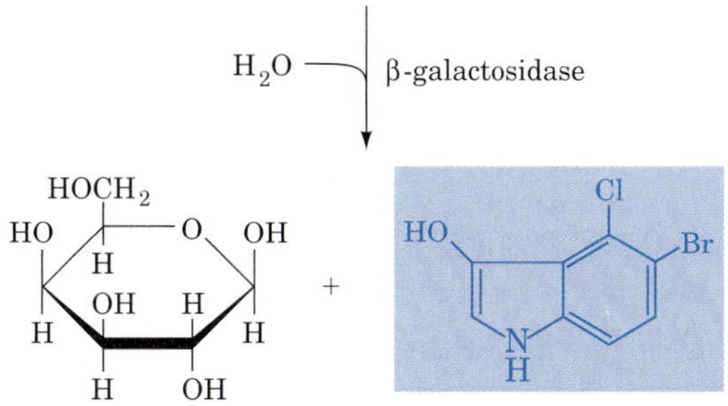

H_2O — β-galactosidase

β-D-Galactose   **5-Bromo-4-chloro-3-hydroxyindole**
(*blue*)

E. coli cells that have been transformed by an unmodified pUC18 plasmid form blue colonies. However, if the plasmid contains a foreign DNA insert in its polylinker region, the colonies are colorless because the insert interrupts the protein-coding sequence of the *lacZ* gene and no functional β-galactosidase is produced. Bacteria that have failed to take up any plasmid are also colorless due to the absence of β-galactosidase, but these cells can be excluded by adding the antibiotic ampicillin to the growth medium (the plasmid includes the gene *amp*^{R}, which confers ampicillin resistance). Thus, successfully transformed cells form colorless colonies in the presence of ampicillin. Genes such as *amp*^{R} are known as **selectable markers.**

Genetically engineered bacteriophage λ vectors contain restriction sites that flank the dispensable central third of the phage genome. This segment can be replaced by foreign DNA, but the chimeric DNA is packaged in phage particles only if its length is from 75 to 105% of the 48.5-kb

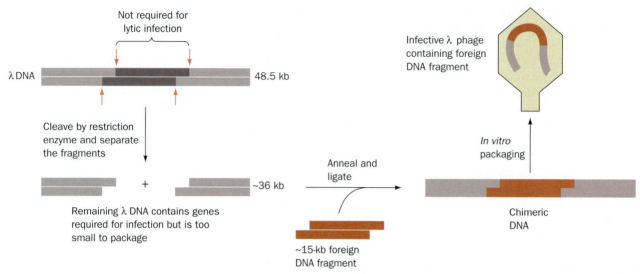

Figure 3-28   Cloning with bacteriophage λ.   Removal of a nonessential portion of the phage genome allows a segment of foreign DNA to be inserted. The DNA insert can be packaged into an infectious phage particle only if the insert DNA has the appropriate size. 🔗 **See the Animated Figures.**

wild-type λ genome (Fig. 3-28). Consequently, λ phage vectors that have failed to acquire a foreign DNA insert are unable to propagate because they are too short to form infectious phage particles. Of course, the production of infectious phage particles results not in a growing bacterial colony but in a **plaque,** a region of lysed bacterial cells, on a culture plate containing a "lawn" of the host bacteria. The recombinant DNA—now much amplified—can be recovered from the phage particles in the plaque.

B   DNA Libraries

In order to clone a particular DNA fragment, it must first be obtained in relatively pure form. The magnitude of this task can be appreciated by considering that, for example, a 1-kb fragment of human DNA represents only 0.000031% of the 3.2 billion-bp human genome. Of course, identifying a particular DNA fragment requires knowing something about its nucleotide sequence or its protein product. In practice, it is usually more difficult to identify a particular DNA fragment from an organism and then clone it than it is to clone all the organism's DNA that might contain the DNA of interest and then identify the clones containing the sequence of interest.

Genomic Libraries.   The cloned set of all DNA fragments from a particular organism is known as its **genomic library.** Genomic libraries are generated by a procedure known as **shotgun cloning.** The chromosomal DNA of the organism is isolated, cleaved to fragments of clonable size, and inserted into a cloning vector. The DNA is usually fragmented by partial rather than exhaustive restriction digestion so that the genomic library contains intact representatives of all the organism's genes, including those that contain restriction sites. DNA in solution can also be mechanically fragmented **(sheared)** by rapid stirring.

Given the large size of the genome relative to a gene, the shotgun cloning method is subject to the laws of probability. The number of randomly generated fragments that must be cloned to ensure a high probability that a desired sequence is represented at least once in the genomic library is cal-

culated as follows: The probability P that a set of N clones contains a fragment that constitutes a fraction f, in bp, of the organism's genome is

$$P = 1 - (1 - f)^N \qquad [3\text{-}1]$$

Consequently,

$$N = \log(1 - P)/\log(1 - f) \qquad [3\text{-}2]$$

Thus, in order for $P = 0.99$ for fragments averaging 10 kb in length, $N = 2162$ for the 4600-kb *E. coli* chromosome and 63,000 for the 137,000-kb *Drosophila* genome. The use of BAC- or YAC-based genomic libraries with their large fragment lengths therefore greatly reduces the effort necessary to obtain a given gene segment from a large genome. After a BAC- or YAC-based clone containing the desired DNA has been identified (see below), its large DNA insert can be further fragmented and cloned again (**subcloned**) to isolate the target DNA.

cDNA Libraries.   A different type of DNA library contains only the expressed sequences from a particular cell type. Such a **cDNA library** is constructed by isolating all the cell's mRNAs and then copying them to DNA using a specialized type of DNA polymerase known as **reverse transcriptase** because it synthesizes DNA on RNA templates (Box 24-2). The **complementary DNA (cDNA)** molecules are then inserted into cloning vectors to form a cDNA library.

Screening.   Once the requisite number of clones is obtained, the genomic library must be **screened** for the presence of the desired gene. This can be done by a process known as **colony** or *in situ* **hybridization** (Latin: *in situ*, in position; Fig. 3-29). The cloned yeast colonies, bacterial colonies, or phage plaques to be tested are transferred, by **replica plating,** from a master plate to a nitrocellulose filter (replica plating is also used to transfer colonies to plates containing different growth media). Next, the filter is treated with NaOH, which lyses the cells or phages and separates the DNA into single strands, which preferentially bind to the nitrocellulose. The filter is then dried to fix the DNA in place and incubated with a labeled **probe.** The probe is a short segment of DNA or RNA whose sequence is complemen-

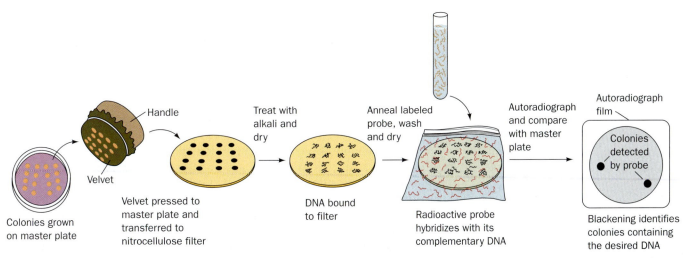

Figure 3-29 Colony (*in situ*) hybridization.   Colonies are transferred from a "master" culture plate by replica plating. Clones containing the DNA of interest are identified by the ability to bind a specific probe. Here, binding is detected by laying X-ray film over the dried filter. Since the colonies on the master plate and on the filter have the same spatial distribution, positive colonies are easily retrieved.

tary to a portion of the DNA of interest. After washing away unbound probe, the presence of the probe on the nitrocellulose is detected by a technique appropriate for the label used (e. g., exposure to X-ray film for a radioactive probe, a process known as **autoradiography,** or illumination with an appropriate wavelength for a fluorescent probe). Only those colonies or plaques containing the desired gene bind the probe and are thereby detected. The corresponding clones can then be retrieved from the master plate. Using this technique, a human genomic library of ~1 million clones can be readily screened for the presence of one particular DNA segment.

Choosing a probe for a gene whose sequence is not known requires some artistry. The corresponding mRNA can be used to do so if it is produced in sufficient quantities to be isolated. Alternatively, if the amino acid sequence of the protein encoded by the gene is known, the probe may be a mixture of the various synthetic oligonucleotides that are complementary to a segment of the gene's inferred base sequence. Several disease-related genes have been isolated using probes specific for nearby markers, such as repeated DNA sequences, that were already known to be genetically linked to the disease genes.

> See Guided Exploration 3
>
> PCR and site-directed mutagenesis.

C DNA Amplification by the Polymerase Chain Reaction

Although molecular cloning techniques are indispensable to modern biochemical research, the **polymerase chain reaction (PCR)** is often a faster and more convenient method for amplifying a specific DNA. Segments of up to 6 kb can be amplified by this technique, which was devised by Kary Mullis in 1985. *In PCR, a DNA sample is separated into single strands and incubated with DNA polymerase, dNTPs, and two oligonucleotide primers whose sequences flank the DNA segment of interest. The primers direct the DNA polymerase to synthesize complementary strands of the target DNA* (Fig. 3-30). Multiple cycles of this process, each doubling the amount of the target DNA, geometrically amplify the DNA starting from as little as a single gene copy. In each cycle, the two strands of the duplex DNA are separated by heating, the primers are annealed to their complementary segments on the DNA, and the DNA polymerase directs the synthesis of the complementary strands. The use of a heat-stable DNA polymerase, such as *Taq* **polymerase** isolated from *Thermus aquaticus,* a bacterium that thrives at 75°C, eliminates the need to add fresh enzyme after each round of heating (heat inactivates most enzymes). Hence, in the presence of sufficient quantities of primers and dNTPs, PCR is carried out simply by cyclically varying the temperature.

Twenty cycles of PCR increase the amount of the target sequence around a millionfold ($\sim2^{20}$) with high specificity. Indeed, PCR can amplify a target DNA present only once in a sample of 10^5 cells, so this method can be used without prior DNA purification. The amplified DNA can then be sequenced or cloned.

PCR amplification has become an indispensable tool. Clinically, it is used to diagnose infectious diseases and to detect rare pathological events such as mutations leading to cancer. Forensically, the DNA from a single hair or sperm can be amplified by PCR so that its RFLPs (Box 3-1) can be used to identify the donor. Traditional ABO blood-type analysis requires a coin-sized drop of blood; PCR is effective on pinhead-sized samples of biological fluids. Most courts now consider DNA sequences as unambiguous identifiers of individuals, as are fingerprints, because the chance of two individuals sharing extended sequences of DNA is typically one in a million or more. In a few cases, PCR has dramatically restored justice to convicts who were

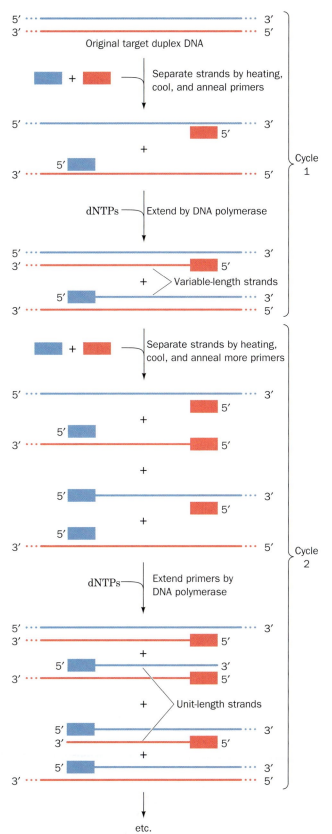

Figure 3-30 The polymerase chain reaction (PCR). In each cycle of the reaction, the strands of the duplex DNA are separated by heating, the reaction mixture is cooled to allow primers to anneal to complementary sequences on each strand, and DNA polymerase extends the primers. The number of "unit-length" strands doubles with every cycle after the second cycle. By choosing primers specific for each end of a gene, the gene can be amplified over a millionfold.

released from prison on the basis of PCR results that proved their innocence—even many years after the crime-scene evidence had been collected.

D Applications of Recombinant DNA Technology

The ability to manipulate DNA sequences allows genes to be altered and expressed in order to obtain proteins with improved functional properties or to correct genetic defects.

Protein Production. The production of large quantities of scarce or novel proteins is relatively straightforward only for bacterial proteins: A cloned gene must be inserted into an **expression vector,** a plasmid that contains properly positioned transcriptional and translational control sequences. The production of a protein of interest may reach 30% of the host's total cellular protein. Such genetically engineered organisms are called **overproducers.** Bacterial cells often sequester large amounts of useless and possibly toxic (to the bacterium) protein as insoluble inclusions, which sometimes simplifies the task of purifying the protein.

Bacteria can produce eukaryotic proteins only if the recombinant DNA that carries the protein-coding sequence also includes bacterial transcriptional and translational control sequences. Synthesis of eukaryotic proteins in bacteria also presents other problems. For example, many eukaryotic genes are large and contain stretches of nucleotides **(introns)** that are transcribed and excised before translation (Section 25-3A); bacteria lack the machinery to excise the introns. In addition, many eukaryotic proteins are posttranslationally modified by the addition of carbohydrates or by other reactions. These problems can be overcome by using expression vectors that propagate in eukaryotic hosts, such as yeast or cultured insect or animal cells.

Table 3-4 lists some recombinant proteins produced for medical and agricultural use. In many cases, purification of these proteins directly from human or animal tissues is unfeasible on ethical or practical grounds. Expression systems permit large-scale, efficient preparation of the proteins while minimizing the risk of contamination by viruses or other pathogens from tissue samples.

Site-Directed Mutagenesis. After isolating a gene, it is possible to modify the nucleotide sequence to alter the amino acid sequence of the encoded protein. **Site-directed mutagenesis,** a technique pioneered by Michael

Table 3-4 Some Proteins Produced by Genetic Engineering

Protein	Use
Human insulin	Treatment of diabetes
Human growth hormone	Treatment of some endocrine disorders
Erythropoietin	Stimulation of red blood cell production
Colony-stimulating factors	Production and activation of white blood cells
Coagulation factors IX and X	Treatment of blood clotting disorders (hemophilia)
Tissue-type plasminogen activator	Lysis of blood clots in heart attack and stroke
Bovine growth hormone	Production of milk in cows

Figure 3-31 Site-directed mutagenesis. A chemically synthesized oligonucleotide incorporating the desired base changes anneals to the DNA encoding the gene to be altered (*green strand*). The mismatched primer is then extended by DNA polymerase, generating the mutated gene (*blue strand*). The altered gene can be inserted into a suitable cloning vector to be amplified, expressed, or used to generate a mutant organism. 🔷 **See the Animated Figures.**

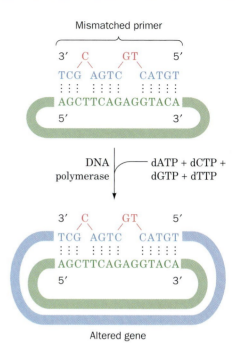

Smith, *mimics the natural process of evolution and allows predictions about the structural and functional roles of particular amino acids in a protein to be rigorously tested in the laboratory.*

Synthetic oligonucleotides are required to specifically alter genes through site-directed mutagenesis. An oligonucleotide whose sequence is identical to a portion of the gene of interest except for the desired base changes is used to direct replication of the gene. The oligonucleotide hybridizes to the corresponding **wild-type** (naturally occurring) sequence if there are no more than a few mismatched base pairs. Extension of the oligonucleotide, called a primer, by DNA polymerase yields the desired altered gene (Fig. 3-31). The altered gene can then be inserted into an appropriate vector. A mutagenized primer can also be used to generate altered genes by PCR.

Transgenic Organisms. For many purposes it is preferable to tailor an intact organism rather than just a protein—true genetic engineering. Multicellular organisms expressing a gene from another organism are said to be **transgenic,** and the transplanted foreign gene is called a **transgene.**

For the change to be permanent, that is, heritable, a transgene must be stably integrated into the organism's germ cells. For mice, this is accomplished by microinjecting cloned DNA encoding the desired altered characteristics into a fertilized egg and implanting it into the uterus of a foster mother. A well-known example of a transgenic mouse contains extra copies of a growth hormone gene (Fig. 3-32).

Transgenic farm animals have also been developed. Ideally, the genes of such animals could be tailored to allow the animals to grow faster on less food or to be resistant to particular diseases. Some transgenic farm animals have been engineered to secrete medically useful proteins into their milk. Harvesting such a substance from milk is much more cost-effective than producing the same substance in bacterial cultures.

Transgenic organisms have greatly enhanced our understanding of gene expression. Animals that have been engineered to contain a defective gene or that lack a gene entirely (a so-called **gene knockout**) also serve as experimental models for human diseases.

One of the most successful transgenic organisms is corn (maize) that has been genetically modified to produce a protein that is toxic to plant-eating insects (but harmless to vertebrates). The toxin is synthesized by the soil microbe *Bacillus thuringiensis.* The toxin gene has been cloned into corn in order to confer protection against the European corn borer, a commercially significant pest that spends much of its life cycle inside the corn plant, where it is largely inaccessible to chemical insecticides. The use of "Bt corn," which is now widely planted in the United States, has greatly reduced the need for such toxic substances.

Transgenic plants have also been engineered for better nutrition. For example, researchers have developed a strain of rice with foreign genes that encode enzymes necessary to synthesize **β-carotene** (an orange pigment that is the precursor of **vitamin A**) and a gene for the iron-storage protein

Figure 3-32 Transgenic mouse. The gigantic mouse on the left was grown from a fertilized ovum that had been microinjected with DNA containing the rat growth hormone gene. He is nearly twice the weight of his normal littermate on the right. [Courtesy of Ralph Brinster, University of Pennsylvania.]

Figure 3-33 Golden rice. The white grains on the right are the wild type. The grains on the left have been engineered to store up to three times more iron and to synthesize β-carotene, which gives them their yellow color. [Courtesy of Ingo Potrykus.]

ferritin. The genetically modified rice, which is named "golden rice" (Fig. 3-33), should help alleviate vitamin A deficiencies (which afflict some 400 million people) and iron deficiencies (an estimated 30% of the world's population suffers from iron deficiency). Other transgenic plants include freeze-tolerant strawberries and slow-ripening tomatoes.

There is presently a widely held popular suspicion, particularly in Europe, that genetically modified or "GM" foods will somehow be harmful. However, extensive research, as well as considerable consumer experience, has failed to reveal any deleterious effects caused by GM foods (see Box 3-3).

Gene Therapy. **Gene therapy** is the transfer of new genetic material to the cells of an individual in order to produce a therapeutic effect. Although the potential benefits of this as yet rudimentary technology are enormous, there are numerous practical obstacles to overcome. For example, the retroviral vectors (RNA-containing viruses) commonly used to directly introduce genes into humans can provoke a fatal immune response.

The only documented success of gene therapy to date has occurred in children with a form of **severe combined immunodeficiency disease (SCID)** known as **SCID-X1,** which without treatment would have required their isolation in a sterile environment to prevent fatal infection. SCID-X1 is caused by a defect in the gene encoding **γc cytokine receptor,** whose ac-

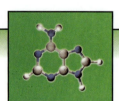

BOX 3-3

Perspectives in Biochemistry

Ethical Aspects of Recombinant DNA Technology

In the early 1970s, when genetic engineering was first discussed, little was known about the safety of the proposed experiments. After considerable debate, during which there was a moratorium on such experiments, regulations for recombinant DNA research were drawn up. The rules prohibit obviously dangerous experiments (e.g., introducing the gene for diphtheria toxin into *E. coli,* which would convert this human symbiont into a deadly pathogen). Other precautions limit the risk of accidentally releasing potentially harmful organisms into the environment. For example, many vectors must be cloned in host organisms with special nutrient requirements. These organisms are unlikely to survive outside the laboratory.

The proven value of recombinant DNA technology has silenced nearly all its early opponents. Certainly, it would not have been possible to study some pathogens, such as the virus that causes AIDS, without cloning. The lack of recombinant-induced genetic catastrophes so far does not guarantee that recombinant organisms won't ever adversely affect the environment. Nevertheless, the techniques used by genetic engineers mimic those used in nature—that is, mutation and selection—so natural and man-made organisms are fundamentally similar. In any case, people have been breeding plants and animals for several millennia already, and for many of the same purposes that guide experiments with recombinant DNA.

There are other ethical considerations to be faced as new genetic engineering techniques become available. Bacterially produced human growth hormone is now routinely prescribed to increase the stature of abnormally short children. However, should athletes be permitted to use this protein, as some reportedly have, to increase their size and strength? Few would dispute the use of gene therapy, if it can be developed, to cure such genetic defects as sickle-cell anemia (Section 7-4) and Lesch–Nyhan syndrome (Section 22-1D). If, however, it becomes possible to alter complex (i.e., multigene) traits such as athletic ability and intelligence, which changes would be considered desirable and who would decide whether to make them? Should gene therapy be used only to correct an individual's defects, or should it also be used to alter genes in the individual's germ cells so that succeeding generations would not inherit the defect? If it becomes easy to determine an individual's genetic makeup, should this information be used in evaluating applicants for educational and employment opportunities or health insurance? These conundrums have led to the creation of a branch of philosophy, named **bioethics,** designed to deal with them.

tion is essential for proper immune system function. Bone marrow cells (the precursors of white blood cells) were removed from the bodies of SCID-X1 victims, incubated with a vector containing a normal γc cytokine receptor gene, and returned to their bodies. The transgenic bone marrow cells restored immune system function.

 ## SUMMARY

1. Nucleotides consist of a purine or pyrimidine base linked to ribose to which at least one phosphate group is attached. RNA is made of ribonucleotides; DNA is made of deoxynucleotides (which contain 2′-deoxyribose).

2. In DNA, two antiparallel chains of nucleotides linked by phosphodiester bonds from a double helix. Bases in opposite strands pair: A with T, and G with C.

3. Single-stranded nucleic acids, such as RNA, can adopt stem–loop structures.

4. DNA carries genetic information in its sequence of nucleotides. When DNA is replicated, each strand acts as a template for the synthesis of a complementary strand.

5. According to the central dogma of molecular biology, one strand of the DNA of a gene is transcribed into mRNA. The RNA is then translated into protein by the ordered addition of amino acids that are bound to tRNA molecules that base-pair with the mRNA at the ribosome.

6. Restriction endonucleases that recognize certain sequences of DNA are used to specifically cleave DNA molecules.

7. Gel electrophoresis is used to separate and measure the sizes of DNA fragments.

8. In the chain-terminator method of DNA sequencing, the sequence of nucleotides in a DNA strand is determined by enzymatically synthesizing complementary polynucleotides that terminate with a dideoxy analog of each of the four nucleotides. Polynucleotide fragments of increasing size are separated by electrophoresis to reconstruct the original sequence.

9. Mutations and other changes to DNA are the basis for the evolution of organisms.

10. In molecular cloning, a fragment of foreign DNA is inserted into a vector for amplification in a host cell. Transformed cells can be identified by selectable markers.

11. Genomic libraries contain all the DNA of an organism. Clones harboring particular DNA sequences are identified by screening procedures.

12. The polymerase chain reaction amplifies selected sequences of DNA.

13. Recombinant DNA methods are used to produce wild-type or selectively mutagenized proteins in cells or entire organisms.

REFERENCES

DNA Structure and Function

Bloomfield, V.A., Crothers, D.M., and Tinoco, I., Jr., *Nucleic Acids. Structures, Properties, and Functions,* University Science Books (2000).

Dickerson, R.E., DNA structure from A to Z, *Methods Enzymol.* **211,** 67–111 (1992). [Describes the various crystallographic forms of DNA.]

Thieffry, D., Forty years under the central dogma, *Trends Biochem. Sci.* **23,** 312–316 (1998). [Traces the origins, acceptance, and shortcomings of the idea that nucleic acids contain biological information.]

Watson, J.D. and Crick, F.H.C., Molecular structure of nucleic acids, *Nature* **171,** 737–738 (1953); *and* Genetical implications of the structure of deoxyribonucleic acid, *Nature* **171,** 964–967 (1953). [The seminal papers that are widely held to mark the origin of modern molecular biology.]

DNA Sequencing

Galperin, M.Y., The molecular biology database collection: 2004 update, *Nucleic Acids Res.* **32,** Database issue D3–D22 (2004). [This article lists 548 databases covering various aspects of molecular biology, biochemistry, and genetics. Additional articles in the same issue provide more information on individual databases. Also available at http://nar.oupjournals.org.]

Graham, C.A. and Hill, A.J.M. (Eds.), *DNA Sequencing Protocols* (2nd ed.), Humana Press (2001).

Higgins, D. and Taylor, W. (Eds.), *Bioinformatics. Sequence, Structure and Databanks,* Oxford University Press (2000).

International Human Genome Sequencing Consortium, Initial sequencing and analysis of the human genome, *Nature* **409,** 860–921 (2001) *and* Venter, J.C., *et al.,* The sequence of the human genome, *Science* **291,** 1304–1351 (2001). [These and other papers in the same issues of *Nature* and *Science* describe the data that constitute the draft sequence of the human genome and discuss how this information can be used in understanding biological function, evolution, and human health.]

Recombinant DNA Technology

Ausubel, F.M., Brent, R., Kingston, R.E., Moore, D.D., Seidman, J.G., Smith, J.A., and Struhl, K., *Short Protocols in Molecular Biology* (4th ed.), Wiley (1999).

Nicholl, D.S.T., *An Introduction to Genetic Engineering* (2nd ed.), Cambridge University Press (2003).

Pingoud, A. and Jeltsch, A., Recognition and cleavage of DNA by type-II restriction endonucleases, *Eur. J. Biochem.* **246,** 1–22 (1997). [Includes an overview of different types of restriction enzymes.]

Sambrook, J., Fritsch, E.F., and Maniatis, T., *Molecular Cloning* (2nd ed.), Cold Spring Harbor Laboratory (1989). [A three-volume "bible" of laboratory protocols with accompanying background explanations.]

 ## KEY TERMS

nucleotide	complementary base pairing	endonuclease	anneal
nucleic acid	genome	exonuclease	ligation
nucleoside	chromosome	palindrome	selectable marker
RNA	diploid	sticky ends	plaque
DNA	haploid	blunt ends	DNA library
polynucleotide	bp	gel electrophoresis	genomic library
phosphodiester bond	kb	restriction map	shotgun cloning
nucleotide residue	stem–loop	dNTP	cDNA
5′ end	gene	primer	colony (*in situ*) hybridization
3′ end	transformation	ddNTP	replica plating
monomer	replication	point mutation	autoradiography
dimer	transcription	recombination	PCR
trimer	translation	transposition	expression vector
tetramer	mRNA	phenotype	overproducer
oligomer	ribosome	vector	site-directed mutagenesis
Chargaff's rules	rRNA	cloning	wild type
tautomer	tRNA	clone	transgenic organism
double helix	gene expression	plasmid	transgene
antiparallel	bacteriophage	recombinant DNA	gene knockout
major groove	modification methylase	BAC	gene therapy
minor groove	restriction endonuclease	YAC	

 ## STUDY EXERCISES

1. Describe the general structure of a nucleotide.

2. Describe the major features of the double-helical structure of DNA.

3. List the chemical and biological differences between DNA and RNA.

4. Summarize the central dogma of molecular biology and explain how a mutation in DNA can alter a protein.

5. Describe the Watson–Crick model of DNA.

6. How does the restriction–modification system operate?

7. Explain the chain-terminator (dideoxy) procedure for sequencing DNA.

8. Summarize the steps involved in cloning a gene.

 ## PROBLEMS

1. Kinases are enzymes that transfer a phosphoryl group from a nucleoside triphosphate. Which of the following are valid kinase-catalyzed reactions?

   (a) ATP + GDP → ADP + GTP

   (b) ATP + GMP → AMP + GTP

   (c) ADP + CMP → AMP + CDP

   (d) AMP + ATP → 2 ADP

2. A diploid organism with a 45,000-kb haploid genome contains 21% G residues. Calculate the number of A, C, G, and T residues in the DNA of each cell in this organism.

3. A segment of DNA containing 20 base pairs includes 7 guanine residues. How many adenine residues are in the segment? How many uracil residues are in the segment?

4. Draw the tautomeric forms of (a) adenine and (b) cytosine.

5. The adenine derivative inosine can base-pair with both cytosine and adenine. Show the structures of these base pairs.

Inosine

6. Explain why the strands of a DNA molecule can be separated more easily at pH > 11.

7. How many different amino acids could theoretically be en-

coded by nucleic acids containing four different nucleotides if (a) each nucleotide coded for one amino acid; (b) consecutive sequences of two nucleotides coded for one amino acid; (c) consecutive sequences of three nucleotides coded for one amino acid; (d) consecutive sequences of four nucleotides coded for one amino acid?

8. The recognition sequence for the restriction enzyme *Taq*I is T↓CGA. Indicate the products of the reaction of *Taq*I with the DNA sequence shown.

5′–ACGTCGAATC–3′
3′–TGCAGCTTAG–5′

9. Using the data in Table 3-2, identify restriction enzymes that (a) produce blunt ends; (b) recognize and cleave the same sequence (called **isoschizomers**); (c) produce identical sticky ends.

10. Construct a restriction map for a circular plasmid from the following data:

Restriction enzyme (kb)	Fragment sizes
*Eco*RI	4.0
*Hae*II	1.6, 2.4
*Pst*I	1.9, 2.1
*Eco*RI and *Hae*II	0.7, 1.6, 1.7
*Eco*RI and *Pst*I	0.8, 1.3, 1.9
*Hae*II and *Pst*I	0.6, 0.9, 1.0, 1.5

11. Describe the outcome of a chain-terminator sequencing pro-

cedure in which (a) too little ddNTP is added; (b) too much ddNTP is added; (c) too few primers are present; (d) too many primers are present.

12. Calculate the number of clones required to obtain with a probability of 0.99 a specific 5-kb fragment from *C. elegans* (Table 3-3).

13. Describe how to select recombinant clones if a foreign DNA is inserted into the polylinker site of pUC18 and then introduced into *E. coli* cells.

14. Describe the possible outcome of a PCR experiment in which (a) one of the primers is inadvertently omitted from the reaction mixture; (b) one of the primers is complementary to several sites in the starting DNA sample; (c) there is a single-stranded break in the target DNA sequence, which is present in only one copy in the starting sample; (d) there is a double-stranded break in the target DNA sequence, which is present in only one copy in the starting sample.

15. Write the sequences of the two 12-residue primers that could be used to amplify the following DNA segment by PCR.

ATAGGCATAGGCCCATATGGCATAAGGCTT-
TATAATATGCGATAGGCGCTGGTCAG

[Problem provided by Bruce Wightman, Muhlenberg College.]

16. (a) Why is a genomic library larger than a cDNA library for a given organism?

(b) Why do cDNA libraries derived from different cell types within the same organism differ from each other?

⬛ BIOINFORMATICS EXERCISES

Bioinformatics Exercises are available for this chapter in the tabbed section at the end of the book and on the text website at www.wiley.com/college/voet.

In addition to the well-known taste sensations of sweet, sour, salty, and bitter is umami, the taste sensation elicited by monosodium glutamate (MSG), an amino acid commonly used as a flavor enhancer. The umami taste receptor on the mammalian tongue recognizes and responds to a variety of amino acids but not to their mirror-image isomers. The evolutionary significance of an amino acid taste receptor is obvious, since amino acids are the building blocks of proteins as well as precursors of other compounds. [Jackson Vereen/ Foodpix/PictureArts Corp.]

Amino Acids

When scientists first turned their attention to nutrition, early in the nineteenth century, they quickly discovered that natural products containing nitrogen were essential for the survival of animals. In 1839, the Dutch chemist G. J. Mulder coined the term **protein** (Greek: *proteios,* primary) for this class of compounds. The physiological chemists of that time did not realize that proteins were actually composed of smaller components, amino acids, although the first amino acids had been isolated in 1830. In fact, for many years, it was believed that substances from plants—including proteins— were incorporated whole into animal tissues. This misconception was laid to rest when the process of digestion came to light. After it became clear that ingested proteins were broken down to smaller compounds containing amino acids, scientists turned their attention to the nutritive qualities of those compounds (Box 4-1).

Modern studies of proteins and amino acids owe a great deal to nineteenth and early twentieth century experiments. We now understand that nitrogen-containing amino acids are essential for life and that they are the building blocks of proteins. The central role of amino acids in biochemistry is perhaps not surprising: Several amino acids are among the organic compounds believed to have appeared early in the earth's history (Section 1-1A). Amino acids, as ancient and ubiquitous molecules, have been co-opted by evolution for a variety of purposes in living systems. We begin this chapter by discussing the structures and chemical properties of the common amino acids, including their stereochemistry, and end with a brief summary of the structures and functions of some related compounds.

1   Amino Acid Structure

The analyses of a vast number of proteins from almost every conceivable source have shown that *all proteins are composed of 20 "standard" amino acids.* Not every protein contains all 20 types of amino acids, but most proteins contain most if not all of the 20 types.

The common amino acids are known as **α-amino acids** because they have a primary amino group (—NH$_2$) as a substituent of the **α carbon** atom, the carbon next to the carboxylic acid group (—COOH; Fig. 4-1). The sole exception is proline, which has a secondary amino group (—NH—), although for uniformity we shall refer to proline as an α-amino acid. The 20 standard amino acids differ in the structures of their side chains **(R groups).** Table 4-1 displays the names and complete chemical structures of the 20 standard amino acids.

Figure 4-1 General structure of an α-amino acid. The R groups differentiate the 20 standard amino acids.

A   General Properties

The amino and carboxylic acid groups of amino acids readily ionize. The pK values of the carboxylic acid groups (represented by pK_1 in Table 4-1) lie in a small range around 2.2, while the pK values of the α-amino groups (pK_2) are near 9.4. *At physiological pH (~7.4), the amino groups are protonated and the carboxylic acid groups are in their conjugate base (carboxylate) form* (Fig. 4-2). An amino acid can therefore act as both an acid and a base. Table 4-1 also lists the pK values for the seven side chains that contain ionizable groups (pK_R).

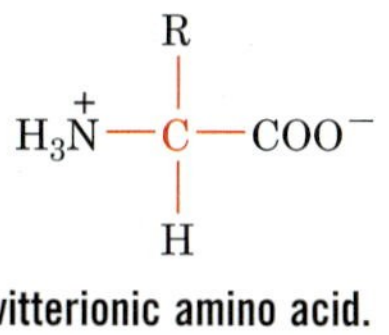

Molecules such as amino acids, which bear charged groups of opposite polarity, are known as **zwitterions** or **dipolar ions.** Amino acids, like other ionic compounds, are more soluble in polar solvents than in nonpolar sol-

Figure 4-2 A zwitterionic amino acid. At physiological pH, the amino group is protonated and the carboxylic acid group is unprotonated.

Table 4-1 *Key to Structure.* Covalent Structures and Abbreviations of the "Standard" Amino Acids of Proteins, Their Occurrence, and the pK Values of Their Ionizable Groups

Name, Three-letter Symbol, and One-letter Symbol	Structural Formula[a]	Residue Mass (D)[b]	Average Occurrence in Proteins (%)[c]	pK_1 α-COOH[d]	pK_2 α-NH$_3^+$[d]	pK_R Side Chain[d]
Amino acids with nonpolar side chains						
Glycine Gly G		57.0	7.2	2.35	9.78	
Alanine Ala A		71.1	7.8	2.35	9.87	
Valine Val V		99.1	6.6	2.29	9.74	
Leucine Leu L		113.2	9.1	2.33	9.74	
Isoleucine Ile I		113.2	5.3	2.32	9.76	
Methionine Met M		131.2	2.2	2.13	9.28	
Proline Pro P		97.1	5.2	1.95	10.64	
Phenylalanine Phe F		147.2	3.9	2.20	9.31	
Tryptophan Trp W		186.2	1.4	2.46	9.41	

[a]The ionic forms shown are those predominating at pH 7.0 (except for that of histidine[f]) although residue mass is given for the neutral compound. The C$_\alpha$ atoms, as well as those atoms marked with an asterisk, are chiral centers with configurations as indicated according to Fischer projection formulas (Section 4-2). The standard organic numbering system is provided for heterocycles.

[b]The residue masses are given for the neutral residues. For the molecular masses of the parent amino acids, add 18.0 D, the molecular mass of H$_2$O, to the residue masses. For side chain masses, subtract 56.0 D, the formula mass of a peptide group, from the residue masses.

[c]Calculated from a database of nonredundant proteins containing 300,688 residues as compiled by Doolittle, R.F. *in* Fasman, G.D. (Ed.), *Predictions of Protein Structure and the Principles of Protein Conformation*, Plenum Press (1989).

[d]Data from Dawson, R.M.C., Elliott, D.C., Elliott, W.H., and Jones, K.M., *Data for Biochemical Research* (3rd ed.), pp. 1–31, Oxford Science Publications (1986).

Table 4-1 (continued)

Name, Three-letter Symbol, and One-letter Symbol	Structural Formula[a]	Residue Mass (D)[b]	Average Occurrence in Proteins (%)[c]	pK_1 α-COOH[d]	pK_2 α-NH$_3^+$[d]	pK_R Side Chain[d]
Amino acids with uncharged polar side chains						
Serine Ser S	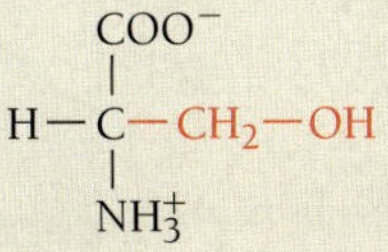 H—C(NH_3^+)(COO$^-$)—CH$_2$—OH	87.1	6.8	2.19	9.21	
Threonine Thr T	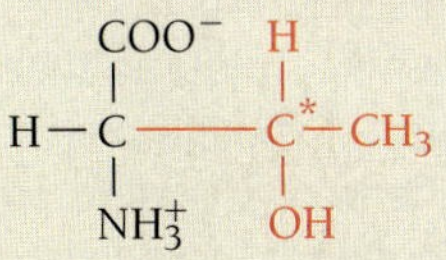 H—C(NH_3^+)(COO$^-$)—C*(H)(OH)—CH$_3$	101.1	5.9	2.09	9.10	
Asparagine[e] Asn N	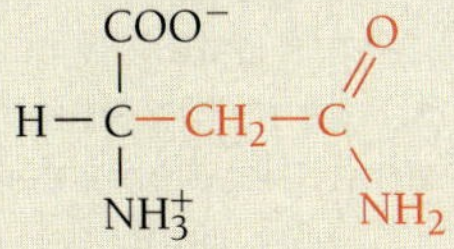 H—C(NH_3^+)(COO$^-$)—CH$_2$—C(=O)—NH$_2$	114.1	4.3	2.14	8.72	
Glutamine[e] Gln Q	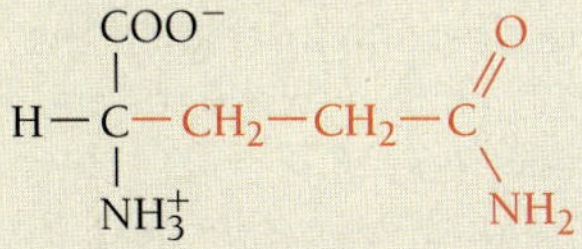 H—C(NH_3^+)(COO$^-$)—CH$_2$—CH$_2$—C(=O)—NH$_2$	128.1	4.3	2.17	9.13	
Tyrosine Tyr Y	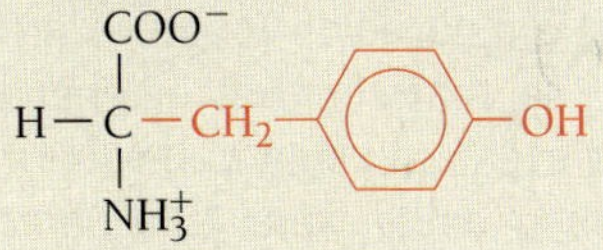 H—C(NH_3^+)(COO$^-$)—CH$_2$—C$_6$H$_4$—OH	163.2	3.2	2.20	9.21	10.46 (phenol)
Cysteine Cys C	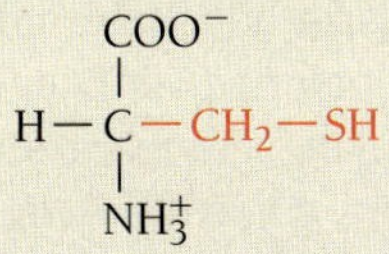 H—C(NH_3^+)(COO$^-$)—CH$_2$—SH	103.1	1.9	1.92	10.70	8.37 (sulfhydryl)
Amino acids with charged polar side chains						
Lysine Lys K	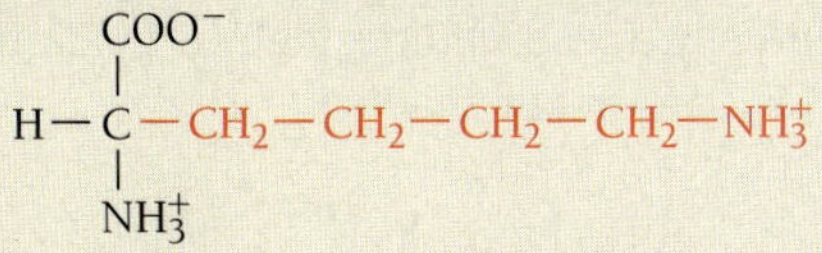 H—C(NH_3^+)(COO$^-$)—CH$_2$—CH$_2$—CH$_2$—CH$_2$—NH$_3^+$	128.2	5.9	2.16	9.06	10.54 (ε-NH$_3^+$)
Arginine Arg R	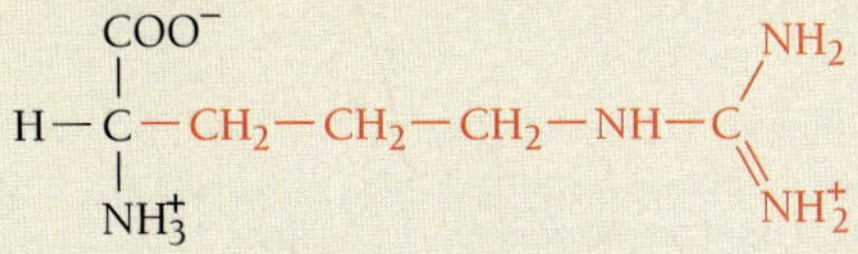 H—C(NH_3^+)(COO$^-$)—CH$_2$—CH$_2$—CH$_2$—NH—C(NH$_2$)(=NH$_2^+$)	156.2	5.1	1.82	8.99	12.48 (guanidino)
Histidine[f] His H	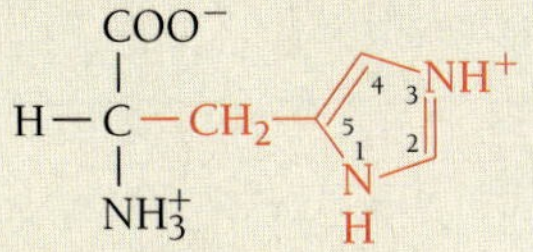 H—C(NH_3^+)(COO$^-$)—CH$_2$—(imidazole)	137.1	2.3	1.80	9.33	6.04 (imidazole)
Aspartic acid[e] Asp D	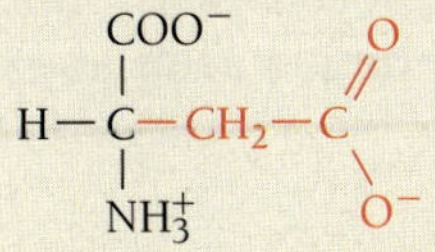 H—C(NH_3^+)(COO$^-$)—CH$_2$—C(=O)—O$^-$	115.1	5.3	1.99	9.90	3.90 (β-COOH)
Glutamic acid[e] Glu E	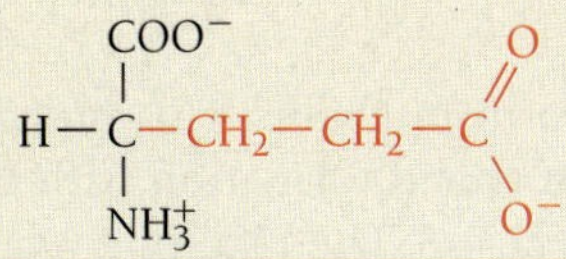 H—C(NH_3^+)(COO$^-$)—CH$_2$—CH$_2$—C(=O)—O$^-$	129.1	6.3	2.10	9.47	4.07 (γ-COOH)

[e]The three- and one-letter symbols for asparagine *or* aspartic acid are Asx and B, whereas for glutamine *or* glutamic acid they are Glx and Z. The one-letter symbol for an undetermined or "nonstandard" amino acid is X.

[f]Both neutral and protonated forms of histidine are present at pH 7.0, since its pK_R is close to 7.0.

BOX 4-1

Pathways of Discovery

William C. Rose and the Discovery of Threonine

Identifying the amino acid constituents of proteins was a scientific challenge that grew out of studies of animal nutrition. At the start of the twentieth century, physical chemists (the term *biochemist* was not yet used) recognized that not all foods provided adequate nutrition. For example, rats fed the corn protein zein as their only source of nitrogen failed to grow unless the amino acids tryptophan and lysine were added to their diet. Knowledge of metabolism at that time was mostly limited to information gleaned from studies in which intake of particular foods in experimental subjects (including humans) was linked to the urinary excretion of various compounds. Results of such studies were consistent with the idea that compounds could be transformed into other compounds, but clearly, nutrients were not wholly interchangeable.

At the University of Illinois, William C. Rose focused his research on nutritional studies to decipher the metabolic relationships of nitrogenous compounds. Among other things, his studies of rat growth and nutrition helped show that purines and pyrimidines were derived from amino acids but that these compounds could not replace dietary amino acids.

In order to examine the nutritional requirements for individual amino acids, Rose hydrolyzed proteins to obtain their component amino acids and then selectively removed certain amino acids. In one of his first experiments, he removed arginine and histidine from a hydrolysate of the milk protein casein. Rats fed on this preparation lost weight unless the amino acid histidine was added back to the food. However, adding back arginine did not compensate for the apparent requirement for histidine. These results prompted Rose to investigate the requirements for all the amino acids. Using similar experimental approaches, Rose demonstrated that cysteine, histidine, and tryptophan could not be replaced by other amino acids.

From preparations based on hydrolyzed proteins, Rose moved to mixtures of pure amino acids. Thirteen of the 19

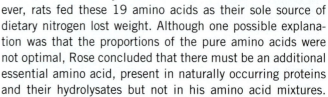

William C. Rose (1887–1985)

known amino acids could be purified, and the other six synthesized. However, rats fed these 19 amino acids as their sole source of dietary nitrogen lost weight. Although one possible explanation was that the proportions of the pure amino acids were not optimal, Rose concluded that there must be an additional essential amino acid, present in naturally occurring proteins and their hydrolysates but not in his amino acid mixtures.

After several years of effort, Rose obtained and identified the missing amino acid. In work published in 1935, Rose showed that adding this amino acid to the other 19 could support rat growth. Thus, the twentieth and last amino acid, threonine, was discovered.

Experiments extending over the next 20 years revealed that 10 of the 20 amino acids found in proteins are nutritionally essential, so that removal of one of these causes growth failure and eventually death in experimental animals. The other 10 amino acids were considered "dispensible" since animals could synthesize adequate amounts of them.

Rose's subsequent work included verifying the amino acid requirements of humans, using graduate students as subjects. Knowing which amino acids were required—and in what amounts—for normal health made it possible to evaluate the potential nutritive value of different types of food proteins. Eventually, these findings helped guide the formulations used for intravenous feeding.

McCoy, R.H., Meyer, C.E., and Rose, W.C., Feeding experiments with mixtures of highly purified amino acids. VIII. Isolation and identification of a new essential amino acid, *J. Biol. Chem.* **112**, 283–302 (1935). [Available at http://www.jbc.org.]

vents. As we shall see, the ionic properties of the side chains influence the physical and chemical properties of free amino acids and amino acids in proteins.

B Peptide Bonds

Amino acids can be polymerized to form chains. This process can be represented as a **condensation reaction** (bond formation with the elimination of a water molecule), as shown in Fig. 4-3. The resulting CO—NH linkage, an amide linkage, is known as a **peptide bond.**

Polymers composed of two, three, a few (3–10), and many amino acid units are known, respectively, as **dipeptides, tripeptides, oligopeptides,** and **polypeptides.** These substances, however, are often referred to simply as "peptides." After they are incorporated into a peptide, the individual amino acids (the monomeric units) are referred to as **amino acid residues.**

Polypeptides are linear polymers rather than branched chains; that is, each amino acid residue participates in two peptide bonds and is linked to its neighbors in a head-to-tail fashion. The residues at the two ends of the polypeptide each participate in just one peptide bond. The residue with a free amino group (by convention, the leftmost residue, as shown in Fig. 4-3) is called the **amino terminus** or **N-terminus.** The residue with a free carboxylate group (at the right) is called the **carboxyl terminus** or **C-terminus.**

Proteins are molecules that contain one or more polypeptide chains. *Variations in the length and the amino acid sequence of polypeptides contribute to the diversity in the shape and biological functions of proteins,* as we shall see in succeeding chapters.

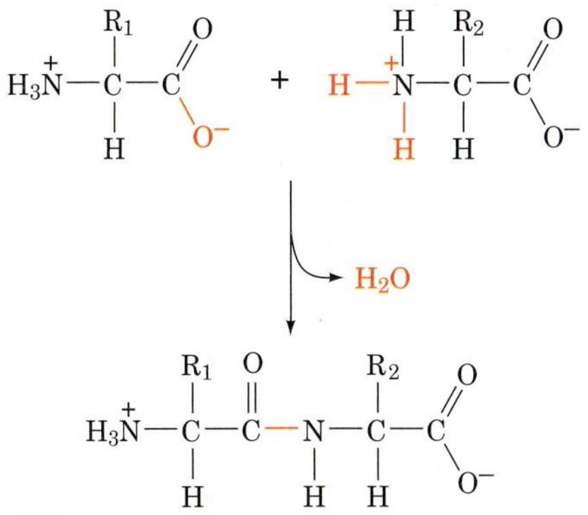

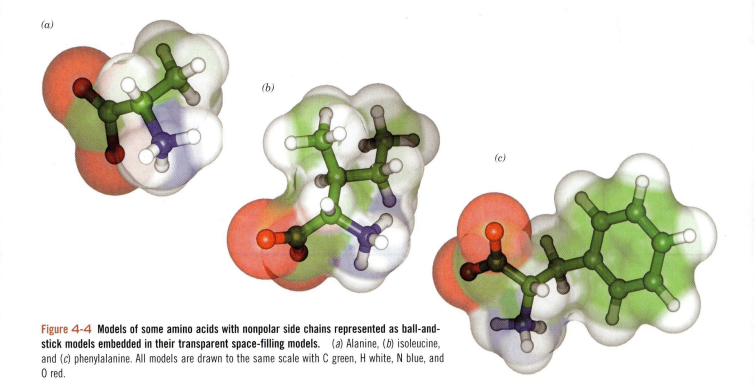

Figure 4-3 **Condensation of two amino acids.** Formation of a CO—NH bond with the elimination of a water molecule produces a dipeptide. The peptide bond is shown in red. The residue with a free amino group is the N-terminus of the peptide, and the residue with a free carboxylate group is the C-terminus.

C Classification and Characteristics

The most useful way to classify the 20 standard amino acids is by the polarities of their side chains. According to the most common classification scheme, there are three major types of amino acids: (1) those with nonpolar R groups, (2) those with uncharged polar R groups, and (3) those with charged polar R groups.

The Nonpolar Amino Acid Side Chains Have a Variety of Shapes and Sizes. Nine amino acids are classified as having nonpolar side chains. The three-dimensional shapes of some of these amino acids are shown in Fig. 4-4. **Glycine** has the smallest possible side chain, an H atom. **Alanine, valine, leucine,** and **isoleucine** have aliphatic hydrocarbon side chains ranging in

(a)

(b)

(c)

Figure 4-4 **Models of some amino acids with nonpolar side chains represented as ball-and-stick models embedded in their transparent space-filling models.** (*a*) Alanine, (*b*) isoleucine, and (*c*) phenylalanine. All models are drawn to the same scale with C green, H white, N blue, and O red.

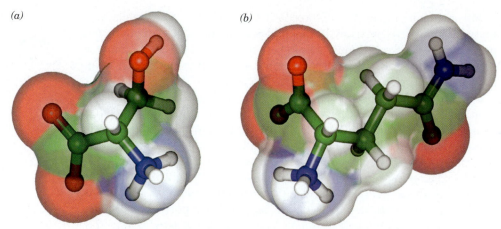

Figure 4-5 Models of some amino acids with uncharged polar side chains.   (*a*) Serine.
(*b*) Glutamine. Atoms are represented and colored as in Fig. 4-4. Note the presence of
electronegative atoms on the side chains.

size from a methyl group for alanine to isomeric butyl groups for leucine
and isoleucine. **Methionine** has a thioether side chain that resembles an
n-butyl group in many of its physical properties (C and S have nearly equal
electronegativities, and S is about the size of a methylene group). **Proline**
has a cyclic pyrrolidine side group. **Phenylalanine** (with its phenyl moiety)
and **tryptophan** (with its indole group) contain aromatic side groups, which
are characterized by bulk as well as nonpolarity.

Uncharged Polar Side Chains Have Hydroxyl, Amide, or Thiol Groups.   Six amino
acids are commonly classified as having uncharged polar side chains
(Table 4-1 and Fig. 4-5). **Serine** and **threonine** bear hydroxylic R groups of
different sizes. **Asparagine** and **glutamine** have amide-bearing side chains
of different sizes. **Tyrosine** has a phenolic group (and, like phenylalanine
and tryptophan, is aromatic). **Cysteine** is unique among the 20 amino acids
in that it has a thiol group that can form a disulfide bond with another cysteine through the oxidation of the two thiol groups (Fig. 4-6). This dimeric
compound was referred to in the older biochemical literature as the amino
acid **cystine,** and cysteine was occasionally called a half-cystine residue.

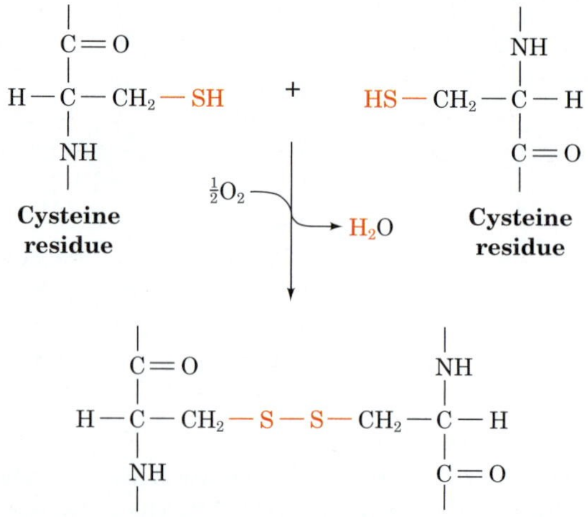

Figure 4-6 Disulfide-bonded cysteine residues.   The disulfide bond forms when the two thiol
groups are oxidized.

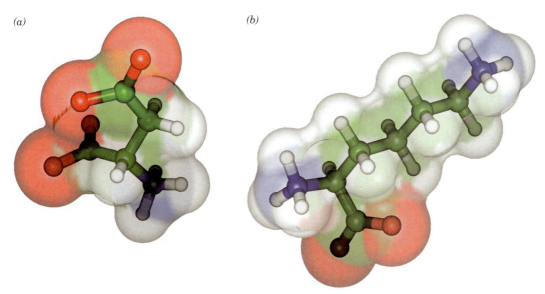

(a)   *(b)*

Figure 4-7 **Models of some amino acids with charged polar side chains.**   *(a)* Aspartate. *(b)* Lysine. Atoms are represented and colored as in Fig. 4-4.

Charged Polar Side Chains Are Positively or Negatively Charged. Five amino acids have charged side chains (Table 4-1 and Fig. 4-7). The side chains of the basic amino acids are positively charged at physiological pH values. **Lysine** has a butylammonium side chain, and **arginine** bears a guanidino group. As shown in Table 4-1, **histidine** carries an imidazolium moiety. Note that only histidine, with a pK_R of 6.04, readily ionizes within the physiological pH range. Consequently, both the neutral and cationic forms occur in proteins. In fact, the protonation–deprotonation of histidine side chains is a feature of numerous enzymatic reaction mechanisms.

The side chains of the acidic amino acids, **aspartic acid** and **glutamic acid,** are negatively charged above pH 3; in their ionized state, they are often referred to as **aspartate** and **glutamate.** Asparagine and glutamine are, respectively, the amides of aspartic acid and glutamic acid.

The allocation of the 20 amino acids among the three different groups is somewhat arbitrary. For example, glycine and alanine, the smallest of the amino acids, and tryptophan, with its heterocyclic ring, might just as well be classified as uncharged polar amino acids. Similarly, tyrosine and cysteine, with their ionizable side chains, might also be thought of as charged polar amino acids, particularly at higher pH values. In fact, the deprotonated side chain of cysteine (which contains the thiolate anion, S^-) occurs in a variety of enzymes, where it actively participates in chemical reactions.

Inclusion of a particular amino acid in one group or another reflects not just the properties of the isolated amino acid, but its behavior when it is part of a polypeptide. The structures of most polypeptides depend on a tendency for polar and ionic side chains to be hydrated and for nonpolar side chains to associate with each other rather than with water. This property of polypeptides is the hydrophobic effect (Section 2-1C) in action. As we shall see, the chemical and physical properties of amino acid side chains also govern the chemical reactivity of the polypeptide. It is worthwhile studying the structures of the 20 standard amino acids in order to appreciate how they vary in polarity, acidity, aromaticity, bulk, conformational flexibility, ability to cross-link, ability to hydrogen bond, and reactivity toward other groups.

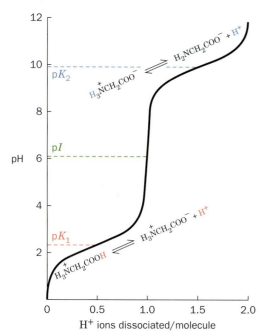

Figure 4-8 Titration of glycine. [After Meister, A., *Biochemistry of Amino Acids* (2nd ed.), Vol. 1, *p.* 30, Academic Press (1965).] *See the Animated Figures.*

D Acid–Base Properties

The α-amino acids have two or, for those with ionizable side chains, three acid–base groups. The titration curve of glycine, the simplest amino acid, is shown in Fig. 4-8. At low pH values, both acid–base groups of glycine are fully protonated, so that the cationic form ($^+H_3NCH_2COOH$) predominates. In the course of titration with a strong base such as NaOH, glycine loses two protons in the stepwise fashion characteristic of a polyprotic acid.

The pK values of glycine's two ionizable groups are sufficiently different so that the Henderson–Hasselbalch equation (Section 2-2B)

$$pH = pK + \log \frac{[A^-]}{[HA]} \qquad [4\text{-}1]$$

adequately describes each leg of the titration curve. Consequently, the pK for each ionization step is the midpoint of the corresponding leg of the titration curve (Section 2-2C). At pH 2.35, the concentrations of the cationic form ($^+H_3NCH_2COOH$) and the zwitterionic form ($^+H_3NCH_2COO^-$) are equal; similarly, at pH 9.78, the concentrations of the zwitterionic form and the anionic form ($H_2NCH_2COO^-$) are equal. Note that *in aqueous solution, the completely neutral form (H_2NCH_2COOH) is present only in vanishingly small quantities.*

The pH at which a molecule carries no net electric charge is known as its **isoelectric point, p*I*.** For the α-amino acids, the application of the Henderson–Hasselbalch equation indicates that, to a high degree of precision,

$$pI = \tfrac{1}{2}(pK_i + pK_j) \qquad [4\text{-}2]$$

where K_i and K_j are the dissociation constants of the two ionizations involving the neutral species. For monoamino, monocarboxylic acids such as glycine, K_i and K_j represent K_1 and K_2. However, for aspartic and glutamic acids, K_i and K_j are K_1 and K_R, whereas for arginine, histidine, and lysine, these quantities are K_R and K_2.

The p*K* Values of Ionizable Groups Depend on Nearby Groups.   The value of pK_1 of glycine (2.35) is much lower than the pK of a simple monocarboxylic acid such as acetic acid (CH_3COOH, pK = 4.76). The large difference in pK values for the same functional group is caused by the electrostatic influence of glycine's positively charged ammonium group. The NH_3^+ group electrostatically stabilizes the COO^- group more than the COOH group. Similarly, the NH_3^+ group of glycine ($pK_2 = 9.78$) is significantly more acidic than are aliphatic amines (pK ≈ 10.7) because of the electron-withdrawing character of glycine's carboxylate group, but it is less acidic than glycine ethyl ester (pK = 7.75), whose carboxyl group is uncharged. Thus, both electronic and electrostatic effects influence the pK of the NH_3^+ group.

The electronic influence of one functional group on another depends on the distance between the groups. For example, the ionization constant of lysine's side chain amino group (separated from the α carbon by four methylene groups) is indistinguishable from that of an aliphatic amine.

Of course, amino acid residues in the interior of a polypeptide chain do not have free α-amino and carboxyl groups that can ionize (these groups are joined in peptide bonds; Fig. 4-3). Furthermore, the pK values of all ionizable groups, including the N- and C-termini, may differ from the pK values listed in Table 4-1 for free amino acids. The pK values of α-carboxyl groups in unfolded proteins range from 3.5 to 4.0, while the pK values for α-amino groups range from 7.5 to 8.5. In addition, the three-dimensional

structure of a folded polypeptide chain may bring polar side chains and the N- and C-termini close together. The resulting electrostatic interactions between these groups may shift their pK values up to several pH units from the values in the corresponding free amino acids.

E A Few Words on Nomenclature

The three-letter abbreviations for the 20 standard amino acids given in Table 4-1 are widely used in the biochemical literature. Most of these abbreviations are taken from the first three letters of the name of the corresponding amino acid and are pronounced as written. The symbol **Glx** indicates Glu or Gln, and similarly, **Asx** means Asp or Asn. This ambiguous notation stems from laboratory experience: Asn and Gln are easily hydrolyzed to Asp and Glu, respectively, under the acidic or basic conditions often used to recover them from proteins (Section 5-3A). Without special precautions, it is impossible to tell whether a detected Glu was originally Glu or Gln, and likewise for Asp and Asn.

The one-letter symbols for the amino acids are also given in Table 4-1. This more compact code is often used when comparing the amino acid sequences of several similar proteins. Note that the one-letter symbol is usually the first letter of the amino acid's name. However, for sets of residues that have the same first letter, this is true only of the most abundant residue of the set.

Amino acid residues in polypeptides are named by dropping the suffix, usually **-ine,** in the name of the amino acid and replacing it by **-yl.** Polypeptide chains are described by starting at the N-terminus and proceeding to the C-terminus. The amino acid at the C-terminus is given the name of its parent amino acid. Thus, the compound

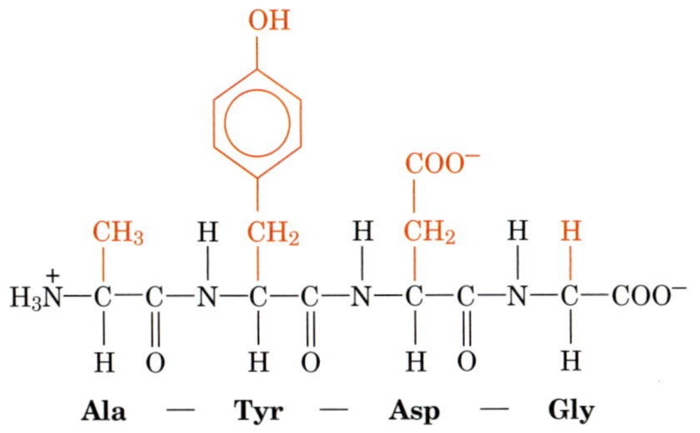

is called alanyltyrosylaspartylglycine. Obviously, such names for polypeptide chains of more than a few residues are extremely cumbersome. The tetrapeptide above can also be written as Ala-Tyr-Asp-Gly using the three-letter abbreviations, or AYDG using the one-letter symbols.

The various atoms of the amino acid side chains are often named in sequence with the Greek alphabet, starting at the carbon atom adjacent to the peptide carbonyl group. Therefore, as Fig. 4-9 indicates, the Lys residue is said to have an ε-amino group and Glu has a γ-carboxyl group. Unfortunately, this labeling system is ambiguous for several amino acids. Consequently, standard numbering schemes for organic molecules are also employed (and are indicated in Table 4-1 for the heterocyclic side chains).

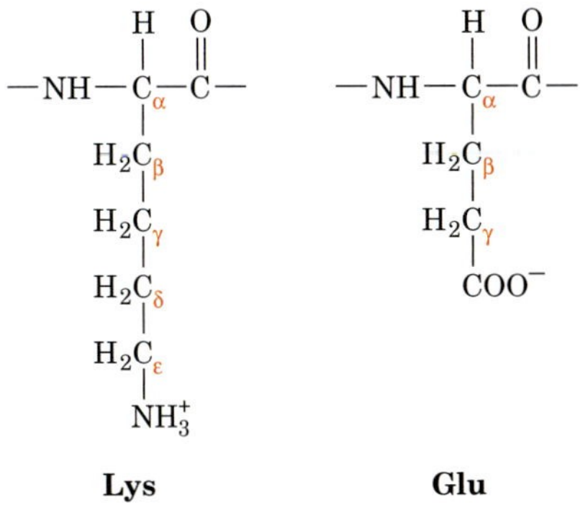

Figure 4-9 Greek nomenclature for amino acids. The carbon atoms are assigned sequential letters in the Greek alphabet, beginning with the carbon next to the carbonyl group.

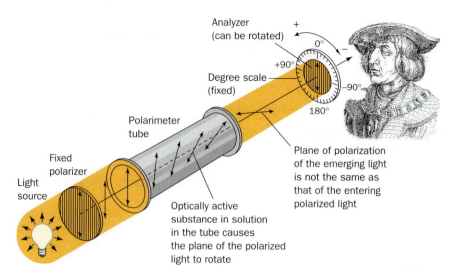

Figure 4-10 Diagram of a polarimeter. This device is used to measure optical rotation.

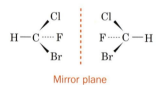

Mirror plane

Figure 4-11 The two enantiomers of fluorochlorobromomethane. The four substituents are tetrahedrally arranged around the central carbon atom. A dotted line indicates that a substituent lies behind the plane of the paper, a wedged line indicates that it lies above the plane of the paper, and a thin line indicates that it lies in the plane of the paper. The mirror plane relating the enantiomers is represented by a vertical dashed line.

2 Stereochemistry

With the exception of glycine, all the amino acids recovered from polypeptides are **optically active;** that is, they rotate the plane of polarized light. The direction and angle of rotation can be measured using an instrument known as a **polarimeter** (Fig. 4-10).

Optically active molecules are asymmetric; that is, they are not superimposable on their mirror image in the same way that a left hand is not superimposable on its mirror image, a right hand. This situation is characteristic of substances containing tetrahedral carbon atoms that have four different substituents. For example, the two molecules depicted in Fig. 4-11 are not superimposable since they are mirror images. The central atoms in such molecules are known as **asymmetric centers** or **chiral centers** and are said to have the property of **chirality** (Greek: *cheir,* hand). The C_α atoms of the amino acids (except glycine) are asymmetric centers. Glycine, which has two H atoms attached to its C_α atom, is superimposable on its mirror image and is therefore not optically active. Many biological molecules in addition to amino acids contain one or more chiral centers.

Chiral Centers Give Rise to Enantiomers. Molecules that are nonsuperimposable mirror images are known as **enantiomers** of one another. *Enantiomeric molecules are physically and chemically indistinguishable by most techniques. Only when probed asymmetrically, for example, by plane-polarized light or by reactants that also contain chiral centers, can they be distinguished or differentially manipulated.*

Unfortunately, there is no clear relationship between the structure of a molecule and the degree or direction to which it rotates the plane of polarized light. For example, leucine isolated from proteins rotates polarized light 10.4° to the left, whereas arginine rotates polarized light 12.5° to the right. (The enantiomers of these compounds rotate polarized light to the same degree but in the opposite direction.) It is not yet possible to predict optical rotation from the structure of a molecule or to derive the **absolute configuration** (spatial arrangement) of chemical groups around a chiral center from optical rotation measurements.

The Fischer Convention Describes the Configuration of Asymmetric Centers. Biochemists commonly use the **Fischer convention** to describe different

forms of chiral molecules. In this system, the configuration of the groups around an asymmetric center is compared to that of **glyceraldehyde,** a molecule with one asymmetric center. In 1891, Emil Fischer proposed that the spatial isomers, or **stereoisomers,** of glyceraldehyde be designated D-glyceraldehyde and L-glyceraldehyde (Fig. 4-12). The prefix L (note the use of a small uppercase letter) signified rotation of polarized light to the left (Greek: *levo,* left), and the prefix D indicated rotation to the right (Greek: *dextro,* right) by the two forms of glyceraldehyde. Fischer assigned the prefixes to the structures shown in Fig. 4-12 without knowing whether the structure on the left and the structure on the right were actually **levorotatory** and **dextrorotatory,** respectively. Only in 1949 did experiments confirm that Fischer's guess was indeed correct.

Fischer also proposed a shorthand notation for molecular configurations, known as **Fischer projections,** which are also given in Fig. 4-12. In the Fischer convention, horizontal bonds extend above the plane of the paper and vertical bonds extend below the plane of the paper.

The configuration of groups around any chiral center can be related to that of glyceraldehyde by chemically converting the groups to those of glyceraldehyde. For α-amino acids, the amino, carboxyl, R, and H groups around the C_α atom correspond to the hydroxyl, aldehyde, CH_2OH, and H groups, respectively, of glyceraldehyde.

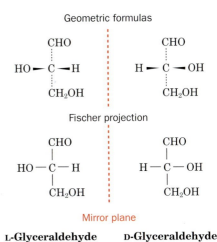

Figure 4-12 The Fischer convention. The enantiomers of glyceraldehyde are shown as geometric formulas (*top*) and as Fischer projections (*bottom*). In a Fischer projection, horizontal lines represent bonds that extend above the page and vertical lines represent bonds that extend below the page (in some Fischer projections, the central chiral carbon atom is not shown explicitly).

Therefore, L-glyceraldehyde and L-α-amino acids are said to have the same **relative configuration.** *All amino acids derived from proteins have the L stereochemical configuration;* that is, they all have the same relative configuration around their C_α atoms. Of course, the L or D designation of an amino acid does not indicate its ability to rotate the plane of polarized light. Many L-amino acids are dextrorotatory.

The Fischer system has some shortcomings, particularly for molecules with multiple asymmetric centers. Each asymmetric center can have two possible configurations, so a molecule with n chiral centers has 2^n different possible stereoisomers. Threonine and isoleucine, for example, each have two chiral carbon atoms, and therefore each has four stereoisomers, or two pairs of enantiomers. [The enantiomers (mirror images) of the L forms are the D forms.] For most purposes, the Fischer system provides an adequate description of biological molecules. A more precise nomenclature system is also occasionally used by biochemists (see Box 4-2).

Life Is Based on Chiral Molecules. Consider the ordinary chemical synthesis of a chiral molecule, which produces a **racemic** mixture (containing equal amounts of each enantiomer). In order to obtain a product with net asymmetry, a chiral process must be employed. One of the most striking characteristics of life is its production of optically active molecules. *Biosynthetic processes almost invariably produce pure stereoisomers.* The fact that the amino acid residues of proteins all have the L configuration is just one example of this phenomenon. Furthermore, because most biological molecules are chiral, a given molecule—present in a single enantiomeric form—will bind to or react with only a single enantiomer of another compound. For example, a protein made of L-amino acid residues that reacts with a

BOX 4-2

Perspectives in Biochemistry

The RS System

A system to unambiguously describe the configurations of molecules with more than one asymmetric center was devised in 1956 by Robert Cahn, Christopher Ingold, and Vladimir Prelog. In the **Cahn–Ingold–Prelog** or ***RS*** system, the four groups surrounding a chiral center are ranked according to a specific although arbitrary priority scheme: Atoms of higher atomic number rank above those of lower atomic number (e.g., —OH ranks above —CH$_3$). If the first substituent atoms are identical, the priority is established by the next atom outward from the chiral center (e.g., —CH$_2$OH takes precedence over —CH$_3$). The order of priority of some common functional groups is

SH > OH > NH$_2$ > COOH > CHO >
CH$_2$OH > C$_6$H$_5$ > CH$_3$ > H

The prioritized groups are assigned the letters W, X, Y, Z such that their order of priority ranking is W > X > Y > Z. To establish the configuration of the chiral center, it is viewed from the asymmetric center toward the Z group (lowest priority). If the order of the groups W → X → Y is clockwise, the configuration is designated *R* (Latin: *rectus,* right). If the

order of W → X → Y is counterclockwise, the configuration is designated *S* (Latin: *sinistrus,* left).

L-Glyceraldehyde is (*S*)-glyceraldehyde because the three highest priority groups are arranged counterclockwise when the H atom (*dashed lines*) is positioned behind the chiral C atom (*large circle*).

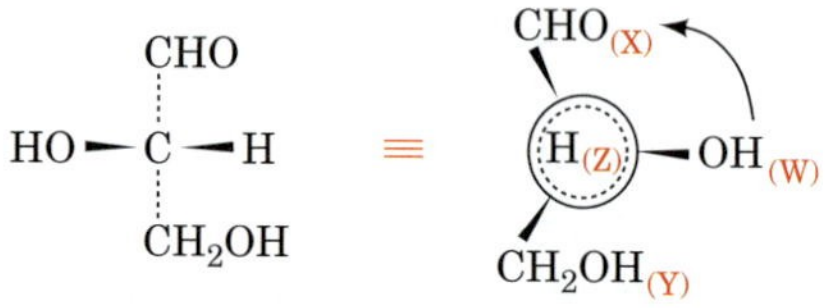

L-Glyceraldehyde   ***(S)*-Glyceraldehyde**

All the L-amino acids in proteins are (*S*)-amino acids except cysteine, which is (*R*)-cysteine because the S in its side chain increases its priority. Other closely related compounds with the same designation under the Fischer DL convention may have different representations under the *RS* system. The *RS* system is particularly useful for describing the chiralities of compounds with multiple asymmetric centers. Thus, L-threonine can also be called (2*S*,3*R*)-threonine.

particular L-amino acid does not readily react with the D form of that amino acid. An otherwise identical synthetic protein made of D-amino acid residues, however, readily reacts only with the corresponding D-amino acid.

D-Amino acid residues are components of some relatively short (<20 residues) bacterial polypeptides. These polypeptides are perhaps most widely distributed as constituents of bacterial cell walls (Section 8-3B). The presence of the D-amino acids renders bacterial cell walls less susceptible to attack by the **peptidases** (enzymes that hydrolyze peptide bonds) that are produced by other organisms to digest bacteria. Likewise, D-amino acids are components of many bacterially produced peptide antibiotics. Most peptides containing D-amino acids are not synthesized by the standard protein synthetic machinery, in which messenger RNA is translated at the ribosome by transfer RNA molecules with attached L-amino acids (Chapter 26). Instead, the D-amino acids are directly joined together by the action of specific bacterial enzymes.

The importance of stereochemistry in living systems is also a concern of the pharmaceutical industry. *Many drugs are chemically synthesized as racemic mixtures, although only one enantiomer has biological activity.* In most cases, the opposite enantiomer is biologically inert and is therefore packaged along with its active counterpart. This is true, for example, of the anti-inflammatory agent **ibuprofen,** only one enantiomer of which is physiologically active (Fig. 4-13). Occasionally, the inactive enantiomer of a useful drug produces harmful effects and must therefore be eliminated from the racemic mixture. The most striking example of this is the drug **thalido-**

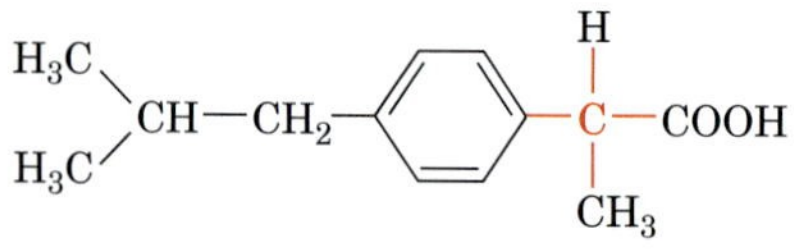

Ibuprofen

Figure 4-13 Ibuprofen.   Only the enantiomer shown has anti-inflammatory action. The chiral carbon is red.

mide (Fig. 4-14), a mild sedative whose inactive enantiomer causes severe birth defects. Partly because of the unanticipated problems caused by inactive drug enantiomers, **chiral organic synthesis** has become an active area of medicinal chemistry.

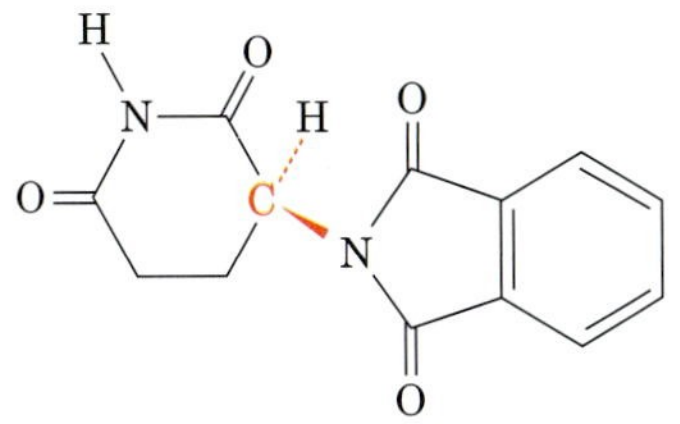

Thalidomide

Figure 4-14 Thalidomide. This drug was widely used in Europe as a mild sedative in the early 1960s. Its inactive enantiomer (not shown), which was present in equal amounts in the formulations used, causes severe birth defects in humans when taken during the first trimester of pregnancy. Thalidomide was often prescribed to alleviate the nausea (morning sickness) that is common during this period.

3 Amino Acid Derivatives

The 20 common amino acids are by no means the only amino acids that occur in biological systems. "Nonstandard" amino acid residues are often important constituents of proteins and biologically active peptides. In addition, many amino acids are not constituents of polypeptides at all but independently play a variety of biological roles.

A Side Chain Modifications in Proteins

The "universal" genetic code, which is nearly identical in all known life-forms (Section 26-1), specifies only the 20 standard amino acids of Table 4-1. Nevertheless, many other amino acids, some of which are shown in Fig. 4-15, are components of certain proteins. *In almost all cases, these unusual amino acids result from the specific modification of an amino acid residue after the polypeptide chain has been synthesized.*

Amino acid modifications include the simple addition of small chemical groups to certain amino acid side chains: hydroxylation, methylation, acetylation, carboxylation, and phosphorylation. Larger groups, including lipids and carbohydrate polymers, are attached to particular amino acid residues of certain proteins. The free amino and carboxyl groups at the N- and C-termini of a polypeptide can also be chemically modified. These modifications are important, if not essential, for the function of the protein. In some cases, several amino acid side chains together form a novel structure (Box 4-3).

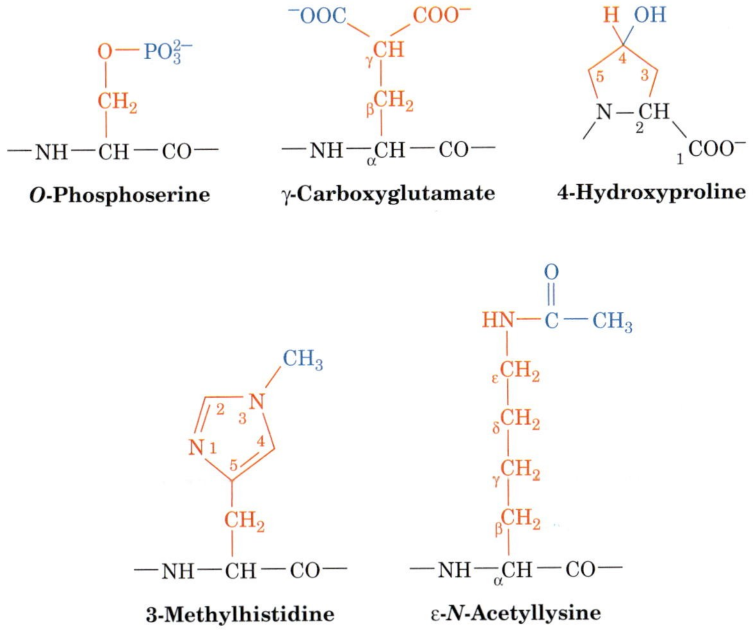

O-Phosphoserine   γ-Carboxyglutamate   4-Hydroxyproline

3-Methylhistidine   ε-*N*-Acetyllysine

Figure 4-15 Some modified amino acid residues in proteins. The side chains of these residues are derived from one of the 20 standard amino acids after the polypeptide has been synthesized. The standard R groups are red, and the modifying groups are blue.

BOX 4-3

Perspectives in Biochemistry

Green Fluorescent Protein

Genetic engineers often link a protein-coding gene to a "reporter gene," for example, the gene for an enzyme that yields a colored reaction product. The intensity of the colored compound can be used to estimate the level of expression of the engineered gene. One of the most useful reporter genes is the one that codes for green fluorescent protein. This protein, from the bioluminescent jellyfish *Aequorea victoria,* fluoresces with a peak wavelength of 508 nm (green light) when irradiated by ultraviolet or blue light (optimally 400 nm). Green fluorescent protein is nontoxic and intrinsically fluorescent; it requires no substrate or small molecule cofactor to fluoresce as do other highly fluorescent proteins. Consequently, when the gene for green fluorescent protein is linked to another gene, the level of expression of the fused genes can be measured noninvasively by fluorescence microscopy.

© 1999 Steven Haddock and Trevor Rivers/Monterey Bay Aquarium Research Institute, Monterey, California.

Green fluorescent protein consists of a chain of 238 amino acid residues. The light-emitting group is a derivative of three consecutive amino acids: Ser, Tyr, and Gly. After the protein has been synthesized, the three amino acids undergo spontaneous cyclization and oxidation. The carbonyl C atom of Ser forms a covalent bond to the amido N atom contributed by Gly, followed by the elimination of water and the oxidation of the C_α—C_β bond of Tyr to a double bond. The resulting structure contains a system of conjugated double bonds that gives the protein its fluorescent properties.

Fluorophore of green fluorescent protein

Cyclization between Ser and Gly is probably rapid, and the oxidation of the Tyr side chain (by O_2) is probably the rate-limiting step of fluorophore generation. Genetic engineering has introduced site-specific mutations that enhance fluorescence intensity and shift the wavelength of the emitted light to different colors, thereby making it possible to simultaneously monitor the expression of two or more different genes.

B   Biologically Active Amino Acids

The 20 standard amino acids undergo a bewildering number of chemical transformations to other amino acids and related compounds as part of their normal cellular synthesis and degradation. In a few cases, the intermediates of amino acid metabolism have functions beyond their immediate use as precursors or degradation products of the 20 standard amino

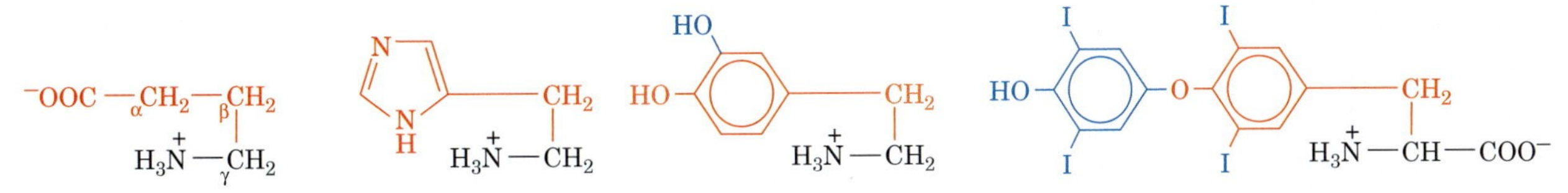

γ-Aminobutyric acid (GABA) Histamine Dopamine Thyroxine

Figure 4-16 Some biologically active amino acid derivatives. The remaining portions of the parent amino acids are black and red, and additional groups are blue.

acids. Moreover, many amino acids are synthesized not to be residues of polypeptides but to function independently. We shall see that many organisms use certain amino acids to transport nitrogen in the form of amino groups (Section 20-2A). Amino acids may also be oxidized as metabolic fuels to provide energy (Section 20-4). In addition, amino acids and their derivatives often function as chemical messengers for communication between cells (Fig. 4-16). For example, glycine, **γ-aminobutyric acid (GABA;** a glutamine decarboxylation product), and **dopamine** (a tyrosine derivative) are **neurotransmitters,** substances released by nerve cells to alter the behavior of their neighbors. **Histamine** (the decarboxylation product of histidine) is a potent local mediator of allergic reactions. **Thyroxine** (another tyrosine derivative) is an iodine-containing thyroid hormone that generally stimulates vertebrate metabolism.

Many peptides containing only a few amino acid residues have important physiological functions as hormones or other regulatory molecules. One nearly ubiquitous tripeptide called **glutathione** plays a role in cellular metabolism. Glutathione is a Glu–Cys–Gly peptide in which the γ-carboxylate group of the glutamate side chain forms an **isopeptide bond** with the amino group of the Cys residue (so called because a peptide bond is taken to be the amide bond formed between an α-carboxylate and an α-amino group of two amino acids). Two of these tripeptides (abbreviated **GSH**) undergo oxidation of their SH groups to form a dimeric disulfide-linked structure called **glutathione disulfide (GSSG):**

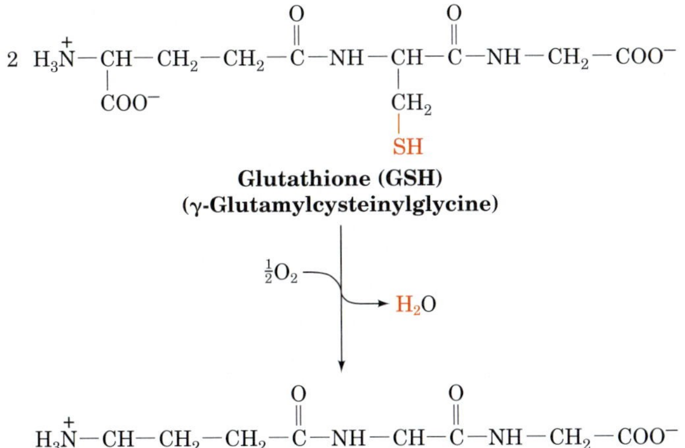

Glutathione (GSH)
(γ-Glutamylcysteinylglycine)

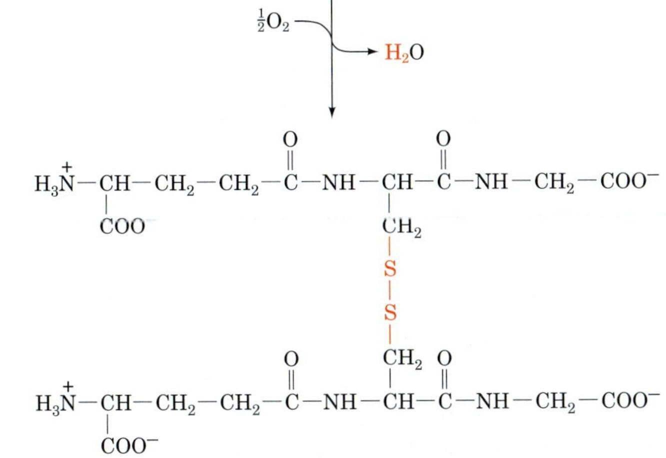

Glutathione disulfide (GSSG)

Glutathione helps inactivate oxidative compounds that could potentially damage cellular structures, since the oxidation of GSH to GSSG is accompanied by the reduction of another compound (shown as O_2 above):

$$2\ GSH + X_{oxidized} \rightarrow GSSG + X_{reduced}$$

GSH must then be regenerated in a separate reduction reaction.

SUMMARY

1. At neutral pH, the amino group of an amino acid is protonated and its carboxylic acid group is ionized.

2. Proteins are polymers of amino acids joined by peptide bonds.

3. The 20 standard amino acids can be classified as nonpolar (Gly, Ala, Val, Leu, Ile, Met, Pro, Phe, Trp), uncharged polar (Ser, Thr, Asn, Gln, Tyr, Cys), and charged (Lys, Arg, His, Asp, Glu).

4. The pK values of the ionizable groups of amino acids are influenced by neighboring groups and may be altered when the amino acid is part of a polypeptide.

5. Amino acids are chiral molecules. Only L-amino acids are found in proteins (some bacterial peptides contain D-amino acids).

6. Amino acids may be covalently modified after they have been incorporated into a polypeptide.

7. Individual amino acids and their derivatives have diverse physiological functions.

REFERENCES

Barrett, G.C. and Elmore, D.T., *Amino Acids and Peptides,* Cambridge University Press (2001). [Includes structures of the common amino acids along with a discussion of their chemical reactivities and information on analytical properties.]

Lamzin, V.S., Dauter, Z., and Wilson, K.S., How nature deals with stereoisomers, *Curr. Opin. Struct. Biol.* **5,** 830–836 (1995). [Discusses proteins synthesized from D-amino acids.]

Solomons, G. and Fryhle, C., *Organic Chemistry* (7th ed.), Chapter 5, Wiley (2000). [A discussion of chirality. Most other organic chemistry textbooks contain similar material.]

KEY TERMS

protein
α-amino acid
α carbon
R group
zwitterion
condensation reaction
peptide bond
dipeptide

tripeptide
oligopeptide
polypeptide
residue
N-terminus
C-terminus
pI
optical activity

chiral center
enantiomers
absolute configuration
Fischer convention
stereoisomers
levorotatory
dextrorotatory
Fischer projection

Cahn–Ingold–Prelog (*RS*) system
racemic mixture
peptidase
neurotransmitter
isopeptide bond

STUDY EXERCISES

1. Draw the structures of the 20 standard amino acids and give their one- and three-letter abbreviations.

2. Classify the 20 standard amino acids by polarity, structure, type of functional group, and acid–base properties.

3. Describe the ionization states of amino acids that have ionizable side chains.

4. Explain how the Fischer convention describes the absolute configuration of a chiral molecule.

5. List some covalent modifications of amino acids in proteins.

PROBLEMS

1. Identify the amino acids that differ from each other by a single methyl or methylene group.

2. The 20 standard amino acids are called α-amino acids. Certain β-amino acids are found in nature. Draw the structure of β-alanine (3-amino-*n*-propionate).

3. Identify the hydrogen bond donor and acceptor groups in asparagine.

4. Draw the dipeptide Asp-His at pH 7.0.

5. Calculate the number of possible pentapeptides that contain one residue each of Ala, Gly, His, Lys, and Val.

6. Determine the net charge of the predominant form of Asp at (a) pH 1.0, (b) pH 3.0, (c) pH 6.0, and (d) pH 11.0.

7. Calculate the pI of (a) Ala, (b) His, and (c) Glu.

8. A sample of the amino acid tyrosine is barely soluble in water. Would a polypeptide containing only Tyr residues, poly(Tyr), be more or less soluble, assuming the total number of Tyr groups remains constant?

9. Circle the chiral carbons in the following compounds:

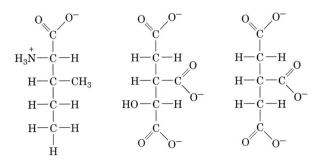

10. Draw the four stereoisomers of threonine.

11. The two C_α H atoms of Gly are said to be prochiral, because when one of them is replaced by another group, C_α becomes chiral. Draw a Fischer projection of Gly and indicate which H must be replaced with CH$_3$ to yield D-Ala.

12. Describe isoleucine (as shown in Table 4-1) using the *RS* system.

13. Some amino acids are synthesized by replacing the keto group (C=O) of an organic acid known as an α-keto acid with an amino group (C—NH$_3^+$). Identify the amino acids that can be produced this way from the following α-keto acids:

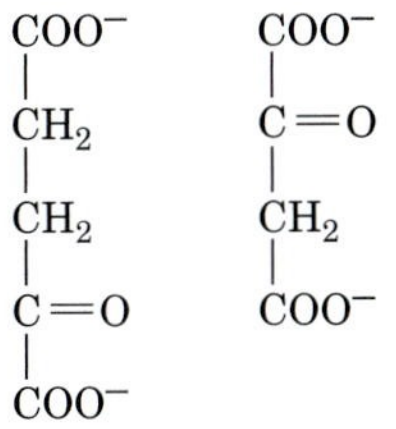

$$
\begin{array}{cc}
\text{COO}^- & \text{COO}^- \\
| & | \\
\text{CH}_2 & \text{C}=\text{O} \\
| & | \\
\text{CH}_2 & \text{CH}_2 \\
| & | \\
\text{C}=\text{O} & \text{COO}^- \\
| & \\
\text{COO}^- &
\end{array}
$$

14. Identify the amino acid residue from which the following compounds are synthesized:

(a) 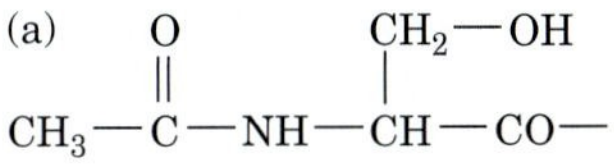
$$
\begin{array}{cc}
\text{O} & \text{CH}_2-\text{OH} \\
\| & | \\
\text{CH}_3-\text{C}-\text{NH}-\text{CH}-\text{CO}- &
\end{array}
$$

(b)
$$
\begin{array}{l}
\text{NH}_3^+ \\
| \\
{}_6\text{CH}_2 \\
| \\
{}_5\text{CH}-\text{OH} \\
| \\
{}_4\text{CH}_2 \\
| \\
{}_3\text{CH}_2 \\
| \\
-\text{NH}-{}_2\text{CH}-{}_1\text{CO}-
\end{array}
$$

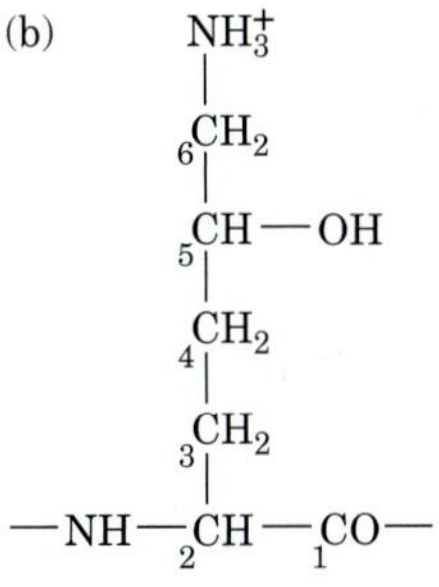

(c)
$$
\begin{array}{l}
\text{S}-\text{CH}_3 \\
| \\
\text{CH}_2 \\
| \\
\text{CH}_2 \\
| \\
\text{HC}-\text{NH}-\text{CH}-\text{CO}-
\end{array}
$$
(with O=)

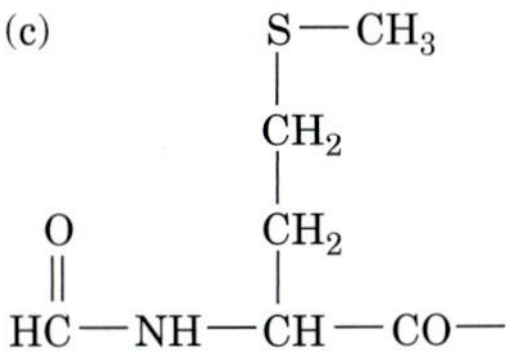

15. Draw the peptide ATLDAK. (a) Calculate its approximate pI. (b) What is its net charge at pH 7.0? [Problem provided by Kathleen Cornely, Providence College.]

16. The protein insulin consists of two polypeptides termed the A and B chains. Insulins from different organisms have been isolated and sequenced. Human and duck insulins have the same amino acid sequence with the exception of six amino acid residues, as shown below. Is the pI of human insulin lower than or higher than that of duck insulin?

Amino acid residue	A8	A9	A10	B1	B2	B27
human	Thr	Ser	Ile	Phe	Val	Thr
duck	Glu	Asn	Pro	Ala	Ala	Ser

[Problem provided by Kathleen Cornely, Providence College.]

The great variation in structure and function among proteins reflects the astronomical variation in the sequences of their component amino acids—there are far more possible amino acid sequences than there are stars in the universe. How can such diversity be assessed in the laboratory and what does it reveal about the evolutionary relationships among proteins? [PhotoDisc, Inc./ Getty Images.]

Proteins: Primary Structure

Proteins are at the center of action in biological processes. Nearly all the molecular transformations that define cellular metabolism are mediated by protein catalysts. Proteins also perform regulatory roles, monitoring extracellular and intracellular conditions and relaying information to other cellular components. In addition, proteins are essential structural components of cells. A complete list of known protein functions would contain many thousands of entries, including proteins that transport other molecules and proteins that generate mechanical and electrochemical forces. And such a list would not account for the thousands of proteins whose functions are not yet fully characterized or, in many cases, are completely unknown.

One of the keys to deciphering the function of a given protein is to understand its structure. Like the other major biological macromolecules, the nucleic acids (Section 3-2) and the polysaccharides (Section 8-2), proteins are polymers of smaller units. But unlike many nucleic acids, proteins do not have uniform, regular structures. This is, in part, because the 20 kinds of amino acid residues from which proteins are made have widely differing chemical and physical properties (Section 4-1C). The sequence in which these amino acids are strung together can be analyzed directly, as we describe in this chapter, or indirectly via DNA sequencing (Section 3-4). In either case, amino acid sequence information provides insights into the chemical and physical properties of proteins, their relationships to other proteins, and ultimately, their mechanisms of action in living organisms.

1 Polypeptide Diversity

Like all polymeric molecules, proteins can be described in terms of levels of organization, in this case, their primary, secondary, tertiary, and quaternary structures. *A protein's **primary structure** is the amino acid sequence of its polypeptide chain,* or chains if the protein consists of more than one polypeptide. An example of an amino acid sequence is given in Fig. 5-1. Each residue is linked to the next via a peptide bond (Fig. 4-3). Higher levels of protein structure—secondary, tertiary, and quaternary— refer to the three-dimensional shapes of folded polypeptide chains and will be described in the next chapter.

Proteins are synthesized *in vivo* by the stepwise polymerization of amino acids in the order specified by the sequence of nucleotides in a gene. The direct correspondence between one linear polymer (DNA) and another (a polypeptide) illustrates the elegant simplicity of living systems and allows us to extract information from one polymer and apply it to the other.

The Theoretical Possibilities for Polypeptides Are Unlimited.   With 20 different choices available for each amino acid residue in a polypeptide chain, it is easy to see that a huge number of different protein molecules are possible. For a protein of n residues, there are 20^n possible sequences. A relatively small protein molecule may consist of a single polypeptide chain of 100

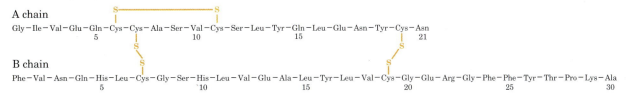

Figure 5-1   The primary structure of bovine insulin.   Note the intrachain and interchain disulfide bond linkages.

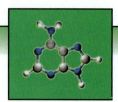

BOX 5-1

Perspectives in Biochemistry

Combinatorial Peptide Libraries

Where nature leaves off, the laboratory picks up, at least as far as peptide diversity is concerned. In pioneering work in the 1980s, Mario Geysen showed that it is possible to chemically synthesize large numbers of oligopeptides whose residues vary systematically at selected positions. For example, a synthetic hexapeptide might have the sequence X-X-A-B-X-X, where X represents any one of the 20 randomly incorporated amino acids and A and B are known. Geysen's approach to hexapeptide synthesis converts an impossible task (synthesizing all possible sequences, which would amount to 20^6, or 64 million peptides) to a manageable one (synthesizing and testing 20^2, or 400, peptides). The so-called **peptide libraries** are therefore of great value in developing new drugs and probes of protein function.

The library of 400 hexapeptides can be rapidly screened for the ability to bind other molecules (antibodies, in Geysen's original experiments) or to produce biological effects. A "positive" X-X-A-B-X-X sequence can then become the starting point for synthesizing a new set of 400 hexapeptides with the sequence X-A-Y-Z-B-X, where Y and Z now represent the

two chosen, invariant residues. Of course, combinatorial peptide libraries are not limited to peptides.

An extended sequence with high biological activity can be constructed by repeated synthesis and screening. In a sense, the chemist exploits the same principles that guide evolution, but on a much smaller time scale. Peptide libraries may include novel peptides—those that don't appear in nature—or those that would be all but impossible to locate and purify from natural sources.

Using similar principles, oligonucleotide libraries have been synthesized *in vitro* and their peptide products tested for their ability to bind other molecules or catalyze chemical reactions. If the oligonucleotides are incorporated into an expression vector, the corresponding oligopeptides can be synthesized in a suitable host organism and evaluated for the desired activity in that *in vivo* system. Oligonucleotide libraries have the added advantage that a single copy of an oligonucleotide can later be amplified by PCR (Section 3-5C) and sequenced.

---

residues. There are $20^{100} \approx 1.27 \times 10^{130}$ possible unique polypeptide chains of this length, a quantity vastly greater than the estimated number of atoms in the universe (9×10^{78}). Clearly, evolution has produced only a tiny fraction of the theoretical possibilities—a fraction that nevertheless represents an astronomical number of different polypeptides (it is possible to generate even more in the laboratory; see Box 5-1).

Actual Polypeptides Are Somewhat Limited in Size and Composition. In general, proteins contain at least 40 residues or so; polypeptides smaller than that are simply called **peptides.** The largest known polypeptide chain belongs to the 26,926-residue **titin,** a giant (2,990 kD) protein that helps arrange the repeating structures of muscle fibers (Section 28-2A). However, *the vast majority of polypeptides contain between 100 and 1000 residues* (Table 5-1). **Multisubunit** proteins contain several identical and/or nonidentical chains called **subunits.** Some proteins are synthesized as single polypeptides that are later cleaved into two or more chains that remain associated; **insulin** is such a protein (Fig. 5-1).

The size range in which most polypeptides fall probably reflects the optimization of several biochemical processes:

1. Forty residues appear to be near the minimum for a polypeptide chain to fold into a discrete and stable shape that allows it to carry out a particular function.

2. Polypeptides with many hundreds of residues may approach the limits of efficiency of the protein synthetic machinery. The longer the polypeptide (and the longer its corresponding mRNA), the greater the likelihood of introducing errors during transcription and translation.

Table 5-1 Compositions of Some Proteins

Protein	Amino Acid Residues	Subunits	Polypeptide Molecular Mass (D)
Proteinase inhibitor III (bitter gourd)	30	1	3,427
Cytochrome c (human)	104	1	11,617
Myoglobin (horse)	153	1	16,951
Interferon-γ (rabbit)	288	2	33,842
Chorismate mutase (*Bacillus subtilis*)	381	3	43,551
Triose phosphate isomerase (*E. coli*)	510	2	53,944
Hemoglobin (human)	574	4	61,986
RNA polymerase (bacteriophage T7)	883	1	98,885
Nucleoside diphosphate kinase (*Dictyostelium discoideum*)	930	6	100,764
Pyruvate decarboxylase (yeast)	2,252	4	245,456
Glutamine synthetase (*E. coli*)	5,616	12	621,264
Titin (human)	26,926	1	2,993,428

In addition to these mild constraints on size, polypeptides are subject to more severe limitations on amino acid composition. The 20 standard amino acids do not appear with equal frequencies in proteins (Table 4-1 lists the average occurrence of each amino acid residue). For example, the most abundant amino acids in proteins are Leu, Ala, Gly, Ser, Val, and Glu; the rarest are Trp, Cys, Met, and His.

Because each amino acid residue has characteristic chemical and physical properties, its presence at a particular position in a protein influences the properties of that protein. In particular, as we shall see, the three-dimensional shape of a folded polypeptide chain is a consequence of the intramolecular forces among its various residues. In general, a protein's hydrophobic residues cluster in its interior, out of contact with water, whereas its hydrophilic side chains tend to occupy the protein's surface.

The characteristics of an individual protein depend more on its amino acid sequence than on its amino acid composition per se, for the same reason that "kitchen" and its anagram "thicken" are quite different words. In addition, many proteins consist of more than just amino acid residues. They may form complexes with metal ions such as Zn^{2+} and Ca^{2+}, they may covalently or noncovalently bind certain small organic molecules, and they may be covalently modified by the posttranslational attachment of groups such as phosphates and carbohydrates.

2 Protein Purification and Analysis

Purification is an all but mandatory step in studying macromolecules, but it is not necessarily easy. Typically, a substance that makes up <0.1% of a tissue's dry weight must be brought to ~98% purity. Purification problems of this magnitude would be considered unreasonably difficult by most synthetic chemists! The following sections outline some of the most common techniques for purifying and, to some extent, characterizing proteins and other macromolecules. Most of these techniques can be used, sometimes in modified form, for nucleic acids and other types of biological molecules.

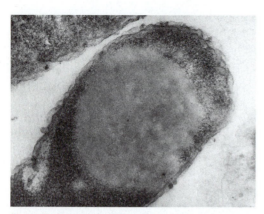

Figure 5-2 Inclusion body.   A genetically engineered organism that produces large amounts of a foreign protein often sequesters it in **inclusion bodies.** This electron micrograph shows an inclusion body of the protein prochymosin in an *E. coli* cell. [Courtesy of Teruhiko Beppu, Nikon University, Japan.]

A General Approach to Purifying Proteins

The task of purifying a protein present in only trace amounts was once so arduous that many of the earliest proteins to be characterized were studied in part because they are abundant and easily isolated. For example, hemoglobin, which accounts for about one-third the weight of red blood cells, has historically been among the most extensively studied proteins. Most of the enzymes that mediate basic metabolic processes or that are involved in the expression and transmission of genetic information are common to all species. For this reason, a given protein is frequently obtained from a source chosen primarily for convenience, for example, tissues from domesticated animals or easily obtained microorganisms such as *E. coli* and *Saccharomyces cerevisiae* (baker's yeast).

The development of molecular cloning techniques (Section 3-5) allows almost any protein-encoding gene to be isolated from its parent organism, specifically altered (genetically engineered) if desired, and expressed at high levels in a microorganism. Indeed, the cloned protein may constitute up to 40% of the microorganism's total cell protein (Fig. 5-2). This high level of protein production generally renders the cloned protein far easier to isolate than it would be from its parent organism (in which it may occur in vanishingly small amounts).

The first step in the isolation of a protein or other biological molecule is to get it out of the cell and into solution. Many cells require some sort of mechanical disruption to release their contents. Most of the procedures for lysing cells use some variation of crushing or grinding followed by filtration or centrifugation to remove large insoluble particles. If the target protein is tightly associated with a lipid membrane, a detergent or organic solvent may be used to solubilize the lipids and recover the protein.

Stabilizing Proteins.   Once a protein has been removed from its natural environment, it becomes exposed to many agents that can irreversibly damage it. These influences must be carefully controlled at all stages of a purification process. The following factors should be considered:

1. *pH.* Biological materials are routinely dissolved in buffer solutions effective in the pH range over which the materials are stable (buffers are described in Section 2-2C). Failure to do so could cause their **denaturation** (structural disruption), if not their chemical degradation.

2. *Temperature.* The thermal stability of proteins varies. Although some proteins denature at low temperatures, most proteins denature at high temperatures, sometimes only a few degrees higher than their native environment. Protein purification is normally carried out at temperatures near 0°C.

3. *Presence of degradative enzymes.* When tissues are destroyed to liberate the molecule of interest, degradative enzymes are also released. These include **proteases** and **nucleases** (enzymes that degrade nucleic acids). Degradative enzymes can be inhibited by adjusting the pH or temperature to values that inactivate them (provided this does not adversely affect the protein of interest) or by adding compounds that specifically block their action.

4. *Adsorption to surfaces.* Many proteins are denatured by contact with the air–water interface or with glass or plastic surfaces. Hence, protein solutions are handled so as to minimize foaming and are kept relatively concentrated.

5. *Long-term storage.* All the factors listed above must be considered when a purified protein sample is to be kept stable. In addition,

processes such as slow oxidation and microbial contamination must be prevented. Protein solutions are sometimes stored under nitrogen or argon gas (rather than under air, which contains ~21% O_2) and/or frozen at $-80°C$ or $-196°C$ (the temperature of liquid nitrogen). However, some so-called cold-labile proteins are unstable below a characteristic temperature.

Assaying Proteins. Purifying a substance requires some means for quantitatively detecting it. Accordingly, an **assay** must be devised that is specific for the target protein, highly sensitive, and convenient to use (especially if it must be repeated at every stage of the purification process).

Among the most straightforward protein assays are those for enzymes that catalyze reactions with readily detected products, because *the rate of product formation is proportional to the amount of enzyme present.* Substances with colored or fluorescent products have been developed for just this purpose. If no such substance is available for the enzyme being assayed, the product of the enzymatic reaction may be converted, by the action of another enzyme, to an easily quantified substance. This is known as a **coupled enzymatic reaction.** Proteins that are not enzymes can be detected by their ability to specifically bind certain substances or to produce observable biological effects.

Immunochemical procedures are among the most sensitive of assay techniques. **Immunoassays** use **antibodies,** proteins produced by an animal's immune system in response to the introduction of a foreign substance (an **antigen**). Antibodies recovered from the blood serum of an immunized animal or from cultures of immortalized antibody-producing cells bind specifically to the original protein antigen.

A protein in a complex mixture can be detected by binding to its corresponding antibodies. In one technique, known as a **radioimmunoassay (RIA),** the protein is indirectly detected by determining the degree to which it competes with a radioactively labeled standard for binding to the antibody. Another technique, the **enzyme-linked immunosorbent assay (ELISA),** has many variations, one of which is diagrammed in Fig. 5-3.

The concentration of a substance in solution can be measured by **absorbance spectroscopy.** A solution containing a solute that absorbs light does so according to the **Beer–Lambert law,**

$$A = \log\left(\frac{I_0}{I}\right) = \varepsilon c l \qquad [5\text{-}1]$$

where A is the solute's **absorbance** (alternatively, its **optical density**), I_0 is the intensity of the incident light at a given wavelength λ, I is its transmitted intensity at λ, ε is the **molar absorptivity** (alternatively, the **molar extinction coefficient**) of the solute at λ, c is its molar concentration, and l is the length of the light path in cm. The value of ε varies with λ; a plot of A or ε vs λ for the solute is called its **absorption spectrum.** If the value of ε for a substance is known, then its concentration can be spectroscopically determined.

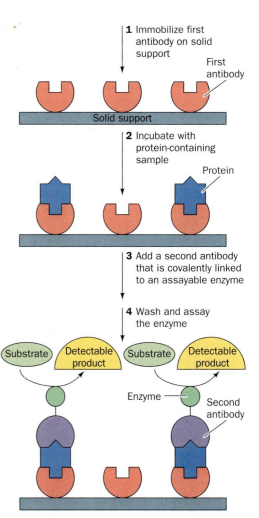

1 Immobilize first antibody on solid support

First antibody

Solid support

2 Incubate with protein-containing sample

Protein

3 Add a second antibody that is covalently linked to an assayable enzyme

4 Wash and assay the enzyme

Substrate | Detectable product | Substrate | Detectable product

Enzyme

Second antibody

Figure 5-3   Enzyme-linked immunosorbent assay. (**1**) An antibody against the protein of interest is immobilized on an inert solid such as polystyrene. (**2**) The solution to be assayed is applied to the antibody-coated surface. The antibody binds the protein of interest, and other proteins are washed away. (**3**) The protein–antibody complex is reacted with a second protein-specific antibody to which an enzyme is attached. (**4**) Binding of the second antibody–enzyme complex is measured by assaying the activity of the enzyme. The amount of substrate converted to product indicates the amount of protein present. ✺ **See the Animated Figures.**

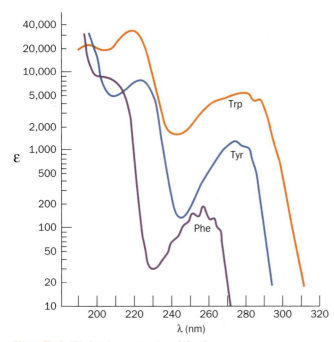

Figure 5-4 UV absorbance spectra of the three aromatic amino acids phenylalanine, tryptophan, and tyrosine.   Note that the molar absorbance, ε, is displayed on a log scale. [After Wetlaufer, D.B., *Adv. Prot. Chem.* **7**, 310 (1962).]

Polypeptides absorb strongly in the ultraviolet (UV) region of the spectrum ($\lambda = 200$ to 400 nm) largely because their aromatic side chains (those of Phe, Trp, and Tyr) have particularly large molar extinction coefficients in this spectral region (ranging into the tens of thousands; Fig. 5-4). However, polypeptides do not absorb visible light ($\lambda = 400$ to 800 nm) so that they are colorless. Nevertheless, if a protein has a **chromophore** that absorbs in the visible region of the spectrum, this absorbance can be used to assay for the presence of this protein in a mixture of other proteins.

In order to follow the purification of a protein, it is important to measure the total amount of protein at every stage of the purification. Absorbance spectroscopy at 280 nm is a convenient way of doing so. However, since many substances that are likely to be present during a protein purification procedure (e.g., nucleic acids) have strong molar absorptivities in the same spectral range as do proteins and proteins vary in their proportion of aromatic residues, spectroscopic measurements in the UV region can only provide estimates of the amount of protein present (although they can accurately determine the amount of a pure protein of known molar absorptivity). Moreover, spectroscopic methods are only moderately sensitive to proteins; they can minimally detect 50 to 100 μg of protein per mL.

Several techniques have been developed to circumvent these difficulties. In the most widely used of these, the **Bradford assay,** the binding of the dye **Coomassie brilliant blue**

R250: R = H
G250: R = CH$_3$

Coomassie brilliant blue

to protein in acidic solution causes the dye's absorption maximum to shift from 465 nm to 595 nm. Hence the absorbance at 595 nm in a Bradford assay provides a direct measure of the amount of protein present. Moreover, the Bradford assay is highly sensitive; it can detect as little as 1 μg of protein per mL.

Separation Techniques.   Proteins are purified by **fractionation procedures.** In a series of independent steps, the various physicochemical properties of the protein of interest are used to separate it progressively from other substances. The idea is not necessarily to minimize the loss of the desired protein, but to *eliminate selectively the other components of the mixture so that only the required substance remains.*

Protein purification is considered as much an art as a science, with many options available at each step. While a trial-and-error approach can work, knowing something about the target protein (or the proteins it is to be separated from) simplifies the selection of separation procedures. Some of the procedures we discuss and the protein characteristics they depend on are as follows:

Protein Characteristic	Purification Procedure
Solubility	Salting out
Ionic Charge	Ion exchange chromatography
	Electrophoresis
	Isoelectric focusing
Polarity	Hydrophobic interaction chromatography
Size	Gel filtration chromatography
	SDS-PAGE
	Ultracentrifugation
Binding specificity	Affinity chromatography

B Protein Solubility

Because a protein contains multiple charged groups, its solubility depends on the concentrations of dissolved salts, the polarity of the solvent, the pH, and the temperature. Some or all of these variables can be manipulated to selectively precipitate certain proteins while others remain soluble.

The solubility of a protein at low ion concentrations increases as salt is added, a phenomenon called **salting in.** The additional ions shield the protein's multiple ionic charges, thereby weakening the attractive forces between individual protein molecules (such forces can lead to aggregation and precipitation). However, as more salt is added, particularly with sulfate salts, the solubility of the protein again decreases. This **salting out** effect is primarily a result of the competition between the added salt ions and the other dissolved solutes for molecules of solvent. At very high salt concentrations, so many of the added ions are solvated that there is significantly less bulk solvent available to dissolve other substances, including proteins.

Since different proteins precipitate at different salt concentrations, salting out is the basis of one of the most commonly used protein purification procedures. Adjusting the salt concentration in a solution containing a mixture of proteins to just below the precipitation point of the protein to be purified eliminates many unwanted proteins from the solution (Fig. 5-5). Then, after removing the precipitated proteins by filtration or centrifugation, the salt concentration of the remaining solution is increased to precipitate the desired protein. This procedure results in a significant purification and concentration of large quantities of protein. Ammonium sulfate, $(NH_4)_2SO_4$, is the most commonly used reagent for salting out proteins because its high solubility (3.9 M in water at 0°C) allows the preparation of solutions with high ionic strength. The pH may be adjusted to approximate the isoelectric point (pI) of the desired protein because a protein is least soluble when its net charge is zero. The pI's of some proteins are listed in Table 5-2.

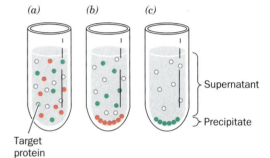

Figure 5-5   Fractionation by salting out.   (*a*) The salt of choice, usually ammonium sulfate, is added to a solution of macromolecules to a concentration just below the precipitation point of the protein of interest. (*b*) After centrifugation, the unwanted precipitated proteins (*red spheres*) are discarded and more salt is added to the supernatant to a concentration sufficient to salt out the desired protein (*green spheres*). (*c*) After a second centrifugation, the protein is recovered as a precipitate, and the supernatant is discarded.

Table 5-2 Isoelectric Points of Several Common Proteins

Protein	pI
Pepsin	<1.0
Ovalbumin (hen)	4.6
Serum albumin (human)	4.9
Tropomyosin	5.1
Insulin (bovine)	5.4
Fibrinogen (human)	5.8
γ-Globulin (human)	6.6
Collagen	6.6
Myoglobin (horse)	7.0
Hemoglobin (human)	7.1
Ribonuclease A (bovine)	9.4
Cytochrome *c* (horse)	10.6
Histone (bovine)	10.8
Lysozyme (hen)	11.0
Salmine (salmon)	12.1

C Chromatography

The process of **chromatography** (Greek: *chroma,* color + *graphein,* to write) was discovered in 1903 by Mikhail Tswett, who separated solubilized plant pigments using solid adsorbents. In most modern chromatographic procedures, a mixture of substances to be fractionated is dissolved in a liquid (the **mobile phase**) and percolated through a column containing a porous solid matrix (the **stationary phase**). As solutes flow through the column, they interact with the stationary phase and are retarded. The retarding force depends on the properties of each solute. If the column is long enough, substances with different rates of migration will be separated. The chromatographic procedures that are most useful for purifying proteins are classified according to the nature of the interaction between the protein and the stationary phase.

One of the earliest chromatographic techniques used strips of filter paper as the stationary phase, a process called **paper chromatography.** Modern column chromatography uses derivatives of cellulose, agarose, or dextran (all carbohydrate polymers) or synthetic substances such as cross-linked polyacrylamide or silica. **High performance liquid chromatography (HPLC)** employs automated systems with precisely applied samples, controlled flow rates at high pressures (up to 5000 psi), a chromatographic matrix of specially fabricated 3- to 300-μm-diameter glass or plastic beads coated with a uniform layer of chromatographic material, and on-line sample detection. This greatly improves the speed, resolution, and reproducibility of the separation—features that are particularly desirable when chromatographic separations are repeated many times or when they are used for analytical rather than preparative purposes.

Ion Exchange Chromatography. In **ion exchange chromatography,** charged molecules bind to oppositely charged groups that have been immobilized on the matrix. Anions bind to cationic groups on **anion exchangers,** and cations bind to anionic groups on **cation exchangers.** Perhaps the most frequently used anion exchanger is a matrix with attached **diethylaminoethyl (DEAE)** groups, and the most frequently used cation exchanger is a matrix bearing **carboxymethyl (CM)** groups

$$\text{DEAE: Matrix} - CH_2 - CH_2 - \overset{+}{N}H(CH_2CH_3)_2$$
$$\text{CM: Matrix} - CH_2 - COO^-$$

Cellulose- and agarose-based resins are among the most frequently used matrix materials in the ion exchange chromatography of proteins.

Proteins and other **polyelectrolytes** (polyionic polymers) that bear both positive and negative charges can bind to both cation and anion exchangers. *The binding affinity of a particular protein depends on the presence of other ions that compete with the protein for binding to the ion exchanger and on the pH of the solution, which influences the net charge of the protein.*

The proteins to be separated are dissolved in a buffer of an appropriate pH and salt concentration and are applied to a column containing the ion exchanger. The column is then washed with the buffer (Fig. 5-6). As the column is washed, proteins with relatively low affinities for the ion exchanger move through the column faster than proteins that bind with higher affinities. The column effluent is collected in a series of fractions. Proteins that bind tightly to the ion exchanger can be **eluted** (washed through the column) by applying a buffer, called the **eluant,** that has a higher salt concentration or a pH that reduces the affinity with which the matrix binds the protein. The column effluent can be monitored for the presence of protein

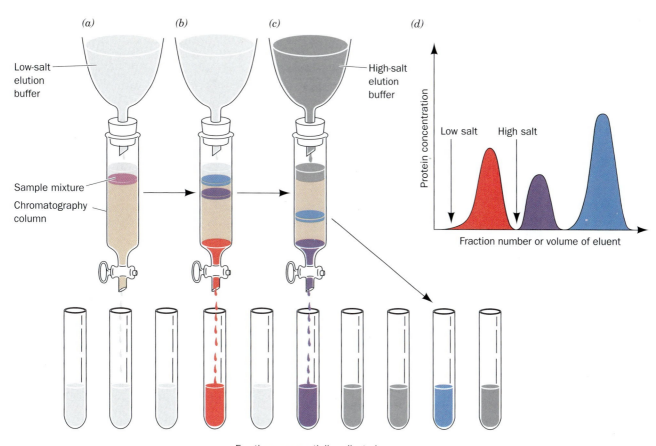

Figure 5-6 Ion exchange chromatography. The tan region of the column represents the ion exchanger and the colored bands represent proteins. (*a*) A mixture of proteins dissolved in a small volume of buffer is applied to the top of the matrix in the column. (*b*) As elution progresses, the proteins separate into discrete bands as a result of their different affinities for the exchanger. In this diagram, the first protein (*red*) has passed through the column and has been isolated as a separate fraction. The other proteins remain near the top of the column. (*c*) The salt concentration in the eluant is increased to elute the remaining proteins. (*d*) The elution diagram of the protein mixture from the column. 🔊 **See the Animated Figures.**

by measuring its absorbance at 280 nm. The eluted fractions can also be tested for the protein of interest using a more specific assay (see above).

Hydrophobic Interaction Chromatography. Hydrophobic interactions between proteins and the chromatographic matrix can be exploited to purify the proteins. In **hydrophobic interaction chromatography,** the matrix material is lightly substituted with octyl or phenyl groups. At high salt concentrations, nonpolar groups on the surface of proteins "interact" with the hydrophobic groups; that is, both types of groups are excluded by the polar solvent (hydrophobic effects are augmented by increased ionic strength). The eluant is typically an aqueous buffer with decreasing salt concentrations, increasing concentrations of detergent (which disrupts hydrophobic interactions), or changes in pH.

Gel Filtration Chromatography. In **gel filtration chromatography** (also called **size exclusion** or **molecular sieve chromatography**), molecules are separated according to their size and shape. The stationary phase consists of gel beads containing pores that span a relatively narrow size range. The pore size is typically determined by the extent of cross-linking between the polymers of the gel material. If an aqueous solution of molecules of various sizes is

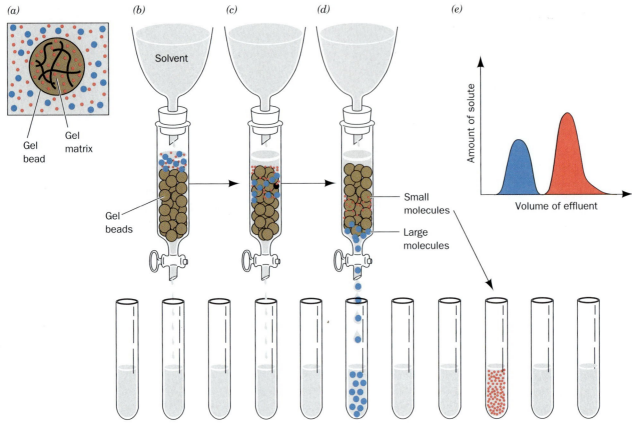

Figure 5-7 Gel filtration chromatography.   (*a*) A gel bead consists of a gel matrix (*wavy solid lines*) that encloses an internal solvent space. Small molecules (*red dots*) can freely enter the internal space of the gel bead. Large molecules (*blue dots*) cannot penetrate the gel pores. (*b*) The sample solution is applied to the top of the column (the gel beads are represented as brown spheres). (*c*) The small molecules can penetrate the gel and consequently migrate through the column more slowly than the large molecules that are excluded from the gel. (*d*) The large molecules elute first and are collected as fractions. Small molecules require a larger volume of solvent to elute. (*e*) The elution diagram, or chromatogram, indicating the complete separation of the two components. ⚙ **See the Animated Figures.**

passed through a column containing such "molecular sieves," the molecules that are too large to pass through the pores are excluded from the solvent volume inside the gel beads. *These large molecules therefore traverse the column more rapidly than small molecules that pass through the pores* (Fig. 5-7). Because the pore size in any gel varies to some degree, gel filtration can be used to separate a range of molecules; larger molecules with access to fewer pores elute sooner (i.e., in a smaller volume of eluant) than smaller molecules that have access to more of the gel's interior volume.

Within the size range of molecules separated by a particular pore size, there is a linear relationship between the relative elution volume of a substance and the logarithm of its molecular mass (assuming the molecules have similar shapes). If a given gel filtration column is calibrated with several proteins of known molecular mass, the mass of an unknown protein can be conveniently estimated by its elution position.

Affinity Chromatography.   A striking characteristic of many proteins is their ability to bind specific molecules tightly but noncovalently. This property can be used to purify such proteins by **affinity chromatography** (Fig. 5-8). In this technique, a molecule (a **ligand**) that specifically binds to the protein of interest (e.g., a nonreactive analog of an enzyme's substrate) is covalently attached to an inert matrix. *When an impure protein solution is passed through this chromatographic material, the desired protein binds to*

the immobilized ligand, whereas other substances are washed through the column with the buffer. The desired protein can then be recovered in highly purified form by changing the elution conditions to release the protein from the matrix. The great advantage of affinity chromatography is its ability to exploit the desired protein's unique biochemical properties rather than the small differences in physicochemical properties between proteins used by other chromatographic methods. Accordingly, the separation power of affinity chromatography for a specific protein is often greater than that of other chromatographic techniques.

Affinity chromatography columns can be constructed by chemically attaching small molecules or proteins to a chromatographic matrix. In **immunoaffinity chromatography,** an antibody is attached to the matrix in order to purify the protein against which the antibody was raised. In all cases, the ligand must have an affinity high enough to capture the protein of interest but not so high as to prevent the protein's subsequent release without denaturing it. The bound protein can be eluted by washing the column with a solution containing a high concentration of free ligand or a solution of different pH or ionic strength.

In **metal chelate affinity chromatography,** a divalent metal ion such as Zn^{2+} or Ni^{2+} is attached to the chromatographic matrix so that proteins bearing metal-chelating groups (e.g., multiple His side chains) can be retained. Recombinant DNA techniques (Section 3-5) can be used to append a segment of six consecutive His residues, known as a **His-Tag,** to the N- or C-terminus of the polypeptide to be isolated. This creates a metal ion-binding site that allows the recombinant protein to be purified by metal chelate chromatography. After the protein has been eluted, usually by altering the pH, the His-Tag can be removed by the action of a specific protease whose recognition sequence separates the $(His)_6$ sequence from the rest of the protein.

D Electrophoresis

Electrophoresis, the migration of ions in an electric field, is described in Section 3-4B. Here, we describe its application to protein purification and analysis.

Polyacrylamide Gel Electrophoresis. The electrophoresis of proteins is typically carried out in agarose or polyacrylamide gels with a characteristic pore size, so *the molecular separations are based on gel filtration (size and shape) as well as electrophoretic mobility (electric charge).* However, electrophoresis differs from gel filtration in that the electrophoretic mobility of smaller molecules is greater than the mobility of larger molecules with the same charge density. The pH of the gel is high enough (usually about pH 9) so that nearly all proteins have net negative charges and move toward the anode when the current is switched on. Molecules of similar size and charge move as a band through the gel.

Following electrophoresis, the separated bands may be visualized in the gel by an appropriate technique, such as soaking the gel in a solution of a stain that binds tightly to proteins (e.g., Coomassie brilliant blue). If the proteins in a sample are radioactive, the gel can be dried and then clamped over a sheet of X-ray film. After a time, the film is developed and the resulting **autoradiograph** shows the positions of the radioactive components by a blackening of the film. If an antibody to a protein of interest is available, it can be used to specifically detect this protein on a gel in the presence of many other proteins, a process called **immunoblotting** or **Western blotting** that is similar to ELISA (Fig. 5-3). Depending on the dimensions of the

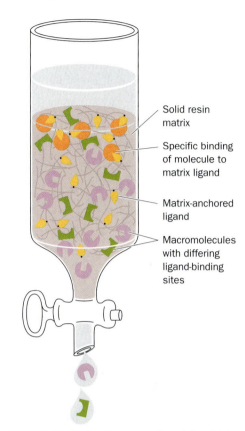

Solid resin matrix

Specific binding of molecule to matrix ligand

Matrix-anchored ligand

Macromolecules with differing ligand-binding sites

Figure 5-8   Affinity chromatography.   A ligand (shown here in yellow) is immobilized by covalently binding it to the chromatographic matrix. The cutout squares, semicircles, and triangles represent ligand-binding sites on macromolecules. Only certain molecules (represented by orange circles) specifically bind to the ligand. The other components are washed through the column.

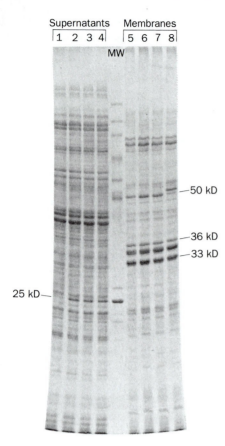

Supernatants Membranes
1 2 3 4 | 5 6 7 8
MW

—50 kD

—36 kD
—33 kD

25 kD—

Figure 5-9 SDS-PAGE. Samples of supernatants (*left*) and membrane fractions (*right*) from a preparation of the bacterium *Salmonella typhimurium* were electrophoresed in parallel lanes on a 35-cm-long × 0.8-mm-thick polyacrylamide slab. The lane marked MW contains molecular weight standards. [Courtesy of Giovanna F. Ames, University of California at Berkley.]

gel and the visualization technique used, samples containing less than a nanogram of protein can be separated and detected by gel electrophoresis.

SDS-PAGE of Proteins.

In one form of **polyacrylamide gel electrophoresis (PAGE),** the detergent sodium dodecyl sulfate (SDS)

$$[CH_3-(CH_2)_{10}-CH_2-O-SO_3^-]\ Na^+$$

is used to denature proteins. Amphiphilic molecules (Section 2-1C) such as SDS interfere with the hydrophobic interactions that normally stabilize proteins. Proteins assume a rodlike shape in the presence of SDS. Furthermore, most proteins bind SDS in a ratio of about 1.4 g SDS per gram protein (about one SDS molecule for every two amino acid residues). The large negative charge that the SDS imparts masks the proteins' intrinsic charge. The net result is that SDS-treated proteins have similar shapes and charge-to-mass ratios. *SDS-PAGE therefore separates proteins purely by gel filtration effects,* that is, according to molecular mass. Figure 5-9 is an example of the resolving power and the reproducibility of SDS-PAGE.

In SDS-PAGE, the relative mobilities of proteins vary approximately linearly with the logarithm of their molecular masses (Fig. 5-10). Consequently, the molecular mass of a protein can be determined with about 5 to 10% accuracy by electrophoresing it together with several "marker" proteins of known molecular masses that bracket that of the protein of interest. Because SDS disrupts noncovalent interactions between polypeptides, SDS-PAGE yields the molecular masses of the subunits of multisubunit proteins. The possibility that subunits are linked by disulfide bonds can be tested by preparing samples for SDS-PAGE in the presence and absence of a reducing agent, such as **2-mercaptoethanol** ($HSCH_2CH_2OH$), that breaks these bonds (Section 5-3A).

Capillary Electrophoresis.

Although gel electrophoresis in its various forms is highly effective at separating charged molecules, it can require up to several hours and is difficult to quantitate and automate. These disadvantages are largely overcome through the use of **capillary electrophoresis (CE),** a technique in which electrophoresis is carried out in very thin capillary tubes (20- to 100-μm inner diameter). Such narrow capillaries rapidly dissipate heat and hence permit the use of very high electric fields, which reduces separation times to a few minutes. The CE techniques have extremely high resolution and can be automated in much the same way as is HPLC, that is, with automatic sample loading and on-line sample detection. Since CE can separate only small amounts of material, it is largely limited to use as an analytical tool.

Isoelectric Focusing and Two-Dimensional Electrophoresis.

A protein has charged groups of both polarities and therefore has an isoelectric point, p*I*, at which it is immobile in an electric field. *If a mixture of proteins is electrophoresed through a solution or gel that has a stable pH gradient in which the pH smoothly increases from anode to cathode, each protein will migrate to the position in the pH gradient corresponding to its pI.* If a protein molecule diffuses away from this position, its net charge will change as it moves into a region of different pH and the resulting electrophoretic forces will move it back to its isoelectric position. Each species of protein is thereby "focused" into a narrow band about its p*I*. This type of electrophoresis is called **isoelectric focusing (IEF).**

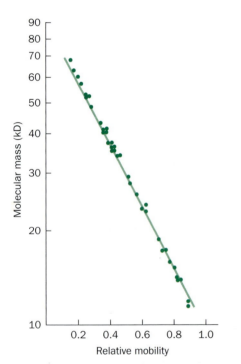

Figure 5-10 Logarithmic relationship between the molecular mass of a protein and its electrophoretic mobility in SDS-PAGE. The masses of 37 proteins ranging from 11 to 70 kD are plotted. [After Weber, K. and Osborn, M., *J. Biol. Chem.* **244,** 4406 (1969).]

IEF can be combined with SDS-PAGE in an extremely powerful two-dimensional separation technique named **two-dimensional (2D) gel electrophoresis.** First, a sample of proteins is subject to IEF in one direction, and then the separated proteins are subjected to SDS-PAGE in the perpendicular direction. This procedure generates an array of spots, each representing a protein (Fig. 5-11). Up to 5000 proteins have been observed on a single two-dimensional electrophoretogram.

Two-dimensional gel electrophoresis is a valuable tool for **proteomics,** a field of study that involves cataloguing all of a cell's expressed proteins with emphasis on their quantitation, localization, modifications, interactions, and activities. Individual protein spots in a stained 2D gel can be excised with a scalpel, destained, and the protein eluted from the gel fragment for identification and/or characterization, often by mass spectrometry (Section 5-3D). 2D electrophoretograms can be analyzed by computer after they have been scanned and digitized. This facilitates the detection of variations in the positions and intensities of protein spots in samples obtained from different tissues or under different growth conditions. Numerous reference 2D gels are publicly available for this purpose in the Web-accessible databases listed at http://us.expasy.org/ch2d/. These databases contain images of 2D gels of a variety of organisms and tissues and identify many of their component proteins.

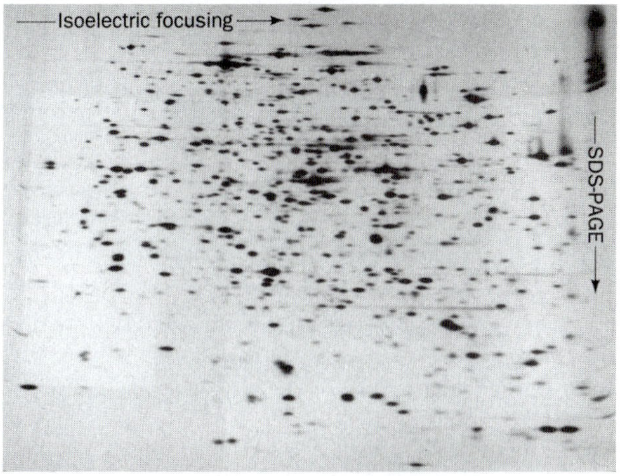

Figure 5-11 Two-dimensional gel electrophoresis. In this example, *E. coli* proteins that had been labeled with [^{14}C]amino acids were subjected to isoelectric focusing (horizontally) followed by SDS-PAGE (vertically). Over 1000 spots can be resolved in the autoradiogram shown here. [Courtesy of Patrick O'Farrell, University of California at San Francisco.]

E Ultracentrifugation

If a container of sand and water is shaken and then allowed to stand quietly, the sand rapidly sediments to the bottom of the container due to the influence of the earth's gravity. Yet macromolecules in solution, which experience the same gravitational field, do not exhibit any perceptible sedimentation because their random thermal (Brownian) motion keeps them uniformly distributed throughout the solution. *Only when subjected to enormous accelerations do macromolecules begin to sediment as do sand grains.*

The **ultracentrifuge,** which was developed around 1923 by the Swedish biochemist The Svedberg, can attain rotational speeds as high as 100,000 rpm so as to generate centrifugal fields in excess of 800,000*g*. Using this instrument, Svedberg first demonstrated that proteins are macromolecules with homogeneous compositions and that many proteins contain subunits.

The rate at which a particle sediments in the ultracentrifuge is related to its mass (the density of the solution and the shape of the particle also affect the sedimentation rate). A protein's sedimentation coefficient (its sedimentation velocity per unit of centrifugal force) is usually expressed in units of 10^{-13} s, which are known as **Svedbergs (S).** The relationship between molecular mass and sedimentation coefficient is not linear; therefore, values of sedimentation coefficients are not additive. The sedimentation coefficients of proteins range from about 1S to about 50S; viruses have sedimentation coefficients in the range of 40S to 1000S. Subcellular particles such as mitochondria have sedimentation coefficients of tens of thousands.

Before about 1970, molecular mass determinations were often made using an **analytical ultracentrifuge,** a device in which the sedimentation of macromolecules can be optically observed. More recently, gel filtration chromatography and SDS-PAGE have proved to be more convenient methods. The analytical ultracentrifuge is still used, however, for characterizing systems of noncovalently associating molecules, including subunits of proteins and other molecules that form macromolecular complexes. In **preparative ultracentrifugation,** sedimentation is carried out in a solution of an inert substance in which the concentration, and therefore the density, of the solution increases from

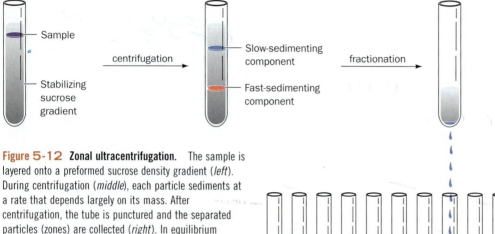

Figure 5-12   Zonal ultracentrifugation. The sample is layered onto a preformed sucrose density gradient (*left*). During centrifugation (*middle*), each particle sediments at a rate that depends largely on its mass. After centrifugation, the tube is punctured and the separated particles (zones) are collected (*right*). In equilibrium density gradient centrifugation, the centrifuge tube is filled with a sample solution containing a substance, usually CsCl, that forms a density gradient as the tube spins.

See Guided Exploration 4

Protein sequence determination

the top to the bottom of the centrifuge tube. The use of such **density gradients** enhances the resolving power of the ultracentrifuge. In **zonal ultracentrifugation,** a macromolecular solution is layered on top of a preformed density gradient, usually made with sucrose. During centrifugation, each species of macromolecule moves through the gradient at a rate largely determined by its sedimentation coefficient and therefore travels as a zone that can be separated from other such zones (Fig. 5-12). After centrifugation, the tube is punctured to collect fractions containing the separated macromolecules.

In **equilibrium density gradient centrifugation,** the sample is dissolved in a relatively concentrated solution of a dense, fast-diffusing substance such as CsCl. Under the high gravitational field produced at high spin rates, the CsCl forms a density gradient. The sample components form bands at positions where their densities are equal to that of the solution. The individual bands can then be separated as in zonal ultracentrifugation.

3 | Protein Sequencing

The first protein sequence, that of bovine insulin, was reported by Frederick Sanger in 1953, thereby definitively establishing that proteins have unique covalent structures (Box 5-2). Since then, many additional proteins have been sequenced, and the sequences of many more proteins have been inferred from their DNA sequences. All told, the amino acid sequences of hundreds of thousands of polypeptides are now known. Such information is valuable for the following reasons:

1. Knowledge of a protein's amino acid sequence is prerequisite for determining its three-dimensional structure and is essential for understanding its molecular mechanism of action.

2. Sequence comparisons among analogous proteins from different species yield insights into protein function and reveal evolutionary relationships among the proteins and the organisms that produce them.

3. Many inherited diseases are caused by mutations that result in an amino acid change in a protein. Amino acid sequence analysis can assist in the development of diagnostic tests and effective therapies.

Sanger's determination of the sequence of insulin's 51 residues (Fig. 5-1) took about 10 years and required ~100 g of protein. Procedures for pri-

BOX 5-2

Pathways of Discovery

Frederick Sanger and Protein Sequencing

At one time, many biochemists believed that proteins were amorphous "colloids" of variable size, shape, and composition. This view was largely abandoned by the 1940s, when it became possible to determine the amino acid composition of a protein, that is, the number of each kind of constituent amino acid. However, such information revealed nothing about the order in which the amino acids were combined. In fact, skeptics still questioned whether proteins even had unique sequences. One plausible theory was that proteins were populations of related molecules that were probably assembled from shorter pieces. Some studies went so far as to describe the rules governing the relative stoichiometries and spacing of the various amino acids in proteins.

During the 1940s, the accuracy of amino acid analysis began to improve as a result of the development of chromatographic techniques for separating amino acids and the use of the reagent **ninhydrin,** which forms colored adducts with amino acids, to chemically quantify them (previous methods used cumbersome biological assays). Frederick Sanger, who began his work on protein sequencing in 1943, used these new techniques as well as classic organic chemistry. Sanger did not actually set out to sequence a protein; his first experiments were directed toward devising a better method for quantifying free amino groups such as the amino groups corresponding to the N-terminus of a polypeptide.

Sanger used a reagent called 2,4-dinitrofluorobenzene, which forms a yellow dinitrophenyl (DNP) derivative at terminal amino groups without breaking any peptide bonds.

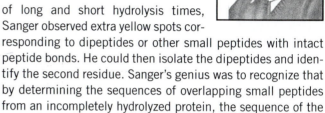

2,4-Dinitrofluoro-benzene (DNFB)   **Polypeptide**   **DNP Polypeptide**

Frederick Sanger (1918–)

of long and short hydrolysis times, Sanger observed extra yellow spots corresponding to dipeptides or other small peptides with intact peptide bonds. He could then isolate the dipeptides and identify the second residue. Sanger's genius was to recognize that by determining the sequences of overlapping small peptides from an incompletely hydrolyzed protein, the sequence of the intact protein could be determined.

An enormous amount of work was required to turn the basic principle into a sound laboratory technique for sequencing a protein. Sanger chose insulin as his subject, because it was one of the smallest known proteins. Insulin contains 51 amino acids in two polypeptide chains (called A and B). The sequence of the 30 residues in the B chain was completed in 1951, and the sequence of the A chain (21 residues) in 1953. Because the two chains are linked through disulfide bonds, Sanger also endeavored to find the optimal procedure for cleaving these bonds and then identifying their positions in the intact protein, a task he completed in 1955.

Sanger's work of over a decade, combining his expertise in organic chemistry (the cleavage and derivatization reactions) with the development of analytical tools for separating and identifying the reaction products, led to his winning a Nobel prize in 1958 [he won a second Nobel prize in 1980, for his invention of the chain-terminator method of nucleic acid sequencing (Section 3-4C)]. A testament to Sanger's brilliance is that his basic approaches to sequencing polypeptides and nucleic acids are still widely used.

Publication of the sequence of insulin in 1955 convinced skeptics that a protein has a unique amino acid sequence. Sanger's work also helped catalyze thinking about the existence of a genetic code that would link the amino acid sequence of a protein to the nucleotide sequence of DNA, a molecule whose structure had just been elucidated in 1953.

When the protein is subsequently hydrolyzed to break its peptide bonds, the N-terminal amino acid retains its DNP label and can be identified when the hydrolysis products are separated by chromatography. Sanger found that some DNP–amino acid derivatives were unstable during hydrolysis, so this step had to be shortened. In comparing the results

Sanger, F., Sequences, sequences, sequences, *Annu. Rev. Biochem.* **57**, 1–28 (1988). [A scientific autobiography that provides a glimpse of the early difficulties in sequencing proteins.]

Sanger, F., Thompson, E.O.P., and Kitai, R., The amide groups of insulin, *Biochem. J.* **59**, 509–518 (1955).

Figure 5-13 Overview of protein sequencing.

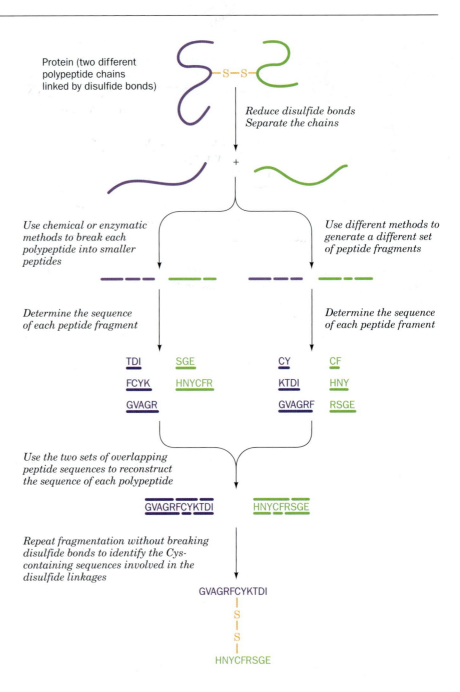

mary structure determination have since been so refined and automated that most proteins can be sequenced within a few hours or days using only a few micrograms of material. Regardless of the technique used, the basic approach for sequencing proteins is similar to the procedure developed by Sanger. *The protein must be broken down into fragments small enough to be individually sequenced, and the primary structure of the intact protein is then reconstructed from the sequences of overlapping fragments* (Fig. 5-13). Such a procedure, as we have seen (Section 3-4C), is also used to sequence DNA.

A   Preliminary Steps

The complete amino acid sequence of a protein includes the sequence of each of its subunits, if any, so the subunits must be identified and isolated before sequencing begins.

N-Terminal Analysis Reveals the Number of Different Types of Subunits.   Each polypeptide chain (if it is not chemically blocked) has an N-terminal residue. *Identifying this **end group** can establish the number of chemically distinct polypeptides in a protein.* For example, insulin has equal amounts of the N-terminal residues Gly and Phe, which indicates that it has equal numbers of two nonidentical polypeptide chains.

The N-terminus of a polypeptide can be determined by several methods. The fluorescent compound **5-dimethylamino-1-naphthalenesulfonyl chloride (dansyl chloride)** reacts with primary amines to yield dansylated polypeptides (Fig. 5-14). The treatment of a dansylated polypeptide with aqueous acid at high temperature hydrolyzes its peptide bonds. This liberates the N-terminal residue, which can then be separated chromatographically from the other amino acids and identified by its intense yellow fluorescence. The N-terminal residue can also be identified by performing the first step of Edman degradation (Section 5-3C), a procedure that liberates amino acids one at a time from the N-terminus of a polypeptide.

Disulfide Bonds between and within Polypeptides Are Cleaved.   Disulfide bonds between Cys residues must be cleaved to separate polypeptide chains—if they are disulfide-linked—and to ensure that polypeptide chains are fully linear (residues in polypeptides that are "knotted" with disulfide bonds may not be accessible to all the enzymes and reagents for sequencing). Disulfide

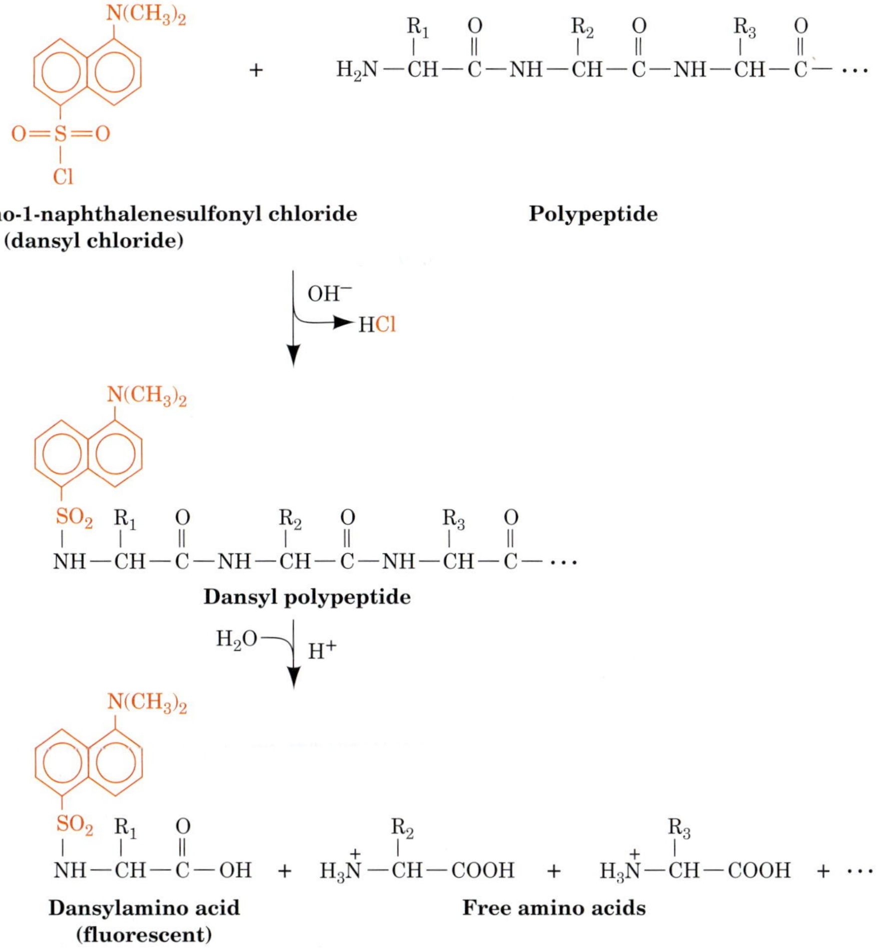

Figure 5-14 The dansyl chloride reaction.   The reaction of dansyl chloride with primary amino groups is used for end group analysis.

bonds can be reductively cleaved by treating them with 2-mercaptoethanol or another **mercaptan** (compounds that contain an —SH group):

Cystine   +   2 HSCH₂CH₂OH ⟶ Cysteine

The resulting free sulfhydryl groups are then alkylated, usually by treatment with **iodoacetate,** to prevent the re-formation of disulfide bonds through oxidation by O_2:

$$Cys—CH_2—SH \;+\; ICH_2COO^- \longrightarrow Cys—CH_2—S—CH_2COO^- \;+\; HI$$

Cysteine   **Iodoacetate**   **S-Carboxymethylcysteine (CM-Cys)**

The Amino Acid Composition of a Polypeptide May Be Determined.

In some cases, it is desirable to know the **amino acid composition** of a polypeptide, that is, the number of each type of amino acid residue present. This information may provide clues to the protein's structure, but it is not required for determining its amino acid sequence.

The amino acid composition of a polypeptide is determined by its complete hydrolysis followed by the analysis of the liberated amino acids. Polypeptide hydrolysis can be accomplished by treating the sample with acid or base, but neither method is fully satisfactory. For example, acid hydrolysis degrades Ser, Thr, Tyr, and Trp and converts Asn and Gln to Asp and Glu. Base hydrolysis destroys Cys, Ser, Thr, and Arg. Enzymatic hydrolysis of polypeptides is often incomplete and is complicated by the fact that the peptidases, being proteins themselves, are subject to proteolytic degradation, thereby contributing to the total amino acid content of the reaction mixture.

The quantitative analysis of the polypeptide hydrolysate is performed by an instrument that separates amino acids by chromatography and derivatizes them (either before or after chromatography) with an easily detected tag. The amino acids are then identified by their characteristic retention times on HPLC and quantified by their absorbance or fluorescence intensities (Fig. 5-15). Modern amino acid analyzers can completely analyze a protein digest containing as little as 1 pmol of each amino acid in <1 h.

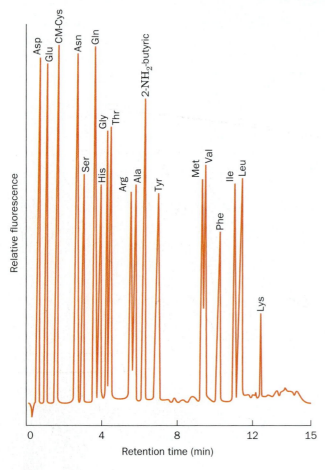

Figure 5-15 Amino acid analysis. The amino acids were derivatized with a fluorescent tag before their separation by HPLC. [After Hunkapiller, M.W., Strickler, J.E., and Wilson, K.J., *Science* **226**, 309 (1984).]

B Polypeptide Cleavage

Polypeptides that are longer than 40 to 100 residues cannot be directly sequenced and must therefore be cleaved, either enzymatically or chemically, to specific fragments that are small enough to be sequenced. Various **endopeptidases** (enzymes that catalyze the hydrolysis of internal peptide bonds, as opposed to **exopeptidases,** which catalyze the hydrolysis of N- or C-terminal residues) can be used to fragment polypeptides. Endopeptidases, as do exopeptidases (which collectively are called **proteases**), have side chain requirements for the residues flanking the scissile peptide bond (that is, the bond that is to be cleaved; Table 5-3). The digestive enzyme **trypsin** has the greatest specificity and is therefore the most valuable member of the arsenal of peptidases used to fragment polypeptides. It cleaves peptide bonds on the C side (toward the carboxyl terminus) of the positively charged residues Arg and Lys if the next residue is not Pro.

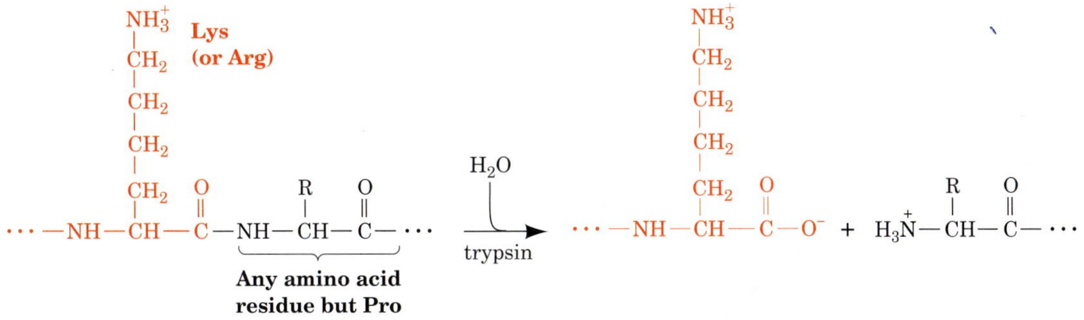

The other endopeptidases listed in Table 5-3 exhibit broader side chain specificities than trypsin and often yield a series of peptide fragments with overlapping sequences. However, through **limited proteolysis,** that is, by adjust-

Table 5-3 Specificities of Various Endopeptidases

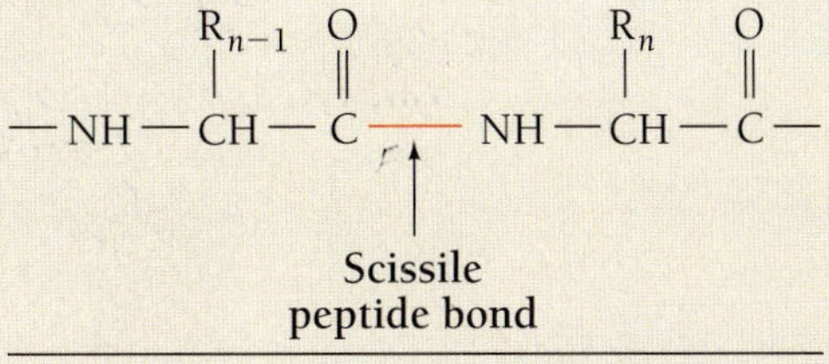

Enzyme	Source	Specificity	Comments
Trypsin	Bovine pancreas	R_{n-1} = positively charged residues: Arg, Lys; $R_n \neq$ Pro	Highly specific
Chymotrypsin	Bovine pancreas	R_{n-1} = bulky hydrophobic residues: Phe, Trp, Tyr; $R_n \neq$ Pro	Cleaves more slowly for R_{n-1} = Asn, His, Met, Leu
Elastase	Bovine pancreas	R_{n-1} = small neutral residues: Ala, Gly, Ser, Val; $R_n \neq$ Pro	
Thermolysin	*Bacillus thermoproteolyticus*	R_n = Ile, Met, Phe, Trp, Tyr, Val; $R_{n-1} \neq$ Pro	Occasionally cleaves at R_n = Ala, Asp, His, Thr; heat stable
Pepsin	Bovine gastric mucosa	R_n = Leu, Phe, Trp, Tyr; $R_{n-1} \neq$ Pro	Also others; quite nonspecific; pH optimum = 2
Endopeptidase V8	*Staphylococcus aureus*	R_{n-1} = Glu	

ing reaction conditions and limiting reaction times, these less specific endopeptidases can yield a set of discrete, nonoverlapping fragments.

Several chemical reagents promote peptide bond cleavage at specific residues. The most useful of these, **cyanogen bromide** (CNBr), cleaves on the C side of Met residues (Fig. 5-16).

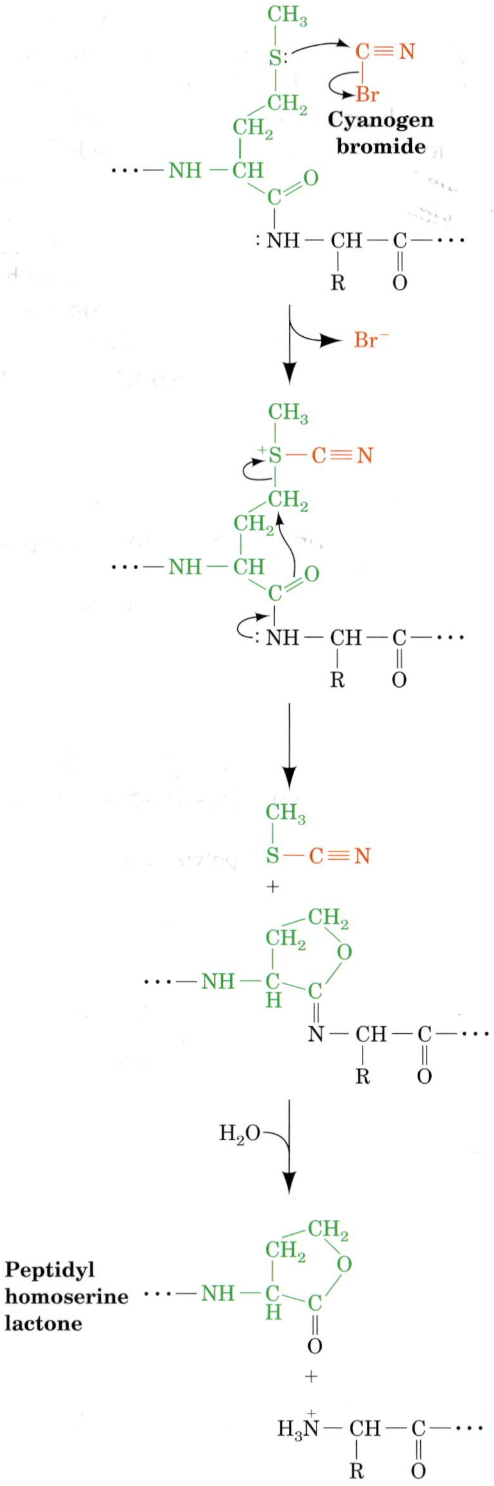

Figure 5-16 Cyanogen bromide cleavage of a polypeptide.   CNBr reacts specifically with Met residues, resulting in cleavage of the peptide bond on their C-terminal side. The newly formed C-terminal residue forms a cyclic structure known as a **peptidyl homoserine lactone.**

C Edman Degradation

Once the peptide fragments formed through specific cleavage reactions have been isolated, their amino acid sequences can be determined. This can be accomplished through repeated cycles of **Edman degradation.** In this process (named after its inventor, Pehr Edman), **phenylisothiocyanate (PITC;** also known as **Edman's reagent)** reacts with the N-terminal amino group of a polypeptide under mildly alkaline conditions to form a **phenylthio-carbamyl (PTC)** adduct (Fig. 5-17). This product is treated with anhydrous **trifluoroacetic acid,** which cleaves the N-terminal residue as a thiazolinone derivative but does not hydrolyze other peptide bonds. Edman degradation therefore releases the N-terminal amino acid residue but leaves intact the rest of the polypeptide chain. The thiazolinone-amino acid is selectively extracted into an organic solvent and is converted to the more stable **phenylthiohydantoin (PTH)** derivative by treatment with aqueous acid. This PTH-amino acid can later be identified by chromatography. Thus, *it is possible to determine the amino acid sequence of a polypeptide chain from the N-terminus inward by subjecting the polypeptide to repeated cycles of*

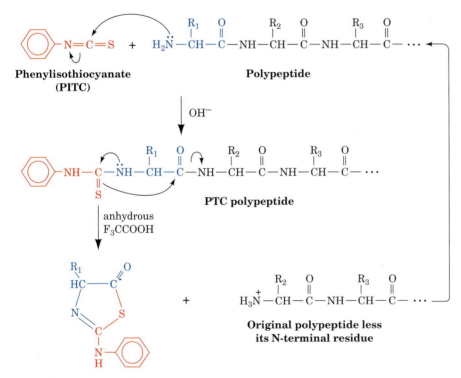

Figure 5-17 Edman degradation.   The reaction occurs in three stages, each requiring different conditions. Amino acid residues can therefore be sequentially removed from the N-terminus of a polypeptide in a controlled stepwise fashion.   See the Animated Figures.

Edman degradation and, after every cycle, identifying the newly liberated PTH-amino acid.

The Edman degradation technique has been automated and refined, resulting in great savings of time and material. In modern instruments, the peptide sample is dried onto a disk of glass fiber paper, and accurately measured quantities of reagents are delivered and products removed as vapors in a stream of argon at programmed intervals. Up to 100 residues can be identified before the cumulative effects of incomplete reactions, side reactions, and peptide loss make further amino acid identification unreliable. Since less than a picomole of a PTH-amino acid can be detected and identified, sequence analysis can be carried out on as little as 5 to 10 pmol of a peptide (<0.1 μg—an invisibly small amount).

D Sequencing by Mass Spectrometry

Mass spectrometry has emerged as an important technique for characterizing and sequencing polypeptides. Mass spectrometry accurately measures the mass-to-charge (m/z) ratio for ions in the gas phase (where m is the ion's mass and z is its charge). Until about 1985, macromolecules such as proteins and nucleic acids could not be analyzed by mass spectrometry. This

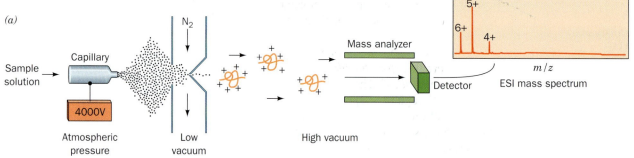

(a)

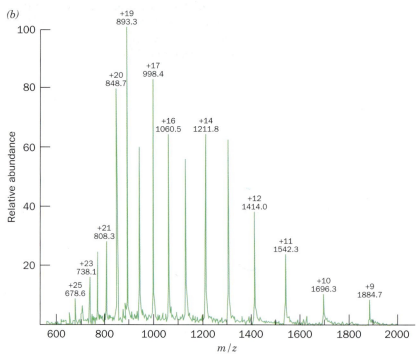

(b)

Figure 5-18 Electrospray ionization mass spectrometry (ESI). (a) Dry N_2 or some other gas promotes the evaporation of solvent from charged droplets containing the protein of interest, leaving gas-phase ions, whose charge is due to the protonation of Arg and Lys residues, thereby yielding so-called $(M + nH)^{n+}$ ions. The mass spectrometer then determines the mass-to-charge ratio of these ions. The resulting mass spectrum consists of a series of peaks corresponding to ions that differ by a single ionic charge and the mass of one proton). [After Fitzgerald, M.C. and Siuzdak, G., *Chem. Biol.* **3,** 708 (1996).] (b) The ESI-MS spectrum of horse heart **apomyoglobin** (myoglobin that lacks its Fe ion). The measured m/z ratios and the inferred charges for most of the peaks are indicated. The data provided by this spectrum permit the mass of the original molecule to be calculated (see Sample Calculation 5-1). The peaks all have shoulders because the polypeptide's component elements contain small admixtures of heavier isotopes (e.g., naturally abundant carbon consists of 98.9% ^{12}C and 1.1% ^{13}C and naturally abundant sulfur consists of 0.8% ^{33}S, 4.2% ^{34}S, and 95.0% ^{35}S). [After Yates, J.R., *Methods Enzymol.* **271,** 353 (1996).]

was because macromolecules were destroyed by the method by which mass spectrometers produced gas-phase ions: vaporization by heating followed by ionization via bombardment with electrons.

Newer techniques have addressed this shortcoming. For example, in the **electrospray ionization (ESI)** technique, a solution of a macromolecule such as a peptide is sprayed from a narrow capillary tube maintained at high voltage (~4000 V), forming fine highly charged droplets from which the solvent rapidly evaporates (Fig. 5-18*a*). This yields a series of gas-phase macromolecular ions that typically have ionic charges in the range +0.5 to +2 per kilodalton. The charges result from the protonation of basic side chains such as Arg and Lys. The ions are directed into the mass spectrometer, which measures their *m/z* values with an accuracy of >0.01% (Fig. 5-18*b*). Consequently, determining an ion's *z* permits its molecular mass to be determined with far greater accuracy than by any other method.

Short polypeptides (<25 residues) can be directly sequenced though the use of a tandem mass spectrometer (Fig. 5-19; two mass spectrometers coupled in series). The first mass spectrometer functions to select and separate the peptide ion of interest from peptide ions of different masses as well as any contaminants that may be present. The selected peptide ion is then passed into a collision cell, where it collides with chemically inert atoms such as helium. The energy thereby imparted to the peptide ion causes it to fragment predominantly at only one of its several peptide bonds, thereby yielding one or two charged fragments per original ion. The molecular masses of the numerous charged fragments so produced are then determined by the second mass spectrometer.

By comparing the molecular masses of successively larger members of a family of fragments, the molecular masses and therefore the identities of the corresponding amino acid residues can be determined. The sequence of an entire polypeptide can thus be elucidated (although mass spectrometry cannot distinguish the isomeric residues Ile and Leu because they have exactly the same mass, and it cannot always reliably distinguish Gln and Lys residues because their molecular masses differ by only 0.036 D).

Computerization of the mass-comparison process has reduced the time required to sequence a short polypeptide to only a few minutes (one cycle of Edman degradation may take an hour). The reliability of this process has been increased through the computerized matching of a measured mass

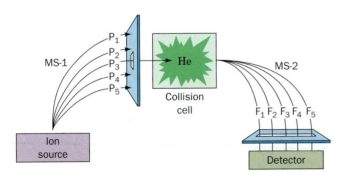

Figure 5-19   Tandem mass spectrometry in amino acid sequencing.   Electrospray ionization (ESI), the ion source, generates gas-phase peptide ions, labeled P_1, P_2, etc., from a digest of the protein to be sequenced. These peptides are separated by the first mass spectrometer (MS-1) according to their *m/z* values, and one of them (here, P_3) is directed into the collision cell, where it collides with helium atoms. This treatment breaks the peptide into fragments (F_1, F_2, etc), which are directed into the second mass spectrometer (MS-2) for determination of their *m/z* values. [After Biemann, K. and Scoble, H.A. *Science* **237**, 992 (1987).]

SAMPLE CALCULATION 5-1

An ESI mass spectrum such as that of apomyoglobin (Fig. 5-18*b*) contains a series of peaks, each corresponding to the *m/z* ratio of an $(M + nH)^{n+}$ ion. Two successive peaks in this mass spectrum have measured *m/z* ratios of 1414.0 and 1542.3. What is the molecular mass of the original apomyoglobin molecule, how does it compare with the value given for it in Table 5-1, and what are the charges of the ions causing these peaks?

The first ($p_1 = 1414.0$) peak arises from an ion with charge *z* and mass M + *z*, where M is the molecular mass of the original protein. Then the adjacent ($p_2 = 1542.3$) peak, which is due to an ion with one less proton, has charge *z* − 1 and mass M + *z* −1. The *m/z* ratios for these ions, p_1 and p_2, are therefore given by the following expressions.

$$p_1 = (M + z)/z$$
$$p_2 = (M + z - 1)/(z - 1)$$

These two linear equations can readily be solved for their unknowns, M and *z*. Solve the first equation for M.

$$M = z(p_1 - 1)$$

Then plug this result into the second equation.

$$p_2 = \frac{z(p_1 - 1) + z - 1}{z - 1} = \frac{zp_1 - 1}{z - 1}$$
$$zp_2 - p_2 = zp_1 - 1$$
$$z = (p_2 - 1)/(p_2 - p_1)$$
$$M = (p_2 - 1)(p_1 - 1)/(p_2 - p_1)$$

Plugging in the values for p_1 and p_2,

M = (1542.3 − 1)(1414.0 − 1)/(1542.3 − 1414.0) = 16,975 D, which is only 0.14% larger than the 16,951 D for horse apomyoglobin given in Table 5-1.

$$z = (1542.3 - 1)/(1542.3 - 1414.0) = 12$$

so that the ionic charge on ion 2 is 12 − 1 = 11.

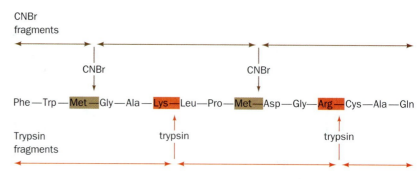

Figure 5-20 **Generating overlapping fragments to determine the amino acid sequence of a polypeptide.** In this example, two sets of overlapping peptide fragments are made by using trypsin to cleave the polypeptide after all its Arg and Lys residues and, in a separate reaction, using CNBr to cleave it after all its Met residues. ☙ **See the Animated Figures.**

spectrum with those of peptides of known sequence as maintained in databases. Mass spectrometry can also be used to sequence peptides with chemically blocked N-termini (which prevents Edman degradation) and to characterize other posttranslational modifications such as phosphorylations and glycosylations.

E Reconstructing the Protein's Sequence

After individual peptide fragments have been sequenced, their order in the original polypeptide must be elucidated. This is accomplished by conducting a second round of protein cleavage with a reagent of different specificity and then comparing the amino acid sequences of the overlapping sets of peptide fragments (Fig. 5-20).

The final step in an amino acid sequence analysis is to determine the positions (if any) of the disulfide bonds. This can be done by cleaving a sample of the protein, with its disulfide bonds intact, to yield pairs of peptide fragments, each containing a single Cys, that are linked by a disulfide bond. After isolating a disulfide-linked polypeptide fragment, the disulfide bond is cleaved and alkylated (Section 5-3A), and the sequences of the two peptides are determined (Fig. 5-21). The various pairs of such polypeptide fragments are identified by comparing their sequences with that of the protein, thereby establishing the locations of the disulfide bonds.

Polypeptide fragment
containing disulfide bond

Reduce disulfide and block
with iodoacetate

$S-CH_2CO_2^-$ + $^-O_2CCH_2-S$

Separate and sequence
the polypeptides

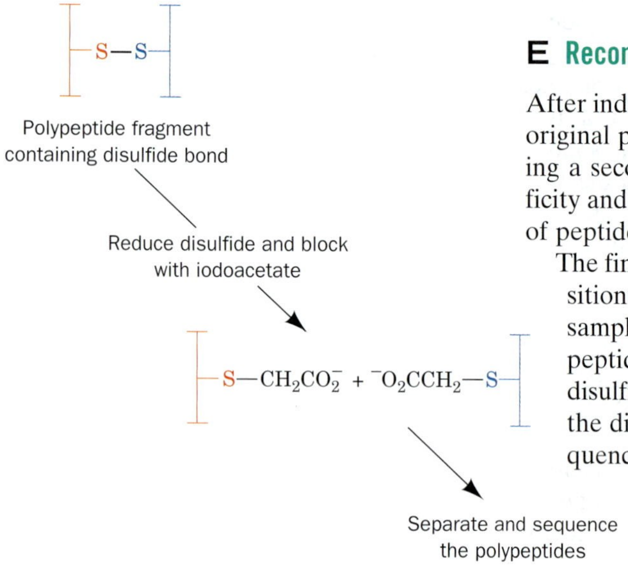

Figure 5-21 **Determining the positions of disulfide bonds.** In this method, disulfide-linked peptide fragments from a protein are reductively cleaved and separately sequenced to identify the positions of the disulfide bonds in the intact protein.

Sequence Databases. After a protein's amino acid sequence has been determined, the information is customarily deposited in a public database. Databases for proteins as well as DNA sequences are accessible via the Internet (Table 5-4). Electronic links between databases allow rapid updates and cross-checking of sequence information.

Table 5-4 **Internet Addresses for the Major Protein and DNA Sequence Data Banks**

Data Banks Containing Protein Sequences

ExPASy Molecular Biology Server (Swiss-Prot): http://au.expasy.org
Protein Information Resource (PIR): http://pir.georgetown.edu/
Protein Research Foundation (PRF): http://www4.prf.or.jp/
UniProt: http://www.ebi.uniprot.org/

Data Banks Containing Gene Sequences

GenBank: http://www.ncbi.nlm.nih.gov/Genbank/GenbankSearch.html
European Bioinformatics Institute (EBI): http://srs.ebi.ac.uk/
DBGET/Integrated Database Retrieval System: http://www.genome.ad.jp/dbget

General information about the UniProt/Swiss-Prot entry	
Entry name	**RSN_HUMAN**
Primary accession number	**Q9HD89**
Entered in Swiss-Prot	Release 40, 16-OCT-2001
Sequence was last modified	Release 40, 16-OCT-2001
Annotations were last modified	Release 44, 05-JUL-2004
Protein description	
Protein name	**Resistin precursor**
Synonyms	Cysteine-rich secreted protein FIZZ3 Adipose tissue-specific secretory factor ADSF C/EBP-epsilon regulated myeloid-specific secreted cysteine-rich protein Cysteine-rich secreted protein A12-alpha-like 2 UNQ407/PRO1199
Origin of the protein	
Gene	Gene name RETN Synonyms RSTN, FIZZ3, HXCP1
From	Homo sapiens (Human)[TaxID:9606]
Taxonomy	Eukaryota; Metazoa; Chordata; Craniata; Vertebrata; Euteleostomi; Mammalia; Eutheria; Primates; Catarrhini; Hominidae; Homo.

Figure 5-22   The initial portion of a Swiss-Prot entry. This information pertains to the protein **resistin**, which is produced by adipose (fat) tissue and may play a role in diabetes. The complete entry includes additional information and references as well as the protein's sequence. [From Swiss-Prot: http://www.uniprot.org/.]

Most sequence databases use similar conventions. For example, consider the annotated protein sequence database named Swiss-Prot. A sequence record in Swiss-Prot begins with the protein's ID code in the form X_Y, where X is a short mnemonic indicating the protein's name (for example, CYC for cytochrome *c* and HBA for hemoglobin α chain) and Y is a five-character identification code indicating the protein's biological source, which usually consists of the first three letters of the genus and the first two letters of the species [e.g., CANFA for *Canis familiaris* (dog)], although for most commonly encountered organisms, Y is self-explanatory (e.g., BOVIN or ECOLI). This is followed by an accession number, which is assigned by the database as a way of identifying an entry even if its ID code must be changed (for example, see Fig. 5-22). The entry continues with the date the sequence was entered into the database and when it was last modified and annotated, a list of pertinent references (which are linked to Med-Line), a description of the protein, and its links to other databases. A Feature Table describes regions or sites of interest in the protein such as disulfide bonds, posttranslational modifications, binding sites, and conflicts between different references. The entry ends with the length of the peptide in residues, its molecular weight, and finally, its sequence using the one-letter code.

Armed with the appropriate software (which is often publicly available at the database sites), a researcher can search a database to find proteins with similar sequences in various organisms. The sequence of even a short peptide fragment may be sufficient to "fish out" the parent protein or its homolog from another species.

Amino acid sequence information is no less valuable when the nucleotide sequence of the corresponding gene is also known, because the protein sequence sometimes provides information about protein structure that is not revealed by nucleic acid sequencing (see Section 3-4). For example, only direct protein sequencing can reveal the locations of disulfide bonds in proteins. In addition, many proteins are modified after they are synthesized. For example, certain residues may be excised to produce the "mature" protein (insulin, shown in Fig. 5-1, is actually synthesized as an 84-residue polypeptide that is proteolytically processed to its smaller two-chain form). Amino acid side chains may also be modified by the addition of carbohydrates, phosphate groups, or acetyl groups, to name only a few. Although some of these modifications occur at characteristic amino acid sequences and are therefore identifiable in nucleotide sequences, only the actual protein sequence can confirm whether and where they occur.

Table 5-5   Amino Acid Sequences of Cytochromes _c_ from 38 Species[a]

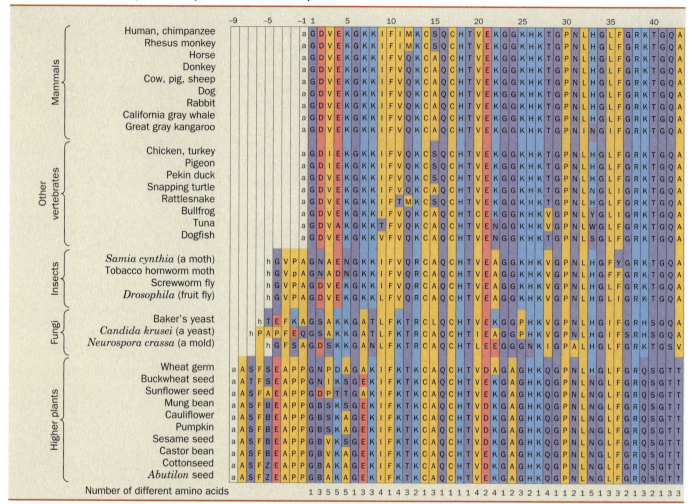

[a]The amino acid side chains have been shaded according to their polarity characteristics so that an invariant or conservatively substituted residue is identified by a vertical band of a single color. The letter a at the beginning of the chain indicates that the N-terminal amino group is acetylated; an h indicates that the acetyl group is absent.

Source: After Dickerson, R.E., _Sci. Am._ **226**(4), 58–72 (1972), with corrections from Dickerson, R.E. and Timkovich, R., _in_ Boyer, P.D. (Ed.), _The Enzymes_ (3rd ed.), Vol. 11, pp. 421–422, Academic Press (1975). Table copyrighted © by Irving Geis.

4 Protein Evolution

Because an organism's genetic material specifies the amino acid sequences of all its proteins, changes in genes due to random mutation can alter a protein's primary structure. A mutation in a protein is propagated only if it somehow increases, or at least does not decrease, the probability that its owner will survive to reproduce. Many mutations are deleterious or produce lethal effects and therefore rapidly die out. On rare occasions, however, a mutation arises that improves the fitness of its host. This is the essence of **Darwinian evolution.**

A Protein Sequence Evolution

The primary structures of a given protein from related species closely resemble one another. Consider **cytochrome _c_,** a protein found in nearly all eukaryotes. Cytochrome _c_ is a component of the mitochondrial electron-

See Guided Exploration 5

Protein evolution

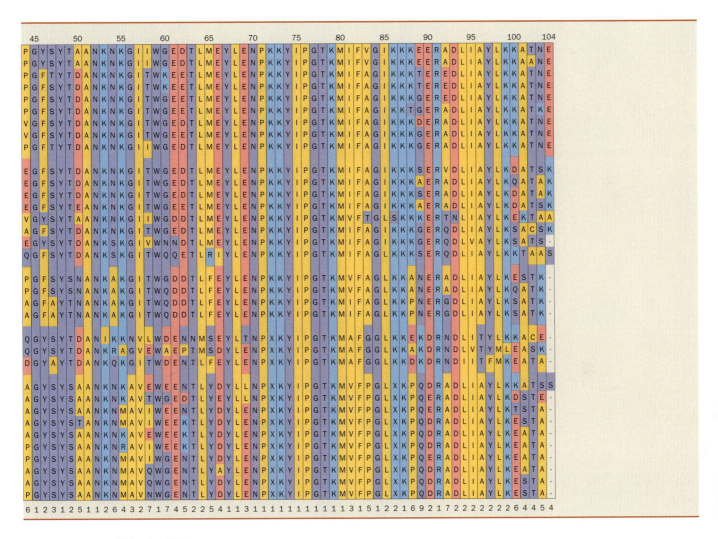

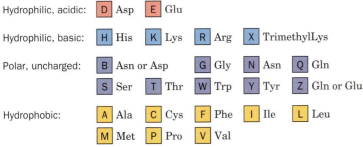

transport system (Section 17-2), which is believed to have taken its present form between 1.5 and 2 billion years ago, when organisms evolved the ability to respire. Emanuel Margoliash, Emil Smith, and others have elucidated the amino acid sequences of the cytochromes *c* from over 100 eukaryotic species ranging in complexity from yeast to humans. The cytochromes *c* from different species are single polypeptides of 104 to 112 residues. The sequences of 38 of these proteins are arranged in Table 5-5 to show the similarities between vertically aligned residues (the residues have been color-coded according to their physical properties). A survey of the aligned sequences (bottom line of Table 5-5) shows that at 38 positions (23 positions in the complete set of >100 sequences), the same amino acid appears

in all species. Most of the remaining positions are occupied by chemically similar residues in different organisms. In only eight positions does the sequence accommodate six or more different residues.

According to evolutionary theory, *related species have evolved from a common ancestor, so it follows that the genes specifying each of their proteins must likewise have evolved from the corresponding gene in that ancestor.* The sequence of the ancestral cytochrome *c* is accessible only indirectly, by examining the sequences of extant proteins. Cytochrome *c* is an evolutionarily **conservative** protein; that is, its sequence has undergone only modest evolutionary changes.

Sequence Comparisons Provide Information on Protein Structure and Function.

*In general, comparisons of the primary structures of **homologous proteins** (evolutionarily related proteins) indicate which of the protein's residues are essential to its function, which are less significant, and which have little specific function.* Finding the same residue at a particular position in the amino acid sequence of a series of related proteins suggests that the chemical or structural properties of that so-called **invariant residue** uniquely suit it to some essential function of the protein. For example, its side chain may be necessary for binding another molecule or participating in a catalytic reaction. Other amino acid positions may have less stringent side chain requirements and can therefore accommodate residues with similar characteristics (e.g., Asp or Glu, Ser or Thr, etc.); such positions are said to be **conservatively substituted.** On the other hand, a particular amino acid position may tolerate many different amino acid residues, indicating that the functional requirements of that position are rather nonspecific. Such a position is said to be **hypervariable.**

Why is cytochrome *c*—an ancient and essential protein—not identical in all species? Even a protein that is well adapted to its function, that is, one that is not subject to physiological improvement, nevertheless continues evolving. *The random nature of mutational processes will, in time, change such a protein in ways that do not significantly affect its function, a process called **neutral drift*** (deleterious mutations are, of course, rapidly rejected through natural selection). Hypervariable residues are apparently particularly subject to neutral drift.

Constructing Phylogenetic Trees.

Far-reaching conclusions about evolutionary relationships can be drawn by comparing the amino acid sequences of homologous proteins. The simplest way to assess evolutionary differences is to count the amino acid differences between proteins. For example, the data in Table 5-5 show that primate cytochromes *c* more nearly resemble those of other mammals than they do those of insects (8–12 differences among mammals versus 26–31 differences between mammals and insects). Similarly, the cytochromes *c* of fungi differ as much from those of mammals (45–51 differences) as they do from those of insects (41–47) or higher plants (47–54). The order of these differences largely parallels that expected from classical taxonomy, which is based primarily on morphological rather than molecular characteristics.

The sequences of homologous proteins can be analyzed by computer to construct a **phylogenetic tree,** a diagram that indicates the ancestral relationships among organisms that produce the protein. The phylogenetic tree for cytochrome *c* is sketched in Fig. 5-23. Similar trees have been derived for other proteins. Each branch point of the tree represents a putative common ancestor for all the organisms above it. The distances between branch points are expressed as the number of amino acid differences per 100

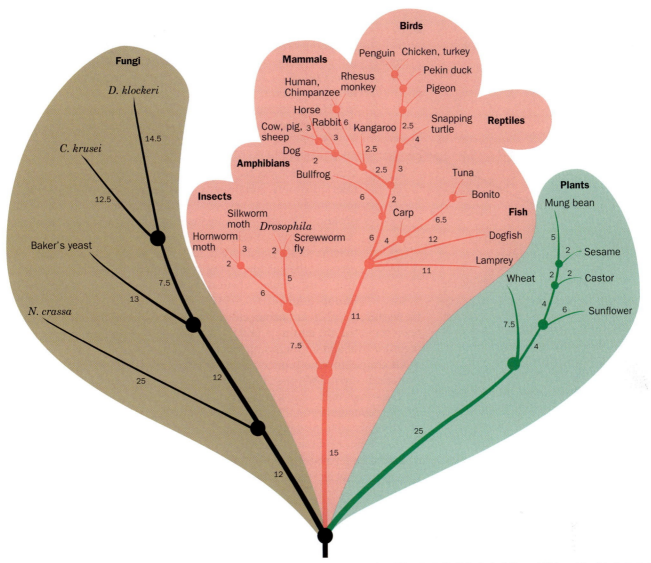

Figure 5-23 Phylogenetic tree of cytochrome *c*. Each branch point represents an organism ancestral to the species connected above it. The number beside each branch indicates the number of inferred differences per 100 residues between the cytochromes *c* of the flanking branch points or species. [After Dayhoff, M.O., Park, C.M., and McLaughlin, P.J., *in* Dayhoff, M.O. (Ed.), *Atlas of Protein Sequence and Structure, p.* 8, National Biomedical Research Foundation (1972).]

residues of the protein. Such trees present a more quantitative measure of the degree of relatedness of the various species than macroscopic taxonomy can provide.

Note that the evolutionary distances from all modern cytochromes *c* to the lowest point, the earliest common ancestor producing this protein, are approximately the same. Thus, "lower" organisms do not represent life-forms that appeared early in history and ceased to evolve further. The cytochromes *c* of all the species included in Fig. 5-23—whether called "primitive" or "advanced"—have evolved to about the same extent.

Proteins Evolve at Characteristic Rates. The protein sequence differences between various species can be plotted against the time when, according to the fossil record, the species diverged. The plot for a given protein is essentially linear, indicating that its mutations accumulate at a constant rate

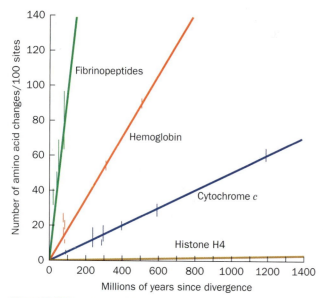

Figure 5-24 **Rates of evolution of four proteins.** The graph was constructed by plotting the number of different amino acid residues in the proteins on two sides of a branch point of a phylogenetic tree versus the time, according to the fossil record, since the corresponding species diverged from their common ancestor. [Illustration, Irving Geis/Geis Archives Trust. Copyright Howard Hughes Medical Institute. Reproduced with permission.]

over a geological time scale. However, rates of evolution vary among proteins (Fig. 5-24). This does not imply that the rates of mutation of the DNAs specifying those proteins differ, but rather that *the rate at which mutations are accepted into a protein depends on the extent to which amino acid changes affect the protein's function.* For example, **histone H4,** a protein that binds to DNA in eukaryotes (Section 23-5A), is among the most highly conserved proteins (the histones H4 from peas and cows, species that diverged 1.2 billion years ago, differ by only two conservative changes in their 102 residues). Evidently, histone H4 is so important for packaging DNA in cells that it is extremely intolerant of any mutations. Cytochrome *c* is only slightly more tolerant. It is a relatively small protein that binds to several other proteins. Hence, any changes in its amino acid sequence must be compatible with all its binding partners. Hemoglobin, which functions as a free-floating molecule, is subject to less selective pressure than histone H4 or cytochrome *c,* so its surface residues are more easily substituted by other amino acids. The **fibrinopeptides** are ~20-residue fragments that are cleaved from the vertebrate protein **fibrinogen** to induce blood clotting. Once they have been removed, the fibrinopeptides are discarded, so there is little selective pressure to maintain their amino acid sequences.

Of course, protein sequences alone cannot reveal the complete story of evolution. Slight differences in protein sequence are not enough to account for the sometimes dramatic morphological features that differentiate even closely related species. Thus, the proteins of humans and chimpanzees are >99% identical on average (e.g., their cytochromes *c* are identical), but the species' anatomical and behavioral differences are so great that they are classified in separate families. Large evolutionary steps that mark the divergence of species appear to be accomplished by mutations in the genes for regulatory proteins that control gene expression, that is, how much of each protein is made, where, and when. Other changes in DNA, such as rearrangements and duplications of genes, give rise to new proteins that are then subject to natural selection.

B Gene Duplication and Protein Families

It is not surprising that proteins with similar functions have similar sequences; such proteins presumably evolved from a common ancestor. Interestingly, protein sequence analysis has revealed that some proteins with widely different physiological functions also have similar sequences of amino acids. In fact, most proteins have extensive sequence similarities with some other proteins from the same organism (we will see in the next chapter that three-dimensional protein structures are likewise highly conserved). Such proteins arose through **gene duplication,** an aberrant genetic recombination event in which one chromosome acquired both copies of the primordial gene (genetic recombination is discussed in Section 24-6). *Gene duplication is a particularly efficient mode of evolution because one copy of the gene evolves a new function through natural selection while its counterpart continues to direct the synthesis of the original protein.* The two independently evolving genes that are derived from a duplication event are said to be **paralogous.**

The **globin family** of proteins provides an excellent example of evolution through gene duplication. **Hemoglobin,** which transports O_2 from the lungs (or gills or skin) to the tissues, is a tetramer with the subunit composition

$\alpha_2\beta_2$ (i.e., two α polypeptides and two β polypeptides). The sequences of the α and β subunits are similar to each other and to the sequence of the protein **myoglobin,** which facilitates oxygen diffusion through muscle tissue (hemoglobin and myoglobin are discussed in more detail in Chapter 7). The primordial globin probably functioned simply as an oxygen-storage protein. Gene duplication allowed one globin to evolve into a monomeric hemoglobin α chain. Duplication of the α chain gene gave rise to the paralogous gene for the β chain. Other members of the globin family include the β-like γ chain that is present in fetal hemoglobin, an $\alpha_2\gamma_2$ tetramer, and the β-like ε and α-like ζ chains that appear together early in embryogenesis as $\zeta_2\varepsilon_2$ hemoglobin. Primates contain a relatively recently duplicated globin, the β-like δ chain, which appears as a minor component ($\sim1\%$) of adult hemoglobin. Although the $\alpha_2\delta_2$ hemoglobin has no known unique function, perhaps it may eventually evolve one. The genealogy of the members of the globin family is diagrammed in Fig. 5-25. The human genome also contains the relics of globin genes that are not expressed. These **pseudogenes** can be considered the dead ends of protein evolution.

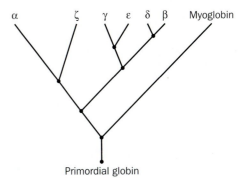

Figure 5-25   Genealogy of the globin family.   Each branch point represents a gene duplication event. Myoglobin is a single-chain protein. The globins identified by Greek letters are subunits of hemoglobins.

C   Protein Modules

Gene duplication is not the only mechanism for generating new proteins. Analysis of protein sequences has revealed that many proteins are mosaics of sequence motifs or **modules** of about 40–100 amino acid residues. In some large proteins, a single module may be repeated many times (Fig. 5-26a); other proteins contain a variety of modules. Some of the proteins involved in blood clotting, for example, are built from a set of smaller modules (Fig. 5-26b). The sequence homology between modules is imperfect, since the sequence of a given protein evolves independently of other sequences. The functions of individual modules are not always known: Some appear to have discrete activities, such as catalyzing a certain type of chemical reaction or binding a particular molecule, but others may merely be spacers or scaffolding for other modules.

Generating new genes by shuffling modules is a much faster process than duplicating an entire gene and allowing it to mutate over time. Neverthe-

(a) Fibronectin

(b) Blood clotting proteins

Factors VII, IX, X, and protein C
Factor XII
Tissue-type plasminogen activator
Protein S

Key

▲ Fibronectin domain 1
■ Fibronectin domain 2
● Fibronectin domain 3
● γ-Carboxyglutamate domain
◆ Epidermal growth factor domain
■ Serine protease domain
▼ Kringle domain
■ Unique domain

Figure 5-26   Modular construction of some proteins.   Each shape represents a segment of $\sim$40–100 residues that appears, with some sequence variation, several times in the same protein or in a number of related proteins, (a) **Fibronectin**, a $\sim$500 kD protein of the extracellular matrix, is composed mostly of repeated modules of three types. (b) Some of the proteins that participate in blood clotting are built from a small set of modules. (The epidermal growth factor domain is so named because it was first observed as a component of **epidermal growth factor.**) [After Baron, M., Norman, D.G., and Campbell, I.D., *Trends Biochem. Sci.* **16**, 14 (1991).]

less, both mechanisms have played important roles in the evolution of species. The appearance of a protein with a novel sequence is a rare event in biology; most proteins are variations on other proteins or parts thereof.

 SUMMARY

1. The properties of proteins depend largely on the sizes and sequences of their component polypeptides.

2. Fractionation procedures are used to purify proteins on the basis of solubility, charge, size, and binding specificity.

3. A purified protein must be stabilized.

4. Differences in solubility permit proteins to be concentrated and purified by salting out.

5. Chromatography, the separation of soluble substances by their rate of movement through an insoluble matrix, is a technique for purifying molecules by charge (ion exchange chromatography), hydrophobicity (hydrophobic interaction chromatography), size (gel filtration chromatography), and binding properties (affinity chromatography). Binding and elution often depend on the salt concentration and pH.

6. Electrophoresis separates molecules by charge and size; SDS-PAGE separates them primarily by size. 2D electrophoresis can resolve thousands of proteins.

7. Proteins can be separated by mass in an ultracentrifuge.

8. Analysis of a protein's sequence begins with end group analysis, to determine the number of different subunits, and the cleavage of disulfide bonds. The amino acid composition may also be determined.

9. Polypeptides are cleaved into fragments suitable for sequencing by Edman degradation, in which residues are removed, one at a time, from the N-terminus. Peptides can also be sequenced by mass spectroscopy.

10. A protein's sequence is reconstructed from the sequences of overlapping peptide fragments and from information about the locations of disulfide bonds. The sequences of numerous proteins are archived in publicly available databases.

11. Proteins evolve through changes in primary structure. Comparisons of proteins in different species may reveal which residues are most essential for a protein's structure and function.

12. Comparisons of polypeptide sequences also reveal the evolutionary relationships between species and between different proteins within a species.

REFERENCES

Protein Purification

Boyer, R.F., *Modern Experimental Biochemistry* (3rd ed.), Benjamin Cummings (2001).

Creighton, T.E. (Ed.), *Protein Structure. A Practical Approach* (2nd ed.), IRL Press (1997). [Includes chapters on electrophoresis, determination of disulfide bonds, mass spectrometry, and immunochemical techniques.]

Hames, B.D. (Ed.), *Gel Electrophoresis of Proteins. A Practical Approach* (3rd ed.), IRL Press (1998).

Janson, J.C. and Rydén, L. (Eds.), *Protein Purification: Principles, High Resolution Methods, and Applications*, Wiley (1998). [Contains detailed discussions of a variety of chromatographic and electrophoretic separation techniques.]

Laue, T.M. and Stafford, W.F., III, Modern applications of analytical ultracentrifugation, *Annu. Rev. Biophys. Biomol. Struct.* **28,** 75–100 (1999).

Roe, S. (Ed.), *Protein Purification Techniques: A Practical Approach* (2nd ed.), Oxford University Press (2001).

Tanford, C. and Reynolds, J., *Nature's Robots: A History of Proteins*, Oxford University Press (2001). [Descriptions of some early discoveries related to the nature of proteins and their purification and analysis.]

Wilson, K. and Walker, J.M. (Eds.), *Principles and Techniques of Practical Biochemistry* (5th ed.), Cambridge University Press (2001). [Includes reviews of centrifugation, spectroscopy, electrophoresis, and chromatography.]

Protein Sequencing

Aebersold, R. and Mann, M., Mass spectrometry-based proteomics, *Nature* **422,** 198–207 (2003). [Describes some of the methods of mass spectrometric analysis of proteins as well as current and potential applications.]

Blackman, D.S., *The Logic of Biochemical Sequencing*, CRC Press (1994). [Question-and-answer format provides overview of strategies and techniques for sequencing proteins and nucleic acids.]

Findlay, J.B.C. and Geisow, M.J. (Eds.), *Protein Sequencing. A Practical Approach*, IRL Press (1989).

Galperin, M.Y., The molecular biology database collection: 2004 update, *Nucleic Acids Res.* **32,** D3–D22 (2004). [This and other articles in the same issue describe the features and potential uses of various protein and DNA sequence databases.]

Protein Evolution

Baxevanis, A.D. and Ouellette, B.F.F. (Eds.), *Bioinformatics, A Practical Guide to the Analysis of Genes and Proteins* (3rd ed.), Wiley-Intescience (2005).

Doolittle, R.F., Feng, D.-F., Tsang, S., Cho, G., and Little, E., Determining divergence times of the major kingdoms of living organisms with a protein clock, *Science* **271,** 470–477 (1996). [Demonstrates how protein sequences can be used to draw phylogenetic trees.]

Henikoff, S., Greene, E.A., Pietrokovski, S., Bork, P., Attwood, T.K., and Hood, L., Gene families: The taxonomy of protein paralogs and chimeras, *Science* **278**, 609–614 (1997). [Explains how gene duplication and rearrangement allow functional innovation in proteins within a species.]

Mount, D.W., *Bioinformatics. Sequence and Genome Analysis.* Cold Spring Harbor Laboratory Press (2001).

KEY TERMS

primary structure	salting in	metal chelate affinity	amino acid composition
peptide	salting out	chromatography	endopeptidase
subunit	chromatography	autoradiograph	exopeptidase
multisubunit protein	mobile phase	immunoblot (Western blot)	limited proteolysis
denaturation	stationary phase	SDS-PAGE	Edman degradation
protease	HPLC	capillary electrophoresis (CE)	mass spectrometry
nuclease	ion exchange	IEF	Darwinian evolution
assay	chromatography	two-dimensional gel	homologous proteins
coupled enzymatic reaction	anion exchanger	electrophoresis	invariant residue
immunoassay	cation exchanger	proteomics	conservative substitution
antibody	polyelectrolyte	ultracentrifugation	hypervariable residue
antigen	elution	Svedberg (S)	neutral drift
RIA	eluant	analytical ultracentrifuge	phylogenetic tree
ELISA	hydrophobic interaction	preparative ultracentrifuge	gene duplication
Beer–Lambert law	chromatography	density gradient	paralogous genes
absorbance (*A*)	gel filtration chromatography	zonal ultracentrifugation	pseudogene
molar absorptivity (ε)	affinity chromatography	equilibrium density gradient	module
chromophore	ligand	centrifugation	
Bradford assay	immunoaffinity	N-terminal analysis	
fractionation procedure	chromatography	mercaptan	

STUDY EXERCISES

1. List the 20 amino acids in order of their abundance in proteins.

2. List some factors that influence the stability of purified proteins.

3. List the separation techniques that exploit the following molecular properties: charge, polarity, size, and specificity.

4. How does ion exchange chromatography differ from affinity chromatography?

5. What kinds of information can be gathered by ultracentrifugation?

6. How can the N-terminal residues of proteins be identified?

7. Why should disulfide bonds be cleaved before protein sequencing?

8. Why might it be helpful to know a protein's amino acid composition before trying to sequence it?

9. Describe the steps of Edman degradation.

10. List some advantages of automating laboratory procedures.

11. What information is provided by comparing the sequences of proteins from different organisms?

12. How is a phylogenetic tree constructed?

PROBLEMS

1. Which peptide has greater absorbance at 280 nm?

    A. Gln–Leu–Glu–Phe–Thr–Leu–Asp–Gly–Tyr

    B. Ser–Val–Trp–Asp–Phe–Gly–Tyr–Trp–Ala

2. (a) In what order would the amino acids Arg, His, and Leu be eluted from a carboxymethyl column at pH 6? (b) In what order would Glu, Lys, and Val be eluted from a diethylaminoethyl column at pH 8?

3. Explain why a certain protein has an apparent molecular mass of 90 kD when determined by gel filtration and 60 kD when determined by SDS-PAGE in the presence or absence of 2-mercaptoethanol. Which molecular mass determination is more accurate?

4. Determine the subunit composition of a protein from the following information:

    Molecular mass by gel filtration: 200 kD

    Molecular mass by SDS-PAGE: 100 kD

    Molecular mass by SDS-PAGE with 2-mercaptoethanol: 40 kD and 60 kD

5. Explain why a protein which has a sedimentation coefficient of 2.6S when ultracentrifuged in a solution containing 0.1 M NaCl has a sedimentation coefficient of 4.3S in a solution containing 1 M NaCl.

6. What fractionation procedure could be used to purify protein 1 from a mixture of three proteins whose amino acid compositions are as follows?

   1. 25% Ala, 20% Gly, 20% Ser, 10% Ile, 10% Val, 5% Asn, 5% Gln, 5% Pro

   2. 30% Gln, 25% Glu, 20% Lys, 15% Ser, 10% Cys

   3. 25% Asn, 20% Gly, 20% Asp, 20% Ser, 10% Lys, 5% Tyr

   All three proteins are similar in size and p*I*, and there is no antibody available for protein 1. [Problem provided by Bruce Wightman, Muhlenberg College.]

7. Explain why the dansyl chloride treatment of a single polypeptide chain followed by its complete acid hydrolysis yields several dansylated amino acids.

8. Identify the first residue obtained by Edman degradation of cytochrome *c* from (a) *Drosophila,* (b) baker's yeast, and (c) wheat germ (see Table 5-5).

9. You must cleave the following peptide into smaller fragments. Which of the proteases listed in Table 5-3 would be likely to yield the most fragments? The fewest?

   NMTQGRCKPVNTFVHEPLVDVQNVCFKE

10. You wish to determine the sequence of a short peptide. Cleavage with trypsin yields three smaller peptides with the sequences Leu–Glu, Gly–Tyr–Asn–Arg, and Gln–Ala–Phe–Val–Lys. Cleavage with chymotrypsin yields three peptides with the sequences Gln–Ala–Phe, Asn–Arg–Leu–Glu, and Val–Lys–Gly–Tyr. What is the sequence of the intact peptide?

11. Separate cleavage reactions of a polypeptide by CNBr and chymotrypsin yield fragments with the following amino acid sequences. What is the sequence of the intact polypeptide?

    CNBr treatment

    1. Arg–Ala–Tyr–Gly–Asn
    2. Leu–Phe–Met
    3. Asp–Met

    Chymotrypsin

    4. Met–Arg–Ala–Tyr
    5. Asp–Met–Leu–Phe
    6. Gly–Asn

12. You wish to determine the sequence of a polypeptide that has the following amino acid composition.

    1 Ala   4 Arg   2 Asn   3 Asp   4 Cys   3 Gly   1 Gln   4 Glu
    1 His   1 Lys   1 Met   1 Phe   2 Pro   4 Ser   2 Tyr   1 Trp

    (a) What is the maximum number of peptides you can expect if you cleave the polypeptide with cyanogen bromide?

    (b) What is the maximum number of peptides you can expect if you cleave the polypeptide with chymotrypsin?

    (c) Analysis of the intact polypeptide reveals that there are no free sulfhydryl groups. How many disulfide bonds are likely to be present?

    (d) How many different arrangements of disulfide bonds are possible?

13. Treatment of a polypeptide with 2-mercaptoethanol yields two polypeptides:

    1. Ala-Val-Cys-Arg-Thr-Gly-Cys-Lys-Asn-Phe-Leu
    2. Tyr-Lys-Cys-Phe-Arg-His-Thr-Lys-Cys-Ser

    Treatment of the intact polypeptide with trypsin yields fragments with the following amino acid compositions:

    3. (Ala, Arg, Cys$_2$, Ser, Val)
    4. (Arg, Cys$_2$, Gly, Lys, Thr, Phe)
    5. (Asn, Leu, Phe)
    6. (His, Lys, Thr)
    7. (Lys, Tyr)

    Indicate the positions of the disulfide bonds in the intact polypeptide

14. You wish to sequence the light chain of a protease inhibitor from the *Brassica nigra* plant. Cleavage of the light chain by trypsin and chymotrypsin yields the following fragments. What is the sequence of the light chain?

    Chymotrypsin

    1. Leu–His–Lys–Gln–Ala–Asn–Gln–Ser–Gly–Gly–Gly–Pro–Ser
    2. Gln–Gln–Ala–Gln–His–Leu–Arg–Ala–Cys–Gln–Gln–Trp
    3. Arg–Ile–Pro–Lys–Cys–Arg–Lys–Phe

    Trypsin

    4. Arg
    5. Ala–Cys–Gln–Gln–Trp–Leu–His–Lys
    6. Cys–Arg
    7. Gln–Ala–Asn–Gln–Ser–Gly–Gly–Gly–Pro–Ser
    8. Phe–Gln–Gln–Ala–Gln–His–Leu–Arg
    9. Ile–Pro–Lys
    10. Lys

    [Problem provided by Kathleen Cornely, Providence College.]

15. In site-directed mutagenesis experiments, Gly is often successfully substituted for Val, but Val can rarely substitute for Gly. Explain.

16. Below is a list of the first 10 residues of the B helix in myoglobin from different organisms.

position	1	2	3	4	5	6	7	8	9	10
human	D	I	P	G	H	G	Q	E	V	L
chicken	D	I	A	G	H	G	H	E	V	L
alligator	K	L	P	E	H	G	H	E	V	I
turtle	D	L	S	A	H	G	Q	E	V	I
tuna	D	Y	T	T	M	G	G	L	V	L
carp	D	F	E	G	T	G	G	E	V	L

    Based on this information, which positions (a) appear unable to tolerate substitutions, (b) can tolerate conservative substitution, and (c) are highly variable?

■ BIOINFORMATICS EXERCISES

Bioinformatics Exercises are available for this chapter in the tabbed section at the end of the book and on the text website at www.wiley.com/college/voet.

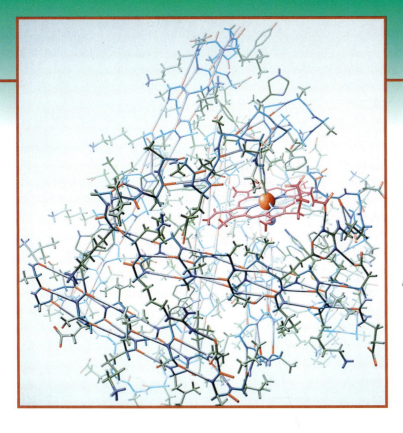

The atomic structure of myoglobin, an oxygen binding protein, is drawn here as a stick model. The overall conformation of a protein such as myoglobin is a function of its amino acid sequence. How do noncovalent forces act on a polypeptide chain to stabilize its unique three-dimensional arrangement of atoms? [Illustration, Irving Geis. Image from the Irving Geis Collection/Howard Hughes Medical Institute. Rights owned by HHMI. Reproduction by permission only.]

Proteins: Three-Dimensional Structure

For many years, it was thought that proteins were colloids of random structure and that the enzymatic activities of certain crystallized proteins were due to unknown entities associated with an inert protein carrier. In 1934, J.D. Bernal and Dorothy Crowfoot Hodgkin showed that a crystal of the protein **pepsin** yielded a discrete diffraction pattern when placed in an X-ray beam. This result provided the first evidence that pepsin was not a random colloid but an ordered array of atoms organized into a large yet uniquely structured molecule.

Even relatively small proteins contain thousands of atoms, almost all of which occupy definite positions in space. The first X-ray structure of a protein, that of sperm whale myoglobin, was reported in 1958 by John Kendrew and co-workers. At the time—only 5 years after James Watson and Francis Crick had elucidated the simple and elegant structure of DNA (Section 3-2B)—protein chemists were chagrined by the complexity and apparent lack of regularity in the structure of myoglobin. In retrospect, such irregularity seems essential for proteins to fulfill their diverse biological roles. However, comparisons of the nearly 30,000 protein structures now known have revealed that proteins actually exhibit a remarkable degree of structural regularity.

As we saw in Section 5-1, the primary structure of a protein is its linear sequence of amino acids. In discussing protein structure, three further levels of structural complexity are customarily invoked:

- **Secondary structure** is the local spatial arrangement of a polypeptide's backbone atoms without regard to the conformations of its side chains.
- **Tertiary structure** refers to the three-dimensional structure of an entire polypeptide, including that of its side chains.
- Many proteins are composed of two or more polypeptide chains, loosely referred to as subunits. A protein's **quaternary structure** refers to the spatial arrangement of its subunits.

The four levels of protein structure are summarized in Fig. 6-1.

In this chapter, we explore secondary through quaternary structure, including examples of proteins that illustrate each of these levels. We also discuss the process of protein folding and the forces that stabilize folded proteins.

1 Secondary Structure

Protein secondary structure includes the regular polypeptide folding patterns such as helices, sheets, and turns. However, before we discuss these basic structural elements, we must consider the geometric properties of peptide groups, which underlie all higher order structures.

A The Peptide Group

Recall from Section 4-1B that a polypeptide is a polymer of amino acid residues linked by amide (peptide) bonds. In the 1930s and 1940s, Linus Pauling and Robert Corey determined the X-ray structures of several amino acids and dipeptides in an effort to elucidate the conformational constraints on a polypeptide chain. These studies indicated that *the peptide group has*

(a) – Lys – Ala – His – Gly – Lys – Lys – Val – Leu – Gly - Ala –
  Primary structure (amino acid sequence in a polypeptide chain)

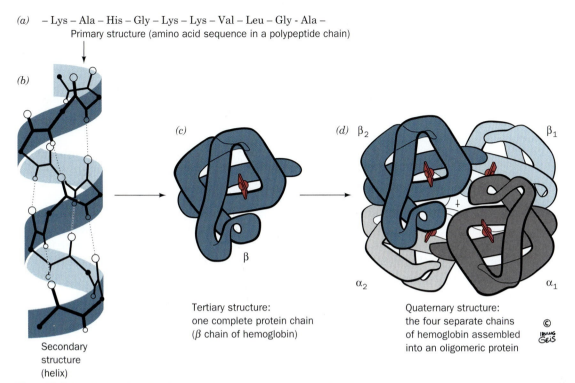

(b)

(c)

(d) β₂   β₁

α₂   α₁

Secondary
structure
(helix)

Tertiary structure:
one complete protein chain
(β chain of hemoglobin)

Quaternary structure:
the four separate chains
of hemoglobin assembled
into an oligomeric protein

© IRVING GEIS

Figure 6-1 Levels of protein structure.   (*a*) Primary structure, (*b*) secondary structure, (*c*) tertiary structure, and (*d*) quaternary structure. [Illustration, Irving Geis. Image from the Irving Geis Collection/Howard Hughes Medical Institute. Rights owned by HHMI. Reproduction by permission only.]

a rigid, planar structure as a consequence of resonance interactions that give the peptide bond ~40% double-bond character:

This explanation is supported by the observations that a peptide group's C—N bond is 0.13 Å shorter than its N—C_α single bond and that its C=O bond is 0.02 Å longer than that of aldehydes and ketones. The planar conformation maximizes π-bonding overlap, which accounts for the peptide group's rigidity.

Peptide groups, with few exceptions, assume the **trans conformation,** in which successive C_α atoms are on opposite sides of the peptide bond joining them (Fig. 6-2). The **cis conformation,** in which successive C_α atoms are on the same side of the peptide bond, is ~8 kJ · mol⁻¹ less stable than the trans conformation because of steric interference between neighboring side chains. However, this steric interference is reduced in peptide bonds to Pro residues, so *~10% of the Pro residues in proteins follow a cis peptide bond.*

Torsion Angles between Peptide Groups Describe Polypeptide Chain Conformations.   The **backbone** or **main chain** of a protein refers to the atoms that participate in peptide bonds, ignoring the side chains of the amino acid residues. The backbone can be drawn as a linked sequence of rigid planar peptide groups (Fig. 6-3). *The conformation of the backbone can therefore*

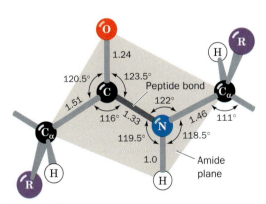

Figure 6-2 The trans peptide group.   The bond lengths (in angstroms) and angles (in degrees) are derived from X-ray crystal structures. [After Marsh, R.E. and Donohue, J., *Adv. Protein Chem.* **22,** 249 (1967).] 🐁 **See Kinemage Exercise 3-1.**

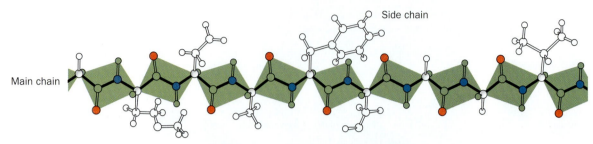

Figure 6-3 Extended conformation of a polypeptide.   The backbone is shown as a series of planar peptide groups. [Illustration, Irving Geis. Image from the Irving Geis Collection/Howard Hughes Medical Institute. Rights owned by HHMI. Reproduction by permission only.]

*be described by the **torsion angles*** (also called **dihedral angles** or rotation angles) *around the C_α—N bond (ϕ) and the C_α—C bond (ψ) of each residue* (Fig. 6-4). These angles, ϕ and ψ, are both defined as 180° when the polypeptide chain is in its fully extended conformation and increase clockwise when viewed from C_α.

The conformational freedom and therefore the torsion angles of a polypeptide backbone are sterically constrained. Rotation around the C_α—N and C_α—C bonds to form certain combinations of ϕ and ψ angles will cause the amide hydrogen, the carbonyl oxygen, or the substituents of C_α of adjacent residues to collide (e.g., Fig. 6-5). Certain conformations of

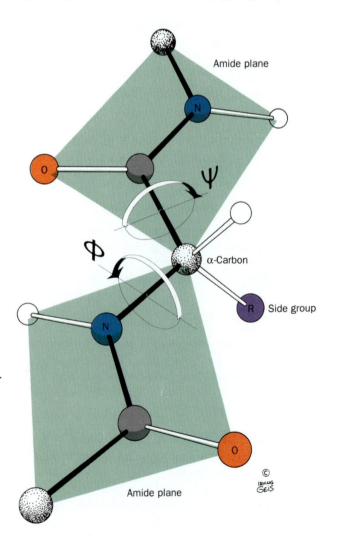

Figure 6-4 Torsion angles of the polypeptide backbone. Two planar peptide groups are shown. The only reasonably free movements are rotations around the C_α—N bond (measured as ϕ) and the C_α—C bond (measured as ψ). By convention, both ϕ and ψ are 180° in the conformation shown and increase, as indicated, when the peptide plane is rotated in the clockwise direction as viewed from C_α. [Illustration, Irving Geis. Image from the Irving Geis Collection/Howard Hughes Medical Institute. Rights owned by HHMI. Reproduction by permission only.] 🖉 **See Kinemage Exercise 3-1.**

longer polypeptides can similarly produce collisions between residues that are far apart in sequence.

The Ramachandran Diagram Indicates Allowed Conformations of Polypeptides.

The sterically allowed values of ϕ and ψ can be calculated. Sterically forbidden conformations, such as the one shown in Fig. 6-5, have ϕ and ψ values that would bring atoms closer than the corresponding van der Waals distance (the distance of closest contact between nonbonded atoms). Such information is summarized in a **Ramachandran diagram** (Fig. 6-6), which is named after its inventor, G. N. Ramachandran.

Most areas of the Ramachandran diagram (most combinations of ϕ and ψ) represent forbidden conformations of a polypeptide chain. Only three small regions of the diagram are physically accessible to most residues. The observed ϕ and ψ values of accurately determined structures nearly always fall within these allowed regions of the Ramachandran plot. There are, however, some notable exceptions:

1. The cyclic side chain of Pro limits its range of ϕ values to angles of around $-60°$, making it, not surprisingly, the most conformationally restricted amino acid residue.

2. Gly, the only residue without a C_β atom, is much less sterically hindered than the other amino acid residues. Hence, its permissible range of ϕ and ψ covers a larger area of the Ramachandran diagram. At Gly residues, polypeptide chains often assume conformations that are forbidden to other residues.

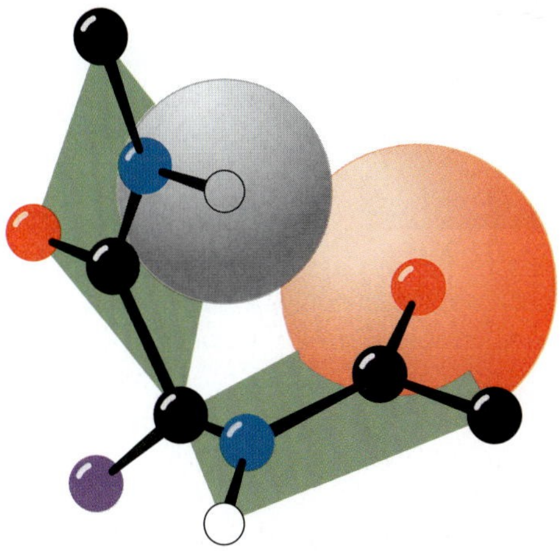

B Regular Secondary Structure: The α Helix and the β Sheet

A few elements of protein secondary structure are so widespread that they are immediately recognizable in proteins with widely differing amino acid sequences. Both the **α helix** and the **β sheet** are such elements; they are called **regular secondary structures** because they are composed of sequences of residues with repeating ϕ and ψ values.

Figure 6-5 Steric interference between adjacent peptide groups.   Rotation can result in a conformation in which the amide hydrogen of one residue and the carbonyl oxygen of the next are closer than their van der Waals distance. [Illustration, Irving Geis. Image from the Irving Geis Collection/Howard Hughes Medical Institute. Rights owned by HHMI. Reproduction by permission only.]

> **See Guided Exploration 6**
> Stable helices in proteins: the α helix.

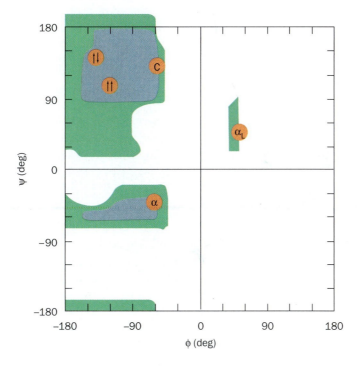

Figure 6-6 The Ramachandran diagram.   The blue-shaded regions indicate the sterically allowed ϕ and ψ angles for all residues except Gly and Pro. The green-shaded regions indicate the more crowded (outer limit) ϕ and ψ angles. The orange circles represent conformational angles of several secondary structures: α, right-handed α helix; ↑↑, parallel β sheet; ↑↓, antiparallel β sheet; C, collagen helix; α_L, left-handed α helix.

BOX 6-1

Pathways of Discovery

Linus Pauling and Structural Biochemistry

(Linus Pauling, 1901–1994)

Linus Pauling, the only person to have been awarded two un-shared Nobel Prizes, is clearly the dominant figure in twentieth-century chemistry and one of the greatest scientific figures of all time. He received his B.Sc. in chemical engineering from Oregon Agricultural College (now Oregon State University) in 1922 and his Ph.D. in chemistry from the California Institute of Technology in 1925, where he spent much of his career.

The major theme throughout Pauling's long scientific life was the study of molecular structures and the nature of the chemical bond. He began this career by using the then recently invented technique of X-ray crystallography to determine the structures of simple minerals and inorganic salts. At that time, methods for solving the phase problem (see Box 7-2) were unknown, so X-ray structures could only be determined using trial-and-error techniques. This limited the possible molecules that could be effectively studied to those with few atoms and high symmetry such that their atomic coordinates could be fully described by only a few parameters (rather than the three-dimensional coordinates of each of its atoms). Pauling realized that the positions of atoms in molecules were governed by fixed atomic radii, bond distances, and bond angles and used this information to make educated guesses about molecular structures. This greatly extended the complexity of the molecules whose structures could be determined.

In his next major contribution, occurring in 1931, Pauling revolutionized the way that chemists viewed molecules by applying the then infant field of quantum mechanics to chemistry. Pauling formulated the theories of orbital hybridization, electron-pair bonding, and resonance and thereby explained the nature of covalent bonds. This work was summarized in his highly influential monograph, *The Nature of the Chemical Bond,* which was first published in 1938.

In the mid-1930s, Pauling turned his attention to biological chemistry. He began these studies in collaboration with his colleague, Robert Corey, by determining the X-ray structures of several amino acids and dipeptides. At that time, the X-ray structural determination of even such small mole-cules required around a year of intense effort, largely because the numerous calculations required to solve a structure had to be made by hand (electronic computers had yet to be invented). Nevertheless, these studies led Pauling and Corey to the conclusions that the peptide bond is planar, which Pauling explained using results from resonance considerations (Section 6-1A), and that hydrogen bonding plays a central role in maintaining macromolecular structures.

In the 1940s, Pauling made several unsuccessful attempts to determine if polypeptides have any preferred conformations. Then, in 1948, while visiting Oxford University, he was confined to bed by a cold. He eventually tired of reading detective stories and science fiction and again turned his attention to proteins. By folding drawings of polypeptides in various ways, he discovered the α helix, whose existence was rapidly confirmed by X-ray studies of α keratin (Section 6-1C). This work was reported in 1951, and later that year Pauling and Corey also proposed both the parallel and antiparallel β pleated sheets. For these ground-breaking insights, Pauling received the Nobel Prize in Chemistry in 1954, although α helices and β sheets were not actually visualized until the first X-ray structures of proteins were determined, five to ten years later.

Pauling made numerous additional pioneering contributions to biological chemistry, most notably that the heme group in hemoglobin changes its electronic state on binding oxygen (Section 7-1A), that vertebrate hemoglobins are $\alpha_2\beta_2$ heterotetramers (Section 7-2A), that the denaturation of proteins is caused by the unfolding of their polypeptide chains, that sickle-cell anemia is caused by a mutation in the β chain of normal adult hemoglobin (the first so-called molecular disease to be characterized; Section 7-4), that molecular complementarity plays an important role in antibody–antigen interactions (Section 28-5C) and by extension all macromolecular interactions, that enzymes catalyze reactions by preferentially binding their transition states (Section 11-3E), and that the

The α Helix.   Only one polypeptide helix has both a favorable hydrogen-bonding pattern and ϕ and ψ values that fall within the fully allowed regions of the Ramachandran diagram: the α helix. Its discovery by Linus Pauling in 1951, through model building, ranks as one of the landmarks of structural biochemistry (Box 6-1).

The α helix (Fig. 6-7) is right-handed; that is, it turns in the direction that the fingers of a right hand curl when its thumb points in the direction that

comparison of the sequences of the corresponding proteins in different organisms yields evolutionary insights (Section 5-4).

Pauling was also a lively and stimulating lecturer who for many years taught a general chemistry course [which one of the authors of this textbook (DV) had the privilege of taking]. His textbook, *General Chemistry*, revolutionized the way that introductory chemistry was taught by presenting it as a subject that could be understood in terms of atomic physics and molecular structure. For a book of such generality, an astounding portion of its subject matter had been elucidated by its author. Pauling's amazing grasp of chemistry was demonstrated by the fact that he dictated each chapter of this textbook in a single sitting.

By the late 1940s, Pauling became convinced that the possibility of nuclear war posed an enormous danger to humanity and calculated that the radioactive fallout from each above-ground test of a nuclear bomb would ultimately cause cancer in thousands of people. He therefore began a campaign to educate the public about the hazards of bomb testing and nuclear war. The political climate in the United States at the time was such that the government considered Pauling to be subversive and his passport was revoked (and only returned two weeks before he was to leave for Sweden to receive his first Nobel Prize). Nevertheless, Pauling persisted in this campaign, which culminated, in 1962, with the signing of the first Nuclear Test Ban Treaty. For his efforts, Pauling was awarded the 1962 Nobel Peace Prize.

Pauling saw science as the search for the truth, which included politics and social causes. In his later years, he became a vociferous promoter of what he called orthomolecular medicine, the notion that large doses of vitamins could ward off and cure many human diseases, including cancer. In the best known manifestation of this concept, Pauling advocated taking large doses of vitamin C to prevent the common cold and lessen its symptoms, advice still followed by millions of people, although the medical evidence supporting this notion is scant. It should be noted, however, that Pauling, who followed his own advice, remained active until he died in 1994 at the age of 93.

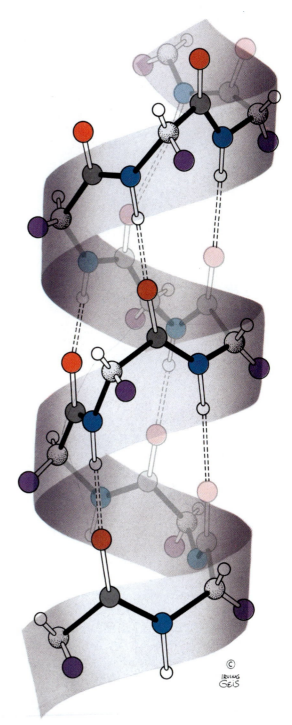

Figure 6-7 *Key to Structure.* **The α helix.** This right-handed helical conformation has 3.6 residues per turn. Dashed lines indicate hydrogen bonds between C=O groups and N—H groups that are four residues farther along the polypeptide chain. [Illustration, Irving Geis. Image from the Irving Geis Collection/Howard Hughes Medical Institute. Rights owned by HHMI. Reproduction by permission only.] 🔴 **See Kinemage Exercise 3-2 and the Animated Figures.**

the helix rises (Fig. 3-7). The α helix has 3.6 residues per turn and a **pitch** (the distance the helix rises along its axis per turn) of 5.4 Å. The α helices of proteins have an average length of ~12 residues, which corresponds to over three helical turns, and a length of ~18 Å.

In the α helix, the backbone hydrogen bonds are arranged such that the peptide C=O bond of the nth residue points along the helix axis toward the peptide N—H group of the (n + 4)th residue. This results in a strong hy-

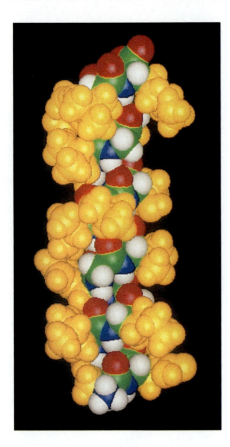

Figure 6-8   Space-filling model of an α helix.   The backbone atoms are colored according to type with C green, N blue, O red, and H white. The side chains (*yellow*) project away from the helix. This α helix is a segment of sperm whale myoglobin.

drogen bond that has the nearly optimum N···O distance of 2.8 Å. Amino acid side chains project outward and downward from the helix (Fig. 6-8), thereby avoiding steric interference with the polypeptide backbone and with each other. The core of the helix is tightly packed; that is, its atoms are in van der Waals contact.

β Sheets.   In 1951, the same year Pauling proposed the α helix, Pauling and Corey postulated the existence of a different polypeptide secondary structure, the β sheet. Like the α helix, the β sheet uses the full hydrogen-bonding capacity of the polypeptide backbone. *In β sheets, however, hydrogen bonding occurs between neighboring polypeptide chains rather than within one* as in an α helix.

Sheets come in two varieties:

1. The **antiparallel β sheet,** in which neighboring hydrogen-bonded polypeptide chains run in opposite directions (Fig. 6-9*a*).
2. The **parallel β sheet,** in which the hydrogen-bonded chains extend in the same direction (Fig. 6-9*b*).

The conformations in which these β structures are optimally hydrogen bonded vary somewhat from that of the fully extended polypeptide shown

> **See Guided Exploration 7**
>
> Hydrogen bonding in β sheets.

(a)   **Antiparallel**

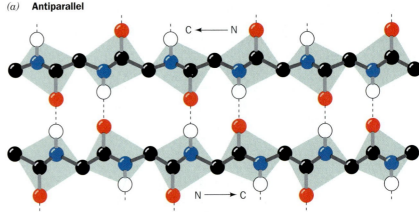

(b)   **Parallel**

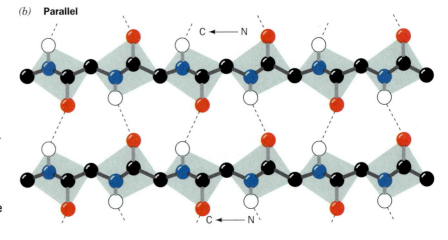

Figure 6-9   *Key to Structure.*   β Sheets.   Dashed lines indicate hydrogen bonds between polypeptide strands. Side chains are omitted for clarity. (*a*) An antiparallel β sheet. (*b*) A parallel β sheet. [Illustration, Irving Geis. Image from the Irving Geis Collection/Howard Hughes Medical Institute. Rights owned by HHMI. Reproduction by permission only.] 🖝 **See Kinemage Exercise 3-3 and the Animated Figures.**

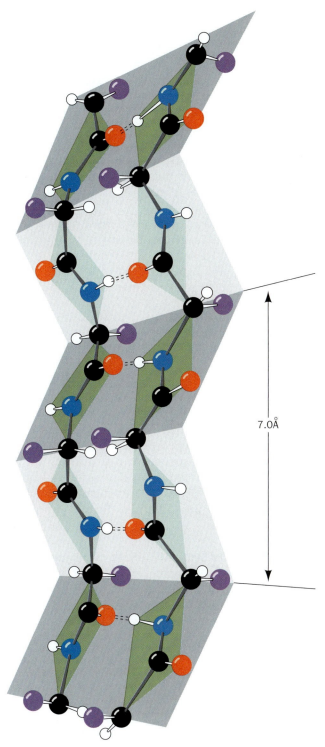

7.0Å

Figure 6-10 Pleated appearance of a β sheet. Dashed lines indicate hydrogen bonds. The R groups (*purple*) on each polypeptide chain alternately extend to opposite sides of the sheet and are in register on adjacent chains. [Illustration, Irving Geis. Image from the Irving Geis Collection/Howard Hughes Medical Institute. Rights owned by HHMI. Reproduction by permission only.] 🔖 **See Kinemage Exercise 3-3.**

in Fig. 6-3. They therefore have a rippled or pleated edge-on appearance (Fig. 6-10) and for that reason are sometimes called "pleated sheets." Successive side chains of a polypeptide chain in a β sheet extend to opposite sides of the sheet with a two-residue repeat distance of 7.0 Å.

β Sheets in proteins contain 2 to as many as 22 polypeptide strands, with an average of 6 strands. Each strand may contain up to 15 residues, the average being 6 residues. A six-stranded antiparallel β sheet is shown in Fig. 6-11.

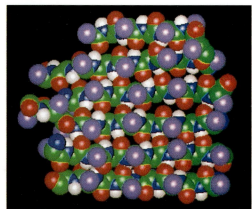

Figure 6-11 Space-filling model of a β sheet. The backbone atoms are colored according to type with C green, N blue, O red, and H white. The R groups are represented by large purple spheres. This six-stranded antiparallel β sheet is from the jack bean protein **concanavalin A.**

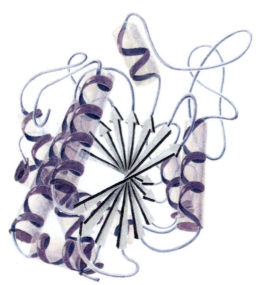

Figure 6-12 Diagram of a β sheet in bovine carboxypeptidase A. The polypeptide backbone is represented by a ribbon with α helices drawn as coils and strands of the β sheet drawn as purple arrows pointing toward the C-terminus. Side chains are not shown. The eight-stranded β sheet forms a saddle-shaped curved surface with a right-handed twist. [After a drawing by Jane Richardson, Duke University. Based on an X-ray structure by William Lipscomb, Harvard University. PDBid 3CPA (for the definition of "PDBid", see Section 6-6).]

(a)   *(b)*

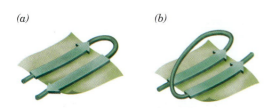

Figure 6-13 Connections between adjacent strands in β sheets. (*a*) Antiparallel strands may be connected by a small loop. (*b*) Parallel strands require a more extensive crossover connection. [After Richardson, J.S., *Adv. Protein Chem.* **34**, 196 (1981).]

Parallel β sheets containing fewer than five strands are rare. This observation suggests that parallel β sheets are less stable than antiparallel β sheets, possibly because the hydrogen bonds of parallel sheets are distorted compared to those of the antiparallel sheets (Fig. 6-9). β Sheets containing mixtures of parallel and antiparallel strands frequently occur.

β Sheets almost invariably exhibit a pronounced right-handed twist when viewed along their polypeptide strands (Fig. 6-12). Conformational energy calculations indicate that the twist is a consequence of interactions between chiral L-amino acid residues in the extended polypeptide chains. The twist distorts and weakens the β sheet's interchain hydrogen bonds. The geometry of a particular β sheet is thus a compromise between optimizing the conformational energies of its polypeptide chains and preserving its hydrogen bonding.

The **topology** (connectivity) of the polypeptide strands in a β sheet can be quite complex. The connection between two antiparallel strands may be just a small loop (Fig. 6-13*a*), but the link between tandem parallel strands must be a crossover connection that is out of the plane of the β sheet (Fig. 6-13*b*). The connecting link in either case can be extensive, often containing helices (e.g., Fig. 6-12).

C Fibrous Proteins

Proteins have historically been classified as either **fibrous** or **globular,** depending on their overall morphology. This dichotomy predates methods for determining protein structure on an atomic scale and does not do justice to proteins that contain both stiff, elongated, fibrous regions as well as more compact, highly folded, globular regions. Nevertheless, the division helps emphasize the properties of fibrous proteins, which often have a protective, connective, or supportive role in living organisms. The two well-characterized fibrous proteins we discuss here—keratin and collagen—are highly elongated molecules whose shapes are dominated by a single type of secondary structure. They are therefore useful examples of these structural elements.

α Keratin—A Coiled Coil. **Keratin** is a mechanically durable and chemically unreactive protein that occurs in all higher vertebrates. It is the principal component of their horny outer epidermal layer and its related appendages such as hair, horn, nails, and feathers. Keratins have been classified as either α keratins, which occur in mammals, or β keratins, which occur in birds and reptiles. Mammals each have over 30 keratin variants that are expressed in a tissue-specific manner.

The X-ray diffraction pattern of α keratin resembles that expected for an α helix (hence the name α keratin). However, α keratin exhibits a 5.1-Å spacing rather than the 5.4-Å distance corresponding to the pitch of the α helix. This discrepancy is the result of *two α keratin polypeptides, each of which forms an α helix, twisting around each other to form a left-handed coil.* The normal 5.4-Å repeat distance of each α helix in the pair is thereby tilted relative to the axis of this assembly, yielding the observed 5.1-Å spacing. The assembly is said to have a **coiled coil** structure because each α helix itself follows a helical path.

The conformation of α keratin's coiled coil is a consequence of its primary structure: The central ~310-residue segment of each polypeptide chain has a 7-residue pseudorepeat, *a-b-c-d-e-f-g*, with nonpolar residues predominating at positions *a* and *d*. Since an α helix has 3.6 residues per turn, α keratin's *a* and *d* residues line up along one side of each α helix (Fig. 6-14*a*). The hydrophobic strip along one helix associates with the hy-

drophobic strip on another helix. Because the 3.5-residue repeat in α keratin is slightly smaller than the 3.6 residues per turn of a standard α helix, the two keratin helices are inclined about 18° relative to one another, resulting in the coiled coil arrangement. This conformation allows the contacting side chains of each helix to interdigitate (Fig. 6-14b). Coiled coils also occur in numerous, not necessarily fibrous, proteins.

The higher order structure of α keratin is not well understood. The N- and C-terminal domains of each polypeptide facilitate the assembly of coiled coils (dimers) into protofilaments, two of which constitute a protofibril (Fig. 6-15). Four protofibrils constitute a microfibril, which associates with other microfibrils to form a macrofibril. A single mammalian hair consists of layers of dead cells, each of which is packed with parallel macrofibrils.

α Keratin is rich in Cys residues, which form disulfide bonds that cross-link adjacent polypeptide chains. The α keratins are classified as "hard" or "soft" according to whether they have a high or low sulfur content. Hard keratins, such as those of hair, horn, and nail, are less pliable than soft keratins, such as those of skin and callus, because the disulfide bonds resist deformation. The disulfide bonds can be reductively cleaved by disulfide interchange with mercaptans (Section 5-3A). Hair so treated can be curled and set in a "permanent wave" by applying an oxidizing agent that reestablishes the disulfide bonds in the new "curled" conformation. Conversely, curly hair can be straightened by the same process.

The springiness of hair and wool fibers is a consequence of the coiled coil's tendency to recover its original conformation after being untwisted by stretching. If some of its disulfide bonds have been cleaved, however, an α keratin fiber can be stretched to over twice its original length. At this point, the polypeptide chains assume a β sheet conformation. β Keratin, such as that in feathers, exhibits a β-like pattern in its native state.

Collagen—A Triple Helix. **Collagen,** which occurs in all multicellular animals, is the most abundant vertebrate protein. Its strong, insoluble fibers are the major stress-bearing components of connective tissues such as bone, teeth, cartilage, tendon, and the fibrous matrices of skin and blood vessels. A single collagen molecule consists of three polypeptide chains. Mammals

(a)

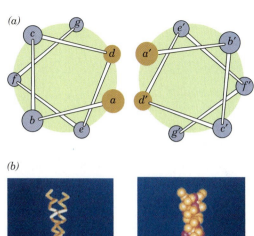

(b)

Figure 6-14 **A coiled coil.** (a) View down the coil axis showing the alignment of nonpolar residues along one side of each α helix. The helices have the pseudorepeating sequence a-b-c-d-e-f-g in which residues a and d are predominately nonpolar. [After McLachlan, A.D. and Stewart, M., *J. Mol. Biol.* **98,** 295 (1975).] (b) Side view of the polypeptide backbones in skeletal (*left*) and space-filling (*right*) forms. Note that the side chains (red spheres in the space-filling model) contact each other. This coiled coil is from the protein tropomyosin. [Courtesy of Carolyn Cohen, Brandeis University.] **See Kinemage Exercises 4-1 and 4-2.**

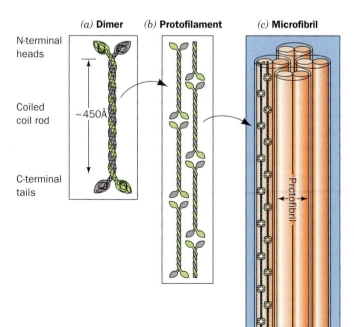

(a) **Dimer** (b) **Protofilament** (c) **Microfibril**

N-terminal heads

Coiled coil rod

~450Å

C-terminal tails

Protofibril

Figure 6-15 **Higher order α keratin structure.** (a) Two keratin polypeptides form a dimeric coiled coil. (b) Protofilaments are formed from two staggered rows of head-to-tail associated coiled coils. (c) Protofilaments dimerize to form a protofibril, four of which form a microfibril. The structures of the latter assemblies are poorly characterized.

have at least 33 genetically distinct chains that are assembled into at least 20 collagen varieties found in different tissues in the same individual. One of the most common collagens, called Type I, consists of two α_1(I) chains and one α_2(I) chain. It has a molecular mass of ~285 kD, a width of ~14 Å, and a length of ~3000 Å.

Collagen has a distinctive amino acid composition: Nearly one-third of its residues are Gly; another 15 to 30% of its residues are Pro and **4-hydroxyprolyl (Hyp)**. **3-Hydroxyprolyl** and **5-hydroxylysyl (Hyl)** residues also occur in collagen, but in smaller amounts.

4-Hydroxyprolyl residue (Hyp)

3-Hydroxyprolyl residue

5-Hydroxylysyl residue (Hyl)

These nonstandard residues are formed after the collagen polypeptides are synthesized. For example, Pro residues are converted to Hyp in a reaction catalyzed by **prolyl hydroxylase.** This enzyme requires **ascorbic acid (vitamin C)** to maintain its activity.

Ascorbic acid (vitamin C)

The disease **scurvy** results from the dietary deficiency of vitamin C (see Box 6-2).

The amino acid sequence of a typical collagen polypeptide consists of monotonously repeating triplets of sequence Gly-X-Y over a segment of ~1000 residues, where X is often Pro and Y is often Hyp. Hyl sometimes appears at the Y position. Collagen's Pro residues prevent it from forming an α helix (Pro residues cannot assume the α-helical backbone conformation and lack the backbone N—H groups that form the intrahelical hydrogen bonds shown in Fig. 6-7). Instead, *the collagen polypeptide assumes a left-handed helical conformation with about three residues per turn. Three parallel chains wind around each other with a gentle, right-handed, ropelike twist to form the triple-helical structure of a collagen molecule* (Fig. 6-16).

This model of the collagen structure has been confirmed by Barbara Brodsky and Helen Berman, who determined the X-ray crystal structure of a collagen-like model polypeptide. Every third residue of each polypeptide chain passes through the center of the triple helix, which is so crowded that only a Gly side chain can fit there. This crowding explains the absolute requirement for a Gly at every third position of a collagen polypeptide chain. The three polypeptide chains are staggered so that a Gly, X, and Y residue occurs at each position along the triple helix axis (Fig. 6-17a). The peptide groups are oriented such that the N—H of each Gly makes a strong hydrogen bond with the carbonyl oxygen of an X (Pro) residue on a neighboring chain (Fig. 6-17b). The bulky and relatively inflexible Pro and Hyp residues confer rigidity on the entire assembly.

Figure 6-16 The collagen triple helix. Left-handed polypeptide helices are twisted together to form a right-handed superhelical structure. [Illustration, Irving Geis. Image from the Irving Geis Collection/Howard Hughes Medical Institute. Rights owned by HHMI. Reproduction by permission only.]

Section 6-1 Secondary Structure   141

BOX 6-2

Biochemistry in Health and Disease

Collagen Diseases

Some collagen diseases have dietary causes. In scurvy (caused by vitamin C deficiency), Hyp production decreases because prolyl hydroxylase requires vitamin C. Thus, in the absence of vitamin C, newly synthesized collagen cannot form fibers properly, resulting in skin lesions, fragile blood vessels, poor wound healing, and, ultimately, death. Scurvy was common in sailors on long voyages whose diets were devoid of fresh foods. The introduction of limes to the diet of the British navy by the renowned explorer Captain James Cook alleviated scurvy and led to the nickname "limey" for the British sailor.

The disease **lathyrism** is caused by regular ingestion of the seeds from the sweet pea *Lathyrus odoratus,* which contain a compound that specifically inactivates lysyl oxidase. The resulting reduced cross-linking of collagen fibers produces serious abnormalities of the bones, joints, and large blood vessels.

Several rare heritable disorders of collagen are known. Mutations of Type I collagen, which constitutes the major structural protein in most human tissues, usually result in **osteogenesis imperfecta** (brittle bone disease). The severity of this disease varies with the nature and position of the

mutation: Even a single amino acid change can have lethal consequences. For example, the central Gly → Ala substitution in the model polypeptide shown in Fig. 6-17 locally distorts the already internally crowded collagen helix. This ruptures the hydrogen bond from the backbone N—H of each Ala (normally Gly) to the carbonyl group of the adjacent Pro in a neighboring chain, thereby reducing the stability of the collagen structure.

Mutations may affect the structure of the collagen molecule or how it forms fibrils. These mutations tend to be dominant because they affect either the folding of the triple helix or fibril formation even when normal chains are also involved.

Many collagen disorders are characterized by deficiencies in the amount of a particular collagen type synthesized, or by abnormal activities of collagen-processing enzymes such as lysyl hydroxylase and lysyl oxidase. One group of at least 10 different collagen-deficiency diseases, the **Ehlers–Danlos syndromes,** are all characterized by the hyperextensibility of the joints and skin. The "India-rubber man" of circus fame had an Ehlers–Danlos syndrome.

(a)   *(b)*

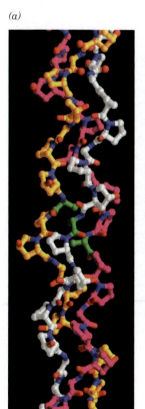

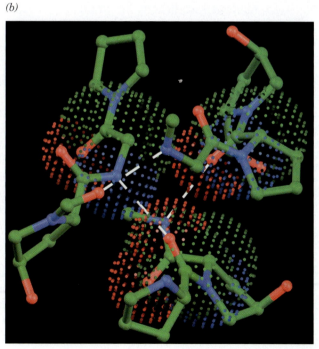

Figure 6-17   Structure of a collagen model peptide.   In this X-ray structure of (Pro-Hyp-Gly)$_{10}$, the fifth Gly of each peptide has been replaced by Ala. *(a)* A ball-and-stick model of the middle portion of the triple helix oriented with the N-termini at the top. The C atoms of the three chains are colored gold, magenta, and white. The N and O atoms on all chains are blue and red. Note how the replacement of Gly with the bulkier Ala (C atoms in green) distorts the triple helix. *(b)* This view from the N-terminus down the helix axis shows the interchain hydrogen-bonding associations. Three consecutive residues from each chain are shown in ball-and-stick form. Hydrogen bonds are represented by dashed lines from Gly N atoms to Pro O atoms in adjacent chains. Dots represent the van der Waals surfaces of the backbone atoms of the central residue in each chain. Note the close packing of the atoms along the triple helix axis. [Based on an X-ray structure by Helen Berman, Rutgers University, and Barbara Brodsky, UMDNJ–Robert Wood Johnson Medical School. PDBid 1CAG.] 🖘 **See Kinemage Exercises 4-3 and 4-4.**

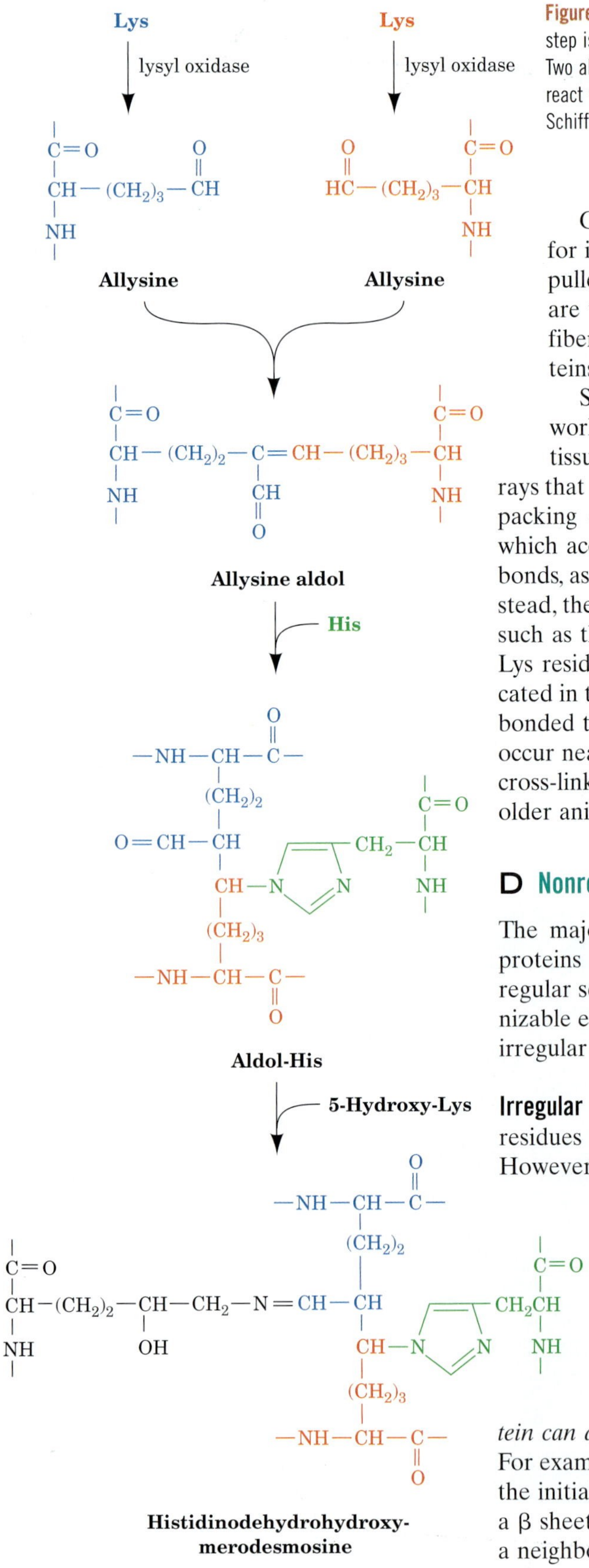

Allysine

Allysine

Allysine aldol

His

Aldol-His

5-Hydroxy-Lys

Histidinodehydrohydroxy-merodesmosine

Figure 6-18   A reaction pathway for cross-linking side chains in collagen.   The first step is the lysyl oxidase–catalyzed oxidative deamination of Lys to form the aldehyde allysine. Two allysines then undergo an aldol condensation to form allysine aldol. This product can react with His to form aldol histidine, which can in turn react with 5-hydroxylysine to form a Schiff base (an imine bond), thereby cross-linking four side chains.

Collagen's well-packed, rigid, triple-helical structure is responsible for its characteristic tensile strength. The twist in the helix cannot be pulled out under tension because its component polypeptide chains are twisted in the opposite direction (Fig. 6-16). Successive levels of fiber bundles in high-quality ropes and cables, as well as in other proteins such as keratin (Fig. 6-14), are likewise oppositely twisted.

Several types of collagen molecules assemble to form loose networks or thick fibrils arranged in bundles or sheets, depending on the tissue. The collagen molecules in fibrils are organized in staggered arrays that are stabilized by hydrophobic interactions resulting from the close packing of triple-helical units. Collagen is also covalently cross-linked, which accounts for its poor solubility. The cross-links cannot be disulfide bonds, as in keratin, because collagen is almost devoid of Cys residues. Instead, the cross-links are derived from Lys and His side chains in reactions such as those shown in Fig. 6-18. **Lysyl oxidase,** the enzyme that converts Lys residues to those of the aldehyde **allysine,** is the only enzyme implicated in this cross-linking process. Up to four side chains can be covalently bonded to each other. The cross-links do not form at random but tend to occur near the N- and C-termini of the collagen molecules. The degree of cross-linking in a particular tissue increases with age. This is why meat from older animals is tougher than meat from younger animals.

D   Nonrepetitive Protein Structure

The majority of proteins are globular proteins that, unlike the fibrous proteins discussed in the preceding section, may contain several types of regular secondary structure, including α helices, β sheets, and other recognizable elements. A significant portion of a protein's structure may also be irregular or unique.

Irregular Structures.   Segments of polypeptide chains whose successive residues do not have similar ϕ and ψ values are sometimes called coils. However, you should not confuse this term with the appellation **random coil,** which refers to the totally disordered and rapidly fluctuating conformations assumed by **denatured** (fully unfolded) proteins in solution. In **native** (folded) proteins, *nonrepetitive structures are no less ordered than are helices or β sheets; they are simply irregular and hence more difficult to describe.*

Variations in Standard Secondary Structure.   *Variations in amino acid sequence as well as the overall structure of the folded protein can distort the regular conformations of secondary structural elements.* For example, the α helix frequently deviates from its ideal conformation in the initial and final turns of the helix. Similarly, a strand of polypeptide in a β sheet may contain an "extra" residue that is not hydrogen bonded to a neighboring strand, producing a distortion known as a **β bulge.**

Many of the limits on amino acid composition and sequence (Section 5-1) may be due in part to conformational constraints in the three-dimensional

structure of proteins. For example, a Pro residue produces a kink in an α helix or β sheet. Similarly, steric clashes between several sequential amino acid residues with large branched side chains (e.g., Ile and Tyr) can destabilize α helices.

Analysis of known protein structures by Peter Chou and Gerald Fasman revealed the propensity P of a residue to occur in an α helix or a β sheet (Table 6-1). Chou and Fasman also discovered that certain residues not only have a high propensity for a particular secondary structure but they tend to disrupt or break other secondary structures. Such data are useful for predicting the secondary structures of proteins with known amino acid sequences.

The presence of certain residues outside of α helices or β sheets may also be nonrandom. For example, α helices are often flanked by residues such as Asn and Gln, whose side chains can fold back to form hydrogen bonds with one of the four terminal residues of the helix, a phenomenon termed **helix capping.** Recall that the four residues at each end of an α helix are not fully hydrogen bonded to neighboring backbone segments (Fig. 6-7).

Turns and Loops.   Segments with regular secondary structure such as α helices or the strands of β sheets are typically joined by stretches of polypeptide that abruptly change direction. Such **reverse turns** or **β bends** (so named because they often connect successive strands of antiparallel β sheets) almost always occur at protein surfaces. Most reverse turns involve four successive amino acid residues more or less arranged in one of two ways, Type I and Type II, that differ by a 180° flip of the peptide unit linking residues 2 and 3 (Fig. 6-19). Both types of turns are stabilized by a hydrogen bond, although deviations from these ideal conformations often disrupt this hydrogen bond. In Type II turns, the oxygen atom of residue 2 crowds the C_β atom of residue 3, which is therefore usually Gly. Residue 2 of either type of turn is often Pro since it can assume the required conformation.

Almost all proteins with more than 60 residues contain one or more loops of 6 to 16 residues, called **Ω loops.** These loops, which have the

Table 6-1 Propensities of Amino Acid Residues for α Helical and β Sheet Conformations

Residue	P_α	P_β
Ala	1.42	0.83
Arg	0.98	0.93
Asn	0.67	0.89
Asp	1.01	0.54
Cys	0.70	1.19
Gln	1.11	1.10
Glu	1.51	0.37
Gly	0.57	0.75
His	1.00	0.87
Ile	1.08	1.60
Leu	1.21	1.30
Lys	1.16	0.74
Met	1.45	1.05
Phe	1.13	1.38
Pro	0.57	0.55
Ser	0.77	0.75
Thr	0.83	1.19
Trp	1.08	1.37
Tyr	0.69	1.47
Val	1.06	1.70

Source: Chou, P.Y. and Fasman, G.D., *Annu. Rev. Biochem.* **47**, 258 (1978).

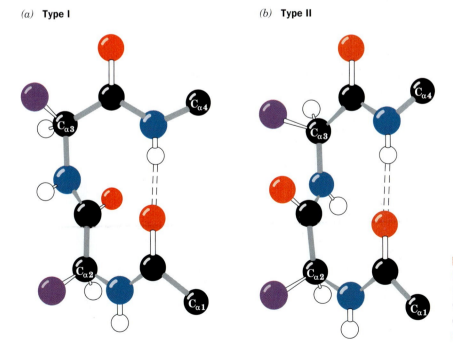

(a) **Type I**   *(b)* **Type II**

Figure 6-19 Reverse turns in polypeptide chains. Dashed lines represent hydrogen bonds. (*a*) Type I. (*b*) Type II. [Illustration, Irving Geis. Image from the Irving Geis Collection/Howard Hughes Medical Institute. Rights owned by HHMI. Reproduction by permission only.]
🖥 **See Kinemage Exercise 3-4.**

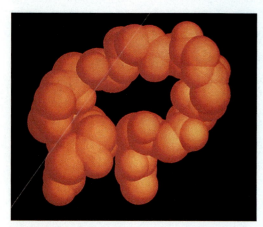

Figure 6-20 Space-filling model of an Ω loop.   Only backbone atoms are shown; the side chains would fill the loop. This structure is residues 40 to 54 from cytochrome *c*. [Courtesy of George Rose, The Johns Hopkins University.]

necked-in shape of the Greek uppercase letter omega (Fig. 6-20), are compact globular entities because their side chains tend to fill in their internal cavities. Since Ω loops are almost invariably located on the protein surface, they may have important roles in biological recognition processes.

2 Tertiary Structure

The tertiary structure of a protein describes the folding of its secondary structural elements and specifies the positions of each atom in the protein, including those of its side chains. The known protein structures have come to light through **X-ray crystallographic** and **nuclear magnetic resonance (NMR)** studies. The atomic coordinates of most of these structures are deposited in a database known as the Protein Data Bank (PDB). These data are readily available via the Internet (http://www.rcsb.org), which allows the tertiary structures of a variety of proteins to be analyzed and compared. The common features of protein tertiary structures reveal much about the biological functions of proteins and their evolutionary origins.

A Determining Protein Structures

X-Ray crystallography is a technique that directly images molecules. X-Rays must be used to do so because, according to optical principles, the uncertainty in locating an object is approximately equal to the wavelength of the radiation used to observe it (covalent bond distances and the wavelengths of the X-rays used in structural studies are both ~1.5 Å; individual molecules cannot be seen in a light microscope because visible light has a minimum wavelength of 4000 Å). There is, however, no such thing as an X-ray microscope because there are no X-ray lenses. Rather, a crystal of the molecule to be imaged (e.g., Fig. 6-21) is exposed to a collimated beam of X-rays and the resulting **diffraction pattern,** which arises from the regularly repeating positions of atoms in the crystal, is recorded by a radiation counter

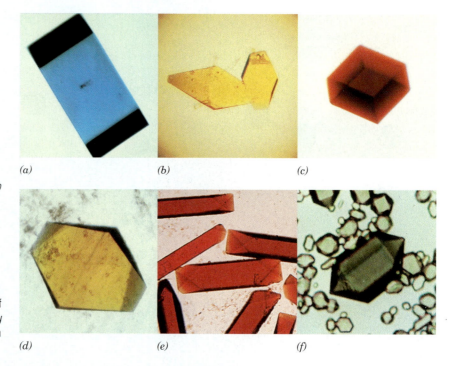

(a)   *(b)*   *(c)*

(d)   *(e)*   *(f)*

Figure 6-21 Protein crystals.   (*a*) Azurin from *Pseudomonas aeruginosa,* (*b*) flavodoxin from *Desulfovibrio vulgaris,* (*c*) rubredoxin from *Clostridium pasteurianum,* (*d*) azidomet myohemerythrin from the marine worm *Siphonosoma funafuti,* (*e*) lamprey hemoglobin, and (*f*) bacteriochlorophyll *a* protein from *Prosthecochloris aestuarii.* These crystals are colored because the proteins contain light-absorbing groups; proteins are colorless in the absence of such groups. [Parts *a–c* courtesy of Larry Siecker, University of Washington; parts *d* and *e* courtesy of Wayne Hendrikson, Columbia University; and part *f* courtesy of John Olsen, Brookhaven National Laboratories, and Brian Matthews, University of Oregon.]

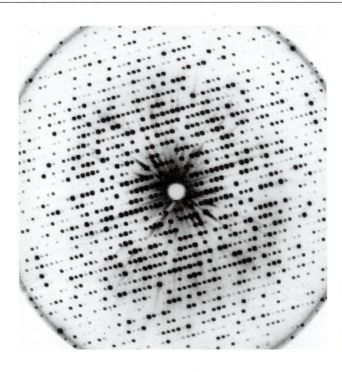

Figure 6-22 An X-ray diffraction photograph of a crystal of sperm whale myoglobin. The intensity of each diffraction maximum (the darkness of each spot) is a function of the crystal's electron density. [Courtesy of John Kendrew, Cambridge University, U.K.]

or, now infrequently, on photographic film (Fig. 6-22). The X-rays used in structural studies are produced by laboratory X-ray generators or, increasingly often, by a **synchrotron,** a type of particle accelerator that produces X-rays of far greater intensity. The intensities of the diffraction maxima (darkness of the spots on a film) are then used to construct mathematically the three-dimensional image of the crystal structure through methods that are beyond the scope of this text. In what follows, we discuss some of the special problems associated with interpreting the X-ray crystal structures of proteins.

X-Rays interact almost exclusively with the electrons in matter, not the nuclei. An X-ray structure is therefore an image of the **electron density** of the object under study. Such **electron density maps** are usually presented with the aid of a graphics computer as one or more sets of contours (e.g., Fig. 6-23) in which a contour represents a specific level of electron density

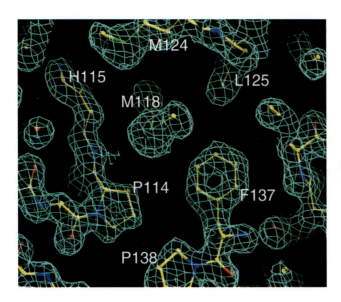

Figure 6-23 A thin section through a 1.5-Å-resolution electron density map of a protein that is contoured in three dimensions. Only a single contour level (*cyan*) is shown, together with a ball-and-stick model of the corresponding polypeptide segments colored according to atom type with C yellow, N blue, and O red. A water molecule is represented by a red sphere. [Courtesy of Xinhua Ji, NCI-Frederick Cancer Research and Development Center, Frederick, Maryland.]

in the same way that a contour on a topographic map indicates locations that have a particular altitude.

Most Protein Crystal Structures Exhibit Less Than Atomic Resolution. The molecules in protein crystals, as in other crystalline substances, are arranged in regularly repeating three-dimensional lattices. Protein crystals, however, differ from those of most small organic and inorganic molecules in being highly hydrated; they are typically 40 to 60% water by volume. The aqueous solvent of crystallization is necessary for the structural integrity of the protein crystals, as J. D. Bernal and Dorothy Crowfoot Hodgkin first noted in 1934 when they carried out the original X-ray studies of protein crystals. This is because water is required for the structural integrity of native proteins themselves (Section 6-4).

The large solvent content of protein crystals gives them a soft, jellylike consistency so that their molecules usually lack the rigid order characteristic of crystals of small molecules such as NaCl or glycine. The molecules in a protein crystal are typically disordered by more than an angstrom so that the corresponding electron density map lacks information concerning structural details of smaller size. The crystal is therefore said to have a resolution limit of that size. Protein crystals typically have resolution limits in the range 1.5 to 3.0 Å, although some are better ordered (have higher resolution, that is, a lesser resolution limit) and many are less ordered (have lower resolution).

Since an electron density map of a protein must be interpreted in terms of its atomic positions, the accuracy and even the feasibility of a crystal structure analysis depends on the crystal's resolution limit. Indeed, the inability to obtain crystals of sufficiently high resolution is a major limiting factor in determining the X-ray crystal structure of a protein or other macromolecule. Figure 6-24 indicates how the quality (degree of focus) of an electron density map varies with its resolution limit. At 6-Å resolution, the presence of a molecule the size of diketopiperazine is difficult to discern. At 2.0-Å resolution, its individual atoms cannot yet be distinguished, although its molecular shape has become reasonably evident. At 1.5-Å resolution, which roughly corresponds to a bond distance, individual atoms become partially resolved. At 1.1-Å resolution, atoms are clearly visible.

Most protein crystal structures are too poorly resolved for their electron density maps to reveal clearly the positions of individual atoms (e.g., Fig. 6-24). Nevertheless, the distinctive shape of the polypeptide backbone usually permits it to be traced, which, in turn, allows the positions and orientations of its side chains to be deduced (e.g., Fig. 6-23). Yet side chains of comparable size and shape, such as those of Leu, Ile, Thr, and Val, cannot always be differentiated with a reasonable degree of confidence (hydrogen atoms, having but one electron, are only visible in the few macromolecular X-ray structures with resolution limits less than ~1.2 Å), so that a protein structure cannot be elucidated from its electron density map alone. Rather, the primary structure of the protein must be known, thereby permitting the sequence of amino acid residues to be fitted to its electron density map. Mathematical refinement can then reduce the uncertainty in the crystal structure's atomic positions to as little as 0.1 Å (in contrast, positional errors in the most accurately determined small molecule X-ray structures are as little as 0.001 Å).

Most Crystalline Proteins Maintain Their Native Conformations. What is the relationship between the structure of a protein in a crystal and that in solu-

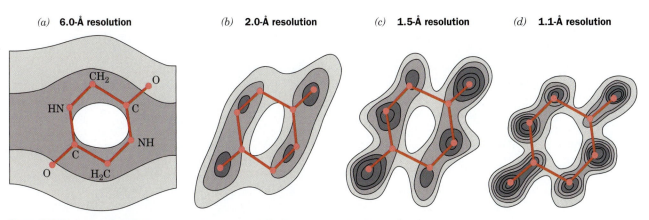

(a) **6.0-Å resolution**   (b) **2.0-Å resolution**   (c) **1.5-Å resolution**   (d) **1.1-Å resolution**

Figure 6-24 **Section through the electron density map of diketopiperazine calculated at the indicated resolution levels.** Hydrogen atoms are not visible in these maps because of their low electron density. [After Hodgkin, D.C., *Nature* **188,** 445 (1960).]

tion, where globular proteins normally function? Several lines of evidence indicate that *crystalline proteins assume very nearly the same structures that they have in solution:*

1. A protein molecule in a crystal is essentially in solution because it is bathed by solvent of crystallization over all of its surface except for the few, generally small patches that contact neighboring protein molecules. In fact, the 40 to 60% water content of typical protein crystals is similar to that of many cells (e.g., see Fig. 1-6).

2. A protein may crystallize in one of several forms or "habits," depending on crystallization conditions, that differ in how the protein molecules are arranged in space relative to each other. In the numerous cases in which different crystal forms of the same protein have been independently analyzed, the molecules have virtually identical conformations. Similarly, in the several cases for which both the X-ray crystal structure and the solution NMR structure of the same protein have been determined, the two structures are, for the most part, identical to within experimental error (see below). Evidently, crystal packing forces do not greatly perturb the structures of protein molecules.

3. The most compelling evidence that crystalline proteins have biologically relevant structures is the observation that many enzymes are catalytically active in the crystalline state. The catalytic activity of an enzyme, as we shall see, is very sensitive to the relative orientations of the groups involved in binding and catalysis (Chapter 11). Active crystalline enzymes must therefore have conformations that closely resemble their solution conformations.

Protein Structure Determination by NMR. The determination of the three-dimensional structures of small globular proteins in aqueous solution has become possible, since the mid-1980s, through the development of **two-dimensional (2D) NMR spectroscopy** (and, more recently, of 3D and 4D techniques), in large part by Kurt Wüthrich. Such NMR measurements, whose description is beyond the scope of this text, yield the interatomic distances between specific protons that are <5 Å apart in a protein of known sequence. The interproton distances may be either through space, as determined by nuclear Overhauser effect spectroscopy (NOESY,

(a)

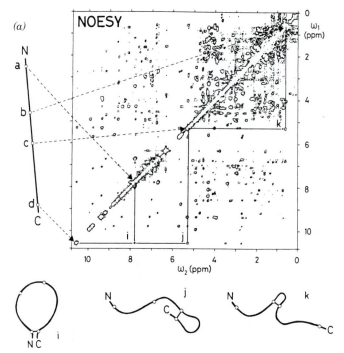

(b)

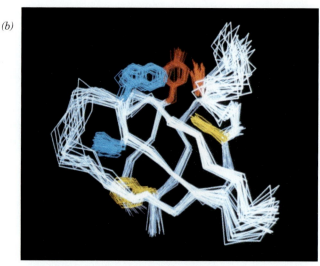

Figure 6-25   2D proton NMR structures of proteins. (*a*) A NOESY spectrum of a protein presented as a contour plot with two frequency axes, ω_1 and ω_2. The conventional 1D-NMR spectrum of the protein, which occurs along the diagonal of the plot ($\omega_1 = \omega_2$), is too crowded with peaks to be directly interpretable (even a small protein has hundreds of protons). The off-diagonal peaks, the so-called cross peaks, each arise from the interaction of two protons that are <5 Å apart in space and whose 1D-NMR peaks are located where the horizontal and vertical lines through the cross peak intersect the diagonal [a **nuclear Overhauser effect (NOE)**]. For example, the line to the left of the spectrum represents the extended polypeptide chain with its N- and C-terminal ends identified by the letters N and C and with the positions of four protons, a to d, represented by small circles. The dashed arrows indicate the diagonal NMR peaks to which these protons give rise. Cross peaks, such as i, j, and k, which are each located at the intersections of the horizontal and vertical lines through two diagonal peaks, are indicative of an NOE between the corresponding two protons, indicating that they are <5 Å apart. These distance relationships are schematically indicated by the three looped structures drawn below the spectrum. Note that the assignment of a distance relationship between two protons in a polypeptide requires that the NMR peaks to which they give rise and their positions in the polypeptide be known, which requires that the polypeptide's amino acid sequence has been previously determined. [After Wüthrich, K., *Science* **243**, 45 (1989).] (*b*) The NMR structure of a 64-residue polypeptide comprising the **Src protein SH3 domain** (Section 21-3D). The drawing represents 20 superimposed structures that are consistent with the 2D- and 3D-NMR spectra of the protein (each calculated from a different, randomly generated starting structure). The polypeptide backbone, as represented by its connected C_α atoms, is white and its Phe, Tyr, and Trp side chains are yellow, red, and blue, respectively. It can be seen that the polypeptide backbone folds into two 3-stranded antiparallel β sheets that form a sandwich. [Courtesy of Stuart Schreiber, Harvard University.]

Fig. 6-25*a*), or through bonds, as determined by correlated spectroscopy (COSY). These distances, together with known geometric constraints such as covalent bond distances and angles, group planarity, chirality, and van der Waals radii, are used to compute the protein's three-dimensional structure. However, since interproton distance measurements are imprecise, they are insufficient to imply a unique structure. Rather, they are consistent with an ensemble of closely related structures. Consequently, an NMR structure of a protein (or any other macromolecule with a well-defined structure) is often presented as a representative sample of structures that are consistent with the constraints (e.g., Figure 6-25*b*). The "tightness" of a bundle of such structures is indicative both of the accuracy with which the structure is known, which in the most favorable cases is roughly comparable to that of an X-ray crystal structure with a resolution of 2 to 2.5 Å, and of the conformational fluctuations that the protein undergoes (Section 6-4B). Although present NMR methods are limited to determining the structures of macromolecules with molecular masses no greater than ~40 kD, recent advances in NMR technology suggest that this limit may soon increase to ~100 kD or more.

In most of the several cases in which both the NMR and X-ray crystal structures of a particular protein have been determined, the two structures

are in good agreement. There are, however, a few instances in which there are real differences between the corresponding X-ray and NMR structures. These, for the most part, involve surface residues that, in the crystal, participate in intermolecular contacts and are thereby perturbed from their solution conformations. NMR methods, besides providing mutual cross-checks with X-ray techniques, can determine the structures of proteins and other macromolecules that fail to crystallize. Moreover, since NMR can probe motions over time scales spanning 10 orders of magnitude, it can also be used to study protein folding and dynamics (Section 6-4).

Visualizing Proteins.   The huge number of atoms in proteins makes it difficult to visualize them using the same sorts of models employed for small organic molecules. Ball-and-stick representations showing all or most atoms in a protein (as in Figs. 6-7 and 6-10) are exceedingly cluttered, and space-filling models (as in Figs. 6-8 and 6-11) obscure the internal details of the protein. Accordingly, computer-generated or artistic renditions (e.g., Fig. 6-12) are often more useful for representing protein structures. The course of the polypeptide chain can be followed by tracing the positions of its C_α atoms or by representing helices as helical ribbons or cylinders and β sheets as sets of flat arrows pointing from the N- to the C-termini.

B Side Chain Location and Polarity

In the years since Kendrew solved the structure of myoglobin, nearly 30,000 protein structures have been reported. No two are exactly alike, but they exhibit remarkable consistencies.

Side Chain Location Varies with Polarity.   The primary structures of globular proteins generally lack the repeating sequences that support the regular conformations seen in fibrous proteins. However, *the amino acid side chains in globular proteins are spatially distributed according to their polarities:*

1. The nonpolar residues Val, Leu, Ile, Met, and Phe occur mostly in the interior of a protein, out of contact with the aqueous solvent. The hydrophobic effects that promote this distribution are largely responsible for the three-dimensional structure of native proteins.

2. The charged polar residues Arg, His, Lys, Asp, and Glu are usually located on the surface of a protein in contact with the aqueous solvent. This is because immersing an ion in the virtually anhydrous interior of a protein is energetically unfavorable.

3. The uncharged polar groups Ser, Thr, Asn, Gln, and Tyr are usually on the protein surface but also occur in the interior of the molecule. When buried in the protein, these residues are almost always hydrogen bonded to other groups; in a sense, the formation of a hydrogen bond "neutralizes" their polarity. This is also the case with the polypeptide backbone.

These general principles of side chain distribution are evident in individual elements of secondary structure (Fig. 6-26) as well as in whole proteins

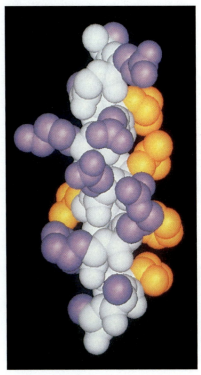

(a)

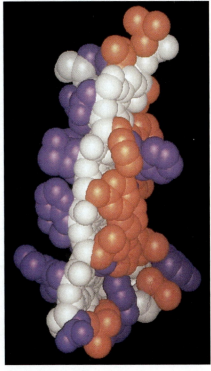

(b)

Figure 6-26   Side chain locations in an α helix and a β sheet.   In these space-filling models, the main chain is white, nonpolar side chains are yellow or brown, and polar side chains are purple. (*a*) An α helix from sperm whale myoglobin. Note that the nonpolar residues are primarily on one side of the helix. (*b*) An antiparallel β sheet from concanavalin A (*side view*). The protein interior is to the right and the exterior is to the left.

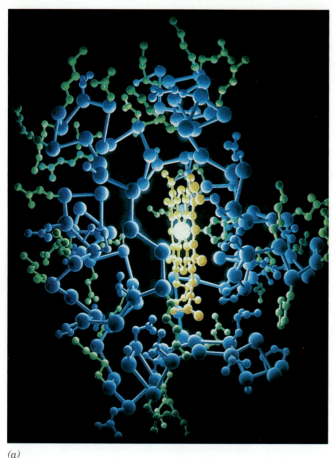

(a)

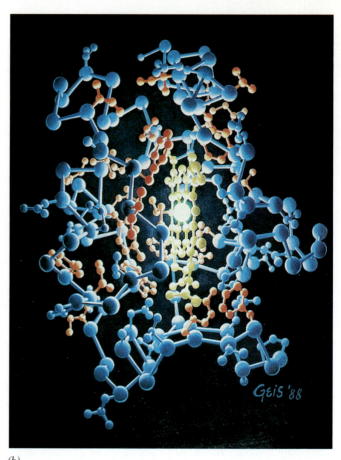

(b)

Figure 6-27 Side chain distribution in horse heart cytochrome *c*. In these paintings, based on an X-ray structure determined by Richard Dickerson, the protein is illuminated by its single iron atom centered in a heme group. Hydrogen atoms are not shown. In (*a*) the hydrophilic side chains are green, and in (*b*) the hydrophobic side chains are orange. [Illustration, Irving Geis. Image from the Irving Geis Collection/Howard Hughes Medical Institute. Rights owned by HHMI. Reproduction by permission only.] 🖐 **See Kinemage Exercise 5.**

See Guided Exploration 8

Secondary structures in proteins

(Fig. 6-27). Polar side chains tend to extend toward—and thereby help form—the protein's surface, whereas nonpolar side chains largely extend toward—and thereby occupy—its interior.

Most proteins are quite compact, with their interior atoms packed together even more efficiently than the atoms in a crystal of small organic molecules. Nevertheless, the atoms of protein side chains almost invariably have low-energy arrangements. Evidently, interior side chains adopt relaxed conformations despite the profusion of intramolecular interactions. Closely packed protein interiors generally exclude water. When water molecules are present, they often occupy specific positions where they can form hydrogen bonds, sometimes acting as a bridge between two hydrogen-bonding protein groups.

C Supersecondary Structures and Domains

The major types of secondary structural elements, α helices and β sheets, occur in globular proteins in varying proportions and combinations. Some proteins, such as *E. coli* **cytochrome b_{562}** (Fig. 6-28*a*), consist only of α helices

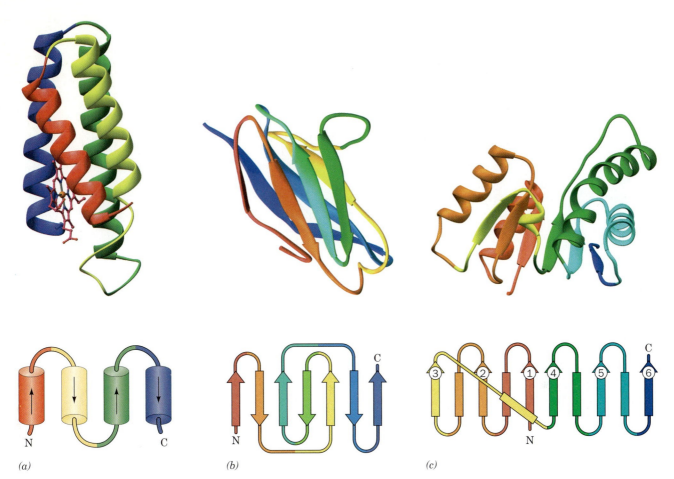

Figure 6-28   A selection of protein structures.   The proteins are represented by their peptide backbones, drawn in ribbon form, with β strands shown as flat arrows pointing from N- to C-terminus. The polypeptide chain is colored, from N- to C-terminus, in rainbow order, from red to blue. Below each drawing is the corresponding topological diagram indicating the connectivity of its helices (represented by cylinders or rectangles) and β strands (represented by flat arrows) (*a*) The X-ray structure of the 106-residue *E. coli* **cytochrome b_{562},** which forms an up–down–up–down 4-helix bundle. Its bound heme group is shown in ball-and-stick form with C magenta, N blue, O red, and Fe orange. (*b*) The X-ray structure of the N-terminal domain of the 103-residue human immunoglobulin fragment **Fab New** showing its immunoglobulin fold. The polypeptide chain is folded into a sandwich of 3- and 4-stranded antiparallel β sheets. (*c*) The X-ray structure of the 163-residue N-terminal domain of dogfish lactate dehydrogenase. It contains a 6-stranded parallel β sheet in which the crossovers between β strands all contain an α helix that forms a right-handed helical turn with its flanking β strands. [Based on X-ray structures by (*a*) F. Scott Matthews, Washington University School of Medicine; (*b*) Roberto Poljak, The Johns Hopkins School of Medicine; and (*c*) Michael Rossmann, Purdue University. PDBids (*a*) 256B, (*b*) 7FAB, and (*c*) 6LDH.]

spanned by short connecting links. Others, such as the **immunoglobulin fold** (Fig. 6-28*b*), have a large proportion of β sheets and are devoid of α helices. Most proteins, such as dogfish **lactate dehydrogenase** (Fig. 6-28*c*) and carboxypeptidase A (Fig. 6-12), have significant amounts of both types of secondary structure (on average, ~31% α helix and ~28% β sheet).

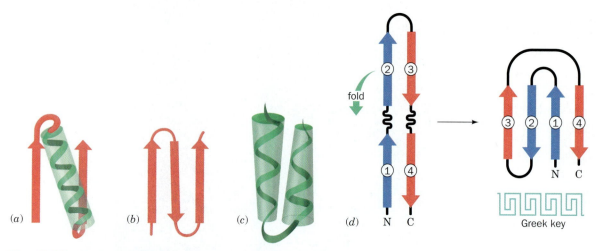

Figure 6-29   Schematic diagrams of supersecondary structures.   (*a*) a βαβ motif, (*b*) a β hairpin motif, (*c*) an αα motif, and (*d*) a Greek key motif, showing how it is constructed from a folded-over β hairpin.

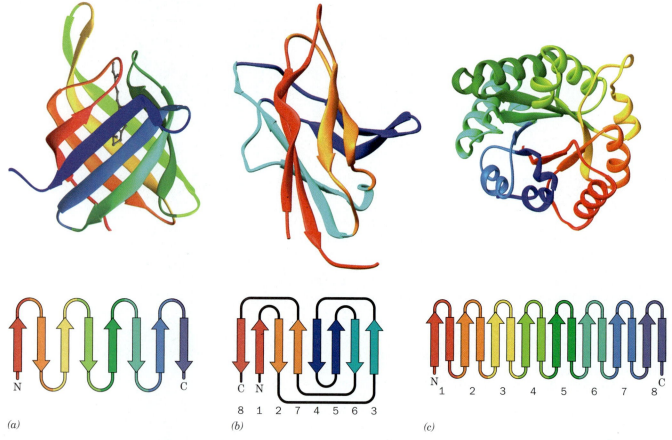

Figure 6-30   X-Ray structures of β barrels.   Each polypeptide is drawn and colored and accompanied by its corresponding topological diagram as is described in the legend to Fig. 6-28. (*a*) Human **retinol binding protein** showing its 8-stranded up-and-down β barrel (residues 1–142 of this 182-residue protein). Note that each β strand is linked via a short loop to its clockwise-adjacent strand as seen from the top. The protein's bound retinol molecule is represented by a gray ball-and-stick model. (*b*) **Peptide-*N*⁴-(*N*-acetyl-β-D-glucosaminyl)asparagine amidase F** from *Flavobacterium meningosepticum* (residues 1–140 of this 340-residue enzyme). Note how its 8-stranded β barrel is formed by rolling up a 4-segment β hairpin. Here the two β strands in each segment of the β hairpin are colored alike with strands 1 and 8 (the N- and C-terminal strands) red, strands

2 and 7 orange, strands 3 and 6 cyan, and strands 4 and 5 blue. This motif, which is known as a **jelly roll** or **Swiss roll barrel,** is so named because of its topological resemblance to these rolled-up pastries. (*c*) Chicken muscle **triose phosphate isomerase** (**TIM;** 247 residues) forms a so-called **α/β barrel** in which 8 pairs of alternating β strands and α helices roll up to form an inner barrel of 8 parallel β strands surrounded by an outer barrel of 8 parallel α helices. The protein is viewed approximately along the axis of the α/β barrel. Note that the α/β barrel is essentially a series of linked βαβ motifs. [Based on X-ray structures by (*a*) T. Alwyn Jones, Biomedical Center, Uppsala, Sweden; (*b*) Patrick Van Roey, New York State Department of Health, Albany, New York; and (*c*) David Phillips, Oxford University, Oxford, U.K. PDBids (*a*) 1RBP, (*b*) 1PNG, and (*c*) 1TIM.]

Certain groupings of secondary structural elements, called **supersecondary structures** or **motifs,** occur in many unrelated globular proteins:

1. The most common form of supersecondary structure is the **βαβ motif,** in which an α helix connects two parallel strands of a β sheet (Fig. 6-29*a*).

2. Another common supersecondary structure, the **β hairpin** motif, consists of antiparallel strands connected by relatively tight reverse turns (Fig. 6-29*b*).

3. In an **αα motif,** two successive antiparallel α helices pack against each other with their axes inclined. This permits energetically favorable intermeshing of their contacting side chains (Fig. 6-29*c*). Similar associations stabilize the coiled coil conformation of α keratin, although its helices are parallel rather than antiparallel (Section 6-1C).

4. In the **Greek key motif** (Fig. 6-29*d;* named after an ornamental design commonly used in ancient Greece; see inset), a β hairpin is folded over to form a 4-stranded antiparallel β sheet. Of the 10 possible ways of connecting the strands of a 4-stranded antiparallel β sheet, the two that form Greek key motifs are, by far, the most common in proteins of known structure.

5. Extended β sheets often roll up to form **β barrels.** Three different types of β barrels are shown in Fig. 6-30.

Motifs may have functional as well as structural significance. For example, Michael Rossmann has shown that a βαβαβ unit, in which the β strands form a parallel sheet with α helical connections, often acts as a nucleotide-binding site. In most proteins that bind dinucleotides [such as nicotinamide adenine dinucleotide (NAD^+); Fig. 11-3], two such βαβαβ units combine to form a motif known as a **dinucleotide-binding fold,** or **Rossmann fold.** Lactate dehydrogenase (Fig. 6-28*c*) is an example of such a protein.

Large Polypeptides Form Domains.   Polypeptide chains containing more than ~200 residues usually fold into two or more globular clusters known as **domains,** which give these proteins a bi- or multilobal appearance. Most domains consist of 100 to 200 amino acid residues and have an average diameter of ~25 Å. Each subunit of the enzyme **glyceraldehyde-3-phosphate dehydrogenase,** for example, has two distinct domains (Fig. 6-31). A polypeptide chain wanders back and forth within a domain, but neighboring domains are usually connected by only one or two polypeptide segments. *Consequently, many domains are structurally independent units that have the characteristics of small globular proteins.* Nevertheless, the domain structure of a protein is not necessarily obvious since its domains may make such extensive contacts with each other that the protein appears to be a single globular entity.

An inspection of the various protein structures diagrammed in this chapter reveals that domains consist of two or more layers of secondary structural elements. The reason for this is clear: At least two such layers are required to seal off a domain's hydrophobic core from its aqueous environment.

Domains often have a specific function such as the binding of a small molecule. In Fig. 6-31, for example, NAD^+ binds to the first domain of glyceraldehyde-3-phosphate dehydrogenase (note its dinucleotide-binding

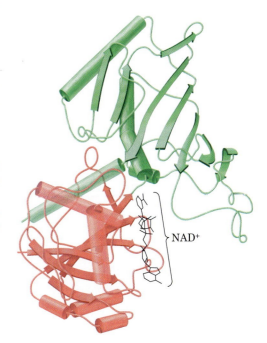

Figure 6-31   The two-domain protein glyceraldehyde-3-phosphate dehydrogenase.   The first domain (*red*) binds NAD^+ (*black*), and the second domain (*green*) binds glyceraldehyde-3-phosphate (not shown). [After Biesecker, G., Harris, J.I., Thierry, J.C., Walker, J.E., and Wonacott, A., *Nature* **266,** 331 (1977).] 🔗 **See the Interactive Exercises.**

fold). In multidomain proteins, binding sites often occupy the clefts between domains; that is, small molecules are bound by groups from two domains. In such cases, the relatively pliant covalent connection between the domains allows flexible interactions between the protein and the small molecule.

D Protein Families

The many thousands of known protein structures, comprising an even greater number of separate domains, can be grouped into families by examining the overall paths followed by their polypeptide chains. When folding patterns are compared without regard to the amino acid sequence or the presence of surface loops, the number of unique structural domains drops to only a few hundred. (Although not all protein structures are known, estimates place an upper limit of about 1000 on the total number of unique protein domains in nature.) Surprisingly, a few dozen folding patterns account for about half of all known protein structures.

There are several possible reasons for the limited number of known domain structures. The numbers may reflect database bias; that is, the collection of known protein structures may not be a representative sample of all protein structures. However, the rapidly increasing number of proteins whose structures have been determined makes this possibility less and less plausible. More likely, the common protein structures may be evolutionary sinks—domains that arose and persisted because of their ability (1) to form stable folding patterns; (2) to tolerate amino acid deletions, substitutions, and insertions, thereby making them more likely to survive evolutionary changes; and/or (3) to support essential biological functions.

Polypeptides with similar sequences tend to adopt similar backbone conformations. This is certainly true for evolutionarily related proteins that carry out similar functions. For example, the cytochromes c of different species are highly conserved proteins with closely similar sequences (see Table 5-5) and three-dimensional structures.

Cytochrome c occurs only in eukaryotes, but prokaryotes contain proteins, known as **c-type cytochromes,** which perform the same general function (that of an electron carrier). The c-type cytochromes from different species exhibit only low degrees of sequence similarity to each other and to eukaryotic cytochromes c. Yet their X-ray structures are clearly similar, particularly in polypeptide chain folding and side chain packing in the protein interior (Fig. 6-32). The major structural differences among c-type

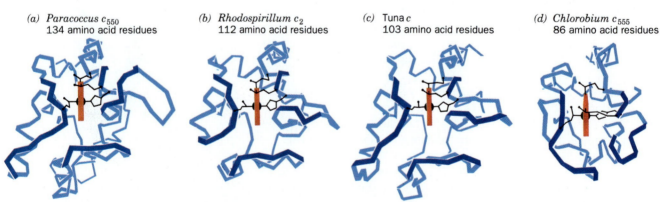

(a) *Paracoccus* c_{550}
134 amino acid residues

(b) *Rhodospirillum* c_2
112 amino acid residues

(c) Tuna c
103 amino acid residues

(d) *Chlorobium* c_{555}
86 amino acid residues

Figure 6-32 Three-dimensional structures of c-type cytochromes. The polypeptide backbones (*blue*) are shown in analogous orientations such that their heme groups (*red*) are viewed edge-on. The Cys, Met, and His side chains that covalently link the heme to the protein are also shown. (*a*) Cytochrome c_{550} from *Paracoccus denitrificans* (134 residues), (*b*) cytochrome c_2 from *Rhodospirillum rubrum* (112 residues), (*c*) cytochrome c from tuna (103 residues), and (*d*) cytochrome c_{555} from *Chlorobium limicola* (86 residues). [Illustration, Irving Geis. Image from the Irving Geis Collection/Howard Hughes Medical Institute. Rights owned by HHMI. Reproduction by permission only.]
 See Kinemage Exercise 5.

cytochromes lie in the various polypeptide loops on their surfaces. The sequences of the *c*-type cytochromes have diverged so far from one another that, in the absence of their X-ray structures, they can be properly aligned only through the use of mathematically sophisticated computer programs. Thus, *it appears that the essential structural and functional elements of proteins, rather than their amino acid residues, are conserved during evolution.*

Structural similarities in proteins with only distantly related functions are commonly observed. For example, many NAD^+-binding enzymes that participate in widely different metabolic pathways contain similar dinucleotide-binding folds (e.g., Fig. 6-31) coupled to diverse domains that carry out specific enzymatic reactions.

3 Quaternary Structure and Symmetry

Most proteins, particularly those with molecular masses >100 kD, consist of more than one polypeptide chain. These polypeptide subunits associate with a specific geometry. The spatial arrangement of these subunits is known as a protein's quaternary structure.

There are several reasons why multisubunit proteins are so common. In large assemblies of proteins, such as collagen fibrils, the advantages of subunit construction over the synthesis of one huge polypeptide chain are analogous to those of using prefabricated components in constructing a building: Defects can be repaired by simply replacing the flawed subunit; the site of subunit manufacture can be different from the site of assembly into the final product; and the only genetic information necessary to specify the entire edifice is the information specifying its few different self-assembling subunits. In the case of enzymes, increasing a protein's size tends to better fix the three-dimensional positions of its reacting groups. *Increasing the size of an enzyme through the association of identical subunits is more efficient than increasing the length of its polypeptide chain since each subunit has an active site. More importantly, the subunit construction of many enzymes provides the structural basis for the regulation of their activities* (Sections 7-3B and 12-3).

Subunits Usually Associate Noncovalently. A multisubunit protein may consist of identical or nonidentical polypeptide chains. Hemoglobin, for example, has the subunit composition $\alpha_2\beta_2$ (Fig. 6-33). Proteins with more than one subunit are called **oligomers,** and their identical units are called **protomers.** A protomer may therefore consist of one polypeptide chain or several unlike polypeptide chains. In this sense, hemoglobin is a dimer of $\alpha\beta$ protomers.

The contact regions between subunits resemble the interior of a single-subunit protein: They contain closely packed nonpolar side chains, hydrogen bonds involving the polypeptide backbones and their side chains, and, in some cases, interchain disulfide bonds. However, the subunit interfaces of proteins that dissociate *in vivo* have lesser hydrophobicities than do permanent interfaces.

Subunits Are Symmetrically Arranged. In the vast majority of oligomeric proteins, the protomers are symmetrically arranged; that is, each protomer occupies a geometrically equivalent position in the oligomer. Proteins cannot have inversion or mirror symmetry, however, because bringing the protomers into coincidence would require converting chiral L residues to D residues. Thus, *proteins can have only **rotational symmetry.***

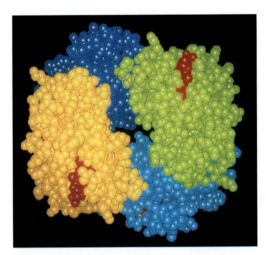

Figure 6-33 Quaternary structure of hemoglobin. In this space-filling model, the α_1, α_2, β_1, and β_2 subunits are colored yellow, green, cyan, and blue, respectively. Heme groups are red. [Based on an X-ray structure by Max Perutz, MRC Laboratory of Molecular Biology, U.K. PDBid 2DHB.]

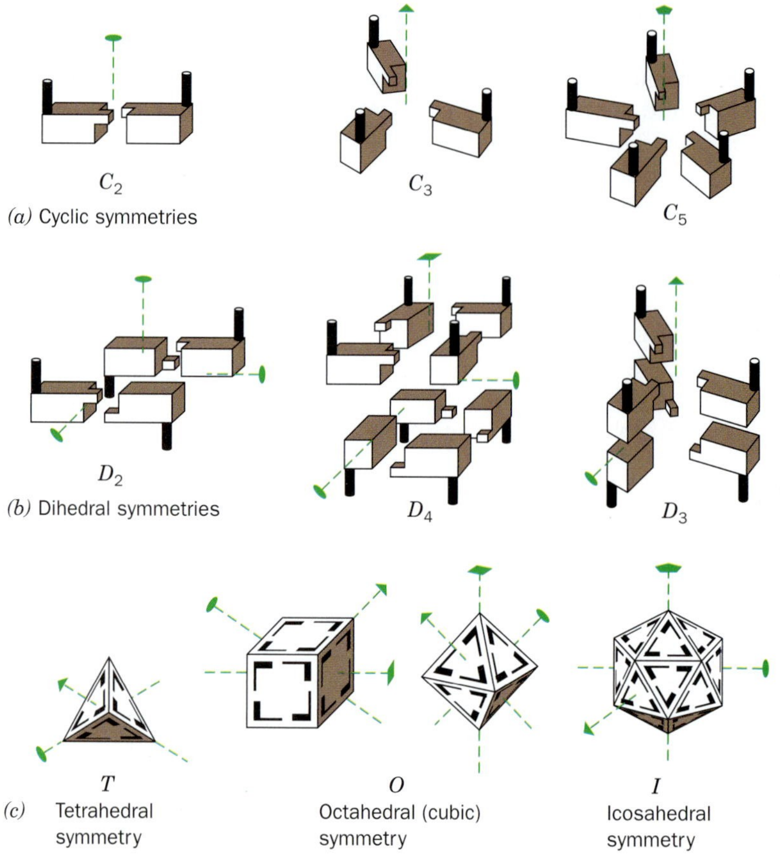

Figure 6-34 **Some symmetries of oligomeric proteins.** The oval, the triangle, the square, and the pentagon at the ends of the dashed green lines indicate, respectively, the unique twofold, threefold, fourfold, and fivefold rotational axes of the objects shown. (*a*) Assemblies with cyclic (*C*) symmetry. (*b*) Assemblies with dihedral (*D*) symmetry. In these objects, a 2-fold axis is perpendicular to another rotational axis. (*c*) Assemblies with the rotational symmetries of a tetrahedron (*T*), a cube or octahedron (*O*), and an icosahedron (*I*). [Illustration, Irving Geis. Image from the Irving Geis Collection/Howard Hughes Medical Institute. Rights owned by HHMI. Reproduction by permission only.] 🖢 **See the Animated Figures.**

In the simplest type of rotational symmetry, **cyclic symmetry,** protomers are related by a single axis of rotation (Fig. 6-34*a*). Objects with two-, three-, or *n*-fold rotational axes are said to have C_2, C_3, or C_n symmetry, respectively. C_2 symmetry is the most common; higher cyclic symmetries are relatively rare.

Dihedral symmetry (D_n), a more complicated type of rotational symmetry, is generated when an *n*-fold rotation axis intersects a twofold rotation axis at right angles (Fig. 6-34*b*). An oligomer with D_n symmetry consists of $2n$ protomers. D_2 symmetry is the most common type of dihedral symmetry in proteins.

Other possible types of rotational symmetry are those of a tetrahedron, cube, and icosahedron (Fig. 6-34*c*). Some multienzyme complexes and spherical viruses are built on these geometric plans.

4 Protein Stability

Incredible as it may seem, thermodynamic measurements indicate that *native proteins are only marginally stable under physiological conditions.* The free energy required to denature them is $\sim$0.4 kJ $\cdot$ mol^{-1} per amino acid residue, so a fully folded 100-residue protein is only about 40 kJ $\cdot$ mol^{-1} more stable than its unfolded form (for comparison, the energy required to break a typical hydrogen bond is $\sim$20 kJ $\cdot$ mol^{-1}). The various noncovalent influences on proteins—hydrophobic effects, electrostatic interactions, and hydrogen bonding—each have energies that may total thousands of kilojoules per mole over an entire protein molecule. Consequently, a pro-

tein structure is the result of a delicate balance among powerful counter-vailing forces. In this section, we discuss the forces that stabilize proteins and the processes by which proteins achieve their most stable folded state.

A Forces That Stabilize Protein Structure

Protein structures are governed primarily by hydrophobic effects and, to a lesser extent, by interactions between polar residues, and by other types of bonds.

The Hydrophobic Effect.   *The hydrophobic effect, which causes nonpolar substances to minimize their contacts with water* (Section 2-1C), *is the major determinant of native protein structure.* The aggregation of nonpolar side chains in the interior of a protein is favored by the increase in entropy of the water molecules that would otherwise form ordered "cages" around the hydrophobic groups. The combined hydrophobic and hydrophilic tendencies of individual amino acid residues in proteins can be expressed as **hydropathies** (Table 6-2). The greater a side chain's hydropathy, the more likely it is to occupy the interior of a protein and vice versa. Hydropathies are good predictors of which portions of a polypeptide chain are inside a protein, out of contact with the aqueous solvent, and which portions are outside (Fig. 6-35).

Site-directed mutagenesis experiments in which individual interior residues have been replaced by a number of others suggest that the factors that affect stability are, in order, the hydrophobicity of the substituted residue, its steric compatibility, and, last, the volume of its side chain.

Electrostatic Interactions.   In the closely packed interiors of native proteins, van der Waals forces, which are relatively weak (Section 2-1A), are nevertheless an important stabilizing influence. This is because these forces act over only short distances and hence are lost when the protein is unfolded.

Perhaps surprisingly, *hydrogen bonds, which are central features of protein structures, make only minor contributions to protein stability.* This is because hydrogen-bonding groups in an unfolded protein form hydrogen bonds with water molecules. Thus the contribution of a hydrogen bond to the stability of a native protein is the small difference in hydrogen bonding free energies between the native and unfolded states (-2 to 8 kJ·mol^{-1} as determined by site-directed mutagenesis studies). Nevertheless, hydrogen bonds are important determinants of native protein structures, because if a protein folded in a way that prevented a hydrogen bond from forming, the stabilizing energy of that hydrogen bond would be lost. Hydrogen bonding therefore fine-tunes tertiary structure by "selecting" the unique native structure of a protein from among a relatively small number of hydrophobically stabilized conformations.

Table 6-2 Hydropathy Scale for Amino Acid Side Chains

Side Chain	Hydropathy
Ile	4.5
Val	4.2
Leu	3.8
Phe	2.8
Cys	2.5
Met	1.9
Ala	1.8
Gly	-0.4
Thr	-0.7
Ser	-0.8
Trp	-0.9
Tyr	-1.3
Pro	-1.6
His	-3.2
Glu	-3.5
Gln	-3.5
Asp	-3.5
Asn	-3.5
Lys	-3.9
Arg	-4.5

Source: Kyte, J. and Doolittle, R.F., *J. Mol. Biol.* **157,** 110 (1982).

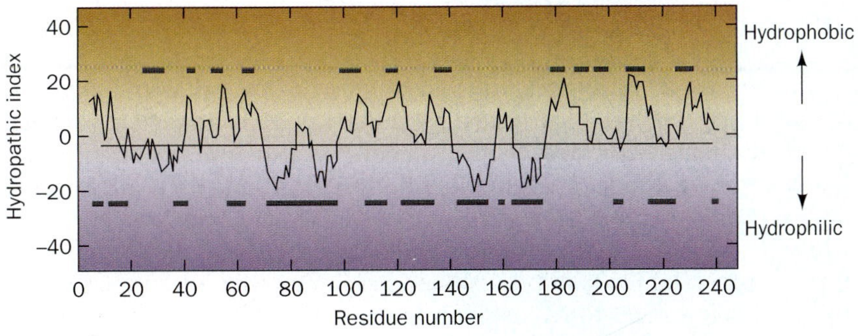

Figure 6-35 A hydropathic index plot for bovine chymotrypsinogen.   The sum of the hydropathies of nine consecutive residues is plotted versus residue sequence number. A large positive hydropathic index indicates a hydrophobic region of the polypeptide, whereas a large negative value indicates a hydrophilic region. The upper bars denote the protein's interior regions, as determined by X-ray crystallography, and the lower bars denote the protein's exterior regions. [After Kyte, J. and Doolittle, R.F., *J. Mol. Biol.* **157,** 111 (1982).]

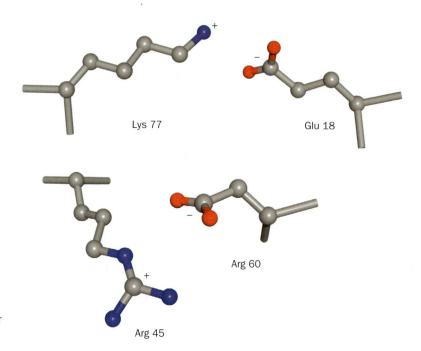

Lys 77

Glu 18

Arg 60

Arg 45

Figure 6-36   Examples of ion pairs in myoglobin.   In each case, oppositely charged side chain groups from residues far apart in sequence closely approach each other through the formation of ion pairs.

The association of two ionic protein groups of opposite charge (e.g., Lys and Asp) is known as an **ion pair** or **salt bridge.** About 75% of the charged residues in proteins are members of ion pairs that are located mostly on the protein surface (Fig. 6-36). Despite the strong electrostatic attraction between the oppositely charged members of an ion pair, these interactions contribute little to the stability of a native protein. This is because the free energy of an ion pair's charge–charge interactions usually fails to compensate for the loss of entropy of the side chains and the loss of solvation free energy when the charged groups form an ion pair. This accounts for the observation that ion pairs are poorly conserved among homologous proteins.

Chemical Cross-Links.   Disulfide bonds (Fig. 4-6) within and between polypeptide chains form as a protein folds to its native conformation. Some polypeptides whose Cys residues have been derivatized or mutagenically replaced to prevent disulfide bond formation can still assume their fully active conformations, suggesting that disulfide bonds are not essential stabilizing forces. They may, however, be important for "locking in" a particular backbone folding pattern as the protein proceeds from its fully extended state to its mature form.

Disulfide bonds are rare in intracellular proteins because the cytoplasm is a reducing environment. Most disulfide bonds occur in proteins that are secreted from the cell into the more oxidizing extracellular environment. The relatively hostile extracellular world (e.g., uncontrolled temperature and pH) apparently requires the additional structural constraints conferred by disulfide bonds.

Metal ions may also function to internally cross-link proteins. For example, at least ten motifs collectively known as **zinc fingers** have been described in nucleic acid–binding proteins. These structures contain about 25–60 residues arranged around one or two Zn^{2+} ions that are tetrahedrally coordinated by the side chains of Cys, His, and occasionally Asp or Glu (Fig. 6-37). The Zn^{2+} ion allows relatively short stretches of polypeptide chain to fold into stable units that can interact with nucleic acids. Zinc fingers are too small to be stable in the absence of Zn^{2+}. Zinc is ideally suited to its structural role in intracellular proteins: Its filled *d* electron shell

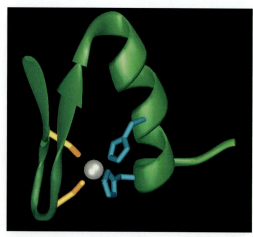

Figure 6-37   A zinc finger motif.   This structure, from the DNA-binding protein **Zif268,** is known as a Cys_2–His_2 zinc finger because the zinc atom (*silver sphere*) is coordinated by two Cys residues (*yellow*) and two His residues (*cyan*). [Based on an X-ray structure by Carl Pabo, MIT. PDBid 1ZAA.]

permits it to interact strongly with a variety of ligands (e.g., sulfur, nitrogen, or oxygen) from different amino acid residues. In addition, zinc has only one stable oxidation state (unlike, for example, copper and iron), so it does not undergo oxidation–reduction reactions in the cell.

B   Protein Dynamics

The plethora of forces acting to stabilize proteins as well as the static way that their structures are usually portrayed may leave the false impression that proteins have fixed and rigid structures. In fact, *proteins are flexible and rapidly fluctuating molecules whose structural mobilities are functionally significant*. Groups ranging in size from individual side chains to entire domains or subunits may be displaced by up to several angstroms through random intramolecular movements or in response to a trigger such as the binding of a small molecule. Extended side chains, such as Lys, and the N- and C-termini of polypeptide chains are especially prone to wave around in solution because there are few forces holding them in place.

Theoretical calculations by Martin Karplus indicate that a protein's native structure probably consists of a large collection of rapidly interconverting conformations that have essentially equal stabilities (Fig. 6-38). Conformational flexibility, or **breathing,** with structural displacement of up to ~2 Å, allows small molecules to diffuse in and out of the interior of certain proteins.

Figure 6-38   Molecular dynamics of myoglobin. Several "snapshots" of the protein calculated at intervals of 5×10^{-12} s are superimposed. The backbone is blue, the heme group is yellow, and the His side chain linking the heme to the protein is orange. [Courtesy of Martin Karplus, Harvard University.]

C   Protein Denaturation and Renaturation

The low conformational stabilities of native proteins make them easily susceptible to denaturation by altering the balance of the weak nonbonding forces that maintain the native conformation. Proteins can be denatured by a variety of conditions and substances:

1. Heating causes a protein's conformationally sensitive properties, such as optical rotation (Section 4-2), viscosity, and UV absorption, to change abruptly over a narrow temperature range. Such a sharp transition indicates that the entire polypeptide unfolds or "melts" **cooperatively,** that is, nearly simultaneously. Most proteins have melting temperatures that are well below 100°C. Among the exceptions are the proteins of thermophilic bacteria (Box 6-3).

2. pH variations alter the ionization states of amino acid side chains, thereby changing protein charge distributions and hydrogen-bonding requirements.

3. Detergents associate with the nonpolar residues of a protein, thereby interfering with the hydrophobic interactions responsible for the protein's native structure.

4. The **chaotropic agents** guanidinium ion and urea,

$$\underset{\textbf{Guanidinium ion}}{H_2N - \underset{\underset{NH_2^+}{\|}}{C} - NH_2} \qquad \underset{\textbf{Urea}}{H_2N - \underset{\underset{O}{\|}}{C} - NH_2}$$

in concentrations in the range 5 to 10 M, are the most commonly used protein denaturants. Chaotropic agents are ions or small organic molecules that increase the solubility of nonpolar substances in water. Their effectiveness as denaturants stems from their ability to disrupt hydrophobic interactions, although their mechanism of action is not well understood.

BOX 6-3

Perspectives in Biochemistry

Thermostable Proteins

Certain species of bacteria known as **hyperthermophiles** grow at temperatures near 100°C. They live in such places as hot springs and submarine hydrothermal vents, with the most extreme, the archeon *Pyrolobus fumarii,* able to grow at temperatures as high as 113°C. These organisms have many of the same metabolic pathways as do **mesophiles** (organisms that grow at "normal" temperatures). Yet most mesophilic proteins denature at temperatures where hyperthermophiles thrive. What is the structural basis for the thermostability of hyperthermophilic proteins?

The difference in the thermal stabilities of the corresponding (hyper)thermophilic and mesophilic proteins does not exceed $\sim$100 kJ·mol^{-1}, the equivalent of a few noncovalent interactions. This is probably why comparisons of the X-ray structures of hyperthermophilic enzymes with their mesophilic counterparts have failed to reveal any striking differences between them. These proteins exhibit some variations in secondary structure but no more than would be expected for homologous proteins from distantly related mesophiles. However, several of these thermostable enzymes have a superabundance of salt bridges on their surfaces, many of which are arranged in extensive networks containing up to 18 side chains.

The idea that salt bridges can stabilize a protein structure appears to contradict the conclusion of Section 6-4A that ion pairs are, at best, marginally stable. The key to this apparent

paradox is that *the salt bridges in thermostable proteins form networks.* Thus, the gain in charge–charge free energy on associating a third charged group with an ion pair is comparable to that between the members of this ion pair, whereas the free energy lost on desolvating and immobilizing the third side chain is only about half that lost in bringing together the first two side chains. The same, of course, is true for the addition of a fourth, fifth, etc., side chain to a salt bridge network.

Not all thermostable proteins have such a high incidence of salt bridges. Structural comparisons suggest that these proteins are stabilized by a combination of small effects, the most important of which are an increased size of the protein's hydrophobic core, an increased size in the interface between its domains and/or subunits, and a more tightly packed core as evidenced by a reduced surface-to-volume ratio.

The fact that the proteins of hyperthermophiles and mesophiles are homologous and carry out much the same functions indicates that mesophilic proteins are by no means maximally stable. This, in turn, strongly suggests *that the marginal stability of most proteins under physiological conditions (averaging $\sim$0.4 kJ·mol^{-1} of amino acid residues) is an essential property that has arisen through evolutionary design.* Perhaps this marginal stability helps confer the structural flexibility that many proteins require to carry out their physiological functions.

Denatured Proteins Can Be Renatured. In 1957, the elegant experiments of Christian Anfinsen on **ribonuclease A (RNase A)** showed that proteins can be denatured reversibly. RNase A, a 124-residue single-chain protein, is completely unfolded and its four disulfide bonds reductively cleaved in an 8 M urea solution containing 2-mercaptoethanol. Dialyzing away the urea and reductant and exposing the resulting solution to O_2 at pH 8 (which oxidizes the SH groups to form disulfides) yields a protein that is virtually 100% enzymatically active and physically indistinguishable from native RNase A (Fig. 6-39). The protein must therefore **renature** spontaneously.

The renaturation of RNase A demands that its four disulfide bonds reform. The probability of one of the eight Cys residues randomly forming a disulfide bond with its proper mate among the other seven Cys residues is 1/7; that of one of the remaining six Cys residues then randomly forming its proper disulfide bond is 1/5; etc. The overall probability of RNase A re-forming its four native disulfide links at random is

$$\frac{1}{7} \times \frac{1}{5} \times \frac{1}{3} \times \frac{1}{1} = \frac{1}{105}$$

Clearly, the disulfide bonds do not randomly re-form under renaturing conditions, since, if they did, only 1% of the refolded protein would be cat-

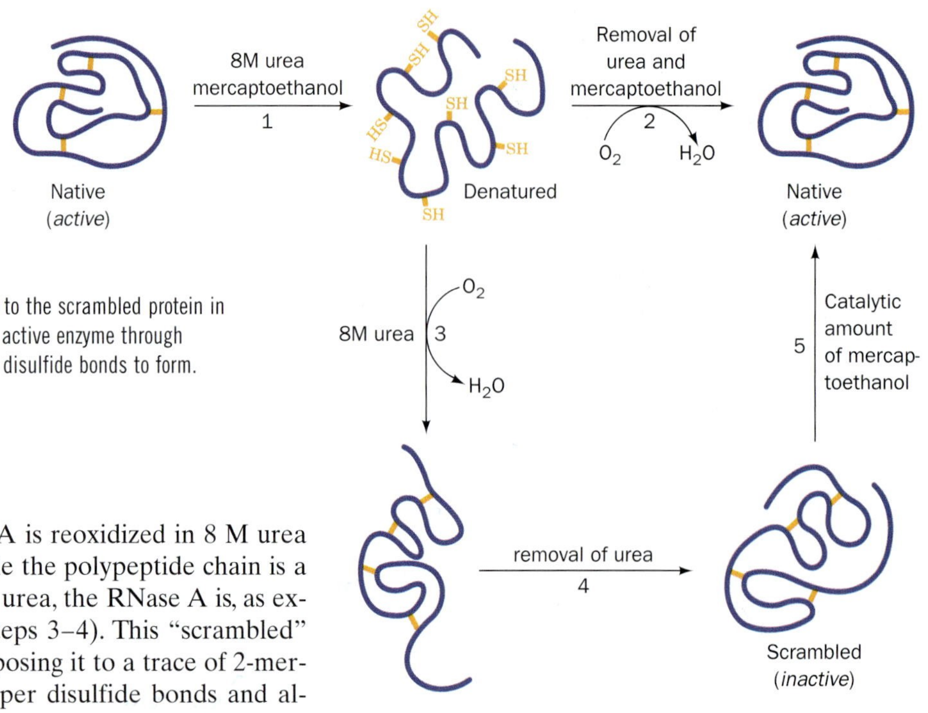

Figure 6-39   Denaturation and renaturation of RNase A. The polypeptide is represented by a purple line, with its disulfide bonds in yellow. After the protein has been denatured and its disulfide bonds cleaved (1), it will renature in the presence of O_2 when the denaturant (urea) and reductant (mercaptoethanol) are removed (2). If the disulfide bonds are allowed to re-form (3) before the urea is removed (4), the bonds form at random and the resulting protein is enzymatically inactive. Adding a small amount of mercaptoethanol to the scrambled protein in the absence of O_2 (5) catalyzes its conversion to the active enzyme through disulfide interchange reactions that allow the native disulfide bonds to form.

alytically active. Indeed, if the RNase A is reoxidized in 8 M urea so that its disulfide bonds re-form while the polypeptide chain is a random coil, then after removal of the urea, the RNase A is, as expected, only ~1% active (Fig. 6-39, Steps 3–4). This "scrambled" protein can be made fully active by exposing it to a trace of 2-mercaptoethanol, which breaks the improper disulfide bonds and allows the proper bonds to form. *Anfinsen's work demonstrated that proteins can fold spontaneously into their native conformations under physiological conditions. This implies that a protein's primary structure dictates its three-dimensional structure.*

5 Protein Folding

Studies of protein stability and renaturation suggest that protein folding is directed largely by the residues that occupy the interior of the folded protein. But *how* does a protein fold to its native conformation? One might guess that this process occurs through the protein's random exploration of all the conformations available to it until it eventually stumbles onto the correct one. A simple calculation first made by Cyrus Levinthal, however, convincingly demonstrates that this cannot possibly be the case: Assume that an n-residue protein's 2^n torsion angles, ϕ and ψ, each have three stable conformations. This yields $3^{2n} \approx 10^n$ possible conformations for the protein (a gross underestimate because we have completely neglected its side chains). Then, if the protein could explore a new conformation every 10^{-13} s (the rate at which single bonds reorient), the time t, in seconds, required for the protein to explore all the conformations available to it is

$$t = \frac{10^n}{10^{13}}$$

For a small protein of 100 residues, $t = 10^{87}$ s, which is immensely greater than the apparent age of the universe (20 billion years, or 6×10^{17} s). Clearly, proteins must fold more rapidly than this.

A Protein Folding Pathways

Experiments have shown that many proteins fold to their native conformations in less than a few seconds. This is because *proteins fold to their native conformations via directed pathways rather than stumbling on them through random conformational searches.* Thus, as a protein folds, its confor-

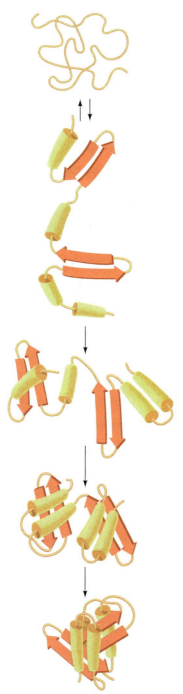

Figure 6-40   Hypothetical protein folding pathway.
This example shows a linear pathway for folding a two-domain protein. [After Goldberg, M.E., *Trends Biochem. Sci.* **10**, 389 (1985).]

mational stability increases sharply (i.e., its free energy decreases sharply), which makes folding a one-way process. A hypothetical folding pathway is diagrammed in Fig. 6-40.

Experimental observations indicate that protein folding begins with the formation of local segments of secondary structure (α helices and β sheets). This early stage of protein folding is extremely rapid, with much of the native secondary structure in small proteins appearing within 5 ms of the initiation of folding. Since native proteins contain compact hydrophobic cores, it is likely that the driving force in protein folding is what has been termed a **hydrophobic collapse.** The collapsed state is known as a **molten globule,** a species that has much of the secondary structure of the native protein but little of its tertiary structure. Theoretical studies suggest that helices and sheets form in part because they are particularly compact ways of folding a polypeptide chain.

Over the next 5 to 1000 ms, the secondary structure becomes stabilized and tertiary structure begins to form. During this intermediate stage, the nativelike elements are thought to take the form of subdomains that are not yet properly docked to form domains. In the final stage of folding, which for small single-domain proteins occurs over the next few seconds, the protein undergoes a series of complex motions in which it attains its relatively stable internal side chain packing and hydrogen bonding while it expels the remaining water molecules from its hydrophobic core.

In multidomain and multisubunit proteins, the respective units then assemble in a similar manner, with a few slight conformational adjustments required to produce the protein's native tertiary or quaternary structure. Thus, *proteins appear to fold in a hierarchical manner, with small local elements of structure forming and then coalescing to yield larger elements, which coalesce with other such elements to form yet larger elements, etc.*

Folding, like denaturation, appears to be a cooperative process, with small elements of structure accelerating the formation of additional structures. A folding protein must proceed from a high-energy, high-entropy state to a low-energy, low-entropy state. This energy–entropy relationship, which is diagrammed in Fig. 6-41, is known as a **folding funnel.** An unfolded polypeptide has many possible conformations (high entropy). As it folds into an ever-decreasing number of possible conformations, its entropy and free energy decrease. The energy–entropy diagram is not a smooth valley but a jagged landscape. Minor clefts and gullies represent conformations that are temporarily trapped until, through random thermal activation, they overcome a slight "uphill" free energy barrier and can then proceed to a lower energy conformation. Evidently, *proteins have evolved to have efficient folding pathways as well as stable native conformations.*

Understanding the process of protein folding as well as the forces that stabilize folded proteins is essential for elucidating the rules that govern the relationship between a protein's amino acid sequence and its three-dimensional structure. Such information will prove useful in predicting the structures of the hundreds of thousands of proteins that are known only from their sequences (Box 6-4).

B   Protein Disulfide Isomerase

Even under optimal experimental conditions, proteins often fold more slowly *in vitro* than they fold *in vivo*. One reason is that folding proteins often form disulfide bonds not present in the native proteins, which then slowly form native disulfide bonds through the process of disulfide interchange. **Protein disulfide isomerase (PDI)** catalyzes this process. Indeed, the observation that RNase A folds so much faster *in vivo* than *in vitro* led Anfinsen to discover this enzyme.

Figure 6-41 Energy–entropy diagram for protein folding. The width of the diagram represents entropy, and the depth, the energy. The unfolded polypeptide proceeds from a high-entropy, disordered state (*wide*) to a single low-entropy (*narrow*), low-energy native conformation. [After Onuchic, J.N., Wolynes, P.G., Luthey-Schulten, Z., and Socci, N.D., *Proc. Natl. Acad. Sci.* **92**, 3626 (1995).]

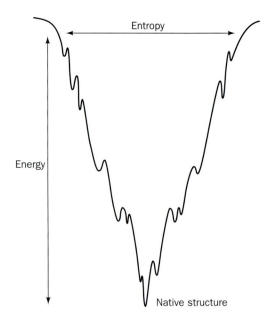

PDI binds to a wide variety of unfolded polypeptides via a hydrophobic patch on its surface. A Cys —SH group on reduced (SH-containing) PDI reacts with a disulfide group on the polypeptide to form a mixed disulfide and a Cys —SH group on the polypeptide (Fig. 6-42*a*). Another disulfide group on the polypeptide, brought into proximity by the spontaneous folding of the polypeptide, is attacked by this Cys —SH group. The newly liberated Cys —SH group then repeats this process with another disulfide bond, and so on, ultimately yielding the polypeptide containing only native disulfide bonds, along with regenerated PDI.

Oxidized (disulfide-containing) PDI also catalyzes the initial formation of a polypeptide's disulfide bonds by a similar mechanism (Fig. 6-42*b*). In this case, the reduced PDI reaction product must be reoxidized by cellular oxidizing agents in order to repeat the process.

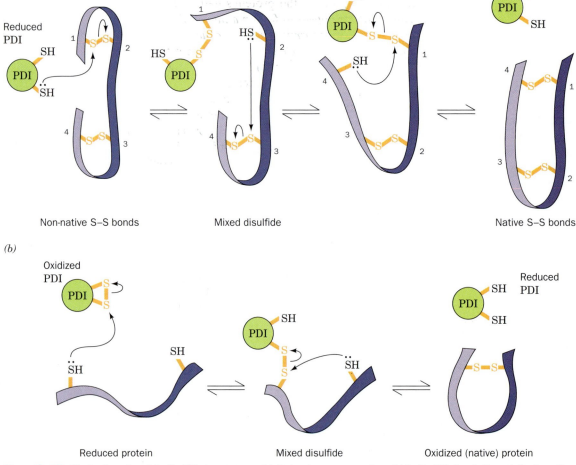

Figure 6-42 Mechanism of protein disulfide isomerase. (*a*) Reduced (SH-containing) PDI catalyzes the rearrangement of a polypeptide's non-native disulfide bonds via disulfide interchange reactions to yield native disulfide bonds. (*b*) Oxidized (disulfide-containing) PDI catalyzes the initial formation of a polypeptide's disulfide bonds through the formation of a mixed disulfide. Reduced PDI can then react with a cellular oxidizing agent to regenerate oxidized PDI. 🐛 **See the Animated Figures.**

BOX 6-4

Perspectives in Biochemistry

Protein Structure Prediction and Protein Design

Hundreds of thousands of protein sequences are known either through direct protein sequencing (Section 5-3) or, more commonly, through genomic DNA sequencing (Section 3-4D). Yet the structures of only ~30,000 of these proteins have been determined by X-ray crystallography or NMR techniques. Consequently, there is a need to develop robust techniques for predicting a protein's structure from its amino acid sequence. This represents a formidable challenge but promises great rewards in terms of understanding protein function, identifying diseases related to abnormal protein sequences, and designing drugs to alter protein structure or function.

There are several major approaches to protein structure prediction. The simplest and most reliable approach, **homology modeling,** aligns the sequence of interest with the sequence of a homologous protein or domain of known structure—compensating for amino acid substitutions, insertions, and deletions—through modeling and energy minimization calculations. This method yields reliable models for proteins that have as little as 25% sequence identity with a protein of known structure, although, of course, the accuracy of the model increases with the degree of sequence identity. The emerging field of **structural genomics,** which seeks to determine the structures of all representative domains, is aimed at expanding this predictive technique. The identification of structural homology is likely to provide clues as to a protein's function even with imperfect structure prediction.

Distantly related proteins may be structurally similar even though they have diverged to such an extent that their sequences show no obvious resemblance. **Threading** is a computational technique that attempts to determine the unknown structure of a protein by ascertaining whether it is consistent with a known protein structure. It does so by placing (threading) the unknown protein's residues along the backbone of a known protein structure and then determining whether the amino acid side chains of the unknown protein are stable in that arrangement. This method is not yet reliable, although it has yielded encouraging results.

Empirical methods based on experimentally determined statistical information such as the α helix and β sheet propensities deduced by Chou and Fasman (Table 6-1) have been moderately successful in predicting the secondary structures of proteins. Their main drawback is that neighboring residues in a polypeptide sometimes exert strong influence on a given residue's tendency to form a particular secondary structure.

Since the native structure of a protein ultimately depends on its amino acid sequence, it should be possible, in principle, to predict the structure of a protein based only on its chemical and physical properties (e.g., the hydrophobicity, size, hydrogen-bonding propensity, and charge of each of its amino acid residues). Such *ab initio* (from the beginning) methods are still only moderately successful in predicting the structures of small polypeptides. Computational molecular biologists periodically test their methods on sequences of proteins whose structures are undergoing conventional structural determination. In many cases, the theoretical results closely approximate the true structures (Fig. 1). This sort of modeling may be enough to provide clues to a protein's function, even if the exact positions of all its side chains are uncertain.

Protein design, the experimental inverse of protein structure prediction, has provided insights into protein folding and stability. Protein design may begin with a target structure such as a simple sandwich of β sheets or a bundle of four α helices. It attempts to construct an amino acid sequence that will form that structure. The designed polypeptide is then chemically or biologically synthesized, and its structure is determined. Fortunately, protein folding seems to be governed more by extended sequences of amino acids than by individual residues, which allows some room for error in designing polypeptides. Experimental results suggest that the greatest challenge of protein design may lie not in getting the polypeptide to fold to the desired conformation but in preventing it from folding into other unwanted conformations. In this respect, science lags far behind nature.

The first wholly successful *de novo* (from the beginning) protein design, accomplished by Stephen Mayo, was for a 28-residue ββα motif that has a backbone conformation designed to resemble a zinc finger (Fig. 6-37) but that

C   Molecular Chaperones

Proteins begin to fold as they are being synthesized, so the renaturation of a denatured protein *in vitro* may not entirely mimic the folding of a protein *in vivo*. In addition, proteins fold *in vivo* in the presence of extremely high concentrations of other proteins with which they can potentially interact. *Molecular chaperones* are essential proteins that bind to unfolded and partially folded polypeptide chains to prevent the improper association of ex-

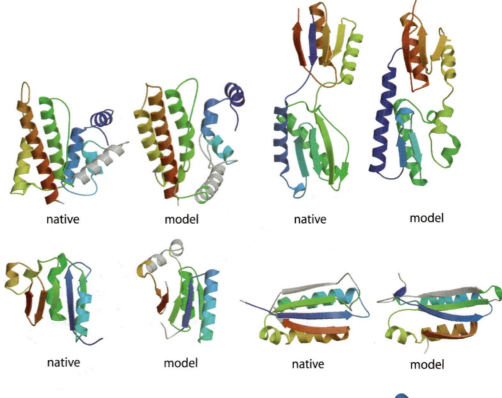

native   model   native   model

Figure 1 Comparison of experimentally determined (native) and predicted (model) folds of polypeptides. The polypeptides are colored in rainbow order from N-terminus (*indigo*) to C-terminus (*red*). [Courtesy of David Baker, University of Washington.]

native   model   native   model

contains no stabilizing metal ions. A computational design process considered the interactions among side chain and backbone atoms, screened all possible amino acid sequences, and, in order to take into account side chain flexibility, tested all sets of energetically allowed torsion angles for each side chain. The number of amino acid sequences to be tested was limited to 1.9×10^{27}, representing 1.1×10^{62} possible conformations! The design process yielded an optimal sequence of 28 residues, which was chemically synthesized and its structure determined by NMR spectroscopy. The designed protein, called FSD-1, closely resembled its predicted structure, and its backbone conformation was nearly superimposable on that of a known zinc finger motif (Fig. 2). Although FSD is relatively small, it folds into a unique stable structure, thereby demonstrating the power of protein design techniques.

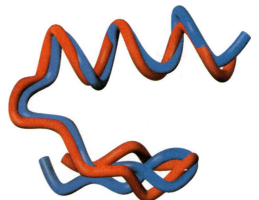

Figure 2 Structure of a designed protein. The structure of the 28-residue FSD-1 was designed to resemble that of a zinc finger motif (from the protein Zif268). The polypeptide backbones of FSD-1 (*blue*) and Zif268 (*red*) are nearly superimposable. [Courtesy of Stephen Mayo, California Institute of Technology.]

posed hydrophobic segments that might lead to non-native folding as well as polypeptide aggregation and precipitation. This is especially important for multidomain and multisubunit proteins, whose components must fold fully before they can properly associate with each other. Molecular chaperones also induce misfolded proteins to refold to their native conformations.

Many molecular chaperones were first described as **heat shock proteins (Hsp)** because their rate of synthesis is increased at elevated temperatures.

Presumably, the additional chaperones are required to recover heat-denatured proteins or to prevent misfolding under conditions of environmental stress.

There are several classes of molecular chaperones in both prokaryotes and eukaryotes, including: (1) the **Hsp70** family of 70-kD proteins, which function as monomers; (2) the **chaperonins,** which are large multisubunit proteins; and (3) the **Hsp90** proteins, which are mainly involved with the folding of proteins involved with signal transduction such as **steroid receptors** (Section 27-3B). All of these molecular chaperones operate by binding to an unfolded or aggregated polypeptide's solvent-exposed hydrophobic surface and subsequently releasing it, often repeatedly, in a manner that facilitates its proper folding. Many molecular chaperones are **ATPases,** that is, enzymes that catalyze the hydrolysis of ATP (adenosine triphosphate) to ADP (adenosine diphosphate) and P_i (inorganic phosphate):

$$ATP + H_2O \rightarrow ADP + P_i$$

The favorable free energy change of ATP hydrolysis drives the chaperone's bind-and-release reaction cycle.

Hsp70 proteins are highly conserved 70-kD monomeric proteins in both prokaryotes and eukaryotes. An Hsp70 chaperone, which functions in association with the **cochaperone** protein **Hsp40,** appears to bind to a newly synthesized polypeptide as it emerges from the ribosome. The binding and release of small hydrophobic regions on the new polypeptide may prevent its premature folding. Other chaperones apparently complete the job begun by the Hsp70 proteins. The Hsp70 proteins also function to unfold proteins in preparation for their transport through membranes (Section 9-4D) and to subsequently refold them.

(a)

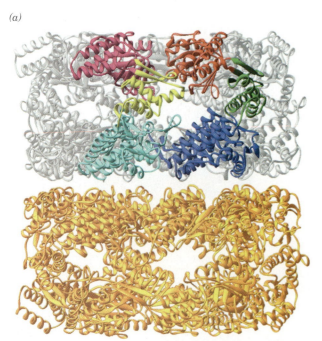

(b)

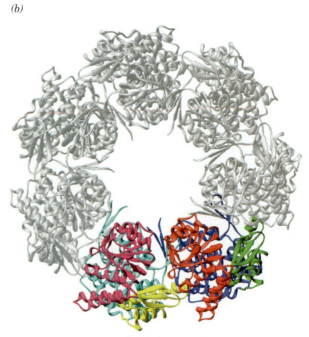

Figure 6-43 X-ray structure of GroEL. (*a*) Side view perpendicular to the 7-fold axis in which the seven identical subunits of the lower ring are gold and those of the upper ring are silver, with the exception of the two subunits nearest the viewer, whose equatorial, intermediate, and apical domains are colored blue, green, and red on the right subunit and cyan, yellow, and magenta on the left subunit. The two rings of the complex are held together through side chain interactions that are not seen in this drawing. (*b*) Top view along the 7-fold axis in which only the upper ring is shown for the sake of clarity. Note the large central channel that appears to run the length of the protein. [Based on an X-ray structure by Axel Brünger, Arthur Horwich, and Paul Sigler, Yale University. PDBid 1OEL.]

The GroEL/ES Chaperones Form a Barrel Structure.   The chaperonins consist of two types of proteins, named **Hsp60** and **Hsp10**. Those in *E. coli,* the best-characterized chaperonins, are called **GroEL** and **GroES.** They are essential for the survival of *E. coli* under all conditions tested. Fourteen identical 549-residue GroEL subunits are arranged in two stacked rings of seven subunits each (Fig. 6-43) to form a complex with D_7 symmetry (with perpendicular 7-fold and 2-fold axes of symmetry). The X-ray structure of GroEL, determined by Arthur Horwich and Paul Sigler, reveals that it forms a porous thick-walled hollow cylinder with an inner diameter of ~45 Å. The central channel forms two chambers in which partially folded proteins fold to their native structures. A constriction in the center prevents a folding protein from passing between the two GroEL rings.

The 97-residue GroES subunits form a domelike heptameric ring with C_7 symmetry (Fig. 6-44) in which the inner surface of the GroES dome is lined with hydrophilic residues. The X-ray structure of a GroEL–GroES–$(ADP)_7$ complex, also determined by Horwich and Sigler, indicates that GroES closes over one GroEL ring like a lid on a pot to form a bullet-shaped structure with C_7 symmetry (Fig. 6-45). The GroEL ring that contacts the GroES heptamer is called the cis ring; the opposing GroEL ring is known as the trans ring.

ATP Binding and Hydrolysis Coordinate the Conformational Changes in GroEL/ES. Each GroEL subunit has a binding pocket for ATP. A conformational change activates GroEL's ATPase activity by completely enclosing the ATP with protein (Fig. 6-45*c*) while shifting a catalytically essential Asp side chain into a productive position. In the structure shown in Fig. 6-45, the cis ring has hydrolyzed its seven molecules of ATP to ADP and has undergone

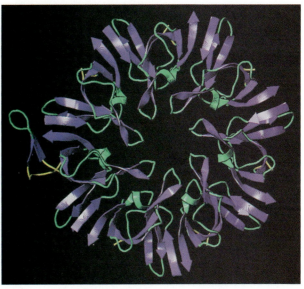

Figure 6-44   X-ray structure of GroES as viewed along its 7-fold axis.   A mobile loop in only one of the seven identical subunits (*left*) is visible in this structure. The polypeptide segments that flank the mobile loop are yellow. [Courtesy of Johann Diesenhofer, University of Texas Southwest Medical Center, Dallas, Texas.]

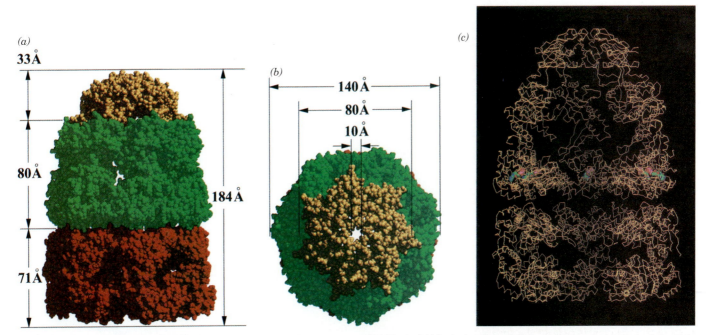

Figure 6-45   X-Ray structure of the GroEL–GroES–(ADP)₇ complex.
(*a*) A space-filling drawing as viewed perpendicularly to the complex's 7-fold axis with the GroES ring gold, the cis ring of GroEL green, and the trans ring of GroEL red. The dimensions of the complex are indicated. Note the different conformations of the two GroEL rings. (*b*) As in Part *a* but viewed along the 7-fold axis. (*c*) The C$_\alpha$ backbone of the complex viewed as in Part *a* but which is cut away along the plane containing the complex's 7-fold axis. The ADPs bound to the cis ring of GroEL are shown in space-filling form. Note the much larger size of the cavity formed by the cis ring and GroES in comparison to that of the trans ring. [Courtesy of Paul Sigler, Yale University. PDBid 1AON.]

conformational changes relative to the trans ring that widen and elongate the cis cavity in a way that more than doubles its volume (from 85,000 Å³ to 175,000 Å³). The enlarged cavity is able to enclose a partially folded substrate protein of at least 70 kD. *All seven subunits of the GroEL ring act in concert; that is, they are mechanically linked such that they can only change their conformations simultaneously.*

The cis and trans GroEL rings undergo conformational changes in a reciprocating fashion, with events in one ring influencing events in the other ring. The entire GroEL/ES chaperonin complex functions as follows (Fig. 6-46):

1. We begin with one GroEL ring binding 7 ATP and an improperly folded substrate protein, which associates with hydrophobic patches on the GroEL apical domains (labeled with an A in Fig. 6-46). The GroEL ring then binds a GroES cap to become the cis ring. GroES binding induces a conformational change in the cis ring that moves the hydrophobic patches to an interior position in GroEL, thereby depriving the substrate protein of its binding sites. This releases the substrate protein into the now enlarged and closed cavity, where the substrate protein commences folding. The cavity, which is now lined only with hydrophilic groups, provides the substrate protein with an

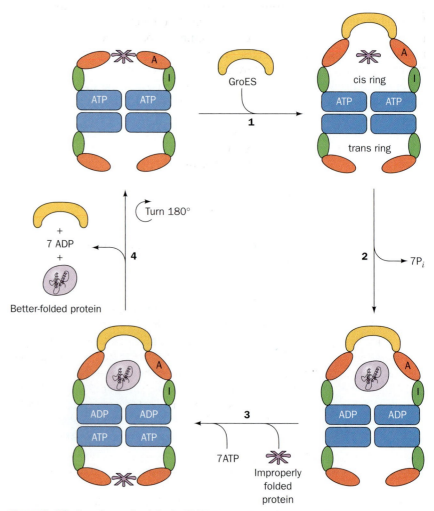

Figure 6-46 Reaction cycle of the GroEL/ES chaperonin. See the text for an explanation.

isolated microenvironment that prevents it from nonspecifically aggregating with other misfolded proteins.

2. Within ~13 s (the time the substrate protein has to fold), the cis ring catalyzes the hydrolysis of its 7 bound ATPs to ADP + P_i and the P_i is released. The absence of ATP's γ phosphate group weakens the interactions that bind GroES to GroEL.

3. A second molecule of substrate protein binds to the trans ring followed by 7 ATP. Conformational linkages between the cis and trans rings prevent the binding of both substrate protein and ATP to the trans ring until the ATP in the cis ring has been hydrolyzed.

4. The binding of substrate protein and ATP to the trans ring conformationally induces the cis ring to release its bound GroES, 7 ADP, and the presumably now better folded substrate protein. This leaves ATP and substrate protein bound only to the trans ring of GroEL, which now becomes the cis ring as it binds GroES.

Steps 1 through 4 are then repeated. The GroEL/ES system expends 7 ATPs per folding cycle. If the released substrate protein has not achieved its native state, it may subsequently rebind to GroEL (a substrate protein that has achieved its native fold lacks exposed hydrophobic groups and hence cannot rebind to GroEL). It requires an average of 24 folding cycles for a protein to attain its native state, which necessitates the hydrolysis of 168 ATPs (which appears to be a profligate use of ATP but constitutes only a small fraction of the thousands of ATPs that must be hydrolyzed to synthesize a typical polypeptide and its component amino acids). Because protein folding occurs alternately in the two GroEL rings, the proper functioning of the chaperonin requires both GroEL rings, even though their two cavities are unconnected.

In addition to forming a protective cage around a folding protein, the chaperonin apparently promotes the refolding of an improperly folded protein by an **iterative annealing** process. In this model, a misfolded protein binds to the hydrophobic patches on two or more of the seven GroEL subunits. The binding of ATP and GroES then triggers the conformational changes that mask these hydrophobic patches, a process that stretches and thereby partially unfolds the bound protein before it is released. This rescues the protein from a local energy minimum in which it had become trapped (Fig. 6-41) and thereby permits it to continue its conformational journey down the folding funnel toward its native state (the state of lowest free energy).

What Kinds of Proteins Are Folded by Chaperonins? *In vivo,* the GroEL/ES system interacts with only a subset of *E. coli* proteins. To identify these proteins, Ulrich Hartl supplied *E. coli* cells with [^{35}S]methionine (^{35}S is a radioactive isotope of S) for the 15 s it takes to synthesize an average-length polypeptide and then added an excess of unlabeled methionine (a **pulse–chase** experiment) for varying lengths of time. The cells were then lysed, the GroEL–GroES–substrate complexes were immunoprecipitated with anti-GroEL antibodics, and the GroEL-bound proteins were separated by electrophoresis. Of the ~2500 cytosolic proteins that could be detected, only ~300 proteins were reproducibly observed to be associated with GroEL. These proteins were isolated and 52 of them were unequivocally identified via the mass spectrometry of their tryptic fragments (see Section 5-3D).

Nearly all of the GroEL substrate proteins that were identified are enzymes that participate in a wide variety of metabolic functions or in tran-

scription or translation. Their molecular masses are mostly in the range 20 to 60 kD. Analysis of these proteins revealed that they tend to contain two or more αβ domains that mainly consist of open β sheets. Such a protein is expected to fold only slowly to its native state because the formation of its hydrophobic β sheets requires the assembly of a large number of specific long-range interactions in their proper orientations. Moreover, such a protein may easily misfold or become kinetically trapped due to the improper packing of its helices and sheets in one domain or, more likely, between domains.

Most of the GroEL-associated proteins dissociated from the chaperonin within a few minutes, after having achieved their native folds. However, ~100 proteins remained partially associated with GroEL, even after 2 hours. Evidently, these proteins repeatedly return to GroEL for conformational maintenance, which strongly suggests that they are structurally labile and/or prone to aggregation.

D Diseases Caused by Protein Misfolding

Most proteins in the body maintain their native conformations or, if they become partially denatured, are either renatured through the auspices of molecular chaperones or are proteolytically degraded (Section 20-1). However, at least 18 different—and usually fatal—human diseases are associated with the extracellular deposition of normally soluble proteins in certain tissues in the form of insoluble fibrous aggregates known as **amyloid** (this term means starchlike; it was originally thought that this material resembled starch). These diseases include **Alzheimer's disease,** the **transmissible spongiform encephalopathies (TSEs),** and the **amyloidoses.** The deposition of amyloid interferes with normal cellular function, resulting in cell death and eventual organ failure.

Alzheimer's Disease. Alzheimer's disease, a neurodegenerative condition that strikes mainly the elderly, causes devastating mental deterioration and eventual death (it affects ~10% of those over 65 and ~50% of those over 85). It is characterized by brain tissue containing abundant amyloid **plaques** (deposits) surrounded by dead and dying neurons (Fig. 6-47). The amyloid plaques consist mainly of fibrils of a 40- to 42-residue protein named **amyloid-β protein (Aβ).** Aβ is a fragment of a 770-residue membrane protein called the **Aβ precursor protein (βPP),** whose normal function is un-

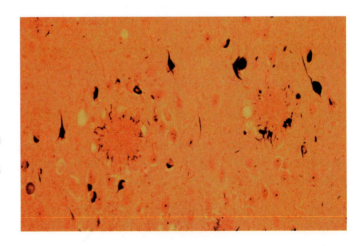

Figure 6-47 Photomicrograph of brain tissue from an individual with Alzheimer's disease. The two circular objects are plaques that consist of amyloid deposits of Aβ protein surrounded by a halo of neurites (axons and dendrites) from dead and dying neurons. [Courtesy of Dennis Selkoe and Marcia Podlisny, Harvard University Medical School.]

known. Aβ is excised from βPP in a multistep process through the actions of two proteolytic enzymes dubbed **β- and γ-secretases.**

It had been hotly debated whether Aβ causes Alzheimer's disease or is merely a product of its neurodegenerative processes. This argument was largely put to rest by the observation that microinjecting 200 pg of fibrillar but not soluble Aβ (the amount of Aβ in an Aβ plaque) into the cerebral cortexes of aged but not young monkeys causes marked neuronal loss and other microscopic changes characteristic of Alzheimer's disease as far as 1.5 mm from the injection site. Evidently, *the Aβ fibrils are neurotoxic even before their deposition in amyloid plaques.*

The age dependence of Alzheimer's disease suggests that Aβ deposition is an ongoing process. Indeed, there are several rare variants of the βPP gene with mutations in their Aβ regions that result in the onset of Alzheimer's disease as early as the fourth decade of life. These mutations affect the proteolytic processing of βPP in a way that increases the rate of Aβ production. A similar phenomenon is seen in individuals with **Down's syndrome** [a condition characterized by mental retardation and a distinctive physical appearance caused by the trisomy (3 copies per cell) of chromosome 21 rather than the normal two copies], who invariably develop Alzheimer's disease by their 40th year. This is because the gene encoding βPP is located on chromosome 21 and hence individuals with Down's syndrome produce βPP and presumably Aβ at an accelerated rate. Consequently, a promising strategy for halting the progression of Alzheimer's disease is to develop drugs that inhibit the action of the β- and/or γ-secretases so as to decrease the rate of Aβ production.

Prion Diseases.   Certain infectious diseases that affect the mammalian central nervous system were originally thought to be caused by "slow viruses" because they take months, years, or even decades to develop. Among them are **scrapie** (a neurological disorder of sheep and goats), **bovine spongiform encephalopathy (BSE** or **mad cow disease),** and **kuru** (a degenerative brain disease in humans that was transmitted by ritual cannibalism among the Fore people of Papua New Guinea; *kuru* means "trembling"). There is also a sporadic (spontaneously arising) human disease with similar symptoms, **Creutzfeldt–Jakob disease (CJD),** which strikes one person per million per year and which may be identical to kuru. In all of these invariably fatal diseases, neurons develop large vacuoles that give brain tissue a spongelike microscopic appearance. Hence the diseases are collectively known as **transmissible spongiform encephalopathies (TSEs).**

Unlike other infectious diseases, *the TSEs are not caused by a virus or microorganism.* Indeed, extensive investigations have failed to show that they are associated with any nucleic acid. Instead, as Stanley Pruisner demonstrated for scrapie, the infectious agent is a protein called a **prion** (for *pro*teinaceous *in*fectious particle that lacks nucleic acid) and hence TSEs are alternatively called **prion diseases.** The scrapie prion, which is named **PrP** (for *Pr*ion *P*rotein), consists of 208 mostly hydrophobic residues. This hydrophobicity causes partially proteolyzed PrP to aggregate as clusters of rodlike particles that closely resemble the amyloid fibrils seen on electron microscopic examination of prion-infected brain tissue (Fig. 6-48). These fibrils presumably form the amyloid plaques that appear to be directly responsible for the neuronal degeneration in TSEs.

How are prion diseases transmitted? PrP is the product of a normal cellular gene that has no known function (mice that mutagenically fail to ex-

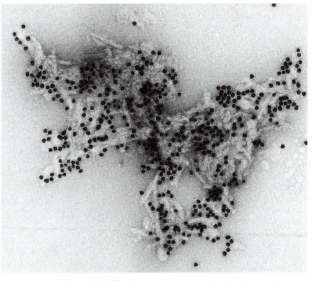

Figure 6-48 Electron micrograph of a cluster of **partially proteolyzed prion rods.**   The black dots are colloidal gold beads that are coupled to anti-PrP antibodies adhering to the PrP. [Courtesy of Stanley Pruisner, University of California at San Francisco Medical Center.]

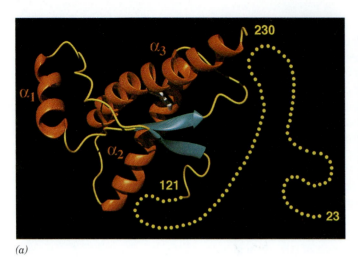

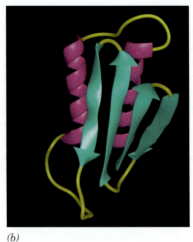

(a)   (b)

Figure 6-49 **Prion protein conformations.** (a) The NMR structure of human prion protein (PrP^C). Its flexibly disordered N-terminal "tail" (residues 23–121) is represented by yellow dots (the protein's N-terminal 23 residues have been posttranslationally excised). (b) A plausible model for the structure of PrP^{Sc}. [Part a courtesy of Kurt Wüthrich, Eidgenössische Technische Hochschule, Zurich, Switzerland. Part b courtesy of Fred Cohen University of California at San Francisco.]

press PrP appear to be normal and have apparently normal progeny). Infection of cells by prions somehow alters the PrP protein (the transcription of the PrP gene itself is not altered). Indeed, various methods have demonstrated that the scrapie form of PrP (PrP^{Sc}) is identical to normal cellular PrP (PrP^C) in sequence and chemical structure but that they differ in their secondary and/or tertiary structures. This suggests that *PrP^{Sc} induces PrP^C to adopt the conformation of PrP^{Sc},* that is, PrP^{Sc} formation is **autocatalytic** (the initially formed PrP^{Sc} induces the formation of additional PrP^{Sc} from PrP^C, etc). This accounts for the observation that mice that do not express the gene encoding PrP cannot be infected with scrapie (which can be transmitted to normal mice by the intracerebral inoculation of PrP^{Sc}).

Human PrP^C consists of a disordered (and hence unseen) 99-residue N-terminal "tail" and a 110-residue C-terminal globular domain containing three α helices and a short two-stranded antiparallel β sheet (Fig. 6-49a). Unfortunately, the insolubility of PrP^{Sc} has precluded its structural determination, but spectroscopic methods indicate that it has a lower α helix content and a higher β sheet content. This suggests that the protein has refolded (Fig. 6-49b). The high β sheet content of PrP^{Sc} presumably facilitates the aggregation of PrP^{Sc} as amyloid fibrils (see below).

Prion diseases can be transmitted by the consumption of nerve tissue from infected individuals, as was first seen in the case of kuru. This has become particularly evident in the case of BSE, which was unknown before 1985 but reached epidemic proportions among cattle in the U.K. in 1993. This is because the process for preparing meat-and-bone meal, which was routinely fed to cattle, was changed in the 1970s from one that fully inactivated prions to one that fails to do so. The BSE epidemic rapidly abated after 1993 due to the banning, in 1988, of the feeding of ruminants with ruminant-derived products other than milk, together with the slaughter of a large number of cattle at risk for having BSE. However, it is now clear that BSE was transmitted to humans who ate meat from BSE-infected cattle: Some 130 cases of so-called **new variant CJD (nvCJD)** have been reported to date, almost entirely in the U.K., many of which occurred in teenagers and young adults. Yet before 1994, CJD under the age of 40 was

extremely rare (its average age of onset is ~64). It should be noted that the transmission of BSE from cattle to humans was unexpected: Scrapie-infected sheep have long been consumed worldwide and yet the incidence of CJD in mainly meat-eating countries such as the U.K. (in which sheep are particularly abundant) was no greater than that in largely vegetarian countries such as India.

Amyloidoses.   Many amyloidogenic proteins are mutant forms of normally occurring proteins. These include **lysozyme** (an enzyme that hydrolyzes bacterial cell walls; Section 11-4) in the disease **familial visceral amyloidosis, transthyretin** (a blood plasma protein that functions as a carrier for water-insoluble hormones including the thyroid hormone **thyroxine;** Fig. 4-16) in **familial amyloid polyneuropathy,** and **fibrinogen** (the precursor of **fibrin,** which forms blood clots; Box 11-4) in **hereditary renal amyloidosis.** Most such diseases do not present (become symptomatic) until the third to seventh decades of life and typically progress over 5 to 15 years, ending in death. In addition, three dominantly inherited neurodegenerative diseases have been traced to mutations in the gene encoding PrP: **familial CJD, Gerstmann–Stråussler–Scheinker syndrome,** and **fatal familial insomnia.** These mutations presumably increase the rate that PrP^C converts to PrP^{Sc}.

Amyloid Fibrils Are β Sheet Structures.   The amyloid fibers that characterize the amyloidoses, Alzheimer's disease, and the TSEs are built from proteins that exhibit no structural or functional similarities in their native states. In contrast, the appearance of their fibrillar forms is strikingly similar. Furthermore, the ability to form amyloid fibrils is not unique to the small set of proteins associated with specific diseases. Under the appropriate conditions, almost any protein can be induced to aggregate. Thus, *the ability to form amyloid may be an intrinsic property of all polypeptide chains.*

Spectroscopic analysis of amyloid fibrils indicates that they are rich in β structure, with individual β strands oriented perpendicular to the fiber axis (Fig. 6-50). Even myoglobin, a globular protein consisting almost entirely of α helices (Fig. 6-38) can be coaxed into a fibrillar form by a pH of 9 and a temperature of 65°C. Results such as these raise the question of how a given amino acid sequence can adopt two very different but well-ordered states. The native structure of myoglobin is defined in large part by interactions between side chains; in the amyloid form, main chain interactions may dominate (side chain interactions are less important in stabilizing β sheets).

A variety of experiments indicate that amyloidogenic mutant proteins are significantly less stable than their wild-type counterparts (e.g., they have significantly lower melting tempera-

Figure 6-50   A model, based on X-ray fiber diffraction measurements, of an amyloid fibril.   (*a*) The model is viewed normal to the fibril axis (*above*) and along the fibril axis (*below*). The arrowheads indicate the path but not necessarily the direction of the β strands. (*b*) An isolated β sheet, which is shown for clarity. The loop regions connecting the β strands have unknown structure. [Courtesy of Colin Blake, Oxford University, Oxford, U.K. and Louise Serpell, University of Cambridge, U.K.]   *(a)*   *(b)*

tures). This suggests that the partially unfolded, aggregation-prone forms are in dynamic equilibrium with the native conformation, even under conditions in which the native state is thermodynamically stable [keep in mind that the equilibrium ratio of unfolded (U) to native (N) protein molecules in the reaction $N \rightleftharpoons U$ is governed by Eq. 1-17: $[U]/[N] = e^{-\Delta G^{\circ\prime}/RT}$, where $\Delta G^{\circ\prime}$ is the standard free energy of unfolding, so that as $\Delta G^{\circ\prime}$ decreases, the equilibrium proportion of U increases]. It is therefore likely that fibrillogenesis is initiated by the association of the β domains of two or more partially unfolded amyloidogenic proteins to form a more extensive β sheet. This would provide a template or nucleus for the recruitment of additional polypeptide chains to form the growing fibril. However, the several decades that most amyloid diseases require to become symptomatic suggests that the spontaneous generation of an amyloid nucleus is a rare event, that is, it has a high free energy of activation (activation barriers and their relationship to reaction rates are discussed in Section 12-1B).

Once it has formed, an amyloid fibril is virtually indestructible under physiological conditions, possibly due to the large number of hydrogen bonds that must be broken in order to separate the individual polypeptide strands. It seems likely that protein folding pathways have evolved not only to allow polypeptides to assume stable native structures but also to avoid forming interchain hydrogen bonds that would lead to fibril formation. The factors that trigger amyloid formation remain obscure, even when mutations (in the case of hereditary amyloidoses) or infection (in the case of TSEs) appear to be the cause. One possibility is that the amyloid diseases result in part from malfunctions in the mechanisms that govern protein folding or the disposal of misfolded proteins.

6 Structural Bioinformatics

The data obtained by X-ray crystallography, NMR spectroscopy, and certain other techniques take the form of three-dimensional coordinates describing the spatial positions of atoms in molecules. This kind of information can be easily stored, displayed, and compared, much like sequence information obtained by nucleotide or protein sequencing methods (see Sections 3-4 and 5-3). **Bioinformatics** is the rapidly growing discipline that deals with the burgeoning amount of information related to molecular sequences and structures. **Structural bioinformatics** is a branch of bioinformatics that is concerned with how macromolecular structures are displayed and compared. Some of the databases and analytical tools that are used in structural bioinformatics are described here.

The Protein Data Bank. The atomic coordinates of nearly all known macromolecular structures are archived in the **Protein Data Bank (PDB).** Indeed, most scientific journals that publish macromolecular structures require that authors deposit their structure's coordinates in the PDB. The PDB contains the coordinates of nearly 30,000 macromolecular structures (proteins, nucleic acids, and carbohydrates as determined by X-ray and other diffraction-based techniques, NMR, and theoretical modeling) and is growing exponentially. The PDB's Web address, from which these coordinates are publicly available, is listed in Table 6-3.

Each independently determined structure in the PDB is assigned a unique four-character identifier (its PDBid). For example, the PDBid for the structure of sperm whale myoglobin is 1MBO. A coordinate file begins with information that identifies the macromolecule, its source (the organ-

Table 6-3 Structural Bioinformatics Internet Addresses

Structural Databases

Protein Data Bank (PDB): http://www.rcsb.org/pdb/
Nucleic Acid Databank: http://ndbserver.rutgers.edu/
Molecular Modeling Database (MMDB): http://www.ncbi.nlm.nih.gov/Structure/index.shtml
Most Representative NMR Structure in an Ensemble: http://pqs.ebi.ac.uk/pqs-nmr.html
PQS Protein Quaternary Structure Query Form at the EBI: http://pqs.ebi.ac.uk/

Molecular Graphics Programs/Plug-Ins

Chime: http://mdli.com/products/framework/chemscape/
Cn3D: http://www.ncbi.nlm.nih.gov/Structure/CN3D/cn3d.shtml
Mage: http://kinemage.biochem.duke.edu/
Protein Explorer: http://www.umass.edu/microbio/chime/explorer/index.htm
RasMol: htttp://www.bernstein-plus-sons.com/software/rasmol/ *and* http://www.umass.edu/microbio/rasmol/
Swiss-PDB Viewer (Deep View): http://us.expasy.org/spdbv/

Structural Classification Algorithms

CATH (*C*lass, *A*rchitecture, *T*opology, and *H*omologous superfamily): http://www.biochem.ucl.ac.uk/bsm/cath/
CE (*C*ombinatorial *E*xtension of optimal pathway): http://cl.sdsc.edu/
FSSP (*F*old classification based on *S*tructure-*S*tructure alignment of *P*roteins): http://www2.ebi.ac.uk/dali/fssp/
SCOP (*S*tructural *C*lassification *O*f *P*roteins): http://scop.mrc-lmb.cam.ac.uk/scop/
VAST (*V*ector *A*lignment *S*earch *T*ool): http://www.ncbi.nlm.nih.gov/Structure/VAST/vast.shtml

ism from which it was obtained), the author(s) who determined the structure, and key journal references. The file continues with a synopsis of how the structure was determined together with indicators of its accuracy. The sequences of the structure's various chains are then listed together with the descriptions and formulas of its so-called HET (for heterogen) groups, which are molecular entities that are not among the "standard" amino acid or nucleotide residues (for example, organic molecules, nonstandard residues such as Hyp, metal ions, and bound water molecules). The positions of the structure's secondary structural elements and its disulfide bonds are then provided.

The bulk of a PDB file consists of a series of lines, each of which provides the coordinates of one ATOM (for a "standard" residue) or HETATM (for a heterogen) in the structure. Each ATOM or HETATM is identified by a serial number and an atom name (for example, C and O for an amino acid residue's carbonyl C and O atoms, CA and CB for C_α and C_β atoms, N1 for atom N1 of a nucleic acid base, C4* for atom C4′ of a ribose or deoxyribose residue). The record also includes the name of the residue (for example, PHE for phenylalanine, G for guanosine, and HEM for a heme group) and a letter to identify the chain to which it belongs (for structures that have more than one chain). The record then continues with the atom's three-dimensional (x, y, z) coordinates in angstroms. For NMR-based structures, the PDB file contains a full set of ATOM and HETATM records for each member of the ensemble of structures that were calculated in solving the structure (the most representative member of such a coordinate set can be obtained from http://pqs.ebi.ac.uk/pqs-nmr.html). PDB files usually end with CONECT (connectivity) records, which denote the nonstandard connectivities between atoms such as disulfide bonds and hydrogen bonds.

A particular PDB file may be located according to its PDBid or, if this is unknown, by searching for a protein's name, its source, the author(s), or

the experimental technique used to determine the structure. Selecting a particular macromolecule in the PDB initially displays a summary page with options for viewing the structure (either statically or interactively), for viewing or downloading the coordinate file, and for classifying or analyzing the structure in terms of its geometric properties and sequence.

The Nucleic Acid Database.   The **Nucleic Acid Database (NDB)** archives the atomic coordinates of structures that contain nucleic acids. Its coordinate files have substantially the same format as PDB files. In fact, the same information is included in the PDB, but the NDB's organization and search algorithms are specialized for dealing with nucleic acids. This is useful because many nucleic acids of known structure are identified only by their sequences—rather than by names, as proteins are—and consequently could easily be overlooked in a search of the PDB.

Viewing Macromolecular Structures in Three Dimensions.   The most informative way to examine a macromolecular structure is through the use of molecular graphics programs. Such programs permit the user to interactively rotate a macromolecule and thereby perceive its three-dimensional structure. This impression may be further enhanced by simultaneously viewing the macromolecule in stereo. Nearly all molecular graphics programs use PDB files as input. The programs described here can be downloaded from the Web addresses listed in Table 6-3, some of which also provide instructions for the program's use.

RasMol, a widely used molecular graphics program written by Roger Sayle, is publicly available for use on a variety of computer platforms (Windows, MacOS, and UNIX). Its Web browser–based counterpart (plug-in) is named **Chime.** RasMol and Chime allow the user to simultaneously display different user-selected portions of a macromolecule in a variety of colors and formats (e.g., wire frame, ball and stick, backbone, space-filling, and cartoons). The Interactive Exercises on the CD-ROM that accompanies this textbook all use Chime. Moreover, the PDB provides the facility for viewing user-selected structures over the Web using several types of viewers including the Chime-based program **Protein Explorer** by Eric Martz. Another molecular graphics program, **Mage,** which was written by David Richardson, displays so-called **Kinemages** on this textbook's accompanying CD-ROM. Mage provides a generally more author-directed user environment than does RasMol or Chime.

The **Swiss-Pdb Viewer** (also called **Deep View**), in addition to displaying molecular structure, provides tools for basic model building, homology modeling, energy minimization, and multiple-sequence alignment. One advantage of the Swiss-Pdb Viewer is that it allows users to easily superimpose two or more models or parts of models. It too is publicly available and works on all major computer platforms.

Structural Classification and Comparison.   Most proteins are structurally related to other proteins. Indeed, *evolution tends to conserve the structures of proteins rather than their sequences* (see Section 6-2D). The computational tools described below facilitate the classification and comparison of protein structures. These programs can be accessed directly via their Web addresses or through the PDB. Studies using these programs yield functional insights, reveal distant evolutionary relationships that are not apparent from sequence comparisons, generate libraries of unique folds for structure prediction, and provide indications as to why certain types of structures are preferred over others.

1. **CATH** (for *C*lass, *A*rchitecture, *T*opology, and *H*omologous superfamily), as its name suggests, categorizes proteins in a four-level structural hierarchy. (1) "Class," the highest level, places the selected protein in one of four categories of gross secondary structure: Mainly α, Mainly β, α/β (having both α helices and β sheets), and Few Secondary Structures. (2) "Architecture" is the description of the gross arrangement of secondary structure independent of topology. (3) "Topology" is indicative of both the overall shape and connectivity of the protein's secondary structures. (4) "Homologous superfamily" is those proteins of known structure that are homologous (share a common ancestor) to the selected protein. A static or an interactive (Chime/RasMol or VRML) drawing of each of these proteins can be displayed.

2. **CE** (for *C*ombinatorial *E*xtension of the optimal path) finds all proteins in the PDB that can be structurally aligned with the query structure to within user-specified geometric criteria. The amino acid sequences of any or all of these proteins can be aligned on the basis of this structural alignment rather than sequence alignment. CE can likewise optimally align and display two user-selected structures.

3. **FSSP** (*F*old classification based on *S*tructure–*S*tructure alignment of *P*roteins) lists the protein structures in the PDB that, at least in part, structurally resemble the query protein based on continuously updated all-against-all comparisons of the protein structures in the PDB. These structural comparisons are made by a program called **Dali** based on the distances between the various atoms in each domain of a protein.

4. **SCOP** (*S*tructural *C*lassification *O*f *P*roteins) classifies protein structures based mainly on manually generated topological considerations according to a six-level hierarchy: Class [all-α, all-β, α/β (having α helices and β strands that are largely interspersed), α + β (having α helices and β strands that are largely segregated), and multidomain (having domains of different classes)], Fold (groups that have similar arrangements of secondary structural elements), Superfamily (indicative of distant evolutionary relationships based on structural criteria and functional features), Family (indicative of near evolutionary relationships based on sequence as well as on structure), Protein, and Species. SCOP permits the user to navigate through its treelike hierarchical organization and lists the known members of any particular branch.

5. **VAST** (*V*ector *A*lignment *S*earch *T*ool), a component of the National Center for Biotechnology Information (NCBI) Entrez system, reports a precomputed list of proteins of known structure that structurally resemble the query protein ("structure neighbors"). The VAST system uses the **Molecular Modeling Database (MMDB),** an NCBI-generated database that is derived from PDB coordinates but in which molecules are represented by connectivity graphs rather than sets of atomic coordinates. VAST displays the superposition of the query protein in its structural alignment with up to five other proteins using **Cn3D** (a molecular graphics program that displays MMDB files and that is publicly available for a variety of computer platforms) or with only one other protein using Mage. VAST also reports a precomputed list of proteins that are similar to the query protein in sequence ("sequence neighbors") and provides links from a selected protein to several bibliographic databases including MedLine.

SUMMARY

1. Four levels of structural complexity are used to describe the three-dimensional shapes of proteins.

2. The conformational flexibility of the peptide group is described by its ϕ and ψ torsion angles.

3. The α helix is a regular secondary structure in which hydrogen bonds form between backbone groups four residues apart. In the β sheet, hydrogen bonds form between the backbones of separate polypeptide segments.

4. Fibrous proteins are characterized by a single type of secondary structure: α keratin is a left-handed coil of two α helices, and collagen is a left-handed triple helix with three residues per turn.

5. Nonrepetitive structures include variations in regular secondary structures, turns, and loops.

6. The structures of proteins can be determined by X-ray crystallography and NMR spectroscopy.

7. The nonpolar side chains of a globular protein tend to occupy the protein's interior, whereas the polar side chains tend to define its surface.

8. Protein structures, which often contain common supersecondary structures (motifs), can be grouped into families according to their folding patterns. Structural elements are more likely to be evolutionarily conserved than are amino acid sequences.

9. The individual subunits of multisubunit proteins are usually symmetrically arranged.

10. Native protein structures are only slightly more stable than their denatured forms. The hydrophobic effect is the primary determinant of protein stability. Hydrogen bonding and ion pairing contribute relatively little to a protein's stability.

11. Studies of protein denaturation and renaturation indicate that the primary structure of a protein determines its three-dimensional structure.

12. Proteins have conformational flexibility that results in small molecular motions.

13. Proteins fold to their native conformations via directed pathways in which small elements of structure coalesce into larger structures.

14. Protein disulfide isomerases and molecular chaperones facilitate protein folding *in vivo*.

15. Diseases caused by protein misfolding include Alzheimer's disease, the transmissable spongiform encephalopathies (TSEs), and the amyloidoses.

16. The field of structural bioinformatics is concerned with the storage, visualization, analysis, and comparison of macromolecular structures.

REFERENCES

General

Branden, C. and Tooze, J., *Introduction to Protein Structure* (2nd ed.), Garland Publishing (1999). [A well-illustrated book with chapters introducing amino acids and protein structure, plus chapters on specific proteins categorized by their structure and function.]

Lesk, A.M., *Introduction to Protein Architecture*, Oxford University Press (2001).

Milner-White, E.J., The partial charge of the nitrogen atom in peptide bonds, *Protein Science* **6,** 2477–2482 (1997). [Discusses the origin of the peptide N atom's partial negative charge.]

Tanford, C. and Reynolds, J., *Nature's Robots*, Oxford University Press (2001). [A history of proteins.]

Fibrous Proteins

Baum, J. and Brodsky, B., Folding of peptide models of collagen and misfolding in disease, *Curr. Opin. Struct. Biol.* **9,** 122–128 (1999).

Kramer, R.Z., Bella, J., Mayville, P., Brodsky, B., and Berman, H.M., Sequence dependent conformational variations of collagen triple-helical structure, *Nature Struct. Biol.* **6,** 454–457 (1999).

Macromolecular Structure Determination

McPherson, A., *Macromolecular Crystallography*, Wiley (2002).

McRee, D.E., *Practical Protein Crystallography* (2nd ed.), Elsevier Science (2002).

Rhodes, G., *Crystallography Made Crystal Clear: A Guide for Users of Macromolecular Models* (2nd ed.), Academic Press (2000). [Includes overviews, methods, and discussions of model quality.]

Wider, G. and Wüthrich, K., NMR spectroscopy of large molecules and multimolecular assemblies in solution, *Curr. Opin. Struct. Biol.* **9,** 594–601 (1999).

Quaternary Structure

Goodsell, D.S. and Olson, J., Structural symmetry and protein function, *Annu. Rev. Biophys. Biomol. Struct.* **29,** 105–153 (2000).

Sheinerman, F.B., Norel, R., and Honig, B., Electrostatic aspects of protein–protein interactions, *Curr. Opin. Struct. Biol.* **10,** 153–159 (2000).

Protein Stability

Fersht, A., *Structure and Mechanism in Protein Science,* Chapter 11, Freeman (1999).

Jaenicke, R. and Böhm, G., The stability of proteins in extreme environments, *Curr. Opin. Struct. Biol.* **8,** 738–748 (1998).

Jones, S. and Thornton, J.M., Principles of protein–protein interactions, *Proc. Natl. Acad. Sci.* **93,** 13–20 (1996).

Protein Folding

Baker, D. and Sali, A., Protein structure prediction and structural genomics, *Science* **294,** 93–96 (2001). [Summarizes the state of the art of protein structure prediction methods and how they can be applied.]

Baldwin, R.L. and Rose, G.D., Is protein folding hierarchic? I. Local structure and peptide folding; *and* II. Folding intermediates and transition states, *Trends Biochem. Sci* **24,** 26–33; *and* 77–83 (1999).

Dobson, C.M. and Karplus, M., The fundamentals of protein folding: bringing together theory and experiment, *Curr. Opin. Struct. Biol.* **9,** 92–101 (1999).

Hartl, F.U. and Hayer-Hartl, M., Molecular chaperones in the cytosol: from nascent chain to folded protein, *Nature* **295,** 1852–1858 (2002). [Discusses the importance of proper folding for newly synthesized proteins and reviews the major chaperone systems.]

Horwich, A.R. (Ed.), Protein folding in the cell, Adv. Prot. Chem. **59** (2002).

Rye, H.S., Roseman, A.M., Chen, S., Furtak, K., Fenton, W.A., Saibil, H.R., and Horwich, A.L., GroEL-GroES cycling: ATP and nonnative polypeptide direct alternation of folding-active rings, *Cell* **97,** 325–338 (1999).

Protein Misfolding Diseases

Bucciantini, M., Giannoni, E., Chiti, F., Baroni, F., Formigli, L., Zurdo, J., Taddei, N., Ramponi, G., Dobson, C.M., and Stefani, M., Inherent toxicity of aggregates implies a common mechanism for protein misfolding diseases, *Nature* **416,** 507–511 (2002). [Provides evidence that a variety of misfolded proteins can form fibrous aggregates that can potentially damage cells.]

Caughey, B., Interactions between prion protein isoforms: the kiss of death? *Trends Biochem. Sci.* **26,** 235–242 (2001).

Jackson, G.S. and Clarke, A.R., Mammalian prion proteins, *Curr. Opin. Struct. Biol.* **10,** 69–74 (2000).

Pruisner, S.B., Scott, M.R., DeArmond, S.J., and Cohen, F.E., Prion protein biology, *Cell* **93,** 337–348 (1998).

Structural Bioinformatics

Bourne, P.E. and Weissig, H. (Eds.), *Structural Bioinformatics,* Wiley-Liss (2003).

Hadley, C. and Jones, J.T., A systematic comparison of protein structure classifications: SCOP, CATH, and FSSP, *Structure* **7,** 1099–1112 (1999).

Orengo, C.A., Todd, A.E., and Thornton, J.M., From protein structure to function, *Curr. Opin. Struct. Biol.* **9,** 374–382 (1999). [Describes how functional information can be derived by examining families of structurally related proteins.]

KEY TERMS

secondary structure	topology	contour map	hydropathy
tertiary structure	fibrous protein	supersecondary structure	ion pair (salt bridge)
quaternary structure	globular protein	(motif)	zinc finger
peptide group	coiled coil	βαβ motif	breathing
trans conformation	denatured	β hairpin	cooperativity
cis conformation	native	αα motif	chaotropic agent
backbone	β bulge	β barrel	renaturation
torsion (dihedral) angle	helix capping	dinucleotide-binding	hydrophobic collapse
φ	reverse turn	(Rossmann) fold	molten globule
ψ	(β bend)	domain	molecular chaperone
Ramachandran diagram	Ω loop	oligomer	heat shock protein
α helix	X-ray crystallography	protomer	ATPase
pitch	NMR	rotational symmetry	amyloid
antiparallel β sheet	diffraction pattern	cyclic symmetry	prion
parallel β sheet	electron density	dihedral symmetry	structural bioinformatics

STUDY EXERCISES

1. Explain why the conformational freedom of peptide bonds is limited.

2. What distinguishes regular and irregular secondary structures?

3. Describe the hydrogen-bonding pattern of an α helix.

4. Why are β sheets pleated?

5. What properties do fibrous proteins confer on substances such as hair, horns, bones, and tendons?

6. Why do turns and loops most often occur on the protein surface?

7. Which side chains usually occur in a protein's interior? On its surface?

8. Give some reasons why the number of possible protein structures is much less than the number of amino acid sequences.

9. List the advantages of multiple subunits in proteins.

10. Why can't proteins have mirror symmetry?

11. Describe the forces that stabilize proteins.

12. Describe the energy and entropy changes that occur during protein folding.

13. How does protein renaturation *in vitro* differ from protein folding *in vivo?*

14. What are amyloid fibrils, what is their origin, and why are they harmful?

15. What is structural bioinformatics?

PROBLEMS

1. Draw a cis peptide bond and identify the groups that experience steric interference.

2. Helices can be described by the notation n_m, where n is the number of residues per helical turn and m is the number of atoms, including H, in the ring that is closed by the hydrogen bond. (a) What is this notation for the α helix? (b) Is the 3_{10} helix steeper or shallower than the α helix?

3. Calculate the length in angstroms of a 100-residue segment of the α keratin coiled coil.

4. Hydrophobic residues usually appear at the first and fourth positions in the seven-residue repeats of polypeptides that form coiled coils. (a) Why do polar or charged residues usually appear in the remaining five positions? (b)Why is the sequence Ile–Gln–Glu–Val–Glu–Arg–Asp more likely than the sequence Trp–Gln–Glu–Tyr–Glu–Arg–Asp to appear in a coiled coil?

5. Globular proteins are typically constructed from several layers of secondary structure, with a hydrophobic core and a hydrophilic surface. Is this true for a fibrous protein such as α keratin?

6. The digestive tract of the larvae of clothes moths is a strongly reducing environment. Why is this beneficial to the larvae?

7. Describe the primary, secondary, tertiary, and quaternary structures of collagen.

8. Explain why gelatin, which is mostly collagen, is nutritionally inferior to other types of protein.

9. Is it possible for a native protein to be entirely irregular, that is, without α helices, β sheets, or other repetitive secondary structure?

10. (a) Is Trp or Gln more likely to be on a protein's surface? (b) Is Ser or Val less likely to be in the protein's interior? (c) Is Leu or Ile less likely to be found in the middle of an α helix? (d) Is Cys or Ser more likely to be in a β sheet?

11. What types of rotational symmetry are possible for a protein with (a) four or (b) six identical subunits?

12. You are performing site-directed mutagenesis to test predictions about which residues are essential for a protein's function. Which of each pair of amino acid substitutions listed below would you expect to disrupt protein structure the most?

Explain.

(a) Val replaced by Ala or Phe.

(b) Lys replaced by Asp or Arg.

(c) Gln replaced by Glu or Asn.

(d) Pro replaced by His or Gly.

13. Laboratory techniques for randomly linking together amino acids typically generate an insoluble polypeptide, yet a naturally occurring polypeptide of the same length is usually soluble. Explain.

14. Given enough time, can all denatured proteins spontaneously renature?

15. Describe the intra- and intermolecular bonds/interactions that are broken or retained when collagen is heated to produce gelatin.

16. Under physiological conditions, polylysine assumes a random coil conformation. Under what conditions might it form an α helix?

17. It is often stated that proteins are quite large compared to the molecules they bind. However, what constitutes a large number depends on your point of view. Calculate the ratio of the volume of a hemoglobin molecule (65 kD) to that of the four O_2 molecules that it binds and the ratio of the volume of a typical office (4 × 4 × 3 m) to that of the typical (70-kg) office worker that occupies it. Assume that the molecular volumes of hemoglobin and O_2 are in equal proportions to their molecular masses and that the office worker has a density of 1.0 g/cm³. Compare these ratios. Is this the result you expected?

18. Which of the following polypeptides is most likely to form an α helix? Which is least likely to form a β strand?

(a) CRAGNRKIVLETY

(b) SEDNFGAPKSILW

(c) QKASVEMAVRNSG

[Problem by Bruce Wightman, Muhlenberg College.]

19. The X-ray crystallographic analysis of a protein often fails to reveal the positions of the first few and/or the last few residues of a polypeptide chain. Explain.

BIOINFORMATICS EXERCISES

Bioinformatics Exercises are available for this chapter in the tabbed section at the end of the book and on the text website at www.wiley.com/college/voet.

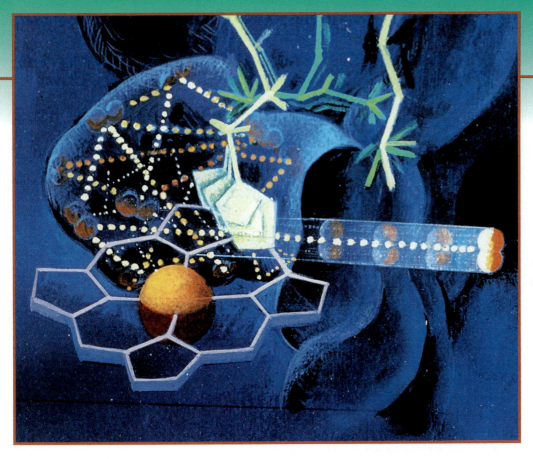

The structure of a protein determines its biological role, whether it is binding a small molecule or interacting with another large molecule. The oxygen binding site of myoglobin, for example, is structured so that O_2 can readily bind to it or, as pictured here, escape from the protein altogether. [Illustration, Irving Geis. Image from the Irving Geis Collection/ Howard Hughes Medical Institute. Rights owned by HHMI. Reproduction by permission only.]

Protein Function: Myoglobin and Hemoglobin

The preceding two chapters have painted a broad picture of the chemical and physical properties of proteins but have not delved deeply into their physiological functions. Nevertheless, it should come as no surprise that the structural complexity and variety of proteins allow them to carry out an enormous array of specialized biological tasks. For example, the enzyme catalysts of virtually all metabolic reactions are proteins (we consider enzymes in detail in Chapters 11 and 12). Genetic information would remain locked in DNA were it not for the proteins that participate in decoding and transmitting that information. Remarkably, the thousands of proteins that participate in building, supporting, recognizing, transporting, and transforming cellular components act with incredible speed and accuracy and in many cases are subject to multiple regulatory mechanisms.

The specialized functions of proteins, from the fibrous proteins we examined in Section 6-1C to the precisely regulated metabolic enzymes we discuss in later chapters, can all be understood in terms of how proteins bind to and interact with other components of living systems. In this chapter, we focus on the oxygen-binding proteins myoglobin and hemoglobin. Their structures can be studied as examples of the principles outlined in the preceding chapters. Both myoglobin and hemoglobin are globular proteins whose primary, secondary, and tertiary structures (and quaternary structure, in the case of hemoglobin) are known in detail. Their functions are vitally important to human health. Most importantly, they serve as models illustrating various relationships between protein structure and function. Studies of myoglobin and hemoglobin have historically represented some of the most significant advances in biochemistry, and these proteins have maintained their position as valuable models for many of the proteins we shall examine when we discuss metabolism and the management of genetic information.

1 Myoglobin

We begin our study of protein function with **myoglobin,** the first protein whose structure was determined by X-ray crystallography. In addition to myoglobin's importance as an oxygen-binding protein, its structure and function provide insights into the structure and function of hemoglobin, which is a tetramer of myoglobin-like polypeptides.

A Myoglobin Structure

Myoglobin is a small intracellular protein in vertebrate muscle. Its X-ray structure, determined by John Kendrew in 1959, revealed that most of myoglobin's 153 residues are members of eight α helices (traditionally labeled A through H) that are arranged to form a globular protein with approximate dimensions $44 \times 44 \times 25$ Å (Fig. 7-1).

Myoglobin, other members of the globin family of proteins (Section 5-4B), and some other proteins such as cytochrome c (Sections 5-4A and 6-2D) all contain a single **heme** group (Fig. 7-2). The heme is tightly wedged in a hydrophobic pocket between the E and F helices in myoglobin. The heterocyclic ring system of heme is a **porphyrin** derivative containing four **pyrrole** groups (labeled A–D) linked by methene bridges (other porphyrins vary in the substituents attached to rings A–D). The Fe(II) atom at the center of heme is coordinated by four porphyrin N atoms and one N from a His side chain (called, in a nomenclature peculiar to myoglobin and hemoglobin, His F8 because it is the eighth residue of the F helix). A molecule

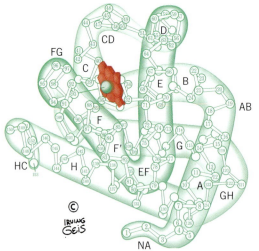

Figure 7-1 Structure of sperm whale myoglobin. This 153-residue monomeric protein consists of eight α helices, labeled A through H, that are connected by short polypeptide links (the last half of what was originally thought to be the EF corner has been shown to form a short helix that is designated the F′ helix). The heme group is shown in red. [Illustration, Irving Geis. Image from the Irving Geis Collection/Howard Hughes Medical Institute. Rights owned by HHMI. Reproduction by permission only.]
🐭 **See Kinemage Exercise 6-1.**

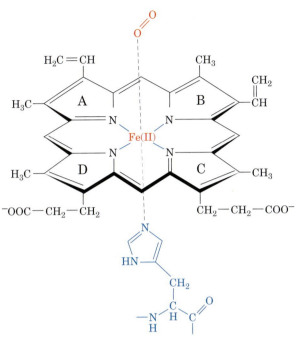

Figure 7-2 The heme group. The central Fe(II) atom is shown liganded to four N atoms of the porphyrin ring, whose pyrrole groups are labeled A–D. The heme is a conjugated system, so all the Fe—N bonds are equivalent. The Fe(II) is also liganded to a His side chain and, when it is present, to O_2. The six ligands are arranged at the corners of an octahedron centered on the Fe ion (octahedral geometry).

of oxygen (O_2) can act as a sixth ligand to the iron atom. His E7 (the seventh residuc of helix E) hydrogen bonds to the O_2 with the geometry shown in Fig. 7-3. Two hydrophobic side chains on the O_2-binding side of the heme, Val E11 and Phe CD1 (the first residue in the segment between helices C and D), help hold the heme in place. These side chains presumably swing aside as the protein "breathes" (Section 6-4B), allowing O_2 to enter and exit.

When exposed to oxygen, the Fe(II) atom of isolated heme is irreversibly oxidized to Fe(III), a form that cannot bind O_2. The protein portion of myoglobin (and of hemoglobin, which contains four heme groups in four globin chains) prevents this oxidation and makes it possible for O_2 to bind reversibly to the heme group. **Oxygenation** alters the electronic state of the Fe(II)–heme complex, as indicated by its color change from dark purple (the color of hemoglobin in venous blood) to brilliant scarlet (the color of hemoglobin in arterial blood). Under some conditions, the Fe(II) of myoglobin or hemoglobin becomes oxidized to Fe(III) to form **metmyoglobin** or **methemoglobin,** respectively; these proteins are responsible for the brown color of old meat and dried blood.

In addition to O_2, certain other small molecules such as CO, NO, and H_2S can bind to heme groups in proteins. These other compounds bind with much higher affinity than O_2, which accounts for their toxicity. CO, for example, has 200-fold greater affinity for hemoglobin than does O_2.

B Myoglobin Function

Although myoglobin was originally thought to be only an oxygen-storage protein, it is now apparent that *its major physiological role is to facilitate oxygen diffusion in muscle* (the most rapidly respiring tissue under condi-

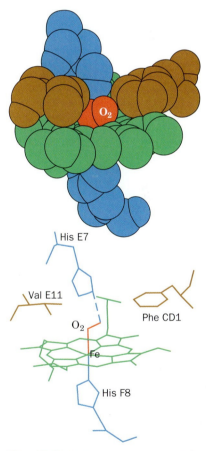

Figure 7-3 The heme complex in myoglobin. In the upper drawing, atoms are represented in space-filling form (H atoms are not shown). The lower drawing shows the corresponding skeletal model with a dashed line representing the hydrogen bond between His E7 and the bound O_2. 🖎 **See Kinemage Exercise 6-1.**

tions of high exertion). The rate at which O_2 can diffuse from the capillaries to the tissues is limited by its low solubility in aqueous solution ($\sim 10^{-4}$ M in blood). Myoglobin increases the effective solubility of O_2 in muscle cells, acting as a kind of molecular bucket brigade to boost the O_2 diffusion rate. The oxygen-storage function of myoglobin is probably significant only in aquatic mammals such as seals and whales, whose muscle myoglobin concentrations are around 10-fold greater than that in terrestrial mammals (which is one reason why Kendrew chose the sperm whale as a source of myoglobin for his X-ray crystallographic studies). Nevertheless, mice in which the gene for myoglobin has been "knocked out" are apparently normal, although their muscles are lighter in color than those of wild-type mice. This experiment suggests that myoglobin is not required by muscles under normal metabolic conditions. In contrast, a recently discovered myoglobin-like protein in the brain, dubbed **neuroglobin,** may be essential for boosting O_2 concentrations in neural tissues, which are metabolically highly active. For example, the brain constitutes only about 2% of the mass of a human body, but it consumes about 20% of the available oxygen.

Myoglobin's Oxygen-Binding Curve Is Hyperbolic.   The reversible binding of O_2 to myoglobin **(Mb)** is described by a simple equilibrium reaction:

$$Mb + O_2 \rightleftharpoons MbO_2$$

The dissociation constant, K, for the reaction is

$$K = \frac{[Mb][O_2]}{[MbO_2]} \tag{7-1}$$

Note that biochemists usually express equilibria in terms of dissociation constants, the reciprocal of the association constants favored by chemists. The O_2 dissociation of myoglobin can be characterized by its **fractional saturation, Y_{O_2},** which is defined as the fraction of O_2-binding sites occupied by O_2:

$$Y_{O_2} = \frac{[MbO_2]}{[Mb] + [MbO_2]} \tag{7-2}$$

Y_{O_2} ranges from zero (when no O_2 is bound to the myoglobin molecules) to one (when the binding sites of all the myoglobin molecules are occupied). Equation 7-1 can be rearranged to

$$[MbO_2] = \frac{[Mb][O_2]}{K} \tag{7-3}$$

When this expression for $[MbO_2]$ is substituted into Eq. 7-2, the fractional saturation becomes

$$Y_{O_2} = \frac{\dfrac{[Mb][O_2]}{K}}{[Mb] + \dfrac{[Mb][O_2]}{K}} \tag{7-4}$$

Factoring out the $[Mb]/K$ term in the numerator and denominator gives

$$Y_{O_2} = \frac{[O_2]}{K + [O_2]} \tag{7-5}$$

Since O_2 is a gas, its concentration is conveniently expressed by its **partial**

pressure, pO_2 (also called the oxygen tension). Equation 7-5 can therefore be expressed as

$$Y_{O_2} = \frac{pO_2}{K + pO_2} \qquad [7\text{-}6]$$

This equation describes a rectangular **hyperbola** *and is identical in form to the equations that describe a hormone binding to its cell-surface receptor or a small molecular substrate binding to the active site of an enzyme.* This hyperbolic function can be represented graphically as shown in Fig. 7-4. At low pO_2, very little O_2 binds to myoglobin (Y_{O_2} is very small). As the pO_2 increases, more O_2 binds to myoglobin. At very high pO_2, virtually all the O_2-binding sites are occupied and myoglobin is said to be **saturated** with O_2.

The steepness of the hyperbola for a simple binding event, such as O_2 binding to myoglobin, increases as the value of K decreases. This means that *the lower the value of K, the tighter is the binding.* K is equivalent to the molar concentration of ligand at which half of the binding sites are occupied. In other words, when $pO_2 = K$, myoglobin is half-saturated with oxygen. This can be shown algebraically by substituting pO_2 for K in Eq. 7-6:

$$Y_{O_2} = \frac{pO_2}{K + pO_2} = \frac{pO_2}{2\,pO_2} = 0.5 \qquad [7\text{-}7]$$

Thus, K can be operationally defined as the value of pO_2 at which $Y = 0.5$ (Fig. 7-4).

It is convenient to define K as p_{50}, that is, the oxygen pressure at which myoglobin is 50% saturated. The p_{50} for myoglobin is 2.8 torr (760 torr = 1 atm). Over the physiological range of pO_2 in the blood (100 torr in arterial blood and 30 torr in venous blood), myoglobin is almost fully saturated with oxygen; for example, $Y_{O_2} = 0.97$ at $pO_2 = 100$ torr and 0.91 at 30 torr. Consequently, *myoglobin efficiently relays oxygen from the capillaries to muscle cells.*

Myoglobin, a single polypeptide chain with one heme group and hence one oxygen-binding site, is a useful model for other binding proteins. Even proteins with multiple binding sites for the same small molecule, or **ligand,** generate hyperbolic binding curves like myoglobin's when the ligands interact with each binding site independently. In practice, the affinity of a ligand for its binding protein may not be known. Constructing a binding curve such as the one shown in Fig. 7-4 may provide this information.

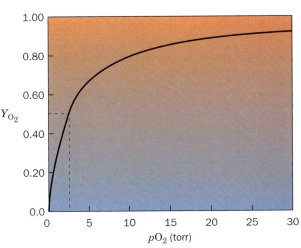

Figure 7-4 **Oxygen-binding curve of myoglobin.** Myoglobin is half-saturated with O_2 ($Y_{O_2} = 0.5$) at an oxygen pressure (pO_2) of 2.8 torr. The hyperbolic shape of myoglobin's binding curve is typical of the simple binding of a small molecule to a protein.

2 | Hemoglobin

Hemoglobin, the intracellular protein that gives red blood cells their color, is one of the best-characterized proteins and was one of the first proteins to be associated with a specific physiological function (oxygen transport). However, hemoglobin is not just a simple oxygen tank; it is a sophisticated delivery system that provides the proper amount of oxygen to the tissues under a wide variety of circumstances. Animals that are too large (>1 mm thick) for simple diffusion to deliver sufficient oxygen to their tissues have circulatory systems containing hemoglobin or a protein of similar function that does so (see Box 7-1).

The efficiency with which hemoglobin binds and releases O_2 is reminiscent of the specificity and efficiency of metabolic enzymes. Because many of the theories formulated to explain O_2 binding to hemoglobin also ex-

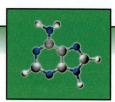

BOX 7-1

Perspectives in Biochemistry

Other Oxygen-Transport Proteins

The presence of O_2 in the earth's atmosphere and its use in the combustion of metabolic fuels have driven the evolution of various mechanisms for storing and transporting oxygen. Small organisms rely on diffusion to supply their respiratory oxygen needs. However, since the rate at which a substance diffuses varies inversely with the square of the distance it must diffuse, organisms of >1-mm thickness overcome the constraints of diffusion with circulatory systems and boost the limited solubility of O_2 in water with specific O_2-transport proteins.

Many invertebrates, and even some plants and bacteria, contain heme-based O_2-binding proteins. Single-subunit and multimeric hemoglobins are found both as intracellular proteins and as extracellular components of blood and other body fluids. The existence of hemoglobin-like proteins in some species of bacteria is evidence of gene transfer from animals to bacteria at one or more points during evolution. In bacteria, these proteins may function as sensors of environmental conditions such as local O_2 concentration. In some leguminous plants, the so-called **leghemoglobins** bind O_2 that would otherwise interfere with nitrogen fixation carried out by bacteria that colonize plant root nodules (Section 20-7). The **chlorocruorins,** which occur in some annelids (e.g., earthworms), contain a somewhat differently derivatized porphyrin than that in hemoglobin, which accounts for the green color of chlorocruorins.

The two other types of O_2-binding proteins, **hemerythrin** and **hemocyanin** (neither of which contains heme groups), occur only in invertebrate animals. Hemerythrin, which occurs in only a few species of marine invertabrates, is an intracellular protein with a subunit mass of ~13 kD. It contains two Fe atoms liganded by His and acidic residues. It is violet-pink when oxygenated and colorless when deoxygenated.

Hemocyanins, which are exclusively extracellular, transport O_2 in mollusks and arthropods. The molluscan and arthropod hemocyanins differ in molecular structure at all levels, although both types are large multimeric proteins. The arthropod hemocyanin subunit is ~75 kD, whereas the molluscan subunit is 350–450 kD and contains seven or eight O_2-binding globular units. Despite the differing tertiary and quaternary structures of hemocyanins, which reflect their dissimilar primary structures, the oxygen-binding sites are highly similar, consisting of a pair of copper atoms, each liganded by three His residues (see figure). The otherwise colorless complex turns blue when it binds O_2.

Hemocyanins must be present at high concentrations in order to function efficiently as oxygen carriers. For example, octopus **hemolymph** (its equivalent of blood) contains about 100 mg/mL hemocyanin. However, this could potentially generate intolerably high osmotic pressures. Arthropod and molluscan hemocyanins minimize this problem by forming multimeric structures with masses as great as 9×10^6 D in some species [vertebrates do so by sequestering hemoglobin in **erythrocytes** (red cells; Greek: *erythrose,* red and *kytos,* a hollow vessel)].

The large extracellular O_2-transport molecules are often the predominant extracellular protein and may therefore have additional functions as buffers against pH changes and osmotic fluctuations. In some invertebrate species, hemocyanins may serve as a nutritional reserve, for example, during metamorphosis or molting.

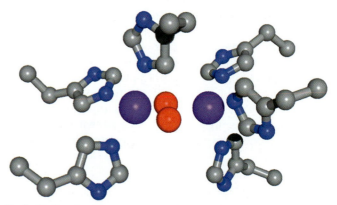

The O_2-binding site of hemocyanin from the horseshoe crab *Limulus polyphemus.* Atoms are colored according to type with C gray, N blue, O red, and Cu purple. [Based on an X-ray structure by Wim Hol, University of Washington School of Medicine. PDBid 1OXY]

plain the control of enzyme activity, hemoglobin has been dubbed an "honorary enzyme." This section includes an examination of hemoglobin's structure and its cooperative oxygen-binding behavior.

A   Hemoglobin Structure

Mammalian hemoglobin, as we saw in Fig. 6-33, is a tetrameric protein with the quaternary structure $\alpha_2\beta_2$ (a dimer of $\alpha\beta$ protomers). The α and β

BOX 7-2

Pathways of Discovery

Max Perutz and the Structure and Function of Hemoglobin

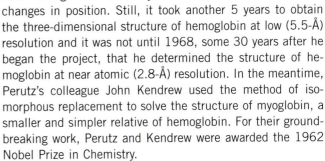

Max Perutz (1914–2002)

The determination of the three-dimensional structures of proteins has become so commonplace that it is difficult to appreciate the challenges that faced the first protein crystallographers. Max Perutz was a pioneer in this area, spending many years determining the structure of hemoglobin at atomic resolution and then using this information to explain the physiological function of the protein.

In 1934, two years before Perutz began his doctoral studies in Cambridge, J.D. Bernal and Dorothy Crowfoot Hodgkin had placed a crystal of the protein pepsin in an X-ray beam and obtained a diffraction pattern. Perutz tried the same experiment with hemoglobin, chosen because of its abundance, ease of crystallization, and obvious physiological importance. Hemoglobin crystals yielded diffraction patterns with thousands of diffraction maxima (called reflections), the result of X-ray scattering by the thousands of atoms in each protein molecule. At the time, X-ray crystallography had been used to determine the structures of molecules containing no more than around 40 atoms, so the prospect of using the technique to determine the atomic structure of hemoglobin seemed impossible. Nevertheless, Perutz took on the challenge and spent the rest of his long career working with hemoglobin.

In X-ray crystallography, the intensities and the positions of the reflections can be readily determined but the values of their phases (the relative positions of the wave peaks, the knowledge of which is as important as wave amplitude for image reconstruction) cannot be directly measured. Although computational techniques for determining the values of the phases had been developed for small molecules, methods for solving this so-called phase problem for such complex entities as proteins seemed hopelessly out of reach. In 1952, Perutz realized that the so-called method of isomorphous replacement might suffice to solve the phase problem for hemoglobin. In this method, a heavy atom such as an Hg^{2+} ion, which is rich in electrons (the particles that scatter X-rays), must bind to specific sites on the protein without significantly disturbing its structure (which would change the positions of the reflections). If this causes measurable changes in the intensities of the reflections, these differences would provide the information to determine their phases. With trepidation followed by jubilation, Perutz observed that Hg-doped hemoglobin crystals indeed yielded reflections with measurable changes in intensity but no changes in position. Still, it took another 5 years to obtain the three-dimensional structure of hemoglobin at low (5.5-Å) resolution and it was not until 1968, some 30 years after he began the project, that he determined the structure of hemoglobin at near atomic (2.8-Å) resolution. In the meantime, Perutz's colleague John Kendrew used the method of isomorphous replacement to solve the structure of myoglobin, a smaller and simpler relative of hemoglobin. For their groundbreaking work, Perutz and Kendrew were awarded the 1962 Nobel Prize in Chemistry.

For Perutz, obtaining the structure of hemoglobin was only part of his goal of understanding hemoglobin. For example, functional studies indicated that the four oxygen-binding sites of hemoglobin interacted, as if they were in close contact, but Perutz's structure showed that the binding sites lay in deep and widely separated pockets. Perutz was also intrigued by the fact that crystals of hemoglobin prepared in the absence of oxygen would crack when they were exposed to air (the result, it turns out, of a dramatic conformational change). Although many other researchers also turned their attention to hemoglobin, Perutz was foremost among them in ascribing oxygen-binding behavior to protein structural features. He also devoted considerable effort to relating functional abnormalities in mutant hemoglobins to structural changes.

Perutz's ground-breaking work on the X-ray crystallography of proteins paved the way for other studies. For example, the first X-ray structure of an enzyme, lysozyme, was determined in 1965. The nearly 30,000 macromolecular structures that have been obtained since then owe a debt to Perutz and his decision to pursue an "impossible" task and to follow through on his structural work to the point where he could use his results to explain biological phenomena.

Perutz, M.F., Rossmann, M.G., Cullis, A.F., Muirhead, H., Will, G., and North, A.C.T., Structure of haemoglobin: A three-dimensional Fourier synthesis at 5.5 Å resolution, obtained by X-ray analysis. *Nature* **185**, 416–422 (1960).

subunits are structurally and evolutionarily related to each other and to myoglobin (the genealogy of the globin polypeptides is discussed in Section 5-4B)

The structure of hemoglobin was determined by Max Perutz (Box 7-2). Only about 18% of the residues are identical in myoglobin and in the α and β subunits of hemoglobin, but the three polypeptides have remarkably similar tertiary structures (hemoglobin subunits follow the myoglobin helix-

labeling system, although the α chain has no D helix). The αβ protomers of hemoglobin are symmetrically related by a 2-fold rotation (i.e., a rotation of 180° brings the protomers into coincidence). In addition, hemoglobin's structurally similar α and β subunits are related by an approximate 2-fold rotation (pseudosymmetry) whose axis is perpendicular to that of the exact 2-fold rotation. Thus, hemoglobin has exact C_2 symmetry and

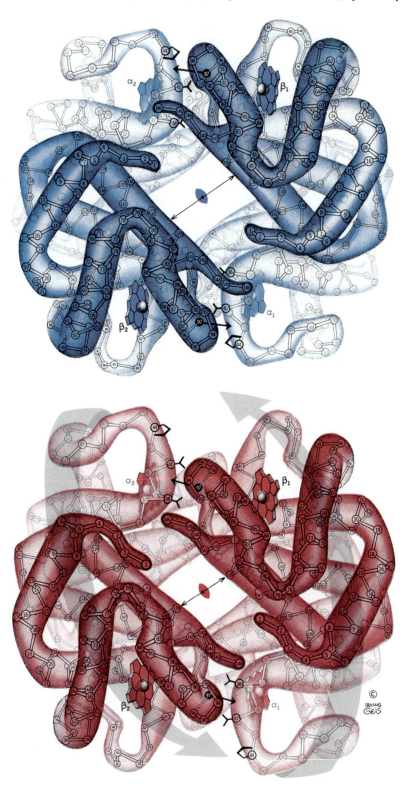

Figure 7-5 Hemoglobin structure.   (*a*) Deoxy-hemoglobin and (*b*) oxyhemoglobin. The $\alpha_1\beta_1$ protomer is related to the $\alpha_2\beta_2$ protomer by a 2-fold axis of symmetry (*lenticular symbol*), which is perpendicular to the page. Oxygenation brings the β chains closer together (compare the lengths of the double-headed arrows) and shifts the contacts between subunits at the $\alpha_1-\beta_2$ and $\alpha_2-\beta_1$ interfaces (some of the relevant side chains are shown in black). The large gray arrows in *b* indicate the molecular movements that accompany oxygenation. [Illustration, Irving Geis. Image from the Irving Geis Collection/Howard Hughes Medical Institute. Rights owned by HHMI. Reproduction by permission only.] 🔗 **See Kinemage Exercises 6-2 and 6-3.**

pseudo-D_2 symmetry (Section 6-3; objects with D_2 symmetry have the rotational symmetry of a tetrahedron). The hemoglobin molecule has overall dimensions of about $64 \times 55 \times 50$ Å.

Oxygen binding alters the structure of the entire hemoglobin tetramer, so the structures of **deoxyhemoglobin** (Fig. 7-5a) and **oxyhemoglobin** (Fig. 7-5b) are noticeably different. In both forms of hemoglobin, the α and β subunits form extensive contacts: Those at the α_1–β_1 interface (and its α_2–β_2 symmetry equivalent) involve 35 residues, and those at the α_1–β_2 (and α_2–β_1) interface involve 19 residues. These associations are predominantly hydrophobic, although numerous hydrogen bonds and several ion pairs are also involved. Note, however, that the α_1–α_2 and β_1–β_2 interactions are tenuous at best because these subunit pairs are separated by an ~20-Å-diameter solvent-filled channel that parallels the 50-Å length of hemoglobin's exact 2-fold axis (Fig. 7-5).

When oxygen binds, the α_1–β_2 (and α_2–β_1) contacts shift, producing a change in quaternary structure. Oxygenation rotates one $\alpha\beta$ dimer ~15° with respect to the other $\alpha\beta$ dimer (Fig. 7-6), which brings the β subunits closer together and narrows the solvent-filled central channel of hemoglobin (Fig. 7-5). Some atoms in the α_1–β_2 and α_2–β_1 interfaces shift by as much as 6 Å (oxygenation causes such extensive quaternary structural changes that crystals of deoxyhemoglobin shatter on exposure to O_2). This structural rearrangement is a crucial element of hemoglobin's oxygen-binding behavior.

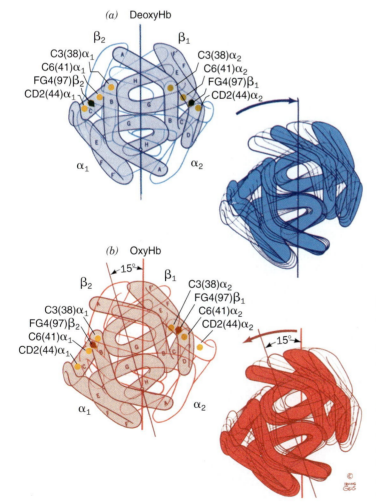

Figure 7-6 The major structural differences between the quaternary conformations of (a) deoxyhemoglobin and (b) oxyhemoglobin. On oxygenation, the $\alpha_1\beta_1$ (*shaded*) and $\alpha_2\beta_2$ (*outlined*) protomers move, as indicated on the right, as rigid units such that there is an ~15° off-center rotation of one protomer relative to the other that preserves the molecule's exact 2-fold symmetry. Note how the position of His FG4β (*pentagons*) changes with respect to Thr C3α, Thr C6α, and Pro CD2α (*yellow dots*) at the α_1–β_2 and α_2–β_1 interfaces. The view is from the right side relative to that in Fig. 7-5. [Illustration, Irving Geis. Image from the Irving Geis Collection/Howard Hughes Medical Institute. Rights owned by HHMI. Reproduction by permission only.]

B Oxygen Binding to Hemoglobin

Hemoglobin has an overall p_{50} of 26 torr (i.e., hemoglobin is half-saturated with O_2 at an oxygen pressure of 26 torr), which is nearly 10 times greater than the p_{50} of myoglobin. Moreover, hemoglobin does not exhibit a myoglobin-like hyperbolic oxygen-binding curve. Instead, O_2 binding to hemoglobin is described by a **sigmoidal** (S-shaped) curve (Fig. 7-7). *This permits the blood to deliver much more O_2 to the tissues than if hemoglobin had a hyperbolic curve with the same p_{50}* (dashed line in Fig. 7-7). For example, hemoglobin is nearly fully saturated with O_2 at arterial oxygen pressures ($Y_{O_2} = 0.95$ at 100 torr) but only about half-saturated at venous oxygen pressures ($Y_{O_2} = 0.55$ at 30 torr). This 0.40 difference in oxygen saturation, a measure of hemoglobin's ability to deliver O_2 from the lungs to the tissues, would be only 0.25 if hemoglobin exhibited hyperbolic binding behavior.

*In any binding system, a sigmoidal curve is diagnostic of a **cooperative** interaction between binding sites.* This means that the binding of a ligand to one site affects the binding of additional ligands to the other sites. In the case of hemoglobin, O_2 binding to one subunit increases the O_2 affinity of the remaining subunits. The initial slope of the oxygen-binding curve (Fig. 7-7) is low, as hemoglobin subunits independently compete for the first O_2. However, an O_2 molecule bound to one of hemoglobin's subunits increases the O_2-binding affinity of its other subunits, thereby accounting for the increasing slope of the middle portion of the sigmoidal curve.

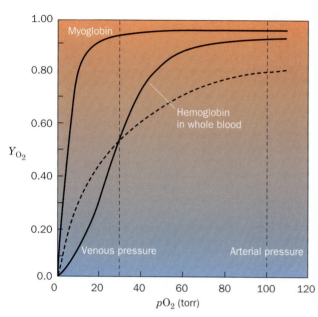

Figure 7-7 *Key to Function.* **Oxygen-binding curve of hemoglobin.** In whole blood, hemoglobin is half-saturated at an oxygen pressure of 26 torr. The normal sea level values of human arterial and venous pO_2 are indicated (atmospheric pO_2 is 160 torr at sea level). The O_2-binding curve for myoglobin is included for comparison. The dashed line is a hyperbolic O_2-binding curve with the same p_{50} as hemoglobin. **See the Animated Figures.**

The Hill Equation Describes Hemoglobin's O_2-Binding Curve. The earliest attempt to analyze hemoglobin's sigmoidal O_2 dissociation curve was formulated by Archibald Hill in 1910. Hill assumed that hemoglobin (Hb) bound n molecules of O_2 in a single step,

$$Hb + n\ O_2 \longrightarrow Hb(O_2)_n$$

that is, with infinite cooperativity. Thus, in analogy with the derivation of Eq. 7-6,

$$Y_{O_2} = \frac{(pO_2)^n}{(p_{50})^n + (pO_2)^n} \qquad [7\text{-}8]$$

which is known as the **Hill equation.** Like Eq. 7-6, it describes the degree of saturation of hemoglobin as a function of pO_2.

Infinite O_2 binding cooperativity, as Hill assumed, is a physical impossibility. Nevertheless, n may be taken to be a nonintegral parameter related to the degree of cooperativity among interacting hemoglobin subunits rather than the number of subunits that bind O_2 in one step. The Hill equation can then be taken as a useful empirical curve-fitting relationship rather than as an indicator of a particular model of ligand binding.

*The quantity n, the **Hill coefficient,** increases with the degree of cooperativity of a reaction and therefore provides a convenient although simplistic characterization of a ligand-binding reaction.* If $n = 1$, Eq. 7-8 describes a hyperbola as does Eq. 7-6 for myoglobin, and the O_2-binding reaction is said to be **noncooperative.** If $n > 1$, the reaction is described as being **positively cooperative,** because O_2 binding increases the affinity of hemoglobin for further O_2 binding (cooperativity is infinite in the limit that $n = 4$, the number of O_2 binding sites in hemoglobin). Conversely, if $n < 1$, the reaction is said to be **negatively cooperative,** because O_2 binding would then reduce the affinity of hemoglobin for subsequent O_2 binding.

The Hill coefficient, n, and the value of p_{50} that best describe hemoglobin's saturation curve can be graphically determined by rearranging Eq. 7-8. First, divide both sides by $1 - Y_{O_2}$:

$$\frac{Y_{O_2}}{1 - Y_{O_2}} = \frac{\dfrac{(pO_2)^n}{(p_{50})^n + (pO_2)^n}}{1 - Y_{O_2}} = \frac{\dfrac{(pO_2)^n}{(p_{50})^n + (pO_2)^n}}{1 - \dfrac{(pO_2)^n}{(p_{50})^n + (pO_2)^n}} \qquad [7\text{-}9]$$

Factoring out the $[(p_{50})^n + (pO_2)^n]$ term gives

$$\frac{Y_{O_2}}{1 - Y_{O_2}} = \frac{(pO_2)^n}{[(p_{50})^n + (pO_2)^n] - (pO_2)^n} = \frac{(pO_2)^n}{(p_{50})^n} \qquad [7\text{-}10]$$

Taking the log of both sides yields a linear equation:

$$\log\left(\frac{Y_{O_2}}{1 - Y_{O_2}}\right) = n \log pO_2 - n \log p_{50} \qquad [7\text{-}11]$$

The linear plot of $\log[Y_{O_2}/(1 - Y_{O_2})]$ versus $\log pO_2$, the **Hill plot,** has a slope of n and an intercept on the $\log pO_2$ axis of $\log p_{50}$ (recall that the linear equation $y = mx + b$ describes a line with a slope of m and an x intercept of $-b/m$).

Figure 7-8 shows the Hill plots for myoglobin and purified hemoglobin. For myoglobin, the plot is linear with a slope of 1, as expected. Although all subunits of hemoglobin do not bind O_2 in a single step as was assumed in deriving the Hill equation, its Hill plot is essentially linear for values of Y_{O_2} between 0.1 and 0.9. When $pO_2 = p_{50}$, $Y_{O_2} = 0.5$, and

$$\frac{Y_{O_2}}{1 - Y_{O_2}} = \frac{0.5}{1 - 0.5} = 1.0 \qquad [7\text{-}12]$$

As can be seen in Fig. 7-8, this is the region of maximum slope, whose value is customarily taken to be the Hill coefficient, n. For normal human hemoglobin, the Hill coefficient is between 2.8 and 3.0; that is, hemoglobin's oxygen binding is highly, but not infinitely, cooperative. Many abnormal hemoglobins exhibit smaller Hill coefficients (Section 7-4), indicating that they have a less than normal degree of cooperativity. Scorpion hemocyanin, a multimer of 24 subunits, each with an oxygen-binding site, has a Hill coefficient of ~20, which is indicative of a high degree of cooperativity.

At Y_{O_2} values near zero, when few hemoglobin molecules have bound even one O_2 molecule, the Hill plot for hemoglobin assumes a slope of 1 (Fig. 7-8, lower asymptote) because the hemoglobin subunits independently compete for O_2 as do molecules of myoglobin. At Y_{O_2} values near 1, when at least three of each of hemoglobin's four O_2-binding sites are occupied, the Hill plot also assumes a slope of 1 (Fig. 7-8, upper asymptote) because the few remaining unoccupied sites are on different molecules and therefore bind O_2 independently.

Extrapolating the lower asymptote in Fig. 7-8 to the horizontal axis indicates, according to Eq. 7-11, that $p_{50} = 30$ torr for binding the first O_2 to

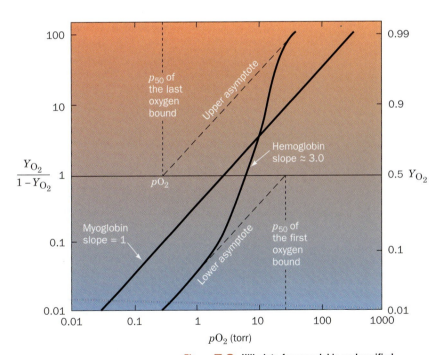

Figure 7-8   Hill plots for myoglobin and purified hemoglobin.   Note that this is a log–log plot. At $pO_2 = p_{50}$, $Y_{O_2}/(1 - Y_{O_2}) = 1$. [The p_{50} for hemoglobin *in vivo* is higher than the p_{50} of purified hemoglobin due to its binding of certain substances present in the red cell (see below)].

purified hemoglobin. Likewise, extrapolating the upper asymptote yields $p_{50} = 0.3$ torr for binding hemoglobin's fourth O_2. Thus, *the fourth O_2 to bind to hemoglobin does so with 100-fold greater affinity than the first*. This difference, as we shall see in Section 7-3A, is entirely due to the influence of the globin chain on the O_2 affinity of heme.

More realistic models than that used in deriving the Hill equation have been developed for analyzing the cooperative binding of O_2 to hemoglobin. We discuss them in Section 7-3A.

3 Cooperativity

Cooperativity among binding sites in macromolecules is a phenomenon that is not limited to oxygen-binding proteins. In fact, cooperative behavior is a common feature of enzymes and provides a sensitive mechanism for regulating their catalytic activity. In this section, we examine the molecular basis for cooperativity in hemoglobin and discuss some of the models used to evaluate cooperative binding behavior in general.

A Mechanism of Cooperativity in Hemoglobin

The cooperativity of oxygen binding to hemoglobin arises from the effect of the ligand-binding state of one heme group on the ligand-binding affinity of another. Yet the hemes are 25 to 37 Å apart—too far to interact electronically. Instead, information about the O_2-binding status of a heme group is mechanically transmitted to the other heme groups by motions of the protein. These movements are responsible for the different quaternary structures of oxy- and deoxyhemoglobin depicted in Fig. 7-5.

Hemoglobin Has Two Conformational States. On the basis of the X-ray structures of oxy- and deoxyhemoglobin, Perutz formulated a mechanism for hemoglobin oxygenation. *In the **Perutz mechanism**, hemoglobin has two stable conformational states, the **T state** (the conformation of deoxyhemoglobin) and the **R state** (the conformation of oxyhemoglobin).* The conformations of all four subunits in T-state hemoglobin differ from those in the R state. Oxygen binding initiates a series of coordinated movements that result in a shift from the T state to the R state within a few microseconds:

1. In the T state, the Fe(II) in each of the four hemes is situated ~0.6 Å out of the heme plane because of a pyramidal doming of the porphyrin group toward His F8 (Fig. 7-9). O_2 binding changes the heme's electronic state, which shortens the Fe—$N_{porphyrin}$ bonds by ~0.1 Å and causes the porphyrin doming to subside. Consequently, during the T → R transition, the Fe(II) moves into the center of the heme plane.

2. The Fe(II) drags the covalently linked His F8 along with it. However, the direct movement of His F8 by 0.6 Å toward the heme plane would cause it to collide with the heme. To avoid this steric clash, the attached F helix tilts and translates by ~1 Å across the heme plane.

3. The changes in tertiary structure are coupled to a shift in the arrangement of hemoglobin's four subunits. The largest change produced by the T → R transition is the result of movements of residues at the α_1–β_2 and α_2–β_1 interfaces; in other words, at the interface between

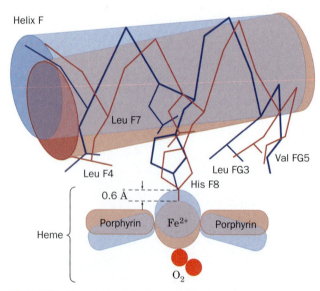

Figure 7-9   Movements of the heme and the F helix during the T → R transition in hemoglobin. In the T form (*blue*), the Fe is 0.6 Å above the center of the domed porphyrin ring. On assuming the R form (*red*), the Fe moves into the plane of the now undomed porphyrin, where it can more tightly bind O_2, and, in doing so, pulls His F8 and its attached F helix with it. ✏ **See Kinemage Exercise 6-4 and the Animated Figures.**

the two protomeric units of hemoglobin. In the T state, His 97 in the β chain contacts Thr 41 in the α chain (Fig. 7-10a). In the R state, His 97 contacts Thr 38, which is positioned one turn back along the C helix (Fig. 7-10b). In both conformations, the "knobs" on one sub-

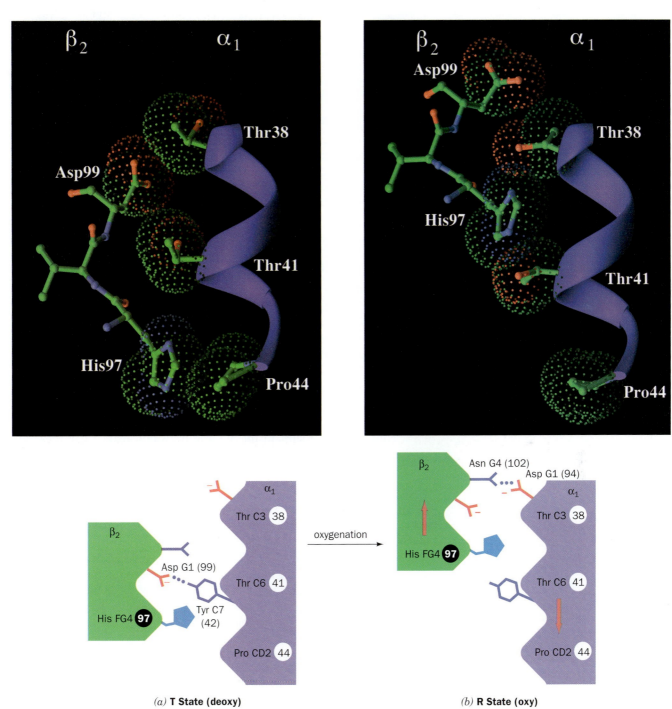

(a) **T State (deoxy)**

(b) **R State (oxy)**

Figure 7-1O   Changes at the α₁–β₂ interface during T → R transition in hemoglobin. (a) The T state, and (b) the R state. In the upper drawings, the C helix is represented by a purple ribbon, the contacting residues forming the α₁C–β₂FG contact are shown in ball-and-stick form colored by atom type (C green, N blue, and O red), and their van der Waals surfaces are outlined by like-colored dots. The lower drawings are the corresponding schematic diagrams of the α₁C–β₂FG contact. Upon a T → R transformation, the β₂FG region shifts by one turn along the α₁C helix with no stable intermediate

(note how in both conformations, the knobs formed by the side chains of His 97β and Asp 99β fit between the grooves on the C helix formed by the side chains of Thr 38α, Thr 41α, and Pro 44α. The subunits are joined by different hydrogen bonds in the two quaternary states. Figs. 7-5 and 7-6 provide additional structural views of these interactions. [Based on X-ray structures by Giulio Fermi, Max Perutz, and Boaz Shaanan, MRC Laboratory of Molecular Biology, Cambridge, U.K. PDBids (a) 2HHB & (b) 1HHO.] 🐁 **See Kinemage Exercise 6-5.**

(a) α Chains

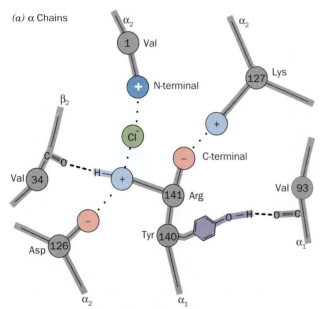

(b) β Chains

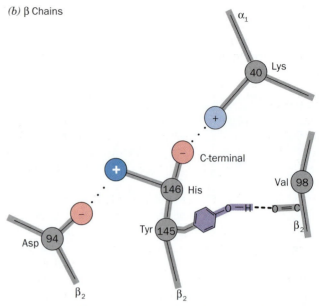

Figure 7-11 Networks of ion pairs and hydrogen bonds in deoxyhemoglobin. These bonds, which involve the last two residues of (a) the α chains and (b) the β chains, are ruptured in the T → R transition. Two groups that become partially deprotonated in the R state (part of the Bohr effect) are indicated by white plus signs. [Illustration, Irving Geis. Image from the Irving Geis Collection/Howard Hughes Medical Institute. Rights owned by HHMI. Reproduction by permission only.]

unit mesh nicely with the "grooves" on the other. An intermediate position would be severely strained because it would bring His 97 and Thr 41 too close together (i.e., knobs on knobs).

4. The C-terminal residues of each subunit (Arg 141α and His 146β) in T-state hemoglobin each participate in a network of intra- and intersubunit ion pairs (Fig. 7-11) that stabilize the T state. However, the conformational shift in the T → R transition tears away these ion pairs in a process that is driven by the energy of formation of the Fe—O_2 bonds.

Considerable experimental evidence supports the sequence of molecular events described here. For example, hemoglobin was engineered so that His F8 was replaced by Gly. The missing imidazole group of the His side chain was then replaced by free imidazole. Without a covalent bond linking the imidazole ring to the rest of the protein, the engineered hemoglobin was able to bind O_2 but, for the most part, did not undergo the O_2-induced conformational changes, starting with the movement of the F helix and culminating in the 15° movement of one αβ protomer relative to the other.

The essential feature of hemoglobin's T → R transition is that *its subunits are so tightly coupled that large tertiary structural changes within one subunit cannot occur without quaternary structural changes in the entire tetrameric protein.* Hemoglobin is limited to only two quaternary forms, T and R, because the intersubunit contacts shown in Fig. 7-10 act as a binary switch that permits only two stable positions of the subunits relative to each other. The inflexibility of the $α_1–β_1$ and $α_2–β_2$ interfaces requires that the T → R shift occur simultaneously at both the $α_1–β_2$ and $α_2–β_1$ interfaces. No one subunit or dimer can greatly change its conformation independently of the others.

We are now in a position to structurally rationalize the cooperativity of oxygen binding to hemoglobin. The T state of hemoglobin has low O_2 affin-

ity, mostly because of the 0.1 Å greater length of its Fe—O_2 bond relative to that of the R state (e.g., the blue structure shown in Fig. 7-9). Experimental evidence indicates that when at least one O_2 has bound to each αβ dimer, the strain in the T-state hemoglobin molecule is sufficient to tear away the C-terminal ion pairs, thereby snapping the protein into the R state. All the subunits are thereby simultaneously converted to the R-state conformation whether or not they have bound O_2. Unliganded subunits in the R-state conformation have increased oxygen affinity because they are already in the O_2-binding conformation. This accounts for the high O_2 affinity of nearly saturated hemoglobin.

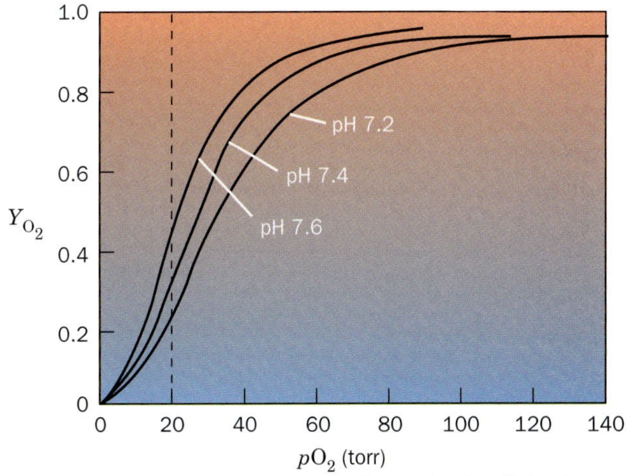

Carbon Dioxide Transport and the Bohr Effect. The conformational changes in hemoglobin that occur on oxygen binding decrease the pK's of several groups. Recall that the tendency for a group to ionize depends on its microenvironment, which may include other ionizable groups. For example, in T-state hemoglobin, the N-terminal amino groups of the α subunits and the C-terminal His of the β subunit are positively charged and participate in ion pairs (see Fig. 7-11). The formation of ion pairs increases the pK values of these groups (makes them less acidic and therefore less likely to give up their protons). In R-state hemoglobin, these ion pairings are absent, and the pK's of the groups decrease (making them more acidic and more likely to give up protons). Consequently, under physiological conditions, hemoglobin releases ~0.6 protons for each O_2 it binds. Conversely, increasing the pH, that is, removing protons, stimulates hemoglobin to bind more O_2 at lower oxygen pressures (Fig. 7-12). This phenomenon is known as the **Bohr effect** after Christian Bohr (father of the physicist Niels Bohr), who first reported it in 1904.

The Bohr effect has important physiological functions in transporting O_2 from the lungs to respiring tissue and in transporting the CO_2 produced by respiration back to the lungs (Fig. 7-13). The CO_2 produced by respiring tissues diffuses from the tissues to the capillaries. This dissolved CO_2 forms bicarbonate (HCO_3^-) only very slowly, by the reaction

$$CO_2 + H_2O \rightleftharpoons H^+ + HCO_3^-$$

Figure 7-12 The Bohr effect. The O_2 affinity of hemoglobin increases with increasing pH. The dashed line indicates the pO_2 in actively respiring muscle. [After Benesch, R.E. and Benesch, R., *Adv. Protein Chem.* **28,** 212 (1974).] ⟳ **See the Animated Figures.**

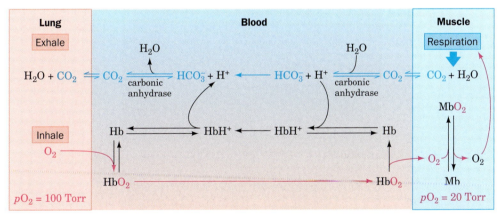

Figure 7-13 *Key to Function.* **The roles of hemoglobin and myoglobin in transporting O_2 from the lungs to respiring tissues and CO_2 (as HCO_3^-) from the tissues to the lungs.** Oxygen is inhaled into the lungs at high pO_2, where it binds to hemoglobin in the blood. The O_2 is then transported to respiring tissue, where the pO_2 is low. The O_2 therefore dissociates from the Hb and diffuses into the tissues, where it is used to oxidize metabolic fuels to CO_2

and H_2O. In rapidly respiring muscle tissue, the O_2 first binds to myoglobin (whose oxygen affinity is higher than that of hemoglobin). This increases the rate at which O_2 can diffuse from the capillaries to the tissues by, in effect, increasing its solubility. The Hb and CO_2 (mostly as HCO_3^-) are then returned to the lungs, where the CO_2 is exhaled.

However, in the erythrocyte, the enzyme **carbonic anhydrase** greatly accelerates this reaction. Accordingly, most of the CO_2 in the blood is carried in the form of bicarbonate.

In the capillaries, where pO_2 is low, the H^+ generated by bicarbonate formation is taken up by hemoglobin in forming the ion pairs of the T state, thereby inducing hemoglobin to unload its bound O_2. This H^+ uptake, moreover, facilitates CO_2 transport by stimulating bicarbonate formation. Conversely, in the lungs, where pO_2 is high, O_2 binding by hemoglobin disrupts the T-state ion pairs to form the R state, thereby releasing the Bohr protons, which recombine with bicarbonate to drive off CO_2. These reactions are closely matched, so they cause very little change in blood pH (see Box 2-2).

The Bohr effect provides a mechanism whereby additional oxygen can be supplied to highly active muscles, where the pO_2 may be <20 torr. Such muscles generate lactic acid (Section 14-3A) so fast that they lower the pH of the blood passing through them from 7.4 to 7.2. At a pO_2 of 20 torr, hemoglobin releases ~10% more O_2 at pH 7.2 than it does at pH 7.4 (Fig. 7-12).

CO_2 also modulates O_2 binding to hemoglobin by combining reversibly with the N-terminal amino groups of blood proteins to form **carbamates:**

$$R{-}NH_2 + CO_2 \rightleftharpoons R{-}NH{-}COO^- + H^+$$

The T (deoxy) form of hemoglobin binds more CO_2 as carbamate than does the R (oxy) form. When the CO_2 concentration is high, as it is in the capillaries, the T state is favored, stimulating hemoglobin to release its bound O_2. The protons released by carbamate formation further promote O_2 release through the Bohr effect. Although the difference in CO_2 binding between the oxy and deoxy states of hemoglobin accounts for only ~5% of

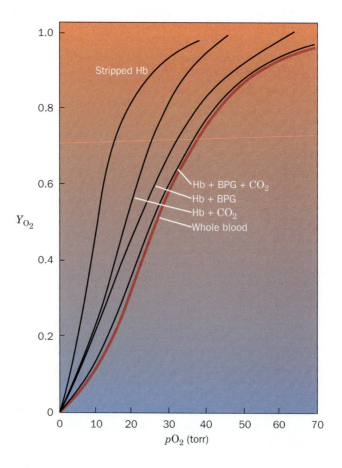

Figure 7-14   The effects of BPG and CO₂ on hemoglobin's O₂ dissociation curve.   Stripped hemoglobin (*left*) has higher O_2 affinity than whole blood (*red curve*). Adding BPG or CO_2 or both to hemoglobin shifts the dissociation curve back to the right (lowers hemoglobin's O_2 affinity). [After Kilmartin, J.V. and Rossi-Bernardi, L., *Physiol. Rev.* **53**, 884 (1973).] 🔊 **See the Animated Figures.**

the total blood CO_2, it is nevertheless responsible for around half the CO_2 transported by the blood. This is because only ~10% of the total blood CO_2 is lost through the lungs in each circulatory cycle.

Bisphosphoglycerate Binds to Deoxyhemoglobin.

Highly purified ("stripped") hemoglobin has a much greater oxygen affinity than hemoglobin in whole blood (Fig. 7-14). This observation led Joseph Barcroft, in 1921, to speculate that blood contains some other substance besides CO_2 that affects oxygen binding to hemoglobin. This compound is D-2,3-bisphosphoglycerate **(BPG)**.

$$\begin{array}{c} {}^-O \diagdown \diagup O \\ C \\ | \\ H-C-OPO_3^{2-} \\ | \\ H-C-OPO_3^{2-} \\ | \\ H \end{array}$$

D-2,3-Bisphosphoglycerate (BPG)

BPG binds tightly to deoxyhemoglobin but only weakly to oxyhemoglobin. *The presence of BPG in mammalian erythrocytes therefore decreases hemoglobin's oxygen affinity by keeping it in the deoxy conformation.* In other vertebrates, different phosphorylated compounds elicit the same effect.

BPG has an indispensable physiological function: In arterial blood, where pO_2 is ~100 torr, hemoglobin is ~95% saturated with O_2, but in venous blood, where pO_2 is ~30 torr, it is only 55% saturated (Fig. 7-7). Consequently, in passing through the capillaries, hemoglobin unloads ~40% of its bound O_2. In the absence of BPG, little of this bound O_2 would be released since hemoglobin's O_2 affinity is increased, thus shifting its O_2 dissociation curve significantly toward lower pO_2 (Fig. 7-14, *left*). BPG also plays an important role in adaptation to high altitudes (see Box 7-3).

The X-ray structure of a BPG–deoxyhemoglobin complex shows that BPG binds in the central cavity of deoxyhemoglobin (Fig. 7-15). The anionic groups of BPG are within hydrogen-bonding and ion-pairing distances of the N-terminal amino groups of both β subunits. The T → R transformation brings the two β H helices together, which narrows the central cavity (compare Figs. 7-5a and 7-5b) and expels the BPG. It also widens the distance between the β N-terminal amino groups from 16 to 20 Å, which prevents their simultaneous hydrogen bonding with BPG's phosphate groups. BPG therefore binds to and stabilizes only the T conformation of hemoglobin by cross-linking its β subunits. This shifts the T ⇌ R equilibrium toward the T state, which lowers hemoglobin's O_2 affinity.

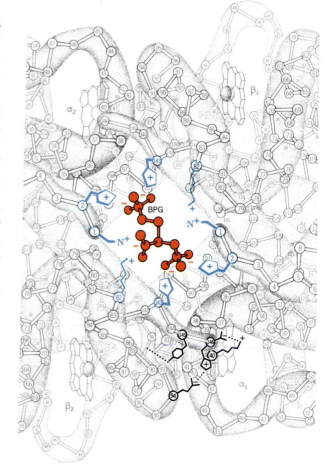

Figure 7-15 Binding of BPG to deoxyhemoglobin. BPG (*red*) binds in hemoglobin's central cavity. The BPG, which has a charge of −5 under physiological conditions, is surrounded by eight cationic groups (*blue*) extending from the two β subunits. In the R state, the central cavity is too narrow to contain BPG. Some of the ion pairs and hydrogen bonds that help stabilize the T state (Fig. 7-11b) are indicated at the lower right. [Illustration, Irving Geis. Image from the Irving Geis Collection/Howard Hughes Medical Institute. Rights owned by HHMI. Reproduction by permission only.] 🐁 **See Kinemage Exercise 6-3.**

BOX 7-3

Biochemistry in Health and Disease

High-Altitude Adaptation

Atmospheric pressure decreases with altitude, so that the oxygen pressure at 3000 m (10,000 feet) is only ~110 torr, 70% of its sea-level pressure. A variety of physiological responses are required to maintain normal oxygen delivery (without adaptation, pO_2 levels of 85 torr or less result in mental impairment).

High-altitude adaptation is a complex process that involves increases in the amount of hemoglobin per erythrocyte and in the number of erythrocytes. It normally requires several weeks to complete. Yet, as is clear to anyone who has climbed to high altitude, even a 1-day stay there results in a noticeable degree of adaptation. This effect results from a rapid increase in the amount of BPG synthesized in erythrocytes (from ~4 mM to ~8 mM; BPG cannot cross the red cell membrane). As illustrated by plots of Y_{O_2} versus pO_2, the high altitude–induced increase in BPG causes the O_2-binding curve of hemoglobin to shift from its sea-level position (*black line*) to a lower affinity position (*red line*). At sea level, the difference between arterial and venous pO_2 is 70 torr (100 torr − 30 torr), and hemoglobin unloads 38% of its bound O_2. However, when the arterial pO_2 drops to 55 torr, as it does at an altitude of 4500 m, hemoglobin would be able to unload only 30% of its O_2. High-altitude adaptation (which decreases the amount of O_2 that hemoglobin can bind in the lungs but, to a greater extent, increases the amount of O_2 it releases at the tissues) allows hemoglobin to deliver a near-normal 37% of its bound O_2. BPG concentrations also increase in individuals suffering from disorders that limit the oxygenation of the blood (**hypoxia**), such as various anemias and cardiopulmonary insufficiency.

The BPG concentration in erythrocytes can be adjusted more rapidly than hemoglobin can be synthesized (Box 14-2; erythrocytes lack nuclei and therefore cannot synthesize proteins). An altered BPG level is also a more sensitive regulator of oxygen delivery than an altered respiratory rate. Hyper-

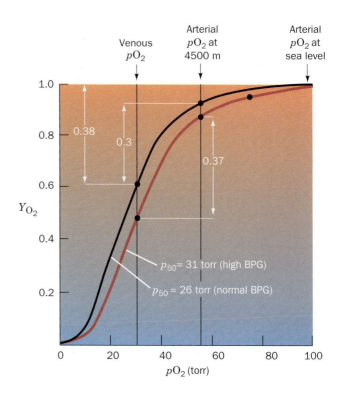

ventilation, another early response to high altitude, may lead to respiratory alkalosis (Box 2-2). Interestingly, individuals in long-established Andean and Himalayan populations exhibit high lung capacity, along with high hemoglobin levels and, often, enlarged right ventricles (reflecting increased cardiac output), compared to individuals from low-altitude populations.

In contrast to the mechanism of human adaptation to high altitude, most mammals that normally live at high altitudes (e.g., the llama) have genetically altered hemoglobins that have higher O_2-binding affinities than do their sea-level cousins. Thus, both raising and lowering hemoglobin's p_{50} can provide high-altitude adaptation.

Fetal Hemoglobin Has Low BPG Affinity.   The effects of BPG also help supply the fetus with oxygen. A fetus obtains its O_2 from the maternal circulation via the placenta. The concentration of BPG is the same in adult and fetal erythrocytes, but BPG binds more tightly to adult hemoglobin than to fetal hemoglobin. The higher oxygen affinity of fetal hemoglobin facilitates the transfer of O_2 to the fetus.

Fetal hemoglobin has the subunit composition $\alpha_2\gamma_2$ in which the γ subunit is a variant of the β chain (Section 5-4B). Residue 143 of the β chain of adult hemoglobin has a cationic His residue, whereas the γ chain has an

uncharged Ser residue. The absence of this His eliminates a pair of inter-actions that stabilize the BPG–deoxyhemoglobin complex (Fig. 7-15).

B Allosteric Proteins

The cooperativity of oxygen binding to hemoglobin is a classic model for the behavior of many other multisubunit proteins (including certain en-zymes) that bind small molecules. In some cases, binding of a ligand to one site increases the affinity of other binding sites on the same protein (as in O_2 binding to hemoglobin). In other cases, a ligand decreases the affinity of other binding sites (as when BPG binding decreases the O_2 affinity of hemoglobin). All these effects are the result of **allosteric interactions** (Greek: *allos*, other + *stereos*, solid or space). *Allosteric effects, in which the binding of a ligand at one site affects the binding of another ligand at an-other site, generally require interactions among subunits of oligomeric pro-teins.* The T → R transition in hemoglobin subunits explains the difference in the oxygen affinities of oxy- and deoxyhemoglobin. Other proteins ex-hibit similar conformational shifts, although the molecular mechanisms that underlie these phenomena are not completely understood.

Two models that account for cooperative ligand binding have received the most attention. One of them, the **symmetry model** of allosterism, formulated in 1965 by Jacques Monod, Jeffries Wyman, and Jean-Pierre Changeux, is defined by the following rules:

1. An allosteric protein is an oligomer of symmetrically related subunits (although the α and β subunits of hemoglobin are only pseudosym-metrically related).

2. Each oligomer can exist in two conformational states, designated R and T; these states are in equilibrium.

3. The ligand can bind to a subunit in either conformation. *Only the con-formational change alters the affinity for the ligand.*

4. *The molecular symmetry of the protein is conserved during the con-formational change.* The subunits must therefore change conforma-tion in a concerted manner; in other words, there are no oligomers that simultaneously contain R- and T-state subunits.

The symmetry model is diagrammed for a tetrameric binding protein in Fig. 7-16. If a ligand binds more tightly to the R state than to the T state, ligand binding will promote the T → R shift, thereby increasing the affin-ity of the unliganded subunits for the ligand.

One major objection to the symmetry model is that it is difficult to be-lieve that oligomeric symmetry is perfectly preserved in all proteins, that is, that the T → R shift occurs simultaneously in all subunits regardless of the number of ligands bound. In addition, the symmetry model can account only for positive cooperativity, although some proteins exhibit negative cooperativity.

An alternative to the symmetry model is the **sequential model** of al-losterism, proposed by Daniel Koshland. According to this model, ligand binding induces a conformational change in the subunit to which it binds,

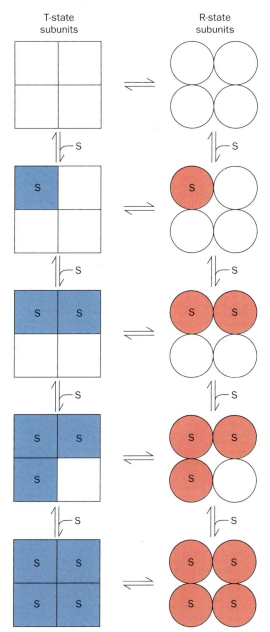

T-state
subunits

R-state
subunits

Figure 7-16   The symmetry model of allosterism.   Squares and circles represent T- and R-state subunits, respectively, of a tetrameric protein. The T and R states are in equilibrium regardless of the number of ligands (represented by S) that have bound to the protein. All the subunits must be in either the T or the R form; the model does not allow combinations of T- and R-state subunits in the same protein.

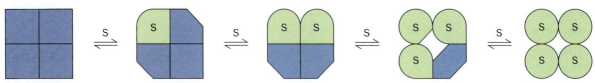

Figure 7-17   The sequential model of allosterism.   Ligand binding progressively induces conformational changes in the subunits, with the greatest changes occurring in those subunits that have bound ligand. The symmetry of the oligomeric protein is not preserved in this process as it is in the symmetry model.

and cooperative interactions arise through the influence of those conformational changes on neighboring subunits. The conformational changes occur sequentially as more ligand-binding sites are occupied (Fig. 7-17). The ligand-binding affinity of a subunit varies with its conformation and may be higher or lower than that of the subunits in the ligand-free protein. Thus, proteins that follow the sequential model of allosterism may be positively or negatively cooperative.

If the mechanical coupling between subunits in the sequential model is particularly strong, the conformational changes occur simultaneously and the oligomer retains its symmetry, as in the symmetry model. Thus, the symmetry model of allosterism may be considered to be an extreme case of the more general sequential model.

Oxygen binding to hemoglobin exhibits features of both models. The quaternary T → R conformational change is concerted, as the symmetry model requires. Yet ligand binding to the T state does cause small tertiary structural changes, as the sequential model predicts. These minor conformational shifts are undoubtedly responsible for the buildup of strain that eventually triggers the T → R transition. It therefore appears that the complexity of ligand–protein interactions in hemoglobin and other proteins allows binding processes to be fine-tuned to the needs of the organism under changing internal and external conditions. We shall revisit allosteric effects when we discuss enzymes in Chapter 12.

4   Abnormal Hemoglobins

Before the advent of recombinant DNA techniques, mutant hemoglobins provided what was essentially a unique opportunity to study structure–function relationships in proteins. This is because, for many years, hemoglobin was the only protein of known structure that had a large number of well-characterized naturally occurring **variants.** The examination of individuals with physiological disabilities, together with the routine electrophoretic screening of human blood samples, has led to the discovery of nearly 900 variant hemoglobins, >90% of which result from single amino acid substitutions in a globin polypeptide chain. Indeed, about 5% of the world's human population are carriers of an inherited variant hemoglobin.

Not all hemoglobin variants produce clinical symptoms, but some abnormal hemoglobin molecules do cause debilitating diseases (~300,000 individuals with serious hemoglobin disorders are born every year; naturally occurring hemoglobin variants that are lethal are, of course, never observed). Table 7-1 lists several of these hemoglobin variants. Mutations that destabilize hemoglobin's tertiary or quaternary structure alter hemoglobin's oxygen-binding affinity (p_{50}) and reduce its cooperativity (Hill coefficient). Moreover, the unstable hemoglobins are degraded by the erythrocytes, and their degradation products often cause the erythrocytes to **lyse**

Table 7-1 Some Hemoglobin Variants

Name[a]	Mutation	Effect
Hammersmith	Phe CD1(42)β → Ser	Weakens heme binding
Bristol	Val E11(67)β → Asp	Weakens heme binding
Bibba	Leu H19(136)α → Pro	Disrupts the H helix
Savannah	Gly B6(24)β → Val	Disrupts the B–E helix interface
Philly	Tyr C1(35)α → Phe	Disrupts hydrogen bonding at the α_1–β_1 interface
Boston	His E7(58)α → Tyr	Promotes methemoglobin formation
Milwaukee	Val E11(67)β → Glu	Promotes methemoglobin formation
Iwate	His F8(87)α → Tyr	Promotes methemoglobin formation
Yakima	Asp G1(99)β → His	Disrupts a hydrogen bond that stabilizes the T conformation
Kansas	Asn G4(102)β → Thr	Disrupts a hydrogen bond that stabilizes the R conformation

[a]Hemoglobin variants are usually named after the place where they were discovered (e.g., hemoglobin Boston).

(break open). The resulting **hemolytic anemia** (anemia is a deficiency of red blood cells) compromises O_2 delivery to tissues.

Certain mutations at the O_2-binding site of either the α or β chain favor the oxidation of Fe(II) to Fe(III). Individuals carrying the resulting methemoglobin subunit exhibit **cyanosis,** a bluish skin color, due to the presence of methemoglobin in their arterial blood. These hemoglobins have reduced cooperativity (Hill coefficient ~1.2 compared to a maximum value of 2, since only two subunits in each of these methemoglobins can bind oxygen).

Mutations that increase hemoglobin's oxygen affinity lead to increased numbers of erythrocytes in order to compensate for the less than normal amount of oxygen released in the tissues. Individuals with this condition, which is named **polycythemia,** often have a ruddy complexion.

Sickle-Cell Anemia.   Most harmful hemoglobin variants occur in only a few individuals, in many of whom the mutation apparently originated. However, ~10% of American blacks and as many as 25% of black Africans carry a single copy of (are **heterozygous** for) the gene for **sickle-cell hemoglobin (hemoglobin S).** Individuals who carry two copies of (are **homozygous** for) the gene for hemoglobin S suffer from **sickle-cell anemia,** in which deoxyhemoglobin S forms insoluble filaments that deform erythrocytes (Fig. 7-18). In this painful, debilitating, and often fatal disease, the rigid, sickle-shaped cells cannot easily pass through the capillaries. Consequently, in a sickle-cell "crisis," the blood flow to some tissues may be completely blocked, resulting in tissue death. In addition, the mechanical fragility of the misshapen cells results in hemolytic anemia. Heterozygotes, whose hemoglobin is ~40% hemoglobin S, usually lead a normal life, although their erythrocytes have a shorter than normal lifetime.

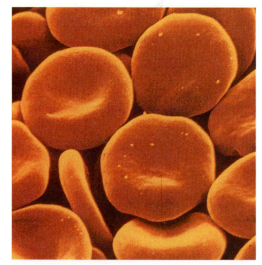

(a)

(b)

Figure 7-18 **Scanning electron micrographs of human erythrocytes.**   (*a*) Normal erythrocytes are flexible biconcave disks that can tolerate slight distortions as they pass through the capillaries (many of which have smaller diameters than erythrocytes). [David M. Phillips/Visuals Unlimited.] (*b*) Sickled erythrocytes from an individual with sickle-cell anemia are elongated and rigid and cannot easily pass through capillaries. [Bill Longcore/Photo Researchers, Inc.]

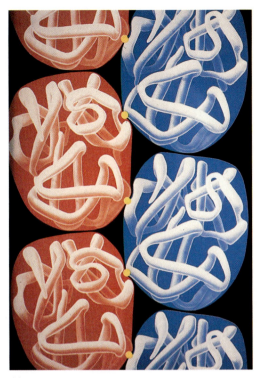

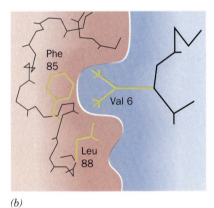

(a)

(b)

Figure 7-19 Structure of a deoxyhemoglobin S fiber.   (a) The arrangement of deoxyhemoglobin S molecules in the fiber. Only three subunits of each deoxyhemoglobin S molecule are shown. (b) The side chain of the mutant Val 6 in the β₂ chain of one hemoglobin S molecule (yellow knob in Part a) binds to a hydrophobic pocket on the β₁ subunit of a neighboring deoxyhemoglobin S molecule. [Illustration, Irving Geis. Image from the Irving Geis Collection/Howard Hughes Medical Institute. Rights owned by HHMI. Reproduction by permission only.]

In 1945, Linus Pauling hypothesized that sickle-cell anemia was the result of a mutant hemoglobin, and in 1949 he showed that the mutant hemoglobin had a less negative ionic charge than normal adult hemoglobin. This was the first evidence that a disease could result from an alteration in the molecular structure of a protein. Furthermore, since sickle-cell anemia is an inherited disease, a defective gene must be responsible for the abnormal protein. Nevertheless, the molecular defect in sickle-cell hemoglobin was not identified until 1956, when Vernon Ingram showed that hemoglobin S contains Val rather than Glu at the sixth position of each β chain. This was the first time an inherited disease was shown to arise from a specific amino acid change in a protein. At present, over 1500 "disease genes" have been identified in which a mutation leads to the production of a defective protein that is directly responsible for the pathological effects of the disease.

The X-ray structure of deoxyhemoglobin S has revealed that one mutant Val side chain in each hemoglobin S tetramer nestles into a hydrophobic pocket on the surface of a β subunit in another hemoglobin tetramer (Fig. 7-19). This intermolecular contact allows hemoglobin S tetramers to form linear polymers. Aggregates of 14 strands that wind around each other form fibers with a diameter of ~220 Å. The fibers extend throughout the length of the erythrocyte (Fig. 7-20). The hydrophobic pocket on the β subunit cannot accommodate the normally occurring Glu side chain, and the pocket is absent in oxyhemoglobin. Consequently, neither normal hemoglobin nor oxyhemoglobin S can polymerize. In fact, hemoglobin S fibers dissolve essentially instantaneously on oxygenation, so none are present in arterial blood. The danger of sickling is greatest when erythrocytes pass through the capillaries, where deoxygenation occurs. The polymerization of hemoglobin S molecules is time and concentration dependent, which explains why blood flow blockage occurs only sporadically (in a sickle-cell "crisis").

Interestingly, many hemoglobin S homozygotes have only a mild form of sickle-cell anemia because they express relatively high levels of fetal hemoglobin, which contains γ chains rather than the defective β chains. The fetal hemoglobin dilutes the hemoglobin S, making it more difficult for

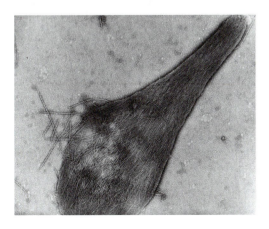

Figure 7-20 Electron micrograph of deoxyhemoglobin S fibers spilling out of a ruptured erythrocyte.   [Courtesy of Robert Josephs, University of Chicago.]

hemoglobin S to aggregate during the 10–20 s it takes for an erythrocyte to travel from the tissues to the lungs for reoxygenation. The administration of **hydroxyurea,**

$$H_2N - \overset{\overset{\displaystyle O}{\|}}{C} - NH - OH$$

Hydroxyurea

the first and as yet the only effective treatment for sickle-cell anemia, ameliorates the symptoms of sickle-cell anemia by increasing the fraction of cells containing fetal hemoglobin (although the mechanism whereby hydroxyurea acts is unknown).

Malaria and Hemoglobin S.   Before the advent of modern palliative therapies, individuals with sickle-cell anemia rarely survived to maturity. Natural selection has not minimized the prevalence of the hemoglobin S variant, however, because heterozygotes are more resistant to **malaria.** *This disease is the most lethal infectious disease that presently affects humanity:* Of the 2.5 billion people living within malaria-endemic areas, 100 million are clinically ill with the disease at any given time and around 1 million, mostly very young children, die from it each year. Malaria is caused by the mosquito-borne protozoan *Plasmodium falciparum,* which resides within an erythrocyte during much of its 48-h life cycle. Infected erythrocytes adhere to capillary walls, causing death when cells impede blood flow to a vital organ.

The regions of equatorial Africa where malaria is a major cause of death coincide closely with those areas where the sickle-cell gene is prevalent (Fig. 7-21), thereby suggesting that the sickle-cell gene confers resistance to malaria. How does it do so? Plasmodia increase the acidity of infected erythrocytes by ~0.4 pH units. The lower pH favors the formation of deoxyhemoglobin via the Bohr effect, thereby increasing the likelihood of sickling in erythrocytes that contain hemoglobin S. Erythrocytes damaged by sickling are normally removed from the circulation by the spleen. During the early stages of a malarial infection, parasite-enhanced sickling probably allows the spleen to preferentially remove infected erythrocytes. In the later stages of infection, when the parasitized erythrocytes attach to the capillary walls (presumably to prevent the spleen from removing them from the circulation), sickling may mechanically disrupt the parasite. Consequently, heterozygous carriers of hemoglobin S in a malarial region have an adaptive advantage: They are more likely to survive to maturity than individuals who are homozygous for normal hemoglobin. Thus, in malarial regions, the fraction of the population who are heterozygotes for the sickle-cell gene increases until their reproductive advantage is balanced by the correspondingly increased proportion of homozygotes (who, without modern medical treatment, die in childhood).

Genetic analysis indicates that the gene for hemoglobin S arose several times, in different parts of the world. Another hemoglobin variant, called **hemoglobin C,** also occurs in malarial regions of West Africa. In hemoglobin C, Glu 6 of the β chain (the same residue mutated in hemoglobin S) is replaced by Lys. The resulting mutant hemoglobin does not polymerize as hemoglobin S does, but it does protect against malaria. Indeed, homozygotes and heterozygotes for hemoglobin C have a 93% and 29% reduced rate of contracting malaria (compared to 73% reduced rate for hemoglobin S heterozygotes). The antimalarial mechanism of hemoglobin C is not understood, but the mutant protein causes only mild hemolytic anemia in the homozygous state, compared to hemoglobin S.

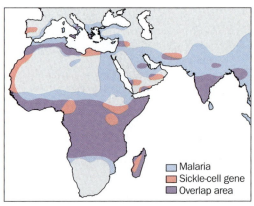

☐ Malaria
☐ Sickle-cell gene
☐ Overlap area

Figure 7-21 Correspondence between malaria and the sickle-cell gene.   The blue areas of the map indicate regions where malaria is or was prevalent. The pink areas represent the distribution of the gene for hemoglobin S. Note the overlap (*purple*) of the distributions.

The mutation that produced hemoglobin C occurred an estimated 1000 years ago, and the gene appears to have replaced that for hemoglobin S in some parts of Africa. However, the selective pressure that would favor the eventual replacement of hemoglobin S by hemoglobin C appears to be diminished by the fact that hemoglobin C protects against malaria in mainly the homozygous state.

 ## SUMMARY

1. Myoglobin, a monomeric heme-containing muscle protein, reversibly binds a single O_2 molecule.

2. Hemoglobin, a tetramer with pseudo-D_2 symmetry, has distinctly different conformations in its oxy and deoxy states.

3. Oxygen binds to hemoglobin in a sigmoidal fashion, indicating cooperative binding.

4. O_2 binding to a heme group induces a conformational change in the entire hemoglobin molecule that includes movements at the subunit interfaces and the disruption of ion pairs. The result is a shift from the T to the R state.

5. CO_2 promotes O_2 dissociation from hemoglobin through the Bohr effect. BPG decreases hemoglobin's O_2 affinity by binding to deoxyhemoglobin.

6. The symmetry and sequential models of allosterism explain how binding of a ligand at one site affects binding of another ligand at a different site.

7. Hemoglobin variants have revealed structure–function relationships. Hemoglobin S produces the symptoms of sickle-cell anemia by forming rigid fibers in the deoxy form.

REFERENCES

General

Dickerson, R.E. and Geis, I., *Hemoglobin*, Benjamin/Cummings (1983). [A beautifully written and lavishly illustrated treatise on the structure, function, and evolution of hemoglobin.]

Hsia, C.C.W., Respiratory function of hemoglobin, *New Engl. J. Med.* **338**, 239–247 (1998). [A short review of hemoglobin's physiological role.]

Judson, H.F., *The Eighth Day of Creation* (Expanded edition), Chapters 9 and 10, Cold Spring Harbor Laboratory Press (1996). [Includes a fascinating historical account of how our present perception of hemoglobin structure and function came about.]

van Holde, K.E., Miller, K.I., and Decker, H., Hemocyanins and invertebrate evolution, *J. Biol. Chem.* **276**, 15563–15566 (2001). [Describes the structures of arthropod and molluscan hemocyanins and their possible evolutionary origins.]

Oxygen Binding to Hemoglobin

Ackers, G.A., Deciphering the molecular code of hemoglobin allostery, *Adv. Prot. Chem.* **51**, 185–253 (1998). [Presents thermodynamic arguments that both of hemoglobin's αβ dimers must be ligated for quaternary switching to occur.]

Barrick, D., Ho, N.T., Simplaceanu, V., Dahlquist, F.W., and Ho, C., A test of the role of the proximal histidines in the Perutz model for cooperativity in haemoglobin, *Nature Struct. Biol.* **4**, 78–83 (1997). [Describes the experiments in which the proximal His is detached from the F helix.]

Eaton, W.A., Henry, E.R., Hofrichter, J., and Mozzarelli, A., Is cooperative oxygen binding by hemoglobin really understood? *Nature Struct. Biol.* **6**, 351–359 (1999). [Discusses the theoretical and experimental underpinnings of the two-state model for the allosteric behavior of hemoglobin.]

Perutz, M.F., Wilkinson, A.J., Paoli, M., and Dodson, G.G., The stereochemical mechanism of the cooperative effects in hemoglobin revisited, *Annu. Rev. Biophys. Biomol. Struct.* **27**, 1–34 (1998).

Abnormal Hemoglobins

Allison, A.C., The discovery of resistance to malaria of sickle-cell heterozygotes, *Biochem. Mol. Biol. Educ.* **30**, 279–287 (2002).

Nagel, R.L., Haemoglobinopathies due to structural mutations, *in* Provan, D. and Gribben, J. (Eds.), *Molecular Haematology*, pp. 121–133, Blackwell Science (2000).

Strasser, B.J., Sickle-cell anemia, a molecular disease, *Science* **286**, 1488–1490 (1999). [A short history of Pauling's characterization of sickle-cell anemia].

 ## KEY TERMS

heme	sigmoidal curve	R state	anemia
oxygenation	cooperative binding	Bohr effect	cyanosis
Y_{O_2}	Hill equation	erythrocyte	polycythemia
pO_2	Hill coefficient	allosteric interaction	heterozygote
hyperbolic curve	positive cooperativity	symmetry model	homozygote
saturation	negative cooperativity	sequential model	sickle-cell anemia
p_{50}	Perutz mechanism	variant	
ligand	T state	lyse	

STUDY EXERCISES

1. Describe the structural and functional differences between myoglobin and hemoglobin.

2. Explain how hemoglobin effectively delivers O_2 to myoglobin in muscles.

3. Explain the structural basis for cooperative O_2 binding to hemoglobin.

4. What is the physiological relevance of the Bohr effect?

5. How does BPG affect O_2 binding to hemoglobin?

6. Differentiate the symmetry and sequential models of allosterism.

7. Explain why mutations can potentially increase or decrease the oxygen affinity of hemoglobin. How can the body compensate for these changes?

PROBLEMS

1. Estimate K from the following data describing ligand binding to a protein.

[Ligand] (mM)	Y
0.25	0.30
0.5	0.45
0.8	0.56
1.4	0.66
2.2	0.80
3.0	0.83
4.5	0.86
6.0	0.93

2. Which set of binding data is likely to represent cooperative ligand binding to an oligomeric protein?

(a) [Ligand] (mM)	Y	(b) [Ligand] (mM)	Y
0.1	0.3	0.2	0.1
0.2	0.5	0.3	0.3
0.4	0.7	0.4	0.6
0.7	0.9	0.6	0.8

3. In active muscles, the pO_2 may be 10 torr at the cell surface and 1 torr at the mitochondria (the organelles where oxidative metabolism occurs). Use Eq. 7-6 to show how myoglobin ($p_{50} = 2.8$ torr) facilitates the diffusion of O_2 through these cells.

4. In humans, the urge to breathe results from high concentrations of CO_2 in the blood; there are no direct physiological sensors of blood pO_2. Skindivers often hyperventilate (breathe rapidly and deeply for several minutes) just before making a dive in the belief that this will increase the O_2 content of their blood. (a) Does it do so? (b) Use your knowledge of hemoglobin function to evaluate whether this practice is useful.

5. Drinking a few drops of a commercial preparation called "vitamin O," which consists of oxygen and sodium chloride dissolved in water, is claimed to increase the concentration of oxygen in the body. (a) Use your knowledge of oxygen transport to evaluate this claim. (b) Would vitamin O be more or less effective if it were infused directly into the bloodstream?

6. Is the p_{50} higher or lower than normal in (a) hemoglobin Yakima and (b) hemoglobin Kansas? Explain.

7. Hemoglobin S homozygotes who are severely anemic often have elevated levels of BPG in their erythrocytes. Is this a beneficial effect?

8. In hemoglobin Rainier, Tyr 145β is replaced by Cys, which forms a disulfide bond with another Cys residue in the same subunit. This prevents the formation of ion pairs that normally stabilize the T state. How does hemoglobin Rainier differ from normal hemoglobin with respect to (a) oxygen affinity, (b) the Bohr effect, and (c) the Hill coefficient?

9. The crocodile, which can remain under water without breathing for up to 1 h, drowns its air-breathing prey and then dines at its leisure. An adaptation that aids the crocodile in doing so is that it can utilize virtually 100% of the O_2 in its blood whereas humans, for example, can extract only ~65% of the O_2 in their blood. Crocodile Hb does not bind BPG. However, crocodile deoxyHb preferentially binds HCO_3^-. How does this help the crocodile obtain its dinner?

10. Some primitive animals have a hemoglobin that consists of two identical subunits.

    (a) Sketch an oxygen-binding curve for this protein.

    (b) What is the likely range of the Hill coefficient for this hemoglobin?

Sugars are relatively simple molecules that can be linked together in various ways to form larger molecules, for example, starch. This storage form of carbohydrate is the primary source of energy in many foods, including bread, rice, and pasta. [Charles D. Winters/Photo Researchers.]

Carbohydrates

Carbohydrates or **saccharides** (Greek: *sakcharon,* sugar) are the most abundant biological molecules. They are chemically simpler than nucleotides or amino acids, containing just three elements—carbon, hydrogen, and oxygen—combined according to the formula $(C \cdot H_2O)_n$, where $n \geq 3$. The basic carbohydrate units are called **monosaccharides.** There are numerous different types of monosaccharides, which, as we discuss below, differ in their number of carbon atoms and in the arrangement of the H and O atoms attached to these carbons. Furthermore, monosaccharides can be strung together in almost limitless ways to form **polysaccharides.**

Until the 1960s, carbohydrates were thought to have only passive roles as energy sources (e.g., glucose and starch) and as structural materials (e.g., cellulose). Carbohydrates, as we shall see, do not catalyze complex chemical reactions as do proteins, nor do carbohydrates replicate themselves as do nucleic acids. And because polysaccharides are not built according to a genetic "blueprint," as are nucleic acids and proteins, they tend to be much more heterogeneous—both in size and in composition—than other biological molecules.

However, it has become clear that the innate structural variation in carbohydrates is fundamental to their biological activity. The apparently haphazard arrangements of carbohydrates on proteins and on the surfaces of cells are the key to many recognition events between proteins and between cells. An understanding of carbohydrate structure, from the simplest monosaccharides to the most complex branched polysaccharides, is essential for appreciating the varied functions of carbohydrates in biological systems.

1 Monosaccharides

Monosaccharides, or simple sugars, are synthesized from smaller precursors that are ultimately derived from CO_2 and H_2O by photosynthesis.

A Classification of Monosaccharides

Monosaccharides are aldehyde or ketone derivatives of straight-chain polyhydroxy alcohols containing at least three carbon atoms. They are classified according to the chemical nature of their carbonyl group and the number of their C atoms. If the carbonyl group is an aldehyde, the sugar is an **aldose.** If the carbonyl group is a ketone, the sugar is a **ketose.** The smallest monosaccharides, those with three carbon atoms, are **trioses.** Those with four, five, six, seven, etc. C atoms are, respectively, **tetroses, pentoses, hexoses, heptoses,** etc.

The aldohexose **D-glucose** has the formula $(C \cdot H_2O)_6$:

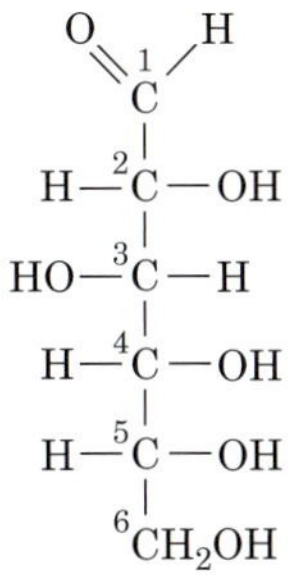

$$
\begin{array}{c}
\overset{1}{C}\diagup^{\displaystyle H}_{\displaystyle }\!\!\diagdown\!\!O \\
| \\
H-\overset{2}{C}-OH \\
| \\
HO-\overset{3}{C}-H \\
| \\
H-\overset{4}{C}-OH \\
| \\
H-\overset{5}{C}-OH \\
| \\
\overset{6}{C}H_2OH
\end{array}
$$

D-Glucose

All but two of its six C atoms, C1 and C6, are chiral centers, so D-glucose is one of $2^4 = 16$ possible stereoisomers. The stereochemistry and nomen-

clature of the D-aldoses are presented in Fig. 8-1. The assignment of D or L is made according to the Fischer convention (Section 4-2): *D sugars have the same absolute configuration at the asymmetric center farthest from their carbonyl group as does D-glyceraldehyde* (i.e., the —OH at C5 of D-glucose is on the right in a Fischer projection). The L sugars are the mirror images of their D counterparts. Because L sugars are biologically much less abundant than D sugars, the D prefix is often omitted.

Sugars that differ only by the configuration around one C atom are known as **epimers** of one another. Thus, D-glucose and **D-mannose** are epimers with respect to C2. The most common aldoses include the six-carbon sugars glucose, mannose, and **galactose.** The pentose **ribose** is a component of the ribonucleotide residues of RNA. The triose **glyceraldehyde** occurs in several metabolic pathways.

The most common ketoses are those with their ketone function at C2 (Fig. 8-2). The position of their carbonyl group gives ketoses one less asymmetric center than their isomeric aldoses, so a ketohexose has only $2^3 = 8$

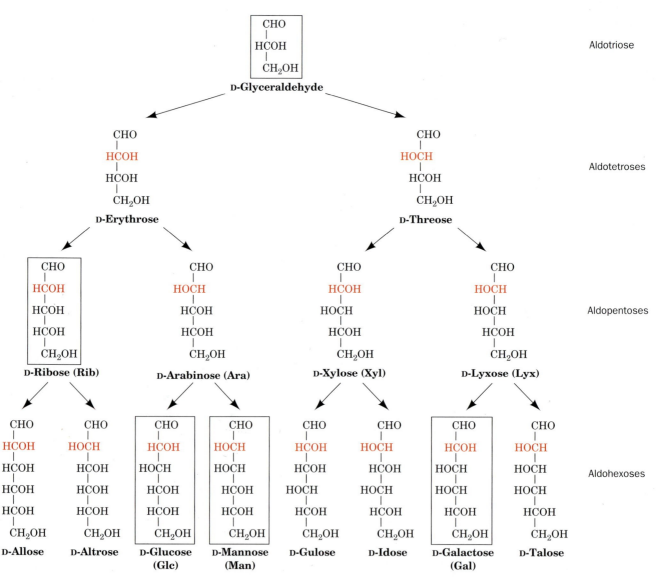

Figure 8-1 The D-aldoses with three to six carbon atoms.   The arrows indicate stereochemical relationships (not biosynthetic pathways). The configuration around C2 (*red*) distinguishes the members of each pair of monosaccharides. The L counterparts of these 15 sugars are their mirror images. The biologically most common aldoses are boxed.

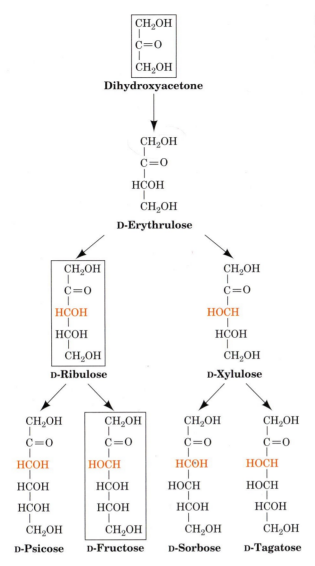

Figure 8-2 The D-ketoses with three to six carbon atoms. The configuration around C3 (*red*) distinguishes the members of each pair. The biologically most common ketoses are boxed.

possible stereoisomers (4 D sugars and 4 L sugars). Note that some ketoses are named by inserting *-ul-* before the suffix *-ose* in the name of the corresponding aldose; thus, D-xylulose is the ketose corresponding to the aldose D-xylose. The most common ketoses are **dihydroxyacetone, ribulose,** and **fructose,** which we shall encounter in our studies of metabolism.

B Configuration and Conformation

Alcohols react with the carbonyl groups of aldehydes and ketones to form **hemiacetals** and **hemiketals,** respectively:

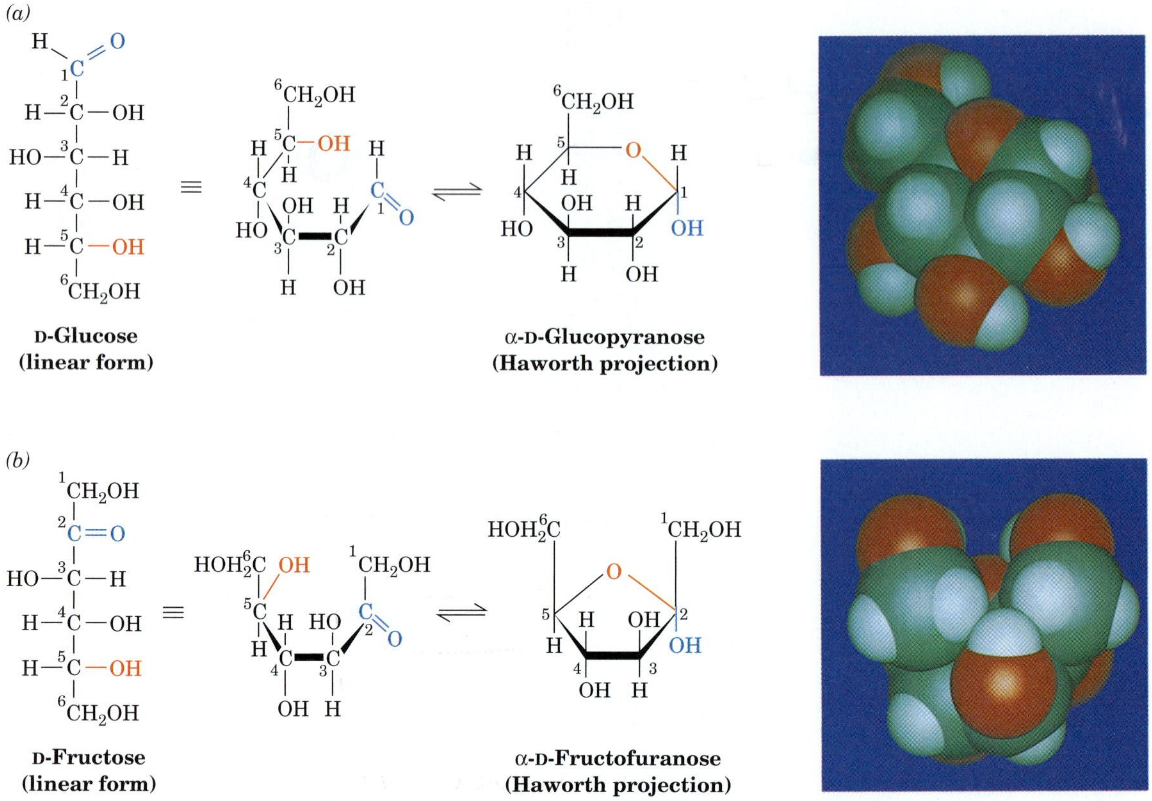

(a)

D-Glucose
(linear form)

α-D-Glucopyranose
(Haworth projection)

(b)

D-Fructose
(linear form)

α-D-Fructofuranose
(Haworth projection)

Figure 8-3 *Key to Structure.* **Cyclization of glucose and fructose.** (*a*) The linear form of D-glucose yields the cyclic hemiacetal α-D-glucopyranose. (*b*) The linear form of D-fructose yields the hemiketal α-D-fructofuranose. The cyclic sugars are shown as both Haworth projections and space-filling models. [Courtesy of Robert Stodola, Fox Chase Cancer Center.]

The hydroxyl and either the aldehyde or the ketone functions of monosaccharides can likewise react intramolecularly to form cyclic hemiacetals and hemiketals (Fig. 8-3). The configurations of the substituents of each carbon atom in these sugar rings are conveniently represented by their **Haworth projections,** in which the heavier ring bonds project in front of the plane of the paper and the lighter ring bonds project behind it.

A sugar with a six-membered ring is known as a **pyranose** in analogy with **pyran,** the simplest compound containing such a ring. Similarly, sugars with five-membered rings are designated **furanoses** in analogy with **furan:**

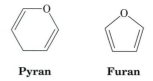

Pyran   **Furan**

The cyclic forms of glucose and fructose with six- and five-membered rings are therefore known as **glucopyranose** and **fructofuranose,** respectively.

Cyclic Sugars Have Two Anomeric Forms. When a monosaccharide cyclizes, the carbonyl carbon, called the **anomeric carbon,** becomes a chiral center with two possible configurations. The pair of stereoisomers that differ in configuration at the anomeric carbon are called **anomers.** In the **α anomer,** the OH substituent of the anomeric carbon is on the opposite side of the sugar ring from the CH_2OH group at the chiral center that designates the

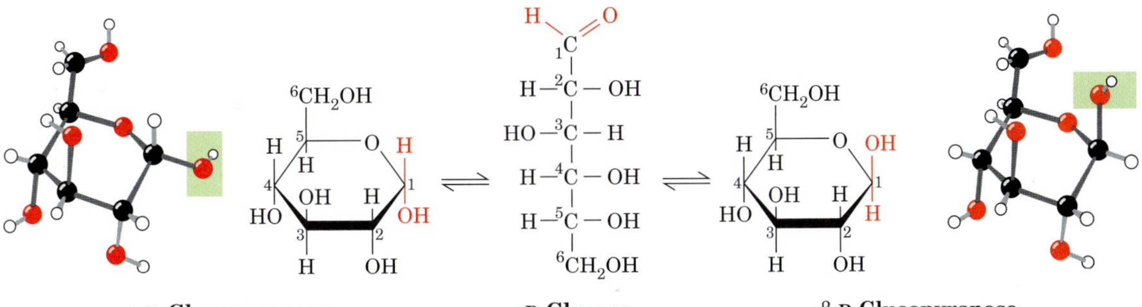

Figure 8-4 α and β anomers. The monosaccharides α-D-glucopyranose and β-D-glucopyranose, drawn as Haworth projections and ball-and-stick models, interconvert through the linear form. They differ only by their configuration about the anomeric carbon, C1. **See Kinemage Exercise 7-1.**

D or L configuration (C5 in hexoses). The other form is known as the **β anomer** (Fig. 8-4).

The two anomers of D-glucose have slightly different physical and chemical properties, including different optical rotations (Section 4-2). *The anomers freely interconvert in aqueous solution,* so at equilibrium, D-glucose is a mixture of the β anomer (63.6%) and the α anomer (36.4%). The linear form is normally present in only minute amounts.

Sugars Are Conformationally Variable. A given hexose or pentose can assume pyranose or furanose forms. In principle, hexoses and larger sugars can form rings of seven or more atoms, but such rings are rarely observed because of the greater stabilities of the five- and six-membered rings. The internal strain of three- and four-membered rings makes them less stable than the linear forms.

The use of Haworth formulas may lead to the erroneous impression that furanose and pyranose rings are planar. This cannot be the case, however, because all the atomic orbitals in the ring atoms are tetrahedrally (sp^3) hybridized. The pyranose ring, like the cyclohexane ring, can assume a chair conformation, in which the substituents of each atom are arranged tetrahedrally. Of the two possible chair conformations, the one that predominates is the one in which the bulkiest ring substituents occupy **equatorial** positions rather than the more crowded **axial** positions (Fig. 8-5). Only β-D-glucose can simultaneously have all five of its non-H substituents in equatorial positions. Perhaps this is why glucose is the most abundant monosaccharide in nature.

Furanose rings can also adopt different conformations, whose stabilities depend on the arrangements of bulky substituents. Note that a monosaccharide can readily shift its *conformation,* because no bonds are broken in

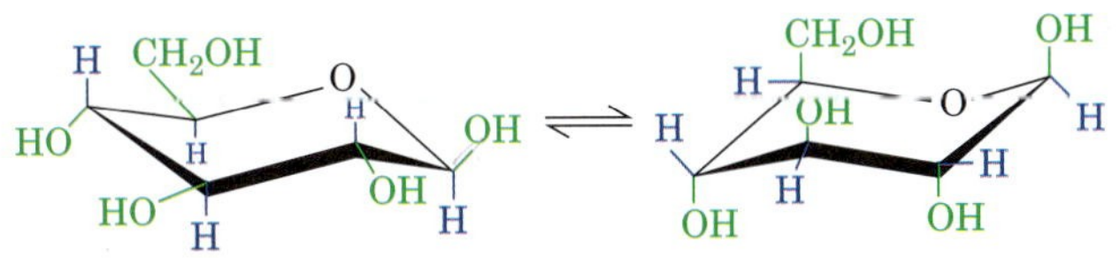

Figure 8-5 The two chair conformations of β-D-glucopyranose. In the conformation on the left, which predominates, the relatively bulky OH and CH₂OH substituents all occupy equatorial positions, where they extend alternately above and below the ring. In the conformation on the right (drawn in ball-and-stick form in Fig. 8-4, *right*), the bulky groups occupy the more crowded axial (vertical) positions. **See Kinemage Exercise 7-1.**

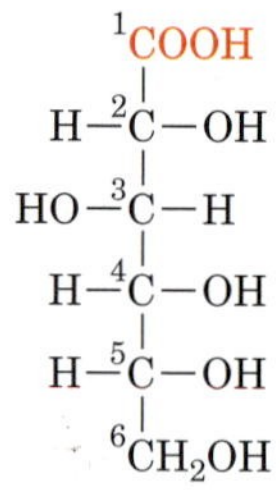

the process. The shift in *configuration* between the α and β anomeric forms or between the pyranose and furanose forms, which requires breaking and re-forming bonds, occurs slowly in aqueous solution. Other changes in configuration, such as **epimerization,** do not occur under physiological conditions without the appropriate enzyme.

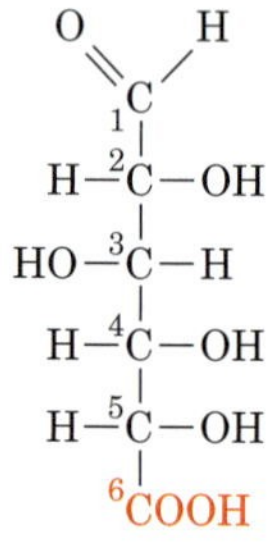

C Sugar Derivatives

Because the cyclic and linear forms of aldoses and ketoses do interconvert, these sugars undergo reactions typical of aldehydes and ketones.

1. Mild chemical or enzymatic oxidation of an aldose converts its aldehyde group to a carboxylic acid group, thereby yielding an **aldonic acid** such as **gluconic acid** (*at left*). Aldonic acids are named by appending the suffix *-onic acid* to the root name of the parent aldose.

2. The specific oxidation of the primary alcohol group of aldoses yields **uronic acids,** which are named by appending *-uronic acid* to the root name of the parent aldose, for example, **D-glucuronic acid** (*at left*). Uronic acids can assume the pyranose, furanose, and linear forms.

3. Aldoses and ketoses can be reduced under mild conditions, for example, by treatment with NaBH₄, to yield acyclic polyhydroxy alcohols known as **alditols,** which are named by appending the suffix *-itol* to the root name of the parent aldose. **Ribitol** is a component of flavin coenzymes (Fig. 13-11), and **glycerol** and the cyclic polyhydroxy alcohol *myo*-**inositol** are important lipid components (Section 9-1). **Xylitol** is a sweetener that is used in "sugarless" gum and candies:

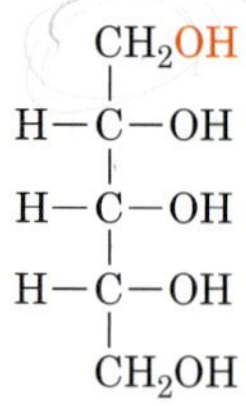

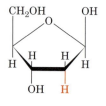

D-Gluconic acid

D-Glucuronic acid

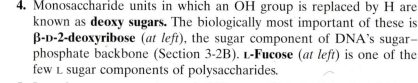

Ribitol 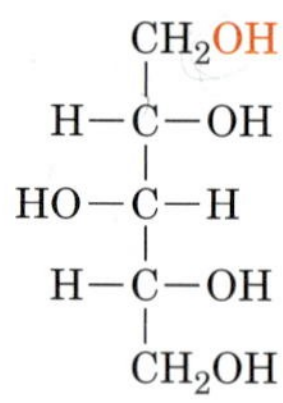   **Xylitol** 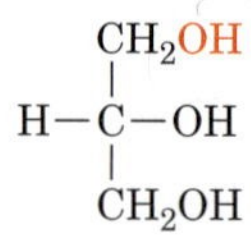   **Glycerol** 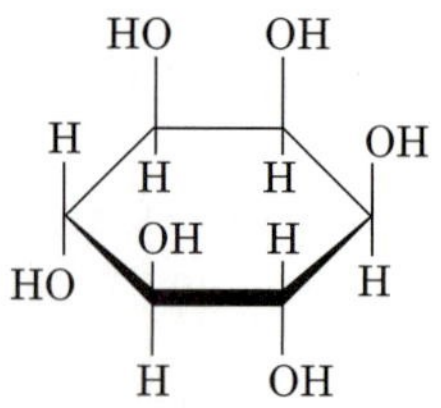   ***myo*-Inositol**

4. Monosaccharide units in which an OH group is replaced by H are known as **deoxy sugars.** The biologically most important of these is **β-D-2-deoxyribose** (*at left*), the sugar component of DNA's sugar–phosphate backbone (Section 3-2B). **L-Fucose** (*at left*) is one of the few L sugar components of polysaccharides.

5. In **amino sugars,** one or more OH groups have been replaced by an amino group, which is often acetylated. **D-Glucosamine** and **D-galactosamine** are the most common:

β-D-2-Deoxyribose

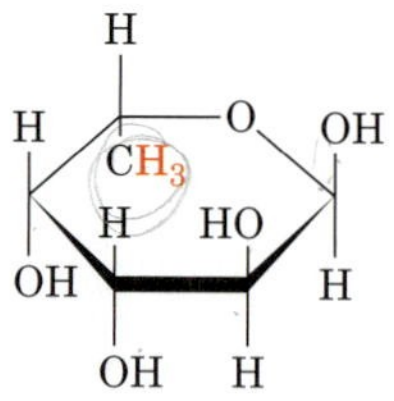

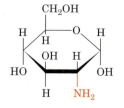

α-L-Fucose

α-D-Glucosamine
(2-amino-2-deoxy-
α-D-glucopyranose)

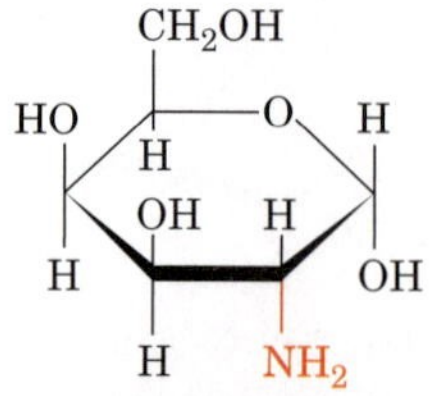

α-D-Galactosamine
(2-amino-2-deoxy-
α-D-galactopyranose)

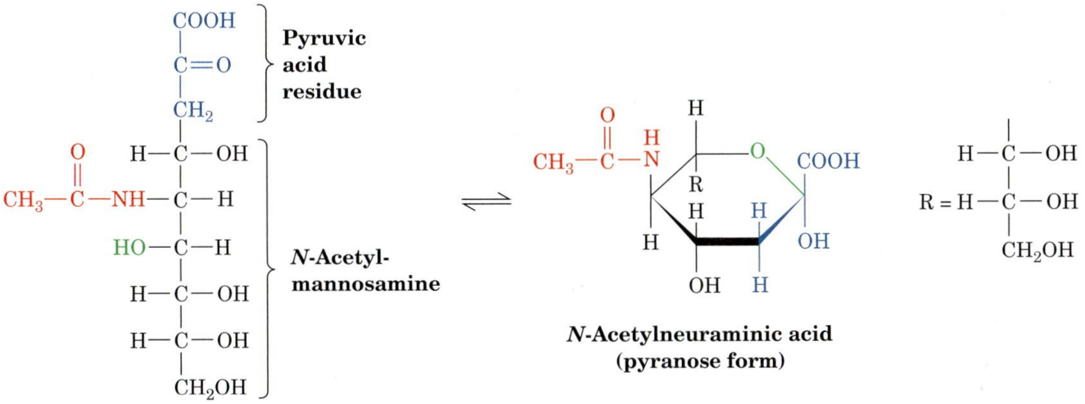

***N*-Acetylneuraminic acid (linear form)**

***N*-Acetylneuraminic acid (pyranose form)**

Figure 8-6 ***N*-Acetylneuraminic acid.** In the cyclic form of this nine-carbon monosaccharide, the pyranose ring incorporates the pyruvic acid residue (*blue*) and part of the mannose moiety.

***N*-Acetylneuraminic acid,** which is derived from ***N*-acetylmannosamine** and **pyruvic acid** (Fig. 8-6), is an important constituent of **glycoproteins** and **glycolipids** (proteins and lipids with covalently attached carbohydrate). *N*-Acetylneuraminic acid and its derivatives are often referred to as **sialic acids.**

Glycosidic Bonds Link the Anomeric Carbon to Other Compounds. The anomeric group of a sugar can condense with an alcohol to form **α- and β-glycosides** (Greek: *glykys*, sweet; Fig. 8-7), which are cyclic **acetals** or **ketals:**

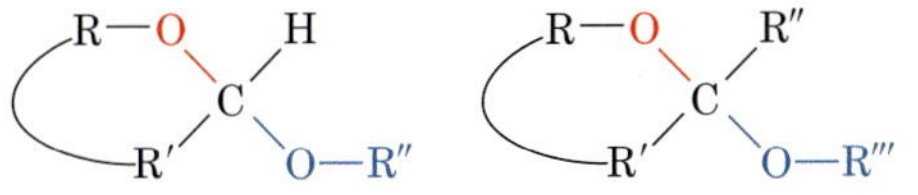

A cyclic acetal **A cyclic ketal**

The bond connecting the anomeric carbon to the alcohol oxygen is termed a **glycosidic bond.** *N*-Glycosidic bonds, which form between the

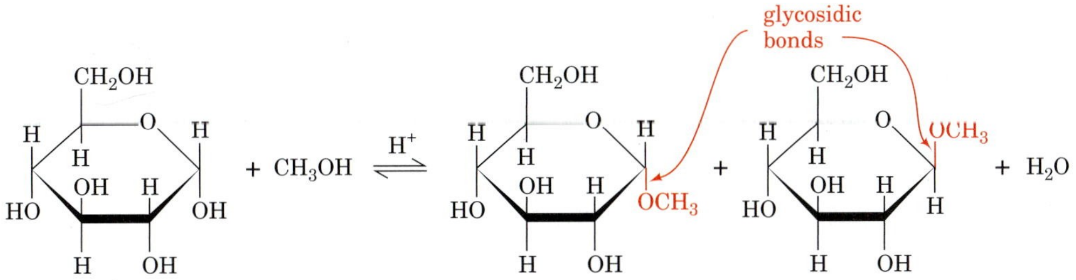

α-D-Glucose **Methyl-α-D-glucoside** **Methyl-β-D-glucoside**

Figure 8-7 **Formation of glycosides.** The acid-catalyzed condensation of α-D-glucose with methanol yields an anomeric pair of **methyl-D-glucosides.**

anomeric carbon and an amine, are the bonds that link D-ribose to purines and pyrimidines in nucleic acids:

N-glycosidic bonds

Like peptide bonds, glycosidic bonds hydrolyze extremely slowly under physiological conditions in the absence of appropriate hydrolytic enzymes. Consequently, an anomeric carbon that is involved in a glycosidic bond cannot freely convert between its α and β anomeric forms. Saccharides bearing anomeric carbons that have not formed glycosides are termed **reducing sugars,** because the free aldehyde group that is in equilibrium with the cyclic form of the sugar reduces mild oxidizing agents. Identification of a sugar as **nonreducing** is evidence that it is a glycoside.

2 Polysaccharides

*Polysaccharides, which are also known as **glycans,** consist of mono-saccharides linked together by glycosidic bonds.* They are classified as **homopolysaccharides** or **heteropolysaccharides** if they consist of one type or more than one type of monosaccharide. Although the monosaccharide sequences of heteropolysaccharides can, in principle, be even more varied than those of proteins, many are composed of only a few types of mono-saccharides that alternate in a repetitive sequence.

Polysaccharides, in contrast to proteins and nucleic acids, form branched as well as linear polymers. This is because glycosidic linkages can be made to any of the hydroxyl groups of a monosaccharide. Fortunately for structural biochemists, most polysaccharides are linear and those that branch do so in only a few well-defined ways.

A complete description of an **oligosaccharide** or polysaccharide includes the identities, anomeric forms, and linkages of all its component monosac-charide units. Some of this information can be gathered through the use of specific **exoglycosidases** and **endoglycosidases,** enzymes that hydrolyze monosaccharide units in much the same way that exopeptidases and en-

dopeptidases cleave amino acid residues from polypeptides (Section 5-3B). NMR measurements are also invaluable in determining both sequences and conformations of polysaccharides.

A Disaccharides

Oligosaccharides containing three or more residues are relatively rare, occurring almost entirely in plants. **Disaccharides,** the simplest polysaccharides, are more common. Many occur as the hydrolysis products of larger molecules. However, two disaccharides are notable in their own right. **Lactose** (*at right*), for example, occurs naturally only in milk, where its concentration ranges from 0 to 7% depending on the species (see Box 8-1). The systematic name for lactose, *O*-β-D-galactopyranosyl-(1→4)-D-glucopyranose, specifies its monosaccharides, their ring types, and how they are linked together. The symbol (1→4) combined with the β in the prefix indicates that the glycosidic bond links C1 of the β anomer of galactose to O4 of glucose. Note that lactose has a free anomeric carbon on its glucose residue and is therefore a reducing sugar.

The most abundant disaccharide is **sucrose** (*at right*), the major form in which carbohydrates are transported in plants. Sucrose is familiar to us as common table sugar (**see Kinemage Exercise 7-2).** The systematic name for sucrose, *O*-α-D-glucopyranosyl-(1→2)-β-D-fructofuranoside, indicates that the anomeric carbon of each sugar (C1 in glucose and C2 in fructose) participates in the glycosidic bond and hence sucrose is not a reducing sugar. Noncarbohydrate molecules that mimic the taste of sucrose are used as sweetening agents in foods and beverages (Box 8-2).

Galactose Glucose

Lactose

Glucose Fructose

Sucrose

B Structural Polysaccharides: Cellulose and Chitin

Plants have rigid cell walls that can withstand osmotic pressure differences between the extracellular and intracellular spaces of up to 20 atm. In large plants, such as trees, the cell walls also have a load-bearing function.

BOX 8-1

Biochemistry in Health and Disease

Lactose Intolerance

In infants, lactose (also known as milk sugar) is hydrolyzed by the intestinal enzyme **β-D-galactosidase** (or **lactase**) to its component monosaccharides for absorption into the bloodstream. The galactose is enzymatically converted (epimerized) to glucose, which is the primary metabolic fuel of many tissues.

Since mammals are unlikely to encounter lactose after they have been weaned, most adult mammals have low levels of β-galactosidase. Consequently, much of the lactose they might ingest moves through their digestive tract to the colon, where bacterial fermentation generates large quantities of CO_2, H_2, and irritating organic acids. These products

cause the embarrassing and often painful digestive upset known as **lactose intolerance.**

Lactose intolerance, which was once considered a metabolic disturbance, is actually the norm in adult humans, particularly those of African and Asian descent. Interestingly, however, β-galactosidase levels decrease only mildly with age in descendants of populations that have historically relied on dairy products for nutrition throughout life. Modern food technology has come to the aid of milk lovers who develop lactose intolerance: Milk in which the lactose has been hydrolyzed enzymatically is widely available.

BOX 8-2

Perspectives in Biochemistry

Artificial Sweeteners

Artificial sweeteners are added to processed foods and beverages to impart a sweet taste without adding calories. This is possible because the compounds mimic sucrose in its interactions with taste receptors but either are not metabolized or contribute very little to energy metabolism because they are used at such low concentrations.

Naturally occurring saccharides, such as fructose, are slightly sweeter than sucrose. Honey, which contains primarily fructose, glucose, and maltose (a glucose disaccharide), is about 1.5 times as sweet as sucrose. How is sweetness measured? There is no substitute for the human sense of taste, so a panel of individuals sample solutions of a compound and compare them to a reference solution containing sucrose. The very sweet compounds listed in the table below must be diluted significantly before testing in this manner.

Compound	Sweetness Relative to Sucrose
Acesulfame	200
Alitame	2000
Aspartame	180
Saccharin	350
Sucralose	600

One of the oldest artificial sweeteners is saccharin (*at right*), discovered in 1879 and commonly consumed as Sweet N Low®. In the 1970s, extremely high doses of saccharin were found to cause cancer in laboratory rats. Such doses are now considered to be so far outside of the range used for sweetening as to be of insignificant concern to users.

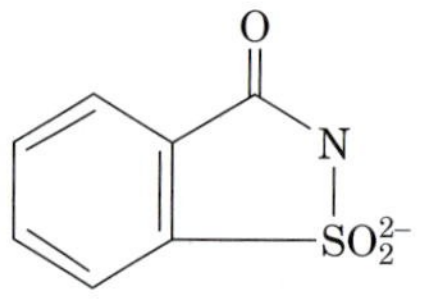

Saccharin

Aspartame, the active ingredient in Nutrasweet® and Equal®, was approved for human use in 1981 and is currently the market leader:

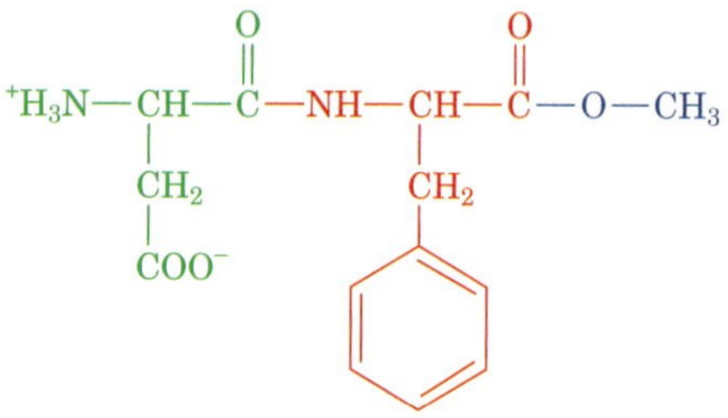

Aspartylphenylalanine methyl ester (aspartame)

Unlike saccharin, which is not metabolized by the human body, aspartame is broken down into its components: aspartate (*green*), phenylalanine (*red*), and methanol (*blue*). The Asp and Phe, like all amino acids, can be metabolized, so aspartame is not calorie-free. Methanol in large amounts is toxic; however, the amount derived from an aspartame-sweetened drink is comparable to the amount naturally present in the same volume of fruit juice. Individuals with the genetic disease **phenylketonuria**, who are unable to metabolize phenylalanine, are advised to avoid ingesting excess Phe in the form of aspartame (or any other polypeptide). The greatest drawback of aspartame may be its instability to heat, which makes it unsuitable for baking. In addition, aspartame in soft drinks hydrolyzes over a period of months and hence loses its flavor.

Acesulfame

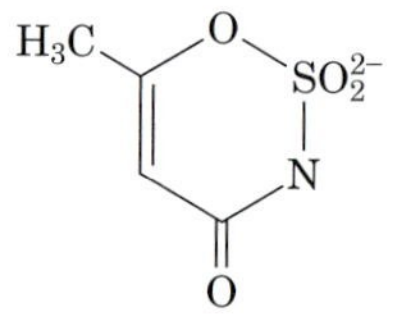

Acesulfame

is sometimes used in combination with aspartame, since the two compounds act synergistically (that is, their sweetness when combined is greater than the sum of their individual sweetnesses). Other artificial sweeteners are derivatives of sugars, such as sucralose (Splenda®; see Problem 8-8), or of aspartame (for example, alitame). Some plant extracts (for example, *Stevia*) are also used as artificial sweeteners.

The market for artificial sweeteners is worth several billion dollars annually. But surprisingly, the most successful sweetening agents have not been the result of dedicated research efforts. Instead, they were discovered by chance or mishap. For example, aspartame was discovered in 1965 by a synthetic chemist who unknowingly got a small amount of the compound on his fingers and happened to lick them. Sucralose came to light when a student was asked to "test" a compound and misunderstood the directions as "taste" the compound.

Cellulose, the primary structural component of plant cell walls (Fig. 8-8), accounts for over half of the carbon in the biosphere: Approximately 10^{15} kg of cellulose is estimated to be synthesized and degraded annually.

Cellulose is a linear polymer of up to 15,000 D-glucose residues linked by β(1→4) glycosidic bonds:

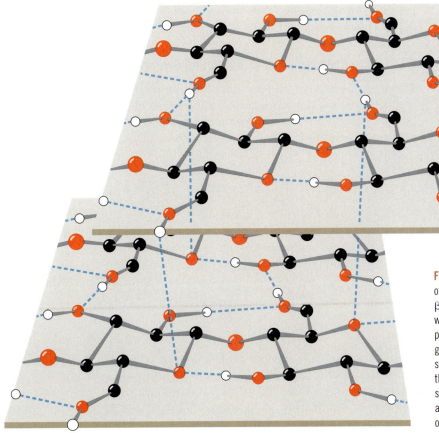

Cellulose

Figure 8-8 Electron micrograph of a plant cell wall. The cellulose fibers in this sample of cell wall from the alga *Chaetomorpha* are arranged in layers. [Biophoto Associates/Photo Researchers.]

X-Ray studies of cellulose fibers led Anatole Sarko to propose the structure diagrammed in Fig. 8-9. This highly cohesive, hydrogen-bonded structure gives cellulose fibers exceptional strength and makes them water insoluble despite their hydrophilicity. In plant cell walls, the cellulose fibers are embedded in and cross-linked by a matrix containing other polysaccharides and **lignin,** a plasticlike phenolic polymer. The resulting composite material can withstand large stresses because the matrix evenly distributes the stresses among the cellulose reinforcing elements.

Although vertebrates themselves do not possess an enzyme capable of hydrolyzing the β(1→4) linkages of cellulose, the digestive tracts of herbi-

Figure 8-9 Model of cellulose. Cellulose fibers consist of ~40 parallel, extended glycan chains. Each of the β(1→4)-linked glucose units in a chain is rotated 180° with respect to its neighboring residues and is held in this position by intrachain hydrogen bonds (*dashed lines*). The glycan chains line up laterally to form sheets, and these sheets stack vertically so that they are staggered by half the length of a glucose unit. The entire assembly is stabilized by intermolecular hydrogen bonds. Hydrogen atoms not participating in hydrogen bonds have been omitted for clarity.

vores contain symbiotic microorganisms that secrete a series of enzymes, collectively known as **cellulases,** that do so. The same is true of termites. Nevertheless, the degradation of cellulose is a slow process because its tightly packed and hydrogen-bonded glycan chains are not easily accessible to cellulase and do not separate readily even after many of their glycosidic bonds have been hydrolyzed. Thus, cows must chew their cud, and the decay of dead trees by fungi and other organisms generally takes many years.

Chitin is the principal structural component of the exoskeletons of invertebrates such as crustaceans, insects, and spiders and is also present in the cell walls of most fungi and many algae. It is therefore almost as abundant as cellulose. Chitin is a homopolymer of $\beta(1 \rightarrow 4)$-linked N-acetyl-D-glucosamine residues:

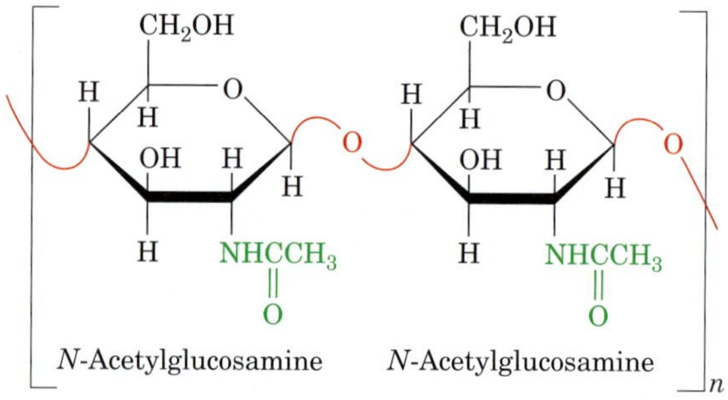

Chitin

It differs chemically from cellulose only in that each C2–OH group is replaced by an acetamido function. X-Ray analysis indicates that chitin and cellulose have similar structures.

C Storage Polysaccharides: Starch and Glycogen

Starch is a mixture of glycans that plants synthesize as their principal energy reserve. It is deposited in the chloroplasts of plant cells as insoluble granules composed of **α-amylose** and **amylopectin.** α-Amylose is a linear polymer of several thousand glucose residues linked by $\alpha(1 \rightarrow 4)$ bonds:

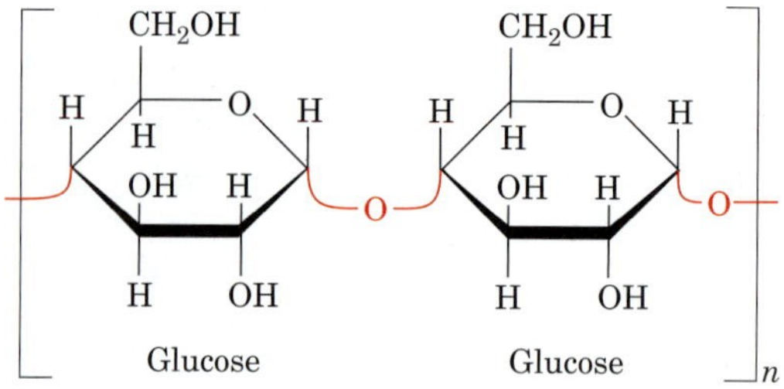

α-Amylose

Note that although α-amylose is an isomer of cellulose, it has very different structural properties. While cellulose's β-glycosidic linkages cause it to assume a tightly packed, fully extended conformation (Fig. 8-9), α-amylose's α-glycosidic bonds cause it to adopt an irregularly aggregating helically coiled conformation (Fig. 8-10).

Amylopectin consists mainly of α(1→4)-linked glucose residues but is a branched molecule with α(1→6) branch points every 24 to 30 glucose residues on average:

Amylopectin

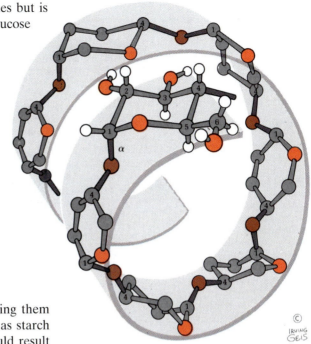

Figure 8-10 α-Amylose. This regularly repeating polymer forms a left-handed helix. Note the great differences in structure and properties that result from changing α-amylose's α(1→4) linkages to the β(1→4) linkages of cellulose (Fig. 8-9). [Illustration, Irving Geis/Geis Archives Trust. Copyright Howard Hughes Medical Institute. Reproduced with permission.]

Amylopectin molecules contain up to 10^6 glucose residues, making them some of the largest molecules in nature. The storage of glucose as starch greatly reduces the large intracellular osmotic pressure that would result from its storage in monomeric form, because osmotic pressure is proportional to the number of solute molecules in a given volume (Section 2-1D). Starch is a reducing sugar, although it has only one residue, called the **reducing end,** that lacks a glycosidic bond.

The digestion of starch, the main carbohydrate source in the human diet, begins in the mouth. Saliva contains an **amylase,** which randomly hydrolyzes the α(1→4) glycosidic bonds of starch. Starch digestion continues in the small intestine under the influence of pancreatic amylase, which degrades starch to a mixture of α(1→4)-linked glucose disaccharides (called **maltose**) and trisaccharides **(maltotriose),** and oligosaccharides known as **dextrins** that contain the α(1→6) branches. These oligosaccharides are hydrolyzed to their component monosaccharides by an **α-glucosidase,** which removes one glucose residue at a time, and a **debranching enzyme,** which hydrolyzes α(1→6) as well as α(1→4) bonds. The resulting monosaccharides are absorbed by the intestine and transported to the bloodstream.

Glycogen, the storage polysaccharide of animals, is present in all cells but is most prevalent in skeletal muscle and in liver, where it occurs as cytoplasmic granules (Fig. 8-11). The primary structure of glycogen resembles that of amylopectin, but glycogen is more highly branched, with branch points occurring every 8 to 14 glucose residues. In the cell, glycogen is degraded for metabolic use by **glycogen phosphorylase,** which phosphorolytically cleaves glycogen's α(1→4) bonds sequentially inward from its nonreducing ends. *Glycogen's highly branched structure, which has many nonreducing ends, permits the rapid mobilization of glucose in times of metabolic need.* The α(1→6) branches of glycogen are cleaved by **glycogen debranching enzyme** (glycogen breakdown is discussed further in Section 15-1).

D Glycosaminoglycans

The extracellular spaces, particularly those of connective tissues such as cartilage, tendon, skin, and blood vessel walls, contain collagen (Section 6-1C) and other proteins embedded in a gel-like matrix that is composed largely

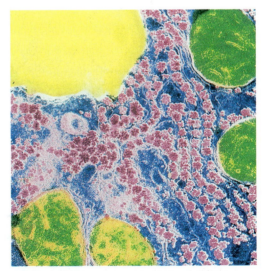

Figure 8-11 Photomicrograph showing the glycogen granules (*pink*) of a liver cell. The greenish objects are mitochondria, and the yellow object is a fat globule. The glycogen content of liver may reach 10% of its net weight. [CNRI/Science Photo Library/Photo Researchers.]

of **glycosaminoglycans.** These unbranched polysaccharides consist of alternating uronic acid and hexosamine residues. Solutions of glycosaminoglycans have a slimy, mucuslike consistency that results from their high viscosity and elasticity.

Hyaluronic acid is an important glycosaminoglycan component of connective tissue, synovial fluid (the fluid that lubricates joints), and the vitreous humor of the eye. Hyaluronic acid molecules are composed of 250 to 25,000 β(1→4)-linked disaccharide units that consist of D-glucuronic acid and ***N*-acetyl-D-glucosamine (GlcNAc)** linked by a β(1→3) bond (Fig. 8-12). The disaccharide units of hyaluronic acid are extended, forming a rigid molecule whose numerous repelling anionic groups bind cations and water molecules. In solution, hyaluronate occupies a volume ~1000 times that in its dry state.

Hyaluronate solutions have a viscosity that is shear dependent (an object under shear stress has equal and opposite forces applied across its opposite faces). At low shear rates, hyaluronate molecules form tangled masses that greatly impede flow; that is, the solution is quite viscous. As the shear stress increases, the stiff hyaluronate molecules tend to line up with the flow and thus offer less resistance to it. This viscoelastic behavior makes hyaluronate solutions excellent biological shock absorbers and lubricants.

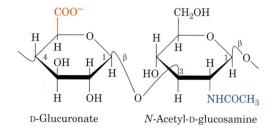

Hyaluronate

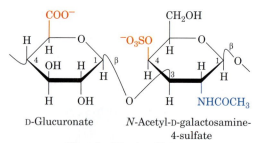

Chondroitin-4-sulfate

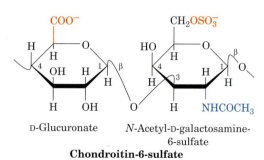

Chondroitin-6-sulfate

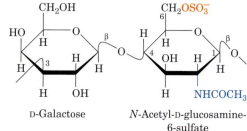

Keratan sulfate

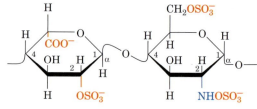

Heparin

Figure 8-12 **Repeating disaccharide units of some glycosaminoglycans.** The anionic groups are shown in red and the *N*-acetylamido groups are shown in blue. 🔴 **See Kinemage Exercise 7-3.**

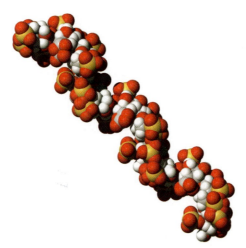

Figure 8-13 NMR structure of heparin containing six pairs of iduronate and glucosamine residues. Atoms are colored according to type with C gray, N blue, O red, and S yellow. Note the high density of anionic sulfate groups. [Based on an NMR structure by Barbara Mulloy and Mark Forster, National Institute for Biological Standards and Control, Herts, U.K. PDBid 1HPN.]

The other common glycosaminoglycans shown in Fig. 8-12 consist of 50 to 1000 sulfated disaccharide units. **Chondroitin-4-sulfate** and **chondroitin-6-sulfate** differ only in the sulfation of their *N*-acetylgalactosamine (GalNAc) residues. **Dermatan sulfate** is derived from chondroitin by enzymatic epimerization of C5 of glucuronate residues to form **iduronate** residues. **Keratan sulfate** (not to be confused with the fibrous protein keratin; Section 6-1C) is the most heterogeneous of the major glycosaminoglycans in that its sulfate content is variable and it contains small amounts of fucose, mannose, GlcNAc, and sialic acid. **Heparin** is also variably sulfated, with an average of 2.5 sulfate residues per disaccharide unit, which makes it the most highly charged polymer in mammalian tissues (Fig. 8-13).

In contrast to the other glycosaminoglycans, heparin is not a constituent of connective tissue but occurs almost exclusively in the intracellular granules of the mast cells that occur in arterial walls. It inhibits the clotting of blood, and its release, through injury, is thought to prevent runaway clot formation. Heparin is therefore in wide clinical use to inhibit blood clotting, for example, in postsurgical patients.

Heparan sulfate (HS), a ubiquitous cell-surface component as well as an extracellular substance in blood vessel walls and brain, resembles heparin but has a far more variable composition with fewer *N*- and *O*-sulfate groups and more *N*-acetyl groups. An important developmental role for HS is beginning to come to light. Various **fibroblast growth factors (FGFs)** and their receptors **(FGFRs)** are essential for many important cellular processes, such as cell differentiation and proliferation. The formation of ternary complexes of HS with FGF and FGFR initiates these signaling processes. Specific sulfation patterns on HS are required for the formation of these ternary complexes.

Plants do not synthesize glycosaminoglycans, but the **pectins,** which are major components of cell walls, may function similarly as shock absorbers. Pectins are heterogeneous polysaccharides with a core of $\alpha(1\rightarrow4)$-linked galacturonate residues interspersed with the hexose **rhamnose:**

Rhamnose

The galacturonate residues may be modified by the addition of methyl and acetyl groups. Other polysaccharide chains, some containing the pentoses arabinose and xylose and other sugars, are attached to the galacturonate. The aggregation of pectin molecules to form bundles requires divalent cations (usually Ca^{2+}), which form cross-links between the anionic carboxylate groups of neighboring galacturonate residues. The tendency for pectin to form highly hydrated gels is exploited in the manufacture of jams and jellies, to which pectin is often added to augment the endogenous pectin content of the fruit.

3 Glycoproteins

Many proteins are actually glycoproteins, with carbohydrate contents varying from <1% to >90% by weight. Glycoproteins occur in all forms of life and have functions that span the entire spectrum of protein activities, including those of enzymes, transport proteins, receptors, hormones, and structural proteins. The polypeptide chains of glycoproteins, like those of all proteins, are synthesized under genetic control. Their carbohydrate chains, in contrast, are enzymatically generated and covalently linked to the polypeptide without the rigid guidance of nucleic acid templates. For this reason, glycoproteins tend to have variable carbohydrate composition, a phenomenon known as **microheterogeneity.**

A Proteoglycans

Proteins and glycosaminoglycans in the extracellular matrix aggregate covalently and noncovalently to form a diverse group of macromolecules known as **proteoglycans.** Electron micrographs (Fig. 8-14a) and other evidence indicate that proteoglycans have a bottlebrush-like molecular architecture, with "bristles" noncovalently attached to a filamentous hyaluronic acid "backbone." The bristles consist of a **core protein** to which glycosaminoglycans, most often keratan sulfate and chondroitin sulfate, are covalently linked (Fig. 8-14b). The interaction between the core protein and the hyaluronate is stabilized by a **link protein.** Smaller oligosaccharides are usually attached to the core protein near its site of attachment to hyaluronate. These oligosaccharides are glycosidically linked to the protein via the amide N of specific Asn residues (and are therefore known as **N-linked oligosaccharides;** Section 8-3C). The keratan sulfate and chondroitin sulfate chains are glycosidically linked to the core protein via oligosaccharides that are covalently bonded to side chain O atoms of specific Ser or Thr residues (i.e., **O-linked oligosaccharides**).

Altogether, a central strand of hyaluronic acid, which varies in length from 4000 to 40,000 Å, can have up to 100 associated core proteins, each of which binds ~50 keratan sulfate chains of up to 250 disaccharide units and ~100 chondroitin sulfate chains of up to 1000 disaccharide units each. This accounts for the enormous molecular masses of many proteoglycans, which range up to tens of millions of daltons.

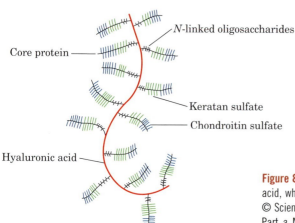

(a)

(b)

Core protein

N-linked oligosaccharides

Keratan sulfate

Chondroitin sulfate

Hyaluronic acid

Figure 8-14   A proteoglycan.   (a) Electron micrograph showing a central strand of hyaluronic acid, which supports numerous projections. [From Caplan, A.I., *Sci. Am.* **251**(4), 87 (1984). Copyright © Scientific American, Inc. Used by permission.] (b) Bottlebrush model of the proteoglycan shown in Part a. Numerous core proteins are noncovalently linked to the central hyaluronic acid strand. Each core protein has three saccharide-binding regions.

The extended brushlike structure of proteoglycans, together with the polyanionic character of their keratan sulfate and chondroitin sulfate components, cause these complexes to be highly hydrated. Cartilage, which consists of a meshwork of collagen fibrils that is filled in by proteoglycans, is characterized by its high resilience: The application of pressure on cartilage squeezes water away from the charged regions of its proteoglycans until charge–charge repulsions prevent further compression. When the pressure is released, the water returns. Indeed, the cartilage in the joints, which lacks blood vessels, is nourished by this flow of liquid brought about by body movements. This explains why long periods of inactivity cause cartilage to become thin and fragile.

B Bacterial Cell Walls

Bacteria are surrounded by rigid cell walls (Fig. 1-6) that give them their characteristic shapes (Fig. 1-7) and permit them to live in **hypotonic** (less than intracellular salt concentration) environments that would otherwise cause them to swell osmotically until their plasma (cell) membranes lysed (burst). Bacterial cell walls are of considerable medical significance because they are responsible for bacterial **virulence** (disease-evoking power). In fact, the symptoms of many bacterial diseases can be elicited in animals merely by the injection of bacterial cell walls. Furthermore, the characteristic antigens of bacteria are components of their cell walls, so injecting bacterial cell wall preparations into an animal often invokes its immunity against these bacteria.

Bacteria are classified as **gram-positive** or **gram-negative** according to whether or not they take up Gram stain (a procedure developed in 1884 by Christian Gram in which heat-fixed cells are successively treated with the dye crystal violet and iodine and then destained by ethanol or acetone). Gram-positive bacteria (Fig. 8-15a) have a thick cell wall (~250 Å) surrounding their plasma membrane, whereas gram-negative bacteria (Fig. 8-15b) have a thin cell wall (~30 Å) covered by a complex outer membrane. This outer membrane functions, in part, to exclude substances toxic to the bacterium, including Gram stain. This accounts for the observation that gram-negative bacteria are more resistant to antibiotics than are gram-positive bacteria.

The cell walls of bacteria consist of covalently linked polysaccharide and polypeptide chains, which form a baglike macromolecule that completely encases the cell. This framework, whose structure was elucidated in large part by Jack Strominger, is known as a **peptidoglycan.** Its polysaccharide

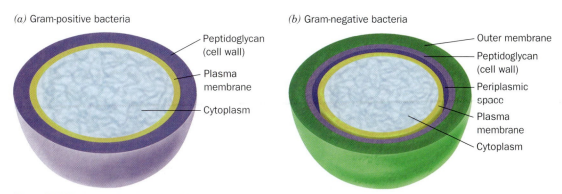

(a) Gram-positive bacteria

- Peptidoglycan (cell wall)
- Plasma membrane
- Cytoplasm

(b) Gram-negative bacteria

- Outer membrane
- Peptidoglycan (cell wall)
- Periplasmic space
- Plasma membrane
- Cytoplasm

Figure 8-15 Bacterial cell walls. This diagram compares the cell envelopes of (a) gram-positive bacteria and (b) gram-negative bacteria.

(a)   *N*-Acetylglucosamine   *N*-Acetylmuramic acid

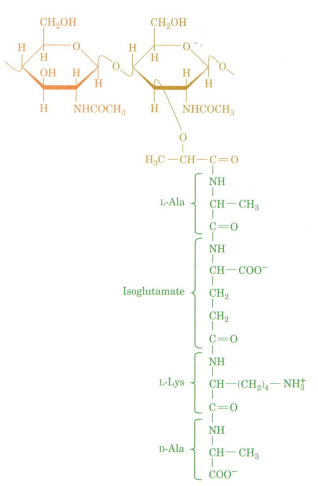

(b)

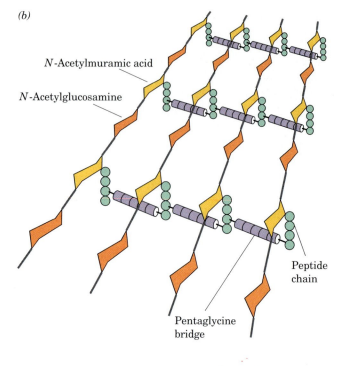

Figure 8-16 Peptidoglycan. (*a*) The repeating unit of peptidoglycan is an *N*-acetylglucosamine–*N*-acetylmuramic acid disaccharide whose lactyl side chain forms an amide bond with a tetrapeptide. The tetrapeptide of *S. aureus* is shown. The isoglutamate is so designated because it forms a peptide link via its γ-carboxyl group. The terminal D-Ala is linked to a second D-Ala in the peptidoglycan precursor. This second D-Ala is replaced by a Gly residue in the crosslinked peptidoglycan. (*b*) The *S. aureus* bacterial cell wall peptidoglycan, showing its pentaglycine connecting bridges (*purple*).

component consists of linear chains of alternating β(1→4)-linked GlcNAc and **N-acetylmuramic acid** (Latin: *murus,* wall). The lactic acid group of *N*-acetylmuramic acid forms an amide bond with a D-amino acid–containing tetrapeptide to form the peptidoglycan repeating unit (Fig. 8-16). Neighboring parallel peptidoglycan chains are covalently cross-linked through their tetrapeptide side chains.

In the bacterium *Staphylococcus aureus,* whose tetrapeptide has the sequence L-Ala-D-isoglutamyl-L-Lys-D-Ala, the cross-link consists of a pentaglycine chain that extends from the terminal carboxyl group of one tetrapeptide to the ε-amino group of the Lys in a neighboring tetrapeptide. The bacterial cell wall consists of several concentric layers of peptidoglycan that are probably cross-linked in the third dimension. The cell walls of gram-positive bacteria also bear additional polysaccharide derivatives that help shield the cell.

The D-amino acids of peptidoglycans render them resistant to proteases, which are mostly specific for L-amino acids. However, **lysozyme,** an enzyme that is present in tears, mucus, and other vertebrate body secretions, as well as in egg whites, catalyzes the hydrolysis of the β(1→4) glycosidic linkage between *N*-acetylmuramic acid and *N*-acetylglucosamine (the structure and mechanism of lysozyme are examined in detail in Section 11-4). The cell wall is also compromised by antibiotics that inhibit bacterial cell wall biosynthesis (see Box 8-3).

BOX 8-3

Biochemistry in Health and Disease

Peptidoglycan-Specific Antibiotics

In 1928, Alexander Fleming noticed that the chance contamination of a bacterial culture plate with the mold *Penicillium notatum* resulted in the lysis of the bacteria in the vicinity of the mold. This was caused by the presence of **penicillin,** an antibiotic secreted by the mold.

Penicillin

Penicillin contains a thiazolidine ring (*red*) fused to a β-lactam ring (*blue*). A variable R group is bonded to the β-lactam ring via a peptide link.

Penicillin specifically binds to and inactivates enzymes that cross-link the peptidoglycan strands of bacterial cell walls. Since cell wall expansion in growing cells requires that their rigid cell walls be opened up for the insertion of new cell wall material, exposure of growing bacteria to penicillin results in cell lysis. However, since no human enzyme binds penicillin specifically, it is not toxic to humans and is therefore therapeutically useful.

Most bacteria that are resistant to penicillin secrete the enzyme **penicillinase** (also called **β-lactamase**), which inactivates penicillin by cleaving the amide bond of its β-lactam ring. Attempts to overcome this resistance have led to the development of β-lactamase inhibitors such as **sulbactam** that are often prescribed as mixtures with penicillin derivatives.

Multiple drug resistant bacteria are a growing problem. For many years, **vancomycin,** the so-called antibiotic of last resort, has been used to treat bacterial infections that do not succumb to other antibiotics.

Vancomycin

Vancomycin acts by inhibiting the transpeptidation (crosslinking) reaction of bacterial cell wall synthesis. The drug binds to the D-Ala—D-Ala sequence in the peptidoglycan precursor, which prevents the replacement of the terminal D-Ala with a pentaglycine bridge, thereby impeding maturation of the cell wall. However, bacteria can become resistant to vancomycin by acquiring a gene that allows cell wall synthesis from a D-Ala—D-lactate precursor sequence, to which vancomycin binds much less effectively.

One limitation of drugs such as vancomycin and penicillin, particularly for slow-growing bacteria, is that the drug may halt bacterial growth without actually killing the cells. For this reason, effective antibacterial treatments may require combinations of antibiotics over a course of several weeks.

C Glycosylated Proteins

Almost all the secreted and membrane-associated proteins of eukaryotic cells are **glycosylated.** Oligosaccharides are covalently attached to proteins by either *N*-glycosidic or *O*-glycosidic bonds.

N-Linked Oligosaccharides. *In N-linked oligosaccharides, GlcNAc is invariably β-linked to the amide nitrogen of an Asn residue in the sequence Asn-*

X-Ser or Asn-X-Thr, where X is any amino acid except possibly Pro or Asp:

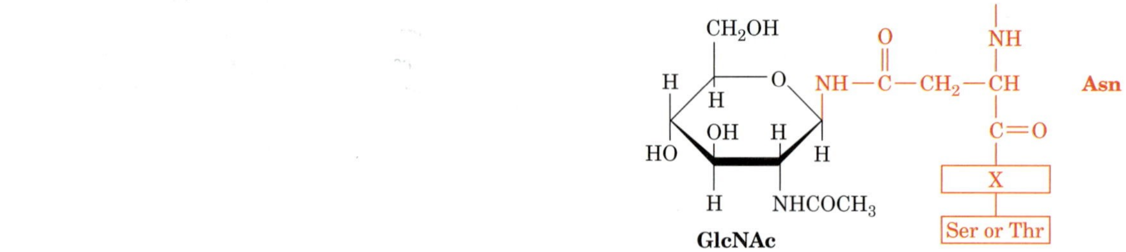

GlcNAc

N-Glycosylation occurs **cotranslationally,** that is, while the polypeptide is being synthesized. Proteins containing *N*-linked oligosaccharides typically are glycosylated and then processed as elucidated, in large part, by Stuart Kornfeld (Fig. 8-17):

1. An oligosaccharide containing 9 mannose residues, 3 glucose residues, and 2 GlcNAc residues is attached to the Asn of a growing polypeptide chain that is being synthesized by a ribosome associated with the endoplasmic reticulum (Section 9-4D).

2. Some of the sugars are removed during processing, which begins in the lumen (internal space) of the endoplasmic reticulum and continues in the Golgi apparatus (Fig. 1-8). Enzymatic trimming is accomplished by glucosidases and mannosidases.

3. Additional monosaccharide residues, including GlcNAc, galactose, fucose, and sialic acid, are added by the action of specific **glycosyltransferases** in the Golgi apparatus.

The exact steps of *N*-linked oligosaccharide processing vary with the identity of the glycoprotein and the battery of endoglycosidases in the cell, but all *N*-linked oligosaccharides have a common core pentasaccharide with the following structure:

$$\left.\begin{array}{l} \text{Man } \alpha(1 \rightarrow 6) \\ \\ \text{Man } \alpha(1 \rightarrow 3) \end{array}\right\rangle \text{Man } \beta(1 \rightarrow 4) \text{ GlcNAc } \beta(1 \rightarrow 4) \text{ GlcNAc}-$$

In some glycoproteins, processing is brief, leaving "high-mannose" oligosaccharides; in other glycoproteins, extensive processing generates large oligosaccharides containing several kinds of sugar residues. *There is enormous diversity among the oligosaccharides of N-linked glycoproteins.* Indeed, even glycoproteins with a given polypeptide chain exhibit considerable microheterogeneity, presumably as a consequence of incomplete glycosylation and lack of absolute specificity on the part of glycosidases and glycosyltransferases.

O-Linked Oligosaccharides.   The most common *O*-glycosidic attachment involves the disaccharide core *β-galactosyl-(1→3)-α-N-acetylgalactosamine linked to the OH group of either Ser or Thr:*

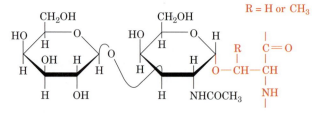

β-Galactosyl-(1 ⟶ 3)-α-N-acetylgalactosaminyl-Ser/Thr

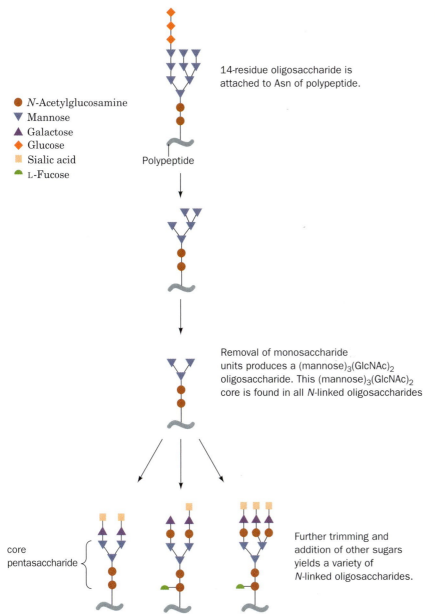

- N-Acetylglucosamine
- Mannose
- Galactose
- Glucose
- Sialic acid
- L-Fucose

Polypeptide

14-residue oligosaccharide is attached to Asn of polypeptide.

Removal of monosaccharide units produces a (mannose)$_3$(GlcNAc)$_2$ oligosaccharide. This (mannose)$_3$(GlcNAc)$_2$ core is found in all N-linked oligosaccharides

core pentasaccharide

Further trimming and addition of other sugars yields a variety of N-linked oligosaccharides.

Figure 8-17 Synthesis of N-linked oligosaccharides. The addition of a (mannose)$_9$(glucose)$_3$(GlcNAc)$_2$ oligosaccharide is followed by removal of monosaccharides as catalyzed by glycosidases, and the addition of other monosaccharides as catalyzed by glycosyl-transferases. The core pentasaccharide occurs in all N-linked oligosaccharides. [Adapted from Kornfeld, R. and Kornfeld, S., *Annu. Rev. Biochem.* **54**, 640 (1985).] 🔬 **See Kinemage Exercise 7-4.**

Less commonly, galactose, mannose, and xylose form O-glycosides with Ser or Thr. Galactose also forms O-glycosidic bonds to the 5-hydroxylysyl residues of collagen (Section 6-1C). O-Linked oligosaccharides vary in size from a single galactose residue in collagen to the chains of up to 1000 disaccharide units in proteoglycans.

O-Linked oligosaccharides are synthesized in the Golgi apparatus by the serial addition of monosaccharide units to a completed polypeptide chain. Synthesis starts with the transfer of GalNAc to a Ser or Thr residue on the polypeptide. N-Linked oligosaccharides are transferred to an Asn in a specific amino acid sequence, but O-glycosylated Ser and Thr residues are not members of any common sequence. Instead, the locations of glycosylation sites are specified only by the secondary or tertiary structure of the polypeptide. O-Glycosylation continues with stepwise addition of sugars by the corresponding glycosyltransferases. The energetics and enzymology of oligosaccharide synthesis are discussed further in Section 15-5.

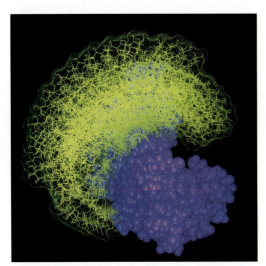

Figure 8-18 Model of oligosaccharide dynamics. The allowed conformations of a (GlcNAc)$_2$(mannose)$_{5-9}$ oligosaccharide (*yellow*) attached to the bovine pancreatic enzyme **ribonuclease B** (*purple*) are shown in superimposed "snapshots." [Courtesy of Raymond Dwek, Oxford University, U.K.]

D Functions of Oligosaccharides

A single protein may contain several *N*- and *O*-linked oligosaccharide chains, although different molecules of the same glycoprotein may differ in the sequences, locations, and numbers of covalently attached carbohydrates (the variant species of a glycoprotein are known as its **glycoforms**). This heterogeneity makes it difficult to assign discrete biological functions to oligosaccharide chains. In fact, certain glycoproteins synthesized by cells that lack particular oligosaccharide-processing enzymes appear to function normally despite abnormal or absent glycosylation. In other cases, however, glycosylation may affect a protein's structure, stability, or activity.

Structural Effects of Oligosaccharides. Oligosaccharides are usually attached to proteins at sequences that form surface loops or turns. Since sugars are hydrophilic, the oligosaccharides tend to project away from the protein surface. Because carbohydrate chains are often conformationally mobile, oligosaccharides attached to proteins can occupy time-averaged volumes of considerable size (Fig. 8-18). In this way, an oligosaccharide can shield a protein's surface, possibly modifying its activity or protecting it from proteolysis.

In addition, some oligosaccharides may play structural roles by limiting the conformational freedom of their attached polypeptide chains. Since *N*-linked oligosaccharides are added as the protein is being synthesized, the attachment of an oligosaccharide may help determine how the protein folds. In addition, the oligosaccharide may help stabilize the folded conformation of a polypeptide by reducing backbone flexibility. In particular, *O*-linked oligosaccharides, which are usually clustered in heavily glycosylated segments of a protein, may help stiffen and extend the polypeptide chain.

During glycosylation, the oligosaccharide may function in quality control and protein sorting. For example, the attachment of the 14-residue oligosaccharide to an Asn residue (Fig. 8-17) verifies that the protein is being successfully translocated from the cytosol to the lumen of the endoplasmic reticulum. The subsequent removal of the two terminal glucose residues allows the protein to interact with a chaperone that recognizes the partially processed oligosaccharide (chaperones are discussed in Section 6-5C). The membrane-bound 572-residue form of the chaperone, called **calnexin,** or its soluble 400-residue homolog **calreticulin,** binds the immature (unfolded) glycoprotein to assist its folding and to protect it from degradation or premature transfer to the Golgi apparatus (where the late stages of oligosaccharide processing take place).

After the folded glycoprotein is released from calnexin/calreticulin, it proceeds through the rest of the oligosaccharide-processing pathway. A glycoprotein that is released before it has adopted its mature conformation can still undergo the next step of processing, which is the removal of another glucose residue. However, when this happens, a glycosyltransferase that recognizes only unfolded glycoproteins reattaches a glucose residue to the immature oligosaccharide. As a result, the glycoprotein can bind again to calnexin/calreticulin for another chance to fold properly (Fig. 8-19). Most glycoproteins undergo reglucosylation at least once.

The glycosylation process also helps direct proteins to their final cellular destinations. Newly synthesized proteins that are destined for **lysosomes** (organelles containing a variety of hydrolytic enzymes and which function as cellular recycling centers) contain *N*-linked oligosaccharides with **mannose-6-phosphate** residues. A mannose-6-phosphate receptor in the Golgi apparatus selects these proteins for transport to lysosomes.

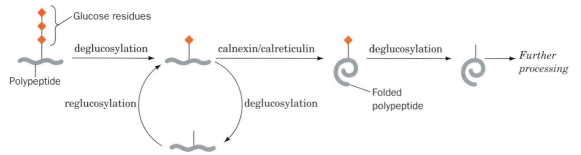

Figure 8-19 Quality control in oligosaccharide processing. A partially processed glycoprotein bearing one terminal glucose residue can bind to the chaperones calnexin or calreticulin. The bound glycoprotein is then deglucosylated and processed further. An improperly folded glycoprotein that becomes deglucosylated is reglucosylated by an enzyme that is specific for unfolded glycoproteins. This mechanism promotes the proper folding and oligosaccharide processing of newly synthesized glycoproteins.

Oligosaccharides may also help identify proteins nearing the end of their life span. For example, circulating glycoproteins that have lost their terminal sialic acid (or sialic acid and galactose) residues are selectively cleared from the blood by liver cells that recognize and bind the newly exposed galactose (or GlcNAc) residues. Thus, the range of glycoforms of a given glycoprotein probably ensures that it has a range of lifetimes in the blood.

Oligosaccharides Mediate Recognition Events. The many possible ways that carbohydrates can be linked together to form branched structures gives them the potential to carry more biological information than either nucleic acids or proteins of similar size. For example, two different nucleotides can make only two distinct dinucleotides, but two different hexoses can combine in 36 different ways (although not all possibilities are necessarily realized in nature).

The first evidence that unique combinations of carbohydrates might be involved in intercellular communication came with the discovery that all cells are coated with sugars in the form of **glycoconjugates** such as glyco-proteins and glycolipids. The oligosaccharides of glycoconjugates form a fuzzy layer up to 1400 Å thick in some cells (Fig. 8-20).

Additional evidence that cell-surface carbohydrates have recognition functions comes from **lectins** (proteins that bind carbohydrates), which are ubiquitous in nature and frequently appear on the surfaces of cells. Lectins are exquisitely specific: They can recognize individual monosaccharides in particular linkages to other sugars in an oligosaccharide (this property also makes lectins useful laboratory tools for isolating glycoproteins and oligosaccharides). Protein–carbohydrate interactions are typically characterized by extensive hydrogen bonding (often including bridging water molecules) and the van der Waals packing of hydrophobic sugar faces against aromatic side chains (Fig. 8-21).

Proteins known as **selectins** mediate the attachment between **leukocytes** (circulating white blood cells) and the surfaces of endothelial cells (the cells

Figure 8-20 Electron micrograph of the erythrocyte surface. Its thick (up to 1400 Å) carbohydrate coat, which is called the **glycocalyx**, consists of closely packed oligosaccharides attached to cell-surface proteins and lipids. [Courtesy of Harrison Latta, UCLA.]

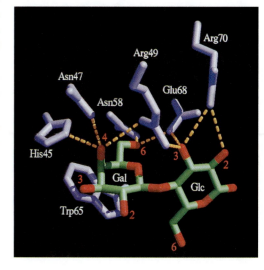

Figure 8-21 Carbohydrate binding by a lectin. Human **galectin-2** binds β-galactosides, such as lactose, primarily through their galactose residue. The galactose and glucose residues are shown in green (with red O atoms), and the lectin amino acid side chains are shown in violet. Hydrogen bonds between the side chains and the sugar residues are shown as dashed yellow lines. [Courtesy of Hakon Leffler, University of California at San Francisco. PDBid 1HLC.]

that line blood vessels). Leukocytes constitutively (continually) express selectins on their surface; endothelial cells transiently display their own selectins in response to tissue damage from infection or mechanical injury. The selectins recognize and bind specific oligosaccharides on cell-surface glycoproteins. Reciprocal selectin–oligosaccharide interactions between the two cell types allow the endothelial cells to "capture" circulating leukocytes, which then crawl past the endothelial cells on their way to eliminate the infection or help repair damaged tissues.

Other cell–cell recognition phenomena depend on oligosaccharides. For example, proteins on the surface of mammalian spermatozoa recognize GlcNAc or galactose residues on the glycoproteins of the ovum as part of the binding and activation events during fertilization. Many viruses, bacteria, and eukaryotic parasites invade their target tissues by first binding to cell-surface carbohydrates.

Oligosaccharides Are Antigenic Determinants. The carbohydrates on cell surfaces are some of the best known immunochemical markers. For example, the **ABO blood group antigens** are oligosaccharide components of glycoproteins and glycolipids on the surfaces of an individual's cells (not just red blood cells). Individuals with type A cells have A antigens on their cell surfaces and carry anti-B antibodies in their blood; those with type B cells, which bear B antigens, carry anti-A antibodies; those with type AB cells, which have both A and B antigens, carry neither anti-A nor anti-B antibodies; and type O individuals, whose cells bear neither antigen, carry both anti-A and anti-B antibodies. Consequently, the transfusion of type A blood into a type B individual, for example, results in an anti-A antibody–A antigen reaction, which agglutinates (clumps together) the transfused erythrocytes, resulting in an often fatal blockage of blood vessels.

Table 8-1 lists the oligosaccharides found in the **A, B,** and **H antigens** (type O individuals have the H antigen). These occur at the nonreducing ends of the oligosaccharide components of glycolipids. The H antigen is the precursor oligosaccharide of A and B antigens. Type A individuals have a 303-residue glycosyltransferase that specifically adds a GalNAc residue to the terminal position of the H antigen. In type B individuals, this enzyme, which differs by four amino acid residues from that of type A individuals, instead adds a galactose residue. In type O individuals, the enzyme is inactive because its synthesis terminates after its 115th residue.

In mammals other than humans and some monkeys, cell-surface carbohydrates include a terminal galactosyl-$\alpha(1\rightarrow3)$-galactose disaccharide. Humans, who lack the galactosyltransferase required for its synthesis, produce antibodies against the disaccharide. This innate immunoreactivity is the primary reason for the rapid (within minutes) rejection of animal tissues that have been transplanted into humans. Even if the recipient's anti-galactose antibodies are removed by affinity adsorption (see Section 5-2C), they eventually return and compromise the long-term survival of the transplanted tissue. Immunosuppressive drugs, which must be taken indefinitely, have their own risks. However, eliminating the organ's antigenic galactose could allow clinically successful **xenotransplants** (from the Greek *xenos,* foreign).

Genetic engineering has produced "transplant-friendly" pigs that lack the gene for α-1,3-galactosyltransferase and hence cannot add the terminal galactose to their cell-surface oligosaccharides (Fig. 8-22). Such animals might become a source of organs for xenotransplantation because pigs are comparable to humans in size and physiology. Of course the development of such technology is leading to many ethical and moral discussions.

Table 8-1 Structures of the A, B, and H Antigenic Determinants in Erythrocytes

Type	Antigen[a]
H	$Gal\beta(1\rightarrow4)GlcNAc\cdots$ $\uparrow 1,2$ L-Fucα
A	$GalNAc\alpha(1\rightarrow3)Gal\beta(1\rightarrow4)GlcNAc\cdots$ $\uparrow 1,2$ L-Fucα
B	$Gal\alpha(1\rightarrow3)Gal\beta(1\rightarrow4)GlcNAc\cdots$ $\uparrow 1,2$ L-Fucα

[a]Gal, Galactose; GalNAc, *N*-acetylgalactosamine; GlcNAc, *N*-acetylglucosamine; L-Fuc, L-fucose.

Figure 8-22 Transplant-friendly pigs. These genetically engineered animals lack the gene for α-1,3-galactosyltransferase, so their organs lack the major antigen responsible for rejection in pig-to-human organ transplants. [© AP/Wide World Photos.]

SUMMARY

1. Monosaccharides, the simplest carbohydrates, are classified as aldoses or ketoses.

2. The cyclic hemiacetal and hemiketal forms of monosaccharides have either the α or β configuration at their anomeric carbon but are conformationally variable.

3. Monosaccharide derivatives include aldonic acids, uronic acids, alditols, deoxy sugars, amino sugars, and α- and β-glycosides.

4. Polysaccharides consist of monosaccharides linked by glycosidic bonds.

5. Cellulose and chitin are polysaccharides whose β(1→4) linkages cause them to adopt rigid and extended structures.

6. The storage polysaccharides starch and glycogen consist of α-glycosidically linked glucose residues.

7. Glycosaminoglycans are unbranched polysaccharides containing uronic acid and amino sugars that are often sulfated.

8. Proteoglycans are enormous molecules consisting of hyaluronic acid with attached core proteins that bear numerous glycosaminoglycans and oligosaccharides.

9. Bacterial cell walls are made of peptidoglycan, a network of polysaccharide and polypeptide chains.

10. Glycosylated proteins may contain N-linked oligosaccharides (attached to Asn) or O-linked oligosaccharides (attached to Ser or Thr) or both. Different molecules of a glycoprotein may contain different sequences and locations of oligosaccharides.

11. Oligosaccharides play important roles in protein processing and in cell-surface recognition phenomena.

REFERENCES

Bernfield, M., Götte, M., Park, P.W., Reizes, O., Fitzgerald, M.L., Linecum, J., and Zako, M., Functions of cell surface heparan sulfate proteoglycans, *Annu. Rev. Biochem.* **68,** 729–777 (1999).

Bush, C.A., Martin-Pastor, M., and Imberty, A., Structure and conformation of complex carbohydrates of glycoproteins, glycolipids, and bacterial polysaccharides, *Annu. Rev. Biophys. Biomol. Struct.* **28,** 269–293 (1999).

Drickamer, K. and Taylor, M.E., Evolving views of protein glycosylation, *Trends Biochem. Sci.* **23,** 321–324 (1998).

Iozzo, R.V., Matrix proteoglycans: from molecular design to cellular function, *Annu. Rev. Biochem.* **67,** 609–652 (1998).

Lehrman, M.A., Oligosaccharide-based information in endoplasmic reticulum quality control and other biological systems, *J. Biol. Chem.* **276,** 8623–8626 (2001). [Summarizes the mechanisms that ensure proper protein folding and oligosaccharide processing.]

Varki, A., Cummings, R., Esko, J., Freeze, H., Hart, G., and Marth, J. (Eds.), *Essentials of Glycobiology,* Cold Spring Harbor Laboratory Press (1999).

Weis, W.I. and Drickamer, K., Structural basis of lectin–carbohydrate recognition, *Annu. Rev. Biochem.* **65,** 441–473 (1996).

Wormald, M.R. and Dwek, R.A., Glycoproteins: glycan presentation and protein-fold stability, *Structure* **7,** R155–R160 (1999).

KEY TERMS

monosaccharide
polysaccharide
aldose
ketose
epimer
hemiacetal
hemiketal
pyranose
furanose
α anomer
β anomer

aldonic acid
uronic acid
alditol
deoxy sugar
amino sugar
glycoprotein
glycolipid
α-glycoside
β-glycoside
glycosidic bond
reducing sugar

glycan
homopolysaccharide
heteropolysaccharide
oligosaccharide
exoglycosidase
endoglycosidase
disaccharide
glycosaminoglycan
microheterogeneity
proteoglycan
N-linked oligosaccharide

O-linked oligosaccharide
gram-positive
gram-negative
peptidoglycan
glycosylation
oligosaccharide processing
glycoforms
glycoconjugate
lectin
xenotransplant

STUDY EXERCISES

1. Show how aldoses and ketoses can form five- and six-membered rings.

2. Draw Fischer and Haworth projections for glucose.

3. Compare and contrast the structures and functions of cellulose, chitin, starch, and glycogen.

4. How do the physical properties of glycosaminoglycans and proteoglycans relate to their biological roles?

5. Explain the differences between N- and O-linked oligosaccharides in glycoproteins.

PROBLEMS

1. How many stereoisomers are possible for (a) a ketopentose, (b) a ketohexose, and (c) a ketoheptose?

2. Which of the following pairs of sugars are epimers of each other?

   (a) D-sorbose and D-psicose

   (b) D-sorbose and D-fructose

   (c) D-fructose and L-fructose

   (d) D-arabinose and D-ribose

   (e) D-ribose and D-ribulose

3. Draw the furanose and pyranose forms of D-ribose.

4. Are (a) D-glucitol, (b) D-galactitol, and (c) D-glycerol optically active?

5. Draw a Fischer projection of L-fucose. L-Fucose is the 6-deoxy form of which L-hexose?

6. Deduce the structure of the disaccharide trehalose from the following information: Complete hydrolysis yields only D-glucose; it is hydrolyzed by α-glucosidase but not β-glucosidase; and it does not reduce Cu^{2+} to Cu^{+}.

7. How many different disaccharides of D-glucopyranose are possible?

8. The artificial sweetener sucralose is a derivative of sucrose with the formal name 1,6-dichloro-1,6-dideoxy-β-D-fructo-furanosyl-4-chloro-4-deoxy-α-D-galactopyranoside. Draw its structure.

9. How many reducing ends are in a molecule of glycogen that contains 10,000 residues with a branch every 10 residues?

10. Is amylose or amylopectin more likely to be a long-term storage polysaccharide in plants?

11. "Nutraceuticals" are products that are believed to have some beneficial effect but are not strictly defined as either food or drug. Why might an individual suffering from osteoarthritis be tempted to consume the nutraceutical glucosamine?

12. Calculate the net charge of a chondroitin-4-sulfate molecule containing 100 disaccharide units.

13. Draw the structure of the O-type oligosaccharide (the H antigen, described in Table 8-1).

14. Glycogen is treated with dimethyl sulfate, which adds a methyl group to every free OH group. Next, the molecule is hydrolyzed to break all the glycosidic bonds between glucose residues. The reaction products are then chemically analyzed.

    (a) How many different types of methylated glucose molecules are obtained?

    (b) Draw the structure of the one that is most abundant.

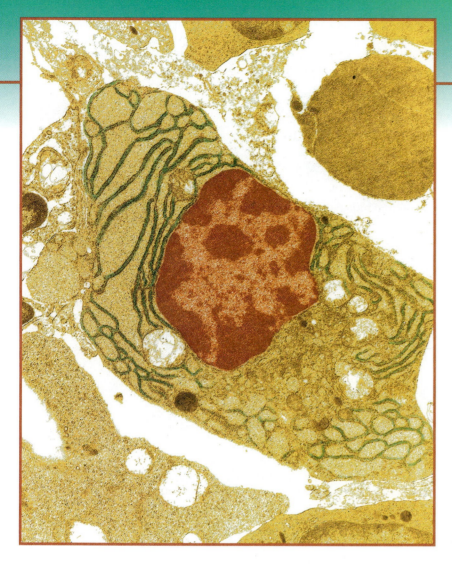

Due to their hydrophobicity, lipids do not mix freely with the aqueous phase but instead can form bilayers. Membranes, which consist of a lipid bilayer and the proteins embedded in it, surround the cytoplasm of all cells and delineate discrete metabolic compartments within cells. [©ISM/Phototake.]

Lipids and Biological Membranes

Lipids (Greek: *lipos,* fat) are the fourth major group of molecules found in all cells. Unlike nucleic acids, proteins, and polysaccharides, lipids are not polymeric. However, they do aggregate, and it is in this state that they perform their most obvious function as the structural matrix of biological membranes.

Lipids exhibit greater structural variety than the other classes of biological molecules. To a certain extent, lipids constitute a catchall category of substances that are similar only in that they are largely hydrophobic and only sparingly soluble in water. In general, lipids perform three biological functions (although certain lipids apparently serve more than one purpose in some cells):

1. Lipid molecules in the form of lipid bilayers are essential components of biological membranes.
2. Lipids containing hydrocarbon chains serve as energy stores.
3. Many intra- and intercellular signaling events involve lipid molecules.

In this chapter we examine the structures and physical properties of the most common types of lipids. Next, we look at the properties of the lipid bilayer and the proteins that are situated within it. Finally, we explore current models of membrane structure. The following chapter examines membrane transport phenomena.

1 Lipid Classification

Lipids are substances of biological origin that are soluble in organic solvents such as chloroform and methanol. Hence, they are easily separated from other biological materials by extraction into organic solvents. Fats, oils, certain vitamins and hormones, and most nonprotein membrane components are lipids. In this section, we discuss the structures and physical properties of the major classes of lipids.

A Fatty Acids

Fatty acids are carboxylic acids with long-chain hydrocarbon side groups (Fig. 9-1). They usually occur in esterified form as major components of the various lipids described in this chapter. The more common biological fatty acids are listed in Table 9-l. In higher plants and animals, the predominant fatty acid residues are those of the C_{16} and C_{18} species: **palmitic, oleic, linoleic,** and **stearic acids.** Fatty acids with <14 or >20 carbon atoms are uncommon. Most fatty acids have an even number of carbon atoms because they are biosynthesized by the concatenation of C_2 units (Section 19-4).

Over half of the fatty acid residues of plant and animal lipids are **unsaturated** (contain double bonds) and are often **polyunsaturated** (contain two or more double bonds). Bacterial fatty acids are rarely polyunsaturated but are commonly branched, hydroxylated, or contain cyclopropane rings.

Table 9-l indicates that the first double bond of an unsaturated fatty acid commonly occurs between its C9 and C10 atoms counting from the carboxyl C atom. This bond is called a Δ^9- or 9-double bond. In polyunsaturated fatty acids, the double bonds tend to occur at every third carbon atom (e.g., —CH=CH—CH$_2$—CH=CH—) and so are not conjugated (as in —CH=CH—CH=CH—). Two important classes of polyunsaturated fatty acids are designated as ω-3 or ω-6 fatty acids, a nomenclature that identities the last double-bonded carbon atom as counted from the methyl terminal (ω) end of the chain. **α-Linolenic acid** and linoleic acid (Fig. 9-1) are examples of such fatty acids.

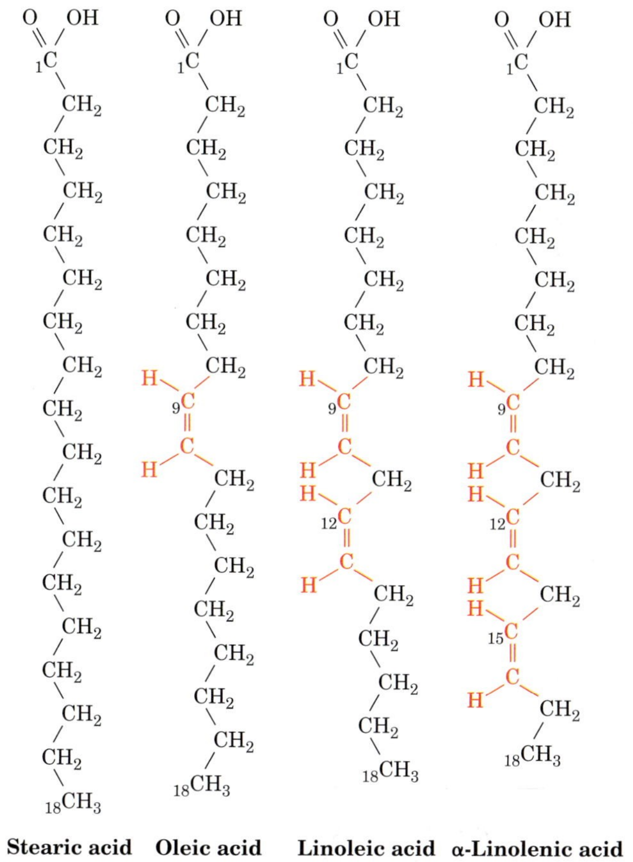

Stearic acid   Oleic acid   Linoleic acid   α-Linolenic acid

Figure 9-1 The structural formulas of some C_{18} fatty acids. The double bonds all have the cis configuration.

Table 9-1 The Common Biological Fatty Acids

Symbol[a]	Common Name	Systematic Name	Structure	mp (°C)
Saturated fatty acids				
12:0	Lauric acid	Dodecanoic acid	$CH_3(CH_2)_{10}COOH$	44.2
14:0	Myristic acid	Tetradecanoic acid	$CH_3(CH_2)_{12}COOH$	52
16:0	Palmitic acid	Hexadecanoic acid	$CH_3(CH_2)_{14}COOH$	63.1
18:0	Stearic acid	Octadecanoic acid	$CH_3(CH_2)_{16}COOH$	69.1
20:0	Arachidic acid	Eicosanoic acid	$CH_3(CH_2)_{18}COOH$	75.4
22:0	Behenic acid	Docosanoic acid	$CH_3(CH_2)_{20}COOH$	81
24:0	Lignoceric acid	Tetracosanoic acid	$CH_3(CH_2)_{22}COOH$	84.2
Unsaturated fatty acids (all double bonds are cis)				
16:1$n-7$	Palmitoleic acid	9-Hexadecenoic acid	$CH_3(CH_2)_5CH=CH(CH_2)_7COOH$	−0.5
18:1$n-9$	Oleic acid	9-Octadecenoic acid	$CH_3(CH_2)_7CH=CH(CH_2)_7COOH$	13.2
18:2$n-6$	Linoleic acid	9,12-Octadecadienoic acid	$CH_3(CH_2)_4(CH=CHCH_2)_2(CH_2)_6COOH$	−9
18:3$n-3$	α-Linolenic acid	9,12,15-Octadecatrienoic acid	$CH_3CH_2(CH=CHCH_2)_3(CH_2)_6COOH$	−17
18:3$n-6$	γ-Linolenic acid	6,9,12-Octadecatrienoic acid	$CH_3(CH_2)_4(CH=CHCH_2)_3(CH_2)_3COOH$	
20:4$n-6$	Arachidonic acid	5,8,11,14-Eicosatetraenoic acid	$CH_3(CH_2)_4(CH=CHCH_2)_4(CH_2)_2COOH$	−49.5
20:5$n-3$	EPA	5,8,11,14,17-Eicosapentaenoic acid	$CH_3CH_2(CH=CHCH_2)_5(CH_2)_2COOH$	−54
22:6$n-3$	DHA	4,7,10,13,16,19-Docosohexenoic acid	$CH_3CH_2(CH=CHCH_2)_6CH_2COOH$	
24:1$n-9$	Nervonic acid	15-Tetracosenoic acid	$CH_3(CH_2)_7CH=CH(CH_2)_{13}COOH$	39

[a]Number of carbon atoms: Number of double bonds. For unsaturated fatty acids, the quantity "$n-x$" indicates the position of the last double bond in the fatty acid, where n is its number of C atoms, and x is the position of the last double-bonded C atom counting from the methyl terminal (ω) end.

Source: Dawson, R.M.C., Elliott, D.C., Elliott, W.H., and Jones, K.M., *Data for Biochemical Research* (3rd ed.), Chapter 8, Clarendon Press (1986).

Saturated fatty acids (which are fully reduced or "saturated" with hydrogen) are highly flexible molecules that can assume a wide range of conformations because there is relatively free rotation around each of their C—C bonds. Nevertheless, their lowest energy conformation is the fully extended conformation, which has the least amount of steric interference between neighboring methylene groups. The melting points (mp) of saturated fatty acids, like those of most substances, increase with their molecular mass (Table 9-1).

Fatty acid double bonds almost always have the cis configuration (Fig. 9-1). This puts a rigid 30° bend in the hydrocarbon chain. Consequently, unsaturated fatty acids pack together less efficiently than saturated fatty acids. The reduced van der Waals interactions of unsaturated fatty acids cause their melting points to decrease with the degree of unsaturation. The fluidity of lipids containing fatty acid residues likewise increases with the degree of unsaturation of the fatty acids. This phenomenon, as we shall see, has important consequences for biological membranes.

B Triacylglycerols

The fats and oils that occur in plants and animals consist largely of mixtures of **triacylglycerols** (also called **triglycerides**). These nonpolar, water-insoluble substances are fatty acid triesters of **glycerol:**

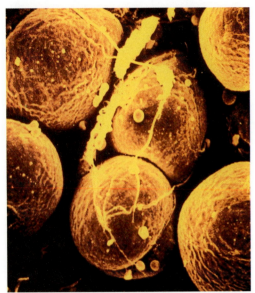

$$^1CH_2-OH$$
$$^2CH-OH$$
$$^3CH_2-OH$$

Glycerol

$$^1CH_2-O-\overset{\overset{O}{\|}}{C}-R_1$$
$$^2CH-O-\overset{\overset{O}{\|}}{C}-R_2$$
$$^3CH_2-O-\overset{\overset{O}{\|}}{C}-R_3$$

Triacylglycerol

Triacylglycerols function as energy reservoirs in animals and are therefore their most abundant class of lipids even though they are not components of cellular membranes.

Triacylglycerols differ according to the identity and placement of their three fatty acid residues. Most triacylglycerols contain two or three different types of fatty acid residues and are named according to their placement on the glycerol moiety, for example, **1-palmitoleoyl-2-linoleoyl-3-stearoyl-glycerol** *(at left)*. Note that the *-ate* ending of the name of the fatty acid becomes *-oyl* in the fatty acid ester. **Fats** and **oils** (which differ only in that fats are solid and oils are liquid at room temperature) are complex mixtures of triacylglycerols whose fatty acid compositions vary with the organism that produced them. Plant oils are usually richer in unsaturated fatty acid residues than animal fats, as the lower melting points of oils imply.

Triacylglycerols Function as Energy Reserves.   Fats are a highly efficient form in which to store metabolic energy. This is because triacylglycerols are less oxidized than carbohydrates or proteins and hence yield significantly more energy per unit mass on complete oxidation. Furthermore, triacylglycerols, which are nonpolar, are stored in anhydrous form, whereas glycogen (Section 8-2C), for example, binds about twice its weight of water under physiological conditions. *Fats therefore provide about six times the metabolic energy of an equal weight of hydrated glycogen.*

In animals, **adipocytes** (fat cells; Fig. 9-2) are specialized for the synthesis and storage of triacylglycerols. Whereas other types of cells have only a

Figure 9-2   Scanning electron micrograph of adipocytes.   Each adipocyte contains a fat globule that occupies nearly the entire cell. [Fred E. Hossler/Visuals Unlimited.]

$$^1CH_2 - {}^2CH - {}^3CH_2$$

| | | |
|O|O|O|

$$C_1{=}O \quad C_1{=}O \quad C_1{=}O$$
$$CH_2 \quad CH_2 \quad CH_2$$
$$CH_2 \quad CH_2 \quad CH_2$$
$$CH_2 \quad CH_2 \quad CH_2$$
$$CH_2 \quad CH_2 \quad CH_2$$
$$CH_2 \quad CH_2 \quad CH_2$$
$$CH_2 \quad CH_2 \quad CH_2$$
$$CH_2 \quad CH_2 \quad CH_2$$
$$CH \quad CH \quad CH_2$$
$$\overset{9}{\|} \quad \overset{9}{\|}$$
$$CH \quad CH \quad CH_2$$
$$CH_2 \quad CH_2 \quad CH_2$$
$$CH_2 \quad CH \quad CH_2$$
$$\quad \overset{12}{\|}$$
$$CH_2 \quad CH \quad CH_2$$
$$CH_2 \quad CH_2 \quad CH_2$$
$$CH_2 \quad CH_2 \quad CH_2$$
$$_{16}CH_3 \quad CH_2 \quad CH_2$$
$$CH_2 \quad CH_2$$
$$_{18}CH_3 \quad _{18}CH_3$$

1-Palmitoleoyl-2-linoleoyl-3-stearoyl-glycerol

few small droplets of fat dispersed in their cytosol, adipocytes may be almost entirely filled with fat globules. **Adipose tissue** is most abundant in a subcutaneous layer and in the abdominal cavity. The fat content of normal humans (21% for men, 26% for women) allows them to survive starvation for 2 or 3 months. In contrast, the body's glycogen supply, which functions as a short-term energy store, can provide for the body's energy needs for less than a day. The subcutaneous fat layer also provides thermal insulation, which is particularly important for warm-blooded aquatic animals, such as whales, seals, geese, and penguins, which are routinely exposed to low temperatures.

C Glycerophospholipids

Glycerophospholipids (or **phosphoglycerides**) are the major lipid components of biological membranes. They consist of **glycerol-3-phosphate** whose C1 and C2 positions are esterified with fatty acids. In addition, the phosphoryl group is linked to another usually polar group, X (Fig. 9-3). *Glycerophospholipids are therefore amphiphilic molecules with nonpolar aliphatic (hydrocarbon) "tails" and polar phosphoryl-X "heads."*

The simplest glycerophospholipids, in which X = H, are **phosphatidic acids;** they are present in only small amounts in biological membranes. In the glycerophospholipids that commonly occur in biological membranes, the head groups are derived from polar alcohols (Table 9-2). Saturated C_{16} or C_{18} fatty acids usually occur at the C1 position of the glycerophospho-

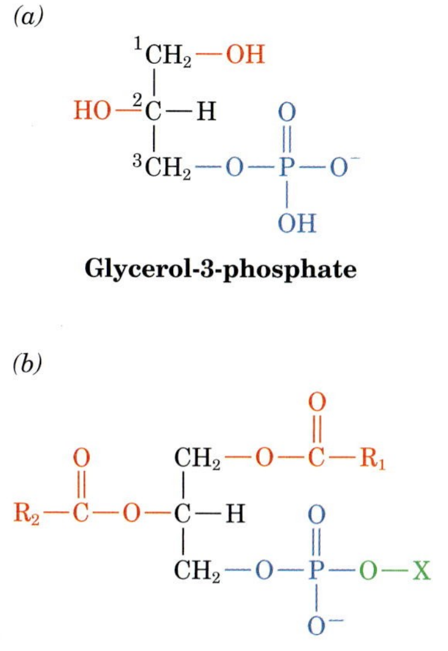

(a)

Glycerol-3-phosphate

(b)

Glycerophospholipid

Figure 9-3 Structure of glycerophospholipids. (a) The backbone, L-glycerol-3-phosphate. (b) The general formula of the glycerophospholipids. R_1 and R_2 are the long-chain hydrocarbon tails of fatty acids, and X is derived from a polar alcohol (Table 9-2). Note that glycerol-3-phosphate and glycerophospholipid are chiral compounds.

Table 9-2 The Common Classes of Glycerophospholipids

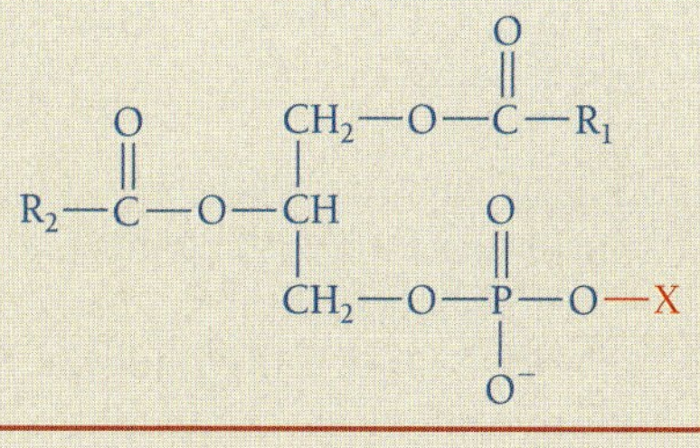

Name of X—OH	Formula of —X	Name of Phospholipid
Water	—H	Phosphatidic acid
Ethanolamine	—$CH_2CH_2NH_3^+$	Phosphatidylethanolamine
Choline	—$CH_2CH_2N(CH_3)_3^+$	Phosphatidylcholine (lecithin)
Serine	—$CH_2CH(NH_3^+)COO^-$	Phosphatidylserine
myo-Inositol	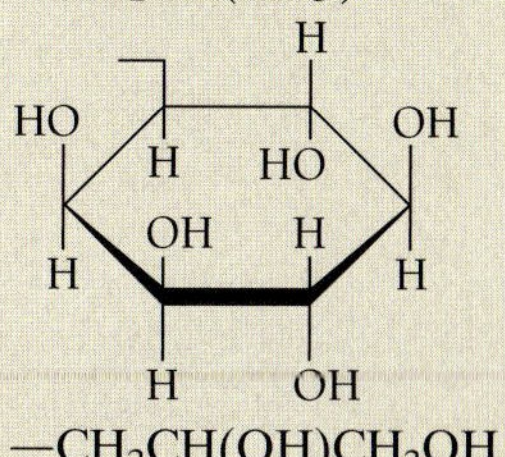 (structure)	Phosphatidylinositol
Glycerol	—$CH_2CH(OH)CH_2OH$	Phosphatidylglycerol
Phosphatidylglycerol	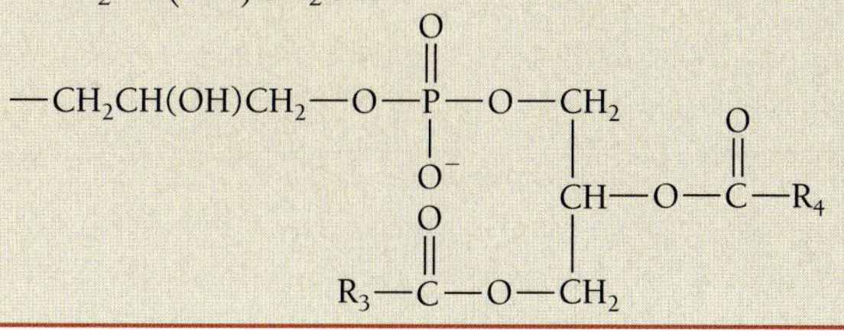 (structure)	Diphosphatidylglycerol (cardiolipin)

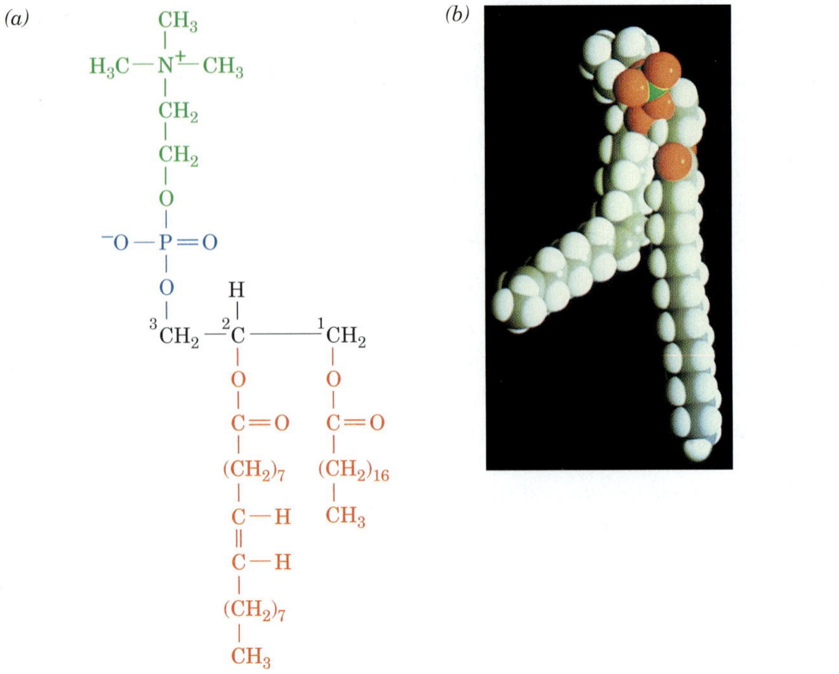

Figure 9-4 **The glycerophospholipid 1-stearoyl-2-oleoyl-3-phosphatidylcholine.** (*a*) Molecular formula in Fischer projection. (*b*) Space-filling model with H white, C gray, O red, and P green. Note how the unsaturated oleoyl chain is bent compared to the saturated stearoyl chain. [Courtesy of Richard Pastor, FDA, Bethesda, Maryland.]

(*a*)

$$CH_3$$
$$|$$
$$H_3C-N^+-CH_3$$
$$|$$
$$CH_2$$
$$|$$
$$CH_2$$
$$|$$
$$O$$
$$|$$
$$^-O-P=O$$
$$|$$
$$O \quad H$$
$$| \quad |$$
$$^3CH_2 \quad —^2C———^1CH_2$$
$$| \quad |$$
$$O \quad O$$
$$| \quad |$$
$$C=O \quad C=O$$
$$| \quad |$$
$$(CH_2)_7 \quad (CH_2)_{16}$$
$$| \quad |$$
$$C-H \quad CH_3$$
$$||$$
$$C-H$$
$$|$$
$$(CH_2)_7$$
$$|$$
$$CH_3$$

(*b*)

1-Stearoyl-2-oleoyl-3-phosphatidylcholine

lipids, and the C2 position is often occupied by an unsaturated C_{16} to C_{20} fatty acid. Individual glycerophospholipids are named according to the identities of these fatty acid residues (e.g., Fig. 9-4). A glycerophospholipid containing two palmitoyl chains is an important component of **lung surfactant** (see Box 9-1).

Phospholipases Hydrolyze Glycerophospholipids. The chemical structures—including fatty acyl chains and head groups—of glycerophospholipids can be determined from the products of the hydrolytic reactions catalyzed by enzymes known as **phospholipases.** For example, **phospholipase A_2** hydrolytically excises the fatty acid residue at C2, leaving a **lysophospholipid** (Fig. 9-5). Lysophospholipids, as their name implies, are powerful detergents that disrupt cell membranes, thereby lysing cells. Bee and snake venoms are rich sources of phospholipase A_2. Other types of phospholipases act at different sites in glycerophospholipids, as shown in Fig. 9-5.

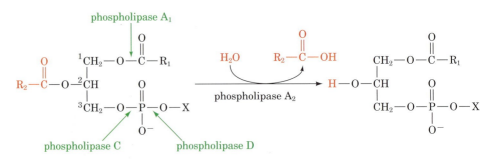

Phospholipid   **Lysophospholipid**

Figure 9-5 **Action of phospholipases.** Phospholipase A_2 hydrolytically excises the C2 fatty acid residue from a triacylglycerol to yield the corresponding lysophospholipid. The bonds hydrolyzed by other types of phospholipases, which are named according to their specificities, are also indicated.

BOX 9-1

Biochemistry in Health and Disease

Lung Surfactant

Dipalmitoyl phosphatidylcholine (DPPC) is the major lipid of lung surfactant, the protein–lipid mixture that is essential for normal pulmonary function. The surfaces of the cells that form the alveoli (small air spaces of the lung) are coated with surfactant, which decreases the alveolar surface tension. Lung surfactant contains 80 to 90% phospholipid by weight, and 70 to 80% of the phospholipid is phosphatidylcholine, mostly the dipalmitoyl species.

Because the palmitoyl chains of DPPC are saturated, they tend to extend straight out without bending. This allows close packing of DPPC molecules, which are oriented in a single layer with their nonpolar tails toward the air and their polar heads toward the alveolar cells. When air is expired from the lungs, the volume and surface area of the alveoli decrease.

The collapse of the alveolar space is prevented by the surfactant, because the closely packed DPPC molecules resist compression. Reopening a collapsed air space requires a much greater force than expanding an already open air space.

Lung surfactant is continuously synthesized, secreted, and recycled by alveolar cells. Because surfactant production is low until just before birth, premature infants are at risk of developing **respiratory distress syndrome,** which is characterized by difficulty in breathing due to alveolar collapse. The syndrome can be treated by introducing exogenous surfactant into the lungs. A related condition in adults **(adult respiratory distress syndrome)** is characterized by insufficient surfactant, usually secondary to other lung injury. This condition, too, can be treated with exogenous surfactant.

Enzymes that act on lipids have fascinated biochemists because the enzymes must gain access to portions of the lipids that are buried in a non-aqueous environment. Phospholipases A_2, which constitute some of the best understood lipid-specific enzymes, are relatively small proteins (~14 kD, ~125 amino acid residues). The X-ray structure of phospholipase A_2 from cobra venom suggests that the enzyme binds a glycerophospholipid molecule such that its polar head group fits into the enzyme's active site, whereas the hydrophobic tails, which extend beyond the active site, interact with several aromatic side chains (Fig. 9-6).

Lipases specific for triacylglycerols and membrane lipids catalyze their degradation *in vivo*. Occasionally, the hydrolysis products are not destined for further degradation but instead serve as intra- and extracellular signal

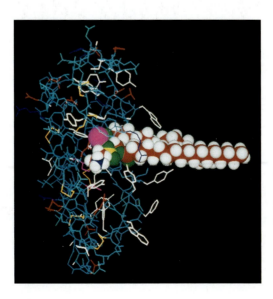

Figure 9-6 Model of phospholipase A_2 and a glycerophospholipid. The X-ray structure of the enzyme from cobra venom is shown with a space-filling model of dimyristoyl phosphatidylethanolamine in its active site as located by NMR methods. A Ca^{2+} ion in the active site is shown in magenta. [Courtesy of Edward A. Dennis, University of California at San Diego.] ✌ **See the Interactive Exercises.**

molecules. For example, **lysophosphatidic acid (1-acyl-glycerol-3-phosphate),** which is not actually lytic since it has a small head group (an unsubstituted phosphate group), is produced by hydrolysis of membrane lipids in blood platelets and injured cells and stimulates cell growth as part of the wound-repair process. **1,2-Diacylglycerol,** derived from membrane lipids by the action of **phospholipase C,** is an intracellular signal molecule that activates a **protein kinase** (Section 21-3F; kinases catalyze ATP-dependent phosphoryl-transfer reactions).

Plasmalogens Contain an Ether Linkage. **Plasmalogens** are glycerophospholipids in which the C1 substituent of the glycerol moiety is linked via an α,β-unsaturated ether linkage in the cis configuration rather than through an ester linkage:

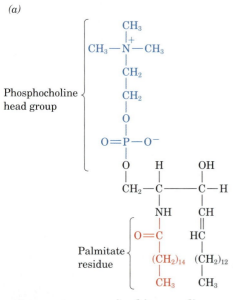

(a)

Phosphocholine head group

Palmitate residue

A sphingomyelin

(b)

Figure 9-7 **A sphingomyelin.** (*a*) Molecular formula. (*b*) Space-filling model with H white, C gray, N blue, and O red. [Courtesy of Richard Pastor, FDA, Bethesda, Maryland.]

A plasmalogen

Ethanolamine, choline, and **serine** (Table 9-2) form the most common plasmalogen head groups. The functions of most plasmalogens are not well understood. Because the vinyl ether group is easily oxidized, plasmalogens may react with oxygen free radicals, by-products of normal metabolism, thereby preventing free-radical damage to other cell constituents.

D Sphingolipids

Sphingolipids are also major membrane components. Their function in cells was at first mysterious, so they were named after the Sphinx. Most sphingolipids are derivatives of the C_{18} amino alcohol **sphingosine,** whose double bond has the trans configuration. The *N*-acyl fatty acid derivatives of sphingosine are known as **ceramides:**

Sphingosine   **A ceramide**

Ceramides are the parent compounds of the more abundant sphingolipids:

1. **Sphingomyelins,** the most common sphingolipids, are ceramides bearing either a phosphocholine (Fig. 9-7) or a phosphoethanolamine

head group, so they can also be classified as **sphingophospholipids.** They typically comprise 10 to 20 mol % of plasma membrane lipids. *Although sphingomyelins differ chemically from phosphatidylcholine and phosphatidylethanolamine, their conformations and charge distributions are quite similar* (compare Figs. 9-4 and 9-7). The membranous myelin sheath that surrounds and electrically insulates many nerve cell axons is particularly rich in sphingomyelins (Fig. 9-8).

2. *Cerebrosides* are ceramides with head groups that consist of a single sugar residue. These lipids are therefore **glycosphingolipids. Galactocerebrosides** and **glucocerebrosides** are the most prevalent. Cerebrosides, in contrast to phospholipids, lack phosphate groups and hence are nonionic.

3. *Gangliosides* are the most complex glycosphingolipids. They are ceramides with attached oligosaccharides that include at least one sialic acid residue. The structures of **gangliosides G$_{M1}$, G$_{M2}$,** and **G$_{M3}$,** three of the over 60 that are known, are shown in Fig. 9-9. Gangliosides are primarily components of cell-surface membranes and constitute a significant fraction (6%) of brain lipids.

Gangliosides have considerable physiological and medical significance. Their complex carbohydrate head groups, which extend beyond the surfaces of cell membranes, act as specific receptors for certain pituitary glycoprotein hormones that regulate a number of important physiological functions. Gangliosides are also receptors for certain bacterial protein toxins such as **cholera toxin.** There is considerable evidence that gangliosides are specific determinants of cell–cell recognition, so they probably have an important role in the growth and differentiation of tissues as well as in car-

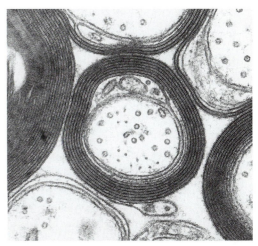

Figure 9-8 **Electron micrograph of myelinated nerve fibers.** This cross-sectional view shows the spirally wrapped membranes around each nerve axon. The myelin sheath may be 10–15 layers thick. Its high lipid content makes it an electrical insulator. [Courtesy of Cedric S. Raine, Albert Einstein College of Medicine.]

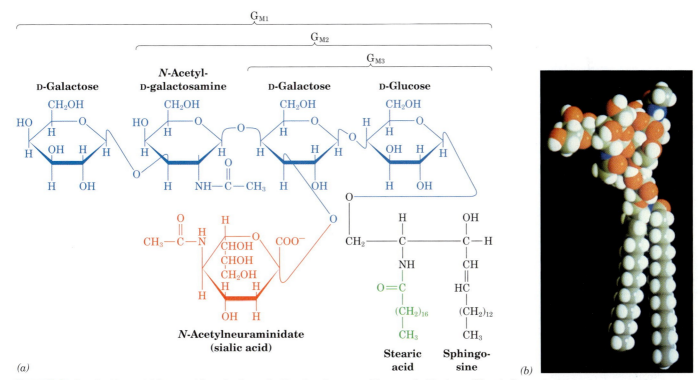

Figure 9-9 **Gangliosides.** (*a*) Structural formula of gangliosides G$_{M1}$, G$_{M2}$, and G$_{M3}$. Gangliosides G$_{M2}$ and G$_{M3}$ differ from G$_{M1}$ only by the sequential absences of the terminal D-galactose and *N*-acetyl-D-galactosamine residues.

Other gangliosides have different oligosaccharide head groups. (*b*) Space-filling model of G$_{M1}$ with H white, C gray, N blue, and O red. [Courtesy of Richard Venable, FDA, Bethesda, Maryland.]

cinogenesis. Disorders of ganglioside breakdown are responsible for several hereditary **sphingolipid storage diseases,** such as **Tay-Sachs disease,** which are characterized by an invariably fatal neurological deterioration in early childhood.

Sphingolipids, like glycerophospholipids, are a source of smaller lipids that have discrete signaling activity. Sphingomyelin itself, as well as the ceramide portions of more complex sphingolipids, appear to specifically modulate the activities of protein kinases and **protein phosphatases** (enzymes that remove phosphoryl groups from proteins) that are involved in regulating cell growth and differentiation.

E Steroids

Steroids, which are mostly of eukaryotic origin, are derivatives of **cyclopentanoperhydrophenanthrene,**

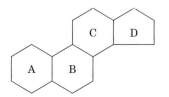

Cyclopentanoperhydrophenanthrene

a compound that consists of four fused, nonplanar rings (labeled A–D). The much maligned **cholesterol,** which is the most abundant steroid in animals, is further classified as a **sterol** because of its C3-OH group (Fig. 9-10). Cholesterol is a major component of animal plasma membranes, typically constituting 30 to 40 mol % of plasma membrane lipids. Its polar OH group gives it a weak amphiphilic character, whereas its fused ring system provides it with greater rigidity than other membrane lipids. Cholesterol can also be esterified to long-chain fatty acids to form **cholesteryl esters,** for example:

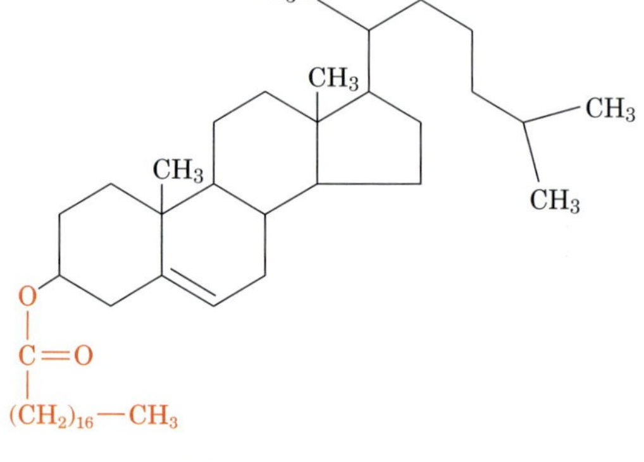

Cholesteryl stearate

Plants contain little cholesterol but synthesize other sterols. Yeast and fungi also synthesize sterols, which differ from cholesterol in their aliphatic side chains and number of double bonds. Prokaryotes contain little, if any, sterol.

In mammals, cholesterol is the metabolic precursor of **steroid hormones,** substances that regulate a great variety of physiological functions. The structures of some steroid hormones are shown in Fig. 9-11. Steroid hormones are classified according to the physiological responses they evoke:

1. The **glucocorticoids,** such as **cortisol** (a C_{21} compound), affect carbohydrate, protein, and lipid metabolism and influence a wide variety

(a)

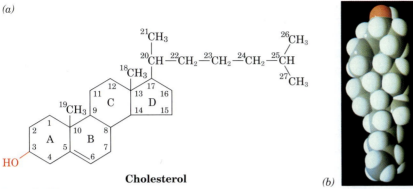

Cholesterol   (b)

Figure 9-10   Cholesterol.   (a) Structural formula with the standard numbering system.
(b) Space-filling model with H white, C gray, and O red. [Courtesy of Richard Pastor, FDA, Bethesda, Maryland.]

of other vital functions, including inflammatory reactions and the capacity to cope with stress.

2. **Aldosterone** and other **mineralocorticoids** regulate the excretion of salt and water by the kidneys.

3. The **androgens** and **estrogens** affect sexual development and function. **Testosterone,** a C_{19} compound, is the prototypic androgen (male sex hormone).

Glucocorticoids and mineralocorticoids are synthesized by the cortex (outer layer) of the adrenal gland. Both androgens and estrogens (female sex hormones) are synthesized by testes and ovaries (although androgens predominate in testes and estrogens predominate in ovaries) and, to a lesser

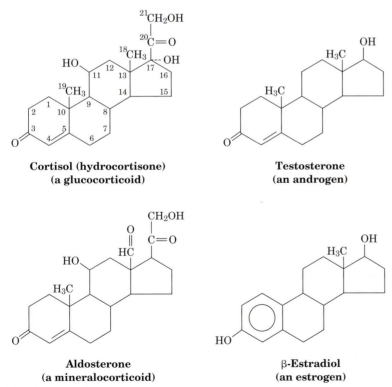

Cortisol (hydrocortisone)
(a glucocorticoid)

Testosterone
(an androgen)

Aldosterone
(a mineralocorticoid)

β-Estradiol
(an estrogen)

Figure 9-11   Some representative steroid hormones.

extent, by the adrenal cortex. Because steroid hormones are water-insoluble, they bind to proteins for transport through the blood to their target tissues.

Impaired adrenocortical function, either through disease or trauma, results in **Addison's disease,** which is characterized by **hypoglycemia** (decreased amounts of glucose in the blood), muscle weakness, Na^+ loss, K^+ retention, impaired cardiac function, and greatly increased susceptibility to stress. The victim, unless treated by the administration of glucocorticoids and mineralocorticoids, slowly languishes and dies without any particular pain or distress.

Vitamin D Regulates Ca^{2+} Metabolism. The various forms of **vitamin D,** which are really hormones, are sterol derivatives in which the steroid B ring is disrupted between C9 and C10:

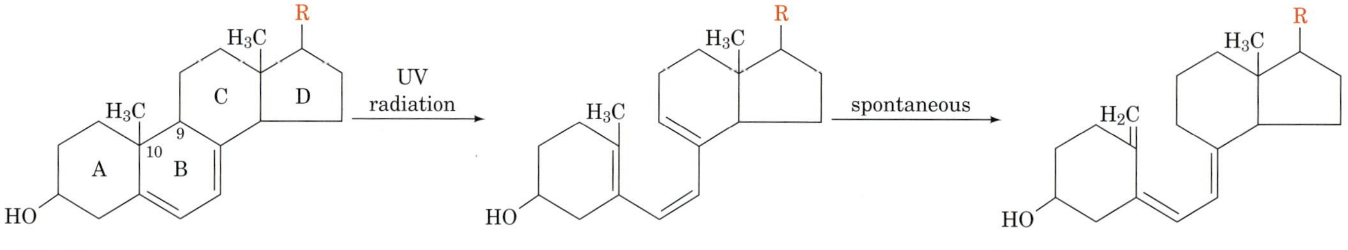

R = X 7-Dehydrocholesterol
R = Y Ergosterol

R = X Vitamin D$_3$ (cholecalciferol)
R = Y Vitamin D$_2$ (ergocalciferol)

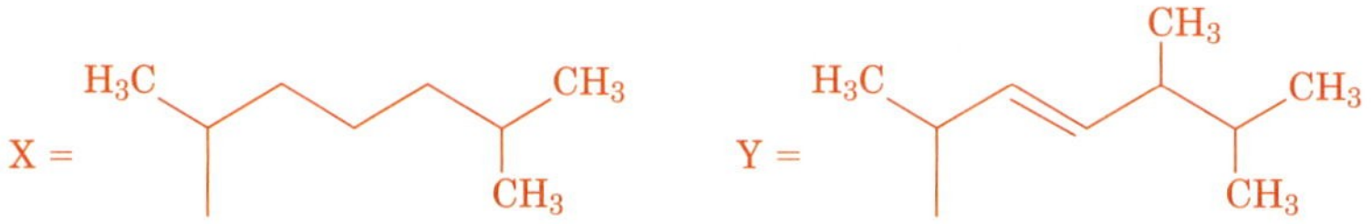

Vitamin D$_2$ (ergocalciferol) is nonenzymatically formed in the skin of animals through the photolytic action of UV light on the plant sterol **ergosterol,** a common milk additive, whereas the closely related **vitamin D$_3$ (cholecalciferol)** is similarly derived from **7-dehydrocholesterol** (hence the saying that sunlight provides vitamin D).

Vitamins D$_2$ and D$_3$ are inactive; the active forms are produced through their enzymatic hydroxylation (addition of an OH group) carried out by the liver (at C25) and by the kidney (at C1) to yield **1α,25-dihydroxycholecalciferol:**

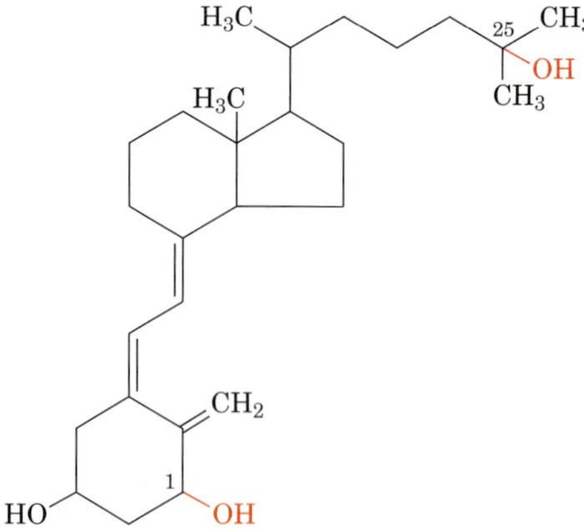

1α,25-Dihydroxycholecalciferol

Active vitamin D increases serum [Ca^{2+}] by promoting the intestinal absorption of dietary Ca^{2+}. This increases the deposition of Ca^{2+} in bones

and teeth. Vitamin D deficiency produces **rickets** in children, a disease characterized by stunted growth and deformed bones caused by insufficient bone mineralization. Although rickets was first described in 1645, it was not until the early twentieth century that eating animal fats, particularly fish liver oils, was shown to prevent this deficiency disease. Rickets can also be prevented by exposing children to sunlight or just UV light in the wavelength range 230 to 313 nm, regardless of their diets.

Since vitamin D is water insoluble, it can accumulate in fatty tissues. Excessive intake of vitamin D over long periods results in **vitamin D intoxication.** The consequent high serum $[Ca^{2+}]$ results in aberrant calcification of soft tissues and in the development of kidney stones, which can cause kidney failure. The observation that the level of skin pigmentation in indigenous human populations tends to increase with their proximity to the equator is explained by the hypothesis that skin pigmentation functions to prevent vitamin D intoxication by filtering out excessive solar radiation.

F Other Lipids

In addition to the well-characterized lipids that are found in large amounts in cellular membranes, many organisms synthesize compounds that are not membrane components but are classified as lipids on the basis of their physical properties. For example, lipids occur in the waxy coatings of plants, where they protect cells from desiccation by creating a water-impermeable barrier.

Isoprenoids Are Built from Five-Carbon Units. Among the compounds that are not structural components of membranes—although they are soluble in the lipid bilayer—are the **isoprenoids,** which are built from five-carbon units with the same carbon skeleton as **isoprene.**

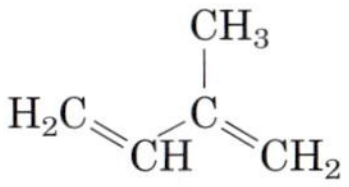

Isoprene

For example, the isoprenoid **ubiquinone** (also known as **coenzyme Q**) is reversibly reduced and oxidized in the mitochondrial membrane (its activity is described in more detail in Section 17-2C). Mammalian ubiquinone consists of 10 isoprenoid units.

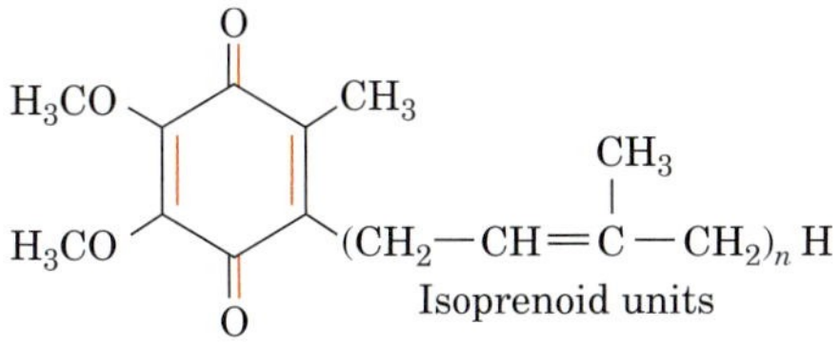

Isoprenoid units

Coenzyme Q (CoQ) or ubiquinone

The plant kingdom is rich in isoprenoid compounds, which serve as pigments, molecular signals (hormones and pheromones), and defensive agents. Indeed, over 25,000 isoprenoids (also known as **terpenoids**), which are mostly of plant, fungal, and bacterial origin, have been characterized. During the course of evolution, vertebrate metabolism has co-opted several of these compounds for other purposes. Some of these compounds (e.g., vitamin D) are known as **fat-soluble vitamins** (vitamins are organic substances that an animal requires in small amounts but cannot synthesize and hence must acquire in its diet).

Vitamin A, or **retinol,**

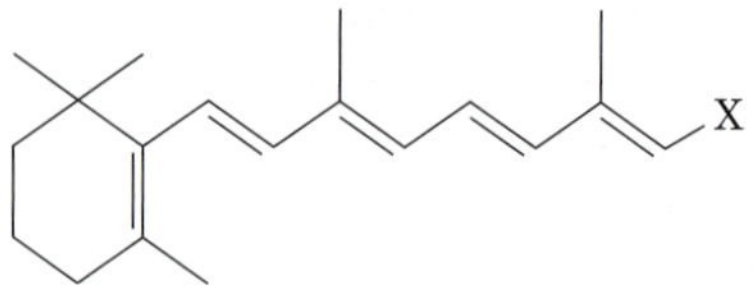

X = CH$_2$OH   **Retinol (vitamin A)**
X = CHO   **Retinal**

is derived mainly from plant products such as **β-carotene** [a red pigment that is present in green vegetables as well as carrots (after which it is named) and tomatoes; Section 18-1B]. Retinol is oxidized to its corresponding aldehyde, **retinal,** which functions as the eye's photoreceptor at low light intensities. Light causes the retinal to isomerize, triggering, via a complex signaling pathway, an impulse through the optic nerve. A severe deficiency of vitamin A can lead to blindness. Retinoic acid also has hormonelike properties in that it stimulates tissue repair. It is used to treat severe acne and skin ulcers and is also used cosmetically to eliminate wrinkles.

Vitamin K is a lipid synthesized by plants (as **phylloquinone**) and bacteria (as **menaquinone**):

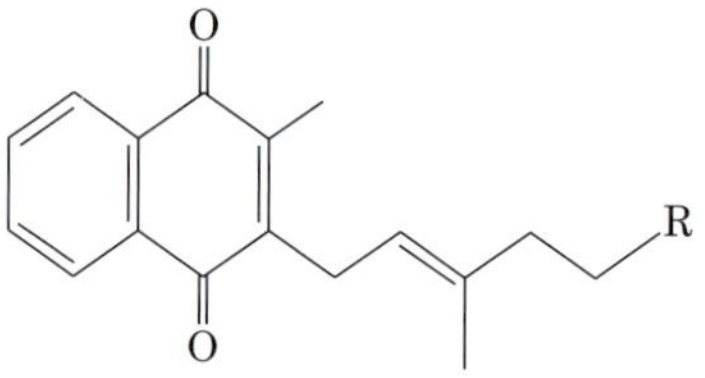

R = 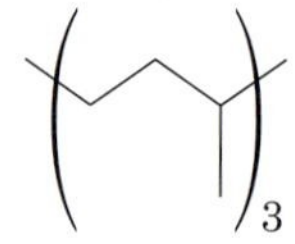   **Phylloquinone**
       (vitamin K$_1$)

R = 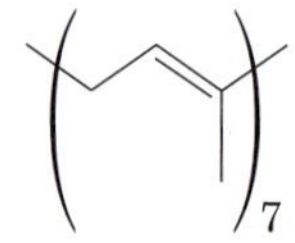   **Menaquinone**
       (vitamin K$_2$)

About half of the daily requirement for humans is supplied by intestinal bacteria. Vitamin K participates in the carboxylation of Glu residues in some of the proteins involved in blood clotting (vitamin K is named for the Danish word *Koagulation*). Vitamin K deficiency prevents this carboxylation, and the resulting inactive clotting proteins lead to excessive bleeding. Compounds that interfere with vitamin K function are the active ingredients in some rodent poisons.

Vitamin E is actually a group of compounds whose most abundant member is **α-tocopherol:**

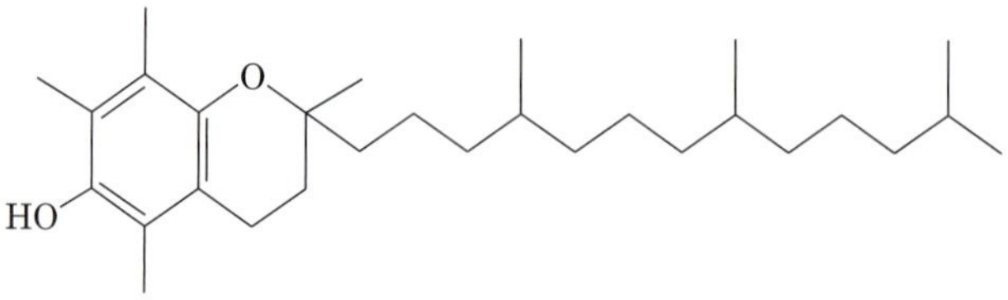

α-Tocopherol
(vitamin E)

This highly hydrophobic molecule is incorporated into cell membranes, where it functions as an antioxidant that prevents oxidative damage to membrane proteins and lipids. A deficiency of vitamin E elicits a variety of nonspecific symptoms, which makes the deficiency difficult to detect. The popularity of vitamin E supplements rests on the hypothesis that vitamin E protects against oxidative damage to cells and hence reduces the effects of aging.

Eicosanoids Are Derived from Arachidonic Acid.   Other less common lipids are derived from relatively abundant membrane lipids. **Prostaglandins** (e.g., Fig. 9-12) were discovered in the 1930s by Ulf von Euler, who thought they were produced by the prostate gland. Prostaglandins and related compounds—**prostacyclins, thromboxanes, leukotrienes,** and **lipoxins**—are known collectively as **eicosanoids** because they are all C_{20} compounds (Greek: *eikosi,* twenty). *The eicosanoids act at very low concentrations and are involved in the production of pain and fever, and in the regulation of blood pressure, blood coagulation, and reproduction.* Unlike hormones, eicosanoids are not transported by the bloodstream to their sites of action but tend to act locally, close to the cells that produced them. In fact, most eicosanoids decompose within seconds or minutes, which limits their effects

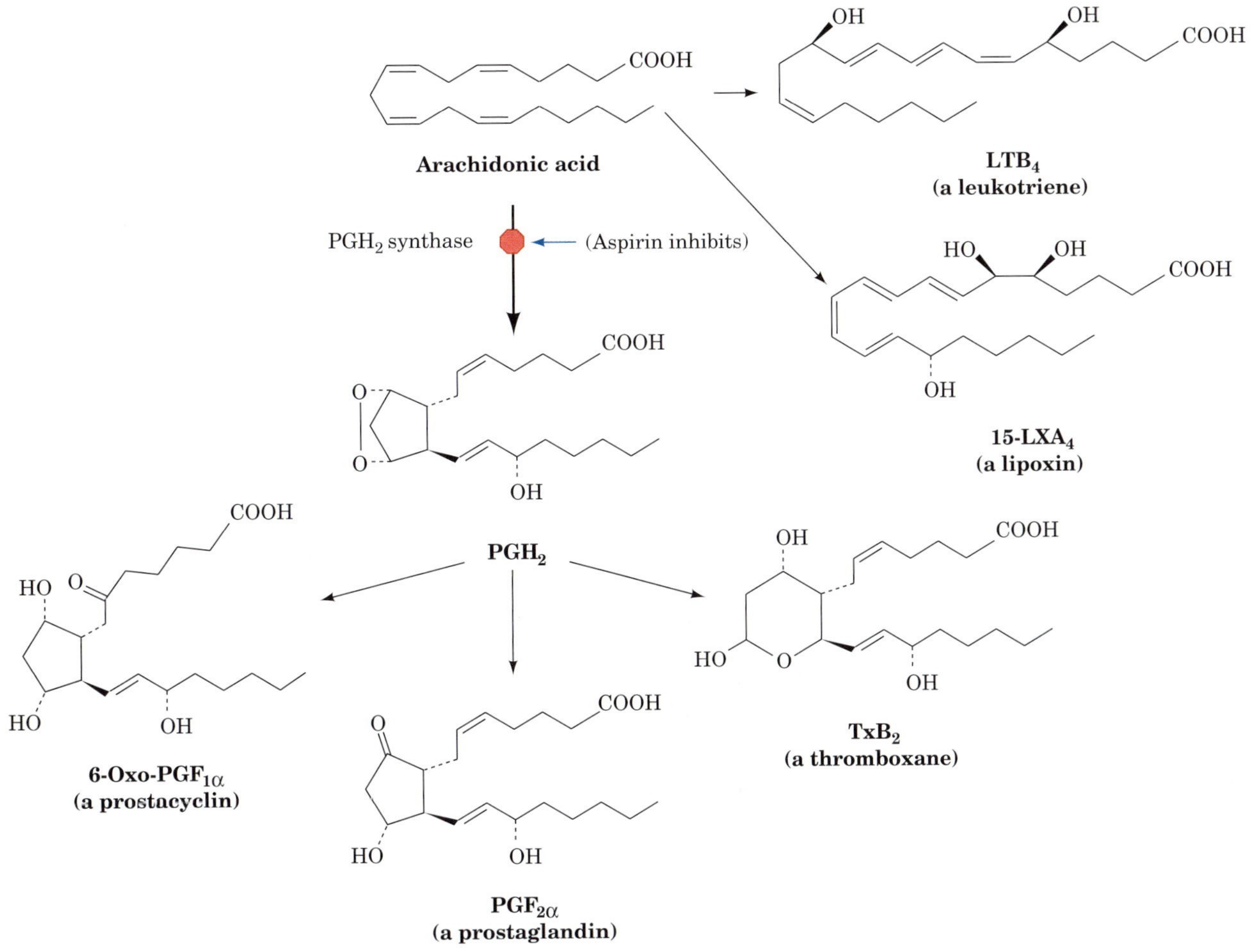

Figure 9-12 Eicosanoids.   Arachidonate is the precursor of prostaglandins (PG), prostacyclins, thromboxanes (Tx), and lipoxins (LX). Arachidonate also leads to leukotrienes. Although only a single example of each type of eicosanoid is shown, each has numerous physiologically significant derivatives, which are designated by letters and subscripts (e.g., **PGH$_2$** for **prostaglandin H$_2$**).

on nearby tissues. The synthesis of the eicosinoids is discussed in Section 19-6C.

In humans, the most important eicosanoid precursor is **arachidonic acid,** a polyunsaturated fatty acid with four double bonds (Table 9-1). Arachidonate is stored in cell membranes as the C2 ester of **phosphatidylinositol** (Table 9-2) and other phospholipids. The fatty acid residue is released by the action of phospholipase A_2 (Fig. 9-5).

The specific products of arachidonate metabolism are tissue-dependent. For example, platelets produce thromboxanes almost exclusively, but endothelial cells (which line the walls of blood vessels) predominantly synthesize prostacyclins. Interestingly, thromboxanes stimulate vasoconstriction and platelet aggregation (which helps initiate blood clotting), while prostacyclins elicit the opposite effects. Thus, the two substances act in opposition to maintain a balance in the cardiovascular system.

2   Lipid Bilayers

In living systems, lipids are seldom found as free molecules but instead associate with other molecules, usually other lipids. In this section, we discuss how lipids aggregate to form micelles and bilayers. We are concerned with the physical properties of lipid bilayers because these aggregates form the structural basis for biological membranes.

A   Why Bilayers Form

In aqueous solutions, amphiphilic molecules such as soaps and detergents form micelles (globular aggregates whose hydrocarbon groups are out of contact with water; Section 2-1C). This molecular arrangement eliminates unfavorable contacts between water and the hydrophobic tails of the amphiphiles and yet permits the solvation of the polar head groups.

The approximate size and shape of a micelle can be predicted from geometrical considerations. Single-tailed amphiphiles, such as soap anions, form spheroidal or ellipsoidal micelles because of their tapered shapes (their hydrated head groups are wider than their tails; Fig. 9-13a,b). The number of molecules in such a micelle depends on the amphiphile, but for many substances it is on the order of several hundred. Too few lipid molecules would expose the hydrophobic core of the micelle to water, whereas too many

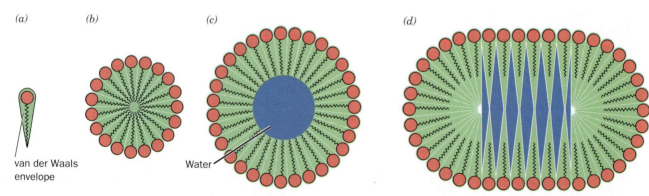

Figure 9-13   Aggregates of single-tailed lipids.   The tapered van der Waals envelope of these lipids (a) permits them to pack efficiently to form a spheroidal micelle (b). The diameter of the micelles depends on the length of the tails. Spheroidal micelles composed of many more lipid molecules than the optimal number (c) would have an unfavorable water-filled center (*blue*). Such micelles could flatten out to collapse the hollow center, but as these ellipsoidal micelles become elongated, they also develop water-filled spaces (d).

would give the micelle an energetically unfavorable hollow center (Fig. 9-13c). Of course, a large micelle could flatten out to eliminate this hollow center, but the resulting decrease of curvature at the flattened surfaces would also generate empty spaces (Fig. 9-13d).

The two hydrocarbon tails of glycerophospholipids and sphingolipids give these amphiphiles a somewhat rectangular cross section (Fig. 9-14a). The steric requirements of packing such molecules together yields large disklike micelles (Fig. 9-14b) that are really extended bimolecular leaflets. These **lipid bilayers** are ~60 Å thick, as measured by electron microscopy and X-ray diffraction techniques, the value expected for more or less fully extended hydrocarbon tails.

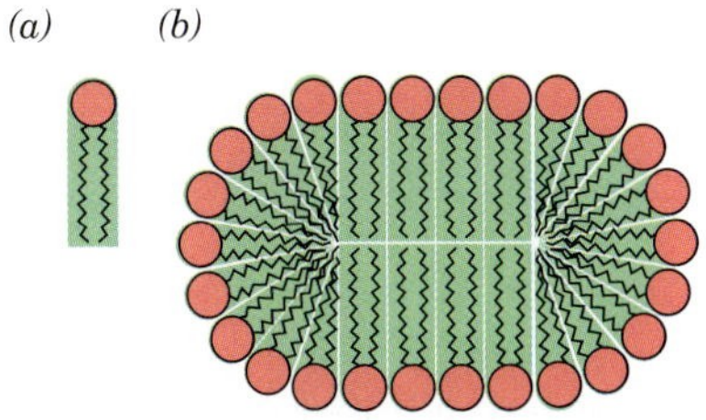

A suspension of phospholipids (glycerophospholipids or sphingomyelins) can form **liposomes**—closed, self-sealing solvent-filled vesicles that are bounded by only a single bilayer (Fig. 9-15). They typically have diameters of several hundred angstroms and, in a given preparation, are rather uniform in size. Once formed, liposomes are quite stable and can be purified by dialysis, gel filtration chromatography, or centrifugation. Liposomes whose internal environment differs from the surrounding solution can therefore be readily prepared. Liposomes serve as models of biological membranes and also hold promise as vehicles for drug delivery since they are absorbed by many cells through fusion with the plasma membrane.

B Lipid Mobility

The transfer of a lipid molecule across a bilayer (Fig. 9-16a), a process termed **transverse diffusion** *or a* **flip-flop,** *is an extremely rare event.* This is because a flip-flop requires the hydrated, polar head group of the lipid to pass through the anhydrous hydrocarbon core of the bilayer. The flip-flop rates of phospholipids have half-times of several days or more. In contrast to their low flip-flop rates, *lipids are highly mobile in the plane of the bilayer* (**lateral diffusion;** Fig. 9-16b). It has been estimated that lipids in a membrane can diffuse the 1-μm length of a bacterial cell in ~1 s. Because of the mobilities of the lipids, the lipid bilayer can be considered to be a two-dimensional fluid.

The interior of the lipid bilayer is in constant motion due to rotations around the C—C bonds of the lipid tails. Various physical measurements suggest that the interior of the bilayer has the viscosity of light machine oil. This feature of the bilayer core is evident in **molecular dynamics simula-**

(a)   (b)

Figure 9-14 Bilayer formation by phospholipids. The cylindrical van der Waals envelope of these lipids (a) causes them to form extended disklike micelles (b) that are better described as lipid bilayers.

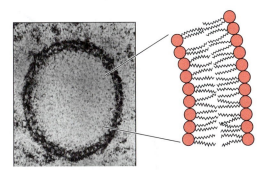

Figure 9-15 Electron micrograph of a liposome. Its wall, as the accompanying diagram indicates, consists of a lipid bilayer. [Courtesy of Walther Stoeckenius, University of California at San Francisco.]

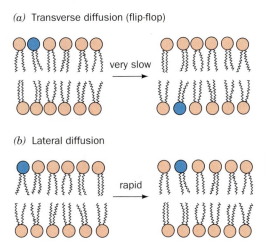

(a) Transverse diffusion (flip-flop)

very slow

(b) Lateral diffusion

rapid

Figure 9-16 Phospholipid diffusion in a lipid bilayer. (a) Transverse diffusion (a flip-flop) is defined as the transfer of a phospholipid molecule from one bilayer leaflet to the other. (b) Lateral diffusion is defined as the pairwise exchange of neighboring phospholipid molecules in the same bilayer leaflet.

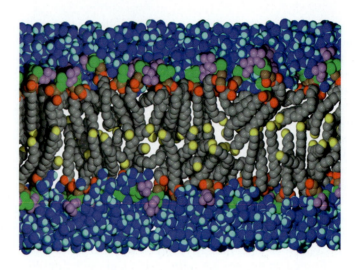

Figure 9-17 Model (snapshot) of a lipid bilayer at a particular instant in time. The conformations of dipalmitoyl phosphatidylcholine molecules in a bilayer surrounded by water were modeled by computer. Atom colors are chain C gray (except terminal methyl C yellow and glycerol C brown), ester O red, phosphate P and O green, choline C and N magenta, water O blue, and water H cyan. Lipid hydrogens have been omitted. [Courtesy of Richard Pastor and Richard Venable, FDA, Bethesda, Maryland.]

tions, in which the time-dependent positions of atoms are predicted from calculations of the forces acting on them (Fig. 9-17). The viscosity of the bilayer increases dramatically closer to the lipid head groups, whose rotation is limited and whose lateral mobility is more constrained by interactions between other polar or charged head groups.

Note that the hydrophobic tails of the lipids shown in Fig. 9-17 are not stiffly regimented as Fig. 9-16 might suggest, but instead bend and interdigitate. A typical biological membrane includes many different lipid molecules, some of whose tails are of different lengths or are kinked due to the presence of double bonds. Under physiological conditions, highly mobile chains fill any gaps that might form between lipids in the bilayer interior.

The model bilayer shown in Fig. 9-17 indicates that the phospholipids bob up and down to some degree. This would also be the case in naturally occurring membranes, which contain a variety of different lipid head groups that must nestle among each other. The polar nature of the outer surface of the bilayer extends from the head groups to the carbonyl groups of the ester and amide bonds that link the fatty acyl chains. Consequently, water molecules penetrate a lipid bilayer to a depth of up to 15 Å, avoiding only the central ~30 Å hydrocarbon core.

The Fluidity of a Lipid Bilayer Is Temperature-Dependent. *As a lipid bilayer cools below a characteristic* **transition temperature,** *it undergoes a sort of phase change in which it becomes a gel-like solid; that is, it loses its fluidity* (Fig. 9-18). Above the transition temperature, the highly mobile lipids are in a state known as a **liquid crystal** because they are ordered in some directions but not in others. The bilayer is thicker in the gel state than in the liquid crystal state due to the stiffening of the hydrocarbon tails at lower temperatures.

The transition temperature of a bilayer increases with the chain length and the degree of saturation of its component fatty acid residues for the same reasons that the melting points of fatty acids increase with these quantities. The transition temperatures of most biological membranes are in the range 10 to 40°C. Bacteria and cold-blooded animals such as fish modify (through lipid synthesis and degradation) the fatty acid compositions of their membrane lipids with ambient temperature so as to maintain a constant level of fluidity. Thus, the fluidity of biological membranes is one of their important physiological attributes.

Cholesterol, which by itself does not form a bilayer, decreases membrane fluidity because its rigid steroid ring system interferes with the motions of the

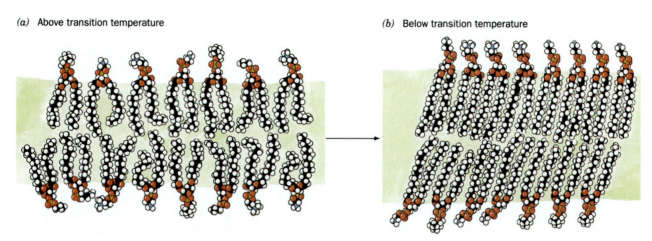

(a) Above transition temperature

(b) Below transition temperature

Figure 9-18 Phase transition in a lipid bilayer.   (a) Above the transition temperature, both the lipid molecules as a whole and their nonpolar tails are highly mobile in the plane of the bilayer. (b) Below the transition temperature, the lipid molecules form a much more orderly array to yield a gel-like solid. [After Robertson, R.N., *The Lively Membranes*, pp. 69–70, Cambridge University Press (1983).]

fatty acid side chains in other membrane lipids. It also broadens the temperature range of the phase transition. This is because cholesterol inhibits the ordering of fatty acid side chains by fitting in between them. Thus, cholesterol functions as a kind of membrane plasticizer.

3 Membrane Proteins

Biological membranes contain proteins as well as lipids. The exact lipid and protein components and the ratio of protein to lipid varies with the identity of the membrane. For example, the lipid-rich myelinated membranes that surround and insulate certain nerve axons (Fig. 9-8) have a protein-to-lipid ratio of 0.23, whereas the protein-rich inner membrane of mitochondria, which mediates numerous chemical reactions, has a protein-to-lipid ratio of 3.2. Eukaryotic plasma membranes are typically ~50% protein.

Membrane proteins catalyze chemical reactions, mediate the flow of nutrients and wastes across the membrane, and participate in relaying information about the extracellular environment to various intracellular components. Such proteins carry out their functions in association with the lipid bilayer. They must therefore interact to some degree with the hydrophobic core and/or the polar surface of the bilayer. In this section, we examine the structures of some membrane proteins, which are classified by their mode of interaction with the membrane.

A Integral Membrane Proteins

Integral or **intrinsic proteins** (Fig. 9-19) associate tightly with membranes through hydrophobic effects and can be separated from membranes only by treatment with agents that disrupt membranes. For example, detergents such as sodium dodecyl sulfate (Section 5-2D) solubilize membrane proteins by taking the place of the membrane lipids that surround the protein. The hydrophobic portions of the detergent molecules coat the hydrophobic regions of the protein, and the polar head groups render the detergent–protein complex soluble in water. Chaotropic agents such as guanidinium ion and urea (Section 6-4C) disrupt water structure, thereby minimizing the hydrophobic effect, the primary force stabilizing the association of the protein with the

Integral membrane protein

Figure 9-19 Model of an integral membrane protein. The protein is solvated by membrane lipids through hydrophobic interactions between the protein and the lipids' nonpolar tails. The polar head groups may also associate with the protein through hydrogen bonding and salt bridges. [After Robertson, R.N., *The Lively Membranes*, p. 56, Cambridge University Press (1983).]

membrane. Some integral proteins bind lipids so tenaciously that they can be freed from them only under denaturing conditions.

Once they have been solubilized, integral proteins can be purified by many of the protein fractionation methods described in Section 5-2. Since these proteins tend to aggregate and precipitate in aqueous solution, their solubility frequently requires the presence of detergents or water-miscible organic solvents such as butanol or glycerol.

Integral Proteins Are Asymmetrically Oriented Amphiphiles. *Integral proteins are amphiphiles; the protein segments immersed in a membrane's nonpolar interior have predominantly hydrophobic surface residues, whereas those portions that extend into the aqueous environment are by and large sheathed with polar residues.* This was demonstrated through **surface labeling,** a technique employing agents that react with proteins but cannot penetrate membranes. For example, the extracellular domain of an integral protein binds antibodies elicited against it, but its cytoplasmic domain will do so only if the membrane has been ruptured. Membrane-impermeable protein-specific reagents that are fluorescent or radioactively labeled can be similarly employed. Alternatively, proteases, which digest only the solvent-exposed portions of an integral protein, may be used to identify the membrane-immersed portions of the protein. These techniques revealed, for example, that the erythrocyte membrane protein **glycophorin A** has three domains (Fig. 9-20): (1) a 72-residue externally located N-terminal

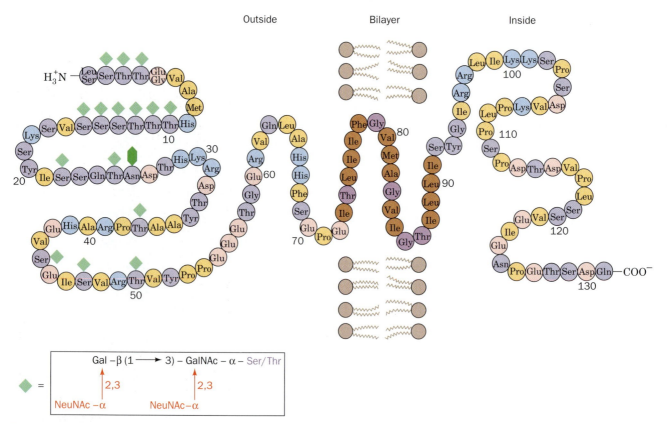

Figure 9-20 Human erythrocyte glycophorin A. The protein bears 15 *O*-linked oligosaccharides (*green diamonds*) and one that is *N*-linked (*dark green hexagon*) on its extracellular domain. The predominant sequence of the *O*-linked oligosaccharides is also shown (NeuNAc = *N*-acetylneuraminic acid). The protein's transmembrane portion (*brown and purple*) consists of 19 sequential predominantly hydrophobic residues. Its C-terminal portion, which is located on the membrane's cytoplasmic face, is rich in anionic (*pink*) and cationic (*blue*) residues. There are two common genetic variants of glycophorin A: Glycophorin A^M has Ser and Gly at positions 1 and 5, whereas glycophorin A^N has Leu and Glu at these positions. [After Marchesi, V.T., *Semin. Hematol.* **16**, 8 (1979).]

domain that bears 16 carbohydrate chains; (2) a 19-residue sequence, consisting almost entirely of hydrophobic residues, that spans the erythrocyte cell membrane; and (3) a 40-residue cytoplasmic C-terminal domain that has a high proportion of charged and polar residues. Thus, glycophorin A is a **transmembrane (TM) protein**; that is, it completely spans the membrane.

Studies of a variety of biological membranes have established that *biological membranes are asymmetric in that a particular membrane protein is invariably located on only one particular face of a membrane, or in the case of a transmembrane protein, oriented in only one direction with respect to the membrane.* However, no protein is known to be completely buried in a membrane; that is, all membrane-associated proteins are at least partially exposed to the aqueous environment.

Transmembrane Proteins May Contain α Helices.

In order for a polypeptide chain to penetrate or span the lipid bilayer, it must have hydrophobic side chains that contact the lipid tails and it must shield its polar backbone groups. This second requirement is met by the formation of secondary structure that satisfies the hydrogen bonding capabilities of the peptide group. Consequently, all known TM segments of integral membrane proteins consist of either α helices or β sheets. For example, glycophorin A's 19-residue TM sequence almost certainly forms an α helix. The existence of such a TM helix can be predicted by comparing the free energy change in transferring α-helical polypeptide segments from the nonpolar interior of a membrane to water (Fig. 9-21). Alternatively, a potential TM sequence can be identified by reference to hydropathy indices such as those in Table 6-2. Methods for predicting the position of a TM α helix are useful because the difficulty in crystallizing integral membrane proteins has permitted relatively few of their X-ray structures to be determined.

Bacteriorhodopsin Contains Seven Transmembrane Helices.

Nigel Unwin and Richard Henderson used **electron crystallography** (Box 9-2) to determine the structure of the integral membrane protein **bacteriorhodopsin**. This 247-residue homotrimeric protein, which is produced by the halophilic (salt-loving) bacterium *Halobacterium salinarium* (it grows best in 4.3 M NaCl), is a light-driven proton pump: It generates a proton concentration gradient across the cell membrane that powers ATP synthesis by a mechanism discussed in Section 17-3C. Bacteriorhodopsin's light-absorbing group, retinal, which is covalently linked to Lys 216 of the protein, is also the light-sensitive element in vision:

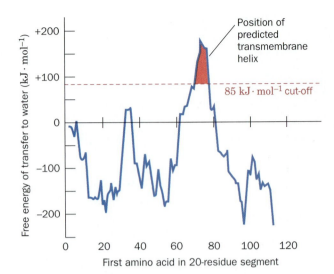

Figure 9-21 Identification of glycophorin A's transmembrane domain. The calculated free energy change in transferring 20-residue-long α-helical segments from the interior of a membrane to water is plotted against the position of the segment's first residue. Peaks higher than +85 kJ · mol^{-1} indicate a transmembrane helix. [After Engleman, D.M., Steitz, T.A., and Goldman, A., *Annu. Rev. Biophys. Biophys. Chem.* **15**, 343 (1986).]

Retinal residue

Bacteriorhodopsin consists largely of a bundle of seven ~25-residue α-helical rods that span the lipid bilayer in directions almost perpendicular to the bilayer plane (Fig. 9-22). As expected, the amino acid side chains that contact the lipid tails are highly hydrophobic. Successive membrane-spanning helices are connected in head-to-tail fashion by hydrophilic loops

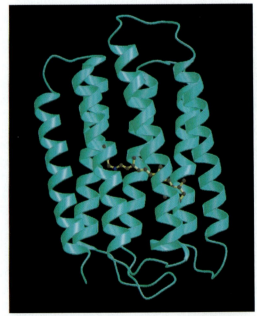

Figure 9-22 The structure of bacteriorhodopsin. The protein is shown in ribbon form (*cyan*) as viewed from within the membrane plane and with its covalently bound retinal shown in ball-and-stick form. [Courtesy of Nikolaus Grigorieff and Richard Henderson, MRC Laboratory of Molecular Biology, Cambridge, U.K.] **See Kinemage Exercise 8-1.**

BOX 9-2

Pathways of Discovery

Richard Henderson and the Structure of Bacteriorhodopsin

Richard Henderson (1945–)

Richard Henderson, like a number of other pioneering structural biologists, began his career as a physicist. He turned his attention to membrane proteins because of their importance in cellular metabolic, transport, and signaling phenomena. Unlike globular proteins, which are soluble in aqueous solution from which they can often be crystallized, integral membrane proteins aggregate in aqueous solution and hence can only be kept in solution by the presence of a suitable detergent. In spite of this, such solubilized proteins rarely crystallize in a manner suitable for X-ray analysis (and none had been made to do so at the time that Henderson began his studies). To circumvent this obstacle, Henderson adapted the technique of electron crystallography to macromolecules, using the membrane protein bacteriorhodopsin as his subject.

Bacteriorhodopsin was discovered in 1967, and its function as a proton pump was described shortly thereafter. The protein is synthesized by halophilic archaebacteria such as *H. salinarum* (formerly known as *H. halobium*). In addition to bacteriorhodopsin, the archaeal rhodopsin family includes chloride-pumping and sensory proteins. Similar proteins have also been identified in eubacteria and in unicellular eukaryotes. What made bacteriorhodopsin attractive as a research subject is its relatively small size (248 residues), its stability, and—most importantly—its unusual proclivity to form ordered two-dimensional arrays in the bacterial cell membrane. In *H. salinarum,* such arrays, which occur as 0.5-μm-wide patches, are known as purple membranes due to the color of the protein's bound retinal molecule. Each purple membrane patch, which is essentially a two-dimensional crystal, consists of 75% bacteriorhodopsin and 25% lipid.

At the time that Henderson began his structural studies of bacteriorhodopsin, the X-ray structures of only around a dozen different globular proteins had been reported. Henderson, working with Nigel Unwin, adapted the principles of X-ray crystallography to the two-dimensional bacteriorhodopsin crystals by measuring the diffraction intensities generated by the electron beam from an electron microscope impinging on a purple membrane patch (as explained by the wave–particle duality, electrons, as do all particles, have wavelike properties with a wavelength, $\lambda = h/mv$, where h is Planck's constant, m is the particle mass, and v is its velocity). It was necessary to use an electron beam rather than X-rays because the electron microscope can focus its electron beam on the microscopically small purple membrane patches. Nevertheless, only very low electron beam intensities could be used because otherwise the resulting radiation damage would destroy the purple membrane. Consequently, to obtain diffraction data of sufficiently high signal-to-noise ratio, the diffraction patterns of around one hundred

or more purple membrane samples had to be averaged.

Working at the limits of the available technology, Henderson and Unwin, in 1975, published a low resolution model for the structure of bacteriorhodopsin. This was the first glimpse of an integral membrane protein. Its seven transmembrane helices were clearly visible as columns of electron density that were approximately perpendicular to the plane of the membrane. However, to obtain three-dimensional diffraction data, a two-dimensional crystal must be systematically tilted relative to the electron beam. Mechanical limitations that prevented the sample from being tilted to the degree necessary to obtain a full three-dimensional data set as well as other technical difficulties therefore yielded a model that had a resolution of 7 Å in the plane of the membrane, but only 14 Å in a perpendicular direction. Consequently, the polypeptide loops connecting the seven helices could not be discerned nor was the protein's associated retinal molecule or any of its side chains visible. However, over the next 15 years, developments in electron microscopy, such as the use of better electron sources and liquid-helium temperatures to minimize radiation damage to the sample, permitted Henderson to extend the resolution of the bacteriorhodopsin structure to 3.5 Å in the plane of the membrane and 10 Å in the perpendicular direction. The resulting electron density map clearly revealed the positions of several bulky aromatic residues and the bound retinal and permitted the loops connecting the transmembrane helices to be visualized. Thus, electron crystallography has become a useful tool for determining the structures of a variety of proteins that can be induced to form two-dimensional arrays as well as those that form very thin three-dimensional crystals.

Henderson's groundbreaking work on bacteriorhodopsin made a seminal contribution to the growing body of biochemical, genetic, spectroscopic, and structural studies of bacteriorhodopsin. This information, together with more recent high-resolution X-ray structures derived from three-dimensional crystals of bacteriorhodopsin, obtained by crystallizing the protein in a lipid matrix, have led to a detailed understanding of the mechanism of the light-induced structural changes through which bacteriorhodopsin pumps protons out of the bacterial cell.

Henderson, R. and Unwin, P.N., Three-dimensional model of purple membrane obtained by electron microscopy, *Nature* **257**, 28–32 (1975).

Henderson, R., Baldwin, J.M., Ceska, T.A., Zemlin, F., Beckmann, E., and Downing, K.H., Model for the structure of bacteriorhodopsin based on high-resolution electron cryo-microscopy, *J. Mol. Biol.* **213**, 899–929 (1990).

of varying size. This arrangement places the protein's charged residues near the surfaces of the membrane in contact with the aqueous environment.

The Photosynthetic Reaction Center Contains Eleven Transmembrane Helices.

The primary photochemical process in purple photosynthetic bacteria is mediated by the so-called **photosynthetic reaction center** (Section 18-2B). This TM protein consists of at least three nonidentical ~300-residue subunits that collectively bind four **chlorophyll** molecules, four other chromophores, and a nonheme Fe atom. The X-ray structure of the 1187-residue photosynthetic reaction center of *Rhodopseudomonas viridis*, which was determined in 1984 by Hartmut Michel, Johann Deisenhofer, and Robert Huber, was the first TM protein to be described in atomic detail (Fig. 9-23). The protein's membrane-spanning portion consists of eleven α helices that form a 45-Å-long cylinder with the expected hydrophobic surface.

Hydrophobic effects, as we saw in Section 6-4A, are the dominant forces stabilizing the three-dimensional structures of water-soluble globular proteins. However, since the TM regions of integral membrane proteins are

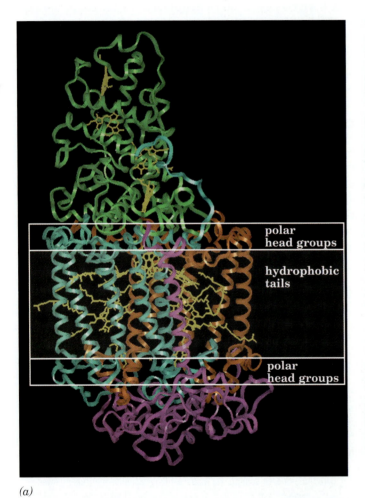

(a)

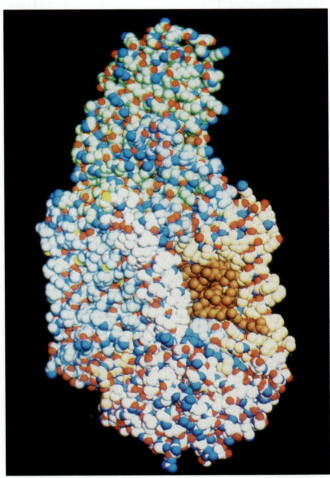

(b)

Figure 9-23   X-Ray structure of the photosynthetic reaction center of *Rhodopseudomonas viridis*.   (*a*) A ribbon diagram showing the C_α backbones of the four subunits in different colors. The various light-absorbing prosthetic groups are yellow. The locations of the polar and hydrophobic regions are indicated schematically. [Based on an X-ray structure by Johann Deisenhofer, Robert Huber, and Hartmut Michel, Max-Planck-Institut für Biochemie, Germany.] (*b*) A space-filling model in which nitrogen atoms are blue, oxygens are red, sulfurs are yellow, and the carbon atoms of the four subunits are tinted pink, blue, orange, and green. Exposed portions of the chromophores are brown. Note how few polar groups (nitrogens and oxygens) are externally exposed in the portion of the protein that is immersed in the nonpolar region of the lipid bilayer. [From Deisenhofer, J. and Michel, H., *Les Prix Nobel* (1989). PDBid 1PRC.]   🔗 **See Kinemage Exercise 8-2.**

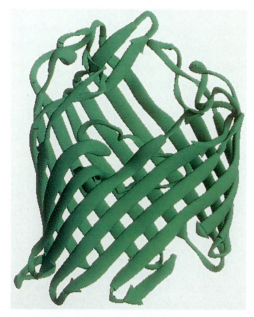

(a)

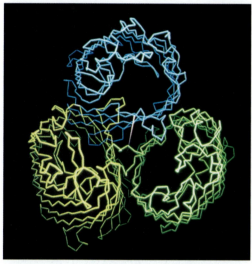

(b)

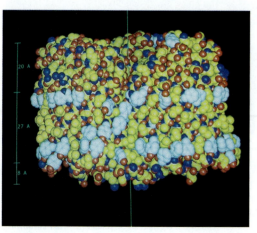

(c)

immersed in nonpolar environments, what stabilizes their structures? Analysis of the photosynthetic reaction center indicates that its interior residues have hydrophobicities comparable to those of water-soluble proteins. However, the membrane-exposed residues of the photosynthetic reaction center, on average, are even more hydrophobic than its interior residues. Thus, *the difference between integral proteins and water-soluble proteins is only skin-deep: Their interiors are similar, but their surface polarities are consistent with the polarities of their environments.*

Some Transmembrane Proteins Contain β Barrels. A protein segment immersed in the nonpolar interior of a membrane must fold so that it satisfies the hydrogen bonding potential of its polypeptide backbone. An α helix can do so as can an antiparallel β sheet that rolls up to form a barrel (a β barrel; Section 6-2C). Transmembrane β barrels of known structure consist of 8 to 22 strands. The number of strands must be even to permit the β sheet to close up on itself.

β Barrels occur in **porins,** which are channel-forming proteins in the outer membrane of gram-negative bacteria (Section 8-3B). The outer membrane protects the bacteria from hostile environments while the porins permit the entry of small polar solutes such as nutrients. Porins also occur in eukaryotes in the outer membranes of mitochondria and chloroplasts (consistent with the descent of these organelles from free-living gram-negative bacteria; Section 1-3B).

Bacterial porins are monomers or trimers of identical 30- to 50-kD subunits. X-Ray structural studies show that most porin subunits consist largely of at least a 16-stranded antiparallel β barrel that forms a solvent-accessible central channel with a length of ~55 Å and a minimum diameter of ~7 Å (Fig. 9-24). As expected, the side chains of the protein's membrane-exposed surface are nonpolar, thereby forming a ~27-Å-high hydrophobic band encircling the trimer (Fig. 9-24c). This band is flanked by more polar aromatic side chains (Table 6-2) that form interfaces with the head groups of the lipid bilayer (Fig. 9-24c). In contrast, the side chains at the solvent-exposed surface of the protein, including those lining the walls of the aqueous channel, are polar. Indeed, even in 8-stranded β barrels, whose central channel is filled in with side chains, these side chains are polar and form a hydrogen bonded network. Possible mechanisms for the solute selectivity of porins are discussed in Section 10-2B.

Figure 9-24   X-Ray structure of the *E. coli* OmpF porin. (*a*) A ribbon diagram of the 16-stranded monomer. (*b*) The C_α backbone of the trimer viewed ~30° from its threefold axis of symmetry, showing the pore through each subunit. (*c*) A space-filling model of the trimer viewed perpendicular to its threefold axis (*vertical green line*). N atoms are blue, O atoms are red, and C atoms are yellow, except those in the side chains of aromatic residues, which are white. The aromatic groups appear to delimit an ~27-Å-high hydrophobic band (*scale at left*) that is immersed in the nonpolar portion of the bacterial outer membrane (with the cell's exterior at the tops of Parts *a* and *c*). Compare this hydrophobic band with that in Fig. 9-23*b*. [Part *a* based on an X-ray structure by and Parts *b* and *c* courtesy of Tilman Schirmer and Johan Jansonius, University of Basel, Switzerland. PDBid 1OPF.] See Kinemage Exercise 8-3.

B   Lipid-Linked Proteins

Some membrane-associated proteins contain covalently attached lipids that anchor the protein to the membrane. The lipid group, like any modifying group, may also mediate protein–protein interactions or modify the structure and activity of the protein to which it is attached. **Lipid-linked proteins** come in three varieties: prenylated proteins, fatty acylated proteins, and glycosylphosphatidylinositol-linked proteins. A single protein may contain more than one covalently linked lipid group.

Prenylated proteins have covalently attached lipids that are built from isoprene units (Section 9-1F). The most common isoprenoid groups are the C_{15} **farnesyl** and C_{20} **geranylgeranyl** residues:

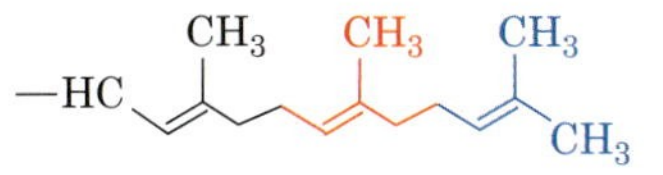

Farnesyl residue

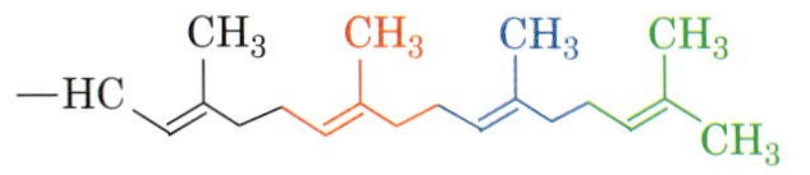

Geranylgeranyl residue

The most common prenylation site in proteins is the C-terminal tetrapeptide C-X-X-Y, where C is Cys and X is often an aliphatic amino acid residue. Residue Y influences the type of prenylation: Proteins are farnesylated when Y is Ala, Met, or Ser and geranylgeranylated when Y is Leu. In both cases, the prenyl group is enzymatically linked to the Cys sulfur atom via a thioether linkage. The X-X-Y tripeptide is then proteolytically excised, and the newly exposed terminal carboxyl group is esterified with a methyl group, producing a C-terminus with the structure

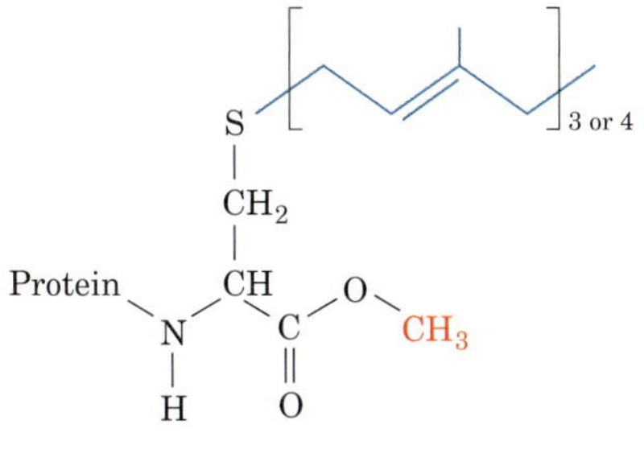

Two kinds of fatty acids, myristic acid and palmitic acid, are linked to membrane proteins. Myristic acid, a biologically rare saturated C_{14} fatty acid, is appended to a protein via an amide linkage to the α-amino group of an N-terminal Gly residue. **Myristoylation** is stable: The fatty acyl group remains attached to the protein throughout its lifetime. Myristoylated proteins are located in a number of subcellular compartments, including the cytosol, endoplasmic reticulum, plasma membrane, and the nucleus.

In **palmitoylation,** the saturated C_{16} fatty acid palmitic acid is joined in thioester linkage to a specific Cys residue. Palmitoylated proteins occur almost exclusively on the cytoplasmic face of the plasma membrane, where many participate in transmembrane signaling. The palmitoyl group can be removed by the action of **palmitoyl thioesterases,** suggesting that reversible palmitoylation may regulate the association of the protein with the membrane and thereby modulate the signaling processes.

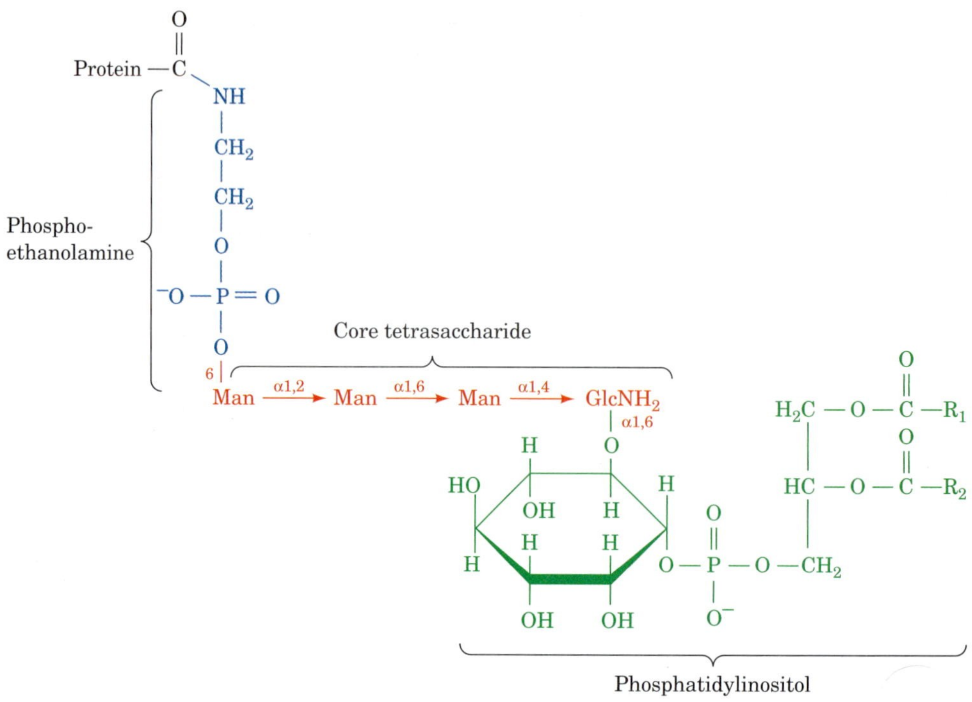

Figure 9-25 **The core structure of the GPI anchors of proteins.** R_1 and R_2 represent fatty acid residues whose identities vary with the protein. The tetrasaccharide may have a variety of attached sugar residues whose identities also vary.

Glycosylphosphatidylinositol-linked proteins (GPI-linked proteins) occur in all eukaryotes but are particularly abundant in some parasitic protozoa, which contain relatively few membrane proteins anchored by transmembrane polypeptide segments. Like glycoproteins and glycolipids, GPI-linked proteins are located only on the exterior surface of the plasma membrane.

The core structure of the GPI group consists of phosphatidylinositol (Table 9-2) glycosidically linked to a linear tetrasaccharide composed of three mannose residues and one glucosaminyl residue (Fig. 9-25). The mannose at the nonreducing end of this assembly forms a phosphodiester bond with a phosphoethanolamine residue that is amide-linked to the protein's C-terminal carboxyl group. The core tetrasaccharide is generally substituted with a variety of sugar residues that vary with the identity of the protein. There is likewise considerable diversity in the fatty acid residues of the phosphatidylinositol group.

C Peripheral Membrane Proteins

Peripheral or **extrinsic proteins,** unlike integral membrane proteins or lipid-linked proteins, can be dissociated from membranes by relatively mild procedures that leave the membrane intact, such as exposure to high ionic strength salt solutions or pH changes. Peripheral proteins do not bind lipid and, once purified, behave like water-soluble proteins. They associate with membranes by binding at their surfaces, most likely to certain lipids or integral proteins, through electrostatic and hydrogen bonding interactions. Cytochrome *c* (Sections 5-4A, 6-2D and 17-2E) is a peripheral membrane protein that is associated with the outer surface of the inner mitochon-

drial membrane. At physiological pH, cytochrome *c* is cationic and can interact with negatively charged phospholipids such as phosphatidylserine and phosphatidylglycerol.

4 **Membrane Structure and Assembly**

Membranes were once thought to consist of a phospholipid bilayer sandwiched between two layers of unfolded polypeptide. This sandwich model, which is improbable on thermodynamic grounds, was further discredited by electron microscopic visualization of membranes and other experimental approaches. More recent studies have revealed insights into the fine structure of membranes, including a surprising degree of heterogeneity.

A The Fluid Mosaic Model

The demonstrated fluidity of artificial lipid bilayers (Section 9-2B) suggests that biological membranes have similar properties. This idea was proposed in 1972 by S. Jonathan Singer and Garth Nicolson in their unifying theory of membrane structure known as the **fluid mosaic model.** In this model, integral proteins are visualized as "icebergs" floating in a two-dimensional lipid "sea" in a random or mosaic distribution (Fig. 9-26). A key element of the model is that integral proteins can diffuse laterally in the lipid matrix unless their movements are restricted by association with other cell components. This model of membrane fluidity explained the earlier experimental results of Michael Edidin, who fused cultured cells and

> See Guided Exploration 9
>
> Membrane structure and the fluid mosaic model.

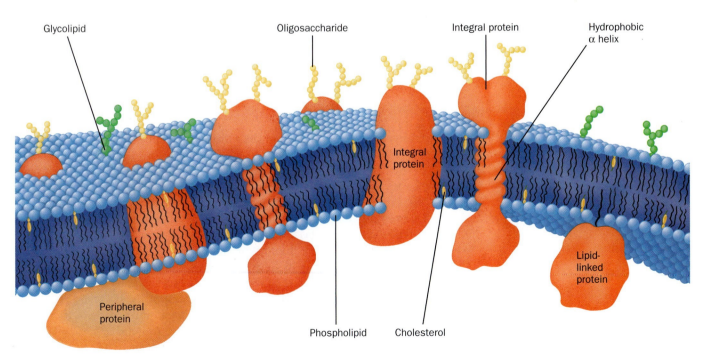

Figure 9-26 Diagram of a plasma membrane.   Integral proteins (*orange*) are embedded in a bilayer composed of phospholipids (*blue head groups attached to wiggly tails*) and cholesterol (*yellow*). The carbohydrate components (*green and yellow beads*) of glycoproteins and glycolipids occur on only the external face of the membrane. Most membranes contain a higher proportion of protein than is depicted here.

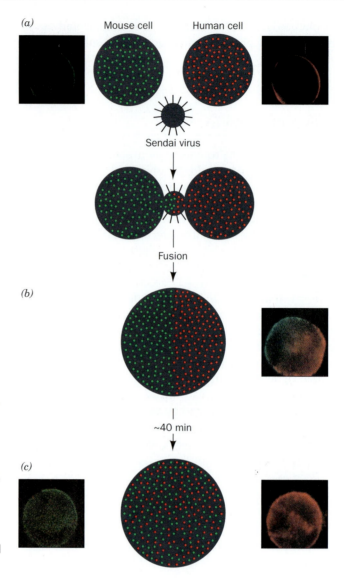

Figure 9-27 **Fusion of mouse and human cells.**
(*a*) The cell-surface proteins of cultured mouse and human cells were labeled with green and red fluorescent markers. The two cell types were fused by treatment with **Sendai virus** to form a hybrid cell known as a **heterokaryon.** (*b*) Immediately after fusion, the mouse and human proteins are segregated. (*c*) However, after 40 min at 37°C, the red and green markers have fully intermixed. The photomicrographs were taken through filters that allowed only red or green light to reach the camera; that in *b* is a double exposure; those in *c* are of the same cell. [Immunofluorescence photomicrographs courtesy of Michael Edidin, The Johns Hopkins University.]

observed the intermingling of their differently labeled cell-surface proteins (Fig. 9-27).

The rates of diffusion of proteins in membranes can be determined from measurements of **fluorescence recovery after photobleaching (FRAP).** In this technique, a **fluorophore** (fluorescent group) is specifically attached to a membrane component in an immobilized cell or in an artificial membrane system. An intense laser pulse focused on a very small area ($\sim$3 μm^2) destroys (bleaches) the fluorophore there (Fig. 9-28). The rate at which the bleached area recovers its fluorescence, as monitored by fluorescence microscopy, indicates the rate at which unbleached and bleached fluorophore-labeled molecules laterally diffuse into and out of the bleached area.

FRAP measurements demonstrate that membrane proteins vary in their lateral diffusion rates. Some 30 to 90% of these proteins are freely mobile; they diffuse at rates only an order of magnitude or so slower than those of the much smaller lipids, so they can diffuse the 20-μm length of a eukaryotic cell within an hour. Other proteins diffuse more slowly, and some are essentially immobile due to submembrane attachments.

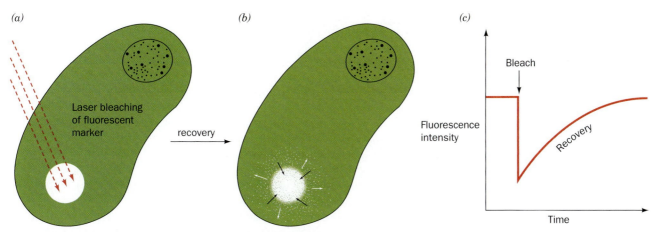

Figure 9-28   The fluorescence recovery after photobleaching (FRAP) technique.   (a) An intense laser light pulse bleaches the fluorescent markers (*green*) from a small region of an immobilized cell that has a fluorophore-labeled membrane component. (b) The fluorescence of the bleached area recovers as the bleached molecules laterally diffuse out of it and intact fluorescent molecules diffuse into it. (c) The fluorescence recovery rate depends on the diffusion rate of the labeled molecule.

B The Membrane Skeleton

Studies of membrane structure and composition often make use of erythrocyte membranes, since these are relatively simple and easily isolated. A mature mammalian erythrocyte lacks organelles and carries out few metabolic processes; it is essentially a membranous bag of hemoglobin. Erythrocyte membranes can be obtained by osmotic lysis, which causes the cell contents to leak out. The resulting membranous particles are known as erythrocyte **ghosts** because, on return to physiological conditions, they reseal to form colorless particles that retain their original shape but are devoid of cytoplasm.

A normal erythrocyte's biconcave disklike shape (Fig. 7-18a) ensures the rapid diffusion of O_2 to its hemoglobin molecules by placing them no further than 1 μm from the cell surface. However, the rim and the dimple regions of an erythrocyte do not occupy fixed positions on the cell membrane. This can be demonstrated by anchoring an erythrocyte to a microscope slide by a small portion of its surface and inducing the cell to move laterally with a gentle flow of buffer. A point originally on the rim of the erythrocyte will move across the dimple to the rim on the opposite side of the cell. Evidently, the membrane rolls across the cell while maintaining its shape, much like the tread of a tractor. This remarkable mechanical property of the erythrocyte membrane results from the presence of a submembranous network of proteins that function as a membrane "skeleton."

The fluidity and flexibility imparted to an erythrocyte by its membrane skeleton have important physiological consequences. A slurry of solid particles of a size and concentration equal to that of red cells in blood has the flow characteristics approximating those of sand. Consequently, in order for blood to flow at all, much less for its erythrocytes to squeeze through capillary blood vessels smaller in diameter than they are, erythrocyte membranes, with their membrane skeletons, must be fluidlike and easily deformable.

The protein **spectrin,** so called because it was discovered in erythrocyte ghosts, accounts for ~75% of the erythrocyte membrane skeleton. It is composed of two similar polypeptide chains, a 280-kD α subunit and a 246-kD

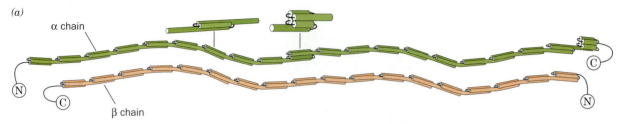

(a)

α chain

β chain

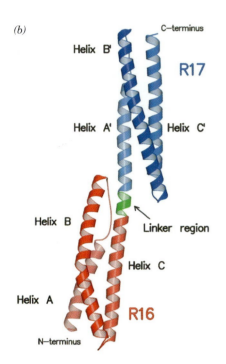

(b)

Helix B'

Helix A'   Helix C'

C–terminus

R17

Helix B   Linker region

Helix C

Helix A   R16

N–terminus

Figure 9-29 Structure of spectrin. (a) Structure of an αβ dimer. Both of these antiparallel polypeptides contain multiple 106-residue repeats, which are thought to form flexibly connected triple-helical bundles. Two of these heterodimers join, head to head, to form an $(\alpha\beta)_2$ heterotetramer. [After Speicher, D.W. and Marchesi, V. *Nature* **311,** 177 (1984).] (b) The X-ray structure of two consecutive repeats of chicken brain α-spectrin. Each of these 106-residue repeats consists of an up-down-up triple helical bundle in which the C-terminal helix of the first repeat (R16; *red*) is continuous, via a 5-residue helical linker (*green*), with the N-terminal helix of the second repeat (R17; *blue*). The helices within each triple helical bundle wrap around each other in a gentle left-handed supercoil that is hydrophobically stabilized by the presence of nonpolar residues at the *a* and *d* positions of heptad repeats on all three of its component α helices (Fig. 6-14). Despite the expected rigidity of α helices, there is considerable evidence that spectrin is a flexible wormlike molecule. [Courtesy of Alfonso Mondragón, Northwestern University. PDBid 1CUN.]

β subunit, which each consist of repeating 106-residue segments that are predicted to fold into triple-stranded α-helical coiled coils (Fig. 9-29). These large polypeptides are loosely intertwined to form a flexible wormlike αβ dimer that is ~1000 Å long. Two such heterodimers further associate in a head-to-head manner to form an $(\alpha\beta)_2$ tetramer. There are ~100,000 spectrin tetramers per cell, and they are cross-linked at both ends by attachments to other cytoskeletal proteins. Together, these proteins form a dense and irregular protein meshwork that underlies the erythrocyte plasma membrane (Fig. 9-30). A defect or deficiency in spectrin synthesis causes **hereditary spherocytosis,** in which erythrocytes are spheroidal and relatively fragile and inflexible. Individuals with the disease suffer from anemia due to erythrocyte lysis and the removal of spherocytic cells by the spleen (which normally functions to filter out aged and hence inflexible erythrocytes from the blood at the end of their ~120-day lifetimes).

Spectrin also associates with an 1880-residue protein known as **ankyrin,** which binds to an integral membrane ion channel protein. This attachment anchors the membrane skeleton to the membrane. Immunochemical studies have revealed spectrin-like and ankyrin-like proteins in a variety of tissues, in addition to erythrocytes.

Ankyrin's N-terminal 798-residue segment consists almost entirely of 24 tandem ~33-residue repeats known as **ankyrin repeats** (Fig. 9-31), which also occur in a variety of other proteins. Each ankyrin repeat consists of two short (8- or 9-residue) antiparallel α helices followed by a long loop, all arranged in a contiguous right-handed helical stack that forms a curved assembly. The loops in this assembly are arranged so as to yield an extended and nearly flat platform, which, together with the adjoining helices, forms an elongated concave surface that is postulated to form the binding sites for various integral proteins as well as spectrin.

The interaction of membrane components with the underlying skeleton helps explain why integral membrane proteins exhibit different degrees of mobility within the membrane: Some integral proteins are firmly attached to elements of the cytoskeleton or are trapped within the spaces defined by those "fences." Other membrane proteins may be able to squeeze

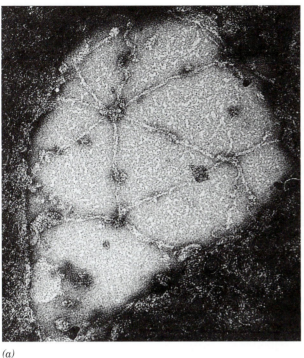

(a)

Figure 9-30   The human erythrocyte membrane skeleton.   (a) An electron micrograph of an erythrocyte membrane skeleton that has been stretched to an area 9 to 10 times greater than that of the native membrane. Stretching makes it possible to obtain clear images of the membrane skeleton, which in its native state is so densely packed and irregularly flexed that it is difficult to pick out individual molecules and to ascertain how they are interconnected. Note the predominantly hexagonal network composed of spectrin tetramers cross-linked by junctions containing **actin** (Section 28-1A) and **band 4.1 protein** (named after its position in an SDS-PAGE electrophoretogram). [Courtesy of Daniel Branton, Harvard University.] (b) A model of the erythrocyte membrane skeleton. The so-called junctional complex, which is magnified in this drawing, contains actin, **tropomyosin** (Section 28-2A), and band 4.1 protein as well as other proteins. [After Goodman, S.R., Krebs, K.E., Whitfield, C.F., Riederer, B.M., and Zagen, I.S., *CRC Crit. Rev. Biochem.* **23**, 196 (1988).]

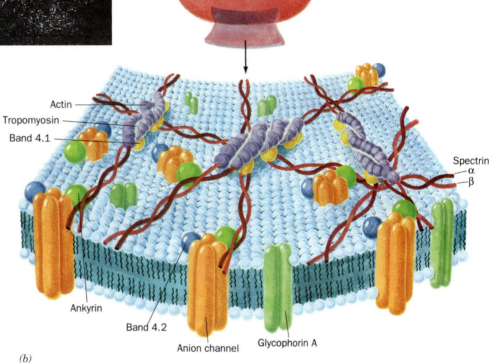

Actin
Tropomyosin
Band 4.1
Spectrin
α
β
Ankyrin
Band 4.2
Anion channel
Glycophorin A

(b)

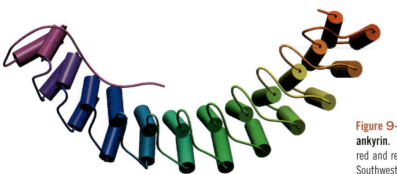

Figure 9-31   The X-ray structure of ankyrin repeats 13 to 24 of human ankyrin.   The individual repeats are colored in rainbow order with repeat 13 red and repeat 24 violet. [Courtesy of Peter Michaely, University of Texas Southwestern Medical Center, Dallas, Texas. PDBid 1N11.]

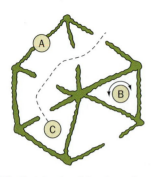

Figure 9-32 **Model rationalizing the various mobilities of membrane proteins.** Protein A, which interacts tightly with the underlying cytoskeleton, is immobile. Protein B is free to rotate within the confines of the cytoskeletal "fences." Protein C diffuses by traveling through "gates" in the cytoskeleton. The diffusion of some membrane proteins is not affected by the cytoskeleton. [After Edidin, M., *Trends Cell Biol.* **2**, 378 (1992).]

through gaps or "gates" between cytoskeletal components, whereas still other proteins can diffuse freely without interacting with the cytoskeleton at all (Fig. 9-32). Support for this **gates and fences model** comes from the finding that partial destruction of the cytoskeleton results in freer protein diffusion.

C Lipid Asymmetry

The lipid and protein components of membranes do not occur in equal proportions on the two sides of biological membranes. For example, *membrane glycoproteins and glycolipids are invariably oriented with their carbohydrate moieties facing the cell's exterior.* The asymmetric distribution of certain membrane lipids between the inner and outer leaflets of a membrane was first established through the use of phospholipases (Section 9-1C). Phospholipases cannot pass through membranes, so phospholipids on only the external surface of intact cells are susceptible to hydrolysis by these enzymes. Such studies reveal that lipids in biological membranes are asymmetrically distributed (e.g., Fig. 9-33). How does this asymmetry arise?

In eukaryotes, the enzymes that synthesize membrane lipids are mostly integral membrane proteins of the **endoplasmic reticulum** (**ER;** the interconnected membranous vesicles that occupy much of the cytosol; Fig. 1-8), whereas in prokaryotes, lipids are synthesized by integral membrane proteins in the plasma membrane. Hence, membrane lipids are fabricated on site. Eugene Kennedy and James Rothman demonstrated this to be the case in bacteria through the use of selective labeling. They gave growing bacteria a 1-minute pulse of $^{32}PO_4^{3-}$ in order to radioactively label the phosphoryl groups of only the newly synthesized phospholipids. Immediately afterward, they added **trinitrobenzenesulfonic acid (TNBS),** a membrane-impermeable reagent that combines with phosphatidylethanolamine (**PE;** Fig. 9-34). Analysis of the resulting doubly labeled membranes showed that none of the TNBS-labeled PE was radioactively labeled. This observation indicates that *newly made PE is synthesized on the cytoplasmic face of the membrane* (Fig. 9-35, *upper right*).

However, if an interval of only 3 minutes was allowed to elapse between the $^{32}PO_4^{3-}$ pulse and the TNBS addition, about half of the ^{32}P-labeled PE was also TNBS labeled (Fig. 9-35, *lower right*). This observation indicates that the flip-flop rate of PE in the bacterial membrane is ~100,000-fold greater than it is in bilayers consisting of only phospholipids (where the flip-flop rates have half-times of many days).

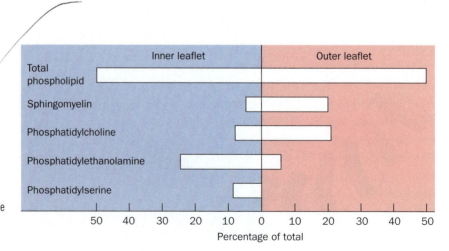

Figure 9-33 **Asymmetric distribution of membrane phospholipids in the human erythrocyte membrane.** The phospholipid content is expressed as mol %. [After Rothman, J.E. and Lenard, J., *Science* **194**, 1744 (1977).]

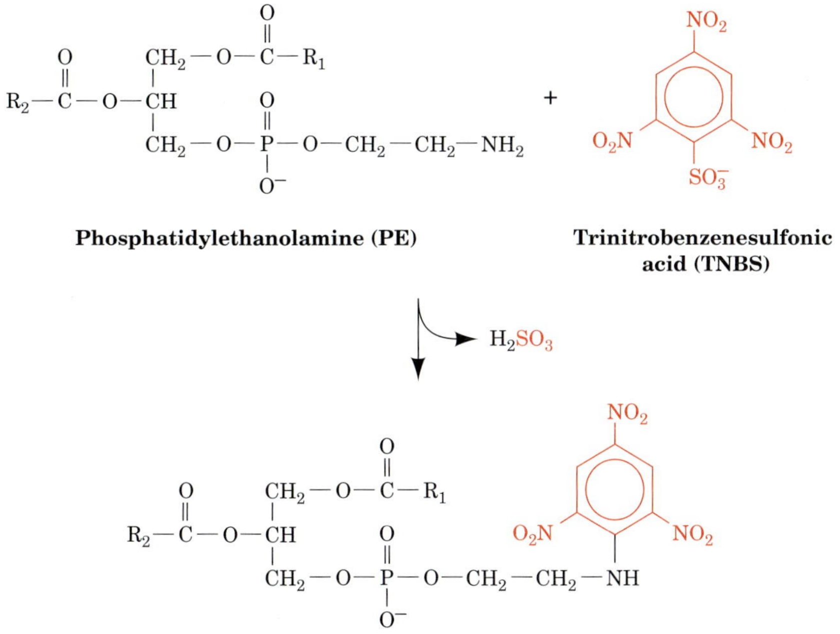

Phosphatidylethanolamine (PE)

Trinitrobenzenesulfonic acid (TNBS)

H_2SO_3

Figure 9-34   The reaction of TNBS with phosphatidylethanolamine.

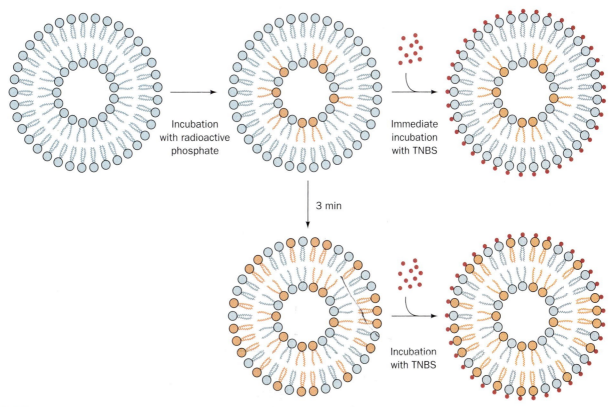

Figure 9-35   The location of lipid synthesis in a bacterial membrane. Newly synthesized PE was labeled by a 1-min pulse of $^{32}PO_4^{3-}$ (*orange head groups*), and the PE on the cell surface was independently labeled by treatment with the membrane-impermeable reagent TNBS. When TNBS labeling (*magenta circles*) occurred immediately after the ^{32}P pulse, none of the ^{32}P- labeled PE was also TNBS labeled (*upper right*), thereby indicating that the PE is synthesized on the cytoplasmic face of the membrane. If, however, there was even a few minutes' delay between the two labeling procedures, much of the TNBS-labeled PE in the external face of the membrane was also ^{32}P labeled (*lower right*).

How do phospholipids synthesized on one side of the membrane reach its other side so quickly? Phospholipid flip-flops in bacteria as well as eukaryotes appear to be facilitated in two ways:

1. Membrane proteins known as **flipases** catalyze the flip-flops of specific phospholipids. These proteins tend to equilibrate the distribution of their corresponding phospholipids across a bilayer; that is, the net transport of a phospholipid is from the side of the bilayer with the higher concentration of the phospholipid to the opposite side. Such a process, as we shall see in Section 10-1, is a form of **facilitated diffusion.**

2. Membrane proteins known as **phospholipid translocases** transport specific phospholipids across a bilayer in a process that is driven by ATP hydrolysis. These proteins can transport certain phospholipids from the side of a bilayer that has the lower concentration of the phospholipid to the opposite side, thereby establishing a nonequilibrium distribution of the phospholipid. Such a process, as we shall see in Section 10-3, is a form of **active transport.**

The observed distribution of phospholipids across membranes (e.g., Fig. 9-33) therefore appears to arise from the membrane orientations of the enzymes that synthesize phospholipids combined with the countervailing tendencies of ATP-dependent phospholipid translocases to generate asymmetric phospholipid distributions and those of flipases to equilibrate these distributions. The importance of these lipid transport systems is demonstrated by the observation that the presence of phosphatidylserine on the exteriors of many cells induces blood clotting (i.e., it is an indication of tissue damage) and, in erythrocytes, marks the cell for removal from the circulation.

In all cells, *new membranes are generated by the expansion of existing membranes.* In eukaryotic cells, lipids synthesized on the cytoplasmic face of the ER are transported to other parts of the cell by membranous vesicles that bud off from the ER and fuse with other cellular membranes. These vesicles also carry membrane proteins.

Lipid Rafts Are Membrane Subdomains. *Lipids and proteins in membranes can also be laterally organized.* Thus the plasma membranes of many eukaryotic cells have two or more distinct domains that have different functions. For example, the plasma membranes of epithelial cells (the cells lining body cavities and free surfaces) have an **apical domain,** which faces the lumen (interior) of the cavity and often has a specialized function (such as the absorption of nutrients in intestinal brush border cells), and a **basolateral domain,** which covers the remainder of the cell. These two domains, which do not intermix, have different compositions of both lipids and proteins.

In addition, the hundreds of different lipids and proteins within a given plasma membrane domain may not be uniformly mixed but instead often segregate to form **microdomains** that are enriched in certain lipids and proteins. This may result from specific interactions between integral membrane proteins and particular types of membrane lipids. Divalent metal ions, notably Ca^{2+}, which ligate negatively charged lipid head groups such as those of phosphatidylserine, may also cause clustering of these lipids.

One type of microdomain, termed a **lipid raft,** appears to consists of closely packed glycosphingolipids (which occur only in the outer leaflet of

the plasma membrane) and cholesterol. By themselves, glycosphingolipids cannot form bilayers because their large head groups prevent the requisite close packing of their predominantly saturated hydrophobic tails. Conversely, cholesterol by itself does not form a bilayer due to its small head group. It is therefore likely that *the glycosphingolipids in lipid rafts associate laterally via weak interactions between their carbohydrate head groups, and the voids between their tails are filled in by cholesterol.*

Owing to the close packing of their component lipids and the long, saturated sphingolipid tails, sphingolipid–cholesterol rafts have a more ordered or crystalline arrangement than other regions of the membrane and are more resistant to solubilization by detergents. The rafts may diffuse laterally within the membrane. Certain proteins preferentially associate with the rafts, including many GPI-linked proteins and some of the proteins that participate in transmembrane signaling processes (Section 21-3). This suggests that lipid rafts, which are probably present in all cell types, function as platforms for the assembly of complex intercellular signaling systems. Several viruses, including influenza virus, measles virus, Ebola virus, and HIV, localize to lipid rafts, which therefore appear to be the sites from which these viruses enter uninfected cells and bud from infected cells. It should be noted that lipid rafts are highly dynamic structures that rapidly exchange both proteins and lipids with their surrounding membrane as a consequence of the weak and transient interactions between membrane components.

Caveolae (Latin for small caves), which are ~75-nm-diameter flask-shaped invaginations on the plasma membrane, are specialized forms of lipid rafts that are associated with ~21-kD integral proteins named **caveolins.** Caveolae, which occur mainly on muscle and epithelial cells, participate in **endocytosis** (the internalization of receptor-bound ligands; Section 19-1B) as well as intercellular signaling.

D   The Secretory Pathway

Membrane proteins, as are all proteins, are ribosomally synthesized under the direction of messenger RNA templates (translation is discussed in Chapter 26). The polypeptide grows from its N-terminus to its C-terminus by the stepwise addition of amino acid residues. Ribosomes may be free in the cytosol or bound to the ER to form the **rough endoplasmic reticulum** (**RER,** so called because of the knobby appearance its bound ribosomes give it; Fig. 1-8). *Free ribosomes synthesize mostly soluble and mitochondrial proteins, whereas membrane-bound ribosomes manufacture transmembrane proteins and proteins destined for secretion, operation within the ER, and incorporation into* **lysosomes** (Fig. 1-8; membranous vesicles containing a battery of hydrolytic enzymes that degrade and recycle cell components). These latter proteins initially appear in the ER.

The Secretory Pathway Accounts for the Targeting of Many Secreted and Membrane Proteins.   How are RER-destined proteins differentiated from other proteins? And how do these large, relatively polar molecules pass through the RER membrane? These processes occur via the **secretory pathway,** which was first described by Günter Blobel, Cesar Milstein, and David Sabatini around 1975. Since ~25% of the various proteins synthesized by all types of cells are integral proteins and many others are secreted, ~40% of *the various types of proteins that a cell synthesizes must be processed via the*

secretory pathway or some other protein targeting pathway. Here we outline the secretory pathway, which is diagrammed in Fig. 9-36:

1. *All secreted, ER-resident, and lysosomal proteins, as well as many TM proteins, are synthesized with leading (N-terminal) 13- to 36-residue signal peptides.* These signal peptides consist of a 6- to 15-residue hydrophobic core flanked by several relatively hydrophilic residues that usually include one or more basic residues near the N-terminus (Fig. 9-37). Signal peptides otherwise have little sequence similarity. However, a variety of evidence indicates they form α helices in nonpolar environments.

2. When the signal peptide first protrudes beyond the ribosomal surface (when the polypeptide is at least ~40 residues long), the **signal recognition particle (SRP),** a 325-kD complex of six different polypeptides and a 300-nucleotide RNA molecule, binds to both the signal peptide and the ribosome accompanied by replacement of the SRP's bound **guanosine diphosphate (GDP;** the guanine analog of ADP) by **guanosine triphosphate (GTP;** the guanine analog of ATP). The SRP's resulting conformational change causes the ribosome to arrest further polypeptide growth, thereby preventing the RER-destined protein from being released into the cytosol.

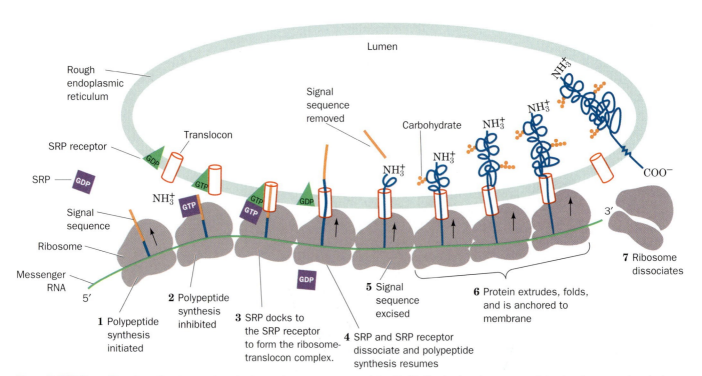

Figure 9-36 Key to Function. The ribosomal synthesis, membrane insertion, and initial glycosylation of an integral protein via the secretory pathway: **(1)** Protein synthesis is initiated at the N-terminus of the polypeptide, which consists of a 13- to 36-residue signal sequence. **(2)** A signal-recognition particle (SRP) binds to the ribosome and the signal sequence emerging from it, thereby arresting polypeptide synthesis. **(3)** The SRP is bound by the transmembrane SRP receptor (SR) in complex with the translocon, thereby bringing together the ribosome and the translocon. **(4)** The SRP and SR hydrolyze their bound GTPs causing them to dissociate from the ribosome–translocon complex. The ribosome then resumes the synthesis of the polypeptide, which passes through the translocon into lumen of the ER. **(5)** Shortly after the entrance of the signal sequence into the lumen of the ER, it is proteolytically excised. **(6)** As the growing polypeptide chain passes into the lumen, it commences folding into its native conformation, a process that is facilitated by its interaction with the chaperone protein Hsp70 (not shown). Simultaneously, enzymes initiate the polypeptide's specific glycosylation. Once the protein has folded, it cannot be pulled out of the membrane. At points determined by its sequence, the polypeptide becomes anchored in the membrane (proteins destined for secretion pass completely into the ER lumen). **(7)** Once polypeptide synthesis is completed, the ribosome dissociates into its two subunits.

☙ **See the Animated Figures.**

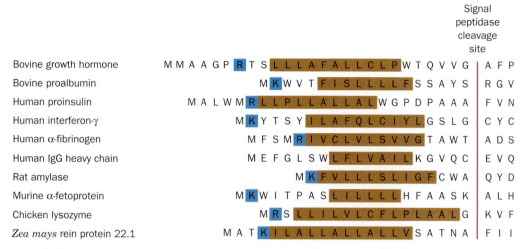

Figure 9-37 **The N-terminal sequences of some eukaryotic preproteins.** The hydrophobic cores (*brown*) of most signal peptides are preceded by basic residues (*blue*). [After Watson, M.E.E., *Nucleic Acids Res.* **12**, 5147–5156 (1984).]

3. The SRP–ribosome complex diffuses to the RER surface, where it is bound by the **SRP receptor (SR;** also called **docking protein)** in complex with the **translocon,** a protein pore in the ER membrane through which the growing polypeptide will be extruded. In forming the SR–translocon complex, the SR's bound GDP is replaced by GTP.

4. The SRP and SR stimulate each other to hydrolyze their bound GTP to GDP (which is energetically equivalent to ATP hydrolysis), resulting in conformational changes that causes them to dissociate from each other and from the ribosome–translocon complex. This permits the bound ribosome to resume polypeptide synthesis such that the growing polypeptide's N-terminus passes through the translocon into the lumen of the ER. Most ribosomal processes, as we shall see in Section 26-4, are driven by GTP hydrolysis.

5. Shortly after the signal peptide enters the ER lumen, it is specifically cleaved from the growing polypeptide by a membrane-bound **signal peptidase** (polypeptide chains with their signal peptide still attached are known as **preproteins;** signal peptides are alternatively called **presequences**).

6. The nascent (growing) polypeptide starts to fold to its native conformation, a process that is facilitated by its interaction with the ER-resident chaperone protein Hsp70 (Section 6-5C). Enzymes in the ER lumen then initiate **posttranslational modification** of the polypeptide, such as the specific attachments of "core" carbohydrates to form glycoproteins (Section 8-3C) and the formation of disulfide bonds as facilitated by protein disulfide isomerase (Section 6-5B).

7. When the synthesis of the polypeptide is completed, it is released from both the ribosome and the translocon and the ribosome dissociates from the RER. Secretory, ER-resident, and lysosomal proteins pass completely through the RER membrane into the lumen. TM proteins, in contrast, contain one or more hydrophobic ~20-residue **membrane anchor** sequences that remain embedded in the membrane.

The secretory pathway also functions in prokaryotes for the insertion of certain proteins into the cell membrane (whose exterior is equivalent to

the ER lumen). Indeed, all forms of life yet tested have homologous SRPs, SRs, and translocons. Nevertheless, it should be noted that cells have several mechanisms for installing proteins in or transporting them through membranes. For example, cytoplasmically synthesized proteins that reside in the mitochondrion reach their destinations via mechanisms that are substantially different from that of the secretory pathway.

The Translocon Is a Multifunctional Transmembrane Pore. How are preproteins transported across or inserted into the RER membrane? In 1975, Blobel postulated that these processes are mediated by an aqueous TM channel. However, it was not until 1991 that he was able to experimentally demonstrate its existence through electrophysiological measurements indicating that the RER membrane contains ion-conducting channels. These channels, now called translocons, enclose aqueous pores that completely span the ER membrane as shown by linking nascent polypeptide chains to fluorescent dyes whose fluorescence is sensitive to the polarity of their environment. The channel-forming component of the translocon is a heterotrimeric protein named **Sec61** in mammals and **SecY** in prokaryotes. This protein is conserved throughout all kingdoms of life and hence is likely to have a similar structure and function in all organisms.

Electron microscopic studies of yeast Sec61 in complex with a ribosome (Fig. 9-38) reveals that Sec61 forms a cylindrical structure that has four attachments to the ribosome. The volume of this cylinder indicates that it consists of three or four Sec61 heterotrimers.

The X-ray structure of the SecY complex from the archaeon *Methanococcus jannaschii,* determined by Stephen Harrison and Tom Rapoport, reveals, as had been predicted, that the α, β, and γ subunits, respectively, have 10, 1, and 1 TM α helices and all have their N-termini in the cytosol (Fig. 9-39). The α subunit's TM helices are wrapped around an hourglass-shaped channel whose minimum diameter is ~3 Å. The channel is blocked at its extracellular end by a short relatively hydrophilic helix

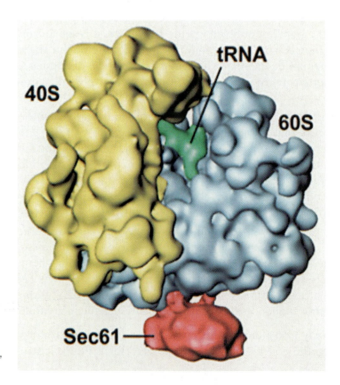

Figure 9-38 **The structure of yeast Sec61 in complex with a translating ribosome as determined by cryoelectron microscopy at 15 Å resolution.** The Sec61 oligomer is red, the ribosome's small and large subunits (Section 26-3) are yellow and blue, and a tRNA that is bound to the ribosome is green. [Courtesy of Joachim Frank, State University of New York at Albany.]

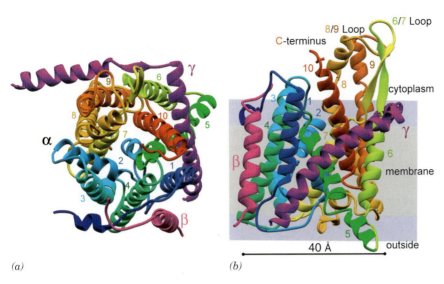

(a)

(b)

Figure 9-39 X-Ray structure of the *M. jannaschii* SecY complex. (a) View from the cytosol in which the α subunit is colored in rainbow order from it N-terminus (*dark blue*) to its C-terminus (*red*), the β subunit is pink, and the γ subunit is magenta. The translocon's putative lateral gate is on the left between helices 2 and 7. (b) View from the right of Part *a* in which the phospholipid head groups and hydrocarbon tail regions of the membrane are indicated by violet and gray background shading. [Courtesy of Stephen Harrison and Tom Rapoport, Harvard Medical School. PDBid 1RH5.]

(blue unnumbered helix in Fig. 9-39*a*). It is proposed that an incoming signal sequence pushes this helix aside and hence that the helix functions as a plug to prevent small molecules from leaking across the membrane in the absence of a translocating polypeptide.

The maximum diameter of an extended polypeptide is ~12 Å and that of an α helix is ~14 Å. Hence, if SecY's central pore indeed functions as the channel through which polypeptides are translocated, it would have to significantly expand to accommodate translocating polypeptides. The SecY structure suggests that this would require relatively simple hingelike motions involving conserved Gly residues.

If the translocon's channel is formed by a single Sec61 or SecY heterotrimer, what is the function of its forming a trimer or tetramer in complex with the ribosome (Fig. 9-38)? Although the answer to this question is as yet unknown, perhaps these multiple protomers are allosterically regulated and/or form binding sites for some of the proteins that are known to interact with the translocon.

The Translocon Inserts Transmembrane Helices into the ER Membrane.

In addition to forming a conduit for soluble proteins to pass through the membrane, *the translocon must mediate the insertion of an integral protein's TM segments into the membrane*. The X-ray structure of SecY suggests that this occurs by the opening of the C-shaped α subunit, as is diagrammed in Fig. 9-40, to permit the lateral installation of the TM segment into the membrane. The basis for the translocon's recognition of TM segments must be more than just an ~20-residue stretch of nonpolar residues since predictions based only on sequence hydrophobicity do not always correctly identify a polypeptide's TM segments.

*The signal sequences of many TM proteins are not cleaved by signal peptidase but, instead, are inserted into the membrane. Such so-called **signal-anchor sequences** may be oriented with either their N- or C-termini in the cytosol (e.g., the SecY subunits or glycophorin A, respectively).* If the N-terminus is installed in the cytosol, then the polypeptide must have looped around before being inserted into the membrane. Moreover, for **polytopic** (multispanning) TM proteins such as the SecY α subunit, this must occur for each successive TM helix. However, since it seems unlikely that the SecY channel could expand to simultaneously accommodate numerous TM helices, these helices are probably installed in the membrane one or two at a time rather than all together.

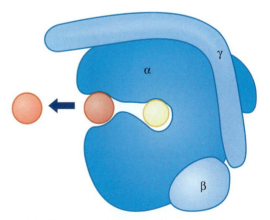

Figure 9-40 Model for the translocon-mediated insertion of a transmembrane helix into a membrane. The translocon (*blue*), which is represented schematically, is viewed as in Fig. 9-39*a*. A polypeptide chain (*yellow*) is shown bound in the translocon's putative pore during its translocation through the membrane, and a signal-anchor sequence (*red*) is shown passing through the translocon's lateral gate (between helices 2 and 7 in Fig. 9-39*a*) and being released into the membrane (*arrow*). [After a drawing by Dobberstein, B. and Sinning, I., *Science* **303**, 320 (2004).]

One might reasonably expect that for a polytopic TM protein, it is the membrane orientation of the N-terminal TM helix that dictates the orientations of the succeeding TM helices (many of which have yet to be synthesized at the time the N-terminal helix is inserted into the membrane). However, the deletion or insertion of a TM helix from/into a polypeptide does not necessarily change the membrane orientations of the succeeding TM helices: In fact, when two successive TM helices have the same preferred orientation, one of them may be forced out of the membrane. Thus, the basis for the orientation of a TM protein in the membrane and the mechanism through which the translocon achieves this orientation are poorly understood.

E Vesicle Trafficking

Shortly after their polypeptide synthesis is completed, the partially processed transmembrane, secretory, and lysosomal proteins appear in the Golgi apparatus (Fig. 1-8), a 0.5- to 1.0-μm-diameter organelle consisting of a stack of three to six or more (depending on the species) flattened and functionally distinct membranous sacs known as **cisternae,** where further posttranslational processing, mainly glycosylation, occurs (Section 8-3C). The Golgi stack (Fig. 9-41) has two distinct faces, each composed of a network of interconnected membranous tubules: the **cis Golgi network (CGN),** which is opposite the ER and is the port through which proteins enter the Golgi apparatus; and the **trans Golgi network (TGN),** through which processed proteins exit to

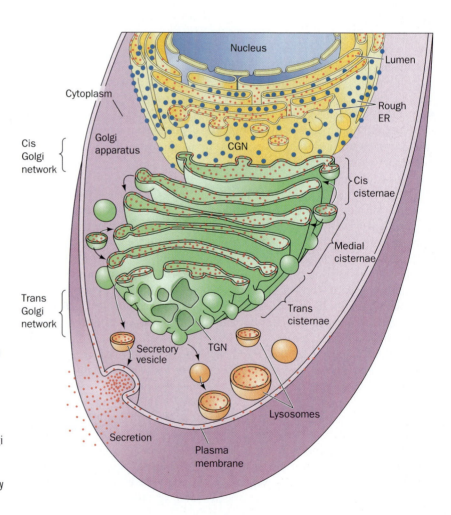

Figure 9-41   The posttranslational processing of proteins.   Proteins destined for secretion, insertion into the plasma membrane, or lysosomes are synthesized by RER-associated ribosomes (*blue dots; top*). As they are synthesized, the proteins (*red dots*) are either injected into the lumen of the ER or inserted into its membrane. After initial processing in the ER, the proteins are encapsulated in vesicles that bud off from the ER membrane and subsequently fuse with the cis Golgi network (CGN). The proteins are progressively processed in the cis-, medial-, and trans- cisternae of the Golgi. Finally, in the trans Golgi network (TGN, *bottom*), the completed glycoproteins are sorted for delivery to their final destinations, the plasma membrane, **secretory vesicles,** or lysosomes, to which they are transported by yet other vesicles.

their final destinations. The intervening Golgi stack contains at least three different types of sacs, the **cis, medial,** and **trans cisternae,** each of which contains different sets of glycoprotein processing enzymes.

Proteins transit from one end of the Golgi stack to the other while being modified in a stepwise manner (Section 8-3C). These proteins are transported via two mechanisms:

1. They are conveyed between successive Golgi compartments in the cis to trans direction as cargo within membranous vesicles that bud off of one compartment and fuse with a successive compartment, a process known as forward or **anterograde transport.**

2. They are carried as passengers in Golgi compartments that transit the Golgi stack, that is, the cis-cisternae eventually become trans-cisternae, a process called **cisternal progression** or **maturation.** This process is mediated through the backward or **retrograde transport** of Golgi-resident proteins from one compartment to the preceding one via membraneous vesicles.

Upon reaching the trans Golgi network, the now mature proteins are sorted and sent to their final cellular destinations.

Membrane, Secretory, and Lysosomal Proteins Are Transported in Coated Vesicles.

The vehicles in which proteins are transported between the RER, the Golgi apparatus, and their final destinations, as well as between the different compartments of the Golgi apparatus, are known as **coated vesicles** (Fig. 9-42).

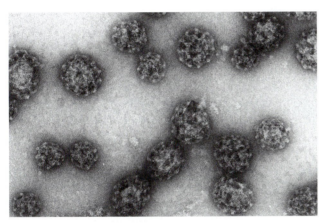

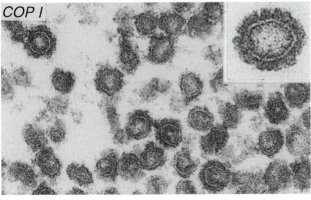

(a)

Figure 9-42   Electron micrographs of coated vesicles.   (a) Clathrin-coated vesicles. Note their polyhedral character. [Courtesy of Barbara Pearse, Medical Research Council, U.K.] (b) COPI-coated vesicles. (c) COPII-coated vesicles. The inserts in Parts b and c show the respective vesicles at higher magnification. [Courtesy of Lelio Orci, University of Geneva, Switzerland.]

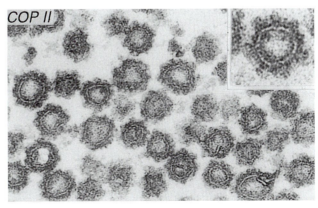

(b)

(c)

This is because these 60- to 150-nm-diameter membranous sacs are initially encased on their outer (cytosolic) faces by specific proteins that act as flexible scaffolding in promoting vesicle formation. A vesicle buds off from its membrane of origin and later fuses to its target membrane. *This process preserves the orientation of the transmembrane protein (Fig. 9-43), so that the lumens of the ER and the Golgi cisternae are topologically equivalent to the outside of the cell. This explains why the carbohydrate moieties of integral glycoproteins and the GPI anchors of GPI-linked proteins occur only on the external surfaces of plasma membranes.*

The three known types of coated vesicles are characterized by their protein coats. These are:

1. **Clathrin** (Fig. 9-42*a*), a protein that forms a polyhedral framework around vesicles that transport TM, GPI-linked, and secreted proteins from the Golgi to the plasma membrane. The clathrin cages can be dissociated to flexible three-legged proteins known as **triskelions**

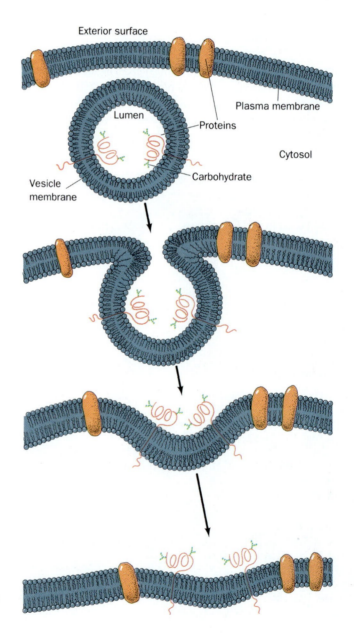

Figure 9-43 The fusion of a vesicle with the plasma membrane preserves the orientation of the integral proteins embedded in the vesicle bilayer. The inside of the vesicle and the exterior of the cell are topologically equivalent because the same side of the protein is always immersed in the cytosol. Note that any soluble proteins contained within the vesicle would be secreted. In fact, proteins destined for secretion are packaged in membranous secretory vesicles that subsequently fuse with the plasma membrane as shown.

(Fig. 9-44) that consist of three so-called heavy chains (**HC,** 190 kD) that each bind one of two homologous light chains, **LCa** or **LCb** (24–27 kD), at random.

2. **COPI** protein (Fig. 9-42*b*; COP for *co*at *p*rotein), which forms a fuzzy rather than a polyhedral coating about vesicles that carry out both the anterograde and retrograde transport of proteins between successive Golgi compartments. In addition, COPI-coated vesicles return ER-resident proteins that have escaped retention from the Golgi back to the ER (see below). The COPI protomer, which contains seven different subunits, is named **coatomer.**

3. **COPII** protein (Fig. 9-42*c*), which transports proteins from the ER to the Golgi. The COPII vesicle components are then recycled by COPI-coated vesicles for participation in another round of vesicle formation (the COPI vesicle components entering the ER are presumably recycled by COPII-coated vesicles). The COPII coat consists of two conserved protein heterodimers.

All of the above coated vesicles also carry receptors, which bind the proteins being transported, as well as the proteins that mediate the fusion of these vesicles with their target membranes (Section 9-4F).

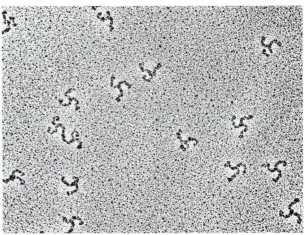

Figure 9-44 An electron micrograph of triskelions. The variable orientations of their legs is indicative of their flexibility. [Courtesy of Daniel Branton, Harvard University.]

Clathrin Cages Are Formed by Overlapping Heavy Chains.   Clathrin-coated vesicles (**CCVs**) are structurally better characterized than those coated with COPI or COPII. Clathrin forms polyhedral cages in which, as Barbara Pearse discovered, each vertex is the center (hub) of a triskelion, and its edges, which are ~150 Å long, are formed by the overlapping legs of four triskelions (Fig. 9-45*a*). Such polyhedra, which have 12 pentagonal faces and a variable number of hexagonal faces, are the most parsimonious way of enclosing spheroidal objects in polyhedral cages. The volume enclosed by a clathrin polyhedron, of course, increases with its number of hexagonal faces.

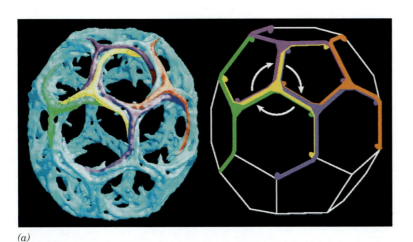

(*a*)

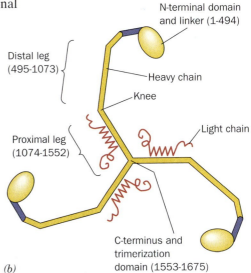

(*b*)

Figure 9-45 The anatomy of a clathrin-coated vesicle.   (*a*) A cryoelectron microscopy–based image of a clathrin cage at 21 Å resolution with its triskelions differently colored. Its receptor-containing core has been removed for clarity. As the accompanying diagram (*right*) indicates, a triskelion is centered on each of this polyhedral cage's 36 vertices, its edges are formed by the antiparallel legs of adjacent triskelions, and the linker and N-terminal domains project inward. Clathrin forms polyhedral cages with a large range of different sizes (number of hexagons): That shown here is only ~600 Å in diameter, whereas clathrin-coated membranous vesicles are typically ~1200 Å in diameter or larger. [Electron micrograph by Barbara Pearse and courtesy of H.T. McMahon, MRC Laboratory for Molecular Biology, Cambridge, U.K.] (*b*) Schematic diagram of a triskelion indicating its structural subdivisions.

The triskelion's ~450-Å-long legs are each formed by the 1675-residue heavy chains (HCs), which trimerize via their C-terminal domains (Fig. 9-45b). Although the X-ray structure of an entire HC has not been determined, those of its N-terminal segment and a portion of its proximal leg have been elucidated:

1. The N-terminal segment (residues 1–494; Figs. 9-46a,b), whose structure was determined by Harrison and Tomas Kirchhausen, consists of two domains: (i) an N-terminal 7-bladed **β propeller** in which each structurally similar propeller blade consists of a 4-stranded antiparallel β sheet (Fig. 9-46b); and (ii) a C-terminal linker that consists of 10 α helices of variable lengths (2–4 turns) connected by short loops and arranged in an irregular right-handed helix (a helix of helices, that is, a **superhelix**) named an **α zigzag.**

2. The proximal leg segment (residues 1210–1516; Fig. 9-46c), whose structure was determined by Peter Hwang and Robert Fletterick, consists of 24 linked α helices that are arranged similarly but more reg-

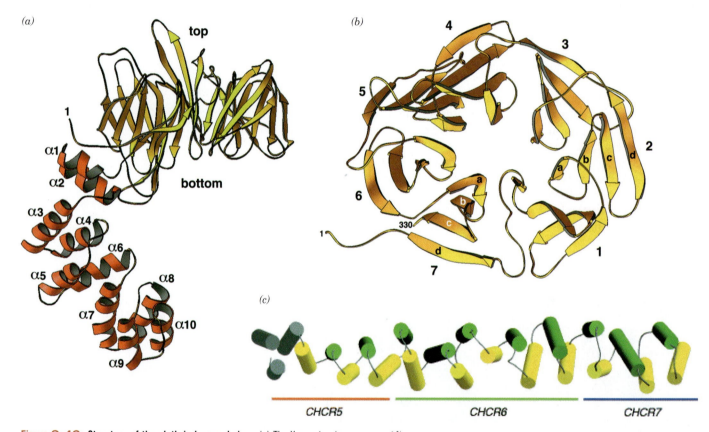

Figure 9-46   Structure of the clathrin heavy chain.   (a) The X-ray structure of the N-terminal domain and part of the linker of rat HC. The N-terminal domain forms a 7-bladed β propeller (*yellow*) that is seen here in side view and the linker (*red*) forms an α zigzag. (b) The β propeller as viewed from the top along its pseudo-7-fold axis. [Parts a and b courtesy of Tomas Kirchhausen, Harvard Medical School. PDBid 1BPO.] (c) The X-ray structure of bovine clathrin HC residues 1210 to 1516 as viewed with its N terminus on the left. The helices are alternately colored yellow and green with the exception of the three N-terminal helices, which are colored gray to indicate that they are poorly resolved. The red, green, and purple bars denote the regions of CHCR5, CHCR6, and CHCR7, respectively. [Courtesy of Peter Hwang, University of California at San Francisco. PDBid 1B89.] (d) Schematic diagram of a single clathrin heavy chain indicating the positions of its N-terminal β propeller (*magenta*), the following α zigzag linker (*blue*), and the proximal leg segment (*cyan*). [Courtesy of Barbara Pearse, MRC Laboratory of Molecular Biology, Cambridge, U.K.]

ularly than the above α zigzag to form a rod-shaped right-handed superhelix. The rigidity of this motif is attributed to its continuous hydrophobic core together with the efficient interdigitation of its side chains where its crossing antiparallel α helices come into contact.

The positions of these fragments in a HC is diagrammed in Fig. 9-46d.

Sequence and structural alignments indicate that HC residues 537 to 1566 consist of seven homologous ~145-residue **clathrin heavy chain repeats (CHCRs)** that are arranged in tandem and which each contain 10 helices (the above proximal leg segment consists of all of CHCR6 together with the C- and N-terminal portions of CHCR5 and CHCR7; Fig. 9-46c). Hence the entire HC leg appears to consist of an extended superhelix of linked α helices. Nevertheless, triskelion legs exhibit considerable flexibility (Fig. 9-44), a functional necessity for the formation of different-sized vesicles as well as for the budding of a vesicle from a membrane surface, which is accompanied by a large change in its curvature. The HC appears to flex mainly along a segment of the knee between its proximal and distal legs that is free of contacts with other molecules in clathrin cages.

The proximal leg segment bears extensive hydrophobic surface patches that follow the grooves between adjacent helices. This suggests that the lengthwise association of two proximal legs in a clathrin cage (Fig. 9-45a) is stabilized by the burial of these hydrophobic patches through the complementary packing of the helices of one proximal leg in the grooves on another.

Light chains (LCs) are not required for clathrin cage assembly. Indeed, LCs inhibit heavy chain polymerization *in vitro*, which suggests that they have a regulatory role in preventing inappropriate clathrin cage assembly in the cytosol. The X-ray structure of the HC proximal leg segment, which encompasses the LC binding site, contains a prominent basic groove in which the highly acidic LCs are postulated to bind. In fact, an 84% identical muscle homolog of clathrin HC, in which three of the basic residues in this groove have been replaced, fails to bind LCs. This suggests that LC binding interferes with the formation of salt bridges between HCs that stabilize cage assembly. The segments of the 60% identical LCa and LCb that differ in sequence are confined to regions that do not participate in HC binding and hence are likely to contain sites for the attachment of cytosolic factors that regulate vesicle uncoating (see below).

Clathrin-Coated Vesicles also Participate in Endocytosis.   CCVs, as we have seen, transport TM and secretory proteins from the trans Golgi network (TGN) to the plasma membrane (Fig. 9-41). In addition, through a process known as **endocytosis** (discussed in Section 19-1B), they act to engulf specific proteins from the extracellular medium by the invagination of a portion of the plasma membrane and to transport them to intracellular destinations.

Proteins Are Directed to the Lysosome by Carbohydrate Recognition Markers.
How are proteins in the ER selected for transport to the Golgi apparatus and from there to their respective membranous destinations? A clue as to the nature of this process is provided by the human hereditary defect known as **I-cell disease** (alternatively, **mucolipidosis II**) which, in homozygotes, is characterized by severe progressive psychomotor retardation, skeletal deformities, and death by age 10. The lysosomes in the connective tissue of I-cell disease victims contain large inclusions (after which the disease is named) of glycosaminoglycans and glycolipids as a result of the absence of

several lysosomal hydrolases. These enzymes are synthesized on the RER with their correct amino acid sequences but rather than being dispatched to the lysosomes, are secreted into the extracellular medium. This misdirection results from the absence of a mannose-6-phosphate recognition marker on the carbohydrate moieties of these hydrolases because of a deficiency of an enzyme required for mannose phosphorylation of the lysosomal proteins. The mannose-6-phosphate residues are normally bound by a receptor in the coated vesicles that transport lysosomal hydrolases from the Golgi apparatus to the lysosomes. No doubt, other glycoproteins are directed to their intracellular destinations by similar carbohydrate markers.

ER-Resident Proteins Have the C-Terminal Sequence KDEL. Most soluble ER-resident proteins in mammals have the C-terminal sequence KDEL (HDEL in yeast), KKXX, or KXKXXX (where X represents any amino acid residue), whose alteration results in the secretion of the resulting protein. By what means are these proteins selectively retained in the ER? Since many ER-resident proteins freely diffuse within the ER, it seems unlikely that they are immobilized by membrane-bound receptors within the ER. Rather, ER-resident proteins, as do secretory and lysosomal proteins, readily leave the ER via COPII-coated vesicles but ER-resident proteins are promptly retrieved from the Golgi and returned to the ER in COPI-coated vesicles. Indeed, coatomer binds the Lys residues in the C-terminal KKXX motif of transmembrane proteins, which presumably permits it to gather these proteins into COPI-coated vesicles. Furthermore, genetically appending KDEL to the lysosomal protease **cathepsin D** causes it to accumulate in the ER, but it nevertheless acquires an N-acetylglucosaminyl-1-phosphate group, a modification that is made in an early Golgi compartment. Presumably, a membrane-bound receptor in a post-ER compartment binds the KDEL signal and the resulting complex is returned to the ER in a COPI-coated vesicle. **KDEL receptors** have, in fact, been identified in yeast and humans. However, the observation that former KDEL proteins whose KDEL sequences have been deleted are, nevertheless, secreted relatively slowly suggests that there are mechanisms for retaining these proteins in the ER by actively withholding them from the bulk flow of proteins through the secretory pathway.

F Vesicle Fusion

In all cells, *new membranes are generated by the expansion of existing membranes.* In eukaryotes, this process occurs mainly via vesicle trafficking in which a vesicle buds off from one membrane (e.g., that of the Golgi apparatus) and fuses to a different membrane (e.g., the plasma membrane or that of the lysosome), thereby transferring both lipids and proteins from the parent to the target membrane.

On arriving at its target membrane, a vesicle fuses with it, thereby releasing its contents on the opposite side of the target membrane (Fig. 9-43). For example (Fig. 9-47), when a nerve impulse in a presynaptic cell reaches a **synapse** [the junction between neurons (nerve cells) or between neurons and muscles], it triggers the fusion of **neurotransmitter**-containing **synaptic vesicles** with the **presynaptic membrane** (a specialized section of the neuron's plasma membrane), thereby releasing the neurotransmitter (a small molecule) into the ~200-Å-wide **synaptic cleft** (the process whereby membranous vesicles fuse with the plasma membrane to release their con-

(a)

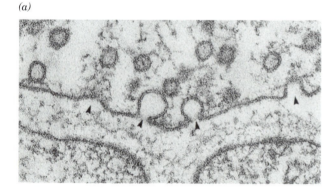

(b)

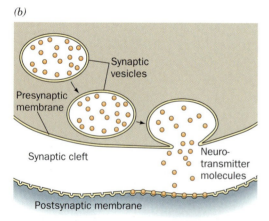

Figure 9-47 Vesicle fusion at a synapse. (*a*) An electron micrograph of a frog neuromuscular synapse. Synaptic vesicles are undergoing fusion (*arrows*) with the presynaptic plasma membrane (*top*). [Courtesy of John Heuser, Washington University School of Medicine, St. Louis, Missouri.] (*b*) This process discharges the neurotransmitter contents of the synaptic vesicles into the synaptic cleft, the space between the neuron and the muscle cell (the postsynaptic cell).

tents outside the cell is called **exocytosis**). The neurotransmitter rapidly (in <0.1 ms) diffuses across the synaptic cleft to the **postsynaptic membrane,** where it binds to specific receptors that thereupon trigger the continuation of the nerve impulse or muscle contraction in the postsynaptic cell. The homogenization of nerve tissue causes its presynaptic endings to pinch off and reseal to form **synaptosomes,** which can be readily isolated by density gradient ultracentrifugation for subsequent study.

How do vesicles fuse and why do they fuse only with their target membranes? Extensive investigations, in large part by James Rothman, have identified numerous proteins that participate in membrane fusion events. One of these is called **NSF** (for **NEM-sensitive fusion protein;** NEM is **N-ethylmaleimide,** which alkylates Cys residues and thereby inactivates NSF). NSF is a cytosolic ATPase that does not function unless a **soluble NSF attachment protein (SNAP)** is also present. SNAPs associate with fusing membranes via **SNAP receptors (SNAREs),** which are integral or lipid-linked proteins.

SNAREs are associated with membranes of both the vesicles **(R-SNAREs)** and their target membranes **(Q-SNAREs),** which are so named because they, respectively, contain conserved Arg and Gln residues (they were originally called **v-** and **t-SNAREs** for *ves*icle and *t*arget). *R-SNARES and Q-SNAREs associate to firmly anchor the vesicle to its target membrane,* a process called docking. The R-SNARE known as **synaptobrevin** and the Q-SNAREs known as **syntaxin** and **SNAP-25** (for *synaptosome-associated protein of 25 kD*) form a highly stable complex (note that SNAP-25 is not a SNAP protein; by coincidence the two independently characterized proteins were assigned the same acronym before it was realized that they are functionally related). Synaptobrevin and syntaxin each have a C-terminal transmembrane helix, and SNAP-25 is anchored to the membrane via the palmitoylation of four Cys residues (Section 9-3B) in the protein's central region.

The X-ray structure of the associating portions of the foregoing SNARE complex, determined by Reinhard Jahn and Axel Brünger, reveals it to be a bundle of four parallel ~65-residue α helices with two of the helices formed by the N- and C-terminal segments of SNAP-25 (Fig. 9-48). Since synaptobrevin is anchored in the vesicle membrane and syntaxin and SNAP-25 are anchored in the target membrane, this so-called core complex firmly ties together the two membranes.

The four helices of the core complex wrap around each other with a gentle left-handed twist. For the most part, the sequence of each helix has the expected 7-residue repeat, $(a\text{-}b\text{-}c\text{-}d\text{-}e\text{-}f\text{-}g)_n$, with residues *a* and *d* hydrophobic (Section 6-1C). However, the central layer of side chains along the length of the 4-helix bundle includes an Arg residue from synaptobrevin that is hydrogen bonded to three Gln side chains, one from syntaxin and one from each of the SNAP-25 helices. These highly conserved polar residues are sealed off from the aqueous environment so that their interactions serve to bring the four helices into proper register. Since cells contain numerous different R-SNAREs and Q-SNAREs, it would seem likely that their interactions are at least partially responsible for the specificity that vesicles exhibit in fusing with their target membranes.

Once the membranes have fused, the SNARE complex is dissociated so that its component proteins can participate in a new round of vesicle fusion. The process is mediated by NSF, an ATP-driven molecular chaperone, that binds to SNAREs through the intermediacy of SNAPs.

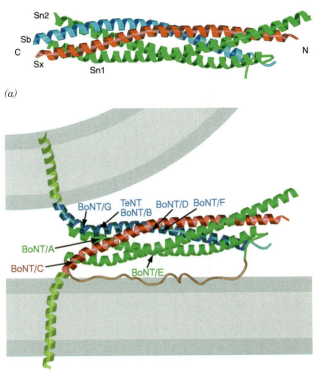

(a)

(b)

Figure 9-48   X-Ray structure of a SNARE complex.
(*a*) Ribbon diagram showing the syntaxin helix (Sx) in red, the synaptobrevin helix (Sb) in blue, and the N- and C-terminal helices of SNAP-25 (Sn1 & Sn2) in green. (*b*) Model of the synaptic fusion complex linking two membranes (*gray*). The helices of the SNARE complex are colored as in Part *a*. The transmembrane C-terminal extensions of syntaxin and synaptobrevin are modeled as helices (*yellow-green*). The peptide segment connecting the N- and C-terminal helices of SNAP-25 is speculatively represented as an unstructured loop (*brown*). This loop is anchored to the membrane via Cys-linked palmitoyl groups (*not shown*). The cleavage sites for the various clostridial neurotoxins are indicated by the arrows. [Courtesy of Axel Brünger, Yale University. PDBid 1SFC.]

Tetanus and Botulinum Toxins Specifically Cleave SNAREs.   The frequently fatal infectious diseases **tetanus** (which arises from wound contamination) and **botulism** (a type of food poisoning) are caused by certain anaerobic bacteria of the genus *Clostridium*. These bacteria produce extremely potent protein neurotoxins that inhibit the release of neurotransmitters into synapses. In fact, botulinum toxins are the most powerful known toxins; they are ~10 million-fold more toxic than cyanide.

There are seven serologically distinct types of botulinum neurotoxins, designated **BoNT/A** through **BoNT/G,** and one type of tetanus neurotoxin, **TeTx.** Each of these homologous proteins is synthesized as a single ~150-kD polypeptide chain that is cleaved by host proteases to yield an ~50-kD light chain and an ~100-kD heavy chain. The heavy chains bind to specific types of neurons and facilitate the uptake of the L chain by endocytosis. *Each light chain is a protease that cleaves its target SNARE at a specific site.* SNARE cleavage prevents the formation of the core complex and thereby halts the exocytosis of synaptic vesicles. The heavy chain of TeTx specifically binds to inhibitory neurons (which function to moderate excitatory nerve impulses) and is thereby responsible for the spastic paralysis characteristic of tetanus. The heavy chains of the BoNTs instead bind to motor neurons (which innervate muscles) and thus cause the flaccid paralysis characteristic of botulism.

The administration of carefully controlled quantities of botulinum toxin ("Botox") is medically useful in relieving the symptoms of certain types of chronic muscle spasms. Moreover, this toxin is being used cosmetically: Its injection into the skin relaxes the small muscles causing wrinkles and hence these wrinkles disappear for ~3 months.

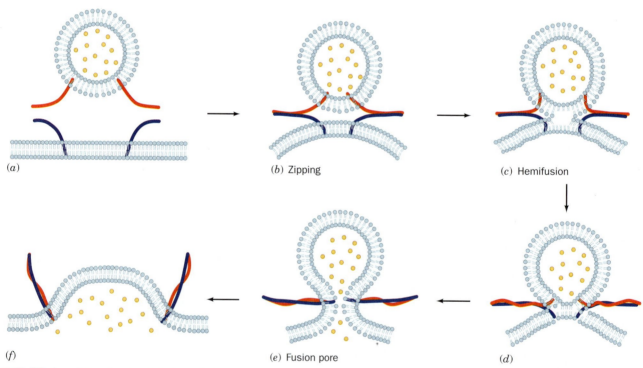

Figure 9-49   A model for SNARE-mediated vesicle fusion.   Here R-SNAREs and Q-SNAREs are schematically represented by red and blue worms. (*a*) The vesicle approaches its target membrane but is not yet in contact with it. (*b*) The SNARE complexes start zipping together (docking) from their N-termini, which draws the two membranes toward each other. (*c*) As docking proceeds, the increased curvature and lateral tension induce the approaching bilayer leaflets to fuse, thereby exposing the bilayer interior. This process is called hemifusion. (*d*) As docking of the SNAREs continues, the two bilayer leaflets that were originally farthest apart are brought together to form a new bilayer. (*e*) The continuing SNARE-induced lateral tension causes membrane breakdown, resulting in the formation of a **fusion pore.** (*f*) The fusion pore then expands as the now fused membrane relaxes. [After a drawing by Chen, Y.A. and Scheller, R.H., *Nature Rev. Mol. Cell Biol.* **2,** 98 (2001).]

Membranes Brought into Close Juxtaposition by SNAREs Spontaneously Fuse. The association of an R-SNARE on a vesicle with Q-SNAREs on its target membrane brings the two bilayers into close proximity (Fig. 9-48*b*). But what induces the fusion of the juxtaposed lipid bilayers? Although the mechanism of membrane fusion is imperfectly understood, in the model that has received the most credence, mechanical stresses arising from the formation of SNARE complexes expel lipid molecules from their bilayer. This, it is hypothesized, results in the formation of a transient structure with lipids from the opposing bilayer, a process that culminates in bilayer fusion (Fig. 9-49). However, *in vitro*, this process takes 30 to 40 minutes, whereas the fusion of a synaptic vesicle with the presynaptic membrane requires ~0.3 ms. This argues that other proteins also participate in inducing bilayer fusion.

SUMMARY

1. Lipids are a diverse group of molecules that are soluble in organic solvents and, in contrast to other major types of biomolecules, do not form polymers.

2. Fatty acids are carboxylic acids whose chain lengths and degrees of unsaturation vary.

3. Adipocytes and other cells contain stores of triacylglycerols, which consist of three fatty acids esterified to glycerol.

4. Glycerophospholipids are amphipathic molecules that contain two fatty acid chains and a polar head group.

5. The sphingolipids include sphingomyelins, cerebrosides, and gangliosides.

6. Cholesterol, steroid hormones, and vitamin D are all based on a four-ring structure.

7. The arachidonic acid derivatives prostaglandins, prostacyclins, thromboxanes, leukotrienes, and lipoxins are signaling molecules that have diverse physiological roles.

8. Glycerophospholipids and sphingolipids form bilayers in which their nonpolar tails associate with each other and their polar head groups are exposed to the aqueous solvent.

9. Although the transverse diffusion of a lipid across a bilayer is extremely slow, lipids rapidly diffuse in the plane of the bilayer. Bilayer fluidity varies with temperature and with the chain lengths and degree of saturation of its component fatty acid residues.

10. The proteins of biological membranes include integral (intrinsic) proteins that contain one or more transmembrane α helices or a β barrel. In all cases, the membrane-exposed surface of the protein is hydrophobic.

11. Other membrane-associated proteins may be anchored to the membrane via isoprenoid, fatty acid, or glycosylphosphatidylinositol (GPI) groups. Peripheral (extrinsic) proteins are loosely associated with the membrane surface.

12. The fluid mosaic model of membrane structure accounts for the lateral diffusion of membrane proteins and lipids.

13. The arrangement of membrane proteins may depend on their interactions with an underlying protein skeleton, as in the erythrocyte.

14. Membrane proteins and lipids are distributed asymmetrically in the two leaflets and may form domains such as lipid rafts.

15. Transmembrane (TM), secretory, and lysosomal proteins are synthesized by means of the secretory pathway. A signal peptide directs a growing polypeptide chain through the RER membrane via a protein pore called the translocon, which also functions to laterally install TM proteins into the RER membrane.

16. Coated vesicles transport membrane-embedded and luminal proteins from the ER to the Golgi apparatus for further processing, and from there to other membranes. The proteins coating these vesicles may consist largely of clathrin, which forms polyhedral cages, or COPI or COPII, which form coats with an amorphous appearance.

17. Membrane fusion occurs via a complex process that involves SNAREs. These form four-helix bundles that bring two membranes into proximity, which in turn, induces membrane fusion.

REFERENCES

Lipids and Membrane Structure

Edidin, M., Shrinking patches and slippery rafts: scale of domains in the plasma membrane, *Trends Cell Biol.* **11,** 492–496 (2001). [Summarizes current views of the structures, dynamics, and functions of membrane domains.]

Gurr, M.I., Harwood, J.L., and Frayn, K.N., *Lipid Biochemistry: An Introduction* (5th ed.), Blackwell Science (2002).

Nagle, J.F. and Tristram-Nagle, S., Lipid bilayer structure, *Curr. Opin. Struct. Biol.* **10,** 474–480 (2000). [Explains why it is difficult to quantitatively describe the structure of the lipid bilayer.]

Membrane Proteins

Grum, V.L., Li, D., MacDonald, R.I., and Mondragón, A., Structures of two repeats of spectrin suggest models of flexibility, *Cell* **98**, 523–535 (1999).

Popot, J.-L. and Engelman, D.M., Helical membrane protein folding, stability, and evolution, *Annu. Rev. Biochem.* **69**, 881–922 (2000). [Shows a number of protein structures and discusses many features of transmembrane proteins.]

Schulz, G.E., β-Barrel membrane proteins, *Curr. Opin. Struct. Biol.* **10**, 443–447 (2000). [Reviews the basic principles of construction for transmembrane β barrels, including the smallest, an 8-stranded barrel.]

Subramaniam, S., The structure of bacteriorhodopsin: an emerging consensus, *Curr. Opin. Struct. Biol.* **9**, 462–468 (1999). [Compares the six structures of bacteriorhodopsin that have been independently determined by electron or X-ray crystallography and finds them to be remarkably similar.]

Udenfriend, S. and Kodukula, K., How glycosyl-phosphatidyl-inositol-anchored membrane proteins are made, *Annu. Rev. Biochem.* **64**, 563–591 (1995).

White, S.H., Ladokhin, A.S., Jayasinghe, S., and Hristova, K., How membranes shape protein structure, *J. Biol. Chem.* **276**, 32395–32398 (2001). [Considers some thermodynamic constraints on membrane proteins.]

Zhang, F.L. and Casey, P.J., Protein prenylation: molecular mechanisms and functional consequences, *Annu. Rev. Biochem.* **65**, 241–269 (1996).

The Secretory Pathway

Johnson, A.E. and van Waes, M.A., The translocon: a dynamic gateway to the ER membrane, *Annu. Rev. Cell Dev. Biol.* **15**, 799–842 (1999).

Keenan, R.J., Fraymann, D.M., Stroud, R.M., and Walter, P., The signal recognition particle, *Annu. Rev. Biochem.* **70**, 755–775 (2001).

van den Berg, B., Clemons, W.M., Jr., Collinson, I., Modis, Y., Hartmann, E., Harrison, S.C., and Rapaport, T.A., X-ray structure of a protein-conducting channel, *Nature* **427**, 36–44 (2004). [The X-ray structure of SecY.]

Vesicle Trafficking

Brodsky, F.M., Chen, C.-Y., Knuehl, C., Towler, M.C., and Wakeham, D.E., Biological basket weaving: formation and function of clathrin-coated vesicles, *Annu. Rev. Cell Dev. Biol.* **17**, 515–568 (2001).

Kirchhausen, T., Clathrin, *Annu. Rev. Biochem.* **69**, 677–706 (2000).

Vesicle Fusion

Brünger, A.T., Structure of proteins involved in synaptic vesicle fusion in neurons, *Annu. Rev. Biophys. Biomol. Struct.* **30**, 151–171 (2001).

Chen, Y.A. and Scheller, R.H., SNARE-mediated membrane fusion, *Nature Rev. Molec. Cell Biol.* **2**, 98–106 (2001). [Reviews SNARE protein structure and the role of SNAREs in lipid bilayer fusion.]

■ KEY TERMS

lipid	sterol	prenylation	SRP receptor (SR)
fatty acid	glucocorticoid	myristoylation	translocon
saturation	mineralocorticoid	palmitoylation	exocytosis
triacylglycerol	androgen	GPI-linked protein	secretory pathway
fats	estrogen	peripheral (extrinsic) protein	signal peptide
oils	isoprenoid	fluid mosaic model	translocon
adipocyte	vitamin	fluorescence recovery after	Golgi apparatus
glycerophospholipid	prostaglandin	photobleaching (FRAP)	coated vesicles
phosphatidic acid	eicosanoid	membrane skeleton	clathrin
phospholipase	lipid bilayer	spectrin	COPI
lysophospholipid	liposome	ankyrin	COPII
plasmalogen	transverse diffusion	gates and fences model	NSF attachment
sphingolipid	lateral diffusion	flipase	protein (SNAP)
ceramide	transition temperature	lipid raft	SNAP receptor (SNARE)
sphingomyelin	integral (intrinsic) protein	secretory pathway	
cerebroside	transmembrane (TM)	signal peptides	
ganglioside	protein	signal recognition	
steroid	electron crystallography	particle (SRP)	

■ STUDY EXERCISES

1. How do lipids differ from the three other major classes of biological molecules?

2. How does unsaturation affect the physical properties of fatty acids or the membrane lipids to which they are esterified?

3. Compare the structures and physical properties of triacyl-glycerols, glycerophospholipids, and sphingolipids.

4. Summarize the functions of steroids and eicosanoids.

5. Explain why lateral diffusion is faster than transverse diffusion in membrane lipids.

6. Explain the differences between integral and peripheral membrane proteins.

7. Describe the covalent modifications of lipid-linked proteins.

8. Describe the fluid mosaic model.

9. How can the cytoskeleton influence membrane protein distribution?

10. Describe the membrane translocation of lysosomal proteins.

11. Trace the route followed by a cell-surface glycoprotein starting from its synthesis on a ribosome.

12. What is the function of vesicles?

13. How do SNARE proteins induce vesicle fusion?

▪ PROBLEMS

1. Does *trans*-oleic acid have a higher or lower melting point than *cis*-oleic acid? Explain.

2. How many different types of triacylglycerols could incorporate the fatty acids shown in Fig. 9-1?

3. Which triacylglycerol yields more energy on oxidation: one containing three residues of linolenic acid or three residues of stearic acid?

4. Draw the structure of a glycerophospholipid that has a saturated C_{16} fatty acyl group at position 1, a monounsaturated C_{18} fatty acyl group at position 2, and an ethanolamine head group.

5. What products are obtained when 1-palmitoyl-2-oleoyl-3-phosphatidylserine is hydrolyzed by (a) phospholipase A_1; (b) phospholipase A_2; (c) phospholipase C; (d) phospholipase D?

6. Which of the glycerophospholipid head groups listed in Table 9-2 can form hydrogen bonds?

7. Does the phosphatidylglycerol "head group" of cardiolipin (Table 9-2) project out of a lipid bilayer like other glycerophospholipid head groups?

8. In some autoimmune diseases, an individual develops antibodies that recognize cell constituents such as DNA and phospholipids. Some of the antibodies react with both DNA and phospholipids. What is the structural basis for this cross-reactivity?

9. Most hormones, such as peptide hormones, exert their effects by binding to cell-surface receptors. However, steroid hormones do so by binding to cytosolic receptors. How is this possible?

10. Animals cannot synthesize linoleic acid (a precursor of arachidonic acid) and therefore must obtain this **essential fatty acid** from their diet. Explain why cultured animal cells can survive in the absence of linoleic acid.

11. Why can't triacylglycerols be significant components of lipid bilayers?

12. Why would a bilayer containing only gangliosides be unstable?

13. When bacteria growing at 20°C are warmed to 30°C, are they more likely to synthesize membrane lipids with (a) saturated or unsaturated fatty acids, and (b) short-chain or long-chain fatty acids? Explain.

14. (a) How many turns of an α helix are required to span a lipid bilayer (~30 Å across)? (b) What is the minimum number of residues required? (c) Why do most transmembrane helices contain more than the minimum number of residues?

15. The distance between the $C_α$ atoms in a β sheet is ~3.5 Å. Can a single 9-residue segment with a β conformation serve as the transmembrane portion of an integral membrane protein?

16. Are the following lipid samples likely to correspond to the inner or outer leaflet of a eukaryotic plasma membrane? (a) 20% Phosphatidylcholine, 15% phosphatidylserine, 65% other lipids. (b) 35% Phosphatidylcholine, 15% gangliosides, 5% cholesterol, 45% other lipids.

17. Describe the labeling pattern of glycophorin A when a membrane-impermeable protein-labeling reagent is added to (a) a preparation of solubilized erythrocyte proteins; (b) intact erythrocyte ghosts; and (c) erythrocyte ghosts that are initially leaky and then immediately sealed and transferred to a solution that does not contain the labeling reagent.

18. Predict the effect of a mutation in signal peptidase that narrows its specificity so that it cleaves only between two Leu residues.

19. Explain why a drug that interferes with the disassembly of a SNARE complex would block neurotransmission.

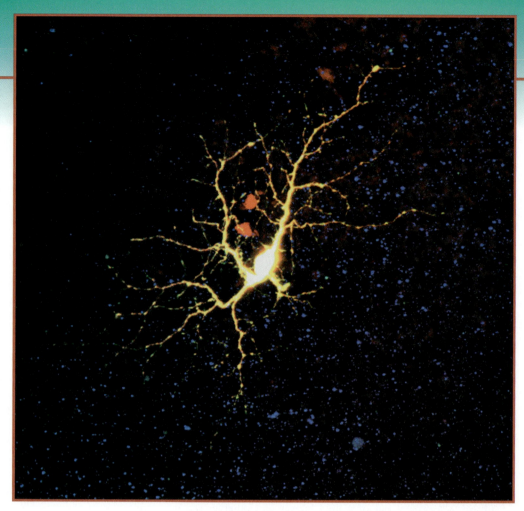

The impermeable membranes of cells, including those of neurons such as that pictured here, permit the establishment of a membrane potential, which changes when certain membrane proteins allow ions to flow into or out of the cell. Different membrane proteins mediate the transmembrane movements of other substances either with or against their concentration gradients. [David Becker/Photo Researchers.]

Membrane Transport

Cells are separated from their environments by plasma membranes. Eukaryotic cells, in addition, are compartmentalized by intracellular membranes that form the boundaries and internal structures of their various organelles. Biological membranes present formidable barriers to the passage of ionic and polar substances so that *these substances can traverse membranes only through the action of specific **transport proteins.*** Such proteins are therefore required to mediate all transmembrane movements of ions, such as Na^+, K^+, Ca^{2+}, and Cl^-, as well as metabolites such as pyruvate, amino acids, sugars, and nucleotides, and even water. Transport proteins are also responsible for all biological electrochemical phenomena such as neurotransmission. More complicated processes (e.g., endocytosis) are required to move larger substances such as proteins and macromolecular aggregates across membranes.

We begin our discussion of membrane transport by considering the thermodynamics of this process. We will then examine the structures and mechanisms of several different types of transport systems.

Thermodynamics of Transport

The diffusion of a substance between two sides of a membrane

$$A(out) \rightleftharpoons A(in)$$

thermodynamically resembles a chemical equilibration. We saw in Section 1-4D that the free energy of a solute, A, varies with its concentration:

$$\overline{G}_A - \overline{G}_A^{\circ\prime} = RT \ln[A] \qquad [10\text{-}1]$$

where $\overline{G}_A$ is the **chemical potential** (partial molar free energy) of A (the bar indicates quantity per mole) and $\overline{G}_A^{\circ\prime}$ is the chemical potential of its standard state. Thus, a difference in the concentrations of the substance on two sides of a membrane generates a **chemical potential difference:**

$$\Delta\overline{G}_A = \overline{G}_A(in) - \overline{G}_A(out) = RT \ln\left(\frac{[A]_{in}}{[A]_{out}}\right) \qquad [10\text{-}2]$$

Consequently, if the concentration of A outside the membrane is greater than that inside, $\Delta\overline{G}_A$ for the transfer of A from outside to inside will be negative and the spontaneous net flow of A will be inward. If, however, [A] is greater inside than outside, $\Delta\overline{G}_A$ is positive and an inward net flow of A can occur only if an exergonic process, such as ATP hydrolysis, is coupled to it to make the overall free energy change negative (see Sample Calculation 10-1).

The transmembrane movement of ions also results in charge differences across the membrane, thereby generating an electrical potential difference, $\Delta\Psi = \Psi(in) - \Psi(out)$, where $\Delta\Psi$ is termed the **membrane potential.** Consequently, if A is ionic, Eq. 10-2 must be amended to include the electrical work required to transfer a mole of A across the membrane from outside to inside:

$$\Delta\overline{G}_A = RT \ln\left(\frac{[A]_{in}}{[A]_{out}}\right) + Z_A \mathscr{F} \Delta\Psi \qquad [10\text{-}3]$$

where Z_A is the ionic charge of A; $\mathscr{F}$, the Faraday constant, is the charge of a mole of electrons (96,485 $C \cdot mol^{-1}$; C is the symbol for coulomb); and $\overline{G}_A$ is now termed the **electrochemical potential** of A. The membrane potentials of living cells are commonly as high as -100 mV (inside negative;

SAMPLE CALCULATION 10-1

Show that $\Delta G < 0$ when Ca^{2+} ions move from the endoplasmic reticulum (where $[Ca^{2+}] = 1$ mM) to the cytosol (where $[Ca^{2+}] = 0.1$ μM). Assume $\Delta\Psi = 0$.

The cytosol is *in* and the endoplasmic reticulum is *out*.

$$\Delta G = RT\ln\frac{[Ca^{2+}]_{in}}{[Ca^{2+}]_{out}} = RT\ln\frac{10^{-7}}{10^{-3}}$$
$$= RT(-9.2)$$

Hence, ΔG is negative.

note that $1 V = 1 J \cdot C^{-1}$). Hence, the last term in Eq. 10-3 is often significant for ionic substances, particularly in mitochondria (Chapter 17) and in neurotransmission (Section 10-2C).

Transport May Be Mediated or Nonmediated.   There are two types of transport processes: **nonmediated transport** and **mediated transport.** Nonmediated transport occurs through simple diffusion. In contrast, mediated transport occurs through the action of specific carrier proteins. The driving force for the nonmediated flow of a substance through a medium is its chemical potential gradient. Thus, *the substance diffuses in the direction that eliminates its concentration gradient, at a rate proportional to the magnitude of this gradient. The rate of diffusion of a substance also depends on its solubility in the membrane's nonpolar core.* Consequently, nonpolar molecules such as steroids and O_2 readily diffuse through biological membranes by nonmediated transport, according to their concentration gradients across the membranes.

Mediated transport is classified into two categories depending on the thermodynamics of the system:

1. **Passive-mediated transport,** or **facilitated diffusion,** in which a specific molecule flows from high concentration to low concentration.
2. **Active transport,** in which a specific molecule is transported from low concentration to high concentration, that is, against its concentration gradient. Such an endergonic process must be coupled to a sufficiently exergonic process to make it favorable (i.e., $\Delta G < 0$).

2 Passive-Mediated Transport

Substances that are too large or too polar to diffuse across lipid bilayers on their own may be conveyed across membranes via proteins or other molecules that are variously called **carriers, permeases, channels,** and **transporters.** These transporters operate under the same thermodynamic principles but vary widely in structure and mechanism, particularly as it relates to their selectivity.

A Ionophores

Ionophores are organic molecules of diverse types, often of bacterial origin, that increase the permeability of membranes to ions. These molecules often exert an antibiotic effect by discharging the vital ion concentration gradients that cells actively maintain.

There are two types of ionophores:

1. *Carrier ionophores, which increase the permeabilities of membranes to their selected ion by binding it, diffusing through the membrane, and releasing the ion on the other side (Fig. 10-1a).* For net transport to occur, the uncomplexed ionophore must then return to the original side of the membrane ready to repeat the process. Carriers therefore share the common property that *their ionic complexes are soluble in nonpolar solvents.*
2. *Channel forming ionophores, which form transmembrane channels or pores through which their selected ions can diffuse (Fig. 10-1b).*

Both types of ionophores transport ions at a remarkable rate. For example, a single molecule of the carrier ionophore **valinomycin** transports up

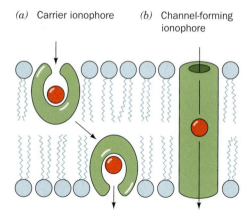

(a) Carrier ionophore   (b) Channel-forming ionophore

Figure 10-1 Ionophore action.   (a) Carrier ionophores transport ions by diffusing through the lipid bilayer. (b) Channel-forming ionophores span the membrane with a channel through which ions can diffuse.

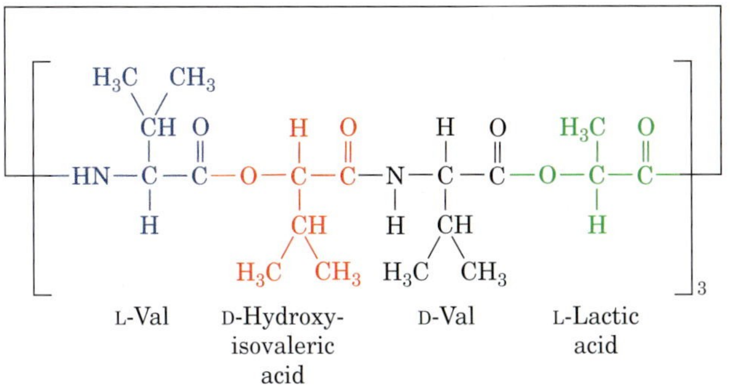

H₃C   CH₃
       CH   O   H   O   H   O   H₃C   O
              | || | || | || | ||
—HN—C—C—O—C—C—N—C—C—O—C—C—
       | | | | |
       H   CH   H CH   H
                               | |
                            H₃C   CH₃   H₃C   CH₃   ⌋₃

L-Val   D-Hydroxy-   D-Val   L-Lactic
          isovaleric   acid
            acid

Valinomycin

Figure 10-2   Valinomycin.   This cyclic ionophore contains ester and amide bonds and D- as well as L-amino acids.

to 10^4 K$^+$ ions per second across a membrane. Channel formers have an even greater ion throughput; for example, each membrane channel composed of the antibiotic **gramicidin A** permits the passage of over 10^7 K$^+$ ions per second. Clearly, the presence of either type of ionophore, even in small amounts, greatly increases the permeability of a membrane toward the specific ions transported. However, *since ionophores passively permit ions to diffuse across a membrane in either direction, their effect can only be to equilibrate the concentrations of their selected ions across the membrane.*

Valinomycin, which is one of the best characterized ionophores, specifically binds K$^+$. It is a cyclic molecule containing D- and L-amino acid residues that participate in ester linkages as well as peptide bonds (Fig. 10-2). The X-ray structure of valinomycin's K$^+$ complex (Fig. 10-3) indicates that the K$^+$ ion is octahedrally coordinated by the carbonyl groups of its six Val residues, and the cyclic valinomycin backbone surrounds the K$^+$ coordination shell. The methyl and isopropyl side chains project outward to provide the complex with a nonpolar exterior that makes it soluble in the hydrophobic cores of lipid bilayers.

The K$^+$ ion (ionic radius, $r = 1.33$ Å) fits snugly into valinomycin's coordination site, but the site is too large for Na$^+$ ($r = 0.95$ Å) or Li$^+$ ($r = 0.60$ Å) to coordinate with all six carbonyl oxygens. Valinomycin therefore has 10,000-fold greater binding affinity for K$^+$ than for Na$^+$. No other known substance discriminates better between Na$^+$ and K$^+$.

Gramicidin A is a 15-residue linear polypeptide consisting of alternating L and D residues, all of which are hydrophobic (Fig. 10-4). NMR and

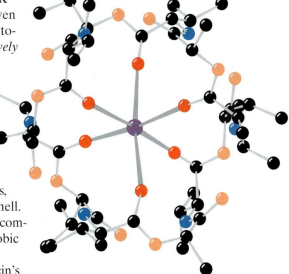

Figure 10-3   X-Ray structure of valinomycin in complex with a K$^+$ ion.   Six oxygen atoms (*dark red*) octahedrally coordinate the K$^+$ ion (*purple*). [After Neupert-Laves, K. and Dobler, M., *Helv. Chim. Acta* **58**, 439 (1975).]

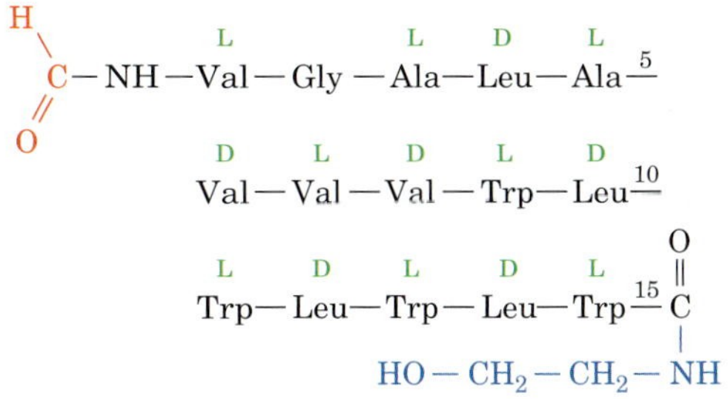

H
 \
  C—NH—Val—Gly—Ala—Leu—Ala—⁵
 //   L   L   D   L
O

          D   L   D   L   D
         Val—Val—Val—Trp—Leu—¹⁰

          L   D   L   D   L   O
         Trp—Leu—Trp—Leu—Trp—¹⁵C
                                    |
              HO—CH₂—CH₂—NH

Gramicidin A

Figure 10-4   Gramicidin A.   This 15-residue peptide of alternating D- and L-amino acids is chemically blocked at both its N- and C-termini.

(a)   *(b)*

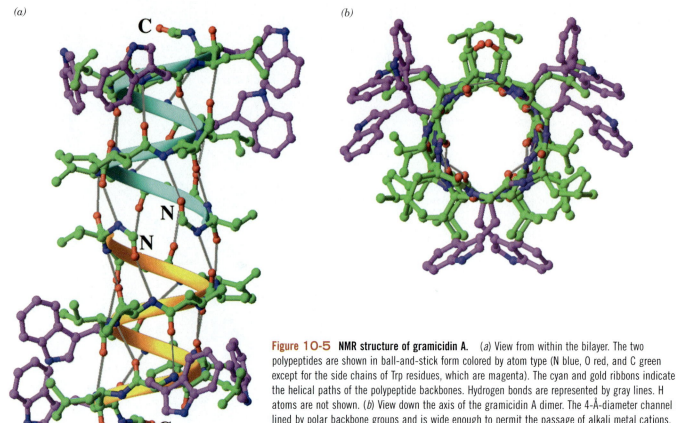

Figure 10-5 **NMR structure of gramicidin A.**   (*a*) View from within the bilayer. The two polypeptides are shown in ball-and-stick form colored by atom type (N blue, O red, and C green except for the side chains of Trp residues, which are magenta). The cyan and gold ribbons indicate the helical paths of the polypeptide backbones. Hydrogen bonds are represented by gray lines. H atoms are not shown. (*b*) View down the axis of the gramicidin A dimer. The 4-Å-diameter channel is lined by polar backbone groups and is wide enough to permit the passage of alkali metal cations. [Based on an NMR structure by Timothy Cross, Florida State University. PDBid 1MAG.]

X-ray crystallographic evidence indicates that gramicidin A dimerizes in a head-to-head fashion to form a transmembrane channel (Fig. 10-5). The alternating L and D residues of gramicidin A allow it to form a helix with a nonpolar exterior and a polar central channel that facilitates the passage of Na^+ and K^+ ions. Recall that a standard α helix contains only L amino acids and does not have a hollow core (Section 6-1B).

B Porins

The porins, introduced in Section 9-3A, are β barrel structures with a central aqueous channel. In the *E. coli* OmpF porin (Fig. 9-24), the channel is constricted to form an elliptical pore with a minimum cross section of 7×11 Å. Consequently, solutes of more than ~600 D are too large to pass through the channel. OmpF is weakly cation selective; other porins are more selective for anions. In general, the size of the channel and the residues that form its walls determine what types of substances can pass through.

Solute selectivity is elegantly illustrated by **maltoporin.** This bacterial outer membrane protein facilitates the diffusion of **maltodextrins,** which are the α(1→4)-linked glucose oligosaccharide degradation products of starch (for example, the glucose disaccharide maltose). The X-ray structure of *E. coli* maltoporin reveals that the protein is structurally similar to OmpF porin but is a homotrimer of 18-stranded rather than 16-stranded antiparallel β barrels. Three long loops from the extracellular face of each maltoporin subunit fold inward into the barrel, thereby constricting the channel near the center of the membrane to a diameter of ~5 Å and giving the channel an hourglass-like cross section. The channel is lined on one side

with a series of six contiguous aromatic side chains arranged in a left-handed helical path that matches the left-hand helical curvature of α-amylose (Fig. 8-10). This so-called greasy slide extends from one end of the channel, through its constriction, to the other end (Fig. 10-6).

How does the greasy slide work? The hydrophobic faces of the maltodextrin glucose residues stack on aromatic side chains, as is often observed in complexes of sugars with proteins. The glucose hydroxyl groups, which are arranged in two strips along opposite edges of the maltodextrins, form numerous hydrogen bonds with polar and charged side chains that line the channel. Tyr 118, which protrudes into the channel opposite the greasy slide, apparently functions as a steric barrier that only permits the passage of near-planar groups such as glucosyl residues. Thus, the hook-shaped sucrose (a glucose–fructose disaccharide) passes only very slowly through the maltoporin channel.

At the start of the translocation process, the entering glucosyl residue interacts with the readily accessible end of the greasy slide in the extracellular vestibule of the channel. Further translocation along the helical channel requires the maltodextrin to follow a screwlike path that maintains the helical structure of the oligosaccharide, much like the movement of a bolt through a nut, thereby excluding molecules of comparable size that have different shapes. *The translocation process is unlikely to encounter any large energy barrier due to the smooth surface of the greasy slide and the multiple polar groups at the channel constriction that would permit the essentially continuous exchange of hydrogen bonds as a maltodextrin moves through the constriction.*

C   Ion Channels

All cells contain ion-specific channels that allow the rapid passage of ions such as Na^+, K^+, and Cl^-. The movement of these ions through such channels, along with their movement through active transporters (discussed in Section 10-3), is essential for maintaining osmotic balance, for signal transduction (Section 21-3), and for effecting changes in membrane potential that are responsible for neurotransmission. Mammalian cells, for example, maintain a nonequilibrium distribution of ions on either side of the plasma membrane: ~150 mM Na^+ and ~4 mM K^+ in the extracellular fluid, and ~12 mM Na^+ and ~140 mM K^+ inside the cell.

Potassium ions passively diffuse from the cytoplasm to the extracellular space through transmembrane proteins known as **K^+ channels.** Although there is a large diversity of K^+ channels, even within a single organism, all of them have similar sequences, exhibit comparable permeability characteristics, and most importantly, are at least 10,000-fold more permeable to K^+ than Na^+. Since this high selectivity (around the same as that of valinomycin; Section 10-2A) implies energetically strong interactions between K^+ and the protein, how can the K^+ channel maintain its observed nearly diffusion-limited throughput rate of up to 10^8 ions per second (a 10^4-fold greater rate than that of valinomycin)?

The X-Ray Structure of KcsA Reveals the Basis of K^+ Channel Selectivity.   One of the best characterized ion channels is a K^+ channel from *Streptomyces lividans* named **KcsA.** This 158-residue integral membrane protein, like all known K^+ channels, functions as a homotetramer. As we shall see, the KcsA channel's three-dimensional structure explains both its functional features and serves as a model for more complicated eukaryotic K^+ channels.

The X-ray structure of KcsA's N-terminal 125-residue segment, determined by Roderick MacKinnon, reveals that each of its subunits forms

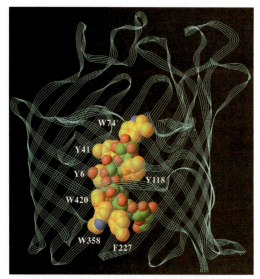

Figure 10-6   Structure of a maltoporin subunit in complex with a maltodextrin of six glycosyl units.   The polypeptide backbone of this *E. coli* protein is represented by a cyan ribbon. Five glucose residues of maltodextrin and the aromatic side chains lining the transport channel are shown in space-filling form with N blue, O red, protein C gold, and glucosyl C green. The "greasy slide" consists of the aromatic side chains of six residues. Tyr 118, which projects into the channel, helps restrict passage to glucosyl residues. [Based on an X-ray structure by Tilman Schirmer, University of Basel, Basel, Switzerland. PDBid 1MPO.]

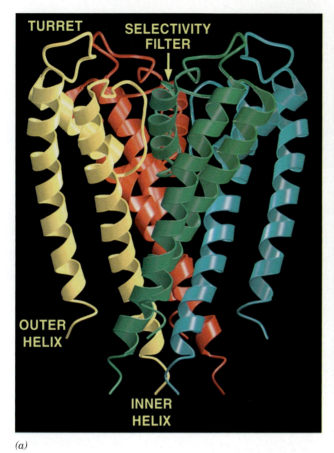

(a)

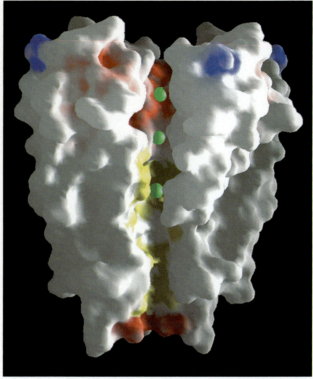

(b)

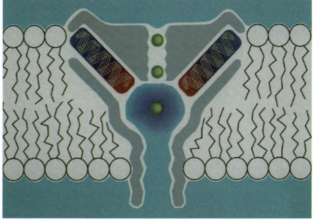

(c)

Figure 10-7 **X-Ray structure of the KscA K$^+$ channel.** (a) Ribbon diagram of the tetramer as viewed from within the plane of the membrane with the cytoplasm below and the extracellular region above. The protein's fourfold axis of rotation is vertical and each of its identical subunits is differently colored. Each subunit has an inner helix that forms part of the central pore, an outer helix that contacts the membrane interior, and a turret that projects out into the extracellular space. The selectivity filter at the extracellular end of the protein allows the passage of K$^+$ ions but not Na$^+$ ions. (b) Cutaway diagram viewed similarly to Part a in which the K$^+$ channel is represented by its solvent-accessible surface. The surface is colored according to its physical properties with negatively charged areas red, uncharged areas white, positively charged areas blue, and hydrophobic areas of the central pore yellow. K$^+$ ions are represented by green spheres. (c) Schematic diagram indicating how the K$^+$ channel stabilizes a cation in the center of the membrane. The central pore's 10-Å-diameter aqueous cavity (which contains ~50 water molecules) stabilizes a K$^+$ ion (green sphere) in the otherwise hydrophobic membrane interior. In addition, the C-terminal ends of the pore helices (red) all point toward the K$^+$ ion, thereby electrostatically stabilizing it via their dipole moments (the alignment of their polar group dipoles makes α helices negative at their C-terminal ends). [Courtesy of Roderick MacKinnon, Rockefeller University. PDBid 1BL8.]

two nearly parallel transmembrane helices that are connected by an ~20-residue pore region (Fig. 10-7a). Four such subunits associate to form a fourfold symmetric assembly surrounding a central pore. The four inner (C-terminal) helices, which largely form the pore, pack against each other near the cytoplasmic side of the membrane much like the poles of an inverted teepee. The four outer helices, which face the lipid bilayer, buttress the inner helices. Several K$^+$ ions and ordered water molecules are seen to occupy the central pore (Figs. 10-7b and 10-8a).

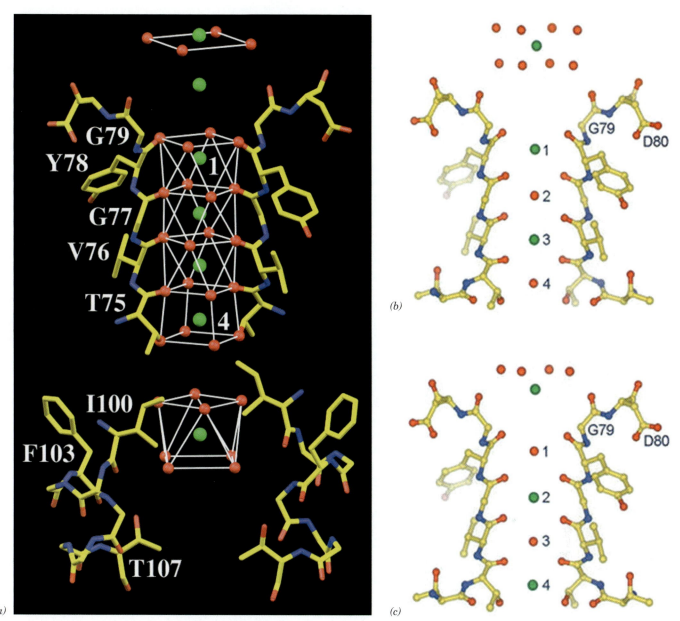

Figure 10-8 Portions of the KcsA K+ channel responsible for its ion selectivity viewed similarly to Fig. 10-7. (*a*) The X-ray structure of the residues forming the cavity (*bottom*) and selectivity filter (*top*) but with the front and back subunits omitted for clarity. Atoms are colored according to type with C yellow, N blue, O red, and K+ ions represented by green spheres. The water and protein O atoms that ligand the K+ ions, including those contributed by the front and back subunits, are represented by red spheres. The coordination polyhedra formed by these O atoms are outlined by thin white lines. (*b* & *c*) Two alternative K+ binding states of the selectivity filter, whose superposition is presumed to be responsible for the electron density observed in the X-ray structure of KcsA. Atoms are colored as in Part *a*. Note that K+ ions occupying the selectivity filter are interspersed with water molecules and that the K+ ion immediately above the selectivity filter in Part *b* is further above the protein than that in Part *c*. Hence these ions maintain a constant spacing while traversing the selectivity filter. [Part *a* based on an X-ray structure by, and Parts *b* and *c* courtesy of, Roderick MacKinnon, Rockefeller University. PDBid 1K4C.] **See the Interactive Exercises.**

The 45-Å-long central pore has variable width: It starts at its cytoplasmic side (Fig. 10-7*b*, *bottom*) as an ~6-Å-diameter and 18-Å-long tunnel, the so-called internal pore, whose entrance is lined with four anionic side chains (red area at the bottom of Fig. 10-7*b*) that presumably attract cations and repel anions. The internal pore then widens to form an ~10-Å-diameter cavity. These regions of the central pore are both wide enough so that a K+

ion could move through them in its hydrated state. However, the upper part of the pore, the so-called selectivity filter, narrows to 3 Å thereby forcing a transiting K^+ ion to shed its waters of hydration. The walls of the internal pore and the cavity are lined with hydrophobic groups that interact minimally with diffusing ions (yellow area of the pore in Fig. 10-7b). However, the selectivity filter (red area of the pore at the top of Fig. 10-7b) is lined with closely spaced main chain carbonyl oxygens of residues (Fig. 10-8a, top) that are highly conserved in all K^+ channels (their so-called signature sequence, TVGYG) and whose mutations disrupt the ability of the channel to discriminate between K^+ and Na^+ ions.

What is the function of the cavity? Energy calculations indicate that an ion moving through a narrow transmembrane pore must surmount an energy barrier that is maximal at the center of the membrane. The existence of the cavity reduces this electrostatic destabilization by surrounding the ion with polarizable water molecules (Fig. 10-7c). Remarkably, the K^+ ion occupying the cavity is liganded by 8 ordered water molecules located at the corners of a square antiprism (a cube with one face twisted by 45° with respect to the opposite face) in which the K^+ ion is centered (Fig. 10-8a, bottom). K^+ in aqueous solution is known to have such an inner hydration shell but it had never before been visualized. The K^+ ion is precisely centered in the cavity but its liganding water molecules are not in van der Waals contact with the walls of the cavity. Indeed, there is room in the cavity for ~40 additional water molecules although they are unseen in the X-ray structure because they are disordered. This disorder arises because the cavity is lined with hydrophobic groups (mainly the side chains of Ile 100 and Phe 103; Fig. 10-8a) that interact but weakly with water molecules, which are thus free to interact with the K^+ ion so as to form an outer hydration shell.

How does the K^+ channel discriminate so acutely between K^+ and Na^+ ions? The main chain O atoms lining the selectivity filter form a stack of rings (Fig. 10-8a; top) that provide a series of closely spaced sites of appropriate dimensions for coordinating dehydrated K^+ ions but not the smaller Na^+ ions. The structure of the protein surrounding the selectivity filter suggests that the diameter of the pore is rigidly maintained, thus making the energy of a dehydrated Na^+ in the selectivity filter considerably higher than that of hydrated Na^+ and thereby accounting for the K^+ channel's high selectivity for K^+ ions.

Since the selectivity filter appears designed to specifically bind K^+ ions, how does it support such a high throughput of these ions (up to 10^8 ions per second)? Figure 10-8a shows what appears to be four K^+ ions in the selectivity filter and two more just outside it on its extracellular (top) side. Such closely spaced positive ions would strongly repel one another and hence represent a high energy situation. However, a variety of evidence suggests that this structure is really a superposition of two sets of K^+ ions, one with K^+ ions at the topmost position in Fig. 10-8a and at positions 1 and 3 in the selectivity filter (Fig. 10-8b) and the second with K^+ ions at the second position from the top in Fig. 10-8a and at positions 2 and 4 in the selectivity filter (Fig. 10-8c; X-ray structures can show overlapping atoms because they are averages of many unit cells). Within the selectivity filter, the positions not occupied by K^+ ions are instead occupied by water molecules that coordinate the neighboring K^+ ions.

The electron density that is represented as the topmost four water molecules in Fig. 10-8a is highly elongated in the vertical direction in this otherwise high resolution (2.0 Å) structure. Hence it is thought to actually arise from eight water molecules that ligand the topmost K^+ ion in Fig. 10-8b to form an inner hydration shell similar to that of the K^+ in the central cav-

ity (Fig. 10-8*a, bottom*). Moreover, the four water molecules liganding the topmost K⁺ ion in Fig. 10-8*c* also contribute to this electron density. This latter ring of four waters provides half of the associated K⁺ ion's eight liganding O atoms. The others are contributed by the carbonyl O atoms of the four Gly 79 residues, which are properly oriented to do so. It therefore appears that a dehydrated K⁺ ion transits the selectivity filter (moves to successive positions in Fig. 10-8*b,c*) by exchanging the properly spaced ligands extending from its walls, and then exits into the extracellular solution by exchanging protein ligands for water molecules to again acquire a hydration shell. These ligands are spaced and oriented such that there is little free energy change (estimated to be <12 kJ·mol^{-1}) as a K⁺ ion transits the selectivity filter and enters the extracellular solution. The rapid dehydration of the K⁺ ion entering the selectivity channel from the cavity is, presumably, similarly managed. The essentially level free energy landscape throughout this process is, of course, conducive to the rapid transit of K⁺ ions through the ion channel and hence must be a product of evolutionary fine-tuning. Energy calculations indicate that mutual electrostatic repulsions between successive K⁺ ions, whose movements are concerted, balances the attractive interactions holding these ions in the selectivity filter and hence further facilitates their rapid transit.

Ion Channels Are Gated. The physiological functions of ion channels depend not only on their exquisite ion specificity and speed of transport but also on their ability to be selectively opened or closed. For example, the ion gradients across cell membranes, which are generated by specific energy-driven pumps (Section 10-3), are discharged through Na⁺ and K⁺ channels. However, the pumps could not keep up with the massive fluxes of ions passing through the open channels, so *ion channels are normally shut and only open transiently to perform some specific task for the cell.* The opening and closing of ion channels, a process known as **gating,** can occur in response to a variety of stimuli:

1. **Mechanosensitive channels** open in response to local deformations in the lipid bilayer. Consequently, they respond to direct physical stimuli such as touch, sound, and changes in osmotic pressure.

2. **Ligand-gated channels** open in response to an extracellular chemical stimulus such as a neurotransmitter.

3. **Signal-gated channels** open on intracellularly binding a Ca^{2+} ion or some other signaling molecule (Section 21-3).

4. **Voltage-gated channels** open in response to a change in membrane potential. Multicellular organisms contain numerous varieties of voltage-gated channels, including those responsible for generating nerve impulses.

Nerve Impulses Are Propagated by Action Potentials. As an example of the functions of voltage-gated channels, let us consider electrical signaling events in neurons (nerve cells). The stimulation of a neuron, a cell specialized for electrical signaling, causes Na⁺ channels to open so that Na⁺ ions spontaneously flow into the cell. The consequent local increase in membrane potential induces neighboring voltage-gated Na⁺ channels to open. The resulting local **depolarization** of the membrane induces nearby voltage-gated K⁺ channels to open. This allows K⁺ ions to spontaneously flow out of the cell in a process called **repolarization** (Fig. 10-9). However, well before the distribution of Na⁺ and K⁺ ions across the membrane equilibrates, the Na⁺ and K⁺ channels spontaneously close. Yet, because depolarization

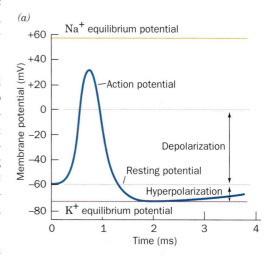

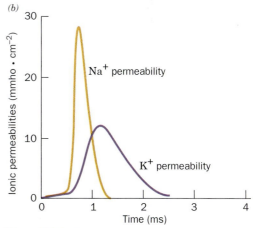

Figure 10-9   Time course of an action potential.
(*a*) The axon membrane undergoes rapid depolarization, followed by a nearly as rapid hyperpolarization and then a slow recovery to its resting potential. (*b*) The depolarization is caused by a transient increase in Na⁺ permeability (conductance), whereas the hyperpolarization results from a more prolonged increase in K⁺ permeability that begins a fraction of a millisecond later. [After Hodgkin, A.L. and Huxley, A.F., *J. Physiol.* **117,** 530 (1952).]

of a membrane segment induces the voltage-gated Na$^+$ channels in a neighboring membrane segment to open, which induces their neighboring voltage-gated Na$^+$ channels to open, etc., a wave of transient change in the membrane potential, called an **action potential,** travels along the length of the nerve cell (which may be over 1 m long). The action potential, which travels at ~10 m/s, propagates in one direction only because, after the ion channels have spontaneously closed, they resist reopening until the membrane potential has regained its resting value, which takes a few milliseconds (Fig. 10-9).

As an action potential is propagated along the length of a nerve cell, it is continuously renewed so that its signal strength remains constant (in contrast, an electrical impulse traveling down a wire dissipates as a consequence of resistive and capacitive effects). Nevertheless, the relative ion imbalance responsible for the resting membrane potential is small; only a tiny fraction of a nerve cell's Na$^+$–K$^+$ gradient (which is generated by ion pumps; Section 10-3A) is discharged by a single nerve impulse (only one K$^+$ ion per 3000–300,000 in the cytosol is exchanged for extracellular Na$^+$ as indicated by measurements with radioactive Na$^+$). A nerve cell can therefore transmit a nerve impulse every few milliseconds without letup. This capacity to fire rapidly is an essential feature of neuronal communications: Since action potentials all have the same amplitude, the magnitude of a stimulus is conveyed by the rate at which a nerve fires.

Voltage Gating in K$_V$ Channels Is Triggered by the Motion of Protein Paddles through the Membrane.

The subunits of all voltage-gated K$^+$ channels contain an ~220-residue N-terminal cytoplasmic domain, an ~250-residue transmembrane domain consisting of six helices, S1 to S6, and an ~150-residue C-terminal cytoplasmic domain (Fig. 10-10). S5 and S6, with their intervening so-called P loop, are homologous to the KcsA channel (Fig. 10-7). In the voltage-gated K$^+$ channels known as **K$_V$ channels,** a conserved ~100-residue loop called the T1 domain precedes the transmembrane domain. In the tetrameric ion channel, the four T1 domains presumably hang from the cytoplasmic face of the K$_V$ channel's transmembrane domain, forming the outer vestibule of the K$^+$ pore.

What is the nature of the gating machinery in voltage-gated ion channels? The ~19-residue S4 helix, which contains approximately five positively charged side chains spaced about every three residues on an otherwise hydrophobic polypeptide, appears to act as a voltage sensor. This was shown by covalently linking a dye, whose fluorescence spectrum varies with the polarity of the environment, to any of several residues in S4. Fluorescence measurements revealed that when the membrane potential increases (the inside becomes less negative), a stretch of at least seven residues at the N-terminal end of S4 moves from a position within the membrane to the extracellular environment. It seems likely that this movement triggers channel opening by displacing a gate that is formed, at least in part, by the cytoplasmic ends of each ion channel's four S6 helices.

Since charged residues are unstable in a lipid environment, it had been widely assumed that the S4 helix remains surrounded by protein throughout its movement. However, this view has been challenged by the X-ray structure, also determined by MacKinnon, of the voltage-dependent K$^+$ channel named **K$_V$AP** from the thermophilic archaebacterium *Aeropyrum pernix*. K$_V$AP is closely similar in sequence and electrophysiological properties to eukaryotic K$_V$ channels. Its S5–P-loop–S6 segment, not surpris-

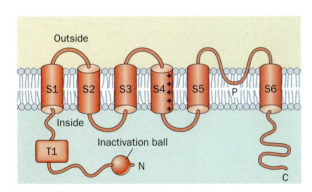

Figure 10-10 **Predicted secondary structure and membrane orientation of voltage-gated K$^+$ channels.** The conserved TVGYG signature sequence is located in the P loop.

Figure 10-11 X-Ray structure of the K$_v$AP channel.   Each of the four subunits is a different color. (*a*) View from the cytoplasmic side of the membrane. The helices of the blue subunit are labeled numerically for S1 to S6 and the pore helix is labeled P. (*b*) View from within the membrane with the cytoplasm below (from the top in Part *a*). Side chains (*green*) of selected residues that participate in voltage-dependent gating are shown for the blue and red subunits. [Courtesy of Roderick MacKinnon, Rockefeller University. PDBid 1ORQ.]

(*a*)

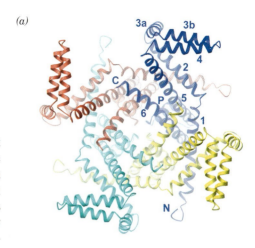

(*b*)

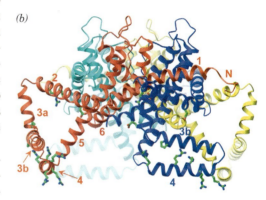

ingly, forms a tetrameric pore structure that is nearly identical to that of KcsA (Fig. 10-7; K$_v$AP's selectivity filter has the signature sequence TVGYG). However, contrary to expectation, helices S3b and S4 form a paddle-shaped assembly that is located on the periphery of the protein such that it extends into the lipid bilayer (Fig. 10-11; the predicted helix S3 is now seen to actually consist of two helices, termed S3a and S3b, joined by a short loop). The paddle has a flexible connection to the rest of the protein as is indicated by comparison to the X-ray structure of segments S1 to S4 alone. This suggests that the K$_v$ channel's four paddles move through the fluid portion of the bilayer toward the cell's exterior in response to an increase in membrane potential (Fig. 10-12) so as to induce a conformational change in the K$^+$ pore that causes it to open. This model is corroborated by experiments that show that the voltage sensor paddles are accessible to tight-binding agents (e.g., antibodies that specifically bind to the loop connecting helices S3b and S4) only from the plasma membrane's cytoplasmic surface when the K$^+$ channel is closed but are only accessible from the membrane's extracellular surface when the channel is open.

Despite the foregoing, a variety of evidence, such as cross-linking studies, indicate that the S4 helix remains in contact with the pore domain during its voltage-induced movement and thus suggests that the observed conformation of the S3b–S4 paddle is an artifact of K$_v$AP crystallization. It is just such discrepancies that spur further research leading to a better understanding of the phenomenon in question.

Ion Channels Have Two Gates.   Electrophysiological measurements indicate that K$_v$ channels spontaneously close a few milliseconds after opening and do not reopen until after the membrane has regained its resting membrane potential. Evidently, *the K$_v$ channel contains two voltage-sensitive gates, one to open the channel on an increase in membrane potential and one to close it a short time later.* This **inactivation** (closing) of the K$_v$ channel is abolished by proteolytically excising its N-terminal 20-residue segment, which

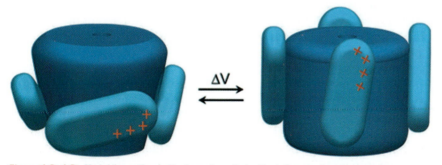

Figure 10-12 Model for gating in K$_v$ channels.   Under the influence of an increase in membrane potential (Na$^+$ flowing into the cell), gating charges (*red plus signs*) are carried through the membrane from inside (*bottom*) to outside (*top*) by the motion of voltage-sensor paddles though the lipid membrane. This motion, it is postulated, opens the pore. [Courtesy of Roderick MacKinnon, Rockefeller University.]

Figure 10-13 Composite model for the closure of the K_V channel. The view is from within the membrane with its extracellular surface above. The bottom portion of the drawing shows a C_α diagram of the X-ray structure of the T1 tetramer with each subunit in a different color (only three of the four subunits are shown for clarity). Only the inactivation peptide that is linked to the red T1 subunit is shown (*yellow*). The upper portion of the drawing represents the K_V channel's transmembrane domain with helices S5 and S6 shown as the X-ray structure of the homologous KcsA channel (Fig. 10-7) and helices S1 to S4 represented schematically. The C-terminal domains, which follow helices S6, have been omitted for clarity. The light green box highlights the selectivity filter formed by the four P loops. The orange box highlights the region occupied by the putative lateral windows through which both the cytoplasmic K^+ ions and the inactivation peptides gain access to the central pore. [Courtesy of Senyon Choe, The Salk Institute, La Jolla, California. PDBids for T1 and for the inactivation peptides: 1EQE and 1ZTN.]

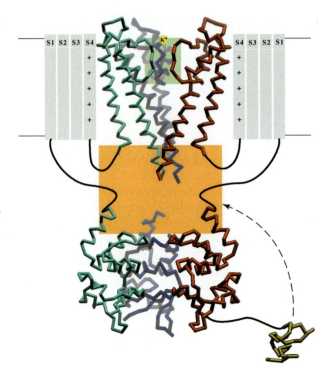

forms a ball-like structure. In the intact K_V channel, this "inactivation ball" is tethered to the end of a flexible 65-residue peptide segment, suggesting that channel inactivation normally occurs when the ball swings around to bind in the mouth of the open K^+ pore, thereby blocking the passage of K^+ ions. This ball-and-chain model for inactivation is shown schematically in Fig. 10-13.

The X-ray structure of isolated T1 domains reveals that they form a rotationally symmetric homotetrameric structure that is presumably coaxial with the tetrameric pore in the intact K_V channel (Fig. 10-13). The T1 domains in a tetramer do not separate far enough to admit the inactivation peptide, as was demonstrated by the observation that cross-linking these T1 domains together such that they cannot separate does not affect the K_V channel's gating properties. Evidently, the inactivation peptide finds its way into the pore through side windows between the T1 tetramer and the central pore (Fig. 10-13).

The cytoplasmic entrance to the K^+ channel pore is only 6 Å in diameter, too narrow to admit the ball. Therefore, it appears that the ball peptide must unfold in order to enter the pore. The first 10 residues of the unfolded ball peptide are predominantly hydrophobic and make contact with the hydrophobic residues lining the K_V channel pore. The next 10 residues, which are largely hydrophilic and contain several basic groups, bind to acidic groups near the pore entrance. Thus, the inactivation peptide acts more like a snake than a ball and chain. A K_V channel engineered so that only one subunit has an inactivation peptide still becomes inactivated but at one-fourth the rate of normal K_V channels. Apparently, any of the normal K_V channel's four inactivation peptides can block the channel and it is simply a matter of chance as to which one does so.

Other Voltage-Gated Cation Channels Contain a Central Pore. Voltage-gated Na^+ and Ca^{2+} channels appear to resemble K^+ channels, although rather than forming homotetramers, they are monomers of four consecutive do-

mains, each of which is homologous to the K^+ channel, separated by often large cytoplasmic loops. These domains presumably assume a pseudotetrameric arrangement about a central pore resembling that of voltage-gated K^+ channels. This structural homology suggests that voltage-gated ion channels share a common architecture in which differences in ion selectivity arise from precise stereochemical variations within the central pore. However, outside of their conserved transmembrane core, voltage-gated ion channels with different ion selectivities are highly divergent. For example, the T1 domain of K_V channels is absent in other types of voltage-gated ion channels.

Cl⁻ Channels Differ from Cation Channels. **Cl⁻ channels,** which occur in all cell types, permit the transmembrane movement of chloride ions along their concentration gradient. In mammals, the extracellular Cl⁻ concentration is ~120 mM and the intracellular concentration is ~4 mM.

ClC Cl⁻ channels form a large family of anion channels that occur widely in both prokaryotes and eukaryotes. The X-ray structures of ClC Cl⁻ channels from two species of bacteria, determined by MacKinnon, reveal, as biophysical measurements had previously suggested, that ClC Cl⁻ channels are homodimers with each subunit forming an anion-selective pore (Fig. 10-14). Each subunit consists mainly of 18 mostly transmembrane α helices that are remarkably tilted with respect to the membrane plane and have variable lengths compared to the transmembrane helices in other integral proteins of known structures (Section 9-3A).

The specificity of the Cl⁻ channel results from an electrostatic field established by basic amino acids on the protein surface, which helps funnel anions toward the pore, and by a selectivity filter formed by the N-terminal ends of several α helices. Because the polar groups of an α helix are all aligned (see Fig. 6-7), it forms a strong electrical dipole with its N-terminal

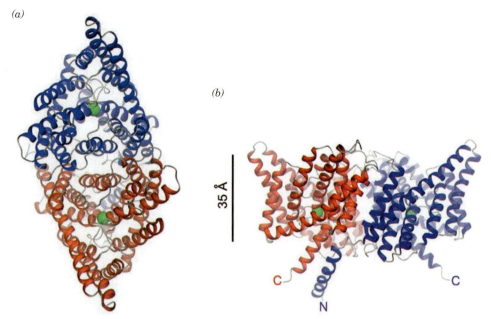

(a)

(b)

35 Å

Figure 10-14   X-Ray structure of the ClC Cl⁻ channel from *Salmonella typhimurium.* Each subunit of the homodimer contains 18 α helices of variable lengths. (*a*) View from the extracellular side of the membrane. The two subunits are colored blue and red. The green spheres represent Cl⁻ ions in the selectivity filter. (*b*) View from within the membrane with the extracellular surface above. The scale bar indicates the thickness of the membrane. [Courtesy of Roderick MacKinnon, Rockefeller University. PDBid 1KPL.]

end positively charged. This feature of the selectivity filter helps attract Cl^- ions, which are specifically coordinated by main chain amide nitrogens and side chain hydroxyls from Ser and Tyr residues (recall that a K^+ ion in the central cavity of a KscA channel interacts with the C-terminal ends of α helices; Fig. 10-7c). A positively charged residue such as Lys or Arg, if it were present in the selectivity filter, would probably bind a Cl^- ion too tightly to facilitate its rapid transit through the channel.

Unlike the K^+ channel, which has a central aqueous cavity (Fig. 10-7c), the Cl^- channel is hourglass-shaped, with its narrowest part in the center of the membrane and flanked by wider aqueous "vestibules." A conserved Glu side chain projects into the pore. This group would repel other anions, suggesting that rapid Cl^- flux requires a protein conformational change in which the Glu side chain moves aside. Another anion could push the Glu away, which explains why some Cl^- channels appear to be activated by Cl^- ions; that is, they open in response to a certain concentration of Cl^- in the extracellular fluid.

D Aquaporins

The observed rapid passage of water molecules across biological membranes had long been assumed to occur via simple diffusion that was made possible by the small size of water molecules and their high concentrations in biological systems. However, certain cells, such as those in the kidney, can sustain particularly rapid rates of water transport, which can be reversibly inhibited by mercuric ions. This suggested the existence of previously unrecognized protein pores that conduct water through biological membranes. The first of these elusive proteins was discovered in 1992 by Peter Agre, who named them **aquaporins.**

Aquaporins are widely distributed in nature; plants may have as many as 50 different aquaporins. The seven mammalian aquaporins that have been identified are expressed at high levels in tissues that rapidly transport water, including kidneys, salivary glands, and lacrimal glands (which produce tears). Aquaporins permit the passage of water molecules at an extremely high rate ($\sim 3 \times 10^9$ per second) but do not permit the transport of solutes (e.g., glycerol or urea) or ions, including, most surprisingly, protons (really hydronium ions; H_3O^+), whose free passage would discharge the cell's membrane potential.

The most extensively characterized member of the aquaporin family, **AQP1,** is a homotetrameric glycoprotein. Its X-ray and electron crystallographic structures reveal that each of its subunits consists mainly of six transmembrane α helices plus two shorter helices that lie within the bilayer (Fig. 10-15). These helices are arranged so as to form an elongated hourglass-shaped central pore that, at its narrowest point, the so-called constriction region, is ~ 2.8 Å wide, which is the van der Waals diameter of a water molecule (Fig. 10-16). Much of the pore is lined with hydrophobic groups whose lack of strong interactions with water molecules facilitates the rapid passage of water through the pore. However, for a water mole-

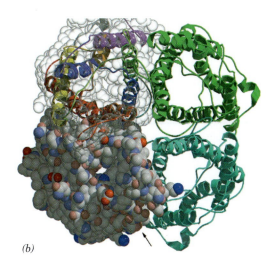

(a)

(b)

Figure 10-15 X-Ray structure of the aquaporin AQP1 from bovine erythrocytes.
(a) Superimposed ribbon and space-filling models of an aquaporin subunit as viewed from within the membrane with its extracellular surface above. The eight helical segments are drawn with different colors. (b) View of the aquaporin tetramer from the extracellular surface. Each subunit forms a water-transport channel, which is visible in the space-filling subunit model at the lower left. [Courtesy of Bing Jap, University of California at Berkeley. PDBid 1J4N.]

cule to transit the constriction region, it must shed its associated waters of hydration. This is facilitated by the side chains of highly conserved Arg and His residues as well as several backbone carbonyl groups, all of which line the constriction region. These groups are oriented so as to form hydrogen bonds to a transiting water molecule and hence readily displace its associated water molecules, much as occurs with K^+ in the selectivity filter of the KcsA channel (Section 10-2C).

If water were to pass through aquaporin as an uninterrupted chain of hydrogen-bonded molecules, this would facilitate the even more rapid passage of protons through the channel via proton jumping (Section 2-2A; in order for more than one such series of proton jumps to occur, each water molecule in the chain must reorient such that one of its protons forms a hydrogen bond to the next water molecule in the chain). However, aquaporin interrupts this process by forming hydrogen bonds from the side chain NH_2 groups of two highly conserved Asn residues to a water molecule that is centrally located in the pore (Fig. 10-16). Consequently, although this central water molecule can readily donate hydrogen bonds to its neighboring water molecules in the hydrogen bonded chain, it cannot accept one from them nor reorient, thereby severing the "proton-conducting wire" (Fig. 2-15).

E   Transport Proteins

Up to this point, we have examined the structures and functions of membrane proteins that form a physical passageway for small molecules, ions, or water. Membrane proteins known as **connexins** also form such channels, in the form of **gap junctions** between cells (see Box 10-1). However, not all membrane transport proteins offer a discrete bilayer-spanning pore. Instead, some proteins undergo conformational changes to move substances from one side of the membrane to the other. The **erythrocyte glucose transporter** (also known as **GLUT1**) is such a protein.

Biochemical evidence indicates that GLUT1 has glucose binding sites on both sides of the membrane. John Barnett showed that adding a propyl group to glucose C1 prevents glucose binding to the outer surface of the membrane, whereas adding a propyl group to C6 prevents binding to the inner surface. He therefore proposed that this transmembrane protein has two alternate conformations: one with the glucose site facing the external cell surface, requiring O1 contact and leaving O6 free, and the other with the glucose site facing the internal cell surface, requiring O6 contact and leaving O1 free. Transport apparently occurs as follows (Fig. 10-17):

1. Glucose binds to the protein on one face of the membrane.
2. A conformational change closes the first binding site and exposes the binding site on the other side of the membrane (transport).
3. Glucose dissociates from the protein.
4. The transport cycle is completed by the reversion of GLUT1 to its initial conformation in the absence of bound glucose (recovery).

This transport cycle can occur in either direction, according to the relative concentrations of intracellular and extracellular glucose. GLUT1 provides

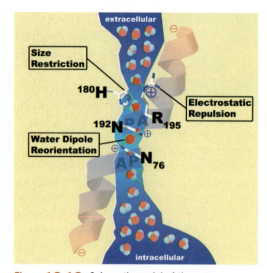

Figure 10-16 Schematic model of the water-conducting pore of aquaporin AQP1 viewed from within the membrane with the extracellular surface above. The positions of residues critical for preventing the passage of protons, other ions, and small molecule solutes are indicated. [Courtesy of Peter Agre, The Johns Hopkins School of Medicine.]

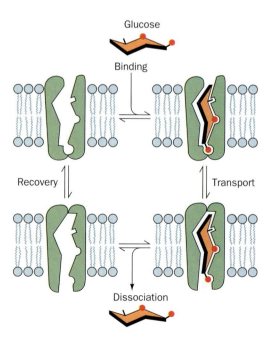

Figure 10-17 *Key to Function.* **Model for glucose transport.** The transport protein alternates between two mutually exclusive conformations. Glucose binds on one side of the membrane and is released on the other side after the protein conformation changes. The glucose molecule (*red*) is not drawn to scale. [After Baldwin, S.A. and Lienhard, G.E., *Trends Biochem. Sci.* **6**, 210 (1981).] *See the Animated Figures.*

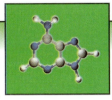

BOX 10-1

Perspectives in Biochemistry

Gap Junctions

Most eukaryotic cells are in metabolic as well as physical contact with neighboring cells. This contact is brought about by tubular particles, named **gap junctions,** that join discrete regions of neighboring plasma membranes much like hollow rivets. The gap junction consists of two apposed plasma membrane–embedded complexes. Small molecules and ions, but not macromolecules, can pass between cells via the gap junction's central channel.

Cytoplasm — Cytoplasm

Ions, amino acids, sugars, nucleotides

Proteins, nucleic acids

Intercellular space

These intercellular channels are so widespread that many whole organs are continuous from within. Thus, *gap junctions are important intercellular communication channels.* For example, the synchronized contraction of heart muscle is brought about by flows of ions through gap junctions, and gap junctions serve as conduits for some of the substances that mediate embryonic development.

Mammalian gap junction channels are 16 to 20 Å in diameter, which Werner Loewenstein established by microinjecting single cells with fluorescent molecules of various sizes and observing with a fluorescence microscope whether the fluorescent probe passed into neighboring cells. The mole-

cules and ions that can pass freely between neighboring cells are limited in molecular mass to a maximum of ~1000 D; macromolecules such as proteins and nucleic acids cannot leave a cell via this route.

The diameter of a gap junction channel varies with Ca^{2+} concentration: The channels are fully open when the Ca^{2+} level is $<10^{-7}$ M and become narrower as the Ca^{2+} concentration increases until, above 5×10^{-5} M, they close. This shutter system is thought to protect communities of interconnected cells from the otherwise catastrophic damage that would result from the death of even one of their members. Cells generally maintain very low cytosolic Ca^{2+} concentrations ($<10^{-7}$ M) by actively pumping Ca^{2+} out of the cell as well as into their mitochondria and endoplasmic reticulum (Section 10-3B). Ca^{2+} floods back into leaky or metabolically depressed cells, thereby inducing closure of their gap junctions and sealing them off from their neighbors.

Gap junctions are constructed from a single sort of protein subunit known as a **connexin.** A single gap junction consists of two hexagonal rings of connexins, called **connexons,** one from each of the adjoining plasma membranes. A given animal expresses numerous genetically distinct connexins, with molecular masses ranging from 25 to 50 kD. At least some connexons may be formed from two or more species of connexins, and the gap junctions joining two cells may consist of two different types of connexons. These various types of gap junctions presumably differ in their selectivities for the substances they transmit.

The structure of a cardiac gap junction, determined by electron crystallography, reveals a symmetrical assembly that has a diameter of ~70 Å, a length of ~150 Å, and encloses a central channel whose diameter varies from ~40 Å at its mouth to ~15 Å in its interior. The transmembrane portions of the gap junction each contain 24 rods of electron density

a means of equilibrating the glucose concentration across the erythrocyte membrane without any accompanying leakage of small molecules or ions (as might occur through an always-open channel such as a porin).

All known transport proteins appear to be asymmetrically situated transmembrane proteins that alternate between two conformational states in which the ligand-binding sites are exposed, in turn, to opposite sides of the membrane. Such a mechanism is analogous to the T → R allosteric transition of proteins such as hemoglobin (Section 7-3). In fact, many of the features of

that are arranged with hexagonal symmetry and which extend normal to the membrane plane.

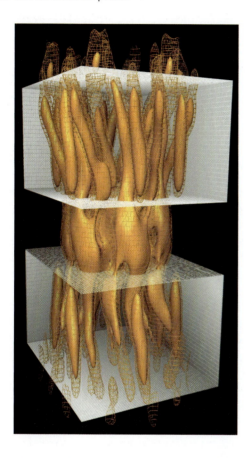

Here, the electron density at two different levels is represented by the solid and mesh contours (*gold*), whereas the white boxes indicate the positions of the cell membranes.

[Electron crystal structure courtesy of Mark Yeager, The Scripps Research Institute, La Jolla, California.]

ligand-binding proteins such as myoglobin and hemoglobin also apply to transport proteins (see Box 10-2).

Sequence analysis indicates that GLUT1 has 12 membrane-spanning α helices. This protein belongs to a large family of transporters whose structures have not been as well characterized as those of channel-type proteins. The 6.5-Å resolution electron crystal structure of a bacterial **oxalate transporter** named **OxlT** reveals that its 12 transmembrane helices are arranged around a central cavity, which presumably represents a substrate-binding

BOX 10-2

Perspectives in Biochemistry

Differentiating Mediated and Nonmediated Transport

Glucose and many other compounds can enter cells by a nonmediated pathway; that is, they slowly diffuse into cells at a rate proportional to their membrane solubility and their concentrations on either side of the membrane. This is a linear process: The **flux** (rate of transport per unit area) of a substance across the membrane increases with the magnitude of its concentration gradient (the difference between its internal and external concentrations). If the same substance, say glucose, moves across a membrane by means of a transport protein, its flux is no longer linear. This is one of four characteristics that distinguish mediated from nonmediated transport:

1. **Speed and specificity.** The solubilities of the chemically similar sugars D-glucose and D-mannitol in a synthetic lipid bilayer are similar. However, the rate at which glucose moves through the erythrocyte membrane is four orders of magnitude faster than that of D-mannitol. The erythrocyte membrane must therefore contain a system that transports glucose and that can distinguish D-glucose from D-mannitol.

2. **Saturation.** The rate of glucose transport into an erythrocyte does not increase infinitely as the external glucose concentration increases: The rate gradually approaches a maximum. Such an observation is evidence that a specific number of sites on the membrane are involved in the transport of glucose. At high [glucose], the transporters become saturated, much like myoglobin becomes saturated with O_2 at high pO_2 (Fig. 7-4). As expected, the following plot of glucose flux versus [glucose] is hyperbolic.

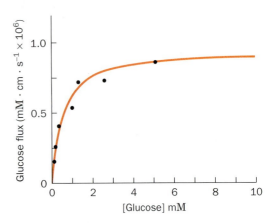

The nonmediated glucose flux increases linearly with [glucose] but would not visibly depart from the baseline on the scale of the above graph.

3. **Competition.** The above curve is shifted to the right in the presence of a substance that competes with glucose for binding to the transporter; for example, 6-*O*-benzyl-D-galactose has this effect. Competition is not a feature of nonmediated transport, since no transport protein is involved.

4. **Inactivation.** Reagents that chemically modify proteins and hence may affect their functions may eliminate the rapid, saturatable flux of glucose into the erythrocyte. The susceptibility of the erythrocyte glucose transport system to protein-modifying reagents is additional proof that it is a protein.

[Graph based on data from Stein, W.D., *Movement of Molecules across Membranes*, p. 134, Academic Press (1967).]

site (Fig. 10-18). The protein exhibits a high degree of symmetry between its cytoplasmic and external halves, which is consistent with its ability to transport substances both into and out of the cell. The structure shown in Fig. 10-18 probably corresponds to a conformational intermediate that is not fully accessible to either the cytoplasmic or extracellular face of the membrane.

Figure 10-18   Electron crystal structure of oxalate transporter OxlT from *Oxalobacter formigenes.* Twelve α helices have been fitted to the electron density map, which has a resolution of 6.5 Å. The green helices are nearly perpendicular to the plane of the membrane; yellow helices include a bend; and magenta helices are both bent and curved to match the electron density. Segments linking the helices are not resolved. [Courtesy of Sriram Subramaniam, NIH, Bethesda, Maryland.]

Some transporters can transport more than one substance. For example, the bacterial oxalate transporter transports **oxalate** into the cell and transports **formate** out.

$$^-OOC{-}COO^- \qquad H{-}COO^-$$
Oxalate   **Formate**

Some transport proteins move more than one substance at a time. Hence, it is useful to categorize mediated transport according to the stoichiometry of the transport process (Fig. 10-19):

1. A **uniport** involves the movement of a single molecule at a time. GLUT1 is a uniport system.
2. A **symport** simultaneously transports two different molecules in the same direction.
3. An **antiport** simultaneously transports two different molecules in opposite directions.

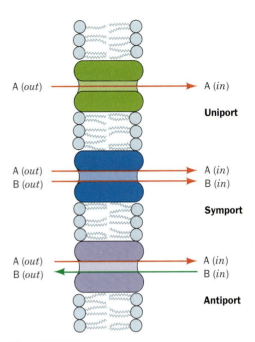

Figure 10-19 Uniport, symport, and antiport translocation systems.

3 Active Transport

Passive-mediated transporters, including porins, ion channels, and proteins such as GLUT1, facilitate the transmembrane movement of substances according to the relative concentrations of the substance on either side of the membrane. For example, the glucose concentration in the blood plasma (~5 mM) is generally higher than in cells, so GLUT1 allows glucose to enter the erythrocyte to be metabolized. Many substances, however, are available on one side of a membrane in lower concentrations than are required on the other side of the membrane. Such substances must be actively and selectively transported across the membrane against their concentration gradients.

Active transport is an endergonic process that, in most cases, is coupled to the hydrolysis of ATP. The elucidation of the mechanism by which the chemical energy released from ATP is used to drive a mechanical process has been a challenging biochemical problem. In this section, we examine membrane-bound ATPases that translocate cations; these proteins carry out **primary active transport.** In **secondary active transport,** the free energy of the electrochemical gradient generated by another mechanism, such as an ion-pumping ATPase, is used to transport a neutral molecule against its concentration gradient.

A $(Na^+{-}K^+){-}ATPase$

One of the most thoroughly studied active transport systems is the **(Na^+–K^+)–ATPase** in the plasma membranes of higher eukaryotes, which was first characterized by Jens Skou. This transmembrane protein consists of two types of subunits: a 110-kD nonglycosylated α subunit that contains the enzyme's catalytic activity and ion-binding sites, and a 55-kD glycoprotein β subunit of unknown function. Sequence analysis suggests that the α subunit has eight transmembrane α-helical segments and two large cytoplasmic domains. The β subunit has a single transmembrane helix and a large extracellular domain. The protein may function as an $(\alpha\beta)_2$ tetramer *in vivo* (Fig. 10-20).

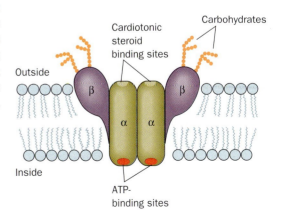

Figure 10-20 (Na^+–K^+)–ATPase. This diagram shows the transporter's putative dimeric structure and its orientation in the plasma membrane. Cardiotonic steroids (Box 10-3) bind to the external surface of the transporter, thereby inhibiting transport.

The $(Na^+–K^+)$–ATPase is often called the **$(Na^+–K^+)$ pump** because it pumps Na^+ out of and K^+ into the cell with the concomitant hydrolysis of intracellular ATP. The overall stoichiometry of the reaction is

$$3\,Na^+(in) + 2\,K^+(out) + ATP + H_2O \rightleftharpoons$$
$$3\,Na^+(out) + 2\,K^+(in) + ADP + P_i$$

The $(Na^+–K^+)$–ATPase is an antiport that generates a charge separation across the membrane, since three positive charges exit the cell for every two that enter. This extrusion of Na^+ enables animal cells to control their water content osmotically; *without functioning $(Na^+–K^+)$–ATPases to maintain a low internal $[Na^+]$, water would osmotically rush in to such an extent that animal cells, which lack cell walls, would swell and burst.* The electrochemical gradient generated by the $(Na^+–K^+)$–ATPase is also responsible for the electrical excitability of nerve cells (Section 10-2C). In fact, all cells expend a large fraction of the ATP they produce (up to 70% in nerve cells) to maintain their required cytosolic Na^+ and K^+ concentrations.

The key to the $(Na^+–K^+)$–ATPase is the phosphorylation of a specific Asp residue of the transport protein. ATP phosphorylates the transporter only in the presence of Na^+, whereas the resulting aspartyl phosphate residue

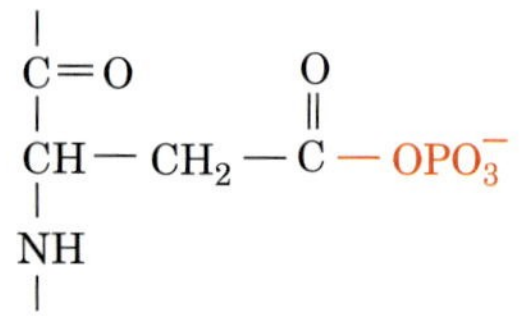

$$\underset{\substack{|\\ NH \\ |}}{\overset{\substack{|\\ C=O \\ |}}{CH}} - CH_2 - \overset{\overset{\displaystyle O}{\|}}{C} - OPO_3^-$$

Aspartyl phosphate residue

is subject to hydrolysis only in the presence of K^+. This suggests that the $(Na^+–K^+)$–ATPase has two conformational states (called E_1 and E_2) with different structures, different catalytic activities, and different ligand specificities. The protein appears to operate in the following manner (Fig. 10-21):

1. The transporter in the E_1 state binds three Na^+ ions inside the cell and then binds ATP to yield an $E_1 \cdot ATP \cdot 3\,Na^+$ complex.
2. ATP hydrolysis produces ADP and a "high-energy" aspartyl phosphate intermediate $E_1{\sim}P \cdot 3\,Na^+$ (here "~" indicates a "high-energy" bond).
3. This "high-energy" intermediate relaxes to its "low-energy" conformation, $E_2—P \cdot 3\,Na^+$, and releases its bound Na^+ outside the cell.
4. $E_2—P$ binds two K^+ ions from outside the cell to form an $E_2—P \cdot 2\,K^+$ complex.
5. The phosphate group is hydrolyzed, yielding $E_2 \cdot 2\,K^+$.
6. $E_2 \cdot 2\,K^+$ changes conformation, releases its two K^+ ions inside the cell, and replaces them with three Na^+ ions, thereby completing the transport cycle.

Although each of the above reaction steps is individually reversible, the cycle, as diagrammed in Fig. 10-21, circulates only in the clockwise direction under normal physiological conditions. This is because ATP hydrolysis and ion transport are coupled vectorial (unidirectional) processes. The vectorial nature of the reaction cycle results from the alternation of some of

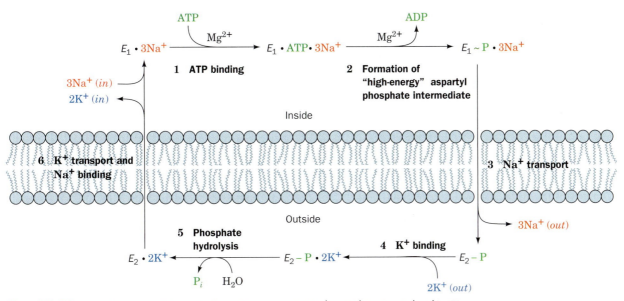

Figure 10-21 *Key to Function.*   Scheme for the active transport of Na$^+$ and K$^+$ by the (Na$^+$–K$^+$)–ATPase.

the steps of the exergonic ATP hydrolysis reaction (Steps 1 + 2 and Step 5) with some of the steps of the endergonic ion transport process (Steps 3 + 4 and Step 6). Thus, *neither reaction can go to completion unless the other one also does.* Study of the (Na$^+$–K$^+$)–ATPase has been greatly facilitated by the use of glycosides that inhibit the transporter (see Box 10-3).

B Ca^{2+}–ATPase

Transient increases in cytosolic [Ca^{2+}] trigger numerous cellular responses including muscle contraction (Section 28-2B), the release of neurotransmitters, and glycogen breakdown (Section 15-3). Moreover, Ca^{2+} is an important activator of oxidative metabolism (Section 17-4).

The [Ca^{2+}] in the cytosol (~0.1 μM) is four orders of magnitude less than it is in the extracellular spaces [~1500 μM; intracellular Ca^{2+} might otherwise combine with phosphate to form Ca$_3$(PO$_4$)$_2$, which has a maximum solubility of only 65 μM]. This large concentration gradient is maintained by the active transport of Ca^{2+} across the plasma membrane and the endoplasmic reticulum (the sarcoplasmic reticulum in muscle) by a **Ca^{2+}–ATPase (Ca^{2+} pump)** that actively pumps two Ca^{2+} ions out of the cytosol at the expense of ATP hydrolysis, while countertransporting two or three protons. The mechanism of the Ca^{2+}–ATPase (Fig. 10-22) resembles that of the (Na$^+$–K$^+$)–ATPase (Fig. 10-21).

The superimposed X-ray structures of the Ca^{2+}–ATPase from rabbit muscle sarcoplasmic reticulum in its E$_1$ and E$_2$ confor-

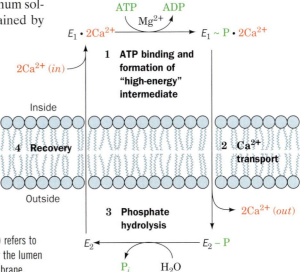

Figure 10-22 **Scheme for the active transport of Ca^{2+} by the Ca^{2+}–ATPase.** Here (*in*) refers to the cytosol and (*out*) refers to the outside of the cell for plasma membrane Ca^{2+}–ATPase or the lumen of the endoplasmic reticulum (or sarcoplasmic reticulum) for the Ca^{2+}–ATPase of that membrane.

BOX 10-3

Biochemistry in Health and Disease

The Action of Cardiac Glycosides

The cardiac glycosides are natural products that increase the intensity of heart muscle contraction. Indeed, **digitalis,** an extract of purple foxglove leaves, which contains a mixture of cardiac glycosides including **digitoxin (digitalin;** see figure below), has been used to treat congestive heart failure for centuries. The cardiac glycoside **ouabain** (pronounced wabane), a product of the East African ouabio tree, has been long used as an arrow poison.

10-21. The resultant increase in intracellular $[Na^+]$ stimulates the cardiac $(Na^+–Ca^{2+})$ antiport system, which pumps Na^+ out of and Ca^{2+} into the cell, ultimately boosting the $[Ca^{2+}]$ in the sarcoplasmic reticulum. Thus, the release of Ca^{2+} to trigger muscle contraction (Section 28-2B) produces a larger than normal increase in cytosolic $[Ca^{2+}]$, thereby intensifying the force of cardiac muscle contraction. Ouabain, which was once thought to be produced only by plants, has

Digitoxin (digitalin)

Ouabain

These two steroids, which are still among the most commonly prescribed cardiac drugs, inhibit the $(Na^+–K^+)$–ATPase by binding strongly to an externally exposed portion of the protein (Fig. 10-20) so as to block Step 5 in Fig.

recently been discovered to be an animal hormone that is secreted by the adrenal cortex and functions to regulate cellular $[Na^+]$ and overall body salt and water balance.

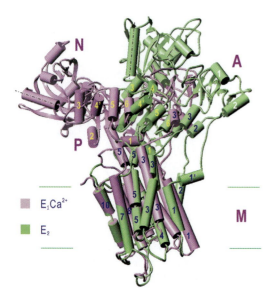

mations, determined by Chikashi Toyoshima, are shown in Fig. 10-23. Two Ca^{2+} ions bind within a bundle of 10 transmembrane helices. Three additional domains form a large structure on the cytoplasmic side of the membrane. The differences between the Ca^{2+}-bound (E_1) and the Ca^{2+}-free (E_2) structures indicate that the transporter undergoes extensive re-

Figure 10-23   X-Ray structures of the Ca^{2+}-free and Ca^{2+}-bound Ca^{2+}–ATPase.   The Ca^{2+}-free form, E_2, is green with black helix numbers, and the Ca^{2+}-bound form, E_1Ca^{2+}, is violet with yellow helix numbers. These proteins, which are superimposed on their transmembrane domains, are viewed from within the membrane with the cytosolic side up. Ten transmembrane helices form the M (for membrane) domain, ATP binds to the N (for nucleotide-binding) domain, the Asp residue that is phosphorylated during the reaction cycle is located on the P (for phosphorylation) domain, and the A (for actuator) domain is so named because it participates in the transmission of major conformational changes. A dashed line highlights the orientation of a helix in the N domain in the two conformations and the horizontal lines delineate the membrane. [Courtesy of Chikashi Toyoshima, University of Tokyo, Japan. PDBids 1EUL and 1WIO.]

arrangements, particularly in the positions of the cytoplasmic domains, but also in the Ca^{2+}-transporting membrane domain, during the reaction cycle. These changes apparently mediate communication between the Ca^{2+} binding sites and the $\sim$80-Å-distant site where bound ATP is hydrolyzed.

C ■ Ion Gradient–Driven Active Transport

Systems such as the $(Na^+–K^+)$–ATPase generate electrochemical gradients across membranes. The free energy stored in an electrochemical gradient (Eq. 10-3) can be harnessed to power various endergonic physiological processes. For example, cells of the intestinal epithelium take up dietary glucose by Na^+-dependent symport (Fig. 10-24). The immediate energy source for this "uphill" transport process is the Na^+ gradient. This process is an example of secondary active transport because *the Na^+ gradient in these cells is maintained by the $(Na^+–K^+)$–ATPase*. The Na^+–glucose transport system concentrates glucose inside the cell. Glucose is then transported into the capillaries through a passive-mediated glucose uniport (which resembles GLUT1; Fig. 10-17). Thus, since glucose enhances Na^+ resorption, which in turn enhances water resorption, glucose, in addition to salt and water, should be fed to individuals suffering from salt and water losses due to diarrhea.

Lactose Permease Requires a Proton Gradient.   Gram-negative bacteria such as *E. coli* contain several active transport systems for concentrating sugars. One extensively studied system, **lactose permease** (also known as **galactoside permease**), *utilizes the proton gradient across the bacterial cell membrane to cotransport H^+ and lactose*. The proton gradient is metabolically generated through oxidative metabolism in a manner similar to that in mitochondria (Section 17-2). The electrochemical potential gradient created by both these systems is used mainly to drive the synthesis of ATP.

Lactose permease is a 417-residue monomer that, like GLUT1 and the oxalate transporter (Section 10-2E), to which it is distantly related, consists largely of 12 transmembrane helices with its N- and C-termini in the cyto-

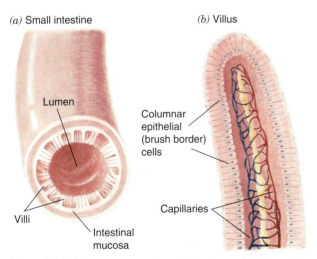

(a) Small intestine

Lumen

Villi

Intestinal mucosa

(b) Villus

Columnar epithelial (brush border) cells

Capillaries

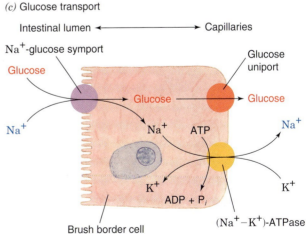

(c) Glucose transport

Intestinal lumen ⟷ Capillaries

Na^+-glucose symport

Glucose

Na^+

Glucose

Glucose uniport

Glucose

Na^+

ATP

ADP + P$_i$

K^+

K^+

$(Na^+–K^+)$-ATPase

Brush border cell

Figure 10-24   Glucose transport in the intestinal epithelium.   The brushlike villi lining the small intestine greatly increase its surface area (*a*), thereby facilitating the absorption of nutrients. The brush border cells from which the villi are formed (*b*) concentrate glucose from the intestinal lumen in symport with Na^+ (*c*), a process that is driven by the $(Na^+–K^+)$–ATPase, which is located on the capillary side of the cell and functions to maintain a low internal $[Na^+]$. The glucose is exported to the bloodstream via a separate passive-mediated uniport system similar to GLUT1.

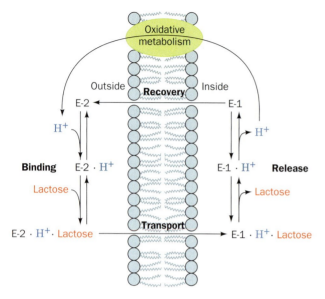

Figure 10-25 **Scheme for the cotransport of H⁺ and lactose by lactose permease in *E. coli*.** H⁺ binds first to E-2 outside the cell, followed by lactose. They are sequentially released from E-1 inside the cell. E-2 must bind to lactose and H⁺ in order to change conformation to E-1, thereby cotransporting these substances into the cell. E-1 changes conformation to E-2 when neither lactose nor H⁺ is bound, thus completing the transport cycle.

plasm. As does (Na^+-K^+)–ATPase, it has two major conformational states (Fig. 10-25):

1. E-1, which has a low-affinity lactose-binding site facing the interior of the cell.

2. E-2, which has a high-affinity lactose-binding site facing the exterior of the cell.

Ronald Kaback established that E-1 and E-2 can interconvert only when their H⁺ and lactose binding sites are either both filled or both empty. This prevents dissipation of the H⁺ gradient without cotransport of lactose into the cell. It also prevents transport of lactose out of the cell since this would require cotransport of H⁺ against its concentration gradient.

The X-ray structure of lactose permease in complex with a tight-binding lactose analog, determined by Kaback and So Iwata, reveals that this protein consists of two structurally similar and twofold symmetrically positioned domains containing six transmembrane helices each (Fig. 10-26*a*). A large internal hydrophilic cavity is open to the cytoplasmic side of the membrane (Fig. 10-26*b*) so that the structure represents the E-1 state of the protein. The lactose analog is bound in the cavity at a position that is approximately equidistant from both sides of the membrane, consistent with the

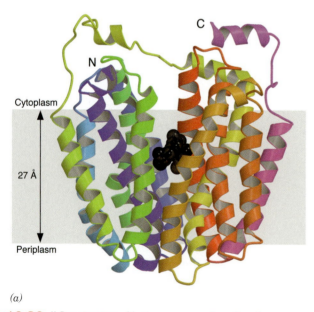

(*a*)

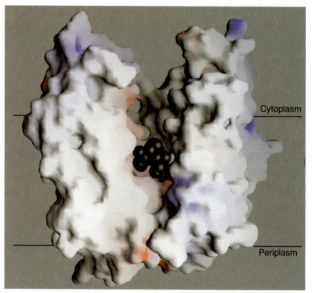

(*b*)

Figure 10-26 **X-Ray structure of lactose permease from *E. coli*.** (*a*) Ribbon diagram as viewed from the membrane with the cytoplasmic side up. The protein's 12 transmembrane helices are colored in rainbow order with the N-terminus purple and the C-terminus pink. The bound lactose analog is represented by black spheres. (*b*) Surface model viewed as in Part *a* but with the two helices closest to the viewer in Part *a* removed to reveal the lactose-binding cavity. The surface is colored according to its electrostatic potential with positively charged areas blue, negatively charged areas red, and neutral areas white. [Courtesy of H. Ronald Kaback, UCLA. PDBid 1PV7.]

model that the lactose binding site is alternately accessible from each side of the membrane (e.g., Fig. 10-17). Arg, His, and Glu residues that mutational studies have implicated in proton translocation are located in the vicinity of the lactose binding site.

SUMMARY

1. The mediated and nonmediated transport of a substance across a membrane is driven by its chemical potential difference.

2. Ionophores facilitate ion diffusion by binding an ion, diffusing through the membrane, and then releasing the ion; or by forming a channel.

3. Porins form β barrel structures about a central channel that is selective for anions, cations, or certain small molecules.

4. Ion channels mediate changes in membrane potential by allowing the rapid and spontaneous transport of ions. Ion channels are highly solute-selective and open and close (gate) in response to various stimuli. Nerve impulses involve ion channels.

5. Aquaporins contain channels that allow the rapid transmembrane diffusion of water but not protons.

6. Transport proteins such as GLUT1 alternate between two conformational states that expose the ligand-binding site to opposite sides of the membrane.

7. Active transport, in most cases, is driven by ATP hydrolysis. In the $(Na^+–K^+)$–ATPase and Ca^{2+}–ATPase, ATP hydrolysis and ion transport are coupled and vectorial.

8. In secondary active transport, an ion gradient maintained by an ATPase drives the transport of another substance. For example, the transport of lactose into a cell by lactose permease is driven by the cotransport of H^+.

REFERENCES

Busch, W. and Saier, M.H., Jr., The transporter classification (TC) system, 2002, *Crit. Rev. Biochem. Mol. Biol.* **37,** 287–337 (2002). [Summarizes the classification of the nearly 400 families of transport systems and their distribution among the three domains of life.]

Dutzler, R., Schirmer, T., Karplus, M., and Fischer, S., Translocation mechanism of long sugar chains across the maltoporin membrane channel, *Structure* **10,** 1273–1284 (2002).

Kovacs, F., Quine, J., and Cross, T.A., Validation of the single-stranded channel conformation of gramicidin A by solid-state NMR, *Proc. Natl. Acad. Sci.* **96,** 7910–7915 (1999).

Unger, V.M., Kumar, N.M., Gilula, N.B., and Yeager, M., Three-dimensional structure of a recombinant gap junction membrane channel, *Science* **283,** 1176–1180 (1999).

Walmsley, A.R., Barrett, M.P., Bringaud, F., and Gould, G.W., Sugar transporters from bacteria, parasites, and mammals: structure–activity relationships, *Trends Biochem. Sci.* **22,** 476–481 (1998). [Provides structure–function analysis of sugar transporters with 12 transmembrane helices.]

Ion Channels

Dutzler, R., Campbell, E.B., Cadene, M., Chait, B.T., and MacKinnon, R., X-ray structure of a ClC chloride channel at 3.0 Å reveals the molecular basis of anion selectivity, *Nature* **415,** 287–294 (2002).

Jiang, Y., Lee, A., Chen, J., Ruta, V., Cadene, M., Chait, B.T., and MacKinnon, R., X-ray structure of a voltage-dependent K^+ channel, *Nature* **423,** 33–41 (2003).

Yellin, G., The voltage-gated potassium channels and their relatives, *Nature* **419,** 35–42 (2002). [Reviews the mechanisms responsible for the ion selectivity, rapid transport, and gating in K^+ channel proteins.]

Zhou, Y., Morais-Cabral, J.H., Kaufman, A., and MacKinnon, R., Chemistry of ion coordination and hydration revealed by a K^+ channel–Fab complex at 2.0 Å resolution, *Nature* **414,** 43–48 (2001).

Aquaporins

Agre, P. and Kozono, D., Aquaporin water channels: Molecular mechanisms for human disease, *FEBS Lett.* **555,** 72–78 (2003).

Sui, H., Han, B.-G., Lee, J.K., and Jap, B.K., Structural basis of water-specific transport through the AQP1 water channel, *Nature* **414,** 872–878 (2001).

Active Transporters

Abramson, J., Smirnova, I., Kasho, V., Verner, G., Kaback, H.R., and Iwata, S., Structure and mechanism of the lactose permease of *Escherichia coli, Science* **301,** 610–615 (2003).

Hille, B., *Ionic Channels of Excitable Membranes* (3rd ed.), Sinauer Associates (2001).

Kaplan, J.H., Biochemistry of Na,K-ATPase, *Annu. Rev. Biochem.* **71,** 511–535 (2002).

Toyoshima, C. and Nomura, H., Structural changes in the calcium pump accompanying the dissociation of calcium, *Nature* **418,** 605–611 (2002); *and* Toyoshima, C., Nomura, H., and Sugita, Y., Structural basis of ion pumping by Ca^{2+}-ATPase of sarcoplasmic reticulum, *FEBS Lett.* **555,** 106–110 (2003).

KEY TERMS

transport protein
chemical potential
$\Delta\Psi$
electrochemical potential
nonmediated transport
mediated transport
passive-mediated transport

ionophore
carrier ionophore
channel-forming ionophore
gating
mechanosensitive channel
ligand-gated channel
signal-gated channel

voltage-gated channel
depolarization
repolarization
action potential
nerve impulse
uniport
symport

antiport
primary active transport
secondary active transport
gap junction
pump
permease

STUDY EXERCISES

1. Explain why the free energy change of membrane transport depends on both the concentration and charge of the transported substance.

2. Explain the differences between mediated and nonmediated transport across membranes.

3. What are the similarities and differences among ionophores, porins, ion channels, and passive-mediated transport proteins?

4. Explain how and why ion channels are gated.

5. Use the terminology of allosteric proteins to discuss the operation of proteins that carry out uniport, symport, and antiport transport processes.

6. Distinguish passive-mediated transport, active transport, and secondary active transport.

7. Explain why the (Na^+-K^+)–ATPase and the Ca^{2+}–ATPase carry out transport in one direction only.

PROBLEMS

1. Indicate whether the following compounds are likely to cross a membrane by nonmediated or mediated transport: (a) ethanol, (b) glycine, (c) cholesterol, (d) ATP.

2. Rank the rate of transmembrane diffusion of the following compounds:

$$CH_3-\overset{\overset{\displaystyle O}{\|}}{C}-NH_2 \qquad CH_3-CH_2-CH_2-\overset{\overset{\displaystyle O}{\|}}{C}-NH_2$$

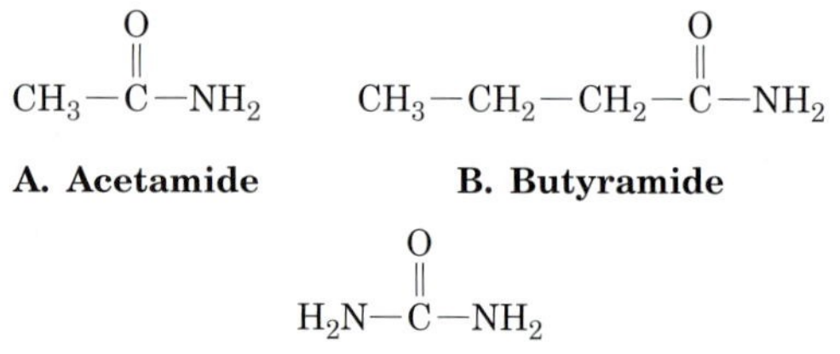

A. Acetamide　　　**B. Butyramide**

$$H_2N-\overset{\overset{\displaystyle O}{\|}}{C}-NH_2$$

C. Urea

3. Calculate the free energy change for glucose entry into cells when the extracellular concentration is 5 mM and the intracellular concentration is 3 mM.

4. (a) Calculate the chemical potential difference when intracellular $[Na^+]$ = 10 mM and extracellular $[Na^+]$ = 150 mM at 37°C. (b) What would the electrochemical potential be if the membrane potential were −60 mV (inside negative)?

5. For the problem in Sample Calculation 10-1, calculate ΔG at 37°C when the membrane potential is (a) −50 mV (cytosol negative) and (b) +150 mV. In which case is Ca^{2+} movement in the indicated direction thermodynamically favorable?

6. (a) What happens to K^+ transport by valinomycin when the membrane is cooled below its transition temperature? (b) The N-terminus of gramicidin A is formylated (Fig. 10-4). Could gramicidin A form a transmembrane channel if its N-terminus were not blocked in this fashion? Explain.

7. How long would it take 100 molecules of valinomycin to transport enough K^+ to change the concentration inside an erythrocyte of volume 100 μm^3 by 10 mM? (Assume that the valinomycin does not also transport any K^+ out of the cell, which it really does, and that the valinomycin molecules inside the cell are always saturated with K^+.)

8. The rate of movement (flux) of a substance X into cells was measured at different concentrations of X to construct the graph below.

   (a) Does this information suggest that the movement of X into the cells is mediated by a protein transporter? Explain.

   (b) What additional experiment could you perform to verify that a transport protein is or is not involved?

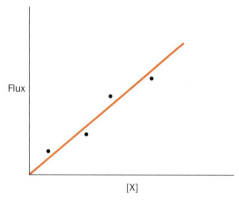

9. If the ATP supply in the cell shown in Fig. 10-24c suddenly vanished, would the intracellular glucose concentration increase, decrease, or remain the same?

10. Endothelial cells and pericytes in the retina of the eye have different mechanisms for glucose uptake. The figure shows the rate of glucose uptake for each type of cell in the presence of increasing amounts of sodium. What do these results reveal about the glucose transporter in each cell type? [Problem provided by Kathleen Cornely, Providence College.]

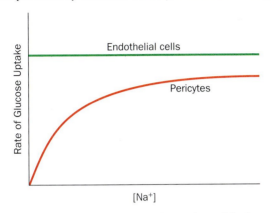

11. The compound shown below is the antiparasitic drug miltefosine.

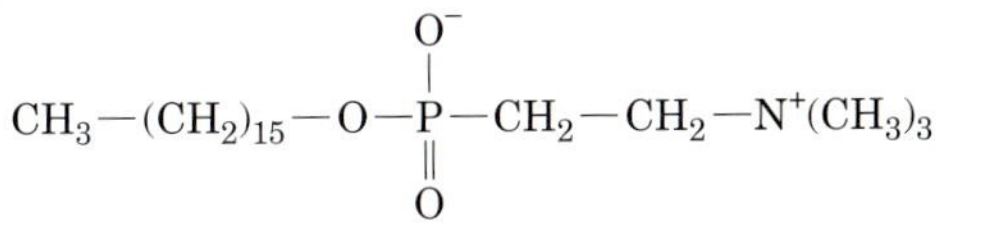

$$CH_3-(CH_2)_{15}-O-\overset{\displaystyle O^-}{\underset{\displaystyle O}{\overset{\displaystyle |}{\underset{\displaystyle ||}{P}}}}-CH_2-CH_2-N^+(CH_3)_3$$

Miltefosine

(a) Is this compound a glycerophospholipid?

(b) How does miltefosine likely cross the parasite cell membrane?

(c) In what part of the cell would the drug tend to accumulate? Explain.

(d) Miltefosine binds to a protein that also binds some sphingolipids and some glycerophospholipids. What feature common to all these compounds is recognized by the protein? The protein does not bind triacylglycerols.

12. In eukaryotes, ribosomes (approximate mass 4×10^6 D) are assembled inside the nucleus, which is enclosed by a double membrane. Protein synthesis occurs in the cytosol. (a) Could a protein similar to a porin or the glucose transporter be responsible for transporting ribosomes into the cytoplasm? Explain. (b) Would free energy be required to move a ribosome from the nucleus to the cytoplasm? Why or why not?

13. In addition to neurons, muscle cells undergo depolarization, although smaller and slower than in the neuron, as a result of the activity of the acetylcholine receptor.

(a) The acetylcholine receptor is also a gated ion channel. What triggers the gate to open?

(b) The acetylcholine receptor/ion channel is specific for Na^+ ions. Would Na^+ ions flow in or out? Why?

(c) How would the Na^+ flow through the ion channel change the membrane potential?

14. Cells in the wall of the mammalian stomach secrete HCl at a concentration of 0.15 M. The secreted protons, which are derived from the intracellular hydration of CO_2 by carbonic anhydrase, are pumped out by an (H^+-K^+)–ATPase antiport. A K^+-Cl^- cotransporter is also required to complete the overall transport process. (a) Calculate the pH of the secreted HCl. How does this compare to the cytosolic pH (7.4)? (b) Write the reaction catalyzed by carbonic anhydrase. (c) Draw a diagram to show how the action of both transport proteins results in the secretion of HCl.

15. Proteins known as multidrug resistance (MDR) transporters use the free energy of ATP hydrolysis to pump a variety of hydrophobic substances out of cells. (a) Write the reaction for ATP hydrolysis and explain, in structural terms, how the transporter might take advantage of this process to drive transport. There is no phosphorylated protein intermediate, as in the (Na^+-K^+)–ATPase. (b) Why would overexpression of an MDR transporter in a cancer cell make the cancer more difficult to treat?

The active site of an enzyme includes functional groups whose arrangement and reactivity allow them to interact with a substrate so as to facilitate its chemical transformation. Like these acrobats, enzymes typically act with great speed and precision, leaving little room for error. [David Madison/ Stone/Getty Images.]

Enzymatic Catalysis

Living systems are shaped by an enormous variety of biochemical reactions, nearly all of which are mediated by a series of remarkable biological catalysts known as enzymes. **Enzymology,** the study of enzymes, has its roots in the early days of biochemistry; both disciplines evolved together from nineteenth century investigations of fermentation and digestion. Initially, the inability to reproduce most biochemical reactions in the laboratory led Louis Pasteur and others to assume that living systems were endowed with a "vital force" that permitted them to evade the laws of nature governing inanimate matter. Some investigators, however, notably Justus von Liebig, argued that biological processes were caused by the action of chemical substances that were then known as "ferments." Indeed, the name "enzyme" (Greek: *en,* in + *zyme,* yeast) was coined in 1878 in an effort to emphasize that there is something *in* yeast, as opposed to the yeast itself, that catalyzes the reactions of fermentation. Eventually, Eduard Buchner showed that a cell-free yeast extract could in fact carry out the synthesis of ethanol from glucose (**alcoholic fermentation;** Section 14-3B):

$$C_6H_{12}O_6 \longrightarrow 2\ CH_3CH_2OH + 2\ CO_2$$

This chemical transformation actually proceeds in 12 enzyme-catalyzed steps.

The chemical composition of enzymes was not firmly established until 1926, when James Sumner crystallized jack bean **urease,** which catalyzes the hydrolysis of urea to NH_3 and CO_2, and demonstrated that these crystals consist of protein. Enzymological experience since then has amply demonstrated that most enzymes are proteins.

Certain species of RNA molecules known as **ribozymes** also have enzymatic activity. These include ribosomal RNA, which catalyzes the formation of peptide bonds between amino acids. In fact, laboratory experiments have produced ribozymes that can catalyze reactions similar to those required for replicating DNA, transcribing it to RNA, and attaching amino acids to transfer RNA. These findings are consistent with a precellular world in which RNA molecules enjoyed a more exalted position as the catalytic workhorses of biochemistry. The present-day RNA catalysts are presumably vestiges of this earlier "RNA world." Proteins have largely eclipsed RNA as cellular catalysts, probably because of the greater chemical versatility of proteins. Whereas nucleic acids are polymers of four types of chemically similar monomeric units, proteins have at their disposal 20 types of amino acids with a greater variety of functional groups.

This chapter is concerned with one of the central questions of biochemistry: How do enzymes work? We shall see that enzymes increase the rates of chemical reactions by lowering the free energy barrier that separates the reactants and products. Enzymes accomplish this feat through various mechanisms that depend on the arrangement of functional groups in the enzyme's **active site,** the region of the enzyme where catalysis occurs. In this chapter, we describe these mechanisms, along with examples that illustrate how enzymes combine several mechanisms to catalyze biological reactions. The following chapter includes a discussion of enzyme kinetics, the study of the rates at which such reactions occur.

1 General Properties of Enzymes

Biochemical research since Pasteur's era has shown that, although enzymes are subject to the same laws of nature that govern the behavior of other substances, enzymes differ from ordinary chemical catalysts in several important respects:

1. *Higher reaction rates.* The rates of enzymatically catalyzed reactions are typically 10^6 to 10^{12} times greater than those of the corresponding uncatalyzed reactions (Table 11-1) and are at least several orders of magnitude greater than those of the corresponding chemically catalyzed reactions.

2. *Milder reaction conditions.* Enzymatically catalyzed reactions occur under relatively mild conditions: temperatures below 100°C, atmospheric pressure, and nearly neutral pH. In contrast, efficient chemical catalysis often requires elevated temperatures and pressures as well as extremes of pH.

3. *Greater reaction specificity.* Enzymes have a vastly greater degree of specificity with respect to the identities of both their **substrates** (reactants) and their products than do chemical catalysts; that is, enzymatic reactions rarely have side products.

4. *Capacity for regulation.* The catalytic activities of many enzymes vary in response to the concentrations of substances other than their substrates. The mechanisms of these regulatory processes include allosteric control, covalent modification of enzymes, and variation of the amounts of enzymes synthesized.

A   Enzyme Nomenclature

Before delving further into the specific properties of enzymes, a word on nomenclature is in order. Enzymes are commonly named by appending the suffix *-ase* to the name of the enzyme's substrate or to a phrase describing the enzyme's catalytic action. Thus, urease catalyzes the hydrolysis of urea, and **alcohol dehydrogenase** catalyzes the oxidation of primary and secondary alcohols to their corresponding aldehydes and ketones by removing hydrogen. Since there were at first no systematic rules for naming enzymes, this practice occasionally resulted in two different names being used for the same enzyme or, conversely, in the same name being used for two different enzymes. Moreover, many enzymes, such as **catalase** (which mediates the dismutation of H_2O_2 to H_2O and O_2), were given names that provide no clue to their function. In an effort to eliminate this confusion and to provide rules for rationally naming the rapidly growing number of newly discovered enzymes, a scheme for the systematic functional classification and nomenclature of enzymes was adopted by the International Union of Biochemistry and Molecular Biology (IUBMB).

Enzymes are classified and named according to the nature of the chemical reactions they catalyze. There are six major classes of enzymatic reactions (Table 11-2), as well as subclasses and sub-subclasses. Each enzyme is assigned two names and a four-part classification number. Its **alternative name** is convenient for everyday use and is often an enzyme's previously

Table 11-1 Catalytic Power of Some Enzymes

Enzyme	Nonenzymatic Reaction Rate (s^{-1})	Enzymatic Reaction Rate (s^{-1})	Rate Enhancement
Carbonic anhydrase	1.3×10^{-1}	1×10^6	7.7×10^6
Chorismate mutase	2.6×10^{-5}	50	1.9×10^6
Triose phosphate isomerase	4.3×10^{-6}	4300	1.0×10^9
Carboxypeptidase A	3.0×10^{-9}	578	1.9×10^{11}
AMP nucleosidase	1.0×10^{-11}	60	6.0×10^{12}
Staphylococcal nuclease	1.7×10^{-13}	95	5.6×10^{14}

Source: Radzicka, A. and Wolfenden, R., *Science* **267,** 91 (1995).

Table 11-2 Enzyme Classification According to Reaction Type

Classification	Type of Reaction Catalyzed
1. Oxidoreductases	Oxidation–reduction reactions
2. Transferases	Transfer of functional groups
3. Hydrolases	Hydrolysis reactions
4. Lyases	Group elimination to form double bonds
5. Isomerases	Isomerization
6. Ligases	Bond formation coupled with ATP hydrolysis

used trivial name. Its **systematic name** is used when ambiguity must be minimized; it is the name of its substrate(s) followed by a word ending in *-ase* specifying the type of reaction the enzyme catalyzes according to its major group classification. For example, the enzyme whose recommended name is carboxypeptidase A (Table 11-1) has the systematic name peptidyl-L-amino acid hydrolase and the **classification number** EC 3.4.17.1 ("EC" stands for Enzyme Commission, and the numbers represent the class, subclass, sub-subclass, and its arbitrarily assigned serial number in its sub-subclass). For our purposes, the recommended name of an enzyme is adequate. Systematic names and EC classification numbers can be obtained via the Internet (http://expasy.org/enzyme/).

B Substrate Specificity

The noncovalent forces through which substrates and other molecules bind to enzymes are similar in character to the forces that dictate the conformations of the proteins themselves (Section 6-4A). Both involve van der Waals, electrostatic, hydrogen bonding, and hydrophobic interactions. In general, a substrate-binding site consists of an indentation or cleft on the surface of an enzyme molecule that is complementary in shape to the substrate **(geometric complementarity).** Moreover, the amino acid residues that form the binding site are arranged to specifically attract the substrate **(electronic complementarity,** Fig. 11-1). Molecules that differ in shape or functional group distribution from the substrate cannot productively bind to the enzyme. X-Ray studies indicate that the substrate-binding sites of most enzymes are largely preformed but undergo some conformational change on substrate binding (a phenomenon called **induced fit**). The complementarity between enzymes and their substrates is the basis of the "lock-and-key" model of enzyme function first proposed by Emil Fischer in 1894. As we shall see, such specific binding is necessary but not sufficient for efficient catalysis.

Enzymes Are Stereospecific. Enzymes are highly specific both in binding chiral substrates and in catalyzing their reactions. This **stereospecificity** arises because enzymes, by virtue of their inherent chirality (proteins consist of only L-amino acids), form asymmetric active sites. For example, the enzyme **aconitase** catalyzes the interconversion of citrate and isocitrate in the citric acid cycle (Section 16-3B):

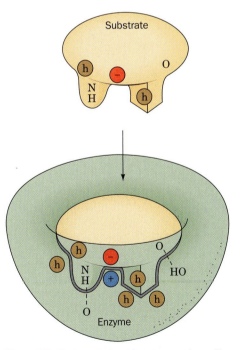

Figure 11-1 An enzyme–substrate complex. The geometric and the electronic complementarity between the enzyme and substrate depend on noncovalent forces. Hydrophobic groups are represented by an h in a brown circle, and dashed lines represent hydrogen bonds.

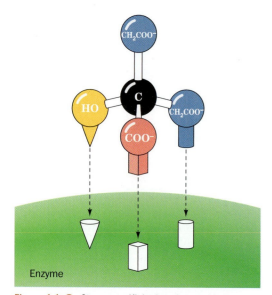

Figure 11-2 Stereospecificity in substrate binding. The specific binding of a prochiral molecule, such as citrate, in an enzyme active site allows the enzyme to differentiate between prochiral groups.

Citrate is a **prochiral** molecule; that is, it can become chiral through the substitution of one of its two carboxymethyl (—CH$_2$COO$^-$) groups (chirality is discussed in Section 4-2). These groups are chemically equivalent but occupy different positions relative to the OH and COO$^-$ groups (likewise, your body is bilaterally symmetric but has distinguishable right and left sides). Aconitase can therefore distinguish between them because citrate interacts asymmetrically with the surface of the enzyme by making a three-point attachment (Fig. 11-2). Because there is only one productive way for citrate to bind to the enzyme, only one of its —CH$_2$COO$^-$ groups reacts to form isocitrate. The stereospecificity of aconitase is by no means unusual. As we consider biochemical reactions, we shall find that *nearly all enzymes that participate in chiral reactions are absolutely stereospecific.*

Enzymes Vary in Geometric Specificity. The stereospecificity of enzymes is not particularly surprising in light of the complementarity of an enzyme's binding site for its substrate. A substance of the wrong chirality will not fit productively into an enzymatic binding site for much the same reason that you cannot fit your right hand into your left glove. In addition to their stereospecificity, however, most enzymes are quite selective about the identities of the chemical groups on their substrates. Indeed, such **geometric specificity** is a more stringent requirement than is stereospecificity.

Enzymes vary considerably in their degree of geometric specificity. A few enzymes are absolutely specific for only one compound. Most enzymes, however, catalyze the reactions of a small range of related compounds although with different efficiencies. For example, alcohol dehydrogenase catalyzes the oxidation of **ethanol** (CH$_3$CH$_2$OH) to **acetaldehyde** (CH$_3$CHO) faster than it oxidizes **methanol** (CH$_3$OH) to **formaldehyde** (H$_2$CO) or **isopropanol** [(CH$_3$)$_2$CHOH] to **acetone** [(CH$_3$)$_2$CO], even though methanol and isopropanol differ from ethanol by only the deletion or addition of a CH$_2$ group.

Some enzymes, particularly digestive enzymes, are so permissive in their ranges of acceptable substrates that their geometric specificities are more accurately described as preferences. Some enzymes are not even very specific in the type of reaction they catalyze. For example, chymotrypsin, in addition to its ability to mediate peptide bond hydrolysis, also catalyzes ester bond hydrolysis.

$$\underset{\textbf{Peptide}}{RC-NHR'} + H_2O \xrightarrow{\text{chymotrypsin}} RC-O^- + H_3\overset{+}{N}R'$$

$$\underset{\textbf{Ester}}{RC-OR'} + H_2O \xrightarrow{\text{chymotrypsin}} RC-O^- + HOR'$$
$$H^+$$

This property makes it convenient to measure chymotrypsin activity using small synthetic esters as substrates. Such permissiveness is much more the exception than the rule. Indeed, most intracellular enzymes function *in vivo* to catalyze a particular reaction on a particular substrate.

C Cofactors and Coenzymes

The functional groups of proteins, as we shall see, can facilely participate in acid–base reactions, form certain types of transient covalent bonds, and take part in charge–charge interactions. They are, however, less suitable for

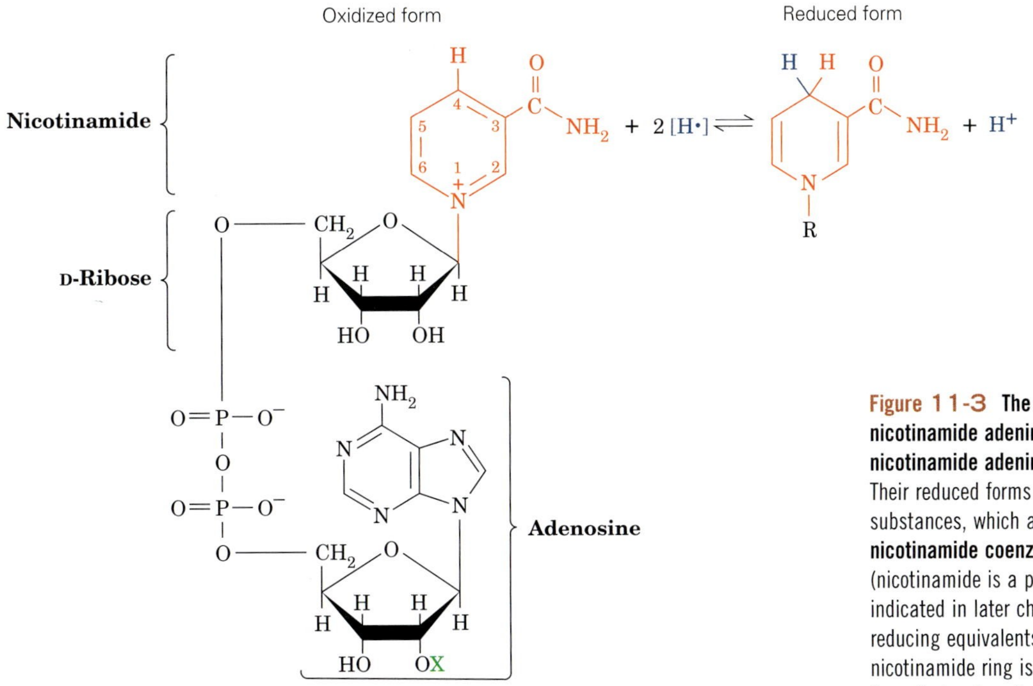

Oxidized form   Reduced form

X = H   Nicotinamide adenine dinucleotide (NAD$^+$)
X = PO$_3^{2-}$ Nicotinamide adenine dinucleotide phosphate (NADP$^+$)

Figure 11-3 The structures and reaction of nicotinamide adenine dinucleotide (NAD$^+$) and nicotinamide adenine dinucleotide phosphate (NADP$^+$). Their reduced forms are **NADH** and **NADPH**. These substances, which are collectively referred to as the **nicotinamide coenzymes** or **pyridine nucleotides** (nicotinamide is a pyridine derivative) function, as is indicated in later chapters, as intracellular carriers of reducing equivalents (electrons). Note that only the nicotinamide ring is changed in the reaction. Reduction formally involves the transfer of two hydrogen atoms (H·), or a hydride ion and a proton (H:$^-$ + H$^+$) although the actual reduction may occur via a different mechanism.

catalyzing oxidation–reduction reactions and many types of group-transfer processes. Although enzymes catalyze such reactions, they can do so only in association with small **cofactors,** which essentially act as the enzymes' "chemical teeth."

Cofactors may be metal ions, such as Cu^{2+}, Fe^{3+}, or Zn^{2+}. The essential nature of these cofactors explains why organisms require trace amounts of certain elements in their diets. It also explains, in part, the toxic effects of certain heavy metals. For example, Cd^{2+} and Hg^{2+} can replace Zn^{2+} (all are in the same group of the periodic table) in the active sites of certain enzymes, including RNA polymerase, and thereby render these enzymes inactive.

Cofactors may also be organic molecules known as **coenzymes.** Some cofactors are only transiently associated with a given enzyme molecule, so that they function as **cosubstrates. Nicotinamide adenine dinucleotide (NAD$^+$)** and **nicotinamide adenine dinucleotide phosphate (NADP$^+$)** are examples of cosubstrates (Fig. 11-3). For instance, NAD$^+$ is an obligatory oxidizing agent in the alcohol dehydrogenase **(ADH)** reaction:

$$CH_3CH_2OH + NAD^+ \xrightarrow{\text{ADH}} CH_3\overset{\overset{\displaystyle O}{\|}}{C}H + NADH + H^+$$

Ethanol   **Acetaldehyde**

The product NADH dissociates from the enzyme for eventual reoxidation to NAD$^+$ in an independent enzymatic reaction.

Other cofactors, known as **prosthetic groups,** are permanently associated with their protein, often by covalent bonds. For example, a heme prosthetic group (Fig. 7-2) is tightly bound to proteins known as **cytochromes** (Box 17-1) through extensive hydrophobic and hydrogen bonding interactions together with covalent bonds between the heme and specific protein side chains.

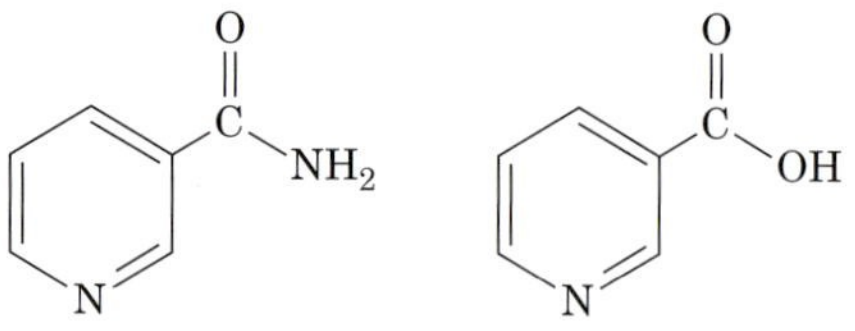

Figure 11-4 **The structures of nicotinamide and nicotinic acid.** These vitamins form the redox-active components of the nicotinamide coenzymes NAD$^+$ and NADP$^+$ (compare with Fig. 11-3).

A catalytically active enzyme–cofactor complex is called a **holoenzyme.** The enzymatically inactive protein resulting from the removal of a holoenzyme's cofactor is referred to as an **apoenzyme;** that is,

apoenzyme *(inactive)* + cofactor $\rightleftharpoons$ holoenzyme *(active)*

Coenzymes Must Be Regenerated.   Coenzymes are chemically changed by the enzymatic reactions in which they participate. *In order to complete the catalytic cycle, the coenzyme must return to its original state.* For a transiently bound coenzyme (cosubstrate), the regeneration reaction may be catalyzed by a different enzyme as we have seen to be the case for NAD$^+$. However, for a prosthetic group, regeneration occurs as part of the enzyme reaction sequence.

Many Vitamins Are Coenzyme Precursors.   Table 11-3 lists the most common coenzymes, along with the types of reactions in which they participate (we shall describe the structures of these substances and their reaction mechanisms in the appropriate sections of the text). Many organisms are unable to synthesize certain portions of essential coenzymes. These substances must be present in the organism's diet and are therefore called **vitamins.** Table 11-3 also lists the vitamin precursors of the common coenzymes.

Many coenzymes were discovered as growth factors for microorganisms or as substances that cure nutritional deficiency diseases in humans and/or animals. For example, the NAD$^+$ component **nicotinamide,** or its carboxylic acid analog **nicotinic acid (niacin;** Fig. 11-4), relieves the ultimately fatal dietary deficiency disease in humans known as **pellagra.** The symptoms of pellagra include diarrhea, dermatitis, and dementia.

The vitamins in the human diet that are coenzyme precursors are all water-soluble vitamins. In contrast, the lipid-soluble vitamins, such as vitamins A and D (Section 9-1), are generally not components of coenzymes, although they are also required in trace amounts in the diets of many higher animals. The distant ancestors of humans probably had the ability to synthesize the various vitamins, as do many modern plants and microorganisms. Yet since vitamins are normally available in the diets of higher animals, which all eat other organisms, or are synthesized by the bacteria that normally inhabit their digestive systems, it seems likely that the superfluous cellular machinery to synthesize them was lost through evolution. For example, vitamin C (ascorbic acid) is required in the diets of only humans, apes, and guinea pigs (Section 6-1C and Box 6-2) because, in what is apparently a recent evolutionary loss, they lack a key enzyme for ascorbic acid biosynthesis.

Table 11-3 Characteristics of Common Coenzymes

Coenzyme	Reaction Mediated	Vitamin Source	Human Deficiency Disease
Biocytin	Carboxylation	Biotin	*a*
Coenzyme A	Acyl transfer	Pantothenate	*a*
Cobalamin coenzymes	Alkylation	Cobalamin (B$_{12}$)	Pernicious anemia
Flavin coenzymes	Oxidation–reduction	Riboflavin (B$_2$)	*a*
Lipoic acid	Acyl transfer	—	*a*
Nicotinamide coenzymes	Oxidation–reduction	Nicotinamide (niacin)	Pellagra
Pyridoxal phosphate	Amino group transfer	Pyridoxine (B$_6$)	*a*
Tetrahydrofolate	One-carbon group transfer	Folic acid	Megaloblastic anemia
Thiamine pyrophosphate	Aldehyde transfer	Thiamine (B$_1$)	Beriberi

[a] No specific name; deficiency in humans is rare or unobserved.

2 Activation Energy and the Reaction Coordinate

Much of our understanding of how enzymes catalyze chemical reactions comes from **transition state theory,** which was developed in the 1930s, principally by Henry Eyring. Consider a bimolecular reaction involving three atoms, such as the reaction of a hydrogen atom with diatomic hydrogen (H_2) to yield a new H_2 molecule and a different hydrogen atom:

$$H_A-H_B + H_C \longrightarrow H_A + H_B-H_C$$

In this reaction, H_C must approach the diatomic molecule H_A-H_B so that, at some point in the reaction, there exists a high-energy (unstable) complex represented as $H_A\cdots H_B\cdots H_C$. In this complex, the H_A-H_B covalent bond is in the process of breaking while the H_B-H_C bond is in the process of forming. The point of highest free energy is called the **transition state** of the system.

Reactants generally approach one another along the path of minimum free energy, their so-called **reaction coordinate.** A plot of free energy versus the reaction coordinate is called a **transition state diagram** or **reaction coordinate diagram** (Fig. 11-5). The reactants and products are states of minimum free energy, and the transition state corresponds to the highest point of the diagram. For the $H + H_2$ reaction, the reactants and products have the same free energy (Fig. 11-5a). If the atoms in the reacting system are of different types, such as in the reaction

$$A + B \longrightarrow X^{\ddagger} \longrightarrow P + Q$$

where A and B are the reactants, P and Q are the products, and $X^{\ddagger}$ represents the transition state, the transition state diagram is no longer symmetrical because there is a free energy difference between the reactants and products (Fig. 11-5b). In either case, $\Delta G^{\ddagger}$, the free energy of the transition state less that of the reactants, is known as the **free energy of activation.**

Passage through the transition state requires only 10^{-13} to 10^{-14} s, so the concentration of the transition state in a reacting system is small. Hence, the decomposition of the transition state to products (or back to reactants) is postulated to be the rate-determining process of the overall reaction.

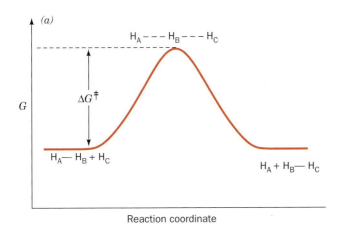

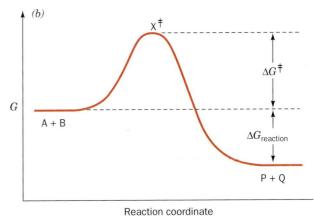

Figure 11-5   Transition state diagrams.   (a) The $H + H_2$ reaction. The reactants and products correspond to low free energy structures. The point of highest free energy is the transition state, in which the reactants are partially converted to products. $\Delta G^{\ddagger}$ is the free energy of activation, the difference in free energy between the reactants and the transition state, $X^{\ddagger}$. (b) Transition state diagram for the reaction $A + B \rightarrow P + Q$. This is a spontaneous reaction; that is, $\Delta G_{\text{reaction}} < 0$ (the free energy of P + Q is less than the free energy of A + B).

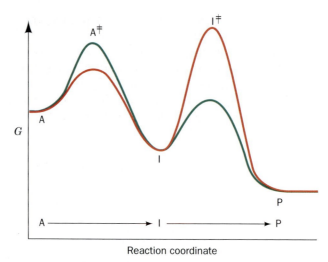

Figure 11-6 Transition state diagram for a two-step reaction.   The blue curve represents a reaction (A → I → P) whose first step is rate determining, and the red curve represents a reaction whose second step is rate determining.

Thermodynamic arguments lead to the conclusion that the reaction rate is proportional to $e^{-\Delta G^{\ddagger}/RT}$, where R is the gas constant and T is the absolute temperature. Thus, *the greater the value of* $\Delta G^{\ddagger}$, *the slower the reaction rate.* This is because the larger the $\Delta G^{\ddagger}$, the smaller the number of reactant molecules that have sufficient thermal energy to achieve the transition state free energy.

Chemical reactions commonly consist of several steps. For a two-step reaction such as

$$A \longrightarrow I \longrightarrow P$$

where I is an intermediate of the reaction, there are two transition states and two activation energy barriers. The shape of the transition state diagram for such a reaction reflects the relative rates of the two steps (Fig. 11-6). If the activation energy of the first step is greater than that of the second step, then the first step is slower than the second step, and conversely, if the activation energy of the second step is greater. In a multistep reaction, the step with the highest transition state free energy acts as a "bottleneck" and is therefore said to be the **rate-determining step** of the reaction.

Catalysts Reduce $\Delta G^{\ddagger}$.   *Catalysts act by providing a reaction pathway with a transition state whose free energy is lower than that in the uncatalyzed reaction* (Fig. 11-7). The difference between the values of $\Delta G^{\ddagger}$ for the uncatalyzed and catalyzed reactions, $\Delta \Delta G^{\ddagger}_{cat}$, indicates the efficiency of the catalyst. The **rate enhancement** (ratio of the rates of the catalyzed and uncatalyzed reactions) is given by $e^{\Delta \Delta G^{\ddagger}_{cat}/RT}$. Hence, at 25°C (298 K), a 10-fold rate enhancement requires a $\Delta \Delta G^{\ddagger}_{cat}$ of only 5.71 kJ · mol^{-1}, which is less than half the free energy of a typical hydrogen bond. Similarly, a million-fold rate acceleration occurs when $\Delta \Delta G^{\ddagger}_{cat} \approx 34$ kJ · mol^{-1}, a small fraction of the free energy of most covalent bonds. Thus, from a theoretical standpoint, tremendous catalytic efficiency seems within reach of the reactive groups that occur in the active sites of enzymes. How these groups actually function at the atomic level is the subject of much of this and other chapters.

Note that a catalyst lowers the free energy barrier by the same amount for both the forward and reverse reactions (Fig. 11-7). Consequently, a cat-

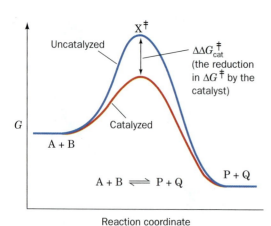

Figure 11-7 Effect of a catalyst on the transition state diagram of a reaction.   Here $\Delta \Delta G^{\ddagger}_{cat} = \Delta G^{\ddagger}(uncat) - \Delta G^{\ddagger}(cat)$.

alyst equally accelerates the forward and reverse reactions. Keep in mind also that while a catalyst can accelerate the conversion of reactants to products (or products back to reactants), the likelihood of the net reaction occurring in one direction or the other depends only on the free energy difference between the reactants and the products. If $\Delta G_{\text{reaction}} < 0$, the reaction proceeds spontaneously from reactants toward products; if $\Delta G_{\text{reaction}} > 0$, the reverse reaction proceeds spontaneously. *An enzyme cannot alter $\Delta G_{\text{reaction}}$; it can only decrease $\Delta G^{\ddagger}$ to allow the reaction to approach equilibrium (where the rates of the forward and reverse reactions are equal) more quickly than it would in the absence of a catalyst.* The actual velocity with which reactants are converted to products is the subject of kinetics (Section 12-1).

3 | Catalytic Mechanisms

Enzymes achieve their enormous rate accelerations via the same catalytic mechanisms used by chemical catalysts. Enzymes have simply been better designed through evolution. Enzymes, like other catalysts, reduce the free energy of the transition state ($\Delta G^{\ddagger}$); that is, *they stabilize the transition state of the catalyzed reaction.* What makes enzymes such effective catalysts is their specificity of substrate binding combined with their arrangement of catalytic groups. As we shall see, however, the distinction between substrate-binding groups and catalytic groups is somewhat arbitrary.

Much can be learned about enzymatic reaction mechanisms by examining the corresponding nonenzymatic reactions of model compounds. Both types of reactions can be described using the **curved arrow convention** to trace the electron pair rearrangements that occur in going from reactants to products. The movement of an electron pair (which may be either a lone pair or a pair forming a covalent bond) is symbolized by a curved arrow emanating from the electron pair and pointing to the electron-deficient center attracting the electron pair. For example, imine **(Schiff base)** formation, a biochemically important reaction between an amine and an aldehyde or ketone, is represented as follows:

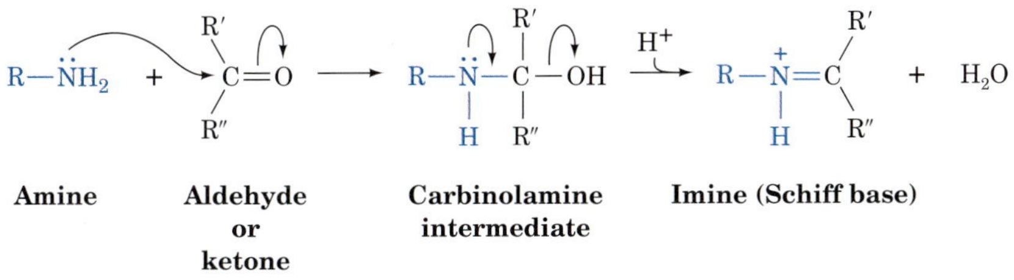

| Amine | Aldehyde or ketone | Carbinolamine intermediate | Imine (Schiff base) |

In the first reaction step, the amine's unshared electron pair adds to the electron-deficient carbonyl carbon while one electron pair from its C=O double bond transfers to the oxygen atom. In the second step, the unshared electron pair on the nitrogen atom adds to the electron-deficient carbon atom with the elimination of water. *At all times, the rules of chemical reason apply to the system:* For example, there are never five bonds to a carbon atom or two bonds to a hydrogen atom.

The types of catalytic mechanisms that enzymes employ have been classified as

1. Acid–base catalysis
2. Covalent catalysis
3. Metal ion catalysis

4. Proximity and orientation effects

5. Preferential binding of the transition state complex

In this section, we consider each of these types of mechanisms in turn.

A Acid–Base Catalysis

***General acid catalysis** is a process in which proton transfer from an acid lowers the free energy of a reaction's transition state.* For example, an uncatalyzed keto–enol tautomerization reaction occurs quite slowly as a result of the high free energy of its carbanion-like transition state (Fig. 11-8*a*; the transition state is drawn in square brackets to indicate its instability). Proton donation to the oxygen atom (Fig. 11-8*b*), however, reduces the carbanion character of the transition state, thereby accelerating the reaction.

A reaction may also be stimulated by **general base catalysis** *if its rate is increased by proton abstraction by a base* (e.g., Fig. 11-8*c*). Some reactions may be simultaneously subject to both processes; these are **concerted acid–base catalyzed reactions.**

Many types of biochemical reactions are susceptible to acid and/or base catalysis. The side chains of the amino acid residues Asp, Glu, His, Cys, Tyr, and Lys have p*K*'s in or near the physiological pH range (Table 4-1), which permits them to act as acid and/or base catalysts. Indeed, *the ability of enzymes to arrange several catalytic groups around their substrates makes concerted acid–base catalysis a common enzymatic mechanism.* The catalytic activity of these enzymes is sensitive to pH, since the pH influences the state of protonation of side chains at the active site (see Box 11-1).

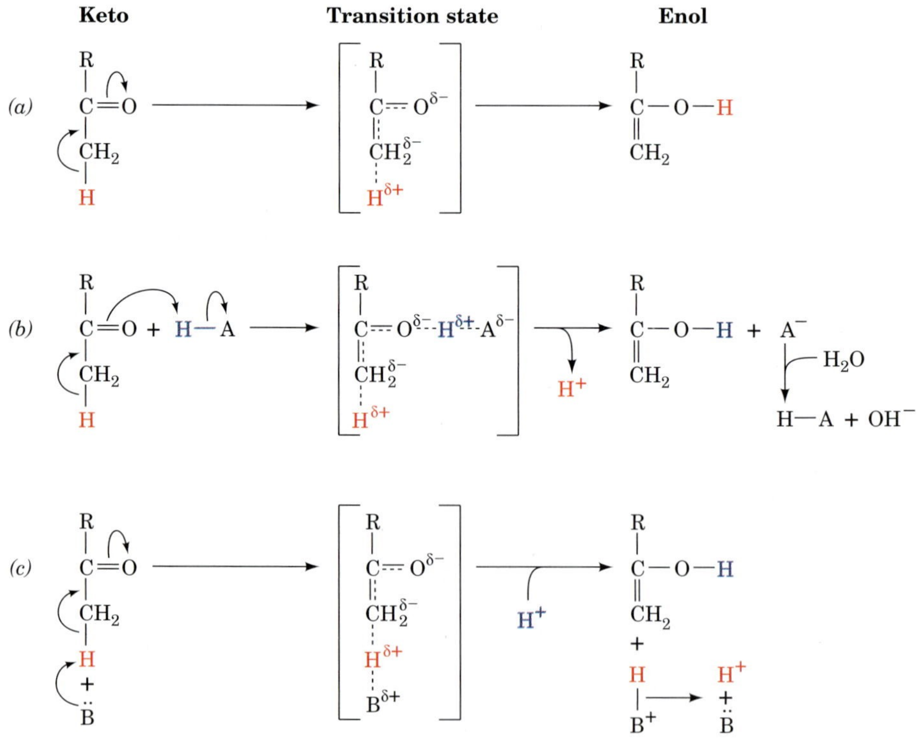

Figure 11-8 Mechanisms of keto–enol tautomerization. (*a*) Uncatalyzed. (*b*) General acid catalyzed. (*c*) General base catalyzed. The acid is represented as H—A and the base as $\ddot{\text{B}}$.

BOX 11-1

Perspectives in Biochemistry

Effects of pH on Enzyme Activity

Most enzymes are active within only a narrow pH range, typically 5 to 9. This is a result of the effects of pH on a combination of factors: (1) the binding of substrate to enzyme, (2) the ionization states of the amino acid residues involved in the catalytic activity of the enzyme, (3) the ionization of the substrate, and (4) the variation of protein structure (usually significant only at extremes of pH).

The rates of many enzymatic reactions exhibit bell-shaped curves as a function of pH. For example, the pH dependence of the rate of the reaction catalyzed by **fumarase** (Section 16-3G) produces the following curve:

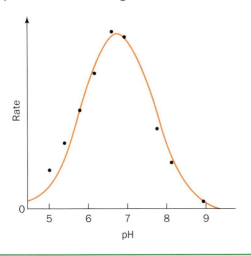

Such curves reflect the ionization of certain amino acid residues that must be in a specific ionization state for enzymatic activity. The observed pK's (the inflection points of the curve) often provide valuable clues to the identities of the amino acid residues essential for enzymatic activity. For example, an observed pK of ~4 suggests that an Asp or Glu residue is essential to the enzyme. Similarly, pK's of ~6 or ~10 suggest the participation of a His or a Lys residue, respectively. However, the pK of a given acid–base group may vary by as much as several pH units from its expected value, depending on its microenvironment (e.g., an Asp residue in a nonpolar environment or in close proximity to another Asp residue would attract protons more strongly than otherwise and hence have a higher pK). Furthermore, pH effects on an enzymatic rate may reflect denaturation of the enzyme rather than protonation or deprotonation of specific catalytic residues. The replacement of a particular residue by site-directed mutagenesis or comparisons of enzyme variants generated by evolution is a more reliable approach to identifying residues that are required for substrate binding or catalysis.

[Figure adapted from Tanford, C., *Physical Chemistry of Macromolecules*, p. 647, Wiley (1961).]

The RNase A Reaction.   **Bovine pancreatic RNaseA (RNase A)** provides an example of enzymatically mediated acid–base catalysis. This digestive enzyme (Fig. 11-9) is secreted by the pancreas into the small intestine, where it hydrolyzes RNA to its component nucleotides. The isolation of 2′,3′-cyclic nucleotides from RNase A digests of RNA indicates that 2′,3′-cyclic nu-

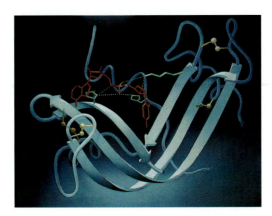

Figure 11-9 **X-Ray structure of bovine pancreatic RNase S.** A nonhydrolyzable substrate analog, the dinucleotide phosphonate UpcA, is bound in the active site. RNase S is a catalytically active form of RNase A in which the peptide bond between residues 20 and 21 has been hydrolyzed. [Illustration Irving Geis/Geis Archives Trust. Copyright Howard Hughes Medical Institute. Reproduced by permission.] **See the Interactive Exercises.**

cleotides are intermediates in the RNase A reaction (Fig. 11-10). The pH dependence of the rate of the RNase A reaction suggests the involvement of two ionizable residues with pK values of 5.4 and 6.4. This information, together with chemical derivatization and X-ray studies, indicates that RNase A has two essential His residues, His 12 and His 119, that act in a concerted manner as general acid and base catalysts. Evidently, the RNase A reaction is a two-step process (Fig. 11-10):

1. His 12, acting as a general base, abstracts a proton from an RNA 2′-OH group, thereby promoting its nucleophilic attack on the adja-

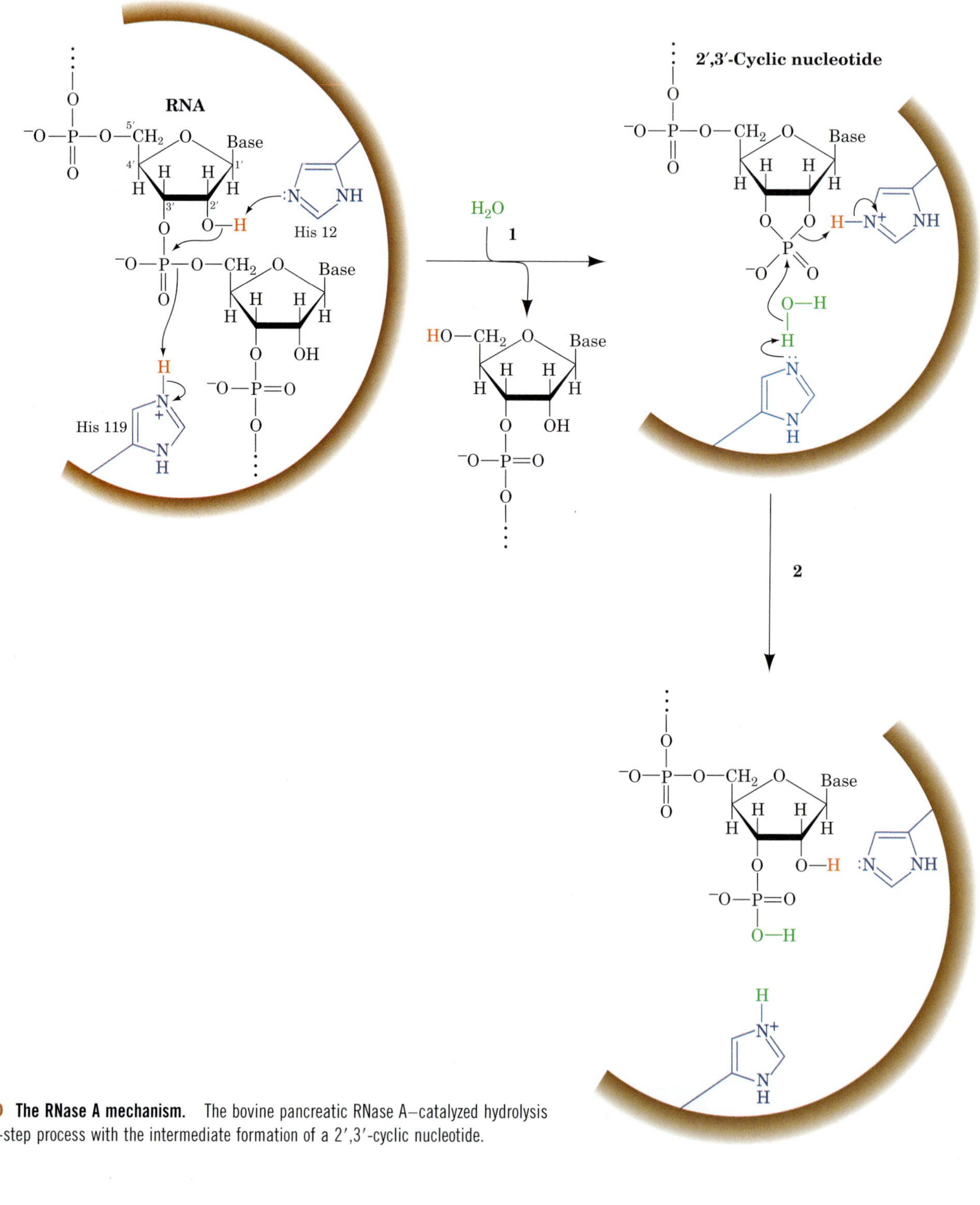

Figure 11-10 **The RNase A mechanism.**   The bovine pancreatic RNase A–catalyzed hydrolysis of RNA is a two-step process with the intermediate formation of a 2′,3′-cyclic nucleotide.

cent phosphorus atom. His 119, acting as a general acid, promotes bond scission by protonating the leaving group.

2. After the leaving group departs, water enters the active site, and the 2′,3′-cyclic intermediate is hydrolyzed through what is essentially the reverse of the first step. Thus, His 12 now acts as a general acid and His 119 as a general base to yield the hydrolyzed RNA and the enzyme in its original state.

B Covalent Catalysis

Covalent catalysis *accelerates reaction rates through the transient formation of a catalyst–substrate covalent bond.* Usually, this covalent bond is formed by the reaction of a nucleophilic group on the catalyst with an electrophilic group on the substrate, and hence this form of catalysis is often also called **nucleophilic catalysis.** The decarboxylation of acetoacetate, as chemically catalyzed by primary amines, is an example of such a process (Fig. 11-11). In the first stage of this reaction, the amine, a nucleophile, attacks the carbonyl group of acetoacetate to form a Schiff base (imine bond):

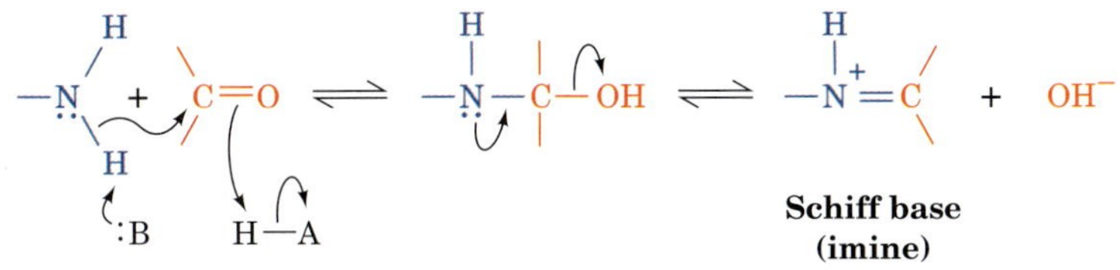

The protonated nitrogen atom of the covalent intermediate then acts as an electron sink (Fig. 11-11, *bottom*) to reduce the high-energy enolate character of the transition state. The formation and decomposition of the Schiff base occurs quite rapidly so that it is not the rate-determining step of this reaction.

Covalent catalysis can be conceptually decomposed into three stages:

1. The nucleophilic reaction between the catalyst and the substrate to form a covalent bond.

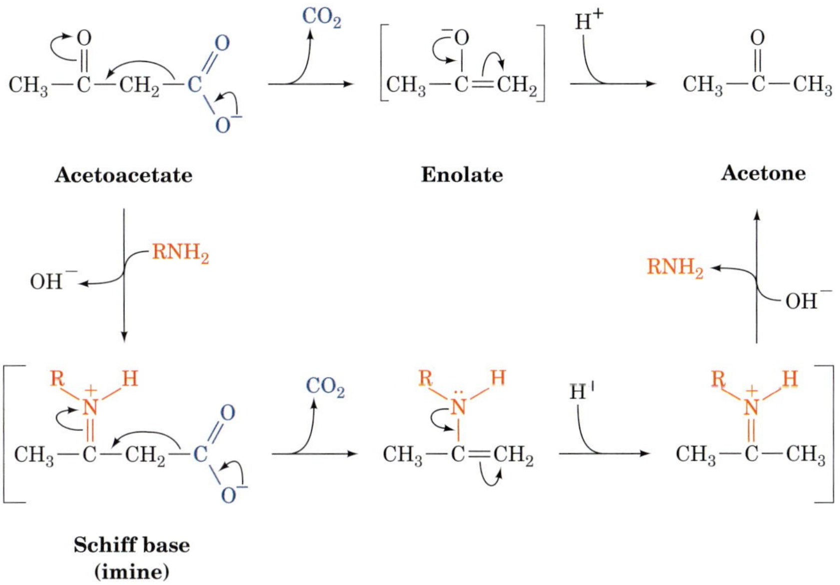

Figure 11-11 The decarboxylation of acetoacetate. The uncatalyzed reaction mechanism is at the top, and the mechanism as catalyzed by primary amines is at the bottom.

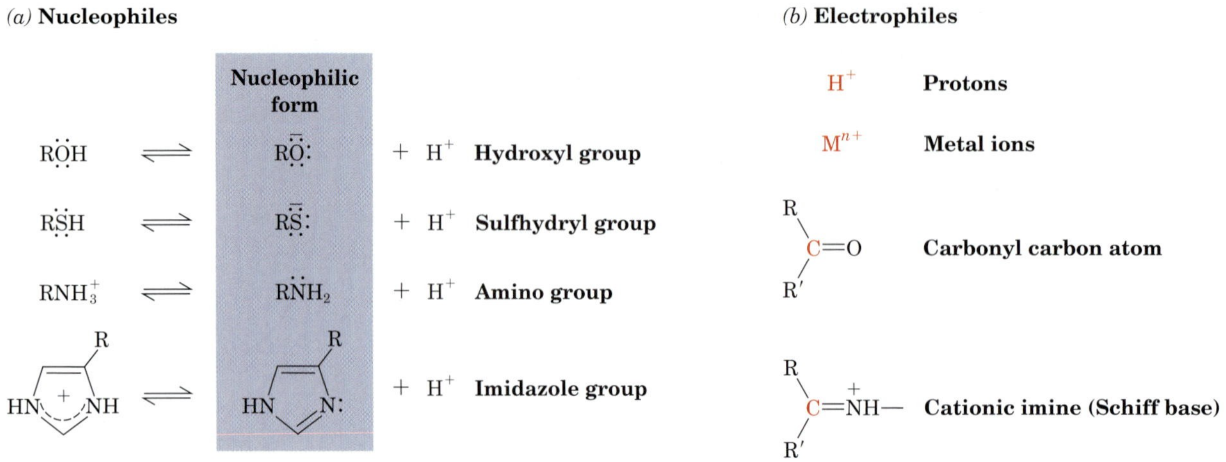

(*a*) **Nucleophiles**

(*b*) **Electrophiles**

H^+	**Protons**
M^{n+}	**Metal ions**

Figure 11-12 Biologically important nucleophilic and electrophilic groups. (*a*) Nucleophilic groups such as hydroxyl, sulfhydryl, amino, and imidazole groups are nucleophiles in their basic forms. (*b*) Electrophilic groups contain an electron-deficient atom (*red*).

2. The withdrawal of electrons from the reaction center by the now electrophilic catalyst.

3. The elimination of the catalyst, a reaction that is essentially the reverse of stage 1.

The nucleophilicity of a substance is closely related to its basicity. Indeed, the mechanism of nucleophilic catalysis resembles that of base catalysis except that, instead of abstracting a proton from the substrate, the catalyst nucleophilically attacks the substrate to form a covalent bond. Biologically important nucleophiles are negatively charged or contain unshared electron pairs that easily form covalent bonds with electron-deficient centers (Fig. 11-12*a*). Electrophiles, in contrast, include groups that are positively charged, contain an unfilled valence electron shell, or contain an electronegative atom (Fig. 11-12*b*).

An important aspect of covalent catalysis is that *the more stable the covalent bond formed, the less easily it can decompose in the final steps of a reaction.* A good covalent catalyst must therefore combine the seemingly contradictory properties of high nucleophilicity and the ability to form a good leaving group, that is, to easily reverse the bond formation step. Groups with high polarizability (highly mobile electrons), such as imidazole and thiol groups, have these properties and hence make good covalent catalysts. Functional groups in proteins that act in this way include the unprotonated ε-amino group of Lys, the imidazole group of His, the thiol group of Cys, the carboxyl group of Asp, and the hydroxyl group of Ser. In addition, several coenzymes, notably **thiamine pyrophosphate** (Section 14-3B) and **pyridoxal phosphate** (Section 20-2A), function in association with their apoenzymes as covalent catalysts. Enzymes commonly employ covalent catalytic mechanisms as is indicated by the large variety of covalently linked enzyme–substrate reaction intermediates that have been isolated.

C Metal Ion Catalysis

Nearly one-third of all known enzymes require the presence of metal ions for catalytic activity. This group of enzymes includes the **metalloenzymes,** which contain tightly bound metal ion cofactors, most commonly transition metal ions such as Fe^{2+}, Fe^{3+}, Cu^{2+}, Mn^{2+}, or Co^{2+}. These catalytically essential metal ions are distinct from ions such as Na^+, K^+, or Ca^{2+}, which

often play a structural rather than a catalytic role in enzymes. Ions such as Mg^{2+} and Zn^{2+} may be either structural or catalytic.

Metal ions participate in the catalytic process in three major ways:

1. By binding to substrates to orient them properly for reaction.
2. By mediating oxidation–reduction reactions through reversible changes in the metal ion's oxidation state.
3. By electrostatically stabilizing or shielding negative charges.

In many metal ion–catalyzed reactions, the metal ion acts in much the same way as a proton to neutralize negative charge. Yet metal ions are often much more effective catalysts than protons because metal ions can be present in high concentrations at neutral pH and may have charges greater than +1.

A metal ion's charge also makes its bound water molecules more acidic than free H_2O and therefore a source of nucleophilic OH^- ions even below neutral pH. An excellent example of this phenomenon occurs in the catalytic mechanism of carbonic anhydrase (Box 2-2), a widely occurring enzyme that catalyzes the reaction

$$CO_2 + H_2O \rightleftharpoons HCO_3^- + H^+$$

Carbonic anhydrase contains an essential Zn^{2+} ion that is implicated in the enzyme's catalytic mechanism as follows:

1. The crystal structure of human carbonic anhydrase (Fig. 11-13) reveals that its Zn^{2+} lies at the bottom of a 15-Å-deep active site cleft, where it is tetrahedrally coordinated by three evolutionarily invariant His side chains and an H_2O molecule. This Zn^{2+}-polarized H_2O ionizes through base catalysis that is facilitated by a fourth His residue (His 64).
2. The resulting Zn^{2+}-bound OH^- nucleophilically attacks the nearby enzymatically bound CO_2, thereby converting it to HCO_3^- (*at right*).
3. The catalytic site is then regenerated by the binding and ionization of another H_2O at the Zn^{2+} (*at right*).

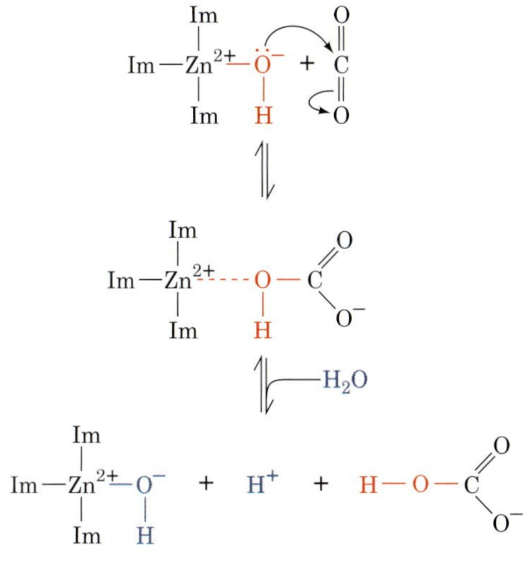

Im = imidazole

D Catalysis through Proximity and Orientation Effects

Although enzymes employ catalytic mechanisms that resemble those of organic model reactions, they are far more catalytically efficient than these models. Such efficiency must arise from the specific physical conditions at

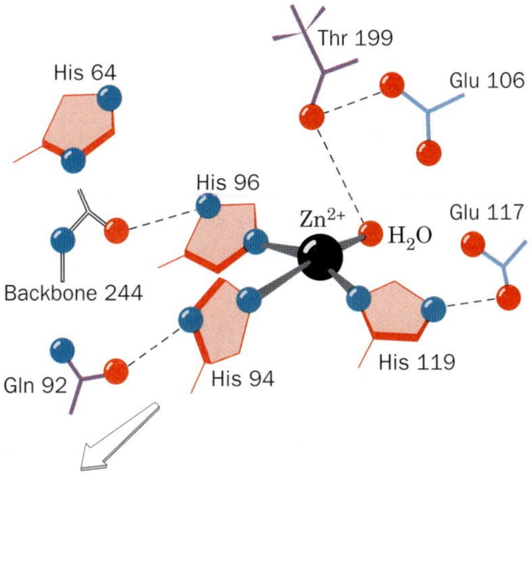

Figure 11-13   The active site of human carbonic anhydrase.   The Zn^{2+} ion is coordinated by the imidazole side chains of three His residues and a water molecule. The arrow points toward the opening of the active site cavity. [After Sheridan, R.P. and Allen, L.C., *J. Am. Chem. Soc.* **103**, 1545 (1981).]   **See the Interactive Exercises.**

enzyme catalytic sites that promote the corresponding chemical reactions. The most obvious effects are **proximity** and **orientation:** Reactants must come together with the proper spatial relationship for a reaction to occur. Consider the bimolecular reaction of imidazole with **p-nitrophenylacetate.**

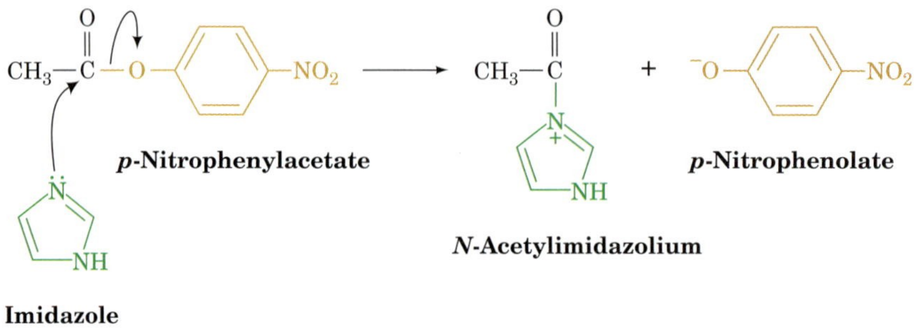

The progress of the reaction is conveniently monitored by the appearance of the intensely yellow **p-nitrophenolate** ion. The related intramolecular reaction

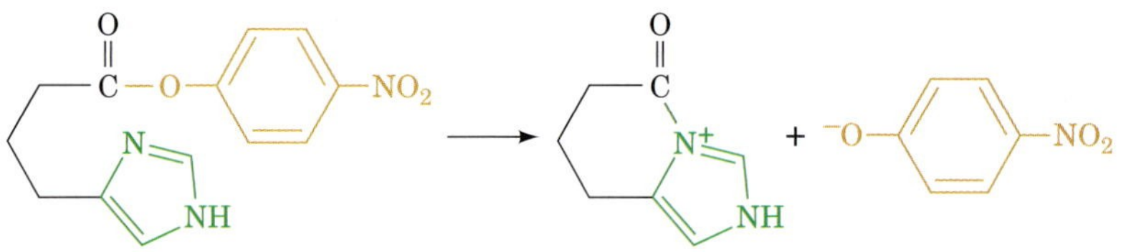

occurs about 24 times faster. Thus, when the imidazole catalyst is covalently attached to the reactant, it is 24 times more effective than when it is free in solution. This rate enhancement results from both proximity and orientation effects.

By simply binding their substrates, enzymes facilitate their catalyzed reactions in four ways:

1. Enzymes bring substrates into contact with their catalytic groups and, in reactions with more than one substrate, with each other. However, calculations based on simple model systems suggest that such proximity effects alone can enhance reaction rates by no more than a factor of ~5.

2. Enzymes bind their substrates in the proper orientations for reaction. Molecules are not equally reactive in all directions. Rather, *they react most readily if they have the proper relative orientation.* For example, in an S$_N$2 (bimolecular nucleophilic substitution) reaction, the incoming nucleophile optimally attacks its target along the direction opposite to that of the bond to the leaving group (Fig. 11-14). Reacting atoms whose approaches deviate by as little as 10° from this optimum direction are significantly less reactive. It is estimated that properly orienting substrates can increase reaction rates by a factor of up to ~100. Enzymes, as we shall see, align their substrates and catalytic groups so as to optimize reactivity.

3. Charged groups may help stabilize the transition state of the reaction, a phenomenon termed **electrostatic catalysis.** The charge distribution around the active sites of enzymes may also guide polar substrates toward their binding site.

4. Enzymes freeze out the relative translational and rotational motions of their substrates and catalytic groups. This is an important aspect of catalysis because, in the transition state, the reacting groups have lit-

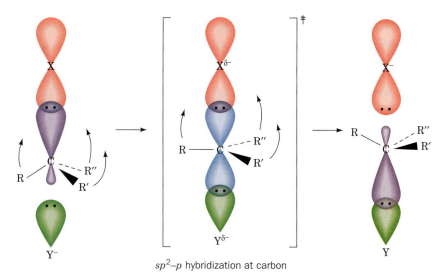

sp²–p hybridization at carbon

Figure 11-14 The geometry of an S$_N$2 reaction. The attacking nucleophile, Y$^-$, must approach the tetrahedrally coordinated and hence *sp³*-hybridized C atom along the direction opposite that of its bond to the leaving group, X, a process called **back side attack**. In the transition state of the reaction, the C atom becomes trigonal bipyramidally coordinated and hence *sp²–p* hybridized, with the *p* orbital (*blue*) forming partial bonds to X and Y. The three *sp²* orbitals form bonds to the C atom's three other substituents (R, R′, and R″), which have shifted their positions into the plane perpendicular to the X—C—Y axis (*curved arrows*). Any deviation from this optimal geometry would increase the free energy of the transition state, $\Delta G^\ddagger$, and hence reduce the rate of the reaction. The transition state then decomposes to products in which the R, R′, and R″ have inverted their positions about the C atom, which has rehybridized to *sp³*, and X$^-$ has been released.

tle relative motion. Indeed, experiments with model compounds suggest that *this effect can promote rate enhancements of up to ~10⁷!*

Bringing substrates and catalytic groups together in a reactive orientation orders them and therefore has a substantial entropic penalty. The free energy required to overcome this entropy loss is supplied by the binding energy of the substrate(s) to the enzyme and contributes to the decreased $\Delta\Delta G^\ddagger$.

E Catalysis by Preferential Transition State Binding

The rate enhancements effected by enzymes are often greater than can be reasonably accounted for by the catalytic mechanisms discussed so far. However, we have not yet considered one of the most important mechanisms of enzymatic catalysis: *An enzyme may bind the transition state of the reaction it catalyzes with greater affinity than its substrates or products.* When taken together with the previously described catalytic mechanisms, preferential transition state binding explains the observed rates of enzyme-catalyzed reactions.

The original concept of transition state binding proposed that enzymes mechanically strained their substrates toward the transition state geometry through binding sites into which undistorted substrates did not properly fit. Such strain promotes many organic reactions. For example, the rate of the reaction

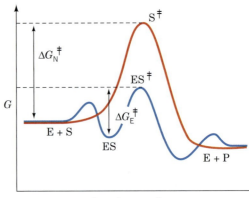

Figure 11-15 *Key to Function.* **Effect of preferential transition state binding.** The reaction coordinate diagram for a hypothetical enzyme-catalyzed reaction involving a single substrate is blue, and the diagram for the corresponding uncatalyzed reaction is red. $\Delta G_N^{\ddagger}$ is the free energy of activation for the nonenzymatic reaction and $\Delta G_E^{\ddagger}$ is the free energy of activation for the enzyme-catalyzed reaction. The small dips in the reaction coordinate diagram for the enzyme-catalyzed reaction arise from the binding of substrate and product to the enzyme. 🐁 **See the Animated Figures.**

is 315 times faster when R is CH_3 rather than H because of the greater steric repulsion between the CH_3 groups and the reacting groups. The strained reactant more closely resembles the transition state of the reaction than does the corresponding unstrained reactant. Thus, as was first suggested by Linus Pauling and further amplified by Richard Wolfenden and Gustav Lienhard, *enzymes that preferentially bind the transition state structure increase its concentration and therefore proportionally increase the reaction rate.*

The more tightly an enzyme binds its reaction's transition state relative to the substrate, the greater is the rate of the catalyzed reaction relative to that of the uncatalyzed reaction; that is, catalysis results from the preferential binding and therefore the stabilization of the transition state relative to the substrate (Fig. 11-15). In other words, the free energy difference between an enzyme–substrate complex (ES) and an enzyme–transition state complex ($ES^{\ddagger}$) is less than the free energy difference between S and $S^{\ddagger}$ in an uncatalyzed reaction.

As we saw in Section 11-2, the rate enhancement of a catalyzed reaction is given by $e^{\Delta\Delta G_{cat}^{\ddagger}/RT}$, where $\Delta\Delta G_{cat}^{\ddagger}$ is the difference in the values of $\Delta G^{\ddagger}$ for the uncatalyzed ($\Delta G_N^{\ddagger}$) and the catalyzed ($\Delta G_E^{\ddagger}$) reactions. Thus, a rate enhancement of 10^6, which requires that an enzyme bind its transition state complex with 10^6-fold higher affinity than its substrate, corresponds to a $34.2 \text{ kJ} \cdot \text{mol}^{-1}$ stabilization at 25°C, roughly the free energy of two hydrogen bonds. Consequently, the enzymatic binding of a transition state by two hydrogen bonds that cannot form when the substrate first binds to the enzyme should result in a rate enhancement of $\sim 10^6$ based on this effect alone.

It is commonly observed that an enzyme binds poor substrates, which have low reaction rates, as well as or even better than good ones, which have high reaction rates. Thus, a good substrate does not necessarily bind to its enzyme with high affinity, but it does so on activation to the transition state.

Transition State Analogs Are Enzyme Inhibitors. *If an enzyme preferentially binds its transition state, then it can be expected that **transition state analogs**, stable molecules that geometrically and electronically resemble the transition state, are potent inhibitors of the enzyme.* For example, the reaction catalyzed

by **proline racemase** from *Clostridium sticklandii* is thought to occur via a planar transition state:

by **proline racemase** from *Clostridium sticklandii* is thought to occur via a planar transition state:

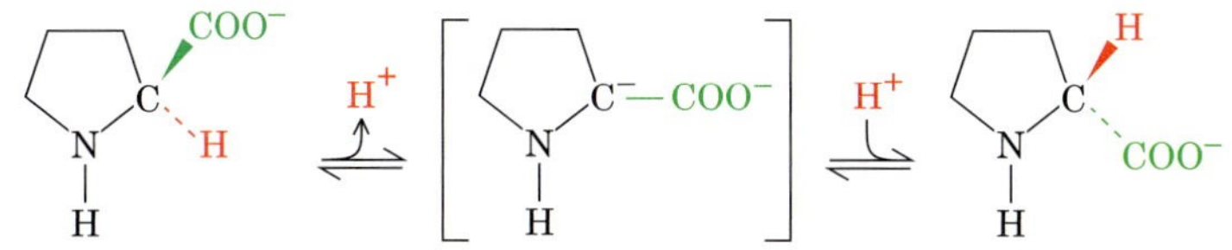

L-Proline   Planar transition state   D-Proline

Proline racemase is inhibited by the planar analogs of proline, **pyrrole-2-carboxylate** and **Δ-1-pyrroline-2-carboxylate**,

Pyrrole-2-carboxylate   Δ-1-Pyrroline-2-carboxylate

both of which bind to the enzyme with 160-fold greater affinity than does proline. These compounds are therefore thought to be analogs of the transition state in the proline racemase reaction.

Hundreds of transition state analogs for various enzymes have been reported. Some are naturally occurring antibiotics. Others were designed to investigate the mechanism of particular enzymes or to act as specific enzyme inhibitors for therapeutic or agricultural use. Indeed, *the theory that enzymes bind transition states with higher affinity than substrates has led to a rational basis for drug design based on the understanding of specific enzyme reaction mechanisms* (Section 12-4).

4 | Lysozyme

In the remainder of this chapter, we investigate the catalytic mechanisms of some well-characterized enzymes. In doing so, we shall see how enzymes apply the catalytic principles described in the preceding section.

Lysozyme is an enzyme that destroys bacterial cell walls. It does so by hydrolyzing the β(1→4) glycosidic linkages from **N-acetylmuramic acid (NAM or MurNAc)** to **N-acetylglucosamine (NAG or GlcNAc)** in cell wall peptidoglycans (Fig. 11-16 and Section 8-3B). It likewise hydrolyzes β(1→4)-linked poly(NAG) (chitin; Section 8-2B), a cell wall component of most fungi

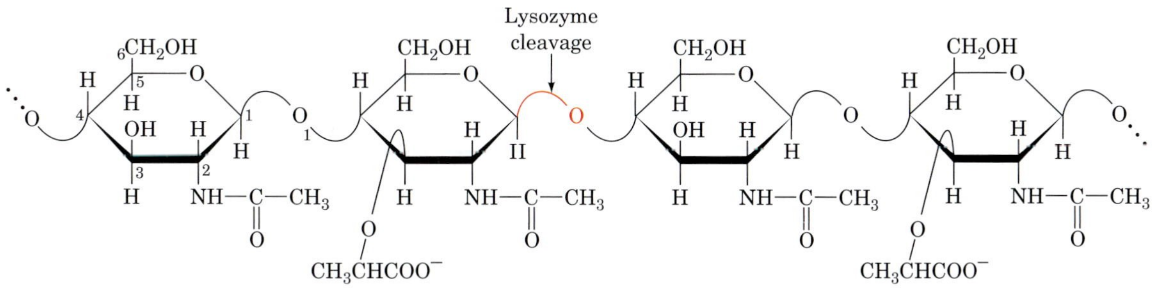

Figure 11-16 The lysozyme cleavage site. The enzyme cleaves after a β(1→4) linkage in the alternating NAG–NAM polysaccharide component of bacterial cell walls.

as well as the major component of the exoskeletons of insects and crustaceans. Lysozyme occurs widely in the cells and secretions of vertebrates, where it probably functions as a bactericidal agent or helps dispose of bacteria after they have been killed by other means.

Hen egg white (HEW) lysozyme is the most widely studied species of lysozyme and is one of the mechanistically best understood enzymes. It is a rather small protein (14.6 kD) whose single polypeptide chain consists of 129 amino acid residues and is internally cross-linked by four disulfide bonds. Lysozyme catalyzes the hydrolysis of its substrate at a rate that is $\sim10^8$-fold greater than that of the uncatalyzed reaction.

A Enzyme Structure

The X-ray structure of HEW lysozyme, which was elucidated by David Phillips in 1965 (and hence was the first enzyme to have its structure determined), shows that the protein molecule is roughly ellipsoidal in shape with dimensions $30 \times 30 \times 45$ Å (Fig. 11-17). *Its most striking feature is a prominent cleft, the substrate-binding site, that traverses one face of the molecule.* The polypeptide chain forms five helical segments as well as a three-stranded antiparallel β sheet that comprises one wall of the binding cleft (Fig. 11-17a). As expected, most of the nonpolar side chains are in the interior of the molecule, out of contact with the aqueous solvent.

The elucidation of an enzyme's mechanism of action requires a knowledge of the structure of its enzyme–substrate complex. This is because, even if the active site residues have been identified through chemical and physical means, their three-dimensional arrangement relative to the substrate as well as to each other must be known in order to understand how the enzyme works. However, an enzyme binds its good substrates only transiently before it catalyzes a reaction and releases products. Consequently, much of our structural knowledge of enzyme–substrate complexes comes from X-ray studies of enzymes in their complexes with nonreactive substrate analogs that remain stably bound to the enzyme for the several hours that it usually takes to measure a protein crystal's X-ray diffraction intensities (but see Box 11-2).

Lysozyme's Catalytic Site Was Identified through Model Building. X-Ray structural analysis revealed that the trisaccharide $(NAG)_3$, which is only slowly hydrolyzed by lysozyme, binds to the enzyme at a site corresponding to the site occupied by residues A, B, and C as drawn in Fig. 11-17a. Lysozyme efficiently catalyzes the hydrolysis of substrates containing at least six saccharide units, so Phillips used model building to investigate how a larger substrate could bind to the enzyme. Lysozyme's active site cleft is long enough to accommodate an oligosaccharide of six residues (designated A to F in Fig. 11-17a). However, the fourth residue (D) appeared unable to bind to the enzyme because its C6 and O6 atoms too closely contacted protein side chains and residue C. This steric interference could be relieved by

Figure 11-17 X-Ray structure of HEW lysozyme (*opposite*). (*a*) The polypeptide chain is shown with a bound $(NAG)_6$ substrate (*green*). The positions of the backbone C_α atoms are indicated together with those of the side chains that line the substrate-binding site and form disulfide bonds. The substrate's sugar rings are designated A, at its nonreducing end (*right*), through F, at its reducing end (*left*). Lysozyme catalyzes the hydrolysis of the glycosidic bond between residues D and E. Rings A, B, and C are observed in the X-ray structure of the complex of $(NAG)_3$ with lysozyme; the positions of rings D, E, and F were inferred from model building studies. [Illustration Irving Geis/Geis Archives Trust. Copyright Howard Hughes Medical Institute. Reproduced with permission.] (*b*) A ribbon diagram of lysozyme highlighting the protein's secondary structure and indicating the positions of its catalytically important side chains. (*c*) A computer-generated model showing the protein's molecular envelope (*purple*) and C_α backbone (*blue*). The side chains of the catalytic residues, Asp 52 (*upper*) and Glu 35 (*lower*), are colored yellow. Note the enzyme's prominent substrate-binding cleft. [Courtesy of Arthur Olson, The Scripps Research Institute, La Jolla, California.] Parts *a*, *b*, and *c* have approximately the same orientation. 🔎 **See the Interactive Exercises and Kinemage Exercise 9.**

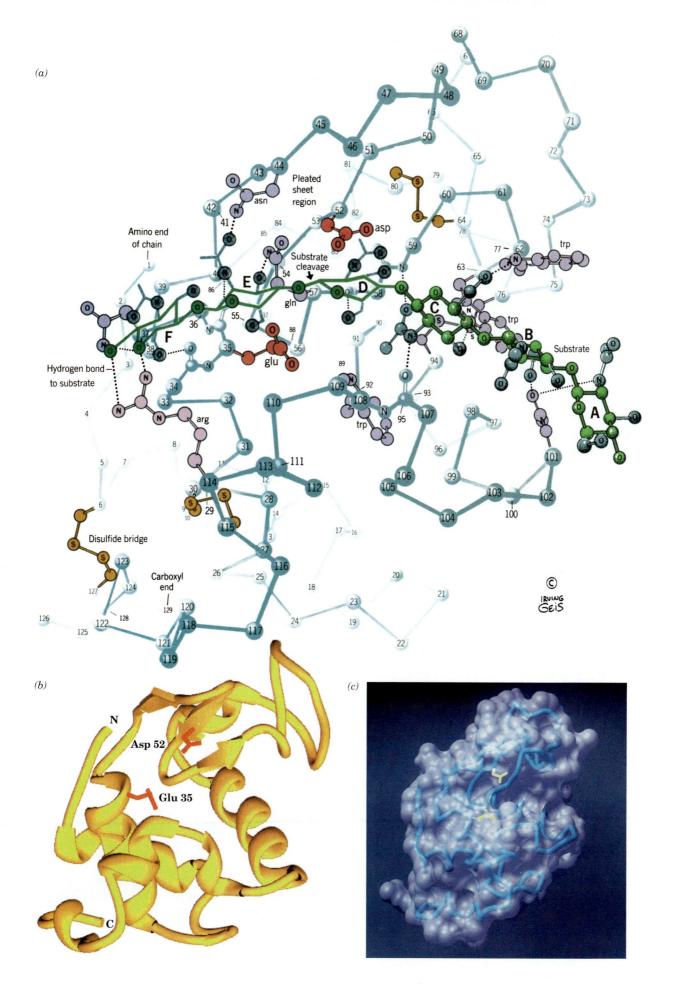

(a)

(b)

(c)

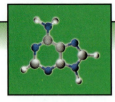

BOX 11-2

Perspectives in Biochemistry

Observing Enzyme Action by X-Ray Crystallography

The atomic rearrangements that occur during catalysis are to some extent observable during X-ray crystallographic analysis. Because protein crystals are mostly solvent, not only are the native structures of proteins preserved in the crystalline state, but often their catalytic functions are also intact. Substrates can diffuse through solvent channels in the crystal into the enzyme's active site. However, an enzyme binds its substrates only transiently before it catalyzes a reaction and releases the products. For this reason, the chemical transformations during an enzymatic reaction would be accessible only from "before and after" snapshots, that is, from the structure of the enzyme in the absence of substrate and, in some cases, the structure of the enzyme with its loosely bound products. Several approaches have been used to get around this limitation.

The X-ray structures of enzymes in complexes with slow-reacting substrates may be determined. In this approach, the substrate must remain stably bound to the enzyme for the several hours that are usually required to measure the crystal's X-ray diffraction intensities. Alternatively, an unreactive substrate analog can be used. In this case, the molecule binds much as a substrate would but is not subject to the catalytic reaction. This approach yields somewhat incomplete information, however, since some of the molecular interactions that allow catalysis to occur are missing or distorted. Nevertheless, such data, along with knowledge of nonenzymatic reaction mechanisms, often allows the enzymatic reaction mechanism to be deduced with a fair degree of certainty. Indeed, most of our present structural knowledge of enzyme–substrate interactions is based on this technique.

Enzymatic reactions, like all chemical reactions, are temperature sensitive. Therefore, cooling a crystallized enzyme can significantly slow the rate at which it reacts with a substrate that diffuses to it. For example, cooling a crystal to less than 50 K can slow reaction times from less than a microsecond to hours or days, long enough to measure a set of X-ray diffraction intensities.

Another approach, which has been successfully used to investigate the atomic events at the heme group of myoglobin, solves two problems inherent in analyzing rapid biochemical events. First, taking a "snapshot" of an enzyme in action requires a short exposure time, in analogy to conventional photography. Accordingly, very intense radiation must be used, in this case an X-ray beam generated by a synchrotron (a type of "atom smasher" in which electrons are accelerated around a circular track to near light speed, thereby emitting intense X-radiation). This radiation is many orders of magnitude more intense than that available from conventional X-ray generators. Second, all the molecules in the crystal must act simultaneously; otherwise, the data will be "blurry." In a study of CO dissociation from myoglobin (the CO binds to myoglobin in much the same way as does O_2 but much more tightly), the molecules were made to dissociate from the heme on cue by a flash of laser light with a duration of a few nanoseconds. Subsequent molecular motions, ultimately leading to the recombination of the CO with myoglobin, were monitored at intervals of microseconds to milliseconds. With refinements, this experimentally complicated technique may eventually prove useful for documenting the operations of enzymes, whose catalytic cycles are complete within nanoseconds.

distorting the glucose ring from its normal chair conformation to that of a half-chair (Fig. 11-18). This distortion, which renders atoms C1, C2, C5, and O5 of residue D coplanar, moves the C6 group from its normal equatorial

Figure 11-18 Chair and half-chair conformations. Hexose rings normally assume the chair conformation. It is postulated, however, that binding by lysozyme distorts the D ring into the half-chair conformation in which atoms C1, C2, C5, and O5 are coplanar. ✎ **See the Animated Figures.**

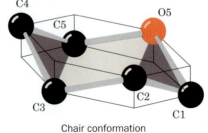

Chair conformation

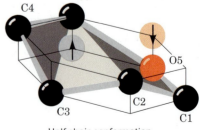

Half-chair conformation

Figure 11-19   The interactions of lysozyme with its substrate.   The view is into the binding cleft with the heavier edges of the rings facing the outside of the enzyme and the lighter ones against the bottom of the cleft. [Modified from a figure by Irving Geis.] 🔖 **See Kinemage Exercise 9.**

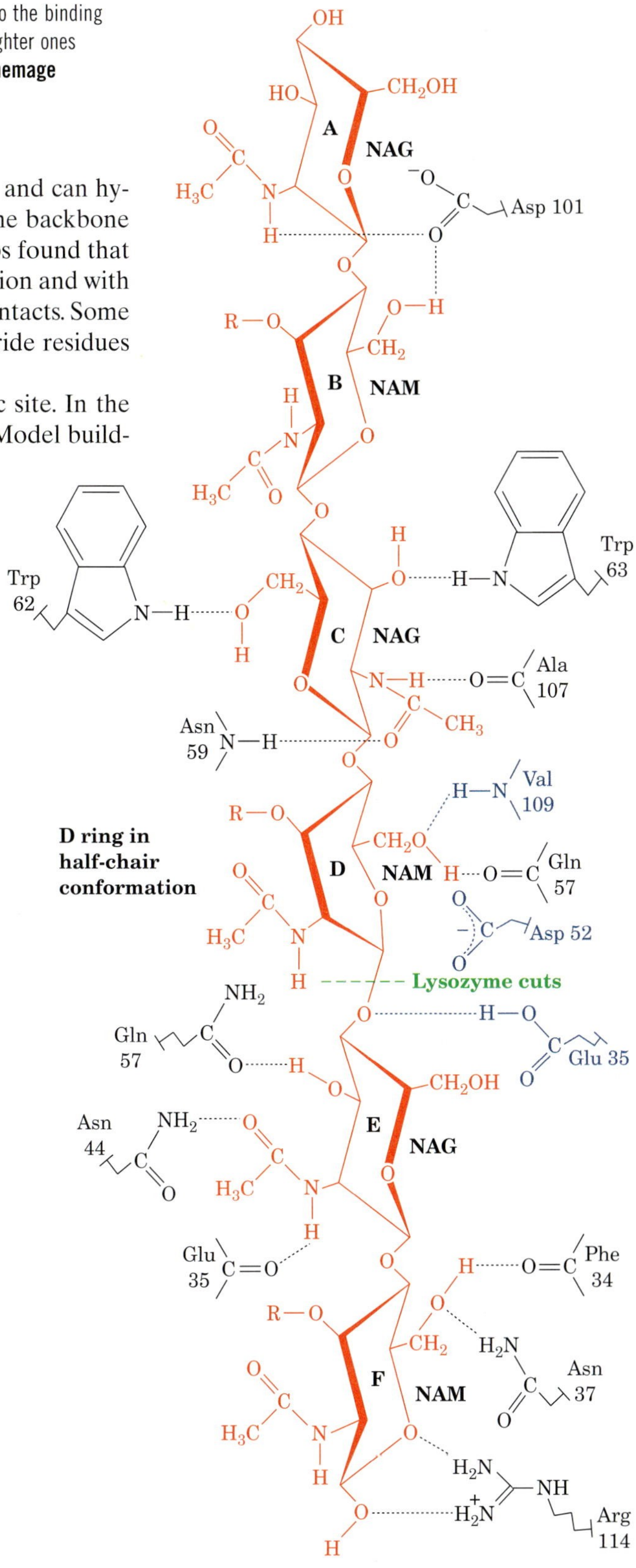

position to an axial position, where it makes no close contacts and can hydrogen bond to the backbone carbonyl group of Gln 57 and the backbone amido group of Val 109. Continuing the model building, Phillips found that residues E and F apparently bind to the enzyme without distortion and with a number of favorable hydrogen bonding and van der Waals contacts. Some of the hydrogen bonds through which the six substrate saccharide residues bind to the enzyme are diagrammed in Fig. 11-19.

We are almost in a position to identify lysozyme's catalytic site. In the enzyme's natural substrate, every second residue is an NAM. Model building, however, indicated that a lactyl side chain cannot be accommodated in the binding subsites of either residues C or E. Hence, the NAM residues must bind at positions B, D, and F.

$$\cdots\text{NAG}-\text{NAM}-\text{NAG}-\text{NAM}-\text{NAG}-\text{NAM}\cdots\longrightarrow \left(\begin{array}{c}\text{reducing}\\ \text{end}\end{array}\right)$$
$$\quad\ \text{A}\qquad\ \text{B}\qquad\ \text{C}\qquad\ \text{D}\qquad\ \text{E}\qquad\ \text{F}$$

The observation that lysozyme hydrolyzes $\beta(1\rightarrow4)$ linkages from NAM to NAG implies that bond cleavage occurs either between residues B and C or between residues D and E. Since $(\text{NAG})_3$ is stably bound but not cleaved while spanning subsites B and C, the probable cleavage site is between residues D and E. Indeed, lysozyme nearly quantitatively hydrolyzes $(\text{NAG})_6$ between the second and third residues from its reducing terminus (the end with a free C1—OH), just as is expected if the enzyme has six saccharide-binding subsites and cleaves its bound substrate between subsites D and E.

The bond that lysozyme cleaves was identified by carrying out the lysozyme-catalyzed hydrolysis of $(\text{NAG})_3$ in $\text{H}_2{}^{18}\text{O}$. The resulting product had ^{18}O bonded to the C1 atom of its newly liberated reducing terminus, thereby demonstrating that bond cleavage occurs between C1 and the bridge oxygen O1:

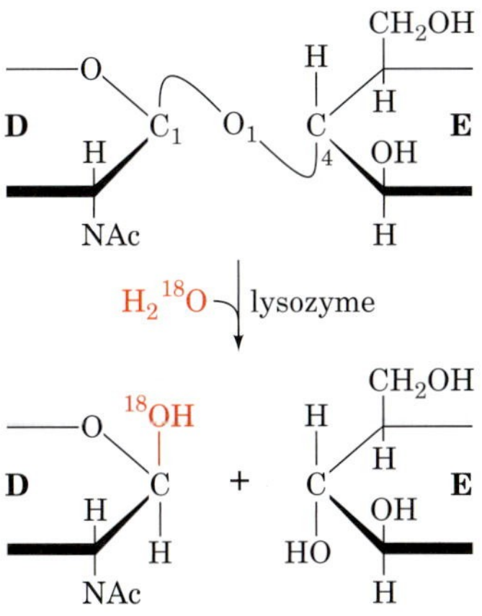

Thus, lysozyme catalyzes the hydrolysis of the C1—O1 bond of a bound substrate's D residue. Moreover, this reaction occurs with retention of configuration so that the D-ring product remains the β anomer.

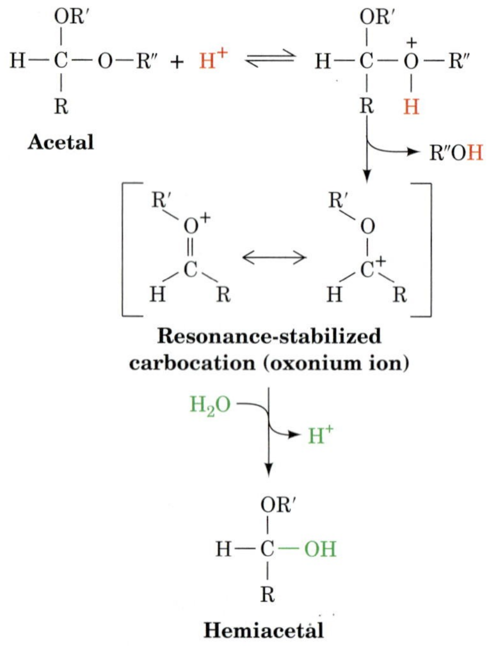

Figure 11-20 The mechanism of the nonenzymatic acid-catalyzed hydrolysis of an acetal to a hemiacetal. The reaction involves the protonation of one of the acetal's oxygen atoms followed by cleavage of its C—O bond to form an alcohol (R″OH) and a resonance-stabilized carbocation (oxonium ion). The addition of water to the oxonium ion forms the hemiacetal and regenerates the H⁺ catalyst. Note that the oxonium ion's C, O, H, R, and R′ atoms all lie in the same plane.

B Catalytic Mechanism

The reaction catalyzed by lysozyme, the hydrolysis of a glycoside, is the conversion of an acetal to a hemiacetal. Nonenzymatic acetal hydrolysis is an acid-catalyzed reaction that involves the protonation of a reactant oxygen atom followed by cleavage of its C—O bond (Fig. 11-20). This results in the transient formation of a resonance-stabilized carbocation that is called an **oxonium ion.** To attain resonance stabilization, the oxonium ion's R and R′ groups must be coplanar with its C, O, and H atoms. The oxonium ion then adds water to yield the hemiacetal and regenerate the acid catalyst. In searching for catalytic groups on an enzyme that mediate acetal hydrolysis, we should therefore seek a potential acid catalyst and possibly a group that can stabilize an oxonium ion transition state.

Glu 35 and Asp 52 Are Lysozyme's Catalytic Residues. The only functional groups in the immediate vicinity of lysozyme's reactive center that have the required catalytic properties are the side chains of Glu 35 and Asp 52. These side chains, which are disposed to either side of the glycosidic linkage to be cleaved (Fig. 11-17), have markedly different environments. Asp 52 is surrounded by several conserved polar residues with which it forms a complex hydrogen bonded network. Asp 52 is therefore predicted to have a normal pK; that is, it should be unprotonated and hence negatively charged throughout the 3 to 8 pH range over which lysozyme is catalytically active. Thus, Asp 52 can function to electrostatically stabilize an oxonium ion. In contrast, the carboxyl group of Glu 35 is nestled in a predominantly nonpolar pocket where it is likely to remain protonated at unusually high pH values for carboxyl groups. This residue can therefore act as an acid catalyst. Studies using protein-modifying reagents and site-directed mutagenesis (e.g., changing Asp 52 to Asn and Glu 35 to Gln) have verified that these residues are catalytically important.

The Lysozyme Reaction Proceeds via a Covalent Intermediate. As originally proposed by Phillips, the lysozyme mechanism involved acid catalysis by Glu 35 and stabilization of an oxonium ion intermediate by the carboxylate group of Asp 52. Formation of a covalent bond between the carboxylate oxygen of Asp 52 and C1 of the substrate D residue did not appear possible because these atoms are separated by ~3 Å (the C—O covalent bond length is ~1.4 Å). However, more recent research has revealed that the lysozyme reaction, in fact, proceeds via a covalent intermediate (see below). The reaction begins when lysozyme attaches to a bacterial cell wall by binding to a hexasaccharide unit. This distorts the D residue toward the half-chair conformation. The rest of the catalytic mechanism occurs as follows (Fig. 11-21):

1. Glu 35 transfers its proton to O1 of the D ring (an example of general acid catalysis). This step involves a resonance stabilized oxonium ion transition state whose formation is facilitated by the strain in the D ring that distorts it to the half-chair conformation in which C1, C2, C5, and O5 are coplanar (catalysis by the preferential binding of the transition state).

2. The carboxylate group of Asp 52 acting as a nucleophile attacks the now electron-poor C1 of the D ring to form a covalent glycosyl–enzyme intermediate (covalent catalysis).

3. Following the departure of the leaving group containing the E ring, water enters the active site to hydrolyze the covalent bond and regenerate the active site groups with the assistance of Glu 35 (general base catalysis).

4. The enzyme then releases the D-ring product, completing the catalytic reaction.

The double-displacement mechanism diagrammed in Fig. 11-21 allows the incoming water molecule to attach to the same face of the D residue as the E residue it replaces. Consequently, the configuration of the D residue is retained. A single-displacement reaction, in which water directly displaces

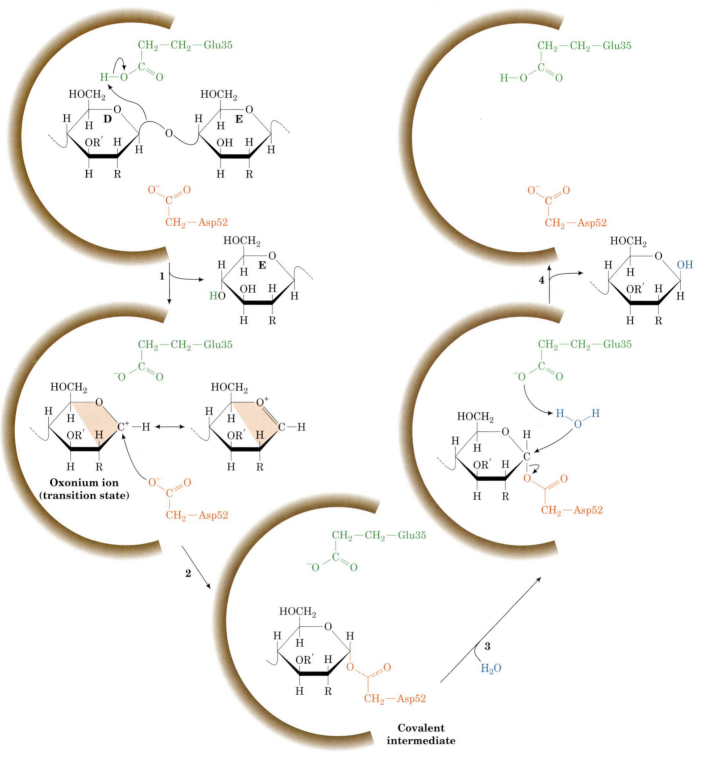

Figure 11-21   The lysozyme reaction mechanism.   Glu 35 acts as an acid catalyst, and Asp 52 acts as a covalent catalyst. Only the substrate D and E rings are shown. R represents the *N*-acetyl group at C2, and R′ represents the CH_3CHCOO^- group at C3. The resonance-stabilized oxonium ion transition state requires that C1, C2, C5 and O5 be coplanar (*shading*) creating a half-chair conformation. Step 4 includes the participation of an oxonium ion transition state that is not shown.   **See Kinemage Exercise 9.**

Table 11-4 Binding Free Energies of HEW Lysozyme Subsites

Site	Bound Saccharide	Binding Free Energy (kJ · mol^{-1})
A	NAG	−7.5
B	NAM	−12.3
C	NAG	−23.8
D	**NAM**	**+12.1**
E	NAG	−7.1
F	NAM	−7.1

Source: Chipman, D.M. and Sharon, N., *Science* **165**, 459 (1969).

the leaving group, would invert the configuration at C1 of the D ring between the substrate and product, a result that is not observed.

Experimental Support for the Lysozyme Mechanism. Many of the mechanistic investigations of lysozyme have had the elusive goal of establishing the catalytic role of strain. Measurements of the binding equilibria of various oligosaccharides to lysozyme indicate that all saccharide residues except that binding to the D subsite contribute energetically toward the binding of substrate to lysozyme; only binding NAM in the D subsite requires a free energy input (Table 11-4). This, according to the above mechanism, represents the energy penalty of straining the D ring from its preferred chair conformation toward the half-chair form.

Experiments with inhibitors also support the above mechanism. As we have discussed in Section 11-3E, an enzyme that catalyzes a reaction by the preferential binding of its transition state has a greater binding affinity for an inhibitor that has the transition state geometry (a transition state analog) than it does for its substrate. The δ-lactone analog of (NAG)$_4$ (Fig. 11-22), which binds tightly to lysozyme, is a transition state analog of lysozyme since *this compound's lactone ring has the half-chair conformation that geometrically resembles the proposed oxonium ion transition state of the substrate's D ring.* X-Ray studies indicate, in accordance with prediction, that this inhibitor binds to lysozyme's A—B—C—D subsites such that the lactone ring occupies the D subsite in a half-chair-like conformation.

Despite the foregoing, *the role of substrate distortion in lysozyme catalysis has been questioned.* Theoretical studies by Michael Levitt and Arieh Warshel on substrate binding by lysozyme suggested that the protein is too flexible to mechanically distort the D ring of a bound substrate. Indeed, Nathan Sharon and David Chipman determined that the NAG lactone inhibitor (Fig. 11-22) binds to the D subsite with only 9.2 kJ · mol^{-1} greater affinity than does NAG. This quantity, as is explained in Section 11-3E, corresponds to no more than an ~40-fold rate enhancement of the lysozyme reaction as a result of strain (recall that the difference in binding energy between a transition state analog and a substrate is indicative of the enzyme's rate enhancement arising from the preferential binding of the transition state). Such an enhancement is hardly a major portion of lysozyme's ~10^8-fold rate enhancement (accounting for only ~20% of the reaction's $\Delta\Delta G_{cat}^{\ddagger}$).

Nevertheless, the X-ray structure of lysozyme in complex with NAM—NAG—NAM shows that the trisaccharide binds, as predicted, to

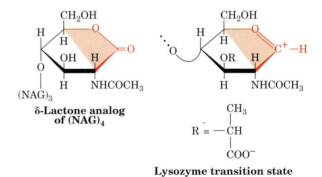

Figure 11-22 Transition state analog inhibition of lysozyme. The δ-lactone analog of (NAG)$_4$ (*left*) resembles the transition state of the lysozyme reaction (*right*). Note that atoms C1, C2, C5, and O5 in each structure are coplanar (as indicated by shading), consistent with the half-chair conformation of the hexose ring.

the B, C, and D subsites of lysozyme. *The NAM in the D subsite is distorted to the half-chair conformation with its —C6H₂OH group in a nearly axial position due to steric clashes that would otherwise occur with the acetamido group of the C-subsite NAG.* This strained conformation is stabilized by a strong hydrogen bond between the D ring O6 and the backbone NH of Val 109 (transition state stabilization). Indeed, the mutation of Val 109 to Pro, which lacks the NH group to make such a hydrogen bond, inactivates the enzyme.

In 1965, when the structure of HEW lysozyme was first determined, the available evidence indicated that the enzyme did not form a covalent intermediate with its substrate. Phillips therefore proposed that the catalytic reaction proceeded noncovalently via an oxonium ion intermediate that was electrostatically stabilized by the nearby negatively charged side chain of Asp 52. The fact that the reaction proceeded with retention of configuration was explained by proposing that the enzyme shielded the back side of the putative oxonium ion intermediate from attack by water. However, since that time all other β-glycosidases of known structure that cleave glycosidic linkages with net retention of configuration (as does HEW lysozyme) have been shown to do so via a covalent glycosyl–enzyme intermediate. The active sites of these so-called **retaining β-glycosidases** structurally resemble that of HEW lysozyme. Moreover, there is no direct evidence indicative of the existence of a long-lived oxonium ion at the active site of any retaining β-glycosidase, including HEW lysozyme (the lifetime of a glucosyl oxonium ion in water is $\sim 10^{-12}$ s, a time period only slightly longer than that of a bond vibration). Consequently, there had been a growing suspicion that the HEW lysozyme reaction also proceeds via a covalent intermediate, as is diagramed in Fig. 11-21, and hence that the oxonium ion is the transition state on the way to this covalent intermediate rather than an intermediate itself.

If, in fact, the HEW lysozyme reaction proceeds via a covalent intermediate, the reason that it had never been observed is that its rate of breakdown must be much faster than its rate of formation. Hence, if this intermediate is to be experimentally observed, its rate of formation must be made significantly greater than its rate of breakdown. To do so, Stephen Withers capitalized on three phenomena. First, if, as postulated, the reaction goes through an oxonium ion transition state, all steps involving its formation should be slowed by the electron withdrawing effects of substituting F (the most electronegative element) at C2 of the D ring. Second, mutating Glu 35 to Gln (E35Q) removes the general acid–base that catalyzes the reaction, further slowing all steps involving the oxonium ion transition state. Third, substituting an additional F atom at C1 of the D ring accelerates the formation of the intermediate because this F is a good leaving group without the need of general acid catalysis. Making all three of these changes should increase the rate of formation of the proposed covalent intermediate relative to its breakdown and hence should result in its accumulation. Withers therefore incubated E35Q HEW lysozyme with NAG-β(1→4)-2-deoxy-2-fluoro-β-D-glucopyranosyl fluoride (**NAG2FGlcF**):

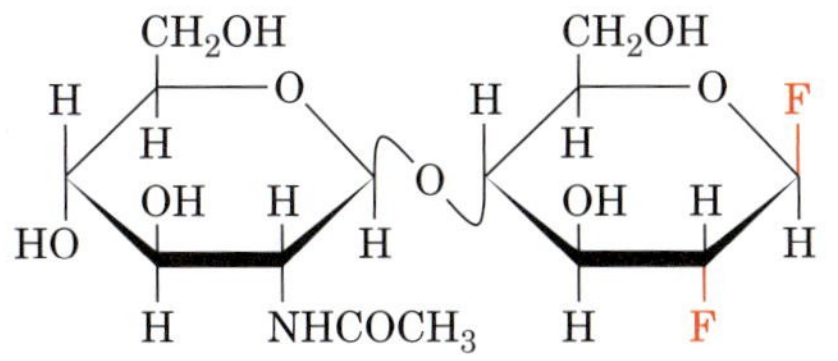

NAG2FGlcF

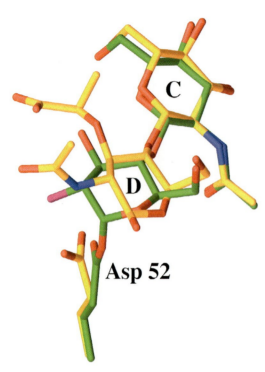

Figure 11-23   Position of the D ring during catalysis by lysozyme.   The position of the substrate's C and D rings and the catalytic Asp 52 are shown in superimposed X-ray structures of the covalent complex between E35Q lysozyme and NAG2FGlcF (C green, N blue, O red, and F magenta) and the noncovalent complex between lysozyme and a NAM—NAG—NAM substrate (C yellow, N blue, and O red). Note that the covalent bond between Asp 52 and C1 of the D ring forms when the D ring in the noncovalent complex relaxes from its distorted half-chair conformation to an undistorted chair conformation. The side chain of Asp 52 also undergoes a 45° rotation. [Based on X-ray structures by David Vocadlo and Stephen Withers, University of British Columbia, Vancouver, Canada; *and* Michael James, University of Alberta, Edmonton, Canada. PDBid 1H6M.]

Electrospray ionization mass spectrometry (ESI-MS; Section 5-3D) of this reaction mixture revealed a sharp peak at 14,683 D, consistent with the formation of the proposed covalent intermediate, but no significant peak at or near the 14,314-D molecular mass of the mutant enzyme alone.

The X-ray structure of this covalent complex, determined in 2001 by Withers, unambiguously reveals the expected ~1.4-Å-long covalent bond between C1 of the D ring and a side chain carboxyl O of Asp 52 (Fig. 11-23). This D ring adopts an undistorted chair conformation, thus indicating that it is a reaction intermediate rather than an approximation of the transition state. The superposition of this covalent complex with that of the above described complex of NAM—NAG—NAM with wild-type HEW lysozyme reveals how this covalent bond forms (Fig. 11-23). The shortening of the 3.2-Å distance between the D ring C1 and the Asp 52 O in the NAM—NAG—NAM complex to ~1.4 Å in the covalent complex is almost entirely a consequence of the relaxation of the D ring from the half-chair to the chair conformation combined with a ~45° rotation of the Asp 52 side chain about its C_β—C_γ bond; the positions of the D ring O4 and O6 atoms are essentially unchanged. Thus, although the noncovalent mechanism for HEW lysozyme proposed by Phillips had been widely accepted for over 35 years, it is now evident that this mechanism must be altered to take into account the transient formation of a covalent glycosyl–enzyme ester intermediate.

5 Serine Proteases

Our next example of enzymatic mechanisms is a diverse and widespread group of proteolytic enzymes known as the **serine proteases,** so named because they have a common catalytic mechanism involving a peculiarly reactive Ser residue. The serine proteases include digestive enzymes from prokaryotes and eukaryotes, as well as more specialized proteins that participate in development, blood coagulation (clotting), inflammation, and numerous other processes. In this section, we focus on some of the best studied serine proteases: chymotrypsin, trypsin, and elastase.

A The Active Site

Chymotrypsin, trypsin, and elastase are digestive enzymes that are synthesized by the pancreas and secreted into the duodenum (the small intestine's upper loop). All these enzymes catalyze the hydrolysis of peptide (amide) bonds but with different specificities for the side chains flanking the scissile (to be cleaved) peptide bond. Chymotrypsin is specific for a bulky hydrophobic residue preceding the scissile bond, trypsin is specific for a positively charged residue, and elastase is specific for a small neutral residue (Table 5-3). Together, they form a potent digestive team.

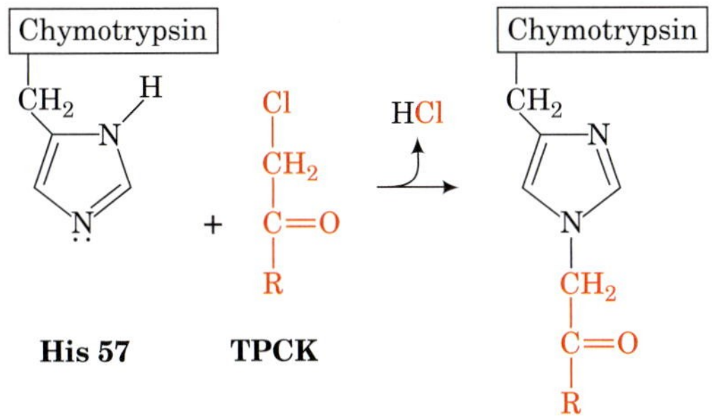

Figure 11-24 Reaction of TPCK with His 57 of chymotrypsin.

Chymotrypsin's catalytically important groups have been identified by chemical labeling studies. A diagnostic test for the presence of the active site Ser of serine proteases is its reaction with **diisopropylphosphofluoridate (DIPF),** which irreversibly inactivates the enzyme (*at right*). Other Ser residues, including those on the same protein, do not react with DIPF. *DIPF reacts only with Ser 195 of chymotrypsin, thereby demonstrating that this residue is the enzyme's active site Ser.* This specificity makes DIPF and related compounds extremely toxic (see Box 11-3).

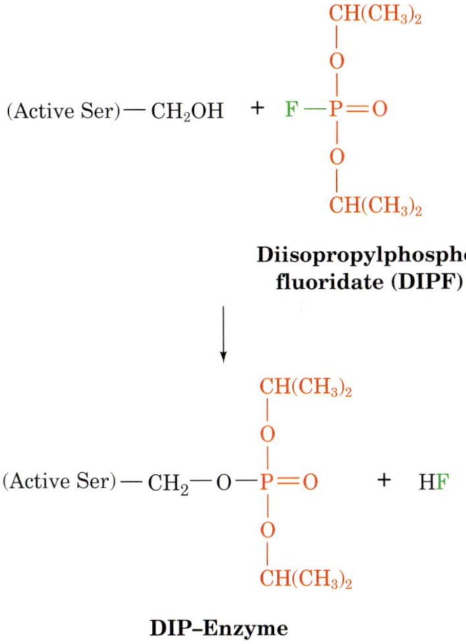

A second catalytically important residue, His 57, was discovered through **affinity labeling.** In this technique, a substrate analog bearing a reactive group specifically binds at the enzyme's active site, where it reacts to form a stable covalent bond with a nearby susceptible group (these reactive substrate analogs have been dubbed the "Trojan horses" of biochemistry). The affinity labeled group(s) can subsequently be isolated and identified.

Chymotrypsin specifically binds **tosyl-L-phenylalanine chloromethylketone (TPCK),**

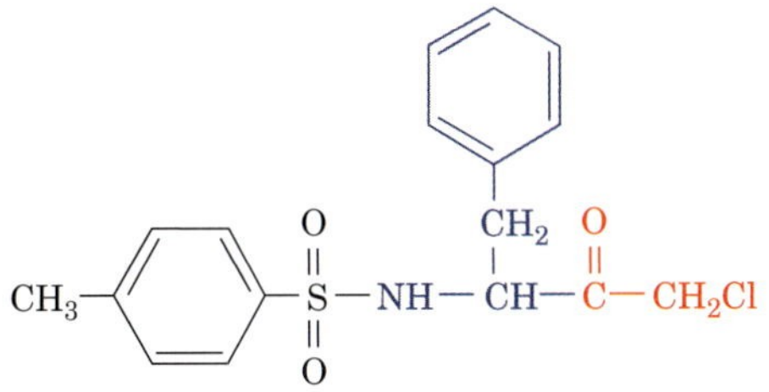

Tosyl-L-phenylalanine chloromethylketone

because of its resemblance to a Phe residue (one of chymotrypsin's preferred substrate residues). Active site–bound TPCK's chloromethylketone group is a strong alkylating agent; it reacts only with His 57 (Fig. 11-24), thereby inactivating the enzyme. Trypsin, which prefers basic residues, is similarly inactivated by **tosyl-L-lysine chloromethylketone:**

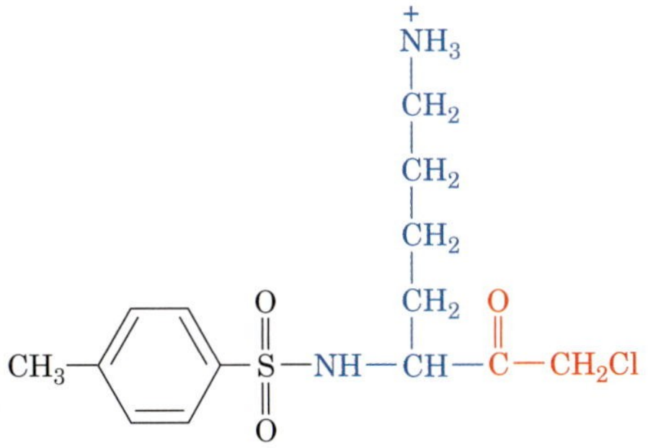

Tosyl-L-lysine chloromethylketone

**Diisopropylphospho-
fluoridate (DIPF)**

DIP–Enzyme

BOX 11-3

Biochemistry in Health and Disease

Nerve Poisons

The use of DIPF as an enzyme-inactivating agent came about through the discovery that organophosphorus compounds such as DIPF are potent nerve poisons. The neurotoxicity of DIPF arises from its ability to inactivate **acetylcholinesterase,** an enzyme that catalyzes the hydrolysis of **acetylcholine:**

$$(CH_3)_3\overset{+}{N}-CH_2-CH_2-O-\overset{\overset{\displaystyle O}{\|}}{C}-CH_3 \ + \ H_2O$$

Acetylcholine

acetylcholinesterase ↓

$$(CH_3)_3\overset{+}{N}-CH_2-CH_2-OH \ + \ \overset{\overset{\displaystyle O}{\|}}{\underset{-O}{C}}-CH_3 \ + \ H^+$$

Choline

The esterase activity of acetylcholinesterase, like that of chymotrypsin (Section 11-1B), requires a reactive Ser residue.

Acetylcholine is a **neurotransmitter:** It transmits nerve impulses across certain types of **synapses** (junctions between nerve cells). Acetylcholinesterase in the synapse normally degrades acetylcholine so that the nerve impulse has a duration of only a millisecond or so. The inactivation of acetylcholinesterase prevents hydrolysis of the neurotransmitter. As a result, the acetylcholine receptor, which is a Na^+–K^+ channel, remains open for longer than normal, thereby interfering with the regular sequence of nerve impulses. DIPF is so toxic to humans (death occurs through the inability to breathe) that it has been used militarily as a nerve gas. Related compounds, such as **parathion** and **malathion,**

Parathion

Malathion

are useful insecticides because they are far more toxic to insects than to mammals. Neurotoxins such as DIPF and **sarin** (which gained notoriety after its release by terrorists in a Tokyo subway in 1995)

Sarin

are inactivated by the enzyme **paraoxonase.** This enzyme occurs as two isoforms (one has Arg at position 192 and the other has Gln) with different activities, and individuals express widely differing levels of the enzyme. These factors may account for the large observed differences in individuals' sensitivity to nerve poisons.

B X-Ray Structures

Chymotrypsin, trypsin, and elastase are strikingly similar: The primary structures of these ~240-residue enzymes are ~40% identical (for comparison, the α and β chains of human hemoglobin have a 44% sequence identity). Furthermore, all these enzymes have a reactive Ser and a catalytically essential His. It therefore came as no surprise when their X-ray structures all proved to be closely related.

The structure of bovine chymotrypsin was elucidated in 1967 by David Blow. This was followed by the determination of the structures of bovine trypsin (Fig. 11-25) by Robert Stroud and Richard Dickerson, and porcine elastase by David Shotton and Herman Watson. Each of these proteins is folded into two domains, both of which have extensive regions of antiparallel β sheets in a barrel-like arrangement but contain little helix. For conven-

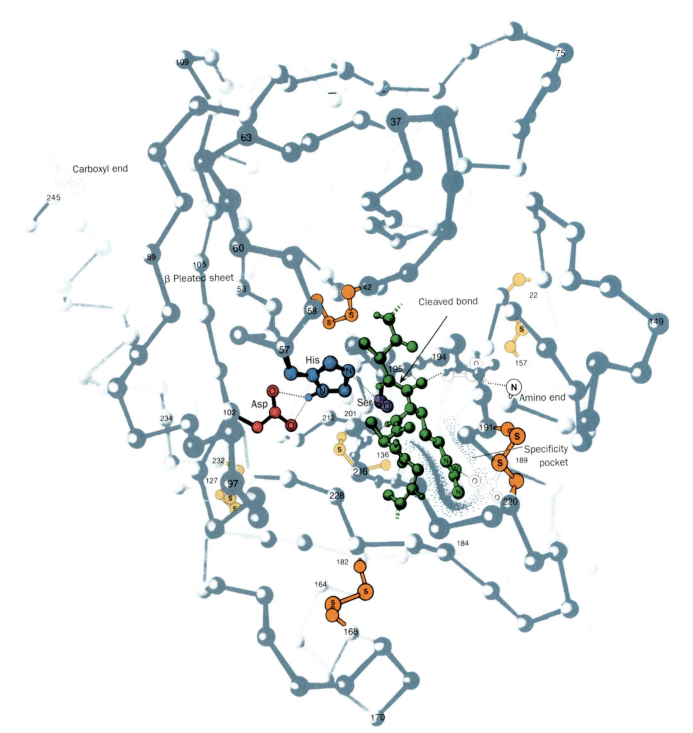

Figure 11-25   X-Ray structure of bovine trypsin.   A drawing of the enzyme showing its disulfide bonds (*orange*) and the side chains of the catalytic triad, Ser 195 (*purple*), His 57 (*blue*), and Asp 102 (*red*). A polypeptide substrate (*green*) is shown with its Arg side chain occupying the enzyme's specificity pocket (*stippling*). The active sites of chymotrypsin and elastase contain almost identically arranged catalytic triads. [Illustration Irving Geis/Geis Archives Trust. Copyright Howard Hughes Medical Institute. Reproduced with permission.]
🖉 **See Kinemage Exercise 10-1.**

ience in comparing the structures of these three enzymes, we shall assign them the same residue numbering system—that of bovine **chymotrypsinogen,** the 245-residue precursor of chymotrypsin (Section 11-5D).

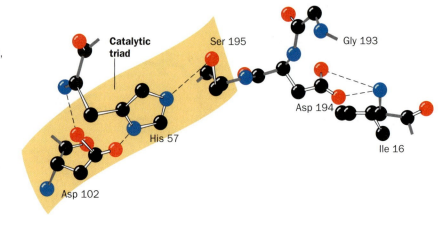

Figure 11-26 **The active site residues of chymotrypsin.** The view is in approximately the same direction as in Fig. 11-25. The catalytic triad consists of Ser 195, His 57, and Asp 102. [After Blow, D.M. and Steitz, T.A., *Annu. Rev. Biochem.* **39,** 86 (1970).]

In all three structures, the catalytically essential His 57 and Ser 195 residues are located in the enzyme's substrate-binding site (center of Fig. 11-25). The X-ray structures also show that Asp 102, which is present in all serine proteases, is buried in a nearby solvent-inaccessible pocket. *These three invariant residues form a hydrogen bonded constellation referred to as the* ***catalytic triad*** (Figs. 11-25 and 11-26).

Substrate Specificities Are Only Partially Rationalized. The X-ray structures of the above three enzymes suggest the basis for their differing substrate specificities (Fig. 11-27):

1. In chymotrypsin, the bulky aromatic side chain of the preferred Phe, Trp, or Tyr residue that contributes the carbonyl group of the scissile peptide fits snugly into a slitlike hydrophobic pocket located near the catalytic groups.

2. In trypsin, the residue corresponding to chymotrypsin Ser 189, which lies at the bottom of the binding pocket, is the anionic residue Asp. The cationic side chains of trypsin's preferred residues, Arg and Lys, can therefore form ion pairs with this Asp residue. The rest of chymotrypsin's specificity pocket is preserved in trypsin so that it can accommodate the bulky side chains of Arg and Lys.

3. Elastase is so named because it rapidly hydrolyzes the otherwise nearly indigestible Ala, Gly, and Val-rich protein **elastin** (a major connective tissue component). Elastase's binding pocket is largely occluded by the side chains of Val and Thr residues that replace the Gly residues lining the specificity pockets in both chymotrypsin and trypsin. Consequently elastase, whose substrate-binding site is better described as merely a depression, specifically cleaves peptide bonds after small neutral residues, particularly Ala. In contrast, chymotrypsin and trypsin hydrolyze such peptide bonds extremely slowly because these small substrates cannot be sufficiently immobilized on the enzyme surface for efficient catalysis to occur.

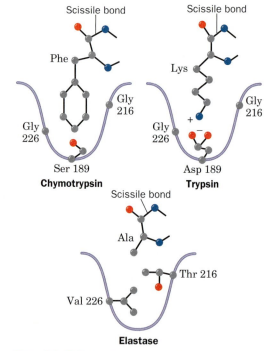

Figure 11-27 **Specificity pockets of three serine proteases.** The side chains of key residues that determine the size and nature of the specificity pocket are shown along with a representative substrate for each enzyme. Chymotrypsin prefers to cleave peptide bonds following large hydrophobic side chains; trypsin prefers Lys or Arg; and elastase prefers Ala, Gly, or Val. [After a drawing in Branden, C. and Tooze, J., *Introduction to Protein Structure* (2nd ed.), Garland Publishing, p. 213 (1999).]

Despite the foregoing, changing trypsin's Asp 189 to Ser by site-directed mutagenesis (Section 3-5D) does not switch its specificity to that of chymotrypsin but instead yields a poor, nonspecific protease. Replacing additional residues in trypsin's specificity pocket with those of chymotrypsin fails to significantly improve this catalytic activity. However, trypsin is converted to a reasonably active chymotrypsin-like enzyme when two surface loops that connect the walls of the specificity pocket (residues 185–188 and 221–225) are also replaced by those of chymotrypsin. These loops, which

are conserved in each enzyme, are apparently necessary not for substrate binding per se but for properly positioning the scissile bond. These results highlight an important caveat for genetic engineers: Enzymes are so exquisitely tailored to their functions that they often respond to mutagenic tinkering in unexpected ways.

Evolutionary Relationships among Serine Proteases.   We have seen that sequence and structural similarities among proteins reveal their evolutionary relationships (Sections 5-4 and 6-2D). *The great similarities among chymotrypsin, trypsin, and elastase indicate that these proteins arose through duplications of an ancestral serine protease gene followed by the divergent evolution of the resulting enzymes.* Indeed, the close structural resemblance of these pancreatic enzymes to certain bacterial proteases indicates that the primordial trypsin gene arose before the divergence of prokaryotes and eukaryotes.

There are several serine proteases whose primary and tertiary structures bear no discernible relationship to each other or to chymotrypsin. Nevertheless, these proteins also contain catalytic triads at their active sites whose structures closely resemble that of chymotrypsin. These enzymes include **subtilisin,** an endopeptidase that was originally isolated from *Bacillus subtilis,* and wheat germ **serine carboxypeptidase II,** an exopeptidase. Since the orders of the corresponding active site residues in the amino acid sequences of these serine proteases are quite different (Fig. 11-28), it seems highly improbable that they could have evolved from a common ancestor protein. These enzymes apparently constitute a remarkable example of **convergent evolution:** *Nature seems to have independently discovered the same catalytic mechanism several times.*

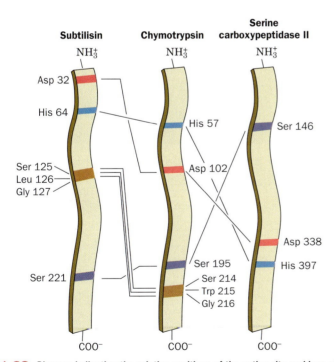

Figure 11-28 **Diagram indicating the relative positions of the active site residues of three unrelated serine proteases.**   The catalytic triads in subtilisin, chymotrypsin, and serine carboxypeptidase II each consist of Ser, His, and Asp residues. The peptide backbones of Ser 214, Trp 215, and Gly 216 in chymotrypsin, and their counterparts in subtilisin, participate in substrate-binding interactions. [After Robertus, J.D., Alden, R.A., Birktoft, J.J., Kraut, J., Powers, J.C., and Wilcox, P.E., *Biochemistry* **11,** 2449 (1972).] 🦎 **See Kinemage Exercise 10-2.**

C Catalytic Mechanism

A catalytic mechanism based on considerable chemical and structural data has been formulated and is given here in terms of chymotrypsin (Fig. 11-29), although it applies to all serine proteases and certain other hydrolytic enzymes:

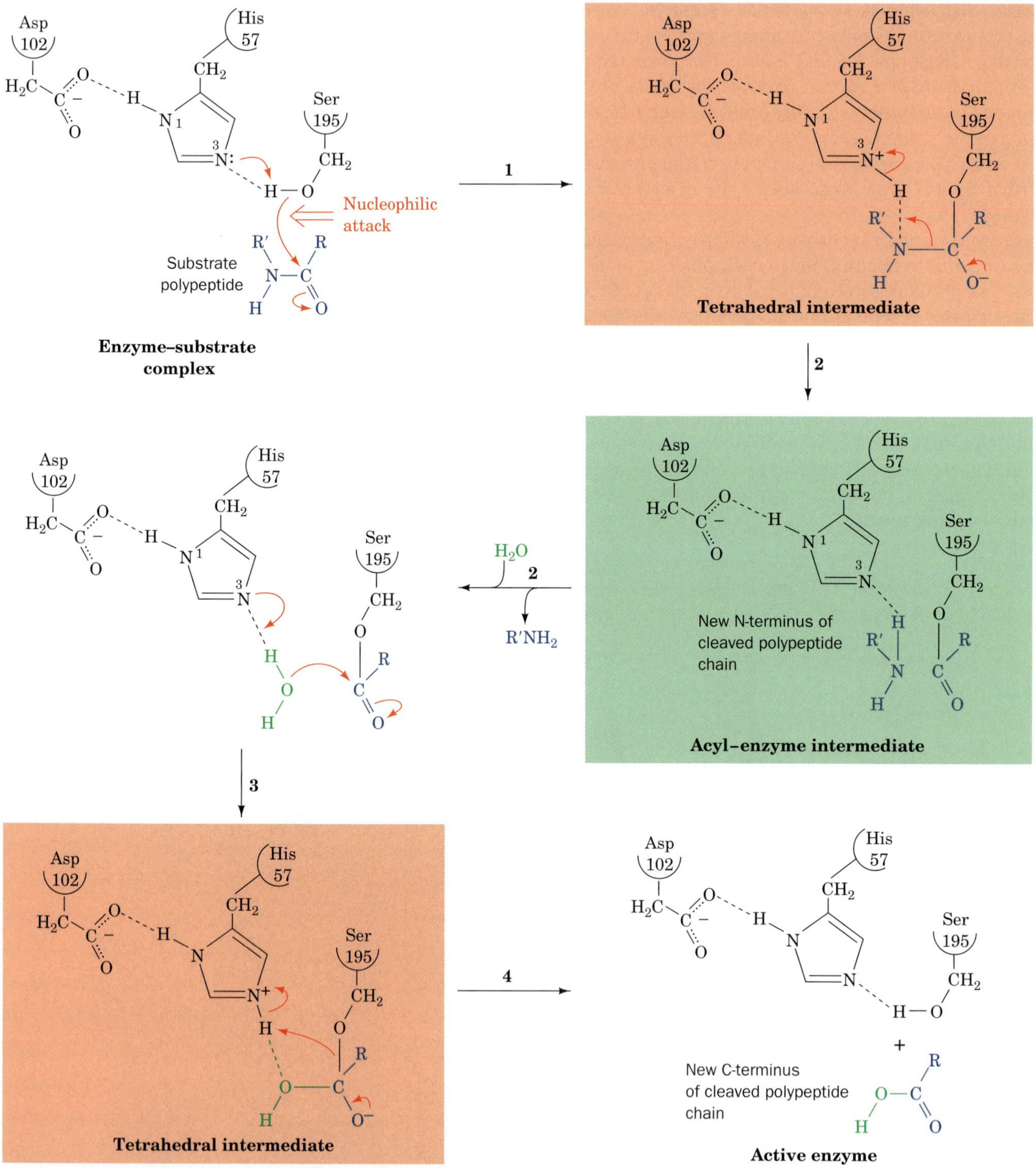

Figure 11-29 Key to Function. The catalytic mechanism of the serine proteases.   The reaction involves (1) the nucleophilic attack of the active site Ser on the carbonyl carbon atom of the scissile peptide bond to form the tetrahedral intermediate; (2) the decomposition of the tetrahedral intermediate to the acyl– enzyme intermediate through general acid catalysis by the active site Asp-polarized His, followed by loss of the amine product and its replacement by a water molecule; (3) the reversal of Step 2 to form a second tetrahedral intermediate; and (4) the reversal of Step 1 to yield the reaction's carboxyl product and the active enzyme.

1. After chymotrypsin has bound a substrate, Ser 195 nucleophilically attacks the scissile peptide's carbonyl group to form the **tetrahedral intermediate,** which resembles the reaction's transition state (covalent catalysis). X-Ray studies indicate that Ser 195 is ideally positioned to carry out this nucleophilic attack (i.e., catalysis also occurs by proximity and orientation effects). This nucleophilic attack involves transfer of a proton to the imidazole ring of His 57, thereby forming an imidazolium ion (general base catalysis). This process is aided by the polarizing effect of the unsolvated carboxylate ion of Asp 102, which is hydrogen bonded to His 57 (electrostatic catalysis). The tetrahedral intermediate has a well-defined, although transient, existence. We shall see that *much of chymotrypsin's catalytic power derives from its preferential binding of the transition state leading to this intermediate (transition state binding catalysis).*

See Guided Exploration 10

The catalytic mechanism of serine proteases.

2. The tetrahedral intermediate decomposes to the **acyl–enzyme intermediate** under the driving force of proton donation from N3 of His 57 (general acid catalysis). The amine leaving group ($R'NH_2$, the new N-terminal portion of the cleaved polypeptide chain) is released from the enzyme and replaced by water from the solvent. The acyl–enzyme intermediate is highly susceptible to hydrolytic cleavage.

3 & 4. The acyl–enzyme intermediate is deacylated by what is essentially the reversal of the previous steps followed by the release of the resulting carboxylate product (the new C-terminal portion of the cleaved polypeptide chain), thereby regenerating the active enzyme. In this process, water is the attacking nucleophile and Ser 195 is the leaving group.

Serine Proteases Preferentially Bind the Transition State. Detailed comparisons of the X-ray structures of several serine protease–inhibitor complexes have revealed a further structural basis for catalysis in these enzymes (Fig. 11-30):

1. The conformational distortion that occurs with the formation of the

(a)

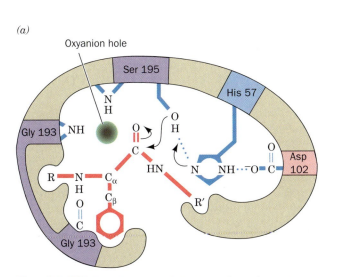

(b)

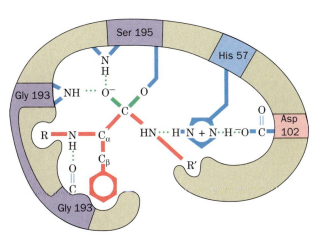

Figure 11-30   Transition state stabilization in the serine proteases.
(*a*) When the substrate binds to the enzyme, the trigonal carbonyl carbon of the scissile peptide is conformationally constrained from binding in the oxyanion hole (*upper left*). (*b*) In the tetrahedral intermediate, the now charged carbonyl oxygen of the scissile peptide (the oxyanion) enters the oxyanion hole and hydrogen bonds to the backbone NH groups of Gly 193 and Ser 195. The

consequent conformational distortion permits the NH group of the residue preceding the scissile peptide bond to form an otherwise unsatisfied hydrogen bond to Gly 193. Serine proteases therefore preferentially bind the tetrahedral intermediate. [After Robertus, J.D., Kraut, J., Alden, R.A., and Birktoft, J.J., *Biochemistry* **11,** 4302 (1972).] 🔍 **See Kinemage Exercise 10-3.**

tetrahedral intermediate causes the now anionic carbonyl oxygen of the scissile peptide to move deeper into the active site so as to occupy a previously unoccupied position called the **oxyanion hole.**

2. There, it forms two hydrogen bonds with the enzyme that cannot form when the carbonyl group is in its normal trigonal conformation. The two enzymatic hydrogen bond donors were first noted by Joseph Kraut to occupy corresponding positions in chymotrypsin and subtilisin. He proposed the existence of the oxyanion hole on the basis of the premise that convergent evolution had made the active sites of these unrelated enzymes functionally identical.

3. The tetrahedral distortion, moreover, permits the formation of an otherwise unsatisfied hydrogen bond between the enzyme and the backbone NH group of the residue preceding the scissile peptide bond.

This preferential binding of the transition state (or the tetrahedral intermediate) over the enzyme–substrate complex or the acyl–enzyme intermediate is responsible for much of the catalytic efficiency of serine proteases. Thus, mutating any or all of the residues in chymotrypsin's catalytic triad yields enzymes that still enhance proteolysis by $\sim 5 \times 10^4$-fold over the uncatalyzed reaction (versus a rate enhancement of $\sim 10^{10}$ for the native enzyme). Similarly, the reason that DIPF is such an effective inhibitor of serine proteases is because its tetrahedral phosphate group makes this compound a transition state analog.

Low-Barrier Hydrogen Bonds May Stabilize the Transition State. The transition state of the chymotrypsin reaction is stabilized not just through the formation of additional hydrogen bonds in the oxyanion hole but possibly also by the formation of an unusually strong hydrogen bond. Proton transfers between hydrogen bonded groups (D—H$\cdots$A) occur at physiologically reasonable rates only when the pK of the proton donor is no more than 2 or 3 pH units greater than that of the protonated form of the proton acceptor. However, when the pK's of the hydrogen bonding donor (D) and acceptor (A) groups are nearly equal, the distinction between them breaks down: *The hydrogen atom becomes more or less equally shared between them* (D$\cdots$H$\cdots$A). Such **low-barrier hydrogen bonds (LBHBs)** are unusually short and strong. Their free energies, as measured in the gas phase, are as high as -40 to -80 kJ$\cdot$mol^{-1} (versus -12 to -30 kJ$\cdot$mol^{-1} for normal hydrogen bonds) and they exhibit a D$\cdots$A bond length of <2.55 Å for O—H$\cdots$O and <2.65 Å for N—H$\cdots$O (versus 2.8 to 3.1 Å for normal hydrogen bonds).

LBHBs are unlikely to exist in dilute aqueous solution because water molecules, which are excellent hydrogen bond donors and acceptors, effectively compete with D—H and A for hydrogen bonding sites. However, LBHBs may exist in the nonaqueous active sites of enzymes. In fact, experimental evidence indicates that in the serine protease catalytic triad, the pK's of the protonated His and Asp are nearly equal, and the hydrogen bond between His and Asp has an unusually short N$\cdots$O distance of 2.62 Å with the H atom nearly centered between the N and O atoms. These findings are consistent with the formation of an LBHB in the transition state. This suggests that the enzyme uses the "strategy" of converting a weak hydrogen bond in the initial enzyme–substrate complex to a strong hydrogen bond in the transition state, thereby facilitating proton transfer while applying the difference in the free energy between the normal and low-barrier hydrogen bonds to preferentially binding the transition state.

Although several studies have revealed the existence of unusually short hydrogen bonds in enzyme active sites, it is far more difficult to demon-

strate experimentally that they are unusually strong, as LBHBs are predicted to be. In fact, several studies of the strengths of unusually short hydrogen bonds in organic model compounds in nonaqueous solutions suggest that these hydrogen bonds are not unusually strong. Consequently, a lively debate has ensued as to the catalytic significance of LBHBs. However, if enzymes do not form LBHBs, it remains to be explained how the conjugate base of an acidic group (e.g., Asp 102) partially abstracts a proton from a far more basic group (e.g., His 57), a feature of numerous enzyme mechanisms.

The Tetrahedral Intermediate Resembles the Complex of Trypsin with Trypsin Inhibitor.

Perhaps the most convincing structural evidence for the existence of the tetrahedral intermediate was provided by Robert Huber in an X-ray study of the complex between **bovine pancreatic trypsin inhibitor (BPTI)** and trypsin. The 58-residue BPTI binds to the active site region of trypsin to form a complex with a tightly packed interface and a network of hydrogen bonded cross-links. This interaction prevents any trypsin that is prematurely activated in the pancreas from digesting that organ (Section 11-5D). The complex's 10^{13} M^{-1} association constant, among the largest of any known protein–protein interaction, emphasizes BPTI's physiological importance.

The portion of BPTI in contact with the trypsin active site resembles bound substrate. A specific Lys side chain of BPTI occupies the trypsin specificity pocket (Fig. 11-31a), and the inhibitor's Lys-Ala peptide bond is positioned as if it were the scissile peptide bond (Fig. 11-31b). What is most remarkable about the BPTI–trypsin complex is that its conformation is well along the reaction coordinate toward the tetrahedral intermediate: The side chain oxygen of trypsin Ser 195, the active Ser, is in closer-than-van der Waals contact (2.6 Å) with the pyramidally distorted carbonyl carbon of BPTI's "scissile" peptide. However, the proteolytic reaction cannot proceed past this point because of the rigidity of the complex and because it is so tightly sealed that the leaving group cannot leave and water cannot enter the reaction site.

Protease inhibitors are common in nature, where they have protective and regulatory functions. For example, certain plants release protease inhibitors in response to insect bites, thereby causing the offending insect to starve by inactivating its digestive enzymes. Protease inhibitors constitute ~10% of the blood plasma proteins. For instance, **α_1-proteinase inhibitor,** which is secreted by the liver, inhibits **leukocyte elastase** (leukocytes are a type of white blood cell; the action of leukocyte elastase is thought to be part of the inflammatory process). Pathological variants of α_1-proteinase inhibitor with reduced activity are associated with **pulmonary emphysema,** a degenerative disease of the lungs resulting from the hydrolysis of its elastic fibers. Smokers also suffer from reduced activity of their α_1-proteinase inhibitor because smoking oxidizes a required Met residue.

(a)

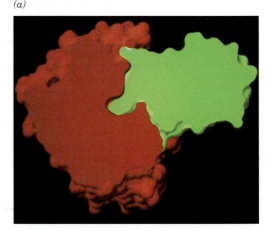

(b)

Figure 11-31 The trypsin–BPTI complex.   (a) The X-ray structure is shown as a computer-generated cutaway drawing indicating how trypsin (*red*) binds BPTI (*green*). The green protrusion extending into the red cavity near the center of the figure represents the inhibitor's Lys 15 side chain occupying trypsin's specificity pocket. Note the close complementary fit of the two proteins. [Courtesy of Michael Connolly, New York University.] (b) Trypsin Ser 195 is in closer-than-van der Waals contact with the carbonyl carbon of BPTI's scissile peptide, which is pyramidally distorted toward Ser 195. The normal proteolytic reaction is apparently arrested somewhere along the reaction coordinate preceding the tetrahedral intermediate. 🐍 **See Kinemage Exercise 10-1.**

The Tetrahedral Intermediate Has Been Directly Observed. Because the tetrahedral intermediate resembles the transition state of the serine protease reaction, it is thought to be short-lived and unstable. Recently, a series of X-ray structures of porcine pancreatic elastase with a peptide substrate were obtained that show the progress of the reaction from the acyl–enzyme intermediate stage to the release of product. This second phase of the proteolysis reaction includes a tetrahedral intermediate (Fig. 11-29).

This acyl–enzyme complex, which is stable at pH 5.0, exhibits the expected structure, with the substrate's C-terminal Ile residue covalently linked via an ester bond to Ser 195 (Fig. 11-32a). In this first view of a serine protease acyl–enzyme intermediate, the acyl group is fully planar, with no distortion toward a tetrahedral geometry. A water molecule is located near the intermediate's ester bond, hydrogen-bonded to His 57, where it appears poised to nucleophilically attack the ester linkage. At pH 5.0, His 57 is protonated and acts as a hydrogen bond donor to water (it cannot function as a base catalyst at this pH).

The hydrolytic reaction was triggered by deprotonation of His 57, accomplished by immersing the acyl–enzyme crystal in a solution of pH 9.0. After a period of 1 or 2 minutes, the crystals were frozen in liquid nitrogen to halt the reaction so that the X-ray structure of the enzyme complex could be determined (see Box 11-2). In this way, the tetrahedral intermediate was trapped and observed (Fig. 11-32b).

During formation of the tetrahedral intermediate, the oxyanion hole does not undergo any change in its structure, but the peptide substrate moves within its binding pocket and becomes distorted toward a tetrahedral geometry. The tetrahedral intermediate has the expected shape, similar to known transition state analog inhibitors. However, it does not bind so tightly to the amide group of the oxyanion hole (Fig. 11-30) that it would not be able to subsequently dissociate.

(a)

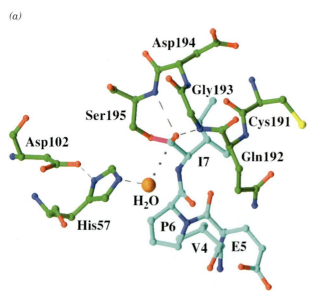

(b)

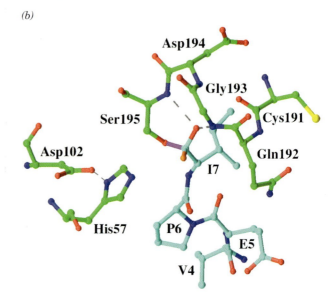

Figure 11-32 Structure of the acyl–enzyme and tetrahedral intermediates. Porcine pancreatic elastase was incubated with a heptapeptide substrate (YPFVEPI, using the one-letter code). Only residues 4–7 are visible. The protease residues are specified by the three-letter code. Atoms are colored according to type with elastase C green, substrate C cyan, N blue, O red, and S yellow. (a) At pH 5.0, a covalent bond (*magenta*) links the Ser 195 O atom to the C-terminal (I7) C atom of the substrate. A water molecule (*orange*) appears poised to nucleophilically attack the acyl–enzyme's carbonyl C atom.

The dashed lines represent catalytically important hydrogen bonds, and the dotted line indicates the trajectory that the water molecule presumably follows in nucleophilically attacking the acyl group's carbonyl C atom. (b) When the complex is brought to pH 9.0 and then rapidly frozen, the water molecule becomes a hydroxyl substituent (*orange*) to the carbonyl C atom, thereby yielding the tetrahedral intermediate, which resembles the transition state. [Based on X-ray structures by Christopher Schofield and Janos Hadju, University of Oxford, U.K. PDBids (a) 1HAX and (b) 1HAZ.]

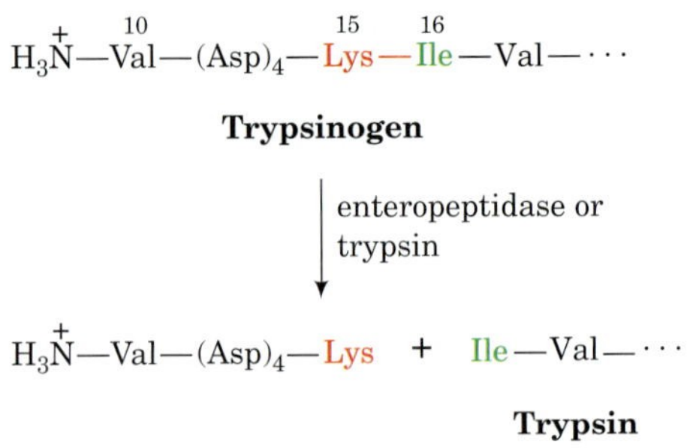

Figure 11-33 The activation of trypsinogen to trypsin. Proteolytic excision of the N-terminal hexapeptide is catalyzed by either enteropeptidase or trypsin. The chymotrypsinogen residue-numbering system is used here; that is, Val 10 is actually trypsinogen's N-terminus and Ile 16 is trypsin's N-terminus.

$$\overset{+}{H_3N}-\overset{10}{Val}-(Asp)_4-\overset{15}{Lys}-\overset{16}{Ile}-Val-\cdots$$

Trypsinogen

enteropeptidase or trypsin

$$\overset{+}{H_3N}-Val-(Asp)_4-Lys \quad + \quad Ile-Val-\cdots$$

Trypsin

D Zymogens

Proteolytic enzymes are usually biosynthesized as somewhat larger inactive precursors known as **zymogens** (enzyme precursors, in general, are known as **proenzymes**). In the case of digestive enzymes, the reason for this is clear: If these enzymes were synthesized in their active forms, they would digest the tissues that synthesized them. Indeed, **acute pancreatitis**, a painful and sometimes fatal condition that can be precipitated by pancreatic trauma, is characterized by the premature activation of the digestive enzymes synthesized by this organ.

The activation of **trypsinogen,** the zymogen of trypsin, occurs when trypsinogen enters the duodenum from the pancreas. **Enteropeptidase,** a serine protease whose secretion from the duodenal mucosa is under hormonal control, excises the N-terminal hexapeptide from trypsinogen by specifically cleaving its Lys 15—Ile 16 peptide bond (Fig. 11-33). Since this activating cleavage occurs at a trypsin-sensitive site (recall that trypsin cleaves after Arg and Lys residues), the small amount of trypsin produced by enteropeptidase also catalyzes trypsinogen activation, generating even more trypsin, etc. Thus, trypsinogen activation is said to be **autocatalytic.** Chymotrypsinogen is then activated by trypsin-catalyzed cleavage of its Arg 15—Ile 16 peptide bond.

Proelastase, the zymogen of elastase, is activated by a single tryptic cleavage that excises a short N-terminal peptide. Trypsin also activates pancreatic **procarboxypeptidases A** and **B** and **prophospholipase A$_2$** (Section 9-1C). The autocatalytic nature of trypsinogen activation and the fact that trypsin activates other hydrolytic enzymes makes it essential that trypsinogen not be activated in the pancreas. We have seen that the all but irreversible binding of trypsin inhibitors such as BPTI to trypsin is a defense against trypsinogen's inappropriate activation.

Sequential proenzyme activation makes it possible to quickly generate large quantities of active enzymes in response to diverse physiological signals. For example, the serine proteases that lead to blood clotting are synthesized as zymogens by the liver and circulate until they are activated by injury to a blood vessel (see Box 11-4).

Zymogens Have Distorted Active Sites. Since the zymogens of trypsin, chymotrypsin, and elastase have all their catalytic residues, why aren't they enzymatically active? Comparisons of the X-ray structures of trypsinogen with that of trypsin, and of chymotrypsinogen with that of chymotrypsin, show that on activation, the newly liberated N-terminal Ile 16 residue moves from the surface of the protein to an internal position, where its free cationic amino group forms an ion pair with the invariant anionic Asp 194, which is close to the catalytic triad (Fig. 11-26). Without this conformational change, the enzyme cannot properly bind its substrate or stabilize the tetrahedral intermediate because its specificity pocket and oxyanion hole are improperly formed. This provides further structural evidence favoring the role of preferential transition state binding in the catalytic mechanism of serine proteases. Nevertheless, because their catalytic triads are structurally intact, the zymogens of serine proteases actually have low levels of enzymatic activity, an observation that was made only after the above structural comparisons suggested that this might be the case.

BOX 11-4

Biochemistry in Health and Disease

The Blood Coagulation Cascade

When a blood vessel is damaged, a clot forms as a result of the aggregation of platelets (small enucleated blood cells) and the formation of an insoluble **fibrin** network that traps additional blood cells.

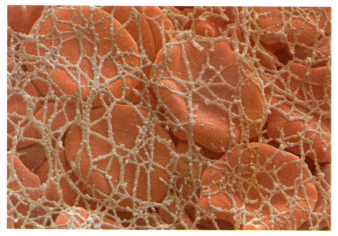

[© Andrew Syred/Photo Researchers]

Fibrin is produced from the soluble circulating protein **fibrinogen** through the action of the serine protease **thrombin.** Thrombin is the last in a series of coagulation enzymes that are sequentially activated by proteolysis of their zymogen forms. The overall process is known as the **coagulation cascade,** although experimental evidence shows that the pathway is not strictly linear, as the waterfall analogy might suggest.

The various components of the coagulation cascade, which include enzymes as well as nonenzymatic protein cofactors, are assigned Roman numerals, largely for historical reasons that do not reflect their order of action *in vivo.* The suffix *a* denotes an active factor.

The catalytic domains of the coagulation proteases resemble trypsin in sequence and mechanism but are much more specific for their substrates. Additional domains mediate interactions with cofactors and help anchor the proteins

to the platelet membrane, which serves as a stage for many of the coagulation reactions.

Coagulation is initiated when a membrane protein **(tissue factor)** exposed to the bloodstream by tissue damage forms a complex with circulating **factor VII** or VIIa (factor VIIa is generated from factor VII by trace amounts of other coagulation proteases, including factor VIIa itself). The tissue factor–VIIa complex proteolytically converts the zymogen **factor X** to factor Xa. Factor Xa then converts **prothrombin** to thrombin, which subsequently generates fibrin from fibrinogen. The tissue factor–dependent steps of coagulation are known as the **extrinsic pathway** because the source of tissue factor is extravascular. The extrinsic pathway is quickly dampened through the action of a protein that inhibits factor VII once factor Xa has been generated.

Sustained thrombin activation requires the activity of the **intrinsic pathway** (so named because all its components are present in the circulation). The intrinsic pathway is stimulated by the tissue factor–VIIa complex, which converts **factor IX** to its active form, factor IXa. The ensuing thrombin activates a number of components of the intrinsic pathway, including **factor XI,** a protease that activates factor IX, to maintain coagulation in the absence of tissue factor or factor VIIa. Thrombin also activates **factors V** and **VIII,** which are cofactors rather than proteases. Factor Va promotes prothrombin activation by factor Xa by as much as 20,000-fold, and factor VIIIa promotes factor X activation by factor IXa by a similar amount. Thus, thrombin promotes its own activation through a feedback mechanism that amplifies the preceding steps of the cascade. **Factor XIII** is also activated by thrombin. Factor XIIIa, which is not a serine protease, chemically cross-links fibrin molecules through formation of peptide bonds between glutamate and lysine side chains, which forms a strong fibrin network.

The intrinsic pathway of coagulation can be triggered by exposure to negatively charged surfaces such as glass. Consequently, blood clots when it is collected in a clean glass test tube. In the absence of tissue factor, a fibrin clot may

---

SUMMARY

1. Enzymes, almost all of which are proteins, are grouped into six mechanistic classes.

2. Enzymes accelerate reactions by factors of up to at least 10^{15}.

3. The substrate specificity of an enzyme depends on the geometric and electronic character of its active site.

4. Some enzymes catalyze reactions with the assistance of metal

not appear for several minutes, but when tissue factor is present, a clot forms within a few seconds. This suggests that rapid blood clotting *in vivo* requires tissue factor as well as the proteins of the intrinsic pathway. Additional evidence for the importance of the extrinsic pathway is that individuals who are deficient in factor VII tend to bleed excessively. Abnormal bleeding also results from congenital defects in factor VIII **(hemophilia a)** or factor IX **(hemophilia b).** Interestingly, a factor XI deficiency causes only a mild bleeding disorder.

The sequential activation of zymogens in the coagulation cascade leads to a burst of thrombin activity, since trace amounts of factors VIIa, IXa, and Xa can activate much larger amounts of their respective substrates. The potential for amplification in the coagulation cascade is reflected in the plasma concentrations of the coagulation proteins (see table).

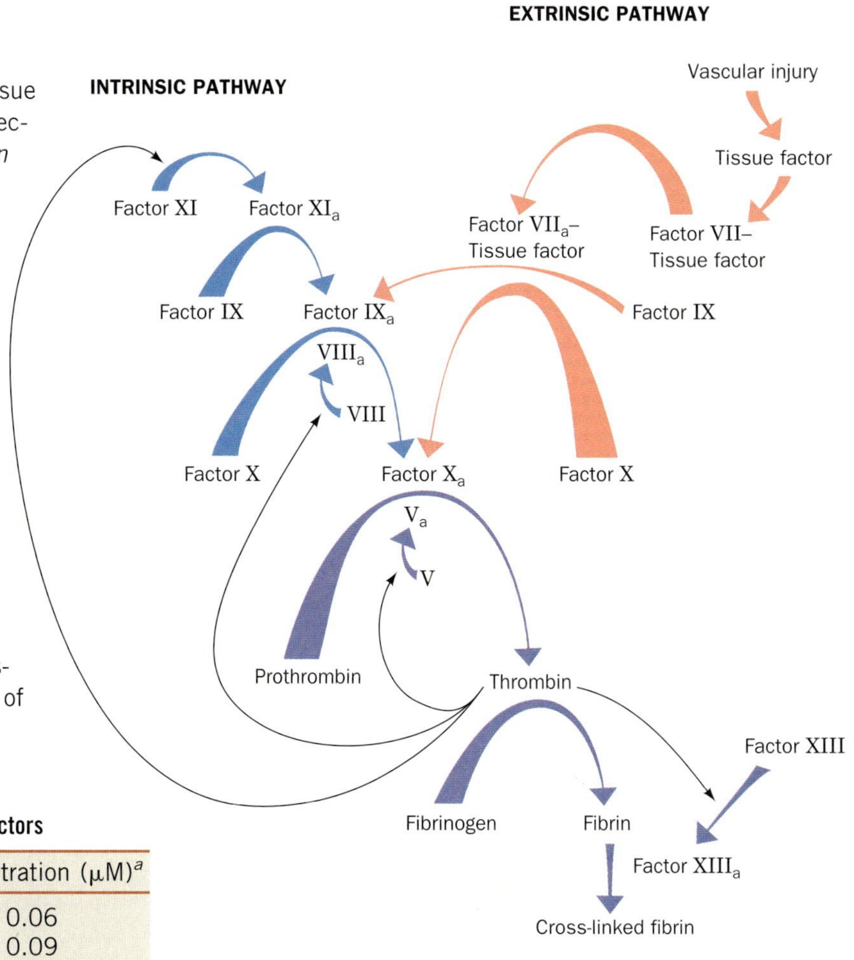

Plasma Concentrations of Some Human Coagulation Factors

Factor	Concentration (μM)[a]
XI	0.06
IX	0.09
VII	0.01
X	0.18
Prothrombin	1.39
Fibrinogen	8.82

[a]Concentrations calculated from data in High, K.A. and Roberts, H.R., (Eds.), *Molecular Basis of Thrombosis and Hemostasis,* Marcel Dekker (1995).

Perhaps not surprisingly, thrombin eventually activates mechanisms that shut down clot formation, thereby limiting the duration of the clotting process and hence the extent of the clot. Such control of clotting is of extreme physiological importance since the formation of even one inappropriate blood clot within an individual's lifetime may have fatal consequences.

[Figure adapted from Davie, E.W., *Thromb. Haemost.* **74,** 2 (1995).]

ion cofactors or organic coenzymes that function as reversibly bound cosubstrates or as permanently associated prosthetic groups. Many coenzymes are derived from vitamins.

5. Enzymes catalyze reactions by decreasing the activation free energy, $\Delta G^{\ddagger}$, which is the free energy required to reach the transition state, the point of highest free energy in the reaction.

6. Enzymes use the same catalytic mechanisms employed by chemical catalysts, including general acid and general base catalysis, covalent catalysis, and metal ion catalysis.

7. The arrangement of functional groups in an enzyme active site allows catalysis by proximity and orientation effects including electrostatic catalysis.

8. A particularly important mechanism of enzyme-mediated catalysis is the preferential binding of the transition state of the catalyzed reaction.

9. In the catalytic mechanism of lysozyme, Glu 35 in its protonated form acts as an acid catalyst to cleave the polysaccharide substrate between its D and E rings, and Asp 52 in its anionic state forms a covalent bond to C1 of the D ring. The reaction is facilitated by the distortion of residue D to the planar half-chair conformation, which resembles the reaction's oxonium ion transition state.

10. Serine proteases contain a Ser–His–Asp catalytic triad near a binding pocket that helps determine the enzymes' substrate specificity.

11. Catalysis in the serine proteases occurs through acid–base catalysis, covalent catalysis, proximity and orientation effects, electrostatic catalysis, and by preferential transition state binding in the oxyanion hole.

12. Synthesis of pancreatic proteases as inactive zymogens protects the pancreas from self-digestion. Zymogens are activated by specific proteolytic cleavages.

REFERENCES

General

Bruice, T.C. and Benkovic, S.J., Chemical basis for enzyme catalysis, *Biochemistry* **39**, 6267–6274 (2000).

Cannon, W.R. and Benkovic, S.J., Solvation, reorganization energy, and biological catalysis, *J. Biol. Chem.* **273**, 26257–26260 (1998). [Summarizes some general features of enzyme function.]

Gerlt, J.A., Protein engineering to study enzyme catalytic mechanisms, *Curr. Opin. Struct. Biol.* **4**, 593–600 (1994). [Describes how information can be gained from mutagenesis and structural analysis of enzymes.]

Hackney, D.D., Binding energy and catalysis, *in* Sigman, D.S. and Boyer, P.D. (Eds.), *The Enzymes* (3rd ed.), Vol. 19, *pp*. 1–36, Academic Press (1990).

Kraut, J., How do enzymes work? *Science* **242**, 533–540 (1988). [A brief and very readable review of transition state theory and applications.]

Schramm, V.L., Enzymatic transition states and transition state analog design, *Annu. Rev. Biochem.* **67**, 693–720 (1998).

Tipton, K.F., The naming of parts, *Trends Biochem. Sci.* **18**, 113–115 (1993). [A discussion of the advantages of a consistent naming scheme for enzymes and the difficulties of formulating one.]

Lysozyme

Kirby, A.J., The lysozyme mechanism sorted—after 50 years, *Nature Struct. Biol.* **8**, 737–739 (2001). [Briefly summarizes the theoretical and experimental evidence for a covalent intermediate in the lysozyme mechanism.]

McKenzie, H.A. and White, F.H., Jr., Lysozyme and α-lactalbumin: Structure, function and interrelationships, *Adv. Protein Chem.* **41**, 173–315 (1991).

Strynadka, N.C.J. and James, M.N.G., Lysozyme revisited: crystallographic evidence for distortion of an *N*-acetylmuramic acid residue bound in site D, *J. Mol. Biol.* **220**, 401–424 (1991).

Serine Proteases

Cleland, W.W., Frey, P.A., and Gerlt, J.A., The low barrier hydrogen bond in enzymatic catalysis, *J. Biol. Chem.* **273**, 25529–25532 (1998).

Davie, E.W., Biochemical and molecular aspects of the coagulation cascade, *Thromb. Haemost.* **74**, 1–6 (1995). [A brief review by one of the pioneers of the cascade hypothesis.]

Fersht, A., *Structure and Mechanism in Protein Science.* Freeman (1999). [Includes detailed reaction mechanisms for chymotrypsin and other enzymes.]

Perona, J.J. and Craik, C.S., Evolutionary divergence of substrate specificity within the chymotrypsin-like protease fold, *J. Biol. Chem.* **272**, 29987–29990 (1997). [Summarizes research identifying the structural basis of substrate specificity in chymotrypsin and related enzymes.]

Wilmouth, R.C., Edman, K., Neutze, R., Wright, P.A., Clifton, I.J., Schneider, T.R., Schofield, C.J., and Hajdu, J., X-Ray snapshots of serine protease catalysis reveal a tetrahedral intermediate, *Nature Struct. Biol.* **8**, 689–694 (2001). [Reports the first structural evidence for a tetrahedral intermediate in the hydrolysis reaction catalyzed by a serine protease.]

KEY TERMS

active site	prosthetic group	general base catalysis	catalytic triad
substrate	holoenzyme	covalent catalysis	convergent evolution
EC classification	apoenzyme	metalloenzyme	tetrahedral intermediate
induced fit	vitamin	electrostatic catalysis	acyl–enzyme intermediate
prochirality	transition state	transition state analog	oxyanion hole
cofactor	$\Delta G^{\ddagger}$	oxonium ion	low-barrier hydrogen bond
coenzyme	rate-determining step	serine protease	zymogen
cosubstrate	general acid catalysis	affinity labeling	

STUDY EXERCISES

1. What properties distinguish enzymes from other catalysts?

2. What factors influence an enzyme's substrate specificity?

3. Why are cofactors required for some enzymatic reactions?

4. Sketch and label the various parts of transition state diagrams for a reaction with and without a catalyst.

5. What is the relationship between ΔG and $\Delta G^{\ddagger}$?

6. Explain how nucleophiles act as covalent catalysts.

7. Why is it unlikely that nonenzymatic catalysts operate by transition state stabilization?

8. Describe the catalytic mechanisms employed in the lysozyme reaction.

9. What is the function of the oxyanion hole in serine proteases?

PROBLEMS

1. Choose the best description of an enzyme:

   (a) It allows a chemical reaction to proceed extremely fast.

   (b) It increases the rate at which a chemical reaction approaches equilibrium relative to its uncatalyzed rate.

   (c) It makes a reaction thermodynamically favorable.

2. Which type of enzyme (Table 11-2) catalyzes the following reactions?

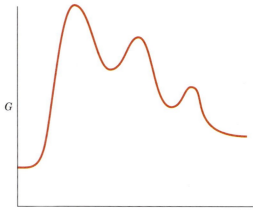

Reaction coordinate

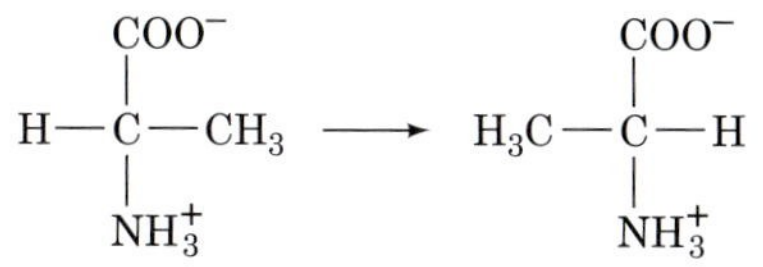

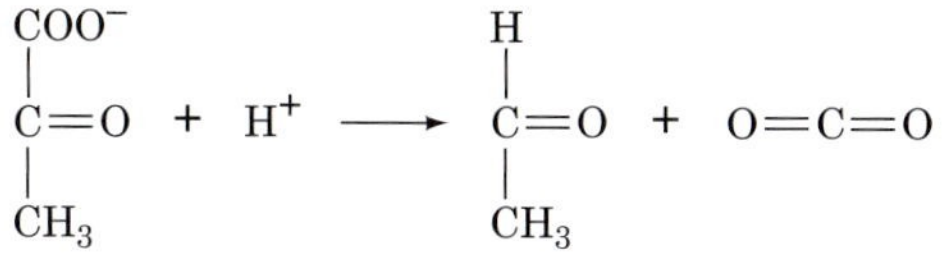

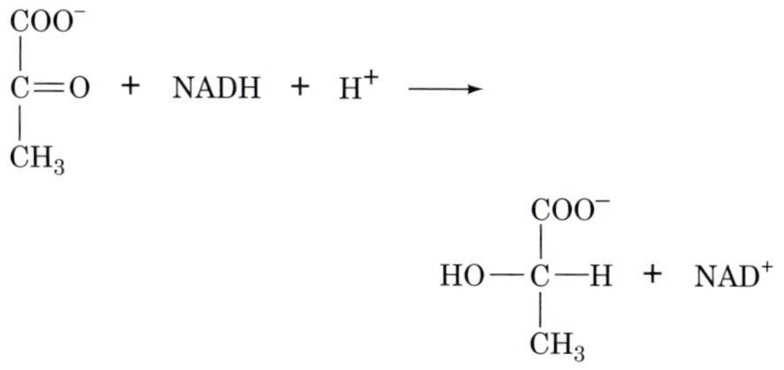

5. Draw a transition state diagram of (a) a nonenzymatic reaction and the corresponding enzyme-catalyzed reaction in which (b) S binds loosely to the enzyme and (c) S binds very tightly to the enzyme. Compare $\Delta G^{\ddagger}$ for each case. Why is tight binding of S not advantageous?

6. Approximately how much does staphylococcal nuclease (Table 11-1) decrease the activation free energy ($\Delta G^{\ddagger}$) of its reaction (the hydrolysis of a phosphodiester bond) at 25°C?

7. Studies at different pH's show that an enzyme has two catalytically important residues whose pK's are ~4 and ~10. Chemical modification experiments indicate that a Glu and a Lys residue are essential for activity. Match the residues to their pK's and explain whether they are likely to act as acid or base catalysts.

8. The covalent catalytic mechanism of an enzyme depends on a single active-site Cys whose pK is 8. A mutation in a nearby residue alters the microenvironment so that this pK increases to 10. Would the mutation cause the reaction rate to increase or decrease? Explain.

9. Explain why RNase A cannot catalyze the hydrolysis of DNA.

10. Suggest a transition state analog for proline racemase that differs from those discussed in the text. Justify your suggestion.

11. Wolfenden has stated that it is meaningless to distinguish between the "binding sites" and the "catalytic sites" of enzymes. Explain.

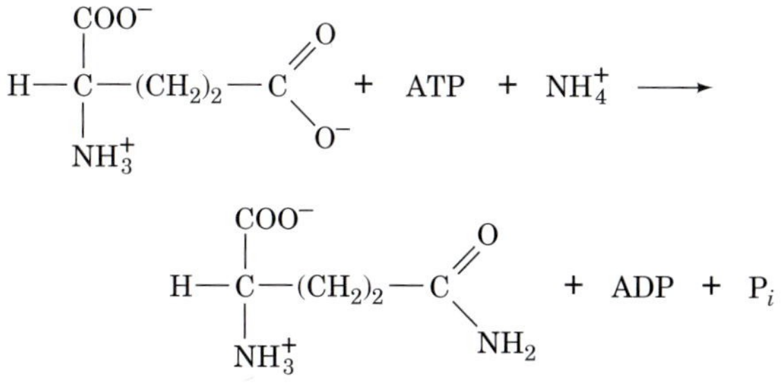

3. What is the relationship between the rate of an enzyme-catalyzed reaction and the rate of the corresponding uncatalyzed reaction? Do enzymes enhance the rates of slow uncatalyzed reactions as much as they enhance the rates of fast uncatalyzed reactions?

4. On the free energy diagram shown, label the intermediate(s) and transition state(s). Is the reaction thermodynamically favorable?

12. Explain why lysozyme cleaves the artificial substrate $(NAG)_4$ ~4000 times more slowly than it cleaves $(NAG)_6$.

13. Lysozyme residues Asp 101 and Arg 114 are required for efficient catalysis, although they are located at some distance from the active site Glu 35 and Asp 52. Substituting Ala for either Asp 101 or Arg 114 does not significantly alter the enzyme's tertiary structure, but it significantly reduces its catalytic activity. Explain. [Problem by Bruce Wightman, Muhlenberg College.]

14. Design a chloromethylketone inhibitor of elastase.

15. Under certain conditions, peptide bond formation rather than peptide bond hydrolysis is thermodynamically favorable. Would you expect chymotrypsin to catalyze peptide bond formation?

16. Diagram the hydrogen bonding interactions of the catalytic triad His–Lys–Ser during catalysis in a hypothetical hydrolytic enzyme.

17. The comparison of the active site geometries of chymotrypsin and subtilisin under the assumption that their similarities have catalytic significance has led to greater mechanistic understanding of both these enzymes. Discuss the validity of this strategy.

18. Predict the effect of mutating Asp 102 of trypsin to Asn (a) on substrate binding and (b) on catalysis.

19. A genetic defect in coagulation factor IX causes hemophilia b, a disease characterized by a tendency to bleed profusely after very minor trauma. However, a genetic defect in coagulation factor XI has no clinical symptoms. Explain this discrepancy in terms of the mechanism for activation of coagulation proteases shown in Box 11-4.

20. Why is the broad substrate specificity of chymotrypsin advantageous *in vivo*? Why would this be a disadvantage for some other proteases?

21. Tofu (bean curd), a high-protein soybean product, is prepared in such a way as to remove the trypsin inhibitor present in soybeans. Explain the reason(s) for this treatment.

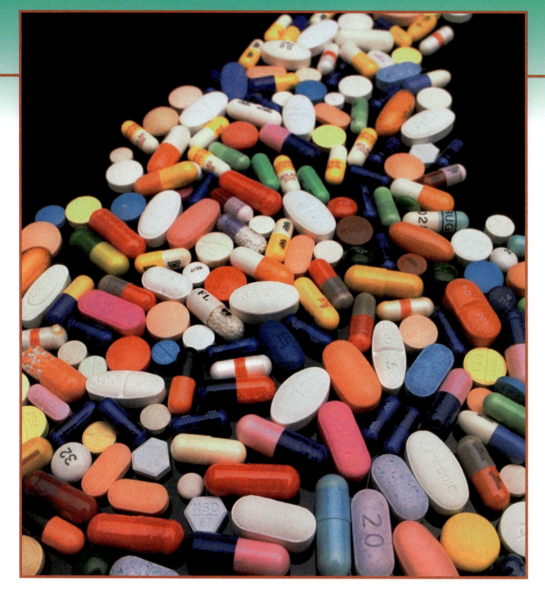

Many natural and synthetic substances are known to inhibit the activities of specific enzymes. The study of inhibitor–enzyme interactions, as quantified through enzyme kinetics, is a mainstay of modern drug development. [Larry Kolvoord/The Image Works.]

Enzyme Kinetics, Inhibition, and Regulation

Early enzymologists, often working with crude preparations of yeast or liver cells, could do little more than observe the conversion of substrates to products catalyzed by as yet unpurified enzymes. Measuring the rates of such reactions therefore came to be a powerful tool for characterizing enzyme activity. The application of simple mathematical models to enzyme activity under varying laboratory conditions, and in the presence of competing substrates or enzyme inhibitors, made it possible to deduce the probable physiological functions and regulatory mechanisms of various enzymes.

The study of enzymatic reaction rates, or **enzyme kinetics,** is no less important now than it was early in the twentieth century. In many cases, the rate of a reaction and how the rate changes in response to different conditions reveal the path followed by the reactants and are therefore indicative of the reaction mechanism. Kinetic data, combined with detailed information about an enzyme's structure and its catalytic mechanisms, provide some of the most powerful clues to the enzyme's biological function and may suggest ways to modify it for therapeutic purposes.

We begin our consideration of enzyme kinetics by reviewing chemical kinetics. Following that, we derive the basic equations of enzyme kinetics and describe the effects of inhibitors on enzymes. We also consider an example of enzyme regulation that highlights several aspects of enzyme function. Finally, we describe some practical applications of enzyme inhibition, the development of enzyme inhibitors as drugs.

1 Reaction Kinetics

Kinetic measurements of enzymatically catalyzed reactions are among the most powerful techniques for elucidating the catalytic mechanisms of enzymes. Enzyme kinetics is a branch of chemical kinetics, so we begin this section by reviewing the principles of chemical kinetics.

A Chemical Kinetics

A reaction of overall stoichiometry

$$A \longrightarrow P$$

where A represents reactants and P represents products, may actually occur through a sequence of **elementary reactions** (simple molecular processes) such as

$$A \longrightarrow I_1 \longrightarrow I_2 \longrightarrow P$$

Here, I_1 and I_2 symbolize **intermediates** in the reaction. Each elementary reaction can be characterized with respect to the number of reacting species and the rate at which they interact. *Descriptions of each elementary reaction collectively constitute the mechanistic description of the overall reaction process.* Even a complicated enzyme-catalyzed reaction can be analyzed in terms of its component elementary reactions.

Reaction Order.   At constant temperature, *the rate of an elementary reaction is proportional to the frequency with which the reacting molecules come together.* The proportionality constant is known as a **rate constant** and is symbolized k. For the elementary reaction $A \rightarrow P$, the instantaneous rate of appearance of product or disappearance of reactant, which is called the **velocity (v)** of the reaction, is

$$v = \frac{d[P]}{dt} = -\frac{d[A]}{dt} = k[A] \qquad [12\text{-}1]$$

In other words, the reaction velocity at any time point is proportional to the concentration of the reactant A. This is an example of a **first-order** reaction. Since the velocity has units of molar per sec ($M \cdot s^{-1}$), the first-order rate constant must have units of reciprocal seconds (s^{-1}). *The **reaction order** of an elementary reaction corresponds to the **molecularity** of the reaction, which is the number of molecules that must simultaneously collide to generate a product.* Thus, a first-order elementary reaction is a **unimolecular** reaction.

Consider the elementary reaction $2A \rightarrow P$. This **bimolecular** reaction is a **second-order** reaction, and its instantaneous velocity is described by

$$v = -\frac{d[A]}{dt} = k[A]^2 \qquad [12\text{-}2]$$

In this case, the reaction velocity is proportional to the square of the concentration of A, and the second-order rate constant k has units of $M^{-1} \cdot s^{-1}$.

The bimolecular reaction $A + B \rightarrow P$ is also a second-order reaction with an instantaneous velocity described by

$$v = -\frac{d[A]}{dt} = -\frac{d[B]}{dt} = k[A][B] \qquad [12\text{-}3]$$

Here, the reaction is said to be first order in [A] and first order in [B] (see Sample Calculation 12-1). Unimolecular and bimolecular reactions are common. **Termolecular** reactions are unusual because the simultaneous collision of three molecules is a rare event. Fourth- and higher-order reactions are unknown.

Rate Equations. *A **rate equation** describes the progress of a reaction as a function of time* and can be derived from the equations that describe the instantaneous reaction velocity. Thus, a first-order rate equation is obtained by rearranging Eq. 12-1

$$\frac{d[A]}{[A]} = d \ln[A] = -k\, dt \qquad [12\text{-}4]$$

and integrating it from $[A]_o$, the initial concentration of A, to [A], the concentration of A at time t:

$$\int_{[A]_o}^{[A]} d \ln[A] = -k \int_0^t dt \qquad [12\text{-}5]$$

This results in

$$\boxed{\ln[A] = \ln[A]_o - kt} \qquad [12\text{-}6]$$

or, taking the antilog of both sides,

$$[A] = [A]_o\, e^{-kt} \qquad [12\text{-}7]$$

Equation 12-6 is a linear equation of the form $y = mx + b$ and can be plotted as in Fig. 12-1. Therefore, if a reaction is first order, a plot of $\ln[A]$ versus t will yield a straight line whose slope is $-k$ (the negative of the first-order rate constant) and whose intercept on the $\ln[A]$ axis is $\ln[A]_o$.

One of the hallmarks of a first-order reaction is that *the time for half of the reactant initially present to decompose, its **half-time** or **half-life**, $t_{1/2}$, is a constant and hence independent of the initial concentration of the reactant.* This is easily demonstrated by substituting the relationship $[A] = [A]_o/2$ when $t = t_{1/2}$ into Eq. 12-6 and rearranging:

$$\ln\left(\frac{[A]_o/2}{[A]_o}\right) = -kt_{1/2} \qquad [12\text{-}8]$$

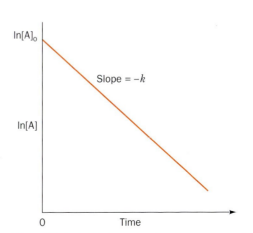

Figure 12-1 A plot of a first-order rate equation. The slope of the line obtained when ln[A] is plotted against time gives the rate constant k.

See Guided Exploration 11

Michaelis–Menten kinetics, Lineweaver–Burk
plots, and enzyme inhibition.

SAMPLE CALCULATION 12-2

The decay of a hypothetical radioisotope has a rate constant of 0.01 s⁻¹. How much time is required for half of a 1-g sample of the isotope to decay?

The units of the rate constant indicate a first-order process. Thus, the half-life is independent of concentration. The half-life of the isotope (the half-time for its decay) is given by Eq. 12-9:

$$t_{1/2} = \frac{\ln 2}{k} = \frac{0.693}{0.01\ \text{s}^{-1}} = 69.3\ \text{s}$$

Thus

$$t_{1/2} = \frac{\ln 2}{k} = \frac{0.693}{k} \qquad [12\text{-}9]$$

Substances that are inherently unstable, such as radioactive nuclei, decompose through first-order reactions (see Box 12-1 and Sample Calculation 12-2).

In a second-order reaction with one type of reactant, $2\ A \rightarrow P$, the variation of [A] with time is quite different from that in a first-order reaction. Rearranging Eq. 12-2 and integrating it over the same limits used for the first-order reaction yields

$$\int_{[A]_o}^{[A]} -\frac{d[A]}{[A]^2} = k \int_0^t dt \qquad [12\text{-}10]$$

so that

$$\boxed{\frac{1}{[A]} = \frac{1}{[A]_o} + kt} \qquad [12\text{-}11]$$

Equation 12-11 is a linear equation in terms of the variables $1/[A]$ and t. *The half-time for a second-order reaction is expressed $t_{1/2} = 1/k[A]_o$, and therefore, in contrast to a first-order reaction, depends on the initial reactant concentration.* Equations 12-6 and 12-11 may be used to distinguish a first-order from a second-order reaction by plotting $\ln[A]$ versus t and $1/[A]$ versus t and observing which, if any, of these plots is a straight line.

To experimentally determine the rate constant for the second-order reaction $A + B \rightarrow P$, it is often convenient to increase the concentration of one reactant relative to the other, for example, $[B] \gg [A]$. Under these conditions, [B] does not change significantly over the course of the reaction. The reaction rate therefore depends only on [A], the concentration of the reactant that is present in limited amounts. Hence, the reaction appears to be first order with respect to A and is therefore said to be a **pseudo-first-order** reaction. The reaction is first order with respect to B when $[A] \gg [B]$.

B Enzyme Kinetics

Enzymes catalyze a tremendous variety of reactions using different combinations of five basic catalytic mechanisms (Section 11-3). Some enzymes act on only a single substrate molecule; others act on two or more different substrate molecules whose order of binding may or may not be obligatory. Some enzymes form covalently bound intermediate complexes with their substrates; others do not. *Yet all enzymes can be analyzed such that their reaction rates as well as their overall efficiency can be quantified.*

The study of enzyme kinetics began in 1902 when Adrian Brown investigated the rate of hydrolysis of sucrose by the yeast enzyme **β-fructofuranosidase**:

$$\text{Sucrose} + \text{H}_2\text{O} \longrightarrow \text{glucose} + \text{fructose}$$

Brown found that when the sucrose concentration is much higher than that of the enzyme, the reaction rate becomes independent of the sucrose concentration; that is, the rate is **zero order** with respect to sucrose. He therefore proposed that the overall reaction is composed of two elementary reactions in which the substrate forms a complex with the enzyme that subsequently decomposes to products, regenerating enzyme:

$$\text{E} + \text{S} \underset{k_{-1}}{\overset{k_1}{\rightleftharpoons}} \text{ES} \overset{k_2}{\longrightarrow} \text{P} + \text{E} \qquad [12\text{-}12]$$

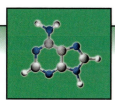

BOX 12-1

Perspectives in Biochemistry

Isotopic Labeling

In the laboratory, it is often useful to label large or small molecules so that they can be easily detected after chromatographic or electrophoretic separation or in various binding assays. One of the most common labeling techniques is to attach a radioactive isotope to a molecule or to synthesize the molecule so that it contains a radioactive isotope in place of a normally occurring isotope. Molecules labeled in this way can be detected in solution or in solid form by measuring the radioactivity emitted by the label. This method is more sensitive than spectroscopic measurements, and it is often easier to carry out than more laborious assays based on chemical or biological activities. Metabolites labeled with NMR-active isotopes such as ^{13}C can also be detected in living tissues by NMR techniques.

Some of the most common radioactive isotopes **(radionuclides)** used in biochemistry are listed below, along with their half-lives and the type of radioactivity emitted by the spontaneously disintegrating atomic nuclei.

Radionuclide	Half-life	Type of Radiation[a]
^{3}H	12 years	β
^{14}C	5715 years	β
^{24}Na	15 hours	β
^{32}P	14 days	β
^{35}S	87 days	β
^{40}K	1.25×10^{9} years	β
^{45}Ca	163 days	β
^{125}I	59 days	γ
^{131}I	8 days	β, γ

[a]β particles are emitted electrons, and γ rays are emitted photons.

Nucleic acids can be easily labeled by attaching a terminal nucleotide that contains ^{32}P in place of the normal nonradioactive ^{31}P. Proteins can be labeled by chemically or enzymatically linking ^{125}I to a Tyr residue. Assays for cell growth and division often measure the uptake of ^{3}H-labeled **thymidine** (thymidine is incorporated exclusively into DNA). Protein synthesis is similarly monitored by the appearance of ^{35}S-labeled Met in proteins. Of course, the choice of a particular isotopic label also depends on the time-course of the experiment and the method for detecting radioactivity.

A **Geiger counter,** which electronically detects the ionization of a gas caused by the passage of radiation, is not sensitive enough to detect low-energy emitters such as ^{3}H and ^{14}C. This limitation is circumvented through **liquid scintillation counting.** In this technique, a β-emitting sample is dissolved in a solvent that contains a fluorescent molecule. The β particles excite this fluor, thereby causing it to emit light that can then be optically detected. Radioactive substances that emit γ rays are detected by **scintillation counters** when the γ rays dislodge electrons from a crystal of NaI in the counter. These electrons induce fluorescence that is measured. In **autoradiography,** a radioactive substance immobilized on paper or in an agarose or polyacrylamide gel is detected by laying X-ray film over the sample followed by incubation and development of the film (dark areas on the developed film correspond to areas exposed to radioactivity). Thin sections of tissue can also be prepared for **microradiography** by covering them with a layer of photographic emulsion and examining the developed emulsion under a microscope (see figure). Instruments that electronically measure radioactivity in solid samples without the use of film **(phosphorimagers)** offer the advantage of digitized results and multiple exposure times (in contrast, film can be developed only once).

The use of radioactive isotopes as molecular labels is not without drawbacks. First and foremost is the danger of working with potentially mutagenic materials (irradiation can cause DNA damage). In addition, radioactive laboratory materials (samples as well as glassware) must be disposed of properly or the resulting contamination can cause errors in subsequent measurements of radioactivity as well as a health hazard. Scintillation fluid presents a particular problem for disposal because of the large volumes required (it also consists largely of organic solvents). The preceding table reveals that while disposal of short-lived radionuclides (such as ^{32}P, ^{35}S, and ^{125}I) can be accomplished mainly by storing the material until the radioactivity has decayed to insignificant levels, the safe disposal of long-lived species (such as ^{3}H and ^{14}C) is a problem that is unlikely to vanish any time soon. This is one reason why molecular labeling techniques that rely on chemical tags or fluorescent compounds have become popular.

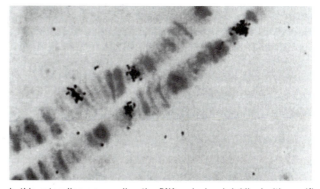

In this autoradiogram, a radioactive RNA probe has hybridized with specific sites in *Drosophila* polytene chromosomes. The black dots reveal the sites where radioactive decay has occurred. [From Loughney, K., Kreber, R., and Ganetzky, B., *Cell* **58**, 1143 (1989) by permission of Cell Press.]

Here E, S, ES, and P symbolize the enzyme, substrate, **enzyme–substrate complex,** and products, respectively. According to this model, when the substrate concentration becomes high enough to entirely convert the enzyme to the ES form, the second step of the reaction becomes rate limiting and the overall reaction rate becomes insensitive to further increases in substrate concentration.

Each of the elementary reactions that make up the above enzymatic reaction is characterized by a rate constant: k_1 and k_{-1} are the forward and reverse rate constants for formation of the ES complex (the first reaction), and k_2 is the rate constant for the decomposition of ES to P (the second reaction). Here we assume, for the sake of mathematical simplicity, that the second reaction is irreversible; that is, no P is converted back to S.

The Michaelis–Menten Equation. The Michaelis–Menten equation describes the rate of the enzymatic reaction represented by Eq. 12-12 as a function of substrate concentration. In this kinetic scheme, the formation of product from ES is a first-order process. Thus, the rate of formation of product can be expressed as the product of the rate constant of the reaction yielding product and the concentration of its immediately preceding intermediate. The general expression for the velocity (rate) of Reaction 12-12 is therefore

$$v = \frac{d[P]}{dt} = k_2[ES] \quad\quad [12\text{-}13]$$

The overall rate of production of ES is the difference between the rates of the elementary reactions leading to its appearance and those resulting in its disappearance:

$$\frac{d[ES]}{dt} = k_1[E][S] - k_{-1}[ES] - k_2[ES] \quad\quad [12\text{-}14]$$

This equation cannot be explicitly integrated, however, without simplifying assumptions. Two possibilities are

1. *Assumption of equilibrium.* In 1913, Leonor Michaelis and Maud Menten, building on the work of Victor Henri, assumed that $k_{-1} \gg k_2$, so that the first step of the reaction reaches equilibrium:

$$K_S = \frac{k_{-1}}{k_1} = \frac{[E][S]}{[ES]} \quad\quad [12\text{-}15]$$

Here K_S is the dissociation constant of the first step in the enzymatic reaction. With this assumption, Eq. 12-14 can be integrated. Although this assumption is often not correct, in recognition of the importance of this pioneering work, the enzyme–substrate complex, ES, is known as the **Michaelis complex.**

2. *Assumption of steady state.* Figure 12-2 illustrates the progress curves of the various participants in Reaction 12-12 under the physiologically common condition that substrate is in great excess over enzyme ([S] ≫ [E]). With the exception of the initial stage of the reaction, which is usually over within milliseconds of mixing E and S, [ES] remains approximately constant until the substrate is nearly exhausted. Hence, the rate of synthesis of

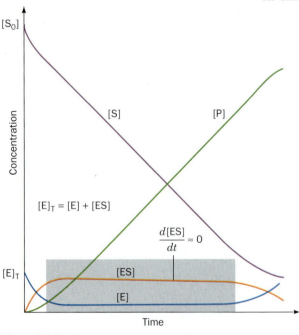

Figure 12-2 The progress curves for a simple enzyme-catalyzed reaction. With the exception of the initial phase of the reaction (before the shaded block), the slopes of the progress curves for [E] and [ES] are essentially zero as long as [S] ≫ [E] (within the shaded block). [After Segel, I.H., *Enzyme Kinetics*, p. 27, Wiley (1975).] 🐚 **See the Animated Figures.**

BOX 12-2

Pathways of Discovery

J.B.S. Haldane and Enzyme Action

J.B.S. Haldane (1892–1964)

John Burdon Sanderson Haldane, the son of a prominent physiologist, was a gifted scientist and writer whose major contributions include the application of mathematics to areas of biology such as genetics and enzyme kinetics. As a scientist as well as a philosopher, he was aware of developments in relativity theory and quantum mechanics and was influenced by a practical philosophy that the natural world obeyed the laws of logic and arithmetic.

When Haldane published his book *Enzymes* in 1930, the idea that enzymes are proteins, rather than small catalysts surrounded by an amorphous protein "colloid," was still controversial. However, even in the absence of structural information, scientists such as Leonor Michaelis and Maud Menten had already applied the principles of thermodynamics to derive some basic equations related to enzyme kinetics. Michaelis and Menten proposed in 1913 that during a reaction, an enzyme and its substrate are in equilibrium with a complex of enzyme and substrate. In 1925, Haldane argued that this was not strictly true, since some enzyme–substrate complex does not dissociate to free enzyme and substrate but instead goes on to form product. When the enzyme and substrate are first mixed together, the concentration of the enzyme–substrate complex increases, but after a time the concentration of the complex levels off because the complex is constantly forming and breaking down to generate product. This principle, the so-called steady state assumption, underlies modern theories of enzyme activity.

Even without knowing what enzymes were made of, Haldane showed great insight in proposing that an enzyme could catalyze a reaction by bringing its substrates into a strained or out-of-equilibrium arrangement. This idea refined Emil Fischer's earlier lock-and-key simile (and was later elaborated further by Linus Pauling). Haldane's idea of strain was not fully appreciated until around 1970, after the X-ray structures of several enzymes, including lysozyme and chymotrypsin (Chapter 11), had been examined.

In addition to his work in enzymology, Haldane articulated the role of genes in heredity and formulated mathematical estimates of mutation rates—many years before the nature of genes or the structure of DNA were known. However, in addition to being a theorist, Haldane was an experimentalist who frequently performed unpleasant or dangerous experiments on himself. For example, he ingested sodium bicarbonate and ammonium chloride in order to investigate their effect on breathing rate. Beyond the laboratory, Haldane was well-known for his efforts to popularize science. In *Daedalus, or, Science and the Future,* Haldane commented on the status of various branches of science circa 1924 and speculated about future developments. His thoughts and his persona are believed to have inspired various plots and characters in other writers' works of science fiction.

Briggs, G.E. and Haldane, J.B.S., A note on the kinetics of enzyme action, *Biochem. J.* **19**, 339 (1925).

ES must equal its rate of consumption over most of the course of the reaction. In other words, ES maintains a **steady state** and [ES] can be treated as having a constant value:

$$\frac{d[\text{ES}]}{dt} = 0 \qquad\qquad [12\text{-}16]$$

This so-called **steady state assumption,** a more general condition than that of equilibrium, was first proposed in 1925 by G.E. Briggs and John B.S. Haldane (see Box 12-2).

In order to be useful, kinetic expressions for overall reactions must be formulated in terms of experimentally measurable quantities. The quantities [ES] and [E] are not, in general, directly measurable, but the total enzyme concentration

$$[\text{E}]_\text{T} = [\text{E}] + [\text{ES}] \qquad\qquad [12\text{-}17]$$

is usually readily determined. The rate equation for the overall enzymatic

reaction as a function of [S] and [E], can then be derived. First, Eq. 12-14 is combined with the steady state assumption (Eq. 12-16) to give

$$k_1[E][S] = k_{-1}[ES] + k_2[ES] \qquad [12\text{-}18]$$

Letting $[E] = [E]_T - [ES]$ and rearranging yields

$$\frac{([E]_T - [ES])[S]}{[ES]} = \frac{k_{-1} + k_2}{k_1} \qquad [12\text{-}19]$$

The **Michaelis constant, K_M,** is defined as

$$K_M = \frac{k_{-1} + k_2}{k_1} \qquad [12\text{-}20]$$

so Eq. 12-19 can then be rearranged to give

$$K_M[ES] = ([E]_T - [ES])[S] \qquad [12\text{-}21]$$

Solving for [ES],

$$[ES] = \frac{[E]_T[S]}{K_M + [S]} \qquad [12\text{-}22]$$

The expression for the **initial velocity (v_o)** of the reaction, the velocity (Eq. 12-13) at $t = 0$, thereby becomes

$$v_o = \left(\frac{d[P]}{dt}\right)_{t=0} = k_2[ES] = \frac{k_2[E]_T[S]}{K_M + [S]} \qquad [12\text{-}23]$$

Both $[E]_T$ and [S] are experimentally measurable quantities. In order to meet the conditions of the steady state assumption, the concentration of the substrate must be much greater than the concentration of the enzyme, which allows each enzyme molecule to repeatedly bind a molecule of substrate and convert it to product, so that [ES] is constant. The use of the initial velocity (operationally taken as the velocity measured before more than ~10% of the substrate has been converted to product)—rather than just the velocity—minimizes such complicating factors as the effects of reversible reactions, inhibition of the enzyme by its product(s), and progressive inactivation of the enzyme. (This is also why the rate of the reverse reaction in Eq. 12-12 can be assumed to be zero.)

The **maximal velocity** of a reaction, V_{max}, occurs at high substrate concentrations when the enzyme is **saturated,** that is, when it is entirely in the ES form:

$$V_{\text{max}} = k_2[E]_T \qquad [12\text{-}24]$$

Therefore, combining Eqs. 12-23 and 12-24, we obtain

$$\boxed{v_o = \frac{V_{\text{max}}[S]}{K_M + [S]}} \qquad [12\text{-}25]$$

This expression, the **Michaelis–Menten equation,** *is the basic equation of enzyme kinetics.* It describes a rectangular hyperbola such as that plotted in Fig. 12-3. The saturation function for oxygen binding to myoglobin (Eq. 7-6) has the same algebraic form.

Significance of the Michaelis Constant. The Michaelis constant, K_M, has a simple operational definition. At the substrate concentration at which [S] = K_M, Eq. 12-25 yields $v_o = V_{\text{max}}/2$ so that K_M *is the substrate concentration at which the reaction velocity is half-maximal.* Therefore, if an enzyme has

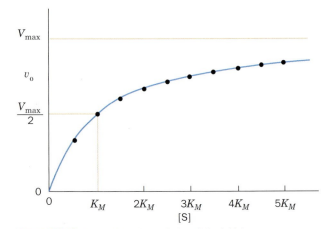

Figure 12-3 *Key to Function.* **A plot of the initial velocity v_o of a simple enzymatic reaction versus the substrate concentration [S].** Points are plotted in 0.5 K_M intervals of substrate concentration between 0.5 K_M and 5 K_M. 🔎 **See the Animated Figures.**

Table 12-1   The Values of K_M, k_{cat}, and k_{cat}/K_M for Some Enzymes and Substrates

Enzyme	Substrate	K_M (M)	k_{cat} (s^{-1})	k_{cat}/K_M ($M^{-1} \cdot s^{-1}$)
Acetylcholinesterase	Acetylcholine	9.5×10^{-5}	1.4×10^4	1.5×10^8
Carbonic anhydrase	CO_2	1.2×10^{-2}	1.0×10^6	8.3×10^7
	HCO_3^-	2.6×10^{-2}	4.0×10^5	1.5×10^7
Catalase	H_2O_2	2.5×10^{-2}	1.0×10^7	4.0×10^8
Chymotrypsin	N-Accetylglycine ethyl ester	4.4×10^{-1}	5.1×10^{-2}	1.2×10^{-1}
	N-Acetylvaline ethyl ester	8.8×10^{-2}	1.7×10^{-1}	1.9
	N-Acetyltyrosine ethyl ester	6.6×10^{-4}	1.9×10^2	2.9×10^5
Fumarase	Fumarate	5.0×10^{-6}	8.0×10^2	1.6×10^8
	Malate	2.5×10^{-5}	9.0×10^2	3.6×10^7
Urease	Urea	2.5×10^{-2}	1.0×10^4	4.0×10^5

a small value of K_M, it achieves maximal catalytic efficiency at low substrate concentrations.

The K_M is unique for each enzyme–substrate pair: Different substrates that react with a given enzyme do so with different K_M values. Likewise, different enzymes that act on a single substrate have different K_M values. The magnitude of K_M varies widely with the identity of the enzyme and the nature of the substrate (Table 12-1). It is also a function of temperature and pH. The Michaelis constant (Eq. 12-20) can be expressed as

$$K_M = \frac{k_{-1}}{k_1} + \frac{k_2}{k_1} = K_S + \frac{k_2}{k_1} \qquad [12\text{-}26]$$

Since K_S is the dissociation constant of the Michaelis complex (Eq. 12-15), as K_S decreases, the enzyme's affinity for substrate increases. K_M is therefore also a measure of the affinity of the enzyme for its substrate, provided k_2/k_1 is small compared to K_S, that is, $k_2 < k_{-1}$ so that the ES $\rightarrow$ P reaction proceeds more slowly than ES reverts to E + S.

k_{cat}/K_M Is a Measure of Catalytic Efficiency.   We can define the **catalytic constant, k_{cat},** of an enzyme as

$$k_{cat} = \frac{V_{max}}{[E]_T} \qquad [12\text{-}27]$$

This quantity is also known as the **turnover number** of an enzyme because it is the number of reaction processes (turnovers) that each active site catalyzes per unit time. The turnover numbers for a selection of enzymes are given in Table 12-l. Note that these quantities vary by over eight orders of magnitude. Equation 12-24 indicates that for a simple system, such as the Michaelis–Menten model reaction (Eq. 12-12), $k_{cat} = k_2$. For enzymes with more complicated mechanisms (e.g., multiple substrates or multiple reaction intermediates), k_{cat} may be a function of several rate constants. Note that whereas k_{cat} is a constant, V_{max} depends on the concentration of the enzyme present in the experimental system. V_{max} increases as $[E]_T$ increases.

When $[S] \ll K_M$, very little ES is formed. Consequently, $[E] \approx [E]_T$, so Eq. 12-23 reduces to a second-order rate equation:

$$v_0 \approx \left(\frac{k_2}{K_M}\right)[E]_T[S] \approx \left(\frac{k_{cat}}{K_M}\right)[E][S] \qquad [12\text{-}28]$$

Here, **k_{cat}/K_M** is the apparent second-order rate constant of the enzymatic reaction; the rate of the reaction varies directly with how often enzyme and

BOX 12-3

Perspectives in Biochemistry

Kinetics and Transition State Theory

How is the rate of a reaction related to its activation energy (Section 11-2)? Consider a bimolecular reaction that proceeds along the following pathway:

$$A + B \underset{}{\overset{K^{\ddagger}}{\rightleftharpoons}} X^{\ddagger} \xrightarrow{k'} P + Q$$

where $X^{\ddagger}$ represents the transition state. The rate of the reaction can be expressed as

$$\frac{d[P]}{dt} = k[A][B] = k'[X^{\ddagger}] \qquad [12\text{-}A]$$

where k is the ordinary rate constant of the elementary reaction and k' is the rate constant for the decomposition of $X^{\ddagger}$ to products.

Although $X^{\ddagger}$ is unstable, it is assumed to be in rapid equilibrium with the reactants; that is,

$$K^{\ddagger} = \frac{[X^{\ddagger}]}{[A][B]} \qquad [12\text{-}B]$$

where $K^{\ddagger}$ is an equilibrium constant. This central assumption of transition state theory permits the powerful formalism of thermodynamics to be applied to the theory of reaction rates.

Since $K^{\ddagger}$ is an equilibrium constant, it can be expressed as

$$-RT \ln K^{\ddagger} = \Delta G^{\ddagger} \qquad [12\text{-}C]$$

where T is the absolute temperature and R (8.3145 J·K^{-1}·mol^{-1}) is the gas constant (this relationship between equilibrium constants and free energy is derived in Section 1-4D). Combining the three preceding equations yields

$$\frac{d[P]}{dt} = k' e^{-\Delta G^{\ddagger}/RT} [A][B] \qquad [12\text{-}D]$$

This equation indicates that the rate of a reaction not only depends on the concentrations of its reactants, but also decreases exponentially with $\Delta G^{\ddagger}$. Thus, *the larger the difference between the free energy of the transition state and that of the reactants (the free energy of activation), that is, the less stable the transition state, the slower the reaction proceeds.*

We must now evaluate k', the rate at which $X^{\ddagger}$ decomposes. The transition state structure is held together by a bond that is assumed to be so weak that it flies apart during its first vibrational excursion. Therefore, k' is expressed

$$k' = \kappa \nu \qquad [12\text{-}E]$$

where ν is the vibrational frequency of the bond that breaks as $X^{\ddagger}$ decomposes to products, and κ, the **transmission coefficient,** is the probability that the breakdown of $X^{\ddagger}$ will be in the direction of product formation rather than back to reactants. For most spontaneous reactions, κ is assumed to be 1.0 (although this number, which must be between 0 and 1, can rarely be calculated with confidence).

Planck's law states that

$$\nu = \varepsilon/h \qquad [12\text{-}F]$$

where, in this case, ε is the average energy of the vibration that leads to the decomposition of $X^{\ddagger}$, and h (= 6.6261 × 10^{-34} J·s) is **Planck's constant.** Statistical mechanics tells us that at a temperature T, the classical energy of an oscillator is

$$\varepsilon = k_B T \qquad [12\text{-}G]$$

where k_B (= 1.3807 × 10^{-23} J·K^{-1}) is the **Boltzmann constant** and $k_B T$ is essentially the available thermal energy. Combining Eqs. 12-E through 12-G gives

$$k' = \frac{k_B T}{h} \qquad [12\text{-}H]$$

Thus, combining Eqs. 12-A, 12-D, and 12-H yields the expression for the rate constant of the elementary reaction:

$$k = \frac{k_B T}{h} e^{-\Delta G^{\ddagger}/RT} \qquad [12\text{-}I]$$

This equation indicates that as the temperature rises, so that there is increased thermal energy available to drive the reacting complex over the activation barrier ($\Delta G^{\ddagger}$), the reaction speeds up.

---

substrate encounter one another in solution. *The quantity k_{cat}/K_M is therefore a measure of an enzyme's catalytic efficiency.*

There is an upper limit to the value of k_{cat}/K_M: It can be no greater than k_1; that is, the decomposition of ES to E + P can occur no more frequently than E and S come together to form ES. The most efficient enzymes have k_{cat}/K_M values near the **diffusion-controlled limit** of 10^8 to 10^9 M^{-1}·s^{-1}.

These enzymes catalyze a reaction almost every time they encounter a substrate molecule and hence have achieved a state of virtual catalytic perfection. The relationship between the catalytic rate and the thermodynamics of the transition state can now be appreciated (see Box 12-3).

C Analysis of Kinetic Data

There are several methods for determining the values of the parameters of the Michaelis–Menten equation (i.e., V_{max} and K_M). At very high values of [S], the initial velocity, v_o, asymptotically approaches V_{max} (see Sample Calculation 12-3). In practice, however, it is very difficult to assess V_{max} accurately from direct plots of v_o versus [S] such as Fig. 12-3, because, even at substrate concentrations as high as [S] = 10 K_M, Eq. 12-25 indicates that v_o is only 91% of V_{max}, so that the value of V_{max} will almost certainly be underestimated.

A better method for determining the values of V_{max} and K_M, which was formulated by Hans Lineweaver and Dean Burk, uses the reciprocal of the Michaelis–Menten equation (Eq. 12-25):

$$\frac{1}{v_o} = \left(\frac{K_M}{V_{max}}\right)\frac{1}{[S]} + \frac{1}{V_{max}} \qquad [12-29]$$

This is a linear equation in $1/v_o$ and $1/[S]$. If these quantities are plotted to obtain the so-called **Lineweaver–Burk** or **double-reciprocal plot,** the slope of the line is K_M/V_{max}, the $1/v_o$ intercept is $1/V_{max}$, and the extrapolated $1/[S]$ intercept is $-1/K_M$ (Fig. 12-4 and Sample Calculation 12-4).

As can be seen in Fig. 12-3, the best estimates of kinetic parameters are obtained by collecting data over a range of [S] from ~0.5 K_M to ~5 K_M. Thus, a disadvantage of the Lineweaver–Burk plots is that most experimental measurements of [S] are crowded onto the left side of the graph (Fig. 12-4). Moreover, for small values of [S], small errors in v_o lead to large errors in $1/v_o$ and hence to large errors in K_M and V_{max}.

Several other types of plots, each with its advantages and disadvantages, can also be used to determine K_M and V_{max} from kinetic data. However, kinetic data are now commonly analyzed by computer using mathemati-

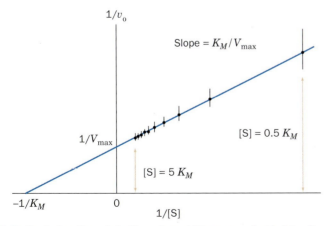

Figure 12-4 *Key to Function.* A double-reciprocal (Lineweaver–Burk) plot.   The error bars represent $\pm 0.05\ V_{max}$. The indicated points are the same as those in Fig. 12-3. Note the large effect of small errors at small [S] (large 1/[S]) and the crowding together of points at large [S]. 🐚 **See the Animated Figures.**

cally sophisticated statistical treatments. Nevertheless, Lineweaver–Burk plots are valuable for the visual presentation of kinetic data.

Steady State Kinetics Cannot Unambiguously Establish a Reaction Mechanism. Although steady state kinetics provides valuable information about the rates of buildup and breakdown of ES, it provides little insight as to the nature of ES. Thus, an enzymatic reaction may, in reality, pass through several more or less stable intermediate states such as

$$E + S \rightleftharpoons ES \rightleftharpoons EX \rightleftharpoons EP \rightleftharpoons E + P$$

or take a more complex path such as

$$E + S \rightleftharpoons ES \begin{array}{c} \nearrow EX \searrow \\ \\ \searrow EY \nearrow \end{array} EP \longrightarrow E + P$$

Unfortunately, steady state kinetic measurements are incapable of revealing the number of intermediates in an enzyme-catalyzed reaction. Thus, such measurements of a multistep reaction can be likened to a "black box" containing a system of water pipes with one inlet and one drain:

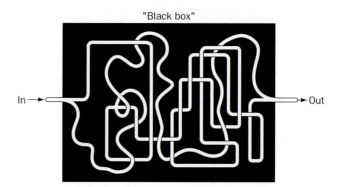

At steady state, that is, after the pipes have filled with water, the relationship between input pressure and output flow can be measured. However, such measurements yield no information concerning the detailed construction of the plumbing connecting the inlet to the drain. This would require additional information, such as opening the box and tracing the pipes. Likewise, steady state kinetic measurements can provide a phenomenological description of enzymatic behavior, but the nature of the intermediates remains indeterminate. The existence of intermediates must be verified independently, for example, by identifying them through the use of spectroscopic techniques.

The foregoing highlights a central principle of enzymology: *The steady state kinetic analysis of a reaction cannot unambiguously establish its mechanism.* This is because no matter how simple, elegant, or rational a postulated mechanism, there are an infinite number of alternative mechanisms that can also account for the kinetic data. Usually, it is the simpler mechanism that turns out to be correct, but this is not always the case. However, *if kinetic data are not compatible with a given mechanism, then that mechanism must be rejected.* Therefore, although kinetics cannot be used to establish a mechanism unambiguously without confirming data, such as the physical demonstration of an intermediate's existence, the steady state ki-

netic analysis of a reaction is of great value because it can be used to eliminate proposed mechanisms.

D Bisubstrate Reactions

We have heretofore been concerned with simple, single-substrate reactions that obey the Michaelis–Menten model (Reaction 12-12). Yet, enzymatic reactions requiring multiple substrates and yielding multiple products are far more common. Indeed, those involving two substrates and yielding two products

$$A + B \underset{}{\overset{E}{\rightleftharpoons}} P + Q$$

account for ~60% of known biochemical reactions. Almost all of these so-called **bisubstrate reactions** are either transfer reactions in which the enzyme catalyzes the transfer of a specific functional group, X, from one of the substrates to the other:

$$P{-}X + B \underset{}{\overset{E}{\rightleftharpoons}} P + B{-}X$$

or oxidation–reduction reactions in which reducing equivalents are transferred between the two substrates. For example, the hydrolysis of a peptide bond by trypsin (Section 11-5) is the transfer of the peptide carbonyl group from the peptide nitrogen atom to water (Fig. 12-5a), whereas in the alcohol dehydrogenase reaction (Section 11-1B), a hydride is formally transferred from ethanol to NAD^+ (Fig. 12-5b). Although bisubstrate reactions could, in principle, occur through a vast variety of mechanisms, only a few types are commonly observed.

Sequential Reactions. *Reactions in which all substrates must combine with the enzyme before a reaction can occur and products be released are known as **Sequential reactions**.* In such reactions, the group being transferred, X, is directly passed from A (= P—X) to B, yielding P and Q (= B—X). Hence, such reactions are also called **single-displacement reactions.**

Sequential reactions can be subclassified into those with a compulsory order of substrate addition to the enzyme, which are said to have an **Ordered mechanism,** and those with no preference for the order of substrate addition, which are described as having a **Random mechanism.** In the Ordered mechanism, the binding of the first substrate is apparently required for the enzyme to form the binding site for the second substrate, whereas in the Random mechanism, both binding sites are present on the free enzyme.

In a notation developed by W.W. Cleland, substrates are designated by the letters A and B in the order that they add to the enzyme, products are designated by P and Q in the order that they leave the enzyme, the enzyme is represented by a horizontal line, and successive additions of substrates and releases of products are denoted by vertical arrows. An Ordered bisubstrate reaction is thereby diagramed:

where A and B are said to be the **leading** and **following** substrates, respectively. Many NAD^+- and $NADP^+$-requiring dehydrogenases follow an Ordered bisubstrate mechanism in which the coenzyme is the leading substrate.

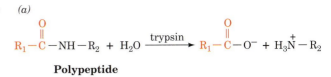

(a)

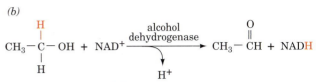

Polypeptide

(b)

Figure 12-5 Some bisubstrate reactions.   (a) In the peptide hydrolysis reaction catalyzed by trypsin, the peptide carbonyl group, with its pendent polypeptide chain, R_1, is transferred from the peptide nitrogen atom to a water molecule. (b) In the alcohol dehydrogenase reaction, a hydride ion is formally transferred from ethanol to NAD^+.

A Random bisubstrate reaction is diagramed:

Some dehydrogenases and kinases operate through Random bisubstrate mechanisms.

Ping Pong Reactions. *Group-transfer reactions in which one or more products are released before all substrates have been added are known as **Ping Pong reactions.*** The Ping Pong bisubstrate reaction is represented by

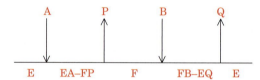

Here, a functional group X of the first substrate A (= P—X) is displaced from the substrate by the enzyme E to yield the first product P and a stable enzyme form F (= E—X) in which X is tightly (often covalently) bound to the enzyme (Ping). In the second stage of the reaction, X is displaced from the enzyme by the second substrate B to yield the second product Q (= B—X), thereby regenerating the original form of the enzyme, E (Pong). Such reactions are therefore also known as **double-displacement reactions.** *Note that in Ping Pong reactions, the substrates A and B do not encounter one another on the surface of the enzyme.* Many enzymes, including trypsin (in which F is the acyl–enzyme intermediate; Section 11-5), transaminases, and some flavoenzymes, react with Ping Pong mechanisms.

Bisubstrate Mechanisms Can Be Distinguished by Kinetic Measurements. The rate equations that describe the foregoing bisubstrate mechanisms are considerably more complicated than the equation for a single-substrate reaction. In fact, the equations for bisubstrate mechanisms (which are beyond the scope of this text) contain as many as four kinetic constants versus two (V_{max} and K_M) for the Michaelis–Menten equation. Nevertheless, steady state kinetic measurements can be used to distinguish among the various bisubstrate mechanisms.

2 Enzyme Inhibition

Many substances alter the activity of an enzyme by combining with it in a way that influences the binding of substrate and/or its turnover number. *Substances that reduce an enzyme's activity in this way are known as **inhibitors.*** A large part of the modern pharmaceutical arsenal consists of enzyme inhibitors. For example, AIDS is treated almost exclusively with drugs that inhibit the activities of certain viral enzymes.

Inhibitors act through a variety of mechanisms. Irreversible enzyme inhibitors, or **inactivators,** bind to the enzyme so tightly that they permanently block the enzyme's activity. Reagents that chemically modify specific amino

acid residues can act as inactivators. For example, the compounds used to identify the catalytic Ser and His residues of serine proteases (Section 11-5A) are inactivators of these enzymes.

Reversible enzyme inhibitors diminish an enzyme's activity by interacting reversibly with it. Some enzyme inhibitors are substances that structurally resemble their enzyme's substrates but either do not react or react very slowly. These substances are commonly used to probe the chemical and conformational nature of an enzyme's active site in an effort to elucidate the enzyme's catalytic mechanism. Other inhibitors affect catalytic activity without interfering with substrate binding. Many do both. In this section, we discuss several of the simplest mechanisms for reversible inhibition and their effects on the kinetic behavior of enzymes that follow the Michaelis–Menten model. As in the preceding discussion, we will base our analysis on a simple one-substrate reaction model.

A Competitive Inhibition

A substance that competes directly with a normal substrate for an enzyme's substrate-binding site is known as a **competitive inhibitor.** Such an inhibitor usually resembles the substrate so that it specifically binds to the active site but differs from the substrate so that it cannot react as the substrate does. For example, **succinate dehydrogenase,** a citric acid cycle enzyme that converts **succinate** to **fumarate** (Section 16-3F), is competitively inhibited by **malonate,** which structurally resembles succinate but cannot be dehydrogenated:

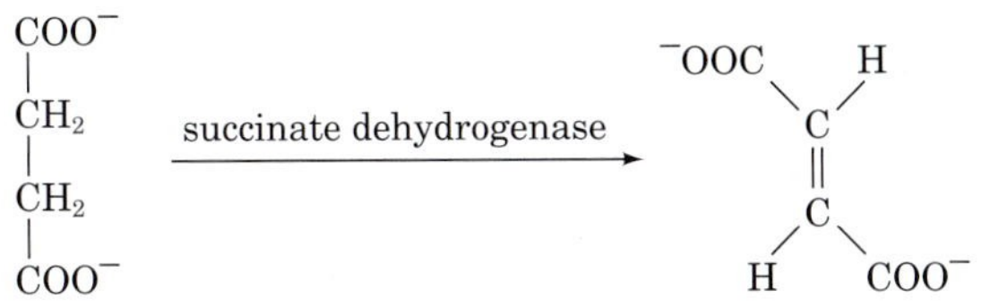

Succinate   **Fumarate**

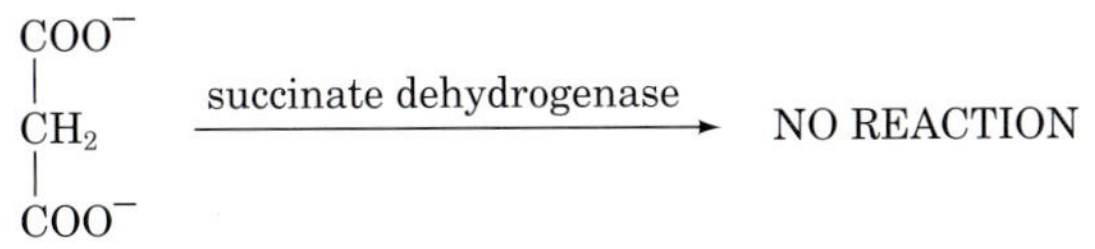

Malonate

The effectiveness of malonate as a competitive inhibitor of succinate dehydrogenase strongly suggests that the enzyme's substrate-binding site is designed to bind both of the substrate's carboxylate groups, presumably through the influence of two appropriately placed positively charged residues.

Similar principles are responsible for **product inhibition.** In this phenomenon, a product of the reaction, which necessarily is able to bind to the enzyme's active site, may accumulate and compete with substrate for binding to the enzyme in subsequent catalytic cycles. Product inhibition is one way in which the cell controls the activities of its enzymes (Section 12-3).

Whereas substrate analogs (including products) are often good competitive inhibitors, **transition state analogs** may be even better inhibitors. This is because effective catalysis depends on the enzyme's ability to bind to and stabilize the reaction's transition state (Section 11-3E). A compound that mimics the transition state can take advantage of these binding interactions

in ways that a substrate analog cannot. For example, **adenosine deaminase** converts the nucleoside adenosine to inosine as follows:

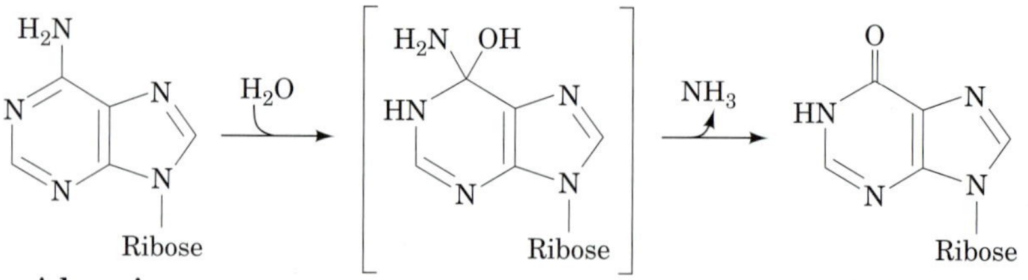

Adenosine   Inosine

The K_M of the enzyme for the substrate adenosine is 3×10^{-5} M. The product inosine acts as an inhibitor of the reaction, with an **inhibition constant** (**K_I**, the dissociation constant for enzyme–inhibitor binding; see below) of 3×10^{-4} M. A transition state analog,

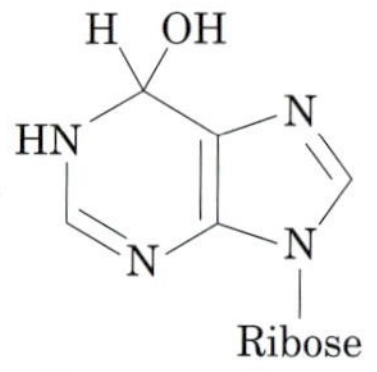

1,6-Dihydroinosine

inhibits the reaction with a K_I of 1.5×10^{-13} M.

The Degree of Competitive Inhibition Depends on the Fraction of Enzyme That Has Bound Inhibitor.

The general model for competitive inhibition is given by the following reaction scheme:

$$E + S \underset{k_{-1}}{\overset{k_1}{\rightleftharpoons}} ES \xrightarrow{k_2} P + E$$

$$+$$

$$I$$

$$K_I$$

$$EI + S \longrightarrow NO\ REACTION$$

Here it is assumed that I, the inhibitor, binds reversibly to the enzyme and is in rapid equilibrium with it so that

$$K_I = \frac{[E][I]}{[EI]} \qquad [12\text{-}30]$$

and EI, the enzyme–inhibitor complex, is catalytically inactive. *A competitive inhibitor therefore reduces the concentration of free enzyme available for substrate binding.*

The Michaelis–Menten equation for a competitively inhibited reaction is derived as before (Section 12-1B), but with an additional term to account for the fraction of $[E]_T$ that binds to I to form EI ($[E]_T = [E] + [ES] + [EI]$). The resulting equation,

$$v_o = \frac{V_{max}[S]}{\alpha K_M + [S]} \qquad [12\text{-}31]$$

is the Michaelis–Menten equation that has been modified by a factor, α, which is defined as

$$\alpha = 1 + \frac{[I]}{K_I} \qquad [12\text{-}32]$$

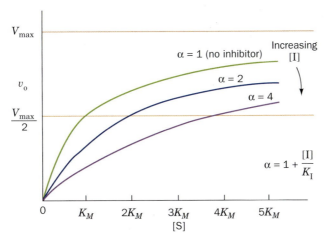

Note that α, a function of the inhibitor's concentration and its affinity for the enzyme, cannot be less than 1. Comparison of Eqs. 12-25 and 12-31 indicates that $K_M^{app} = \alpha K_M$, where K_M^{app} is the apparent K_M, that is, the K_M value that would be measured in the absence of the knowledge that inhibitor is present. Figure 12-6 shows the hyperbolic plots of Eq. 12-31 for increasing values of α. The presence of I makes [S] appear to be less than it really is (makes K_M appear to be larger than it really is), a consequence of the binding of I and S to E being mutually exclusive. However, increasing [S] can overwhelm a competitive inhibitor. In fact, α *is the factor by which [S] must be increased in order to overcome the effect of the presence of inhibitor.* As [S] approaches infinity, v_0 approaches V_{max} for any value of α (that is, for any concentration of inhibitor). Thus, the inhibitor does not affect the enzyme's turnover number.

Competitive inhibition is the principle behind the use of ethanol to treat methanol poisoning. Methanol itself is only mildly toxic. However, the liver enzyme alcohol dehydrogenase converts methanol to the highly toxic formaldehyde, only small amounts of which cause blindness and death. Ethanol competes with methanol for binding to the active site of liver alcohol dehydrogenase, thereby slowing the production of formaldehyde from methanol (the ethanol is converted to the readily metabolized acetaldehyde):

$$
\begin{array}{ccc}
\text{H} & & \text{O} \\
| & & \| \\
\text{H}-\text{C}-\text{OH} & \longrightarrow & \text{H}-\text{C}-\text{H} \\
| & & \\
\text{H} & & \\
\textbf{Methanol} & & \textbf{Formaldehyde}
\end{array}
$$

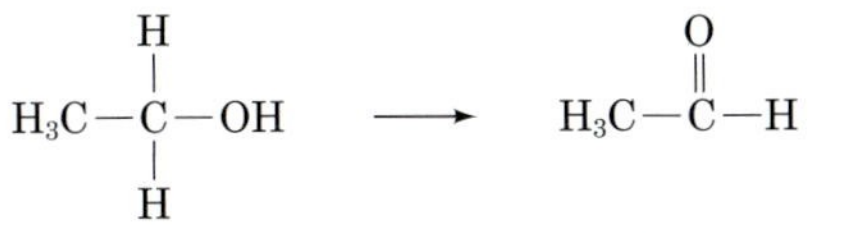

$$
\begin{array}{ccc}
\text{H} & & \text{O} \\
| & & \| \\
\text{H}_3\text{C}-\text{C}-\text{OH} & \longrightarrow & \text{H}_3\text{C}-\text{C}-\text{H} \\
| & & \\
\text{H} & & \\
\textbf{Ethanol} & & \textbf{Acetaldehyde}
\end{array}
$$

Thus, through the administration of ethanol, a large portion of the methanol will be harmlessly excreted from the body in the urine before it can be converted to formaldehyde. The same principle underlies the use of ethanol to treat antifreeze (ethylene glycol, $HOCH_2CH_2OH$) poisoning, which often occurs in cats and dogs.

K_I Can Be Measured. Recasting Eq. 12-31 in the double-reciprocal form yields

$$\frac{1}{v_0} = \left(\frac{\alpha K_M}{V_{max}}\right)\frac{1}{[S]} + \frac{1}{V_{max}} \qquad [12\text{-}33]$$

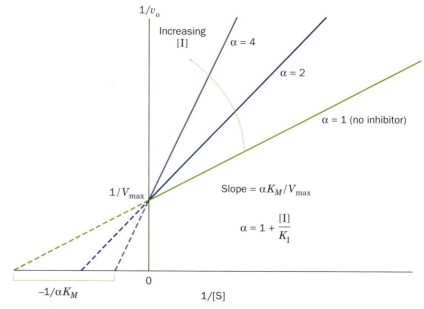

Figure 12-7 A Lineweaver–Burk plot of the competitively inhibited Michaelis–Menten enzyme described by Fig. 12-6. Note that all lines intersect on the $1/v_o$ axis at $1/V_{max}$. The varying slopes indicate the effect of the inhibitor on $\alpha K_M = K_M^{app}$. 🔁 **See the Animated Figures.**

SAMPLE CALCULATION 12-5

An enzyme has a K_M of 8 μM in the absence of a competitive inhibitor and a K_M^{app} of 12 μM in the presence of 3 μM of the inhibitor. Calculate K_I.

First calculate the value of α when $K_M = 8$ μM and $K_M^{app} = 12$ μM:

$$K_M^{app} = \alpha K_M$$
$$\alpha = \frac{K_M^{app}}{K_M}$$
$$\alpha = \frac{12\ \mu M}{8\ \mu M} = 1.5$$

Next, calculate K_I from Eq. 12-32:

$$\alpha = 1 + \frac{[I]}{K_I}$$
$$K_I = \frac{[I]}{\alpha - 1}$$
$$K_I = \frac{3\ \mu M}{1.5 - 1} = 6\ \mu M$$

A plot of this equation is linear and has a slope of $\alpha K_M/V_{max}$, a $1/[S]$ intercept of $-1/\alpha K_M$, and a $1/v_o$ intercept of $1/V_{max}$ (Fig. 12-7). The double-reciprocal plots for a competitive inhibitor at various concentrations of I intersect at $1/V_{max}$ on the $1/v_o$ axis, a property that is diagnostic of competitive inhibition.

The value of K_I for a competitive inhibitor can be determined from the plot of $K_M^{app} = (1 + [I]/K_I)K_M$ vs. $[I]$; its intercept on the $[I]$ axis is $-K_I$. K_I can also be calculated from Eq. 12-32 if $K_M^{app} = \alpha K_M$ is determined at a known $[I]$ for an enzyme of known K_M (see Sample Calculation 12-5). Comparing the K_I values of competitive inhibitors with different structures can provide information about the binding properties of an enzyme's active site and hence its catalytic mechanism. For example, to ascertain the importance of the various segments of an ATP molecule

for binding to the active site of an ATP-requiring enzyme, one might determine the K_I, say, for ADP, AMP, ribose, triphosphate, etc. Since many of these ATP components are unreactive, inhibition studies are the most convenient method of monitoring their binding to the enzyme.

Competitive inhibition studies are also used to determine the affinities of transition state analogs for an enzyme's active site (Section 11-3E). For example, the HIV protease inhibitors (Box 12-4) have been designed to mimic this enzyme's transition state and thus bind to the enzyme with high affinity. Inhibitor studies are the mainstays of such drug development, as is described in Section 12-4.

B Uncompetitive Inhibition

In **uncompetitive inhibition,** the inhibitor binds directly to the enzyme–substrate complex but not to the free enzyme:

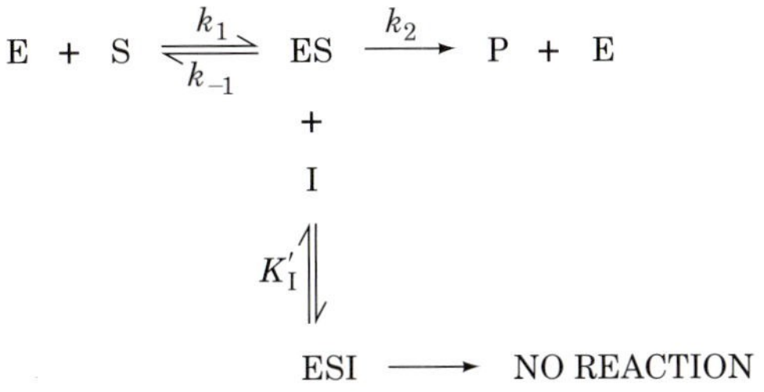

$$E + S \underset{k_{-1}}{\overset{k_1}{\rightleftharpoons}} ES \xrightarrow{k_2} P + E$$
$$+$$
$$I$$
$$K_I' \updownarrow$$
$$ESI \longrightarrow NO\ REACTION$$

In this case, the inhibitor binding step has the dissociation constant

$$K_I' = \frac{[ES][I]}{[ESI]} \qquad [12\text{-}34]$$

The uncompetitive inhibitor, which need not resemble the substrate, presumably *distorts the active site, thereby rendering the enzyme catalytically inactive.*

The Michaelis–Menten equation for uncompetitive inhibition and the equation for its double-reciprocal plot are given in Table 12-2. The double-

Table 12-2 Effects of Inhibitors on Michaelis–Menten Reactions[a]

Type of Inhibition	Michaelis–Menten Equation	Lineweaver–Burk Equation	Effect of Inhibitor
None	$v_o = \dfrac{V_{max}[S]}{K_M + [S]}$	$\dfrac{1}{v_o} = \dfrac{K_M}{V_{max}}\dfrac{1}{[S]} + \dfrac{1}{V_{max}}$	None
Competitive	$v_o = \dfrac{V_{max}[S]}{\alpha K_M + [S]}$	$\dfrac{1}{v_o} = \dfrac{\alpha K_M}{V_{max}}\dfrac{1}{[S]} + \dfrac{1}{V_{max}}$	Increases K_M^{app}
Uncompetitive	$v_o = \dfrac{V_{max}[S]}{K_M + \alpha'[S]} = \dfrac{(V_{max}/\alpha')[S]}{K_M/\alpha' + [S]}$	$\dfrac{1}{v_o} = \dfrac{K_M}{V_{max}}\dfrac{1}{[S]} + \dfrac{\alpha'}{V_{max}}$	Decreases K_M^{app} and V_{max}^{app}
Mixed (noncompetitive)	$v_o = \dfrac{V_{max}[S]}{\alpha K_M + \alpha'[S]} = \dfrac{(V_{max}/\alpha')[S]}{(\alpha/\alpha')K_M + [S]}$	$\dfrac{1}{v_o} = \dfrac{\alpha K_M}{V_{max}}\dfrac{1}{[S]} + \dfrac{\alpha'}{V_{max}}$	Decreases V_{max}^{app}; may increase or decrease K_M^{app}

[a] $\alpha = 1 + \dfrac{[I]}{K_I}$ and $\alpha' = 1 + \dfrac{[I]}{K_I'}$

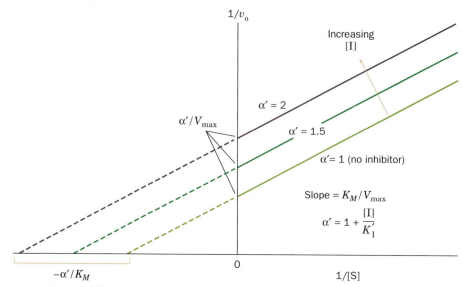

Figure 12-8   A Lineweaver–Burk plot of Michaelis–Menten enzyme in the presence of an uncompetitive inhibitor.   Note that all lines have identical slopes of K_M/V_{max}. 🐾 **See the Animated Figures.**

reciprocal plot consists of a family of parallel lines (Fig. 12-8) with slope K_M/V_{max}, $1/v_o$ intercepts of α'/V_{max}, and $1/[S]$ intercepts of $-\alpha'/K_M$. Note that in uncompetitive inhibition, both $K_M^{app} = K_M/\alpha'$ and $V_{max}^{app} = V_{max}/\alpha'$ are decreased, but that $K_M^{app}/V_{max}^{app} = K_M/V_{max}$. In contrast to the case for competitive inhibition, adding substrate does not reverse the effect of an uncompetitive inhibitor because the binding of substrate does not interfere with the binding of uncompetitive inhibitor.

Uncompetitive inhibition requires that the inhibitor affect the catalytic function of the enzyme but not its substrate binding. This is difficult to envision for single-substrate enzymes. In actuality, uncompetitive inhibition is significant only for multisubstrate enzymes.

C   Mixed Inhibition

Many reversible inhibitors interact with the enzyme in a way that affects substrate binding as well as catalytic activity. In other words, both the enzyme and the enzyme–substrate complex bind inhibitor, resulting in the following model:

$$E + S \underset{k_{-1}}{\overset{k_1}{\rightleftharpoons}} ES \xrightarrow{k_2} P + E$$

$$+ \qquad\qquad +$$
$$I \qquad\qquad I$$

$$K_I \Updownarrow \qquad\qquad K_I' \Updownarrow$$

$$EI \qquad\qquad ESI \longrightarrow \text{NO REACTION}$$

This phenomenon is known as **mixed inhibition** (alternatively, **noncompetitive inhibition**). Presumably, *a mixed inhibitor binds to enzyme sites that participate in both substrate binding and catalysis.* For example, metal ions, which do not compete directly with substrates for binding to an enzyme ac-

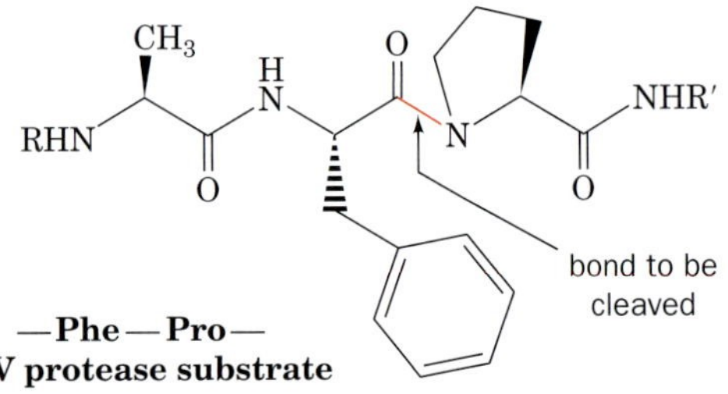

—Phe—Pro—
HIV protease substrate

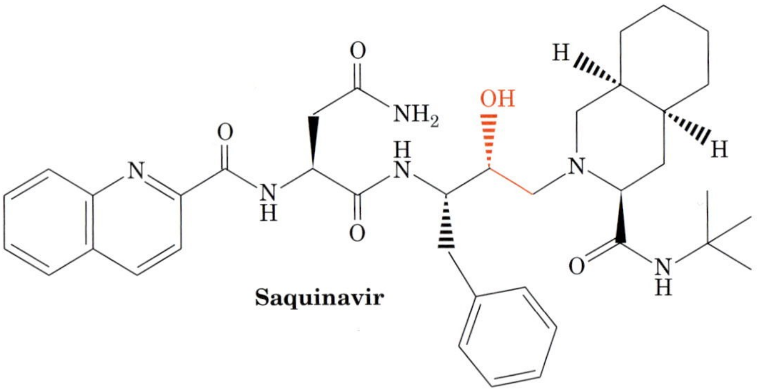

Saquinavir

$K_I = 0.40$ nM

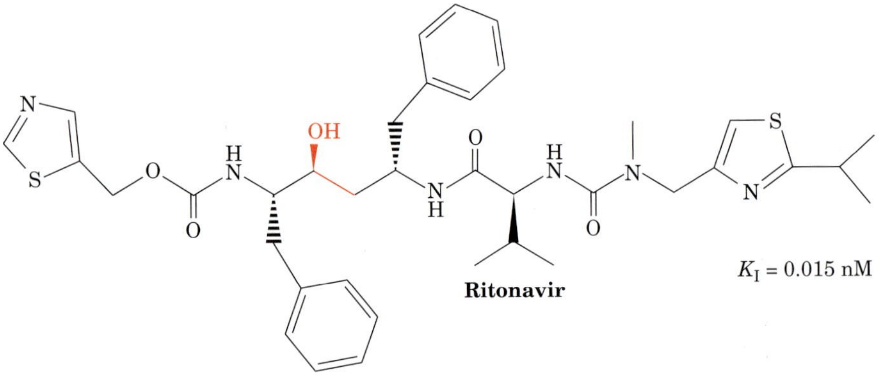

Ritonavir

$K_I = 0.015$ nM

active site. Of perhaps even greater importance, these drugs have the geometry of the catalyzed reaction's tetrahedral transition state (*red*). The enzyme's peptide substrate is shown for comparison. **See the Interactive Exercises.**

The efficacy of anti-HIV agents, like that of many drugs, is limited by their side effects. Despite their preference for viral enzymes, anti-HIV drugs also interfere with normal cellular processes. For example, the inhibition of DNA synthesis by reverse transcriptase inhibitors in rapidly dividing cells, such as the bone marrow cells that give rise to erythrocytes, can lead to severe anemia. Other side effects include nausea, kidney stones, and rashes. Side effects are particularly problematic in HIV infection, because drugs must be taken several times daily for many years, if not for a lifetime.

Acquired resistance also limits the effectiveness of antiviral drugs. This is a significant problem in HIV infection because the error-prone reverse transcriptase allows HIV to mutate rapidly. Numerous mutations in HIV are known to be associated with drug resistance.

In the case of HIV infection, it seems unlikely that a single drug will prove to be a "magic bullet," in part because HIV infects many cell types, but mostly because of its ability to rapidly evolve resistance against any one drug. The outstanding success of anti-HIV therapy rests on combination therapy, in which several different drugs are administered simultaneously. This successfully keeps AIDS at bay by reducing levels of HIV, in some cases to undetectable levels, thereby reducing the probability that HIV will evolve a drug-resistant variant. The advantages of using an inhibitor "cocktail" containing inhibitors of reverse transcriptase and HIV protease include (1) decreasing the likelihood that a viral strain will simultaneously develop resistance to every compound in the mix and (2) decreasing the doses and hence the side effects of the individual compounds.

operates at low pH). Comparisons among the active sites of the aspartic proteases were instrumental in designing HIV protease inhibitors. HIV protease cleaves a number of specific peptide bonds, including Phe–Pro and Tyr–Pro peptide bonds in its physiological substrates, the HIV proteins (*see above*). Inhibitors based on these sequences should therefore selectively inhibit the viral protease. The **peptidomimetic** (peptide-imitating) drugs **ritonavir** and **saquinavir** contain phenyl and other bulky groups that bind in the HIV protease

The kinetics of an enzyme inactivator (an irreversible inhibitor) resembles that of a pure noncompetitive inhibitor because the inactivator reduces the concentration of functional enzyme at all substrate concentrations. Consequently, V_{max} decreases and K_M is unchanged. The double-reciprocal plots for irreversible inactivation therefore resemble those for pure noncompetitive inhibition (the lines intersect on the 1/[S] axis).

3 Allosteric Regulation of Enzyme Activity

An organism must be able to regulate the catalytic activities of its component enzymes so that it can coordinate its numerous metabolic processes, respond to changes in its environment, and grow and differentiate, all in an orderly manner. There are two ways that this may occur:

1. **Control of enzyme availability.** The amount of a given enzyme in a cell depends on both its rate of synthesis and its rate of degradation. Each of these rates is directly controlled by the cell and is subject to dramatic changes over time spans of minutes (in bacteria) to hours (in higher organisms).

2. **Control of enzyme activity.** An enzyme's catalytic activity can be directly regulated through structural alterations that influence the enzyme's substrate-binding affinity or turnover number. Just as hemoglobin's oxygen affinity is allosterically regulated by the binding of ligands such as O_2, CO_2, H^+, and BPG (Section 7-3A), an enzyme's substrate-binding affinity may likewise vary with the binding of small molecules, called **allosteric effectors**. *Allosteric mechanisms can cause large changes in enzymatic activity.* The activities of some enzymes are similarly regulated by covalent modification, usually phosphorylation and dephosphorylation of specific Ser, Thr, or Tyr residues.

In this section, we consider the allosteric control of enzymatic activity by considering one example—**aspartate transcarbamoylase (ATCase)** from *E. coli*. We shall examine other examples of allosteric control as well as covalent modification in later chapters.

The Feedback Inhibition of ATCase Regulates Pyrimidine Synthesis. Aspartate transcarbamoylase catalyzes the formation of *N*-**carbamoyl aspartate** from **carbamoyl phosphate** and aspartate:

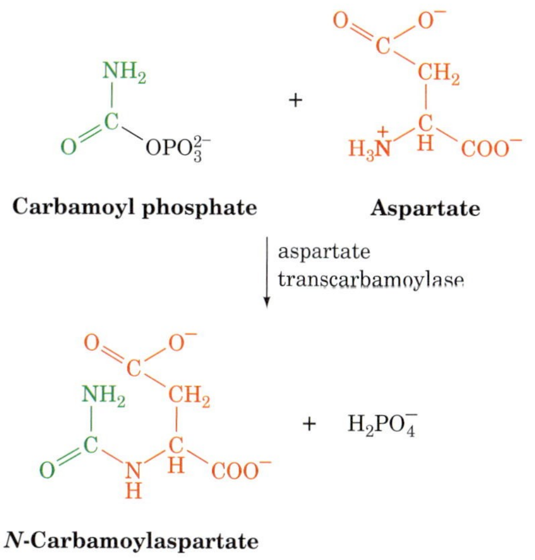

Carbamoyl phosphate   Aspartate

aspartate transcarbamoylase

N-Carbamoylaspartate

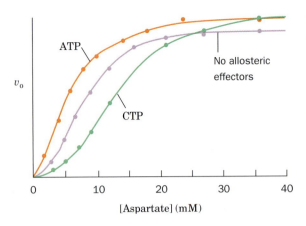

Figure 12-10   Plot of v_o versus [Aspartate] for the ATCase reaction.   Reaction velocity was measured in the absence of allosteric effectors, in the presence of 0.4 mM CTP (an inhibitor), and in the presence of 2.0 mM ATP (an activator). [After Kantrowitz, E.R., Pastra-Landis, S.C., and Lipscomb, W.N., *Trends Biochem. Sci.* **5**, 125 (1980).] 🎵 See the Animated Figures.

This reaction is the first step unique to the biosynthesis of pyrimidines (Section 22-2A). The allosteric behavior of *E. coli* ATCase has been investigated by John Gerhart and Howard Schachman, who demonstrated that both of its substrates bind cooperatively to the enzyme. Moreover, ATCase is allosterically inhibited by **cytidine triphosphate (CTP),** a pyrimidine nucleotide, and is allosterically activated by adenosine triphosphate (ATP), a purine nucleotide.

The v_o versus [S] curve for ATCase (Fig. 12-10) is sigmoidal, rather than hyperbolic as it is in enzymes that follow the Michaelis–Menten model. This is consistent with cooperative substrate binding (recall that hemoglobin's O_2-binding curve is also sigmoidal; Fig. 7-7). ATCase's allosteric effectors shift the entire curve to the right or the left: At a given substrate concentration, CTP decreases the enzyme's catalytic rate, whereas ATP increases it.

CTP, which is a product of the pyrimidine biosynthetic pathway, is an example of a **feedback inhibitor,** since *it inhibits an earlier step in its own biosynthesis* (Fig. 12-11). Thus, when CTP levels are high, CTP binds to

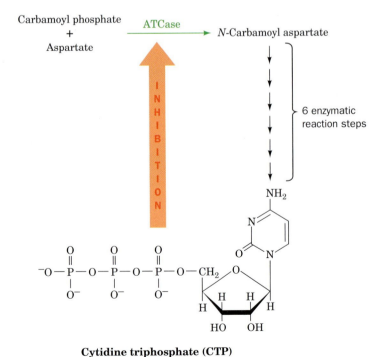

Cytidine triphosphate (CTP)

Figure 12-11   **A schematic representation of the pyrimidine biosynthesis pathway.**   CTP, the end product of the pathway, inhibits ATCase, which catalyzes the pathway's first step.

ATCase, thereby reducing the rate of CTP synthesis. Conversely, when cellular [CTP] decreases, CTP dissociates from ATCase and CTP synthesis accelerates.

The metabolic significance of the ATP activation of ATCase is that it tends to coordinate the rates of synthesis of purine and pyrimidine nucleotides, which are required in roughly equal amounts in nucleic acid biosynthesis. For instance, if the ATP concentration is much greater than that of CTP, ATCase is activated to synthesize pyrimidine nucleotides until the concentrations of ATP and CTP become balanced. Conversely, if the CTP concentration is greater than that of ATP, CTP inhibition of ATCase permits purine nucleotide biosynthesis to balance the ATP and CTP concentrations.

Allosteric Changes Alter ATCase's Substrate-Binding Sites. *E. coli* ATCase (300 kD) has the subunit composition c_6r_6, where c and r represent its catalytic and regulatory subunits. The X-ray structure of ATCase (Fig. 12-12), determined by William Lipscomb, reveals that the catalytic subunits are arranged as two sets of trimers (c_3) in complex with three sets of regulatory dimers (r_2). Each regulatory dimer joins two catalytic subunits in different c_3 trimers.

The isolated catalytic trimers are catalytically active, have a maximum catalytic rate greater than that of intact ATCase, exhibit a noncooperative (hyperbolic) substrate saturation curve, and are unaffected by the presence of ATP or CTP. The isolated regulatory dimers bind these allosteric effectors but are devoid of enzymatic activity. Evidently, *the regulatory subunits allosterically reduce the activity of the catalytic subunits in the intact enzyme.*

As allosteric theory predicts (Section 7-3B), the activator ATP preferentially binds to ATCase's active (R or high substrate affinity) state, whereas the inhibitor CTP preferentially binds to the enzyme's inactive (T or low substrate affinity) state. Similarly, the unreactive bisubstrate analog **N-(phosphonacetyl)-L-aspartate (PALA)**

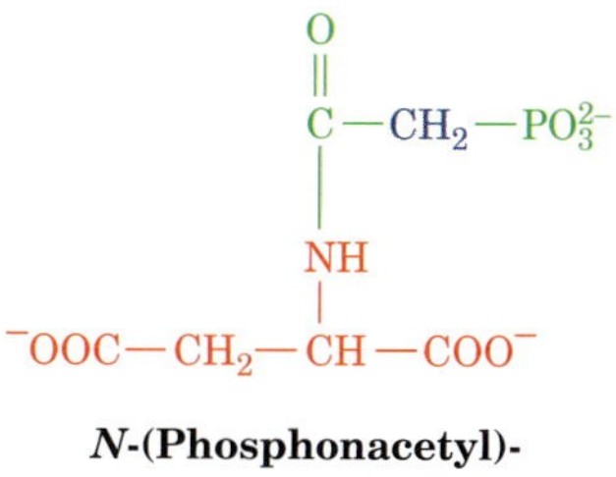

N-(Phosphonacetyl)-
L-aspartate (PALA)

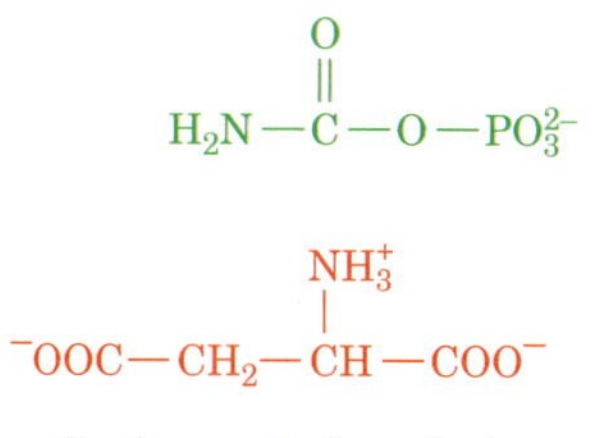

Carbamoyl phosphate
+
Aspartate

binds tightly to R-state but not to T-state ATCase.

X-Ray structures have been determined for the T-state ATCase–CTP complex and the R-state ATCase–PALA complex (as a rule, unreactive substrate analogs, such as PALA, form complexes with an enzyme that are more amenable to structural analysis than complexes of the enzyme with rapidly reacting substrates; Box 11-2). Structural studies reveal that in the T → R transition, the enzyme's catalytic trimers separate along the molecular threefold axis by ~11 Å and reorient around this axis relative to each other by 12° (Fig. 12-12b,c). In addition, the regulatory dimers rotate clockwise by 15° around their twofold axes and separate by ~4 Å along

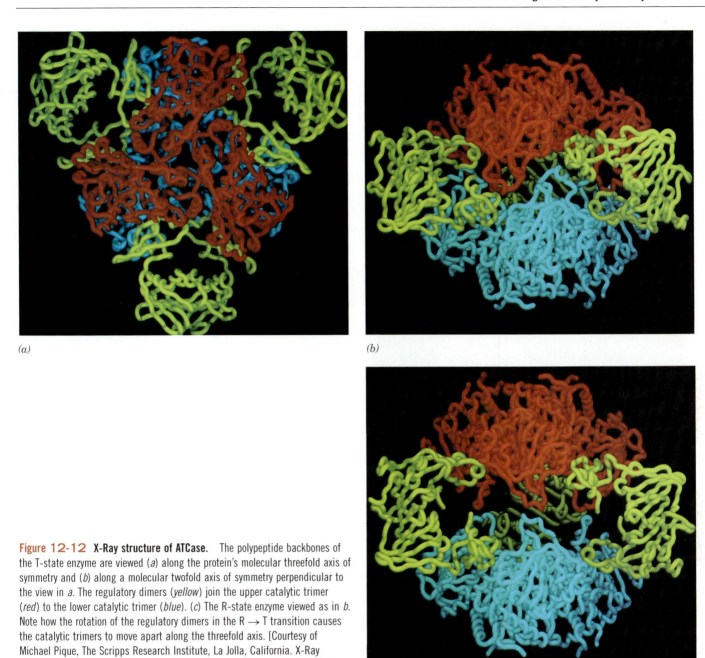

(a)

(b)

(c)

Figure 12-12 X-Ray structure of ATCase.   The polypeptide backbones of the T-state enzyme are viewed (a) along the protein's molecular threefold axis of symmetry and (b) along a molecular twofold axis of symmetry perpendicular to the view in a. The regulatory dimers (yellow) join the upper catalytic trimer (red) to the lower catalytic trimer (blue). (c) The R-state enzyme viewed as in b. Note how the rotation of the regulatory dimers in the R → T transition causes the catalytic trimers to move apart along the threefold axis. [Courtesy of Michael Pique, The Scripps Research Institute, La Jolla, California. X-Ray structures by William Lipscomb, Harvard University. PDBids 4AT1 and 8ATC.]
See Kinemage Exercise 11-1.

the threefold axis. Such large quaternary shifts are reminiscent of those in hemoglobin (Section 7-3A).

Each catalytic subunit of ATCase consists of a carbamoyl phosphate–binding domain and an aspartate-binding domain. The binding of PALA to the enzyme, which presumably mimics the binding of both substrates, induces a conformational change that swings the two domains together such

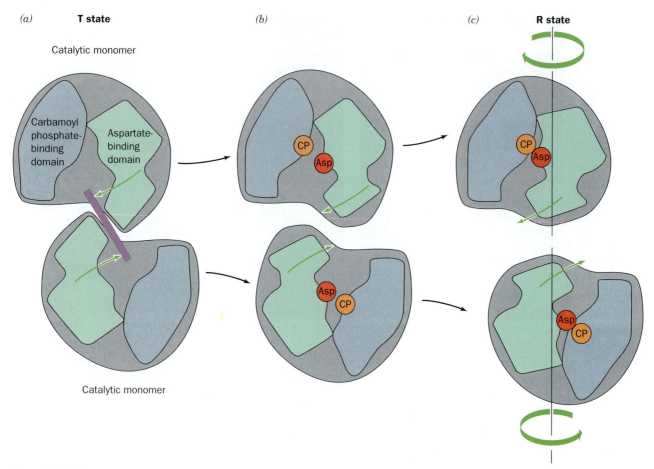

Figure 12-13 **Schematic diagram of tertiary and quaternary conformational changes in ATCase.** Two of the six vertically interacting catalytic ATCase subunits are shown. (*a*) In the absence of bound substrate, the protein is held in the T state because the motions that bring together the two domains of each subunit (*green arrows*) are prevented by steric interference (*purple bar*) between the contacting aspartic acid–binding domains. (*b*) The binding of carbamoyl phosphate (CP) followed by aspartic acid (Asp) to their respective binding sites causes the subunits to move apart and rotate with respect to each other so as to permit the T → R transition. (*c*) In the R state, the two domains of each subunit come together to promote the reaction of their bound substrates to form products. [Illustration, Irving Geis/Geis Archives Trust. Copyright Howard Hughes Medical Institute. Reproduced with permission.]

 See Kinemage Exercises 11-1 and 11-2.

that their two bound substrates can react to form product (Fig. 12-13). The conformational changes—movements of up to 8 Å for some residues—in a single catalytic subunit trigger ATCase's T → R quaternary shift. ATCase's tertiary and quaternary shifts are tightly coupled (i.e., ATCase closely follows the symmetry model of allosterism; Section 7-3B). *The binding of substrate to one catalytic subunit therefore increases the substrate-binding affinity and catalytic activity of the other five catalytic subunits* and hence accounts for the enzyme's cooperative substrate binding.

The Structural Basis of Allosterism in ATCase. The structural basis for the effects of CTP and ATP on ATCase activity is gradually being unveiled. Both the inhibitor CTP and the activator ATP bind to the same site on the outer edge of the regulatory subunit, about 60 Å away from the nearest catalytic site. CTP binds preferentially to the T state, increasing its stability, whereas ATP binds preferentially to the R state, increasing its stability.

The binding of CTP and ATP to their less favored enzyme states also has structural consequences. When CTP binds to R-state ATCase, it induces

a contraction in the regulatory dimer that causes the catalytic trimers to come together by 0.5 Å (become more T-like, that is, less active). This, in turn, reorients key residues in the enzyme's active sites, thereby decreasing the enzyme's catalytic activity. ATP has essentially opposite effects when binding to the T-state enzyme: It causes the catalytic trimers to move apart by 0.4 Å (become more R-like, that is, more active), thereby reorienting key residues in the enzyme's active sites so as to increase the enzyme's catalytic activity.

Allosteric Transitions in Other Enzymes Resemble Those of Hemoglobin and ATCase.   Allosteric enzymes are widely distributed in nature and tend to occupy key regulatory positions in metabolic pathways. Such enzymes are symmetrical proteins containing at least two subunits. In all known cases, quaternary structural changes communicate binding and catalytic effects among all active sites in the enzyme. The quaternary shifts are primarily rotations of subunits relative to one another. Secondary structures are largely preserved in T → R transitions, which is probably important for mechanically transmitting allosteric effects over distances of tens of angstroms.

4 Drug Design

The use of drugs to treat various maladies has a long history, but the modern pharmaceutical industry, which is based on science (rather than tradition or superstition), is a product of the twentieth century. For example, at the beginning of the twentieth century, only a handful of drugs, apart from folk medicines, were known, including **digitalis,** a heart stimulant from the foxglove plant (see Box 10-3); **quinine** (Fig. 12-14), obtained from the bark and roots of the *Cinchona* tree, which was used to treat malaria; and

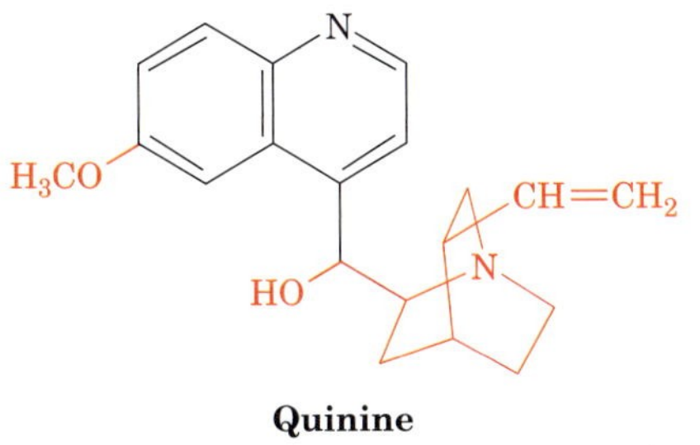

Quinine

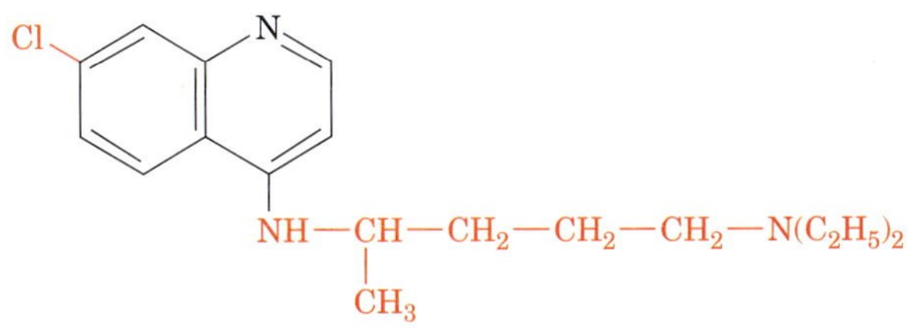

Chloroquine

Figure 12-14   Quinine and chloroquine.   These two compounds, which share a quinoline ring system, are effective antimalarial agents. The *Plasmodium* parasite multiplies within red blood cells, where it proteolyzes hemoglobin to meet its nutritional needs. This process releases heme, which in its soluble form, is toxic to the parasite. The parasite sequesters the heme in a crystalline (nontoxic) form. Quinine and chloroquine pass through cell membranes and inhibit heme crystallization in the parasite.

mercury, which was used to treat syphilis—a cure that was often worse than the disease. Almost all drugs in use today were discovered and developed in the past three decades. The majority of drugs act by modifying the activity of a receptor protein, with enzyme inhibitors constituting the second largest class of drugs. Indeed, the techniques of enzyme kinetics have proven to be invaluable for evaluating drug candidates.

A Drug Discovery

How are new drugs discovered? Nearly all drugs that have been in use for over a decade were discovered by screening large numbers of compounds—either synthetic compounds or those derived from natural products (often plants used in folk remedies). Initial screening involves *in vitro* assessment of, for example, the degree of binding of a drug candidate to an enzyme that is implicated in a disease of interest, that is, determining its K_I value. Later, as the number of drug candidates is winnowed down, more sensitive screens such as testing in animals are employed.

A drug candidate that exhibits a desired effect is called a **lead compound.** A good lead compound binds to its target protein with a dissociation constant (for an enzyme, an inhibition constant) of less than 1 μM. Such a high affinity is necessary to minimize a drug's less specific binding to other macromolecules in the body and to ensure that only low doses of the drug need be taken.

A lead compound is used as a point of departure to design more efficacious compounds. Even minor modifications to a drug candidate can result in major changes in its pharmacological properties. Thus, substitution of methyl, chloro, hydroxyl, or benzyl groups at various places on a lead compound may improve its action. For most drugs in use today, 5000 to 10,000 related compounds were typically synthesized and tested. Drug development is a systematic and iterative process in which the most promising derivatives of the lead compound serve as the starting points for the next round of derivatization and testing.

Structure-Based Drug Design. Since the mid 1980s, dramatic advances in the speed and precision with which a macromolecular structure can be determined by X-ray crystallography and NMR (see Section 6-2A) have enabled **structure-based drug design,** a process that greatly reduces the number of compounds that need be synthesized in a drug discovery program. As its name implies, structure-based drug design (also called **rational drug design**) uses the structure of a receptor or enzyme in complex with a drug candidate to guide the development of more efficacious compounds. Such a structure will reveal, for example, the positions of the hydrogen bonding donors and acceptors in the binding site as well as cavities in the binding site into which substituents might be placed on a drug candidate to increase its binding affinity. These direct visualization techniques are usually supplemented with molecular modeling tools such as the computation of the minimum energy conformation of a proposed derivative, quantum mechanical calculations that determine its charge distribution and hence how it would interact electrostatically with the protein, and docking simulations in which an inhibitor candidate is computationally modeled into the binding site on the receptor to assess potential interactions. A structure-based approach was used to develop the pain relievers Celebrex and Vioxx (Section 19-6C) and to develop drugs to treat HIV infection (Box 12-4).

Combinatorial Chemistry and High-Throughput Screening. As structure-based methods were developed, it appeared that they would become the domi-

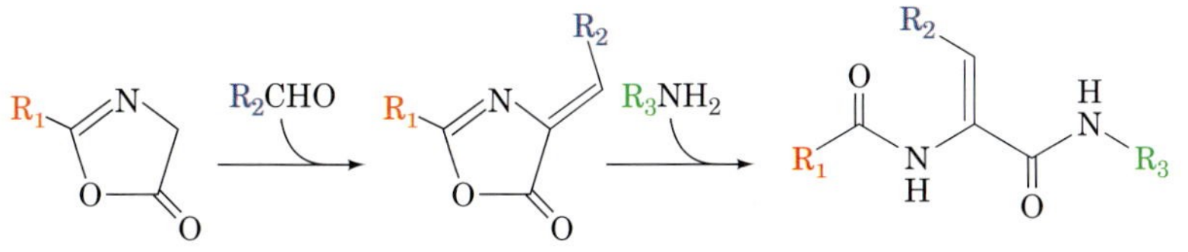

Figure 12-15 **The combinatorial synthesis of arylidene diamides.**   If ten different variants of each R group are used in the synthesis, then 1000 different derivatives will be synthesized.

nant mode of drug discovery. However, the recent advent of **combinatorial chemistry** techniques (see Box 5-1) to rapidly and inexpensively synthesize large numbers of related compounds combined with the development of robotic **high-throughput screening** techniques has caused the drug discovery "pendulum" to again swing toward the "make-many-compounds-and-see-what-they-do" approach. If a lead compound can be synthesized in a stepwise manner from several smaller modules, then the substituents on each of these modules can be varied in parallel to produce a library of related compounds (Fig. 12-15).

A variety of synthetic techniques have been developed that permit the combinatorial synthesis of thousands of related compounds in a single procedure. Thus, whereas investigations into the importance of a hydrophobic group at a particular position in a lead compound might previously have prompted the individual syntheses of only the ethyl, propyl, and benzyl derivatives of the compound, the use of combinatorial synthesis would permit the generation of perhaps 100 different groups at that position. This would far more effectively map out the potential range of the substituent and possibly identify an unexpectedly active analog.

B Bioavailability and Toxicity

The *in vitro* development of an effective drug candidate is only the first step in the drug development process. *Besides causing the desired response in its isolated target protein, a useful drug must be delivered in sufficiently high concentration to this protein where it resides in the human body.* For example, a drug that is administered orally (the most convenient route) must surmount a series of formidable barriers: (1) The drug must be chemically stable in the highly acidic environment of the stomach and must not be degraded by digestive enzymes; (2) it must be absorbed from the gastrointestinal tract into the bloodstream, that is, it must pass through several cell membranes; (3) it must not bind too tightly to other substances in the body (e.g., lipophilic substances tend to be absorbed by certain plasma proteins and by fat tissue); (4) it must survive derivatization by the battery of enzymes, mainly in the liver, that detoxify **xenobiotics** (foreign compounds; note that the intestinal blood flow drains directly into the liver via the portal vein so that the liver processes all orally ingested substances before they reach the rest of the body); (5) it must avoid rapid excretion by the kidneys; (6) it must pass from the capillaries to its target tissue; (7) if it is targeted to the brain, it must cross the **blood–brain barrier,** which blocks the passage of most polar substances; and (8) if it is targeted to an intracellular protein, it must pass through the plasma membrane and possibly other intracellular membranes.

The ways in which a drug interacts with the barriers listed above is known as its **pharmacokinetics.** Thus, the **bioavailability** of a drug (the extent to

which it reaches its site of action, which is usually taken to be the systemic circulation) depends on both the dose given and its pharmacokinetics. *The most effective drugs are usually a compromise; they are neither too lipophilic nor too hydrophilic.* In addition, their pK values are usually in the range 6 to 8 so that they can readily assume both their ionized and unionized forms at physiological pH's. This permits them to cross cell membranes in their unionized form and to bind to their target protein in their ionized form.

C Clinical Trials

Above all else, a successful drug candidate must be safe and efficacious in humans. Tests for these properties are initially carried out in animals, but since humans and animals often react quite differently to a drug, it must ultimately be tested in humans through **clinical trials.** In the United States, clinical trials are monitored by the Food and Drug Administration (FDA) and have three increasingly detailed (and expensive) phases:

Phase I. This phase is primarily designed to test the safety of a drug candidate but is also used to determine its dosage range and the optimal dosage method (e.g., oral vs injected) and frequency. It is usually carried out on a small number (20–100) of normal, healthy volunteers, but in the case of a drug candidate known to be highly toxic (e.g., a cancer chemotherapeutic agent), it is carried out on volunteer patients with the target disease.

Phase II. This phase mainly tests the efficacy of the drug against the target disease in 100 to 500 volunteer patients but also refines the dosage range and checks for side effects. The effects of the drug candidate are usually assessed via **single blind tests,** procedures in which the patient is unaware of whether he/she has received the drug or a control substance. Usually the control substance is a **placebo** (an inert substance with the same physical appearance, taste, etc., as the drug being tested) but, in the case of a life-threatening disease, it is an ethical necessity that the control substance be the best available treatment against the disease.

Phase III. This phase monitors adverse reactions from long-term use as well as confirming efficacy in 1000 to 5000 patients. It pits the drug candidate against control substances through the statistical analysis of carefully designed **double blind tests,** procedures in which neither the patients nor the clinical investigators know whether a given patient has received the drug or a control substance. This is done to minimize bias in the subjective judgments the investigators must make.

Currently, only about 5 drug candidates in 5000 that enter preclinical trials reach clinical trials. Of these, only one, on average, is ultimately approved for clinical use, with the majority failing in Phase II trials. In recent years, the preclinical portion of a drug discovery process has averaged ~3 years to complete, whereas successful clinical trials have usually required an additional 7 to 10 years. These successive stages of the drug discovery process are increasingly expensive so that successfully bringing a drug to market costs, on average, around $300 million.

The most time-consuming and expensive aspect of a drug development program is identifying a drug candidate's rare adverse reactions. Nevertheless, it is not an uncommon experience for a drug to be brought to market only to be withdrawn some months or years later when it is found to

have caused unanticipated life-threatening side effects in as few as 1 in 10,000 individuals (the post-marketing surveillance of a drug is known as its Phase IV clinical trial). For example, in 1997, the FDA withdrew its approval of the drug **fenfluramine (fen),**

Fenfluramine

Phentermine

which it had approved in 1973 for use as an appetite suppressant in short-term (a few weeks) weight-loss programs. Fenfluramine had become widely prescribed, often for extended periods, together with another appetite suppressant, **phentermine (phen;** approved in 1959), a combination known as **fen-phen** (although the FDA had not approved of the use of the two drugs in combination, once it approves a drug for some purpose, a physician may prescribe it for any other purpose). The withdrawal of fenfluramine was prompted by over 100 reports of heart valve damage in individuals (mostly women) who had taken fen-phen for an average of 12 months (phentermine was not withdrawn because the evidence indicated that fenfluramine was the responsible agent). This rare side effect had not been observed in the clinical trials of fenfluramine, in part because it is such an unusual type of drug reaction that it had not been screened for.

D Cytochromes P450 and Adverse Drug Reactions

Why is it that a drug that is well tolerated by the majority of patients can pose a danger to others? *Differences in reactions to drugs arise from genetic differences among individuals as well as differences in their disease states, other drugs they are taking, age, sex, and environmental factors.* The **cytochromes P450,** which function in large part to detoxify xenobiotics and participate in the metabolic clearance of the majority of drugs in use, provide instructive examples of these phenomena.

The cytochromes P450 constitute a superfamily of heme-containing enzymes that occur in nearly all living organisms, from bacteria to mammals [their name arises from the characteristic 450-nm peak in their absorption spectra when reacted in their Fe(II) state with CO]. Humans express ~100 **isozymes** (catalytically and structurally similar but genetically distinct enzymes from the same organism; also called **isoforms**) of cytochrome P450, mainly in the liver but also in other tissues (its various isozymes are named by the letters "CYP" followed by a number designating its family, an upper case letter designating its subfamily, and often another number; e.g., CYP2D6). These **monooxygenases** (Fig. 12-16), which in animals are embedded in the endoplasmic reticulum membrane, catalyze reactions of the sort

$$RH + O_2 + 2H^+ + 2e^- \rightarrow ROH + H_2O$$

The electrons (e^-) are supplied by NADPH, which passes them to cytochrome P450's heme prosthetic group via the intermediacy of the enzyme **cytochrome P450 reductase.** Here RH represents a wide variety of usually lipophilic compounds for which the different cytochromes P450 are specific. They include polycyclic aromatic hydrocarbons [PAHs; frequently car-

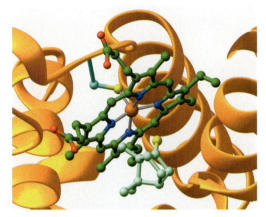

Figure 12-16 The X-ray structure of the cytochrome P450 from *Pseudomonas putida* showing its active site region. The heme group, the Cys side chain that axially ligands its Fe atom, and the enzyme's lipophilic substrate **thiocamphor** are shown in ball-and-stick form with N blue, O red, S yellow, Fe orange, and the C atoms of the heme, its liganding Cys side chain, and the thiocamphor green, cyan, and pale blue-green, respectively. The bonds ligading the Fe are gray. [Based on an X-ray structure by Thomas Poulos, University of California at Irvine. PDBid 8CPP.]

cinogenic (cancer-causing) compounds that are present in tobacco smoke, broiled meats, and other pyrolysis products], polycyclic biphenyls (PCBs; which were widely used in electrical insulators and as plasticizers and are also carcinogenic), steroids (in whose syntheses cytochromes P450 participate), and many different types of drugs. The xenobiotics are thereby converted to a more water-soluble form, which aids in their excretion by the kidneys. Moreover, the newly generated hydroxyl groups are often enzymatically conjugated (covalently linked) to polar substances such as glucuronic acid (Section 8-1C), glycine, sulfate, and acetate, which further enhances aqueous solubility. The many types of cytochromes P450 in animals, which have different substrate specificities (although these specificities tend to be broad and hence often overlap), are thought to have arisen in response to the numerous toxins that plants produce, presumably to discourage animals from eating them.

Drug–drug interactions *are often mediated by cytochromes P450.* For example, if drug A is metabolized by or otherwise inhibits a cytochrome P450 isozyme that metabolizes drug B, then coadministering drugs A and B will cause the bioavailability of drug B to increase above the value it would have had if it alone had been administered. This phenomenon is of particular concern if drug B has a low **therapeutic index** (the ratio of the dose of the drug that produces toxicity to that which produces the desired effect). Conversely, if, as is often the case, drug A induces the increased expression of the cytochrome P450 isozyme that metabolizes it and drug B, then coadministering drugs A and B will reduce drug B's bioavailability, a phenomenon that was first noted when certain antibiotics caused oral contraceptives to lose their efficacy. Moreover, if drug B is metabolized to a toxic product, its increased rate of reaction may result in an adverse reaction. Environmental pollutants such as PAHs and PCBs are also known to induce the expression of specific cytochrome P450 isozymes and thereby alter the rates that certain drugs are metabolized. Finally, some of these same effects may occur in patients with liver disease, as well as arising from age-based, gender-based, and individual differences in liver physiology.

Although the cytochromes P450 presumably evolved to detoxify and/or help eliminate harmful substances, in several cases they have been shown to participate in converting relatively innocuous compounds to toxic agents. For example, **acetaminophen** (Fig. 12-17), a widely used analgesic and antipyretic (fever reducer) is quite safe when taken in therapeutic doses (1.2 g/day for an adult) but, in large doses (>10 g), is highly toxic. This is because, in therapeutic amounts, 95% of the acetaminophen present is enzymatically glucuronidated or sulfated at its —OH group to the corresponding conjugates, which are readily excreted. The remaining 5% is converted, through the action of a cytochrome P450 (CYP2E1), to **acetimidoquinone** (Fig. 12-17), which is then conjugated with glutathione (Section 4-3B). However, when acetaminophen is taken in large amounts, the glucuronidation and sulfation pathways become saturated and hence the cytochrome P450-mediated pathway becomes increasingly important. If hepatic (liver) glutathione is depleted faster than it can be replaced, acetimidoquinone, a reactive compound, instead conjugates with the sulfhydryl groups of cellular proteins resulting in often fatal hepatotoxicity.

Many of the cytochromes P450 in humans are unusually **polymorphic,** that is, there are several common alleles (variants) of the genes encoding each of these enzymes in the human population. Alleles that cause diminished, enhanced, and qualitatively altered rates of drug metabolism have been characterized for many of the cytochromes P450. The distributions of these various alleles differs markedly among ethnic groups and hence prob-

ably arose to permit each group to cope with the toxins in its particular diet.

Polymorphism in a given cytochrome P450 results in differences between individuals in the rates that they metabolize certain drugs. For instance, in cases that a cytochrome P450 variant has absent or diminished activity, otherwise standard doses of a drug that the enzyme normally metabolizes may cause the bioavailability of the drug to reach toxic levels. Conversely, if a particular P450 enzyme has enhanced activity (usually because the gene encoding it has been duplicated one or more times), higher than normal doses of a drug that the enzyme metabolizes would have to be administered to obtain the required therapeutic effect. However, if the drug is metabolized to a toxic product, this may result in an adverse reaction. Several known P450 variants have altered substrate specificities and hence produce unusual metabolites, which also may cause harmful side effects.

Experience has amply demonstrated that *there is no such thing as a drug that is entirely free of adverse reactions.* However, as the enzymes and their variants that participate in drug metabolism are characterized and rapid and inexpensive genotyping methods are developed, it is becoming possible to tailor drug treatment to an individual's genetic makeup rather than to the population as a whole. This rapidly developing area of study is called **pharmacogenomics.**

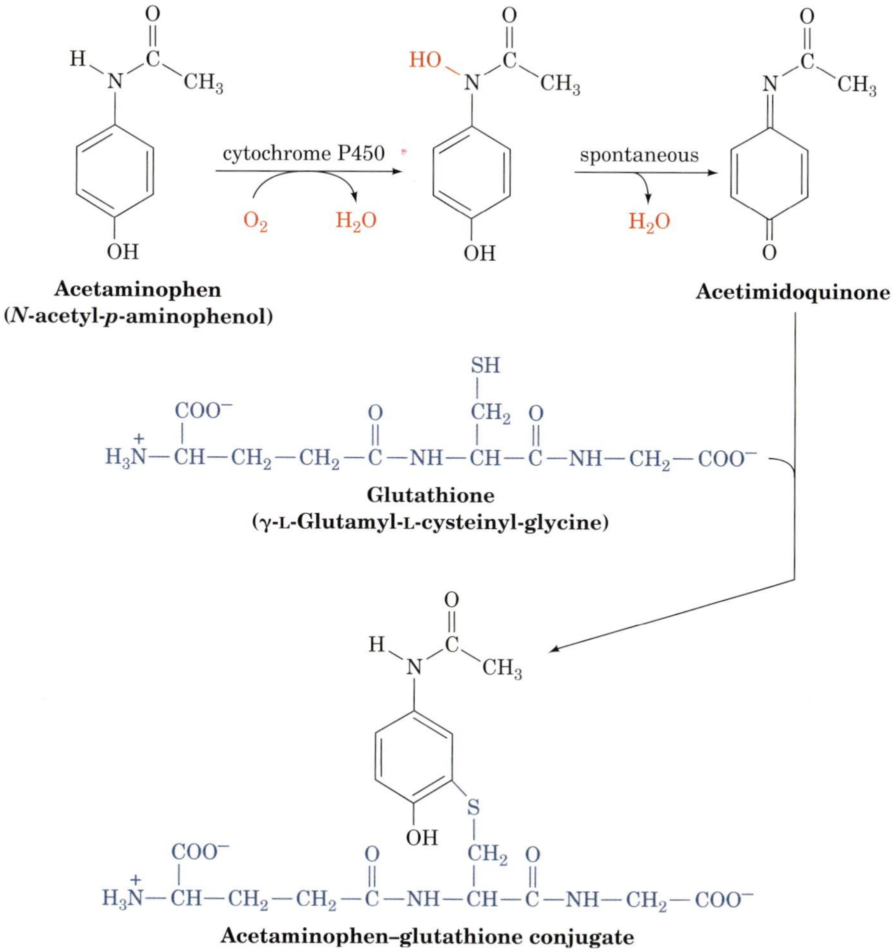

Figure 12-17   The metabolic reactions of acetaminophen that convert it to its conjugate with glutathione.

SUMMARY

1. Elementary chemical reactions may be first order, second order, or rarely, third order. In each case, a rate equation describes the progress of the reaction as a function of time.

2. The Michaelis–Menten equation describes the relationship between initial reaction velocity and substrate concentration under steady state conditions.

3. K_M is the substrate concentration at which the reaction velocity is half-maximal. The value of k_{cat}/K_M indicates an enzyme's catalytic efficiency.

4. Kinetic data can be plotted in double-reciprocal form to determine K_M and V_{max}.

5. Bisubstrate reactions are classified as Sequential (single displacement) or Ping Pong (double displacement). A Sequential reaction may proceed by an Ordered or Random mechanism.

6. Reversible inhibitors reduce an enzyme's activity by binding to the substrate-binding site (competitive inhibition), to the enzyme–substrate complex (uncompetitive inhibition), or to both the enzyme and the enzyme–substrate complex (mixed inhibition).

7. Enzyme activity may be regulated by allosteric effectors.

8. The activity of ATCase is increased by ATP and decreased by CTP, which alter the conformation of the catalytic sites by stabilizing the R and the T states of the enzyme, respectively.

9. An enzyme inhibitor can be developed for use as a drug through structure-based or combinatorial methods. It must then be tested for safety and efficacy in clinical trials. Adverse reactions to drugs and drug–drug interactions are often mediated by a cytochrome P450.

REFERENCES

Kinetics

Cornish-Bowden, A., *Fundamentals of Enzyme Kinetics* (revised ed.), Portland Press (1995). [A lucid and detailed account of enzyme kinetics.]

Fersht, A., *Structure and Mechanism in Protein Science: A Guide to Enzyme Catalysis and Protein Folding*, W.H. Freeman (1999).

Gutfreund, H., *Kinetics for the Life Sciences: Receptors, Transmitters, and Catalysts*, Cambridge University Press (1995).

Segel, I.H., *Enzyme Kinetics*, Wiley-Interscience (1993). [A detailed and understandable treatise providing full explanations of many aspects of enzyme kinetics.]

Wood, W.B., Wilson, J.H., Benbow, R.M., and Hood, L.E., *Biochemistry. A Problems Approach* (2nd ed.), Chapter 8, Benjamin/Cummings (1981). [Contains instructive problems on enzyme kinetics with answers worked out in detail.]

Allosteric Regulation

Jin, L., Stec, B., Lipscomb, W.N., and Kantrowitz, E.R., Insights into the mechanisms of catalysis and heterotropic regulation of *Escherichia coli* aspartate transcarbamoylase based upon a structure of the enzyme complexed with the bisubstrate analogue *N*-phosphonacetyl-L-aspartate at 2.1 Å, *Proteins* **37**, 729–742 (1999).

Lipscomb, W.N., Structure and function of allosteric enzymes, *Chemtracts—Biochem. Mol. Biol.* **2**, 1–15 (1991).

Perutz, M., *Mechanisms of Cooperativity and Allosteric Regulation in Proteins*, Cambridge University Press (1990).

Drug Design

Ingelman-Sundberg, M., Oscarson, M., and McLellan, R.A., Polymorphic human cytochrome P450 enzymes: An opportunity for individualized drug treatment, *Trends Pharmacol. Sci.* **20**, 342–349 (1999).

Marrone, T.J., Briggs, J.M., and McCammon, J.A., Structure-based drug design: Computational advances, *Annu. Rev. Pharmacol. Toxicol.* **37**, 71–90 (1997).

Ohlstein, E.H., Ruffolo, R.R., Jr., and Elliott, J.D., Drug discovery in the next millennium, *Annu. Rev. Pharmacol. Toxicol.* **40**, 177–191 (2000).

Smith, D.A. and van der Waterbeemd, H., Pharmacokinetics and metabolism in early drug design, *Curr. Opin. Chem. Biol.* **3**, 373–378 (1999).

White, R.E., High-throughput screening in drug metabolism and pharmokinetic support of drug discovery, *Annu. Rev. Pharmacol. Toxicol.* **40**, 133–157 (2000).

Wlodawer, A. and Vondrasek, J., Inhibitors of HIV-1 protease: A major success of structure-assisted drug design, *Annu. Rev. Biophys. Biomol. Struct.* **27**, 249–284 (1998). [Reviews the development of some HIV-1 protease inhibitors.]

KEY TERMS

elementary reaction	$t_{1/2}$	enzyme saturation	Ordered mechanism
k	ES complex	Michaelis–Menten equation	Random mechanism
v	k_{-1}	k_{cat}	Ping Pong reaction
reaction order	k_2	turnover number	single-displacement reaction
first-order reaction	Michaelis complex	k_{cat}/K_M	double-displacement reaction
second-order reaction	steady state assumption	diffusion-controlled limit	
pseudo-first-order reaction	K_M	Lineweaver–Burk (double-reciprocal) plot	inhibitor
molecularity	v_o	Sequential reaction	inactivator
rate equation	V_{max}		competitive inhibition

uncompetitive inhibition	K_M^{app}	structure-based (rational)	bioavailability
mixed (noncompetitive)	V_{max}^{app}	drug design	clinical trials
inhibition	allosteric effector	combinatorial chemistry	cytochrome P450
product inhibition	feedback inhibition	xenobiotic	drug–drug interactions
transition state analog	lead compound	pharmacokinetics	pharmacogenomics
K_I			

STUDY EXERCISES

1. Write the rate equations for a first-order and a second-order reaction.

2. What are the differences between instantaneous velocity, initial velocity, and maximal velocity for an enzymatic reaction?

3. Derive the Michaelis–Menten equation.

4. What do the values of K_M and k_{cat}/K_M reveal about an enzyme?

5. Write the Lineweaver–Burk (double-reciprocal) equation and describe the features of a Lineweaver–Burk plot.

6. Use Cleland notation to describe Ordered and Random Sequential reactions and a Ping Pong reaction.

7. Describe the effects of competitive and mixed inhibitors on K_M and V_{max}.

8. What distinguishes an inhibitor from an inactivator?

9. List some mechanisms for regulating enzyme activity.

10. Explain the structural basis for cooperative substrate binding and allosteric regulation in ATCase.

11. Summarize the chemical and biological features of effective drugs.

12. Indicate how drugs that are well tolerated by the majority of the population cause adverse reactions in certain individuals.

PROBLEMS

1. Consider the nonenzymatic elementary reaction A → B. When the concentration of A is 20 mM, the reaction velocity is measured as 5 μM B produced per minute. (a) Calculate the rate constant for this reaction. (b) What is the molecularity of the reaction?

2. If there are 10 μmol of the radioactive isotope ^{32}P (half-life 14 days) at $t = 0$, how much ^{32}P will remain at (a) 7 days, (b) 14 days, (c) 21 days, and (d) 70 days?

3. The hypothetical elementary reaction 2 A → B + C has a rate constant of 10^{-6} $M^{-1} \cdot s^{-1}$. What is the reaction velocity when the concentration of A is 10 mM?

4. For each reaction below, determine whether the reaction is first order or second order and calculate the rate constant.

Time (s)	Reaction A reactant (mM)	Reaction B reactant (mM)
0	6.2	5.4
1	3.1	4.6
2	2.1	3.9
3	1.6	3.2
4	1.3	2.7
5	1.1	2.3

5. For an enzymatic reaction, draw curves that show the appropriate relationships between the variables in each plot below.

6. Explain why it is usually easier to calculate an enzyme's reaction velocity from the rate of appearance of product rather than the rate of disappearance of a substrate.

7. At what concentration of S (expressed as a multiple of K_M) will $v_o = 0.95$ V_{max}?

8. Identify the enzymes in Table 12-1 whose catalytic efficiencies are near the diffusion-controlled limit.

9. Explain why each of the following data sets from a Lineweaver–Burk plot are not individually ideal for determining K_M for an enzyme-catalyzed reaction that follows Michaelis–Menten kinetics.

Set A	$1/[S]$ (mM^{-1})	$1/v_o$ $(\mu M^{-1} \cdot s)$
	0.5	2.4
	1.0	2.6
	1.5	2.9
	2.0	3.1

Set B	$1/[S]$ (mM^{-1})	$1/v_o$ $(\mu M^{-1} \cdot s)$
	8	5.9
	10	6.8
	12	7.8
	14	8.7

10. Calculate K_M and V_{max} from the following data:

$[S]$ (μM)	v_o $(mM \cdot s^{-1})$
0.1	0.34
0.2	0.53
0.4	0.74
0.8	0.91
1.6	1.04

11. You are trying to determine the K_M for an enzyme. Due to a lab mishap, you have only two usable data points:

Substrate concentration (μM)	Reaction velocity (μM·s^{-1})
1	5
100	50

Use these data to calculate an approximate value for K_M. Is this value likely to be an overestimate or an underestimate of the true value? Explain.

12. You are attempting to determine K_M by measuring the reaction velocity at different substrate concentrations, but you do not realize that the substrate tends to precipitate under the experimental conditions you have chosen. How would this affect your measurement of K_M?

13. You are constructing a velocity versus [substrate] curve for an enzyme whose K_M is believed to be about 2 μM. The enzyme concentration is 200 nM and the substrate concentrations range from 0.1 μM to 10 μM. What is wrong with this experimental setup and how could you fix it?

14. Is it necessary for measurements of reaction velocity to be expressed in units of concentration per time (M·s^{-1}, for example) in order to calculate an enzyme's K_M?

15. Is it necessary to know $[E]_T$ in order to determine (a) K_M, (b) V_{max}, or (c) k_{cat}?

16. The K_M for the reaction of chymotrypsin with N-acetylvaline ethyl ester is 8.8×10^{-2} M, and the K_M for the reaction of chymotrypsin with N-acetyltyrosine ethyl ester is 6.6×10^{-4} M. (a) Which substrate has the higher apparent affinity for the enzyme? (b) Which substrate is likely to give a higher value for V_{max}?

17. Enzyme A catalyzes the reaction S → P and has a K_M of 50 μM and a V_{max} of 100 nM·s^{-1}. Enzyme B catalyzes the reaction S → Q and has a K_M of 5 mM and a V_{max} of 120 nM·s^{-1}. When 100 μM of S is added to a mixture containing equivalent amounts of enzymes A and B, after one minute which reaction product will be more abundant: P or Q?

18. In a bisubstrate reaction, a small amount of the first product P is isotopically labeled (P*) and added to the enzyme and the first substrate A. No B or Q is present. Will A (= P—X) become isotopically labeled (A*) if the reaction follows (a) a Ping Pong mechanism or (b) a Sequential mechanism?

19. Determine the type of inhibition of an enzymatic reaction from the following data collected in the presence and absence of the inhibitor.

[S] (mM)	v_o (mM·min^{-1})	v_o with I present (mM·min^{-1})
1	1.3	0.8
2	2.0	1.2
4	2.8	1.7
8	3.6	2.2
12	4.0	2.4

20. Estimate K_I for a competitive inhibitor when [I] = 5 mM gives an apparent value of K_M that is three times the K_M for the uninhibited reaction.

21. For an enzyme-catalyzed reaction, the presence of 5 nM of a reversible inhibitor yields a V_{max} value that is 80% of the value in the absence of the inhibitor. The K_M value is unchanged. (a) What type of inhibition is likely occurring?

(b) What proportion of the enzyme molecules have bound inhibitor? (c) Calculate the inhibition constant.

22. How would diisopropylphosphofluoridate (DIPF; Section 11-5A) affect the apparent K_M and V_{max} of a sample of chymotrypsin?

23. Based on some preliminary measurements, you suspect that a sample of enzyme contains an irreversible enzyme inhibitor. You decide to dilute the sample 100-fold and remeasure the enzyme's activity. What would your results show if the inhibitor in the sample is (a) irreversible or (b) reversible?

24. Enzyme X and enzyme Y catalyze the same reaction and exhibit the v_o versus [S] curves shown below. Which enzyme is more efficient at low [S]? Which is more efficient at high [S]?

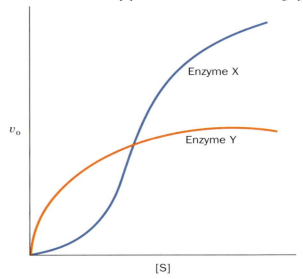

25. Sphingosine-1-phosphate (SPP) is important for cell survival. The synthesis of SPP from sphingosine and ATP is catalyzed by the enzyme sphingosine kinase. An understanding of the kinetics of the sphingosine kinase reaction may be important in the development of drugs to treat cancer. The velocity of the sphingosine kinase reaction was measured in the presence and absence of *threo*-sphingosine, a stereoisomer of sphingosine that inhibits the enzyme. The results are shown below.

[Sphingosine] (μM)	v_o (mg·min^{-1}) (no inhibitor)	v_o (mg·min^{-1}) (with *threo*-sphingosine)
2.5	32.3	8.5
3.5	40	11.5
5	50.8	14.6
10	72	25.4
20	87.7	43.9
50	115.4	70.8

Construct a Lineweaver–Burk plot to answer the following questions:

(a) What are the apparent K_M and V_{max} values in the presence and absence of the inhibitor?

(b) What kind of an inhibitor is *threo*-sphingosine? Explain. [Problem provided by Kathleen Cornely, Providence College.]

▪ BIOINFORMATICS EXERCISES

Bioinformatics Exercises are available for this chapter in the tabbed section at the end of the book and on the text website at www.wiley.com/college/voet.

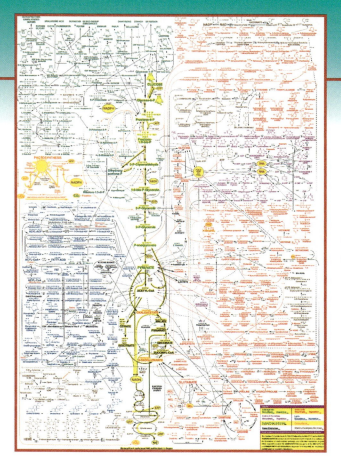

The process by which biological molecules are broken down and resynthesized form a complex, yet highly regulated, network of interdependent enzymatic reactions that are collectively known as life. [Designed by Donald E. Nicholson, Department of Biochemistry and Molecular Biology, The University of Leeds, England, and Sigma.]

Introduction to Metabolism

1. Overview of Metabolism
   - A. *Trophic Strategies*
   - B. *Metabolic Pathways*
   - C. *Thermodynamic Considerations*
   - D. *Control of Metabolic Flux*

2. "High-Energy" Compounds
   - A. *ATP and Phosphoryl Group Transfer*
   - B. *Coupled Reactions*
   - C. *Other Phosphorylated Compounds*
   - D. *Thioesters*

3. Oxidation–Reduction Reactions
   - A. *NAD^+ and FAD*
   - B. *The Nernst Equation*
   - C. *Measurements of Reduction Potential Differences*

4. Experimental Approaches to the Study of Metabolism
   - A. *Tracing Metabolic Fates*
   - B. *Perturbing the System*
   - C. *DNA Microarrays*
   - D. *Proteomics*

Understanding the chemical compositions and three-dimensional structures of biological molecules is not sufficient to understand how they are assembled into organisms or how they function to sustain life. We must therefore examine the reactions in which biological molecules are built and broken down. We must also consider how free energy is consumed in building cellular materials and carrying out cellular work and how free energy is generated from organic or other sources. **Metabolism,** the overall process through which living systems acquire and use free energy to carry out their various functions, is traditionally divided into two parts:

1. **Catabolism,** or degradation, in which nutrients and cell constituents are broken down to salvage their components and/or to generate energy.
2. **Anabolism,** or biosynthesis, in which biomolecules are synthesized from simpler components.

In general, catabolic reactions carry out the exergonic oxidation of nutrient molecules. The free energy thereby released is used to drive such endergonic processes as anabolic reactions, the performance of mechanical work, and the active transport of molecules against concentration gradients. Exergonic and endergonic processes are often coupled through the intermediate synthesis of a "high-energy" compound such as ATP. This simple principle underlies many of the chemical reactions presented in the following chapters. In this chapter, we introduce the general features of metabolic reactions and the roles of ATP and other compounds as energy carriers. Because many metabolic reactions are also oxidation–reduction reactions, we review the thermodynamics of these processes. Finally, we examine some approaches to studying metabolic reactions.

1 Overview of Metabolism

A bewildering array of chemical reactions occur in any living cell. Yet the principles that govern metabolism are the same in all organisms, a result of their common evolutionary origin and the constraints of the laws of thermodynamics. In fact, many of the specific reactions of metabolism are common to all organisms, with variations due primarily to differences in the source of the free energy that supports them.

A Trophic Strategies

The nutritional requirements of an organism reflect its source of metabolic energy. For example, some prokaryotes are **autotrophs** (Greek: *autos,* self + *trophos,* feeder), which can synthesize all their cellular constituents from simple molecules such as H_2O, CO_2, NH_3, and H_2S. There are two possible free energy sources for this process. **Chemolithotrophs** (Greek: *lithos,* stone) obtain their energy through the oxidation of inorganic compounds such as NH_3, H_2S, or even Fe^{2+}:

$$2\ NH_3 + 4\ O_2 \rightarrow 2\ HNO_3 + 2\ H_2O$$
$$H_2S + 2\ O_2 \rightarrow H_2SO_4$$
$$4\ FeCO_3 + O_2 + 6\ H_2O \rightarrow 4\ Fe(OH)_3 + 4\ CO_2$$

Photoautotrophs do so via photosynthesis, a process in which light energy powers the transfer of electrons from inorganic donors to CO_2 to produce carbohydrates, $(CH_2O)_n$, which are later oxidized to release free energy.

Heterotrophs (Greek: *hetero,* other) obtain free energy through the oxidation of organic compounds (carbohydrates, lipids, and proteins) and hence ultimately depend on autotrophs for these substances.

Organisms can be further classified by the identity of the oxidizing agent for nutrient breakdown. **Obligate aerobes** (which include animals) must use O_2, whereas **anaerobes** employ oxidizing agents such as sulfate or nitrate. **Facultative anaerobes,** such as *E. coli,* can grow in either the presence or the absence of O_2. **Obligate anaerobes,** in contrast, are poisoned by the presence of O_2. Their metabolisms are thought to resemble those of the earliest life forms, which arose over 3.5 billion years ago when the earth's atmosphere lacked O_2. Most of our discussion of metabolism will focus on aerobic processes.

B Metabolic Pathways

Metabolic pathways are series of connected enzymatic reactions that produce specific products. Their reactants, intermediates, and products are referred to as **metabolites.** There are over 2000 known metabolic reactions, each catalyzed by a distinct enzyme. The types of enzymes and metabolites in a given cell vary with the identity of the organism, the cell type, its nutritional status, and its developmental stage. Many metabolic pathways are branched and interconnected, so delineating a pathway from a network of thousands of reactions is somewhat arbitrary and is driven by tradition as much as by chemical logic.

In general, catabolic and anabolic pathways are related as follows (Fig. 13-1): In catabolic pathways, complex metabolites are exergonically broken down into simpler products, in many cases, a two-carbon acetyl unit linked to **coenzyme A** to form **acetyl-coenzyme A (acetyl-CoA;** Section 13-2D). The free energy released in this degradative process is conserved by the synthesis of ATP from ADP + P_i or by the reduction of the coenzyme $NADP^+$ (Fig. 11-3) to NADPH. ATP and NADPH are the major free energy sources for anabolic reactions. We shall take a closer look at the thermodynamic properties of acetyl-CoA, ATP, and NADPH later in this chapter.

A striking characteristic of degradative metabolism is that *the pathways for the catabolism of a large number of diverse substances (carbohydrates, lipids, and proteins) converge on a few common intermediates.* These intermediates are then further metabolized in a central oxidative pathway. Figure 13-2 outlines the breakdown of various foodstuffs to their monomeric units and then to acetyl-CoA. This is followed by the oxidation of the acetyl carbons to CO_2 by the **citric acid cycle** (Chapter 16). When one substance is **oxidized** (loses electrons), another must be **reduced** (gain electrons; see Box 13-1). The citric acid cycle thus produces the reduced coenzymes **NADH** and **FADH₂** (Section 13-3A),

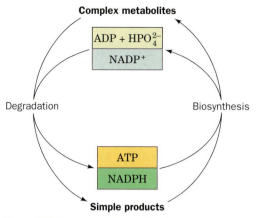

Figure 13-1 Roles of ATP and NADP⁺ in metabolism. ATP and NADPH generated through the degradation of complex metabolites are sources of free energy for biosynthetic and other reactions.

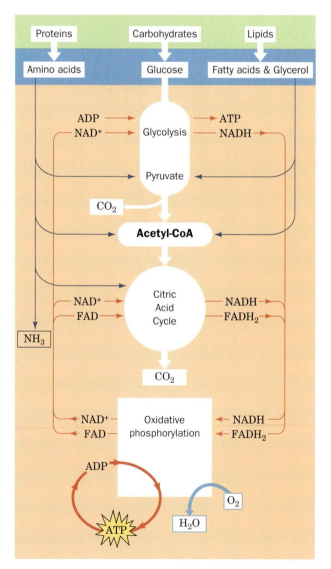

Figure 13-2 *Key to Metabolism.* Overview of catabolism. Complex metabolites such as carbohydrates, proteins, and lipids are degraded first to their monomeric units, chiefly glucose, amino acids, fatty acids, and glycerol, and then to the common intermediate, acetyl-CoA. The acetyl group is oxidized to CO_2 via the citric acid cycle with concomitant reduction of NAD^+ and FAD. Reoxidation of NADH and FADH₂ by O_2 during electron transport and oxidative phosphorylation yields H_2O and ATP.

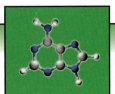

BOX 13-1

Perspectives in Biochemistry

Oxidation States of Carbon

The carbon atoms in biological molecules can assume different oxidation states depending on the atom to which they are bonded. For example, a carbon atom bonded to less electronegative hydrogen atoms is more reduced than a carbon atom bonded to highly electronegative oxygen atoms.

The simplest way to determined the oxidation number (and hence the oxidation state) of a particular carbon atom is to examine each of its bonds and assign the electrons to the more electronegative atom. In a C—O bond, both electrons "belong" to O; in a C—H bond, both electrons "belong" to C; and in a C—C bond, each carbon "owns" one electron. An atom's oxidation number is the number of valence electrons on the free atom (4 for carbon) minus the number of its lone pair and assigned electrons. For example, the oxidation number of carbon in CO_2 is $4 - (0 + 0) = +4$, and the oxidation number of carbon in CH_4 is $4 - (0 + 8) = -4$. Keep in mind, however, that oxidation numbers are only accounting devices; actual atomic charges are much closer to neutrality.

The following compounds are listed according to the oxidation state of the highlighted carbon atom. In general, the more oxidized compounds have fewer electrons per C atom and are richer in oxygen, and the more reduced compounds have more electrons per C atom and are richer in hydrogen. But note that not all reduction events (gain of electrons) or oxidation events (loss of electrons) are associated with bonding to oxygen. For example, when an alkane is converted to an alkene, the formation of a carbon–carbon double bond involves the loss of electrons and therefore is an oxidation reaction although no oxygen is involved. Knowing the oxidation number of a carbon atom is seldom required. However, it is useful to be able to determine whether the oxidation state of a given atom increases or decreases during a chemical reaction.

Compound	Formula	Oxidation Number
Carbon dioxide	O=C=O	4 (most oxidized)
Acetic acid	$H_3C-C\overset{O}{\underset{OH}{}}$	3
Carbon monoxide	:C≡O:	2
Formic acid	$H-C\overset{O}{\underset{OH}{}}$	2
Acetone	$H_3C-\overset{O}{\overset{\|}{C}}-CH_3$	2
Acetaldehyde	$H_3C-\overset{O}{\overset{\|}{C}}-H$	1
Formaldehyde	$H-\overset{O}{\overset{\|}{C}}-H$	0
Acetylene	HC≡CH	−1
Ethanol	$H_3C-\overset{H}{\underset{H}{\overset{\|}{\underset{\|}{C}}}}-OH$	−1
Ethene	$H_2C=C\overset{H}{\underset{H}{}}$	−2
Ethane	$H_3C-\overset{H}{\underset{H}{\overset{\|}{\underset{\|}{C}}}}-H$	−3
Methane	$H-\overset{H}{\underset{H}{\overset{\|}{\underset{\|}{C}}}}-H$	−4 (least oxidized)

which then pass their electrons to O_2 to produce H_2O in the process of **oxidative phosphorylation** (Chapter 17).

Biosynthetic pathways carry out the opposite process. *Relatively few metabolites serve as starting materials for a host of varied products.* In the next several chapters, we discuss many catabolic and anabolic pathways in detail.

Enzymes Catalyze the Reactions of Metabolic Pathways. With a few exceptions, the interconversions of metabolites in degradative and biosynthetic pathways are catalyzed by enzymes. In the absence of enzymes, the reactions would occur far too slowly to support life. In addition, the specificity of enzymes guarantees the efficiency of metabolic reactions by preventing

BOX 13-2

Perspectives in Biochemistry

Mapping Metabolic Pathways

The task of cataloging all the enzymatic reactions that occur in a given organism is formidable and, in many cases, far from complete. However, the metabolic reactions that constitute the major catabolic and anabolic pathways can be organized in a diagram. Typically, the structures or names of the intermediates are shown, and details about the enzymes that catalyze their interconversions can be called up. The following figure, which diagrams a portion of the photosynthetic pathway in plants, shows the interconversions of several three- and four-carbon compounds. The boxes represent enzymes, identified by their EC number (Section 11-1A).

Various Internet-accessible databases contain "universal" maps of all possible metabolic reactions or more specialized maps devoted to a single species or focused on a specific pathway. Such databases may link individual enzymes in a pathway to gene sequences and to the three-dimensional structures of the proteins, if known. Information about an enzyme's activators and inhibitors may also be included.

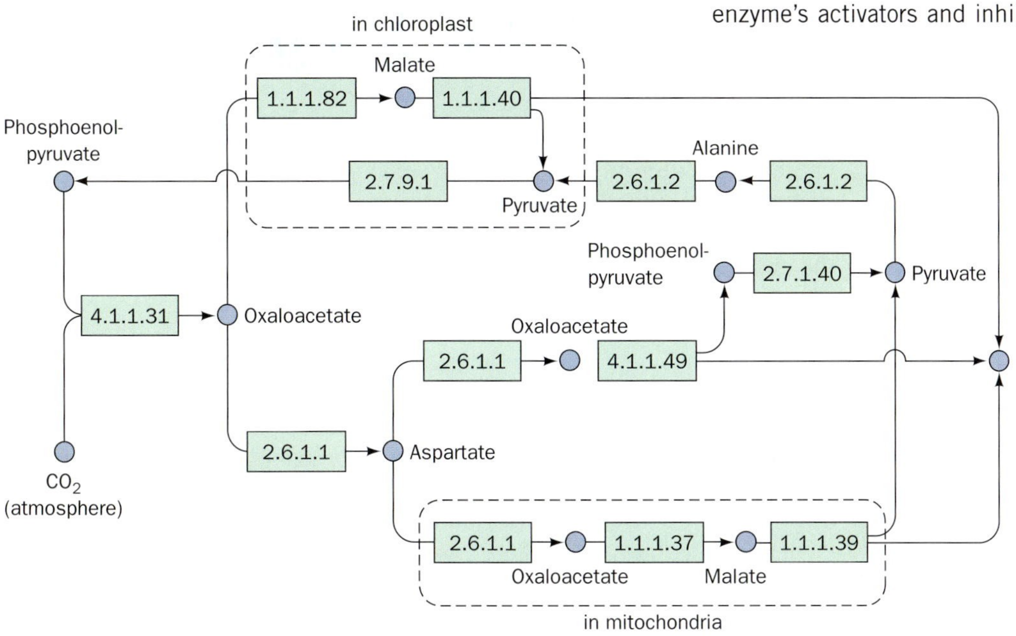

Understanding the chemical logic of metabolic pathways requires more than listing the substrates and products of each step. The ability to recognize the types of reactions that occur in a pathway provides insights into the overall metabolic capabilities of an organism and makes it easier to see the similarities and differences between pathways. At a deeper level, understanding the mechanisms and regulation of metabolic enzymes may lead to advances in treating metabolic and other diseases.

[Redrawn from the Kyoto Encyclopedia of Genes and Genomes (KEGG) Metabolic Pathways: http://www.genome.jp/kegg/metabolism.html.]

the formation of useless or toxic by-products. Most importantly, enzymes provide a mechanism for coupling an endergonic chemical reaction (which would not occur on its own) with an energetically favorable reaction, as discussed below.

We will see examples of reactions catalyzed by all six classes of enzymes introduced in Section 11-1A. These reactions fall into four major types: **oxidations and reductions** (catalyzed by oxidoreductases), **group-transfer reactions** (catalyzed by transferases and hydrolases), **eliminations, isomerizations, and rearrangements** (catalyzed by isomerases and mutases), and **reactions that make or break carbon–carbon bonds** (catalyzed by hydrolases, lyases, and ligases). Details about the enzymes that catalyze individual steps of metabolic pathways are available from Internet-accessible databases (Box 13-2).

Metabolic Pathways Occur in Specific Cellular Locations. The compartmentation of the eukaryotic cytoplasm allows different metabolic pathways to

Table 13-1 Metabolic Functions of Eukaryotic Organelles

Organelle	Major functions
Mitochondrion	Citric acid cycle, oxidative phosphorylation, fatty acid oxidation, amino acid breakdown
Cytosol	Glycolysis, pentose phosphate pathway, fatty acid biosynthesis, many reactions of gluconeogenesis
Lysosomes	Enzymatic digestion of cell components and ingested matter
Nucleus	DNA replication and transcription, RNA processing
Golgi apparatus	Posttranslational processing of membrane and secretory proteins; formation of plasma membrane and secretory vesicles
Rough endoplasmic reticulum	Synthesis of membrane-bound and secretory proteins
Smooth endoplasmic reticulum	Lipid and steroid biosynthesis
Peroxisomes (glyoxysomes in plants)	Oxidative reactions catalyzed by amino acid oxidases and catalase; glyoxylate cycle reactions in plants

operate in different locations. For example, oxidative phosphorylation occurs in the mitochondria, while **glycolysis** (a carbohydrate degradation pathway) and fatty acid biosynthesis occur in the cytosol. Table 13-1 lists the major metabolic features of eukaryotic organelles. Metabolic processes in prokaryotes, which lack organelles, may be localized to particular areas of the cytosol.

The synthesis of metabolites in specific membrane-bounded compartments in eukaryotic cells requires mechanisms to transport these substances between compartments. Accordingly, transport proteins (Chapter 10) are essential components of many metabolic processes. For example, a transport protein is required to move ATP, which is generated in the mitochondria, to the cytosol, where most of it is consumed.

In multicellular organisms, compartmentation is carried a step further to the level of tissues and organs. The mammalian liver, for example, is largely responsible for the synthesis of glucose from noncarbohydrate precursors (**gluconeogenesis;** Section 15-4) so as to maintain a relatively constant level of glucose in the circulation, whereas adipose tissue is specialized for storage of triacylglycerols. The interdependence of the metabolic functions of the various organs is the subject of Chapter 21.

An intriguing manifestation of specialization of tissues and subcellular compartments is the existence of **isozymes,** enzymes that catalyze the same reaction but are encoded by different genes and have different kinetic or regulatory properties. For example, vertebrates possess two homologs of the enzyme **lactate dehydrogenase:** the M type, which predominates in tissues subject to anaerobic conditions such as skeletal muscle and liver, and the H type, which predominates in aerobic tissues such as heart muscle. Lactate dehydrogenase catalyzes the interconversion of **pyruvate,** a product of glycolysis, and **lactate** (Section 14-3). The M-type isozyme appears to mainly function in the reduction by NADH of pyruvate to lactate, whereas the H-type enzyme appears to be better adapted to catalyze the reverse reaction. The existence of isozymes allows for the testing of various illnesses. For example, heart attacks cause the death of heart muscle cells, which consequently rupture and release H-type LDH into the blood. A blood test indicating the presence of H-type LDH is therefore diagnostic of a heart attack.

C Thermodynamic Considerations

Knowing the location of a metabolic pathway and enumerating its substrates and products does not necessarily reveal how that pathway functions as part of a larger network of interrelated biochemical processes. It is also necessary to appreciate how fast end product can be generated by the pathway as well as how pathway activity is regulated as the cell's needs change. Conclusions about a pathway's output and its potential for regulation can be gleaned from information about the thermodynamics of each enzyme-catalyzed step.

Recall from Section 1-4D that the free energy change (ΔG) of a biochemical process, such as the reaction

$$A + B \rightleftharpoons C + D$$

is related to the standard free energy change ($\Delta G^{\circ\prime}$) and the concentrations of the reactants and products (Eq. 1-15):

$$\Delta G = \Delta G^{\circ\prime} + RT \ln\left(\frac{[C][D]}{[A][B]}\right) \qquad [13\text{-}1]$$

At equilibrium $\Delta G = 0$, and the equation becomes

$$\Delta G^{\circ\prime} = -RT \ln K_{eq} \qquad [13\text{-}2]$$

Thus, the value of $\Delta G^{\circ\prime}$ can be calculated from the equilibrium constant and vice versa (see Sample Calculation 13-1).

When the reactants are present at values close to their equilibrium values, $[C]_{eq}[D]_{eq}/[A]_{eq}[B]_{eq} \approx K_{eq}$, and $\Delta G \approx 0$. This is the case for many metabolic reactions, which are said to be **near-equilibrium reactions.** Because their ΔG values are close to zero, they can be relatively easily reversed by changing the ratio of products to reactants. When the reactants are in excess of their equilibrium concentrations, the net reaction proceeds in the forward direction until the excess reactants have been converted to products and equilibrium is attained. Conversely, when products are in excess, the net reaction proceeds in the reverse direction so as to convert products to reactants until the equilibrium concentration ratio is again achieved. *Enzymes that catalyze near-equilibrium reactions tend to act quickly to restore equilibrium concentrations, and the net rates of such reactions are effectively regulated by the relative concentrations of substrates and products.*

Other metabolic reactions function far from equilibrium; that is, they are irreversible. This is because an enzyme catalyzing such a reaction has insufficient catalytic activity (the rate of the reaction it catalyzes is too slow) to allow the reaction to come to equilibrium. Reactants therefore accumulate in large excess of their equilibrium amounts, making $\Delta G \ll 0$. Changes in substrate concentrations therefore have relatively little effect on the rate of an irreversible reaction; the enzyme is essentially saturated. Only changes in the activity of the enzyme, through allosteric interactions, for example, can significantly alter this rate. The enzyme is therefore analogous to a dam on a river: *It controls the flow of substrate through the reaction by varying its activity, much as a dam controls the flow of a river by varying the opening of its flood gates.*

Understanding the **flux** (rate of flow) of metabolites through a metabolic pathway requires knowledge of which reactions are functioning near equilibrium and which are far from it. Most enzymes in a metabolic pathway operate near equilibrium and therefore have net rates that vary with their substrate concentrations. However, certain enzymes that operate far from equilibrium are strategically located in metabolic pathways. This has several important implications:

S A M P L E   C A L C U L A T I O N   1 3 - 1

Calculate the equilibrium constant for the hydrolysis of glucose-1-phosphate at 37°C.

$\Delta G^{\circ\prime}$ for the reaction

Glucose-1-phosphate + $H_2O \rightarrow$ glucose + P_i

is -20.9 kJ $\cdot$ mol^{-1} (Table 13-2). At equilibrium, $\Delta G = 0$ and Eq. 13-1 becomes

$$\Delta G^{\circ\prime} = -RT \ln K \quad \text{(Eq. 13-2)}$$

Therefore,

$K = e^{-\Delta G^{\circ\prime}/RT}$
$K = e^{-(-20,900 \text{ J} \cdot \text{mol}^{-1})/(8.3145 \text{ J} \cdot \text{K}^{-1} \cdot \text{mol}^{-1})(310 \text{ K})}$
$K = 3.3 \times 10^3$

1. ***Metabolic pathways are irreversible.*** A highly exergonic reaction (one with $\Delta G \ll 0$) is irreversible; that is, it goes to completion. If such a reaction is part of a multistep pathway, it confers directionality on the pathway; that is, it makes the entire pathway irreversible.

2. ***Every metabolic pathway has a first committed step.*** Although most reactions in a metabolic pathway function close to equilibrium, there is generally an irreversible (exergonic) reaction early in the pathway that "commits" its product to continue down the pathway (likewise, water that has gone over a dam cannot return).

3. ***Catabolic and anabolic pathways differ.*** If a metabolite is converted to another metabolite by an exergonic process, free energy must be supplied to convert the second metabolite back to the first. This energetically "uphill" process requires a different pathway for at least one of the reaction steps.

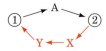

The existence of independent interconversion routes, as we shall see, is an important property of metabolic pathways because it allows independent control of the two processes. If metabolite 2 is required by the cell, it is necessary to "turn off" the pathway from 2 to 1 while "turning on" the pathway from 1 to 2. Such independent control would be impossible without different pathways.

D Control of Metabolic Flux

Living organisms are thermodynamically open systems that tend to maintain a steady state rather than reaching equilibrium (Section 1-4E). This is strikingly demonstrated by the observation that, over a 40-year time span, a normal human adult consumes literally tons of nutrients and imbibes over 20,000 L of water but does so without significant weight change. *The flux of intermediates through a metabolic pathway in a steady state is more or less constant; that is, the rates of synthesis and breakdown of each pathway intermediate maintain it at a constant concentration.* A steady state far from equilibrium is thermodynamically efficient, because only a nonequilibrium process ($\Delta G \neq 0$) can perform useful work. Indeed, living systems that have reached equilibrium are dead.

Since a metabolic pathway is a series of enzyme-catalyzed reactions, it is easiest to describe the flux of metabolites through the pathway by considering its reaction steps individually. The flux of metabolites, J, through each reaction step is the rate of the forward reaction, v_f, less that of the reverse reaction, v_r:

$$J = v_f - v_r \qquad \text{[13-3]}$$

At equilibrium, by definition, there is no flux ($J = 0$), although v_f and v_r may be quite large. In reactions that are far from equilibrium, $v_f \gg v_r$, so the flux is essentially equal to the rate of the forward reaction ($J \approx v_f$).

For the pathway as a whole, flux is set by the rate-determining step of the pathway. By definition, this step is the pathway's slowest step, which is often the first committed step of the pathway. In some pathways, flux control is distributed over several enzymes, all of which help determine the overall rate of flow of metabolites through the pathway. Because a rate-determining step is slow relative to other steps in the pathway, its product is removed by succeeding steps in the pathway before it can equilibrate

with reactant. Thus, *the rate-determining step functions far from equilibrium and has a large negative free energy change.* In an analogous manner, a dam creates a difference in water levels between its upstream and downstream sides, and a large negative free energy change results from the hydrostatic pressure difference. The dam can release water to generate electricity, varying the water flow according to the need for electrical power.

Reactions that function near equilibrium respond rapidly to changes in substrate concentration. For example, upon a sudden increase in the concentration of a reactant for a near equilibrium reaction, the enzyme catalyzing it would increase the net reaction rate so as to rapidly achieve the new equilibrium level. Thus, a series of near-equilibrium reactions downstream from the rate-determining step all have the same flux.

In practice, it is often possible to identify flux control points for a pathway by identifying reactions that have large negative free energy changes. The relative insensitivity of the rates of these nonequilibrium reactions to variations in the concentrations of their substrates permits the establishment of a steady state flux of metabolites through the pathway. Of course, flux through a pathway must vary in response to the organism's requirements so as to reach a new steady state. Altering the rates of the rate-determining steps can alter the flux of material through the entire pathway, often by an order of magnitude or more.

Cells use several mechanisms to control flux through the rate-determining steps of metabolic pathways:

1. **Allosteric control.** Many enzymes are allosterically regulated (Section 12-3) by effectors that are often substrates, products, or coenzymes of the pathway but not necessarily of the enzyme in question. For example, in negative feedback regulation, the product of a pathway inhibits an earlier step in the pathway:

Thus, as we have seen, CTP, a product of pyrimidine biosynthesis, inhibits ATCase, which catalyzes the rate-determining step in this pathway (Fig. 12-11).

2. **Covalent modification.** Many enzymes that control pathway fluxes have specific sites that may be enzymatically phosphorylated and dephosphorylated (the attachment and removal of a phosphoryl group) or covalently modified in some other way.

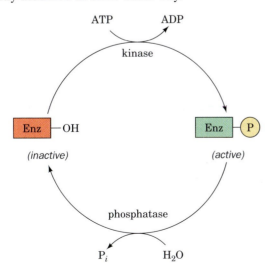

Such enzymatic modification processes, which are themselves subject to control, greatly alter the activities of the modified enzymes. This flux control mechanism is discussed in Section 15-3B.

3. ***Substrate cycles.*** If v_f and v_r represent the rates of two opposing nonequilibrium reactions that are catalyzed by different enzymes, v_f and v_r may be independently varied.

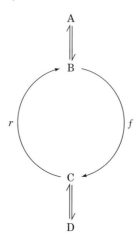

For example, flux ($v_f - v_r$) can be increased not just by accelerating the forward reaction but by slowing the reverse reaction. The flux through such a **substrate cycle,** as we shall see in Section 14-4, is more sensitive to the concentrations of allosteric effectors than is the flux through a single unopposed nonequilibrium reaction.

4. ***Genetic control.*** Enzyme concentrations, and hence enzyme activities, may be altered by protein synthesis in response to metabolic needs. Genetic control of enzyme concentrations is a major concern of Part V of this text.

Mechanisms 1 to 3 can respond rapidly (within seconds or minutes) to external stimuli and are therefore classified as "short-term" control mechanisms. Mechanism 4 responds more slowly to changing conditions (within hours or days in higher organisms) and is therefore regarded as a "long-term" control mechanism.

Control of most metabolic pathways involves several nonequilibrium steps. Hence, the flux of material through a pathway that supplies intermediates for use by an organism may depend on multiple effectors whose relative importance reflects the overall metabolic demands of the organism at a given time. Thus, a metabolic pathway is part of a **supply-demand process.**

2 "High-Energy" Compounds

The complete oxidation of a metabolic fuel such as glucose

$$C_6H_{12}O_6 + 6\ O_2 \rightarrow 6\ CO_2 + 6\ H_2O$$

releases considerable energy ($\Delta G^{\circ\prime} = -2850$ kJ · mol^{-1}). The complete oxidation of palmitate, a typical fatty acid,

$$C_{16}H_{32}O_2 + 23\ O_2 \rightarrow 16\ CO_2 + 16\ H_2O$$

is even more exergonic ($\Delta G^{\circ\prime} = -9781$ kJ · mol^{-1}). Oxidative metabolism proceeds in a stepwise fashion, so the released free energy can be recov-

BOX 13-3

Pathways of Discovery

Fritz Lipmann and "High-energy" Compounds

Fritz Albert Lipmann (1899–1986)

Among the many scientists who fled Europe for the United States in the 1930s was Fritz Lipmann, a German-born physician-turned-biochemist. During the first part of the century, scientists were primarily interested in the structures and compositions of biological molecules, and not much was known about their biosynthesis. Lipmann's contribution to this field centers on his understanding of "energy-rich" phosphates and other "active" compounds.

Lipmann began his research career by studying creatine phosphate, a compound that could provide energy for muscle contraction. He, like many of his contemporaries, was puzzled by the absence of an obvious link between the phosphorylated compound and the known metabolic activity of a contracting muscle, namely, converting glucose to lactate. One link was discovered by Otto Warburg (see Box 14-1), who showed that one of the steps of glycolysis was accompanied by the incorporation of inorganic phosphate. The resulting acyl phosphate (1,3-bisphosphoglycerate) could then react with ADP to form ATP.

Lipmann wondered whether other phosphorylated compounds might behave in a similar manner. Because the purification of such labile (prone to degradation) compounds from whole cells was impractical, Lipmann synthesized them himself. He was able to show that cell extracts used synthetic acetyl phosphate to produce ATP. Lipmann went on to propose that cells contain two classes of phosphorylated compounds, which he termed "energy-poor" and "energy-rich", by which he meant compounds with low and high negative free energies of hydrolysis (the "squiggle," $\sim$, which is still used, was his symbol for an "energy-rich" bond). Lipmann described a sort of "phosphate current" in which photosynthesis or breakdown of food molecules generates "energy-rich" phosphates that lead to the synthesis of ATP. The ATP,

in turn, can power mechanical work such as muscle contraction or drive biosynthetic reactions.

Until this point (1941), biochemists studying biosynthetic processes were largely limited to working with whole animals or relatively intact tissue slices. Lipmann's insight regarding the role of ATP freed researchers from their cumbersome and poorly reproducible experimental systems. Biochemists could simply add ATP to their cell-free preparations to reconstitute the biosynthetic process.

Lipmann was intrigued by the discovery that a two-carbon group, an "active acetate," served as a precursor for the synthesis of fatty acids and steroids. Was acetyl phosphate also the "active acetate"? This proved not to be the case, although Lipmann was able to show that the addition of a two-carbon unit to another molecule (acetylation) required acetate, ATP, and a heat-stable factor present in pigeon liver extracts. He isolated and determined the structure of this factor, which he named coenzyme A. For this seminal discovery, Lipmann was awarded the 1953 Nobel Prize for Physiology or Medicine.

Even after "high-energy" thioesters (as in acetyl-CoA) came on the scene, Lipmann remained a staunch advocate of "high-energy" phosphates. For example, he realized that carbamoyl phosphate ($H_2N-COO-PO_3^{2-}$) could function as an "active" carbamoyl group donor in biosynthetic reactions. He also helped identify more obscure compounds, mixed anhydrides between phosphate and sulfate, as "active" sulfates that function as sulfate group donors.

Kleinkauf, H., von Döhren, H., and Jaenicke, L. (Eds.), *The Roots of Modern Biochemistry. Fritz Lipmann's Squiggle and its Consequences,* Walter de Gruyter (1988).

---

ered in a manageable form at each exergonic step of the overall process. *These "packets" of energy are conserved by the synthesis of a few types of* **"high-energy" intermediates** *whose subsequent exergonic breakdown drives endergonic processes.* These intermediates therefore form a sort of free energy "currency" through which free energy–producing reactions such as glucose oxidation or fatty acid oxidation "pay for" the free energy–consuming processes in biological systems (Box 13-3).

The cell uses several forms of energy currency, including phosphorylated compounds such as the nucleotide ATP (the cell's primary energy currency), compounds that contain thioester bonds, and reduced coenzymes such as NADH. Each of these represents a source of free energy that the cell can

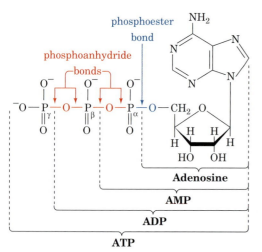

Figure 13-3 *Key to Structure.* The structure of ATP indicating its relationship to ADP, AMP, and adenosine. The phosphoryl groups, starting from AMP, are referred to as the α-, β-, and γ-phosphates. Note the differences between phosphoester and phosphoanhydride bonds.

Table 13-2 Standard Free Energies of Phosphate Hydrolysis of Some Compounds of Biological Interest

Compound	$\Delta G^{\circ\prime}$ (kJ · mol^{-1})
Phosphoenolpyruvate	−61.9
1,3-Bisphosphoglycerate	−49.4
ATP (→ AMP + PP$_i$)	−45.6
Acetyl phosphate	−43.1
Phosphocreatine	−43.1
ATP (→ ADP + P$_i$)	−30.5
Glucose-1-phosphate	−20.9
PP$_i$	−19.2
Fructose-6-phosphate	−13.8
Glucose-6-phosphate	−13.8
Glycerol-3-phosphate	−9.2

Source: Mostly from Jencks, W.P., *in* Fasman, G.D. (Ed.), *Handbook of Biochemistry and Molecular Biology* (3rd ed.), Physical and Chemical Data, Vol. I, pp. 296–304, CRC Press (1976).

use in various ways, including the synthesis of ATP. We will first examine ATP and then discuss the properties of other forms of energy currency.

A ATP and Phosphoryl Group Transfer

The "high-energy" intermediate adenosine triphosphate (ATP; Fig. 13-3) occurs in all known life forms. ATP consists of an **adenosine** moiety (adenine + ribose) to which three phosphoryl (—PO$_3^{2-}$) groups are sequentially linked via a **phosphoester** bond followed by two **phosphoanhydride** bonds.

The biological importance of ATP rests in the large free energy change that accompanies cleavage of its phosphoanhydride bonds. This occurs when either a phosphoryl group is transferred to another compound, leaving ADP, or a nucleotidyl (AMP) group is transferred, leaving **pyrophosphate** (P$_2$O$_7^{4-}$; **PP$_i$**). When the acceptor is water, the process is known as hydrolysis:

$$ATP + H_2O \rightleftharpoons ADP + P_i$$
$$ATP + H_2O \rightleftharpoons AMP + PP_i$$

Most biological group-transfer reactions involve acceptors other than water. However, knowing the free energy of hydrolysis of various phosphoryl compounds allows us to calculate the free energy of transfer of phosphoryl groups to other acceptors by determining the difference in free energy of hydrolysis of the phosphoryl donor and acceptor.

The $\Delta G^{\circ\prime}$ values for hydrolysis of several phosphorylated compounds of biochemical importance are tabulated in Table 13-2. The negatives of these values are often referred to as **phosphoryl group-transfer potentials;** they are a measure of the tendency of phosphorylated compounds to transfer their phosphoryl groups to water. Note that ATP has an intermediate phosphate group-transfer potential. Under standard conditions, the compounds above ATP in Table 13-2 can spontaneously transfer a phosphoryl group to ADP to form ATP, which can, in turn, spontaneously transfer a phosphoryl group to the appropriate groups to form the compounds listed below it. Note that a favorable free energy change for a reaction does not indicate how quickly the reaction occurs. Despite their high group-transfer potentials, ATP and related phosphoryl compounds are **kinetically stable** and do not react at a significant rate unless acted upon by an appropriate enzyme.

Rationalizing the "Energy" in "High-Energy" Compounds. Bonds whose hydrolysis proceeds with large negative values of $\Delta G^{\circ\prime}$ (customarily more than −25 kJ · mol^{-1}) are often referred to as **"high-energy" bonds** or **"energy-rich" bonds** and are frequently symbolized by the squiggle (~). Thus, ATP can be represented as AR—P~P~P, where A, R, and P symbolize adenyl, ribosyl, and phosphoryl groups, respectively. Yet the phosphoester bond joining the adenosyl group of ATP to its α-phosphoryl group appears to be not greatly different in electronic character from the so-called "high-energy" bonds bridging its α- and β- and its β- and γ-phosphoryl groups. In fact, none of these bonds has any unusual properties, so the term "high-energy" bond is somewhat of a misnomer (in any case, it should not be confused with the term "bond energy," which is defined as the energy required to break, not hydrolyze, a covalent bond). Why, then, are the phosphoryl group-transfer reactions of ATP so exergonic? Several factors appear to be responsible for the "high-energy" character of phosphoanhydride bonds such as those in ATP (Fig. 13-4):

1. The resonance stabilization of a phosphoanhydride bond is less than that of its hydrolysis products. This is because a phosphoanhydride's

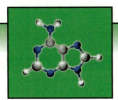

BOX 13-4

Perspectives in Biochemistry

ATP and ΔG

The standard conditions reflected in $\Delta G°'$ values never occur in living organisms. Furthermore, other compounds that are present at high concentrations and that can potentially interact with the substrates and products of a metabolic reaction may dramatically affect ΔG values. For example, Mg^{2+} ions in cells partially neutralize the negative charges on the phosphate groups in ATP and its hydrolysis products, thereby diminishing the electrostatic repulsions that make ATP hydrolysis so exergonic. Similarly, changes in pH alter the ionic character of phosphorylated compounds and therefore alter their free energies.

In a given cell, the concentrations of many ions, coenzymes, and metabolites vary with both location and time, often by several orders of magnitude. Intracellular ATP concentrations are maintained within a relatively narrow range, usually 2–10 mM, but the concentrations of ADP and P_i are more variable. Consider a typical cell with [ATP] = 3.0 mM, [ADP] = 0.8 mM, and [P_i] = 4.0 mM. Using Eq. 13-1, the actual free energy of ATP hydrolysis at 37°C is calculated below.

$$\Delta G = \Delta G°' + RT \ln\left(\frac{[ADP][P_i]}{[ATP]}\right)$$
$$= -30.5 \text{ kJ} \cdot \text{mol}^{-1} + (8.3145 \text{ J} \cdot \text{K}^{-1} \cdot \text{mol}^{-1})(310 \text{ K})$$
$$\ln\left(\frac{(0.8 \times 10^{-3} \text{ M})(4.0 \times 10^{-3} \text{ M})}{(3.0 \times 10^{-3} \text{ M})}\right)$$
$$= -30.5 \text{ kJ} \cdot \text{mol}^{-1} - 17.6 \text{ kJ} \cdot \text{mol}^{-1}$$
$$= -48.1 \text{ kJ} \cdot \text{mol}^{-1}$$

This value is even greater than the standard free energy of ATP hydrolysis. However, because of the difficulty in accurately measuring the concentrations of particular chemical species in a cell or organelle, the ΔG's for most *in vivo* reactions are little more than estimates. For the sake of consistency, we shall, for the most part, use $\Delta G°'$ values in this textbook.

---

two strongly electron-withdrawing groups must compete for the π electrons of its bridging oxygen atom, whereas this competition is absent in the hydrolysis products. In other words, the electronic requirements of the phosphoryl groups are less satisfied in a phosphoanhydride than in its hydrolysis products.

2. Of perhaps greater importance is the destabilizing effect of the electrostatic repulsions between the charged groups of a phosphoanhydride compared to those of its hydrolysis products. In the physiological pH range, ATP has three to four negative charges whose mutual electrostatic repulsions are partially relieved by ATP hydrolysis.

3. Another destabilizing influence, which is difficult to assess, is the smaller solvation energy of a phosphoanhydride compared to that of its hydrolysis products. Some estimates suggest that this factor provides the dominant thermodynamic driving force for the hydrolysis of phosphoanhydrides.

Of course, the free energy change for any reaction, including phosphoryl group transfer from a "high-energy" compound, depends in part on the concentrations of the reactants and products (Eq. 13-1). Furthermore, because ATP and its hydrolysis products are ions, ΔG also depends on pH and ionic strength (see Box 13-4).

Figure 13-4 Resonance and electrostatic stabilization in a phosphoanhydride and its hydrolytic products. The competing resonances (*curved arrows* from the central O) and charge–charge repulsions (*zigzag lines*) between phosphoryl groups decrease the stability of a phosphoanhydride relative to its hydrolysis products.

B Coupled Reactions

The exergonic reactions of "high-energy" compounds can be coupled to endergonic processes to drive them to completion. The thermodynamic explanation for the coupling of an exergonic and an endergonic process is

based on the additivity of free energy. Consider the following two-step reaction pathway:

$$(1) \quad A + B \rightleftharpoons C + D \quad \Delta G_1$$
$$(2) \quad D + E \rightleftharpoons F + G \quad \Delta G_2$$

If $\Delta G_1 \geq 0$, Reaction 1 will not occur spontaneously. However, if ΔG_2 is sufficiently exergonic so that $\Delta G_1 + \Delta G_2 < 0$, then although the equilibrium concentration of D in Reaction 1 will be relatively small, it will be larger than that in Reaction 2. As Reaction 2 converts D to products, Reaction 1 will operate in the forward direction to replenish the equilibrium concentration of D. The highly exergonic Reaction 2 therefore "drives" or "pulls" the endergonic Reaction 1, and the two reactions are said to be coupled through their common intermediate, D. That these coupled reactions proceed spontaneously can also be seen by summing Reactions 1 and 2 to yield the overall reaction

$$(1 + 2) \quad A + B + E \rightleftharpoons C + F + G \quad \Delta G_3$$

where $\Delta G_3 = \Delta G_1 + \Delta G_2 < 0$. *As long as the overall pathway is exergonic, it will operate in the forward direction.*

To illustrate this concept, let us consider two examples of phosphoryl group-transfer reactions. The initial step in the metabolism of glucose is its conversion to **glucose-6-phosphate** (Section 14-2A). Yet the direct reaction of glucose and P_i is thermodynamically unfavorable ($\Delta G^{\circ\prime} = +13.8\ kJ \cdot mol^{-1}$; Fig. 13-5*a*). In cells, however, this reaction is coupled to

(a)

				$\Delta G^{\circ\prime}$ (kJ • mol^{-1})
Endergonic half-reaction 1	P_i + glucose	$\rightleftharpoons$	glucose-6-P + H_2O	+13.8
Exergonic half-reaction 2	ATP + H_2O	$\rightleftharpoons$	ADP + P_i	–30.5
Overall coupled reaction	ATP + glucose	$\rightleftharpoons$	ADP + glucose-6-P	–16.7

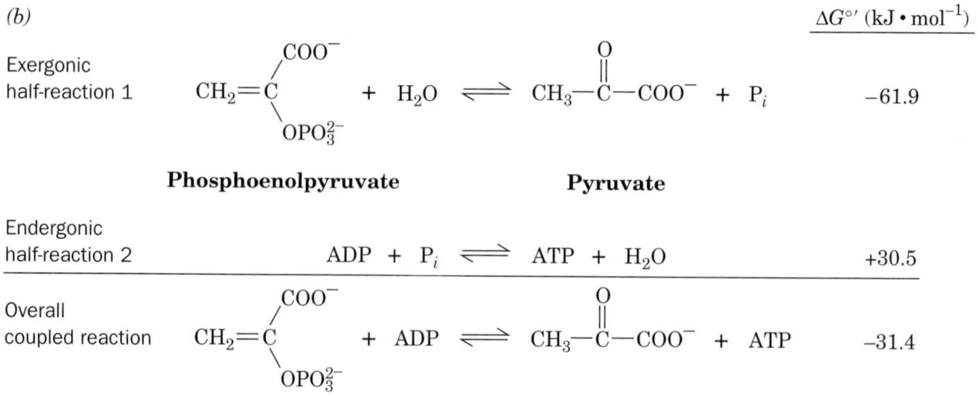

Figure 13-5 Some coupled reactions involving ATP. (*a*) The phosphorylation of glucose to form glucose-6-phosphate and ADP. (*b*) The phosphorylation of ADP by phosphoenolpyruvate to form ATP and pyruvate. Each reaction has been conceptually decomposed into a direct phosphorylation step (half-reaction 1) and a step in which ATP is hydrolyzed (half-reaction 2). Both half-reactions proceed in the direction that makes the overall reaction exergonic ($\Delta G < 0$).

the exergonic cleavage of ATP (for ATP hydrolysis, $\Delta G^{\circ\prime} = -30.5$ kJ · mol^{-1}), so the overall reaction is thermodynamically favorable ($\Delta G^{\circ\prime} = +13.8 - 30.5 = -16.7$ kJ · mol^{-1}). ATP can be similarly regenerated ($\Delta G^{\circ\prime} = +30.5$ kJ · mol^{-1}) by coupling its synthesis from ADP and P_i to the even more exergonic cleavage of **phosphoenolpyruvate** ($\Delta G^{\circ\prime} = -61.9$ kJ · mol^{-1}; Fig. 13-5*b* and Section 14-2J).

Note that the half-reactions shown in Fig. 13-5 do not actually occur as written in an enzyme active site. **Hexokinase,** the enzyme that catalyzes the formation of glucose-6-phosphate (Fig. 13-5*a*), does not catalyze ATP hydrolysis but instead catalyzes the transfer of a phosphoryl group from ATP directly to glucose. Likewise, **pyruvate kinase,** the enzyme that catalyzes the reaction shown in Fig. 13-5*b*, does not add a free phosphoryl group to ADP but transfers a phosphoryl group from phosphoenolpyruvate to ADP to form ATP.

Phosphoanhydride Hydrolysis Drives Some Biochemical Processes.

The free energy of the phosphoanhydride bonds of "high-energy" compounds such as ATP can be used to drive reactions even when the phosphoryl groups are not transferred to another organic compound. For example, ATP hydrolysis (i.e., phosphoryl group transfer directly to H_2O) provides the free energy for the operation of molecular chaperones (Section 6-5C), muscle contraction (Section 28-2B), and transmembrane active transport (Section 10-3). In these processes, proteins undergo conformational changes in response to binding ATP. *The exergonic hydrolysis of ATP and release of ADP and P_i renders these changes irreversible and thereby drives the processes forward.* GTP hydrolysis functions similarly to drive some of the reactions of signal transduction (Section 21-3B) and protein synthesis (Section 26-4).

In the absence of an appropriate enzyme, phosphoanhydride bonds are stable; that is, they hydrolyze quite slowly, despite the large amount of free energy released by these reactions. This is because these hydrolysis reactions have unusually high free energies of activation ($\Delta G^{\ddagger}$; Section 11-2). Consequently, *ATP hydrolysis is thermodynamically favored but kinetically disfavored.* For example, consider the reaction of glucose with ATP that yields glucose-6-phosphate (Fig. 13-5*a*). $\Delta G^{\ddagger}$ for the nonenzymatic transfer of a phosphoryl group from ATP to glucose is greater than that for ATP hydrolysis, so the hydrolysis reaction predominates (although neither reaction occurs at a biologically significant rate). However, in the presence of the appropriate enzyme, **hexokinase** (Section 14-2A), glucose-6-phosphate is formed far more rapidly than ATP is hydrolyzed. This is because the catalytic influence of the enzyme reduces the activation energy for phosphoryl group transfer from ATP to glucose to less than the activation energy for ATP hydrolysis. This example underscores the point that even a thermodynamically favored reaction ($\Delta G < 0$) may not occur in a living system in the absence of a specific enzyme that catalyzes the reaction (i.e., lowers $\Delta G^{\ddagger}$ to increase the rate of product formation; Box 12-3).

Inorganic Pyrophosphatase Catalyzes Additional Phosphoanhydride Bond Cleavage.

Although many reactions involving ATP yield ADP and P_i (**orthophosphate cleavage**), others yield AMP and PP_i (**pyrophosphate cleavage**). In these latter cases, the PP_i is rapidly hydrolyzed to 2 P_i by **inorganic pyrophosphatase** ($\Delta G^{\circ\prime} = -19.2$ kJ · mol^{-1}) so that *the pyrophosphate cleavage of ATP ultimately consumes two "high-energy" phosphoanhydride bonds.* The attachment of amino acids to tRNA molecules for protein synthesis is an

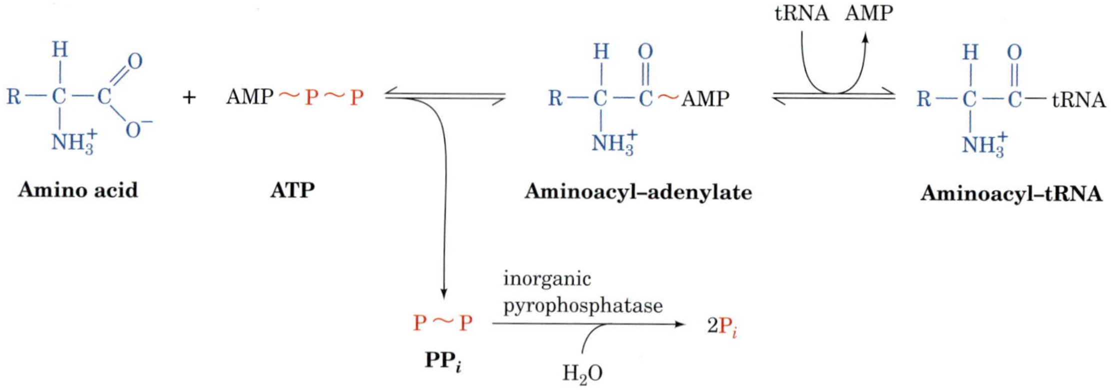

Figure 13-6   Pyrophosphate cleavage in the synthesis of an aminoacyl–tRNA.   In the first reaction step, the amino acid is **adenylylated** by ATP. In the second step, a tRNA molecule displaces the AMP moiety to form an aminoacyl–tRNA. The exergonic hydrolysis of pyrophosphate ($\Delta G^{\circ\prime} = -19.2$ kJ · mol^{-1}) drives the reaction forward.

example of this phenomenon (Fig. 13-6 and Section 26-2B). The two steps of the reaction are readily reversible because the free energies of hydrolysis of the bonds formed are comparable to that of ATP hydrolysis. The overall reaction is driven to completion by the irreversible hydrolysis of PP$_i$. Nucleic acid biosynthesis from nucleoside triphosphates also releases PP$_i$ (Sections 24-1 and 25-1). The standard free energy changes of these reactions are around 0, so the subsequent hydrolysis of PP$_i$ is also essential for the synthesis of nucleic acids.

C   Other Phosphorylated Compounds

"High-energy" compounds other than ATP are essential for energy metabolism, in part because they help maintain a relatively constant level of cellular ATP. *ATP is continually being hydrolyzed and regenerated.* Indeed, experimental evidence indicates that the metabolic half-life of an ATP molecule varies from seconds to minutes depending on the cell type and its metabolic activity. For instance, brain cells have only a few seconds supply of ATP (which partly accounts for the rapid deterioration of brain tissue by oxygen deprivation). An average person at rest consumes and regenerates ATP at a rate of ~3 mol (1.5 kg) per hour and as much as an order of magnitude faster during strenuous activity.

Just as ATP drives endergonic reactions through the exergonic process of phosphoryl group transfer and phosphoanhydride hydrolysis, *ATP itself can be regenerated by coupling its formation to a more highly exergonic metabolic process.* As Table 13-2 indicates, in the thermodynamic hierarchy of phosphoryl-transfer agents, ATP occupies the middle rank. ATP can therefore be formed from ADP by direct transfer of a phosphoryl group from a "high-energy" compound (e.g., phosphoenolpyruvate; Fig. 13-5*b* and Section 14-2J). Such a reaction is referred to as a **substrate-level phosphorylation.** Other mechanisms generate ATP indirectly, using the energy supplied by transmembrane proton concentration gradients. In oxidative metabolism, this process is called **oxidative phosphorylation** (Section 17-3), whereas in photosynthesis, it is termed **photophosphorylation** (Section 18-2D).

The flow of energy from "high-energy" phosphate compounds to ATP and from ATP to "low-energy" phosphate compounds is diagrammed in Fig. 13-7. These reactions are catalyzed by enzymes known as **kinases,** which transfer phosphoryl groups from ATP to other compounds or from phos-

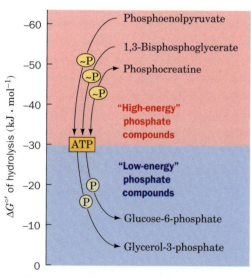

Figure 13-7   Position of ATP relative to "high-energy" and "low-energy" phosphate compounds.   Phosphoryl groups flow from the "high-energy" donors, via the ATP–ADP system, to "low-energy" acceptors.

phorylated compounds to ADP. We shall revisit these processes in our discussions of carbohydrate metabolism in Chapters 14 and 15.

The compounds whose phosphoryl group-transfer potentials are greater than that of ATP have additional stabilizing effects. For example, the hydrolysis of **acyl phosphates** (mixed phosphoric–carboxylic anhydrides), such as **acetyl phosphate** and **1,3-bisphosphoglycerate,**

$$CH_3 - \overset{\overset{\textstyle O}{\|}}{C} \sim OPO_3^{2-} \qquad ^{-2}O_3POCH_2 - \overset{\overset{\textstyle OH}{|}}{CH} - \overset{\overset{\textstyle O}{\|}}{C} \sim OPO_3^{2-}$$

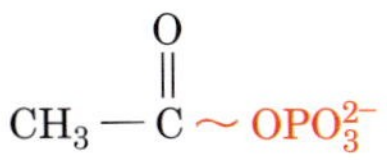

Acetyl phosphate 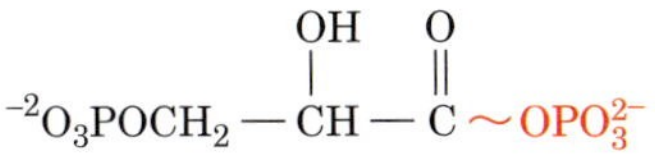   **1,3-Bisphosphoglycerate**

is driven by the same competing resonance and differential solvation effects that influence the hydrolysis of phosphoanhydrides (Fig. 13-4). Apparently, these effects are more pronounced for acyl phosphates than for phosphoanhydrides, as the rankings in Table 13-2 indicate.

In contrast, compounds such as glucose-6-phosphate and **glycerol-3-phosphate,**

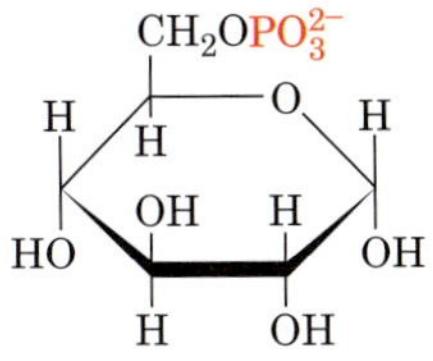

α-D-Glucose-6-phosphate 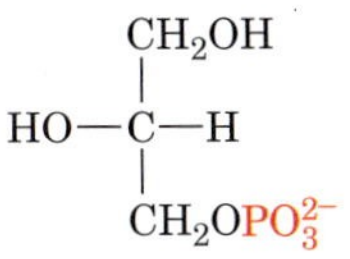   **L-Glycerol-3-phosphate**

which are below ATP in Table 13-2, have no significantly different resonance stabilization or charge separation compared to their hydrolysis products. Their free energies of hydrolysis are therefore much less than those of the preceding "high-energy" compounds.

The high phosphoryl group-transfer potentials of **phosphoguanidines,** such as **phosphocreatine** and **phosphoarginine,** largely result from the competing resonances in the **guanidino** group, which are even more pronounced than they are in the phosphate group of phosphoanhydrides:

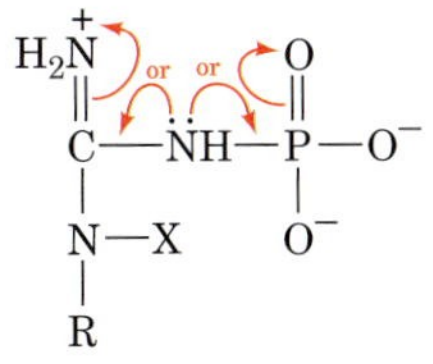

$$R = CH_2 - CO_2^- \; ; \; X = CH_3 \qquad \textbf{Phosphocreatine}$$

$$R = CH_2 - CH_2 - CH_2 - \overset{\overset{\textstyle NH_3^+}{|}}{CH} - CO_2^- \; ; \; X = H \qquad \textbf{Phosphoarginine}$$

Consequently, phosphocreatine can transfer its phosphoryl group to ADP to form ATP.

Phosphocreatine Provides a "High-Energy" Reservoir for ATP Formation.

Muscle and nerve cells, which have a high ATP turnover, rely on phosphoguanidines to regenerate ATP rapidly. In vertebrates, phosphocreatine is synthesized by the reversible phosphorylation of creatine by ATP catalyzed by **creatine kinase:**

ATP + creatine $\rightleftharpoons$ phosphocreatine + ADP

$$\Delta G^{\circ\prime} = +12.6 \; kJ \cdot mol^{-1}$$

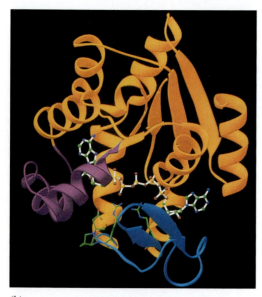

(a)

(b)

Figure 13-8 Conformational changes in *E. coli* adenylate kinase on binding substrate. (*a*) The unliganded enzyme. (*b*) The enzyme with the bound bisubstrate analog Ap₅A. The Ap₅A is shown in ball-and-stick form (C green, N blue, O red, and P yellow). Several of the protein's side chains that have been implicated in substrate binding are shown in stick form. The protein's magenta and blue domains undergo extensive conformational changes on ligand binding, whereas the remainder of the protein (*gold*), whose orientation is the same in *a* and *b*, largely maintains its conformation. [Based on X-ray structures by Georg Schulz, Institut für Organische Chemie und Biochemie, Freiburg, Germany. PDBids (*a*) 4AKE and (*b*) 1AKE.]   **See the Interactive Exercises.**

Note that this reaction is endergonic under standard conditions; however, *the intracellular concentrations of its reactants and products are such that it operates close to equilibrium* ($\Delta G \approx 0$). Accordingly, when the cell is in a resting state, so that [ATP] is relatively high, the reaction proceeds with net synthesis of phosphocreatine, whereas at times of high metabolic activity, when [ATP] is low, the equilibrium shifts so as to yield net synthesis of ATP from phosphocreatine and ADP. *Phosphocreatine thereby acts as an ATP "buffer" in cells that contain creatine kinase.* A resting vertebrate skeletal muscle normally has sufficient phosphocreatine to supply its free energy needs for several minutes (but for only a few seconds at maximum exertion). In the muscles of some invertebrates, such as lobsters, phosphoarginine performs the same function. These phosphoguanidines are collectively named **phosphagens.**

Nucleoside Triphosphates Are Freely Interconverted. Many biosynthetic processes, such as the synthesis of proteins and nucleic acids, require nucleoside triphosphates other than ATP. For example, RNA synthesis requires the ribonucleotides CTP, GTP, and UTP, along with ATP, and DNA synthesis requires dCTP, dGTP, dTTP, and dATP (Section 3-1). All these nucleoside triphosphates **(NTPs)** are synthesized from ATP and the corresponding nucleoside diphosphate **(NDP)** in a reaction catalyzed by the nonspecific enzyme **nucleoside diphosphate kinase:**

$$\text{ATP} + \text{NDP} \rightleftharpoons \text{ADP} + \text{NTP}$$

The $\Delta G°'$ values for these reactions are nearly 0, as might be expected from the structural similarities among the NTPs. These reactions are driven by the depletion of the NTPs through their exergonic utilization in subsequent reactions.

Other kinases reversibly convert nucleoside monophosphates to their diphosphate forms at the expense of ATP. One of these phosphoryl group-transfer reactions is catalyzed by **adenylate kinase:**

$$\text{AMP} + \text{ATP} \rightleftharpoons 2\ \text{ADP}$$

This enzyme is present in all tissues, where it functions to maintain equilibrium concentrations of the three nucleotides. When AMP accumulates, it is converted to ADP, which can be used to synthesize ATP through substrate-level phosphorylation, oxidative phosphorylation, or photophosphorylation. The reverse reaction helps restore cellular ATP as rapid consumption of ATP increases the level of ADP.

The X-ray structure of adenylate kinase, determined by Georg Schulz, reveals that, in the reaction catalyzed by this enzyme, two ~30-residue domains of the enzyme close over the substrates (Fig. 13-8), thereby tightly binding them and preventing water from entering the active site (which would lead to hydrolysis rather than phosphoryl group transfer). The movement of one of these domains depends on the presence of four invariant charged residues. Interactions between these groups and the bound substrates apparently trigger the rearrangements around the substrate-binding site (Fig. 13-8*b*).

Once the adenylate kinase reaction is complete, the tightly bound products must be rapidly released to maintain the enzyme's catalytic efficiency. Yet since the reaction is energetically neutral (the net number of phosphoanhydride bonds is unchanged), another source of free energy is required for rapid product release. The comparison of the X-ray structures of unliganded adenylate kinase and adenylate kinase in complex with the bisubstrate model compound **Ap₅A** (AMP and ATP connected by a fifth

phosphate) show how the enzyme avoids the kinetic trap of tight-binding substrates and products: On binding substrate, a portion of the protein remote from the active site increases its chain mobility and thereby consumes some of the free energy of substrate binding. The region "resolidifies" when the binding site is opened and the products are released. This mechanism is thought to act as an "energetic counterweight" to help adenylate kinase maintain a high reaction rate.

D Thioesters

The ubiquity of phosphorylated compounds in metabolism is consistent with their early evolutionary appearance. Yet phosphate is (and was) scarce in the abiotic world, which suggests that other kinds of molecules might have served as energy-rich compounds even before metabolic pathways became specialized for phosphorylated compounds. One candidate for a primitive "high-energy" compound is the **thioester,** which offers as its main recommendation its occurrence in the central metabolic pathways of all known organisms. Notably, the thioester bond is involved in substrate-level phosphorylation, an ATP-generating process that is independent of—and presumably arose before—oxidative phosphorylation.

The thioester bond appears in modern metabolic pathways as a reaction intermediate (involving a Cys residue in an enzyme active site) and in the form of acetyl-CoA (Fig. 13-9), the common product of carbohydrate, fatty acid, and amino acid catabolism. **Coenzyme A (CoASH or CoA)** consists of a β-mercaptoethylamine group bonded through an amide linkage to the vitamin **pantothenic acid,** which, in turn, is attached to a 3′-phosphoadeno-

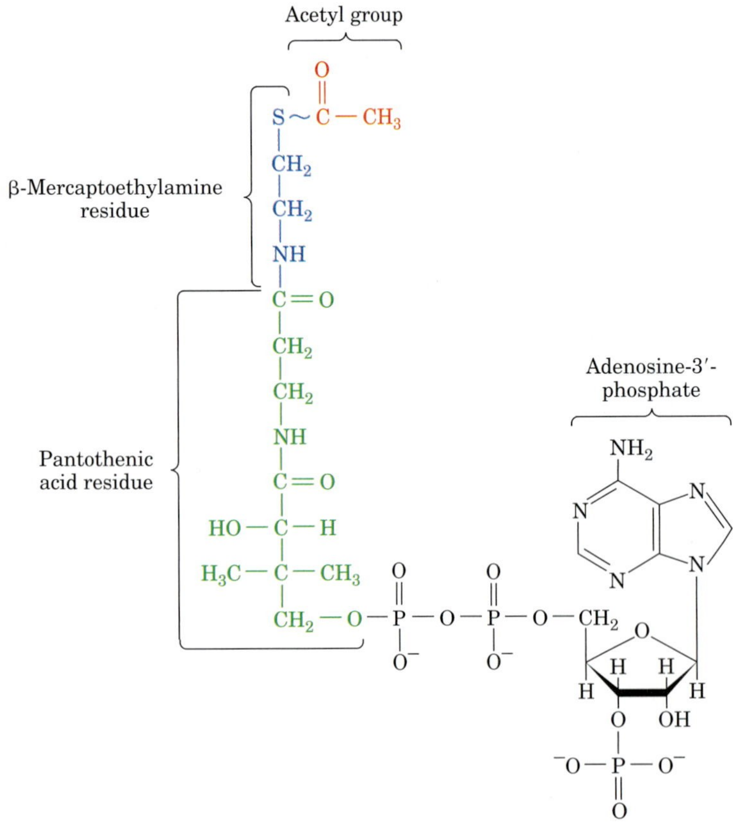

Acetyl-coenzyme A (acetyl-CoA)

Figure 13-9   The chemical structure of acetyl-CoA. The thioester bond is drawn with a ~ to indicate that it is a "high-energy" bond (has a high negative free energy of hydrolysis). In CoA, the acetyl group is replaced by hydrogen.

sine moiety via a pyrophosphate bridge. The acetyl group of acetyl-CoA is bonded as a thioester to the sulfhydryl portion of the β-mercaptoethylamine group. *CoA thereby functions as a carrier of acetyl and other acyl groups (the A of CoA stands for "Acetylation")*. Thioesters also take the form of acyl chains bonded to a phosphopantetheine residue that is linked to a Ser OH group in a protein rather than to 3'-phospho-AMP, as in CoA.

Acetyl-CoA is a "high-energy" compound. The $\Delta G^{\circ\prime}$ for the hydrolysis of its thioester bond is $-31.5 \text{ kJ} \cdot \text{mol}^{-1}$, which makes this reaction slightly $(1 \text{ kJ} \cdot \text{mol}^{-1})$ more exergonic than ATP hydrolysis. The hydrolysis of thioesters is more exergonic than that of ordinary esters because the thioester is less stabilized by resonance. This destabilization is a result of the large atomic radius of S, which reduces the electronic overlap between C and S compared to that between C and O.

The formation of a thioester bond in a metabolic intermediate conserves a portion of the free energy of oxidation of a metabolic fuel. That free energy can then be used to drive an exergonic process. In the citric acid cycle, for example, cleavage of a thioester **(succinyl-CoA)** releases sufficient free energy to synthesize GTP from GDP and P_i (Section 16-3E).

3 Oxidation–Reduction Reactions

As metabolic fuels are oxidized to CO_2, electrons are transferred to molecular carriers that, in aerobic organisms, ultimately transfer the electrons to molecular oxygen. The process of electron transport results in a transmembrane proton concentration gradient that drives ATP synthesis (oxidative phosphorylation; Section 17-3). Even obligate anaerobes, which do not carry out oxidative phosphorylation, rely on the oxidation of substrates to drive ATP synthesis. In fact, oxidation–reduction reactions (also known as **redox reactions**) supply living things with most of their free energy. In this section, we examine the thermodynamic basis for the conservation of free energy during substrate oxidation.

A NAD⁺ and FAD

Two of the most widely occurring electron carriers are the nucleotide coenzymes nicotinamide adenine dinucleotide (NAD^+) and **flavin adenine dinucleotide (FAD)**. The nicotinamide portion of NAD^+ (and its phosphorylated counterpart $NADP^+$; Fig. 11-3) is the site of reversible reduction, which formally occurs as the transfer of a hydride ion (H^-; a proton with two electrons) as shown in Fig. 13-10. The terminal electron acceptor in aerobic organisms, O_2, can accept only unpaired electrons; that is, elec-

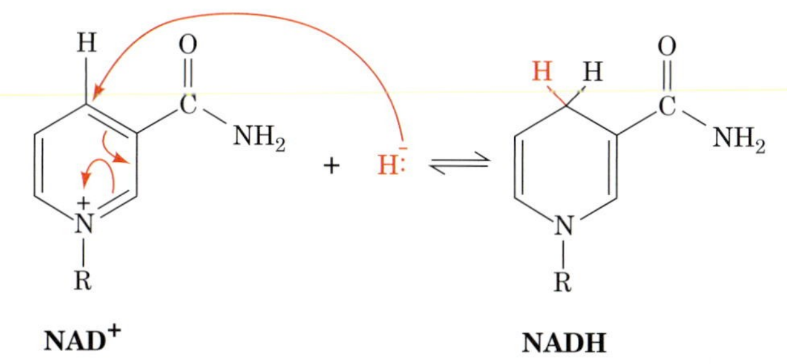

Figure 13-10 Reduction of NAD⁺ to NADH. R represents the ribose–pyrophosphoryl–adenosine portion of the coenzyme. Only the nicotinamide ring is affected by reduction, which is formally represented here as occurring by hydride transfer.

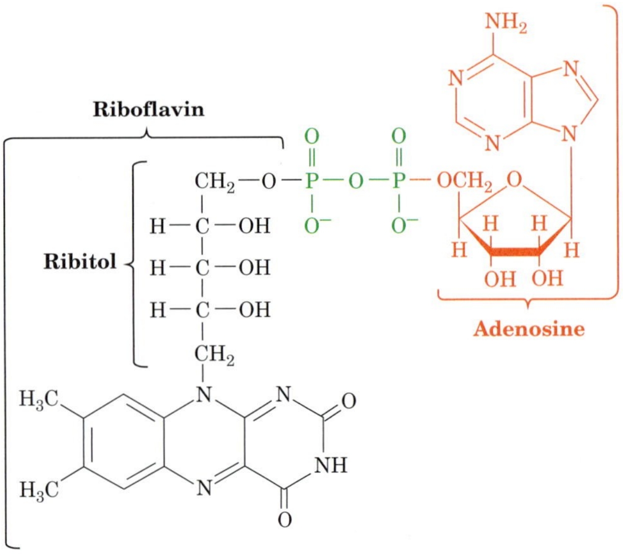

Figure 13-11 Flavin adenine dinucleotide (FAD). Adenosine (*red*) is linked to **riboflavin** (*black*) by a pyrophosphoryl group (*green*). The riboflavin portion of FAD is also known as **vitamin B$_2$**.

trons must be transferred to O$_2$ one at a time. Electrons that are removed from metabolites as pairs (e.g., with the two-electron reduction of NAD$^+$) must be transferred to other carriers that can undergo one-electron redox reactions. FAD (Fig. 13-11) is one such coenzyme; it can undergo both one-electron and two-electron transfers.

The conjugated ring system of FAD can accept one or two electrons to produce the stable radical (semiquinone) FADH· or the fully reduced (hydroquinone) FADH$_2$ (Fig. 13-12). The change in the electronic state of the ring system on reduction is reflected in a color change from brilliant yellow (in FAD) to pale yellow (in FADH$_2$). The metabolic functions of NAD$^+$ and FAD demand that they undergo reversible reduction so that they can accept electrons, pass them on to other electron carriers, and thereby be regenerated to participate in additional cycles of oxidation and reduction.

Humans cannot synthesize the flavin moiety of FAD but, rather, must obtain it from their diets, for example, in the form of riboflavin (vitamin B$_2$; Fig. 13-11). Nevertheless, riboflavin deficiency is quite rare in humans, in part because of the tight binding of flavin prosthetic groups to their apoenzymes. The symptoms of riboflavin deficiency, which are associated with general malnutrition or bizarre diets, include an inflamed tongue, lesions in the corner of the mouth, and dermatitis.

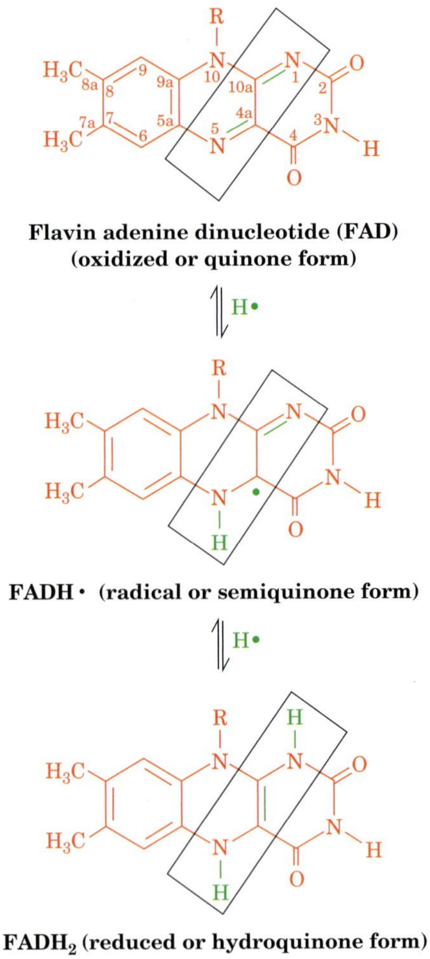

B The Nernst Equation

Oxidation–reduction reactions resemble other types of group-transfer reactions except that the "groups" transferred are electrons, which are passed from an **electron donor (reductant** or **reducing agent)** to an **electron acceptor (oxidant** or **oxidizing agent).** For example, in the reaction

$$Fe^{3+} + Cu^+ \rightleftharpoons Fe^{2+} + Cu^{2+}$$

Cu$^+$, the reductant, is oxidized to Cu^{2+} while Fe^{3+}, the oxidant, is reduced to Fe^{2+}.

Redox reactions can be divided into two **half-reactions,** such as

$$Fe^{3+} + e^- \rightleftharpoons Fe^{2+} \text{ (reduction)}$$
$$Cu^+ \rightleftharpoons Cu^{2+} + e^- \text{ (oxidation)}$$

Figure 13-12 Reduction of FAD to FADH$_2$. R represents the ribitol–pyrophosphoryl–adenosine portion of the coenzyme. The conjugated ring system of FAD undergoes two sequential one-electron reductions or a two-electron transfer that bypasses the **semiquinone** state.

whose sum is the whole reaction above. These particular half-reactions occur during the oxidation of cytochrome c oxidase in the mitochondrion (Section 17-2F). Note that for electrons to be transferred, both half-reactions must occur simultaneously. In fact, the electrons are the two half-reactions' common intermediate.

Electrochemical Cells.   A half-reaction consists of an electron donor and its conjugate electron acceptor; in the oxidative half-reaction shown above, Cu^+ is the electron donor and Cu^{2+} is its conjugate electron acceptor. Together these constitute a **redox couple** or **conjugate redox pair** analogous to a conjugate acid–base pair (HA and A^-; Section 2-2B). An important difference between redox pairs and acid–base pairs, however, is that *the two half-reactions of a redox reaction, each consisting of a conjugate redox pair, can be physically separated to form an **electrochemical cell*** (Fig. 13-13). In such a device, each half-reaction takes place in its separate **half-cell,** and electrons are passed between half-cells as an electric current in the wire connecting their two electrodes. A salt bridge is necessary to complete the electrical circuit by providing a conduit for ions to migrate and thereby maintain electrical neutrality.

The free energy of an oxidation–reduction reaction is particularly easy to determine by simply measuring the voltage difference between its two half-cells. Consider the general reaction

$$A_{ox}^{n+} + B_{red} \rightleftharpoons A_{red} + B_{ox}^{n+}$$

in which n electrons per mole of reactants are transferred from reductant (B_{red}) to oxidant (A_{ox}^{n+}). The free energy of this reaction is expressed as

$$\Delta G = \Delta G^{\circ\prime} + RT \ln\left(\frac{[A_{red}][B_{ox}^{n+}]}{[A_{ox}^{n+}][B_{red}]}\right) \qquad [13\text{-}4]$$

Under reversible conditions,

$$\Delta G = -w' = -w_{el} \qquad [13\text{-}5]$$

where w' is non-pressure–volume work. In this case, w' is equivalent to w_{el}, the electrical work required to transfer the n moles of electrons through the **electrical potential difference, $\Delta \mathscr{E}$** [where the units of $\mathscr{E}$ are volts (V), the number of joules (J) of work required to transfer 1 coulomb (C) of charge]. This, according to the laws of electrostatics, is

$$w_{el} = n\mathscr{F}\Delta\mathscr{E} \qquad [13\text{-}6]$$

where $\mathscr{F}$, the **faraday,** is the electrical charge of 1 mol of electrons (1 $\mathscr{F} = 96{,}485 \ C \cdot mol^{-1} = 96{,}485 \ J \cdot V^{-1} \cdot mol^{-1}$), and n is the number of moles of electrons transferred per mole or reactant converted. Thus, substituting Eq. 13-6 into Eq. 13-5,

$$\Delta G = -n\mathscr{F}\Delta\mathscr{E} \qquad [13\text{-}7]$$

Combining Eqs. 13-4 and 13-7, and making the analogous substitution for $\Delta G^{\circ\prime}$, yields the **Nernst equation:**

$$\boxed{\Delta\mathscr{E} = \Delta\mathscr{E}^{\circ\prime} - \frac{RT}{n\mathscr{F}} \ln\left(\frac{[A_{red}][B_{ox}^{n+}]}{[A_{ox}^{n+}][B_{red}]}\right)} \qquad [13\text{-}8]$$

which was originally formulated in 1881 by Walther Nernst. Here $\mathscr{E}$ is the **reduction potential,** the tendency for a substance to undergo reduction (gain electrons). $\Delta\mathscr{E}$, the **electromotive force (emf),** can be described as the

"electron pressure" that the electrochemical cell exerts. The quantity $\mathcal{E}^\circ$, the reduction potential when all components are in their standard states, is called the **standard reduction potential.** If these standard states refer to biochemical standard states (Section 1-4D), then $\Delta\mathcal{E}^\circ$ is replaced by $\Delta\mathcal{E}^{\circ\prime}$. Note that a positive $\Delta\mathcal{E}$ in Eq. 13-7 results in a negative ΔG; in other words, *a positive $\Delta\mathcal{E}$ indicates a spontaneous reaction, one that can do work.*

C Measurements of Reduction Potential Differences

Equation 13-7 shows that the free energy change of a redox reaction can be determined by directly measuring its change in reduction potential with a voltmeter (Fig. 13-13). Such measurements make it possible to determine the order of spontaneous electron transfers among a set of electron carriers such as those of the electron transport pathway that mediates oxidative phosphorylation in cells.

Any redox reaction can be divided into its component half-reactions:

$$A_{ox}^{n+} + n\,e^- \rightleftharpoons A_{red}$$
$$B_{ox}^{n+} + n\,e^- \rightleftharpoons B_{red}$$

where, by convention, both half-reactions are written as reductions. These half-reactions can be assigned reduction potentials, $\mathcal{E}_A$ and $\mathcal{E}_B$, in accordance with the Nernst equation:

$$\mathcal{E}_A = \mathcal{E}_A^{\circ\prime} - \frac{RT}{n\mathcal{F}} \ln\left(\frac{[A_{red}]}{[A_{ox}^{n+}]}\right) \qquad [13\text{-}9]$$

$$\mathcal{E}_B = \mathcal{E}_B^{\circ\prime} - \frac{RT}{n\mathcal{F}} \ln\left(\frac{[B_{red}]}{[B_{ox}^{n+}]}\right) \qquad [13\text{-}10]$$

For the overall redox reaction involving the two half-reactions, the difference in reduction potential, $\Delta\mathcal{E}^{\circ\prime}$, is defined as

$$\Delta\mathcal{E}^{\circ\prime} = \mathcal{E}^{\circ\prime}_{(e^- \text{ acceptor})} - \mathcal{E}^{\circ\prime}_{(e^- \text{ donor})} \qquad [13\text{-}11]$$

Thus, when the reaction proceeds with A as the electron acceptor and B as the electron donor, $\Delta\mathcal{E}^{\circ\prime} = \mathcal{E}_A^{\circ\prime} - \mathcal{E}_B^{\circ\prime}$, and $\Delta\mathcal{E} = \mathcal{E}_A - \mathcal{E}_B$.

Standard Reduction Potentials. Reduction potentials, like free energies, must be defined with respect to some arbitrary standard, in this case, the hydrogen half-reaction

$$2\,H^+ + 2\,e^- \rightleftharpoons H_2\,(g)$$

in which H^+ is in equilibrium with $H_2\,(g)$ that is in contact with a Pt electrode. This half-cell is arbitrarily assigned a standard reduction potential $\mathcal{E}^\circ$ of 0 V (1 V = 1 J·C^{-1}) at pH 0, 25°C, and 1 atm. Under the biochemical convention, where the standard state is pH 7.0, the hydrogen half-reaction has a standard reduction potential $\mathcal{E}^{\circ\prime}$ of −0.421 V.

When $\Delta\mathcal{E}$ is positive, ΔG is negative (Eq. 13-7), indicating a spontaneous process. In combining two half-reactions under standard conditions, the direction of spontaneity therefore involves the reduction of the redox couple with the more positive standard reduction potential. In other words, *the more positive the standard reduction potential, the higher the affinity of the redox couple's oxidized form for electrons, that is, the greater the tendency for the redox couple's oxidized form to accept electrons and thus become reduced.*

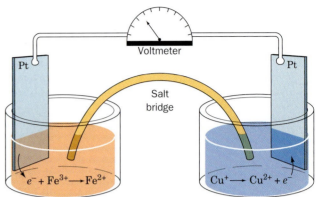

Figure 13-13 **An electrochemical cell.** The half-cell undergoing oxidation (here $Cu^+ \rightarrow Cu^{2+} + e^-$) passes the liberated electrons through the wire to the half-cell undergoing reduction (here $e^- + Fe^{3+} \rightarrow Fe^{2+}$). Electroneutrality in the two half-cells is maintained by the transfer of ions through the electrolyte-containing salt bridge.

Biochemical Half-Reactions Are Physiologically Significant. The biochemical standard reduction potentials ($\mathcal{E}^{\circ\prime}$) of some biochemically important half-reactions are listed in Table 13-3. The oxidized form of a redox couple with a large positive standard reduction potential has a high affinity for electrons and is a strong electron acceptor (oxidizing agent), whereas its conjugate reductant is a weak electron donor (reducing agent). For example, O_2 is the strongest oxidizing agent in Table 13-3, whereas H_2O, which tightly holds its electrons, is the table's weakest reducing agent. The converse is true of half-reactions with large negative standard reduction potentials.

Since electrons spontaneously flow from low to high reduction potentials, they are transferred, under standard conditions, from the reduced products in any half-reaction in Table 13-3 to the oxidized reactants of any half-reaction above it (see Sample Calculation 13-2). However, such a reaction may not occur at a measurable rate in the absence of a suitable enzyme. Note that Fe^{3+} ions of the various cytochromes listed in Table 13-3 have significantly different reduction potentials. This indicates that *the protein components of redox enzymes play active roles in electron-transfer reactions by modulating the reduction potentials of their bound redox-active centers.*

Electron-transfer reactions are of great biological importance. For example, in the mitochondrial electron-transport chain (Section 17-2), electrons are passed from NADH along a series of electron acceptors of increasing reduction potential (including FAD and others listed in Table 13-3) to O_2. ATP is generated from ADP and P_i by coupling its synthesis to this free energy cascade. *NADH thereby functions as an energy-rich elec-*

Table 13-3   Standard Reduction Potentials of Some Biochemically Important Half-Reactions

Half-Reaction	$\mathcal{E}^{\circ\prime}$ (V)
$\frac{1}{2} O_2 + 2\,H^+ + 2\,e^- \rightleftharpoons H_2O$	0.815
$SO_4^{2-} + 2\,H^+ + 2\,e^- \rightleftharpoons SO_3^{2-} + H_2O$	0.48
$NO_3^- + 2\,H^+ + 2\,e^- \rightleftharpoons NO_2^- + H_2O$	0.42
Cytochrome a_3 (Fe^{3+}) $+ e^- \rightleftharpoons$ cytochrome a_3 (Fe^{2+})	0.385
$O_2(g) + 2\,H^+ + 2\,e^- \rightleftharpoons H_2O_2$	0.295
Cytochrome a (Fe^{3+}) $+ e^- \rightleftharpoons$ cytochrome a (Fe^{2+})	0.29
Cytochrome c (Fe^{3+}) $+ e^- \rightleftharpoons$ cytochrome c (Fe^{2+})	0.235
Cytochrome c_1 (Fe^{3+}) $+ e^- \rightleftharpoons$ cytochrome c_1 (Fe^{2+})	0.22
Cytochrome b (Fe^{3+}) $+ e^- \rightleftharpoons$ cytochrome b (Fe^{2+}) (*mitochondrial*)	0.077
Ubiquinone $+ 2\,H^+ + 2\,e^- \rightleftharpoons$ ubiquinol	0.045
Fumarate$^-$ $+ 2\,H^+ + 2\,e^- \rightleftharpoons$ succinate$^-$	0.031
FAD $+ 2\,H^+ + 2\,e^- \rightleftharpoons$ FADH$_2$ (*in flavoproteins*)	~0.
Oxaloacetate$^-$ $+ 2\,H^+ + 2\,e^- \rightleftharpoons$ malate$^-$	−0.166
Pyruvate$^-$ $+ 2\,H^+ + 2\,e^- \rightleftharpoons$ lactate$^-$	−0.185
Acetaldehyde $+ 2\,H^+ + 2\,e^- \rightleftharpoons$ ethanol	−0.197
FAD $+ 2\,H^+ + 2\,e^- \rightleftharpoons$ FADH$_2$ (*free coenzyme*)	−0.219
$S + 2\,H^+ + 2\,e^- \rightleftharpoons H_2S$	−0.23
Lipoic acid $+ 2\,H^+ + 2\,e^- \rightleftharpoons$ dihydrolipoic acid	−0.29
$NAD^+ + H^+ + 2\,e^- \rightleftharpoons$ NADH	−0.315
$NADP^+ + H^+ + 2\,e^- \rightleftharpoons$ NADPH	−0.320
Cystine $+ 2\,H^+ + 2\,e^- \rightleftharpoons$ 2 cysteine	−0.340
Acetoacetate$^-$ $+ 2\,H^+ + 2\,e^- \rightleftharpoons$ β-hydroxybutyrate$^-$	−0.346
$H^+ + e^- \rightleftharpoons \frac{1}{2} H_2$	−0.421
Acetate$^-$ $+ 3\,H^+ + 2\,e^- \rightleftharpoons$ acetaldehyde $+ H_2O$	−0.581

Source: Mostly from Loach, P.A., *In* Fasman, G.D. (Ed.), *Handbook of Biochemistry and Molecular Biology* (3rd ed.), Physical and Chemical Data, Vol. I, pp. 123–130, CRC Press (1976).

SAMPLE CALCULATION 13-2

Calculate $\Delta G^{\circ\prime}$ for the oxidation of NADH by FAD.

Combining the relevant half-reactions gives

$$NADH + FAD + H^+ \rightarrow NAD^+ + FADH_2$$

Next, calculate the electromotive force ($\Delta \mathscr{E}^{\circ\prime}$) from the standard reduction potentials given in Table 13-3, using one of the following methods:

Method 1
According to Eq. 13-11,

$$\Delta \mathscr{E}^{\circ\prime} = \mathscr{E}^{\circ\prime}_{(e^-\ acceptor)} - \mathscr{E}^{\circ\prime}_{(e^-\ donor)}$$

Since FAD ($\mathscr{E}^{\circ\prime} = -0.219$ V) is the electron acceptor, and NADH ($\mathscr{E}^{\circ\prime} = -0.315$ V) is the electron donor,

$$\Delta \mathscr{E}^{\circ\prime} = (-0.219\text{ V}) - (-0.315\text{ V}) = +0.096\text{ V}$$

Method 2
Write the net reaction as a sum of the two relevant half-reactions. For FAD, the half-reaction is the same as the reductive half-reaction given in Table 13-3, and its $\mathscr{E}^{\circ\prime}$ value is -0.219 V. For NADH, which undergoes oxidation rather than reduction, the half-reaction is the reverse of the one given in Table 13-3, and its $\mathscr{E}^{\circ\prime}$ value is $+0.315$ V, the reverse of the reduction potential given in the table. The two half-reactions are added to give the net oxidation–reduction reaction, and the $\mathscr{E}^{\circ\prime}$ values are also added:

$FAD + 2\ H^+ + 2\ e^- \rightarrow FADH_2$	$\mathscr{E}^{\circ\prime} = -0.219$ V
$NADH \rightarrow NAD^+ + H^+ + 2\ e^-$	$\mathscr{E}^{\circ\prime} = +0.315$ V
$NADH + FAD + H^+ \rightarrow NAD^+ + FADH_2$	$\Delta\mathscr{E}^{\circ\prime} = +0.096$ V

Next, use Eq. 13-8 to calculate $\Delta G^{\circ\prime}$. Because two moles of electrons are transferred for every mole of NADH oxidized to NAD^+, $n = 2$.

$$\Delta G^{\circ\prime} = -n\mathscr{F}\Delta\mathscr{E}^{\circ\prime}$$
$$\Delta G^{\circ\prime} = -(2)(96{,}485\text{ J} \cdot \text{V}^{-1} \cdot \text{mol}^{-1})(0.096\text{ V})$$

tron-transfer coenzyme. In fact, the oxidation by O_2 of one NADH to NAD^+ supplies sufficient free energy to generate three ATPs. NAD^+ is an electron acceptor in many exergonic metabolite oxidations. In serving as the electron donor in ATP synthesis, it fulfills its cyclic role as a free energy conduit in a manner analogous to ATP (Fig. 13-7).

4 Experimental Approaches to the Study of Metabolism

A metabolic pathway can be understood at several levels:

1. In terms of the sequence of reactions by which a specific nutrient is converted to end products, and the energetics of these conversions.
2. In terms of the mechanisms by which each intermediate is converted to its successor. Such an analysis requires the isolation and characterization of the specific enzymes that catalyze each reaction.
3. In terms of the control mechanisms that regulate the flow of metabolites through the pathway. These mechanisms include the interorgan relationships that adjust metabolic activity to the needs of the entire organism.

Elucidating a metabolic pathway on all these levels is a complex process, often requiring contributions from a variety of disciplines.

The outlines of the major metabolic pathways have been known for decades, although in many cases, the enzymology behind various steps of the pathways remains obscure. Likewise, the mechanisms that regulate pathway activity under different physiological conditions are not entirely understood. These areas are of great interest because of their potential to yield information that could be useful in improving human health and curing metabolic diseases. In addition, the unexplored metabolisms of unusual organisms, including recently discovered "extremophiles," hold the promise of novel biological materials and enzymatic processes that can be exploited for the environmentally sensitive production of industrial materials, foods, and therapeutic drugs.

Early metabolic studies used whole organisms, often yeast, but also mammals. For example, Frederick Banting and Charles Best established the role of the pancreas in diabetes in 1921; they surgically removing that organ from dogs and observed that the animals then developed the disease. Techniques for studying metabolic processes have since become more refined, progressing from whole-organ preparations and thin tissue slices to cultured cells and organelles isolated by ultracentrifugation (Section 5-2E). The most recent approaches include identifying active genes and cataloguing their protein products.

A   Tracing Metabolic Fates

A metabolic pathway in which one compound is converted to another can be followed by tracing a specifically labeled metabolite. Franz Knoop formulated this technique in 1904 to study fatty acid oxidation. He fed dogs fatty acids chemically labeled with phenyl groups and isolated the phenyl-substituted end products from the dogs' urine. From the differences in these products, depending on whether the phenyl-substituted starting material contained odd or even numbers of carbon atoms, Knoop deduced that fatty acids are degraded in two-carbon units (Section 19-2).

Chemical labeling has the disadvantage that the chemical properties of labeled metabolites differ from those of normal metabolites. This problem is eliminated by labeling molecules with isotopes. *The fate of an isotopically labeled atom in a metabolite can therefore be elucidated by following its progress through the metabolic pathway of interest.* The advent of isotopic labeling and tracing techniques in the 1940s revolutionized the study of metabolism.

One of the early advances in metabolic understanding resulting from the use of isotopic tracers was the demonstration, by David Shemin and David Rittenberg in 1945, that the nitrogen atoms of heme (Fig. 7-2) are derived from glycine rather than from ammonia, glutamic acid, proline, or leucine (Section 20-6A). They showed this by feeding rats these ^{15}N-labeled nutrients, isolating the heme in their blood, and analyzing it for ^{15}N content. Only when the rats were fed [^{15}N]glycine did the heme contain ^{15}N. This technique was also used with the radioactive isotope ^{14}C to demonstrate that all of cholesterol's carbon atoms are derived from acetyl-CoA (Section 19-7A). Radioactive isotopes (Box 12-1) have become virtually indispensable for establishing the metabolic origins of complex metabolites.

Another method for tracing the fates of labeled metabolites is nuclear magnetic resonance (NMR), which detects specific isotopes, including ^{1}H, ^{13}C, ^{15}N, and ^{31}P, by their characteristic nuclear spins. Since the NMR spectrum of a particular nucleus varies with its immediate environment, it is

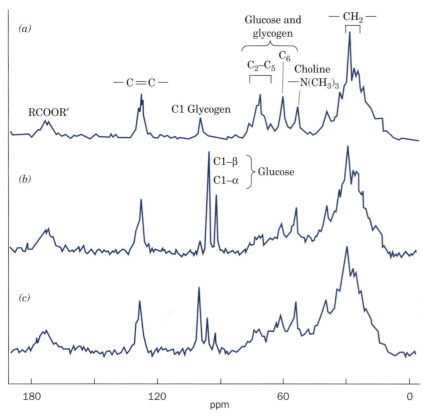

Figure 13-14 **The conversion of [1-¹³C]glucose to glycogen as observed by localized *in vivo* ¹³C NMR.** (*a*) The natural abundance ¹³C NMR spectrum of the liver of a live rat. Note the resonance corresponding to C1 of glycogen. (*b*) The ¹³C NMR spectrum of the liver of the same rat ~5 min after it was intravenously injected with 100 mg of [1-¹³C]glucose (90% enriched). The resonances of the C1 atom of both the α and β anomers of glucose are clearly distinguishable from each other and from the resonance of the C1 atom of glycogen. (*c*) The ¹³C NMR spectrum of the liver of the same rat ~30 min after the [1-¹³C]glucose injection. The C1 resonances of both the α- and β-glucose anomers are much reduced while the C1 resonance of glycogen has increased. [After Reo, N.V., Siegfried, B.A., and Acherman, J.J.H., *J. Biol. Chem.* **259**, 13665 (1984).]

possible to identify the peaks corresponding to specific atoms even in relatively complex mixtures. The development of magnets large enough to accommodate animals and humans, and to localize spectra to specific organs, has made it possible to study metabolic pathways noninvasively by NMR techniques. For example, ³¹P NMR can be used to study energy metabolism in muscle by monitoring the levels of phosphorylated compounds such as ATP, ADP, and phosphocreatine.

Isotopically labeling specific atoms of metabolites with ¹³C (which is only 1.10% naturally abundant) permits the metabolic progress of the labeled atoms to be followed by ¹³C NMR. Figure 13-14 shows *in vivo* ¹³C NMR spectra of a rat liver before and after an injection of D-[1-¹³C]glucose. The ¹³C can be seen entering the liver and then being incorporated into glycogen (the storage form of glucose; Section 15-2).

B Perturbing the System

Many of the techniques used to elucidate the intermediates and enzymes of metabolic pathways involve perturbing the system in some way and observing how this affects the activity of the pathway. One way to perturb a

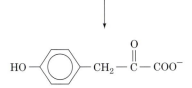

Phenylalanine

Tyrosine

p-Hydroxyphenylpyruvate

Homogentisate

defective in
alcaptonuria

$H_2O + CO_2$

Figure 13-15 **Pathway for phenylalanine degradation.**
Alcaptonurics lack the enzyme that breaks down
homogentisate; therefore, this intermediate accumulates
and is excreted in the urine.

pathway is to add certain substances, called **metabolic inhibitors,** that block the pathway at specific points, thereby causing the preceding intermediates to build up. This approach was used in elucidating the conversion of glucose to ethanol in yeast by glycolysis (Section 14-2). Similarly, the addition of substances that block electron transfer at different sites was used to deduce the sequence of electron carriers in the mitochondrial electron-transport chain (Section 17-2B).

Genetic Defects also Cause Metabolic Intermediates to Accumulate. Archibald Garrod's realization, in the early 1900s, that human genetic diseases are the consequence of deficiencies in specific enzymes also contributed to the elucidation of metabolic pathways. For example, upon the ingestion of either phenylalanine or tyrosine, individuals with the largely harmless inherited condition known as **alcaptonuria,** but not normal subjects, excrete **homogentisic acid** in their urine (Box 20-2). This is because the liver of alcaptonurics lacks an enzyme that catalyzes the breakdown of homogentisic acid (Fig. 13-15).

Genetic Manipulation Alters Metabolic Processes. Early studies of metabolism led to the astounding discovery that *the basic metabolic pathways in most organisms are essentially identical.* This metabolic uniformity has greatly facilitated the study of metabolic reactions. Thus, although a mutation that inactivates or deletes an enzyme in a pathway of interest may be unknown in higher organisms, it can be readily generated in a rapidly reproducing microorganism through the use of **mutagens** (chemical agents that induce genetic changes; Section 24-4A), X-rays, or, more recently, through genetic engineering techniques (Section 3-5). The desired mutants, which cannot synthesize the pathway's end product, can be identified by their requirement for that product in their culture medium.

Higher organisms that have been engineered to lack particular genes (i.e., gene "knockouts"; Section 3-5D) are useful, particularly in cases in which the absence of a single gene product results in a metabolic defect but is not lethal. Genetic engineering techniques have advanced to the point that it is possible to selectively "knock out" a gene only in a particular tissue. This approach is necessary in cases in which a gene product is required for development and therefore cannot be entirely deleted. In the opposite approach, techniques for constructing transgenic animals make it possible to express genes in tissues in which they were not originally present.

C DNA Microarrays

The overall metabolic capabilities of an organism are encoded by its genes. In theory, it should be possible to reconstruct a cell's metabolic activities from its DNA sequences. At present, this can be done only in a general sense. For example, the sequenced genome of *Vibrio cholerae,* the bacterium that causes cholera, reveals a large repertoire of genes encoding transport proteins and enzymes for catabolizing a wide range of nutrients. This is consistent with the complicated lifestyle of *V. cholerae,* which can live on its own or in association with zooplankton, as well as in the human gastrointestinal tract.

Of course, a simple catalog of an organism's genes does not reveal how the genes function. For example, some genes are expressed continuously at high levels, whereas others are expressed rarely, for example, only when the organism encounters a particular metabolite. Creating an accurate picture of gene expression is the goal of **transcriptomics,** the study of a cell's **trans-**

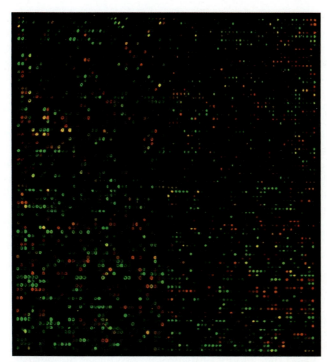

Figure 13-16 A DNA chip. This ~6000-gene array contains most of the genes from baker's yeast, one per spot. The chip had been hybridized to the cDNAs derived from mRNAs extracted from yeast (a cDNA molecule is a DNA copy of an RNA molecule; see Section 3-5B). The cDNAs derived from cells that were grown in glucose were labeled with a red-fluorescing dye, whereas the cDNAs derived from cells grown in the absence of glucose were labeled with a green-fluorescing dye. Thus the red and green spots, respectively, reveal those genes that are transcriptionally activated by the presence or absence of glucose, whereas the yellow spots (*red plus green*) indicate genes whose expression is unaffected by the level of glucose. [Courtesy of Patrick Brown, Stanford University School of Medicine.]

criptome (which, in analogy with the word "genome," is the entire collection of RNAs that the cell transcribes). Identifying and quantifying all the mRNA transcripts from a single cell type reveals which genes are active. This can be done by depositing numerous different DNA segments of known sequences in a precise array on a solid support such as a coated glass surface. The DNAs in these so-called **DNA microarrays** or **DNA chips** (which contain up to hundreds of thousands of DNAs in an area of ~1 cm^2) are often PCR-amplified cDNA clones derived from mRNAs (PCR is discussed in Section 3-5C) or their robotically synthesized counterparts. The mRNAs extracted from cells, tissues, or other biological sources are then labeled with a fluorescent dye, allowed to hybridize with the DNAs on the DNA microarray, and the unhybridized mRNA is washed away. The resulting fluorescence intensity at each site on the DNA microarray then indicates how much mRNA has bound to a particular complementary DNA sequence.

The DNA on a microarray may represent an entire genome or just a few selected genes. Because the detection and quantification of DNA–RNA hybrids is easily automated, it is possible to examine multiple preparations from cells at different developmental stages or from cells grown under different conditions (Fig. 13-16). Differences in the expression of particular genes can then be correlated with developmental processes or growth patterns. For example, DNA microarrays have been used to profile the patterns of gene expression in tumor cells because different types of tu-

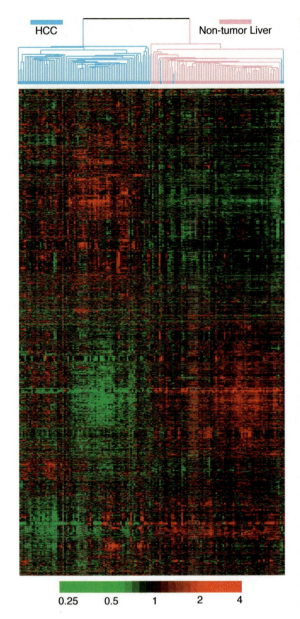

0.25   0.5   1   2   4

Figure 13-17 **The relative transcriptional activities of the genes in hepatocellular carcinoma (HCC) tumors as determined using DNA microarrays.** The data are presented in matrix form with each column representing one of 156 tissue samples [82 HCC tumors (the most common human liver cancer and among the five leading causes of cancer deaths in the world) and 74 nontumor liver tissues] and each row representing one of 3180 genes (those of the ~17,400 genes on the DNA microarray with the greatest variation in transcriptional activity among the various tissue samples). The data are arranged so as to group the genes as well as the tissue samples on the basis of similarities of their expression patterns. The color of each cell indicates the expression level of the corresponding gene in the corresponding tissue relative to its mean expression level in all the tissue samples with bright red, black, and bright green indicating expression levels of 4, 1, and 1/4 times that of the mean for that gene (as indicated on the scale below). The dendrogram at the top of the matrix indicates the similarities in expression patterns among the various tissue samples. [Courtesy of David Botstein and Patrick Brown, Stanford University School of Medicine.]

mors synthesize different proteins (Fig. 13-17). This information is useful in choosing how best to treat a cancer.

D Proteomics

Unfortunately, the correlation between the amount of a particular mRNA and the amount of its protein product is not perfect. The various mRNAs are both translated and degraded at different rates. Furthermore, many proteins are posttranslationally modified, sometimes in several different ways (e.g., by alternative splicing). Consequently, the number of proteins in a cell exceeds the number of mRNAs.

A more reliable way than transcriptomics to assess gene expression is to examine a cell's **proteome,** the complete set of proteins that the cell synthesizes. This approach requires that the proteins first be separated, usually by **two-dimensional (2D) gel electrophoresis,** a technique that separates proteins by mass in one direction and by isoelectric point in the perpendicular direction (Section 5-2D).

Protein-cataloging efforts are limited mostly by the technical problems of detecting minute quantities of thousands of different proteins using techniques that have not yet been fully automated, as they have been for DNA microarrays. In addition, there is no procedure comparable to PCR for amplifying trace amounts of protein. Most likely, a profile of a cell's metabolic activities will come to light through a combination of transcriptomics, **proteomics** (the study of proteomes), and more traditional approaches.

SUMMARY

1. The free energy released from catabolic oxidation reactions is used to drive endergonic anabolic reactions.

2. Heterotrophic organisms obtain their free energy from compounds synthesized by chemolithotrophic or photoautotrophic organisms.

3. Metabolic pathways are sequences of enzyme-catalyzed reactions that occur in different cellular locations.

4. Near-equilibrium reactions are freely reversible, whereas reactions that function far from equilibrium serve as regulatory points and render metabolic pathways irreversible.

5. Flux through a metabolic pathway is controlled by regulating the activities of the enzymes that catalyze its rate-determining steps.

6. The free energy of the "high-energy" compound ATP is made available through cleavage of one or both of its phosphoanhydride bonds.

7. An exergonic reaction such as ATP or PP_i hydrolysis can be coupled to an endergonic reaction to make it more favorable.

8. Substrate-level phosphorylation is the synthesis of ATP from ADP by phosphoryl group transfer from another compound.

9. The common product of carbohydrate, lipid, and protein catabolism, acetyl-CoA, is a "high-energy" thioester.

10. The coenzymes NAD^+ and FAD are reversibly reduced during the oxidation of metabolites.

11. The Nernst equation relates the electromotive force of a redox reaction to the standard reduction potentials and concentrations of the electron donors and acceptors.

12. Electrons flow spontaneously from the reduced member of a redox couple with the more negative reduction potential to the oxidized member of a redox couple with the more positive reduction potential.

13. Studies of metabolic pathways endeavor to determine the order of metabolic transformations, their enzymatic mechanisms, their regulation, and their relationships to metabolic processes in other tissues.

14. Metabolic pathways are studied using isotopic tracers, enzyme inhibitors, natural and engineered mutations, DNA microarrays, and proteomics techniques.

REFERENCES

Alberty, R.A., Calculating apparent equilibrium constants of enzyme-catalyzed reactions at pH 7, *Biochem. Ed.* **28,** 12–17 (2000).

Goodridge, A.G., The new metabolism: molecular genetics in the analysis of metabolic regulation, *FASEB J.* **4,** 3099–3110 (1990).

Hanson, R.W., The role of ATP in metabolism, *Biochem. Ed.* **17,** 86–92 (1989). [Provides an excellent explanation of why ATP is an energy transducer rather than an energy store.]

Harold, F.M., *The Vital Force: A Study of Bioenergetics,* Chapters 1 and 2, Freeman (1986).

Schena, M., *Microarray Analysis,* Wiley-Liss (2003).

Schulman, R.G. and Rothman, D.L., ^{13}C NMR of intermediary metabolism: Implications for systematic physiology, *Annu. Rev. Physiol.* **63,** 15–48 (2001).

Scriver, C.R., Beaudet, A.L., Sly, W.S., and Valle, D., (Eds.), *The Metabolic & Molecular Bases of Inherited Disease* (8th ed.), McGraw-Hill (2001). [Most chapters in this encyclopedic work include a review of a normal metabolic process that is disrupted by disease.]

Westheimer, F.H., Why nature chose phosphates, *Science* **235,** 1173–1178 (1987).

Young, R., Biomedical discovery with DNA arrays, *Cell* **102,** 9–15 (2000).

KEY TERMS

metabolism	oxidation	substrate-level phosphorylation	conjugate redox pair
catabolism	reduction	oxidative phosphorylation	electrochemical cell
anabolism	isozyme	photophosphorylation	$\Delta\mathscr{E}$
autotroph	near-equilibrium reaction	kinase	$\mathscr{F}$
chemolithotroph	substrate cycle	phosphagen	Nernst equation
photoautotroph	flux	reducing agent	$\mathscr{E}°'$
heterotroph	"high-energy" intermediate	oxidizing agent	transcriptomics
aerobic		half-reaction	DNA microarray
anaerobic	orthophosphate cleavage	redox couple	proteomics
metabolite	pyrophosphate cleavage		

STUDY EXERCISES

1. Describe the differences between autotrophs and heterotrophs.

2. Explain the metabolic significance of reactions that function near equilibrium and reactions that function far from equilibrium.

3. Why is ATP a "high-energy" compound?

4. Describe the ways an exergonic process can drive an endergonic process.

5. What is the metabolic role of reduced coenzymes?

6. Explain the terms of the Nernst equation.

7. How is $\Delta\mathscr{E}$ related to ΔG?

8. Describe how information about an organism's genome can be used to assess and manipulate its metabolic activities.

PROBLEMS

1. Rank the following compounds in order of increasing oxidation state.

$$\underset{\textbf{A}}{H_3C-\overset{\displaystyle OH}{\underset{\displaystyle |}{CH}}-CH_2OH} \qquad \underset{\textbf{B}}{^-OOC-CH_2-COO^-}$$

$$\underset{\textbf{C}}{H_3C-CH_2-CH_3} \qquad \underset{\textbf{D}}{H_3C-CH=CH_2} \qquad \underset{\textbf{E}}{H_3C-\overset{\displaystyle O}{\underset{\displaystyle ||}{C}}-COO^-}$$

2. A certain metabolic reaction takes the form A → B. Its standard free energy change is 7.5 kJ · mol^{-1}. (a) Calculate the equilibrium constant for the reaction at 25°C. (b) Calculate ΔG at 37°C when the concentration of A is 0.5 mM and the concentration of B is 0.1 mM. Is the reaction spontaneous under these conditions? (c) How might the reaction proceed in the cell?

3. Choose the best definition for a near-equilibrium reaction:

   (a) always operates with a favorable free energy change.

   (b) has a free energy change near zero.

   (c) is usually a control point in a metabolic pathway.

   (d) operates very slowly *in vivo*.

4. Assuming 100% efficiency of energy conservation, how many moles of ATP can be synthesized under standard conditions by the complete oxidation of (a) 1 mol of glucose and (b) 1 mol of palmitate?

5. Does the magnitude of the free energy change for ATP hydrolysis increase or decrease as the pH increases from 5 to 6?

6. The reaction for "activation" of a fatty acid (RCOO$^-$),

   $$ATP + CoA + RCOO^- \rightleftharpoons RCO-CoA + AMP + PP_i$$

   has $\Delta G^{\circ\prime} = +4.6$ kJ · mol^{-1}. What is the thermodynamic driving force for this reaction?

7. Predict whether creatine kinase will operate in the direction of ATP synthesis or phosphocreatine synthesis at 25°C when [ATP] = 4 mM, [ADP] = 0.15 mM, [phosphocreatine] = 2.5 mM, and [creatine] = 1 mM.

8. If intracellular [ATP] = 5 mM, [ADP] = 0.5 mM, and [P$_i$] = 1.0 mM, calculate the concentration of AMP at pH 7 and 25°C under the condition that the adenylate kinase reaction is at equilibrium.

9. List the following substances in order of their decreasing reducing power: (a) acetoacetate, (b) cytochrome b (Fe^{3+}), (c) NAD$^+$, (d) SO$_4^{2-}$, and (e) pyruvate.

10. Write a balanced equation for the oxidation of ubiquinol by cytochrome c. Calculate $\Delta G^{\circ\prime}$ and $\Delta \mathscr{E}^{\circ\prime}$ for the reaction.

11. Under standard conditions, will the following reactions proceed spontaneously as written?

    (a) Fumarate + NADH + H$^+$ $\rightleftharpoons$ succinate + NAD$^+$

    (b) Cyto a (Fe^{2+}) + cyto b (Fe^{3+}) $\rightleftharpoons$
    cyto a (Fe^{3+}) + cyto b (Fe^{2+})

12. Under standard conditions, is the oxidation of free FADH$_2$ by ubiquinone sufficiently exergonic to drive the synthesis of ATP?

13. A hypothetical three-step metabolic pathway consists of intermediates W, X, Y, and Z and enzymes A, B, and C. Deduce the order of the enzymatic steps in the pathway from the following information:

    1. Compound Q, a metabolic inhibitor of enzyme B, causes Z to build up.

    2. A mutant in enzyme C requires Y for growth.

    3. An inhibitor of enzyme A causes W, Y, and Z to accumulate.

    4. Compound P, a metabolic inhibitor of enzyme C, causes W and Z to build up.

14. A certain metabolic pathway can be diagramed as

    $$A \xrightarrow{\;X\;} B \xrightarrow{\;Y\;} C \xrightarrow{\;Z\;} D$$

    where A, B, C, and D are the intermediates, and X, Y, and Z are the enzymes that catalyze the reactions. The physiological free energy changes for the reactions are

X	−0.2 kJ · mol^{-1}
Y	−12.3 kJ · mol^{-1}
Z	−1.2 kJ · mol^{-1}

    (a) Which reaction is likely to be a major regulatory point for the pathway? (b) If your answer in Part a was in fact the case, in the presence of an inhibitor that blocks the activity of enzyme Z, would the concentrations of A, B, C, and D increase, decrease, or not be affected?

BIOINFORMATICS EXERCISES

Bioinformatics Exercises are available for this chapter in the tabbed section at the end of the book and on the text website at www.wiley.com/college/voet.

During a short race, for example, of 100 to 200 m, a major source of power for a runner's muscles is provided by the free energy produced through anaerobic glycolysis, a catabolic pathway that breaks down carbohydrates and produces ATP but does not depend on the presence of oxygen. Even under aerobic conditions, glycolysis is the major starting point for carbohydrate metabolism. [AFLO Foto/Alamy Images.]

Glucose Catabolism

The fermentation (anaerobic breakdown) of glucose to ethanol and CO_2 by yeast has been exploited for many centuries in baking and winemaking. However, scientific investigation of the chemistry of this catabolic pathway began only in the mid-nineteenth century, with the experiments of Louis Pasteur and others. Nearly a century would pass before the complete pathway was elucidated. During that interval, several important features of the pathway came to light:

1. In 1905, Arthur Harden and William Young discovered that phosphate is required for glucose fermentation.

2. Certain reagents, such as iodoacetic acid and fluoride ion, inhibit the formation of pathway products, thereby causing pathway intermediates to accumulate. Different substances caused the buildup of different intermediates and thereby revealed the sequence of molecular interconversions.

3. Studies of how different organisms break down glucose indicated that, with few exceptions, all of them do so the same way.

The efforts of many investigators came to fruition in 1940, when the complete pathway of glucose breakdown was described. This pathway, which is named **glycolysis** (Greek: *glykus,* sweet + *lysis,* loosening), is alternately known as the **Embden–Meyerhof–Parnas pathway** to commemorate the work of Gustav Embden, Otto Meyerhof, and Jacob Parnas in its elucidation. The discovery of glycolysis came at a time when other significant inroads were being made in the area of metabolism (Box 14-1).

Glycolysis, which is probably the most completely understood biochemical pathway, is a sequence of 10 enzymatic reactions in which one molecule of glucose is converted to two molecules of the three-carbon compound pyruvate with the concomitant generation of 2 ATP. It plays a key role in energy metabolism by providing a significant portion of the free energy used by most organisms and by preparing glucose and other compounds for further oxidative degradation. Thus, it is fitting that we begin our discussion of specific metabolic pathways by considering glycolysis. We shall examine the sequence of reactions by which glucose is degraded, along with some of the relevant enzyme mechanisms. We will then examine the features that influence glycolytic flux and the ultimate fate of its products. Finally, we will discuss the catabolism of other hexoses and the **pentose phosphate pathway,** an alternative pathway for glucose catabolism that functions to provide biosynthetic precursors.

See Guided Exploration 12

Glycolysis overview.

---

1 Overview of Glycolysis

Before beginning our detailed discussion of glycolysis, let us first take a moment to survey the overall pathway as it fits in with animal metabolism as a whole. Glucose usually appears in the blood as a result of the breakdown of polysaccharides (e.g., liver glycogen or dietary starch and glycogen) or from its synthesis from noncarbohydrate precursors (**gluconeogenesis;** Section 15-4). Glucose enters most cells by a specific carrier that transports it from the exterior of the cell into the cytosol (Section 10-2E). The enzymes of glycolysis are located in the cytosol, where they are only loosely associated, if at all, with each other or with other cell structures.

Glycolysis converts glucose to two C_3 units (pyruvate). The free energy released in this process is harvested to synthesize ATP from ADP and P_i. Thus, glycolysis is a pathway of chemically coupled phosphorylation reactions (Section 13-2B). The 10 reactions of glycolysis are diagrammed in

BOX 14-1

Pathways of Discovery

Otto Warburg and Studies of Metabolism

Otto Warburg (1883–1970)

One of the great figures in biochemistry—by virtue of his own contributions and his influence on younger researchers—is the German biochemist Otto Warburg. His long career spanned a period during which studies of whole organisms and crude extracts gave way to molecular explanations of biological structure and function. Like others of his generation, he earned a doctorate in chemistry at an early age and went on to obtain a medical degree, although he spent the remainder of his career in scientific research rather than in patient care. He became interested primarily in three subjects related to the chemistry of oxygen and carbon dioxide: respiration, photosynthesis, and cancer.

One of Warburg's first accomplishments was to develop a technique for studying metabolic reactions in thin slices of animal tissue. This method produced more reliable results than the alternative practice of chopping or mincing tissues (such manipulations tend to release lysosomal enzymes that degrade enzymes and other macromolecules). Warburg was also largely responsible for refining manometry, the measurement of gas pressure, as a technique for analyzing the consumption and production of O_2 and CO_2 by living tissues.

Warburg received a Nobel prize in 1931 for his discovery of the catalytic role of iron porphyrins (heme groups) in biological oxidation (the subject was the reaction carried out by the enzyme complex now known as cytochrome oxidase; Section 17-2F). Warburg also identified nicotinamide as an active part of some enzymes. In 1944, he was offered a second Nobel prize for his work with enzymes, but he was unable to accept the award, owing to Hitler's decree that Germans could not accept Nobel prizes. In fact, Warburg's apparent allegiance to the Nazi regime incensed some of his colleagues in other countries and may have contributed to their resistance to some of his more controversial scientific pronouncements. In any case, Warburg was not known for his warm personality. He was never a teacher and tended to recruit younger research assistants who were expected to move on after a few years.

In addition to the techniques he developed, which were widely adopted, and a number of insights into enzyme action, Warburg formulated some wide-reaching theories about the growth of cancer cells. He showed that cancer cells could live and develop even in the absence of oxygen. Moreover, he came to believe that anaerobiosis triggered the development of cancer, and he rejected the notion that viruses could cause cancer, a principle that had already been demonstrated in animals but not in humans. In the eyes of many, Warburg was guilty of equating the absence of evidence with the evidence of absence in the matter of virus-induced human cancer. Nevertheless, Warburg's observations of cancer cell metabolism, which is generally characterized by a high rate of glycolysis, were sound. Even today, the oddities of tumor metabolism offer opportunities for chemotherapy. Warburg's dedication to his research in cancer and other areas is revealed by the fact that he continued working in his laboratory until just a few days before his death at age 87.

Warburg, O., On the origin of cancer cells, *Science* **123**, 309–314 (1956).

Fig. 14-1. Note that ATP is used early in the pathway to synthesize phosphorylated compounds (Reactions 1 and 3) but is later resynthesized twice over (Reactions 7 and 10). Glycolysis can therefore be divided into two stages:

Stage I   Energy investment (Reactions 1–5). In this preparatory stage, the hexose glucose is phosphorylated and cleaved to yield two molecules of the triose **glyceraldehyde-3-phosphate.** This process consumes 2 ATP.

Stage II   Energy recovery (Reactions 6–10). The two molecules of glyceraldehyde-3-phosphate are converted to pyruvate, with concomitant generation of 4 ATP. Glycolysis therefore has a net "profit" of 2 ATP per glucose: Stage I consumes 2 ATP; Stage II produces 4 ATP.

The phosphoryl groups that are initially transferred from ATP to the hexose do not immediately result in "high-energy" compounds. However, subsequent enzymatic transformations convert these "low-energy" prod-

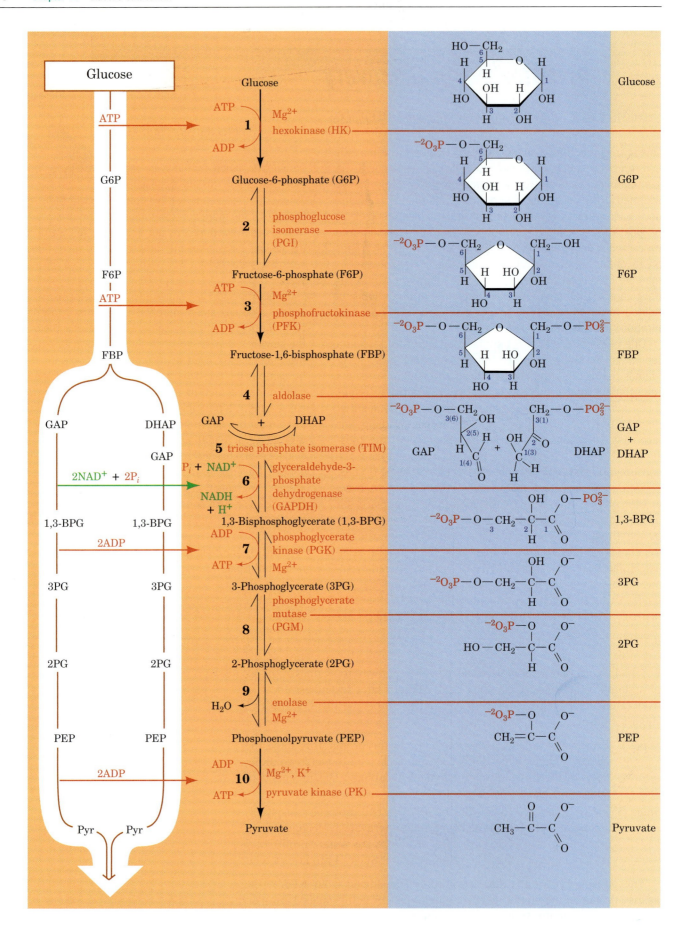

Figure 14-1 *(opposite)* ***Key to Metabolism.*** **Glycolysis.** In its first stage (Reactions 1–5), one molecule of glucose is converted to two glyceraldehyde-3-phosphate molecules in a series of reactions that consumes 2 ATP. In the second stage of glycolysis (Reactions 6–10), the two glyceraldehyde-3-phosphate molecules are converted to two pyruvate molecules, generating 4 ATP and 2 NADH. *See the Animated Figures.*

ucts to compounds with high phosphoryl group-transfer potentials, which are capable of phosphorylating ADP to form ATP. The overall reaction is

$$\text{Glucose} + 2 \text{ NAD}^+ + 2 \text{ ADP} + 2 \text{ P}_i \longrightarrow$$
$$2 \text{ pyruvate} + 2 \text{ NADH} + 2 \text{ ATP} + 2 \text{ H}_2\text{O} + 4 \text{ H}^+$$

Hence, the NADH formed in the process must be continually reoxidized to keep the pathway supplied with its primary oxidizing agent, NAD^+. In Section 14-3, we shall examine how organisms do so under aerobic or anaerobic conditions.

2 The Reactions of Glycolysis

In this section, we examine the reactions of glycolysis more closely, describing the properties of the individual enzymes and their mechanisms. As we study the individual glycolytic enzymes, we shall encounter many of the catalytic mechanisms described in Section 11-3.

A Hexokinase: First Use of ATP

Reaction 1 of glycolysis is the transfer of a phosphoryl group from ATP to glucose to form **glucose-6-phosphate (G6P)** in a reaction catalyzed by **hexokinase** *(at right)*.

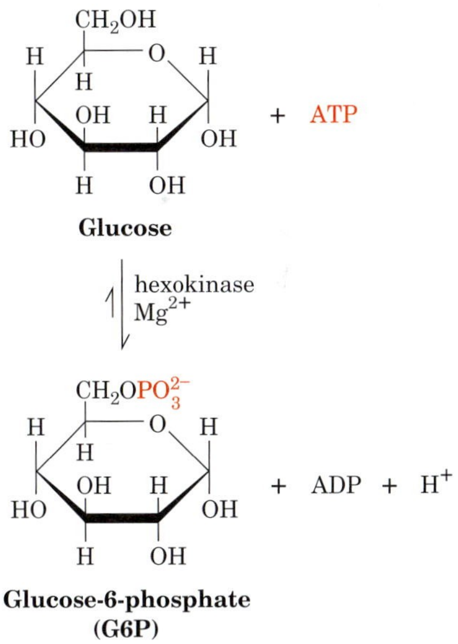

A kinase is an enzyme that transfers phosphoryl groups between ATP and a metabolite (Section 13-2C). The metabolite that serves as the phosphoryl group acceptor is indicated in the prefix of the kinase name. Hexokinase is a ubiquitous, relatively nonspecific enzyme that catalyzes the phosphorylation of hexoses such as D-glucose, D-mannose, and D-fructose. Liver cells also contain the isozyme **glucokinase,** which catalyzes the same reaction but which is primarily involved in maintaining blood glucose levels (Section 21-1D).

The second substrate for hexokinase, as for other kinases, is an Mg^{2+}–ATP complex. In fact, uncomplexed ATP is a potent competitive inhibitor of hexokinase. Although we do not always explicitly mention the participation of Mg^{2+}, it is essential for kinase activity. The Mg^{2+} shields the negative charges of the ATP's α- and β- or β- and γ-phosphate oxygen atoms, making the γ-phosphorus atom more accessible for nucleophilic attack by the C6-OH group of glucose:

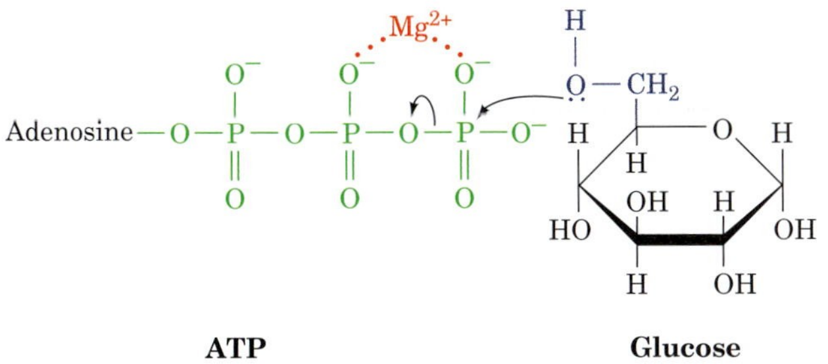

Glucose

hexokinase
Mg^{2+}

Glucose-6-phosphate (G6P)

ATP   **Glucose**

Comparison of the X-ray structures of yeast hexokinase and the glucose–hexokinase complex indicates that *glucose induces a large conformational*

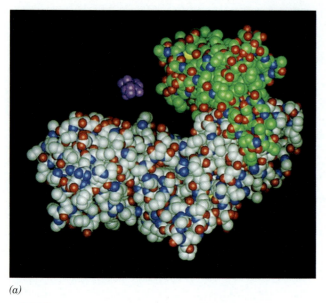

(a)

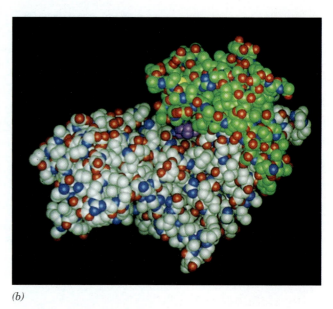

(b)

Figure 14-2 Substrate-induced conformational changes in yeast hexokinase. (*a*) Space-filling model of a hexokinase subunit showing the prominent bilobal appearance of the free enzyme (the C atoms in the small lobe are shaded green, and those in the large lobe are light gray; the N and O atoms are blue and red). (*b*) Model of the hexokinase complex with glucose (*magenta*). The lobes have swung together to engulf the substrate. [Based on X-ray structures by Thomas Steitz, Yale University. PDBids (*a*) 2 YHX and (*b*) 1 HKG.] 🔗 **See the Interactive Exercises.**

change in hexokinase (Fig. 14-2). The two lobes that form its active site cleft swing together by up to 8 Å so as to engulf the glucose in a manner that suggests the closing of jaws. *This movement places the ATP close to the —C6H₂OH group of glucose and excludes water from the active site (catalysis by proximity effects;* Section 11-3D). If the catalytic and reacting groups were in the proper position for reaction while the enzyme was in the open position (Fig. 14-2*a*), ATP hydrolysis (i.e., phosphoryl group transfer to water, which is thermodynamically favored; Fig. 13-5*a*) would almost certainly be the dominant reaction.

Clearly, the substrate-induced conformational change in hexokinase is responsible for the enzyme's specificity. In addition, the active site polarity is reduced by exclusion of water, thereby expediting the nucleophilic reaction process. Other kinases have the same deeply clefted structure as hexokinase and undergo conformational changes on binding their substrates (e.g., adenylate kinase; Fig. 13-8).

B Phosphoglucose Isomerase

Reaction 2 of glycolysis is the conversion of G6P to **fructose-6-phosphate (F6P)** by **phosphoglucose isomerase (PGI;** *at left*). This is the isomerization of an aldose to a ketose.

Since G6P and F6P both exist predominantly in their cyclic forms, the reaction requires ring opening followed by isomerization and subsequent ring closure (the interconversions of cyclic and linear forms of hexoses are shown in Fig. 8-3).

A proposed reaction mechanism for the PGI reaction involves general acid–base catalysis by the enzyme (Fig. 14-3):

Step 1 The substrate binds.

Step 2 An enzymatic acid, probably the ε-amino group of a conserved Lys residue, catalyzes ring opening.

Step 3 A base, thought to be a conserved His residue, abstracts the acidic

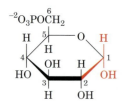

Glucose-6-phosphate (G6P)

phosphoglucose isomerase (PGI)

$^{-2}O_3POCH_2$ ... CH_2OH

Fructose-6-phosphate (F6P)

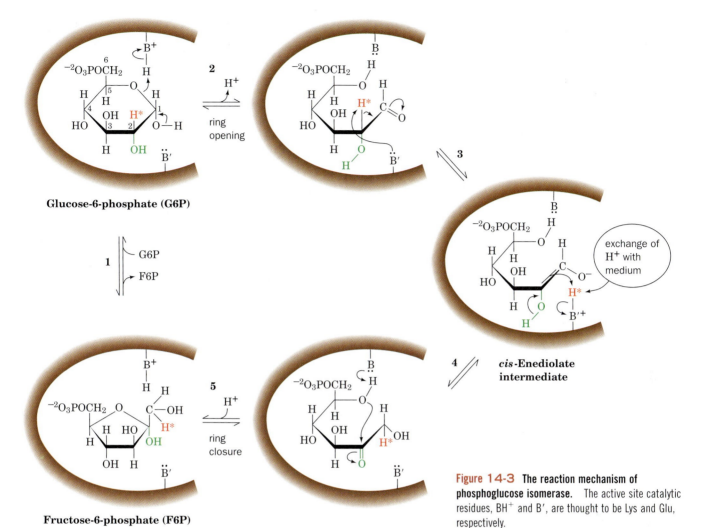

Figure 14-3   The reaction mechanism of phosphoglucose isomerase.   The active site catalytic residues, BH$^+$ and B′, are thought to be Lys and Glu, respectively.

proton from C2 to form a *cis*-enediolate intermediate (this proton is acidic because it is α to a carbonyl group).

Step 4   The proton is replaced on C1 in an overall proton transfer. Protons abstracted by bases rapidly exchange with solvent protons. Nevertheless, Irwin Rose confirmed this step by demonstrating that 2-[^{3}H]G6P is occasionally converted to 1-[^{3}H]F6P by intramolecular proton transfer before the ^{3}H has had a chance to exchange with the medium.

Step 5   The ring closes to form the product, which is subsequently released to yield free enzyme, thereby completing the catalytic cycle.

C   Phosphofructokinase: Second Use of ATP

In Reaction 3 of glycolysis, **phosphofructokinase (PFK)** phosphorylates F6P to yield **fructose-1,6-bisphosphate** (**FBP** or **F1,6P**; *at right*). (The product is a *bis*phosphate rather than a *di*phosphate because its two phosphate groups are not attached directly to each other.)

The PFK reaction is similar to the hexokinase reaction. The enzyme catalyzes the nucleophilic attack by the C1-OH group of F6P on the electrophilic γ-phosphorus atom of the Mg^{2+}–ATP complex.

Phosphofructokinase plays a central role in control of glycolysis because it catalyzes one of the pathway's rate-determining reactions. In many organisms, the activity of PFK is enhanced allosterically by several substances, including

Fructose-6-phosphate
(F6P)

Fructose-1,6-bisphosphate
(FBP)

AMP, and inhibited allosterically by several other substances, including ATP and citrate. The regulatory properties of PFK are examined in Section 14-4A.

D Aldolase

Aldolase catalyzes Reaction 4 of glycolysis, the cleavage of FBP to form the two trioses **glyceraldehyde-3-phosphate (GAP)** and **dihydroxyacetone phosphate (DHAP):**

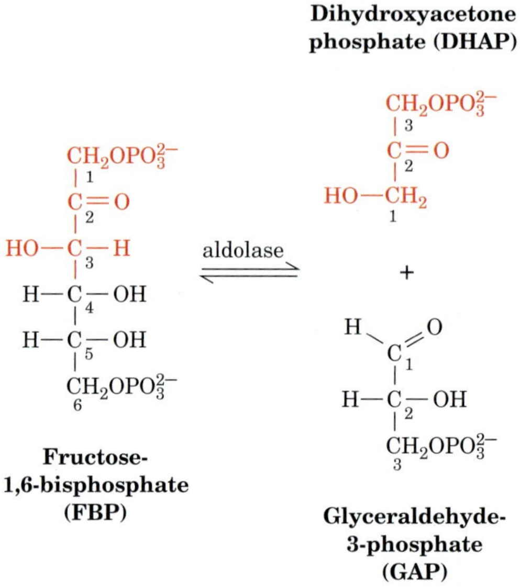

Note that at this point in the pathway, the atom numbering system changes. Atoms 1, 2, and 3 of glucose become atoms 3, 2, and 1 of DHAP, thus reversing order. Atoms 4, 5, and 6 become atoms 1, 2, and 3 of GAP.

Reaction 4 is an **aldol cleavage (retro aldol condensation)** whose non-enzymatic base-catalyzed mechanism is shown in Fig. 14-4. The **enolate** intermediate is stabilized by resonance, as a result of the electron-withdrawing character of the carbonyl oxygen atom. Note that aldol cleavage between C3 and C4 of FBP requires a carbonyl at C2 and a hydroxyl at C4. Hence, the "logic" of Reaction 2 in the glycolytic pathway, the isomerization of G6P to F6P, is clear. Aldol cleavage of G6P would yield products of unequal carbon chain length, while *aldol cleavage of FBP results in two interconvertible C$_3$ compounds that can therefore enter a common degradative pathway.*

Aldol cleavage is catalyzed by stabilizing its enolate intermediate through increased electron delocalization. In animals and plants, the reaction occurs as follows (Fig. 14-5):

Step 1 Substrate binding.

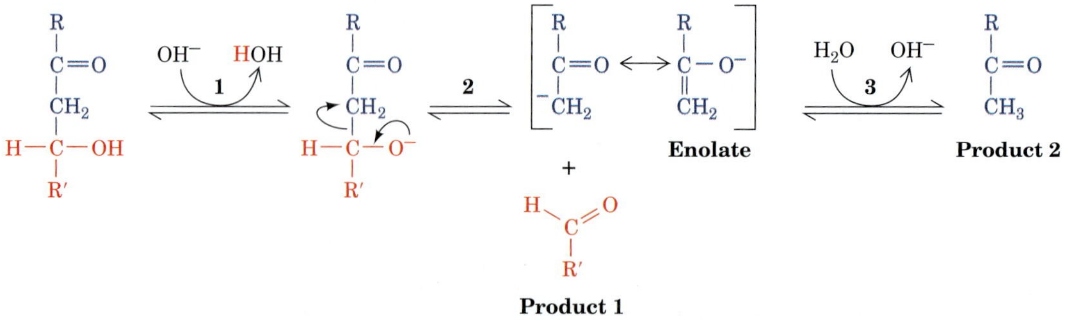

Product 1

Figure 14-4 The mechanism of base-catalyzed aldol cleavage.   Aldol condensation occurs by the reverse mechanism.

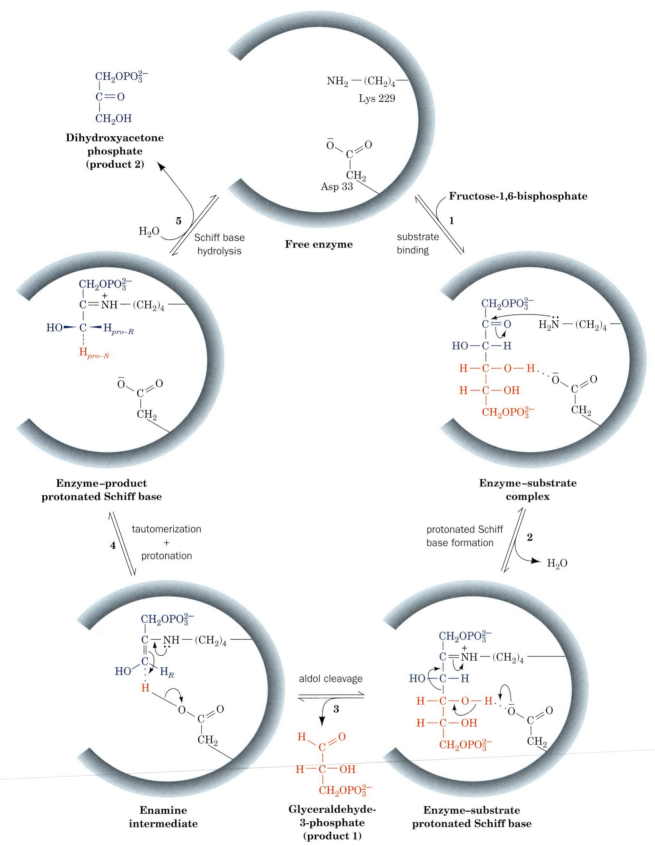

Figure 14-5 The enzymatic mechanism of aldolase.   The reaction involves (1) substrate binding; (2) Schiff base (imine) formation between the enzyme's active site Lys residue and the open-chain form of FBP; (3) aldol cleavage to form an enamine intermediate of the enzyme and DHAP, with release of GAP; (4) tautomerization and protonation to the iminium form of the Schiff base; and (5) hydrolysis of the Schiff base with release of DHAP. ✐ **See the Animated Figures.**

Step 2 Reaction of the FBP carbonyl group with the ε-amino group of the active site Lys to form an iminium cation, that is, a protonated Schiff base.

Step 3 C3—C4 bond cleavage resulting in enamine formation and the release of GAP. The iminium ion is a better electron-withdrawing group than the oxygen atom of the precursor carbonyl group. Thus, catalysis occurs because the enamine intermediate (Fig. 14-5, Step 3) is more stable than the corresponding enolate intermediate of the base-catalyzed aldol cleavage reaction (Fig. 14-4, Step 2).

Step 4 Protonation of the enamine to an iminium cation.

Step 5 Hydrolysis of this iminium cation to release DHAP, with regeneration of the free enzyme.

E Triose Phosphate Isomerase

Only one of the products of the aldol cleavage reaction, GAP, continues along the glycolytic pathway (Fig. 14-1). However, DHAP and GAP are ketose–aldose isomers (like F6P and G6P). They are interconverted by an isomerization reaction with an **enediol** (or **enediolate**) **intermediate. Triose phosphate isomerase (TIM)** catalyzes this process in Reaction 5 of glycolysis, the final reaction of Stage I:

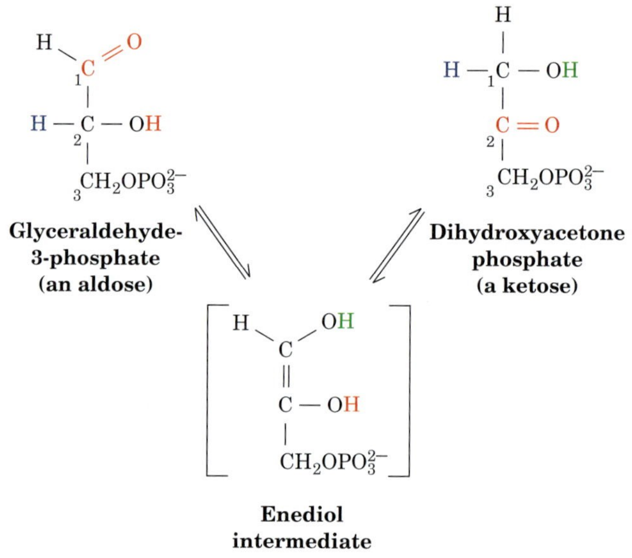

Glyceraldehyde-3-phosphate (an aldose)

Dihydroxyacetone phosphate (a ketose)

Enediol intermediate

Support for this reaction scheme comes from the use of the transition state analogs **phosphoglycohydroxamate** and **2-phosphoglycolate,** stable compounds whose geometry resembles that of the proposed enediol or enediolate intermediate:

Phosphoglyco-hydroxamate

Proposed enediolate intermediate

2-Phosphoglycolate

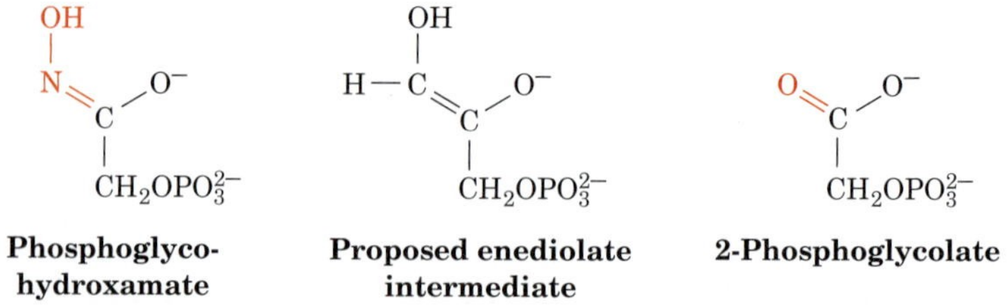

Enzymes catalyze reactions by binding the transition state complex more tightly than the substrate (Section 11-3E), and, in fact, phosphoglycohy-

droxamate and 2-phosphoglycolate bind 155- and 100-fold more tightly to TIM than does either GAP or DHAP.

Glu 165 and His 95 Act as General Acids and Bases.
Mechanistic considerations suggest that the conversion of GAP to the enediol intermediate is catalyzed by a general base, which abstracts a proton from C2 of GAP, and by a general acid, which protonates its carbonyl oxygen atom. X-Ray studies reveal that the Glu 165 side chain is ideally situated to abstract the C2 proton from GAP (Fig. 14-6). In fact, the mutagenic replacement of Glu 165 by Asp, which X-ray studies show withdraws the carboxylate group only ~1 Å farther away from the substrate than its position in the wild-type enzyme, reduces TIM's catalytic activity 1000-fold. X-Ray studies similarly indicate that His 95 is hydrogen bonded to and hence is properly positioned to protonate GAP's carbonyl oxygen. The positively charged side chain of Lys 12 is thought to electrostatically stabilize the negatively charged transition state in the reaction. In the conversion of the enediol intermediate to DHAP, Glu 165 acts as a general acid to protonate C1 and His 95 acts as a general base to abstract the proton from the OH group, thereby restoring these catalytic groups to their initial protonation states.

A Flexible Loop Closes over the Active Site.
The comparison of the X-ray structure of TIM (Fig. 6-30c) with that of the enzyme–phosphoglycohydroxamate complex reveals that when substrate binds to TIM, a conserved 10-residue loop closes over the active site like a hinged lid, in a movement that involves main chain shifts of >7 Å (Fig. 14-6). A four-residue segment of this loop makes a hydrogen bond with the phosphate group of the substrate. Mutagenic excision of these four residues does not significantly distort the protein, so substrate binding is not greatly impaired. However, the catalytic power of the mutant enzyme is reduced 10^5-fold, and it only weakly binds phosphoglycohydroxamate. Evidently, loop closure preferentially stabilizes the enzymatic reaction's enediol-like transition state.

Loop closure in the TIM reaction also supplies a striking example of the so-called **stereoelectronic control** that enzymes can exert on a reaction. In solution, the enediol intermediate readily breaks down with the elimination of the phosphate at C3 to form the toxic compound **methylglyoxal:**

$$
\begin{array}{c}
\text{O} \\
\parallel \\
\text{H}-\text{C} \\
\diagdown \\
\text{C}=\text{O} \\
\diagup \\
\text{H}_3\text{C}
\end{array}
$$

Methylglyoxal

On the enzyme's surface, however, this reaction is prevented because the phosphate group is held by the flexible loop in a position that disfavors phosphate elimination. In the mutant enzyme lacking the flexible loop, the enediol is able to escape: ~85% of the enediol intermediate is released into solution where it rapidly decomposes to methylglyoxal and P_i. Thus, the flexible loop closure assures that substrate is efficiently transformed to product.

α/β Barrel Enzymes May Have Evolved by Divergent Evolution.
TIM was the first protein found to contain an α/β barrel (also known as a TIM barrel), a cylinder of eight parallel β strands surrounded by eight parallel α helices (Fig. 6-30c). This striking structural motif has since been found in numerous different proteins, essentially all of which are enzymes (including the

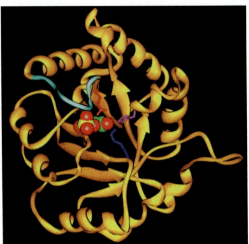

Figure 14-6   A ribbon diagram of yeast TIM in complex with its transition state analog 2-phosphoglycolate. A single subunit of this homodimeric enzyme is viewed roughly along the axis of its α/β barrel. The enzyme's flexible loop is cyan, and the side chains of the catalytic Lys, His, and Glu residues are blue, magenta, and red, respectively. The 2-phosphoglycolate is represented by a space-filling model colored according to atom type (C, green; O, red; P, yellow). [Based on an X-ray structure by Gregory Petsko, Brandeis University. PDBid 2YPI.]
See the Interactive Exercises and Kinemage Exercises 12-1 and 12-2.

glycolytic enzymes aldolase, enolase, and pyruvate kinase). Intriguingly, the active sites of all known α/β barrel enzymes are located in the mouth of the barrel at the end that contains the C-terminal ends of the β strands, although there is no obvious structural rationale for this. Despite the fact that few of these proteins exhibit significant sequence similarity, it has been postulated that all of them have evolved from a common ancestor (divergent evolution). However, it has also been argued that the α/β barrel is a particularly stable arrangement that nature has independently discovered on several occasions (convergent evolution).

Triose Phosphate Isomerase Is a Catalytically Perfect Enzyme.
Jeremy Knowles has demonstrated that TIM has achieved **catalytic perfection.** This means that the rate of the bimolecular reaction between enzyme and substrate is diffusion controlled, so product formation occurs as rapidly as enzyme and substrate can collide in solution. Any increase in TIM's catalytic efficiency therefore would not increase its reaction rate.

GAP and DHAP are interconverted so efficiently that the concentrations of these two metabolites are maintained at their equilibrium values: $K = [GAP]/[DHAP] = 4.73 \times 10^{-2}$. At equilibrium, $[DHAP] \gg [GAP]$. However, under the steady state conditions in a cell, GAP is consumed in the succeeding reactions of the glycolytic pathway. *As GAP is siphoned off in this manner, more DHAP is converted to GAP to maintain the equilibrium ratio.* In effect, DHAP follows GAP into the second stage of glycolysis, so a single pathway accounts for the metabolism of both products of the aldolase reaction.

Taking Stock of Glycolysis So Far.
At this point in the glycolytic pathway, one molecule of glucose has been transformed into two molecules of GAP. This completes the first stage of glycolysis (Fig. 14-7). Note that 2 ATP have been consumed in generating the phosphorylated intermediates. This energy investment has not yet paid off, but with a little chemical artistry, the

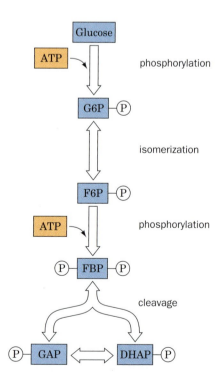

Figure 14-7 Schematic diagram of the first stage of glycolysis.   In this series of reactions, a hexose is phosphorylated, isomerized, phosphorylated again, and then cleaved to two interconvertible triose phosphates. Two ATP are consumed in the process.

"low-energy" GAP can be converted to "high-energy" compounds whose free energies of hydrolysis can be coupled to ATP synthesis in the second stage of glycolysis.

F   Glyceraldehyde-3-Phosphate Dehydrogenase: First "High-Energy" Intermediate Formation

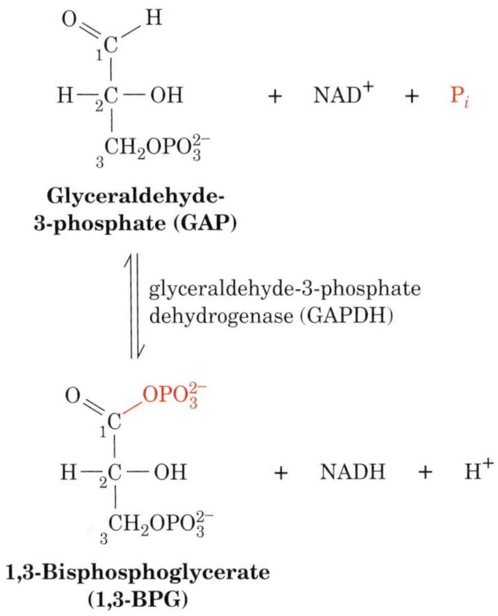

Reaction 6 of glycolysis is the oxidation and phosphorylation of GAP by NAD$^+$ and P$_i$ as catalyzed by **glyceraldehyde-3-phosphate dehydrogenase** (**GAPDH;** *at right;* Fig. 6-31). This is the first instance of the chemical artistry alluded to above. *In this reaction, aldehyde oxidation, an exergonic reaction, drives the synthesis of the "high-energy" acyl phosphate **1,3-bisphosphogly-cerate (1,3-BPG)**.* Recall that acyl phosphates are compounds with high phosphoryl group-transfer potential (Section 13-2C).

Mechanistic Studies. Several key enzymological experiments have contributed to the elucidation of the GAPDH reaction mechanism:

1. GAPDH is inactivated by alkylation with stoichiometric amounts of iodoacetate. The presence of **carboxymethylcysteine** in the hydrolysate of the resulting alkylated enzyme (Fig. 14-8*a*) suggests that GAPDH has an active site Cys sulfhydryl group.
2. GAPDH quantitatively transfers ^{3}H from C1 of GAP to NAD$^+$ (Fig. 14-8*b*), thereby establishing that this reaction occurs via direct hydride transfer.

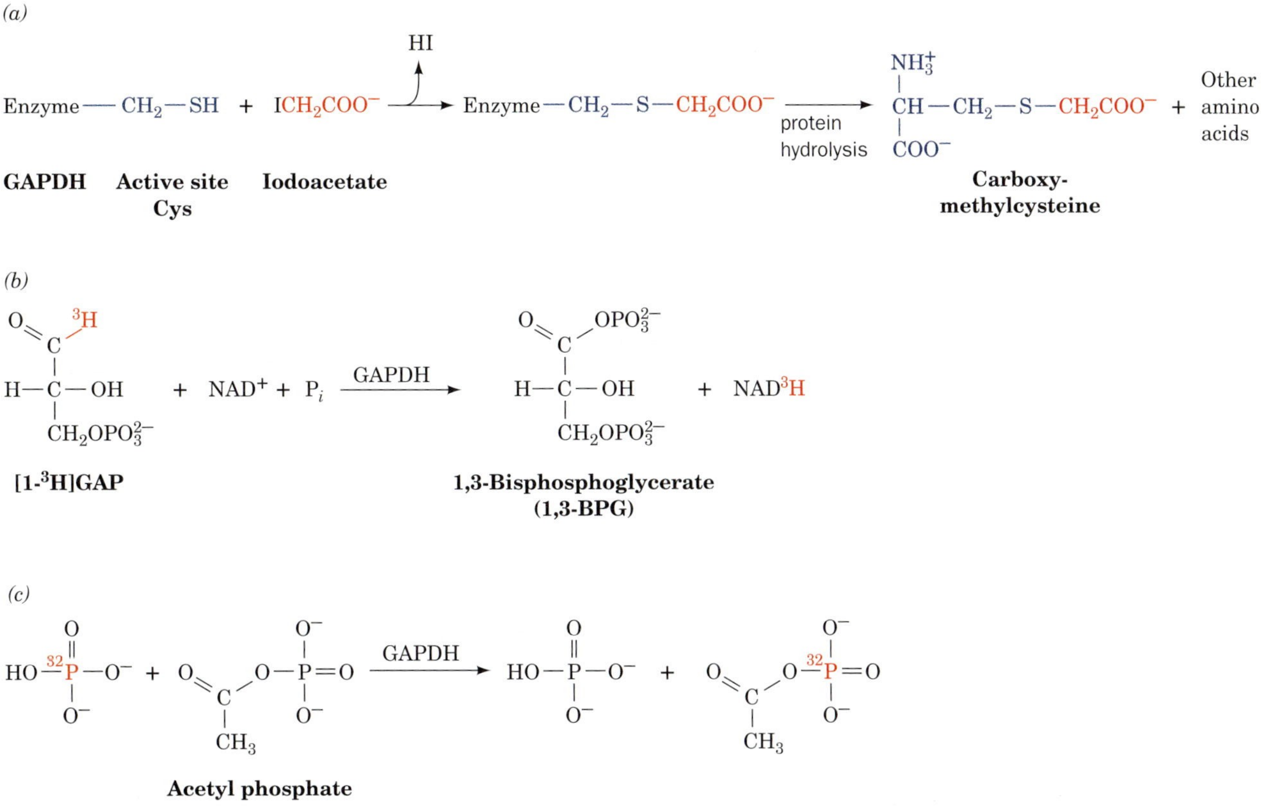

Glyceraldehyde-3-phosphate (GAP)

glyceraldehyde-3-phosphate dehydrogenase (GAPDH)

1,3-Bisphosphoglycerate (1,3-BPG)

(a)

(b)

(c)

Acetyl phosphate

Figure 14-8   Reactions that were used to elucidate the enzymatic mechanism of GAPDH. (*a*) The reaction of iodoacetate with an active site Cys residue. (*b*) Quantitative tritium transfer from substrate to NAD$^+$. (*c*) The enzyme-catalyzed exchange of ^{32}P from phosphate to acetyl phosphate.

3. GAPDH catalyzes exchange of ^{32}P between P_i and the product analog **acetyl phosphate** (Fig. 14-8c). Such isotope exchange reactions are indicative of an acyl–enzyme intermediate; that is, the acetyl group forms a covalent complex with the enzyme, similar to the acyl–enzyme intermediate in the serine protease reaction mechanism (Section 11-5C).

David Trentham has proposed a mechanism for GAPDH based on this information and the results of kinetic studies (Fig. 14-9):

Step 1 GAP binds to the enzyme.

Step 2 The essential sulfhydryl group, acting as a nucleophile, attacks the aldehyde to form a **thiohemiacetal.**

Step 3 The thiohemiacetal undergoes oxidation to an **acyl thioester** by direct hydride transfer to NAD$^+$. This intermediate, which has been isolated, has a large free energy of hydrolysis. Thus, *the energy of aldehyde oxidation has not been dissipated but has been conserved through the synthesis of the thioester and the reduction of NAD$^+$ to NADH.*

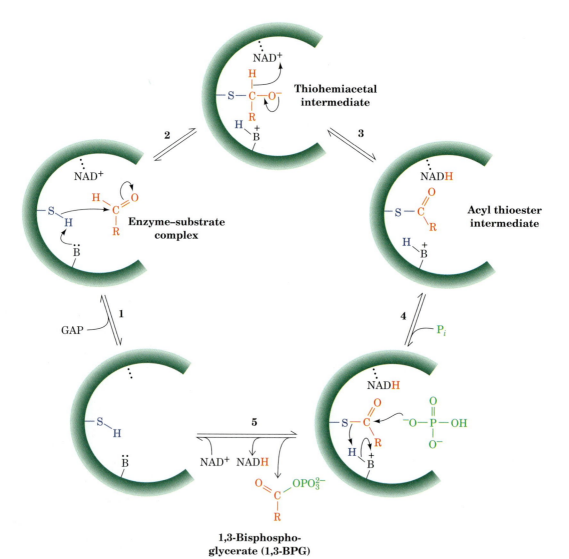

Figure 14-9 The enzymatic mechanism of GAPDH. (1) GAP binds to the enzyme; (2) the active site sulfhydryl group forms a thiohemiacetal with the substrate; (3) NAD$^+$ oxidizes the thiohemiacetal to a thioester; (4) P_i binds to the enzyme; and (5) attacks the thioester, forming the acyl phosphate product, 1,3-BPG, which dissociates from the enzyme followed by the replacement of the newly formed NADH by NAD$^+$, thereby regenerating the active enzyme. ♪ **See the Animated Figures.**

Step 4 P_i binds to the enzyme–thioester–NADH complex.

Step 5 The thioester intermediate undergoes nucleophilic attack by P_i to form the "high-energy" mixed anhydride 1,3-BPG, which then dissociates from the enzyme followed by replacement NADH by another molecule of NAD^+ to regenerate the active enzyme.

G Phosphoglycerate Kinase: First ATP Generation

Reaction 7 of the glycolytic pathway yields ATP together with **3-phosphoglycerate (3PG)** in a reaction catalyzed by **phosphoglycerate kinase (PGK;** *at right*). (Note that this enzyme is called a "kinase" because the reverse reaction is phosphoryl group transfer from ATP to 3PG.)

PGK (Fig. 14-10) is conspicuously bilobal in appearance. The Mg^{2+}–ADP binding site is located on one domain, ~10 Å from the 1,3-BPG binding site, which is on the other domain. Physical measurements suggest that, on substrate binding, the two domains of PGK swing together to permit the substrates to react in a water-free environment, as occurs in hexokinase (Section 14-2A). Indeed, the appearance of PGK is remarkably similar to that of hexokinase (Fig. 14-2), even though the structures of these proteins are otherwise unrelated.

Coupling between the GAPDH and PGK Reactions. As described in Section 13-2B, a slightly unfavorable reaction can be coupled to a highly favorable reaction so that both reactions proceed in the forward direction. In the case of the sixth and seventh reactions of glycolysis, *1,3-BPG is the common intermediate whose consumption in the PGK reaction "pulls" the GAPDH reaction forward.* The energetics of the overall reaction pair are

$$GAP + P_i + NAD^+ \longrightarrow 1,3\text{-BPG} + NADH \qquad \Delta G^{\circ\prime} = +6.7 \text{ kJ} \cdot \text{mol}^{-1}$$

$$1,3\text{-BPG} + ADP \longrightarrow 3PG + ATP \qquad \Delta G^{\circ\prime} = -18.8 \text{ kJ} \cdot \text{mol}^{-1}$$

$$GAP + P_i + NAD^+ + ADP \longrightarrow 3PG + NADH + ATP$$
$$\Delta G^{\circ\prime} = -12.1 \text{ kJ} \cdot \text{mol}^{-1}$$

Although the GAPDH reaction is endergonic, the strongly exergonic nature of the transfer of a phosphoryl group from 1,3-BPG to ADP makes the overall synthesis of NADH and ATP from GAP, P_i, NAD^+, and ADP favorable. *This production of ATP, which does not involve O_2, is an example of substrate-level phosphorylation.* The subsequent oxidation of the NADH produced in this reaction by O_2 generates additional ATP by oxidative phosphorylation, as we shall see in Section 17-3.

H Phosphoglycerate Mutase

In Reaction 8 of glycolysis, 3PG is converted to **2-phosphoglycerate (2PG)** by **phosphoglycerate mutase (PGM):**

1,3-Bisphosphoglycerate
(1,3-BPG)

$$Mg^{2+} \qquad \text{phosphoglycerate kinase (PGK)}$$

3-Phosphoglycerate
(3PG)

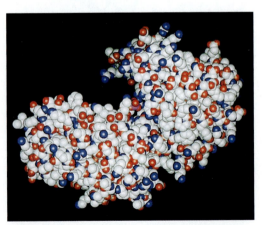

Figure 14-10 A space-filling model of yeast phosphoglycerate kinase. The substrate-binding site is at the bottom of a deep cleft between the two lobes of the protein. This site is marked by the P atom (*magenta*) of 3PG. Compare this structure with that of hexokinase (Fig. 14-2*a*). [Based on an X-ray structure by Herman Watson, University of Bristol, U.K. PDBid 3PGK.]

Phospho-His residue

A **mutase** catalyzes the transfer of a functional group from one position to another on a molecule. This more or less energetically neutral reaction is necessary preparation for the next reaction in glycolysis, which generates a "high-energy" phosphoryl compound.

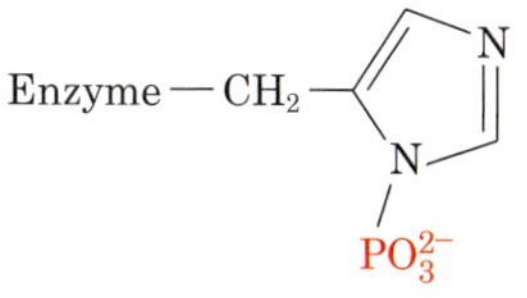

Reaction Mechanism of Phosphoglycerate Mutase.   At first sight, the reaction catalyzed by phosphoglycerate mutase appears to be a simple intramolecular phosphoryl group transfer. This is not the case, however. The active enzyme has a phosphoryl group at its active site, attached to His 8 *(at left)*. The phosphoryl group is transferred to the substrate to form a bisphospho intermediate. This intermediate then rephosphorylates the enzyme to form the product and regenerate the active phosphoenzyme. The enzyme's X-ray structure shows the proximity of His 8 to the substrate (Fig. 14-11).

Catalysis by phosphoglycerate mutase occurs as follows (Fig. 14-12):

Step 1   3PG binds to the phosphoenzyme in which His 8 is phosphorylated.

Step 2   The enzyme's phosphoryl group is transferred to the substrate, resulting in an intermediate 2,3-bisphosphoglycerate–enzyme complex.

Steps 3 & 4   The complex decomposes to form the product 2PG and regenerate the phosphoenzyme.

The phosphoryl group of 3PG therefore ends up on C2 of the next 3PG to undergo reaction.

Occasionally, 2,3-bisphosphoglycerate (2,3-BPG) formed in Step 2 of the reaction dissociates from the dephosphoenzyme, leaving it in an inactive form. Trace amounts of 2,3-BPG must therefore always be available to regenerate the active phosphoenzyme by the reverse reaction. 2,3-BPG also specifically binds to deoxyhemoglobin, thereby decreasing its oxygen affinity (Section 7-3A). Consequently, erythrocytes require much more 2,3-BPG (5 mM) than the trace amounts that are used to prime phosphoglycerate mutase (see Box 14-2).

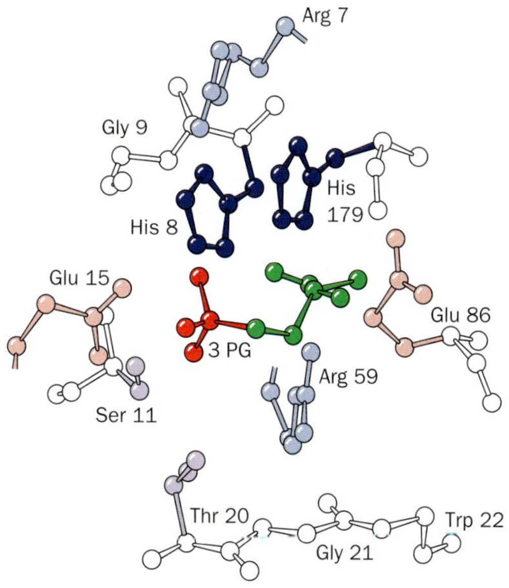

Enolase: Second "High-Energy" Intermediate Formation

In Reaction 9 of glycolysis, 2PG is dehydrated to **phosphoenolpyruvate (PEP)** in a reaction catalyzed by **enolase:**

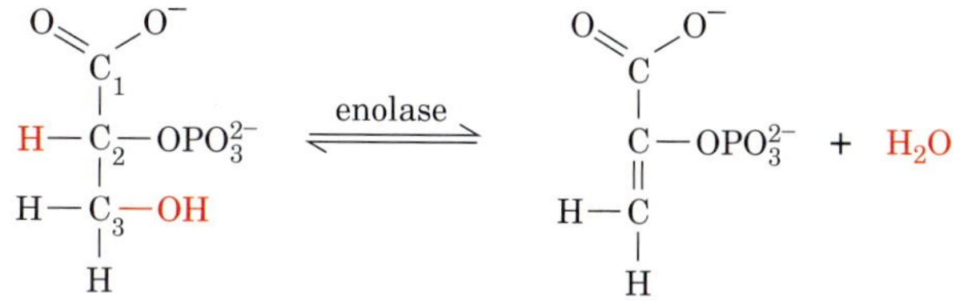

2-Phosphoglycerate (2PG)   **Phosphoenolpyruvate (PEP)**

The enzyme forms a complex with a divalent cation such as Mg^{2+} before the substrate binds. Fluoride ion inhibits glycolysis by blocking enolase activity (F^- was one of the metabolic inhibitors used in elucidating the glycolytic pathway). In the presence of P_i, F^- blocks substrate binding to enolase by forming a bound complex with Mg^{2+} at the enzyme's active site. Enolase's substrate, 2PG, therefore builds up, and, through the action of PGM, 3PG also builds up.

Figure 14-11   The active site region of yeast phosphoglycerate mutase (dephospho form).   The substrate, 3PG, binds to an ionic pocket. His 8 is phosphorylated in the active enzyme. [After Winn, S.I., Watson, H.I., Harkins, R.N., and Fothergill, L.A., *Phil. Trans. R. Soc. London Ser. B* **293**, 126 (1981). PDBid 3PGM.]

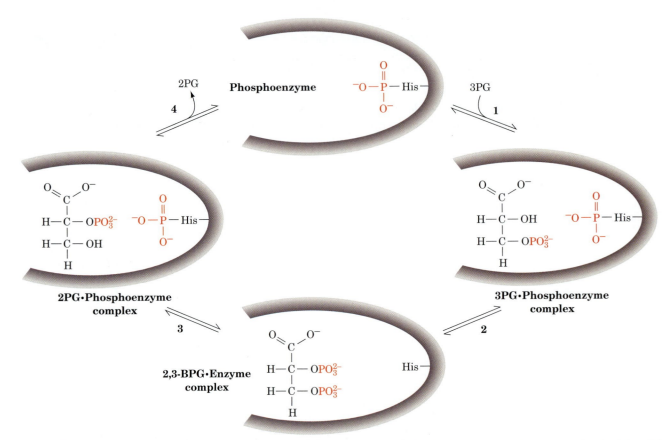

Figure 14-12 A proposed reaction mechanism for phosphoglycerate mutase.   The active form of the enzyme contains a phospho-His residue at the active site. (1) Formation of an enzyme–substrate complex; (2) transfer of the enzyme-bound phosphoryl group to the substrate; (3) rephosphorylation of the enzyme by the other phosphoryl group of the substrate; and (4) release of product, regenerating the active phosphoenzyme.

J Pyruvate Kinase: Second ATP Generation

In Reaction 10 of glycolysis, its final reaction, **pyruvate kinase (PK)** couples the free energy of PEP cleavage to the synthesis of ATP during the formation of pyruvate:

Phosphoenolpyruvate (PEP) $+$ ADP $+$ H$^+$

pyruvate kinase (PK)

Pyruvate $+$ ATP

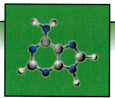

BOX 14-2

Perspectives in Biochemistry

Synthesis of 2,3-Bisphosphoglycerate in Erythrocytes and Its Effect on the Oxygen Carrying Capacity of the Blood

The specific binding of 2,3-bisphosphoglycerate (2,3-BPG) to deoxyhemoglobin decreases the oxygen affinity of hemoglobin (Section 7-3A). Erythrocytes synthesize and degrade 2,3-BPG by a detour from the glycolytic pathway.

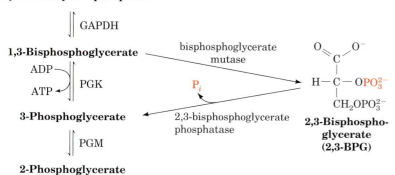

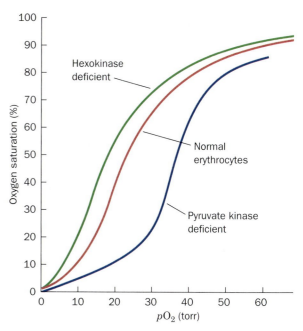

Bisphosphoglycerate mutase catalyzes the transfer of a phosphoryl group from C1 to C2 of 1,3-BPG. The resulting 2,3-BPG is hydrolyzed to 3PG by **2,3-bisphosphoglycerate phosphatase.** The 3PG then continues through the glycolytic pathway.

The level of available 2,3-BPG regulates hemoglobin's oxygen affinity. Consequently, inherited defects of glycolysis in erythrocytes alter the ability of the blood to carry oxygen as is indicated by the oxygen-saturation curve of its hemoglobin.

For example, in hexokinase-deficient erythrocytes, the concentrations of all the glycolytic intermediates are low (since hexokinase catalyzes the first step of glycolysis), thereby resulting in a diminished 2,3-BPG concentration and an increased hemoglobin oxygen affinity (*green curve*). Conversely, a deficiency in pyruvate kinase (which catalyzes the final reaction of glycolysis; Fig. 14-1) decreases hemoglobin's oxygen affinity (*purple curve*) through an increase in 2,3-BPG concentration resulting from this blockade. Thus, although erythrocytes, which lack nuclei and other organelles, have only a minimal metabolism, this metabolism is physiologically significant.

[Oxygen-saturation curves after Deliveria-Papadopoulos, M., Oski, F.A., and Gottlieb, A.J., *Science* **165**, 601 (1969).]

Catalytic Mechanism of Pyruvate Kinase. The PK reaction, which requires both monovalent (K^+) and divalent (Mg^{2+}) cations, occurs as follows (Fig. 14-13):

Step 1 A β-phosphoryl oxygen of ADP nucleophilically attacks the PEP phosphorus atom, thereby displacing **enolpyruvate** and forming ATP.

Step 2 Enolpyruvate tautomerizes to pyruvate.

The PK reaction is highly exergonic, supplying more than enough free energy to drive ATP synthesis (another example of substrate-level

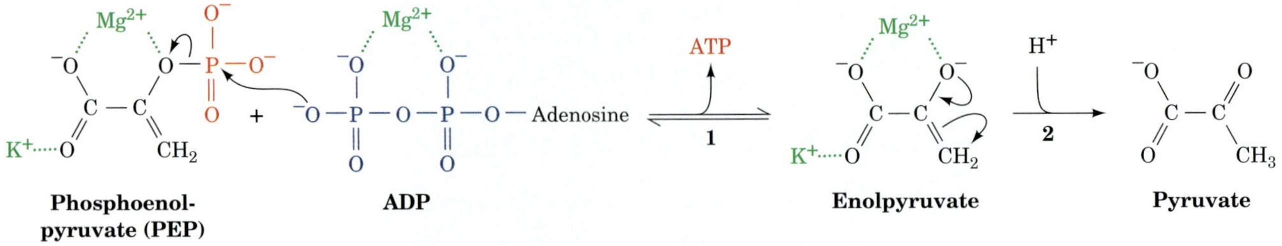

Figure 14-13 The mechanism of the reaction catalyzed by pyruvate kinase. (1) Nucleophilic attack of an ADP β-phosphoryl oxygen atom on the phosphorus atom of PEP to form ATP and enolpyruvate; and (2) tautomerization of enolpyruvate to pyruvate.

phosphorylation). At this point, the "logic" of the enolase reaction becomes clear. The standard free energy of hydrolysis of 2PG is only $-16 \text{ kJ} \cdot \text{mol}^{-1}$, which is insufficient to drive ATP synthesis from ADP ($\Delta G^{\circ\prime} = 30.5 \text{ kJ} \cdot \text{mol}^{-1}$). However, the dehydration of 2PG results in the formation of a "high-energy" compound capable of such synthesis. *The high phosphoryl group-transfer potential of PEP reflects the large release of free energy on converting the product enolpyruvate to its keto tautomer.* Consider the hydrolysis of PEP as a two-step reaction (Fig. 14-14). The tautomerization step supplies considerably more free energy than the phosphoryl group-transfer step.

Assessing Stage II of Glycolysis. The energy investment of the first stage of glycolysis (2 ATP consumed) is doubly repaid in the second stage of glycolysis because two phosphorylated C_3 units are transformed to two pyru-

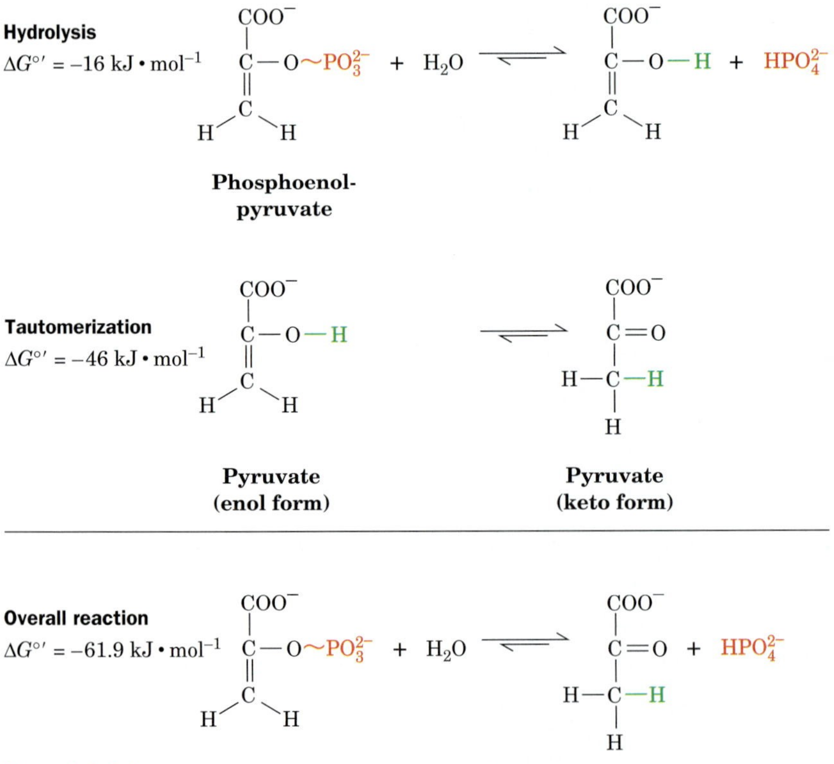

Figure 14-14 The hydrolysis of PEP. The reaction is broken down into two steps, hydrolysis and tautomerization. The overall $\Delta G^{\circ\prime}$ value is much more negative than that required to provide the $\Delta G^{\circ\prime}$ for ATP synthesis from ADP and P_i.

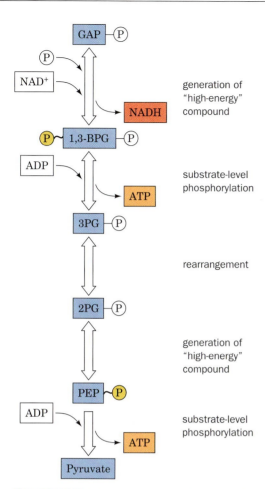

Figure 14-15 **Schematic diagram of the second stage of glycolysis.** In this series of reactions, GAP undergoes phosphorylation and oxidation, followed by molecular rearrangements so that both phosphoryl groups have sufficient free energy to be transferred to ADP to produce ATP. Two molecules of GAP are converted to pyruvate for every molecule of glucose that enters Stage I of glycolysis.

vates with the coupled synthesis of 4 ATP. This process is shown schematically in Fig. 14-15.

The overall reaction of glycolysis, as we have seen, is

$$\text{Glucose} + 2\,\text{NAD}^+ + 2\,\text{ADP} + 2\,\text{P}_i \longrightarrow$$
$$2\,\text{pyruvate} + 2\,\text{NADH} + 2\,\text{ATP} + 2\,\text{H}_2\text{O} + 4\,\text{H}^+$$

Let us consider each of the three products of glycolysis:

1. **ATP.** The initial investment of 2 ATP per glucose in Stage I and the subsequent generation of 4 ATP by substrate-level phosphorylation (two for each GAP that proceeds through Stage II) gives a net yield of 2 ATP per glucose. In some tissues and organisms for which glucose is the primary metabolic fuel, ATP produced by glycolysis satisfies most of the cell's energy needs. For example, the parasitic protozoans *Trypanosoma* and *Leishmania* (which are the causes of several human diseases, including **African sleeping sickness** and **leishmaniasis**) rely almost entirely on glycolysis. Some of the glycolytic enzymes in these organisms differ structurally from their mammalian counterparts, which makes them attractive targets for rational drug design (see Section 12-4).

2. **NADH.** During its catabolism by the glycolytic pathway, glucose is oxidized to the extent that two NAD^+ are reduced to two NADH. As described in Section 13-3C, reduced coenzymes such as NADH represent a source of free energy than can be recovered by subsequent oxidation. Under aerobic conditions, electrons pass from reduced coenzymes through a series of electron carriers to the final oxidizing agent, O_2, in a process known as **electron transport** (Section 17-2). The free energy of electron transport drives the synthesis of ATP from ADP (oxidative phosphorylation; Section 17-3). In aerobic organisms, this sequence of events also serves to regenerate oxidized NAD^+ that can participate in further rounds of catalysis mediated by GAPDH. Under anaerobic conditions, NADH must be reoxidized by other means in order to keep the glycolytic pathway supplied with NAD^+ (Section 14-3).

3. **Pyruvate.** The two pyruvate molecules produced through the partial oxidation of each glucose are still relatively reduced molecules. Under aerobic conditions, complete oxidation of the pyruvate carbon atoms to CO_2 is mediated by the citric acid cycle (Chapter 16). The energy released in that process drives the synthesis of much more ATP than is generated by the limited oxidation of glucose by the glycolytic pathway alone. In anaerobic metabolism, pyruvate is metabolized to a lesser extent to regenerate NAD^+, as we shall see in the following section.

3 Fermentation: The Anaerobic Fate of Pyruvate

The three common metabolic fates of pyruvate produced by glycolysis are outlined in Fig. 14-16.

1. *Under aerobic conditions, the pyruvate is completely oxidized via the citric acid cycle to CO_2 and H_2O.*

2. *Under anaerobic conditions, pyruvate must be converted to a reduced end product in order to reoxidize the NADH produced by the GAPDH reaction. This occurs in two ways:*

Figure 14-16 **Metabolic fate of pyruvate.**   Under aerobic conditions (*left*), the pyruvate carbons are oxidized to CO_2 by the citric acid cycle and the electrons are eventually transferred to O_2 to yield H_2O in oxidative phosphorylation. Under anaerobic conditions in muscle, pyruvate is reversibly converted to lactate (*middle*), whereas in yeast, it is converted to CO_2 and ethanol (*right*).

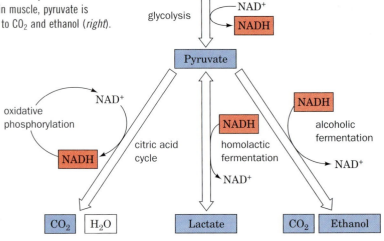

(a) Under anaerobic conditions in muscle, pyruvate is reduced to **lactate** to regenerate NAD^+ in a process known as **homolactic fermentation** (a fermentation is an anaerobic biological reaction process).

(b) In yeast, pyruvate is decarboxylated to yield CO_2 and **acetaldehyde,** which is then reduced by NADH to yield NAD^+ and ethanol. This process is known as **alcoholic fermentation.**

Thus, in aerobic glycolysis, NADH acts as a "high-energy" compound, whereas in anaerobic glycolysis, its free energy of oxidation is dissipated as heat.

A   Homolactic Fermentation

In muscle, during vigorous activity, when the demand for ATP is high and oxygen is in short supply, ATP is largely synthesized via anaerobic glycolysis, which rapidly generates ATP, rather than through the slower process of oxidative phosphorylation. Under these conditions, **lactate dehydrogenase (LDH)** catalyzes the oxidation of NADH by pyruvate to yield NAD^+ and lactate:

Pyruvate   **NADH**

lactate
dehydrogenase (LDH)

L-Lactate   **NAD⁺**

This reaction is often classified as Reaction 11 of glycolysis. The lactate dehydrogenase reaction is freely reversible, so *pyruvate and lactate concentrations are readily equilibrated.*

In the proposed mechanism for pyruvate reduction by LDH, a hydride ion is stereospecifically transferred from C4 of NADH to C2 of pyruvate

with concomitant transfer of a proton from the imidazolium moiety of His 195:

NADH   **Pyruvate**   Arg 109   CH$_2$—His 195   Arg 171

L-Lactate

Both His 195 and Arg 171 interact electrostatically with the substrate to orient pyruvate (or lactate, in the reverse reaction) in the enzyme active site.

The overall process of anaerobic glycolysis in muscle can be represented as

$$\text{Glucose} + 2\ \text{ADP} + 2\ \text{P}_i \rightarrow 2\ \text{lactate} + 2\ \text{ATP} + 2\ \text{H}_2\text{O} + 2\ \text{H}^+$$

Lactate represents a sort of dead end for anaerobic glucose metabolism. The lactate can either be exported from the cell or converted back to pyruvate. Much of the lactate produced in skeletal muscle cells is carried by the blood to the liver, where it is used to synthesize glucose (Section 21-1F).

Contrary to widely held belief, it is not lactate buildup in the muscle per se that causes muscle fatigue and soreness but the accumulation of glycolytically generated acid (muscles can maintain their workload in the presence of high lactate concentrations if the pH is kept constant).

B Alcoholic Fermentation

Under anaerobic conditions in yeast, NAD$^+$ for glycolysis is regenerated in a process that has been valued for thousands of years: the conversion of pyruvate to ethanol and CO$_2$. Ethanol is, of course, the active ingredient of wine and spirits; CO$_2$ so produced leavens bread.

Yeast (Fig. 14-17) produces ethanol and CO$_2$ via two consecutive reactions (Fig. 14-18):

1. The decarboxylation of pyruvate to form acetaldehyde and CO$_2$ as catalyzed by **pyruvate decarboxylase** (an enzyme not present in animals).

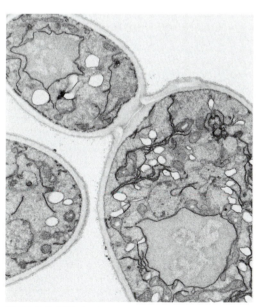

Figure 14-17   An electron micrograph of yeast cells.
[Biophoto Associates Photo Researchers.]

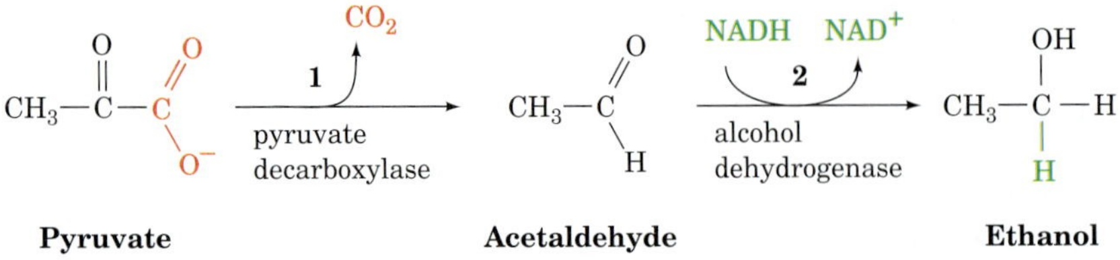

Figure 14-18 The two reactions of alcoholic fermentation. (1) Decarboxylation of pyruvate to form acetaldehyde; and (2) reduction of acetaldehyde to ethanol by NADH.

2. The reduction of acetaldehyde to ethanol by NADH as catalyzed by alcohol dehydrogenase (Section 11-1B), thereby regenerating NAD^+ for use in the GAPDH reaction of glycolysis.

TPP Is an Essential Cofactor of Pyruvate Decarboxylase. Pyruvate decarboxylase contains the coenzyme **thiamine pyrophosphate (TPP;** also called **thiamin diphosphate, ThDP):**

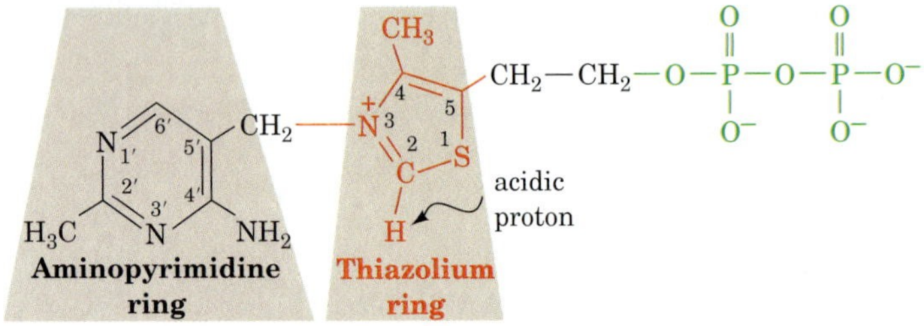

Thiamine pyrophosphate (TPP)

TPP, which is synthesized from thiamine **(vitamin B_1),** binds tightly but noncovalently to pyruvate decarboxylase (Fig. 14-19).

The enzyme uses TPP because uncatalyzed decarboxylation of an α-keto acid such as pyruvate requires the buildup of negative charge on the carbonyl carbon atom in the transition state, an unstable situation:

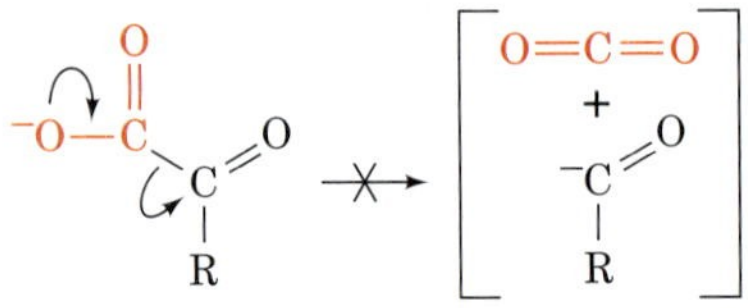

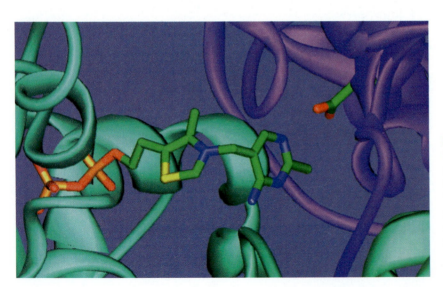

Figure 14-19 TPP binding to pyruvate decarboxylase from *Saccharomyces uvarum* (brewer's yeast). The TPP and the side chain of Glu 51 are shown in skeletal form with C green, N blue, O red, S yellow, and P orange. The TPP binds in a cavity situated between the dimer's two subunits (*cyan and magenta*) where it hydrogen bonds to Glu 51. [Based on an X-ray structure by William Furey and Martin Sax, Veterans Administration Medical Center and University of Pittsburgh. PDBid 1PYD.] ⚛️ **See the Interactive Exercises.**

This transition state can be stabilized by delocalizing the developing negative charge into a suitable "electron sink." The amino acid residues of proteins function poorly in this capacity but TPP does so easily.

*TPP's catalytically active functional group is the **thiazolium ring.*** The C2-H atom of this group is relatively acidic because of the adjacent positively charged quaternary nitrogen atom, which electrostatically stabilizes the carbanion formed when the proton dissociates. This dipolar carbanion (or **ylid**) is the active form of the coenzyme. Pyruvate decarboxylase operates as follows (Fig. 14-20):

Step 1 Nucleophilic attack by the ylid form of TPP on the carbonyl carbon of pyruvate.

Step 2 Departure of CO_2 to generate a resonance-stabilized carbanion adduct in which the thiazolium ring of the coenzyme acts as an electron sink.

Step 3 Protonation of the carbanion.

Step 4 Elimination of the TPP ylid to form acetaldehyde and regenerate the active enzyme.

This mechanism has been corroborated by the isolation of the **hydroxyethylthiamine pyrophosphate** intermediate.

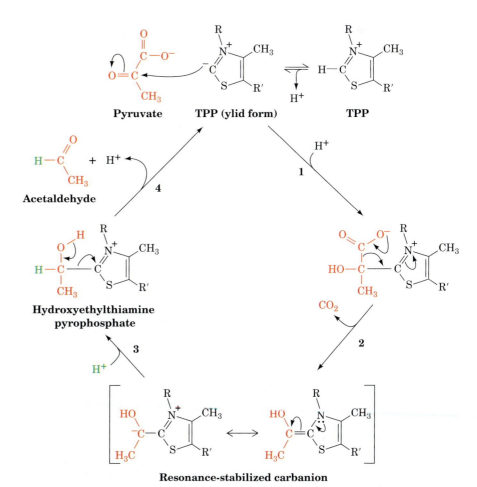

Figure 14-20   The reaction mechanism of pyruvate decarboxylase.   (1) Nucleophilic attack by the ylid form of TPP on the carbonyl carbon of pyruvate; (2) departure of CO_2 to generate a resonance-stabilized carbanion; (3) protonation of the carbanion; and (4) elimination of the TPP ylid and release of product.

Vitamin B₁ Deficiency Causes Beriberi. The ability of TPP's thiazolium ring to add to carbonyl groups and act as an electron sink makes it the coenzyme most utilized in α-keto acid decarboxylation reactions. Consequently, thiamine (vitamin B_1), which is neither synthesized nor stored in significant quantities by the tissues of most vertebrates, is required in their diets. Thiamine deficiency in humans results in an ultimately fatal condition known as **beriberi** (Singhalese for weakness) that is characterized by neurological disturbances causing pain, paralysis, and atrophy (wasting) of the limbs and/or edema (accumulation of fluid in tissues and body cavities). Beriberi was particularly prevalent in the late eighteenth and early nineteenth centuries in the rice-consuming areas of Asia after the introduction of steam-powered milling machines that polished the rice grains to remove their coarse but thiamine-containing outer layers (the previously used milling procedures were less efficient and hence left sufficient thiamine on the grains). Parboiling rice before milling, a process common in India, causes the rice kernels to absorb nutrients from their outer layers, thereby decreasing the incidence of beriberi. Once thiamine deficiency was recognized as the cause of beriberi, enrichment procedures were instituted so that today it has ceased to be a problem except in areas undergoing famine. However, beriberi occasionally develops in alcoholics due to their penchant for drinking but not eating.

Reduction of Acetaldehyde and Regeneration of NAD⁺. Yeast alcohol dehydrogenase **(YADH)**, the enzyme that converts acetaldehyde to ethanol, is a tetramer, each subunit of which binds one Zn^{2+} ion. The Zn^{2+} polarizes the carbonyl group of acetaldehyde to stabilize the developing negative charge in the transition state of the reaction *(at right)*. This facilitates the stereospecific transfer of a hydrogen from NADH to acetaldehyde.

Mammalian liver alcohol dehydrogenase **(LADH)** metabolizes the alcohols anaerobically produced by the intestinal flora as well as those from external sources (the direction of the alcohol dehydrogenase reaction varies with the relative concentrations of ethanol and acetaldehyde). Mammalian LADH is a dimer with significant amino acid sequence similarity to YADH, although LADH subunits each contain a second Zn^{2+} ion that presumably has a structural role.

C Energetics of Fermentation

Thermodynamics permits us to dissect the process of fermentation into its component parts and to account for the free energy changes that occur. This enables us to calculate the efficiency with which the free energy of glucose catabolism is used in the synthesis of ATP. For homolactic fermentation,

$$\text{Glucose} \longrightarrow 2 \text{ lactate} + 2 \text{ H}^+ \qquad \Delta G^{\circ\prime} = -196 \text{ kJ} \cdot \text{mol}^{-1}$$

For alcoholic fermentation,

$$\text{Glucose} \longrightarrow 2 \text{ CO}_2 + 2 \text{ ethanol} \qquad \Delta G^{\circ\prime} = -235 \text{ kJ} \cdot \text{mol}^{-1}$$

Each of these processes is coupled to the net formation of 2 ATP, which requires $\Delta G^{\circ\prime} = +61 \text{ kJ} \cdot \text{mol}^{-1}$ of glucose consumed. Dividing $\Delta G^{\circ\prime}$ of ATP formation by that of lactate formation indicates that homolactic fermentation is 31% "efficient"; that is, 31% of the free energy released by this process under standard biochemical conditions is sequestered in the form of ATP. The rest is dissipated as heat, thereby making the process

irreversible. Likewise, alcoholic fermentation is 26% efficient under biochemical standard state conditions. *Under physiological conditions, where the concentrations of reactants and products differ from those of the standard state, these reactions have thermodynamic efficiencies of >50%.*

Anaerobic fermentation uses glucose in a profligate manner compared to oxidative phosphorylation: Fermentation results in the production of 2 ATP per glucose, whereas oxidative phosphorylation yields up to 38 ATP per glucose (Section 17-3C). This accounts for Pasteur's observation that yeast consume far more sugar when growing anaerobically than when growing aerobically (the **Pasteur effect**). However, *the rate of ATP production by anaerobic glycolysis can be up to 100 times faster than that of oxidative phosphorylation. Consequently, when tissues such as muscle are rapidly consuming ATP, they regenerate it almost entirely by anaerobic glycolysis.* (Homolactic fermentation does not really "waste" glucose since the lactate can be aerobically reconverted to glucose by the liver; Section 21-1F.) Certain muscles are specialized for the rapid production of ATP by glycolysis (see Box 14-3).

4 Control of Glycolysis

Under steady state conditions, glycolysis operates continuously in most tissues, although the glycolytic flux must vary to meet the needs of the organism. Elucidation of the flux control mechanisms of a given pathway, such as glycolysis, commonly involves three steps:

1. Identification of the rate-determining step(s) of the pathway by measuring the *in vivo* ΔG for each reaction. Enzymes that operate far from equilibrium are potential control points (Section 13-1C).

2. *In vitro* identification of allosteric modifiers of the enzymes catalyzing the rate-determining reactions. The mechanisms by which these compounds act are determined from their effects on the enzymes' kinetics.

3. Measurement of the *in vivo* levels of the proposed regulators under various conditions to establish whether these concentration changes are consistent with the proposed control mechanism.

Let us examine the thermodynamics of glycolysis in muscle tissue with an eye toward understanding its control mechanisms (keep in mind that different tissues control glycolysis in different ways). Table 14-1 lists the

Table 14-1 $\Delta G^{\circ\prime}$ and ΔG for the Reactions of Glycolysis in Heart Muscle[a]

Reaction	Enzyme	$\Delta G^{\circ\prime}$ $(kJ \cdot mol^{-1})$	ΔG $(kJ \cdot mol^{-1})$
1	Hexokinase	−20.9	−27.2
2	PGI	+2.2	−1.4
3	PFK	−17.2	−25.9
4	Aldolase	+22.8	−5.9
5	TIM	+7.9	~0
6 + 7	GAPDH + PGK	−16.7	−1.1
8	PGM	+4.7	−0.6
9	Enolase	−3.2	−2.4
10	PK	−23.0	−13.9

[a]Calculated from data in Newsholme, E.A. and Start, C., *Regulation in Metabolism, p.* 97, Wiley (1973).

BOX 14-3

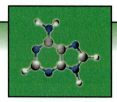

Perspectives in Biochemistry

Glycolytic ATP Production in Muscle

Skeletal muscle consists of both **slow-twitch** (Type I) and **fast-twitch** (Type II) **fibers.** Fast-twitch fibers, so called because they predominate in muscles capable of short bursts of rapid activity, are nearly devoid of mitochondria (where oxidative phosphorylation occurs). Consequently, they must obtain nearly all of their ATP through anaerobic glycolysis, for which they have a particularly large capacity. Muscles designed to contract slowly and steadily, in contrast, are enriched in slow-twitch fibers that are rich in mitochondria and obtain most of their ATP through oxidative phosphorylation.

Fast- and slow-twitch fibers were originally known as white and red fibers, respectively, because otherwise pale-colored muscle tissue, when enriched with mitochondria, takes on the red color charac-

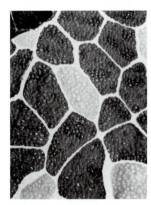

Slow-twitch muscle fiber

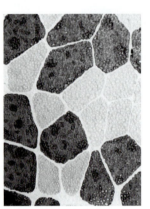

Fast-twitch muscle fiber

teristic of their heme-containing cytochromes. However, fiber color is an imperfect indictor of muscle physiology.

In a familiar example, the flight muscles of migratory birds such as ducks and geese, which need a continuous energy supply, are rich in slow-twitch fibers. Therefore, these birds have dark breast meat. In contrast, the flight muscles of less ambitious fliers, such as chickens and turkeys, which are used only for short bursts (often to escape danger), consist mainly of fast-twitch fibers that form white meat. In humans, the muscles of sprinters are relatively rich in fast-twitch fibers, whereas distance runners have a greater proportion of slow-twitch fibers (although their muscles have the same color).

[Photo of muscle courtesy of J.D. MacDougall, McMaster University, Canada.]

---

standard free energy changes ($\Delta G^{\circ\prime}$) and the actual physiological free energy change (ΔG) associated with each reaction in the pathway. It is important to realize that the free energy changes associated with the reactions under standard conditions may differ dramatically from the actual values *in vivo.*

Only three reactions of glycolysis, those catalyzed by hexokinase, phosphofructokinase, and pyruvate kinase, function with large negative free energy changes in heart muscle under physiological conditions (Fig. 14-21). These nonequilibrium reactions of glycolysis are candidates for flux-control points. The other glycolytic reactions function near equilibrium: Their forward and reverse rates are much faster than the actual flux through the pathway. Consequently, these equilibrium reactions are very sensitive to changes in the concentration of pathway intermediates and readily accommodate changes in flux generated at the rate-determining step(s) of the pathway.

A Phosphofructokinase: The Major Flux-Controlling Enzyme of Glycolysis in Muscle

In vitro studies of hexokinase, phosphofructokinase, and pyruvate kinase indicate that each is controlled by a variety of compounds. Yet when the G6P source for glycolysis is glycogen, rather than glucose, as is often the case in skeletal muscle, the hexokinase reaction is not required (Section

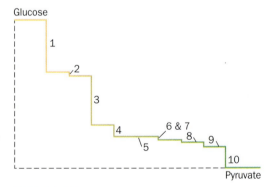

Figure 14-21   Diagram of free energy changes in glycolysis. This "waterfall" diagram illustrates the actual free energy changes for the glycolytic reactions in heart muscle (see Table 14-1). Reactions 1, 3, and 10 are irreversible. The other reactions operate near equilibrium and can mediate flux in either direction.

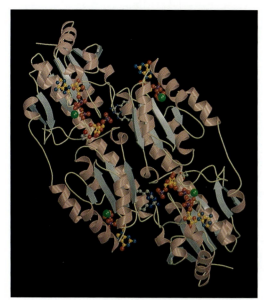

Figure 14-22   X-Ray structure of PFK from *E. coli*. Two subunits of the tetrameric enzyme are shown in ribbon form with helices pink, β strands gray, and the remaining chain segments white. Each subunit binds its substrates F6P (*near the center of each subunit*) and Mg^{2+}–ATP (*lower right and upper left;* the green balls represent Mg^{2+}), along with the activator Mg^{2+}–ADP (*top right and lower left, in the rear*). [Courtesy of Philip Evans, Cambridge University. PDBid 1PFK.] 🐚 **See Kinemage Exercise 13-1.**

15-1). Pyruvate kinase catalyzes the last reaction of glycolysis and is therefore unlikely to be the primary point for regulating flux through the entire pathway. Evidently, PFK, an elaborately regulated enzyme functioning far from equilibrium, is the major control point for glycolysis in muscle under most conditions.

PFK (Fig. 14-22) is a tetrameric enzyme with two conformational states, R and T, that are in equilibrium. ATP is both a substrate and an allosteric inhibitor of phosphofructokinase. Other compounds, including ADP, AMP, and **fructose-2,6-bisphosphate (F2,6P),** reverse the inhibitory effects of ATP and are therefore activators of PFK. Each PFK subunit has two binding sites for ATP: a substrate site and an inhibitor site. The substrate site binds ATP equally well in either conformation, but the inhibitor site binds ATP almost exclusively in the T state. The other substrate of PFK, F6P, preferentially binds to the R state. Consequently, at high concentrations, ATP acts as an allosteric inhibitor of PFK by binding to the T state, thereby shifting the T ⇌ R equilibrium in favor of the T state and thus decreasing PFK's affinity for F6P (this is similar to the action of 2,3-BPG in decreasing the affinity of hemoglobin for O$_2$; Section 7-3A).

In graphical terms, high concentrations of ATP shift the curve of PFK activity versus [F6P] to the right and make it even more sigmoidal (cooperative) (Fig. 14-23). For example, when [F6P] = 0.5 mM (the dashed line in Fig. 14-23), the enzyme is nearly maximally active, but in the presence of 1 mM ATP, the activity drops to 15% of its original level, a nearly sevenfold decrease. An activator such as AMP or ADP counters the effect of ATP by binding to R-state PFK, thereby shifting the T ⇌ R equilibrium toward the R state. (Actually, the most potent allosteric effector of PFK is F2,6P, which we will discuss in Section 15-4C.)

Structural Basis for Allosterism in Phosphofructokinase.

The X-ray structures of PFK from several organisms have been determined in both the R and the T states by Philip Evans. The R state of PFK is stabilized by the binding of its substrate F6P. In the R state of *Bacillus stearothermophilus* PFK, the side chain of Arg 162 forms an ion pair with the phosphoryl group of an F6P bound in the active site of another subunit (Fig. 14-24). However, Arg 162 is located at the end of a helical turn that unwinds on transition to the T state. The positively charged side chain of Arg 162 thereby swings away and is replaced by the negatively charged side chain of Glu 161. As a consequence, the doubly negative phosphoryl group of F6P has a greatly diminished affinity for the T-state enzyme. The unwinding of this helical turn, which is obligatory for the R → T transition, is prevented by the binding of the activator ADP to its effector site on the enzyme. Presumably, ATP can bind to this site only when the helical turn is in its unwound conformation (the T state).

AMP Overcomes the ATP Inhibition of PFK.

Direct allosteric regulation of PFK by ATP may at first appear to be the means by which glycolytic flux is controlled. After all, when [ATP] is high as a result of low metabolic demand, PFK is inhibited and flux through glycolysis is low; conversely when [ATP] is low, flux through the pathway is high and ATP is synthesized to replenish the pool. Consideration of the physiological variation in ATP concentration, however, indicates that the situation must be more complex. The metabolic flux through glycolysis may vary by 100-fold or more, depending on the metabolic demand for ATP. However, *measurements of [ATP] in vivo at various levels of metabolic activity indicate that [ATP] varies <10% between rest and vigorous exertion.* Yet there is no known allosteric mecha-

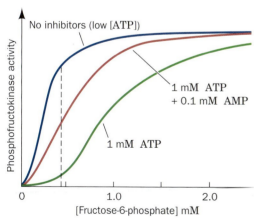

Figure 14-23   PFK activity versus F6P concentration. The various conditions are as follows: purple, no inhibitors or activators; green, 1 mM ATP; and red, 1 mM ATP + 0.1 mM AMP. [After data from Mansour, T.E. and Ahlfors, C.E., *J. Biol. Chem.* **243**, 2523–2533 (1968).] 🐚 **See the Animated Figures.**

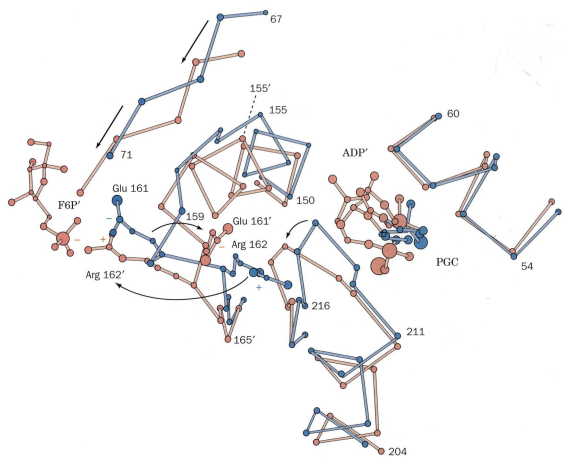

Figure 14-24 Allosteric changes in PFK from *Bacillus stearothermophilus*. Segments of the T state (*blue*) are superimposed on segments of the R state (*pink*) that undergo a large conformational rearrangement on the T → R allosteric transition (indicated by the arrows). Residues of the R-state structure are marked by a prime. Note that in the R state, Arg 162′ forms an attractive ionic interaction with F6P′, whereas in the T state, F6P′ is repelled by Glu 161. Also shown are bound ligands: the nonphysiological inhibitor 2-phosphoglycolate (PGC; a PEP analog) for the T state, and the cooperative substrate F6P and the activator ADP for the R state. [After Schirmer, T. and Evans, P.R., *Nature* **343,** 142 (1990). PDBids 4PFK and 6PFK.] **See Kinemage Exercise 13-2.**

nism that can account for a 100-fold change in flux of a nonequilibrium reaction with only a 10% change in effector concentration. Thus, some other mechanism(s) must be responsible for controlling glycolytic flux.

The inhibition of PFK by ATP is relieved by AMP as well as ADP. This results from AMP's preferential binding to the R state of PFK. If a PFK solution containing 1 mM ATP and 0.5 mM F6P is brought to 0.1 mM in AMP, the activity of PFK rises from 15 to 50% of its maximal activity, a threefold increase (Fig. 14-23).

The [ATP] decreases by only 10% in going from a resting state to one of vigorous activity because it is buffered by the action of two enzymes: creatine kinase and adenylate kinase (Section 13-2C). Adenylate kinase catalyzes the reaction

$$2 \text{ ADP} \rightleftharpoons \text{ATP} + \text{AMP} \qquad K = \frac{[\text{ATP}][\text{AMP}]}{[\text{ADP}]^2} = 0.44$$

which rapidly equilibrates the ADP resulting from ATP hydrolysis in muscle contraction with ATP and AMP.

In muscle, [ATP] is ~50 times greater than [AMP] and ~10 times greater than [ADP]. Consequently, *a change in [ATP] from, for example, 1 to*

0.9 mM, a 10% decrease, can result in a 100% increase in [ADP] (from 0.1 to 0.2 mM) as a result of the adenylate kinase reaction, and a >400% increase in [AMP] (from 0.02 to ~0.1 mM). Therefore, a metabolic signal consisting of a decrease in [ATP] too small to relieve PFK inhibition is amplified significantly by the adenylate kinase reaction, which increases [AMP] by an amount that produces a much larger increase in PFK activity.

B Substrate Cycling

Even a finely tuned allosteric mechanism like that of PFK can account for only a fraction of the 100-fold alterations in glycolytic flux. Additional control may be achieved by substrate cycling. Recall from Section 13-1C that only a near-equilibrium reaction can undergo large changes in flux because, in a near-equilibrium reaction, $v_f - v_r \approx 0$ (where v_f and v_r are the forward and reverse reaction rates) and hence a small change in v_f will result in a large fractional change in $v_f - v_r$. However, this is not the case for the PFK reaction because, for such nonequilibrium reactions, v_r is negligible.

Nevertheless, *such equilibrium-like conditions may be imposed on a nonequilibrium reaction if a second enzyme (or series of enzymes) catalyzes the regeneration of its sub-strate from its product in a thermodynamically favorable manner.* This can be diagrammed as

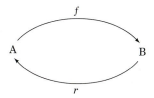

Since two different enzymes catalyze the forward (f) and reverse (r) reactions, v_f and v_r may be independently varied and v_r is no longer negligible compared to v_f. Note that the forward process (e.g., formation of FBP from F6P) and the reverse process (e.g., breakdown of FBP to F6P) *must* be carried out by different enzymes since the laws of thermodynamics would otherwise be violated (i.e., for a single reaction, the forward and reverse reactions cannot simultaneously be favorable).

Under physiological conditions, the reaction catalyzed by PFK:

$$F6P + ATP \longrightarrow FBP + ADP$$

is highly exergonic ($\Delta G = -25.9$ kJ · mol^{-1}). Consequently, the back reaction has a negligible rate compared to the forward reaction. **Fructose-1,6-bisphosphatase (FBPase),** however, which is present in many mammalian tissues (and which is an essential enzyme in gluconeogenesis; Section 15-4B), catalyzes the exergonic hydrolysis of FBP ($\Delta G = -8.6$ kJ · mol^{-1}):

$$FBP + H_2O \longrightarrow F6P + P_i$$

Note that the combined reactions catalyzed by PFK and FBPase result in net ATP hydrolysis:

$$ATP + H_2O \Longleftrightarrow ADP + P_i$$

Such a set of opposing reactions (Section 13-1D) is known as a **substrate cycle** because it cycles a substrate to an intermediate and back again. When this set of reactions was discovered, it was referred to as a **futile cycle** since its net result seemed to be the useless consumption of ATP.

Eric Newsholme has proposed that substrate cycles are not at all "futile" but, rather, have a regulatory function. *The combined effects of allosteric*

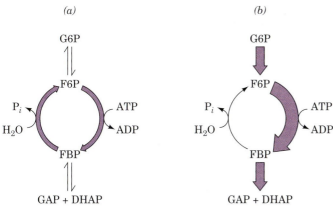

Figure 14-25   Substrate cycling in the regulation of PFK.   (*a*) In resting muscle, both enzymes in the F6P/FBP substrate cycle are active, and glycolytic flux is low. (*b*) In active muscle, PFK activity increases while FBPase activity decreases. This dramatically increases the flux through PFK and therefore results in high glycolytic flux.

effectors on the opposing reactions of a substrate cycle can produce a much greater fractional effect on pathway flux ($v_f - v_r$) than is possible through allosteric regulation of a single enzyme. For example, the allosteric effector F2,6P activates the PFK reaction while inhibiting the FBPase reaction (this regulatory mechanism is important for balancing glycolysis and gluconeogenesis in liver cells; Section 15-4C).

Substrate cycling does not increase the maximum flux through a pathway. On the contrary, it functions to decrease the minimum flux. In a sense, the substrate is put into a "holding pattern." In the PFK/FBPase example (Fig. 14-25), the cycling of substrate appears to be the energetic "price" that a muscle must pay to be able to change rapidly from a resting state (where $v_f - v_r$ is small), in which substrate cycling is maximal, to one of sustained high activity (where $v_f - v_r$ is large). The rate of substrate cycling itself may be under hormonal or neuronal control so as to increase the sensitivity of the metabolic system under conditions when high activity (fight or flight) is anticipated.

Substrate cycling and other mechanisms that control PFK activity *in vivo* are part of larger systems that regulate all the cell's metabolic activities. At one time, it was believed that because PFK is the controlling enzyme of glycolysis, increasing its level of expression via genetic engineering would increase flux through glycolysis. However, this is not the case, because the *activity* of PFK, whatever its concentration, is ultimately controlled by factors that reflect the cell's demand for the products supplied by glycolysis and all other metabolic pathways.

Substrate Cycling, Thermogenesis, and Obesity.   Many animals, including adult humans, are thought to generate much of their body heat, particularly when it is cold, through substrate cycling in muscle and liver, a process known as **nonshivering thermogenesis** (the muscle contractions of shivering or any other movement also produce heat). Substrate cycling is stimulated by thyroid hormones (which stimulate metabolism in most tissues) as is indicated, for example, by the observation that rats lacking a functional thyroid gland do not survive at 5°C. Chronically obese individuals tend to have lower than normal metabolic rates, which is probably due, in part, to a reduced rate of nonshivering thermogenesis. Such individuals therefore tend to be cold sen

sitive. Indeed, whereas normal individuals increase their rate of thyroid hormone activation on exposure to cold, genetically obese animals and obese humans fail to do so.

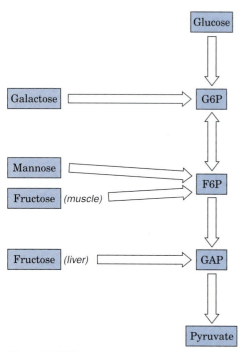

Figure 14-26 Entry of other hexoses into glycolysis. Fructose (in muscle) and mannose are converted to F6P; liver fructose is converted to GAP; and galactose is converted to G6P.

5 Metabolism of Hexoses Other than Glucose

Together with glucose, the hexoses fructose, galactose, and mannose are prominent metabolic fuels. After digestion, these monosaccharides enter the bloodstream, which carries them to various tissues. Fructose, galactose, and mannose are converted to glycolytic intermediates that are then metabolized by the glycolytic pathway (Fig. 14-26).

A Fructose

Fructose is a major fuel source in diets that contain large amounts of fruit or sucrose (a disaccharide of fructose and glucose; Section 8-2A). There are two pathways for the metabolism of fructose; one occurs in muscle and the other occurs in liver. This dichotomy results from the different enzymes present in these tissues.

Fructose metabolism in muscle differs little from that of glucose. Hexokinase (Section 14-2A), which converts glucose to G6P, also phosphorylates fructose, yielding F6P (Fig. 14-27, *left*). The entry of fructose into glycolysis therefore involves only one reaction step.

Liver contains a hexokinase known as **glucokinase,** which has a low affinity for hexoses, including fructose (Section 21-1D). Fructose metabolism in liver must therefore differ from that in muscle. In fact, liver converts fructose to glycolytic intermediates through a pathway that involves seven enzymes (Fig. 14-27, *right*):

1. **Fructokinase** catalyzes the phosphorylation of fructose by ATP at C1 to form **fructose-1-phosphate.** Neither hexokinase nor PFK can phosphorylate fructose-1-phosphate at C6 to form the glycolytic intermediate FBP.

2. Aldolase (Section 14-2D) has several isozymic forms. Muscle contains Type A aldolase, which is specific for FBP. Liver, however, contains Type B aldolase, for which fructose-1-phosphate is also a substrate (Type B aldolase is sometimes called **fructose-1-phosphate aldolase**). In liver, fructose-1-phosphate therefore undergoes an aldol cleavage:

    Fructose-1-phosphate $\rightleftharpoons$
    dihydroxyacetone phosphate + glyceraldehyde

3. Direct phosphorylation of **glyceraldehyde** by ATP through the action of **glyceraldehyde kinase** forms the glycolytic intermediate GAP.

4–7. Alternatively, glyceraldehyde is converted to the glycolytic intermediate DHAP by its NADH-dependent reduction to glycerol as catalyzed by alcohol dehydrogenase (Reaction 4), phosphorylation to **glycerol-3-phosphate** through the action of **glycerol kinase** (Reaction 5), and NAD$^+$-dependent reoxidation to DHAP catalyzed by **glycerol phosphate dehydrogenase** (Reaction 6). The DHAP is then converted to GAP by triose phosphate isomerase (Reaction 7).

The two pathways leading from glyceraldehyde to GAP have the same net cost: Both consume ATP, and although NADH is oxidized in Reaction 4,

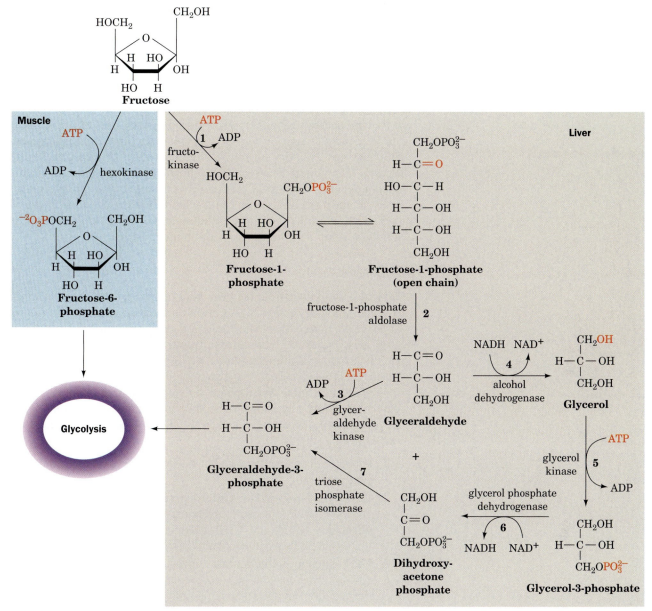

Figure 14-27   The metabolism of fructose.   In muscle (*left*), the conversion of fructose to the glycolytic intermediate F6P involves only one enzyme, hexokinase. In liver (*right*), seven enzymes participate in the conversion of fructose to glycolytic intermediates: (1) fructokinase, (2) fructose-1-phosphate aldolase, (3) glyceraldehyde kinase, (4) alcohol dehydrogenase, (5) glycerol kinase, (6) glycerol phosphate dehydrogenase, and (7) triose phosphate isomerase.

it is reduced again in Reaction 6. The longer pathway, however, produces glycerol-3-phosphate, which (along with DHAP) can become the glycerol backbone of glycerophospholipids and triacylglycerols (Section 19-6A).

Is Excess Fructose Harmful?   The consumption of fructose in the United States has increased at least 10-fold in the last quarter century, in large part due to the use of high-fructose corn syrup as a sweetener in soft drinks and other foods. Fructose has a sweeter taste than sucrose (see Box 8-2) and is inexpensive to produce. One possible hazard of excessive fructose intake is that fructose catabolism in liver bypasses the PFK-catalyzed step of glycolysis and thereby avoids a major metabolic control point. This could po-

tentially disrupt fuel metabolism so that glycolytic flux is directed toward lipid synthesis in the absence of a need for ATP production. This hypothesis suggests a link between the increase in both fructose consumption and the incidence of obesity in the United States.

At the opposite extreme are individuals with **fructose intolerance,** which results from a deficiency in Type B aldolase. In the absence of the aldolase, fructose-1-phosphate may accumulate enough to deplete the liver's store of P_i. Under these conditions, [ATP] drops, which causes liver damage. In addition, the increased [fructose-1-phosphate] inhibits both **glycogen phosphorylase** (an essential enzyme in the breakdown of glycogen to glucose; Section 15-1A) and fructose-1,6-bisphosphatase (an essential enzyme in gluconeogenesis; Section 15-4B), thereby causing severe **hypoglycemia** (low levels of blood glucose), which can reach life-threatening proportions. However, frutose intolerance is self-limiting: Individuals with this condition rapidly develop a strong distaste for anything sweet.

B Galactose

Galactose is obtained from the hydrolysis of lactose (a disaccharide of galactose and glucose; Section 8-2A) in dairy products. Galactose and glucose *(at left)* are epimers that differ only in their configuration at C4. Although hexokinase phosphorylates glucose, fructose, and mannose, it does not recognize galactose. An epimerization reaction must therefore occur before galactose enters glycolysis. This reaction takes place after the conversion of galactose to its **uridine diphosphate** derivative (the role of UDP–sugars and other nucleotidyl–sugars is discussed in more detail in Section 15-5). The entire pathway converting galactose to a glycolytic intermediate requires four reactions (Fig. 14-28):

1. Galactose is phosphorylated at C1 by ATP in a reaction catalyzed by **galactokinase.**
2. **Galactose-1-phosphate uridylyl transferase** transfers the uridylyl group of UDP–glucose to **galactose-1-phosphate** to yield **glucose-1-phosphate (G1P)** and **UDP–galactose** by the reversible cleavage of UDP–glucose's pyrophosphoryl bond.
3. **UDP–galactose-4-epimerase** converts UDP–galactose back to UDP–glucose. This enzyme has an associated NAD^+, which suggests that the reaction involves the sequential oxidation and reduction of the hexose C4 atom:

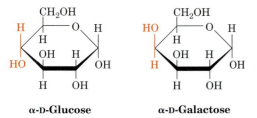

α-D-Glucose   α-D-Galactose

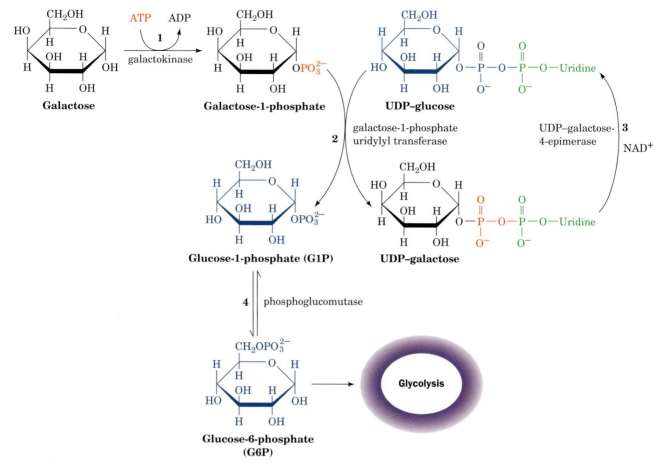

Figure 14-28 **The metabolism of galactose.**   Four enzymes participate in the conversion of galactose to the glycolytic intermediate G6P:

(1) galactokinase, (2) galactose-1-phosphate uridylyl transferase, (3) UDP–galactose-4-epimerase, and (4) phosphoglucomutase.

4. G1P is converted to the glycolytic intermediate G6P by the action of **phosphoglucomutase.**

Galactosemia.   Galactosemia is a genetic disease characterized by the inability to convert galactose to glucose. Its symptoms include failure to thrive, mental retardation, and, in some instances, death from liver damage. Most cases of galactosemia involve a deficiency in the enzyme catalyzing Reaction 2 of the interconversion, galactose-1-phosphate uridylyl transferase. Formation of UDP–galactose from galactose-1-phosphate is thus prevented, leading to a buildup of toxic metabolic by-products. For example, the increased galactose concentration in the blood results in a higher galactose concentration in the lens of the eye, where this sugar is reduced to **galactitol:**

$$
\begin{array}{c}
\text{CH}_2\text{OH} \\
|\\
\text{H} - \text{C} - \text{OH} \\
|\\
\text{HO} - \text{C} - \text{H} \\
|\\
\text{HO} - \text{C} - \text{H} \\
|\\
\text{H} - \text{C} - \text{OH} \\
|\\
\text{CH}_2\text{OH}
\end{array}
$$

D-Galactitol

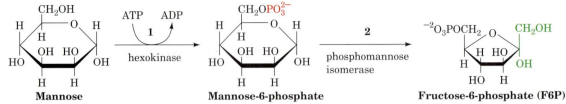

Figure 14-29 The metabolism of mannose. Two enzymes are required to convert mannose to the glycolytic intermediate F6P: (1) hexokinase and (2) phosphomannose isomerase.

The presence of this sugar alcohol in the lens eventually causes cataract formation (clouding of the lens).

Galactosemia is treated by a galactose-free diet. Except for the mental retardation, this reverses all symptoms of the disease. The galactosyl units that are essential for the synthesis of glycoproteins (Section 8-3C) and glycolipids (Section 9-1D) can be synthesized from glucose by a reversal of the epimerase reaction. These syntheses therefore do not require dietary galactose.

C Mannose

Mannose, a product of digestion of polysaccharides and glycoproteins, is the C2 epimer of glucose:

α-D-Glucose α-D-Mannose

Mannose enters the glycolytic pathway after its conversion to F6P via a two-reaction pathway (Fig. 14-29):

1. Hexokinase recognizes mannose and converts it to **mannose-6-phosphate.**

2. **Phosphomannose isomerase** then converts this aldose to the glycolytic intermediate F6P in a reaction whose mechanism resembles that of phosphoglucose isomerase (Section 14-2B).

6 The Pentose Phosphate Pathway

ATP is the cell's "energy currency"; its exergonic cleavage is coupled to many otherwise endergonic cell functions. *Cells also have a second currency, reducing power.* Many endergonic reactions, notably the reductive biosynthesis of fatty acids (Section 19-4) and cholesterol (Section 19-7A), require NADPH in addition to ATP. Despite their close chemical resemblance, *NADPH and NADH are not metabolically interchangeable.* Whereas NADH uses the free energy of metabolite oxidation to synthesize ATP (oxidative phosphorylation), NADPH uses the free energy of metabolite oxidation for reductive biosynthesis. This differentiation is possible because the dehydrogenases involved in oxidative and reductive metabolism are highly specific for their respective coenzymes. Indeed, cells normally maintain their $[NAD^+]/[NADH]$ ratio near 1000, which favors metabolite oxi-

dation, while keeping their $[NADP^+]/[NADPH]$ ratio near 0.01, which favors reductive biosynthesis.

NADPH is generated by the oxidation of glucose-6-phosphate via an alternative pathway to glycolysis, the pentose phosphate pathway (also called the **hexose monophosphate shunt;** Fig. 14-30). Tissues most heavily involved in lipid biosynthesis (liver, mammary gland, adipose tissue, and adrenal cor-

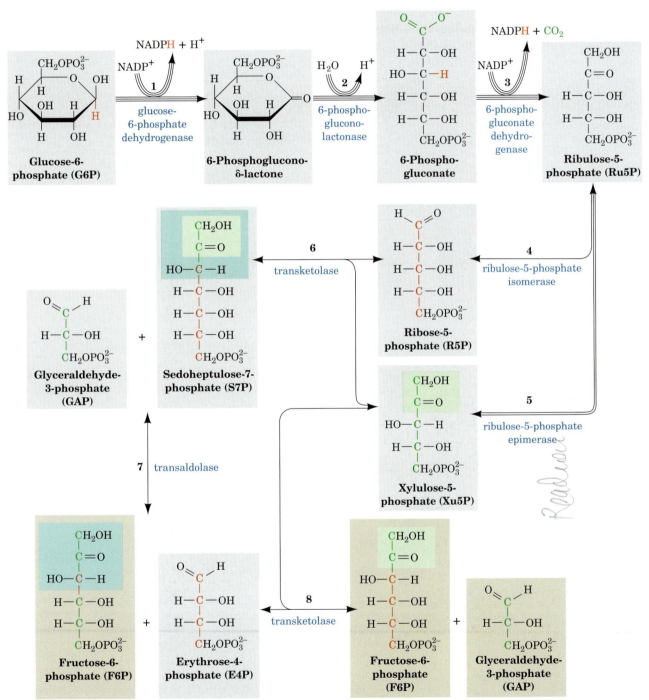

Figure 14-30 *Key to Metabolism.*   **The pentose phosphate pathway.**   The number of lines in an arrow represents the number of molecules reacting in one turn of the pathway so as to convert 3 G6P to 3 CO$_2$, 2 F6P, and 1 GAP. For the sake of clarity, sugars from Reaction 3 onward are shown in their linear forms.

The carbon skeleton of R5P and the atoms derived from it are drawn in red, and those from Xu5P are drawn in green. The C$_2$ units transferred by transketolase are shaded in green, and the C$_3$ units transferred by transaldolase are shaded in blue. Double-headed arrows indicate reversible reactions.

tex) are rich in pentose phosphate pathway enzymes. Indeed, some 30% of the glucose oxidation in liver occurs via the pentose phosphate pathway rather than glycolysis.

The overall reaction of the pentose phosphate pathway is

$$3 \text{ G6P} + 6 \text{ NADP}^+ + 3 \text{ H}_2\text{O} \rightleftharpoons$$
$$6 \text{ NADPH} + 6 \text{ H}^+ + 3 \text{ CO}_2 + 2 \text{ F6P} + \text{GAP}$$

However, the pathway can be considered to have three stages:

Stage 1 Oxidative reactions (Fig. 14-30, Reactions 1–3), which yield NADPH and **ribulose-5-phosphate (Ru5P):**

$$3 \text{ G6P} + 6 \text{ NADP}^+ + 3 \text{ H}_2\text{O} \longrightarrow$$
$$6 \text{ NADPH} + 6 \text{ H}^+ + 3 \text{ CO}_2 + 3 \text{ Ru5P}$$

Stage 2 Isomerization and epimerization reactions (Fig. 14-30, Reactions 4 and 5), which transform Ru5P either to **ribose-5-phosphate (R5P)** or to **xylulose-5-phosphate (Xu5P):**

$$3 \text{ Ru5P} \rightleftharpoons \text{R5P} + 2 \text{ Xu5P}$$

Stage 3 A series of C—C bond cleavage and formation reactions (Fig. 14-30, Reactions 6–8) that convert two molecules of Xu5P and one molecule of R5P to two molecules of F6P and one molecule of GAP.

The reactions of Stages 2 and 3 are freely reversible, so the products of the pathway vary with the needs of the cell (see below). In this section, we discuss the three stages of the pentose phosphate pathway and how this pathway is controlled.

A   Stage 1: Oxidative Reactions of NADPH Production

G6P is considered the starting point of the pentose phosphate pathway. This metabolite may arise through the action of hexokinase on glucose (Reaction 1 of glycolysis; Section 14-2A) or from glycogen breakdown (which produces G6P directly; Section 15-1). Only the first three reactions of the pentose phosphate pathway are involved in NADPH production (Fig. 14-30):

1. **Glucose-6-phosphate dehydrogenase (G6PD)** catalyzes net transfer of a hydride ion to NADP$^+$ from C1 of G6P to form **6-phospho-glucono-δ-lactone:**

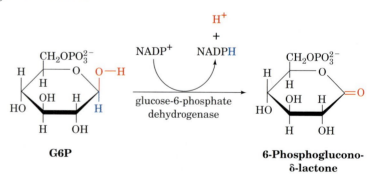

G6P   **6-Phosphoglucono-δ-lactone**

G6P, a cyclic hemiacetal with C1 in the aldehyde oxidation state, is thereby oxidized to a cyclic ester (lactone). The enzyme is specific for NADP$^+$ and is strongly inhibited by NADPH.

2. **6-Phosphogluconolactonase** increases the rate of hydrolysis of 6-phosphoglucono-δ-lactone to **6-phosphogluconate** (the nonenzymatic reaction occurs at a significant rate).

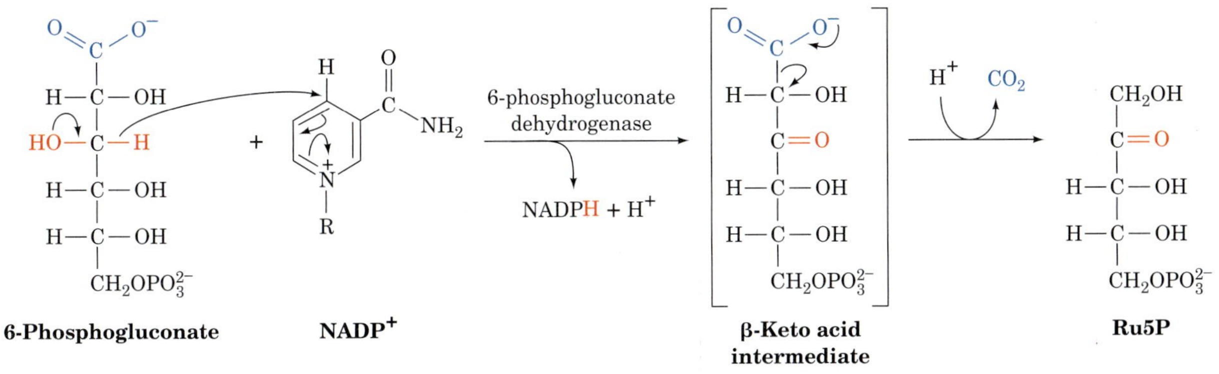

Figure 14-31   The 6-phosphogluconate dehydrogenase reaction.   Oxidation of the OH group forms an easily decarboxylated β-keto acid (although the proposed intermediate has not been isolated).

3. **6-Phosphogluconate dehydrogenase** catalyzes the oxidative decarboxylation of 6-phosphogluconate, a β-hydroxy acid, to Ru5P and CO_2 (Fig. 14-31). This reaction is thought to proceed via the formation of a β-keto acid intermediate. The keto group presumably facilitates decarboxylation by acting as an electron sink.

Formation of Ru5P completes the oxidative portion of the pentose phosphate pathway. *It generates two molecules of NADPH for each molecule of G6P that enters the pathway.*

B Stage 2: Isomerization and Epimerization of Ribulose-5-Phosphate

Ru5P is converted to R5P by **ribulose-5-phosphate isomerase** (Fig. 14-30, Reaction 4) or to Xu5P by **ribulose-5-phosphate epimerase** (Fig. 14-30, Reaction 5). These isomerization and epimerization reactions, like the reaction catalyzed by triose phosphate isomerase (Section 14-2E), are thought to occur via enediolate intermediates.

The relative amounts of R5P and Xu5P produced from Ru5P depend on the needs of the cell. For example, R5P is an essential precursor in the biosynthesis of nucleotides (Chapter 22). Accordingly, R5P production is relatively high (in fact, the entire pentose phosphate pathway activity may be elevated) in rapidly dividing cells, in which the rate of DNA synthesis is increased. If the pathway is being used solely for NADPH production, Xu5P and R5P are produced in a 2:1 ratio for conversion to glycolytic intermediates in the third stage of the pentose phosphate pathway as is discussed below.

C Stage 3: Carbon–Carbon Bond Cleavage and Formation Reactions

How is a five-carbon sugar transformed to a six-carbon sugar such as F6P? The rearrangements of carbon atoms in the third stage of the pentose phosphate pathway are easier to follow by considering the stoichiometry of the pathway. Every three G6P molecules that enter the pathway yield three Ru5P molecules in Stage 1. These three pentoses are then converted to one R5P and two Xu5P (Fig. 14-30, Reactions 4 and 5). The conversion of these three C_5 sugars to two C_6 sugars and one C_3 sugar involves a remarkable "juggling act" catalyzed by two enzymes, **transaldolase** and **transketolase.** These enzymes have mechanisms that involve the generation of stabilized carbanions and their addition to the electrophilic centers of aldehydes.

Transketolase Catalyzes the Transfer of C_2 Units. Transketolase, which has a thiamine pyrophosphate cofactor (TPP; Section 14-3B), catalyzes the transfer of a C_2 unit from Xu5P to R5P, yielding GAP and **sedoheptulose-7-phosphate** (**S7P**; Fig. 14-30, Reaction 6). The reaction intermediate is a covalent adduct between Xu5P and TPP (Fig. 14-32). The X-ray structure of the dimeric enzyme shows that the TPP binds in a deep cleft between the subunits so that residues from both subunits participate in its binding, just as in pyruvate decarboxylase (another TPP-requiring enzyme; Fig. 14-19). In fact, the structures are so similar that they likely diverged from a common ancestor.

Transaldolase Catalyzes the Transfer of C_3 Units. Transaldolase catalyzes the transfer of a C_3 unit from S7P to GAP yielding **erythrose-4-phosphate** (**E4P**) and F6P (Fig. 14-30, Reaction 7). The reaction occurs by aldol cleavage (Section 14-2D), which begins with the formation of a Schiff base between an ε-amino group of an essential Lys residue and the carbonyl group of S7P (Fig. 14-33).

A Second Transketolase Reaction Yields Glyceraldehyde-3-Phosphate and a Second Fructose-6-Phosphate Molecule. In a second transketolase reaction, a C_2 unit is transferred from a second molecule of Xu5P to E4P to form GAP and another molecule of F6P (Fig. 14-30, Reaction 8). The third stage of the pentose phosphate pathway thus transforms two molecules of Xu5P and one of R5P to two molecules of F6P and one molecule of GAP. These carbon skeleton transformations (Fig. 14-30, Reactions 6–8) are summarized in Fig. 14-34.

D Control of the Pentose Phosphate Pathway

The principal products of the pentose phosphate pathway are R5P and NADPH. The transaldolase and transketolase reactions convert excess R5P to glycolytic intermediates when the metabolic need for NADPH exceeds that of R5P in nucleotide biosynthesis. The resulting GAP and F6P can be consumed through glycolysis and oxidative phosphorylation or recycled by gluconeogenesis (Section 15-4) to form G6P.

When the need for R5P outstrips the need for NADPH, F6P and GAP can be diverted from the glycolytic pathway for use in the synthesis of R5P by reversal of the transaldolase and transketolase reactions. The relationship between glycolysis and the pentose phosphate pathway is diagrammed in Fig. 14-35.

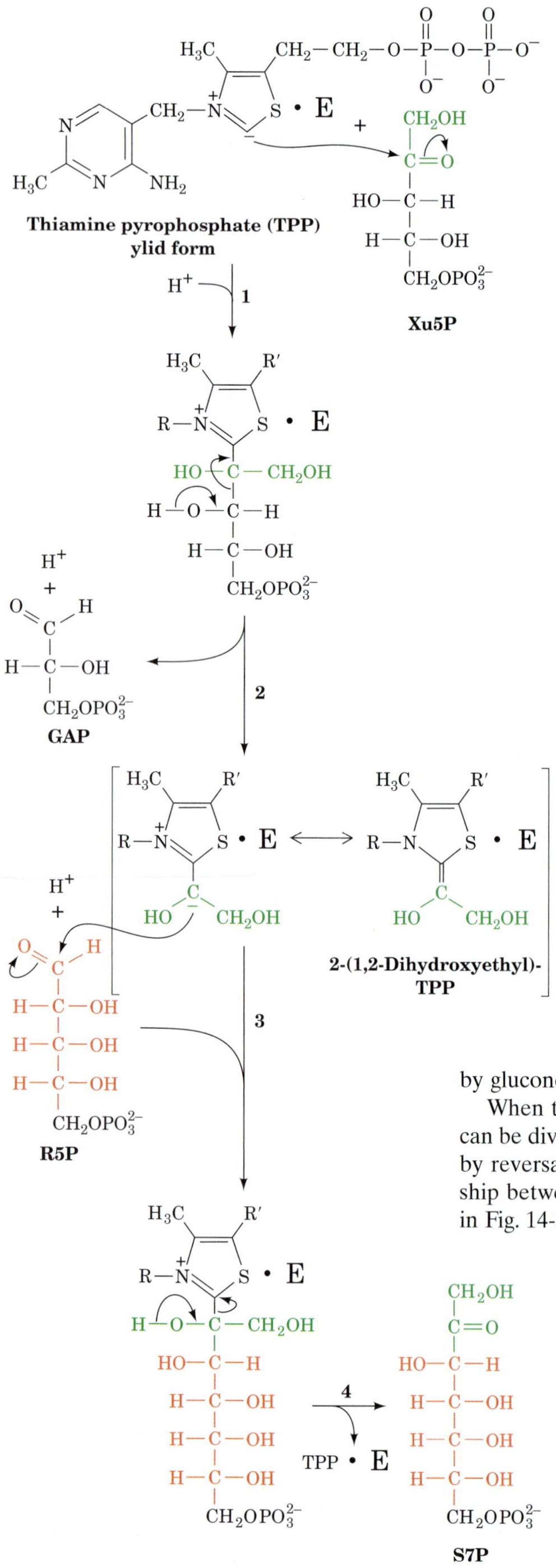

Figure 14-32 Mechanism of transketolase. Transketolase (represented by E) uses the coenzyme TPP to stabilize the carbanion formed on cleavage of the C2—C3 bond of Xu5P. The reaction occurs as follows: (1) The TPP ylid attacks the carbonyl group of the Xu5P; (2) C2—C3 bond cleavage yields GAP and enzyme-bound 2-(1,2-dihydroxyethyl)-TPP, a resonance-stabilized carbanion; (3) the C2 carbanion attacks the aldehyde carbon of R5P to form an S7P–TPP adduct; (4) TPP is eliminated, yielding S7P and the regenerated TPP–enzyme.

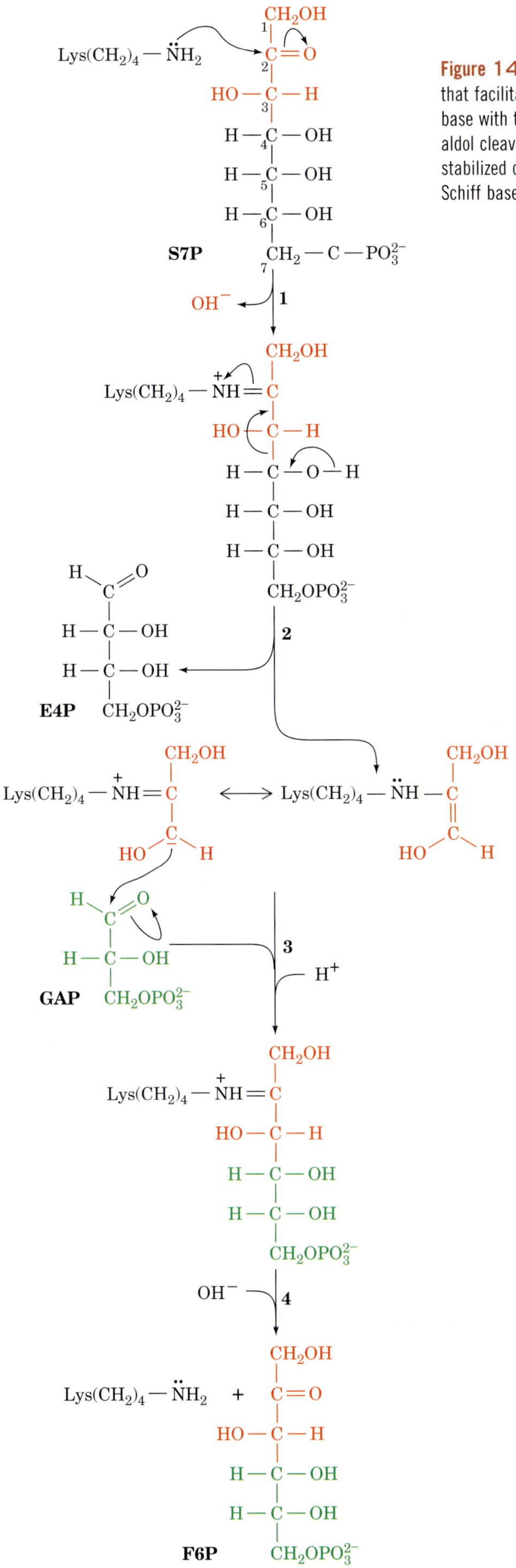

Figure 14-33 Mechanism of transaldolase.   Transaldolase contains an essential Lys residue that facilitates an aldol cleavage reaction as follows: (**1**) The ε-amino group of Lys forms a Schiff base with the carbonyl group of S7P; (**2**) a Schiff base–stabilized C3 carbanion is formed in an aldol cleavage reaction between C3 and C4 that eliminates E4P; (**3**) the enzyme-bound resonance-stabilized carbanion adds to the carbonyl C atom of GAP, forming F6P linked to the enzyme via a Schiff base; (**4**) the Schiff base hydrolyzes, regenerating active enzyme and releasing F6P.

$$(\textbf{6}) \quad C_5 + C_5 \rightleftharpoons C_7 + C_3$$

$$(\textbf{7}) \quad C_7 + C_3 \rightleftharpoons C_6 + C_4$$

$$(\textbf{8}) \quad \underline{C_5 + C_4 \rightleftharpoons C_6 + C_3}$$

$$(\text{Sum}) \quad 3\,C_5 \rightleftharpoons 2\,C_6 + C_3$$

Figure 14-34 Summary of carbon skeleton rearrangements in the pentose phosphate pathway. A series of carbon–carbon bond formations and cleavages convert three C_5 sugars to two C_6 and one C_3 sugar. The number to the left of each reaction is keyed to the corresponding reaction in Fig. 14-30.

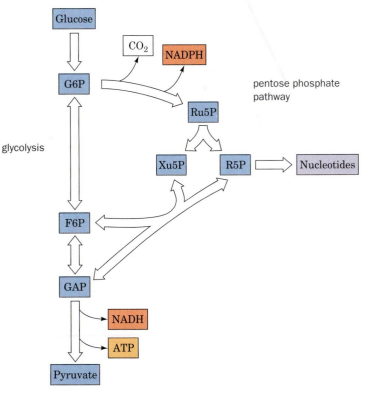

Figure 14-35 Relationship between glycolysis and the pentose phosphate pathway.   The pentose phosphate pathway, which begins with G6P produced in Step 2 of glycolysis, generates NADPH for use in reductive reactions and R5P for nucleotide synthesis. Excess R5P is converted to glycolytic intermediates by a sequence of reactions that can operate in reverse to generate additional R5P, if needed.

BOX 14-4

Biochemistry in Health and Disease

Glucose-6-Phosphate Dehydrogenase Deficiency

NADPH is required for several reductive processes in addition to biosynthesis. For example, erythrocytes require a plentiful supply of reduced **glutathione (GSH),** a Cys-containing tripeptide (see Section 4-3B).

$$H_3\overset{+}{N}-\underset{\underset{COO^-}{|}}{CH}-CH_2-CH_2-\overset{\overset{O}{\|}}{C}-NH-\underset{\underset{\underset{SH}{|}}{CH_2}}{CH}-\overset{\overset{O}{\|}}{C}-NH-CH_2-COO^-$$

Glutathione (GSH)
(γ-L-glutamyl-L-cysteinylglycine)

A major function of GSH in the erythrocyte is to reductively eliminate H_2O_2 and organic hydroperoxides, which are reactive oxygen metabolites that can irreversibly damage hemoglobin and cleave the C—C bonds in the phospholipid tails of cell membranes. The unchecked buildup of peroxides results in premature cell lysis. Peroxides are eliminated by reaction with glutathione, catalyzed by **glutathione peroxidase:**

$$2\,GSH + R-O-O-H \xrightarrow{\substack{\text{glutathione} \\ \text{peroxidase}}} GSSG + ROH + H_2O$$

Organic
hydroperoxide

GSSG represents oxidized glutathione (two GSH molecules linked through a disulfide bond between their sulfhydryl groups). Reduced GSH is subsequently regenerated by the reduction of GSSG by NADPH as catalyzed by **glutathione reductase:**

$$GSSG + NADPH + H^+ \xrightarrow{\substack{\text{glutathione} \\ \text{reductase}}} 2\,GSH + NADP^+$$

A steady supply of NADPH is therefore vital for erythrocyte integrity.

The erythrocytes in individuals who are deficient in glucose-6-phosphate dehydrogenase (G6PD) are particularly sensitive to oxidative damage, although clinical symptoms may be absent. This enzyme deficiency, which is common in African, Asian, and Mediterranean populations, came to light through investigations of the hemolytic anemia that is induced in these individuals when they ingest drugs such as the antimalarial compound **primaquine**

Primaquine

or eat **fava beans (broad beans,** *Vicia faba*), a staple Middle Eastern vegetable. Primaquine stimulates peroxide formation, thereby increasing the demand for NADPH to a level that the mutant cells cannot meet. Certain toxic glycosides present in small amounts in fava beans have the same effect, producing a condition known as **favism.**

The major reason for low enzymatic activity in affected cells appears to be an accelerated rate of breakdown of the

Flux through the pentose phosphate pathway and thus the rate of NADPH production is controlled by the rate of the glucose-6-phosphate dehydrogenase reaction (Fig. 14-30, Reaction 1). The activity of this enzyme, which catalyzes the pathway's first committed step ($\Delta G = -17.6$ kJ · mol^{-1} in liver), is regulated by the $NADP^+$ concentration (i.e., regulation by substrate availability). When the cell consumes NADPH, the $NADP^+$ concentration rises, increasing the rate of the G6PD reaction and thereby stimulating NADPH regeneration. In some tissues, the amount of enzyme synthesized also appears to be under hormonal control. A deficiency in G6PD is the most common clinically significant enzyme defect of the pentose phosphate pathway (see Box 14-4).

mutant enzyme. This explains why patients with relatively mild forms of G6PD deficiency react to primaquine with hemolytic anemia but recover within a week despite continued primaquine treatment. Mature erythrocytes lack a nucleus and protein synthesizing machinery and therefore cannot synthesize new enzyme molecules to replace degraded ones (they likewise cannot synthesize new membrane components, which is why they are so sensitive to membrane damage in the first place). The initial primaquine treatments result in the lysis of old red blood cells whose defective G6PD has been largely degraded. Lysis products stimulate the release of young cells that contain more enzyme and are therefore better able to cope with primaquine stress.

It is estimated that ~400 million people are deficient in G6PD, which makes this condition the most common human enzyme deficiency. Indeed, ~400 G6PD variants have been reported and at least 125 of them have been characterized at the molecular level. G6PD is active in a dimer-tetramer equilibrium. Many of the mutation sites in individuals with the most severe G6PD deficiency are at the dimer interface, shifting the equilibrium toward the inactive and unstable monomer.

The high prevalence of defective G6PD in malarial areas of the world suggests that such mutations confer resistance to the malarial parasite, *Plasmodium falciparum*. Indeed, erythrocytes with G6PD deficiency appear to be less suitable hosts for plasmodia than normal cells. Thus, like the sickle-cell trait (Section 7-4), *a defective G6PD confers a selective advantage on individuals living where malaria is endemic.*

The G6PD deficiency primarily affects erythrocytes, in which the lack of a nucleus prevents replacement of the unstable mutant enzyme. However, the importance of NADPH in cells other than erythrocytes has been demonstrated through the development of mice in which the G6PD gene has been knocked out. All the cells in these animals are extremely sensitive to oxidative stress, even though they contain other mechanisms for eliminating reactive oxygen species.

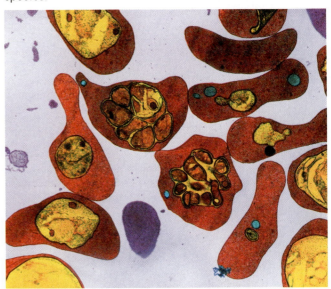

Photo of red blood cells showing intracellular *Plasmodium falciparum* (the malaria parasite). [© Dr. Gopal Murte/Photo Researchers.]

SUMMARY

1. Glycolysis is a sequence of 10 enzyme-catalyzed reactions by which one molecule of glucose is converted to two molecules of pyruvate, with the net production of 2 ATP and the reduction of 2 NAD$^+$ to 2 NADH.

2. In the first stage of glycolysis, glucose is phosphorylated by hexokinase, isomerized by phosphoglucose isomerase (PGI), phosphorylated by phosphofructokinase (PFK), and cleaved by aldolase to yield the trioses glyceraldehyde-3-phosphate (GAP) and dihydroxyacetone phosphate (DHAP), which are interconverted by triose phosphate isomerase (TIM). These reactions consume 2 ATP per glucose.

3. In the second stage of glycolysis, GAP is oxidatively phosphorylated by glyceraldehyde-3-phosphate dehydrogenase (GAPDH), dephosphorylated by phosphoglycerate kinase (PGK) to produce ATP, isomerized by phosphoglycerate mutase (PGM), dehydrated by enolase, and dephosphorylated by pyruvate kinase to produce a second ATP and pyruvate. This stage produces 4 ATP per glucose for a net yield of 2 ATP per glucose.

4. Under anaerobic conditions, pyruvate is reduced to regenerate NAD$^+$ for glycolysis. In homolactic fermentation, pyruvate is reversibly reduced to lactate.

5. In alcoholic fermentation, pyruvate is decarboxylated by a thiamine pyrophosphate (TPP)-dependent mechanism, and the resulting acetaldehyde is reduced to ethanol.

6. The glycolytic reactions catalyzed by hexokinase, phosphofructokinase, and pyruvate kinase are metabolically irreversible.

7. Phosphofructokinase is the primary flux control point for glycolysis. ATP inhibition of this allosteric enzyme is relieved by AMP and ADP, whose concentrations change more dramatically than those of ATP.

8. The opposing reactions of the fructose-6-phosphate (F6P)/fructose-1,6-bisphosphate (FBP) substrate cycle allow large changes in glycolytic flux.

9. Fructose, galactose, and mannose are enzymatically converted to glycolytic intermediates for catabolism.

10. In the pentose phosphate pathway, glucose-6-phosphate (G6P) is oxidized and decarboxylated to produce two NADPH, CO_2, and ribulose-5-phosphate (Ru5P).

11. Depending on the cell's needs, ribulose-5-phosphate may be isomerized to ribose-5-phosphate (R5P) for nucleotide synthesis or converted, via ribose-5-phosphate and xylulose-5-phosphate (Xu5P), to fructose-6-phosphate and glyceraldehyde-3-phosphate, which can re-enter the glycolytic pathway.

REFERENCES

Berstein, B.E., Michels, P.A.M., and Hol, W.G.J., Synergistic effects of substrate-induced conformational changes in phosphoglycerate activation, *Nature* **385,** 275–278 (1997).

Dalby, A., Dauter, Z., and Littlechild, J.A., Crystal structure of human muscle aldolase complexed with fructose 1,6-bisphosphate: Mechanistic implications, *Protein Science* **8,** 291–297 (1999).

Depre, C., Rider, M.H., and Hue, L., Mechanisms of control of heart glycolysis, *Eur. J. Biochem.* **258,** 277–290 (1998). [Discusses how the control of glycolysis in heart muscle is distributed among several enzymes, transporters, and other pathways.]

Frey, P.A., The Leloir pathway: a mechanistic imperative for three enzymes to change the stereochemical configuration of a single carbon in galactose, *FASEB J.* **10,** 461–470 (1996).

Gefflaut, T., Blonski, C., Perie, J., and Wilson, M., Class I aldolases: substrate specificity, mechanism, inhibitors and structural aspects, *Prog. Biophys. Molec. Biol.* **63,** 301–340 (1995).

Hofmeyr, J.-H.S. and Cornish-Bowden, A., Regulating the cellular economy of supply and demand, *FEBS Lett.* **476,** 47–51 (2000).

Lindqvist, Y. and Schneider, G., Thiamin diphosphate dependent enzymes: transketolase, pyruvate oxidase and pyruvate decarboxylase, *Curr. Opin. Struct. Biol.* **3,** 896–901 (1993).

Muirhead, H. and Watson, H., Glycolytic enzymes; from hexose to pyruvate, *Curr. Opin. Struct. Biol.* **2,** 870–876 (1992). [A brief summary of the structures of glycolytic enzymes.]

Schirmer, T. and Evans, P.R., Structural basis of the allosteric behaviour of phosphofructokinase, *Nature* **343,** 140–145 (1990).

Scriver, C.R., Beaudet, A., Sly, W.S., and Valle, D. (Eds.), *The Metabolic & Molecular Bases of Inherited Disease* (8th ed.), pp. 4517–4553, McGraw-Hill (2001). [Chapters 70 and 72 discuss fructose and galactose metabolism and their genetic disorders. Chapter 179 discusses glucose-6-phosphate dehydrogenase deficiency.]

KEY TERMS

glycolysis	enediol intermediate	homolactic fermentation	Pasteur effect
pentose phosphate pathway	catalytic perfection	alcoholic fermentation	substrate cycle
aldol cleavage	mutase	TPP	

STUDY EXERCISES

1. Write the reactions of glycolysis, showing the structural formulas of the intermediates and the names of the enzymes that catalyze the reactions.

2. Describe the three possible fates of pyruvate.

3. Describe the mechanisms that regulate phosphofructokinase activity.

4. What is the metabolic advantage of a substrate cycle?

5. Describe how fructose, galactose, and mannose enter the glycolytic pathway.

6. Outline the reactions of the pentose phosphate pathway.

7. How does flux through the pentose phosphate pathway change in response to the need for NADPH or ribose-5-phosphate?

PROBLEMS

1. Which of the 10 reactions of glycolysis are (a) phosphorylations, (b) isomerizations, (c) oxidation–reductions, (d) dehydrations, and (e) carbon–carbon bond cleavages?

2. The aldolase reaction can proceed in reverse as an enzymatic aldol condensation. If the enzyme were not stereospecific, how many different products would be obtained?

3. Bacterial aldolase does not form a Schiff base with the substrate. Instead, it has a divalent Zn^{2+} ion in the active site. How does this ion facilitate the aldolase reaction?

4. Arsenate (AsO_4^{3-}), a structural analog of phosphate, can act as a substrate for any reaction in which phosphate is a substrate. Arsenate esters, unlike phosphate esters, are kinetically as well as thermodynamically unstable and hydrolyze almost instantaneously. Write a balanced overall equation for the conversion of glucose to pyruvate in the presence of ATP, ADP, NAD^+, and either (a) phosphate or (b) arsenate. (c) Why is arsenate a poison?

5. Draw the enediolate intermediates of the ribulose-5-phosphate isomerase reaction (Ru5P → R5P) and the ribulose-5-phosphate epimerase reaction (Ru5P → Xu5P).

6. (a) Why is it possible for the ΔG values in Table 14-1 to differ from the $\Delta G^{\circ\prime}$ values? (b) If a reaction has a $\Delta G^{\circ\prime}$ value of at least -30.5 kJ · mol^{-1}, sufficient to drive the synthesis of ATP ($\Delta G^{\circ\prime} = 30.5$ kJ · mol^{-1}), can it still drive the synthesis of ATP *in vivo* when its ΔG is only -10 kJ · mol^{-1}? Explain.

7. $\Delta G^{\circ\prime}$ for the aldolase reaction is 22.8 kJ · mol^{-1}. In the cell at 37°C, [DHAP]/[GAP] = 5.5. Calculate the equilibrium ratio of [FBP]/[GAP] when [GAP] = 10^{-4} M.

8. The half-reactions involved in the lactate dehydrogenase (LDH) reaction and their standard reduction potentials are

$$Pyruvate + 2\ H^+ + 2\ e^- \longrightarrow lactate \qquad \mathscr{E}^{\circ\prime} = -0.185\ V$$
$$NAD^+ + 2\ H^+ + 2\ e^- \longrightarrow NADH + H^+ \quad \mathscr{E}^{\circ\prime} = -0.315\ V$$

Calculate ΔG at pH 7.0 for the LDH-catalyzed reduction of pyruvate under the following conditions:

(a) [lactate]/[pyruvate] = 1 and [NAD^+]/[NADH] = 1

(b) [lactate]/[pyruvate] = 160 and [NAD^+]/[NADH] = 160

(c) [lactate]/[pyruvate] = 1000 and [NAD^+]/[NADH] = 1000

(d) Discuss the effect of the concentration ratios in parts a–c on the direction of the reaction.

9. Although it is not the primary flux-control point for glycolysis, pyruvate kinase is subject to allosteric regulation. (a) What is the metabolic importance of regulating flux through the pyruvate kinase reaction? (b) What is the advantage of activating pyruvate kinase with fructose-1,6-bisphosphate?

10. Compare the ATP yield of three glucose molecules that enter glycolysis and are converted to pyruvate with that of three glucose molecules that proceed through the pentose phosphate pathway such that their carbon skeletons (as two F6P and one GAP) re-enter glycolysis and are metabolized to pyruvate.

11. If G6P is labeled at its C2 position, where will the label appear in the products of the pentose phosphate pathway?

12. (a) Describe the lengths of the products of the transketolase reaction when the two substrates are both five-carbon sugars. (b) Describe the products of the reaction when the substrates are a five-carbon aldose and a six-carbon ketose. Does it matter which of the substrates binds to the enzyme first?

13. Explain why some tissues continue to produce CO_2 in the presence of high concentrations of fluoride ion, which inhibits glycolysis.

14. The catalytic behavior of liver and brain phosphofructokinase-1 (PFK-1) was observed in the presence of AMP, phosphate, and fructose-2,6-bisphosphate. The following table lists the concentrations of each effector required to achieve 50% of the maximal velocity. Compare the response of the two isozymes to the three effectors and discuss the possible implications of their different responses.

PFK-1 isozyme	Phosphate	AMP	F2,6P
Liver	200 μM	10 μM	0.05 μM
Brain	350 μM	75 μM	4.5 μM

[Problem provided by Kathleen Cornely, Providence College.]

15. Some bacteria catabolize glucose by the Entner–Doudoroff pathway, a variant of glycolysis in which glucose-6-phosphate is converted to 6-phosphogluconate (as in the pentose phosphate pathway) and then to **2-keto-3-deoxy-6-phosphogluconate (KDPG).**

$$
\begin{array}{c}
COO^- \\
| \\
C=O \\
| \\
H-C-H \\
| \\
H-C-OH \\
| \\
H-C-OH \\
| \\
CH_2OPO_3^{2-}
\end{array}
$$

KDPG

Next, an aldolase acts on KDPG. (a) Draw the structures of the products of the KDPG aldolase reaction. (b) Describe how these reaction products are further metabolized by glycolytic enzymes. (c) What is the ATP yield when glucose is metabolized to pyruvate by the Entner–Doudoroff pathway? How does this compare to the ATP yield of glycolysis?

The muscles of animals contain glycogen, a storage form of the metabolic fuel glucose. In living animals, the balance between glycogen synthesis and utilization is carefully regulated. Measurement of glycogen levels in lobster has been used to determine environmental conditions, since animals that have been undernourished or stressed have depleted glycogen stores. [Andrew J. Martinez/Photo Researchers.]

Glycogen Metabolism and Gluconeogenesis

Glycogen (in animals, fungi, and bacteria) and starch (in plants) can function to stockpile glucose for later metabolic use. In animals, a constant supply of glucose is essential for tissues such as the brain and red blood cells, which depend almost entirely on glucose as an energy source (other tissues can also oxidize fatty acids for energy; Section 19-2). The mobilization of glucose from glycogen stores, primarily in the liver, provides a constant supply of glucose (~ 5 mM in blood) to all tissues. When glucose is plentiful, such as immediately after a meal, glycogen synthesis accelerates. Yet the liver's capacity to store glycogen is sufficient to supply the brain with glucose for about half a day. Under fasting conditions, most of the body's glucose needs are met by **gluconeogenesis** (literally, new glucose synthesis) from noncarbohydrate precursors such as amino acids. Not surprisingly, the regulation of glucose synthesis, storage, mobilization, and catabolism by glycolysis (Section 14-2) or the pentose phosphate pathway (Section 14-6) is elaborate and is sensitive to the immediate and long-term energy needs of the organism.

The importance of glycogen for glucose storage is plainly illustrated by the effects of deficiencies of the enzymes that release stored glucose. **McArdle's disease,** for example, is an inherited condition whose major symptom is painful muscle cramps on exertion. The muscles in afflicted individuals lack the enzyme required for glycogen breakdown to yield glucose. Although glycogen is synthesized normally, it cannot supply fuel for glycolysis to keep up with the demand for ATP.

Figure 15-1 summarizes the metabolic uses of glucose. Glucose-6-phosphate (G6P), a key branch point, is derived from free glucose through the action of hexokinase (Section 14-2A) or is the product of glycogen breakdown or gluconeogenesis. G6P has several possible fates: It can be used to synthesize glycogen; it can be catabolized via glycolysis to yield ATP and carbon atoms (as acetyl-CoA) that are further oxidized by the citric acid cycle; and it can be shunted through the pentose phosphate pathway to generate NADPH and/or ribose-5-phosphate. In the liver, G6P can be converted to glucose for export to other tissues via the bloodstream.

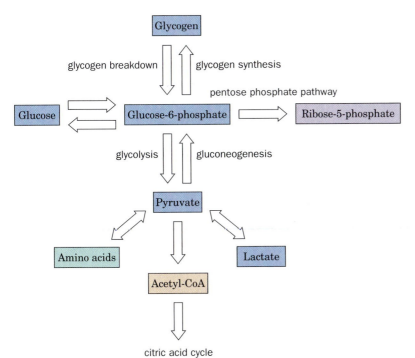

Figure 15-1 Overview of glucose metabolism. Glucose-6-phosphate (G6P) is produced by the phosphorylation of free glucose, by glycogen degradation, and by gluconeogenesis. It is also a precursor for glycogen synthesis and the pentose phosphate pathway. The liver can hydrolyze G6P to glucose. Glucose is metabolized by glycolysis to pyruvate, which can be further broken down to acetyl-CoA for oxidation by the citric acid cycle. Lactate and amino acids, which are reversibly converted to pyruvate, are precursors for gluconeogenesis. 🔎 **See the Animated Figures.**

(a)

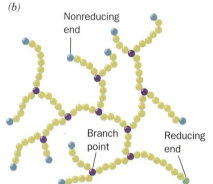

(b)

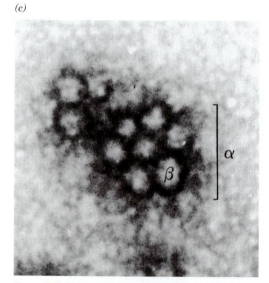

(c)

Figure 15-2 **The structure of glycogen.** (*a*) Molecular formula. In the actual molecule, there are ~12 residues per chain. (*b*) Schematic diagram of glycogen's branched structure. Note that the molecule has many nonreducing ends but only one reducing end. (*c*) Electron micrograph of a glycogen granule from rat skeletal muscle. Each granule (labeled α) consists of several spherical glycogen molecules (β) and associated proteins. [From Calder, P.C., *Int. J. Biochem.* **23**, 1339 (1991). Copyright Elsevier Science. Used with permission.]

The opposing processes of glycogen synthesis and degradation, and of glycolysis and gluconeogenesis, are reciprocally regulated; that is, one is largely turned on while the other is largely turned off. In this chapter, we examine the enzymatic steps of glycogen metabolism and gluconeogenesis, paying particular attention to the regulatory mechanisms that ensure efficient operation of opposing metabolic pathways.

1 Glycogen Breakdown

Glycogen is a polymer of α(1→4)-linked D-glucose with α(1→6)-linked branches every 8–14 residues (Fig. 15-2*a,b* and Section 8-2C). Glycogen occurs as intracellular granules of 100- to 400-Å-diameter spheroidal molecules that each contain up to 120,000 glucose units (Fig. 15-2*c*). The granules are especially prominent in the cells that make the greatest use of glycogen: muscle (up to 1–2% glycogen by weight) and liver cells (up to 10% glycogen by weight; Fig. 8-11). Glycogen granules also contain the enzymes that catalyze glycogen synthesis and degradation as well as many of the proteins that regulate these processes.

Glucose units are mobilized by their sequential removal from the nonreducing ends of glycogen (the ends lacking a C1-OH group). Whereas glycogen has only one reducing end, there is a nonreducing end on every branch. *Glycogen's highly branched structure therefore permits rapid glucose mobilization through the simultaneous release of the glucose units at the end of every branch.*

Glycogen breakdown, or **glycogenolysis,** requires three enzymes:

1. **Glycogen phosphorylase** (or simply **phosphorylase**) catalyzes glycogen **phosphorolysis** (bond cleavage by the substitution of a phosphate group) to yield **glucose-1-phosphate (G1P):**

$$\text{Glycogen} + \text{P}_i \rightleftharpoons \text{glycogen} + \text{G1P}$$
$$(n \text{ residues}) \qquad (n-1 \text{ residues})$$

This enzyme releases a glucose unit only if it is at least five units away from a branch point.

2. **Glycogen debranching enzyme** removes glycogen's branches, thereby making additional glucose residues accessible to glycogen phosphorylase.

BOX 15-1

Pathways of Discovery

Carl and Gerty Cori and Glucose Metabolism

Carl F. Cori (1896–1984)
Gerti T. Cori (1896–1957)

A lifelong collaboration commenced with the marriage of Carl and Gerty Cori in 1920. Although the Coris began their professional work in Austria, they fled the economic and social hardships of Europe in 1922 and moved to Buffalo, New York. They later made their way to Washington University School of Medicine in St. Louis, where Carl served as chair of the Pharmacology department and later, chair of the Biochemistry department. Despite her role as an equal partner in their research work, Gerty officially remained a Research Associate.

The Coris' research focused primarily on the metabolism of glucose. One of their first discoveries was the connection between glucose metabolism in muscle and glycogen metabolism in the liver. The "Cori cycle" (Section 21-1F) describes how lactate produced by glycolysis in active muscle is transported to the liver, where it is used to synthesize glucose that is stored as glycogen until needed.

After describing the interorgan movement of glucose in intact animals, the Coris turned their attention to the metabolic fate of glucose, specifically, the intermediates and enzymes of glucose metabolism. In 1936, using a preparation of minced frog muscle, the Coris found glucose in the form of a phosphate ester (called the Cori ester, now known as glucose-1-phosphate). They traced the presence of the Cori ester to the activity of a phosphorylase (glycogen phosphorylase). This was a notable discovery, because the enzyme used phosphate, rather than water, to split glucose residues from the ends of glycogen chains. Even more remarkably, the enzyme could be made to work in reverse to elongate a glycogen polymer by adding glucose residues (from glucose-1-phosphate). For the first time, a large biological molecule could be synthesized *in vitro*.

During the 1940s, the Coris unraveled many of the secrets of glycogen phosphorylase. For example, they found that the enzyme exists in two forms, one that requires the activator AMP and one that is active in the absence of an allosteric activator. Although it was not immediately appreciated that the differences between the two forms resulted from the presence of covalently bound phosphate, this work laid the foundation for subsequent research on enzyme regulation through phosphorylation and dephosphorylation.

Carl and Gerty Cori also described phosphoglucomutase, the enzyme that converts glucose-1-phosphate to glucose-6-phosphate so that it can participate in other pathways of glucose metabolism. Over time, the Cori lab became a magnet for scientists interested in purifying and characterizing other enzymes of glucose metabolism.

Perhaps because of their experience with discrimination and—especially for Gerty—the lack of equal opportunity, the Cori lab welcomed a more diverse group of scientists than was typical of labs of that era. The Coris received the 1947 Nobel prize in Medicine or Physiology. Several of their junior colleagues, Arthur Kornberg (see Box 24-1), Severo Ochoa, Luis Leloir, Earl Sutherland, Christian de Duve, and Edwin G. Krebs, later earned Nobel prizes of their own, quite possibly reflecting the work ethic, broad view of science and medicine, and meticulous work habits instilled by Carl and Gerty Cori.

Cori, G.T., Colowick, S.P., and Cori, C.F., The activity of the phosphorylating enzyme in muscle extracts, *J. Biol. Chem.* **127**, 771–782 (1939).

Kornberg, A., Remembering our teachers, *J. Biol. Chem. Reflections*, www.jbc.org.

3. **Phosphoglucomutase** converts G1P to G6P, which has several metabolic fates (Fig. 15-1).

Several key features of glycogen metabolism were discovered by the team of Carl and Gerty Cori (Box 15-1).

A Glycogen Phosphorylase

Glycogen phosphorylase is a dimer of identical 842-residue (97-kD) subunits that catalyzes the rate-controlling step in glycogen breakdown. It is regulated both by allosteric interactions and by **covalent modification** (phosphorylation and dephosphorylation). The phosphorylated form of the

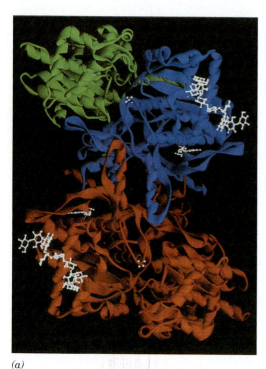

(a)

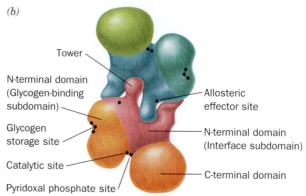

(b)

Figure 15-3 **The X-ray structure of rabbit muscle glycogen phosphorylase.** (a) Ribbon diagram of the phosphorylase a dimer viewed along its molecular twofold axis of symmetry. The bottom subunit is colored orange, and the top subunit's N-terminal and C-terminal domains are colored blue and green, respectively. The various bound ligands are white: The phosphate group at the center of each subunit marks the enzyme's catalytic site; maltoheptose (a glucose heptamer) is bound at each glycogen storage site; and the AMPs at the "back" of the protein identify the allosteric effector sites. [Courtesy of Stephen Sprang, University of Texas Southwestern Medical Center.] (b) An interpretive drawing of the structure in a showing the enzyme's various ligand-binding sites. 🔮 **See Kinemage Exercise 14-1.**

enzyme, **phosphorylase *a*,** has a phosphoryl group esterified to Ser 14. The dephospho form is called **phosphorylase *b*.** Phosphorylase's allosteric inhibitors (ATP, G6P, and glucose) and its allosteric activator (AMP) interact differently with the phospho- and dephosphoenzymes, resulting in an extremely sensitive regulation process (Section 15-3).

The X-ray structures of phosphorylase *a* and phosphorylase *b*, respectively determined by Robert Fletterick and Louise Johnson, are similar. Both structures have a large N-terminal domain (484 residues; the largest known domain) and a smaller C-terminal domain (Fig. 15-3). The N-terminal domain includes the phosphorylation site (Ser 14), the allosteric effector site, a glycogen-binding site (called the glycogen storage site), and all the intersubunit contacts in the dimer. The catalytic site is located at the center of the subunit.

An ~30-Å-long crevice on the surface of the phosphorylase monomer connects the glycogen storage site to the active site. *Since this crevice can accommodate four or five sugar residues in a chain but is too narrow to admit branched oligosaccharides, it provides a clear physical rationale for the inability of phosphorylase to cleave glycosyl residues closer than five units from a branch point.* Presumably, the glycogen storage site increases the catalytic efficiency of phosphorylase by permitting it to phosphorylize many glucose residues on the same glycogen particle without having to dissociate and reassociate completely between catalytic cycles.

Phosphorylase binds the cofactor **pyridoxal-5′-phosphate (PLP;** *at left*), which it requires for activity. This prosthetic group, a **vitamin B$_6$** derivative, is covalently linked to the enzyme via a Schiff base (imine) formed between its aldehyde group and the ε-amino group of Lys 680. PLP also occurs in a variety of enzymes involved in amino acid metabolism, where PLP's conjugated ring system functions catalytically to delocalize electrons (Sections 20-2A and 20-4A). In phosphorylase, however, only the phosphate group participates in catalysis, where it acts as a general acid–base catalyst. Phosphorolysis of glycogen proceeds by a Random mechanism (Section 12-1D) involving an enzyme · P$_i$ · glycogen ternary complex. An oxonium ion in-

Pyridoxal-5′-phosphate (PLP)

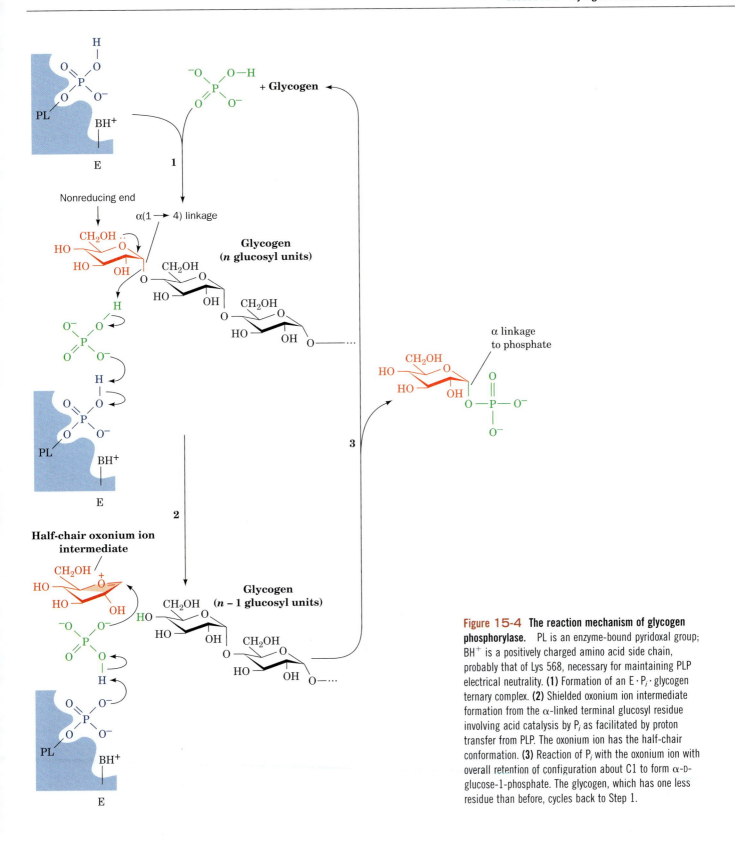

Figure 15-4 **The reaction mechanism of glycogen phosphorylase.** PL is an enzyme-bound pyridoxal group; BH$^+$ is a positively charged amino acid side chain, probably that of Lys 568, necessary for maintaining PLP electrical neutrality. **(1)** Formation of an E · P$_i$ · glycogen ternary complex. **(2)** Shielded oxonium ion intermediate formation from the α-linked terminal glucosyl residue involving acid catalysis by P$_i$ as facilitated by proton transfer from PLP. The oxonium ion has the half-chair conformation. **(3)** Reaction of P$_i$ with the oxonium ion with overall retention of configuration about C1 to form α-D-glucose-1-phosphate. The glycogen, which has one less residue than before, cycles back to Step 1.

termediate forms during C1—O1 bond cleavage, similar to the transition state that forms in the reaction catalyzed by lysozyme (Section 11-4B). The reaction mechanism is diagrammed in Fig. 15-4, which shows the participation of PLP's phosphate group as a general acid–base catalyst.

Conformational Changes in Glycogen Phosphorylase. The structural differences between the active (R) and inactive (T) conformations of phosphorylase (Fig. 15-5) are fairly well understood in terms of the symmetry model of allosterism (Section 7-3B). The T-state enzyme has a buried active site and hence a low affinity for its substrates, whereas the R-state enzyme has an accessible catalytic site and a high-affinity phosphate-binding site.

AMP promotes phosphorylase's T (*inactive*) → R (*active*) conformational shift by binding to the R state of the enzyme at its allosteric effector site. In doing so, AMP's adenine, ribose, and phosphate groups bind to separate segments of the polypeptide chain to link the active site, the subunit interface, and the N-terminal region, the latter having undergone a large conformational shift (involving a 36-Å movement of Ser 14) from its position in the T-state enzyme. AMP binding also causes glycogen phosphorylase's tower helices (Figs. 15-3 and 15-5) to tilt and pull apart so as to pack more favorably. These tertiary movements trigger a concerted T → R transition, which largely consists of an ~10° relative rotation of the two subunits.

The movement of the tower helices displaces and disorders a loop that covers the T-state active site so as to prevent substrate access. Tower movement also causes the Arg 569 side chain, which is located in the active site near the PLP and the P$_i$-binding site, to rotate in a way that increases the enzyme's binding affinity for its anionic P$_i$ substrate (Fig. 15-5).

ATP also binds to the allosteric effector site, but in the T state, so that it inhibits rather than promotes the T → R conformational shift. This is because the β- and γ-phosphate groups of ATP prevent the proper alignment

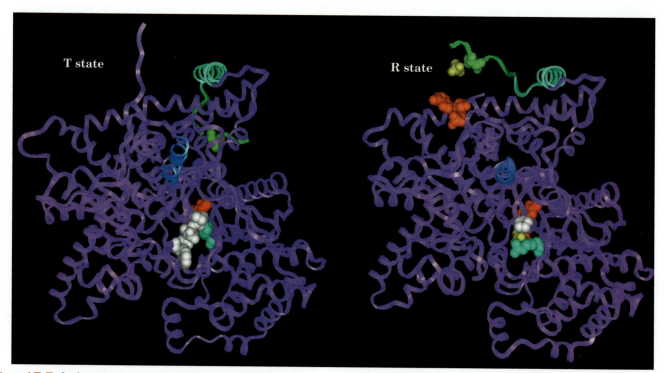

Figure 15-5 Conformational changes in glycogen phosphorylase. One subunit of the dimeric phosphorylase *b* is shown (*left*) in the T state in the absence of allosteric effectors and (*right*) in the R state with bound AMP. The view is of the lower (*orange*) subunit in Fig. 15-3 as seen from the top of the page. The tower helix is blue, the N-terminal helix is cyan, and the N-terminal residues that change conformation on AMP binding are green. Of the groups that are shown in space-filling representation, Ser 14, the phosphorylation site, is light green; AMP is orange; the active site PLP is red; the Arg 569 side chain, which reorients in the T → R transition so as to interact with the substrate phosphate, is cyan; loop residues 282 to 284, which in the R state are mostly disordered and hence not seen, are white; and the phosphates, both at the active site and at the R state Ser 14 phosphorylation site (not present in phosphorylase *b* but shown for position), are yellow. [X-ray structure coordinates courtesy of Stephan Sprang, University of Texas Southwestern Medical Center.]
🖱 **See Kinemage Exercises 14-2 and 14-3.**

of its ribose and α-phosphate groups that result in the conformational changes elicited by AMP.

Phosphorylation and dephosphorylation can alter enzymatic activity in a manner reminiscent of allosteric regulation. The phosphate group has a double negative charge (a property not shared by naturally occurring amino acid residues) and its covalent attachment to a protein can induce dramatic conformational changes. In phosphorylase, the phosphorylation of Ser 14 causes tertiary and quaternary changes as the N-terminal segment moves to allow the phospho-Ser to ion pair with two cationic Arg residues. *The presence of the Ser 14–phosphoryl group causes similar conformational changes as does the binding of AMP, thereby shifting the enzyme's T ⇌ R equilibrium in favor of the R state.* This accounts for the observation that phosphorylase *b* requires AMP for activity and that the *a* form is active without AMP. We shall return to the regulation of phosphorylase activity when we discuss the mechanisms that balance glycogen synthesis against glycogen degradation (Section 15-3).

B Glycogen Debranching Enzyme

Phosphorolysis proceeds along a glycogen branch until it approaches to within four or five residues of an α(1→6) branch point, leaving a "limit branch." Glycogen debranching enzyme acts as an **α(1→4) transglycosylase** (glycosyltransferase) by transferring an α(1→4)-linked trisaccharide unit from a limit branch of glycogen to the nonreducing end of another branch (Fig. 15-6). This reaction forms a new α(1→4) linkage with three more units available for phosphorylase-catalyzed phosphorolysis. The

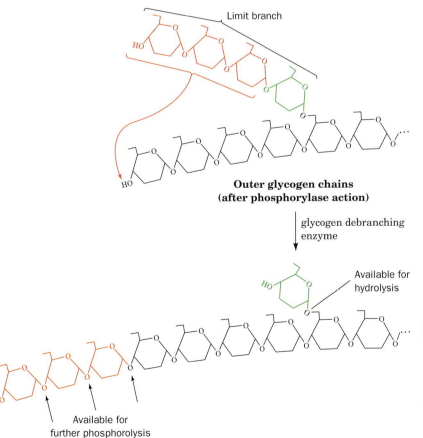

Figure 15-6 The reactions catalyzed by debranching enzyme. The enzyme transfers the terminal three α(1 → 4)-linked glucose residues from a "limit branch" of glycogen to the nonreducing end of another branch. The α(1 → 6) bond of the residue remaining at the branch point is hydrolyzed by further action of debranching enzyme to yield free glucose. The newly elongated branch is subject to degradation by glycogen phosphorylase.

α(1→6) bond linking the remaining glycosyl residue in the branch to the main chain is hydrolyzed (not phosphorylyzed) by the same debranching enzyme to yield glucose and debranched glycogen. About 10% of the residues in glycogen (those at the branch points) are therefore converted to glucose rather than G1P. *Debranching enzyme has separate active sites for the transferase and the α(1→6)-glucosidase reactions.* The presence of two independent catalytic activities on the same enzyme no doubt improves the efficiency of the debranching process.

The maximal rate of the glycogen phosphorylase reaction is much greater than that of the glycogen debranching reaction. Consequently, the outermost branches of glycogen, which constitute nearly half of its residues, are degraded in muscle in a few seconds under conditions of high metabolic demand. Glycogen degradation beyond this point requires debranching and hence occurs more slowly. This, in part, accounts for the fact that a muscle can sustain its maximum exertion for only a few seconds.

C Phosphoglucomutase

Phosphorylase converts the glucosyl units of glycogen to G1P, which, in turn, is converted by phosphoglucomutase to G6P. The phosphoglucomutase reaction is similar to that catalyzed by phosphoglycerate mutase (Section 14-2H). A phosphoryl group is transferred from the active phosphoenzyme to G1P, forming **glucose-1,6-bisphosphate (G1,6P),** which then rephosphorylates the enzyme to yield G6P (Fig. 15-7; this near-equilibrium reaction also functions in reverse). An important difference between this enzyme and phosphoglycerate mutase is that the phosphoryl group in phosphoglucomutase is covalently bound to a Ser hydroxyl group rather than to a His imidazole nitrogen.

Glucose-6-Phosphatase Generates Glucose in the Liver. The G6P produced by glycogen breakdown can continue along the glycolytic pathway or the pentose phosphate pathway (note that the glucose is already phosphorylated, so that the ATP-consuming hexokinase-catalyzed phosphorylation of glu-

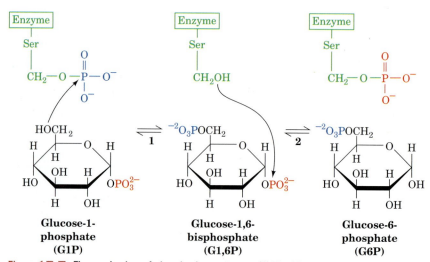

Figure 15-7 The mechanism of phosphoglucomutase. (1) The OH group at C6 of G1P attacks the phosphoenzyme to form a dephosphoenzyme–G1,6P intermediate. (2) The Ser OH group on the dephosphoenzyme attacks the phosphoryl group at C1 to regenerate the phosphoenzyme with the formation of G6P.

cose is bypassed). In the liver, G6P is also made available for use by other tissues. Because G6P cannot pass through the cell membrane, it is first hydrolyzed by **glucose-6-phosphatase (G6Pase):**

$$G6P + H_2O \rightarrow glucose + P_i$$

Although G6P is produced in the cytosol, G6Pase resides in the endoplasmic reticulum (ER) membrane. Consequently G6P must be imported into the ER by a **G6P translocase** before it can be hydrolyzed. The resulting glucose and P_i are then returned to the cytosol via specific transport proteins. A defect in any of the components of this G6P hydrolysis system results in **type I glycogen storage disease** (Box 15-2). Glucose leaves the liver cell via a specific glucose transporter named **GLUT2** and is carried by the blood to other tissues. Muscle and other tissues lack G6Pase and therefore retain their G6P.

2 Glycogen Synthesis

The $\Delta G^{\circ\prime}$ for the glycogen phosphorylase reaction is $+3.1 \text{ kJ} \cdot \text{mol}^{-1}$, but under physiological conditions, glycogen breakdown is exergonic ($\Delta G^{\circ\prime} = -5$ to $-8 \text{ kJ} \cdot \text{mol}^{-1}$). The synthesis of glycogen from G1P under physiological conditions is therefore thermodynamically unfavorable without free energy input. Consequently, *glycogen synthesis and breakdown must occur by separate pathways*. This recurrent metabolic strategy—that biosynthetic and degradative pathways of metabolism are different—is particularly important when both pathways must operate under similar physiological conditions. This situation is thermodynamically impossible if one pathway is just the reverse of the other.

It was not thermodynamics, however, that led to recognition of the separation of synthetic and degradative pathways for glycogen, but McArdle's disease. Individuals with the disease lack muscle glycogen phosphorylase activity and therefore cannot break down glycogen. Yet their muscles contain moderately high quantities of normal glycogen. Clearly, glycogen synthesis does not require glycogen phosphorylase. In this section, we describe the three enzymes that participate in glycogen synthesis: **UDP–glucose pyrophosphorylase, glycogen synthase,** and **glycogen branching enzyme.** The opposing reactions of glycogen synthesis and degradation are diagramed in Fig. 15-8.

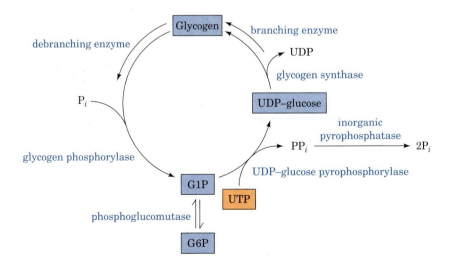

Figure 15-8 Opposing pathways of glycogen synthesis and degradation. The exergonic process of glycogen breakdown is reversed by a process that uses UTP to generate a UDP–glucose intermediate.

BOX 15-2

Biochemistry in Health and Disease

Glycogen Storage Diseases

Glycogen storage diseases are inherited disorders that affect glycogen metabolism, producing glycogen that is abnormal in either quantity or quality. Studies of the genetic defects that underlie these diseases have helped elucidate the complexities of glycogen metabolism (e.g., McArdle's disease). Conversely, the biochemical characterization of the pathways affected by a genetic disease often leads to useful strategies for its treatment. The table on p. 483 lists the enzyme deficiencies associated with each type of glycogen storage disease.

Glycogen storage diseases that mainly affect the liver generally produce **hepatomegaly** (enlarged liver) and **hypoglycemia** (low blood sugar), whereas glycogen storage diseases that affect the muscles cause muscle cramps and weakness. Both types of disease may also cause cardiovascular and renal disturbances.

Type I: Glucose-6-Phosphatase Deficiency (von Gierke's Disease).
Glucose-6-phosphatase catalyzes the final step leading to the release of glucose into the bloodstream by the liver. Deficiency of this enzyme results in an increase of intracellular [G6P], which leads to a large accumulation of glycogen in the liver and kidney (recall that G6P activates glycogen synthase) and an inability to increase blood glucose concentration in response to the hormones glucagon or epinephrine. The symptoms of Type I glycogen storage disease include severe hepatomegaly and hypoglycemia and a general failure to thrive. Treatment of the disease has included drug-induced inhibition of glucose uptake by the liver (to increase blood [glucose]), continuous intragastric feeding overnight (again to increase blood [glucose]), surgical transposition of the portal vein, which ordinarily feeds the liver directly from the intestines (to allow this glucose-rich blood to reach peripheral tissues before it reaches the liver), and liver transplantation.

Type II: α-1,4-Glucosidase Deficiency (Pompe's Disease).
α-1,4-Glucosidase deficiency is the most devastating of the glycogen storage diseases. It results in a large accumulation of glycogen of normal structure in the lysosomes of all cells and causes death by cardiorespiratory failure, usually before the age of 1 year. α-1,4-Glucosidase is not involved in the main pathways of glycogen metabolism. It occurs in lysosomes, where it hydrolyzes maltose (a glucose disaccharide) and other linear oligosaccharides, as well as the outer branches of glycogen, thereby yielding free glucose. Normally, this alternative pathway of glycogen metabolism is not quantitatively important, and its physiological significance is not known.

Type III: Amylo-1,6-Glucosidase (Debranching Enzyme) Deficiency (Cori's Disease).
In Cori's disease, glycogen of abnormal structure containing very short outer chains accumulates in both liver and muscle since, in the absence of debranching enzyme, the glycogen cannot be further degraded. The resulting hypoglycemia is not as severe as in von Gierke's disease (Type I) and can be treated with frequent feedings and a high-protein diet (to offset the loss of amino acids used for gluconeogenesis). For unknown reasons, the symptoms of Cori's disease often disappear at puberty.

Type IV: Amylo-(1,4→1,6)-Transglycosylase (Branching Enzyme) Deficiency (Andersen's Disease).
Andersen's disease is one of the most severe glycogen storage diseases; victims rarely survive past the age of 4 years because of liver dysfunction. Liver glycogen is present in normal concentrations, but it contains long unbranched chains that greatly reduce its solubility. The liver dysfunction may be caused by a "foreign body" immune reaction to the abnormal glycogen.

Type V: Muscle Phosphorylase Deficiency (McArdle's Disease).
The symptoms of McArdle's disease, painful muscle cramps on exertion, typically do not appear until early adulthood and can be prevented by avoiding strenuous exercise. This condition affects glycogen metabolism in muscle but not in liver, which contains normal amounts of a different phosphorylase isozyme.

Type VI: Liver Phosphorylase Deficiency (Hers' Disease).
Patients with a deficiency of liver glycogen phosphorylase have symptoms similar to those with mild forms of Type I glycogen storage disease. The hypoglycemia in this case results from

the inability of liver glycogen phosphorylase to respond to the need for circulating glucose.

Type VII: Muscle Phosphofructokinase Deficiency (Tarui's Disease).

The result of a deficiency of the glycolytic enzyme PFK in muscle is an abnormal buildup of the glycolytic metabolites G6P and F6P. High concentrations of G6P increase the activities of glycogen synthase and UDP–glucose pyrophosphorylase (G6P is in equilibrium with G1P, a substrate for UDP–glucose pyrophosphorylase) so that glycogen accumulates in muscle. Other symptoms are similar to those of muscle phosphorylase deficiency, since PFK deficiency prevents glycolysis from keeping up with the ATP demand in contracting muscle.

Type VIII: X-Linked Phosphorylase Kinase Deficiency.

Some individuals with symptoms of Type VI glycogen storage disease have normal phosphorylase enzymes but a defective phosphorylase kinase, which results in their inability to convert phosphorylase *b* to phosphorylase *a*. The α subunit of phosphorylase kinase is encoded by a gene on the X chromosome, so Type VIII disease is X-linked rather than autosomal recessive, as are the other glycogen storage diseases.

Type IX: Phosphorylase Kinase Deficiency.

Phosphorylase kinase deficiency, an autosomal recessive disease, results from a mutation in one of the genes that encode the β, γ, and δ subunits of phosphorylase kinase. Because different tissues contain different phosphorylase kinase isozymes, the symptoms and severity of the disease vary according to the affected organs. Techniques for identifying genetic lesions are therefore more reliable than clinical symptoms for diagnosing a particular glycogen storage disease.

Type 0: Liver Glycogen Synthase Deficiency.

Liver glycogen synthase deficiency is the only disease of glycogen metabolism in which there is a deficiency rather than an overabundance of glycogen. The activity of liver glycogen synthase is extremely low in individuals with Type 0 disease, who exhibit hyperglycemia after meals and hypoglycemia at other times. Some individuals, however, are asymptomatic, which suggests that there may be multiple forms of this autosomal recessive disorder.

Hereditary Glycogen Storage Diseases

Type	Enzyme Deficiency	Tissue	Common Name	Glycogen Structure
I	Glucose-6-phosphatase	Liver	von Gierke's disease	Normal
II	α-1,4-Glucosidase	All lysosomes	Pompe's disease	Normal
III	Amylo-1,6-glucosidase (debranching enzyme)	All organs	Cori's disease	Outer chains missing or very short
IV	Amylo-(1,4→1,6)-transglycosylase (branching enzyme)	Liver, probably all organs	Andersen's disease	Very long unbranched chains
V	Glycogen phosphorylase	Muscle	McArdle's disease	Normal
VI	Glycogen phosphorylase	Liver	Hers' disease	Normal
VII	Phosphofructokinase	Muscle	Tarui's disease	Normal
VIII	Phosphorylase kinase	Liver	X-linked phosphorylase kinase deficiency	Normal
IX	Phosphorylase kinase	All organs		Normal
0	Glycogen synthase	Liver		Normal, deficient in quantity

A UDP–Glucose Pyrophosphorylase

Since the direct conversion of G1P to glycogen and P_i is thermodynamically unfavorable (positive ΔG) under physiological conditions, glycogen biosynthesis requires an exergonic step. This is accomplished, as Luis Leloir discovered in 1957, by combining G1P with uridine triphosphate (UTP) in a reaction catalyzed by UDP–glucose pyrophosphorylase (Fig. 15-9). The product of this reaction, **uridine diphosphate glucose (UDP–glucose** or **UDPG),** is an "activated" compound that can donate a glucosyl unit to the growing glycogen chain. The formation of UDPG itself has $\Delta G^{\circ\prime} \approx 0$ (it is a phosphoanhydride exchange reaction), but the subsequent exergonic hydrolysis of PP_i by the omnipresent enzyme inorganic pyrophosphatase makes the overall reaction exergonic.

	$\Delta G^{\circ\prime}$ (kJ · mol^{-1})
G1P + UTP $\rightleftharpoons$ UDPG + PP_i	~ 0
H_2O + PP_i $\rightarrow$ 2 P_i	-19.2
Overall G1P + UTP $\rightarrow$ UDPG + 2 P_i	-19.2

This is an example of the common biosynthetic strategy of cleaving a nucleoside triphosphate to form PP_i. The free energy of PP_i hydrolysis can then be used to drive an otherwise unfavorable reaction to completion (Section 13-2B); the near total elimination of the PP_i by the highly exergonic (irreversible) pyrophosphatase reaction prevents the reverse of the PP_i-producing reaction from occurring.

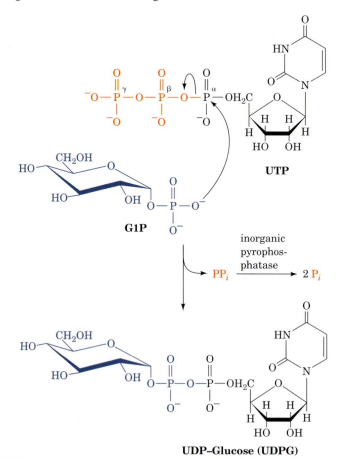

Figure 15-9 The reaction catalyzed by UDP–glucose pyrophosphorylase.   In this phosphoanhydride exchange reaction, the phosphoryl oxygen of G1P attacks the α-phosphorus atom of UTP to form UDP–glucose and PP_i. The PP_i is rapidly hydrolyzed by inorganic pyrophosphatase.

B Glycogen Synthase

In the next step of glycogen synthesis, the glycogen synthase reaction, the glucosyl unit of UDPG is transferred to the C4-OH group on one of glycogen's nonreducing ends to form an $\alpha(1\rightarrow4)$ glycosidic bond. The $\Delta G^{\circ\prime}$ for the glycogen synthase reaction

$$\text{UDPG} + \text{glycogen} \rightarrow \text{UDP} + \text{glycogen}$$
$$(n \text{ residues}) \qquad\qquad (n + 1 \text{ residues})$$

is $-13.4\,\text{kJ}\cdot\text{mol}^{-1}$, making the overall reaction spontaneous under the same conditions that glycogen breakdown by glycogen phosphorylase is also spontaneous. However, glycogen synthesis does have an energetic price. Combining the first two reactions of glycogen synthesis gives

$$\text{Glycogen} + \text{G1P} + \text{UTP} \rightarrow \text{glycogen} + \text{UDP} + 2\,\text{P}_i$$
$$(n \text{ residues}) \qquad\qquad (n + 1 \text{ residues})$$

Thus, *one molecule of UTP is cleaved to UDP for each glucose residue incorporated into glycogen*. The UTP is replenished through a phosphoryl-transfer reaction mediated by nucleoside diphosphate kinase (Section 13-2C):

$$\text{UDP} + \text{ATP} \rightleftharpoons \text{UTP} + \text{ADP}$$

so that UTP consumption is energetically equivalent to ATP consumption.

The transfer of a glucosyl unit from UDPG to a growing glycogen chain involves the formation of a glycosyl oxonium ion by the elimination of UDP, a good leaving group (Fig. 15-10). The enzyme is inhibited by **1,5-gluconolactone** (*at right*), an analog that mimics the oxonium ion's half-chair geometry. The same analog inhibits both glycogen phosphorylase (Section 15-1A) and lysozyme (Section 11-4), which have similar mechanisms.

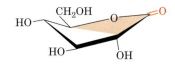

1,5-Gluconolactone

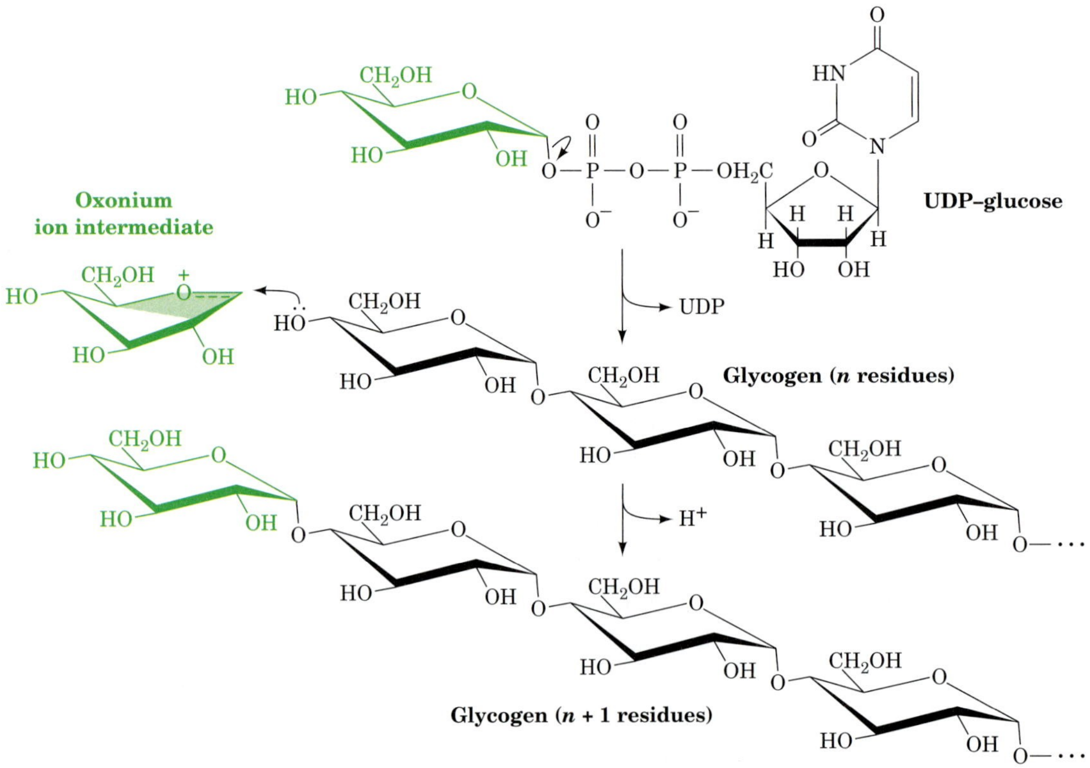

Figure 15-10 The reaction catalyzed by glycogen synthase. This reaction involves a glucosyl oxonium ion intermediate.

Human muscle glycogen synthase is a homotetramer of 737-residue subunits (the liver isozyme has 703-residue subunits). Like glycogen phosphorylase, it has two enzymatically interconvertible forms; in this case, however, the phosphorylated *b* form is less active, and the original (dephosphorylated) *a* form is more active. (Note: For enzymes subject to covalent modification, "*a*" refers to the more active form and "*b*" refers to the less active form.)

Glycogen synthase is under allosteric control; it is strongly inhibited by physiological concentrations of ATP, ADP, and P_i. In fact, the phosphorylated enzyme is almost totally inactive *in vivo*. The dephosphorylated enzyme, however, can be activated by G6P, so the cell's glycogen synthase activity varies with [G6P] and the fraction of the enzyme in its dephosphorylated form. The mechanistic details of the interconversion of phosphorylated and dephosphorylated forms of glycogen synthase are complex and are not as well understood as those of glycogen phosphorylase (for one thing, glycogen synthase has multiple phosphorylation sites). We shall discuss the regulation of glycogen synthase further in Section 15-3B.

Glycogenin Primes Glycogen Synthesis. Glycogen synthase cannot simply link together two glucose residues; it can only extend an already existing $\alpha(1\rightarrow4)$-linked glucan chain. How, then, is glycogen synthesis initiated? In the first step of this process, a 349-residue protein named **glycogenin,** acting as a glycosyltransferase, attaches a glucose residue donated by UDPG to the OH group of its Tyr 194. Glycogenin then extends the glucose chain by up to seven additional UDPG-donated glucose residues to form a glyco-

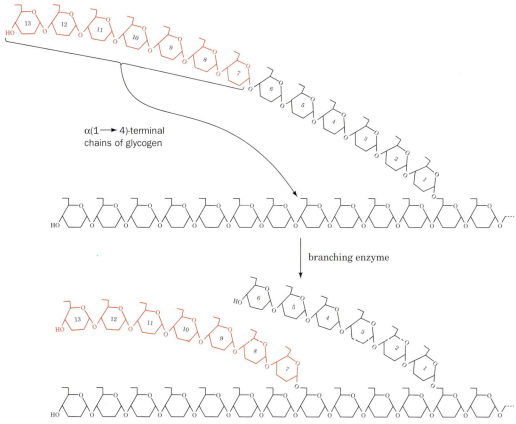

$\alpha(1\longrightarrow4)$-terminal chains of glycogen

branching enzyme

Figure 15-11 The branching of glycogen. Branches are formed by transferring a 7-residue terminal segment from an $\alpha(1\rightarrow4)$-linked glucan chain to the C6-OH group of a glucose residue on the same or another chain.

gen "primer." Only at this point does glycogen synthase commence glycogen synthesis by extending the primer. Analysis of glycogen granules suggests that each glycogen molecule is associated with only one molecule each of glycogenin and glycogen synthase.

C Glycogen Branching Enzyme

Glycogen synthase generates only α(1→4)-linkages to yield α-amylose. Branching to form glycogen is accomplished by a separate enzyme, **amylo-(1,4→1,6)-transglycosylase (branching enzyme),** which is distinct from glycogen debranching enzyme (Section 15-1B). A branch is created by transferring a 7-residue segment from the end of a chain to the C6-OH group of a glucose residue on the same or another glycogen chain (Fig. 15-11). Each transferred segment must come from a chain of at least 11 residues, and the new branch point must be at least 4 residues away from other branch points. The branching pattern of glycogen has been optimized by evolution for the efficient storage and mobilization of glucose (see Box 15-3).

3 Control of Glycogen Metabolism

If glycogen synthesis and breakdown proceed simultaneously, all that is accomplished is the wasteful hydrolysis of UTP. Glycogen metabolism must therefore be controlled according to cellular needs. *The regulation of glycogen metabolism involves allosteric control as well as hormonal control by covalent modification of the pathway's regulatory enzymes.*

A Direct Allosteric Control of Glycogen Phosphorylase and Glycogen Synthase

As we saw in Sections 13-1D and 14-4B, the net flux, J, of reactants through a step in a metabolic pathway is the difference between the forward and reverse reaction velocities, v_f and v_r. However, the flux varies dramatically with substrate concentration as the reaction approaches equilibrium ($v_f \approx v_r$). The flux through a near-equilibrium reaction is therefore all but uncontrollable. *Precise flux control of a pathway is possible when an enzyme functioning far from equilibrium is opposed by a separately controlled enzyme. Then, v_f and v_r vary independently and v_r can be larger or smaller than v_f, allowing control of both rate and direction.* Exactly this situation occurs in glycogen metabolism through the opposition of the glycogen phosphorylase and glycogen synthase reactions.

Both glycogen phosphorylase and glycogen synthase are under allosteric control by effectors that include ATP, G6P, and AMP. Muscle glycogen phosphorylase is activated by AMP and inhibited by ATP and G6P. Glycogen synthase, on the other hand, is activated by G6P. This suggests that when there is high demand for ATP (low [ATP], low [G6P], and high [AMP]), glycogen phosphorylase is stimulated and glycogen synthase is inhibited, which favors glycogen breakdown. Conversely, when [ATP] and [G6P] are high, glycogen synthesis is favored.

In vivo, this allosteric scheme is superimposed on an additional control system based on covalent modification. For example, phosphorylase *a* is active even without AMP stimulation (Section 15-1A), and glycogen synthase is essentially inactive (Section 15-2B) unless it is dephosphorylated and G6P is present. *Thus, covalent modification (phosphorylation and dephosphorylation) of glycogen phosphorylase and glycogen synthase provides a more so-*

BOX 15-3

Perspectives in Biochemistry

Optimizing Glycogen Structure

The function of glycogen in animal cells is to store the metabolic fuel glucose and to release it rapidly when needed. Glucose must be stored as a polymer, because glucose itself could not be stored without a drastic increase in intracellular osmotic pressure (Section 2-1D). It has been estimated that the total concentration of glucose residues stored as glycogen in a liver cell is ~0.4 M, whereas the concentration of glycogen is only ~10 nM. This huge difference mitigates osmotic stress.

To fulfill its biological function, the glycogen polymer must store the largest amount of glucose in the smallest possible volume while maximizing both the amount of glucose available for release by glycogen phosphorylase and the number of nonreducing ends (to maximize the rate at which glucose residues can be mobilized). All these criteria must be met by optimizing just two variables: the degree of branching and chain length.

In a glycogen molecule, shown schematically here,

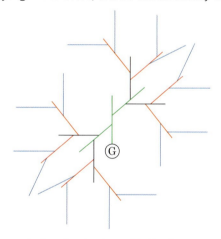

the glycogen chains, beginning with the innermost chain attached to glycogenin (G), have two branches (the outermost chains are unbranched). The entire molecule is roughly spherical and is organized in tiers. There are an estimated 12 tiers in mature glycogen (only 4 are shown above).

With two branches per chain, the number of chains in a given tier is twice the number of the preceding tier, and the outermost tier contains about half of the total glucose residues (regardless of the number of tiers). When the degree of branching increases, for example, to three branches per chain, the proportion of residues in the outermost tier increases, but so does the density of glucose residues. This severely limits the maximum size of the glycogen particle and the number of glucose residues it can accommodate. Thus, glycogen has around two branches per chain.

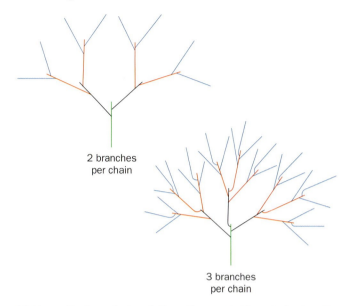

2 branches
per chain

3 branches
per chain

Mathematical analysis of the other variable, chain length, yields an optimal value of 13, which is in good agreement with the actual length of glycogen chains in cells (8–14 residues). Consider the two simplified glycogen molecules shown below, which contain the same number of glucose residues (the same total length of line segments) and the same branching pattern:

The molecule with the shorter chains packs more glucose in a given volume and has more points for phosphorylase attack, but only about half the amount of glucose can be released before debranching must occur (debranching is much slower than phosphorolysis). In the less dense molecule, the longer chains increase the number of residues that can be continuously phosphorylyzed; however, there are fewer points of attack. Thirteen residues is apparently a compromise for mobilizing the largest amount of glucose in the shortest time.

Amylopectin (Section 8-2C), which is chemically similar to glycogen, is a much larger molecule and has longer chains. Amylose lacks branches altogether. Evidently, starch, unlike glycogen, is not designed for rapid mobilization of metabolic fuel.

[Figures adapted from Meléndez-Hevia, E., Waddell, T.G., and Shelton, E.D., *Biochem. J.* **295**, 477–483 (1993).]

phisticated control system that modulates the responsiveness of these enzymes to their allosteric effectors.

B Covalent Modification of Glycogen Phosphorylase and Glycogen Synthase

See Guided Exploration 13

Control of glycogen breakdown.

The interconversion of the *a* and *b* forms of glycogen synthase and glycogen phosphorylase is accomplished through enzyme-catalyzed phosphorylation and dephosphorylation, a process that is under hormonal control (Section 15-3C). Enzymatically interconvertible enzyme systems can therefore respond to a greater number of effectors than simple allosteric systems. Furthermore, if the enzymes that covalently modify a target enzyme are themselves under allosteric control, it is possible for a small change in the concentration of an allosteric effector of a modifying enzyme (resulting from hormonal stimulation, for example) to cause a large change in the activity of its modifiable target enzyme. A set of kinases and phosphatases linked in cascade fashion therefore has enormous potential for signal amplification and flexibility in response to different metabolic signals. Note that the correlation between phosphorylation and enzyme activity varies with the enzyme. For example, glycogen phosphorylase is activated by phosphorylation ($b \rightarrow a$), whereas glycogen synthase is inactivated by phosphorylation ($a \rightarrow b$). Conversely, dephosphorylation inactivates glycogen phosphorylase and activates glycogen synthase.

Glycogen Phosphorylase Is Activated by Phosphorylation.   The cascade that governs enzymatic interconversion of glycogen phosphorylase involves three enzymes (Fig. 15-12):

1. *Phosphorylase kinase,* which specifically phosphorylates Ser 14 of glycogen phosphorylase *b.*

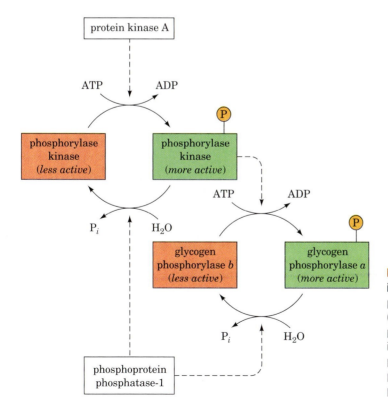

Figure 15-12 The glycogen phosphorylase interconvertible enzyme system.   Conversion of phosphorylase *b* (the less active form) to phosphorylase *a* (the more active form) is accomplished through phosphorylation catalyzed by phosphorylase kinase, which is itself subject to activation through phosphorylation by protein kinase A (PKA). Both glycogen phosphorylase *a* and phosphorylase kinase are dephosphorylated by phosphoprotein phosphatase-1.

> **2. *Protein kinase A (PKA),*** which phosphorylates and thereby activates phosphorylase kinase.
>
> **3. *Phosphoprotein phosphatase-1,*** which dephosphorylates and thereby deactivates both glycogen phosphorylase *a* and phosphorylase kinase.

Phosphorylase *b* is sensitive to allosteric effectors, but phosphorylase *a* is much less so (Fig. 15-13). In the resting cell, the concentrations of ATP and G6P are high enough to inhibit phosphorylase *b*. The level of phosphorylase activity is therefore largely determined by the fraction of enzyme present as phosphorylase *a*. The steady state fraction of phosphorylated enzyme depends on the relative activities of phosphorylase kinase, PKA, and phosphoprotein phosphatase-1. Let us examine the factors that regulate the activities of these enzymes before we return to the regulation of glycogen synthase activity.

Protein Kinase A.   The primary intracellular signal for glycogen phosphorylase activation by phosphorylase kinase is **adenosine-3′,5′-cyclic monophosphate (cyclic AMP** or **cAMP).** The cAMP concentration in a cell is a function of the ratio of its rate of synthesis from ATP by **adenylate cyclase** and its rate of breakdown to AMP by a specific **phosphodiesterase:**

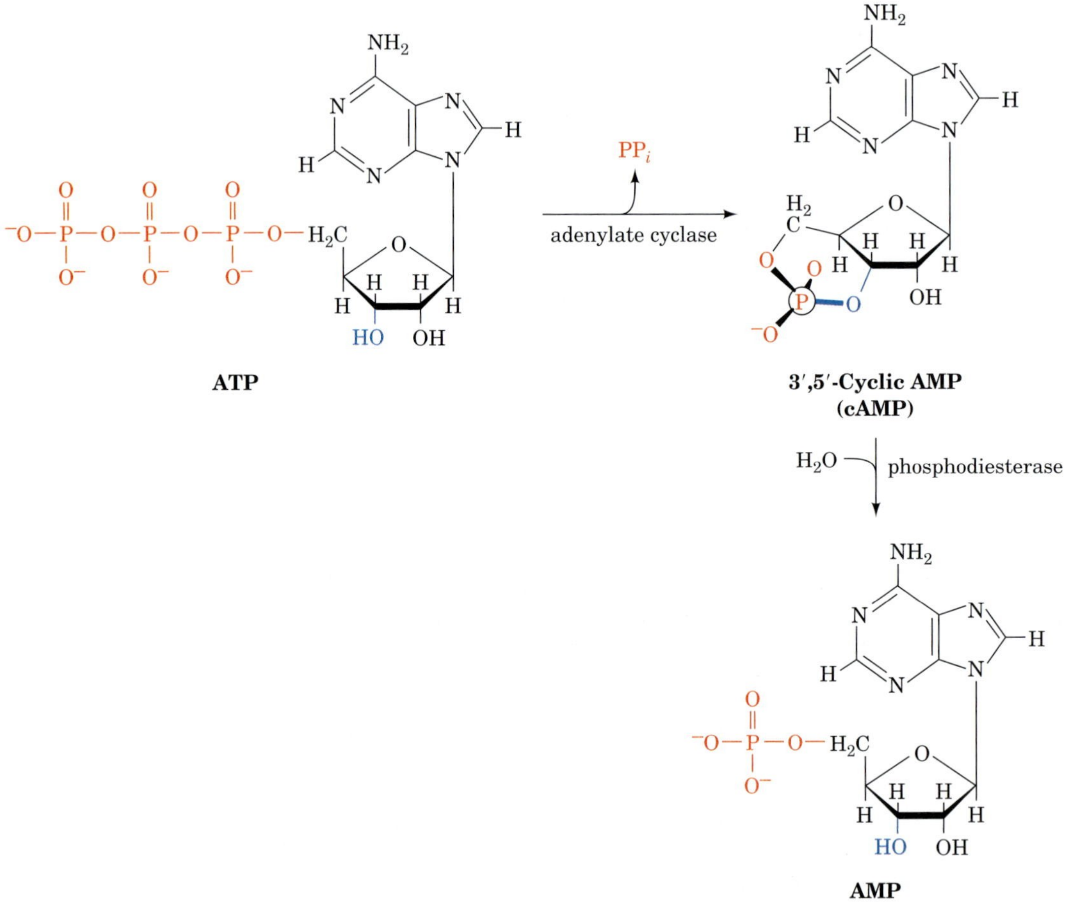

ATP

3′,5′-Cyclic AMP (cAMP)

AMP

Adenylate cyclase, a transmembrane protein, is stimulated by the binding of certain hormones (Section 15-3C) to their cell-surface receptors, to catalyze the synthesis of cAMP inside the cell.

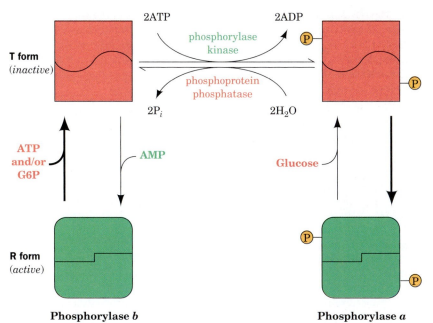

Figure 15-13   The control of glycogen phosphorylase activity.   The enzyme may assume the enzymatically inactive T conformation (*above*) or the catalytically active R form (*below*). The conformation of phosphorylase *b* is allosterically controlled by effectors such as AMP, ATP, and G6P and is mostly in the T state under physiological conditions. In contrast, the modified form of the enzyme, phosphorylase *a,* is largely unresponsive to these effectors and is mostly in the R state unless there is a high level of glucose. Under usual physiological conditions, the enzymatic activity of glycogen phosphorylase is essentially determined by its rates of modification and demodification. Note that only the T-state enzyme is subject to phosphorylation and dephosphorylation, so effector binding influences the rates of these modification/demodification events.

cAMP is absolutely required for the activity of PKA (also known as **cAMP-dependent protein kinase or cAPK),** an enzyme that phosphorylates specific Ser or Thr residues of numerous cellular proteins, including phosphorylase kinase and glycogen synthase. These proteins all contain a consensus kinase-recognition sequence, Arg-Arg-X-Ser/Thr-Y, where Ser/Thr is the phosphorylation site, X is any small residue, and Y is a large hydrophobic residue. In the absence of cAMP, PKA is an inactive tetramer consisting of two regulatory and two catalytic subunits, R_2C_2. The cAMP binds to the regulatory subunit to cause the dissociation of active catalytic monomers:

$$\underset{(\textit{inactive})}{R_2C_2} + 4\ cAMP \rightleftharpoons \underset{(\textit{active})}{2\ C} + R_2(cAMP)_4$$

The intracellular concentration of cAMP therefore determines the fraction of PKA in its active form and thus the rate at which it phosphorylates its substrates.

The X-ray structure of the 350-residue C subunit of mouse PKA in complex with ATP and a 20-residue inhibitor peptide, which was determined by Susan Taylor and Janusz Sowadski, is shown in Fig. 15-14. The C subunit, like other kinases (Figs. 14-2 and 14-10), is bilobal. In the PKA struc-

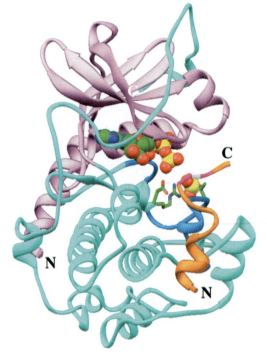

Figure 15-14   X-ray structure of the catalytic (C) subunit of mouse protein kinase A (PKA). The protein is in complex with ATP and a 20-residue peptide segment of a naturally occurring protein kinase inhibitor. The N-terminal domain is pink, the C-terminal domain is cyan, and the "activation loop" containing Thr 197 is light blue. The polypeptide inhibitor is orange, and its pseudo-target sequence, Arg-Arg-Asn-Ala-Ile, is magenta (the Ala, which replaces the Ser or Thr of a true substrate, is white). The substrate ATP and the phosphoryl group of phospho-Thr 197 are shown in space-filling form and the side chains of the catalytically essential Arg 165, Asp 166, and Thr 197 are shown in stick form, all colored according to atom type (C green, N blue, O red, and P yellow). Note that the inhibitor's pseudo-target sequence is close to ATP's γ phosphate group, the group that the enzyme transfers to the Ser or Thr of the target sequence. [Based on an X-ray structure by Susan Taylor and Janusz Sowadski, University of California at San Diego. PDBid 1ATP.]

🔖 See the Interactive Exercises and Kinemage Exercise 15.

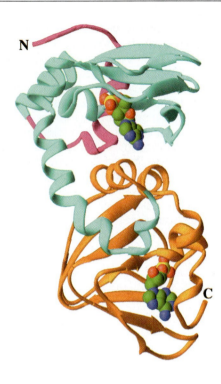

Figure 15-15 X-ray structure of the regulatory (R) subunit of bovine protein kinase A (PKA) in complex with cAMP. Domain A is cyan, and domain B is orange. The cAMP molecules are shown in space-filling form colored according to atom type (C green, N blue, O red, and P yellow). The region containing the autoinhibitor segment is magenta. The N-terminal 91 residues of the R subunit, which mediate dimerization, are absent from this polypeptide. [Based on an X-ray structure by Susan Taylor, University of California at San Diego. PDBid 1RGS.]

ture, the deep cleft between the lobes is occupied by ATP and a segment of the inhibitor peptide that resembles the 5-residue consensus sequence for phosphorylation except that the phosphorylated Ser/Thr is replaced by Ala. The C subunit of PKA itself must be phosphorylated at Thr 197 for maximal activity. The phospho-Thr side chain helps orient the kinase's active site residues.

A large family of protein kinases play key roles in the signaling pathways by which many hormones, growth factors, neurotransmitters, and toxins affect the functions of their target cells (Section 21-3). Indeed, analysis of the human genome has revealed the sequences of several hundred kinases. The activities of such enzymes are reflected in the observation that up to 30% of the proteins in mammalian cells are phosphorylated.

The R subunit of protein kinase A competitively inhibits its C subunit. The R subunit has a well-defined structure containing two homologous cAMP-binding domains, A and B, and a so-called **autoinhibitor segment** (Fig. 15-15). In the inactive R_2C_2 complex, the autoinhibitor segment, which resembles the C subunit's substrate, binds in the C subunit's active site (as does the inhibitory peptide in Fig. 15-14) so as to block substrate binding. The binding of two cAMP molecules to each of its subunits causes the R_2 dimer to release its two bound C subunits.

Phosphorylase Kinase. Phosphorylase kinase is a 1300-kD protein with four nonidentical subunits, known as α, β, γ, and δ. The γ subunit contains the catalytic site, and the other three subunits have regulatory functions. *Phosphorylase kinase is maximally activated by Ca^{2+} and by the phosphorylation of its α and β subunits by PKA.*

The γ subunit of phosphorylase kinase contains a 386-residue kinase domain, which is 36% identical in sequence to the PKA C subunit and has a similar structure (Fig. 15-16). The γ subunit is not subject to phosphorylation, as are many other protein kinases, because the Ser, Thr, or Tyr residue that is phosphorylated to activate these other kinases is replaced by a Glu residue in the γ subunit. The negative charge of the Glu is thought to mimic the presence of a phosphate group and interact with a conserved Arg residue near the active site. However, full catalytic activity of the γ subunit is prevented by an autoinhibitory C-terminal segment, which binds to and blocks the kinase's active site, much like the R subunit blocks the activity of the C subunit of protein kinase A. An inhibitory segment in the β subunit may also block the activity of the γ subunit.

Autoinhibition of phosphorylase kinase is relieved by PKA–catalyzed phosphorylation of both the α and β subunits. This presumably causes the β inhibitor segment to move aside (the way in which phosphorylation of the α subunit modulates the enzyme's behavior is not understood). However, full activity of the γ subunit also requires Ca^{2+} binding to the δ subunit, which is also known as **calmodulin (CaM;** see below). Ca^{2+} concentrations as low as 10^{-7} M activate phosphorylase kinase by inducing a conformational change in CaM that causes it to bind to and extract the γ subunit's autoinhibitor segment from its catalytic site.

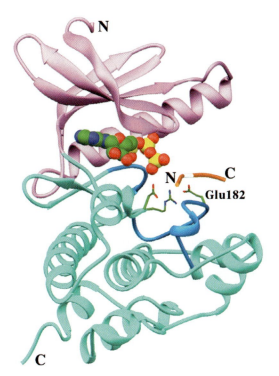

Figure 15-16 **X-ray structure of the γ subunit of rabbit muscle phosphorylase kinase in complex with ATP and a heptapeptide analog of the enzyme's natural substrate.** The N-terminal domain is pink, the C-terminal domain is cyan, the activation loop is light blue, and the heptapeptide is orange, with its residue to be phosphorylated (Ser) white. The ATP is shown in space-filling form and the side chains of the catalytically essential Arg 148, Asp 149, and Glu 182 are shown in stick form, all colored according to atom type (C green, N blue, O red, and P yellow). Note the structural similarities and differences between this protein and the homologous C subunit of protein kinase A (Fig. 15-14). [After an X-ray structure by Louise Johnson, Oxford University, U.K., PDBid 2PHK.]

The conversion of glycogen phosphorylase *b* to glycogen phosphorylase *a* through the action of phosphorylase kinase increases the rate of glycogen breakdown. The physiological significance of the Ca^{2+} trigger for this activation is that muscle contraction is also triggered by a transient increase in the level of cytosolic Ca^{2+} (Section 28-2B). The rate of glycogen breakdown is thereby linked to the rate of muscle contraction. This is critical because glycogen breakdown provides fuel for glycolysis to generate the ATP required for muscle contraction. Since Ca^{2+} release occurs in response to nerve impulses, whereas the phosphorylation of phosphorylase kinase ultimately occurs in response to the presence of certain hormones, these two signals act synergistically in muscle cells to stimulate glycogenolysis.

Calmodulin. CaM is a ubiquitous, often free-floating eukaryotic Ca^{2+}-binding protein that participates in numerous cellular regulatory processes. The X-ray structure of this highly conserved 148-residue protein has a curious dumbbell-like shape in which two structurally similar globular domains are connected by a seven-turn α helix (Fig. 15-17).

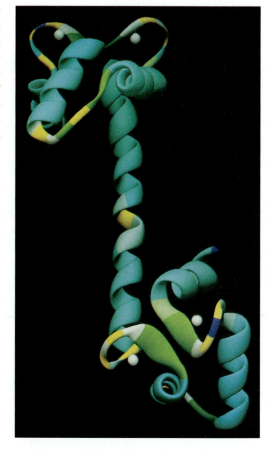

Figure 15-17 **The X-ray structure of rat testis calmodulin.** A seven-turn α helix separates two remarkably similar globular domains. The residues are color-coded according to their backbone conformation angles (φ and ψ; Fig. 6-6): cyan, α helical; green, β sheet; yellow, between helix and sheet; and purple, left-handed helix. The Gly residues are white and the N-terminus is blue. The two Ca^{2+} ions bound to each domain are represented by white spheres. NMR studies of CaM show that in solution the middle portion of the central helix is actually nonhelical. [Courtesy of Mike Carson, University of Alabama at Birmingham. X-Ray structure determined by Charles Bugg, University of Alabama at Birmingham. PDBid 3CLN.] 🐚 **See Kinemage Exercise 16-1.**

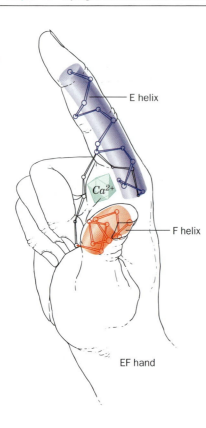

Figure 15-18 The EF hand. The Ca^{2+}-binding sites in many proteins that sense the level of Ca^{2+} are formed by helix–loop–helix motifs named EF hands. [After Kretsinger, R.H., *Annu. Rev. Biochem.* **45**, 241 (1976).] 🔖 **See Kinemage Exercise 16-1.**

E helix

Ca^{2+}

F helix

EF hand

CaM's two globular domains each contain two high-affinity Ca^{2+}-binding sites. The Ca^{2+} ion in each of these sites is octahedrally coordinated by oxygen atoms from the backbone and side chains as well as from a protein-associated water molecule. Each of these Ca^{2+}-binding sites is formed by nearly superimposable helix–loop–helix motifs known as **EF hands** (Fig. 15-18) that form the Ca^{2+}-binding sites in numerous other Ca^{2+}-sensing proteins of known structure.

The binding of Ca^{2+} to either domain of CaM induces a conformational change in that domain, which exposes an otherwise buried Met-rich hydrophobic patch. This patch, in turn, binds with high affinity to the CaM-binding domain of the phosphorylase kinase γ subunit (which contains the kinase inhibitor segment) and to the CaM-binding domains of many other Ca^{2+}-regulated proteins. These CaM-binding domains have little mutual sequence homology but are all basic amphiphilic α helices.

Despite uncomplexed CaM's extended appearance (Fig. 15-17), a variety of studies indicate that both of its globular domains bind to a single target helix. This was confirmed by both the NMR and X-ray structures of Ca^{2+}-CaM in complex with target proteins (Fig. 15-19). Thus, CaM's central α helix serves as a flexible tether rather than a rigid spacer, a property that probably extends the range of sequences to which CaM can bind. In-

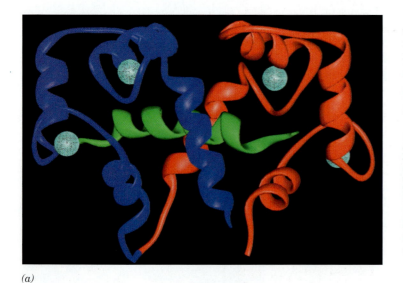

(a)

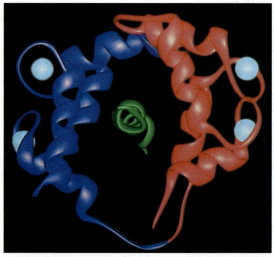

(b)

Figure 15-19 NMR structure of calmodulin in complex with a 26-residue target polypeptide. The N-terminal domain of CaM (from the fruit fly *Drosophila melanogaster*) is blue, its C-terminal domain is red, the target polypeptide is green, and the Ca^{2+} ions are represented by cyan spheres. (*a*) A view of the complex in which the N-terminus of the target polypeptide is on the right. (*b*) The perpendicular view as seen from the right side of the structure shown in Part *a*. In both views, the pseudo–twofold axis relating the N- and C-terminal domains of CaM is approximately vertical. Note how the segment that joins the two domains is unwound and bent (bottom loop in Part *b*) so that CaM forms a globular protein that largely encloses the helical target polypeptide within a hydrophobic tunnel in a manner resembling two hands holding a rope. [Based on an NMR structure by Marius Clore, Angela Gronenborn, and Ad Bax, National Institutes of Health. PDBid 2BBM.] 🔖 **See Kinemage Exercise 16-2.**

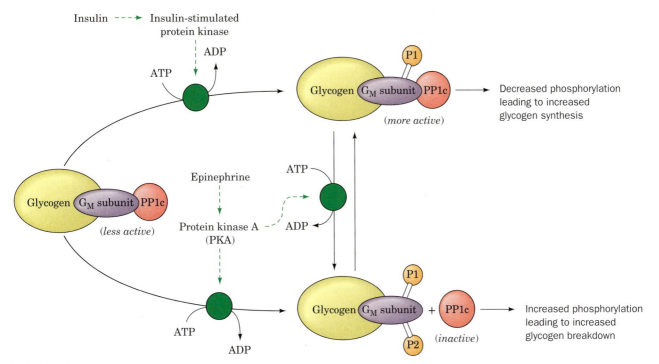

Figure 15-20 Regulation of phosphoprotein phosphatase-1 in muscle. The antagonistic effects of insulin and epinephrine on glycogen metabolism in muscle occur through their effects on the phosphoprotein phosphatase-1 catalytic subunit, PP1c, via its glycogen-bound G_M subunit. Green circles and dashed arrows indicate activation.

deed, both of CaM's globular domains are required for CaM to activate its targets: Domains that have been separated by tryptic cleavage bind to their target peptides but do not cause enzyme activation.

Phosphoprotein Phosphatase-1. A steady state for many phosphorylated enzymes is maintained by a balance between phosphorylation, as catalyzed by a corresponding kinase, and hydrolytic dephosphorylation, as catalyzed by a phosphatase. Phosphoprotein phosphatase-1 removes the phosphoryl groups from glycogen phosphorylase a and the α and β subunits of phosphorylase kinase (Fig. 15-12), as well as those of other proteins involved in glycogen metabolism (see below).

Phosphoprotein phosphatase-1 is controlled differently in muscle and in liver. In muscle, the catalytic subunit of phosphoprotein phosphatase-1 (called **PP1c**) is active only when it is bound to glycogen through its glycogen-binding G_M **subunit.** The activity of PP1c and its affinity for the G_M subunit are regulated by phosphorylation of the G_M subunit at two separate sites (Fig. 15-20). Phosphorylation of site 1 by an **insulin-stimulated protein kinase** (a homolog of PKA and the γ subunit of phosphorylase kinase) activates PP1c, whereas phosphorylation of site 2 by PKA (which can also phosphorylate site 1) causes PP1c to be released into the cytoplasm, where it cannot dephosphorylate the glycogen-bound enzymes of glycogen metabolism.

In the cytosol, phosphoprotein phosphatase-1 is also inhibited by its binding to the protein **phosphoprotein phosphatase inhibitor 1 (inhibitor-1).** This latter protein provides yet another example of control by covalent modification: It too is activated by PKA and deactivated by phosphopro-

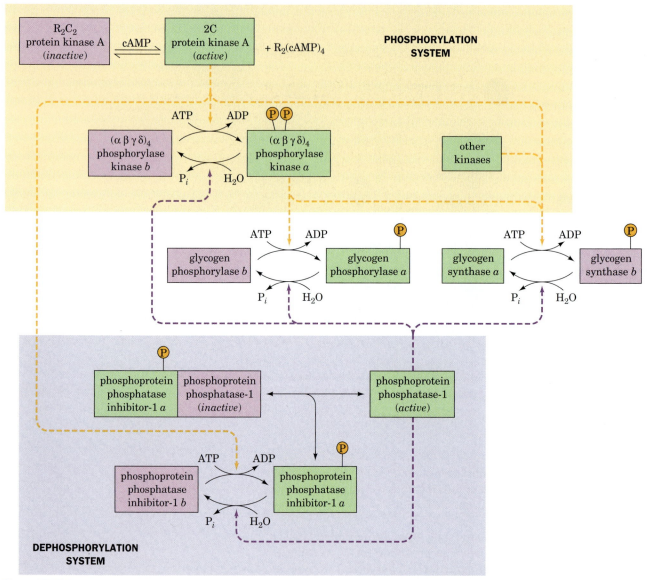

Figure 15-21 *Key to Metabolism.* The major phosphorylation and dephosphorylation systems that regulate glycogen metabolism in muscle. Activated enzymes are shaded green, and deactivated enzymes are shaded pink. Dashed arrows indicate facilitation of a phosphorylation or dephosphorylation reaction. 🎞️ **See the Animated Figures.**

tein phosphatase-1 (Fig. 15-21, *lower left*). *The concentration of cAMP therefore controls the fraction of an enzyme in its phosphorylated form, not only by increasing the rate at which it is phosphorylated, but also by decreasing the rate at which it is dephosphorylated.* In the case of glycogen phosphorylase, an increase in [cAMP] not only increases this enzyme's rate of activation, but also decreases its rate of deactivation.

In liver, phosphoprotein phosphatase-1 is also bound to glycogen, but through the intermediacy of a glycogen-binding subunit named **G_L.** In contrast to G_M, G_L is not subject to control via phosphorylation. The activity of phosphoprotein phosphatase-1 · G_L complex is controlled by its binding to phosphorylase *a*. Both the R and T forms of phosphorylase *a* strongly bind phosphoprotein phosphatase-1, but only in the T state is the Ser 14 phosphoryl group accessible for hydrolysis (in the R state, the Ser 14 phosphoryl group is buried at the dimer interface; Fig. 15-5). Consequently,

when phosphorylase *a* is in its active R form, it effectively sequesters phosphoprotein phosphatase-1. However, under conditions at which phosphorylase *a* converts to the T state (see below), phosphoprotein phosphatase-1 hydrolyzes the now exposed Ser 14 phosphoryl group, thereby converting phosphorylase *a* to phosphorylase *b,* which has only a low affinity for phosphoprotein phosphatase-1 · G_L complex. One effect of phosphorylase *a* dephosphorylation, therefore, is to relieve the inhibition of phosphoprotein phosphatase-1. Since liver cells contain 10 times more glycogen phosphorylase than phosphoprotein phosphatase-1, the phosphatase is not released until more than ~90% of the glycogen phosphorylase is in the *b* form. Only then can phosphoprotein phosphatase-1 dephosphorylate its other target proteins, including glycogen synthase.

Glucose is an allosteric inhibitor of phosphorylase *a* (Fig. 15-13). Consequently, when the concentration of glucose is high, phosphorylase *a* converts to its T form, thereby leading to its dephosphorylation and the dephosphorylation of glycogen synthase. Glucose is therefore thought to be important in the control of glycogen metabolism in the liver.

Glycogen Synthase.   Phosphorylase kinase, which activates glycogen phosphorylase, also phosphorylates and thereby inactivates glycogen synthase. Six other protein kinases, including PKA and phosphorylase kinase, are known to at least partially deactivate human muscle glycogen synthase by phosphorylating one or more of the nine Ser residues on each of its subunits (Fig. 15-21). The reason for this elaborate regulation of glycogen synthase is unclear.

The balance between net synthesis and degradation of glycogen as well as the rates of these processes depend on the relative activities of glycogen synthase and glycogen phosphorylase. To a large extent, the rates of the phosphorylation and dephosphorylation of these enzymes control glycogen synthesis and breakdown. The two processes are linked by PKA and phosphorylase kinase, which, through phosphorylation, activate glycogen phosphorylase as they inactivate glycogen synthase (Fig. 15-21). They are also linked by phosphoprotein phosphatase-1, which in liver is inhibited by phosphorylase *a* and therefore unable to activate (dephosphorylate) glycogen synthase unless it first inactivates (also by dephosphorylation) phosphorylase *a.* Of course, control by allosteric effectors is superimposed on control by covalent modification so that, for example, the availability of the substrate G6P (which activates glycogen synthase) also influences the rate at which glucose residues are incorporated into glycogen. Inherited deficiencies of enzymes can disrupt the fine control of glycogen metabolism, leading to various diseases (see Box 15-2).

C   Hormonal Effects on Glycogen Metabolism

Glycogen metabolism in the liver is largely controlled by the polypeptide hormones insulin (Fig. 5-1) and **glucagon** acting in opposition. Glucagon,

$$H_3\overset{+}{N}—His—Ser   Glu   Gly —Thr—Phe—Thr —Ser —Asp—Tyr— 10$$

$$Ser—Lys —Tyr—Leu—Asp—Ser —Arg —Arg—Ala—Gln— 20$$

$$Asp—Phe—Val —Gln —Trp—Leu—Met—Asn—Thr—COO^- 29$$

Glucagon

like insulin, is synthesized by the pancreas in response to the concentration of glucose in the blood. In muscles and various tissues, control is exerted

by insulin and by the adrenal hormones **epinephrine (adrenalin)** and **norepinephrine (noradrenalin):**

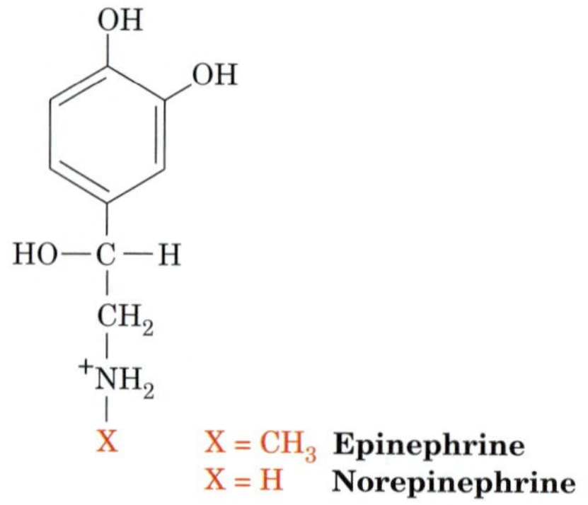

$$\text{X} = CH_3 \quad \textbf{Epinephrine}$$
$$\text{X} = H \quad \textbf{Norepinephrine}$$

These hormones affect metabolism in their target tissues by ultimately stimulating covalent modification (phosphorylation) of regulatory enzymes. They do so by binding to transmembrane **receptors** on the surface of cells. Different cell types have different complements of receptors and therefore respond to different sets of hormones. The responses involve the release inside the cell of molecules collectively known as **second messengers,** that is, intracellular mediators of the externally received hormonal message. Different receptors cause the release of different second messengers. cAMP, identified by Earl Sutherland in the 1950s, was the first second messenger discovered. Ca^{2+}, as released from intracellular reservoirs into the cytosol, is also a common second messenger. Receptors and second messengers are discussed in greater depth in Section 21-3.

When hormonal stimulation increases the intracellular cAMP concentration, PKA activity increases, increasing the rates of phosphorylation of many proteins and decreasing their dephosphorylation rates as well. Because of the cascade nature of the regulatory system diagramed in Fig. 15-21, *a small change in [cAMP] results in a large change in the fraction of phosphorylated enzymes.* When a large fraction of the glycogen metabolism enzymes are phosphorylated, the metabolic flux is in the direction of glycogen breakdown, since glycogen phosphorylase is active and glycogen synthase is inactive. When [cAMP] decreases, phosphorylation rates decrease, dephosphorylation rates increase, and the fraction of enzymes in their dephospho forms increases. The resulting activation of glycogen synthase and inhibition of glycogen phosphorylase cause the flux to shift to net glycogen synthesis.

Glucagon binding to its receptor on liver cells, which generates intracellular cAMP, results in glucose mobilization from stored glycogen (Fig. 15-22). Glucagon is released from the pancreas when the concentration of circulating glucose decreases to less than ~5 mM, such as during exercise or several hours after a meal has been digested. Glucagon is therefore critical for the liver's function in supplying glucose to tissues that depend primarily on glycolysis for their energy needs. Muscle cells do not respond to glucagon because they lack the appropriate receptor.

Epinephrine and norepinephrine, which are often called the "fight or flight" hormones, are released into the bloodstream by the adrenal glands in response to stress. There are two types of receptors for these hormones: the **β-adrenergic receptor,** which is linked to the adenylate cyclase system, and the **α-adrenergic receptor,** whose second messenger causes intracellular $[Ca^{2+}]$ to increase (Section 21-3). Muscle cells, which have the β-adrenergic receptor (Fig. 15-22), respond to epinephrine by breaking down glycogen for glycolysis, thereby generating ATP and helping the muscles cope with the stress that triggered the epinephrine release.

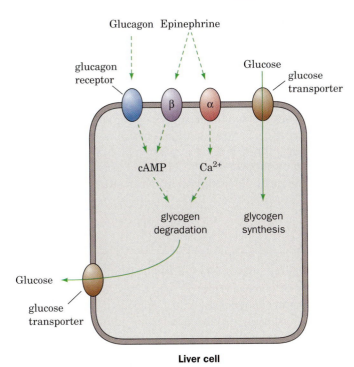

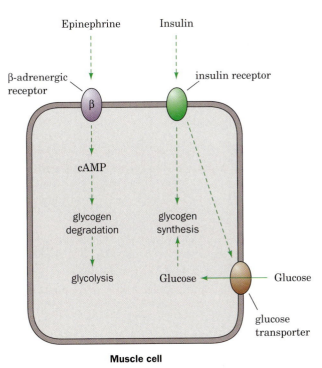

Liver cell   **Muscle cell**

Figure 15-22 Hormonal control of glycogen metabolism. Epinephrine binding to β-adrenergic receptors on liver and muscle cells increases intracellular [cAMP], which promotes glycogen degradation to G6P for glycolysis (in muscle) or to glucose for export (in liver). The liver responds similarly to glucagon. Epinephrine binding to α-adrenergic receptors on liver cells leads to increased cytosolic [Ca^{2+}], which also promotes glycogen degradation. When circulating glucose is plentiful, insulin stimulates glucose uptake and glycogen synthesis in muscle cells. The liver responds directly to increased glucose by increasing glycogen synthesis. 🐁 **See the Animated Figures.**

Liver cells respond to epinephrine directly and indirectly. Epinephrine promotes the release of glucagon from the pancreas, and glucagon binding to its receptor on liver cells stimulates glycogen breakdown as described above. Epinephrine also binds directly to both α- and β-adrenergic receptors on the surfaces of liver cells (Fig. 15-22). Binding to the β-adrenergic receptor results in increased intracellular cAMP, which leads to glycogen breakdown. Epinephrine binding to the α-adrenergic receptor stimulates an increase in intracellular [Ca^{2+}], which reinforces the cells' response to cAMP (recall that phosphorylase kinase, which activates glycogen phosphorylase and inactivates glycogen synthase, is fully active only when phosphorylated and in the presence of increased [Ca^{2+}]). In addition, glycogen synthase is inactivated through phosphorylation catalyzed by several Ca^{2+}-dependent protein kinases.

Insulin and Epinephrine Are Antagonists. Insulin is released from the pancreas in response to high levels of circulating glucose (e.g., immediately after a meal). Hormonal stimulation by insulin increases the rate of glucose transport into the many types of cells that have insulin receptors on their surfaces (e.g., muscle and fat cells, but not liver and brain cells). In addition, [cAMP] decreases, causing glycogen metabolism to shift from glycogen breakdown to glycogen synthesis (Fig. 15-22). The mechanism of insulin action is very complex (Section 21-3F), but one of its target enzymes appears to be phosphoprotein phosphatase-1. As outlined in Fig. 15-20, insulin activates insulin-stimulated protein kinase in muscle to phosphorylate site 1 on the glycogen-binding G_M subunit of phosphoprotein phosphatase-1 so as to activate this protein and thus dephosphorylate the enzymes of glycogen metabolism. The storage of glucose as glycogen is thereby promoted

through the inhibition of glycogen breakdown and the stimulation of glycogen synthesis.

In liver, it is thought that glucose itself, rather than insulin, may be the messenger to which the glycogen metabolism system responds. *Glucose inhibits phosphorylase a by binding to the enzyme's inactive T state and thereby shifting the $T \rightleftharpoons R$ equilibrium toward the T state* (Fig. 15-13). This conformational shift exposes the Ser 14 phosphoryl group to dephosphorylation. An increase in glucose concentration therefore promotes inactivation of glycogen phosphorylase *a* through its conversion to phosphorylase *b*. The subsequent release of phosphoprotein phosphatase-1 activates glycogen synthase. Thus when glucose is plentiful, the liver can store the excess as glycogen.

4 Gluconeogenesis

When dietary sources of glucose are not available and when the liver has exhausted its supply of glycogen, glucose is synthesized from noncarbohydrate precursors by **gluconeogenesis.** In fact, gluconeogenesis provides a substantial fraction of the glucose produced in fasting humans, even within a few hours of eating. Gluconeogenesis occurs in liver, and to a lesser extent, in kidney.

The noncarbohydrate precursors that can be converted to glucose include the glycolysis products lactate and pyruvate, citric acid cycle intermediates, and the carbon skeletons of most amino acids. First, however, all these substances must be converted to the four-carbon compound **oxaloacetate** (*at left*), which itself is a citric acid cycle intermediate (Section 16-1). The only amino acids that cannot be converted to oxaloacetate in animals are leucine and lysine because their breakdown yields only acetyl-CoA (Section 20-4E) and because *there is no pathway in animals for the net conversion of acetyl-CoA to oxaloacetate.* Likewise, fatty acids cannot serve as glucose precursors in animals because most fatty acids are degraded completely to acetyl-CoA (Section 19-2).

For convenience, we consider gluconeogenesis to be the pathway by which pyruvate is converted to glucose. Most of the reactions of gluconeogenesis are glycolytic reactions that proceed in reverse (Fig. 15-23). However, the glycolytic enzymes hexokinase, phosphofructokinase, and pyruvate kinase catalyze reactions with large negative free energy changes. These reactions must therefore be replaced in gluconeogenesis by reactions that make glucose synthesis thermodynamically favorable.

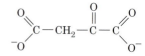

Oxaloacetate

A Pyruvate to Phosphoenolpyruvate

We begin our examination of the reactions unique to gluconeogenesis with the conversion of pyruvate to phosphoenolpyruvate (PEP). Because this step is the reverse of the highly exergonic reaction catalyzed by pyruvate kinase (Section 14-2J), it requires free energy input. This is accomplished by first converting the pyruvate to oxaloacetate. Oxaloacetate is a "high-energy" intermediate because its exergonic decarboxylation provides the free energy necessary for PEP synthesis. The process requires two enzymes (Fig. 15-24):

1. **Pyruvate carboxylase** catalyzes the ATP-driven formation of oxaloacetate from pyruvate and HCO_3^-.
2. **PEP Carboxykinase (PEPCK)** converts oxaloacetate to PEP in a reaction that uses GTP as a phosphoryl-group donor.

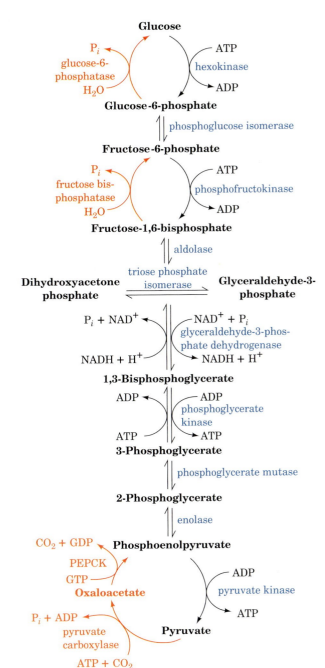

Figure 15-23 *Key to Metabolism.* **Comparison of the pathways of gluconeogenesis and glycolysis.** The red arrows represent the steps that are catalyzed by different enzymes in gluconeogenesis. The other seven reaction steps of gluconeogenesis are catalyzed by glycolytic enzymes that function near equilibrium. **⌘ See the Animated Figures.**

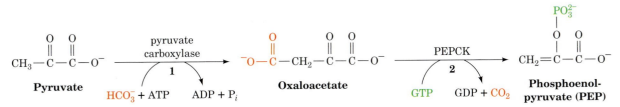

Figure 15-24 **The conversion of pyruvate to phosphoenolpyruvate (PEP).** This process requires **(1)** pyruvate carboxylase to convert pyruvate to oxaloacetate and **(2)** PEP carboxykinase (PEPCK) to convert oxaloacetate to PEP.

Pyruvate Carboxylase Has a Biotin Prosthetic Group. Pyruvate carboxylase is a tetrameric protein of identical ~1160-residue subunits, each of which has a **biotin** prosthetic group. Biotin (Fig. 15-25a) functions as a CO_2 carrier by forming a carboxyl substituent at its **ureido group** (Fig. 15-25b). Biotin is covalently bound to an enzyme Lys residue to form a **biocytin** (alternatively, **biotinyllysine**) residue (Fig. 15-25b). The biotin ring system is therefore at the end of a 14-Å-long flexible arm. Biotin, which was first identified in 1935 as a growth factor in yeast, is an essential human nutrient. Its nutritional deficiency is rare, however, because it occurs in many foods and is synthesized by intestinal bacteria.

The pyruvate carboxylase reaction occurs in two phases (Fig. 15-26):

Phase I The cleavage of ATP to ADP acts to dehydrate bicarbonate via the formation of a "high-energy" carboxyphosphate intermediate. The reaction of the resulting CO_2 with biotin is exergonic. The biotin-bound carboxyl group is therefore "activated" relative to bicarbonate and can be transferred to another molecule without further free energy input.

Phase II The activated carboxyl group is transferred from carboxybiotin to pyruvate in a three-step reaction to form oxaloacetate.

These two reaction phases occur on different subsites of the same enzyme; the 14-Å arm of biocytin transfers the biotin ring between the two sites.

Oxaloacetate is both a precursor for gluconeogenesis and an intermediate of the citric acid cycle (Section 16-3). When the citric acid cycle substrate acetyl-CoA accumulates, it allosterically activates pyruvate carboxylase, thereby increasing the amount of oxaloacetate that can participate in the citric acid cycle. When citric acid cycle activity is low, oxaloacetate instead enters the gluconeogenic pathway.

PEP Carboxykinase. PEPCK, a monomeric ~610-residue enzyme, catalyzes the GTP-requiring decarboxylation/phosphorylation of oxaloacetate to form PEP and GDP (Fig. 15-27). Note that the CO_2 that carboxylates pyruvate to yield oxaloacetate is eliminated in the formation of PEP. The favorable decarboxylation reaction drives the formation of the enol that GTP phosphorylates. *Oxaloacetate can therefore be considered as "activated" pyruvate, with CO_2 and biotin facilitating the activation at the expense of ATP.*

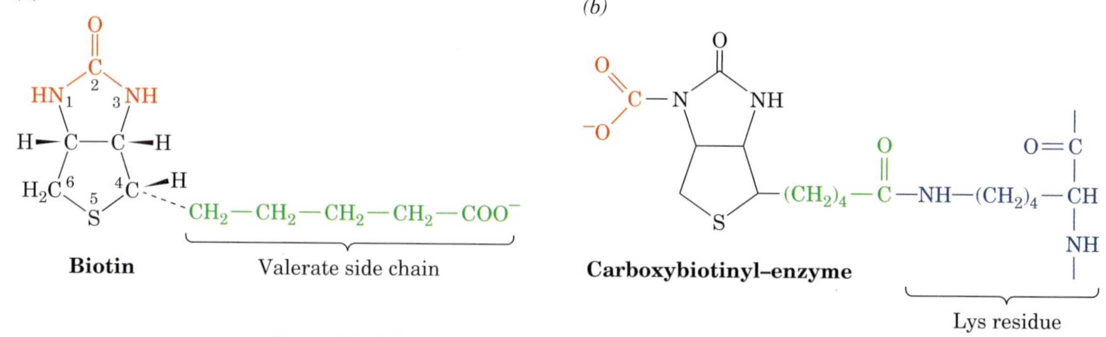

(a)

Biotin   Valerate side chain

(b)

Carboxybiotinyl–enzyme   Lys residue

Figure 15-25 Biotin and carboxybiotinyl–enzyme. (*a*) Biotin consists of an imidazoline ring that is cis-fused to a tetrahydrothiophene ring bearing a valerate side chain. Positions 1, 2, and 3 constitute a ureido group. (*b*) Biotin is covalently attached to carboxylases by an amide linkage between its valeryl carboxyl group and an ε-amino group of an enzyme Lys side chain. The carboxybiotinyl–enzyme forms when N1 of the biotin ureido group is carboxylated.

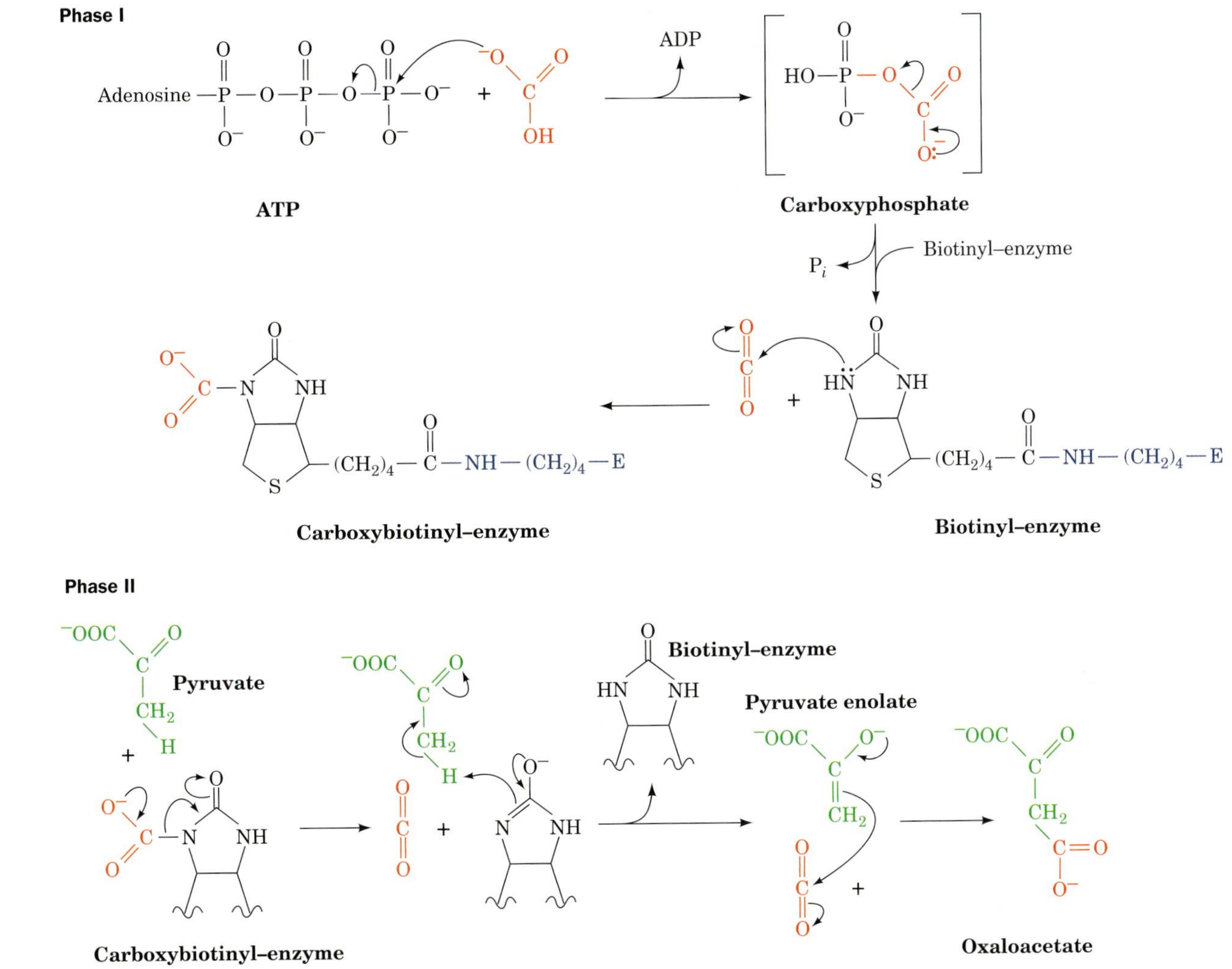

Phase I

ATP **Carboxyphosphate**

Carboxybiotinyl–enzyme **Biotinyl–enzyme**

Phase II

Carboxybiotinyl–enzyme **Oxaloacetate**

Figure 15-26 The two-phase reaction mechanism of pyruvate carboxylase. Phase I is a three-step reaction in which carboxyphosphate is formed from bicarbonate and ATP, followed by the generation of CO_2, which then carboxylates biotin. Phase II is a three-step reaction in which CO_2 is produced at the active site via the elimination of the biotinyl–enzyme, which accepts a proton from pyruvate to generate pyruvate enolate. This enolate, in turn, nucleophilically attacks the CO_2, yielding oxaloacetate. [After Knowles, J.R., *Annu. Rev. Biochem.* **58**, 217 (1989).]

Gluconeogenesis Requires Metabolite Transport between Mitochondria and Cytosol. The generation of oxaloacetate from pyruvate or citric acid cycle intermediates occurs only in the mitochondrion, whereas the enzymes that convert PEP to glucose are cytosolic. The cellular location of PEPCK varies:

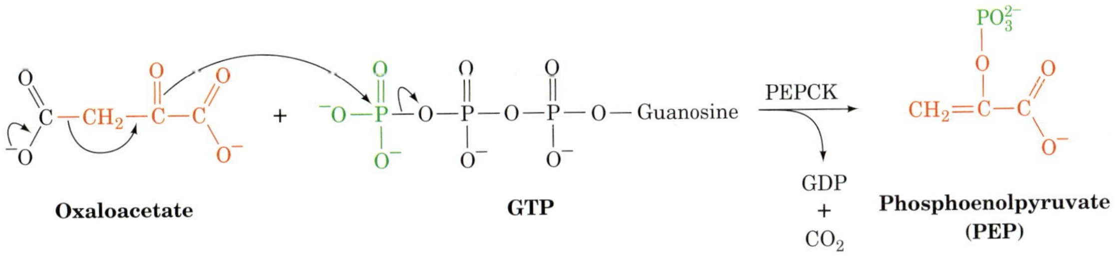

Oxaloacetate **GTP** **Phosphoenolpyruvate (PEP)**

GDP + CO_2

Figure 15-27 The PEPCK mechanism. Decarboxylation of oxaloacetate (a β-keto acid) forms a resonance-stabilized enolate anion whose oxygen atom attacks the γ-phosphoryl group of GTP, forming PEP and GDP.

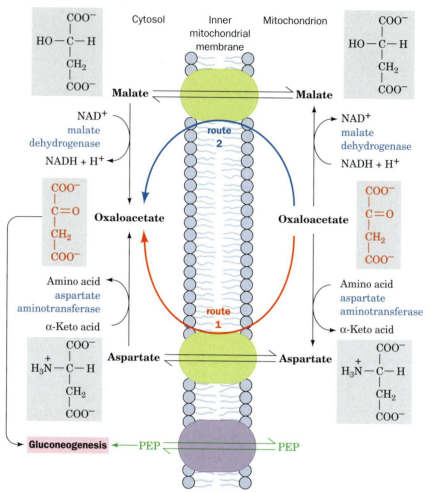

Figure 15-28 The transport of PEP and oxaloacetate from the mitochondrion to the cytosol.
PEP is directly transported between these compartments. Oxaloacetate, however, must first be converted to either aspartate through the action of aspartate aminotransferase (Route 1) or to malate by malate dehydrogenase (Route 2). Route 2 involves the mitochondrial oxidation of NADH followed by the cytosolic reduction of NAD$^+$ and therefore also transfers NADH reducing equivalents from the mitochondrion to the cytosol. 🐎 See the Animated Figures.

In some species, it is mitochondrial; in some, it is cytosolic; and in some (including humans) it is equally distributed between the two compartments. In order for gluconeogenesis to occur, either oxaloacetate must leave the mitochondrion for conversion to PEP or the PEP formed there must enter the cytosol.

PEP is transported across the mitochondrial membrane by specific membrane transport proteins. There is, however, no such transport system for oxaloacetate. *In species with cytosolic PEPCK, oxaloacetate must first be converted either to aspartate* (Fig. 15-28, Route 1) *or to **malate*** (Fig. 15-28, Route 2), for which mitochondrial transport systems exist. The difference between these two routes involves the transport of NADH **reducing equivalents** (in the transport of reducing equivalents, the electrons—but not the electron carrier—cross the membrane). The **malate dehydrogenase** route (Route 2) results in the transport of reducing equivalents from the mitochondrion to the cytosol, since it uses mitochondrial NADH and produces cytosolic NADH. The **aspartate aminotransferase** route (Route 1) does not

involve NADH. Cytosolic NADH is required for gluconeogenesis, so, under most conditions, the route through malate is a necessity. However, when the gluconeogenic precursor is lactate, its oxidation to pyruvate generates cytosolic NADH, and either transport system can then be used. All the reactions shown in Fig. 15-28 are freely reversible, so that under appropriate conditions, the malate–aspartate shuttle system also operates to transport NADH reducing equivalents into the mitochondrion for oxidative phosphorylation (Section 17-1B). Liver has a variation of Route 1 in which aspartate entering the cytosol is deaminated via the urea cycle before undergoing a series of reactions that yield oxaloacetate (Section 20-3A).

B Hydrolytic Reactions

The route from PEP to fructose-1,6-bisphosphate (FBP) is catalyzed by the enzymes of glycolysis operating in reverse. However, *the glycolytic reactions catalyzed by phosphofructokinase (PFK) and hexokinase are endergonic in the gluconeogenesis direction and hence must be bypassed by different gluconeogenic enzymes.* FBP is hydrolyzed by fructose-1,6-bisphosphatase (FBPase). The resulting fructose-6-phosphate (F6P) is isomerized to G6P, which is then hydrolyzed by glucose-6-phosphatase, the same enzyme that converts glycogen-derived G6P to glucose (Section 15-1C) and which is present only in liver and kidney. Note that these two hydrolytic reactions release P_i rather than reversing the ATP $\rightarrow$ ADP reactions that occur at this point in the glycolytic pathway.

The net energetic cost of converting two pyruvate molecules to one glucose molecule by gluconeogenesis is six ATP equivalents: two each at the steps catalyzed by pyruvate carboxylase, PEPCK, and phosphoglycerate kinase. Since the energetic profit of converting one glucose molecule to two pyruvate molecules via glycolysis is two ATP, (Section 14-1), the energetic cost of the futile cycle in which glucose is converted to pyruvate and then resynthesized is four ATP equivalents. Such free energy losses are the thermodynamic price that must be paid to maintain the independent regulation of two opposing pathways.

Although glucose is considered the endpoint of the gluconeogenic pathway, it is possible for pathway intermediates to be directed elsewhere, for example, through the transketolase and transaldolase reactions of the pentose phosphate pathway (Section 14-6C) to produce ribose-5-phosphate. The G6P produced by gluconeogenesis may not be hydrolyzed to glucose but may instead be converted to G1P for incorporation into glycogen.

C Regulation of Gluconeogenesis

The opposing pathways of gluconeogenesis and glycolysis, like glycogen synthesis and degradation, do not proceed simultaneously *in vivo*. Instead, these pathways are reciprocally regulated to meet the needs of the organism. There are three substrate cycles and therefore three potential points for regulating glycolytic versus gluconeogenic flux (Fig. 15-29).

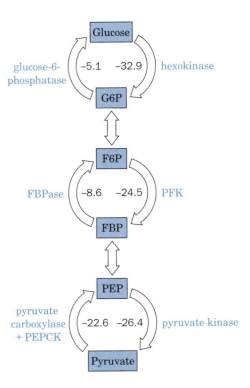

Figure 15-29 Substrate cycles in glucose metabolism. The interconversions of glucose and G6P, F6P and FBP, and PEP and pyruvate are catalyzed by different enzymes in the forward and reverse directions so that all reactions are exergonic (the ΔG values for the reactions in liver are given in kJ · mol^{-1}). [ΔG's obtained from Newsholme, E.A. and Leech, A.R., *Biochemistry for the Medical Sciences*, p. 448, Wiley (1983).]

Fructose-2,6-Bisphosphate Activates Phosphofructokinase and Inhibits Fructose-1,6,-Bisphosphatase.

The net flux through the substrate cycle created by the opposing actions of PFK and FBPase (described in Section 14-4B) is determined by the concentration of fructose-2,6-bisphosphate (F2,6P).

β-D-Fructose-2,6-bisphosphate
(F2,6P)

F2,6P, which is not a glycolytic intermediate, is an extremely potent allosteric activator of PFK and an inhibitor of FBPase.

The concentration of F2,6P in the cell depends on the balance between its rates of synthesis and degradation by **phosphofructokinase-2 (PFK-2)** and **fructose bisphosphatase-2 (FBPase-2),** respectively (Fig. 15-30). These enzyme activities are located on different domains of the same ~100-kD homodimeric protein. The bifunctional enzyme is regulated by a variety of allosteric effectors and by phosphorylation and dephosphorylation as catalyzed by PKA and a phosphoprotein phosphatase. Thus, the balance between gluconeogenesis and glycolysis is under hormonal control.

For example, when [glucose] is low, glucagon stimulates the production of cAMP in liver cells. This activates PKA to phosphorylate the bifunctional enzyme at a specific Ser residue, which inactivates the enzyme's PFK-2 activity and activates its FBPase-2 activity. The net result is a decrease in [F2,6P], which shifts the balance between the PFK and FBPase reactions in favor of FBP hydrolysis and hence increases gluconeogenic flux (Fig. 15-31). The concurrent increases in gluconeogenesis and glycogen breakdown allow the liver to release glucose into the circulation. Conversely, when the blood [glucose] is high, cAMP levels decrease, and the resulting increase in [F2,6P] promotes glycolysis.

In muscle, which is not a gluconeogenic tissue, the F2,6P control system functions quite differently from that in liver due to the presence of different PFK-2/FBPase-2 isozymes. For example, hormones that stimulate glycogen breakdown in heart muscle lead to phosphorylation of a site on the bifunctional enzyme that activates rather than inhibits PFK-2. The resulting increase in F2,6P stimulates glycolysis so that glycogen breakdown and

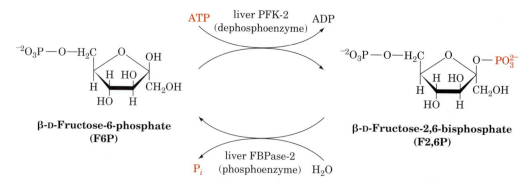

Figure 15-30   The formation and degradation of β-D-fructose-2,6-bisphosphate (F2,6P).   The enzymatic activities of phosphofructokinase-2 (PFK-2) and fructose bisphosphatase-2 (FBPase-2) occur on different domains of the same protein molecule. The phosphorylation of the liver enzyme inactivates PFK-2 while activating FBPase-2.

glycolysis are coordinated. The skeletal muscle isozyme lacks a phosphorylation site altogether and is therefore not subject to cAMP-dependent control.

Other Allosteric Effectors Influence Gluconeogenic Flux. Acetyl-CoA activates pyruvate carboxylase (Section 15-4A), but there are no known allosteric effectors of PEPCK, which together with pyruvate carboxylase reverses the pyruvate kinase reaction. Pyruvate kinase, however, is allosterically inhibited in the liver by alanine, a major gluconeogenic precursor. Alanine is converted to pyruvate by the transfer of its amino group to an α-keto acid to yield a new amino acid and the α-keto acid pyruvate,

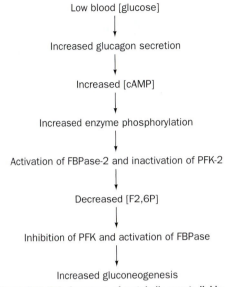

Figure 15-31 Sequence of metabolic events linking low blood [glucose] to gluconeogenesis in liver.

a process termed **transamination** (which is discussed in Section 20-2A). Liver pyruvate kinase is also inactivated by phosphorylation, further increasing gluconeogenic flux. Since phosphorylation also activates glycogen phosphorylase, the pathways of gluconeogenesis and glycogen breakdown both flow toward G6P, which is converted to glucose for export from the liver.

The activity of hexokinase (or glucokinase, the liver isozyme) is also controlled, as we shall see in Section 21-1D. The activity of glucose-6-phosphatase is controlled as well but this process is complex and poorly understood.

The regulation of glucose metabolism occurs not only through allosteric effectors, but also through long-term changes in the amounts of enzymes synthesized. Pancreatic and adrenal hormones influence the rates of transcription and the stabilities of the mRNAs encoding many of the regulatory proteins of glucose metabolism. For example, insulin inhibits transcription of the gene for PEPCK, whereas high concentrations of intracellular cAMP promote the transcription of the genes for PEPCK, FBPase, and glucose-6-phosphatase, and repress transcription of the genes for glucokinase, PFK, and the PFK-2/FBPase-2 bifunctional enzyme.

5 Other Carbohydrate Biosynthetic Pathways

The liver, by virtue of its mass and its metabolic machinery, is primarily responsible for maintaining a constant level of glucose in the circulation. Glucose produced by gluconeogenesis or from glycogen breakdown is released from the liver for use by other tissues as an energy source. Of course, glucose has other uses in the liver and elsewhere, for example, in the synthesis of lactose (see Box 15-4).

Nucleotide Sugars Power the Formation of Glycosidic Bonds. Glucose and other monosaccharides (principally mannose, *N*-acetylglucosamine, fucose, galactose, *N*-acetylneuraminic acid, and *N*-acetylgalactosamine) occur in glycoproteins and glycolipids. Formation of the glycosidic bonds that link sugars to each other and to other molecules requires free energy input under physiological conditions, ($\Delta G^{\circ\prime} = 16 \ kJ \cdot mol^{-1}$). This free energy, as we

BOX 15-4

Perspectives in Biochemistry

Lactose Synthesis

Like sucrose in plants, lactose is a disaccharide that is synthesized for later use as a metabolic fuel, in this case, after digestion by very young mammals. Lactose, or milk sugar, is produced in the mammary gland by **lactose synthase**. In this reaction, the donor sugar is UDP–galactose, which is formed by the epimerization of UDP–glucose (Fig. 14-28). The acceptor sugar is glucose:

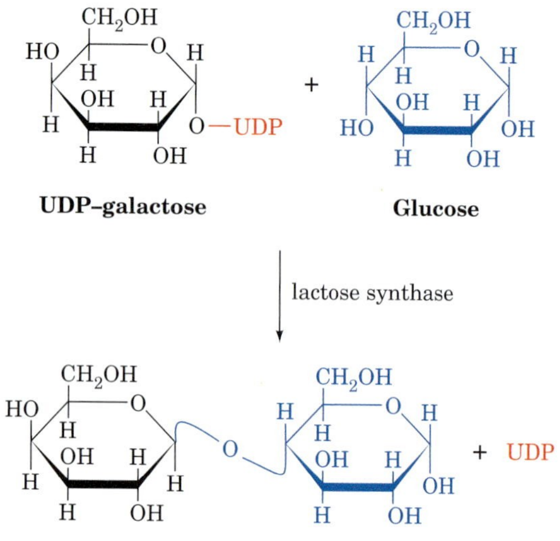

UDP–galactose   **Glucose**

lactose synthase

Lactose
[β-galactosyl-(1 ⟶ 4)-glucose]

Thus, both saccharide units of lactose are ultimately derived from glucose. Lactose synthase consists of two subunits:

1. **Galactosyltransferase,** the catalytic subunit, occurs in many tissues, where it catalyzes the reaction of UDP–galactose and *N*-acetylglucosamine to yield **N-acetyllactosamine,**

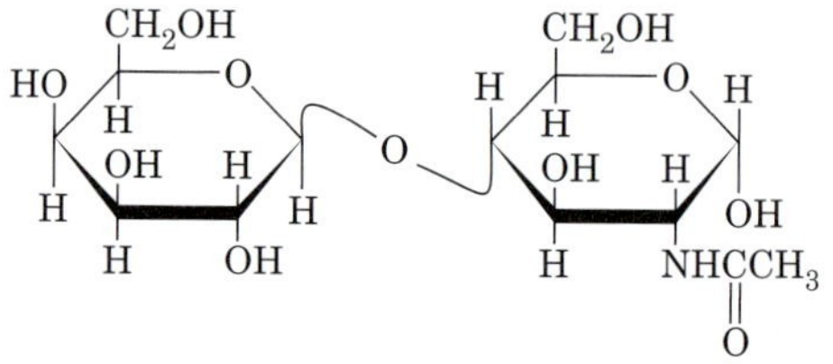

N-Acetyllactosamine

   a constituent of many complex oligosaccharides.

2. **α-Lactalbumin,** a mammary gland protein with no catalytic activity, alters the specificity of galactosyltransferase so that it uses glucose as an acceptor, rather than *N*-acetylglucosamine, to form lactose instead of *N*-acetyllactosamine.

Synthesis of α-lactalbumin, whose sequence is ~37% identical to that of lysozyme (which also participates in reactions involving sugars), is triggered by hormonal changes at parturition (birth), thereby promoting lactose synthesis for milk production.

have seen in glycogen synthesis (Section 15-2A), is acquired through the synthesis of a nucleotide sugar from a nucleoside triphosphate and a monosaccharide, thereby releasing PP_i, whose exergonic hydrolysis drives the reaction. The nucleoside diphosphate at the sugar's anomeric carbon atom is a good leaving group and thereby facilitates formation of a glycosidic bond to a second sugar in a reaction catalyzed by a glycosyltransferase (Fig. 15-32). In mammals, most glycosyl groups are donated by UDP–sugars, but fucose and mannose are carried by GDP, and sialic acid by CMP. In plants, starch is built from glucose units donated by **ADP–glucose,** and cellulose synthesis relies on ADP–glucose or **CDP–glucose.**

O-linked Oligosaccharides Are Posttranslationally Formed.   Nucleotide sugars are the donors in the synthesis of O-linked oligosaccharides and in the processing the N-linked oligosaccharides of glycoproteins (Section 8-3C). O-linked oligosaccharides are synthesized in the Golgi apparatus by the serial addition of monosaccharide units to a completed polypeptide chain

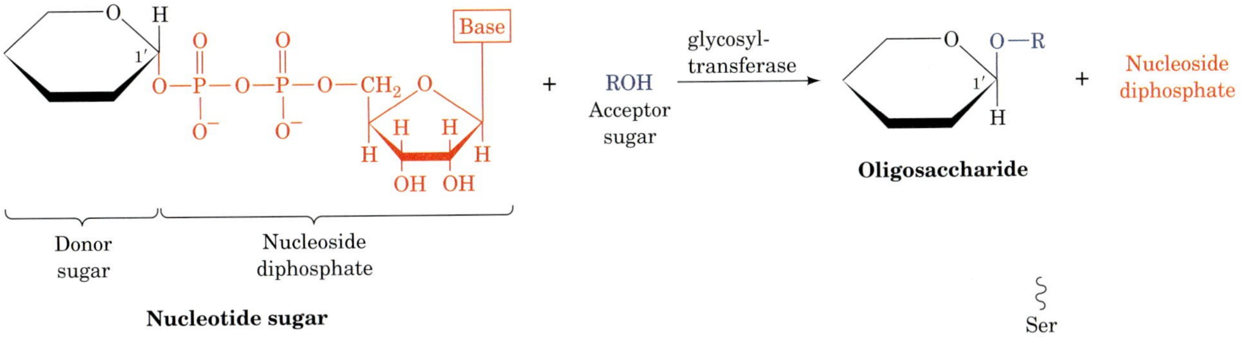

Figure 15-32 Role of nucleotide sugars. These compounds are the glycosyl donors in oligosaccharide biosynthetic reactions as catalyzed by glycosyltransferases.

(Fig. 15-33). Synthesis begins with the transfer, as catalyzed by **GalNAc transferase,** of N-acetylgalactosamine (GalNAc) from UDP–GalNAc to a Ser or Thr residue on the polypeptide. The location of the glycosylation site is thought to be specified only by the secondary or tertiary structure of the polypeptide. Glycosylation continues with the stepwise addition of sugars such as galactose, sialic acid, N-acetylglucosamine, and fucose. In each case, the sugar residue is transferred from its nucleotide sugar derivative by a corresponding glycosyltransferase.

N-linked Oligosaccharides Are Constructed on Dolichol Carriers.

The synthesis of N-linked oligosaccharides is more complicated than that of O-linked oligosaccharides. In the early stages of N-linked oligosaccharide synthesis, sugar residues are sequentially added to a lipid carrier, **dolichol pyrophosphate** (Fig. 15-34). Dolichol is a long-chain polyisoprenoid containing 17 to 21 isoprene units in animals and 14 to 24 units in fungi and plants. It anchors the growing oligosaccharide to the endoplasmic reticulum membrane, where the initial glycosylation reactions take place.

Although nucleotide sugars are the most common monosaccharide donors in glycosyltransferase reactions, several mannosyl and glucosyl residues are transferred to growing dolichol-PP-oligosaccharides from their corresponding dolichol-P derivatives. Dolichol phosphate "activates" a sugar residue for subsequent transfer, as does a nucleoside diphosphate.

The construction of an N-linked oligosaccharide begins, as is described in Section 8-3C, by the synthesis of an oligosaccharide with the composition (N-acetylglucosamine)$_2$(mannose)$_9$(glucose)$_3$. This occurs on a dolichol carrier in a 12-step process catalyzed by a series of specific glycosyltrans-

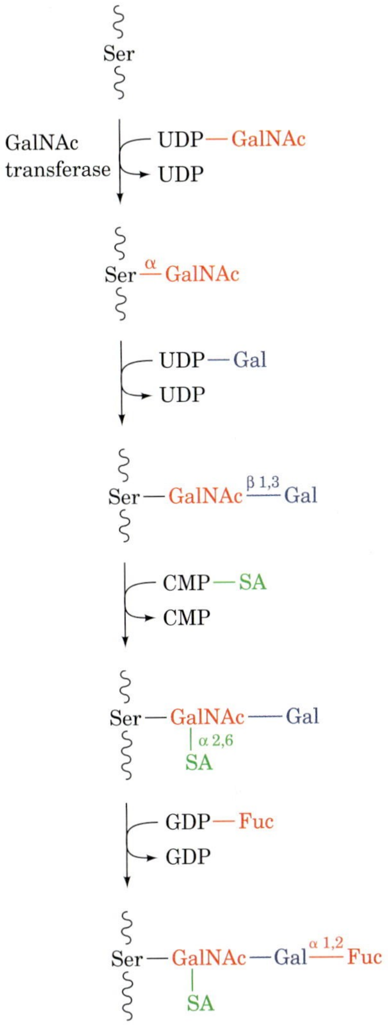

Figure 15-33 Synthesis of an O-linked oligosaccharide chain. This pathway shows the proposed steps in the assembly of a carbohydrate moiety in canine submaxillary mucin. SA is sialic acid.

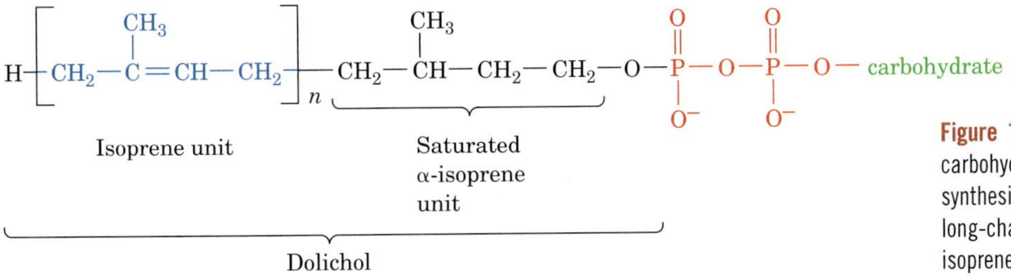

Figure 15-34 Dolichol pyrophosphate glycoside. The carbohydrate precursors of N-linked glycosides are synthesized as oligosaccharides attached to dolichol, a long-chain polyisoprenol ($n = 14$–24) in which the α-isoprene unit is saturated.

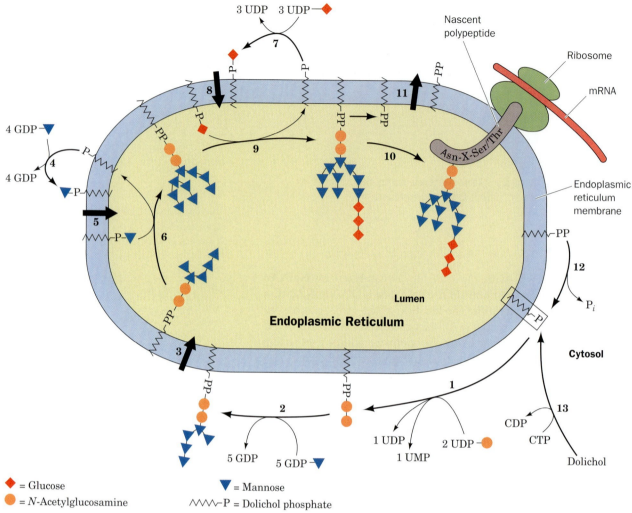

◆ = Glucose   ▼ = Mannose
● = N-Acetylglucosamine   〰〰〰–P = Dolichol phosphate

Figure 15-35   The pathway of dolichol-PP-oligosaccharide synthesis.
(1) Addition of *N*-acetylglucosamine-1-P and a second *N*-acetylglucosamine to dolichol-P. **(2)** Addition of five mannosyl residues from GDP–mannose in reactions catalyzed by five different mannosyltransferases. **(3)** Membrane translocation of dolichol-PP-(*N*-acetylglucosamine)₂(mannose)₅ to the lumen of the endoplasmic reticulum (ER). **(4)** Cytosolic synthesis of dolichol-P-mannose from GDP–mannose and dolichol-P. **(5)** Membrane translocation of dolichol-P-mannose to the lumen of the ER. **(6)** Addition of four mannosyl residues from dolichol-P-mannose in reactions catalyzed by four different mannosyl-

transferases. **(7)** Cytosolic synthesis of dolichol-P-glucose from UDPG and dolichol-P. **(8)** Membrane translocation of dolichol-P-glucose to the lumen of the ER. **(9)** Addition of three glucosyl residues from dolichol-P-glucose. **(10)** Transfer of the oligosaccharide from dolichol-PP to the polypeptide chain at an Asn residue in the sequence Asn-X-Ser/Thr, releasing dolichol-PP. **(11)** Translocation of dolichol-PP to the cytoplasmic surface of the ER membrane. **(12)** Hydrolysis of dolichol-PP to dolichol-P. **(13)** Dolichol-P can also be formed by phosphorylation of dolichol by CTP. [Modified from Abeijon, C. and Hirschberg, C.B., *Trends Biochem. Sci.* **17**, 34 (1992).] ✷ **See the Animated Figures.**

feases (Fig. 15-35). Note that some of these reactions take place on the lumenal surface of the endoplasmic reticulum, whereas others occur on its cytoplasmic surface. Hence, on four occasions (Reactions 3, 5, 8, and 11 in Fig. 15-35), dolichol and its attached hydrophilic group are translocated, via unknown mechanisms, across the endoplasmic reticulum membrane. In the final steps of this process, the oligosaccharide is transferred to the Asn residue in a segment of sequence Asn-X-Ser/Thr (where X is any residue except Pro and possibly Asp) on a growing polypeptide chain. The resulting dolichol pyrophosphate is hydrolyzed to dolichol phosphate and P_i, a process similar to the pyrophosphatase cleavage of PP_i to 2 P_i. Further processing of the oligosaccharide takes place, as described in Section 8-3C, first in the endoplasmic reticulum and then in the Golgi apparatus (Fig. 8-17), where certain monosaccharide residues are trimmed away by specific

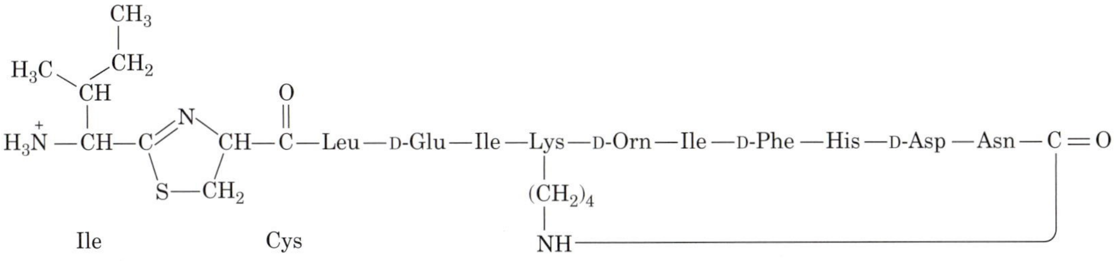

$$H_3\overset{+}{N}-\overset{\displaystyle CH}{\underset{\displaystyle \underset{\displaystyle S-CH_2}{|}}{C}} \overset{\displaystyle N}{=}\overset{\displaystyle CH-C}{\underset{\displaystyle \overset{\displaystyle O}{||}}{}}-Leu-\text{D-Glu}-Ile-Lys-\text{D-Orn}-Ile-\text{D-Phe}-His-\text{D-Asp}-Asn-C=O$$

Ile   Cys $(CH_2)_4$
                                                      |
                                                     NH

Bacitracin

Figure 15-36 The chemical structure of bacitracin. Note that this dodecapeptide has four D-amino acid residues and two unusual intrachain linkages. "Orn" represents the nonstandard amino acid ornithine (Fig. 20-9).

glycosylases and others are added by specific nucleotide sugar–requiring glycosyltransferases.

Bacitracin Interferes with the Dephosphorylation of Dolichol Pyrophosphate. A number of compounds block the actions of specific glycosylation enzymes, including **bacitracin** (Fig. 15-36), a cyclic polypeptide that is a widely used antibiotic. Bacitracin forms a complex with dolichol pyrophosphate that inhibits its dephosphorylation (Fig. 15-35, Reaction 12), thereby preventing the synthesis of glycoproteins from dolichol-linked oligosaccharide precursors. Bacitracin is clinically useful because it inhibits bacterial cell wall synthesis (which also involves dolichol-linked oligosaccharides) but does not affect animal cells since it cannot cross cell membranes (bacterial cell wall synthesis is an extracellular process).

SUMMARY

1. Glycogen breakdown requires three enzymes. Glycogen phosphorylase converts the glucosyl units at the nonreducing ends of glycogen to glucose-1-phosphate (G1P). Debranching enzyme transfers an $\alpha(1{\rightarrow}4)$-linked trisaccharide to a nonreducing end and hydrolyzes the $\alpha(1{\rightarrow}6)$ linkage. Phosphoglucomutase converts G1P to glucose-6-phosphate (G6P). In liver, G6P is hydrolyzed by glucose-6-phosphatase to glucose for export to the tissues.

2. Glycogen synthesis requires a different pathway in which G1P is activated by reaction with UTP to form UDP–glucose. Glycogen synthase adds glucosyl units to the nonreducing ends of a growing glycogen molecule that has been primed by glycogenin. Branching enzyme removes an $\alpha(1{\rightarrow}4)$-linked 7-residue segment and reattaches it through an $\alpha(1{\rightarrow}6)$ linkage to form a branched chain.

3. Glycogen metabolism is controlled in part by allosteric effectors such as AMP, ATP, and G6P. Covalent modification of glycogen phosphorylase and glycogen synthase shifts their $T \rightleftharpoons R$ equilibria and therefore alters their sensitivity to allosteric effectors.

4. The ratio of phosphorylase *a* (more active) to phosphorylase *b* (less active) depends on the activity of phosphorylase kinase, which is regulated by the activity of protein kinase A (PKA), a cAMP-dependent enzyme, and on the activity of phosphoprotein phosphatase-1. Glycogen phosphorylase is activated by phosphorylation, whereas glycogen synthase is activated by dephosphorylation.

5. Hormones such as glucagon, epinephrine, and insulin control glycogen metabolism. Hormone signals that generate cAMP as a second messenger or that elevate intracellular Ca^{2+}, which binds to the calmodulin subunit of phosphorylase kinase, promote glycogen breakdown. Insulin stimulates glycogen synthesis in part by activating phosphoprotein phosphatase-1.

6. Compounds that can be converted to oxaloacetate can subsequently be converted to glucose. The conversion of pyruvate to glucose by gluconeogenesis requires enzymes that bypass the three exergonic steps of glycolysis: Pyruvate carboxylase and PEP carboxykinase (PEPCK) bypass pyruvate kinase, fructose-1,6-bisphosphatase (FBPase) bypasses phosphofructokinase, and glucose-6-phosphatase bypasses hexokinase.

7. Gluconeogenesis is regulated by changes in enzyme synthesis and by allosteric effectors, including fructose-2,6-bisphosphate (F2,6P), which inhibits FBPase and activates phosphofructokinase (PFK) and whose synthesis depends on the phosphorylation state of the bifunctional enzyme phosphofructokinase-2/fructose bisphosphatase-2 (PFK-2/FBPase-2).

8. Formation of glycosidic bonds requires nucleotide sugars.

SELECTED READINGS

Bollen, M., Keppens, S., and Stalmans, W., Specific features of glycogen metabolism in the liver, *Biochem. J.* **336,** 19–31 (1998). [Describes the activities of the enzymes involved in glycogen synthesis and degradation and discusses the mechanisms for regulating these processes.]

Brosnan, J.T., Comments on metabolic needs for glucose and the role of gluconeogenesis, *Eur. J. Clin. Nutr.* **53,** S107–S111 (1999). [A very readable review that discusses possible reasons why carbohydrates are used universally as metabolic fuels and why glucose is stored as glycogen.]

Browner, M.F. and Fletterick, R.J., Phosphorylase: a biological transducer, *Trends Biochem. Sci.* **17,** 66–71 (1992).

Burda, P. and Aebi, M., The dolichol pathway of N-linked glycosylation, *Biochim. Biophys. Acta* **1426,** 239–257 (1999).

Chen, Y.-T., Glycogen storage diseases, *in* Scriver, C.R., Beaudet, A.L., Sly, W.S., and Valle, D. (Eds.), *The Metabolic & Molecular Bases of Inherited Disease* (8th ed.), pp. 1521–1552, McGraw-Hill (2001). [Begins with a review of glycogen metabolism.]

Croniger, C.M., Olswang, Y., Reshef, L., Kalhan, S.C., Tilghman, S.M., and Hanson, R.W., Phosphoenolpyruvate carboxykinase revisited. Insights into its metabolic role, *Biochem. Mol. Biol. Educ.* **30,** 14–20 (2002); *and* Croniger, C.M., Chakravarty, K., Olswang, Y., Cassuto, H., Reshef, L., and Hanson, R.W., Phosphoenolpyruvate carboxykinase revisited. II. Control of PEPCK-C gene expression, *Biochem. Mol. Biol. Educ.* **30,** 353–362 (2002).

Johnson, L.N., Lowe, E.D., Noble, M.E.M., and Owen, D.J., The structural basis for substrate recognition and control by protein kinases, *FEBS Lett.* **430,** 1–11 (1998).

Meléndez-Hevia, E., Waddell, T.G., and Shelton, E.D., Optimization of molecular design in the evolution of metabolism: the glycogen molecule, *Biochem. J.* **295,** 477–483 (1993).

Nordlie, R.C., Foster, J.D., and Lange, A.J., Regulation of glucose production by the liver, *Annu. Rev. Nutr.* **19,** 379–406 (1999).

Okar, D.A., Manzano, À., Navarro-Sabatè, A., Riera, L., Bartrons, R., and Lange, A.J., PFK-2/FBPase-2: Maker and breaker of the essential biofactor fructose-2,6-bisphosphate, *Trends Biochem. Sci.* **26,** 30–35 (2001).

Roach, P.J. and Skurat, A.V., Self-glucosylating initiator proteins and their role in glycogen biosynthesis, *Prog. Nucl. Acid Res. Mol. Biol.* **57,** 289–316 (1997). [Discusses glycogenin.]

Smith, C.M., Radzio-Andzelm, E., Akamine, M.P., Madhusudan, Akamine, P., and Taylor, S.S., The catalytic subunit of cAMP-dependent protein kinase: prototype for an extended network of communication, *Prog. Biophys. Mol. Biol.* **71,** 313–341 (1999).

KEY TERMS

glycogenolysis

phosphorolysis

debranching

nucleotide sugar

interconvertible enzymes

protein kinase A (PKA)

phophoprotein phosphatase

phosphorylase kinase

cAMP

calmodulin (CaM)

glycogen storage disease

second messenger

gluconeogenesis

reducing equivalent

dolichol

STUDY EXERCISES

1. List the metabolic sources and products of G6P.

2. How does the structure of glycogen relate to its metabolic function?

3. Describe the enzymatic degradation and synthesis of glycogen.

4. Why must opposing biosynthetic and degradative pathways differ in at least one enzyme?

5. Why does a phosphorylation/dephosphorylation system allow more sensitive regulation of a metabolic process than a simple allosteric system?

6. How does regulation of glycogen metabolism differ between liver and muscle?

7. Describe the reactions of gluconeogenesis.

8. Why is the malate–aspartate shuttle system important for gluconeogenesis?

9. Describe the role of fructose-2,6-bisphosphate in regulating gluconeogenesis.

PROBLEMS

1. Indicate the energy yield or cost, in ATP equivalents, for the following processes:

    (a) glycogen (3 residues) → 6 pyruvate

    (b) 3 glucose → 6 pyruvate

    (c) 6 pyruvate → 3 glucose

2. Write the balanced equation for (a) the sequential conversion of glucose to pyruvate and of pyruvate to glucose and (b) the catabolism of six molecules of G6P by the pentose phosphate pathway followed by conversion of ribulose-5-phosphate back to G6P by gluconeogenesis.

3. Phosphoglucokinase catalyzes the phosphorylation of the C6-OH group of G1P. Why is this enzyme important for the normal function of phosphoglucomutase?

4. The free energy of hydrolysis of an $\alpha(1 \rightarrow 4)$ glycosidic bond is -15.5 kJ $\cdot$ mol^{-1}, whereas that of an $\alpha(1 \rightarrow 6)$ glycosidic bond is -7.1 kJ $\cdot$ mol^{-1}. Use these data to explain why glycogen debranching includes three reactions [breaking and reforming $\alpha(1 \rightarrow 4)$ bonds and hydrolyzing $\alpha(1 \rightarrow 6)$ bonds], while glycogen branching requires only two reactions [breaking $\alpha(1 \rightarrow 4)$ bonds and forming $\alpha(1 \rightarrow 6)$ bonds].

5. Calculations based on the volume of a glucose residue and the branching pattern of cellular glycogen indicate that a glycogen molecule could have up to 28 branching tiers before becoming impossibly dense. What are the advantages of such a molecule and why is it not found *in vivo*?

6. One molecule of dietary glucose can be oxidized through glycolysis and the citric acid cycle to generate a maximum of 38 molecules of ATP. Calculate the fraction of this energy that is lost when the glucose is stored as glycogen before it is catabolized.

7. Glucose binds to glycogen phosphorylase and competitively inhibits the enzyme. What is the physiological advantage of this?

8. Many diabetics do not respond to insulin because of a deficiency of insulin receptors on their cells. How does this affect (a) the levels of circulating glucose immediately after a meal and (b) the rate of glycogen synthesis in muscle?

9. Glucose-6-phosphatase is located inside the endoplasmic reticulum. Describe the probable symptoms of a defect in G6P transport across the endoplasmic reticulum membrane.

10. Individuals with McArdle's disease often experience a "second wind" resulting from cardiovascular adjustments that allow glucose mobilized from liver glycogen to fuel muscle contraction. Explain why the amount of ATP derived in the muscle from circulating glucose is less than the amount of ATP that would be obtained by mobilizing the same amount of glucose from muscle glycogen.

11. A sample of glycogen from a patient with liver disease is incubated with P_i, normal glycogen phosphorylase, and normal debranching enzyme. The ratio of G1P to glucose formed in this reaction mixture is 100. What is the patient's most probable enzymatic deficiency?

The synthesis and degradation of numerous biological materials depends on the flow of molecules and energy through the citric acid cycle, which has been likened to a metabolic water wheel. [Al Zwiazek/ SUPERSTOCK.]

Citric Acid Cycle

In the preceding two chapters, we examined the catabolism of glucose and its biosynthesis, storage, and mobilization. Although glucose is a source of energy for nearly all cells, it is not the only metabolic fuel, nor is glycolysis the only energy-yielding catabolic pathway. Cells that rely exclusively on glycolysis to meet their energy requirements actually waste most of the chemical potential energy of carbohydrates. When glucose is converted to lactate or ethanol, a relatively reduced product leaves the cell. If the end product of glycolysis is instead further oxidized, the cell can recover considerably more energy.

The oxidation of an organic compound requires an electron acceptor, such as NO_3^-, SO_4^{2-}, Fe^{3+}, or O_2, all of which are exploited as oxidants in different organisms. In aerobic organisms, the electrons produced by oxidative metabolism are ultimately transferred to O_2. Oxidation of metabolic fuels is carried out by the citric acid cycle, a sequence of reactions that arose sometime after levels of atmospheric oxygen became significant, about 3 billion years ago. As the reduced carbon atoms of metabolic fuels are oxidized to CO_2, electrons are transferred to electron carriers that are subsequently reoxidized by O_2. In this chapter, we examine the oxidation reactions of the citric acid cycle itself. In the following chapter, we examine the fate of the electrons and see how their energy is used to drive the synthesis of ATP.

It is sometimes convenient to think of the citric acid cycle as an addendum to glycolysis. Pyruvate derived from glucose can be split into CO_2 and a two-carbon fragment that enters the cycle for oxidation as acetyl-CoA (Fig. 16-1). However, it is really misleading to think of the citric acid cycle as merely a continuation of carbohydrate catabolism. *The citric acid cycle is a central pathway for recovering energy from several metabolic fuels, including carbohydrates, fatty acids, and amino acids, that are broken down to acetyl-CoA for oxidation.* In fact, under some conditions, the principal function of the citric acid cycle is to recover energy from fatty acids. We shall also see that the citric acid cycle supplies the reactants for a variety of biosynthetic pathways.

We begin this chapter with an overview of the citric acid cycle. Next, we explore how acetyl-CoA, its starting compound, is formed from pyruvate. After discussing the reactions catalyzed by each of the enzymes of the cycle, we consider the regulation of these enzymes. Finally, we examine the links between citric acid cycle intermediates and other metabolic processes.

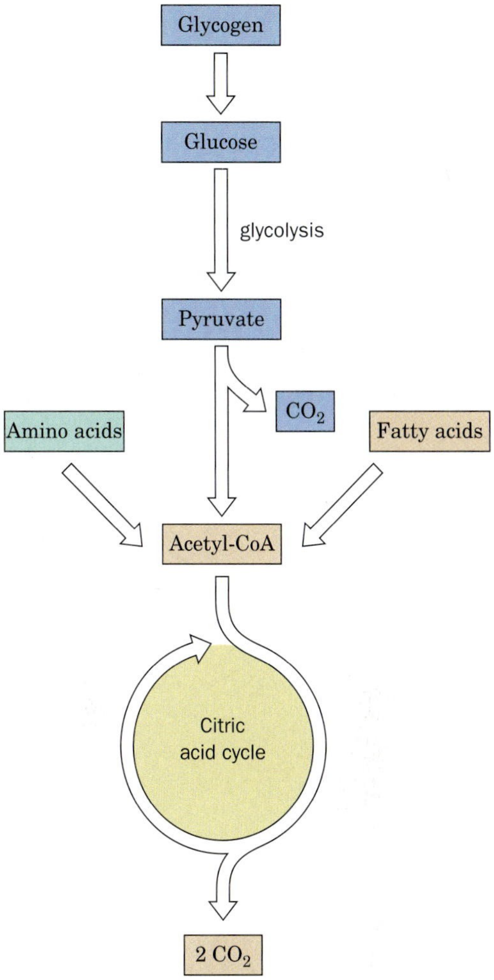

Figure 16-1 Overview of oxidative fuel metabolism. Acetyl groups derived from carbohydrates, amino acids, and fatty acids enter the citric acid cycle, where they are oxidized to CO_2. ⟲ See the Animated Figures.

1 Overview of the Citric Acid Cycle

The citric acid cycle (Fig. 16-2) is an ingenious series of eight reactions that oxidizes the acetyl group of acetyl-CoA to two molecules of CO_2 in a manner that conserves the liberated free energy in the reduced compounds NADH and $FADH_2$. The cycle is named after the product of its first reaction, **citrate.** One complete round of the cycle yields two molecules of CO_2, three NADH, one $FADH_2$, and one "high-energy" compound (GTP or ATP).

The citric acid cycle first came to light in the 1930s, when Hans Krebs, building on the work of others, proposed a circular reaction scheme for the interconversion of certain compounds containing two or three carboxylic acid groups (that is, di- and tricarboxylates). At the time, many of the citric acid cycle intermediates were already well known as plant products: **citrate** from citrus fruit, **aconitate** from monkshood (*Aconitum*), **succinate** from amber (*Succinum*), **fumarate** from the herb *Fumaria,* and **malate** from apple (*Malus*). Two other intermediates, **α-ketoglutarate** and **oxaloacetate,** are

| See Guided Exploration 14 |
| Citric acid cycle overview. |

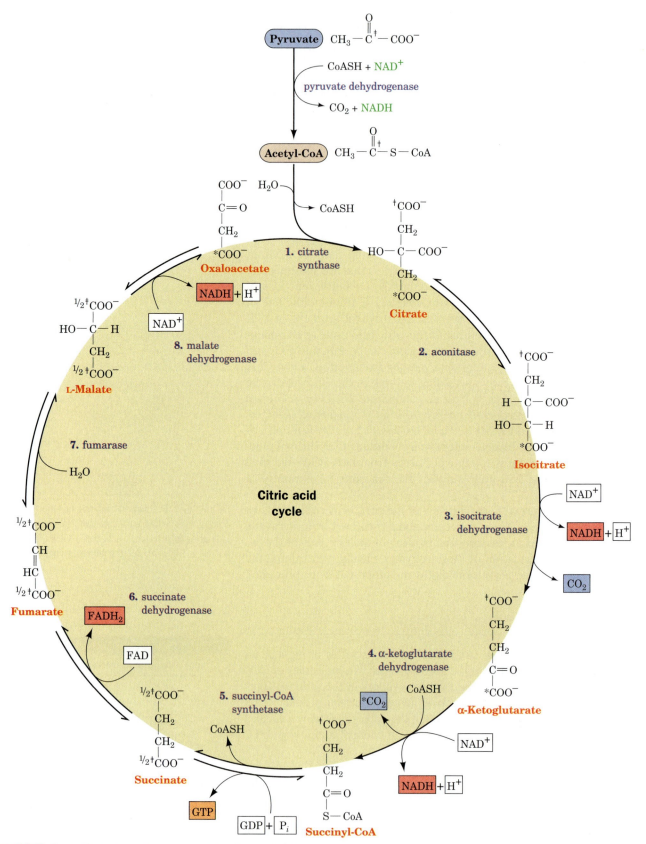

Figure 16-2 *Key to Metabolism.*   The reactions of the citric acid cycle. The reactants and products of this catalytic cycle are boxed. The pyruvate → acetyl-CoA reaction (*top*) supplies the cycle's substrate via carbohydrate metabolism but is not considered to be part of the cycle.

An isotopic label at C4 of oxaloacetate (*) becomes C1 of α-ketoglutarate and is released as CO_2 in Reaction 4. An isotopic label at C1 of acetyl-CoA ($\ddagger$) becomes C5 of α-ketoglutarate and is scrambled in Reaction 5 between C1 and C4 of succinate ($\frac{1}{2}\ddagger$). **See the Animated Figures.**

known by their chemical names because they were synthesized before they were identified in living organisms. Krebs was the first to show how the metabolism of these compounds was linked to the oxidation of metabolic fuels. His discovery of the citric acid cycle, in 1937, ranks as one of the most important achievements of metabolic chemistry (Box 16-1). Although the enzymes and intermediates of the citric acid cycle are now well established, many investigators continue to explore the molecular mechanisms of the enzymes and how the enzymes are regulated for optimal performance under varying metabolic conditions in different organisms.

Before we examine each of the reactions in detail, we should emphasize some general features of the citric acid cycle:

1. The circular pathway, which is also called the **Krebs cycle** or the **tricarboxylic acid (TCA) cycle,** oxidizes acetyl groups from many sources, not just pyruvate. Because it accounts for the major portion of carbohydrate, fatty acid, and amino acid oxidation, the citric acid cycle is often considered the "hub" of cellular metabolism.

2. The net reaction of the citric acid cycle is

$$3 \ NAD^+ + FAD + GDP + P_i + \text{acetyl-CoA} \longrightarrow$$
$$3 \ NADH + FADH_2 + GTP + CoA + 2 \ CO_2$$

   The oxaloacetate that is consumed in the first step of the citric acid cycle is regenerated in the last step of the cycle. Thus, *the citric acid cycle acts as a multistep catalyst that can oxidize an unlimited number of acetyl groups.*

3. In eukaryotes, all the enzymes of the citric acid cycle are located in the mitochondrion, so all substrates, including NAD^+ and GDP, must be generated in the mitochondria or be transported into mitochondria from the cytosol. Similarly, all the products of the citric acid cycle must be consumed in the mitochondria or transported into the cytosol.

4. The carbon atoms of the two molecules of CO_2 produced in one round of the cycle are not the two carbons of the acetyl group that began the round (Fig. 16-2). These acetyl carbon atoms are lost in subsequent rounds of the cycle. However, the net effect of each round of the cycle is the oxidation of one acetyl group to 2 CO_2.

5. Citric acid cycle intermediates are precursors for the biosynthesis of other compounds (e.g., oxaloacetate for gluconeogenesis; Section 15-4).

6. The oxidation of an acetyl group to 2 CO_2 requires the transfer of four pairs of electrons. The reduction of 3 NAD^+ to 3 NADH accounts for three pairs of electrons; the reduction of FAD to $FADH_2$ accounts for the fourth pair. Much of the free energy of oxidation of the acetyl group is conserved in these reduced coenzymes. Energy is also recovered in GTP (or ATP). In Section 17-3C, we shall see that 11 ATP are formed when the four pairs of electrons are eventually transferred to O_2.

2 Synthesis of Acetyl-Coenzyme A

Acetyl groups enter the citric acid cycle as part of the "high-energy" compound acetyl-CoA (recall that thioesters have high free energies of hydrolysis; Section 13-2D). Although acetyl-CoA can also be derived from fatty acids (Section 19-2) and some amino acids (Section 20-4), we shall focus here on the production of acetyl-CoA from pyruvate derived from carbohydrates.

BOX 16-1

Pathways of Discovery

Hans Krebs and the Citric Acid Cycle

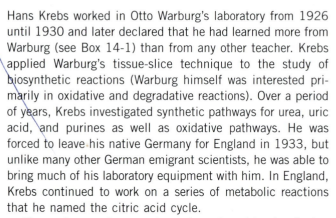

Hans Krebs (1900–1981)

Hans Krebs worked in Otto Warburg's laboratory from 1926 until 1930 and later declared that he had learned more from Warburg (see Box 14-1) than from any other teacher. Krebs applied Warburg's tissue-slice technique to the study of biosynthetic reactions (Warburg himself was interested primarily in oxidative and degradative reactions). Over a period of years, Krebs investigated synthetic pathways for urea, uric acid, and purines as well as oxidative pathways. He was forced to leave his native Germany for England in 1933, but unlike many other German emigrant scientists, he was able to bring much of his laboratory equipment with him. In England, Krebs continued to work on a series of metabolic reactions that he named the citric acid cycle.

The cycle was discovered not through sudden inspiration but through a series of careful experiments spanning the years 1932 to 1937. Krebs became interested in the "combustion" phase of fuel use, namely, what occurs after the fermentation of glucose to lactate. Until the 1930s, the mechanism of glucose oxidation and its relationship to cellular respiration (oxygen uptake) was a mystery. Krebs understood that the stoichiometry of the overall process (glucose + 6 O_2 → 6 CO_2 + 6 H_2O) required a multistep pathway. He was also aware that other researchers had examined the ability of muscle tissue to rapidly oxidize various dicarboxylates (α-ketoglutarate, succinate, and malate) and a tricarboxylate (citrate), but none of these substances had a clear relationship to any foodstuffs.

In 1935, Albert Szent-Györgyi found that cellular respiration was dramatically accelerated by small amounts of succinate, fumarate, malate, or oxaloacetate. In fact, the addition of any of these compounds stimulated O_2 uptake and CO_2 production far in excess of what would be expected for their oxidation. In other words, the compounds acted catalytically to boost the combustion of other compounds in the cell. At about the same time, Carl Martius and Franz Knoop showed that citrate could be converted to α-ketoglutarate. Soon, Krebs had an entire sequence of reactions for converting citrate to oxaloacetate: citrate → aconitate → isocitrate → α-ketoglutarate → succinate → fumarate → malate → oxaloacetate. However, in order for this sequence of reactions to work catalytically, it must repeatedly return to its starting point; that is, the first compound must be regenerated. And there was still no obvious link to glucose metabolism!

Krebs believed that citrate and the other intermediates were involved in glucose combustion because they appeared to burn at the same rate as foodstuffs and were the only substances that did so. In addition, earlier work had shown that the three-carbon compound malonate not only blocks the conversion of succinate to fumarate, it blocks all combustion by living cells. In 1937, Martius and Knoop provided Krebs with a key piece of information: oxaloacetate and pyruvate could be converted to citrate in the presence of hydrogen peroxide.

Krebs now had the missing link: Pyruvate is a product of glucose metabolism, and its reaction with oxaloacetate to form citrate closed off the linear series of reactions to form a cycle. The idea of a circular pathway was not new to Krebs. He and Kurt Henseleit had elucidated the four-step urea cycle in 1932 (Section 20-3). Krebs quickly showed that the reaction of pyruvate with oxaloacetate to form citrate took place in living tissue and that the rates of citrate synthesis and breakdown were high enough to account for the observed fuel combustion in a variety of tissue types.

Remarkably, Krebs' initial report on the citric acid cycle was rejected by *Nature,* a leading journal, before being accepted for publication in the less prestigious *Enzymologia.* In his report, Krebs established the major outlines of the pathway, although some details were later revised. For example, the mechanism of citrate formation (which involves acetyl-CoA rather than pyruvate) and the participation of succinyl-CoA in the cycle were not immediately appreciated. Coenzyme A was not discovered until 1945, and only in 1951 was acetyl-CoA shown to be the intermediate that condenses with oxaloacetate to form citrate. Work by Krebs and others established that the citric acid cycle plays a major role in the oxidation of amino acids and fatty acids. In fact, the pathway accounts for approximately two-thirds of the energy derived from metabolic fuels. Krebs also recognized the role of the citric acid cycle in supplying precursors for synthetic reactions.

[Krebs, H.A. and Johnson, W.A., The role of citric acid in intermediate metabolism in animal tissues, *Enzymologia* **4**, 148–156 (1937).]

As we saw in Section 14-3, the end product of glycolysis under anaerobic conditions is lactate or ethanol. However, under aerobic conditions, when the NADH generated by glycolysis is reoxidized in the mitochondria,

the final product is pyruvate. A transport protein imports pyruvate along with H^+ (i.e., a pyruvate–H^+ symport) into the mitochondrion for further oxidation.

A The Pyruvate Dehydrogenase Multienzyme Complex

Multienzyme complexes are groups of noncovalently associated enzymes that catalyze two or more sequential steps in a metabolic pathway. Virtually all organisms contain multienzyme complexes, which represent a step forward in the evolution of catalytic efficiency because they offer the following advantages:

1. Enzymatic reaction rates are limited by the frequency with which enzymes collide with their substrates (Section 11-3D). When a series of reactions occurs within a multienzyme complex, the distance that substrates must diffuse between active sites is minimized, thereby enhancing the reaction rate.

2. The channeling of metabolic intermediates between successive enzymes in a metabolic pathway reduces the opportunity for these intermediates to react with other molecules, thereby minimizing side reactions.

3. The reactions catalyzed by a multienzyme complex can be coordinately controlled.

Acetyl-CoA is formed from pyruvate through oxidative decarboxylation by a multienzyme complex named **pyruvate dehydrogenase.** This complex contains multiple copies of three enzymes: **pyruvate dehydrogenase (E_1), dihydrolipoyl transacetylase (E_2),** and **dihydrolipoyl dehydrogenase (E_3).** The *E. coli* pyruvate dehydrogenase complex is an ~4600-kD particle with a diameter of about 300 Å (Fig. 16-3a). The core of the particle is made

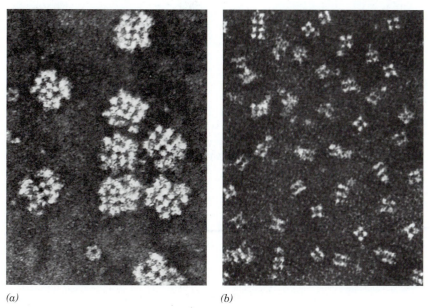

(a)   *(b)*

Figure 16-3 Electron micrographs of the *E. coli* pyruvate dehydrogenase multienzyme complex. (*a*) The intact complex. (*b*) The dihydrolipoyl transacetylase (E_2) core complex. [Courtesy of Lester Reed, University of Texas at Austin.]

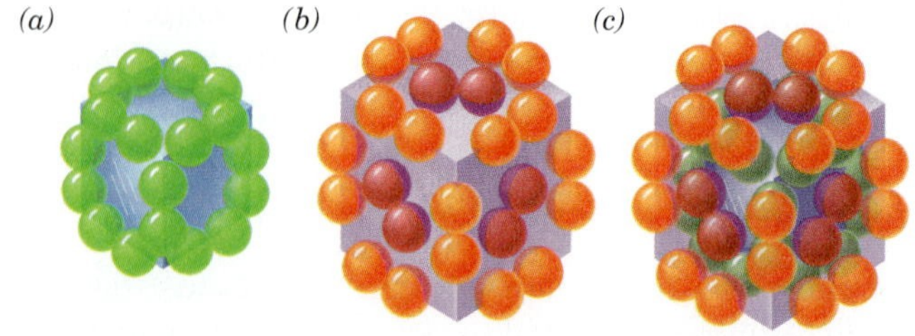

(a)   (b)   (c)

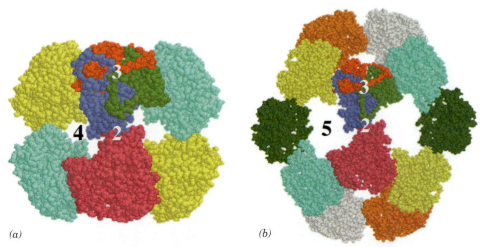

Figure 16-4   Structural organization of the *E. coli* pyruvate dehydrogenase multienzyme complex.   (*a*) The dihydrolipoyl transacetylase (E$_2$) core. The 24 E$_2$ proteins (*green spheres*) associate as trimers at the corners of a cube. (*b*) The 24 pyruvate dehydrogenase (E$_1$) proteins (*orange spheres*) form dimers that associate with the E$_2$ core (*shaded cube*) along its 12 edges. The 12 dihydrolipoyl dehydrogenase (E$_3$) proteins (*purple spheres*) form dimers that attach to the six faces of the E$_2$ cube. (*c*) Parts *a* and *b* combined form the entire 60-subunit complex.

of 24 E$_2$ proteins arranged in a cube (Figs. 16-3*b*, 4*a*, and 5*a*), which is surrounded by 24 E$_1$ proteins and 12 E$_3$ proteins (Fig. 16-4*b,c*). In mammals, yeast, and some bacteria, the pyruvate dehydrogenase complex is even larger and more complicated, although it catalyzes the same reactions using homologous enzymes and similar mechanisms. In these larger complexes, the E$_2$ core consists of 60 subunits (20 trimers) arranged with dodecahedral symmetry [Fig. 16-5*b*; a dodecahedron is a regular polyhedron with *I* symmetry (Fig. 6-34*c*) that has 20 vertices, each lying on a three-fold axis, and 12 pentagonal faces having an aggregate of 30 edges]. Thus, the

(a)   (b)

Figure 16-5   Comparison of the X-ray structures of the cubic and dodecahedral dihydrolipoyl transacetylase (E$_2$) cores of pyruvate dehydrogenase multienzyme complexes.   The structures are shown in space-filling representation and viewed along their twofold axes of symmetry. The rear portion of each complex, which is almost entirely eclipsed by the forward portion, has been deleted for clarity. (*a*) The cubic *Azobacter vinelandii* E$_2$ core. It consists of 24 subunits that form 8 trimers, shown here in different colors. The positions of a twofold, a threefold, and a fourfold axis of symmetry are indicated. Its height is ~125 Å. (*b*) The dodecahedral *B. stearothermophilus* E$_2$ core. It consists of 60 subunits that form 20 trimers, shown here in different colors. The positions of a twofold, a threefold, and a fivefold axis of symmetry are indicated. Its outer diameter is ~237 Å. The subunits forming the trimer above the center of each drawing are individually colored. Note that these subunits are extensively associated, but that the interactions between contacting trimers in both types of complexes are relatively tenuous. Also note that these contacting trimers form 4- and 5-membered rings, respectively, that comprise the square and pentagonal faces of the cubic and dodecahedral complexes. [Based on X-ray structures by Wim Hol, University of Washington. PDBids 1EAB and 1B5S.]

mitochondrially located ~10,000-kD eukaryotic complex, the largest known multienzyme complex, contains a dodecahedral core of 20 E_2 trimers (one centered on every vertex) surrounded by 30 E_1 $\alpha_2\beta_2$ heterotetramers (one centered on every edge) and 12 E_3 dimers (one centered in every face), together with 12 or so copies of **E_3 binding protein,** which facilitates the binding of E_3 to the E_2 core. Mammalian complexes also contain several copies of a kinase and a phosphatase that function to regulate the activity of the complex (see Section 16-4A).

B The Reactions of the Pyruvate Dehydrogenase Complex

The pyruvate dehydrogenase complex catalyzes five sequential reactions with the overall stoichiometry

$$\text{Pyruvate} + \text{CoA} + \text{NAD}^+ \longrightarrow \text{acetyl-CoA} + \text{CO}_2 + \text{NADH}$$

Five different coenzymes are required: thiamine pyrophosphate (TPP; page 449), **lipoamide,** coenzyme A (Fig. 13-9), FAD (Fig. 13-11), and NAD^+ (Fig. 11-3). The coenzymes and their mechanistic functions are listed in Table 16-1. The sequence of reactions catalyzed by the pyruvate dehydrogenase complex is as follows (Fig. 16-6):

1. Pyruvate dehydrogenase (E_1), a TPP-requiring enzyme, decarboxylates pyruvate with the formation of a hydroxyethyl-TPP intermediate:

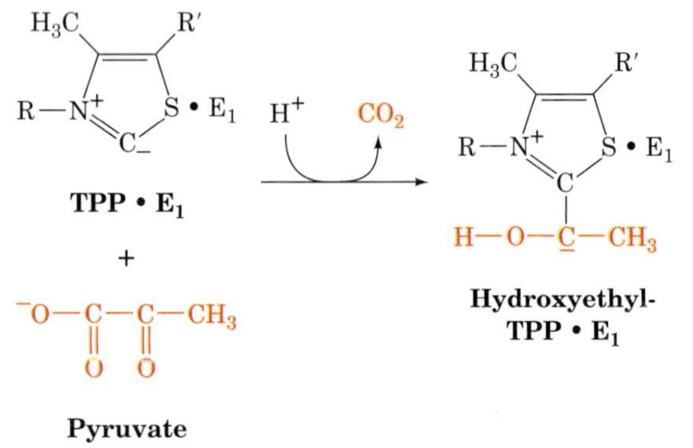

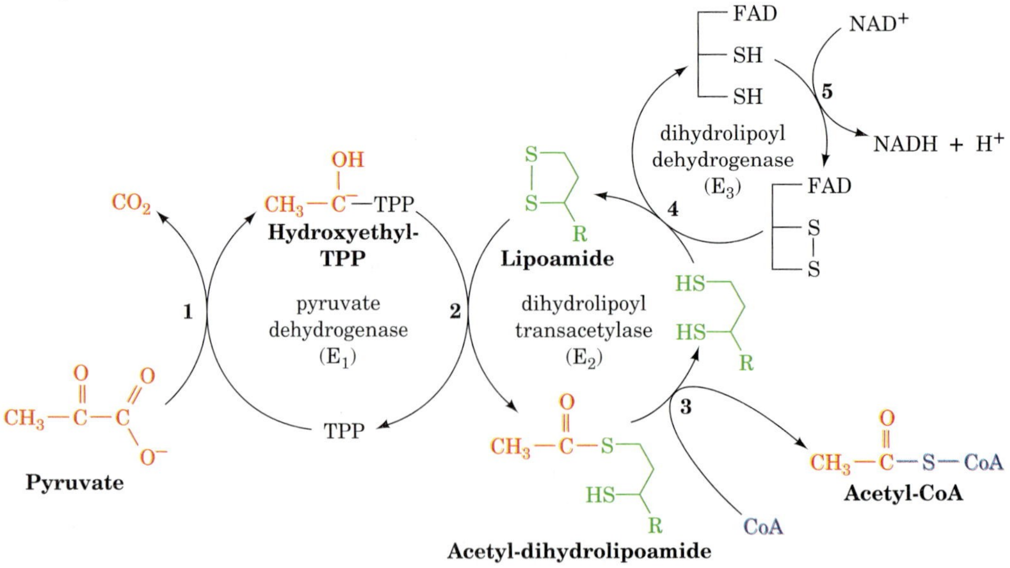

Figure 16-6 The five reactions of the pyruvate dehydrogenase multienzyme complex. E_1 (pyruvate dehydrogenase) contains TPP and catalyzes Reactions 1 and 2. E_2 (dihydrolipoyl transacetylase) contains lipoamide and catalyzes Reaction 3. E_3 (dihydrolipoyl dehydrogenase) contains FAD and a redox-active disulfide and catalyzes Reactions 4 and 5.

Table 16-1 The Coenzymes and Prosthetic Groups of Pyruvate Dehydrogenase

Cofactor	Location	Function
Thiamine pyrophosphate (TPP)	Bound to E_1	Decarboxylates pyruvate yielding a hydroxyethyl-TPP carbanion
Lipoic acid	Covalently linked to a Lys on E_2 (lipoamide)	Accepts the hydroxyethyl carbanion from TPP as an acetyl group
Coenzyme A (CoA)	Substrate for E_2	Accepts the acetyl group from lipoamide
Flavin adenine dinucleotide (FAD)	Bound to E_3	Reduced by lipoamide
Nicotinamide adenine dinucleotide (NAD^+)	Substrate for E_3	Reduced by $FADH_2$

This reaction is identical to that catalyzed by yeast pyruvate decarboxylase (Fig. 14-20).

Recall (Section 14-3B) that the ability of TPP's thiazolium ring to add to carbonyl groups and act as an electron sink makes it the coenzyme most utilized in α-keto acid decarboxylation reactions.

2. The hydroxyethyl group is transferred to the next enzyme, dihydrolipoyl transacetylase (E_2), which contains a lipoamide group. Lipoamide consists of **lipoic acid** linked via an amide bond to the ε-amino group of a Lys residue (Fig. 16-7). The reactive center of lipoamide is a cyclic disulfide that can be reversibly reduced to yield **dihydrolipoamide.** The hydroxyethyl group derived from pyruvate attacks the lipoamide disulfide, and TPP is eliminated. The hydroxyethyl

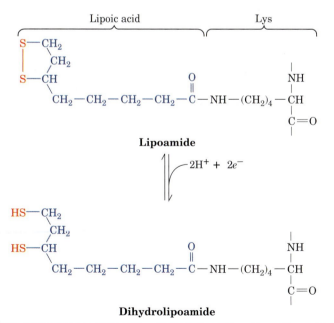

Figure 16-7 Interconversion of lipoamide and dihydrolipoamide. Lipoamide consists of lipoic acid covalently joined to the ε-amino group of a Lys residue via an amide bond.

carbanion is thereby oxidized to an acetyl group as the lipoamide disulfide is reduced:

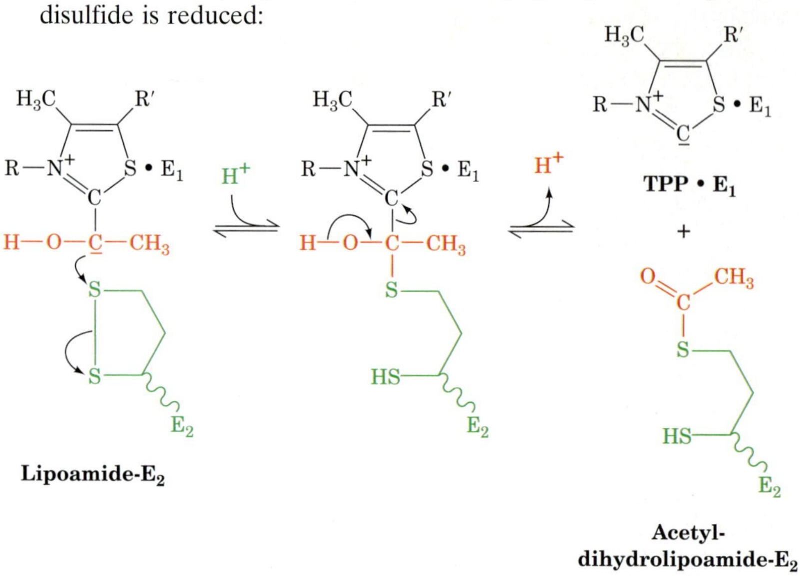

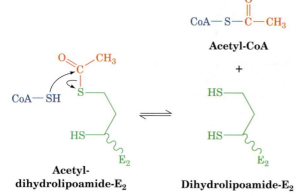

Lipoamide-E$_2$

TPP • E$_1$

+

Acetyl-dihydrolipoamide-E$_2$

3. E$_2$ then catalyzes a transesterification reaction in which the acetyl group is transferred to CoA, yielding acetyl-CoA and dihydrolipoamide-E$_2$:

$$CoA-S-\overset{\overset{\textstyle O}{\|}}{C}-CH_3$$

Acetyl-CoA

+

Acetyl-dihydrolipoamide-E$_2$ **Dihydrolipoamide-E$_2$**

4. Acetyl-CoA has now been formed, but the lipoamide group of E$_2$ must be regenerated. Dihydrolipoyl dehydrogenase (E$_3$) reoxidizes dihydrolipoamide to complete the catalytic cycle of E$_2$. Oxidized E$_3$ contains a reactive Cys—Cys disulfide group and a tightly bound FAD. The oxidation of dihydrolipoamide is a disulfide interchange reaction:

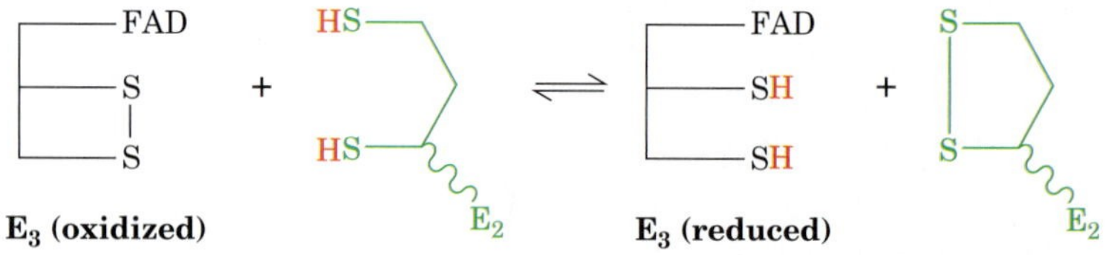

E$_3$ (oxidized)   **E$_2$**   **E$_3$ (reduced)**   **E$_2$**

5. Finally, reduced E$_3$ is reoxidized. The sulfhydryl groups are reoxidized by a mechanism in which FAD funnels electrons to NAD$^+$ yielding NADH:

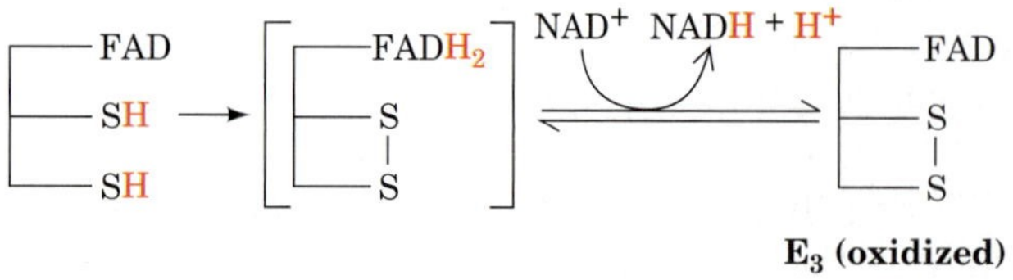

NAD$^+$ NADH + H$^+$

E$_3$ (oxidized)

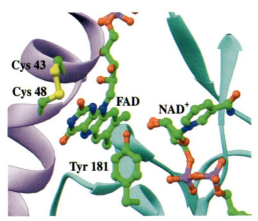

Figure 16-8 **Active site of dihydrolipoamide dehydrogenase (E₃).** In this X-ray structure of the enzyme from *Pseudomonas putida,* the redox-active portions of the bound NAD⁺ and FAD cofactors, the side chains of Cys 43 and Cys 48 forming the redox-active disulfide bond, and the side chain of Tyr 181 are shown in ball-and-stick form with C green, N blue, O red, P magenta, and S yellow. Note that the side chain of Tyr 181 is interposed between the flavin and the nicotinamide rings. [Based on an X-ray structure by Wim Hol, University of Washington. PDBid 1LVL.]

The X-ray structure of dihydrolipoyl dehydrogenase together with mechanistic information indicate that the reaction catalyzed by dihydrolipoamide dehydrogenase (E₃) is more complex than Reactions 4 and 5 in Fig. 16-6 suggest. The enzyme's redox-active disulfide bond occurs between Cys 43 and Cys 48, which reside on a highly conserved segment of the enzyme's polypeptide chain. The disulfide bond links successive turns in a distorted segment of an α helix (in an undistorted helix, the C_α atoms of Cys 43 and Cys 48 would be too far apart to permit the disulfide bond to form). The enzyme's flavin group is almost completely buried in the protein, which prevents the surrounding solution from interfering with the electron-transfer reaction catalyzed by the enzyme. The nicotinamide ring of NAD⁺ binds on the side of the flavin opposite the disulfide. In the absence of NAD⁺, the phenol side chain of Tyr 181 covers the nicotinamide-binding pocket so as to shield the flavin from contact with the solution (Fig. 16-8). The Tyr side chain apparently moves aside to allow the nicotinamide ring to bind near the flavin ring.

FAD prosthetic groups in proteins have reduction potentials of around 0 V (see Table 13-3), which makes $FADH_2$ unsuitable for donating electrons to NAD⁺ ($\mathscr{E}°' = -0.315$ V). Evidence suggests that the FAD group in dihydrolipoamide dehydrogenase never becomes fully reduced as $FADH_2$. Due to the precise positioning of the flavin and nicotinamide ring, electrons are rapidly transferred from the enzyme disulfide through FAD to NAD⁺, so a reduced flavin anion (FADH⁻) has but a transient existence. Thus, *FAD appears to function more as an electron conduit than as a source or sink of electrons.*

A Swinging Arm Transfers Intermediates. How are reaction intermediates channeled between E₂ (the core of the pyruvate dehydrogenase complex) and the E₁ and E₃ proteins on the outside? The key is the lipoamide group of E₂. The lipoic acid residue and the side chain of the Lys residue to which it is attached have a combined length of about 14 Å. This **lipoyllysyl arm** *(at left)* apparently acts as a long tether that swings the disulfide group from E₁ (where it picks up a hydroxyethyl group), to the E₂ active site (where the hydroxyethyl group is transferred to form acetyl-CoA), and from there to E₃ (where the reduced disulfide is reoxidized). The domains of E₂ that carry the lipoyllysyl arms are linked to the rest of the E₂ protein by a highly flexible Pro- and Ala-rich segment that contributes to the mobility of the lipoyllysyl arm. Because of the flexibility and reach of the lipoyllysyl arms, one E₁ protein can acetylate numerous E₂ proteins, and one E₃ protein can reoxidize several dihydrolipoamide groups. The lipoyllysyl arms probably protrude into the hollow interior of the E₂ core (Fig. 16-5) and swing around in order to "visit" the active sites of E₁, E₂, and E₃. The entire pyruvate dehydrogenase complex can be inactivated by the reaction of the lipoamide group with certain arsenic-containing compounds (see Box 16-2).

Lipoyllysyl arm (fully extended)

14 Å

3 Enzymes of the Citric Acid Cycle

In this section, we discuss the eight enzymes of the citric acid cycle. The elucidation of the mechanisms for each of these enzymes is the result of an enormous amount of experimental work. Even so, there remain questions about the mechanistic details of the enzymes and their regulatory properties.

BOX 16-2

Biochemistry in Health and Disease

Arsenic Poisoning

The toxicity of arsenic has been known since ancient times. As(III) compounds such as **arsenite** (AsO_3^{3-}) and **organic arsenicals** are toxic because they bind to sulfhydryl compounds (including lipoamide) that can form bidentate adducts:

Arsenite + **Dihydro-lipoamide** → + $2 H_2O$

Organic arsenical + → + H_2O

The inactivation of lipoamide-containing enzymes by arsenite, especially the pyruvate dehydrogenase and α-keto-glutarate dehydrogenase complexes, brings respiration to a halt. However, organic arsenicals are more toxic to microorganisms than they are to humans, apparently because of differences in the sensitivities of their various enzymes to these compounds. This differential toxicity is the basis for the early twentieth century use of organic arsenicals in the treatment of **syphilis** (a bacterial disease) and trypanosomiasis (a parasitic disease). These compounds were actually the first antibiotics, although, not surprisingly, they produced severe side effects.

Arsenic is often suspected as a poison in untimely deaths. It was long thought that Napoleon Bonaparte died from arsenic poisoning while in exile on the island of St. Helena, a suspicion that is strongly supported by the recent finding that a lock of his hair contains high levels of arsenic. But was it murder or environmental pollution? Arsenic-containing dyes were used in wallpaper at the time, and it was eventually determined that in damp weather, fungi convert the arsenic to a volatile compound. Surviving samples of the wallpaper from Napoleon's room in fact contain arsenic. Napoleon's arsenic poisoning may therefore have been unintentional.

Napoleon Bonaparte [© Victoria and Albert Museum/Art Resource].

Charles Darwin may also have been an unwitting victim of chronic arsenic poisoning. In the years following his epic voyage on the *Beagle,* Darwin was plagued by eczema, vertigo, headaches, gout, and nausea—all symptoms of arsenic poisoning. Fowler's solution, a widely used nineteenth century "tonic," contained 10 mg of arsenite per mL. Many individuals, quite possibly Darwin himself, took this "medication" for years.

Charles Darwin [© Photo Researchers].

A Citrate Synthase

Citrate synthase catalyzes the condensation of acetyl-CoA and oxaloacetate. This initial reaction of the citric acid cycle is the point at which carbon atoms (from carbohydrates, fatty acids, and amino acids) are "fed into the furnace" as acetyl-CoA. The citrate synthase reaction proceeds with an Ordered Sequential kinetic mechanism in which oxaloacetate binds before acetyl-CoA.

X-Ray studies show that the free enzyme (a dimer) is in an "open" form, with two domains that form a cleft containing the oxaloacetate binding site

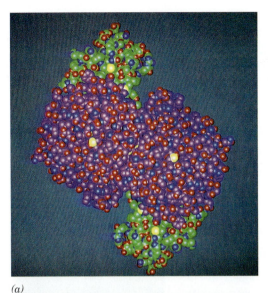

(a)

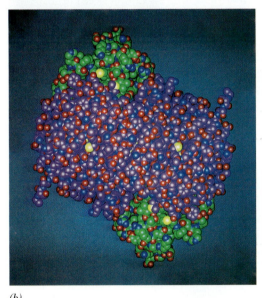

(b)

Figure 16-9 **Conformational changes in citrate synthase.** (*a*) The open conformation. (*b*) The closed, substrate-binding conformation. The C atoms of the small domain in each subunit of the enzyme are green, and those of the large domain are magenta. N, O, and S atoms in both domains are blue, red, and yellow. The large conformational shift between the open and closed forms entails relative interatomic movements of up to 15 Å. [Courtesy of Anne Dallas, University of Pennsylvania; and Helen Berman, Rutgers University. Based on X-ray structures determined by James Remington and Robert Huber, Max-Planck-Institut für Biochemie, Martinsried, Germany. PDBids 1CTS and 2CTS.] **See the Interactive Exercises.**

(Fig. 16-9*a*). When oxaloacetate binds, the smaller domain undergoes a remarkable 18° rotation, which closes the cleft (Fig. 16-9*b*). The existence of the "open" and "closed" forms explains the enzyme's Ordered Sequential kinetic behavior. *The conformational change generates the acetyl-CoA binding site and seals the oxaloacetate binding site so that solvent cannot reach the bound substrate.* A similar conformational change occurs in adenylate kinase (Fig. 13-8) and hexokinase (Fig. 14-2) to prevent ATP hydrolysis.

In the reaction mechanism proposed by James Remington, three ionizable side chains of citrate synthase participate in catalysis (Fig. 16-10):

1. The enol of acetyl-CoA is generated in the rate-limiting step of the reaction when Asp 375 (a base) removes a proton from the methyl group. His 274 forms a hydrogen bond with the enolate oxygen.

2. **Citryl-CoA** is formed in a concerted acid–base catalyzed step, in which the acetyl-CoA enolate (a nucleophile) attacks oxaloacetate. His 320 (an acid) donates a proton to oxaloacetate's carbonyl group. The citryl-CoA intermediate remains bound to the enzyme. Citrate synthase is one of the few enzymes that can directly form a carbon–carbon bond without the assistance of a metal ion cofactor.

3. Citryl-CoA is hydrolyzed to citrate and CoA. This hydrolysis provides the reaction's thermodynamic driving force ($\Delta G^{\circ\prime} = -31.5 \text{ kJ} \cdot \text{mol}^{-1}$). We shall see later why this reaction requires such a large, seemingly wasteful, expenditure of free energy.

B Aconitase

Aconitase catalyzes the reversible isomerization of citrate and **isocitrate,** with **cis-aconitate** as an intermediate:

$$
\begin{array}{ccccc}
\text{COO}^- & & \left[\text{COO}^-\right. & & \text{COO}^- \\
| & & | & & | \\
\text{CH}_2 & & \text{CH}_2 & & \text{CH}_2 \\
| & \xrightarrow{\text{H}_2\text{O}} & | & \xrightarrow{\text{H}_2\text{O}} & | \\
\text{HO}-\text{C}-\text{COO}^- & \rightleftharpoons & \text{C}-\text{COO}^- & \rightleftharpoons & \text{H}-\text{C}-\text{COO}^- \\
| & & \| & & | \\
\text{CH}_2 & & \text{CH} & & \text{HO}-\text{C}-\text{H} \\
| & & | & & | \\
\text{COO}^- & & \left.\text{COO}^-\right] & & \text{COO}^- \\
\textbf{Citrate} & & \textit{cis-}\textbf{Aconitate} & & \textbf{Isocitrate}
\end{array}
$$

The reaction begins with a dehydration step in which a proton and an OH group are removed. Since citrate has two carboxymethyl groups substituent to its central C atom, it is prochiral rather than chiral. Thus, although water might conceivably be eliminated from either of the two carboxymethyl arms, aconitase removes water only from citrate's lower (*pro-R*) arm (i.e., such that the product molecule has the *R* configuration; see Box 4-2).

Aconitase contains a **[4Fe–4S] iron–sulfur cluster** (an arrangement of four iron atoms and four sulfur atoms, Section 17-2C) that presumably coordinates the OH group of citrate to facilitate its elimination. Iron–sulfur clusters normally participate in redox processes (Section 17-2C); aconitase is an intriguing exception.

The second stage of the aconitase reaction is rehydration of the double bond of *cis*-aconitate to form isocitrate. Although addition of water across the double bond of *cis*-aconitate could potentially yield four stereoisomers, aconitase catalyzes the stereospecific addition of OH^- and H^+ to produce only one isocitrate stereoisomer. The ability of an enzyme to differentiate its substrate's *pro-R* and *pro-S* groups was not appreciated until 1948, when Alexander Ogston pointed out that aconitase can distinguish between the two —CH_2COO^- groups of citrate when it is bound to the enzyme (see Section 11-1B).

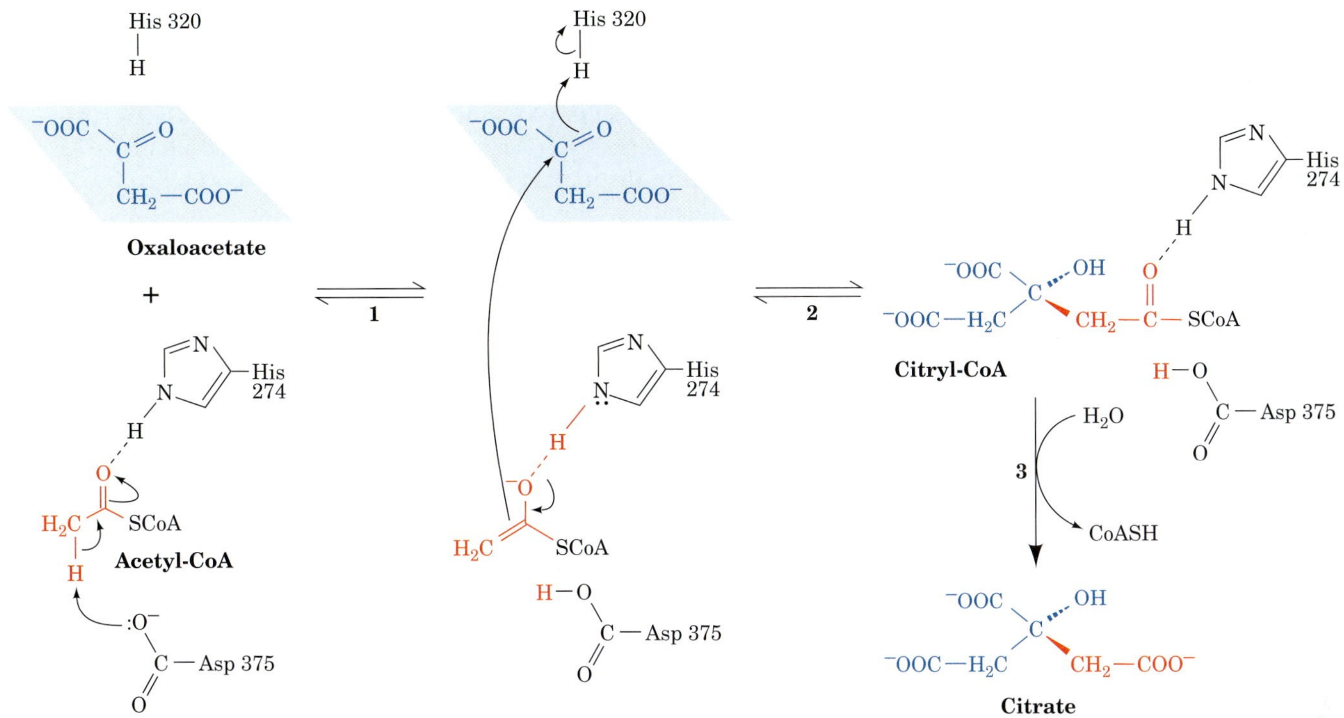

Figure 16-10   The mechanism of the citrate synthase reaction. His 274 and His 320 in their neutral forms and Asp 375 have been implicated as general acid–base catalysts. The rate-limiting step is the formation of the acetyl-CoA enolate, which is stabilized by a hydrogen bond to the His 274.

The acetyl-CoA enolate then nucleophilically attacks oxaloacetate's carbonyl carbon. The resulting intermediate, citryl-CoA, is hydrolyzed to yield citrate and CoA. [Mostly after Remington, J.S., *Curr. Opin. Struct. Biol.* **2,** 732 (1992).]

C   NAD⁺-Dependent Isocitrate Dehydrogenase

Isocitrate dehydrogenase catalyzes the oxidative decarboxylation of isocitrate to α-ketoglutarate. This reaction produces the first CO_2 and NADH of the citric acid cycle. Note that this CO_2 began the citric acid cycle as a component of oxaloacetate, not of acetyl-CoA (Fig. 16-2). (Mammalian tissues also contain an isocitrate dehydrogenase isozyme that uses $NADP^+$ as a cofactor.)

NAD⁺-dependent isocitrate dehydrogenase, which also requires a Mn^{2+} or Mg^{2+} cofactor, catalyzes the oxidation of a secondary alcohol (isocitrate) to a ketone **(oxalosuccinate)** followed by the decarboxylation of the carboxyl group β to the ketone (Fig. 16-11). Mn^{2+} helps polarize the newly formed carbonyl group. The isocitrate dehydrogenase reaction mechanism is similar to that of phosphogluconate dehydrogenase in the pentose phosphate pathway (Section 14-6A).

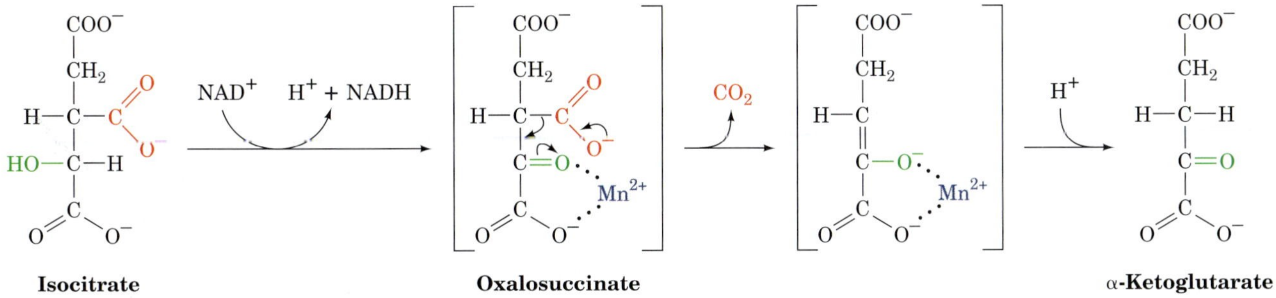

Figure 16-11   The reaction mechanism of isocitrate dehydrogenase.   Oxalosuccinate is shown in brackets because it does not dissociate from the enzyme.

The oxalosuccinate intermediate of the isocitrate dehydrogenase reaction exists only transiently, and its existence was therefore difficult to confirm. However, an enzymatic reaction can be slowed by mutating catalytically important residues—in this case, Tyr 160 and Lys 230—to create kinetic "bottlenecks" so that reaction intermediates accumulate. Accordingly, crystals of the mutant isocitrate dehydrogenase were exposed to the substrate isocitrate and immediately visualized via X-ray crystallography using rapid X-ray intensity measurement techniques that require the highly intense X-rays generated by a synchrotron. These studies revealed the oxalosuccinate intermediate in the active site of the enzyme.

D α-Ketoglutarate Dehydrogenase

α-Ketoglutarate dehydrogenase catalyzes the oxidative decarboxylation of an α-keto acid (α-ketoglutarate). This reaction produces the second CO_2 and NADH of the citric acid cycle:

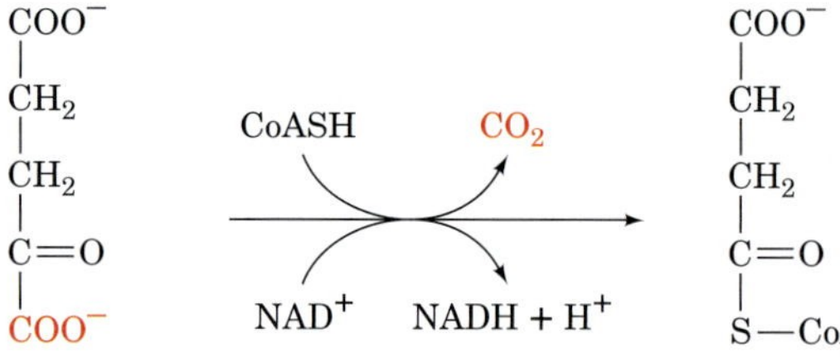

The reaction converts **α-Ketoglutarate** to **Succinyl-CoA**, with CoASH and NAD^+ as reactants, producing CO_2 and $NADH + H^+$.

Again, this CO_2 entered the citric acid cycle as a component of oxaloacetate rather than of acetyl-CoA (Fig. 16-2). Thus, although each round of the citric acid cycle oxidizes two C atoms to CO_2, the C atoms of the entering acetyl groups are not oxidized to CO_2 until subsequent rounds of the cycle.

The α-ketoglutarate dehydrogenase reaction chemically resembles the reaction catalyzed by the pyruvate dehydrogenase multienzyme complex. α-Ketoglutarate dehydrogenase is a multienzyme complex containing **α-ketoglutarate dehydrogenase (E_1), dihydrolipoyl transsuccinylase (E_2),** and **dihydrolipoyl dehydrogenase (E_3).** Indeed, this E_3 is identical to the E_3 of the pyruvate dehydrogenase complex (a third member of the **2-keto acid dehydrogenase** family of multienzyme complexes is **branched-chain α-keto acid dehydrogenase,** which participates in the degradation of isoleucine, leucine, and valine; Section 20-4). The reactions catalyzed by the α-ketoglutarate dehydrogenase complex occur by mechanisms identical to those of the pyruvate dehydrogenase complex. Again, the product is a "high-energy" thioester, in this case, **succinyl-CoA.**

E Succinyl-CoA Synthetase

Succinyl-CoA synthetase (also called **succinate thiokinase**) couples the cleavage of the "high-energy" succinyl-CoA to the synthesis of a "high-energy" nucleoside triphosphate (both names for the enzyme reflect the reverse reaction). GTP is usually synthesized from GDP + P_i by the mammalian enzyme; plant and bacterial enzymes usually use ADP + P_i to form ATP. These reactions are nevertheless energetically equivalent since ATP and GTP are rapidly interconverted through the action of nucleoside diphosphate kinase (Section 13-2C):

$$GTP + ADP \rightleftharpoons GDP + ATP \qquad \Delta G^{\circ\prime} = 0$$

How does succinyl-CoA synthetase couple the exergonic cleavage of succinyl-CoA ($\Delta G^{\circ\prime} = -32.6$ kJ·mol^{-1}) to the endergonic formation of a nucleoside triphosphate ($\Delta G^{\circ\prime} = 30.5$ kJ·mol^{-1}) from the corresponding nucleoside diphosphate and P_i? This question was answered by an experiment with isotopically labeled ADP. In the absence of succinyl-CoA, the spinach enzyme catalyzes the transfer of the γ-phosphoryl group from ATP to [^{14}C]ADP, producing [^{14}C]ATP. Such an isotope-exchange reaction suggests the participation of a phosphoryl-enzyme intermediate that mediates the reaction sequence

This information led to the isolation of a kinetically active phosphoryl-enzyme in which the phosphoryl group is covalently linked to the N3 position of a His residue. A three-step mechanism for succinyl-CoA synthetase is shown in Fig. 16-12.

1. Succinyl-CoA reacts with P_i to form **succinyl-phosphate** and CoA.

2. The phosphoryl group is then transferred from succinyl-phosphate to a His residue on the enzyme, releasing succinate.

3. The phosphoryl group on the enzyme is transferred to GDP, forming GTP.

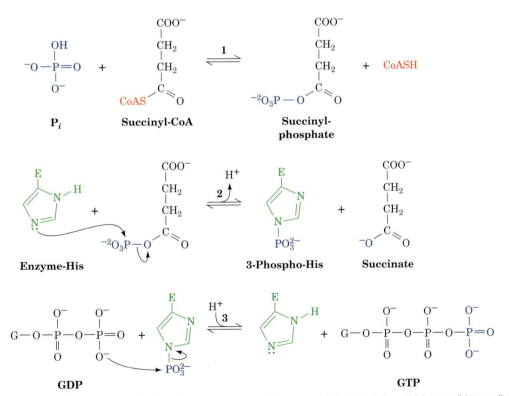

Figure 16-12 The reaction catalyzed by succinyl-CoA synthetase. (1) Formation of succinyl-phosphate, a "high-energy" acyl phosphate.

(2) Formation of phosphoryl-His, a "high-energy" intermediate. (3) Transfer of the phosphoryl group to GDP, forming GTP.

Note that in each of these steps, *the energy of succinyl-CoA is conserved through the formation of "high-energy" compounds: first, succinyl-phosphate, then a 3-phospho-His residue, and finally GTP.* The process is reminiscent of passing a hot potato. The reaction catalyzed by succinyl-CoA synthetase is another example of substrate-level phosphorylation (ATP synthesis that does not directly depend on the presence of oxygen).

By this point in the citric acid cycle, one acetyl equivalent has been completely oxidized to two CO_2. Two NADH and one GTP (equivalent to one ATP) have also been generated. In order to complete the cycle, succinate must be converted back to oxaloacetate. This is accomplished by the cycle's remaining three reactions.

F Succinate Dehydrogenase

Succinate dehydrogenase catalyzes the stereospecific dehydrogenation of succinate to fumarate:

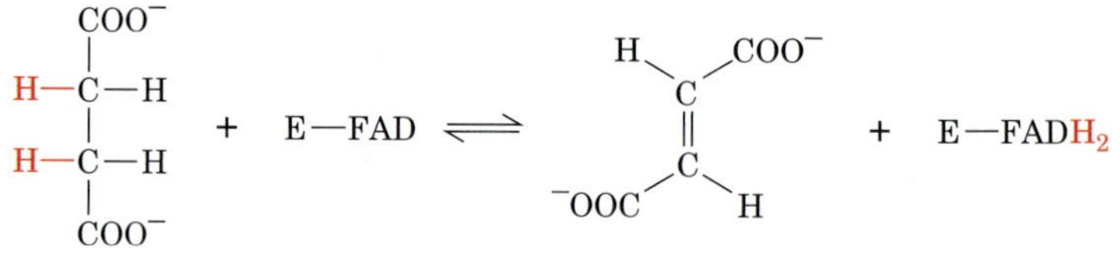

This enzyme is strongly inhibited by malonate,

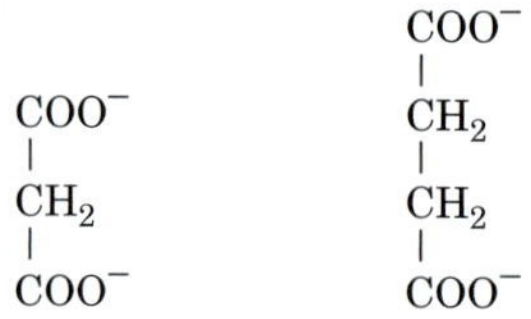

a structural analog of succinate and a classic example of a competitive inhibitor. When Krebs was formulating his theory of the citric acid cycle, the inhibition of cellular respiration by malonate provided one of the clues that succinate plays a catalytic role in oxidizing substrates and is not just another substrate.

Succinate dehydrogenase contains an FAD prosthetic group that is covalently linked to the enzyme via a His residue (Fig. 16-13); in most other FAD-containing enzymes, the FAD is held tightly but noncovalently). In general, FAD functions biochemically to oxidize alkanes (such as succinate) to alkenes (such as fumarate), whereas NAD^+ participates in the more exergonic oxidation of alcohols to aldehydes or ketones (e.g., in the reaction catalyzed by isocitrate dehydrogenase). The dehydrogenation of succinate produces $FADH_2$, which must be reoxidized before succinate dehydrogenase can undertake another catalytic cycle. The reoxidation of $FADH_2$ occurs when its electrons are passed to the mitochondrial electron transport chain, which we shall examine in Section 17-2. Succinate dehydrogenase is the only membrane-bound enzyme of the citric acid cycle (the others are components of the mitochondrial matrix), so it is positioned to funnel electrons directly into the electron transport machinery of the mitochondrial membrane.

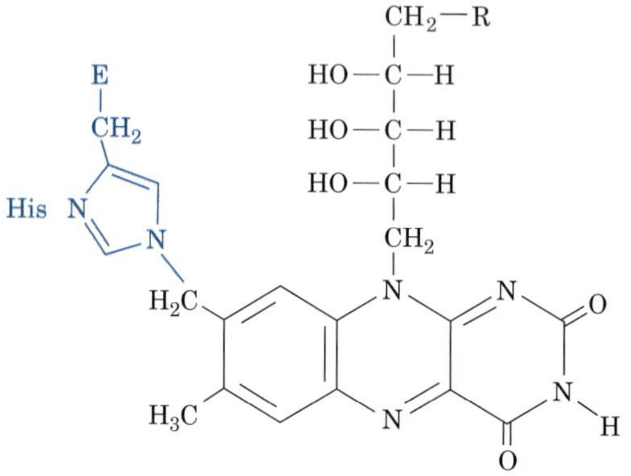

Figure 16-13 The covalent attachment of FAD to a His residue of succinate dehydrogenase. R represents the ADP moiety.

G Fumarase

Fumarase (fumarate hydratase) catalyzes the hydration of the double bond of fumarate to form malate. The hydration reaction proceeds via a carbanion transition state. OH^- addition occurs before H^+ addition:

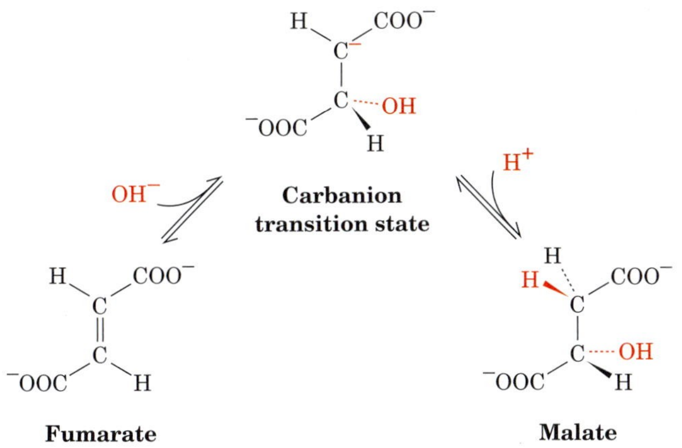

Fumarate

Carbanion transition state

Malate

H Malate Dehydrogenase

Malate dehydrogenase catalyzes the final reaction of the citric acid cycle, the regeneration of oxaloacetate. The hydroxyl group of malate is oxidized in an NAD^+-dependent reaction:

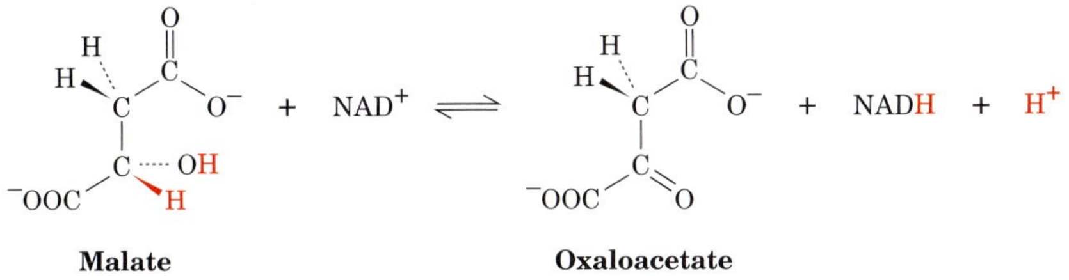

Malate   Oxaloacetate

Transfer of the hydride ion to NAD^+ occurs by the same mechanism used for hydride ion transfer in lactate dehydrogenase and alcohol dehydrogenase (Section 14-3). X-Ray crystallographic comparisons of the NAD^+-binding domains of these three enzymes indicate that they are remarkably similar, consistent with the proposal that all NAD^+-binding domains evolved from a common ancestor.

The $\Delta G^{\circ\prime}$ value for the malate dehydrogenase reaction is $+29.7 \text{ kJ} \cdot \text{mol}^{-1}$; therefore, the concentration of oxaloacetate at equilibrium (and under cellular conditions) is very low relative to malate. Recall, however, that the reaction catalyzed by citrate synthase, the first reaction of the citric acid cycle, is highly exergonic ($\Delta G^{\circ\prime} = -31.5 \text{ kJ} \cdot \text{mol}^{-1}$) because of the cleavage of the thioester bond of citryl-CoA. We can now understand the necessity for such a seemingly wasteful process. It allows citrate formation to be exergonic even at the low oxaloacetate concentrations present in cells and thus helps keep the citric acid cycle rolling.

4 Regulation of the Citric Acid Cycle

The capacity of the citric acid cycle to generate energy for cellular needs is closely regulated. The availability of substrates, the need for citric acid cycle intermediates as biosynthetic precursors, and the demand for ATP all influence the operation of the cycle. There is some evidence that

the enzymes of the citric acid cycle are physically associated, which might contribute to their coordinated regulation. Before we examine the various mechanisms for regulating the citric acid cycle, let us briefly consider the energy-generating capacity of the cycle.

The oxidation of one acetyl group to two molecules of CO_2 is a four-electron pair process (but keep in mind that it is not the carbon atoms of the incoming acetyl group that are oxidized). For every acetyl-CoA that enters the cycle, three molecules of NAD^+ are reduced to NADH, which accounts for three of the electron pairs, and one molecule of FAD is reduced to $FADH_2$, which accounts for the fourth electron pair. In addition, one GTP (or ATP) is produced.

The electrons carried by NADH and $FADH_2$ are funneled into the electron-transport chain, which culminates with the reduction of O_2 to H_2O. The energy of electron transport is conserved in the synthesis of ATP by oxidative phosphorylation (Section 17-3). For every NADH that passes its electrons on, approximately 3 ATP are produced from ADP + P_i. For every $FADH_2$, approximately 2 ATP are produced. Thus, one turn of the citric acid cycle ultimately generates approximately 12 ATP. We will see in Section 17-3C why these values are only approximations.

When glucose is converted to two molecules of pyruvate by glycolysis, two molecules of ATP are generated and two molecules of NAD^+ are reduced (Section 14-1). The NADH molecules yield approximately 6 molecules of ATP on passing their electrons to the electron-transport chain. When the two pyruvate molecules are converted to two acetyl-CoA by the pyruvate dehydrogenase complex, the two molecules of NADH produced in that process also eventually give rise to ~6 ATP. Two turns of the citric acid cycle (one for each acetyl group) generate ~24 ATP. Thus, one molecule of glucose can potentially yield ~38 molecules of ATP under aerobic conditions, when the citric acid cycle is operating. In contrast, only 2 molecules of ATP are produced per glucose molecule under anaerobic conditions.

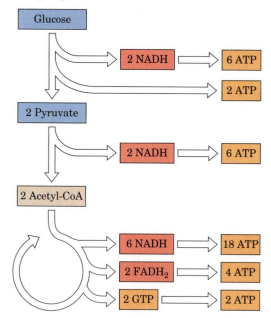

A Regulation of Pyruvate Dehydrogenase

Given the large amount of ATP that can potentially be generated from carbohydrate catabolism via the citric acid cycle, it is not surprising that the entry of acetyl units derived from carbohydrate sources is regulated. The

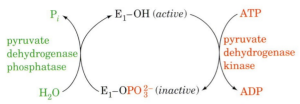

Figure 16-14   Covalent modification of eukaryotic pyruvate dehydrogenase. E_1 is inactivated by the specific phosphorylation of one of its Ser residues in a reaction catalyzed by pyruvate dehydrogenase kinase. This phosphoryl group is hydrolyzed through the action of pyruvate dehydrogenase phosphatase, thereby reactivating E_1.

decarboxylation of pyruvate by the pyruvate dehydrogenase complex is irreversible, and since there are no other pathways in mammals for the synthesis of acetyl-CoA from pyruvate, it is crucial that the reaction be precisely controlled. Two regulatory systems are used:

1. *Product inhibition by NADH and acetyl-CoA.* These compounds compete with NAD^+ and CoA for binding sites on their respective enzymes. They also drive the reversible transacetylase (E_2) and dihydrolipoyl dehydrogenase (E_3) reactions backward (Fig. 16-6). High [NADH]/[NAD^+] and [acetyl-CoA]/[CoA] ratios therefore maintain E_2 in the acetylated form, incapable of accepting the hydroxyethyl group from the TPP on E_1. This, in turn, ties up the TPP on the E_1 subunit in its hydroxyethyl form, decreasing the rate of pyruvate decarboxylation.

2. *Covalent modification by phosphorylation/dephosphorylation of E_1.* In eukaryotes, the products of the pyruvate dehydrogenase reaction, NADH and acetyl-CoA, also activate the pyruvate dehydrogenase kinase associated with the enzyme complex. The resulting phosphorylation of a specific dehydrogenase Ser residue inactivates the pyruvate dehydrogenase complex (Fig. 16-14). Insulin, the hormone that signals fuel abundance, reverses the inactivation by activating pyruvate dehydrogenase phosphatase, which removes the phosphate groups from pyruvate dehydrogenase. Recall that insulin also activates glycogen synthesis by activating phosphoprotein phosphatase (Section 15-3B). Thus, in response to increases in blood [glucose], insulin promotes the synthesis of acetyl-CoA as well as glycogen.

Other regulators of the pyruvate dehydrogenase system include pyruvate and ADP, which inhibit pyruvate dehydrogenase kinase, and Ca^{2+}, which inhibits pyruvate dehydrogenase kinase and activates pyruvate dehydrogenase phosphatase. In contrast to the glycogen metabolism control system (Section 15-3B), pyruvate dehydrogenase activity is unaffected by cAMP.

B   The Rate-Controlling Enzymes of the Citric Acid Cycle

To understand how a metabolic pathway is controlled, we must identify the enzymes that catalyze its rate-determining steps, the *in vitro* effectors of the enzymes, and the *in vivo* concentrations of these substances. *A proposed mechanism of flux control must operate within the physiological concentration range of the effector.*

Identifying the rate-determining steps of the citric acid cycle is more difficult than it is for glycolysis because most of the cycle's metabolites are present in both mitochondria and cytosol and we do not know their distri-

Table 16-2 Standard Free Energy Changes ($\Delta G°'$) and Physiological Free Energy Changes (ΔG) of Citric Acid Cycle Reactions

Reaction	Enzyme	$\Delta G°'$ (kJ·mol^{-1})	ΔG (kJ·mol^{-1})
1	Citrate synthase	−31.5	Negative
2	Aconitase	~5	~0
3	Isocitrate dehydrogenase	−21	Negative
4	α-Ketoglutarate dehydrogenase multienzyme complex	−33	Negative
5	Succinyl-CoA synthetase	−2.1	~0
6	Succinate dehydrogenase	+6	~0
7	Fumarase	−3.4	~0
8	Malate dehydrogenase	+29.7	~0

bution between these two compartments (recall that identifying a pathway's rate-determining steps requires determining the ΔG of each of its reactions from the concentrations of its substrates and products). However, we shall assume that the compartments are in equilibrium and use the total cell concentrations of these substances to estimate their mitochondrial concentrations. Table 16-2 gives the standard free energy changes for the eight citric acid cycle enzymes and estimates of the physiological free energy changes for the reactions in heart muscle or liver tissue. We can see that *three of the enzymes are likely to function far from equilibrium under physiological conditions (negative ΔG): citrate synthase, NAD$^+$-dependent isocitrate dehydrogenase, and α-ketoglutarate dehydrogenase.* These are therefore the rate-determining enzymes of the cycle.

In heart muscle, where the citric acid cycle is active, the flux of metabolites through the citric acid cycle is proportional to the rate of cellular oxygen consumption. *Because oxygen consumption, NADH reoxidation, and ATP production are tightly coupled (Section 17-3), the citric acid cycle must be regulated by feedback mechanisms that coordinate NADH production with energy expenditure.* Unlike the rate-limiting enzymes of glycolysis and glycogen metabolism, which regulate flux by elaborate systems of allosteric control, substrate cycles, and covalent modification, the regulatory enzymes of the citric acid cycle seem to control flux primarily by three simple mechanisms: (1) substrate availability, (2) product inhibition, and (3) competitive feedback inhibition by intermediates further along the cycle. Some of the major regulatory mechanisms are diagramed in Fig. 16-15. There is no single flux-control point in the citric acid cycle; rather, flux control is distributed among several enzymes.

Perhaps the most crucial regulators of the citric acid cycle are its substrates, acetyl-CoA and oxaloacetate, and its product, NADH. Both acetyl-CoA and oxaloacetate are present in mitochondria at concentrations that do not saturate citrate

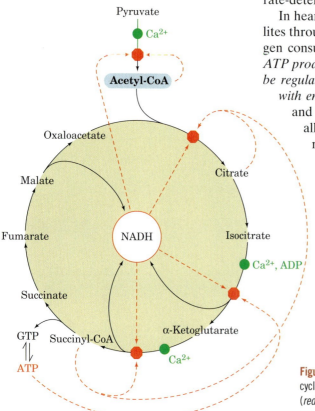

Figure 16-15 Regulation of the citric acid cycle. This diagram of the citric acid cycle, which includes the pyruvate dehydrogenase reaction, indicates points of inhibition (*red octagons*) and the pathway intermediates that function as inhibitors (*dashed red arrows*). ADP and Ca^{2+} (*green dots*) are activators. 🔴 **See the Animated Figures.**

synthase. The metabolic flux through the enzyme therefore varies with substrate concentration and is controlled by substrate availability. We have already seen that the production of acetyl-CoA from pyruvate is regulated by the activity of pyruvate dehydrogenase. The concentration of oxaloacetate, which is in equilibrium with malate, fluctuates with the $[NADH]/[NAD^+]$ ratio according to the equilibrium expression

$$K = \frac{[\text{oxaloacetate}][\text{NADH}]}{[\text{malate}][\text{NAD}^+]}$$

If, for example, the muscle workload and respiration rate increase, mitochondrial [NADH] decreases. The consequent increase in [oxaloacetate] stimulates the citrate synthase reaction, which controls the rate of citrate formation.

Aconitase functions close to equilibrium, so the rate of citrate consumption depends on the activity of NAD^+-dependent isocitrate dehydrogenase, which is strongly inhibited *in vitro* by its product NADH. Citrate synthase is also inhibited by NADH but is less sensitive than isocitrate dehydrogenase to changes in [NADH].

Other instances of product inhibition in the citric acid cycle are the inhibition of citrate synthase by citrate (citrate competes with oxaloacetate) and the inhibition of α-ketoglutarate dehydrogenase by NADH and succinyl-CoA. Succinyl-CoA also competes with acetyl-CoA in the citrate synthase reaction (competitive feedback inhibition). This interlocking system helps keep the citric acid cycle coordinately regulated and the concentrations of its intermediates within reasonable bounds.

Additional Regulatory Mechanisms.   *In vitro* studies of citric acid cycle enzymes have identified a few allosteric activators and inhibitors. ADP is an allosteric activator of isocitrate dehydrogenase, whereas ATP inhibits this enzyme. Ca^{2+}, in addition to its many other cellular functions, regulates the citric acid cycle at several points. It activates pyruvate dehydrogenase phosphatase (Fig. 16-14), which in turn activates the pyruvate dehydrogenase complex to produce acetyl-CoA. Ca^{2+} also activates both isocitrate dehydrogenase and α-ketoglutarate dehydrogenase (Fig. 16-15). Thus Ca^{2+}, the signal that stimulates muscle contraction, also stimulates the production of the ATP to fuel it.

5 | Reactions Related to the Citric Acid Cycle

At first glance, a metabolic pathway appears to be either catabolic, with the release and conservation of free energy, or anabolic, with a requirement for free energy. The citric acid cycle is catabolic, of course, because it involves degradation and is a major free-energy conservation system in most organisms. Cycle intermediates are required in only catalytic amounts to maintain the degradative function of the cycle. However, several biosynthetic pathways use citric acid cycle intermediates as starting materials for anabolic reactions. The citric acid cycle is therefore **amphibolic** (both anabolic and catabolic). In this section, we examine some of the reactions that feed intermediates into the citric acid cycle or draw them off; we also examine the **glyoxylate pathway,** a variation of the citric acid cycle that occurs only in plants and converts acetyl-CoA to oxaloacetate. Some

Figure 16-16   Amphibolic functions of the citric acid cycle.   The diagram indicates the positions at which intermediates are drawn off by cataplerotic reactions for use in anabolic pathways (*red arrows*) and the points where anaplerotic reactions replenish cycle intermediates (*green arrows*). Reactions involving amino acid transamination and deamination are reversible, so their direction varies with metabolic demand. ✍ **See the Animated Figures.**

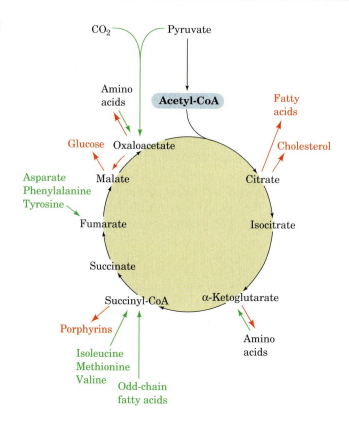

of the reactions that use and replenish citric acid cycle intermediates are summarized in Fig. 16-16.

A   Pathways That Use Citric Acid Cycle Intermediates

Reactions that utilize and therefore drain citric acid cycle intermediates are called **cataplerotic reactions** (emptying; Greek: *cata*, down + *plerotikos*, to fill). These reactions serve not only to synthesize important products but also to avoid the inappropriate buildup of citric acid cycle intermediates in the mitochondrion, for example, when there is a high rate of breakdown of amino acids to citric acid cycle intermediates. Cataplerotic reactions occur in the following pathways:

1. **Glucose biosynthesis** (gluconeogenesis) utilizes oxaloacetate (Section 15-4). Because gluconeogenesis takes place in the cytosol, oxaloacetate must be converted to malate or aspartate for transport out of the mitochondrion (Fig. 15-28). Since the citric acid cycle is a cyclical pathway, any of its intermediates can be converted to oxaloacetate and used for gluconeogenesis.

2. **Fatty acid biosynthesis** is a cytosolic process that requires acetyl-CoA. Acetyl-CoA is generated in the mitochondria and is not transported across the mitochondrial membrane. *Cytosolic acetyl-CoA is therefore generated by the breakdown of citrate, which can cross the membrane, in a reaction catalyzed by* **ATP-citrate lyase** (Section 19-4A). This reaction uses the free energy of ATP to "undo" the citrate synthase reaction:

$$\text{ATP} + \text{citrate} + \text{CoA} \longrightarrow \text{ADP} + \text{P}_i + \text{oxaloacetate} + \text{acetyl-CoA}$$

3. **Amino acid biosynthesis** uses α-ketoglutarate and oxaloacetate as starting materials. For example, α-ketoglutarate is converted to glutamate by reductive amination catalyzed by a **glutamate dehydrogenase** that utilizes either NADH or NADPH:

$$
\begin{array}{c}
\text{COO}^- \\
| \\
\text{CH}_2 \\
| \\
\text{CH}_2 \\
| \\
\text{C}=\text{O} \\
| \\
\text{COO}^-
\end{array}
+ \text{NADH} + \text{H}^+ + \text{NH}_4^+ \;\rightleftharpoons\;
\begin{array}{c}
\text{COO}^- \\
| \\
\text{CH}_2 \\
| \\
\text{CH}_2 \\
| \\
\text{H}-\text{C}-\text{NH}_3^+ \\
| \\
\text{COO}^-
\end{array}
+ \text{NAD}^+ + \text{H}_2\text{O}
$$

α-Ketoglutarate   **Glutamate**

Oxaloacetate undergoes transamination with alanine to produce aspartate and pyruvate (Section 20-2A):

$$
\begin{array}{c}
\text{COO}^- \\
| \\
\text{C}=\text{O} \\
| \\
\text{CH}_2 \\
| \\
\text{COO}^-
\end{array}
+
\begin{array}{c}
\text{COO}^- \\
| \\
\text{H}_3\overset{+}{\text{N}}-\text{C}-\text{H} \\
| \\
\text{CH}_3
\end{array}
\;\rightleftharpoons\;
\begin{array}{c}
\text{COO}^- \\
| \\
\text{H}_3\overset{+}{\text{N}}-\text{C}-\text{H} \\
| \\
\text{CH}_2 \\
| \\
\text{COO}^-
\end{array}
+
\begin{array}{c}
\text{COO}^- \\
| \\
\text{C}=\text{O} \\
| \\
\text{CH}_3
\end{array}
$$

Oxaloacetate   **Alanine**   **Aspartate**   **Pyruvate**

B   Reactions That Replenish Citric Acid Cycle Intermediates

In aerobic organisms, the citric acid cycle is the major source of free energy, and hence the catabolic function of the citric acid cycle cannot be interrupted: Cycle intermediates that have been siphoned off must be replenished. The replenishing reactions are called **anaplerotic reactions** (filling up, Greek: *ana,* up + *plerotikos,* to fill). The most important of these reactions is catalyzed by pyruvate carboxylase, which produces oxaloacetate from pyruvate:

$$\text{Pyruvate} + \text{CO}_2 + \text{ATP} + \text{H}_2\text{O} \longrightarrow \text{oxaloacetate} + \text{ADP} + \text{P}_i$$

(This is also one of the first steps of gluconeogenesis; Section 15-4A). Pyruvate carboxylase "senses" the need for more citric acid cycle intermediates through its activator, acetyl-CoA. *Any decrease in the rate of the cycle caused by insufficient oxaloacetate or other intermediates allows the concentration of acetyl-CoA to rise.* This activates pyruvate carboxylase, which replenishes oxaloacetate. The reactions of the citric acid cycle convert the oxaloacetate to citrate, α-ketoglutarate, succinyl-CoA, and so on, until all the intermediates are restored to appropriate levels.

An increase in the concentrations of citric acid cycle intermediates supports increased flux of acetyl groups through the cycle. For example, flux through the citric acid cycle may increase as much as 60- to 100-fold in muscle cells during intense exercise. Not all of this increase is due to elevated concentrations of cycle intermediates (which only increase about fourfold), because other regulatory mechanisms (as described in Section 16-4B) also promote flux through the rate-controlling steps of the cycle.

During exercise, some of the pyruvate generated by increased glycolytic flux is directed toward oxaloacetate synthesis as catalyzed by pyruvate carboxylase. Pyruvate can also accept an amino group from glutamate (a transamination reaction) to generate alanine (the amino acid counterpart

of pyruvate) and the citric acid cycle intermediate α-ketoglutarate (the ketone counterpart of glutamate). Both of these mechanisms help the citric acid cycle efficiently catabolize the acetyl groups derived—also from pyruvate—by the reactions of the pyruvate dehydrogenase complex.

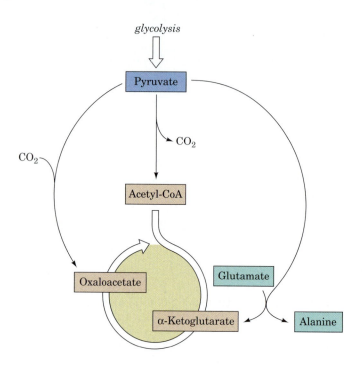

The end result is increased production of ATP to power muscle contraction.

Other metabolites that feed into the citric acid cycle are succinyl-CoA, a product of the degradation of odd-chain fatty acids (Section 19-2E) and certain amino acids (Section 20-4), and α-ketoglutarate and oxaloacetate, which are formed by the reversible transamination of certain amino acids, as indicated above. The links between the citric acid cycle and other metabolic pathways offer some clues to its evolution (Box 16-3).

C The Glyoxylate Cycle

Plants, bacteria, and fungi, but not animals, possess enzymes that mediate the net conversion of acetyl-CoA to oxaloacetate, which can be used for gluconeogenesis. In plants, these enzymes constitute the glyoxylate cycle (Fig. 16-17), which operates in two cellular compartments: the mitochondrion and the **glyoxysome,** a membrane-bounded plant organelle that is a specialized peroxisome. Most of the enzymes of the glyoxylate cycle are the same as those of the citric acid cycle.

The glyoxalate cycle consists of five reactions (Fig. 16-17):

Reactions 1 and 2. Glyoxysomal oxaloacetate is condensed with acetyl-CoA to form citrate, which is isomerized to isocitrate as in the citric acid cycle. Since the glyoxysome contains no aconitase, Reaction 2 presumably takes place in the cytosol.

Reaction 3. Glyoxysomal **isocitrate lyase** cleaves the isocitrate to succinate and **glyoxylate** (hence the cycle's name).

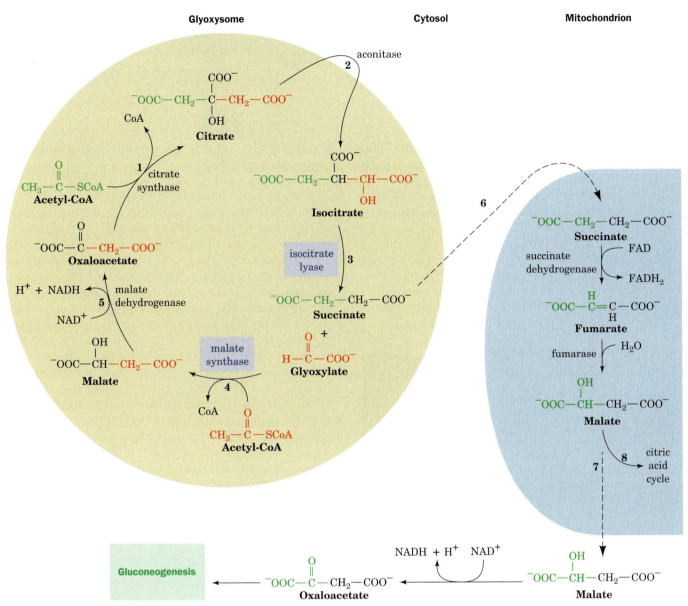

Figure 16-17 The glyoxylate cycle. The cycle results in the net conversion of two acetyl-CoA to succinate in the glyoxysome, which can be converted to malate in the mitochondrion for use in gluconeogenesis. Isocitrate lyase and malate synthase, enzymes unique to glyoxysomes (which occur only in plants), are boxed in blue. (1) Glyoxysomal citrate synthase catalyzes the condensation of oxaloacetate with acetyl-CoA to form citrate. (2) Cytosolic aconitase catalyzes the conversion of citrate to isocitrate. (3) Isocitrate lyase catalyzes the cleavage of isocitrate to succinate and glyoxylate. (4) Malate synthase catalyzes the condensation of glyoxylate with acetyl-CoA to form malate. (5) Glyoxysomal malate dehydrogenase catalyzes the oxidation of malate to oxaloacetate, completing the cycle. (6) Succinate is transported to the mitochondrion, where it is converted to malate via the citric acid cycle. (7) Malate is transported to the cytosol, where malate dehydrogenase catalyzes its oxidation to oxaloacetate, which can then be used in gluconeogenesis. (8) Alternatively, malate can continue in the citric acid cycle, making the glyoxylate cycle anaplerotic.

Reaction 4.   **Malate synthase,** a glyoxysomal enzyme, condenses glyoxylate with a second molecule of acetyl-CoA to form malate.

Reaction 5.   Glyoxysomal malate dehydrogenase catalyzes the oxidation of malate to oxaloacetate by NAD^+.

The glyoxylate cycle therefore results in the net conversion of two acetyl-CoA to succinate instead of to four molecules of CO_2 as would occur in the cit-

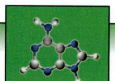

BOX 16-3

Perspectives in Biochemistry

Evolution of the Citric Acid Cycle

The citric acid cycle is ubiquitous in aerobic organisms and plays a central role in energy metabolism in these cells. However, an eight-step catalytic cycle such as the citric acid cycle is unlikely to have arisen all at once and must have evolved from a simpler set of enzyme-catalyzed reactions. Clues to its origins can be found by examining the metabolism of cells that resemble early life-forms. Such organisms emerged before significant quantities of atmospheric oxygen became available some 3 billion years ago. These cells may have used sulfur as their ultimate oxidizing agent, reducing it to H_2S. Their modern-day counterparts are anaerobic autotrophs that harvest free energy by pathways that are independent of the pathways that oxidize carbon-containing compounds.

These organisms therefore do not use the citric acid cycle to generate reduced cofactors that are subsequently oxidized by molecular oxygen. However, all organisms must synthesize the small molecules from which they can build proteins, nucleic acids, carbohydrates, lipids, and so on.

The task of divining an organism's metabolic capabilities has been facilitated through bioinformatics. By comparing the sequences of prokaryotic genomes and assigning functions to various homologous genes, it is possible to reconstruct the central metabolic pathways for the organisms. This approach has been fruitful because many "housekeeping" genes, which encode enzymes that make free energy and molecular building blocks available to the cell, are highly conserved among different species and hence are relatively easy to recognize.

Genomic analysis reveals that many prokaryotes lack the citric acid cycle. However, these organisms do contain genes for some citric acid cycle enzymes. The last four reactions of the cycle, leading from succinate to oxaloacetate, appear to be the most highly conserved. This pathway fragment constitutes a mechanism for accepting electrons that are released during sugar fermentation. For example, the reverse of this pathway could regenerate NAD^+ from the NADH produced by the glyceraldehyde-3-phosphate dehydrogenase step of glycolysis.

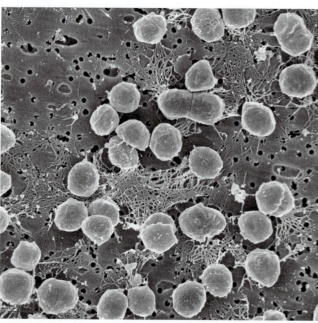

Methanococcus jannaschii, an organism without a citric acid cycle. [Courtesy of Boonyaratanakornkit, B., Clark, D.S., and Vrdoljak, G., University of California at Berkeley.]

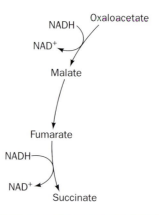

ric acid cycle. The succinate produced in Reaction 3 is transported to the mitochondrion where it enters the citric acid cycle and is converted to malate, which has two alternative fates: (1) It can be converted to oxaloacetate in the mitochondrion, continuing the citric acid cycle and thereby making the glyoxylate cycle an anaplerotic process (Section 16-5B); or (2) it can be transported to the cytosol, where it is converted to oxaloacetate for entry into gluconeogenesis.

The resulting succinate could then be used as a starting material for the biosynthesis of other compounds.

Many archaeal cells have a pyruvate:ferredoxin oxidoreductase that converts pyruvate to acetyl-CoA (but without producing NADH). In a primitive cell, the resulting acetyl groups could have condensed with oxaloacetate (by the action of a citrate synthase), eventually giving rise to an oxidative sequence of reactions resembling the first few steps of the modern citric acid cycle.

Interestingly, a primitive citric acid cycle that operated in the reverse (counterclockwise) direction could have provided a route for fixing CO_2 (that is, incorporating CO_2 into biological molecules).

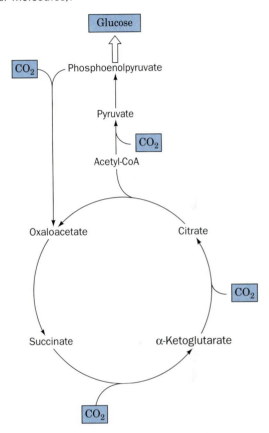

The α-ketoglutarate produced in this way can be converted to glutamate and other amino acids.

The reductive and oxidative branches of the citric acid cycle outlined so far function in modern bacterial cells such as *E. coli* cells when they are growing anaerobically, suggesting that similar pathways could have filled the metabolic needs of early cells. The evolution of a complete citric acid cycle in which the two branches are linked and both proceed in an oxidative direction (clockwise) would have required an enzyme such as α-ketoglutarate:ferredoxin reductase (a homolog of pyruvate:ferredoxin oxidoreductase) to link α-ketoglutarate and succinate.

The genes encoding enzymes that catalyze the steps of such a pathway have been identified in several modern autotrophic bacteria. This reductive pathway, which occurs in some deeply rooted archaeal species, possibly predates the CO_2-fixing pathway used in some photosynthetic bacteria and in the chloroplasts of green plants (Section 18-3A).

The overall reaction of the glyoxylate cycle can be considered to be the formation of oxaloacetate from two molecules of acetyl-CoA:

$$2\,\text{Acetyl-CoA} + 2\,\text{NAD}^+ + \text{FAD} \rightarrow$$
$$\text{oxaloacetate} + 2\,\text{CoA} + 2\,\text{NADH} + \text{FADH}_2 + 2\,\text{H}^+$$

Isocitrate lyase and malate synthase occur only in plants. These enzymes enable germinating seeds to convert their stored triacylglycerols, through

acetyl-CoA, to glucose. It had long been assumed that this was a requirement of germination. However, a mutant of *Arabidopsis thaliana* (an oilseed plant) lacking isocitrate lyase, and hence unable to convert lipids to carbohydrate, nevertheless germinated. This process was only inhibited when the mutant plants were subjected to low light conditions. Therefore, it now appears that the glyoxylate cycle's importance in seedling growth is its anaplerotic function in providing four-carbon units to the citric acid cycle, which can then oxidize the triacylglycerol-derived acetyl-CoA.

Organisms that lack the glyoxylate pathway cannot undertake the net synthesis of glucose from acetyl-CoA. This is the reason humans cannot convert fats (that is, fatty acids, which are catabolized to acetyl-CoA) to carbohydrates (that is, glucose).

Some human pathogens use the glyoxylate cycle, sometimes to great advantage. For example, *Mycobacterium tuberculosis,* which causes tuberculosis, can persist for years in the lung without being attacked by the immune system. During this period, the bacterium subsists largely on lipids, using the citric acid cycle to produce precursors for amino acid synthesis and using the glyoxylate cycle to produce carbohydrate precursors. Drugs that are designed to inhibit the bacterial isocitrate lyase can therefore potentially limit the pathogen's survival. The virulence of the yeast *Candida albicans,* which often infects immunosuppressed individuals, may also depend on activation of the glyoxylate cycle when the yeast cells take up residence inside macrophages.

SUMMARY

1. The eight enzymes of the citric acid cycle function in a multistep catalytic cycle to oxidize an acetyl group to two CO_2 molecules with the concomitant generation of three NADH, one $FADH_2$, and one GTP. The free energy released when the reduced coenzymes ultimately reduce O_2 is used to generate ATP.

2. Acetyl groups enter the citric acid cycle as acetyl-CoA. The pyruvate dehydrogenase multienzyme complex, which contains three types of enzymes and five types of coenzymes, generates acetyl-CoA from the glycolytic product pyruvate. The lipoyllysyl arm of E_2 acts as a tether that swings reactive groups between enzymes in the complex.

3. Citrate synthase catalyzes the condensation of acetyl-CoA and oxaloacetate in a highly exergonic reaction.

4. Aconitase catalyzes the isomerization of citrate to isocitrate, and isocitrate dehydrogenase catalyzes the oxidative decarboxylation of isocitrate to α-ketoglutarate to produce the citric acid cycle's first CO_2 and NADH.

5. α-Ketoglutarate dehydrogenase catalyzes the oxidative decarboxylation of α-ketoglutarate to produce succinyl-CoA and the citric acid cycle's second CO_2 and NADH.

6. Succinyl-CoA synthetase couples the cleavage of succinyl-CoA to the synthesis of GTP (or in some organisms, ATP) via a phosphoryl-enzyme intermediate.

7. The citric acid cycle's remaining three reactions, catalyzed by succinate dehydrogenase, fumarase, and malate dehydrogenase, regenerate oxaloacetate to continue the citric acid cycle.

8. Entry of glucose-derived acetyl-CoA into the citric acid cycle is regulated at the pyruvate dehydrogenase step by product inhibition (by NADH and acetyl-CoA) and by covalent modification.

9. The citric acid cycle itself is regulated at the steps catalyzed by citrate synthase, NAD^+-dependent isocitrate dehydrogenase, and α-ketoglutarate dehydrogenase. Regulation is accomplished mainly by substrate availability, product inhibition, and feedback inhibition.

10. Cataplerotic reactions deplete citric acid cycle intermediates. Some citric acid cycle intermediates are substrates for gluconeogenesis, fatty acid biosynthesis, and amino acid biosynthesis.

11. Anaplerotic reactions such as the pyruvate carboxylase reaction replenish citric acid cycle intermediates.

12. The glyoxylate cycle, which operates only in plants, bacteria, and fungi, requires the glyoxysomal enzymes isocitrate lyase and malate synthase. This variation of the citric acid cycle permits net synthesis of glucose from acetyl-CoA.

REFERENCES

Eastmond, P.J. and Graham, I.A., Re-examining the role of the glyoxylate cycle in oilseeds, *Trends Plant Sci.* **6,** 72–77 (2001).

Huynen M.A., Dandekar, T., and Bork, P., Variation and evolution of the citric-acid cycle: a genomic perspective, *Trends Microbiol.* **7,** 281–291 (1999). [Discusses how genome studies can allow reconstruction of metabolic pathways, even when some enzymes appear to be missing.]

Izard, T., Ævarsson, A., Allen, M.D., Westphal, A.H., Perham, R.N., de Kok, A., and Hol, W.G.J., Principles of quasi-equivalence and euclidean geometry govern the assembly of cubic and dodecahedral cores of pyruvate dehydrogenase, *Proc. Natl. Acad. Sci.* **96,** 1240–1245 (1999).

Owen, O.E., Kalhan, S.C., and Hanson, R.W., The key role of anaplerosis and cataplerosis for citric acid cycle function, *J. Biol. Chem.* **277,** 30409–30412 (2002). [Describes the influx (anaplerosis) and efflux (cataplerosis) of citric acid cycle intermediates in different organ systems.]

Perham, R.N., Swinging arms and swinging domains in multifunctional enzymes: catalytic machines for multistep reactions, *Annu. Rev. Biochem.* **69,** 961–1004 (2000). [An authoritative review on multienzyme complexes.]

KEY TERMS

multienzyme complex	amphibolic pathway	anaplerotic reaction	glyoxylate cycle
lipoyllysyl arm	cataplerotic reaction		

STUDY EXERCISES

1. Describe the five reactions of the pyruvate dehydrogenase multienzyme complex.

2. Draw the structures of the eight intermediates of the citric acid cycle and name the enzymes that catalyze their interconversion.

3. Write the net equations for oxidation of pyruvate, acetyl-CoA, and glucose to CO_2.

4. Which steps of the citric acid cycle regulate flux through the cycle?

5. Describe the role of Ca^{2+}, acetyl-CoA, and NADH in regulating pyruvate dehydrogenase and the citric acid cycle.

6. Explain how a catalytic cycle can supply precursors for other metabolic pathways without depleting its own intermediates.

7. Describe the reactions of the glyoxylate cycle.

PROBLEMS

1. (a) Explain why obligate anaerobes contain some citric acid cycle enzymes. (b) Why don't these organisms have a complete citric acid cycle?

2. The first organisms on earth may have been chemoautotrophs in which the citric acid cycle operated in reverse to "fix" atmospheric CO_2 in organic compounds. Complete a catalytic cycle that begins with the hypothetical overall reaction succinate + 2 $CO_2 \rightarrow$ citrate.

3. Which one of the five steps of the pyruvate dehydrogenase complex reaction is most likely to be metabolically irreversible? Explain.

4. The CO_2 produced in one round of the citric acid cycle does not originate in the acetyl carbons that entered that round. (a) If acetyl-CoA is labeled with ^{14}C at its carbonyl carbon, how many rounds of the cycle are required before $^{14}CO_2$ is released? (b) How many rounds are required if acetyl-CoA is labeled at its methyl group?

5. The branched-chain α-keto acid dehydrogenase complex, which participates in amino acid catabolism, contains the same three types of enzymes as are in the pyruvate dehy-drogenase and the α-ketoglutarate dehydrogenase complexes. Draw the reaction product when valine is deaminated as in the glutamate $\rightleftharpoons$ α-ketoglutarate reaction (Section 16-5A) and then is acted on by the branched-chain α-keto acid dehydrogenase.

6. Refer to Table 13-3 to explain why FAD rather than NAD^+ is used in the succinate dehydrogenase reaction.

7. Malonate is a competitive inhibitor of succinate in the succinate dehydrogenase reaction. Explain why increasing the oxaloacetate concentration can overcome malonate inhibition.

8. Why is it advantageous for citrate, the product of Reaction 1 of the citric acid cycle, to inhibit phosphofructokinase, which catalyzes the third reaction of glycolysis?

9. Anaplerotic reactions permit the citric acid cycle to supply intermediates to biosynthetic pathways while maintaining the proper levels of cycle intermediates. Write the equation for the net synthesis of citrate from pyruvate.

10. Many amino acids are broken down to intermediates of the citric acid cycle. (a) Why can't these amino acid "remnants" be directly oxidized to CO_2 by the citric acid cycle?

(b) Explain why amino acids that are broken down to pyruvate can be completely oxidized by the citric acid cycle.

11. Certain microorganisms with an incomplete citric acid cycle decarboxylate α-ketoglutarate to produce **succinate semialdehyde.** A dehydrogenase then converts succinate semialdehyde to succinate.

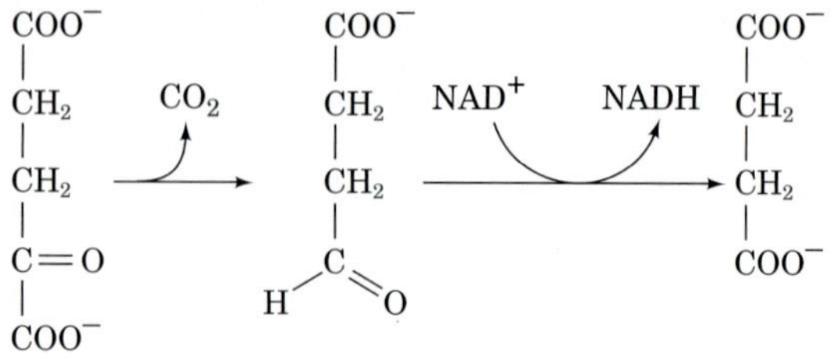

α-Ketoglutarate   **Succinate**   **Succinate**
                 semialdehyde

These reactions can be combined with other standard citric acid cycle reactions to create a pathway from citrate to oxaloacetate. Compare the ATP and reduced cofactor yield of the standard and alternate pathways.

12. Given the following information, calculate the physiological ΔG of the isocitrate dehydrogenase reaction at 25°C and pH 7.0: $[NAD^+]/[NADH] = 8$, [α-ketoglutarate] = 0.1 mM; and [isocitrate] = 0.02 mM. Assume standard conditions for CO_2 ($\Delta G^{\circ\prime}$ is given in Table 16-2). Is this reaction a likely site for metabolic control?

13. Although animals cannot synthesize glucose from acetyl-CoA, if a rat is fed ^{14}C-labeled acetate, some of the label appears in glycogen extracted from its muscles. Explain.

A resting human body consumes approximately 420 kilojoules of energy per hour, a power requirement that is only slightly greater than that of a 100-watt light bulb. The body's energy needs are largely met by the electrochemical events in mitochondria, which support a relatively modest voltage of approximately 0.2 V (a household power outlet in the United States supplies 110 V) but a current of about 500 amps, representing the transmembrane movement of approximately 3×10^{21} protons per second. It is this movement that powers ATP synthesis. [Image State/ Alamy Images.]

Electron Transport and Oxidative Phosphorylation

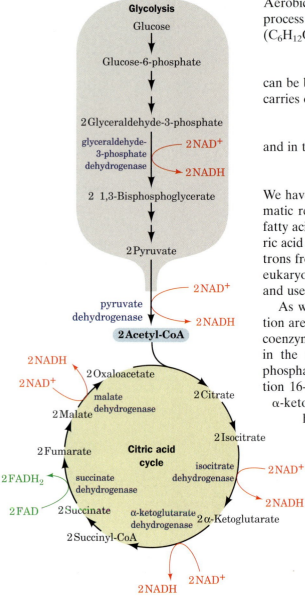

Figure 17-1 The sites of electron transfer that form NADH and FADH$_2$ in glycolysis and the citric acid cycle.

Aerobic organisms consume oxygen and generate carbon dioxide in the process of oxidizing metabolic fuels. The complete oxidation of glucose ($C_6H_{12}O_6$), for example, by molecular oxygen

$$C_6H_{12}O_6 + 6\,O_2 \rightarrow 6\,CO_2 + 6\,H_2O$$

can be broken down into two half-reactions that the metabolic machinery carries out. In the first, glucose carbon atoms are oxidized:

$$C_6H_{12}O_6 + 6\,H_2O \rightarrow 6\,CO_2 + 24\,H^+ + 24\,e^-$$

and in the second, molecular oxygen is reduced:

$$6\,O_2 + 24\,H^+ + 24\,e^- \rightarrow 12\,H_2O$$

We have already seen that the first half-reaction is mediated by the enzymatic reactions of glycolysis and the citric acid cycle (the breakdown of fatty acids—the other major type of metabolic fuel—also requires the citric acid cycle). In this chapter, we describe the pathway by which the electrons from reduced fuel molecules are transferred to molecular oxygen in eukaryotes. We also examine how the energy of fuel oxidation is conserved and used to synthesize ATP.

As we have seen, the 12 electron pairs released during glucose oxidation are not transferred directly to O_2. Rather, they are transferred to the coenzymes NAD$^+$ and FAD to form 10 NADH and 2 FADH$_2$ (Fig. 17-1) in the reactions catalyzed by the glycolytic enzyme glyceraldehyde-3-phosphate dehydrogenase (Section 14-2F), pyruvate dehydrogenase (Section 16-2B), and the citric acid cycle enzymes isocitrate dehydrogenase, α-ketoglutarate dehydrogenase, succinate dehydrogenase, and malate dehydrogenase (Section 16-3). *The electrons then pass into the **mitochondrial electron-transport chain,** a system of linked electron carriers.* The following events occur during the electron-transport process:

1. By transferring their electrons to other substances, the NADH and FADH$_2$ are reoxidized to NAD$^+$ and FAD so that they can participate in additional substrate oxidation reactions.

2. The transferred electrons participate in the sequential oxidation–reduction of multiple **redox centers** (groups that undergo oxidation–reduction reactions) in four enzyme complexes before reducing O_2 to H_2O.

3. During electron transfer, protons are expelled from the mitochondrion, producing a proton gradient across the mitochondrial membrane. *The free energy stored in this electrochemical gradient drives the synthesis of ATP from ADP and P$_i$ through **oxidative phosphorylation.***

1 The Mitochondrion

The mitochondrion (Greek: *mitos,* thread + *chondros,* granule) is the site of eukaryotic oxidative metabolism. Mitochondria contain pyruvate dehydrogenase, the citric acid cycle enzymes, the enzymes catalyzing fatty acid oxidation (Section 19-2), and the enzymes and redox proteins involved in electron transport and oxidative phosphorylation. It is therefore with good reason that the mitochondrion is often described as the cell's "power plant."

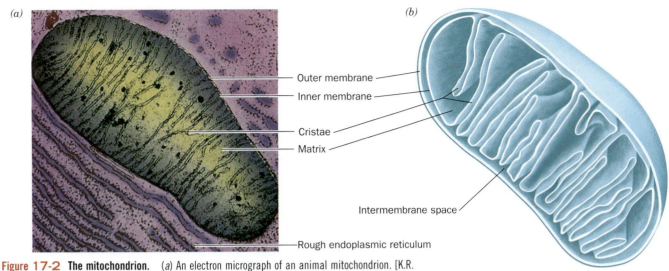

Figure 17-2 The mitochondrion. (*a*) An electron micrograph of an animal mitochondrion. [K.R. Porter/Photo Researchers, Inc.] (*b*) Cutaway diagram of a mitochondrion. [© K.R. Porter/Photo Researchers.]

A Mitochondrial Anatomy

Mitochondria vary in size and shape, depending on their source and metabolic state, but they are often ellipsoidal with dimensions of around 0.5 × 1.0 μm, about the size of a bacterium. A eukaryotic cell typically contains ~2000 mitochondria, which occupy roughly one-fifth of its total cell volume. A mitochondrion is bounded by a smooth outer membrane and contains an extensively invaginated inner membrane (Fig. 17-2). The number of invaginations, called **cristae** (Latin: crests), reflects the respiratory activity of the cell. The proteins mediating electron transport and oxidative phosphorylation are bound in the inner mitochondrial membrane, so the respiration rate varies with membrane surface area.

The inner membrane divides the mitochondrion into two compartments, the **intermembrane space** and the internal **matrix.** The matrix is a gel-like solution that contains extremely high concentrations of the soluble enzymes of oxidative metabolism as well as substrates, nucleotide cofactors, and inorganic ions. The matrix also contains the mitochondrial genetic machinery—DNA, RNA, and ribosomes—that generates several (but by no means all) mitochondrial proteins.

Two-dimensional electron micrographs of mitochondria such as Fig. 17-2*a* suggest that mitochondria are discrete kidney-shaped organelles. In fact, some mitochondria adopt a tubular shape that extends throughout the cytosol. Furthermore, mitochondria are highly variable structures. For example, the cristae may not resemble baffles and the intercristal spaces may not communicate freely with the mitochondrion's intermembrane space. Electron microscopy–based three-dimensional image reconstruction methods have revealed that cristae can range in shape from simple tubular entities to more complicated lamellar assemblies that merge with the inner membrane via narrow tubular structures (Fig. 17-3). Evidently, cristae form microcompartments that restrict the diffusion of substrates and ions between the intercristal and inter-

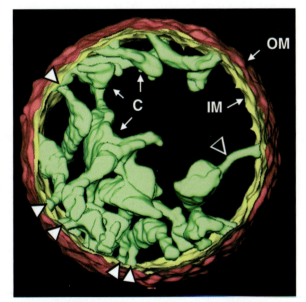

Figure 17-3 Electron microscopy–based three-dimensional image reconstruction of a rat liver mitochondrion. The outer membrane (OM) is red, the inner membrane (IM) is yellow, and the cristae (C) are green. The arrowheads point to tubular regions of the cristae that connect them to the inner membrane and to each other. [Courtesy of Carmen Mannella, Wadsworth Center, Albany, New York.]

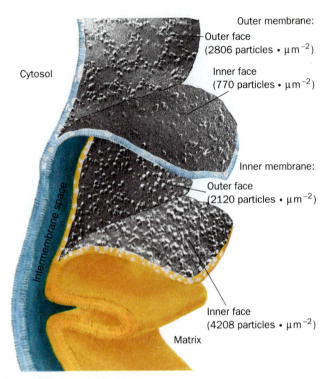

Outer membrane:
Outer face
(2806 particles · μm^{-2})

Inner face
(770 particles · μm^{-2})

Inner membrane:

Outer face
(2120 particles · μm^{-2})

Inner face
(4208 particles · μm^{-2})

Cytosol

Intermembrane space

Matrix

Figure 17-4 Electron micrographs of the inner and outer mitochondrial membranes that have been split to expose the inner surfaces of their bilayer leaflets. Note that the inner membrane contains about twice the density of embedded particles as does the outer membrane. The particles are the portions of integral membrane proteins that were exposed when the bilayers were split. [Courtesy of Lester Packer, University of California at Berkeley.]

membrane spaces. This has important functional implications because it would result in a locally greater pH gradient across cristal membranes than across inner membranes that are not part of cristae, thereby significantly influencing the rate of oxidative phosphorylation (Section 17-3).

B Mitochondrial Transport Systems

Like bacterial outer membranes, the outer mitochondrial membrane contains porins, proteins that permit the free diffusion of molecules of up to 10 kD (Section 10-2B). *The intermembrane space is therefore equivalent to the cytosol in its concentrations of metabolites and ions.* The inner membrane, which is ~75% protein by mass, is considerably richer in proteins than is the outer membrane (Fig. 17-4). It is freely permeable only to O_2, CO_2, and H_2O and contains, in addition to respiratory chain proteins, numerous transport proteins that control the passage of metabolites such as ATP, ADP, pyruvate, Ca^{2+}, and phosphate. *The controlled impermeability of the inner mitochondrial membrane to most ions and metabolites permits the generation of ion gradients across this barrier and results in the compartmentalization of metabolic functions between cytosol and mitochondria.*

Cytosolic Reducing Equivalents Are "Transported" into Mitochondria. The NADH produced in the cytosol by glycolysis must gain access to the mitochondrial electron-transport chain for aerobic oxidation. However, the inner mitochondrial membrane lacks an NADH transport protein. *Only the electrons from cytosolic NADH are transported into the mitochondrion by one of several ingenious "shuttle" systems.* We have already discussed the **malate–aspartate shuttle** (Fig. 15-28), in which, when run in reverse, cytosolic oxaloacetate is reduced to malate for transport into the mitochondrion. When malate is reoxidized in the matrix, it gives up the reducing equivalents that originated in the cytosol.

In the **glycerophosphate shuttle** (Fig. 17-5) of insect flight muscle (the tissue with the largest known sustained power output—about the same power-to-weight ratio as a small automobile engine), **3-phosphoglycerol dehydrogenase** catalyzes the oxidation of cytosolic NADH by dihydroxyacetone phosphate to yield NAD^+, which re-enters glycolysis. The electrons of the resulting **3-phosphoglycerol** are transferred to **flavoprotein dehydrogenase** to form $FADH_2$. This enzyme, which is situated on the inner mitochondrial membrane's outer surface, supplies electrons directly to the electron-transport chain.

The ADP–ATP Translocator. Most of the ATP generated in the mitochondrial matrix through oxidative phosphorylation is used

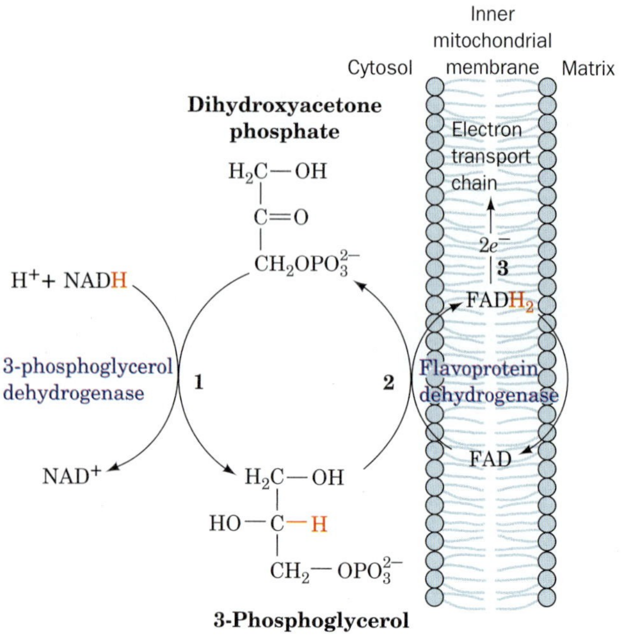

Figure 17-5 **The glycerophosphate shuttle.** The electrons of cytosolic NADH are transported to the mitochondrial electron-transport chain in three steps (shown in red as hydride transfers): **(1)** Cytosolic oxidation of NADH by dihydroxyacetone phosphate catalyzed by 3-phosphoglycerol dehydrogenase. **(2)** Oxidation of 3-phosphoglycerol by flavoprotein dehydrogenase with reduction of FAD to $FADH_2$. **(3)** Reoxidation of $FADH_2$ with passage of electrons into the electron-transport chain.

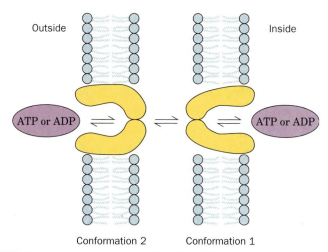

Figure 17-6 Conformational mechanism of the ADP–ATP translocator. An adenine nucleotide–binding site located in the intersubunit contact area of the translocator dimer is alternately exposed to the two sides of the membrane.

in the cytosol. The inner mitochondrial membrane contains an **ADP–ATP translocator** (also called the **adenine nucleotide translocase**) that transports ATP out of the matrix in exchange for ADP produced in the cytosol by ATP hydrolysis.

The ADP–ATP translocator, a dimer of identical 30-kD subunits, has one binding site for which ADP and ATP compete. It has two major conformations: one with its ATP–ADP binding site facing the inside of the mitochondrion, and the other with this site facing outward (Fig. 17-6). The translocator must bind ligand to change from one conformation to the other at a physiologically reasonable rate. Thus it functions as an exchanger by importing one ADP for every ATP that is exported. In this respect, it differs from the glucose transporter (Fig. 10-17), which can change its conformation in the absence of ligand. Note that the export of ATP (net charge -4) and the import of ADP (net charge -3) results in the export of one negative charge per transport cycle. This **electrogenic** antiport is driven by the membrane potential difference, $\Delta\Psi$, across the inner mitochondrial membrane (positive outside), which is a consequence of the transmembrane proton gradient.

Phosphate Transport.   ATP is synthesized from ADP $+$ P$_i$ in the mitochondrion but is utilized in the cytosol. The P$_i$ is returned to the mitochondrion by the **phosphate carrier,** an electroneutral P$_i$–H symport that is driven by ΔpH. The transmembrane proton gradient generated by the electron-transport machinery of the inner mitochondrial membrane thus not only provides the thermodynamic driving force for ATP synthesis (Section 17-3), it also motivates the transport of the raw materials—ADP and P$_i$—required for this process.

2 Electron Transport

The electron carriers that ferry electrons from NADH and FADH$_2$ to O$_2$ are associated with the inner mitochondrial membrane. Some of these redox centers are mobile, and others are components of integral membrane protein complexes. The sequence of electron carriers roughly reflects their

relative reduction potentials, so that the overall process of electron transport is exergonic. We begin this section by examining the thermodynamics of electron transport. We then consider the molecular characteristics of the various electron carriers.

A Thermodynamics of Electron Transport

We can estimate the thermodynamic efficiency of electron transport by inspecting the standard reduction potentials of the redox centers. As we saw in our thermodynamic considerations of oxidation–reduction reactions (Section 13-3), an oxidized substrate's affinity for electrons increases with its standard reduction potential, $\mathscr{E}°'$ (Table 13-3 lists the standard reduction potentials of some biologically important half-reactions). The standard reduction potential difference, $\Delta\mathscr{E}°'$ for a redox reaction involving any two half-reactions is expressed

$$\Delta\mathscr{E}°' = \mathscr{E}°'_{(e^-\ acceptor)} - \mathscr{E}°'_{(e^-\ donor)}$$

NADH Oxidation Is Highly Exergonic. The half-reactions for oxidation of NADH by O_2 are

$$NAD^+ + H^+ + 2\ e^- \rightleftharpoons NADH \qquad \mathscr{E}°' = -0.315\ V$$

and

$$\tfrac{1}{2}O_2 + 2\ H^+ + 2\ e^- \rightleftharpoons H_2O \qquad \mathscr{E}°' = 0.815\ V$$

Since the O_2/H_2O half-reaction has the greater standard reduction potential and therefore the higher affinity for electrons, the NADH half-reaction is reversed so that NADH is the electron donor in this couple and O_2 the electron acceptor. The overall reaction is

$$\tfrac{1}{2}O_2 + NADH + H^+ \rightleftharpoons H_2O + NAD^+$$

so that

$$\Delta\mathscr{E}°' = 0.815\ V - (-0.315\ V) = 1.130\ V$$

The standard free energy change for the reaction can then be calculated from Eq. 13-7:

$$\Delta G°' = -n\mathscr{F}\Delta\mathscr{E}°'$$

For NADH oxidation, $\Delta G°' = -218\ kJ \cdot mol^{-1}$. In other words, the oxidation of 1 mol of NADH by O_2 (the transfer of 2 mol e^-) under standard biochemical conditions is associated with the release of 218 kJ of free energy.

Because the standard free energy required to synthesize 1 mol of ATP from ADP + P_i is 30.5 kJ $\cdot$ mol^{-1}, the oxidation of NADH by O_2 is theoretically able to drive the formation of several moles of ATP. In mitochondria, the coupling of NADH oxidation to ATP synthesis is achieved by an electron-transport chain in which electrons pass through three protein complexes. *This allows the overall free energy change to be broken into three smaller parcels, each of which contributes to ATP synthesis by oxidative phosphorylation. Oxidation of one NADH results in the synthesis of approximately three ATP* (we shall see later why the relationship is not strictly stoichiometric). The thermodynamic efficiency of oxidative phosphorylation is therefore 3 $\times$ 30.5 kJ $\cdot$ mol^{-1} $\times$ 100/218 kJ $\cdot$ mol^{-1} = 42% under standard biochemical conditions. However, under physiological conditions in active mitochondria (where the reactant and product concentrations as well as the pH deviate from standard conditions), this thermodynamic

efficiency is thought to be ~70%. In comparison, the energy efficiency of a typical automobile engine is <30%.

B The Sequence of Electron Transport

Oxidation of NADH and FADH$_2$ is carried out by the electron-transport chain, a set of protein complexes containing redox centers with progressively greater affinities for electrons (increasing standard reduction potentials). Electrons travel through this chain from lower to higher standard reduction potentials (Fig. 17-7). Electrons are carried from **Complexes I** and **II** to **Complex III** by **coenzyme Q** (**CoQ** or **ubiquinone;** so named because of its ubiquity in respiring organisms), and from Complex III to **Complex IV** by the peripheral membrane protein **cytochrome c.**

Complex I catalyzes oxidation of NADH by CoQ:

$$\text{NADH} + \text{CoQ } (oxidized) \rightarrow \text{NAD}^+ + \text{CoQ } (reduced)$$
$$\Delta \mathscr{E}^{\circ\prime} = 0.360 \text{ V} \qquad \Delta G^{\circ\prime} = -69.5 \text{ kJ} \cdot \text{mol}^{-1}$$

Complex III catalyzes oxidation of CoQ (reduced) by cytochrome c:

$$\text{CoQ } (reduced) + 2 \text{ cytochrome } c \ (oxidized) \rightarrow$$
$$\text{CoQ } (oxidized) + 2 \text{ cytochrome } c \ (reduced)$$
$$\Delta \mathscr{E}^{\circ\prime} = 0.190 \text{ V} \qquad \Delta G^{\circ\prime} = -36.7 \text{ kJ} \cdot \text{mol}^{-1}$$

Complex IV catalyzes oxidation of reduced cytochrome c by O$_2$, the terminal electron acceptor of the electron-transport process.

$$2 \text{ Cytochrome } c \ (reduced) + \tfrac{1}{2} \text{O}_2 \rightarrow 2 \text{ cytochrome } c \ (oxidized) + \text{H}_2\text{O}$$
$$\Delta \mathscr{E}^{\circ\prime} = 0.580 \text{ V} \qquad \Delta G^{\circ\prime} = -112 \text{ kJ} \cdot \text{mol}^{-1}$$

> **See Guided Exploration 15**
>
> Electron transport and oxidative phosphorylation overview.

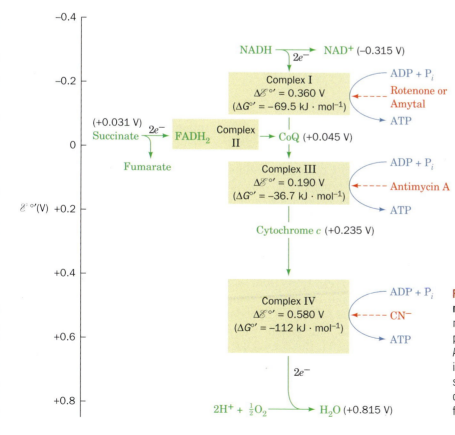

Figure 17-7   Overview of electron transport in the mitochondrion.   The standard reduction potentials of its most mobile components (*green*) are indicated, as are the points where sufficient free energy is released to synthesize ATP (*blue*) and the sites of action of several respiratory inhibitors (*red*). Complexes I, III, and IV do not directly synthesize ATP but sequester the free energy necessary to do so by pumping protons outside the mitochondrion to form a proton gradient.

As an electron pair successively traverses Complexes I, III, and IV, sufficient free energy is released at each step to power the synthesis of an ATP molecule.

Complex II catalyzes the oxidation of FADH₂ by CoQ:

$$FADH_2 + CoQ \ (oxidized) \rightarrow FAD + CoQ \ (reduced)$$
$$\Delta \mathscr{E}^{\circ\prime} = 0.085 \ V \qquad \Delta G^{\circ\prime} = -16.4 \ kJ \cdot mol^{-1}$$

This redox reaction does not release sufficient free energy to synthesize ATP; it functions only to inject the electrons from FADH₂ into the electron-transport chain.

Inhibitors Reveal the Workings of the Electron-Transport Chain. The sequence of events in electron transport was elucidated largely through the use of specific inhibitors and later corroborated by measurements of the standard reduction potentials of the redox components. The rate at which O_2 is consumed by a suspension of mitochondria is a sensitive measure of the activity of the electron-transport chain. Compounds that inhibit electron transport, as judged by their effect on O_2 consumption, include **rotenone** (a plant toxin used by Amazonian Indians to poison fish and which is also used as an insecticide), **amytal** (a barbiturate), **antimycin A** (an antibiotic), and **cyanide.**

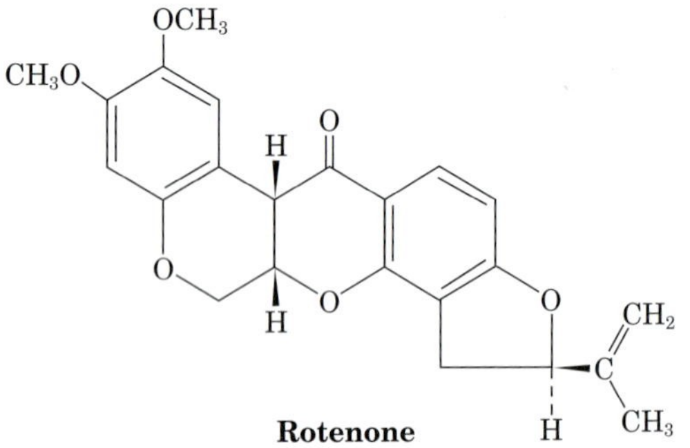

Rotenone

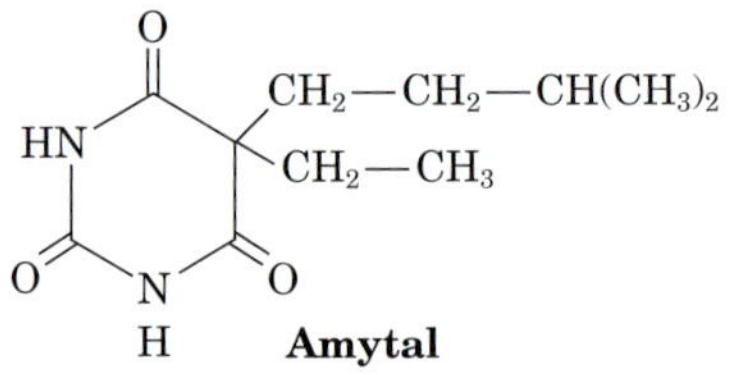

Amytal

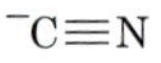

Cyanide

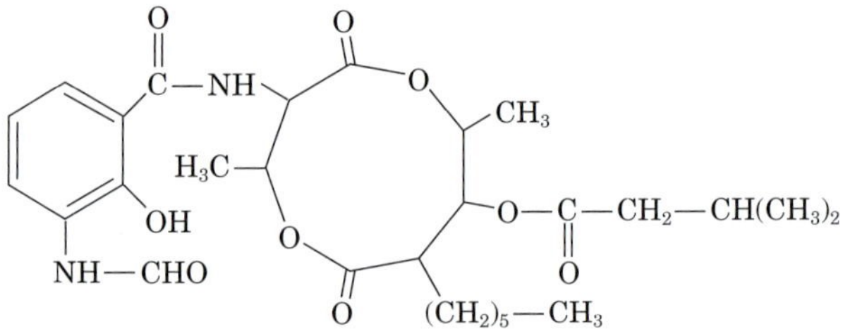

Antimycin A

Adding rotenone or amytal to a suspension of mitochondria blocks electron transport in Complex I; antimycin A blocks Complex III, and CN^- blocks electron transport in Complex IV (Fig. 17-7). Each of these inhibitors also halts O_2 consumption. Oxygen consumption resumes following addition of a substance whose electrons enter the electron-transport chain "downstream" of the block. For example, the addition of succinate to rotenone-blocked mitochondria restores electron transport and O_2 consumption. Experiments with inhibitors of electron transport thus reveal the points of entry of electrons from various substrates.

Each of the four respiratory complexes of the electron-transport chain consists of several protein components that are associated with a variety of redox-active prosthetic groups with successively increasing reduction potentials (Table 17-1). The complexes are all laterally mobile within the inner mitochondrial membrane; they do not appear to form any stable higher structures. Indeed, they are not even present in equimolar amounts. In the following sections, we examine their structures and the molecules that transfer electrons between them. Their relationships are summarized in Fig. 17-8.

Table 17-1 **Reduction Potentials of Electron-Transport Chain Components in Resting Mitochondria**

Component	$\mathscr{E}^{\circ\prime}$ (V)
NADH	−0.315
Complex I (NADH–CoQ oxidoreductase; ~900 kD, 43 subunits):	
FMN	?
(Fe–S)N-1a	−0.380
(Fe–S)N-1b	−0.250
(Fe–S)N-2	−0.030
(Fe–S)N-3,4	−0.245
(Fe–S)N-5,6	−0.270
Succinate	0.031
Complex II (succinate–CoQ oxidoreductase; ~120 kD, 4 subunits):	
FAD	−0.040
[2Fe–2S]	−0.030
[4Fe–4S]	−0.245
[3Fe–4S]	0.060
Heme b_{560}	−0.080
Coenzyme Q	0.045
Complex III (CoQ–cytochrome c oxidoreductase; ~240 kD, 9–11 subunits):	
Heme b_H (b_{562})	0.030
Heme b_L (b_{566})	−0.030
[2Fe–2S]	0.280
Heme c_1	0.215
Cytochrome c	0.235
Complex IV (cytochrome c oxidase; ~205 kD, 8–13 subunits):	
Heme a	0.210
Cu_A	0.245
Cu_B	0.340
Heme a_3	0.385
O_2	0.815

Source: Mainly Wilson, D.F., Erecinska, M., and Dutton, P.L., *Annu. Rev. Biophys. Bioeng.* **3**, 205 and 208 (1974); *and* Wilson, D.F., *in* Bittar, E.E. (Ed.), *Membrane Structure and Function,* Vol. 1, p. 160, Wiley (1980).

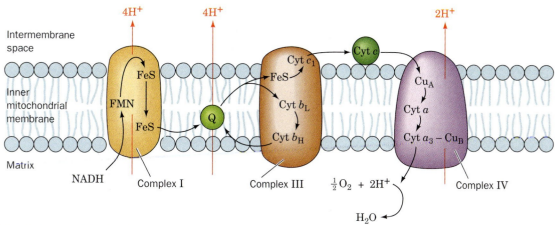

Figure 17-8 *Key to Function.* **The mitochondrial electron-transport chain.** This diagram indicates the pathways of electron transfer (*black*) and proton translocation (*red*). Electrons are transferred between Complexes I and III by the membrane-soluble coenzyme Q (Q) and between Complexes III and IV by the peripheral membrane protein cytochrome *c*. Complex II (not shown) transfers electrons from succinate to coenzyme Q. 🔶 **See the Animated Figures.**

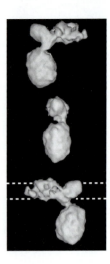

Figure 17-9 Cryoelectron microscopy–based image of bovine Complex I at 22 Å resolution. Successive views, top to bottom, are rotated by 90° about the vertical axis. The boundaries of the lipid bilayer in which the complex is immersed are indicated by the dashed lines. The vertical arm protrudes into the mitochondrial matrix. [Courtesy of Nikolaus Grigorieff, Brandeis University.]

C Complex I (NADH–Coenzyme Q Oxidoreductase)

Complex I, which passes electrons from NADH to CoQ, may be the largest protein complex in the inner mitochondrial membrane, containing, in mammals, 43 subunits with a total mass of ~900 kD. Electron microscopy reveals it to be an L-shaped protein with one arm embedded in the inner mitochondrial membrane and the other extending into the matrix (Fig. 17-9). Complex I contains one molecule of **flavin mononucleotide** (**FMN,** a redox-active prosthetic group that differs from FAD only by the absence of the AMP group) and six to seven **iron–sulfur clusters** that participate in electron transport.

The Coenzymes of Complex I. Iron–sulfur clusters occur as the prosthetic groups of **iron–sulfur proteins** (also called **nonheme iron proteins**). The two most common types, designated **[2Fe–2S]** and **[4Fe–4S] clusters** (*at left*), consist of equal numbers of iron and sulfide ions and are both coordinated to four protein Cys sulfhydryl groups. Note that the Fe atoms in both types of clusters are each coordinated by four S atoms, which are more or less

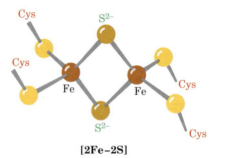

[2Fe–2S]

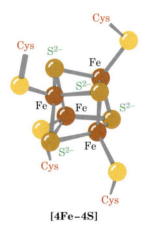

[4Fe–4S]

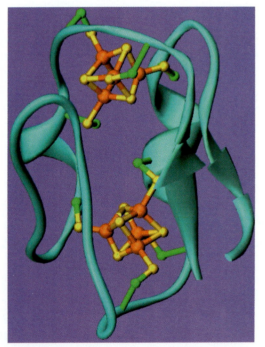

Figure 17-10 X-Ray structure of ferredoxin from *Peptococcus aerogenes*. This monomeric 54-residue protein contains two [4Fe–4S] clusters. The C_β atoms of the four Cys residues liganding each cluster are green, the Fe atoms are orange, and the S atoms are yellow. [Based on an X-ray structure by Elinor Adman, Larry Sieker, and Lyle Jensen, University of Washington. PDBid 1FDX.] 🐁 See the Interactive Exercises.

tetrahedrally disposed around the Fe. The protein **ferredoxin** from bacteria contains two [4Fe–4S] clusters (Fig. 17-10).

Iron–sulfur clusters can undergo one-electron oxidation and reduction. *The oxidized and reduced states of all iron–sulfur clusters differ by one formal charge regardless of their number of Fe atoms.* This is because the Fe atoms in each cluster form a conjugated system and thus can have oxidation states between the normal $+2$ and $+3$ values for individual Fe ions.

FMN and CoQ can each adopt three oxidation states (Fig. 17-11). They are capable of accepting and donating either one or two electrons because their semiquinone forms are stable (these semiquinones are stable **free radicals,** molecules with an unpaired electron). FMN is tightly bound to proteins; however, CoQ has a hydrophobic tail that makes it soluble in the inner mitochondrial membrane's lipid bilayer. In mammals, this tail consists of 10 C_5 isoprenoid units (Section 9-1F) and hence the coenzyme is

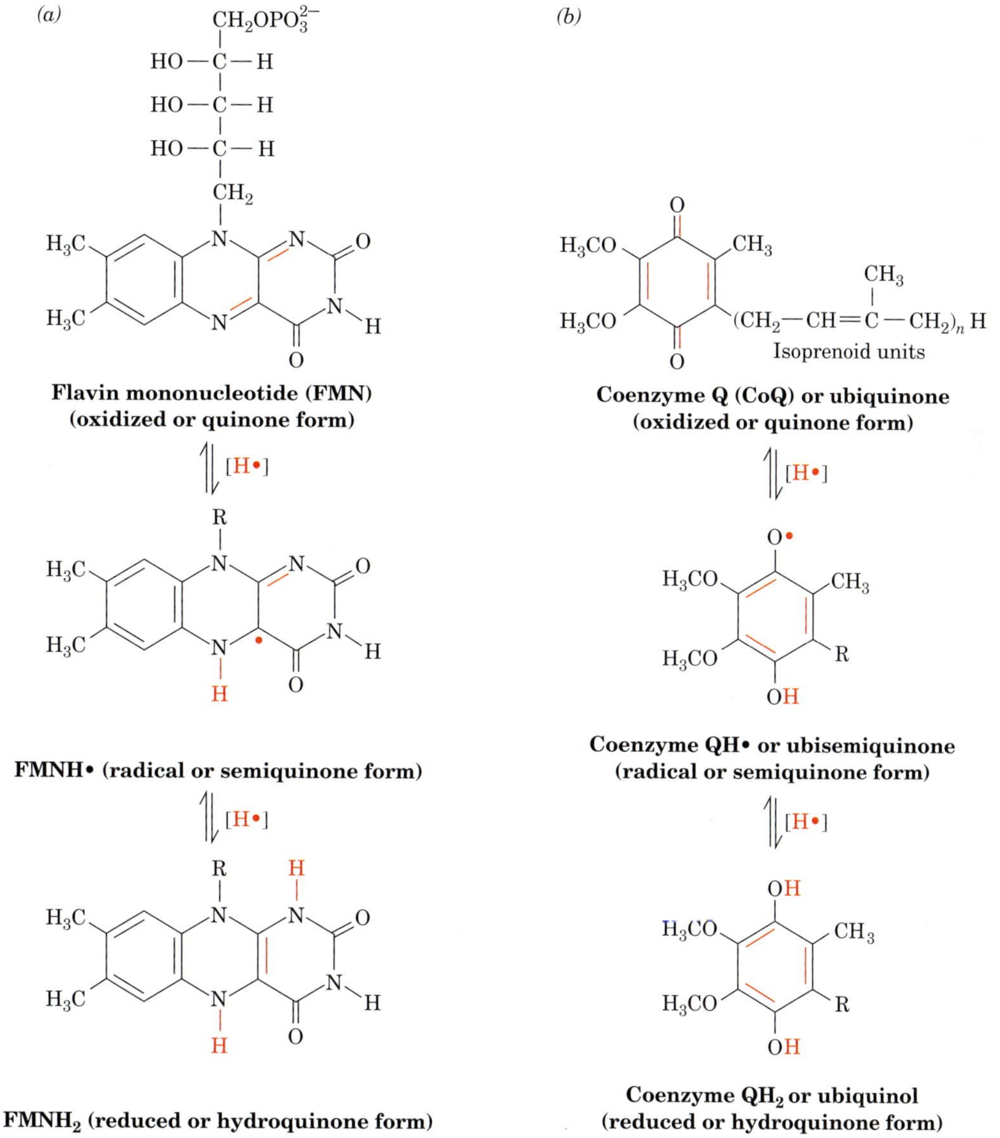

Figure 17-11 The oxidation states of FMN and coenzyme Q. Both (*a*) FMN and (*b*) coenzyme Q form stable semiquinone free-radical states.

designated $\mathbf{Q_{10}}$. In other organisms, CoQ may have only 6 $(\mathbf{Q_6})$ or 8 $(\mathbf{Q_8})$ isoprenoid units.

Electron Transfer in Complex I. The transit of electrons from NADH to CoQ presumably occurs by a stepwise mechanism, according to the reduction potential of the various redox centers in Complex I (Table 17-1). This process involves the transient reduction of each group as it binds electrons and its reoxidation when it passes the electrons to the next group. The exact sequence of electron movements is not known, in part because reducing equivalents appear to distribute among the iron–sulfur clusters faster than they are donated by NADH. Note that redox centers do not need to come into contact in order to transfer an electron. An electron's quantum mechanical properties enable it to "tunnel" (jump) between protein-embedded redox groups that are separated by <14 Å at physiological rates. This is because electron transfer rates decrease exponentially with the distance between redox centers (they exhibit an ~10-fold decrease for each 1.7 Å increase in distance). Hence electron transfers over distances longer than 14 Å always involve chains of redox centers.

NADH can participate in only a two-electron transfer reaction. In contrast, the cytochromes of Complex III (see below), to which reduced CoQ passes its electrons, are capable of only one-electron reactions. *FMN and CoQ, which can transfer one or two electrons at a time, therefore provide an electron conduit between the two-electron donor NADH and the one-electron acceptors, the cytochromes.*

Proton Translocation. *As electrons are transferred between the redox centers of Complex I, four protons are translocated from the matrix to the intermembrane space.* The fate of protons donated by NADH is uncertain: They may be among those pumped across the membrane or used in the reduction of CoQ to its hydroquinone form, $CoQH_2$. The proton-pumping mechanism of Complex I is not well understood but almost certainly is driven by conformational changes induced by changes in the redox state of the protein. These conformational changes alter the pK values of ionizable side chains so that protons are taken up or released as electrons are transferred.

Because a proton is simply an atomic nucleus, it cannot be transported across a membrane in the same way as ions such as Na^+ and K^+. However, a proton can be translocated by "hopping" along a chain of hydrogen-bonded groups in a transmembrane channel, just as it "jumps" between hydrogen-bonded water molecules in solution (Fig. 2-15). The arrangement of hydrogen-bonded groups in the protein has been described as a **proton wire** and may include water molecules. Presumably, mitochondrial Complex I includes a proton wire that allows four protons to move across the membrane for every pair of electrons that passes from NADH to CoQ.

Bacteriorhodopsin: A Model Proton Pump. A useful model for proton-translocating complexes is bacteriorhodopsin, an integral membrane protein of *Halobacter halobium* that contains seven transmembrane helical segments surrounding a central polar channel (Fig. 9-22). Bacteriorhodopsin is a light-driven proton pump: It obtains the free energy required for pumping protons through the absorbance of light by its retinal prosthetic group. The retinal is linked to the protein via a protonated Schiff base to the side chain of Lys 216.

On absorbing light, the all-*trans*-retinal isomerizes to its 13-*cis* configuration:

all-*trans*-Retinal

| light

13-*cis*-Retinal

This structural change initiates a sequence of protein conformational adjustments that restore the system to its ground state over a period of ~10 ms. These conformational changes alter the pK's of several amino acid side chains (Fig. 17-12). Specifically, the pK of Asp 85 increases so that it can receive a proton from the Schiff base. Asp 85 then transfers the proton to the extracellular medium via a hydrogen-bonded network that includes Arg 82, Glu 194, Glu 204, and several water molecules. Water molecules also move into position to form a hydrogen-bonded network that reprotonates the Schiff base with an intracellular proton via Asp 96, whose pK decreases. The net result is that a proton appears to move from the cytosol to the cell exterior (the proton that leaves the cytosol is not the same proton that enters the extracellular space).

The various amino acid side chains involved in proton transport in bacteriorhodopsin move by ~1 Å or less, but this is enough to alter their pK values and to sequentially make and break hydrogen bonds so as to allow a proton to pass along the proton wire. The vectorial (one-way) nature of this process arises from the unidirectional series of conformational changes made by the photoexcited retinal as it relaxes to its ground state. Electron transfers between the various redox cofactors of mitochondrial Complex I likely motivate a similar sequence of conformational and pK changes.

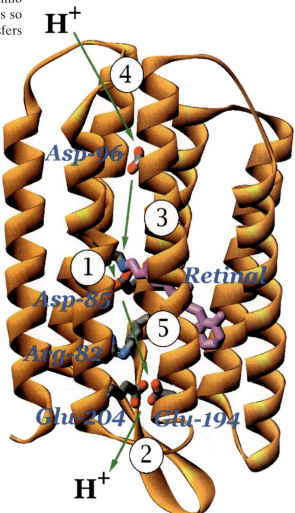

Figure 17-12 Proton translocation in bacteriorhodopsin. The retinal prosthetic group in Schiff base linkage to Lys 216 of the seven-transmembrane helix protein is shown in purple. The side chains of amino acids that participate in light-driven proton translocation are shown in stick form with C gray, N blue, and O red. The arrows with their associated numbers indicate the order of proton transfer steps during the photochemical cycle: (**1**) deprotonation of the Schiff base and protonation of Asp 85; (**2**) proton release to the extracellular surface; (**3**) reprotonation of the Schiff base and deprotonation of Asp 96; (**4**) reprotonation of Asp 96 from the cytoplasmic surface; and (**5**) deprotonation of Asp 85 and reprotonation of the proton release site. [Courtesy of Janos Lanyi, University of California at Irvine. PDBid 1C3W.]

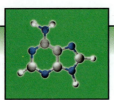

BOX 17-1

Perspectives in Biochemistry

Cytochromes Are Electron-Transport Heme Proteins

Cytochromes, whose function was elucidated in 1925 by David Keilin, are redox-active proteins that occur in all organisms except a few types of obligate anaerobes. These proteins contain heme groups that alternate between their Fe(II) and Fe(III) oxidation states during electron transport.

The heme groups of the reduced Fe(II) cytochromes have prominent visible absorption spectra consisting of three peaks: the α, β, and γ **(Soret)** bands. The spectrum for cytochrome *c* is shown in Fig. *a*.

The wavelength of the α peak, which varies characteristically with the reduced cytochrome species (it is absent in oxidized cytochromes), is used to differentiate the various cytochromes in mitochondrial membranes (*top right of Fig. a and Fig. b*).

Each group of cytochromes contains a differently substituted heme group coordinated with the redox-active iron atom (*right*). The *b*-type cytochromes contain **protoporphyrin IX,** which also occurs in myoglobin and hemoglobin (Section 7-1A). The heme group of *c*-type cytochromes differs from protoporphyrin IX in that its vinyl groups have added Cys sulfhydryls across their double bonds to form thioether linkages to the protein. Heme *a* contains a long hydrophobic tail of isoprene units attached to the porphyrin, as well as a formyl group in place of a methyl substituent in hemes *b* and *c*. The axial ligands of the heme iron also vary with the cytochrome type. In cytochromes *a* and *b*, both ligands are His residues, whereas in cytochromes *c*, one is His and the other is the S atom of Met.

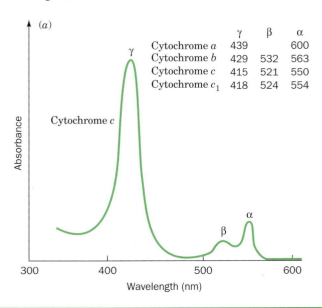

	γ	β	α
Cytochrome *a*	439		600
Cytochrome *b*	429	532	563
Cytochrome *c*	415	521	550
Cytochrome *c*$_1$	418	524	554

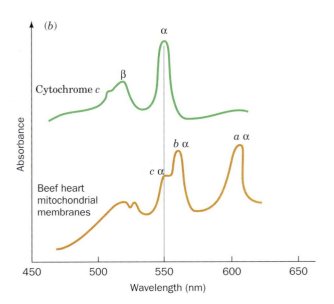

D Complex II (Succinate–Coenzyme Q Oxidoreductase)

Complex II, which contains the citric acid cycle enzyme succinate dehydrogenase (Section 16-3F), passes electrons from succinate to CoQ. Its redox groups include succinate dehydrogenase's covalently bound FAD (Fig. 16-13) to which electrons are initially passed, one [4Fe–4S] cluster, a [3Fe–4S] cluster (essentially a [4Fe–4S] complex that lacks one Fe atom), one [2Fe–2S] cluster, and one **cytochrome b_{560}** (cytochromes are discussed in Box 17-1).

The free energy for electron transfer from succinate to CoQ (Fig. 17-7) is insufficient to drive ATP synthesis. The complex is nevertheless important because it allows relatively high-potential electrons to enter the electron-transport chain by bypassing Complex I.

Within each group of cytochromes, different heme group environments may be characterized by slightly different α peak wavelengths. For this reason, it is convenient to identify cytochromes by the wavelength (in nm) at which its α band absorbance is maximal (e.g., cytochrome b_{560} in Complex II). Cytochromes are also identified nondescriptively with either numbers or letters.

Reduced heme groups are highly reactive entities; they can transfer electrons over distances of 10 to 20 Å at physiologically significant rates. Hence cytochromes, in a sense, have the opposite function of enzymes: Instead of persuading unreactive substrates to react, they must prevent their hemes from transferring electrons nonspecifically to other cellular components. This, no doubt, is why these hemes are almost entirely enveloped by protein. However, cytochromes must also provide a path for electron transfer to an appropriate partner. Since electron transfer occurs far more efficiently through bonds than through space, protein structure appears to be an important determinant of the rate of electron transfer between proteins.

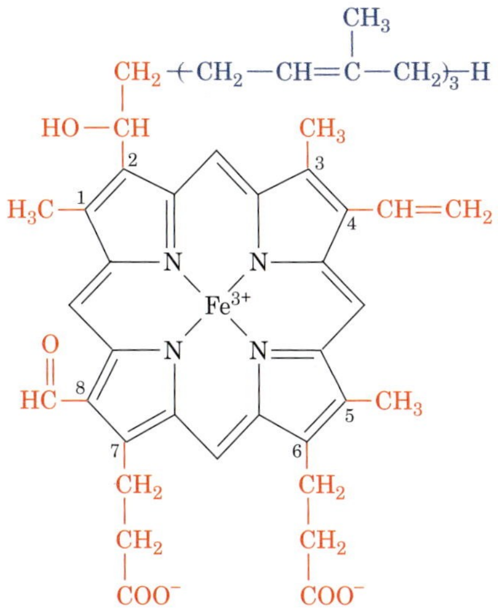

Heme a

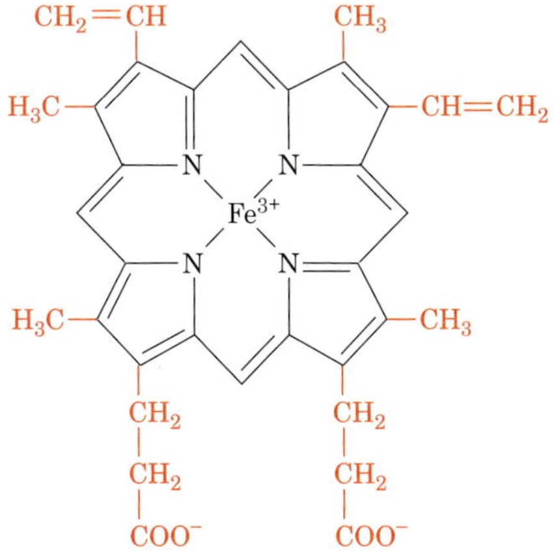

Heme b
(iron–protoporphyrin IX)

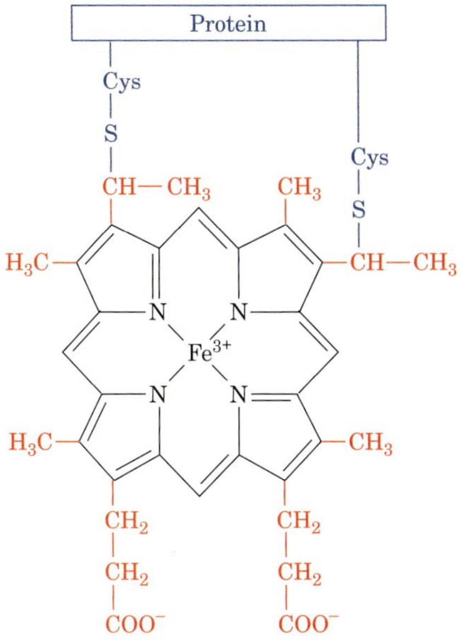

Heme c

Note that Complexes I and II, despite their names, do not operate in series. But both accomplish the same result: the transfer of electrons to CoQ from reduced substrates (NADH or succinate). *CoQ, which diffuses in the lipid bilayer among the respiratory complexes, therefore serves as a sort of collection point for electrons.* As we shall see in Section 19-2C, the first step in fatty acid oxidation generates electrons that enter the electron-transport chain at the level of CoQ. CoQ also collects electrons from the $FADH_2$ produced by the glycerophosphate shuttle (Fig. 17-5).

Complex II Contains a Linear Chain of Redox Cofactors. Although the X-ray structure of mitochondrial Complex II is unknown, that of the closely

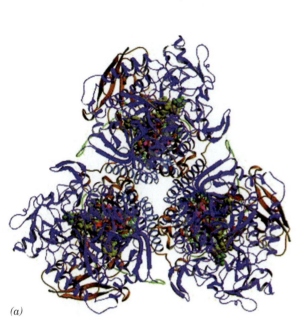

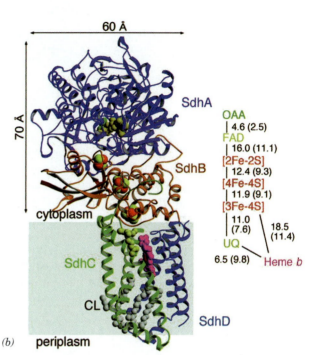

Figure 17-13 X-Ray structure of *E. coli* Complex II. (*a*) The homotrimeric complex as viewed from the cytoplasm perpendicular to the membrane. (*b*) A protomer as viewed parallel to the membrane with the cytoplasm above. SdhA, SdhB, SdhC, and SdhD are respectively purple, brown, green, and blue. The bound oxaloacetate inhibitor (*green*), the FAD (*yellow-green*), the Fe–S clusters (*Fe red and S green*), the ubiquinone (*green*), and the heme *b* (*magenta*) are drawn in space-filling form as are bound cardiolipin molecules (*gray*). The center-to-center and edge-to-edge (in parentheses) distances between redox centers are shown. The inferred position of the membrane is indicated by the light blue shading. [Courtesy of So Iwata, Imperial College London, U.K. PDBid 1NEK.]

related *E. coli* Complex II together with its inhibitor oxaloacetate has been determined by So Iwata (Fig. 17-13). *E. coli* Complex II is a 360-kD mushroom-shaped homotrimer whose protomers each consist of two hydrophilic subunits, a flavoprotein **(SdhA)** and an iron–sulfur subunit **(SdhB),** that occupy the cytoplasm (the equivalent of the mitochondrial matrix), and two hydrophobic membrane-anchor subunits, **SdhC** and **SdhD,** which each have three transmembrane helices and which collectively bind one *b*-type heme and one ubiquinone. SdhA binds both the substrate (whose binding site is occupied by oxaloacetate in the X-ray structure) and the FAD prosthetic group, whereas SdhB binds the complex's three iron–sulfur clusters. The substrate and ubiquinone binding sites are connected by a >40-Å-long chain of redox centers with the sequence substrate–FAD—[2Fe–2S]—[4Fe–4S]—[3Fe–4S]—Q (top to bottom in Fig. 17-13). These redox centers are separated by edge-to-edge distances of 2.5 to 11.1 Å, which are common cofactor separations in electron-transfer chains. The heme *b*, which is not located in this direct electron transfer pathway, apparently fine tunes the system's electronic properties so as to suppress side reactions that form damaging **reactive oxygen species** such as H_2O_2 (Section 17-4C).

E Complex III (Coenzyme Q–Cytochrome *c* Oxidoreductase)

Complex III (also known as **cytochrome *bc*₁**) passes electrons from reduced CoQ to cytochrome *c*. It contains two ***b*-type cytochromes,** one **cytochrome *c*₁,** and one [2Fe–2S] cluster in which one of the Fe atoms is coordinated by two His residues rather than two Cys residues (and which is therefore

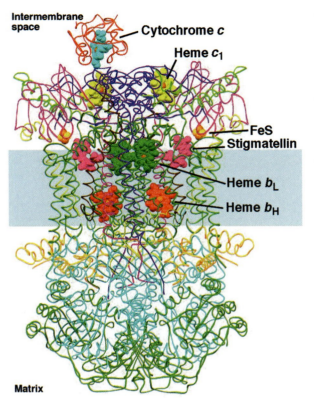

Figure 17-14 X-Ray structure of yeast Complex III in complex with cytochrome *c* and the inhibitor stigmatellin. The homodimeric complex is viewed perpendicular to its twofold axis from within the plane of the membrane with the intermembrane space above. The nine different subunits in each protomer, which collectively have 12 transmembrane helices, are differently colored with cytochrome *b* (8 transmembrane helices) green, cytochrome c_1 (1 transmembrane helix) purple, the ISP (1 transmembrane helix) magenta, and cytochrome *c* red. The four different heme groups, the [2Fe–2S] cluster, and stigmatellin are drawn in space-filling form in different colors but with all Fe atoms orange. The inferred position of the membrane is indicated by the blue shading. Note that only one cytochrome *c* is bound to the homodimeric Complex III. [Based on an X-ray structure by Carola Hunte, Max Planck Institute for Biophysics, Frankfurt am Main, Germany. PDBid 1KYO.] **See the Interactive Exercises.**

known as a **Rieske center** after its discoverer, John Rieske). Complex III from yeast mitochondria is a 419-kD homodimer with 9 subunits (11 in the 485-kD bovine heart mitochondrial Complex III). Its X-ray structure (Fig. 17-14) reveals a pear-shaped dimer whose widest part extends ~75 Å into the mitochondrial matrix. The ~40-Å-thick transmembrane portion consists of 12 transmembrane helices per protomer (14 in bovine heart Complex III), most of which are tilted with respect to the plane of the membrane. Eight of these helices belong to the **cytochrome *b*** subunit, which binds both *b*-type cytochrome hemes, b_{562} (or b_H, for high potential, which lies near the intermembrane space) and b_{566} (or b_L, for low potential, which lies near the matrix). The cytochrome c_1 subunit is anchored by a single transmembrane helix, with its globular head, which contains a *c*-type heme, extending into the intermembrane space. The **iron–sulfur protein (ISP),** which contains the Rieske center, is similarly anchored by a single transmembrane helix and extends into the intermembrane space. The two ISPs of the dimeric complex are intertwined so that the [2Fe–2S] cluster in the ISP of one protomer interacts with the cytochrome *b* and cytochrome c_1 subunits of the other protomer.

Electron Transport and Proton Pumping in Complex III: The Q Cycle.

Complex III functions to permit one molecule of $CoQH_2$, a two-electron carrier, to reduce two molecules of cytochrome *c*, a one-electron carrier. This occurs by a surprising bifurcation of the flow of electrons from $CoQH_2$ to cytochrome c_1 and to cytochrome *b* (in which the flow is cyclic). It is this so-called **Q cycle** that permits Complex III to pump protons from the matrix to the intermembrane space.

The essence of the Q cycle is that *$CoQH_2$ undergoes a two-cycle reoxidation in which the semiquinone, CoQ^-, is a stable intermediate.* This involves

See Guided Exploration 16

The Q cycle.

Figure 17-15 The Q cycle. The Q cycle, which is mediated by Complex III, results in the translocation of H$^+$ from the matrix to the intermembrane space as driven by the transport of electrons from cytochrome b to cytochrome c. The overall cycle is actually two cycles, the first requiring Reactions 1 through 7 and the second requiring Reactions 1 through 6 and 8. (**1**) Coenzyme QH$_2$ is supplied by Complex I on the matrix side of the membrane. (**2**) QH$_2$ diffuses to the cytosolic side of the membrane, where it binds in the Q$_o$ site on the cytochrome b subunit of Complex III. (**3**) QH$_2$ reduces the Rieske iron–sulfur protein (ISP), forming Q$^{\bar{\cdot}}$ semiquinone and releasing 2 H$^+$. The ISP goes on to reduce heme c_1. (**4**) Q$^{\bar{\cdot}}$ reduces heme b_L to form coenzyme Q. (**5**) Q diffuses to the matrix side, where, in cycle 1 only, it binds in the Q$_i$ site on cytochrome b. (**6**) Heme b_L reduces heme b_H. (**7**, Cycle 1 only) Q is reduced to Q$^{\bar{\cdot}}$ by heme b_H. (**8**, Cycle 2 only) Q$^{\bar{\cdot}}$ bound in the Q$_i$ site is reduced to QH$_2$ by heme b_H. The net reaction is the transfer of two electrons from QH$_2$ to cytochrome c_1 and the translocation of four protons from the matrix to the intermembrane space. [After Trumpower, B.L., *J. Biol. Chem.* **265**, 11410 (1990).]

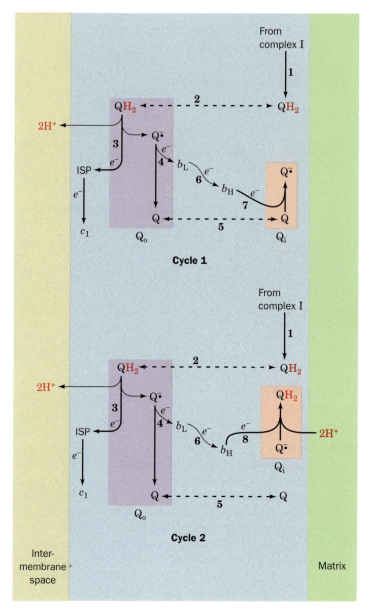

two independent binding sites for coenzyme Q: Q$_o$, which binds CoQH$_2$ and is located between the Rieske [2Fe–2S] center and heme b_L in proximity to the intermembrane space; and Q$_i$, which binds both CoQ$^{\bar{\cdot}}$ and CoQ and is located near heme b_H in proximity to the matrix. In the first cycle (Fig. 17-15, *top*), CoQH$_2$ from Complex I (**1** and **2**) binds to the Q$_o$ site, where it transfers one of its electrons to the ISP (**3**), releasing its two protons into the intermembrane space and yielding CoQ$^{\bar{\cdot}}$. The ISP goes on to reduce cytochrome c_1, whereas the CoQ$^{\bar{\cdot}}$ transfers its remaining electron to cytochrome b_L (**4**), yielding fully oxidized CoQ. Cytochrome b_L then reduces cytochrome b_H (**6**). The CoQ from Step 4 is released from the Q$_o$ site and rebinds to the Q$_i$ site (**5**), where it picks up the electron from cytochrome b_H (**7**), reverting to the semiquinone form, CoQ$^{\bar{\cdot}}$. Thus, the reaction for this first cycle is

$$\text{CoQH}_2 + \text{cytochrome } c_1 \text{ (Fe}^{3+}) \rightarrow$$
$$\text{CoQ}^{\bar{\cdot}} + \text{cytochrome } c_1 \text{ (Fe}^{2+}) + 2 \text{ H}^+ \text{ (}\textit{intermembrane}\text{)}$$

In the second cycle (Fig. 17-15, *bottom*), another $CoQH_2$ from Complex I repeats Steps 1 through 6: One electron reduces the ISP and then cytochrome c_1, and the other electron sequentially reduces cytochrome b_L and then cytochrome b_H. This second electron then reduces the CoQ^- at the Q_i site produced in the first cycle (8), yielding $CoQH_2$. The protons consumed in this last step originate in the mitochondrial matrix. The reaction for the second cycle is therefore

$$CoQH_2 + CoQ^- + \text{cytochrome } c_1 \text{ (Fe}^{3+}) + 2 H^+ \text{ (\textit{matrix}) } \rightarrow$$
$$CoQ + CoQH_2 + \text{cytochrome } c_1 \text{ (Fe}^{2+}) + 2 H^+ \text{ (\textit{intermembrane})}$$

For every two $CoQH_2$ that enter the Q cycle, one $CoQH_2$ is regenerated. The combination of both cycles, in which two electrons are transferred from $CoQH_2$ to cytochrome c_1, results in the overall reaction

$$CoQH_2 + 2 \text{ cytochrome } c_1 \text{ (Fe}^{3+}) + 2 H^+ \text{ (\textit{matrix}) } \rightarrow$$
$$CoQ + 2 \text{ cytochrome } c_1 \text{ (Fe}^{2+}) + 4 H^+ \text{ (\textit{intermembrane})}$$

How does the structure of Complex III support the operation of the Q cycle? First, X-ray studies provide direct evidence for the independent existence of the Q_o and Q_i sites. The antifungal agent **stigmatellin,**

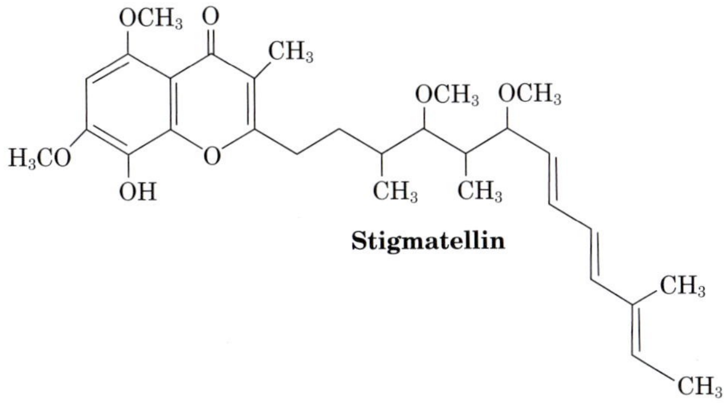

Stigmatellin

which is known to inhibit electron flow from $CoQH_2$ to the ISP and to heme b_L (Steps 3 and 4 of both cycles), binds in a pocket within cytochrome b midway between the iron positions of the Rieske [2Fe–2S] center and heme b_L. Thus, this binding pocket is likely to overlap the Q_o site. Similarly, antimycin A, which has been shown to block electron flow from heme b_H to CoQ or CoQ^- (Step 7 of Cycle 1 and Step 8 of Cycle 2), binds in a pocket near heme b_H, thereby identifying this pocket as site Q_i.

The circuitous route of electron transfer in Complex III is tied to the ability of coenzyme Q to diffuse within the hydrophobic core of the membrane in order to bind to both the Q_o and Q_i sites. In fact, the mitochondrial membrane likely contains a pool of CoQ, CoQ^-, and $CoQH_2$, so that the ubiquinone molecule released from the Q_o site may not be the same one that rebinds to the Q_i site in Cycle 1 (Fig. 17-15).

X-Ray structures of cytochrome bc_1 also explain why Q_o-bound CoQ^- exclusively reduces heme b_L rather than the Rieske [2Fe–2S] cluster of the ISP, despite the greater reduction potential difference ($\Delta\mathscr{E}$) favoring the latter reaction (Table 17-1). The globular domain of the ISP can swing via a ~20-Å hinge motion between the Q_o site and cytochrome c_1. *Consequently, the ISP acquires an electron from $CoQH_2$ in the Q_o site and mechanically delivers it to the heme c_1 group.* The CoQ^- product cannot reduce the ISP (after it has reduced cytochrome c_1) because the ISP has moved too far away for this to occur.

The net reaction for the Q cycle indicates that *when $CoQH_2$ is oxidized, two reduced cytochrome c molecules and four protons appear on the outer*

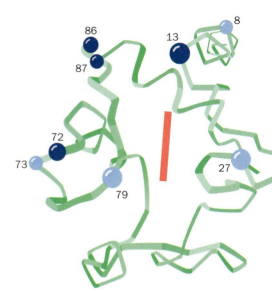

Figure 17-16 Ribbon diagram of cytochrome *c* showing the Lys residues involved in intermolecular complex formation.   Dark and light blue balls, respectively, mark the position of Lys residues whose ε-amino groups are strongly and less strongly protected by cytochrome c_1 or cytochrome *c* oxidase against acetylation. Note that these Lys residues form a ring around the heme (*solid bar*) on one face of the protein. [After Mathews, F.S., *Prog. Biophys. Mol. Biol.* **45**, 45 (1986).] 🐛 **See the Interactive Exercises and Kinemage Exercise 5.**

side of the membrane. Proton transport by the Q cycle thus differs from the proton-pumping mechanism of Complexes I and IV (see below): In the Q cycle, a redox center itself (CoQ) is the proton carrier.

Cytochrome *c*: A Soluble Electron Carrier. The electrons that flow to cytochrome c_1 are transferred to cytochrome *c*, which, unlike the other cytochromes of the respiratory electron-transport chain, is a peripheral membrane protein. It shuttles electrons between Complexes III and IV on the outer surface of the inner mitochondrial membrane. The structure and evolution of cytochrome *c* are discussed in Sections 5-4A and 6-2D. Several invariant Lys residues in cytochrome *c* lie in a ring around the exposed edge of its otherwise buried heme group (Fig. 17-16). These residues constitute a binding site that was identified by **differential labeling:** Treatment of cytochrome *c* with acetic anhydride (which acetylates Lys residues) in the presence and absence of cytochrome c_1 demonstrated that cytochrome c_1 completely shields these cytochrome *c* Lys residues. The reactivities of other cytochrome *c* Lys residues that are distant from the exposed heme edge are unaffected by the presence of cytochrome c_1. Nearly identical results were obtained when cytochrome c_1 was replaced by cytochrome *c* oxidase. This suggests that both these complexes have negatively charged sites that are complementary to the ring of positively charged Lys residues on cytochrome *c*. These interactions presumably serve to align redox groups for optimal electron transfer.

The X-ray structure of yeast cytochrome bc_1 in complex with cytochrome *c* (Fig. 17-14) reveals, as expected, that cytochrome *c* binds to the cytochrome c_1 subunit of cytochrome bc_1. This association appears to be particularly tenuous because its interfacial area (880 Å²) is significantly less than that exhibited by protein–protein complexes known to have low stability (typically <1600 Å²). Such a small interface is well suited for fast binding and release. This interface involves only two cytochrome *c* Lys residues, Lys 86 and Lys 79, which respectively contact Glu 235 and Ala 164 of cytochrome c_1. Other pairs of charged and often conserved residues surround the contact site but they are not close enough for direct polar interactions. Perhaps these interactions are mediated by water molecules that are not seen in the X-ray structure. The closest approach between the heme groups of the contacting proteins is 4.5 Å between atoms of their respective vinyl side chains, which accounts for the rapid rate of electron transfer between these two redox centers.

F Complex IV (Cytochrome *c* Oxidase)

Cytochrome *c* oxidase catalyzes the one-electron oxidations of four consecutive reduced cytochrome *c* molecules and the concomitant four-electron reduction of one O_2 molecule:

$$4 \text{ Cytochrome } c \text{ (Fe}^{2+}) + 4 \text{ H}^+ + O_2 \rightarrow 4 \text{ cytochrome } c \text{ (Fe}^{3+}) + 2 \text{ H}_2\text{O}$$

Mammalian Complex IV is a ~410-kD homodimer whose component protomers are each composed of 13 subunits. The X-ray structure of Complex IV from bovine heart mitochondria, determined by Shinya Yoshikawa, reveals that ten of its subunits are transmembrane proteins that contain a total of 28 membrane-spanning α helices (Fig. 17-17). The core of Complex IV consists of its three largest and most hydrophobic subunits, I, II, and III, which are encoded by mitochondrial DNA (the remaining subunits are nuclearly encoded and must be transported into the mitochondrion). A con-

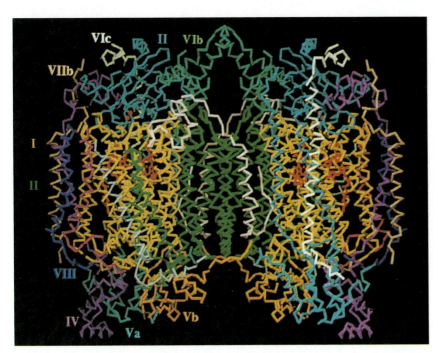

Figure 17-17 X-Ray structure of the bovine heart cytochrome *c* oxidase homodimer. Each of its 13 subunits is shown as a C_α trace of a different color. The view is from within the membrane, with the intermembrane space at the top. [Courtesy of Shinya Yoshikawa, Himeji Institute of Technology, Hyogo, Japan. PDBid 1OCC.]
☙ See the Interactive Exercises.

cave area on the surface of the protein that faces the intermembrane space contains numerous acidic amino acids that can potentially interact with the ring of Lys residues on cytochrome *c*, the electron donor for Complex IV.

Complex IV contains four redox centers: **cytochrome *a*, cytochrome *a*₃, a copper atom known as Cu_B,** and a pair of copper atoms known as the **Cu_A center** (Fig. 17-18). The Cu_A center, which is bound to Subunit II, lies 8 Å above the membrane surface. Its two copper ions are bridged by the sulfur atoms of two Cys residues, giving it a geometry similar to that of a [2Fe–2S] cluster. The other redox groups—Cu_B and cytochromes *a* and *a*₃—all bind to Subunit I and lie ~13 Å below the membrane surface.

Spectroscopic studies have shown that electron transfer in Complex IV is linear, proceeding from cytochrome *c* to the Cu_A center, then to heme *a*, and finally to heme *a*₃ and Cu_B. The Fe of heme *a*₃ lies only 4.9 Å from Cu_B; these redox groups really form a single binuclear complex. Electrons appear to travel between the redox centers of Complex IV via a hydrogen bonded network, involving amino acid side chains, the polypeptide backbone, and the propionate side chains of the heme groups.

Reduction of O₂ by Cytochrome *c* Oxidase. The reduction of O_2 to $2 H_2O$ by cytochrome *c* oxidase takes place at the cytochrome *a*₃–Cu_B binuclear complex and requires the nearly simultaneous input of four electrons. However, the fully reduced Fe(II)–Cu(I) binuclear complex can readily contribute

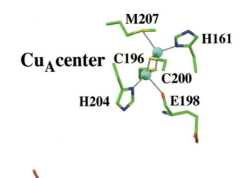

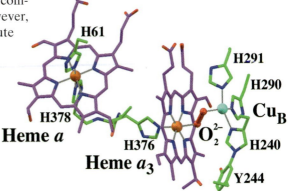

Figure 17-18 The redox centers of bovine heart cytochrome *c* oxidase. The Fe and Cu ions are represented by orange and cyan spheres. Their liganding heme and protein groups are drawn in stick form colored according to atom type (heme C magenta, protein C green, N blue, O red, and S yellow). The peroxy group that bridges the Cu_B and heme *a*₃ Fe ions is shown in ball-and-stick form in red. Coordination bonds are drawn as gray lines. Note that the side chains of His 240 and Tyr 244 are joined by a covalent bond (*lower right*). [Based on an X-ray structure by Shinya Yoshikawa, Himeji Institute of Technology, Hyogo, Japan. PDBid 2OCC.]

only three electrons to its bound O_2 in reaching its fully oxidized Fe(IV)–Cu(II) state [cytochrome a_3 assumes its Fe(IV) or **ferryl** oxidation state during the reduction of O_2; see below]. What is the source of the fourth electron?

X-Ray structures of cytochrome *c* oxidase clearly indicate that the His 240 ligand of Cu_B is covalently bonded to the side chain of a conserved Tyr residue (Tyr 244; Fig. 17-18, *lower right*). This places the Tyr phenolic —OH group close to the heme a_3–ligated O_2 such that *Tyr 244 can supply the fourth electron by transiently forming a tyrosyl radical* (TyrO·). Tyrosyl radicals have been implicated in several other enzyme-mediated redox processes, including the generation of O_2 from H_2O in photosynthesis (Section 18-2C) and in the **ribonucleotide reductase** reaction (which converts NDP to dNDP; Section 22-3A). In cytochrome *c* oxidase, the Tyr phenolic —OH group is within hydrogen bonding distance of the enzyme-bound O_2 and hence is a likely H^+ donor during O—O bond cleavage. The formation of the covalent cross-link is expected to lower both the reduction potential and the pK of Tyr 244, thereby facilitating both radical formation and proton donation.

A proposed reaction sequence for cytochrome *c* oxidase, which was elucidated through the use of a variety of spectroscopic techniques, is shown in Fig. 17-19:

1 and 2. The oxidized binuclear complex [Fe(III)$_{a3}$—OH⁻ Cu(II)$_B$] is reduced to its [Fe(II)$_{a3}$ Cu(I)$_B$] state by two consecutive one-electron transfers from cytochrome *c* via cytochrome *a* and Cu_A. A proton from the matrix is concomitantly acquired and an H_2O is released in this process. Tyr 244 (Y—OH) is in its phenolic state.

3. O_2 binds to the reduced binuclear complex so as to ligand its Fe(II)$_{a3}$ atom. It binds to the heme with much the same configuration it has in oxymyoglobin (Fig. 7-3).

4. Internal electron redistribution rapidly yields the oxyferryl complex [Fe(IV)=O²⁻ HO⁻—Cu(II)] in which Tyr 244 has donated an electron and a proton to the complex and thereby

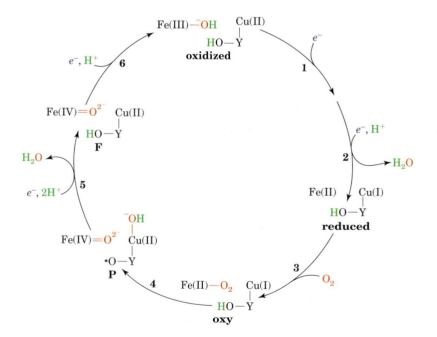

Figure 17-19 Proposed reaction sequence for cytochrome *c* oxidase. A total of four electrons ultimately donated by four cytochrome *c* molecules, together with four protons, are required to reduce O_2 to H_2O at the cytochrome a_3–Cu_B binuclear complex. The numbered steps are discussed in the text. The entire reaction is extremely fast; it goes to completion in ∼1 ms at room temperature. [Modified from Babcock, G.T., *Proc. Natl. Acad. Sci.* **96**, 12971 (1999).]

assumed its neutral radical state $(Y\!-\!O\cdot)$. This is known as compound P because it was once thought to be a peroxy compound.

5. A third one-electron transfer from cytochrome c together with the acquisition of two protons reconverts Tyr 244 to its phenolic state, yielding compound F (for ferryl) and releasing an H_2O.

6. A fourth and final electron transfer and proton acquisition yields the oxidized $[Fe(III)_{a3}\!-\!OH^-\ Cu(II)_B]$ complex, thereby completing the catalytic cycle.

Cytochrome c Oxidase Has Two Proton-Translocating Channels. Four so-called **chemical** or **scalar** protons are taken up from the matrix during the reduction of O_2 by cytochrome c oxidase to yield 2 H_2O. This four-electron process is coupled to the translocation of up to four so-called **pumped** or **vectorial** protons from the matrix to the intermembrane space, thereby contributing to the proton gradient that powers ATP synthesis (Section 17-3). Note that for each turnover of the enzyme,

$$8\ H^+_{matrix} + O_2 + 4\ \text{cytochrome}\ c\ (Fe^{2+}) \rightarrow$$
$$4\ \text{cytochrome}\ c\ (Fe^{3+}) + 2\ H_2O + 4\ H^+_{intermembrane}$$

a total of eight positive charges are transported across the inner mitochondrial membrane, thus contributing to its membrane potential.

X-Ray structures of cytochrome c oxidase reveal the presence of two channels that lead from the matrix to the vicinity of the O_2-reducing center, which could potentially transport protons via a proton wire mechanism. The **K-channel** (so named because it contains an essential Lys residue) leads from the matrix side of the protein to Tyr 244, the residue that forms a free radical as described above. Since this channel does not appear to be connected to the intermembrane space, it is thought to supply the chemical protons for O_2 reduction. The **D-channel** (named for a key Asp residue) extends from the matrix to the vicinity of the heme a_3–Cu_B center, where it connects to the so-called **exit channel,** which communicates with the intermembrane space. Apparently, the D-channel, in series with the exit channel, functions to pump vectorial protons from the matrix to the intermembrane space. Moreover, there are indications that the D-channel also functions as a conduit for the chemical protons that are required for the second part of the reaction cycle (steps 5 and 6 of Fig. 17-19). Despite the foregoing, the mechanism that couples O_2 reduction to proton pumping in Complex IV remains an enigma.

3   Oxidative Phosphorylation

The endergonic synthesis of ATP from ADP and P_i in mitochondria is catalyzed by an **ATP synthase** (also known as **Complex V**) that is driven by the electron-transport process. *The free energy released by electron transport through Complexes I–IV must be conserved in a form that ATP synthase can use.* Such energy conservation is referred to as **energy coupling.**

The physical characterization of energy coupling proved to be surprisingly elusive; many sensible and often ingenious ideas failed to withstand the test of experimental scrutiny. For example, one theory—now abandoned—was that electron transport yields a "high-energy" intermediate, such as phosphoenolpyruvate (PEP) in glycolysis (Section 14-2J), whose subsequent breakdown drives ATP synthesis. No such intermediate has ever been identified. In fact, ATP synthesis is coupled to electron transport through the formation of a transmembrane proton gradient during electron

transport by Complexes I, III, and IV. In this section, we explore this coupling mechanism and the operation of ATP synthase.

A The Chemiosmotic Theory

The **chemiosmotic theory,** proposed in 1961 by Peter Mitchell, spurred considerable controversy before becoming widely accepted (Box 17-2). Mitchell's theory states that *the free energy of electron transport is conserved by pumping H^+ from the mitochondrial matrix to the intermembrane space to create an electrochemical H^+ gradient across the inner mitochondrial membrane. The electrochemical potential of this gradient is harnessed to synthesize ATP* (Fig. 17-20). Several key observations are explained by the chemiosmotic theory:

1. Oxidative phosphorylation requires an intact inner mitochondrial membrane.

2. The inner mitochondrial membrane is impermeable to ions such as H^+, OH^-, K^+, and Cl^-, whose free diffusion would discharge an electrochemical gradient.

3. Electron transport results in the transport of H^+ out of intact mitochondria (the intermembrane space is equivalent to the cytosol), thereby creating a measurable electrochemical gradient across the inner mitochondrial membrane.

4. Compounds that increase the permeability of the inner mitochondrial membrane to protons, and thereby dissipate the electrochemical gradient, allow electron transport (from NADH and succinate oxidation) to continue but inhibit ATP synthesis; that is, they "uncouple" electron transport from oxidative phosphorylation. Conversely, increasing the acidity outside the inner mitochondrial membrane stimulates ATP synthesis.

An entirely analogous process occurs in bacteria, whose electron-transporting machinery is located in their plasma membranes (see Box 17-3).

Electron Transport Generates a Proton Gradient. *Electron transport, as we have seen, causes Complexes I, III, and IV to transport protons across the inner mitochondrial membrane from the matrix, a region of low $[H^+]$, to the inter-*

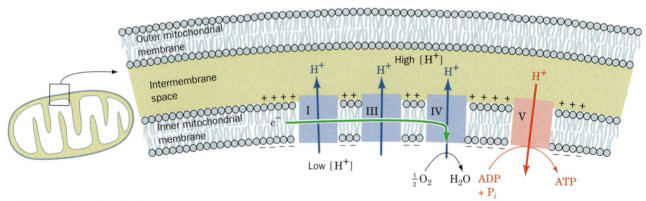

Figure 17-20 **The coupling of electron transport and ATP synthesis.** Electron transport (*green arrow*) generates a proton electrochemical gradient across the inner mitochondrial membrane. H^+ is pumped out of the mitochondrion during electron transport (*blue arrows*) and its exergonic return powers the synthesis of ATP (*red arrows*). Note that the intermembrane space is topologically equivalent to the cytosol because the outer mitochondrial membrane is permeable to H^+. 🞲 **See the Animated Figures.**

BOX 17-2

Pathways of Discovery

Peter Mitchell and the Chemiosmotic Theory

Peter Mitchell (1920–1992)

One of the most dramatic paradigm shifts in biochemistry came about through the work of Peter Mitchell, whose chemiosmotic hypothesis linked biological electron transport to ATP synthesis. Mitchell was primarily a theoretical biochemist, although he also generated experimental data to support his hypothesis. He once compared the human mind to a garden planted with facts and ideas that are constantly being rearranged. However, he promoted his own ideas, which were highly controversial, with great tenacity and very little flexibility.

Mitchell graduated from the University of Cambridge in 1942 and conducted research there, laying the groundwork for the chemiosmotic theory. Mitchell continued his work after 1955 at the University of Edinburgh. He was captivated by the idea of compartmentation in living cells and by the vectorial, or one-way, aspect of metabolic processes. Initially, he chose to study phosphate transport in bacteria because this process was linked both to metabolism and to transmembrane transport. He realized that the vectorial nature of membrane transport must be due to the presence of membrane-associated systems that were driven by chemical forces.

Mitchell also became interested in the respiratory chain, an idea formulated by David Keilin at Cambridge, because this too was a clearly vectorial biological phenomenon. Mitchell reasoned that there must be enzymes that, like transporters, convert a substrate on one side of a membrane to a product on the other side. Other researchers had already discovered that the activity of the respiratory chain generated a pH gradient. Mitchell's genius was to explain how this pH gradient could drive ATP synthesis. In his seminal paper of 1961, Mitchell proposed that the respiratory chain, associated with the cristae in the mitochondrion, generates a protonmotive force due to electrical and pH differences across the membrane. This force drives an ATPase, working in reverse, to catalyze the condensation of phosphate with ADP to produce ATP.

Mitchell's hypothesis, elegant as it was, met strong resistance from other biochemists for several reasons. First, chemiosmosis was a theoretical notion without direct experimental evidence. Second, the study of oxidative phosphorylation was dominated by a few powerful laboratories that were not inclined to welcome new theories. In particular, metabolic studies since the mid-1940s had been focused on the activities of soluble enzymes. Mitchell's theory was not a product of this classic biochemical approach but instead came through an understanding of membrane physiology. Furthermore, the prevailing theories about the connection between electron transport and ATP synthesis centered on a phosphorylated compound as a high-energy intermediate. Fritz Lipmann (see Box 13-3) had proposed that a high-energy phosphate group might become attached to some component of the respiratory chain. This hypothesis was later amended to invoke a soluble phosphorylated intermediate. The search for the elusive compound lasted some 20 years, and the investment of time and money may have made some researchers hesitant to abandon this theory in favor of Mitchell's seemingly outrageous proposal. In the meantime, a third theory was proposed, in which electron transport was coupled to ATP synthesis through protein conformational changes, with the observed pH gradient presumed to be simply a by-product of this process.

Mitchell's chemiosmotic hypothesis did not garner significant support until about 10 years after its publication. During this period of sometimes acrimonious debate, Mitchell became ill, moved to Cornwall, and renovated a manor house, part of which became a private laboratory, known as Glynn Research, that was funded by Mitchell's family fortune. Here, he and his life-long collaborator, Jennifer Moyle, produced experimental evidence to support the chemiosmotic theory. Ultimately, Mitchell was proven correct by other researchers who demonstrated the proton-pumping activity of purified mitochondrial components reconstituted in liposomes. Mitchell was awarded a Nobel Prize in Physiology or Medicine in 1978.

Mitchell succeeded in changing the prevailing views of a central feature of aerobic metabolism, although it was a long battle. He later expressed sadness that his work was taken for granted, as if it had been "self-evident from the beginning." Oddly, Mitchell stubbornly resisted altering any of his own ideas. For example, he never wavered in his belief that protons participate directly in ADP phosphorylation in the active site of ATP synthase. And for many years, Mitchell refused to acknowledge proton pumping (as we now know occurs in Complexes I and IV). Instead, he insisted that the source of the proton gradient was a "redox loop" in which two electrons are transferred from the positive to the negative side of the membrane and combine with two protons to reduce a quinone to a quinol. The quinol then diffuses back across the bilayer to be reoxidized at the positive side, where the protons are released. The redox loop mechanism, which requires two "active sites," does occur during the Q cycle in mitochondrial Complex III and in certain bacterial systems, but it cannot account for the entire protonmotive force that is the heart of the chemiosmotic mechanism.

Mitchell, P., Coupling of phosphorylation to electron and hydrogen transfer by a chemiosmotic type of mechanism, *Nature* **191**, 144–148 (1961).

Prebble, J., Peter Mitchell and the ox phos wars, *Trends Biochem. Sci.* **27**, 209–212 (2002).

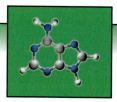

BOX 17-3

Perspectives in Biochemistry

Bacterial Electron Transport and Oxidative Phosphorylation

It comes as no surprise that aerobic bacteria (such as those pictured), whose ancestors gave rise to mitochondria, use similar machinery to oxidize reduced coenzymes and conserve their energy in ATP synthesis. In bacteria, the components of the respiratory electron-transport chain are located in the plasma membrane, and protons are pumped from the cytosol to the outside of the plasma membrane. Protons flow back into the cell via an ATP synthase, whose catalytic component is oriented toward the cytosol. This is exactly the arrangement expected if bacteria and mitochondria are evolutionarily related.

The oxidation of $CoQH_2$ is universal in aerobic organisms. In mitochondria, CoQ collects electrons donated by NADH (via Complex I), succinate (via Complex II), and fatty acids. In aerobic bacteria, CoQ is the collection point for electrons extracted by dehydrogenases specific for a wide variety of substrates. In bacteria, as in mitochondria, electrons flow from CoQ through cytochrome-based oxidoreductases before reaching O_2. In some species, two protein complexes (analogous to mitochondrial Complexes III and IV) carry out this process. In other species, including *E. coli,* a single type of enzyme, **quinol oxidase,** uses the electrons donated by CoQ to reduce O_2 (*below, right*).

The advantage of a multicomplex electron-transport pathway is that it affords more opportunities for proton translocation across the bacterial membrane, so the ATP yield per electron is greater. However, the shorter electron-transport pathways may confer a selective advantage in the presence of toxins that inactivate the bacterial counterpart of mitochondrial Complex III. Multiple routes for electron transport probably also allow bacteria to adjust oxidative phosphorylation to the availability of different energy sources and to balance ATP synthesis against the regeneration of various reduced coenzymes. For example, in facultative anaerobic bacteria (which can grow in either the absence or presence of O_2), when energy needs are met through anaerobic fermentation, electron transport can be adjusted to regenerate NAD^+ without synthesizing ATP by oxidative phosphorylation.

A variety of cytochrome-containing protein complexes occur in bacterial plasma membranes. Some of these proteins represent more streamlined versions of the mitochondrial complexes since they lack the additional subunits encoded by the nuclear genome of eukaryotes. However, this is not a universal feature of respiratory complexes, and many bacterial proteins (e.g., **cytochrome *d***) have no counterparts encoded by either the mitochondrial or nuclear genomes.

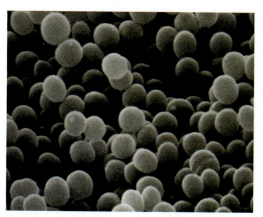

Staphylococcus aureus. [© Tony Brain/Photo Researchers.]

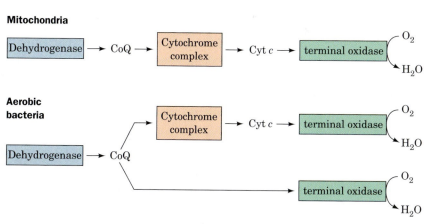

membrane space (which is in contact with the cytosol), a region of high [H$^+$] (Fig. 17-8). The free energy sequestered by the resulting electrochemical gradient (also called the **protonmotive force, pmf**) powers ATP synthesis.

The free energy change of transporting a proton out of the mitochondrion has a chemical as well as an electrical component, since H$^+$ is an ion (see Section 10-1). ΔG is therefore expressed by Eq. 10-3, which in terms of pH is

$$\Delta G = 2.3\, RT[\text{pH }(in) - \text{pH }(out)] + Z\mathscr{F}\Delta\Psi \qquad [17\text{-}1]$$

where Z is the charge on the proton (including sign), $\mathscr{F}$ is the Faraday constant, and $\Delta\Psi$ is the membrane potential. The sign convention for $\Delta\Psi$ is that when a proton is transported from negative to positive, $\Delta\Psi$ is positive. Since pH (*out*) is less than pH (*in*), *the export of protons from the mitochondrial matrix (against the proton gradient) is an endergonic process.*

The measured membrane potential across the inner membrane of a liver mitochondrion, for example, is 0.168 V (inside negative). The pH of its matrix is 0.75 units higher than that of its intermembrane space. ΔG for proton transport out of this mitochondrial matrix is therefore 21.5 kJ · mol^{-1}. Because formation of the proton gradient is an endergonic process, discharge of the gradient is exergonic. This free energy is harnessed by ATP synthase to drive the phosphorylation of ADP.

An ATP molecule's estimated physiological free energy of synthesis, around +40 to +50 kJ · mol^{-1}, is too large for ATP synthesis to be driven by the passage of a single proton back into the mitochondrial matrix; at least two protons are required. In fact, most experimental measurements (which are difficult to precisely quantitate) indicate that around three protons are required per ATP synthesized.

B   ATP Synthase

ATP synthase, also known as **proton-pumping ATP synthase** and **F_1F_0-ATPase,** is a multisubunit transmembrane protein with a total molecular mass of 450 kD. Efraim Racker discovered that mitochondrial ATP synthase is composed of two functional units, **F_0** and **F_1.** F_0 is a water-insoluble transmembrane protein containing as many as eight different types of subunits. F_1 is a water-soluble peripheral membrane protein, composed of five types of subunits, that is easily and reversibly dissociated from F_0 by treatment with urea. Solubilized F_1 hydrolyzes ATP but cannot synthesize it (hence the name ATPase).

Electron micrographs of the inner mitochondrial membrane reveal that its matrix surface is studded with molecules of ATP synthase whose F_1 component is connected to the membrane-embedded F_0 component by a protein stalk, thereby giving F_1 a lollipop-like appearance (Fig. 17-21a). Similar entities have been observed lining the inner surface of the bacterial plasma membrane and in chloroplasts (Section 18-2D). Higher resolution electron micrographs of the ATP synthase from *E. coli* reveal that its F_1 and F_0 components are joined by both an ~45-Å-long central stalk and a less substantial peripherally located connector (Fig. 17-21b).

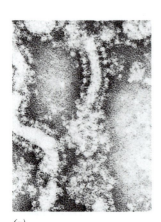
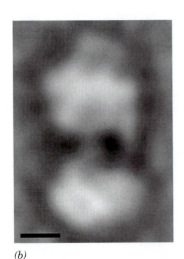
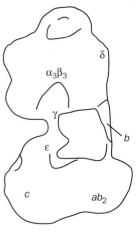

(a)   (b)

Figure 17-21 Structure of ATP synthase. (*a*) An electron micrograph of cristae from a mitochondrion showing their F_1 "lollipops" projecting into the matrix. [From Parsons, D.F., *Science* **140**, 985 (1963). Copyright © 1963 American Association for the Advancement of Science. Used by permission.] (*b*) Electron microscopy–based image of *E. coli* F_1F_0-ATPase. The accompanying interpretive drawing indicates the positions of its component subunits, which are described in the text. [Courtesy of Roderick Capaldi, University of Oregon.]

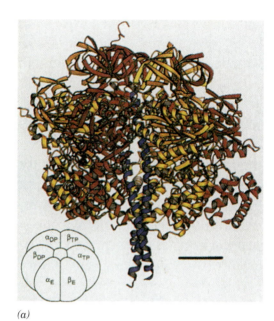

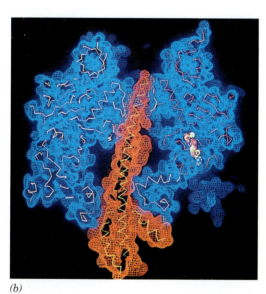

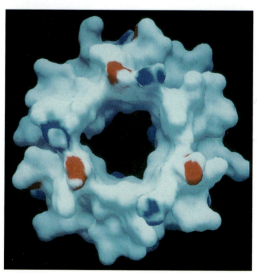

(a)

(b)

(c)

The F₁ Component. The F_1 component of mitochondrial ATP synthase has the subunit composition $\alpha_3\beta_3\gamma\delta\varepsilon$. The X-ray structure of the 3440-residue (371-kD) F_1 subunit from bovine heart mitochondria, determined by John Walker and Andrew Leslie, reveals that it consists of an 80-Å-high and 100-Å-wide spheroid that is mounted on a 30-Å-long stem (Fig. 17-22a). The α and β subunits, which are 20% identical in sequence and have nearly identical folds, are arranged alternately, like the segments of an orange, around the upper portion of a 90-Å-long α helix formed by the C-terminal segment of the γ subunit (Fig. 17-22b). The lower portion of the helix forms a bent left-handed antiparallel coiled coil with the N-terminal segment of the γ subunit. This coiled coil is part of the stalk that links F_1 to F_0. The X-ray structure of bovine F_1F_0-ATPase that was crystallized under different conditions than those used to determine the structure in Fig. 17-22 reveals that the δ and ε subunits are wrapped around the base of the γ subunit's coiled coil. (Note that, in an unfortunate confusion of nomenclature, the *E. coli* ε subunit is the homolog of the mitochondrial δ subunit, the *E. coli* δ subunit is the homolog of the mitochondrial subunit known as **OSCP,** and the mitochondrial ε subunit has no counterpart in either bacterial or chloroplast ATP synthases.)

The cyclic arrangement and structural similarities of F_1's α and β subunits give it both pseudo-threefold and pseudo-sixfold rotational symmetry (Fig. 17-22c). Nevertheless, the protein is asymmetric due to the presence of the γ subunit but, more importantly, because each pair of α and β subunits adopts a different conformation, each with a different substrate affinity. Thus one β subunit (designated β_{TP} in Fig. 17-22a) binds a molecule of a nonhydrolyzable ATP analog, the second (β_{DP}) binds ADP, and the third (β_E) has an empty and distorted binding site. Only the β subunits catalyze the ATP synthesis reaction although the α subunits also bind ATP.

The F₀ Component. The F_0 component of bacterial and mitochondrial F_1F_0-ATPases consists of multiple subunits. In *E. coli*, three transmembrane subunits—*a*, *b*, and *c*—form an $a_1b_2c_{9-12}$ complex. Mitochondrial F_0 contains additional subunits whose functions are unclear. A variety of evidence indicates that the *c* subunits associate to form a ring that is embedded in the membrane.

The NMR structure of the 79-residue *E. coli* *c* subunit consists of two α helices of different lengths that are connected by a four-residue polar loop and that are arranged in a banana-shaped antiparallel coiled coil (Fig. 17-23). The conformation of this assembly varies with pH.

The low-resolution X-ray structure of yeast F_1 in complex with its *c*-ring oligomer (Fig. 17-24) reveals that its α and β subunits have conformations and bound nucleotides similar to those in bovine F_1 (Fig. 17-22). The yeast

Figure 17-22   X-Ray structure of F₁-ATPase from bovine heart mitochondria.   (*a*) A ribbon diagram in which the α, β, γ subunits are red, yellow, and blue, respectively. The inset drawing indicates the orientation of these subunits in this view. The bar is 20 Å long. (*b*) Cross section through the electron density map of the protein (the α and β subunits are blue, and the γ subunit is orange). The superimposed C_α backbones of these subunits are yellow, and a bound ATP analog is represented in space-filling form (C yellow, N blue, O red). (*c*) Pseudosymmetrical arrangement of the $\alpha_3\beta_3$ assembly as viewed from the top of Parts *a* and *b*. The surface is colored according to its electrical potential, with positive potentials blue, negative potentials red, and neutral potentials white. Note the absence of charge on the inner surface of this sleeve. The portion of the γ subunit's C-terminal helix that contacts the sleeve is similarly devoid of charge. [From Abrahams, J.P., Leslie, A.G.W., Lutter, R., and Walker, J.E., *Nature* **370,** 623 and 627 (1994). PDBid 1BMF.] 🖱 **See the Interactive Exercises.**

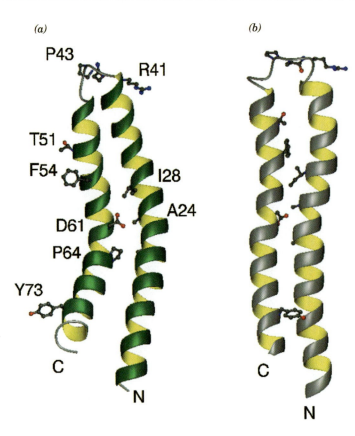

(a)

P43
R41
T51
F54
I28
D61
A24
P64
Y73
C
N

(b)

C
N

Figure 17-23 NMR structure of the c subunit of E. coli F₁F₀-ATPase. (a) At pH 8, Asp 61 (D61) is deprotonated. (b) At pH 5, D61 is protonated. Selected side chains are shown to aid in the comparison of the two structures. Note that the C-terminal helix in the pH 8 structure is rotated 140° clockwise, as viewed from the top of the drawing, relative to that in the pH 5 structure. [Courtesy of Mark Girvin, Albert Einstein College of Medicine. PDBid 1COV.]

c oligomer consists of 10 subunits (a number which may differ from that in E. coli) that each resemble the NMR structure of the isolated E. coli subunit (Fig. 17-23). The c subunits associate side by side so as to form two concentric rings of α helices. The N-terminal domain of the bovine δ subunit (the homolog of the E. coli ε subunit) fits into a region of the yeast F₁ electron density, where it contacts both the base of the γ subunit and the c-ring. This provides a footlike interface between F₀ and F₁ such that about two-thirds of the top surface of the c-ring contacts the base of the F₁ stalk (Fig. 17-24).

The sequence of the a subunit suggests that this highly hydrophobic 271-residue protein forms five transmembrane helices. The 156-residue b subunit consists of a single transmembrane helix anchoring a polar domain that homodimerizes to form a parallel α helical coiled coil that extends from the pe-

Figure 17-24 X-Ray structure of the yeast mitochondrial F₁–c₁₀ complex. This low-resolution (3.9 Å) electron density map (*pink*) shows the complex as viewed from within the inner mitochondrial membrane with the matrix above. The Cα backbone of bovine F₁ (with α orange, β yellow, and γ green) is superimposed on the electron density map. The inset indicates the location of the subunits of the complex, with the dashed lines indicating the presumed position of the inner mitochondrial membrane (M) and with the c subunits numbered. [Courtesy of Andrew Leslie and John Walker, Medical Research Council, Cambridge, U.K. PDBid 1Q01.]

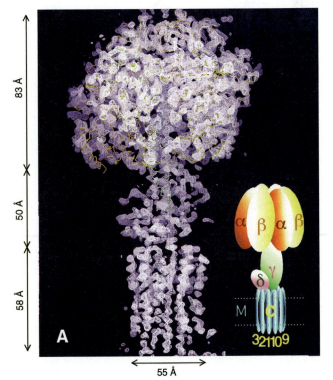

83 Å

50 Å

58 Å

α β α β

δ γ

M   c

3 2 1 10 9

A

55 Å

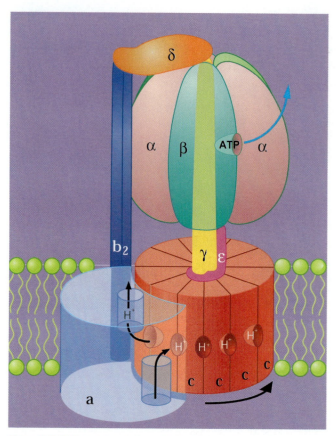

Figure 17-25   Model of the *E. coli* F₁F₀-ATPase.   The $\gamma\varepsilon-c_{12}$ ring complex is the rotor and the $ab_2-\alpha_3\beta_3\delta$ complex is the stator. Rotational motion is imparted to the rotor by the passage of protons from the outside (periplasmic space) to the inside (cytoplasm). Protons entering from the outside bind to a *c* subunit where it interacts with the *a* subunit, and exit to the inside after the *c*-ring has made a nearly full rotation as indicated (*black arrow*), so that the *c* subunit again contacts the *a* subunit. The $b_2\delta$ complex presumably functions to prevent the $\alpha_3\beta_3$ assembly from rotating with the γ subunit. [Courtesy of Richard Cross, State University of New York, Syracuse, New York.]

riphery of the *c*-ring into the matrix where it contacts the $\delta\alpha_3\beta_3$ assembly (the peripherally located connector in Fig. 17-21*b*). A model of the entire *E. coli* F₁F₀-ATPase is shown in Fig. 17-25.

The Binding Change Mechanism.   The mechanism of ATP synthesis by proton-translocating ATP synthase can be conceptually broken down into three phases:

1. Translocation of protons carried out by F₀.
2. Catalysis of formation of the phosphoanhydride bond of ATP carried out by F₁.
3. Coupling of the dissipation of the proton gradient with ATP synthesis, which requires interaction of F₁ and F₀.

Considerable evidence supports a mechanism for ATP formation proposed by Paul Boyer. According to this **binding change mechanism,** F₁ has three interacting catalytic protomers ($\alpha\beta$ units), each in a different conformational state: one that binds substrates and products loosely (L state),

one that binds them tightly (T state), and one that does not bind them at all (open or O state). *The free energy released on proton translocation is harnessed to interconvert these three states.* The phosphoanhydride bond of ATP is synthesized only in the T state, and ATP is released only in the O state. The reaction involves three steps (Fig. 17-26):

1. ADP and P_i bind to the loose (L) binding site (β_{DP} in Fig. 17-22*a*).

2. A free energy–driven conformational change converts the L site to a tight (T) binding site (β_{TP}) that catalyzes the formation of ATP. This step also involves conformational changes of the other two protomers that convert the ATP-containing T site to an open (O) site (β_E) and convert the O site to an L site.

3. ATP is synthesized at the T site on one subunit while ATP dissociates from the O site on another subunit. The reaction forming ATP is essentially at equilibrium under the conditions at the enzyme's active site. The free energy supplied by the proton flow primarily facilitates the release of the newly synthesized ATP from the enzyme; that is, it drives the T → O transition, thereby disrupting the enzyme–ATP interactions that had previously promoted the spontaneous formation of ATP from ADP + P_i in the T site.

How is the free energy of proton transfer coupled to the synthesis of ATP? The cyclic nature of the binding change mechanism led Boyer to propose that *the binding changes are driven by the rotation of the catalytic assembly, $\alpha_3\beta_3$, with respect to other portions of the F_1F_0-ATPase.* This hypothesis is supported by the X-ray structure of F_1. Thus, the closely fitting nearly circular arrangement of the α and β subunits' inner surface about the γ subunit's helical C-terminus is reminiscent of a cylindrical bearing rotating in a sleeve (Figs. 17-22*b,c*). Indeed, the contacting hydrophobic surfaces in this assembly are devoid of the hydrogen bonding and ionic interactions that would interface with their free rotation. (Fig. 17-22*c*); that is, the bearing and sleeve appear to be "lubricated." Moreover, the central cavity in the $\alpha_3\beta_3$ assembly (Fig. 17-22*b*) would permit the passage of the γ subunit's N-terminal helix within the core of this particle during rotation. Finally, the conformational differences between F_1's three catalytic sites appear to be correlated with the position of the γ subunit. *Apparently the γ subunit, which rotates within the fixed $\alpha_3\beta_3$ assembly, acts as a molecular cam shaft in linking the proton gradient–driven rotational motor to the conformational changes in the catalytic sites of F_1.*

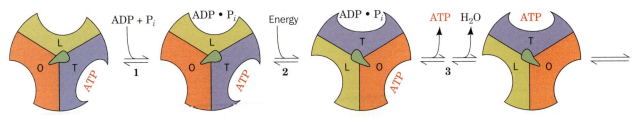

Figure 17-26   *Key to Function.*   The binding change mechanism for ATP synthase. F_1 has three chemically identical but conformationally distinct interacting $\alpha\beta$ protomers: O, the open conformation, has very low affinity for ligands and is catalytically inactive; L binds ligands loosely and is catalytically inactive; T binds ligands tightly and is catalytically active. ATP synthesis occurs in three steps: (1) ADP and P_i bind to site L. (2) An energy-dependent conformational change converts binding site L to T, T to O, and O to L. (3) ATP is synthesized at site T and ATP is released from site O. The enzyme returns to its initial state after two more passes of this reaction sequence. The free energy that drives the conformational change is transmitted to the catalytic $\alpha_3\beta_3$ assembly via the rotation of the $\gamma\varepsilon$ assembly (in *E. coli*; $\gamma\delta$ in mitochondria), here represented by the centrally located asymmetric object (*green*). [After Cross, R.L., *Annu. Rev. Biochem.* **50**, 687 (1980).] ✐ **See the Animated Figures.**

The F₁F₀-ATPase Is a Rotary Engine. The proposed rotation of the $\alpha_3\beta_3$ assembly with respect to the γ subunit engendered by the binding change mechanism has led to the model of the F₁F₀-ATPase diagrammed in Fig. 17-25. A rotational engine must have both a rotor (which rotates) and a stator (which is stationary). In the F₁F₀-ATPase, the rotor is proposed to be an assembly of the c-ring with the γ and (*E. coli*) ε subunits, whereas the ab_2 unit and the (*E. coli*) δ subunit together with the $\alpha_3\beta_3$ spheroid form the stator. The rotation of the c-ring in the membrane relative to the stationary a subunit is driven by the migration of protons from the outside to the inside as we discuss below (here "outside" refers to the mitochondrial intermembrane space or the bacterial exterior, whereas "inside" refers to the mitochondrial matrix or the bacterial cytoplasm). The $b_2\delta$ assembly presumably functions to hold the $\alpha_3\beta_3$ spheroid in position while the γ subunit rotates inside it.

In the model for proton-driven rotation of the F₀ subunit that is diagrammed in Fig. 17-25, protons from the outside enter a hydrophilic channel between the a subunit and the c-ring, where they bind to a c subunit. The c-ring then rotates nearly a full turn (while protons bind to successive c subunits as they pass this input channel) until the subunit reaches a second hydrophilic channel between the a subunit and the c-ring that opens into the inside, where the proton is released (in an alternative model, the protons are released through putative channels between the C-terminal helices of adjacent c-subunits, channels that are occluded when the a and c subunits are in contact). Thus, the F₁F₀-ATPase, which generates 3 ATP per turn and (at least in yeast) has 10 c subunits in its F₀ assembly, ideally forms $3/10 = 0.3$ ATP for every proton it passes from outside to inside. But how does the passage of protons through this system induce the rotation of the c-ring and hence the synthesis of ATP? Protons most likely bind to Asp 61 of each c subunit, an invariant residue whose protonation and deprotonation alter the subunit's conformation (Fig. 17-23). Evidently, when a c subunit binds a proton as it passes the input channel, its conformation changes, which causes it to mechanically push against the a subunit so as to induce the c-ring to rotate in the direction indicated in Fig. 17-25. This process is augmented by the interaction between Asp 61 on the c subunit and the invariant Arg 210 on the a subunit. These two residues have been shown to become juxtaposed at some point during the c-ring's rotation cycle. It has therefore been proposed that the electrostatic attraction between the cationic Arg 210 and the anionic Asp 61 helps rotate the c-ring so as to bring these two residues into opposition but, as this occurs, Asp 61 becomes protonated, thereby permitting the c-ring to continue its rotation.

The rotation of the $\gamma\varepsilon$–c-ring rotor with respect to the ab_2–$\alpha_3\beta_3\delta$ stator has been ingeniously demonstrated by Masamitsu Futai (Fig. 17-27a). The $\alpha_3\beta_3$ spheroid of *E. coli* F₁F₀-ATPase was fixed, head down, to a glass surface as follows. Six consecutive His residues (a so-called **His tag**) were mutagenically appended to the N-terminus of the α subunit, which is located

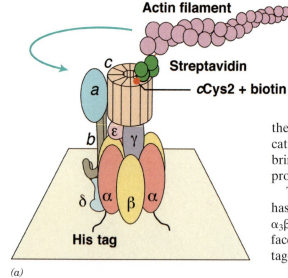

Actin filament

c

a

Streptavidin

cCys2 + biotin

b

ε γ

δ α β α

His tag

(*a*)

(*b*)

Figure 17-27   The rotation of the c-ring in *E. coli* F₁F₀-ATPase. (*a*) The experimental system used to observe the rotation. See the text for details. The blue arrow indicates the observed direction of rotation of the fluorescently labeled actin filament that was linked to the c-ring. (*b*) The rotation of a 3.6-µm-long actin filament in the presence of 5 mM MgATP as seen in successive video images taken through a fluorescence microscope. [Courtesy of Masamitsu Futai, Osaka University, Osaka, Japan.]

at the top of the $\alpha_3\beta_3$ spheroid as it is drawn in Fig. 17-22a. The His-tagged assembly was applied to a glass surface coated with horseradish peroxidase (which, like most proteins, sticks to glass) conjugated with **Ni^{2+}-nitriloacetic acid** [Ni^{2+}-N(CH$_2$COOH)$_3$, which tightly binds His tags], thereby binding the F$_1$F$_0$-ATPase to the surface with its F$_0$ side facing away from the surface. The Glu 2 residues of this assembly's c subunits, which are located on the side of the c-ring facing away from F$_1$, had been mutagenically replaced by Cys residues, which were then covalently linked to biotin (Section 15-4A). A fluorescently labeled and biotinylated (at one end) filament of the muscle protein **actin** (Section 28-1A) was then attached to the c subunit through the addition of a bridging molecule of **streptavidin,** a bacterial protein that avidly binds biotin to each of four binding sites.

The *E. coli* F$_1$F$_0$-ATPase can work in reverse, that is, it can pump protons from the inside to the outside at the expense of ATP hydrolysis (this enables the bacterium to maintain its proton gradient under anaerobic conditions, which it uses to drive various processes). Thus, the foregoing preparation was observed under a fluorescence microscope as a 5 mM MgATP solution was infused over it. *Many of the actin filaments were seen to rotate (Fig. 17-27b), and always in a counterclockwise direction when viewed looking down on the glass surface (from the outside).* This would permit the γ subunit to sequentially interact with the β subunits in the direction

$$\beta_E(\text{O-state}) \rightarrow \beta_{DP}(\text{L-state}) \rightarrow \beta_{TP}(\text{T-state})$$

(Figs. 17-22a and 17-26), the direction expected for ATP hydrolysis. Similar experiments revealed that the γ subunit rotates mostly in increments of 120°. Presumably, electrostatic interactions between the γ and β subunits of F$_1$ act as a catch that temporarily holds the γ subunit in place. As the c-ring rotates, strain builds up and causes the γ subunit to snap to the next β subunit.

C The P/O Ratio

ATP synthesis is tightly coupled to the proton gradient; that is, ATP synthesis requires the discharge of the proton gradient, and the proton gradient cannot be discharged without the synthesis of ATP. The proton gradient is established through the activity of the electron-transporting complexes of the inner mitochondrial membrane. Therefore, it is possible to express the amount of ATP synthesized in terms of substrate molecules oxidized. Experiments with isolated mitochondria show that the oxidation of NADH is associated with the synthesis of approximately 3 ATP, and the oxidation of FADH$_2$ with approximately 2 ATP. Oxidation of the nonphysiological compound **tetramethyl-*p*-phenylenediamine,**

Tetramethyl-*p*-phenylenediamine

which donates an electron pair directly to Complex IV, yields approximately 1 ATP. These stoichiometric relationships are called **P/O ratios** because they relate the amount of ATP synthesized (P) to the amount of oxygen reduced (O).

Experimentally determined P/O ratios are compatible with the chemiosmotic theory and the known structure of ATP synthase. The flow of two

electrons through Complexes I, III, and IV results in the translocation of 10 protons into the intermembrane space (Fig. 17-8). Influx of these 10 protons through the F_1F_0-ATPase is sufficient to drive the synthesis of ~3 ATP. Because eukaryotic F_0 appears to contain 10 c subunits (Fig. 17-24), 10 proton translocations result in one complete rotation of the c-ring–γ subunit rotor relative to the $\alpha_3\beta_3$ spheroid. Electrons that enter the electron-transport chain as $FADH_2$ at Complex II bypass Complex I and therefore lead to the transmembrane movement of only 6 protons, enough to synthesize ~2 ATP (corresponding to approximately two-thirds of a full rotation of the ATP synthase rotary engine). The transit of two electrons through Complex IV alone contributes 2 protons to the gradient, enough for ~1 ATP (around one-third of a rotation).

In actively respiring mitochondria, *P/O ratios are almost certainly not integral numbers.* This is also consistent with the chemiosmotic theory. Oxidation of a physiological substrate contributes to the transmembrane proton gradient at several points, but the gradient is tapped at only a single point, the F_1F_0-ATPase. Therefore, the number of protons translocated out of the mitochondrion by any component of the electron-transport chain need not be an integral multiple of the number of protons required to synthesize ATP from ADP + P_i. Moreover, the proton gradient is dissipated to some extent by the nonspecific leakage of protons back into the matrix and by the consumption of protons for other purposes, such as the transport of P_i into the matrix (Section 17-1B). Taking P_i transport into account gives a stoichiometry of four protons consumed per ATP synthesized from ADP + P_i.

Peter Hinkle's re-examination of P/O ratios suggests that the values given above are actually closer to 2.5, 1.5, and 1. If these values are correct, then the number of ATPs that are synthesized per molecule of glucose oxidized is 2.5 ATP/NADH $\times$ 10 NADH/glucose + 1.5 ATP/$FADH_2$ $\times$ 2 $FADH_2$/glucose + 2 ATP/glucose from the citric acid cycle + 2 ATP/glucose from glycolysis = 32 ATP/glucose rather than the conventional value of 38 implied by P/O ratios of 3, 2, and 1. For the sake of consistency, we shall use the value of 38 ATP/glucose throughout this textbook, but keep in mind that this value is disputed.

D   Uncoupling Oxidative Phosphorylation

Electron transport (the oxidation of NADH and $FADH_2$ by O_2) and oxidative phosphorylation (the proton gradient–driven synthesis of ATP) are normally tightly coupled. *This coupling depends on the impermeability of the inner mitochondrial membrane, which allows an electrochemical gradient to be established across this membrane by H^+ translocation during electron transport.* Virtually the only way for H^+ to re-enter the matrix is through the F_0 portion of ATP synthase. In the resting state, when oxidative phosphorylation is minimal, the electrochemical gradient across the inner mitochondrial membrane builds up to the extent that it prevents further proton pumping and therefore inhibits electron transport. When ATP synthesis increases, the electrochemical gradient dissipates, allowing electron transport to resume.

Over the years, compounds such as **2,4-dinitrophenol (DNP)** have been found to "uncouple" electron transport and ATP synthesis. DNP is a lipophilic weak acid that readily passes through membranes in its neutral, protonated state. In a pH gradient, it binds protons on the acidic side of the membrane, diffuses through the membrane, and releases the protons on the membrane's alkaline side, thereby acting as a proton-transporting ionophore (Section 10-2A) and dissipating the gradient (Fig. 17-28). The

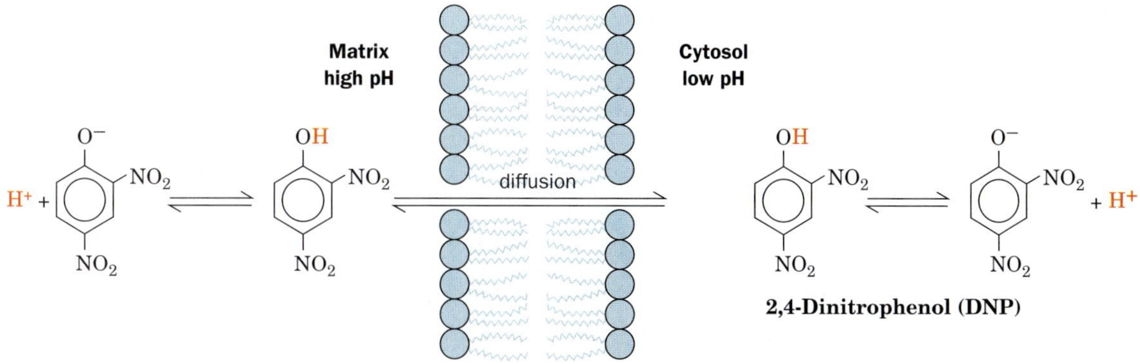

Figure 17-28 Action of 2,4-dinitrophenol. A proton-transporting ionophore such as DNP uncouples oxidative phosphorylation from electron transport by discharging the electrochemical proton gradient generated by electron transport.

chemiosmotic theory provides a rationale for understanding the action of such **uncouplers.** *The presence in the inner mitochondrial membrane of an agent that increases its permeability to H^+ uncouples oxidative phosphorylation from electron transport by providing a route for the dissipation of the proton electrochemical gradient that does not require ATP synthesis.* Uncoupling therefore allows electron transport to proceed unchecked even when ATP synthesis is inhibited. Consequently, in the 1920s, DNP was used as a "diet pill," a practice that was effective in inducing weight loss but often caused fatal side effects. Under physiological conditions, *the dissipation of an electrochemical H^+ gradient, which is generated by electron transport and uncoupled from ATP synthesis, produces heat* (see Box 17-4).

4 Control of Oxidative Metabolism

An adult woman requires some 1500 to 1800 kcal (6300–7500 kJ) of metabolic energy per day. This corresponds to the free energy of hydrolysis of over 200 mol of ATP to ADP and P_i. Yet the total amount of ATP present in the body at any one time is <0.1 mol; obviously, this sparse supply of ATP must be continually recycled. As we have seen, when carbohydrates serve as the energy supply and aerobic conditions prevail, this recycling involves glycogenolysis, glycolysis, the citric acid cycle, and oxidative phosphorylation.

Of course, the need for ATP is not constant. There is a 100-fold change in the rate of ATP consumption between sleep and vigorous activity. *The activities of the pathways that produce ATP are under strict coordinated control so that ATP is never produced more rapidly than necessary.* We have already discussed the control mechanisms of glycolysis, and the citric acid cycle (Sections 14-4, 15-3, and 16-4). In this section, we discuss the mechanisms that control the rate of oxidative phosphorylation.

A Control of Oxidative Phosphorylation

In our discussions of metabolic pathways, we have seen that most of their reactions function close to equilibrium. The few irreversible reactions constitute potential control points of the pathways and usually are catalyzed by regulatory enzymes that are under allosteric control. In the case of ox-

BOX 17-4

Perspectives in Biochemistry

Uncoupling in Brown Adipose Tissue Generates Heat

Heat generation is the physiological function of **brown adipose tissue (brown fat).** This tissue is unlike typical (white) adipose tissue in that it contains numerous mitochondria whose cytochromes cause its brown color. Newborn mammals that lack fur, such as humans, as well as hibernating mammals, all contain brown fat in their neck and upper back that generates heat by **nonshivering thermogenesis.** Other sources of heat are the ATP hydrolysis that occurs during muscle contraction (in shivering or any other

[© K.M. Highfill/Photo Researchers.]

movement) and the operation of ATP-hydrolyzing substrate cycles (see Section 14-4B).

The mechanism of heat generation in brown fat involves the regulated uncoupling of oxidative phosphorylation. Brown fat mitochondria contain a proton channel known as **uncoupling protein (UCP1;** also called **thermogenin;** see below). In cold-adapted animals, UCP1 constitutes up to 15% of the protein in the inner mitochondrial membranes of brown fat. The flow of protons through UCP1 is in-

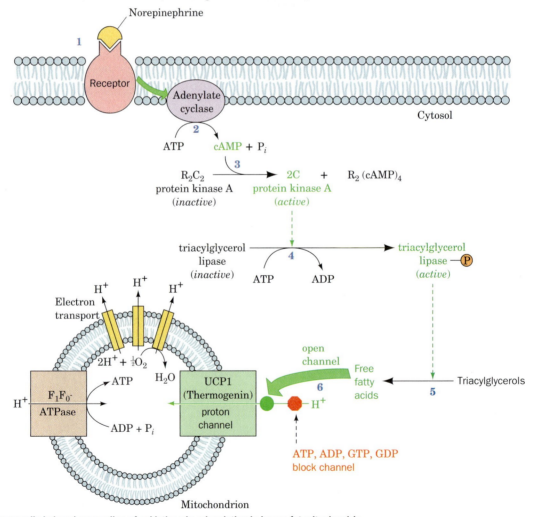

Mechanism of hormonally induced uncoupling of oxidative phosphorylation in brown fat mitochondria.

hibited by physiological concentrations of purine nucleotides (ADP, ATP, GDP, GTP), but this inhibition can be overcome by free fatty acids.

Thermogenesis in brown fat mitochondria is under hormonal control (see diagram). Norepinephrine (**1**; noradrenaline) induces the production of the second messenger cAMP (**2**) and thereby activates protein kinase A (**3**; Section 15-3B). The kinase then activates **hormone-sensitive triacylglycerol lipase** (**4**) by phosphorylating it. The activated lipase hydrolyzes triacylglycerols (**5**) to yield free fatty acids that counteract the inhibitory effect of the purine nucleotides on UCP1 (**6**). The resulting flow of protons through UCP1 dissipates the proton gradient across the inner mitochondrial membrane. This allows substrate oxidation to proceed (and generate heat) without the synthesis of ATP.

Adult humans lack brown fat, but the mitochondria of ordinary adipose tissue and muscle appear to contain uncoupling proteins known as **UCP2** and **UCP3**. These proteins may help regulate metabolic rates, and variations in UCP levels or activity might explain why some people seem to have a "fast" or "slow" metabolism. UCPs are being studied as targets for treating obesity, since increasing the activity of UCPs could uncouple respiration from ATP synthesis, thus permitting stored metabolic fuels (especially fat) to be metabolized.

Uncoupling proteins may also play a role in maintaining body temperature, and their function may not be limited to the animal kingdom. Some plants express uncoupling proteins in response to cold stress or to increase flower temperature, possibly to enhance the vaporization of scent to attract pollinators.

idative phosphorylation, the pathway from NADH to cytochrome c functions near equilibrium:

$$\tfrac{1}{2} \text{NADH} + \text{cytochrome } c(\text{Fe}^{3+}) + \text{ADP} + \text{P}_i \Longrightarrow$$
$$\tfrac{1}{2} \text{NAD}^+ + \text{cytochrome } c(\text{Fe}^{2+}) + \text{ATP} \qquad \Delta G' \approx 0$$

and hence

$$K_{eq} = \left(\frac{[\text{NAD}^+]}{[\text{NADH}]}\right)^{\!\!\frac{1}{2}} \frac{[c^{2+}]}{[c^{3+}]} \frac{[\text{ATP}]}{[\text{ADP}][\text{P}_i]}$$

This pathway is therefore readily reversed by the addition of its product, ATP. However, *the cytochrome c oxidase reaction (the terminal step of the electron-transport chain) is irreversible and is therefore a potential control site.* Cytochrome c oxidase, in contrast to most regulatory enzyme systems, appears to be controlled primarily by the availability of one of its substrates, reduced cytochrome c (c^{2+}). Since c^{2+} is in equilibrium with the rest of the coupled oxidative phosphorylation system, the concentration of c^{2+} ultimately depends on the intramitochondrial ratios of [NADH]/[NAD$^+$] and [ATP]/[ADP][P$_i$] (this latter quantity is known as the **ATP mass action ratio**). We can see by rearranging the foregoing equilibrium expression:

$$\frac{[c^{2+}]}{[c^{3+}]} = \left(\frac{[\text{NADH}]}{[\text{NAD}^+]}\right)^{\!\!\frac{1}{2}} \frac{[\text{ADP}][\text{P}_i]}{[\text{ATP}]} K_{eq}$$

The higher the [NADH]/[NAD$^+$] ratio and the lower the ATP mass action ratio, the higher the concentration of reduced cytochrome c and thus the higher the cytochrome c oxidase activity.

How is this system affected by changes in physical activity? In an individual at rest, the rate of ATP hydrolysis to ADP and P$_i$ is minimal and the ATP mass action ratio is high; the concentration of reduced cytochrome c is therefore low and the rate of oxidative phosphorylation is minimal.

Increased activity results in hydrolysis of ATP to ADP and P_i, thereby decreasing the ATP mass action ratio and increasing the concentration of reduced cytochrome c. This results in an increase in the rate of electron transport and ADP phosphorylation.

The concentrations of ATP, ADP, and P_i in the mitochondrial matrix depend on the activities of the transport proteins that import these substances from the cytosol. Thus, the ADP–ATP translocator and the P_i transporter may play a part in regulating oxidative phosphorylation. There is also some evidence that Ca^{2+} stimulates the electron-transport complexes and possibly ATP synthase itself. This is consistent with the many other instances in which Ca^{2+} directly stimulates oxidative metabolic processes.

Mitochondria contain an 84-residue protein, called **IF_1** that functions to regulate ATP synthase. In actively respiring mitochondria, in which the matrix pH is relatively high, IF_1 exists as an inactive tetramer. However, below pH 6.5, the protein dissociates into dimers and in this form inhibits the ATPase activity of the F_1 component by binding to the interface between its α_{DP} and β_{DP} subunits so as to trap the ATP bound to the β_{DP} subunit. This appears to be a mechanism to prevent ATP hydrolysis when respiratory activity (and therefore the proton gradient) is temporarily interrupted by lack of O_2. Otherwise, the F_1F_0-ATPase would reverse its direction of rotation as driven by the hydrolysis of ATP (then generated by glycolysis), thus depriving the cell of its remaining energy resources.

B   Coordinated Control of Oxidative Metabolism

The primary sources of the electrons that enter the mitochondrial electron-transport chain are glycolysis, fatty acid degradation, and the citric acid cycle. For example, 10 molecules of NAD^+ are converted to NADH per molecule of glucose oxidized (Fig. 17-1). Not surprisingly, the control of glycolysis and the citric acid cycle is coordinated with the demand for oxidative phosphorylation. An adequate supply of electrons to feed the electron-transport chain is provided by regulation of the control points of glycolysis and the citric acid cycle (phosphofructokinase, pyruvate dehydrogenase, citrate synthase, isocitrate dehydrogenase, and α-ketoglutarate dehydrogenase) by adenine nucleotides or NADH or both, as well as by certain metabolites (Fig. 17-29).

One particularly interesting regulatory effect is the inhibition of phosphofructokinase (PFK) by citrate. When demand for ATP decreases, [ATP] increases and [ADP] decreases. Because isocitrate dehydrogenase is activated by ADP and α-ketoglutarate dehydrogenase is inhibited by ATP, the citric acid cycle slows down. This causes the citrate concentration to build up. Citrate leaves the mitochondrion via a specific transport system and, *once in the cytosol, acts to restrain further carbohydrate breakdown by inhibiting PFK.*

C   Physiological Implications of Aerobic Metabolism

Not all organisms carry out oxidative phosphorylation. However, those that do are able to extract considerably more energy from a given amount of a metabolic fuel. This principle is illustrated by the Pasteur effect (Section 14-3C): When anaerobically growing yeast are exposed to oxygen, their glucose consumption drops precipitously. An analogous effect is observed in mammalian muscle; the concentration of lactic acid (the anaerobic product of muscle glycolysis; Section 14-3A) drops dramatically when cells switch to aerobic metabolism. These effects are easily understood by ex-

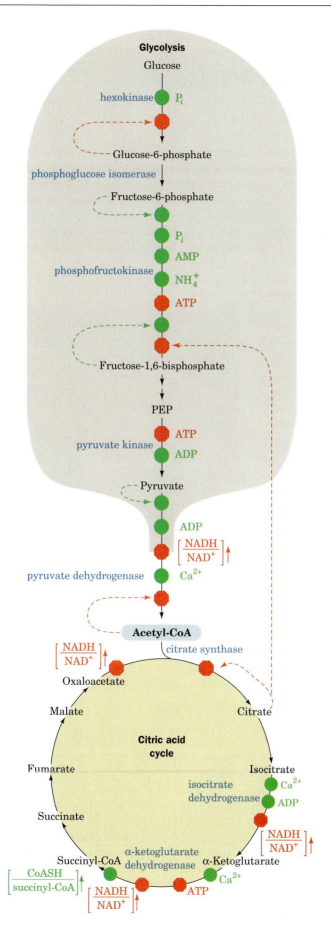

Figure 17-29 The coordinated control of glycolysis and the citric acid cycle. The diagram shows the effects of ATP, ADP, AMP, P$_i$, Ca^{2+}, and the [NADH]/[NAD$^+$] ratio (the vertical arrows indicate increases in this ratio). Here a green dot signifies activation and a red octagon represents inhibition. [After Newsholme, E.A. and Leech, *A.R.*, *Biochemistry for the Medical Sciences*, pp. 316, 320, Wiley (1983).] **See the Animated Figures.**

BOX 17-5

Biochemistry in Health and Disease

Oxygen Deprivation in Heart Attack and Stroke

As outlined in Boxes 2-1 and 7-1, organisms larger than 1 mm thick require circulatory systems to deliver nutrients to cells and dispose of cellular wastes. In addition, the circulatory fluid in most larger organisms contains proteins specialized for oxygen transport (e.g., hemoglobin). The sophistication of oxygen-delivery systems and their elaborate regulation are consistent with their essential nature and their long period of evolution.

What happens during oxygen deprivation? Consider two common causes of human death, **myocardial infarction** (heart attack) and **stroke,** which result from interruption of the blood (O_2) supply to a portion of the heart or the brain, respectively. In the absence of O_2, a cell, which must then rely only on glycolysis for ATP production, rapidly depletes its stores of phosphocreatine (a source of rapid ATP production; Section 13-2C) and glycogen. As the rate of ATP production falls below the level required by membrane ion pumps for maintaining proper intracellular ion concentrations, osmotic balance is disrupted so that the cell and its membrane-enveloped organelles begin to swell. The resulting over-stretched membranes become permeable, thereby leaking their enclosed contents. For this reason, a useful diagnostic criterion for myocardial infarction is the presence in the blood of heart-specific enzymes, such as the H-type isozyme of lactate dehydrogenase (Section 13-1B), which leak out of necrotic (dead) heart tissue. Moreover, the decreased intra-cellular pH that accompanies anaerobic glycolysis (because of lactic acid production) permits the released lysosomal enzymes (which are active only at acidic pHs) to degrade the cell contents. Thus, O_2 deprivation leads not only to cessation of cellular activity but to irreversible cell damage and cell death. Rapidly respiring tissues, such as the heart and brain, are particularly susceptible to damage by oxygen deprivation.

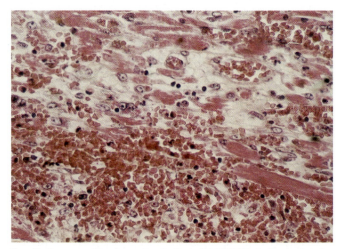

Necrotic tissue resulting from a heart attack. [© CNRI//Photo Researchers.]

amining the stoichiometries of anaerobic and aerobic breakdown of glucose (Section 16-4):

Anaerobic glycolysis:

$$C_6H_{12}O_6 + 2\ ADP + 2\ P_i \rightarrow 2\ lactate + 2\ H^+ + 2\ H_2O + 2\ ATP$$

Aerobic metabolism of glucose:

$$C_6H_{12}O_6 + 38\ ADP + 38\ P_i + 6\ O_2 \rightarrow 6\ CO_2 + 44\ H_2O + 38\ ATP$$

Thus, *aerobic metabolism is up to 19 times more efficient than anaerobic glycolysis in producing ATP.*

Aerobic metabolism has its drawbacks, however. Many organisms and tissues depend exclusively on aerobic metabolism and suffer irreversible damage during oxygen deprivation (see Box 17-5). Oxidative metabolism is also accompanied by the production of low levels of reactive oxygen metabolites that, over time, may damage cellular components. Evidently, the organisms that have existed during the last 3 billion years (the period in which the earth's atmosphere has contained significant amounts of O_2) exhibit physiological and biochemical adaptations that permit them to take

advantage of the oxidizing power of O_2 while minimizing the potential dangers of oxygen itself.

Reactive Oxygen Species (ROS).   Although the four-electron reduction of O_2 by cytochrome c oxidase is nearly always orchestrated with great rapidity and precision, *O_2 is occasionally only partially reduced, yielding oxygen species that readily react with a variety of cellular components.* The best known reactive oxygen species is the **superoxide radical:**

$$O_2 + e^- \rightarrow O_2^- \cdot$$

Superoxide radical is a precursor of other reactive species. Protonation of $O_2^- \cdot$ yields $HO_2 \cdot$, a much stronger oxidant than $O_2^- \cdot$. The most potent oxygen species in biological systems is probably the hydroxyl radical, which forms from the relatively harmless hydrogen peroxide (H_2O_2):

$$H_2O_2 + Fe^{2+} \rightarrow \cdot OH + OH^- + Fe^{3+}$$

The hydroxyl radical also forms through the reaction of superoxide with H_2O_2:

$$O_2^- \cdot + H_2O_2 \rightarrow O_2 + H_2O + \cdot OH$$

Although most free radicals are extremely short-lived (the half-life of $O_2^- \cdot$ is 1×10^{-6} s, and that of $\cdot OH$ is 1×10^{-9} s), they readily extract electrons from other molecules, converting them to free radicals and thereby initiating a chain reaction.

The random nature of free-radical attacks makes it difficult to characterize their reaction products, but all classes of biological molecules are susceptible to oxidative damage caused by free radicals. The oxidation of polyunsaturated lipids in cells may disrupt the structures of biological membranes, and oxidative damage to DNA may result in point mutations. Enzyme function may also be compromised through radical reactions with amino acid side chains. Because the mitochondrion is the site of the bulk of the cell's oxidative metabolism, its lipids, DNA, and proteins probably bear the brunt of free radical–related damage.

Several degenerative diseases, including **Parkinson's, Alzheimer's,** and **Huntington's diseases,** are associated with oxidative damage to mitochondria. Such observations have led to the free-radical theory of aging, which holds that *free-radical reactions arising during the course of normal oxidative metabolism are at least partially responsible for the aging process.* In fact, individuals with congenital defects in their mitochondrial DNA suffer from a variety of symptoms typical of old age, including neuromotor difficulties, deafness, and dementia. Their genetic defects may make their mitochondria all the more susceptible to the reactive oxygen species generated by the electron-transport machinery.

Antioxidant Mechanisms.   **Antioxidants** destroy oxidative free radicals such as $O_2^- \cdot$ and $\cdot OH$. In 1969, Irwin Fridovich discovered that the enzyme **superoxide dismutase (SOD),** which is present in nearly all cells, catalyzes the conversion of $O_2^- \cdot$ to H_2O_2.

$$2\, O_2^- \cdot + 2\, H^+ \rightarrow H_2O_2 + O_2$$

Mitochondrial and bacterial SOD are both Mn-containing tetramers; eukaryotic cytosolic SOD is a dimer containing copper and zinc ions. The rate of nonenzymatic superoxide breakdown is $\sim 2 \times 10^5\, M^{-1} \cdot s^{-1}$, whereas the rate of the Cu,Zn-SOD–catalyzed reaction is $\sim 2 \times 10^9\, M^{-1} \cdot s^{-1}$. This

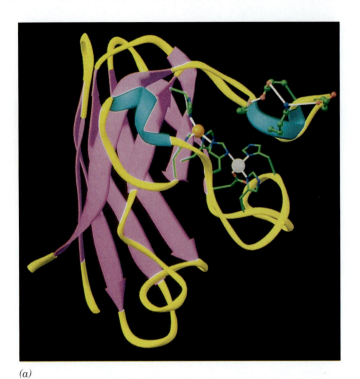

(a)

(b)

Figure 17-30 Electrostatic effects in human Cu,Zn-superoxide dismutase. (*a*) The X-ray structure of a subunit of the dimeric enzyme. The polypeptide backbone is colored according to its secondary structure, the Cu and Zn ions are represented by orange and silver spheres, the side chains that ligand these metal ions are drawn in stick form, and the side chains forming the hydrogen-bonded network at the entrance to the active site pocket are shown in ball-and-stick form. Metal–ligand bonds and hydrogen bonds are represented by thin white lines. [Based on an X-ray structure by John Tainer, The Scripps Research Institute, La Jolla, California. PDBid 1SOS.] (*b*) Cross section of the active site channel of Cu,Zn-SOD. Its molecular surface is represented by a dot surface that is colored according to charge: red, most negative; yellow, negative; green, neutral; light blue, positive; dark blue, most positive. The electrostatic field vectors are represented by similarly colored arrows. The O_2^{-}·-binding site is located between the Cu ion and the side chain of Arg 143. [Courtesy of Elizabeth Getzoff, The Scripps Research Institute, La Jolla, California.]

rate enhancement, which is close to the diffusion-controlled limit (Section 12-1B), is apparently accomplished by electrostatic guidance of the negatively charged superoxide substrate into the enzyme's active site (Fig. 17-30). The active-site Cu ion lies at the bottom of a deep pocket in each enzyme subunit. A hydrogen-bonded network of Glu 123, Glu 133, Lys 136, and Thr 137 at the entrance to the pocket facilitates the diffusion of O_2^{-}· to a site between the Cu ion and an Arg residue.

SOD is considered a first-line defense against reactive oxygen species. The H_2O_2 produced in the reaction, which can potentially react to yield other reactive oxygen species, is degraded to water and oxygen by enzymes such as **catalase,** which catalyzes the reaction

$$2\ H_2O_2 \rightarrow 2\ H_2O + O_2$$

and glutathione peroxidase (Box 14-4), which uses glutathione (GSH) as the reducing agent:

$$2\ GSH + H_2O_2 \rightarrow GSSG + 2\ H_2O$$

The latter enzyme also catalyzes the breakdown of organic hydroperoxides. Some types of glutathione peroxidase require Se for activity; this is one reason why Se appears to have antioxidant activity.

Other potential antioxidants are plant-derived compounds such as ascorbate (vitamin C; Section 6-1C) and α-tocopherol (vitamin E; Section 9-1F).

These compounds may help protect plants from oxidative damage during photosynthesis, a process in which H_2O is oxidized to O_2. Their efficacy as antioxidants in humans, however, has not been proven.

 ## SUMMARY

1. Electrons from the reduced coenzymes NADH and $FADH_2$ pass through a series of redox centers in the electron-transport chain before reducing O_2. During electron transfer, protons are translocated out of the mitochondrion to form an electrochemical gradient whose free energy drives ATP synthesis.

2. The mitochondrion contains soluble and membrane-bound enzymes for oxidative metabolism. Reducing equivalents are imported from the cytosol via a shuttle system. Specific transporters mediate the transmembrane movements of ADP, ATP, and P_i.

3. Electrons flow from redox centers with more negative reduction potentials to those with more positive reduction potentials. Inhibitors have been used to reveal the sequence of electron carriers and the points of entry of electrons into the electron transport chain.

4. Electron transport is mediated by one-electron carriers (Fe–S clusters, cytochromes, and Cu ions) and two-electron carriers (CoQ, FMN, FAD).

5. Complex I transfers two electrons from NADH to CoQ while translocating four protons to the intermembrane space.

6. Complex II transfers electrons from succinate through FAD to CoQ.

7. Complex III transfers two electrons from $CoQH_2$ to two molecules of cytochrome c. The concomitant operation of the Q cycle translocates four protons to the intermembrane space.

8. Complex IV reduces O_2 to 2 H_2O using four electrons donated by four cytochrome c and four protons from the matrix. Two protons are translocated to the intermembrane space for every two electrons that reduce oxygen.

9. As explained by the chemiosmotic theory, protons translocated into the intermembrane space during electron transport through Complexes I, III, and IV establish an electrochemical gradient across the inner mitochondrial membrane.

10. The influx of protons through the F_0 component of ATP synthase (F_1F_0-ATPase) drives its F_1 component to synthesize ATP from ADP + P_i via the binding change mechanism, a process that is mechanically driven by the F_0-mediated rotation of F_1's γ subunit with respect to its catalytic $\alpha_3\beta_3$ assembly.

11. The P/O ratio, the number of ATPs synthesized per oxygen reduced, need not be an integral number.

12. Agents that discharge the proton gradient can uncouple oxidative phosphorylation from electron transport.

13. Oxidative phosphorylation is controlled by the ratio $[NADH]/[NAD^+]$ and by the ATP mass action ratio. Glycolysis and the citric acid cycle are coordinately regulated according to the need for oxidative phosphorylation.

14. Aerobic metabolism is more efficient than anaerobic metabolism. However, aerobic organisms must guard against damage caused by reactive oxygen species.

 ## REFERENCES

Beinert, H., Holm, R.H., and Münck, E., Iron-sulfur clusters: Nature's modular, multipurpose structures, *Science* **277**, 653–659 (1997).

Boyer, P.D., Catalytic site forms and controls in ATP synthase catalysis, *Biochim. Biophys. Acta* **1458**, 252–262 (2000). [A description of the steps of ATP synthesis and hydrolysis, along with experimental evidence and alternative explanations, by the author of the binding change mechanism.]

Crofts, A.R., The cytochrome bc_1 complex: Function in the context of structure, *Annu. Rev. Physiol.* **66**, 689–733 (2004).

Frey, T.G. and Mannella, C.A., The internal structure of mitochondria, *Trends Biochem. Sci.* **23**, 319–324 (2000).

Gennis, R.B., Multiple proton-conducting pathways in cytochrome oxidase and a proposed role for the active-site tyrosine, *Biochim. Biophys. Acta* **1458**, 241–248 (2000).

Kühlbrandt, W., Bacteriorhodopsin—the movie, *Nature* **406**, 569–570 (2000).

Lange, C. and Hunte, C., Crystal structure of the yeast cytochrome bc_1 complex with its bound substrate cytochrome c, *Proc. Natl. Acad. Sci.* **99**, 2800–2805 (2002).

Lanyi, J.K., Bacteriorhodopsin, *Annu. Rev. Physiol.* **66**, 665–688 (2004).

Nicholls, D.G. and Ferguson, S.J., *Bioenergetics 3*, Academic Press (2002). [An authoritative monograph devoted almost entirely to the mechanism of oxidative phosphorylation and the techniques used to elucidate it.]

Noji, H. and Yoshida, M., The rotary machine in the cell, ATP synthase, *J. Biol. Chem.* **276**, 1665–1668 (2001).

Pebay-Peyroula, E., Dahout-Gonzalez, C., Kahn, R., Trézéguet, V., Lauquin, G.J.-M., and Brandolin, G., Structure of mitochondrial ADP/ATP carrier in complex with carboxyatractyloside, *Nature* **426**, 39–44 (2003). [Reports the 2.2-Å-resolution structure of the bovine carrier monomer and proposes a mechanism in which conformational changes in each monomer of

the dimeric complex allow simultaneous transport of ADP into and ATP out of the matrix.]

Saraste, M., Oxidative phosphorylation at the *fin de siècle, Science* **283,** 1488–1493 (1999). [Provides a succinct overview of the entire process of electron transport and ATP synthesis.]

Schultz, B.E. and Chan, S.I., Structures and proton-pumping strategies of mitochondrial respiratory enzymes, *Annu. Rev. Biophys. Biomol. Struct.* **30,** 23–65 (2001).

Stock, D., Gibbons, C., Arechaga, I., Leslie, A.G.W., and Walker, J.E., The rotary mechanism of ATP synthase, *Curr. Opin. Struct. Biol.* **10,** 692–679 (2000).

Walker, J.E. (Ed.), *The Mechanism of F₁F₀-ATPase, Biochim. Biophys. Acta* **1458,** 221–514 (2002). [A series of authoritative reviews.]

Yankovskaya, V., Horsefield, R., Törnroth, S., Luna-Chavez, C., Miyoshi, H., Léger, C., Byrne, B., Cecchini, G., and Iwata, S., Architecture of succinate dehydrogenase and reactive oxygen species generation, *Science* **299,** 700–704 (2003). [X-Ray structure of *E. coli* Complex II.]

 ## KEY TERMS

redox center
cristae
intermembrane space
mitochondrial matrix
malate–aspartate shuttle
glycerophosphate shuttle

ADP–ATP translocator
electrogenic transport
coenzyme Q
iron–sulfur protein
free radical
proton wire

cytochrome
Q cycle
differential labeling
energy coupling
protonmotive force
F₁F₀-ATPase

P/O ratio
uncoupler
ATP mass action ratio
superoxide radical
antioxidant

 ## STUDY EXERCISES

1. Describe the route followed by electrons from glucose to O₂.

2. Describe how a pair of electrons from NADH is transferred to the one-electron carrier cytochrome *c*.

3. How does the mechanism of proton translocation in Complexes I and IV differ from that in Complex III?

4. Summarize the chemiosmotic theory.

5. Describe how protons move from the intermembrane space into the matrix. How is proton translocation linked to ATP synthesis?

6. Describe the binding change mechanism of F₁F₀-ATP synthase.

7. Explain why the P/O ratio for a given substrate is not necessarily an integer.

8. Explain how oxidative phosphorylation is linked to electron transport and how the two processes can be uncoupled.

9. What are the advantages and disadvantages of O₂-based metabolism?

 ## PROBLEMS

1. Explain why a liver cell mitochondrion contains fewer cristae than a mitochondrion from a heart muscle cell.

2. How many ATPs are synthesized for every cytoplasmic NADH that participates in the glycerophosphate shuttle in insect flight muscle? How does this compare to the ATP yield when NADH reducing equivalents are transferred into the matrix via the malate–aspartate shuttle?

3. Calculate ΔG°′ for the oxidation of free FADH₂ by O₂. What is the maximum number of ATPs that can be synthesized, assuming standard conditions and 100% conservation of energy?

4. The O₂-consumption curve of a dilute, well-buffered suspen-

sion of mitochondria containing an excess of ADP and Pᵢ takes the form

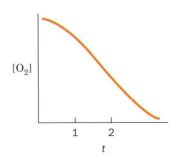

Sketch the curves obtained when (a) amytal is added at time $t = 1$; (b) amytal is added at $t = 1$ and succinate is added at $t = 2$; (c) CN^- is added at $t = 1$ and succinate is added at $t = 2$; (d) oligomycin (which binds to F_0 and prevents ATP synthesis) is added at $t = 1$ and DNP is added at $t = 2$.

5. Why is it possible for electrons to flow from a redox center with a more positive $\mathscr{E}°'$ to one with a more negative $\mathscr{E}°'$ within an electron-transfer complex?

6. Bombarding a suspension of mitochondria with high frequency sound waves **(sonication)** produces **submitochondrial particles** derived from the inner mitochondrial membrane. These membranous vesicles seal inside out, so that the intermembrane space of the mitochondrion becomes the lumen of the submitochondrial particle. (a) Diagram the process of electron transfer and oxidative phosphorylation in these particles. (b) Assuming all the substrates for oxidative phosphorylation are present in excess, does ATP synthesis increase or decrease with an increase in the pH of the fluid in which the submitochondrial particles are suspended?

7. The difference in pH between the internal and external surfaces of the inner mitochondrial membrane is 1.4 pH units (external side acidic). If the membrane potential is 0.06 V (inside negative) what is the free energy released on transporting 1 mol of protons back across the membrane? How many protons must be transported to provide enough free energy for the synthesis of 1 mol of ATP (assuming standard biochemical conditions)?

8. Consider the mitochondrial ADP–ATP translocator and the P_i–H^+ symport protein. (a) How do the activities of these two transporters affect the electrochemical gradient across the mitochondrial membrane? (b) What thermodynamic force drives the two transport systems?

9. Dicyclohexylcarbodiimide (DCCD) is a reagent that reacts with Asp or Glu residues.

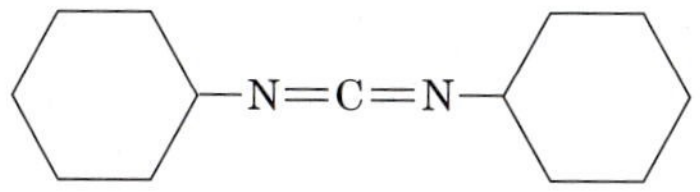

**Dicyclohexylcarbodiimide
(DCCD)**

Explain why the reaction of DCCD with the c subunits of F_1F_0-ATPase blocks its ATP-synthesizing activity.

10. How do the P/O ratios for NADH differ in ATP synthases that contain 9 and 12 c subunits?

11. Explain why compounds such as DNP increase metabolic rates.

12. What is the advantage of hormones activating a lipase to stimulate nonshivering thermogenesis in brown fat rather than activating UCP1 directly (see Box 17-4)?

13. Describe the changes in [NADH]/[NAD$^+$] and [ATP]/[ADP] that occur during the switch from anaerobic to aerobic metabolism. How do these ratios influence the activity of glycolysis and the citric acid cycle?

14. Activated neutrophils and macrophages (types of white blood cells) fight invading bacteria by releasing superoxide. These cells contain an **NADPH oxidase** that catalyzes the reaction

$$2\ O_2 + NADPH \rightarrow 2\ O_2^{-}\!\cdot + NADP^+ + H^+$$

Explain why flux through the glucose-6-phosphate dehydrogenase reaction increases in these cells.

Plants, such as these trees, can survive literally on water, air, and light. They oxidize H_2O to O_2 and convert CO_2 from air into carbohydrates in a process that is driven by light. [Michael Busselle/Stone/Getty Images.]

Photosynthesis

The notion that plants obtain nourishment from such insubstantial things as light and air was not validated until the eighteenth century. Evidence that plants produce a vital substance — O_2 — was not obtained until Joseph Priestly noted that the air in a jar in which a candle had burnt out could be "restored" by introducing a small plant into the jar. In the presence of sunlight, plants and cyanobacteria consume CO_2 and H_2O and produce O_2 and "fixed" carbon in the form of carbohydrate:

$$CO_2 + H_2O \xrightarrow{\text{light}} (CH_2O) + O_2$$

Photosynthesis, in which light energy drives the reduction of carbon, is essentially the reverse of oxidative carbohydrate metabolism. Photosynthetically produced carbohydrates therefore serve as an energy source for the organism that produces them as well as for nonphotosynthetic organisms that directly or indirectly consume photosynthetic organisms. It is estimated that photosynthesis annually fixes $\sim 10^{11}$ tons of carbon, which represents the storage of over 10^{18} kJ of energy. The process by which light energy is converted to chemical energy has its roots early in evolution, and its complexity is consistent with its long history. Our discussion focuses first on purple photosynthetic bacteria, because of the relative simplicity of their photosynthetic machinery, and then on plants, whose chloroplasts are the site of photosynthesis.

Early in the twentieth century, it was mistakenly thought that light absorbed by photosynthetic pigments directly reduced CO_2, which then combined with water to form carbohydrate. In fact, photosynthesis in plants is a two-stage process in which light energy is harnessed to oxidize H_2O:

$$2 H_2O \xrightarrow{\text{light}} O_2 + 4 [H \cdot]$$

The electrons thereby obtained subsequently reduce CO_2:

$$4 [H \cdot] + CO_2 \rightarrow (CH_2O) + H_2O$$

The two stages of photosynthesis are traditionally referred to as the **light reactions** and **dark reactions:**

1. In the light reactions, specialized pigment molecules capture light energy and are thereby oxidized. A series of electron-transfer reactions, which culminate with the reduction of NADP to NADPH, generate a transmembrane proton gradient whose energy is tapped to synthesize ATP from ADP + P_i. The oxidized pigment molecules are reduced by H_2O, thereby generating O_2.

2. The dark reactions use NADPH and ATP to reduce CO_2 and incorporate it into the three-carbon precursors of carbohydrates.

As we shall see, both processes occur in the light and are therefore better described as light-dependent and light-independent reactions. After describing the chloroplast and its contents, we shall consider the light reactions and dark reactions in turn.

1 Chloroplasts

The site of photosynthesis in eukaryotes (algae and higher plants) is the ***chloroplast.*** Cells contain 1 to 1000 chloroplasts, which vary considerably in size and shape but are typically ~ 5-μm-long ellipsoids. These organelles presumably evolved from photosynthetic bacteria.

A   Chloroplast Anatomy

Like mitochondria, which they resemble in many ways, chloroplasts have a highly permeable outer membrane and a nearly impermeable inner membrane separated by a narrow intermembrane space (Fig. 18-1). The inner membrane encloses the **stroma,** a concentrated solution of enzymes, including those required for carbohydrate synthesis. The stroma also contains the DNA, RNA, and ribosomes involved in the synthesis of several chloroplast proteins.

The stroma, in turn, surrounds a third membranous compartment, the **thylakoid** (Greek: *thylakos,* a sac or pouch). The thylakoid is probably a single highly folded vesicle, although in most organisms it appears to consist of stacks of disklike sacs named **grana,** which are interconnected by unstacked **stromal lamellae.** A chloroplast usually contains 10 to 100 grana. Thylakoid membranes arise from invaginations in the inner membrane of developing chloroplasts and therefore resemble mitochondrial cristae. The thylakoid membrane contains protein complexes involved in harvesting light energy, transporting electrons, and synthesizing ATP. In photosynthetic bacteria, the machinery for the light reactions is located in the plasma membrane, which often forms invaginations or multilammelar structures that resemble grana.

B   Light-Absorbing Pigments

The principal photoreceptor in photosynthesis is **chlorophyll.** This cyclic tetrapyrrole, like the heme group of globins and cytochromes (Section 7-1A and Box 17-1), is derived biosynthetically from protoporphyrin IX. Chlorophyll molecules, however, differ from heme in several respects (Fig. 18-2). In chlorophyll, the central metal ion is Mg^{2+} rather than Fe(II) or Fe(III), and a cyclopentanone ring, Ring V, is fused to pyrrole Ring III. The major chlorophyll forms in plants and cyanobacteria, **chlorophyll *a* (Chl *a*)** and **chlorophyll *b* (Chl *b*),** and the major forms in photosynthetic bacteria, **bacteriochlorophyll *a* (BChl *a*)** and **bacteriochlorophyll *b* (BChl *b*),** also differ from heme and from each other in the degree of saturation of Rings II and IV and in the substituents of Rings I, II, and IV.

The highly conjugated chlorophyll molecules, along with other photosynthetic pigments, strongly absorb visible light (the most intense form of the solar radiation reaching the earth's surface; Fig. 18-3). The relatively

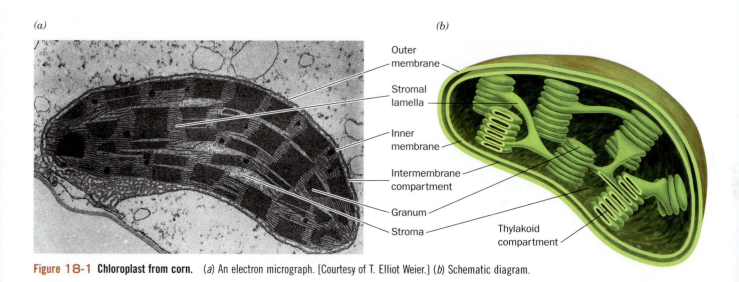

(a) *(b)*

Outer membrane
Stromal lamella
Inner membrane
Intermembrane compartment
Granum
Stroma
Thylakoid compartment

Figure 18-1   Chloroplast from corn.   (*a*) An electron micrograph. [Courtesy of T. Elliot Weier.] (*b*) Schematic diagram.

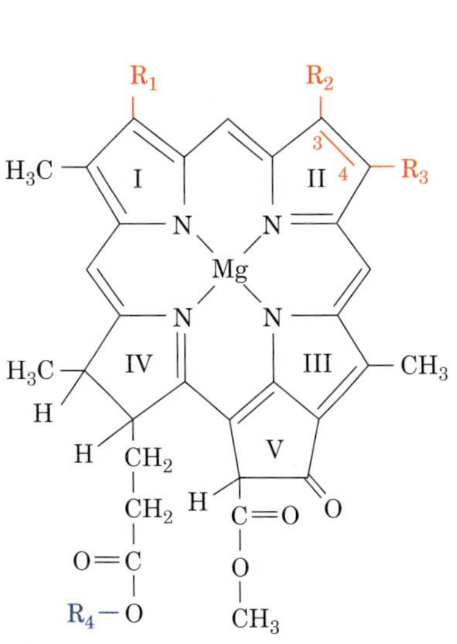

Chlorophyll

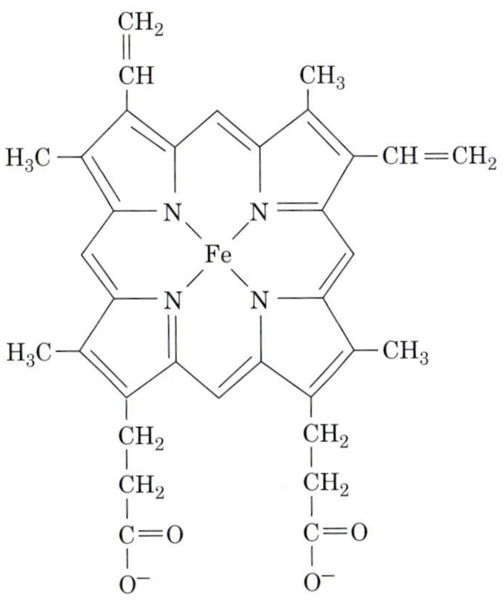

Iron–protoporphyrin IX

Figure 18-2 Chlorophyll structures. The molecular formulas of chlorophylls *a* and *b* and bacteriochlorophylls *a* and *b* are compared to that of iron–protoporphyrin IX (heme). The isoprenoid phytyl and geranylgeranyl tails presumably increase the chlorophylls' solubility in nonpolar media.

	R_1	R_2	R_3	R_4
Chlorophyll *a*	$-CH=CH_2$	$-CH_3$	$-CH_2-CH_3$	P
Chlorophyll *b*	$-CH=CH_2$	$\overset{O}{\overset{\|}{-C}}-H$	$-CH_2-CH_3$	P
Bacteriochlorophyll *a*	$\overset{O}{\overset{\|}{-C}}-CH_3$	$-CH_3^{\,a}$	$-CH_2-CH_3^{\,a}$	P or G
Bacteriochlorophyll *b*	$\overset{O}{\overset{\|}{-C}}-CH_3$	$-CH_3^{\,a}$	$=CH-CH_3^{\,a}$	P

^{a} No double bond between positions C3 and C4.

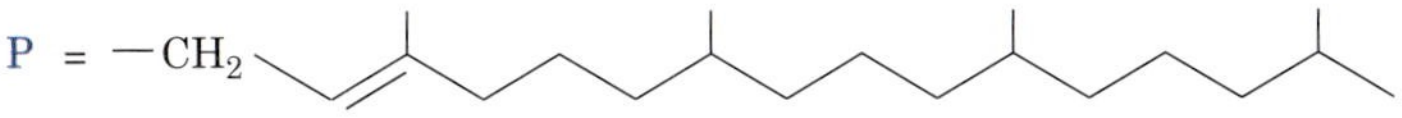

P = $-CH_2$

Phytyl side chain

G = $-CH_2$

Geranylgeranyl side chain

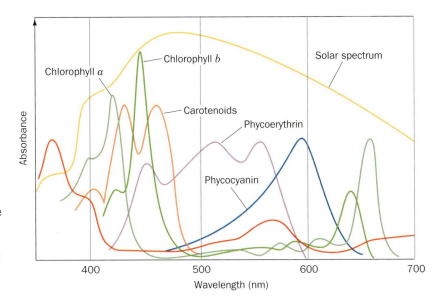

Figure 18-3 Absorption spectra of various photosynthetic pigments. The chlorophylls have two absorption bands, one in the red (long wavelength) and one in the blue (short wavelength). Phycoerythrin absorbs blue and green light, whereas phycocyanin absorbs yellow light. Together, these pigments absorb most of the visible light in the solar spectrum. [After a drawing by Govindjee, University of Illinois.]

small chemical differences among the various chlorophylls greatly affect their absorption spectra.

Light-Harvesting Complexes Contain Multiple Pigments.

The primary reactions of photosynthesis, as is explained in Section 18-2B, take place at **photosynthetic reaction centers.** *Yet photosynthetic assemblies contain far more chlorophyll molecules than are contained in reaction centers.* This is because most chlorophyll molecules do not participate directly in photochemical reactions but function to gather light; that is, *they act as light-harvesting antennas.* These **antenna chlorophylls** pass the energy of absorbed **photons** (units of light) from molecule to molecule until it reaches a photosynthetic reaction center (Fig. 18-4).

Transfer of energy from the antenna system to a reaction center **(RC)** occurs in $<10^{-10}$ s with an efficiency of $>90\%$. This high efficiency depends on the chlorophyll molecules having appropriate spacings and relative orientations. Even in bright sunlight, an RC directly intercepts only ~1 photon per second, a metabolically insignificant rate. Hence, a complex of antenna pigments, or **light-harvesting complex (LHC),** is essential.

LHCs consist of arrays of membrane-bound hydrophobic proteins that each contain numerous, often symmetrically arranged pigment molecules. For example, **LH-2** from the purple photosynthetic bacterium *Rhodospirillum molischianum* is an integral membrane protein that consists of eight α subunits and eight β subunits arranged in two eightfold symmetric concentric rings between which are sandwiched 32 pigment molecules (Fig. 18-5). Other LHCs vary widely in their structure and complement of light-harvesting pigments. The number and arrangement of pigment molecules in each LHC have presumably been optimized for efficient energy transfer throughout the LHC. Indeed, the ring of 16 BChl *a* molecules in LH-2 (lower part of Fig. 18-5*b*) are so strongly coupled that they absorb radiation almost as a single unit.

Most LHCs contain other light-absorbing substances besides chlorophyll. These **accessory pigments** "fill in" the absorption spectra of the antenna complexes, covering the spectral regions where chlorophylls do not absorb strongly (Fig. 18-3). For example, **carotenoids,** which are linear polyenes such as **β-carotene,**

β-Carotene

are components of all green plants and many photosynthetic bacteria and are therefore the most common accessory pigments. They are largely responsible for the brilliant fall colors of deciduous trees as well as for the orange color of carrots (after which they are named).

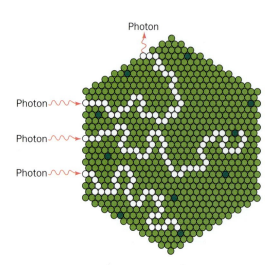

Figure 18-4 Flow of energy through a photosynthetic antenna complex. The energy of an absorbed photon randomly migrates among the molecules of the antenna complex (*light green circles*) until it reaches a reaction center chlorophyll (*dark green circles*) or, less frequently, is re-emitted (fluorescence).

(a)

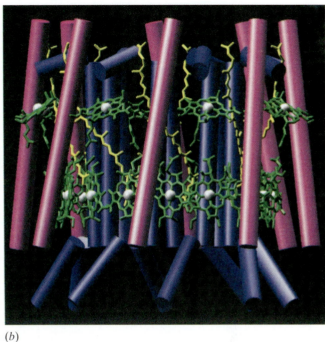

(b)

Figure 18-5 **X-Ray structure of the light-harvesting complex LH-2 from *Rs. molischianum.*** (*a*) View perpendicular to the photosynthetic membrane showing that the α subunits (*blue;* 56 residues) and the β subunits (*magenta;* 45 residues), as represented by their Cα backbones, are arranged in two concentric eightfold symmetric rings. Twenty-four bacteriochlorophyll *a* (BChl *a; green*) and eight **lycopene** (a carotenoid; *yellow*) molecules are sandwiched between the protein rings. (*b*) View from the plane of the membrane, using the same colors as in Part *a,* in which the α helical portions of the proteins are represented by cylinders and the Mg²⁺ ions are represented by white spheres.

Note that 8 of the BChl *a* molecules are bound near the top of the complex with their ring systems nearly parallel to the plane of the membrane, whereas the remaining 16 BChl *a* molecules are bound near the bottom of the complex with their ring systems approximately perpendicular to the plane of the membrane. This arrangement, together with that of the lycopene molecules, presumably optimizes the light-absorbing and excitation-transmitting capability of this antenna system. [Courtesy of Juergen Koepke and Hartmut Michel, Max-Planck Institut für Biochemie, Germany. PDBid 1LGH.] 🔗 **See the Interactive Exercises.**

Water-dwelling photosynthetic organisms, which carry out nearly half of the photosynthesis on earth, additionally contain other types of accessory pigments. This is because light outside the wavelengths 450 to 550 nm (blue and green light) is absorbed almost completely by passage through more than 10 m of water. In red algae and cyanobacteria, Chl *a* therefore is replaced as an antenna pigment by a set of linear tetrapyrroles, notably the red **phycoerythrobilin** and the blue **phycocyanobilin** (their spectra are shown in Fig. 18-3).

Phycoerythrobilin and Phycocyanobilin

2 | The Light Reactions

Photosynthesis is a process in which electrons from excited chlorophyll molecules are passed through a series of acceptors that convert electronic energy to chemical energy. We can thus ask two questions: (1) What is the mechanism of energy transduction; and (2) How do photooxidized chlorophyll molecules regain their lost electrons?

A | The Interaction of Light and Matter

Electromagnetic radiation is propagated as discrete **quanta** (photons) whose energy E is given by **Planck's law:**

$$E = h\nu = \frac{hc}{\lambda}$$

where h is **Planck's constant** (6.626×10^{-34} J·s), c is the speed of light (2.998×10^8 m·s^{-1} in vacuum), ν is the frequency of the radiation, and λ is its wavelength.

When a molecule absorbs a photon, one of its electrons is promoted from its ground (lowest energy) state molecular orbital to one of higher energy. However, *a given molecule can absorb photons of only certain wavelengths because, as is required by the law of conservation of energy, the energy difference between the two states must exactly match the energy of the absorbed photon.*

An electronically excited molecule can dissipate its excitation energy in several ways (Fig. 18-6):

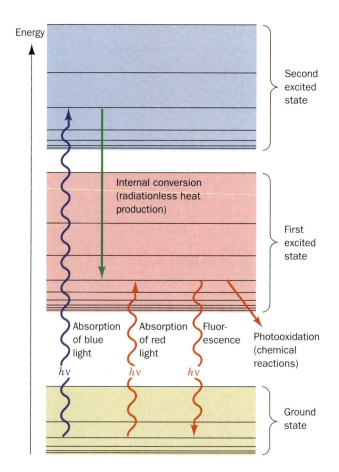

Figure 18-6 An energy diagram indicating the electronic states of chlorophyll and their most important modes of interconversion. The wiggly arrows represent the absorption of photons or their fluorescent emission. Excitation energy may also be dissipated in radiationless processes such as internal conversion (heat production) and chemical reactions. *See the Animated Figures.*

1. **Internal conversion,** a common mode of decay in which electronic energy is converted to the kinetic energy of molecular motion, that is, to heat. Many molecules relax in this manner to their ground states. Chlorophyll molecules, however, usually relax only to their lowest excited states. Consequently, the photosynthetically applicable excitation energy of a chlorophyll molecule that has absorbed a photon in its short-wavelength band, which corresponds to its second excited state, is no different than if it had absorbed a photon in its less energetic long-wavelength band.

2. **Fluorescence,** in which an electronically excited molecule decays to its ground state by emitting a photon. A fluorescently emitted photon generally has a longer wavelength (lower energy) than that initially absorbed. Fluorescence accounts for the dissipation of only 3 to 6% of the light energy absorbed by living plants.

3. **Exciton transfer** (also known as **resonance energy transfer**), in which an excited molecule directly transfers its excitation energy to nearby unexcited molecules with similar electronic properties. This process occurs through interactions between the molecular orbitals of the participating molecules. *Light energy is funneled to RCs through exciton transfer among antenna pigments.* The energy (excitation) is trapped at the RC chlorophylls because they have slightly lower excited state energies than the antenna chlorophylls (Fig. 18-7). This energy difference is lost as heat.

4. **Photooxidation,** in which a light-excited donor molecule is oxidized by transferring an electron to an acceptor molecule, which is thereby reduced. This process occurs because the transferred electron is less tightly bound to the donor in its excited state than it is to the ground state. In photosynthesis, excited chlorophyll (Chl*) is such a donor. *The energy of the absorbed photon is thereby chemically transferred to the photosynthetic reaction system.* Photooxidized chlorophyll, Chl^+, a cationic free radical, eventually returns to its reduced state by oxidizing some other molecule.

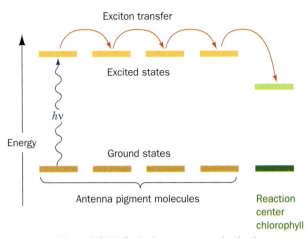

Figure 18-7   Excitation energy trapping by the photosynthetic reaction center.   Light energy that has been passed among pigment molecules by exciton transfer is trapped by the reaction center chlorophyll because its lowest excited state has a lower energy than those of the antenna pigment molecules.

B   Electron Transport in Photosynthetic Bacteria

In purple photosynthetic bacteria, a membrane-bound bacteriochlorophyll complex undergoes photooxidation when illuminated with red light. The excited electron is transferred along a series of carriers until it returns to the original bacteriochlorophyll complex. During the electron-transfer process, cytoplasmic protons are translocated across the plasma membrane. Dissipation of the resulting proton gradient drives ATP synthesis. The relatively simple RC of purple photosynthetic bacteria **(PbRC)** illustrates some general principles of the photochemical events that occur in the more complicated photosynthetic apparatus of plants and cyanobacteria (Section 18-2C).

The PbRC Is a Transmembrane Protein.   The RCs from several species of purple photosynthetic bacteria each contain three hydrophobic subunits known as H, L, and M. The L and M subunits collectively bind four molecules of bacteriochlorophyll, two molecules of **bacteriopheophytin (BPheo;** bacteriochlorophyll in which the Mg^{2+} ion is replaced by two protons), one Fe(II)

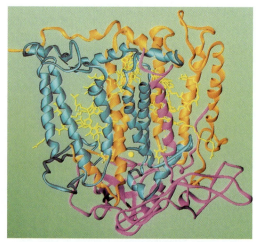

ion, and two molecules of the redox coenzyme ubiquinone (Fig. 17-11*b*) or one molecule of ubiquinone and one of the related **menaquinone:**

$$(CH_2{-}CH{=}C{-}CH_2)_8H$$

Menaquinone

The PbRC from *Rhodopseudomonas (Rps.) viridis* was the first transmembrane protein to be described in atomic detail (Fig. 9-23). Its arrangement of 11 membrane-spanning helices closely resembles that of the PbRC from *Rhodobacter (Rb.) sphaeroides* (Fig. 18-8), although the *Rb. sphaeroides* protein lacks the bound *c*-type cytochrome of the *Rps. viridis* protein.

The disposition of prosthetic groups in the *Rps. viridis* protein is shown in Fig. 18-9. *The most striking aspect of the PbRC is that these groups are arranged with nearly perfect twofold symmetry.* Two of the BChl molecules, the so-called **special pair,** are closely associated; they are nearly parallel and have an Mg—Mg distance of ~7 Å. The special pair is named for its wavelength (in nm) of maximum absorbance [**P870** or **P960** depending on whether it consists of Bchl *a* or BChl *b;* photosynthetic bacteria tend to inhabit murky stagnant ponds where visible light (400–800 nm) does not penetrate; they require a near infrared–absorbing species of chlorophyll]. Each member of the special pair, here P960, contacts another BChl molecule that, in turn, associates with a BPheo molecule. The menaquinone is close to the L subunit BPheo (Fig. 18-9, *right*), whereas the ubiquinone associates with the M subunit BPheo *b* (Fig. 18-9, *left*). The Fe(II) is positioned between the menaquinone and the ubiquinone rings. Curiously, the two symmetry-related sets of prosthetic groups are not functionally equivalent; electrons are almost exclusively transferred through the L subunit (the right sides of Figs. 18-8 and 18-9). This effect is generally attributed to subtle structural and electronic differences between the L and M subunits.

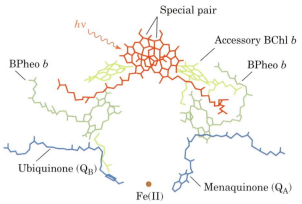

Figure 18-9   Disposition of prosthetic groups in the photosynthetic reaction center of *Rps. viridis.* Note that their rings, but not their aliphatic side chains, are arranged with close to twofold symmetry. The prosthetic groups bound by the L subunit are on the right and those bound by the M subunit are on the left. Photons are absorbed by the special pair of BChl *b* molecules (*red*). ✍ **See Kinemage Exercise 8-2.**

Photon Absorption Rapidly Photooxidizes the Special Pair. The photochemical events mediated by the PbRC occur as follows:

1. The primary photochemical event of bacterial photosynthesis is the absorption of a photon by the special pair (e.g., P960). The excited electron is delocalized over both its BChl molecules.

2. P960*, the excited state of P960, has but a fleeting existence. Within ~3 picoseconds (ps; 10^{-12} s), P960* transfers an electron to the BPheo on the right in Fig. 18-9 to yield $P960^+$ BPheo b^- (the intervening BChl group probably plays a role in conveying electrons, although it is not itself reduced; it is therefore known as the **accessory BChl**).

3. During the next 200 ps, the electron migrates to the menaquinone (or, in many species, the second ubiquinone), designated Q_A, to form the anionic semiquinone radical $Q_A^-\cdot$. All these electron transfers, as diagrammed in Fig. 18-10, are to progressively lower energy states, which makes this process all but irreversible.

Rapid removal of the excited electron from the vicinity of $P960^+$ is an essential feature of the PbRC; this prevents return of the electron to $P960^+$, which would lead to the wasteful internal conversion of its excitation energy to heat. In fact, *electron transfer in the PbRC is so efficient that its overall **quantum yield** (ratio of molecules reacted to photons absorbed) is virtually 100%.* No man-made device has yet approached this level of efficiency.

Electrons Are Returned to the Photooxidized Special Pair via an Electron-Transport Chain. $Q_A^-\cdot$, which occupies a hydrophobic pocket in the PbRC, transfers its excited electron to the more solvent-exposed ubiquinone, Q_B, to form $Q_B^-\cdot$ (the Fe ion positioned between Q_A and Q_B does not directly participate in these redox reactions). Q_A never becomes fully reduced; it shuttles between its oxidized and semiquinone forms.

When the PbRC is again excited, it transfers a second electron to $Q_B^-\cdot$ to form the fully reduced Q_B^{2-}. This anionic quinol takes up two protons from the cytoplasmic side of the plasma membrane to form Q_BH_2. Thus, *Q_B is a molecular transducer that converts two light-driven one-electron excitations to a two-electron chemical reduction.*

The electrons taken up by Q_BH_2 are eventually returned to $P960^+$ via an electron-transport chain (Fig. 18-10). The details of this process are species-dependent. The available redox carriers include a membrane-bound pool of ubiquinone molecules, a **cytochrome bc_1 complex,** and **cytochrome c_2**

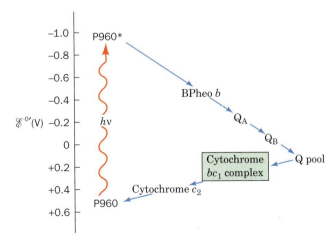

Figure 18-10   The photosynthetic electron-transport system of purple photosynthetic bacteria. Electrons liberated by the absorption of photons by P960 pass through BPheo b and Q_A before reaching Q_B, which exchanges with a pool of free ubiquinone. Electrons from QH_2 pass through cytochrome bc_1 to cytochrome c_2, which then reduces $P960^+$. Note that two photons are required for the two-electron reduction of Q to QH_2 and that cytochrome c_2 carries one electron at a time back to the PbRC. The overall process is essentially irreversible because electrons are transferred to progressively lower energy states (more positive standard reduction potentials).

(whose structure is drawn in Fig. 6-32*b*). The electron-transport pathway leads from Q_BH_2 through the ubiquinone pool, with which Q_BH_2 exchanges, to cytochrome bc_1, and then to cytochrome c_2. The reduced cytochrome c_2, which closely resembles mitochondrial cytochrome *c*, carries an electron back to $P960^+$. The PbRC is thereby reduced and prepared to absorb another photon.

Since electron transport in purple photosynthetic bacteria is a cyclic process (Fig. 18-10), *it results in no net oxidation–reduction.* However, when QH_2 transfers its electrons to cytochrome bc_1, its protons are translocated across the plasma membrane. Cytochrome bc_1 is a transmembrane protein complex containing a [2Fe–2S] iron–sulfur protein, cytochrome c_1, and a cytochrome *b* that contains two hemes, b_H and b_L (H and L for high and low potential). Note that cytochrome bc_1 is strikingly similar to the proton-translocating Complex III of mitochondria (which is also called cytochrome bc_1; Section 17-2E). In fact, electron transfer from QH_2 (a two-electron carrier) to the one-electron acceptor cytochrome c_2 occurs in a two-stage Q cycle, exactly as occurs in mitochondrial electron transport (Fig. 17-15). The net result is that for every two electrons transferred from QH_2 to cytochrome c_2, four protons enter the periplasmic space. Thus, photon absorption by the PbRC generates a transmembrane H^+ gradient. Light-dependent synthesis of ATP, a process known as **photophosphorylation,** is driven by the dissipation of this gradient (Section 18-2D).

See Guided Exploration 18

Two-center photosynthesis (Z-scheme) overview.

C Two-Center Electron Transport

In plants and cyanobacteria, *photosynthesis is a noncyclic process that uses the reducing power generated by the light-driven oxidation of H_2O to produce NADPH.* This multistep process involves two photosynthetic reaction centers (RCs) that each bear considerable resemblance to PbRCs. These RCs are **photosystem II (PSII),** which oxidizes H_2O, and **photosystem I (PSI),** which reduces $NADP^+$. Each photosystem is independently activated by light, with electrons flowing from PSII to PSI. *PSII and PSI therefore operate in electrical series to couple H_2O oxidation with $NADP^+$ reduction.* Evidence for the existence of two photosystems came from observations that in the presence of both red light (which activates only PSI) and yellow-green light (which also activates PSII), plants produce O_2 (i.e., oxidize H_2O) at a greater rate than the sum of the rates for each light acting alone. The herbicide **3-(3,4-dichlorophenyl)-1,1-dimethylurea (DCMU)**

**3-(3,4-Dichlorophenyl)-1,1-dimethylurea
(DCMU)**

blocks electron flow from PSII to PSI so that even with adequate illumination (i.e., activation of both PSI and PSII), PSI is not supplied with electrons, PSII cannot be reoxidized, and photosynthetic oxygen production ceases.

The pathway of electron transport in the chloroplast is more elaborate than in photosynthetic bacteria. *The components involved in electron transport from H_2O to NADPH are largely organized into three thylakoid membrane-bound particles (Fig. 18-11): PSII, a* **cytochrome b_6f complex,** *and PSI.* Electrons are transferred between these complexes via mobile

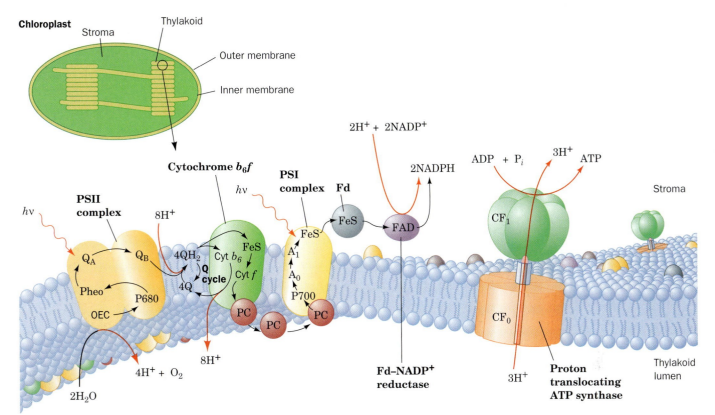

Figure 18-11 *Key to Function.* **A model of the thylakoid membrane.** The electron-transport system consists of three protein complexes: PSII, the cytochrome $b_6 f$ complex, and PSI, which are electrically "connected" by the diffusion of the electron carriers plastoquinone (Q) and plastocyanin (PC). Light-driven transport of electrons (*black arrows*) from H_2O to $NADP^+$ motivates the transport of protons (*red arrows*) into the thylakoid lumen. Additional protons are split off from water by the oxygen-evolving center (OEC), yielding O_2. The resulting proton gradient powers the synthesis of ATP by the $CF_1 CF_0$ proton-translocating ATP synthase. The membrane also contains light-harvesting complexes (not shown) whose component pigments transfer their excitations to PSII and PSI. Fd represents ferredoxin. [After Ort, D.R. and Good, N.E., *Trends Biochem. Sci.* **13**, 469 (1988).]

electron carriers, much as occurs in the respiratory electron-transport chain. The ubiquinone analog **plastoquinone (Q),** via its reduction to **plastoquinol (QH$_2$),**

Plastoquinone

2 [H•]

Plastoquinol

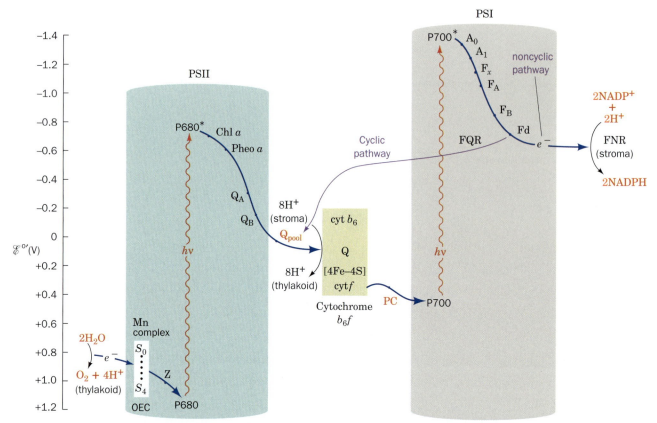

Figure 18-12   The Z-scheme of photosynthesis.   Electrons ejected from P680 in PSII by the absorption of photons are replaced with electrons abstracted from H_2O by an Mn complex (the oxygen-evolving center; OEC), thereby forming O_2 and 4 H^+. Each ejected electron passes through a chain of electron carriers to a pool of plastoquinone molecules (Q). The resulting plastoquinol, in turn, reduces the cytochrome b_6f complex (*yellow box*) with the concomitant translocation of protons into the thylakoid lumen. Cytochrome b_6f then transfers the electrons to plastocyanin (PC). The plastocyanin regenerates photooxidized P700 in PSI. The electron ejected from P700, through the intermediacy of a chain of electron carriers, reduces $NADP^+$ to NADPH in noncyclic electron transport. Alternatively, the electron may return to the cytochrome b_6f complex in a cyclic process that translocates additional protons into the thylakoid lumen. The reduction potentials increase downward so that electrons flow spontaneously in this direction.

links PSII to the cytochrome b_6f complex, which, in turn, interacts with PSI through the mobile peripheral membrane protein **plastocyanin (PC).** Electrons eventually reach **ferredoxin–NADP$^+$ reductase (FNR),** where they are used to reduce $NADP^+$. The oxidation of water and the passage of electrons through a Q cycle generate a transmembrane proton gradient, with the greater $[H^+]$ in the thylakoid lumen. The free energy of this proton gradient is tapped by chloroplast ATP synthase.

The various prosthetic groups of the photosynthetic apparatus of plants can be arranged in a diagram known as the **Z-scheme** (Fig. 18-12). As in other electron-transport systems, electrons flow from low to high reduction potentials. The zig-zag nature of the Z-scheme reflects the two loci for photochemical events (one at PSII, one at PSI) that are required to drive electrons from H_2O to $NADP^+$.

O_2 Is Generated by a Five-Stage Water-Splitting Reaction. The **oxygen-evolving center (OEC)** of PSII is also known as the **water-splitting enzyme** because it breaks down two water molecules to O_2, four protons, and four electrons. Insight into this process was garnered by Pierre Joliet and Bessel Kok, who analyzed the production of O_2 by dark-adapted chloroplasts that

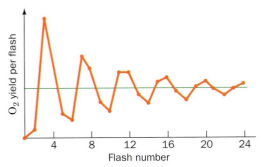

Figure 18-13   The O$_2$ yield per flash in dark-adapted spinach chloroplasts.   Note that the yield peaks on the third flash and then on every fourth flash thereafter until the curve eventually damps out to its average value. [After Forbush, B., Kok, B., and McGloin, M.P., *Photochem. Photobiol.* **14,** 309 (1971).]

were exposed to a series of short flashes of light. O$_2$ was evolved with a peculiar oscillatory pattern (Fig. 18-13). There is virtually no O$_2$ evolved by the first two flashes. The third flash results in the maximum O$_2$ yield. Thereafter, the amount of O$_2$ produced peaks with every fourth flash until the oscillations damp out to a steady state. This periodicity indicates that each OEC must undergo four light-dependent reactions—that is, four electron transfers—before releasing O$_2$.

The OEC is thought to cycle through five different states, S_0 through S_4 (Fig. 18-14). O$_2$ is released in the transition between S_4 and S_0. The observation that O$_2$ evolution peaks at the third rather than the fourth flash indicates that the OEC's resting state is predominantly S_1 rather than S_0. The oscillations gradually damp out because a small fraction of the RCs fail to be excited or become doubly excited by a given flash of light, so that the RCs eventually lose synchrony. The complete reaction sequence releases a total of four water-derived protons into the inner thylakoid space in a stepwise manner. These protons contribute to the transmembrane proton gradient.

Since the OEC abstracts electrons from H$_2$O, its five states must have extraordinarily high reduction potentials (recall from Table 13-3 that the O$_2$/H$_2$O half-reaction has a standard reduction potential of 0.815 V). PSII must also stabilize the highly reactive intermediates for extended periods (as much as minutes) in close proximity to water. As we shall see below, we are just beginning to understand how this occurs.

PSII Resembles the PbRC.   PSII from the thermophilic cyanobacterium *Synechococcus elongatus* consists of 19 subunits, 14 of which occupy the photosynthetic membrane. These transmembrane subunits include the reaction center proteins **D1 (PsbA)** and **D2 (PsbD),** the chlorophyll-containing inner-antenna subunits **CP43 (PsbC)** and **CP47 (PsbB),** and **cytochrome b_{559}.** The X-ray structure of this PSII (Fig. 18-15), determined by James Barber and So Iwata, reveals that this ~340-kD protein is a symmetric dimer, whose protomeric units each contain 35 transmembrane helices, 22 of which are portions of D1, D2, CP43, and CP47. Each protomer binds 36 Chl a's, 2 **pheophytin a's** (**Pheo a's;** Chl a with its Mg^{2+} replaced by two protons), one heme b, one heme c, 2 plastoquinones, one nonheme Fe, 7 all-trans carotenoids presumed to be β-carotene, 2 HCO$_3^-$ ions, and one OEC. Each protomer has pseudo-twofold symmetry that relates its D1,

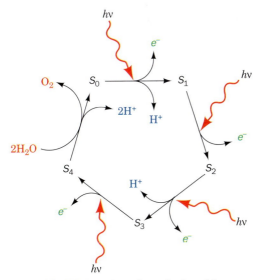

Figure 18-14   The schematic mechanism of O$_2$ generation in chloroplasts.   Four electrons are stripped, one at a time, in light-driven reactions ($S_0 \rightarrow S_4$), from two bound H$_2$O molecules. In the recovery step ($S_4 \rightarrow S_0$), which is light independent, O$_2$ is released and two more H$_2$O molecules are bound. Three of these five steps release protons into the thylakoid lumen.

(a)

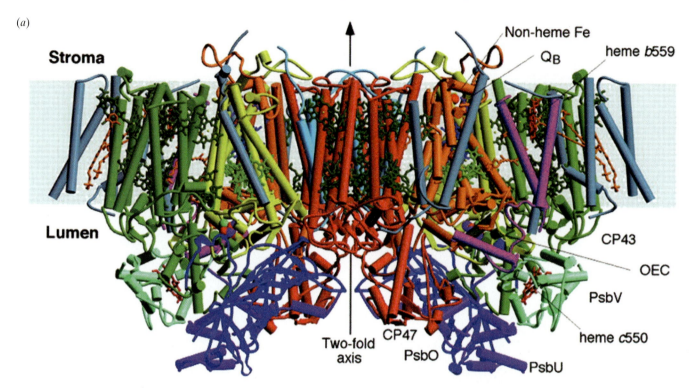

Figure 18-15 X-Ray structure of PSII from *S. elongatus*.   (a) The PSII dimer as viewed from within the plane of the membrane. Its transmembrane subunits are D1 (*yellow*); D2 (*orange*); CP47 (*red*); CP43 (*green*); cytochrome b_{559} (*magenta*); PsbL, PsbM, and PsbT (*light blue*); and PsbH, PsbI, PsbJ, PsbK, PsbN, PsbX, and PsbZ (*blue-gray*). Its extrinsic proteins are PsbO (*dark blue*), PsbU (*purple*), and PsbV (*cyan*). The various cofactors are drawn in stick form with the chlorophylls of the D1/D2 reaction center light green, those of the antenna complexes dark green, pheophytins dark blue, hemes red, β-carotenes orange, Q_A and Q_B purple, and the nonheme Fe represented by a red sphere. The OEC is shown in space-filling form with Mn ions purple, Ca^{2+} ion cyan, and O atoms red. The inferred position of the membrane is indicated by the light blue band. (b) View of a PSII protomer perpendicular to the membrane from the thylakoid lumen showing only the transmembrane portions of the complex and colored as in Part *a*. A portion of the other protomer in the PSII dimer is shown in muted colors with the dashed line indicating the region of monomer–monomer interactions. The pseudo-twofold axis, which is perpendicular to the membrane and passes through the nonheme Fe, relates the transmembrane helices of the D1/D2 heterodimer, CP43 and CP47, and PsbI and PsbX as emphasized by the black lines encircling these subunits. [Courtesy of James Barber and So Iwata, Imperial College London, U.K. PDBid 1S5L.]

(b)

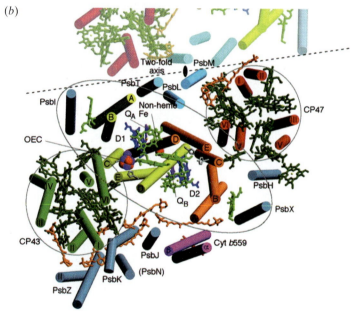

CP47, and **PsbI** subunits to its D2, CP43, and **PsbX** subunits. (In higher plants, the PSII protomer contains ~25 subunits and forms a ~1000-kD transmembrane supercomplex with several antenna proteins.)

The arrangement of the 5 transmembrane helices in both D1 and D2 resembles that in the L and M subunits of the PbRC (Fig. 18-8). Indeed, these two sets of subunits have similar sequences, thereby indicating that they arose from a common ancestor. CP43 and CP47 each contain 6 helices arranged as a trimer of dimers and respectively bind 12 and 14 antenna Chl *a*'s.

The cofactors of PSII's RC (Fig. 18-16) are organized similarly to those of the bacterial system (Fig. 18-9): They have essentially the same compo-

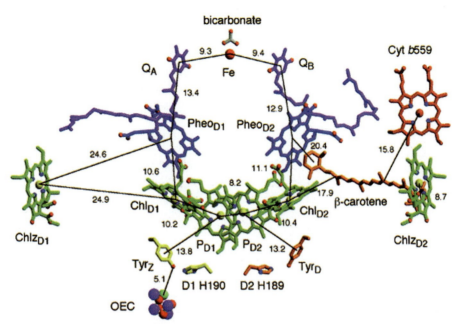

bicarbonate

Figure 18-16 The arrangement of electron transfer cofactors in PSII from *S. elongatus.* The complex is viewed along the membrane plane with the thylakoid lumen below. The cofactors are colored as in Fig. 18-15 but with Mg^{2+} ions yellow, N atoms blue, and O atoms red. The phytyl tails of the chlorophylls and pheophytins have been removed for clarity. The side chains of Tyr_Z (D1 Tyr 161) and D1 His 190 are yellow, and those of Tyr_D (D2 Tyr 160) and D2 His 189 are orange. The numbers indicate the center-to-center distances, in Å, between the cofactors spanned by the accompanying thin black lines. Compare this figure to Fig. 18-9 (which is drawn upside down relative to this figure). [Courtesy of James Barber and So Iwata, Imperial College London, U.K. PDBid 1S5L.]

nents (with Chl *a*, Pheo *a*, and plastoquinone replacing BChl *b*, BPheo *b*, and menaquinone, respectively) and are symmetrically organized along the complex's pseudo-twofold axis. The two Chl *a* rings labeled P_{D1} and P_{D2} in Fig. 18-16 are positioned analogously to the BChl *b*'s of P960's "special pair" and are therefore presumed to form PSII's primary electron donor, **P680** (named after the wavelength of its absorption maximum). The electron ejected from P680 follows a similar asymmetric course as that in the PbRC even though the two systems operate over different ranges of reduction potential (compare Figs. 18-10 and 18-12). As indicated in the central part of Fig. 18-12, the electron is transferred to a molecule of Pheo *a* ($Pheo_{D1}$ in Fig. 18-16), probably via a Chl *a* molecule (Chl_{D1}), and then to a bound plastoquinone (Q_A). The electron is subsequently transferred to a second plastoquinone molecule, Q_B, which after it receives a second electron in a like manner, takes up two protons at the stromal (cytoplasmic in cyanobacteria) surface of the thylakoid membrane. The resulting plastoquinol, Q_BH_2, then exchanges with a membrane-bound pool of plastoquinone molecules. DCMU as well as many other commonly used herbicides compete with plastoquinone for the Q_B-binding site on PSII, which explains how they inhibit photosynthesis.

The Mechanism of Water Oxidation by the OEC Is Poorly Understood.
The OEC, which is located at the lumenal surface of the D1 subunit (Fig. 18-16), is a Mn_4CaO_4 complex that consists of a tetrahedral constellation of three Mn ions and a Ca^{2+} ion, all bridged by O atoms to form a cubanelike Mn_3CaO_4 cluster, in which one of the O atoms that ligands the Ca^{2+} ion also ligands an additional Mn ion (Fig. 18-17). Each Mn ion is further liganded by one or two carboxylate groups from Asp or Glu side chains. In

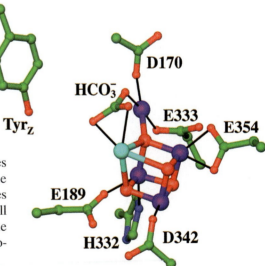

Figure 18-17 Structure of the OEC and its ligands. The complex is shown in ball-and-stick form with Mn ions purple, the Ca^{2+} ion cyan, C green, O red, and N blue. Metal coordination bonds are represented by thin black lines. All protein side chains shown are members of the D1 subunit of PSII except E354, which is a member of the D2 subunit. [Based on an X-ray structure by James Barber and So Iwata, Imperial College London, U.K. PDBid 1S5L.]

addition, the Ca^{2+} ion is liganded by an HCO_3^- ion, which also ligands the Mn ion that is not part of the Mn_3CaO_4 cluster. Several water molecules and/or hydroxide ions, which are not visible in the relatively poorly resolved X-ray structure of PSII but for which there is room, may also participate in liganding the Mn and Ca^{2+} ions so as to fill out their coordination numbers to the usual six or seven for a Ca^{2+} ion and six for a Mn ion.

The water-splitting reaction is driven by the excitation of the PSII RC. A variety of evidence indicates that the Mn ions in the OEC's various S states (Fig. 18-14) cycle through specific combinations of Mn(II), Mn(III), Mn(IV), and Mn(V) while abstracting protons and electrons from two H_2O molecules to yield O_2, which is released into the thylakoid lumen. Nevertheless, the mechanism whereby this occurs, that is, the nature of the five S states, remains unclear due to a lack of structural information concerning these states.

The electrons abstracted from water by the OEC are relayed, one at a time, to $P680^+$ by an entity originally named Z (Fig. 18-12). Spectroscopic measurements have identified Z as a transient neutral tyrosyl radical (TyrO·) located on D1 Tyr 161 (known as Tyr_Z), which is situated between the OEC and P680 (Fig. 18-16). Note that a tyrosyl radical has also been implicated in the reduction of O_2 to $2 H_2O$ by cytochrome c oxidase (Complex IV) in the respiratory electron transport chain (Section 17-2F).

Electron Transport through the Cytochrome b_6f Complex Generates a Proton Gradient. From the plastoquinone pool, electrons pass through the cytochrome b_6f complex. This integral membrane assembly resembles

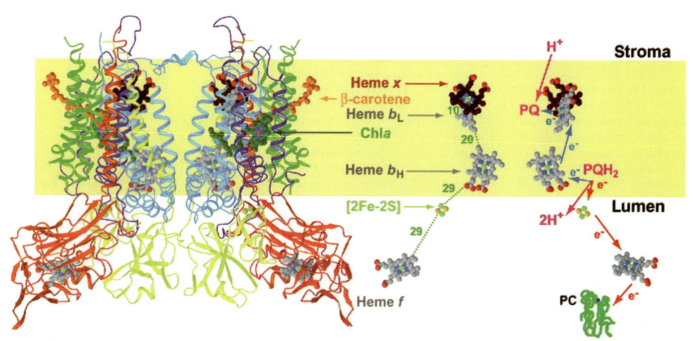

Figure 18-18 X-Ray structure of the cytochrome b_6f complex from the thermophilic cyanobacterium *Mastigocladus laminosus*.   A ribbon diagram of the dimeric complex is drawn on the left with cytochrome b_6 blue, subunit IV purple, cytochrome f red, the iron–sulfur protein (ISP) yellow, and the PetG, -L, -M, and -N subunits green. The inferred position of the lipid bilayer is indicated by a yellow band. Compare this figure to Fig. 17-14 (which is upside down relative to this figure). The paths of electron and proton transfer through the complex and the distances, in Å, between redox centers are shown on the right. [Modified from a drawing by William A. Cramer and Janet Smith, Purdue University. PDBid 1UM3.]

cytochrome bc_1, its purple bacterial counterpart (Section 18-2B), as well as Complex III of the mitochondrial electron-transport chain (Section 17-2E). Electron flow through the cytochrome b_6f complex occurs through a Q cycle (Fig. 17-15). Accordingly, two protons are translocated across the thylakoid membrane for every electron transported. The four electrons abstracted from 2 H_2O by the OEC therefore lead to the translocation of eight H^+ from the stroma to the thylakoid lumen. *Electron transport via the cytochrome b_6f complex generates much of the electrochemical proton gradient that drives the synthesis of ATP in chloroplasts.*

The X-ray structure of cytochrome b_6f (Fig. 18-18) was independently determined by Janet Smith and William Cramer and by Jean-Luc Polpot and Daniel Picot. Cytochrome b_6f is a dimer of ~109-kD protomers that each contain four large subunits (18–32 kD) that have counterparts in cytochrome bc_1: **cytochrome b_6,** a homolog of the N-terminal half of cytochrome b; **subunit IV,** a homolog of the C-terminal half of cytochrome b; **cytochrome f** (f for *feuille,* French for leaf), a c-type cytochrome that is a functional analog of cytochrome c_1, although the two are unrelated in structure or sequence; and a Rieske iron–sulfur protein (ISP), which is also present in cytochrome bc_1. In addition, cytochrome b_6f has four small hydrophobic subunits, **PetG, -L, -M,** and **-N,** that have no equivalents in cytochrome bc_1. Each protomer contains 13 transmembrane helices, four in cytochrome b_6, three in subunit IV, and one each in the remaining subunits. Cytochrome b_6f binds cofactors that are the equivalents of all of those in cytochrome bc_1: **heme f,** a c-type heme bound by cytochrome f; a [2Fe–2S] cluster bound by the ISP; hemes b_H and b_L, both of which are doubly axially liganded by His side chains of cytochrome b_6; a plastoquinone molecule that occupies the Q_i site (the quinone-binding site at which fully reduced quinone is regenerated during the Q cycle; Section 17-2E); and (in the X-ray structure) a quinone-analog inhibitor that occupies the Q_o site. In addition, cytochrome b_6f binds several cofactors that have no counterparts in cytochrome bc_1: a Chl a, a β-carotene, and, unexpectedly, a novel heme named **heme x** (alternatively, **heme c_i**), which is covalently linked to the protein via a single thioether bond to Cys 35 of cytochrome b_6, and whose only axial ligand is a water molecule (compare with hemes a, b, and c in Box 17-1).

Plastocyanin Transports Electrons from Cytochrome b_6f to PSI. Electron transfer between cytochrome f, the terminal electron carrier of the cytochrome b_6f complex, and PSI is mediated by plastocyanin (PC), a peripheral membrane protein located on the thylakoid lumenal surface (Fig. 18-11). The Cu-containing redox center of this mobile monomer cycles between its Cu(I) and Cu(II) oxidation states. The X-ray structure of PC from poplar leaves shows that the Cu atom is coordinated with distorted tetrahedral geometry by a Cys, a Met, and two His residues (Fig. 18-19). Cu(II) complexes with four ligands normally adopt a square planar coordination geometry, whereas those of Cu(I) are usually tetrahedral. Evidently, the strain of Cu(II)'s protein-imposed tetrahedral coordination in PC promotes its reduction to Cu(I). This hypothesis accounts for PC's high standard reduction potential (0.370 V) compared to that of the normal Cu(II)/Cu(I) half-reaction (0.158 V) and illustrates how proteins can modulate the reduction potentials of their redox centers. In the case of plastocyanin, this facilitates electron transfer from the cytochrome b_6f complex to PSI.

The structures of cytochrome f and PC suggest how these proteins associate. Cytochrome f's Lys 187, a member of a conserved group of five positively charged residues on the protein's surface, can be cross-linked to

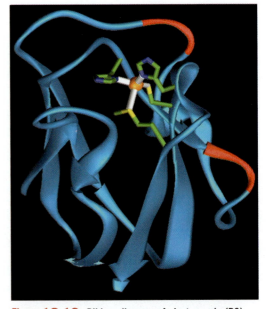

Figure 18-19   Ribbon diagram of plastocyanin (PC) from poplar leaves. This 99-residue monomeric protein, a member of the family of **blue copper proteins,** folds into a β sandwich. Its Cu atom (*orange sphere*), which alternates between its Cu(I) and Cu(II) oxidation states, is tetrahedrally coordinated by the side chains of His 37, Cys 84, His 87, and Met 92, which are shown in stick form with their C, N, and S atoms green, blue, and yellow. Six conserved Asp and Glu residues that form a negatively charged patch on the protein's surface are red. [Based on an X-ray structure by Mitchell Guss and Hans Freeman, University of Sydney, Australia. PDBid 1PLC.]

Asp 44 on PC, which occupies a conserved negatively charged surface patch. Quite possibly the two proteins associate through electrostatic interactions, much like PC's functional analog cytochrome *c* interacts with its redox partners in the mitochondrial electron transport chain (Section 17-2E).

The PSI RC Resembles Both the PSII RC and the PbRC. Cyanobacterial PSIs are trimers of protomers that each consist of at least 11 different protein subunits coordinating >100 cofactors. The X-ray structure of PSI from *S. elongatus* (Fig. 18-20), determined by Wolfram Saenger, reveals that each of its 356-kD protomers contains nine transmembrane subunits (**PsaA, PsbB, PsaF, PsaI–M,** and **PsaX**) and three stromal (cytoplasmic in cyanobacteria) subunits (**PsaC–E),** which collectively bind 127 cofactors that comprise 30% of PSI's mass. The cofactors forming the PSI RC are all bound by the homologous subunits PsaA (755 residues) and PsaB (740 residues), whose 11 transmembrane helices each are arranged in a manner resembling those in the L and M subunits of the PbRC (Fig. 18-8) and the D1 and D2 subunits of PSII (Fig. 18-15), thus supporting the notion that all RCs arose from a common ancestor. PsaA and PsaB, together with other transmembrane subunits, also bind the cofactors of the core antenna system (see below).

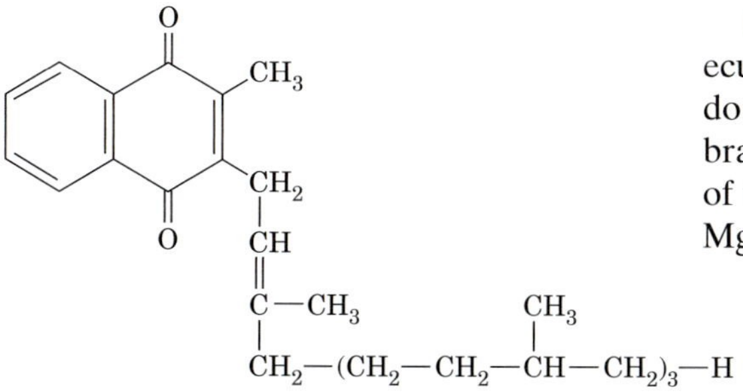

Figure 18-21 indicates that PSI's RC consists of 6 Chl *a*'s and two molecules of **phylloquinone** (*at left*), which has the same phytyl side chain as do chlorophylls (Fig. 18-2), all arranged in two pseudosymmetrically related branches, followed by three [4Fe–4S] clusters. The primary electron donor of this system, **P700,** consists of a pair of parallel Chl *a*'s, A1 and B1, whose Mg^{2+} ions are separated by 6.3 Å, and thus resembles the "special pair" in the PbRC. A1 is followed in the left branch of Fig. 18-21 by two more Chl *a* rings, B2 and A3, and B1 is followed by A2 and B3 in the right branch. One or both of the third pair of Chl *a* molecules, A3 and B3, probably form the spectroscopically identified primary electron acceptor A_0 (right side of Fig. 18-12). The Mg^{2+} ions of A3 and B3 are each axially liganded by the S atoms of a Met residue rather than by His side chains (thereby forming the only known biological examples of Mg^{2+}—S coordination). Electrons are passed from A3 and B3 to the phylloquinones, Q_K-A and Q_K-B, which almost certainly correspond to the spectroscopically identified electron acceptor A_1. Spectroscopic investigations indicate that, in contrast to the case for the PbRC, electrons pass through both branches of the PSI RC, although at different rates. Indeed, the PSI RC is most closely related to the RC of **green sulfur bacteria** (a second class of photosynthetic bacteria), which is a true homodimer.

Up until this point, PSI's RC resembles those of PSII and purple photosynthetic bacteria. However, rather than the reduced forms of either Q_K-A or Q_K-B dissociating from PSI, both of these quinones directly pass their photoexcited electron to a chain of three spectroscopically identified [4Fe–4S] clusters designated F_X, F_A, and F_B (right side of Fig. 18-12). F_X, which lies on the pseudo-twofold axis relating PsaA and PsaB, is coordinated by two Cys residues from each of these subunits. F_A and F_B are bound to the stromal subunit PsaC, which structurally resembles bacterial 2[4Fe–4S] ferredoxins (e.g., Fig. 17-10). The observation that both branches of PSI's electron transfer pathways are active, in contrast to only one active branch in PSII and the PbRC, is rationalized by the fact that the two quinones at the ends of each branch are functionally equivalent in PSI but functionally different in PSII and the PbRC.

PSI's core antenna system consists of 90 Chl *a* molecules and 22 carotenoids (Fig. 18-20*a*). The spatial distribution of these antenna Chl *a*'s

Phylloquinone

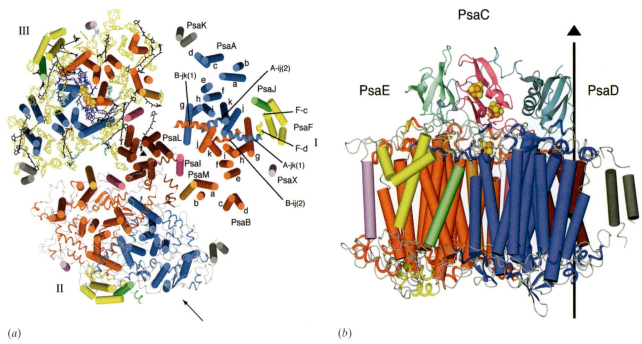

(a)

(b)

Figure 18-20   X-Ray structure of PSI from *S. elongatus*.   (a) View of the trimeric complex perpendicular to the membrane from its stromal side. The stromal subunits have been removed for clarity. PSI's threefold axis of symmetry is represented by the small black triangle. Different structural elements are shown for each of the three protomers (I, II, and III). I shows the arrangement of transmembrane helices (*cylinders*), which are differently colored for each subunit. The transmembrane helices of both PsaA (*blue*) and PsaB (*red*) are named a through k from their N- to C-termini. The six helices in extramembranous loop regions are drawn as spirals. II shows the transmembrane helices as cylinders with the stromal and luminal loop regions drawn in ribbon form. III shows the transmembrane helices as cylinders

together with all cofactors. The RC Chl *a*'s and quinones, drawn in stick form, are blue, the Fe and S atoms of the [4Fe–4S] clusters are drawn as orange and yellow spheres, the antenna system Chl *a*'s (whose side chains have been removed for clarity) are yellow, the carotenoids are black, and the bound lipids are light green. (b) One protomer as viewed parallel to the membrane along the arrow in Part *a* with the stroma above. The transmembrane subunits are colored as in Part *a* with the stromal subunits PsaC, PsaD, and PsaE pink, cyan, and light green. The vertical line and triangle mark the trimer's threefold axis of symmetry. [Courtesy of Wolfram Saenger, Freie Universität Berlin, Germany. PDBid 1JB0.]

Figure 18-21   The cofactors of the PSI RC and PsaC as viewed parallel to the membrane plane with the stroma above.   The Chl *a* and phylloquinone molecules are arranged in two branches that are related by PSI's twofold axis of pseudosymmetry, which is vertical in this drawing. The Chl *a*'s are labeled A or B to indicate that their Mg^{2+} ions are liganded by the side chains of PsaA or PsaB, respectively, and, from the luminal side upward, by different colors and numbers, 1 to 3. The phylloquinones are named Q_K-A and Q_K-B. PsaC is shown in ribbon form with those portions resembling segments in bacterial 2[4Fe–4S] ferredoxins pink and with insertions and extensions green. The three [4Fe–4S] clusters are shown in ball-and-stick form and are labeled according to their spectroscopic identities F_X, F_A, and F_B. The center-to-center distances between cofactors (*vertical black lines*) are given in Å. Compare this figure with Figs. 18-9 and 18-16. [Courtesy of Wolfram Saenger, Freie Universität Berlin, Germany. PDBid 1JB0.]

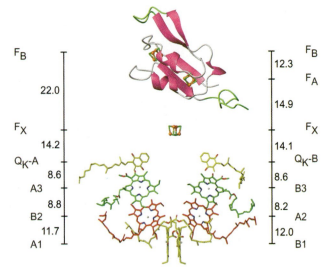

resembles that in the core antenna subunits CP43 and CP47 of PSII (Fig. 18-15). Indeed, the N-terminal domains of PsaA and PsaB are similar in sequence to those of CP43 and CP47 and fold into similar structures containing six transmembrane helices each. The carotenoids, which are mostly β-carotenes, are deeply buried in the membrane, where they are in van der Waals contact with Chl *a* rings. This permits efficient energy transfer from photoexcited carotenoids to Chl *a* as well as protecting PSI from photooxidative damage.

PSIs from higher plants are monomers rather than trimers as are cyanobacterial PSIs. Nevertheless, the X-ray structure of PSI from peas reveals that the positions and orientations of the chlorophylls in both species of PSIs are nearly identical, a remarkable finding considering the >1 billion years since chloroplasts diverged from their cyanobacterial ancestors. However, pea PSI has four antenna proteins not present in cyanobacterial PSI that are arranged in a crescent-shaped transmembrane belt around one side of its RC and which collectively bind 56 chlorophyll molecules.

PSI-Activated Electrons May Reduce NADP⁺ or Motivate Proton Gradient Formation. *Electrons ejected from F_B in PSI may follow either of two alternative pathways (Fig. 18-12):*

1. Most electrons follow a noncyclic pathway by reducing a ~100-residue, [2Fe–2S]-containing, soluble protein called **ferredoxin (Fd)** that is located in the stroma. Reduced Fd, in turn, reduces NADP⁺ in a reaction mediated by the ~310-residue, monomeric, FAD-containing ferredoxin–NADP⁺ reductase (FNR, Fig. 18-22), to yield the final product of the chloroplast light reactions, NADPH. Two reduced Fd molecules successively deliver one electron each to the FAD of FNR, which thereby sequentially assumes the neutral semiquinone and fully reduced states before transferring the two electrons and a proton to the NADP⁺ via what is formally a hydride ion transfer.

2. Some electrons are returned from PSI, via cytochrome b_6, to the plastoquinone pool, thereby traversing a cyclic pathway that translocates protons across the thylakoid membrane. A mechanism that has been proposed for this process is that Fd transfers an electron to heme *x* of cytochrome b_6 (Fig. 18-18) rather than to FNR. Since heme *x* contacts heme b_L at the periphery of cytochrome $b_6 f$'s Q_i site, an electron injected into heme *x* would be expected to reduce plastoquinone via a Q cycle-like mechanism (Fig. 17-15). Note that the cyclic pathway is independent of the action of PSII and hence does not result in the evolution of O_2. This accounts for the observation that chloroplasts absorb more than eight photons per O_2 molecule evolved.

   The cyclic electron flow presumably functions to increase the amount of ATP produced relative to that of NADPH and thus per-

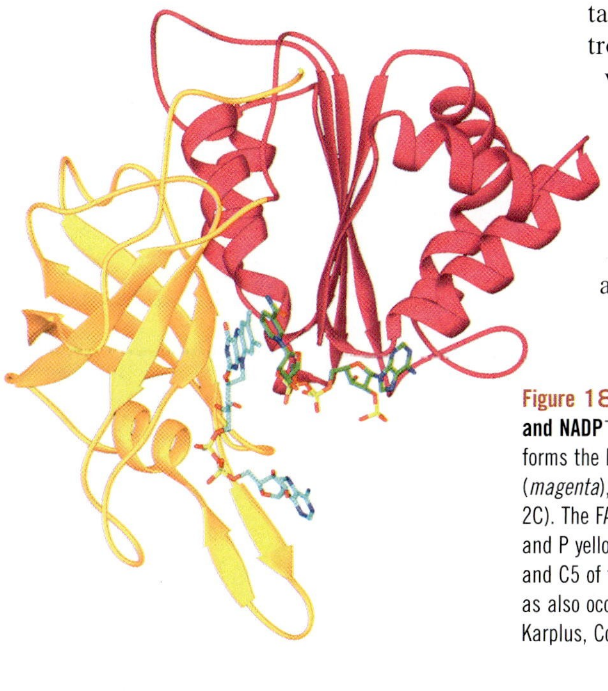

Figure 18-22 X-Ray structure of pea ferredoxin–NADP⁺ reductase (FNR) in complex with FAD and NADP⁺. This 308-residue protein has two domains: The N-terminal domain (*gold*), which forms the FAD binding site, folds into an antiparallel β barrel, whereas the C-terminal domain (*magenta*), which provides the NADP⁺ binding site, forms a dinucleotide-binding fold (Section 6-2C). The FAD and NADP⁺ are shown in stick form with NADP⁺ C green, FAD C cyan, N blue, O red, and P yellow. The flavin and nicotinamide rings are in opposition with C4 of the nicotinamide ring and C5 of the flavin ring 3.0 Å apart, an arrangement that is consistent with direct hydride transfer as also occurs in dihydrolipoyl dehydrogenase (Fig. 16-8). [Based on an X-ray structure by Andrew Karplus, Cornell University. PDBid 1QFY.] ✪ **See the Interactive Exercises.**

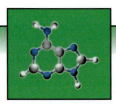

BOX 18-1

Perspectives in Biochemistry

Segregation of PSI and PSII

Electron microscopy has revealed that the protein complexes of the thylakoid membrane have characteristic distributions (*see figure*).

1. PSI occurs mainly in the unstacked stromal lamellae, in contact with the stroma, where it has access to $NADP^+$.
2. PSII is located almost exclusively between the closely stacked grana, out of direct contact with the stroma.
3. Cytochrome b_6f is uniformly distributed throughout the membrane.

The high mobilities of plastoquinone and plastocyanin, the electron carriers that shuttle electrons between these particles, permit photosynthesis to proceed at a reasonable rate.

What function is served by the segregation of PSI and PSII? If these two photosystems were in close proximity, the higher excitation energy of PSII (P680 versus P700) would cause it to pass a large fraction of its absorbed photons to PSI via exciton transfer; that is, PSII would act as a light-harvesting antenna for PSI. The separation of the particles by around 100 Å eliminates this difficulty.

The physical separation of PSI and PSII also permits the chloroplast to respond to changes in illumination. The relative amounts of light absorbed by the two photosystems vary with how the light-harvesting complexes are distributed between the stacked and unstacked portions of the thylakoid membrane. Under high illumination (normally direct sunlight, which contains a high proportion of short-wavelength blue light), PSII absorbs more light than PSI. PSI is then unable to take up electrons as fast as PSII can supply them, so the plastoquinone is predominantly in its reduced state. The reduced plastoquinone activates a protein kinase to phosphorylate specific Thr residues of the LHCs, which, in response, migrate to the unstacked regions of the thylakoid membrane, where they associate with PSI. A greater fraction of the incident light is thereby funneled to PSI.

Under low illumination (normally shady light, which contains a high proportion of long-wavelength red light), PSI takes up electrons faster than PSII can provide them so that plastoquinone predominantly assumes its oxidized form. The LHCs are consequently dephosphorylated and migrate to the stacked portions of the thylakoid membrane, where they associate with PSII. The chloroplast therefore maintains the balance between its two photosystems by a light-activated feedback mechanism.

[Figure based on Anderson, J.M. and Anderson, B., *Trends Biochem. Sci.* **7**, 291 (1982).]

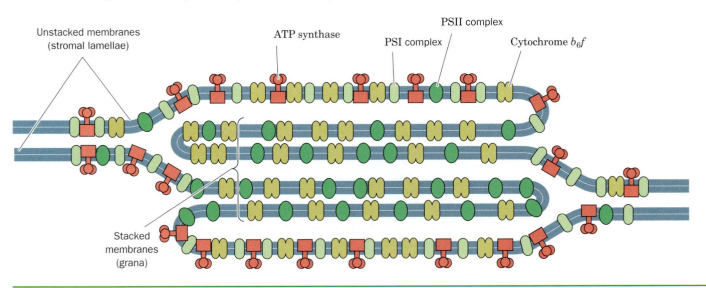

Unstacked membranes (stromal lamellae)
ATP synthase
PSI complex
PSII complex
Cytochrome b_6f
Stacked membranes (grana)

mits the cell to adjust the relative amounts of these two substances produced according to its needs. However, the mechanism that apportions electrons between the cyclic and noncyclic pathways is unknown. Fine tuning of the light reactions also depends on the segregation of PSI and PSII in distinct portions of the thylakoid membrane (see Box 18-1).

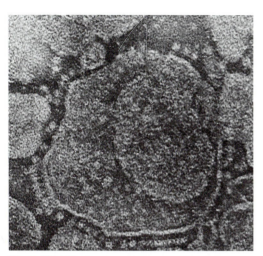

Figure 18-23 Electron micrograph of thylakoids showing the CF1 "lollipops" of their ATP synthases projecting from their stromal surfaces.   Compare this with Fig. 17-21a. [Courtesy of Peter Hinkle, Cornell University.]

D   Photophosphorylation

Chloroplasts generate ATP in much the same way as mitochondria, that is, by coupling the dissipation of a proton gradient to the enzymatic synthesis of ATP (Section 17-3). This light-dependent process is known as **photophosphorylation**. Like oxidative phosphorylation, it requires an intact thylakoid membrane and can be uncoupled from light-driven electron transport by compounds such as 2,4-dinitrophenol (Fig. 17-28).

Electron micrographs of thylakoid membrane stromal surfaces and bacterial plasma membrane inner surfaces reveal lollipop-shaped structures (Fig. 18-23). These closely resemble the F_1 units of the proton-translocating ATP synthase in mitochondria (Fig. 17-21). In fact, the chloroplast ATP synthase, which is also known as the **CF_1CF_0 complex** (C for chloroplast), is remarkably similar to the mitochondrial F_1F_0 complex. For example,

1. Both the F_0 and the CF_0 units are hydrophobic transmembrane proteins that contain a proton-translocating channel.
2. Both the F_1 and the CF_1 units are hydrophilic peripheral membrane proteins of subunit composition $\alpha_3\beta_3\gamma\delta\varepsilon$, of which β is a reversible ATPase.
3. Both ATP synthases are inhibited by oligomycin and by dicyclohexylcarbodiimide (DCCD).

Clearly, proton-translocating ATP synthase must have evolved very early in the history of cellular life. Note, however, that whereas chloroplast ATP synthase translocates protons out of the thylakoid space into the stroma (Fig. 18-11), mitochondrial ATP synthase conducts them from the intermembrane space into the matrix space (Fig. 17-20). This is because the stroma is topologically analogous to the mitochondrial matrix.

Photosynthesis with Noncyclic Electron Transport Produces Around 1.25 ATP per Absorbed Photon. At saturating light intensities, chloroplasts generate proton gradients of ~3.5 pH units across their thylakoid membranes as a result of two processes:

1. The evolution of a molecule of O_2 from two H_2O molecules releases four protons into the thylakoid lumen.
2. The transport of the liberated four electrons through the cytochrome b_6f complex occurs with the translocation of eight protons from the stroma to the thylakoid lumen.

Altogether, ~12 protons enter the lumen per molecule of O_2 produced by noncyclic electron transport.

The thylakoid membrane, in contrast to the inner mitochondrial membrane, is permeable to ions such as Mg^{2+} and Cl^-. Translocation of protons and electrons across the thylakoid membrane is consequently accompanied by the passage of these ions so as to maintain electrical neutrality (Mg^{2+} out and Cl^- in). This all but eliminates the membrane potential. *The electrochemical gradient in chloroplasts is therefore almost entirely a result of the pH (concentration) gradient.*

Chloroplast ATP synthase, according to most estimates, produces one ATP for every three protons it transports from the thylakoid lumen to the stroma. Noncyclic electron transport in chloroplasts therefore results in the production of ~12/3 = 4 molecules of ATP per molecule of O_2 evolved (cyclic electron transport generates more ATP because more protons are translocated to the thylakoid lumen via the Q cycle mediated by cytochrome b_6f).

Noncyclic electron transport, of course, also yields NADPH (2 NADPH for every 4 electrons liberated from 2 H_2O by the OEC). Each NADPH has the free energy to produce 3 ATP (Section 17-3C; although NADPH is not used to drive ATP synthesis), for a total of 6 more ATP equivalents per O_2 produced. Consequently, a total of 10 ATP equivalents are generated per O_2 produced. A minimum of two photons is required for each electron traversing the system from H_2O to NADPH, that is, eight photons per O_2 produced. This is confirmed by experimental measurements which indicate that plants and algae require 8 to 10 photons of visible light to produce one molecule of O_2. Thus, the overall efficiency of the light reactions is 10 ATP/ 8–10 photons, or approximately 1.25 ATP per absorbed photon.

3 The Dark Reactions

In the previous section we saw how plants harness light energy to generate ATP and NADPH. In this section we discuss how these products are used to synthesize carbohydrates and other substances from CO_2.

A The Calvin Cycle

The metabolic pathway by which plants incorporate CO_2 into carbohydrates was elucidated between 1946 and 1953 by Melvin Calvin, James Bassham, and Andrew Benson. They did so by tracing the metabolic fate of the radioactive label from $^{14}CO_2$ in cultures of algal cells. Some of Calvin's earliest experiments indicated that algae exposed to $^{14}CO_2$ for a minute or more synthesize a complex mixture of labeled metabolites, including sugars and amino acids. Analysis of the algae within 5 s of their exposure to $^{14}CO_2$, however, showed that *the first stable radioactive compound formed is 3-phosphoglycerate (3PG),* which is initially labeled only in its carboxyl group. This result immediately suggested that the 3PG was formed by the carboxylation of a C_2 compound. Yet no such precursor was found. The actual carboxylation reaction involves the pentose **ribulose-5-phosphate (Ru5P):**

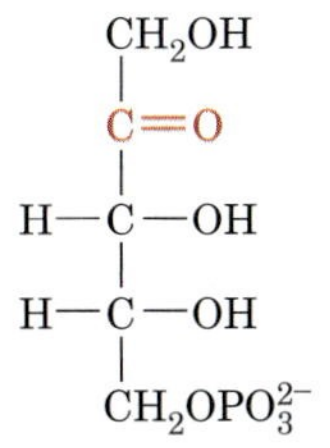

$$
\begin{array}{c}
CH_2OH \\
| \\
C{=}O \\
| \\
H-C-OH \\
| \\
H-C-OH \\
| \\
CH_2OPO_3^{2-}
\end{array}
$$

Ribulose-5-phosphate (Ru5P)

The resulting C_6 product splits into two C_3 compounds, both of which turn out to be 3PG. The overall pathway, diagrammed in Fig. 18-24, is known as the **Calvin cycle** or the **reductive pentose phosphate cycle.** It involves the carboxylation of a pentose, the formation of carbohydrate products, and the regeneration of the pentose.

During the search for the carboxylation substrate, several other photosynthetic intermediates had been identified and their labeling patterns elucidated. For example, the hexose fructose-1,6-bisphosphate (FBP) is initially labeled only at its C3 and C4 positions but later becomes labeled to a lesser degree at its other atoms. A consideration of the flow of labeled carbon through the various tetrose, pentose, hexose, and heptose phos-

Figure 18-24 (*Opposite*) *Key to Metabolism.* **The Calvin cycle.** The number of lines in an arrow indicates the number of molecules reacting in that step for a single turn of the cycle that converts three CO_2 molecules to one GAP molecule. For the sake of clarity, the sugars are all shown in their linear forms, although the hexoses and heptoses predominantly exist in their cyclic forms. The ^{14}C-labeling patterns generated in one turn of the cycle through the use of $^{14}CO_2$ are indicated in red. Note that two of the product Ru5Ps are labeled only at C3, whereas the third Ru5P is equally labeled at C1, C2, and C3. 🔊 **See the Animated Figures.**

phates led, in what is a milestone of metabolic biochemistry, to the deduction of the Calvin cycle as is diagrammed in Fig. 18-24. The existence of many of its postulated reactions was eventually confirmed by *in vitro* studies using purified enzymes.

The Calvin Cycle Generates GAP from CO_2 via a Two-Stage Process. The Calvin cycle can be considered to have two stages:

Stage 1 The production phase (top line of Fig. 18-24), in which three molecules of Ru5P react with three molecules of CO_2 to yield six molecules of glyceraldehyde-3-phosphate (GAP) at the expense of nine ATP and six NADPH molecules. *The cyclic nature of the pathway makes this process equivalent to the synthesis of one GAP from three CO_2 molecules.* At this point, GAP can be bled off from the cycle for use in biosynthesis.

Stage 2 The recovery phase (bottom lines of Fig. 18-24), in which the carbon atoms of the remaining five GAPs are shuffled in a remarkable series of reactions, similar to those of the pentose phosphate pathway (Section 14-6), to re-form the three Ru5Ps with which the cycle began. This stage can be conceptually decomposed into four sets of reactions (with the numbers keyed to the corresponding reactions in Fig. 18-24):

   6. $C_3 + C_3 \rightarrow C_6$
   8. $C_3 + C_6 \rightarrow C_5 + C_4$
   9. $C_3 + C_4 \rightarrow C_7$
   11. $C_3 + C_7 \rightarrow C_5 + C_5$

The overall stoichiometry for this process is therefore

$$5\ C_3 \rightarrow 3\ C_5$$

Note that this stage of the Calvin cycle occurs without further input of free energy (ATP) or reducing power (NADPH).

The first reaction of the Calvin cycle is the phosphorylation of Ru5P by **phosphoribulokinase** to form **ribulose-1,5-bisphosphate (RuBP).** Following the carboxylation of RuBP (Reaction 2; discussed below), the resulting 3PG is converted first to 1,3-bisphosphoglycerate (BPG) and then to GAP. This latter sequence is the reverse of two consecutive glycolytic reactions (Section 14-2G and 14-2F) except that the Calvin cycle reaction uses NADPH rather than NADH.

The second stage of the Calvin cycle begins with the reverse of a familiar glycolytic reaction, the isomerization of GAP to dihydroxyacetone phosphate (DHAP) by triose phosphate isomerase (Section 14-2E). Following this, DHAP is directed along two analogous paths: Reactions 6 to 8 or Reactions 9 to 11. Reactions 6 and 9 are aldolase-catalyzed aldol condensations in which DHAP is linked to an aldehyde. Reaction 6 is also the reverse of a glycolytic reaction (Section 14-2D). Reactions 7 and 10 are phos-

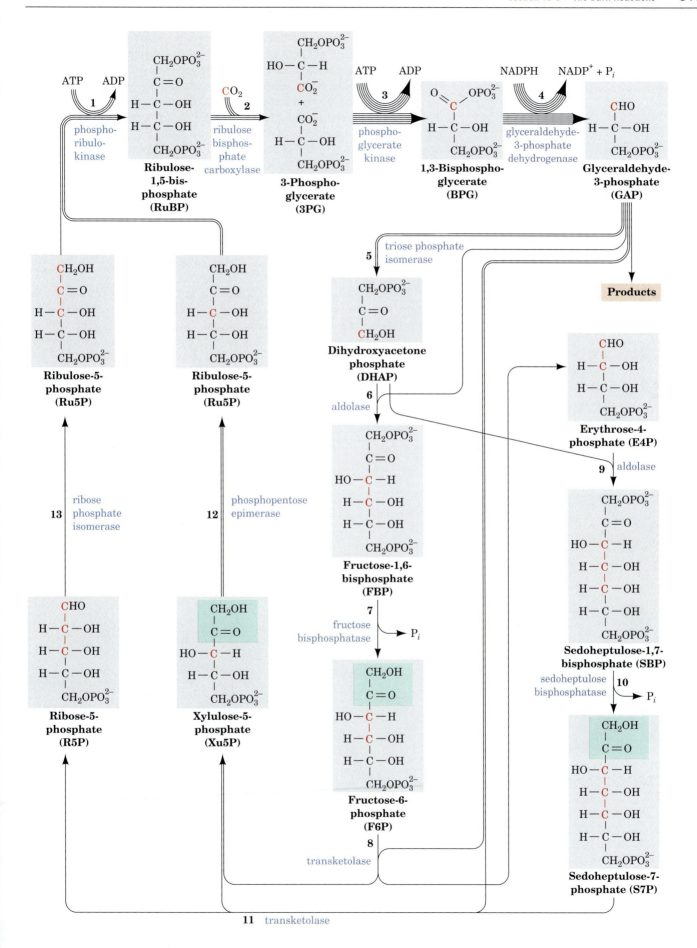

phate hydrolysis reactions that are catalyzed, respectively, by fructose bisphosphatase (FBPase; Section 14-4B) and **sedoheptulose bisphosphatase (SBPase).** The remaining Calvin cycle reactions are catalyzed by enzymes that also participate in the pentose phosphate pathway. In Reactions 8 and 11, both catalyzed by transketolase, a C_2 keto unit (shaded green in Fig. 18-24) is transferred from a ketose to GAP to form xylulose-5-phosphate (Xu5P), leaving the aldoses erythrose-4-phosphate (E4P) in Reaction 8 and ribose-5-phosphate (R5P) in Reaction 11. The E4P produced by Reaction 8 feeds into Reaction 9. The Xu5Ps produced by Reactions 8 and 11 are converted to Ru5P by **phosphopentose epimerase** in Reaction 12. The R5P from Reaction 11 is also converted to Ru5P by **ribose phosphate isomerase** in Reaction 13, thereby completing a turn of the Calvin cycle. Only 3 of the 11 Calvin cycle enzymes—phosphoribulokinase, the carboxylation enzyme ribulose bisphosphate carboxylase, and SBPase—have no equivalents in animal tissues.

RuBP Carboxylase Catalyzes CO$_2$ Fixation. The enzyme that catalyzes CO_2 fixation, ribulose bisphosphate carboxylase **(RuBP carboxylase),** is arguably the world's most important enzyme since nearly all life on earth ultimately depends on its action. This protein, presumably as a consequence of its low catalytic efficiency ($k_{cat} \approx 3\ s^{-1}$), accounts for up to 50% of leaf proteins and is therefore the most abundant protein in the biosphere. RuBP carboxylase from higher plants and most photosynthetic microorganisms

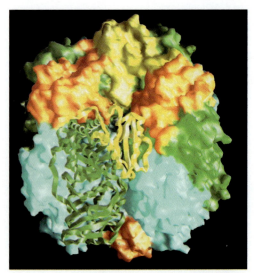

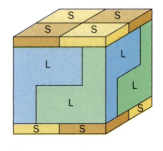

(a)

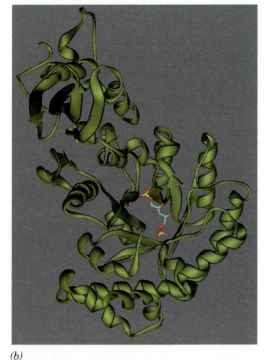

(b)

Figure 18-25 X-Ray structure of RuBP carboxylase. (a) The quaternary structure of the L_8S_8 protein. One L and one S subunit are drawn as ribbons with the remainder represented by their solvent-accessible surfaces. The protein, which has D_4 symmetry (the symmetry of a square prism; Fig. 6-34b), is viewed with its fourfold axis tipped toward the viewer. As the accompanying diagram schematically indicates, the elongated L subunits (six are clearly visible in the structural drawing) can be considered to associate as two interdigitated tetramers, with that extending from the top green and that extending from the bottom cyan. The subunits of the S_4 tetramers that cap the top and bottom of the complex are alternately colored yellow and orange (only one subunit of the lower S_4 tetramer is visible). [Based on an X-ray structure by Yasushi Kai, Osaka University, Japan. PDBid 1BUR.] (b) An L subunit. The transition state inhibitor 2-carboxyarabinitol-1,5-bisphosphate, represented in stick form, is positioned in the substrate-binding site in the mouth of the enzyme's α/β barrel. The subunit is oriented by a rotation about the vertical axis relative to that drawn in ribbon form in Part a. [Based on an X-ray structure by David Eisenberg, UCLA. PDBid 1RLC.]

consists of eight large (L) subunits (477 residues in tobacco leaves) encoded by chloroplast DNA, and eight small (S) subunits (123 residues) specified by a nuclear gene (the RuBP carboxylase from certain photosynthetic bacteria is an L_2 dimer whose L subunit has 28% sequence identity with and is structurally similar to that of the L_8S_8 enzyme). X-Ray studies by Carl-Ivar Brändén and by David Eisenberg demonstrated that the L_8S_8 enzyme has the symmetry of a square prism (Fig. 18-25a). The L subunit is made of a β sheet domain and an α/β barrel domain that contains the enzyme's catalytic site (Fig. 18-25b). The function of the S subunit is unknown; attempts to show that is has a regulatory role, in analogy with other enzymes, have been unsuccessful.

The accepted mechanism of RuBP carboxylase, which was largely formulated by Calvin, is indicated in Fig. 18-26. Abstraction of the C3 proton of RuBP, the reaction's rate-determining step, generates an enediolate that nucleophilically attacks CO_2. The resulting β-keto acid is rapidly attacked at its C3 position by H_2O to yield an adduct that splits, by a reaction similar to aldol cleavage, to yield the two product 3PG molecules. *The driving force for the overall reaction, which is highly exergonic ($\Delta G^{\circ\prime} = -35.1$ kJ · mol^{-1}), is provided by the cleavage of the β-keto acid intermediate to yield an additional resonance-stabilized carboxylate group.*

RuBP carboxylase activity requires Mg^{2+}, which probably stabilizes developing negative charges during catalysis. The Mg^{2+} is, in part, bound to the enzyme by a catalytically important carbamate group ($-NH-COO^-$) that is generated by the reaction of a nonsubstrate CO_2 with the ε-amino group of Lys 201. This essential reaction is catalyzed *in vivo* by the enzyme **RuBP carboxylase activase** in an ATP-driven process.

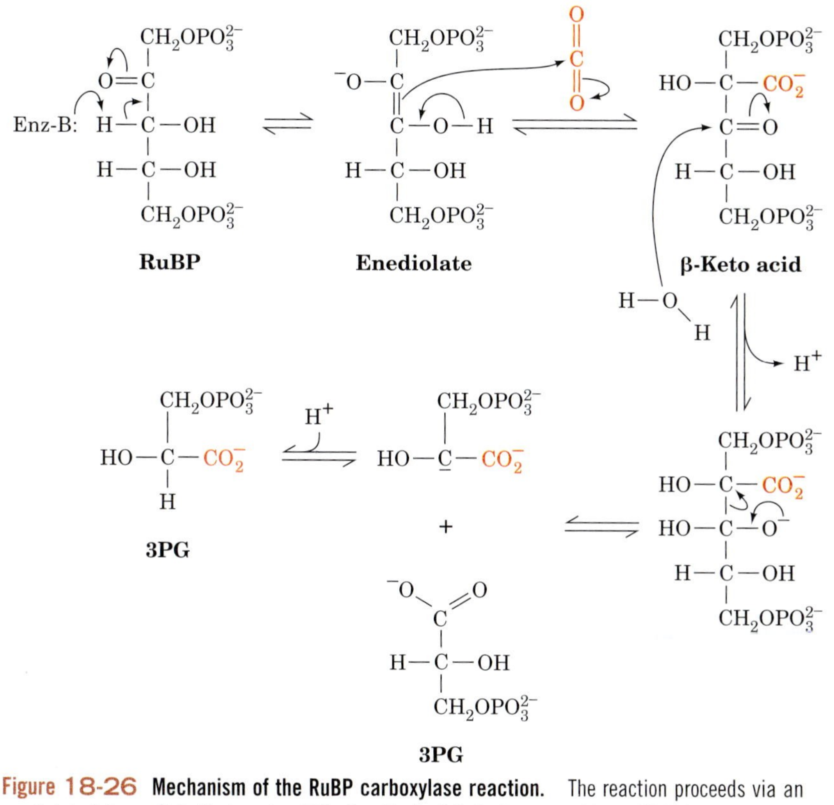

Figure 18-26 Mechanism of the RuBP carboxylase reaction. The reaction proceeds via an enediolate intermediate that nucleophilically attacks CO_2 to form a β-keto acid. This intermediate reacts with water to yield two molecules of 3PG. *See the Animated Figures.*

B Carbohydrate Synthesis

The overall stoichiometry of the Calvin cycle is

$$3 \text{ CO}_2 + 9 \text{ ATP} + 6 \text{ NADH} \rightarrow \text{GAP} + 9 \text{ ADP} + 8 \text{ P}_i + 6 \text{ NADP}^+$$

GAP, the primary product of photosynthesis, is used in a variety of biosynthetic pathways, both inside and outside the chloroplast. For example, it can be converted to fructose-6-phosphate by the further action of Calvin cycle enzymes and then to glucose-1-phosphate (G1P) by phosphoglucose isomerase and phosphoglucomutase (Section 15-1C). *G1P is the precursor of the higher order carbohydrates characteristic of plants.*

The polysaccharide α-amylose, a major component of starch (Section 8-2C), is synthesized in the chloroplast stroma as a temporary storage depot for glucose. It is also synthesized as a long-term storage molecule elsewhere in the plant, including leaves, seeds, and roots. G1P is first activated by its reaction with ATP to form ADP–glucose as catalyzed by **ADP–glucose pyrophosphorylase. Starch synthase** then transfers the glucose residue to the nonreducing end of an α-amylose molecule, forming a new glycosidic linkage (Fig. 18-27). The overall reaction is driven by the exergonic hydrolysis of the PP$_i$ released in the formation of ADP–glucose. A similar reaction sequence occurs in glycogen synthesis, which uses UDP–glucose (Section 15-2).

Sucrose, a disaccharide of glucose and fructose (Section 8-2A), is the major transport sugar for delivering carbohydrates to nonphotosynthesizing cells and hence is the major photosynthetic product of green leaves. Since sucrose is synthesized in the cytosol, either glyceraldehyde-3-phosphate or dihydroxyacetone phosphate is transported out of the chloroplast by an antiporter that exchanges phosphate for a triose phosphate. Two trioses combine to form fructose-6-phosphate (F6P) and subsequently glucose-1-phosphate (G1P), which is then activated by UTP to form UDP–glucose. Next, sucrose-6-phosphate is produced in a reaction catalyzed **by sucrose-phosphate synthase.** Finally, sucrose-6-phosphate is hydrolyzed by **sucrose-phosphate phosphatase** to yield sucrose,

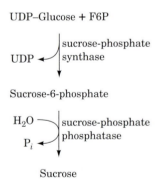

which is then exported to other plant tissues.

Cellulose, which consists of long chains of β(1→4)-linked glucose units and is the major polysaccharide of plants, is also synthesized from UDP–glucose. Plant cell walls consist of almost-crystalline cables containing ~36 cellulose chains, all embedded in an amorphous matrix of other polysaccharides and lignin (Section 8-2B). Unlike starch in plants or glycogen in mammals, cellulose is synthesized by multisubunit enzyme complexes in the plant plasma membrane and extruded into the extracellular space.

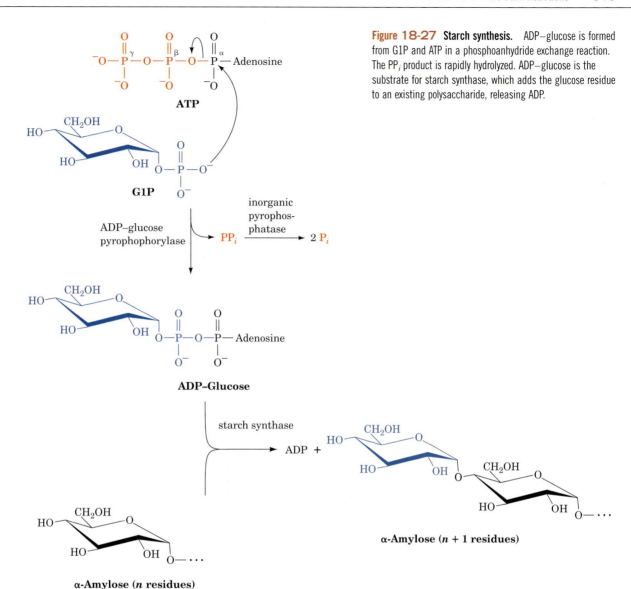

Figure 18-27 Starch synthesis. ADP–glucose is formed from G1P and ATP in a phosphoanhydride exchange reaction. The PP_i product is rapidly hydrolyzed. ADP–glucose is the substrate for starch synthase, which adds the glucose residue to an existing polysaccharide, releasing ADP.

C Control of the Calvin Cycle

During the day, plants satisfy their energy needs via the light and dark reactions of photosynthesis. At night, however, like other organisms, they must use their nutritional reserves to generate ATP and NADPH through glycolysis, oxidative phosphorylation, and the pentose phosphate pathway. Since the stroma contains the enzymes of glycolysis and the pentose phosphate pathway as well as those of the Calvin cycle, *plants must have a light-sensitive control mechanism to prevent the Calvin cycle from consuming this catabolically produced ATP and NADPH in a wasteful futile cycle.*

As we have seen, the control of flux in a metabolic pathway occurs at enzymatic steps that are far from equilibrium (large negative value of ΔG).

Table 18-1 Standard and Physiological Free Energy Changes for the Reactions of the Calvin Cycle

Step[a]	Enzyme	$\Delta G^{\circ\prime}$ ($kJ \cdot mol^{-1}$)	ΔG ($kJ \cdot mol^{-1}$)
1	Phosphoribulokinase	−21.8	−15.9
2	Ribulose bisphosphate carboxylase	−35.1	−41.0
3 + 4	Phosphoglycerate kinase + glyceraldehyde-3-phosphate dehydrogenase	+18.0	−6.7
5	Triose phosphate isomerase	−7.5	−0.8
6	Aldolase	−21.8	−1.7
7	Fructose bisphosphatase	−14.2	−27.2
8	Transketolase	+6.3	−3.8
9	Aldolase	−23.4	−0.8
10	Sedoheptulose bisphosphatase	−14.2	−29.7
11	Transketolase	+0.4	−5.9
12	Phosphopentose epimerase	+0.8	−0.4
13	Ribose phosphate isomerase	+2.1	−0.4

[a]Refer to Fig. 18-24.

Source: Bassham, J.A. and Buchanan, B.B., *in* Govindjee (Ed.), *Photosynthesis,* Vol. II, p. 155, Academic Press (1982).

Inspection of Table 18-1 indicates that the three best candidates for flux control in the Calvin cycle are the reactions catalyzed by RuBP carboxylase, FBPase, and SBPase (Reactions 2, 7, and 10 of Fig. 18-24). In fact, the catalytic efficiencies of these three enzymes all vary *in vivo* with the level of illumination.

The activity of RuBP carboxylase responds to three light-dependent factors:

1. pH. On illumination, the pH of the stroma increases from ~7.0 to ~8.0 as protons are pumped from the stroma into the thylakoid lumen. RuBP carboxylase has a sharp pH optimum near pH 8.0.

2. $[Mg^{2+}]$. Recall that the light-induced influx of protons to the thylakoid lumen is accompanied by the efflux of Mg^{2+} to the stroma (Section 18-2D). This Mg^{2+} stimulates RuBP carboxylase.

3. The transition state analog **2-carboxyarabinitol-1-phosphate (CA1P).**

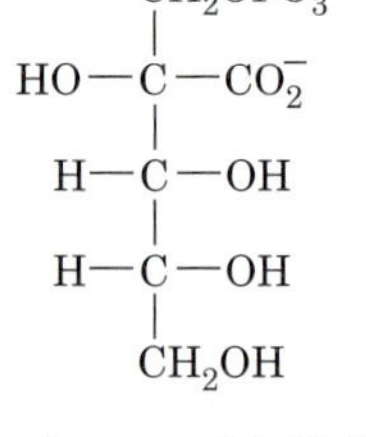

$$
\begin{array}{c}
CH_2OPO_3^{2-} \\
| \\
HO-C-CO_2^- \\
| \\
H-C-OH \\
| \\
H-C-OH \\
| \\
CH_2OH
\end{array}
$$

**2-Carboxyarabinitol-
1-phosphate
(CA1P)**

Many plants synthesize this compound, which inhibits RuBP carboxylase, only in the dark. RuBP carboxylase activase facilitates the release of the tight-binding CA1P from RuBP carboxylase as well as catalyzing its carbamoylation (Section 18-3A).

FBPase and SBPase are also activated by increased pH and $[Mg^{2+}]$, and by NADPH as well. The action of these factors is complemented by a sec-

Figure 18-28   The light-activation mechanism of FBPase and SBPase.   Photoactivated PSI reduces soluble ferredoxin (Fd), which reduces ferredoxin–thioredoxin reductase, which, in turn, reduces the disulfide linkage of thioredoxin. Reduced thioredoxin reacts with the inactive bisphosphatases by disulfide interchange, thereby activating these flux-controlling Calvin cycle enzymes.

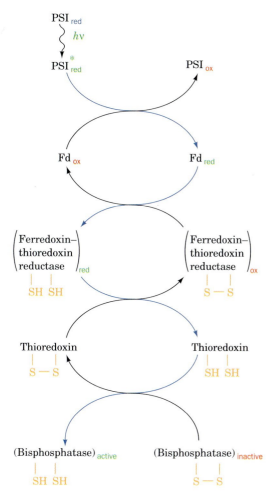

ond regulatory system that responds to the redox potential of the stroma. **Thioredoxin,** an ∼105-residue protein that occurs in many types of cells, contains a reversibly reducible cystine disulfide group. Reduced thioredoxin activates both FBPase and SBPase by a disulfide interchange reaction (Fig. 18-28). The redox level of thioredoxin is maintained by a second disulfide-containing enzyme, **ferredoxin–thioredoxin reductase,** which directly responds to the redox state of the soluble ferredoxin in the stroma. This in turn varies with the illumination level. The thioredoxin system also deactivates phosphofructokinase (PFK), the main flux-generating enzyme of glycolysis (Section 14-4A). Thus, in plants, *light stimulates the Calvin cycle while deactivating glycolysis, whereas darkness has the opposite effect* (that is, the so-called dark reactions do not occur in the dark).

D   Photorespiration

It has been known since the 1960s that *illuminated plants consume O_2 and evolve CO_2 in a pathway distinct from oxidative phosphorylation. In fact, at low CO_2 and high O_2 levels, this* **photorespiration** *process can outstrip photosynthetic CO_2 fixation.* The basis of photorespiration was unexpected: O_2 competes with CO_2 as a substrate for RuBP carboxylase (RuBP carboxylase is therefore also called **RuBP carboxylase–oxygenase** or **RuBisCO**). In the oxygenase reaction, O_2 reacts with the enzyme's other substrate, RuBP, to form 3PG and **2-phosphoglycolate** (Fig. 18-29). The 2-phosphoglycolate is hydrolyzed to **glycolate** by **glycolate phosphatase** and, as described below, is partially oxidized to yield CO_2 by a series of enzymatic reactions that occur in the **peroxisome** and the mitochondrion. Thus, photorespiration is a seemingly wasteful process that undoes some of the work of photosynthesis. In this section we discuss the biochemical basis of photorespiration and how certain plants manage to evade its deleterious effects.

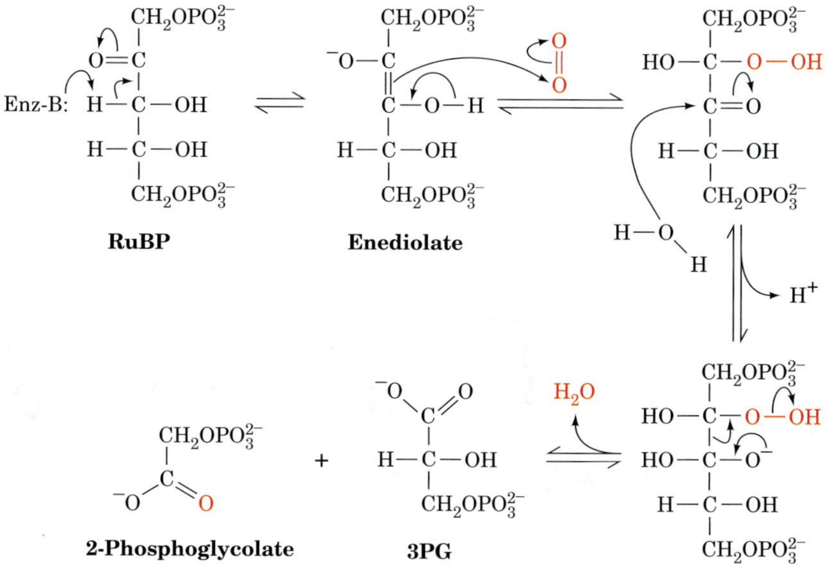

RuBP   **Enediolate**

2-Phosphoglycolate   **3PG**

Figure 18-29   Probable mechanism of the oxygenase reaction catalyzed by RuBP carboxylase–oxygenase. Note the similarity of this mechanism to that of the carboxylase reaction catalyzed by the same enzyme (Fig. 18-26).

Photorespiration Dissipates ATP and NADPH. The photorespiration pathway is outlined in Fig. 18-30. Glycolate is exported from the chloroplast to the peroxisome (also called the glyoxysome; Section 16-5C), where it is oxidized by **glycolate oxidase** to glyoxylate and H_2O_2. The H_2O_2, a potentially harmful oxidizing agent, is converted to H_2O and O_2 by the heme-containing enzyme catalase (Section 17-4C). The glyoxylate can be converted to glycine in a transamination reaction, as is discussed in Section 20-2A, and exported

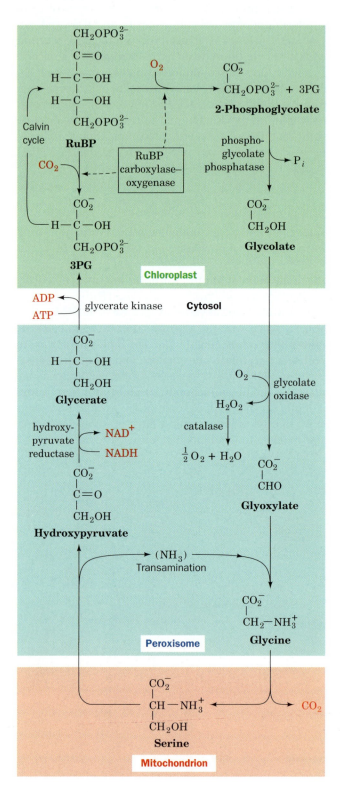

Figure 18-30 Photorespiration. This pathway metabolizes the phosphoglycolate produced by the RuBP carboxylase–catalyzed oxidation of RuBP. The reactions occur, as indicated, in the chloroplast, the peroxisome, the mitochondrion, and the cytosol. Note that two glycines are required to form serine + CO_2.

to the mitochondrion. There, two molecules of glycine are converted to one molecule of serine and one of CO_2. *This is the origin of the CO_2 generated by photorespiration.* The serine is transported back to the peroxisome, where a transamination reaction converts it to **hydroxypyruvate.** This substance is reduced to **glycerate** and phosphorylated in the cytosol to 3PG, which re-enters the chloroplast and is reconverted to RuBP in the Calvin cycle. *The net result of this complex photorespiration cycle is that some of the ATP and NADPH generated by the light reactions is uselessly dissipated.*

Although photorespiration has no known metabolic function, the RuBP carboxylases from the great variety of photosynthetic organisms so far tested all exhibit oxygenase activity. Yet, over the eons, the forces of evolution must have optimized the function of this important enzyme. Photorespiration may confer a selective advantage by protecting the photosynthetic apparatus from photooxidative damage when insufficient CO_2 is available to otherwise dissipate its absorbed light energy. This hypothesis is supported by the observation that when chloroplasts or leaf cells are brightly illuminated in the absence of both CO_2 and O_2, their photosynthetic capacity is rapidly and irreversibly lost.

C_4 Plants Concentrate CO_2.   On a hot, bright day, when photosynthesis has depleted the level of CO_2 at the chloroplast and raised that of O_2, the rate of photorespiration approaches the rate of photosynthesis. This phenomenon is a major limitation on the growth of many plants (and is therefore an important agricultural problem that is being attacked through genetic engineering studies—none of which has yet been successful). However, *certain species of plants, such as sugarcane, corn, and most important weeds, have a metabolic cycle that concentrates CO_2 in their photosynthetic cells, thereby almost totally preventing photorespiration.* The leaves of plants that have this so-called **C_4 cycle** have a characteristic anatomy. Their fine veins are concentrically surrounded by a single layer of so-called **bundle-sheath cells,** which in turn are surrounded by a layer of **mesophyll cells.**

The C_4 cycle (Fig. 18-31) was elucidated in the 1960s by Marshall Hatch and Rodger Slack. It begins when mesophyll cells, which lack RuBP car-

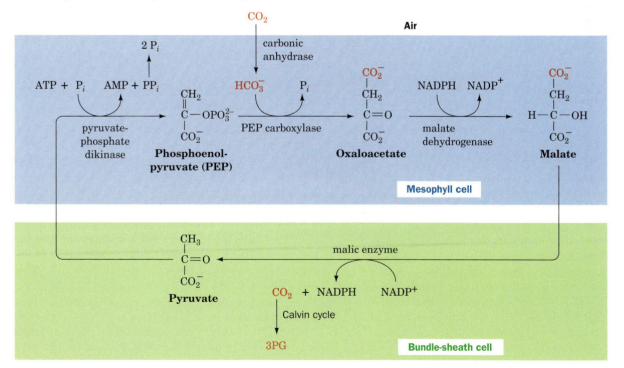

Figure 18-31   The C_4 pathway. CO_2 is concentrated in the mesophyll cells and transported to the bundle-sheath cells for entry into the Calvin cycle.

boxylase, take up atmospheric CO_2 by condensing it as HCO_3^- with phosphoenolpyruvate (PEP) to yield oxaloacetate. The oxaloacetate is reduced by NADPH to malate, which is exported to the bundle-sheath cells (the name C_4 refers to these four-carbon acids). There, the malate is oxidatively decarboxylated by NADP to form CO_2, pyruvate, and NADPH. The CO_2, which has been concentrated by this process, enters the Calvin cycle. The pyruvate is returned to the mesophyll cells, where it is phosphorylated to regenerate PEP. The enzyme that mediates this reaction, **pyruvate-phosphate dikinase,** has the unusual action of activating a phosphate group through the hydrolysis of ATP to AMP + PP_i. This PP_i is further hydrolyzed to two P_i, which is tantamount to the consumption of a second ATP. *CO_2 is thereby concentrated in the bundle-sheath cells at the expense of 2 ATP per CO_2. Photosynthesis in C_4 plants therefore consumes a total of 5 ATP per CO_2 fixed versus the 3 ATP required by the Calvin cycle alone.*

C_4 plants occur largely in tropical regions because they grow faster under hot and sunny conditions than other, so-called **C_3 plants** (so named because they initially fix CO_2 in the form of three-carbon acids). In cooler climates, where photorespiration is less of a burden, C_3 plants have the advantage because they require less energy to fix CO_2.

CAM Plants Store CO_2 through a Variant of the C_4 Cycle. A variant of the C_4 cycle that separates CO_2 acquisition and the Calvin cycle in time rather than in space occurs in many desert-dwelling succulent plants. If these plants opened their stomata (pores in the leaves) by day to acquire CO_2, as most plants do, they would lose an unacceptable amount of water by evaporation. To minimize this loss, these succulents absorb CO_2 only at night and use the reactions of the C_4 pathway (Fig. 18-31) to store it as malate. This process is known as **crassulacean acid metabolism (CAM),** so named because it was first discovered in plants of the family Crassulaceae. The large amount of PEP necessary to store a day's supply of CO_2 is obtained by the breakdown of starch via glycolysis. During the course of the day, the malate is broken down to CO_2, which enters the Calvin cycle, and pyruvate, which is used to resynthesize starch. CAM plants are thus able to carry out photosynthesis with minimal water loss.

SUMMARY

1. Photosynthesis is the process whereby light energy drives the reduction of CO_2 to yield carbohydrates. In plants and cyanobacteria, photosynthesis oxidizes water to O_2.

2. In plants, the photosynthetic machinery consists of protein complexes embedded in the thylakoid membrane and enzymes dissolved in the stroma of chloroplasts.

3. Chlorophyll and other light-absorbing pigments are organized in light-harvesting complexes that funnel light energy to photosynthetic reaction centers (RCs).

4. The purple bacterial photosynthetic reaction center (PbRC) undergoes photooxidation when it absorbs a photon. The excited electron passes through a series of electron carriers before reducing ubiquinone. The reduced ubiquinone is reoxidized by cytochrome bc_1, which in the process translocates four protons from the cytosol to the periplasmic space via a Q cycle. The electron is then returned to the PbRC

via an electron-transport chain resulting in no net oxidation–reduction.

5. In plants and cyanobacteria, photosystems I and II (PSI and PSII) operate in electrical series in an arrangement known as the Z-scheme. The oxidation of water by the Mn-containing oxygen-evolving complex (OEC) is driven by the photooxidation of PSII.

6. The electrons released by the photooxidation of PSII are transferred, via plastoquinone, to the cytochrome b_6f complex, which mediates a proton-translocating Q cycle while passing the electrons to plastocyanin.

7. The photooxidation of PSI drives the electrons obtained from plastocyanin to ferredoxin and then to NADP to produce NADPH. However, in cyclic electron flow, electrons return to cytochrome b_6f, thereby bypassing the need for PSII photooxidation.

8. The reaction centers of the PbRC, PSII, and PSI have similar structures and mechanisms and therefore appear to have arisen from a common ancestor.

9. In photophosphorylation, the protons released by the oxidation of H_2O and proton translocation into the thylakoid lumen generate a transmembrane proton gradient that is tapped by chloroplast ATP synthase to drive the phosphorylation of ADP. A similar process occurs in purple photosynthetic bacteria.

10. The dark reactions use the ATP and NADPH produced in the light reactions to power the synthesis of carbohydrates from CO_2. In the first phase of the Calvin cycle, CO_2 reacts with ribulose-1,5-bisphosphate (RuBP) to ultimately yield glyceraldehyde-3-phosphate (GAP). The remaining reactions of the cycle regenerate the RuBP acceptor of CO_2.

11. RuBP carboxylase, the key enzyme of the dark reactions, is regulated by pH, $[Mg^{2+}]$, and the inhibitory compound 2-carboxyarabinitol-1-phosphate (CA1P). The two bisphosphatases of the Calvin cycle are controlled by the redox state of the chloroplast via disulfide interchange reactions mediated in part by thioredoxin.

12. Photorespiration, in which plants consume O_2 and evolve CO_2, uses the ATP and NADPH produced by the light reactions. C_4 plants minimize the oxygenase activity of RuBP carboxylase (RuBisCO) by concentrating CO_2 in their photosynthetic cells. CAM plants use a related mechanism to conserve water.

 ## REFERENCES

Barber, J., Photosystem II: a multisubunit membrane protein that oxidises water, *Curr. Opin. Struct. Biol.* **12,** 523–530 (2002).

Ben-Shem, A., Frolow, F., and Nelson, N., Crystal structure of plant Photosystem I, *Nature* **426,** 630–635 (2003).

Blankenship, R.E., *Molecular Mechanisms of Photosynthesis,* Blackwell Science (2002).

Chitnis, P.R., Photosystem I: Function and physiology, *Annu. Rev. Plant Physiol. Plant Biol.* **52,** 593–626 (2001).

Ferreira, K.N., Iverson, T.M., Maghlaoui, K., Barber, J., and Iwata, S., Architecture of the photosynthetic oxygen-evolving center, *Science* **303,** 1831–1838 (2004). [The X-ray structure of PSII.]

Heathcote, P., Fyfe, P.K., and Jones, M.R., Reaction centres: the structure and evolution of biological solar power, *Trends Biochem. Sci.* **27,** 79–87 (2002).

Jordan, P., Fromme, P., Witt, H.T., Klukas, O., Saenger, W., and Krauss, N., Three-dimensional structure of cyanobacterial photosystem I at 2.5 Å resolution, *Nature* **411,** 909–917 (2001).

Koepke, J., Hu, X., Muenke, C., Schulen, K., and Michel, H., The crystal structure of the light-harvesting complex II (B800–850) from *Rhodospirillum molischianum, Structure* **4,** 581–597 (1996).

Kurisu, G., Zhang, H., Smith, J.L., and Cramer, W.A., Structure of the cytochrome $b_6 f$ complex of oxygenic photosynthesis: Tuning the cavity, *Science* **302,** 1009–1014 (2003); *and* Stroebel, D., Choquet, Y., Popot, J.-L., and Picot, D., An atypical heam in the cytochrome $b_6 f$ complex, *Nature* **426,** 413–418 (2003).

Spreitzer, R.J. and Salvucci, M.E., Rubisco: structure, regulatory interactions, and possibilities for a better enzyme, *Annu. Rev. Plant Biochem.* **53,** 449–475 (2001).

 ## KEY TERMS

photosynthesis
light reactions
dark reactions
stroma
thylakoid
grana
stromal lamella

photosynthetic reaction center
antenna chlorophyll
LHC
accessory pigment
Planck's law
internal conversion

fluorescence
exciton transfer (resonance energy transfer)
photooxidation
special pair
PSI
PSII

Z-scheme
photophosphorylation
Calvin cycle
photorespiration
C_4 plant
C_3 plant
CAM

 ## STUDY EXERCISES

1. Explain the relationship between the light and dark reactions.

2. Why are light-harvesting complexes important?

3. How do molecules dissipate absorbed light energy?

4. Describe the Z-scheme.

5. What are the implications of cyclic and noncyclic electron transfer in PSI?

6. Compare and contrast photophosphorylation and oxidative phosphorylation.

7. Summarize the two stages of the Calvin cycle.

8. How is photosynthesis regulated?

9. How do plants minimize photorespiration?

PROBLEMS

1. The "red tide" is a massive proliferation of certain algal species that causes seawater to become visibly red. Describe the spectral characteristics of the dominant photosynthetic pigments in these algae.

2. The net equation for oxidative phosphorylation can be written as

$$2\ NADH + 2\ H^+ + O_2 \rightarrow 2\ H_2O + 2\ NAD^+$$

Write an analogous equation for the light reactions of photosynthesis.

3. The three electron-transporting complexes of the thylakoid membrane can be called plastocyanin–ferredoxin oxidoreductase, plastoquinone–plastocyanin oxidoreductase, and water–plastoquinone oxidoreductase. What are the common names of these enzymes and in what order do they act?

4. $H_2^{18}O$ is added to a suspension of chloroplasts capable of photosynthesis. Where does the label appear when the suspension is exposed to sunlight?

5. (a) Calculate the energy of one mole of photons of red light ($\lambda = 700$ nm). (b) How many moles of ATP could theoretically be synthesized using this energy?

6. (a) Calculate $\Delta\mathscr{E}^{\circ\prime}$ and $\Delta G^{\circ\prime}$ for the light reactions in plants, that is, the four-electron oxidation of H_2O by NADP. (b) Use the solution of Problem 18-5 to calculate how many moles of photons of red light ($\lambda = 700$ nm) are theoretically required to drive this process. (c) How many moles of photons of UV light ($\lambda = 220$ nm) would be required?

7. Describe the functional similarities between the purple bacterial photosynthetic reaction center and PSI.

8. Why is it possible for chloroplasts to absorb much more than 8–10 photons per O_2 molecule evolved?

9. Under conditions of very high light intensity, excess absorbed solar energy is dissipated by the action of photoprotective proteins in the thylakoid membrane. (a) Explain why it is advantageous for these proteins to be activated by buildup of the proton gradient across the membrane. (b) Which of the mechanisms for dissipating light energy shown in Fig. 18-6 would best protect the photosystems from excess light energy?

10. Predict the effect of an uncoupler such as dinitrophenol (Fig. 17-28) on production of (a) ATP and (b) NADPH in a chloroplast.

11. Describe the effects of an increase in oxygen pressure on the dark reactions of photosynthesis.

12. Chloroplasts are illuminated until the levels of the Calvin cycle intermediates reach a steady state. The light is then turned off. How do the levels of RuBP and 3PG vary after this point?

13. Calculate the energy cost of the Calvin cycle combined with glycolysis and oxidative phosphorylation, that is, the ratio of the energy spent synthesizing starch from CO_2 and photosynthetically produced NADPH and ATP to the energy generated by the complete oxidation of starch. Assume that each NADPH is energetically equivalent to 3 ATP and that starch biosynthesis and breakdown are mechanistically identical to glycogen synthesis and breakdown.

14. The leaves of some species of desert plants taste sour in the early morning, but, as the day wears on, they become tasteless and then bitter. Explain.

Lipids act as an organism's major storage form of metabolic energy. In hibernating mammals, such as this marmot, these fat reserves are synthesized during the warmer months and are oxidized during the winter to keep the animal alive. [Jeff Foote/Bruce Coleman, Inc.]

Lipid Metabolism

Most cells contain a wide variety of lipids, but many of these structurally distinct molecules are functionally similar. For example, most cells can tolerate variations in the lipid composition of their membranes, provided that membrane fluidity, which is largely a property of their component fatty acid chains, is maintained (Section 9-2B).

Even greater variation is exhibited in the cellular content of lipids stored as energy reserves. Triacylglycerol stores are built up and gradually depleted in response to changing physiological demands. The camel's hump is a well-known example of a fat depot that supplies energy (the catabolism of the fat also generates water). Other organisms that undergo dramatic changes in body fat content are hibernating mammals and birds that migrate long distances without refueling. Lipid metabolism in these cases is notable for the sheer amount of material that flows through the few relatively simple biosynthetic and degradative pathways.

In this chapter, we acknowledge the central function of lipids in energy metabolism by first examining the absorption and transport of their component fatty acids and the oxidation of these fatty acids to produce energy. The second part of this chapter examines the synthesis of fatty acids and other lipids, including glycerophospholipids, sphingolipids, and cholesterol.

1 Lipid Digestion, Absorption, and Transport

Triacylglycerols (also called fats or triglycerides) constitute ~90% of the dietary lipid and are the major form of metabolic energy storage in humans. Triacylglycerols consist of glycerol triesters of fatty acids such as palmitic and oleic acids (*at left*; the names and structural formulas of some biologically common fatty acids are listed in Table 9-1). The mechanisms for digesting, absorbing, and transporting triacylglycerols from the intestine to the tissues must accommodate their inherent hydrophobicity.

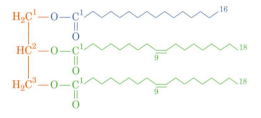

1-Palmitoyl-2,3-dioleoyl-glycerol

A Digestion and Absorption

Since triacylglycerols are water insoluble, whereas digestive enzymes are water soluble, triacylglycerol digestion takes place at lipid–water interfaces. The rate of triacylglycerol digestion therefore depends on the surface area of the interface, which is greatly increased by the churning peristaltic movements of the intestine combined with the emulsifying action of **bile acids.** The bile acids (also called **bile salts**) are amphipathic detergent-like molecules that act to solubilize fat globules. Bile acids are cholesterol derivatives that are synthesized by the liver and secreted as glycine or **taurine** conjugates (Fig. 19-1) into the gallbladder for storage. From there, they are secreted into the small intestine, where lipid digestion and absorption mainly take place.

Lipases Act at the Lipid–Water Interface. Pancreatic **lipase (triacylglycerol lipase)** catalyzes the hydrolysis of triacylglycerols at their 1 and 3 positions to form sequentially **1,2-diacylglycerols** and **2-acylglycerols,** together with the Na^+ and K^+ salts of fatty acids (soaps). The enzymatic activity of pancreatic lipase greatly increases when it contacts the lipid–water interface, a phenomenon known as **interfacial activation.** Binding to the lipid–water interface requires mixed micelles of phosphatidylcholine and bile acids, as well as pancreatic **colipase,** a 90-residue protein that forms a 1:1 complex with lipase. The X-ray structures, determined by Christian Cambillau, of pancreatic lipase–colipase complexes reveal the structural basis of the

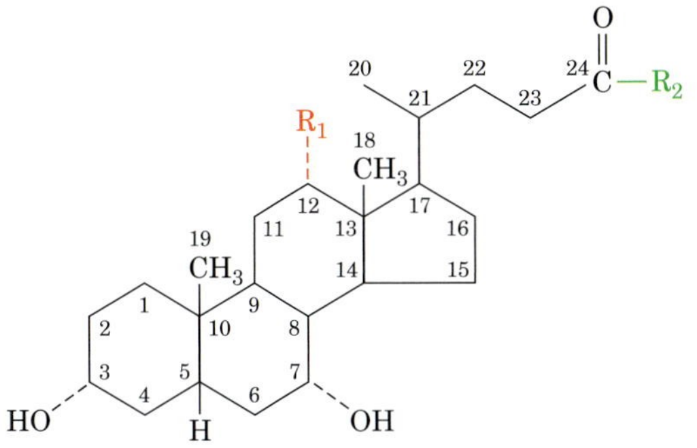

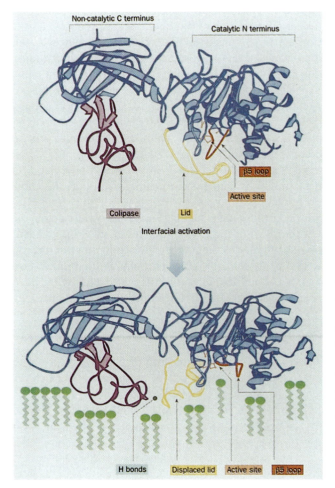

$R_1 = OH$ $R_1 = H$

	$R_1 = OH$	$R_1 = H$
$R_2 = OH$	**Cholic acid**	**Chenodeoxycholic acid**
$R_2 = NH—CH_2—COOH$	**Glycocholic acid**	**Glycochenodeoxycholic acid**
$R_2 = NH—CH_2—CH_2—SO_3H$	**Taurocholic acid**	**Taurochenodeoxycholic acid**

Figure 19-1 Structures of the major bile acids and their glycine and taurine conjugates.

interfacial activation of lipase as well as how colipase and micelles help lipase bind to the lipid–water interface (Fig. 19-2).

The active site of the 449-residue pancreatic lipase, which is contained in the enzyme's N-terminal domain (residues 1–336), contains a catalytic triad that closely resembles that in serine proteases (Section 11-5B; recall that ester hydrolysis is mechanistically similar to peptide hydrolysis). In the

Figure 19-2 The mechanism of interfacial activation of triacylglycerol lipase.   The enzyme is in complex with procolipase (the precursor of colipase; *magenta*). On binding to a phospholipid micelle (*green*), the 25-residue lid (*yellow*) covering the enzyme's active site (*tan*) changes conformation so as to expose its hydrophobic residues, thereby uncovering the active site. This causes the 10-residue β5 loop (*brown*) to move aside in a way that forms the enzyme's oxyanion hole. The procolipase also changes its conformation so as to hydrogen bond to the "open" lid, thereby stabilizing it in this conformation and, together with lipase, forming an extended hydrophobic surface. [From *Nature* **362**, 793 (1993). Reproduced with permission. PDBid 1LPA.]

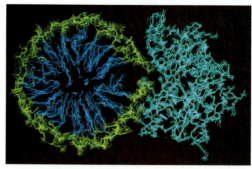

(a)

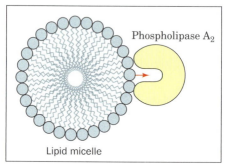

(b)

Figure 19-3   Substrate binding to phospholipase A₂.
(a) Hypothetical model of phospholipase A₂ in complex with a micelle of lysophosphatidylethanolamine as shown in cross section. The protein is drawn in cyan, the phospholipid head groups are yellow, and their hydrocarbon tails are blue. The calculated atomic motions of the assembly are indicated through a series of superimposed images taken at 5-ps intervals. [Courtesy of Raymond Salemme, E. I. du Pont de Nemours & Company.] (b) Schematic diagram of a phospholipid contained in a micelle entering the hydrophobic phospholipid channel in phospholipase A₂ (*red arrow*).

absence of lipid micelles, lipase's active site is covered by a 25-residue helical "lid." However, in the presence of the micelles, the lid undergoes a complex structural reorganization that exposes the active site. Simultaneously, a 10-residue loop, called the β5 loop, changes conformation in a way that forms the active enzyme's oxyanion hole and generates a hydrophobic surface near the entrance to the active site.

Colipase binds to the C-terminal domain of lipase (residues 337–449) such that the hydrophobic tips of its three loops extend from the complex. This creates a continuous hydrophobic plateau, extending >50 Å past the lipase active site, that presumably helps bind the complex to the lipid surface. Colipase also forms three hydrogen bonds to the opened lid, thereby stabilizing it in this conformation.

Other lipases, such as phospholipase A₂ (Fig. 9-6), also preferentially catalyze reactions at interfaces. Instead of changing its conformation, however, phospholipase A₂ contains a hydrophobic channel that provides the substrate with direct access from the phospholipid aggregate (micelle or membrane) surface to the bound enzyme's active site (Fig. 19-3). Hence, on leaving its micelle to bind to the enzyme, the substrate need not become solvated and then desolvated. In contrast, soluble and dispersed phospholipids must first surmount these significant kinetic barriers in order to bind to the enzyme.

Bile Acids and Fatty Acid–Binding Protein Facilitate the Intestinal Absorption of Lipids.

The mixture of fatty acids and mono- and diacylglycerols produced by lipid digestion is absorbed by the cells lining the small intestine (the intestinal **mucosa**). *Bile acids not only aid lipid digestion; they are essential for the absorption of the digestion products.* The micelles formed by the bile acids take up the nonpolar lipid degradation products so as to permit their transport across the unstirred aqueous boundary layer at the intestinal wall. The importance of this process is demonstrated in individuals with obstructed bile ducts: They absorb little of their ingested lipids but, rather, eliminate them in hydrolyzed form in their feces. Bile acids are likewise required for the efficient intestinal absorption of the lipid-soluble vitamins A, D, E, and K.

Inside the intestinal cells, fatty acids form complexes with **intestinal fatty acid–binding protein (I-FABP),** a cytoplasmic protein that increases the effective solubility of these water-insoluble substances and also protects the cell from their detergent-like effects. The X-ray structure of rat I-FABP, determined by James Sacchettini, shows that this monomeric, 131-residue protein consists largely of 10 antiparallel β strands stacked in two approximately orthogonal β sheets (Fig. 19-4). A fatty acid molecule occupies a gap between two of the β strands, lying between the β sheets so that it is more or less parallel to the gapped β strands (a structure that has been dubbed a "β-clam"). The fatty acid's carboxyl group interacts with Arg 106, Gln 115, and two bound water molecules, whereas its tail in encased by the side chains of several hydrophobic, mostly aromatic residues.

B Lipid Transport

The fatty acid products of lipid digestion that are absorbed by the intestinal mucosa make their way to other tissues for catabolism or storage. But

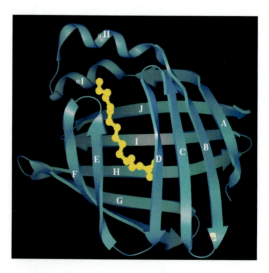

Figure 19-4   X-Ray structure of rat intestinal fatty acid–binding protein.   The structure is shown in ribbon form (*blue*), in complex with palmitate, shown in ball-and-stick form (*yellow*). [Courtesy of James Sacchettini, Texas A&M University. PDBid 2IFB.]

Table 19-1 Characteristics of the Major Classes of Lipoproteins in Human Plasma.

	Chylomicrons	VLDL	IDL	LDL	HDL
Density (g·cm^{-3})	<0.95	<1.006	1.006–1.019	1.019–1.063	1.063–1.210
Particle diameter (Å)	750–12,000	300–800	250–350	180–250	50–120
Particle mass (kD)	400,000	10,000–80,000	5000–10,000	2300	175–360
% Protein[a]	1.5–2.5	5–10	15–20	20–25	40–55
% Phospholipids[a]	7–9	15–20	22	15–20	20–35
% Free cholesterol[a]	1–3	5–10	8	7–10	3–4
% Triacylglycerols[b]	84–89	50–65	22	7–10	3–5
% Cholesteryl esters[b]	3–5	10–15	30	35–40	12
Major apolipoproteins	A-I, A-II, B-48, C-I, C-II, C-III, E	B-100, C-I, C-II, C-III, E	B-100, C-I, C-II, C-III, E	B-100	A-I, A-II, C-I, C-II, C-III, D, E

[a]Surface components.

[b]Core lipids.

because they are only sparingly soluble in aqueous solution, lipids are transported by the circulation in complex with proteins.

Lipoproteins Are Complexes of Lipid and Protein. *Lipoproteins are globular micelle-like particles that consist of a nonpolar core of triacylglycerols and cholesteryl esters surrounded by an amphiphilic coating of protein, phospholipid, and cholesterol.* There are five classes of lipoproteins (Table 19-1), which vary in composition and physiological function.

Intestinal mucosal cells convert dietary fatty acids to triacylglycerols and package them, along with dietary cholesterol, into lipoproteins called **chylomicrons.** These particles are released into the intestinal lymph and are transported through the lymphatic vessels before draining into the large veins. The bloodstream then delivers chylomicrons throughout the body. Other lipoproteins known as **very low density lipoproteins (VLDL), intermediate density lipoproteins (IDL),** and **low density lipoproteins (LDL)** are synthesized by the liver to transport endogenous (internally produced) triacylglycerols and cholesterol from the liver to the tissues. **High density lipoproteins (HDL)** transport cholesterol and other lipids from the tissues back to the liver.

Each lipoprotein contains just enough protein, phospholipid, and cholesterol to form an ~20-Å-thick monolayer of these substances on the particle surface (Fig. 19-5). Lipoprotein densities increase with decreasing particle diameter because the density of their outer coating is greater than

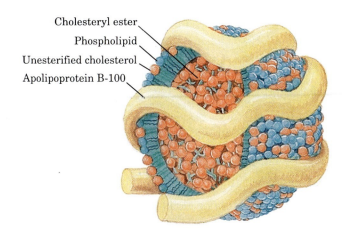

Cholesteryl ester
Phospholipid
Unesterified cholesterol
Apolipoprotein B-100

Figure 19-5 Diagram of LDL, the major cholesterol carrier of the bloodstream. This spheroidal particle consists of some 1500 cholesteryl ester molecules surrounded by an amphiphilic coat of ~800 phospholipid molecules, ~500 cholesterol molecules, and a single 4536-residue molecule of apolipoprotein B-100.

that of their inner core. Thus, the HDL, which are the most dense of the lipoproteins, are also the smallest.

Apolipoproteins Coat Lipoprotein Surfaces. The protein components of lipoproteins are known as **apolipoproteins** or just **apoproteins.** At least nine apolipoproteins are distributed in different amounts in the human lipoproteins (Table 19-1). For example, LDL contain **apolipoprotein B-100 (apoB-100).** This protein, a 4536-residue monomer (and thus one of the largest monomeric proteins known), has a hydrophobicity approaching that of integral membrane proteins. Each LDL particle contains only one molecule of apoB-100, which appears to cover at least half of the particle surface (Fig. 19-5).

Unlike apoB-100, the other apolipoproteins are water soluble and associate rather weakly with lipoproteins. These apolipoproteins also have a high helix content, which increases when they are incorporated into lipoproteins. Contact with a hydrophobic surface apparently favors formation of helices, which satisfy the hydrogen bonding potential of the protein's polar backbone groups. In addition, *the helices in apolipoproteins have hydrophilic and hydrophobic side chains on opposite sides of the helical cylinder, suggesting that lipoprotein α helices are amphipathic and float on phospholipid surfaces, much like logs on water.* The charged head groups of the lipids presumably bind to oppositely charged residues on the helix while the first few methylene groups of their fatty acyl chains associate with the nonpolar face of the helix.

Apolipoprotein A-I (apoA-I), which occurs in chylomicrons and HDL, is a 243-residue, 29-kD polypeptide. It consists largely of tandem 22-residue segments of similar sequence. The X-ray structure of a truncated apolipoprotein A-I (lacking residues 1–43) reveals that the polypeptide chain forms a pseudocontinuous α helix that is punctuated by kinks at Pro residues spaced about every 22 residues. Four monomers associate to form the structure shown in Fig. 19-6a. The size and twisted elliptical shape of this complex seem ideal for wrapping around an HDL particle. Figure 19-6b contains a helical wheel representation of a portion of apoA-I, illustrating the amphipathic nature of the helix.

Chylomicrons Are Delipidated in the Capillaries of Peripheral Tissues.

Chylomicrons adhere to binding sites on the inner surface (endothelium) of the capillaries in skeletal muscle and adipose tissue. The chylomicron's component triacylglycerols are hydrolyzed through the action of the extracellular enzyme

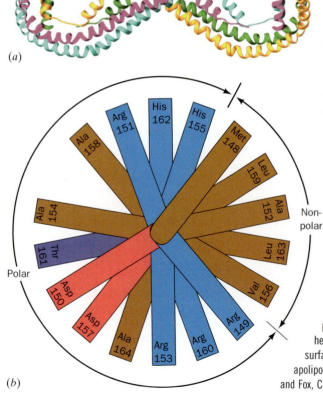

(a)

(b)

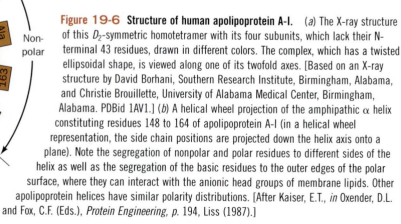

Figure 19-6 Structure of human apolipoprotein A-I. (a) The X-ray structure of this D_2-symmetric homotetramer with its four subunits, which lack their N-terminal 43 residues, drawn in different colors. The complex, which has a twisted ellipsoidal shape, is viewed along one of its twofold axes. [Based on an X-ray structure by David Borhani, Southern Research Institute, Birmingham, Alabama, and Christie Brouillette, University of Alabama Medical Center, Birmingham, Alabama. PDBid 1AV1.] (b) A helical wheel projection of the amphipathic α helix constituting residues 148 to 164 of apolipoprotein A-I (in a helical wheel representation, the side chain positions are projected down the helix axis onto a plane). Note the segregation of nonpolar and polar residues to different sides of the helix as well as the segregation of the basic residues to the outer edges of the polar surface, where they can interact with the anionic head groups of membrane lipids. Other apolipoprotein helices have similar polarity distributions. [After Kaiser, E.T., *in* Oxender, D.L. and Fox, C.F. (Eds.), *Protein Engineering, p.* 194, Liss (1987).]

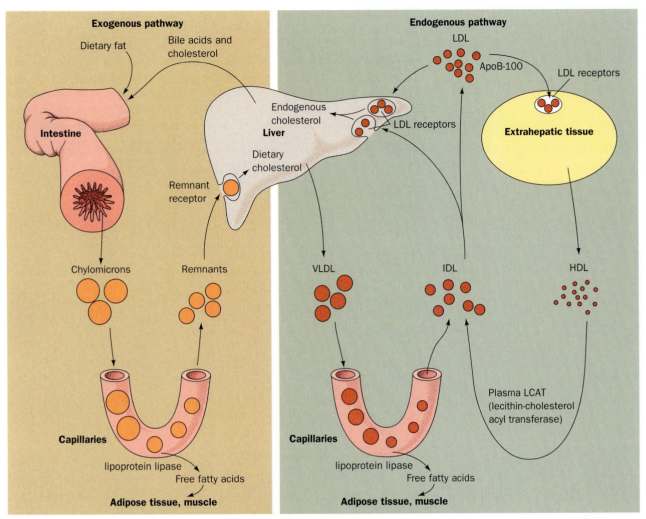

Figure 19-7 Model for plasma triacylglycerol and cholesterol transport in humans. [After Brown, M.S. and Goldstein, J.L., *in* Brunwald, E., Isselbacher, K.J., Petersdorf, R.G., Wilson, J.D., Martin, J.B., and Fauci, A.S. (Eds.), *Harrison's Principles of Internal Medicine* (11th ed.), p. 1652, McGraw-Hill (1987).]
🔁 **See the Animated Figures.**

lipoprotein lipase. The tissues then take up the liberated monoacylglycerol and fatty acids. The chylomicrons shrink as their triacylglycerols are progressively hydrolyzed until they are reduced to cholesterol-enriched **chylomicron remnants.** The remnants dissociate from the capillary endothelium and re-enter the circulation to be taken up by the liver. *Chylomicrons therefore deliver dietary triacylglycerols to muscle and adipose tissue, and dietary cholesterol to the liver* (Fig. 19-7, *left*).

VLDL Are Gradually Degraded.   Very low density lipoproteins (VLDL), which transport endogenous triacylglycerols and cholesterol, are also degraded by lipoprotein lipase in the capillaries of adipose tissue and muscle (Fig. 19-7, *right*). The released fatty acids are taken up by the cells and oxidized for energy or used to resynthesize triacylglycerols. The glycerol backbone of triacylglycerols is transported to the liver or kidneys and converted to the glycolytic intermediate dihydroxyacetone phosphate. Oxidation of this three-carbon unit, however, yields only a small fraction of the energy available from oxidizing the three fatty acyl chains of a triacylglycerol.

After giving up their triacylglycerols, the VLDL remnants, which have also lost some of their apolipoproteins, appear in the circulation first as

IDL and then as LDL. About half of the VLDL, after degradation to IDL and LDL, are taken up by the liver.

Cells Take Up LDL by Receptor-Mediated Endocytosis. Animal cells acquire cholesterol, an essential component of cell membranes (Section 9-2B), either by synthesizing it or by taking up LDL, which are rich in cholesterol and cholesteryl esters. The latter process, as Michael Brown and Joseph Goldstein have demonstrated, occurs by **receptor-mediated endocytosis** (engulfment) of the LDL (Fig. 19-8). The LDL particles are sequestered by

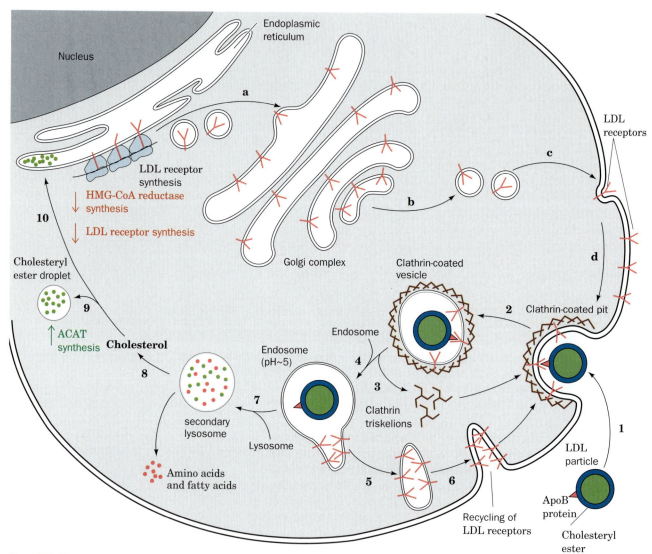

Figure 19-8 *Key to Function.* **Receptor-mediated endocytosis of LDL.** LDL receptor is synthesized on the endoplasmic reticulum (**a**), processed in the Golgi apparatus (**b**), and inserted into the plasma membrane (**c**), where it becomes a component of clathrin-coated pits (**d**). The apolipoprotein B-100 (apoB-100) component of LDL specifically binds to LDL receptors on clathrin-coated pits (**1**). These bud into the cell (**2**) to form clathrin-coated vesicles, whose clathrin coats depolymerize to form triskelions (**3**), which are recycled to the cell surface. The uncoated vesicles fuse with vesicles called endosomes (**4**), which have an internal pH of ~5.0. The acidity induces the LDL particle to dissociate from its receptor. The LDL accumulates in the vesicular portion of the endosome, whereas the LDL receptors concentrate in the membrane of an attached tubular structure, which then separates from the endosome (**5**) and subsequently recycles the LDL receptors to the plasma membrane (**6**). The vesicular portion of the endosome fuses with a lysosome (**7**), yielding a **secondary lysosome** wherein the apoB-100 component of the LDL and the cholesteryl esters are hydrolyzed. This releases cholesterol (**8**), which either is converted to cholesteryl esters through the action of **acyl-CoA:cholesterol acyltransferase (ACAT)** and sequestered in a droplet (**9**) or proceeds to the endoplasmic reticulum (**10**). Increased cholesterol concentration in the endoplasmic reticulum decreases the rate of synthesis of **HMG-CoA reductase** (the enzyme that catalyzes the rate-limiting step of cholesterol biosynthesis; Section 19-7B) and LDL receptors (*down arrows*), while increasing that of ACAT (*up arrow*). [After Brown, M.S. and Goldstein, J.L., *Curr. Top. Cell. Reg.* **26**, 7 (1985).] *See the Animated Figures.*

LDL receptors, cell-surface transmembrane glycoproteins that specifically bind apoB-100. LDL receptors cluster into **clathrin-coated pits,** which gather the cell-surface receptors that are destined for endocytosis while excluding other cell-surface proteins. The coated pits invaginate from the plasma membrane to form **clathrin-coated vesicles** (Section 9-4E). Then, after divesting themselves of their clathrin coats, the vesicles fuse with vesicles known as **endosomes,** whose internal pH is ~5.0. Under these conditions, the LDL particle dissociates from its receptor. The receptors are recycled back to the cell surface, while the endosome with its enclosed LDL fuses with a lysosome. In the lysosome, LDL's apoB-100 is rapidly degraded to its component amino acids, and the cholesteryl esters are hydrolyzed to yield cholesterol and fatty acids.

Receptor-mediated endocytosis is a general mechanism whereby cells take up large molecules, each through a corresponding specific receptor. Many cell-surface receptors recycle between the plasma membrane and the endosomal compartment as does the LDL receptor, even in the absence of ligand. The LDL receptor cycles in and out of the cell about every 10 minutes. New receptors are synthesized on the endoplasmic reticulum and travel via vesicles through the Golgi complex to the cell surface (Fig. 19-8; Section 9-4E). The intracellular concentration of free cholesterol controls the rate of LDL receptor synthesis (Sections 19-7A and 19-7B). However, defects in the LDL receptor system lead to abnormally high levels of circulating cholesterol with the attendant increased risk of heart disease (Section 19-7C).

HDL Transports Cholesterol from the Tissues to the Liver. HDL have essentially the opposite function of LDL: *They remove cholesterol from the tissues.* HDL are assembled in the plasma from components largely obtained through the degradation of other lipoproteins. *A circulating HDL particle acquires its cholesterol by extracting it from cell-surface membranes.* The cholesterol is then converted to cholesteryl esters by the HDL-associated enzyme **lecithin–cholesterol acyltransferase (LCAT),** which is activated by apoA-I. HDL therefore function as cholesterol scavengers.

The liver is the only organ capable of disposing of significant quantities of cholesterol (by its conversion to bile acids; Fig. 19-1). About half of the VLDL, after their degradation IDL and LDL, are taken up by the liver via LDL receptor-mediated endocytosis (Fig. 19-7, *right*). However, liver cells take up HDL by an entirely different mechanism: Rather than being engulfed and degraded, an HDL particle binds to a cell-surface receptor named **SR-BI** (for *scavenger receptor class B type I*) and selectively transfers its component lipids to the cell. The lipid-depleted HDL particle then dissociates from the cell and re-enters the circulation.

2 Fatty Acid Oxidation

The triacylglycerols stored in adipocytes are mobilized in times of metabolic need by the action of **hormone-sensitive lipase** (Section 19-5). The free fatty acids are released into the bloodstream, where they bind to albumin, a soluble 66-kD monomeric protein. In the absence of albumin, the maximum solubility of fatty acids is ~10^{-6} M; the effective solubility of fatty acids in complex with albumin is as high as 2 mM. Nevertheless, those rare individuals with **analbuminemia** (severely depressed levels of albumin) suffer no apparent adverse symptoms; evidently, their fatty acids are transported in complex with other serum proteins.

Fatty acids are catabolized by an oxidative process that releases free energy. The biochemical strategy of fatty acid oxidation was understood long

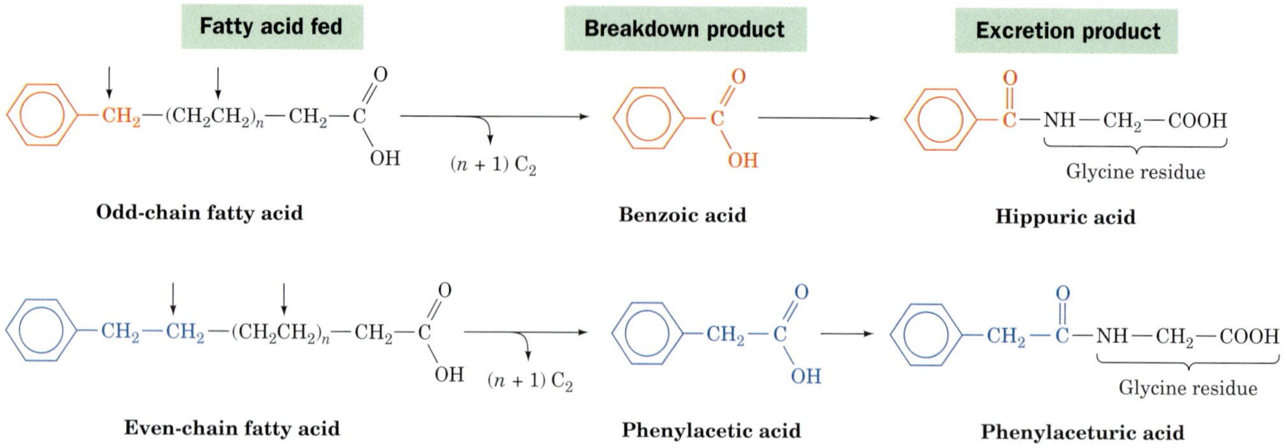

Fatty acid fed   **Breakdown product**   **Excretion product**

Odd-chain fatty acid   Benzoic acid   Hippuric acid

Even-chain fatty acid   Phenylacetic acid   Phenylaceturic acid

Figure 19-9 Franz Knoop's classic experiment indicating that fatty acids are metabolically oxidized at their β-carbon atom.   ω-Phenyl-labeled fatty acids containing an odd number of carbon atoms are oxidized to the phenyl-labeled C_1 product, benzoic acid, whereas those with an even number of carbon atoms are oxidized to the phenyl-labeled C_2 product, phenylacetic acid. These products are excreted as their respective glycine amides, hippuric and phenylaceturic acids. The vertical arrows indicate the deduced sites of carbon oxidation. The intermediate C_2 products are oxidized to CO_2 and H_2O and were therefore not isolated.

before the oxidative enzymes were purified. In 1904, Franz Knoop, in the first use of chemical labels to trace metabolic pathways, fed dogs fatty acids labeled at their ω (last) carbon atom by a benzene ring and isolated the phenyl-containing metabolic products from their urine. Dogs fed labeled odd-chain fatty acids excreted **hippuric acid,** the glycine amide of **benzoic acid,** whereas those fed labeled even-chain fatty acids excreted **phenylaceturic acid,** the glycine amide of **phenylacetic acid** (Fig. 19-9). Knoop therefore deduced that fatty acids are progressively degraded by two-carbon units and that this process involves the oxidation of the carbon atom β to the carboxyl group. Otherwise, the phenylacetic acid would be further oxidized to benzoic acid. Knoop's **β oxidation** hypothesis was only confirmed in the 1950s. The β-oxidation pathway is a series of enzyme-catalyzed reactions that operates in a repetitive fashion to progressively degrade fatty acids by removing two-carbon units.

A Fatty Acid Activation

Before fatty acids can be oxidized, they must be "primed" for reaction in an ATP-dependent acylation reaction to form fatty acyl-CoA. The activation process is catalyzed by a family of at least three **acyl-CoA synthetases** (also called **thiokinases**) that differ in their chain-length specificities. These enzymes, which are associated with either the endoplasmic reticulum or the outer mitochondrial membrane, all catalyze the reaction

$$\text{Fatty acid} + \text{CoA} + \text{ATP} \rightleftharpoons \text{acyl-CoA} + \text{AMP} + \text{PP}_i$$

This reaction proceeds via an acyladenylate mixed anhydride intermediate that is attacked by the sulfhydryl group of CoA to form the thioester product (Fig. 19-10), thereby preserving the free energy of ATP hydrolysis in the "high-energy" thioester bond (Section 13-2D). The overall reaction is driven to completion by the exergonic hydrolysis of pyrophosphate catalyzed by inorganic pyrophosphatase.

B Transport across the Mitochondrial Membrane

Although fatty acids are activated for oxidation in the cytosol, they are oxidized in the mitochondrion, as Eugene Kennedy and Albert Lehninger

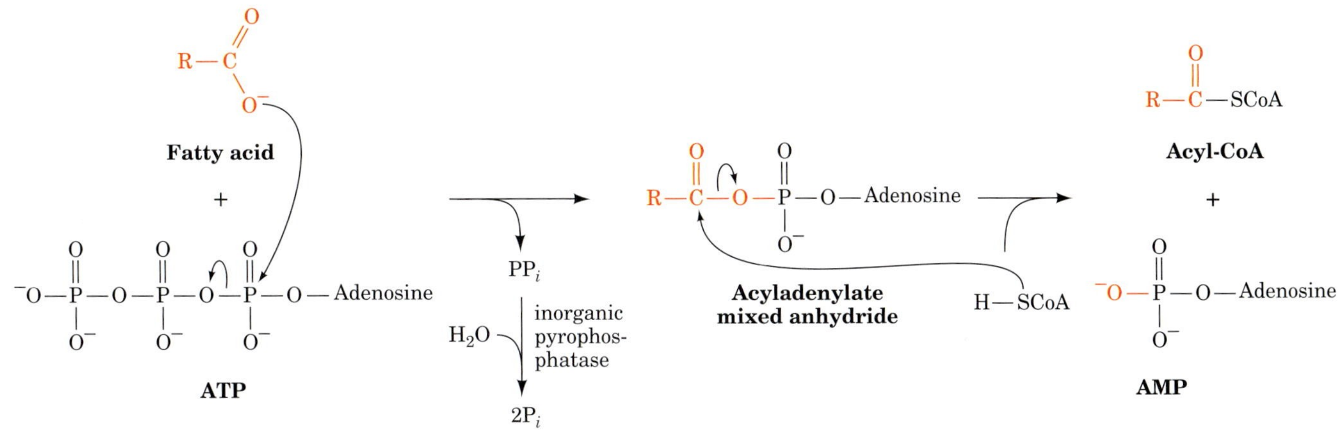

Figure 19-10 The mechanism of fatty acid activation catalyzed by acyl-CoA synthetase.
Formation of acyl-CoA involves an intermediate acyladenylate mixed anhydride.

established in 1950. We must therefore consider how fatty acyl-CoA is transported across the inner mitochondrial membrane. A long-chain fatty acyl-CoA cannot directly cross the inner mitochondrial membrane. Instead, its acyl portion is first transferred to **carnitine,** a compound that occurs in both plant and animal tissues.

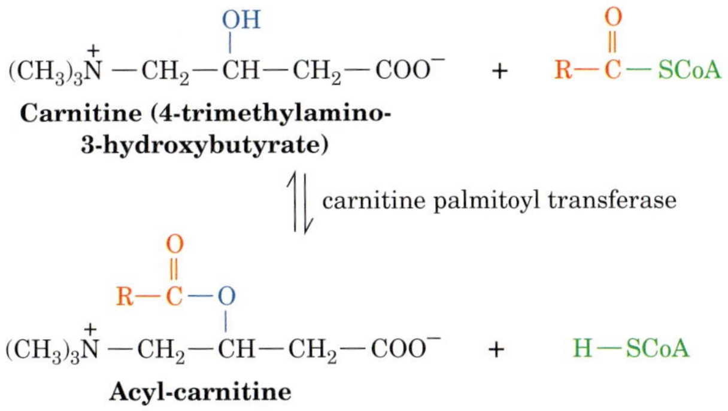

This transesterification reaction has an equilibrium constant close to 1, which indicates that the *O*-acyl bond of **acyl-carnitine** has a free energy of hydrolysis similar to that of acyl-CoA's thioester bond. **Carnitine palmitoyl transferases I** and **II,** which can transfer a variety of acyl groups (not just palmitoyl groups), are located, respectively, on the external and internal surfaces of the inner mitochondrial membrane. The translocation process itself is mediated by a specific carrier protein that transports acyl-carnitine into the mitochondrion while transporting free carnitine in the opposite direction. The acyl-CoA transport system is diagrammed in Fig. 19-11.

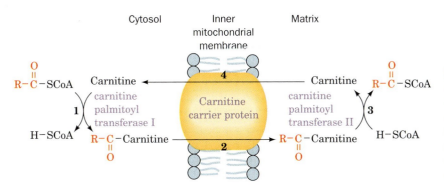

Figure 19-11 The transport of fatty acids into the mitochondrion. (1) The acyl group of a cytosolic acyl-CoA is transferred to carnitine, thereby releasing the CoA to its cytosolic pool. (2) The resulting acyl-carnitine is transported into the mitochondrial matrix by the carrier protein. (3) The acyl group is transferred to a CoA molecule from the mitochondrial pool. (4) The product carnitine is returned to the cytosol.

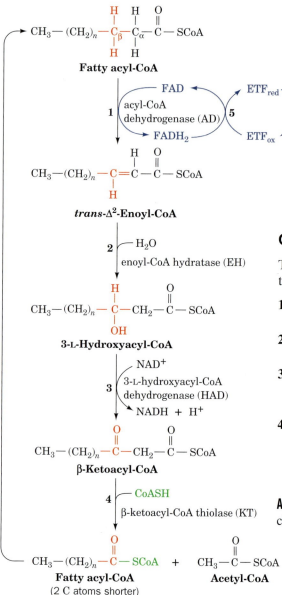

Figure 19-12 *Key to Metabolism.* The β-oxidation pathway of fatty acyl-CoA. 🐚 See the Animated Figures.

C β Oxidation

The degradation of fatty acyl-CoA via β oxidation occurs in four reactions (Fig. 19-12):

1. Formation of a trans-α,β double bond through dehydrogenation by the flavoenzyme **acyl-CoA dehydrogenase (AD).**

2. Hydration of the double bond by **enoyl-CoA hydratase (EH)** to form a **3-L-hydroxyacyl-CoA.**

3. NAD$^+$-dependent dehydrogenation of this β-hydroxyacyl-CoA by **3-L-hydroxyacyl-CoA dehydrogenase (HAD)** to form the corresponding β-ketoacyl-CoA.

4. C_α—C_β cleavage in a thiolysis reaction with CoA as catalyzed by **β-ketoacyl-CoA thiolase (KT;** also called just **thiolase)** to form acetyl-CoA and a new acyl-CoA containing two fewer C atoms than the original one.

Acyl-CoA Dehydrogenase Is Linked to the Electron-Transport Chain. Mitochondria contain four acyl-CoA dehydrogenases, with specificities for short (C_4 to C_6), medium (C_6 to C_{10}), long (between medium and very long), and very long chain (C_{12} to C_{18}) fatty acyl-CoAs. The reaction catalyzed by these enzymes is thought to involve removal of a proton at C_α and transfer of a hydride ion equivalent from C_β to FAD (Fig. 19-12, Reaction 1). The X-ray structure of **medium-chain acyl-CoA dehydrogenase (MCAD)** in complex with **octanoyl-CoA,** determined by Jung-Ja Kim, clearly shows how the enzyme orients a basic group (Glu 376), the substrate C_α—C_β bond, and the FAD prosthetic group for reaction (Fig. 19-13).

A deficiency of MCAD has been identified in ~10% of cases of **sudden infant death syndrome (SIDS).** Glucose is the principal energy metabolism substrate just after eating, but when the glucose level later decreases, the rate of fatty acid oxidation must correspondingly increase. The sudden death of infants lacking MCAD may be caused by the imbalance between glucose and fatty acid oxidation.

The FADH$_2$ resulting from the oxidation of the fatty acyl-CoA substrate is reoxidized by the mitochondrial electron-transport chain through a series of electron-transfer reactions. **Electron-transfer flavoprotein (ETF)** transfers an electron pair from FADH$_2$ to the flavo-iron–sulfur protein **ETF:ubiquinone oxidoreductase,** which in turn transfers an electron pair to the mitochondrial electron-transport chain by reducing coenzyme Q (CoQ; Fig. 19-12, Reactions 5–8). Reduction of O$_2$ to H$_2$O by the electron-

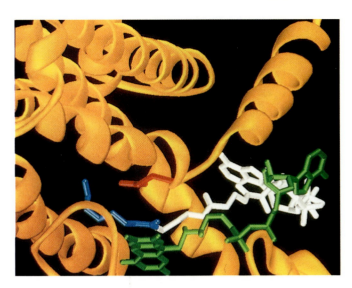

Figure 19-13 Ribbon diagram of the active site region of medium-chain acyl-CoA dehydrogenase. The enzyme, from pig liver mitochondria, is a tetramer of identical 385-residue subunits, each of which binds an FAD prosthetic group (*green*) and an octanoyl-CoA substrate (whose octanoyl and CoA moieties are blue and white) in largely extended conformations. The octanoyl-CoA binds such that its C_α—C_β bond is sandwiched between the carboxylate group of Glu 376 (*red*) and the flavin ring (*green*), consistent with the proposal that Glu 376 is the general base that abstracts the α proton in the α,β dehydrogenation reaction catalyzed by the enzyme. [Based on an X-ray structure by Jung-Ja Kim, Medical College of Wisconsin. PDBid 3MDE.] **See the Interactive Exercises.**

transport chain beginning at the CoQ stage results in the synthesis of 2 ATP per electron pair transferred (Section 17-3C).

Long-Chain Enoyl-CoAs Are Converted to Acetyl-CoA and a Shorter Acyl-CoA by Mitochondrial Trifunctional Protein.

The product of acyl-CoA dehydrogenases are 2-enoyl-CoAs. Depending on their chain lengths, their processing is continued by one of three systems (Fig. 19-12): the short chain, medium chain, or long chain 2-enoyl-CoA hydratases (EHs), hydroxyacyl-CoA dehydrogenases (HADs), and 3-ketoacyl-CoA thiolases (KTs). The long chain (LC) versions of these enzymes are contained on one $\alpha_4\beta_4$ heterooctameric, trifunctional protein, located in the inner mitochondrial membrane. LCEH and LCHAD are contained on the α subunits while LCKT is located on the β subunits. The protein is therefore both a multifunctional protein (more than one enzyme activity on a single polypeptide chain) and a multienzyme complex (a complex of polypeptides catalyzing more than one reaction). The advantage of such a trifunctional enzyme is the ability to channel the intermediates toward the final product. Indeed, no long-chain hydroxyacyl-CoA or ketoacyl-CoA intermediates are released into solution by this system.

The Thiolase Reaction Occurs via Claisen Ester Cleavage.

The fourth step of β oxidation is the thiolase reaction (Fig. 19-12, Reaction 4), which yields acetyl-CoA and a new acyl-CoA that is two carbon atoms shorter than the one that began the cycle (Fig. 19-14):

1. An active site thiol group adds to the β-keto group of the substrate acyl-CoA.

2. Carbon–carbon bond cleavage yields a thioester between the acyl-CoA substrate and the active site thiol group, together with an acetyl-CoA carbanion intermediate that is stabilized by electron withdrawal into this thioester's carbonyl group. This type of reaction is known as a Claisen ester cleavage (the reverse of

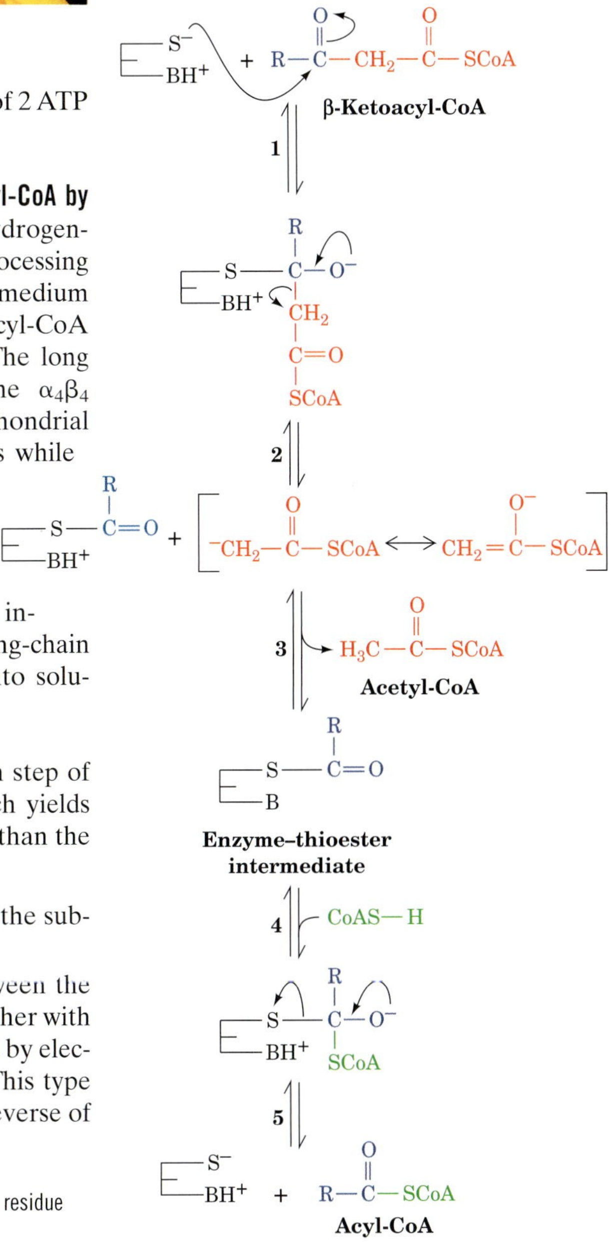

Figure 19-14 Mechanism of action of β-ketoacyl-CoA thiolase. An active site Cys residue participates in the formation of an enzyme–thioester intermediate.

a Claisen condensation). The citric acid cycle enzyme citrate synthase also catalyzes a reaction that involves a stabilized acetyl-CoA carbanion intermediate (Section 16-3A).

3.   An enzyme acidic group protonates the acetyl-CoA carbanion, yielding acetyl-CoA.

4 and 5.   Finally, CoA displaces the enzyme thiol group from the enzyme–thioester intermediate, yielding an acyl-CoA that is shortened by two C atoms.

Fatty Acid Oxidation Is Highly Exergonic.   The function of fatty acid oxidation is, of course, to generate metabolic energy. Each round of β oxidation produces one NADH, one $FADH_2$, and one acetyl-CoA. Oxidation of acetyl-CoA via the citric acid cycle generates an additional $FADH_2$ and 3 NADH, which are reoxidized through oxidative phosphorylation to form ATP. Complete oxidation of a fatty acid molecule is therefore a highly exergonic process that yields numerous ATPs. For example, oxidation of palmitoyl-CoA (which has a C_{16} fatty acyl group) involves seven rounds of β oxidation, yielding 7 $FADH_2$, 7 NADH, and 8 acetyl-CoA. Oxidation of the 8 acetyl-CoA, in turn, yields 8 GTP, 24 NADH, and 8 $FADH_2$. Since oxidative phosphorylation of the 31 NADH molecules yields 93 ATP and that of the 15 $FADH_2$ yields 30 ATP, subtracting the 2 ATP equivalents required for fatty acyl-CoA formation (Section 19-2A), *the oxidation of one palmitate molecule has a net yield of 129 ATP.*

D   Oxidation of Unsaturated Fatty Acids

Almost all unsaturated fatty acids of biological origin contain only cis double bonds, which most often begin between C9 and C10 (referred to as a Δ^9 or 9-double bond; Table 9-1). Additional double bonds, if any, occur at three-carbon intervals and are therefore never conjugated. Two examples of unsaturated fatty acids are oleic acid and linoleic acid (*at left*).

Note that one of the double bonds in linoleic acid is at an odd-numbered carbon atom and the other is at an even-numbered carbon atom. The double bonds in fatty acids such as linoleic acid pose three problems for the β-oxidation pathway that are solved through the actions of four additional enzymes (Fig. 19-15).

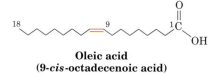

Oleic acid
(9-*cis*-octadecenoic acid)

Linoleic acid
(9,12-*cis*-octadecadienoic acid)

> **Problem 1: A β,γ Double Bond.** The first enzymatic difficulty occurs after the third round of β oxidation: The resulting cis-β,γ double bond–containing enoyl-CoA is not a substrate for enoyl-CoA hydratase. **Enoyl-CoA isomerase,** however, converts the cis-Δ^3 double bond to the trans-Δ^2 form. The Δ^2 compound is the normal substrate of enoyl-CoA hydratase, so β oxidation can continue.
>
> **Problem 2: A Δ^4 Double Bond Inhibits Enoyl-CoA Hydratase.** The next difficulty arises in the fifth round of β oxidation: The presence of a double bond at an even-numbered carbon atom results in the formation of 2,4-dienoyl-CoA, which is a poor substrate for enoyl-CoA hydratase. However, NADPH-dependent **2,4-dienoyl-CoA reductase** reduces the Δ^4 double bond. The *E. coli* reductase produces *trans*-2-enoyl-CoA, a normal substrate of β oxidation. The mammalian reductase, however, yields *trans*-3-enoyl-CoA, which, to proceed along the β-oxidation pathway, must first be isomerized to *trans*-2-enoyl-CoA by **3,2-enoyl-CoA isomerase.**

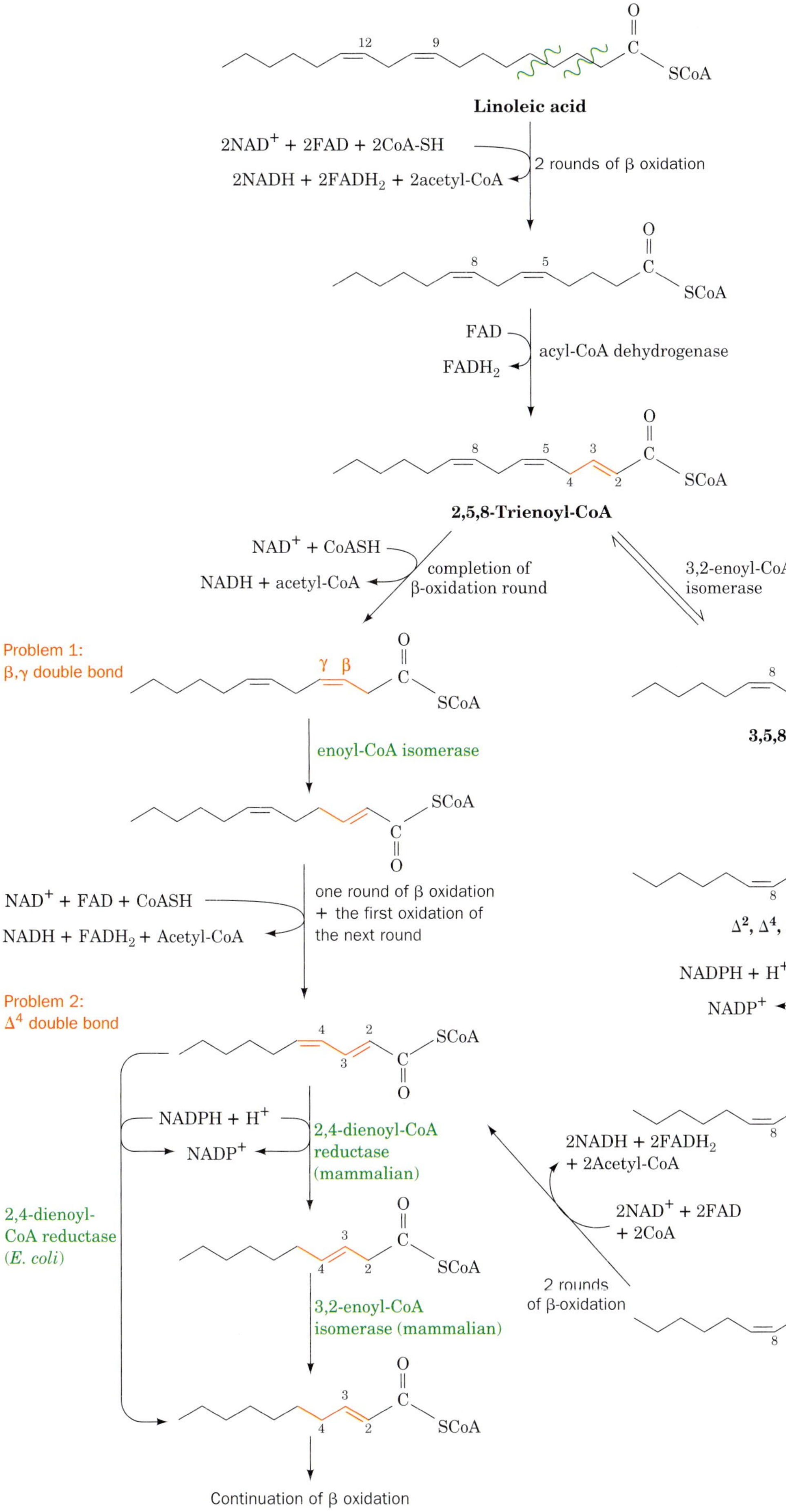

Figure 19-15 The oxidation of unsaturated fatty acids. β oxidation of an unsaturated fatty acid such as linoleic acid presents three problems. The first problem, the presence of a β,γ double bond, seen in the left-hand pathway, is solved by the bond's conversion to a trans-α,β double bond. The second problem in the left-hand pathway, that a 2,4-dienoyl-CoA is a poor substrate for enoyl-CoA hydratase, is eliminated by the NADPH-dependent reduction of the Δ^4 bond by 2,4-dienoyl-CoA reductase to yield the β-oxidation substrate *trans*-2-enoyl-CoA. This step requires one enzyme in *E. coli* and two in mammals. The third problem, the isomerization of 2,5 dienoyl-CoA (originating from the oxidation of unsaturated fatty acids with double bonds at odd-numbered C atoms) to 3,5-dienoyl-CoA, is solved by converting the 3,5-dienoyl-CoA to 2,4-dienoyl-CoA, a substrate for 2,4-dienoyl-CoA reductase.

641

Problem 3: The Unanticipated Isomerization of 2,5-enoyl-CoA by 3,2-enoyl-CoA Isomerase. Mammalian 3,2-enoyl-CoA isomerase catalyzes a reversible reaction that interconverts Δ^2 and Δ^3 double bonds. A carbonyl group is stabilized by being conjugated to a Δ^2 double bond. However, the presence of a Δ^5 double bond (originating from an unsaturated fatty acid with a double bond at an odd-numbered C atom such as the Δ^9 double bond of linoleic acid) is likewise stabilized by being conjugated with a Δ^3 double bond. If a 2,5-enoyl-CoA is converted by 3,2-enoyl-CoA isomerase to 3,5-enoyl CoA, which occurs up to 20% of the time, another enzyme is necessary to continue the oxidation: **3,5–2,4-Dienoyl-CoA isomerase** isomerizes the 3,5 diene to a 2,4 diene, which is then reduced by 2,4-dienoyl-CoA reductase and isomerized by 3,2-enoyl-CoA isomerase as in Problem 2 above. After two more rounds of β oxidation, the cis-Δ^4 double bond originating from the cis-Δ^{12} double bond of linoleic acid is also dealt with as in Problem 2.

E Oxidation of Odd-Chain Fatty Acids

Most fatty acids, for reasons explained in Section 19-4, have even numbers of carbon atoms and are therefore completely converted to acetyl-CoA. Some plants and marine organisms, however, synthesize fatty acids with an odd number of carbon atoms. *The final round of β oxidation of these fatty acids yields propionyl-CoA, which is converted to succinyl-CoA for entry into the citric acid cycle.* Propionate and propionyl-CoA are also produced by the oxidation of the amino acids isoleucine, valine, and methionine (Section 20-4D).

The conversion of propionyl-CoA to succinyl-CoA involves three enzymes (Fig. 19-16). The first reaction, catalyzed by **propionyl-CoA carboxylase,** requires a biotin prosthetic group and is driven by the hydrolysis of ATP to ADP + P_i. This reaction resembles that of pyruvate carboxylase (Fig. 15-26).

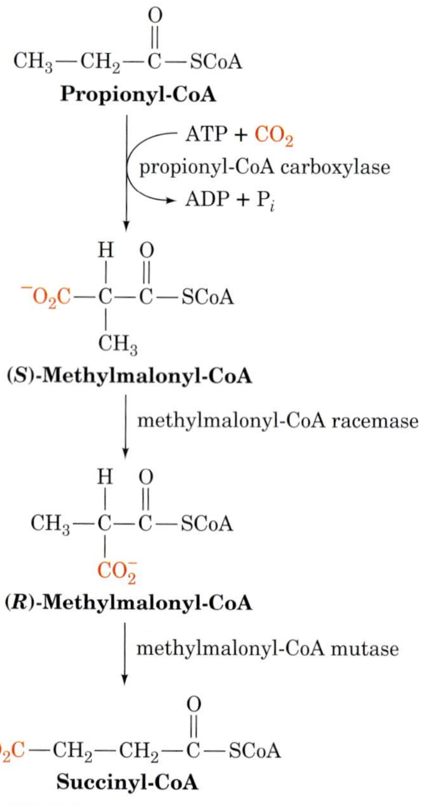

The (S)-methylmalonyl-CoA product of the carboxylase reaction is converted to the R form by **methylmalonyl-CoA racemase.** (R)-Methylmalonyl-CoA is a substrate for **methylmalonyl-CoA mutase,** which catalyzes the third reaction in Fig. 19-16.

Methylmalonyl-CoA mutase catalyzes an unusual carbon skeleton rearrangement.

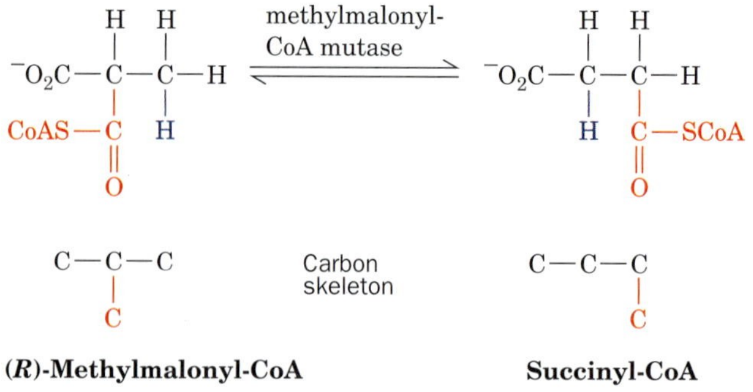

(R)-Methylmalonyl-CoA   **Succinyl-CoA**

The enzyme uses a **5′-deoxyadenosylcobalamin** prosthetic group (**AdoCbl;** also called coenzyme B_{12}, a derivative of **cobalamin,** or **vitamin B_{12};** see Box 19-1). Dorothy Hodgkin determined the structure of this complex molecule (Fig. 19-17) in 1956 through X-ray crystallographic analysis combined

Figure 19-16 Conversion of propionyl-CoA to succinyl-CoA.

with chemical degradation studies, a landmark achievement (Box 19-2).

5′-Deoxyadenosylcobalamin contains a hemelike **corrin** ring whose four pyrrole N atoms each ligand a six-coordinate Co ion. The fifth Co ligand is an N atom of a **5,6-dimethylbenzimidazole (DMB)** nucleotide that is co-valently linked to the corrin D ring. The sixth ligand is a 5′-deoxyadenosyl group in which the deoxyribose C5′atom forms a covalent C—Co bond, *one of only two carbon–metal bonds known in biology* (the other is a C—Ni bond in the bacterial enzyme **carbon monoxide dehydrogenase**). In some cobalamin-dependent enzymes, the sixth ligand instead is a CH_3 group that likewise forms a C—Co bond. There are only about a dozen known cobalamin-dependent enzymes, which catalyze molecular rearrangements or methyl-group transfer reactions.

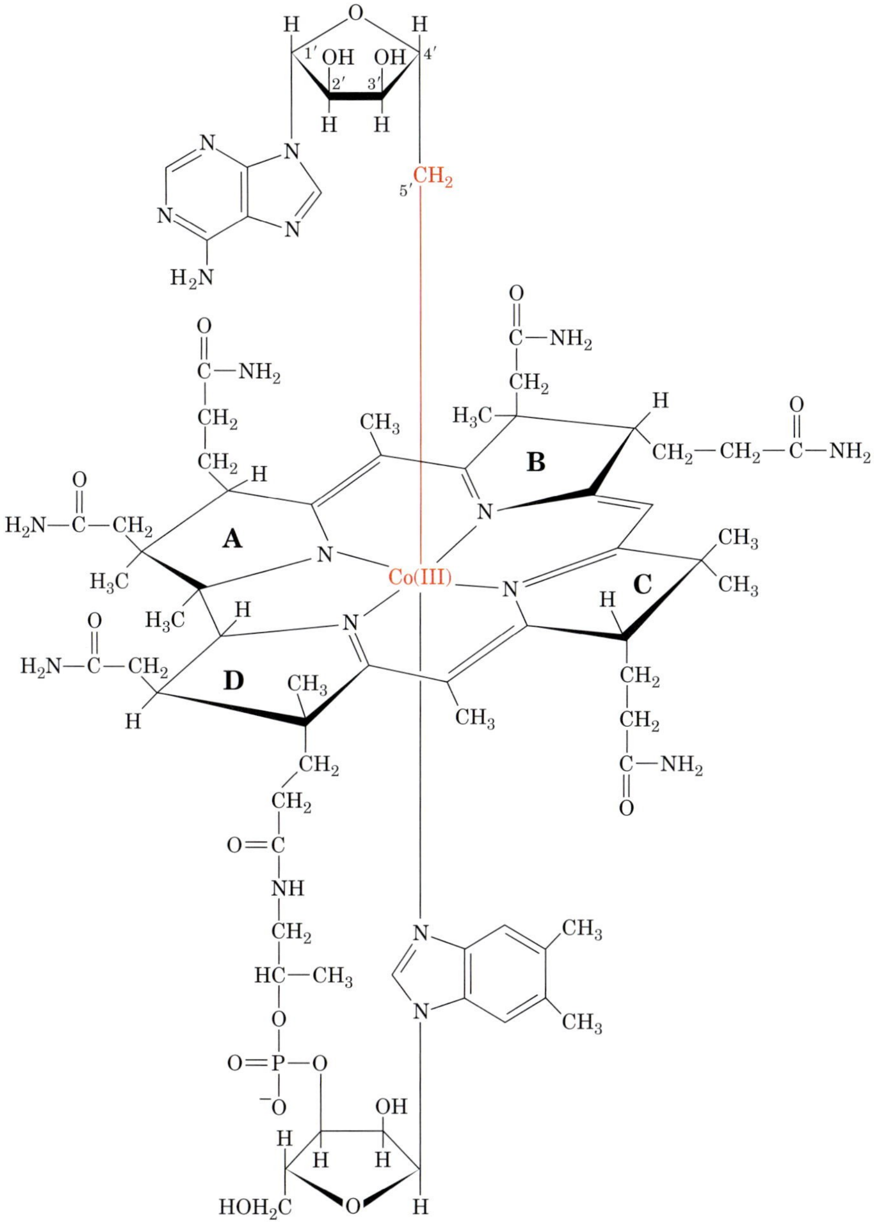

5′-Deoxyadenosylcobalamin (coenzyme B_{12})

Figure 19-17 Structure of 5′-deoxyadenosylcobalamin.

BOX 19-1

Biochemistry in Health and Disease

Vitamin B₁₂ Deficiency

The existence of vitamin B_{12} came to light in 1926 when George Minot and William Murphy discovered that **pernicious anemia,** an often fatal disease of the elderly characterized by decreased numbers of red blood cells, low hemoglobin levels, and progressive neurological deterioration, can be treated by the daily consumption of large amounts of raw liver. Nevertheless, the antipernicious anemia factor—vitamin B_{12}— was not isolated until 1948.

Vitamin B_{12} is synthesized by neither plants nor animals but only by a few species of bacteria. Herbivores obtain their vitamin B_{12} from the bacteria that inhabit their gut (in fact, some animals, such as rabbits, must periodically eat some of their feces to obtain sufficient amounts of this essential substance). Humans, however, obtain almost all their vitamin B_{12} directly from their diet, particularly from meat. In the intestine, the glycoprotein **intrinsic factor,** which is secreted by the stomach, specifically binds vitamin B_{12}, and the protein–vitamin complex is absorbed via a receptor in the intestinal mucosa. The complex dissociates and the liberated vitamin B_{12} is transported to the bloodstream. At least three different plasma proteins, called **transcobalamins,** bind the vitamin and facilitate its uptake by the tissues.

Pernicious anemia is not usually a dietary deficiency disease but, rather, results from insufficient secretion of intrinsic factor. The normal human requirement for cobalamin is very small, $\sim3\ \mu g \cdot day^{-1}$, and the liver stores a 3- to 5-year supply of this vitamin. This accounts for the insidious onset of pernicious anemia and the fact that true dietary deficiency of vitamin B_{12}, even among strict vegetarians, is extremely rare.

Methylmalonyl-CoA Mutase Has a Unique α/β Barrel. The X-ray structure of methylmalonyl-CoA mutase from *Propionibacterium shermanii*, an αβ heterodimer, in complex with the substrate analog **2-carboxypropyl-CoA** (which lacks methylmalonyl-CoA's thioester oxygen atom) was determined by Philip Evans. Its AdoCbl cofactor is sandwiched between the catalytically active α subunit's two domains: a 559-residue N-terminal α/β barrel (TIM barrel, the most common enzymatic motif; Sections 6-2C and 14-2) and a 169-residue C-terminal α/β domain that resembles a Rossmann fold (Section 6-2C). The structure of the α/β barrel contains several surprising features (Fig. 19-18):

1. The active sites of nearly all α/β barrel enzymes are located at the C-terminal ends of the barrel's β strands. However, in methylmalonyl-CoA mutase, the AdoCbl is packed against the N-terminal ends of the barrel's β strands.

2. In free AdoCbl, the Co atom is axially liganded by an N atom of its DMB group and by the adenosyl residue's 5'-CH₂ group (Fig. 19-17). However, in the enzyme, the DMB has swung aside to bind in a separate pocket and has been replaced by the side chain of His 610 from the C-terminal domain. The adenosyl group is not visible in the structure due to disorder and hence has probably also swung aside.

3. In nearly all other α/β barrel-containing enzymes, the center of the barrel is occluded by large, often branched, hydrophobic side chains. However, in methylmalonyl-CoA mutase, the 2-carboxypropyl-CoA's pantetheine group binds in a narrow tunnel along the center of the α/β barrel so as to put the methylmalonyl group of an intact substrate in close proximity to the unliganded face of the cobalamin ring. This tunnel provides the only direct access to the active site cavity, thereby protecting the reactive free radical intermediates that are produced

(a)

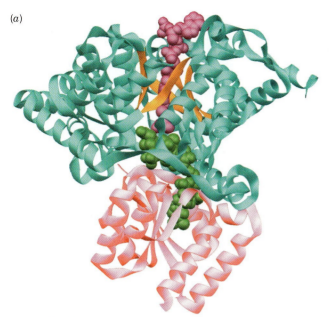

(b)

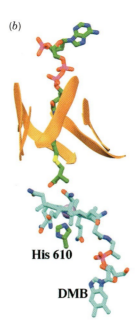

His 610

DMB

Figure 19-18 X-Ray structure of *P. shermanii* methylmalonyl-CoA mutase in complex with 2-carboxypropyl-CoA and AdoCbl. (*a*) The catalytically active α subunit in which the N-terminal domain is cyan, the β strands of its α/β barrel are orange, and the C-terminal domain is pink. The 2-carboxypropyl-CoA (*magenta*) and AdoCbl (*green*) are drawn in space-filling form. The 2-carboxypropyl-CoA passes through the center of the α/β barrel and is oriented such that the methylmalonyl group of methylmalonyl-CoA would contact the corrin ring of the AdoCbl, which is sandwiched between the enzyme's N- and C-terminal domains. (*b*) The arrangement of the AdoCbl and 2-carboxypropyl-CoA molecules which, together with the side chain of His 610, are represented in stick form colored according to atom type (2-carboxypropyl-CoA and His C green, AdoCbl C cyan, N blue, O red, P magenta, and S yellow). The corrin ring's Co atom is represented by a lavender sphere and the α/β barrel's β strands are represented by orange ribbons. The view is similar to that in Part *a*. Note that the DMB group (*bottom*) has swung away from the corrin ring (seen edgewise) to be replaced by the side chain of His 610 from the C-terminal domain and that the 5′-deoxyadenosyl group is unseen (due to disorder). [Based on an X-ray structure by Philip Evans, MRC Laboratory of Molecular Biology, Cambridge, U.K. PDBid 7REQ.] 🔁 **See the Interactive Exercises.**

in the catalytic reaction from side reactions (see below). The tunnel is lined by small hydrophilic residues (Ser and Thr).

Methylmalonyl-CoA mutase's substrate binding mode resembles that of several other AdoCbl-containing enzymes of known structure, which are collectively unique among α/β barrel-containing enzymes.

Methylmalonyl-CoA Mutase Stabilizes and Protects Free Radical Intermediates.

The proposed methylmalonyl-CoA mutase reaction mechanism (Fig. 19-19) begins with the **homolytic cleavage** of the cobalamin C—Co bond; in other words, the C and Co atoms each acquire one of the electrons that formed the cleaved electron pair bond. (Note that a homolytic cleavage reaction is unusual in biology; most other biological bond-cleavage reactions occur via **heterolytic cleavage** in which the electron pair forming the cleaved bond is fully acquired by one of the separating atoms.) The Co ion therefore alternates between its Co(III) and Co(II) oxidation states and hence *functions as a reversible free radical generator.* The C—Co(III) bond is well suited to this function because it is inherently weak (dissociation energy 109 kJ · mol^{-1}) and is further weakened through steric interactions with the enzyme.

Spectroscopic measurements indicate that the Co atom in methylmalonyl-CoA mutase is in the Co(II) state, thereby confirming that it has no sixth ligand (as occurs during its catalytic cycle; Fig. 19-19). Protein-induced strain makes the His N—Co bond extremely long (2.5 Å versus

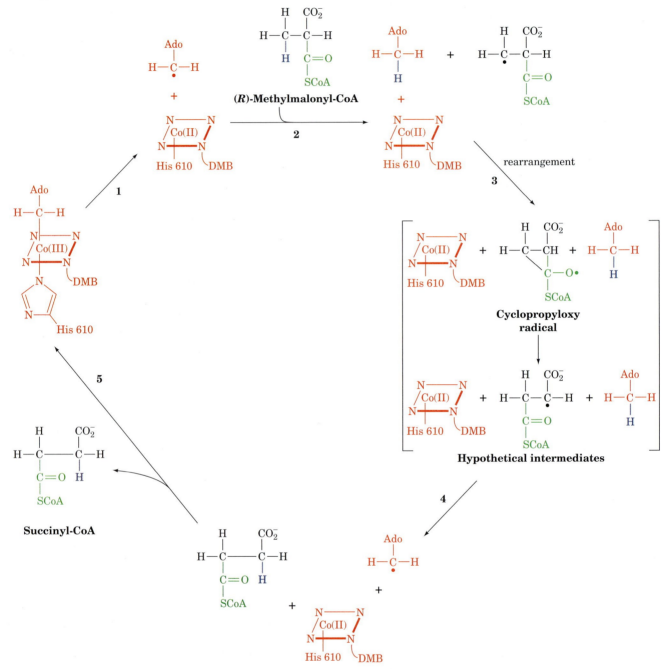

Figure 19-19 Proposed mechanism of methylmalonyl-CoA mutase.
(**1**) The homolytic cleavage of the C—Co(III) bond yields a 5′-deoxyadenosyl (Ado) radical and cobalamin in its Co(II) oxidation state. (**2**) The 5′-deoxyadenosyl radical abstracts a hydrogen atom from methylmalonyl-CoA, thereby generating a methylmalonyl-CoA radical. (**3**) Carbon skeleton rearrangement yields a succinyl-CoA radical via a proposed cyclopropyloxy radical intermediate. (**4**) The succinyl-CoA radical abstracts a hydrogen atom from 5′-deoxyadenosine to regenerate the 5′-deoxyadenosyl radical. (**5**) The release of succinyl-CoA re-forms the coenzyme.

1.9–2.0 Å in various other B_{12}-containing structures), stabilizing the Co(II) species relative to Co(III) and thereby favoring the formation of the adenosyl radical. This radical abstracts a hydrogen atom from the methylmalonyl-CoA substrate, thereby facilitating its rearrangement to succinyl-CoA through the intermediate formation of a cyclopropyloxy radical.

BOX 19-2

Pathways of Discovery

Dorothy Crowfoot Hodgkin and the Structure of Vitamin B₁₂

Dorothy Crowfoot Hodgkin (1910–1994)

The third woman to receive a Nobel Prize in Chemistry, after Marie Curie in 1911 and her daughter Irene Joliet-Curie in 1935, was Dorothy Crowfoot Hodgkin, who received the award in 1964 for determining the structures of biological molecules by X-ray crystallography. Hodgkin's career as a scientist is notable not only for her accomplishments—elucidating the structures of sterols, penicillin, vitamin B_{12}, and insulin—but also for her contributions to the methodology of X-ray crystallography.

Hodgkin's fascination with crystals reportedly began at age 10 with a simple experiment involving alum. Her scientific tendencies were encouraged by her parents, particularly her mother, who was an accomplished amateur botanist. Hodgkin attributed her independent spirit to the fact that her parents were absent for long periods, as her father worked as an archaeologist in Egypt and the Middle East.

Hodgkin's formal training in crystallography took place at Oxford during her undergraduate years and continued as she earned a doctorate under the guidance of J.D. Bernal at Cambridge. Unlike many of his colleagues at the time, Bernal was generous in giving credit to his junior associates. Consequently, Hodgkin received recognition along with Bernal in 1934 when they reported the first X-ray diffraction pattern of a protein, pepsin. Bernal and Hodgkin had thereby shown that proteins are not amorphous colloids but have discrete structures. They also advanced the field of macromolecular crystallography by noting the necessity of examining crystals surrounded by their mother liquor (the solution from which the molecules crystallize) rather than air-dried, as was standard procedure at the time.

After returning to Oxford, Hodgkin established her own laboratory, a challenging task for someone striking out in a new field, and even more difficult for a woman trying to balance work with marriage and motherhood in the late 1930s. At Oxford, Hodgkin undertook studies of crystallized cholesterol and other steroids, showing that calculations based on X-ray diffraction in three rather than two dimensions could disclose considerable information about molecular structure, including the stereochemistry of each carbon atom. Hodgkin also began studying crystals of insulin, although an atomic model of the 777-atom protein took 34 years to complete (the sequence of insulin, determined by Frederick Sanger, was not known until 1955; Box 5-2). During World War II, Hodgkin turned her attention to the structure of penicillin and used X-ray crystallography to elucidate its unexpected ring structure (three carbons and a nitrogen; Box 8-3). This discovery helped pave the way for the synthesis of penicillin and its derivatives. Hodgkin's work also marked the beginning of an electronic age in biochemical research, as she used one of the first analog computers from IBM to help with the required calculations.

In 1955, a visitor brought Hodgkin crystals of cyanocobalamin, whose deep-red color may have enticed Hodgkin to immediately examine their X-ray diffraction. The cobalamin structure, four times larger than that of penicillin, had not yet yielded its structural secrets to conventional chemical approaches. Hodgkin's success in determining the structure of cobalamin was the result of experience as well as intuition. She was reportedly able to discern features of molecular structure simply by examining the diffraction pattern. However, she also took advantage of powerful computers available to her collaborator Kenneth Trueblood in Los Angeles, who, due to the great distance between them, communicated with her through letters and telegrams. Hodgkin bravely decided to use the cobalt atom naturally present in cobalamin to solve the "phase problem" that prevented the straightforward determination of X-ray structures (Box 7-2). Although others believed that the cobalt atom would not scatter X-rays strongly enough relative to the lighter atoms in the structure, Hodgkin was able to derive the necessary phase information and thus was spared having to produce cobalamin crystals containing a heavier atom.

Hodgkin and her collaborators identified the unusual porphyrin structure of cyanocobalamin, in which rings A and D are directly linked and ring A is fully saturated (Fig. 19-17). They subsequently identified cobalamin, with its Co—C bond, as the first known biological organometallic compound. Hodgkin later determined the structure of the physiological form of vitamin B_{12}, adenosylcobalamin, in 1961. Now that the structure of the vitamin was known, it could be synthesized and used to treat pernicious anemia (Box 19-1). In addition to its obvious implications for human health, Hodgkin's work with cobalamin encouraged other researchers to pursue the structures of large compounds previously thought to be too complicated for crystallographic analysis. Hodgkin herself finally completed a three-dimensional model of insulin in 1969.

After receiving the Nobel prize in 1964, along with many other honors, Hodgkin became increasingly involved in international organizations, including the International Union of Crystallography and other groups dedicated to promoting scientific cooperation as a means for decreasing international tensions during the Cold War. Her own laboratory over the years hosted scientists from around the globe. By the end of her life, her efforts to foster international goodwill were no less notable than her accomplishments as a crystallographer.

Ferry, G., *Dorothy Hodgkin: A Life,* Cold Spring Harbor Laboratory Press (2000).

Hodgkin, D.C., Kamper, J., Mackay, M., Pickworth, J., Trueblood, K.N. and White, J.G., Structure of vitamin B-12, *Nature* **178,** 64 (1956).

Succinyl-CoA Is Not Directly Consumed by the Citric Acid Cycle. Methylmalonyl-CoA mutase catalyzes the conversion of a metabolite to a citric acid cycle intermediate other than acetyl-CoA. However, such C_4 intermediates are actually catalysts, not substrates, of the citric acid cycle. *In order for succinyl-CoA to undergo net oxidation by the citric acid cycle, it must first be converted to pyruvate and thence to acetyl-CoA.* This is accomplished by converting succinyl-CoA to malate (Reactions 5–7 of the citric acid cycle; Fig. 16-2) followed by its transport to the cytosol and oxidative decarboxylation to pyruvate and CO_2 by **malic enzyme**

$$
\underset{\textbf{Malate}}{
\begin{array}{c}
CO_2^- \\
| \\
HO-C-H \\
| \\
CH_2 \\
| \\
\underset{O}{\overset{\displaystyle\parallel}{C}}\!-\!O^-
\end{array}}
\xrightarrow[\text{NADP}^+\ \ \text{NADPH}]{\text{H}^+ +}
\left[
\begin{array}{c}
CO_2^- \\
| \\
C=O \\
| \\
CH_2 \\
| \\
\underset{O}{\overset{\displaystyle\parallel}{C}}\!-\!O^-
\end{array}
\right]
\xrightarrow{\ CO_2\ }
\underset{\textbf{Pyruvate}}{
\begin{array}{c}
CO_2^- \\
| \\
C=O \\
| \\
CH_3
\end{array}}
$$

(this enzyme also functions in the C_4 cycle of photosynthesis; Fig. 18-31). Pyruvate is then completely oxidized via pyruvate dehydrogenase and the citric acid cycle.

F Peroxisomal β Oxidation

In mammalian cells, the bulk of β oxidation occurs in the mitochondria, but peroxisomes (Fig. 19-20) also oxidize fatty acids, particularly those with very long chains or branched chains. In plants, fatty acid oxidation occurs exclusively in the peroxisomes and glyoxysomes (which are specialized peroxisomes). In addition to lipid catabolism, mammalian peroxisomes participate in the synthesis of certain lipids, including bile acids. A variety of human diseases result from defects in peroxisomal enzymes or proteins that transport intermediates across the peroxisomal membrane.

Peroxisomal β oxidation, which differs only slightly from mitochondrial β oxidation, shortens very long chain fatty acids (>22 carbon atoms), which are then fully degraded by the mitochondrial pathway. Very long chain fatty acids are transported into the peroxisomes by a mechanism that does not require carnitine, and are activated by a long chain acyl-CoA synthetase. Peroxisomal β oxidation results in the same chemical changes to fatty acids as in the mitochondrial pathway but requires only three enzymes:

1. **Acyl-CoA oxidase** catalyzes the reaction

   Fatty acyl-CoA + O_2 → *trans*-Δ^2-enoyl-CoA + H_2O_2

   This enzyme uses an FAD cofactor, but the abstracted electrons are transferred directly to O_2 rather than passing through the electron-transport chain with its concomitant oxidative phosphorylation (Fig. 19-12, Reactions 5–8). Peroxisomal fatty acid oxidation therefore generates two fewer ATP per C_2 cycle than mitochondrial fatty acid oxidation. Catalase converts the H_2O_2 produced in the oxidase reaction to $H_2O + O_2$.

2. Peroxisomal enoyl-CoA hydratase and 3-L-hydroxyacyl-CoA dehydrogenase activities occur on a single polypeptide. The reactions catalyzed are identical to those of the mitochondrial system (Fig. 19-12, Reactions 2 and 3).

3. Peroxisomal thiolase catalyzes the final step of oxidation. This enzyme is almost inactive with acyl-CoAs of length C_8 or less, so peroxisomes incompletely oxidize fatty acids.

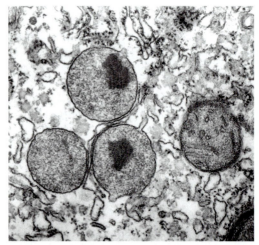

Figure 19-20 Peroxisomes. These membrane-bounded organelles perform a variety of metabolic functions, including the oxidation of very long chain fatty acids. [© Donald Fawcett/Visuals Unlimited.]

The peroxisome contains both a carnitine acetyltransferase and a transferase specific for longer chain acyl groups. Acyl-CoAs that have been chain-shortened by peroxisomal β oxidation are thereby converted to their carnitine esters. These substances, for the most part, passively diffuse out of the peroxisome to the mitochondrion, where they are oxidized further.

3 Ketone Bodies

The acetyl-CoA produced by oxidation of fatty acids can be further oxidized via the citric acid cycle. In liver mitochondria, however, a significant fraction of this acetyl-CoA has another fate. By a process known as **ketogenesis,** acetyl-CoA is converted to **acetoacetate** or **D-β-hydroxybutyrate.** These compounds together with acetone are somewhat inaccurately referred to as **ketone bodies:**

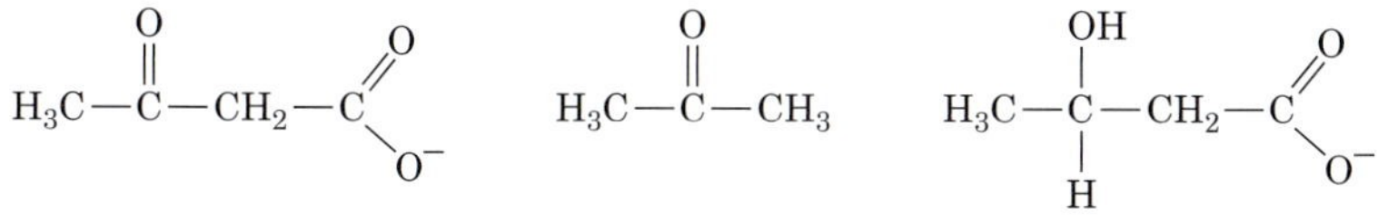

$$H_3C-\overset{\overset{\displaystyle O}{\|}}{C}-CH_2-\overset{\overset{\displaystyle O}{\|}}{C}\diagdown O^- \qquad H_3C-\overset{\overset{\displaystyle O}{\|}}{C}-CH_3 \qquad H_3C-\overset{\overset{\displaystyle OH}{|}}{\underset{\underset{\displaystyle H}{|}}{C}}-CH_2-\overset{\overset{\displaystyle O}{\|}}{C}\diagdown O^-$$

Acetoacetate   **Acetone**   **D-β-Hydroxybutyrate**

Ketone bodies are important metabolic fuels for many peripheral tissues, particularly heart and skeletal muscle. The brain, under normal circumstances, uses only glucose as its energy source (fatty acids are unable to pass through the blood–brain barrier), but during starvation, the small, water-soluble ketone bodies become the brain's major fuel source (Section 21-4A).

Acetoacetate formation occurs in three reactions (Fig. 19-21):

1. Two molecules of acetyl-CoA are condensed to **acetoacetyl-CoA** by thiolase (also called **acetyl-CoA acetyltransferase**) working in the

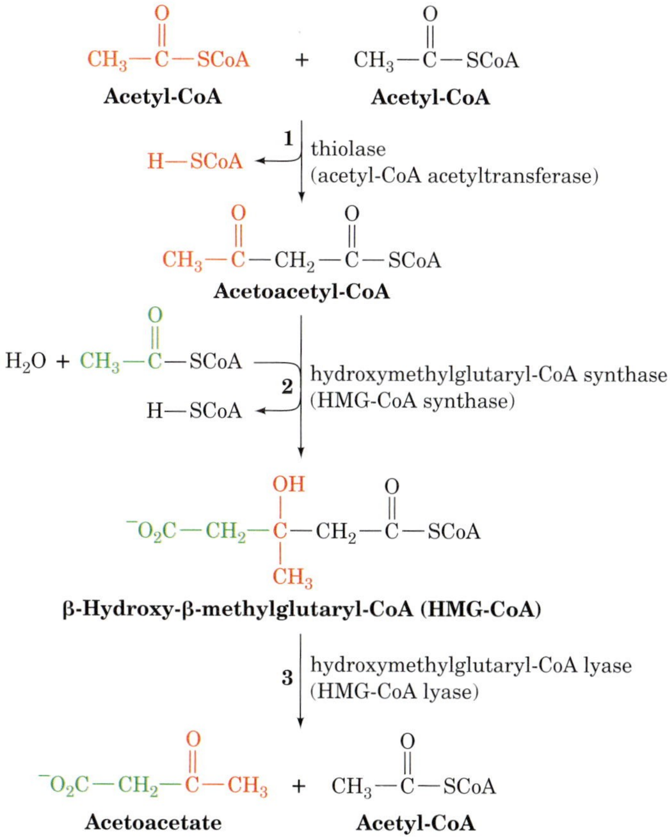

Figure 19-21 Ketogenesis. Acetoacetate is formed from acetyl-CoA in three steps. (**1**) Two molecules of acetyl-CoA condense to form acetoacetyl-CoA. (**2**) A Claisen ester condensation of the acetoacetyl-CoA with a third acetyl-CoA forms β-hydroxy-β-methylglutaryl-CoA (HMG-CoA). (**3**) HMG-CoA is degraded to acetoacetate and acetyl-CoA in a mixed aldol–Claisen ester cleavage.

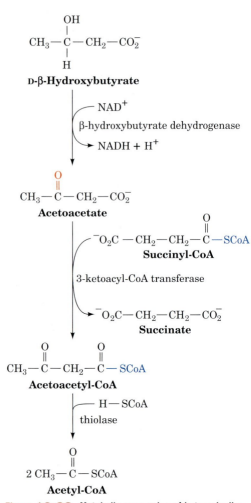

Figure 19-22 Metabolic conversion of ketone bodies to acetyl-CoA.

reverse direction from the way it does in the final step of β oxidation (Fig. 19-12, Reaction 4).

2. Condensation of the acetoacetyl-CoA with a third acetyl-CoA by **HMG-CoA synthase** forms **β-hydroxy-β-methylglutaryl-CoA (HMG-CoA).** The mechanism of this reaction resembles the reverse of the thiolase reaction (Fig. 19-14) in that an active site thiol group forms an acyl-thioester intermediate.

3. HMG-CoA is degraded to acetoacetate and acetyl-CoA in a mixed aldol–Claisen ester cleavage by **HMG-CoA lyase.** The mechanism of this reaction is analogous to the reverse of the citrate synthase reaction (Fig. 16-10). HMG-CoA is also a precursor in cholesterol biosynthesis (Section 19-7A). HMG-CoA lyase is present only in liver mitochondria and therefore does not interfere with cholesterol synthesis in the cytoplasm.

Acetoacetate may be reduced to D-β-hydroxybutyrate by **β-hydroxybutyrate dehydrogenase:**

$$\begin{array}{ccc} CH_3 & & CH_3 \\ | & H^+ + & | \\ C=O & NADH \quad NAD^+ & HO-C-H \\ | & \curvearrowright & | \\ CH_2 & \text{β-hydroxybutyrate} & CH_2 \\ | & \text{dehydrogenase} & | \\ CO_2^- & & CO_2^- \end{array}$$

Acetoacetate   **D-β-Hydroxybutyrate**

Acetoacetate, a β-keto acid, also undergoes relatively facile nonenzymatic decarboxylation to acetone and CO_2. Indeed, in individuals with **ketosis,** a pathological condition in which acetoacetate is produced faster than it is metabolized (a symptom of diabetes; Section 21-4B), the breath has the characteristic sweet smell of acetone.

The liver releases acetoacetate and β-hydroxybutyrate, which are carried by the bloodstream to the peripheral tissues for use as alternative fuels. There, these products are converted to two acetyl-CoA as in diagrammed in Fig. 19-22. Succinyl-CoA, which acts as the CoA donor in this process, can also be converted to succinate with the coupled synthesis of GTP in the succinyl-CoA synthetase reaction of the citric acid cycle (Section 16-3E). The "activation" of acetoacetate bypasses this step and therefore "costs" the free energy of GTP hydrolysis.

4 Fatty Acid Biosynthesis

Fatty acid biosynthesis occurs through condensation of C_2 units, the reverse of the β-oxidation process. Through isotopic labeling techniques, David Rittenberg and Konrad Bloch demonstrated, in 1945, that these condensation units are derived from acetic acid. Subsequent research showed that both acetyl-CoA and bicarbonate are required, and that a C_3 unit, **malonyl-CoA,** is an intermediate of fatty acid biosynthesis.

The pathway of fatty acid synthesis differs from that of fatty acid oxidation. This situation, as we saw in Section 15-3, is typical of opposing biosynthetic and degradative pathways because it permits them both to be thermodynamically favorable and independently regulated under similar physiological conditions. Figure 19-23 outlines fatty acid oxidation and synthesis with emphasis on the differences between these pathways, including the cellular locations of the pathways, the redox coenzymes, and the manner in which C_2 units are removed or added to the fatty acyl chain.

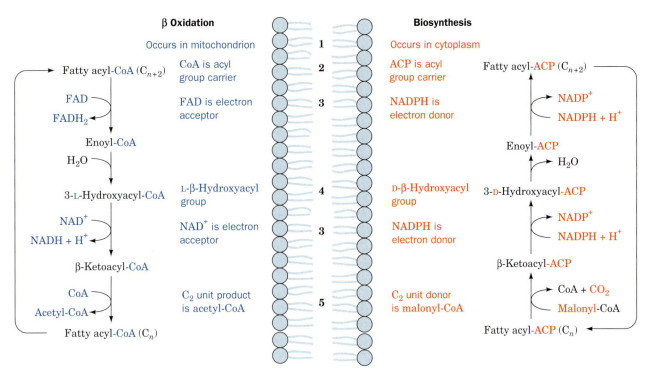

Figure 19-23 Comparison of fatty acid β oxidation and fatty acid biosynthesis.   Differences occur in (**1**) cellular location, (**2**) acyl group carrier, (**3**) electron acceptor/donor, (**4**) stereochemistry of the hydration/ dehydration reaction, and (**5**) the form in which C_2 units are produced/donated. 🐾 **See the Animated Figures.**

A   Transport of Mitochondrial Acetyl-CoA into the Cytosol

Acetyl-CoA, the starting material for fatty acid synthesis, is generated in the mitochondrion by the oxidative decarboxylation of pyruvate as catalyzed by pyruvate dehydrogenase (Section 16-2B) as well as by the oxidation of fatty acids. When the demand for ATP is low, so that the oxidation of acetyl-CoA via the citric acid cycle and oxidative phosphorylation is minimal, this mitochondrial acetyl-CoA may be stored for future use as fat. Fatty acid biosynthesis occurs in the cytosol, however, and the mitochondrial membrane is essentially impermeable to acetyl-CoA. *Acetyl-CoA enters the cytosol in the form of citrate via the* **tricarboxylate transport system** (Fig. 19-24). **ATP-citrate lyase** then catalyzes the reaction

$$\text{Citrate} + \text{CoA} + \text{ATP} \rightarrow \text{acetyl-CoA} + \text{oxaloacetate} + \text{ADP} + P_i$$

which resembles the reverse of the citrate synthase reaction (Fig. 16-10) except that ATP hydrolysis is required to drive the synthesis of the thioester bond. Oxaloacetate is then reduced to malate by malate dehydrogenase. Malate is oxidatively decarboxylated to pyruvate by malic enzyme and returned in this form to the mitochondrion. This reaction involves the reoxidation of malate to oxaloacetate, a β-keto acid, which is then decarboxylated, a reaction reminiscent of the isocitrate dehydrogenase reaction in the citric acid cycle (Section 16-3C). The NADPH produced is used in the reductive reactions of fatty acid biosynthesis.

B   Acetyl-CoA Carboxylase

Acetyl-CoA carboxylase (ACC) catalyzes the first committed step of fatty acid biosynthesis and one of its rate-controlling steps. The mechanism of

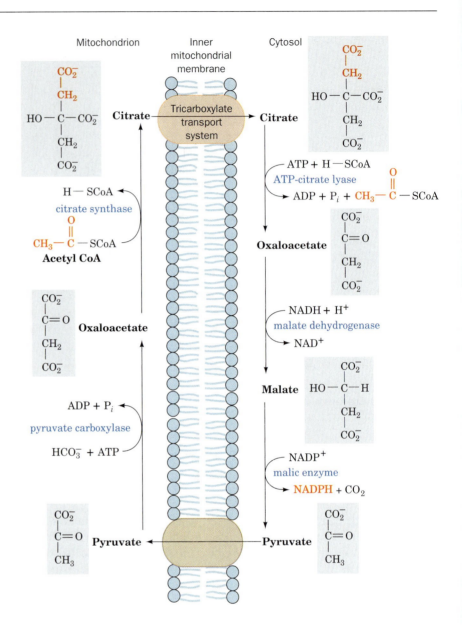

Figure 19-24 **The tricarboxylate transport system.** The overall effect of this sequence of reactions is the transfer of acetyl groups from the mitochondrion to the cytosol.

this biotin-dependent enzyme is similar to those of propionyl-CoA carboxylase (Section 19-2E) and pyruvate carboxylase (Fig. 15-26). The reaction occurs in two steps, a CO_2 activation and a carboxylation:

$$
\begin{array}{ccc}
\text{E—biotin} & \xrightarrow[\text{HCO}_3^- + \text{ATP} \quad \text{ADP} + \text{P}_i]{} & \text{E—biotin—CO}_2^- \\
\textbf{Biotinyl-enzyme} & & \textbf{Carboxybiotinyl-enzyme}
\end{array}
$$

$$
\text{CH}_3-\overset{\overset{\text{O}}{\|}}{\text{C}}-\text{SCoA} \quad \textbf{Acetyl-CoA}
$$

$$
\xrightarrow{\hspace{2cm}} \quad {}^-\text{O}_2\text{C}-\text{CH}_2-\overset{\overset{\text{O}}{\|}}{\text{C}}-\text{SCoA} + \text{E—biotin}
$$

Malonyl-CoA

The result is a three-carbon (malonyl) group linked as a thioester to CoA.

Mammalian acetyl-CoA carboxylase, a 230-kD polypeptide, is subject to allosteric and hormonal control. For example, citrate stimulates acetyl-CoA carboxylase and long chain fatty acyl-CoAs are feedback inhibitors of the enzyme. Fine-tuning of enzyme activity is accomplished through covalent modification. Acetyl-CoA carboxylase is a substrate for several kinases. It

has six phosphorylation sites, but phosphorylation of only one (Ser 79) is clearly associated with enzyme inactivation. Ser 79 is phosphorylated by **AMP-dependent protein kinase (AMPK)** in a cAMP-independent pathway. However, glucagon as well as epinephrine, which act through protein kinase A (PKA; Section 15-3B), promote the phosphorylation of Ser 79, possibly by inhibiting its dephosphorylation (recall that this occurs in glycogen metabolism when the PKA-mediated phosphorylation of phosphoprotein phosphatase inhibitor-1 inhibits dephosphorylation; Fig. 15-21). Insulin, on the other hand, stimulates dephosphorylation of acetyl-CoA carboxylase and thereby activates the enzyme.

Mammalian Acetyl-CoA Carboxylase Has Two Major Isoforms. There are two major isoforms of ACC. **α-ACC** occurs in adipose tissue and **β-ACC** occurs in tissues that oxidize but do not synthesize fatty acids, such as heart muscle. Tissues that both synthesize and oxidize fatty acids, such as liver, contain both isoforms, which are homologous although the genes encoding them are located on different chromosomes. What is the function of β-ACC? The product of the ACC-catalyzed reaction, malonyl-CoA, strongly inhibits the mitochondrial import of fatty acyl-CoA for fatty acid oxidation, the major control point for this process. Thus it appears that β-ACC has a regulatory function (Section 19-5).

The *E. coli* acetyl-CoA carboxylase, which is a multisubunit protein, is regulated by guanine nucleotides so that fatty acid synthesis is coordinated with cell growth. In prokaryotes, fatty acids serve primarily as phospholipid precursors, since these organisms do not synthesize triacylglycerols for energy storage.

C Fatty Acid Synthase

The synthesis of fatty acids, mainly palmitic acid, from acetyl-CoA and malonyl-CoA involves seven enzymatic reactions. These reactions were first studied in cell-free extracts of *E. coli,* in which they are catalyzed by independent enzymes. Individual enzymes with these activities also occur in chloroplasts (plant fatty acid synthesis does not occur in the cytosol). In yeast, **fatty acid synthase** is a cytosolic, 2500-kD multifunctional enzyme with the composition $\alpha_6\beta_6$, whereas in animals it is a 534-kD multifunctional enzyme consisting of two identical polypeptide chains. Presumably, such proteins evolved by the joining of previously independent genes for the enzymes.

Although fatty acid synthesis begins with the synthesis of a CoA ester, malonyl-CoA, the growing fatty acid is anchored to **acyl-carrier-protein** (**ACP;** Fig. 19-25). ACP, like CoA, contains a phosphopantetheine group

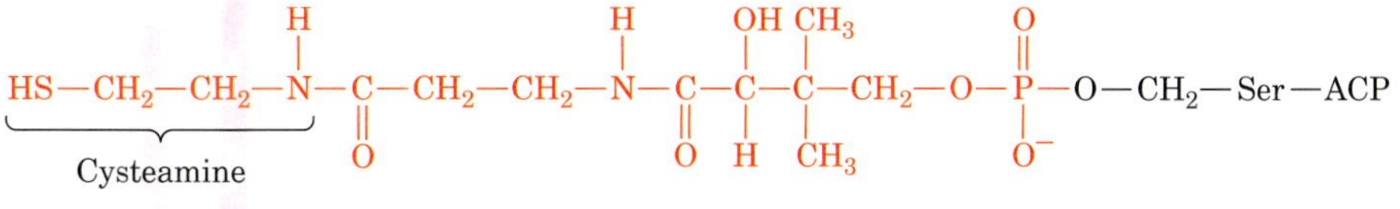

Phosphopantetheine prosthetic group of ACP

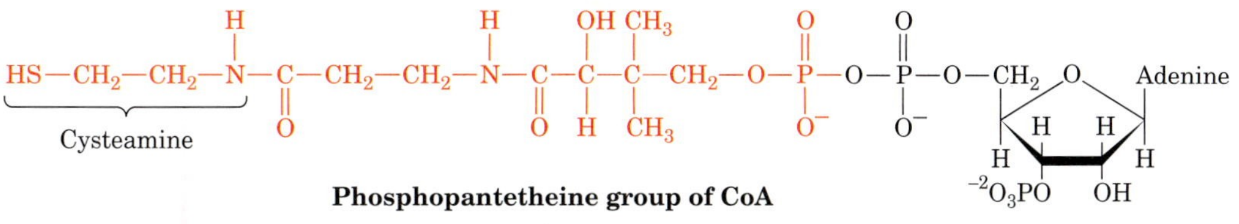

Phosphopantetheine group of CoA

Figure 19-25 **The phosphopantetheine group in acyl-carrier protein (ACP) and in CoA.**

that forms a thioester with an acyl group. The phosphopantetheine phosphoryl group is esterified with a Ser OH group of ACP, whereas in CoA it is linked to AMP. In *E. coli,* ACP is a 10-kD polypeptide, whereas in animals it is part of the multifunctional fatty acid synthase.

The reactions catalyzed by mammalian fatty acid synthase are diagrammed in Fig. 19-26:

1.   These are priming reactions in which the synthase is "loaded" with the precursors for the condensation reaction. In mammals, **malonyl/ acetyl-CoA-ACP transacetylase (MAT)** catalyzes two similar reactions at a single active site: An acetyl group originally linked as a thioester in acetyl-CoA is transferred to ACP **(1a),** and a malonyl group is transferred from malonyl-CoA to ACP **(1b).**

2.   The **β-ketoacyl-ACP synthase (KS;** also known as **condensing enzyme)** first transfers the acetyl group from ACP to an enzyme Cys residue **(2a).** In the condensation reaction **(2b),** the malonyl-ACP is decarboxylated, and the resulting carbanion attacks the acetyl-thioester to form a four-carbon β-ketoacyl-ACP. The decarboxylation reaction drives the condensation reaction:

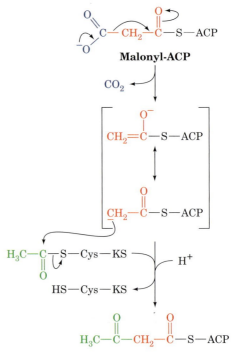

3–5. Two reductions and a dehydration convert the β-keto group to an alkyl group. The coenzyme in both reductive steps is NADPH. In β oxidation, the analogs of Reactions 3 and 5, respectively, use NAD^+ and FAD (Fig. 19-12, Reactions 3 and 1). Moreover, Reaction 4 requires a D-β-hydroxyacyl substrate, whereas the analogous reaction in β oxidation forms the corresponding L isomer.

At this point, the acyl group, originally an acetyl group, has been elongated by a C_2 unit. This butyryl group is then transferred from ACP to the Cys-SH of the enzyme (a repeat of Reaction 2a) so that it can be extended by additional rounds of the fatty acid synthase reaction sequence. Note that the malonyl-CoA synthesized by the acetyl-CoA carboxylase reaction is decarboxylated in the condensation reaction. The formation of a C—C bond

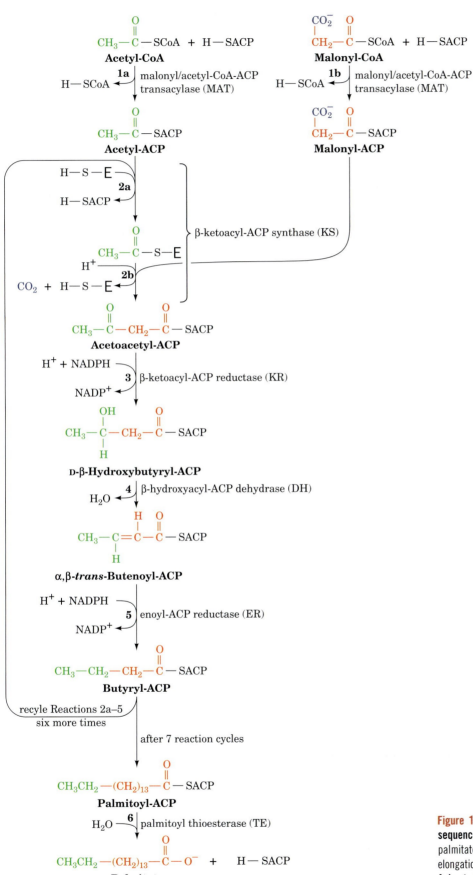

Figure 19-26 *Key to Metabolism.* **Reaction sequence for the biosynthesis of fatty acids.** In forming palmitate, the pathway is repeated for seven cycles of C_2 elongation followed by a final hydrolysis step. 🐚 **See the Animated Figures.**

is an endergonic process requiring an activated precursor. Malonyl-CoA is a β-keto ester whose exergonic decarboxylation yields the acetyl-CoA carbanion required for C—C bond formation. The free energy required for the overall reaction is supplied by the ATP hydrolysis in the acetyl-CoA carboxylase reaction, which generates the malonyl-CoA. This carboxylation–decarboxylation sequence is similar to the activation of pyruvate to oxaloacetate for conversion to phosphoenolpyruvate in gluconeogenesis (Section 15-4A).

In each reaction cycle, the ACP is "reloaded" with a malonyl group, and the acyl chain grows by two carbon atoms. Note that each new acetyl unit donated by malonyl-CoA adds to the growing acyl chain at its point of attachment to the enzyme. Thus, the fatty acid grows from its thioester end, not from its methyl end (likewise, fatty acids are catabolized from their thioester ends).

Seven cycles of elongation are required to form **palmitoyl-ACP.** The thioester bond is then hydrolyzed by **palmitoyl thioesterase (TE;** Fig. 19-26, Reaction 6), yielding palmitate, the normal product of the fatty acid synthase pathway, and regenerating the enzyme for a new round of synthesis.

The stoichiometry of palmitate synthesis is therefore

$$\text{Acetyl-CoA} + 7 \text{ malonyl-CoA} + 14 \text{ NADPH} + 7 \text{ H}^+ \rightarrow$$
$$\text{palmitate} + 7 \text{ CO}_2 + 14 \text{ NADP}^+ + 8 \text{ CoA} + 6 \text{ H}_2\text{O}$$

Since the 7 malonyl-CoA are derived from acetyl-CoA as follows:

$$7 \text{ Acetyl-CoA} + 7 \text{ CO}_2 + 7 \text{ ATP} \rightarrow$$
$$7 \text{ malonyl-CoA} + 7 \text{ ADP} + 7 \text{ P}_i + 7 \text{ H}^+$$

the overall stoichiometry for palmitate biosynthesis is

$$8 \text{ Acetyl-CoA} + 14 \text{ NADPH} + 7 \text{ ATP} \rightarrow$$
$$\text{palmitate} + 14 \text{ NADP}^+ + 8 \text{ CoA} + 6 \text{ H}_2\text{O} + 7 \text{ ADP} + 7 \text{ P}_i$$

The Two Halves of Fatty Acid Synthase Operate in Concert.   In animals, the seven reactions of fatty acid synthesis are localized to six discrete active sites (MAT carries out two reactions, 1a and 1b). The enzymatic activities are arranged along the polypeptide chain as indicated in Fig. 19-27. These contiguous stretches of polypeptide fold to form a series of domains, which when separated by limited proteolysis, exhibit many of the enzymatic activities of the intact protein. Several other enzymes exhibit similar multifunctionality, but none has as many separate catalytic activities as animal fatty acid synthase.

The condensation reaction (2b) requires the juxtaposition of the sulfhydryl group of an ACP phosphopantetheine and the active site Cys residue, which are located at opposite ends of the fatty acid synthase monomer (Fig. 19-27). Experiments with fatty acid synthase mutants indicate that these groups can interact in the monomer. However, in native fatty acid synthase, which is a dimer, the interacting groups more often belong to different monomers. This is consistent with the known structure of the animal enzyme, in which the two monomers are arranged head to tail

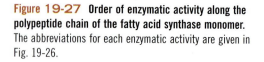

Figure 19-27 Order of enzymatic activity along the polypeptide chain of the fatty acid synthase monomer. The abbreviations for each enzymatic activity are given in Fig. 19-26.

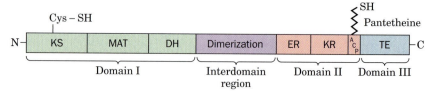

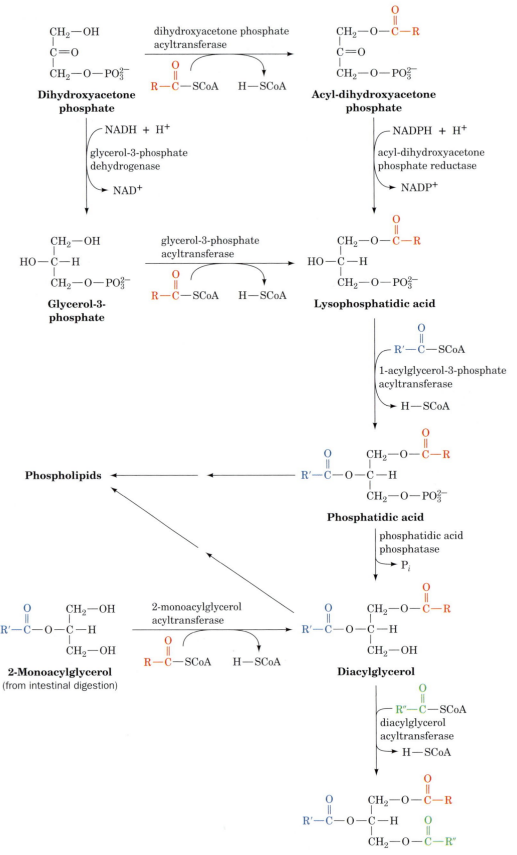

Figure 19-30 The reactions of triacylglycerol biosynthesis.

latter case, the product acyl-dihydroxyacetone phosphate is reduced to the corresponding **lysophosphatidic acid** by an NADPH-dependent reductase. The lysophosphatidic acid is converted to a triacylglycerol by the successive actions of **1-acylglycerol-3-phosphate acyltransferase, phosphatidic acid phosphatase,** and **diacylglycerol acyltransferase.** The intermediate phosphatidic acid and 1,2-diacylglycerol can also be converted to phospholipids by the pathways described in Section 19-6A. The acyltransferases are not completely specific for particular fatty acyl-CoAs, either in chain length or degree of unsaturation, but in human adipose tissue triacylglycerols, palmitate tends to be concentrated at position 1 and oleate at position 2.

Glyceroneogenesis Is Important for Triacylglycerol Biosynthesis. The dihydroxyacetone phosphate used to make glycerol-3-phosphate for triacylglycerol synthesis comes either from glucose via the glycolytic pathway (Fig. 14-1) or from oxaloacetate via an abbreviated version of gluconeogenesis (Fig. 15-23) termed **glyceroneogenesis.** Glyceroneogenesis is necessary in times of starvation, since approximately 30% of the fatty acids that enter the liver during a fast are re-esterified to triacylglycerol and exported in VLDL (Section 19-1). Adipocytes also carry out glyceroneogenesis in times of starvation. They do not carry out gluconeogenesis but contain the gluconeogenic enzyme phosphoenolpyruvate carboxykinase (PEPCK), which is upregulated when glucose concentration is low, and participates in the glyceroneogenesis required for triacylglycerol biosynthesis.

At this point, we can appreciate how triacylglycerols synthesized from fatty acids built from two-carbon acetyl units can be broken back down into acetyl units. In the liver, the resulting acetyl-CoA may be shunted to the formation of ketone bodies and later converted back to acetyl-CoA by another tissue. The acetyl-CoA can then either be used to build fatty acids that are stored as triacylglycerols or be oxidized by the citric acid cycle to generate considerable ATP by oxidative phosphorylation. As we shall see, the flux of material in the direction of triacylglycerol synthesis or triacylglycerol degradation depends on the metabolic energy needs of the organism and the need for synthesis of other compounds, such as membrane lipids (Section 19-6) and cholesterol (Section 19-7A). These key features of lipid metabolism are summarized in Fig. 19-31.

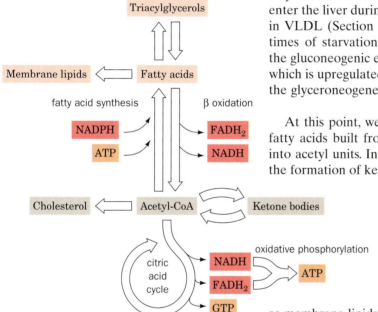

Figure 19-31   A summary of lipid metabolism.

5   Regulation of Fatty Acid Metabolism

Discussions of metabolic control are usually concerned with the regulation of metabolite flow through a pathway in response to the differing energy needs and dietary states of an organism. In mammals, glycogen and triacylglycerols serve as primary fuels for energy-requiring processes and are synthesized in times of plenty for future use. *Synthesis and breakdown of glycogen and triacylglycerols are processes that concern the whole organism, with its organs and tissues forming an interdependent network connected by the bloodstream.* The blood carries the metabolites responsible for energy production: triacylglycerols in the form of chylomicrons and VLDL, fatty acids as their albumin complexes, ketone bodies, amino acids, lactate, and glucose. As in glycogen metabolism (Section 15-3), *hormones such as insulin and glucagon regulate the rates of the opposing pathways of lipid metabolism and thereby control whether fatty acids will be oxidized or synthesized.* The major control mechanisms are summarized in Fig. 19-32.

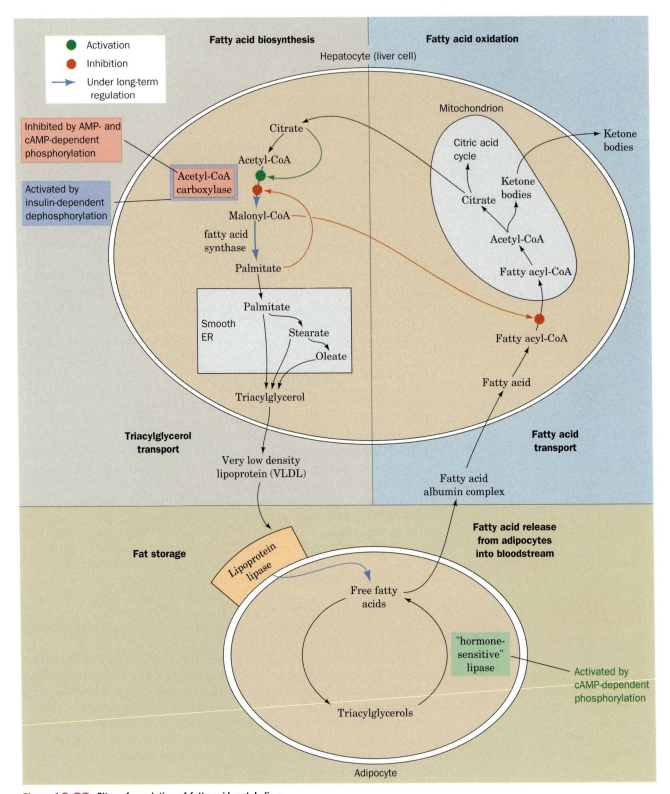

Figure 19-32 Sites of regulation of fatty acid metabolism.

Fatty acid oxidation is regulated largely by the concentration of fatty acids in the blood, which is, in turn, controlled by the hydrolysis rate of tri-acylglycerols in adipose tissue by **hormone-sensitive triacylglycerol lipase.** This enzyme is so named because it is susceptible to regulation by phos-phorylation and dephosphorylation in response to hormonally controlled

cAMP levels. Glucagon, epinephrine, and norepinephrine, which are released in times of metabolic need, increase adipose tissue cAMP concentrations. cAMP allosterically activates PKA (Fig. 15-21), which in turn phosphorylates certain enzymes. Phosphorylation activates hormone-sensitive lipase, thereby stimulating lipolysis in adipose tissue, raising blood fatty acid levels, and ultimately activating the β-oxidation pathway in other tissues such as liver and muscle. In liver, this process leads to the production of ketone bodies that are secreted into the bloodstream for use as an alternative fuel to glucose by peripheral tissues. PKA also inactivates acetyl-CoA carboxylase (Section 19-4B), so *cAMP-dependent phosphorylation simultaneously stimulates fatty acid oxidation and inhibits fatty acid synthesis.*

Insulin, a pancreatic hormone released in response to high blood glucose concentrations (the fed state), has the opposite effect of glucagon and epinephrine: It stimulates the formation of glycogen and triacylglycerols. Insulin decreases cAMP levels, leading to the dephosphorylation and thus the inactivation of hormone-sensitive lipase. This reduces the amount of fatty acid available for oxidation. Insulin also activates acetyl-CoA carboxylase (Section 19-4B). *The glucagon:insulin ratio therefore determines the rate and direction of fatty acid metabolism.*

Another mechanism that inhibits fatty acid oxidation when fatty acid synthesis is stimulated is the inhibition of carnitine palmitoyl transferase I by malonyl-CoA. This inhibition keeps the newly synthesized fatty acids out of the mitochondria (Section 19-2B) and thus away from the β-oxidation system. In fact, heart muscle, an oxidative tissue that does not carry out fatty acid biosynthesis, contains an isoform of acetyl-CoA carboxylase, β-ACC (Section 19-4B), whose sole function appears to be the synthesis of malonyl-CoA to regulate fatty acid oxidation.

AMP-dependent protein kinase (AMPK), which phosphorylates (inactivates) ACC, may itself be an important regulator of fatty acid metabolism. This enzyme is activated by AMP and inhibited by ATP and thus has been proposed to serve as a fuel gauge for the cell. When ATP levels are high, signaling the fed and rested state, this kinase is inhibited, allowing ACC to become dephosphorylated (activated) so as to stimulate malonyl-CoA production for fatty acid synthesis in adipose tissue and for inhibition of fatty acid oxidation in muscle cells. When activity levels increase, causing ATP levels to decrease with a concomitant increase in AMP levels, AMPK is activated to phosphorylate (inactivate) ACC. The resulting decrease in malonyl-CoA levels causes fatty acid biosynthesis to decrease in adipose tissue while fatty acid oxidation increases in muscle to provide the ATP for continued activity.

Factors such as substrate availability, allosteric interactions, and covalent modification (phosphorylation) regulate enzyme activity with response times of minutes or less. Such **short-term regulation** is complemented by **long-term regulation,** which requires hours or days and governs a pathway's regulatory enzyme by altering the amount of enzyme present. This is accomplished through changes in the rates of protein synthesis and/or breakdown.

The long-term regulation of lipid metabolism includes stimulation by insulin and inhibition by starvation of the synthesis of acetyl-CoA carboxylase and fatty acid synthase. The amount of adipose tissue lipoprotein lipase, the enzyme that initiates the entry of lipoprotein-packaged fatty acids into adipose tissue for storage (Section 19-1B), is also increased by insulin and decreased by starvation. Thus, an abundance of glucose, reflected in the level of insulin, promotes fatty acid synthesis and the storage of fatty acids by adipocytes, whereas starvation, when glucose is unavailable, decreases fatty acid synthesis and the uptake of fatty acids by adipocytes.

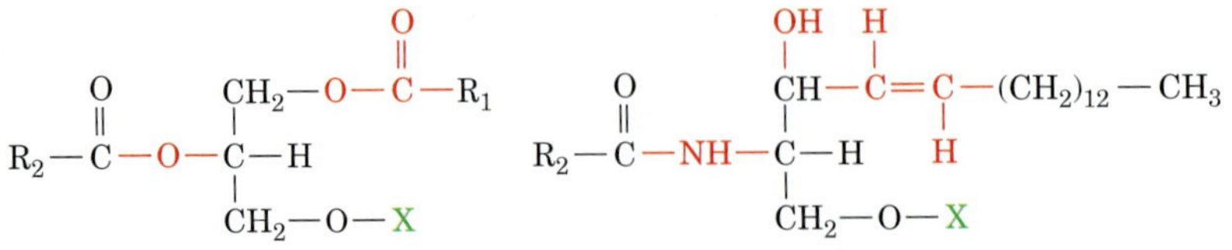

Glycerolipid **Sphingolipid**

X = H	**1,2-Diacylglycerol**	***N*-Acylsphingosine (ceramide)**
X = Carbohydrate	**Glyceroglycolipid**	**Sphingoglycolipid (glycosphingolipid)**
X = Phosphate ester	**Glycerophospholipid**	**Sphingophospholipid**

Figure 19-33 Glycerolipids and sphingolipids. The structures of the common head groups, X, are presented in Table 9-2. Plant membranes are particularly rich in glyceroglycolipids.

6 Synthesis of Other Lipids

Fatty acids are the precursors not just of triacylglycerols but of a variety of other compounds, including membrane lipids and certain signaling molecules. Most membrane lipids are dual-tailed amphipathic molecules composed of either 1,2-diacylglycerol or *N*-acylsphingosine (ceramide) linked to a polar head group that is either a carbohydrate or a phosphate ester (Fig. 19-33). In this section, we describe the biosynthesis of these complex lipids from their simpler components. These lipids are synthesized in membranes, mostly on the cytosolic face of the endoplasmic reticulum, and from there are transported in vesicles to their final cellular destinations (Section 9-4E). Arachidonate groups released from membrane lipids give rise to prostaglandins of various types.

A Glycerophospholipids

Glycerophospholipids have significant asymmetry in their C1- and C2-linked fatty acyl groups: C1 substituents are mostly saturated fatty acids, whereas those at C2 are by and large unsaturated fatty acids.

Biosynthesis of Diacylglycerophospholipids. The triacylglycerol precursors 1,2-diacylglycerol and phosphatidic acid are also the precursors of glycerophospholipids (Fig. 19-30). The polar head groups of glycerophospholipids are linked to C3 of the glycerol via a phosphodiester bond. In mammals, the head groups **ethanolamine** and **choline** are activated before being attached to the lipid (Fig. 19-34):

1. ATP phosphorylates the OH group of choline or ethanolamine.
2. The phosphoryl group of the resulting **phosphoethanolamine** or **phosphocholine** then attacks CTP, displacing PP_i, to form the corresponding CDP derivatives, which are activated phosphate esters of the polar head group.
3. The C3-OH group of 1,2-diacylglycerol attacks the phosphoryl group of the activated CDP–ethanolamine or CDP–choline, displacing CMP to yield the corresponding glycerophospholipid.

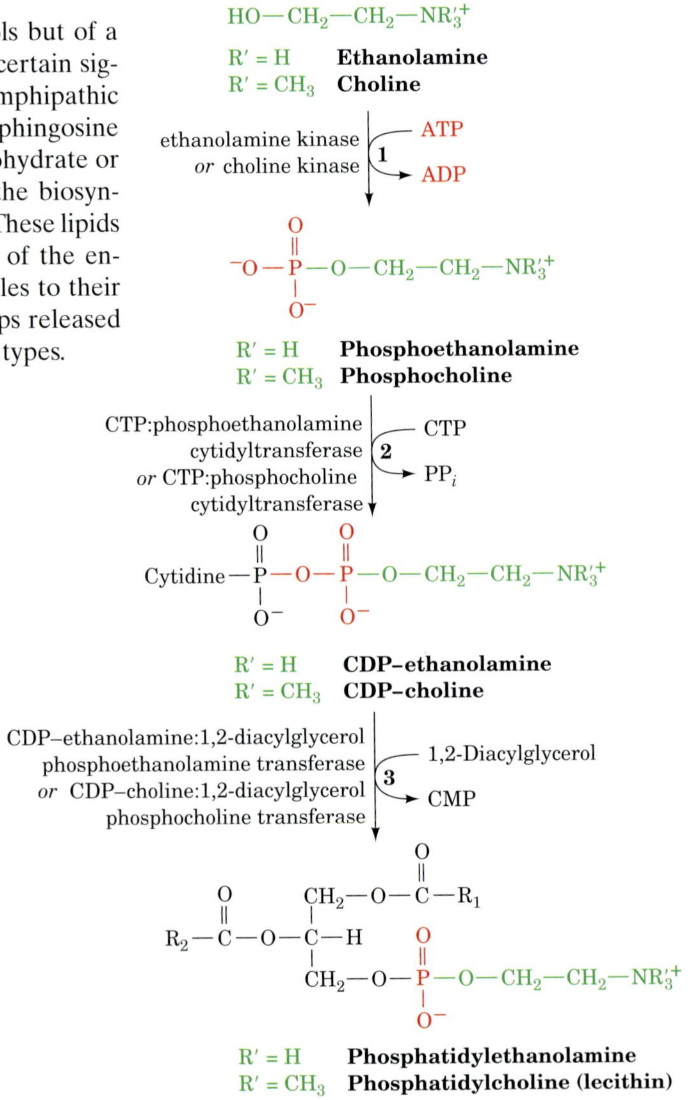

Figure 19-34 Biosynthesis of phosphatidylethanolamine and phosphatidylcholine. In mammals, CDP–ethanolamine and CDP–choline are the precursors of the head groups.

In liver, phosphatidylethanolamine can also be converted to phosphatidylcholine by the addition of methyl groups.

Phosphatidylserine is synthesized from phosphatidylethanolamine by a head group exchange reaction catalyzed by **phosphatidylethanolamine transferase** in which serine's OH group attacks the donor's phosphoryl group. The original head group is then eliminated, forming phosphatidylserine:

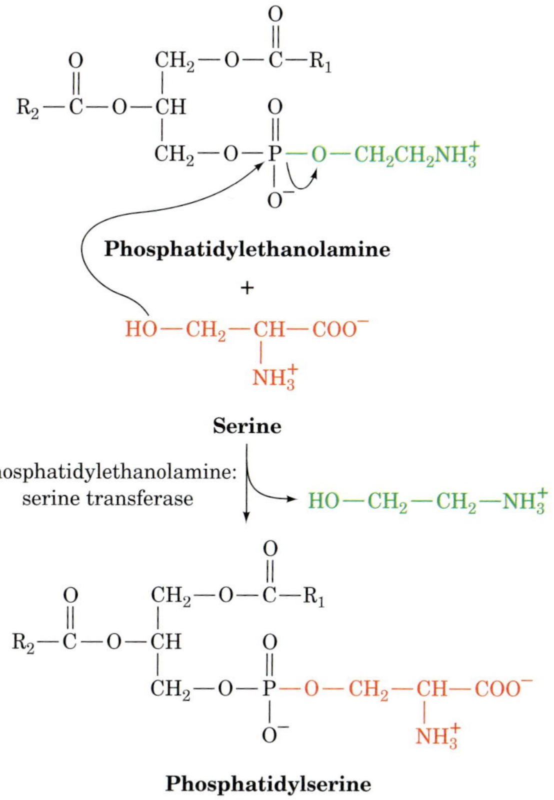

Phosphatidylethanolamine

+

Serine

phosphatidylethanolamine:
serine transferase

Phosphatidylserine

In the synthesis of **phosphatidylinositol** and **phosphatidylglycerol,** the hydrophobic tail is activated rather than the polar head group. Phosphatidic acid, the precursor of 1,2-diacylglycerol (Fig. 19-30), attacks the α-phosphoryl group of CTP to form the activated **CDP–diacylglycerol** and PP$_i$ (Fig. 19-35). Phosphatidylinositol results from the attack of inositol on CDP–diacylglycerol. Phosphatidylglycerol is formed in two reactions: (1) attack of the C1-OH group of glycerol-3-phosphate on CDP–diacylglycerol, yielding **phosphatidylglycerol phosphate;** and (2) hydrolysis of the phosphoryl group to form phosphatidylglycerol.

Cardiolipin forms by the condensation of two molecules of phosphatidylglycerol with the elimination of one molecule of glycerol:

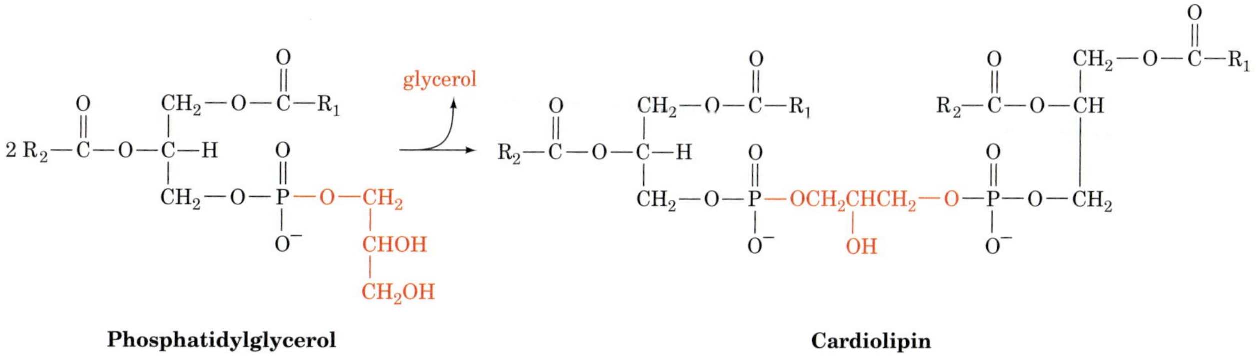

Phosphatidylglycerol   **Cardiolipin**

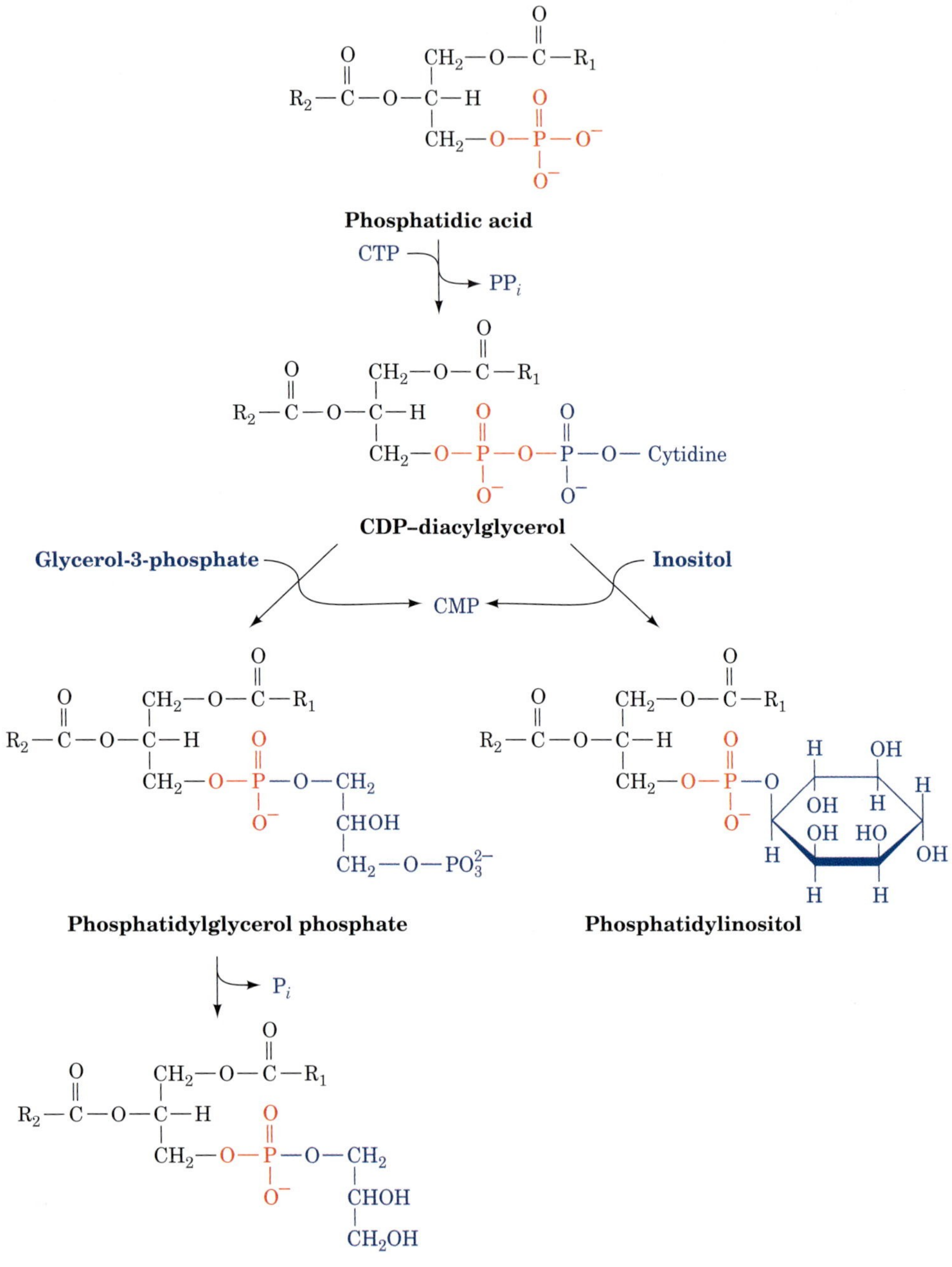

Figure 19-35 Biosynthesis of phosphatidylinositol and phosphatidylglycerol.   In mammals, this process involves a CDP–diacylglycerol intermediate.

Enzymes that synthesize phosphatidic acid have a general preference for saturated fatty acids at C1 and for unsaturated fatty acids at C2. Yet this general preference cannot account, for example, for the observations that ~80% of brain phosphatidylinositol has a stearoyl group (18:0) at C1 and an arachidonoyl group (20:4) at C2, and that ~40% of lung phosphatidyl-choline has palmitoyl groups (16:0) at both positions (this latter substance is the major component of the surfactant that prevents the lung from col-lapsing when air is expelled; its deficiency in premature infants is respon-sible for respiratory distress syndrome; see Box 9-1). *William Lands showed that such side chain specificity results from "remodeling" reactions in which*

specific acyl groups of individual glycerophospholipids are exchanged by specific phospholipases and acyltransferases.

Biosynthesis of Plasmalogens and Alkylacylglycerophospholipids. Eukaryotic membranes contain significant amounts of two other types of glycerophospholipids: **plasmalogens,** which contain a hydrocarbon chain linked to glycerol C1 via a vinyl ether linkage, and **alkylacylglycerophospholipids,** in which the alkyl substituent at glycerol C1 is attached via an ether linkage.

A plasmalogen

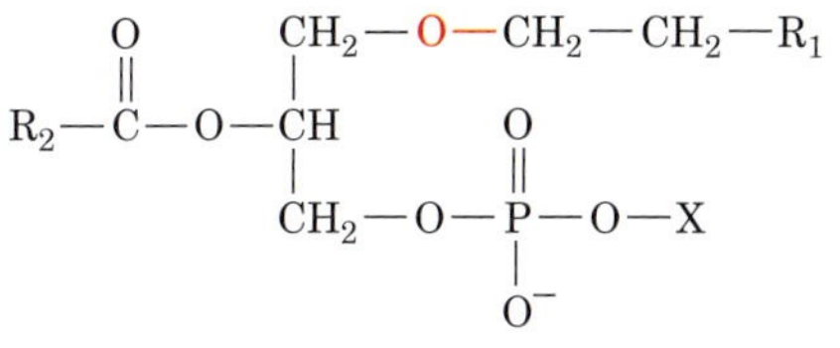

An alkylacyl-glycerophospholipid

About 20% of mammalian glycerophospholipids are plasmalogens, but the exact percentage varies both among species and among tissues within a given organism. For example, plasmalogens account for only 0.8% of the phospholipids in human liver but 23% of those in human nervous tissue. The alkylacylglycerophospholipids are much less abundant than the plasmalogens. The biosynthetic pathway for these lipids yields the alkylacylglycerophospholipid, and this ether is then oxidized to a vinyl ether by the action of a desaturase to yield the plasmalogen.

B Sphingolipids

Most sphingolipids are **sphingoglycolipids;** that is, their polar head groups consist of carbohydrate units. **Cerebrosides** are ceramide monosaccharides, whereas **gangliosides** are sialic acid–containing ceramide oligosaccharides (Section 9-1D). These lipids are synthesized by attaching carbohydrate units to the C1-OH group of ceramide (*N*-acylsphingosine).

N-Acylsphingosine is synthesized in four reactions from the precursors palmitoyl-CoA and serine (Fig. 19-36):

1. **3-Ketosphinganine synthase** catalyzes condensation of palmitoyl-CoA with serine, yielding **3-ketosphinganine.**
2. **3-Ketosphinganine reductase** catalyzes the NADPH-dependent reduction of 3-ketosphinganine's keto group to form **sphinganine (dihydrosphingosine).**
3. **Dihydroceramide** is formed by transfer of an acyl group from an acyl-CoA to sphinganine's 2-amino group, forming an amide bond.
4. **Dihydroceramide reductase** converts dihydroceramide to ceramide by an FAD-dependent oxidation reaction.

The nonglycosylated lipid sphingomyelin, an important structural lipid of nerve cell membranes,

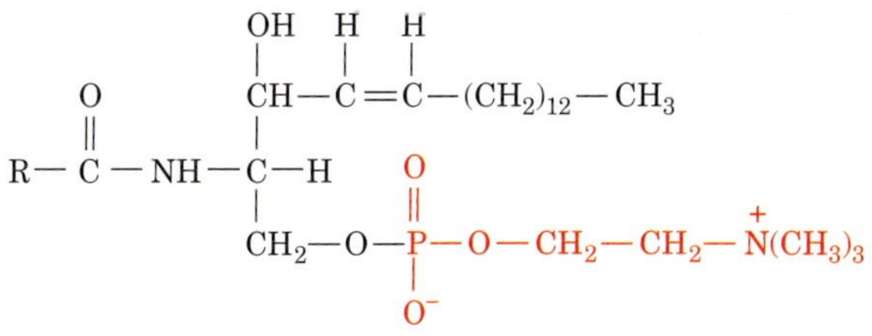

Sphingomyelin

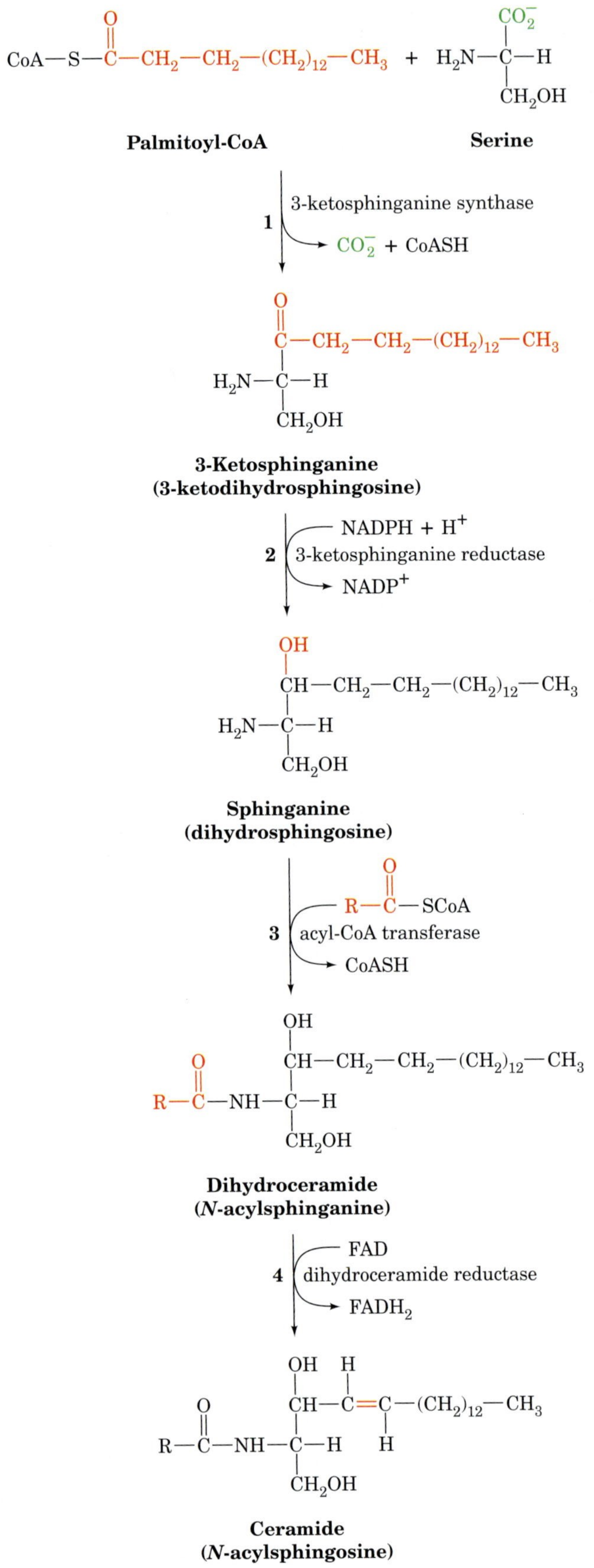

$$CoA-S-\overset{\overset{\displaystyle O}{\|}}{C}-CH_2-CH_2-(CH_2)_{12}-CH_3 \quad + \quad H_2N-\overset{\overset{\displaystyle CO_2^-}{|}}{\underset{\underset{\displaystyle CH_2OH}{|}}{C}}-H$$

Palmitoyl-CoA   **Serine**

1 — 3-ketosphinganine synthase
$\longrightarrow CO_2^- + $ CoASH

$$\overset{\overset{\displaystyle O}{\|}}{C}-CH_2-CH_2-(CH_2)_{12}-CH_3$$
$$H_2N-\overset{|}{\underset{\underset{\displaystyle CH_2OH}{|}}{C}}-H$$

3-Ketosphinganine
(3-ketodihydrosphingosine)

2 — NADPH + H$^+$
3-ketosphinganine reductase
$\longrightarrow$ NADP$^+$

$$\overset{\overset{\displaystyle OH}{|}}{CH}-CH_2-CH_2-(CH_2)_{12}-CH_3$$
$$H_2N-\overset{|}{\underset{\underset{\displaystyle CH_2OH}{|}}{C}}-H$$

Sphinganine
(dihydrosphingosine)

3 — R$-\overset{\overset{\displaystyle O}{\|}}{C}-$SCoA
acyl-CoA transferase
$\longrightarrow$ CoASH

$$\overset{\overset{\displaystyle OH}{|}}{CH}-CH_2-CH_2-(CH_2)_{12}-CH_3$$
$$R-\overset{\overset{\displaystyle O}{\|}}{C}-NH-\overset{|}{\underset{\underset{\displaystyle CH_2OH}{|}}{C}}-H$$

Dihydroceramide
(*N*-acylsphinganine)

4 — FAD
dihydroceramide reductase
$\longrightarrow$ FADH$_2$

$$\overset{\overset{\displaystyle OH}{|}}{CH}-\overset{\overset{\displaystyle H}{|}}{C}=\overset{|}{\underset{\underset{\displaystyle H}{|}}{C}}-(CH_2)_{12}-CH_3$$
$$R-\overset{\overset{\displaystyle O}{\|}}{C}-NH-\overset{|}{\underset{\underset{\displaystyle CH_2OH}{|}}{C}}-H$$

Ceramide
(*N*-acylsphingosine)

Figure 19-36 Biosynthesis of ceramide (*N*-acylsphingosine).

BOX 19-4

Biochemistry in Health and Disease

Sphingolipid Degradation and Lipid Storage Diseases

Sphingoglycolipids are lysosomally degraded by a series of enzymatically mediated hydrolytic reactions. Below is shown the pathway for the degradation of ganglioside G_{M1} and a related globoside and sulfatide. The abbreviations for the monosaccharide residues are Gal, galactose; GalNAc, N-acetylgalactosamine; Glc, glucose; NANA, N-acetylneuraminic acid (sialic acid). Cer represents ceramide.

A hereditary defect in one of these enzymes (indicated by a red bar) results in a **sphingolipid storage disease.** The substrate of the missing enzyme therefore accumulates, often with disastrous consequences. In many cases, affected individuals suffer from mental retardation and die in infancy or early childhood. One of the most common lipid storage dis-

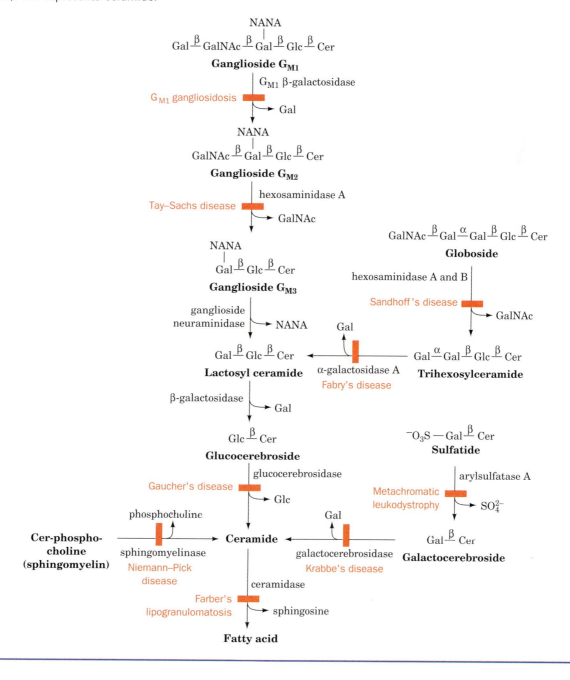

eases is **Tay–Sachs disease,** an autosomal recessive deficiency in **hexosaminidase A,** which hydrolyzes *N*-acetylgalactosamine from ganglioside G_{M2}. The absence of hexosaminidase A activity results in the accumulation of G_{M2} as shell-like inclusions in neuronal cells as shown below:

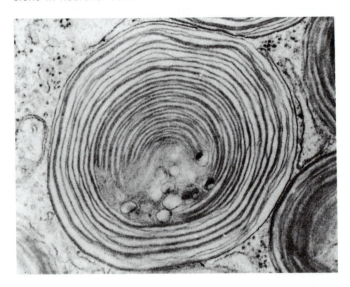

Although infants born with Tay–Sachs disease at first appear normal, by ~1 year of age, when sufficient G_{M2} has ac-

cumulated to interfere with neuronal function, they become progressively weaker, retarded, and blinded until they die, usually by the age of 3 years. It is possible, however, to screen potential carriers of this disease by a simple serum assay for hexosaminidase A.

Experiments using a mouse model of Tay–Sachs disease suggest that G_{M2} accumulation can be reduced by inhibiting its synthesis. This is accomplished in mice by administering the imino sugar ***N*-butyldeoxynojirimycin,**

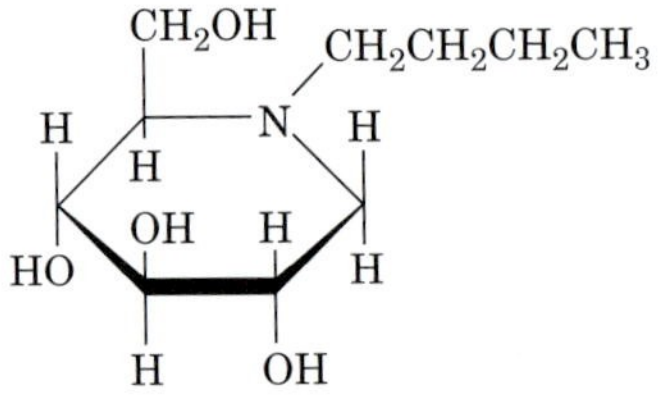

***N*-Butyldeoxynojirimycin**

which inhibits the glycosyltransferase in the first step of glucoceramide synthesis. Similar "substrate deprivation" approaches may be effective in treating other lipid storage diseases, particularly when the defective but essential enzyme has some residual activity.

[Photo courtesy of John S. O'Brien, University of California at San Diego.]

---

is the product of a reaction in which phosphatidylcholine donates its phosphocholine group to the C1-OH group of *N*-acylsphingosine.

Cerebrosides, which are most commonly 1-β-galactoceramide or 1-β-glucoceramide, are synthesized from ceramide by the addition of the glycosyl unit from the corresponding UDP–hexose to ceramide's C1-OH group. The more elaborate oligosaccharide head groups of gangliosides (Fig. 9-9) are constructed through the action of a series of glycosyltransferases. Defects in the pathways for degrading these complex lipids are responsible for certain **lipid storage diseases** (see Box 19-4).

C   Prostaglandins

Prostaglandins and related compounds (Fig. 9-12) are derivatives of C_{20} fatty acids such as arachidonate, which is released from membrane phospholipids in response to hormones and other signals. The functions of prostaglandins vary in a tissue-specific manner, but several of them trigger pain, fever, or inflammation. The production of prostaglandins begins with the formation of a cyclopentane ring in the linear fatty acid, as catalyzed

by **prostaglandin H₂ synthase** (Fig. 19-37). This heme-containing enzyme contains two catalytic activities: a **cyclooxygenase** that adds two molecules of O_2 to arachidonate, and a **peroxidase** that converts the resulting hydroperoxy group to an OH group. The enzyme is commonly called **COX,** after its cyclooxygenase activity (not to be confused with cytochrome *c* oxidase, which is also called COX).

The use of **aspirin** as an analgesic (pain-relieving), antipyretic (fever-reducing), and anti-inflammatory agent has been widespread since the nineteenth century. Yet it was not until 1971 that John Vane discovered its mechanism of action: Aspirin inhibits the synthesis of prostaglandins by acetylating a specific Ser residue of prostaglandin H₂ synthase, which prevents arachidonate from reaching the cyclooxygenase active site. Other **nonsteroidal anti-inflammatory drugs (NSAIDs)** such as **ibuprofen** and **acetaminophen** noncovalently bind to the enzyme so as to similarly block its active site.

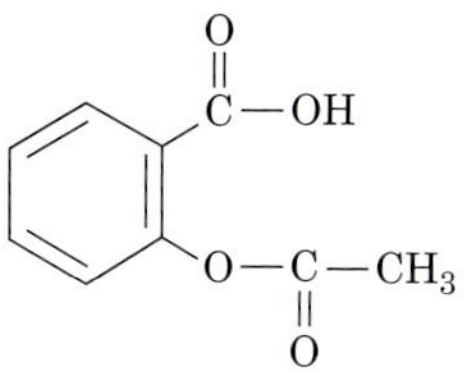

Aspirin
(acetylsalicylic acid) 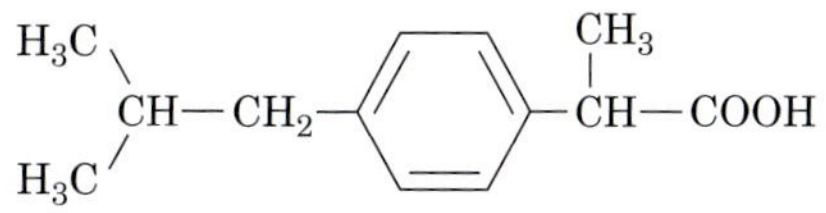   **Ibuprofen** 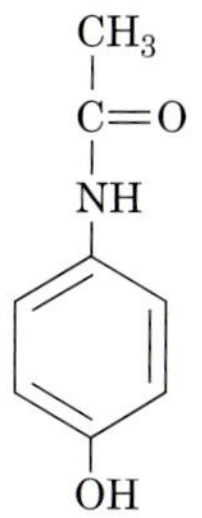   **Acetaminophen**

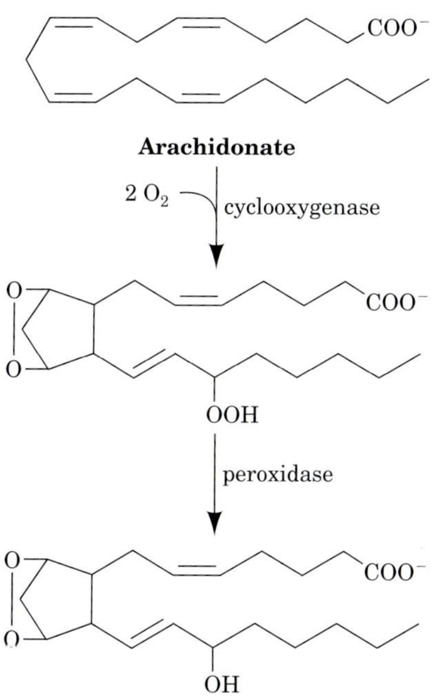

Arachidonate

$2 O_2$ ⟍ cyclooxygenase

OOH

peroxidase

OH

PGH₂

Figure 19-37 The prostaglandin H₂ synthase reaction.
A cyclooxygenase activity catalyzes the additions and rearrangements that generate the cyclopentane ring. The enzyme's peroxidase activity converts the peroxide intermediate to prostaglandin H₂ (PGH₂), which is the precursor of other prostaglandins.

COX-2 Inhibitors Lack the Side Effects of Other NSAIDs. Prostaglandin H₂ synthase has two isoforms, **COX-1** and **COX-2,** that share a high degree (60%) of sequence identity and structural homology. COX-1 is constitutively (without regulation) expressed in most, if not all, mammalian tissues, thereby supporting levels of prostaglandin synthesis necessary to maintain organ and tissue homeostasis. In contrast, COX-2 is only expressed in certain tissues in response to inflammatory stimuli and hence is responsible for the elevated prostaglandin levels that cause inflammation.

The NSAIDs shown above are relatively nonspecific and therefore can have adverse side effects (e.g., gastrointestinal ulceration) when used to treat inflammation or fever. A structure-based drug design program (Section 12-4) was therefore instituted to create inhibitors that would target COX-2 but not COX-1. The three-dimensional structures of COX-1 and COX-2 are almost identical. However, their amino acid differences make COX-2's active site channel ~20% larger in volume than that of COX-1. Chemists therefore synthesized inhibitors, collectively known as **coxibs,** that could enter the COX-2 channel but are excluded from that of COX-1. Two of these inhibitors, **rofecoxib (Vioxx)** and **celecoxib (Celebrex),**

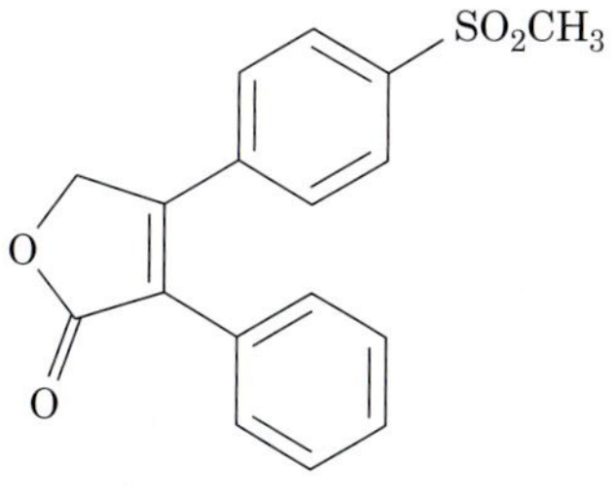

Rofecoxib (Vioxx) 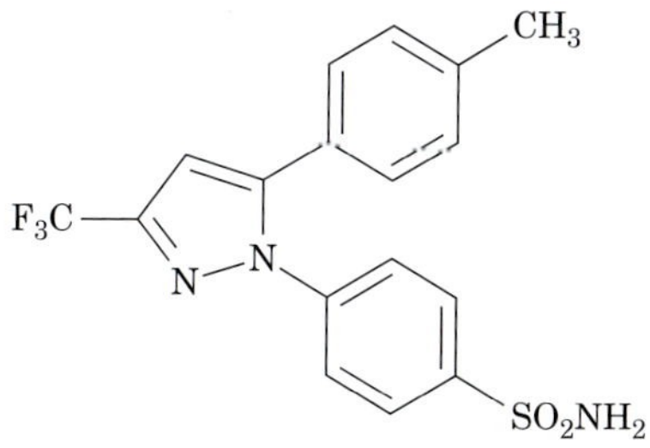   **Celecoxib (Celebrex)**

became major drugs for the treatment of inflammatory diseases such as arthritis because they lack the major side effects of the nonspecific NSAIDs. However, in 2004, Vioxx was withdrawn from the market because of unanticipated cardiac side effects.

Interestingly, acetaminophen, the most commonly used analgesic/antipyretic, binds poorly to both COX-1 and COX-2 and is not (despite its inclusion in this category) an anti-inflammatory agent. Its mechanism of action remained a mystery until Daniel Simmons' recent discovery of a third COX isozyme, **COX-3**, which is expressed at high levels in the central nervous system and is apparently the target of drugs that decrease pain and fever.

7 Cholesterol Metabolism

Cholesterol is a vital constituent of cell membranes and the precursor of steroid hormones and bile acids. It is clearly essential to life, yet its deposition in arteries is associated with cardiovascular disease and stroke, two leading causes of death in humans. *In a healthy organism, an intricate balance is maintained between the biosynthesis, utilization, and transport of cholesterol, keeping its harmful deposition to a minimum.* In this section, we study the pathways of cholesterol biosynthesis and transport and how they are controlled.

A Cholesterol Biosynthesis

Cholesterol biosynthesis follows a lengthy pathway, first outlined by Konrad Bloch, in which acetate (from acetyl-CoA) is converted to **isoprene units** that have the carbon skeleton of **isoprene**:

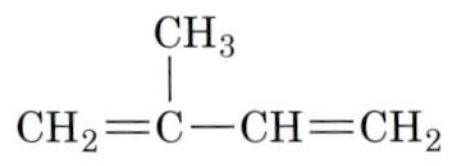

$$CH_2\!=\!\overset{\overset{\displaystyle CH_3}{|}}{C}\!-\!CH\!=\!CH_2 \qquad \overset{\overset{\displaystyle C}{|}}{C}\!-\!\overset{\overset{\displaystyle C}{|}}{C}\!-\!C\!-\!C$$

Isoprene | **An isoprene unit**
(2-methyl-1,3-butadiene)

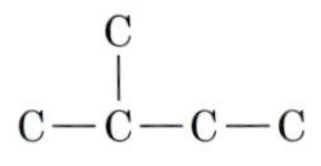

The isoprene units then condense to form a linear molecule with 30 carbons that cyclizes to form the four-ring structure of cholesterol.

HMG-CoA Is a Key Cholesterol Precursor. Acetyl-CoA is converted to isoprene units by a series of reactions that begins with formation of hydroxymethylglutaryl-CoA (HMG-CoA; this compound is also an intermediate in ketone body synthesis; Fig. 19-21). HMG-CoA synthesis requires thiolase and HMG-CoA synthase. In mitochondria, these two enzymes form HMG-CoA for ketone body synthesis. Cytosolic isozymes of these two proteins generate the HMG-CoA that is used in cholesterol biosynthesis. Four additional reactions convert HMG-CoA to the isoprenoid intermediate **isopentenyl pyrophosphate** (Fig. 19-38):

1. The CoA thioester group of HMG-CoA is reduced to an alcohol in an NADPH-dependent four-electron reduction catalyzed by **HMG-CoA reductase**, yielding **mevalonate**, a C_6 compound. This is the rate-determining step of cholesterol biosynthesis.

2. The new OH group is phosphorylated by **mevalonate-5-phosphotransferase**.

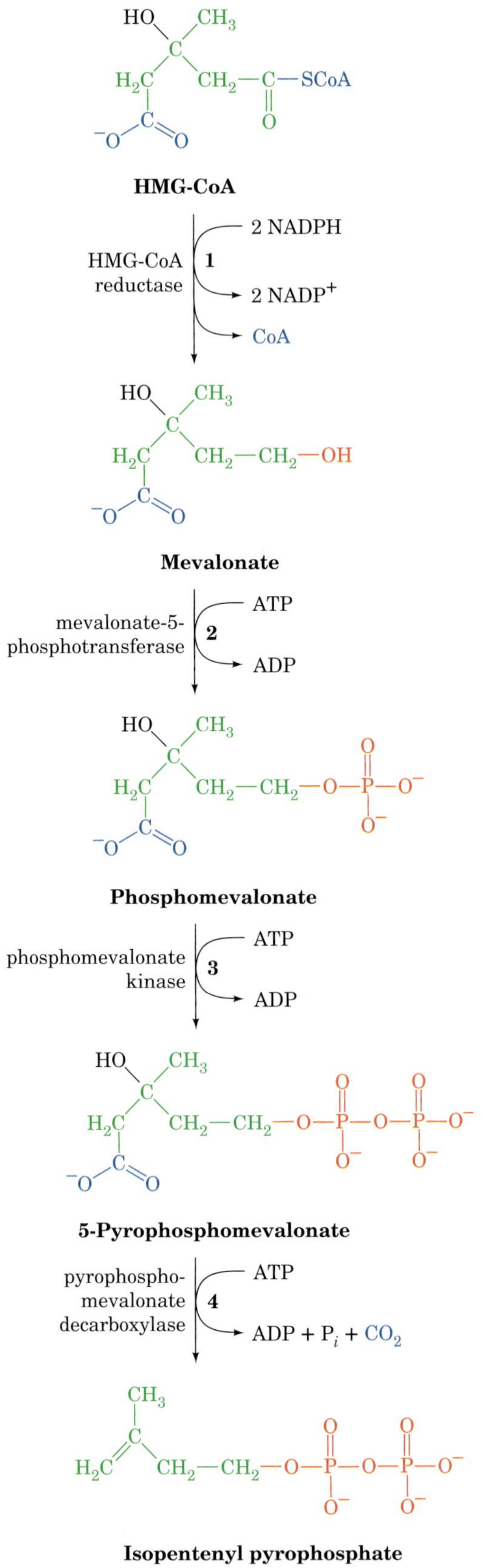

Figure 19-38 Formation of isopentenyl pyrophosphate from HMG-CoA.

3. The phosphate group is converted to a pyrophosphate by **phosphomevalonate kinase.**

4. The molecule undergoes an ATP-dependent decarboxylation–dehydration reaction catalyzed by **pyrophosphomevalonate decarboxylase:**

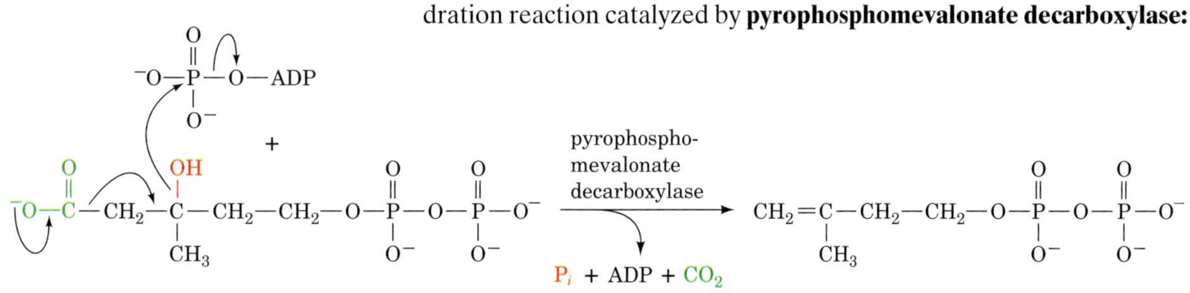

5-Pyrophosphomevalonate **Isopentenyl pyrophosphate**

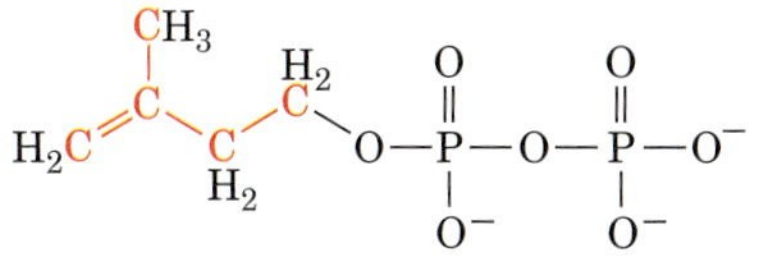

Isopentenyl pyrophosphate

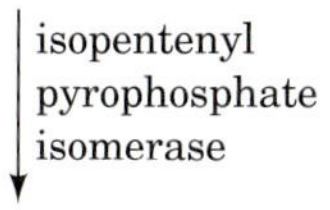

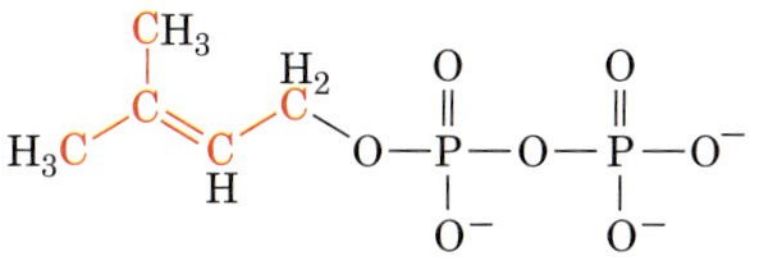

isopentenyl pyrophosphate isomerase

Dimethylallyl pyrophosphate

Squalene Is Formed by the Condensation of Six Isoprene Units. Isopentenyl pyrophosphate is converted to **dimethylallyl pyrophosphate** by **isopentenyl pyrophosphate isomerase** (*at left*). Four isopentenyl pyrophosphates and two dimethylallyl pyrophosphates condense to form the C_{30} cholesterol precursor **squalene** in three reactions catalyzed by two enzymes (Fig. 19-39):

1. **Prenyltransferase** catalyzes the head-to-tail condensation of dimethylallyl pyrophosphate and isopentenyl pyrophosphate to yield the C_{10} compound **geranyl pyrophosphate.**

2. Prenyltransferase catalyzes a second head-to-tail condensation of geranyl pyrophosphate and isopentenyl pyrophosphate to yield the C_{15} compound **farnesyl pyrophosphate.** The prenyltransferase catalyzes an S_N1 reaction to form a carbocation intermediate with an ionization–condensation–elimination mechanism:

Ionization–condensation–elimination

S_N1

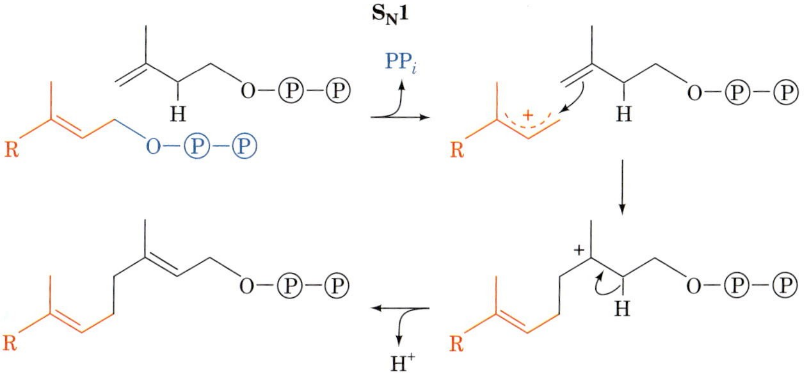

3. **Squalene synthase** then catalyzes the head-to-head condensation of two farnesyl pyrophosphate molecules to form squalene.

Farnesyl pyrophosphate is also the precursor of other isoprenoid compounds in mammals, including ubiquinone (Section 9-1F) and the isoprenoid tails of some lipid-linked membrane proteins (Section 9-3B).

Squalene Cyclization Eventually Yields Cholesterol. Squalene, a linear hydrocarbon, cyclizes to form the tetracyclic steroid skeleton in two steps. **Squalene epoxidase** catalyzes oxidation of squalene to form **2,3-oxidosqualene** (Fig. 19-40). **Oxidosqualene cyclase** converts this epoxide to the steroid **lanosterol.** The reaction is a chemically complex process involving cycliza-

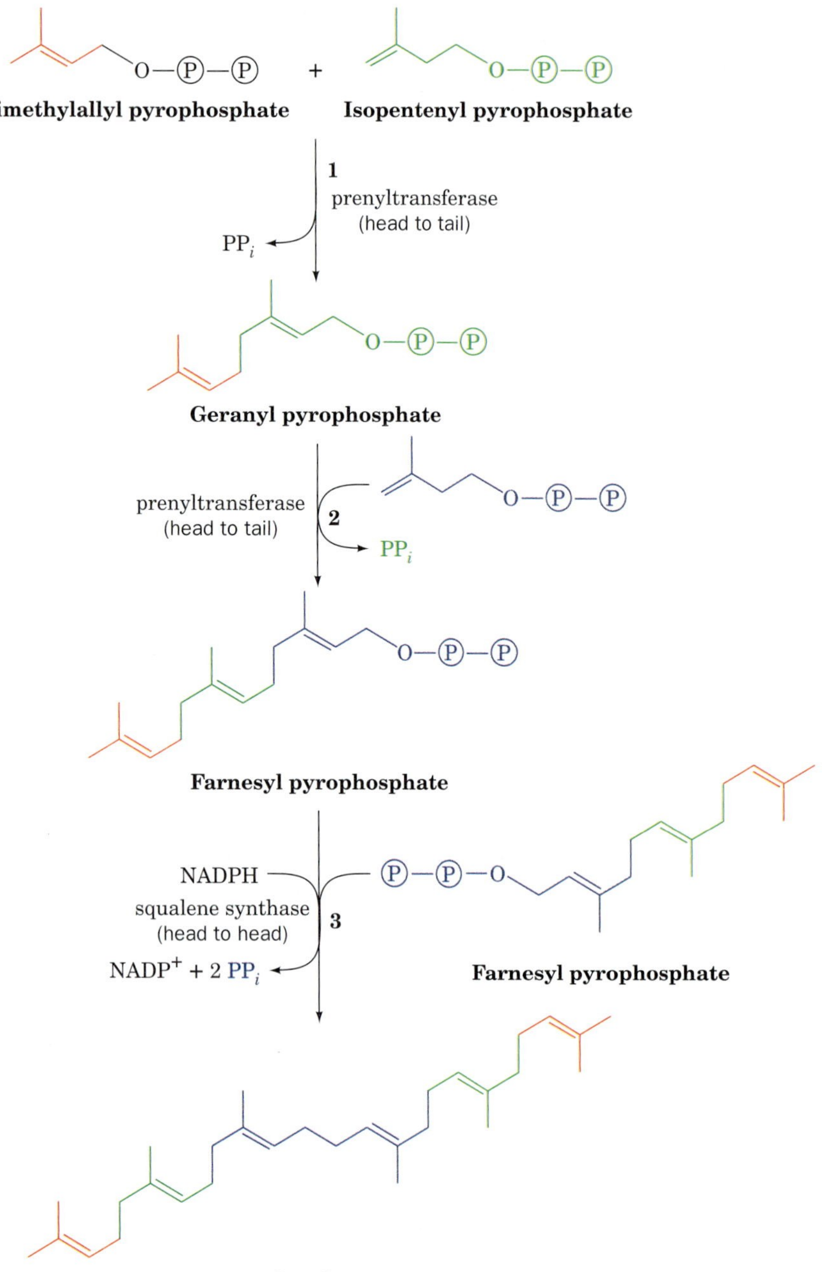

Dimethylallyl pyrophosphate + **Isopentenyl pyrophosphate**

1 prenyltransferase (head to tail)

PP$_i$

Geranyl pyrophosphate

prenyltransferase (head to tail) 2 PP$_i$

Farnesyl pyrophosphate

NADPH squalene synthase (head to head) 3 NADP$^+$ + 2 PP$_i$

Farnesyl pyrophosphate

Squalene

Figure 19-39 Formation of squalene from isopentenyl pyrophosphate and dimethylallyl pyrophosphate. The pathway involves two head-to-tail condensations catalyzed by prenyltransferase and a head-to-head condensation catalyzed by squalene synthase.

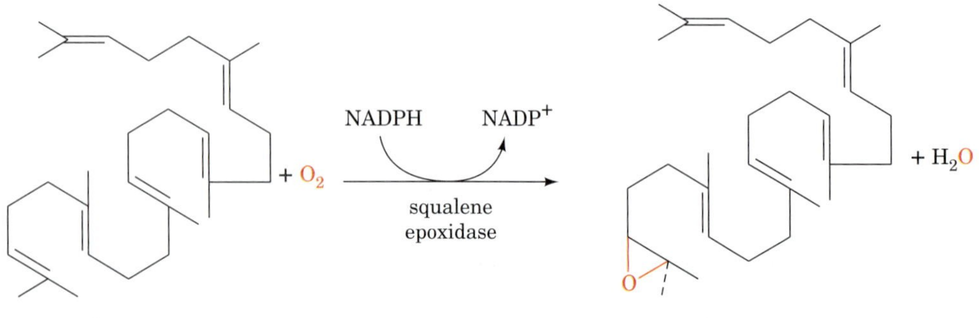

Squalene + O$_2$

NADPH → NADP$^+$

squalene epoxidase

2,3-Oxidosqualene + H$_2$O

Figure 19-40 The squalene epoxidase reaction.

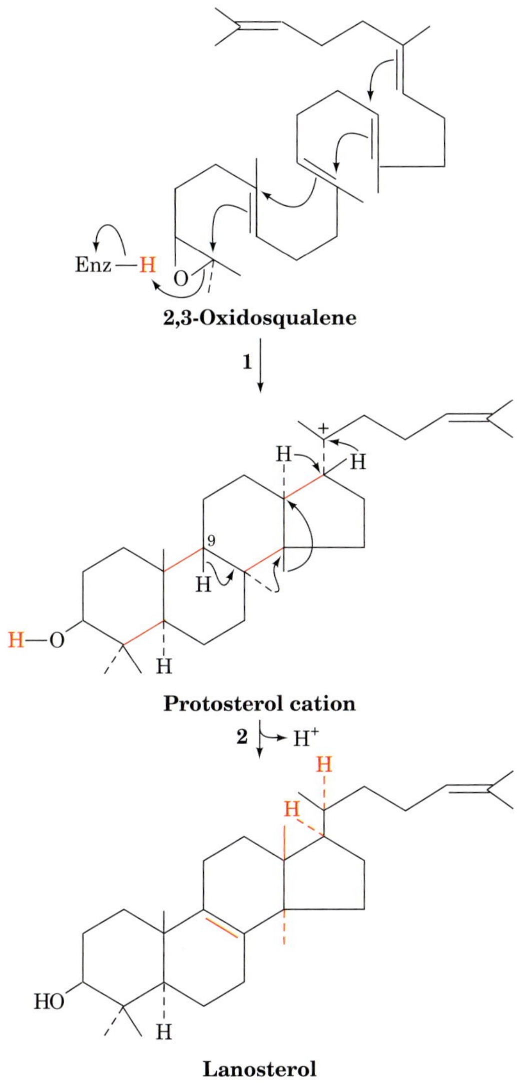

2,3-Oxidosqualene

1 ↓

Protosterol cation

2 → H⁺

Lanosterol

Figure 19-41 **The oxidosqualene cyclase reaction.** (**1**) 2,3-Oxidosqualene is cyclized to the protosterol cation in a process that is initiated by the enzyme-mediated protonation of the squalene epoxide oxygen. The opening of the epoxide leaves an electron-deficient center whose migration drives the series of cyclizations that form the protosterol cation. (**2**) A series of methyl and hydride migrations followed by the elimination of a proton from C9 of the sterol to form a double bond ultimately yields neutral lanosterol.

tion of 2,3-oxidosqualene to a **protosterol** cation and rearrangement of this cation to lanosterol by a series of 1,2 hydride and methyl shifts (Fig. 19-41).

Conversion of lanosterol to cholesterol

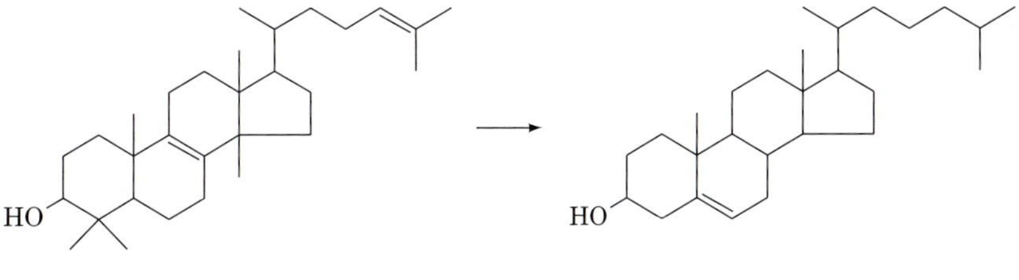

Lanosterol   **Cholesterol**

is a 19-step process that involves an oxidation and the loss of three methyl groups. The enzymes required for this process are embedded in the endoplasmic reticulum membrane.

Uses of Cholesterol. Cholesterol is the precursor of steroid hormones such as cortisol, androgens, and estrogens (Section 9-1E). The liver converts cholesterol to bile acids (Fig. 19-1), which act as emulsifying agents in the digestion and absorption of fats (Section 19-1A). An efficient recycling system allows the bile acids to re-enter the bloodstream and return to the liver for reuse several times each day. The bile acids that escape this recycling are further metabolized by intestinal microorganisms and excreted. *This is the only route for cholesterol excretion.*

Cholesterol synthesized by the liver may be esterified by **acyl-CoA:cholesterol acyltransferase (ACAT)** to form cholesteryl esters.

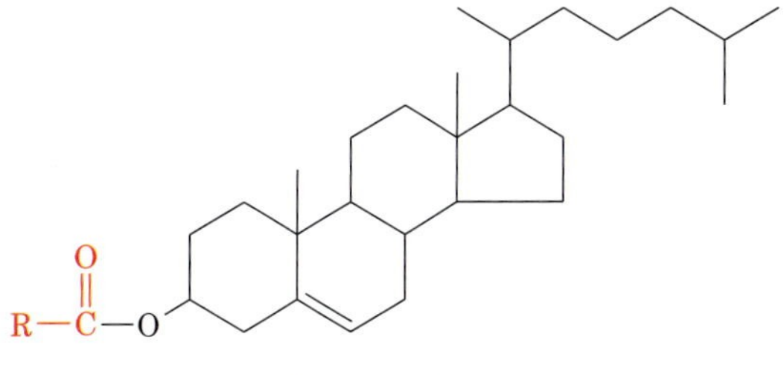

Cholesteryl ester

These highly hydrophobic compounds are transported throughout the body in lipoprotein complexes (Section 19-1B).

B Regulation of Cholesterol Synthesis

HMG-CoA reductase, which catalyzes the rate-limiting step of cholesterol biosynthesis (Fig. 19-38), is the pathway's main regulatory site. The enzyme is subject to short-term regulation by competitive inhibition, allosteric effects, and covalent modification involving reversible phosphorylation. Like glycogen phosphorylase, glycogen synthase, and other enzymes, HMG-CoA

reductase exists in interconvertible more active and less active forms. When phosphorylated at Ser 871, the enzyme is less active. Phosphorylation is carried out by AMP-dependent protein kinase (AMPK), the same enzyme that inactivates acetyl-CoA carboxylase (Section 19-4B). It appears that this control mechanism conserves energy when ATP levels fall and AMP levels rise, by generally inhibiting biosynthetic pathways.

The primary regulatory mechanism for HMG-CoA reductase activity is long-term feedback control of the amount of enzyme present in the cell. The amount of enzyme can rise as much as 200-fold, due to an increase in enzyme synthesis combined with a decrease in its degradation. In fact, cholesterol itself regulates the expression of the HMG-CoA reductase gene along with more than 20 other genes involved in its biosynthesis and uptake, including the gene encoding the LDL receptor. These genes all contain a specific recognition sequence called a **sterol regulatory element (SRE;** control of eukaryotic gene expression is discussed in detail in Chapter 27). The transcription of these genes, as Brown and Goldstein elucidated, requires the binding to their SRE of a portion of **sterol regulatory element binding protein (SREBP).** However, when cholesterol levels are sufficiently high, SREBP resides in the endoplasmic reticulum (ER) membrane as an inactive 1160-residue precursor that binds to **SREBP cleavage-activating protein (SCAP).** SCAP is an ~1276-residue intracellular cholesterol sensor that contains two domains: an N-terminal transmembrane domain called the **sterol-sensing domain** that interacts with sterols, and a C-terminal domain containing five copies of a protein–protein interaction motif known as a **WD repeat** that interacts with the C-terminal so-called regulatory domain of SREBP (Fig. 19-42). When cholesterol in the ER membrane is depleted, SCAP changes conformation and escorts its bound SREPB to the Golgi apparatus via membranous vesicles. SREBP is then sequentially cleaved by two Golgi proteases: **site-1 protease (S1P),** a serine protease

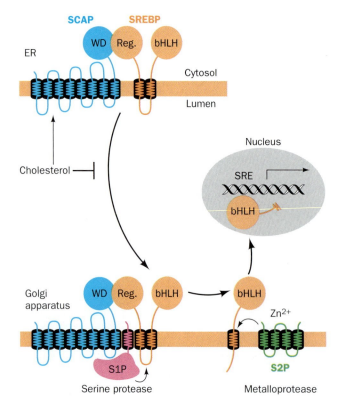

Figure 19-42 The cholesterol-mediated proteolytic activation of SREBP. When cholesterol levels in the cell are high, the SREBP–SCAP complex resides in the ER. When cholesterol levels are low, SCAP escorts SREBP via membranous vesicles to the Golgi apparatus, where SREBP undergoes sequential proteolytic cleavage by the membrane-bound proteases S1P and S2P. This releases SREBP's N-terminal domain, which enters the nucleus where it binds to the SREs of its target genes, thereby inducing their transcription. [After Goldstein, J., Rawson, R.B., and Brown, M., *Arch. Biochem. Biophys.* **397,** 139 (2002).]

that cleaves SREBP only when it is associated with SCAP, and **site-2 protease (S2P),** a zinc metalloprotease that cleaves SREBP at a peptide bond exposed by the S1P cleavage. This results in the release of the active fragment, a soluble 480-residue protein that travels to the nucleus, where it activates the transcription of genes containing an SRE by binding to the SRE via a DNA-binding motif known as a **basic helix–loop–helix (bHLH;** Section 23-4C; Fig. 19-42). The cholesterol level in the cell thereby rises until SCAP no longer induces the translocation of SREBP to the Golgi apparatus, a classic case of feedback inhibition.

Statins Inhibit HMG-CoA Reductase. High levels of circulating cholesterol, a condition known as **hypercholesterolemia,** can be treated with drugs called **statins** that inhibit HMG-CoA reductase (Fig. 19-43). These compounds all contain an HMG-like group that acts as a competitive inhibitor of HMG-CoA binding to the enzyme. The statins bind extremely tightly, with K_I values in the nanomolar range, whereas the substrate, HMG-CoA, has a K_M of ~4 μM.

The X-ray structures of HMG-CoA reductase in its complexes with six different statins reveal that the bulky hydrophobic groups of the inhibitors play a major role in interfering with enzyme activity. The HMG-CoA reductase active site normally accommodates NADH as well as the pantothenate portion of CoA. Statin binding does not interfere with NADH

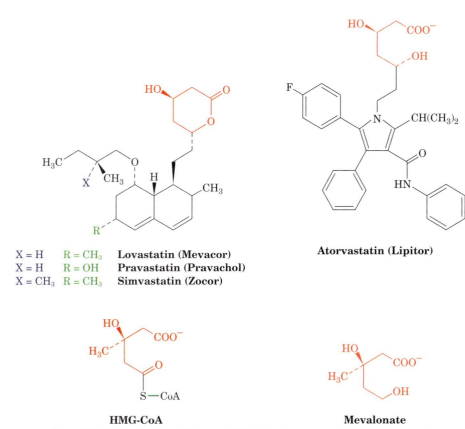

X = H	R = CH$_3$	**Lovastatin (Mevacor)**
X = H	R = OH	**Pravastatin (Pravachol)**
X = CH$_3$	R = CH$_3$	**Simvastatin (Zocor)**

Atorvastatin (Lipitor)

HMG-CoA

Mevalonate

Figure 19-43 Competitive inhibitors of HMG-CoA reductase used for the treatment of hypercholesterolemia. The molecular formulas of lovastatin (Mevacor), pravastatin (Pravachol), simvastatin (Zocor), and atorvastatin (Lipitor), which are known as statins, are given. The structures of HMG-CoA and the HMG-CoA reductase product mevalonate are shown for comparison. Note that lovastatin, pravastatin, and simvastatin are lactones while atorvastatin and mevalonate are hydroxy acids. The lactones are hydrolyzed enzymatically *in vivo* to the active hydroxy-acid forms.

binding. However, the enzyme alters its conformation in order to accommodate the large hydrophobic groups of the statins. It appears that these inhibitors exploit the inherent flexibility of the enzyme. Furthermore, despite the structural variation among the different statins, they all experience extensive van der Waals contacts with the enzyme. *This high degree of complementarity between the statins and the active site, a result of the enzyme's flexibility, accounts for the extremely tight inhibition constants.*

The initial decreased cellular cholesterol supply caused by the presence of statins is met by the induction of the LDL receptor and HMG-CoA reductase (Sections 19-1B and 19-7C; Fig. 19-8) so that at the new steady state, the HMG-CoA level is almost that of the predrug state. However, the increased number of LDL receptors causes increased removal from the blood of both LDL and IDL (the apoB-100 containing precursor to LDL), decreasing serum LDL-cholesterol levels appreciably.

C   Cholesterol Transport and Atherosclerosis

As mentioned above, cellular cholesterol concentrations depend not only on the rate of cholesterol synthesis but also on the ability of cells to absorb cholesterol from circulating lipoproteins (Section 19-1B). The role of lipoproteins in cholesterol metabolism has been studied extensively because an elevated cholesterol level in the blood, primarily in the form of LDL, is a strong risk factor for cardiovascular disease.

Atherosclerosis Results from Accumulation of Lipid in Vessel Walls.   Approximately half of all deaths in the United States are linked to the vascular disease **atherosclerosis** (Greek: *athera,* mush + *sclerosis,* hardness). Atherosclerosis is a slow progressive disease that begins with the deposition of lipids in the walls of large blood vessels, particularly the coronary arteries. The initial event appears to be the association of lipoproteins with vessel wall proteoglycans (Section 8-3A). The trapped lipids trigger inflammation by inducing the endothelial cells lining the vessels to express adhesion molecules specific for monocytes, a type of white blood cell. Once these cells burrow into the vessel wall, they differentiate into macrophages that take up the accumulated lipids, becoming so engorged that they are known as "foam cells." Macrophages do not normally take up lipids but do so when the lipids are oxidized, as apparently occurs when LDL particles are trapped for an extended period in blood vessel walls. Factors released by the foam cells, and possibly the oxidized lipids themselves, recruit more white blood cells, perpetuating a state of inflammation.

The damaged vessel wall forms a **plaque** with a core of cholesterol, cholesteryl esters, and remnants of dead macrophages, surrounded by proliferating smooth muscle cells that may undergo calcification, as occurs in bone formation (hence the "hardening" of the arteries). Although a very large plaque can occlude the lumen of the artery (Fig. 19-44), blood flow is usually not completely blocked unless the plaque ruptures. This triggers formation of a blood clot that can prevent circulation to the heart, causing **myocardial infarction** (heart attack). Stoppage of blood flow to the brain causes **stroke.**

The development of atherosclerosis is strongly correlated with the concentration of circulating LDL (often referred to as "bad cholesterol"). Some

Figure 19-44   An atherosclerotic plaque in a coronary artery.   The vessel wall is dramatically thickened as a result of lipid accumulation and activation of inflammatory processes. [© Eye of Science/Photo Researchers.]

high-fat diets may contribute to atherosclerosis by boosting LDL levels, but genetic factors and infection also increase the risk of atherosclerosis. Smoking contributes to the disease because cigarette smoke oxidizes LDL, which promotes their uptake by macrophages. Atherosclerosis is less likely to occur in individuals who maintain low total cholesterol levels in their blood and who have high levels of HDL (or "good cholesterol"). These lipoproteins transport excess cholesterol to the liver for disposal as bile acids. Women have more HDL than men and a lower risk of heart disease. Statins also reduce the risk of cardiovascular disease by decreasing blood cholesterol levels (Section 19-7B).

Role of the LDL Receptor. LDL receptors clearly play an important role in the maintenance of plasma LDL levels. Circulating LDL and IDL (which both contain apolipoproteins that specifically bind to the LDL receptor) re-enter the liver through receptor-mediated endocytosis (Fig. 19-8). *Individuals with the inherited disease **familial hypercholesterolemia (FH)** are deficient in functional LDL receptors.* FH homozygotes, who lack the receptors entirely, have such high levels of cholesterol-rich LDL in their plasma that their cholesterol levels are three to five times greater than the average level of ~200 mg/dL. This situation results in the deposition of cholesterol in their skin and tendons as yellow nodules known as **xanthomas.** However, far greater damage is caused by the development of atherosclerosis, which causes death from heart attacks as early as age 5. FH heterozygotes (about 1 person in 500) have about half the normal number of functional LDL receptors and exhibit plasma LDL levels of about twice the average. They typically develop symptoms of cardiovascular disease after age 30.

The long-term ingestion of a high-fat/high-cholesterol diet has an effect similar to although not as extreme as FH. Cholesterol regulates the synthesis of the LDL receptor by the same mechanism that regulates HMG-CoA reductase synthesis (Section 19-7A). Consequently, high intracellular concentrations of cholesterol suppress LDL receptor synthesis so that more LDL particles remain in the circulation. Excessive dietary cholesterol, delivered to the tissues via chylomicrons, therefore contributes to high plasma LDL levels.

Cholesterol Efflux from Cells. Most cells do not consume cholesterol by converting it to steroid hormones or bile acids, for example, but all cells require cholesterol to maintain membrane fluidity. Cholesterol in excess of this requirement can be esterified by the action of ACAT and stored as cholesteryl esters in intracellular deposits. Cholesterol can also be eliminated from cells by a mechanism illuminated through studies of individuals with **Tangier disease.** In this recessive inherited disorder, almost no HDL are produced, because cells have a defective transport protein, known as **ATP-cassette binding protein A1 (ABCA1).**

In normal individuals, ABCA1 apparently acts as a flipase (Section 9-4C) to transfer cholesterol, cholesteryl esters, and other lipids from the inner to the outer leaflet of the plasma membrane, from which they can be captured by apolipoprotein A-I to form HDL. Cells lacking ABCA1 are unable to offload cholesterol and accumulate cholesteryl esters in the cytoplasm. Macrophages thus engorged with lipids contribute to the development of atherosclerosis, so that individuals with Tangier disease exhibit symptoms similar to those with FH. The functions of the LDL receptor and ABCA1 in cholesterol homeostasis are presented schematically in Fig. 19-45.

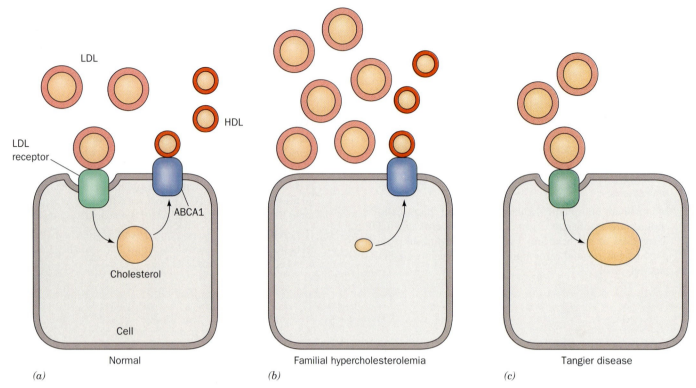

Figure 19-45 The role of LDL and HDL in cholesterol metabolism.
(*a*) Cells acquire cholesterol and cholesteryl esters via endocytosis of LDL as mediated by the LDL receptor (*green*). Cholesterol efflux to form HDL is assisted by the ABCA1 transporter (*blue*). (*b*) In familial hypercholesterolemia, a lack of functional LDL receptors results in high levels of circulating LDL. (*c*) In Tangier disease, cells become laden with cholesterol and cholesteryl esters and few HDL are formed because efflux is prevented due to lack of functional ABCA1.

SUMMARY

1. Triacylglycerol digestion depends on the emulsifying activity of bile acids and the activation of lipases at the lipid–water interface.

2. Lipoproteins, complexes of nonpolar lipids surrounded by a coat of amphipathic lipids and apolipoproteins, transport lipids in the bloodstream. Cells take up cholesterol and other lipids by the receptor-mediated endocytosis of LDL.

3. Fatty acid oxidation begins with the activation of the acyl group by formation of a thioester with CoA. The acyl group is transferred to carnitine for transport into the mitochondria, where it is re-esterified to CoA.

4. β oxidation occurs in four reactions: (1) formation of an α, β double bond, (2) hydration of the double bond, (3) dehydrogenation to form a β-ketoacyl-CoA, and (4) thiolysis by CoA to produce acetyl-CoA and an acyl-CoA shortened by two carbons. This process is repeated until fatty acids with even numbers of carbon atoms are converted to acetyl-CoA and the fatty acids with odd numbers of carbon atoms are converted to acetyl-CoA and one molecule of propionyl-CoA. The acetyl-CoA is oxidized by the citric acid cycle and oxidative phosphorylation to generate ATP.

5. The oxidation of unsaturated fatty acids requires an isomerase to convert Δ^3 double bonds to Δ^2 double bonds and

a reductase to remove Δ^4 double bonds. The oxidation of odd-chain fatty acids yields propionyl-CoA, which is converted to succinyl-CoA through a cobalamin (B_{12})-dependent pathway. Very long chain fatty acids are partially oxidized by a three-enzyme system in peroxisomes.

6. The liver uses acetyl-CoA to synthesize the ketone bodies acetoacetate and β-hydroxybutyrate, which are released into the bloodstream. Tissues that use these ketone bodies for fuel convert them back to acetyl-CoA.

7. In fatty acid synthesis, mitochondrial acetyl-CoA is shuttled to the cytosol via the tricarboxylate transport system and activated to malonyl-CoA by the action of acetyl-CoA carboxylase.

8. A series of seven enzymatic activities, which in mammals are contained in a multifunctional homodimeric enzyme, extend acyl-ACP chains by two carbons at a time. An enzyme-bound acyl group and malonyl-CoA condense to form a β-ketoacyl intermediate and CO_2. Two reductions and a dehydration yield an acyl-ACP in a series of reactions that resemble the reverse of β oxidation but are catalyzed by separate enzymes in the cytosol. Palmitate (C_{16}), the normal product of fatty acid biosynthesis, is synthesized in seven such reaction cycles and is then cleaved from ACP by a thioesterase.

9. Other fatty acids are synthesized from palmitate through the action of elongases and desaturases. Human triacylglycerols synthesized from fatty acyl-CoA and glyceraldehyde-3-phosphate or dihydroxyacetone phosphate tend to contain saturated fatty acids at C1 and unsaturated fatty acids at C2.

10. The opposing pathways of fatty acid degradation and synthesis are hormonally regulated. Glucagon and epinephrine activate hormone-sensitive lipase in adipose tissue, thereby increasing the supply of fatty acids for oxidation in other tissues, and inactivate acyl-CoA carboxylase. Insulin has the opposite effect. Insulin also regulates the levels of acetyl-CoA carboxylase and fatty acid synthase by controlling their rates of synthesis.

11. Mammalian phosphatidylethanolamine and phosphatidylcholine are synthesized from 1,2-diacylglycerol and CDP derivatives of the head groups. Phosphatidylinositol, phosphatidylglycerol, and cardiolipin syntheses begin with CDP–diacylglycerol.

12. Sphingoglycolipids are synthesized from ceramide (N-acyl-sphingosine, a derivative of palmitate and serine) by the addition of glycosyl units donated by nucleotide sugars.

13. Arachidonate is the precursor of prostaglandins. Certain drugs, including aspirin, block prostaglandin synthesis by inhibiting cyclooxygenase.

14. Cholesterol is synthesized from acetyl units that pass through HMG-CoA and mevalonate intermediates on the way to being converted to a C_5 isoprene unit. Six isoprene units condense to form the C_{30} compound squalene, which cyclizes to yield lanosterol, the steroid precursor of cholesterol.

15. A cholesterol sensing system in the endoplasmic reticulum regulates the synthesis of HMG-CoA reductase and LDL receptor in the cell. HMG-CoA reductase is sensitive to statin drugs.

16. Hypercholesterolemia, associated with certain genetic defects or a high-cholesterol diet, contributes to the development of atherosclerosis.

REFERENCES

Lipoproteins
Borhani, D.W., Rogers, D.P., Engler, J.A., and Brouillette, C.G., Crystal structure of truncated human apolipoprotein A-I suggests a lipid-bound conformation, *Proc. Natl. Acad. Sci.* **94,** 12291–12296 (1997).

Fatty Acid Metabolism
Eaton, S., Bartlett, K., and Pourfarzam, M., Mammalian mitochondrial β-oxidation, *Biochem. J.* **320,** 345–357 (1996). [Discusses the reactions and enzymes of mitochondrial β oxidation as well as their regulation and diseases associated with their deficiencies.]

Kent, C., Eukaryotic phospholipid synthesis, *Annu. Rev. Biochem.* **64,** 315–342 (1995).

Kim, J.-J.P. and Battaile, K.P., Burning fat: the structural basis of fatty acid β-oxidation, *Curr. Opin. Struct. Biol.* **12,** 721–728 (2002).

Marsh, E.N.G. and Drennan, C.L., Adenosylcobalamin-dependent isomerases: new insights into structure and mechanism, *Curr. Opin. Chem. Biol.* **5,** 499–505 (2001).

Metabolism of Other Lipids
Kurumbail, R.G., Kiefer, J.R., and Marnett, L.J., Cyclooxygenase enzymes: catalysis and inhibition, *Curr. Opin. Struct. Biol.* **11,** 752–760 (2001).

Cholesterol Metabolism
Istvan, E.S. and Deisenhofer, J., Structural mechanism for statin inhibition of HMG-CoA reductase, *Science* **292,** 1160–1164 (2001).

Scriver, C.R., Beaudet, A.C., Sly, W.S., and Valle, D. (Eds.), *The Metabolic and Molecular Bases of Inherited Disease* (8th ed.), McGraw-Hill (2001). [Volumes 2 and 3 contain numerous chapters on defects in lipid metabolism.]

Simons, K. and Ikonen, E., How cells handle cholesterol, *Science* **290,** 1721–1726 (2000). [Discusses mechanisms of cholesterol influx and efflux from cells.]

Steinberg, D., Atherogenesis in perspective: hypercholesterolemia and inflammation as partners in crime, *Nature Medicine* **8,** 1211–1216 (2002). [Summarizes current hypotheses linking cholesterol levels to the development of atherosclerosis.]

KEY TERMS

interfacial activation	receptor-mediated	acyl-carrier protein	lipid storage disease
chylomicron	endocytosis	elongase	isoprene unit
VLDL	β oxidation	desaturase	hypercholesterolemia
IDL	homolytic cleavage	essential fatty acid	atherosclerosis
LDL	heterolytic cleavage	hormone-sensitive lipase	
HDL	ketogenesis	short-term regulation	
apolipoprotein	ketosis	long-term regulation	

STUDY EXERCISES

1. How do bile acids aid in the digestion and absorption of lipids?

2. Describe the roles of the lipoproteins in humans.

3. Summarize the chemical transformations that occur during fatty acid activation and its degradation to acetyl-CoA.

4. What additional steps are required to oxidize unsaturated and odd-chain fatty acids?

5. How are ketone bodies synthesized and degraded?

6. Describe the shuttle systems for transporting fatty acids into the mitochondria and acetyl-CoA into the cytosol.

7. Summarize the similarities and differences between fatty acid oxidation and biosynthesis.

8. Describe the major mechanisms of regulating fatty acid metabolism in humans.

9. How are triacylglycerols, glycerophospholipids, and sphingolipids synthesized?

10. Summarize the chemical events in cholesterol biosynthesis.

11. How does cholesterol regulate its own synthesis?

12. What do familial hypercholesterolemia and Tangier disease reveal about the development of atherosclerosis?

PROBLEMS

1. Explain why individuals with a hereditary deficiency of carnitine palmitoyl transferase II have muscle weakness. Why are these symptoms more severe during fasting?

2. The removal of fatty acids from triacylglycerols leaves glycerol. Show how the actions of glycerol kinase and glycerol-3-phosphate dehydrogenase on glycerol produce an intermediate of glycolysis.

3. The first three steps of β oxidation (Fig. 19-12) chemically resemble three successive steps of the citric acid cycle. Which steps are these?

4. Certain branched-chain fatty acids such as **pristanic acid**

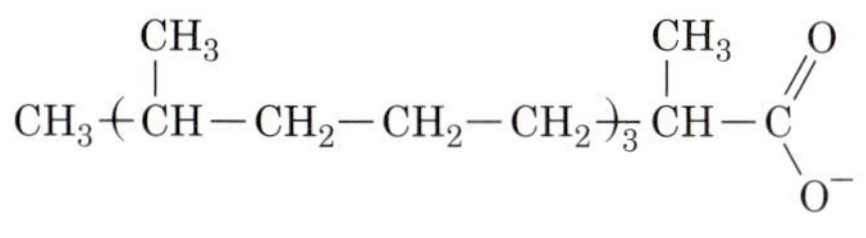

$$CH_3 \overset{\underset{|}{CH_3}}{+} CH-CH_2-CH_2-CH_2 \overset{}{)_3} \overset{\underset{|}{CH_3}}{CH}-C\overset{O}{\underset{O^-}{\diagup}}$$

Pristanic acid

can undergo β oxidation. (a) How many cycles of β oxidation are required to completely degrade this fatty acid? (b) How many C_2 (acetyl-CoA), C_3 (propionyl-CoA), and C_4 (methylpropionyl-CoA) products are obtained?

5. Why are unsaturated fats preferable to saturated fats for an individual whose caloric intake must be limited?

6. Explain why the degradation of odd-chain fatty acids can boost the activity of the citric acid cycle.

7. The complete combustion of palmitate and glucose yield 9781 kJ · mol^{-1} and 2850 kJ · mol^{-1} of free energy, respectively. Compare these values to the free energy (as ATP) obtained

though cellular catabolism of palmitate and glucose. Which process is more efficient?

8. Why is it important that liver cells lack 3-ketoacyl-CoA transferase (Fig. 19-22)?

9. The tricarboxylate transport system supplies cytosolic acetyl-CoA for palmitate synthesis. What percentage of the NADPH required for palmitate synthesis is thereby provided?

10. On what carbon atoms does the $^{14}CO_2$ used to synthesize malonyl-CoA from acetyl-CoA appear in palmitate?

11. Explain why adipocytes need glucose as well as fatty acids in order to synthesize triacylglycerols.

12. Is the fatty acid shown below likely to be synthesized in animals? Explain.

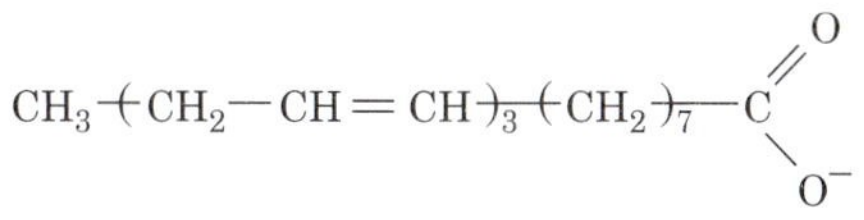

$$CH_3 + CH_2-CH=CH \overset{}{)_3} + CH_2 \overset{}{)_7}-C\overset{O}{\underset{O^-}{\diagup}}$$

13. Compare the energy cost, in ATP equivalents, of synthesizing stearate from mitochondrial acetyl-CoA to the energy recovered by degrading stearate (a) to acetyl-CoA and (b) to CO_2.

14. An animal is fed palmitate with a ^{14}C-labeled carboxyl group. (a) Under ketogenic conditions, where would the label appear in acetoacetate? (b) Under conditions of membrane lipid synthesis, where would the label appear in sphinganine?

15. Hypercholesterolemic individuals taking statins are sometimes advised to take supplements of coenzyme Q. Explain.

All organisms need a source of nitrogen. Complex metabolic pathways convert a few nitrogen-containing compounds into many others, including amino acids for protein synthesis. These pea plants grow in symbiotic association with microorganisms that make nitrogen available to them; other crops require a nitrogen-containing fertilizer. [Maurice Nimmo/Corbis Images.]

Amino Acid Metabolism

The metabolism of amino acids comprises a wide array of synthetic and degradative reactions by which amino acids are assembled as precursors of polypeptides or other compounds and broken down to recover metabolic energy. The chemical transformations of amino acids are distinct from those of carbohydrates or lipids in that they involve the element nitrogen. We must therefore examine the origin of nitrogen in biological systems and its disposal.

The bulk of the cell's amino acids are incorporated into proteins, which are constantly being synthesized and degraded. Aside from this dynamic pool of polymerized amino acids, there is no true storage form of amino acids analogous to glycogen or triacylglycerols. Mammals synthesize certain amino acids and obtain the rest from their diet. *Excess dietary amino acids are not simply excreted but are converted to common metabolites that are precursors of glucose, fatty acids, and ketone bodies and are therefore metabolic fuels.*

In this chapter, we consider the pathways of amino acid metabolism, beginning with the degradation of proteins and the **deamination** (amino group removal) of their component amino acids. We then examine the incorporation of nitrogen into urea for excretion. Next, we examine the pathways by which the carbon skeletons of individual amino acids are broken down and synthesized. We conclude with a brief examination of some other biosynthetic pathways involving amino acids and **nitrogen fixation,** a process that converts atmospheric N_2 to a biologically useful form.

1 Protein Degradation

The components of living cells are constantly turning over. Proteins have lifetimes that range from as short as a few minutes to weeks or more. In any case, *cells continuously synthesize proteins from and degrade them to amino acids.* This seemingly wasteful process has three functions: (1) to store nutrients in the form of proteins and to break them down in times of metabolic need, processes that are most significant in muscle tissue; (2) to eliminate abnormal proteins whose accumulation would be harmful to the cell; and (3) to permit the regulation of cellular metabolism by eliminating superfluous enzymes and regulatory proteins. *Controlling a protein's rate of degradation is therefore as important to the cellular and organismal economy as is controlling its rate of synthesis.*

The half-lives of different enzymes in a given tissue vary substantially, as is indicated for rat liver in Table 20-1. Remarkably, *the most rapidly degraded enzymes all occupy important metabolic control points, whereas the relatively stable enzymes have nearly constant catalytic activities under all physiological conditions.* The susceptibilities of enzymes to degradation have evidently evolved along with their catalytic and allosteric properties so that cells can efficiently respond to environmental changes and metabolic requirements. The rate of protein degradation in a cell also varies with its nutritional and hormonal state. For example, under conditions of nutritional deprivation, cells increase their rate of protein degradation so as to provide the necessary nutrients for indispensable metabolic processes.

A Lysosomal Degradation

Lysosomes contain ~50 hydrolytic enzymes, including a variety of proteases known as **cathepsins.** The lysosome maintains an internal pH of ~5, and its enzymes have acidic pH optima. This situation presumably protects the cell

Table 20-1 Half-Lives of Some Rat Liver Enzymes

Enzyme	Half-Life (h)
Short-Lived Enzymes	
Ornithine decarboxylase	0.2
RNA polymerase I	1.3
Tyrosine aminotransferase	2.0
Serine dehydratase	4.0
PEP carboxylase	5.0
Long-Lived Enzymes	
Aldolase	118
GAPDH	130
Cytochrome b	130
LDH	130
Cytochrome c	150

Source: Dice, J.F. and Goldberg, A.L., *Arch. Biochem. Biophys.* **170,** 214 (1975).

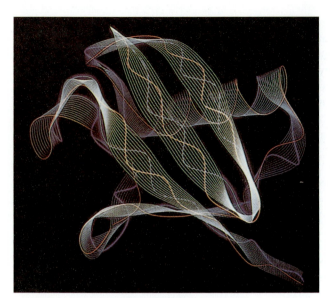

Figure 20-1 X-Ray structure of ubiquitin.   The white ribbon represents the polypeptide backbone and the red and blue curves, respectively, indicate the edges of the β strands toward which the carbonyl O's and the amide H's point. [Courtesy of Mike Carson, University of Alabama at Birmingham. X-Ray structure determined by Charles Bugg, University of Alabama at Birmingham. PDBid 1UBQ.]
🖱 **See the Interactive Exercises.**

against accidental lysosomal leakage since lysosomal enzymes are largely inactive at cytosolic pH's.

Lysosomes degrade substances that the cell takes up via endocytosis (Section 19-1B). They also recycle intracellular constituents that are enclosed within vacuoles that fuse with lysosomes. *In well nourished cells, lysosomal protein degradation is nonselective.* In starving cells, however, this degradation would deplete essential enzymes and regulatory proteins. Lysosomes therefore also have a selective pathway, which is activated only after a prolonged fast, that imports and degrades cytosolic proteins containing the pentapeptide Lys-Phe-Glu-Arg-Gln (KFERQ) or a closely related sequence. Such **KFERQ proteins** are selectively lost from tissues that atrophy in response to fasting (e.g., liver and kidney) but not from tissues that do not do so (e.g., brain and testes). Many normal and pathological processes are associated with increased lysosomal activity, for example, the muscle wastage caused by disuse, denervation, or traumatic injury. The regression of the uterus after childbirth, in which this muscular organ reduces its mass from 2 kg to 50 g in nine days, is a striking example of this process. Many chronic inflammatory diseases such as **rheumatoid arthritis** involve the extracellular release of lysosomal enzymes, which break down surrounding tissues.

B Ubiquitin

Protein breakdown in eukaryotic cells also occurs in an ATP-requiring process that is independent of lysosomes. This process involves **ubiquitin** (Fig. 20-1), a 76-residue monomeric protein named for its ubiquity and abundance. It is one of the most highly conserved eukaryotic proteins known (it is identical in such diverse organisms as humans, trout, and *Drosophila*), suggesting that it is uniquely suited for some essential cellular function(s).

Proteins are marked for degradation by covalently linking them to ubiquitin. This process occurs in three steps elucidated notably by Avram Hershko, Aaron Ciechanover, and Irwin Rose (Fig. 20-2):

1. In an ATP-requiring reaction, ubiquitin's terminal carboxyl group is conjugated, via a thioester bond, to **ubiquitin-activating enzyme (E1).** Most organisms have only one type of E1.

2. The ubiquitin is then transferred to a specific Cys sulfhydryl group on one of numerous homologous proteins named **ubiquitin-conjugating enzymes (E2s;** 11 in yeast and >20 in mammals). The various E2s are characterized by a ~150-residue catalytic core containing the active site Cys.

3. **Ubiquitin-protein ligase (E3)** transfers the activated ubiquitin from E2 to a Lys ε-amino group of a previously bound protein, thereby forming an **isopeptide bond.** Cells contain many species of E3s, each

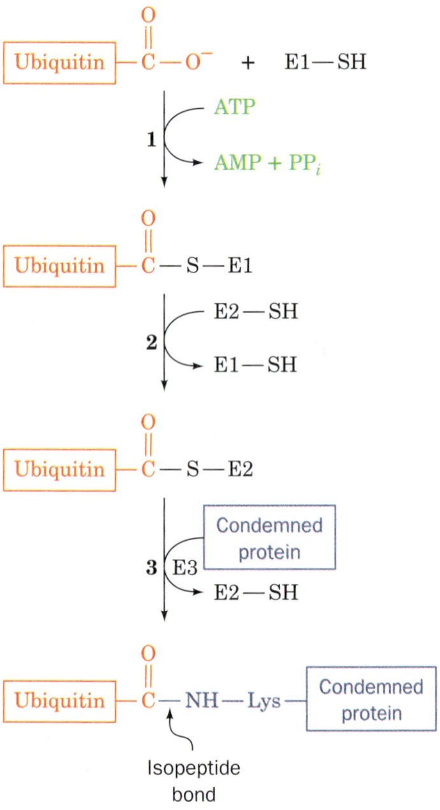

Figure 20-2 Reactions involved in protein ubiquitination.   Ubiquitin's terminal carboxyl group is first joined, via a thioester linkage, to E1 in a reaction driven by ATP hydrolysis. The activated ubiquitin is subsequently transferred to a sulfhydryl group of E2 and then, in a reaction catalyzed by E3, to a Lys ε-amino group on a condemned protein, thereby marking the protein for proteolytic degradation.

of which mediates the ubiquitination (alternatively, ubiquitylation) of a specific set of proteins and thereby marks them for degradation. Each E3 is served by one or a few specific E2s. The known E3s are members of two unrelated families, those containing a **HECT** domain (HECT for *homologous to E6AP C-t*erminus) and those containing a so-called **RING finger** (RING for *really interesting new gene*), although some E2s react well with members of both families.

In order for a protein to be efficiently degraded, it must be linked to a chain of at least four tandemly linked ubiquitin molecules in which Lys 48 of each ubiquitin forms an isopeptide bond with the C-terminal carboxyl group of the following ubiquitin. These **polyubiquitin** chains may contain 50 or more ubiquitin units. Ubiquitinated proteins are dynamic entities, with ubiquitin molecules being rapidly attached and removed (the latter by **ubiquitin isopeptidases**).

The Ubiquitin System Has Both Housekeeping and Regulatory Functions. Until the mid-1990s, it appeared that the ubiquitin system functioned mainly in a "housekeeping" capacity to maintain the proper balance among metabolic proteins and to eliminate damaged proteins. Indeed, as Alexander Varshavsky discovered, the half-lives of many cytoplasmic proteins vary with the identities of their N-terminal residues via the so-called **N-end rule:** Proteins with the "destabilizing" N-terminal residues Asp, Arg, Leu, Lys, and Phe have half-lives of only 2 to 3 minutes, whereas those with the "stabilizing" N-terminal residues Ala, Gly, Met, Ser, Thr, and Val have half-lives of >10 hours in prokaryotes and >20 hours in eukaryotes. The N-end rule applies in both eukaryotes and prokaryotes, which suggests the system that selects proteins for degradation is conserved in eukaryotes and prokaryotes, even though prokaryotes lack ubiquitin.

In eukaryotes, the N-end rule results from the actions of a RING finger E3 named **E3α** whose ubiquitination signals are the destabilizing N-terminal residues. However, it is now clear that the ubiquitin system is far more sophisticated than a simple garbage disposal system. Thus, the growing list of known E3s have a variety of ubiquitination signals that often occur on a quite limited range of target proteins, many of which have regulatory functions. For example, it has long been known that proteins with segments rich in Pro (P), Glu (E), Ser (S), and Thr (T), the so-called **PEST proteins,** are rapidly degraded. It is now realized that this is because these PEST elements often contain phosphorylation sites that target their proteins for ubiquitination. The resulting destruction of these regulatory proteins, of course, has important regulatory consequences.

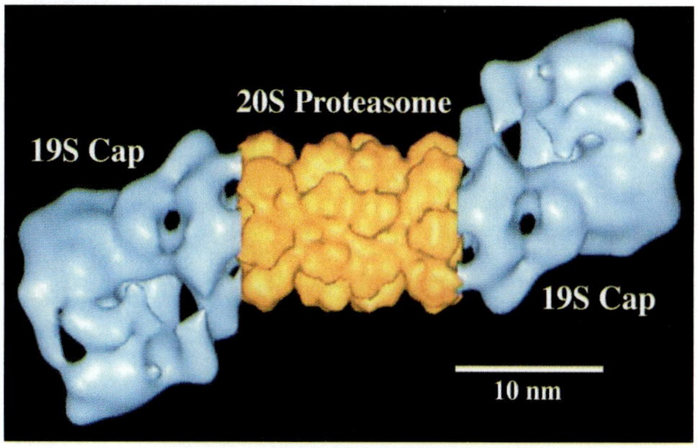

C   The Proteasome

Ubiquitinated proteins are proteolytically degraded in an ATP-dependent process mediated by a large (~2100 kD, 26S) multiprotein complex named the **26S proteasome** (Fig. 20-3). The 26S proteasome consists of a hollow cylindrical core, known as the **20S proteasome,** which is covered at each end by a **19S cap.**

The yeast 20S proteasome, which is closely similar to other eukaryotic 20S proteasomes, is composed of seven different types of α-like subunits and seven different types of β-like subunits. The X-ray structure of this enormous (6182-residue, ~670-kD) protein complex, determined by Robert

Figure 20-3 Electron microscopy–based image of the *Drosophila melanogaster* 26S proteasome. The complex is around 450 × 190 Å. The central portion of this twofold symmetric multiprotein complex (*yellow*), the 20S proteasome, consists of four stacked seven-membered rings of subunits that form a hollow barrel in which the proteolysis of ubiquitin-linked proteins occurs. The 19S caps (*blue*), which may attach to one or both ends of the 20S proteasome, control the access of condemned proteins to the 20S proteasome (see text). [Courtesy of Wolfgang Baumeister, Max-Planck-Institut für Biochemie, Martinsried, Germany.]

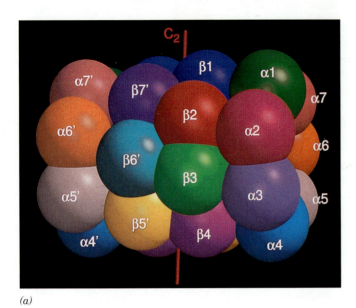

(a)

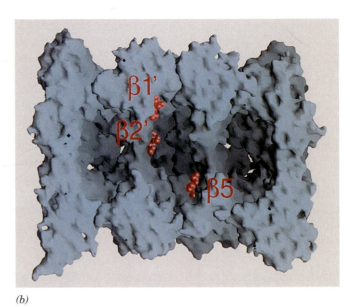

(b)

Figure 20-4 **X-Ray structure of the yeast 20S proteasome.**   (a) The arrangement of the 28 subunits shown as spheres. Four rings of seven subunits each are stacked to form a barrel (the ends of the barrel are at the right and left) with the α-type and β-type subunits forming the outer and inner rings, respectively. The complex's twofold (C_2) axis of symmetry is represented by the vertical red line. (b) Surface view of the proteasome core cut along its cylindrical axis. Three bound protease inhibitor molecules marking the three active β subunits are shown in red as space-filling models. [Courtesy of Robert Huber, Max-Planck-Institut für Biochemie, Martinsried, Germany. PDBid 1RYP.]

Huber, reveals that it consists of four stacked rings of subunits with its outer and inner rings, respectively, consisting of seven different α-type subunits and seven different β-type subunits (Fig. 20-4a). The various α-type subunits have folds that are similar to one another and to the various β-type subunits. Consequently, this 28-subunit complex has exact twofold rotational symmetry relating the two pairs of rings, but only pseudo-sevenfold rotational symmetry relating the subunits within each ring. The 20S proteasome's hollow core consists of three large chambers (Fig. 20-4b): Two are located at the interfaces between adjoining rings of α and β subunits, with the third, larger chamber centrally located between the two rings of β subunits.

Although the α-type subunits and the β-type subunits are structurally similar, only three of the β-type subunits have proteolytic activity. The X-ray structure together with enzymological studies reveal that the three active sites catalyze peptide bond hydrolysis via a novel mechanism in which the β subunits' N-terminal Thr residues function as catalytic nucleophiles. The active sites are located inside the central chamber of the 20S proteasome, thereby preventing this omnivorous protein-dismantling machine from indiscriminately hydrolyzing the proteins in its vicinity. It appears that polypeptide substrates must enter the central chamber of the barrel through the narrow axially located apertures in the α rings that are lined with hydrophobic residues so that only unfolded proteins can enter the central chamber. Nevertheless, in the X-ray structure of the yeast 20S proteasome (Fig. 20-4b), these apertures are blocked by a plug formed by the interdigitation of the α subunit's N-terminal tails.

The three active β-type subunits have different substrate specificities, cleaving after acidic residues (the β1 subunit), basic residues (the β2 subunit), and hydrophobic residues (the β5 subunit). As a result, the 20S proteasome cleaves its polypeptide substrates into ~8-residue fragments, which

then diffuse out of the proteasome. Cytosolic peptidases degrade these peptides to their component amino acids. The ubiquitin molecules attached to the target protein are not degraded, however, but are returned to the cell and reused.

19S Caps Control the Access of Ubiquitinated Proteins to the 20S Proteasome.

The 20S proteasome probably does not exist alone *in vivo;* it is most often in complex with two 19S caps that function to recognize ubiquitinated proteins, unfold them, and feed them into the 20S proteasome in an ATP-dependent manner. The 19S cap, which consists of ~18 different subunits, is poorly characterized due in large part to its low intrinsic stability. Its so-called base complex consists of 9 different subunits, 6 of which are ATPases that form a ring that abuts the α ring of the 20S proteasome (Fig. 20-3). Cecile Pickart has demonstrated, via cross-linking experiments, that one of these ATPases, named **S6'**, contacts the polyubiquitin signal that targets a condemned protein to the 26S proteasome. This suggests that the recognition of the polyubiquitin chain as well as substrate protein unfolding are ATP-driven processes. Moreover, the ring of ATPases must function to open (gate) the otherwise closed axial aperture of the 20S proteasome so as to permit the entry of the unfolded substrate protein.

Eight additional subunits form the so-called lid complex, the portion of the 19S cap that is more distant from the 20S proteasome. The functions of the lid subunits are largely unknown, although a truncated 26S proteasome that lacks the lid subunits is unable to degrade polyubiquitinated substrates. Several other subunits may be transiently associated with the 19S cap and/or with the 20S proteasome.

Eubacteria Also Contain Self-Compartmentalized Proteases.

Although 20S proteasomes occur in all eukaryotes and archaebacteria yet examined, they are absent in nearly all eubacteria (which provides further evidence that eukaryotes arose from archaea; Fig. 1-9). Nevertheless, eubacteria have ATP-dependent proteolytic assemblies that share a barrel-shaped architecture with proteasomes and carry out similar functions. For example, in *E. coli,* two proteins known as **Lon** and **Clp** account for up to 80% of the bacterium's protein degradation. Thus, *all cells appear to contain proteases whose active sites are only available from the inner cavity of a hollow particle to which access is controlled.* These so-called **self-compartmentalized proteases** appear to have arisen early in the history of cellular life, before the advent of eukaryotic membrane-bound organelles such as the lysosome, which similarly carry out degradative processes in a way that protects the cell contents from indiscriminant destruction.

Clp protease consists of two components, the proteolytically active **ClpP** and one of several ATPases, which in *E. coli* are **ClpA** and **ClpX.** The X-ray structure of ClpP reveals that it oligomerizes to form a hollow barrel ~90 Å long and wide that consists of two back-to-back sevenfold symmetric rings of 193-residue subunits and thereby has the same rotational symmetry as does the 20S proteasome (Fig. 20-5). Nevertheless, the ClpP subunit has a novel fold that is entirely different from that of the 20S proteasome's homologous α and β subunits. The ClpP active site, which is only exposed

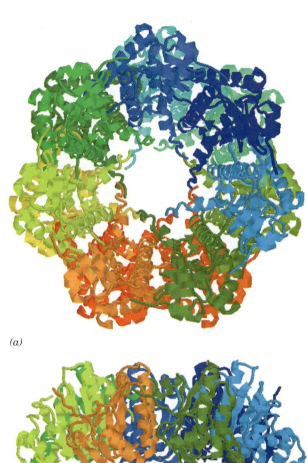

(a)

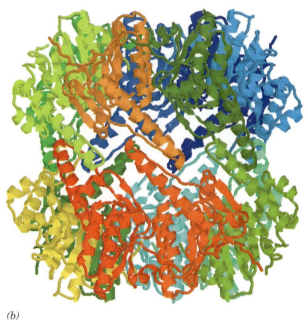

(b)

Figure 20-5   X-Ray structure of *E. coli* ClpP. (*a*) View of the heptameric complex along its sevenfold axis with each of its subunits differently colored. (*b*) View along the complex's twofold axis (rotated 90° about a horizontal axis with respect to Part *a*). [Based on an X-ray structure by John Flannagan, Brookhaven National Laboratory, Upton, New York. PDBid 1TYF.]

on the inside of the barrel, contains a catalytic triad composed of its Ser 97, His 122, and Asp 171, and hence is a serine protease (Fig. 11-26).

2 Amino Acid Deamination

Free amino acids originate from the degradation of cellular proteins and from the digestion of dietary proteins. The gastric protease **pepsin,** the pancreatic enzymes trypsin, chymotrypsin, and elastase (discussed in Sections 5-3B and 11-5), and a host of other endo- and exopeptidases degrade polypeptides to oligopeptides and amino acids. These substances are absorbed by the intestinal mucosa and transported via the bloodstream to be absorbed by other tissues.

The further degradation of amino acids takes place intracellularly and includes a step in which the α-amino group is removed. In many cases, the amino group is converted to ammonia, which is then incorporated into urea for excretion (Section 20-3). The remaining carbon skeleton (α-keto acid) of the amino acid can be broken down to other compounds (Section 20-4). This metabolic theme is outlined in Fig. 20-6.

A Transamination

Most amino acids are deaminated by **transamination,** the transfer of their amino group to an α-keto acid to yield the α-keto acid of the original amino acid and a new amino acid. The predominant amino group acceptor is α-ketoglutarate, producing glutamate and the new α-keto acid:

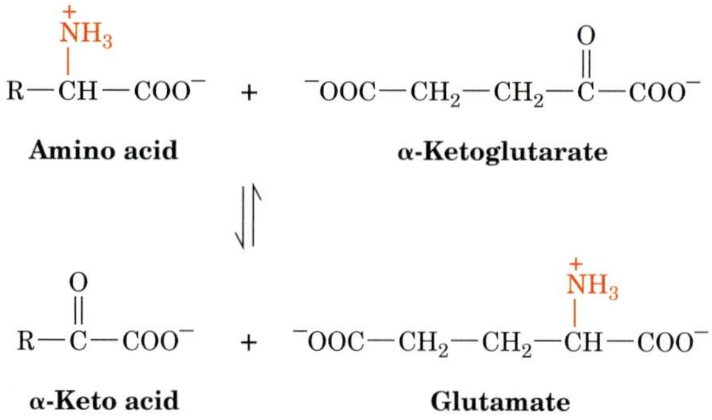

$$\overset{\overset{+}{N}H_3}{\underset{}{R-\overset{|}{C}H-COO^-}} + \ ^-OOC-CH_2-CH_2-\overset{\overset{O}{\|}}{C}-COO^-$$

Amino acid   **α-Ketoglutarate**

$$\overset{\overset{O}{\|}}{R-C-COO^-} + \ ^-OOC-CH_2-CH_2-\overset{\overset{+}{N}H_3}{\underset{}{\overset{|}{C}H-COO^-}}$$

α-Keto acid   **Glutamate**

Glutamate's amino group, in turn, can be transferred to oxaloacetate in a second transamination reaction, yielding aspartate and reforming α-ketoglutarate:

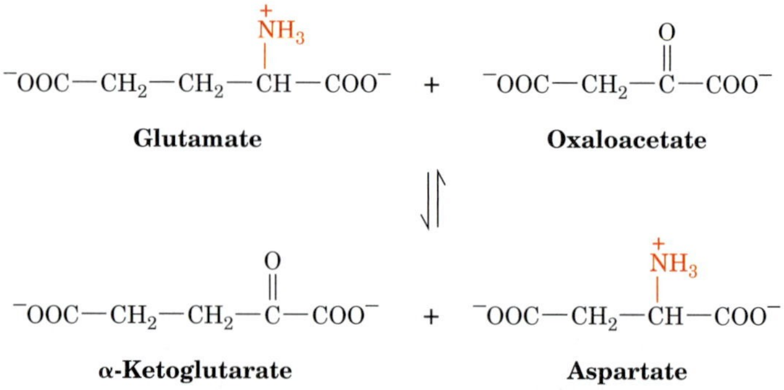

$$^-OOC-CH_2-CH_2-\overset{\overset{+}{N}H_3}{\underset{}{\overset{|}{C}H-COO^-}} + \ ^-OOC-CH_2-\overset{\overset{O}{\|}}{C}-COO^-$$

Glutamate   **Oxaloacetate**

$$^-OOC-CH_2-CH_2-\overset{\overset{O}{\|}}{C}-COO^- + \ ^-OOC-CH_2-\overset{\overset{+}{N}H_3}{\underset{}{\overset{|}{C}H-COO^-}}$$

α-Ketoglutarate   **Aspartate**

The enzymes that catalyze transamination, called **aminotransferases** or **transaminases,** require the coenzyme **pyridoxal-5′-phosphate (PLP;** Fig.

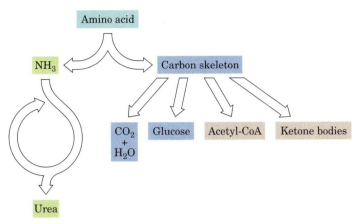

Figure 20-6 Overview of amino acid catabolism. The amino group is removed and incorporated into urea for disposal. The remaining carbon skeleton (α-keto acid) can be broken down to CO_2 and H_2O or converted to glucose, acetyl-CoA, or ketone bodies.

20-7a). PLP is a derivative of **pyridoxine** (**vitamin B$_6$;** Fig. 20-7b). The coenzyme is covalently attached to the enzyme via a Schiff base (imine) linkage formed by the condensation of its aldehyde group with the ε-amino group of an enzyme Lys residue (Fig. 20-7c). The Schiff base, which is conjugated to the pyridinium ring, is the center of the coenzyme's activity. When PLP accepts the amino group from an amino acid as described below, it becomes **pyridoxamine-5'-phosphate** (**PMP;** Fig. 20-7d).

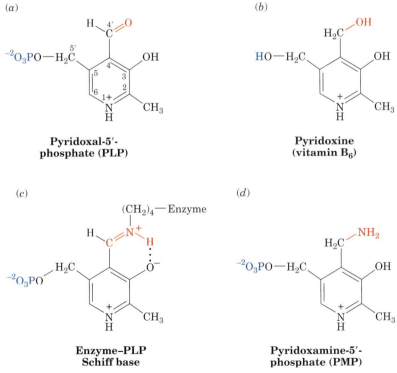

Figure 20-7 Forms of pyridoxal-5'-phosphate. (a) The coenzyme pyridoxal-5'-phosphate (PLP). (b) Pyridoxine (vitamin B$_6$). (c) The Schiff base that forms between PLP and an enzyme ε-amino group. (d) Pyridoxamine-5'-phosphate (PMP).

Esmond Snell, Alexander Braunstein, and David Metzler demonstrated that the aminotransferase reaction occurs via a Ping Pong mechanism (Section 12-1D) whose two stages consist of three steps each (Fig. 20-8):

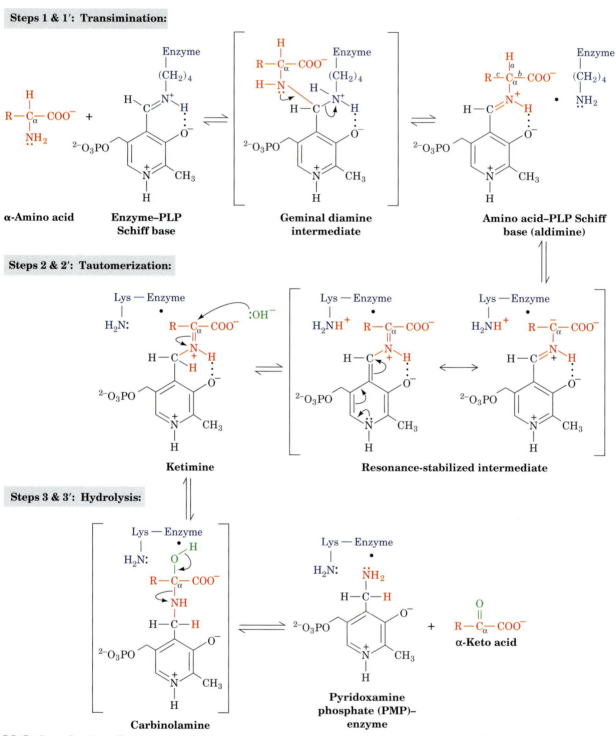

Steps 1 & 1′: Transimination:

α-Amino acid   Enzyme–PLP Schiff base   Geminal diamine intermediate   Amino acid–PLP Schiff base (aldimine)

Steps 2 & 2′: Tautomerization:

Ketimine   Resonance-stabilized intermediate

Steps 3 & 3′: Hydrolysis:

Carbinolamine   Pyridoxamine phosphate (PMP)–enzyme   α-Keto acid

Figure 20-8 *Key to Function.* **The mechanism of PLP-dependent enzyme-catalyzed transamination.**   The first stage of the reaction, in which the α-amino group of an amino acid is transferred to PLP yielding an α-keto acid and PMP, consists of three steps: (**1**) transimination, (**2**) tautomerization, in which the Lys released during the transimination reaction acts as a general acid–base catalyst, and (**3**) hydrolysis. The second stage of the reaction, in which the amino group of PMP is transferred to a different α-keto acid to yield a new α-amino acid and PLP, is essentially the reverse of the first stage: Steps 3′, 2′, and 1′ are, respectively, the reverse of Steps 3, 2, and 1. 🖭 **See the Animated Figures.**

Stage I: Conversion of an Amino Acid to a Keto Acid.

1. The amino acid's nucleophilic amino group attacks the enzyme–PLP Schiff base carbon atom in a **transimination** reaction to form an amino acid–PLP Schiff base (an aldimine), with concomitant release of the enzyme's Lys amino group. This Lys is then free to act as a general base at the active site.

2. The amino acid–PLP Schiff base tautomerizes to an α-keto acid–PMP Schiff base (a ketimine) by the active site Lys–catalyzed removal of the amino acid α-hydrogen and protonation of PLP atom C4′ via a resonance-stabilized carbanion intermediate. This resonance stabilization facilitates the cleavage of the C_α—H bond.

3. The α-keto acid–PMP Schiff base is hydrolyzed to PMP and an α-keto acid.

Stage II: Conversion of an α-Keto Acid to an Amino Acid.

To complete the aminotransferase's catalytic cycle, the coenzyme must be converted from PMP back to the enzyme–PLP Schiff base. This involves the same three steps as above, but in reverse order:

3′. PMP reacts with an α-keto acid to form a Schiff base.

2′. The α-keto acid–PMP Schiff base tautomerizes to form an amino acid–PLP Schiff base.

1′. The ε-amino group of the active site Lys residue attacks the amino acid–PLP Schiff base in a transimination reaction to regenerate the active enzyme–PLP Schiff base and release the newly formed amino acid.

Note that removal of the substrate amino acid's amino group produces a resonance-stabilized C_α carbanion whose electrons are delocalized all the way to the coenzyme's protonated pyridinium nitrogen atom; that is, *PLP functions as an electron sink.* For transamination reactions, this electron-withdrawing capacity facilitates removal of the α proton (*a* bond cleavage, top right of Fig. 20-8) during tautomerization. PLP functions similarly in enzymatic reactions involving *b* and *c* bond cleavage.

Aminotransferases differ in their specificity for amino acid substrates in the first stage of the transamination reaction, thereby producing the correspondingly different α-keto acid products. Most aminotransferases, however, accept only α-ketoglutarate or (to a lesser extent) oxaloacetate as the α-keto acid substrate in the second stage of the reaction, thereby yielding glutamate or aspartate as their only amino acid product. *The amino groups from most amino acids are consequently funneled into the formation of glutamate or aspartate.* The transaminase reaction is freely reversible, so transaminases participate in pathways for amino acid synthesis as well as degradation. Lysine is the only amino acid that is not transaminated.

The presence of transaminases in muscle and liver cells makes them useful markers of tissue damage. Assays of the enzymes' activities in the blood are the basis of the commonly used clinical measurements known as **SGOT (serum glutamate-oxaloacetate transaminase,** also known as **aspartate transaminase, AST)** and **SGPT (serum glutamate-pyruvate transaminase, or alanine transaminase, ALT).** The concentrations of these enzymes in the blood increase after a heart attack, when damaged heart muscle leaks its intracellular contents. Liver damage is also monitored by SGOT and SGPT levels.

B Oxidative Deamination

Transamination, of course, does not result in any net deamination. Glutamate, however, is oxidatively deaminated by **glutamate dehydrogenase (GDH),** yielding ammonia and regenerating α-ketoglutarate for use in additional transamination reactions.

Glutamate dehydrogenase, a mitochondrial enzyme, is the only known enzyme that can accept either NAD^+ or $NADP^+$ as its redox coenzyme. Oxidation is thought to occur with transfer of a hydride ion from glutamate's C_α to $NAD(P)^+$, thereby forming α-iminoglutarate, which is hydrolyzed to α-ketoglutarate and ammonia:

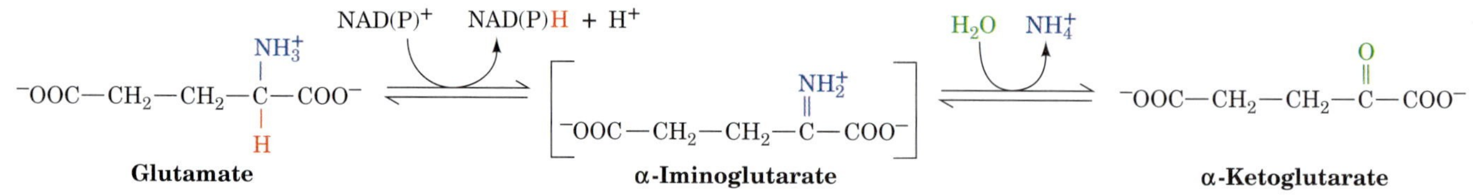

The enzyme is allosterically inhibited by GTP and NADH (signaling abundant metabolic energy) and activated by ADP and NAD^+ (signaling the need to generate ATP). Because the product of the reaction, α-ketoglutarate, is an intermediate of the citric acid cycle, activation of glutamate dehydrogenase can stimulate flux through the citric acid cycle, leading to increased ATP production by oxidative phosphorylation.

The equilibrium position of the glutamate dehydrogenase reaction ($\Delta G^{\circ\prime} \approx 30$ kJ·mol^{-1}) favors glutamate synthesis, the reverse of the reaction written above. At one time, it was believed that this reaction represented a route for the body to remove free ammonia, which is toxic at high concentrations. Under physiological conditions, the enzyme was thought to function close to equilibrium so changes in ammonia concentration could, in principle, cause a shift in equilibrium toward glutamate synthesis to remove the excess ammonia. However, a form of hyperinsulinism that is characterized by hypoglycemia and **hyperammonemia (HI/HA;** hyperammonemia is elevated levels of ammonia in the blood) is caused by mutations in GDH resulting in decreased sensitivity to GTP inhibition and therefore increased GDH activity. Since HI/HA patients have increased GDH activity but higher levels of NH_3 than normal, this accepted role of GDH functioning close to equilibrium and preventing ammonia toxicity cannot be correct. Indeed, if GDH functioned close to equilibrium, changes in its activity resulting from allosteric interactions would not result in significant flux changes (recall that enzymes controlling flux must function far from equilibrium).

The ammonia liberated in the GDH reaction as written above is eventually excreted in the form of urea. Thus, *the glutamate dehydrogenase reaction functions to eliminate amino groups from amino acids that undergo transamination reactions with α-ketoglutarate.*

3 The Urea Cycle

Living organisms excrete the excess nitrogen arising from the metabolic breakdown of amino acids in one of three ways. Many aquatic animals simply excrete ammonia. Where water is less plentiful, however, processes

have evolved that convert ammonia to less toxic waste products that require less water for excretion. One such product is urea, which is produced by most terrestrial vertebrates; another is **uric acid,** which is excreted by birds and terrestrial reptiles:

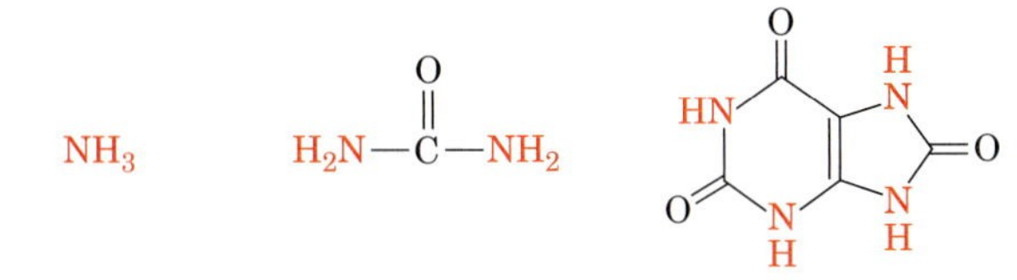

$$NH_3 \qquad H_2N-\overset{\overset{\textstyle O}{\|}}{C}-NH_2$$

Ammonia   **Urea**   **Uric acid**

In this section, we focus our attention on urea formation. Uric acid biosynthesis is discussed in Section 22-4.

Urea is synthesized in the liver by the enzymes of the **urea cycle.** It is then secreted into the bloodstream and sequestered by the kidneys for excretion in the urine. The urea cycle was outlined in 1932 by Hans Krebs and Kurt Henseleit (the first known metabolic cycle; Krebs did not elucidate the citric acid cycle until 1937; Box 16-1). Its individual reactions were later described in detail by Sarah Ratner and Philip Cohen. The overall urea cycle reaction is

$$NH_3 + HCO_3^- + {}^-OOC-CH_2-\overset{\overset{\textstyle NH_3^+}{|}}{C}H-COO^-$$

Aspartate

$$\begin{array}{c} \text{3 ATP} \\ \downarrow \\ \text{2 ADP} + \text{2 P}_i + \text{AMP} + \text{PP}_i \end{array}$$

$$H_2N-\overset{\overset{\textstyle O}{\|}}{C}-NH_2 + {}^-OOC-CH=CH-COO^-$$

Urea   **Fumarate**

Thus, urea's two nitrogen atoms are contributed by ammonia and aspartate, whereas its carbon atom comes from HCO_3^-.

A   Reactions of the Urea Cycle

Five enzymatic reactions are involved in the urea cycle, two of which are mitochondrial and three cytosolic (Fig. 20-9).

1. Carbamoyl Phosphate Synthetase: Acquisition of the First Urea Nitrogen Atom.

Carbamoyl phosphate synthetase (CPS) is technically not a member of the urea cycle. It catalyzes the condensation and activation of NH_3 and HCO_3^- to form **carbamoyl phosphate,** the first of the cycle's two nitrogen-containing substrates, with the concomitant cleavage of 2 ATP. Eukaryotes have two forms of CPS: Mitochondrial **CPS I** uses ammonia as its nitrogen donor and participates in urea biosynthesis, whereas cytosolic **CPS II** uses glutamine as its nitrogen donor and is involved in pyrimidine biosynthesis

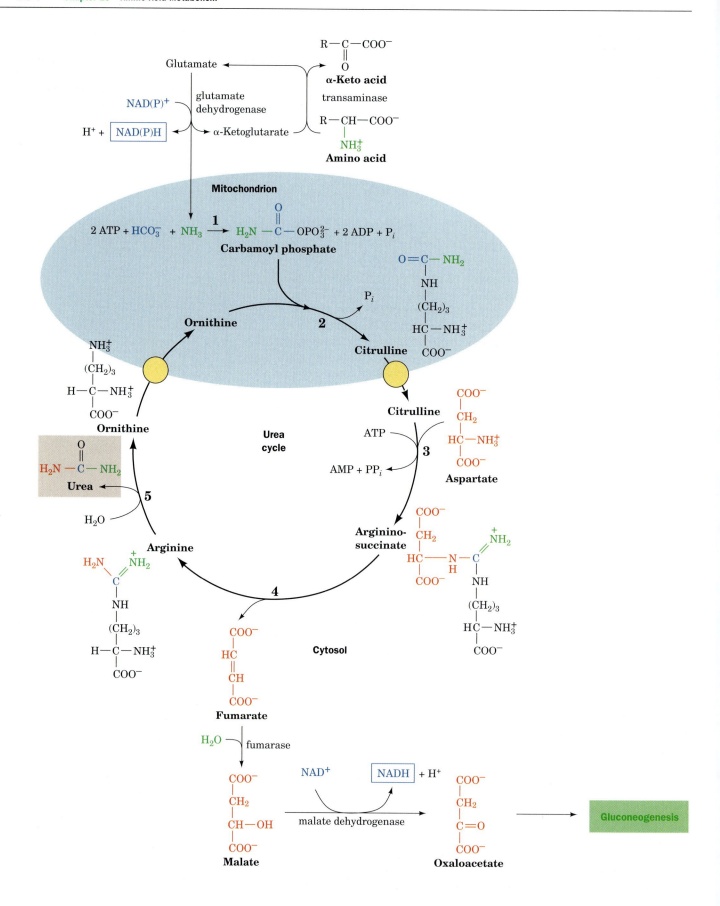

Figure 20-9 (*opposite*) *Key to Metabolism.* **The urea cycle.** Five enzymes participate in the urea cycle: (**1**) carbamoyl phosphate synthetase, (**2**) ornithine transcarbamoylase, (**3**) argininosuccinate synthetase, (**4**) argininosuccinase, and (**5**) arginase. Enzymes 1 and 2 are mitochondrial and enzymes 3–5 are cytosolic. Ornithine and citrulline must therefore be transported across the mitochondrial membrane by specific transport systems (*yellow circles*). The urea amino groups arise from the deamination of amino acids. One amino group (*green*) originates as ammonia that is generated by the glutamate dehydrogenase reaction (*top*). The other amino group (*red*) is obtained from aspartate (*right*), which is formed via the transamination of oxaloacetate by an amino acid. The fumarate product of Reaction 4 is converted to the gluconeogenic precursor oxaloacetate by the action of cytosolic fumarase and malate dehydrogenase (*bottom*). 🔁 **See the Animated Figures.**

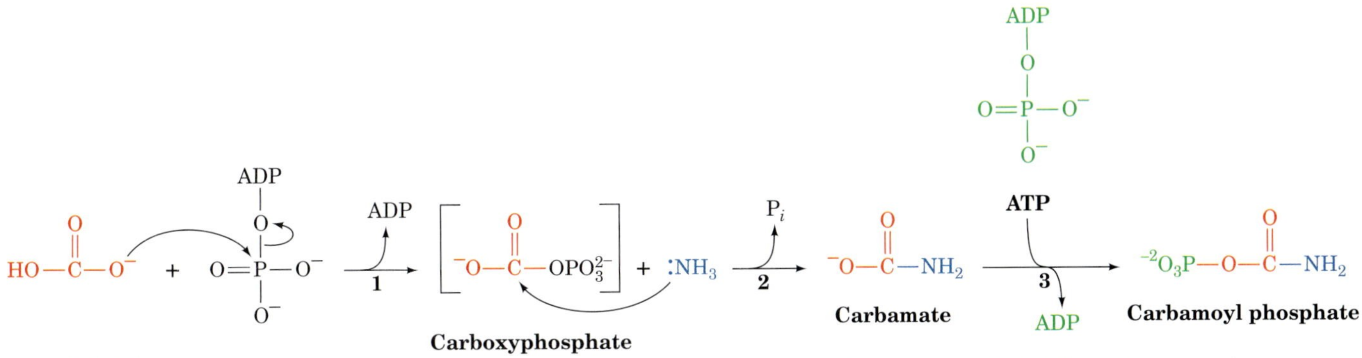

Figure 20-10 **The mechanism of action of CPS I.** (**1**) Phosphorylation activates HCO_3^- to form the intermediate carboxyphosphate. (**2**) NH_3 attacks carboxyphosphate to form carbamate. (**3**) ATP phosphorylates carbamate, yielding carbamoyl phosphate.

(Section 22-2A). CPS I catalyzes an essentially irreversible reaction that is the rate-limiting step of the urea cycle (Fig. 20-10):

1. ATP activates HCO_3^- to form **carboxyphosphate** and ADP.

2. Ammonia attacks carboxyphosphate, displacing the phosphate to form **carbamate** and P_i.

3. A second ATP phosphorylates carbamate to form carbamoyl phosphate and ADP.

In *E. coli*, a single CPS with the subunit structure $(\alpha\beta)_4$ generates carbamoyl phosphate, using glutamine as the nitrogen donor. The enzyme's small subunit is a **glutaminase** that hydrolyzes glutamine, and its large subunit catalyzes Reactions 1, 2, and 3 of Fig. 20-10. The X-ray structure of *E. coli* CPS, determined by Hazel Holden and Ivan Rayment, reveals the astonishing fact that although the three active sites are widely separated in space, they are connected by a narrow 96-Å-long tunnel that runs nearly the entire length of the protein molecule (Fig. 20-11). The NH_3 produced by the glutaminase active site travels ~45 Å in order to react with carboxyphosphate. The resulting carbamate then travels ~35 Å to reach the carbamoyl phosphate synthesis site. This phenomenon, in which the intermediate of two reactions is directly transferred from one enzyme active site to another, is called **channeling.**

Channeling increases the rate of a metabolic pathway by preventing the loss of its intermediate products as well as protecting the intermediates from degradation. Channeling is critical for CPS because the intermediates

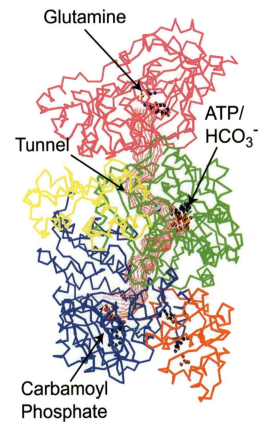

Figure 20-11 **X-Ray structure of *E. coli* carbamoyl phosphate synthetase.** The protein is represented by its C_α backbone. The small subunit (*magenta*) contains the glutamine-binding site where NH_3 is produced. The large subunit consists of several domains (*green, yellow, blue,* and *orange*) and contains the other two active sites. The 96-Å-long tunnel connecting the three active sites is outlined in red. [Courtesy of Hazel Holden and Ivan Rayment, University of Wisconsin. PDBid 1JDB.]

carboxyphosphate and carbamate are extremely reactive, having half-lives of 28 and 70 ms, respectively, at neutral pH. Also, channeling allows the local concentration of NH_3 to reach a higher value than is present in the cellular medium. We shall encounter several other examples of channeling in our studies of metabolic enzymes, but the CPS tunnel is among the longest known.

2. Ornithine Transcarbamoylase.   **Ornithine transcarbamoylase** transfers the carbamoyl group of carbamoyl phosphate to **ornithine,** yielding **citrulline** (Fig. 20-9, Reaction 2). Note that both of these latter compounds are "nonstandard" α-amino acids that do not occur in proteins. The transcarbamoylase reaction occurs in the mitochondrion, so ornithine, which is produced in the cytosol, must enter the mitochondrion via a specific transport system. Likewise, since the remaining urea cycle reactions occur in the cytosol, citrulline must be exported from the mitochondrion.

3. Argininosuccinate Synthetase: Acquisition of the Second Urea Nitrogen Atom. Urea's second nitrogen atom is introduced by the condensation of citrulline's ureido group with an aspartate amino group by **argininosuccinate synthetase** (Fig. 20-12). ATP activates the ureido oxygen atom as a leaving group through formation of a citrullyl–AMP intermediate, and AMP is subsequently displaced by the aspartate amino group. The PP_i formed in this reaction is hydrolyzed to 2 P_i, so the reaction consumes two ATP equivalents.

4. Argininosuccinase.   With the formation of argininosuccinate, all of the urea molecule components have been assembled. However, the amino group donated by aspartate is still attached to the aspartate carbon skeleton. This situation is remedied by the **argininosuccinase**-catalyzed elimination of fumarate, leaving arginine (Fig. 20-9, Reaction 4). Arginine is urea's immediate precursor. The fumarate produced in the argininosuccinase reaction is converted to oxaloacetate by the action of fumarase and malate dehydrogenase. These two reactions are the same as those that occur in the citric acid cycle, although they take place in the cytosol rather than in the mitochondrion. The oxaloacetate is then used for gluconeogenesis (Section 15-4).

5. Arginase.   The urea cycle's final reaction is the **arginase**-catalyzed hydrolysis of arginine to yield urea and regenerate ornithine (Fig. 20-9, Reaction 5). Ornithine is then returned to the mitochondrion for another round of the cycle.

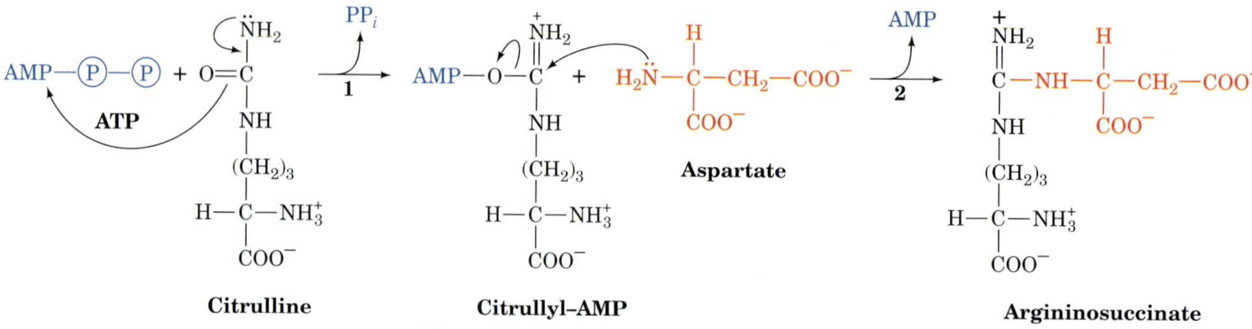

Figure 20-12   The mechanism of action of argininosuccinate synthetase. (1) The formation of citrullyl–AMP activates the ureido oxygen of citrulline. (2) The α-amino group of aspartate displaces AMP.

The urea cycle thus converts two amino groups, one from ammonia and one from aspartate, and a carbon atom from HCO_3^- to the relatively non-toxic product, urea, at the cost of four "high-energy" phosphate bonds. The energy spent is more than recovered, however, by the oxidation of the carbon skeletons of the amino acids that have donated their amino groups, via transamination, to glutamate and aspartate. Indeed, half the oxygen that the liver consumes is used to provide this energy.

B Regulation of the Urea Cycle

Carbamoyl phosphate synthetase I, which catalyzes the first committed step of the urea cycle, is allosterically activated by **N-acetylglutamate** (*at right*). This metabolite is synthesized from glutamate and acetyl-CoA by **N-acetyl-glutamate synthase.** When amino acid breakdown rates increase, the concentration of glutamate increases as a result of transamination. The increased glutamate stimulates *N*-acetylglutamate synthesis. The resulting activation of carbamoyl phosphate synthetase increases the rate of urea production. Thus, the excess nitrogen produced by amino acid breakdown is efficiently excreted. Note that the urea cycle, like gluconeogenesis (Section 15-4) and ketogenesis (Section 19-3), is a pathway that occurs in the liver but serves the needs of the body as a whole.

The remaining enzymes of the urea cycle are controlled by the concentrations of their substrates. In individuals with inherited deficiencies in urea cycle enzymes other than arginase, the corresponding substrate builds up, increasing the rate of the deficient reaction so that the rate of urea production is normal (the total lack of a urea cycle enzyme, however, is lethal). The anomalous substrate buildup is not without cost, however. The substrate concentrations become elevated all the way back up the cycle to ammonia, resulting in hyperammonemia. Although the root cause of ammonia toxicity is not completely understood, it is clear that the brain is particularly sensitive to high ammonia concentrations (symptoms of urea cycle enzyme deficiencies include mental retardation and lethargy).

4 Breakdown of Amino Acids

Amino acids are degraded to compounds that can be metabolized to CO_2 and H_2O or used in gluconeogenesis. Indeed, oxidative breakdown of amino acids typically accounts for 10 to 15% of the metabolic energy generated by animals. In this section we consider how the carbon skeletons of the 20 "standard" amino acids are catabolized. We shall not describe in detail all of the many reactions involved. Rather, we shall consider how these pathways are organized and focus on a few reactions of chemical and/or medical interest.

"Standard" amino acids are degraded to one of seven metabolic intermediates: pyruvate, α-ketoglutarate, succinyl-CoA, fumarate, oxaloacetate, acetyl-CoA, or acetoacetate (Fig. 20-13). The amino acids can therefore be divided into two groups based on their catabolic pathways:

1. **Glucogenic amino acids,** which are degraded to pyruvate, α-ketoglutarate, succinyl-CoA, fumarate, or oxaloacetate and are therefore glucose precursors (Section 15-4).

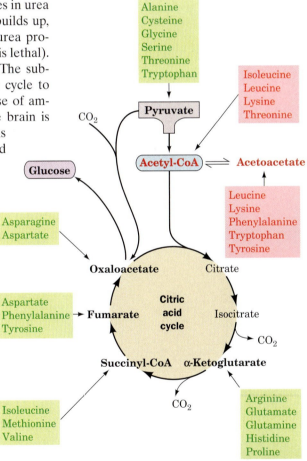

N-Acetylglutamate

Figure 20-13 Degradation of amino acids to one of seven common metabolic intermediates. Glucogenic and ketogenic degradations are indicated in green and red, respectively.

2. **Ketogenic amino acids,** which are broken down to acetyl-CoA or ace-toacetate and can thus be converted to fatty acids or ketone bodies (Section 19-3).

Some amino acids are precursors of both carbohydrates and ketone bodies. Since animals lack any metabolic pathways for the net conversion of acetyl-CoA or acetoacetate to gluconeogenic precursors, *no net synthesis of carbohydrates is possible from the purely ketogenic amino acids Lys and Leu.*

In studying the specific pathways of amino acid breakdown, we shall organize the amino acids into groups that are degraded to each of the seven metabolites mentioned above.

A   Alanine, Cysteine, Glycine, Serine, and Threonine Are Degraded to Pyruvate

Five amino acids—alanine, cysteine, glycine, serine, and threonine—are broken down to yield pyruvate (Fig. 20-14). Alanine is straightforwardly transaminated to pyruvate. Serine is converted to pyruvate through dehydration by **serine dehydratase.** This PLP-dependent enzyme, like the amino-

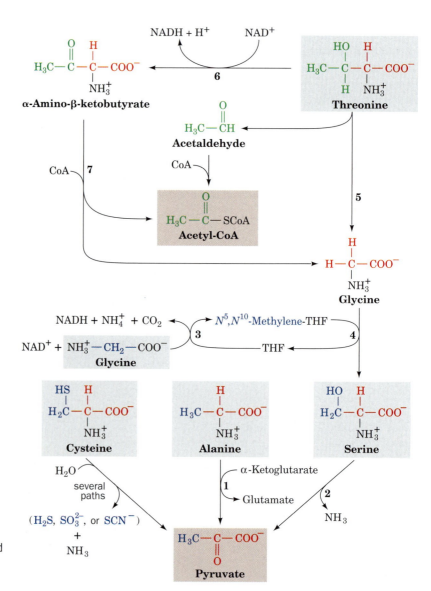

Figure 20-14   The pathways converting alanine, cysteine, glycine, serine, and threonine to pyruvate. The enzymes involved are (**1**) alanine aminotransferase, (**2**) serine dehydratase, (**3**) glycine cleavage system, (**4** and **5**) serine hydroxymethyltransferase, (**6**) threonine dehydrogenase, and (**7**) α-amino-β-ketobutyrate lyase.

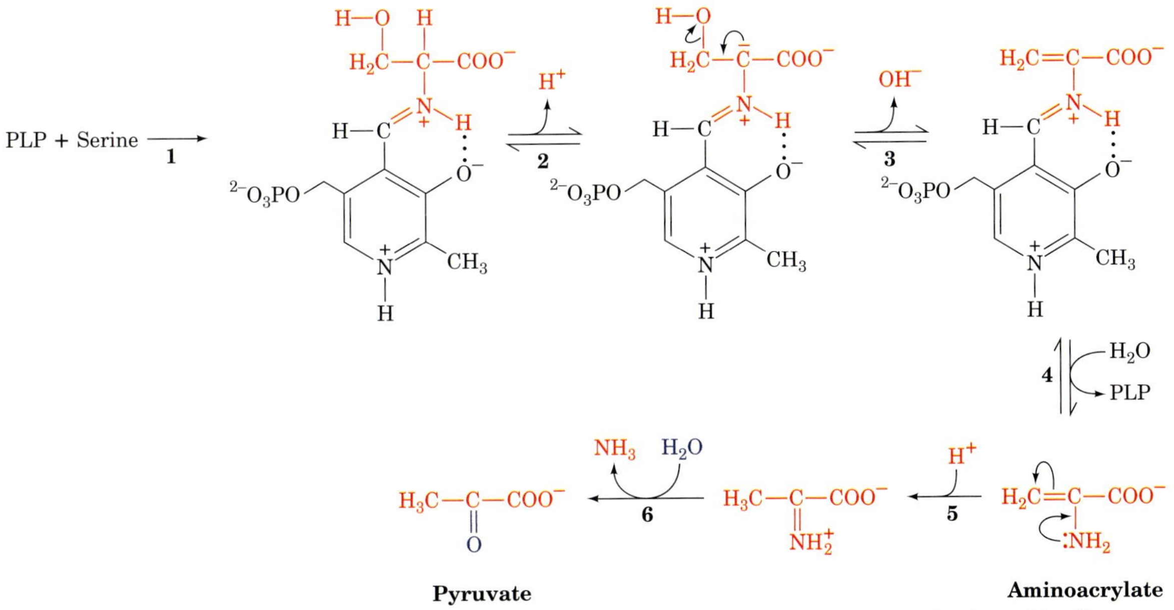

Figure 20-15 The serine dehydratase reaction. This PLP-dependent enzyme catalyzes the elimination of water from serine in six steps: (**1**) formation of a serine–PLP Schiff base, (**2**) removal of the α-H atom of serine to form a resonance-stabilized carbanion, (**3**) β elimination of OH$^-$, (**4**) hydrolysis of the Schiff base to yield the PLP–enzyme and aminoacrylate, (**5**) nonenzymatic tautomerization to the imine, and (**6**) nonenzymatic hydrolysis to form pyruvate and ammonia.

transferases (Section 20-2A), forms a PLP–amino acid Schiff base which facilitates the removal of the amino acid's α-hydrogen atom. In the serine dehydratase reaction, however, the C_α carbanion breaks down with the elimination of the amino acid's C_β OH, rather than with tautomerization (Fig. 20-8, Step 2), so that the substrate undergoes α,β elimination of H_2O rather than deamination (Fig. 20-15). The product of the dehydration, the enamine **aminoacrylate,** tautomerizes nonenzymatically to the corresponding imine, which spontaneously hydrolyzes to pyruvate and ammonia.

Cysteine can be converted to pyruvate via several routes in which the sulfhydryl group is released as H_2S, SO_3^{2-}, or SCN$^-$.

Glycine is converted to pyruvate by first being converted to serine by the enzyme **serine hydroxymethyltransferase,** another PLP-containing enzyme (Fig. 20-14, Reaction 4). This enzyme uses N^5,N^{10}-**methylene-tetrahydrofolate (N^5,N^{10}-methylene-THF)** as a one-carbon donor (the structure and chemistry of THF cofactors are described in Section 20-4D). The methylene group of the THF cofactor is obtained from a second glycine in Reaction 3 of Fig. 20-14, which is catalyzed by the **glycine cleavage system** (called the **glycine decarboxylase multienzyme system** in plants). This enzyme is a multiprotein complex that resembles pyruvate dehydrogenase (Section 16-2). The glycine cleavage system mediates the major route of glycine degradation in mammalian tissues. An inherited deficiency of the glycine cleavage system causes the disease **nonketotic hyperglycinemia,** which is characterized by mental retardation and accumulation of large amounts of glycine in body fluids.

Threonine is both glucogenic and ketogenic since it generates both pyruvate and acetyl-CoA. Its major route of breakdown is through **threonine dehydrogenase** (Fig. 20-14, Reaction 6), producing **α-amino-β-ketobutyrate,** which is converted to acetyl-CoA and glycine by **α-amino-β-ketobutyrate lyase** (Fig. 20-14, Reaction 7). The glycine can be converted, through serine, to pyruvate.

Serine Hydroxymethyltransferase Catalyzes PLP-Dependent C_α—C_β Bond Formation and Cleavage. Threonine can also be converted directly to glycine and acetaldehyde (which is subsequently oxidized to acetyl-CoA) via Reaction 5 of Fig. 20-14, which breaks threonine's C_α—C_β bond. This PLP-dependent reaction is catalyzed by serine hydroxymethyltransferase, the same enzyme that adds a hydroxymethyl group to glycine to produce serine (Fig. 20-14, Reaction 4). In the glycine → serine reaction, the amino acid's C_α—H bond is cleaved (as occurs in transamination; Fig. 20-8) and a C_α—C_β bond is formed. In contrast, the degradation of threonine to glycine by serine hydroxymethyltransferase acts in reverse, beginning with C_α—C_β bond cleavage:

With the cleavage of any of the bonds to C_α, the PLP group delocalizes the electrons of the resulting carbanion. This feature of PLP action is the key to understanding how the same amino acid–PLP Schiff base can undergo cleavage of different bonds to C_α in different enzymes (bonds *a*, *b*, or *c* in the upper right of Fig. 20-8). The bond that is cleaved is the one that lies in the plane perpendicular to that of the π-orbital system of the PLP (Fig. 20-16). This arrangement allows the PLP π-orbital system to overlap the bonding orbital containing the electron pair being delocalized. Any other geometry would reduce or even eliminate this orbital overlap (the newly formed double bond would be twisted out of planarity) yielding a higher

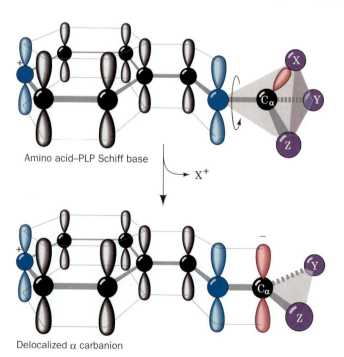

Figure 20-16 The π-orbital framework of a PLP–amino acid Schiff base. The bond from X to C_α is in a plane perpendicular to the plane of the PLP π-orbital system (*top*) and is therefore labile. The broken bond's electron pair (*bottom*) is delocalized over the conjugated molecule.

Amino acid–PLP Schiff base

X⁺

Delocalized α carbanion

energy arrangement. Different bonds to C_α can be positioned for cleavage by rotation around the C_α—N bond. Evidently, *each enzyme binds its amino acid–PLP Schiff base adduct with the appropriate geometry for bond cleavage.*

B Asparagine and Aspartate Are Degraded to Oxaloacetate

Transamination of aspartate leads directly to oxaloacetate:

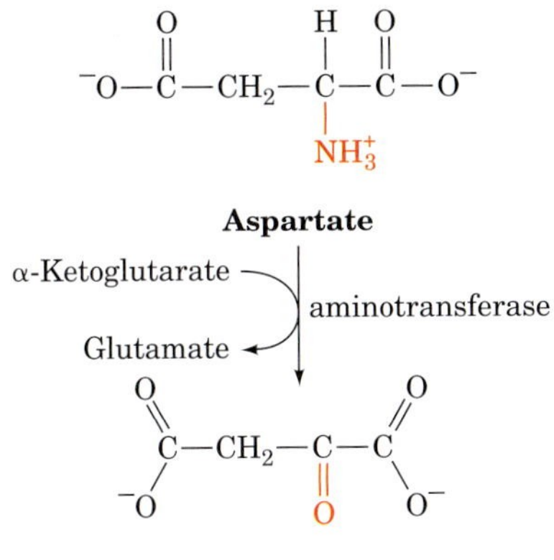

Aspartate

α-Ketoglutarate

Glutamate

aminotransferase

Oxaloacetate

Asparagine is also converted to oxaloacetate in this manner after its hydrolysis to aspartate by **L-asparaginase:**

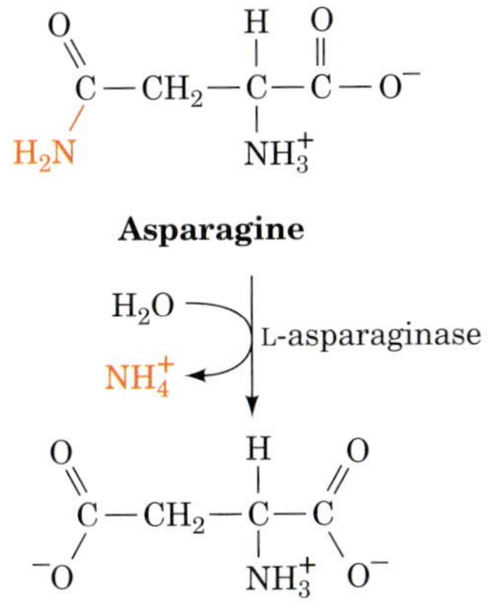

Asparagine

H_2O

NH_4^+

L-asparaginase

Aspartate

Interestingly, L-asparaginase is an effective chemotherapeutic agent in the treatment of cancers that must obtain asparagine from the blood, particularly **acute lymphoblastic leukemia.**

C Arginine, Glutamate, Glutamine, Histidine, and Proline Are Degraded to α-Ketoglutarate

Arginine, glutamine, histidine, and proline are all degraded by conversion to glutamate (Fig. 20-17), which in turn is oxidized to α-ketoglutarate by glutamate dehydrogenase (Section 20-2B). Conversion of glutamine to glutamate involves only one reaction: hydrolysis by **glutaminase.** In the kidney, the action of glutaminase produces ammonia, which combines with a proton to form the ammonium ion (NH_4^+) and is excreted. During metabolic acidosis (see Box 2-2), kidney glutaminase helps eliminate excess acid. Although free NH_3 in the blood could serve the same acid-absorbing purpose, ammonia is toxic. It is therefore converted to glutamine by glutamine synthetase in the liver (Section 20-5A). Glutamine therefore acts as an ammonia transport system between the liver, where much of it is synthesized, and the kidneys, where it is hydrolyzed by glutaminase.

Figure 20-17 The degradation of arginine, glutamate, glutamine, histidine, and proline to α-ketoglutarate. The enzymes catalyzing the reactions are (**1**) glutamate dehydrogenase, (**2**) glutaminase, (**3**) arginase, (**4**) ornithine-δ-aminotransferase, (**5**) glutamate-5-semialdehyde dehydrogenase, (**6**) proline oxidase, (**7**) spontaneous, (**8**) histidine ammonia lyase, (**9**) urocanate hydratase, (**10**) imidazolone propionase, and (**11**) glutamate formiminotransferase.

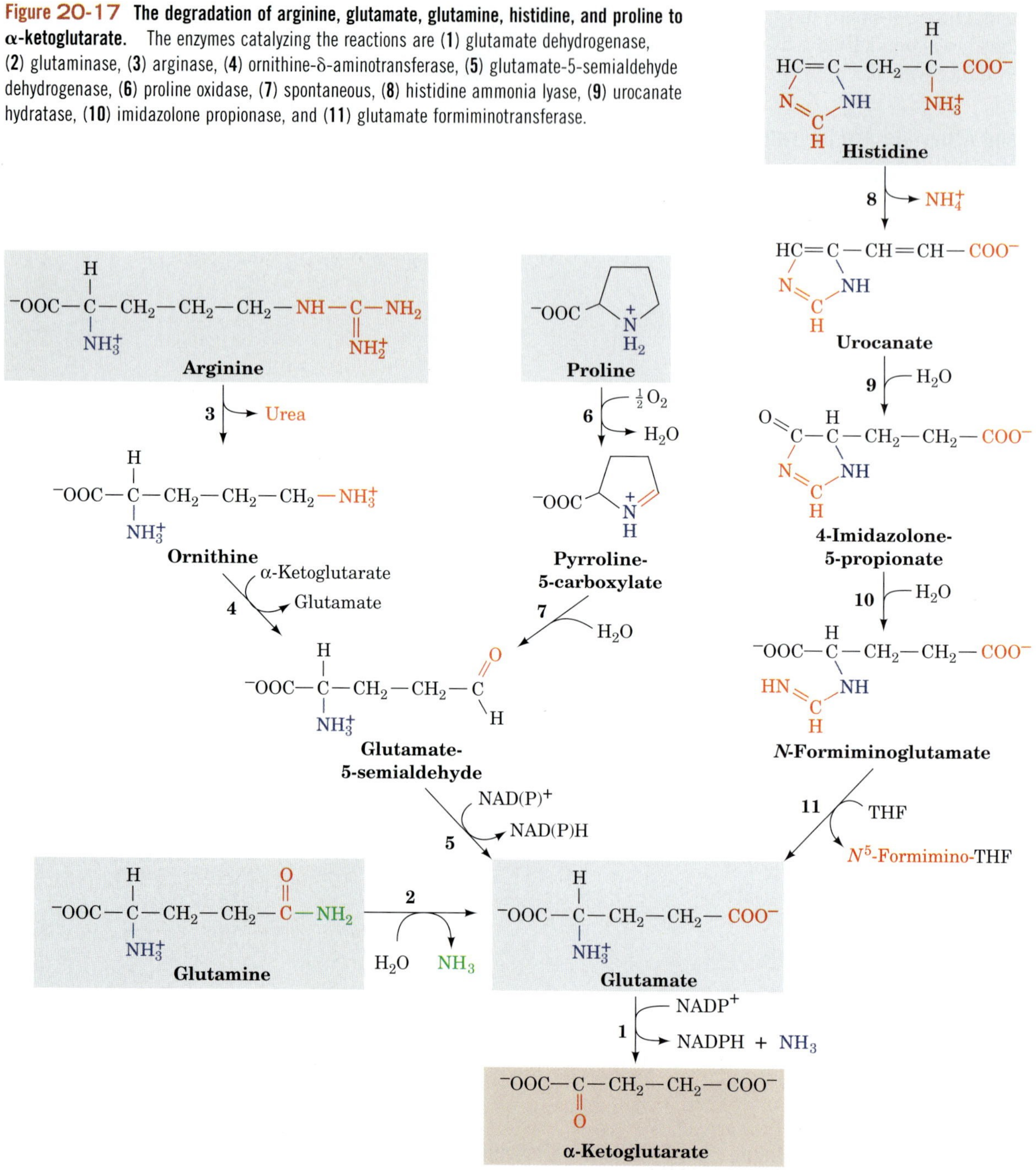

Histidine's conversion to glutamate is more complicated: It is nonoxidatively deaminated, then it is hydrated, and its imidazole ring is cleaved to form **N-formiminoglutamate.** The formimino group is then transferred to tetrahydrofolate forming glutamate and **N^5-formimino-tetrahydrofolate** (Section 20-4D). Both arginine and proline are converted to glutamate through the intermediate formation of **glutamate-5-semialdehyde.**

D Isoleucine, Methionine, and Valine Are Degraded to Succinyl-CoA

Isoleucine, methionine, and valine have complex degradative pathways that all yield propionyl-CoA, which is also a product of odd-chain fatty acid

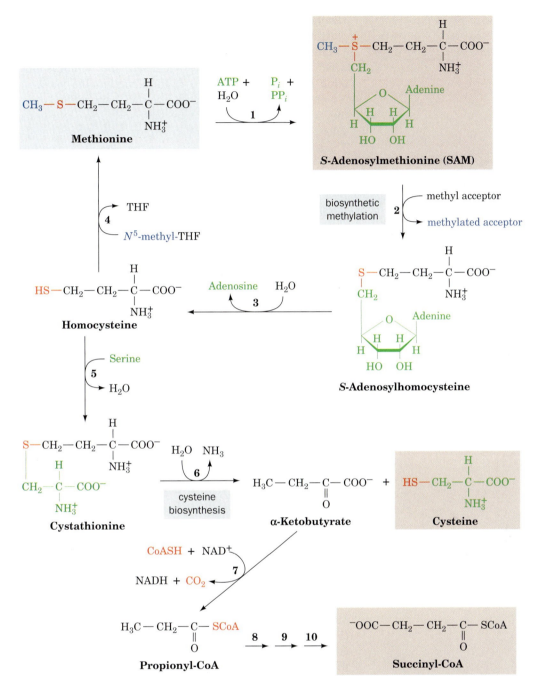

Figure 20-18   Methionine degradation.   This pathway yields cysteine and succinyl-CoA. The enzymes are (**1**) methionine adenosyltransferase in a reaction that yields the biological methylating agent *S*-adenosylmethionine (SAM), (**2**) methyltransferase, (**3**) adenosylhomocysteinase, (**4**) methionine synthase (a coenzyme B$_{12}$–dependent enzyme), (**5**) cystathionine β-synthase (a PLP- dependent enzyme), (**6**) cystathione γ-lyase, (**7**) α-keto acid dehydrogenase, (**8**) propionyl-CoA carboxylase, (**9**) methylmalonyl-CoA racemase, and (**10**) methylmalonyl-CoA mutase (a coenzyme B$_{12}$–dependent enzyme). Reac- tions 8–10 are discussed in Section 19-2E.

degradation. Propionyl-CoA is converted to succinyl-CoA by a series of reactions requiring biotin and coenzyme B$_{12}$ (Section 19-2E).

Methionine Breakdown Involves Synthesis of *S*-Adenosylmethionine and Cysteine.

Methionine degradation (Fig. 20-18) begins with its reaction with ATP to form **S-adenosylmethionine (SAM;** alternatively **AdoMet).** *This sulfonium ion's highly reactive methyl group makes it an important biological meth-*

ylating agent. For instance, SAM is the methyl donor in the synthesis of phosphatidylcholine from phosphatidylethanolamine (Section 19-6A).

Donation of a methyl group from SAM leaves **S-adenosylhomocysteine,** which is then hydrolyzed to adenosine and **homocysteine.** The homocysteine can be methylated to re-form methionine via a reaction in which N^5-**methyl-tetrahydrofolate** (see below) is the methyl donor. Alternatively, the homocysteine can combine with serine to yield **cystathionine,** which subsequently forms cysteine (cysteine biosynthesis) and **α-ketobutyrate.** The α-ketobutyrate continues along the degradative pathway to propionyl-CoA and then succinyl-CoA. High homocysteine levels are associated with disease (see Box 20-1).

Tetrahydrofolates Are One-Carbon Carriers. Many biosynthetic processes involve the addition of a C_1 unit to a metabolic precursor. In most carboxylation reactions (e.g., pyruvate carboxylase; Fig. 15-26), the enzyme uses a biotin cofactor. In some reactions, *S*-adenosylmethionine (Fig. 20-18) functions as a methylating agent. However, tetrahydrofolate (THF) is more versatile than either of these cofactors because it can transfer C_1 units in several oxidation states.

THF is a 6-methylpterin derivative linked in sequence to a **p-aminobenzoic acid** and a Glu residue:

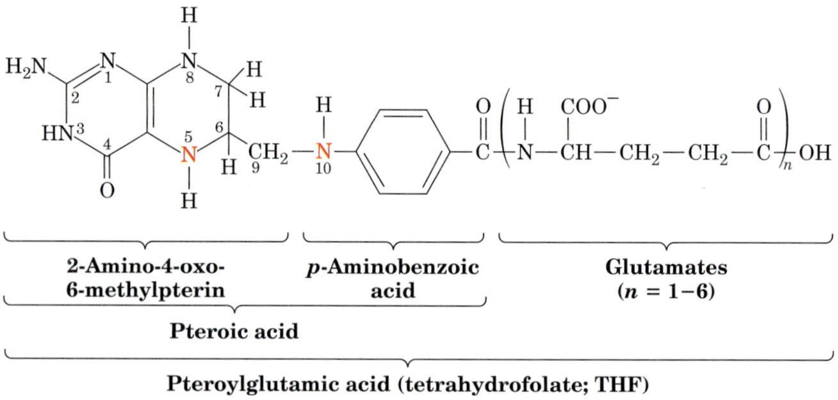

Pteroylglutamic acid (tetrahydrofolate; THF)

Up to five additional Glu residues are linked to the first glutamate via isopeptide bonds to form a polyglutamyl tail. THF is derived from the vitamin **folic acid** (Latin: *folium,* leaf), a doubly oxidized form of THF that must be enzymatically reduced before it becomes an active coenzyme (Fig. 20-19). Both reductions are catalyzed by **dihydrofolate reductase (DHFR).** Mammals cannot synthesize folic acid, so it must be provided in the diet or by intestinal microorganisms.

C_1 units are covalently attached to THF at positions N5, N10, or both N5 and N10. These C_1 units, which may be at the oxidation levels of formate, formaldehyde, or methanol (Table 20-2), are all interconvertible by enzymatic redox reactions (Fig. 20-20).

Table 20-2 Oxidation Levels of C_1 Groups Carried by THF

Oxidation Level	Group Carried	THF Derivative(s)
Methanol	Methyl ($-CH_3$)	N^5-Methyl-THF
Formaldehyde	Methylene ($-CH_2-$)	N^5,N^{10}-Methylene-THF
Formate	Formyl ($-CH=O$)	N^5-Formyl-THF, N^{10}-formyl-THF
	Formimino ($-CH=NH$)	N^5-Formimino-THF
	Methenyl ($-CH=$)	N^5,N^{10}-Methenyl-THF

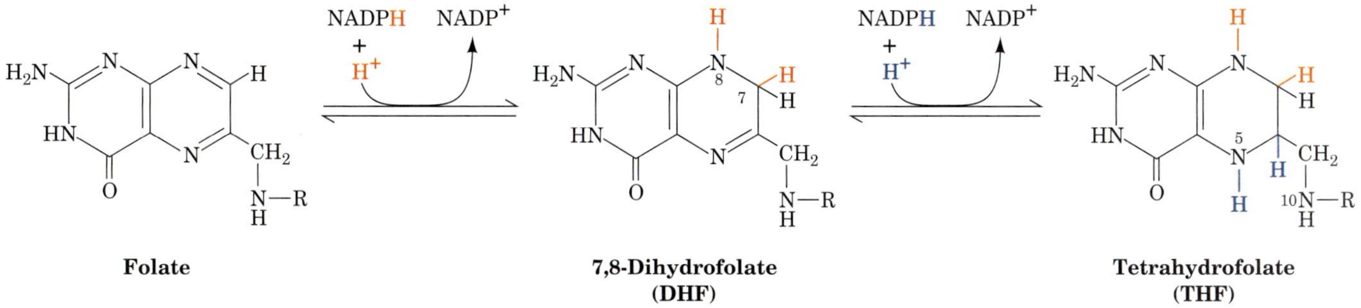

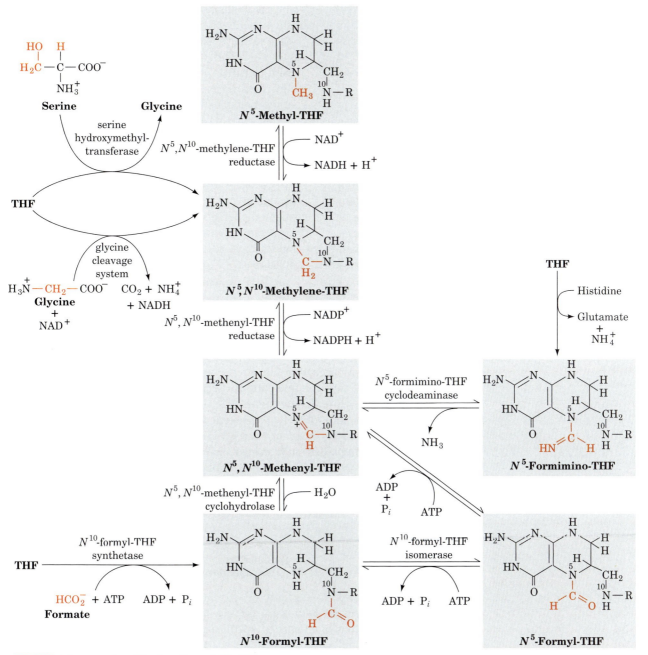

Figure 20-19   The two-stage reduction of folate to THF.   Both reactions are catalyzed by dihydrofolate reductase.

Figure 20-20   Interconversion of the C_1 units carried by THF.

BOX 20-1

Biochemistry in Health and Disease

Homocysteine, a Marker of Disease

The cellular level of homocysteine depends on its rate of synthesis through methylation reactions utilizing SAM (Fig. 20-18, Reactions 2 and 3) and its rate of utilization through remethylation to form methionine (Fig. 20-18, Reaction 4) and reaction with serine to form cystathionine in the cysteine biosynthetic pathway (Fig. 20-18, Reaction 5). An increase in homocysteine levels leads to **hyperhomocysteinemia,** elevated concentrations of homocysteine in the blood, which is associated with cardiovascular disease. The link was first discovered in individuals with **homocysteinuria,** a disorder in which excess homocysteine is excreted in the urine. These individuals develop atherosclerosis as children, possibly because homocysteine causes oxidative damage to the walls of blood vessels even in the absence of elevated LDL levels (Section 19-7C).

Hyperhomocysteinemia is also associated with **neural tube defects,** the cause of a variety of severe birth defects including **spina bifida** (defects in the spinal column that often result in paralysis) and **anencephaly** (the invariably fatal failure of the brain to develop, which is the leading cause of infant death due to congenital anomalies). Hyperhomocysteinemia is readily controlled by ingesting the vitamin precursors of the coenzymes that participate in homocysteine

breakdown, namely, B_6 (pyridoxine, the PLP precursor; Fig. 20-7), B_{12} (Fig. 19-17), and folate (Section 20-4D). Folate, especially, alleviates hyperhomocysteinemia; its administration to pregnant women dramatically reduces the incidence of neural tube defects in newborns. Because neural tube development is one of the earliest steps of embryogenesis, women of childbearing age are encouraged to consume adequate amounts of folate even before they become pregnant.

Around 10% of the population is homozygous for an Ala → Val mutation in N^5,N^{10}-**methylene-tetrahydrofolate reductase (MTHFR),** which catalyzes the conversion of N^5,N^{10}-methylene-THF to N^5-methyl-THF (Fig. 20-20, *top center*). This reaction generates the N^5-methyl-THF required to convert homocysteine to methionine (Fig. 20-18, Reaction 4). The mutation does not affect the enzyme's reaction kinetics but instead increases the rate at which its essential flavin cofactor dissociates. Folate derivatives that bind to the enzyme decrease the rate of flavin loss, thus increasing the enzyme's overall activity and decreasing the homocysteine concentration. The prevalence of the MTHFR mutation in the human population suggests that it has (or once had) some selective advantage; however, this is as yet a matter of speculation.

THF acquires C_1 units in the conversion of serine to glycine by serine hydroxymethyltransferase (the reverse of Reaction 4, Fig. 20-14), in the cleavage of glycine (Fig. 20-14, Reaction 3), and in histidine breakdown (Fig. 20-17, Reaction 11). The C_1 units carried by THF are used in the synthesis of thymine nucleotides (Section 22-3B) and in the synthesis of methionine from homocysteine (Fig. 20-18). By promoting this latter process, supplemental folate helps prevent diseases associated with abnormally high levels of homocysteine (Box 20-1).

Sulfonamides (sulfa drugs) such as **sulfanilamide** are antibiotics that are structural analogs of the *p*-aminobenzoic acid constituent of THF:

$$\text{H}_2\text{N} - \!\!\bigcirc\!\!- \overset{\displaystyle \text{O}}{\underset{\displaystyle \text{O}}{\overset{\|}{\underset{\|}{\text{S}}}}} - \text{NH} - \text{R} \qquad \text{H}_2\text{N} - \!\!\bigcirc\!\!- \overset{\displaystyle \text{O}}{\overset{\|}{\text{C}}} - \text{OH}$$

Sulfonamides   ***p*-Aminobenzoic acid**
(R = H, sulfanilamide)

They competitively inhibit bacterial synthesis of THF at the *p*-aminobenzoic acid incorporation step, thereby blocking THF-requiring reactions. The inability of mammals to synthesize folic acid leaves them unaffected by sulfonamides, which accounts for the medical utility of these widely used antibacterial agents.

Branched-Chain Amino Acid Degradation Involves Acyl-CoA Oxidation.

Degradation of the branched-chain amino acids isoleucine, leucine, and valine begins with three reactions that employ common enzymes (Fig. 20-21):

1. Transamination to the corresponding α-keto acid.
2. Oxidative decarboxylation to the corresponding acyl-CoA.
3. Dehydrogenation by FAD to form a double bond.

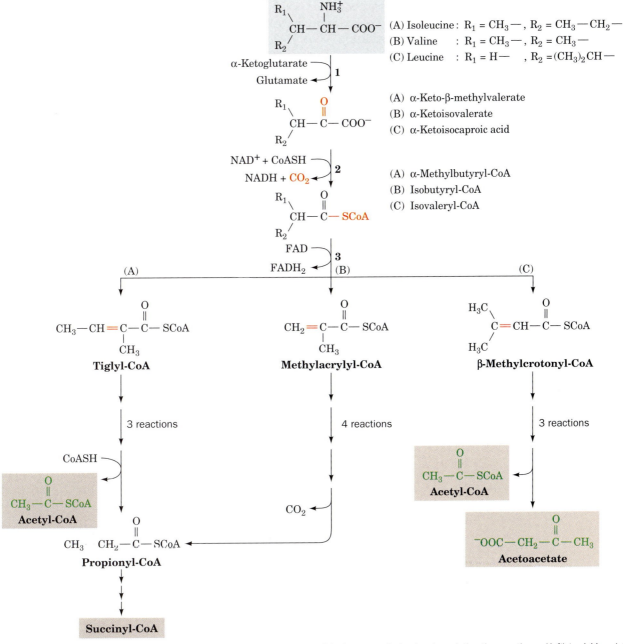

Figure 20-21 The degradation of the branched-chain amino acids. Isoleucine (A), valine (B), and leucine (C) follow an initial common pathway utilizing three enzymes: (**1**) branched-chain amino acid aminotransferase, (**2**) branched-chain α-keto acid dehydrogenase (BCKDH), and (**3**) acyl-CoA dehydrogenase. Isoleucine degradation then continues (*left*) to yield acetyl-CoA and succinyl-CoA; valine degradation continues (*center*) to yield succinyl-CoA; and leucine degradation continues (*right*) to yield acetyl-CoA and acetoacetate.

The remaining reactions of the isoleucine degradation pathway are analogous to those of fatty acid oxidation (Section 19-2C), thereby yielding acetyl-CoA and propionyl-CoA, which is subsequently converted via three reactions to succinyl-CoA. The degradation of valine, which contains one less carbon than isoleucine, yields CO_2 and propionyl-CoA, which is then converted to succinyl-CoA. The further degradation of leucine, which yields acetoacetate instead of propionyl-CoA, is considered in Section 20-4E.

Branched-chain α-keto acid dehydrogenase (BCKDH), which catalyzes Reaction 2 of Fig. 20-21, is a multienzyme complex that closely resembles the pyruvate dehydrogenase and α-ketoglutarate dehydrogenase complexes (Sections 16-2 and 16-3D). Indeed, all three multienzyme complexes share a common subunit, E_3 (dihydrolipoamide dehydrogenase), and employ the coenzymes TPP, lipoamide, and FAD in addition to their terminal oxidizing agent, NAD^+.

A genetic deficiency in BCKDH causes **maple syrup urine disease,** so named because the consequent buildup of branched-chain α-keto acids imparts the urine with the characteristic odor of maple syrup. Unless promptly treated by a diet low in branched-chain amino acids, maple syrup urine disease is rapidly fatal.

E Leucine and Lysine Are Degraded Only to Acetyl-CoA and/or Acetoacetate

Leucine degradation begins in the same manner as isoleucine and valine degradation (Fig. 20-21), but the dehydrogenated CoA adduct β-methylcrotonyl-CoA is converted to acetyl-CoA and acetoacetate, a ketone body.

The predominant pathway for lysine degradation in mammalian liver produces acetoacetate and 2 CO_2 via the initial formation of the α-ketoglutarate–lysine adduct **saccharopine** (Fig. 20-22). This pathway is worth examining in detail because we have encountered 7 of its 11 reactions in other pathways. Reaction 4 is a PLP-dependent transamination. Reaction 5 is the oxidative decarboxylation of an α-keto acid by a multienzyme complex similar to pyruvate dehydrogenase (Section 16-2). Reactions 6, 8, and 9 are standard reactions of fatty acyl-CoA oxidation: dehydrogenation by FAD, hydration, and dehydrogenation by NAD^+. Reactions 10 and 11 are standard reactions in ketone body formation. Two moles of CO_2 are produced at Reactions 5 and 7 of the pathway.

The saccharopine pathway is thought to predominate in mammals because a genetic defect in the enzyme that catalyzes Reaction 1 in the sequence results in **hyperlysinemia** and **hyperlysinuria** (elevated levels of lysine in the blood and urine, respectively) along with mental and physical retardation. This is yet another example of how the study of rare inherited disorders has helped to trace metabolic pathways.

F Tryptophan Is Degraded to Alanine and Acetoacetate

The complexity of the major tryptophan degradation pathway (outlined in Fig. 20-23) precludes a detailed discussion of all its reactions. However, one reaction is of particular interest: The fourth reaction is catalyzed by **kynureninase,** whose PLP group facilitates cleavage of the C_β—C_γ bond to release alanine. The kynureninase reaction follows the same initial steps as transamination (Fig. 20-8), but an enzyme nucleophilic group then attacks C_γ of the resonance-stabilized intermediate, resulting in C_β—C_γ bond cleavage. The remainder of the tryptophan skeleton is converted in five reac-

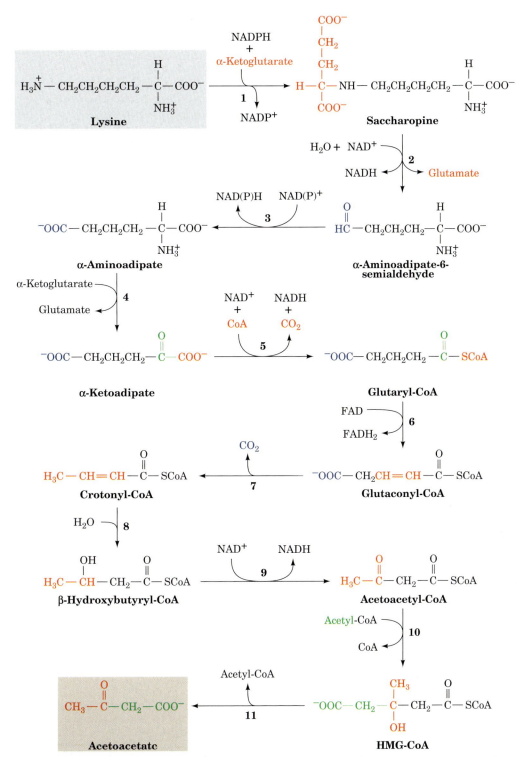

Figure 20-22 The pathway of lysine degradation in mammalian liver.
The enzymes are (**1**) saccharopine dehydrogenase (NADP$^+$, lysine forming),
(**2**) saccharopine dehydrogenase (NAD$^+$, glutamate forming), (**3**) aminoadipate
semialdehyde dehydrogenase, (**4**) aminoadipate aminotransferase (a PLP-
dependent enzyme), (**5**) α-keto acid dehydrogenase, (**6**) glutaryl-CoA
dehydrogenase, (**7**) decarboxylase, (**8**) enoyl-CoA hydratase, (**9**) β-hydroxyacyl-
CoA dehydrogenase, (**10**) HMG-CoA synthase, and (**11**) HMG-CoA lyase.
Reactions 10 and 11 are discussed in Section 19-3.

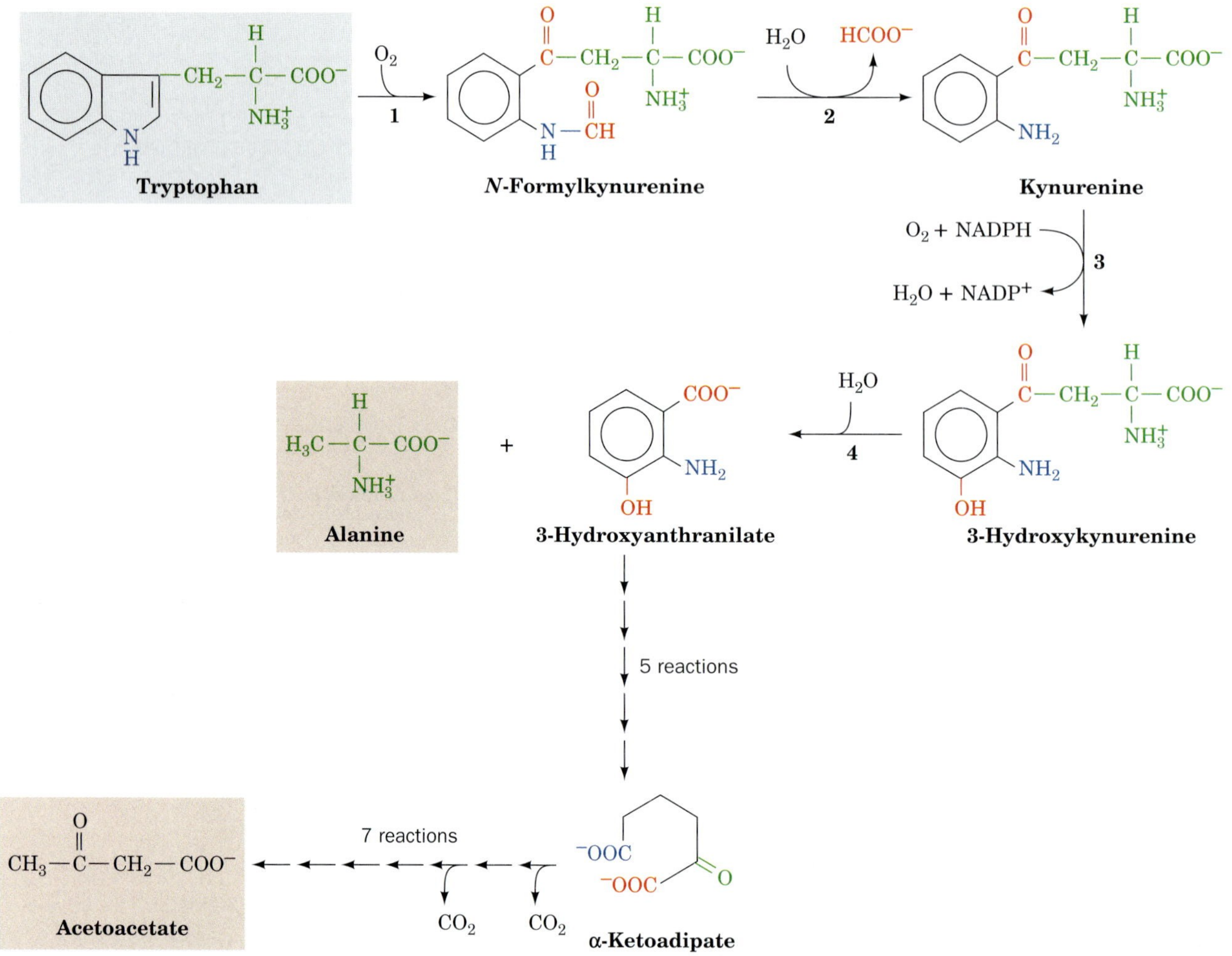

Figure 20-23 The pathway of tryptophan degradation.   The enzymatic reactions shown are (**1**) tryptophan-2,3-dioxygenase, (**2**) formamidase, (**3**) kynurenine-3-monooxygenase, and (**4**) kynureninase (a PLP-dependent enzyme). Five reactions convert 3-hydroxyanthranilate to α-ketoadipate, which is converted to acetyl-CoA and acetoacetate in seven reactions as shown in Fig. 20-22, Reactions 5–11.

tions to α-ketoadipate, which is also an intermediate in lysine degradation. α-Ketoadipate is broken down to 2 CO_2 and acetoacetate in seven reactions, as shown in Fig. 20-22.

G Phenylalanine and Tyrosine Are Degraded to Fumarate and Acetoacetate

Since the first reaction in phenylalanine degradation is its hydroxylation to tyrosine, a single pathway (Fig. 20-24) is responsible for the breakdown of both of these amino acids. The final products of the six-reaction degradation are fumarate, a citric acid cycle intermediate, and acetoacetate, a ketone body. Defects in the enzymes that catalyze Reactions 1 and 4 cause disease (see Box 20-2).

Pterins Are Redox Cofactors

The hydroxylation of phenylalanine by the Fe(III)-containing enzyme **phenylalanine hydroxylase** (Fig. 20-24, Reaction 1) requires the cofactor **biopterin,** a **pterin** derivative. Pterins are compounds that contain the **pteridine** ring (Fig. 20-25). Note the resemblance between the pteridine ring

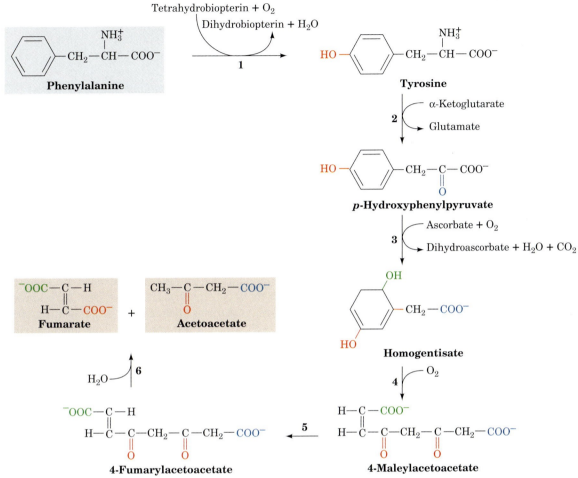

Figure 20-24 The pathway of phenylalanine degradation. The enzymes involved are (**1**) phenylalanine hydroxylase, (**2**) aminotransferase, (**3**) *p*-hydroxyphenylpyruvate dioxygenase, (**4**) homogentisate dioxygenase, (**5**) maleylacetoacetate isomerase, and (**6**) fumarylacetoacetase.

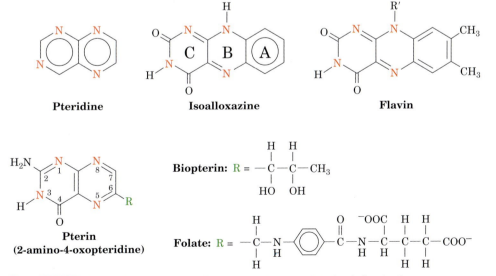

Figure 20-25 The pteridine ring nucleus of biopterin and folate. Note the similar structures of pteridine and the isoalloxazine ring of flavin coenzymes.

BOX 20-2

Biochemistry in Health and Disease

Phenylketonuria and Alcaptonuria Result from Defects in Phenylalanine Degradation

Archibald Garrod realized in the early 1900s that human genetic diseases result from specific enzyme deficiencies. We have repeatedly seen how this realization has contributed to the elucidation of metabolic pathways. The first such disease Garrod recognized was **alcaptonuria,** which results in the excretion of large quantities of **homogentisic acid** (Section 13-4). This condition results from deficiency of **homogentisate dioxygenase** (Fig. 20-24, Reaction 4). Alcaptonurics suffer no ill effects other than arthritis later in life (although their urine darkens alarmingly because of the rapid air oxidation of the homogentisate they excrete).

Individuals suffering from **phenylketonuria (PKU)** are not so fortunate. Severe mental retardation occurs within a few months of birth if the disease is not detected and treated immediately. PKU is caused by the inability to hydroxylate phenylalanine (Fig. 20-24, Reaction 1) and therefore results in increased blood levels of phenylalanine **(hyperphenylalaninemia).** The excess phenylalanine is transaminated to **phenylpyruvate**

$$\text{(phenyl)}-CH_2-\overset{\overset{\displaystyle O}{\|}}{C}-COO^-$$

Phenylpyruvate

by an otherwise minor pathway. The "spillover" of phenylpyruvate (a phenylketone) into the urine was the first observation connected with the disease and gave the disease its name.

All babies born in the United States are now screened for PKU immediately after birth by testing for elevated levels of phenylalanine in the blood.

Classical PKU results from a deficiency in phenylalanine hydroxylase. When this was established in 1947, it was the first inborn error of metabolism whose basic biochemical defect had been identified. Since all of the tyrosine breakdown enzymes are normal, treatment consists in providing the patient with a low-phenylalanine diet and monitoring the blood level of phenylalanine to ensure that it remains within normal limits for the first 5 to 10 years of life (the adverse effects of hyperphenylalaninemia seem to disappear after that age). **Aspartame (Nutrasweet),** an often used sweetening ingredient in diet soft drinks and many other dietetic food products, is Asp-Phe-methyl ester (see Box 8-2) and is therefore a source of dietary phenylalanine. Consequently, a warning label for phenylketonurics appears on all these products.

Phenylalanine hydroxylase deficiency also accounts for another common symptom of PKU: Its victims have lighter hair and skin color than their siblings. This is because elevated phenylalanine levels inhibit tyrosine hydroxylation, the first reaction in the formation of the skin pigment **melanin** (Fig. 20-39).

Other causes of hyperphenylalaninemia have been discovered since the introduction of infant screening techniques. These are caused by deficiencies in the enzymes catalyzing the formation or regeneration of 5,6,7,8-tetrahydrobiopterin, the phenylalanine hydroxylase cofactor (Fig. 20-26).

and the isoalloxazine ring of the flavin coenzymes (Fig. 13-11); the positions of the nitrogen atoms in pteridine are identical to those of the B and C rings of isoalloxazine. Folate derivatives also contain the pterin ring (Section 20-4D).

Pterins, like flavins, participate in biological oxidations. The active form of biopterin is the fully reduced form, **5,6,7,8-tetrahydrobiopterin.** It is produced from **7,8-dihydrobiopterin** and NADPH, in what may be considered a priming reaction, by dihydrofolate reductase (Fig. 20-26), which simultaneously reduces dihydrofolate to tetrahydrofolate (Fig. 20-19). In the phenylalanine hydroxylase reaction, 5,6,7,8-tetrahydrobiopterin is hydroxylated to **pterin-4a-carbinolamine** (Fig. 20-26), which is then converted to 7,8-dihydrobiopterin **(quinoid form)** by **pterin-4a-carbinolamine dehydratase.** This quinoid is subsequently reduced by the NAD(P)H-requiring enzyme **dihydropteridine reductase** to regenerate the active cofactor. Although dihydrofolate reductase and dihydropteridine reductase produce the same product, they use different tautomers of the substrate.

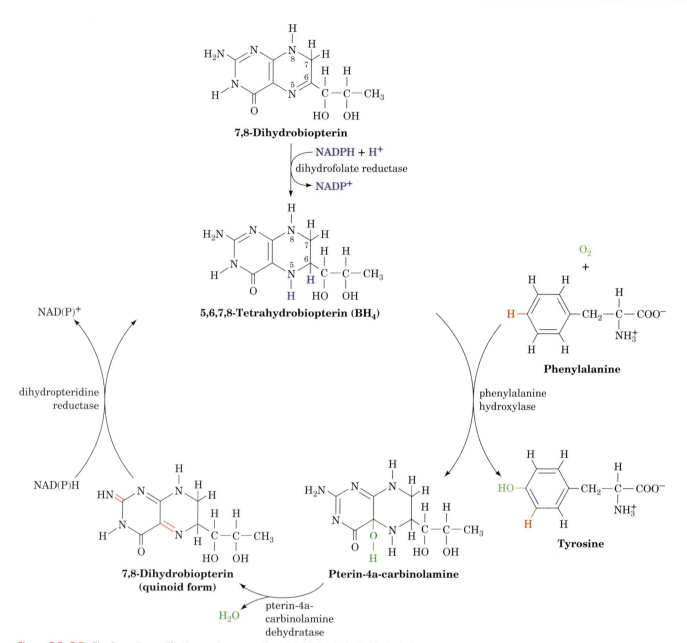

Figure 20-26 The formation, utilization, and regeneration of 5,6,7,8-tetrahydrobiopterin in the phenylalanine hydroxylase reaction.

5 | Amino Acid Biosynthesis

Many amino acids are synthesized by pathways that are present only in plants and microorganisms. Since mammals must obtain these amino acids in their diets, these substances are known as **essential amino acids.** The other amino acids, which can be synthesized by mammals from common intermediates, are termed **nonessential amino acids.** Their α-keto acid carbon skeletons are converted to amino acids by transamination reactions (Section 20-2A) utilizing the preformed α-amino nitrogen of another amino acid, usually glutamate. Although it was originally presumed that glutamate

Table 20-3 Essential and Nonessential Amino Acids in Humans

Essential	Nonessential
Arginine[a]	Alanine
Histidine	Asparagine
Isoleucine	Aspartate
Leucine	Cysteine
Lysine	Glutamate
Methionine	Glutamine
Phenylalanine	Glycine
Threonine	Proline
Tryptophan	Serine
Valine	Tyrosine

[a]Although mammals synthesize arginine, they cleave most of it to form urea (Section 20-3A).

can be synthesized from ammonia and α-ketoglutarate by GDH acting in reverse, it now appears that the predominant physiological direction of this enzyme is glutamate breakdown (Section 20-2B). Consequently, *preformed α-amino nitrogen should also be considered to be an essential nutrient.* In this context, it is interesting to note that, in addition to the four well-known taste receptors, those for sweet, sour, salty, and bitter tastes, a fifth taste receptor has recently been characterized, that for the meaty taste of **monosodium glutamate (MSG),** which is known as **umami** (a Japanese name).

The essential and nonessential amino acids for humans are listed in Table 20-3. Arginine is classified as essential, even though it is synthesized by the urea cycle (Section 20-3A), because it is required in greater amounts than can be produced by this route during the normal growth and development of children (but not adults). The essential amino acids occur in animal and vegetable proteins. Different proteins, however, contain different proportions of the essential amino acids. Milk proteins, for example, contain them all in the proportions required for proper human nutrition. Bean protein, on the other hand, contains an abundance of lysine but is deficient in methionine, whereas wheat is deficient in lysine but contains ample methionine. A balanced protein diet therefore must contain a variety of different protein sources that complement each other to supply the proper proportions of all the essential amino acids.

In this section we study the pathways involved in the formation of the nonessential amino acids. We also briefly consider such pathways for the essential amino acids as they occur in plants and microorganisms. Keep in mind that *there is considerable variation in these pathways among different species. In contrast, the basic pathways of carbohydrate and lipid metabolism are all but universal.*

A Biosynthesis of the Nonessential Amino Acids

All the nonessential amino acids except tyrosine are synthesized by simple pathways leading from one of four common metabolic intermediates: pyruvate, oxaloacetate, α-ketoglutarate, and 3-phosphoglycerate. Tyrosine, which is really misclassified as being nonessential, is synthesized by the one-step hydroxylation of the essential amino acid phenylalanine (Fig. 20-24). Indeed, the dietary requirement for phenylalanine reflects the need for tyrosine as well. The presence of dietary tyrosine therefore decreases the need for phenylalanine.

Alanine, Asparagine, Aspartate, Glutamate, and Glutamine Are Synthesized from Pyruvate, Oxaloacetate, and α-Ketoglutarate. Pyruvate, oxaloacetate, and α-ketoglutarate are the α-keto acids (the so-called carbon skeletons) that correspond to alanine, aspartate, and glutamate, respectively. Indeed, the synthesis of each of these amino acids is a one-step transamination reaction (Fig. 20-27, Reactions 1–3). The ultimate source of the α-amino group in these transamination reactions is glutamate, which is synthesized in microorganisms, plants, and lower eukaryotes by glutamate synthase (Section 20-7), an enzyme that is absent in vertebrates.

Asparagine and glutamine are, respectively, synthesized from aspartate and glutamate by ATP-dependent amidation. In the **glutamine synthetase** reaction (Fig. 20-27, Reaction 5), glutamate is first activated by reaction with ATP to form a γ-**glutamylphosphate** intermediate. NH_4^+ then displaces the phosphate group to produce glutamine. Curiously, aspartate amidation

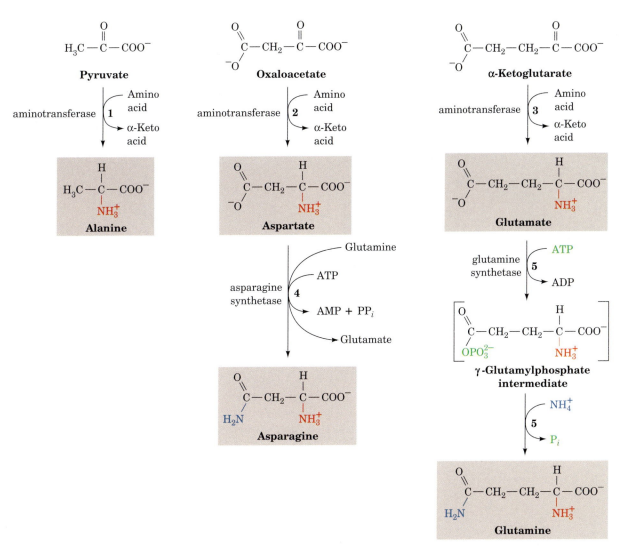

Figure 20-27 The syntheses of alanine, aspartate, glutamate, asparagine, and glutamine. These reactions involve, respectively, transamination of (1) pyruvate, (2) oxaloacetate, and (3) α-ketoglutarate, and amidation of (4) aspartate and (5) glutamate.

by **asparagine synthetase** to form asparagine follows a different route; it uses glutamine as its amino group donor and cleaves ATP to AMP + PP$_i$ (Fig. 20-27, Reaction 4).

Glutamine Synthetase Is a Central Control Point in Nitrogen Metabolism.

Glutamine is the amino group donor in the formation of many biosynthetic products as well as being a storage form of ammonia. The control of glutamine synthetase is therefore vital for regulating nitrogen metabolism. Mammalian glutamine synthetases are activated by α-ketoglutarate, the product of glutamate's oxidative deamination (Section 20-2B). This control presumably helps prevent the accumulation of the ammonia produced by that reaction.

Bacterial glutamine synthetase, as Earl Stadtman showed, has a much more elaborate control system. This enzyme, which consists of 12 identical 469-residue subunits arranged at the corners of a hexagonal prism (Fig.

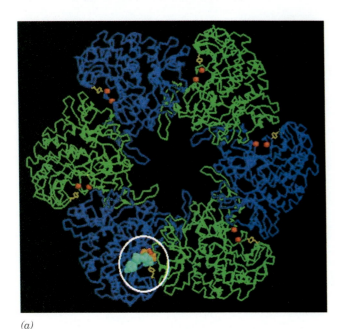

(a)

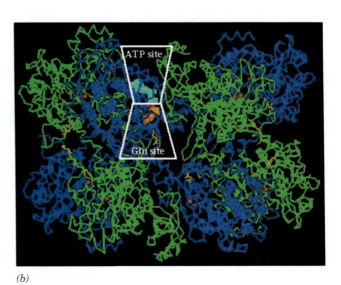

(b)

Figure 20-28 X-Ray structure of glutamine synthetase from the bacterium *Salmonella typhimurium.* The enzyme consists of 12 identical subunits, here represented by their C_α backbones, arranged with D_6 symmetry (the symmetry of a hexagonal prism). (*a*) View down the sixfold axis of symmetry showing only the six subunits of the upper ring in alternating blue and green. The subunits of the lower ring are roughly directly below those of the upper ring. The protein, including its side chains (not shown) has a diameter of

143 Å. Pairs of Mg^{2+} ions (*red spheres*) that are required for enzymatic activity are shown in each active site. Each adenylylation site, Tyr 397 (*yellow*), lies between two subunits. Also drawn in one active site are ADP (*cyan*) and phosphinothricin (*orange*), a competitive inhibitor of glutamate. (*b*) Side view along one of the twofold axes showing only the six nearest subunits. The molecule extends 103 Å along the sixfold axis, which is vertical in this view. [Based on an X-ray structure by David Eisenberg, UCLA. PDBid 1FPY.]

20-28), is regulated by several allosteric effectors as well as by covalent modification. Several aspects of its control system bear note. *Nine allosteric feedback inhibitors, each with its own binding site, control the activity of bacterial glutamine synthetase in a cumulative manner.* Six of these effectors— histidine, tryptophan, carbamoyl phosphate (as synthesized by carbamoyl phosphate synthetase II), glucosamine-6-phosphate, AMP, and CTP—are all end products of pathways leading from glutamine. The other three— alanine, serine, and glycine—reflect the cell's nitrogen level.

E. coli glutamine synthetase is covalently modified by **adenylylation** (addition of an AMP group) of a specific Tyr residue (Fig. 20-29). The enzyme's susceptibility to cumulative feedback inhibition increases, and its activity therefore decreases, with its degree of adenylylation. The level of adenylylation is controlled by a complex metabolic cascade that is conceptually similar to that controlling glycogen phosphorylase (Section 15-3B). Both adenylylation and deadenylylation of glutamine synthetase are catalyzed by **adenylyltransferase** in complex with a tetrameric regulatory protein, **P_{II}.** This complex deadenylylates glutamine synthetase when P_{II} is **uridylylated** (also at a Tyr residue) and adenylylates glutamine synthetase when P_{II} lacks UMP residues. The level of P_{II} uridylylation, in turn, depends on the relative levels of two enzymatic activities located on the same protein: a **uridylyltransferase** that uridylylates P_{II} and a **uridylyl-removing enzyme** that hydrolytically excises the attached UMP groups of P_{II}. The uridylyltransferase is activated by α-ketoglutarate and ATP and inhibited by glutamine and P_i, whereas uridylyl-removing enzyme is insensitive to these metabolites. This intricate metabolic cascade therefore renders the activity of *E. coli* glutamine synthetase extremely responsive to the cell's nitrogen requirements.

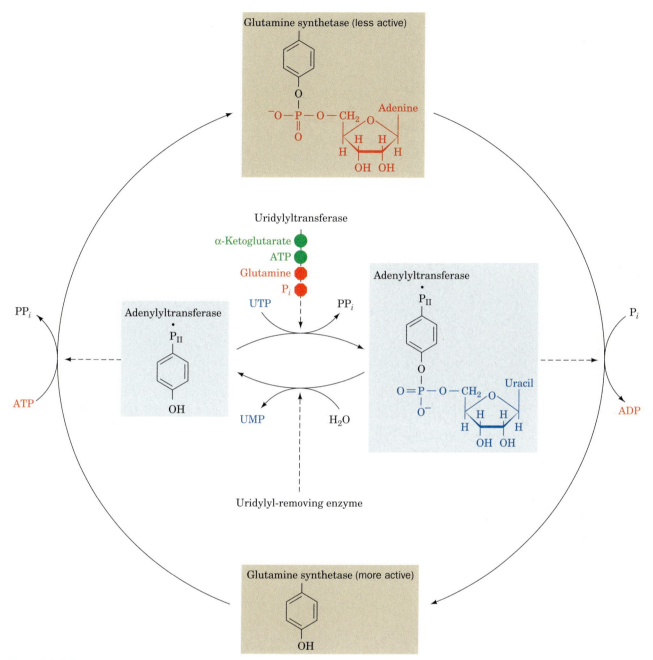

Figure 20-29 The regulation of bacterial glutamine synthetase. The adenylylation/deadenylylation of a specific Tyr residue is controlled by the level of uridylylation of a specific adenylyltransferase·P_{II} Tyr residue. This uridylylation level, in turn, is controlled by the relative activities of uridylyltransferase, which is sensitive to the levels of a variety of nitrogen metabolites, and uridylyl-removing enzyme, whose activity is independent of these metabolite levels.

Glutamate Is the Precursor of Proline, Ornithine, and Arginine. Conversion of glutamate to proline (Fig. 20-30, Reactions 1–4) involves the reduction of the γ-carboxyl group to an aldehyde followed by the formation of an internal Schiff base whose further reduction yields proline. Reduction of the glutamate γ-carboxyl group to an aldehyde is an endergonic process that is facilitated by first phosphorylating the carboxyl group in a reaction catalyzed by **γ-glutamyl kinase.** The unstable product, **glutamate-5-phosphate,** has not been isolated from reaction mixtures but is presumed to be the sub-

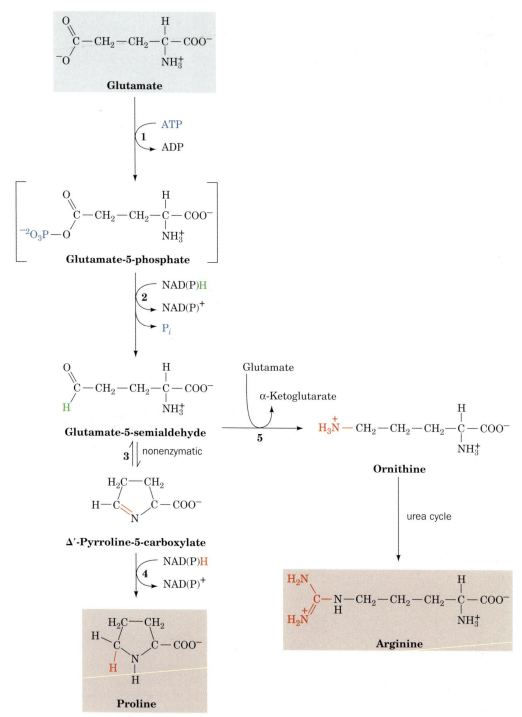

Figure 20-30 **The biosynthesis of the glutamate family of amino acids: arginine, ornithine, and proline.** The catalysts for proline biosynthesis are (**1**) γ-glutamyl kinase, (**2**) dehydrogenase, (**3**) nonenzymatic, and (**4**) pyrroline- 5-carboxylate reductase. In mammals, ornithine is produced from glutamate-5-semialdehyde by the action of ornithine-δ-aminotransferase (**5**). Ornithine is converted to arginine via the urea cycle (Section 20-3A).

strate for the reduction that follows. The resulting **glutamate-5-semialdehyde** (which is also a product of arginine and proline degradation; Fig. 20-17) cyclizes spontaneously to form the internal Schiff base **Δ¹-pyrroline-5-carboxylate.** The final reduction to proline is catalyzed by **pyrroline-**

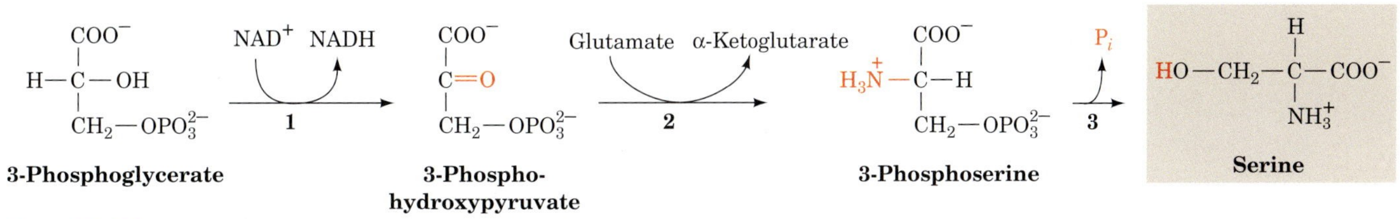

Figure 20-31 The conversion of 3-phosphoglycerate to serine. The pathway enzymes are (**1**) 3-phosphoglycerate dehydrogenase, (**2**) a PLP-dependent aminotransferase, and (**3**) phosphoserine phosphatase.

5-carboxylate reductase. Whether the enzyme requires NADH or NADPH is unclear.

In humans, a three-step pathway leads from glutamate to ornithine via a branch from proline biosynthesis after Step 2 (Fig. 20-30). Glutamate-5-semialdehyde, which is in equilibrium with Δ^1-pyrroline-5-carboxylate, is directly transaminated to yield ornithine in a reaction catalyzed by **ornithine-δ-amino transferase** (Fig. 20-30, Reaction 5). Ornithine is converted to arginine by the reactions of the urea cycle (Fig. 20-9).

Serine, Cysteine, and Glycine Are Derived from 3-Phosphoglycerate.

Serine is formed from the glycolytic intermediate 3-phosphoglycerate in a three-reaction pathway (Fig. 20-31):

1. Conversion of 3-phosphoglycerate's 2-OH group to a ketone, yielding **3-phosphohydroxypyruvate,** serine's phosphorylated keto acid analog.
2. Transamination of 3-phosphohydroxypyruvate to phosphoserine.
3. Hydrolysis of phosphoserine to serine.

Serine participates in glycine synthesis in two ways:

1. Direct conversion of serine to glycine by serine hydroxymethyltransferase in a reaction that also yields N^5,N^{10}-methylene-THF (Fig. 20-14, Reaction 4 in reverse).
2. Condensation of the N^5,N^{10}-methylene-THF with CO_2 and NH_4^+ by glycine synthase (Fig. 20-14, Reaction 3 in reverse).

In animals, cysteine is synthesized from serine and homocysteine, a breakdown product of methionine (Fig. 20-18, Reactions 5 and 6). Homocysteine combines with serine to yield cystathionine, which subsequently forms cysteine and α-ketobutyrate. Since cysteine's sulfhydryl group is derived from the essential amino acid methionine, cysteine can be considered to be an essential amino acid.

B Biosynthesis of the Essential Amino Acids

Essential amino acids, like nonessential amino acids, are synthesized from familiar metabolic precursors. Their synthetic pathways are present only in microorganisms and plants, however, and usually involve more steps than those of the nonessential amino acids. The enzymes that synthesize essential amino acids were apparently lost early in animal evolution, possibly because of the ready availability of these amino acids in the diet. We shall focus on only a few of the many reactions in the biosynthesis of essential amino acids.

The Aspartate Family: Lysine, Methionine, and Threonine.   In bacteria, aspartate is the common precursor of lysine, methionine, and threonine (Fig. 20-32). The biosyntheses of these essential amino acids all begin with the **aspartokinase**-catalyzed phosphorylation of aspartate to yield **aspartyl-β-phosphate.** We have seen that the control of metabolic pathways commonly occurs at the first committed step of the pathway. One might therefore expect lysine, methionine, and threonine biosynthesis to be controlled as a group. Each of these pathways is, in fact, independently controlled. *E. coli* has three isozymes of aspartokinase that respond differently to the three amino acids in terms both of feedback inhibition of enzyme activity and repression of enzyme synthesis. In addition, the pathway direction is controlled by feedback inhibition at the branch points by the amino acid products of the branches.

Methionine synthase (alternatively **homocysteine methyltransferase**) catalyzes the methylation of homocysteine to form methionine using N^5-methyl-THF as its methyl group donor (Reaction 4 in both Figs. 20-18 and

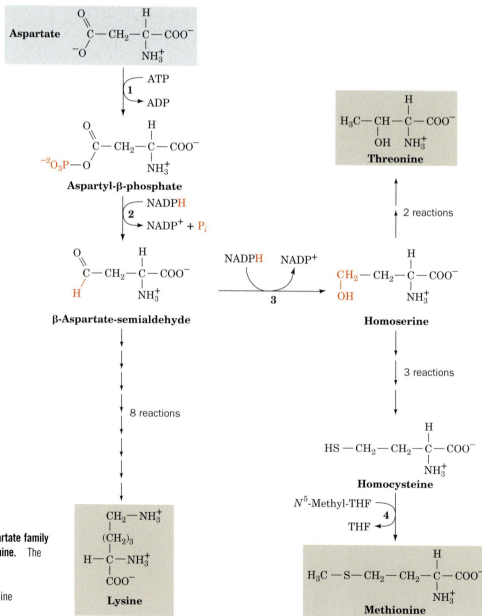

Figure 20-32 The biosynthesis of the aspartate family of amino acids: lysine, methionine, and threonine.   The pathway enzymes shown are (**1**) aspartokinase, (**2**) β-aspartate semialdehyde dehydrogenase, (**3**) homoserine dehydrogenase, and (**4**) methionine synthase (a coenzyme B_{12}–dependent enzyme).

20-32). Methionine synthase is the only coenzyme B_{12}–associated enzyme in mammals besides methylmalonyl-CoA mutase (Section 19-2E). However, the coenzyme B_{12}'s Co ion in methionine synthase is axially liganded by a methyl group to form **methylcobalamin** rather than by a 5'-adenosyl group as in methylmalonyl-CoA mutase (Fig. 19-17). In mammals, the primary function of methionine synthase is not *de novo* methionine synthesis, as Met is an essential amino acid. Instead, it functions in the cyclic synthesis of SAM for use in biological methylations (Fig. 20-18).

The Pyruvate Family: Leucine, Isoleucine, and Valine. Valine and isoleucine follow the same biosynthetic pathway utilizing pyruvate as a starting reactant, the only difference being in the first step of the series (Fig. 20-33). In this thiamine pyrophosphate–dependent reaction, which resembles those

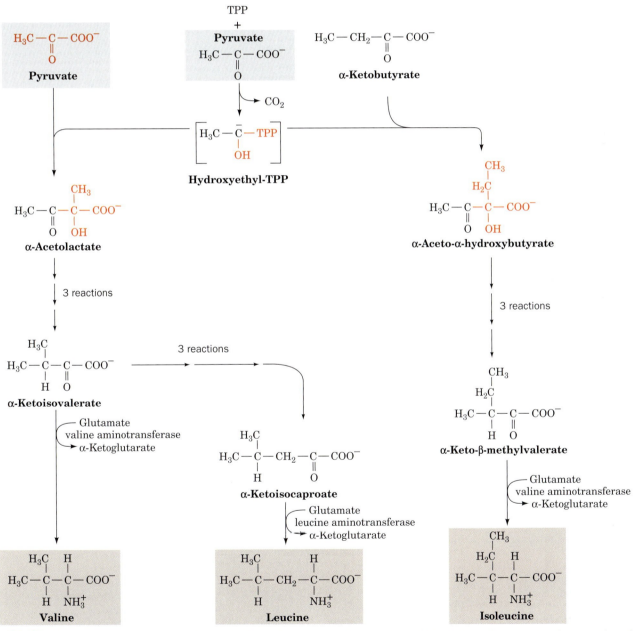

Figure 20-33 **The biosynthesis of the pyruvate family of amino acids: isoleucine, leucine, and valine.** The first enzyme, **acetolactate synthase** (a TPP enzyme), catalyzes two reactions, one leading to valine and leucine, and the other to isoleucine. Note also that **valine aminotransferase** catalyzes the formation of both valine and isoleucine from their respective α-keto acids.

catalyzed by pyruvate decarboxylase (Fig. 14-20) and transketolase (Fig. 14-32), pyruvate forms an adduct with TPP that is decarboxylated to hydroxyethyl-TPP. This resonance-stabilized carbanion adds either to the keto group of a second pyruvate to form **acetolactate** on the way to valine, or to the keto group of **α-ketobutyrate** to form **α-aceto-α-hydroxybutyrate** on the way to isoleucine. The leucine biosynthetic pathway branches off

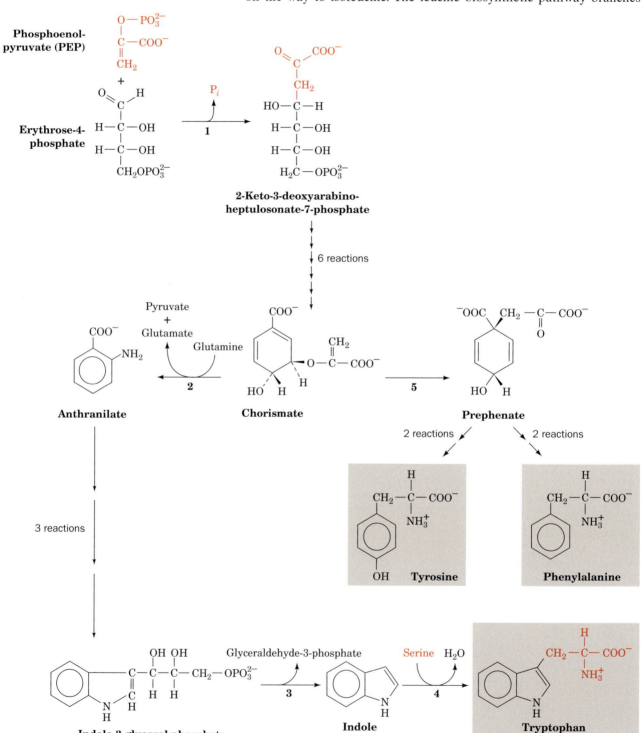

Figure 20-34   The biosynthesis of phenylalanine, tryptophan, and tyrosine. The enzymes shown are (**1**) 2-keto-3-deoxy-D-arabinoheptulosonate-7-phosphate synthase, (**2**) anthranilate synthase, (**3**) tryptophan synthase, α subunit, (**4**) tryptophan synthase, β subunit (a PLP-dependent enzyme), and (**5**) chorismate mutase.

from the valine pathway. The final step in each of the three pathways, which begin with pyruvate rather than an amino acid, is the PLP-dependent transfer of an amino group from glutamate to form the amino acid.

The Aromatic Amino Acids: Phenylalanine, Tyrosine, and Tryptophan.   The precursors of the aromatic amino acids are the glycolytic intermediate phosphoenolpyruvate (PEP) and erythrose-4-phosphate (an intermediate in the pentose phosphate pathway; Fig. 14-30). Their condensation forms **2-keto-3-deoxy-D-arabinoheptulosonate-7-phosphate** (Fig. 20-34). This C_7 compound cyclizes and is ultimately converted to **chorismate,** the branch point for tryptophan synthesis. Chorismate is converted to either **anthranilate** and then to tryptophan, or to **prephenate** and on to tyrosine or phenylalanine. Although mammals synthesize tyrosine by the hydroxylation of phenylalanine (Fig. 20-24), many microorganisms synthesize it directly from prephenate. The last step in the synthesis of tyrosine and phenylalanine is the addition of an amino group through transamination. In tryptophan synthesis, the amino group is part of the serine molecule that is added to **indole.**

Indole Is Channeled between Two Active Sites in Tryptophan Synthase.   The final two reactions of tryptophan biosynthesis (Reactions 3 and 4 in Fig. 20-34) are both catalyzed by **tryptophan synthase:**

1. The α subunit (268 residues) of this $\alpha_2\beta_2$ bifunctional enzyme cleaves **indole-3-glycerol phosphate,** yielding indole and glyceraldehyde-3-phosphate.

2. The β subunit (396 residues) joins indole with serine in a PLP-dependent reaction to form tryptophan.

Either subunit alone is enzymatically active, but when they are joined in the $\alpha_2\beta_2$ tetramer, the rates of both reactions and their substrate affinities increase by 1 to 2 orders of magnitude. Indole, the intermediate product, does not appear free in solution because it is channeled between the two subunits.

The X-ray structure of tryptophan synthase from *Salmonella typhimurium,* determined by Craig Hyde, Edith Miles, and David Davies, reveals that the protein forms a 150-Å-long, twofold symmetric α-β-β-α complex in which the active sites of neighboring α and β subunits are separated by ~25 Å (Fig. 20-35). *These active sites are joined by a solvent-filled tunnel that is wide enough to permit the passage of the intermediate substrate, indole.* This structure, the first in which a tunnel connecting two active sites was observed, suggests the following series of events: The indole-3-

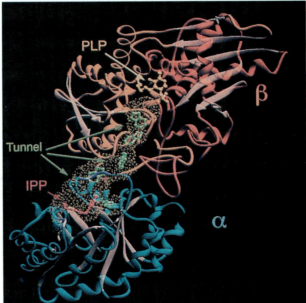

Figure 20-35 A ribbon diagram of the bifunctional enzyme tryptophan synthase from *S. typhimurium.*   Only one αβ subunit of the bifunctional αββα heterotetramer is shown. The α subunit is blue, the β subunit's N-terminal domain is gold, its C-terminal domain is red-orange, and all β sheets are tan. The active site of the α subunit is located by its bound competitive inhibitor, **indolepropanol phosphate** (IPP; *red ball-and-stick model*), whereas that of the β subunit is marked by its PLP coenzyme (*yellow ball-and-stick model*). The solvent-accessible surface of the ~25-Å-long tunnel connecting the α and β active sites is outlined by yellow dots. Several indole molecules (*green ball-and-stick models*) have been modeled into the tunnel to show that it is wide enough for indole to pass from one active site to the other. [Courtesy of Craig Hyde, Edith Miles, and David Davies, NIH.] ☙ **See the Interactive Exercises.**

glycerol phosphate substrate binds to the α subunit through an opening into its active site, its "front door," and the glyceraldehyde-3-phosphate product leaves via the same route. Similarly, the β subunit active site has a "front door" opening to the solvent through which serine enters and tryptophan leaves. Both active sites also have "back doors" that are connected by the tunnel. *The indole intermediate presumably diffuses between the two active sites via the tunnel and hence does not escape to the solvent.* Channeling may be particularly important for indole since this nonpolar molecule otherwise can escape the bacterial cell by diffusing through its plasma and outer membranes.

In order for channeling to increase tryptophan synthase's catalytic efficiency, (1) its connected active sites must be coupled such that their catalyzed reactions occur in phase, and (2) after substrate has bound to the α subunit, its active site ("front door") must close off to ensure that the product indole passes through the tunnel ("back door") to the β subunit rather than escaping into solution. A variety of experimental evidence indicates that this series of events is facilitated through allosteric signals derived from covalent transformations at the β subunit's active site. These switch the enzyme between an open, low-activity conformation to which substrates bind, and a closed, high-activity conformation from which indole cannot escape.

Histidine Biosynthesis. Five of histidine's six C atoms are derived from **5-phosphoribosyl-α-pyrophosphate** (**PRPP;** Fig. 20-36), a phospho-sugar intermediate that is also involved in the biosynthesis of purine and pyrimidine nucleotides (Sections 22-1A and 22-2A). The histidine's sixth carbon originates from ATP. The ATP atoms that are not incorporated into histidine are eliminated as **5-aminoimidazole-4-carboxamide ribonucleotide** (Fig. 20-36, Reaction 2), which is also an intermediate in purine biosynthesis.

The unusual biosynthesis of histidine from a purine (N^1-**5′-phosphoribosyl ATP,** the product of Reaction 1 in Fig. 20-36) has been cited as evidence supporting the hypothesis that life was originally RNA-based. His residues, as we have seen, are often components of enzyme active sites, where they act as nucleophiles and/or general acid–base catalysts. The discovery that RNA can have catalytic properties therefore suggests that the imidazole moiety of purines plays a similar role in these RNA enzymes. This further suggests that the histidine biosynthetic pathway is a "fossil" of the transition to more efficient protein-based life-forms.

6 Other Products of Amino Acid Metabolism

Certain amino acids, in addition to their major function as protein building blocks, are essential precursors of a variety of important biomolecules, including nucleotides and nucleotide coenzymes, heme, and various hormones and neurotransmitters. In this section, we consider the pathways leading to some of these substances. The biosynthesis of nucleotides is considered in Chapter 22.

A Heme Biosynthesis and Degradation

Heme, as we have seen, is an Fe-containing prosthetic group that is an essential component of many proteins, notably hemoglobin, myoglobin, and the cytochromes. The initial reactions of heme biosynthesis are common to the formation of other tetrapyrroles including chlorophyll in plants and bacteria (Fig. 18-2) and coenzyme B_{12} in bacteria (Fig. 19-17).

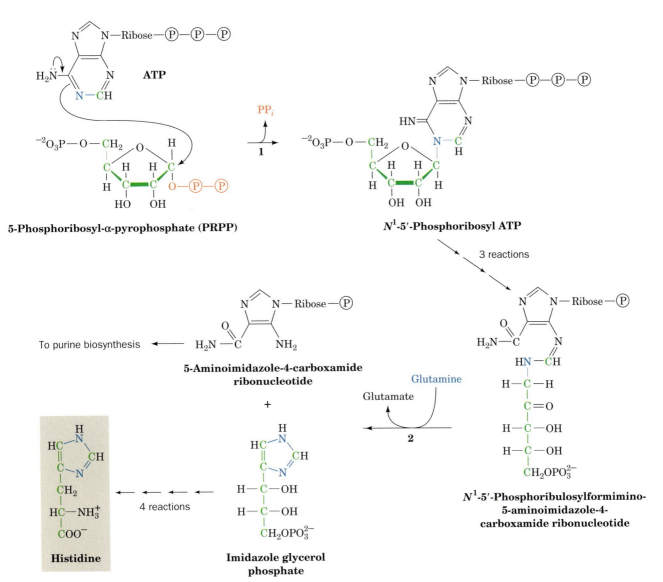

Figure 20-36   The biosynthesis of histidine.   The enzymes shown are (1) ATP phosphoribosyltransferase and (2) glutamine amidotransferase.

Elucidation of the heme biosynthetic pathway involved some interesting detective work. David Shemin and David Rittenberg, who were among the first to use isotopic tracers in the elucidation of metabolic pathways, demonstrated, in 1945, that *all of heme's C and N atoms can be derived from acetate and glycine.* Heme biosynthesis takes place partly in the mitochondrion and partly in the cytosol (Fig. 20-37). Mitochondrial acetate is metabolized via the citric acid cycle to succinyl-CoA, which condenses with glycine in a reaction that produces CO_2 and **δ-aminolevulinic acid (ALA).** ALA is transported to the cytosol, where it combines with a second ALA to yield **porphobilinogen (PBG).** The reaction is catalyzed by the Zn-requiring enzyme **porphobilinogen synthase.**

Inhibition of PBG synthase by lead is one of the major manifestations of acute lead poisoning. Indeed, it has been suggested that the accumulation, in the blood, of ALA, which resembles the neurotransmitter **γ-aminobutyric acid** (Section 20-6B), is responsible for the psychosis that often accompanies lead poisoning.

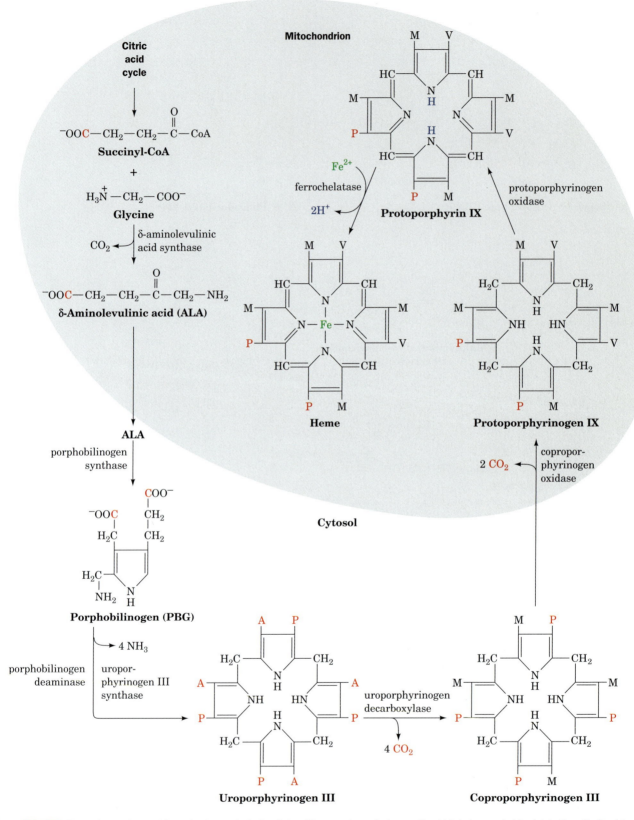

Figure 20-37 The pathway of heme biosynthesis.   δ-Aminolevulinic acid (ALA) is synthesized in the mitochondrion from succinyl-CoA and glycine by ALA synthase. ALA is transported to the cytosol, where two molecules condense to form PBG, four molecules of which condense to form a porphyrin ring. The next three reactions involve oxidation of the pyrrole ring substituents, yielding protoporphyrinogen IX, which is transported back into the mitochondrion during its formation. After oxidation of the methylene groups, ferrochelatase catalyzes the insertion of Fe^{2+} to yield heme. A, P, M, and V, respectively, represent acetyl, propionyl, methyl, and vinyl (—CH=CH$_2$) groups. C atoms originating as the carboxyl group of acetate are red.

The next phase of heme biosynthesis is the condensation of four PBG molecules to form **uroporphyrinogen III,** the porphyrin nucleus, in a series of reactions catalyzed by **porphobilinogen deaminase** (also called **uroporphyrinogen synthase**) and **uroporphyrinogen III cosynthase.** The initial product, **hydroxymethylbilane,**

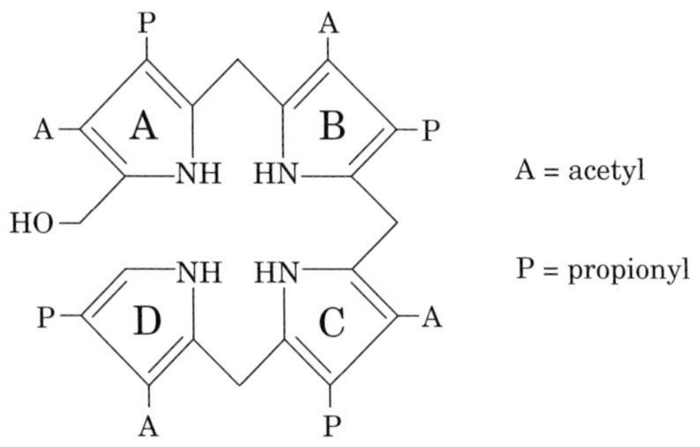

A = acetyl

P = propionyl

Hydroxymethylbilane

is a linear tetrapyrrole that cyclizes. **Protoporphyrin IX,** to which Fe is added to form heme, is produced from uroporphyrinogen III in a series of reactions catalyzed by (1) **uroporphyrinogen decarboxylase,** which decarboxylates all four acetate side chains (A) to form methyl groups (M); (2) **coproporphyrinogen oxidase,** which oxidatively decarboxylates two of the propionate side chains (P) to vinyl groups (V); and (3) **protoporphyrinogen oxidase,** which oxidizes the methylene groups linking the pyrrole rings to methenyl groups. During the coproporphyrinogen oxidase reaction, the porphyrin is transported back into the mitochondrion. In the final reaction of heme biosynthesis, **ferrochelatase** inserts Fe(II) into protoporphyrin IX.

Regulation of Heme Biosynthesis.

The two major sites of heme biosynthesis are erythroid cells, which synthesize ~85% of the body's heme groups, and the liver, which synthesizes most of the remainder. In liver, the level of heme synthesis must be adjusted according to metabolic conditions. For example, the synthesis of the heme-containing cytochrome P450 (Section 12-4D) fluctuates with the need for detoxification. In contrast, heme synthesis in erythroid cells is a one-time event; heme and protein synthesis cease when the cell matures, so that the hemoglobin must last the erythrocyte's lifetime (~120 days).

In liver, the main control target in heme biosynthesis is ALA synthase, the enzyme catalyzing the pathway's first committed step. Heme, or its Fe(III) oxidation product **hemin,** controls this enzyme's activity through feedback inhibition, inhibition of the transport of ALA synthase from its site of synthesis in the cytosol to its reaction site in the mitochondrion (Fig. 20-37), and repression of ALA synthase synthesis.

In erythroid cells, heme exerts quite a different effect on its biosynthesis. Heme induces, rather than represses, protein synthesis in **reticulocytes** (immature erythrocytes). Although the vast majority of the protein synthesized by reticulocytes is globin, heme may also induce these cells to synthesize the enzymes of heme biosynthesis. Moreover, the rate-determining steps of heme biosynthesis in erythroid cells may be the ferrochelatase and porphobilinogen deaminase reactions rather than the ALA synthase reaction. This is consistent with the supposition that when erythroid heme biosynthesis is "switched on," all of its steps function at their maximal rates rather than any one step limiting the flow through the pathway. Heme-

BOX 20-3

Biochemistry in Health and Disease

The Porphyrias

Defects in heme biosynthesis in liver or erythroid cells result in the accumulation of porphyrin and/or its precursors and are therefore known as porphyrias. Two such defects are known to affect erythroid cells: uroporphyrinogen III cosynthase deficiency (**congenital erythropoietic porphyria**) and ferrochelatase deficiency (**erythropoietic protoporphyria**). The former results in accumulation of uroporphyrinogen derivatives. Excretion of these compounds colors the urine red; their deposition in the teeth turns them reddish brown; and their accumulation in the skin renders it extremely photosensitive, so that it ulcerates and forms disfiguring scars. Increased hair growth is also observed in afflicted individuals; fine hair may cover much of the face and extremities. These symptoms have prompted speculation that the werewolf legend has a biochemical basis.

The most common porphyria that primarily affects liver is porphobilinogen deaminase deficiency (**acute intermittent porphyria**). This disease is marked by intermittent attacks of abdominal pain and neurological dysfunction. Excessive amounts of ALA and PBG are excreted in the urine during and after such attacks. The urine may become red resulting from the excretion of excess porphyrins synthesized from PBG in nonhepatic cells, although the skin does not become unusually photosensitive. King George III, who ruled England during the American Revolution, and who has been widely portrayed as being mad, in fact, had attacks characteristic of acute intermittent porphyria; he was reported to have urine the color of port wine, and had several descendants who were diagnosed as having this disease. American history might have been quite different had George III not inherited this metabolic defect.

stimulated synthesis of globin also ensures that heme and globin are synthesized in the correct ratio for assembly into hemoglobin (Section 27-3C). Genetic defects in heme biosynthesis cause conditions known as **porphyrias** (see Box 20-3).

Heme Degradation. At the end of their lifetime, red cells are removed from the circulation and their components degraded. Heme catabolism (Fig. 20-38) begins with oxidative cleavage, by heme oxygenase, of the porphyrin between rings A and B to form **biliverdin,** a green linear tetrapyrrole. Biliverdin's central methenyl bridge (between rings C and D) is then reduced to form the red-orange **bilirubin.** The changing colors of a healing bruise are a visible manifestation of heme degradation.

In the reaction forming biliverdin, the methenyl bridge carbon between rings A and B is released as CO, which is a tenacious heme ligand (with 200-fold greater affinity for hemoglobin than O_2; Section 7-1A). Consequently, ~1% of hemoglobin's binding sites are blocked by CO even in the absence of air pollution.

The highly lipophilic bilirubin is insoluble in aqueous solutions. Like other lipophilic metabolites, such as free fatty acids, it is transported in the blood in complex with serum albumin. Bilirubin derivatives are secreted in the bile and for the most part are further degraded by bacterial enzymes in the large intestine. Some of the resulting **urobilinogen** is reabsorbed and transported via the bloodstream to the kidney, where it is converted to the yellow **urobilin** and excreted, thus giving urine its characteristic color. Most urobilinogen, however, is microbially converted to the deeply red-brown **stercobilin,** the major pigment of feces.

When the blood contains excessive amounts of bilirubin, the deposition of this highly insoluble substance colors the skin and the whites of the eyes yellow. This condition, called **jaundice** (French: *jaune,* yellow), signals either an abnormally high rate of red cell destruction, liver dysfunction, or bile duct obstruction. Newborn infants, particularly when premature, often

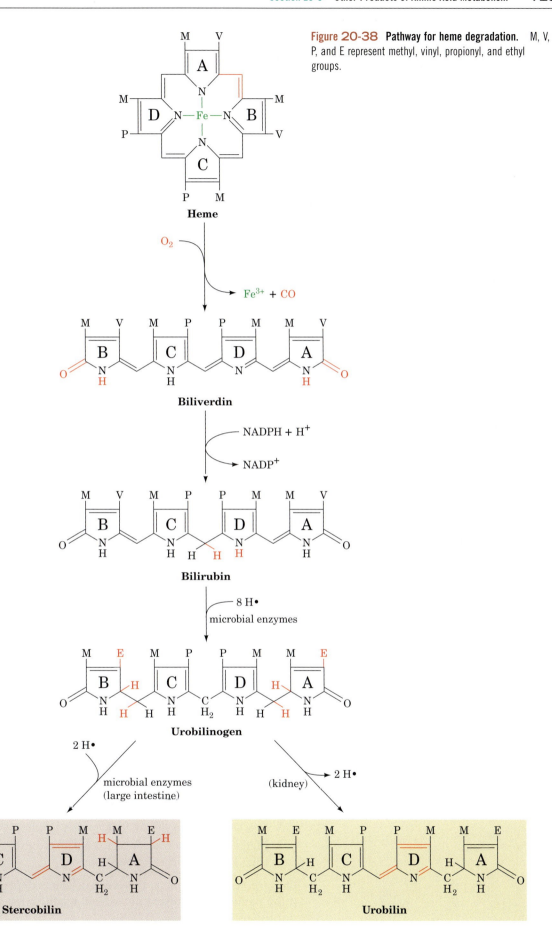

Figure 20-38   Pathway for heme degradation.   M, V, P, and E represent methyl, vinyl, propionyl, and ethyl groups.

become jaundiced because they lack an enzyme that degrades bilirubin. Jaundiced infants are treated by bathing them with light from a fluorescent lamp; this photochemically converts bilirubin to more soluble isomers that the infant can degrade and excrete.

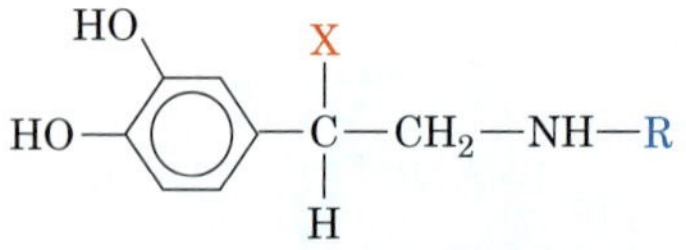

B Biosynthesis of Physiologically Active Amines

Epinephrine (adrenalin), norepinephrine, dopamine, serotonin (5-hydroxytryptamine), γ-aminobutyric acid (GABA), and **histamine** *(at left)* are hormones and/or neurotransmitters derived from amino acids. Epinephrine, as we have seen, activates muscle adenylate cyclase, thereby stimulating glycogen breakdown (Section 15-3B); deficiency in dopamine production is associated with **Parkinson's disease,** a degenerative condition causing "shaking palsy"; serotonin causes smooth muscle contraction; GABA is one of the brain's major inhibitory neurotransmitters; and histamine is involved in allergic responses (as allergy sufferers who take antihistamines will realize), as well as in the control of acid secretion by the stomach.

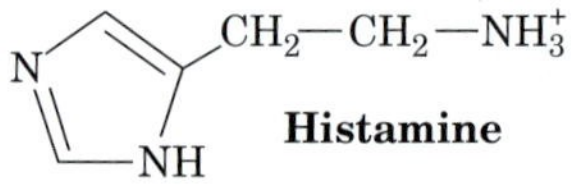

The biosynthesis of each of these physiologically active amines involves decarboxylation of the corresponding precursor amino acid. Amino acid decarboxylases are PLP-dependent enzymes that form a PLP–Schiff base with the substrate so as to stabilize (by delocalization) the C_α carbanion formed on C_α—COO^- bond cleavage (Section 20-2A):

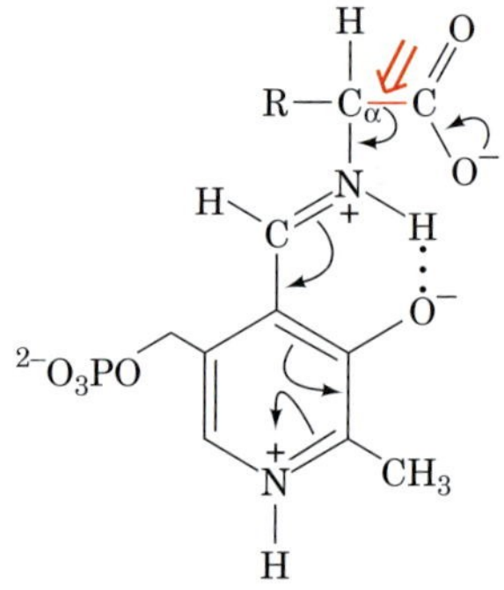

Formation of histamine (from histidine) and formation of GABA (from glutamate) are one-step processes; the synthesis of serotonin from tryptophan requires a hydroxylation step as well as decarboxylation. The various **catecholamines**—dopamine, norepinephrine, and epinephrine—are related to **catechol**

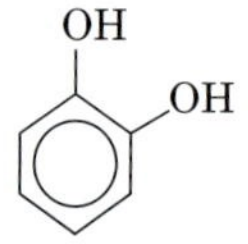

Catechol

and are sequentially synthesized from tyrosine (Fig. 20-39):

1. Tyrosine is hydroxylated to **3,4-dihydroxyphenylalanine (L-DOPA)** in a reaction that requires 5,6,7,8-tetrahydrobiopterin (Fig. 20-26).
2. L-DOPA is decarboxylated to dopamine.
3. A second hydroxylation yields norepinephrine.
4. Methylation of norepinephrine's amino group by *S*-adenosylmethionine (Fig. 20-18) produces epinephrine.

The specific catecholamine that a cell produces depends on which enzymes of the pathway are present. In adrenal medulla, for example, epinephrine

X = OH, R = CH₃   **Epinephrine (Adrenalin)**
X = OH, R = H   **Norepinephrine**
X = H,   R = H   **Dopamine**

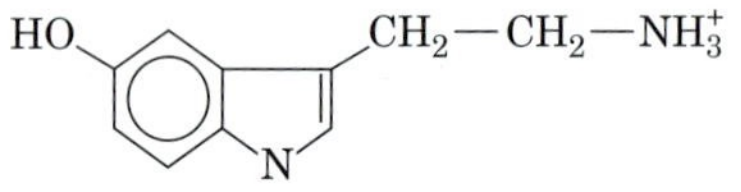

Serotonin
(5-hydroxytryptamine)

$^-OOC—CH_2—CH_2—CH_2—NH_3^+$

γ-Aminobutyric acid (GABA)

Histamine

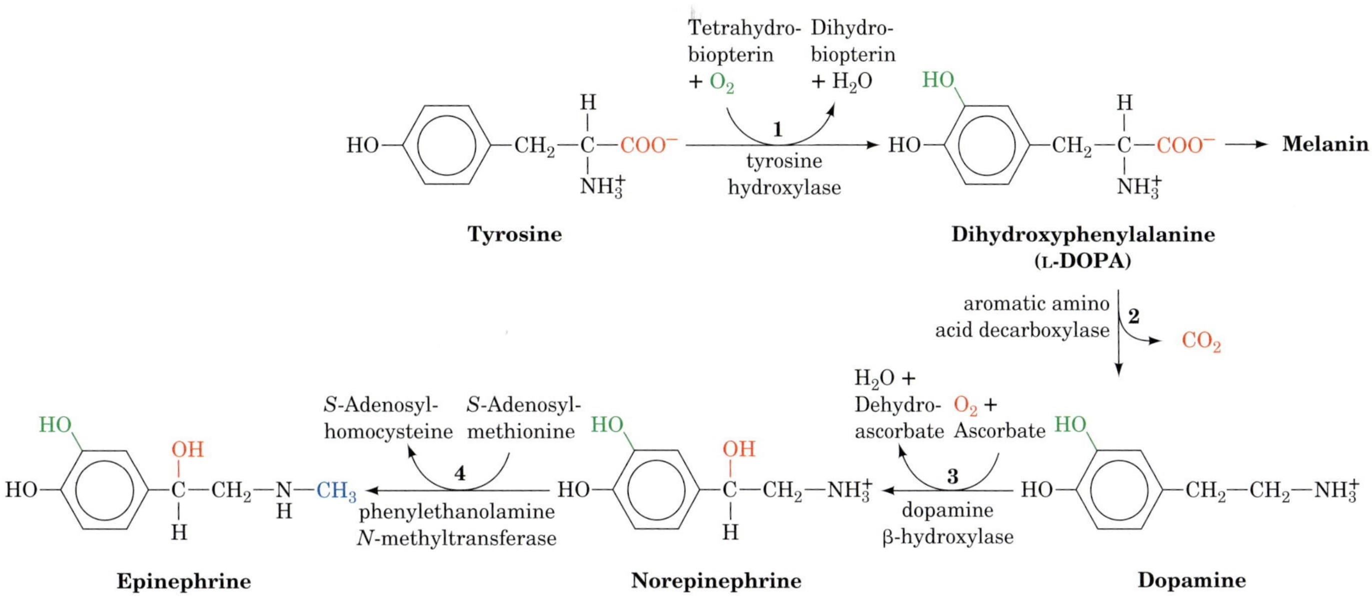

Figure 20-39 The sequential synthesis of L-DOPA, dopamine, norepinephrine, and epinephrine from tyrosine. L-DOPA is also the precursor of the skin pigment melanin.

is the predominant product. In some areas of the brain, norepinephrine is more common. In other areas, the pathway stops at dopamine. In melanocytes, L-DOPA is the precursor of red and black melanins, which are irregular cross-linked polymers that give hair and skin much of their color.

C Nitric Oxide

Arginine is the precursor of a substance that was originally called **endothelium-derived relaxing factor (EDRF)** because it was synthesized by vascular endothelial cells and caused the underlying smooth muscle to relax. The signal for vasodilation was not a peptide, as expected, but the stable free radical **nitric oxide, NO.** The identification of NO as a vasodilator came in part from studies that identified NO as the decomposition product that mediates the vasodilating effects of compounds such as **nitroglycerin** *(at right)*. Nitroglycerin is often administered to individuals suffering from **angina pectoris,** a disease caused by insufficient blood flow to the heart muscle, to rapidly but temporarily relieve their chest pain.

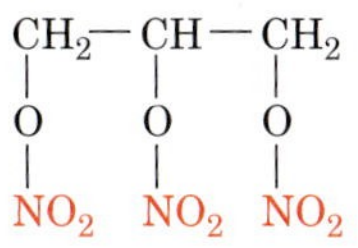

Nitroglycerin

The reaction that converts arginine to NO and citrulline is catalyzed by **nitric oxide synthase (NOS):**

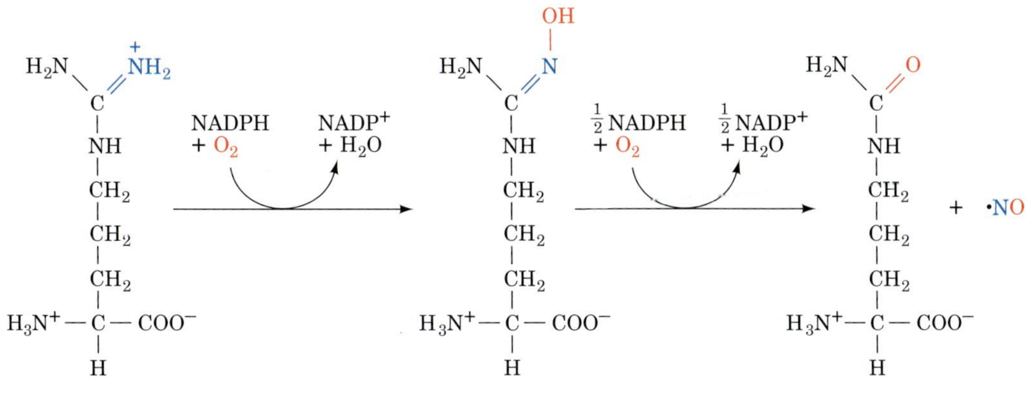

L-Arginine **L-Hydroxyarginine** **L-Citrulline**

The reaction proceeds via an enzyme-bound hydroxyarginine intermediate and requires an array of redox coenzymes. NOS is a homodimeric protein of 125- to 160-kD subunits, and each subunit contains one FMN, one FAD, one tetrahydrobiopterin (Fig. 20-26), and one Fe(III)-heme. These cofactors facilitate the five-electron oxidation of arginine to produce NO.

Because NO is a gas, it rapidly diffuses across cell membranes, although its high reactivity (half-life ~5 s) prevents it from acting much further than ~1 mm from its site of synthesis. NO is produced by endothelial cells in response to a wide variety of agents and physiological conditions. Neuronal cells also synthesize NO (neuronal NOS is ~55% homologous to endothelial NOS). This endothelium-independent NO synthesis dilates cerebral and other arteries and is responsible for penile erection. The brain contains more NOS than any other tissue in the body, suggesting that NO is essential for the function of the central nervous system. A third type of NOS is found in leukocytes (white blood cells). These cells produce NO as part of their cytotoxic arsenal. NO combines with superoxide (Section 17-4C) to produce the highly reactive hydroxyl radical, which kills invading bacteria. The sustained release of NO has been implicated in **endotoxic shock** (an often fatal immune system overreaction to bacterial infection), in inflammation-related tissue damage, and in the damage to neurons in the vicinity of but not directly killed by a stroke (which often does greater harm than the stroke itself).

7 Nitrogen Fixation

The most prominent chemical elements in living systems are O, H, C, N, and P. The elements O, H, and P occur widely in metabolically available forms (H_2O, O_2, and P_i). However, the major forms of C and N, CO_2 and N_2, are extremely stable (unreactive); for example, the $N{\equiv}N$ triple bond has a bond energy of 945 kJ·mol^{-1} (versus 351 kJ·mol^{-1} for a C—O single bond). CO_2, with minor exceptions, is metabolized (fixed) only by photosynthetic organisms (Chapter 18). *N_2 fixation is even less common; this element is converted to metabolically useful forms by only a few strains of bacteria, called **diazotrophs.*** These organisms include certain marine cyanobacteria and bacteria that colonize the root nodules of legumes (plants belonging to the pea family, including beans, clover, and alfalfa; Fig. 20-40).

Figure 20-40 Root nodules of a pea plant. Bacteria of the genus *Rhizobium*, which carry out nitrogen fixation, live symbiotically with the root nodules of such legumes. [Vu/Cabisco/Visuals Unlimited.]

Diazotrophs produce the enzyme **nitrogenase,** which catalyzes the reduction of N_2 to NH_3:

$$N_2 + 8\ H^+ + 8\ e^- + 16\ ATP + 16\ H_2O \rightarrow$$
$$2\ NH_3 + H_2 + 16\ ADP + 16\ P_i$$

In legumes, this nitrogen-fixing system produces more metabolically useful nitrogen than the legume needs; the excess is excreted into the soil, enriching it. It is therefore common agricultural practice to plant a field with alfalfa every few years to build up the supply of usable nitrogen in the soil for later use in growing other crops.

Nitrogenase Contains Novel Redox Centers.   **Nitrogenase,** which catalyzes the reduction of N_2 to NH_3, is a complex of two proteins:

1. The **Fe-protein,** a homodimer that contains one [4Fe–4S] cluster and two ATP binding sites.
2. The **MoFe-protein,** an $\alpha_2\beta_2$ heterotetramer that contains Fe and Mo.

The X-ray structures of *Azotobacter vinelandii* nitrogenase in complex with the inhibitor $ADP \cdot AlF_4^-$ (which mimics the transition state in ATP hydrolysis), determined by Douglas Rees, reveals that each MoFe-protein associates with two molecules of Fe-protein (Fig. 20-41).

Each Fe-protein dimer's single [4Fe–4S] cluster is located in a solvent-exposed cleft between the two subunits and is symmetrically linked to Cys 97 and Cys 132 from both subunits such that an Fe-protein resembles an

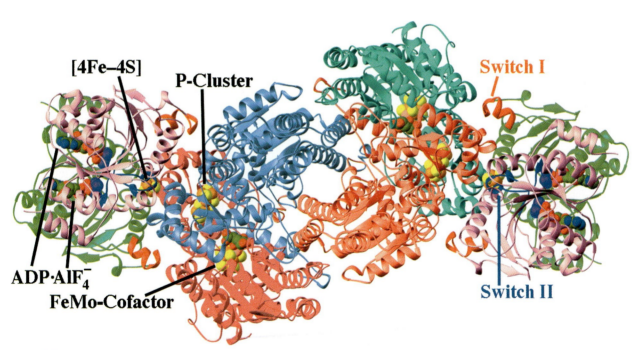

Figure 20-41 X-Ray structure of the *A. vinelandii* nitrogenase in complex with ADP·AlF$_4^-$.   The enzyme, which is viewed along its molecular twofold axis, is an $(\alpha\beta\gamma_2)_2$ heterooctamer in which the β-α-α-β assembly, the MoFe-protein, is flanked by two γ_2 Fe-proteins whose 289-residue subunits are related by local twofold symmetry. The homologous α subunits (*cyan and red;* 491 residues) and β subunits (*light red and light blue;* 522 residues) are related by pseudo-twofold symmetry. The two γ subunits forming each Fe-protein (*pink and green with their Switch I and Switch II segments red and blue*) bind to the MoFe-protein with the twofold axis relating them coincident with the pseudo-twofold axis relating the MoFe-protein's α and β subunits. The $ADP \cdot AlF_4^-$, [4Fe–4S] cluster, FeMo-cofactor, and P-cluster are drawn in space-filling form with C green, N blue, O red, S yellow, Fe orange, Mo pink, and the AlF_4^- ion purple. [Based on an X-ray structure by Douglas Rees, California Institute of Technology. PDBid 1N2C.]

"iron butterfly" with the [4Fe–4S] cluster at its head. Its nucleotide binding sites are located at the interface between its two subunits.

The MoFe-protein's α and β subunits assume similar folds and extensively associate to form a pseudo-twofold symmetric αβ dimer, two of which more loosely associate to form the twofold symmetric $\alpha_2\beta_2$ tetramer (Fig. 20-41). Each αβ dimer has two bound redox centers:

1. The **P-cluster** (Figs. 20-42a,b), which consists of two [4Fe–3S] clusters linked through an additional sulfide ion forming the eighth corner of both of the clusters to make cubane-like structures, and bridged by two Cys thiol ligands, each coordinating one Fe from each cluster. Four additional Cys thiols coordinate the remaining four Fe atoms. The positions of two of the Fe atoms in one of the [4Fe–3S] clusters change on oxidation, rupturing the bonds from these Fe atoms to the linking sulfide ion. These bonds are replaced in the oxidized state by a Ser oxygen ligand to one of the Fe atoms, and by a bond to the amide N of a Cys from the other Fe atom.

2. The **FeMo-cofactor** (Fig. 20-42c), which consists of a [4Fe–3S] cluster and a [1Mo–3Fe–3S] cluster bridged by three sulfide ions. The FeMo-cofactor's Mo atom is approximately octahedrally coordinated by three cofactor sulfurs, a His imidazole nitrogen, and two oxygens from a bound **homocitrate** ion:

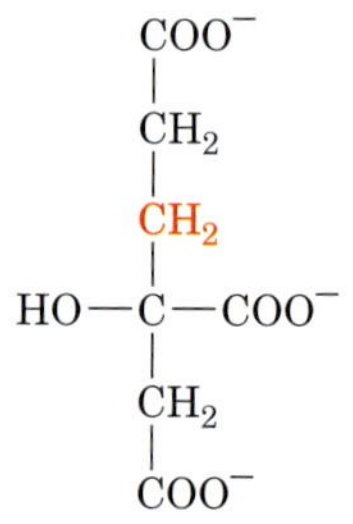

$$
\begin{array}{c}
COO^- \\
| \\
CH_2 \\
| \\
CH_2 \\
| \\
HO-C-COO^- \\
| \\
CH_2 \\
| \\
COO^-
\end{array}
$$

Homocitrate

(an essential component of the FeMo-cofactor). The FeMo-cofactor contains a central cavity that a high resolution X-ray structure of *A. vinelandii* MoFe-protein, also determined by Rees, reveals contains what most probably is a nitrogen atom (although a C or an O atom cannot be ruled out). This putative N atom is liganded to the FeMo-cofactor's central six Fe atoms such that it completes the approximate tetrahedral coordination environment of each of these Fe atoms.

ATP Hydrolysis Triggers Conformational Changes in Nitrogenase. Nitrogen fixation by nitrogenase requires a source of electrons. These are generated either oxidatively or photosynthetically, depending on the organism. The electrons are transferred to **ferredoxin,** a [4Fe–4S]-containing electron carrier that transfers an electron to the Fe-protein of nitrogenase, beginning the nitrogen fixation process (Fig. 20-43).

The transfer of electrons during the nitrogenase reaction requires ATP-dependent protein conformational changes and the dissociation of the Fe-protein from the MoFe-protein after each electron transfer. During the reaction cycle, two molecules of ATP bind to the Fe-protein and are hydrolyzed as an electron passes to the MoFe-protein. ATP hydrolysis induces a conformational change in the Fe-protein that alters its redox potential from -0.29 to -0.40 V, making the electron capable of N_2 reduction ($\mathscr{E}°'$ for the reaction $N_2 + 6\,H^+ + 6\,e^- \rightleftharpoons 2\,NH_3$ is -0.34 V).

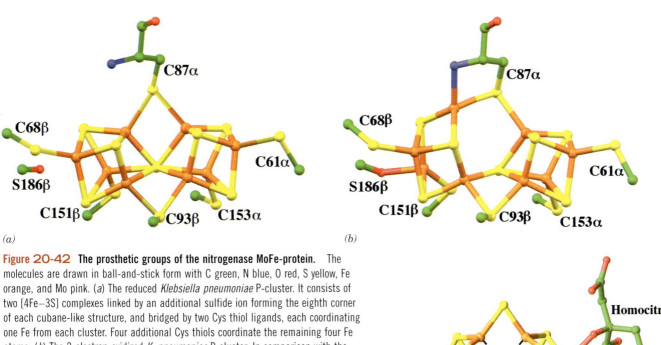

(a)   *(b)*

Figure 20-42 **The prosthetic groups of the nitrogenase MoFe-protein.**   The molecules are drawn in ball-and-stick form with C green, N blue, O red, S yellow, Fe orange, and Mo pink. (*a*) The reduced *Klebsiella pneumoniae* P-cluster. It consists of two [4Fe–3S] complexes linked by an additional sulfide ion forming the eighth corner of each cubane-like structure, and bridged by two Cys thiol ligands, each coordinating one Fe from each cluster. Four additional Cys thiols coordinate the remaining four Fe atoms. (*b*) The 2-electron-oxidized *K. pneumoniae* P-cluster. In comparison with the reduced complex in Part *a*, two of the Fe—S bonds from the centrally located sulfide ion that bridges the two [4Fe–3S] clusters have been replaced by ligands from the Cys 87α amide N and the Ser 186β side chain O yielding a [4Fe–3S] cluster (*left*) and a [4Fe–4S] cluster (*right*) that remain linked by a direct Fe–S bond and two bridging Cys thiols. (*c*) The *A. vinelandii* FeMo-cofactor. It consists of a [4Fe–3S] cluster and a [1Mo–3Fe–3S] cluster that are bridged by three sulfide ions. The FeMo-cofactor is linked to the protein by only two ligands at its opposite ends, one from His 442α to the Mo atom and the other from Cys 275α to an Fe atom. The Mo atom is additionally doubly liganded by homocitrate. What is most likely an N atom (*blue sphere*) is liganded to the FeMo-cluster's six central Fe atoms (*dashed black lines*). [Parts *a* and *b* based on X-ray structures by David Lawson, John Innes Centre, Norwich, U.K. Part *c* based on an X-ray structure by Douglas Rees, California Institute of Technology. PDBids (*a*) 1QGU, (*b*) 1QH1, and (*c*) 1M1N.]

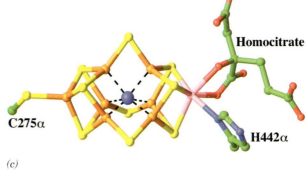

(c)

How are events at the ATP-binding site linked to electron transfer? ATP hydrolysis occurs at a significant rate only when the Fe-protein is in complex with the MoFe-protein (and, of course, electron transfer can occur only when the two proteins are associated). Intriguingly, the binding of $ADP \cdot AlF_4^-$ to the Fe-protein induces conformational changes in two re-

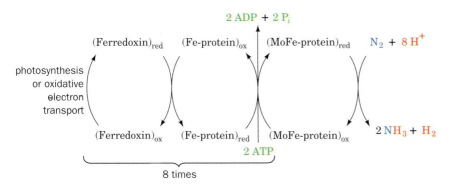

Figure 20-43 **The flow of electrons in the nitrogenase-catalyzed reduction of N_2.**

gions of the Fe-protein, designated Switch I and Switch II (Fig. 20-41), that are homologous to segments of signal-transducing G proteins in which nucleotide hydrolysis is coupled to protein conformational changes (Section 21-3B). At Switch I, these conformational changes affect the interactions between the Fe-protein and the MoFe-protein, whereas at Switch II they affect the environment of the [4Fe–4S] cluster. X-Ray structures of the Fe-protein alone and with bound ADP or ATP indicate that ATP hydrolysis triggers a structural change including rotation of the two Fe-protein subunits by about 13° toward each other. This conformational change brings the [4Fe–4S] cluster closer to the P-cluster of the adjacent MoFe-protein (from 18 to 14 Å apart), thereby promoting electron transfer from the Fe-protein to the MoFe-protein.

Kinetic studies of nitrogenase indicate that the rate-limiting step of N_2 reduction is the dissociation of the Fe-protein from the MoFe-protein. If the Fe-protein functions in the same way as a G protein, then nucleotide hydrolysis might trigger its dissociation from the MoFe-protein. Reassociation could occur when the ADP exits the active site and is replaced by ATP. ATP hydrolysis would not occur again until the Fe-protein was docked with the MoFe-protein and poised to transfer electrons.

N_2 Reduction Is Energetically Costly. The actual reduction of N_2 occurs on the MoFe-protein in three discrete steps, each involving an electron pair:

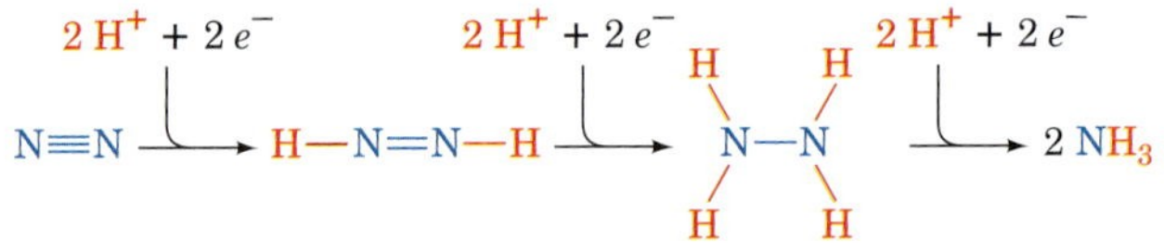

Diimine   **Hydrazine**

An electron transfer must occur six times per N_2 molecule fixed, so a total of 12 ATP are required to fix one N_2 molecule. However, nitrogenase also reduces H_2O to H_2, which in turn reacts with **diimine** to re-form N_2.

$$HN{=}NH + H_2 \rightarrow N_2 + 2\,H_2$$

This futile cycle is favored when the ATP level is low and/or the reduction of the Fe-protein is sluggish. Even when ATP is plentiful, however, the cycle cannot be suppressed beyond about one H_2 molecule produced per N_2 reduced and hence appears to be a requirement for the nitrogenase reaction. The total cost of N_2 reduction is therefore 8 electrons transferred and 16 ATP hydrolyzed. Under cellular conditions, the cost is closer to 20–30 ATP. Consequently, nitrogen fixation is an energetically expensive process.

Although atmospheric N_2 is the ultimate nitrogen source for all living things, most plants do not support the symbiotic growth of nitrogen-fixing bacteria. They must therefore depend on a source of "prefixed" nitrogen such as nitrate or ammonia. These nutrients come from lightning discharges (the source of ~10% of naturally fixed N_2), decaying organic matter in the soil, or from fertilizer applied to it (~50% of the nitrogen fixed is now generated by the **Haber-Bosch process,** which is used in the industrial production of ammonia from N_2 and H_2). One major long-term goal of genetic engineering is to induce agriculturally useful nonleguminous plants to fix their own nitrogen. This would free farmers, particularly those in developing countries, from either purchasing fertilizers, periodically letting their fields lie fallow (giving legumes the opportunity to grow), or following the slash-and-burn techniques that are rapidly destroying the world's tropical forests.

The Nitrogen Cycle Describes the Interconversion of Nitrogen in the Biosphere.

The ammonia produced by the nitrogenase reaction is recycled in the biosphere as described by the **nitrogen cycle** (Fig. 20-44). Nitrate is produced by certain bacteria that oxidize NH_3 to NO_2^- and then NO_3^-, a process called **nitrification.** Still other organisms convert nitrate back to N_2, which is known as **denitrification.** In addition, nitrate is reduced to NH_3 by plants, fungi, and many bacteria, a process called **ammonification** in which **nitrate reductase** catalyzes the two-electron reduction of nitrate to nitrite (NO_2^-):

$$NO_3^- + 2\ H^+ + 2\ e^- \rightarrow NO_2^- + H_2O$$

and then **nitrite reductase** converts nitrite to ammonia,

$$NO_2^- + 7\ H^+ + 6\ e^- \rightarrow NH_3 + 2\ H_2O$$

The direct anaerobic oxidation of NH_3 back to N_2 without the intermediacy of nitrate, the reverse of nitrogen fixation, has recently been discovered in certain bacteria (Box 20-4).

Fixed Nitrogen Must Be Assimilated.

After atmospheric N_2 has been converted to a biologically useful form (e.g., ammonia), it must be **assimilated** (incorporated) into a cell's biomolecules. Once nitrogen has been introduced into an amino acid, the amino group can be transferred to other compounds by transamination. Because most organisms do not fix nitrogen and hence must rely on prefixed nitrogen, nitrogen assimilation reactions are critical for conserving this essential element.

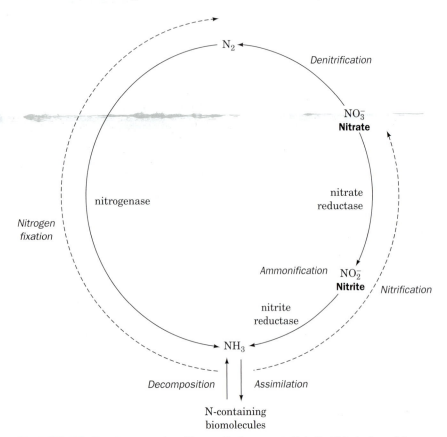

Figure 20-44 The nitrogen cycle. Nitrogen fixation converts N_2 to the biologically useful ammonia. Nitrate can also be converted to ammonia. Ammonia is transformed to N_2 by nitrification followed by denitrification. Ammonia may be assimilated into nitrogen-containing biomolecules, which may be decomposed back to ammonia.

BOX 20-4

Perspectives in Biochemistry

Anammox Bacteria

Because nitrogen fixation, the conversion of N_2 to NH_3, is so energetically expensive, the reverse process must be much more favorable. In fact, nitrification followed by denitrification (Fig. 20-44) is a well-known pathway that transforms fixed nitrogen back to N_2. This pathway occurs in certain bacteria that decompose organic matter. Under anaerobic conditions, denitrification can occur, but the oxidation of ammonia to nitrate (nitrification) cannot take place due to the lack of oxygen. However, in oxygen-deficient environments such as the depths of the Black Sea, the expected ammonia does not appear to accumulate. This observation suggested that there must be a class of microorganisms capable of carrying out the anaerobic oxidation of ammonia to N_2, a process known as **anammox.**

An anaerobic ammonia-oxidizing organism was isolated in 1999 from a wastewater treatment system. The bacterium was identified as a member of the *Planctomycetales* division of the Bacteria. Since then, several species of anammox bacteria have been described. They appear to play a significant and previously unrecognized role in the nitrogen cycle in anaerobic environments, including the depths of the Black Sea. The organisms grow slowly and are difficult to culture, which helps explain why they went undetected for so long.

Planctomycetes are unusual among bacteria in lacking peptidoglycan in their cell wall (see Section 8-3B) and in having a membrane-bounded intracellular compartment. In the anammox bacteria, the compartment is known as the **anammoxosome** (*see figure*) and is the site of the anammox reaction:

$$NH_4^+ + NO_2^- \rightarrow N_2 + 2\ H_2O$$

The nitrite (NO_2^-) functions as an electron acceptor in this process, which is a major energy-generating pathway that drives carbon fixation in these lithotrophic organisms. The intermediates of the anammox reaction include hydroxylamine

(NH_2OH) and hydrazine (N_2H_4), both of which are highly reactive and toxic to cells. The anammoxosome presumably sequesters these compounds to protect the rest of the cell, much as the eukaryotic lysosome protects the cytoplasm from the enclosed hydrolytic enzymes.

Bacteria are known to contain some unusual lipids, and the anammoxosome membrane exhibits some strikingly novel structures. Among these are lipids consisting of a glycerol backbone to which C_{20} alkyl chains are attached via ether linkages. This sort of structure is not unprecedented, but the hydrocarbon chains include a hitherto unobserved set of fused cyclobutane rings.

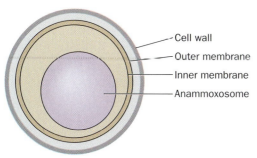

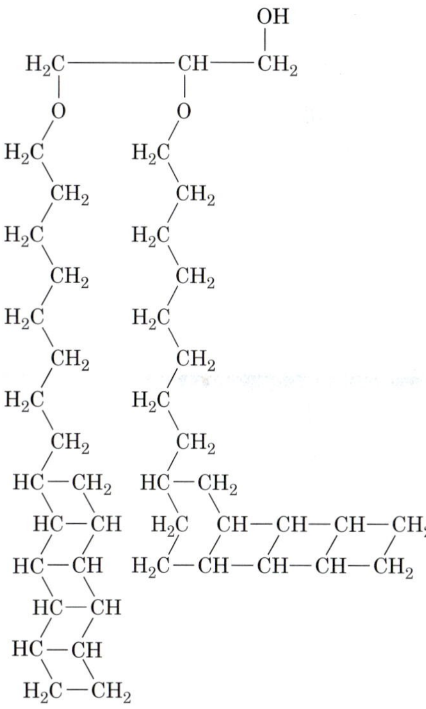

The staircase-like side chains have been dubbed "ladderanes" (compare the lipid shown here to the conventional membrane lipids described in Section 9-1). The ladderane lipids apparently make the anammoxosome membrane more dense and rigid, thereby decreasing its permeability to substances such as hydroxylamine and hydrazine, which can diffuse relatively freely through ordinary biological membranes. An impermeable anammoxosome membrane might be necessary for the bacterium, not only to protect the rest of the cell from the anammox intermediates, but also to maintain the concentrations of the intermediates in these very slow-growing organisms.

Cell wall
Outer membrane
Inner membrane
Anammoxosome

Cross section of an anammox bacterium.

We have already considered the reaction catalyzed by glutamine synthetase (Section 20-5A), which in microorganisms represents a metabolic entry point for fixed nitrogen (in animals, this reaction helps "mop up" excess ammonia). The glutamine synthetase reaction requires the nitrogen-containing compound glutamate as a substrate. So what is the source of the amino group in glutamate? In bacteria and plants, but not animals, the enzyme **glutamate synthase** converts α-ketoglutarate and glutamine to two molecules of glutamate:

α-Ketoglutarate + glutamine + NADPH + H$^+$ →

2 glutamate + NADP$^+$

This reductive amination reaction requires electrons from NADPH and takes place at three distinct active sites in the $\alpha_2\beta_2$ heterotetramer (Fig. 20-45). X-Ray structures of the enzyme reveal that the substrate binding sites are widely separated. As a result, the ammonia that is transferred from glutamine to α-ketoglutarate must travel through a 31-Å-long tunnel in the

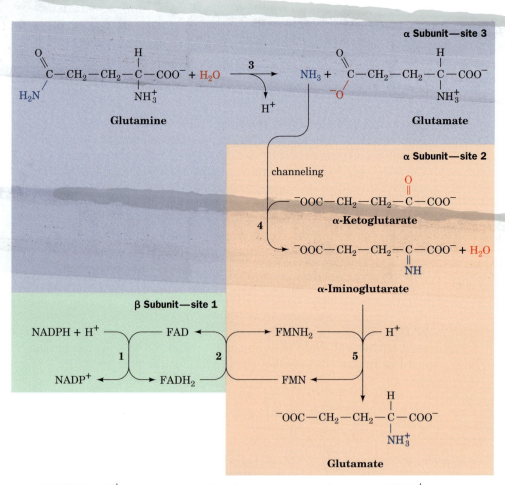

Overall: NADPH + H$^+$ + glutamine + α-ketoglutarate ⟶ 2 glutamate + NADP$^+$

Figure 20-45 The glutamate synthase reaction. (1) Electrons are transferred from NADPH to FAD at active site 1 on the β subunit to yield FADH$_2$. (2) Electrons travel from the FADH$_2$ to FMN at site 2 on an α subunit to yield FMNH$_2$. (3) Glutamine is hydrolyzed to glutamate and ammonia at site 3. (4) The ammonia produced in Step 3 moves to site 2 by channeling, where it reacts with α-ketoglutarate. (5) The α-iminoglutarate product is reduced by FMNH$_2$ to form glutamate.

Figure 20-46 **X-Ray structure of the α subunit of glutamate synthase from the nitrogen-fixing bacterium *Azospirillum brasilense*.**   The subunit, which is represented by its C_α backbone, consists of four domains: an N-terminal glutamine amidotransferase domain (*blue*) to which methionine sulfone (methionine with its S atom additionally forming two S=O bonds, which is a glutamine tetrahedral transition state analog) is bound; a central domain (*red*); an FMN-binding domain (*green*) to which an FMN, a [3Fe–4S] cluster, and an α-ketoglutarate are bound; and a so-called β helix domain (*purple*). The foregoing ligands are drawn in black in ball-and-stick form. The 31-Å-long tunnel from the methyl group on the methionine sulfone (analogous to the amido group of glutamine) to the α-keto group of the α-ketoglutarate is outlined by a gray surface. The tunnel is blocked in this structure (it is divided into two cavities) by the main chain atoms of four residues that protrude into the tunnel. [Courtesy of Andrea Mattevi, Universitá degli Studi di Pavia, Italy. PDBid 1EA0.]

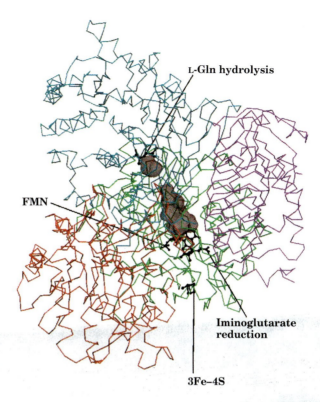

L-Gln hydrolysis

FMN

Iminoglutarate reduction

3Fe–4S

protein (Fig. 20-46). This channeling probably helps prevent the loss of NH_3 to the cytosol and maintains this intermediate in its more reactive deprotonated state (in the cytosol, NH_3 would immediately acquire a proton to become NH_4^+). Furthermore, the tunnel is blocked by the side chains of several residues and opens only when α-ketoglutarate is bound to the enzyme and NADPH is available. This mechanism apparently prevents the wasteful hydrolysis of glutamine.

The net result of the glutamine synthetase and glutamate synthase reactions is

$$\alpha\text{-Ketoglutarate} + NH_4^+ + NADPH + ATP \rightarrow$$
$$\text{glutamate} + NADP^+ + ADP + P_i$$

Thus, *the combined action of these two enzymes assimilates fixed nitrogen (NH_4^+) into an organic compound (α-ketoglutarate) to produce an amino acid (glutamate).* Once the nitrogen is assimilated as glutamate, it can be used in the synthesis of other amino acids by transamination.

■ SUMMARY

1. Intracellular proteins are degraded by lysosomal proteases or, after being ubiquitinated, by the action of proteasomes.

2. The degradation of an amino acid almost always begins with the removal of its amino group in a PLP-facilitated transamination reaction.

3. In the urea cycle, a nitrogen atom from ammonia (a product of the oxidative deamination of glutamate) and a nitrogen atom from aspartate combine with HCO_3^- to form urea for

excretion. The rate-limiting step of this process is catalyzed by carbamoyl phosphate synthetase.

4. The 20 "standard" amino acids are degraded to compounds that give rise either to glucose or to ketone bodies or fatty acids: pyruvate, α-ketoglutarate, succinyl-CoA, fumarate, oxaloacetate, acetyl-CoA, or acetoacetate.

5. The nonessential amino acids are synthesized in simple pathways from pyruvate, oxaloacetate, α-ketoglutarate, and

3-phosphoglycerate. The pathways for the syntheses of essential amino acids are more complicated and vary among organisms.

6. Amino acids are the precursors of various biomolecules. Heme is synthesized from glycine and acetate. Various hormones and neurotransmitters are synthesized by the decarboxylation and hydroxylation of histidine, glutamate, trypto-phan, and tyrosine. The five-electron oxidation of arginine yields the bioactive stable radical nitric oxide.

7. Nitrogen fixation in bacteria, which requires 8 electrons and at least 16 ATP, is catalyzed by nitrogenase, a multisubunit protein with redox centers containing Fe, S, and Mo. Fixed nitrogen is incorporated into amino acids via the glutamine synthetase and glutamate synthase reactions.

 ## REFERENCES

Protein Degradation

Groll, M. and Clausen, M., Molecular shredders: How proteasomes fulfill their roles. *Curr. Opin. Struct. Biol.* **13,** 665–673 (2003).

Hartmann-Peterson, R., Seeger, M., and Gordon, C., Transferring substrates to the 26S proteasome, *Trends Biochem. Sci.* **28,** 26–31 (2003).

Urea Cycle

Withers, P.C., Urea: Diverse functions of a 'waste product,' *Clin. Exp. Pharm. Physiol.* **25,** 722–727 (1998). [Discusses the various roles of urea, contrasting it with ammonia with respect to toxicity, acid–base balance, and nitrogen transport.]

Amino Acid Catabolism and Synthesis

Brosnan, J.T., Glutamate, at the interface between amino acid and carbohydrate metabolism, *J. Nutr.* **130,** 988S–990S (2000). [Summarizes the metabolic roles of glutamate as a fuel and as a participant in deamination and transamination reactions.]

Eisenberg, D., Gill, H.S., Pfluegl, M.U., and Rotstein, S.H., Structure–function relationships of glutamine synthetases, *Biochim. Biophys. Acta* **1477,** 122–145 (2000).

Huang, X., Holden, H.M., and Raushel, F.M., Channeling of substrates and intermediates in enzyme-catalyzed reactions, *Annu. Rev. Biochem.* **70,** 149–180 (2001).

Katagiri, M. and Nakamura, M., Animals are dependent on preformed α-amino nitrogen as an essential nutrient, *Life* **53,** 125–129 (2002).

Medina, M.Á., Urdiales, J.L., and Amores-Sánchez, M.I., Roles of homocysteine in cell metabolism: Old and new functions, *Eur. J. Biochem.* **268,** 3871–3882 (2001).

Scriver, C.R., Beaudet, A.L., Sly, S.S., and Vale, D. (Eds.), *Metabolic & Inherited Bases of Inherited Disease* (8th ed.), McGraw-Hill (2001). [Volume 2, Part 8 contains numerous chapters on amino acid metabolism and its defects.]

Nitrogen Fixation

Igarashi, R.Y. and Seefeldt, L.C., Nitrogen fixation: The mechanism of Mo-dependent nitrogenase, *Crit. Rev. Biochem. Mol. Biol.* **38,** 351–384 (2003).

Jang, S.B., Seefeldt, L.C., and Peters, J.W., Insights into nucleotide signal transduction in nitrogenase: structure of an iron protein with MgADP, *Biochemistry* **39,** 14745–14752 (2000).

Rees, D.C. and Howard, J.B., Nitrogenase: standing at the crossroads, *Curr. Opin. Chem. Biol.* **4,** 559–566 (2000).

 ## KEY TERMS

ubiquitin	hyperammonemia	biopterin	catecholamine
isopeptide bond	urea cycle	essential amino acid	nitrogen fixation
N-end rule	channeling	nonessential amino acid	anammox
proteasome	glucogenic amino acid	adenylylation	nitrogen assimilation
transamination	ketogenic amino acid	uridylylation	
PLP	THF	porphyria	
deamination	SAM	jaundice	

 ## STUDY EXERCISES

1. Describe the pathway for proteasome-mediated protein degradation, including the roles of ubiquitin and ATP.

2. How does PLP facilitate amino acid deamination?

3. Summarize the steps of the urea cycle. How do the amino groups of amino acids enter the cycle?

4. Describe the two general metabolic fates of the carbon skeletons of amino acids.

5. What are the metabolic precursors of the nonessential amino acids?

6. Why is nitrogen fixation so energetically costly?

7. Describe the reactions by which fixed nitrogen is introduced into amino acids.

PROBLEMS

1. Explain why protein degradation by proteasomes requires ATP even though proteolysis is an exergonic process.

2. Explain why the symptoms of a partial deficiency in a urea cycle enzyme can be attenuated by a low-protein diet.

3. Production of the enzymes that catalyze the reactions of the urea cycle can increase or decrease according to the metabolic needs of the organism. High levels of these enzymes are associated with high-protein diets as well as starvation. Explain this apparent paradox.

4. Which three mammalian enzymes can potentially react with and thereby decrease the concentration of free NH_4^+?

5. *Helicobacter pylori,* the bacterium responsible for gastric ulcers, can survive in the stomach (where the pH is as low as 1.5) in part because it synthesizes large amounts of the enzyme urease. (a) Write the reaction for urea hydrolysis by urease. (b) Explain why this reaction could help establish a more hospitable environment for *H. pylori,* which tolerates acid but prefers to grow at near-neutral pH.

6. In the degradation pathway for isoleucine (Fig. 20-21), draw the reactions that convert tiglyl-CoA to acetyl-CoA and propionyl-CoA.

7. Draw the amino acid–Schiff base that forms in the breakdown of 3-hydroxykynurenine to form 3-hydroxyanthranilate in the tryptophan degradation pathway (Fig. 20-23, Reaction 4) and indicate which bond is to be cleaved.

8. Which of the 20 "standard" amino acids are (a) purely glucogenic, (b) purely ketogenic, and (c) both glucogenic and ketogenic?

9. Alanine, cysteine, glycine, serine, and threonine are amino acids whose breakdown yields pyruvate. Which, if any, of the remaining 15 amino acids also do so?

10. What are the metabolic consequences of a defective uridylyl-removing enzyme in *E. coli?*

11. Many of the most widely used herbicides inhibit the synthesis of aromatic amino acids. Explain why these compounds are safe to use near animals.

12. One of the symptoms of **kwashiorkor,** the dietary protein deficiency disease in children, is the depigmentation of the skin and hair. Explain the biochemical basis of this symptom.

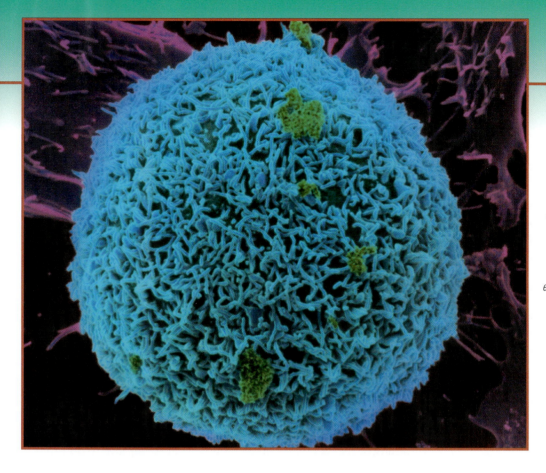

Mammalian Fuel Metabolism: Integration and Regulation

In even the simplest prokaryotic cell, metabolic processes must be coordinated so that opposing pathways do not operate simultaneously and so that the organism can respond to changing external conditions such as the availability of nutrients. In addition, the organism's metabolic activities must meet the demands set by genetically programmed growth and reproduction. The challenges of coordinating energy acquisition and expenditure are markedly more complex in multicellular organisms, in which individual cells must cooperate. In animals and plants, this task is simplified by the division of metabolic labor among tissues.

In animals, the interconnectedness of various tissues is ensured by neuronal circuits and by hormones. Such regulatory systems do not simply switch cells on and off but elicit an almost infinite array of responses. The exact response of a cell to a given regulatory signal depends on the cell's ability to recognize the signal and on the presence of synergistic or antagonistic signals.

Our examination of mammalian carbohydrate, lipid, and amino acid metabolism (Chapters 14–20) would be incomplete without a discussion of how such processes are coordinated at the molecular level and how their malfunctions produce disease. In this chapter, we summarize the specialized metabolism of different organs and the pathways that link them. We also examine the mechanisms by which extracellular hormones influence intracellular events. We conclude with a discussion of disruptions in mammalian fuel metabolism.

1 Organ Specialization

Many of the metabolic pathways discussed so far have to do with the oxidation of metabolic fuels for the production of ATP. These pathways, which encompass the synthesis and breakdown of glucose, fatty acids, and amino acids, are summarized in Fig. 21-1.

1. **Glycolysis.** The metabolic degradation of glucose begins with its conversion to two molecules of pyruvate with the net generation of two molecules of ATP (Section 14-1).

2. **Gluconeogenesis.** Mammals can synthesize glucose from a variety of precursors, such as pyruvate, via a series of reactions that largely reverse the path of glycolysis (Section 15-4).

3. **Glycogen degradation and synthesis.** The opposing processes catalyzed by glycogen phosphorylase and glycogen synthase are reciprocally regulated by hormonally controlled phosphorylation and dephosphorylation (Section 15-3).

4. **Fatty acid synthesis and degradation.** Fatty acids are broken down through β oxidation to form acetyl-CoA (Section 19-2), which, through its conversion to malonyl-CoA, is also the substrate for fatty acid synthesis (Section 19-4).

5. **The citric acid cycle.** The citric acid cycle (Section 16-1) oxidizes acetyl-CoA to CO_2 and H_2O with the concomitant production of reduced coenzymes whose reoxidation drives ATP synthesis. Many glucogenic amino acids can be oxidized via the citric acid cycle following their breakdown to one of its intermediates (Section 20-4), which, in turn, are broken down to pyruvate and then to acetyl-CoA, the cycle's only substrate.

6. **Oxidative phosphorylation.** This mitochondrial pathway couples the oxidation of NADH and $FADH_2$ produced by glycolysis, β oxidation, and the citric acid cycle to the phosphorylation of ADP (Section 17-3).

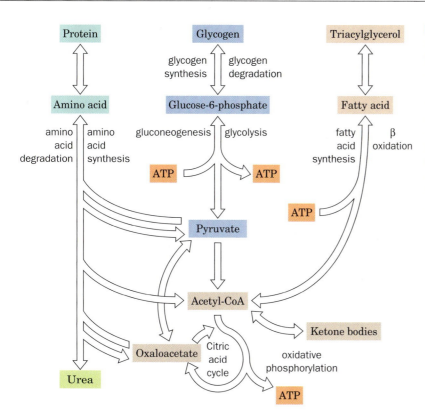

Figure 21-1 The major pathways of fuel metabolism in mammals. Proteins, glycogen, and triacylglycerols are built up from and broken down to smaller units: amino acids, glucose-6-phosphate, and fatty acids. Oxidation of these fuels yields metabolic energy in the form of ATP. Pyruvate (a product of glucose and amino acid degradation) and acetyl-CoA (a product of glucose, amino acid, and fatty acid degradation) occupy central positions in mammalian fuel metabolism. Compounds that give rise to pyruvate, such as oxaloacetate, can be used for gluconeogenesis; acetyl-CoA can give rise to ketone bodies but not glucose. Not all the pathways shown here occur in all cells or occur simultaneously in a given cell.

7. Amino acid synthesis and degradation. Excess amino acids are degraded to metabolic intermediates of glycolysis and the citric acid cycle (Section 20-4). The amino group is disposed of through urea synthesis (Section 20-3). Nonessential amino acids are synthesized via pathways that begin with common metabolites (Section 20-5A).

Two compounds lie at the crossroads of the major metabolic pathways: acetyl-CoA and pyruvate (Fig. 21-1). Acetyl-CoA is the common degradation product of glucose, fatty acids, and ketogenic amino acids. Its acetyl group can be oxidized to CO_2 and H_2O via the citric acid cycle and oxidative phosphorylation or used to synthesize ketone bodies or fatty acids. Pyruvate is the product of glycolysis and the breakdown of glucogenic amino acids. It can be oxidatively decarboxylated to yield acetyl-CoA, thereby committing its atoms either to oxidation or to the biosynthesis of fatty acids. Alternatively, pyruvate can be carboxylated via the pyruvate carboxylase reaction to form oxaloacetate, which can either replenish citric acid cycle intermediates or give rise to glucose or certain amino acids.

All of the above pathways are affected in one way or another by the need for ATP, as is indicated by the cell's AMP level. Several enzymes are either activated or inhibited allosterically by AMP, and several others are phosphorylated by **AMP-dependent protein kinase** (**AMPK,** which is known as the cell's fuel gauge). AMPK activates metabolic breakdown pathways that generate ATP while inhibiting biosynthetic pathways so as to conserve ATP for more vital processes. AMPK's targets include the heart and inducible isozymes of the bifunctional enzyme PFK-2/FBPase-2, which control the F2,6P concentration (Section 15-4C). The phosphorylation of these isozymes activates their PFK-2 activity, increasing [F2,6P], which in turn activates PFK-1 and glycolysis. Consequently, when there is insufficient oxygen for oxidative phosphorylation to maintain adequate concentrations of ATP, the resulting AMP buildup causes the cell to switch to anaerobic glycolysis for ATP production.

AMPK also phosphorylates and thereby activates hormone-sensitive triacylglycerol lipase (Section 19-5). This increases the amount of fatty acid available for oxidation and hence provides another means of generating ATP. AMPK-mediated phosphorylation also inhibits certain enzymes including acetyl-CoA carboxylase (ACC, which catalyzes the first committed step of fatty acid synthesis; Section 19-4B), hydroxymethylglutaryl-CoA reductase (HMG-CoA reductase; which catalyzes the rate-determining step in cholesterol biosynthesis; Section 19-7B), and glycogen synthase (which catalyzes the rate-limiting reaction in glycogen synthesis; Section 15-3B). Consequently, when the rate of ATP production is inadequate, these biosynthetic pathways are turned off, thereby conserving ATP for the most vital cellular functions.

Only a few tissues, such as liver, can carry out all the reactions shown in Fig. 21-1, and in a given cell only a small portion of all possible metabolic reactions occur at a significant rate. The flux through any sequence of reactions depends on the presence of the appropriate enzyme catalysts and on the organism's need for the reaction products.

We shall consider the metabolism of five mammalian organs: brain, muscle, adipose tissue, liver, and kidney. Metabolites flow between these organs in well-defined pathways in which flux varies with the nutritional state of the animal (Fig. 21-2). For example, immediately following a meal, glucose, amino acids, and fatty acids are directly available from the intestine. Later, when these fuels have been exhausted, the liver supplies other tissues with glucose and ketone bodies, whereas adipose tissue provides them with fatty acids. All these organs are connected via the bloodstream.

A The Brain

Brain tissue has a remarkably high respiration rate. Although the human brain constitutes only ~2% of the adult body mass, it is responsible for ~20% of its resting O_2 consumption. Most of the brain's energy production powers the plasma membrane (Na^+–K^+)–ATPase (Section 10-3A), which maintains the membrane potential required for nerve impulse transmission.

Under usual conditions, glucose is the brain's primary fuel (although during an extended fast, the brain gradually switches to ketone bodies; Section 21-4A). Since brain cells store very little glycogen, *they require a steady supply of glucose from the blood.* A blood glucose concentration of less than half the normal value of ~5 mM results in brain dysfunction. Levels much below this result in coma, irreversible damage, and ultimately death.

B Muscle

Muscle's major fuels are glucose (from glycogen), fatty acids, and ketone bodies. Rested, well-fed muscle synthesizes a glycogen store comprising 1 to 2% of its mass. Although triacylglycerols are a more efficient form of energy storage (Section 9-1B), the metabolic effort of synthesizing glycogen is cost-effective because glycogen can be mobilized more rapidly than fat and because glucose can be metabolized anaerobically, whereas fatty acids cannot.

In muscle, glycogen is converted to glucose-6-phosphate (G6P) for entry into glycolysis. Muscle cannot export glucose, however, because it lacks glucose-6-phosphatase. Furthermore, although muscle can synthesize glycogen from glucose, it does not participate in gluconeogenesis because it lacks the required enzymatic machinery. Consequently, *muscle carbohydrate metabolism serves only muscle.*

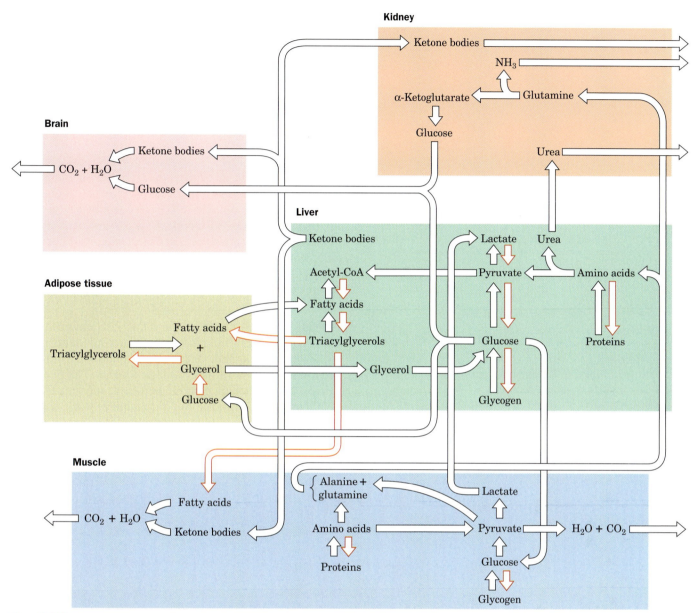

Figure 21-2 *Key to Metabolism.* The metabolic interrelationships among brain, adipose tissue, muscle, liver, and kidney. The red arrows indicate pathways that predominate in the well-fed state.

Muscle Contraction Is Anaerobic under Conditions of High Exertion.

Muscle contraction is driven by ATP hydrolysis (Section 28-2) and therefore requires either an aerobic or an anaerobic ATP regeneration system. Respiration (the citric acid cycle and oxidative phosphorylation) is the body's major source of ATP resupply. Skeletal muscle at rest uses ~30% of the O_2 consumed by the human body. A muscle's respiration rate may increase in response to a heavy workload by as much as 25-fold. Yet, its rate of ATP hydrolysis can increase by a much greater amount. The ATP is initially regenerated by the reaction of phosphocreatine with ADP (Section 13-2C):

$$\text{Phosphocreatine} + \text{ADP} \rightleftharpoons \text{creatine} + \text{ATP}$$

(phosphocreatine is resynthesized in resting muscle by the reversal of this

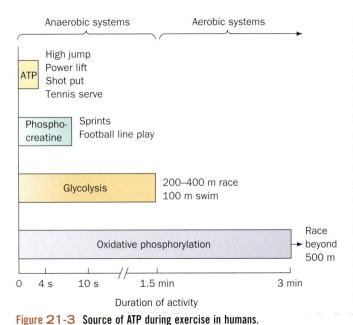

Anaerobic systems   Aerobic systems

ATP — High jump / Power lift / Shot put / Tennis serve

Phospho-creatine — Sprints / Football line play

Glycolysis — 200–400 m race / 100 m swim

Oxidative phosphorylation — Race beyond 500 m

| 0 | 4 s | 10 s | // | 1.5 min | 3 min |

Duration of activity

Figure 21-3   Source of ATP during exercise in humans. The supply of endogenous ATP is extended for a few seconds by phosphocreatine, after which anaerobic glycolysis generates ATP. The shift from anaerobic to aerobic metabolism (oxidative phosphorylation) occurs after about 90 sec, or slightly later in trained athletes. [Adapted from McArdle, W.D., Katch, F.I., and Katch, V.L., *Exercise Physiology*, 2nd ed., Lea & Febiger (1986), p. 348.]

reaction). Under conditions of maximum exertion, however, such as during a sprint, a muscle has only about a 4-s supply of phosphocreatine. It must then shift to ATP production via glycolysis of G6P, a process whose maximum flux greatly exceeds those of the citric acid cycle and oxidative phosphorylation. Much of the G6P is therefore degraded anaerobically to lactate (Section 14-3A). As we shall see in Section 21-1F, export of lactate relieves much of the muscle's respiratory burden. Muscle fatigue, which occurs after ~20 s of maximal exertion, is not caused by exhaustion of the muscle's glycogen supply but by the drop in pH that results from the buildup of lactate (Section 14-3A). This phenomenon may be an adaptation that prevents muscle cells from committing suicide by fully depleting their ATP supply. Exercise lasting more than a minute or two is fueled primarily by oxidative phosphorylation, which generates ATP more slowly but much more efficiently than glycolysis alone. The source of ATP during exercise of varying duration is summarized in Fig. 21-3.

The Heart Is Largely Aerobic. The heart is a muscular organ that acts continuously rather than intermittently. Therefore, heart muscle relies entirely on aerobic metabolism and is richly endowed with mitochondria; they occupy up to 40% of its cytoplasmic space. The heart can metabolize fatty acids, ketone bodies, glucose, pyruvate, and lactate. Fatty acids are the resting heart's fuel of choice, but during heavy work, the heart greatly increases its consumption of glucose, which is derived mostly from its relatively limited glycogen store. Some individuals with atherosclerosis suffer from **angina** (heart pain) due to an insufficient O_2 supply to the heart. The drug **ranolazine**

Ranolazine

relieves angina by partially inhibiting fatty acid oxidation so that the heart muscle burns relatively more glucose, a process that requires less oxygen for the same amount of ATP production.

C Adipose Tissue

The function of adipose tissue is to store and release fatty acids as needed for fuel. Adipose tissue is widely distributed throughout the body but occurs most prominently under the skin, in the abdominal cavity, and in skeletal muscle. The adipose tissue of a normal 70-kg man contains ~15 kg of fat. This amount represents some 590,000 kJ of energy (141,000 dieter's Calories), which is sufficient to maintain life for ~3 months.

Adipose tissue obtains most of its fatty acids for storage from circulating lipoproteins as described in Section 19-1B. Fatty acids are activated by the formation of the corresponding fatty acyl-CoA and then esterified with glycerol-3-phosphate to form the stored triacylglycerols. The glycerol-3-phosphate

arises from the reduction of dihydroxyacetone phosphate, which must be glycolytically generated from glucose.

In times of metabolic need, adipocytes hydrolyze triacylglycerols to fatty acids and glycerol through the action of hormone-sensitive lipase (Section 19-5). If glycerol-3-phosphate is abundant, many of the fatty acids so formed are reesterified to triacylglycerols. If glycerol-3-phosphate is in short supply, the fatty acids are released into the bloodstream. Thus, *fatty acid mobilization depends in part on the rate of glucose uptake since glucose is the precursor of glycerol-3-phosphate.* Metabolic need is signaled directly by a decrease in [glucose] as well as by hormonal stimulation.

D Liver

The liver is the body's central metabolic clearinghouse. It maintains the proper levels of circulating fuels for use by the brain, muscles, and other tissues. The liver is uniquely situated to carry out this task because all the nutrients absorbed by the intestines except fatty acids are released into the portal vein, which drains directly into the liver.

Glucokinase Converts Blood Glucose to Glucose-6-Phosphate. *One of the liver's major functions is to act as a blood glucose "buffer."* It does so by taking up and releasing glucose in response to hormones and to the concentration of glucose itself. After a carbohydrate-containing meal, when the blood glucose concentration reaches ~6 mM, the liver takes up glucose by converting it to G6P. This reaction is catalyzed by **glucokinase,** a liver isozyme of hexokinase. The hexokinases in most cells obey Michaelis–Menten kinetics, have a high glucose affinity ($K_M < 0.1$ mM), and are inhibited by their reaction product (G6P). Glucokinase, in contrast, has much lower glucose affinity (reaching half-maximal velocity at ~5 mM) and displays sigmoidal kinetics. Consequently, *glucokinase activity increases rapidly with blood [glucose] over the normal physiological range* (Fig. 21-4). Glucokinase, moreover, is not inhibited by physiological concentrations of G6P. Therefore, the higher the blood [glucose], the faster the liver converts glucose to G6P. At low blood [glucose], the liver does not compete with other tissues for the available glucose, whereas at high blood [glucose], when the glucose needs of these tissues are met, the liver can take up the excess glucose at a rate roughly proportional to the blood glucose concentration.

Glucokinase is a monomeric enzyme, so its sigmoidal kinetic behavior is somewhat puzzling (models of allosteric interactions do not explain cooperative behavior in a monomeric protein; Section 7-3B). Glucokinase is subject to metabolic control, however. Emile Van Schaftingen has isolated a **glucokinase regulatory protein** from rat liver, which, in the presence of the glycolytic intermediate fructose-6-phosphate (F6P), is a competitive inhibitor of glucokinase. Since F6P and the glucokinase product G6P are equilibrated in liver cells by phosphoglucose isomerase, glucokinase is, in effect, inhibited by its product. Fructose-1-phosphate (F1P), an

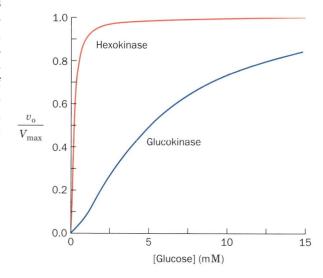

Figure 21-4 Relative enzymatic activities of hexokinase and glucokinase over the physiological blood glucose range. Glucokinase has much lower affinity for glucose ($K_M \approx 5$ mM) than does hexokinase ($K_M = 0.1$ mM) and exhibits sigmoidal rather than hyperbolic variation with [glucose]. [The glucokinase curve was generated using the Hill equation (Eq. 7-8) with $K = 10$ mM and $n = 1.5$ as obtained from Cardenas, M.L., Rabajille, E., and Niemeyer, H., *Eur. J. Biochem.* **145,** 163–171 (1984).]

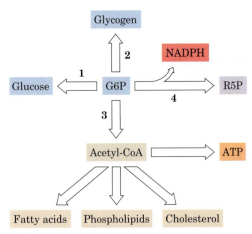

Figure 21-5 **Metabolic fate of glucose-6-phosphate in liver.** G6P can be converted (**1**) to glucose for export or (**2**) to glycogen for storage. Acetyl-CoA derived from G6P degradation (**3**) is the starting material for lipid biosynthesis. It is also consumed in generating ATP by respiration. Degradation of G6P via the pentose phosphate pathway (**4**) yields NADPH.

intermediate in liver fructose metabolism (Section 14-5A), overcomes this inhibition. Since fructose is normally available only from dietary sources, fructose may be the signal that triggers the uptake of dietary glucose by the liver.

Glucose-6-Phosphate Is at the Crossroads of Carbohydrate Metabolism. G6P has several alternative fates in the liver, depending on the glucose demand (Fig. 21-5):

1. G6P can be converted to glucose, by the action of glucose-6-phosphatase, for transport via the bloodstream to the peripheral organs. This occurs only when the blood [glucose] drops below ~5 mM. During exercise or fasting, low concentrations of blood glucose cause the pancreas to secrete glucagon. Glucagon receptors on the liver cell surface respond by activating adenylate cyclase. The resulting increase in intracellular [cAMP] triggers glycogen breakdown (Section 15-3).

2. G6P can be converted to glycogen (Section 15-2) when the body's demand for glucose is low.

3. G6P can be converted to acetyl-CoA via glycolysis and the action of pyruvate dehydrogenase. This glucose-derived acetyl-CoA, if it is not oxidized via the citric acid cycle and oxidative phosphorylation to generate ATP, can be used to synthesize fatty acids (Section 19-4), phospholipids (Section 19-6), and cholesterol (Section 19-7A).

4. G6P can be degraded via the pentose phosphate pathway (Section 14-6) to generate the NADPH required for the biosynthesis of fatty acids and other compounds.

The Liver Can Synthesize or Degrade Triacylglycerols. Fatty acids are also subject to alternative metabolic fates in the liver. When the demand for metabolic fuels is high, fatty acids are degraded to acetyl-CoA and then to ketone bodies for export to the peripheral tissues. The liver itself cannot use ketone bodies as fuel, because liver cells lack 3-ketoacyl-CoA transferase, an enzyme required to convert ketone bodies back to acetyl-CoA (Section 19-3). Fatty acids rather than glucose or ketone bodies are therefore the liver's major acetyl-CoA source under conditions of high metabolic demand. The liver generates its ATP from this acetyl-CoA through the citric acid cycle and oxidative phosphorylation.

When the demand for metabolic fuels is low, fatty acids are incorporated into triacylglycerols that are secreted into the bloodstream as VLDL for uptake by adipose tissue. Fatty acids can also be incorporated into phospholipids (Section 19-6). Under these conditions, fatty acids synthesized in the liver are not oxidized to acetyl-CoA because fatty acid synthesis (in the cytosol) is separated from fatty acid oxidation (in the mitochondria) and because malonyl-CoA, an intermediate in fatty acid synthesis, inhibits the transport of fatty acids into the mitochondria.

Amino Acids Are Metabolic Fuels. The liver degrades amino acids to a variety of metabolic intermediates that can be completely oxidized to CO_2 and H_2O or converted to glucose or ketone bodies (Section 20-4). Oxidation of amino acids provides a significant fraction of metabolic energy immediately after feeding, when the amino acids are present in relatively high concentrations in the blood. During a fast, when other fuels become scarce, glucose is produced from amino acids arising mostly from muscle protein degradation to alanine and glutamine. Thus, *proteins, in addition to their structural and functional roles, are important fuel reserves.*

E Kidney

The kidney filters urea and other waste products from the blood while it recovers important metabolites such as glucose. In addition, the kidney maintains the blood's pH by regenerating depleted blood buffers such as bicarbonate (lost by the exhalation of CO_2) and by excreting excess H^+ together with the conjugate bases of excess metabolic acids such as the ketone bodies acetoacetate and β-hydroxybutyrate. Protons are also excreted in the form of NH_4^+ with the ammonia derived from glutamine or glutamate. The remaining amino acid skeleton, α-ketoglutarate, can be converted to glucose by gluconeogenesis (the kidney is the only tissue besides liver that can undertake glucose synthesis). During starvation, the kidneys generate as much as 50% of the body's glucose supply.

F Interorgan Metabolic Pathways

The ability of the liver to supply other tissues with glucose or ketone bodies, or the ability of adipocytes to make fatty acids available to other tissues, depends, of course, on the circulatory system, which transports metabolic fuels, intermediates, and waste products among tissues. In addition, several important metabolic pathways are composed of reactions occurring in multiple tissues. In this section, we describe two well-known interorgan pathways.

The Cori Cycle. The ATP that powers muscle contraction is generated through oxidative phosphorylation (in mitochondrion-rich slow-twitch muscle fibers; Box 14-3) or by rapid catabolism of glucose to lactate (in fast-twitch muscle fibers). Slow-twitch fibers also produce lactate when ATP demand exceeds oxidative flux. The lactate is transferred via the bloodstream to the liver, where it is reconverted to pyruvate by lactate dehydrogenase and then to glucose by gluconeogenesis. Thus, liver and muscle are linked by the bloodstream in a metabolic cycle known as the **Cori cycle** (Fig. 21-6) in honor of Carl and Gerty Cori (Box 15-1), who first described it.

The ATP-consuming glycolysis/gluconeogenesis cycle would be a futile cycle if it occurred within a single cell. In this case, the two halves of the pathway occur in different organs. Liver ATP is used to resynthesize glucose from lactate produced in muscle. The resynthesized glucose returns to the muscle, where it may be stored as glycogen or catabolized immediately to generate ATP for muscle contraction.

The ATP consumed by the liver during the operation of the Cori cycle is regenerated by oxidative phosphorylation. After vigorous exertion, it may take at least 30 min for the oxygen consumption rate to decrease to its resting level. The elevated O_2 consumption pays off the **oxygen debt** created by the demand for ATP to drive gluconeogenesis.

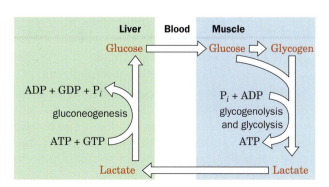

Figure 21-6 The Cori cycle. Lactate produced by muscle glycolysis is transported by the bloodstream to the liver, where it is converted to glucose by gluconeogenesis. The bloodstream carries the glucose back to the muscle, where it may be stored as glycogen. 🔊 **See the Animated Figures.**

Figure 21-7 The glucose–alanine cycle. Pyruvate produced by muscle glycolysis is the amino-group acceptor for muscle aminotransferases. The resulting alanine is transported by the bloodstream to the liver, where it is converted back to pyruvate (its amino group is disposed of via urea synthesis). The pyruvate is a substrate for gluconeogenesis, and the bloodstream carries the resulting glucose back to the muscles. ♻ **See the Animated Figures.**

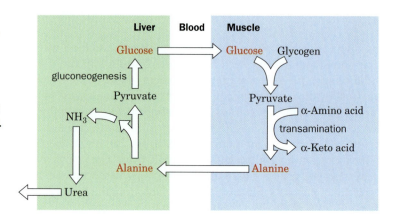

The Glucose–Alanine Cycle. In a pathway similar to the Cori cycle, alanine rather than lactate travels from muscle to the liver. In muscle, certain aminotransferases use pyruvate as their α-keto acid substrate rather than α-ketoglutarate or oxaloacetate (Section 20-2A):

$$\underset{\textbf{Amino acid}}{R-\underset{\overset{|}{\overset{+}{NH_3}}}{CH}-COO^-} + \underset{\textbf{Pyruvate}}{H_3C-\overset{\overset{O}{||}}{C}-COO^-} \rightleftharpoons \underset{\alpha\textbf{-Keto acid}}{R-\overset{\overset{O}{||}}{C}-COO^-} + \underset{\textbf{Alanine}}{H_3C-\underset{\overset{|}{\overset{+}{NH_3}}}{CH}-COO^-}$$

The product amino acid, alanine, is released into the bloodstream and transported to the liver, where it undergoes transamination back to pyruvate. This pyruvate is a substrate for gluconeogenesis, and the resulting glucose can be returned to the muscles to be glycolytically degraded. This is the **glucose–alanine cycle** (Fig. 21-7). The amino group carried by alanine ends up in either ammonia or aspartate and can be used for urea biosynthesis (which occurs only in the liver). Thus, *the glucose–alanine cycle is a mechanism for transporting nitrogen from muscle to liver.*

During fasting, the glucose formed in the liver by this route is also used by other tissues, breaking the cycle. Because the pyruvate originates from muscle protein degradation, muscle supplies glucose to other tissues even though it does not carry out gluconeogenesis.

2 Hormonal Control of Fuel Metabolism

Living things coordinate their activities at every level of their organization through complex signaling systems involving chemical messengers known as **hormones**. In higher animals, **endocrine glands** synthesize and release hormones, which are carried by the bloodstream to their target cells (Fig. 21-8). The human endocrine system secretes a wide variety of hormones that enable the body to

1. Maintain **homeostasis** (a steady state; e.g., insulin and glucagon maintain the blood glucose level within rigid limits during feast or famine).
2. Respond to a wide variety of external stimuli (such as the preparation for "fight-or-flight" by epinephrine and norepinephrine).
3. Follow various cyclic and developmental programs (for instance, sex hormones regulate sexual differentiation, maturation, the menstrual cycle, and pregnancy).

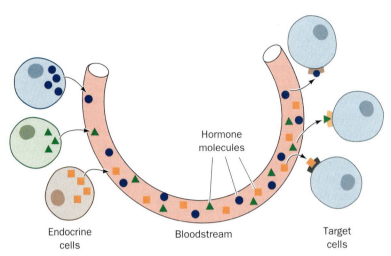

Figure 21-8 Endocrine signaling. Hormones produced by endocrine cells reach their target cells via the bloodstream. Only cells that display the appropriate receptors can respond to the hormones.

We have already discussed steroid hormones (Section 9-1E), the peptide hormones insulin and glucagon (Section 15-3C), and the catecholamines epinephrine and norepinephrine (Section 20-6B). With the exception of steroids, which can diffuse through cell membranes and interact directly with intracellular components (and which we shall discuss further in Section 27-3B), these extracellular signals must first bind to a cell-surface receptor. A **receptor** in this context is defined as a binding protein that is specific for its ligand and behaves in such a way that it elicits a discrete biochemical effect when its ligand is bound, a feature that differentiates a receptor from a simple binding or transport protein such as myoglobin (Section 7-1) or a glucose transporter (Section 10-2E). In this section, we outline the effects of hormones on fuel metabolism in different mammalian tissues. In the following section, we will examine the pathways by which these hormones exert their effects on cells.

Pancreatic and Adrenal Hormones. The pancreas and the adrenal glands synthesize the hormones that play the largest roles in regulating the metabolism of fuels. The pancreas is a large glandular organ, the bulk of which is dedicated to producing digestive enzymes such as trypsin, RNase A, α-amylase, and phospholipase A_2. These proteins are secreted via the pancreatic duct into the small intestine. However, ~1 to 2% of pancreatic tissue consists of scattered clumps of cells known as **islets of Langerhans** (Fig. 21-9), which secrete polypeptide hormones into the bloodstream. These hormones, like other proteins destined for secretion, are ribosomally synthesized as inactive precursors, processed in the rough endoplasmic reticulum and Golgi apparatus to form the mature hormones, and then packaged in secretory granules to await the signal for their release by exocytosis (Section 9-4D). The **β cells** of the pancreatic islets secrete insulin (51 residues; Fig. 5-1) in response to high blood glucose levels. The **α cells** of the pancreatic islets secrete glucagon (29 residues; p. 497) in response to low blood glucose.

In contrast to the pancreatic islets, the adrenal glands release hormones in response to neuronal signals. The adrenal glands consist of two distinct types of tissue: the **medulla** (core), which is really an extension of the nervous system, and the more typically glandular **cortex** (outer layer), which

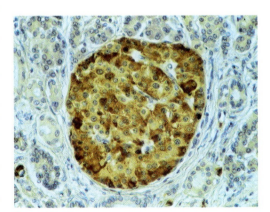

Figure 21-9 Pancreatic islet cells. The hormone-producing islet cells (*brown stain*) constitute only a small fraction of the pancreatic cells, most of which synthesize digestive enzymes. [© Parviz M. Pour/Photo Researchers, Inc.]

synthesizes and secretes steroid hormones. The adrenal medulla synthesizes norepinephrine and its methyl derivative epinephrine (p. 498) from tyrosine as described in Section 20-6B. These catecholamines are stored in granules to await their exocytotic release.

Insulin Release Is Triggered by Glucose.   The pancreas responds to increases in the concentration of blood glucose by secreting insulin, which therefore serves as a signal for plentiful metabolic fuel. Pancreatic β cells are most sensitive to glucose at concentrations of 5.5 to 6.0 mM (normal blood glucose concentrations range from 3.6 to 5.8 mM). There is no evidence for a cell-surface glucose "receptor" that might relay a signal to the secretory machinery in the β cell. In fact, the metabolism of glucose by β cells, which it enters via a passive transporter, generates the signal for insulin secretion.

The rate-limiting step of glucose metabolism in β cells is the reaction catalyzed by glucokinase (the same enzyme that occurs in hepatocytes). Consequently, glucokinase is considered the β cell's glucose "sensor." Glucokinase's G6P product is not used to synthesize glycogen, and the activity of the pentose phosphate pathway is minor. Furthermore, lactate dehydrogenase activity is low. As a result, essentially all the G6P produced in β cells is degraded to pyruvate and then converted to acetyl-CoA for oxidation by the citric acid cycle. This one-way, linear catabolic pathway for glucose directly links the β cell's rate of oxidative phosphorylation to the amount of available glucose. By mechanisms that are not entirely understood, the overall level of the β cell's respiratory activity regulates insulin synthesis and secretion.

Insulin Promotes Fuel Storage in Muscle and Adipose Tissue.   Insulin acts as the primary regulator of blood glucose concentration by promoting glucose uptake in muscles and adipose tissue and by inhibiting hepatic glucose production. Insulin also stimulates cell growth and differentiation by increasing the synthesis of glycogen, proteins, and triacylglycerols.

Muscle cells and adipocytes express an insulin-sensitive glucose transporter known as **GLUT4.** Insulin stimulates GLUT4 activity. Such an increase can occur through an increase in the intrinsic activity of the transporter molecules, but in the case of GLUT4, the increase is accomplished through the appearance of additional transporter molecules in the plasma membrane (Fig. 21-10). In the absence of insulin, GLUT4 is localized in intracellular vesicles and tubular structures. Insulin promotes the fusion of these vesicles to the plasma membrane in a process that is mediated by SNAREs (Section 9-4F). GLUT4 appears on the cell surface only a few minutes after insulin stimulation. GLUT4 has a relatively low K_M for glucose (2–5 mM), so cells containing this transporter can rapidly take up glucose from the blood. On insulin withdrawal, the glucose transporters are gradually sequestered through endocytosis.

The insulin-dependent GLUT4 transport system allows muscle and adipose tissue to quickly stockpile metabolic fuel immediately after a meal. Tissues such as the brain, which uses glucose

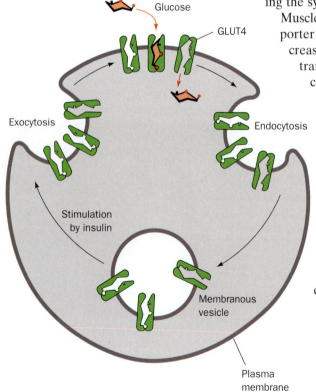

Glucose

GLUT4

Exocytosis

Endocytosis

Stimulation
by insulin

Membranous
vesicle

Plasma
membrane

Figure 21-10   GLUT4 activity.   Glucose uptake in muscle and fat cells is regulated by the insulin-stimulated exocytosis of membranous vesicles containing GLUT4 (*left*). On insulin withdrawal, the process reverses itself through endocytosis (*right*). 🎵 **See the Animated Figures.**

almost exclusively as a fuel, constitutively express an insulin-insensitive glucose transporter. Consequently, the central nervous system does not experience large fluctuations in glucose absorption. Significantly, the liver also lacks GLUT4 and therefore does not respond to increases in insulin levels by increasing its rate of glucose uptake.

Once glucose enters cells, it can be used to synthesize glycogen (in muscle) and triacylglycerols (in adipocytes; the glucose must first be metabolized to acetyl-CoA for fatty acid synthesis). Insulin specifically promotes these metabolic activities. As discussed in Section 15-3, insulin activates glycogen synthase by promoting its dephosphorylation. In adipocytes, insulin activates the pyruvate dehydrogenase complex (by activating the associated phosphatase; Section 16-4A), and it activates acetyl-CoA carboxylase and increases levels of fatty acid synthase (Section 19-5). At the same time, insulin inhibits lipolysis by inhibiting hormone-sensitive lipase.

Insulin Blocks Liver Gluconeogenesis and Glycogenolysis. Although the liver does not respond to insulin by increasing its rate of uptake of glucose, insulin binding to its receptor on hepatocytes has several consequences. The inactivation of phosphorylase kinase decreases the rate of glycogenolysis, and the activation of glycogen synthase promotes glycogen synthesis. Insulin also inhibits transcription of the genes encoding the gluconeogenic enzymes phosphoenolpyruvate carboxykinase, fructose-1,6-bisphosphatase, and glucose-6-phosphatase (see Section 15-4) and stimulates transcription of the genes for the glycolytic enzymes glucokinase and pyruvate kinase. The expression of lipogenic enzymes such as acetyl-CoA carboxylase and fatty acid synthase also increases. The result of these regulatory changes is that *the liver stores glucose (as glycogen and as triacylglycerols) rather than producing glucose by glycogenolysis or gluconeogenesis*. The major metabolic effects of insulin are summarized in Table 21-1.

Glucagon and Catecholamines Counter the Effects of Insulin. As we discussed in Section 15-3C, the peptide hormone glucagon activates a series of intracellular events that lead to glycogenolysis in the liver. This control mechanism helps make glucose available to other tissues when the concentration of circulating glucose drops. Muscle cells, which lack a glucagon receptor and cannot respond directly to the hormone, benefit indirectly from the glucose released by the liver. Glucagon also stimulates fatty acid mobilization from adipose tissue by activating hormone-sensitive lipase.

The catecholamines elicit responses similar to glucagon's. Epinephrine and norepinephrine, which are released during times of stress, bind to two

Table 21-1   Hormonal Effects on Fuel Metabolism

Tissue	Insulin	Glucagon	Epinephrine
Muscle	↑ Glucose uptake ↑ Glycogen synthesis	No effect	↑ Glycogenolysis
Adipose tissue	↑ Glucose uptake ↑ Lipogenesis ↓ Lipolysis	↑ Lipolysis	↑ Lipolysis
Liver	↑ Glycogen synthesis ↑ Lipogenesis ↓ Gluconeogenesis	↓ Glycogen synthesis ↑ Glycogenolysis	↓ Glycogen synthesis ↑ Glycogenolysis ↑ Gluconeogenesis

different types of receptors: the **β-adrenergic receptor,** which is linked to the adenylate cyclase system, and the **α-adrenergic receptor,** whose second messenger (discussed in Section 21-3F) causes intracellular Ca^{2+} concentrations to increase.

Liver cells respond to epinephrine directly and indirectly. Epinephrine promotes the release of glucagon from the pancreas, and glucagon binding to its receptor on liver cells stimulates glycogen breakdown. Epinephrine also binds directly to both α- and β-adrenergic receptors on the surfaces of liver cells (Fig. 15-22, *left*). Binding to the β-adrenergic receptor results in increased intracellular cAMP, which leads to glycogen breakdown and gluconeogenesis. Epinephrine binding to the α-adrenergic receptor stimulates an increase in intracellular $[Ca^{2+}]$, which reinforces the cells' response to cAMP (recall that phosphorylase kinase, which activates glycogen phosphorylase and inactivates glycogen synthase, is fully active only when phosphorylated and in the presence of increased $[Ca^{2+}]$; Section 15-3B). In addition, glycogen synthase is inactivated through phosphorylation catalyzed by several Ca^{2+}-dependent protein kinases.

Epinephrine binding to the β-adrenergic receptor on muscle cells similarly promotes glycogen degradation, thereby mobilizing glucose that can be metabolized by glycolysis to produce ATP. In adipose tissue, epinephrine binding to several α- and β-type receptors leads to activation of hormone-sensitive lipase, which results in the mobilization of fatty acids that can be used as fuels by other tissues. In addition, epinephrine stimulates smooth muscle relaxation in the bronchi and blood vessels supplying skeletal muscle, while it stimulates constriction of the blood vessels that supply skin and other peripheral organs. Most of these diverse effects are directed toward a common end: *to mobilize energy reserves and shunt them to where they are most needed to prepare the body for sudden action.* The major responses to glucagon and epinephrine are summarized in Table 21-1, and the metabolic effects of insulin, glucagon, and epinephrine under different conditions are presented schematically in Fig. 21-11.

3 Signal Transduction

As we saw in previous chapters, the cellular response to a hormone signal depends on a set of components that includes a specific receptor and other proteins that can transduce the signal to the cell interior. The hormones described in Section 21-2, as well as many other hormones, signaling molecules, and growth factors, use a variety of **signal-transduction pathways** to set in motion a series of biochemical reactions that produce a biological response such as altered metabolism, cell differentiation, or cell growth and division. The exact nature of the response depends on many factors. Cells respond to a hormone, generally referred to as a **ligand,** only if they display the appropriate receptor. Specific intracellular responses are modulated by the number, type, and cellular locations of the elements of the signaling systems. Furthermore, a given cell typically contains receptors for many different ligands, so the response to one particular ligand may depend on the level of engagement of the other ligands with the cell's signal-transduction machinery. As a result, *cells can react to discrete signals or combinations of signals with variations in the magnitude and duration of the cellular response.*

In general, every signal-transduction pathway consists of a receptor, a mechanism for transmitting the ligand-binding event to the cell interior, and a series of intracellular responses that may involve a second messenger

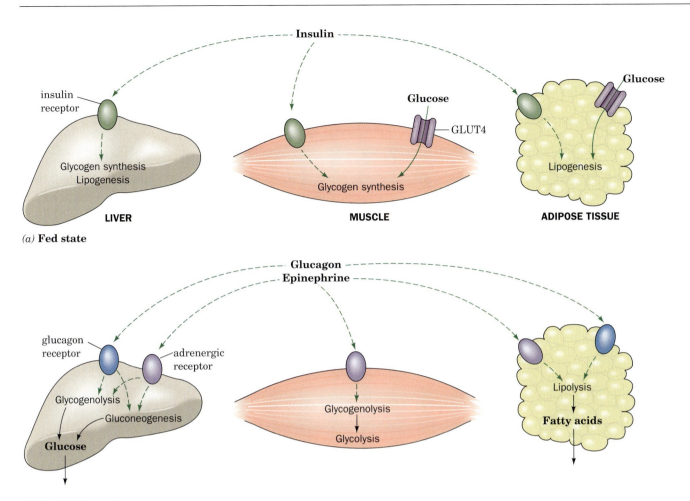

Figure 21-11 Overview of hormonal control of fuel metabolism.
(*a*) Immediately after a meal, when glucose and fatty acids are abundant, insulin signals tissues to store fuel as glycogen and triacylglycerols. Insulin also stimulates tissues other than liver to take up glucose via the GLUT4 transporter. (*b*) When dietary fuels are not available, glucagon stimulates the liver to release glucose and adipose tissue to release fatty acids. During stress, epinephrine elicits similar responses.

and/or chemical changes catalyzed by kinases and phosphatases. As we explore the various pathways below, try to identify these "standard" features.

A G Protein–Coupled Receptors

We begin our discussion of signal-transduction pathways with a common class of receptors that includes the glucagon receptor, the β-adrenergic receptor (to which epinephrine binds), and a host of other proteins that bind peptide hormones, odorant molecules, eicosanoids (Section 19-6C), and other compounds. All of these receptors are integral membrane proteins with seven transmembrane α helices. They are known collectively as **G protein–coupled receptors (GPCRs)** because they interact with heterotrimeric G proteins, which are described below. The mammalian genome (with ~30,000 genes) is estimated to contain over 1000 different GPCRs. The importance of these receptors is also evident in the fact that some 60% of therapeutic drugs target specific GPCRs.

The first GPCR to be structurally characterized at the atomic level was **rhodopsin,** a light-sensing protein in the retina. Rhodopsin consists of the

Figure 21-12   X-Ray structure of bovine rhodopsin.
The structure is viewed parallel to the plane of the membrane with the cytoplasm above. The transparent surface represents the protein's solvent-accessible surface. The polypeptide backbone is shown in tube-and-arrow representation (*blue*). Note its bundle of seven nearly parallel transmembrane helices. The protein's retinal prosthetic group is shown in space-filling form (*red*), and its two N-linked oligosaccharides (*dark blue*) and its two covalently attached palmitoyl groups (*green*) are shown in ball-and-stick form. Detergent molecules, which facilitated the crystallization of this integral protein and which are associated with its hydrophobic transmembrane surface, are drawn in ball-and-stick form in yellow. [Courtesy of Ronald Stenkamp, University of Washington. PDBid 1HZX.]

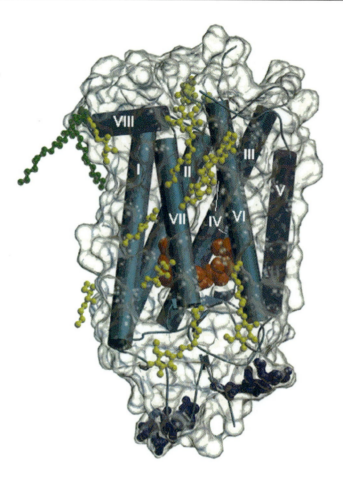

348-residue protein **opsin** and a covalently linked chromophore **retinal** (Fig. 21-12), as in the homologous protein bacteriorhodopsin (Section 9-3A). The absorption of a photon causes the rhodopsin-bound retinal to isomerize from its ground-state 11-cis form to its all-trans form. The isomerization causes opsin to undergo a transient conformational change that activates its associated G protein.

The transmembrane helices of GPCRs are generally uniform in size: 20 to 27 residues, which is sufficient to span a lipid bilayer. However, their N- and C-terminal segments and the loops connecting their transmembrane helices vary widely in length. These are the portions of the protein that bind ligands (on the extracellular side) and G proteins (on the cytoplasmic side). Rhodopsin is posttranslationally modified by oligosaccharides, two of which are positioned on extracellular loops, and palmitoyl groups, at least one of which interacts with membrane lipids.

GPCRs and other cell-surface receptors function much like allosteric proteins such as hemoglobin (Section 7-3). *By alternating between two discrete conformations, one with ligand bound and one without, the receptor can transmit an extracellular signal to the cell interior.* This model of receptor action is similar to the operation of membrane transport proteins (e.g., Fig. 10-17); in fact, some membrane-bound receptors are ion channels that switch between the open and closed conformations in response to ligand binding.

One hallmark of biological signaling systems is that they adapt to long-term stimuli by reducing their response to them, a process named **desensitization.** These signaling systems therefore respond to changes in stimulation levels rather than to their absolute values. In the case of the β-adrenergic receptor, continuous exposure to epinephrine leads to the phosphorylation of one or more of the receptor's Ser residues. This

phosphorylation, which is catalyzed by a specific kinase that acts on the hormone–receptor complex but not on the receptor alone, reduces the receptor's affinity for epinephrine. If the epinephrine level is reduced, the receptor is slowly dephosphorylated, thereby restoring the cell's epinephrine sensitivity.

B Heterotrimeric G Proteins

The class of proteins known as **G proteins,** which were first characterized by Alfred Gilman and Martin Rodbell, are named for their ability to bind the guanine nucleotides GTP and GDP and to hydrolyze GTP to GDP + P_i. As we shall see in Chapter 26, several of the proteins that assist in protein synthesis by ribosomes are G proteins, but in this chapter we will focus on the G proteins that participate in signal transduction. **Heterotrimeric G proteins,** as their name implies, are G proteins that consist of an α, β, and γ subunit (45, 37, and 9 kD, respectively).

The X-ray structures of entire heterotrimeric G proteins were independently determined by Gilman and Stephan Sprang (Fig. 21-13) and by Heidi Hamm and Paul Sigler. The large α subunit, designated G_α, contains the nucleotide binding site in a deep cleft and is anchored to the membrane by a myristoyl or palmitoyl group, or both, covalently attached near its N-terminus. The β subunit includes a seven-blade propeller-like structure built of four-stranded β sheets. The γ subunit, which is anchored to the membrane via prenylation of its C-terminus, binds to the G_β subunit with such high affinity that they only dissociate under denaturing conditions. Consequently, we shall henceforth refer to their complex as $G_{\beta\gamma}$.

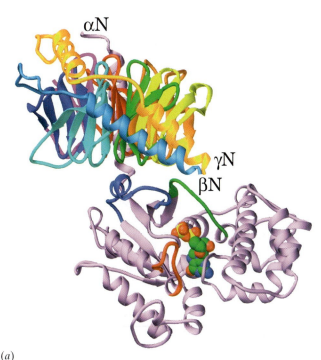

(a)

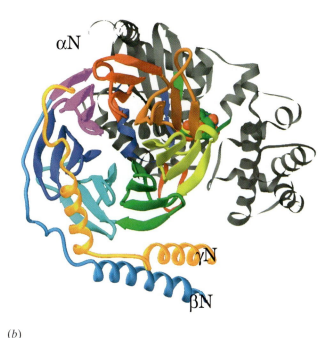

(b)

Figure 21-13   X-Ray structure of a heterotrimeric G protein.   (a) The G_α subunit is violet with the segments known as Switch I, II, and III green, blue, and red, respectively. A bound GDP is shown in space-filling form with C green, N blue, O red, and P yellow. The G_β subunit's N-terminal segment is light blue and each blade of its β-propeller has a different color. The G_γ subunit is gold. The plasma membrane is probably at the top of the drawing, as inferred from the positions of the N terminus of G_α and the neighboring C terminus of G_γ, which are lipid-linked *in vivo,* although the orientation of the protein relative to the membrane is unknown. (b) View related to that in Part *a* by a 90° rotation about the horizontal axis, that is, looking from the general direction of the plasma membrane. The protein is colored as in Part *a* except that the G_α subunit is mainly gray. [Based on an X-ray structure by Alfred Gilman and Stephan Sprang, University of Texas Southwestern Medical Center. PDBid 1GP2.] 🔍 **See the Interactive Exercises.**

In its unactivated state, a heterotrimeric G protein maintains its heterotrimeric state and its G_α subunit binds GDP. However, the binding of such a $G_\alpha \cdot GDP$–$G_{\beta\gamma}$ complex to its cognate GPCR in complex with its ligand induces the G_α subunit to exchange its bound GDP for GTP. The presence of GTP's γ phosphate group, in turn, promotes conformational changes in three so-called **switch regions** of G_α (Fig. 21-13a) that cause G_α to dissociate from $G_{\beta\gamma}$. This is because the γ phosphoryl group of GTP hydrogen bonds with side chains in Switches I and II so as to prevent these segments from interacting with the loops and turns at the bottom of the β-propeller of $G_{\beta\gamma}$. Comparison of the X-ray structures of the $G_\alpha \cdot GDP$–$G_{\beta\gamma}$ complex and $G_{\beta\gamma}$ alone indicates that the structure of $G_{\beta\gamma}$ is unchanged by its association with $G_\alpha \cdot GDP$. Nevertheless, both G_α and $G_{\beta\gamma}$ are active in signal transduction; they interact with additional cellular components, as we discuss below.

The effect of G protein activation is short-lived, because G_α is also a **GTPase** that catalyzes the hydrolysis of its bound GTP to GDP + P_i, although at the relatively sluggish rate of 2 to 3 min^{-1}. GTP hydrolysis causes the heterotrimeric G protein to reassemble as the inactive $G_\alpha \cdot GDP$–$G_{\beta\gamma}$ complex. This prevents a runaway response to ligand binding to a GPCR. Nevertheless, ligand binding to a single GPCR molecule can activate several G proteins, each of which can act on several other proteins (see below).

G Proteins Often Require Accessory Proteins to Function. The proper physiological functioning of a G protein often requires the participation of other proteins:

1. A **GTPase-activating protein (GAP),** as the name implies, stimulates its corresponding G protein to hydrolyze its bound GTP. This rate enhancement can be >2000-fold. Some downstream effectors of G proteins exhibit GTPase-activating activities toward their G proteins, which otherwise would hydrolyze GTP at physiologically insignificant rates.

2. A **guanine nucleotide exchange factor (GEF)** induces its corresponding G protein to release its bound GDP. The G protein subsequently binds another guanine nucleotide. This could be either GTP or GDP, which most G proteins bind with approximately equal affinity. However, since cells maintain a GTP concentration that is 10-fold higher than that of GDP, the overall effect is to exchange the bound GDP for GTP. For heterotrimeric G proteins, the GPCR–ligand complex functions as a GEF.

Heterotrimeric G Proteins Activate Other Proteins. A mammalian cell can contain several different kinds of heterotrimeric G proteins, since there are 20 different α subunits, 6 different β subunits, and 12 different γ subunits. This heterozygosity presumably permits various cell types to respond in different ways to a variety of stimuli. One of the major targets of the G-protein system is the enzyme **adenylate cyclase** (described in the next section). For example, when a $G_\alpha \cdot GTP$ complex dissociates from $G_{\beta\gamma}$, it may bind with high affinity to adenylate cyclase, thereby activating the enzyme. Such a G_α protein is known as a stimulatory G protein, $\mathbf{G_{s\alpha}}$. Other G_α proteins, known as inhibitory G proteins, $\mathbf{G_{i\alpha}}$, inhibit adenylate cyclase activity. The heterotrimeric $\mathbf{G_s}$ and $\mathbf{G_i}$ proteins, which differ in their α subunits, may actually contain the same β and γ subunits. Other types of heterotrimeric G proteins—acting through their G_α or $G_{\beta\gamma}$ units—stimulate the opening of ion channels, participate in the phosphoinositide signaling system (Section 21-3F), activate phosphodiesterases, and activate protein kinases.

Because a single ligand–receptor interaction can activate more than one G protein, this step of the signal-transduction pathway serves to amplify the original extracellular signal. In addition, several types of ligand–receptor complexes may activate the same G protein so that different extracellular signals elicit the same cellular response.

C   Adenylate Cyclase

The adenylate cyclase target of heterotrimeric G proteins catalyzes the reaction

$$ATP \longrightarrow cAMP + PP_i$$

cAMP (3′,5′-cyclic AMP) is a second messenger that activates a specific protein kinase, among other things (Section 15-3B). Mammals have 10 different isoforms of adenylate cyclase, which are each expressed in a tissue-specific manner and differ in their regulatory properties. These ~120-kD transmembrane glycoproteins each consist of a small N-terminal domain (N), followed by two repeats of a unit consisting of a transmembrane domain (M) followed by two consecutive cytoplasmic domains (C), thus forming the sequence $NM_1C_{1a}C_{1b}M_2C_{2a}C_{2b}$ (Fig. 21-14). The 40% identical C_{1a} and C_{2a} domains associate to form the enzyme's catalytic core, whereas C_{1b}, as well as C_{1a} and C_{2a}, bind regulatory molecules. For example, $G_{s\alpha}$ binds to C_{2a} to activate adenylate cyclase, and $G_{i\alpha}$ binds to C_{1a} to inhibit the enzyme. Other regulators of adenylate cyclase activity include Ca^{2+}, calmodulin (see Section 15-3B), protein kinase A (PKA), and protein kinase C (PKC). Clearly, *cells can respond to a great variety of stimuli in determining their cAMP levels.*

The structure of intact adenylate cyclase is not known, but X-ray structural studies of the catalytic domains indicate that $G_{s\alpha} \cdot GTP$ binds to the $C_{1a} \cdot C_{2a}$ complex via its Switch II region. This binding alters the orientation of the C_{1a} and C_{2a} domains so as to position their catalytic residues for the efficient conversion of ATP to cAMP. When $G_{s\alpha}$ hydrolyzes its bound GTP, the Switch II region reorients so that it can no longer bind to C_{2a}, and the adenylate cyclase reverts to its inactive conformation.

cAMP Is a Second Messenger.   cAMP, a polar, freely diffusing cytoplasmic molecule, is called a second messenger since it mediates the hormonal (primary) message within a cell. As described in Section 15-3B, cAMP binds to

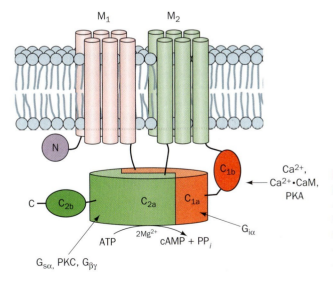

Figure 21-14   Schematic diagram of a typical mammalian adenylate cyclase.   The M_1 and M_2 domains are each predicted to contain six transmembrane helices. C_{1a} and C_{2a} form the enzyme's pseudosymmetric catalytic core. The domains with which various regulatory proteins are known to interact are indicated. [After Tesmer, J.J.G. and Sprang, S.R., *Curr. Opin. Struct. Biol.* **8**, 713 (1998).]

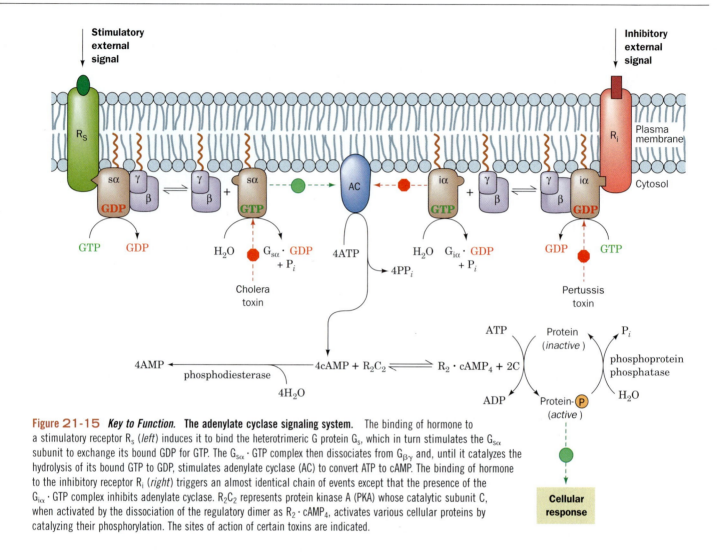

Figure 21-15 *Key to Function.* **The adenylate cyclase signaling system.** The binding of hormone to a stimulatory receptor R_s (*left*) induces it to bind the heterotrimeric G protein G_s, which in turn stimulates the $G_{s\alpha}$ subunit to exchange its bound GDP for GTP. The $G_{s\alpha} \cdot$ GTP complex then dissociates from $G_{\beta\gamma}$ and, until it catalyzes the hydrolysis of its bound GTP to GDP, stimulates adenylate cyclase (AC) to convert ATP to cAMP. The binding of hormone to the inhibitory receptor R_i (*right*) triggers an almost identical chain of events except that the presence of the $G_{i\alpha} \cdot$ GTP complex inhibits adenylate cyclase. R_2C_2 represents protein kinase A (PKA) whose catalytic subunit C, when activated by the dissociation of the regulatory dimer as $R_2 \cdot$ cAMP$_4$, activates various cellular proteins by catalyzing their phosphorylation. The sites of action of certain toxins are indicated.

the regulatory subunits of the cAMP-dependent PKA, thereby activating the kinase to phosphorylate specific Ser or Thr residues on target proteins.

The targets of PKA include enzymes involved in glycogen metabolism. Thus, when epinephrine binds to the β-adrenergic receptor of a muscle cell, for example, the sequential activation of a heterotrimeric G protein, adenylate cyclase, and PKA leads to the activation of glycogen phosphorylase, thereby making glucose-6-phosphate available for glycolysis in a "fight-or-flight" response.

Each step of a signal-transduction pathway can potentially be regulated, so *the nature and magnitude of the cellular response ultimately reflect the presence and degree of activation or inhibition of all the preceding components of the pathway.* For example, the adenylate cyclase signaling pathway can be limited or reversed through ligand activation of a receptor coupled to an inhibitory G protein. The activity of the cAMP second messenger can be attenuated by the action of a phosphodiesterase that hydrolyzes cAMP to AMP. In addition, reactions catalyzed by PKA are reversed by protein phosphatases that dephosphorylate proteins containing phospho-Ser and phospho-Thr. Some of these features of the adenylate cyclase signaling pathway are illustrated in Fig. 21-15. Many drugs and toxins exert their effects by modifying components of the adenylate cyclase system (see Box 21-1).

BOX 21-1

Biochemistry in Health and Disease

Drugs and Toxins That Affect Cell Signaling

Complex processes such as the adenylate cyclase signaling system can be sabotaged by a variety of agents. For example, the methylated purine derivatives **caffeine** (an ingredient of coffee and tea), **theophylline** (an asthma treatment), and **theobromine** (found in chocolate)

R = CH$_3$ X = CH$_3$ **Caffeine (1,3,7-trimethylxanthine)**
R = H   X = CH$_3$ **Theophylline (1,3-dimethylxanthine)**
R = CH$_3$ X = H   **Theobromine (1,7-dimethylxanthine)**

are stimulants because they antagonize adenosine receptors that act through inhibitory G proteins. This antagonism results in an increase in cAMP concentration.

Deadlier effects result from certain bacterial toxins that interfere with heterotrimeric G protein function. The toxin released by *Vibrio cholerae* (the bacterium causing cholera) causes massive fluid loss of over a liter per hour from diarrhea. Victims die from dehydration unless their lost water and salts are replaced. **Cholera toxin,** an 87-kD protein of subunit composition AB$_5$, binds to ganglioside G$_{M1}$ (Fig. 9-9) on the surface of intestinal cells via its B subunits. This permits the toxin to enter the cell, probably via receptor-mediated endocytosis, where an ~195-residue proteolytic fragment of its A subunit is released. This fragment catalyzes the transfer of the ADP–ribose unit from NAD$^+$ to a specific Arg side chain of G$_{s\alpha}$ (*below*).

ADP-ribosylated G$_{s\alpha}$·GTP can activate adenylate cyclase but cannot hydrolyze its bound GTP (Fig. 21-15). As a consequence, the adenylate cyclase is locked in its active state and cellular cAMP levels increase ~100-fold. Intestinal cells, which normally respond to small increases in cAMP by secreting digestive fluid (an HCO$_3^-$-rich salt solution), pour out enormous quantities of this fluid in response to the elevated cAMP concentrations.

Other bacterial toxins act similarly. Certain strains of *E. coli* cause a diarrheal disease similar to but less serious than cholera through their production of **heat-labile enterotoxin,** a protein that is closely similar to cholera toxin (their A and B subunits are >80% identical) and has the same mechanism of action. **Pertussis toxin** (secreted by *Bordetella pertussis,* the bacterium that causes **pertussis,** or whooping cough, which is responsible for ~400,000 infant deaths per year worldwide) is an AB$_5$ protein homologous to cholera toxin that ADP-ribosylates a specific Cys residue of G$_{i\alpha}$. The modified G$_{i\alpha}$ cannot exchange its bound GDP for GTP and therefore cannot inhibit adenylate cyclase (Fig. 21-15).

G$_{s\alpha}$

(CH$_2$)$_3$

NH   **Arg**

C=NH$_2^+$

NH$_2$

+

NAD$^+$

Adenosine—O—P—O—P—O—CH$_2$

Nicotinamide

cholera toxin →

+

ADP-ribosylated G$_{s\alpha}$

G$_{s\alpha}$

(CH$_2$)$_3$

NH

C=NH$_2^+$

NH

Adenosine—O—P—O—P—O—CH$_2$

See Guided Exploration 20

Mechanisms of hormone signaling involving the receptor tyrosine kinase system.

D   Receptor Tyrosine Kinases

Insulin and many other protein hormones known as **growth factors** do not act via GPCRs and cAMP-dependent pathways. Instead, these hormones bind to receptors whose C-terminal domains have **tyrosine kinase** activity. A tyrosine kinase catalyzes the ATP-dependent phosphorylation of a Tyr side chain:

$$\text{Protein}-CH_2-\text{—}OH \xrightarrow[\ \ -ADP\ \ ]{ATP} \text{Protein}-CH_2-\text{—}O-\overset{\displaystyle O}{\underset{\displaystyle O^-}{\overset{\displaystyle \|}{\underset{\displaystyle |}{P}}}}-O^-$$

Receptor tyrosine kinases (RTKs) typically contain only a single transmembrane segment and are monomers in the unliganded state. These structural features make it unlikely that ligand binding to an extracellular domain manifests itself as a conformational change in an intracellular domain [such a conformational shift seems more likely to occur in receptors with multiple transmembrane segments, such as the GPCRs (e.g., rhodopsin; Fig. 21-12)]. Indeed, the most common mechanism for activating RTKs appears to be ligand-induced dimerization of receptor proteins (i.e., ligand binding causes two monomeric receptors to form a dimer). The insulin receptor is unusual in that it is a dimer in the unliganded state. In this case, ligand binding apparently does induce a conformational change in the receptor.

Receptor Tyrosine Kinases Undergo Autophosphorylation. When an RTK dimerizes (or its conformation changes on ligand binding, in the case of the insulin receptor), the cytoplasmic tyrosine kinase domains are brought close together so that they cross-phosphorylate each other on specific tyrosine residues. This **autophosphorylation** activates the tyrosine kinase so that it can phosphorylate other protein substrates (likewise, phosphorylation of protein kinase A activates it as a kinase; Section 15-3B).

How does autophosphorylation activate the tyrosine kinase activity of the insulin receptor? The protein is synthesized as a single 1382-residue precursor peptide that is proteolytically processed to yield the disulfide-linked α and β subunits of the mature receptor (Fig. 21-16). Insulin binds to the α subunits, which are entirely extracellular. The X-ray structure of a 306-residue portion of the β subunit reveals that the tyrosine kinase domain (Fig. 21-17a) structurally resembles Ser/Thr kinases such as PKA (Fig. 15-14) and the phosphorylase kinase γ subunit (Fig. 15-16).

On ligand binding and phosphorylation, the tyrosine kinase's N-terminal lobe undergoes a nearly rigid 21° rotation relative to the C-terminal lobe (Fig. 21-17b). This dramatic conformational change closes the active site

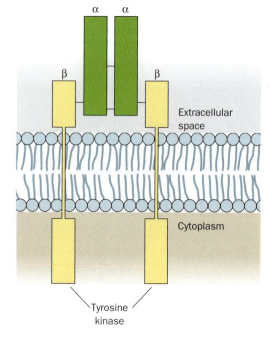

Figure 21-16   Schematic diagram of the insulin receptor. The subunits of the $\alpha_2\beta_2$ heterotetramer are linked by disulfide bonds (short horizontal bars). The extracellular α subunits form the insulin-binding site. The cytoplasmic portions of the β subunits are tyrosine kinases that are activated by insulin binding.

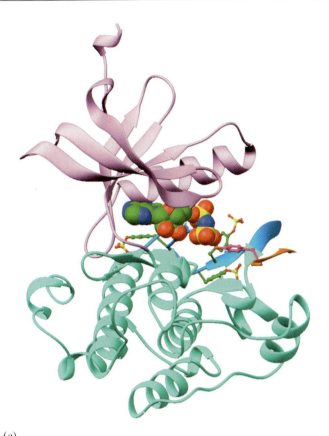

(a)

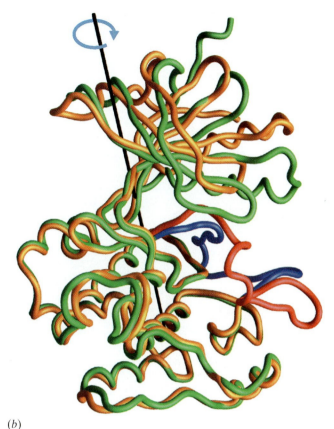

(b)

Figure 21-17   X-Ray structure of the tyrosine kinase domain of the insulin receptor.   (a) The tyrosine kinase domain is shown in the "standard" protein kinase orientation with its N-terminal domain lavender, its C-terminal domain cyan, and its activation loop light blue. Its three phosphorylated Tyr side chains are shown in ball-and-stick form with C green, N blue, O red, and P yellow. An ATP analog is shown in space-filling form. The substrate polypeptide is orange (only six of its residues are visible) and its phosphorylatable Tyr residue is shown with C magenta and O red. Compare this structure with those of PKA and phosphorylase kinase (Figs. 15-14 and 15-16). (b) The polypeptide backbones of the phosphorylated and unphosphorylated forms of the insulin receptor tyrosine kinase domain are shown superimposed on their C-terminal lobes. The phosphorylated protein is green with its activation loop blue, and the unphosphorylated protein is yellow with its activation loop red. The blue arrow and black axis indicate the rotation required to align the two N-terminal lobes. [Part a based on an X-ray structure by and Part b courtesy of Stevan Hubbard, New York University Medical School. PDBids 1IR3 and 1IRK.] 🖱 **See the Interactive Exercises.**

cleft, presumably positioning critical residues for substrate binding and catalysis. Three tyrosine residues that are phosphorylated are all located on the "activation loop." The unphosphorylated activation loop threads through the tyrosine kinase active site so as to prevent the binding of both ATP and protein substrates. On phosphorylation, however, the activation loop assumes a conformation that does not occlude the active site (Fig. 21-17b) but instead forms part of the substrate recognition site. In fact, experiments indicate that the tyrosine kinase activity of the insulin receptor increases with the degree of phosphorylation at its three autophosphorylatable Tyr side chains. The specificity of the protein for phosphorylating Tyr rather than Ser or Thr is explained by the observation that the side chain of Tyr, but not those of Ser or Thr, is long enough to reach the active site.

The main targets of the insulin receptor tyrosine kinase are known as **insulin receptor substrates 1** and **2 (IRS-1 and IRS-2).** When phosphorylated, these proteins can interact with yet another set of proteins that contain one or two conserved ~100-residue modules known as **Src homology 2 (SH2) domains** (because they are similar to the sequence of a domain in the protein known as **Src**). SH2 domains bind phospho-Tyr residues with high affinity but do not bind the far more abundant phospho-Ser and

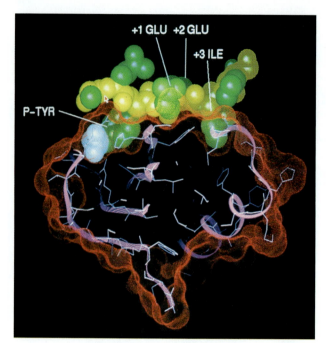

Figure 21-18 **X-Ray structure of the Src SH2 domain.**
An 11-residue polypeptide containing the protein's
phospho-Tyr-Glu-Glu-Ile target tetrapeptide is bound to the
SH2 domain. In this cutaway view, the protein surface is
represented by red dots, the protein backbone (*pink*) is
shown in ribbon form with its side chains in stick form,
and the bound polypeptide's N-terminal 8-residue segment
is shown in space-filling form with its backbone yellow, its
side chains green, and its phosphate group white.
[Courtesy of John Kuriyan, The Rockefeller University.]

phospho-Thr residues. This specificity has a simple explanation.
X-Ray structural studies reveal that phospho-Tyr interacts with
an Arg at the bottom of a deep pocket (Fig. 21-18). The side
chains of Ser and Thr are too short to interact with this residue.
There is growing evidence that analogous protein domains recognize phospho-Ser and phospho-Thr groups as part of other signaling pathways.

The SH2-containing proteins that interact with the IRSs
and other substrates have varied functions: Some activate G proteins, some are kinases, and some are phosphatases. To further
complicate things, IRS-1 undergoes serine phosphorylation,
which attenuates the effects of insulin-stimulated tyrosine phosphorylation.

Like other intracellular signals, cytoplasmic tyrosine kinases
must be "turned off" after the system has delivered its message.
The off switch is provided by **protein tyrosine phosphatases,** enzymes that dephosphorylate phospho-Tyr residues.

The complex web of interactions in signal-transduction pathways explains why different extracellular signals seem to converge on the same intracellular pathways. In the case of insulin
signaling, the intersecting pathways mediate changes in vesicle
trafficking (e.g., translocation of GLUT4), enzyme activation and
inactivation (e.g., phosphorylation and dephosphorylation), and changes in
gene expression. The multiple steps by which insulin exerts its effects on carbohydrate and lipid metabolism are not completely understood. We discuss
the present state of our understanding of this complex system after looking
at the details of some other RTK-mediated pathways that are better known.

Growth Factor Receptor Tyrosine Kinase Initiates a Kinase Cascade. Molecular
genetic analysis of signaling in a variety of distantly related organisms has
revealed a remarkably conserved pathway that regulates such essential
functions as cell growth and differentiation. Briefly, growth factor binding
to a receptor tyrosine kinase activates a monomeric G protein named
Ras, which in turn activates a "kinase cascade" that relays the signal to the
transcriptional apparatus in the nucleus.

The binding of a growth factor to its RTK leads to autophosphorylation
of the RTK, which then interacts with an SH2-containing protein (Fig. 21-19,
left). Many proteins that contain SH2 domains also have one or more unrelated 50- to 75-residue **SH3 domains.** SH3 domains, which bind Pro-rich
sequences of 9 or 10 residues, are also present in some proteins that lack
SH2 domains. Both types of domains mediate the interactions between kinases and regulatory proteins. In the signaling cascade shown in Fig. 21-19,
a 217-residue mammalian protein known as **Grb2** (**Sem-5** in *Caenorhabditis
elegans*), which consists almost entirely of an SH2 domain flanked by two
SH3 domains, forms a complex with the 1596-residue **Sos protein.** Sos contains a Pro-rich sequence that binds specifically to SH3 domains. The
Grb2–Sos complex bridges the activated RTK and Ras in a way that induces Ras to exchange its bound GDP for GTP, thereby activating it. Only
Ras · GTP is capable of further relaying the signal from the RTK. However,
Ras, like other G proteins, hydrolyzes GTP to GDP + P_i, thereby limiting
the magnitude of the cell's response to the growth factor. GTP hydrolysis
is accelerated by a GTPase-activating protein (GAP).

The signaling pathway downstream of Ras consists of a linear cascade of
protein kinases (Fig. 21-19, *right*). The Ser/Thr kinase **Raf,** which is activated
by direct interaction with Ras · GTP, phosphorylates a protein alternatively

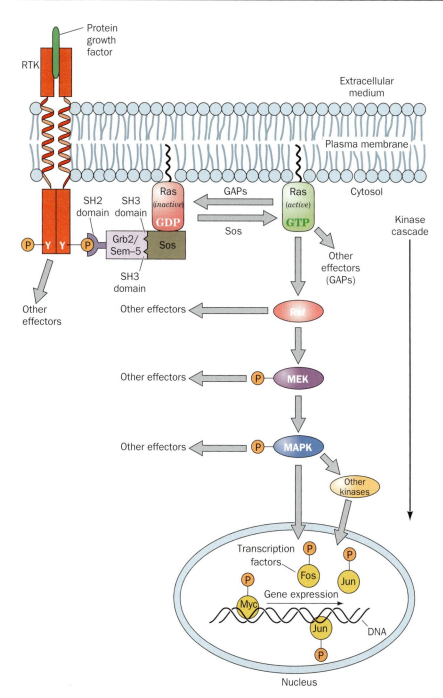

Figure 21-19 **The Ras signaling cascade.**
RTK binding to its cognate growth factor induces the autophosphorylation of the RTK's cytosolic domain. Grb2/Sem-5 binds to the resulting phospho-Tyr–containing peptide segment via its SH2 domain and simultaneously binds to Pro-rich segments on Sos via its two SH3 domains. This activates Sos to exchange Ras's bound GDP for GTP, which activates Ras to bind to Raf. Then, in a so-called kinase cascade, Raf, a Ser/Thr kinase, phosphorylates MEK, which in turn phosphorylates MAPK, which then migrates to the nucleus, where it phosphorylates transcription factors such as Fos, Jun, and Myc, thereby modulating gene expression. Ras is inactivated by GTP hydrolysis, a process that is accelerated by GTPase-activating proteins (GAPs). [After Egan, S.E. and Weinberg, R.A., *Nature* **365,** 782 (1993).]
🔊 **See the Animated Figures.**

known as **MEK** or **MAP kinase kinase,** thereby activating it as a kinase. Activated MEK phosphorylates a family of proteins variously termed **mitogen-activated protein kinases (MAPKs)** or **extracellular-signal-regulated kinases (ERKs).** A MAPK must be phosphorylated at both its Thr and Tyr residues in the sequence Thr-Glu-Tyr for full activity. MEK (which stands for *M*AP kinase/*E*RK kinase-activating *k*inase) catalyzes both phosphorylations; it is therefore a Ser/Thr kinase as well as a Tyr kinase.

The activated MAP kinases migrate from the cytosol to the nucleus, where they phosphorylate a variety of proteins, including **Fos, Jun,** and **Myc.** These proteins are **transcription factors** (proteins that induce the transcription of their target genes; Section 25-2C): In their activated forms they

BOX 21-2

Biochemistry in Health and Disease

Oncogenes and Cancer

The growth and differentiation of cells in the body are normally strictly controlled. Thus, with few exceptions (e.g., blood-forming cells and hair follicles), cells in the adult body are largely quiescent. However, for a variety of reasons, a cell may be made to proliferate uncontrollably to form a tumor.

Malignant tumors (cancers) grow in an invasive manner and are almost invariably life threatening. They are responsible for 20% of the mortalities in the United States.

Among the many causes of cancer are viruses that carry **oncogenes** (Greek: *onkos*, mass or tumor). For example, the **Rous sarcoma virus (RSV),** which induces the formation of **sarcomas** (cancers arising from connective tissues) in chickens, contains four genes. Three of these genes are essential for viral replication, whereas the fourth, **v-*src*** (v for viral, *src* for sarcoma), an oncogene, induces tumor formation. What is the origin of v-*src*, and how does it function? Hybridization studies by Michael Bishop and Harold Varmus in 1976 led to the remarkable discovery that uninfected chicken cells contain a gene, **c-*src*** (c for cellular), that is homologous to v-*src* and that is highly conserved in a wide variety of eukaryotes, suggesting that it is an essential cellular gene. Apparently, v-*src* was originally acquired from a cellular source by a non-tumor-forming ancestor of RSV. Both v-*src* and c-*src* encode a 60-kD tyrosine kinase. However, whereas the

activity of c-*src* is strictly regulated, that of v-*src* is under no such control and hence its presence maintains the host cell in a proliferative state. Since cells are not killed by an RSV infection, this presumably enhances the viral replication rate.

Other oncogenes have been similarly linked to processes that regulate cell growth. For example, the **v-*erbB*** oncogene specifies a truncated version of the **epidermal growth factor (EGF) receptor,** which lacks the EGF-binding domain but retains its transmembrane segment and its tyrosine kinase domain. This kinase phosphorylates its target proteins in the absence of an extracellular signal, thereby driving uncontrolled cell proliferation.

The **v-*ras*** oncogene encodes a 21-kD protein, **v-Ras,** that resembles cellular Ras but hydrolyzes GTP much more slowly. The reduced braking effect of GTP hydrolysis on the rate of protein phosphorylation leads to increased activation of the kinases downstream of Ras (Fig. 21-19).

The transcription factors that respond to Ras-mediated signaling (e.g., Fos and Jun) are also encoded by **proto-oncogenes,** the normal cellular analogs of oncogenes. The viral genes **v-*fos*** and **v-*jun*** encode proteins that are nearly identical to their cellular counterparts and mimic their effects on host cells but in an uncontrolled fashion.

Oncogenes are not necessarily of viral origin. Indeed, few human cancers are caused by viruses. Rather, they are caused by proto-oncogenes that have mutated to form oncogenes. For example, a mutation in the **c-*ras*** gene, which converts Gly 12 of Ras to Val, reduces Ras's GTPase activity without affecting its ability to stimulate protein phosphorylation. This prolongs the time that Ras is in the "on" state, thereby inducing uncontrolled cell proliferation. In fact, oncogenic versions of c-*ras* are among the most commonly implicated oncogenes in human cancers.

To date, over 50 oncogenes have been identified. The subversive effects of oncogene products arise through their differences from the corresponding normal cellular proteins: They may have different rates of synthesis and/or degradation; they may have altered cellular functions; or they may resist control by cellular regulatory mechanisms. However, in order for a normal cell to undergo a **malignant transformation** (become a cancer cell), it must undergo several (an average of five) independent oncogenic events. This is a reflection of the complexity of cellular signaling networks (cells respond to a variety of hormones, growth factors, and transcription factors in partially overlapping ways) and explains why the incidence of cancer increases with age.

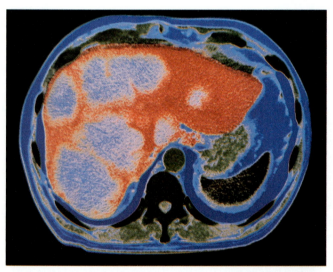

An X-ray based false color image showing an axial section through a human abdomen that has cancer of the liver. The liver is the large red mass occupying much of the abdomen; the light patches on the liver are cancerous tumors. A vertebra (*dark green*) can be seen at the lower center of the image. [Salisbury/Photo Researchers].

stimulate various genes to produce the effects commissioned by the extra-cellular presence of the growth factor that initiated the signaling cascade. When insulin activates the Ras signaling pathway, the result is an increase in protein synthesis that supports cell growth and differentiation, a response consistent with insulin's function as a signal of fuel abundance. Variant proteins encoded by **oncogenes** subvert such signaling pathways so as to induce uncontrolled cell growth (see Box 21-2).

E Protein Phosphatases

Examination of the human genome reveals the presence of ~500 genes en-coding protein kinases. These enzymes participate in signal transduction, regulation of central metabolic pathways, cell cycle control, and cell growth and differentiation. One of the challenges in understanding the function of a particular kinase is identifying its physiological target proteins. This is no small task, as an estimated one-third of eukaryotic proteins are subject to reversible phosphorylation. In many cases, the kinase's target is another ki-nase (e.g., Fig. 21-19). The advantage of such a kinase cascade is that *a small signal can be amplified manyfold inside the cell.* In addition, phosphoryla-tion of more than one target protein can lead to the simultaneous activa-tion of several intracellular processes.

The activity of protein kinases is balanced by the activity of protein phos-phatases that hydrolyze the phosphoryl groups attached to Ser, Thr, or Tyr side chains and thereby reverse the effects of the signal that originally acti-vated the kinase. Although kinases have traditionally garnered more atten-tion, mammalian cells contain a large number of phosphatases with sub-strate specificities comparable to those of kinases.

Protein Tyrosine Phosphatases.   The enzymes that dephosphorylate Tyr residues, the **protein tyrosine phosphatases (PTPs),** are not just simple housekeeping enzymes but are signal transducers in their own right. These enzymes form a large family of diverse proteins that are present in all eukaryotes. Each tyrosine phosphatase contains at least one conserved ~240-residue phosphatase domain that has the 11-residue signature se-quence [(I/V)HCXAGXGR(S/T)G], the so-called CX_5R motif, which con-tains the enzyme's catalytically essential Cys and Arg residues. During the reaction, the phosphoryl group is transferred from the tyrosyl residue of the substrate protein to the essential Cys on the enzyme, forming a cova-lent Cys–phosphate intermediate that is subsequently hydrolyzed.

Some tyrosine phosphatases are constructed much like the receptor ty-rosine kinases; that is, they have an extracellular domain, a single trans-membrane helix, and a cytoplasmic domain consisting of a catalytically ac-tive PTP domain that, in most cases, is followed by a second PTP domain with little or no catalytic activity. These inactive PTP domains are, never-theless, highly conserved, which suggests that they have an important although as yet unknown function. Biochemical and structural analyses indicate that ligand-induced dimerization of a receptor-like PTP reduces its catalytic activity, probably by blocking its active sites.

A second group of PTPs, intracellular PTPs, contain only one tyrosine phosphatase domain, which is flanked by regions containing motifs, such as SH2 domains (e.g., Fig. 21-18), that participate in protein–protein interac-tions. The PTP known as **SHP-2,** which is expressed in all mammalian cells, binds to a variety of phosphorylated (that is, ligand-activated) receptor tyrosine kinases. The X-ray structure of SHP-2 lacking its C-terminal tail reveals two SH2 domains, followed by a tyrosine phosphatase domain

Figure 21-20 **X-Ray structure of the protein tyrosine phosphatase SHP-2.** In this drawing, its N-SH2 domain is gold with its D′E loop red, its C-SH2 domain is green, and its PTP domain is cyan, with its 11-residue signature sequence, its CX_5R motif, blue and the side chain of its catalytically essential Cys residue shown in ball-and-stick form with C green and S yellow. [Based on an X-ray structure by Michael Eck and Steven Shoelson, Harvard Medical School. PDBid 2SHP.]

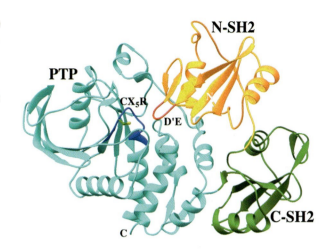

(Fig. 21-20). The N-terminal SH2 domain (N-SH2) functions as an autoinhibitor by inserting a protein loop (labeled D′E in Fig. 21-20) into the PTP's catalytic cleft. When N-SH2 recognizes and binds a phospho-Tyr group on a substrate protein, its conformation changes, unmasking the PTP catalytic site so that the phosphatase can hydrolyze another phospho-Tyr group on the target protein (activated RTKs typically bear multiple phosphorylated Tyr residues).

The active-site cleft of intracellular tyrosine phosphatases such as SHP-2 is too deep to bind phospho-Ser/Thr side chains. However, the active site pockets of a third group of PTPs, the so-called dual-specificity tyrosine phosphatases, are sufficiently shallow to bind both phospho-Tyr and phospho-Ser/Thr residues.

Protein Ser/Thr Phosphatases. The **protein Ser/Thr phosphatases** in mammalian cells belong to two protein families: the **PPP family** and the **PPM family.** The PPP and PPM families are unrelated to each other or to the protein tyrosine phosphatases. X-Ray structures have shown that PPP catalytic centers each contain an Fe^{2+} (or possibly an Fe^{3+}) ion and a Zn^{2+} (or possibly an Mn^{2+}) ion, whereas PPM catalytic centers each contain two Mn^{2+} ions. These binuclear metal ion centers nucleophilically activate water molecules to dephosphorylate substrates in a single reaction step.

We have already considered phosphoprotein phosphatase-1 (PP1, a member of the PPP family) in connection with its role in dephosphorylating the proteins that regulate glycogen metabolism (Section 15-3B). The PPP member known as **PP2A** participates in a wide variety of regulatory processes including those governing metabolism, DNA replication, transcription, and development. In addition to its catalytic subunit and a set of regulatory subunits, PP2A contains a **scaffold subunit** (the A subunit) with an unusual structure (Fig. 21-21). Scaffold subunits bind the other subunits of a protein so as to ensure that these subunits interact only with one another.

The PPP family of phosphatases also includes **calcineurin** (also called **PP2B**), a Ser/Thr phosphatase that is activated by Ca^{2+}. Calcineurin plays an essential role in T cell proliferation. It is inhibited by the action of drugs

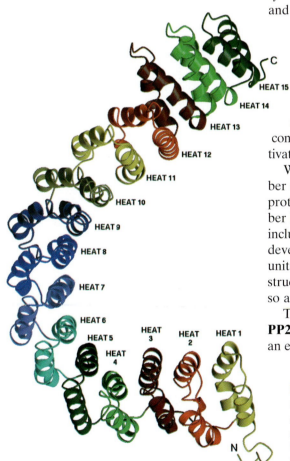

Figure 21-21 **X-Ray structure of the A subunit of the protein phosphatase PP2A.** This solenoid-shaped protein is built from 15 imperfect repeats, each shown in a different color, of a 39-residue sequence termed HEAT for this characteristic amino acid sequence. The entire structure is ~100 Å long. [Courtesy of Bostjan Kobe, St. Vincent's Institute of Medical Research, Fitzroy, Victoria, Australia. X-Ray structure by David Barford, University of Oxford, U.K. PDBid 1B3U.]

such as **cyclosporin,** which is used to suppress immune system function following organ transplants.

F   The Phosphoinositide Pathway

A discussion of signal-transduction pathways would not be complete without the **phosphoinositide pathway,** which leads to an increase in intracellular Ca^{2+} concentrations. This signaling pathway requires a receptor with seven transmembrane segments, a heterotrimeric G protein, a specific kinase, and a phosphorylated glycerophospholipid that is a minor component of the plasma membrane's inner leaflet.

Ligand binding to its receptor, such as epinephrine binding to the α_1-adrenergic receptor, activates a heterotrimeric G protein, G_q, whose membrane-anchored α subunit in complex with GTP diffuses laterally along the plasma membrane to activate the membrane-bound enzyme **phospholipase C (PLC;** Fig. 21-22, *upper left*). Activated PLC catalyzes the hydrolysis of **phosphatidylinositol-4,5-bisphosphate (PIP$_2$)** at its glycero-phospho

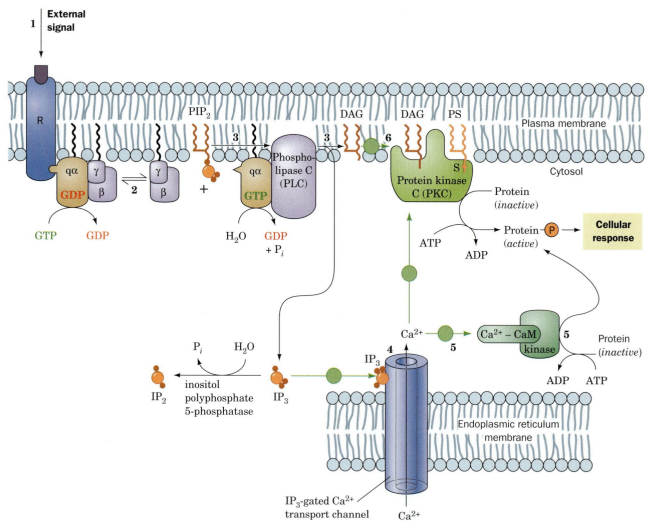

Figure 21-22   *Key to Function*.   The phosphoinositide signaling system. Ligand binding to a cell-surface receptor R (**1**) activates phospholipase C through the heterotrimeric G protein G_q (**2**). Phospholipase C catalyzes the hydrolysis of PIP$_2$ to IP$_3$ and DAG (**3**). The water-soluble IP$_3$ stimulates the release of Ca^{2+} sequestered in the endoplasmic reticulum (**4**), which in turn activates numerous cellular processes through the intermediacy of calmodulin (CaM; **5**). The nonpolar DAG remains associated with the membrane, where it activates protein kinase C (PKC) to phosphorylate and thereby modulate the activities of a number of cellular proteins (**6**). PKC activation also requires the presence of the membrane lipid phosphatidylserine (PS) and Ca^{2+}.   *See the Animated Figures.*

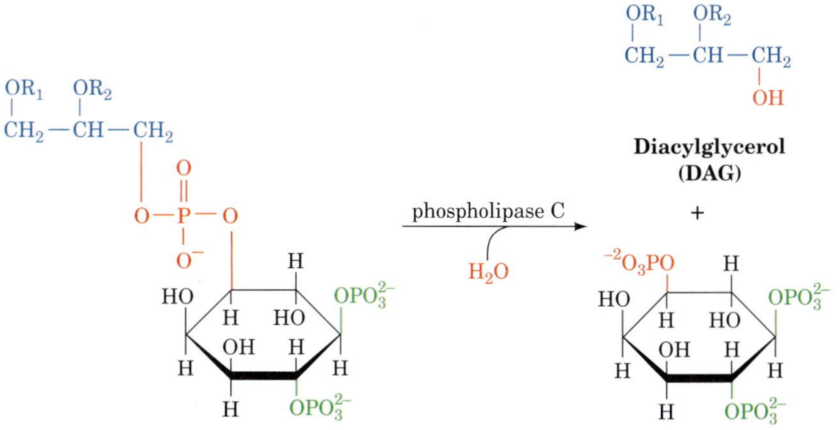

Phosphatidylinositol-4,5-bisphosphate (PIP_2)

Diacylglycerol (DAG)

Inositol-1,4,5-trisphosphate (IP_3)

Figure 21-23 Phosphatidylinositol-4,5-bisphosphate (PIP_2) and its hydrolysis products. PIP_2 is cleaved by phospholipase C to produce diacylglycerol (DAG) and inositol-1,4,5-trisphosphate (IP_3), both of which are second messengers. (The *bis* and *tris* prefixes denote, respectively, two and three phosphoryl groups that are linked separately to the inositol; in di- and triphosphates, the phosphoryl groups are linked sequentially.)

bond (Section 9-1C), yielding **inositol-1,4,5-trisphosphate (IP_3)** and **1,2-diacylglycerol (DAG;** Fig. 21-23). PLC, which in mammals is actually a set of 11 isozymes, has a hydrophobic ridge consisting of three protein loops that is postulated to penetrate into the membrane's nonpolar region during catalysis. This would explain how the enzyme can catalyze hydrolysis of the membrane-bound PIP_2, leaving the DAG product associated with the membrane.

The charged IP_3 molecule is a water-soluble second messenger that diffuses through the cytoplasm to the endoplasmic reticulum. There, it binds to and induces the opening of a Ca^{2+} transport channel (an example of a receptor that is also an ion channel), thereby allowing the efflux of Ca^{2+} from the endoplasmic reticulum (ER). This causes the cytosolic $[Ca^{2+}]$ to increase from ~0.1 μM to as much as 10 μM, which triggers such diverse cellular processes as glucose mobilization and muscle contraction through the intermediacy of calmodulin (Section 15-3B) and its homologs. The ER contains embedded Ca^{2+}–ATPases that actively pump Ca^{2+} from the cytosol to the ER (Section 10-3B) so that in the absence of IP_3, the cytosolic $[Ca^{2+}]$ rapidly returns to its resting state.

The nonpolar diacylglycerol product of phospholipase C action is a lipid-soluble second messenger. It remains embedded in the plasma membrane, where it activates **protein kinase C (PKC)** to phosphorylate and thereby modulate the activities of several different cellular proteins. The phosphoinositide signaling system is diagrammed in Fig. 21-22.

Multiple PKC enzymes are known; they differ in tissue expression, intracellular location, and their requirement for the diacylglycerol that activates them. PKC is a phosphorylated, cytosolic protein in its resting state. Diacylglycerol increases the membrane affinity of PKC and also helps stabilize its active conformation.

The X-ray structure of the DAG-bound segment of PKC shows that the 50-residue motif is largely knit together by two Zn^{2+} ions, each of which is tetrahedrally liganded by one His and three Cys side chains (Fig. 21-24). A DAG analog, **phorbol-13-acetate,**

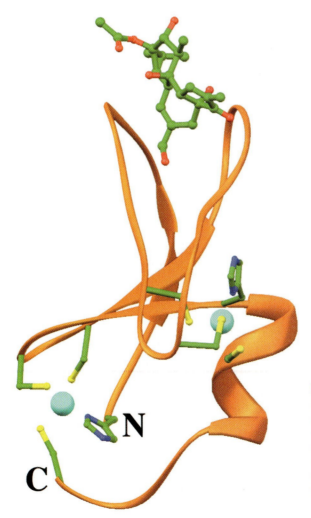

Phorbol-13-acetate

binds in a narrow groove between two long nonpolar loops. Very few soluble proteins have such a large continuous nonpolar region, suggesting that this portion of PKC inserts into the membrane. Full activation of PKC requires phosphatidylserine (which is present only in the cytoplasmic leaflet of the plasma membrane) and, in some cases, Ca^{2+} (presumably made available through the action of the IP_3 second messenger). Like other signaling systems, the phosphoinositide system is limited by the destruction of its second messengers, for example, through the action of **inositol polyphosphate 5-phosphatase** (Fig. 21-22, *lower left*).

Variations in Lipid Signaling. Choline-containing phospholipids hydrolyzed by phospholipase C yield diacylglycerols that differ not only from those

Figure 21-24 X-Ray structure of a portion of protein kinase C in complex with phorbol-13-acetate.   The protein tetrahedrally ligands two Zn^{2+} ions (*cyan spheres*), each via His and Cys side chains (shown in ball-and-stick form). Phorbol-13-acetate (*top*), which mimics the natural diacylglycerol ligand, binds between two nonpolar protein loops. Atoms are colored: C green, N blue, and O red. [Based on an X-ray structure by James Hurley, NIH. PDBid 1PTR.]

released from PIP$_2$ but also in their effects on PKC. Another lipid second messenger, sphingosine released from sphingolipids, inhibits PKC. The catalytic activities of PKC and PKA are similar: Both kinases phosphorylate Ser and Thr residues.

The phosphoinositide signaling pathway in some cells yields a diacylglycerol that is predominantly 1-stearoyl-2-arachidonoyl-glycerol. This molecule is further degraded to yield arachidonate, the precursor of the bioactive eicosanoids (prostaglandins and thromboxanes; Section 19-6C). In other cells, IP$_3$ and diacylglycerol are rapidly recycled to re-form PIP$_2$ in the inner leaflet of the membrane. Some receptor tyrosine kinases activate a form of phospholipase C that contains two SH2 domains. This is one example of **cross talk,** the interactions of different signal-transduction pathways.

The insulin signaling pathway involves both the tyrosine kinase signaling cascade and phosphoinositides. Phosphorylation of IRS proteins (Section 21-3D) leads to activation of an enzyme known as a **phosphoinositide 3-kinase (PI3K).** This enzyme adds a phosphoryl group to the 3′-OH group of a phosphatidylinositol, often the 4,5-bisphosphate shown in Fig. 21-23. The 3-phosphorylated lipid activates **phosphoinositide-dependent protein kinase-1 (PDK1)** that in turn activates a PKC and initiates cascades leading to glycogen synthesis and the translocation of GLUT4 to the surface of insulin-responsive cells, as well as affecting cell growth and differentiation.

Figure 21-25 illustrates the multiple pathways of action of insulin signaling. The reaction of such a complex signaling system to ligand stimulation is extremely difficult to predict from a knowledge of its individual steps. However, it is clear that such a system is responsive to a variety of inputs in a graded manner. Moreover, by activating multiple pathways, a hormone such as insulin can trigger a variety of physiological effects that would not be possible in a one hormone–one target regulatory system. Similarly, during *Bacillus anthracis* infection, the anthrax toxin interferes with more than one signaling pathway (Box 21-3).

4 Disturbances in Fuel Metabolism

The complexity of the mechanisms that regulate mammalian fuel metabolism permit the body to respond efficiently to changing energy demands and to accommodate changes in the availability of various fuels, maintaining reasonable **metabolic homeostasis** (balance between energy inflow and output). Such complex systems can also malfunction, producing acute or chronic diseases of variable severity. Considerable effort has been directed at elucidating the molecular basis of conditions such as diabetes and obesity, both of which are essentially disorders of fuel metabolism. In this section, we examine the metabolic changes that occur in starvation, diabetes, and obesity.

A Starvation

Because humans do not eat continuously, the disposition of dietary fuels and the mobilization of fuel stores shifts dramatically during the few hours between meals. Yet humans can survive fasts of up to a few months by adjusting their fuel metabolism. Such metabolic flexibility certainly evolved before modern humans became accustomed to thrice-daily meals.

Absorbed Fuels Are Allocated Immediately. When a meal is digested, nutrients are broken down to small, usually monomeric, units for absorption by the

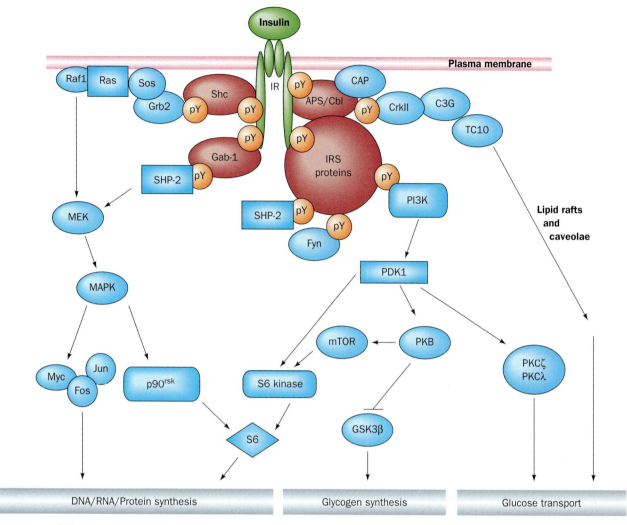

Figure 21-25 Insulin signal transduction. The binding of insulin to the insulin receptor **(IR)** induces tyrosine phosphorylations (pY) that lead to the activation of the MAPK and PI3K phosphorylation cascades as well as a lipid raft and caveolae-associated regulation process. The MAPK cascade regulates the expression of genes involved in cellular growth and differentiation. The PI3K cascade leads to changes in the phosphorylation states of several enzymes, so as to stimulate glycogen synthesis, as well as other metabolic pathways. The PI3K cascade also participates in the control of vesicle trafficking, leading to the translocation of the GLUT4 glucose transporter to the cell surface and thus increasing the rate of glucose transport into the cell. Glucose transport control is also exerted by the APS/Cbl system in a PI3K-independent manner involving lipid rafts and caveolae (Section 9-4C). Other symbols: Myc, Fos, and Jun (transcription factors), SHP-2 (an SH2-containing PTP), **CAP** (Cbl-associated protein), **C3G** [a guanine nucleotide exchange factor (GEF)], **CrkII** (an SH2/SH3-containing adapter protein), PDK1 (phosphoinositide-dependent protein kinase-1), PKB (protein kinase B, also named Akt), **mTOR** [for *mammalian target of rapamycin*, a PI3K-related protein kinase; **rapamycin** is an immunosuppressant similar to FK506; mTOR is also known as **FKBP12-rapamycin-associated protein (FRAP)**], **S6** (a protein subunit of the eukaryotic ribosome's small subunit whose phosphorylation stimulates translation), and PKCζ and PKCλ (atypical isoforms of protein kinase C). [After Zick, Y. *Trends Cell Biol.* **11,** 437 (2001).]

intestinal mucosa. From there, the products of digestion pass through the circulation to the rest of the body. Dietary proteins, for example, are broken down to amino acids for absorption. The small intestine uses amino acids for fuel, but most travel to the liver via the portal vein. There, they are used for protein synthesis or, if present in excess, oxidized to produce energy or converted to glycogen for storage. They may also be converted to glucose or triacylglycerols for export, depending on conditions. There is no dedicated storage depot for amino acids; whatever the liver does not metabolize circulates to peripheral tissues to be catabolized or used for protein synthesis.

BOX 21-3

Biochemistry in Health and Disease

Anthrax

Anthrax is a bacterial disease that is widespread among herbivorous animals. It is rare in humans but potentially deadly, leading to its exploitation as an agent of biological warfare. The effectiveness of anthrax as a weapon was first confirmed through the accidental release of anthrax spores from a laboratory in the Soviet Union in 1979, when 68 people died.

Bacillus anthracis is a nonmotile aerobic bacterium that quickly dies outside of host tissues. However, it forms spores, particles about 1 μm in diameter, that can survive for decades.

Such spores are naturally present in soils worldwide and are ingested by herbivores. The mechanism whereby the spores germinate to form full-sized bacterial cells is not well understood. If released into the air, the odorless and invisible spores can travel long distances and easily find their way indoors. These spores can cause inhalation anthrax in humans, which is an extremely rare disease. A cluster of inhalation anthrax cases almost certainly indicates that the spores have been specifically targeted to humans, for example, in letters or packages. As recent events in the United States have shown, spores from one piece of mail can easily contaminate many individuals.

Historically, anthrax infections in humans have been of the cutaneous variety and are easily recognized by the black lesions that result (cutaneous anthrax was understood to be an occupational hazard for woolsorters and others who worked with animal hides). Unfortunately, the early symptoms of inhalation anthrax are nonspecific and resemble the flu. The later stages of the disease are rapidly fatal, with an average interval of only 3 days between the onset of symptoms and death.

Like many deadly microbes, *B. anthracis* synthesizes a toxin (see below), which is particularly lethal to cells of the immune system. Thus, the toxin prevents the immune system from destroying the bacteria. Early identification of anthrax infection—or even just exposure—is essential to prevent death. Most naturally occurring strains of *B. anthracis* are sensitive to penicillin, but it is feared that "weaponized" anthrax may have been engineered for resistance to common antibiotics. The drug of choice is therefore the newer, broad-spectrum antibiotic **ciprofloxacin** (which inhibits DNA gyrase, a bacterial enzyme that helps maintain the proper degree of DNA supercoiling; Section 23-1C and Box 23-2). Treatment for about 60 days is required to prevent infection by spores that have delayed germination. The good news is that anthrax, unlike smallpox and bubonic plague, for example, is not highly contagious.

The Anthrax Toxin   Even if *B. anthracis* cells can be eliminated, the toxin they have already synthesized can continue to damage the host. Anthrax toxin consists of three proteins that act in concert: **protective antigen (PA), edema factor (EF),** and **lethal factor (LF).** PA, which is named for its use in vaccines, is a 735-residue protein that binds to a host cell-surface protein. This receptor is a 368-residue membrane protein with a single bilayer-spanning α helix. Its normal cellular function is not known. After PA has bound, a cell-surface protease cleaves it, and its N-terminal fragment diffuses away. The remaining membrane-bound portions of

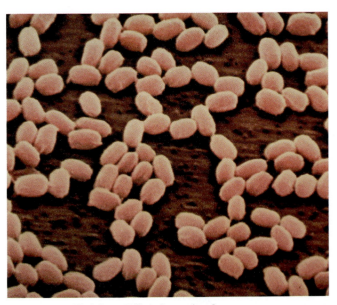

Anthrax spores. [© A. Dowsett/Photo Researchers.]

Dietary carbohydrates, like proteins, are degraded in the intestine, and the absorbed monomeric products (e.g., glucose derived from dietary starch) are delivered to the liver via the portal vein. As much as one-third of the dietary glucose is immediately converted to glycogen in the liver; at

PA form a heptameric complex that can bind the other two toxin proteins, EF and LF. The resulting toxin complex is then internalized by receptor-mediated endocytosis (Section 19-1B). Once it is inside the low-pH endosomal compartment, the PA subunits form a 14-stranded β-barrel pore that allows EF and LF to enter the cytosol.

EF is an adenylate cyclase that interferes with normal intracellular signaling, especially in cells such as macrophages, whose functions include the engulfment and destruction of pathogenic bacteria. *The toxic effects of EF may reflect the ability of the resulting increased concentrations of cAMP to inhibit a cellular signaling pathway necessary to maintain an effective immune response.* EF also upsets water homeostasis and hence is probably responsible for the massive edema (abnormal buildup of intercellular fluid) seen in cutaneous anthrax infection.

The mechanism of EF action is notable because it requires the host protein calmodulin (bacteria lack calmodulin, so this feature may protect *B. anthracis* from its own toxin). The two-lobed calmodulin (see Figs. 15-17 to 15-19) binds to EF by wrapping around it. This induces a conformational change that stabilizes the substrate-binding site of adenylate cyclase, thereby activating it. In addition, the bound calmodulin can no longer carry out its normal cellular duties, which are to bind Ca^{2+} ions and interact with specific intracellular target proteins such as cAMP phosphodiesterase (Section 15-3B). Thus, by "soaking up" calmodulin, EF enhances its ability to produce cAMP.

The other component of the anthrax toxin, LF, is a relatively large (90 kD) protein. Its X-ray structure (*at right*) reveals that the protein consists of four domains. Domain I binds to the protective antigen. Domain II resembles another bacterial toxin but has a mutated and nonfunctional active site. Domain III appears to be a duplicated version of Domain II. Domains II and III function to hold the substrate for Domain IV, which is a protease with an active site zinc ion. The sequence of Domain IV exhibits no homology to known zinc proteases; its activity was identified on the basis of its three-dimensional structure and the position of its catalytic zinc ion.

The substrates for lethal factor are the members of the mitogen-activated protein kinase kinase (MAPKK) family (e.g., MEK; Fig. 21-19). LF is an extremely specific protease because it interacts extensively with its substrates: The substrate-binding groove extends for about 40 Å and accommodates a 16-residue sequence of a MAPKK. Cleavage of the kinase by LF excises its docking sequence for downstream MAPKs and thereby blocks its signaling activity. Low levels of LF, which occur early in *B. anthracis* infection, inhibit the ability of macrophages to release inflammatory mediators such as cytokines and nitric oxide (Section 20-6C). This has the effect of reducing or delaying an immune response to the bacteria. Later in infection, when concentrations of the toxin are high, LF triggers macrophage lysis, causing the sudden release of inflammatory mediators. This probably results in the massive septic shock that causes death.

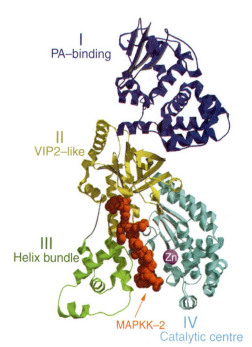

X-Ray structure of anthrax lethal factor. [Courtesy of Robert Liddington, The Burnham Institute, La Jolla, California. PDBid 1J7N.]

least half of the remainder is converted to glycogen in muscle cells, and the rest is oxidized by these and other tissues for immediate energy needs. Excess glucose is converted to triacylglycerol in the liver and exported for storage in adipose tissue. Both glucose uptake and glycogen and fatty acid

biosynthesis are stimulated by insulin, whose concentration in the blood increases in response to high blood [glucose].

Dietary fatty acids are packaged as triacylglycerols in chylomicrons (Section 19-1), which circulate first in the lymph and then in the bloodstream and therefore are not delivered directly to the liver as are absorbed amino acids and carbohydrates. Instead, a significant portion of the dietary fatty acids are taken up by adipose tissue. Lipoprotein lipase first hydrolyzes the triacylglycerols, and the released fatty acids are absorbed and esterified in the adipocytes.

Blood Glucose Remains Constant.   As tissues take up and metabolize glucose, blood [glucose] drops, thereby causing the pancreatic α cells to release glucagon. This hormone stimulates glycogen breakdown and the release of glucose from the liver. It also promotes gluconeogenesis from amino acids and lactate. *The reciprocal effects of insulin and glucagon, which both respond to and regulate blood [glucose], ensure that the concentration of glucose available to extrahepatic tissues remains relatively constant.*

However, the body stores less than a day's supply of carbohydrate (Table 21-2). After an overnight fast, the combination of increased glucagon secretion and decreased insulin secretion promotes the mobilization of fatty acids from adipose tissue (Section 19-5). The diminished insulin also inhibits glucose uptake by muscle tissue. Muscles therefore switch from glucose to fatty acid metabolism for energy production. This adaptation spares glucose for use by tissues, such as the brain, that cannot utilize fatty acids.

Gluconeogenesis Supplies Glucose during Starvation.   After a lengthy fast, the liver's store of glycogen becomes depleted (Fig. 21-26). Under these conditions, the rate of gluconeogenesis increases. Gluconeogenesis supplies ~96% of the glucose produced by the liver after 40 hours of fasting. The kidney also is active in gluconeogenesis under these conditions. In animals, glucose cannot be synthesized from fatty acids. This is because neither pyruvate nor oxaloacetate, the precursors of glucose in gluconeogenesis, can be synthesized in a net manner from acetyl-CoA. During starvation, glucose must therefore be synthesized from the glycerol product of triacylglycerol breakdown and, more importantly, from the amino acids derived from the proteolytic degradation of proteins, the major source of which is muscle. The breakdown of muscle cannot continue indefinitely, however, since loss

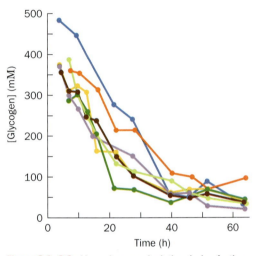

Figure 21-26 Liver glycogen depletion during fasting. The liver glycogen content in seven subjects was measured using ^{13}C NMR over the course of a 64-hour fast. [After Rothman, D.L., Magnusson, I., Katz, L.D., Shulman, R.G., and Shulman, G.I., *Science* **254**, 575 (1991).]

Table 21-2 Fuel Reserves for a Normal 70-kg Man

Fuel	Mass (kg)	Calories[a]
Tissues		
Fat (adipose triacyglycerols)	15	141,000
Protein (mainly muscle)	6	24,000
Glycogen (muscle)	0.150	600
Glycogen (liver)	0.075	300
Circulating fuels		
Glucose (extracellular fluid)	0.020	80
Free fatty acids (plasma)	0.0003	3
Triacylglycerols (plasma)	0.003	30
Total		**166,000**

[a]1 (dieter's) Calorie = 1 kcal = 4.184 kJ.

Source: Cahill, G.E., Jr., *New Engl. J. Med.* **282**, 669 (1970).

of muscle mass would eventually prevent an animal from moving about in search of food. The organism must therefore make alternate metabolic arrangements.

After several days of starvation, the liver directs acetyl-CoA, which is derived from fatty acid β oxidation, to the synthesis of ketone bodies (Section 19-3). These fuels are then released into the blood. The brain gradually adapts to using ketone bodies as fuel through the synthesis of the appropriate enzymes: After a 3-day fast, only about one-third of the brain's energy requirements are satisfied by ketone bodies, and after 40 days of starvation, ~70% of its energy needs are so met. The rate of muscle breakdown during prolonged starvation consequently decreases to ~25% of its rate after a several-day fast. *The survival time of a starving individual therefore depends much more on the size of fat reserves than on muscle mass.* Indeed, highly obese individuals can survive a year or more without eating (and have done so in clinically supervised weight-reduction programs).

B   Diabetes Mellitus

In the disease **diabetes mellitus** (often called just **diabetes**), which is the third leading cause of death in the United States after heart disease and cancer, insulin either is not secreted in sufficient amounts or does not efficiently stimulate its target cells. As a consequence, blood glucose levels become so elevated that the glucose "spills over" into the urine, providing a convenient diagnostic test for the disease. Yet, despite these high blood glucose levels, cells "starve" since insulin-stimulated glucose entry into cells is impaired. Triacylglycerol hydrolysis, fatty acid oxidation, gluconeogenesis, and ketone body formation are accelerated and, in a condition known as **ketosis,** ketone body levels in the blood become abnormally high. Since ketone bodies are acids, their high concentration puts a strain on the buffering capacity of the blood and on the kidney, which controls blood pH by excreting the excess H^+ into the urine. This H^+ excretion is accompanied by NH_4^+, Na^+, K^+, P_i, and H_2O excretion, causing severe dehydration (excessive thirst is a classic symptom of diabetes) and a decrease in blood volume—ultimately life-threatening situations.

There are two major forms of diabetes mellitus:

1. **Insulin-dependent** or **juvenile-onset diabetes mellitus,** which most often strikes suddenly in childhood.

2. **Non-insulin-dependent** or **maturity-onset diabetes mellitus,** which usually develops gradually after the age of 40.

Insulin-Dependent Diabetes Is Caused by a Deficiency of Pancreatic β Cells.   In insulin-dependent (type I) diabetes mellitus, insulin is absent or nearly so because the pancreas lacks or has defective β cells. This condition usually results from an autoimmune response that selectively destroys pancreatic β cells. Individuals with insulin-dependent diabetes, as Frederick Banting and Charles Best first demonstrated in 1921 (Box 21-4), require daily insulin injections to survive and must follow carefully balanced diet and exercise regimens. Their life spans are, nevertheless, reduced by up to one-third as a result of degenerative complications such as kidney malfunction, nerve impairment, and cardiovascular disease that apparently arise from the imprecise metabolic control provided by periodic insulin injections. The **hyperglycemia** (high blood [glucose]) of diabetes mellitus also leads to blindness through retinal degeneration and the glucosylation of lens proteins, which causes cataracts (Fig. 21-27).

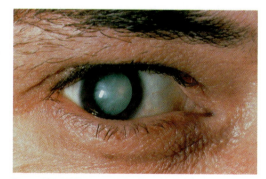

Figure 21-27   Photo of a diabetic cataract.   The accumulation of glucose in the lens leads to swelling and precipitation of lens proteins. The resulting opacification causes blurred vision and ultimately complete loss of sight. [© Sue Ford/Photo Researchers.]

BOX 21-4

Pathways of Discovery

Frederick Banting and Charles Best and the Discovery of Insulin

Frederick Banting (1891–1941)

Charles Best (1899–1978)

One of the most celebrated clinical research stories is the discovery and therapeutic use of insulin, the hormone missing in type I diabetes. Since the end of the nineteenth century, clinical scientists understood that there was a connection between diabetes and secretions from the pancreas, specifically, from the clumps of cells known as the islets of Langerhans. Several unsuccessful attempts were made to extract the secretions. In 1920, Frederick Banting, reading an account of such work, wondered whether an active extract could be purified if the pancreatic ducts, which collect potentially destructive digestive enzymes, were first ligated (closed off).

Banting, a young surgeon whose medical practice had not yet gotten off the ground, was working as a demonstrator at the University of Western Ontario. He took his idea to John J.R. Macleod, the head of the Department of Physiology at the University of Toronto. A skeptical Macleod gave Banting the use of laboratory space for the summer of 1921, access to experimental dogs, and the help of assistant Charles Best. Banting, despite his medical training and experience with the Canadian Army Medical Service, was not much of an experimentalist. Best, however, had just completed a degree in Physiology and Biochemistry and had already worked as an assistant to Macleod. Best quickly learned surgical techniques from Banting, as Banting picked up analytical techniques, such as measuring sugar levels in dogs' blood and urine, from Best.

Banting's and Best's experimental protocol required pancreatectomized dogs, but the surgery was difficult, and many dogs died of infection (it was rumored that strays from the streets of Toronto were gathered as replacements). Successfully pancreatectomized dogs, who developed the fatal symptoms of diabetes, were injected with a pancreatic extract that Banting and Best named insulin, after the Latin word *insula* (island). Finally, at the end of July, Banting and Best achieved their desired result, when the insulin injection dramatically reduced the dog's blood sugar level. The experiment was repeated, with similar results, in other dogs.

Although Banting's temporary position was to have been terminated at the end of the summer, his results were promising enough to secure him additional funding, the supervision of a more enthusiastic Macleod, and the continued assistance of Best, who had originally intended to work with Banting for only 2 months. By fall, Banting's and Best's success at prolonging the lives of diabetic dogs was putting enormous pressure on their ability to produce insulin in large amounts and of consistent purity. Assistance was offered by James Collip, a biochemist on sabbatical leave from the University of Alberta. Collip's insulin preparations, which were made without any clue as to the proteinaceous nature of the hormone, proved good enough for experimental use in humans. At that time, a diagnosis of diabetes was a virtual death sentence, with the only treatment being a severely limited diet that probably extended life for only a few months (many patients died of malnutrition).

In January 1923, Banting injected Collip's extract into Leonard Thompson, a 14-year-old diabetic who was near death. Thompson's blood sugar level immediately returned to normal, and his strength increased. Such a dramatic clinical outcome did not go unnoticed, and within months, Banting opened a clinic to begin treating diabetics desperate for a treatment that restored their hopes for a near-normal life. He enlisted the help of the Connaught Antitoxin Laboratories at the University of Toronto and the Eli Lilly Pharmaceutical Company to produce insulin on a scale suitable for clinical testing. Best, just 23, was put in charge of insulin production for all of Canada. It eventually became clear that Banting's original assumption—that the presence of digestive enzymes compromised the purification of insulin—was mistaken, and insulin could be successfully isolated from an intact pancreas.

Amid some controversy, the 1923 Nobel Prize for Physiology or Medicine was awarded to Banting and Macleod. Banting, annoyed by the omission of Best, split his share of the prize with his assistant. Macleod, who had not offered much support during the initial research, split his share with Collip. Banting subsequently held a variety of administrative posts and pursued research in silicosis, but he never accomplished much. He was killed in a plane crash en route to Britain for a wartime mission in 1941. Best, who replaced Macleod as head of Physiology at the age of 29, continued his research into the biological action of insulin. He also supervised efforts to purify the anticoagulant glycosaminoglycan heparin and to produce dried human serum for medical use by the military.

Banting, F.G. and Best, C.H., The internal secretion of the pancreas, *J. Lab. Clin. Med.* **7**, 251–266 (1922).

The usually rapid onset of the symptoms of insulin-dependent diabetes had suggested that the autoimmune attack on the pancreatic β cells is of short duration. Typically, however, the disease develops over several years as the immune system slowly destroys the β cells. Only when >80% of these cells have been eliminated do the classic symptoms of diabetes suddenly emerge. Consequently, one of the most successful treatments for insulin-dependent diabetes is a β-cell transplant, a procedure that became possible with the development of relatively benign immunosuppressive drugs.

Non-Insulin-Dependent Diabetes May Be Caused by a Deficiency of Insulin Receptors or Insulin Signal Transduction.

Non-insulin-dependent (type II) diabetes mellitus, which accounts for over 90% of the diagnosed cases of diabetes and affects 18% of the population over 65 years of age, usually occurs in obese individuals with a genetic predisposition for this condition. These individuals have normal or even greatly elevated insulin levels, but their cells are not responsive to insulin and are therefore said to be **insulin resistant.** As a result, blood glucose concentrations are much higher than normal, particularly after a meal (Fig. 21-28).

The hyperglycemia that accompanies insulin resistance induces the pancreatic β cells to increase their production of insulin. Yet the high basal level of insulin secretion diminishes the ability of the β cells to respond to further increases in blood glucose. Consequently, the hyperglycemia and its attendant complications tend to worsen over time. A small percentage of cases of type II diabetes result from mutations in the insulin receptor that affect its insulin-binding ability or tyrosine kinase activity. However, a clear genetic cause has not been identified in the vast majority of cases. It is therefore likely that many factors play a role in the development of this disease. For example, the increased insulin production resulting from overeating may eventually suppress the synthesis of insulin receptors. This hypothesis accounts for the observation that diet alone often decreases the severity of the disease.

Another hypothesis, put forward by Gerald Shulman, is that the elevated

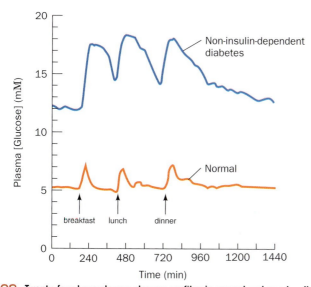

Figure 21-28   Twenty-four-hour plasma glucose profiles in normal and non-insulin-dependent diabetic subjects.   The basal level of glucose and the peaks following meals are higher in the diabetic individuals. [After Bell, G.I., Pilkis, S.J., Weber, I.T., and Polonsky, K.S., *Annu. Rev. Physiol.* **58**, 178 (1996).]

concentration of free fatty acids in the blood caused by obesity decreases insulin signal transduction. This high concentration ultimately results in an activation of a PKC isoform that phosphorylates insulin receptor substrates (IRSs) on Ser and Thr, thereby inhibiting the Tyr phosphorylation that activates them. The failure to activate IRSs decreases the cell's response to insulin (Fig. 21-25).

Other treatments for non-insulin-dependent diabetes are drugs such as **metformin** and the **thiazolidinediones (TZDs),**

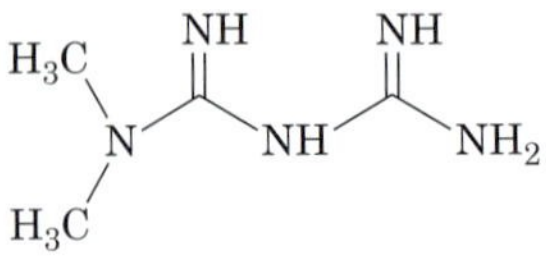

Metformin

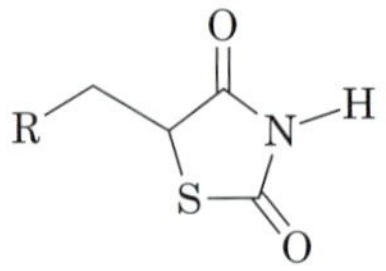

A thiazolidinedione (TZD)

which decrease insulin resistance by either decreasing glucose release by the liver (metformin) or increasing insulin-stimulated glucose disposal in muscle (TZDs). These drugs both target Complex I of the mitochondrial electron transport system, whose inhibition decreases respiratory ATP production, thereby increasing [AMP]. The resulting increase in AMPK activity is thought to account for the effects of both drugs, by decreasing gluconeogenesis in liver and increasing glucose utilization in muscle. In addition, the TZDs decrease insulin resistance by binding to and activating a transcription factor known as a **peroxisome proliferator–activated receptor-γ (PPAR-γ),** primarily in adipose tissue. Among other things, PPAR-γ activation leads to an increase in fatty acid uptake by adipocytes, which may indirectly help muscle and liver make better use of circulating glucose.

C Obesity

The human body regulates glycogen and protein levels within relatively narrow limits, but fat reserves, which are much larger, can become enormous. The accumulation of fatty acids as triacylglycerols in adipose tissue is largely a result of excess fat or carbohydrate intake compared to energy expenditure. Fat synthesis from carbohydrates occurs when the carbohydrate intake is high enough that glycogen stores, to which excess carbohydrate is normally directed, approach their maximum capacity.

A chronic imbalance between fat and carbohydrate consumption and utilization increases the mass of adipose tissue through an increase in the number of adipocytes or their size (once formed, adipocytes are not lost, although their size may increase or decrease). The increase in adipose tissue mass increases the pool of fatty acids that can be mobilized to generate metabolic energy. Eventually, a steady state is achieved in which the mass of adipose tissue no longer increases and fat storage is balanced by fat mobilization. This phenomenon explains in part the high incidence of obesity in affluent societies, where fat- and carbohydrate-rich foods are plentiful and physical activity is not a requirement for survival. Considerable evidence suggests that behavior (e.g., eating habits and levels of physical activity) influences an individual's body composition. Yet some cases of obesity are clearly also the result of innate disturbances in an individual's capacity to metabolize fuels.

A Genetic Basis for Obesity.   Most animals, including humans, tend to have stable weights; that is, if they are given free access to food, they eat just enough to maintain their "set-point" weight. However, a strain of mice that are homozygous for defects in the *obese* gene (known as *ob/ob* mice) are over twice the weight of normal (*OB/OB*) mice (Fig. 21-29) and overeat when given access to unlimited quantities of food. The obese mice lack the protein **leptin** (Greek: *leptos,* thin), the product of the *obese* gene. Leptin is a 146-residue polypeptide that is normally produced by adipocytes (Fig. 21-30). When leptin is injected into *ob/ob* mice, they eat less and lose weight. Leptin has therefore been considered a "satiety" signal that affects the appetite control system of the brain. Leptin has also been found to cause higher energy expenditure.

The simple genetic basis for obesity in *ob/ob* mice does not appear to apply to most obese humans. Leptin levels in humans increase with the percentage of body fat, consistent with the synthesis of leptin by adipocytes. Thus, obesity in humans is apparently the result not of faulty leptin production but of "leptin resistance," perhaps due to a decrease in the level of a leptin receptor in the brain or the saturation of the receptor that transports leptin across the blood–brain barrier into the central nervous system.

A diminished response to leptin leads to high concentrations of **neuropeptide Y,**

Figure **21-29** Normal (*OB/OB, left*) and obese (*ob/ob, right*) mice.   [Courtesy of Richard D. Palmiter, University of Washington.]

$$^1 \quad\quad\quad\quad\quad\quad\quad\quad\quad ^{10}$$
Tyr—Pro—Ser—Lys—Pro—Asp—Asn—Pro—Gly—Glu—Asp—Ala—

^{20}
Pro—Ala—Glu—Asp—Met—Ala—Arg—Tyr—Tyr—Ser—Ala—Leu—

$$^{30} \quad\quad\quad\quad\quad\quad\quad\quad\quad\quad\quad ^{36}$$
Arg—His—Tyr—Ile—Asn—Leu—Ile—Thr—Arg—Gln—Arg—Tyr—NH$_2$

Neuropeptide Y
(The C-terminal carboxyl is amidated)

a 36-residue peptide released by the **hypothalamus,** a part of the brain that controls many physiological functions. Neuropeptide Y stimulates appetite, which in turn leads to fat accumulation.

Additional Peptide Hormones Play a Role in Obesity.   It is not surprising that fuel metabolism, body weight, and appetite are linked. In addition to leptin, other peptide hormones are known to be involved in appetite control. Insulin receptors are present in the hypothalamus. These, along with leptin receptors, act, on hormone binding, to inhibit neuropeptide Y secretion.

Ghrelin is an appetite-stimulating peptide secreted by the empty stomach.

^{10}
Gly—Ser—X—Phe—Leu—Ser—Pro—Glu—His—Gln—

^{20}
Arg—Val—Gln—Gln—Arg—Lys—Glu—Ser—Lys—Lys—

^{28}
Pro—Pro—Ala—Lys—Leu—Gln—Pro—Arg

Ghrelin
(X = Ser modified with *n*-octanoic acid)

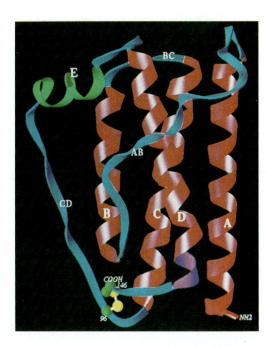

Figure **21-30** **X-Ray structure of human leptin-E100.**   This mutant form of leptin (Trp 100 → Glu) has comparable biological activity to the wild-type protein but crystallizes more readily. Leptin's residues pack into a bundle of four α helices (labeled A–D). A disulfide bond links Cys 146 to Cys 96 (atoms shown in green and yellow). [Courtesy of Faming Zhang, Eli Lilly & Co., Indianapolis, Indiana. PDBid 1AX8.] **See the Interactive Exercises.**

Ghrelin appears to boost levels of neuropeptide Y, and most likely is part of a short-term appetite-control system, since ghrelin levels fluctuate on the order of hours, increasing before meals and decreasing immediately afterward.

The gastrointestinal tract secretes an appetite-suppressing hormone called **PYY$_{3-36}$**:

$$
\overset{3}{\text{Ile}} - \text{Lys} - \text{Pro} - \text{Glu} - \text{Ala} - \text{Pro} - \text{Gly} - \overset{10}{\text{Glu}} -
$$

$$
\text{Asp} - \text{Ala} - \text{Ser} - \text{Pro} - \text{Glu} - \text{Glu} - \text{Leu} - \text{Asn} - \text{Arg} - \overset{20}{\text{Tyr}} -
$$

$$
\text{Tyr} - \text{Ala} - \text{Ser} - \text{Leu} - \text{Arg} - \text{His} - \text{Thr} - \text{Leu} - \text{Asn} - \overset{30}{\text{Leu}} -
$$

$$
\text{Val} - \text{Thr} - \text{Arg} - \text{Gln} - \text{Arg} - \overset{36}{\text{Tyr}}
$$

PYY$_{3-36}$

In animal and human experiments, an infusion of this peptide decreases food intake by inhibiting neuropeptide Y secretion. In summary, ghrelin stimulates appetite while leptin, insulin, and PYY$_{3-36}$ suppress it, all by interacting to regulate the secretion of neuropeptide Y from the hypothalamus.

In addition to leptin, adipocytes secrete a 108-residue polypeptide known as **resistin.** The hormone is named for its ability to block the action of insulin on adipocytes. In fact, resistin production is decreased by the antidiabetic thiazolidinediones, a feature that led to its discovery. Intriguingly, resistin synthesis seems to be higher in abdominal fat than in subcutaneous fat, which is consistent with the correlation between abdominal fat and the risk of developing type II diabetes. The role of resistin in regulating the insulin responsiveness of other tissues is not known.

The hormonal mechanisms that contribute to obesity and its attendant health problems, such as diabetes, may have conferred a selective advantage in premodern times. For example, the ability to gain weight easily would have protected against famine. One theory proposes that as humans evolved, the regulation of fuel metabolism was actually geared toward overeating large quantities of food when available, obtaining essential nutrients while storing excess nonessential energy as fat for use during famine. In modern Western cultures, where famines are rare and nutritious food is readily available, human genetic heritage has apparently contributed to an epidemic of obesity: An estimated 30% of U.S. adults are obese and another 35% are overweight.

SUMMARY

1. The pathways for the synthesis and degradation of the major metabolic fuels (glucose, fatty acids, and amino acids) converge on acetyl-CoA and pyruvate. In mammals, flux through these pathways is tissue specific. AMP-dependent protein kinase (AMPK), the cell's fuel gauge, senses the cell's need for ATP and activates metabolic breakdown pathways while inhibiting biosynthetic pathways.

2. The brain uses glucose as its primary metabolic fuel. Muscle can oxidize a variety of fuels but depends on anaerobic glycolysis for maximum exertion. Adipose tissue stores excess fatty acids as triacylglycerols and mobilizes them as needed.

3. The liver maintains the concentrations of circulating fuels. The action of glucokinase allows liver to take up excess glucose, which can then be directed to several metabolic fates. The liver also converts fatty acids to ketone bodies and metabolizes amino acids derived from the diet or from protein breakdown. Both the liver and kidney carry out gluconeogenesis.

4. The Cori cycle and the glucose–alanine cycle are multiorgan pathways through which the liver and muscle exchange metabolic intermediates.

5. Hormones such as insulin, glucagon, and epinephrine transmit regulatory signals to target tissues by binding to recep-

tors that transduce the signal to responses in the interior of the cell.

6. The G protein–coupled receptors (GPCRs) have seven transmembrane helices and, on ligand binding, activate an associated heterotrimeric G protein. The G_α and $G_{\beta\gamma}$ units may activate or inhibit targets such as adenylate cyclase, which produces the cAMP activator of protein kinase A.

7. On ligand binding, receptor tyrosine kinases such as the insulin receptor undergo autophosphorylation. This activates them to phosphorylate their target proteins, in some cases triggering a kinase cascade. The activity of protein kinases is reversed by the activity of protein phosphatases.

8. In the phosphoinositide pathway, hormone binding leads to the hydrolysis of phosphatidylinositol-4,5-bisphosphate (PIP_2) to yield inositol-1,4,5-trisphosphate (IP_3), which opens Ca^{2+} channels, and diacylglycerol (DAG), which activates protein kinase C (PKC).

9. During starvation, when dietary fuels are unavailable, the liver releases glucose first by glycogen breakdown and then by gluconeogenesis from amino acid precursors. Eventually, ketone bodies supplied by fatty acid breakdown meet most of the body's energy needs.

10. Diabetes mellitus causes hyperglycemia and other physiological difficulties resulting from destruction of insulin-producing pancreatic β cells or from insulin resistance (from loss of receptors or their insensitivity to insulin).

11. Obesity, an imbalance between food intake and energy expenditure, may result from abnormal regulation by peptide hormones produced in different tissues.

REFERENCES

Fuel Metabolism

Bollen, M., Keppens, S., and Stalmans, W., Specific features of glycogen metabolism in the liver. *Biochem. J.* **336,** 19–31 (1998).

Bryant, N.J., Govers, R., and James, D.E., Regulated transport of the glucose transporter GLUT4, *Nature Rev. Mol. Cell Biol.* **3,** 267–273(2002).

Obesity, *Science* **299,** 845–860 (2003). [A series of informative reports on the origins and treatments of obesity.]

Saltiel, A.R. and Kahn, C.R., Insulin signaling and the regulation of glucose and lipid metabolism, *Nature* **414,** 799–806 (2001). [Summarizes the mechanisms of insulin action and insulin resistance.]

Spiegelman, B.M. and Flier, J.S., Obesity and the regulation of energy balance, *Cell* **104,** 531–543 (2001).

Signal Transduction

Chang, L. and Karin, M., Mammalian MAP kinase signaling cascades, *Nature* **410,** 37–40 (2001).

Marinissen, M.J. and Gutkind, J.S., G-protein-coupled receptors and signaling networks: emerging paradigms, *Trends Pharm. Sci.* **22,** 368–375 (2001). [Discusses some of the pathways in which G proteins mediate signal transduction.]

Neves, S.R., Ram, P.T., and Iyengar, R., G protein pathways, *Science* **296,** 1636–1639 (2002). [A brief introduction to the four families of G proteins.]

Ottensmeyer, F.P., Beniac, D.R., Luo, R.Z.-T., and Yip, C.C., Mechanism of transmembrane signaling: insulin binding and the insulin receptor, *Biochemistry* **39,** 12103–12111 (2000).

Science's Signal Transduction Knowledge Environment (STKE). http://stke.sciencemag.org/. [A database on signaling molecules and their relationships to each other. Full access to the database requires an individual or institutional subscription.]

Simonds, W.F., G protein regulation of adenylate cyclase, *Trends Pharm. Sci.* **20,** 66–73 (1999).

Teller, D.C., Okada, T., Behnke, C.A., Palczewski, K., and Stenkamp, R.E., Advances in determination of a high-resolution three-dimensional structure of rhodopsin, a model of G-protein-coupled receptors (GPCRs), *Biochemistry* **40,** 7761–7772 (2001).

Yaffe, M.B., Phosphotyrosine-binding domains in signal transduction, *Nature Rev. Mol. Cell Biol.* **3,** 177–186 (2002).

KEY TERMS

angina	receptor	heterotrimeric G protein	cross talk
Cori cycle	adrenergic receptor	GTPase	diabetes
oxygen debt	signal transduction	tyrosine kinase	hyperglycemia
glucose–alanine cycle	ligand	autophosphorylation	insulin resistance
hormone	GPCR	phosphatase	
homeostasis	desensitization	phosphoinositide pathway	

STUDY EXERCISES

1. Summarize the major features of fuel metabolism in the brain, muscle, adipose tissue, liver, and kidney.

2. Explain why the high K_M of glucokinase is important for the role of the liver in buffering blood glucose.

3. Describe the conditions under which the Cori cycle and the glucose–alanine cycle operate.

4. Summarize the role of insulin, glucagon, and epinephrine in regulating mammalian fuel metabolism.

5. Describe the activity of each component of the signaling pathways based on adenylate cyclase, receptor tyrosine kinases, and phosphoinositides.

6. How do heterotrimeric G proteins work?

7. Describe the metabolic changes that occur during starvation.

8. Distinguish insulin-dependent and non-insulin-dependent diabetes mellitus.

9. How is obesity related to non-insulin-dependent diabetes mellitus?

PROBLEMS

1. Predict the effect of an overdose of insulin on brain function in a normal person.

2. Why would oxidative metabolism, which generates ATP, cease when a cell's ATP supply is exhausted?

3. The passive glucose transporter, called GLUT1 (Fig. 10-17), is present in the membranes of many cells, but not in the liver. Instead, liver cells express the GLUT2 transporter, which exhibits different transport kinetics. (a) Given what you know about the role of the liver in buffering blood glucose, compare the K_M values of GLUT1 and GLUT2. (b) Explain why a deficiency of GLUT2 produces symptoms resembling those of Type I glycogen storage disease (Box 15-3).

4. Pancreatic β cells have been found to express a receptor for fatty acids. Fatty acid binding to this protein appears to stimulate insulin secretion. (a) Does this phenomenon make metabolic sense? (b) Fatty acids appear to stimulate insulin secretion to a much greater extent when glucose is also present. Why is this significant?

5. Would you expect insulin to increase or decrease the activity of the enzyme ATP-citrate lyase?

6. The sense of taste is mediated by GPCRs on the tongue. There are five classes of these receptors, which detect the tastes known as sweet, salty, sour, bitter, and umami. The umami receptor responds to amino acids such as glutamate (i.e., the flavor enhancer monosodium glutamate, MSG). Given what you know about glutamate metabolism, what might be the selective advantage of having an umami receptor?

7. How does the presence of the poorly hydrolyzable GTP analog **GTPγS** (in which an O atom on the terminal phosphate is replaced by an S atom) affect cAMP production by adenylate cyclase?

8. Phosphatidylethanolamine and PIP₂ containing identical fatty acyl residues can be hydrolyzed with the same efficiency by a certain phospholipase C. Will the hydrolysis products of the two lipids have the same effect on protein kinase C? Explain.

9. A growth factor that acts through a receptor tyrosine kinase stimulates cell division. Predict the effect of a viral protein that inhibits the corresponding protein tyrosine phosphatase.

10. Explain why insulin is required for adipocytes to synthesize triacylglycerols from fatty acids.

11. Experienced runners know that it is poor practice to ingest large amounts of glucose immediately before running a marathon. What is the metabolic basis for this apparent paradox?

12. After several days of starvation, the capacity of the liver to metabolize acetyl-CoA via the citric acid cycle is greatly diminished. Explain.

13. If the circulatory system of an *ob/ob* mouse is surgically joined to that of a normal mouse, what will be the effect on the appetite and weight of the *ob/ob* mouse?

14. Why can adipose tissue be considered to be an endocrine organ?

15. In experiments to test the appetite-suppressing effects of PYY₃₋₃₆, why must the hormone be administered intravenously rather than orally?

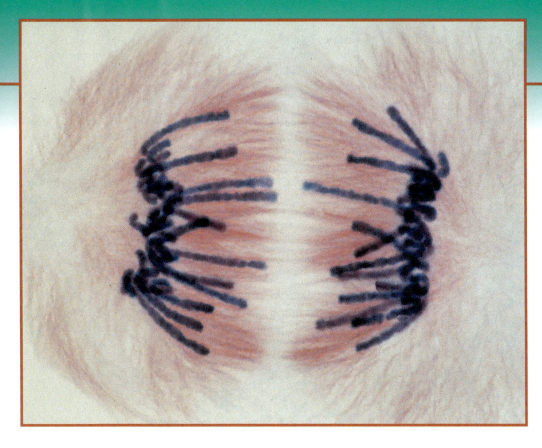

The synthesis of two complete sets of chromosomes requires large amounts of newly synthesized nucleotides. During cell division (mitosis), sets of identical chromosomes (purple) are pulled apart. [Courtesy of Andrew Bajer, University of Oregon.]

Nucleotide Metabolism

Nucleotides are phosphate esters of a pentose (ribose or deoxyribose) in which a purine or pyrimidine base is linked to C1′ of the sugar (Section 3-1). Nucleoside triphosphates are the monomeric units that act as precursors of nucleic acids; nucleotides also perform a wide range of other biochemical functions. For example, we have seen how the cleavage of "high-energy" compounds such as ATP provides the free energy that makes various reactions thermodynamically favorable. We have also seen that nucleotides are components of some of the central cofactors of metabolism, including FAD, NAD$^+$, and coenzyme A. The importance of nucleotides in cellular metabolism is indicated by the observation that nearly all cells can synthesize them both *de novo* (anew) and from the degradation products of nucleic acids. However, unlike carbohydrates, amino acids, and fatty acids, nucleotides do not provide a significant source of metabolic energy.

In this chapter, we consider the nature of the nucleotide biosynthetic pathways. In doing so, we shall examine how they are regulated and the consequences of their blockade, both by genetic defects and through the administration of chemotherapeutic agents. We then discuss how nucleotides are degraded. In following the general chemical themes of nucleotide metabolism, we shall break our discussion into sections on purines, pyrimidines, and deoxynucleotides (including thymidylate). The structures and nomenclature of the major purines and pyrimidines are given in Table 3-1.

1 Synthesis of Purine Ribonucleotides

In 1948, John Buchanan obtained the first clues to the *de novo* synthesis of purine nucleotides by feeding a variety of isotopically labeled compounds to pigeons and chemically determining the positions of the labeled atoms in their excreted **uric acid** (a purine).

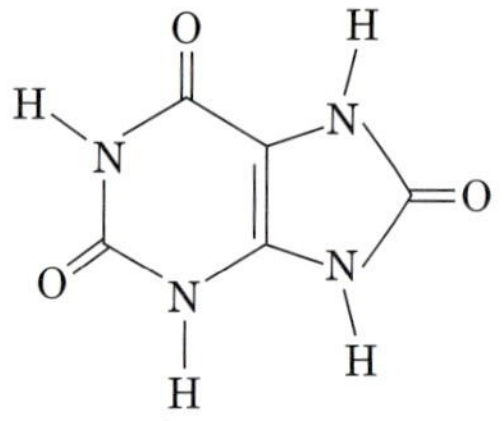

Uric acid

The results of his studies demonstrated that N1 of purines arises from the amino group of aspartate; C2 and C8 originate from formate; N3 and N9 are contributed by the amide group of glutamine; C4, C5, and N7 are derived from glycine (strongly suggesting that this molecule is wholly incorporated into the purine ring); and C6 comes from HCO$_3^-$.

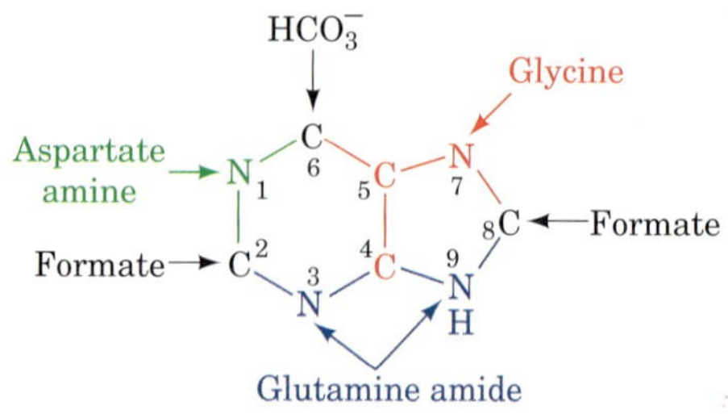

The actual pathway by which these precursors are incorporated into the purine ring was elucidated in subsequent investigations performed largely

by Buchanan and G. Robert Greenberg. The initially synthesized purine derivative is **inosine monophosphate (IMP),**

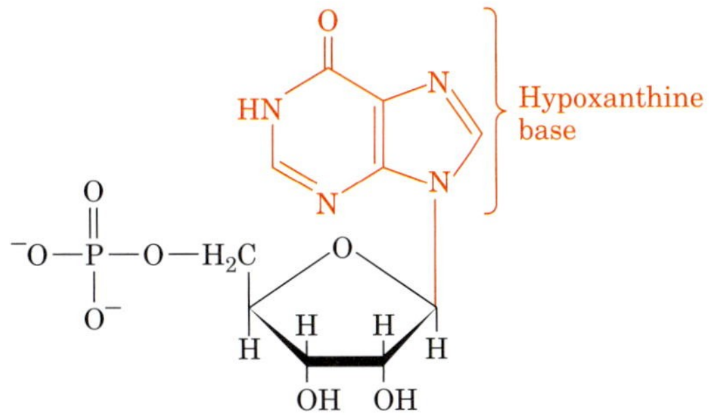

Inosine monophosphate (IMP)

the nucleotide of the base **hypoxanthine.** IMP is the precursor of both AMP and GMP. Thus, contrary to expectation, *purines are initially formed as ribonucleotides rather than as free bases.* Additional studies have demonstrated that such widely divergent organisms as *E. coli,* yeast, pigeons, and humans have virtually identical pathways for the biosynthesis of purine nucleotides, thereby further demonstrating the biochemical unity of life.

A Synthesis of Inosine Monophosphate

IMP is synthesized in a pathway composed of 11 reactions (Fig. 22-1):

1. **Activation of ribose-5-phosphate.** The starting material for purine biosynthesis is α-D-ribose-5-phosphate, a product of the pentose phosphate pathway (Section 14-6). In the first step of purine biosynthesis, **ribose phosphate pyrophosphokinase** activates the ribose by reacting it with ATP to form **5-phosphoribosyl-α-pyrophosphate (PRPP).** This compound is also a precursor in the biosynthesis of pyrimidine nucleotides (Section 22-2A) and the amino acids histidine and tryptophan (Section 20-5B). As is expected for an enzyme at such an important biosynthetic crossroads, the activity of ribose phosphate pyrophosphokinase is carefully regulated.

2. **Acquisition of purine atom N9.** In the first reaction unique to purine biosynthesis, **amidophosphoribosyl transferase** catalyzes the displacement of PRPP's pyrophosphate group by glutamine's amide nitrogen. The reaction occurs with inversion of the α configuration at C1 of PRPP, thereby forming **β-5-phosphoribosylamine** and establishing the anomeric form of the future nucleotide. The reaction, which is driven to completion by the subsequent hydrolysis of the released PP_i, is the pathway's flux-controlling step.

3. **Acquisition of purine atoms C4, C5, and N7.** Glycine's carboxyl group forms an amide with the amino group of phosphoribosylamine, yielding **glycinamide ribotide (GAR).** This reaction is reversible, despite its concomitant hydrolysis of ATP to ADP + P_i. It is the only step of the purine biosynthetic pathway in which more than one purine ring atom is acquired.

4. **Acquisition of purine atom C8.** GAR's free α-amino group is formylated to yield **formylglycinamide ribotide (FGAR).** The formyl donor in this reaction is N^{10}-formyltetrahydrofolate (N^{10}-formyl-THF), a coenzyme that transfers C_1 units (THF cofactors are described in Section 20-4D). The X-ray structure of the enzyme catalyzing this

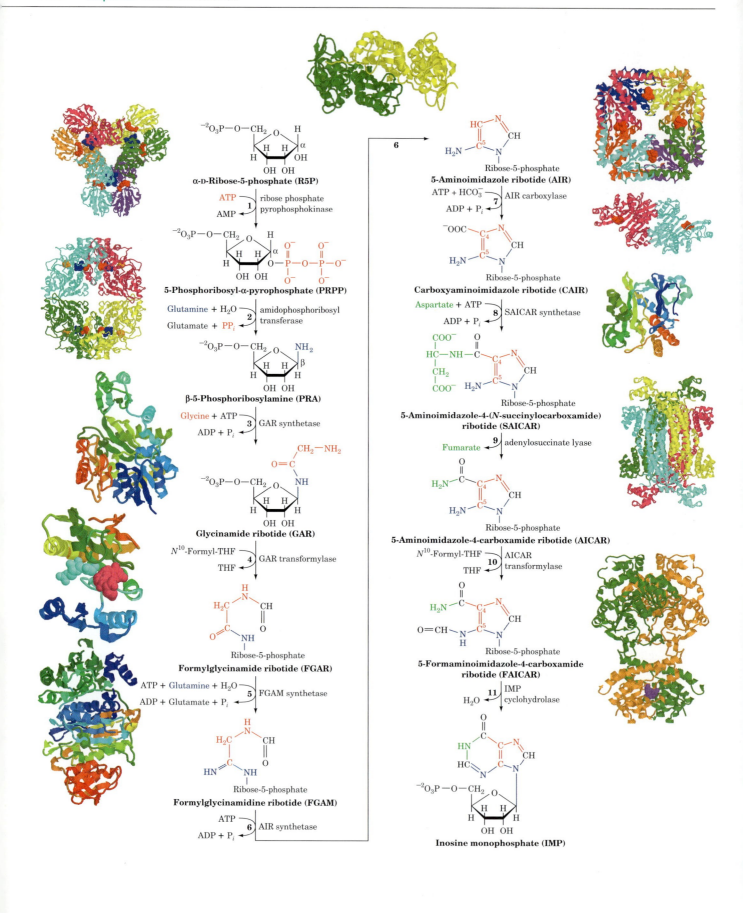

α-D-**Ribose-5-phosphate (R5P)**

1 · ribose phosphate pyrophosphokinase · ATP → AMP

5-**Phosphoribosyl-α-pyrophosphate (PRPP)**

2 · amidophosphoribosyl transferase · Glutamine + H_2O → Glutamate + PP_i

β-5-**Phosphoribosylamine (PRA)**

3 · GAR synthetase · Glycine + ATP → ADP + P_i

Glycinamide ribotide (GAR)

4 · GAR transformylase · N^{10}-Formyl-THF → THF

Formylglycinamide ribotide (FGAR)

5 · FGAM synthetase · ATP + Glutamine + H_2O → ADP + Glutamate + P_i

Formylglycinamidine ribotide (FGAM)

6 · AIR synthetase · ATP → ADP + P_i

5-**Aminoimidazole ribotide (AIR)**

7 · AIR carboxylase · ATP + HCO_3^- → ADP + P_i

Carboxyaminoimidazole ribotide (CAIR)

8 · SAICAR synthetase · Aspartate + ATP → ADP + P_i

5-**Aminoimidazole-4-(N-succinylocarboxamide) ribotide (SAICAR)**

9 · adenylosuccinate lyase · Fumarate

5-**Aminoimidazole-4-carboxamide ribotide (AICAR)**

10 · AICAR transformylase · N^{10}-Formyl-THF → THF

5-**Formaminoimidazole-4-carboxamide ribotide (FAICAR)**

11 · IMP cyclohydrolase · H_2O

Inosine monophosphate (IMP)

Figure 22-1 (*Opposite*) *Key to Metabolism.* The metabolic pathway for the *de novo* biosynthesis of IMP. Here the purine residue is built up on a ribose ring in 11 enzyme-catalyzed reactions. The X-ray structures for all these enzymes are shown to the outside of the corresponding reaction arrow. The peptide chains of monomeric enzymes are color-ramped from N-terminus (*blue*) to C-terminus (*red*). The oligomeric enzymes, all of which consist of identical polypeptide chains, are viewed along a rotation axis with their various chains differently colored. Bound ligands are shown in space-filling form. PDBids: enzyme 1, 1DKU; enzyme 2, 1AOO; enzyme 3, 1GSO; enzyme 4, 1CDE; enzyme 5, 1VK3; enzyme 6, 1CLI; enzyme 7, 1D7A (PurE) and 1B6S (PurK); enzyme 8, 1A48; enzyme 9, 1C3U; enzymes 10 and 11, 1G8M. 🔁 **See the Animated Figures.**

reaction, **GAR transformylase,** in complex with GAR and the THF analog **5-deazatetrahydrofolate (5dTHF)** was determined by Robert Almassy (Fig. 22-2). Note the proximity of the GAR amino group to N10 of 5dTHF. This supports enzymatic studies suggesting that the GAR transformylase reaction proceeds via the nucleophilic attack of the GAR amine group on the formyl carbon of N^{10}-formyl-THF to yield a tetrahedral intermediate.

5. **Acquisition of purine atom N3.** The amide amino group of a second glutamine is transferred to the growing purine ring to form **formylglycinamidine ribotide (FGAM).** This reaction is driven by the coupled hydrolysis of ATP to ADP + P_i.

6. **Formation of the purine imidazole ring.** The purine imidazole ring is closed in an ATP-requiring intramolecular condensation that yields **5-aminoimidazole ribotide (AIR).** The aromatization of the imidazole ring is facilitated by the tautomeric shift of the reactant from its imine to its enamine form.

7. **Acquisition of C6.** Purine C6 is introduced as HCO_3^- (CO_2) in a reaction catalyzed by **AIR carboxylase** that yields **carboxyaminoimidazole ribotide (CAIR).** In yeast, plants, and most prokaryotes (including *E. coli*), AIR carboxylase consists of two proteins called **PurE** and **PurK.** Although PurE alone can catalyze the carboxylation reaction, its K_M for HCO_3^- is ~110 mM, so the reaction would require an unphysiologically high HCO_3^- concentration (~100 mM) to proceed. PurK decreases the HCO_3^- concentration required for the PurE reaction by >1000-fold but at the expense of ATP hydrolysis.

8. **Acquisition of N1.** Purine atom N1 is contributed by aspartate in an amide-forming condensation reaction yielding **5-aminoimidazole-4-(*N*-succinylocarboxamide) ribotide (SACAIR).** This reaction, which is driven by the hydrolysis of ATP, chemically resembles Reaction 3.

9. **Elimination of fumarate.** SACAIR is cleaved with the release of fumarate, yielding **5-aminoimidazole-4-carboxamide ribotide (AICAR).** Reactions 8 and 9 chemically resemble the reactions in the urea cycle in which citrulline is aminated to form arginine (Section 20-3A). In both pathways, aspartate's amino group is transferred to an acceptor through an ATP-driven coupling reaction followed by the elimination of the aspartate carbon skeleton as fumarate.

10. **Acquisition of C2.** The final purine ring atom is acquired through formylation by N^{10}-formyl-THF, yielding **5-formaminoimidazole-4-carboxamide ribotide (FAICAR).** This reaction and Reaction 4 of purine biosynthesis are inhibited indirectly by **sulfonamides,**

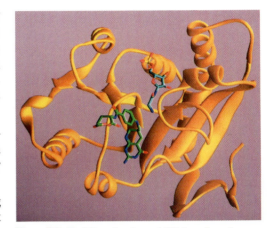

Figure 22-2 **X-Ray structure of GAR transformylase in complex with GAR and 5dTHF.** The C atoms of GAR (*upper right*) and 5dTHF (*lower left*) are cyan and green, whereas their N, C, and P atoms are blue, red, and yellow, respectively. Note the proximity of GAR to 5dTHF. [Based on an X-ray structure by Robert Almassy, Agouron Pharmaceuticals, San Diego, California. PDBid 1CDE.]

structural analogs of the *p*-aminobenzoic acid constituent of THF (Section 20-4D).

11. **Cyclization to form IMP.** The final reaction in the purine biosynthetic pathway, ring closure to form IMP, occurs through the elimination of water. In contrast to Reaction 6, the cyclization that forms the imidazole ring, this reaction does not require ATP hydrolysis.

In animals, Reactions 10 and 11 are catalyzed by a bifunctional enzyme, as are Reactions 7 and 8. Reactions 3, 4, and 6 also take place on a single protein. *The intermediate products of these multifunctional enzymes are not readily released to the medium but are channeled to the succeeding enzymatic activities of the pathway.* As in the reactions catalyzed by the pyruvate dehydrogenase complex (Section 16-2), fatty acid synthase (Section 19-4C), bacterial glutamate synthase (Section 20-7), and tryptophan synthase (Section 20-5B), channeling in the nucleotide synthetic pathways increases the overall rate of these multistep processes and protects intermediates from degradation by other cellular enzymes.

B Synthesis of Adenine and Guanine Ribonucleotides

IMP does not accumulate in the cell but is rapidly converted to AMP and GMP. AMP, which differs from IMP only in the replacement of its 6-keto group by an amino group, is synthesized in a two-reaction pathway (Fig. 22-3, *left*). In the first reaction, aspartate's amino group is linked to IMP in a reaction powered by the hydrolysis of GTP to GDP + P_i to yield **adenylosuccinate.** In the second reaction, **adenylosuccinate lyase** eliminates fumarate from adenylosuccinate to form AMP. The same enzyme catalyzes Reaction 9 of the IMP pathway (Fig. 22-1). Both reactions add a nitrogen with the elimination of fumarate.

GMP is also synthesized from IMP in a two-reaction pathway (Fig. 22-3, *right*). In the first reaction, IMP is dehydrogenated via the reduction of NAD^+ to form **xanthosine monophosphate (XMP; the ribonucleotide of the base xanthine).** XMP is then converted to GMP by the transfer of the glutamine amide nitrogen in a reaction driven by the hydrolysis of ATP to AMP + PP_i (and subsequently to 2 P_i). In *B* and *T* lymphocytes, which mediate the immune response, IMP dehydrogenase activity is high in order to supply the guanosine these cells need for proliferation. The fungal compound **mycophenolic acid**

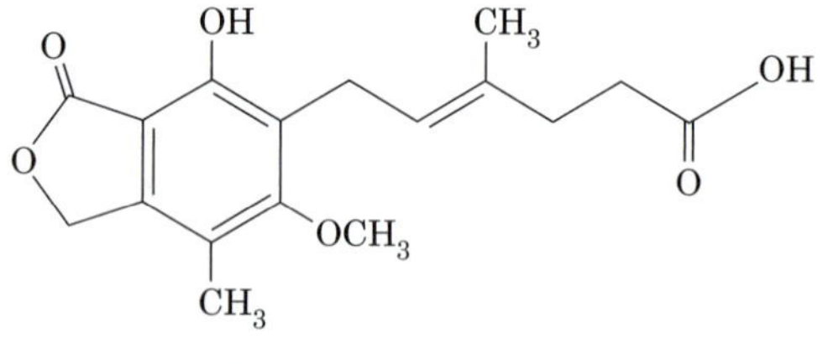

Mycophenolic acid

inhibits the enzyme and is used as an immunosuppressant following kidney transplants.

Nucleoside Diphosphates and Triphosphates Are Synthesized by the Phosphorylation of Nucleoside Monophosphates. *In order to participate in nucleic acid synthesis, nucleoside monophosphates must first be converted to the corresponding nucleoside triphosphates.* First, nucleoside diphosphates are synthesized from the corresponding nucleoside monophosphates by base-specific

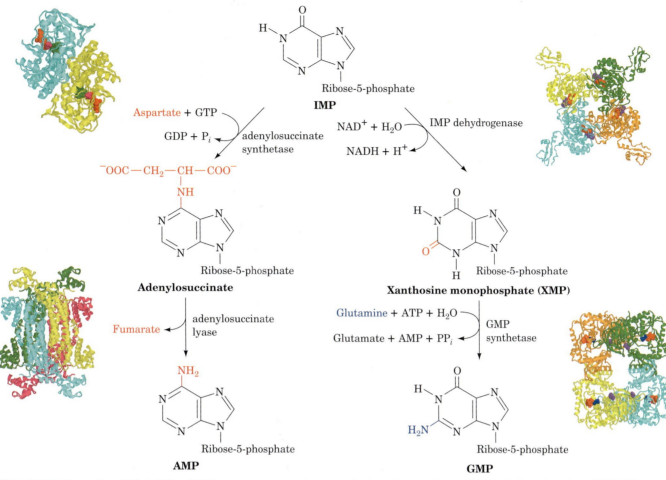

Figure 22-3 Conversion of IMP to AMP or GMP in separate two-reaction pathways. The X-ray structures of the homo-oligomeric enzymes catalyzing these reactions are shown as is described in the legend for Fig. 22-1. PDBids: adenylosuccinate synthetase, 1G1M; adenylosuccinate lyase, 1C3U; IMP dehydrogenase, 1JR1; GMP synthetase, 1GPM.

nucleoside monophosphate kinases. For example, adenylate kinase (Section 13-2C) catalyzes the phosphorylation of AMP to ADP:

$$AMP + ATP \rightleftharpoons 2\ ADP$$

Similarly, GDP is produced by **guanylate kinase:**

$$GMP + ATP \rightleftharpoons GDP + ADP$$

These nucleoside monophosphate kinases do not discriminate between ribose and deoxyribose in the substrate.

Nucleoside diphosphates are converted to the corresponding triphosphates by **nucleoside diphosphate kinase;** for instance,

$$GDP + ATP \rightleftharpoons GTP + ADP$$

Although the reaction is written with ATP as the phosphoryl donor, this enzyme exhibits no preference for the bases of its substrates or for ribose over deoxyribose. Furthermore, the nucleoside diphosphate kinase reaction, as might be expected from the nearly identical structures of its substrates and products, normally operates close to equilibrium ($\Delta G \approx 0$). ADP is, of course, also converted to ATP by a variety of energy-releasing

reactions such as those of glycolysis and oxidative phosphorylation. Indeed, it is these reactions that ultimately drive the foregoing kinase reactions.

C   Regulation of Purine Nucleotide Biosynthesis

The pathways synthesizing IMP, ATP, and GTP are individually regulated in most cells so as to control the total amounts of purine nucleotides available for nucleic acid synthesis, as well as the relative amounts of ATP and GTP. This control network is diagrammed in Fig. 22-4.

The IMP pathway is regulated at its first two reactions: those catalyzing the synthesis of PRPP and 5-phosphoribosylamine. Ribose phosphate pyrophosphokinase, the enzyme catalyzing Reaction 1 of the IMP pathway (Fig. 22-1), is inhibited by both ADP and GDP. Amidophosphoribosyl transferase, the enzyme catalyzing the first committed step of the IMP pathway (Reaction 2) is likewise subject to feedback inhibition. In this case, the enzyme binds ATP, ADP, and AMP at one inhibitory site and GTP, GDP, and GMP at another. *The rate of IMP production is therefore independently but synergistically controlled by the levels of adenine nucleotides and guanine nucleotides.* Moreover, amidophosphoribosyl transferase is allosterically stimulated by PRPP **(feedforward activation).**

A second level of regulation occurs immediately below the branch point leading from IMP to AMP and GMP. AMP and GMP are each competitive inhibitors of IMP in their own synthesis, which prevents excessive buildup of the pathway products. In addition, the rates of adenine and guanine nucleotide synthesis are coordinated. Recall that GTP powers the synthesis of AMP from IMP, whereas ATP powers the synthesis of GMP from IMP (Fig. 22-3). This reciprocity balances the production of AMP and GMP (which are required in roughly equal amounts in nucleic acid biosynthesis):

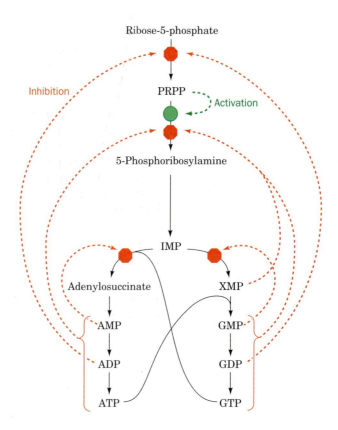

Figure 22-4   Control of the purine biosynthesis pathway.   Red octagons and green dots indicate control points. Feedback inhibition is indicated by dashed red arrows, and feedforward activation is represented by a dashed green arrow.   **See the Animated Figures.**

The rate of synthesis of GMP increases with [ATP], whereas that of AMP increases with [GTP].

D Salvage of Purines

In most cells, the turnover of nucleic acids, particularly some types of RNA, releases adenine, guanine, and hypoxanthine (Section 22-4A). These free purines are reconverted to their corresponding nucleotides through **salvage pathways.** In contrast to the *de novo* purine nucleotide synthetic pathway, which is virtually identical in all cells, salvage pathways are diverse in character and distribution. In mammals, purines are mostly salvaged by two different enzymes. **Adenine phosphoribosyltransferase (APRT)** mediates AMP formation using PRPP:

$$\text{Adenine} + \text{PRPP} \rightleftharpoons \text{AMP} + \text{PP}_i$$

Hypoxanthine–guanine phosphoribosyltransferase (HGPRT) catalyzes the analogous reaction for both hypoxanthine and guanine:

$$\text{Hypoxanthine} + \text{PRPP} \rightleftharpoons \text{IMP} + \text{PP}_i$$
$$\text{Guanine} + \text{PRPP} \rightleftharpoons \text{GMP} + \text{PP}_i$$

Lesch–Nyhan Syndrome Results from HGPRT Deficiency. The symptoms of **Lesch–Nyhan syndrome,** which is caused by a severe HGPRT deficiency, indicate that purine salvage reactions have functions other than conservation of the energy required for *de novo* purine biosynthesis. This sex-linked congenital defect (it affects mostly males) results in excessive uric acid production (uric acid is a purine degradation product; Section 22-4A) and neurological abnormalities such as spasticity, mental retardation, and highly aggressive and destructive behavior, including a bizarre compulsion toward self-mutilation. For example, many children with Lesch–Nyhan syndrome have such an irresistible urge to bite their lips and fingers that they must be restrained. If the restraints are removed, communicative patients will plead that the restraints be replaced, even as they attempt to injure themselves.

The excessive uric acid production in patients with Lesch–Nyhan syndrome is readily explained. The lack of HGPRT activity leads to an accumulation of the PRPP that would normally be used to salvage hypoxanthine and guanine. The excess PRPP activates amidophosphoribosyl transferase (which catalyzes Reaction 2 of the IMP biosynthetic pathway), thereby greatly accelerating the synthesis of purine nucleotides and thus the formation of their degradation product, uric acid. Yet the physiological basis of the associated neurological abnormalities remains obscure. That a defect in a single enzyme can cause such profound but well-defined behavioral changes nevertheless has important psychiatric implications.

2 Synthesis of Pyrimidine Ribonucleotides

The biosynthesis of pyrimidines is simpler than that of purines. Isotopic labeling experiments have shown that atoms N1, C4, C5, and C6 of the pyrimidine ring are all derived from aspartic acid, C2 arises from HCO_3^-, and N3 is contributed by glutamine.

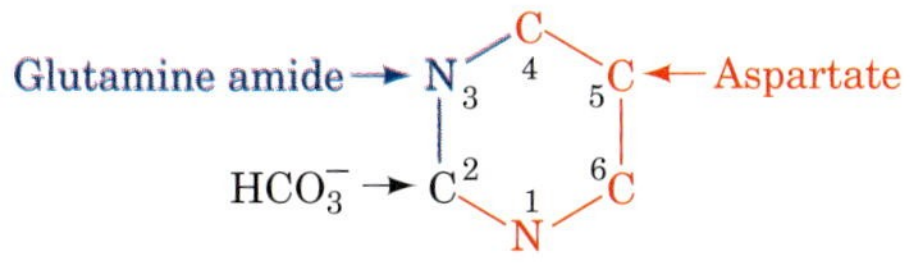

A Synthesis of UMP

UMP, which is also the precursor of CMP, is synthesized in a six-reaction pathway (Fig. 22-5). In contrast to purine nucleotide synthesis, the pyrimidine ring is coupled to the ribose-5-phosphate moiety *after* the ring has been synthesized.

1. **Synthesis of carbamoyl phosphate.** The first reaction of pyrimidine biosynthesis is the synthesis of **carbamoyl phosphate** from HCO_3^- and the amide nitrogen of glutamine by the cytosolic enzyme **carbamoyl phosphate synthetase II.** This reaction consumes two molecules of ATP: One provides a phosphate group and the other energizes the reaction. Carbamoyl phosphate is also synthesized in the urea cycle (Section 20-3A). In that reaction, catalyzed by the mitochondrial enzyme carbamoyl phosphate synthetase I, ammonia is the nitrogen source.

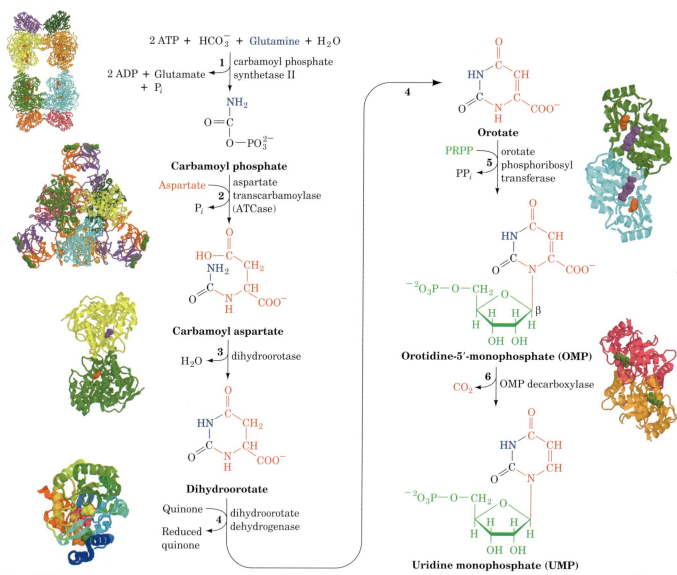

Figure 22-5 *Key to Metabolism.* **The *de novo* synthesis of UMP.** The X-ray structures of the six enzymes catalyzing these reactions are shown as is described in the legend for Fig. 22-1. PDBids: enzyme 1, 1BXR; enzyme 2, 5AT1; enzyme 3, 1J79; enzyme 4, 1D3H; enzyme 5, 1OPR; enzyme 6, 1DBT. 🔊 **See the Animated Figures.**

2. **Synthesis of carbamoyl aspartate.** Condensation of carbamoyl phosphate with aspartate to form **carbamoyl aspartate** is catalyzed by **aspartate transcarbamoylase (ATCase).** This reaction proceeds without ATP hydrolysis because carbamoyl phosphate is already "activated." The structure and regulation of *E. coli* ATCase are discussed in Section 12-3.

3. **Ring closure to form dihydroorotate.** The third reaction of the pathway is an intramolecular condensation catalyzed by **dihydroorotase** to yield **dihydroorotate.**

4. **Oxidation of dihydroorotate.** Dihydroorotate is irreversibly oxidized to **orotate** by **dihydroorotate dehydrogenase.** The eukaryotic enzyme, which contains FMN and nonheme Fe, is located on the outer surface of the inner mitochondrial membrane, where quinones supply its oxidizing power. The other five enzymes of pyrimidine nucleotide biosynthesis are cytosolic in animal cells. Inhibition of dihydroorotate dehydrogenase blocks pyrimidine synthesis in *T* lymphocytes, thereby attenuating the autoimmune disease rheumatoid arthritis.

5. **Acquisition of the ribose phosphate moiety.** Orotate reacts with PRPP to yield **orotidine-5′-monophosphate (OMP)** in a reaction catalyzed by **orotate phosphoribosyl transferase.** This reaction, which is driven by the hydrolysis of the eliminated PP_i, fixes the anomeric form of pyrimidine nucleotides in the β configuration. Orotate phosphoribosyl transferase also salvages other pyrimidine bases, such as uracil and cytosine, by converting them to their corresponding nucleotides.

6. **Decarboxylation to form UMP.** The final reaction of the pathway is the decarboxylation of OMP by **OMP decarboxylase (ODCase)** to form UMP. ODCase enhances the rate (k_{cat}/K_M) by a factor of 2×10^{23} over that of the uncatalyzed reaction, making it the most catalytically proficient enzyme known. Nevertheless, this reaction requires no cofactors to help stabilize its putative carbanion intermediate. Although the mechanism of the ODCase reaction is not fully understood, the removal of OMP's phosphate group, which is quite distant from the C6 carboxyl group, deceases the reaction rate by a factor of 7×10^7, thus providing a striking example of how binding energy can be applied to catalysis (preferential transition state binding).

In bacteria, the six enzymes of UMP biosynthesis occur as independent proteins. In animals, however, as Mary Ellen Jones demonstrated, the first three enzymatic activities of the pathway—carbamoyl phosphate synthetase II, ATCase, and dihydroorotase—occur on a single 210-kD polypeptide chain.

The pyrimidine biosynthetic pathway is a target for antiparasitic drugs. For example, the parasite *Toxoplasma gondii* (Fig. 22-6), which infects most mammals, causes **toxoplasmosis,** a disease whose complications include blindness, neurological dysfunction, and death in immunocompromised individuals (e.g., those with AIDS). Most parasites have evolved to take advantage of nutrients supplied by their hosts, but *T. gondii* is unable to meet its needs exclusively through nucleotide salvage pathways and retains the ability to synthesize uracil *de novo*. Drugs that target the parasite's carbamoyl phosphate synthetase II (an enzyme whose structure and kinetics distinguish it from its mammalian counterpart) could therefore prevent *T. gondii* growth. Moreover, there is evidence that *T. gondii* strains that have been engineered to lack carbamoyl phosphate synthetase II are avirulent and could be useful as vaccines in humans and livestock.

Figure 22-6 *Toxoplasma gondii.* This intracellular parasite (*yellow*) causes toxoplasmosis. [© Dennis Kunkel/Phototake.]

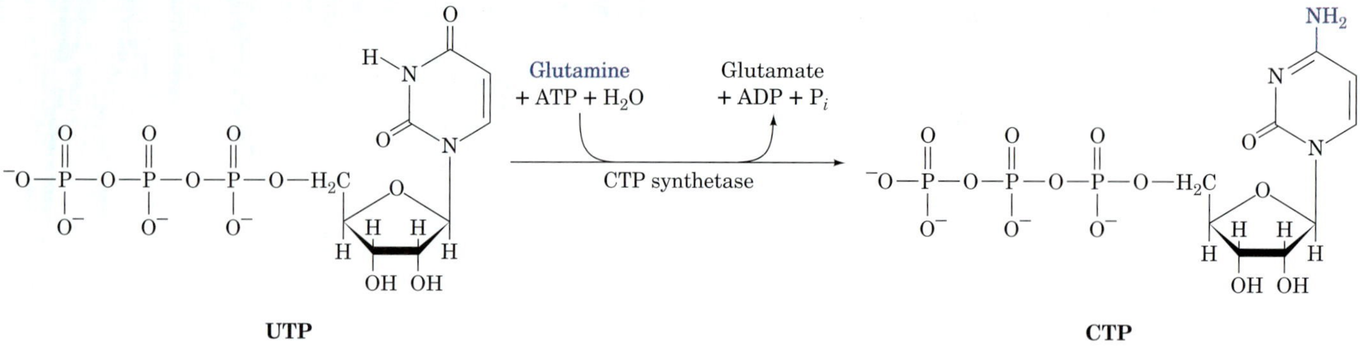

Figure 22-7   The synthesis of CTP from UTP.

B Synthesis of UTP and CTP

The synthesis of UTP from UMP is analogous to the synthesis of purine nucleoside triphosphates (Section 22-1B). The process occurs by the sequential actions of a nucleoside monophosphate kinase and nucleoside diphosphate kinase:

$$UMP + ATP \rightleftharpoons UDP + ADP$$
$$UDP + ATP \rightleftharpoons UTP + ADP$$

CTP is formed by amination of UTP by **CTP synthetase** (Fig. 22-7). In animals, the amino group is donated by glutamine, whereas in bacteria it is supplied directly by ammonia.

C Regulation of Pyrimidine Nucleotide Biosynthesis

In bacteria, the pyrimidine biosynthetic pathway is primarily regulated at Reaction 2, the ATCase reaction (Fig. 22-8a). In *E. coli,* control is exerted through the allosteric stimulation of ATCase by ATP and its inhibition by CTP (Section 12-3). In many bacteria, however, UTP is the major ATCase inhibitor.

In animals, ATCase is not a regulatory enzyme. Rather, pyrimidine biosynthesis is controlled by the activity of carbamoyl phosphate synthetase II, which is inhibited by UDP and UTP and activated by ATP and PRPP (Fig. 22-8b). A second level of control in the mammalian pathway occurs at OMP decarboxylase, for which UMP and to a lesser extent CMP are competitive inhibitors. In all organisms, the rate of OMP production varies with the availability of its precursor, PRPP. Recall that the PRPP level depends on the activity of ribose phosphate pyrophosphokinase (Fig. 22-1, Reaction 1), which is inhibited by ADP and GDP (Section 22-1C).

Orotic Aciduria Results from an Inherited Enzyme Deficiency.   **Orotic aciduria,** an inherited human disease, is characterized by the urinary excretion of large amounts of orotic acid, retarded growth, and severe anemia. It results from a deficiency in the bifunctional enzyme catalyzing Reactions 5 and 6 of pyrimidine nucleotide biosynthesis. Consideration of the biochemistry of this situation led to its effective treatment: the administration of uridine and/or cytidine. The UMP formed through the phosphorylation of these nucleosides, besides replacing that normally synthesized, inhibits carbamoyl phosphate synthetase II so as to attenuate the rate of orotic acid synthesis. No other genetic deficiency in pyrimidine nucleotide biosynthesis is known in humans, presumably because such defects are lethal *in utero.*

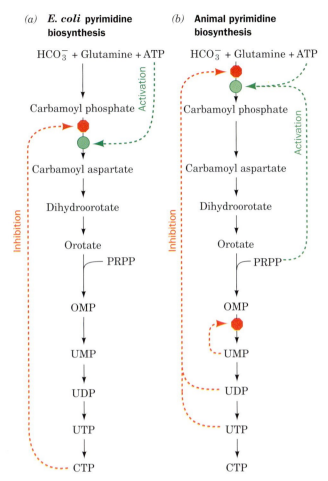

(a)   **E. coli pyrimidine biosynthesis**

(b)   **Animal pyrimidine biosynthesis**

Figure 22-8 Regulation of pyrimidine biosynthesis. The control networks are shown for (a) *E. coli* and (b) animals. Red octagons and green dots indicate control points. Feedback inhibition is represented by dashed red arrows, and activation is indicated by dashed green arrows. 🐚 See the Animated Figures.

---

3 Formation of Deoxyribonucleotides

DNA differs chemically from RNA in two major respects: (1) Its nucleotides contain 2′-deoxyribose residues rather than ribose residues, and (2) it contains the base thymine (5-methyluracil) rather than uracil. In this section, we consider the biosynthesis of these DNA components.

A Production of Deoxyribose Residues

Deoxyribonucleotides are synthesized from their corresponding ribonucleotides by the reduction of their C2′ position rather than by their de novo synthesis from deoxyribose-containing precursors (at right).

Enzymes that catalyze the formation of deoxyribonucleotides by the reduction of the corresponding ribonucleotides are named **ribonucleotide reductases (RNRs).** There are three classes of RNRs, which differ in their prosthetic groups, although they all replace the 2′-OH group of ribose with H via a free-radical mechanism. We shall discuss the mechanism of the so-called Class I RNRs, which contain an Fe prosthetic group and which occur in most eukaryotes and aerobic prokaryotes.

Class I RNRs reduce ribonucleoside diphosphates (NDPs) to the corresponding deoxyribonucleoside diphosphates (dNDPs). *E. coli* ribonucleotide reductase, as Peter Reichard demonstrated, is mainly present *in vitro* as a heterotetramer that can be decomposed to two catalytically

NDP

dNDP

(a)

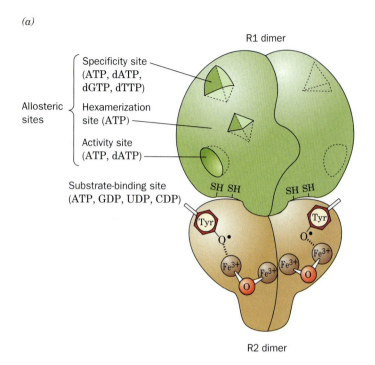

(b)

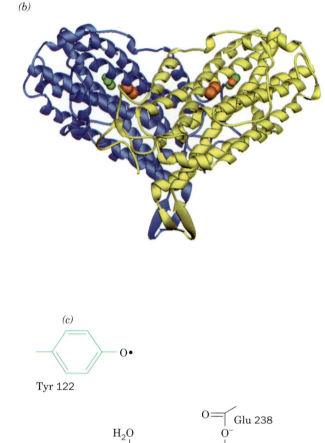

(c)

Figure 22-9   Class I ribonucleotide reductase from *E. coli*.   (*a*) Schematic diagram of the quaternary structure. The enzyme consists of two identical pairs of subunits, R1$_2$ and R2$_2$. Each R2 subunit contains a binuclear Fe(III) complex that generates a phenoxy radical at Tyr 122. The R1 subunits each contain three different allosteric effector sites and five catalytically important Cys residues. The enzyme's two active sites occur at the interface between the R1 and R2 subunits. (*b*) A ribbon diagram of R2$_2$ viewed perpendicularly to its twofold axis with the subunits drawn in blue and yellow. The Fe(III) ions are represented by orange spheres, and the radical-harboring Tyr 122 side chains are shown in space-filling representation with their C and O atoms green and red. (*c*) The binuclear Fe(III) complex of R2. Each Fe(III) ion is octahedrally coordinated by a His N atom and five O atoms, including those of the O^{2-} ion and the Glu carboxyl group that bridge the two Fe(III) ions. [Part *b* based on an X-ray structure by Hans Eklund, Swedish University of Agricultural Sciences, Uppsala, Sweden. PDBid 1RIB.] ✑⌇ **See the Interactive Exercises.**

inactive homodimers, R1$_2$ and R2$_2$ (Fig. 22-9*a*). Each R1 subunit contains a substrate-binding site that includes several redox-active thiol groups. The R1 subunits also contain three effector-binding sites that control the enzyme's catalytic activity as well as its substrate specificity (see below).

The X-ray structure of R2$_2$ (Fig. 22-9*b*), which was determined by Hans Eklund, reveals that the subunits are bundles of eight unusually long helices. Each subunit contains a novel binuclear Fe(III) prosthetic group whose Fe(III) ions are liganded by a variety of groups including an O^{2-} ion (Fig. 22-9*c*). The Fe(III) complex interacts with Tyr 122 to form an unusual tyrosyl free radical. Tyrosine radicals also take part in the reactions catalyzed by cytochrome *c* oxidase (Section 17-2F) and plant Photosystem II (Section 18-2C).

JoAnne Stubbe has proposed the following catalytic mechanism for *E. coli* ribonucleotide reductase (Fig. 22-10):

1.   Ribonucleotide reductase's free radical (X·) abstracts an H atom from C3′ of the substrate in the reaction's rate-determining step.

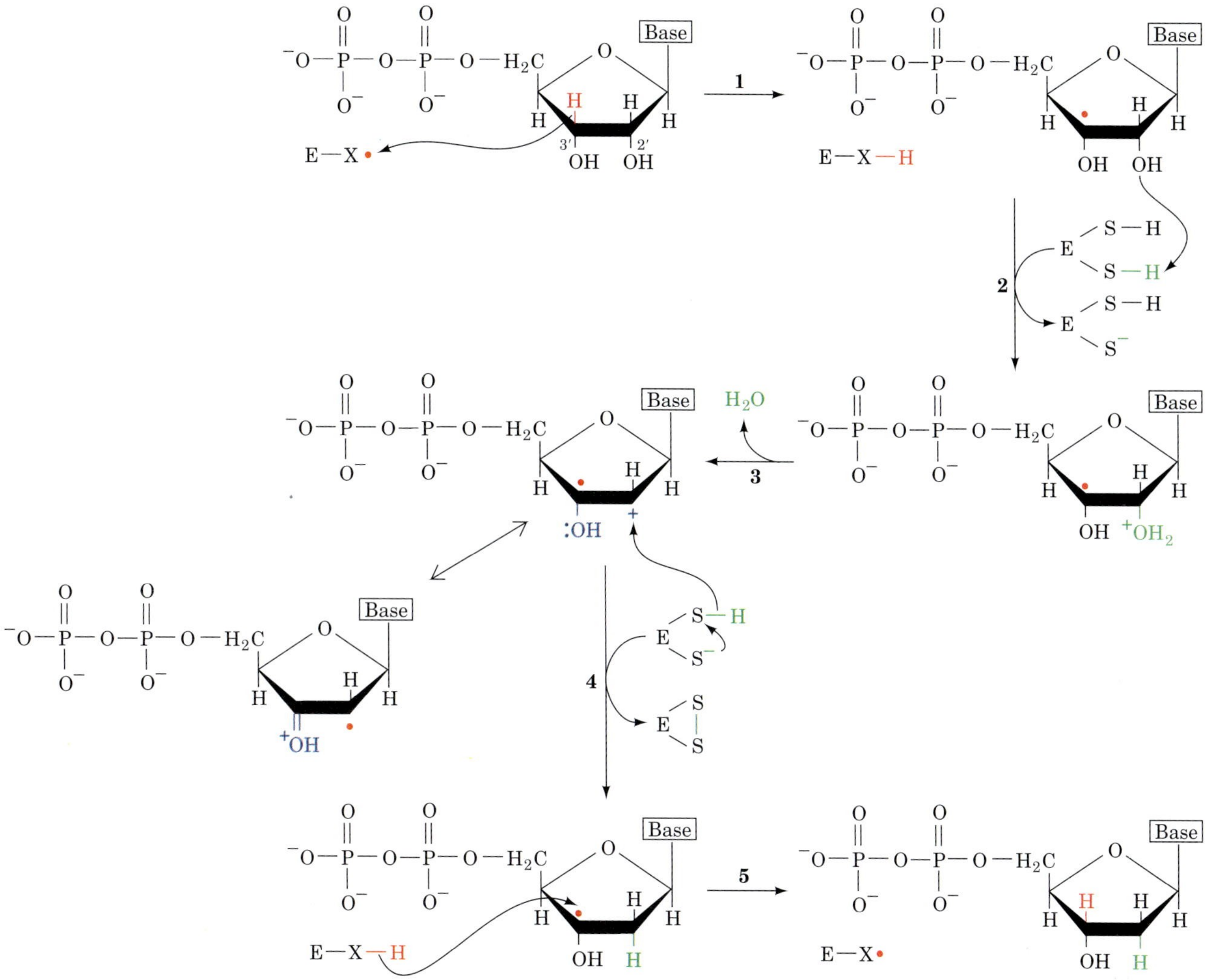

Figure 22-10 Enzymatic mechanism of ribonucleotide reductase. The reaction occurs via a free radical–mediated process in which reducing equivalents are supplied by the formation of an enzyme disulfide bond. [After Stubbe, J.A., *J. Biol. Chem.* **265**, 5330 (1990).]

2 and 3. Acid-catalyzed cleavage of the C2'—OH bond releases H_2O to yield a radical–cation intermediate. The C3'-OH group's unshared electron pair stabilizes the C2' cation. This accounts for the radical's catalytic role.

4. The radical–cation intermediate is reduced by the enzyme's redox-active sulfhydryl pair to yield a 3'-deoxynucleotide radical and a protein disulfide group (this group must eventually be oxidized to regenerate the enzyme's activity).

5. The 3' radical abstracts an H atom from the protein to yield the product deoxynucleoside diphosphate and restore the enzyme to its radical state.

The Tyr 122 radical in R2 is buried 10 Å beneath the surface of the protein, too far for the enzyme's catalytic site to abstract an electron directly from the substrate. Evidently, the protein mediates electron transfer from this tyrosyl radical to another group that is closer to the substrate, probably the thiyl radical (—S·) form of Cys 439 in R1 (represented as X· in Fig. 22-10). Two other R1 Cys residues probably form the redox-active

sulfhydryl pair that directly reduces the substrate. The resulting disulfide bond is reduced via disulfide interchange with yet two other Cys residues, which are positioned to accept electrons from external reducing agents to regenerate the active enzyme. Thus, each R1 subunit contains at least five Cys residues that chemically participate in nucleotide reduction.

The Inability of Oxidized Ribonucleotide Reductase to Bind Substrate Serves an Essential Protective Function.

Comparison of the X-ray structures of reduced R1 (in which the redox-active Cys 225 and Cys 462 residues are in their SH forms) and oxidized R1 (in which Cys 225 and Cys 462 are disulfide-linked) reveals that Cys 462 in reduced R1 has rotated away from its position in oxidized R1 to become buried in a hydrophobic pocket, whereas Cys 225 moves into the region formerly occupied by Cys 462. The distance between the formerly disulfide-linked S atoms thereby increases from 2.0 Å to 5.7 Å. These movements are accompanied by small shifts of the surrounding polypeptide chain. R1 Cys 225 in oxidized ribonucleotide reductase prevents the binding of substrate through steric interference of its S atom with the substrate dNDP's O2′ atom.

The inability of oxidized ribonucleotide reductase to bind substrate has functional significance. In the absence of substrate, the enzyme's free radical is stored in the interior of the R2 subunit, close to its dinuclear iron center. When substrate is bound, the radical is presumably transferred to it via a series of protein side chains in both R2 and R1. If the substrate is unable to properly react after accepting this free radical, as would be the case if the enzyme was in its oxidized state, the free radical could potentially destroy both the substrate and the enzyme. Thus, *an important role of the enzyme is to control the release of the radical's powerful oxidizing capability. It does so in part by preventing the binding of substrate while the enzyme is in its oxidized form.*

Thioredoxin Reduces Ribonucleotide Reductase.

The final step in the ribonucleotide reductase catalytic cycle is reduction of the enzyme's newly formed disulfide bond to re-form its redox-active sulfhydryl pair. One of the enzyme's physiological reducing agents is **thioredoxin,** a ubiquitous monomeric protein with a pair of neighboring Cys residues (and which also participates in regulating the Calvin cycle; Section 18-3C). Thioredoxin reduces oxidized ribonucleotide reductase via disulfide interchange.

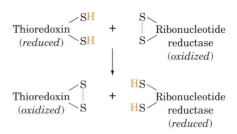

The X-ray structure of thioredoxin (Fig. 22-11) reveals that its redox-active disulfide group is located on a molecular protrusion, making this protein the only known example of a "male" enzyme.

Oxidized thioredoxin is, in turn, reduced in a reaction mediated by **thioredoxin reductase,** which contains redox-active thiol groups and an FAD prosthetic group. This enzyme is a homolog of glutathione reductase (Box 14-4) and catalyzes a similar reaction: the NADPH-mediated reduction of a substrate disulfide bond. NADPH therefore serves as the terminal reducing agent in the ribonucleotide reductase–catalyzed reduction of NDPs to dNDPs (Fig. 22-12).

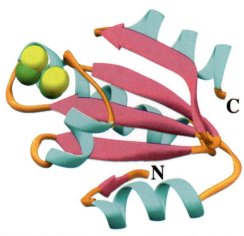

Figure 22-11 X-Ray structure of human thioredoxin in its reduced (sulfhydryl) state. The backbone of this 105-residue monomer is colored according to secondary structure with helices cyan, sheets magenta, and the remaining portions orange. The side chains of the redox-active Cys residues are shown in space-filling form with C green and S yellow. [Based on an X-ray structure by William Montfort, University of Arizona. PDBid 1ERT.]

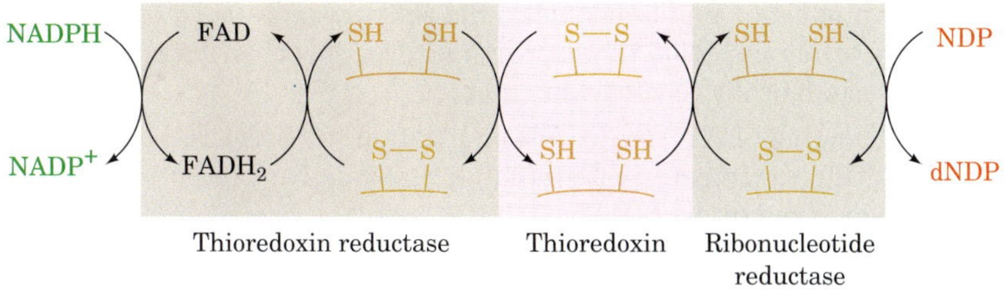

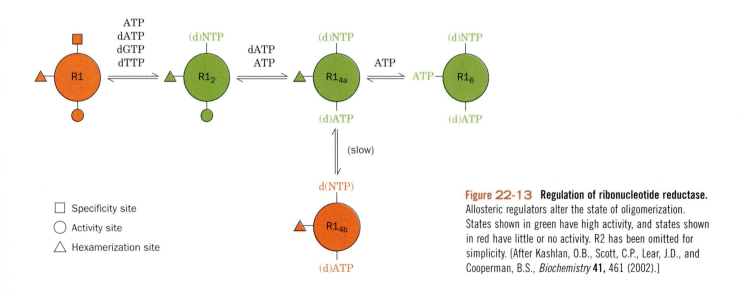

Figure 22-12 An electron-transfer pathway for nucleoside diphosphate (NDP) reduction. NADPH provides the reducing equivalents for this process through the intermediacy of thioredoxin reductase, thioredoxin, and ribonucleotide reductase.

Ribonucleotide Reductase Is Regulated by a Complex Feedback Network. The synthesis of the four dNTPs in the amounts required for DNA synthesis is accomplished through feedback control. Maintaining the proper intracellular ratios of dNTPs is essential for normal growth. Indeed, *a deficiency of any dNTP is lethal, whereas an excess is mutagenic because the probability that a given dNTP will be erroneously incorporated into a growing DNA strand increases with its concentration relative to those of the other dNTPs.*

The activities of both *E. coli* and mammalian ribonucleotide reductases are remarkably responsive to the levels of nucleotides in the cell. For example, ATP induces the reduction of CDP and UDP; dTTP induces the reduction of GDP and inhibits the reduction of CDP and UDP; and dATP inhibits the reduction of all NDPs. Barry Cooperman has shown that the catalytic activity of mouse ribonucleotide reductase varies with its state of oligomerization, which in turn is governed by the binding of nucleotide effectors to three independent allosteric sites on R1 (Fig. 22-9*a*): (1) the specificity site, which binds ATP, dATP, dGTP, and dTTP; (2) the activity site, which binds ATP and dATP; and (3) the hexamerization site, which binds only ATP. Cooperman's model for the allosteric regulation of Class I ribonucleotide reductase, which quantitatively accounts for the enzyme's regulatory properties, has the following features (Fig. 22-13):

1. The binding of ATP, dATP, dGTP, or dTTP to the specificity site induces the catalytically inactive R1 monomers to form a catalytically active dimer, R1$_2$.

Figure 22-13 Regulation of ribonucleotide reductase. Allosteric regulators alter the state of oligomerization. States shown in green have high activity, and states shown in red have little or no activity. R2 has been omitted for simplicity. [After Kashlan, O.B., Scott, C.P., Lear, J.D., and Cooperman, B.S., *Biochemistry* **41,** 461 (2002).]

2. The binding of dATP or ATP to the activity site causes the dimers to form tetramers, R1$_{4a}$, that slowly but reversibly change conformation to a catalytically inactive state, R1$_{4b}$.

3. The binding of ATP to the hexamerization site induces the tetramers to further aggregate to form catalytically active hexamers, R1$_6$, the enzyme's major active form.

The concentration of ATP in a cell is such that, *in vivo*, R1 is almost entirely in its tetrameric or hexameric forms. As a consequence, ATP couples the overall rate of DNA synthesis to the cell's energy state.

dNTPs Are Produced by Phosphorylation of dNDPs.   *The final step in the production of all dNTPs is the phosphorylation of the corresponding dNDPs:*

$$\text{dNDP} + \text{ATP} \rightleftharpoons \text{dNTP} + \text{ADP}$$

This reaction is catalyzed by nucleoside diphosphate kinase, the same enzyme that phosphorylates NDPs (Section 22-1B). As before, the reaction is written with ATP as the phosphoryl donor, although any NTP or dNTP can function in this capacity.

B Origin of Thymine

The dTMP component of DNA is synthesized by methylation of dUMP. The dUMP is generated through the hydrolysis of dUTP by **dUTP diphosphohydrolase (dUTPase):**

$$\text{dUTP} + \text{H}_2\text{O} \rightarrow \text{dUMP} + \text{PP}_i$$

dTMP, once it is formed, is phosphorylated to form dTTP. The apparent reason for the energetically wasteful process of dephosphorylating dUTP and rephosphorylating dTMP is that cells must minimize their concentration

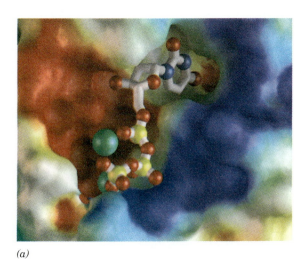

(a)

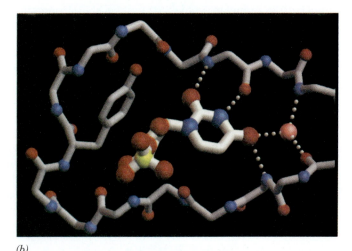

(b)

Figure 22-14 X-Ray structure of human dUTPase.   (*a*) The active site region of dUTPase in complex with dUTP. The protein is represented by its molecular surface colored according to its electrostatic potential (negative, red; positive, blue; and near neutral, white). The dUTP is shown in ball-and-stick form with its N, O, and P atoms blue, red, and yellow. Mg^{2+} ions that have been modeled into the structure are represented by green spheres. Note the complementary fit of the uracil ring into its binding pocket, particularly the close contacts that discriminate against a methyl group on C5 of the pyrimidine ring and a 2'-OH group on the ribose ring. (*b*) The binding site of dUMP, showing the hydrogen bonding system responsible for the enzyme's specific binding of a uracil ring. The dUMP and the polypeptide backbone binding it are shown in ball-and-stick form with atoms colored as in Part *a*; hydrogen bonds are indicated by white dotted lines; and a conserved water molecule is represented by a pink sphere. The side chain of a conserved Tyr is tightly packed against the ribose ring so as to discriminate against the presence of a 2'-OH group. [Courtesy of John Tainer and Clifford Mol, The Scripps Research Institute, La Jolla, California.]

of dUTP in order to prevent incorporation of uracil into their DNA (the enzyme system that synthesizes DNA from dNTPs does not efficiently discriminate between dUTP and dTTP; Section 24-2A).

Human dUTPase is a homotrimer of 141-residue subunits. Its X-ray structure, determined by John Tainer, reveals the basis for this enzyme's exquisite specificity for dUTP. Each subunit binds dUTP in a snug-fitting cavity that sterically excludes thymine's C5 methyl group via the side chains of conserved residues (Fig. 22-14a). The enzyme differentiates uracil from the similarly shaped cytosine via a set of hydrogen bonds that in part mimic adenine's base pairing interactions (Fig. 22-14b). The 2'-OH group of ribose is likewise sterically excluded by the side chain of a conserved Tyr.

Thymidylate Synthase. Thymidylate (dTMP) is synthesized from dUMP by **thymidylate synthase** with N^5,N^{10}-methylenetetrahydrofolate (N^5,N^{10}-methylene-THF) as the methyl donor:

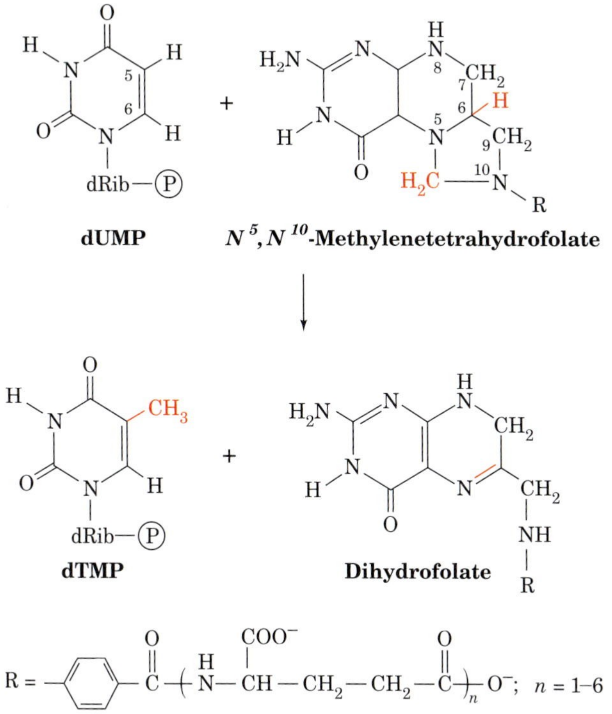

Note that the transferred methylene group (in which the carbon has the oxidation state of formaldehyde) is reduced to a methyl group (which has the oxidation state of methanol) at the expense of the oxidation of the THF cofactor to **dihydrofolate (DHF).**

Thymidylate synthase, a highly conserved 70-kD dimeric protein, follows a mechanistic scheme proposed by Daniel Santi (Fig. 22-15):

1. An enzyme nucleophile, identified as the thiolate group of Cys 146, attacks C6 of dUMP to form a covalent adduct.

2. C5 of the resulting enolate ion attacks the CH_2 group of the iminium cation in equilibrium with N^5,N^{10}-methylene-THF to form an enzyme–dUMP–THF ternary covalent complex.

3. An enzyme base abstracts the acidic proton at the C5 position of the enzyme-bound dUMP, forming an exocyclic methylene group and eliminating the THF cofactor. The abstracted proton subsequently exchanges with solvent.

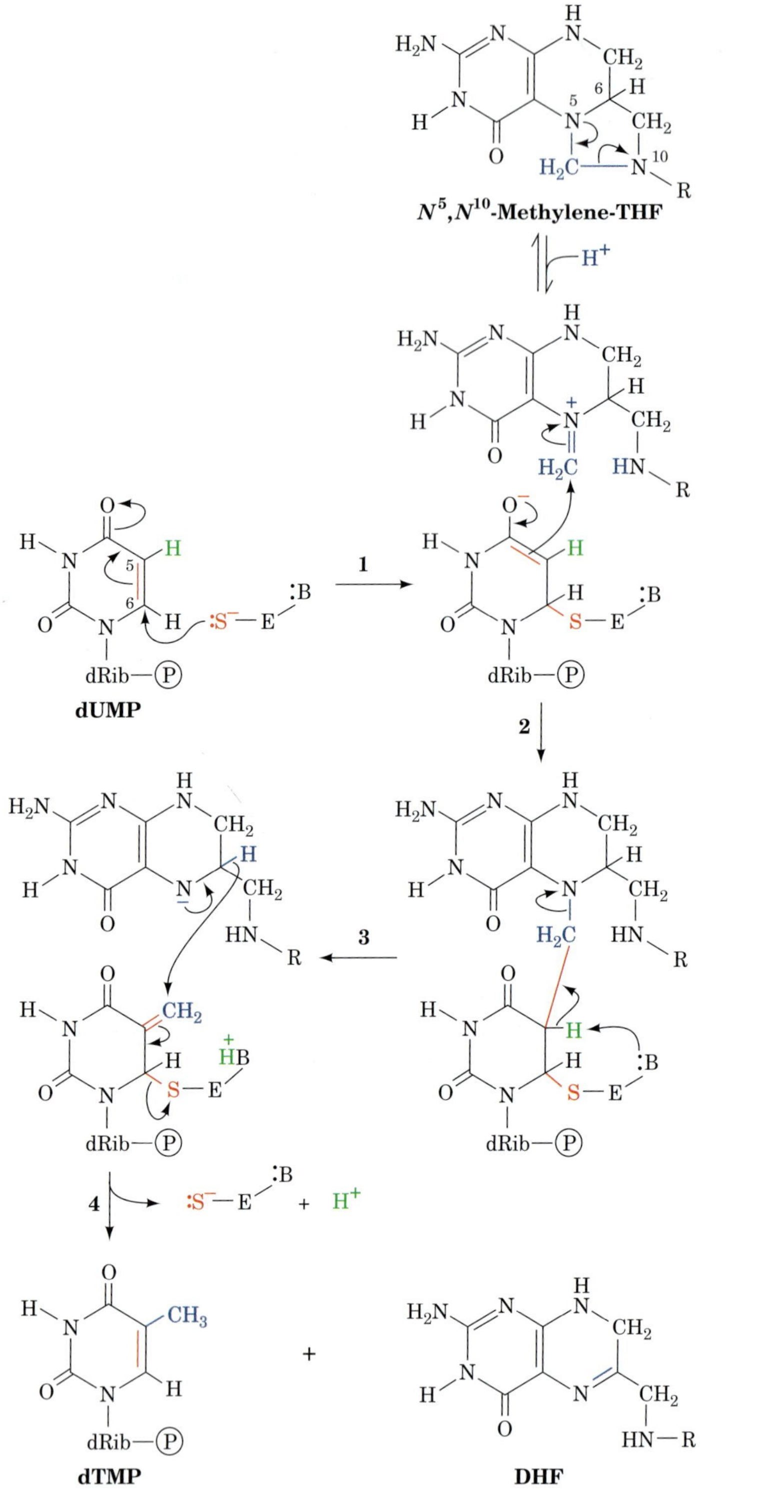

Figure 22-15 Catalytic mechanism of thymidylate synthase. The methyl group is supplied by N^5,N^{10}-methylene-THF, which is concomitantly oxidized to dihydrofolate.

4. The redox change occurs via the migration of the C6-H atom of THF as a hydride ion to the exocyclic methylene group, converting it to a methyl group and yielding DHF. This reduction promotes the displacement of the Cys thiolate group from the intermediate to release product, dTMP, and re-form the active enzyme.

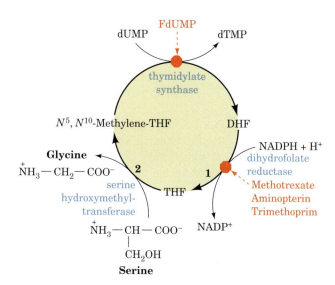

Figure 22-16 Regeneration of N^5,N^{10}-methylene-tetrahydrofolate. The DHF product of the thymidylate synthase reaction is converted back to N^5,N^{10}-methylene-THF by the sequential actions of (1) dihydrofolate reductase and (2) serine hydroxymethyltransferase. The sites of action of some inhibitors are indicated by red octagons. Thymidylate synthase is inhibited by FdUMP, whereas dihydrofolate reductase is inhibited by the antifolates methotrexate, aminopterin, and trimethoprim (see Box 22-1).

Tetrahydrofolate Is Regenerated in Two Reactions.

The thymidylate synthase reaction is biochemically unique in that it oxidizes THF to DHF; no other enzymatic reaction employing a THF cofactor alters this coenzyme's net oxidation state. The DHF product of the thymidylate synthase reaction is recycled back to N^5,N^{10}-methylene-THF through two sequential reactions (Fig. 22-16):

1. DHF is reduced to THF by NADPH as catalyzed by **dihydrofolate reductase (DHFR;** Fig. 22-17). Although in most organisms DHFR is a monomeric, monofunctional enzyme, in protozoa and some plants DHFR and thymidylate synthase occur on the same polypeptide chain to form a bifunctional enzyme that has been shown to channel DHF from its thymidylate synthase to its DHFR active sites.

2. Serine hydroxymethyltransferase (Section 20-4A) transfers the hydroxymethyl group of serine to THF yielding N^5,N^{10}-methylene-THF and glycine.

Inhibition of thymidylate synthase or DHFR blocks dTMP synthesis and is therefore the basis of cancer chemotherapies (see Box 22-1).

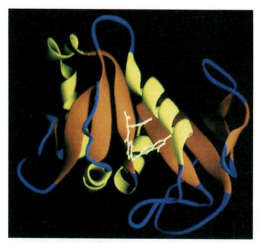

Figure 22-17 X-Ray structure of human dihydrofolate reductase in complex with folate. The helices of this monomeric enzyme are drawn in yellow, the β sheets in orange, and the other polypeptide segments in blue. [Courtesy of Jay F. Davies II and Joseph Kraut, University of California at San Diego. PDBid 1DHF.] See the Interactive Exercises.

BOX 22-1

Biochemistry in Health and Disease

Inhibition of Thymidylate Synthesis in Cancer Therapy

dTMP synthesis is a critical process for rapidly proliferating cells, such as cancer cells, which require a steady supply of dTMP for DNA synthesis. Interruption of dTMP synthesis can therefore kill these cells. Most normal mammalian cells, which grow slowly if at all, require less dTMP and so are less sensitive to agents that inhibit thymidylate synthase or dihydrofolate reductase (notable exceptions are the bone marrow cells that constitute the blood-forming tissue and much of the immune system, the intestinal mucosa, and hair follicles).

(continued on next page)

5-Fluorodeoxyuridylate (FdUMP)

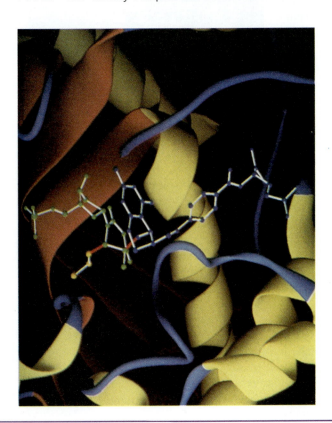

5-Fluorodeoxyuridylate (FdUMP)

is an irreversible inhibitor of thymidylate synthase. This substance, like dUMP, binds to the enzyme (an F atom is not much larger than an H atom) and undergoes the first two steps of the normal enzymatic reaction (Fig. 22-15). In Step 3, however, the enzyme cannot abstract the F atom as F$^+$ (F is the most electronegative element) so that the enzyme is frozen in an enzyme–FdUMP–THF ternary covalent complex. The X-ray structure of this covalent thymidylate synthase–FdUMP–THF ternary complex has been determined.

Here, its active site region is shown with helices yellow, β strands orange, and other polypeptide segments blue. The C5 and C6 atoms of FdUMP (*green spheres*) form covalent bonds (*red*) with the CH$_2$ group substituent to N5 of THF (*blue spheres*) and the S atom of Cys 146 (*yellow spheres*).

Enzyme inhibitors such as FdUMP, which inactivate an enzyme only after undergoing part or all of its normal catalytic reaction, are called **mechanism-based inhibitors** (alternatively, **suicide substrates** because they cause the enzyme to "commit suicide"). Because of their extremely high specificity, mechanism-based inhibitors are among the most useful therapeutic agents.

Inhibition of DHFR blocks dTMP synthesis as well as all other THF-dependent biological reactions, because the THF converted to DHF by the thymidylate synthase reaction cannot be regenerated. **Methotrexate (amethopterin), aminopterin,** and **trimethoprim**

R = H   **Aminopterin**
R = CH$_3$ **Methotrexate (amethopterin)**

Trimethoprim

are DHF analogs that competitively although nearly irreversibly bind to DHFR with an ~1000-fold greater affinity than does DHF. These **antifolates** (substances that interfere with the action of folate cofactors) are effective anticancer agents, particularly against childhood leukemias. In fact, a successful chemotherapeutic strategy is to treat a cancer victim with a lethal dose of methotrexate and some hours later "rescue" the patient (but hopefully not the cancer) by administering massive doses of 5-formyl-THF and/or thymidine. Trimethoprim, which was discovered by George Hitchings and Gertrude Elion, binds much more tightly to bacterial DHFRs than to those of mammals and is therefore a clinically useful antibacterial agent.

[X-ray structure courtesy of Jesus Villafranca and David Matthews, Agouron Pharmaceuticals, La Jolla, California.]

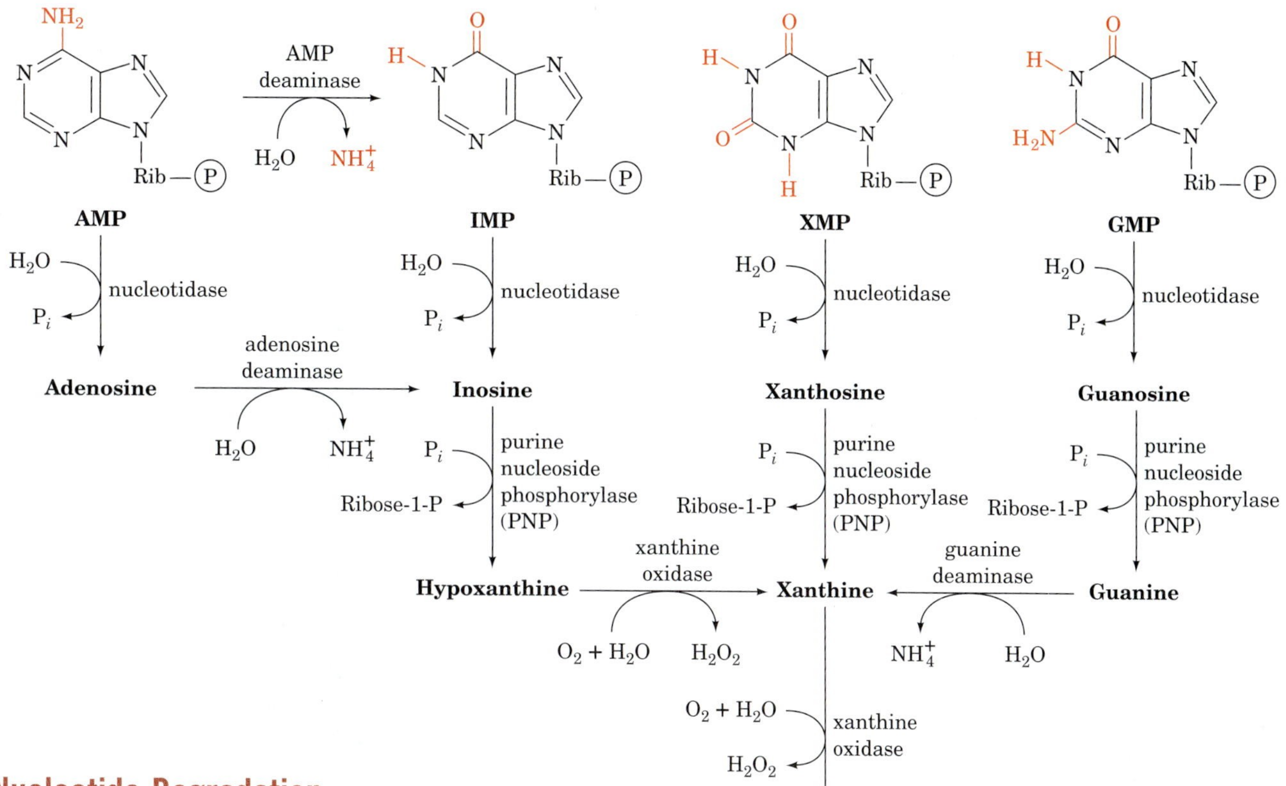

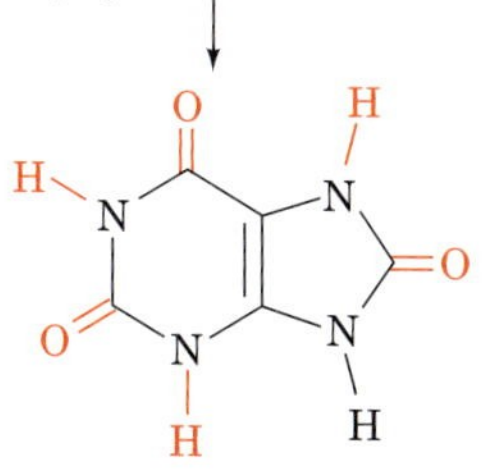

Figure 22-18 The major pathways of purine catabolism in animals. The various purine nucleotides are all degraded to uric acid.

4 | Nucleotide Degradation

Most foodstuffs, being of cellular origin, contain nucleic acids. Dietary nucleic acids survive the acidic medium of the stomach; they are degraded to their component nucleotides, mainly in the intestine, by pancreatic nucleases and intestinal phosphodiesterases. The ionic nucleotides, which cannot pass through cell membranes, are then hydrolyzed to nucleosides by a variety of group-specific nucleotidases and nonspecific phosphatases. Nucleosides may be directly absorbed by the intestinal mucosa or further degraded to free bases and ribose or ribose-1-phosphate through the action of **nucleosidases** and **nucleoside phosphorylases:**

$$\text{Nucleoside} + H_2O \xrightarrow{\text{nucleosidase}} \text{base} + \text{ribose}$$

$$\text{Nucleoside} + P_i \xrightarrow[\text{phosphorylase}]{\text{nucleoside}} \text{base} + \text{ribose-1-P}$$

Radioactive labeling experiments have demonstrated that only a small fraction of the bases of ingested nucleic acids are incorporated into tissue nucleic acids. Evidently, *the de novo pathways of nucleotide biosynthesis largely satisfy an organism's need for nucleotides.* Consequently, ingested bases are mostly degraded and excreted. Cellular nucleic acids are also subject to degradation as part of the continual turnover of nearly all cellular components. In this section, we outline these catabolic pathways and discuss the consequences of several of their inherited defects.

A Catabolism of Purines

The major pathways of purine nucleotide and deoxynucleotide catabolism in animals are diagrammed in Fig. 22-18. The pathways in other organisms differ somewhat, but all the pathways lead to uric acid. Of course, the pathway

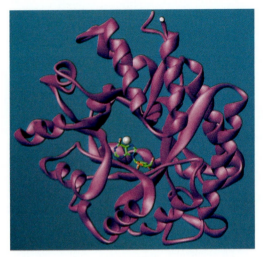

Figure 22-19 **X-Ray structure of murine adenosine deaminase.** The enzyme is viewed approximately down the axis of its α/β barrel from the N-terminal ends of its β strands. A transition state analog, **6-hydroxyl-1,6-dihydropurine ribonucleoside (HDPR),** is shown in skeletal form with its C, N, and O atoms green, blue, and red. The enzyme-bound Zn^{2+} ion, which is coordinated by HDPR's 6-hydroxyl group, is represented by a silver sphere. [Based on an X-ray structure by Florante Quiocho, Baylor College of Medicine. PDBid 1ADA.] 🔖 **See the Interactive Exercises.**

intermediates may be directed to purine nucleotide synthesis via salvage reactions. In addition, ribose-1-phosphate, a product of the reaction catalyzed by **purine nucleoside phosphorylase (PNP),** is a precursor of PRPP.

Adenosine and deoxyadenosine are not degraded by mammalian PNP. Rather, adenine nucleosides and nucleotides are deaminated by **adenosine deaminase (ADA)** and **AMP deaminase** to their corresponding inosine derivatives, which can then be further degraded.

ADA is an eight-stranded α/β barrel (Fig. 22-19) with its active site in a pocket at the C-terminal end of the β barrel, as in all known α/β barrel enzymes. A catalytically essential zinc ion is bound in the deepest part of the active site pocket. Mutations that affect the active site of ADA selectively kill lymphocytes, causing the disease **severe combined immunodeficiency disease (SCID).** Without special protective measures, the disease is invariably fatal in infancy because of overwhelming infection.

Biochemical considerations provide a plausible explanation of SCID's etiology (causes). In the absence of active ADA, deoxyadenosine is phosphorylated to yield levels of dATP that are 50-fold greater than normal. This high concentration of dATP inhibits ribonucleotide reductase (Section 22-3A), thereby preventing the synthesis of the other dNTPs, choking off DNA synthesis and thus cell proliferation. The tissue-specific effect of ADA deficiency on the immune system can be explained by the observation that lymphoid tissue is particularly active in deoxyadenosine phosphorylation. ADA deficiency was one of the first genetic diseases to be successfully treated by gene therapy.

The Purine Nucleotide Cycle. The deamination of AMP to IMP, when combined with the synthesis of AMP from IMP (Fig. 22-3, *left*), has the net effect of deaminating aspartate to yield fumarate (Fig. 22-20). John Lowenstein demonstrated that this **purine nucleotide cycle** has an important metabolic role in skeletal muscle. An increase in muscle activity

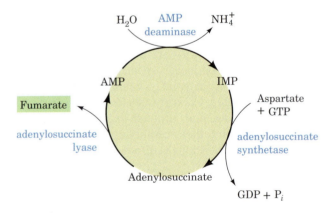

Figure 22-20 **The purine nucleotide cycle.** This pathway functions in muscle to prime the citric acid cycle by generating fumarate.

Net: H_2O + Aspartate + GTP ⟶ NH_4^+ + GDP + P_i + fumarate

requires an increase in the activity of the citric acid cycle. This process usually occurs through the generation of additional citric acid cycle intermediates (Section 16-5B). Muscles, however, lack most of the enzymes that catalyze these anaplerotic (filling up) reactions in other tissues. Instead, muscle replenishes its citric acid cycle intermediates with fumarate generated in the purine nucleotide cycle.

The importance of the purine nucleotide cycle in muscle metabolism is indicated by the observation that the activities of the three enzymes involved are all severalfold higher in muscle than in other tissues. In fact, individuals with an inherited deficiency in muscle AMP deaminase **(myo-adenylate deaminase deficiency)** are easily fatigued and usually suffer from cramps after exercise.

Xanthine Oxidase Is a Mini-Electron-Transport System.   **Xanthine oxidase (XO)** converts hypoxanthine (the base of IMP) to xanthine, and xanthine to uric acid (Fig. 22-18, *bottom*). The reaction product is an enol (which has a pK of 5.4; hence the name uric *acid*). The enol tautomerizes to the more stable keto form:

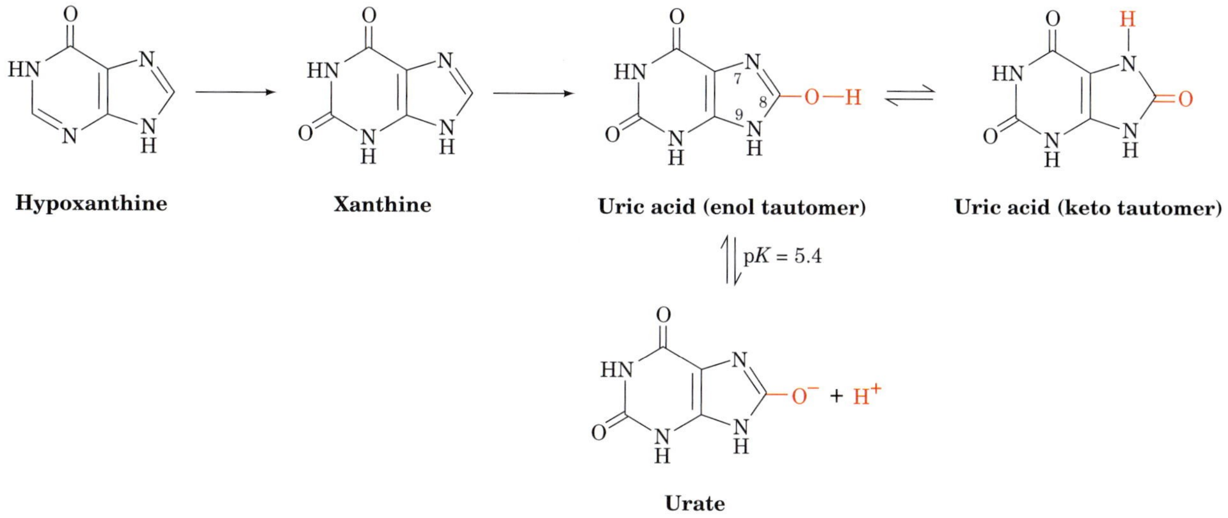

| Hypoxanthine | Xanthine | Uric acid (enol tautomer) | Uric acid (keto tautomer) |

$$pK = 5.4$$

Urate

In mammals, xanthine oxidase occurs almost exclusively in the liver and the small intestinal mucosa. It is a dimeric protein of identical 130-kD subunits, each of which contains an entire "zoo" of electron-transfer agents: an FAD, a Mo complex that cycles between its Mo(VI) and Mo(IV) oxidation states, and two different Fe–S clusters. The final electron acceptor is O_2, which is converted to H_2O_2, a potentially harmful oxidizing agent (Section 17-4C) that is subsequently converted to H_2O and O_2 by catalase.

B Fate of Uric Acid

In humans and other primates, the final product of purine degradation is uric acid, which is excreted in the urine. The same is true of birds, terrestrial reptiles, and many insects, but these organisms, which do not excrete urea, also catabolize their excess amino acid nitrogen to uric acid via purine biosynthesis. This complicated system of nitrogen excretion has a straightforward function: *It conserves water.* Uric acid is only sparingly soluble in water, so its excretion as a paste of uric acid crystals is accompanied by very little water. In contrast, the excretion of an equivalent amount of the much more water-soluble urea osmotically sequesters a significant amount of water.

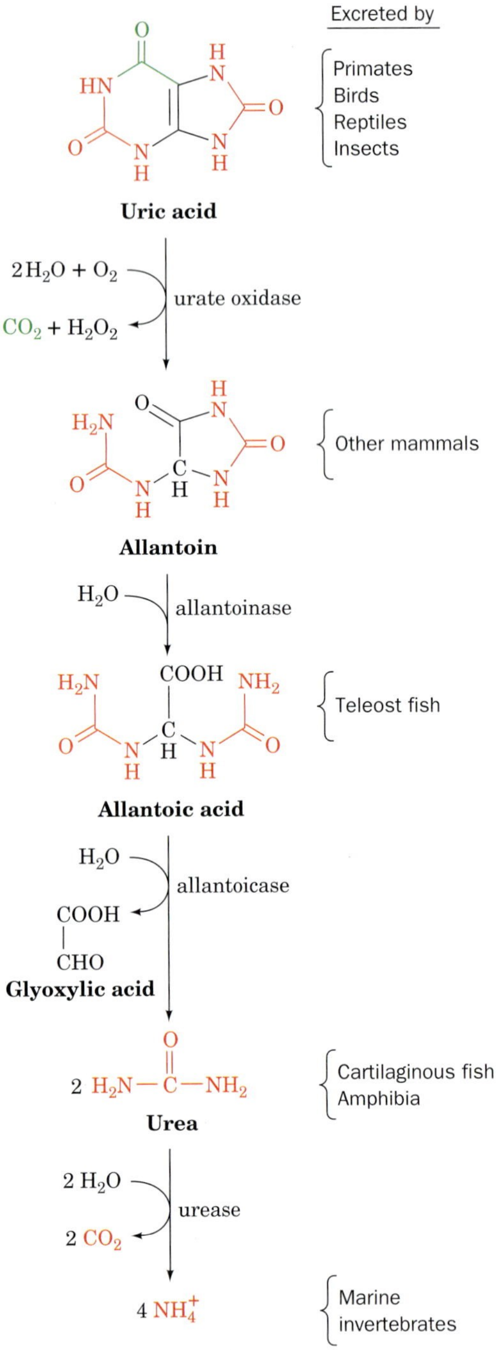

Figure 22-21 Degradation of uric acid to ammonia.
The process is arrested at different stages in the indicated species, and the resulting nitrogen-containing product is excreted.

In all other organisms, uric acid is further processed before excretion (Fig. 22-21). Mammals other than primates oxidize it to their excretory product, **allantoin,** in a reaction catalyzed by the Cu-containing enzyme **urate oxidase.** A further degradation product, **allantoic acid,** is excreted by teleost (bony) fish. Cartilaginous fish and amphibia further degrade allantoic acid to urea prior to excretion. Finally, marine invertebrates decompose urea to NH_4^+.

Gout Is Caused by an Excess of Uric Acid.   **Gout** is a disease characterized by elevated levels of uric acid in body fluids. Its most common manifestation is excruciatingly painful arthritic joint inflammation of sudden onset, most

Figure 22-22 *The Gout,* a cartoon by James Gillray **(1799).** [Courtesy of Yale University Medical Historical Library.]

often of the big toe (Fig. 22-22), caused by deposition of nearly insoluble crystals of sodium urate. Sodium urate and/or uric acid may also precipitate in the kidneys and ureters as stones, resulting in renal damage and urinary tract obstruction.

Gout, which affects ~3 per 1000 persons, predominantly males, has been traditionally, although inaccurately, associated with overindulgent eating and drinking. The probable origin of this association is that in previous centuries, when wine was often contaminated with lead during its manufacture and storage, heavy drinking resulted in chronic lead poisoning that, among other things, decreases the kidney's ability to excrete uric acid.

The most prevalent cause of gout is impaired uric acid excretion (although usually for reasons other than lead poisoning). Gout may also result from a number of metabolic insufficiencies, most of which are not well characterized. One well-understood cause is HGPRT deficiency (Lesch–Nyhan syndrome in severe cases), which leads to excessive uric acid production through PRPP accumulation (Section 22-1D).

Gout can be treated by administering the xanthine oxidase inhibitor **allopurinol,** a hypoxanthine analog with interchanged N7 and C8 positions.

Allopurinol   **Hypoxanthine**

Xanthine oxidase hydroxylates allopurinol, as it does hypoxanthine, yielding **alloxanthine,**

Alloxanthine

which remains tightly bound to the reduced form of the enzyme, thereby inactivating it. Allopurinol consequently alleviates the symptoms of gout by decreasing the rate of uric acid production while increasing the levels

of the more soluble hypoxanthine and xanthine. Although allopurinol controls the gouty symptoms of Lesch–Nyhan syndrome, it has no effect on its neurological symptoms.

C Catabolism of Pyrimidines

Animal cells degrade pyrimidine nucleotides to their component bases (Fig. 22-23, *top*). These reactions, like those of purine nucleotides, occur through dephosphorylation, deamination, and glycosidic bond cleavages. The resulting uracil and thymine are then broken down in the liver through reduction (Fig. 22-23, *middle*) rather than by oxidation as occurs in purine

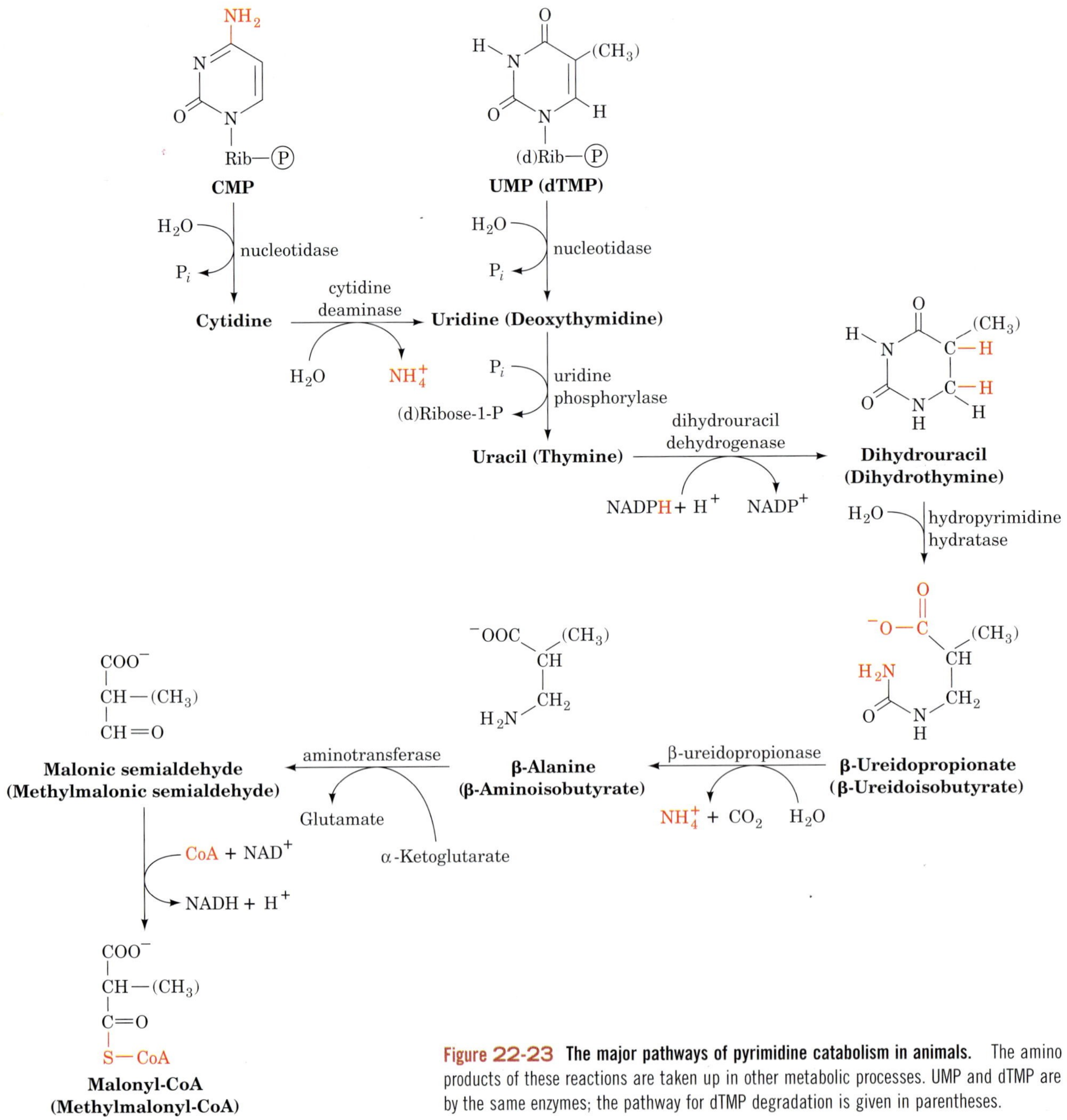

Figure 22-23 The major pathways of pyrimidine catabolism in animals. The amino acid products of these reactions are taken up in other metabolic processes. UMP and dTMP are degraded by the same enzymes; the pathway for dTMP degradation is given in parentheses.

catabolism. The end products of pyrimidine catabolism, **β-alanine** and **β-aminoisobutyrate,** are amino acids and are metabolized as such. They are converted, through transamination and activation reactions, to malonyl-CoA and methylmalonyl-CoA (Fig. 22-23, *bottom left*). Malonyl-CoA is a precursor of fatty acid synthesis (Fig. 19-26), and methylmalonyl-CoA is converted to the citric acid cycle intermediate succinyl-CoA (Fig. 19-16). Thus, *to a limited extent, catabolism of pyrimidine nucleotides contributes to the energy metabolism of the cell.*

SUMMARY

1. The purine nucleotide IMP is synthesized in 11 steps from ribose-5-phosphate, aspartate, fumarate, glutamine, glycine, and HCO_3^-. Purine nucleotide synthesis is regulated at its first and second steps.

2. IMP is the precursor of AMP and GMP, which are phosphorylated to produce the corresponding di- and triphosphates.

3. The pyrimidine nucleotide UMP is synthesized from 5-phosphoribosyl pyrophosphate, aspartate, glutamine, and HCO_3^- in six reactions. UMP is converted to UTP and CTP by phosphorylation and amination.

4. Pyrimidine nucleotide synthesis is regulated in bacteria at the ATCase step and in animals at the step catalyzed by carbamoyl phosphate synthetase II.

5. Deoxyribonucleoside diphosphates are synthesized from the corresponding NDP by the action of ribonucleotide reductase, which contains a binuclear Fe(III) prosthetic group, a tyrosyl radical, and several redox-active sulfhydryl groups. Enzyme activity is regenerated through disulfide interchange with thioredoxin.

6. Ribonucleotide reductase is regulated by allosteric effectors, which ensure that deoxynucleotides are synthesized in the amounts required for DNA synthesis.

7. dTMP is synthesized from dUMP by thymidylate synthase. The dihydrofolate produced in this reaction is converted back to tetrahydrofolate by dihydrofolate reductase (DHFR).

8. Purine nucleotides are degraded by nucleosidases and purine nucleoside phosphorylase (PNP). Adenine nucleotides are deaminated by adenosine deaminase and AMP deaminase. The synthesis and degradation of AMP in the purine nucleotide cycle yield the citric acid cycle intermediate fumarate in muscles. Xanthine oxidase catalyzes the oxidation of hypoxanthine to xanthine and of xanthine to uric acid.

9. In humans, the ultimate product of purine degradation is uric acid, which is excreted. Other organisms degrade urate further.

10. Pyrimidines are broken down to intermediates of fatty acid metabolism.

REFERENCES

Carreras, C.W. and Santi, D.V., The catalytic mechanism and structure of thymidylate synthase, *Annu. Rev. Biochem.* **64,** 721–762 (1995).

Finer-Moore, J.S., Santi, D.V., and Stroud, R.M., Lessons and conclusions from dissecting the mechanism of a bisubstrate enzyme: thymidylate synthase mutagenesis, function, and structure, *Biochemistry* **42,** 248–256 (2003).

Greasley, S.E., Horton, P., Ramcharan, J., Beardsley, G.P., Benkovic, S.J., and Wilson, I.A., Crystal structure of a bifunctional transformylase and cyclohydrolase enzyme in purine biosynthesis, *Nature Struct. Biol.* **8,** 402–406 (2001).

Kappock, T.J., Ealick, S.E., and Stubbe, J., Modular evolution of the purine biosynthetic pathway, *Curr. Opin. Chem. Biol.* **4,** 567–572 (2000).

Liu, S., Neidhardt, E.A., Grossman, T.H., Ocain, T., and Clardy, J., Structures of human dihydroorotate dehydrogenase in complex with antiproliferative agents, *Structure* **8,** 25–33 (1999).

Scriver, C.R., Beaudet, A.L., Sly, W.S., and Valle, D. (Eds.), *The Metabolic and Molecular Bases of Inherited Disease* (8th ed.), Chapters 106–114, McGraw-Hill (2001). [These chapters describe normal and abnormal pathways of nucleotide metabolism.]

Stubbe, J., Ge, J., and Yee, C.S., The evolution of ribonucleotide reduction revisited, *Trends Biochem. Sci.* **26,** 93–99 (2001).

KEY TERMS

PRPP	salvage pathway	purine nucleotide cycle	mechanism-based inhibitor
feedforward activation			

STUDY EXERCISES

1. Review the nomenclature of bases, nucleosides, and nucleotides.

2. Compare the pathways of purine and pyrimidine nucleotide synthesis with respect to (a) precursors, (b) energy cost, (c) acquisition of the ribose moiety, and (d) number of enzymatic steps.

3. How do PRPP levels influence purine and pyrimidine nucleotide synthesis?

4. How are folate cofactors involved in nucleotide metabolism?

5. Why are antifolates effective drugs?

6. Describe the metabolic defects of Lesch–Nyhan syndrome, orotic aciduria, SCID, and gout.

7. How does the cell balance the production of (a) purine and pyrimidine nucleotides and (b) ribonucleotides and deoxyribonucleotides?

8. What compounds are produced by the degradation of purines and pyrimidines?

 PROBLEMS

1. Calculate the cost, in ATP equivalents, of synthesizing *de novo* (a) IMP, (b) AMP, and (c) CTP. Assume all substrates (e.g., ribose-5-phosphate and glutamine) and cofactors are available.

2. Certain glutamine analogs irreversibly inactivate enzymes that bind glutamine. Identify the nucleotide biosynthetic intermediates that accumulate in the presence of these compounds.

3. Rats are given cytidine that is ^{14}C-labeled at both its base and ribose components. Their DNA is then extracted and degraded with nucleases. Describe the labeling pattern of the recovered deoxycytidylate residues if deoxycytidylate production in the cell followed a pathway in which (a) intact CDP is reduced to dCDP, and (b) CDP is broken down to cytosine and ribose before reduction.

4. Explain why hydroxyurea,

$$H_2N-\overset{\overset{\displaystyle O}{\|}}{C}-NH-OH$$

Hydroxyurea

which destroys tyrosyl radicals, is useful as an antitumor agent.

5. Why is deoxyadenosine toxic to mammalian cells?

6. Why do individuals who are undergoing chemotherapy with FdUMP or methotrexate often temporarily go bald?

7. Normal cells die in a nutrient medium containing thymidine and methotrexate, whereas mutant cells defective in thymidylate synthase survive and grow. Explain.

8. Explain why methotrexate inhibits the synthesis of histidine and methionine.

9. Some microorganisms lack DHFR activity, but their thymidylate synthase has an FAD cofactor. What is the function of the FAD?

10. Explain whether the following are mechanism-based inhibitors: (a) trimethoprim with bacterial dihydrofolate reductase, and (b) allopurinol with xanthine oxidase.

11. Why does von Gierke's glycogen storage disease (Box 15-3) cause symptoms of gout?

12. In animals, one pathway for NAD$^+$ synthesis begins with nicotinamide. Draw the structures generated by the reactions shown.

Nicotinamide

PRPP — | nicotinamide
PP$_i$ — | phosphoribosyl
| transferase

ATP — | NAD$^+$
PP$_i$ — | pyrophosphorylase

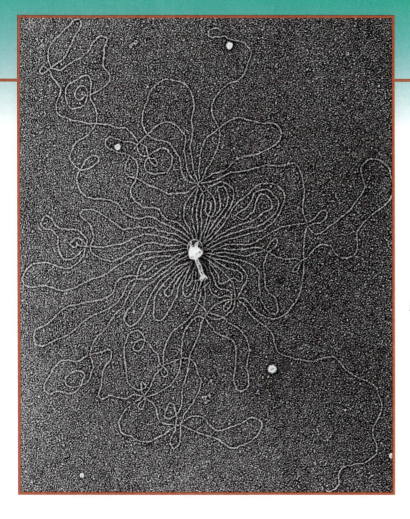

DNA molecules, such as that spilling out of the osmotically lysed bacteriophage T2 shown here, are enormous. A cell must be able to efficiently package and safely store its DNA. However, despite DNA's highly condensed structure, a cell must be able to read and interpret the meaning of the encoded genetic informtion. [From Roth, M.B., and Gall, J.G., J. Cell Biol. **105,** 1049 (1987), by copyright permission of The Rockefeller University Press.]

Nucleic Acid Structure

In every organism, the ultimate source of biological information is nucleic acid. The shapes and activities of individual cells are, to a large extent, determined by genetic instructions contained in DNA (or RNA, in some viruses). According to the central dogma of molecular biology (Section 3-3B), sequences of nucleotide bases in DNA encode the amino acid sequences of proteins. Many of the cell's proteins are enzymes that carry out the metabolic processes we have discussed in Chapters 14–22. Other proteins have a structural or regulatory role or participate in maintaining and transmitting genetic information.

Two kinds of nucleic acids, DNA and RNA, store information and make it available to the cell. The structures of these molecules must be consistent with the following:

1. Genetic information must be stored in a form that is manageable in size and stable over a long period.

2. Genetic information must be decoded—often many times—in order to be used. **Transcription** is the process by which nucleotide sequences in DNA are copied onto RNA so that they can direct protein synthesis, a process known as **translation.**

3. Information contained in DNA or RNA must be accessible to proteins and other nucleic acids. These agents must recognize nucleic acids (in many cases, in a sequence-specific fashion) and bind to them in a way that alters their function.

4. The progeny of an organism must be equipped with the same set of instructions as in the parent. Thus, DNA is **replicated** (an exact copy made) so that each daughter cell receives the same information.

As we shall see, many cellular components are required to execute all the functions of nucleic acids. Yet nucleic acids are hardly inert "read-only" entities. RNA in particular, owing to its single-stranded nature, is a dynamic molecule that provides structural scaffolding as well as catalytic proficiency in a number of processes that decode genetic information. In this chapter, we shall focus on the structural properties of nucleic acids, including their interactions with proteins, that allow them to carry out their duties. In subsequent chapters, we shall examine the processes of replication (Chapter 24), transcription (Chapter 25), and translation (Chapter 26).

1 The DNA Helix

We begin our discussion of nucleic acid structure by examining the various forms of DNA, with an eye toward understanding how this molecule safeguards genetic information while leaving it accessible for replication and transcription.

A The Geometry of DNA

DNA is a two-stranded polymer of deoxynucleotides linked by phosphodiester bonds (Figs. 3-3 and 3-8). The biologically most common form of DNA is known as **B-DNA,** which has the structural features first noted by James Watson and Francis Crick together with Rosalind Franklin and others (Section 3-2B and Box 23-1):

1. The two antiparallel polynucleotide strands wind in a right-handed manner around a common axis to produce an ~20-Å-diameter double helix.

See Guided Exploration 21

DNA structures.

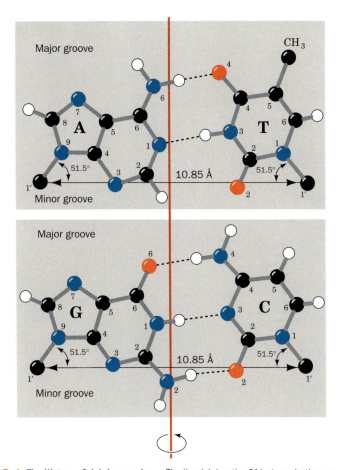

Figure 23-1 The Watson–Crick base pairs.   The line joining the C1′ atoms is the same length in both A · T and G · C base pairs and makes equal angles with the glycosidic bonds to the bases. This gives DNA a series of pseudo-twofold symmetry axes that pass through the center of each base pair (*red line*) and are perpendicular to the helix axis. [After Arnott, S., Dover, S.D., and Wonacott, A.J., *Acta Cryst.* **B25**, 2196 (1969).] ♻ **See Kinemage Exercise 17-2.**

2. The planes of the nucleotide bases, which form hydrogen-bonded pairs, are nearly perpendicular to the helix axis. In B-DNA, the bases occupy the core of the helix while the sugar–phosphate backbones wind around the outside, forming the major and minor grooves. Only the edges of the base pairs are exposed to solvent.

3. Each base pair has approximately the same width (Fig. 23-1), which accounts for the near-perfect symmetry of the DNA molecule, regardless of base composition. A · T and G · C base pairs are interchangeable: *They can replace each other in the double helix without altering the positions of the sugar–phosphate backbones' C1′ atoms.* Likewise, the partners of a Watson–Crick base pair can be switched (i.e., by changing a G · C to a C · G or an A · T to a T · A). In contrast, any other combination of bases would significantly distort the double helix.

4. The "ideal" B-DNA helix has 10 base pairs (bp) per turn (a helical twist of 36° per bp) and, since the aromatic bases have van der Waals thicknesses of 3.4 Å and are partially stacked on each other, the helix has a pitch (rise per turn) of 34 Å.

Double-helical DNA can assume several distinct structures depending on the solvent composition and base sequence. The major structural variants

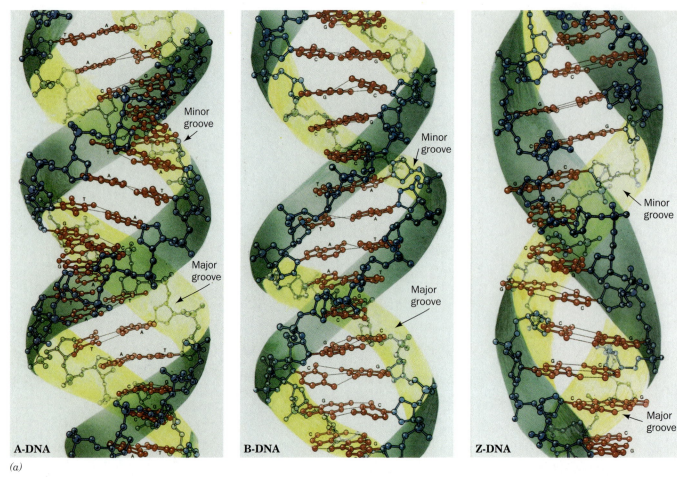

(a)

Figure 23-2   Structures of A-, B-, and Z-DNA.   (*a*) View perpendicular to the helix axis. The sugar–phosphate backbones are outlined by a green ribbon and the bases are red. (*b*) (*opposite*) Space-filling models, with C, N, O, and P atoms colored white, blue, red, and green, respectively. H atoms have been omitted for clarity. (*c*) (*opposite*) View down the helix axis. The ribose ring O atoms are red and the nearest base is white. [Molecular models based on structures determined by (*a*) Olga Kennard, Dov Rabinovitch, Zippora Shakked, and Mysore Viswamitra (Nucleic Acid Database ID ADH010); (*b*) Richard Dickerson and Horace Drew, PDBid 1BNA; (*c*) Andrew Wang and Alexander Rich, PDBid 2DCG). Illustrations in Part *a*, Irving Geis/Geis Archives Trust. Copyright Howard Hughes Medical Institute. Reproduced with permission.] 🔬 **See Kinemage Exercises 17-1, 17-4, 17-5, and 17-6.**

Table 23-1 *Key to Structure.* Structural Features of Ideal A-, B-, and Z-DNA

	A	B	Z
Helical sense	Right handed	Right handed	Left handed
Diameter	~26 Å	~20 Å	~18 Å
Base pairs per helical turn	11.6	10	12 (6 dimers)
Helical twist per base pair	31°	36°	60° (per dimer)
Helix pitch (rise per turn)	34 Å	34 Å	44 Å
Helix rise per base pair	2.9 Å	3.4 Å	7.4 Å per dimer
Base tilt normal to the helix axis	20°	6°	7°
Major groove	Narrow and deep	Wide and deep	Flat
Minor groove	Wide and shallow	Narrow and deep	Narrow and deep
Sugar pucker	C3'-*endo*	C2'-*endo*	C2'-*endo* for pyrimidines; C3'-*endo* for purines
Glycosidic bond conformation	Anti	Anti	Anti for pyrimidines; syn for purines

of DNA are **A-DNA** and **Z-DNA.** The geometries of these molecules are summarized in Table 23-1 and Fig. 23-2.

A-DNA's Base Pairs Are Inclined to the Helix Axis.   Under dehydrating conditions, B-DNA undergoes a reversible conformational change to A-DNA,

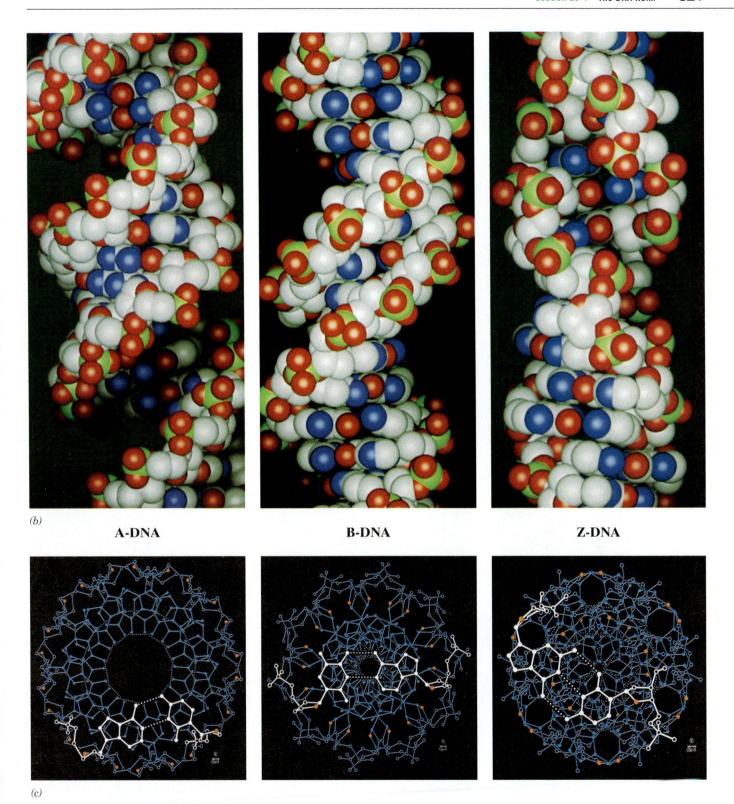

(b)

A-DNA   **B-DNA**   **Z-DNA**

(c)

Figure 23-2 *(Continued)*

which forms a wider and flatter right-handed helix than does B-DNA. A-DNA has 11.6 bp per turn and a pitch of 34 Å, which gives it an axial hole (Fig. 23-2*c, left*). A-DNA's most striking feature, however, is that the planes of its base pairs are tilted 20° with respect to the helix axis. Since the axis does not pass through its base pairs, A-DNA has a deep major

BOX 23-1

Pathways of Discovery

Rosalind Franklin and the Structure of DNA

Rosalind Franklin (1920–1958)

James Watson and Francis Crick were the first to publish an accurate model of the structure of DNA. This seminal discovery was not only based on their own insights but, like all scientific discoveries, built on the work of others. One of the major contributors to this process was Rosalind Franklin, who was probably never fully aware of her role in this discovery and did not share the 1962 Nobel Prize in Physiology or Medicine awarded to Watson, Crick, and Maurice Wilkins.

Franklin was born in England in an intellectually oriented and well-to-do family. She excelled in mathematics, and although career opportunities for women with her talents were few, she obtained a doctorate in physical chemistry at the University of Cambridge in 1945. From 1947 to 1950 she worked in a French government laboratory, where she became an authority in applying X-ray diffraction techniques to imperfectly crystalline substances such as coal; her numerous publications on the structures of coal and other forms of carbon changed the way that these substances were viewed.

In 1951, she returned to England at the invitation of John Randall, the head of the Medical Research Council's (MRC's) Biophysics Research Unit at King's College London, to investigate the structure of DNA. Randall had indicated to Franklin that she would be given an independent position, whereas Wilkins, who had been working on DNA at King's for some time, was given the impression that Franklin would be working under his direction. This misunderstanding, together with their sharply contrasting personalities (Franklin was quick, assertive, and confrontational, whereas Wilkins was deliberate, shy, and indirect) led to a falling out between the two such that they were barely on speaking terms. Hence,

they worked independently. Some popular accounts of Franklin's life and work have implied that the atmosphere at King's College was inhospitable to women, but Franklin's correspondence as well as firsthand accounts indicate that the atmosphere was in fact congenial. Still, she was unhappy at King's, apparently for personal reasons, and in the spring of 1953, departed for Birkbeck College in London. There, until her untimely death of ovarian cancer in 1958, she carried out groundbreaking investigations on the structures of viruses, most notably tobacco mosaic virus (see Box 23-4).

Early in her tenure at King's, Franklin discovered, through the analysis of X-ray fiber diffraction patterns, that DNA (which she obtained from Wilkins, who had gotten it from Rudoph Signer in Switzerland) exists in two distinct conformations, which she called the A and B forms. Prior to this discovery, the X-ray diffraction patterns of DNA that had been obtained were confusing because they were of mixtures of the A and B forms. Through careful control of the humidity, Franklin obtained an X-ray fiber diffraction photograph of B-DNA of unprecedented clarity (Fig. 3-5) that strikingly indicated the DNA's helical character. Analysis of this photograph permitted Franklin to determine that B-DNA is double helical, it has a diameter of 20 Å, and each turn of the helix is 34 Å long and contains 10 base pairs, each separated by 3.4 Å. Further analysis suggested that the hydrophilic sugar–phosphate chains were on the outside of the helix and the relatively hydrophobic bases were on the inside. Although Franklin was aware of Chargaff's rules (Section 3-2A) and

groove and a very shallow minor groove; it can be described as a flat ribbon wound around a 6-Å-diameter cylindrical hole.

Z-DNA Forms a Left-Handed Helix.   Occasionally, a familiar system exhibits quite unexpected properties. Over 25 years after the discovery of the Watson–Crick DNA structure, the crystal structure determination of d(CGCGCG) by Andrew Wang and Alexander Rich revealed, quite surprisingly, a left-handed double helix. This helix, which was dubbed Z-DNA, has 12 Watson–Crick base pairs per turn, a pitch of 44 Å, a deep minor groove, and no discernible major groove. Z-DNA therefore resembles a left-handed drill bit in appearance (Fig. 23-2, *right*).

Fiber diffraction and NMR studies have shown that complementary polynucleotides with alternating purines and pyrimidines, such as poly d(GC)·poly d(GC) or poly d(AC)·poly d(GT), assume the Z conformation at high salt concentrations. The salt stabilizes Z-DNA relative to B-DNA

Jerry Donohue's work concerning the tautomeric forms of the bases (Section 3-2B), she did not deduce the existence of base pairs in double-stranded DNA.

In January 1953, Wilkins showed Franklin's X-ray photograph of B-DNA to Watson, when he visited King's College. Moreover, in February 1953, Max Perutz (Box 7-2), Crick's thesis advisor at Cambridge University, showed Watson and Crick his copy of the 1952 Report of the MRC, which summarized the work of all of its principal investigators, including that of Franklin. Within a week (and after 13 months of inactivity on the project), Watson and Crick began building a model of DNA with a backbone structure compatible with Franklin's data [in earlier modeling attempts, they had placed the bases on the outside of the helix (as did a model published by Linus Pauling; Box 6-1) because they assumed that the bases could transmit genetic information only if they were externally accessible]. On several occasions Crick acknowledged that Franklin's findings were crucial to this enterprise.

Watson and Crick published their model of B-DNA in *Nature* in April 1953. This paper was followed, back-to-back, by papers by Wilkins and by Franklin on their structural studies of DNA. Franklin's manuscript had been written in March 1953, before she knew about Watson and Crick's work. The only change that Franklin made to her manuscript when she became aware of Watson and Crick's model was the addition of a single sentence, "Thus our general ideas are not inconsistent with the model proposed by Watson and Crick in the preceding communication." She was apparently unaware that the Watson–Crick model was, to a significant extent, based on her work. Since Watson and Crick did not acknowledge Franklin in their 1953 *Nature* paper, her paper was widely taken as data that supported the Watson–Crick model rather than being an important element in its formulation. Only after her death did Watson and Crick indicate the crucial role of Franklin's contributions.

Interestingly, Watson, Crick, and Franklin developed a close friendship. Starting in 1954 they maintained a correspondence and commented on each other's work. In the summer of 1954, Watson offered to drive Franklin across the United States from Woods Hole, Massachusetts, to her destination at Caltech, where he was also going. In the spring of 1956, she toured Spain with Crick and his wife Odile and, later, stayed with them at their house in Cambridge when she was recovering from her treatments for ovarian cancer.

If Watson and Crick had not formulated their model of B-DNA, would Franklin, who had a distaste for speculative modeling, have eventually done so? We, of course, shall never know. And, had she lived, would Franklin have received the Nobel prize together with Watson, Crick, and Wilkins (the prize cannot be awarded posthumously)? Those who argue that she was an indispensable participant in the discovery process would say yes. However, the Nobel committee typically awards those who initiate the research (i.e., Wilkins), and the prize cannot be shared by more than three individuals.

Franklin, R.E. and Gosling, R.G., Molecular configuration in sodium thymonucleate, *Nature* **171**, 740–741 (1953).

Maddox, B., *Rosalind Franklin: The Dark Lady of DNA*, HarperCollins (2002).

by reducing the electrostatic repulsions between closest approaching phosphate groups on opposite strands (which are 8 Å apart in Z-DNA and 12 Å apart in B-DNA).

Does Z-DNA have any biological function? Rich has proposed that the reversible conversion of specific segments of B-DNA to Z-DNA under appropriate circumstances acts as a kind of switch in regulating gene expression. Yet the *in vivo* existence of Z-DNA has been difficult to prove. A major problem is demonstrating that a particular probe for detecting Z-DNA, for example, a Z-DNA-specific antibody, does not in itself cause what would otherwise be B-DNA to assume the Z conformation—a kind of biological uncertainty principle (i.e., the act of measurement inevitably disturbs the system being measured). Recently, however, Rich has discovered a family of Z-DNA-binding protein domains named **Zα,** whose existence strongly suggests that Z-DNA does, in fact, exist *in vivo.* The X-ray structure of a Zα domain in complex with double-stranded DNA reveals that one Zα

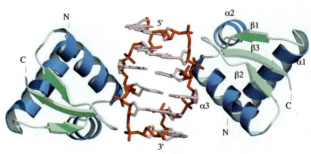

Figure 23-3 X-Ray structure of Zα in complex with Z-DNA. The DNA sequence d(CGCGCG) is shown in stick form with its backbone red. The Zα domains are drawn with α helices blue and β sheets cyan. Note that each Zα domain contacts only one strand of Z-DNA and that none of the bases of the Z-DNA interact directly with the protein. [Courtesy of Alexander Rich, MIT. PDBid 1QBJ.]

domain binds to each strand of Z-DNA (Fig. 23-3). The protein–DNA interactions primarily take the form of hydrogen bonds and ionic interactions between polar and basic side chains and the sugar–phosphate backbone of the DNA; none of the DNA's bases participate in these associations. The protein's DNA-binding surface is complementary in shape to the Z-DNA and is positively charged, as is expected for a protein that interacts with closely spaced anionic phosphate groups.

RNA Can Form an A Helix. Double-stranded RNA is the genetic material of certain viruses, but it is synthesized only as a single strand. Nevertheless, single-stranded RNA can fold back on itself so that complementary sequences base pair to form double-stranded stems with single-stranded loops (Fig. 3-9). Moreover, short segments of double-stranded RNA have been implicated in the control of gene expression (Section 27-3C). Double-stranded RNA is unable to assume a B-DNA-like conformation because of steric clashes involving its 2'-OH groups. Rather, it usually assumes a conformation resembling A-DNA (Fig. 23-2) that ideally has 11.0 bp per helical turn, a pitch of 30.9 Å, and base pairs that, on average, are inclined to the helix axis by 16.7°.

Hybrid double helices, which consist of one strand each of RNA and DNA, also have an A-DNA-like conformation (Fig. 23-4), although hybrid helices also have B-DNA-like qualities in that their overall conformation is intermediate to those of A-DNA and B-DNA. Short stretches of RNA–DNA hybrid helices occur during the initiation of DNA replication by small segments of RNA (Section 24-1) and during the transcription of RNA on DNA templates (Section 25-1C).

B Flexibility of DNA

The structurally distinct A, B, and Z forms of DNA are not thought to freely interconvert *in vivo*. Rather, the transition from one form to another requires unusual physical conditions (e.g., low humidity) or the influence of DNA-binding proteins. In addition, real

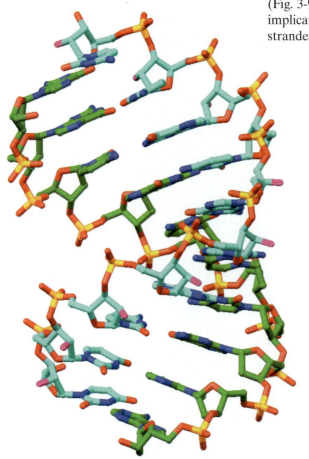

Figure 23-4 X-Ray structure of a 10-bp RNA–DNA hybrid helix. The complex consists of the RNA UUCGGGCGCC that is base paired to its DNA complement. The structure is shown in stick form with RNA C atoms cyan, DNA C atoms green, N blue, O red except for RNA O2' atoms, which are magenta, and P gold. [Based on an X-ray structure by Barry Finzel, Pharmacia & Upjohn, Inc., Kalamazoo, Michigan. PDBid 1FIX.] 🔗 **See the Interactive Exercises.**

DNA molecules deviate from the ideal structures described in the preceding section. X-Ray structures of B-DNA segments reveal that *individual residues significantly depart from the average conformation in a sequence-dependent manner.* For example, the helical twist per base pair may range from 26° to 43°. Each base pair can also deviate from its ideal conformation by rolling or twisting like the blade of a propeller. Such conformational variation appears to be important for the sequence-specific recognition of DNA by the proteins that process genetic information.

B-DNA molecules, which are 20 Å thick and many times as long, are not perfectly rigid rods. In fact, it is imperative that these molecules be somewhat flexible so that they can be packaged in cells. DNA helices can adopt different degrees of curvature ranging from gentle arcs to sharp bends. The more severe distortions from linearity generally occur in response to the binding of specific proteins.

The Conformational Flexibility of DNA Is Limited.

The conformation of a nucleotide unit, as Fig. 23-5 indicates, is specified by the six torsion angles of the sugar–phosphate backbone and the torsion angle describing the orientation of the base around the glycosidic bond (the bond joining C1′ to the base). It would seem that these seven degrees of freedom per nucleotide would render polynucleotides highly flexible. Yet these torsion angles are subject to a variety of internal constraints that greatly restrict their rotational freedom.

The rotation of a base around its glycosidic bond (angle χ) is greatly hindered. Purine residues have two sterically permissible orientations known as the **syn** (Greek: with) and **anti** (Greek: against) conformations (Fig. 23-6). Only the anti conformation of pyrimidines is stable, because, in the syn conformation, the sugar residue sterically interferes with the pyrimidine's C2 substituent. *In most double helical nucleic acids, all bases are in the anti conformation.* The exception is Z-DNA (Section 23-1A), in which the alternating pyrimidine and purine residues are anti and syn, respectively (this is one reason why the repeating unit of Z-DNA is considered to be a dinucleotide).

The flexibility of the ribose ring itself is also limited. The vertex angles of a regular pentagon are 108°, a value quite close to the tetrahedral angle (109.5°), so one might expect the ribofuranose ring to be nearly flat. However, the ring substituents are eclipsed when the ring is planar. To relieve this crowding, which occurs even between hydrogen atoms, the ring puckers; that is, it becomes slightly nonplanar. In the great majority of known nucleoside and nucleotide X-ray structures, four of the ring atoms are coplanar to within a few hundredths of an angstrom and the remaining atom is

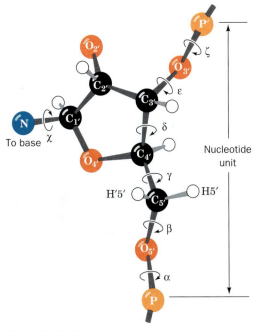

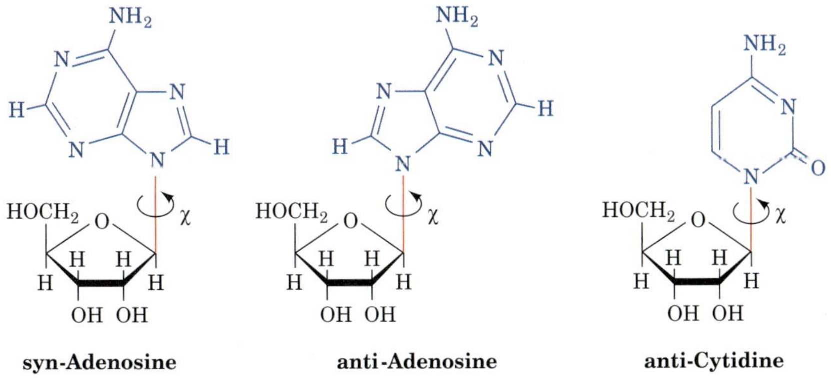

Figure 23-5 The seven torsion angles that determine the conformation of a nucleotide unit.

syn-**Adenosine**   anti-**Adenosine**   anti-**Cytidine**

Figure 23-6 The sterically allowed orientations of purine and pyrimidine bases with respect to their attached ribose units.   In B-DNA, the nucleotide residues all have the anti conformation.

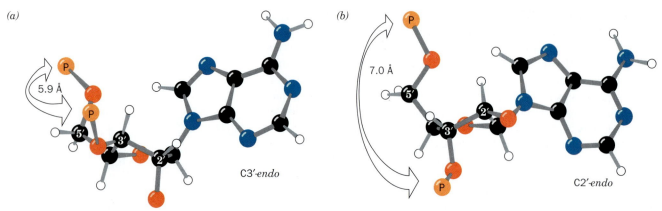

(a)

5.9 Å

C3′-*endo*

(b)

7.0 Å

C2′-*endo*

Figure 23-7 Nucleotide sugar conformations. (*a*) The C3′-*endo* conformation (C3′ is displaced to the same side of the ring as C5′), which occurs in A-RNA. (*b*) The C2′-*endo* conformation, which occurs in B-DNA. The distances between adjacent P atoms in the sugar–phosphate backbone are indicated. [After Saenger, W., *Principles of Nucleic Acid Structure*, p. 237, Springer-Verlag (1983).] 🔬 **See Kinemage Exercise 17-3.**

out of this plane by several tenths of an angstrom. The out-of-plane atom is almost always C2′ or C3′ (Fig. 23-7). The two most common ribose conformations are known as **C3′-*endo*** and **C2′-*endo*;** "*endo*" (Greek: *endon,* within) indicates that the displaced atom is on the same side of the ring as C5′; whereas "*exo*" (Greek: *exo,* out of) indicates displacement toward the opposite side of the ring from C5′.

The ribose pucker is conformationally important in nucleic acids because it governs the relative orientations of the phosphate substituents to each ribose residue. In fact, B-DNA has the C2′-*endo* conformation, whereas A-DNA is C3′-*endo*. In Z-DNA, the purine nucleotides are all C3′-*endo* and the pyrimidine nucleotides are C2′-*endo*.

The Sugar–Phosphate Backbone Is Conformationally Constrained. Finally, if the torsion angles of the sugar–phosphate chain (angles α to ζ in Fig. 23-5) were completely free to rotate, there could probably be no stable nucleic acid structure. However, these angles are actually quite restricted. This is because of noncovalent interactions between the ribose ring and the phosphate groups and, in polynucleotides, steric interferences between residues. The overall result is that *the sugar–phosphate chains of the double helix are stiff, although the sugar–phosphate conformational angles are reasonably strain-free.*

See Guided Exploration 22

DNA supercoiling.

C Supercoiled DNA

The chromosomes of many viruses and bacteria are circular molecules of duplex DNA. In electron micrographs (e.g., Fig. 23-8), some of these molecules have a peculiar twisted appearance, a phenomenon known as **supercoiling** or

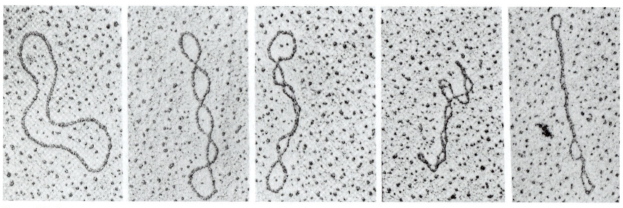

Figure 23-8 Electron micrographs of circular duplex DNAs. Their conformations vary from no supercoiling (*left*) to tightly supercoiled (*right*).

[Electron micrographs by Laurien Polder. From Kornberg, A. and Baker, T.A., *DNA Replication* (2nd ed.), *p.* 36, W.H. Freeman (1992). Used by permission.]

superhelicity. Supercoiled DNA molecules are more compact than "relaxed" molecules with the same number of nucleotides. This has important consequences for packaging DNA in cells (Section 23-5) and for the unwinding events that occur as part of DNA replication and RNA transcription.

Superhelix Topology Can Be Simply Described. Consider a double-helical DNA molecule in which each of its polynucleotide strands forms a covalently closed circle, thus forming a circular duplex molecule. A geometric property of such an assembly is that *its number of coils cannot be altered without first cleaving at least one of its strands.* You can easily demonstrate this with a buckled belt in which each edge of the belt represents a strand of DNA. The number of times the belt is twisted before it is buckled cannot be changed without unbuckling the belt (cutting a polynucleotide strand). This phenomenon is mathematically expressed

$$L = T + W \qquad [23\text{-}1]$$

in which:

1. *L*, the **linking number,** is the number of times that one DNA strand winds around the other. This integer quantity is most easily counted when the molecule is made to lie flat on a plane. The linking number cannot be changed by twisting or distorting the molecule, as long as both its polynucleotide strands remain covalently intact.

2. *T*, the **twist,** is the number of complete revolutions that one polynucleotide strand makes around the duplex axis. By convention, *T* is positive for right-handed duplex turns so that, for B-DNA, the twist is normally the number of base pairs divided by 10.4 (the observed number of base pairs per turn of the B-DNA double helix in aqueous solution).

3. *W*, the **writhing number,** is the number of turns that the duplex axis makes around the superhelix axis. *It is a measure of the DNA's superhelicity.* The difference between writhing and twisting is illustrated in Fig. 23-9. When a circular DNA is constrained to lie in a plane, *W* = 0.

The two DNA conformations diagrammed on the right of Fig. 23-10 are topologically equivalent; that is, they have the same linking number, *L*, but differ in their twists and writhing numbers (topology is the study of the geometric properties of objects that are unaltered by deformation but not by cutting).

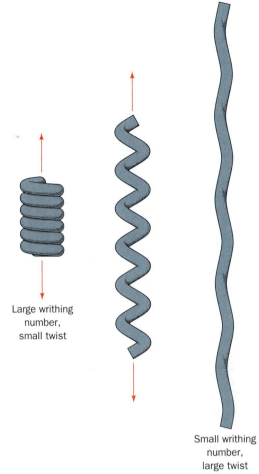

Large writhing number, small twist

Small writhing number, large twist

Figure 23-9   The difference between writhing and twist as demonstrated by a coiled telephone cord.   In its relaxed state (*left*), the cord is in a helical form that has a large writhing number and a small twist. As the coil is pulled out (*center*) until it is nearly straight (*right*), its writhing number becomes small and its twist becomes large.

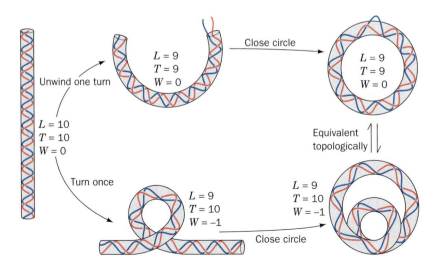

Unwind one turn

Close circle

$L = 9$
$T = 9$
$W = 0$

$L = 9$
$T = 9$
$W = 0$

$L = 10$
$T = 10$
$W = 0$

Turn once

$L = 9$
$T = 10$
$W = -1$

Close circle

Equivalent topologically

$L = 9$
$T = 10$
$W = -1$

Figure 23-10   Two ways of introducing one supercoil into a DNA that has 10 duplex turns.   The two closed circular forms shown (*right*) are topologically equivalent; that is, they are interconvertible without breaking any covalent bonds. The linking number *L*, twist *T*, and writhing number *W* are indicated for each form. Strictly speaking, the linking number is defined only for a covalently closed circle.

Since L is a constant in an intact duplex DNA circle, for every new double-helical twist, ΔT, there must be an equal and opposite superhelical twist; that is, $\Delta W = -\Delta T$. For example, a closed circular DNA without supercoils (Fig. 23-10, *upper right*) can be converted to a negatively supercoiled conformation (Fig. 23-10, *lower right*) by winding the duplex helix the same number of positive (right-handed) turns.

Supercoiled DNA Is Relaxed by Nicking One Strand.

Supercoiled DNA may be converted to **relaxed circles** (as appears in the leftmost panel of Fig. 23-8) by treatment with **pancreatic DNase I,** an **endonuclease** (an enzyme that cleaves phosphodiester bonds within a polynucleotide strand), which cleaves only one strand of a duplex DNA. *One single-strand nick is sufficient to relax a supercoiled DNA.* This is because the sugar–phosphate chain opposite the nick is free to swivel about its backbone bonds (Fig. 23-5) so as to change the molecule's linking number and thereby alter its superhelicity. Supercoiling builds up elastic strain in a DNA circle, much as it does in a rubber band. This is why the relaxed state of a DNA circle is not supercoiled.

Naturally Occurring DNA Circles Are Underwound.

The linking numbers of natural DNA circles are less than those of their corresponding relaxed circles; that is, they are underwound. However, because DNA tends to adopt an overall conformation that maintains its normal twist of 1 turn/10.4 bp, the molecule is negatively supercoiled ($W < 0$; Fig. 23-11, *left*). If the duplex is unwound (if T decreases), then W increases (L must remain constant). At first, this reduces the superhelicity of an underwound circle. However, with continued unwinding, the value of W passes through zero (a relaxed circle; Fig. 23-11, *center*) and then becomes positive, yielding a positively coiled superhelix (Fig. 23-11, *right*).

Topoisomerases Control DNA Supercoiling.

DNA functions normally only if it is in the proper topological state. In such basic biological processes as replication and transcription, complementary polynucleotide strands must separate. The negative supercoiling of naturally occurring DNAs in both prokaryotes and eukaryotes promotes such separations since it tends to unwind the duplex helix (an increase in W must be accompanied by a

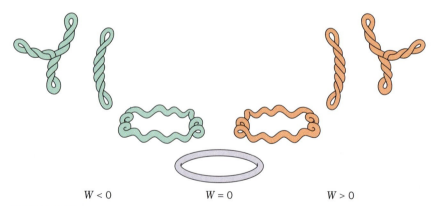

$W < 0$ $\qquad\qquad\qquad$ $W = 0$ $\qquad\qquad\qquad$ $W > 0$

Figure 23-11 Progressive unwinding of a negatively supercoiled DNA molecule. As the double helix in a negatively supercoiled circle ($W < 0$) is unwound without breaking covalent bonds (T decreases), W increases until it reaches 0. Further unwinding of the double helix then causes the DNA to supercoil in the opposite direction, yielding a positively coiled superhelix ($W > 0$).

decrease in T). *If DNA lacks the proper superhelical tension, the above vital processes cannot occur.*

The supercoiling of DNA is controlled by a remarkable group of enzymes known as **topoisomerases.** They are so named because they alter the topological state (linking number) of circular DNA but not its covalent structure. There are two classes of topoisomerases in prokaryotes and eukaryotes:

1. **Type I topoisomerases** act by creating transient single-strand breaks in DNA. Type I enzymes are further classified as **type IA** and **type IB topoisomerases,** which differ in sequence and reaction mechanism.

2. **Type II topoisomerases** act by making transient double-strand breaks in DNA.

Type IA Topoisomerases Relax Negatively Supercoiled DNA.

Type I topoisomerases catalyze the relaxation of supercoiled DNA by changing the linking number in increments of one. Type IA enzymes, which are present in all cells, relax only negatively supercoiled DNA. A clue to the mechanism of type IA topoisomerases was provided by the observation that they reversibly **catenate** (interlink) single-stranded circles (Fig. 23-12*a*). Apparently the enzyme operates by cutting a single strand, passing a single-strand loop through the resulting gap, and then resealing the break (Fig. 23-12*b*), thereby twisting double-helical DNA by one turn. This increases the linking number, thereby making the writhing number more positive (fewer negative supercoils). For duplex DNA, this process can be repeated until all of its supercoils have been removed ($W = 0$). This **strand-passage** mechanism is supported by the observation that the denaturation of type IA topoisomerase that has been incubated with single-stranded circular DNA yields a linear DNA that has its 5′-terminal phosphoryl group linked to the enzyme via a phospho-Tyr diester linkage:

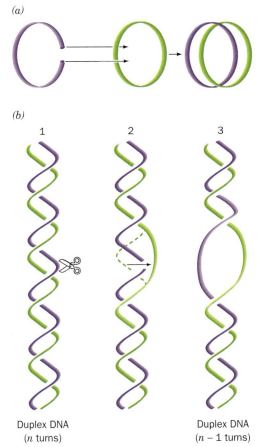

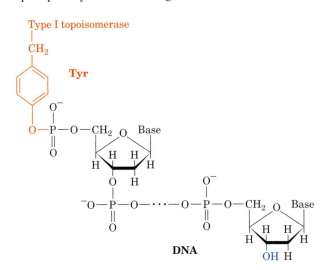

Formation of this covalent enzyme–DNA intermediate conserves the free energy of the cleaved phosphodiester bond so that no free energy input is needed to later reseal the nick in the DNA.

E. coli **topoisomerase III,** a type IA enzyme, is a 659-residue monomer. The X-ray structure of its catalytically inactive Y328F mutant in complex with the single-stranded octanucleotide d(CGCAACTT) was determined by Alfonso Mondragón. The enzyme folds into four domains that enclose an ~20 by 28 Å hole which is large enough to contain duplex DNA and which

(a)

(b)

1	2	3

Duplex DNA   Duplex DNA
(n turns)   ($n - 1$ turns)

Figure 23-12 Type IA topoisomerase action.   By cutting a single-stranded DNA, passing a loop of a second strand through the break, and then resealing the break, a type IA topoisomerase can (*a*) catenate two single-stranded circles or (*b*) unwind duplex DNA by one turn.

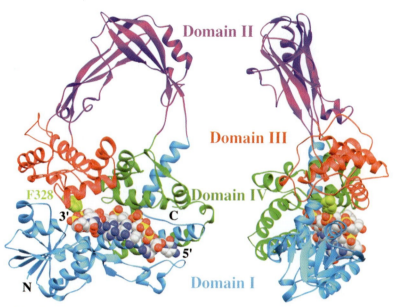

Figure 23-13 X-Ray structure of the Y328F mutant of *E. coli* topoisomerase III in complex with the single-stranded octanucleotide d(CGCAACTT). The protein's four domains are drawn in different colors and the two views shown are related by a 90° rotation about a vertical axis. The DNA is drawn in space-filling form with C gray, N blue, O red, and P yellow. This type IA topoisomerase's active site is marked by the side chain of Phe 328, which is shown in space-filling form in yellow-green. [Based on an X-ray structure by Alfonso Mondragón, Northwestern University. PDBid 1I7D.]

is lined with numerous Arg and Lys residues (Fig. 23-13). The octanucleotide binds in a groove that is also lined with Arg and Lys side chains with its sugar–phosphate backbone in contact with the protein and with most of its bases exposed for possible base pairing. This single-stranded DNA assumes a B-DNA-like conformation even though its complementary strand would be sterically excluded from the groove. The DNA strand is oriented with its 3′ end near the active site where, if the mutant Phe 328 were the wild-type Tyr, its side chain would be properly positioned to nucleophilically attack the phosphate group bridging the DNA's C-6 and T-7 nucleotides to form a 5′-phosphoTyr linkage with T-7 and release C-6 with a free 3′-OH. This structure suggests the mechanism for the type IA topoisomerase-catalyzed strand-passage reaction that is diagrammed in Fig. 23-14.

Type IB Topoisomerases Relax Supercoiled DNA via Controlled Rotation. Type IB topoisomerases can relax both negative and positive supercoils. In doing so, they transiently cleave one strand of a duplex DNA through the nucleophilic attack of an active site Tyr on a DNA P atom to yield a 3′-linked phospho-Tyr intermediate and a free 5′-OH group on the succeeding nucleotide (in contrast to the 5′-linked phospho-Tyr and free 3′-OH group formed by type IA topoisomerases). The X-ray structure of the catalytically inactive Y723F mutant of human **topoisomerase I**, a type IB topoisomerase, in complex with a 22-bp duplex DNA was determined by Wim Hol. The core domain of this bilobal protein is wrapped around the DNA in a tight embrace (Fig. 23-15). If the mutant Phe 723 were the wild-type Tyr, its OH group would be ideally positioned to nucleophilically attack the P on the scissile P—O5′ bond so as to form a covalent linkage with the 3′ end of the cleaved strand. The protein interacts to a much greater extent with the five base pairs of the DNA's "downstream" segment (which would contain the cleaved strand's

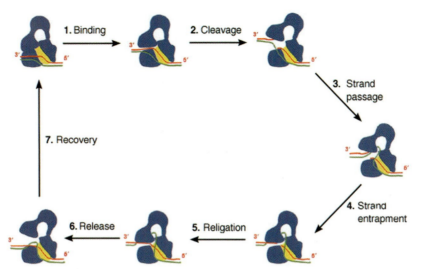

Figure 23-14   Proposed mechanism for type IA topoisomerases.   The enzyme is shown in blue with the yellow patch representing the binding groove for single-stranded (ss) DNA. The two DNA strands, which are drawn in red and green, could represent the two strands of a covalently closed circular duplex or two ss circles. **(1)** The enzyme recognizes a ss region of the DNA, here the red strand, and binds it in its binding groove. This is followed by or occurs simultaneously with the opening of a gap between domains I and III. **(2)** The DNA is cleaved, with the newly formed 5′ end becoming covalently linked to the active site Tyr and the segment with the newly formed 3′ end remaining tightly but noncovalently bound in the binding groove. **(3)** The unbroken (*green*) strand is passed through the opening or gate formed by the cleaved (*red*) strand to enter the protein's central hole. **(4)** The unbroken strand is trapped by the partial closing of the gap. **(5)** The two cleaved ends of the red strand are rejoined in what is probably a reversal of the cleavage reaction. **(6)** The gap between domains I and III reopens to permit the escape of the red strand, yielding the reaction product in which the green strand has been passed through a transient break in the red strand. **(7)** The enzyme returns to its initial state. If the two strands form a negatively supercoiled duplex DNA, its linking number, L, has increased by 1; if they are separate ss circles, they have been catenated or decatenated. For negatively supercoiled duplex DNA, this process can be repeated until all of its supercoils have been removed ($W = 0$). [After a drawing by Alfonso Mondragón, Northwestern University.]

newly formed 5′ end) than it does with the base pairs of the DNA's "up-stream" segment (to which Tyr 723 would be covalently linked), and in both cases it does so in a largely sequence-independent manner.

Topoisomerase I does not appear sterically capable of unwinding super-coiled DNA via the strand-passage mechanism that type IA topoisomerases

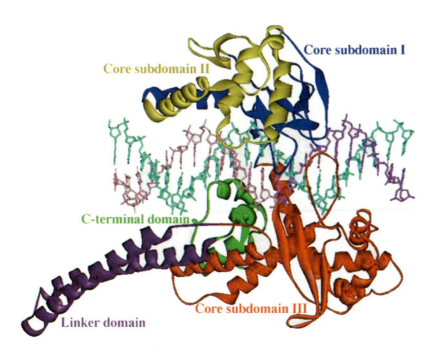

Figure 23-15   X-Ray structure of the Y723F mutant of human topoisomerase I in complex with a 22-bp duplex DNA.   This type IB topoisomerase consists of several domains and subdomains that are drawn here in different colors. Tyr 723 is on the C-terminal domain (*green*). The DNA's uncleaved strand is cyan, and the upstream and downstream portions of the scissile strand are magenta and pink, respectively. [Courtesy of Wim Hol, University of Washington. PDBid 1A36.]

Figure 23-16 **The controlled rotation mechanism for type IB topoisomerases.** A highly negatively supercoiled DNA (*red, with a right-handed writhe*) is converted, via stages (*a*) through (*g*), to a less supercoiled form (*green*). Topoisomerase I is drawn as a bilobal space-filling structure (*cyan and magenta*). The structure shown in (*d*), which is expanded by a factor of 2, shows the downstream portion of the rotating DNA (that containing the cleaved strand's new 5' end) at 30° intervals, all differently colored. Since the enzyme is not always in direct contact with the rotating DNA, small rocking motions of the protein (*small curved arrows*) may accompany the controlled rotation. [Courtesy of Wim Hol, University of Washington.]

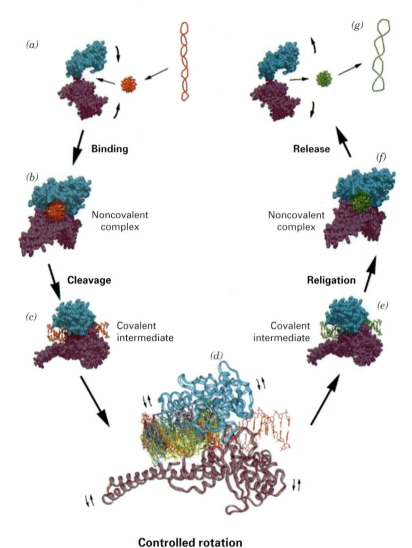

(a)

Binding

(b)

Noncovalent complex

Cleavage

(c)

Covalent intermediate

(d)

Controlled rotation

(g)

Release

(f)

Noncovalent complex

Religation

(e)

Covalent intermediate

follow (Fig. 23-14). Rather, as is diagrammed in Fig. 23-16, it is likely that topoisomerase I relaxes DNA supercoils by permitting the cleaved duplex DNA's loosely held downstream segment to rotate relative to the tightly held upstream segment. This rotation can only occur about the sugar–phosphate bonds in the uncleaved strand (α, β, γ, ε, and ζ in Fig. 23-5) that are opposite the cleavage site because the cleavage frees these bonds to rotate. In support of this mechanism, the protein region surrounding the downstream segment contains 16 conserved, positively charged residues that form a ring about this duplex DNA, which apparently holds the DNA in the ring but not in any specific orientation. Nevertheless, the downstream segment is unlikely to rotate freely because the cavity containing it is shaped so as to interact with the downstream segment during some portions of its rotation. Hence, type IB topoisomerases are said to mediate a **controlled rotation** mechanism in relaxing supercoiled DNA. This unwinding is driven by the superhelical tension in the DNA and hence requires no other energy input. Eventually, the DNA is religated by a reversal of the cleavage reaction and the now less supercoiled DNA is released.

Type II Topoisomerases Hydrolyze ATP. Type II topoisomerases are multimeric enzymes that require ATP hydrolysis to complete a reaction cycle in which

two DNA strands are cleaved, duplex DNA is passed through the break, and the break is resealed. Type II topoisomerases therefore change the linking number in increments of two rather than one, as in type I topoisomerases. Both prokaryotic and eukaryotic type II enzymes relax negative and positive supercoils, but only the prokaryotic enzyme (also known as **DNA gyrase**) can introduce negative supercoils. The negative supercoils in eukaryotic chromosomes result primarily from its packaging in nucleosomes (Section 23-5B) rather than from topoisomerase action.

Type II topoisomerases superficially resemble the type I enzymes: They have a pair of catalytic Tyr residues, which form transient covalent intermediates with the 5′ ends of duplex DNA. They thereby mediate the cleavage of the two DNA strands at staggered sites to produce 4-nucleotide "sticky ends." Stephen Harrison and James Wang determined the X-ray structure of a large homodimeric fragment of yeast **topoisomerase II,** a type II topoisomerase, that can cleave duplex DNA but cannot transport it through the break because it lacks the intact protein's N-terminal ATPase domain. The heart-shaped dimer has a triangular hole that is 55 Å wide at its base and 60 Å high (Fig. 23-17), far larger than the diameter of B-DNA. The two active-site Tyr residues are located 27 Å apart, so they must move a considerable distance to link to the 5′ ends of a cleaved duplex DNA.

A proposed mechanism for type II topoisomerases, based on this X-ray structure, is diagrammed in Fig. 23-18. The DNA to be cleaved first binds to the enzyme and is clamped in place. In the presence of ATP, the bound DNA is cleaved and a second duplex DNA is passed through the opening into the central hole of the protein. The cleaved DNA is then resealed and the transported DNA exits the complex at a point opposite its point of entry. ATP hydrolysis to ADP + P_i prepares the enzyme for an additional catalytic cycle. The importance of topoisomerases in maintaining DNA in its proper topological state is indicated by the fact that many antibiotics and chemotherapeutic agents are inhibitors of topoisomerases (see Box 23-2).

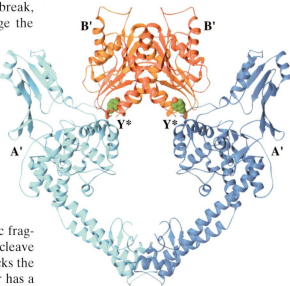

Figure 23-17 X-Ray structure of yeast topoisomerase II. This homodimeric type II enzyme (residues 410–1202 of the 1429-residue protein) is viewed with its twofold axis vertical. The A′ and B′ subfragments of one subunit are blue and red, and those of the other subunit are cyan and orange. The active-site Tyr side chains (Y*) are shown in space-filling form with C green and O red. [Based on an X-ray structure by James Berger, Stephen Harrison, and James Wang, Harvard University. PDBid 1BGW.] ♻ **See the Interactive Exercises.**

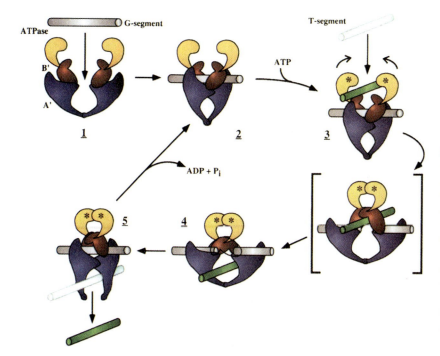

Figure 23-18 Proposed mechanism for Type II topoisomerase. The protein's ATPase, B′, and A′ domains are colored yellow, red, and purple. (**1**) An unliganded enzyme binds duplex DNA (the so-called G-segment; *gray*), and a conformational change (**2**) clamps the DNA in place. (**3**) ATP binding (represented by asterisks) promotes cleavage of the G-segment DNA so that another duplex DNA (the T-segment; *green*) can pass into the central hole of the enzyme (**4**). This DNA transport step is shown as proceeding through a hypothetical intermediate shown in square brackets. The G-segment is then resealed (**5**), and the T-segment exits the enzyme. ATP hydrolysis and release regenerate state **2**. [Courtesy of Stephen Harrison and James Wang, Harvard University.]

BOX 23-2

Biochemistry in Health and Disease

Inhibitors of Topoisomerases as Antibiotics and Anticancer Chemotherapeutic Agents

Type II topoisomerases are inhibited by a variety of compounds. For example, **ciprofloxacin** and **novobiocin** specifically inhibit DNA gyrase but not eukaryotic topoisomerase II and are therefore antibiotics. In fact, ciprofloxacin is the most efficacious oral antibiotic presently in clinical use (e.g., it is the preferred antibiotic in the treatment of anthrax). In contrast, novobiocin's adverse side effects and the rapid generation of bacterial resistance to its presence have resulted in the discontinuation of its use in the treatment of human infections. A number of substances, including **doxorubicin** and **etoposide,** inhibit eukaryotic type II topoisomerases and are therefore widely used in cancer chemotherapy.

Different type II topoisomerase inhibitors act in one of two ways. Many of these agents, including novobiocin, inhibit their target enzyme's ATPase activity. They therefore kill cells by blocking topoisomerase activity, which results in the arrest of DNA replication and RNA transcription since the cell can no longer control the level of supercoiling in its DNA. However, other substances, including ciprofloxacin, doxorubicin, and etoposide, enhance the rate at which their target type II topoisomerases cleave double-stranded DNA and/or reduce the rate at which these breaks are resealed. Consequently, these substances induce higher than normal levels of transient protein-bridged breaks in the DNA of treated cells. These protein bridges are easily ruptured by the passage of the replication and transcription machinery, thereby rendering the breaks permanent. Although all cells have extensive enzymatic machinery for repairing damaged DNA (Section 24-5), a sufficiently high level of DNA damage can overwhelm the repair mechanisms and trigger cell death. Consequently, since rapidly replicating cells such as cancer cells have elevated levels of type II topoisomerases, they are far more likely to incur lethal DNA damage through the inhibition of their type II topoisomerases than are slow-growing or quiescent cells.

Camptothecin and its derivatives,

Camptothecin

the only known inhibitors of type IB topoisomerases, act by prolonging the lifetime of the covalent enzyme–DNA intermediate. These substances bind to the complex between the ends of the severed DNA strand, positioning the 5'-OH groups over 4.5 Å away from the phosphate groups that must be reattached in the religation reaction. Because the nicked DNA cannot be replicated, camptothecin prevents the proliferation of rapidly growing cells such as cancer cells and hence is a potent anticancer agent.

Ciprofloxacin

Doxorubicin

Novobiocin

Etoposide

2 | Forces Stabilizing Nucleic Acid Structures

DNA does not exhibit the structural complexity of proteins because it has only a limited repertoire of secondary structures and no comparable tertiary or quaternary structures. This is perhaps to be expected since the 20 amino acid residues of proteins have a far greater range of chemical and physical properties than do the four DNA bases. Nevertheless, many RNAs have well-defined tertiary structures. In this section, we examine the forces that give rise to the structures of nucleic acids.

A   Denaturation and Renaturation

When a solution of duplex DNA is heated above a characteristic temperature, its native structure collapses and its two complementary strands separate and assume random conformations (Fig. 23-19). This denaturation process is accompanied by a qualitative change in the DNA's physical properties. For example, the characteristic high viscosity of native DNA solutions, which arises from the resistance to deformation of its rigid and rodlike duplex molecules, drastically decreases when the DNA decomposes to the conformationally flexible single chains. Likewise, DNA's ultraviolet absorbance, which is almost entirely due to its aromatic bases, increases by ~40% on denaturation (Fig. 23-20) as a consequence of the disruption of the electronic interactions among neighboring bases. The increase in absorbance is known as the **hyperchromic effect.**

Monitoring the changes in absorbance at a single wavelength (usually 260 nm) as the temperature increases reveals that the increase in absorbance occurs over a narrow temperature range (Fig. 23-21). This indicates that *the denaturation of DNA is a cooperative phenomenon in which*

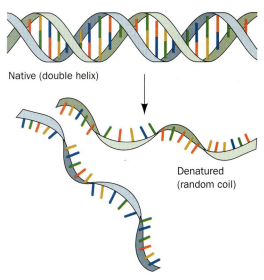

Figure 23-19 A schematic representation of DNA denaturation.

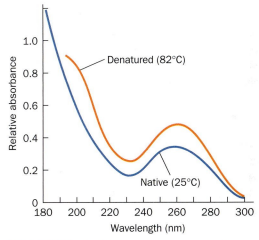

Figure 23-20 **The UV absorbance spectra of native and heat-denatured *E. coli* DNA.** Note that denaturation does not change the general shape of the absorbance curve but only increases its intensity. [After Voet, D., Gratzer, W.B., Cox, R.A., and Doty, P., *Biopolymers* **1**, 205 (1963).] See the Animated Figures.

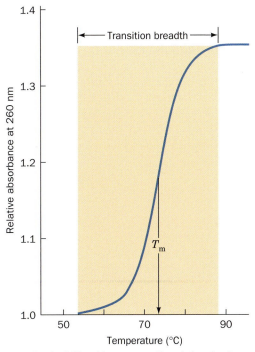

Figure 23-21 An example of a DNA melting curve.   The relative absorbance is the ratio of the absorbance (customarily measured at 260 nm) at the indicated temperature to that at 25°C. The melting temperature, T_m, is the temperature at which half of the maximum absorbance increase is attained. See the Animated Figures.

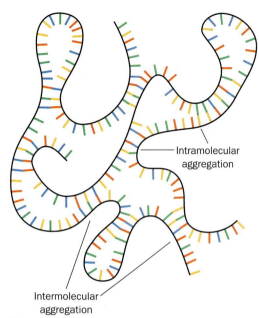

Figure 23-22   Partially renatured DNA. This schematic representation shows the imperfectly base-paired structures assumed by DNA that has been heat denatured and then rapidly cooled. Note that both intramolecular and intermolecular aggregation can occur.

the collapse of one part of the structure destabilizes the remainder. In analogy with the melting of a solid, Fig. 23-21 is referred to as a **melting curve,** and the temperature at its midpoint is defined as its **melting temperature, T_m.**

The stability of the DNA double helix, and hence its T_m, depends on several factors, including the nature of the solvent, the identities and concentrations of the ions in solution, and the pH. T_m also increases linearly with the mole fraction of G·C base pairs, although this is not, as one might expect, entirely because G·C base pairs contain one more hydrogen bond than A·T base pairs (see below).

DNA Can Be Renatured. If a solution of denatured DNA is rapidly cooled below its T_m, the resulting DNA will be only partially base paired (Fig. 23-22), because the complementary strands will not have had sufficient time to find each other before the randomly base-paired structure becomes effectively "frozen in." If, however, the temperature is maintained ~25°C below the T_m, enough thermal energy is available for short base-paired regions to rearrange by melting and re-forming. Under such **annealing** conditions, as Julius Marmur discovered in 1960, denatured DNA eventually completely renatures. Likewise, complementary strands of RNA and DNA, in a process known as **hybridization,** form RNA–DNA hybrid double helices that are only slightly less stable than the corresponding DNA double helices.

B   Base Pairing

Base pairing is apparently a "glue" that holds together double-stranded nucleic acids. Only Watson–Crick pairs occur in the crystal structures of self-complementary oligonucleotides. However, other hydrogen-bonded base pairs with reasonable geometries are known (Fig. 23-23). For example, when monomeric adenine and thymine derivatives are cocrystallized, the A·T base pairs that form invariably have adenine N7 as the hydrogen bond acceptor (**Hoogsteen** geometry; Fig. 23-23b) rather than N1 (Watson–Crick geometry; Fig. 23-1).

These and other observations indicate that *Watson–Crick geometry is the most stable mode of base pairing in the double helix, even though non-Watson–Crick base pairs are theoretically possible.* Initially, it was believed that the geometrical constraints of the double helix precluded other types of base pairing. Recall that because A·T, T·A, G·C, and C·G base pairs are geometrically similar, they can be interchanged without altering the conformation of the sugar–phosphate chains. However, experimental measurements

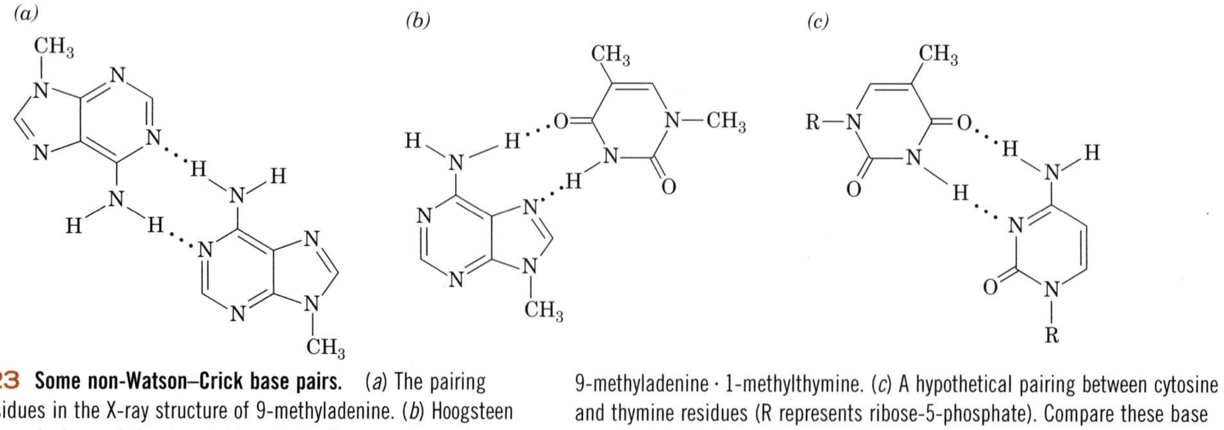

(a)

(b)

(c)

Figure 23-23   Some non-Watson–Crick base pairs. (a) The pairing of adenine residues in the X-ray structure of 9-methyladenine. (b) Hoogsteen pairing between adenine and thymine residues in the X-ray structure of 9-methyladenine · 1-methylthymine. (c) A hypothetical pairing between cytosine and thymine residues (R represents ribose-5-phosphate). Compare these base pairs to those shown in Fig. 23-1.

show that a major reason that other base pairs do not appear in the double helix is that *Watson–Crick base pairs have an intrinsic stability that non-Watson–Crick base pairs lack; that is, the bases in a Watson–Crick pair have a higher mutual affinity than those in a non-Watson–Crick pair.* Nevertheless, as we shall see, the double-helical segments of many RNAs contain unusual base pairs, such as G · U, that help stabilize their tertiary structures.

Hydrogen Bonds Only Weakly Stabilize Nucleic Acid Structures. It is clear that hydrogen bonding is required for the specificity of base pairing in DNA. Yet, as is also true for proteins (Section 6-4A), *hydrogen bonding contributes little to the stability of nucleic acid structures.* This is because, on denaturation, the hydrogen bonds between the base pairs of a native nucleic acid are replaced by energetically similar hydrogen bonds between the bases and water. Other types of forces must therefore play an important role in stabilizing nucleic acid structures.

C Base Stacking and Hydrophobic Interactions

Purines and pyrimidines tend to form extended stacks of planar parallel molecules. This has been observed in the structures of nucleic acids (Fig. 23-2c) and in the structures of crystallized nucleic acid bases (e.g., Fig. 23-24). These **stacking interactions** are a form of van der Waals interaction (Section 2-1A). Interactions between stacked G and C bases are greater than those between stacked A and T bases (Table 23-2), which largely accounts for the greater thermal stability of DNAs with a high G + C content. Note also that differing sets of base pairs in a stack have different stacking energies. Thus, *the stacking energy of a double helix is sequence-dependent.*

Stacking Interactions in Aqueous Solution Result from Hydrophobic Forces. One might assume that hydrophobic interactions in nucleic acids are qualitatively similar to those that stabilize protein structures. This is not the case. Recall that folded proteins are stabilized primarily by the increase in entropy of the solvent water molecules (the hydrophobic effect; Section 6-4A); in other words, protein folding is enthalpically opposed and entropically driven. In contrast, thermodynamic measurements reveal that the base stacking in nucleic acids is enthalpically driven and entropically opposed, although the theoretical basis for this observation is not well understood. The difference between hydrophobic forces in proteins and in nucleic acids may reflect the fact that nucleic acid bases are much more polar than most protein side chains (e.g., adenine versus the side chain of Phe or Leu). Whatever their origin, hydrophobic forces are of central importance in determining nucleic acid structures, as is clear from the denaturing effect of adding nonpolar solvents to aqueous solutions of DNA.

D Ionic Interactions

Any theory of the stability of nucleic acid structures must take into account the electrostatic interactions of their charged phosphate groups. For example, the melting temperature of duplex DNA increases with the Na^+ concentration because these ions electrostatically shield the anionic phosphate groups from each other. Other monovalent cations such as Li^+ and K^+ have similar nonspecific interactions with phosphate groups. Divalent cations, such as Mg^{2+}, Mn^{2+}, and Co^{2+}, in contrast, specifically bind to phosphate groups, so *they are far more effective shielding agents for nucleic acids than*

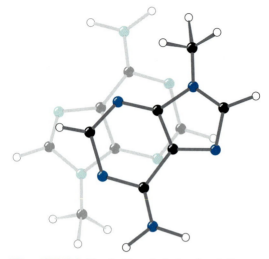

Figure 23-24 **The stacking of adenine rings in the X-ray structure of 9-methyladenine.** The partial overlap of the rings is typical of the association between bases in crystal structures and in double-helical nucleic acids. [After Stewart, R.F. and Jensen, L.H., *J. Chem. Phys.* **40**, 2071 (1964).]

Table 23-2 Stacking Energies for the Ten Possible Dimers in B-DNA

Stacked Dimer	Stacking Energy (kJ · mol⁻¹)
C · G G · C	−61.0
C · G A · T	−44.0
C · G T · A	−41.0
G · C C · G	−40.5
G · C G · C	−34.6
G · C A · T	−28.4
T · A A · T	−27.5
G · C T · A	−27.5
A · T A · T	−22.5
A · T T · A	−16.0

Source: Ornstein, R.L., Rein, R., Breen, D.L., and MacElroy, R.D., *Biopolymers* **17**, 2356 (1978).

are monovalent cations. For example, an Mg^{2+} ion has an influence on the DNA double helix comparable to that of 100 to 1000 Na$^+$ ions. Indeed, enzymes that mediate reactions with nucleic acids or nucleotides almost always require Mg^{2+} for activity. Mg^{2+} ions also play an essential role in stabilizing the complex structures assumed by many RNAs.

E RNA Structure

The structure of RNA is stabilized by the same forces that stabilize DNA, and its conformational flexibility is limited by many of the same features that limit DNA conformation. In fact, RNA may be even more rigid than DNA owing to the presence of a greater number of water molecules that form hydrogen bonds to the 2'-OH groups of RNA. Even so, RNA comes in a greater variety of shapes and sizes than DNA and, in some cases, has catalytic activity.

RNAs May Contain Double-Stranded Segments.   Bacterial ribosomes, which are two-thirds RNA and one-third protein, contain three highly conserved RNA molecules (Section 26-3). The smallest of these is the 120-nucleotide **5S RNA** [so named because it sediments in the ultracentrifuge at the rate of 5 Svedbergs (S); Section 5-2E]. 5S RNA has a secondary structure consisting of several base-paired regions connected by loops of various kinds (Fig. 23-25a). These structural elements can be discerned in the X-ray structure of this RNA molecule, in which the base-paired segments form mostly A-type helices (Fig. 23-25b). Approximately two-thirds of the bases in 5S rRNA participate in base pairing interactions; the remaining nucleotides, which are located in loops and at the 3' and 5' ends, are free to interact with ribosomal proteins or with unpaired nucleotides in other rRNA molecules.

Large RNA molecules presumably fold in stages, as do multidomain proteins (Section 6-5). RNA folding is almost certainly a cooperative process,

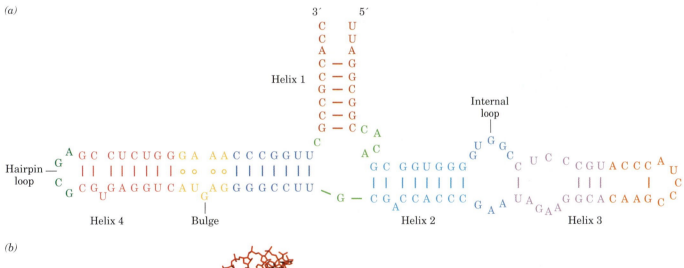

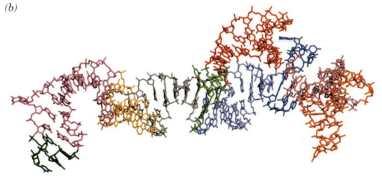

Figure 23-25 Secondary and tertiary structure of 5S RNA from *Haloarcula marismortui.* (*a*) Secondary structure diagram. This single-stranded RNA molecule contains four double-stranded base-paired helices, as well as hairpin loops, internal loops, and bulges. The dashes indicate Watson–Crick base pairing, and circles indicate other modes of base pairing. (*b*) The X-ray structure in stick form colored according to its secondary structure as in Part *a*. [Based on an X-ray structure by Peter Moore and Thomas Steitz, Yale University. PDBid 1JJ2.]

with the rapid formation of short duplex regions preceding the collapse of the structure into its mature conformation. The complete folding process may take several minutes, it may involve relatively stable intermediates, and it may require the assistance of proteins.

RNA Molecules Are Stabilized by Stacking Interactions.

The three-dimensional structure of the yeast transfer RNA that forms a covalent complex with Phe (**tRNA^{Phe}**) was elucidated in 1974 by Alexander Rich in collaboration with Sung-Hou Kim and, in a different crystal form, by Aaron Klug. The molecule is compact and L-shaped, with each leg of the L ~60 Å long (Fig. 23-26). The structural complexity of yeast tRNA^{Phe} is reminiscent of that of a protein. Although only 42 of its 76 bases occur in double-helical stems, 71 of them participate in stacking associations.

tRNA structures are characterized by the presence of covalently modified bases and unusual base pairs, including hydrogen-bonding associations involving three bases. These tertiary interactions contribute to the compact structure of the tRNAs (tRNA structure is discussed in greater detail in Section 26-2A).

Some RNAs Are Catalysts.

Although the discovery of catalytic RNA in 1982 was initially greeted with skepticism, subsequent studies have demonstrated that the catalytic potential of RNA is virtually unlimited. *At least nine naturally occurring types of catalytic RNAs have been described, and many more* **ribozymes** *have been developed in the laboratory.* Most naturally occurring

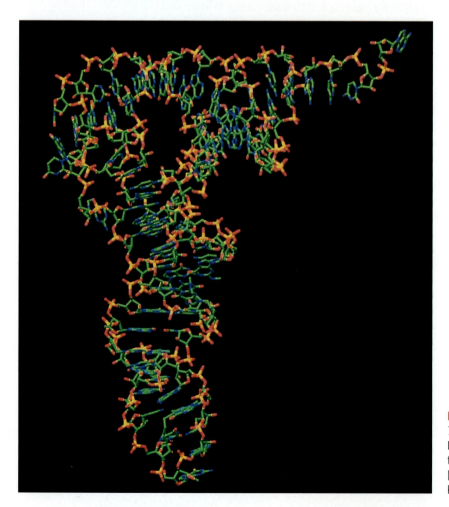

Figure 23-26   Structure of yeast tRNA^{Phe}. The 76-nucleotide RNA is shown in stick form with C green, N blue, O red, and P gold. Nearly all the bases are stacked, thereby stabilizing the compact structure of the tRNA. [After an X-ray structure by Alexander Rich and Sung-Hou Kim, MIT. PDBid 6TNA.]

RNA catalysts participate in aspects of RNA metabolism such as mRNA splicing (Section 25-3A) and hydrolytic RNA processing (Section 25-3B). The formation of peptide bonds during protein synthesis is catalyzed by ribosomal RNA (Section 26-4). *In vitro,* RNA molecules can catalyze some of the reactions required for their own production, such as the synthesis of glycosidic bonds and nucleotide polymerization. This functional versatility has led to the hypothesis that RNA once carried out many of the basic activities of early life, before DNA or proteins had evolved. This scenario has been dubbed the RNA world (see Box 23-3).

One of the best-characterized ribozymes is the **hammerhead ribozyme,** an ~40-nucleotide molecule that participates in the replication of certain virus-like RNAs that infect plants. The hammerhead ribozyme catalyzes site-specific cleavage of one of its own phosphodiester bonds. The secondary structures of the hammerhead ribozyme, which cleaves itself between its C-17 and A-1.1 nucleotides, has three duplex stems and a conserved core of two nonhelical segments (Fig. 23-27a). The X-ray structure of this hammerhead ribozyme (Fig. 23-27b), determined by William Scott and Klug, reveals that it forms three A-type helices, although its overall shape more closely resembles a wishbone than a hammerhead. The nucleotides in the helical stems form normal Watson–Crick base pairs, whereas nucleotides U-7 through A-9 form non-Watson–Crick base pairs with nucleotides G-12 through A-14 in which ribose oxygens participate as both hydrogen bond donors and acceptors.

In the catalytic core of the hammerhead ribozyme, the absolutely conserved CUGA tetranucleotide loop forms a pocket into which C-17 inserts. This region of the molecule has sufficient flexibility to position the O2′ atom of C-17 for nucleophilic attack on the P atom linking the O3′ atom of C-17 to the O5′ atom of A-1.1, thereby yielding a cyclic 2′,3′-phosphodiester on C-17 together with a free 5′-OH on A-1.1. An Mg^{2+} ion bound to the ribozyme appears to stabilize the RNA structure rather than participate directly in catalysis.

The hammerhead ribozyme is not a true catalyst, since it is its own substrate and hence cannot return to its original state. However, other

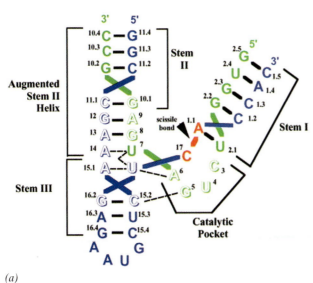

(a)

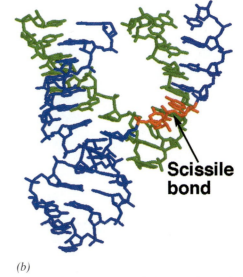

(b)

Figure 23-27   Structure of a hammerhead ribozyme.   (a) The sequence and schematic structural representation of the ribozyme drawn with its 16-nucleotide enzyme strand green, its 25-nucleotide substrate strand purple, and nucleotides spanning the cleavage site red. Essential and highly conserved residues are represented by hollow letters, and the universal numbering system is provided. Watson–Crick and nonstandard G · A base pairs are shown as black dashes, and single hydrogen bonds are shown by dashed lines. (b) A stick model of the ribozyme colored as in Part a. [Courtesy of William Scott, University of California at Santa Cruz. PDBid 1MME.]   See the Interactive Exercises.

BOX 23-3

Perspectives in Biochemistry

The RNA World

The observations that nucleic acids but not proteins can direct their own synthesis, that cells contain batteries of protein-based enzymes for manipulating DNA (Chapter 24) but few for processing RNA, and that many coenzymes in biosynthetic processes and energy-harvesting pathways are ribonucleotides (e.g., ATP, NAD^+, FAD, and coenzyme A) led to the hypothesis that *RNAs were the original biological catalysts in precellular times—the so-called* **RNA world.** Indeed, RNA remains the carrier of genetic information in viruses with single- or double-stranded RNA genomes. The fact that RNA-based functions are distributed across all three domains of life provides additional evidence that the role of RNA in modern cells reflects its establishment very early in evolution.

Assuming that an RNA world did once exist, we must ask two questions: Why did protein catalysts take the place of most ribozymes, and why did DNA become the dominant molecule of heredity? The first question is readily answered by observing that the 20 amino acid residues of proteins contain functional groups—hydroxyl, sulfhydryl, amide, and carboxylate groups—that are lacking in RNA, whose four nucleotide residues are more uniform and less reactive. Thus, proteins probably replaced RNA as cellular workhorses by virtue of their greater chemical virtuosity. Nevertheless, RNA continues to play the dominant role in the synthesis of proteins: Ribosomes are predominantly RNA, which is now known to catalyze peptide bond formation (Section 26-4B).

As for DNA, it is more stable than RNA because RNA is highly susceptible to base-catalyzed hydrolysis by the reaction mechanism at the right. The base-induced deprotonation of the 2′-OH group facilitates its nucleophilic attack on the adjacent phosphorus atom, thereby cleaving the RNA backbone. The resulting 2′,3′-cyclic phosphate group subsequently hydrolyzes to produce a 2′- or 3′-nucleotide product (the RNase A–catalyzed hydrolysis of RNA follows a nearly identical reaction sequence but generates only 3′-nucleotides; Section 11-3A). DNA is not susceptible to such degradation because it lacks a 2′-OH group. This greater chemical stability of DNA makes it more suitable than RNA for the long-term storage of genetic information.

RNA

2′,3′-Cyclic nucleotide

2′-Nucleotide   or   **3′-Nucleotide**

ribozymes catalyze multiple turnovers without themselves undergoing alteration. Ribozymes are comparable to protein enzymes in their rate enhancement (10^9-fold for the hammerhead ribozyme), in their ability to use cofactors such as metal ions or imidazole groups, and in their regulation by small allosteric effectors. Although RNA molecules lack the repertoire of functional groups possessed by protein catalysts, RNA molecules can assume conformations that specifically bind substrates, orient them for reaction, and stabilize the transition state—all hallmarks of enzyme catalysts.

3 Fractionation of Nucleic Acids

In Section 5-2 we considered the most common procedures for isolating and characterizing proteins. Most of these methods, often with some modifications, are also used to fractionate nucleic acids according to size, composition, and sequence. There are also many techniques that apply only to nucleic acids. In this section, we outline some of the most useful nucleic acid fractionation procedures.

Nucleic acids in cells are invariably associated with proteins. Once cells have been broken open, their nucleic acids are usually deproteinized. This can be accomplished by shaking the protein–nucleic acid mixture with a phenol solution so that the protein precipitates and can be removed by centrifugation. Alternatively, the protein can be dissociated from the nucleic acids by detergents, guanidinium chloride, or high salt concentration, or it can be enzymatically degraded by proteases. In all cases, the nucleic acids, a mixture of RNA and DNA, can then be isolated by precipitation with ethanol. The RNA can be recovered from such precipitates by treating them with pancreatic DNase to eliminate the DNA. Conversely, the DNA can be freed of RNA by treatment with RNase.

In all these and subsequent manipulations, the nucleic acids must be protected from degradation by nucleases that occur both in the experimental material and on human hands. Nucleases can be inhibited by chelating agents such as EDTA, which sequester the divalent metal ions they require for activity. Laboratory glassware can also be autoclaved to heat denature the nucleases. Nevertheless, nucleic acids are generally easier to handle than proteins because most lack a complex tertiary structure and are therefore relatively tolerant of extreme conditions.

A Chromatography

Many of the chromatographic techniques that are used to separate proteins (Section 5-2C) also apply to nucleic acids. However, **hydroxyapatite,** a form of calcium phosphate [$Ca_5(PO_4)_3OH$], is particularly useful in the chromatographic purification and fractionation of DNA. Double-stranded DNA binds to hydroxyapatite more tightly than do most other molecules. Consequently, DNA can be rapidly isolated by passing a cell lysate through a hydroxyapatite column, washing the column with a phosphate buffer of concentration low enough to release only the RNA and protein, and then eluting the DNA with a concentrated phosphate solution.

Affinity chromatography is used to isolate specific nucleic acids. For example, most eukaryotic messenger RNAs (mRNAs) have a poly(A) sequence at their 3' ends (Section 25-3A). They can be isolated on agarose or cellulose to which poly(U) is covalently attached. The poly(A) sequences specifically bind to the complementary poly(U) in high salt and at low temperature and can later be released by altering these conditions.

B Electrophoresis

Nucleic acids of a given type can be separated by polyacrylamide gel electrophoresis (Section 3-4B) because their electrophoretic mobilities in such gels vary inversely with their molecular masses. However, DNAs of more than a few thousand base pairs cannot penetrate even a weakly cross-linked polyacrylamide gel and so must be separated in agarose gels. Yet conventional gel electrophoresis is limited to DNAs of <100,000 bp, because larger DNA molecules tend to worm their way through the agarose at a rate

independent of their size. Charles Cantor and Cassandra Smith overcame this limitation by developing **pulsed-field gel electrophoresis (PFGE),** which can resolve DNAs of up to 10 million bp (6.6 million kD). In the simplest type of PFGE apparatus, the polarity of the electrodes is periodically reversed, with the duration of each pulse varying from 0.1 to 1000 s, depending on the sizes of the DNAs being separated. With each change in polarity, the migrating DNA must reorient to the new electrical field before it can resume movement. A DNA molecule may actually migrate backward for part of the time. Because small molecules reorient more quickly than large molecules, different-sized DNA molecules gradually separate (Fig. 23-28).

Intercalation Agents Stain Duplex DNA. The various DNA bands in a gel can be stained by planar aromatic cations such as **ethidium ion, acridine orange,** or **proflavin:**

Ethidium

Proflavin   **Acridine orange**

These dyes bind to double-stranded DNA by **intercalation** (slipping in between the stacked bases; Fig. 23-29), where they exhibit a fluorescence under UV light that is far more intense than that of the free dye. As little as 50 ng of DNA can be detected in a gel by staining it with ethidium bromide. Single-stranded DNA and RNA also stimulate the fluorescence of ethidium but to a lesser extent than does duplex DNA.

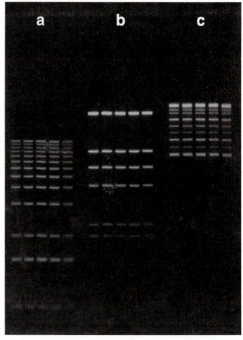

Figure 23-28 Pulsed-field gel electrophoresis. The three samples contain DNA fragments ranging from 0.5 to 48.5 kb. The smallest (fastest migrating) molecules are at the bottom. [Courtesy of Hoefer Scientific Instruments.]

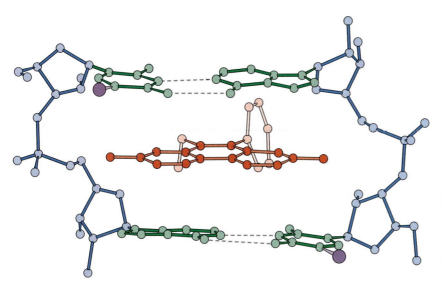

Figure 23-29 X-Ray structure of a complex of ethidium with 5-iodo-UpA. Ethidium (*red*) intercalates between the base pairs of the double-helically paired dinucleotide and thereby provides a model of the binding of ethidium to duplex DNA. [After Tsai, C.-C., Jain, S.C., and Sobell, H.M., *Proc. Natl. Acad. Sci.* **72,** 629 (1975).]

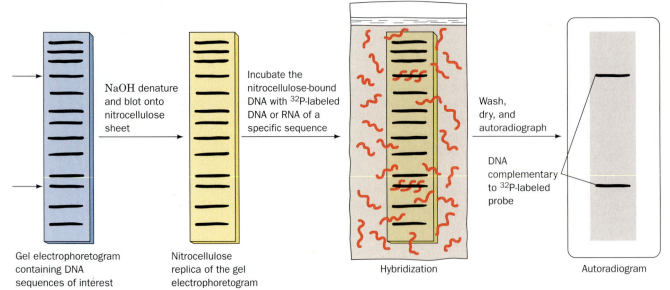

Figure 23-30 The detection of DNAs containing specific base sequences by Southern blotting.

Gel electrophoretogram containing DNA sequences of interest

NaOH denature and blot onto nitrocellulose sheet

Nitrocellulose replica of the gel electrophoretogram

Incubate the nitrocellulose-bound DNA with ^{32}P-labeled DNA or RNA of a specific sequence

Hybridization

Wash, dry, and autoradiograph

DNA complementary to ^{32}P-labeled probe

Autoradiogram

Southern Blotting Identifies DNAs with Specific Sequences. DNA with a specific base sequence can be identified through a procedure developed by Edwin Southern known as **Southern blotting** (Fig. 23-30). This procedure takes advantage of the ability of nitrocellulose to tenaciously bind single-stranded but not duplex DNA. Following gel electrophoresis of double-stranded DNA, the gel is soaked in 0.5 M NaOH to convert the DNA to its single-stranded form. The gel is then overlaid by a sheet of nitrocellulose paper. Molecules in the gel are forced through the nitrocellulose by drawing out the liquid with a stack of absorbent towels compressed against the far side of the nitrocellulose, or by using an electrophoretic process **(electroblotting).** The single-stranded DNA binds to the nitrocellulose at the same position it had in the gel. After drying at 80°C, which permanently fixes the DNA in place, the nitrocellulose sheet is moistened with a minimal quantity of solution containing a ^{32}P-labeled single-stranded DNA or RNA probe that is complementary in sequence to the DNA of interest. The moistened nitrocellulose is held at a suitable renaturation temperature for several hours to permit the probe to hybridize to its target sequence(s), washed to remove the unbound radioactive probe, dried, and then autoradiographed by placing it for a time over a sheet of X-ray film. The positions of the molecules that are complementary to the radioactive probe are indicated by a blackening of the developed film.

Specific RNA sequences can be detected through a variation of the Southern blot, punningly named a **Northern blot,** in which the RNA is immobilized on nitrocellulose paper and probed with a complementary radioactive RNA or DNA. A specific protein can be analogously detected in an **immunoblot** or **Western blot** through the use of antibodies directed against the protein in a procedure similar to that used in an ELISA (Fig. 5-3).

C Ultracentrifugation

Equilibrium density gradient ultracentrifugation (Section 5-2E) in CsCl is one of the most commonly used DNA separation procedures. The buoyant density of double-stranded Cs$^+$DNA depends on its base composition, with DNAs of higher G + C content having a greater density (Fig. 23-31).

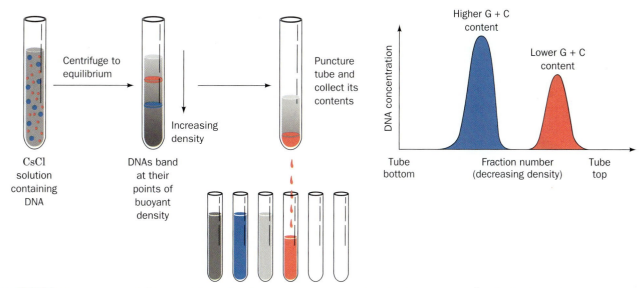

Figure 23-31 The separation of DNAs by equilibrium density gradient ultracentrifugation in CsCl solution.   An initially 8 M CsCl solution forms a density gradient that varies linearly from $\sim$1.80 g$\cdot$cm^{-3} at the bottom of the centrifuge tube to $\sim$1.55 g$\cdot$cm^{-3} at the top. The sedimentation rates of the DNAs depend on their base composition. The amount of DNA in each fraction is estimated from its UV absorbance, usually at 260 nm.

Single-stranded DNA is $\sim$0.015 g$\cdot$cm^{-3} denser than the corresponding double-stranded DNA, so the two can be separated by equilibrium density gradient ultracentrifugation. Circular DNA can also be separated from linear DNA in this manner, as can circular DNAs that are supercoiled to different extents.

RNA is too dense to band in CsCl but does so in Cs_2SO_4 solutions. RNA–DNA hybrids band in CsCl but at a higher density than the corresponding duplex DNA. RNA can also be fractionated by zonal ultracentrifugation (Fig. 5-12) through a sucrose gradient. RNAs are separated by this technique largely on the basis of their size.

4 DNA–Protein Interactions

The accessibility of genetic information depends on the ability of proteins to recognize and interact with DNA in a manner that allows the encoded information to be copied as DNA (in replication) or as RNA (in transcription). Even the most basic steps of these processes require many proteins that interact with each other and with the nucleic acids. In addition, organisms regulate the expression of most genes, which requires yet other proteins that act as repressors or activators of transcription.

Many proteins bind DNA nonspecifically, that is, without regard to the sequence of nucleotides. For example, histones (which are involved in packaging DNA; Section 23-5A) and certain DNA replication proteins (which must potentially interact with all the sequences of an organism's genome) bind to DNA primarily through interactions between protein functional groups and the sugar–phosphate backbone of DNA. Proteins that recognize specific DNA sequences presumably also bind nonspecifically but loosely to DNA so that they can scan the polynucleotide chain for their target sequences to which they then bind specifically and tightly. Sequence-specific DNA–protein interactions must be extremely precise so that the proteins can exert their effects at sites selected from among—in the human

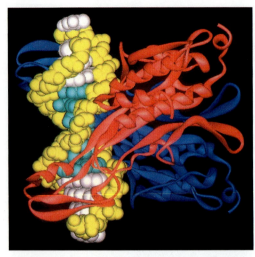

(a)

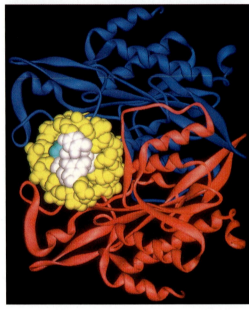

(b)

Figure 23-32 **X-Ray structure of *Eco*RI endonuclease in complex with DNA.** The segment of duplex DNA has the self-complementary sequence TCGCGAATTCGCG (12 bp with an overhanging T at both 5′ ends; the enzyme's 6-bp target sequence is underlined) and is drawn as a space-filling model with its sugar–phosphate chains yellow, its recognition sequence bases cyan, and its other bases white. The protein is drawn in ribbon form with its two identical subunits red and blue. The complex is shown (*a*) with its DNA helix axis vertical and (*b*) in end view. The complex's twofold axis is horizontal and the DNA's major groove faces right (toward the protein) in both views. [Based on an X-ray structure by John Rosenberg, University of Pittsburgh. PDBid 1ERI.] **See Kinemage Exercise 18-1.**

genome—billions of base pairs. How do such proteins interact with their target sites on DNA?

Sequence-specific DNA-binding proteins generally do not disrupt the base pairs of the duplex DNA to which they bind. They do, however, *discriminate among the four base pairs (A · T, T · A, G · C, and C · G) according to the functional groups of the base pairs that project into DNA's major and minor grooves.* As can be seen in Fig. 23-2, these groups are more exposed in the major groove of B-DNA than in its narrower minor groove. Moreover, the major groove contains more sequence-specific functional groups than does the minor groove (Fig. 23-1). However, as revealed in a systematic survey of 129 protein–DNA complexes, only about one-third of the interactions between the binding protein and the DNA involve hydrogen bonds, either directly or via intervening water molecules. Approximately two-thirds of all interactions are van der Waals contacts. Ionic interactions with backbone phosphate groups also occur. The complementarity of the binding partners in many cases is augmented by an "induced fit" phenomenon in which the protein and the nucleic acid change their conformations for greater stability. In some cases, these changes allow the binding proteins to interact with other proteins or alter the accessibility of the DNA to other molecules.

The forces that stabilize DNA–protein interactions must necessarily derive from those that stabilize proteins (Section 6-4A) and nucleic acids (Section 23-2). But because DNA stability is imperfectly understood, the interactions that underlie DNA–protein complexes are also somewhat murky. There is no doubt, however, that these forces are substantial. Most DNA–protein complexes have dissociation constants ranging from 10^{-9} to 10^{-12} M; these values represent binding that is 10^3 to 10^7 times stronger than nonspecific binding. Below, we examine some examples of specific DNA–protein binding. We shall encounter many more examples of nucleic acid–protein interactions in subsequent chapters.

A Restriction Endonucleases

Type II restriction endonucleases rid bacterial cells of foreign DNA by cleaving the DNA at specific sites that have not yet been methylated by the host's modification methylase (Section 3-4A). Restriction enzymes recognize palindromic DNA sequences of ~4 to 8 base pairs with such remarkable specificity that a single base change can reduce their activity by a millionfold. This degree of specificity is necessary to prevent accidental cleavage of other sites in a DNA sequence.

The X-ray structure of *Eco*RI endonuclease in complex with a segment of B-DNA containing the enzyme's recognition sequence was determined by John Rosenberg. The DNA binds in the symmetric cleft between the two identical 276-residue subunits of the dimeric enzyme (Fig. 23-32),

Figure 23-33 X-Ray structure of *Eco*RV endonuclease in complex with DNA.   The 10-bp segment of DNA has the self-complementary sequence GGGATATCCC (the enzyme's 6-bp target sequence is underlined). The DNA and protein are colored as in Fig. 23-32. (*a*) The complex as viewed along its twofold axis, facing the DNA's major groove. The two symmetry-related protein loops that overlie the major groove (composed of residues 182 to 186) are the only parts of the enzyme that make base-specific contacts with the DNA. (*b*) The complex as viewed from the right in *a* (the DNA's major groove faces left). Note how the protein kinks the DNA toward its major groove. [Based on an X-ray structure by Fritz Winkler, Hoffman-LaRoche Ltd., Basle, Switzerland. PDBid 4RVE.] 🔖 **See Kinemage Exercise 18-2.**

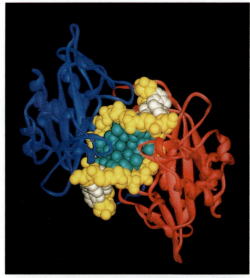

(*a*)

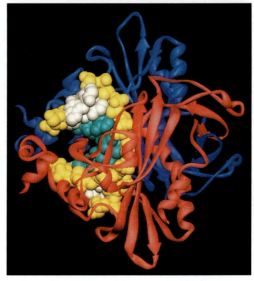

(*b*)

thereby accounting for the DNA's palindromic recognition sequence. Protein binding causes the dihedral angle between the recognition sequence's central two base pairs to open up by ~50° toward the minor groove. These base pairs thereby become unstacked, but the DNA remains nearly straight due to compensating bends at the adjacent base pairs. Nevertheless, this unwinds the DNA by 28° and widens the major groove by 3.5 Å at the recognition site. The N-terminal ends of a pair of parallel helices from each protein subunit are inserted into the widened major groove, where they participate in a hydrogen-bonded network with the bases of the recognition sequence. The phosphodiester cleavage points are located two bases from the center of the palindrome (Table 3-2).

Certain other restriction endonucleases, including *Eco*RV (Fig. 23-33), also induce kinks in the DNA, in some cases by opening up the minor groove and compressing the major groove. However, *complementary hydrogen bonding between the nucleotide bases and protein side chain and backbone groups, rather than the DNA distortion per se, is the primary prerequisite for the formation of a sequence-specific endonuclease–DNA complex.* For example, in the *Bam*HI–DNA complex, every potential hydrogen bond donor and acceptor in the major groove of the recognition site takes part in direct or water-mediated hydrogen bonds with the protein. No other DNA sequence could support this degree of complementarity with *Bam*HI.

B Prokaryotic Transcriptional Control Motifs

In prokaryotes, the expression of many genes is governed at least in part by **repressors,** proteins that bind at or near the gene so as to prevent its transcription (Section 27-2). These repressors often contain ~20-residue polypeptide segments that form a **helix–turn–helix (HTH) motif** containing two α helices that cross at an angle of ~120°. HTH motifs, which are apparently evolutionarily related, occur as components of domains that otherwise have widely varying structures, although all of them bind DNA. Note that HTH motifs are structurally stable only when they are components of larger proteins.

Like restriction endonuclease recognition sites, the sequences to which repressors bind (called **operators**) exhibit palindromic symmetry or nearly so. Typically, the repressors are dimeric, although their interactions with DNA may not be perfectly symmetrical. Binding involves interactions between the amino acid side chains extending from the second helix of the HTH motif (the "recognition" helix) and the bases and sugar–phosphate chains of the DNA.

The X-ray structure of a dimeric repressor from **bacteriophage 434** bound to its 20-bp recognition sequence was determined by Stephen Harrison. The **434 repressor** associates with the DNA in a twofold symmetric manner with a recognition helix from each subunit bound in successive

See Guided Exploration 23

Transcription factor–DNA interactions.

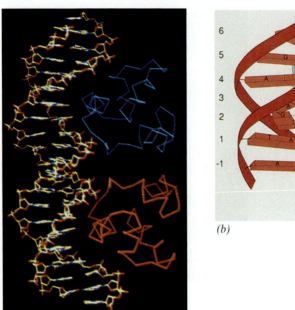

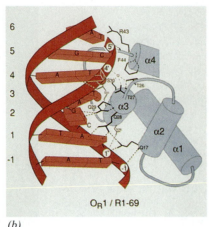

(b)

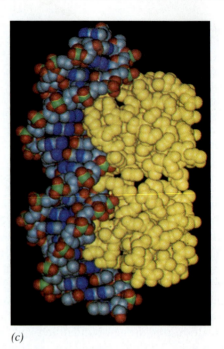

(a)   *(c)*

Figure 23-34 X-Ray structure of a portion of the 434 phage repressor in complex with its target DNA. One strand of the 20-bp DNA (*left*) has the sequence d(TATACAAGAAAGTTTGTACT). (*a*) A skeletal model showing the DNA and the first 63 residues of the protein's two identical subunits (*blue and red*, C_α backbone only). (*b*) A schematic drawing indicating how the helix–turn–helix motif, which encompasses helices α2 and α3, interacts with the DNA. Residues are identified by single letter codes. Short bars emanating from the polypeptide chain represent peptide NH groups, dashed lines represent hydrogen bonds, and numbered circles represent DNA phosphates. The small circle is a water molecule. (*c*) A space-filling model corresponding to *a*. All the protein's non-H atoms are drawn in yellow. [Courtesy of Aneel Aggarwal, John Anderson, and Stephen Harrison, Harvard University. PDBid 2OR1.] **See the Interactive Exercises and Kinemage Exercise 19.**

turns of the DNA's major groove (Fig. 23-34). The repressor closely conforms to the DNA surface and interacts with its paired bases and sugar–phosphate chains through elaborate networks of hydrogen bonds, salt bridges, and van der Waals contacts. In the repressor–DNA complex, the DNA bends around the protein in an arc of radius ~65 Å, which compresses the minor groove by ~2.5 Å near its center (between the two protein monomers) and widens it by ~2.5 Å toward its ends.

The *E. coli trp* Repressor Binds DNA Indirectly. The *E. coli* **trp** **repressor** regulates the transcription of genes required for tryptophan biosynthesis (Section 27-2C). Paul Sigler determined the X-ray structure of this protein in complex with a DNA containing an 18-bp palindrome (of single-strand sequence TGTACTAGTTAACTAGTAC, where the *trp* repressor's target sequence is underlined) that closely resembles the *trp* operator. The homodimeric repressor protein also has HTH motifs whose recognition helices bind, as expected, in successive major grooves of the DNA, each in contact with half of the operator sequence (ACTAGT; Fig. 23-35). There are numerous hydrogen-bonding contacts between the *trp* repressor and the DNA's nonesterified phosphate oxygens. Astoundingly, however, there are no direct hydrogen bonds or nonpolar contacts that can explain the repressor's specificity for its operator. Rather, all but one of the side chain–base hydrogen-bonding interactions are mediated by bridging water molecules. In addition, the operator contains several base pairs that are not in contact with the repressor but whose mutation nevertheless greatly decreases repressor binding affinity. This suggests that *the operator assumes*

a sequence-specific conformation that makes favorable contacts with the repressor. Other DNA sequences could conceivably assume the same conformation but at too high an energy cost to form a stable complex with the repressor. This phenomenon, in which a protein senses the base sequence of DNA through the DNA's backbone conformation and/or flexibility, is referred to as **indirect readout.** This finding puts to rest the notion that proteins recognize nucleic acid sequences exclusively via particular sets of pairings between amino acid side chains and nucleotide bases analogous to Watson–Crick base pairing.

The *met* Repressor Binds DNA via a Two-Stranded β Sheet.

Simon Phillips first determined the X-ray structure of the *E. coli* **met repressor** in the absence of DNA (the *met* repressor regulates the transcription of genes involved in methionine biosynthesis). The homodimeric protein lacks an HTH motif, but model-building studies suggested that the repressor might bind to its palindromic target DNA via a symmetry-related pair of protruding α helices, reminiscent of the way the recognition helices of HTH motifs interact with DNA. However, the subsequently determined X-ray structure of the *met* repressor–operator complex showed that the protein actually binds its target DNA sequence through a pair of symmetrically related β strands (located on the opposite side of the protein from the protruding α helices) that form a two-stranded antiparallel β sheet that inserts into the DNA's major groove (Fig. 23-36). The β strands make sequence-specific contacts with the DNA via hydrogen bonding and, probably, indirect readout. This result indicates that the conclusions of even what appear to be straightforward model-building studies should be viewed with skepticism. In the case of the *met* repressor–operator complex, the model-building study favored the incorrect model because it could not take into account the small conformational adjustments that both the protein and the DNA make on binding one another.

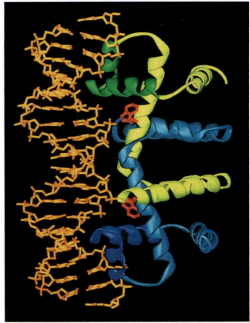

Figure 23-35 X-Ray structure of an *E. coli* trp repressor–operator complex. The molecular twofold axis is horizontal and in the plane of the paper. The protein's two identical subunits are shown in ribbon form in green and blue with the HTH motifs more deeply colored. The 18-bp self-complementary DNA is yellow. *trp* repressor binds its operator only when L-tryptophan (*red*) is also bound. Note that the protein's recognition helices bind, as expected, in successive major grooves of the DNA but extend approximately perpendicular to the DNA helix axis, whereas those of 434 phage repressor are nearly parallel to the major grooves of its bound DNA (Fig. 23-34). [Based on an X-ray structure by Paul Sigler, Yale University. PDBid 1TRO.] 🔗 **See the Interactive Exercises.**

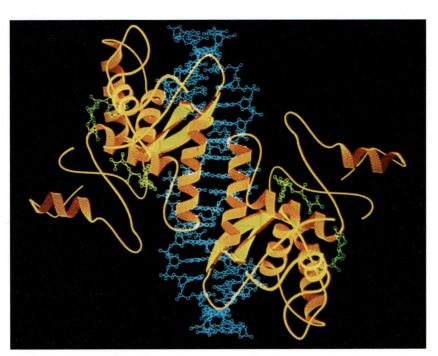

Figure 23-36 X-Ray structure of the *E. coli* met repressor–operator complex. The 104-residue repressor subunits are shown in gold. The self-complementary 19-bp DNA is shown as a blue ball-and-stick model. The methionine derivative *S*-adenosylmethionine, shown in green, must be bound to the repressor for it to bind DNA. Note that the DNA has four bound repressor subunits: Pairs of subunits form symmetric dimers in which each subunit donates one strand of the two-stranded β sheet that is inserted in the DNA's major groove (*upper left and lower right*). Two such dimers pair across the complex's twofold axis via their antiparallel N-terminal helices, which contact each other over the DNA's minor groove. [Courtesy of Simon Phillips, University of Leeds, Leeds, U.K. PDBid 1CMA.] 🔗 **See the Interactive Exercises.**

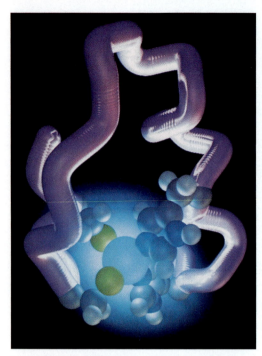

Figure 23-37 NMR structure of a single zinc finger from the *Xenopus* protein Xfin. The Zn^{2+} ion together with the atoms of its His and Cys ligands are represented as spheres with Zn^{2+} cyan, C gray, N blue, S yellow, and H white. [Courtesy of Michael Pique, The Scripps Research Institute, La Jolla, California. Based on an NMR structure by Peter E. Wright, The Scripps Research Institute. PDBid 1ZNF.]

C Eukaryotic Transcription Factors

In eukaryotes, genes are selectively expressed in different cell types; this requires more complicated regulatory machinery than in prokaryotes. Prokaryotic repressors of known structure either contain an HTH motif or resemble the *met* repressor. However, eukaryotic DNA-binding proteins employ a much wider variety of structural motifs to bind DNA. A number of proteins known as **transcription factors** (Section 25-2C) promote the transcription of genes by binding to DNA sequences at or near those genes. In this section, we describe a variety of the DNA-binding motifs in eukaryotic transcription factors.

Zinc Finger DNA-Binding Motifs. The first of the predominantly eukaryotic DNA-binding motifs, the **zinc finger,** was discovered by Klug in **transcription factor IIIA (TFIIIA)** from *Xenopus laevis* (an African clawed toad). The 344-residue TFIIIA contains nine similar, tandemly repeated, ~30-residue modules, each of which contains two invariant Cys residues and two invariant His residues. Each of these units binds a Zn^{2+} ion, which is tetrahedrally liganded by the Cys and His residues (Fig. 23-37). In some zinc fingers, the two Zn^{2+}-liganding His residues are replaced by two additional Cys residues, whereas others have six Cys residues liganding two Zn^{2+} ions. Indeed, structural diversity is a hallmark of zinc finger proteins. In all cases, however, the Zn^{2+} ions appear to knit together relatively small globular domains, thereby eliminating the need for much larger hydrophobic protein cores (Section 6-4A).

The **Cys₂–His₂ zinc finger** (Figs. 6-37 and 23-37) contains a two-stranded antiparallel β sheet and one α helix. Three of these motifs are incorporated into a 72-residue segment of the mouse protein **Zif268,** whose X-ray structure in complex with a target DNA was elucidated by Carl Pabo (Fig. 23-38). The three zinc fingers are arranged as separate domains in a C-shaped structure that fits snugly into the DNA's major groove. Each zinc finger interacts in a conformationally identical manner with successive 3-bp segments of the DNA, predominantly through hydrogen bonds between the zinc finger's α helix and one strand of the DNA. Each zinc finger specifically hydrogen bonds to two bases in the major groove. Interestingly, five of these six associations involve Arg–guanine pairs. In addition to these sequence-specific interactions, each zinc finger hydrogen bonds with the DNA's phosphate groups via conserved Arg and His residues.

Figure 23-38 X-Ray structure of a three–zinc finger segment of Zif268 in complex with a 10-bp DNA. The protein and DNA (with a single nucleotide overhang at each end) are shown in stick form, with superimposed cylinders and ribbons marking the protein's α helices and β sheets. Finger 1 is orange, finger 2 is yellow, finger 3 is pink, the DNA is blue, and the Zn^{2+} ions are represented by white spheres. Note how the N-terminal (lower) end of each zinc finger's helix extends into the DNA's major groove to contact three base pairs. [Courtesy of Carl Pabo, MIT. PDBid 1ZAA.]
🐾 See the Interactive Exercises.

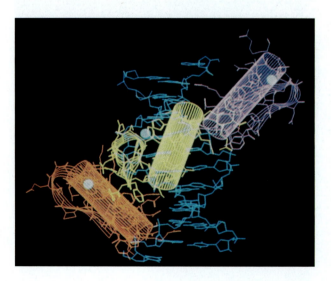

The Cys$_2$–His$_2$ zinc finger broadly resembles the prokaryotic HTH motif as well as most other DNA-binding motifs we shall encounter (including other types of zinc fingers). All these DNA-binding motifs provide a platform for inserting an α helix into the major groove of B-DNA. However, the Cys$_2$–His$_2$ zinc fingers, unlike other DNA-binding motifs, occur as modules that each contact successive DNA segments (some transcription factors contain over 60 zinc fingers). *Such a modular system can recognize extended asymmetric base sequences.*

A binuclear **Cys$_6$ zinc finger** mediates the DNA binding of the yeast protein **GAL4,** a transcriptional activator of several genes that encode galactose-metabolizing enzymes. Residues 1 to 65 of this 881-residue protein include six Cys residues that collectively bind two Zn^{2+} ions (Fig. 23-39*a*). Each Zn^{2+} ion is tetrahedrally coordinated by four Cys residues, with two of these residues ligating both metal ions. GAL4 binds to its 17-bp target DNA as a symmetric dimer (Fig. 23-39*b*), although in the absence of DNA it is a monomer. Each subunit includes a compact zinc finger (residues 8–40) that binds DNA, an extended linker (residues 41–49), and an α helix (residues 50–64) that assists in GAL4 dimerization. The N-terminal helix of the zinc finger inserts into the DNA's major groove, making sequence-specific contacts with a highly conserved CCG sequence at each end of the recognition sequence. The bound DNA retains its B conformation.

The dimerization helices of GAL4 (center of Fig. 23-39*b*) are positioned over the minor groove of the DNA. The linkers connecting these helices to the zinc fingers wrap around the DNA, largely following its minor groove. The two symmetrically related DNA-binding zinc fingers thereby approach the major groove from opposite sides of the DNA, ~1.5 helical turns apart,

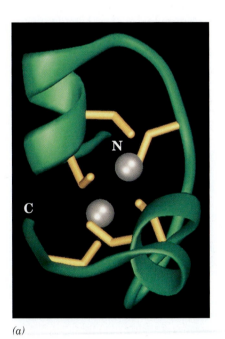

(a)

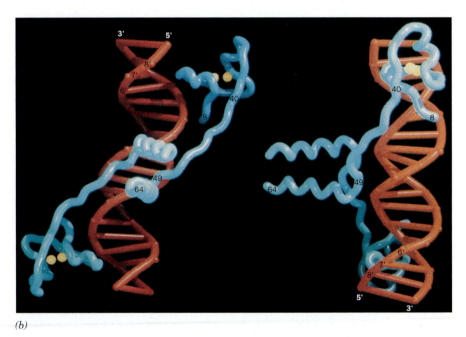

(b)

Figure 23-39 X-Ray structure of the GAL4 DNA-binding domain in complex with DNA. (*a*) A ribbon model of the protein's zinc finger domain (residues 8–40) with its six Cys side chains in stick form (*yellow*) and its Zn^{2+} ions shown as silver spheres. Compare this structure with Figs. 6-37 and 23-37. (*b*) The complex of the dimeric GAL4 protein with a palindromic 19-bp DNA (except for the central base pair) containing the protein's binding sequence. The structure is shown in tube form with the DNA red, the protein backbone cyan, and the Zn^{2+} ions represented by yellow spheres. The views are along the complex's twofold axis (*left*) and turned 90° with the twofold axis horizontal (*right*). Note how the C-terminal end of each subunit's N-terminal helix extends into the DNA's major groove. [Part *b* courtesy of and Part *a* based on an X-Ray structure by Ronen Mamorstein and Stephen Harrison, Harvard University. PDBid 1D66.] 🔎 **See the Interactive Exercises.**

rather than from the same side the DNA, ~1 turn apart, as do HTH motifs. The resulting relatively open structure could permit other proteins to bind simultaneously to the DNA.

Transcription Factors with Leucine Zippers.   Segments of certain eukaryotic transcription factors, such as the yeast protein **GCN4,** contain a Leu at every seventh position. We have already seen that α helices with the seven-residue pseudorepeating sequence $(a\text{-}b\text{-}c\text{-}d\text{-}e\text{-}f\text{-}g)_n$, in which the a and d residues are hydrophobic, have a hydrophobic strip along one side that allows them to dimerize as a coiled coil (e.g., α-keratin; Section 6-1C). Steven McKnight suggested that DNA-binding proteins containing such **heptad repeats** also form coiled coils in which the Leu side chains interdigitate, much like the teeth of a zipper (Fig. 23-40). In fact, *these **leucine zippers** mediate the dimerization of certain DNA-binding proteins but are not themselves DNA-binding motifs.*

The X-ray structure of the 33-residue polypeptide corresponding to the leucine zipper of GCN4 was determined by Peter Kim and Thomas Alber. The first 30 residues, which contain ~3.6 heptad repeats (Fig. 23-40a), coil into an ~8-turn α helix that dimerizes as McKnight predicted to form ~1/4 turn of a parallel left-handed coiled coil (Fig. 23-40b). The dimer can be envisioned as a twisted ladder whose sides consist of the helix backbones and whose rungs are formed by the interacting hydrophobic side chains. The conserved Leu residues in the heptad position d, which corresponds to every second rung, are not interdigitated as McKnight originally suggested, but instead make side-to-side contacts. The alternate rungs are likewise formed by the a residues of the heptad repeat (which are mostly Val). These contacts form an extensive hydrophobic interface between the two helices.

In many leucine zipper–containing proteins, a DNA-binding region that is rich in basic residues is immediately N-terminal to the leucine zipper and hence these proteins are known as **basic region leucine zipper (bZIP) proteins.** For example, in GCN4, the C-terminal 56 residues form an extended α helix. The last 25 residues of two such helices associate as a leucine zipper. The N-terminal portions of the helices smoothly diverge to bind in the major grooves on opposite sides of the DNA, thereby clamping the

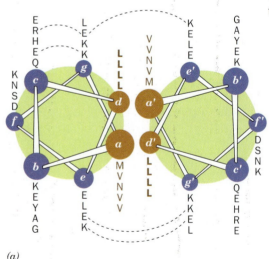

(a)

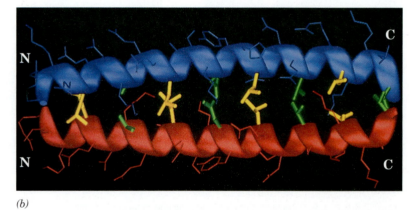

(b)

Figure 23-40   The GCN4 leucine zipper motif.   (*a*) A helical wheel representation of the motif's two helices as viewed from their N-termini. The sequences of residues at each position are indicated by the adjacent column of one-letter codes. Residues that form ion pairs in the crystal structure are connected by dashed lines. Note that all residues at positions d and d' are Leu (L), those at positions a and a' are mostly Val (V), and those at other positions are mostly polar. [After O'Shea, E.K., Klemm, J.D., Kim, P.S., and Alber, T., *Science* **254,** 540 (1991).] (*b*) The X-ray structure, in side view, in which the helices are shown in ribbon form. Side chains are shown in stick form with the contacting Leu residues at positions d and d' yellow and residues at positions a and a' green. [Based on an X-ray structure by Peter Kim, MIT, and Tom Alber, University of Utah School of Medicine. PDBid 2ZTA.] 🔖 **See Kinemage Exercise 4-1.**

Figure 23-41 X-Ray structure of the GCN4 bZIP region in complex with its target DNA.   The DNA (*red*), shown in stick form, consists of a 19-bp segment with a single nucleotide overhang at each end and contains the protein's palindromic (except for the central base pair) 7-bp target sequence. The two identical GCN4 subunits, shown in ribbon form, each contain a continuous 52-residue α helix. At their C-terminal ends, (*yellow*), the two subunits associate in a parallel coiled coil (a leucine zipper), and at their basic regions (*green*), they smoothly diverge to each engage the DNA in its major groove at the target sequence. The N-terminal ends are white. [Based on an X-ray structure by Stephen Harrison, Harvard University. PDBid 1YSA.] 🔁 **See the Interactive Exercises.**

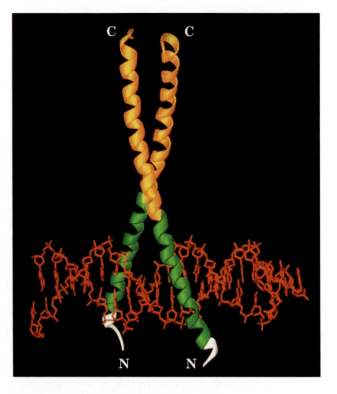

DNA in a sort of scissors grip (Fig. 23-41). The basic region residues that are conserved in bZIP proteins make numerous contacts with both the bases and the phosphate oxygens of their DNA target sequence without distorting its conformation.

Many eukaryotic transcription factors contain a conserved DNA-binding basic region that is immediately followed by two amphipathic helices connected by a loop to form a so-called **basic helix–loop–helix (bHLH) motif.** The basic region, together with the N-terminal portion of the first (H1) helix of the bHLH motif, binds in the major groove of its target DNA. The C-terminal (H2) helix of the bHLH motif mediates the dimerization of the protein via the formation of a coiled coil. In many proteins, the bHLH motif is continuous with a leucine zipper (Z) that presumably augments the dimerization. Thus, as is shown in Fig. 23-42 for the protein **Max,** the dimeric **bHLH/Z** protein grips the DNA in a manner reminiscent of a pair of forceps. Each basic region contacts specific bases of the DNA as well as phosphate groups. Side chains of both the loop and the N-terminal end of the H2 helix also contact DNA phosphate groups.

5 Eukaryotic Chromosome Structure

Although each base pair of B-DNA contributes only ~3.4 Å to its **contour length** (the end-to-end length of a stretched-out native molecule), DNA molecules are generally enormous (as shown on page 817). Indeed, the 23 chromosomes of the 3.2 billion-bp human genome have a total contour length of almost 1 m. One of the enduring questions of molecular biology is how such vast quantities of genetic information can be scanned and decoded in a reasonable time while stored in a small portion of the cell's volume.

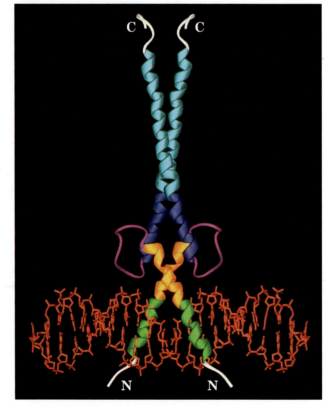

Figure 23-42 X-Ray structure of Max binding to DNA.   Residues 22–113 of the 160-residue transcription factor Max form a complex with a 22-bp DNA containing the protein's palindromic 6-bp target sequence. The DNA (*red*) is shown in stick form, and the homodimeric protein is shown in ribbon form. The protein's N-terminal basic region (*green*) forms an α helix that engages its target sequence in the DNA's major groove and then merges smoothly with the H1 helix (*yellow*) of the basic helix–loop–helix (bHLH) motif. Following the loop (*magenta*), the protein's two H2 helices (*purple*) of the bHLH motif form a short parallel four-helix bundle with the N-terminal ends of the two H1 helices. Each H2 helix then merges smoothly with the leucine zipper (Z) motif (*cyan*) to form a parallel coiled coil. The protein's N- and C-terminal ends are white. [Based on an X-ray structure by Edward Ziff and Stephen Burley, The Rockefeller University. PDBid 1AN2.] 🔁 **See the Interactive Exercises.**

Figure 23-43 Electron micrograph of a human metaphase chromosome. [Courtesy of Gunther Bahr, Armed Forces Institute of Pathology.]

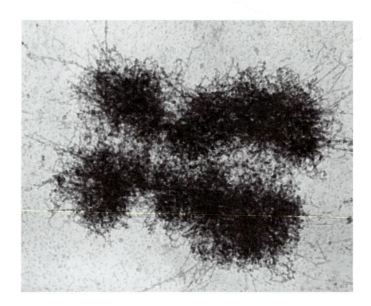

The elongated shape of duplex DNA (its diameter is only 20 Å) and its relative stiffness make it susceptible to mechanical damage when outside the protective environment of the cell. For example, a *Drosophila* chromosome, if expanded by a factor of 500,000, would have the shape and some of the mechanical properties of a 6-km long strand of uncooked spaghetti. In fact, the **shear degradation** of DNA by stirring, shaking, or pipetting a DNA solution is a standard laboratory method for preparing DNA fragments.

Prokaryotic genomes typically comprise a single circular DNA molecule. However, most eukaryotes condense and package their genome in several **chromosomes.** Each chromosome is a complex of DNA and protein, a material known as **chromatin,** and is a dynamic entity whose appearance varies dramatically with the stage of the **cell cycle** (the general sequence of events that occur in the lifetime of a eukaryotic cell; Section 27-3D). For example, chromosomes assume their most condensed forms only during the metaphase stage of cell division (Fig. 23-43). During the remainder of the cell cycle, when the DNA is transcribed and replicated, the chromosomes of most cells become so highly dispersed that they cannot be distinguished. Yet the DNA of these chromosomes is still compacted relative to its free B-helix form. Human chromosomes have contour lengths between 1.5 and 8.4 cm but in their most condensed state are only 1.3 to 10 μm long. In this section, we examine how DNA is packaged in cells to achieve this degree of condensation.

A Histones

Chromatin is about one-half protein by mass, and most of this protein consists of **histones.** To understand how DNA is packaged, we must first examine these proteins. The five major classes of histones, **H1, H2A, H2B, H3,** and **H4,** all have a large proportion of positively charged residues (Arg and Lys; Table 23-3). These proteins can therefore bind DNA's negatively charged phosphate groups through electrostatic interactions.

The amino acid sequences of histones H2A, H2B, H3, and H4 are remarkably conserved. For example, histones H4 from cows and peas, species that diverged 1.2 billion years ago, differ by only two conservative residue changes, which makes this protein among the most evolutionarily conserved

Table 23-3 Calf Thymus Histones

Histone	Number of Residues	Mass (kD)	% Arg	% Lys
H1	215	23.0	1	29
H2A	129	14.0	9	11
H2B	125	13.8	6	16
H3	135	15.3	13	10
H4	102	11.3	14	11

proteins known (Section 5-4A). *Such evolutionary stability implies that the histones have critical functions to which their structures are so well tuned that they are all but intolerant to change.* The fifth histone, H1, is more variable than the other histones; we shall see below that it also has a somewhat different role.

Histones are subject to posttranslational modifications that include methylation, acetylation, and phosphorylation of specific Arg, His, Lys, Ser, and Thr residues. *These modifications, many of which are reversible, all decrease the histones' positive charges, thereby significantly altering histone–DNA interactions.* Despite the histones' great evolutionary stability, their degree of modification varies enormously with the species, the tissue, and the stage of the cell cycle. As we shall see (Section 27-3A), modification of histones has been linked to transcriptional activation. The modifications probably create new protein binding sites on the histones themselves or on the DNA with which they associate.

B Nucleosomes

The first level of chromatin organization was pointed out by Roger Kornberg in 1974 from several lines of evidence:

1. Chromatin contains roughly equal numbers of histones H2A, H2B, H3, and H4, and no more than half that number of histone H1.

2. Electron micrographs of chromatin preparations at low ionic strength reveal ~100-Å-diameter particles connected by thin strands of apparently naked DNA, rather like beads on a string (Fig. 23-44).

3. Brief digestion of chromatin by **micrococcal nuclease** (which hydrolyzes double-stranded DNA) cleaves the DNA only between the above particles; apparently the particles protect the DNA closely associated with them from nuclease digestion. Gel electrophoresis indicates that each particle contains ~200 bp of DNA.

Kornberg proposed that the chromatin particles, which are called **nucleosomes,** consist of the octamer $(H2A)_2(H2B)_2(H3)_2(H4)_2$ in association with ~200 bp of DNA. The fifth histone, H1, was postulated to be associated in some other manner with the nucleosome (see below).

See Guided Exploration 24

Nucleosome structure.

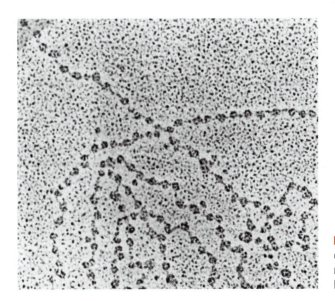

Figure 23-44 Electron micrograph of *D. melanogaster* chromatin showing strings of closely spaced nucleosomes.   [Courtesy of Oscar L. Miller, Jr., and Steven McKnight, University of Virginia.]

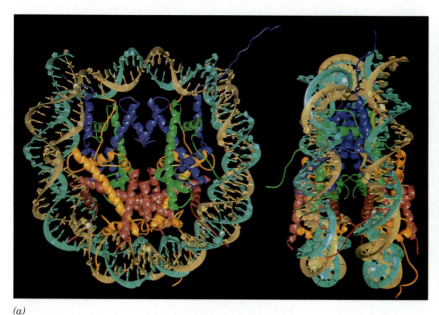

(a)

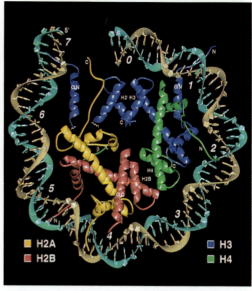

(b)

Figure 23-45 **X-Ray structure of the nucleosome core particle.**
(*a*) The entire core particle as viewed (*left*) along its superhelical axis and
(*right*) rotated 90° about the vertical axis. The proteins of the histone
octamer are drawn in ribbon form with H3 blue, H4 green, H2A yellow, and
H2B red. The sugar–phosphate backbones of the 146-bp DNA are drawn as
tan and turquoise ribbons whose attached bases are represented by
polygons of the same color. In both views, the pseudo-twofold axis is vertical
and passes through the DNA center at the top. (*b*) The top half of the

nucleosome core particle as viewed in Part *a, left* and identically colored. The
numbers 0 through 7 arranged about the inside of the 73-bp DNA superhelix
mark the positions of sequential double-helical turns. Those histones that are
drawn in their entirety are primarily associated with this DNA segment, whereas
only fragments of H3 and H2B from the other half of the particle are shown. The
two four-helix bundles shown are labeled H3′ H3 and H2B H4. [Courtesy of
Timothy Richmond, Eidgenössische Technische Hochschule, Zürich, Switzerland.
PDBid 1AOI.]

DNA Coils around a Histone Octamer to Form the Nucleosome Core Particle.
Micrococcal nuclease initially degrades chromatin to single nucleosomes in
complex with histone H1. Further digestion trims away additional DNA,
releasing histone H1. This leaves the so-called **nucleosome core particle,**
which consists of a 146-bp strand of DNA associated with the histone octamer. A segment of **linker DNA,** the DNA that is removed by the nuclease, joins neighboring nucleosomes. Its length varies between 8 and 114 bp
among organisms and tissues, although it is usually ~55 bp.

The X-ray structure of the nucleosome core particle, independently
determined by Timothy Richmond and Gerard Bunick, reveals a nearly
twofold symmetric complex in which B-DNA is wrapped around the outside of the histone octamer in 1.65 turns of a left-handed superhelix
(Fig. 23-45). This is the origin of supercoiling in eukaryotic DNA. Despite
having only weak sequence similarity, all four types of histone share a similar ~70-residue fold in which a long central helix is flanked on each side
by a loop and a shorter helix (Fig. 23-46). Pairs of histones interdigitate in
a sort of "molecular handshake" to form the crescent-shaped heterodimers
H2A–H2B and H3–H4, each of which binds 2.5 turns of duplex DNA that
curves around it in a 140° bend. The H3–H4 pairs interact, via a bundle of
four helices from the two H3 histones, to form an (H3–H4)$_2$ tetramer with
which each H2A–H2B pair interacts, via a similar four-helix bundle between H2A and H4, to form the histone octamer (Fig. 23-45*b*).

The histones bind exclusively to the inner face of the DNA, primarily
via its sugar–phosphate backbones, through hydrogen bonds, salt bridges,
and helix dipoles (their positive N-terminal ends), all interacting with phosphate oxygens, as well as through hydrophobic interactions with the deoxyribose rings. There are few contacts between the histones and the bases,

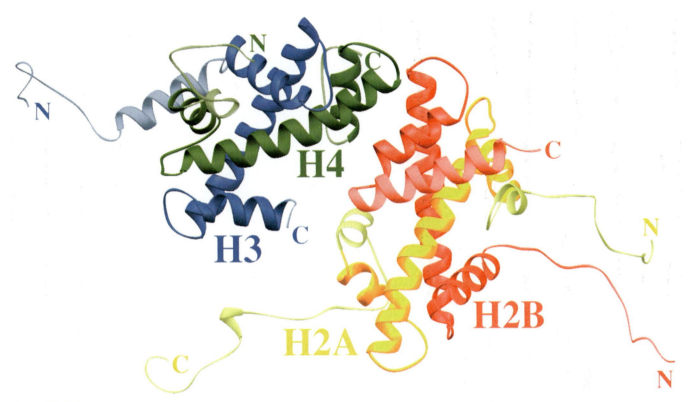

Figure 23-46 X-Ray structure of half of a histone octamer within the nucleosome core particle. Those portions of H2A, H2B, H3, and H4 that form the histone folds are yellow, red, blue, and green, respectively, with their N- and C-terminal tails colored in lighter shades. [Based on an X-ray structure by Gerard Bunick, University of Tennessee and Oak Ridge National Laboratory, Oak Ridge, Tennessee. PDBid 1EQZ.]

in accord with the nucleosome's lack of sequence specificity. However, an Arg side chain is inserted into the DNA's minor groove at each of the 14 positions at which it faces the histone octamer. The DNA superhelix has a radius of 42 Å and a pitch (rise per turn) of 26 Å. The DNA does not follow a uniform superhelical path but, rather, is bent fairly sharply at several locations due to outward bulges of the histone core. Moreover, the DNA double helix exhibits considerable conformational variation along its length such that its twist, for example, varies from 7.5 to 15.2 bp/turn with an average value of 10.4 bp/turn (versus 10.4 bp/turn for DNA in solution). Approximately 75% of the DNA surface is accessible to solvent and hence appears to be available for interactions with DNA-binding proteins.

Linker Histones Bring Nucleosomes Together. In the micrococcal nuclease digestion of chromatin fibers, the ~200-bp DNA is first degraded to 166 bp. Then there is a pause before histone H1 is released and the DNA is further shortened to 146 bp. Since the 146-bp DNA of the core particle makes 1.65 superhelical turns, the 166-bp intermediate should make nearly two full superhelical turns, which would bring its two ends close together. Klug has proposed that histone H1 binds to nucleosomal DNA at this point, where the DNA segments enter and leave the core particle (Fig. 23-47).

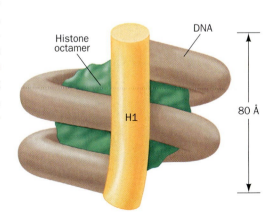

Figure 23-47 Model of histone H1 binding to the DNA of the 166-bp nucleosome. The DNA's two complete superhelical turns enable H1 to bind to the DNA's two ends and its middle. Here the histone octamer is represented by the green central spheroid and the H1 molecule is represented by the yellow cylinder.

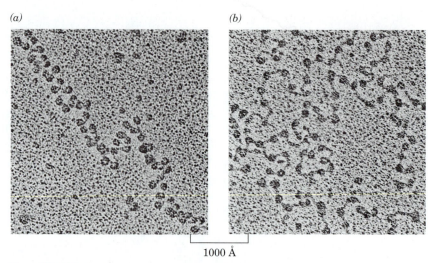

(a)   (b)

1000 Å

Figure 23-48   Electron micrographs of chromatin.   Both H1-containing chromatin (a) and H1-depleted chromatin (b) are in 5 to 15 mM salt. [Courtesy of Fritz Thoma, Eidgenössische Technische Hochschule, Zürich, Switzerland.]

Chromatin fibers containing H1 have closely spaced nucleosomes (Fig. 23-48a), consistent with DNA entering and exiting the nucleosome on the same side. In H1-depleted chromatin, the entry and exit points tend to occur on opposite sites of the nucleosome, producing a more dispersed arrangement (Fig. 23-48b). Evidently, linker histones have a relatively active role in condensing chromatin fibers and regulating the access of other proteins to the DNA.

C   Higher Levels of Chromatin Organization

Winding the DNA helix around a nucleosome reduces its contour length sevenfold (the 560-Å length of 166 bp is compressed to an ~80-Å-high supercoil). At physiological salt concentrations, chromatin condenses further by folding in a zig-zag fashion to form a filament with a diameter of ~30 nm (Fig. 23-49).

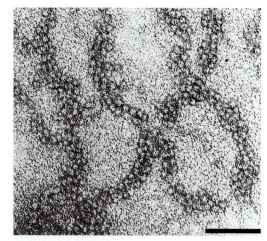

Figure 23-49   Electron micrograph of 30-nm-diameter chromatin filaments.   Note that the filaments are two to three nucleosomes across. The bar represents 100 nm. [Courtesy of Jerome B. Rattner, University of Calgary, Canada.]

Klug proposed that the 30-nm filament is a solenoid with ~6 nucleosomes per turn and a pitch of 110 Å (the diameter of a nucleosome; Fig. 23-50). The solenoid is presumed to be stabilized by interactions between histone H1 molecules in adjacent nucleosomes. Note, however, that several other plausible models for the 30-nm chromatin filament have also been formulated. Indeed, Kensal van Holde has argued that the 30-nm filament does not have a regular structure but rather, because of the varying lengths of the presumably straight and stiff linker DNAs connecting the nucleosome cores (with each additional nucleotide adding a 36° rotation between these cores), has an irregular helix-like structure that simulations indicate forms a filament with an average diameter of 30 nm. This would account for the difficulty in experimentally determining the structure of the 30-nm filament despite numerous attempts to do so over nearly three decades.

In the X-ray structure of the nucleosome, the N-terminal tails of histones H2B, H3, and H4 pass between the gyres (turns) of the DNA superhelix (Fig. 23-45a). The N-terminal segments of these tails are highly basic and contain the histones' acetylation sites. It is therefore likely that these tails stabilize the formation of chromatin filaments through their interactions with neighboring nucleosomes. Indeed, in the X-ray structure of the nucleosome, an H4's positively charged N-terminal tail binds, in an extended

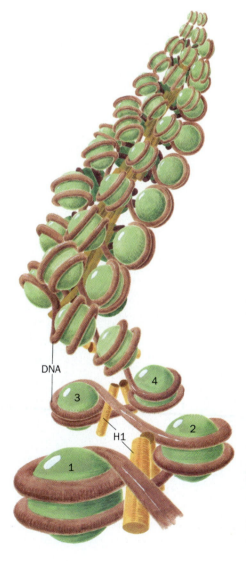

Figure 23-50 Proposed model of the 30-nm-diameter chromatin fiber. The zigzag pattern of nucleosomes (1, 2, 3, 4) closes up to form a solenoid with ~6 nucleosomes per turn. In this model, the H1 molecules (*yellow*) run along the center of the solenoid.

conformation, to an intensely negatively charged region on an exposed face of an H2A–H2B dimer from a neighboring nucleosome. Acetylation of H4's N-terminal tail reduces its positive charge and would therefore weaken this interaction. Since histone tails contain numerous acetylation sites, the disruption of higher order chromatin structure is likely to be controlled, at least in part, by the acetylation of certain residues but not others.

Loops of DNA Are Attached to a Scaffold.   Histone-depleted metaphase chromosomes exhibit a central fibrous protein "scaffold" surrounded by an extensive halo of DNA (Fig. 23-51a). The strands of DNA that can be followed form loops that enter and exit the scaffold at nearly the same point (Fig. 23-51b). Most of these loops are 15 to 30 μm long (which corresponds to 45–90 kb), so when condensed as 30-nm filaments, they would be ~0.6 μm long. Electron micrographs of metaphase chromosomes in cross section, such as Fig. 23-52a, strongly suggest that the chromatin fibers of metaphase chromosomes are radially arranged. If the observed loops correspond to these radial fibers, they would each contribute 0.3 μm to the diameter of the chromosome (a fiber must double back on itself to form a loop). Taking into account the 0.4-μm width of the scaffold, this model predicts the diameter of the metaphase chromosome to be 1.0 μm, in agreement with observations (Fig. 23-52b). A typical human chromosome, which contains ~140 million bp, would therefore have ~2000 of these ~70-kb radial loops. The 0.4-μm-diameter scaffold of such a chromosome has sufficient surface area along its 6-μm length to bind this number of radial loops. The radial loop model therefore accounts for DNA's observed packing ratio in metaphase chromosomes.

(a)

(b)

Figure 23-51 Electron micrographs of a histone-depleted metaphase human chromosome.   (a) The fibrous central protein matrix (scaffold) anchors the surrounding DNA. (b) Higher magnification reveals that the DNA is attached to the scaffold in loops. [Courtesy of Ulrich Laemmli, University of Geneva, Switzerland.]

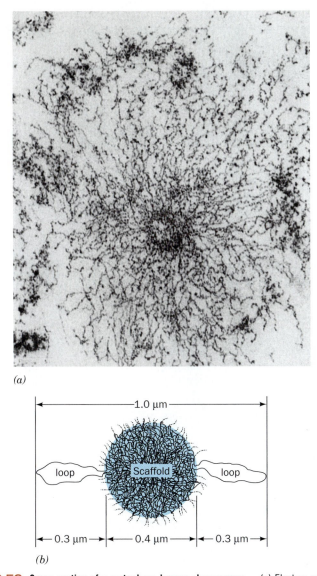

(a)

(b)

Figure 23-52 **Cross section of a metaphase human chromosome.** (a) Electron micrograph. Note the mass of chromatin fibers radially projecting from the central scaffold. [Courtesy of Ulrich Laemmli, University of Geneva, Switzerland.] (b) Interpretive diagram indicating how the 0.3-μm-long radial loops are thought to combine with the 0.4-μm-wide scaffold to form the 1.0-μm-diameter metaphase chromosome.

Almost nothing is known about how the 30-nm filaments organize to form radial loops or about how metaphase chromosomes and the far more dispersed chromosomes of nondividing cells interconvert. Certainly, nonhistone proteins, which constitute ~10% of the chromosomal proteins, must be involved in these processes.

In prokaryotes, DNA is also packaged through its association with highly basic proteins that functionally resemble histones. Nucleosome-like particles condense to form large loops that are attached to a protein scaffold, yielding a relatively compact chromosome. The demands of packaging nucleic acid molecules in small volumes are particularly acute in viruses, which are essentially DNA or RNA molecules surrounded by protein coats (see Box 23-4).

BOX 23-4

Perspectives in Biochemistry

Packaging Viral Nucleic Acids

Viruses are parasites that consist of a nucleic acid molecule encased by a protein **capsid** and, in some cases, an additional lipid bilayer and glycoprotein envelope. Since most viruses contain only a few genes, they depend on the metabolic apparatus of their host cells for replication (and hence viruses are not considered to be alive). The small size of the viral "genome" limits the number of proteins that it can encode. Therefore, its capsid must be built from one or a few kinds of protein subunits that are often arranged in a symmetrical fashion. In the **helical viruses,** the coat protein subunits associate to form helical tubes, whereas in the **icosahedral viruses,** coat proteins aggregate as closed polyhedral shells that have icosahedral symmetry [an icosahedron is a regular polyhedron with 20 faces and 12 vertices (Fig. 6-34c)]. In both cases, the viral nucleic acid occupies the capsid's core.

The most extensively characterized helical virus is **tobacco mosaic virus (TMV),** a rod-shaped particle ~3000 Å long and 180 Å in diameter. The coat consists of ~2130 identical copies of a 158-residue protein. Its assembly begins when 34 of these wedge-shaped proteins spontaneously form a shallow protohelix (the yellow structure in Part a below). The viral genome, an ~6400-nucleotide RNA molecule (*red*) forms a hairpin loop that inserts into the protohelix's central cavity so that three nucleotides bind to each protein subunit (*b*). As additional protohelices associate with the growing viral particle, the RNA is pulled up into the central cavity to form a

coil with a radius of ~40 Å (*c*). The model below, of a portion of TMV, illustrates the helical arrangement of its coat protein subunits and the RNA (exposed at the top of the viral helix).

[Image of TMV assembly courtesy of Hong Wang and Gerald Stubbs, Vanderbilt University. TMV model courtesy of Gerald Stubbs and Keiichi Namba, Vanderbilt University; and Donald Caspar, Brandeis University.]

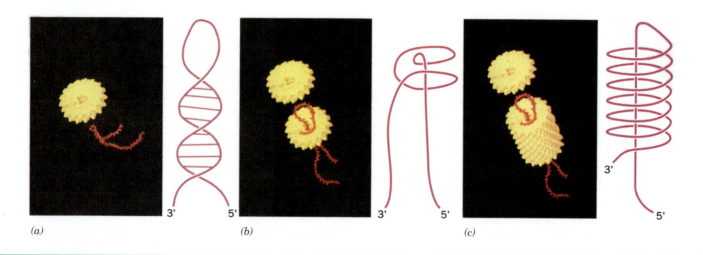

(*a*)   (*b*)   (*c*)

Icosahedral viruses have icosahedral capsids, which require a minimum of 60 protein subunits. A pseudosymmetrical arrangement of protein subunits on each triangular face of the icosahedron permits the capsid to contain more than 60 proteins. For example, the ~340-Å-diameter **tomato bushy stunt virus (TBSV)** has 180 identical coat proteins arranged, as schematically illustrated, around a single ~4800-nucleotide RNA molecule.

Satellite tobacco mosaic virus (STMV), one of the smallest known viruses (it can replicate only in cells that are co-infected with TMV), is an icosahedral virus with a diameter of ~170 Å. It consists of 60 identical protein subunits that enclose an RNA molecule of only 1058 nucleotides. This RNA (*yellow ribbons*), which is extensively base paired, nestles just below the surface of the viral capsid (whose coat proteins are shown in different colors in this cutaway view).

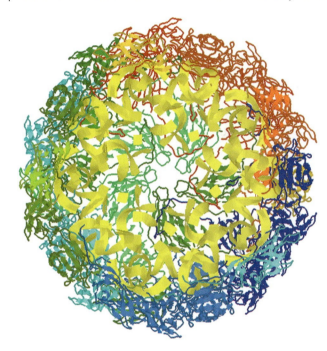

The capsid proteins of TBSV actually form two layers, and the RNA is sandwiched in between them. The volume constraints imposed by this arrangement require that the RNA be tightly packed. The negative charges of the RNA phosphate groups are presumably neutralized by the numerous positively charged Arg and Lys residues of the capsid proteins.

[Drawing of TBSV by Irving Geis/Geis Archives Trust; copyright Howard Hughes Medical Institute; reproduced with permission. X-Ray structure of STMV courtesy of Alexander McPherson, University of California at Irvine. PDBid 1A34.]

SUMMARY

1. The most common form of DNA is B-DNA, which is a right-handed double helix containing A · T and G · C base pairs of similar geometry. The A-DNA helix, which also occurs in double-stranded RNA, is wider and flatter than the B-DNA helix. The left-handed Z-DNA helix may occur in sequences of alternating purines and pyrimidines.

2. The flexibility of nucleotides in nucleic acids is constrained by the allowed rotation angles around the glycosidic bond, the puckering of the ribose ring, and the torsion angles of the sugar–phosphate backbone.

3. Naturally occurring DNA is negatively supercoiled (underwound). Topoisomerases relax supercoils by cleaving one or both strands of the DNA, passing the DNA through the break (types IA and II topoisomerases) or allowing controlled rotation of the uncleaved strand (type IB topoisomerase), and resealing the broken strand(s).

4. Nucleic acids can be denatured by increasing the temperature above their T_m and renatured by lowering the temperature to ~25°C below their T_m.

5. The structures of nucleic acids are stabilized by Watson–Crick base pairing, by hydrophobic interactions between stacked base pairs, and by divalent cations that shield adjacent phosphate groups.

6. RNA molecules assume a variety of structures consisting of double-stranded stems and single-stranded loops. Some RNAs have catalytic activity.

7. Nucleic acids are fractionated by methods similar to those used to fractionate proteins, including solubilization, affinity chromatography, electrophoresis, and ultracentrifugation.

8. Sequence-specific DNA-binding proteins interact primarily with bases in the major groove and with phosphate groups through direct and indirect hydrogen bonds, van der Waals interactions, and ionic interactions. The conformations of both the protein and the DNA may change on binding.

9. Common structural motifs in DNA-binding proteins include the helix–turn–helix (HTH) motif in prokaryotic repressors, and zinc fingers, leucine zippers, and basic helix–loop–helix (bHLH) motifs in eukaryotic transcription factors.

10. The DNA of eukaryotic chromatin winds around histone octamers to form nucleosome core particles that further condense in the presence of linker histones. Additional condensation is accomplished by folding chromatin into 30-nm-diameter filaments, which are then attached in loops to a fibrous protein scaffold to form a condensed (metaphase) chromosome.

 ## REFERENCES

Nucleic Acid Structure

Dickerson, R.E., DNA structures from A to Z, *Methods Enzymol.* **211,** 67–111 (1992).

Doudna, J.A. and Cech, T.R., The chemical repertoire of natural ribozymes, *Nature* **418,** 222–228 (2002). [Describes some of the major ribozymes and how they relate to the RNA world.]

Snustad, D.P. and Simmons, M.J., *Principles of Genetics,* Wiley (2003). [This and other textbooks review DNA structure and function.]

The double helix—50 years, *Nature* **421,** 395–453 (2003). [A supplement containing a series of articles on the historical, cultural, and scientific influences of the DNA double helix on the fiftieth anniversary of its discovery.]

Topoisomerases

Champoux, J.J., DNA topoisomerases: Structure, function, and mechanism, *Annu. Rev. Biochem.* **70,** 369–413 (2001).

Wang, J.C., Cellular roles of DNA topoisomerases: a molecular perspective, *Nature Rev. Mol. Cell Biol.* **3,** 430–440 (2002).

Chromatin

Gruss, C. and Knippers, R., Structure of replicating chromatin, *Prog. Nucl. Acid Res. Mol. Biol.* **52,** 337–365 (1996).

Luger, K., Mäder, A.W., Richmond, R.K., Sargent, D.F., and Richmond, T.J., Crystal structure of the nucleosome core particle at 2.8 Å resolution, *Nature* **389,** 251–260 (1997); *and* Harp, J.M., Hanson, B.L., Timm, D.E., and Bunick, G.J., Asymmetries in the nucleosome core particle at 2.5 Å resolution, *Acta Cryst.* **D56,** 1513–1534 (2000).

van Holde, K. and Zlatanova, J., Chromatin higher order structure: Chasing a mirage? *J. Biol. Chem.* **270,** 8373–8376 (1995). [Presents the arguments that the 30-nm-diameter chromatin filament has an irregular structure.]

Widom, J., Structure, dynamics, and function of chromatin in vitro, *Annu. Rev. Biophys. Biomol. Struct.* **27,** 285–327 (1998).

DNA–Protein Interactions

Aggarwal, A.K., Structure and function of restriction endonucleases, *Curr. Opin. Struct. Biol.* **5,** 11–19 (1995).

Berg, J.M. and Shi, Y., The galvanization of biology: a growing appreciation of the roles of zinc, *Science* **271,** 1081–1085 (1996). [Summarizes different types of zinc fingers and discusses why zinc is suitable for stabilizing small protein domains.]

Ellenberger, T.E., Getting a grip on DNA recognition: structures of the basic region leucine zipper, and the basic region helix-loop-helix DNA-binding domains, *Curr. Opin. Struct. Biol.* **4,** 12–21 (1994).

Luscombe, N.M., Austin, S.E., Berman, H.M., and Thornton, J.M., An overview of the structures of protein-DNA complexes, *Genome Biol.* **1,** reviews 001.1–001.37 (2000). [This review, which contains many molecular models, is available online at http://genomebiology.com/2000/1/1/reviews/001.]

Marmorstein, R., Carey, M., Ptashne, M., and Harrison, S.C., DNA recognition by GAL4: structure of a protein–DNA complex, *Nature* **356,** 408–414 (1992).

Somers, W.S. and Phillips, S.E.V., Crystal structure of the *met* repressor–operator complex at 2.8 Å resolution reveals DNA recognition by β-strands, *Nature* **359,** 387–393 (1992).

 ## KEY TERMS

syn conformation	twist	anneal	pulsed-field gel
anti conformation	writhing number	hybridize	electrophoresis
C3'-*endo*	topoisomerase	Watson–Crick base pair	intercalation agent
C2'-*endo*	hyperchromic effect	Hoogsteen base pair	Southern blotting
supercoiling	melting curve	stacking interactions	repressor
linking number	T_m	ribozyme	HTH motif

operator	heptad repeat	shear degradation	nucleosome
indirect readout	leucine zipper	chromosome	nucleosome core particle
transcription factor	bHLH motif	chromatin	linker DNA
zinc finger	contour length	histone	

STUDY EXERCISES

1. Summarize the differences between A-, B-, and Z-DNA.

2. How do the structures of RNA and DNA differ?

3. How do type IA, type IB, and type II topoisomerases alter DNA topology?

4. Explain the molecular events of nucleic acid denaturation and renaturation.

5. Describe the forces that stabilize nucleic acid structure.

6. What properties allow RNA molecules to act as catalysts?

7. Describe the types of interactions between nucleic acids and proteins.

8. How does the DNA binding of a zinc finger–containing transcription factor differ from that of a prokaryotic repressor?

9. Describe how DNA is packaged in eukaryotic cells.

PROBLEMS

1. Amino acid residues in proteins are each specified by three contiguous bases. What is the contour length of a segment of B-DNA that encodes a 50-kD protein? Calculate the contour length for the same gene if it assumed an A-DNA conformation.

2. Unusual bases and non-Watson–Crick base pairs frequently appear in tRNA molecules. (a) Which base is most likely to pair with hypoxanthine (Section 22-1)? Draw this base pair. (b) Draw the structure of a G · U base pair.

3. The ends of eukaryotic chromosomes terminate in a G-rich single-stranded overhang that can fold up on itself to form a four-stranded structure. In this structure, four guanine residues assume a hydrogen-bonded planar arrangement with an overall geometry that can be represented as

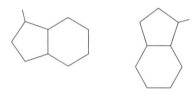

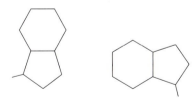

   (a) Draw the complete structure of this "G quartet," including the hydrogen bonds between the purine bases.

   (b) Show schematically how a single strand of four repeating TTAGGG sequences can fold to generate a structure with three stacked G quartets linked by TTA loops.

4. The degree of supercoiling of a circular DNA molecule can be assessed by using an ultracentrifuge to measure its sedimentation velocity relative to the corresponding relaxed circular DNA molecule. (a) Does the supercoiled circle sediment faster or slower than the relaxed circle? (b) Can the ultracentrifuge distinguish between negatively and positively supercoiled molecules? Explain.

5. You have discovered an enzyme secreted by a particularly virulent bacterium that cleaves the C2′—C3′ bond in the deoxyribose residues of duplex DNA. What is the effect of this enzyme on supercoiled DNA?

6. A closed circular duplex DNA has a 100-bp segment of alternating C and G residues. On transfer to a high salt solution, this segment undergoes a transition from the B conformation to the Z conformation. What is the change in its linking number, writhing number, and twist?

7. Compare the melting temperature of a 1-kb segment of DNA containing 20% A residues to that of a 1-kb segment containing 30% A residues under the same conditions.

8. How is the melting curve of duplex DNA affected by (a) decreasing the ionic strength of the solution, and (b) adding a small amount of ethanol?

9. The melting curve for the polyribonucleotide poly(A) is shown below. (a) Explain why absorbance increases with increasing temperature. (b) Why does the shape of the curve differ from the one shown in Fig. 23-21?

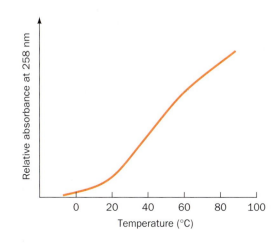

10. For the RNA sequence AUUGGCAUCCGAUAA, draw the secondary structure that maximizes its base pairing.

11. *E. coli* ribosomes contain three RNA molecules named for their sedimentation behavior: 5S, 18S, and 28S. Draw a diagram of a centrifuge tube showing the approximate positions of the three RNA species following ultracentrifugation in a sucrose density gradient.

12. What is the probability that the palindromic symmetry of the *trp* repressor target DNA sequence (Section 23-4B) is merely accidental?

13. For a linear B-DNA molecule of 50,000 kb, calculate (a) the contour length, (b) the length of the DNA as packaged in nucleosomes with linker histones present, and (c) the length of the DNA in a 30-nm-diameter filament.

14. Mouse genomic DNA is treated with a restriction endonuclease and electrophoresed in an agarose gel. A radioactive probe made from the human gene *rxr-1* is used to perform a Southern blot. The experiment was repeated three times. Explain the results of these repeated experiments:

Experiment 1. The autoradiogram shows a large smudge at a position corresponding to the top of the lane in which the mouse DNA was electrophoresed.

Experiment 2. The autoradiogram shows a smudge over the entire lane containing the mouse DNA.

Experiment 3. The autoradiogram shows three bands of varying intensity.

[Problem provided by Bruce Wightman, Muhlenberg College.]

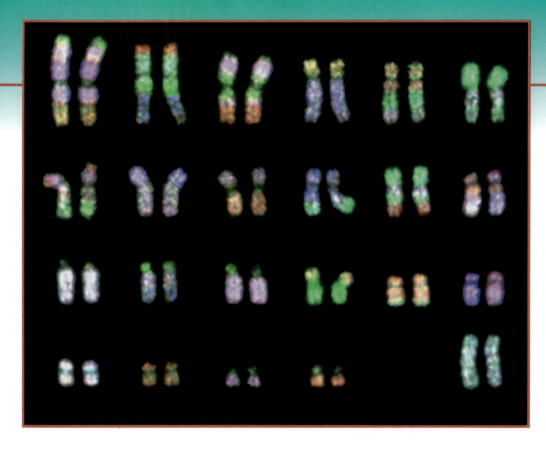

Each time a human cell divides, the DNA in each of its 46 chromosomes must be replicated rapidly and accurately. Errors that do occur may be repaired to ensure the faithful transmission of genetic information.
[L. Willatt/Photo Researchers.]

DNA Replication, Repair, and Recombination

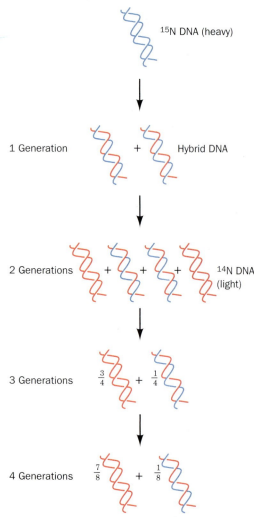

15N DNA (heavy)

1 Generation   +   Hybrid DNA

2 Generations   +   +   +   14N DNA (light)

3 Generations $\frac{3}{4}$ + $\frac{1}{4}$

4 Generations $\frac{7}{8}$ + $\frac{1}{8}$

Figure 24-1 The Meselson and Stahl experiment.
Parental (^{15}N-labeled or "heavy") DNA is replicated
semiconservatively so that in the first generation, DNA
molecules contain one parental (*blue*) strand and one newly
synthesized (*red*) strand. In succeeding generations, the
proportion of ^{14}N-labeled ("light") strands increases, but
hybrid molecules containing one heavy and one light strand
persist. ♻ See the Animated Figures.

Watson and Crick's seminal paper describing the DNA double helix ended
with the statement: "It has not escaped our notice that the specific pairing
we have postulated immediately suggests a possible copying mechanism
for the genetic material." As they predicted, when DNA replicates, each
polynucleotide strand acts as a template for the formation of a complementary strand through base pairing interactions. The two strands of the
parent molecule must therefore separate so that a complementary daughter strand can be enzymatically synthesized on the surface of each parent
strand. This results in two molecules of duplex DNA, each consisting of one
polynucleotide strand from the parent molecule and a newly synthesized
complementary strand (Fig. 3-11). Such a mode of replication is termed
semiconservative (in **conservative** replication, the parental DNA would
remain intact and both strands of the daughter duplex would be newly
synthesized).

The semiconservative nature of DNA replication was elegantly demonstrated in 1958 by Matthew Meselson and Franklin Stahl. The density of
DNA was increased by labeling it with ^{15}N, a heavy isotope of nitrogen (^{14}N
is the naturally abundant isotope). This was accomplished by growing *E. coli*
in a medium that contained ^{15}NH$_4$Cl as its only nitrogen source. The labeled bacteria were then abruptly transferred to a ^{14}N-containing medium
and the density of their DNA was monitored over several generations by
equilibrium density gradient ultracentrifugation (Section 5-2E).

Meselson and Stahl found that after one generation (one doubling of
the cell population), all the DNA had a density exactly halfway between
the densities of fully ^{15}N-labeled DNA and unlabeled DNA. This DNA
must therefore contain equal amounts of ^{14}N and ^{15}N as is expected after
one generation of semiconservative replication (Fig. 24-1). Conservative
DNA replication, in contrast, would preserve the fully ^{15}N-labeled parental
strand and generate an equal amount of unlabeled DNA. After two generations, half of the DNA molecules were unlabeled and the remainder
were ^{14}N–^{15}N hybrids. In succeeding generations, the amount of unlabeled
DNA increased relative to the amount of hybrid DNA although the hybrid
never totally disappeared. This is in accord with semiconservative replication but at odds with conservative replication, in which hybrid DNA never
forms.

The details of DNA replication, including the unwinding of the parental
strands and the assembly of complementary strands from nucleoside
triphosphates, have emerged gradually since 1958. DNA replication is far
more complex than the overall chemistry of this process might suggest, in
large part because replication must be extremely accurate in order to preserve the integrity of the genome from generation to generation. In this
chapter, we examine the protein assemblies that mediate DNA replication
in prokaryotes and eukaryotes. We also discuss the mechanisms for ensuring fidelity during replication and for correcting polymerization errors and
other types of DNA damage.

1 Overview of DNA Replication

DNA is replicated by enzymes known as **DNA-directed DNA polymerases** or simply **DNA polymerases.** These enzymes use single-stranded
DNA **(ssDNA)** as templates on which to catalyze the synthesis of the
complementary strand from the appropriate deoxynucleoside triphosphates
(Fig. 24-2). The reaction occurs through the nucleophilic attack of the
growing DNA chain's 3'-OH group on the α-phosphoryl of an incoming

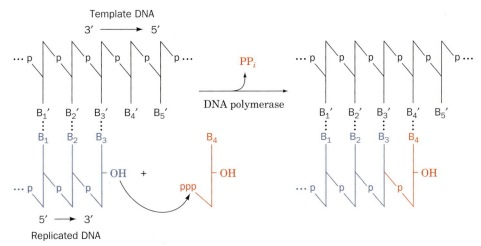

Template DNA

Replicated DNA

Figure 24-2   Action of DNA polymerases.   These enzymes assemble incoming deoxynucleoside triphosphates on single-stranded DNA templates. The OH group at the 3' end of a growing polynucleotide strand is a nucleophile that attacks the α-phosphate group of an incoming dNTP that base pairs with the template DNA strand, and hence the growing strand is elongated in its 5' → 3' direction. Although formation of a new phosphodiester bond is reversible, the subsequent hydrolysis of the PP$_i$ product makes the overall reaction irreversible.

nucleoside triphosphate. The otherwise reversible reaction is driven by the subsequent hydrolysis of the eliminated PP$_i$. The incoming nucleotides are selected by their ability to form Watson–Crick base pairs with the template DNA so that the newly synthesized DNA strand forms a double helix with the template strand. Nearly all known DNA polymerases can add a nucleotide only to the free 3'-OH group of a base-paired polynucleotide so that *DNA chains are extended only in the 5' → 3' direction.*

DNA Replication Occurs at Replication Forks.   John Cairns observed DNA replication through the autoradiography of chromosomes from *E. coli* grown in a medium containing [³H]thymidine. These circular chromosomes contain replication "eyes" or "bubbles" (Fig. 24-3), called **θ structures** (after their resemblance to the Greek letter theta), that form when the two parental strands of DNA separate to allow the synthesis of their complementary daughter strands. DNA replication involving θ structures is known as **θ replication.**

A branch point in a replication eye at which DNA synthesis occurs is called a **replication fork.** A replication bubble may contain one or two replication forks **(unidirectional** or **bidirectional replication).** Autoradiographic studies have demonstrated that θ replication is almost always bidirectional

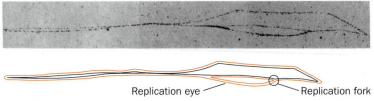

Replication eye   Replication fork

Figure 24-3   An autoradiogram and its interpretive drawing of a replicating *E. coli* chromosome.   The bacterium had been grown in a medium containing [³H]thymidine, thereby labeling the subsequently synthesized DNA so that it appears as a line of dark grains in the photographic emulsion (*red lines in the drawing*). The size of the replication eye indicates that this circular chromosome is about one-eighth duplicated. [Courtesy of John Cairns, Cold Spring Harbor Laboratory.]

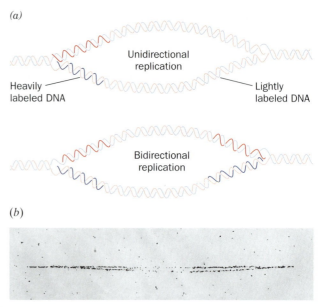

(a)

Unidirectional replication

Heavily labeled DNA

Lightly labeled DNA

Bidirectional replication

(b)

Figure 24-4 **The autoradiographic differentiation of unidirectional and bidirectional θ replication of DNA.** (a) An organism is grown for several generations in a medium that is lightly labeled with [³H]thymidine so that all of its DNA will be visible in an autoradiogram. A large amount of [³H]thymidine is then added to the medium for a few seconds before the DNA is isolated (**pulse labeling**) in order to heavily label bases near the replication fork(s). Unidirectional DNA replication will exhibit only one heavily labeled branch point, whereas bidirectional DNA replication will exhibit two such branch points. (b) An autoradiogram of *E. coli* DNA demonstrating that it is bidirectionally replicated. [Courtesy of David M. Prescott, University of Colorado.]

(Fig. 24-4). In other words, *DNA synthesis proceeds in both directions from the point where replication is initiated.*

Replication Is Semidiscontinuous. The low-resolution images provided by autoradiograms such as Figs. 24-3 and 24-4*b* suggest that duplex DNA's two antiparallel strands are simultaneously replicated at an advancing replication fork. Yet since DNA polymerases extend DNA strands only in the $5' \rightarrow 3'$ direction, how can they copy the parent strand that extends in the $5' \rightarrow 3'$ direction past the replication fork? This question was answered in 1968 by Reiji Okazaki through the following experiment. If a growing *E. coli* culture is pulse labeled for 30 s with [³H]thymidine, much of the radioactive and hence newly synthesized DNA consists of 1000- to 2000-nucleotide (**nt**) fragments (in eukaryotes, these so-called **Okazaki fragments** are 100–200 nt long). When the cells are transferred to an unlabeled medium after the [³H]thymidine pulse, the size of the labeled fragments increases over time. *The Okazaki fragments must therefore become covalently incorporated into larger DNA molecules.*

Okazaki interpreted his experimental results in terms of the **semidiscontinuous replication** model (Fig. 24-5). The two parent strands are replicated in different ways. The newly synthesized DNA strand that extends $5' \rightarrow 3'$ in the direction of replication fork movement, the **leading strand,** is continuously synthesized in its $5' \rightarrow 3'$ direction as the replication fork advances. The other new strand, the **lagging strand,** is also synthesized in its $5' \rightarrow 3'$ direction. However, it can only be made discontinuously, as Okazaki fragments, as single-stranded parental

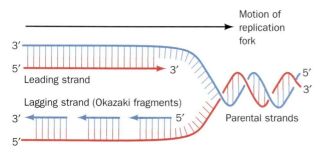

Motion of replication fork

3′
5′
Leading strand
3′
Lagging strand (Okazaki fragments)
3′   5′
5′
5′
3′
Parental strands

Figure 24-5 **Semidiscontinuous DNA replication.** Both daughter strands (*leading strand red, lagging strand blue*) are synthesized in their $5' \rightarrow 3'$ direction. The leading strand is synthesized continuously, whereas the lagging strand is synthesized discontinuously. The lagging strand segments are known as Okazaki fragments.

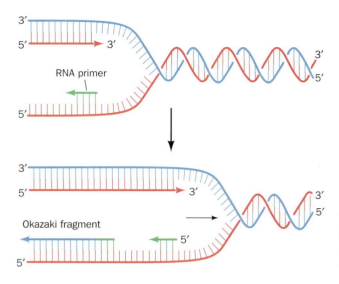

Figure 24-6   Priming of DNA synthesis by short RNA segments. Each Okazaki fragment consists of an RNA primer (*green*) that has been extended by DNA polymerase. The RNA primer is later removed.

DNA becomes newly exposed at the replication fork. The Okazaki fragments are later covalently joined together by the enzyme **DNA ligase.**

DNA Synthesis Extends RNA Primers.   Given that DNA polymerases require a free 3′-OH group to extend a DNA chain, how is DNA synthesis initiated? Careful analysis of Okazaki fragments revealed that their 5′ ends consist of RNA segments of 1 to 60 nt (the length is species dependent) that are complementary to the template DNA chain (Fig. 24-6). In *E. coli*, these **RNA primers** are synthesized by the enzyme **primase.** Multiple primers are required for lagging strand synthesis, but only one primer is required to initiate synthesis of the leading strand. Mature DNA, however, does not contain RNA. *The RNA primers are eventually replaced with DNA.*

2 Prokaryotic DNA Replication

DNA replication involves a great variety of enzymes in addition to those mentioned above. Many of the required enzymes were first isolated from prokaryotes and are therefore better understood than their eukaryotic counterparts. Accordingly, we begin with a detailed consideration of prokaryotic DNA replication. Replication in eukaryotes is discussed in Section 24-3.

A DNA Polymerases

In 1957, Arthur Kornberg discovered an *E. coli* enzyme that catalyzes the synthesis of DNA, based on its ability to incorporate the radioactive label from [^{14}C]thymidine triphosphate into DNA (Box 24-1). This enzyme, which is now known as **DNA polymerase I** or **Pol I,** consists of a single 928-residue polypeptide. Pol I is said to be a **processive** enzyme because *it catalyzes a series of successive nucleotide polymerization steps, typically 20 or more, without releasing the single-stranded template.*

Pol I Has Exonuclease Activity.   In addition to its polymerase activity, Pol I has two independent hydrolytic activities that occupy separate active sites: a 3′ → 5′ exonuclease and a 5′ → 3′ exonuclease. *The 3′ → 5′ exonuclease*

BOX 24-1

Pathways of Discovery

Arthur Kornberg and DNA Polymerase I

Like a number of his contemporaries during the Great Depression, Arthur Kornberg entered medical school because of the lack of jobs in teaching or industry. He subsequently served in the U.S. Coast Guard as a ship's doctor but quickly realized that he was better suited to the laboratory than the sea. When he began his work at the National Institutes of Health in 1942, the classic studies of nutrition and vitamins were giving way to the new science of enzymology. After purifying aconitase from rat hearts, Kornberg stated that he found enzymes intoxicating. He likened purifying an enzyme to ascending an uncharted mountain, with the reward being the view from the top and the satisfaction of being the first one there.

Kornberg spent 6 months at the Cori laboratory (see Box 15-1) and in 1948 began working on enzymes involved in the synthesis of nucleotide cofactors. He found that the enzyme nucleotide pyrophosphatase catalyzed a reaction in which a nucleotide was incorporated into a coenzyme:

$$\text{Nicotinate ribonucleotide} + \text{ATP} \rightarrow \text{NAD}^+ + \text{PP}_i$$

Next, Kornberg began to search for an enzyme that could assemble many nucleotides to make a nucleic acid chain. Some people have mistakenly assumed that Kornberg was inspired to look for DNA polymerase by the publication of the Watson–Crick model of DNA in 1953. In fact, Kornberg was following his instincts as an enzymologist, and his curiosity stemmed from his familiarity with other enzymes, including glycogen phosphorylase, which was capable of synthesizing the polymer glycogen *in vitro*. Kornberg had also experimented with enzymes involved in the synthesis of phospholipids, which—like DNA—contain phosphodiester bonds (he abandoned this line of investigation because he didn't like working with "greasy" molecules).

To purify a DNA-synthesizing enzyme, Kornberg started with extracts of fast-growing *E. coli* cells, which had replaced slower growing yeast cells as a laboratory subject. He added radioactive thymidine to the extracts and measured the production of radioactive DNA. Disappointingly, the incorporation of thymidine was extremely low, but thymidine phosphate (TMP) worked better, and thymidine triphosphate (TTP) better still. Kornberg also discovered that the amount of newly synthesized DNA increased when a small amount of DNA was

Arthur Kornberg (1918–)

included in the reaction mixture. This was not entirely unexpected, as Kornberg already knew that a small amount of glycogen could serve as a primer for additional glycogen synthesis. However, at first, Kornberg believed that the DNA added to his reaction mixture acted as a substrate for nucleases that were present in cell extracts and thereby protected the newly synthesized radioactive DNA from degradation. Later, he realized that the added DNA functioned as a template for the synthesis of a new strand and through its partial degradation also supplied the other, nonradioactive nucleotides required for polymerization. The idea of a template was, at the time, foreign to most enzymologists and other biochemists, but biologists seemed more receptive to the role of a template in DNA synthesis. For his discovery and characterization of DNA polymerase (later known as DNA polymerase I or Pol I), Kornberg was awarded the 1959 Nobel Prize in Physiology or Medicine.

Even with purified Pol I in hand, Kornberg was faced with the need to prove that the reaction product was biologically active. He therefore used Pol I to synthesize the 5386-bp DNA of bacteriophage ϕX174 on viral DNA templates and then DNA ligase to close up the synthetic DNA molecule to yield a circular DNA that was infectious. To Kornberg's chagrin, the popular press misunderstood this work, hailing it as the creation of life in a test tube.

Kornberg later confessed that he was amazed by the virtuosity of the DNA polymerase he had isolated from *E. coli*: The enzyme could synthesize a chain of thousands of nucleotides with an accuracy that exceeded chemical predictions. However, during the 1970s, genetic studies and other evidence clearly indicated that other proteins were responsible for DNA replication in *E. coli,* and DNA polymerases II and III were soon discovered and characterized. For the next two decades, Kornberg led the effort to determine the mechanism of DNA replication. Indeed, many of the leading researchers in this field were trained in his laboratory.

Kornberg, A., Active center of DNA polymerase, *Science* **163**, 1410–1418 (1969).

Kornberg, A., *For Love of Enzymes: The Odyssey of a Biochemist,* Harvard University Press (1989). [A scientific autobiography.]

activity allows Pol I to edit its mistakes. If Pol I erroneously incorporates a mispaired nucleotide at the end of a growing DNA chain, the polymerase activity is inhibited and the $3' \rightarrow 5'$ exonuclease hydrolytically excises the

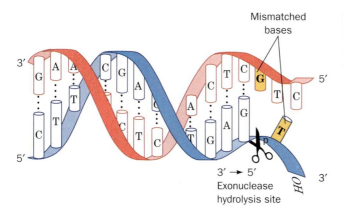

Figure 24-7 The 3′ → 5′ exonuclease function of DNA polymerase I. This enzymatic activity excises mispaired nucleotides from the 3′ end of the growing DNA strand (*blue*).

offending nucleotide (Fig. 24-7). The polymerase activity then resumes DNA replication. This **proofreading** mechanism explains the high fidelity of DNA replication by Pol I.

The Pol I 5′ → 3′ exonuclease binds to duplex DNA at single-strand nicks (breaks). It cleaves the nicked DNA strand in a base-paired region beyond the nick to excise the DNA as either mononucleotides or oligonucleotides of up to 10 residues (Fig. 24-8).

Although Pol I was the first of the *E. coli* DNA polymerases to be discovered, it is not *E. coli*'s primary replicase. Rather, its most important (and only essential) function is in lagging strand synthesis, in which it removes the RNA primers and replaces them with DNA. This process involves the 5′ → 3′ exonuclease and polymerase activities of Pol I working in concert to excise the ribonucleotides on the 5′ end of the single-strand nick between the new and old (previously synthesized) Okazaki fragments and to replace them with deoxynucleotides that are appended to the 3′ end of the old Okazaki fragment (Fig. 24-9). The nick is thereby translated (moved) toward the DNA strand's 3′ end, a process known as **nick translation.** When the RNA has been entirely excised, the nick is sealed by the action of DNA ligase (Section 24-2C), thereby linking the new and old Okazaki fragments.

Biochemists use nick translation to prepare radioactive DNA. Double-stranded DNA **(dsDNA)** is nicked in only a few places by treating it with small amounts of pancreatic **DNase I.** Radioactively labeled dNTPs are then added and Pol I translates the nicks, thereby replacing unlabeled deoxynucleotides with labeled deoxynucleotides.

Pol I also functions in the repair of damaged DNA. As we discuss in Section 24-5, damaged DNA is detected by a variety of DNA repair systems, many of which endonucleolytically cleave the damaged DNA on the 5′ side of the lesion. Pol I's 5′ → 3′ exonuclease activity then excises the damaged DNA while its polymerase activity fills in the resulting single-strand gap in the same way it replaces the RNA primers of Okazaki fragments. Thus, Pol I has indispensable roles in *E. coli* DNA replication and repair although it is not, as was first supposed, responsible for the bulk of DNA synthesis.

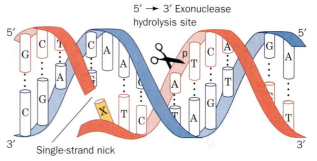

Figure 24-8 The 5′ → 3′ exonuclease function of DNA polymerase I. This enzymatic activity excises up to 10 nucleotides from the 5′ end of a single-strand nick. The nucleotide immediately past the nick (X) may or may not be paired.

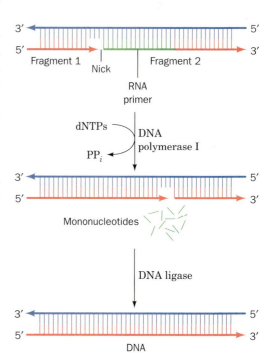

Figure 24-9 The replacement of RNA primers by DNA in lagging strand synthesis. The RNA primer at the 5′ end of a newly synthesized Okazaki fragment (*Fragment 2*) is excised through the action of Pol I's 5′ → 3′ exonuclease function and replaced through its polymerase function, which adds deoxynucleotides to the 3′ end of the previously synthesized Okazaki fragment (*Fragment 1*). This translates the nick originally at the 5′ end of the RNA to the position that was occupied by its 3′ end (nick translation). The nick is sealed by the action of DNA ligase.

Figure 24-10 X-Ray structure of *E. coli* DNA polymerase I Klenow fragment in complex with a double-helical DNA. The solvent-accessible surface of the Klenow fragment is drawn in yellow with a 12-nt DNA template strand in cyan and a 14-nt primer strand in red. [Courtesy of Thomas Steitz, Yale University. PDBid 1KLN.] ✌ **See the Interactive Exercises.**

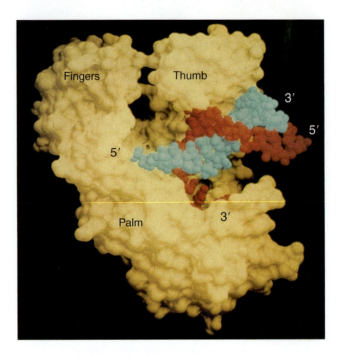

Structure of Klenow Fragment. *E. coli* DNA polymerase I, a protein whose three enzymatic activities occupy three separate active sites, can be proteolytically cleaved to a large or **"Klenow" fragment** (residues 324–928), which contains both the polymerase and the $3' \rightarrow 5'$ exonuclease activities, and a small fragment (residues 1–323), which contains the $5' \rightarrow 3'$ exonuclease activity. The X-ray structure of the Klenow fragment, determined by Thomas Steitz, reveals that it consists of two domains (Fig. 24-10). The smaller domain (residues 324–517; the lower portion of the structure shown in Fig. 24-10) contains the $3' \rightarrow 5'$ exonuclease site. The larger domain (residues 521–928) contains the polymerase active site at the bottom of a prominent cleft, a surprisingly large distance ($\sim$25 Å) from the $3' \rightarrow 5'$ exonuclease site. The cleft, which is lined with positively charged residues, has the appropriate size and shape ($\sim$22 Å $\times$ $\sim$30 Å) to bind a B-DNA molecule in a manner resembling a right hand grasping a rod (note the thumb, fingers, and palm structures in Fig. 24-10). The active sites of nearly all DNA and RNA polymerases of known structure are located at the bottoms of similarly shaped clefts.

Steitz cocrystallized the Klenow fragment with a short DNA "template" strand and a complementary "primer" strand. The protein contacts only the phosphate backbone of the duplex DNA, consistent with Pol I's lack of sequence specificity in binding DNA. The separation of the polymerase and $3' \rightarrow 5'$ exonuclease active sites suggests that the bound DNA undergoes a large conformational shift in shuttling between these sites.

DNA Polymerase Senses Watson–Crick Base Pairs via Sequence-Independent Interactions. The C-terminal domain of the thermostable *Thermus aquaticus (Taq)* DNA polymerase I **(Klentaq1)** is 50% identical in sequence and closely similar in structure to the large domain of Klenow fragment (Klentaq1 lacks a functional $3' \rightarrow 5'$ exonuclease site). Gabriel Waksman crystallized Klentaq1 in complex with an 11-bp DNA that has a GGAAA-5' overhang at the 5' end of its template strand and the crystals were incubated with 2',3'-dideoxy-CTP (ddCTP; which lacks a 3'-OH group). The

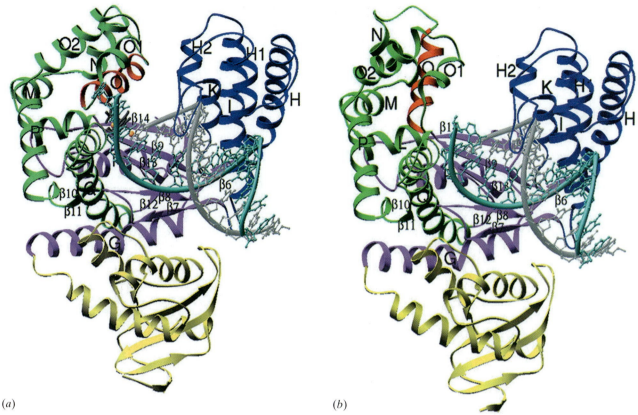

(a)

(b)

Figure 24-11 X-Ray structure of Klentaq1 in complex with DNA with and without ddCTP. (*a*) The closed conformation (with ddCTP). (*b*) The open conformation (ddCTP depleted). The protein, which is viewed similarly to that in Fig. 24-10, is represented in ribbon form with its N-terminal, palm, fingers, and thumb domains colored yellow, magenta, green, and dark blue, respectively, and with the O helix in the fingers domain red. The DNA is shown in stick form, and its sugar–phosphate backbone is also represented in tube form with the template strand cyan and the primer strand silver. In Part *a*, the bound ddCTP, which is base paired with a template G residue at the active site, is shown in stick form in black, and its two bound metal ions are represented by orange spheres. [Courtesy of Gabriel Waksman, Washington University School of Medicine. PDBids 3KTQ and 2KTQ.]

X-ray structure of these crystals (Fig. 24-11*a*) revealed that a ddC residue had been covalently linked to the 3′ end of the primer where it formed a Watson–Crick pair with the template overhang's 3′ G. Moreover, a ddCTP molecule (to which the primer's new 3′-terminal ddC residue is incapable of forming a covalent bond) occupied the enzyme's active site where it formed a Watson–Crick pair with the template's next G. Clearly, Klentaq1 retained its catalytic activity in this crystal.

A DNA polymerase must distinguish correctly paired bases from mismatches and yet do so via sequence-independent interactions with the incoming dNTP. The foregoing X-ray structure reveals that this occurs through an active site pocket that is complementary in shape to Watson–Crick base pairs. This pocket is formed by the stacking of a conserved Tyr side chain on the template base, as well as by van der Waals interactions with the protein and with the preceding base pair. In addition, although the dsDNA is mainly in the B conformation, the 3 base pairs nearest the active site assume the A conformation, as has also been observed in the X-ray structures of several other DNA polymerases in their complexes with DNA. The resulting wider and shallower minor groove (Section 23-1A) permits protein side chains to form hydrogen bonds with the otherwise inaccessible N3 atoms of the purine bases and O2 atoms of the pyrimidine bases. The positions of these hydrogen bond acceptors are sequence-independent,

as can be seen from an inspection of Fig. 23-1 (in contrast, the positions of the hydrogen bonding acceptors in the major groove vary with the sequence). However, with a non-Watson–Crick pairing—that is, with a mismatched dNTP in the active site—these hydrogen bonds would be greatly distorted if not completely disrupted. The polymerase also makes extensive sequence-independent hydrogen bonding and van der Waals interactions with the DNA's sugar–phosphate backbone.

The above Klentaq1 · DNA · ddCTP crystals were partially depleted of ddCTP by soaking them in a stabilizing solution that lacks ddCTP. The X-ray structure of the ddCTP-depleted crystals (Fig. 24-11b) revealed that Klentaq1's fingers domain assumed a so-called open conformation, which differs significantly from that in the so-called closed conformation described above. Evidently, the O, O1, and O2 helices in the open conformation move via a hingelike motion in the direction of the active site so as to bury the bound ddCTP, thereby assembling the productive ternary closed complex (Fig. 24-11a). These observations are consistent with kinetic measurements on Pol I indicating that the binding of the correct dNTP to the enzyme induces a rate-limiting conformational change that yields a tight ternary complex. It therefore appears that the enzyme rapidly samples the available dNTPs in its open conformation but only when it binds the correct dNTP in a Watson–Crick pairing with the template base does it form the catalytically competent closed conformation. The subsequent reaction steps then rapidly yield the product complex which, following a second conformational change, releases the product PP_i. Finally, the DNA is translocated in the active site so as to position it for the next reaction cycle.

The DNA Polymerase Catalytic Mechanism Involves Two Metal Ions. The X-ray structures of a variety of DNA polymerases suggest that they share a common catalytic mechanism for nucleotidyl transfer (Fig. 24-12). Their active sites all contain two metal ions, usually Mg^{2+}, that are liganded by two invariant Asp side chains in the palm domain. Metal ion B in Fig. 24-12 is liganded by all three phosphate groups of the bound dNTP, whereas metal ion A bridges the α-phosphate group of this dNTP and the primer's 3'-OH group. Metal ion A presumably activates the primer's 3'-OH group for a nucleophilic attack on the α-phosphate group, whereas metal ion B functions to orient its bound triphosphate group and to electrostatically shield its negative charges as well as the additional negative charge on the transition state leading to the release of the PP_i ion.

DNA Polymerase III Is *E. coli's* DNA Replicase. The discovery of normally growing *E. coli* mutants that have very little (but not entirely absent) Pol I activity stimulated the search for additional DNA polymerizing activities. This effort was rewarded by the discovery of two more enzymes, designated, in the order they were discovered, **DNA polymerase II (Pol II)** and **DNA**

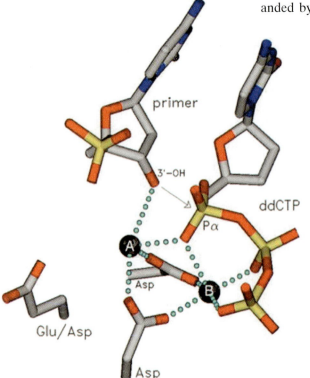

Figure 24-12 Diagram for the nucleotidyl transferase mechanism of DNA polymerases. A and B represent enzyme-bound metal ions that usually are Mg^{2+}. Atoms are colored according to type (C gray, N blue, O red, and P yellow), and metal ion coordination is represented by green dotted lines. Metal ion A activates the primer's 3'-OH group for nucleophilic attack on the incoming dNTP's α-phosphate group (*arrow*), whereas metal ion B acts to orient and electrostatically stabilize the negatively charged triphosphate group. [Courtesy of Tom Ellenberger, Harvard Medical School.]

Table 24-1 Properties of *E. coli* DNA Polymerases

	Pol I	Pol II	Pol III
Mass (kD)	103	90	130
Molecules/cell	400	?	10–20
Turnover number[a]	600	30	9000
Structural gene	*polA*	*polB*	*polC*
Conditionally lethal mutant	+	−	+
Polymerization: 5′ → 3′	+	+	+
Exonuclease: 3′ → 5′	+	+	+
Exonuclease: 5′ → 3′	+	−	−

[a]Nucleotides polymerized $\text{min}^{-1} \cdot \text{molecule}^{-1}$ at 37°C.

Source: Kornberg, A. and Baker, T.A., *DNA Replication* (2nd ed.), p. 167, Freeman (1992).

polymerase III (Pol III). The properties of these enzymes are compared with those of Pol I in Table 24-1. Pol II and Pol III had not previously been detected because their combined activities in the assays used are normally <5% that of Pol I. Pol II participates in DNA repair; mutant cells lacking Pol II can therefore grow normally. The absence of Pol III, however, is lethal, demonstrating that it is *E. coli's* DNA replicase.

The catalytic core of Pol III consists of three subunits: α, which contains the complex's DNA polymerase activity; ε, its 3′ → 5′ exonuclease; and θ. However, at least seven other subunits (τ, γ, δ, δ′, χ, ψ, and β) combine with them to form a labile multisubunit enzyme known as the **Pol III holoenzyme.** The catalytic properties of the Pol III core resemble those of Pol I except that Pol III lacks 5′ → 3′ exonuclease activity on dsDNA. Thus, *Pol III can synthesize a DNA strand complementary to a single-stranded template and can edit the polymerization reaction to increase replication fidelity, but it cannot catalyze nick translation.*

B Initiation of Replication

The *E. coli* chromosome is a supercoiled DNA molecule of 4.6×10^6 bp. Since DNA polymerase requires a single-stranded template, other proteins participate in DNA replication by locating the replication initiation site, unwinding the DNA, and preventing the single strands from reannealing. Replication in *E. coli* begins at a 245-bp region known as *oriC.* Elements of this sequence are highly conserved among gram-negative bacteria. Multiple copies of a 52-kD protein known as **DnaA** bind to *oriC* and cause ~45 bp of an AT-rich segment of the DNA to separate into single strands. This melting requires the free energy of ATP hydrolysis and is probably also facilitated by both the AT-rich nature of the DNA segment and the negative supercoiling (underwinding) of the circular DNA chromosome [the latter being generated by DNA gyrase, a type II topoisomerase (Section 23-1C) whose activity is required for prokaryotic DNA replication].

Helicases Unwind DNA. DnaA bound to *oriC* recruits two hexameric complexes of **DnaB,** one to each end of the melted region. DnaB is a **helicase** that further separates the DNA strands. Helicases are a diverse group of enzymes that unwind DNA during replication, transcription, and a variety of other processes. DnaB is one of 12 helicases expressed by *E. coli.* Helicases function by translocating along one strand of a double-helical nucleic acid so as to mechanically unwind the helix in their path, a process that is driven by the free energy of NTP hydrolysis.

See Guided Exploration 25

The replication of DNA in *E. coli*

Figure 24-13 X-Ray structure of T7 helicase. Each subunit of the cyclic hexamer is drawn in a different color. Four bound ATP analogs are shown in ball-and-stick form. Note that the conformations of adjacent subunits are not identical. This structure is part of a larger protein complex that also contains primase activity. [Courtesy of Dale Wigley, Cancer Research U.K., London Research Institute. PDBid 1E0J.]

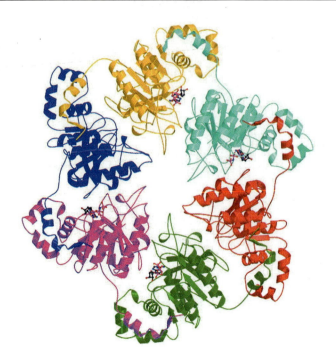

E. coli DnaB, a hexamer of identical 471-residue subunits, separates the two strands of the parental DNA by translocating along the lagging strand template in the $5' \rightarrow 3'$ direction, while hydrolyzing ATP (it can also use GTP and CTP but not UTP). Some helicases move in the $3' \rightarrow 5'$ direction, and some are dimers rather than hexamers. The X-ray structure of a hexameric helicase from bacteriophage T7, which infects *E. coli,* reveals that the protein forms a ring with a central channel large enough to accommodate a DNA strand (Fig. 24-13). Adjacent helicase subunits have different conformations, as is also the case in the F_1F_0-ATPase (Section 17-3B). This suggests that when a helicase subunit binds and hydrolyzes an NTP, it undergoes a conformational change that alters its interaction with the ssDNA in the center of the ring. The cumulative effect of each subunit sequentially binding NTP, hydrolyzing it, and releasing the products is that *the protein pulls itself along a DNA strand, mechanically separating the dsDNA ahead of it.*

Single-Strand Binding Protein Prevents DNA from Reannealing. The separated DNA strands behind an advancing helicase do not reanneal to form dsDNA because they become coated with **single-strand binding protein (SSB).** The SSB coat also prevents ssDNA from forming secondary structures (such as stem-loops) and protects it from nucleases. Note that the DNA must be stripped of SSB before it can be replicated by DNA polymerase.

E. coli SSB is a tetramer of 177-residue subunits that can bind to DNA in several different ways. In the major binding mode (Fig. 24-14), only two of the four subunits interact with ssDNA, with the two ends of the bound DNA emerging from opposite ends of the tetramer. This would permit an unlimited series of SSB tetramers to interact end-to-end along the length of a ssDNA. The DNA-binding cleft of SSB, which is contained in its N-terminal 115 residues, is positively charged so that the protein can interact electrostatically with DNA phosphate groups. The cleft is too narrow to accommodate dsDNA.

The Primosome Synthesizes RNA Primers. All DNA synthesis, both of leading and lagging strands, requires the prior synthesis of an RNA primer. Primer synthesis in *E. coli* is mediated by an ~600-kD protein assembly

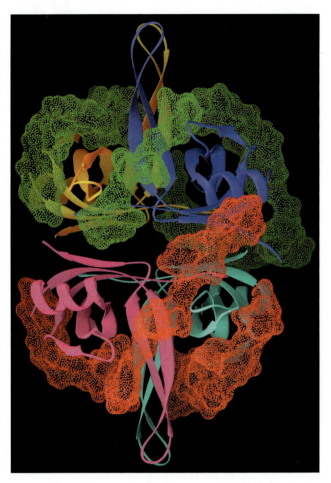

Figure 24-14 X-Ray structure of SSB in complex with dC(pC)$_{34}$.   The homotetramer, which has D_2 symmetry, is viewed along one of its twofold axes with its other twofold axes horizontal and vertical. Each of its subunits (which include the N-terminal 134 residues of the 177-residue polypeptide) are differently colored. Its two bound ssDNA molecules are represented by green and red dot surfaces (the red strand is partially disordered and not fully visible). [Based on an X-ray structure by Timothy Lohman and Gabriel Waksman, Washington University School of Medicine. PDBid 1EYG.]

known as a **primosome,** which includes the DnaB helicase and an RNA-synthesizing **primase** called **DnaG,** as well as five other types of subunits. *E. coli* DnaG is a monomeric protein whose catalytic domain does not resemble the typical hand-shaped DNA and RNA polymerases. Nevertheless, it catalyzes the same polymerization reaction (Fig. 24-2 using NTPs rather than dNTPs) to produce an RNA segment of ~11 nucleotides.

The primosome is propelled in the $5' \rightarrow 3'$ direction along the DNA template for the lagging strand (i.e., toward the replication fork) in part by DnaB-catalyzed ATP hydrolysis. This motion, which displaces the SSB in its path, is opposite in direction to that of template reading during DNA chain synthesis. Consequently, the primosome reverses its migration momentarily to allow primase to synthesize an RNA primer in the $5' \rightarrow 3'$ direction (Fig. 24-6).

The primosome is required to initiate each Okazaki fragment. The single RNA segment that primes the synthesis of the leading strand can be synthesized, at least *in vitro,* by either primase or **RNA polymerase** (the enzyme that synthesizes RNA transcripts from a DNA template; Section 25-1), but its rate of synthesis is greatly enhanced when both enzymes are present.

C Synthesis of the Leading and Lagging Strands

In *E. coli,* the Pol III holoenzyme catalyzes the synthesis of both the leading and lagging strands. This occurs in a single multiprotein particle, the **replisome,** which contains two Pol III enzymes. In some other prokaryotes

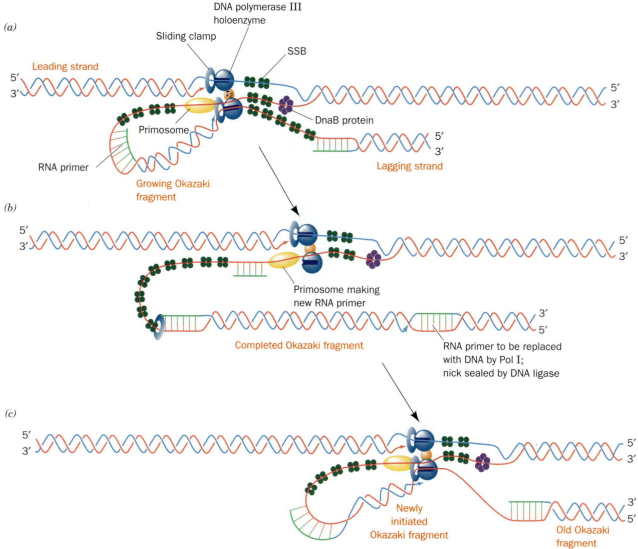

Figure 24-15 *Key to Function.* **The replication of *E. coli* DNA.** (*a*) The replisome, which contains two DNA polymerase III holoenzymes, synthesizes both the leading and the lagging strands. The lagging strand template must loop around to permit the holoenzyme to extend the primed lagging strand. (*b*) The holoenzyme releases the lagging strand template when it encounters the previously synthesized Okazaki fragment. This may signal the primosome to initiate synthesis of a lagging strand RNA primer. (*c*) The holoenzyme rebinds the lagging strand template and extends the RNA primer to form a new Okazaki fragment. Note that in this model, leading strand synthesis is always ahead of lagging strand synthesis.

and in eukaryotes, two different polymerases synthesize the leading and lagging strands, but like Pol III, they are part of a multiprotein replisome. *In order for the replisome to move as a single unit in the 5′ → 3′ direction along the leading strand, the lagging strand template must loop around* (Fig. 24-15). After completing the synthesis of an Okazaki fragment, the lagging strand holoenzyme relocates to a new primer near the replication fork and resumes synthesis. *The result of this process is a continuous leading strand and a series of RNA-primed Okazaki fragments separated by single-strand nicks.* The RNA primers are replaced with DNA through Pol I–catalyzed nick translation, and the nicks in the lagging strand are then sealed through the action of DNA ligase (see below).

A Sliding Clamp Promotes Pol III Processivity. The Pol III core enzyme dissociates from the template DNA after replicating only ~12 residues; that

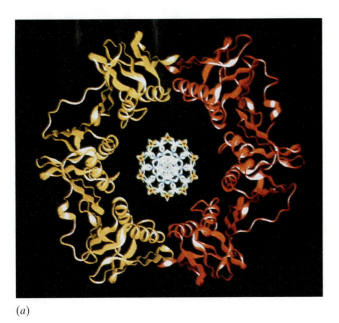

(a)

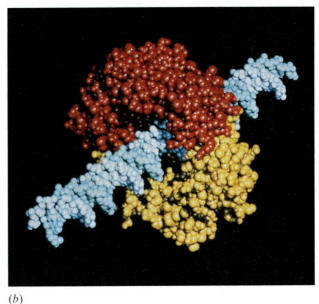

(b)

Figure 24-16 X-Ray structure of the β clamp of *E. coli* Pol III holoenzyme. (*a*) A ribbon drawing showing the two monomeric units of the dimeric protein in yellow and red as viewed along the dimer's twofold axis. A stick model of B-DNA is placed with its helix axis coincident with the protein dimer's twofold axis. (*b*) A space-filling model of the protein, colored as in Part *a*, in a hypothetical complex with B-DNA (*cyan*). [Courtesy of John Kuriyan, The Rockefeller University. PDBid 2POL.]

is, it has a processivity of ~12 residues. However, the Pol III holoenzyme has a processivity of >5000 residues, due to the presence of its **β subunit.** β Subunit bound to a cut circular DNA slides to the break and falls off. This suggests that β subunit *forms a ring around the DNA that functions as a **sliding clamp** (alternatively, **β clamp**) that can move along it, thereby keeping the Pol III holoenzyme from diffusing away.*

The X-ray structure of the β clamp (Fig. 24-16), determined by John Kuriyan, reveals that it is a dimer of C-shaped monomers that form an ~80-Å-diameter donut-shaped structure. The central ~35-Å-diameter hole is larger than the 20- and 26-Å diameters of B- and A-DNAs (the hybrid helices containing RNA primers and DNA have an A-DNA-like conformation; Section 23-1A). Each β subunit forms three domains of similar structure so that the dimeric ring is a pseudo-symmetrical six-pointed star. The interior surface of the ring is positively charged, whereas its outer surface is negatively charged.

Model-building studies in which a B-DNA helix is threaded through the central hole (Fig. 24-16) indicate that the protein's α helices span the major and minor grooves of the DNA rather than entering into them as do, for example, the recognition helices of helix–turn–helix motifs (Section 23-4B). It appears that the β subunit is designed to minimize its associations with DNA. This presumably permits the protein to freely slide along the DNA helix.

E. coli DNA is replicated at a rate of ~1000 nt/s. Thus, in lagging strand synthesis, the DNA polymerase holoenzyme must be reloaded onto the template strand every ~1 second (Okazaki fragments are ~1000 nt long). This requires that a new β clamp, which promotes Pol III's processivity, be installed around the lagging strand template every ~1 sec. The **γ complex** of the Pol III holoenzyme (subunit composition $\gamma\tau_2\delta\delta'\chi\psi$) is also known as the **clamp loader** because it opens the dimeric β clamp to load it onto the DNA template in an ATP-dependent manner. The γ complex bridges the

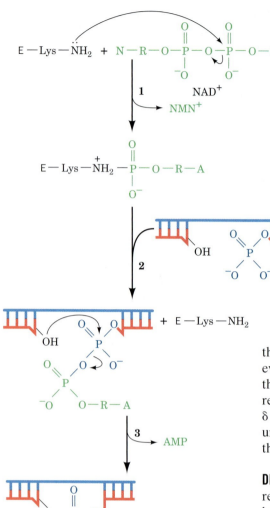

Figure 24-17 **The reactions catalyzed by *E. coli* DNA ligase.** In eukaryotic and T4 ligases, NAD⁺ is replaced by ATP so that PP$_i$ rather than NMN⁺ is eliminated in the first reaction step. Here A, R, and N represent the adenine, ribose, and nicotinamide residues, respectively.

replisome's two Pol III cores (αεθ) via the C-terminal segments of its two τ subunits (Fig. 24-15), which also bind the DnaB helicase.

Once the β clamp has been loaded onto the DNA, the Pol III core binds to the β clamp more tightly than does the γ complex, thereby displacing it and permitting processive DNA replication to proceed. When the polymerase encounters the previously synthesized Okazaki fragment, that is, when the gap between the two successively synthesized Okazaki fragments has been reduced to a nick, the Pol III core releases the DNA and loses its affinity for the β clamp. The γ subunit and the Pol III core can then rapidly initiate the synthesis of a new Okazaki fragment because they are held in the vicinity of the lagging strand template through their linkage to the Pol III core engaged in leading strand synthesis (which remains tethered to the DNA by its associated β clamp). This, of course, must be preceded by the loading of a new β clamp about the template DNA in the vicinity of the newly synthesized primer.

The sliding clamp that remains around the completed Okazaki fragment probably functions to recruit Pol I and DNA ligase so as to replace the RNA primer on the previously synthesized Okazaki fragment with DNA and seal the remaining nick. However, the sliding clamp must eventually be recycled. It was initially assumed that this was the job of the clamp loader. However, it is now clear that the release of the sliding clamp from its associated DNA is carried out by free δ subunit (the "wrench" in the clamp loader that cracks apart the β subunits forming the sliding clamp), which is synthesized in 5-fold excess over that required to populate the cell's few clamp loaders.

DNA Ligase Is Activated by NAD⁺ or ATP.   The free energy for the DNA ligase reaction is obtained, in a species-dependent manner, through the coupled hydrolysis of either NAD⁺ to **nicotinamide mononucleotide (NMN⁺)** + AMP, or ATP to PP$_i$ + AMP. *E. coli* DNA ligase, a 77-kD monomer that uses NAD⁺, catalyzes a three-step reaction (Fig. 24-17):

1. The adenylyl group of NAD⁺ is transferred to the ε-amino group of an enzyme Lys residue to form an unusual phosphoamide adduct.

2. The adenylyl group of this activated enzyme is transferred to the 5′-phosphoryl terminus of the nick to form an adenylylated DNA. Here, AMP is linked to the 5′-nucleotide via a pyrophosphate rather than the usual phosphodiester bond.

3. DNA ligase catalyzes the formation of a phosphodiester bond by attack of the 3′-OH on the 5′-phosphoryl group, thereby sealing the nick and releasing AMP.

ATP-requiring DNA ligases, such as those of eukaryotes, release PP$_i$ in the first step of the reaction rather than NMN⁺. The DNA ligase from the bacteriophage T4 is notable because it can link together two duplex DNAs that lack complementary single-stranded ends **(blunt end ligation)** in a reaction that is a boon to genetic engineering (Section 3-5).

D   Termination of Replication

The *E. coli* replication terminus is a large (350-kb) region flanked by seven nearly identical nonpalindromic ~23-bp terminator sites, **TerE, TerD,** and

TerA on one side and ***TerG, TerF, TerB,*** and ***TerC*** on the other (Fig. 24-18; note that *oriC* is directly opposite the termination region on the *E. coli* chromosome). A replication fork, traveling counterclockwise as drawn in Fig. 24-18 passes through *TerG, TerF, TerB,* and *TerC* but stops on encountering either *TerA, TerD,* or *TerE* (*TerD* and *TerE* are presumably backup sites for *TerA*). Similarly, a clockwise-traveling replication fork passes *TerE, TerD,* and *TerA* but halts at *TerC* or, failing that, *TerB, TerF,* or *TerG.* Thus, these termination sites are polar; they act as one-way valves that allow replication forks to enter the terminus region but not to leave it. *This arrangement guarantees that the two replication forks generated by bidirectional initiation at oriC will meet in the replication terminus even if one of them arrives there well ahead of its counterpart.*

The arrest of replication fork motion at *Ter* sites requires the action of **Tus protein,** a 309-residue monomer that is the product of the ***tus*** gene (for *t*erminator *u*tilization *s*ubstance). Tus protein specifically binds to a *Ter* site, where it prevents strand displacement by DnaB helicase, thereby arresting replication fork advancement.

The X-ray structure of Tus in complex with a 15-bp *Ter* fragment (Fig. 24-19) reveals that the protein forms a deep positively charged cleft in which the DNA binds. A 5-bp segment of the DNA near the side of Tus that permits the passage of the replication fork is deformed and underwound relative to canonical (normal) DNA: Its major groove is deeper, and its minor groove is significantly expanded. Protein side chains at the bottom of the cleft penetrate the DNA's widened major groove to make sequence-specific contacts such that the protein cannot release the bound DNA without a large conformational change. Nevertheless, the mechanism through which Tus prevents replication fork advancement from one side of a *Ter* site but not the other is unclear. Curiously, however, this termination system is not essential for termination. When the replication terminus is deleted, replication simply stops, apparently through the collision of opposing replication forks. Nevertheless, this termination system is highly conserved in gram-negative bacteria.

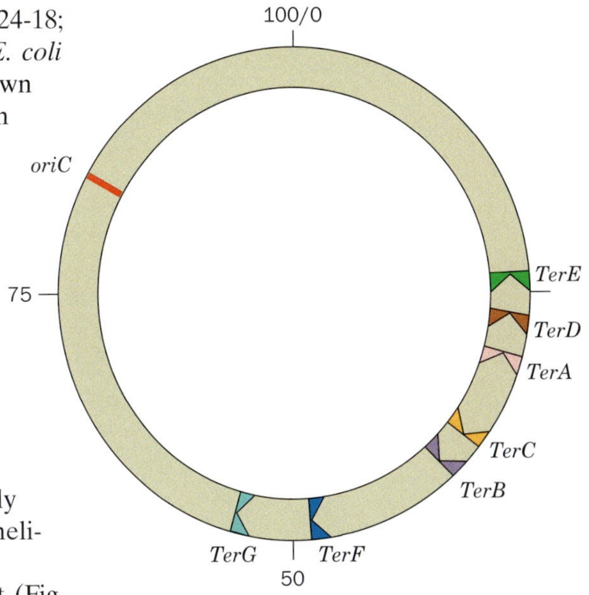

Figure 24-18 Map of the *E. coli* chromosome showing the positions of the *Ter* sites. The *TerC, TerB, TerF,* and *TerG* sites, in combination with Tus protein, allow a counterclockwise-moving replisome to pass but not a clockwise-moving replisome. The opposite is true of the *TerA, TerD,* and *TerE* sites. Consequently, two replication forks that initiate bidirectional DNA replication at *oriC* will meet between the oppositely facing *Ter* sites.

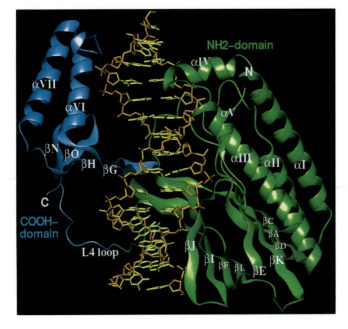

Figure 24-19 X-Ray structure of *E. coli* Tus in complex with a 15-bp *Ter*-containing DNA. The protein is shown in ribbon form with its N- and C-terminal domains colored green and blue. The DNA is represented in stick form with its bases yellow and its sugar–phosphate backbone gold. [From Kamada, K., Horiuchi, T., Ohsumi, K.; Shimamoto, N., and Morikawa, K., *Nature* **383,** 599 (1996). Used with permission. PDBid 1ECR.] **See the Interactive Exercises.**

The final step in *E. coli* DNA replication is the topological unlinking of the catenated (interlocked) parental DNA strands, thereby separating the two replication products. This reaction is probably catalyzed by one or more topoisomerases.

E Fidelity of Replication

Since a single polypeptide as small as the Pol I Klenow fragment can replicate DNA by itself, why does *E. coli* maintain a battery of over 20 intricately coordinated proteins to replicate its chromosome? The answer apparently is *to ensure the nearly perfect fidelity of DNA replication required to accurately transmit genetic information.*

The rates of reversion of mutant *E. coli* or T4 phages to the wild type indicates that only one mispairing occurs per 10^8 to 10^{10} base pairs replicated. This corresponds to ~1 error per 1000 bacteria per generation. Such high replication accuracy arises from four sources:

1. Cells maintain balanced levels of dNTPs through the mechanisms discussed in Sections 22-1C and 22-2C. This is important because a dNTP present at aberrantly high levels is more likely to be misincorporated and, conversely, one present at low levels is more likely to be replaced by one of the dNTPs present at higher levels.

2. The polymerase reaction itself has extraordinary fidelity because it occurs in two stages. First, the incoming dNTP base pairs with the template while the enzyme is in an open, catalytically inactive conformation. Polymerization occurs only after the polymerase has closed around the newly formed base pair, which properly positions the catalytic residues (induced fit; see Fig. 24-11). *The protein conformational change constitutes a double check for correct Watson–Crick base pairing between the dNTP and the template.*

3. The $3' \rightarrow 5'$ exonuclease functions of Pol I and Pol III detect and eliminate the occasional errors made by their polymerase functions.

4. A remarkable set of enzyme systems in all cells repairs residual errors in the newly synthesized DNA as well as any damage that may occur after its synthesis through chemical and/or physical insults. We discuss these DNA repair systems in Section 24-5.

In addition, the inability of a DNA polymerase to initiate chain elongation without a primer increases DNA replication fidelity. The first few nucleotides of a chain are those most likely to be mispaired because of the cooperative nature of base-pairing interactions (Section 23-2). The use of RNA primers eliminates this source of error since the RNA is eventually replaced by DNA under conditions that permit more accurate base pairing.

3 Eukaryotic DNA Replication

Eukaryotic and prokaryotic DNA replication mechanisms are remarkably similar, although the eukaryotic system is vastly more complex in terms of the amount of DNA to be replicated and the number of proteins required (estimated at >27 in yeast and mammals). Several different modes of DNA replication occur in eukaryotic cells, which contain nuclear DNA as well as mitochondrial and, in plants, chloroplast DNA. In this section, we consider some of the proteins of eukaryotic DNA replication as well as the challenge of replicating the ends of linear chromosomes.

Table 24-2 Properties of Some Eukaryotic DNA Polymerases

	α	δ	ε
$3' \rightarrow 5'$ Exonuclease	no	yes	yes
Associates with primase	yes	no	no
Processivity	moderate	high	high
Requires PCNA	no	yes	no

A Eukaryotic DNA Polymerases

Animal cells contain at least 13 distinct DNA polymerases, which have been named with Greek letters according to their order of discovery. A newer classification scheme uses sequence homology to group eukaryotic as well as prokaryotic polymerases into six families: A, B, C, D, X, and Y. In this section, we describe the three main enzymes involved in replicating eukaryotic nuclear DNA: **polymerases α, δ,** and **ε,** which are all members of the B-family of polymerases (Table 24-2).

DNA polymerase α **(pol α),** like all DNA polymerases, replicates DNA by extending a primer in the $5' \rightarrow 3'$ direction under the direction of a ssDNA template. This enzyme has no exonuclease activity and therefore cannot proofread its polymerization product. Pol α is only moderately processive (polymerizing ~100 nucleotides at a time) and associates tightly with a primase, indicating that it is involved in initiating DNA replication. The pol α/primase complex synthesizes a 7- to 10-nt RNA primer and extends it by an additional 15 or so deoxynucleotides. Its lack of proofreading activity is not problematic, since the first few residues of newly synthesized DNA are typically removed and replaced along with the RNA primer.

DNA polymerase δ **(pol δ)** does not associate with a primase and contains a $3' \rightarrow 5'$ exonuclease active site. In addition, the processivity of pol δ is essentially unlimited (it can replicate the entire length of a template DNA), but only when it is in complex with a sliding-clamp protein named **proliferating cell nuclear antigen (PCNA).** The X-ray structure of PCNA (Fig. 24-20), also determined by Kuriyan, reveals that it forms a trimeric ring with almost identical structure (and presumably function) as the *E. coli* β$_2$ sliding clamp (Fig. 24-16). Intriguingly, PCNA and the β clamp exhibit no significant sequence identity, even when their structurally similar portions are aligned.

Pol δ in complex with PCNA is required for both leading and lagging strand synthesis. During replication, the eukaryotic counterpart of the *E. coli* γ complex (the clamp loader) loads PCNA onto the template strand near the primer. This displaces pol α, allowing pol δ to bind and processively extend the new DNA strand.

DNA polymerase ε **(pol ε)** superficially resembles pol δ but is highly processive in the absence of PCNA and has a $3' \rightarrow 5'$ exonuclease activity that degrades ssDNA to 6- or 7-residue oligonucleotides rather than to mononucleotides, as does the exonuclease activity of pol δ. Although pol ε is required for the viability of yeast, its essential function can be carried out by only the noncatalytic C-terminal half of its 256-kD catalytic subunit, which is unique among B-family DNA polymerases. It therefore appears that, at least in yeast, pol ε has an essential control function but not a catalytic function.

DNA polymerase γ (pol γ), an A-family enzyme, occurs exclusively in the mitochondrion, where it presumably replicates the mitochondrial genome. Chloroplasts contain a similar enzyme. An additional member of

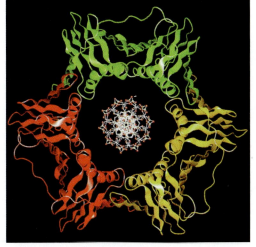

Figure 24-20 X-Ray structure of PCNA. The three protein monomers (*red, green,* and *yellow*) form a threefold symmetric ring structure. A model of duplex B-DNA (viewed along its helix axis) has been placed in the center of the PCNA ring. Compare this structure to that of the β clamp of the *E. coli* Pol III holoenzyme (Fig. 24-16). [Courtesy of John Kuriyan, The Rockefeller University. PDBid 1PLQ.]

🔗 **See the Interactive Exercises.**

Perspectives in Biochemistry

Reverse Transcriptase

Reverse transcriptase (**RT**) is an essential enzyme of **retroviruses,** which are RNA-containing eukaryotic viruses such as **human immunodeficiency virus** (**HIV,** the causative agent of AIDS). RT, which was independently discovered in 1970 by Howard Temin and David Baltimore, synthesizes DNA in the $5' \rightarrow 3'$ direction from an RNA template. Although the activity of this enzyme was initially considered antithetical to the central dogma of molecular biology (Section 3-3B), there is no thermodynamic prohibition to the RT reaction (in fact, under certain conditions, Pol I can copy RNA templates). RT catalyzes the first step in the conversion of the virus' single-stranded RNA genome to a double-stranded DNA.

After the virus enters a cell, its RT uses the viral RNA as a template to synthesize a complementary DNA strand, yielding an RNA–DNA hybrid helix. The DNA synthesis is primed by a host cell tRNA whose 3' end unfolds to base pair with a complementary segment of viral RNA. The viral RNA strand is then nucleolytically degraded by an **RNase H** (an RNase activity that hydrolyzes the RNA of an RNA–DNA hybrid helix). Finally, the DNA strand acts as a template for the synthesis of its complementary DNA, yielding dsDNA that is then integrated into a host cell chromosome (*at right*).

RT has been a particularly useful tool in genetic engineering because it can transcribe mRNAs to complementary strands of DNA (**cDNA**). mRNA-derived cDNAs can be used, for example, to express eukaryotic structural genes in *E. coli* (Section 3-5D). Since *E. coli* lacks the machinery to splice out introns (Section 25-3A), the use of genomic DNA to express a eukaryotic structural gene in *E. coli* would require the prior excision of its introns—a technically difficult feat.

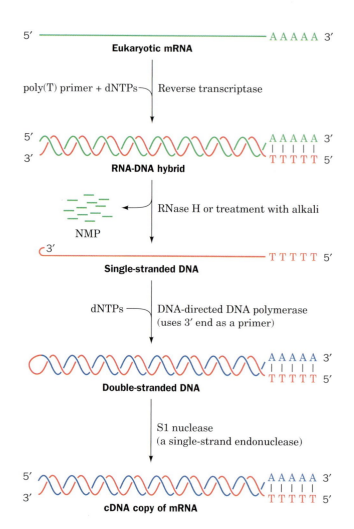

the polymerase family of proteins is the viral enzyme **reverse transcriptase,** an RNA-directed DNA polymerase (see Box 24-2).

B Initiation and Elongation of Eukaryotic DNA Replication

DNA polymerase α synthesizes DNA at the rate of ~50 nt/s (~20 times slower than prokaryotic DNA polymerases). Since a eukaryotic chromosome typically contains 60 times more DNA than does a prokaryotic chromosome, its bidirectional replication from a single origin, as in prokaryotes, would require ~1 month. Electron micrographs such as Fig. 24-21, however, show that *eukaryotic chromosomes contain multiple origins*, one every 3 to 300 kb, depending on both the species and the tissue, so replication is complete in a few hours.

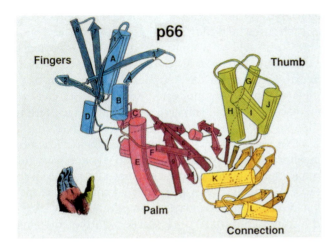

HIV-1 reverse transcriptase is a dimeric protein whose subunits are synthesized as identical 66-kD polypeptides, known as **p66,** that each contain a polymerase domain and an RNase H domain. However, the RNase H domain of one of the two subunits is proteolytically excised, thereby yielding a 51-kD polypeptide named **p51.** Thus, RT is a dimer of p66 and p51.

The X-ray structure of HIV-1 RT shows that the two subunits have different structures, although each has a fingers, palm, and thumb domain as well as a "connection" domain. The RNase H domain of the p66 subunit follows the connection domain. The p66 and p51 subunits are not related by twofold molecular symmetry (a rare but not unprecedented phenomenon) but instead associate in a sort of head-to-tail arrangement. Consequently, RT has only one polymerase active site.

Reverse transcriptase lacks a proofreading exonuclease function and hence is highly error prone. Indeed, it is HIV's capacity to rapidly evolve, even within a single host, that presents a major obstacle to the development of an anti-HIV vaccine. This rapid rate of mutation is also the main contributor to the ability of HIV to rapidly develop resistance to drugs that inhibit virally encoded enzymes, including RT. **See the Interactive Exercises.**

[Structure of RT courtesy of Thomas Steitz, Yale University. PDBid 3HVT.]

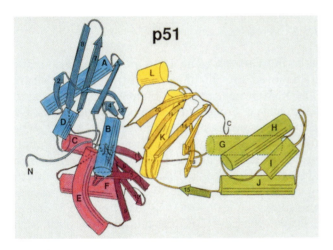

Cytological observations indicate that the various chromosomal regions are not all replicated simultaneously. Rather, clusters of 20 to 80 adjacent **replicons** (replicating units; DNA segments that are each served by a

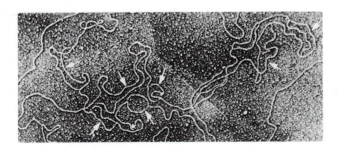

Figure 24-21 Electron micrograph of a fragment of replicating *Drosophila* DNA. The arrows indicate the multiple replication eyes. [From Kreigstein, H.J. and Hogness, D.S., *Proc. Natl. Acad. Sci.* **71,** 136 (1974).]

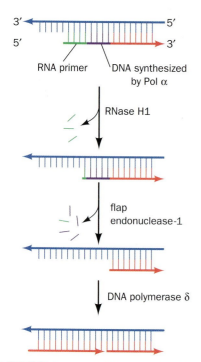

Figure 24-22   Removal of RNA primers in eukaryotes.
RNase H1 excises all but the 5'-ribonucleotide of the RNA primer. FEN1, a 5' → 3' endonuclease, then removes the remaining ribonucleotide along with a segment of adjoining DNA if it contains mismatches. The excised nucleotides are replaced as DNA polymerase δ completes the synthesis of the next Okazaki fragment (on the left in this diagram). The nick is eventually sealed by DNA ligase.

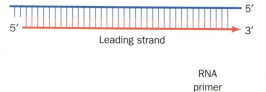

Figure 24-23   Replication of a linear chromosome.
Leading strand synthesis can proceed to the end of the chromosome (*top*). However, DNA polymerase cannot synthesize the extreme 5' end of the lagging strand because it can only extend an RNA primer that is paired with the 3' end of a template strand (*bottom*). Removal of the primer and degradation of the remaining single-stranded extension would cause the chromosome to shorten with each round of replication.

replication origin) are activated simultaneously. New sets of replicons are activated until the entire chromosome has been replicated. During this process, replicons that have already been replicated are distinguished from those that have not; that is, *a cell's chromosomal DNA is replicated once and only once per cell cycle.*

In yeast cells, the initiation of DNA replication occurs at **autonomously replicating sequences (ARS),** which are conserved 11-bp sequences adjacent to easily unwound DNA. As in prokaryotes, a helicase is required to prepare the DNA for replication, and the resulting ssDNA is coated with a eukaryotic analog of SSB. Aided by a multitude of accessory proteins, DNA pol α/primase begins to synthesize a new DNA strand, then is replaced by pol δ, which extends the chain. DNA replication proceeds in each direction from the origin of replication until each replication fork collides with a fork from the adjacent replicon. Eukaryotes appear to lack termination sequences analogous to the *Ter* sites in *E. coli.*

Unlike prokaryotic DNA, eukaryotic DNA is packaged in nucleosomes (Section 23-5B). Some alteration of this structure is probably necessary for initiation, but once replication is under way, nucleosomes do not seem to impede the progress of DNA polymerases. Experiments with labeled histones indicate that nucleosomes just ahead of the replication fork disassemble and the freed histones immediately reassociate with the emerging daughter duplexes. The parental histones randomly associate with the leading and lagging duplexes. DNA replication (which occurs in the nucleus) is coordinated with histone protein synthesis in the cytosol so that new histones are available in the required amounts.

Primer Removal Requires Two Enzymes.   The RNA primers of eukaryotic Okazaki fragments are removed through the actions of two enzymes: **RNase H1** removes most of the RNA, leaving only a 5'-ribonucleotide adjacent to the DNA, which is then removed through the action of **flap endonuclease-1 (FEN1).** However, as we have seen, pol α extends the RNA primer by ~15 nt of DNA before it is displaced by pol δ. Since pol α lacks proofreading ability, this primer extension is more likely to contain errors than the DNA synthesized by pol δ. However, FEN1 provides what is, in effect, pol α's proofreading function: It is also an endonuclease that excises mismatch-containing oligonucleotides up to 15 nt long from the 5' end of an annealed DNA strand. Moreover, FEN1 can make several such excisions in succession to remove more distant mismatches. The excised segment is later replaced by pol δ as it synthesizes the succeeding Okazaki fragment (Fig. 24-22).

C   Telomeres and Telomerase

The ends of linear chromosomes present a problem for the replication machinery. Specifically, *DNA polymerase cannot synthesize the extreme 5' end of the lagging strand* (Fig. 24-23). Even if an RNA primer were paired with the 3' end of the DNA template, it could not be replaced with DNA (recall that DNA polymerase operates only in the 5' → 3' direction; it can only extend an existing primer; and the primer must be bound to its complementary strand). Consequently, *in the absence of a mechanism for completing the lagging strand, linear chromosomes would be shortened at both ends by at least the length of an RNA primer with each round of replication.*

Telomeres Are Synthesized by Telomerase.   The ends of eukaryotic chromosomes, the **telomeres** (Greek: *telos,* end), have an unusual structure. Telomeric

DNA consists of 1000 or more tandem repeats of a short G-rich sequence (TTGGGG in the protozoan *Tetrahymena* and TTAGGG in humans) on the 3'-ending strand of each chromosome end. Moreover, this strand has a 12- to 16-nt single-strand overhang. *This 3' extension can serve as a template for the primer that initiates the final Okazaki fragment of the lagging strand.*

Elizabeth Blackburn has shown that telomeric DNA is synthesized and maintained by an enzyme named **telomerase.** *Tetrahymena* telomerase, for example, adds tandem repeats of the telomeric sequence TTGGGG to the 3' end of any G-rich telomeric oligonucleotide independently of any exogenously added template. A clue as to how this occurs came from the discovery that telomerases are **ribonucleoproteins** (complexes of protein and RNA) whose RNA components contain a segment that is complementary to the repeating telomeric sequence. This RNA apparently acts as a template for a reaction in which nucleotides are added to the 3' end of the DNA (Fig. 24-24, *top*). Telomerase thus functions similarly to reverse transcriptase (Box 24-2); in fact, its protein component is homologous to reverse transcriptase. Telomerase repeatedly translocates to the new 3' end of the DNA strand, thereby adding multiple telomeric sequences to the DNA (Fig. 24-24, *bottom*). The DNA strand complementary to the telomeric G-rich strand is apparently synthesized by the normal cellular machinery for lagging strand synthesis, which necessarily leaves a 3' overhang on the G-rich strand.

The absence of telomerase, which allows the gradual truncation of chromosomes with each round of DNA replication, contributes to the normal senescence of cells. Conversely, enhanced telomerase activity permits the uncontrolled replication and cell growth that occur in cancer (see Box 24-3).

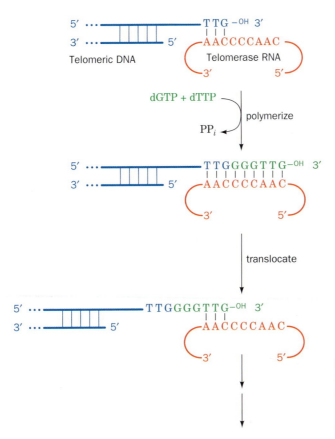

Figure 24-24 Mechanism for the synthesis of telomeric DNA by *Tetrahymena* telomerase. The telomere's 5'-ending strand is later extended by normal lagging strand synthesis. [After Greider, C.W. and Blackburn, E.H., *Nature* **337**, 336 (1989).]

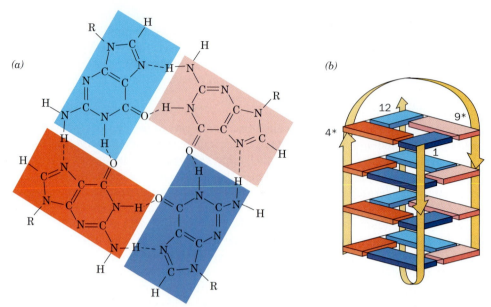

Figure 24-25 **Structure of the telomeric oligonucleotide d(GGGGTTTTGGGG).** (*a*) The base-pairing interactions in the G-quartet. (*b*) Schematic diagram of the NMR structure in which the strand directions are indicated by arrows. The nucleotides are numbered 1 to 12 in one strand and 1* to 12* in the symmetry-related strand. Guanine residues G1 to G4 are represented by blue rectangles, G9 to G12 are cyan, G1* to G4* are red, and G9* to G12* are pink. [After Schultze, P., Smith, F.W., and Feigon, J., *Structure* **2**, 227 (1994). PDBid 156D.]

Telomeres Form G-Quartets. Guanine-rich polynucleotides are notoriously difficult to work with. This is because of their propensity to aggregate via Hoogsteen-type base pairing (Section 23-2B) to form cyclic tetramers known as **G-quartets** (Fig. 24-25*a*). Indeed, the G-rich overhanging strands of telomeres fold back on themselves to form a hairpin, two of which associate in an antiparallel fashion to form stable complexes of stacked G-quartets (Fig. 24-25*b*). Such structures presumably serve as binding sites for capping proteins, some of which mediate the formation of loops at the ends of otherwise linear telomeric DNA. These so-called **T-loops** are visible by electron microscopy (Fig. 24-26). Loop formation may help regulate telomere length and prevent activation of DNA repair mechanisms that recognize the ends of broken DNA molecules.

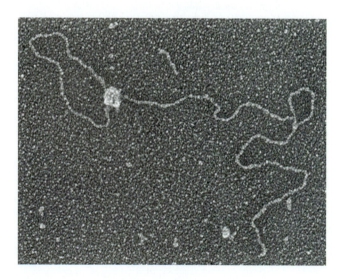

Figure 24-26 **Electron micrograph of a T-loop.** Proteins known as **telomere repeat-binding factors** are required for loop formation. [Courtesy of Jack Griffith, University of North Carolina at Chapel Hill.]

BOX 24-3

Biochemistry in Health and Disease

Telomerase, Aging, and Cancer

Without the action of telomerase, a chromosome would be shortened at both ends by at least the length of an RNA primer with every cycle of DNA replication and cell division. It was therefore initially assumed that, in the absence of active telomerase, essential genes located near the ends of chromosomes would eventually be lost, thereby killing the descendents of the originally affected cells. However, it is now evident that telomeres serve a vital chromosomal function that is compromised before this can happen. Free DNA ends trigger DNA damage repair systems that normally function to rejoin the ends of broken chromosomes (Section 24-5E). Thus exposed telomeric DNA would result in the end-to-end fusion of chromosomes, a process that leads to chromosomal instability and eventual cell death (fused chromosomes often break in mitosis; their two centromeres cause them to be pulled in opposite directions). However, in a process known as **capping,** telomeric DNA is specifically bound by proteins that hide the DNA ends. There is mounting evidence that capping is a dynamic process in which the probability of a telomere spontaneously upcapping increases as telomere length decreases.

The somatic cells of multicellular organisms lack telomerase activity. This explains why such cells in culture can only undergo a limited number of doublings (20–60) before they reach **senescence** (a stage in which they cease dividing) and eventually die. Indeed, otherwise immortal *Tetrahymena* cultures with mutationally impaired telomerases exhibit characteristics reminiscent of senescent mammalian cells before

dying off. Apparently, *the loss of telomerase function in somatic cells is a basis for aging in multicellular organisms.*

Despite the foregoing, there is only a weak correlation between the proliferative capacity of a cell culture and the age of its donor. There is, however, a strong correlation between the initial telomere length in a cell culture and its proliferative capacity. Cells that initially have relatively short telomeres undergo significantly fewer doublings than cells with longer telomeres. Moreover, fibroblasts from individuals with **progeria** (a rare disease characterized by rapid and premature aging resulting in childhood death) have short telomeres, an observation that is consistent with their known reduced proliferative capacity in culture. In contrast, sperm (which are essentially immortal) have telomeres that do not vary in length with donor age, which indicates that telomerase is active during germ-cell growth. Likewise, those few cells in culture that become immortal (capable of unlimited proliferation) exhibit an active telomerase and a telomere of stable length, as do the cells of unicellular eukaryotes (which are also immortal).

What selective advantage might multicellular organisms gain by eliminating the telomerase activity in their somatic cells? An intriguing possibility is that cellular senescence is a mechanism that protects multicellular organisms from cancer. Indeed, cancer cells, which are immortal and grow uncontrollably, contain active telomerase. For example, the enzyme is active in ovarian cancer cells but not in normal ovarian tissue. This hypothesis makes telomerase inhibitors an attractive target for antitumor drug development.

4 DNA Damage

The fidelity of DNA replication carried out by DNA polymerases and their attendant proofreading functions is essential for the accurate transmission of genetic information during cell division. Yet errors in polymerization occasionally occur and, if not corrected, may alter the nucleotide sequences of genes. DNA can also be chemically altered by agents that are naturally present in the cell or in the cell's external environment.

In many cases, damaged DNA can be repaired, as discussed in Section 24-5. Severe lesions, however, may be irreversible, leading to the loss of genetic information and, often, cell death. Even when damaged DNA can be mended, the restoration may be imperfect, producing a **mutation,** a heritable alteration of genetic information. In multicellular organisms, genetic changes are usually notable only when they occur in germ-line cells so that the change is passed on to all the cells of the organism's offspring. Damage to the DNA of a somatic cell, in contrast, rarely has an effect beyond that cell unless the mutation contributes to a malignant transformation (cancer).

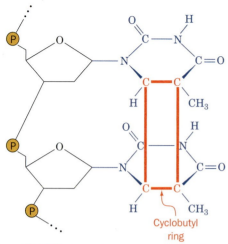

Figure 24-27 The cyclobutylthymine dimer. The dimer forms on UV irradiation of two adjacent thymine residues on a DNA strand. The ~1.6-Å-long covalent bonds joining the thymine rings (*red*) are much shorter than the normal 3.4-Å spacing between stacked rings in B-DNA, thereby locally distorting the DNA.

A Mutagenesis

Environmental agents such as ultraviolet light, ionizing radiation, and certain chemical agents can physically damage DNA. For example, UV radiation (200–300 nm) promotes the formation of a cyclobutyl ring between adjacent thymine residues on the same DNA strand to form an intrastrand **thymine dimer** (Fig. 24-27). Similar cytosine and thymine–cytosine dimers also form but less frequently. Such **pyrimidine dimers** locally distort DNA's base-paired structure, interfering with transcription and replication. Ionizing radiation also damages DNA either through its direct action on the DNA molecule or indirectly by inducing the formation of free radicals, particularly the hydroxyl radical (HO·), in the surrounding aqueous medium. This can lead to strand breakage.

The DNA damage produced by **chemical mutagens,** substances that induce mutations, falls into two major classes:

1. **Point mutations,** in which one base pair replaces another. These are subclassified as:
   (a) **Transitions,** in which one purine (or pyrimidine) is replaced by another.
   (b) **Transversions,** in which a purine is replaced by a pyrimidine or vice versa.
2. **Insertion/deletion mutations,** in which one or more nucleotide pairs are inserted in or deleted from DNA.

Point Mutations Are Generated by Altered Bases. One cause of point mutations is treatment of DNA with nitrous acid (HNO_2), which oxidatively deaminates aromatic primary amines. Cytosine is thereby converted to uracil and adenine to the guanine-like hypoxanthine (which forms two of guanine's three hydrogen bonds with cytosine; Fig. 24-28). Hence, treatment of DNA with nitrous acid results in both $A \cdot T \rightarrow G \cdot C$ and $G \cdot C \rightarrow A \cdot T$ transitions. Despite its potential mutagenic activity, nitrite (the conjugate base of nitrous acid) is used as a preservative in prepared meats such as

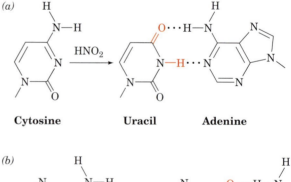

Figure 24-28 Oxidative deamination by nitrous acid. (*a*) Cytosine is converted to uracil, which base pairs with adenine. (*b*) Adenine is converted to hypoxanthine, a guanine derivative (it lacks guanine's 2-amino group) that base pairs with cytosine.

frankfurters because it also prevents the growth of *Clostridium botulinum,* the organism that causes botulism. Base deamination reactions also occur spontaneously in the absence of nitrous acid.

Cellular metabolism itself exposes DNA to the damaging effects of reactive oxygen species (e.g., the superoxide ion $O_2^-\cdot$, the hydroxyl radical, and H_2O_2) that are normal by-products of oxidative metabolism (Section 17-4C). Over 100 different oxidative modifications to DNA have been catalogued. For example, guanine can be oxidized to **8-oxoguanine (oxoG):**

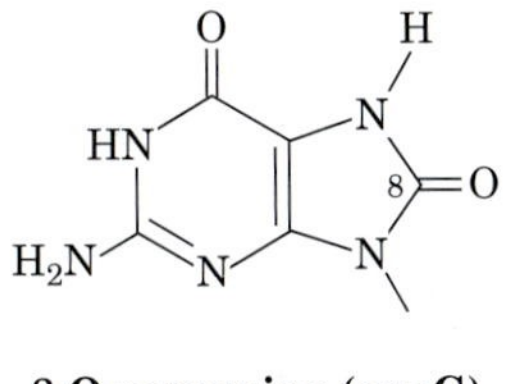

8-Oxoguanine (oxoG)

When the modified DNA strand is replicated, the oxoG can base pair with either C or A, causing a $G\cdot C \rightarrow T\cdot A$ transversion.

Alkylating agents such as dimethyl sulfate, **nitrogen mustard, ethylnitrosourea,** and *N*-**methyl-***N'*-**nitro-***N*-**nitrosoguanidine (MNNG)**

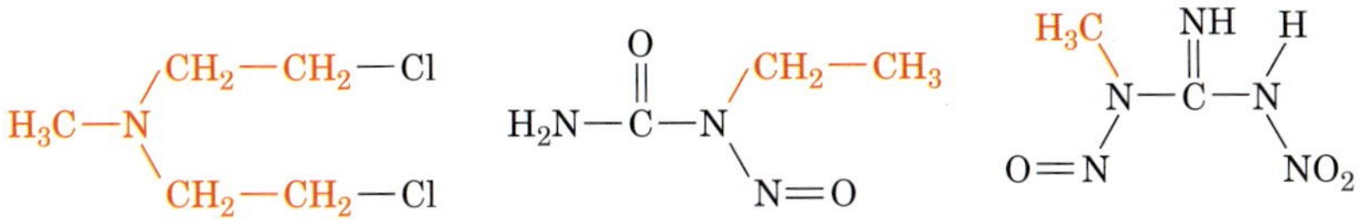

Nitrogen mustard   **Ethylnitrosourea**   *N*-**Methyl-***N'*-**nitro-***N*-**nitrosoguanidine (MNNG)**

can also generate transversions. For example, the exposure of DNA to MNNG yields, among other products, *O*⁶-**methylguanine** residues,

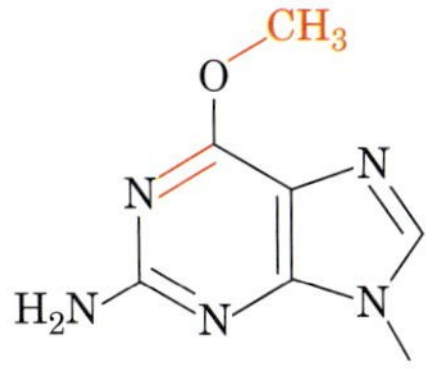

*O*⁶-**Methylguanine residue**

which can base pair with either C or T. The metabolic methylating agent *S*-adenosylmethionine (Section 20-4D) occasionally nonenzymatically methylates a base to form derivatives such as 3-methyladenine and 7-methylguanine residues. However, base methylations also have normal physiological functions (see Box 24-4).

The alkylation of the N7 position of a purine nucleotide renders its glycosidic bond susceptible to hydrolysis, leading to loss of the base. The resulting gap in the sequence is filled in by an error-prone enzymatic repair system. Transversions arise when the missing purine is replaced by a pyrimidine. Even in the absence of alkylating agents, the glycosidic bonds of an estimated 10,000 of the 3.2 billion purine nucleotides in the human genome spontaneously hydrolyze every day.

Insertion/Deletion Mutations Are Generated by Intercalating Agents.

Insertion/deletion mutations may arise from the treatment of DNA with intercalating agents such as acridine orange or proflavin (Section 23-3B). The distance between two consecutive base pairs is roughly doubled by the intercalation

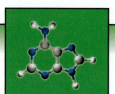

BOX 24-4

Perspectives in Biochemistry

DNA Methylation

Not all DNA modifications are detrimental. For example, the A and C residues of DNA may be methylated, in a species-specific pattern, to form **N^6-methyladenine (m^{6}A)**, **N^4-methylcytosine (m^{4}C)**, and **5-methylcytosine (m^{5}C)** residues:

N^6-**Methyladenine (m^{6}A)**
residue

5-Methylcytosine (m^{5}C)
residue

N^4-**Methylcytosine (m^{4}C)**
residue

These methyl groups project into B-DNA's major groove, where they can interact with DNA-binding proteins. In most cells, only a few percent of the susceptible bases are methylated, although this figure rises to >30% of the C residues in some plants.

Bacterial DNAs are methylated at their own particular restriction sites by modification methylases, thereby preventing the corresponding restriction endonuclease from cleaving the DNA (Section 3-4A). Other **methyltransferases** also modify DNA bases in a sequence-specific manner. For example, in *E.coli*, the **Dam methyltransferse** methylates the A residue in all GATC sequences and the **Dcm methyltransferse** methylates both C residues in CC$_T^A$TGG at their C5 positions. Note that both of these sequences are palindromic.

In addition to its role in restriction–modification systems, DNA methylation in prokaryotes functions most conspicuously as a marker of parental DNA in the repair of mismatched base pairs. Any replicational mispairing that has eluded the editing functions of Pol I and Pol III may still be corrected by a process known as mismatch repair (Section 24-5D). However, if this system is to correct errors rather than perpetuate them, it must distinguish the parental DNA, which has the correct base, from the daughter strand, which has the incorrect although normal base. The observation that *E. coli* with a deficient Dam methyltransferase have higher mutation rates than wild-type bacteria suggests how this distinction is made. A newly replicated daughter strand is undermethylated compared to the parental strand because DNA methylation lags behind DNA synthesis.

5-Methylcytosine is the only methylated base in most eukaryotic DNAs, including those of vertebrates. This modification occurs largely in the CG dinucleotide of various palindromic sequences. CG is present in the vertebrate genome at only about one-fifth its randomly expected frequency. The upstream regions of many genes, however, have normal CG frequencies and are therefore known as **CpG islands.**

There is clear evidence that *DNA methylation switches off eukaryotic gene expression, particularly when it occurs in the promoter regions upstream of a gene's transcribed sequence (Section 27-3A).* For example, globin genes are less methylated in erythroid cells than they are in nonerythroid cells and, in fact, the specific methylation of the control region in a recombinant globin gene inhibits its transcription. Moreover, the methylation pattern of a parental DNA strand directs the methylation of its daughter strand (a methylated CG sequence directs a methyltransferase to methylate its complementary CG sequence), so that the "inheritance" of a methylation pattern in a cell line permits all the cells to have the same differentiated phenotype. Variations in methylation are responsible for **genomic imprinting** in mammals, the phenomenon in which certain maternal and paternal genes are differentially expressed in the offspring (Section 27-3A).

of such a molecule between them. The replication of this distorted DNA occasionally results in the insertion or deletion of one or more nucleotides in the newly synthesized polynucleotide. (Insertions and deletions of large segments generally arise from aberrant crossover events; Section 24-6A.)

All Mutations Are Random.   The bulk of the scientific data regarding mutagenesis is that mutations, whether the result of polymerase errors, spontaneous

modification, or chemical damage to DNA, occur at random. This paradigm was challenged by John Cairns, who demonstrated that bacteria unable to digest lactose preferentially acquired the mutations they needed in order to use lactose when it was the only nutrient available. This observation, which suggests that bacteria can "direct" mutations that benefit them, more likely reflects a nonspecific adaptive response in which the overall rate of mutation—useful as well as nonuseful—increases when the cells are under metabolic stress. *The hypermutable state appears to reflect the activation of error-prone DNA repair and recombination systems that are relatively inactive in normally growing cells.*

B Carcinogens

Not all alterations to DNA have phenotypic consequences. For example, mutations in noncoding segments of DNA are often invisible. Similarly, the redundancy of the genetic code (more than one trinucleotide may specify a particular amino acid; Section 26-1C) can mask point mutations. Even when a protein's amino acid sequence is altered, its function may be preserved if the substitution is conservative (Section 5-4A) or occurs on a surface loop. Nevertheless, even a single point mutation, if appropriately located, can irreversibly alter cellular metabolism, for example, by causing cancer. As many as 80% of human cancers may be caused by **carcinogens** that damage DNA or interfere with its replication or repair. Consequently, many mutagens are also carcinogens.

There are presently over 60,000 man-made chemicals of commercial importance, and ~1000 new ones are introduced every year. The standard animal tests for carcinogenesis, exposing rats or mice to high levels of the suspected carcinogen and checking for cancer, are expensive and require ~3 years to complete. Thus, relatively few substances have been tested in this manner. Likewise, epidemiological studies in humans are costly, time-consuming, and often inconclusive.

Bruce Ames has devised a rapid and effective bacterial assay for carcinogenicity that is based on the high correlation between carcinogenesis and mutagenesis. He constructed special tester strains of the bacterium *Salmonella typhimurium* that are his^- (cannot synthesize histidine and therefore cannot grow in its absence). Mutagenesis in these strains is indicated by their reversion to the his^+ phenotype.

In the **Ames test,** ~10^9 tester strain bacteria are spread on a culture plate that lacks histidine. A mutagen placed in the culture medium causes some of these his^- bacteria to become his^+, which is detected by their growth into visible colonies after 2 days at 37°C (Fig. 24-29). The mutagenicity of a substance is scored as the number of such colonies minus the few spontaneously revertant colonies that occur in the absence of the mutagen.

Many noncarcinogens are converted to carcinogens in the liver or in other tissues via a variety of detoxification reactions (e.g., those catalyzed by the cytochromes P450; Section 12-4D). A small amount of rat liver homogenate is therefore included in the Ames test medium in order to approximate the effects of mammalian metabolism.

About 80% of the compounds determined to be carcinogens in whole-animal experiments are also mutagenic by the Ames test. Dose–response curves, which are generated by testing a given compound at a number of concentrations, are almost always linear, indicating that *there is no threshold concentration for mutagenesis.* Several compounds to which humans have been extensively exposed that were found to be mutagenic by the Ames test were later found to be carcinogenic in animal tests. These include

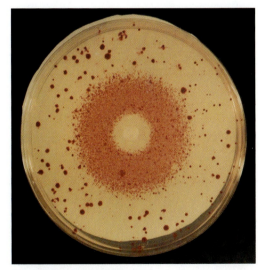

Figure 24-29   The Ames test for mutagenesis. A filter paper disk containing a mutagen, in this case the alkylating agent ethyl methanesulfonate, is centered on a culture plate containing *his⁻* tester strains of *Salmonella typhimurium* in a medium that lacks histidine. A dense halo of revertant bacterial colonies appears around the disk from which the mutagen diffused. The larger colonies distributed around the culture plate are spontaneous revertants. The bacteria near the disk have been killed by the toxic mutagen's high concentration. [Courtesy of Raymond Devoret, Institut Curie, Orsay, France.]

tris(2,3-dibromopropyl)phosphate, which was used as a flame retardant in children's sleepwear in the mid-1970s and can be absorbed through the skin; and **furylfuramide,** which was used in Japan in the 1960s and 1970s as an antibacterial additive in many prepared foods (and which had passed two animal tests before it was found to be mutagenic).

5 DNA Repair

DNA damage must be repaired to maintain the integrity of genetic information. The biological importance of DNA repair is indicated by the great variety of repair mechanisms in even simple organisms such as *E. coli.* These systems include enzymes that simply reverse the chemical modification of nucleotide bases as well as more complicated multienzyme systems that depend on the inherent redundancy of the information in duplex DNA to restore the damaged molecule.

A Direct Reversal of Damage

Several enzymes recognize and reverse certain types of DNA damage. For example, pyrimidine dimers (Fig. 24-27) may be restored to their monomeric forms by **photoreactivation** catalyzed by light-absorbing enzymes known as **DNA photolyases.** These 55- to 65-kD monomeric enzymes are found in many prokaryotes and eukaryotes but not in humans. Photolyases contain two prosthetic groups: a light-absorbing cofactor and FADH⁻. In the *E. coli* enzyme, the cofactor N^5,N^{10}-methenyltetrahydrofolate (Section 20-4D) absorbs UV–visible light (300–500 nm) and transfers the excitation energy to the FADH⁻, which then transfers an electron to the pyrimidine dimer, thereby splitting it. The resulting pyrimidine anion reduces the FADH· to regenerate the enzyme.

E. coli DNA photolyase binds ssDNA or dsDNA without regard to base sequence. Its X-ray structure reveals its DNA binding site to be a positively charged flat surface with a hole that has a size and polarity complementary to that of a pyrimidine dimer (Fig. 24-30). In order to bind in this site and contact the isoalloxazine ring of the FADH⁻ for electron transfer, the pyrimidine dimer must flip out of the double helix. This conformational change is probably facilitated by the relatively weak base pairing of the pyrimidine dimer and the distortion it imposes on the double helix. **Base-flipping** is a common feature of enzymes that chemically alters DNA bases as part of repair pathways or normal metabolism (e.g., cytosine methylation; Section 27-3A).

Another type of direct DNA repair is the reversal of base methylation by **alkyltransferases.** For example, O^6-methylguanine and O^6-ethylguanine lesions of DNA are repaired by O^6**-alkylguanine–DNA alkyltransferase,** which directly transfers the offending methyl or ethyl group to one of its own Cys residues. This reaction inactivates the protein, which therefore cannot be strictly classified as an enzyme. Apparently, the cost of sacrificing the alkyltransferase is justified by the highly mutagenic nature of the modified guanine residue.

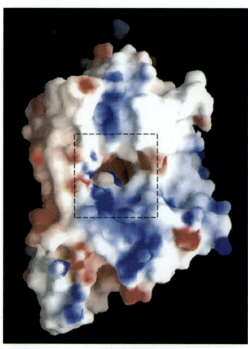

Figure 24-30   X-Ray structure of *E. coli* DNA photolyase.   The solvent-accessible surface of the enzyme is shaded according to its electrostatic potential (*blue,* most positive; *red,* most negative; and *white,* neutral). The dashed lines surround the hole in the protein surface where the pyrimidine dimer is thought to bind. [Courtesy of Johann Deisenhofer, University of Texas Southwestern Medical Center, Dallas, Texas. PDBid 1DNP.]

B Base Excision Repair

Damaged bases that cannot be directly repaired may be removed and replaced in a process known as **base excision repair (BER).** This pathway, as its name implies, begins with removal of the damaged base. Cells contain

a variety of **DNA glycosylases** that each cleave the glycosidic bond of a corresponding type of altered nucleotide, leaving a deoxyribose residue with no attached base (Fig. 24-31). Such **apurinic** or **apyrimidinic sites (AP** or **abasic sites)** also result from the occasional spontaneous hydrolysis of glycosidic bonds. The deoxyribose residue is then cleaved on one side by an **AP endonuclease,** the deoxyribose and several adjacent residues are removed by the action of a cellular exonuclease (possibly associated with a DNA polymerase), and the gap is filled in and sealed by a DNA polymerase and DNA ligase.

The enzymes of BER, which correct the most frequent type of DNA damage, include a glycosylase that recognizes 8-oxoguanine and the enzyme **uracil–DNA glycosylase (UDG),** which excises uracil residues. The latter arise from cytosine deamination as well as the occasional misincorporation of uracil instead of thymine into DNA (Box 24-5).

The X-ray structure of human UDG in complex with a 10-bp DNA containing a U·G mismatch (which forms a doubly hydrogen-bonded base pair whose shape differs from that of Watson–Crick base pairs; Section 23-1A), determined by John Tainer, reveals that the enzyme has bound the DNA with its uridine nucleotide flipped out of the double helix (Fig. 24-32). Moreover, the enzyme has hydrolyzed uridine's glycosidic bond, yielding the free uracil base and an AP site on the DNA, although both products remain bound to the enzyme. The cavity in the DNA's base stack that would otherwise be occupied by the flipped-out uracil is filled by the side chain of Leu 272, which intercalates into the DNA from its minor groove side.

How does UDG detect a base-paired uracil in the center of DNA and how does it discriminate so acutely between uracil and other bases, particularly the closely similar thymine? The X-ray structures indicate that the phosphate groups flanking the flipped-out base are 4 Å closer together than they are in B-DNA (8 Å versus 12 Å), which causes the DNA to kink by ~45° in

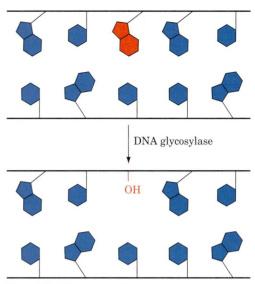

Figure 24-31 Action of DNA glycosylases. These enzymes hydrolyze the glycosidic bond of their corresponding altered base (*red*) to yield an AP site.

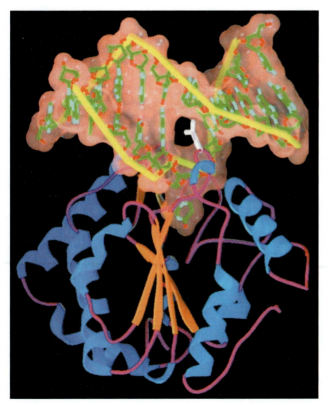

Figure 24-32 X-Ray structure of a complex of human uracil–DNA glycosylase with a 10-bp DNA containing a U·G base pair.   The protein (the C-terminal 223 residues of the 304-residue monomer) is colored according to its secondary structure (helices blue, β strands orange, and other segments magenta). The DNA, viewed looking into its minor groove, is drawn in stick form colored according to atom type (C green, N cyan, O red) and with its phosphate backbone traced by yellow tubes (phosphate O atoms have been omitted for clarity). The DNA's transparent solvent-accessible surface is pink and the side chain of Leu 272 is white. The uridine nucleotide has flipped out of the double helix (below the DNA) and has been hydrolyzed to yield an AP nucleotide and uracil, which remains bound in the enzyme's binding pocket. The side chain of Leu 272 has intercalated into the DNA base stack to fill the space vacated by the flipped-out uracil base. [Courtesy of John Tainer, The Scripps Research Institute, La Jolla, California. PDBid 4SKN.]

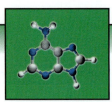

BOX 24-5

Perspectives in Biochemistry

Why Doesn't DNA Contain Uracil?

Three of the deoxynucleotide bases in DNA (adenine, guanine, and cytosine) also occur as ribonucleotide bases in RNA. The fourth deoxynucleotide base, thymine, is synthesized—at considerable metabolic effort (Section 22-3B)—from uracil, which occurs in RNA. Since uracil and thymine have identical base-pairing properties, why do cells bother to synthesize thymine at all?

This enigma was solved by the discovery of cytosine's penchant for conversion to uracil by deamination, either spontaneously or by reaction with nitrites (Section 24-4A). If U were a normal DNA base, the deamination of C would be highly mutagenic because there would be no indication of whether the resulting mismatched G·U base pair had initially been G·C or A·U. Since T is DNA's normal base, however, any U in DNA is almost certainly a deaminated C and can be removed by uracil–DNA glycosylase.

Uracil–DNA glycosylase also has an important function in DNA replication. dUTP, an intermediate in dTTP synthesis, is present in all cells in small amounts. DNA polymerases do not discriminate well between dUTP and dTTP, both of which can base pair with the template C. Consequently, newly synthesized DNA contains an occasional U. These U's are rapidly replaced by T through base excision repair.

the direction parallel to the view in Fig. 24-32. Tainer has postulated that UDG rapidly scans a DNA molecule for uracil by periodically binding to it so as to compress and thereby slightly bend the DNA backbone. The DNA bends more readily at a uracil-containing site (a U·G base pair is smaller than C·G and hence leaves a space in the base stack, whereas a U·A base pair is even weaker than T·A), permitting the enzyme to flip out the uracil by inserting Leu 272 into the minor groove. The exquisite specificity of the UDG's binding pocket for uracil prevents the binding and hydrolysis of any other base that the enzyme might have induced to flip. Thus the overall shapes of adenine and guanine exclude them from this pocket, whereas thymine's 5-methyl group is sterically blocked by the rigidly held side chain of Tyr 147. Cytosine, which has approximately the same shape as uracil, is excluded through a set of hydrogen bonds emanating from the protein that mimic those made by adenine in a Watson–Crick A·U base pair (dUTPase discriminates similarly between uracil and other bases; Fig. 22-14).

AP sites in mammalian DNA are highly cytotoxic because they irreversibly trap mammalian topoisomerase I in its covalent complex with DNA (Section 23-1C). Moreover, since the ribose at the AP site lacks a glycosidic bond, it can readily convert to its linear form (Section 8-1B), whose reactive aldehyde group can cross-link to other cell components. This rationalizes why AP sites remain tightly bound to UDG in solution. UDG activity is enhanced by AP endonuclease, the next enzyme in the base excision repair pathway, but the two enzymes do not interact in the absence of DNA. This suggests that UDG remains bound to an AP site it generated until it is displaced by the more tightly binding AP endonuclease, thereby protecting the cell from the AP site's cytotoxic effects. Its seems likely that other damage-specific DNA glycosylases function similarly.

C Nucleotide Excision Repair

All cells have a more elaborate pathway, **nucleotide excision repair (NER),** to correct pyrimidine dimers and other DNA lesions in which the bases are displaced from their normal position or have bulky substituents. The NER

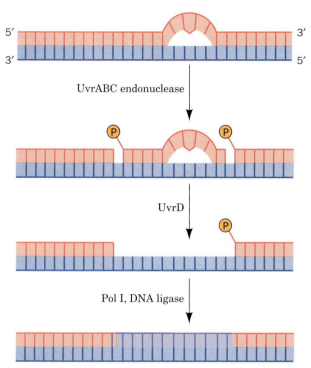

Figure 24-33 The mechanism of nucleotide excision repair (NER) of pyrimidine dimers.

system appears to respond to helix distortions rather than by the recognition of any particular group. In humans, NER is the major defense against two important carcinogens, sunlight and tobacco smoke.

In *E. coli*, NER is carried out in an ATP-dependent process through the actions of the **UvrA, UvrB,** and **UvrC** proteins (the products of the *uvrA, uvrB,* and *uvrC* genes). This system, which is often referred to as the **UvrABC endonuclease** (although there is no complex that contains all three subunits), cleaves the damaged DNA strand at the seventh and at the third or fourth phosphodiester bonds from the lesion's 5′ and 3′ sides, respectively (Fig. 24-33). The excised 11- or 12-nt oligonucleotide is then displaced by the binding of **UvrD** (also called **helicase II**) and is replaced through the actions of Pol I and DNA ligase.

Xeroderma Pigmentosum and Cockayne Syndrome Are Caused by Genetically Defective NER.

In humans, NER requires at least 16 proteins, and removes oligonucleotides of ~30 residues. Many of the enzymes involved in this pathway have been identified through mutations that are manifested as two genetic diseases. The inherited disease **xeroderma pigmentosum** (**XP;** Greek: *xeros*, dry + *derma*, skin) is mainly characterized by the inability of skin cells to repair UV-induced DNA lesions. Individuals suffering from this autosomal recessive condition are extremely sensitive to sunlight. During infancy they develop marked skin changes such as dryness, excessive freckling, and keratoses (a type of skin tumor; the skin of these children is described as resembling that of farmers with many years of sun exposure), together with eye damage, such as opacification and ulceration of the cornea. Moreover, they develop often fatal skin cancers at an ~2000-fold greater rate than normal and internal cancers at a 10- to 20-fold increased rate. Curiously, many individuals with XP also have a bewildering variety of seemingly unrelated symptoms including progressive neurological degeneration and developmental deficits.

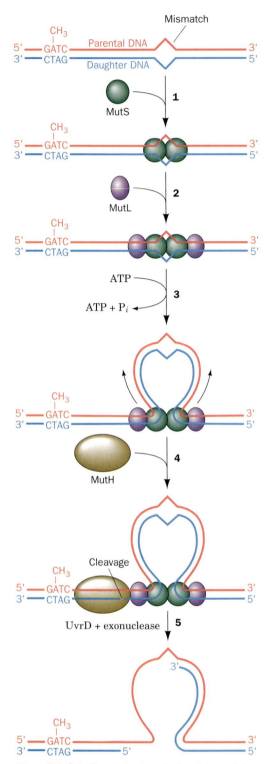

Figure 24-34 The mechanism of mismatch repair in *E. coli.*

Cockayne syndrome (CS), a rare inherited disease that is also associated with defective NER, arises from defects in three of the same genes that are defective in XP as well as in two additional genes. Individuals with CS are hypersensitive to UV radiation and exhibit stunted growth, neurological dysfunction due to neuron demyelination, and the appearance of premature aging but, intriguingly, have a normal incidence of skin cancer. The defective proteins in CS participate in recognizing an RNA polymerase whose progress has been halted by a damaged or distorted DNA template. In order for transcription to resume, the stalled RNA polymerase must be removed so that the DNA damage can be repaired by the NER system. In CS, the DNA cannot be repaired, which causes the cell to undergo **apoptosis** (programmed cell death; Section 27-3D). The death of transcriptionally active cells may account for the developmental symptoms of Cockayne syndrome.

D Mismatch Repair

Any replicational mispairing that has eluded the editing functions of the various participating DNA polymerases may still be corrected by a process known as **mismatch repair (MMR).** The MMR system can also correct insertions or deletions of up to four nucleotides (which arise from the slippage of one strand relative to the other in the active site of DNA polymerase). The importance of mismatch repair is indicated by the fact that defects in the human mismatch repair system result in a high incidence of cancer, most notably **hereditary nonpolyposis colorectal cancer syndrome,** which affects several organs and may be the most common inherited predisposition to cancer.

Mismatch repair in *E. coli* is carried out by three proteins (Fig. 24-34). A homodimer of **MutS** binds to a mismatched base pair, or to unpaired bases and then binds a homodimer of **MutL.** Then, in an ATP-dependent process, the resulting $MutS_2MutL_2$ complex translocates along the DNA in both directions, thereby causing the duplex DNA to form a mismatch-containing loop that is closed by the $MutS_2MutL_2$ complex. The mismatch repair system can distinguish the two DNA strands because newly replicated DNA remains hemimethylated until methyltransferases have had sufficient time to methylate the daughter strand (Box 24-4). On encountering a hemimethylated GATC palindrome, the $MutS_2MutL_2$ complex recruits **MutH** and activates this endonuclease to make a nick on the 5′ side of the unmethylated GATC sequence. This site may be located on either side of the mismatch and over 1000 bp distant from it. The UvrD helicase, which also participates in NER, separates the parental and daughter strands. An exonuclease completely removes the defective daughter segment, which is then replaced by the action of DNA polymerase III.

Eukaryotic cells contain several homologs of MutS and MutL but not of MutH and must therefore use some other cue than methylation status to differentiate the daughter and parental strands. One possibility is that a newly synthesized daughter strand is identified by its as yet unsealed nicks.

E Error-Prone Repair

Individuals with a variant form of XP exhibit increased rates of skin cancer even though their NER proteins are normal. The defect in this disorder has been pinpointed to an enzyme known as **DNA polymerase η (pol η)** When functioning normally, pol η can bypass DNA lesions such as UV-induced thymine dimers, incorporating two adenine bases in the new strand. Although it is useful as a translesion polymerase, pol η is relatively

inaccurate and has no proofreading exonuclease activity: It inserts an incorrect base on average every 30 nucleotides.

The existence of error-prone polymerases provides a fail-safe mechanism for replicating stretches of DNA that cannot be navigated by the standard replication machinery. In fact, such alternative DNA polymerases typically outnumber the polymerases devoted to normal replication. For example, eukaryotes contain at least five such "bypass" polymerases that act with different degrees of accuracy. However, the errors generated by bypass polymerases may still be corrected by the mismatch repair system.

End-Joining Repairs Double-Strand Breaks. Both the base excision repair and nucleotide excision repair pathways act when the lesion affects only one strand of DNA. However, DNA is susceptible to double-strand breaks **(DSBs),** often caused by ionizing radiation or the free radical by-products of oxidative metabolism. Unrepaired or misrepaired DSBs can be lethal to cells or cause chromosomal aberrations that may lead to cancer. Hence the efficient repair of DSBs is essential for cell viability and genomic integrity.

Cells have two general mechanisms to repair DSBs: **recombination repair** and **nonhomologous end-joining (NHEJ).** Here we discuss NHEJ, a process which, as its name implies, directly rejoins DSBs. The recombination repair of DSBs is discussed in Section 24-6B.

In NHEJ, the ends of the broken DNA must be aligned, frayed ends trimmed off or filled in, and the strands ligated. The core of the end-joining machinery in eukaryotes includes the protein **Ku** (a heterodimer of homologous 70- and 83-kD subunits, **Ku70** and **Ku80**), which appears to be the cell's broken-DNA sensor. The X-ray structure of Ku in complex with a 14-bp DNA, determined by Jonathan Goldberg, reveals that the protein cradles the dsDNA segment along its entire length and encircles its central ~3 bp segment (Fig. 24-35). Ku makes no specific contacts with the DNA's bases and few with its backbone but instead fits snugly into the major and minor grooves. Ku–DNA complexes dimerize so as to align both halves of

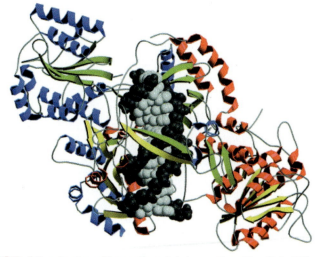

Figure 24-35 X-Ray structure of human Ku protein in complex with a 14-bp DNA. The subunits of Ku70 (*red helices and yellow β strands*) and Ku80 (*blue helices and green β strands*) are viewed along the pseudo-twofold axis relating them. The DNA is drawn in space-filling form with its sugar–phosphate backbone dark gray and its base pairs light gray. Note that the DNA is surrounded by a ring of protein. [Courtesy of John Tainer, The Scripps Research Institute, La Jolla, California. Based on an X-ray structure by Jonathan Goldberg, Memorial Sloan-Kettering Cancer Center, New York, New York. PDBid 1JEY.]

a double-strand break while leaving the strand ends accessible to nucleases, polymerases, and ligases. Nucleotide trimming, of course, generates mutations, but an unrepaired double-strand break would be even more detrimental to the cell.

The *E. coli* SOS Response Introduces Errors. In *E. coli*, agents that damage DNA induce a complex system of cellular changes known as the **SOS response.** Cells undergoing the SOS response cease dividing and increase their capacity to repair damaged DNA. **LexA,** a repressor, and **RecA,** a DNA-binding protein, regulate the activity of this system. (RecA is also a key player in homologous recombination, discussed in the following section, which offers a means for repairing damaged DNA after it has been replicated.) During normal growth, LexA represses SOS gene expression. However, when DNA is damaged (and cannot fully replicate), the resulting single strands bind to RecA to form a complex that activates LexA to cleave and thereby inactivate itself. The SOS genes, which include *recA* and *lexA* as well as the NER genes *uvrA* and *uvrB* (Section 24-5C), are thereby expressed. On the repair of the DNA, the DNA · RecA complex is no longer present, so the newly synthesized LexA again represses the expression of the SOS genes.

Among the 43 genes controlled by the SOS system are **DNA polymerases IV** and **V,** both of which replicate DNA with low fidelity and lack proofreading exonuclease activity. Because these translesion polymerases can synthesize DNA even when there is no information as to which bases were originally present, *the SOS repair system is error prone and consequently mutagenic.* It is therefore a process of last resort that is only initiated ~50 min after SOS induction if the DNA has not already been repaired by other means. Hence, the SOS repair system is a testimonial to the proposition that survival with a chance of loss of function (and the possible gain of a new one) is advantageous, in the Darwinian sense, over death—although only a small fraction of cells survive this process.

6 Recombination

Over the years, genetic studies have shown that genes are not immutable. In higher organisms, pairs of genes may exchange by crossing-over when homologous chromosomes are aligned (Fig. 24-36). Bacteria, which

(a)

(b)

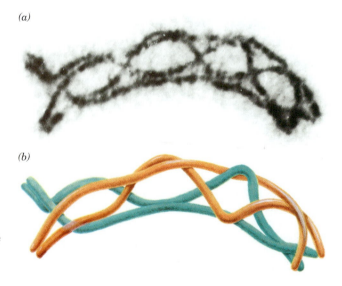

Figure 24-36 Crossing-over. (*a*) An electron micrograph and (*b*) its interpretive drawing of two homologous pairs of chromatids during meiosis in the grasshopper *Chorthippus parallelus.* Nonsister chromatids (*different colors*) may recombine at any of the points where they cross over. [Courtesy of Bernard John, The Australian National University, Canberra, Australia.]

do not contain duplicate chromosomes, also have an elaborate mechanism for recombining genetic information. In addition, foreign DNA can be installed in a host's chromosome through recombination. In this section, we examine the molecular events of recombination and discuss the biochemistry of **transposons,** which are mobile genetic elements.

A The Mechanism of Homologous Recombination

Homologous recombination (also called **general recombination**) occurs between DNA segments with extensive homology; **site-specific recombination** occurs between two short, specific DNA sequences. The prototypical model for homologous recombination (Fig. 24-37) was proposed by Robin Holliday in 1964. The corresponding strands of two aligned homologous DNA duplexes are nicked, and the nicked strands cross over to pair with the nearly complementary strands on the homologous duplex, thereby forming a segment of **heterologous DNA,** after which the nicks are sealed (Fig. 24-37a–e). The crossover point is a four-stranded structure known as a

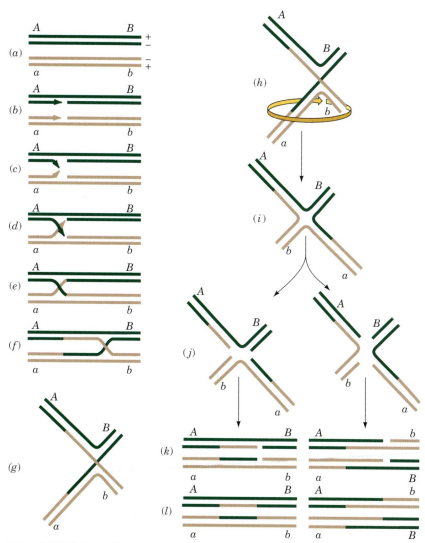

Figure 24-37 Key to Function.   The Holliday model of general recombination between homologous DNA duplexes.   **See the Animated Figures.**

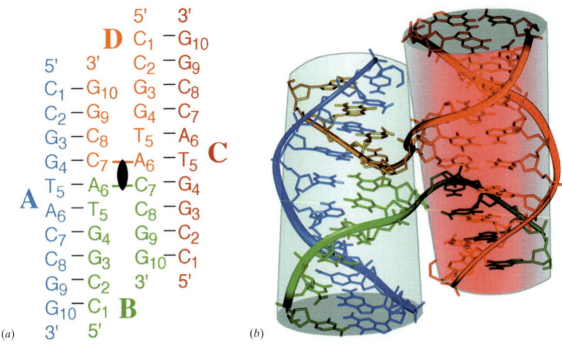

(a)

(b)

Figure 24-38   X-Ray structure of a Holliday junction.   (a) The secondary structure of the four-stranded Holliday junction formed by the palindromic sequence d(CCGGTACCGG) in which the four strands, A, B, C, and D, are individually colored, their nucleotides are numbered 1 to 10 from their 5′ to 3′ termini, and Watson–Crick base-pairing interactions are represented by black dashes. The twofold axis relating the two helices is represented by the black lenticular symbol. (b) The three-dimensional structure of the Holliday junction, as viewed along its twofold axis, in which the oligonucleotides are represented in stick form with their backbones traced by ribbons, all colored as in Part a. With the exception of the backbones of strands B and D at the crossovers, the two arms of this structure each form an undistorted B-DNA helix, including the stacking of the base pairs flanking the crossovers. Note that Fig. 24-37g is a schematic representation of this structure as viewed from the side in this drawing. [Courtesy of Shing Ho, Oregon State University. PDBid 1DCW.]

Holliday junction. A Holliday junction has, in fact, been observed in the X-ray structure of the self-complementary d(CCGGTACCGG), determined by Shing Ho (Fig. 24-38). The crossover point can move in either direction, often thousands of nucleotides, in a process known as **branch migration** (Fig. 24-37e and f).

The Holliday junction can be "resolved" into two duplex DNAs in two equally probable ways (Fig. 24-37g–l):

1. Cleavage of the strands that did not cross over exchanges the ends of the original duplexes to form, after nick sealing, the traditional recombinant DNA molecule (right branch of Fig. 24-37j–l).

2. Cleavage of the strands that crossed over exchanges a pair of homologous single-stranded segments (left branch of Fig. 24-37j–l).

RecA Promotes Recombination in *E. coli*.   The 352-residue RecA protein polymerizes on ssDNA or dsDNA that contains a single-strand gap. Electron microscopy reveals that RecA filaments bound to DNA form a right-handed helix with ~6.2 RecA monomers per turn (Fig. 24-39). Three nucleotides (or base pairs) bind to each RecA monomer so that the DNA assumes an extended conformation with ~18.6 nt (or bp) per turn (B-DNA has 10.4 bp per turn). The X-ray structure of RecA confirms that the central cavity of the RecA filament has a diameter of 25 Å, large enough to accommodate DNA.

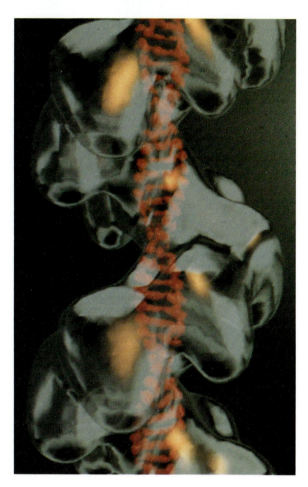

Figure 24-39   Model of a RecA–DNA complex.   In this electron microscopy–based image, the transparent surface represents an *E. coli* RecA filament. The extended and untwisted duplex DNA (*red*) has been modeled into this image. [Courtesy of Edward Egelman, University of Minnesota Medical School.]

How does RecA mediate DNA strand exchange between ssDNA and dsDNA? On encountering a duplex DNA with a strand that is complementary to its bound ssDNA, RecA partially unwinds the duplex and, in an ATP-driven reaction, exchanges the ssDNA with the corresponding strand on the dsDNA. ATP hydrolysis is thought to drive the rearrangement of a three-stranded DNA intermediate bound to the protein (Fig. 24-40) in a process that tolerates only a limited degree of mispairing. As

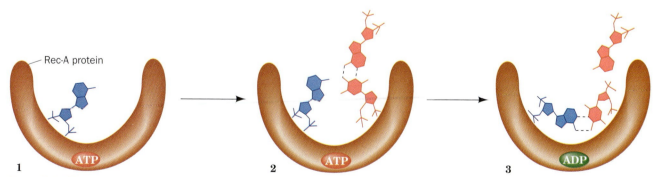

Figure 24-40   A model for RecA-mediated pairing.   (**1**) A ssDNA binds to RecA to form an initiation complex. (**2**) dsDNA binds to the initiation complex so as to transiently form a three-stranded helix. (**3**) RecA rotates the bases of the aligned homologous strands to effect strand exchange in an ATP-driven process. [After West, S.C., *Annu. Rev. Biochem.* **61**, 618 (1992).]

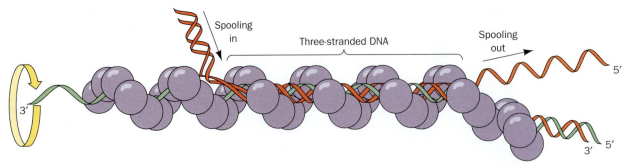

Figure 24-41 **Model for RecA-mediated strand exchange.** Homologous DNA molecules are paired in advance of strand exchange in a three-stranded helix. The ATP-driven rotation of the RecA filament (*purple*) around its helix axis causes duplex DNA to be "spooled in" to the filament, left to right as drawn. [After West, S.C., *Annu. Rev. Biochem.* **61,** 617 (1992).]

the RecA filament rotates about its axis, the duplex DNA is spooled in (Fig. 24-41). Two such strand-exchange processes must occur simultaneously in a Holliday junction. Eukaryotes contain proteins that are homologous to *E. coli* RecA and that apparently function in a similar ATP-dependent manner to mediate DNA recombination.

RecBCD Initiates Recombination by Making Single-Strand Nicks. The single-strand nicks to which RecA binds are made by the **RecBCD** protein, the 330-kD heterotrimeric product of the SOS genes *recB, recC,* and *recD,* which has both helicase and nuclease activities (Fig. 24-42). The process begins with RecBCD binding to the end of a dsDNA and then unwinding it via its ATP-driven helicase function. As it does so, it nucleolytically degrades the unwound single strands behind it, with the 3'-ending strand being cleaved more often and hence broken down to smaller fragments than the 5'-ending strand. However, on encountering the sequence GCTGGTGG from its 3' end (the so-called **Chi sequence,** which occurs about every 5 kb in the *E. coli* genome), it stops cleaving the 3'-ending strand and increases the rate at which it cleaves the 5'-ending strand, thereby yielding the 3'-ending single-strand segment to which RecA binds. This explains the observation that regions containing Chi sequences have elevated rates of recombination.

RecBCD can only commence unwinding DNA at a free duplex end. Such ends are not normally present in *E. coli,* which has a circular genome, but become available during recombinational processes.

Figure 24-42 **The generation of a 3'-ending single-strand DNA segment by RecBCD to initiate recombination.** (**1**) RecBCD binds to a free end of a dsDNA and, in an ATP-driven process, advances along the helix, unwinding the DNA and degrading the resulting single strands behind it, with the 3'-ending strand cleaved more often than the 5'-ending strand. (**2**) When RecBCD encounters a properly oriented Chi sequence, it increases the frequency at which it cleaves the 5'-ending strand but stops cleaving the 3'-ending strand, thereby generating the 3'-ending strand segment to which RecA binds.

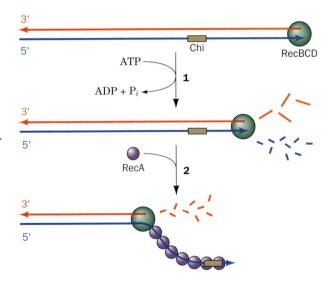

RuvABC Mediates the Branch Migration and the Resolution of the Holliday Junction.   Branch migration requires the proteins **RuvA** and **RuvB**. The X-ray structure of RuvA in complex with a Holliday junction, determined by Kosuke Morikawa, indicates that RuvA forms a homotetramer with the appearance of a four-petaled flower (C_4 symmetry) and is relatively flat ($80 \times 80 \times 45$ Å) with one face concave and the other convex (Fig. 24-43). The concave face (that facing the viewer in Fig. 24-43), which is highly positively charged and is studded with numerous conserved residues, has four symmetry-related grooves that bind the Holliday junction's four arms. This face's four centrally located projections or "pins" are negatively charged and hence the repulsive forces between them and the Holliday junction's anionic phosphate groups probably facilitate the separation of the ssDNA segments and guide them from one double helix to another.

However, this is not the entire story. The X-ray structure of a RuvA–Holliday junction complex crystallized under different conditions than that in Fig. 24-43 resembles the complex in Fig. 24-43 but with a second RuvA tetramer in face-to-face contact with the first. Here the Holliday junction is contained in two intersecting tunnels running through the resulting D_2-symmetric RuvA octamer. In addition, the X-ray structure of RuvA in complex with RuvB shows this octamer in complex with four noncontacting RuvB monomers. However, in the presence of dsDNA, RuvB, an ATPase, oligomerizes to form a pseudohexameric ring with a 30-Å-diameter hole through which a dsDNA can be threaded. A model for RuvAB action, based on these X-ray structures together with electron micrographs of the RuvAB–Holliday junction complex (Fig. 24-44a), places two RuvB hexamers on opposite sides of a RuvA octamer–Holliday junction complex (Fig. 24-44b). The two RuvB rings were originally postulated to counter-rotate as driven by ATP hydrolysis so as to screw the single strands of the vertical

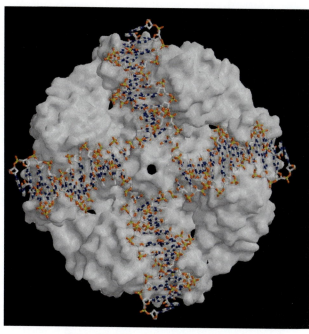

Figure 24-43   X-Ray structure of a RuvA–Holliday junction complex.   The protein tetramer, which is viewed along its fourfold axis, is represented by its molecular surface (*gray*). The DNA is drawn in stick form colored according to atom type (C white, N blue, O red, and P yellow). [Courtesy of Kosuke Morikawa, Biomolecular Engineering Research Institute, Osaka, Japan. PDBid 1C7Y.]

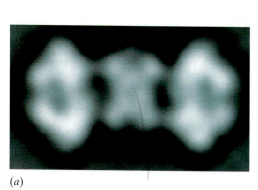

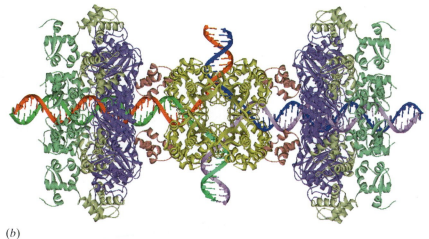

(a)   (b)

Figure 24-44   RuvAB–Holliday junction complex.   (a) An electron microcopy–based image of the complex. (b) A model of the complex based on fitting the X-ray structures of the RuvA octamer and the RuvB hexameric ring to the image in Part a. The RuvA octamer is yellow with its RuvB-contacting domains orange. The N, M, and C domains of the RuvB subunits are colored purple, light yellow, and light green, respectively. The DNA of the Holliday junction is drawn in ladder form with its homologous red and magenta strands complementary to its green and blue strands. The complex is postulated to drive branch migration via the ATP-driven "counter-rotation" of two oppositely located dsDNA strands by their associated RuvB hexamers. This unidirectionally pumps (screws) the vertical dsDNA strands in through the center of the Holliday junction on RuvA and out through the RuvB hexamers, a process in which the strands of the vertical dsDNAs separate and then rejoin, via base pairing, to form the horizontal dsDNAs. [Courtesy of Kazuhiro Yamada, Tomoko Miyata, and Kosuke Morikawa, Biomolecular Engineering Research Institute, Osaka, Japan.]

dsDNAs through the center of the junction and into the horizontal dsDNAs, thereby effecting branch migration. However, the large area of contact between RuvA and RuvB suggests that the RuvB hexamer does not rotate relative to the RuvA octamer. Instead, it appears that RuvB's DNA-contacting loops unidirectionally "walk" along the horizontal dsDNA's grooves (much like the hexagonal helicases; Section 24-2B), which screws these dsDNAs out through RuvB. This pulls the single strands of the vertical dsDNAs through the center of RuvA, where they rejoin to form the horizontal dsDNAs. The direction of branch migration depends on which pair of opposing arms of the Holliday junction the RuvB hexamers are loaded onto.

The final stage of homologous recombination is the resolution of the Holliday junction into its two homologous dsDNAs. This process is carried out by **RuvC,** a homodimeric nuclease. This requires that one of the RuvA tetramers dissociate from the complex shown in Fig. 24-44*b* to expose the Holliday junction DNA. RuvC presumably sits down on the open face of the resulting RuvA–Holliday junction complex (the side facing the viewer in Fig. 24-43) to cleave oppositely located strands at the Holliday junction (Figs. 24-37*i* and *j*). The resulting single-strand nicks are then sealed by DNA ligase (Figs. 24-37*k* and *l*).

B Recombination Repair

In haploid organisms such as bacteria, homologous recombination between chromosomal DNA and exogenously supplied DNA as occurs, for example, in transformation (Section 3-3A) is such a rare event that the vast majority of such cells never participate in this process. Similarly, in multicellular organisms, the only time that gene shuffling through homologous recombination occurs is during meiosis (Fig. 24-36), which takes place only in germ cells. Why then do nearly all cells have elaborate systems for mediating homologous recombination? This is because damaged replication forks occur at a frequency of at least once per bacterial cell generation and perhaps ten times per eukaryotic cell cycle. The DNA lesions that damaged the replication forks can be corrected via homologous recombination in a process named recombination repair [error-prone repair (Section 24-5E), which is highly mutagenic, is a process of last resort]. Indeed, the rates of synthesis of RuvA and RuvB are greatly enhanced by the SOS response. Thus, as Michael Cox pointed out, *the primary function of homologous recombination is to repair damaged replication forks.* In what follows, we describe recombination repair as it occurs in *E. coli.*

Recombination Repair Reconstitutes Damaged Replication Forks. Recombination repair is called into play when a replication fork encounters an unrepaired single-strand nick (Fig. 24-45):

1. When this occurs, the replication fork collapses (falls apart).
2. The repair process begins via the RecBCD + RecA–mediated invasion of the newly synthesized and undamaged 3′-ending strand into the homologous dsDNA starting at its broken end.
3. Branch migration, as mediated by RuvAB, then yields a Holliday junction, which exchanges the replication fork's 3′-ending strands.
4. RuvC then resolves the Holliday junction yielding a reconstituted replication fork ready for replication restart.

Thus, the 5′-ending strand of the nick has, in effect, become the 5′ end of an Okazaki fragment.

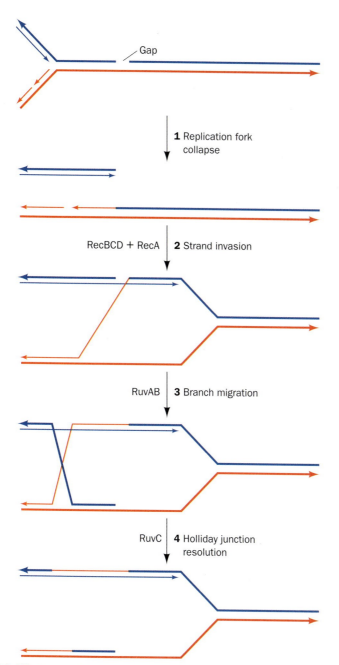

Figure 24-45   **The recombination repair of a replication fork that has encountered a single-strand nick.**   Thick lines indicate parental DNA, thin lines indicate newly synthesized DNA, and the arrows point in the 5' → 3' direction. [After Cox, M.M., *Annu. Rev. Genet.* **35**, 53 (2001).]

The final step in the recombination repair process is the restart of DNA replication. This process is, of necessity, distinct from the replication initiation that occurs at *oriC* (Section 24-2B). **Origin-independent replication restart** is mediated by a specialized seven-subunit primosome, which has therefore been named the **restart primosome.**

Recombination Repair Reconstitutes Double-Strand Breaks.   We have seen that double-strand breaks (DSBs) in DNA can be rejoined, often mutagenically, by nonhomologous end-joining (NHEJ; Section 24-5E). DSBs may also be

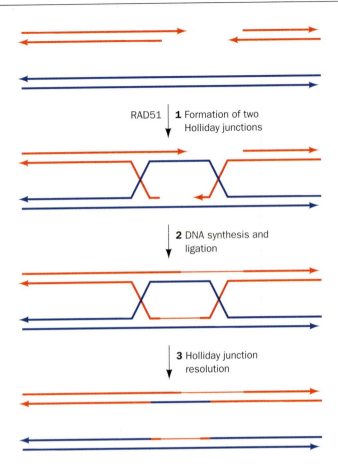

Figure 24-46 The repair of a double-strand break in DNA by homologous end-joining. Thick lines indicate parental DNA, thin lines indicate newly synthesized DNA, and the arrows point in the 5′ → 3′ direction. [After Haber, J.E., *Trends Genet.* **16**, 259 (2000).]

nonmutagenically repaired through a recombination repair process known as **homologous end-joining,** which occurs via two Holliday junctions (Fig. 24-46):

1. The DSB's double-stranded ends are resected to produce single-stranded ends. One of the 3′-ending strands invades the corresponding sequence of a homologous chromosome to form a Holliday junction, a process that, in eukaryotes, is mediated by the RecA homolog **RAD51.** The other 3′-ending strand pairs with the displaced strand segment on the homologous chromosome to form a second Holliday junction.

2. DNA synthesis and ligation fills in the gaps and seals the joints.

3. Both Holliday junctions are resolved to yield two intact double strands.

Thus, the sequences that may have been lost in the formation of the DSB are copied from the homologous chromosome. Of course, a limitation of homologous end-joining, particularly in haploid cells, is that a homologous chromosomal segment may not be available.

The importance of recombination repair in humans is demonstrated by the observation that defects in the proteins **BRCA1** (1863 residues) and **BRCA2** (3418 residues), both of which interact with RAD51, are associated with a greatly increased incidence of breast, ovarian, prostate, and pancreatic cancers. Indeed, individuals with mutant *BRCA1* or *BRCA2* genes have up to an 80% lifetime risk of developing cancer.

C Transposition

In the early 1950s, Barbara McClintock reported that the variegated pigmentation pattern of maize (Indian corn) kernels results from the action of genetic elements that can move within the maize genome. This proposal

was resoundingly ignored because it was contrary to the then-held ortho-doxy that chromosomes consist of genes linked in fixed order. Another 20 years were to pass before evidence of mobile genetic elements was found in another organism, *E. coli.*

Transposons Move Genes between Unrelated Sites   It is now known that **transposable elements,** or **transposons,** are common in both prokaryotes and eukaryotes, where they influence the variation of phenotypic expression over the short term and evolutionary development over the long term. Each transposon codes for the enzymes that insert it into the recipient DNA. This process differs from homologous recombination in that it requires no homology between donor and recipient DNA and occurs at a rate of only one event in every 10^4 to 10^7 cell divisions.

Prokaryotic transposons with three levels of complexity have been characterized:

1. The simplest transposons are named **insertion sequences** or **IS elements.** They are normal constituents of bacterial chromosomes and **plasmids** (autonomously replicating circular DNA molecules that usually consist of several thousand base pairs; Section 3-1A). For example, a common *E. coli* strain has eight copies of **IS1** and five copies of **IS2.** IS elements generally consist of <2000 bp, comprising a so-called **transposase** gene and, in some cases, a regulatory gene, flanked by short inverted (having opposite orientation) terminal repeats. An inserted IS element is flanked by a directly (having the same orientation) repeated segment of host DNA (Fig. 24-47). This suggests that *an IS element is inserted in the host DNA at a staggered cut that is later filled in* (Fig. 24-48). The length of this target sequence (most commonly 5–9 bp), but not its sequence, is characteristic of the IS element.

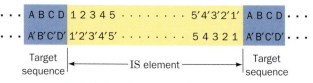

Figure 24-47   Structure of IS elements.   These and other transposons have inverted terminal repeats (*numerals*) and are flanked by direct repeats of host DNA sequences (*letters*).

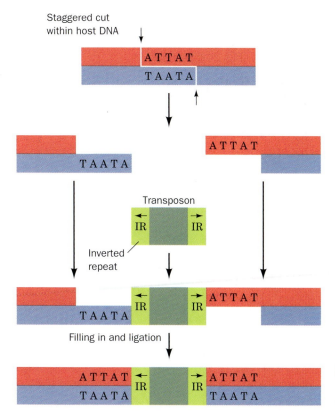

Figure 24-48   A model for transposon insertion.   A staggered cut that is later filled in generates direct repeats of the target sequence.

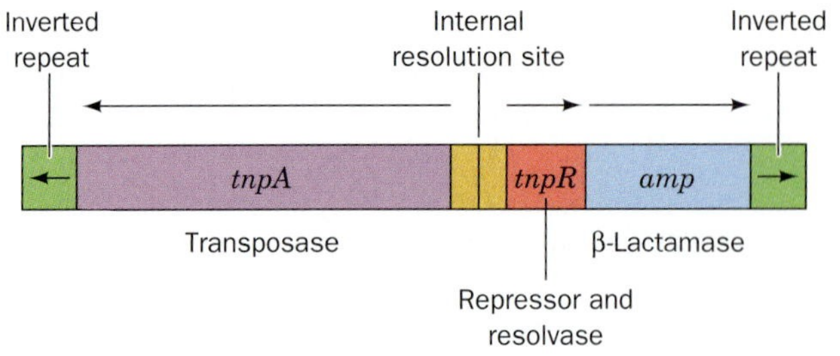

Inverted
repeat

Internal
resolution site

Inverted
repeat

| | *tnpA* | | *tnpR* | *amp* | |

Transposase

β-Lactamase

Repressor and
resolvase

Figure 24-49   A map of transposon Tn3.

2. *More complex transposons carry genes not involved in the transposition process,* such as antibiotic resistance genes. For example, **Tn3** (Fig. 24-49) consists of 4957 bp and has inverted terminal repeats of 38 bp each. The central region of Tn3 codes for three proteins: (1) a 1015-residue transposase named **TnpA;** (2) a 185-residue protein known as **TnpR,** which represses the expression of both *tnpA* and *tnpR* and mediates the site-specific recombination reaction necessary for transposition (see below); and (3) a **β-lactamase** (encoded by *amp*) that inactivates ampicillin (Box 8-3). The site-specific recombination occurs in an AT-rich region, the **internal resolution site,** between *tnpA* and *tnpR*.

3. The so-called **composite transposons** (Fig. 24-50) consist of a gene-containing central region flanked by two identical or nearly identical IS-like modules that have either the same or an inverted relative orientation. Composite transposons apparently arose by the association of two originally independent IS elements. Experiments demonstrate that *composite transposons can transpose any sequence of DNA in their central region.*

A Proposed Transposition Mechanism.   Transposons do not simply jump from point to point within a genome, as their name implies. Instead, *the mechanism of transposition involves the replication of the transposon.* A model for the movement of a transposon between two plasmids consists of the following steps (Fig. 24-51):

1. A pair of staggered single-strand cuts (such as in Fig. 24-48) is made at the target sequence of the recipient plasmid. Similarly, single-strand cuts are made on opposite strands on either side of the transposon.

2. Each of the transposon's free ends is ligated to a protruding single

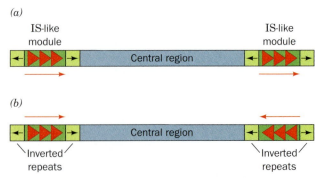

(a)

IS-like
module

IS-like
module

Central region

(b)

Central region

Inverted
repeats

Inverted
repeats

Figure 24-50   A composite transposon.   This element consists of two identical or nearly identical IS-like modules (*green*) flanking a central region carrying various genes. The IS-like modules may have either (*a*) direct or (*b*) inverted relative orientations.

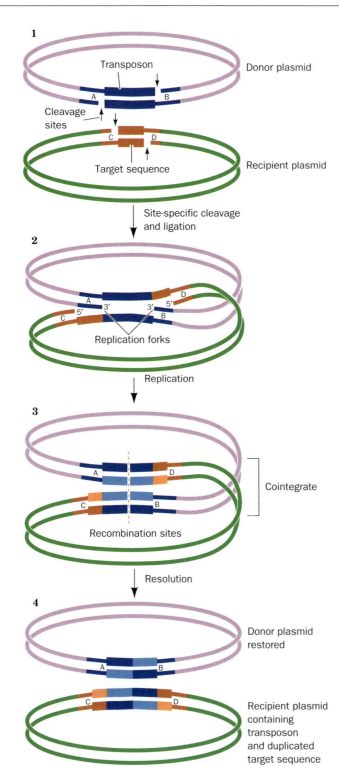

Figure 24-51 A model for transposition involving the intermediacy of a cointegrate. More lightly shaded bars represent newly synthesized DNA. [After Shapiro, J.A., *Proc. Natl. Acad. Sci.* **76**, 1934 (1979).]

strand at the insertion site. This forms a replication fork at each end of the transposon.

3. The transposon is replicated, thereby yielding a **cointegrate** (the fusion of the two plasmids). Such cointegrates have been isolated.

4. Through a site-specific crossover between the internal resolution sites of the two transposons, the cointegrate is resolved into two separate

plasmids, each of which contains the transposon. This recombination process is catalyzed by a transposon-coded **resolvase** (TnpR in Tn3) rather than by RecA.

Transposition Is Responsible for Much Genetic Rearrangement. In addition to mediating their own insertion into DNA, *transposons promote inversions, deletions, and rearrangements of the host DNA.* Inversions can occur when the host DNA contains two copies of a transposon in inverted orientation. The recombination of these transposons inverts the region between them (Fig. 24-52*a*). If, instead, the two transposons have the same orientation, recombination deletes the segment between them (Fig. 24-52*b*). The deletion of a chromosomal segment in this manner, followed by its integration into the chromosome at a different site by a separate recombination event, results in chromosomal rearrangement.

Transposons can be considered nature's genetic engineering "tools." For example, the rapid evolution, since antibiotics came into common use, of plasmids that confer resistance to several antibiotics (Section 3-5A) has resulted from the accumulation of the corresponding antibiotic-resistance transposons in these plasmids. Transposon-mediated rearrangements may also have been responsible for forming new proteins by linking two formerly independent gene segments. Moreover, transposons can apparently mediate the transfer of genetic information between unrelated species.

Most Eukaryotic Transposons Resemble Retroviruses. Transposons occur in such distantly related eukaryotes as yeast, maize, fruit flies, and humans. In fact, ~3% of the human genome consists of DNA-based transposons although, in most cases, their sequences have mutated so as to render them inactive, that is, these transposons are evolutionary fossils.

Despite the foregoing, many eukaryotic transposons exhibit little similarity to those of prokaryotes. Rather, their base sequences resemble those of retroviruses (Box 24-2), which suggests that these transposons are degenerate

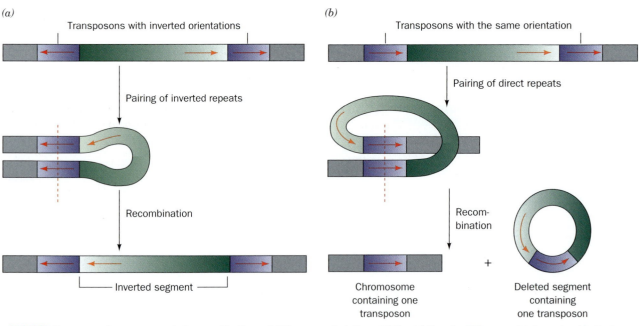

(a) Transposons with inverted orientations

Pairing of inverted repeats

Recombination

Inverted segment

(b) Transposons with the same orientation

Pairing of direct repeats

Recombination

Chromosome containing one transposon

+

Deleted segment containing one transposon

Figure 24-52 Chromosomal rearrangement via recombination. (*a*) The inversion of a DNA segment between two identical transposons with inverted orientations. (*b*) The deletion of a DNA segment between two identical transposons with the same orientation.

retroviruses. The transposition of these so-called **retrotransposons** occurs via a pathway that resembles the replication of retroviral DNA: (1) their transcription to RNA, (2) the reverse transcriptase–mediated copying of this RNA to cDNA, and (3) the largely random insertion of this DNA into the host organism's genome as mediated by enzymes known as **integrases** (which catalyze reactions similar to and structurally resemble DNA transposases).

Vertebrate genomes also contain retrotransposons that replicate via a different mechanism from that of retroviruses. A common family of these **nonviral retrotransposons** is the 6- to 8-kb **long interspersed nuclear elements (LINEs).** Around 1.4 million LINEs or LINE fragments are present in the human genome and comprise ~20% of its 3.2 billion bp. The great majority of these molecular parasites have mutated to the point of inactivity but a few still appear capable of further transposition. Indeed, several hereditary diseases are caused by the insertion of a LINE into a gene. Several other types of retrotransposons collectively comprise another 22% of the human genome, for a total of ~45% transposon content.

SUMMARY

1. DNA is replicated semiconservatively through the action of DNA polymerases that use the separated parental strands as templates for the synthesis of complementary daughter strands.

2. Replication is semidiscontinuous: The leading strand is synthesized continuously while the lagging strand is synthesized as RNA-primed Okazaki fragments that are later joined.

3. *E. coli* DNA polymerase I (Pol I), which has $3' \rightarrow 5'$ and $5' \rightarrow 3'$ exonuclease activities in addition to its $5' \rightarrow 3'$ polymerase activity, excises RNA primers and replaces them with DNA. Pol III is the primary polymerase in *E. coli*.

4. To initiate replication, parental strands are first melted apart at a specific site named *oriC* and further unwound by a helicase. Single-strand binding protein (SSB) prevents the resulting single strands from reannealing. A primase-containing primosome synthesizes an RNA primer.

5. Because DNA polymerases operate only in the $5' \rightarrow 3'$ direction, the lagging strand template must loop back to the replisome, which contains two Pol III units. A sliding clamp increases Pol III processivity. DNA ligase seals the nicks between Okazaki fragments. Replication proceeds until the two replication forks meet between oppositely facing *Ter* sequences.

6. The high fidelity of DNA replication is achieved by the regulation of dNTP levels, by the requirement for RNA primers, by $3' \rightarrow 5'$ proofreading, and by DNA repair mechanisms.

7. Eukaryotes contain a number of DNA polymerases. Pol α extends primers, and the highly processive pol δ is the main polymerase. Eukaryotic replication has multiple origins and proceeds through nucleosomes.

8. In order to replicate the 5' end of the lagging strand, eukaryotic chromosomes end with repeated telomeric sequences added by the ribonucleoprotein telomerase. The 3' extension at the end of each chromosome serves as a template for primer synthesis. Somatic cells lack telomerase activity, which may prevent them from transforming to cancer cells.

9. Mutations in nucleotide sequences arise spontaneously from replication errors and from alterations triggered by agents such as UV light, radiation, and chemical mutagens. Many compounds that are mutagenic in the Ames test are also carcinogenic.

10. Some forms of DNA damage (e.g., alkylated bases and pyrimidine dimers) may be reversed in a single step. In base excision repair, a glycosylase removes damaged bases. In nucleotide excision repair, an oligonucleotide containing the lesion is removed and replaced. Repair pathways that use error-prone polymerases, nonhomologous end-joining, and the *E. coli* SOS response are mutagenic.

11. Homologous recombination, in which strands of homologous DNA segments are exchanged, involves a crossover structure (Holliday junction). Recombination requires proteins to unwind DNA, mediate strand exchange, drive branch migration, and resolve the Holliday junction. Damaged DNA can be repaired through recombination.

12. Transposons are genetic elements that move within a genome by mechanisms involving their replication. Transposons mediate rearrangement of the host DNA.

REFERENCES

DNA Replication

Bell, S.P. and Dutta, A., DNA replication in eukaryotic cells, *Annu. Rev. Biochem.* **71,** 333–374 (2002).

Benkovic, J.J., Valentine, A.M., and Salinas, F., Replisome-mediated DNA replication, *Annu. Rev. Biochem.* **70,** 181–208 (2001).

Blackburn, E.H., Telomere states and cell fates, *Nature* **408,** 53–56 (2000).

Caruthers, J.M. and McKay, D.B., Helicase structure and mechanism, *Curr. Opin. Struct. Biol.* **12,** 123–133 (2002).

Hübscher, U., Maga, G., and Spadari, S., Eukaryotic DNA polymerases, *Annu. Rev. Biochem.* **71,** 133–163 (2002).

Jeruzalmi, D., O'Donnell, M., and Kuriyan, J., Clamp loaders and sliding clamps, *Curr. Opin. Struct. Biol.* **12,** 217–224 (2002).

Kelleher, C., Teixeira, M.T., Förstemann, K., and Lingner, J., Telomerase: Biochemical considerations for enzyme and substrate, *Trends Biochem. Sci.* **27,** 572–579 (2002).

Smogorzewska, A. and de Lange, T., Regulation of telomerase by telomeric proteins, *Annu. Rev. Biochem.* **73,** 177–208 (2004).

Steitz, T.A., DNA polymerases: structural diversity and common mechanisms, *J. Biol. Chem.* **274,** 17395–17398 (1999).

DNA Repair

Friedberg, E.C., Wagner, R., and Radman, M., Specialized DNA polymerases, cellular survival, and the genesis of mutations, *Science* **296,** 1627–1630 (2002). [Reviews the roles of error-prone DNA polymerases in DNA repair and antibody generation.]

Rouse, J. and Jackson, S.P., Interfaces between the detection, signaling, and repair of DNA damage, *Science* **297,** 547–551 (2002). [Summarizes current understanding of how signaling pathways triggered by DNA damage lead to DNA repair.]

Sancar, A., Lindsey-Bolz, L.A., Ünsal-Kaçmaz, K., and Linn, S., Molecular mechanisms of mammalian DNA repair and the DNA damage checkpoints, *Annu. Rev. Biochem.* **73,** 39–85 (2004).

Scriver, C.R., Beaudet, A.L., Sly, W.S., and Valle, D. (Eds.), *The Metabolic and Molecular Bases of Inherited Disease* (8th ed.), Chaps. 28 and 32, McGraw-Hill (2001). [Discussions of xeroderma pigmentosum, Cockayne syndrome, and hereditary nonpolyposis colorectal cancer.]

Recombination and Transposition

Cox, M.M., Recombinational DNA repair of damaged replication forks in *Escherichia coli:* questions, *Annu. Rev. Genet.* **35,** 53–82 (2001).

Craig, N.L., Craigie, R., Gellert, M., and Lambowitz, A.M. (Eds.), *Mobile DNA II,* ASM Press (2002). [A compendium of authoritative articles.]

Grindley, N.D.F. and Leschziner, A.E., DNA transpositions: from a black box to a color monitor, *Cell* **83,** 1063–1066 (1995). [Describes the mechanism of transposition and the structure and function of transposases.]

Haber, J.E, DNA recombination: the replication connection, *Trends Biochem. Sci.* **24,** 271–275 (1999).

Kowalczykowski, S.C., Initiation of genetic recombination and recombination-dependent replication, *Trends Biochem. Sci.* **25,** 156–165 (2000). [Presents models for and describes the proteins involved in recombination/replication events in bacteria.]

Yamada, K., Ariyoshi, M., and Morikawa, K., Three-dimensional structural views of branch migration and resolution in DNA homologous recombination, *Curr. Opin. Struct. Biol.* **14,** 130–137 (2004).

 KEY TERMS

θ structure	primase	insertion/deletion mutation	heterologous DNA
replication fork	replisome	carcinogen	Holliday junction
Okazaki fragment	blunt end ligation	Ames test	branch migration
pulse-labeling	replicon	photoreactivation	recombination repair
leading strand	telomere	base excision repair (BER)	homologous end-joining
lagging strand	telomerase	DNA glycosylase	transposon
ligase	ribonucleoprotein	AP site	IS element
primer	G-quartet	nucleotide excision repair (NER)	internal resolution site
processivity	mutation	mismatch repair (MMR)	composite transposon
proofreading	pyrimidine dimer	nonhomologous end-joining (NHEJ)	cointegrate
nick translation	mutagen		retrovirus
helicase	point mutation	SOS response	retrotransposon
SSB	transition	homologous recombination	
primosome	transversion		

 STUDY EXERCISES

1. Explain how DNA replication is semiconservative, bidirectional, and semidiscontinuous.

2. Summarize the role of RNA in replicating the lagging strand and the ends of linear chromosomes.

3. Describe the functions of the three catalytic activities of *E. coli* DNA Pol I.

4. What is the purpose of the β clamp and PCNA?

5. Which reactions required for DNA replication are driven

by ATP hydrolysis? What drives the non-ATP-dependent reactions?

6. What are the sources of high DNA replication fidelity?

7. How does DNA replication differ in eukaryotes and prokaryotes?

8. Describe the structure and function of telomeres.

9. List some of the ways mutations can arise.

10. Summarize the steps of base excision repair and nucleotide excision repair.

11. What are the advantages and disadvantages of error-prone DNA polymerases?

12. Describe the protein activities required for homologous recombination. What is the role of recombination in repairing damaged DNA?

13. What protein activities are required to support recombination?

14. How do transposons mediate genetic rearrangements?

PROBLEMS

1. Approximately how many Okazaki fragments are synthesized in the replication of the *E. coli* chromosome?

2. Explain why a DNA polymerase that could synthesize DNA in the $3' \to 5'$ direction would have a selective disadvantage even if it had $5' \to 3'$ proofreading activity.

3. You have discovered a drug that inhibits the activity of inorganic pyrophosphatase. What effect would this drug have on DNA synthesis?

4. A reaction mixture contains DNA polymerase, the four dNTPs, and one of the DNA molecules whose structure is represented below. Which reaction mixtures generate PP$_i$?

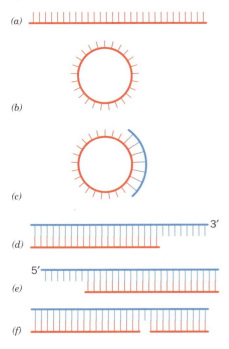

(a)

(b)

(c)

(d)   3'

(e) 5'

(f)

5. Why are there no Pol I mutants that completely lack $5' \to 3'$ exonuclease activity?

6. Why is it advantageous to use only the Klenow fragment, rather than intact *E. coli* Pol I, in DNA sequencing reactions (Section 3-4C)?

7. *E. coli oriC* is rich in A · T base pairs. Why is this advantageous?

8. Explain why DNA gyrase is required for efficient unwinding of DNA by helicase at the replication fork.

9. Why is the observed mutation rate of *E. coli* 10^{-8} to 10^{-10} per base pair replicated, even though the error rates of Pol I and Pol III are 10^{-6} to 10^{-7} per base pair replicated?

10. Why can't a linear duplex DNA, such as that of bacteriophage T7, be fully replicated by just *E. coli*-encoded proteins?

11. The base analog **5-bromouracil (5BU),**

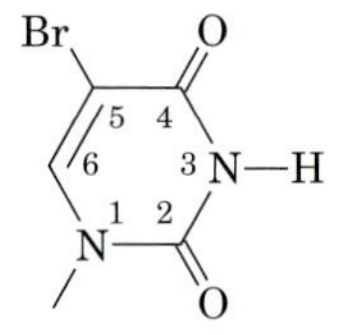

5-Bromouracil (5BU)

which sterically resembles thymine, more readily undergoes tautomerization from its keto form to its enol form than does thymine. 5BU can be incorporated into newly synthesized DNA when it pairs with adenine on the template strand. However, the enol form of 5BU can pair with guanine rather than adenine. (a) Draw the 5BU · G base pair. (b) What type of mutation results?

12. The deamination of adenine yields the base hypoxanthine, which can pair with cytosine. What type of repair system would cells most likely use to repair DNA with a deaminated adenine?

13. **Hydroxylamine** (NH$_2$OH) converts cytosine to the compound shown below.

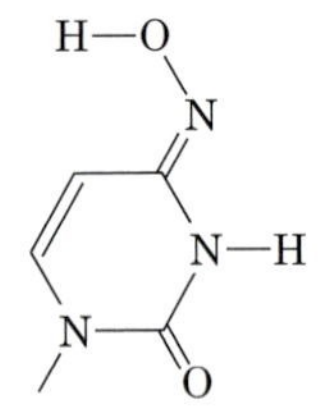

With which base does this modified cytosine pair? Does this generate a transition or a transversion mutation?

14. In eukaryotic cells, a specific triphosphatase cleaves deoxy-8-oxoguanosine triphosphate (oxo-dGTP) to oxo-dGMP + PP$_i$. What is the advantage of this reaction?

15. Certain sites in the *E. coli* chromosome are known as **hot spots** because they have unusually high rates of point mutations. Many of these sites contain a 5-methylcytosine residue. Explain the existence of such hot spots.

16. Predict whether loss of the following *E. coli* genes would be lethal or not: (a) *dnaB* (which encodes DnaB), (b) *polA* (which encodes Pol I), (c) *ssb*, (d) *recA*.

17. In *E. coli*, all newly synthesized DNA appears to be fragmented (an observation that could be interpreted to mean that the leading strand as well as the lagging strand is synthesized discontinuously). However, in *E. coli* mutants that are defective in uracil–DNA glycosylase, only about half the newly synthesized DNA is fragmented. Explain.

18. *E. coli* DNA polymerase V has the ability to bypass thymine dimers. However, Pol V tends to incorporate G rather than A opposite the damaged T bases. Would you expect Pol V to be more or less processive than Pol III? Explain.

Transcription and RNA Processing

DNA is confined almost exclusively to the nucleus of eukaryotic cells, as shown by microscopists in the 1930s. By the 1950s, the site of protein synthesis was identified by showing that radioactively labeled amino acids that had been incorporated into proteins were associated with cytosolic RNA–protein complexes called ribosomes. Thus, *protein synthesis is not immediately directed by DNA because, at least in eukaryotes, DNA and ribosomes are never in contact.* The intermediary between DNA and the protein-biosynthesis machinery, as outlined in Francis Crick's central dogma of molecular biology (Section 3-3B), is RNA.

Cells contain three major types of RNA: **ribosomal RNA (rRNA),** which constitutes two-thirds of the ribosomal mass; **transfer RNA (tRNA),** a set of small compact molecules that deliver amino acids to the ribosomes for assembly into proteins; and **messenger RNA (mRNA),** whose nucleotide sequences direct protein synthesis. In addition, a host of other small RNA species plays various roles in the processing of newly transcribed RNA molecules. All types of RNA can be shown to hybridize with complementary sequences on DNA from the same organism. Thus, *all cellular RNAs are transcribed from DNA templates.*

The transcription of DNA to RNA is carried out by **RNA polymerases (RNAPs)** that operate as multisubunit complexes, as do the DNA polymerases that catalyze DNA replication. In this chapter, we examine the catalytic properties of RNAPs and discuss how these proteins—unlike DNA polymerases—are targeted to specific genes. We shall also see how newly synthesized RNA is processed to become fully functional.

1   RNA Polymerase

RNAP, the enzyme responsible for the DNA-directed synthesis of RNA, was discovered independently in 1960 by Samuel Weiss and Jerard Hurwitz. The enzyme couples together the ribonucleoside triphosphates (NTPs) ATP, CTP, GTP, and UTP on DNA templates in a reaction that is driven by the release and subsequent hydrolysis of PP_i:

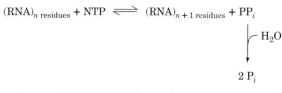

$$(RNA)_{n \text{ residues}} + NTP \rightleftharpoons (RNA)_{n+1 \text{ residues}} + PP_i$$

$$\downarrow H_2O$$

$$2\,P_i$$

All cells contain RNAP. In bacteria, one enzyme synthesizes all of the cell's RNA except the short RNA primers employed in DNA replication (Section 24-2B). Eukaryotic cells contain four or five RNAPs that each synthesize a different class of RNA. We shall first consider the bacterial enzyme because it is smaller and simpler than the eukaryotic enzymes.

A   Enzyme Structure

The *E. coli* RNAP **holoenzyme** is an ~449-kD protein with subunit composition $\alpha_2\beta\beta'\omega\sigma$. Once RNA synthesis has been initiated, however, the σ subunit (also called the **σ factor**) dissociates from the **core enzyme** $\alpha_2\beta\beta'\omega$, which carries out the actual polymerization process. RNAP is large enough to be clearly visible in electron micrographs (Fig. 25-1).

Low-resolution structural studies indicate that bacterial RNAP exhibits an overall "hand" shape similar to that of many DNA polymerases (Sec-

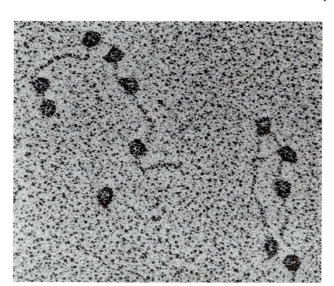

Figure 25-1 An electron micrograph of *E. coli* RNA polymerase holoenzyme. This soluble enzyme, one of the largest known, is attached to various promoter sites on bacteriophage T7 DNA. [From Williams, R.C., *Proc. Natl. Acad. Sci.* **74**, 2313 (1977).]

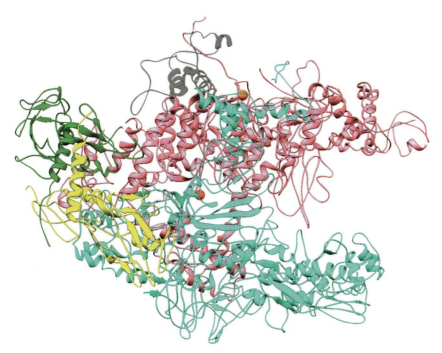

Figure 25-2 **X-Ray structure of *Taq* core RNA polymerase.** The two α subunits are yellow and green, the β subunit is cyan, the β′ subunit (part of which is disordered and therefore not visible in this model) is pink, and the ω subunit is white. Bound Mg^{2+} and Zn^{2+} ions are represented by red and orange spheres, respectively. [Based on an X-ray structure by Seth Darst, The Rockefeller University.]

tion 24-2A) and reverse transcriptase (Box 24-2). The X-ray structure of *E. coli* RNAP has not been determined. However, Seth Darst has elucidated the X-ray structure of the closely related enzyme from *Thermus aquaticus (Taq)*. The core enzyme is shaped like a crab claw whose two pincers are formed by the β and β′ subunits (Fig. 25-2). The channel or space between the pincers is ~27 Å high. The active site, which includes a Mg^{2+} ion, is located at the base of the channel. In the *Taq* holoenzyme, the σ subunit extends across the top of the core enzyme and causes the pincers to come together so as to narrow the channel between them by about 10 Å. The outer surface of the holoenzyme is almost uniformly negatively charged, whereas those surfaces presumed to interact with nucleic acids, particularly the inner walls of the channel, are positively charged.

B Template Binding

RNA synthesis is normally initiated only at specific sites on the DNA template. In contrast to replication, which requires that both strands of the chromosome be entirely copied, the regulated expression of genetic information involves much smaller, single-strand portions of the genome. The DNA strand that serves as a template during transcription is known as the **antisense** or **noncoding strand** since its sequence is complementary to that of the RNA. The other DNA strand, which has the same sequence (except for the replacement of U with T) as the transcribed RNA, is known as the **sense** or **coding strand** (Fig. 25-3). The two strands of DNA in an organism's chromosome can therefore contain different sets of genes.

Keep in mind that "gene" is a relatively loose term that refers to sequences that encode polypeptides, as well as those that correspond to the sequences of rRNA, tRNA, and other RNA species. Furthermore, a gene typically includes sequences that participate in initiating and terminating transcription (and translation) that are not actually transcribed (or translated). The expression of many genes also depends on

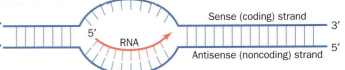

Figure 25-3 Sense and antisense DNA strands. The template strand of duplex DNA is known as its antisense or noncoding strand. Its complementary sense or coding strand has the same nucleotide sequence and orientation as the transcribed RNA.

Figure 25-4 **The *E. coli lac* operon.** This DNA includes genes encoding the proteins mediating lactose metabolism and the genetic sites that control their expression. The *Z, Y,* and *A* genes, respectively, specify the proteins **β-galactosidase, galactoside permease,** and **thiogalactoside transacetylase.** The closely linked regulatory gene, *I*, which is not part of the *lac* operon, encodes a repressor that inhibits transcription of the *lac* operon.

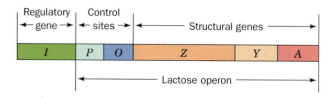

regulatory sequences that do not directly flank the coding regions but may be located a considerable distance away.

Most protein-coding genes (called **structural genes**) in eukaryotes are transcribed individually. In prokaryotic genomes, however, genes are frequently arranged in tandem along a single strand so that they can be transcribed together. These genetic units, called **operons,** typically contain genes with related functions. For example, *E. coli's* three different rRNA genes occur in single operons (Section 25-3B). The *E. coli **lac* operon** whose expression is described in detail in Section 27-2A, contains three genes encoding proteins involved in lactose metabolism as well as sequences that control their transcription (Fig. 25-4). Other operons contain genes encoding proteins required for biosynthetic pathways, for example, the ***trp* operon,** whose six **gene products** (proteins are often referred to as gene products) catalyze tryptophan synthesis. An operon is transcribed as a single unit, giving rise to a **polycistronic mRNA** that directs the more-or-less simultaneous synthesis of each of the encoded polypeptides (the term **cistron** is a somewhat archaic synonym for gene). In contrast, eukaryotic structural genes, which are not part of operons, give rise to **monocistronic mRNAs.**

RNA Polymerase Holoenzyme Binds to Promoters.   How does RNAP recognize the correct DNA strand and initiate RNA synthesis at the beginning of a gene (or operon)? *RNAP binds to its initiation sites through base sequences known as **promoters** that are recognized by the corresponding σ factor.* The existence of promoters was first revealed through mutations that enhanced or diminished the transcription rates of certain genes. Promoters consist of ~40-bp sequences that are located on the 5′ side of the transcription start site. By convention, the sequence of this DNA is represented by its sense (nontemplate) strand so that it will have the same sequence and directionality as the transcribed RNA. A base pair in a promoter region is assigned a negative or positive number that indicates its position, upstream or downstream in the direction of RNAP travel, from the first nucleotide that is transcribed to RNA; this start site is +1 and there is no 0. Because RNA is synthesized in the 5′ → 3′ direction (Section 25-1C), the promoter is said to lie upstream of the RNA's starting nucleotide.

The holoenzyme forms tight complexes with promoters (dissociation constant $K \approx 10^{-14}$ M). This tight binding can be demonstrated by showing that the holoenzyme protects the bound DNA segments from digestion *in vitro* by the endonuclease DNase I. Sequence determinations of the protected regions from numerous *E. coli* genes have identified the "consensus" sequence of *E. coli* promoters (Fig. 25-5). Their most conserved sequence is a hexamer centered at about the −10 position (sometimes called the **Pribnow box,** after David Pribnow, who described it in 1975). It has a consensus sequence of TATAAT in which the leading TA and the final T are highly conserved. Upstream sequences around position −35 also have a region of sequence similarity, TTGACA. The initiating (+1) nucleotide, which is nearly always A or G, is centered in a poorly conserved CAT or CGT sequence. Most promoter sequences vary considerably from the consensus sequence (Fig. 25-5). Nevertheless, a mutation in one of the partially

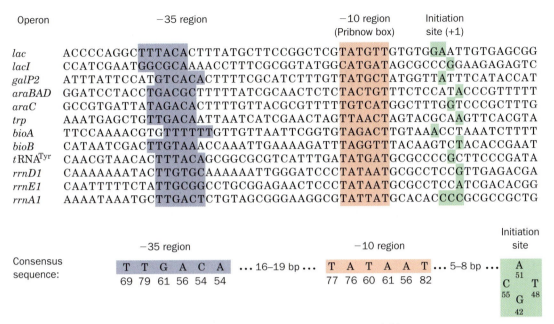

Operon	−35 region		−10 region (Pribnow box)	Initiation site (+1)
lac	ACCCCAGGC**TTTACA**CTTTATGCTTCCGGCTCG**TATGTT**GTGTGG**A**ATTGTGAGCGG
lacI	CCATCGAAT**GGCGCA**AAACCTTTCGCGGTATGG**CATGAT**AGCGCCC**G**GAAGAGAGTC
galP2	ATTTATTCCAT**GTCACA**CTTTTCGCATCTTTGT**TATGCT**ATGGTT**A**TTTCATACCAT
araBAD	GGATCCTACC**TGACGC**TTTTTATCGCAACTCTC**TACTGT**TTCTCCAT**A**CCCGTTTTT
araC	GCCGTGATTA**TAGACA**CTTTTGTTACGCGTTTTT**GTCAT**GGCTTTGG**T**CCCGCTTTG
trp	AAATGAGCTG**TTGACA**ATTAATCATCGAACTAGT**TAACT**AGTACGCAAGTTCACGTA
bioA	TTCCAAAACGTG**TTTTTT**GTTGTTAATTCGGTG**TAGACT**TGTAAA**CC**TAAATCTTTT
bioB	CATAATCGAC**TTGTAAA**CCAAATTGAAAAGATT**TAGGTT**TACAAGTC**T**ACACCGAAT
tRNA^Tyr	CAACGTAACAC**TTTACA**GCGGCGCGTCATTTGA**TATGAT**GCGCCCC**G**CTTCCCGATA
rrnD1	CAAAAAAT**ACTTGTGC**AAAAAATTGGGATCCC**TATAAT**GCGCCTCC**G**TTGAGACGA
rrnE1	CAATTTTTCTA**TTGCGG**CCTGCGGAGAACTCCC**TATAAT**GCGCCTCC**A**TCGACACGG
rrnA1	AAAATAAATGC**TTGACT**CTGTAGCGGGAAGGCG**TATTAT**GCACACCCCGCGCCGCTG

	−35 region		−10 region		Initiation site
Consensus sequence:	T T G A C A	... 16–19 bp ...	T A T A A T	... 5–8 bp ...	A^51 C^55 T^48 G^42
	69 79 61 56 54 54		77 76 60 61 56 82		

Figure 25-5   The sense (coding) strand sequences of selected *E. coli* promoters. A 6-bp region centered around the −10 position (*red shading*) and a 6-bp sequence around the −35 region (*blue shading*) are both conserved. The transcription initiation sites (+1), which in most promoters occur at a single purine nucleotide, are shaded in green. The bottom row shows the consensus sequence of 298 *E. coli* promoters with the number below each base indicating its percentage occurrence. [After Rosenberg, M. and Court, D., *Annu. Rev. Genet.* **13**, 321–323 (1979). Consensus sequence from Lisser, D. and Margalit, H., *Nucleic Acids Res.* **21**, 1512 (1993).]

conserved regions can greatly increase or decrease a promoter's initiation efficiency. This is because *the rates at which E. coli genes are transcribed vary directly with the rates at which their promoters form stable initiation complexes with the holoenzyme.*

Initiation Requires the Formation of an Open Complex. The promoter regions in contact with the RNAP holoenzyme have been identified by a procedure named **footprinting.** In this procedure, DNA is incubated with a protein to which it binds and is then treated with an alkylation agent such as dimethyl sulfate (DMS). This results in alkylation of the DNA's bases followed by backbone cleavage at the alkylated positions. However, those portions of the DNA that bind proteins are protected from alkylation and hence cleavage. The resulting pattern of protection is called the protein's footprint. The footprint of RNAP holoenzyme indicates that it contacts the promoter primarily at its −10 and −35 regions. In some genes, additional upstream sequences may also influence RNAP binding to DNA.

DMS methylates G residues at N7 and A residues at N3 in both double- and single-stranded DNA. DMS also methylates N1 of A and N3 of C, but only if these latter positions are not involved in base-pairing interactions. The pattern of DMS methylation therefore reveals whether the DNA is single or double stranded. Footprinting studies indicate that holoenzyme binding "melts" (separates) ~11 bp of DNA (from −9 to +2). The resulting **open complex** is analogous to the region of unwound DNA at the replication origin (Section 24-2B).

The core enzyme, which does not specifically bind promoters, tightly binds duplex DNA (the complex's dissociation constant is $K \approx 5 \times 10^{-12}$ M, and its half-life is ~60 min). *The holoenzyme, in contrast, binds to nonpromoter*

DNA comparatively loosely $K \approx 10^{-7}$ M and a half-life of >1 s. Evidently, the σ subunit allows the holoenzyme to move rapidly along a DNA strand in search of the σ subunit's corresponding promoter. Once transcription has been initiated and the σ subunit jettisoned, the tight binding of the core enzyme to DNA apparently stabilizes the ternary enzyme–DNA–RNA complex.

Gene Expression Is Controlled by Different σ Factors.   Because different σ factors recognize different promoters, *a cell's complement of σ factors determines which genes are transcribed.* Development and differentiation, which involve the temporally ordered expression of sets of genes, can be orchestrated through a "cascade" of σ factors. For example, infection of *Bacillus subtilis* by **bacteriophage SP01** requires the expression of different sets of phage genes at different times. The first set, known as the **early genes,** are transcribed using the bacterial σ factor. One of the early phage gene products is a σ subunit known as σ^{gp28}, which displaces the host σ factor and thereby permits the RNAP to recognize only the phage **middle gene** promoters. The phage middle genes, in turn, specify $\sigma^{\text{gp33/34}}$, which promotes transcription of only phage **late genes.**

Several bacteria, including *E. coli* and *Bacillus subtilis*, likewise have several different σ factors. These are not necessarily used in a sequential manner. For example, σ factors in *E. coli* that differ from its primary σ factor (which is named σ^{70} because its molecular mass is 70 kD) control the transcription of coordinately expressed groups of special-purpose genes whose promoters are quite different from those recognized by σ^{70}.

C   Chain Elongation

Because RNA synthesis, like DNA synthesis, proceeds in the $5' \rightarrow 3'$ direction (Fig. 25-6), the growing RNA molecule has a 5'-triphosphate group. Mature RNA molecules, as we shall see, may also be chemically modified at one or both ends.

A portion of the double-stranded DNA template remains opened up at the point of RNA synthesis. This allows the antisense strand to be transcribed onto its complementary RNA strand. The RNA chain transiently forms a short length of RNA–DNA hybrid duplex. The unpaired "bubble" of DNA in the open initiation complex apparently travels along the DNA with the RNAP (Fig. 25-7).

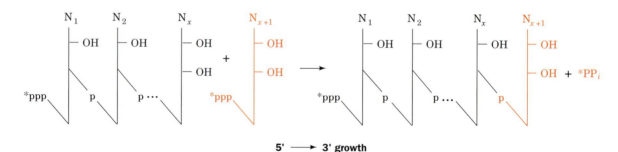

$5' \longrightarrow 3'$ growth

Figure 25-6 $5' \rightarrow 3'$ RNA chain growth.   Nucleotides are added to the 3' end of the growing RNA chain via attack of the 3'-OH group on the incoming nucleoside triphosphate. A radioactive label at the γ position of an NTP (indicated by an asterisk) is retained in the initial nucleotide of the RNA (the 5' end) but is lost as PP$_i$ during the polymerization of subsequent nucleotides.

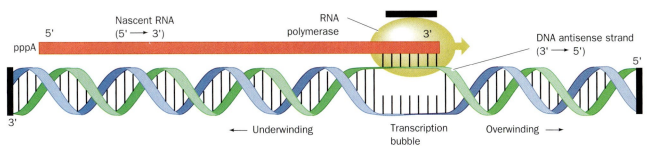

Figure 25-7 DNA supercoiling during transcription.   In the region being transcribed, the DNA double helix is unwound by about a turn to permit the DNA's antisense strand to form a short segment of DNA–RNA hybrid double helix. As the RNA polymerase advances along the DNA template (here to the right), the DNA unwinds ahead of the RNA's growing 3′ end and rewinds behind it, thereby stripping the newly synthesized RNA from the template strand. Because the ends of the DNA as well as the RNA polymerase are apparently prevented from rotating by attachments within the cell (*black bars*), the DNA becomes overwound ahead of the advancing transcription bubble and underwound behind it (consider the consequences of placing your finger between the twisted DNA strands in this model and pushing toward the right). [After Futcher, B., *Trends Genet.* **4**, 272 (1988).]

As DNA's helical turns are pushed ahead of the advancing transcription bubble, they become more tightly wound (more positively supercoiled) while the DNA behind the bubble becomes equivalently unwound (more negatively supercoiled). This scenario is supported by the observation that the transcription of plasmids in *E. coli* induces their positive supercoiling in gyrase mutants (which cannot relax positive supercoils; Section 23-1C) and their negative supercoiling in topoisomerase I mutants (which cannot relax negative supercoils). Inappropriate superhelicity in the DNA being transcribed halts transcription (Section 23-1C). Quite possibly the torsional tension in the DNA generated by negative superhelicity behind the transcription bubble is required to help drive the transcriptional process, whereas too much such tension prevents the opening and maintenance of the transcription bubble.

RNA Polymerase Is Processive.   Once the open complex has been formed, transcription proceeds without dissociation of the enzyme from the template. Processivity is accomplished without an obvious clamplike structure (e.g., the β clamp of *E. coli* DNA polymerase III; Fig. 24-16), although the RNAP itself apparently functions as a sliding clamp by binding tightly but flexibly to the DNA–RNA complex. In experiments in which the RNAP was immobilized and a magnetic bead was attached to the DNA, up to 180 rotations (representing nearly two thousand base pairs at 10.4 bp per turn) were observed before the polymerase slipped. Such processivity is essential, since genes are often thousands (in eukaryotes, sometimes millions) of nucleotides in length.

The tight association between RNAP and DNA and their multiple attachment sites may explain why a transcription complex does not completely dissociate from a DNA template even when transcription is interrupted by the DNA replication machinery proceeding along the same strand of DNA (see Box 25-1).

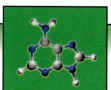

BOX 25-1

Perspectives in Biochemistry

Collisions between DNA Polymerase and RNA Polymerase

In rapidly proliferating bacterial cells, DNA synthesis is likely to occur even as genes are being transcribed. The DNA replication machinery moves along the circular chromosome at a rate many times faster than the movement of the transcription machinery. Collisions between DNA polymerase and RNA polymerase seem unavoidable. What happens when the two enzyme complexes collide? Using *in vitro* model systems, Bruce Alberts has shown that when both enzymes are moving in the same direction, the replication fork passes the RNA polymerase without displacing it, leaving it fully competent to resume RNA chain elongation.

When the replication fork collides head-on with a transcription complex, however, the replication fork pauses briefly before moving past the RNA polymerase. Surprisingly, this causes the RNA polymerase to switch its template strand.

The growing RNA chain dissociates from the original template DNA strand and hybridizes with the newly synthesized daughter DNA strand of the same sequence before RNA elongation resumes.

Head-on collisions are disadvantageous because (1) replication slows when DNA polymerase pauses, and (2) dissociation of the RNA polymerase during the jump from one template strand to the other could abort the transcription process. Indeed, in many bacterial and phage genomes, the most heavily transcribed genes are oriented so that replication and transcription complexes move in the same direction. It remains to be seen whether a similar arrangement holds in eukaryotic genomes, which contain multiple replication origins and genes that are much larger than are prokaryotic genes.

Transcription Is Rapid.   In *E. coli*, the *in vivo* rate of transcription is 20 to 50 nucleotides per second at 37°C (but still several times slower than the DNA replication rate; Section 24-2C). The error frequency in RNA synthesis is one wrong base incorporated for every $\sim 10^4$ transcribed. This frequency, which is 10^4 to 10^6 times higher than that for DNA synthesis, is tolerable because of the repeated transcription of most genes, because the genetic code contains numerous synonyms (Section 26-1C), and because amino acid substitutions in proteins are often functionally innocuous.

Once an RNAP molecule has initiated transcription and moved away from the promoter, another RNAP can follow suit. The synthesis of RNAs that are needed in large quantities, rRNAs, for example, is initiated as often as is sterically possible, about once per second. This gives rise to an arrowhead appearance of the transcribed DNA (Fig. 25-8). mRNAs encoding proteins are generally synthesized at less frequent intervals, and there is enormous variation in the amounts of different polypeptides produced. For example, an *E. coli* cell may contain 10,000 copies of a ribosomal protein, whereas a regulatory protein may be present in only a few copies per cell. Many enzymes, particularly those involved in basic cellular "housekeeping" functions, are synthesized at a more or less constant rate; they are called

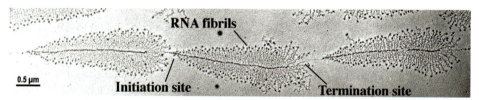

0.5 μm

RNA fibrils

Initiation site

Termination site

Figure 25-8   **An electron micrograph of three contiguous ribosomal genes undergoing transcription.**   The "arrowhead" structures result from the increasing lengths of the nascent RNA chains as the RNA polymerases synthesizing them move from the initiation site on the DNA to the termination site. [Courtesy of Ulrich Scheer, University of Würzburg, Germany.]

constitutive enzymes. Other enzymes, termed **inducible enzymes,** are synthesized at rates that vary with the cell's circumstances. To a large extent, the regulation of gene expression relies on mechanisms that govern the rate of transcription, as we shall see in Section 27-2. The products of transcription also vary in their stability. Ribosomal RNA turns over much more slowly than mRNA, which is rapidly synthesized and rapidly degraded (sometimes so fast that the 5' end of an mRNA is degraded before its 3' end has been synthesized).

D Chain Termination

Electron micrographs such as Fig. 25-8 suggest that DNA contains specific sites at which transcription is terminated. The transcription termination sequences of many *E. coli* genes share two common features (Fig. 25-9a):

1. A series of 4 to 10 consecutive A · T base pairs, with the A's on the template strand. The transcribed RNA is terminated in or just past this sequence.

2. A G+C-rich region with a palindromic sequence that immediately precedes the series of A · T's.

The RNA transcript of this region can therefore form a self-complementary "hairpin" structure that is terminated by several U residues (Fig. 25-9b).

The structural stability of an RNA transcript at its terminator's G+C-rich hairpin and the weak base pairing of its oligo(U) tail to template DNA appear to be important factors in ensuring proper chain termination. The formation of the G+C-rich hairpin causes RNAP to pause for several seconds at the termination site. This probably induces a conformational change in the RNAP that permits the nontemplate DNA strand to displace the weakly bound oligo(U) tail from the template strand, thereby terminating transcription.

Despite the foregoing, experiments by Michael Chamberlin indicate that the RNA-terminator hairpin and U-rich 3' tail do not function independently of their upstream and downstream flanking regions. Indeed, terminators that lack a U-rich segment can be highly efficient when joined to the appropriate sequence immediately downstream from the termination site. Termination efficiency also varies with the concentrations of nucleoside triphosphates, with the level of supercoiling in the DNA template, with

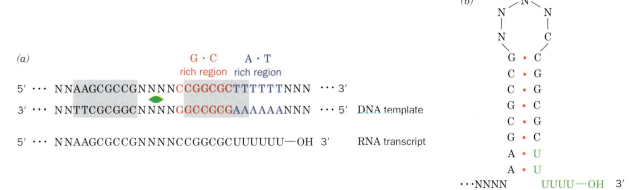

Figure 25-9   A hypothetical strong (efficient) *E. coli* terminator. The base sequence was deduced from the sequences of several transcripts. (*a*) The DNA sequence together with its corresponding RNA. The A · T-rich and G · C-rich sequences are shown in blue and red, respectively. The twofold symmetry axis (*green symbol*) relates the flanking shaded segments that form an inverted repeat. (*b*) The RNA hairpin structure and poly(U) tail that triggers transcription termination. [After Pribnow, D., *in* Goldberger, R.F. (Ed.), *Biological Regulation and Development,* Vol. 1, p. 253, Plenum Press (1979).]

changes in the salt concentration, and with the sequence of the terminator. These results suggest that *termination is a complex multistep process.*

Termination Often Requires Rho Factor.

The termination sequences described above induce the spontaneous termination of transcription. Other termination sites, however, lack any obvious similarities and are unable to form strong hairpins; *they require a protein known as **Rho factor** to terminate transcription.* Rho factor, a hexamer of identical 419-residue subunits, enhances the termination efficiency of spontaneously terminating transcripts and induces termination of nonspontaneously terminating transcripts.

Several key observations have led to a model of Rho-dependent termination:

1. Rho factor is a helicase that catalyzes the unwinding of RNA–DNA and RNA–RNA double helices. This process is powered by the hydrolysis of nucleoside triphosphates to nucleoside diphosphates + P_i with little preference for the identity of the base.

2. Genetic manipulations indicate that Rho-dependent termination requires a specific recognition sequence on the nascent (still being synthesized) RNA upstream of the termination site.

These observations suggest that Rho factor attaches to the RNA at its recognition sequence and then migrates along the RNA in the $5' \rightarrow 3'$ direction until it encounters an RNAP paused at the termination site (without the pause, Rho might not be able to overtake the RNAP). There, Rho unwinds the RNA–DNA duplex at the transcription bubble, thereby releasing the RNA transcript. Rho-terminated transcripts have 3′ ends that typically vary over a range of ~50 nt. This suggests that Rho gradually pries the RNA away from its template DNA rather than liberating the RNA at a specific point.

Each Rho subunit consists of two domains: Its N-terminal domain binds single-stranded polynucleotides and its C-terminal domain, which is homologous to the α and β subunits of the F_1-ATPase (Section 17-3B), binds an NTP. The X-ray structure of Rho in complex with an ATP analog and an 8-nt RNA has been determined by James Berger. Rather than forming a closed hexameric ring, as does F_1-ATPase, Rho forms a lock washer–shaped helix that is 120 Å in diameter with an ~30-Å-diameter central hole and whose first and sixth subunits are separated by a 12-Å gap and a rise of 45 Å along the helix axis (Fig. 25-10). The RNA binds along the interior of the helix to the so-called primary RNA binding sites on the N-terminal domains and to the so-called secondary RNA binding sites on the C-terminal domains that have been implicated in mRNA translocation and unwinding. Since electron microscopic images of Rho show both closed as well as notched hexameric rings, the X-ray structure probably represents an open state that is poised to bind mRNA that has entered its central cavity through the notch. Presumably, mRNA binding would cause the hexameric ring to close. Ensuing cycles of ATP hydrolysis, with its attendant conformational changes, would then propel Rho along the mRNA in the $5' \rightarrow 3'$ direction.

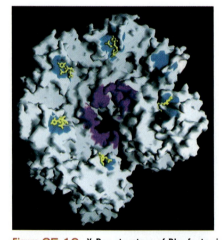

Figure 25-10   X-Ray structure of Rho factor in complex with RNA.   The solvent-accessible surface of the Rho hexamer is viewed from the top of the lock washer–shaped helix. The primary RNA binding sites are cyan and the secondary RNA binding sites are magenta. The RNA, which is drawn in yellow stick form, is bound to the primary RNA binding sites. It is only partially visible in the X-ray structure. [Courtesy of James Berger, University of California at Berkeley. PDBid 1PVO.]

2 Transcription in Eukaryotes

Although the fundamental principles of transcription are similar in prokaryotes and eukaryotes, *eukaryotic transcription is distinguished by having multiple RNAPs and by relatively complicated control sequences.* Moreover, the eukaryotic transcription machinery, as we shall see, is far more complex than that of prokaryotes, requiring well over 100 polypep-

tides that form assemblies with molecular masses of several million daltons to recognize the control sequences and initiate transcription.

A Eukaryotic RNA Polymerases

Eukaryotic nuclei contain three distinct types of RNA polymerase that differ in the RNAs they synthesize:

1. **RNA polymerase I (RNAP I),** which is located in the **nucleoli** (dark-staining nuclear bodies where ribosomes are assembled; Fig. 1-8), synthesizes the precursors of most rRNAs.
2. **RNA polymerase II (RNAP II),** which occurs in the nucleoplasm, synthesizes the mRNA precursors.
3. **RNA polymerase III (RNAP III),** which also occurs in the nucleoplasm, synthesizes the precursors of 5S rRNA, the tRNAs, and a variety of other small nuclear and cytosolic RNAs.

In addition to these nuclear enzymes, eukaryotic cells contain separate mitochondrial and (in plants) chloroplast RNAPs. The essential function of RNAPs in all cells makes them attractive targets for antibiotics and other drugs (see Box 25-2).

Eukaryotic RNAPs, which have molecular masses of as much as 600 kD, have considerably greater subunit complexity than those of prokaryotes. Each type of enzyme contains two nonidentical "large" (>120 kD) subunits, which are homologs of the prokaryotic β and β' subunits, and an array of up to 13 different "small" (<50 kD) subunits, two of which are homologs of the prokaryotic α subunit and one of which is a homolog of the ω subunit. Five of the small subunits, including the ω homolog, are identical in all three eukaryotic enzymes and the α homologs are identical in RNAPs I and III.

In a crystallographic tour de force, Roger Kornberg determined the X-ray structure of yeast RNAP II (Fig. 25-11). This enzyme, as expected,

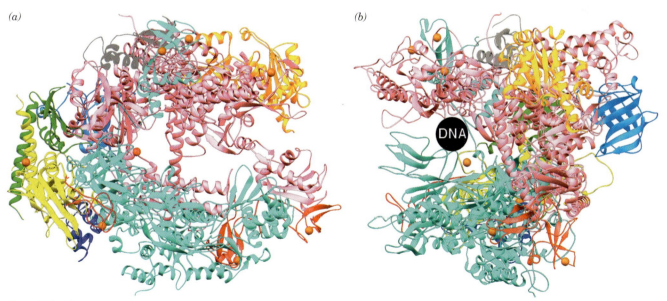

(a)   *(b)*

Figure 25-11 X-Ray structure of yeast RNA polymerase II.   (*a*) The enzyme is oriented similarly to the *Taq* RNAP in Fig. 25-2 and its subunits are differently colored, with subunits homologous to those in *Taq* RNA polymerase given the same colors. The position of the active site Mg^{2+} ion is marked by a red sphere, and eight Zn^{2+} ions are represented by orange spheres. Two nonessential polymerase subunits and the C-terminus of the β' homolog are not visible in this structure. (*b*) View of the enzyme from the right of Part *a*, showing the DNA binding cleft. The circle has the approximate diameter of B-DNA. [Based on an X-ray structure by Roger Kornberg, Stanford University. PDBid 1I50.]

BOX 25-2

Biochemistry in Health and Disease

Inhibitors of Transcription

A wide variety of compounds inhibit transcription in prokaryotes and eukaryotes. These agents are therefore toxic to susceptible organisms; that is, they function as antibiotics. Such compounds are also useful research tools since they arrest the transcription process at well-defined points.

Two related antibiotics, **rifamycin B,** which is produced by *Streptomyces mediterranei,* and its semisynthetic derivative **rifampicin,**

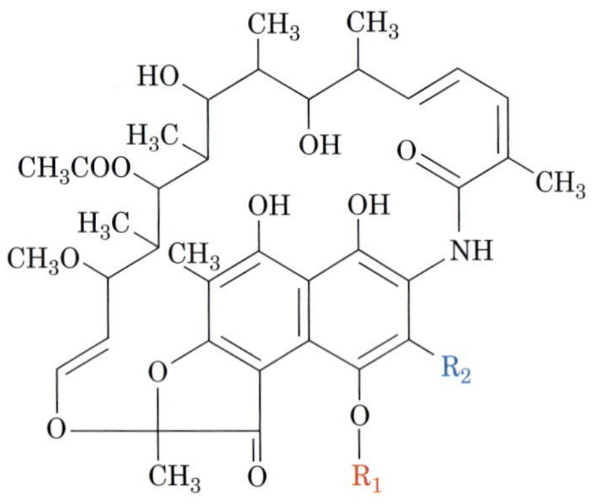

Rifamycin B $R_1 = CH_2COO^-$; $R_2 = H$

Rifampicin $R_1 = H$; $R_2 = CH=\overset{+}{N}$ ⬭ $N-CH_3$

specifically inhibit transcription by prokaryotic but not eukaryotic RNA polymerases. The selectivity and high potency of rifampicin (2×10^{-8} M results in 50% inhibition of bacterial RNA polymerase) makes it a medically useful bactericidal agent. Rifamycins inhibit neither the binding of RNA polymerase to the promoter nor the formation of the first phosphodiester bonds, but they prevent further chain elongation. The inactivated RNA polymerase remains bound to the promoter, thereby blocking initiation by uninhibited enzyme. Once RNA chain initiation has occurred, however, rifamycins have no effect on the subsequent elongation process. The rifamycins can therefore be used in the laboratory to dissect transcriptional initiation and elongation.

Actinomycin D,

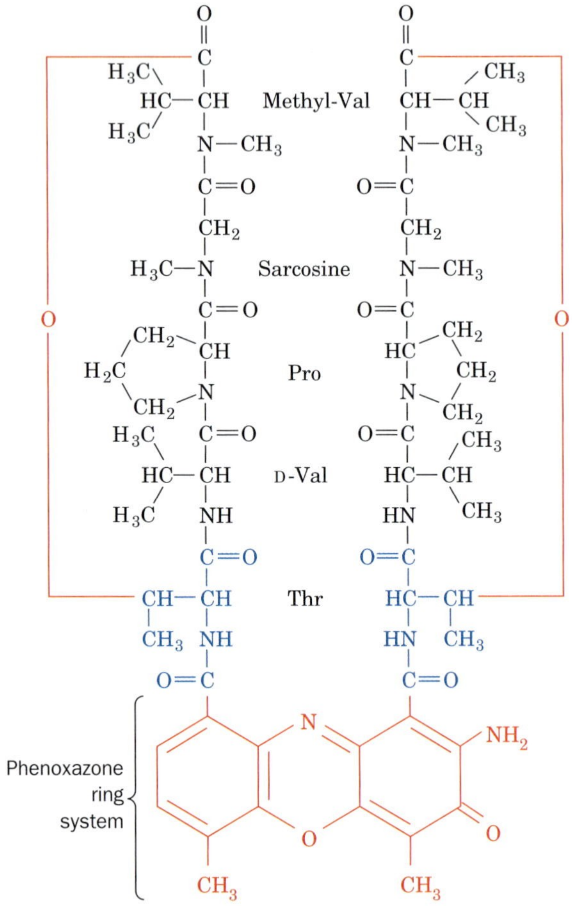

Actinomycin D

a useful antineoplastic (anticancer) agent produced by *Streptomyces antibioticus,* tightly binds to duplex DNA and, in doing so, strongly inhibits both transcription and DNA replication, presumably by interfering with the passage of RNA and DNA polymerases. The X-ray structure of actinomycin D in complex with an 8-bp DNA is shown below as two vertically stacked complexes in space-filling form with the DNA's sugar–phosphate backbone yellow, its bases white, and with the actinomycins colored according to atom type (C green, N blue, and O red).

resembles *Taq* RNAP (Fig. 25-2) in its overall crab claw–like shape and in the positions and core folds of their homologous subunits, although RNAP II is somewhat larger than and has several subunits that have no counterpart in *Taq* RNAP. RNAP II binds two Mg^{2+} ions at its active site in the vicinity of five conserved acidic residues, which suggests that RNAPs

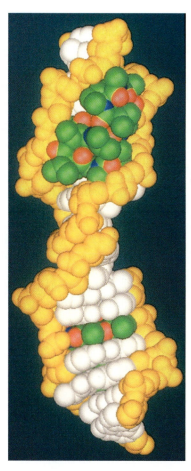

including a series of unusual bicyclic octapeptides known as **amatoxins**. α-**Amanitin,**

α-Amanitin

which is representative of the amatoxins, forms a tight 1:1 complex with RNAP II ($K = 10^{-8}$ M) and a looser one with RNAP III ($K = 10^{-6}$ M) so as to specifically block their elongation steps. The X-ray structure of RNAP II in complex with α-amanitin reveals that α-amanitin binds in the funnel beneath the protein's bridge helix (Fig. 25-12b) such that it interacts almost exclusively with the residues of the bridge helix and the adjacent part of Rpb1. The α-amanitin binding site is too far away from the enzyme active site to directly interfere with NTP entry or RNA synthesis, consistent with the observation that α-amanitin does not influence the affinity of RNAP II for NTPs. Most probably, α-amanitin binding impedes the conformational change of the bridge helix postulated to motivate the RNAP translocation step (Section 25-2A), which further supports this mechanism. RNAP I as well as mitochondrial, chloroplast, and prokaryotic RNA polymerases are insensitive to α-amanitin.

Despite the amatoxins' high toxicity (5–6 mg, contained in ~40 g of fresh mushrooms, is sufficient to kill a human adult), they act slowly. Death, usually from liver dysfunction, occurs no earlier than several days after mushroom ingestion (and after recovery from the effects of other mushroom toxins). This, in part, reflects the slow turnover rate of eukaryotic mRNAs and proteins.

[Structure of actinomycin D–DNA complex based on an X-ray structure by Fusao Takusagawa, University of Kansas. PDBid 172D.]

The actinomycin's phenoxazone ring system (which is visible in the lower molecule) intercalates between the DNA's base pairs, thereby unwinding the DNA helix by 23° and separating the neighboring base pairs by 7.0 Å. Actinomycin's chemically identical cyclic **depsipeptides** (which are seen in the upper molecule; depsipeptides have both peptide bonds and ester linkages) extend in opposite directions from the intercalation site along the minor groove of the DNA. Other intercalation agents, including ethidium and proflavin (Section 23-3B), also inhibit nucleic acid synthesis, presumably by similar mechanisms.

The poisonous mushroom *Amanita phalloides* (death cap), which is responsible for the majority of fatal mushroom poisonings in Europe, contains several types of toxic substances,

catalyze RNA elongation via a two-metal ion mechanism similar to that employed by DNA polymerases (Section 24-2A). As is the case with *Taq* RNAP, the surface of RNAP II is almost entirely negatively charged except for the DNA-binding cleft and the region about the active site, which are positively charged.

RNAP II's **Rpb1** subunit (pink in Fig. 25-11), the homolog of the β′ subunit in prokaryotic RNAPs, has an extraordinary C-terminal domain **(CTD).** In mammals, the CTD contains 52 highly conserved repeats with the consensus sequence Pro-Thr-Ser-Pro-Ser-Tyr-Ser (26 repeats in yeast, with other eukaryotes having intermediate values). As many as 50 Ser residues in this hydroxyl-rich protein segment are subject to reversible phosphorylation. RNAP II initiates transcription only when the CTD is unphosphorylated but commences elongation only after the CTD has been phosphorylated. Charge–charge repulsions between nearby phosphate groups probably cause a highly phosphorylated CTD to project as far as 500 Å from the globular portion of the polymerase. As we shall see, the phosphorylated CTD provides the binding sites for numerous protein factors that are essential for transcription.

Structure of a Transcribing RNA Polymerase.   To produce a snapshot of RNAP II in action, Kornberg incubated the enzyme with a dsDNA molecule bearing a 3′ single-stranded tail on one strand (the template strand) together with all the NTP substrates except UTP. As a consequence, the polymerase synthesized a short (14-nt) RNA strand before pausing at the first template A residue (when the crystals of this complex were soaked in UTP, transcription resumed, demonstrating that the complex was active). The X-ray structure and a cutaway diagram of this paused RNAP II are shown in Fig. 25-12.

In the RNAP–DNA–RNA complex, a massive (~50-kD) portion of the **Rpb2** subunit (the β homolog; cyan in Fig. 25-11), named the "clamp," has swung downward over the DNA to trap it in the cleft, in large part accounting for the enzyme's essentially infinite processivity.

(a)

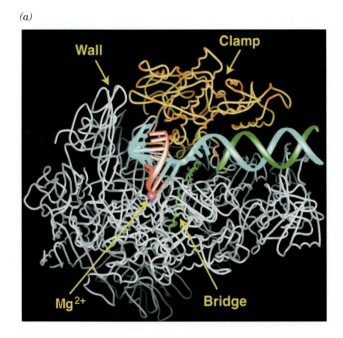

(b)

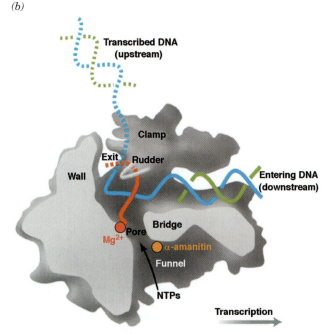

Figure 25-12 The RNA polymerase II elongation complex.   (*a*) X-Ray structure of the complex as viewed from the bottom of Fig. 25-11*a* (portions of Rpb2 that form the near side of the cleft have been removed to expose the bound RNA · DNA complex). The protein is represented by its backbone in which the clamp, which is closed over the downstream DNA duplex, is yellow, the bridge helix is green, and the remaining portions of the protein are gray. The template DNA strand (*cyan*), the nontemplate DNA strand (*green*), and the newly synthesized RNA (*red*) are drawn with their well-ordered portions in ladder form and their less ordered portions drawn in backbone form. The active site Mg^{2+} ion is represented by a magenta sphere. (*b*) Cutaway diagram of the transcribing complex in Part *a* in which the cut surfaces of the protein are light gray, its remaining surfaces are darker gray, and several of its functionally important structural features are labeled. The DNA, RNA, and active site Mg^{2+} ion are colored as in Part *a* with portions of the DNA and RNA that are not visible in the X-ray structure represented by dashed lines. α-Amanitin (*orange ball*) is discussed in Box 25-2. [Modified from diagrams by Roger Kornberg, Stanford University. PDBid 1I6H.]

The DNA unwinds by three nucleotides before entering the active site (which is contained on Rpb1). Past this point, however, a portion of Rpb2 dubbed the "wall" directs the template strand out of the cleft in an ~90° turn. As a consequence, the template base at the active site (+1) points toward the floor of the active site where it can be read out by the polymerase. This base is paired with the ribonucleotide at the 3′ end of the RNA, which is positioned above a "pore" at the end of a "funnel" to the protein exterior through which NTPs presumably gain access to the otherwise sealed-off active site.

The hybrid helix adopts a nonstandard conformation intermediate between those of A- and B-DNAs, which is underwound relative to that in the X-ray structure of an RNA · DNA hybrid helix alone (Fig. 23-4). Nearly all contacts that the protein makes with the RNA and DNA are to their sugar–phosphate backbones; none are with the edges of their bases. The specificity of the enzyme for a ribonucleotide rather than a deoxyribonucleotide appears due to the enzyme's recognition of both the incoming ribose sugar and the RNA · DNA hybrid helix. After about one turn of hybrid helix, a loop extending from the clamp known as the "rudder" separates the RNA and template DNA strands, thereby permitting the DNA double helix to re-form as it exits the enzyme (although the unpaired 5′ tail of the nontemplate strand and the 3′ tail of the template strand are disordered in the X-ray structure).

How does RNAP translocate its bound DNA–RNA assembly at the end of each catalytic cycle? A highly conserved helical segment of Rpb1 called the "bridge" (because it bridges the two pincers forming the enzyme's cleft) nonspecifically contacts the template DNA base at the +1 position. The bridge is straight in the X-ray structures of RNAP II (Figs. 25-11a and 25-12a) but bent in the X-ray structure of *Taq* RNAP (Fig. 25-2). If the bridge helix, in fact, alternates between its straight and bent conformations, it would move by 3 to 4 Å. Kornberg has therefore speculated that translocation occurs through the bending of the bridge helix so as to push the paired nucleotides at position +1 to position −1. The recovery of the bridge helix to its straight conformation would then yield an empty site at position +1 for entry of the next NTP, thereby preparing the enzyme for a new round of nucleotide addition.

RNAP II, like DNA polymerase I, edits its work. If a deoxynucleotide or a mispaired ribonucleotide is mistakenly incorporated into RNA, the DNA–RNA hybrid helix becomes distorted. This causes polymerization to cease, and the newly synthesized RNA backs out of the active site through the pore and funnel by which ribonucleotides enter the active site. The exposed RNA containing the error is then trimmed by a nuclease, which excises between 1 and 11 residues. Transcription may resume if the 3′ end of the truncated transcript is then repositioned at the active site.

B Eukaryotic Promoters

In eukaryotes, RNAP does not include a removable σ factor. Instead, a number of accessory proteins identify promoters and recruit RNAP to the transcription start site (as described more fully in Section 25-2C). As expected, eukaryotic promoters are more complex and diverse than prokaryotic promoters. In addition, the three eukaryotic RNAPs recognize different types of promoters.

Mammalian RNA Polymerase I Has a Bipartite Promoter. Both prokaryotic and eukaryotic genomes contain multiple copies of their rRNA genes in order to meet the enormous demand for these rRNAs (which comprise, e.g., 80%

of an *E. coli* cell's RNA content). Since the numerous rRNA genes in a given eukaryotic cell have essentially identical sequences, its RNAP I recognizes only one promoter. Yet in contrast to RNAP II and III promoters, RNAP I promoters are species specific; that is, an RNAP I recognizes only its own promoter and those of closely related species.

RNAP I promoters were identified by determining how the transcription rate of an rRNA gene is affected by a series of increasingly longer deletions approaching its transcription start site from either its upstream or its downstream side. Such studies have indicated, for example, that mammalian RNAP I requires a so-called **core promoter element,** which spans positions -31 to $+6$ and hence overlaps the transcribed region. However, efficient transcription also requires an **upstream promoter element,** which is located between residues -187 and -107. These elements, which are G+C-rich and ~85% identical, are bound by specific transcription factors which then recruit RNAP I to the transcription start site.

RNA Polymerase II Promoters Are Complex and Diverse.   The promoters recognized by RNAP II are considerably longer and more diverse than those of prokaryotic genes. The structural genes expressed in all tissues (the housekeeping genes, which are thought to be constitutively transcribed) have one or more copies of the sequence GGGCGG or its complement (the **GC box**) located upstream from their transcription start sites. The analysis of deletion and point mutations indicates that *GC boxes function analogously to prokaryotic promoters.* On the other hand, structural genes that are selectively expressed in one or a few cell types often lack these GC-rich sequences. Instead, *they contain a conserved AT-rich sequence located 25 to 30 bp upstream from their transcription start sites* (Fig. 25-13). Note that this so-called **TATA box** resembles the -10 region of a prokaryotic promoter (TATAAT), although it differs in its location relative to the transcription start site (-27 versus -10). The functions of the two promoter elements are not strictly analogous, however, since the deletion of the TATA box does not necessarily eliminate transcription. Rather, TATA box deletion or mutation generates heterogeneities in the transcription start site, thereby indicating that the TATA box participates in selecting this site.

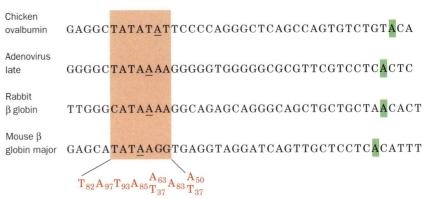

Figure 25-13 The promoter sequences of selected eukaryotic structural genes.   The homologous segment, the TATA box, is shaded in red with the base at position -27 underlined. The initial nucleotide to be transcribed ($+1$) is shaded in green. The bottom row indicates the consensus sequence of several such promoters, with the subscripts indicating the percentage occurrence of the corresponding base. [After Gannon, F., O'Hare, K., Perrin, F., Le Pennec, J.P., Benoist, C., Cochet, M., Breathnach, R., Royal, A., Garapin, A., Cami, B., and Chambon, P., *Nature* **278,** 433 (1978).]

The gene region extending between about −50 and −110 also contains promoter elements. For instance, many eukaryotic structural genes have a conserved consensus sequence of CCAAT (the **CCAAT box**) located between about −70 and −90 whose alteration greatly reduces the gene's transcription rate. Evidently, *the promoter sequences upstream of the TATA box constitute DNA-binding sites for RNAP II and other proteins involved in transcription initiation.*

Enhancers Are Transcriptional Activators That Can Have Variable Positions and Orientations.

Perhaps the most surprising aspect of eukaryotic transcriptional control elements is that some of them need not have fixed positions and orientations relative to their corresponding transcribed sequences. For example, the genome of **simian virus 40 (SV40),** in which such elements were first discovered, contains two repeated sequences of 72 bp each that are located upstream from the promoter for early gene expression. Transcription is unaffected if one of these repeats is deleted but is nearly eliminated when both are absent. The analysis of a series of SV40 mutants containing only one of these repeats demonstrated that its ability to stimulate transcription from its corresponding promoter is all but independent of its position and orientation. Indeed, transcription is unimpaired when this segment is several thousand base pairs upstream or downstream from the transcription start site. Gene segments with such properties are named **enhancers** to indicate that they differ from promoters, with which they must be associated in order to trigger site-specific and strand-specific transcription initiation (although the characterization of numerous promoters and enhancers indicates that their functional properties are similar). Enhancers occur in cellular genes as well as eukaryotic viruses.

Enhancers are required for the full activities of their cognate promoters. It was originally thought that enhancers somehow acted as entry points on DNA for RNAP II (perhaps by altering DNA's local conformation or through a lack of binding affinity for the histones that normally coat eukaryotic DNA; Section 23-5). However, it is now clear that *enhancers are recognized by specific transcription factors that stimulate RNAP II to bind to the corresponding but distant promoter.* This requires that the DNA between the enhancer and promoter loop around so that the transcription factor can simultaneously contact the enhancer and RNAP II at the promoter. Most cellular enhancers are associated with genes that are selectively expressed in specific tissues. It therefore seems, as we discuss in Section 27-3B, that *enhancers mediate much of the selective gene expression in eukaryotes.*

RNA Polymerase III Promoters Can Be Located Downstream from Their Transcription Start Sites.

The promoters of some genes transcribed by RNAP III are located entirely within the genes' transcribed regions. In a gene for the 5S RNA of *Xenopus borealis,* deletions of base sequences that start from outside one or the other end of the transcribed portion of the gene prevent transcription only if they extend into the segment between nucleotides +40 and +80. This portion of the gene is effective as a promoter because it contains the binding site for a transcription factor that stimulates upstream binding of RNAP III. Further studies have shown, however, that the promoters of other RNAP III–transcribed genes lie entirely upstream of their start sites. These upstream sites also bind transcription factors that recruit RNAP III.

C Transcription Factors

Differentiated eukaryotic cells possess a remarkable capacity for the selective expression of specific genes. The synthesis rates of a particular protein in two cells of the same organism may differ by as much as a factor of 10^9. Thus, for example, reticulocytes (immature red blood cells) synthesize large amounts of hemoglobin but no detectable insulin, whereas the pancreatic β cells produce large quantities of insulin but no hemoglobin. In contrast, prokaryotic systems generally exhibit no more than a thousand-fold range in their transcription rates so that at least a few copies of all the proteins they encode are present in any cell. Nevertheless, *the basic mechanism for initiating transcription of structural genes is the same in eukaryotes and prokaryotes: Protein factors bind selectively to the promoter regions of DNA*. In eukaryotes, a complex of at least six **general transcription factors** (**GTFs;** Table 25-1) operate as a formal equivalent of a prokaryotic σ factor. The structures of several eukaryotic transcription factors are described in Section 23-4C.

The six GTFs, which are highly conserved from yeast to humans, are required for the synthesis of all mRNAs, even those with strong promoters. The GTFs allow a low (basal) level of transcription that can be augmented by the participation of other gene-specific factors. We will examine the action of some of these other proteins in Section 27-3. The GTFs, whose names begin with TF (for transcription factor) followed by the Roman numeral II to indicate that they are involved in transcription by RNAP II, combine with the enzyme and promoter DNA in an ordered pathway to form a **preinitiation complex (PIC).**

PIC Formation Begins with TATA-Binding Protein Binding to the TATA Box.   As indicated in Section 25-2B, the promoters of eukaryotic structural genes often contain a TATA box at position -27. The GTFs are targeted to this and nearby sequences so that the entire eukaryotic promoter typically extends over ~100 bp. The first transcription factor to bind to the promoter is the **TATA-binding protein (TBP),** which as its name indicates, binds to the TATA box and thereby helps identify the transcription start site. TBP is subsequently joined on the promoter by additional subunits to form the ~770-kD multisubunit complex **TFIID.**

The highly conserved C-terminal domain of TBP contains two ~40% identical direct repeats of 66 residues separated by a highly basic segment. Its X-ray structure, which was independently determined by Roger

Table 25-1 Properties and Functions of the Eukaryotic General Transcription Factors

Factor	Number of Subunits	Mass (kD)[a]	Function
TFIID			
TBP	1	27	Bending TATA box DNA around TFIIB and RNAP II
TAFs	14	749	Promoter recognition
TFIIA	2	46	Stabilizes binding of TBP on DNA
TFIIB	1	38	Start site determination
TFIIE	2	92	Couples RNAP II–promoter interaction to recruitment of TFIIH
TFIIF	3	156	Interaction with nontemplate DNA strand
TFIIH	9	525	Promoter opening and RNAP II phosphorylation

[a]Mass data are for yeast proteins.

Source: Mainly Bushnell, D.A., Westover, K.D., Davis, R.E., and Kornberg, R.D., *Science* **303,** 983 (2004).

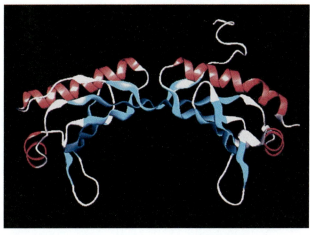

(a)

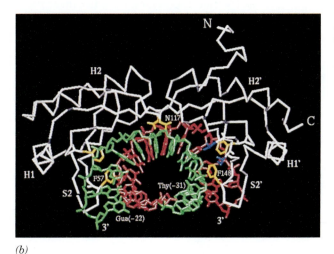

(b)

Figure 25-14 X-Ray structure of TATA-binding protein (TBP) from the plant *Arabidopsis thaliana.* (*a*) A ribbon diagram of the protein in the absence of DNA, in which α helices are red, β strands are blue, and the remainder of the polypeptide backbone is white. The protein's pseudo-twofold axis of symmetry is vertical. (*b*) TBP in complex with a 14-bp TATA box–containing segment of DNA. The DNA, which is largely in the B form, is drawn in stick form with its sense and antisense strands green and red, respectively. It enters its binding site with the 5′ end of the sense strand on the right and exits on the left with its helix axis nearly perpendicular to the page. The protein is represented by its C_α backbone (*white*); together with the side chains of Phe residues 57, 74, 148, and 165 (*yellow*), which induce sharp kinks in the DNA; Asn residues 27 and 117 (*also yellow*), which make hydrogen bonds in the minor groove; and Ile 52 and Leu 163 (*blue*), which are implicated in specific DNA recognition. Between the kinks, which are located at each end of the TATA box, the DNA is partially unwound with the protein's eight-stranded β sheet inserted into the DNA's greatly widened minor groove. TBP does not contact the DNA's major groove. [Courtesy of Stephen Burley, Structural GenomiX, Inc., San Diego, California.] *🖉* **See the Interactive Exercises.**

Kornberg and Stephen Burley, reveals a saddle-shaped molecule that consists of two structurally similar domains arranged with pseudo-twofold symmetry (Fig. 25-14*a*). TBP's overall structure suggests that it could fit snugly astride an undistorted B-DNA helix. However, the X-ray structures of TBP–DNA complexes, independently determined by Burley and Paul Sigler, reveal a quite different interaction. The DNA indeed binds to the concave surface of TBP but with its duplex axis nearly perpendicular rather than parallel to the saddle's cylindrical axis (Fig. 25-14*b*). The bound DNA is kinked by ~45° between the first two and the last two base pairs of its 8-bp TATA element, thereby assuming a cranklike shape. The TBP, which undergoes little conformational change on binding DNA, does so via hydrogen bonding and van der Waals interactions. The kinked and partially unwound DNA is stabilized by a wedge of two Phe side chains on each side of the saddle structure that pry apart the two base pairs flanking each kink from their minor groove sides. The bent conformation of DNA creates a stage for the assembly of other proteins to form the PIC.

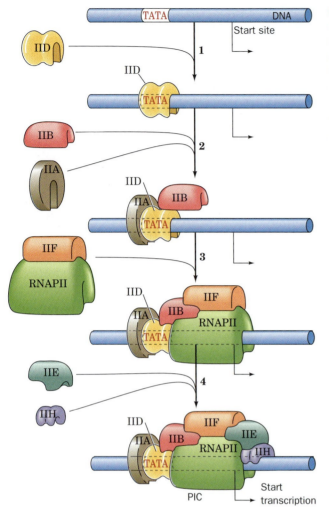

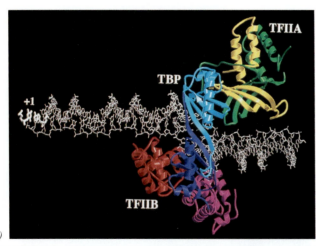

(a)

Figure 25-16 **Structural models for the interactions of TFIIB.** (a) Model of the TFIIA–TBP–TFIIB_C–promoter complex based on the independently determined X-ray crystal structures of the TFIIA–TBP–promoter and TFIIB_C–TBP–promoter complexes. The DNA (*white sticks*) has been extended in both directions beyond the TATA box. The pseudo-symmetrical TBP (*cyan and purple*), viewed from above, induces two sharp kinks in the DNA, giving it a cranklike shape. Both TFIIA (*yellow and green*) and TFIIB_C (*magenta and red*) interact with TBP, but each interacts with independent sites on the DNA. The transcription start site (+1) is at the left. [Courtesy of Stephen Burley, Structural GenomiX,

Figure 25-15 *Key to Function.* **The assembly of the preinitiation complex (PIC) on a TATA box–containing promoter.** (1) The TBP component of TFIID binds to the TATA box of the promoter. (2) TFIIA and TFIIB then bind. (3) TFIIF binds to RNAP II and escorts it to the complex. (4) Finally, TFIIE and TFIIH are sequentially recruited, thereby completing the PIC. [After Zawel, L. and Reinberg, D. *Curr. Opin. Cell Biol.* **4,** 490 (1992).]

TFIIA, TFIIB, and TAFs Interact with TBP and RNAP II. The other GTFs required for basal transcription assemble as is diagrammed in Fig. 25-15. The PIC requires, at a minimum, TBP, **TFIIB, TFIIE, TFIIF,** and **TFIIH.** TFIIB consists of two domains, an N-terminal domain **(TFIIB_N),** which interacts with RNAP II, and a C-terminal domain **(TFIIB_C),** which binds DNA and interacts with TBP. The X-ray structures of TFIIA–TBP–DNA and TFIIB_C–TBP–DNA complexes have permitted the generation of a plausible model for the TFIIA–TFIIB_C–TBP–DNA complex (Fig. 25-16a). The three proteins bind to the DNA just upstream from the transcription start site, leaving ample room for additional proteins and RNAP II to bind. Since the pseudosymmetric TBP has been shown to bind to the TATA box in either orientation, it appears that base-specific interactions between TFIIB and the promoter function to position TFIIB to properly orient the TBP on the promoter.

The X-ray structure of the TFIIB_N–RNAP II complex, also determined by Kornberg, reveals that TFIIB_N contacts the "dock" domain of RNAP II near its RNA exit channel and that TFIIB_N inserts its "finger" domain into the active center of RNAP II. A model of the TFIIB–TBP–RNAP II–DNA com-

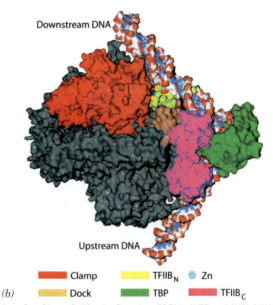

(b)

🟥 Clamp	🟨 TFIIB_N	🔵 Zn
🟧 Dock	🟩 TBP	🟥 TFIIB_C

Inc., San Diego, California. Based on PDBids 1YTF and 1VOL.] (b) Model of the TFIIB–TBP–RNAP II–promoter complex based on the X-ray structures of the TFIIB_N–RNAP II and TFIIB_C–TBP–promoter complexes. The view of the RNAP II is from the bottom of Fig. 25-12 after it has been turned 180° in the plane of the paper. The protein and the DNA are represented by their solvent-accessible surfaces with the protein colored grey except as indicated in the accompanying key and the DNA colored according to atom type (C white, N blue, O red, and P yellow). The DNA has been extended by 20 bp at both ends. [Courtesy of Roger Kornberg, Stanford University. Based on PDBids 1R5U and 1VOL.]

plex based on this X-ray structure together with that of the TFIIB$_C$–TBP–DNA complex is shown in Fig. 25-16*b*.

The remaining components of TFIID, which are known as **TBP-associated factors (TAFs),** form a horseshoe-shaped complex to which TFIIA and TFIIB are bound (Fig. 25-17). TBP is located at the top of the cavity, and the promoter DNA passes through the cavity. Portions of 9 of the 14 known TAFs, which are highly conserved from yeast to humans, are homologous to nonlinker histones (Section 23-5A). Indeed, X-ray and other studies suggest that four of these TAFs associate to form a nucleosome-like heterooctamer. Nevertheless, it is unlikely that the DNA wraps around the TAFs as it does in a nucleosome because many of the histone residues that make critical contacts with the DNA in the nucleosome have not been conserved in the histone-like TAFs.

In the final steps of PIC formation (Fig. 25-15), TFIIF recruits RNAP II to the promoter in a manner reminiscent of the way that σ factor interacts with bacterial RNAP. In fact, the second largest of TFIIF's three subunits is homologous to σ^{70}, the predominant bacterial σ factor, and, moreover, can specifically interact with bacterial RNAPs (although it does not participate in promoter recognition). Finally, TFIIE and TFIIH join the assembly. Once this complex has been assembled, the ATP-dependent helicase activity of TFIIH induces the formation of the open complex so that RNA synthesis can commence. Note that the yeast PIC, exclusive of the ~12-subunit, ~600-kD RNAP II, contains 32 subunits with an aggregate mass of 1633 kD (Table 25-1). Many of the proteins in the PIC are the targets of transcriptional regulators.

Many Class II Core Promoters Lack a TATA Box.

The core promoters of ~65% of the genes transcribed by RNAP II (class II genes) lack TATA boxes. They are often "housekeeping" genes, that is, genes that are constitutively expressed in all cells at relatively low rates. How can RNAP II properly initiate transcription at these TATA-less promoters? Investigations have shown that TATA-less promoters often contain a so-called **initiator (Inr)** element that extends between positions −6 and +11 and has the loose consensus sequence YYAN^A_TYY, where Y is a pyrimidine (C or T), N is any nucleotide, and A is the initiating (+1) nucleotide. The presence of an Inr element is sufficient to direct RNAP II to the correct start site. These systems require the participation of many of the same GTFs that initiate transcription from TATA box–containing promoters. Surprisingly, they also require TBP. This suggests that with TATA-less promoters, Inr recruits TFIID such that its component TBP binds to the −30 region in a sequence-nonspecific manner. Indeed, in Inr-containing promoters that also contain a TATA box, the two elements act synergistically to promote transcriptional initiation. Nevertheless, a mutant TBP that is defective in TATA box binding will support efficient transcription from some TATA-less promoters although not from others. This suggests that the former promoters do not require a stable interaction with TBP. Consequently, the scheme outlined in Fig. 25-15 should be taken as a flexible framework for transcription initiation in eukaryotes, with the exact protein requirements depending on the nature of the promoter and the presence of additional protein factors.

TBP Is a Universal Transcription Factor.

RNAP I and RNAP III require different sets of GTFs from each other and from RNAP II to initiate transcription at their respective promoters. This is not unexpected considering the very different organizations of these three classes of promoters (Section 25-2B). Indeed, the promoters recognized by RNAP I (class I promoters) and nearly all those recognized by RNAP III (class III

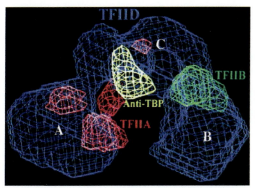

Figure 25-17 Electron microscopy–based image of the human TFIID–TFIIA–TFIIB complex at 35 Å resolution. The complex consists of three domains, A, B, and C, arranged in a horseshoe shape roughly 200 Å wide, 135 Å high, and 110 Å thick. The red and green meshes indicate the positions of TFIIA and TFIIB as determined by comparison with the EM-based images of the TFIID–TFIIA complex and TFIID alone. The yellow mesh indicates the binding position of an anti-TBP antibody. [Courtesy of Eva Nogales, University of California at Berkeley.]

promoters) lack TATA boxes. Thus, it came as a surprise when it was demonstrated that *TBP is required for initiation by both RNAP I and RNAP III.* TBP participates by combining with different sets of TAFs to form the GTFs **SLI** (with class I promoters) and **TFIIIB** (with class III promoters). As with certain class II TATA-less promoters, a TBP mutant that is defective for TATA box binding can still support *in vitro* transcriptional initiation by both RNAP I and RNAP III. Clearly, TBP, the only known universal transcription factor, is an unusually versatile protein.

Elongation Requires Different Transcription Factors.   After RNAP II initiates RNA synthesis and successfully produces a short transcript, the transcription machinery undergoes a transition to the elongation mode. The switch appears to involve phosphorylation of the C-terminal domain (CTD) of RNAP II's Rpb1 subunit. Phosphorylated RNAP II releases some of the transcription-initiating factors and advances beyond the promoter region. In fact, when RNAP II moves away from ("clears") the promoter, it leaves behind some GTFs, including TFIID. These proteins can reinitiate transcription by recruiting another RNAP II to the promoter. Consequently, *the first RNAP to transcribe a gene may act as a "pioneer" polymerase that helps pave the way for additional rounds of transcription.*

During elongation, a six-protein complex called **Elongator** binds to the phosphorylated CTD of Rpb1, taking the place of the jettisoned transcription factors. Although Elongator is not essential for transcription by RNAP II *in vitro*, its presence accelerates transcription. Interestingly, TFIIF and TFIIH remain associated with the polymerase during elongation.

Eukaryotes Lack Precise Transcription Termination Sites.   The sequences signaling transcriptional termination in eukaryotes have not been identified. This is largely because the termination process is imprecise; that is, the primary transcripts of a given structural gene have heterogeneous 3′ sequences. *However, a precise termination site is not required because the transcript undergoes processing that includes endonucleolytic cleavage at a specific site* (see below). The endonuclease may act even while the polymerase is still transcribing, so RNA cleavage itself may signal the polymerase to stop.

3 Posttranscriptional Processing

The immediate products of transcription, the **primary transcripts,** are not necessarily functional. In order to acquire biological activity, many of them must be specifically altered: (1) by the exo- and endonucleolytic removal of polynucleotide segments; (2) by appending nucleotide sequences to their 3′ and 5′ ends; and/or (3) by the modification of specific nucleotide residues. The three major classes of RNA—mRNA, rRNA, and tRNA—are altered in different ways in prokaryotes and in eukaryotes. In this section, we shall outline these **posttranscriptional modification** processes.

A   Messenger RNA Processing

In prokaryotes, most primary mRNA transcripts are translated without further modification. Indeed, protein synthesis usually begins before transcription is complete (Section 26-4). In eukaryotes, however, mRNAs are synthesized in the cell nucleus, whereas translation occurs in the cytosol. Eukaryotic mRNA transcripts can therefore undergo extensive posttranscriptional processing while still in the nucleus.

Eukaryotic mRNAs Have 5′ Caps. *Eukaryotic mRNAs have a **cap structure** consisting of a **7-methylguanosine (m⁷G)** residue joined to the transcript's initial (5′) nucleotide via a 5′–5′ triphosphate bridge* (Fig. 25-18). The cap, which adds to the growing transcript when it is ~30 nt long, identifies the eukaryotic translation start site (Section 26-4A). Capping involves several enzymatic reactions: (1) the removal of the leading phosphate group from the mRNA's 5′ terminal triphosphate group by an **RNA triphosphatase;** (2) the guanylation of the mRNA by **capping enzyme,** which requires GTP and yields the 5′–5′ triphosphate bridge and PP$_i$; and (3) the methylation of guanine by **guanine-7-methyltransferase,** in which the methyl group is supplied by *S*-adenosylmethionine (SAM). In addition, the cap may be $O^{2'}$-methylated at the first and second nucleotides of the transcript by a SAM-requiring **2′-O-methyltransferase.** Capping enzyme binds RNAP II's phosphorylated CTD, and hence it appears that capping marks the completion of RNAP II's switch from transcription initiation to elongation. Capped mRNAs are resistant to 5′-exonucleolytic degradation.

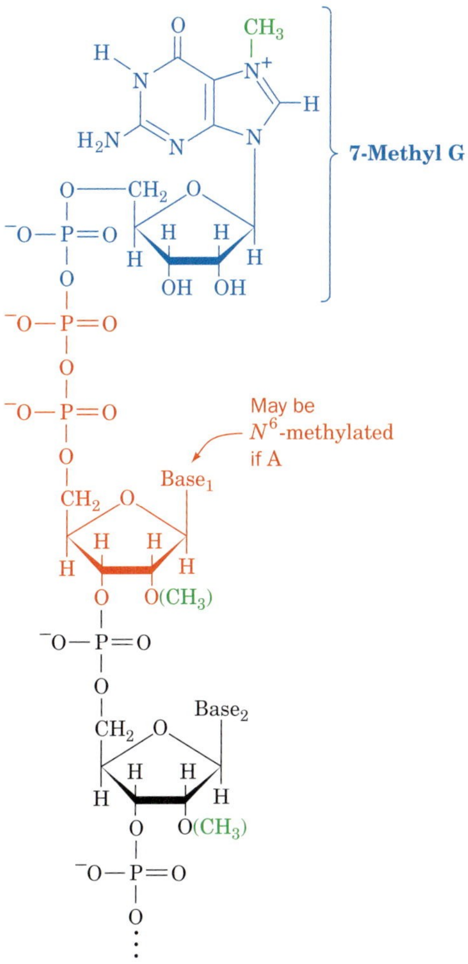

Eukaryotic mRNAs Have Poly(A) Tails. Eukaryotic RNA transcripts have heterogeneous 3′ sequences because the transcription termination process is imprecise. *Mature eukaryotic mRNAs, however, have well-defined 3′ ends terminating in **poly(A) tails** of ~250 nt (~80 nt in yeast).* The poly(A) tails are enzymatically appended to the primary transcripts in two reactions:

1. A transcript is cleaved 15 to 25 nt past a highly conserved AAUAAA sequence and less than 50 nt before a less-conserved U-rich or GU-rich sequence. The precision of this cleavage reaction has apparently eliminated the need for accurate transcription termination.

2. The poly(A) tail is subsequently generated from ATP through the stepwise action of **poly(A) polymerase (PAP),** a template-independent RNA polymerase that elongates an mRNA primer with a free 3′-OH group. PAP is activated by **cleavage and polyadenylation specificity factor (CPSF)** when the latter protein recognizes the AAUAAA sequence. Once the poly(A) tail has grown to ~10 residues, the AAUAAA sequence is no longer required for further chain elongation. This suggests that CPSF becomes disengaged from its recognition site in a manner reminiscent of the way σ factor is released from the transcription initiation site during elongation of prokaryotic RNA.

CPSF binds to the phosphorylated RNAP II CTD (Section 25-2A); deleting the CTD inhibits polyadenylation. Evidently, the CTD couples polyadenylation to transcription.

PAP is part of a 500- to 1000-kD complex that also contains the proteins required for mRNA cleavage. Consequently, the cleaved transcript is polyadenylated before it can dissociate and be digested by cellular nucleases (see below). The maximum length of the poly(A) tail may be determined by the stoichiometric binding of multiple copies of **poly(A) binding protein II (PAB II).**

In vitro studies indicate that a poly(A) tail is not required for mRNA translation. Rather, the observation that an mRNA's poly(A) tail shortens as it ages in the cytosol suggests that poly(A) tails have a protective role. In fact, the only mature mRNAs that lack poly(A) tails, those of histones (which are required in large quantities only during the relatively short period during the cell cycle when DNA is being replicated), have cytosolic lifetimes of <30 min versus hours or days for most other mRNAs. The poly(A) tails are specifically complexed in the cytosol by **poly(A)-binding protein (PABP;** not related to PAB II), which organizes poly(A)-bearing

Figure 25-18 Structure of the 5′ cap of eukaryotic mRNAs. A cap may be $O^{2'}$-methylated at the transcript's leading nucleoside (the predominant cap in multicellular organisms), at its first two nucleosides, or at neither of these positions (the predominant cap in unicellular eukaryotes). If the first nucleoside is adenosine (it is usually a purine), it may also be N^6-methylated.

mRNAs into ribonucleoprotein particles. PABP is thought to protect mRNA from degradation as is suggested, for example, by the observation that the addition of PABP to a cell-free system containing mRNA and mRNA-degrading nucleases greatly reduces the rate at which the mRNAs are degraded and the rate at which their poly(A) tails are shortened.

Eukaryotic Genes Consist of Alternating Exons and Introns. The most striking difference between eukaryotic and prokaryotic structural genes is that *the coding sequences of most eukaryotic genes are interspersed with unexpressed regions*. The primary transcripts, also called **pre-mRNAs** or **heterogeneous nuclear RNAs (hnRNAs),** are variable in length and are much larger (~2000 to >20,000 nt) than expected from the known sizes of eukaryotic proteins. Rapid labeling experiments demonstrated that little of the hnRNA is ever transported to the cytosol; most of it is quickly degraded in the nucleus. Yet the hnRNA's 5′ caps and 3′ tails eventually appear in cytosolic mRNAs. The straightforward explanation of these observations, that pre-mRNAs are processed by the excision of internal sequences, seemed so bizarre that it came as a great surprise in 1977 when Phillip Sharp and Richard Roberts independently demonstrated that this is actually the case (Box 25-3). Thus, *pre-mRNAs are processed by the excision of nonexpressed **intervening sequences (introns),** following which the flanking **expressed sequences (exons)** are **spliced** (joined) together.*

A pre-mRNA typically contains eight introns whose aggregate length averages 4 to 10 times that of its exons. This situation is graphically illustrated in Fig. 25-19, which is an electron micrograph of chicken **ovalbumin** mRNA hybridized to the antisense (template) strand of the ovalbumin gene (ovalbumin is the major protein component of egg white). In humans, the number of introns in a gene varies from none to 234 (in the gene encoding the 26,926-residue muscle protein **titin,** the largest known single-chain

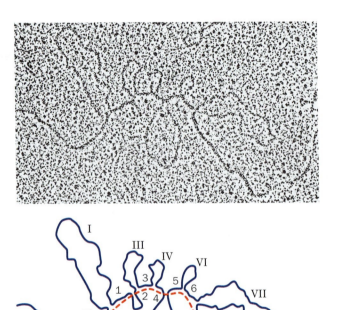

Figure 25-19 The chicken ovalbumin gene and its mRNA. The electron micrograph and its interpretive drawing show the hybridization of the antisense (template) strand of the chicken ovalbumin gene and its corresponding mRNA. The complementary segments of the DNA (*purple line in drawing*) and mRNA (*red line*) have annealed to reveal the exon positions (*L, 1–7*). The looped-out segments (I–VII), which have no complementary sequences in the mRNA, are the introns. [From Chambon, P., *Sci. Am.* **244**(5), 61 (1981).]

BOX 25-3

Pathways of Discovery

Richard Roberts and Phillip Sharp and the Discovery of Introns

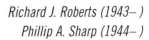
Richard J. Roberts (1943–)
Phillip A. Sharp (1944–)

For several decades following the discovery that DNA is the genetic material in all organisms, it was believed that genes were continuous sequences of nucleic acid. A messenger RNA molecule was thought to be a faithful copy of the gene, aligning exactly with the DNA sequence. In fact, this is largely true for simple genetic systems such as bacteria and bacteriophages, which were widely used in early studies of molecular biology. However, Richard Roberts and Phillip Sharp, working independently, showed in 1977 that genes could be discontinuous. We now understand that the mRNA segments corresponding to these "split genes" must be spliced together in order for the gene to be properly expressed.

Richard Roberts began his scientific career as an organic chemist. While working on his thesis, he read a book by John Kendrew (who had determined the X-ray structure of myoglobin; Section 7-1A) and became hooked on the new field of molecular biology. His first project in the world of nucleic acids was to determine the sequence of nucleotides in a tRNA molecule. He was able to take advantage of new sequencing techniques devised by Frederick Sanger (who subsequently developed the dideoxy sequencing method; Section 3-4C). Roberts next turned his attention to restriction endonucleases. He recognized that these enzymes could be invaluable tools for cutting large DNA molecules down to size, including the adenovirus-2 genome that he began to characterize.

During roughly the same period, Phillip Sharp completed a thesis describing his use of statistical and physical theories to describe the DNA polymer. By his own admission, he was not a skilled experimenter. His outlook changed, however, when he joined a molecular biology laboratory that made extensive use of electron microscopy to examine DNA–RNA heteroduplexes. When Sharp subsequently turned to the study of eukaryotic gene expression, the only practical experimental systems were animal viruses with DNA genomes, such as adenovirus-2. Because this virus, which is a cause of the common cold, infects mammalian cells, its genes were thought to resemble those of its host.

Thus, in the mid-1970s, both Roberts (at the Cold Spring Harbor Laboratory on Long Island, New York) and Sharp (at the Massachusetts Institute of Technology) came to be mapping the adenovirus genome. They located viral genes by obtaining expressed mRNA molecules and hybridizing them to segments of the viral DNA. Sharp observed that the nuclei of virus-infected cells accumulated viral mRNAs that were not transported to the cytoplasm, and he speculated that the

nuclear mRNAs were processed in order to generate the cytoplasmic mRNAs. Meanwhile, Roberts observed that "late" (mature) mRNAs began with an oligonucleotide that was not encoded in the DNA next to the main body of the mRNA.

Both scientists prepared samples in which the mRNA was allowed to hybridize with the DNA, and then visualized the hybrid by electron microscopy. The results were as exciting as they were unexpected: A single mRNA molecule hybridized with as many as four well-separated segments of the DNA molecule, so that the unpaired DNA sequences between the hybrid segments looped out (as in Fig. 25-19). The inescapable conclusion was that viral genetic information was organized discontinuously, a notion that contradicted commonly held views about the nature of genes. Nevertheless, other researchers were eager to see whether split genes occurred in other viruses and animal cells. Within a year, similar results were confirmed for a handful of other genes. Subsequent research showed that most animal genes are discontinuous. The sequences that are ultimately expressed were termed exons, and the intervening sequences that did not appear in the mature mRNA were named introns. For their discoveries of split genes, Roberts and Sharp shared the 1993 Nobel Prize in Physiology or Medicine.

The existence of split genes solved some biochemical puzzles but, as is always the case with important discoveries, introduced new ones: How are introns cut out and the remaining exons joined together? Are the same exons spliced together in all cells? Could evolution occur more rapidly through exon shuffling? Answers to these questions are still being refined.

Chow, L.T., Gelinas, R.E., Broker, T.R., and Roberts, R.J., An amazing sequence arrangement at the 5′ ends of adenovirus 2 messenger RNA, *Cell* **12**, 1–8 (1977).

Berget, S.M., Moore, C., and Sharp, P.A., Spliced segments at the 5′ terminus of adenovirus 2 late mRNA, *Proc. Natl. Acad. Sci.* **74**, 3171–3175 (1977).

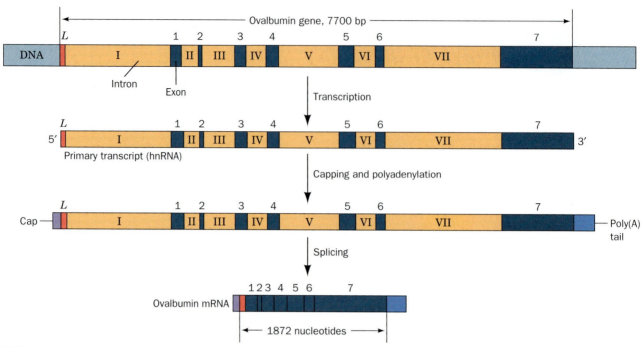

Figure 25-20   The sequence of steps in the production of mature eukaryotic mRNA.   This example shows the chicken ovalbumin gene. Following transcription, the primary transcript is capped and polyadenylated. The introns are then excised and the exons spliced together to form the mature mRNA. However, splicing may also occur cotranscriptionally.

protein), with intron lengths ranging from ~65 to 2.4 million nt (in the gene encoding the muscle protein **dystrophin**) and averaging ~3500 nt (exons, in contrast, have lengths that average ~150 nt and range up to 17,106 nt in the gene encoding titin). The introns from corresponding genes in two vertebrate species rarely vary in number and position, but often differ extensively in length and sequence so as to bear little resemblance to one another.

The production of a translation-competent eukaryotic mRNA begins with the transcription of the entire gene, including its introns (Fig. 25-20). Capping occurs soon after initiation, and splicing commences during the elongation phase of transcription. The mature mRNA emerges only after splicing is complete and the RNA has been polyadenylated. The mRNA is then transported to the cytosol, where the ribosomes are located, for translation into protein.

Exons Are Spliced in a Two-Stage Reaction.   Sequence comparisons of exon–intron junctions from a diverse group of eukaryotes indicate that they have a high degree of homology (Fig. 25-21), with most of them having an invariant GU at the intron's 5′ boundary and an invariant AG at its 3′ boundary. *These sequences are necessary and sufficient to define a splice junction.* The splicing reaction occurs via two transesterification reactions (Fig. 25-22):

Figure 25-21   The consensus sequences at the exon–intron junctions of eukaryotic pre-mRNAs.   The subscripts indicate the percentage of pre-mRNAs in which the specified base(s) occurs. Note that the 3′ splice site is preceded by a tract of 11 predominantly pyrimidine nucleotides. [Based on data from Padgett, R.A., Grabowski, P.J., Konarska, M.M., Seiler, S.S., and Sharp, P.A., *Annu. Rev. Biochem.* **55**, 1123 (1986).]

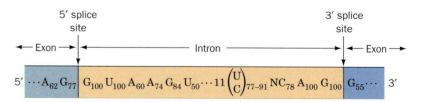

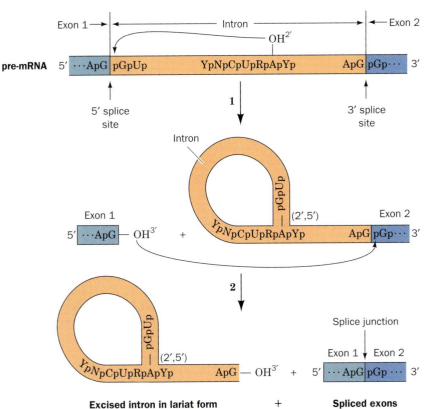

Figure 25-22 *Key to Function.*   **The splicing reaction.**   Two transesterification reactions splice together the exons of eukaryotic pre-mRNAs. The exons and introns are drawn in blue and orange, and R and Y represent purine and pyrimidine residues. (**1**) The 2′-OH group of a specific intron A residue nucleophilically attacks the 5′-phosphate at the 5′ intron boundary to displace the 3′ end of the 5′ exon, thereby yielding a 2′,5′-phosphodiester bond and forming a lariat structure. (**2**) The liberated 3′-OH group attacks the 5′-phosphate of the 5′-terminal residue of the 3′ exon, forming a 3′,5′-phosphodiester bond, thereby displacing the intron in lariat form and splicing the two exons together.

1. A 2′,5′-phosphodiester bond forms between an intron adenosine residue and the intron's 5′-terminal phosphate group. The 5′ exon is thereby released and the intron assumes a novel **lariat structure** (so called because of its shape). The adenosine at the lariat branch point is typically located in a conserved sequence 20 to 50 residues upstream of the 3′ splice site. Mutations that change this branch point A residue abolish splicing at that site.

2. The 5′ exon's free 3′-OH group displaces the 3′ end of the intron, forming a phosphodiester bond with the 5′-terminal phosphate of the 3′ exon and yielding the spliced product. The intron is thereby eliminated in its lariat form with a free 3′-OH group. Mutations that alter the conserved AG at the 3′ splice junction block this second step, although they do not interfere with lariat formation. The lariat is eventually debranched (linearized) and, *in vivo,* is rapidly degraded.

Note that the splicing process proceeds without free energy input; its transesterification reactions preserve the free energy of each cleaved phosphodiester bond through the concomitant formation of a new one.

The sequences required for splicing are the short consensus sequences at the 3′ and 5′ splice sites and at the branch site. Nevertheless, these sequences are poorly conserved. However, other short sequence elements within exons that are known as **exonic sequence enhancers (ESEs)** also play important roles in splice site selection although their characteristics are poorly understood (even highly sophisticated computer programs are only ~50% successful in predicting actual splice sites over apparently equally good candidates that are not). In contrast, large portions of most introns can be deleted without impeding splicing.

A simplistic interpretation of Fig. 25-22 suggests that any 5' splice site could be joined with any following 3' splice site, thereby eliminating all the intervening exons together with the introns joining them. However, such **exon skipping** does not normally take place (but see below). Rather, all of a pre-mRNA's introns are individually excised in what appears to be a largely fixed order that more or less proceeds in the 5' → 3' direction. This occurs, at least in part, because splicing occurs cotranslationally. Thus, as a newly synthesized exon emerges from an RNAP II, it is bound by splicing factors (see below) that are also bound to the RNAP II's highly phosphorylated CTD. This tethers the exon and its associated splicing machinery to the CTD so as to ensure that splicing occurs when the next exon emerges from the RNAP II.

Splicing Is Mediated by snRNPs in the Spliceosome. How are splice junctions recognized and how are the two exons to be joined brought together in the splicing process? Part of the answer to this question was established by Joan Steitz going on the assumption that one nucleic acid is best recognized by another. The eukaryotic nucleus contains numerous copies of highly conserved 60- to 300-nt RNAs called **small nuclear RNAs (snRNAs),** which form protein complexes termed **small nuclear ribonucleoproteins (snRNPs;** pronounced "snurps"). Steitz noted that the 5' end of one of these snRNAs, **U1-snRNA** (which is so named because of its high uridine content), is partially complementary to the consensus sequence of 5' splice junctions. Apparently, *U1-snRNP recognizes the 5' splice junction.* Other snRNPs that participate in splicing are **U2-snRNP, U4–U6-snRNP** (in which the **U4-** and **U6-snRNAs** associate via base pairing), and **U5-snRNP.**

Splicing takes place in an ~45S particle dubbed the **spliceosome.** The spliceosome brings together a pre-mRNA, the snRNPs, and a variety of pre-mRNA binding proteins. The spliceosome, which consists of 5 RNAs and ~65 polypeptides, is comparable in size and complexity to the ribosome (Section 26-3). A pre-assembled spliceosome appears to associate with the pre-mRNA and then undergoes significant structural changes, including rearrangement of base-paired stems among the snRNAs, as it carries out the two transesterification reactions. The U6 snRNA coordinates a catalytically essential metal ion that enhances the nucleophilicity of the attacking OH group and stabilizes the leaving group. The existence of self-splicing introns (see below) and the results of experiments with protein-free snRNAs strongly suggest that the RNA components of the spliceosome, rather than the proteins, catalyze splicing.

All four snRNPs involved in mRNA splicing contain the same so-called **snRNP core protein,** which consists of seven **Sm proteins,** named **B, D_1, D_2, D_3, E, F,** and **G proteins.** These proteins collectively bind to a conserved AAUUUGUGG sequence known as the **Sm RNA motif,** which occurs in U1-, U2-, U4-, and U5-snRNAs. The X-ray structures of the D_3B and D_1D_2 heterodimers reveal that these four proteins share a common fold, which consists of an N-terminal helix followed by a five-stranded antiparallel β sheet that is strongly bent so as to form a hydrophobic core. Model building based on these X-ray structures together with biochemical and mutagenic evidence suggest that the seven Sm proteins form a heptameric ring in which the fourth β strand of one Sm protein interacts with the fifth β strand of an adjacent Sm protein through main-chain hydrogen bonding (Fig. 25-23). The funnel-shaped central hole of the Sm ring, which is lined with positive charges, is large enough to accommodate single-stranded snRNA. This model is corroborated by the X-ray structure of an Sm-like protein from the hypertherm-

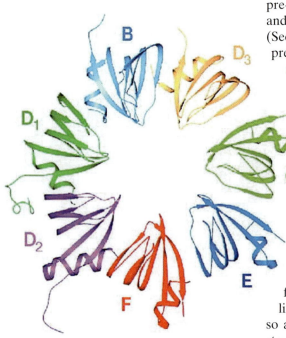

Figure 25-23 **A model of the snRNP core protein.** Its seven Sm proteins, which are each differently colored, are arranged in a seven-membered ring. This heptameric ring has an outer diameter of 70 Å and a central hole with a diameter of 20 Å when only the main chain atoms are considered. [Courtesy of Kiyoshi Nagai, MRC Laboratory of Molecular Biology, Cambridge, U.K.]

ophilic archeon *Pyrobaculum aerophilum* that forms a homoheptameric ring that is structurally similar to the heteroheptameric model. Evidently, the seven eukaryotic Sm proteins arose through a series of duplications of an archaeal Sm-like protein gene.

Mammalian U1-snRNP consists of U1-snRNA and ten proteins, namely, the seven Sm proteins that are common to all U-snRNPs as well as three that are specific to U1-snRNP: **U1-70K, U1-A,** and **U1-C.** The predicted secondary structure of the 165-nt U1-snRNA (Fig. 25-24*a*) contains five double-helical stems, four of which come together at a four-way junction. U1-70K and U1-A bind directly to RNA stem–loops I and II, respectively, whereas U1-C is bound by other proteins.

An electron microscopy–based image of U1-snRNP (Fig. 25-24*b*), elucidated by Holgar Stark and Reinhard Lührmann, contains as its most obvious feature a ring-shaped body that is 70 to 80 Å in diameter with a funnel-shaped central hole that closely matches the model of the snRNP core protein in Fig. 25-23. Proteins were assigned to U1-snRNP's various protuberances based on cross-linking and binding studies as well electron micrographs of U1-snRNP lacking U1-A or U1-70K. The positions of the U1-snRNA's various structural elements, which were identified on the basis of known protein–RNA interactions, indicate that the Sm RNA motif, in fact, passes through the central hole in the snRNP core protein.

The Significance of Gene Splicing.   Analysis of the large body of known DNA sequences reveals that introns are rare in prokaryotic structural genes, uncommon in lower eukaryotes such as yeast (which has a total of 239 introns in its ~6000 genes and, with two exceptions, only one intron per polypeptide), and abundant in higher eukaryotes (the only known vertebrate structural genes lacking introns are those encoding histones and the antiviral proteins known as **interferons**). Pre-mRNA introns, as we have

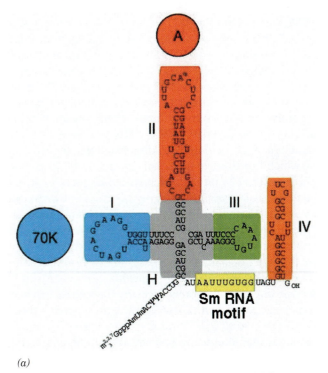

(a)

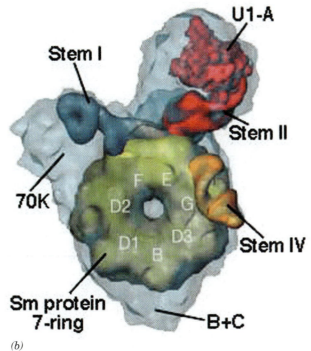

(b)

Figure 25-24   Electron microscopy–based structure of U1-snRNP at 10 Å resolution.   (*a*) The predicted secondary structure of U1-snRNA with the positions at which the proteins U1-70K and U1-A bind the RNA indicated. (*b*) The molecular outline of U1-snRNP in light gray with its component ring-shaped Sm core protein yellow (and viewed oppositely from Fig. 25-23) and U1-snRNA colored as in Part *a*. [Courtesy of Holgar Stark, Max-Planck-Institut für biophysikalische Chemie, Göttingen, Germany.]

seen, can be quite long and many genes contain large numbers of them. Consequently, unexpressed sequences constitute ~80% of a typical vertebrate structural gene and >99% of a few of them.

The argument that introns are only molecular parasites (**junk DNA**) seems untenable since it would then be difficult to rationalize why the evolution of complex splicing machinery offered any selective advantage over the elimination of the split genes. What then is the function of gene splicing? Although, since its discovery, the significance of gene splicing has been debated, often vehemently, two important roles for it have emerged:

1. Gene splicing is an agent for rapid protein evolution. Many eukaryotic proteins consist of modules that also occur in other proteins (Section 5-4C). For example, SH2 and SH3 domains occur in many of the proteins involved in signal transduction (Section 21-3D). *It therefore appears that the genes encoding these modular proteins arose by the stepwise collection of exons that were assembled by (aberrant) recombination between their neighboring introns.*

2. Through **alternative splicing,** which we discuss below, gene splicing permits a single gene to encode several (sometimes many) proteins that may have significantly different functions.

Alternative mRNA Splicing Yields Multiple Proteins from a Single Gene. *The expression of numerous cellular genes is modulated by the selection of alternative splice sites.* Thus, genes containing multiple exons may give rise to transcripts containing mutually exclusive exons. In effect, certain exons in one type of cell may be introns in another. For example, a single rat gene encodes seven tissue-specific variants of the muscle protein **α-tropomyosin** through the selection of alternative splice sites (Fig. 25-25).

Alternative splicing occurs in all multicellular organisms and is especially prevalent in vertebrates. In fact, an estimated 60% of human structural genes are subject to alternative splicing. This helps rationalize the discrepancy between the ~30,000 genes identified in the human genome and earlier estimates that it contains over 100,000 structural genes. The variations in spliced mRNA sequences can take several different forms: Exons can be retained or skipped; introns may be excised or retained; and the positions of 5' and 3' splice sites can be shifted to make exons longer or shorter. Alterations in the transcriptional start site and/or the polyadenylation site can further contribute to the diversity of the mRNAs that are transcribed from a single gene. In a particularly striking example, the *Drosophila* **DSCAM** protein, which functions in neuronal development, is encoded by 24 exons of which there are 12 mutually exclusive variants of exon 4, 48 of exon 6, 33 of exon 9, and 2 of exon 17 (which are therefore known as **cassette exons**) for a total of 38,016 possible variants of this protein (compared to ~13,000 identified genes in the *Drosophila* genome). Although it is unknown whether all possible DSCAM variants are produced, experimental evidence suggests that the *Dscam* gene expresses many thousands of them. Clearly, *the number of genes in an organism's genome does not by itself provide an adequate assessment of its protein diversity.* Indeed, it has been estimated that, on average, each human structural gene encodes three different proteins.

The types of changes that alternative splicing confers on expressed proteins span the entire spectrum of protein properties and functions. Entire functional domains or even single amino acid residues may be inserted into or deleted from a protein, and the insertion of a **Stop codon** may truncate the expressed polypeptide [a codon is a 3-nt sequence that specifies an

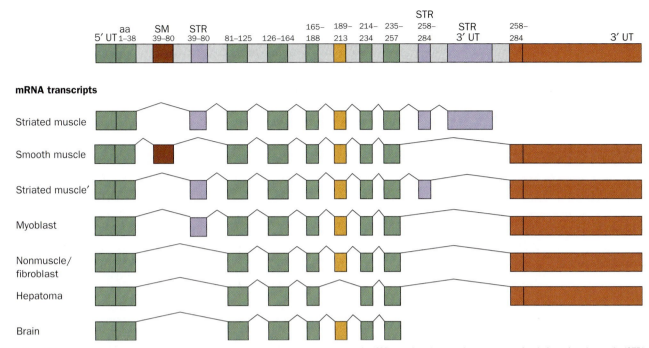

Figure 25-25 Alternative splicing in the rat α-tropomyosin gene. Seven alternative splicing pathways give rise to cell-specific α-tropomyosin variants. The thin kinked lines indicate the positions occupied by the introns before they are spliced out to form the mature mRNAs. Tissue-specific exons are indicated together with the amino acid (aa) residues they encode: "constitutive" exons (those expressed in all tissues) are green; those expressed only in smooth muscle (SM) are dark brown; those expressed only in striated muscle (STR) are purple; and those variably expressed are yellow. Note that the smooth and striated muscle exons encoding amino acid residues 39 to 80 are mutually exclusive and, likewise, there are alternative 3′ untranslated (UT) exons. [After Breitbart, R.E., Andreadis, A., and Nadal-Ginard, B., *Annu. Rev. Biochem.* **56,** 481 (1987).]

amino acid in a polypeptide; a Stop codon is a 3-nt sequence that instructs the ribosome to terminate polypeptide synthesis (Section 26-1C)]. Splice variations may, for example, control whether a protein is soluble or membrane bound, whether it is phosphorylated by a specific kinase, the subcellular location to which it is targeted, whether an enzyme binds a particular allosteric effector, the affinity with which a receptor binds a ligand, etc. Changes in an mRNA, particularly in its noncoding regions, may also influence the rate at which it is transcribed and its susceptibility to degradation. Since the selection of alternative splice sites is both tissue- and developmental stage-specific, splice site choice must be tightly regulated in both space and time. In fact, ~15% of human genetic diseases are caused by point mutations that result in pre-mRNA splicing defects. Some of these mutations delete functional splice sites, thereby activating nearby preexisting **cryptic splice sites.** Others generate new splice sites that are used instead of the normal ones. In addition, tumor progression is correlated with changes in levels of proteins implicated in alternative splice site selection.

How are alternative splice sites selected? The best understood examples of such processes occur in the pathway responsible for sex determination in *Drosophila,* two of which we discuss here:

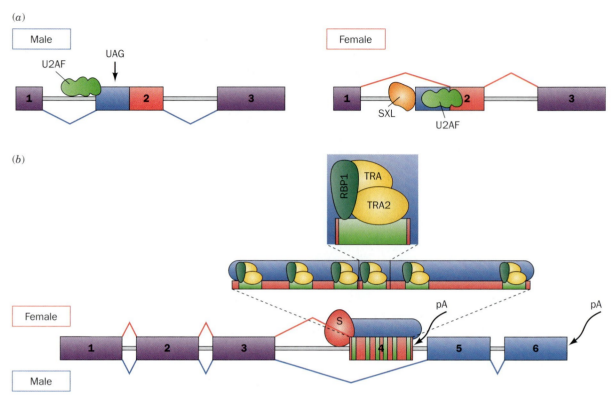

Figure 25-26 Alternative splice site selection in the *Drosophila* sex-determination pathway.   The mechanisms are described in the text. In all panels, exons are represented by colored rectangles and introns are shown as gray bands with those to be excised flanked by kinked lines. (*a*) Alternative splicing in *tra* pre-mRNA. UAG is a Stop codon. (*b*) Alternative splicing in *dsx* pre-mRNA. The six exonic splice enhancers (ESEs) in exon 4 are indicated by green stripes and S represents the splicing machinery. In females, polyadenylation (pA) of *dsx* mRNA occurs downstream of exon 4, whereas in males, it occurs downstream of exon 6. [After a drawing by Maniatis, T. and Tasic, B., *Nature* **418,** 236 (2002).]

1. Exon 2 of *transformer (tra)* pre-mRNA contains two alternative 3′ splice sites, with the proximal (close; to exon 1) site used in males and the distal (far) site used in females (Fig. 25-26*a*). The region between these two sites contains a Stop codon with the sequence UAG. In males, the splicing factor **U2-snRNP auxiliary factor (U2AF)** binds to the proximal 3′ splice site to yield an mRNA containing this premature Stop codon, which thereby directs the synthesis of a truncated and hence nonfunctional **TRA** protein. However, in females, the proximal 3′ splice site is bound by the female-specific **SXL** protein, the product of the *sex-lethal (sxl)* gene (which is only expressed in females), so as to block the binding of U2AF, which then binds to the distal 3′ splice site, thereby inducing the expression of functional TRA protein.

2. In *doublesex (dsx)* pre-mRNA, the first three exons are constitutively spliced in both males and females. However, the branch site immediately upstream of exon 4 has a suboptimal pyrimidine tract to which U2AF does not bind (Fig. 25-26*b*). Hence in males, exon 4 is not included in *dsx* mRNA, leading to the synthesis of male-specific **DSX-M** protein that functions as a repressor of female-specific genes. However, in females, TRA protein promotes the cooperative binding of the splicing factors **RBP1** and **TRA2** [the product of the *transformer-2 (tra-2)* gene] to six copies of an exonic splice enhancer (ESE) within exon 4. This heterotrimeric complex recruits the splicing machinery to the upstream 3′ splice site of exon 4, leading to its inclu-

sion in *dsx* mRNA. The resulting female-specific **DSX-F** protein is a repressor of male-specific genes.

Thus, the synthesis of functional TRA protein involves the repression of a splice site, whereas the synthesis of female-specific DSX-F protein involves the activation of a splice site. Similar mechanisms of alternative splice site selection have been identified in vertebrates.

mRNA May Be Edited. Additional posttranscriptional modifications of eukaryotic mRNAs include the methylation of certain A residues. In a few cases, mRNAs undergo **RNA editing,** which involves base changes, deletions, or insertions. In the most extreme examples of this phenomenon, which occur in the trypanosomes and related protozoa, several hundred U residues may be added and removed to produce a translatable mRNA. Not surprisingly, the mRNA editing machinery includes RNAs known as **guide RNAs (gRNAs)** that base pair with the immature mRNA to direct its alteration.

In a few mRNAs, specific bases are chemically altered, for example, by enzyme-catalyzed deamination of C to produce U and/or deamination of A to produce inosine (which is read as G by the ribosome), a process called **substitutional editing.** For example, humans express two forms of **apolipoprotein B (apoB): apoB-48,** which is made only in the small intestine and functions in chylomicrons to transport triacylglycerols from the intestine to the liver and peripheral tissues; and **apoB-100,** which is made only in the liver and functions with VLDL, IDL, and LDL to transport cholesterol from the liver to the peripheral tissues (Table 19-1). ApoB-100 is an enormous 4536-residue protein, whereas apoB-48 consists of apoB-100's N-terminal 2152 residues and therefore lacks the C-terminal domain of apoB-100 that mediates LDL receptor binding. Although apoB-48 and apoB-100 are expressed from the same gene, the mRNAs encoding these two proteins differ by a single C $\rightarrow$ U change: The codon for Gln 2153 (CAA) in apoB-100 mRNA is, in apoB-48 mRNA, a UAA Stop codon. The activity that catalyzes this conversion is a protein: It is destroyed by proteases but not by nucleases. Evidently, the editing activity is a site-specific **cytidine deaminase.**

Substitutional editing may contribute to protein diversity. For example, *Drosophila cacophony* pre-mRNA, which encodes a voltage-gated Ca^{2+} channel subunit, contains 10 different substitutional editing sites and hence has the potential of generating $2^{10} \approx 1000$ different isoforms in the absence of alternative splicing. Substitutional editing can also generate alternative splice sites. For example, the enzyme **ADAR2** (ADAR for *a*denosine *d*eaminase *a*cting on *R*NA) edits its own pre-mRNA by converting an intronic AA dinucleotide to AI, which mimics the AG normally found at 3′ splice sites (Fig. 25-22). The consequent new splice site adds 47 nucleotides near the 5′ end of the *ADAR2* mRNA so as to generate a new translational initiation site. The resulting ADAR2 isozyme is catalytically active but is produced in smaller amounts than that from unedited transcripts, perhaps due to a less efficient translational initiation site. Thus, ADAR2 appears to regulate its own rate of expression.

B Ribosomal RNA Processing

E. coli has three types of rRNAs, the **5S, 16S,** and **23S rRNAs.** These are specified by seven operons, each of which contains one nearly identical copy of each of the three rRNA genes. The polycistronic primary transcripts of these operons are >5500-nt long and contain, in addition to the rRNAs,

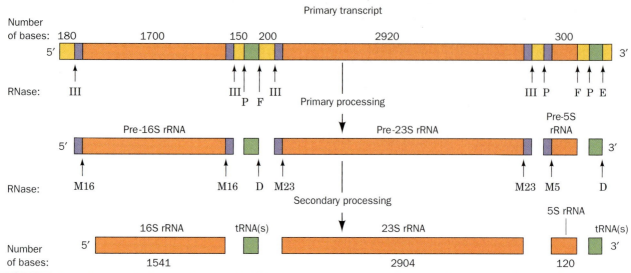

Figure 25-27 **The posttranscriptional processing of *E. coli* rRNA.** The transcriptional map is shown approximately to scale. The labeled arrows indicate the positions of the various nucleolytic cuts and the nucleases that generate them. [After Apiron, D., Ghora, B.K., Plantz, G., Misra, T.K., and Gegenheimer, P., *in* Söll, D., Abelson, J.N., and Schimmel, P.R. (Eds.), *Transfer RNA: Biological Aspects, p. 148*, Cold Spring Harbor Laboratory (1980).]

transcripts for as many as four tRNAs (Fig. 25-27). The steps in processing these primary transcripts to mature rRNAs were elucidated with the aid of mutants defective in one or more of the processing enzymes.

The initial processing, which yields products known as **pre-rRNAs,** commences while the primary transcript is still being synthesized. It consists of specific endonucleolytic cleavages by **RNase III, RNase P, RNase E,** and **RNase F** at the sites indicated in Fig. 25-27. The cleavage sites are probably recognized on the basis of their secondary structures.

The 5′ and 3′ ends of the pre-rRNAs are trimmed away in secondary processing steps through the action of **RNases D, M16, M23,** and **M5** to produce the mature rRNAs. These final cleavages occur only after the pre-rRNAs become associated with ribosomal proteins. During ribosomal assembly, specific rRNA residues are methylated, which may help protect them from inappropriate nuclease digestion.

Eukaryotic rRNA Processing. The eukaryotic genome typically has several hundred tandemly repeated copies of rRNA genes. These genes are transcribed and processed in the nucleolus. The primary eukaryotic rRNA transcript is an ~7500-nt **45S RNA** that contains the **18S, 5.8S,** and **28S rRNAs** separated by spacer sequences (Fig. 25-28). In the first stage of its processing, 45S RNA is specifically methylated at numerous sites (106 in humans) that occur mostly in its rRNA sequences. About 80% of these modifications yield $O^{2'}$**-methylribose** residues, and the remainder yield methylated bases such as N^6,N^6**-dimethyladenine** and **2-methylguanine.** The subsequent cleavage and trimming of the 45S RNA superficially resemble those of prokaryotic rRNAs. In fact, enzymes exhibiting RNase III- and RNase P-like activities occur in eukaryotes. Eukaryotic ribosomes contain four different rRNAs (Section 26-3). The fourth type, 5S eukaryotic rRNA, is separately processed in a manner resembling that of tRNA (Section 25-3C).

The methylation sites in yeast and vertebrate rRNAs generally occur in invariant sequences, although the methylation sites do not appear to have a consensus structure that might be recognized by a single methyltransferase. How are the methylation sites targeted? Pre-rRNAs interact with members of a large family of **small nucleolar RNAs**

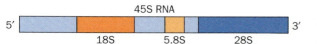

Figure 25-28 **The organization of the 45S primary transcript of eukaryotic rRNA.**

(snoRNAs). Mammals have ~200 snoRNAs, most of which are encoded by the introns of structural genes (and hence not all excised introns are discarded). The snoRNAs, whose lengths vary from 70 to 100 nt, contain segments of 10 to 21 nt that are precisely complementary to segments of the mature rRNAs that contain $O^{2'}$-methylation sites. The snoRNAs appear to direct a protein complex containing a methyltransferase to the methylation site. Presumably, without the snoRNAs, the cell would need to synthesize a different methyltransferase to recognize each methylation sequence. Complexes of small RNAs and proteins also catalyze the conversion of certain rRNA uridine residues to **pseudouridine (ψ):**

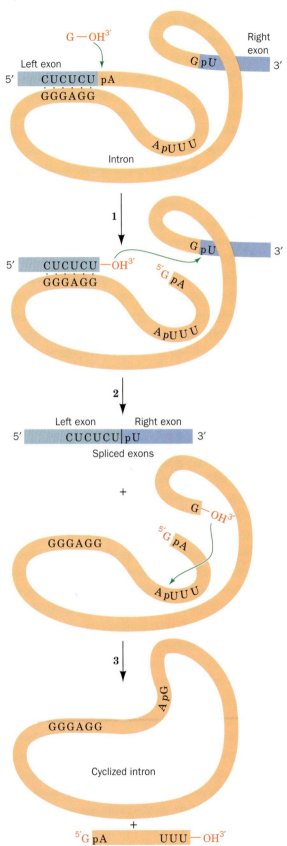

Uridine → Pseudouridine (ψ)

Some Eukaryotic rRNAs Are Self-Splicing. A few eukaryotic rRNA genes contain introns. In fact, it was Thomas Cech's study of how these introns are spliced out that led to the astonishing discovery, in 1982, that *RNA can act as an enzyme* (a ribozyme; Section 23-2E). Cech showed that when the isolated pre-rRNA of the ciliated protozoan *Tetrahymena thermophila* is incubated with guanosine or a free guanine nucleotide (GMP, GDP, or GTP), but in the absence of protein, its single 413-nt intron excises itself and splices together its flanking exons; that is, *the pre-rRNA is self-splicing.* The three-step reaction sequence of this process (Fig. 25-29) resembles that of mRNA splicing:

1. The 3'-OH group of the guanosine attacks the intron's 5' end, displacing the 3'-OH group of the 5' exon and thereby forming a new phosphodiester linkage with the 5' end of the intron.

2. The 3'-terminal OH group of the newly liberated 5' exon attacks the 5'-phosphate of the 3' exon, forming a new phosphodiester bond, thereby splicing together the two exons and displacing the intron.

3. The 3'-terminal OH group of the intron attacks a phosphate of the nucleotide 15 residues from the intron's end, displacing the 5' terminal fragment and yielding the 3' terminal fragment in cyclic form.

The self-splicing process consists of a series of transesterifications and therefore does not require free energy input.

Self-splicing RNAs that react as shown in Fig. 25-29 are known as **group I introns** and occur in the nuclei, mitochondria, and chloro-

Figure 25-29 The sequence of reactions in the self-splicing of *Tetrahymena* pre-rRNA. (1) The 3'-OH group of a guanine nucleotide attacks the intron's 5'-terminal phosphate to form a phosphodiester bond, displacing the 5' exon. (2) The newly generated 3'-OH group of the 5' exon attacks the 5'-terminal phosphate of the 3' exon, thereby splicing the two exons and displacing the intron. (3) The 3'-OH group of the intron attacks the phosphate of the nucleotide that is 15 residues from the 5' end so as to cyclize the intron and displace its 5'-terminal fragment. Throughout this process, the RNA maintains a folded, internally hydrogen-bonded conformation that permits the precise excision of the intron.

plasts of diverse eukaryotes (but not vertebrates), and in some bacteria. **Group II introns,** which occur in the mitochondria and chloroplasts of fungi and plants, react via a lariat intermediate and do not require an external nucleotide. The chemical similarities of the pre-mRNA and group II intron splicing reactions therefore suggest that *spliceosomes are ribozymal systems whose RNA components have evolved from primordial self-splicing RNAs and that their protein components serve mainly to fine-tune ribozymal structure and function.* This, of course, is consistent with the hypothesis that RNAs were the original biological catalysts in precellular times (the so-called RNA world; Box 23-3).

The X-Ray Structure of a Group I Ribozyme. In order to function as a catalyst, the self-splicing RNA must be able to assume a stable tertiary structure, as does a protein enzyme. The sequence of the *Tetrahymena* group I intron contains nine double-stranded segments, designated P1 through P9 (Fig. 25-30a). Jennifer Doudna and Cech determined the

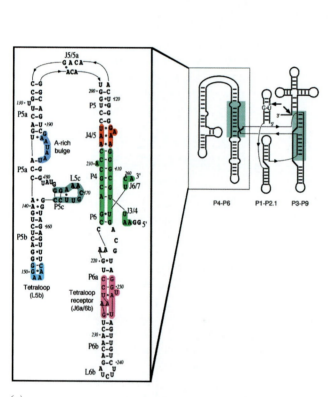

(a)

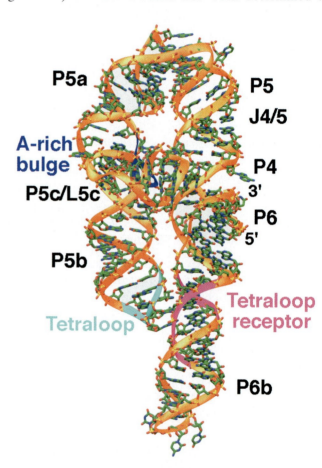

(b)

Figure 25-30 The self-splicing group I intron from *Tetrahymena thermophila.* (a) The secondary structure of the entire 413-nt intron is shown on the right with its phylogenetically conserved catalytic core shaded in gray. Helical regions are numbered sequentially along the intron's sequence with P for base-paired segment, J for joining region, and L for loop; the short arrows indicate the 5' and 3' splice sites. The sequence of the independently folded 160-nt P4–P6 domain is enlarged on the left. Segments of functional interest are highlighted as follows: the GAAA tetraloop is cyan; the conserved tetraloop receptor is magenta; the A-rich bulge, which is required for the proper folding of P4–P6, is blue; segments of the conserved core are green and red; and P5c

is gray. Watson–Crick and non-Watson–Crick base-pairing interactions are represented by short horizontal lines and small circles. (b) The X-ray structure of P4–P6 viewed as in Part a. The structure is drawn in stick form with C green, N blue, O red, and P yellow. The sugar–phosphate backbone is traced by a ribbon that is colored as in Part a for the tetraloop, tetraloop receptor, and A-rich bulge and is gold elsewhere. Note the numerous interactions between the various segments of this RNA molecule. [Part a based on a drawing by and Part b based on an X-ray structure by Jennifer Doudna, Yale University. PDBid 1GID.]
🖱 **See the Interactive Exercises.**

X-ray structure of the 160-nt P4–P6 domain (Fig. 25-30*b*), which contains the active site. The structure consists mostly of two coaxially stacked sets of A-RNA helices, one of 29 bp and the other of 23 bp. Extensive hydrogen bonding and Mg^{2+} ions coordinated by the oxygens of two or more phosphate groups contribute to a structure whose interior is densely packed and solvent inaccessible, much like the interior of a protein enzyme. Of particular note are its so-called A-rich bulge, a 7-nt sequence about halfway along the short arm of the U-shaped macromolecule, and the 6-nt sequence at the tip of the short arm of the U, whose central GAAA assumes a characteristic conformation known as a **tetraloop.** In both of these substructures, the bases are splayed outward so as to stack on each other and to associate in the minor groove of specific segments of the long arm of the U via hydrogen-bonding interactions involving ribose residues as well as bases. In the interaction involving the A-rich bulge, the close packing of phosphates from adjacent helices is mediated by hydrated Mg^{2+} ions. Throughout this structure, the defining characteristic of RNA, its 2'-OH group, is both a donor and acceptor of hydrogen bonds to phosphate, bases, and other 2'-OH groups (which may explain why "deoxyribozymes" are unknown in biology, although synthetic ssDNAs with catalytic properties are known).

C   Transfer RNA Processing

A tRNA molecule, as discussed in Section 23-2E, consists of ~80 nucleotides, many of which are chemically modified, that assume a cloverleaf-shaped secondary structure with four base-paired stems (Fig. 25-31). The *E. coli* chromosome contains ~60 tRNA genes. Some of them are components of rRNA operons; the others are distributed, often in clusters, throughout the chromosome. The primary tRNA transcripts, which contain as many as five identical tRNA species, have extra nucleotides at the 3' and 5' ends of each tRNA sequence. The excision and trimming of these tRNA sequences resemble *E. coli* rRNA processing (Fig. 25-27) in that the two processes employ some of the same nucleases.

E. coli RNase P, which processes rRNA and generates the 5' ends of tRNAs, is a particularly interesting enzyme because it is a ribozyme with a catalytically essential 377-nt (125 kD) RNA and a 119-residue (14 kD) polypeptide. At first, the RNA was believed to recognize the substrate through base pairing and thereby guide the protein subunit, which was presumed to be the nuclease, to the cleavage site. However, Sidney Altman has shown that *the RNA component of RNase P is, in fact, the enzyme's catalytic subunit by demonstrating that free RNase P RNA catalyzes the cleavage of substrate RNA at high salt concentrations.* The RNase P protein, which is basic, evidently functions to electrostatically reduce the repulsions between the polyanionic ribozyme and its substrate RNAs. RNase P activity also occurs in eukaryotes, although the enzyme includes 9 or 10 protein subunits. Indeed, RNase P mediates one of the two ribozymal activities that occur in all cellular life, the other being associated with ribosomes (Section 26-4B).

The RNase P from *T. thermophilus* contains a 393-residue RNA whose predicted secondary structure includes a catalytic and a specificity domain

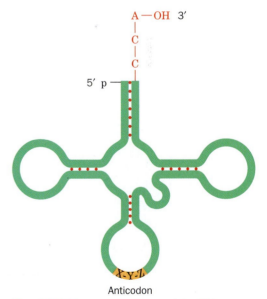

Figure 25-31 Schematic diagram of the tRNA cloverleaf secondary structure.   Each red dot indicates a base pair in the hydrogen-bonded stems. The positions of the **anticodon** (the 3-nt sequence that binds to mRNA during translation) and the 3'-terminal —CCA are indicated.

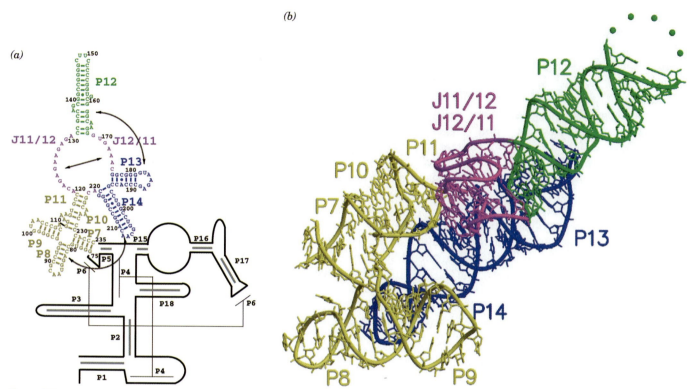

Figure 25-32 **The structure of the RNA of *T. thermophilus* RNase P.** (*a*) Its predicted secondary structure in which the specificity domain is drawn in various colors and the catalytic domain is black. (*b*) The X-ray structure of the specificity domain in which its various segments are colored as in Part *a*. [Courtesy of Alfonso Mondragón, Northwestern University. PDBid 1U9S.]

(Fig. 25-32*a*). The X-ray structure of the specificity domain (Fig. 25-32*b*), determined by Alfonso Mondragón, confirms the predicted secondary structure. The domain consists mainly of stacked helical stems (P7–P10–P11, P8–P9, P12, and P13–14 in Fig. 25-32) together with an unusual nonhelical module (J11/12 and J12/11) that sits over a four-way junction formed by stems P7 through P11.

Many Eukaryotic Pre-tRNAs Have Introns. Eukaryotic genomes contain from several hundred to several thousand tRNA genes. Many eukaryotic primary tRNA transcripts, for example, that of yeast **tRNA^Tyr** (Fig. 25-33), contain a small intron as well as extra nucleotides at their 5′ and 3′ ends. tRNA processing therefore includes nucleolytic removal of these extra nucleotides. The three nucleotides, CCA, at the 3′ termini of all tRNAs, the sites at which amino acids are attached (Section 26-2B), are lacking in the immature tRNA transcripts. This trinucleotide is appended by the enzyme **tRNA nucleotidyltransferase,** which sequentially adds two C's and an A to tRNA using CTP and ATP as substrates (prokaryotic tRNA primary transcripts include the CCA sequence).

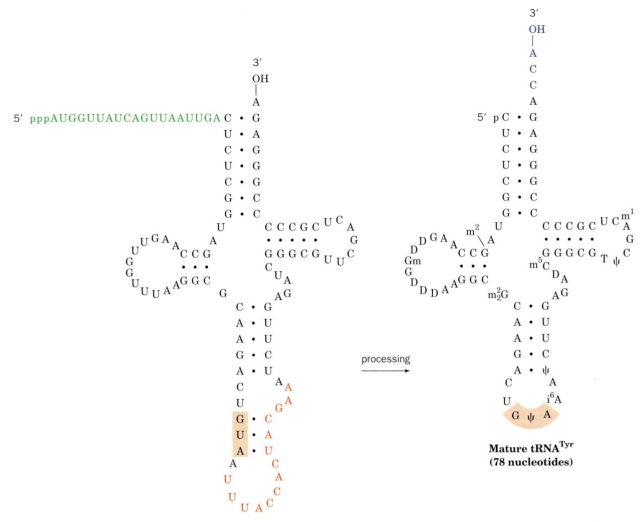

Figure 25-33 **The posttranscriptional processing of yeast tRNA^Tyr.** A 14-nt intervening sequence (*red*) and a 19-nt 5'-terminal sequence (*green*) are excised from the primary transcript, a —CCA (*blue*) is appended to its 3' end, and several of the bases are modified to form the mature tRNA

(m²G, *N²*-methylguanosine; D, dihydrouridine; Gm, 2'-methylguanosine; m²₂G, *N²,N²*-dimethylguanosine; ψ pseudouridine; i⁶A, *N⁶*-isopentenyladenosine; m⁵C, 5-methylcytosine; m¹A, 1-methyladenosine; see Fig. 26-4). The anticodon is shaded. [After DeRobertis, E.M. and Olsen, M.V., *Nature* **278**, 142 (1989).]

SUMMARY

1. RNA polymerase synthesizes a polynucleotide chain from ribonucleoside triphosphates using a single strand (the antisense or noncoding strand) of DNA as a template.

2. The σ factor of *E. coli* RNA polymerase holoenzyme recognizes and binds to a promoter sequence to position the enzyme to initiate transcription.

3. RNA synthesis requires the formation of an open complex. The transcription bubble travels along the DNA as the RNA chain is elongated by the processive activity of RNA polymerase.

4. In *E. coli*, RNA synthesis is terminated in response to specific secondary structural elements in the transcript and may require the action of Rho factor.

5. Eukaryotes contain three nuclear RNA polymerases that synthesize different types of RNAs.

6. Eukaryotic promoters are diverse: They vary in position relative to the transcription start site and may consist of multiple upstream sequences. Enhancers form the binding sites for regulatory proteins that function as activators of transcription.

7. Basal transcription of eukaryotic genes requires six general transcription factors, including TBP, that form a preinitiation complex with RNA polymerase at the promoter. Some of these proteins remain with the polymerase when it is phosphorylated and shifts to elongation.

8. The primary transcripts of most eukaryotic structural genes are posttranscriptionally modified by the addition of a 5′ cap and a 3′ poly(A) tail. mRNAs that contain introns undergo splicing, in which the introns are excised and the flanking exons are joined together via two transesterification reactions mediated by an snRNA-containing spliceosome.

9. The processing of pre-rRNAs includes nucleolytic cleavage and snoRNA-assisted methylation. Some eukaryotic rRNA transcripts undergo splicing catalyzed by the intron itself.

10. tRNA transcripts may be processed by the addition, removal, and modification of nucleotides.

 ## REFERENCES

RNA Polymerase

Cramer, P., Multisubunit RNA polymerases, *Curr. Opin. Struct. Biol.* **12**, 89–97 (2002).

Kadonaga, K.T., The RNA polymerase II core promoter, *Annu. Rev. Biochem.* **72**, 449–479 (2003).

Marakami, K.S. and Darst, S.A., Bacterial RNA polymerases: the whole story. *Curr. Opin. Struct. Biol.* **13**, 31–39 (2003).

Shilatifard, A., Conway, R.C., and Conway, J.W., The RNA polymerase II elongation complex, *Annu. Rev. Biochem.* **72**, 693–715 (2003).

Skordalakes, E. and Berger, J.M., The structure of Rho transcription terminator: Mechanism of mRNA recognition and helicase loading. *Cell* **114**, 135–146 (2003).

RNA Processing

Black, D.L., Mechanism of alternative pre-messenger RNA splicing, *Annu. Rev. Biochem.* **72**, 291–336 (2003).

Decatur, W.A. and Fournier, M.J., RNA-guided nucleotide modification of ribosomal and other RNAs, *J. Biol. Chem.* **278**, 695–698 (2003).

Doherty, E.A. and Doudna, J.A., Ribozyme structures and mechanisms, *Annu. Rev. Biophys. Biomol. Struct.* **30**, 457–475 (2001).

Frank, D.N. and Pace, N.R., Ribonuclease P: Unity and diversity in a tRNA processing ribozyme, *Annu. Rev. Biochem.* **67**, 153–180 (1998).

Maniatis, T. and Reed, R., An extensive network of coupling among gene expression machines, *Nature* **416**, 499–506 (2002). [Describes why efficient gene expression requires coupling between transcription and RNA processing.]

Maniatis, T. and Tasic, B., Alternative pre-mRNA splicing and proteome expansion in metazoans, *Nature* **418**, 236–243 (2002).

Proudfoot, N.J., Furger, A., and Dye, M.J., Integrating mRNA processing with transcription, *Cell* **108**, 501–512 (2002).

 ## KEY TERMS

rRNA	cistron	CCAAT box	exon
tRNA	polycistronic RNA	enhancer	splicing
mRNA	monocistronic RNA	general transcription factor	snRNA
RNAP	promoter	preinitiation complex	snRNP
holoenzyme	Pribnow box	TBP	spliceosome
σ factor	footprinting	TAFs	alternative splicing
core enzyme	open complex	primary transcript	RNA editing
antisense strand	constitutive enzyme	posttranscriptional	gRNA
noncoding strand	inducible enzyme	modification	snoRNA
sense strand	Rho factor	cap structure	group I intron
coding strand	nucleolus	poly(A) tail	group II intron
structural gene	GC box	hnRNA	
operon	TATA box	intron	

 ## STUDY EXERCISES

1. Why is it difficult to precisely define the term "gene"?

2. Compare DNA and RNA polymerases with respect to overall structure, substrates, mechanism of action, error rate, and template specificity.

3. What are the advantages and disadvantages of arranging genes in operons?

4. What is a consensus sequence?

5. What is the significance of the different DNA-binding properties of prokaryotic RNA polymerase core enzyme and holoenzyme?

6. What are the functions of the three eukaryotic RNA polymerases?

7. Describe the assembly of the eukaryotic preinitiation complex.

8. Summarize the posttranscriptional modification of eukaryotic mRNA, rRNA, and tRNA.

9. What is the advantage of starting RNA processing before transcription is complete?

10. Compare the mechanism of splicing in mRNA processing and in group I introns.

 PROBLEMS

1. The antibiotic **cordycepin** inhibits bacterial RNA synthesis.

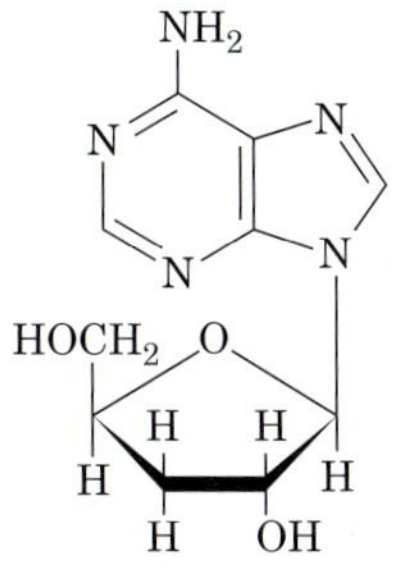

Cordycepin

(a) Of which nucleoside is cordycepin a derivative?

(b) Explain cordycepin's mechanism of action.

2. Indicate the -10 region, the -35 region, and the initiating nucleotide on the sense strand of the *E. coli* tRNA^{Tyr} promoter shown below.

5' CAACGTAACACTTTACAGCGGCGCGTCATTTGATATGATGCGCCCCGCTTCCCGATA 3'
3' GTTGCATTGTGAAATGTCGCCGCGCAGTAAACTATACTACGCGGGGCGAAGGGCTAT 5'

3. Design a six-residue nucleic acid probe that would hybridize with the greatest number of *E. coli* gene promoters.

4. Explain why inserting 5 bp of DNA at the -50 position of a eukaryotic gene decreases the rate of RNA polymerase II transcription initiation to a greater extent than inserting 10 bp at the same site.

5. Why does promoter efficiency tend to decrease with the number of $G \cdot C$ base pairs in the -10 region of a prokaryotic gene?

6. A eukaryotic ribosome contains 4 different rRNA molecules and ~82 different proteins. Why does a cell contain many more copies of the rRNA genes than the ribosomal protein genes?

7. Design an oligonucleotide-based affinity chromatography system for purifying mature mRNAs from eukaryotic cell lysates.

8. A eukaryotic cell carrying out transcription and RNA processing is incubated with ^{32}P-labeled ATP. Where will the radioactive isotope appear in mature mRNA if the ATP is labeled at the (a) α position, (b) β position, and (c) γ position?

9. Explain why the active site of poly(A) polymerase is much narrower than that of DNA and RNA polymerases.

10. Explain why the $O^{2'}$-methylation of ribose residues protects rRNA from RNases.

11. Would you expect spliceosome-catalyzed intron removal to be reversible in a highly purified *in vitro* system and *in vivo?* Explain.

12. Introns in eukaryotic protein-coding genes may be quite large, but almost none are smaller than about 65 bp. What is the reason for this minimum intron size?

13. Infection with certain viruses inhibits snRNA processing in eukaryotic cells. Explain why this favors the expression of viral genes in the host cell.

Proteins are assembled from amino acids that are joined together in the order specified by a messenger RNA molecule. A ribosome, consisting of RNA as well as protein, can translate a given mRNA molecule many times to produce a large number of protein molecules. Similarly, a printing press can generate numerous copies of a page. [Erich Lessing/Art Resource.]

Translation

How is the genetic information encoded in DNA decoded? In the preceding chapter, we saw how the base sequence of DNA is transcribed into that of RNA. In this chapter, we shall consider the remainder of the decoding process by examining how the base sequences of RNAs are translated into the amino acid sequences of proteins. This second part of the central dogma of molecular biology (DNA → RNA → protein) shares a number of features with both DNA replication and transcription, the other major events of nucleic acid metabolism. First, all three processes are carried out by large, complicated protein-containing macromolecular machines whose proper functioning depends on a variety of specific and nonspecific protein–nucleic acid interactions. Accessory factors may also be required for the initiation, elongation, and termination phases of these processes. Furthermore, translation, like replication and transcription, must be executed with accuracy. Although translation involves base pairing between complementary nucleotides, it is amino acids, rather than nucleotides, that are ultimately joined to generate a polymeric product. This process, like replication and transcription, is endergonic and requires the cleavage of "high-energy" phosphoanhydride bonds.

An understanding of translation requires not only a knowledge of the macromolecules that participate in polypeptide synthesis, but an appreciation for the mechanisms that produce a chain of linked amino acids in the exact order specified by mRNA. Accordingly, we begin this chapter by examining the **genetic code,** the correspondence between nucleic acid sequences and polypeptide sequences. We then consider in turn the structures and properties of tRNAs and ribosomes. Next, we examine how the translation machinery operates as a coordinated whole. Finally, we take a brief look at some posttranslational steps of protein synthesis.

1 The Genetic Code

One of the most fascinating puzzles in molecular biology is how a sequence of nucleotides composed of only four types of residues can specify the sequence of up to 20 types of amino acids in a polypeptide chain. Clearly, a one-to-one correspondence between nucleotides and amino acids is not possible. *A group of several bases, termed a **codon,** is necessary to specify a single amino acid.* A triplet code, that is, one with 3 bases per codon, is more than sufficient to specify all the amino acids since there are $4^3 = 64$ different triplets of bases. A doublet code with 2 bases per codon ($4^2 = 16$ possible doublets) would be inadequate. The triplet code allows many amino acids to be specified by more than one codon. Such a code, in a term borrowed from mathematics, is said to be **degenerate.**

How does the polypeptide-synthesizing apparatus group DNA's continuous sequences of bases into codons? For example, the code might be overlapping; that is, in the sequence

ABCDEFGHIJ · · ·

ABC might code for one amino acid, BCD for a second, CDE for a third, etc. Alternatively, the code might be nonoverlapping, so that ABC specifies one amino acid, DEF a second, GHI a third, etc. In fact, *the genetic code is a nonoverlapping, degenerate, triplet code.* A number of elegant experiments, some of which are outlined below, revealed the workings of the genetic code.

A   Codons Are Triplets That Are Read Sequentially

In genetic experiments on bacteriophage T4, Francis Crick and Sydney Brenner found that a mutation that resulted in the deletion of a nucleotide could abolish the function of a specific gene. However, a second mutation, in which a nucleotide was inserted at a different but nearby position, could restore gene function. These two mutations are said to be **suppressors** of one another; that is, they cancel each other's mutant properties. On the basis of these experiments, Crick and Brenner concluded that *the genetic code is read in a sequential manner starting from a fixed point in the gene.* The insertion or deletion of a nucleotide shifts the **reading frame** (grouping) in which succeeding nucleotides are read as codons. Insertions or deletions of nucleotides are therefore known as **frameshift mutations.**

In further experiments, Crick and Brenner found that whereas two closely spaced deletions or two closely spaced insertions could not suppress each other (restore gene function), three closely spaced deletions or insertions could do so. These observations clearly established that *the genetic code is a triplet code.*

The foregoing principles are illustrated by the following analogy. Consider a sentence (gene) in which the words (codons) each consist of three letters (bases)

THE   BIG   RED   FOX   ATE   THE   EGG

Here the spaces separating the words have no physical significance; they are present only to indicate the reading frame. The deletion of the fourth letter, which shifts the reading frame, changes the sentence to

THE   IGR   EDF   OXA   TET   HEE   GG

so that all words past the point of deletion are unintelligible (specify the wrong amino acids). An insertion of any letter, however, say an X in the ninth position,

THE   IGR   EDX   FOX   ATE   THE   EGG

restores the original reading frame. Consequently, only the words between the two changes (mutations) are altered. As in this example, such a sentence might still be intelligible (the gene could still specify a functional protein), particularly if the changes are close together. Two deletions or two insertions, no matter how close together, would not suppress each other but just shift the reading frame. However, three insertions, say X, Y, and Z in the fifth, eighth, and twelfth positions, respectively, would change the sentence to

THE   BXI   GYR   EDZ   FOX   ATE   THE   EGG

which, after the third insertion, restores the original reading frame. The same would be true of three deletions. As before, if all three changes were close together, the sentence might still retain its meaning. Like this textual analogy, *the genetic code has no internal punctuation to indicate the reading frame; instead, the nucleotide sequence is read sequentially, triplet by triplet.*

Since any nucleotide sequence may have three reading frames, it is possible, at least in principle, for a polynucleotide to encode two or even three different polypeptides. In fact, some single-stranded DNA bacteriophages (which presumably must make maximal use of their small complement of DNA) contain completely overlapping genes that have different reading frames. A similar form of coding economy is exhibited by bacteria, in which the ribosomal initiation sequence of one gene in a polycistronic mRNA often overlaps the end of the preceding gene.

B Deciphering the Genetic Code

In order to understand how the genetic code dictionary was elucidated, we must first review how proteins are synthesized. An mRNA does not directly recognize amino acids. Rather, *it specifically binds molecules of tRNA that each carry a corresponding amino acid* (Fig. 26-1). Each tRNA contains a trinucleotide sequence, its **anticodon,** that is complementary to an mRNA codon specifying the tRNA's amino acid. During translation, amino acids carried by tRNAs are joined together according to the order in which the tRNA anticodons bind to the mRNA codons at the ribosome (Fig. 3-14).

The genetic code could, in principle, be determined by simply comparing the base sequence of an mRNA with the amino acid sequence of the polypeptide it specifies. In the 1960s, however, techniques for isolating and sequencing mRNAs had not yet been developed. Moreover, techniques for synthesizing RNAs were quite rudimentary. They utilized **polynucleotide phosphorylase,** an enzyme from *Azotobacter vinelandii* that links together nucleotides without the use of a template.

$$(RNA)_n + NDP \rightleftharpoons (RNA)_{n+1} + P_i$$

Thus, the NDPs are linked together at random so that the base composition of the product RNA reflects that of the reactant NDP mixture. The elucidation of the genetic code therefore proved to be a difficult task, even with the development of cell-free translation systems.

E. coli cells that have been gently broken open and centrifuged to remove cell walls and membranes yield an extract containing DNA, mRNA, ribosomes, enzymes, and other cell constituents necessary for protein synthesis. When fortified with ATP, GTP, and amino acids, this system synthesizes small amounts of protein. A cell-free translation system, of course, produces proteins specified by the cell's DNA. Adding DNase halts protein synthesis after a few minutes because the system can no longer synthesize mRNA, and the mRNA originally present is rapidly degraded. At this point, purified mRNA or synthetic mRNA can be added to the system and the resulting polypeptide products can be subsequently recovered.

In 1961, Marshall Nirenberg and Heinrich Matthaei added the synthetic polyribonucleotide poly(U) to a cell-free translation system containing isotopically labeled amino acids and recovered a labeled poly(Phe) polypeptide. They concluded that *UUU must be the codon that specifies Phe.* Similar experiments with poly(A) and poly(C) yielded poly(Lys) and poly(Pro), respectively, thereby identifying AAA as a codon for Lys and CCC as a codon for Pro.

The Genetic Code Was Elucidated through Triplet Binding Assays and the Use of Polyribonucleotides with Known Sequences.   In another series of experiments, different trinucleotides were tested for their ability to promote tRNA binding to ribosomes. The ribosomes, with their bound tRNAs, are retained by a nitrocellulose filter, but free tRNAs are not. The bound tRNA can be identified by the radioactive amino acid attached to it. This simple binding assay revealed that, for instance, UUU stimulates the ribosomal binding of only Phe tRNA. Likewise, UUG, UGU, and GUU stimulate the binding of Leu, Cys, and Val tRNAs, respectively. Hence UUG, UGU, and GUU must be codons that specify Leu, Cys, and Val, respectively. In this way, the amino acids specified by some 50 codons were identified. For the remaining codons, the binding assay was either negative (no tRNA bound) or ambiguous.

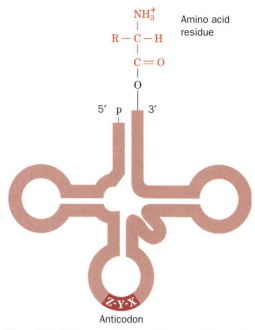

Figure 26-1 Transfer RNA in its "cloverleaf" form.   Its covalently linked amino acid residue is at the top, and its anticodon (a trinucleotide segment that base pairs with the complementary mRNA codon during translation) is at the bottom.

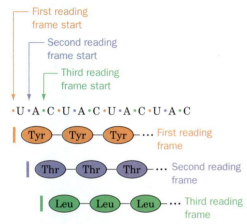

First reading
frame start

Second reading
frame start

Third reading
frame start

·U·A·C·U·A·C·U·A·C·U·A·C

| Tyr — Tyr — Tyr — ... First reading frame

| Thr — Thr — Thr — ... Second reading frame

| Leu — Leu — Leu — ... Third reading frame

Figure 26-2 **The three potential reading frames of an mRNA.** Each reading frame would yield a different polypeptide.

The genetic code dictionary was completed and previous results confirmed through H. Gobind Khorana's chemical synthesis of polynucleotides with specified repeating sequences (at the time, an extremely laborious process). In a cell-free translation system, UCUCUCUC · · · , for example, is read

$$\text{UCU} \quad \text{CUC} \quad \text{UCU} \quad \text{CUC} \quad \text{UCU} \quad \text{C}\cdots$$

so that it specifies a polypeptide chain of two alternating amino acid residues. This particular mRNA stimulated the production of

$$\text{Ser}-\text{Leu}-\text{Ser}-\text{Leu}-\text{Ser}-\text{Leu}-\cdots$$

since UCU codes for Ser and CUC codes for Leu.

Alternating sequences of three nucleotides, such as poly(UAC), specify three different homopolypeptides because ribosomes may initiate polypeptide synthesis on these synthetic mRNAs in any of the three possible reading frames (Fig. 26-2). Analyses of the polypeptides specified by various alternating sequences of two and three nucleotides confirmed the identify of many codons and filled out missing portions of the genetic code.

C The Nature of the Genetic Code

The genetic code, presented in Table 26-1, has several remarkable features:

1. *The code is highly degenerate.* Three amino acids—Arg, Leu, and Ser—are each specified by six different codons, and most of the rest are specified by either four, three, or two codons. Only Met and Trp, two of the least common amino acids in proteins (Table 4-1), are represented by a single codon. Codons that specify the same amino acid are termed **synonyms.**

2. *The arrangement of the code table is nonrandom.* Most synonyms occupy the same box in Table 26-1; that is, they differ only in their third nucleotide. XYU and XYC always specify the same amino acid; XYA and XYG do so in all but two cases. Moreover, changes in the first codon position tend to specify similar (if not the same) amino acids, whereas codons with second position pyrimidines encode mostly hydrophobic amino acids (tan in Table 26-1), and those with second position purines encode mostly polar amino acids (blue, red, and purple in Table 26-1). These observations suggest a nonrandom origin of the genetic code and indicate that the code evolved so as to minimize the deleterious effects of mutations (see Box 26-1).

3. *UAG, UAA, and UGA are* **Stop codons.** These three codons (also known as **nonsense codons**) do not specify amino acids but signal the ribosome to terminate polypeptide chain elongation. UAG, UAA, and UGA are often referred to as **amber, ochre,** and **opal codons** (these names are the result of a laboratory joke: The German word for *amber* is Bernstein, the name of an individual who helped discover amber mutations, which change some other codon to UAG; *ochre* and *opal* are puns on *amber*).

4. *AUG and GUG are chain initiation codons.* The codons AUG and, less frequently, GUG specify the starting point for polypeptide chain synthesis. However, they also specify the amino acids Met and Val, respectively, at internal positions in polypeptide chains. We shall see in Section 26-4A how ribosomes differentiate the two types of codons.

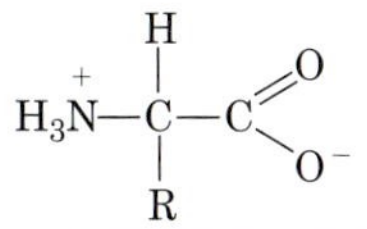

Table 26-1. Key to Function. The "Standard" Genetic Code[a]

First position (5′ end)	Second position				Third position (3′ end)
	U	**C**	**A**	**G**	
U	UUU Phe	UCU	UAU Tyr	UGU Cys	U
	UUC	UCC	UAC	UGC	C
	UUA Leu	UCA Ser	UAA STOP	UGA STOP	A
	UUG	UCG	UAG	UGG Trp	G
C	CUU	CCU	CAU His	CGU	U
	CUC	CCC	CAC	CGC	C
	CUA Leu	CCA Pro	CAA Gln	CGA Arg	A
	CUG	CCG	CAG	CGG	G
A	AUU	ACU	AAU Asn	AGU Ser	U
	AUC Ile	ACC	AAC	AGC	C
	AUA	ACA Thr	AAA Lys	AGA Arg	A
	AUG Met[b]	ACG	AAG	AGG	G
G	GUU	GCU	GAU Asp	GGU	U
	GUC	GCC	GAC	GGC	C
	GUA Val	GCA Ala	GAA Glu	GGA Gly	A
	GUG	GCG	GAG	GGG	G

[a]Nonpolar amino acid residues are tan, basic residues are blue, acidic residues are red, and polar uncharged residues are purple.
[b]AUG forms part of the initiation signal as well as coding for internal Met residues.

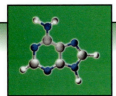

BOX 26-1

Perspectives in Biochemistry

Evolution of the Genetic Code

Because of the degeneracy of the genetic code, a point mutation in the third position of a codon seldom alters the specified amino acid. For example, a GUU → GUA transversion still codes for Val and is therefore said to be phenotypically silent. Other point mutations, even at the first or second codon positions, producing AUU (Ile) or GCU (Ala), for instance, result in the substitution of a chemically similar amino acid and may have minimal impact on the encoded protein's overall structure or function. This built-in protection against mutation may be more than accidental.

In the 1960s, the perceived universality of the genetic code led Francis Crick to propose the "frozen accident" theory, which holds that codons were allocated to different amino acids entirely by chance. Once assigned, the meaning of a codon could not change because of the high probability of disrupting the structure of the encoded protein. Thus, once established, the genetic code was thought to have ceased evolving.

However, the distribution of codons as presented in Table 26-1 suggests an alternative evolutionary history of the genetic code, in which a few simple codons corresponding to a handful of amino acids gradually became more complex. One scenario begins with an RNA-based world containing only A and U nucleotides. Uracil is almost certainly a primordial base since the pyrimidine biosynthetic pathway yields uracil nucleotides before cytosine or thymine nucleotides (Section 22-2). Adenine would have been required as uracil's complement.

Assuming that a triplet-based genetic code was established at the outset (and it is difficult to envision how any other arrangement could have given rise to the present-day

triplet code), the two bases could have coded for $2^3 = 8$ amino acids. In fact, the contemporary genetic code assigns these all-U/A codons to six amino acids and a stop signal:

UUU = Phe	AAA = Lys
UUA = Leu	AAU = Asn
UAU = Tyr	AUA = Ile
AUU = Ile	UAA = Stop

The AUA codon may well have originally specified the initiating Met (now encoded by AUG), bringing the total to seven amino acids.

When G and C appeared in evolving life-forms, these nucleotides were incorporated into RNA. Codons containing three or four types of bases could specify additional amino acids, but because of selective pressure against introducing disruptive mutations into proteins, the level of codon redundancy increased. An inspection of Table 26-1 shows that triplet codons made entirely of G and C specify only four different amino acids, which is half the theoretical maximum of eight:

GGG, GGC	= Gly
GCG, GCC	= Ala
CGG, CGC	= Arg
CCC, CCG	= Pro

This nonrandom allocation of codons to amino acids argues against a completely random origin for the genetic code. The gradual introduction of two new bases (G and C) to a primitive genetic code based on U and A must have allowed greater information capacity (i.e., coding for 20 amino acids) while minimizing the rate of deleterious substitutions.

The "Standard" Genetic Code Is Not Universal. For many years, it was thought that the "standard" genetic code (that given in Table 26-1) was universal. This assumption was based in part on the observation that one kind of organism (e.g., *E. coli*) can accurately translate the genes for quite different organisms (e.g., humans). Indeed, this phenomenon is the basis of genetic engineering. DNA studies in 1981 nevertheless revealed that *the genetic codes of certain mitochondria are variants of the "standard" genetic code.* For example, in mammalian mitochondria, AUA, as well as the standard AUG, is a Met/initiation codon; UGA specifies Trp rather than "Stop"; and AGA and AGG are "Stop" rather than Arg. Apparently, mitochondria, which contain their own genes and protein synthesizing systems, are not subject to the same evolutionary constraints as are nuclear genomes. An alternate genetic code also appears to have evolved in ciliated protozoa,

which branched off very early in eukaryotic evolution. Thus, the "standard" genetic code, although very widely utilized, is not universal.

2 Transfer RNA and Its Aminoacylation

See Guided Exploration 26

The structure of tRNA.

Cells must translate the language of RNA base sequences into the language of polypeptides. In 1955, Francis Crick hypothesized that translation occurs through the mediation of "adaptor" molecules, which we now know are tRNAs, that carry a specific amino acid and recognize the corresponding mRNA codon.

A tRNA Structure

In 1965, after a 7-year effort, Robert Holley reported the first known base sequence of a biologically significant nucleic acid, that of the 76-residue yeast **alanine tRNA (tRNA^{Ala}).** Currently, the base sequences of over 4000 tRNAs from more than 200 organisms and organelles are known (most from their DNA sequences). They vary in length from 54 to 100 nucleotides (18–28 kD) although most have ~76 nucleotides.

Almost all known tRNAs can be schematically arranged in the so-called cloverleaf secondary structure (Fig. 26-3). Starting from the 5′ end, they have the following common features:

1. A 5′-terminal phosphate group.

2. A 7-bp stem that includes the 5′-terminal nucleotide and that may contain non-Watson–Crick base pairs such as G · U. This assembly is known as the **acceptor** or **amino acid stem** because the amino acid residue carried by the tRNA is covalently attached to its 3′-terminal OH group.

3. A 3- or 4-bp stem ending in a 5- to 7-nt loop that frequently contains the modified base **dihydrouridine (D).** This stem and loop are therefore collectively termed the **D arm.**

4. A 5-bp stem ending in a loop that contains the anticodon. These features are known as the **anticodon arm.**

5. A 5-bp stem ending in a loop that usually contains the sequence TψC (where ψ is the symbol for **pseudouridine;** Section 25-3B). This assembly is called the **TψC** or **T arm.**

6. A 3′ CCA sequence with a free 3′-OH group. The —CCA may be genetically specified or enzymatically added to immature tRNA, depending on the species (Section 25-3C).

tRNAs have 15 invariant positions (always have the same base) and 8 **semi-invariant** positions (only a purine or only a pyrimidine) that occur mostly in the loop regions. The purine on the 3′ side of the anticodon is invariably modified. The site of greatest variability among the known tRNAs occurs in the so-called **variable arm.** It has from 3 to 21 nucleotides and may have a stem consisting of up to 7 bp.

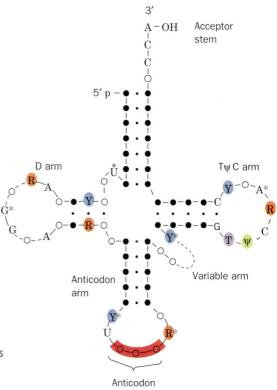

Figure 26-3 The cloverleaf secondary structure of tRNA.   Filled circles connected by dots represent Watson–Crick base pairs, and open circles indicate bases involved in non-Watson–Crick base pairing. Invariant positions are indicated: R and Y represent invariant purines and pyrimidines, and ψ represents pseudouracil. The starred nucleotides are often modified. The D and variable arms contain different numbers of nucleotides in the various tRNAs.

tRNAs Have Numerous Modified Bases.   One of the most striking characteristics of tRNAs is their large proportion, up to 25%, of posttranscriptionally modified bases. Nearly 80 such bases, found at >60 different tRNA positions, have been characterized. A few of them, together with their standard abbreviations, are indicated in Fig. 26-4. None of these modifications are essential for maintaining a tRNA's structural integrity or for its proper binding to the ribosome. However, base modifications may help promote attachment of the proper amino acid to the acceptor stem or strengthen codon–anticodon interactions.

tRNA Has a Complex Tertiary Structure.   As described in Section 23-2E, tRNA molecules have an L-shape in which the acceptor and T stems form one leg and the D and anticodon stems form the other (Fig. 26-5). Each leg of the L is ~60 Å long and the anticodon and amino acid acceptor sites are at opposite ends of the molecule, some 76 Å apart. The narrow 20- to 25-Å width of tRNA is essential to its biological function: During protein synthesis, three tRNA molecules must simultaneously bind in close proximity at adjacent codons on mRNA (Section 26-4B).

tRNA's complex tertiary structure is maintained by extensive stacking interactions and base pairing within and between the helical stems. Many of the tertiary base-pairing interactions are non-Watson–Crick associations. Moreover, most of the bases involved in these interactions are either invariant or semi-invariant, consistent with the notion that *all tRNAs have similar conformations.* The compact structure of yeast tRNA^{Phe} renders most of its bases inaccessible to solvent. The most notable exceptions are the anticodon bases and those of the amino acid–bearing —CCA terminus. Both of these groupings must be accessible in order to carry out their biological functions.

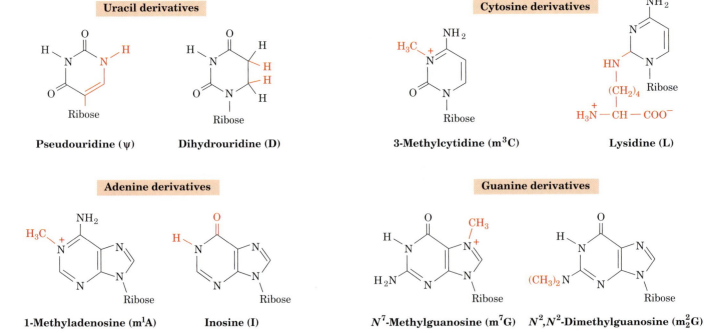

Figure 26-4   A few of the modified nucleosides that occur in tRNAs.   Note that although inosine chemically resembles guanosine, it is biochemically derived by the deamination of adenosine. Nucleosides may also be methylated at their ribose 2′ positions to form residues symbolized, for instance, by Cm, Gm, and Um.   **See Kinemage Exercise 20-2.**

(a)

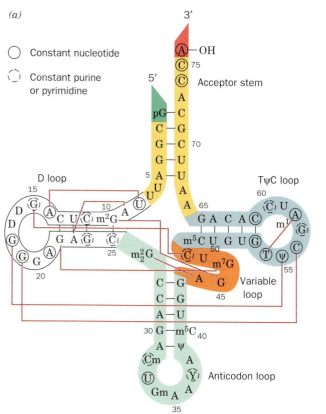

○ Constant nucleotide

⟨ ⟩ Constant purine or pyrimidine

(b)

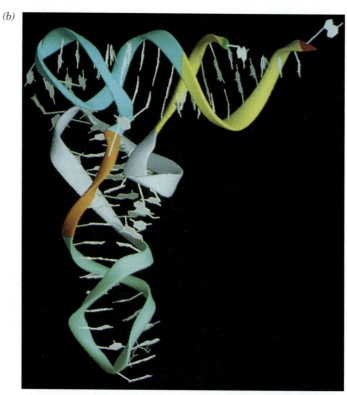

Figure 26-5   *Key to Structure.*   The structure of yeast tRNA^Phe.   (*a*) The base sequence drawn in cloverleaf form. Tertiary base-pairing interactions are represented by thin red lines connecting the participating bases. Bases that are conserved or semiconserved in all tRNAs are circled by solid and dashed lines, respectively. The 5′ terminus is colored bright green, the acceptor stem is yellow, the D arm is white, the anticodon arm is light green, the variable arm is orange, the TψC arm is cyan, and the 3′ terminus is red. (*b*) The X-ray structure drawn to show how its base-paired stems are arranged to form the L-shaped molecule. The sugar–phosphate backbone is represented by a ribbon with the same color scheme as in Part *a*. [Courtesy of Mike Carson, University of Alabama at Birmingham. PDBid 6TNA.] 🐭 **See Kinemage Exercise 20-1.**

B   Aminoacyl–tRNA Synthetases

Accurate translation requires two equally important recognition steps:

1. The correct amino acid must be selected for covalent attachment to a tRNA by an **aminoacyl–tRNA synthetase** (discussed below).

2. The correct **aminoacyl–tRNA (aa–tRNA)** must pair with an mRNA codon at the ribosome (discussed in Section 26-4B).

An amino acid–specific aminoacyl–tRNA synthetase (**aaRS**) appends an amino acid to the 3′-terminal ribose residue of its cognate tRNA to form an aa–tRNA (Fig. 26-6). Aminoacylation occurs in two sequential reactions that are catalyzed by a single enzyme.

1. The amino acid is first "activated" by its reaction with ATP to form an **aminoacyl–adenylate,**

$$R-\overset{\overset{\displaystyle H}{|}}{\underset{\underset{\displaystyle NH_3^+}{|}}{C}}-\overset{\overset{\displaystyle O}{\|}}{\underset{\underset{\displaystyle O^-}{|}}{C}} + ATP \rightleftharpoons R-\overset{\overset{\displaystyle H}{|}}{\underset{\underset{\displaystyle NH_3^+}{|}}{C}}-\overset{\overset{\displaystyle O}{\|}}{C}-O-\overset{\overset{\displaystyle O}{\|}}{\underset{\underset{\displaystyle O^-}{|}}{P}}-O-\text{Ribose–Adenine} + PP_i$$

Amino acid   **Aminoacyl–adenylate (aminoacyl–AMP)**

See Guided Exploration 27
The structures of aminoacyl–tRNA synthetases and their interactions with tRNAs.

Aminoacyl–tRNA

Figure 26-6   An aminoacyl–tRNA.   The amino acid residue is esterified to the tRNA's 3′-terminal nucleotide at either its 3′-OH group, as shown here, or its 2′-OH group.

which, with all but three aaRSs, can occur in the absence of tRNA. Indeed, this intermediate can be isolated, although it normally remains tightly bound to the enzyme.

2. This mixed anhydride then reacts with tRNA to form the aa–tRNA:

$$\text{Aminoacyl–AMP} + \text{tRNA} \rightleftharpoons \text{aminoacyl–tRNA} + \text{AMP}$$

The overall aminoacylation reaction

$$\text{Amino acid} + \text{tRNA} + \text{ATP} \longrightarrow \text{aminoacyl–tRNA} + \text{AMP} + \text{PP}_i$$

is driven to completion by the hydrolysis of the PP_i generated in the first reaction step. The aa–tRNA product is a "high-energy" compound (Section 13-2A); for this reason, the amino acid is said to be "activated" and the tRNA is said to be "charged." Amino acid activation resembles fatty acid activation (Section 19-2A); the major difference is that tRNA is the acyl acceptor in amino acid activation whereas CoA performs this function in fatty acid activation.

Table 26-2 Classification of *E. coli* Aminoacyl–tRNA Synthetases

Class I Amino Acid	Class II Amino Acid
Arg	Ala
Cys	Asn
Gln	Asp
Glu	Gly
Ile	His
Leu	Lys
Met	Pro
Trp	Phe
Tyr	Ser
Val	Thr

There Are Two Classes of Aminoacyl–tRNA Synthetases. Cells must have at least one aaRS for each of the 20 amino acids. The similarity of the reaction catalyzed by these enzymes and the structural similarities among tRNAs suggest that all aaRSs evolved from a common ancestor and should therefore be structurally related. This is not the case. In fact, *the aaRSs form a diverse group of enzymes with different sizes and quaternary structures and little sequence similarity.* Nevertheless, these enzymes can be grouped into two classes that each have the same 10 members in nearly all organisms (Table 26-2). **Class I** and **Class II aminoacyl–tRNA synthetases** differ in several ways:

1. **Structural motifs.** The Class I enzymes share two homologous polypeptide segments that are components of a dinucleotide-binding fold (Rossmann fold, which is also present in many NAD^+- and ATP-binding proteins; Section 6-2C). The Class II synthetases lack the foregoing sequences but have three other sequences in common.

2. **Anticodon recognition.** Many Class I aaRSs must recognize the anticodon to aminoacylate their cognate tRNAs. In contrast, several Class II enzymes do not interact with their bound tRNA's anticodon.

3. **Site of aminoacylation.** All Class I enzymes aminoacylate their bound tRNA's 3′-terminal 2′-OH group, whereas Class II enzymes charge the 3′-OH group. Nevertheless, an aminoacyl group attached at the 2′ position rapidly equilibrates between the 2′ and 3′ positions (it must be at the 3′ position to take part in protein synthesis).

4. **Amino acid specificity.** The amino acids for which the Class I synthetases are specific tend to be larger and more hydrophobic than those for Class II synthetases.

Aminoacyl–tRNA Synthetases Recognize Unique Structural Features of tRNA. How does an aaRS recognize a tRNA so that it can be charged with the proper amino acid? First, all tRNAs have similar structures, so the features that differentiate them must be subtle variations in sequence or local structure. On the other hand, since the genetic code is degenerate, more than one tRNA may carry a given amino acid. The members of each set of these so-called **isoaccepting tRNAs** must all be recognized by their cognate aaRS. Finally, the tRNA must be charged with only the amino acid that corresponds to its anticodon, and not any of the 19 other amino acids.

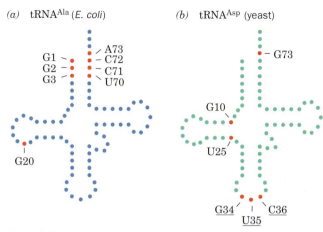

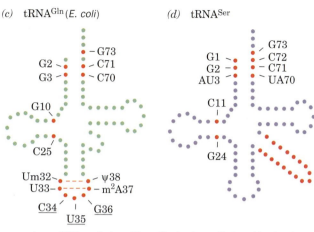

(a) tRNA^Ala (*E. coli*)   *(b)* tRNA^Asp (yeast)   *(c)* tRNA^Gln (*E. coli*)   *(d)* tRNA^Ser

Figure 26-7   Major identity elements in four tRNAs. Each base in the tRNA is represented by a filled circle. Red circles indicate positions that have been shown to be identity elements for the recognition of the tRNA by its cognate aa–tRNA synthetase. The anticodon bases that are identity elements are underlined.

Clues to the specificity of synthetase–tRNA interactions have been gleaned from studies using tRNA fragments, mutationally altered tRNAs, chemical cross-linking agents, computerized sequence comparisons, and X-ray crystallography. The most common synthetase contact sites on tRNA occur on the inner (concave) face of the L and, of course, include the acceptor stem. Other than that, there appears to be little regularity in how the synthetases recognize their cognate tRNAs.

The elements in four tRNAs that are recognized by different aaRSs are shown in Fig. 26-7. In all of these cases, *the acceptor stem is critical for the enzyme–tRNA interaction, as are other sites, not necessarily including the anticodon.* A single base or base pair may constitute an identity element. When the experimentally determined identity elements for a variety of tRNAs are mapped onto a three-dimensional model of a tRNA molecule, they cluster in the acceptor stem and the anticodon loop (Fig. 26-8). These sites lie at opposite ends of the molecule, so that aaRSs that interact with both of them must have a size and structure adequate to bind both legs of the L-shaped tRNA.

The X-ray structures of synthetase–tRNA complexes reveal extensive contacts between the protein and the inside face of the tRNA L. The structure of *E. coli* **glutaminyl–tRNA synthetase (GlnRS)** in complex with **tRNA^Gln** and ATP (Fig. 26-9), determined by Thomas Steitz, was the first to be elucidated. GlnRS, a 553-residue monomeric Class I enzyme, has an elongated shape so that it binds the anticodon near one end of the protein and the acceptor stem near the other. Genetic and biochemical data indicate that the identity elements for tRNA^Gln include all seven bases of the anticodon loop (Fig. 26-7c). The bases of the anticodon itself are unstacked and splay outward so as to bind in separate recognition pockets of GlnRS. The 3′ end of tRNA^Gln plunges deeply into a protein pocket that also binds the enzyme's ATP and glutamine substrates.

Yeast **AspRS,** a Class II enzyme, is an α_2 dimer of 557-residue subunits. Its X-ray structure in complex with **tRNA^Asp,** determined by Dino Moras,

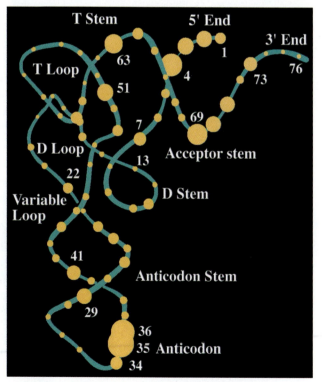

Figure 26-8   Experimentally observed identity elements of tRNAs. The tRNA backbone is cyan, and each of its nucleotides is represented by a yellow circle whose diameter is proportional to the fraction of the various tRNA acceptor types for which the nucleotide is an observed determinant. [Courtesy of William McClain, University of Wisconsin.]

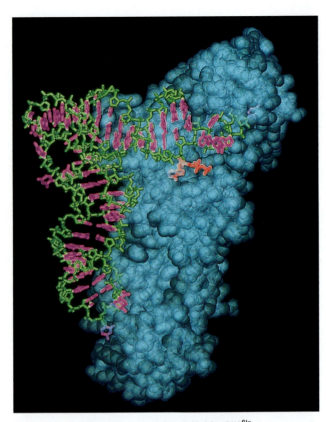

Figure 26-9 X-Ray structure of *E. coli* GlnRS · tRNA^Gln · ATP. The tRNA and ATP are shown in skeletal form with the tRNA sugar–phosphate backbone green, its bases magenta, and the ATP red. The protein is represented by a translucent cyan space-filling model that reveals the buried portions of the tRNA and ATP. Note that both the 3′ end of the tRNA (*top right*) and its anticodon bases (*bottom*) are inserted into deep pockets in the protein. [Based on an X-ray structure by Thomas Steitz, Yale University. PDBid 1GSG.] ✎ **See Kinemage Exercise 21.**

reveals that the protein symmetrically binds two tRNA molecules by contacting them principally at their acceptor stem and anticodon regions (Fig. 26-10). The anticodon arm of tRNA^Asp is bent by as much as 20 Å toward the inside of the L relative to that in the X-ray structure of uncomplexed tRNA^Asp, and its anticodon bases are unstacked. The hinge point for this bend is a G30 · U40 base pair in the anticodon stem (nearly all other species of tRNA contain a Watson–Crick base pair at this point). The anticodon bases of tRNA^Gln are also unstacked in contacting GlnRS but with a backbone conformation that differs from that in tRNA^Asp. Evidently, the conformation of a tRNA in complex with its cognate synthetase is dictated more by its interactions with the protein (induced fit) than by its sequence. This is perhaps one reason why *the members of each set of isoaccepting tRNAs in a cell are recognized by a single aaRS.*

The different modes of tRNA binding by GlnRS and AspRS are highlighted in Fig. 26-11. Although both tRNAs approach their synthetases along the inside of the L shapes, tRNA^Gln does so from the direction of the minor groove of its acceptor stem, whereas tRNA^Asp does so from the direction of its major groove. The 3′ end of tRNA^Asp thereby continues its helical track as it plunges into AspRS's catalytic site, whereas the 3′ end of tRNA^Gln bends backward into a hairpin turn as it enters its active site. These structural differences account for the observation that Class I and Class II enzymes aminoacylate different OH groups on the 3′ terminal ribose of tRNA.

Proofreading Enhances the Fidelity of Amino Acid Attachment to tRNA. The charging of a tRNA with its cognate amino acid is a remarkably accurate process. Experimental measurements indicate, for example, that at equal concentrations of isoleucine and valine, **IleRS** transfers ~40,000 isoleucines to tRNA^Ile for every valine it so transfers. This high degree of accuracy is surprising because valine, which differs from isoleucine only by the lack of

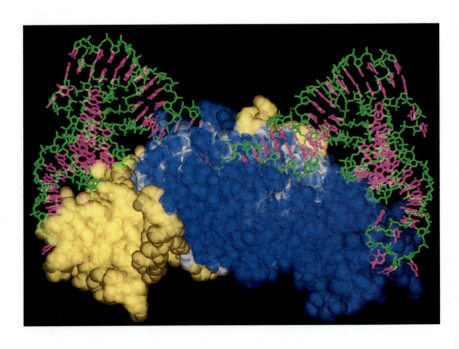

Figure 26-10 X-Ray structure of yeast AspRS · tRNA^Asp · ATP. The homodimeric enzyme with its two symmetrically bound tRNAs is viewed with its twofold axis approximately vertical. The tRNAs are shown in skeletal form with their sugar–phosphate backbones green and their bases magenta. The two protein subunits are represented by translucent yellow and blue space-filling models that reveal buried portions of the tRNAs. [Based on an X-ray structure by Dino Moras, CNRS/INSERM/ULP, Illkirch Cédex, France. PDBid 1ASY.]

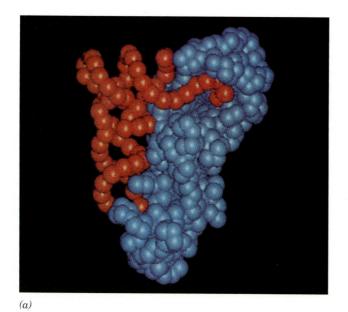

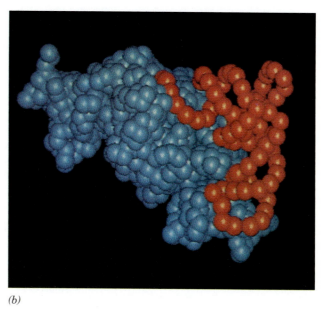

(a)

(b)

Figure 26-11   Comparison of the binding of GlnRS and AspRS to their cognate tRNAs.   The proteins and tRNAs are represented by blue and red spheres marking the C_α and P atom positions. Note how GlnRS (*a*), a Class I synthetase, binds tRNA^{Gln} from the minor groove side of its acceptor stem so as to bend its 3′ end into a hairpin conformation. In contrast, AspRS (*b*), a Class II synthetase, binds tRNA^{Asp} from the major groove side of its acceptor stem so that its 3′ end continues its helical path on entering the active site. [Courtesy of Dino Moras, CNRS/INSERM/ULP, Illkirch Cédex, France.]

a single methylene group, should fit easily into the isoleucine-binding site of IleRS. The binding free energy of a methylene group is estimated to be ~12 kJ · mol^{-1}. Equation 1-16 indicates that the ratio f of the equilibrium constants, K_1 and K_2, with which two substances bind to a given binding site is given by

$$f = \frac{K_1}{K_2} = \frac{e^{-\Delta G_1^{\circ\prime}/RT}}{e^{-\Delta G_2^{\circ\prime}/RT}} = e^{-\Delta\Delta G^{\circ\prime}/RT} \qquad [26\text{-}1]$$

where $\Delta\Delta G^{\circ\prime} = \Delta G_1^{\circ\prime} - \Delta G_2^{\circ\prime}$ is the difference between the free energies of binding of the two substances. It is therefore estimated that isoleucyl–tRNA synthetase could discriminate between isoleucine and valine by no more than a factor of ~100.

Paul Berg resolved this apparent paradox by demonstrating that, in the presence of tRNA^{Ile}, IleRS catalyzes the quantitative hydrolysis of valine–adenylate, the intermediate of the aminoacylation reaction, to valine + AMP rather than forming Val–tRNA^{Ile}. Thus, *isoleucyl–tRNA synthetase subjects aminoacyl–adenylates to a proofreading or editing step*, a process reminiscent of that carried out by DNA polymerase I (Section 24-2A).

IleRS has two active sites that operate as a double sieve. The first site activates isoleucine and the chemically similar valine. The second active site, which hydrolyzes aminoacylated tRNA^{Ile}, admits only tRNAs with aminoacyl groups that are smaller than isoleucine (i.e., Val–tRNA^{Ile}). The X-ray structure of a bacterial IleRS, a Class I enzyme, determined by Steitz, reveals that the protein contains an additional editing domain inserted into its dinucleotide-binding fold domain (Fig. 26-12). The 3′ terminus of tRNA^{Ile} appears to shuttle between the aminoacylation and editing sites by a conformational change. *The combined selectivities of*

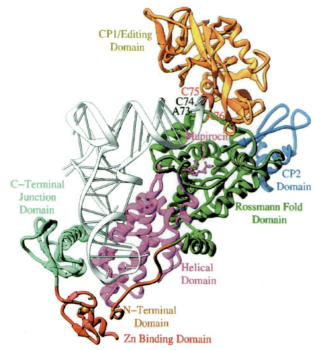

Figure 26-12   X-Ray structure of *Staphylococcus aureus* isoleucyl–tRNA synthetase in complex with tRNA^{Ile}.   The tRNA is white, the protein is colored by domain, and **mupirocin,** an antibiotic that binds to IleRS, is drawn in stick form in pink. The four terminal residues of the tRNA are labeled. The 3′ terminus of the tRNA is positioned near the editing active site. [Courtesy of Thomas Steitz, Yale University. PDBid 1QU2.]

the aminoacylation and editing steps are responsible for the high fidelity of translation, a phenomenon that occurs at the expense of ATP hydrolysis.

Other aminoacyl–tRNA synthetases discriminate against noncognate amino acids through a variety of noncovalent interactions. In **ValRS,** the side chain of Asp 279 protrudes into the editing active site, where it hydrogen bonds with the hydroxyl group of threonine. The isosteric (having the same shape) valine lacks this hydroxyl group and is thereby excluded from the editing pocket. **ThrRS** has the opposite problem: It must synthesize Thr–tRNA^{Thr} but not Val–tRNA^{Thr}. Specificity is conferred by the aminoacylation site, which contains a Zn^{2+} ion that is coordinated by the side chain OH group of threonine. Valine cannot coordinate the Zn^{2+} in this way and hence does not undergo adenylylation by ThrRS. A separate editing site deals with misacylated Ser–tRNA^{Thr}. **TyrRS** distinguishes between tyrosine and phenylalanine through hydrogen bonding with tyrosine's OH group. Since no other amino acid resembles tyrosine, the enzyme can do without an editing function.

Some Organisms Lack Certain aaRSs. Many bacteria lack the expected complement of 20 aminoacyl–tRNA synthetases. For example, GlnRS is absent in gram-positive bacteria, archaebacteria, cyanobacteria, mitochondria, and chloroplasts. Instead, glutamate is linked to tRNA^{Gln} by the same GluRS that synthesizes Glu–tRNA^{Glu}. The resulting Glu–tRNA^{Gln} is then converted to Gln–tRNA^{Gln} by **Glu–tRNA^{Gln} amidotransferase** in an ATP-requiring reaction in which glutamine is the amide donor. Some microorganisms that lack **AsnRS** use a similar transamidation pathway for the synthesis of Asn–tRNA^{Asn} from Asp–tRNA^{Asn}.

Certain archaebacteria lack a gene for **CysRS.** In these cells, Cys–tRNA^{Cys} is produced by a ProRS (called **ProCysRS**). Despite its unusual dual specificity, ProCysRS does not generate either Cys–tRNA^{Pro} or Pro–tRNA^{Cys}. This is because adenylylation of Pro does not require the presence of tRNA^{Pro}, but the activation of Cys requires tRNA^{Cys}. Presumably, the binding of Pro–AMP or tRNA^{Cys} elicits conformational changes that are mutually exclusive.

C Codon–Anticodon Interactions

In protein synthesis, the proper tRNA is selected only through codon–anticodon interactions; the aminoacyl group does not participate in this process (this is one reason why accurate aminoacylation is critical for protein synthesis). The three nucleotides of an mRNA codon pair with the three nucleotides of a complementary tRNA anticodon in an antiparallel fashion. One might naively guess that each of the 61 codons specifying an amino acid would be read by a different tRNA. Yet even though most cells contain numerous groups of isoaccepting tRNAs, *many tRNAs bind to two or three of the codons specifying their cognate amino acids.* For example, yeast tRNA^{Phe}, which has the anticodon GmAA (where Gm indicates G with a 2′-methyl group), recognizes the codons UUC and UUU,

	3′ 5′	3′ 5′
Anticodon:	—A—A—Gm—	—A—A—Gm—
	• • •	• • •
	5′ • • • 3′	5′ • • • 3′
Codon:	—U—U—C—	—U—U—U—

and yeast tRNA^{Ala}, which has the anticodon IGC (where I is inosine), recognizes the codons GCU, GCC, and GCA:

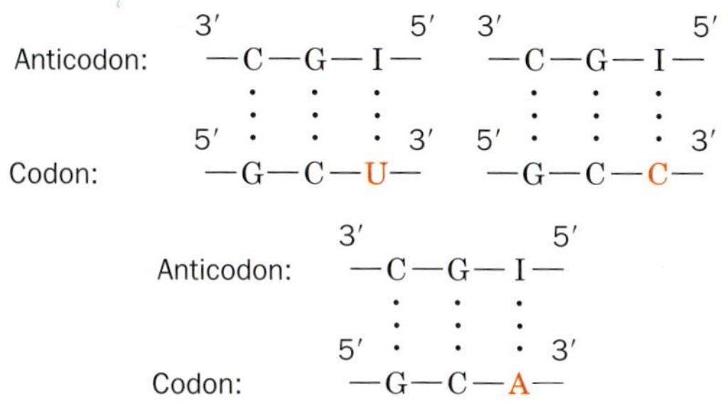

```
                3'        5'   3'        5'
Anticodon:    —C—G—I—      —C—G—I—
               · ·  ·       · ·  ·
               · ·  ·  3'  5' · ·  ·  3'
Codon:        —G—C—U—      —G—C—C—

                    3'        5'
      Anticodon:  —C—G—I—
                   · ·  ·
                   · ·  ·  3'
      Codon:    —G—C—A—
```

It therefore seems that non-Watson–Crick base pairing can occur at the third codon–anticodon position (the anticodon's first position is defined as its 3′ nucleotide), the site of most codon degeneracy (Table 26-1). Note that the third (5′) anticodon position commonly contains a modified base such as Gm or I.

The Wobble Hypothesis Accounts for Codon Degeneracy.

By combining structural insight with logical deduction, Crick proposed the **wobble hypothesis** to explain how a tRNA can recognize several degenerate codons. He assumed that *the first two codon–anticodon pairings have normal Watson–Crick geometry and that there could be a small amount of play or "wobble" in the third anticodon position to allow limited conformational adjustments in its pairing geometry*. This permits the formation of several non-Watson–Crick pairs such as U · G and I · A (Fig. 26-13). The allowed pairings for the third codon–anticodon position are listed in Table 26-3. An anticodon with C or A in its third position can potentially pair only with its Watson–Crick complementary codon (although, in fact, there is no known instance of a tRNA with an A in its third anticodon position). If U or G occupies the third anticodon position, two codons can potentially be recognized. I at the third anticodon position can pair with U, C, or A.

A consideration of the various wobble pairings indicates that at least 31 tRNAs are required to translate all 61 coding triplets of the genetic code (there are 32 tRNAs in the minimal set because translation initiation requires a separate tRNA; Section 26-4A). Most cells have >32 tRNAs, some of which have identical anticodons. In fact, mammalian cells have >150 tRNAs. Some organisms contain unique tRNAs that are charged with unusual amino acids (Box 26-2).

Frequently Used Codons Are Complementary to the Most Abundant tRNA Species.

The analysis of the base sequences of several highly expressed structural genes of baker's yeast, *Saccharomyces cerevisiae,* has revealed a remarkable bias in their codon usage. Only 25 of the 61 coding triplets are com-

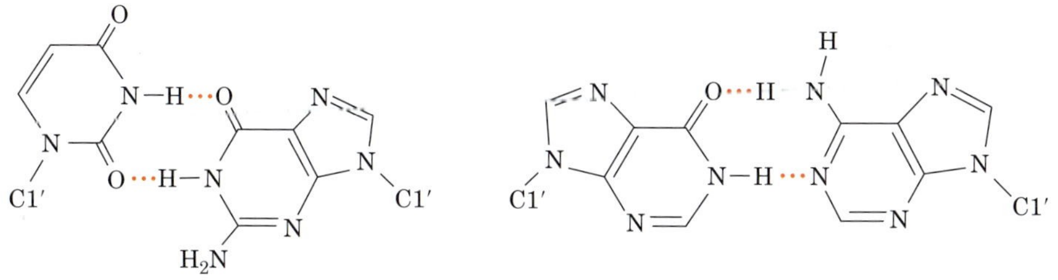

Table 26-3 Allowed Wobble Pairing Combinations in the Third Codon–Anticodon Position

5′-Anticodon Base	3′-Codon Base
C	G
A	U
U	A or G
G	U or C
I	U, C, or A

Figure 26-13 U · G and I · A wobble pairs. Both have been observed in X-ray structures.

Perspectives in Biochemistry

Expanding the Genetic Code

It is widely stated that proteins are synthesized from 20 "standard" amino acids, but two other amino acids are known to be incorporated into proteins during translation (other nonstandard amino acid residues in proteins are the result of posttranslational modifications). In both cases, the amino acid is attached to a unique tRNA that recognizes a Stop codon.

Several enzymes in eukaryotes and prokaryotes contain the nonstandard amino acid **selenocysteine (Sec):**

$$\begin{array}{c} | \\ NH \\ | \\ CH-CH_2-Se-H \\ | \\ C{=}O \\ | \end{array}$$

**The selenocysteine
(Sec) residue**

The Sec residues of **selenoproteins** are thought to participate in redox reactions such as those catalyzed by mammalian glutathione peroxidase (Box 14-4) and thioredoxin reductase (Section 22-3A). For this reason, selenium is an essential trace element. The human genome contains an estimated 25 selenoproteins.

Selenocysteine, sometimes called the "twenty-first amino acid," is incorporated into proteins with the aid of a tRNA that interprets the UGA Stop codon as a Sec codon. **tRNA^{Sec}** is initially charged with serine in a reaction catalyzed by the same SerRS that charges tRNA^{Ser}. The resulting seryl–tRNA^{Sec} is enzymatically selenylated to produce selenocysteinyl–tRNA^{Sec}. Although tRNA^{Sec} must resemble tRNA^{Ser} enough to interact with the same SerRS, its acceptor stem has 8 bp (rather than 7), its D arm has a 6-bp stem and a 4-base loop (rather than a 4-bp stem and a 7–8-base loop),

its TψC stem has 4 bp rather than 5, and its anticodon, UCA, recognizes a UGA Stop codon rather than a Ser codon. In addition, several of the invariant residues of other tRNAs are altered in tRNA^{Sec}. These changes explain why tRNA^{Sec} is not recognized by EF-Tu·GTP, which escorts other aminoacyl–tRNAs to the ribosome (Section 26-4B). Instead, a dedicated protein (a special elongation factor) named **SELB** in complex with GTP is required to deliver Sec–tRNA^{Sec} to the ribosome. SELB·GTP·Sec–tRNA^{Sec} reads the UGA codon as "Sec" rather than "Stop," provided that the ribosomally bound mRNA has a hairpin loop on the 3′ side of the UGA specifying Sec.

The archaeal protein **methylamine methyltransferase** includes the amino acid **pyrrolysine (Pyl),** a lysine with its ε-nitrogen in amide linkage to a pyrroline group:

$$\begin{array}{c} | \\ NH \\ | \\ CH-CH_2-CH_2-CH_2-CH_2-NH-C \\ | \\ C{=}O \\ | \end{array}$$

The pyrrolysine (Pyl) residue

Pyl is specified by the codon UAG (normally a Stop codon). **tRNA^{Pyl}** differs from typical tRNAs in having an anticodon stem with 6 bp rather than 5. The tRNA is charged by a specific aminoacyl–tRNA synthetase that differs from known lysyl–tRNA synthetases. Because the PylRS readily generates Lys–tRNA^{Pyl}, it is not clear whether the substrate for aminoacylation is lysine, which is later modified, or preformed pyrrolysine. The mRNA signals that allow UAG to be read as a Pyl codon rather than as a Stop codon have not yet been elucidated.

monly used. *The preferred codons are those that are most nearly complementary, in the Watson–Crick sense, to the anticodons in the most abundant species in each set of isoaccepting tRNAs.* A similar phenomenon occurs in *E. coli,* although several of its 22 preferred codons differ from those in yeast. The degree with which the preferred codons occur in a given gene is strongly correlated, in both organisms, with the gene's level of expression. This probably permits the mRNAs of proteins that are required in high abundance to be rapidly and smoothly translated.

3 | Ribosomes

Ribosomes, small organelles that were once thought to be artifacts of cell disruption, were identified as the site of protein synthesis in 1955 by Paul Zamecnik, who demonstrated that ^{14}C-labeled amino acids are transiently associated with ribosomes before they appear in free proteins. The ribosome is both enormous (~2.5 × 10^6 D in bacteria and 3.9 to 4.5 × 10^6 D in eukaryotes) and complex (ribosomes contain several large RNA molecules and dozens of different proteins). This complexity is necessary for the ribosome to carry out the following vital functions:

1. The ribosome binds mRNA such that its codons can be read with high fidelity.

2. The ribosome includes specific binding sites for tRNA molecules.

3. The ribosome mediates the interactions of nonribosomal protein factors that promote polypeptide chain initiation, elongation, and termination.

4. The ribosome catalyzes peptide bond formation.

5. The ribosome undergoes movement so that it can translate sequential codons.

In this section, we discuss the structure of the ribosome, beginning with that of the smaller and simpler prokaryotic ribosome and ending with that of the larger and more complicated eukaryotic ribosome.

A The Prokaryotic Ribosome

Ribosomal components are traditionally described in terms of their rate of sedimentation in an ultracentrifuge, which correlates roughly with their size (Section 5-2E). Thus, the intact *E. coli* ribosome has a sedimentation coefficient of 70S. As James Watson discovered, the ribosome can be dissociated into two unequal subunits (Table 26-4). The small **(30S)** subunit consists of a **16S rRNA** molecule and 21 different proteins, whereas the large **(50S)** subunit contains a **5S** and a **23S rRNA** together with 31 different proteins. By convention, ribosomal proteins from the small and large subunits are designated with the prefixes S and L, respectively, followed by a number that roughly increases from the largest to the smallest. These proteins, which range in size from 46 to 557 residues, occur in only one copy per ribosome with the exception of L12, which is present in four copies. Most

Table 26-4   Components of *E. coli* Ribosomes

	Ribosome	Small Subunit	Large Subunit
Sedimentation coefficient	70S	30S	50S
Mass (kD)	2520	930	1590
RNA			
Major		16S, 1542 nucleotides	23S, 2904 nucleotides
Minor			5S, 120 nucleotides
RNA mass (kD)	1664	560	1104
Proportion of mass	66%	60%	70%
Proteins		21 polypeptides	31 polypeptides
Protein mass (kD)	857	370	487
Proportion of mass	34%	40%	30%

of these proteins, which exhibit little sequence similarity with one another, are rich in the basic amino acids Lys and Arg and contain few aromatic residues, as is expected for proteins that are closely associated with polyanionic RNA molecules. The up to 20,000 ribosomes in an *E. coli* cell account for ~80% of its RNA content and 10% of its protein.

Ribosomal RNAs Have Complicated Secondary Structures. The *E. coli* 16S rRNA, which was sequenced by Harry Noller, consists of 1542 nucleotides. A computerized search of this sequence for stable double-helical segments yielded many plausible but often mutually exclusive secondary structures. However, the comparison of the sequences of 16S rRNAs from several prokaryotes, under the assumption that their structures have been evolutionarily conserved, led to the flowerlike secondary structure for 16S rRNA seen in Fig. 26-14a. In this four-domain structure, which is 54% base paired, the double-helical stems tend to be short (<8 bp) and many of them are imperfect. Intriguingly, electron micrographs of the 16S rRNA resemble those of the complete 30S subunit, thereby suggesting that the 30S subunit's overall shape is largely determined by the 16S rRNA. The large ribosomal subunit's 5S and 23S rRNAs, which consist of 120 and 2904 nucleotides, respectively, have also been sequenced. As with the 16S rRNA, they have extensive secondary structures (Fig. 26-14b).

The Ribosome Has a Highly Complicated Three-Dimensional Structure. The structure of the ribosome began to come into focus through electron microscopy (EM; Fig. 26-15a) and later through **cryoelectron microscopy** (**cryo-EM;** Fig. 26-15b). In the latter technique, the sample is cooled to near liquid N_2 temperatures (−196°C) so rapidly (in a few milliseconds) that the water in the sample does not have time to crystallize but, rather, assumes a vitreous (glasslike) state. Consequently, the sample remains hydrated and hence retains its native shape to a greater extent than in conventional electron microscopy. Cryo-EM studies, carried out largely by Joachim Frank,

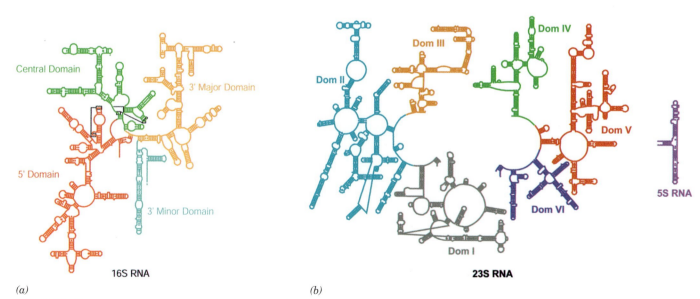

(a) *(b)*

Figure 26-14 Secondary structures of the *E. coli* ribosomal RNAs. (*a*) 16S rRNA and (*b*) 23S and 5S rRNAs. The rRNAs are colored by domain with short lines spanning a stem representing Watson–Crick base pairs, small dots representing G · U base pairs, and large dots representing other non-Watson–Crick base pairs. Note the flowerlike series of stems and loops forming each domain. [Courtesy of V. Ramakrishnan, MRC Laboratory of Molecular Biology, Cambridge, U.K., and Peter Moore, Yale University. Adapted from diagrams in http://www.rna.icmb.utexas.edu.]

(a)

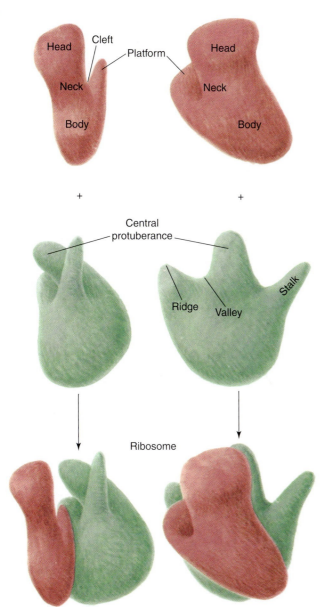

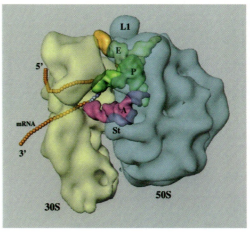

(b)

Figure 26-15   Structure of the *E. coli* ribosome. (*a*) Two views of a low resolution three-dimensional, electron microscopy–based model. The small (16S) subunit (*top*) combines with the large (30S) subunit (*middle*) to form the complete ribosome (*bottom*).
(*b*) Cryoelectron microscopy–based image at ~25 Å resolution. In this semitransparent three-dimensional model, the 30S subunit (*yellow*) is on the left and the 50S subunit (*cyan*) is on the right. The tRNAs that occupy the A, P, and E sites (Section 26-4B) are colored magenta, green, and gold. The inferred path of the mRNA is represented by a chain of orange beads with the six nucleotides contacting the A and P sites blue and purple, respectively. The exit of the tunnel in the 50S subunit through which the growing polypeptide chain is extruded is visible (*center right*). [Courtesy of Joachim Frank, State University of New York at Albany.]

revealed that the ribosome has an irregular shape, about 250 Å across, with numerous lobes and bulges as well as channels and tunnels.

The fine structure of the ribosome was determined by X-ray crystallography, a landmark achievement owing to the enormous size of the particle. Ribosomal subunits were first crystallized by Ada Yonath in 1980 although they diffracted X-rays poorly. Over the course of several years the quality of these crystals was incrementally improved. However, it was not until the late 1990s that technology was up to the task of determining the X-ray structures of these gargantuan molecular complexes. In 2000, Peter Moore and Steitz reported the X-ray structure of the ~100,000-atom 50S ribosomal subunit of the halophilic (salt-loving) bacterium *Haloarcula marismortui* at atomic (2.4 Å) resolution, and shortly thereafter V. Ramakrishnan and Yonath independently reported the X-ray structure of the 30S subunit of *T. thermophilus* at ~3 Å resolution. In 2001, Noller reported the low (5.5 Å) resolution structure of the entire *T. thermophilus* ribosome in complex with three tRNAs and a 36-nt mRNA fragment.

Several generalizations can be made about prokaryotic ribosomal architecture based on the structures of the 30S and 50S subunits:

1. Both the 16S and 23S rRNAs are assemblies of helical elements connected by loops, most of which are irregular extensions of helices (Fig. 26-16). These structures, which are in close accord with previous secondary structure predictions (Fig. 26-14), are stabilized by interactions between helices such as minor groove to minor groove packing, which has also been seen in the structure of the group I intron (Section 25-3B; recall that A-form RNA has a very shallow minor groove); the insertion of a phosphate ridge into a minor groove; and adenines which are distant in sequence but often highly conserved that are inserted into minor grooves.

2. Each of the 16S rRNA's four domains, which extend out from a central junction (Fig. 26-14a), forms a morphologically distinct portion of the 30S subunit (Fig. 26-16a): The 5' domain forms most of the body (Fig. 26-15a), the central domain forms the platform, the 3' major domain forms the entire head, and the 3' minor domain, which consists of just 2 helices, is located at the interface between the 30S and 50S subunits. In contrast, the 23S rRNA's six domains (Fig. 26-14b) are intricately intertwined in the 50S subunit (Fig. 26-16b). Since the ribosomal proteins are embedded in the RNA (see below), this suggests that the domains of the 30S subunit can move relative to one another during protein synthesis, whereas the 50S subunit appears to be rigid.

3. The distribution of the proteins in the two ribosomal subunits is not uniform (Fig. 26-17). The vast majority of the ribosomal proteins are located on the back and sides of their subunits. In contrast, the face of each subunit that forms the interface between the two subunits,

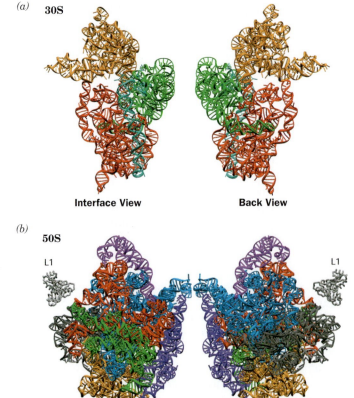

Figure 26-16 Tertiary structures of the ribosomal **RNAs.** (a) The 16S rRNA of *T. thermophilus*. (b) The 23S rRNA of *H. marismortui*. The rRNAs are colored according to domain as in Fig. 26-14. The interface view of a ribosomal subunit (*left*) is toward its surface that associates with the other subunit in the whole ribosome and the back view (*right*) is from the opposite (solvent-exposed) side. Note that the secondary structure domains of the 16S rRNA fold as separate tertiary structure domains, whereas in the 23S rRNA, the secondary structure domains are convoluted together. The L1 protein, which forms the large subunit's ridge (Fig. 26-15a), is shown for purposes of orientation. [Courtesy of V. Ramakrishnan, MRC Laboratory of Molecular Biology, Cambridge, U.K., and Peter Moore, Yale University. PDBids 1J5E and 1JJ2.]

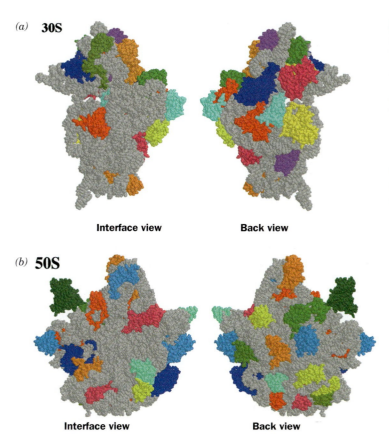

(a) **30S**

Interface view   Back view

(b) **50S**

Interface view   Back view

Figure 26-17 Distribution of protein and RNA in the ribosomal subunits. (*a*) The 30S subunit of *T. thermophilus.* (*b*) The 50S subunit of *H. marismortui.* The subunits are drawn in space-filling form with their RNAs gray and their proteins in various colors. Note that the interface side of each subunit is largely free of protein. The globular portions are exposed on the surface of their associated subunit (Fig. 26-18), whereas the extended segments are largely buried in the RNA. [Part *a* based on an X-ray structure by V. Ramakrishnan, MRC Laboratory of Molecular Biology, Cambridge, U.K. Part *b* based on an X-ray structure by Peter Moore and Thomas Steitz, Yale University. PDBids 1J5E and 1JJ2.]

particularly those regions that bind the tRNAs and mRNA (see below), is largely devoid of proteins.

4. Most ribosomal proteins consist of a globular domain, which is, for the most part, located on a subunit surface (Fig. 26-17), and a long segment that is largely devoid of secondary structure and unusually rich in basic residues that infiltrates between the RNA helices into the subunit interior (Fig. 26-18). Indeed, a few ribosomal proteins lack

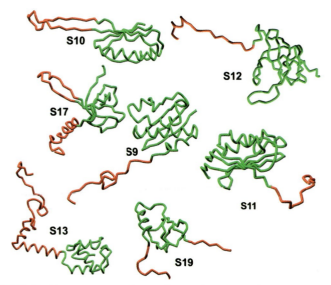

S10

S12

S17

S9

S11

S13

S19

Figure 26-18 Backbone structures of some ribosomal proteins. Globular portions of the proteins are green, and extended segments are red. [Courtesy of V. Ramakrishnan, MRC Laboratory of Molecular Biology, Cambridge, U.K. PDBid 1J5E.]

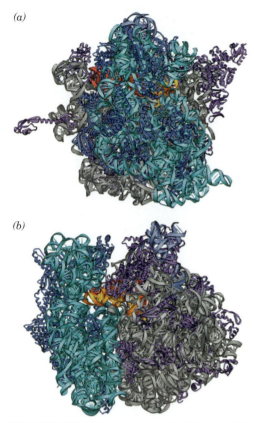

(a)

(b)

Figure 26-19 **X-Ray structure of the *T. thermophilus* 70S ribosome in complex with three tRNAs and an mRNA fragment.** Here the 16S RNA is cyan, the 23S RNA is gray, the 5S RNA is light blue, the small subunit proteins are dark blue, the large subunit proteins are violet, and the tRNAs bound to the A, P, and E sites (which are largely occluded) are gold, orange, and red, respectively. (*a*) View similar to that on the lower right of Fig. 26-15*a* in which the small subunit is in front of the large subunit. (*b*) A view rotated 90° around the vertical axis relative to Part *a* which resembles that in Fig. 26-15*b*. Here the A-site tRNA is more clearly visible at the bottom of a funnel in which elongation factors bind (Section 26-4B). [Courtesy of Harry Noller, University of California at Santa Cruz. PDBids 1G1X and 1G1Y.] 🔁 **See the Interactive Exercises.**

a globular domain altogether. These protein tails make far fewer base-specific interactions than do other known RNA-binding proteins. They tend to interact with the RNA through salt bridges between their positively charged side chains and the RNAs' negatively charged phosphate oxygen atoms, thereby neutralizing the repulsive charge–charge interactions between nearby RNA segments. Moreover, the sequences of these proteins' tails are more conserved than their attached globular domains. This is consistent with the hypothesis that the primordial ribosome consisted entirely of RNA (the RNA world) and that the proteins that were eventually acquired stabilized its structure and fine-tuned its function.

Although the X-ray structure of the entire *T. thermophilus* ribosome has only been determined at 5.5 Å resolution, its RNA backbones can be confidently traced and its proteins of known structure can be properly positioned (Fig. 26-19). The structures of the associated 30S and 50S subunits closely resemble those of the isolated subunits. Nevertheless, there are several regions at the subunit interface that exhibit significant conformational shifts (between 3.5 and 10 Å), which suggests that these changes occur as a consequence of subunit association. The two subunits in the intact ribosome contact each other at 12 positions via RNA–RNA, protein–protein, and RNA–protein bridges (Fig. 26-20). These intersubunit bridges have a distinct distribution: The RNA–RNA bridges are centrally located adjacent to the three bound tRNAs, whereas the protein–protein and RNA–protein bridges are peripherally located away from the ribosome's functional sites.

Ribosomes, as we shall see in Section 26-4B, have three functionally distinct tRNA-binding sites: the **A** or **aminoacyl site** (it accommodates the incoming aminoacyl–tRNA), the **P** or **peptidyl site** (it accommodates the **peptidyl–tRNA,** the tRNA to which the growing peptide chain is attached), and the **E** or **exit site** (it accommodates a deacylated tRNA that is about to exit the ribosome). The ribosome binds all three tRNAs in a similar manner with their anticodon arms bound to the 30S subunit and their remaining portions bound to the 50S subunit. These interactions, which mainly consist of RNA–RNA contacts, are made to the tRNAs' universally conserved segments, thereby permitting the ribosome to bind different species of tRNAs in the same way. Nevertheless, the three bound tRNAs have slightly different conformations, with the A-site tRNA most closely resembling the X-ray structure of tRNA^{Phe} (Fig. 26-5*b*), and the E-site tRNA most distorted.

The growing polypeptide fits into a tunnel on the 50S subunit that extends from the P site to the outer ribosomal surface (the tunnel exit is approximately in the center of the 50S subunit as is seen in Fig. 26-15*b*). The ~100-Å-long tunnel is lined with mostly hydrophilic residues and has an average diameter of ~15 Å. This is barely large enough for an α helix, so significant protein folding probably cannot occur until the polypeptide exits the ribosome.

We discuss the path of the mRNAs and how it interacts with the tRNAs in Section 26-4. There we shall see that *the large subunit is mainly involved in mediating biochemical tasks such as catalyzing the reactions of polypeptide elongation, whereas the small subunit is the major actor in ribosomal recognition processes such as mRNA and tRNA binding* (although, as we have seen, the large subunit also participates in tRNA binding). We shall also see that *rRNA has the major functional role in ribosomal processes* (recall that RNA has demonstrated catalytic properties; Sections 23-2E and 25-3A).

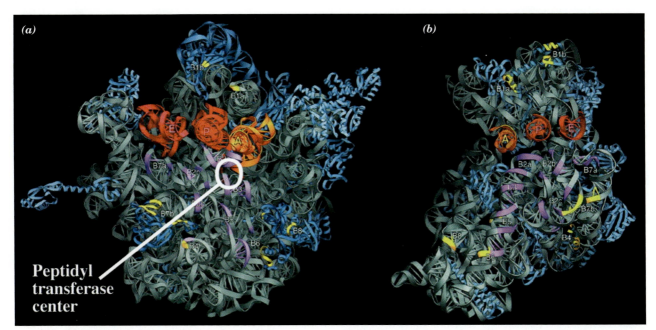

Figure 26-20 Ribosomal subunits in the X-ray structure of the *T. thermophilus* ribosome in complex with three tRNAs and an mRNA. (*a*) Interface view of the large subunit (similar to Fig. 26-17*b, left*). (*b*) Interface view of the small subunit (similar to Fig. 26-17*a, left*). The RNA is gray with segments that participate in intersubunit contacts magenta, and the proteins are blue with segments that participate in intersubunit contacts yellow. The tRNAs bound in the A, P, and E sites are gold, orange, and red, respectively. The active site where peptide bond formation occurs is circled. [Courtesy of Harry Noller, University of California at Santa Cruz. PDBids 1GIX and 1GIY.] **See the Interactive Exercises.**

B The Eukaryotic Ribosome

Although eukaryotic and prokaryotic ribosomes resemble each other in both structure and function, they differ in nearly all details. Eukaryotic ribosomes have particle masses in the range 3.9 to 4.5×10^6 D and have a nominal sedimentation coefficient of 80S. They dissociate into two unequal subunits with compositions that are distinctly different from those of prokaryotes (Table 26-5; compare with Table 26-4). The small (**40S**) subunit of the rat liver cytoplasmic ribosome, the best characterized eukaryotic ribosome, consists of 33 unique polypeptides and an **18S rRNA.** Its large (**60S**) subunit contains 49 different polypeptides and three rRNAs of 28S,

Table 26-5 Components of Rat Liver Cytoplasmic Ribosomes

	Ribosome	Small Subunit	Large Subunit
Sedimentation coefficient	80S	40S	60S
Mass (kD)	4220	1400	2820
RNA			
Major		18S, 1874 nucleotides	28S, 4718 nucleotides
Minor			5.8S, 160 nucleotides
			5S, 120 nucleotides
RNA mass (kD)	2520	700	1820
Proportion of mass	60%	50%	65%
Proteins		33 polypeptides	49 polypeptides
Protein mass (kD)	1700	700	1000
Proportion of mass	40%	50%	35%

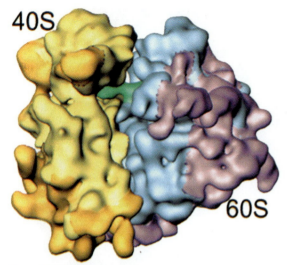

Figure 26-21 Cryo-EM–based image of the yeast 80S ribosome at 15 Å resolution. In this side view, the small (40S) subunit is yellow, the large (60S) subunit is cyan, and the tRNA that is bound in the ribosomal P site is green. Portions of this ribosome that are not homologous to the RNA or proteins of the *E. coli* ribosome are shown in gold for the small subunit and magenta for the large subunit. [Courtesy of Joachim Frank, State University of New York at Albany.]

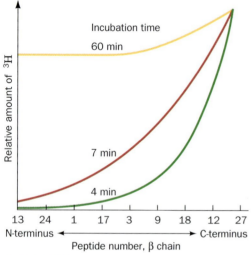

Figure 26-22 Demonstration that polypeptide synthesis proceeds from the N- to the C-terminus. Rabbit reticulocytes were incubated with [³H]leucine for the indicated times. The curves show the distribution of [³H]Leu among the tryptic peptides from the β subunit of soluble rabbit hemoglobin. The numbers on the horizontal axis are peptide identifiers, arranged from the N-terminus to the C-terminus. [After Dintzis, H.M., *Proc. Natl. Acad. Sci.* **47**, 255 (1961).]

5.8S, and 5S. The additional complexity of the eukaryotic ribosome relative to its prokaryotic counterpart is presumably due to the eukaryotic ribosome's additional functions: Its mechanism of translational initiation is more complex (Section 26-4A); it must be transported from the nucleus, where it is formed, to the cytoplasm, where translation occurs; and the machinery with which it participates in the secretory pathway is more complicated (Section 9-4D).

Sequence comparisons of the corresponding rRNAs from various species indicate that evolution has conserved their secondary structures rather than their base sequences (Fig. 26-14). For example, a $G \cdot C$ in a base-paired stem of *E. coli* 16S rRNA has been replaced by an $A \cdot U$ in the analogous stem of yeast 18S rRNA. The **5.8S rRNA,** which occurs in the large eukaryotic subunit in base-paired complex with the **28S rRNA,** is homologous in sequence to the 5′ end of prokaryotic 23S rRNA. Apparently 5.8S RNA arose through mutations that altered rRNA's posttranscriptional processing to produce a fourth rRNA.

The cryo-EM–based image of the yeast 80S ribosome (Fig. 26-21), determined at 15 Å resolution by Andrej Sali, Günter Blobel, and Frank, reveals that there is a high degree of structural conservation between eukaryotic and prokaryotic ribosomes (compare Fig. 26-21 with Fig. 26-15b). Although the yeast 40S subunit (which consists of a 1798-nt 18S rRNA and 32 proteins) contains an additional 256 nt of RNA and 11 proteins relative to the *E. coli* 30S subunit (Table 26-4; 15 of the yeast proteins are homologous to those of *E. coli*), both exhibit similar structures. Many of the differences between these two small ribosomal subunits are accounted for by the 40S subunit's additional RNA and proteins, although their homologous portions exhibit several distinct conformational differences. Similarly, the yeast 60S subunit (which consists of an aggregate of 3671 nt and 45 proteins) structurally resembles the considerably smaller prokaryotic 50S subunit. The yeast ribosome exhibits 16 intersubunit bridges, 12 of which match the 12 that were observed in the X-ray structure of the *T. thermophilus* ribosome (Fig. 26-20), a remarkable evolutionary conservation that indicates the importance of these bridges.

4 Translation

In order to appreciate the manner in which the ribosome orchestrates the translation of mRNA to synthesize polypeptides, it is helpful to assimilate the following points:

1. *Polypeptide synthesis proceeds from the N-terminus to the C-terminus;* that is, a **peptidyl transferase** activity appends an incoming amino acid to a growing polypeptide's C-terminus. This was shown to be the case in 1961 by Howard Dintzis, who exposed reticulocytes (immature red blood cells) that were actively synthesizing hemoglobin to ³H-labeled leucine for less time than it takes to synthesize an entire polypeptide. The extent to which the tryptic peptides from the soluble (completed) hemoglobin molecules were labeled increased with their proximity to the C-terminus (Fig. 26-22), thereby indicating that incoming amino acids are appended to the growing polypeptide's C-terminus.

2. *Chain elongation occurs by linking the growing polypeptide to the incoming tRNA's amino acid residue.* If the growing polypeptide is

released from the ribosome by treatment with high salt concentrations, its C-terminal residue is esterified to a tRNA molecule as a peptidyl–tRNA (*at right*). The nascent (growing) polypeptide must therefore grow by being transferred from the peptidyl–tRNA in the P site to the incoming aa–tRNA in the A site to form a peptidyl–tRNA with one more residue (Fig. 26-23). After the peptide bond has formed, the new peptidyl–tRNA, which now occupies the A site, is translocated to the P site so that a new aa–tRNA can enter the A site. The uncharged tRNA in the P site moves to the E site before it dissociates from the ribosome (we discuss the details of chain elongation in Section 26-4B).

3. *Ribosomes read mRNA in the 5′ → 3′ direction.* This was shown through the use of a cell-free protein-synthesizing system in which the mRNA was poly(A) with a 3′-terminal C:

$$5′ \ A—A—A—\cdots—A—A—A—C \ 3′$$

Such a system synthesizes a poly(Lys) that has a C-terminal Asn:

$$\overset{+}{H_3N}—Lys—Lys—Lys—\cdots—Lys—Lys—Asn—COO^-$$

Together with the knowledge that AAA and AAC code for Lys and Asn (Table 26-1) and the polarity of peptide synthesis, this establishes that the mRNA is read in the 5′ → 3′ direction. Because mRNA is also synthesized in the 5′ → 3′ direction, prokaryotic ribosomes can commence translation as soon as a nascent mRNA emerges from RNA polymerase. This, however, is not possible in eukaryotes because the nuclear membrane separates the site of transcription (the nucleus) from the site of translation (the cytosol).

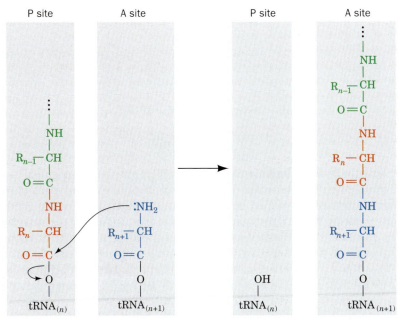

Figure 26-23 The ribosomal peptidyl transferase reaction forming a peptide bond. The amino group of the aminoacyl–tRNA in the A site nucleophilically displaces the tRNA of the peptidyl–tRNA ester in the P site, thereby forming a new peptide bond and transferring the nascent polypeptide to the A-site tRNA.

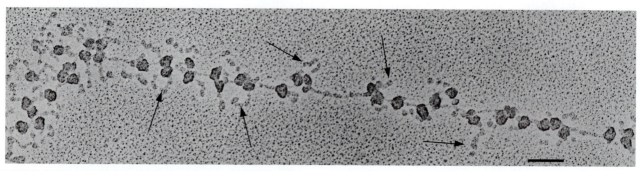

Figure 26-24   Electron micrograph of polysomes from silk gland cells of the silkworm *Bombyx mori.*   The 3′ end of the mRNA is on the left. Arrows point to the silk fibroin polypeptides. The bar represents 0.1 μm. [Courtesy of Oscar L. Miller, Jr., and Steven L. McKnight, University of Virginia.]

4. *Active translation occurs on polysomes.* In both prokaryotes and eukaryotes, multiple ribosomes can bind to a single mRNA transcript, giving rise to a beads-on-a-string structure called a **polyribosome (polysome;** Fig. 26-24). Individual ribosomes are separated by gaps of 50 to 150 Å so that they have a maximum density on the mRNA of ~1 ribosome per 80 nt. Polysomes arise because once an active ribosome has cleared its initiation site on mRNA, a second ribosome can initiate translation at that site.

See Guided Exploration 28

Translational initiation.

A   Chain Initiation

The first indication of how ribosomes initiate polypeptide synthesis was the observation that almost half of the *E. coli* proteins begin with the otherwise uncommon amino acid residue Met. In fact, the tRNA that initiates translation is a peculiar form of Met–tRNA^{Met} in which the Met residue is *N*-formylated:

$$\text{HC}\overset{\displaystyle O}{\underset{\displaystyle \|}{}}-\text{NH}-\text{CH}-\overset{\displaystyle \text{CH}_2}{\underset{\displaystyle |}{\overset{|}{\text{CH}_2}}}\cdots$$

N-Formylmethionine–tRNA$_f^{Met}$
(fMet–tRNA$_f^{Met}$)

Because the ***N*-formylmethionine** residue **(fMet)** already has an amide bond, it can only be the N-terminal residue of a polypeptide. *E. coli* proteins are posttranslationally modified by deformylation of their fMet residue and, in many proteins, by the subsequent removal of the resulting N-terminal Met. This processing usually occurs on the nascent polypeptide, which accounts for the observation that mature *E. coli* proteins all lack fMet.

The tRNA that recognizes the initiation codon, **tRNA$_f^{Met}$,** differs from the tRNA that carries internal Met residues, **tRNA$_m^{Met}$,** although they both recognize the same AUG codon. Presumably, the conformations of these tRNAs are different enough to permit them to be distinguished in the reactions of chain initiation and elongation.

In *E. coli*, uncharged tRNA$_f^{Met}$ is aminoacylated with Met by the same MetRS that charges tRNA$_m^{Met}$. The resulting Met–tRNA$_f^{Met}$ is specifically *N*-formylated to yield fMet–tRNA$_f^{Met}$ by a transformylase that employs N^{10}-formyltetrahydrofolate (Section 20-4D) as its formyl donor. This transformylase does not recognize Met–tRNA$_m^{Met}$.

Initiation
codon

araB	– U U U G G A U G G A G U G A A A C G A U G G C G A U U –
galE	– A G C C U A A U G G A G C G A A U U A U G A G A G U U –
lacI	– C A A U U C A G G G U G G U G A U U G U G A A A C C A –
lacZ	– U U C A C A C A G G A A A C A G C U A U G A C C A U G –
Qβ phage replicase	– U A A C U A A G G A U G A A A U G C A U G U C U A A G –
φX174 phage A protein	– A A U C U U G G A G G C U U U U U U A U G G U U C G U –
R17 phage coat protein	– U C A A C C G G G G U U U G A A G C A U G G C U U C U –
Ribosomal S12	– A A A A C C A G G A G C U A U U U A A U G G C A A C A –
Ribosomal L10	– C U A C C A G G A G C A A A G C U A A U G G C U U U A –
trpE	– C A A A A U U A G A G A A U A A C A A U G C A A A C A –
trp leader	– G U A A A A A G G G U A U C G A C A A U G A A A G C A –

3′ end of 16S rRNA   3′ $_{HO}$A U U C C U C C A C U A G – 5′

Figure 26-25   Some translation initiation sequences recognized by *E. coli* ribosomes.   The RNAs are aligned according to their initiation codons (*blue shading*). Their Shine–Dalgarno sequences (*red shading*) are complementary, counting G·U pairs, to a portion of the 16S rRNA's 3′ end (*below*). [After Steitz, J.A., *in* Chambliss, G., Craven, G. R., Davies, J., Davis, K., Kahan, L., and Nomura, M. (Eds.), *Ribosomes. Structure, Function and Genetics*, pp. 481–482, University Park Press (1979).]

Base Pairing between mRNA and the 16S rRNA Helps Select the Translation Initiation Site.

AUG codes for internal Met residues as well as the initiating Met residue of a polypeptide. Moreover, mRNAs usually contain many AUGs (and GUGs) in different reading frames. Clearly, *a translation initiation site must be specified by more than just an initiation codon.*

In *E. coli*, the 16S rRNA contains a pyrimidine-rich sequence at its 3′ end. This sequence, as John Shine and Lynn Dalgarno pointed out in 1974, is partially complementary to a purine-rich tract of 3 to 10 nucleotides, the **Shine–Dalgarno sequence,** that is centered ~10 nucleotides upstream from the start codon of nearly all known prokaryotic mRNAs (Fig. 26-25). *Base-pairing interactions between an mRNA's Shine–Dalgarno sequence and the 16S rRNA apparently permit the ribosome to select the proper initiation codon.*

The X-ray structure of the 70S ribosome reveals, in agreement with Fig. 26-15*b*, that an ~30-nt segment of the mRNA is wrapped in a groove that encircles the neck of the 30S subunit (Fig. 26-26). The mRNA codons in the A and P sites are exposed on the interface side of the 30S subunit, whereas its 5′ and 3′ ends are bound in tunnels composed of RNA and protein. The mRNA's Shine–Dalgarno sequence, which is located near its 5′ end, is base paired, as expected, with the 16S rRNA's anti-Shine–Dalgarno sequence, which is situated close to the E site. The proteins that in part form the tunnel through which the mRNA enters the ribosome (green in Fig. 26-26) probably function as a helicase to remove secondary structures from the mRNA that would otherwise interfere with tRNA binding.

Initiation Requires Soluble Protein Factors.

Translation initiation in *E. coli* is a complex process in which the two ribosomal subunits and fMet–tRNA$_f^{Met}$ assemble on a properly aligned mRNA to form a complex that can commence chain elongation. This process also requires **initiation factors** that are not permanently associated with the ribosome, designated **IF-1, IF-2,** and **IF-3** in *E. coli* (Table 26-6).

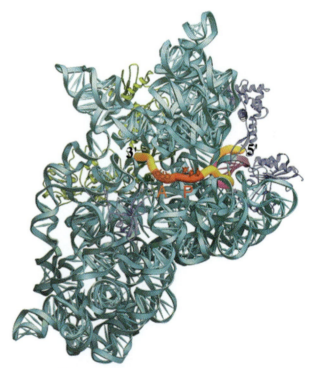

Figure 26-26   Path of mRNA through the ribosomal 30S subunit as viewed from its interface side.   The 16S rRNA is cyan, the mRNA is represented in worm form with its A- and P-site codons orange and red, the Shine–Dalgarno helix (which includes a segment of 16S rRNA) magenta, and its remaining segments yellow. The ribosomal proteins that in part form the entry and exit tunnels for the mRNA are drawn in green and purple. The remaining ribosomal proteins have been omitted for clarity. [Courtesy of Gloria Culver, Iowa State University. Based on an X-ray structure by Harry Noller, University of California at Santa Cruz. PDBid 1JGO.]

Table 26-6 The Soluble Protein Factors of *E. coli* Protein Synthesis

Factor	Number of Residues[a]	Function
Initiation Factors		
IF-1	71	Assists IF-3 binding
IF-2	890	Binds initiator tRNA and GTP
IF-3	180	Releases mRNA and tRNA from recycled 30S subunit and aids new mRNA binding
Elongation Factors		
EF-Tu	393	Binds aminoacyl–tRNA and GTP
EF-Ts	282	Displaces GDP from EF-Tu
EF-G	703	Promotes translocation through GTP binding and hydrolysis
Release Factors		
RF-1	360	Recognizes UAA and UAG Stop codons
RF-2	365	Recognizes UAA and UGA Stop codons
RF-3	528	Stimulates RF-1/RF-2 release via GTP hydrolysis
RRF	185	Together with EF-G, induces ribosomal dissociation to small and large subunits

[a]All *E. coli* translational factors are monomeric proteins.

Translation initiation in *E. coli* occurs in three stages (Fig. 26-27):

1. On completing a cycle of polypeptide synthesis, the 30S and 50S subunits remain associated as an inactive 70S ribosome. IF-3 binds to the 30S subunit so as to promote the dissociation of this complex.

2. mRNA and IF-2 in a ternary complex with GTP and fMet–tRNA$_f^{Met}$, along with IF-1, subsequently bind to the 30S subunit in either order. Since the ternary complex containing fMet–tRNA$_f^{Met}$ can bind to the ribosome before mRNA, fMet–tRNA$_f^{Met}$ binding must not be mediated by a codon–anticodon interaction; it is the only tRNA–ribosome association that does not require one, although this interaction helps bind fMet–tRNA$_f^{Met}$ to the ribosome. IF-1 binds in the A site, where it may prevent the inappropriate binding of a tRNA. IF-3 also functions at this stage of the initiation process by preventing the binding of tRNAs other than tRNA$_f^{Met}$.

3. Last, in a process that is preceded by IF-1 and IF-3 release, the 50S subunit joins the 30S initiation complex in a manner that stimulates IF-2 to hydrolyze its bound GTP to GDP + P$_i$. This irreversible reaction conformationally rearranges the 30S subunit and releases IF-2 for participation in further initiation reactions.

Initiation results in the formation of an fMet–tRNA$_f^{Met}$ · mRNA · ribosome complex in which the fMet–tRNA$_f^{Met}$ occupies the ribosome's P site while its A site is poised to accept an incoming aa–tRNA (an arrangement analogous to that at the conclusion of a round of elongation; Section 26-4B). Note that tRNA$_f^{Met}$ is the only tRNA that directly enters the P site. All other tRNAs must first enter the A site during chain elongation.

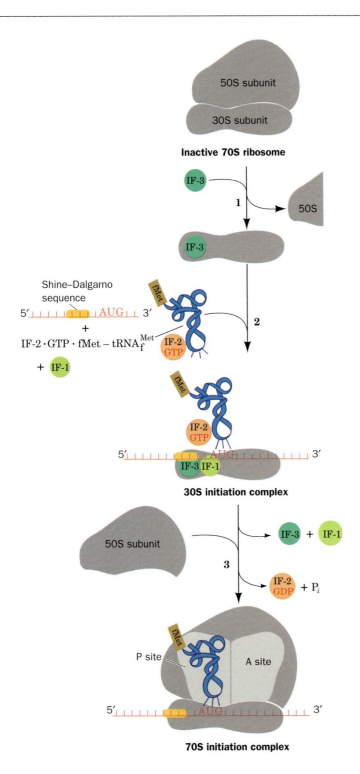

Figure 26-27   Translation initiation pathway in *E. coli*.

Eukaryotic Initiation Is Far More Complicated than That of Prokaryotes.

Ribosomal initiation in eukaryotes requires the assistance of at least 11 initiation factors (designated eIF*n;* "e" for *e*ukaryotic) that consist of 26 polypeptide chains. Nevertheless, Steps 1 and 3 of the prokaryotic process (Fig. 26-27) are superficially similar in eukaryotes. However, the way in which the mRNA's initiating codon is identified in eukaryotes is fundamentally different from that in prokaryotes.

The pairing of the initiator tRNA with the initiating AUG in the ribosomal P site begins with **eIF2,** a heterotrimer in complex with GTP, escorting the initiator tRNA to the complex of the 40S subunit with several other initiation factors, thereby forming the **43S preinitiation complex** (a process that resembles Step 2 in prokaryotic initiation but without the mRNA). The initiator tRNA is **tRNA$_i^{Met}$** ("i" for *i*nitiator). Its appended Met residue is not formylated as in prokaryotes. Nevertheless, both species of initiator tRNAs are readily interchangeable *in vitro*.

Eukaryotic mRNAs lack the complementary sequences to bind to the 18S rRNA in the Shine–Dalgarno manner. Rather, they have an entirely different mechanism for recognizing the mRNA's initiating AUG codon. *Eukaryotic mRNAs, nearly all of which have a 7-methylguanosine (m^{7}G) cap and a poly(A) tail (Section 25-3A), are invariably monocistronic and almost always initiate translation at their leading AUG.* This AUG, which occurs at the end of a 5′-untranslated region of 50 to 70 nt, is embedded in the consensus sequence GCCRCC**AUG**G. Changes in the purine (R) 3 nt before the AUG and in the G immediately following it each reduce translational efficiency by ~10-fold. The recognition of the initiation site begins by the binding of **eIF4F** to the mRNA's m^{7}G cap. eIF4F is a heterotrimeric complex of **eIF4E, eIF4G,** and **eIF4A** (all monomers), in which eIF4E (also known as **cap-binding protein**) recognizes the mRNA's m^{7}G cap and eIF4G serves as a scaffold to join eIF4E with eIF4A.

The structure of eIF4E in complex with **m^{7}GDP** reveals that the protein binds the m^{7}G base by intercalating it between two highly conserved Trp residues in a region that is adjacent to a positively charged cleft that presumably forms the mRNA-binding site (Fig. 26-28). The m^{7}G base is specifically recognized by hydrogen bonding to protein side chains in a manner reminiscent of G · C base pairing. eIF4G also binds poly(A)-binding protein (PABP; Section 25-3A), which coats the mRNA's poly(A) tail, thereby circularizing the mRNA. Although this explains the synergism between an mRNA's m^{7}G cap and its poly(A) tail in stimulating translational initiation, the function of the circle is unclear. However, an attractive hypothesis is that it enables a ribosome that has finished translating the mRNA to reinitiate translation without having to disassemble and then reassemble. Another possibility is that it prevents the translation of incomplete (broken) mRNAs.

The eIF4F–mRNA complex is subsequently joined by several additional initiation factors. The resulting complex joins the 43S preinitiation complex, which then scans down the mRNA in an ATP-driven process until it encounters the mRNA's initiating AUG codon, thereby forming the **48S initiation complex.** The recognition of the AUG occurs mainly through base pairing with the CUA anticodon on the bound Met–tRNA$_i^{Met}$, as was demonstrated by the observation that mutating this anticodon results in the recognition of the new cognate codon instead of AUG.

In the analog of Step 3 of prokaryotic initiation (Fig. 26-27), the eIF2-catalyzed hydrolysis of its bound GTP induces the release of all initiation factors from the 48S initiation complex. The resulting 40S subunit–Met-tRNA$_i^{Met}$ complex is joined by the 60S subunit in a GTP-dependent reaction mediated by **eIF5B** (a monomer and homolog of prokaryotic IF-2), thereby forming the 80S ribosomal initiation complex.

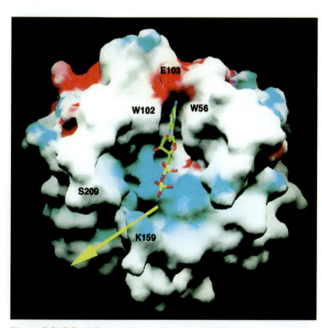

Figure 26-28 **X-Ray structure of murine eIF4E in complex with the m^{7}G cap analog m^{7}GDP.** The protein is shown as its solvent-accessible surface colored according to its electrostatic potential (red negative, blue positive, and white neutral). The m^{7}GDP is drawn in stick form with C green, N blue, O red, and P yellow. The two Tyr residues that bind the m^{7}G base are indicated, as are a Lys and Ser residue that flank the putative mRNA-binding cleft (*yellow arrow*). [Courtesy of Nahum Sonenberg, McGill University, Montreal, Quebec, Canada. PDBid 1EJ1.]

B Chain Elongation

Ribosomes elongate polypeptide chains in a three-stage reaction cycle (Fig. 26-29):

See Guided Exploration 29
Translational elongation.

1. **Decoding,** in which the ribosome selects and binds an aminoacyl–tRNA whose anticodon is complementary to the mRNA codon in the A site.
2. **Transpeptidation,** or peptide bond formation, in which the peptidyl group in the P-site tRNA is transferred to the aminoacyl group in the A site.
3. **Translocation,** in which the A-site and P-site tRNAs are respectively transferred to the P and E sites, accompanied by their bound mRNA; that is, the mRNA, together with its base-paired tRNAs, is ratcheted through the ribosome by one codon.

This process, which occurs at a rate of 10 to 20 amino acid residues per second, requires several nonribosomal proteins known as **elongation factors** (Table 26-6).

Decoding: An Aminoacyl–tRNA Binds to the Ribosomal A Site.   In the first stage of the *E. coli* elongation cycle, a binary complex of GTP and the elonga-

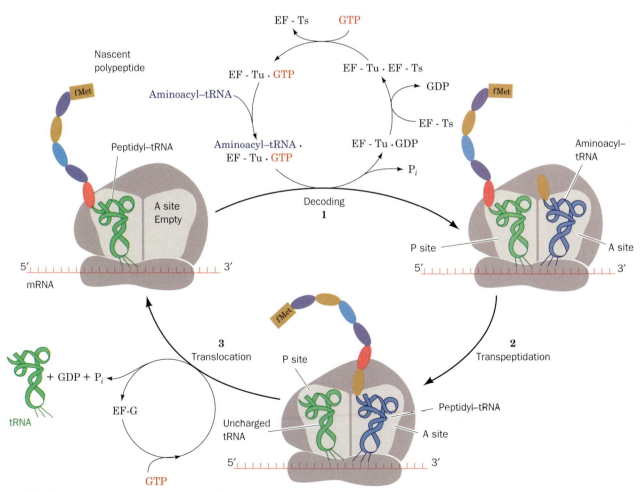

Figure 26-29 *Key to Function.* **Elongation cycle in *E. coli* ribosomes.** The E site is not shown. Eukaryotic elongation follows a similar cycle, but EF-Tu and EF-Ts are replaced by a single multisubunit protein, eEF1, and EF-G is replaced by eEF2.

tion factor **EF-Tu** combines with an aa–tRNA. The resulting ternary complex binds to the ribosome. Binding of the aa–tRNA in a codon–anticodon complex to the ribosomal A site is accompanied by the hydrolysis of GTP to GDP so that EF-Tu · GDP and P$_i$ are released. The EF-Tu · GTP complex is regenerated when GDP is displaced from EF-Tu · GDP by the elongation factor **EF-Ts,** which in turn is displaced by GTP.

Aminoacyl–tRNAs can bind to the ribosomal A site without EF-Tu but at a rate too slow to support cell growth. The importance of EF-Tu is indicated by the fact that it is the most abundant *E. coli* protein; it is present in ~100,000 copies per cell (>5% of the cell's protein), which is approximately the number of tRNA molecules in the cell. Consequently, *the cell's entire complement of aa–tRNAs is essentially sequestered by EF-Tu.*

The X-ray structure of EF-Tu in complex with Phe–tRNA^{Phe} and the nonhydrolyzable GTP analog **guanosine-5′-(β,γ-imido)triphosphate (GMPPNP; alternatively GDPNP),**

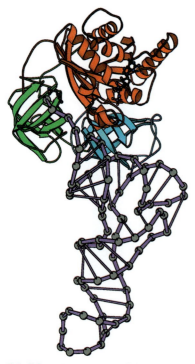

Guanosine-5′-(β,γ-imido)triphosphate (GMPPNP)

was determined by Brian Clark and Jens Nyborg (Fig. 26-30). This complex has a corkscrew-shaped structure in which the EF-Tu and the tRNA's acceptor stem form a knoblike handle and the tRNA's anticodon helix forms the screw. The macromolecules appear to associate rather tenuously via three major regions: (1) The 3′-CCA–Phe segment of the Phe–tRNA^{Phe} binds in the cleft between domains 1 and 2 of EF-Tu · GMPPNP (red and green in Fig. 26-30); (2) the 5′-phosphate of the tRNA binds in a depression at the junction of EF-Tu's three domains; and (3) one side of the TψC stem of the tRNA makes contacts with exposed main chain and side chains of EF-Tu domain 3 (cyan in Fig. 26-30).

EF-Tu binds all aminoacylated tRNAs, but not uncharged tRNAs or initiator tRNAs. Evidently, the tight association of the aminoacyl group with EF-Tu greatly increases the affinity of EF-Tu for the otherwise loosely bound tRNA. tRNA$_f^{Met}$ does not bind to EF-Tu because it has a 3′ overhang of 5 nt versus 4 nt in an elongator tRNA. This, together with the formyl group attached to the fMet residue, apparently prevents fMet–tRNA$_f^{Met}$ from binding to EF-Tu, thereby explaining why the initiator tRNA never translates internal AUG codons.

EF-Tu is a G protein that undergoes a large conformational change on hydrolyzing GTP. Its N-terminal domain 1 (red in Fig. 26-30) resembles other G proteins (Section 21-3B), with Switch I and Switch II regions signaling the state of the bound nucleotide (GTP or GDP). GTP hydrolysis also causes domain 1 to reorient with respect to domains 2 and 3 by a dramatic 91° rotation (Fig. 26-31), a conformational change that eliminates the tRNA binding site. Since EF-Tu is a G protein, the ribosome can be classified as its GTPase-activating protein (GAP) and EF-Ts functions as its guanine nucleotide exchange factor (GEF).

The Ribosome Monitors Correct Codon–Anticodon Pairing.   When EF-Tu · GTP delivers an aa–tRNA to the ribosome, it first inserts the anticodon end of the tRNA into the ribosome. The tRNA's aminoacyl end moves fully into

Figure 26-30 X-Ray structure of the ternary complex of yeast Phe–tRNA^{Phe}, *Thermus aquaticus* EF-Tu, and GMPPNP.   EF-Tu domains 1, 2, and 3 (N- to C-terminal) are red, green, and cyan, and the tRNA is shown in ladder form in purple. The GMPPNP is drawn in ball-and-stick form (*black*). [Courtesy of Jens Nyborg, University of Aarhus, Århus, Denmark. PDBid 1TTT.]

residue. Yet the binding energy loss arising from a single base mismatch in a codon–anticodon interaction is estimated to be $\sim$12 kJ $\cdot$ mol^{-1}, which, according to Eq. 26-1, cannot account for a ribosomal decoding accuracy of less than $\sim$10^{-2} errors per codon. Evidently, the ribosome has some sort of proofreading mechanism that increases its overall decoding accuracy.

A proofreading step must be entirely independent of the initial selection step. Only then can the overall probability of error be equal to the product of the probabilities of error of the individual selection steps. We have seen that DNA polymerases and aminoacyl–tRNA synthetases maintain the independence of their two selection steps by carrying them out at separate active sites (Sections 24-2A and 26-2B). Yet the ribosome recognizes the incoming aa–tRNA only according to its anticodon's complementarity to the codon in the A site. Consequently, the ribosome must somehow examine this codon–anticodon interaction in two separate ways. How does this occur?

Cryo-EM studies by Frank indicate that in the ribosome $\cdot$ aa–tRNA $\cdot$ EF-Tu $\cdot$ GTP complex, the aa–tRNA assumes an otherwise energetically unfavorable conformation in which the anticodon arm has moved via a hinge motion toward the inside of the tRNA's L. The formation of a correct codon–anticodon complex triggers EF-Tu to hydrolyze its bound GTP and thereby dissociate from the aa–tRNA, which then assumes its energetically more favorable conformation while maintaining its codon–anticodon interaction. This movement swings the acceptor stem with its attached aminoacyl group into the peptidyl transferase center on the 50S subunit (see below). A noncognate (mispaired) aa–tRNA would presumably have insufficient strength in its codon–anticodon interaction to hold it in place during this conformational change and would therefore dissociate from the ribosome. EF-Tu's irreversible GTPase reaction must precede this proofreading step because otherwise the dissociation of a noncognate aa–tRNA (the release of its anticodon from the codon) would simply be the reverse of the initial binding step; that is, it would be part of the initial selection step rather than proofreading. *GTP hydrolysis therefore provides the second context necessary for proofreading; it is the entropic price the system must pay for accurate tRNA selection.*

Transpeptidation: The Ribosome Is a Ribozyme. In the second (transpeptidation) stage of the elongation cycle (Fig. 26-29), the peptide bond is formed through the nucleophilic displacement of the P site tRNA by the amino group of the 3'-linked aa–tRNA in the A site (Fig. 26-23). The nascent polypeptide chain is thereby lengthened at its C-terminus by one residue and transferred to the A-site tRNA. The reaction occurs without the need of activating cofactors such as ATP because the ester linkage between the nascent polypeptide and the P-site tRNA is a "high-energy" bond.

*The **peptidyl transferase** center that catalyzes peptide bond formation is located entirely on the large subunit.* This was demonstrated by the observation that in high concentrations of organic solvents such as ethanol, the large subunit alone catalyzes peptide bond formation. The organic solvent apparently distorts the large subunit in a way that mimics the effect of small subunit binding.

What is the nature of the peptidyl transferase center; that is, does it consist of RNA, protein, or both? Since all proteins, including those associated with ribosomes, are ribosomally synthesized, the primordial ribosome must have preceded the primordial proteins and hence consisted entirely of RNA. Despite this (in hindsight) obvious evolutionary argument, the idea that rRNA functions catalytically was not seriously entertained until after it had been discovered that RNA can, in fact, act as a catalyst (Section

25-3B). Several other observations further indicate that the ribosome is a ribozyme:

1. The absence of any one of the 50S subunit's 31 proteins but **L2, L3,** and **L4** does not abolish its peptidyl transferase function.

2. rRNAs are more highly conserved throughout evolution than are ribosomal proteins.

3. Most mutations that confer resistance to antibiotics that inhibit protein synthesis occur in genes encoding rRNAs rather than ribosomal proteins.

Nevertheless, the unambiguous demonstration that rRNA functions catalytically in polypeptide synthesis proved to be surprisingly elusive until the X-ray structure of the large ribosomal subunit was determined.

The peptidyl transferase center was unequivocally identified in the X-ray structure of the 50S subunit that contained a bound analog of the reaction's tetrahedral intermediate (see below). The ribosome's active site lies in a highly conserved region in domain V of the 23S rRNA (Figs. 26-14b and 26-34) and, most significantly, the nearest protein side chain is ~18 Å away from the newly formed peptide bond—too far to be involved in catalysis. Moreover, all the nucleotides that contact the analog are >95% conserved among all three kingdoms of life. Clearly, *the ribosomal transpeptidase reaction is catalyzed solely by RNA.*

The ribosomal transpeptidase reaction has a 2×10^7-fold greater rate than that of the uncatalyzed reaction. How does the ribosome catalyze this reaction? Ribosomally mediated peptide bond formation proceeds via the nucleophilic attack of the amino group on the carbonyl group of an ester to form a tetrahedral intermediate that collapses to an amide and an alcohol (Fig. 26-23). However, in the physiological pH range, the attacking amino group is predominantly in its ammonium form (RNH_3^+), which lacks the lone pair necessary to undertake a nucleophilic attack. This suggests that the transpeptidase reaction is catalyzed in part by a general base that abstracts a proton from the ammonium group to generate the required free amino group (RNH_2).

Inspection of the peptidyl transferase active site reveals that the only basic group within 5 Å of the inferred position of the attacking amino group is atom N3 of the invariant rRNA base A2486 (A2451 in *E. coli*). It is ~3 Å from and hence hydrogen bonded to the attacking amino group (Fig. 26-35). This further suggests that the protonated A2486-N3 electrostatically stabilizes the oxyanion of the tetrahedral reaction intermediate and then donates the proton to the leaving group of the P-site tRNA to yield a 3'-OH group (general acid catalysis). However, in order for A2486-N3 to act as a general base in abstracting the proton from an ammonium group (whose pK is ~10), it must have a pK of at least 7 (recall that proton transfers between hydrogen-bonded groups occur at physiologically significant rates only when the pK of the proton donor is no more than 2 or 3 pH units greater than that of the proton acceptor; Section 11-5C). Yet, the pK of N3 in AMP is <3.5. Moreover, the model displayed in Fig. 26-35 indicates that the tetrahedral intermediate's oxyanion would point away from and hence could not be stabilized by protonated A2486-N3.

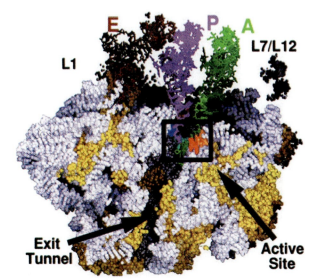

Figure 26-34 X-Ray structure of the large ribosomal subunit with three tRNAs superimposed on the positions they occupy in the A, P, and E sites of the 70S ribosome. The subunit is shown in cutaway view so as to expose the lumen of the peptide exit tunnel and the peptidyl transferase site (*boxed*). The complex is drawn in space-filling form with its rRNA white, its proteins yellow, the tRNAs in the A, P, and E sites green, maroon, and brown and some of the nucleotides that interact with the A-site tRNA red and orange. [Courtesy of Peter Moore and Thomas Steitz, Yale University.]

The Ribosome Is an Entropy Trap. The resolution to this conundrum was provided by Marina Rodnina and Richard Wolfenden who noted that the uncatalyzed reaction of esters with amines to form amides occurs quite facilely in aqueous solution. They therefore measured the rates of both uncatalyzed

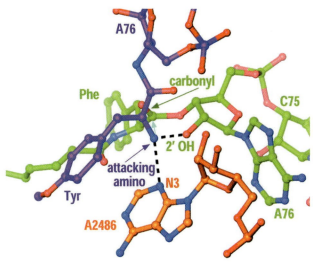

Figure 26-35 Model of the substrate complex of the 50S ribosomal subunit.   Atoms are colored according to type with the A-site substrate (Tyr-tRNA^{Tyr}) C purple, P-site substrate C green, 23S rRNA C orange, N blue, and O red. The attacking amino group of the A-site aminoacyl residue is held in position for nucleophilic attack (*cyan arrow*) on the carbonyl C of the P-site aminoacyl ester through hydrogen bonds (*dashed black lines*) to A2486-N3 and the 2′-O of the P-site A76. [Courtesy of Peter Moore and Thomas Steitz, Yale University.]

peptide bond formation by model compounds and peptidyl transfer by the ribosome at several different temperatures. This provided values of $\Delta\Delta H_{cat}^{\ddagger}$ and $\Delta\Delta S_{cat}^{\ddagger}$, the reaction's change in the enthalpy and entropy of activation by the ribosome relative to the uncatalyzed reaction. Here $\Delta\Delta H_{cat}^{\ddagger} - T\Delta\Delta S_{cat}^{\ddagger} = \Delta\Delta G_{cat}^{\ddagger} = \Delta G^{\ddagger}(\text{uncat}) - \Delta G^{\ddagger}(\text{cat})$, where $\Delta\Delta G_{cat}^{\ddagger}$ is the change in the reaction's free energy of activation by the ribosome relative to the uncatalyzed reaction, and $\Delta G^{\ddagger}(\text{uncat})$ and $\Delta G^{\ddagger}(\text{cat})$ are the free energies of activation of the uncatalyzed and catalyzed (ribosomal) reactions (Sections 1-4D and 11-2). The measured value of $\Delta\Delta H_{cat}^{\ddagger}$ is -19 kJ·mol^{-1}, a quantity that would be positive, not negative, if the ribosomal reaction had a significant component of chemical catalysis such as acid–base catalysis and/or the formation of new hydrogen bonds. In contrast, the value of $T\Delta\Delta S_{cat}^{\ddagger}$ is $+52$ kJ·mol^{-1}, which indicates that the Michaelis complex (E·S; Section 12-1B) in the ribosomal reaction is significantly more ordered relative to the transition state than is the uncatalyzed reaction. This value of $T\Delta\Delta S_{cat}^{\ddagger}$ largely accounts for the observed 2×10^{7}-fold rate enhancement of the ribosomal reaction relative to the uncatalyzed reaction (the rate enhancement by the ribosome is given by $e^{\Delta\Delta G_{cat}^{\ddagger}/RT}$; Section 11-2). Evidently, *the ribosome enhances the rate of peptide bond formation by properly positioning and orienting its substrates and/or excluding water from the active site (a form of reactant ordering) rather than by chemical catalysis.*

Translocation: The Ribosome Moves to the Next Codon.   Following transpeptidation, the now uncharged P-site tRNA moves to the E site (not shown in Fig. 26-29), and the new peptidyl–tRNA in the A site, together with its bound mRNA, moves to the P site. These events, which leave the A site vacant, prepare the ribosome for the next elongation cycle. During translocation, peptidyl–tRNA's interaction with the codon, which is no longer necessary for amino acid specification, acts as a placekeeper so that the ribosome can advance by exactly three nucleotides along the mRNA, as required to preserve the reading frame. Indeed, certain mutant tRNA molecules that cause frameshifting induce the ribosome to translocate by four nucleotides, thereby demonstrating that mRNA movement is directly coupled to tRNA movement.

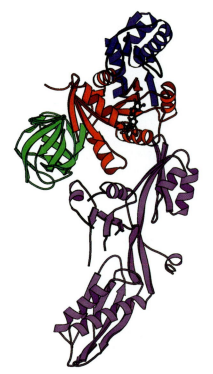

Figure 26-36 X-Ray structure of EF-G in complex with GDP. Domain 1 is red with its α-helical insert dark purple, domain 2 is green, and domains 3, 4, and 5 are magenta. The GDP is drawn in ball-and-stick form (*black*). Portions of the structure are not visible. Note the remarkable resemblance in shape between this structure and that of Phe–tRNA^{Phe} · EF-Tu · GMPPNP (Fig. 26-30). [Courtesy of Jens Nyborg, University of Aarhus, Århus, Denmark. Based on an X-ray structure by Thomas Steitz and Peter Moore, Yale University. PDBid 2EFG.]

The translocation process requires an elongation factor, **EF-G** in *E. coli*, that binds to the ribosome together with GTP and is released only on hydrolysis of the GTP to GDP + P$_i$ (Fig. 26-29). EF-G release is prerequisite for beginning the next elongation cycle because the ribosomal binding sites of EF-G and EF-Tu partially or completely overlap and hence their ribosomal binding is mutually exclusive. GTP hydrolysis, which precedes translocation, provides free energy for tRNA movement.

EF-G Structurally Mimics the EF-Tu · tRNA Complex. The X-ray structure of EF-G (Fig. 26-36), determined by Steitz and Moore, reveals that this protein has an elongated shape comprising five domains, the first two of which are arranged similarly to the first two domains of the EF-Tu · GMPPNP complex (Fig. 26-30). Most intriguingly, the remaining three domains of EF-G have a conformation reminiscent of the shape of tRNA bound to EF-Tu. Such molecular mimicry raises the possibility that EF-G drives translocation not just by inducing a conformational change in the ribosome but by actively displacing the peptidyl–tRNA from the A site. The structural similarity between EF-G domains 3 to 5 and tRNA also suggests that in the earliest cells, whose functions were based on RNA, proteins evolved by mimicking shapes already used successfully by RNA.

Translocation Occurs Via Intermediate States. Variations in chemical footprinting patterns during the elongation cycle, together with X-ray and cryo-EM studies, indicate that the translocation of tRNA through the ribosome occurs in several discete steps (Fig. 26-37):

1. In the **posttranslocational state,** a deacylated tRNA occupies the E site of both the 30S and 50S subunits (the E/E state), a peptidyl–tRNA occupies both P sites (the P/P state), and the A site is vacant. An aa–tRNA in complex with EF-Tu · GTP binds to the ribosome accompanied by the release of the E-site tRNA (see below). This yields a complex in which the incoming aa–tRNA is bound in the 30S subunit's A site via a codon–anticodon interaction (recall that the mRNA is bound to the 30S subunit) but with the EF-Tu · GTP preventing the entry of the tRNA's aminoacyl end into the 50S subunit's A site, an arrangement termed the A/T state (T for EF-*T*u).

2. As we discussed above, EF-Tu hydrolyzes its bound GTP to GDP + P$_i$ and is released from the ribosome, thereby permitting the aa–tRNA to fully bind to the A site (the A/A state).

3. The transpeptidation reaction occurs, yielding the **pretranslocational state.**

4. The acceptor end of the new peptidyl–tRNA shifts to the P site of the 50S subunit, while the tRNA's anticodon end remains associated with the A site of the 30S subunit (yielding the A/P state). The acceptor end of the newly deacylated tRNA simultaneously moves to the E site of the 50S subunit while its anticodon end remains associated with the P site of the 30S subunit (the P/E state).

5. EF-G · GTP binding to the ribosome and the resulting GTP hydrolysis impel the anticodon ends of the two tRNAs, together with their bound mRNA, to move relative to the small ribosomal subunit such that the peptidyl–tRNA assumes the P/P state and the deacylated tRNA assumes the E/E state (the posttranslocational state), thereby completing the elongation cycle.

The binding of tRNAs to the A and E sites of the ribosome, as Knud Nierhaus has shown, exhibits negative allosteric cooperativity. In the pre-

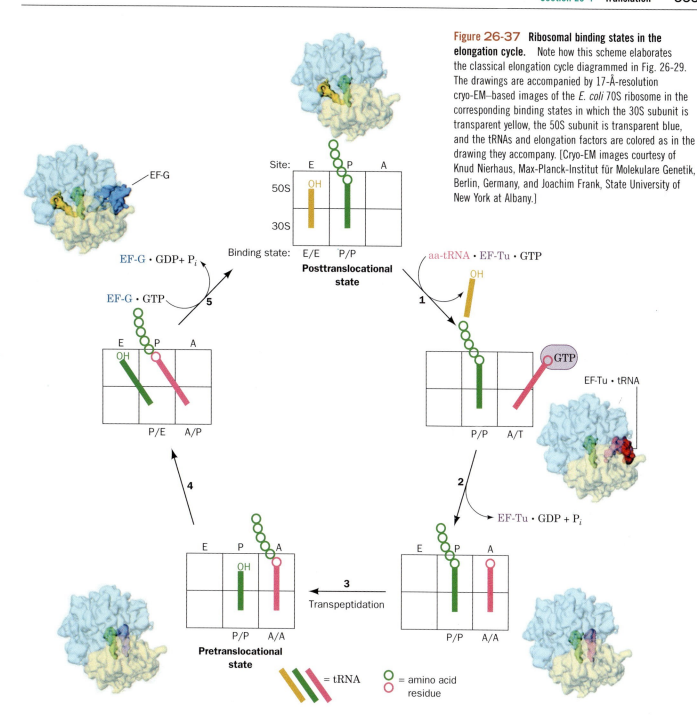

Figure 26-37 Ribosomal binding states in the elongation cycle. Note how this scheme elaborates the classical elongation cycle diagrammed in Fig. 26-29. The drawings are accompanied by 17-Å-resolution cryo-EM–based images of the *E. coli* 70S ribosome in the corresponding binding states in which the 30S subunit is transparent yellow, the 50S subunit is transparent blue, and the tRNAs and elongation factors are colored as in the drawing they accompany. [Cryo-EM images courtesy of Knud Nierhaus, Max-Planck-Institut für Molekulare Genetik, Berlin, Germany, and Joachim Frank, State University of New York at Albany.]

translocational state, the E site binds the newly deacylated tRNA with high affinity, whereas the now empty A site has low affinity for aa–tRNA. The binding of a new aa–tRNA · EF-Tu · GTP complex induces the ribosome to undergo a conformational change that converts the A site to a high-affinity state and the E site to a low-affinity state that consequently releases the deacylated tRNA. Thus, the E site is not simply a passive holding site for spent tRNAs but performs an essential function in the translation process.

The requirement that the GTP-hydrolyzing proteins EF-Tu and EF-G alternate in their activities ensures that the ribosome cycles unidirectionally through the transpeptidation and translocation stages of translation. Translocation can occur in the absence of GTP, which indicates that the free energy

BOX 26-3

Biochemistry in Health and Disease

The Effects of Antibiotics on Protein Synthesis

The majority of known antibiotics, including a great variety of medically useful substances, block translation. This situation is presumably a consequence of the translation machinery's enormous complexity, which makes it vulnerable to disruption in many ways. Antibiotics have also been useful in analyzing ribosomal mechanisms because the blockade of a specific function often permits its biochemical dissection into its component steps. For example, the ribosomal elongation cycle was originally characterized through the use of the antibiotic **puromycin,** which resembles the 3′ end of Tyr–tRNA:

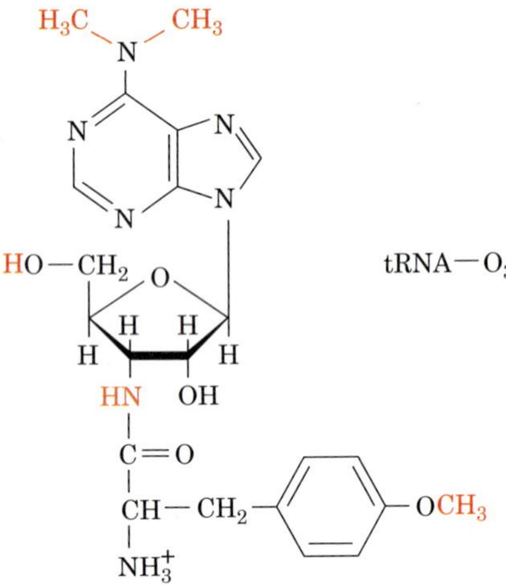

Puromycin

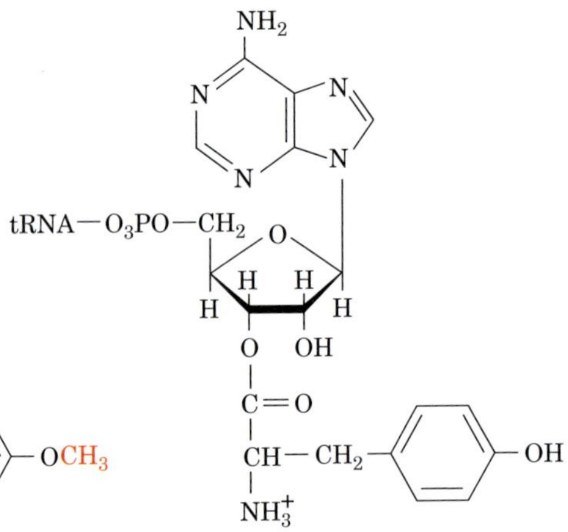

Tyrosyl–tRNA

Puromycin binds to the ribosomal A site without the need of elongation factors. The transpeptidation reaction yields a peptidyl–puromycin in which puromycin's "amino acid residue" is linked to its "tRNA" via an amide rather than an ester bond. The ribosome therefore cannot catalyze further transpeptidation, and polypeptide synthesis is aborted.

Streptomycin is a medically important member of a family

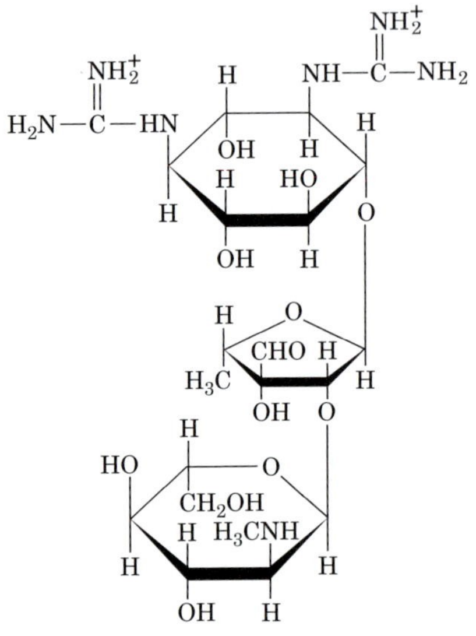

Streptomycin

of antibiotics known as **aminoglycosides** that inhibit prokaryotic ribosomes in a variety of ways. At low concentrations, streptomycin induces the ribosome to characteristically misread mRNA: One pyrimidine may be mistaken for the other in the first and second codon

of the transpeptidation reaction is sufficient to drive the entire translational process. However, the GTP hydrolysis catalyzed by EF-Tu and EF-G increases the overall rate of translation. *Because GTP hydrolysis is irreversible, the accompanying conformational changes in the ribosome are also irreversible and hence unidirectional.*

Several of the many antibiotics that disrupt ribosomal functioning are discussed in Box 26-3.

The Eukaryotic Elongation Cycle Resembles That of Prokaryotes. Elongation in eukaryotes closely resembles that in prokaryotes. In eukaryotes, the functions of EF-Tu and EF-Ts are assumed by the eukaryotic elongation factors **eEF1A** and **eEF1B.** Likewise, **eEF2** functions in a manner analogous to prokaryotic EF-G. However, the corresponding eukaryotic and prokaryotic elongation factors are not interchangeable.

positions, and either pyrimidine may be mistaken for adenine in the first position. This inhibits the growth of susceptible cells but does not kill them. At higher concentrations, however, streptomycin prevents proper chain initiation and thereby causes cell death.

Chloramphenicol,

Chloramphenicol

the first of the "broad-spectrum" antibiotics, inhibits the peptidyl transferase activity of prokaryotic ribosomes. However, its clinical uses are limited to severe infections because of its toxic side effects, which are caused in part by the chloramphenicol sensitivity of mitochondrial ribosomes. Chloramphenicol binds to the large subunit near the A site, which explains why it competes for binding with puromycin and the 3' end of aminoacyl–tRNA but not with peptidyl–tRNAs.

Tetracycline

Tetracycline

and its derivatives are broad-spectrum antibiotics that bind to the small subunit of prokaryotic ribosomes. Tetracycline binding prevents the entry of aminoacyl–tRNAs into the A site but allows EF-Tu to hydrolyze its GTP. As a result, protein synthesis cannot proceed, and GTP hydrolysis, which occurs every time another aminoacyl–tRNA attempts to enter the ribosome, presents an enormous energetic drain on the cell.

Tetracycline-resistant bacterial strains have become quite common, thereby precipitating a serious clinical problem. Most often resistance is conferred by a decrease in bacterial cell membrane permeability to tetracycline rather than any alteration of ribosomal components that would overcome the inhibitory effect on translation.

Although it is a protein rather than a small molecule, the enzyme **ricin** can be considered an antibiotic. Ricin is an *N*-glycosidase from castor beans, the source of castor oil. The enzyme inactivates the large subunit of eukaryotic ribosomes by hydrolytically removing the adenine base of a highly conserved residue of 28S rRNA. The modified ribosome is unable to bind elongation factors, and translation ceases. Because it acts catalytically rather than stoichiometrically, a single ricin molecule can inactivate tens of thousands of ribosomes. Its deadliness in minute amounts makes ricin attractive to bioterrorists, but methods for selectively targeting ricin to particular cells, such as cancer cells, raise the possibility of its therapeutic use.

C   Chain Termination

Polypeptide synthesis under the direction of synthetic mRNAs such as poly(U) results in a peptidyl–tRNA "stuck" in the ribosome. However, the translation of natural mRNAs, which contain the Stop codons UAA, UGA, or UAG, yields free polypeptides. In *E. coli*, the Stop codons, which normally have no corresponding tRNAs, are recognized by protein **release factors** (Table 26-6): **RF-1** recognizes UAA and UAG, whereas **RF-2** recognizes UAA and UGA. In eukaryotes, a single release factor, **eRF1,** recognizes all three termination codons.

Termination, like initiation and elongation, has several stages. The sequence of events in *E. coli* is diagrammed in Fig. 26-38:

1. RF-1 or RF-2 recognizes a corresponding Stop codon in the ribosome's A site.

Figure 26-38 The translation termination pathway in *E. coli* ribosomes. RF-1 recognizes the Stop codons UAA and UAG, whereas RF-2 (not shown) recognizes UAA and UGA. Eukaryotic termination follows an analogous pathway but requires only a single release factor, eRF1, that recognizes all three Stop codons.

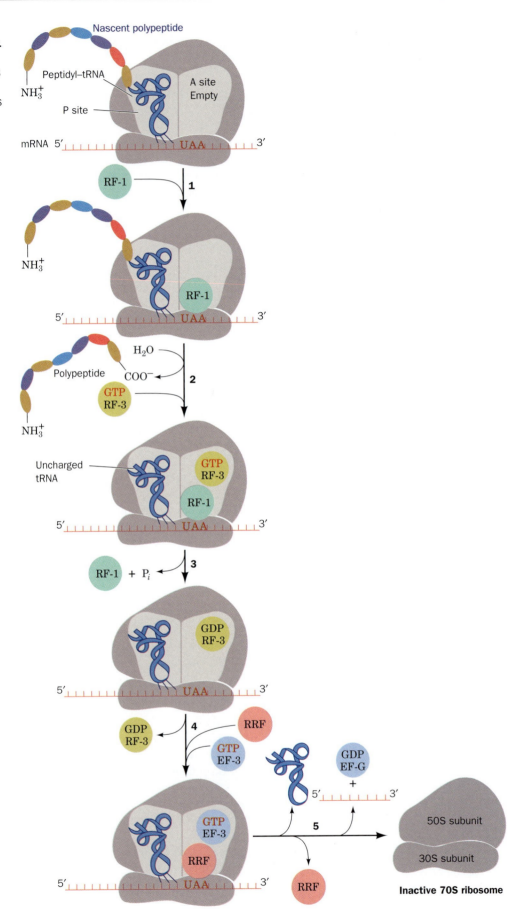

2. The release factor induces the transfer of the peptidyl group from the peptidyl–tRNA to water, rather than to another aa–tRNA, to yield an uncharged tRNA in the P site and a free polypeptide that dissociates from the ribosome.

3. A third factor, **RF-3,** a G protein, binds to the ribosome in complex with GTP. The subsequent hydrolysis of this GTP induces the ribosome to release its bound RF-1 or RF-2.

4. **Ribosome recycling factor (RRF),** which was largely characterized by Akira Kaji, binds in the A site, followed by EF-G · GTP.

5. EF-G hydrolyzes its bound GTP, which causes RRF to move to the P site and the tRNAs occupying the P and E sites (the latter not shown in Fig. 26-38) to be released. Finally, the RRF and then the EF-G · GDP and mRNA dissociate, yielding an inactive 70S ribosome ready for reinitiation (Fig. 26-27).

The release factors' functional mimicry of tRNAs suggests that they structurally resemble tRNAs. Indeed, the X-ray structures of human eRF1 (Fig. 26-39a) and *E. coli* RF-2 (Fig. 26-39b) indicate that these two unrelated proteins both structurally resemble tRNA. However, cryo-EM studies reveal that RF-2 undergoes a large conformational change on binding to the ribosome so that it no longer mimics tRNA. The X-ray structure of *Thermatoga maritima* RRF (Fig. 26-39c) also has an L-shaped structure that could potentially fit one of the ribosome's tRNA binding sites. However, footprinting studies indicate that RRF binds to the A site in an orientation that differs markedly from any previously observed for a tRNA. Evidently, structural mimicry can be misleading.

Nonsense Suppressors Prevent Termination. A mutation that converts an aminoacyl-coding ("sense") codon to a Stop codon is known as a **nonsense mutation** and leads to the premature termination of translation. An organism with such a mutation may be "rescued" by a second mutation in

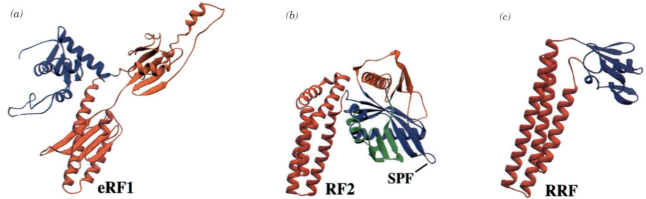

(a)   *(b)*   *(c)*

eRF1   **RF2**   SPF   **RRF**

Figure 26-39 X-Ray structures of putative tRNA mimics that participate in translation termination. (*a*) Human eRF1, (*b*) *E. coli* RF-2, and (*c*) *Thermatoga maritima* RRF. The various domains in these proteins are differently colored with the domains that appear to mimic the tRNA anticodon stem drawn in red. The SPF tripeptide in RF-2 is believed to act as an anticodon. Compare these structures to Figs. 26-30 and 26-36. [Courtesy of V. Ramakrishnan, MRC Laboratory of Molecular Biology, Cambridge, U.K. Part *a* is based on an X-ray structure by David Barford, Institute of Cancer Research, London, U.K.; Part *b* is based on an X-ray structure by Richard Buckingham, CNRS, Paris, France, and Jens Nyborg and Morten Kjeldgaard, University of Aarhus, Århus, Denmark; and Part *c* is based on an X-ray structure by Akira Kaji, University of Pennsylvania, and Anders Liljas, Lund University, Lund, Sweden. PDBids (*a*) 1DT9, (*b*) 1GQE, and (*c*) 1DD5.]

a tRNA gene that causes the tRNA to recognize a Stop (nonsense) codon. This **nonsense suppressor** tRNA carries the same amino acid as its wild-type progenitor and appends it to the growing polypeptide at the Stop codon, thereby preventing chain termination. For example, the *E. coli* **amber suppressor** known as *su3* is a tRNA^{Tyr} whose anticodon has mutated from the wild-type GUA (which reads the Tyr codons UAU and UAC) to CUA (which recognizes the amber Stop codon UAG). An *su3$^+$ E. coli* cell with an otherwise lethal amber mutation in a gene coding for an essential protein would be viable if the replacement of the wild-type amino acid residue by Tyr does not inactivate the protein.

How do cells tolerate a mutation that both eliminates a normal tRNA and prevents the termination of polypeptide synthesis? They survive because the mutated tRNA is usually a minor member of a set of isoaccepting tRNAs and because nonsense suppressor tRNAs must compete with release factors for binding to Stop codons. Consequently, many suppressor-rescued mutants grow more slowly than wild-type cells.

5 Posttranslational Processing

The tunnel that leads from the peptidyl transferase active site to the exterior of the bacterial ribosome (Fig. 26-34) is ~100 Å long, enough to shelter a polypeptide of ~30 residues. Once it emerges from the ribosome, the polypeptide faces two challenges: how to fold properly and how to reach its final cellular destination. Both processes occur with the aid of other proteins. In addition, a nascent polypeptide may be covalently modified before it achieves its mature form. In this section we review some of the events of **posttranslational protein processing.**

A Protein Folding

The ribosome's peptide exit tunnel is too narrow (~15 Å wide) to permit the formation of secondary structure other than perhaps helices, so a nascent polypeptide cannot begin to fold until after it emerges from the ribosome. However, folding commences well before translation is complete. Molecular chaperones (Section 6-5C) bind to the N-terminus of nascent polypeptides to prevent their aggregation, to facilitate their folding, and to promote their proper association with other subunits.

A growing body of evidence suggests that ribosomal proteins play a role in protein folding by recruiting chaperones. For example, in *E. coli*, a protein known as **trigger factor** associates with the ribosomal protein **L23,** which is located at the outlet of the peptide exit tunnel (Fig. 26-40). Trigger factor recognizes relatively short hydrophobic protein segments and may be the first chaperone on the scene. **DnaK,** a member of the Hsp70 family of chaperones, together with **DnaJ,** an Hsp40 cochaperonin (Section 6-5C), can bind to and protect longer polypeptides (DnaK and DnaJ were so named because they were discovered through the isolation of mutants that do not support the growth of bacteriophage λ and hence were initially assumed to participate in DNA replication). Ultimately, the chaperonins GroEL and GroES facilitate protein folding after translation termination (Section 6-5C). Eukaryotic cells lack a homolog of trigger factor but contain other small chaperones that may operate in the same way.

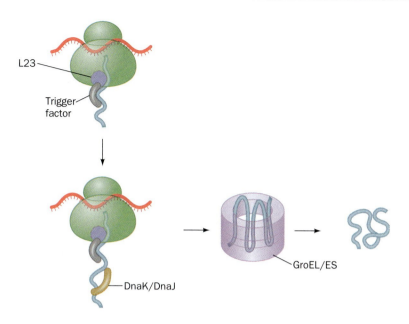

Figure 26-40   Chaperone binding to newly synthesized bacterial proteins. The ribosomal protein L23 first recruits trigger factor, which interacts with the emerging polypeptide at the ribosomal exit site. DnaK and DnaJ then assist protein folding. Folding mediated by the GroEL/ES complex occurs posttranslationally.

B Covalent Modification

The list of known posttranslational modifications is long and includes the removal and/or derivatization of specific residues. For example, a polypeptide's N-terminal Met or fMet residue is usually excised. Other common posttranslational modifications are the hydroxylation of Pro and Lys residues in collagen (Section 6-1C), glycosylation (Sections 8-3C and 15-5), and the prenylation and acylation of membrane proteins (Section 9-3B). Many covalent modifications, such as phosphorylation (Section 15-3B) and palmitoylation (Section 9-3B), are reversible. Over 150 different types of side chain modifications are known; these involve all side chains except those of Ala, Gly, Ile, Leu, Met, and Val. The functions of such modifications are varied and in many cases remain enigmatic.

A growing number of proteins are known to be ubiquitinated. Recall from Section 20-1B that the attachment of the 76-residue protein ubiquitin marks a protein for degradation by the proteasome. A similar protein known as **small ubiquitin-related modifier (SUMO)** can be ligated to a Lys residue, as is ubiquitin, to regulate protein function. But whereas ubiquitination is usually associated with protein degradation, sumoylation appears to play a role in determining protein localization inside cells.

Proteins that are synthesized as inactive precursors, called **proproteins** or proenzymes, are activated by limited proteolysis. The zymogens of serine proteases are converted to active enzymes in this fashion (Section 11-5D). Proteins that are translocated into the endoplasmic reticulum for export from the cell typically contain a signal peptide that must be excised (Section 9-4D). Proteins bearing a signal peptide are known as **preproteins,** or **preproproteins** if they undergo additional proteolysis during their maturation. For example, the hormone insulin (Section 21-2) is synthesized as a preproprotein, which is converted to the proprotein **proinsulin,** a single 84-residue polypeptide. Proteolysis at two sites generates the mature hormone, whose A and B chains remain linked by disulfide bonds (Fig. 26-41). The 33-residue C chain is discarded.

Figure 26-41   Conversion of proinsulin to insulin. The prohormone, with three disulfide bonds, is proteolyzed at two sites (*arrows*) to eliminate the C chain. The mature hormone insulin consists of the disulfide-linked A and B chains.

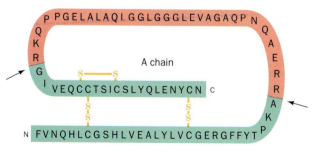

The Signal Recognition Particle Is a Ribonucleoprotein.

Many transmembrane and secretory proteins are translocated into or completely through cellular membranes [the endoplasmic reticulum (ER) in eukaryotes and the plasma membrane in prokaryotes] via a series of events known as the secretory pathway (Section 9-4D). As their N-terminal signal sequences emerge from the ribosome, these proteins associate with the signal recognition particle (SRP). Peptide chain elongation ceases on SRP binding to the ribosome but resumes after the SRP docks with its receptor on the ER membrane. The SRP then dissociates from the ribosome, allowing the nascent polypeptide to pass into or through the membrane via a protein pore called the translocon.

Like the ribosome and the spliceosome (Section 25-3A), the SRP is a ribonucleoprotein. The eukaryotic SRP contains six different polypeptides and a 300-residue RNA. The *E. coli* SRP, which consists of a single polypeptide called **Ffh** and a 114-nt RNA, appears to be a structurally minimized version of the eukaryotic SRP.

Jennifer Doudna determined the structure of the highly conserved core of the SRP, represented by the so-called M domain of Ffh and domain IV of the RNA. The 49-nt RNA segment forms a 70-Å-long double-helical rod in which the RNA chain doubles back on itself (Fig. 26-42a). The RNA and the protein interact mainly via a dense hydrogen-bonded network involving the RNA's unpaired bases and α helices from the protein (Fig. 26-42b). Ffh forms a deep groove that is lined almost entirely with hydrophobic residues, including 11 of the protein's 14 Met residues. This 15-Å-wide by 25-Å-long groove apparently binds the signal peptide as an α helix. The Met side chains, which have flexibility and polarity similar to *n*-butyl groups, provide some structural plasticity, so that *the SRP can accommodate different signal sequences as long as they are hydrophobic and form an α helix.*

What is the function of the SRP RNA? The X-ray structure reveals that the RNA seamlessly continues the protein's hydrophobic binding groove onto a ledge formed by the RNA's sugar–phosphate backbone. The basic residues of the signal sequence (Fig. 9-37) may bind to this anionic surface.

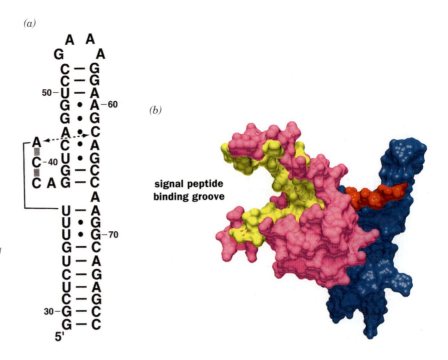

Figure 26-42 X-Ray structure of the SRP core from *E. coli.* (*a*) The secondary structure of the 49-nt RNA. (*b*) The molecular surface of the complex oriented to show the putative signal peptide binding groove. The double-helical RNA is dark blue, the protein is pink, the hydrophobic residues lining the signal peptide binding groove are yellow, and the RNA's adjoining phosphate groups are red. [Courtesy of Robert Batey and Jennifer Doudna, Yale University. PDBid 1DUL.]

SUMMARY

1. The genetic code, by which nucleic acid sequences are translated into amino acid sequences, is composed of three-nucleotide codons that do not overlap and are read sequentially by the protein-synthesizing machinery.

2. The standard genetic code of 64 codons includes numerous synonyms, three Stop codons, and one initiation codon.

3. All tRNAs have numerous chemically modified bases and a similar cloverleaf secondary structure comprising an acceptor stem, D arm, TψC arm, anticodon arm, and variable arm. The three-dimensional structures of tRNAs are likewise similar and are maintained by stacking interactions and non-Watson–Crick hydrogen-bonded cross-links.

4. Aminoacyl–tRNA synthetases (aaRSs) catalyze the ATP-dependent attachment of an amino acid to the appropriate tRNA to yield an aminoacyl–tRNA (aa–tRNA). The acceptor stem and anticodon loop are common identity elements for tRNA–aaRS interactions. The fidelity of aminoacylation is enhanced by proofreading.

5. Wobble pairing between mRNA codons and tRNA anticodons at the third position accounts for much of the degeneracy of the genetic code. Within a species, the preferred codons correspond to the most abundant species in each set of isoaccepting tRNAs.

6. Ribosomes, which are large complexes of RNA and protein, have a structure that is determined by their RNA components. A small subunit and large subunit associate to form the intact ribosome, which accommodates an aa–tRNA in the A site, a peptidyl–tRNA in the P site, and a deacylated tRNA in the E site.

7. During translation initiation, an initiator tRNA charged with fMet (in prokaryotes) or Met (in eukaryotes), an mRNA with an AUG initiation codon, and the ribosomal subunits assemble. Initiation requires GTP-hydrolyzing initiation factors. The initiating AUG codon is identified in prokaryotes via the mRNA's Shine–Dalgarno sequence. In eukaryotes it is the mRNA's initiating AUG that is identified in a complex process by its proximity to the mRNA's 5′ cap.

8. A polypeptide chain is elongated from its N- to its C-terminus. An aa–tRNA in complex with a GTP-hydrolyzing elongation factor binds to the A site, where the ribosome senses correct codon–anticodon pairing, a process whose accuracy is enhanced by proofreading. The peptidyl transferase activity of the large subunit's RNA catalyzes the attack of the A-site bound aa–tRNA's amino group on the peptidyl–tRNA in the P site. Following transpeptidation, a second GTP-hydrolyzing elongation factor promotes the translocation of the new peptidyl–tRNA to the P site.

9. Translation termination requires release factors that recognize Stop codons.

10. As a polypeptide emerges from the ribosome, it begins to fold with the assistance of chaperones. Proteins may undergo posttranslational processing that includes proteolysis, covalent modification, and translocation through a membrane.

REFERENCES

Genetic code

Judson, J.F., *The Eighth Day of Creation,* Expanded edition, Part II, Cold Spring Harbor Laboratory Press (1996). [A fascinating historical narrative on the elucidation of the genetic code.]

Knight, R.D., Freeland, S.J., and Landweber, L.F., Selection, history and chemistry: the three faces of the genetic code, *Trends Biochem. Sci.* **24,** 241–247 (1999).

Transfer RNA structure and aminoacylation

Ibba, M., Becker, H.D., Stathopoulos, C., Tumbula, D.L., and Söll, D., The adaptor hypothesis revisited, *Trends Biochem. Sci.* **25,** 311–316 (1999). [Describes exceptions to the 20 aminoacyl–tRNA synthetase rule.]

Nureki, O., Vassylyev, D.G., Tateno, M., Shimada, A., Nakama, T., Fukai, S., Konno, M., Hendrickson, T.L., Schimmel, P., and Yokoyama, S., Enzyme structure with two catalytic sites for double-sieve selection of substrate, *Science* **280,** 578–582 (1998). [The X-ray structures of IleRS in complexes with isoleucine and valine.]

Ribosomes and translation

Culver, G.M., Meanderings of the mRNA through the ribosome, *Structure* **9,** 751–758 (2001).

Green, R. and Lorsch, J.R., The path to perdition is paved with protons, *Cell* **110,** 665–668 (2002). [Discusses the difficulties in characterizing the catalytic mechanism of the ribosomal peptidyl transferase.]

Moore, P.B. and Steitz, T.A., RNA, the first macromolecular catalyst: the ribosome is a ribozyme, *Trends Biochem. Sci.* **28,** 411–418 (2003); *and* The structural basis of large ribosomal subunit function, *Annu. Rev. Biochem.* **72,** 813–850 (2003).

Nissen, P., Kjeldgaard, M., and Nyborg, J., Macromolecular mimicry, *EMBO J.* **19,** 489–495 (2000). [Describes how translation factors and other proteins function by adopting the same structures and interacting with the same binding sites as nucleic acids.]

Ogle, J.M., Brodersen, D.E., Clemons, W.M., Jr., Tarry, M.J., Carter, A.P., and Ramakrishnan, V., Recognition of cognate transfer RNA by the 30S ribosomal subunit, *Science* **292,** 897–902 (2001). [Presents X-ray structural evidence for an rRNA sensor that verifies correct codon–anticodon pairing in the ribosome.]

Sievers, A., Beringer, M., Rodnina, M.V., and Wolfenden, R., The ribosome as an entropy trap, *Proc. Natl. Acad. Sci.* **101,** 7897–7901 (2004).

Valle, M., Zavialov, A., Li, W., Stagg, S.M., Sengupta, J., Nielsen, R.C., Nissen, P., Harvey, S.C., Ehrenberg, M., and Frank, J., Incorporation of aminoacyl-tRNA into the ribosome as seen

by cryo-electron microscopy, *Nature Struct. Biol.* **10**, 899–906 (2003).

Yusupov, M.M., Yusupova, G.Zh., Baucom, A., Lieberman, K., Earnest, T.N., Cate, J.H.D., and Noller, H.F., Crystal structure of the ribosome at 5.5 Å resolution, *Science* **292**, 883–896 (2001). [Describes the structure of the intact 70S bacterial ribosome with bound mRNA and three tRNA molecules.]

Posttranslational processing

Batey, R.T., Rambo, R.P., Lucast, L., Rha, B., and Doudna, J.A., Crystal structure of the ribonuclear core protein of the signal recognition particle, *Science* **287**, 1232–1239 (2000).

Hartl, F.U. and Hayer-Hartl, M., Molecular chaperones in the cytosol: from nascent chain to folded protein, *Science* **295**, 1852–1858 (2002). [Describes the structures, functions, and potential interactions of the various chaperone systems in eukaryotes, eubacteria, and archaea.]

KEY TERMS

genetic code	D arm	cryoelectron microscopy	initiation factor
codon	anticodon arm	(cryo-EM)	transpeptidation
degeneracy	TψC arm	A site	translocation
suppressor	variable arm	P site	elongation factor
reading frame	aminoacyl–tRNA synthetase	peptidyl–tRNA	release factor
frameshift mutation	(aaRS)	E site	nonsense suppressor
anticodon	aminoacyl–tRNA	peptidyl transferase	posttranslational processing
synonym	(aa–tRNA)	polysome	proprotein
Stop codon	isoaccepting tRNA	fMet	preprotein
acceptor stem	wobble hypothesis	Shine–Dalgarno sequence	preproprotein

STUDY EXERCISES

1. List the overall similarities and differences among replication, transcription, and translation.

2. Explain why codons must consist of at least three nucleotides.

3. Describe the major structural features of tRNA.

4. Describe how the double-sieve mechanism of IleRS promotes accurate tRNA aminoacylation.

5. Describe the functions of the three tRNA binding sites in the ribosome.

6. Summarize the roles of initiation, elongation, and release factors in translation.

7. Summarize the role of GTP hydrolysis in promoting the efficiency of translation initiation, decoding, translocation, and chain termination.

8. How does the ribosome verify correct tRNA–mRNA pairing?

PROBLEMS

1. Could a single nucleotide deletion restore the function of a protein-coding gene interrupted by the insertion of a 4-nt sequence? Explain.

2. List all possible codons present in a ribonucleotide polymer containing U and G in random sequence. Which amino acids are encoded by this RNA?

3. Which amino acids are specified by codons that can be changed to an amber codon by a single point mutation?

4. A double-stranded fragment of viral DNA, one of whose strands is shown below, encodes two peptides, called *vir-1* and *vir-2*. Adding this double-stranded DNA fragment to an *in vitro* transcription and translation system yields peptides of 10 residues (*vir-1*) and 5 residues (*vir-2*).

AGATCGGATGCTCAACTATATGTGATTAACAGAG-CATGCGGCATAAACT

(a) Identify the DNA sequence that encodes each peptide.

(b) Determine the amino acid sequence of each peptide.

(c) In a mutant viral strain, the T at position 23 has been replaced with G. Determine the amino acid sequences of the two peptides encoded by the mutant virus.

[Problem provided by Bruce Wightman, Muhlenberg College.]

5. Shown on the facing page is the sequence of the sense strand of a mammalian gene. Determine the sequences of the mature RNA and the encoded protein. Assume that transcription initiates approximately 25 bp downstream of the TATAAT sequence, that each 5′ splice site has the sequence AG/GUAAGU, and that each 3′ splice site has the sequence AGG/N, where N stands for any of the four RNA bases and the / marks the location of the splice.

TATAATACGCGCAATACAATCTACAGCTTCGCGTA
AATCGTAGGTAAGTTGTAATAAATATAAGTGAGT
ATGATAGGGCTTTGGACCGATAGATGCGACCCTG
GAGGTAAGTATAGATTAATTAAGCACAGGCATGCA
GGGATATCCTCCAAATAGGTAAGTAACCTTACGG
TCAATTAATTAGGCAGTAGATGAATAAACGATAT
CGATCGGTTAGGTAAGTCTGAT

[Problem provided by Bruce Wightman, Muhlenberg College.]

6. IleRS uses a double-sieve mechanism to accurately produce Ile–tRNA^{Ile} and prevent the synthesis of Val–tRNA^{Ile}. Which other pairs of amino acids differ in structure by a single carbon and might have aaRSs that use a similar double-sieve proofreading mechanism?

7. In eukaryotes, the primary rRNA transcript is a 45S rRNA that includes the sequences of the 18S, 5.8S, and 28S rRNAs separated by short spacers. What is the advantage of this operon-like arrangement of rRNA genes?

8. Explain why the translation of a given mRNA can be inhibited by a segment of its complementary sequence, a so-called **antisense RNA.**

9. Explain the significance of the observation that peptides such as fMet-Leu-Phe "activate" the phagocytotic (particle-engulfing) functions of mammalian leukocytes (white blood cells).

10. Explain why prokaryotic ribosomes can translate a circular mRNA molecule, whereas eukaryotic ribosomes normally cannot, even in the presence of the required cofactors.

11. EF-Tu binds all aminoacyl–tRNAs with approximately equal affinity so that it can deliver them to the ribosome with the same efficiency. Based on the experimentally determined binding constants for EF-Tu and correctly charged and mischarged aminoacyl–tRNAs (see Table), explain how the tRNA–EF-Tu recognition system could prevent the incorporation of the wrong amino acid during translation.

Aminoacyl–tRNA	Dissociation constant (nM)
Ala–tRNA^{Ala}	6.2
Gln–tRNA^{Ala}	0.05
Gln–tRNA^{Gln}	4.4
Ala–tRNA^{Gln}	260

[Source: LaRiviere, F.J., Wolfson, A.D., and Uhlenbeck, O.C., *Science* **294,** 167 (2001).]

12. The antibiotic **paromomycin** binds to a ribosome and induces the same conformational changes in 16S rRNA residues A1492 and A1493 as are induced by codon–anticodon pairing (Fig. 26-33). Propose an explanation for the antibiotic effect of paromomycin.

13. The rate of the peptidyl transferase reaction increases as the pH increases from 6 to 8. (a) Explain these results in terms of its reaction mechanism. (b) It has been proposed that residue A2486 is protonated and therefore stabilizes the tetrahedral reaction intermediate. Is this mechanistic embellishment consistent with the observed pH effect?

14. All cells contain an enzyme called **peptidyl–tRNA hydrolase,** and cells that are deficient in the enzyme grow very slowly. What is the probable function of the enzyme and why is it necessary?

15. Calculate the energy required, in ATP equivalents, to synthesize a 100-residue protein from free amino acids in *E. coli* (assume that the N-terminal Met remains attached to the polypeptide and that no ribosomal proofreading occurs).

16. Design an mRNA with the necessary prokaryotic control sites that codes for the octapeptide Lys-Pro-Ala-Gly-Thr-Glu-Asn-Ser.

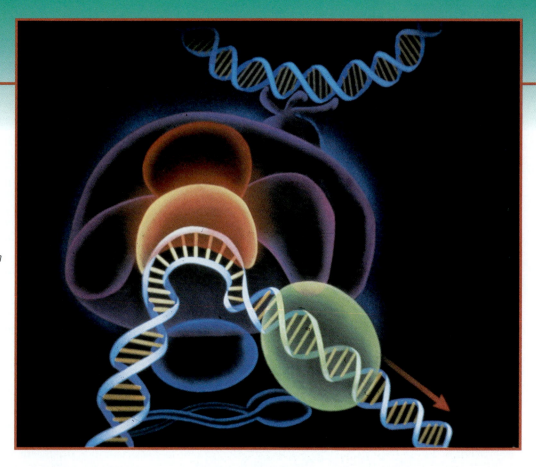

The controlled expression of genetic information requires a multitude of factors that interact specifically and nonspecifically with DNA. This painting schematically diagrams transcription factors in the preinitiation complex whose formation in eukaryotes must precede the transcription of DNA to mRNA. [Illustration, Irving Geis. Image from the Irving Geis Collection/Howard Hughes Medical Institute. Rights owned by HHMI. Reproduction by permission only.]

Regulation of Gene Expression

The faithful replication of DNA ensures that all the descendants of a single cell contain virtually identical sets of genetic instructions. Yet individual cells may differ—sometimes dramatically—from their progenitors, depending on how those instructions are read. The **expression** of genetic information in a given cell or organism, that is, the synthesis of RNA and proteins specified by the DNA sequence, is neither random nor fully preprogrammed. Rather, *the information in an organism's genome must be tapped in an orderly fashion during development and yet must also be available to direct the organism's responses to changes in internal or external conditions.*

Much of the mystery surrounding the regulation of gene expression has to do with how genetic information is organized and how it can be located and accessed on an appropriate time scale. The complexity of the mechanisms for transcribing and translating genes suggests the potential for even more complicated systems for enhancing or inhibiting these processes. Indeed, we have already seen numerous examples of how accessory protein factors influence the rates or specificities of transcription and translation. In this chapter, we consider some additional aspects of gene expression by examining a variety of strategies used by prokaryotes and eukaryotes to control how genetic information specifies cell structures and metabolic functions.

1 Genome Organization

Genomics, the study of organisms' genomes, was established as a discipline with the advent of techniques for rapidly sequencing enormous tracts of DNA, such as make up the chromosomes of living things. Studies that once relied on the hybridization of oligonucleotide probes to identify genes can now be conducted by comparing nucleotide sequences deposited in databases (Section 5-3E). In fact, a computerized database is the only practical format for storing the vast amount of information provided by a whole-genome sequence. Even a simplified diagram of the genome of the relatively simple bacterium *Helicobacter pylori* reveals a bewildering array of genes (Fig. 27-1). Nevertheless, the ability to sequence and map the en-

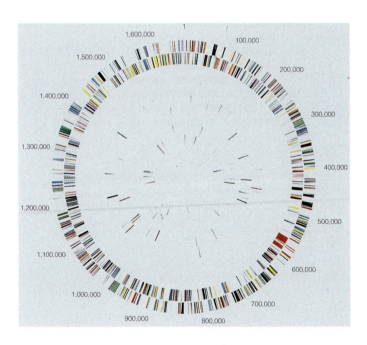

Figure 27-1   Map of the 1670-kb *H. pylori* circular chromosome.   The 1590 predicted protein-coding sequences (91% of the genome) are indicated in different colors on the two outermost rings, which represent the two DNA strands. The third and fourth rings indicate intervening sequence elements and other repeating sequences (2.3% of the genome). The fifth and sixth rings show the genes for tRNAs, rRNAs, and other small RNAs (0.7% of the genome). Intergenic sequences account for 6% of the genome. [Courtesy of J. Craig Venter, The Institute for Genomic Research, Rockville, Maryland.]

tire genome of an organism makes it possible to draw far-reaching conclusions about how many genes an organism contains, how they are organized, and how they serve the organism.

A   Gene Number

The rough correlation between the quantity of an organism's unique genetic material (its **C value**) and the complexity of its morphology and metabolism (Table 3-3) has numerous exceptions, known as the **C-value paradox.** For example, the genomes of lungfishes are 10 to 15 times larger than those of mammals (Fig. 27-2). Some algae have genomes 10 times larger still. We know that much of this "extra" DNA is unexpressed, but its function is largely a matter of conjecture. The complete genomic sequences of numerous prokaryotes and eukaryotes, including data from some large complicated genomes, nevertheless indicate that *the apparent number of genes, like the overall quantity of DNA, roughly parallels the organism's complexity* (Table 27-1). Thus, humans have ~30,000 genes compared to *E. coli's* ~4300 genes.

Identifying Genes.   About 2,900,000 kb of the 3,200,000-kb human genome has been fully sequenced (the remaining DNA includes gene-poor regions near the centromeres and telomeres). *No more than about 1.4% of the genome encodes proteins,* and the 30,000 or so protein-coding genes are dis-

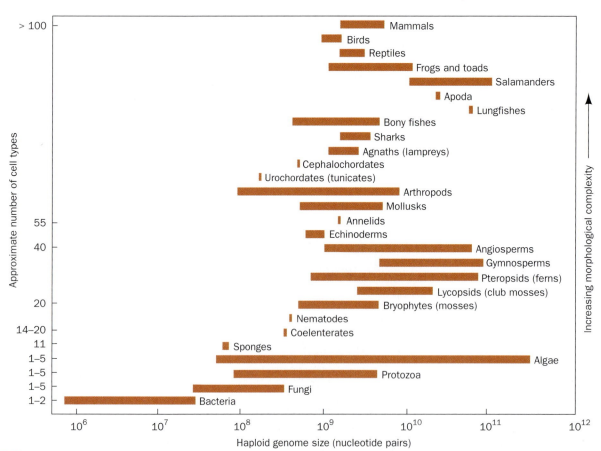

Figure 27-2   Range of haploid genome DNA contents in various categories of organisms.   The morphological complexity of the organisms, as estimated from their number of different cell types, increases from bottom to top. Haploid genome size roughly correlates with complexity; exceptions are described by the C-value paradox. [After Raff, R.A. and Kaufman, T.C., *Embryos, Genes, and Evolution*, p. 314, Macmillan (1983).]

Table 27-1   Genome Size and Gene Number in Some Organisms

Organism	Genome Size (kb)	Number of Genes
Haemophilus influenzae (bacterium)	1,830	1,740
Escherichia coli (bacterium)	4,639	4,289
Saccharomyces cerevisiae (yeast)	11,700	6,034
Caenorhabditis elegans (nematode)	97,000	19,099
Oryza sativa (rice)	430,000	~35,000
Arabidopsis thaliana (mustard weed)	117,000	~26,000
Drosophila melanogaster (fruit fly)	137,000	13,061
Mus musculus (mouse)	2,500,000	~30,000
Homo sapiens (human)	3,200,000	~30,000

tributed throughout the genome, sometimes in clusters. Identifying the genes in human or other genomes involves several approaches.

A gene can be identified by its homology to a previously described mRNA or protein sequence. A protein-coding gene may also be identified as an **open reading frame (ORF),** a sequence that is not interrupted by Stop codons and that exhibits the same codon-usage preferences as other genes in the organism. However, our incomplete knowledge of the features through which cells recognize genes, combined with the fact that human genes consist of relatively short exons (averaging ~150 nt) separated by much longer introns (averaging ~3500 nt and often much longer), has limited the success of computer-based gene identification algorithms. Such algorithms therefore rely on sequence alignments with **expressed sequence tags** (**ESTs;** cDNAs that have been reverse-transcribed from mRNAs) together with alignments with known genes from other organisms. Genes may also be revealed by the presence of a **CpG island** (Box 24-4). CpG dinucleotides are present in vertebrate genomes at only about one-fifth their randomly expected frequency (because the spontaneous deamination of m^5C yields a normal T and therefore often results in a CG → TA mutation; Box 24-5), but they appear in clusters, or islands, at near-normal frequencies in the upstream regions of many genes (~56% of human genes).

Around 30,000 putative genes have been identified in the human genome. The discrepancy between this number and previous estimates of 50,000 to 140,000 genes is largely attributed to a much greater prevalence of alternative splicing than had previously been surmised (Section 25-3A). Genes may be classified as those that encode proteins (structural genes) and those that are transcribed to RNAs that are not translated. These latter so-called **noncoding RNAs (ncRNAs)** consist of tRNAs, rRNAs, and other small RNAs.

A total of 26,383 predicted structural genes in the human genome have been classified according to molecular function through sequence comparisons at the level both of protein families and of domains (Fig. 27-3). Note that nearly 42% of these genes have unknown functions, as is likewise the case with most other genomes of known sequence, including those of prokaryotes. Genes with no known function are called **orphan genes.** Some of these sequences may represent novel genes whose protein products have not yet been discovered. Others may simply be counterparts of known genes but are too different in sequence to be recognized as such.

About three-quarters of all known human genes appear to have counterparts in other species. About one-quarter are present only in other vertebrates, and one-quarter are present in prokaryotes as well as eukaryotes. As expected, the human genome contains approximately the same number

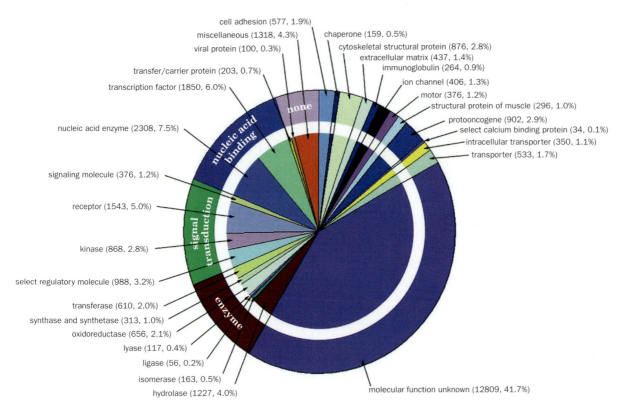

cell adhesion (577, 1.9%)
miscellaneous (1318, 4.3%)
viral protein (100, 0.3%)
chaperone (159, 0.5%)
cytoskeletal structural protein (876, 2.8%)
extracellular matrix (437, 1.4%)
transfer/carrier protein (203, 0.7%)
immunoglobulin (264, 0.9%)
transcription factor (1850, 6.0%)
ion channel (406, 1.3%)
motor (376, 1.2%)
structural protein of muscle (296, 1.0%)
nucleic acid enzyme (2308, 7.5%)
protooncogene (902, 2.9%)
select calcium binding protein (34, 0.1%)
intracellular transporter (350, 1.1%)
transporter (533, 1.7%)
signaling molecule (376, 1.2%)
receptor (1543, 5.0%)
kinase (868, 2.8%)
select regulatory molecule (988, 3.2%)
transferase (610, 2.0%)
synthase and synthetase (313, 1.0%)
oxidoreductase (656, 2.1%)
lyase (117, 0.4%)
ligase (56, 0.2%)
isomerase (163, 0.5%)
hydrolase (1227, 4.0%)
molecular function unknown (12809, 41.7%)

none
nucleic acid binding
signal transduction
enzyme

Figure 27-3 Distribution of molecular functions of
26,383 putative structural genes in the human genome.
Each wedge of this pie chart lists, in parentheses, the
number and percentage of the genes assigned to the
indicated category of molecular function. The outer circle
indicates the general functional categories whereas the
inner circle provides a more detailed breakdown of these
categories. [Courtesy of J. Craig Venter, Celera Genomics,
Rockville, Maryland.]

of "housekeeping" genes (genes required for the most fundamental cellular activities) as other eukaryotes. But it includes relatively more genes for vertebrate-specific activities, such as those relevant to the immune system and neuronal and hormonal signaling pathways.

Practical Aspects of Genomics. Because many human genes have a homolog in a model organism such as yeast or mouse, it is worthwhile to alter or delete that gene in the model organism and observe whether any function is lost. It is also possible to systematically mutate every gene in a small organism such as *S. cerevisiae* in order to assign a function to each gene. These findings can then be extrapolated to the human genome, which may contain a family of similar genes.

In addition, the availability of human genome sequence information should make it easier to identify genes associated with diseases. Currently, about 1000 such genes have been catalogued, including those for cystic fibrosis (see Box 3-2) and some hereditary forms of cancer. Unfortunately, monogenetic diseases are relatively rare. Most diseases result from interactions among several genes and from environmental factors. One goal of genomics is to identify genetic features that can be linked to susceptibility to disease or infection. To that end, a catalog of human DNA sequence variations is being compiled as a database of **single nucleotide polymorphisms (SNPs;** pronounced "snips"). A SNP, or single-base difference between individuals, occurs about every 1250 bp on average. Nearly 2 million SNPs have been described. Although less than 1% of them result in protein variants, and probably fewer still have functional consequences, it is becoming increasingly apparent that SNPs are largely responsible for an individual's susceptibility to many diseases as well as to adverse reactions to drugs (side effects; Section 12-4). Moreover, SNPs can potentially serve as genetic markers for nearby disease-related genes.

B Gene Clusters

Genes are not necessarily randomly distributed throughout an organism's genome. For example, the average gene frequency in the human genome is ~1 gene per 100 kb but ranges from 0 to 64 genes per 100 kb. Moreover, some protein-coding genes and other chromosomal elements exhibit a certain degree of organization. Prokaryotic genomes, for example, contain numerous operons, in which, as we have seen (Section 25-1B), genes with related functions (e.g., encoding the proteins involved in a particular metabolic pathway) occur close together, sometimes in the same order in which their encoded proteins act in a metabolic reaction sequence, and are transcribed onto a single polycistronic mRNA.

Gene clusters also occur in both prokaryotes and eukaryotes. Although most genes occur only once in an organism's haploid genome, genes such as those for rRNA and tRNA, whose products are required in relatively large amounts, may occur in multiple copies. As we saw in Section 25-3B, large rRNA transcripts are cleaved to yield the mature rRNA molecules. Furthermore, the transcribed blocks of 18S, 5.8S, and 28S eukaryotic rRNA genes are arranged in tandem repeats that are separated by nontranscribed spacers (Fig. 27-4). The rRNA genes, which may be distributed among several chromosomes, vary in haploid number from less than 50 to over 10,000, depending on the species. Humans, for example, have 50 to 200 blocks of rRNA genes spread over five chromosomes. tRNA genes are similarly reiterated and clustered.

Protein-coding genes almost never occur in multiple copies, presumably because the repeated translation of a few mRNA transcripts provides adequate amounts of most proteins. One exception is histone proteins, which are required in large amounts during the short time when eukaryotic DNA synthesis occurs. Not only are histone genes reiterated (up to ~100 times in *Drosophila*), they often occur as sets of each of the five different histone genes separated by nontranscribed sequences (Fig. 27-5). The gene order and the direction of transcription in these quintets are preserved over large evolutionary distances. The spacer sequences vary widely among species and, to a limited extent, among the repeating quintets within a genome. In birds and mammals, which contain 10 to 20 copies of each of the five histone genes, the genes occur in clusters but in no particular order.

Other gene clusters contain genes of similar but not identical sequence. For example, human globin genes are arranged in two clusters on separate

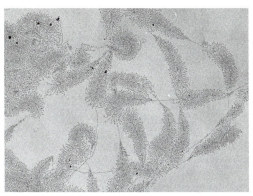

Figure 27-4 An electron micrograph of tandem arrays of actively transcribing 18S, 5.8S, and 28S rRNA genes from the newt *Notophthalmus viridescens*. The axial fibers are DNA. The fibrillar "Christmas tree" matrices, which consist of newly synthesized RNA strands in complex with proteins, outline each transcriptional unit. Note that the longest ribonucleoprotein branches are only ~10% of the length of their corresponding DNA. Apparently, the RNA strands are compacted through secondary structure interactions and/or protein associations. The matrix-free segments of DNA are the untranscribed spacers. [Courtesy of Oscar L. Miller, Jr., and Barbara R. Beatty, University of Virginia.]

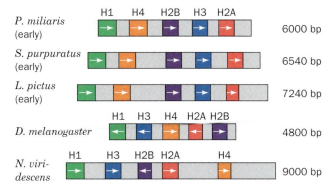

Figure 27-5 The organization and lengths of the histone gene cluster repeating units in a variety of organisms. Coding regions are indicated in color, and spacers are gray. The arrows denote the direction of transcription (the top three organisms are distantly related sea urchins).

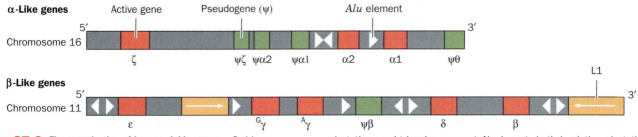

Figure 27-6 The organization of human globin genes.   Red boxes represent active genes; green boxes represent pseudogenes; yellow boxes represent repetitive sequences, with the arrows indicating their relative orientations; and triangles represent *Alu* elements in their relative orientations. [After Karlsson, S. and Nienhuis, A.W., *Annu. Rev. Biochem.* **54,** 1074 (1985).]

chromosomes (Fig. 27-6). The various genes are transcribed at different developmental stages. Adult hemoglobin is an $\alpha_2\beta_2$ tetramer, whereas the first hemoglobin made by the human embryo is a $\zeta_2\varepsilon_2$ tetramer in which ζ and ε are α- and β-like subunits, respectively (Fig. 27-7). At approximately 8 weeks after conception, fetal hemoglobin containing α and γ subunits appears. The γ subunit is gradually supplanted by β starting a few weeks before birth. Adult human blood normally contains ~97% $\alpha_2\beta_2$ hemoglobin, 2% $\alpha_2\delta_2$ (in which δ is a β variant), and 1% $\alpha_2\gamma_2$.

The **α-globin gene cluster** (Fig. 27-6, *top*), which spans 28 kb, contains three functional genes: the embryonic ζ gene and two slightly different α genes, $\alpha1$ and $\alpha2$, which encode identical polypeptides. The α cluster also contains four pseudogenes (nontranscribed relics of ancient gene duplications): $\psi\zeta$, $\psi\alpha2$, $\psi\alpha1$, and $\psi\theta$. The **β-globin gene cluster** (Fig. 27-6, *bottom*), which spans >60 kb, contains five functional genes: the embryonic ε gene, the fetal genes $^G\gamma$ and $^A\gamma$ (duplicated genes encoding polypeptides that differ only by having either Gly or Ala at position 136), and the adult genes δ and β. The β-globin cluster also contains one pseudogene, $\psi\beta$. Both α and β gene clusters also include copies of **repetitive DNA sequences** (see below).

C Repetitive DNA Sequences

Approximately 11% of the *E. coli* genome consists of nontranscribed regions, including the regulatory sequences that separate individual genes and sites that govern the origin and termination of replication. In addition, bac-

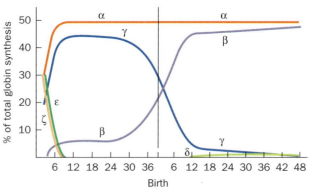

Figure 27-7 The progression of human globin chain synthesis with fetal development.   Note that any red blood cell contains only one type each of α- and β-like subunits. [After Weatherall, D.J. and Clegg, J.B., *The Thalassaemia Syndromes* (3rd ed.), *p.* 64, Blackwell Scientific Publications (1981).]

terial genomes typically contain insertion sequences (Section 24-6C) and the remnants of integrated bacteriophages.

The small sizes of prokaryotic genomes probably exert selective pressure against the accumulation of useless DNA. Eukaryotic genomes, however, which are usually much larger than prokaryotic genomes, apparently are not subject to the same selective forces. About 30% of the yeast genome, for example, consists of nonexpressed sequences. The proportion is far greater in higher eukaryotes (as much as 98% in humans). Much of this DNA consists of repetitive sequences. High concentrations of repetitive DNA are located at the **centromeres** of eukaryotic chromosomes, the regions attached to the microtubular spindle during mitosis. These sequences may help align chromosomes and facilitate recombination. The telomeres also are composed of repeating DNA sequences (Section 24-3C). Over a dozen human diseases result from excessively reiterated trinucleotide sequences (see Box 27-1).

Highly repetitive sequences (present at $>10^6$ copies per haploid genome) are clusters of nearly identical sequences up to 10 bp long that are tandemly repeated thousands of times. These DNA regions are also known as **satellite DNAs** because their distinctive base compositions cause them to form bands separate from the main DNA band in density gradient ultracentrifugation of fragmented DNA (Fig. 27-8). The three satellites of *Drosophila virilis* DNA are closely related heptanucleotide repeats:

$$5'—ACAAACT—3' \qquad 5'—ATAAACT—3' \qquad 5'—ACAAATT—3'$$
$$3'—TGTTTGA—5' \qquad 3'—TATTTGA—5' \qquad 3'—TGTTTAA—5'$$

$$\textbf{Satellite I} \qquad\qquad \textbf{Satellite II} \qquad\qquad \textbf{Satellite III}$$

Approximately 3% of the human genome consists of highly repetitive DNA sequences (also known as **short tandem repeats; STRs**), with the greatest contribution (0.5%) provided by dinucleotide repeats, most frequently $(CA)_n$ and $(TA)_n$. STRs appear to have arisen by template slippage during DNA replication. This occurs more frequently with short repeats, which

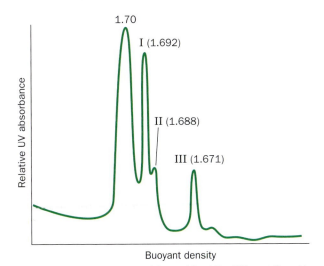

Figure 27-8 The buoyant density pattern of *Drosophila virilus* DNA centrifuged to equilibrium in neutral CsCl. Three prominent bands of satellite DNA (with densities of 1.692, 1.688, and 1.671 $g \cdot cm^{-3}$) are present, in addition to the main DNA band (density of 1.70 $g \cdot cm^{-3}$). [After Gall, J.G., Cohen, E.H., and Atherton, D.D., *Cold Spring Harbor Symp. Quant. Biol.* **38**, 417 (1973).]

BOX 27-1

Biochemistry in Health and Disease

Trinucleotide Repeat Diseases

At least 14 human diseases are associated with repeated trinucleotides in certain genes. These **trinucleotide repeats** exhibit an unusual genetic instability: Above a threshold of about 35 to 50 copies (100–150 bp), the repeats tend to expand with successive generations. Because the overall length of the repeat typically correlates with the age of onset of the disease, descendants of an individual with a trinucleotide repeat disease tend to be more severely affected and at an earlier age. The disease is therefore said to exhibit **genetic anticipation.**

Some types of trinucleotide repeat diseases are caused by massive expansion (usually to hundreds of copies) of a trinucleotide in the noncoding region of a gene, for example, in a region upstream of the transcription start site, in a 5' or 3' untranslated region **(UTR),** or in an intron (see table). These expansions generally affect gene expression. For example, **myotonic dystrophy** results from aberrant expression of a protein kinase. The severity of the symptoms, progressive muscle weakness and wasting, correlate with the number of CTG repeats (>2000 in some cases) in its gene's 5' UTR.

Fragile X syndrome, the most common cause of mental retardation after Down's syndrome, is so named because the tip of the X chromosome's long arm is connected to the rest of the chromosome by a slender thread that is easily broken. As in many trinucleotide repeat diseases, the genetics of fragile X syndrome are bizarre. The maternal grandfathers of individuals having fragile X syndrome may be asymptomatic. Their daughters are likewise asymptomatic, but these daughters' children of either sex may have the syndrome. Evidently, the fragile X defect is activated by passage through a female.

The affected gene in fragile X syndrome, *FMR1* (for *fragile X mental retardation 1*), encodes a 632-residue RNA-binding protein named **FMRP** (for *FMR protein*), which apparently functions in the transport of certain mRNAs from the nucleus to the cytoplasm. FMRP, which is highly conserved in vertebrates, is heavily expressed in brain neurons, where a variety of evidence indicates that its participation is required for the proper formation and/or function of synapses.

In the general population, the 5' untranslated region of *FMR1* contains a $(CGG)_n$ sequence with n ranging from 6 to 60 (such a gene is said to be **polymorphic**). However, in individuals with fragile X syndrome, this triplet repeat has undergone an astonishing expansion to values of n ranging from >200 to several thousand. Moreover, these triplet repeats differ in size among siblings and often exhibit heterogeneity within an individual, suggesting that they are somatically generated.

Some trinucleotide repeat diseases result from the moderate expansion of a CAG triplet (which codes for Gln) in the protein-coding region of a gene. For example, in normal individuals, **huntingtin,** a polymorphic, ~3150-residue protein of unknown function, contains a stretch of 11 to 34 consecutive Gln residues beginning 17 residues from its N-terminus. However, in **Huntington's disease (HD),** this poly(Gln) tract has expanded to between 37 and 86 repeats. Synthetic poly(Gln) aggregates as β sheets that are linked by hydrogen bonds involving both their main chain and side chain amide groups.

Individuals with HD, an autosomal dominant condition, suffer progressive choreic (jerky and disordered) movements, cognitive decline, and emotional disturbances over a 15 to 20 year course that is invariably fatal. This devastating disease typically develops late in life, around age 40, or earlier in individuals with high numbers of trinucleotide repeats. One of the hallmarks of HD, as well as of other neurodegenerative diseases such as Alzheimer's disease, is the deposition of insoluble protein aggregates in the cytosol and nuclei of neurons (Section 6-5D). The appearance of these intracellular aggregates (presumably huntingtin and/or its proteolytic products in HD) often coincides with the onset of neurological symptoms, strongly suggesting that protein aggregation directly causes neuronal death and its attendant symptoms. The long incubation period before the symptoms of HD become evident is attributed to the lengthy nucleation time for aggregate formation, much like what we have seen occurs in the formation of amyloid fibrils (Section 6-5D).

therefore have a high degree of length polymorphism in the human population. The determination of n for several well-characterized STRs (those whose numbers of repeats have been determined in numerous individuals of multiple ethnicities) can unambiguously identify the DNA's donor.

Some Diseases Associated with Trinucleotide Repeats

Disease	Repeat	Site of Repeat
Fragile X syndrome	CGG	5' UTR
Myotonic dystrophy	CTG	Upstream region, 3' UTR
Friedrich's ataxia	GAA	Intron
Spinobulbar muscular atrophy	CAG	Exon
Huntington's disease	CAG	Exon

The expansion of trinucleotide repeats occurs by an unknown mechanism. One theory, that additional nucleotides are introduced by the slippage of DNA polymerases during replication, is inconsistent with the gradual accumulation of trinucleotide repeats over time in cells such as neurons that normally do not divide. Another possibility is that the additional nucleotides are introduced during DNA repair processes (which are ongoing and independent of DNA replication) and may result from hairpin formation within the trinucleotide repeat region. These intrastrand base-paired structures may lead to misalignment of DNA strands and the polymerization of additional nucleotides to fill in the gaps.

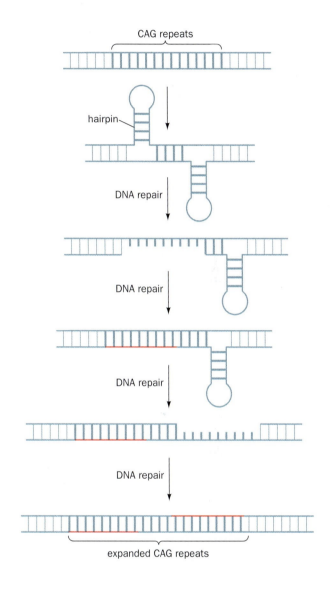

STRs are also widely used to prove or disprove familial relationships. For example, oral tradition suggests that Thomas Jefferson, the third American president, fathered a son, Eston Hemmings (born in 1808), with his slave Sally Hemmings (Eston Hemmings was said to bear a striking

physical resemblance to Jefferson). Only the tips of the Y chromosome undergo recombination (with the X chromosome) and the rest of it is passed unchanged from father to son (except for occasional mutations). The finding that the Y chromosomes of male-line descendents of both Eston Hemmings and Jefferson's father's brother (Jefferson had no surviving legitimate sons) had STRs of identical lengths indicates that Thomas Jefferson was probably Eston Hemmings' father (although this could also be true of any of Jefferson's contemporary male-line relatives).

Moderately repetitive sequences ($<10^6$ copies per haploid genome) occur in segments of 100 to several thousand base pairs that are interspersed with larger blocks of unique DNA. Most of the repetitive sequences are retrotransposons (transposable elements that propagate through the intermediate synthesis of RNA; Section 24-6C). Around 42% of the human genome consists of three types of retrotransposons:

1. **Long interspersed nuclear elements (LINEs)** are 6- to 8-kb long segments (Section 24-6C). The great majority of these have accumulated mutations that render them transcriptionally inactive.

2. **Short interspersed nuclear elements (SINEs)** consist of 100- to 400-bp elements. The most common SINEs in the human genome are members of the *Alu* **family,** which are so named because most of their ~300-bp segments contain a cleavage site for the restriction endonuclease *Alu*I (AGCT; Table 3-2). The globin gene cluster contains several *Alu* elements (Fig. 27-6).

3. Retrotransposons with **long terminal repeats (LTRs).**

In addition, the human genome contains DNA transposons that resemble bacterial transposons. Overall, ~45% of the human genome consists of widely dispersed and almost entirely inactive transposable elements (Table 27-2).

No function has been unequivocally assigned to moderately repetitive DNA, which therefore has been termed **selfish** or **junk DNA.** This DNA apparently is a molecular parasite that, over many generations, has disseminated itself throughout the genome through transposition. The theory of natural selection predicts that the increased metabolic burden imposed by the replication of an otherwise harmless selfish DNA would eventually lead to its elimination. Yet for slowly growing eukaryotes, the relative disadvantage of replicating an additional 100 bp of selfish DNA in an ~1-billion-bp genome would be so slight that its rate of elimination would be balanced by its rate of propagation. *Because unexpressed sequences are subject to little selective pressure, they accumulate mutations at a greater rate than do expressed sequences.* The resulting variations can be used to trace evolutionary relationships (see Box 27-2).

Table 27-2   Moderately Repetitive Sequences in the Human Genome

Type of Repeat	Length (bp)	Number of Copies ($\times$ 1000)	Percentage of Genome
LINEs	6000–8000	868	20.4
SINEs	100–300	1558	13.1
LTR retrotransposons	1500–11,000	443	8.3
DNA transposons	80–3000	294	2.8
Total			44.8

Source: International Human Genome Sequencing Consortium, *Nature* **409,** 800 (2001).

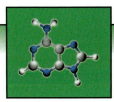

BOX 27-2

Perspectives in Biochemistry

Inferring Genealogy from DNA Sequences

Mutations in DNA sequences can create or obliterate restriction sites, giving rise to different sized fragments when the DNA is digested by prokaryotic restriction endonucleases (Section 3-4A and Box 3-1). The genealogy of several human populations has been inferred from such restriction fragment length polymorphisms in five segments of their β-globin gene clusters. This study has led to the construction of a "family tree" in which the length of the horizontal axis indicates the genetic distance between the populations and hence the times between their divergence.

This tree suggests that Eurasian populations are much more closely related to each other than they are to African populations. Fossil evidence indicates that anatomically modern humans arose in Africa about 100,000 years ago and rapidly spread throughout that continent. This family tree therefore suggests that all Eurasian populations are descended from a surprisingly small "founder population" (perhaps only a few hundred individuals) that left Africa ~50,000 years ago.

[Figure after Wainscoat, J.S., Hill, A.V.S., Boyce, A.L., Flint, J., Hernandez, M., Thein, S.L., Old, J.M., Lynch, J.R., Falusi, A.G., Weatherall, D.J., and Clegg, J.B., *Nature* ***319,*** 493 (1986).]

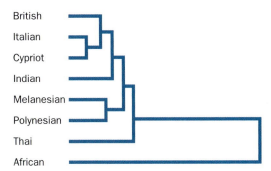

2 Regulation of Prokaryotic Gene Expression

> **See Guided Exploration 30**
>
> The regulation of gene expression by the *lac* repressor system.

A complete genome sequence reveals the metabolic capabilities of an organism, but *gene sequences alone do not necessarily indicate when or where the encoded molecules are produced.* The route from gene sequence to fully functional gene product offers many potential points for regulation, but *in prokaryotes, gene expression is primarily controlled at the level of transcription.* This is perhaps because prokaryotic mRNAs have lifetimes of only a few minutes, so translational control is less necessary. In this section, we examine a few well-documented examples of gene regulation in prokaryotes. In the following section, we shall consider how eukaryotic cells regulate gene expression.

A The *lac* Repressor

Bacteria adapt to their environments by producing enzymes that metabolize certain nutrients only when those substances are available. For example, *E. coli* cells grown in the absence of lactose are initially unable to metabolize this disaccharide. To do so they require two proteins: **β-galactosidase,** which catalyzes the hydrolysis of lactose to its component monosaccharides,

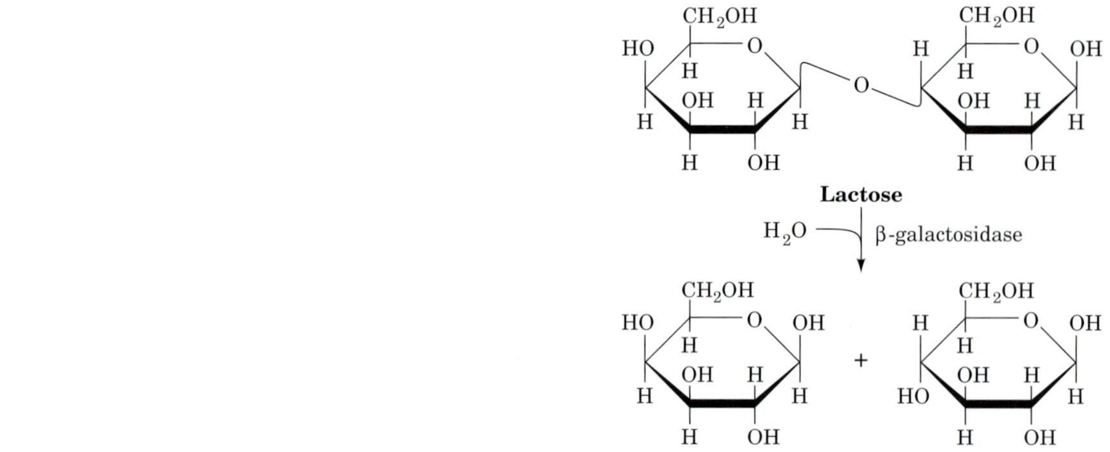

Lactose

H_2O → β-galactosidase

Galactose   +   **Glucose**

and **galactoside permease** (also known as **lactose permease**), which transports lactose into the cell. Cells grown in the absence of lactose contain only a few molecules of these proteins. Yet a few minutes after lactose is introduced into their medium, the cells increase the rate at which they synthesize these proteins by ~1000-fold and maintain this pace until lactose is no longer available. *This ability to produce a series of proteins only when the substances they metabolize are present permits the bacteria to adapt to their environment without the debilitating need to continuously synthesize large quantities of otherwise unnecessary enzymes.*

Lactose or one of its metabolic products must somehow act as an **inducer** to trigger the synthesis of the above proteins. The physiological inducer of the lactose system, the lactose isomer **1,6-allolactose,**

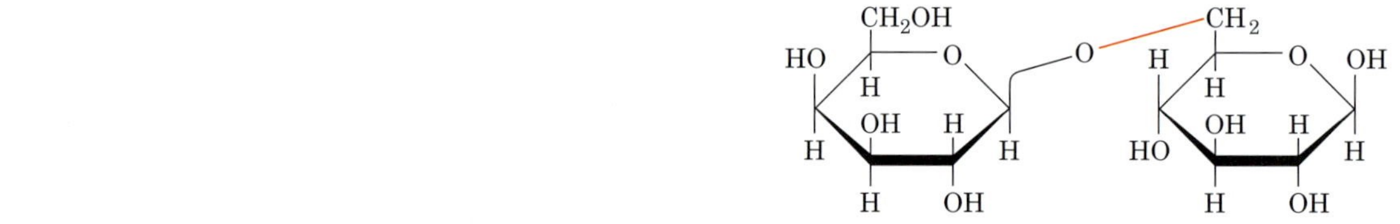

1,6-Allolactose

arises from lactose's occasional transglycosylation by β-galactosidase. Most *in vitro* studies of lactose metabolism use **isopropylthiogalactoside (IPTG),**

Isopropylthiogalactoside (IPTG)

a synthetic inducer that structurally resembles allolactose but is not degraded by β-galactosidase. Natural and synthetic inducers also stimulate the synthesis of **thiogalactoside transacetylase,** an enzyme whose physiological role is unknown.

The genes specifying β-galactosidase, galactoside permease, and thiogalactoside transacetylase, designated *Z, Y,* and *A,* respectively, are contiguously arranged in the *lac* operon (Fig. 25-4). All three structural genes are translated from a single mRNA transcript. A nearby gene, *I,* encodes the ***lac* repressor,** a protein that inhibits the synthesis of the three *lac* proteins.

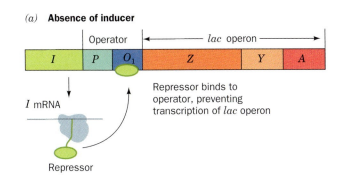

(a) **Absence of inducer**

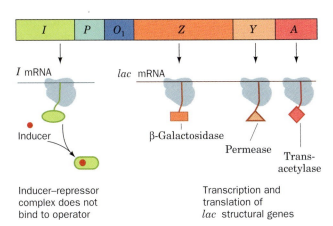

(b) **Presence of inducer**

Figure 27-9 *Key to Function. The expression of the lac operon.* (*a*) In the absence of inducer, the repressor (the product of the *I* gene) binds to the operator (O_1), thereby preventing transcription of the *lac* operon from the promoter (*P*). (*b*) On binding inducer, the repressor dissociates from the operator, which permits the transcription and subsequent translation of the *lac* structural genes (*Z*, *Y*, and *A*, which respectively encode β-galactosidase, lactose permease, and thiogalactoside transacetylase).

lac Repressor Recognizes Operator Sequences. The target of the *lac* repressor is a region of the *lac* operon known as its **operator**, which lies near the beginning of the β-galactosidase gene. *In the absence of inducer, lac repressor specifically binds to the operator to prevent the transcription of mRNA (Fig. 27-9). On binding inducer, the repressor dissociates from the operator, thereby permitting the transcription and subsequent translation of the lac enzymes (Fig. 27-9b).*

The *lac* operator actually contains three operator sequences to which *lac* repressor binds with high affinity, known as O_1, O_2, and O_3. O_1, the primary repressor-binding site, was identified through its protection by *lac* repressor from nuclease digestion. The 26-bp protected sequence lies within a nearly twofold symmetrical sequence of 35 bp (Fig. 27-10). O_1 overlaps the transcription start site of the *lacZ* gene. The operator sequence O_2 is centered 401 bp downstream, fully within the *lacZ* gene, and O_3 is centered 93 bp upstream of O_1, at the end of the *lacI* gene. Genetic engineering experiments show that all three operator sequences must be present for maximum repression *in vivo*.

The observed rate constant for the binding of *lac* repressor to *lac* operator is $k \approx 10^{10}$ $M^{-1} \cdot s^{-1}$. This "on" rate is much greater than that calculated for the diffusion-controlled process in solution: $k \approx 10^7$ $M^{-1} \cdot s^{-1}$ for molecules the size of *lac* repressor. Since it is impossible for a reaction to proceed faster than its diffusion-controlled rate, *lac* repressor must not encounter operator from solution in a random three-dimensional search. Rather, it appears that *lac repressor finds its operator by nonspecifically binding to DNA and sliding along it in a far more efficient one-dimensional search.*

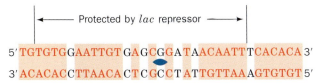

Figure 27-10 **The base sequence of the *lac* operator O_1.** Its symmetry-related regions, which comprise 28 of its 35 bp, are shaded in red.

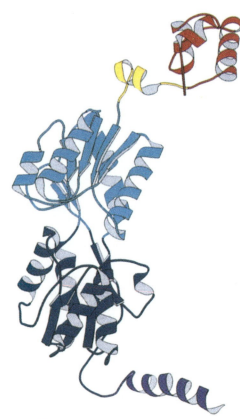

Figure 27-11 **Ribbon diagram of the *lac* repressor monomer.** The DNA-binding domain containing its helix–turn–helix motif is red; the hinge helix is yellow; the inducer-binding core is light and dark blue; and the tetramerization helix is purple. [Courtesy of Ponzy Lu and Mitchell Lewis, University of Pennsylvania. PDBid 1LBI.]

***lac* Repressor Binds Two DNA Segments Simultaneously.** The isolation of *lac* repressor, by Beno Müller-Hill and Walter Gilbert in 1966, was exceedingly difficult because the repressor constitutes only ~0.002% of the protein in wild-type *E. coli.* Molecular cloning techniques (Section 3-5) later made it possible to produce large quantities of *lac* repressor. Nevertheless, it was not until 1996 that Ponzy Lu and Mitchell Lewis reported the complete three-dimensional structure of this protein. Each 360-residue subunit of the repressor homotetramer has four functional units (Fig. 27-11):

1. An N-terminal "headpiece," which contains a helix–turn–helix (HTH) motif that resembles those in other prokaryotic DNA-binding proteins (Section 23-4B) and which specifically binds operator DNA sequences.

2. A linker, which contains a short α helix that acts as a hinge and also binds DNA. In the absence of DNA, the hinge helices of the *lac* repressor tetramer are disordered, allowing the headpieces to move freely.

3. A two-domain core, which binds inducers such as IPTG.

4. A C-terminal α helix, which is required for the quaternary structure of *lac* repressor. In the tetramer, all four C-terminal helices associate. Surprisingly, the repressor homotetramer does not exhibit the D_2 symmetry of nearly all homotetrameric proteins of known structure (three mutually perpendicular twofold axes; Section 6-3) but is instead V-shaped (with only twofold symmetry) and is therefore best considered to be a dimer of dimers (Fig. 27-12).

The X-ray structure of a complex between *lac* repressor and a 21-bp synthetic DNA containing a high-affinity binding sequence reveals that each repressor tetramer binds two DNA segments (Fig. 27-12). The repressor's HTH motif fits snugly into the major groove, bending the DNA so that it has a radius of curvature of 60 Å. The two bound DNA segments are laterally separated by ~25 Å.

IPTG binds to the repressor core at the interface between the two domains colored light and dark blue in Fig. 27-11. This binding induces a conformational change in the repressor dimer that is communicated through the hinge helices to the headpieces. This causes the two DNA-binding do-

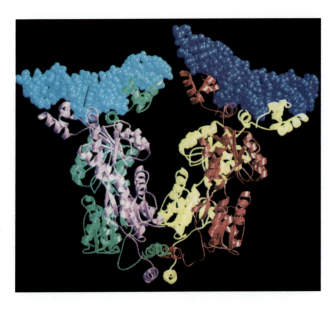

Figure 27-12 **X-Ray structure of *lac* repressor tetramer bound to two 21-bp segments of DNA.** The monomeric units of the repressor, shown in ribbon form, are green, pink, yellow, and red, and the DNA segments, shown in space-filling form, are cyan and blue. [Courtesy of Ponzy Lu and Mitchell Lewis, University of Pennsylvania. PDBid 1LBG.]

mains in each dimer to separate by ~3.5 Å so that they can no longer simultaneously bind DNA, thereby causing the repressor to dissociate from the DNA.

The *lac* repressor is an allosteric protein: IPTG binding to one subunit alters the DNA-binding activity of its dimeric partner (but not of the other dimer in the repressor tetramer). Since the allosteric transition occurs within the dimer, why does full repressor activity require a tetramer? Model-building studies provide a plausible answer to this puzzle. *A lac repressor tetramer simultaneously binds to two operators so that it brings them together, forming a loop of DNA either 93 or 401 bp long, depending on whether the repressor binds O_1 and O_3 or O_1 and O_2.* The formation of a stable looped structure may require additional DNA-binding proteins; one candidate is **CAP** (Section 27-2B), a DNA-binding protein that binds to the DNA between O_1 and O_3 (Fig. 27-13). In the model shown in Fig. 27-13, the *lac* promoter is part of the looped DNA.

It was widely assumed for years that *lac* repressor simply physically obstructs the binding of RNA polymerase (RNAP) to the *lac* promoter. However, experiments have demonstrated that RNAP can bind to the promoter in the presence of the repressor but cannot properly initiate transcription. Dissociation of the repressor in response to an inducer would allow unimpeded transcription. If RNAP were already bound to the *lac* promoter, transcription could begin immediately. Nevertheless, in the model shown in Fig. 27-13, the contact surface for RNAP is on the inside of the DNA loop, which presumably would preclude the binding of RNAP. Further studies are needed to resolve this apparent contradiction.

B Catabolite Repression: An Example of Gene Activation

Glucose is *E. coli*'s metabolic fuel of choice; adequate amounts of glucose prevent the full expression of genes specifying proteins involved in the fermentation of numerous other catabolites, including lactose, arabinose, and galactose, even when they are present in high concentrations. This phenomenon, which is known as **catabolite repression,** prevents the wasteful duplication of energy-producing enzyme systems. Catabolite repression is overcome in the absence of glucose by a cAMP-dependent mechanism. cAMP levels are low in the presence of glucose but rise when glucose becomes scarce.

CAP–cAMP Complex Stimulates the Transcription of Catabolite Repressed Operons.

Certain *E. coli* mutants, in which the absence of glucose does not relieve catabolite repression, are missing a cAMP-binding protein that is synonymously named **catabolite gene activator protein (CAP)** or **cAMP receptor protein (CRP).** CAP is a dimeric protein of identical 210-residue subunits that undergoes a large conformational change on binding cAMP. The CAP–cAMP complex, but not CAP alone, binds to the promoter region of the *lac* operon (among others) and stimulates transcription in the absence of repressor. CAP is therefore a **positive regulator** (it turns transcription on), in contrast to *lac* repressor, which is a **negative regulator** (it turns transcription off).

How does the CAP–cAMP complex operate? The *lac* operon has a weak (low-efficiency) promoter. One possibility is that CAP–cAMP binding enhances the ability of RNAP to transcribe the *lac* operon by inducing a conformational change in the promoter DNA. In fact, the X-ray structure of CAP–cAMP in complex with a 30-bp segment of DNA, whose sequence resembles the CAP-binding site, reveals that the CAP dimer binds in

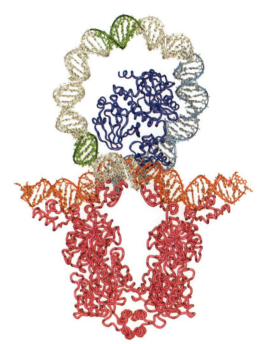

Figure 27-13 Model of the 93-bp loop formed when the *lac* repressor tetramer binds to O_1 and O_3. The proteins are represented by their C_α backbones, and the DNA is shown in skeletal form with its sugar–phosphate backbones traced by helical ribbons. The model was constructed from the X-ray structure of *lac* repressor (*red*) in complex with two 21-bp operator DNA segments (*orange*) and the X-ray structure of CAP (*purple*) in complex with its 30-bp target DNA (*cyan*). The remainder of the DNA loop was generated by applying a smooth curvature to B-DNA (*white*) with the −10 and −35 regions of the *lac* promoter highlighted in green. [Courtesy of Ponzy Lu and Mitchell Lewis, University of Pennsylvania.]

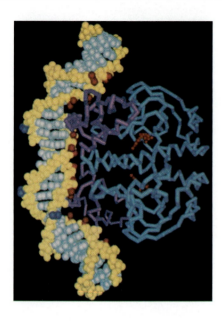

Figure 27-14 **X-Ray structure of the CAP–cAMP dimer in complex with DNA.** The protein, viewed with its twofold axis of symmetry horizontal, is represented by its C_α backbone, with its N-terminal cAMP-binding domain blue and its C-terminal DNA-binding domain magenta. The 30-bp self-complementary DNA is shown in space-filling form with its sugar–phosphate backbone yellow and its bases white. The DNA phosphates, whose ethylation interferes with CAP binding, are red. Those in the complex that are hypersensitive to DNase I are blue (these phosphates bridge the CAP-induced kinks where the minor groove has been dramatically widened). The bound cAMPs are shown in ball-and-stick form in red. [Courtesy of Thomas Steitz, Yale University. PDBid 1CGP.]
🖱 **See the Interactive Exercises.**

successive turns of DNA's major groove via its two HTH motifs so as to bend the DNA by ~90° around the protein dimer (Fig. 27-14). The bend arises from two ~45° kinks in the DNA between the fifth and sixth bases out from the complex's twofold axis in both directions and results in a closing of the major groove and in an enormous widening of the minor groove at each kink. The distorted DNA may be a more efficient substrate for transcription initiation than linear DNA. A second possibility is that CAP–cAMP stimulates transcription initiation through direct contact with RNAP. Indeed, the CAP–cAMP binding site on the *lac* operon overlaps the *lac* promoter.

The model for *lac* repressor binding (Fig. 27-13) paradoxically includes CAP, which is an activator. This dual binding may be a mechanism for conserving cellular energy in the absence of both glucose and lactose. If lactose became available, *lac* repressor would dissociate, and CAP–cAMP would be poised to promote transcription of the *lac* operon.

C Attenuation

The *E. coli* ***trp* operon** encodes five polypeptides (which form three enzymes) that mediate the synthesis of tryptophan from chorismate (Section 20-5B). These five *trp* operon genes (*A–E*; Fig. 27-15) are coordinately ex-

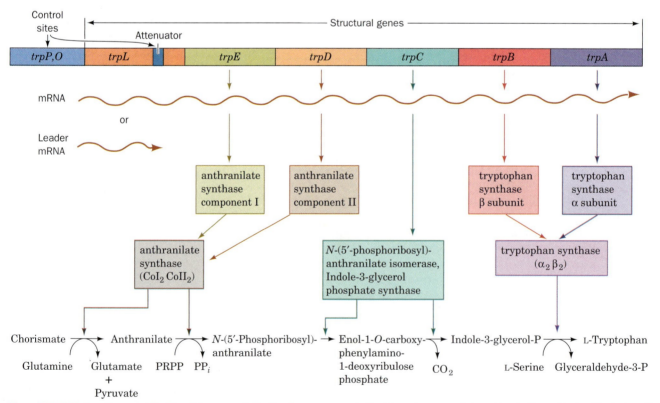

Figure 27-15 **Genetic map of the *E. coli trp* operon indicating the enzymes it specifies and the reactions they catalyze.** The gene product of *trpC* catalyzes two sequential reactions in the synthesis of tryptophan (Section 20-5C). [After Yanofsky, C., *J. Am. Med. Assoc.* **218**, 1027 (1971).]

pressed under the control of the *trp* repressor, which binds L-tryptophan to form a complex that specifically binds to the *trp* operator to reduce the rate of *trp* operon transcription 70-fold (Section 23-4B). In this system, tryptophan acts as a **corepressor;** its presence prevents what would be superfluous tryptophan biosynthesis.

The *trp* repressor–operator system was at first thought to fully account for the regulation of tryptophan biosynthesis in *E. coli*. However, the discovery of *trp* deletion mutants located downstream from the operator (*trpO*) that increase *trp* operon expression sixfold indicated the existence of an additional transcriptional control element. This element is located ~30 to 60 nucleotides upstream of the structural gene *trpE* in a 162-nucleotide **leader sequence** (*trpL;* Fig. 27-15).

When tryptophan is scarce, the entire 6720-nucleotide polycistronic *trp* mRNA, including the *trpL* sequence, is synthesized. When the encoded enzymes begin synthesizing tryptophan, the rate of *trp* operon transcription decreases as tryptophan binds to *trp* repressor. Of the *trp* mRNA that is transcribed, however, an increasing proportion consists of only a 140-nucleotide segment corresponding to the 5′ end of *trpL*. *The availability of tryptophan therefore results in the premature termination of trp operon transcription.* The control element responsible for this effect is consequently termed an **attenuator.**

The *trp* Attenuator's Transcription Terminator Is Masked when Tryptophan Is Scarce.

The attenuator transcript contains four complementary segments that can form one of two sets of mutually exclusive base-paired hairpins (Fig. 27-16). Segments 3 and 4 together with the succeeding residues constitute a transcription terminator (Section 25-1D): a G + C-rich hairpin followed by several sequential U's (compare with Fig. 25-9). Transcription rarely proceeds beyond this termination site unless tryptophan is in short supply.

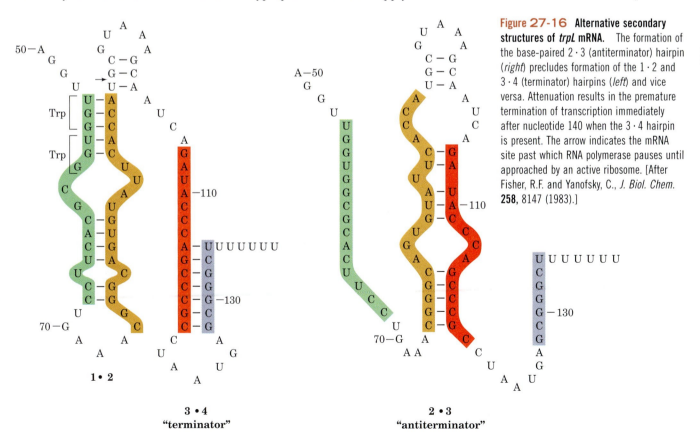

Figure 27-16   Alternative secondary structures of *trpL* mRNA.   The formation of the base-paired 2·3 (antiterminator) hairpin (*right*) precludes formation of the 1·2 and 3·4 (terminator) hairpins (*left*) and vice versa. Attenuation results in the premature termination of transcription immediately after nucleotide 140 when the 3·4 hairpin is present. The arrow indicates the mRNA site past which RNA polymerase pauses until approached by an active ribosome. [After Fisher, R.F. and Yanofsky, C., *J. Biol. Chem.* **258,** 8147 (1983).]

1·2

3·4
"terminator"

2·3
"antiterminator"

How does a shortage of tryptophan cause transcription to proceed past this terminator? A section of the leader sequence, which includes segment 1 of the attenuator, is translated to form a 14-residue polypeptide that contains two consecutive Trp residues (Fig. 27-16, *left*). The position of this particularly rare dipeptide (~1% of the residues in *E. coli* proteins are Trp) provided an important clue to the mechanism of attenuation, as proposed by Charles Yanofsky (Fig. 27-17): An RNA polymerase that has escaped repression initiates *trp* operon transcription. Soon after the ribosomal initiation site of the *trpL* sequence has been transcribed, a ribosome attaches to it and begins translating the leader peptide. When tryptophan is abundant (i.e., there is a plentiful supply of Trp–tRNA^{Trp}), the ribosome follows closely behind the transcribing RNA polymerase. Indeed, RNA polymerase pauses past position 92 of the transcript and continues transcribing only on the approach of a ribosome, thereby ensuring the close coupling of transcription and translation. The progress of the ribosome prevents the formation of the 2·3 hairpin and permits the formation of the 3·4 hairpin, which terminates transcription (Fig. 27-17a). When tryptophan is scarce, however, the ribosome stalls at the tandem UGG codons (which specify Trp) because of the lack of Trp–tRNA^{Trp}. As transcription continues, the newly synthesized segments 2 and 3 form a hairpin because the stalled ribosome prevents the otherwise competitive formation of the 1·2 hairpin (Fig. 27-17b). The 3·4 hairpin does not form, allowing transcription to proceed past this region and into the remainder of the *trp* operon. Thus, attenuation regulates *trp* operon transcription according to the tryptophan supply.

The leader peptides of the five other amino acid–biosynthesizing operons known to be regulated by attenuation are all rich in their corresponding amino acid residues. For example, the *E. coli* **his operon,** which speci-

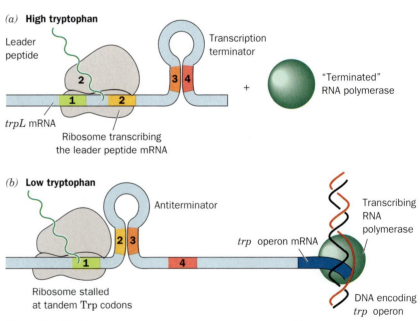

(a) **High tryptophan**

Leader peptide

Transcription terminator

"Terminated" RNA polymerase

trpL mRNA

Ribosome transcribing the leader peptide mRNA

(b) **Low tryptophan**

Antiterminator

Transcribing RNA polymerase

trp operon mRNA

Ribosome stalled at tandem Trp codons

DNA encoding *trp* operon

Figure 27-17 Attenuation in the *trp* operon. (*a*) When Trp–tRNA^{Trp} is abundant, the ribosome translates *trpL* mRNA. The presence of the ribosome on segment 2 prevents the formation of the base-paired 2·3 hairpin. The 3·4 hairpin, an essential component of the transcriptional terminator, can then form, thus aborting transcription. (*b*) When Trp–tRNA^{Trp} is scarce, the ribosome stalls on the tandem Trp codons of segment 1. This situation permits the formation of the 2·3 hairpin, which precludes the formation of the 3·4 hairpin. RNA polymerase therefore transcribes through this unformed terminator and continues transcribing the *trp* operon.

fies enzymes synthesizing histidine, has seven tandem His residues in its 16-residue leader peptide, whereas the ***ilv* operon,** which specifies enzymes participating in isoleucine, leucine, and valine biosynthesis, has five Ile, three Leu, and six Val residues in its 32-residue leader peptide. The leader transcripts of these operons resemble that of the *trp* operon in their capacity to form two alternative secondary structures, one of which contains a trailing transcription terminator.

D Riboswitches

We have just seen how the formation of secondary structure in a growing RNA transcript can regulate gene expression though attenuation. The conformational flexibility of mRNA also allows it to regulate genes by directly interacting with certain cellular metabolites, thereby eliminating the need for sensor proteins such as the *lac* repressor, CAP, and the *trp* repressor.

In *E. coli*, the biosynthesis of thiamine pyrophosphate (TPP; Section 14-3B) requires the action of several proteins whose levels vary according to the cell's need for TPP. In at least two of the relevant genes the untranslated regions at the 5′ end of the mRNA include a highly conserved sequence called the ***thi* box.** The susceptibility of the *thi* box to chemical or enzymatic cleavage, as Ronald Breaker showed, differs in the presence and absence of TPP, suggesting that the RNA changes its secondary structure when TPP binds to it (the binding of a metabolite by RNA is not unprecedented; synthetic oligonucleotides known as **aptamers** bind specific molecules with high specificity and affinity). The predicted secondary structures of the mRNA are shown in Fig. 27-18. In the absence of TPP, the mRNA assumes a conformation that allows a ribosome to begin translation. In the presence of TPP, an alternative secondary structure masks its

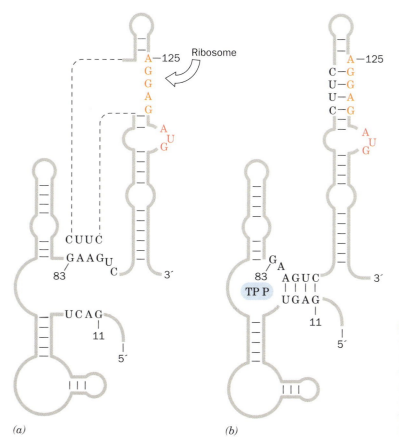

(a)   (b)

Figure 27-18 TPP-dependent mRNA conformations. The predicted secondary structure of a 165-residue segment at the 5′ end of the *E. coli thiM* gene is shown in the absence (*a*) and presence (*b*) of TPP. The TPP-binding conformation masks the Shine–Dalgarno sequence (*orange*) required by the ribosome to initiate translation at the AUG sequence (*red*) just downstream. [After Winkler, W., Nahvi, A., and Breaker, R.R., *Nature* **419**, 955 (2002).]

Shine–Dalgarno sequence (Section 26-4A) so that the ribosome cannot initiate the mRNA's translation. Thus, *the concentration of a metabolite can regulate the expression of genes required for its synthesis.*

The TPP-sensing mRNA element has been dubbed a **riboswitch.** At least eight bacterial riboswitches have been identified, including those that regulate the expression of enzymes involved in the synthesis of coenzyme B_{12} (Fig. 19-17), riboflavin (Fig. 13-11), and *S*-adenosylmethionine (SAM; Fig. 20-18). In some cases, metabolite binding to the mRNA controls the formation of an internal transcription termination site so that transcription beyond this site proceeds only when the metabolite is absent. In others, a ribozyme that forms a portion of an mRNA is activated by the binding of a metabolite to self-cleave, thereby inactivating the mRNA. Plants and fungi also appear to contain riboswitches. The fact that the interaction of riboswitches with their effectors does not require the participation of proteins suggests that they are relics of the RNA world (Box 23-3) and hence among the oldest regulatory systems.

3 | Regulation of Eukaryotic Gene Expression

The general principles that govern the expression of prokaryotic genes apply also to eukaryotic genes: *The expression of specific genes may be actively inhibited or stimulated through the effects of proteins that bind to DNA or RNA.* As in prokaryotes, the majority of known regulatory mechanisms act at the level of gene transcription, but unlike prokaryotic control systems, the eukaryotic mechanisms must contend with much larger amounts of DNA that is packaged in seemingly inaccessible structures. In this section, we describe a few of the strategies whereby eukaryotic cells manage their genetic information.

A Chromatin Structure and Gene Expression

The vast majority of DNA in multicellular organisms is not expressed. This includes the large portion of the genome that does not encode protein or RNA, as well as the genes whose expression is inappropriate for a particular cell type. Although nearly all the cells in an organism contain identical sets of DNA, genes are expressed in a highly tissue-specific manner. For example, most pancreatic cells synthesize and secrete digestive enzymes, but the pancreatic islet cells synthesize insulin or glucagon instead (Section 21-2).

Nonexpressed DNA is typically highly condensed in a form known as **heterochromatin.** An extreme example of this is the complete inactivation of one of the two X chromosomes in female mammals (Box 27-3). Transcriptionally active DNA, which is known as **euchromatin,** is less condensed, presumably to provide access to the transcription machinery. In fact, Harold Weintraub demonstrated that transcriptionally active chromatin is more susceptible to digestion by pancreatic DNase I than transcriptionally inactive chromatin. Yet nuclease sensitivity apparently reflects a gene's potential for transcription rather than transcription itself.

The banding pattern of **polytene chromosomes** found in certain secretory cells of dipteran (two-winged) flies reflects the selective condensation of nontranscribed DNA (Fig. 27-19). Polytene chromosomes result from the multiple replications of chromosomes such that the replicas remain attached to each other and in register. The light bands of such chromosomes correspond to specific genes; when undergoing transcription, these bands

BOX 27-3

Perspectives in Biochemistry

X Chromosome Inactivation

Female mammalian cells contain two X chromosomes, whereas male cells have one X and one Y chromosome. *Female somatic cells, however, maintain only one of their X chromosomes in a transcriptionally active state.* Consequently, both males and females make approximately equal amounts of X chromosome-encoded gene products, a phenomenon known as **dosage compensation.**

The inactive X chromosome is visible as a highly condensed and darkly staining **Barr body** at the periphery of the cell nucleus.

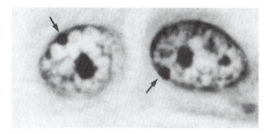

[From Moore, K.L. and Barr, M.L., *Lancet* **2,** 57 (1955).]

In marsupials (pouched mammals such as kangaroos), the Barr body is always the paternally inherited X chromosome, but in placental mammals, one randomly selected X chromosome in every somatic cell is inactivated when the embryo consists of only a few cells. The progeny of each of these cells maintain the same inactive X chromosome. Female placental animals are therefore mosaics of cloned groups of cells in which the active X chromosome is either paternally or maternally inherited. The calico cat, for example, with its patches of black and yellow fur, is almost always a female

cat whose two X chromosomes specify the different coat colors.

[© Hank Delespinasse/Age Fotostock America, Inc.]

The mechanism of chromosome inactivation is only beginning to come to light. Inactivation appears to be triggered by the transcription of the *Xist* gene in the inactive chromosome only. The resulting *Xist* RNA coats the inactive X chromosome over its entire length but does not bind to the active X chromosome. The bound *Xist* RNA appears to recruit specific DNA binding proteins that repress transcription, as well as variant histones. The CpG islands (Section 27-1A) within many promoters on the inactive X chromosome become methylated, probably as part of the mechanism that maintains the chromosome's inactive state in subsequent cell generations.

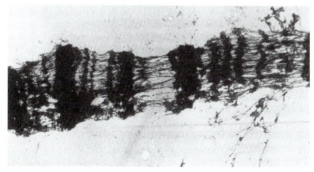

Figure 27-19   Electron micrograph of a segment of a polytene chromosome from *D. melanogaster*.   Note that its interband regions consist of chromatin fibers that are more or less parallel to the long axis of the chromosome, whereas its bands, which contain ~95% of the chromosome's DNA, are much more highly condensed. [Courtesy of Gary Burkholder, University of Saskatchewan, Canada.]

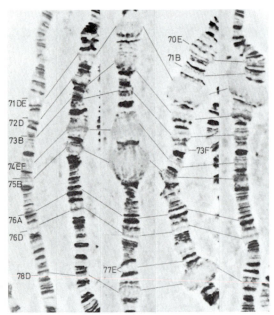

Figure 27-20 Formation and regression of chromosome puffs *(lines).* This series of photomicrographs shows a *D. melanogaster* polytene chromosome over a 22-h period of larval development. [Courtesy of Michael Ashburner, Cambridge University, UK.]

are further decondensed to form **chromosome puffs** (Fig. 27-20). In *Drosophila,* these puffs reproducibly form and regress as part of the normal larval development program and in response to physiological stimuli such as hormones and heat.

In order for genetic information to be expressed, the transcription machinery must gain access to the DNA, which is packaged in nucleosomes (Section 23-5B). Not surprisingly, nucleosomes are not fixed structures but can undergo remodeling. In addition, histones can be covalently modified to alter their ability to interact with other cellular components.

Chromatin-Remodeling Complexes Shuffle Nucleosomes. Sequence-specific DNA-binding proteins must gain access to their target DNAs before they can bind to them. Yet nearly all DNA in eukaryotes is sequestered by nucleosomes if not by higher order chromatin. How then do the proteins that bind to DNA segments gain access to their target DNAs? The answer, which has only become apparent since the mid-1990s, is that *chromatin contains ATP-driven complexes that remodel nucleosomes,* that is, they somehow disrupt the interactions between histones and DNA in nucleosomes to make the DNA more accessible. This may cause the histone octamer to slide along the DNA strand to a new location or even relocate to a different DNA strand. Thus, *these **chromatin-remodeling complexes** impose a "fluid" state on chromatin that maintains its DNA's overall packaging but transiently exposes individual sequences to interacting factors.*

Chromatin-remodeling complexes consist of multiple subunits. The first of them to be characterized was the yeast **SWI/SNF** complex, so called because it is essential for mating type switching (SWI for *swi*tching defective) and for growth on sucrose (SNF for *s*ucrose *n*on*f*ermenter). SWI/SNF, an 1150-kD complex of 11 different types of subunits, is essential for the expression of only ~3% of yeast genes and is not required for cell viability. However, a related complex named **RSC** (for *r*emodels the *s*tructure of *c*hromatin) is ~100 times more abundant in yeast and is required for cell viability. RSC shares two subunits with SWI/SNF, and many of their remaining subunits are homologs, including their ATPase subunits. All eukaryotes contain multiple chromatin-remodeling complexes.

An electron microscopy–based image of yeast RSC, determined by Francisco Asturias and Roger Kornberg, reveals that RSC consists of four modules surrounding a central cavity (Fig. 27-21*a*). Biochemical studies indicate that RSC binds tightly to nucleosomes in a 1:1 complex. Indeed, the size and shape of RSC's central cavity appear to be appropriate for binding a single nucleosome core particle as Fig. 27-21*b* indicates.

The simultaneous release of all of the many interactions holding DNA to a histone octamer would require an enormous free energy input and hence is unlikely to occur. How, then, do chromatin-remodeling complexes function? Their various ATPase subunits share a region of homology with helicases (Section 24-2B), although they lack helicase activity. Nevertheless, it seems plausible that, like helicases, chromatin-remodeling complexes "walk" up DNA strands as driven by ATP hydrolysis. If such a complex was somehow tethered to a histone, this would put torsional strain on the DNA in the nucleosome, thereby decreasing its local twist (DNA supercoiling is discussed in Section 23-1C). The region of decreased twist could diffuse along the DNA wrapped around the nucleosome, thereby transiently loosening the histone octamer's grip on a segment of DNA. The torsional strain might also be partially accommodated as a writhe, which would lift a seg-

Figure 27-21   Electron microscopy–based image of yeast RSC. (*a*) Two views of the structure (*front and back*) at ~28 Å resolution showing four modules surrounding a central cavity. (*b*) A model made by manually fitting the X-ray structure of the core nucleosome (Fig. 23-45) reduced to 25 Å resolution into the central cavity of the RSC structure in Part *a*. The complex is shown in mesh outline with RSC red and the nucleosome, which is viewed edgewise, yellow. The nucleosome fits snugly into the cavity with no steric clash. The scale bar in both parts represents 100 Å. [Courtesy of Francisco Asturias, The Scripps Research Institute, La Jolla, California.]

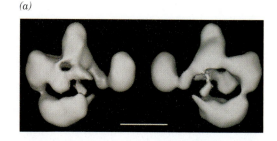

(*a*)

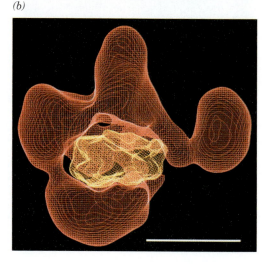

(*b*)

ment of DNA off the nucleosome's surface. In either case, the resulting DNA distortion could diffuse around the surface of the nucleosome in a wave that would locally and transiently release the DNA from the histone octamer as it passed (Fig. 27-22) and hence permit the DNA to bind to its cognate DNA-binding factors. Multiple cycles of ATP hydrolysis would send multiple DNA-loosening waves around the nucleosome, thereby sliding the nucleosome along the DNA.

HMG Proteins Are Architectural Proteins That Help Regulate Gene Expression. The variation of a given gene's transcriptional activity according to cell type indicates that chromosomal proteins participate in the gene activation process. Yet histones' chromosomal abundance and lack of variety make it highly unlikely that they have the specificity required for this role. Among the most common nonhistone proteins are the members of the **high mobility group (HMG),** so named because of their high electrophoretic mobilities in polyacrylamide gels. These highly conserved, low molecular mass

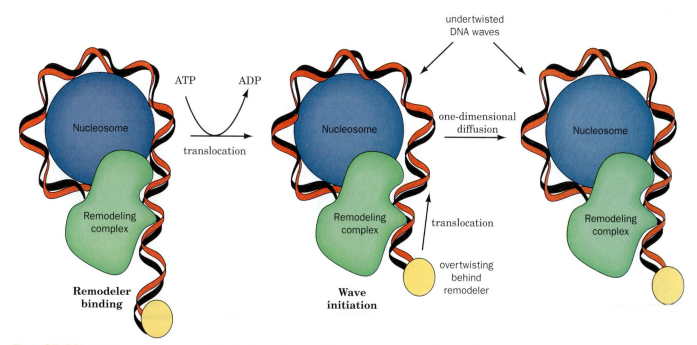

Figure 27-22   Model for nucleosome remodeling by chromatin-remodeling complexes. The chromatin-remodeling complex (*green*) couples the free energy of ATP hydrolysis to the translocation and concomitant twisting of the DNA in the nucleosome (*blue,* only half of which is shown for clarity) as depicted by the movement of a fixed point on the DNA (*yellow ball*). This locally breaks the contacts between the histones and the DNA. The position of the undertwisted and/or bulged DNA propagates around the nucleosome in a one-dimensional wave that transiently releases the DNA from the histone as it passes, thereby providing DNA-binding factors access to the DNA. [After a drawing by Saha, A., Wittmeyer, J., and Cairns, B.R., *Genes Dev.* **16**, 2120 (2002).]

(<30 kD) proteins, which have the unusual amino acid composition of ~25% basic side chains and 30% acidic side chains, are relatively abundant with ~1 HMG molecule per 10 to 15 nucleosomes.

The yeast HMG protein known as **NHP6A** contains an ~80-residue structural motif called an **HMG box.** Its NMR structure in complex with DNA, determined by Juli Feigon, reveals that it causes the DNA to bend by as much as 130° toward its major groove (Fig. 27-23). Such proteins presumably facilitate the binding of other regulatory proteins to the DNA and are therefore described as **architectural proteins.** Other HMG proteins differ in structure but can also alter DNA conformation by inducing it to bend, straighten, unwind, or form loops. Some HMG proteins compete with histones for DNA, thereby altering nucleosomal structure.

Histones Are Covalently Modified. The posttranslational modifications to which core histones are subject include the acetylation/deacetylation of specific Lys side chains, the methylation of specific Lys and Arg side chains, the phosphorylation/dephosphorylation of specific Ser and possibly Thr side chains, and the ubiquitination of specific Lys side chains (Fig. 27-24). Moreover, Lys side chains can be mono-, di-, and trimethylated and Arg side chains can be mono- and both symmetrically and asymmetrically dimethylated. The core histones' N-terminal tails, as we have seen (Section 23-5C), are implicated in stabilizing the structures of both core nucleosomes and higher order chromatin. All of these modifications but methylations reduce (make more negative) the electronic charge of the side chains to which they are appended and hence are likely to weaken histone–DNA interactions

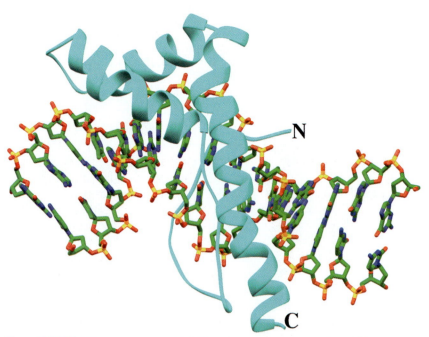

Figure 27-23 NMR structure of yeast NHP6A protein in complex with a 15-bp DNA. The protein is drawn in ribbon form (*cyan*) and the DNA is drawn in stick form colored according to atom type (C green, N blue, O red, and P yellow). The three helices of the HMG box form an L shape, with the inside of the L inserted into the DNA's minor groove so as to bend the DNA by ~70° toward its major groove. [Based on an NMR structure by Juli Feigon, University of California at Los Angeles. PDBid 1J5N.]

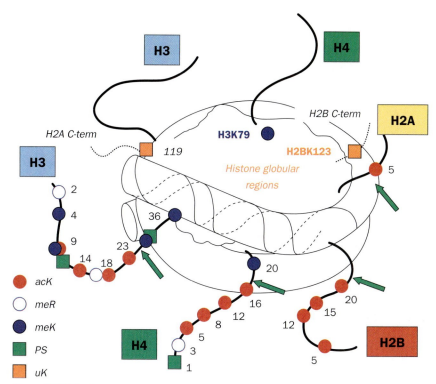

Figure 27-24 Histone modifications on the nucleosome core particle.   Posttranslational modification sites are indicated by the residue numbers and the colored symbols, which are defined in the key at the lower left (acK = acetyl-Lys, meR = methyl-Arg, meK = methyl-Lys, PS = phospho-Ser, and uK = ubiquitinated Lys). Note that H3 Lys 9 can be either methylated or acetylated. The N-terminal tail modifications are shown on only one of the two molecules of H3 and H4, and only one molecule each of H2A and H2B is shown. The C-terminal tails of one H2A and one H2B are represented by dashed lines. The green arrows indicate the sites in intact nucleosomes that are susceptible to trypsin cleavage. This cartoon summarizes data from several organisms, some of which may lack particular modifications. [Courtesy of Bryan Turner, University of Birmingham School of Medicine, U.K.]

so as to promote chromatin decondensation, although as we shall see, this is not always the case. Methyl groups, in contrast, increase the basicity and hydrophobicity of the side chains to which they are linked and hence tend to stabilize chromatin structure. Modified histone tails also interact with specific chromatin-associated nonhistone proteins in a way that changes the transcriptional accessibility of their associated genes.

The characterization of a variety of histone tail modifications led David Allis to hypothesize that *there is a* **histone code** *in which specific modifications evoke certain chromatin-based functions and that these modifications act sequentially or in combination to generate unique biological outcomes.* For example, uncondensed and hence transcriptionally active chromatin is associated with the acetylation of histone H3 Lys 9 and 14 and histone H4 Lys 5 and the methylation of H3 Lys 4 and H4 Arg 3; condensed and hence transcriptionally inactive chromatin is associated with the acetylation of H4 Lys 12 and the methylation of H3 Lys 9; and nucleosome deposition is associated with the phosphorylation of H3 Ser 10 and 28. It can be seen from Fig. 27-24 that there are a vast number of possible combinations of histone modifications.

Histone Acetyltransferases Are Components of Multisubunit Transcriptional Co-activators. Histone Lys side chains are acetylated in a sequence-specific manner by enzymes known as **histone acetyltransferases (HATs),** all of which employ acetyl-CoA as their acetyl group donor:

$$\text{CoA}-\text{S}-\overset{\overset{\textstyle O}{\|}}{\text{C}}-\text{CH}_3 \quad + \quad \text{Lys}-(\text{CH}_2)_4-\overset{+}{\text{N}}\text{H}_3$$

Acetyl-CoA   **Histone Lys**

$$\downarrow \text{HAT}$$

$$\text{CoA}-\text{SH} \quad + \quad \text{Lys}-(\text{CH}_2)_4-\text{NH}-\overset{\overset{\textstyle O}{\|}}{\text{C}}-\text{CH}_3$$

Histone Acetyl-Lys

Most if not all HATs function *in vivo* as members of often large (10–20 subunits) multisubunit complexes, many of which were initially characterized as transcriptional regulators. For example, **TAF1,** the largest subunit of the general transcription factor TFIID (TAF for *T*BP-*a*ssociated *f*actor, where TBP is TATA-binding protein; Section 25-2C), is a HAT. Moreover, many **HAT complexes** share subunits. Thus, the HAT complex named **SAGA** contains **TAF5, TAF6, TAF9, TAF10,** and **TAF12** as does the **PCAF complex** with the exception that TAF5 and TAF6 are replaced in the PCAF complex by their close homologs **PAF65β** and **PAF65α.** Portions of TAF6, TAF9, and TAF12 are structural homologs of histones H3, H4, and H2B, respectively (Section 23-5). Consequently, these TAFs probably associate to form an architectural element that is common to TFIID, SAGA, and the PCAF complex and hence these complexes are likely to interact with TBP in a similar manner. The various HAT complexes presumably target their component HATs to the promoters of active genes.

The X-ray structure of the HAT domain of *Tetrahymena thermophila* **GCN5** (residues 48–210 of the 418-residue protein) in complex with a bisubstrate inhibitor was determined by Ronen Marmorstein. The bisubstrate inhibitor (Fig. 27-25*a*) consists of CoA covalently linked from its S atom via an isopropionyl group (which mimics an acetyl group) to the side chain of Lys 14 (the acetyl group acceptor) of the 20-residue N-terminal segment of histone H3. The enzyme (Fig. 27-25*b*) is deeply clefted and contains a core region common to all HATs of known structure (magenta in Fig. 27-25*b*) that consists of a three-stranded antiparallel β sheet connected via an α helix to a fourth β strand that forms a parallel interaction with the β sheet. Only 6 residues of the histone tail, Gly 12 through Arg 17, are visible in the X-ray structure. The CoA moiety binds in the enzyme's cleft such that it is mainly contacted by core residues. The comparison of this structure with other GCN5-containing structures indicates that the cleft has closed down about the CoA moiety.

Bromodomains Recruit Coactivators to Acetylated Lys Residues in Histone Tails. The different patterns of histone acetylation required for different functions (the histone code) suggest that the function of histone acetylation is more complex than merely attenuating the charge–charge interactions between the cationic histone N-terminal tails and anionic DNA. In fact, there is growing evidence that specific acetylation patterns are recognized by protein modules of transcriptional coactivators in much the same way that specific phosphorylated sequences are recognized by protein modules such as the SH2 domain that mediates signal transduction via protein kinase cascades (Section 21-3D). Thus, nearly all HAT-associated transcriptional coactivators contain ~110-residue modules known as **bromodomains** that

(a)

Histone H3 N-terminal peptide

$$Ac-A_1-R-T-K-Q_5-T-A-R-K-S_{10}-T-G-G-K-A_{15}-P-R-K-Q-L_{20}-COO^-$$

CH₂

CH₂

CH₂

CH₃   O   CH₂

CoA—S—CH—C—NH

Isopropionyl
group

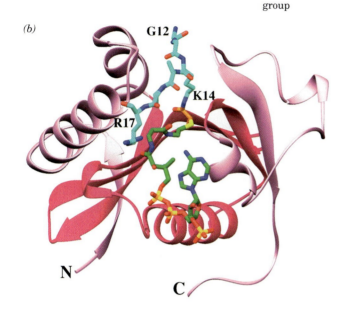

(b)

G12

K14

R17

N

C

Figure 27-25 X-Ray structure of *Tetrahymena* GCN5 in complex with an inhibitor. (*a*) The bisubstrate inhibitor, a peptide–CoA conjugate, consists of CoA covalently linked from its S atom via an isopropyl group to the side chain of Lys 14 of the 20-residue N-terminal segment of histone H3. (*b*) The protein is drawn in ribbon form in lavender with its core region magenta. The peptide–CoA conjugate is drawn in stick form colored according to atom type (histone C blue, isopropionyl group C orange, CoA C green, N blue, O red, S yellow, and P gold). [Part *b* based on an X-ray structure by Ronen Marmorstein, The Wistar Institute, Philadelphia, Pennsylvania. PDBid 1MID.]

Figure 27-26 X-Ray structure of the human TAF1 double bromodomain. Each bromodomain consists of an antiparallel four-helix bundle whose helices are colored, from N- to C-termini, red, yellow, green, and blue, with the remaining portions of the protein orange. The acetyl-Lys binding sites occupy deep hydrophobic pockets at the end of each four-helix bundle opposite its N- and C-termini. [Based on an X-ray structure by Robert Tjian, University of California at Berkeley. PDBid 1EQF.]

specifically bind acetylated Lys residues on histones. For example, GCN5 essentially consists of a HAT domain followed by a bromodomain, whereas TAF1 consists mainly of an N-terminal kinase domain followed by a HAT domain and two tandem bromodomains.

The X-ray structure of human TAF1's double bromodomain (residues 1359–1638 of the 1872-residue protein), determined by Robert Tjian, reveals that it consists of two nearly identical antiparallel four-helix bundles (Fig. 27-26). A variety of evidence, including NMR structures of single bromodomains in complex with their target acetyl-Lys–containing peptides, indicates that the acetyl-Lys binding site of each bromodomain occurs in a deep hydrophobic pocket that is located at the end of its four-helix bundle opposite its N- and C-termini. The double bromodomain's two binding pockets are separated by ~25 Å, which makes them ideally positioned to bind two acetyl-Lys residues that are separated by 7 or 8 residues. In fact, the N-terminal tail of histone H4 contains Lys residues at its positions 5, 8, 12, and 16 (Fig. 27-24), whose acetylation is correlated with increased transcriptional activity. Moreover, the 36-residue N-terminal peptide of histone H4, when fully acetylated, binds to the TAF1 double bromodomain in 1:1 ratio with 70-fold higher affinity than to single bromodomains but fails to bind when it is unacetylated.

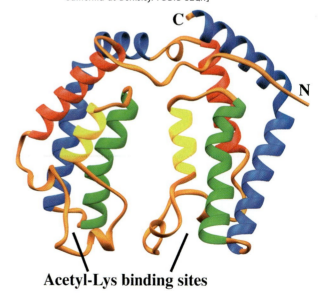

C

N

Acetyl-Lys binding sites

The foregoing structure suggests that the TAF1 bromodomains serve to target TFIID to promoters that are within or near nucleosomes [in contrast to the widely held notion that TFIID targets PICs (preinitiation complexes) to nucleosome-free regions]. Tjian has therefore postulated that the transcriptional initiation process begins with the recruitment of a HAT-containing coactivator complex by an upstream DNA-binding protein (Fig. 27-27). The HAT could then acetylate the N-terminal histone tails of nearby nucleosomes, which would recruit TFIID to an appropriately located promoter via the binding of its TAF1 bromodomains to the acetyl-Lys residues. Moreover, the TAF1 HAT activity could acetylate other nearby nucleosomes, thereby initiating a cascade of acetylation events that would render the DNA template competent for transcriptional initiation.

Histone acetylation is a reversible process. The enzymes that remove the acetyl groups from histones, the **histone deacetylases (HDACs),** promote transcriptional repression and gene silencing. Eukaryotic cells from yeast to humans typically contain numerous different HDACs; 10 HDACs have been identified in yeast and 17 in humans.

Methylated Histones Are Bound by Chromodomains. Histone methylation at both the Lys and Arg side chains of histone H3 and H4 N-terminal tails (Fig. 27-24) tends to silence the associated genes by inducing the formation of heterochromatin. The enzymes mediating these methylations, the **histone methyltransferases (HMTs),** all utilize S-adenosylmethionine (SAM) as their methyl donor. Thus, the lysine HMTs, the most extensively characterized HMTs, catalyze the following reaction:

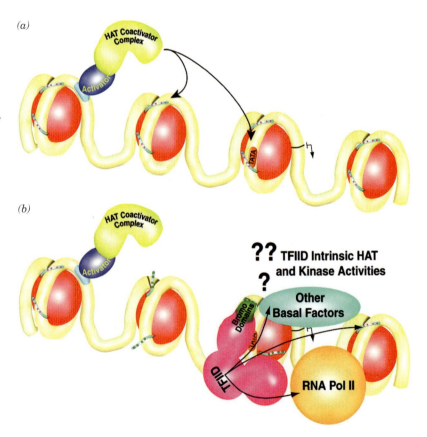

Figure 27-27   Simplified model for the assembly of a transcriptional initiation complex on chromatin-bound templates. Here the DNA is represented by a yellow worm, the histone octamers around which the DNA wraps to form nucleosomes are shown as red spheres, and their N-terminal histone tails are drawn as short cyan rods with the red and green dots representing unacetylated and acetylated Lys residues. The transcription initiation site is represented by the black ring about the DNA from which the squared-off arrow points downstream. (*a*) The process begins by the recruitment of a HAT-containing transcriptional coactivator complex (*yellow-green*) through its interactions with a DNA-binding activator protein (*purple*) that is bound to an upstream enhancer (*light blue*). The HAT coactivator complex is thereby positioned to acetylate the N-terminal tail on nearby nucleosomes (*curved arrows*). (*b*) The binding of its TAF1's bromodomains to the acetylated histone tails could then help recruit TFIID (*magenta*) to a nearby TATA box (*orange patch*). Further acetylation of nearby histone tails by the TAF1's HAT domain could help recruit other basal factors (*cyan*) and RNAP II (*orange*) to the promoter, thus stimulating PIC formation. [Modified from a drawing by Robert Tjian, University of California at Berkeley.]

Histone Lys   **S-Adenosylmethionine**

lysine histone methyltransferase

S-Adenosylhomocysteine

These enzymes all have a so-called **SET domain,** which contains their catalytic sites.

The human lysine HMT named **SET7/9** monomethylates Lys 4 of histone H3. The X-ray structure of the SET domain of SET7/9 (residues 108–366 of the 366-residue protein) in complex with SAM and the N-terminal decapeptide of histone H3 in which Lys 4 is monomethylated was determined by Steven Gamblin. Interestingly, SAM and the peptide substrate bind to opposite sides of the protein (Fig. 27-28). However, there is a narrow tunnel through the protein into which the Lys 4 side chain is inserted such that its amine group is properly positioned for methylation by SAM. The arrangement of the hydrogen bonding acceptors for the Lys amino group stabilize the methyl-Lys side chain in its observed orientation about the C_ε—N_ζ bond, thus sterically precluding the methyl-Lys group from assuming a conformation in which it could be further methylated by SAM.

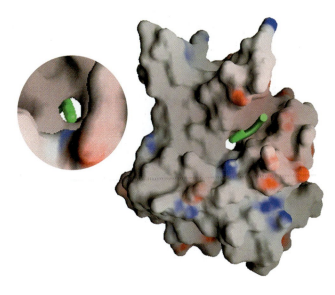

Figure 27-28 X-Ray structure of the human histone methyltransferase SET7/9 in complex with SAM and the histone H3 N-terminal decapeptide with its Lys 4 monomethylated. The protein is represented by its surface diagram colored according to charge (blue most positive, red most negative, and gray neutral) and the H3 decapeptide is represented by a green ribbon. Note the narrow tunnel through the protein in which the methyl-Lys side chain is inserted. The inset on the left shows a close-up of this Lys access channel containing the methyl-Lys side chain (*green*) as viewed from the opposite side of the protein from the SAM-binding site. [Courtesy of Steven Gamblin, National Institute for Medical Research, London, U.K.]

Methylated histones are recognized by so-called **chromodomains.** For example, methylated H3 Lys 9 is bound by the chromodomain-containing **heterochromatin protein 1 (HP1),** which thereon recruits proteins that control chromatin structure and gene expression. The NMR structure of the mouse HP1 chromodomain (residues 8–80 of the 185-residue protein) in complex with the N-terminal 18-residue tail of H3 in which Lys 4 and 9 are dimethylated, determined by Natalia Murzina and Earnest Laue, reveals that HP1 binds the H3 tail in an extended β-strand-like conformation in a groove on its surface (Fig. 27-29). The chromodomain buries the side chain of H3 Lys 9 (but not that of H3 Lys 4) such that its two methyl groups are contained in a hydrophobic box formed by three conserved aromatic residues. In contrast, the unmethylated H3 tail does not bind to HP1.

Unlike acetylation, methylation of histones is not readily reversed and appears to correlate with the formation of heterochromatin. In transcriptionally silent chromatin, the DNA itself may also be methylated (see below).

Heterochromatin has a tendency to spread, thus silencing the newly heterochromatized genes. One way that this appears to occur is via the binding of HP1 to nucleosomes whose H3 Lys 9 residues have been methylated (which is associated with transcriptionally inactive chromatin). The bound HP1 recruits the HMT **Suv39h,** which methylates nearby nucleosomes at their H3 Lys 9 residues, thereby recruiting additional HP1, etc.

Methyltransferases Flip Their Target Bases Out of the DNA Double Helix. The methylation of DNA, as we have seen (Box 24-4), switches off eukaryotic gene expression. Eukaryotic DNAs are methylated only on their cytosine residues to form 5-methylcytosine (m^5C) residues, a modification that occurs largely on the palindromic CpG islands that commonly occur in the upstream promoter regions of genes (Section 27-1A). The enzymes that mediate these reactions, the **DNA methyltransferases (DNA MTases),** all use SAM as their methyl donor. Cells also contain **demethylases,** which catalyze the reaction

$$\text{5-Methylcytosine} + H_2O \longrightarrow \text{cytosine} + HO{-}CH_3$$

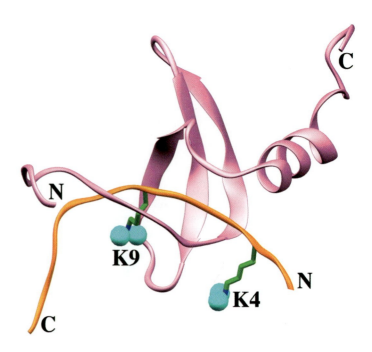

Figure 27-29 X-Ray structure of the mouse HP1 chromodomain in complex with the 18-residue N-terminal tail of histone H3 in which Lys 4 and Lys 9 are dimethylated.   The 80-residue chromodomain is lavender, the H3 N-terminal tail is orange, and its two dimethyl-Lys side chains (K9 and K4) are drawn in stick form with C green and N blue, with their methyl groups represented by cyan spheres. The side chain of H3 Lys 9, but not that of H3 Lys 4, is buried by the chromodomain. [Based on an X-ray structure by Natalia Murzina and Ernest Laue, University of Cambridge, U.K. PDBid 1GUW.]

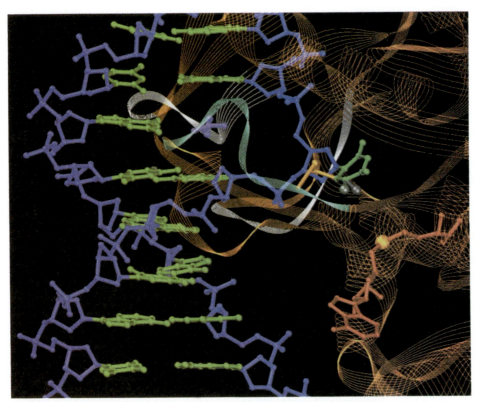

Figure 27-30   X-Ray structure of the M.HhaI DNA methyltransferase in complex with *S*-adenosylhomocysteine and a dsDNA containing a methylated 5-fluorocytosine base at the enzyme's target site.   The DNA is shown with its bases green and its backbone purple. The protein backbone is represented by a multiline orange ribbon with its active site loop cyan and its recognition loops white. The methylated 5-fluorocytosine residue has swung out of the DNA helix into the enzyme's active site pocket, where its C6 forms a covalent bond with the S atom of an enzyme Cys residue (*yellow*). The methyl group and a fluorine atom at C5 (which prevents the methylation reaction from going to completion) are both represented by silver spheres because their positions cannot be differentiated in this X-ray structure. The position of the flipped-out cytosine base in the DNA double helix is occupied by the side chain of a Gln residue (*magenta*), which hydrogen bonds to the "orphaned" guanine base. The *S*-adenosylmethionine, which has given up its methyl group, is shown in red with its S atom represented by a yellow sphere. [Based on an X-ray structure by Richard Roberts, New England Biolabs, Beverly, Massachusetts, and Xiaodong Cheng, Cold Spring Harbor Laboratory, Cold Spring Harbor, New York. PDBid 1MHT.]

In DNA MTase reactions, SAM reacts with the enzyme's targeted cytosine base to form a tetrahedral intermediate in which the cytosine C5 is substituted by both its original hydrogen atom and the donated methyl group. This requires that the attacking methyl group approach C5 from above or below the cytosine ring. Yet the faces of nucleic acid bases are inaccessible within the DNA double helix. Consequently, a DNA MTase must induce its target cytosine base to flip out of the double helix to where it can be methylated by SAM. Such a flipped-out intermediate is observed in the X-ray structure of the DNA MTase from *Haemophilus haemolyticus* (**M.HhaI**), determined by Richard Roberts and Xiaodong Cheng, in which the target cytosine base has been replaced by 5-fluorocytosine (Fig. 27-30). Although the normally present C5—H atom is eliminated from the foregoing tetrahedral intermediate as H^+ to yield the methylation reaction's m^5C product, the high electronegativity of fluorine prevents the C5—F

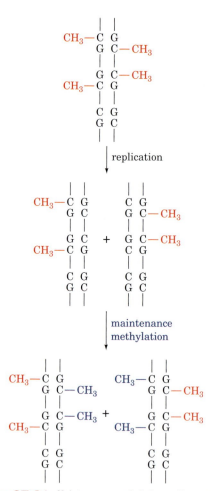

Figure 27-31 Maintenance methylation.   The pattern of methylation on a parental DNA strand induces the corresponding methylation pattern in the complementary strand. In this way, a stable methylation pattern may be maintained in a cell line.

atom from being eliminated as F^+, thereby trapping the tetrahedral intermediate on the enzyme. Certain DNA repair enzymes that act on individual DNA bases also operate by flipping their target base out the DNA helix (Section 24-5A).

DNA Methylation in Eukaryotes Is Self-Perpetuating.   The palindromic nature of DNA methylation sites in eukaryotes permits the methylation pattern on a parental DNA strand to direct the generation of the same pattern in its daughter strand (Fig. 27-31). This **maintenance methylation** would result in the stable "inheritance" of a methylation pattern in a cell line and hence cause these cells to all have the same differentiated phenotype. Such changes to the genome are described as being **epigenetic** (Greek: *epi*, on or beside) because they provide an additional layer of information that specifies when and where specific portions of the otherwise fixed genome are expressed [epigenetic changes that we have already encountered are the lengthening of telomeres in germ cells (Section 24-3C) and the inactivation of a specific X chromosome in female mammals (Box 27-3)]. Epigenetic characteristics, as we shall see, are not bound by the laws of Mendelian inheritance.

There is considerable experimental evidence favoring the existence of maintenance methylation, including the observation that artificially methylated viral DNA, when taken up by eukaryotic cells, maintains its methylation pattern for at least 30 cell generations. Maintenance methylation in mammals appears to be mediated mainly by **DNMT1** protein, which has a strong preference for methylating hemimethylated substrate DNAs. In contrast, prokaryotic DNA MTases such as M.HhaI do not differentiate between hemimethylated and fully methylated substrate DNAs. The importance of maintenance methylation is demonstrated by the observation that mice that are homozygous for the deletion of the *DNMT1* gene die early in embryonic development.

The pattern of DNA methylation in mammals varies in early embryonic development. DNA methylation levels are high in mature gametes (sperm and ova) but are nearly eliminated by the time a fertilized ovum has become a **blastocyst** (a hollow ball of cells, the stage at which the embryo implants into the uterine wall; embryonic development is discussed in Section 27-3E). After that stage, however, the embryo's DNA methylation levels globally rise until, by the time the embryo has reached the developmental stage known as a **gastrula,** its DNA methylation levels have risen to adult levels, where they remain for the lifetime of the animal. This *de novo* (new) methylation appears to be mediated by two DNA MTases distinct from DNMT1 named **DNMT3a** and **DNMT3b.** An important exception to this remethylation process is that the CpG islands of germline cells (cells that give rise to sperm or ova) remain unmethylated. This ensures the faithful transmission of the CpG islands to the succeeding generation in the face of the strong mutagenic pressure of m^5C deamination (which yields T, a mutation that mismatch repair occasionally fails to correct; Box 24-5). In female mammals, the inactivation of the same X-chromosome from cell generation to cell generation is, at least in part, preserved by maintenance methylation (Box 27-3).

The change in DNA methylation levels (epigenetic reprogramming) during embryonic development suggests that the pattern of genetic expression differs in embryonic and somatic cells. This explains the observed high failure rate in cloning mammals (sheep, mice, cattle, etc.) by transferring the nucleus of an adult cell into an enucleated oocyte (immature ovum). Few of these animals survive to birth, many of those that do so die shortly there-

after, and most of the ~1% that do survive have a variety of abnormalities, most prominently an unusually large size. However, the survival of any embryos at all is indicative that the oocyte has the remarkable capacity to epigenetically reprogram somatic chromosomes (although it is rarely entirely successful in doing so) and that mammalian embryos are relatively tolerant of epigenetic abnormalities. Presumably, the reproductive cloning of humans from adult nuclei would result in similar abnormalities and for this reason (in addition to social and ethical prohibitions) should not be attempted.

Genomic Imprinting Results from Differential DNA Methylation.   It has been known for thousands of years that maternal and paternal inheritance can differ. For example, a mule (the offspring of a mare and a male donkey) and a hinny (the offspring of a stallion and a female donkey) have obviously different physical characteristics, a hinny having shorter ears, a thicker mane and tail, and stronger legs than a mule. This is because, in mammals only, certain maternally and paternally supplied genes are differentially expressed, a phenomenon termed **genomic imprinting.** The genes that are subject to genomic imprinting are, as Rudolph Jaenisch has shown, differentially methylated in the two parents during gametogenesis and the resulting different methylation patterns are resistant to the wave of demethylation that occurs during the formation of the blastocyst and to the wave of *de novo* methylation that occurs thereafter.

The importance of genomic imprinting is demonstrated by the observation that an embryo derived from the transplantation of two male or two female pronuclei into an ovum fails to develop (pronuclei are the nuclei of mature sperm and ova before they fuse during fertilization). Inappropriate imprinting is also associated with certain diseases. For example, **Prader-Willi syndrome (PWS),** which is characterized by the failure to thrive in infancy, small hands and feet, marked obesity, and variable mental retardation, is caused by a >5000-kb deletion in a specific region of the paternally inherited chromosome 15. In contrast, **Angelman syndrome (AS),** which is manifested by severe mental retardation, a puppetlike ataxic (uncoordinated) gait, and bouts of inappropriate laughter, is caused by a deletion of the same region from the maternally inherited chromosome 15. These syndromes are also exhibited by those rare individuals who inherit both their chromosomes 15 from their mothers for PWS and from their fathers for AS. Evidently, certain genes on the deleted chromosomal region must be paternally inherited to avoid PWS and others must be maternally inherited to avoid AS. Several other human diseases are also associated with either maternal or paternal inheritance or lack of it. Aberrant epigenetic programming may also play a role in tumor growth, as cancer cells often exhibit abnormal patterns of DNA methylation.

B   Control of Transcription in Eukaryotes

In addition to the six general transcription factors described in Section 25-2C, eukaryotic cells contain a host of other proteins that interact with DNA and/or other transcription factors to stimulate or repress the transcription of class II genes (genes that are transcribed by RNAP II). Some of these proteins bind to any available DNA containing their target sequences, whereas others must be activated and deactivated, often as part of a signal transduction pathway (Section 21-3). These regulatory proteins are known as **activators** and **repressors,** and their target DNA sites are known as **enhancers** and **silencers,** respectively.

An enhancer typically is not essential for transcription but greatly increases the rate of transcription. Unlike promoters, which are necessarily located a short distance from the transcription start site, *enhancers (or silencers) need not have fixed positions and orientations* (Section 25-2B). For example, William Rutter has linked the upstream (5′-flanking) sequences of either the insulin or the chymotrypsin gene to the sequence encoding **chloramphenicol acetyltransferase (CAT),** an easily assayed enzyme not normally present in eukaryotic cells. A plasmid containing the insulin gene sequences elicits expression of the CAT gene only when introduced into cultured cells that normally produce insulin. Likewise, the chymotrypsin recombinants are active only in chymotrypsin-producing cells. Dissection of the insulin gene control sequence indicates that the enhancer lies between positions -103 and -333 and, in insulin-producing cells only, it stimulates the transcription of the CAT gene with little regard to its position and orientation relative to the promoter. Because so many enhancers are located upstream of transcribed sequences, the proteins that bind to them are often called **upstream transcription factors.**

Upstream Transcription Factors Act Cooperatively with Each Other and the PIC.

How do upstream transcription factors stimulate (or inhibit) transcription? *Evidently, when these proteins bind to their target DNA sites in the vicinity of a PIC (in some cases, many thousands of base pairs distant), they somehow activate (or repress) its component RNAP II to initiate transcription.* Transcription factors may bind cooperatively to each other and/or the PIC, thereby synergistically stimulating (or repressing) transcriptional initiation. Indeed, molecular cloning experiments indicate that many enhancers and silencers consist of segments (modules) whose individual deletion reduces but does not eliminate enhancer/silencer activity. *Such complex arrangements presumably permit transcriptional control systems to respond to a variety of stimuli in a graded manner.* In some cases, however, several transcription factors together with architectural proteins cooperatively assemble on an $\sim$100-bp enhancer to form a multisubunit complex known as an **enhanceosome,** in which the absence of a single subunit all but eliminates its ability to stimulate transcriptional initiation at the associated promoter. Thus, enhanceosomes function more like on/off switches rather than providing a graded response. Enhanceosomes may also contain **coactivators** and/or **corepressors,** proteins that do not bind to DNA but, rather, interact with proteins that do so to activate or repress transcription.

The functional properties of many upstream transcription factors are surprisingly simple. They typically consist of (at least) two domains:

1. A DNA-binding domain that specifically binds to the protein's target DNA sequence (several such domains are described in Section 23-4C).

2. A domain containing the transcription factor's activation function. Sequence analysis indicates that many of these **activation domains** have conspicuously acidic surface regions whose negative charges, if mutationally increased or decreased, respectively raise or lower the transcription factor's activity. This suggests that the associations between these transcription factors and a PIC are mediated by relatively nonspecific electrostatic interactions rather than by conformationally more demanding hydrogen bonds. Other types of activation domains have also been characterized, including those with Gln-rich regions and those with Pro-rich regions.

The DNA-binding and activation functions of eukaryotic transcription factors can be physically separated (which is why they are thought to occur on different domains). In fact, a genetically engineered hybrid protein, containing the DNA-binding domain of one transcription factor and the activation domain of a second, activates the same genes as the first transcription factor. Moreover, it makes little functional difference as to whether the activation domain is placed on the N-terminal side of the DNA-binding domain or on its C-terminal side. This geometric permissiveness in the binding between the activation domain and its target protein is also indicated by the observation that transcription factors are largely insensitive to the orientations and positions of their corresponding enhancers relative to the transcriptional start site. Of course, *the DNA between an enhancer and its distant transcriptional start site must be looped around for an enhancer-bound transcription factor to interact with the promoter-bound PIC* (Section 25-2C).

The synergy (cooperativity) of multiple transcription factors in initiating transcription may be understood in terms of a simple recruitment model. Suppose an enhancer-bound transcription factor increases the affinity with which a PIC binds to the enhancer's associated promoter so as to increase the rate at which the PIC initiates transcription there by a factor of 10. Then, if another transcription factor binding to a different enhancer subsite likewise increases the initiation rate by a factor of 20, both transcription factors acting together will increase the initiation rate by a factor of 200. *In this way, a limited number of transcription factors can support a much larger number of transcription patterns.* Transcriptional activation, according to this model, is essentially a mass action effect: The binding of a transcription factor to an enhancer increases the transcription factor's effective concentration at the associated promoter (the DNA holds the transcription factor in the vicinity of the promoter), which consequently increases the rate at which the PIC binds to the promoter. This explains why a transcription factor that is not bound to DNA (or even lacks a DNA-binding domain) inhibits transcriptional initiation. Such unbound transcription factors compete with DNA-bound transcription factors for their target sites and thereby reduce the rate at which the PIC is recruited to the associated promoter. This phenomenon, which is known as **squelching,** is apparently why transcription factors in the nucleus are almost always bound to inhibitors unless they are actively engaged in transcriptional initiation.

Mediator Provides the Interface between Transcriptional Activators and RNAP II.

Eukaryotic genomes encode as many as several thousand transcriptional regulators for class II genes (e.g., Fig. 27-3). However, activators fail to stimulate transcription by a reconstituted PIC *in vitro*. Evidently, an additional factor is required to do so. *Indeed, genetic studies in yeast by Kornberg led him to discover an ~20-subunit, ~1000-kD complex named **Mediator,** whose presence is required for transcription from nearly all class II gene promoters in yeast.* Mediator, which is therefore considered to be a coactivator, binds to the C-terminal domain (CTD) of RNAP II's β′ subunit (Section 25-2A) to form the so-called **RNAP II holoenzyme.** Further investigations revealed that multicellular organisms contain several multisubunit complexes that function similarly to yeast Mediator and which share many of their numerous subunits. Moreover, many of their subunits are related, albeit distantly, to those of yeast Mediator. *Mediators apparently function as adaptors that bridge DNA-bound transcriptional regulators and RNAP II so as to influence (induce or inhibit) the formation of a stable PIC at the*

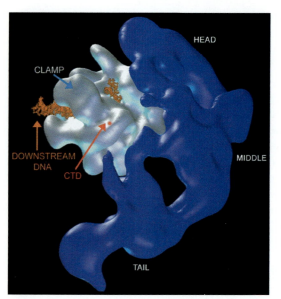

Figure 27-32 Electron microscopy-based projection of the yeast RNA polymerase II holoenzyme at 35 Å resolution. The head, middle, and tail domains of Mediator (*blue*) are clearly distinguishable. The independently determined EM-based image of RNAP II (*white*) is oriented to best match the polymerase density in the lower resolution holoenzyme image. Promoter DNA (*orange*) was modeled in, based on the structure of the yeast RNAP II elongation complex (Fig. 25-12*b*). Note that the RNAP II's DNA-binding cleft remains fully accessible in the holoenzyme complex. The red dot labeled CTD indicates the position where RNAP II's C-terminal domain extends from the enzyme. [Courtesy of Francisco Asturias, The Scripps Research Institute, La Jolla, California, and Roger Kornberg, Stanford University School of Medicine.]

associated promoter. They thereby function to integrate the various signals implied by the binding of these transcriptional regulators to their target DNAs. The different mediators in multicellular organisms presumably relay signals from different sets of transcriptional regulators.

The EM-based image of yeast RNAP II holoenzyme (Fig. 27-32), determined by Kornberg and Asturias, reveals that Mediator consists of three domains known as the head, middle, and tail domains. The head domain interacts closely with the RNAP II, although >75% of RNAP II's surface remains accessible for interaction with other components of the PIC. However, the tail domain appears not to contact RNAP II at all.

Altogether, then, the transcriptional machinery for class II genes comprises nearly 60 polypeptides with an aggregate molecular mass of ~3 million D. Nevertheless, this ribosome-sized assembly (the eukaryotic ribosome has a molecular mass of ~4.2 million D; Table 26-5) requires considerable assistance from yet other large macromolecular assemblies to gain access to the DNA in chromatin.

Insulators Limit the Effects of Enhancers and Stop Heterochromatin Spreading. Enhancers function independently of their position relative to the promoter. So what prevents an enhancer from affecting the transcription of all the genes in its chromosome? Conversely, heterochromatin, as we have seen (Section 27-3A), appears to be self-nucleating. What prevents heterochromatin from spreading into neighboring segments of euchromatin so as to prevent the transcription of their component genes? Experiments in which DNA sequences near *Drosophila* genes were rearranged revealed that short (<2 kb) segments of DNA known as **insulators** define the boundaries of functional units for transcription. Inserting an insulator between a gene and its upstream enhancer blocks the effect of the enhancer on transcription. Insulators may also prevent heterochromatin from spreading. For example, the **HS4** insulator in the chicken β-globin gene cluster recruits HATs that acetylate H3 Lys 9 on nearby nucleosomes (which is associated with transcriptional activity), thereby blocking their methylation (which induces heterochromatin formation). The mechanism of action of insulators is enigmatic. Presumably, it is not the insulators themselves but the proteins that bind to them that form the active insulator elements. Indeed, the HAT-recruiting activity of HS4 is distinct from its enhancer-blocking function, which is mediated by the binding of an 11-zinc finger protein named **CTCF** to a different subsite of HS4 than that to which HATs bind.

Many Signal Transduction Pathways Activate Transcription Factors. A variety of signaling pathways, including some involving steroid hormones, heterotrimeric G proteins, receptor tyrosine kinases (RTKs), and phosphoinositide cascades (Section 21-3), result in the activation (or inactivation) of transcription factors. In this way, extracellular factors such as hormones can influence gene expression inside a cell. We have already seen that the Ras signaling cascade (Fig. 21-19) results in the phosphorylation of several transcription factors, including Fos, Jun, and Myc, thereby modulating their activities. The sterol regulatory element binding protein (SREBP), which regulates the expression of the genes involved in cholesterol biosynthesis by binding to their sterol regulatory elements (SREs; Section 19-7B), is also an inducible transcription factor. In the following paragraphs, we discuss two additional examples of signaling pathways that activate transcription factors.

The JAK-STAT Pathway Relays Cytokine-Based Signals.   The protein growth factors that regulate the differentiation, proliferation, and activities of numerous types of cells, most conspicuously blood cells, are known as **cytokines.** The signal that certain cytokines have been extracellularly bound by their cognate receptors is transmitted within the cell, as James Darnell elucidated, by the **JAK-STAT pathway. Cytokine receptors** form complexes with proteins of the **Janus kinase (JAK)** family of **nonreceptor tyrosine kinases (NRTKs),** so named because each of its four ~1150-residue members has two tyrosine kinase domains (Janus is the two-faced Roman god of gates and doorways), although only the C-terminal domain is functional. **STATs** (for *s*ignal *t*ransducers and *a*ctivators of *t*ranscription) comprise a family of seven ~800-residue proteins that are the only known transcription factors whose activities are regulated by Tyr phosphorylation and that have SH2 domains.

The JAK-STAT pathway functions as is diagrammed in Fig. 27-33:

1. Ligand binding induces the cytokine receptor to dimerize (or, in some cases, to trimerize or even tetramerize).

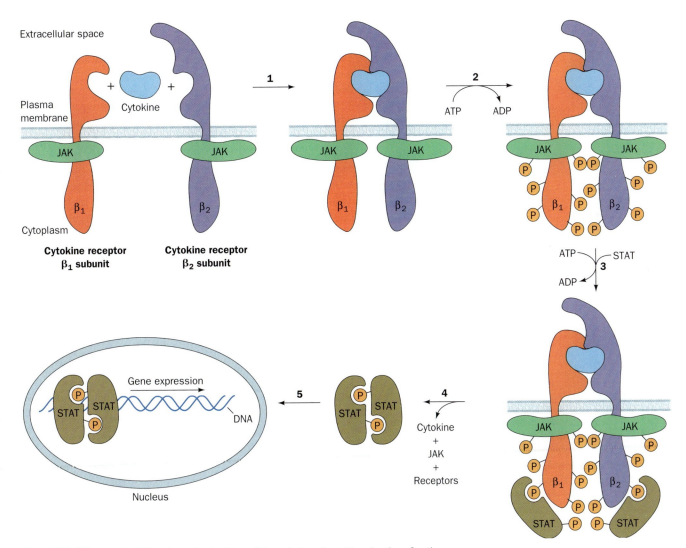

Figure 27-33   The JAK-STAT pathway for the intracellular relaying of cytokine signals.   See the text for details. [After Carpenter, L.R., Yancopoulos, G.D., and Stahl, N., *Adv. Prot. Chem.* **52,** 109 (1999).]

2. The cytokine receptor's two associated JAKs are thereby brought into apposition, whereon they reciprocally phosphorylate each other and then their associated receptors, a process resembling the autophosphorylation of dimerized RTKs (Section 21-3D).

3. STATs bind to the phospho-Tyr group on their cognate activated receptor via their SH2 domain and are then phosphorylated on a conserved Tyr residue by the associated JAK.

4. Following their dissociation from the receptor, the phosphorylated STATs homo- or heterodimerize via the association of their phospho-Tyr residue with the SH2 domain on the opposing subunit.

5. The STAT dimers are then translocated to the nucleus, where these now functional transcription factors specifically bind to 9-bp DNA segments with the consensus sequence TTCCGGGAA, thereby inducing the transcription of their associated genes.

Like many other components of signal transduction pathways, the STAT signal is inactivated through the action of phosphatases so that a STAT typically remains active for only a few minutes. The X-ray structure of the STAT named **Stat3β,** determined by Christoph Müller, is shown in Fig. 27-34.

Nuclear Receptors Are Activated by Hormones.   The **nuclear receptor superfamily,** which occurs in animals ranging from worms to humans, is composed of >150 proteins that bind a variety of hormones including steroids (glucocorticoids, mineralocorticoids, estrogens, and androgens; Section 9-1E), **thyroid hormones** such as thyroxine (Fig. 4-16) that stimulate metabolism, and vitamin D (Section 9-1E), all of which are nonpolar molecules. The nuclear receptors, many of which activate distinct but overlapping sets of genes, share a conserved modular organization that includes, from N- to C-terminus, a poorly conserved activation domain, a highly conserved DNA-binding domain, a connecting hinge region, and a ligand-binding domain. The DNA-binding domains each contain eight Cys residues that, in groups of four, tetrahedrally coordinate two Zn^{2+} ions. Many members of the nuclear receptor superfamily recognize specific DNA segments known as **hormone response elements (HREs)** that have the half-site consensus sequences 5′-AGAACA-3′ for steroid receptors and 5′-AGGTCA-3′ for

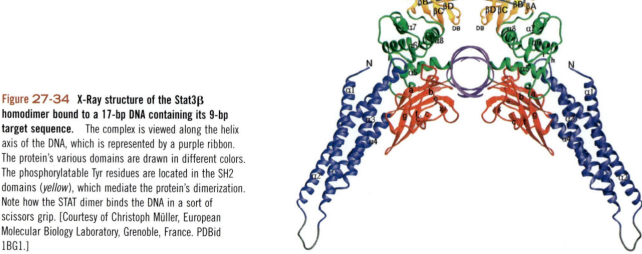

Figure 27-34   X-Ray structure of the Stat3β homodimer bound to a 17-bp DNA containing its 9-bp target sequence.   The complex is viewed along the helix axis of the DNA, which is represented by a purple ribbon. The protein's various domains are drawn in different colors. The phosphorylatable Tyr residues are located in the SH2 domains (*yellow*), which mediate the protein's dimerization. Note how the STAT dimer binds the DNA in a sort of scissors grip. [Courtesy of Christoph Müller, European Molecular Biology Laboratory, Grenoble, France. PDBid 1BG1.]

other nuclear receptors. These sequences are arranged in direct repeats ($\rightarrow n \rightarrow$), inverted repeats ($\rightarrow n \leftarrow$), and everted repeats ($\leftarrow n \rightarrow$), where n represents a 0- to 8-bp spacer (usually 1–5 bp) to whose length a specific receptor is targeted. **Steroid receptors** bind to their hormone response elements as homodimers, whereas other nuclear receptors do so as homodimers, as heterodimers, and in a few cases as monomers.

Steroid hormones, being nonpolar, readily pass through cell membranes into the cytosol or nucleus, where they bind to their receptors (although in some cases, they may also interact with cell surface receptors). In the absence of their cognate steroid, these receptors are part of large multiprotein complexes. Steroid binding releases these receptors, whereon they dimerize and, if they are cytoplasmic, enter the nucleus, where they bind to their target HREs so as to induce, or in some cases repress, the transcription of the associated genes.

The X-ray structure of the 86-residue DNA-binding domain of rat **glucocorticoid receptor** (**GR;** which regulates carbohydrate, protein, and lipid metabolism) in complex with DNA containing two ideal 6-bp **glucocorticoid response element (GRE)** half-sites arranged in inverted repeats was determined by Paul Sigler and Keith Yamamoto (Fig. 27-35). The protein forms a symmetric dimer involving protein–protein contacts even though it exhibits no tendency to dimerize in the absence of DNA (NMR measurements indicate that the contact region is flexible in solution). Each protein subunit consists of two structurally distinct modules, each nucleated by a Zn^{2+} coordination center, that closely associate to form a compact globular fold. The C-terminal module provides the entire dimerization interface and also makes several contacts with the phosphate groups of the DNA backbone. The N-terminal module, which is also anchored to the phosphate backbone, makes all of the GR's sequence-specific interactions with the GRE via three side chains that extend from the N-terminal α helix, its recognition helix, which is inserted into the GRE's major groove.

Afterword. As we have seen, eukaryotic transcriptional initiation is an astoundingly complex process that involves the synergistic participation of numerous multisubunit complexes comprising several hundred often loosely or sequentially interacting polypeptides (i.e., histones of various types and subtypes; the PIC; Mediator-like complexes; a variety of transcription factors, architectural factors, coactivators, and corepressors that in some cases form enhanceosomes; chromatin-remodeling complexes, and several types of histone modification complexes) as well as large segments of DNA. Intensive investigations in many laboratories over the past two decades have, as we discussed above, identified many of these complexes, characterized their component polypeptides, and in many cases elucidated their general functions. However, we are far from having a more than rudimentary understanding of how these various components interact *in vivo* to transcribe only those genes required by their cell under its particular circumstances in the appropriate amounts and with the proper timing. Nevertheless, an increase in the complexity of transcriptional regulation—rather than an increase in gene number per se—is likely to be at least partially responsible for the vast differences in morphology and behavior among nematodes, fruit flies, and humans, which have comparable numbers of genes (Table 27-1).

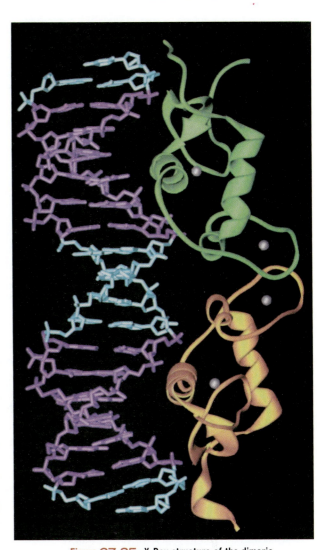

Figure 27-35 X-Ray structure of the dimeric glucocorticoid receptor DNA-binding domain in complex with DNA. The 18-bp DNA, which has a single nucleotide overhang at each of its 5′ ends, contains two symmetrical 6-bp glucocorticoid response element half-sites (*magenta*) separated by a 4-bp spacer (*cyan*). The protein subunits, which are drawn as green and gold ribbons, each contain two zinc fingers (Section 23-4C) in which each Zn^{2+} ion (*silver spheres*) is tetrahedrally liganded by four Cys side chains. Note how the glucocorticoid receptor's two N-terminal helices are inserted into successive major grooves in a manner reminiscent of the recognition helices of prokaryotic helix–turn–helix (HTH) domains (Section 23-4B). [Based on an X-ray structure by Paul Sigler, Yale University, and Keith Yamamoto, University of California at San Francisco. PDBid 1GLU.] 🐁 **See the Interactive Exercises.**

C Posttranscriptional Control Mechanisms

Although most regulation of gene expression in eukaryotes occurs at the level of transcription, additional control mechanisms act after an RNA transcript has been synthesized. In this section, we consider several of these mechanisms, including mRNA degradation, RNA interference, and control of translation initiation. We have already discussed alternative mRNA splicing (Section 25-3A) as a posttranscriptional mechanism for modulating gene expression through the selection of different exons.

mRNAs Are Degraded at Different Rates.   The range of mRNA stability in eukaryotic cells, measured in half-lives, varies from a few minutes to many hours or days. The mRNA molecules themselves appear to contain elements that dictate their decay rates. These elements include the poly(A) tail, the 5′ cap, and sequences that are located within the coding region.

A major route for mRNA degradation begins with the progressive removal of its poly(A) tail, a process catalyzed by **deadenylases** that appear to be located throughout the cytosol. When the residual poly(A) tail is less than ~10 nt long and hence no longer capable of interacting with poly(A) binding protein (Section 25-3A), the mRNA becomes a substrate for a **decapping enzyme,** which hydrolytically excises the mRNA's m^{7}GDP cap. The resulting RNA is then degraded by exonucleases. In yeast and other cells, a decapping enzyme, 5′→3′ exonucleases, and accessory proteins appear to form a complex called a **processing body,** or **P body,** that may function either to degrade mRNA or to store it in an inactive form.

How does degradation of the 3′ end of the mRNA stimulate decapping at the 5′ end? *In vivo,* as we have seen (Section 26-4A), the translational initiation factor eIF4G interacts with both poly(A) binding protein and cap binding protein, thereby circularizing the mRNA so that events at the 3′ end can be coupled to events at the 5′ end.

Proteins that bind to AU-rich sequences in the 3′ untranslated region of the mRNA also appear to increase or decrease the rate of mRNA degradation, although their exact action is not understood. In some cases, RNA secondary structure and RNA-binding proteins—which may be susceptible to modification by cellular signaling pathways—play a role in regulating mRNA stability. mRNAs that cannot be translated due to the presence of a premature Stop codon are specifically targeted for degradation (Box 27-4).

RNA Interference Is a Type of Posttranscriptional Gene Silencing.   In recent years it has become increasingly clear that noncoding RNAs can have important roles in controlling gene expression. One of the first indications of this phenomenon occurred in Richard Jorgensen's attempt to genetically engineer more vividly purple petunias by introducing extra copies of the gene that directs the synthesis of the purple pigment. Unexpectedly, the resulting transgenic plants had variegated or entirely white flowers. Apparently, the purple-making genes switched each other off. This result was at first attributed to the well-known phenomenon in which **antisense RNA** (RNA that is complementary to a portion of an mRNA) prevents the translation of the corresponding mRNA because the ribosome cannot translate double-stranded RNA. However, injecting **sense RNA** (RNA with the same sequence as the mRNA) into experimental organisms such as the nematode *C. elegans* also blocked protein production. Since the added RNA somehow interferes with gene expression, this phenomenon is known as **RNA interference (RNAi).** RNAi is now known to occur in all eukaryotes except perhaps yeast.

BOX 27-4

Perspectives in Biochemistry

Nonsense-Mediated Decay

Errors during DNA replication, transcription, or mRNA splicing can give rise to a Stop codon either through substitution of one nucleotide for another or through frameshifting. "Premature" Stop codons that interrupt a coding sequence may account for as many as one-third of human genetic diseases. Interestingly, eukaryotic mRNAs with premature Stop codons rarely produce the corresponding truncated polypeptide because the mRNA is destroyed through **nonsense-mediated decay (NMD)** soon after its synthesis.

How do cells distinguish a premature Stop codon from a normal Stop codon at the end of the coding sequence? Some experiments suggest that the signal for NMD is an intron downstream of the Stop codon. However, a translation-ready mRNA has already had its introns spliced out (Section 25-3A). One possible explanation for this paradox is that after mRNA splicing, an **exon-junction protein complex (EJC)** containing splicing factors and other components remains associated with each exon–exon junction, even after the mRNA has been exported to the cytosol for translation. An EJC following a Stop codon could therefore mark the absence of a normally present coding sequence, and the cell could then selectively destroy that mRNA.

A revised model for NMD proposes that quality control occurs before the mRNA leaves the nucleus. Although the bulk of translation occurs in the cytosol, *an estimated 10 to 15% of total translation in mammalian cells takes place inside the nucleus.* During nuclear translation, a ribosome paused at a Stop codon may be converted to a "surveillance" complex that then scans the rest of the mRNA for the presence of an EJC. If an EJC is detected, the mRNA is degraded by removal of its 5′ cap and 3′ poly(A) tail and by the action of exonucleases.

Further experiments by Andrew Fire and Craig Mello revealed that double-stranded RNA is even more effective than either of its component strands alone at interfering with gene expression. In fact, RNAi can be induced by just a few molecules of double-stranded RNA, suggesting that RNAi is a catalytic rather than a stoichiometric phenomenon.

RNA interference is not merely an artifact of genetic engineering. In many cells, naturally occurring small RNA molecules, called **short interfering RNAs (siRNAs)** or **micro RNAs (miRNAs),** depending on their origin, down-regulate gene expression by binding to complementary mRNA molecules. Several hundred miRNAs, ranging in length from 18 to 25 nucleotides, have been identified in mammals, although their mRNA targets are only just beginning to be characterized.

Several Nucleases Participate in RNAi.   Work with exogenous double-stranded RNA in *C. elegans* and *Drosophila* has led to the elucidation of the following pathway for RNAi (Fig. 27-36):

1. The trigger double-stranded RNA, as Phillip Zamore discovered, is chopped up into ~21- to 23-nt fragments (that is, siRNAs), each of whose strands has a 2-nt overhang at its 3′ end and a 5′ phosphate. The cleavage reaction is catalyzed by an ATP-dependent RNase named **Dicer.** Model-building studies based on homologous RNases suggest that Dicer functions as a homodimer whose active sites are separated by a distance that corresponds to ~22 nucleotides of an RNA helix.

2. An siRNA is transferred to a 250- to 500-kD enzyme complex known as the **RNA-induced silencing complex (RISC).** RISC has at least four protein components, one of which is an ATP-dependent RNA heli-

Figure 27-36 A mechanism of RNA interference.
See the text for details. ATP is required for Dicer-catalyzed cleavage of RNA and for RISC-associated helicase unwinding of double-stranded RNA. Depending on the species, the mRNA may not be completely degraded.

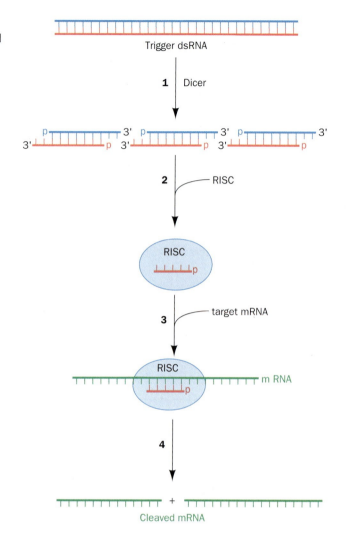

case that is probably responsible for generating single-stranded siRNA. In some species, but apparently not in humans, the original siRNA signal is amplified by the action of an **RNA-dependent RNA polymerase.**

3. The antisense strand of the siRNA guides the RISC to an mRNA with the complementary sequence.

4. An endonuclease component of RISC cleaves the mRNA, probably opposite the bound siRNA. The cleaved mRNA is then further degraded by cellular nucleases, thereby preventing its translation.

Although the molecular machinery involved in RNAi is highly conserved from yeast to humans, variations abound, even within a single organism. For example, different Dicer enzymes can potentially generate different lengths of siRNAs, which may differ in their gene-silencing activities. The exact length of the siRNA, or the degree to which its sequence exactly complements the mRNA sequence, may determine whether a corresponding mRNA is completely degraded or just rendered nontranslatable.

The ease with which RNAi can be induced by exogenous RNAs suggests that the pathway may have arisen as a defense against viruses. In many eukaryotes, the major source of double-stranded RNA is RNA viruses, many of which encode their own RNA-dependent RNA polymerase to convert

their single-stranded genome into double-stranded RNA. In fact, many plant viruses contain genes that suppress various steps of RNAi and that are therefore essential for pathogenesis.

RNAi Has Numerous Applications. RNAi has become the method of choice for "knocking out" specific genes in plants and invertebrates. For example, in *C. elegans*, RNAi has been used to systematically inactivate over 16,000 of its ~19,000 protein-coding genes in an attempt to assign a function to each gene. *C. elegans* is particularly amenable to the RNAi approach, since these worms eat *E. coli* cells, and it is relatively easy to genetically engineer the bacterial cells to express double-stranded RNA that becomes part of the worms' diet. One limitation of the RNAi method is that only gene inactivation—rather than gene activation—can be examined.

Manipulating the RNAi pathway in mammals is more difficult, mainly due to the difficulty of delivering double-stranded RNA to cells and eliciting more than transient gene silencing. To complicate matters, mammalian cells have additional pathways for dealing with foreign RNA, which result in nonspecific degradation of RNA and cessation of all translation (these responses probably also help prevent infection by RNA viruses). Nevertheless, experiments have demonstrated that it is possible to use RNAi to block the liver's inflammatory response to a hepatitis virus, at least in mice, and to prevent HIV replication in cultured human cells. One challenge for the future is to devise protocols for longer-lasting gene silencing that would make it possible to prevent viral infections or to block the effects of disease-causing mutant genes.

Translational Control. In some cells, altering the rates of mRNA production or degradation does not provide the necessary level of control. For example, the early embryonic development of sea urchins, insects, and frogs depends on the rapid translation of mRNA that has been stockpiled in the oocyte. The mRNA is stored in inactive form in association with proteins but on fertilization becomes available for translation. This permits embryogenesis to commence immediately, without waiting for mRNAs from paternally supplied genes to be synthesized.

Globin synthesis in reticulocytes (immature red blood cells) also proceeds rapidly, but only if heme is available. The inhibition of globin synthesis occurs at the level of translation initiation. In the absence of heme, reticulocytes accumulate a protein, **heme-regulated inhibitor (HRI).** HRI is a kinase that phosphorylates a specific Ser residue, Ser 51, on the α subunit of eIF2 (the initiation factor that delivers GTP and Met–tRNA$_i^{Met}$ to the ribosome; Section 26-4A).

Phosphorylated eIF2 participates in translation initiation in much the same way as unphosphorylated eIF2, but it is not regenerated normally. At the completion of the initiation process, unmodified eIF2 exchanges its bound GDP for GTP in a reaction mediated by another initiation factor, **eIF2B:**

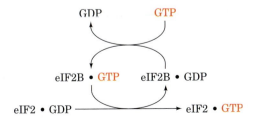

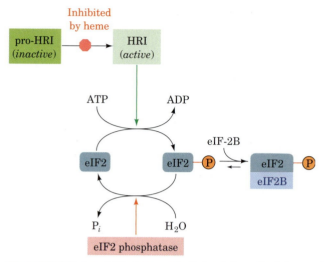

Figure 27-37 A model for heme-controlled protein synthesis in reticulocytes.

Phosphorylated eIF2 forms a much tighter complex with eIF2B than does unphosphorylated eIF2. This sequesters eIF2B (Fig. 27-37), which is present in lesser amounts than is eIF2, thereby preventing regeneration of the eIF2·GTP required for translation.

In the presence of heme, the heme-binding sites in HRI are occupied, which inactivates the kinase. The eIF2 molecules that are already phosphorylated are reactivated through the action of **eIF2 phosphatase,** which is unaffected by heme. The reticulocyte thereby coordinates its synthesis of globin and heme.

D The Cell Cycle, Cancer, and Apoptosis

The **cell cycle,** the general sequence of events that occur during the lifetime of a eukaryotic cell, is divided into four distinct phases (Fig. 27-38):

1. Mitosis and cell division occur during the relatively brief **M phase** (for mitosis).
2. This is followed by the **G₁ phase** (for gap), which covers the longest part of the cell cycle.

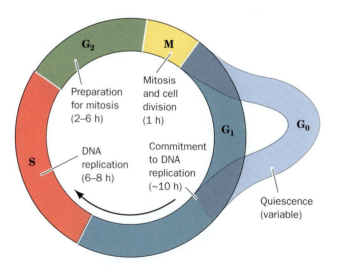

Figure 27-38 **The eukaryotic cell cycle.** Cells may enter a quiescent phase (G₀) rather than continuing about the cycle.

3. G_1 gives way to the **S phase** (for synthesis), which, in contrast to events in prokaryotes, *is the only period in the cell cycle when DNA is synthesized.*

4. During the relatively short **G_2 phase,** the now tetraploid cell prepares for mitosis. It then enters M phase once again and thereby commences a new round of the cell cycle.

The cell cycle for cells in culture typically occupies a 16- to 24-h period. In contrast, cell cycle times for the different types of cells in a multicellular organism may vary from as little as 8 h to >100 days. Most of this variation occurs in the G_1 phase. Moreover, many terminally differentiated cells, such as neurons or muscle cells, never divide; they assume a quiescent state known as the **G_0 phase.**

Progression through the cell cycle is triggered by external as well as internal signals. In addition, the cell cycle has a series of **checkpoints** that monitor its progress and the health of the cell and arrest the cell cycle if certain conditions have not been satisfied. For example, G_2 has a checkpoint that prevents the initiation of M until all of the cell's DNA has been replicated, thereby ensuring that both daughter cells will receive a full complement of DNA. Similarly, a checkpoint in M prevents mitosis until all chromosomes have properly attached to the mitotic spindle (if this were not the case, even for one chromosome, one daughter cell would lack this chromosome and the other would have two, both deleterious if not lethal conditions). Checkpoints in G_1 and S also arrest the cell cycle in response to damaged DNA so as the give the cell time to repair the damage (Section 24-5). In the cells of multicellular organisms, if after a time the checkpoint conditions have not been met, the cell may be directed to commit suicide, a process named **apoptosis** (see below), thereby preventing the proliferation of an irreparably damaged and hence dangerous (e.g., cancerous) cell.

The Cell Cycle Is Controlled by Cyclins and Cyclin-Dependent Protein Kinases.

The progression of a cell through the cell cycle is regulated by proteins known as **cyclins** and **cyclin-dependent protein kinases (Cdks).** Cyclins are so named because they are synthesized during one phase of the cell cycle and are completely degraded during a succeeding phase (protein degradation is discussed in Section 20-1). A particular cyclin specifically binds to and thereby activates its corresponding Cdk(s), which are Ser/Thr protein kinases, to phosphorylate their target nuclear proteins. These proteins, which include histone H1, several oncogene proteins (see below), and proteins involved in nuclear disassembly and cytoskeletal rearrangement, are thereby activated to carry out the processes making up that phase of the cell cycle.

The human Cdk named **Cdk2** is activated by the binding of **cyclin A** and by phosphorylation of its Thr 160 as catalyzed by **Cdk-activating kinase (CAK;** this kinase is itself a complex of **cyclin H** and **Cdk7).** Phosphorylation of Cdk2 residues Thr 14 and Tyr 15 negatively regulates Cdk2 activity. The X-ray structure of unphosphorylated Cdk2 in complex with ATP but in the absence of cyclin A (Fig. 27-39*a*) closely resembles that of the catalytic subunit of protein kinase A (PKA; Fig. 15-14). However, Cdk2 is inactive as a kinase in part because access of protein substrates to the γ-phosphate of the bound ATP is blocked by a 19-residue protein loop dubbed the **T loop** (which contains Thr 160). Phosphorylation of Thr 160 and binding of cyclin A alter the conformation of Cdk2 (Fig. 27-39*b*). In particular, the N-terminal α helix of Cdk2, which contains the PSTAIRE

(a)

(b)

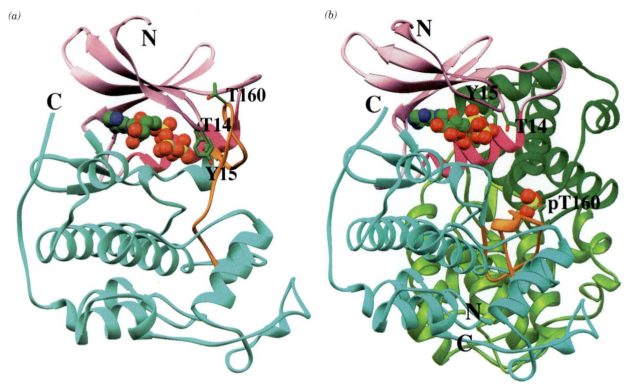

Figure 27-39 X-Ray structure of human cyclin-dependent kinase 2 (Cdk2). (a) Cdk2 in complex with ATP. The protein is shown in the "standard" protein kinase orientation with its N-terminal lobe pink, its C-terminal lobe cyan, its PSTAIRE helix (residues 45–56) magenta, and its T loop (residues 152–170) orange. The ATP is shown in space-filling form and the phosphorylatable side chains of Thr 14, Tyr 15, and Thr 160 are shown in stick form, all colored according to atom type (C green, N blue, O red, and P yellow). Compare this structure with that of protein kinase A (PKA; Fig. 15-14). [Based on an X-ray structure by Sung-Hou Kim, University of California at Berkeley. PDBid 1HCK.] (b) The complex of Thr 160–phosphorylated Cdk2 with cyclin A

and ATP. The Cdk2 and ATP are represented as in Part *a* and viewed similarly. The cyclin A is colored light and dark green. The phosphoryl group at Thr 160 of Cdk2 is drawn in space-filling form. Note how the binding of cyclin A together with the phosphorylation of Thr 160 has caused a major structural reorganization of the T loop and the Cdk2 N-terminal lobe, including its PSTAIRE helix. Also note the different conformations of the ATP triphosphate group in the two structures. [Based on an X-ray structure by Nikola Pavletich, Memorial Sloan-Kettering Cancer Center, New York, New York. PDBid 1JST.] 🙡 **See the Interactive Exercises.**

sequence motif characteristic of the Cdk family, rotates about its axis by 90° and moves several angstroms in order to position several residues in a catalytically active arrangement. In addition, Cdk2's T loop undergoes a dramatic reorganization, involving position shifts of up to 21 Å, that allows protein substrates access to the active site. The phosphate group on Thr 160 fits snugly into a positively charged pocket composed of three Arg residues that forms, in part, on cyclin A binding.

In addition to their control by phosphorylation/dephosphorylation and by the binding of the appropriate cyclin, *Cdk activities are regulated by cyclin-dependent kinase inhibitors (CKIs), which arrest the cell cycle in response to such antiproliferative signals as contact with other cells, DNA damage, terminal differentiation, and senescence (in which cell cycle arrest is permanent).* The importance of CKIs is indicated by their frequent alterations in cancer, which is manifest as uncontrolled cell division. The pathways by which sensor proteins monitor cellular conditions and activate or inactivate Cdks are not yet fully understood. However, considerable information has been provided by studies of proteins, including p53 and pRb (see below), whose mutations are associated with loss of cell cycle control.

p53 Is a Tumor Suppressor That Arrests the Cell Cycle in G₂. Individuals with the rare inherited condition known as **Li–Fraumeni syndrome** are highly

susceptible to a variety of malignant tumors, particularly breast cancer, which they often develop before their thirtieth birthdays. These individuals have germ-line mutations in their *p53* gene, which suggests that its normal protein product, **p53** (a *p*olypeptide with a nominal mass of *53* kD), is a **tumor suppressor,** that is, *p53 functions to restrain the uninhibited cell proliferation that is characteristic of cancer.* Indeed, the *p53* gene is the most commonly altered gene in human cancers; ~50% of human cancers contain a mutation in *p53,* and many other oncogenic (cancer-causing) mutations occur in genes that encode proteins that directly or indirectly interact with p53. Evidently, p53 functions as a "molecular policeman" in monitoring genome integrity.

Despite the central role of p53 in preventing tumor formation, the way it does so has only gradually come to light. p53 is specifically bound by **Mdm2** protein, a ubiquitin-protein ligase (E3) that specifically ubiquitinates p53, thereby marking it for proteolytic degradation by the proteasome (Section 20-1). Consequently, the amplification (generation of several copies) of the *mdm2* gene, which occurs in >35% of human **sarcomas** (none of which have a mutated *p53* gene; sarcomas are malignancies of connective tissues such as muscle, tendon, and bone), results in an increased rate of degradation of p53, thereby predisposing cells to malignant transformation. Thus, *an additional way that oncogenes (cancer-causing genes; Box 21-2) can cause cancer is by inactivating normal tumor suppressors.*

p53 is an efficient transcriptional activator. Indeed, all point-mutated forms of p53 that are implicated in cancer have lost their sequence-specific DNA-binding properties. How, then, does p53 function as a tumor suppressor? A clue to this riddle came from the observation that the treatment of cells with DNA-damaging ionizing radiation results in the accumulation of normal p53. This led to the discovery that the activated protein kinases **ATM** and **Chk2** both phosphorylate p53 [ATM for *a*taxia *t*elangiectasia *m*utated (**ataxia telangiectasia** is a rare genetic disease characterized by a progressive loss of motor control, growth retardation, premature aging, and a greatly increased risk of cancer); Chk for *c*heckpoint *k*inase]. ATM and Chk2 had previously been shown to be activated by and participate in a phosphorylation cascade that induces cell cycle arrest at the G_2 checkpoint on the detection of damaged or unreplicated DNA. The phosphorylation of p53 prevents its binding by Mdm2 and hence increases the otherwise low level of p53 in the nucleus. *Although p53 does not initiate cell cycle arrest in G_2, its presence is required to prolong this process. It does so by activating the transcription of the gene encoding the cyclin-dependent kinase inhibitor (CKI) named $p21^{Cip1}$,* which binds to several Cdk–cyclin complexes so as to inhibit both the G_1/S and G_2/M transitions.

Cells that are irreparably damaged synthesize excessive levels of p53, which in turn induces these cells to commit suicide by activating the expression of several of the proteins that participate in apoptosis (see below). In the absence of p53 activation, cells control the level of p53 through a feedback loop in which p53 stimulates the transcription of the *mdm2* gene.

The X-Ray Structure of p53 Explains Its Oncogenic Mutations.

p53 is a tetramer of identical 393-residue subunits, which each contain a sequence-specific DNA-binding core. The X-ray structure of this domain (residues 102–313) in complex with a 21-bp target DNA sequence, determined by Nikola Pavletich, is shown in Fig. 27-40. The p53 DNA-binding motif does not re-

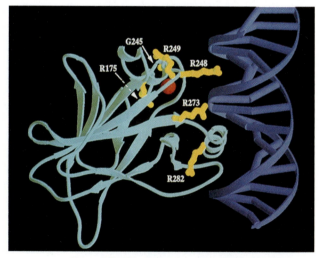

Figure 27-40 X-Ray structure of the DNA-binding domain of human p53 in complex with its target DNA. The protein is shown in ribbon form (*cyan*) and the DNA in ladder form with its bases represented by cylinders (*blue*). A tetrahedrally liganded Zn^{2+} ion is shown as a red sphere, and the side chains of the six most frequently mutated residues in human tumors are shown in stick form (*yellow*) and identified with their one-letter codes. [Courtesy of Nikola Pavletich, Memorial Sloan-Kettering Cancer Center, New York. PDBid 1TSR.] **See the Interactive Exercises.**

semble any other that has previously been characterized (Sections 23-4B and 23-4C). It makes sequence-specific contacts with the bases of the DNA in its major groove (lower right of Fig. 27-40). In addition, the side chain of Arg 248 extends into the DNA's minor groove (upper right of Fig. 27-40). The protein also contacts the DNA backbone between the major and minor grooves in this region (notably with Arg 273).

The structure's most striking feature is that *its DNA-binding motif consists of conserved regions comprising the most frequently mutated residues in the >1000 p53 variants found in human tumors.* Among them are one Gly and five Arg residues (highlighted in yellow in Fig. 27-40) whose mutations collectively account for over 40% of the *p53* variants in tumors. The two most frequently mutated residues, Arg 248 and Arg 273, as we saw, directly contact the DNA. The other four "mutational hotspot" residues appear to play a critical role in structurally stabilizing p53's DNA-binding surface. The relatively sparse secondary structure in the polypeptide segments forming this surface (one helix and three loops) accounts for this high mutational sensitivity: Its structural integrity mostly relies on specific side chain–side chain and side chain–backbone interactions.

p53 Is a Sensor That Integrates Information from Several Pathways. p53 may be activated by several other pathways. For example, aberrant growth signals, including those generated by oncogenic variants of Ras signaling cascade components (Fig. 21-20) such as Ras, cause the inappropriate activation of a variety of transcription factors. One of them, **Myc,** activates the transcription of the gene encoding **p14^ARF,** which binds to Mdm2 and thereby inhibits its activity. This prevents the degradation of p53 and hence triggers the p53-dependent transcriptional programs leading to cell cycle arrest as well as apoptosis. Evidently, p14^ARF acts as part of a p53-dependent fail-safe system to counteract hyperproliferative signals.

A third activation pathway for p53 is induced by a wide variety of DNA-damaging chemotherapeutic agents, protein kinase inhibitors, and UV radiation. These activate a protein kinase named **ATR** to phosphorylate p53 so as to reduce its affinity for Mdm2 in much the same way as do ATM and Chk2. p53 is also subject to a rich variety of reversible posttranslational modifications that markedly influence the expression of its target genes, including acetylation at several Lys residues, glycosylation, and sumoylation (Section 26-5B), in addition to its phosphorylation at multiple Ser/Thr residues and ubiquitination.

p53, as we have only glimpsed, is the recipient of a vast number of intracellular signals and, in turn, controls the activities of a large number of downstream regulators. One way to understand the operation of this highly complex and interconnected network is in analogy with the Internet. In the Internet (cell), a small number of highly connected servers or hubs ("master" proteins) transmit information to/from a large number of computers or nodes (other proteins) that directly interact with only a few other nodes (proteins). In such a network, overall performance is largely unperturbed by the inactivation of one of the nodes (other proteins). However, the inactivation of a hub ("master" protein) will greatly impact system performance. p53 is a "master" protein, that is, it is analogous to a hub. Inactivation of one of the many proteins that influences its performance or one of the many proteins whose activity it influences usually has little effect on cellular events due to the cell's redundant and highly interconnected components. However, the inactivation of p53 or several of its most closely associated proteins (e.g., Mdm2) disrupts the cell's responses to DNA damage and tumor-predisposing stresses, thereby leading to tumor formation.

Loss of pRb Protein Leads to Cancer.   **Retinoblastoma,** a cancer of the developing retina that affects infants and young children, is associated with the loss of the *Rb* gene, which encodes the tumor suppressor **pRb.** This 928-residue DNA-binding protein interacts with the **E2F** family of transcription factors, which has six members in mammals. E2F proteins induce the transcription of genes that encode proteins required for entry into S phase. pRb can be phosphorylated at as many as 16 of its Ser/Thr residues by various Cdk–cyclin complexes (the various complexes phosphorylate different sets of sites on pRb). In non-proliferating cells (those in early G_1), pRb is hypophosphory-lated. In this state, it binds to E2F so as to prevent it from activating transcription at the promoters to which it is bound. In response to a mitogenic signal (a signal that induces mitosis), the levels of D-type cyclins increase, which triggers phosphorylation of pRb by **Cdk4/6–cyclin D** complexes. Hyperphosphorylated pRb releases E2F, which thereon induces the expression of genes that promote cell cycle progression, including genes for additional cyclins and Cdks. The E2F-binding site of pRb is the major site of *Rb* gene alterations in tumors.

Apoptosis Is an Essential Process.   **Programmed cell death** or **apoptosis** (Greek: falling off, as leaves from a tree), which was first described by John Kerr in the late 1960s, is a normal part of development as well as maintenance and defense of the adult animal body. For example, in many vertebrates, the digits of the developing hands and feet are initially connected by webbing that is eliminated by programmed cell death (Fig. 27-41). In the adult human body, which consists of nearly 10^{14} cells, an estimated 10^{11} cells are eliminated each day through programmed cell death (which closely matches the number of new cells produced by mitosis). The immune system eliminates virus-infected cells, in part by inducing them to undergo apoptosis, in order to prevent viral replication. Cells with irreparably damaged DNA and hence at risk for malignant transformation undergo apoptosis, thereby protecting the entire organism from cancer. In fact, as Martin Raff pointed out, *apoptosis appears to be the default option for animal cells: Unless they continually receive external hormonal and/or neuronal signals not to commit suicide, they will do so.* Thus, adult organs maintain their constant size by balancing cell proliferation with apoptosis. Not surprisingly, therefore, inappropriate apoptosis has been implicated in several neurodegenerative diseases including Alzheimer's disease (Section 6-5D), Parkinson's disease (Section 20-6B), and Huntington's disease (Box 27-1), as well as much of the damage caused by stroke and heart attack.

Apoptosis is qualitatively different from **necrosis,** the type of cell death caused by trauma (e.g., lack of oxygen, extremes of temperature, and mechanical injury). Cells undergoing necrosis essentially explode: They and their membrane-enclosed organelles swell as water rushes in through their compromised membranes, releasing lytic enzymes that digest the cell contents until the cell lyses, spilling its contents into the surrounding region (Box 17-5). The cytokines that the cell releases often induce an inflammatory response (which can damage surrounding cells). In contrast, apoptosis begins with the loss of intercellular contacts by an apparently healthy cell followed by

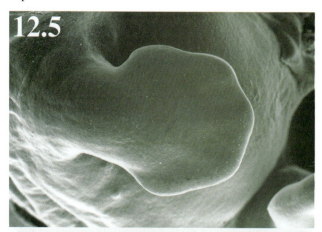

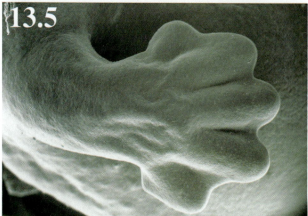

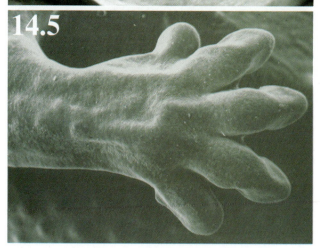

Figure 27-41   Programmed cell death in the embryonic mouse paw.   At day 12.5 of development, its digits are fully connected by webbing. At day 13.5, the webbing has begun to die. By day 14.5, this apoptotic process is complete. [Courtesy of Paul Martin, University College of London, U.K.]

its shrinkage, the condensation of its chromatin at the nuclear periphery, the collapse of its cytoskeleton, the dissolution of its nuclear envelope, the fragmentation of its DNA, and violent blebbing (blistering) of its plasma membrane. Eventually, the cell disintegrates into numerous membrane-enclosed **apoptotic bodies** that are phagocytosed (engulfed) by neighboring cells as well as by roving macrophages without spilling the cell contents and hence not inducing an inflammatory response.

Caspases Participate in Apoptosis.   Apoptosis involves a family of proteases known as **caspases** (for *c*ysteinyl *asp*artate-specific prote*ases*), which are **cysteine proteases** whose mechanism resembles that of serine proteases (Section 11-5) but with Cys replacing the active site Ser. Caspases cleave target polypeptides after an Asp residue.

Caspases are $\alpha_2\beta_2$ heterotetramers that consist of two large α subunits (~300 residues) and two small β subunits (~100 residues). They are expressed as single-chained zymogens **(procaspases)** that are activated by proteolytic excision of their N-terminal prodomains and proteolytic separation of their α and β subunits. The activating cleavage sites all follow Asp residues and are, in fact, targets for caspases, suggesting that caspase activation may either be autocatalytic or be catalyzed by another caspase.

The X-ray structure of human **caspase-7** (Fig. 27-42), determined by Keith Wilson and Paul Charifson, reveals that each $\alpha\beta$ heterodimer contains a six-stranded β sheet flanked by five α helices that are approximately parallel to the β strands. The β sheet is continued across the enzyme's twofold axis to form a twisted 12-stranded β sheet. The active site of each $\alpha\beta$ heterodimer is located at the C-terminal ends of its parallel β strands. The structures of other caspases differ mainly in the conformations of the four loops forming their active sites. In the inactive **procaspase-7,** these loops are folded so as to obliterate the active site.

Over 60 cellular proteins have been identified as caspase substrates. These include cytoskeletal proteins, proteins involved in cell cycle regula-

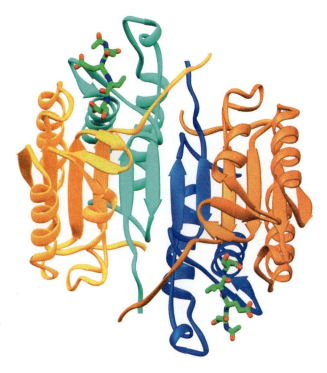

Figure 27-42 **X-Ray structure of caspase-7 in complex with a tetrapeptide aldehyde inhibitor.**   The $\alpha_2\beta_2$ heterotetrameric enzyme is viewed along its twofold axis with its large (α) subunits orange and gold and its small (β) subunits cyan and light blue. The acetyl-Asp-Glu-Val-Asp-CHO inhibitor is drawn in stick form with C green, N blue, and O red. [Based on an X-ray structure by Keith Wilson and Paul Charifson, Vertex Pharmaceuticals, Cambridge, Massachusetts. PDBid 1F1J.]

Figure 27-43   The extrinsic pathway of apoptosis.   Flat gray arrows indicate activation. The binding of a trimeric death ligand (e.g., FasL) on the inducing cell to the death receptor (e.g., Fas) on the apoptotic cell causes the death receptor's cytoplasmic death domains (DDs) to trimerize. This recruits adaptors (e.g., FADD), which bind via their DDs to the DDs of the death receptor. The adaptors, in turn, recruit initiator procaspases (e.g., procaspase-8) via the interactions between the death effector domains (DEDs) on the adaptors and the initiator procaspases, which induces the autoactivation of the initiator procaspases to form the corresponding heterotetrameric initiator caspases (e.g., caspase-8). The initiator caspases then proteolytically activate effector procaspases (e.g., procaspase-3) to yield the heterotetrameric effector caspases (e.g., caspase-3), which catalyze the proteolytic cleavages resulting in apoptosis.

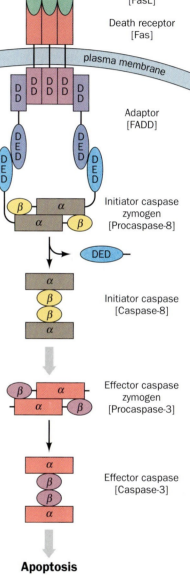

tion (including cyclin A, p21^{Cip1}, ATM, and pRb), proteins that participate in DNA replication, transcription factors, and proteins that participate in signal transduction. Nevertheless, how the cleavage of these numerous proteins causes the morphological changes that cells undergo during apoptosis is unclear. The induction of apoptosis also causes the rapid degradation of DNA by the action of **caspase-activated DNase.** Presumably, DNA degradation prevents the genetic transformation of other cells that subsequently phagocytose apoptotic bodies containing viral DNA or damaged chromosomal DNA.

Apoptosis Is Triggered Extracellularly or Intracellularly.   *Apoptosis in a given cell may be induced either by externally supplied signals in the so-called* **extrinsic pathway** *(death by commission) or by the absence of external signals that inhibit apoptosis in the so-called* **intrinsic pathway** *(death by omission).* The extrinsic pathway is initiated by the association of a cell destined to undergo apoptosis with a cell that has selected it to do so. In what is perhaps the best characterized such pathway (Fig. 27-43), a transmembrane protein named **Fas ligand (FasL)** that projects from the plasma membrane of the inducing cell, a so-called **death ligand,** binds to a transmembrane protein known as **Fas** that projects from the plasma membrane of the apoptotic cell, a so-called **death receptor.** Fas ligand is a homotrimeric protein whose binding to three Fas molecules causes the Fas cytoplasmic domains to trimerize. The trimerized Fas recruits three molecules of a 208-residue adaptor protein named **FADD** (for *F*as-*a*ssociating *d*eath *d*omain-containing protein), which in turn recruits **procaspase-8** and **procaspase-10.** The consequent clustering of procaspase-8 and procaspase-10 results in the proteolytic autoactivation of these zymogens, thereby generating **caspases-8 and -10,** which are termed **initiator caspases.** This is because these enzymes then activate **caspase-3,** which is known as an **effector (executioner) caspase** because its actions cause the cell to undergo apoptosis.

The intrinsic pathway for initiating apoptosis follows a slightly different route to caspase-3 activation. Most animal cells are continuously bathed in an extracellular soup, generated in part by neighboring cells, that contains a wide variety of substances that regulate the cell's growth, differentiation, activity, and survival. The withdrawal of this chemical support for its survival or the loss of direct cell–cell interactions induces a cell to undergo apoptosis via the intrinsic pathway. The initial step of this pathway appears to be the activation of one or more members of the **Bcl-2** family (so named because its founding member is involved in *B* cell *l*ymphoma). Association of certain Bcl-2 proteins with the mitochondrion causes it to release cy-

Figure 27-44 Cryoelectron microscopy–based image of the apoptosome at 27 Å resolution. In its top view, the particle is viewed along its sevenfold axis of symmetry. The side view reveals the flattened nature of this wheel-like particle. The scale bar represents 100 Å. [Courtesy of Christopher Akey, Boston University School of Medicine.]

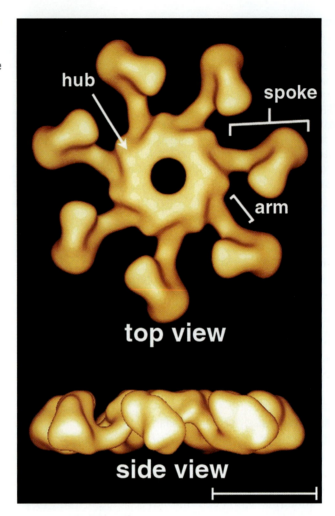

tochrome c (Section 17-2E) from the intermembrane space into the cytosol. It is not clear how cytochrome c exits the mitochondrion. It may traverse the outer membrane via a newly formed pore or via an existing pore whose conformation is altered to accommodate the ~12-kD cytochrome, or the outer membrane may rupture as a result of Bcl-2 activity.

The cytochrome c combines with a 1248-residue protein called **Apaf-1** (for *apoptosis protease-activating factor-1*) and ATP or dATP to form an ~1100-kD wheel-shaped complex named the **apoptosome** (Fig. 27-44). The apoptosome binds several molecules of **procaspase-9** in a manner that induces their autoactivation. The resulting **caspase-9,** still bound to the apoptosome, then activates procaspase-3 to instigate cell death.

E The Molecular Basis of Development

Perhaps the most awe-inspiring event in biology is the growth and development of a fertilized ovum to form an extensively differentiated multicellular organism. No outside instruction is required to do so; *fertilized ova contain all the information necessary to form complex multicellular organisms such as human beings.* Much of what we know about the molecular basis of cell differentiation is based on studies of the fruit fly *Drosophila melanogaster.* We therefore begin this section with a synopsis of *Drosophila* embryogenesis.

***Drosophila* Development.** Almost immediately after the *Drosophila* egg (Fig. 27-45*a*) is laid (which, rather than the earlier fertilization, triggers development), it commences a series of rapid, synchronized nuclear divisions, one every 6 to 10 min. Here, the nuclear division process is not accompanied by the formation of new cell membranes; the nuclei continue sharing their common cytoplasm to form a so-called **syncytium** (Fig. 27-45*b*). After the eighth round of nuclear division, the ~256 nuclei begin to migrate toward the cortex (outer layer) of the egg where, by around the eleventh nuclear division, they have formed a single layer surrounding a yolk-rich core (Fig. 27-45*c;* the germ-cell progenitors, the pole cells, are set aside after the ninth division). At this stage, the mitotic cycle time begins to lengthen while the nuclear genes, which have heretofore been fully engaged in DNA replication, become transcriptionally active (a freshly laid egg contains an enormous store of mRNA that has been contributed by the developing oocyte's surrounding "nurse" cells). In the fourteenth nuclear division cycle, which lasts ~60 min, the egg's plasma membrane invaginates around each of the ~6000 nuclei to yield a cellular monolayer called a **blastoderm** (Fig. 27-45*d*). At this point, after ~2.5 h of development, transcriptional activity reaches its maximum and mitotic synchrony is lost.

During the next few hours, the embryo undergoes **gastrulation** (migration of cells to form a triple-layered structure) and organogenesis. A striking aspect of this remarkable process, in *Drosophila* as well as in higher animals, is the division of the embryo into a series of segments corresponding to the adult organism's organization (Fig. 27-44*e*). The *Drosophila* embryo has at least three segments that eventually merge to form its head (Md, Mx, and Lb for mandibulary, maxillary, and labial), three thoracic segments (T1–T3), and eight abdominal segments (A1–A8). As development continues, the embryo elongates and several of its abdominal segments fold over its thoracic segments (Fig. 27-45*f*). At this stage, the segments become subdivided into anterior (forward) and posterior (rear) compartments. The embryo then shortens and unfolds to form a larva that hatches 1 day after beginning development (Fig. 27-45*g*). Over the next 5 days, the larva feeds, grows, molts twice, pupates, and commences metamorphosis to form an adult (Fig. 27-45*h*). In this latter process, the larval epidermis is almost entirely replaced (through apoptosis) by the outgrowth of apparently undifferentiated patches of larval epithelium known as **imaginal disks** that are committed to their developmental fates as early as the blastoderm stage. These structures, which maintain the larva's segmental boundaries, form the adult's legs, wings, antennae, eyes, etc. About 10 days after commencing development, the adult emerges and, within a few hours, initiates a new reproductive cycle.

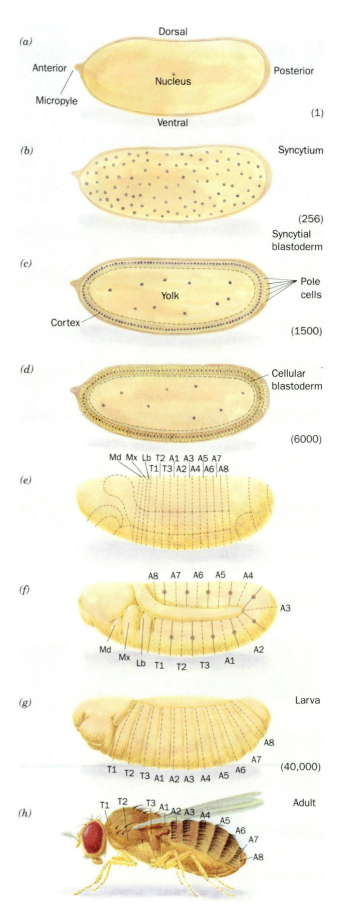

Figure 27-45   Development in Drosophila.   The various stages are explained in the text. Note that the embryos and newly hatched larva are all the same size, ~0.5 mm long. The adult is, of course, much larger. The approximate numbers of cells in the early stages of development are given in parentheses.

Developmental Patterns Are Genetically Mediated. What is the mechanism of embryonic pattern formation? Much of what we know about this process stems from genetic analyses of a series of bizarre mutations in three classes of *Drosophila* genes that normally specify progressively finer regions of cellular specialization in the developing embryo:

1. *Maternal-effect genes, which define the embryo's polarity,* that is, its anteroposterior (head to tail) and dorsoventral (back to belly) axes. Mutations of these genes globally alter the embryonic body pattern, producing, for example, nonviable embryos with two anterior or two posterior ends pointing in opposite directions.

2. *Segmentation genes, which specify the correct number and polarity of embryonic body segments.* Investigations by Christiane Nüsslein-Volhard and Eric Wieschaus led to their subclassification as follows:

   (a) **Gap genes,** the first of a developing embryo's to be transcribed, are so named because their mutations result in gaps in the embryo's segmentation pattern. Embryos with defective **hunchback (hb)** genes, for example, lack mouthparts and thorax structures.

   (b) **Pair-rule genes** specify the division of the embryo's broad gap domains into segments. These genes are so named because their mutations usually delete portions of every second segment.

   (c) **Segment polarity genes** specify the polarities of the developing segments. Thus, homozygous **engrailed (en)** mutants lack the posterior compartment of each segment.

3. *Homeotic selector genes, which specify segmental identity.* Mutations of homeotic selector genes transform one body part into another. For instance, **Antennapedia (antp,** antenna-foot) mutants have legs in place of antennae (Fig. 27-46*a*), whereas the mutations **bithorax (bx), anteriorbithorax (abx),** and **postbithorax (pbx)** each transform sections of halteres (vestigial wings that function as balancers), which normally occur only on segment T3, to the corresponding sections of wings, which normally occur only on segment T2 (Fig. 27-46*b*).

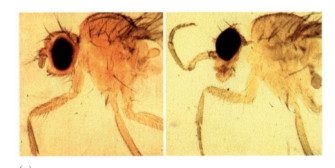

(a)

(b)

Figure 27-46 Developmental mutants of *Drosophila*.   (*a*) Head and thorax of a wild-type adult fly (*left*) and one that is homozygous for a mutant form of the homeotic *Antennapedia* (*antp*) gene (*right*). The mutant gene is inappropriately expressed in the imaginal disks that normally form antennae (where the wild-type *antp* gene is not expressed) so that they develop as the legs that normally occur only on segment T2. [Courtesy of Ginés Morata, Universidad Autónoma de Madrid, Spain.] (*b*) A four-winged *Drosophila* (it normally has two wings) that results from the presence of three mutations in the bithorax complex. These mutations cause the normally haltere-bearing segment T3 to develop as if it were the wing-bearing segment T2. [Courtesy of Edward B. Lewis, Caltech.]

The properties of maternal-effect gene mutants suggest that maternal-effect genes specify substances known as **morphogens** *whose distributions in the egg cytoplasm define the future embryo's spatial coordinate system.* Indeed, immunofluorescence studies by Nüsslein-Volhard have demonstrated that the product of the **bicoid (bcd)** gene is distributed in a gradient that decreases toward the posterior end of the normal embryo (Fig. 27-47a), whereas embryos with *bcd*-deficient mothers lack this gradient. The gradient arises through the secretion, by ovarian nurse cells, of *bcd* mRNA into the anterior end of the oocyte during oogenesis. The **nanos** gene mRNA is similarly deposited near the egg's posterior pole. The *bcd* and *nanos* gene products regulate the expression of specific gap genes. Some other maternal-effect genes specify proteins that trap the localized mRNAs in their area of deposition. This explains why early embryos produced by females homozygous for maternal-effect mutations can often be "rescued" by the injection of cytoplasm, or sometimes just the mRNA, from early wild-type embryos.

The mRNA of the gap gene *hunchback (hb)* is deposited uniformly in the unfertilized egg (Fig. 27-47a). However, **Bicoid protein** activates the transcription of the embryonic *hb* gene, whereas **Nanos protein** inhibits the translation of *hb* mRNA. Consequently, **Hunchback protein** becomes distributed in a gradient that decreases from anterior to posterior (Fig. 27-47b). Footprinting studies have demonstrated that Bicoid protein binds to five homologous sites (consensus sequence TCTAATCCC) in the *hb* gene's upstream promoter region.

Hunchback protein controls the expression of several other gap genes (Fig. 27-47c, d): High levels of Hunchback protein induce **giant** expression; **Krüppel** (German: cripple) is expressed where the level of Hunchback protein begins to decline; **knirps** (German: pigmy) is expressed at even lower levels of Hunchback protein; and *giant* is again activated in regions where Hunchback protein is undetectable. These patterns of gene expression are stabilized and maintained by additional interactions. For example, **Krüppel protein** binds to the promoters of the *hb* gene, which it activates, and the *knirps* gene, which it represses. Conversely, **Knirps protein** represses the *Krüppel* gene.

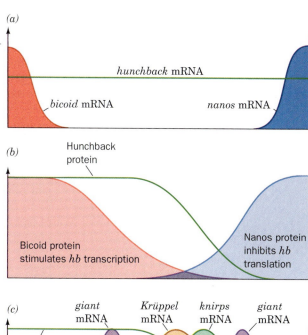

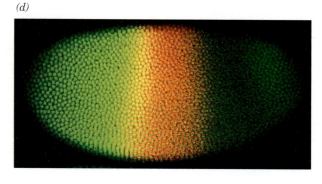

Figure 27-47 **The formation and effects of the Hunchback protein gradient in *Drosophila* embryos.** (*a*) The unfertilized egg contains maternally supplied *bicoid* and *nanos* mRNAs placed at its anterior and posterior poles, together with a uniform distribution of *hunchback* mRNA. (*b*) On fertilization, the three mRNAs are translated. Bicoid and Nanos proteins are not bound in place as are their mRNAs and hence their gradients are broader than those of the mRNAs. Bicoid protein stimulates the translation of *hunchback* mRNA whereas Nanos protein inhibits its translation, resulting in a gradient of Hunchback protein that decreases nonlinearly from anterior to posterior. (*c*) Specific concentrations of Hunchback protein induce the transcription of the *giant*, *Krüppel*, and *knirps* genes. The gradient of Hunchback protein thereby specifies the positions at which these latter mRNAs are synthesized. (*d*) A photomicrograph of a *Drosophila* embryo (*anterior end left*) that has been immunofluorescently stained for both Hunchback (*green*) and Krüppel proteins (*red*). The region where these proteins overlap is yellow. [Parts *a, b*, and *c* after Gilbert, S. F., *Developmental Biology* (5th ed.), pp. 550 and 565, Sinauer Associates (1997); Part *d* courtesy of Jim Langeland, Stephen Paddock, and Sean Carroll, Howard Hughes Medical Institute, University of Wisconsin—Madison.]

Figure 27-48 *Drosophila* embryos stained for pair-rule genes. The **Fushi tarazu** protein **(Ftz)** is brown, and the **Eve** protein is gray. These proteins are each expressed in seven stripes. [Courtesy of Peter Lawrence, MRC Laboratory of Molecular Biology, Cambridge, U.K.]

This mutual repression is thought to be responsible for the sharp boundaries between the various gap domains.

Pair-rule genes are expressed in sets of seven stripes, each just a few nuclei wide, along the early embryo's anterior–posterior axis (Fig. 27-48). The gap gene products directly control three **primary pair-rule genes: *hairy, even-skipped (eve),*** and ***runt.*** The promoters of most primary pair-rule genes consist of a series of modules, each of which contains a particular arrangement of activating and inhibitory binding sites for the various gap gene proteins. As a result, the expression of a pair-rule gene reflects the combination of gap gene proteins present, giving rise to a "zebra stripe" pattern. As with the gap genes, the patterns of expression of the primary pair-rule genes become stabilized through interactions among themselves. The primary pair-rule gene products also induce or inhibit the expression of five **secondary pair-rule genes** including ***fushi tarazu*** (***ftz;*** Japanese for not enough segments). Thus, as Walter Gehring demonstrated, *ftz* transcripts first appear in the nuclei lining the cortical cytoplasm during the embryo's tenth nuclear division cycle. By the fourteenth division cycle, when the cellular blastoderm forms, *ftz* is expressed in a pattern of seven belts around the blastoderm, each 3 or 4 cells wide (Fig. 27-48).

The expression of eight known segment polarity genes is initiated by pair-rule gene products. For example, by the thirteenth nuclear division cycle, as Thomas Kornberg demonstrated, *engrailed (en)* transcripts become detectable but are more or less evenly distributed throughout the embryonic cortex. However, since *en* is preferentially expressed in nuclei containing high concentrations of either Eve or Ftz proteins, by the fourteenth cycle they form a striking pattern of 14 stripes around the blastoderm (half the spacing of *ftz* expression). The *en* gene product thereby induces the posterior half of each segment to develop in a different fashion from its anterior half.

Homeotic Genes Direct Development of Individual Body Parts. The structural components of developmentally analogous body parts, say *Drosophila* antennae and legs, are nearly identical; only their organizations differ (Fig. 27-49). *Consequently, developmental genes must control the pattern of structural gene expression rather than simply turning these genes on or off.*

The *Drosophila* homeotic selector genes map into two large gene families: the **bithorax complex (BX-C),** which controls differentiation in the thoracic and abdominal segments, and the **antennapedia complex (ANT-C),** which primarily affects head and thoracic segments. *Homozygous mutations in BX-C cause one or more segments to develop as if they were more ante-*

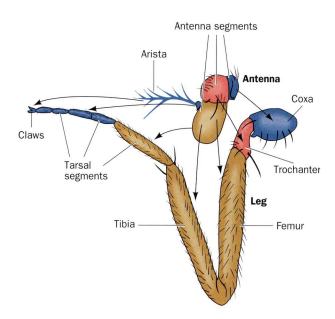

Figure 27-49 **The correspondence between** *Drosophila* **antennae and legs.**   [After Postlethwait, J. H. and Schneiderman, H. A., *Dev. Biol.* **25**, 622 (1971).]

rior segments (e.g., segment T3 develops as if it were segment T2; Fig. 27-46*b*). The entire deletion of *BX-C* causes all segments posterior to T2 to resemble T2; apparently T2 is the developmental "ground state" of the more distal segments. The evolution of homeotic gene families, it is thought, permitted arthropods (the phylum containing insects) to arise from the more primitive annelids (segmented worms) in which all segments are nearly alike.

Detailed genetic analysis of *BX-C* led Edward B. Lewis to formulate a model for segmental differentiation (Fig. 27-50): *BX-C*, Lewis proposed, contains at least one gene for each segment from T3 to A8 (numbered 0 to 8 in Fig. 27-50). Starting with segment T3, progressively more posterior segments express successively more *BX-C* genes until, in segment A8, all of these genes are expressed. Such a pattern of gene expression may result from a gradient in the concentration of a *BX-C* repressor that decreases from the anterior to the posterior end. The developmental fate of a segment is thereby determined by its position in the embryo. Subsequent sequence analysis revealed that the nine "genes" in the Lewis model are actually enhancer elements on three *BX-C* genes.

In characterizing the *Antennapedia (antp)* gene, Gehring and Matthew Scott independently discovered that *antp* cDNA hybridizes to both the *antp* and the *ftz* genes, indicating that *these genes share a common base sequence.* Subsequent experiments revealed that a similar sequence, called a **homeodomain** or **homeobox,** occurs in many *Drosophila* homeotic genes. These sequences,

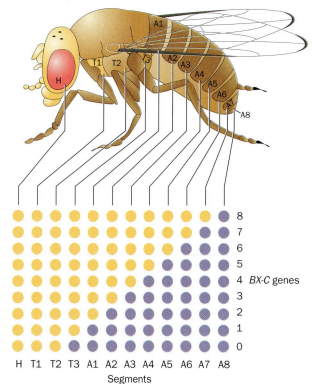

Figure 27-50 **Model for the differentiation of embryonic segments in** *Drosophila.* Segments T2, T3, and A1–8, as the lower drawing indicates, are each characterized by a unique combination of active (*purple circles*) and inactive (*yellow circles*) *BX-C* "genes." These "genes" (which are really enhancer elements), here numbered 0 to 8, are thought to be sequentially activated from anterior to posterior in the embryo so that segment T2, the developmentally most primitive segment, has no active *BX-C* genes, while in segment A8, all of them are active. [After Ingham, P., *Trends Genet.* **1**, 113 (1985).]

which are 70 to 90% identical to one another, encode even more identical 60-residue polypeptide segments.

Further hybridization studies using homeodomain probes led to the truly astonishing finding that *homeodomains are also present in the genomes of many animals.* Homeodomain-containing genes have collectively become known as **Hox genes.** In vertebrates, they are organized in four clusters of 9 to 11 genes, each located on a separate chromosome and spanning more than 100 kb. In contrast, *Drosophila,* as we saw, has two *Hox* clusters, whereas nematodes, which are evolutionarily more primitive than insects, have only one *Hox* cluster. The various *Hox* clusters, as well as their component genes, almost certainly arose through a series of gene duplications.

Hox Genes Encode Transcription Factors.   Some *Hox* genes are remarkably homologous; for example, the homeodomains of the *Drosophila antp* gene and the frog **MM3 gene** encode polypeptides that have 59 of their 60 amino acids in common. Since vertebrates and invertebrates diverged over 600 million years ago, this strongly suggests that the product of the homeodomain has an essential function.

The polypeptide encoded by the homeodomain of the *Drosophila engrailed* gene specifically binds to the DNA sequences just upstream from the transcription start sites of both the *en* and the *ftz* genes. Moreover, fusing the *ftz* gene's upstream sequence to other genes imposes *ftz*'s pattern of stripes (Fig. 27-48) on the expression of these genes in *Drosophila* embryos. *These observations suggest that homeodomain-containing genes encode transcription factors that regulate the expression of other genes.*

Thomas Kornberg and Carl Pabo have determined the X-ray structure of the 61-residue homeodomain from the *Drosophila* Engrailed protein in complex with a 21-bp DNA (Fig. 27-51). The homeodomain consists

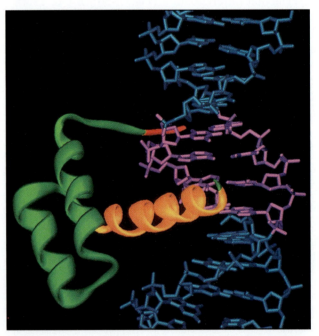

Figure 27-51   X-Ray structure of the Engrailed protein homeodomain in complex with its target DNA.   The protein is shown in ribbon form (*green*) with its recognition helix (helix 3; *gold*) bound in the DNA's major groove. The N-terminus (*red*) binds in the minor groove. The DNA is shown in stick form (*blue*) with the base pairs containing its TAAT subsite highlighted in magenta. [Based on an X-ray structure by Carl Pabo, MIT. PDBid 1HDD.]   **See the Interactive Exercises.**

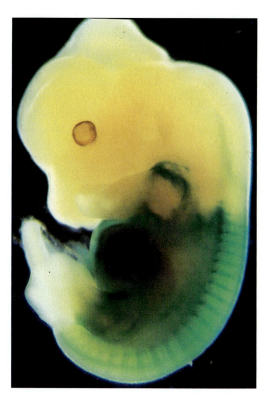

Figure 27-52 Pattern of expression of the *Hox-3.1* gene in a 12.5-day-old mouse embryo. The protein-coding portion of the *Hox-3.1* gene was replaced by the *lacZ* gene. The regions of this transgenic embryo in which *Hox-3.1* is expressed are revealed by soaking the embryo in a buffer containing a substance that turns blue when hydrolyzed by the *lacZ* gene product, β-galactosidase. [Courtesy of Phillipe Brûlet, Collège de France and the Institut Pasteur, France.]

largely of three α helices, the last two of which form an HTH motif that is closely superimposable on the HTH motifs of prokaryotic repressors (Section 23-4B).

Vertebrate *Hox* genes, like those of *Drosophila*, are expressed in specific patterns and at particular stages during embryogenesis. That the *Hox* genes directly specify the identities and fates of embryonic cells was shown, for example, by the following experiment. Mouse embryos were made transgenic for the *Hox-1.1* gene that had been placed under the control of a promoter that is active throughout the body even though *Hox-1.1* is normally expressed only below the neck. The resulting mice had severe craniofacial abnormalities such as a cleft palate and an extra vertebra and an intervertebral disk at the base of the skull. Some also had an extra pair of ribs in the neck region. Thus, the altered expression of the *Hox-1.1* gene induced a homeotic mutation, that is, a change in the development pattern, analogous to those observed in *Drosophila* (Fig. 27-46).

Homozygotic mice whose *Hox-3.1* coding sequence has been replaced with that of *lacZ* are born alive but usually die within a few days. They exhibit skeletal deformities in their trunk regions in which several skeletal segments are transformed into the likenesses of more anterior segments. The pattern of β-galactosidase activity, as colorimetrically detected through the use of a substrate analog whose hydrolysis products are blue (Fig. 27-52), indicates that *Hox-3.1* deletion modifies the properties but not the positions of the embryonic cells that normally express *Hox-3.1*.

SUMMARY

1. Genomic DNA sequence data reveals the total number of genes, their probable functions, and possible links to disease.

2. Certain genes are found in clusters, for example, those in bacterial operons, rRNA and tRNA genes, and eukaryotic histone genes. The human globin gene clusters contain genes expressed at different developmental stages.

3. Prokaryotic genomes contain small amounts of nontranscribed DNA, which include control regions for replication and transcription. The genomes of higher eukaryotes contain much larger proportions of nontranscribed DNA in the form of repetitive sequences, many of them the remnants of transposons.

4. Prokaryotic gene expression is controlled primarily at the level of transcription. Regulation of the *lac* operon is mediated by the binding of the *lac* repressor to its operator sequences. This binding, which prevents transcription of the operon, is reversed by the binding to the *lac* repressor of an inducer whose presence signals the availability of lactose, the substrate for the *lac* operon–encoded enzymes.

5. In catabolite repression, a complex of CAP and cAMP, which signals the scarcity of glucose, binds to its target DNA to stimulate the transcription of genes encoding proteins that participate in the metabolism of other sugars.

6. Attenuation is a mechanism whereby the translation-dependent formation of alternate mRNA secondary structures in an operon's leader sequence determines whether transcription proceeds or terminates. In a riboswitch, metabolite binding to an mRNA regulates gene expression.

7. The expression of eukaryotic genetic information involves the repositioning of nucleosomes and the covalent modification of histones as part of the histone code that is read by regulatory proteins. DNA methylation allows epigenetic inheritance.

8. In eukaryotes, activators and repressors, which bind to DNA enhancers and silencers, act cooperatively to regulate the rate of transcription initiation. Some of these transcription factors are activated by hormone signaling.

9. Eukaryotic gene expression may also be controlled by variable rates of mRNA degradation and regulation of translation initiation. RNA interference is a posttranscriptional gene silencing pathway in which double-stranded RNA directs the selective degradation of complementary mRNA.

10. The eukaryotic cell cycle is governed by cyclin-dependent kinases, whose targets include the tumor suppressors p53 and pRb. p53 functions as a transcriptional activator that detects a variety of pathological states such as DNA damage and thereon helps maintain cell cycle arrest and, if the damage cannot be repaired, apoptosis (programmed cell death).

11. Extracellular or intracellular signals may trigger apoptosis, a cell-suicide pathway that requires the activity of various caspases.

12. The development of the *Drosophila* embryo is controlled by maternal-effect genes, which define the embryo's polarity; gap, pair-rule, and segment polarity genes, which specify the number and polarity of embryonic body segments; and homeotic selector genes (*Hox* genes), which encode transcription factors that regulate the expression of genes and therefore govern cell differentiation. *Hox* genes similarly regulate vertebrate development.

REFERENCES

Genome Organization

Cummings, C.J. and Zoghbi, H.Y., Trinucleotide repeats: mechanisms and pathophysiology, *Annu. Rev. Genomics Hum. Genet.* **1**, 281–328 (2002).

Jurka, J., Repeats in genomic DNA: mining and meaning, *Curr. Opin. Struct. Biol.* **8**, 333–337 (1998). [Reviews the evolutionary history and function of repetitive DNA sequences.]

The International HapMap Consortium, The International HapMap Project, *Nature* **426**, 789–796 (2003). [Describes how patterns of single-nucleotide polymorphism can be mapped and used as markers for genetic diseases.]

Prokaryotic Gene Expression

Bell, C.E. and Lewis, M., The Lac repressor: a second generation of structural and functional studies, *Curr. Opin Struct. Biol.* **11**, 19–25 (2001).

Kolb, A., Busby, S., Buc, H., Garges, S., and Adhya, S., Transcriptional regulation by cAMP and its receptor protein, *Annu. Rev. Biochem.* **62**, 749–795 (1993).

Nudler, E. and Mironov, A.S., The riboswitch control of bacterial metabolism, *Trends Biochem. Sci.* **29**, 11–17 (2004).

Yanofsky, C., Transcription attenuation, *J. Biol. Chem.* **263**, 609–612 (1988). [A general discussion of attenuation.]

Eukaryotic Gene Expression

Becker, P.B. and Hörz, W., ATP-dependent nucleosome remodeling, *Annu. Rev. Biochem.* **71**, 247–273 (2002).

Brivanlou, A.H. and Darnell, J.E., Jr., Signal transduction and the control of gene expression, *Science* **295**, 813–818 (2002).

Cohen, D.E. and Lee, J.T., X-Chromosome inactivation and the search for chromosome-wide silencers, *Curr. Opin. Genet. Dev.* **12**, 219–224 (2002).

Denli, A.M. and Hannon, G.J., RNAi: an ever-growing puzzle, *Trends Biochem. Sci.* **28**, 196–201 (2003).

Felsenfeld, G. and Groudine, M., Controlling the double helix, *Nature* **421**, 448–453 (2003). [A brief review of chromatin packaging, including histone modifications and epigenetics.]

Gilbert, S.F., *Developmental Biology* (7th ed.), Sinauer Associates (2003).

Gustafsson, C.M. and Samuelsson, T., Mediator—a universal complex in transcription regulation, *Mol. Microbiol.* **41**, 1–8 (2001).

Hickman, E.S., Moroni, M.C., and Helin, K., The role of p53 and pRB in apoptosis and cancer, *Curr. Opin. Genet. Dev.* **12**, 60–66 (2002).

Iizuka, M. and Smith, M.M., Functional consequences of histone modifications, *Curr. Opin. Genet. Dev.* **13**, 154–160 (2003).

Jiang, X. and Wang, X., Cytochrome *c*-mediated apoptosis, *Annu. Rev. Biochem.* **73**, 87–106 (2004).

Jones, P.A. and Takai, D., The role of DNA methylation in mammalian epigenetics, *Science* **293**, 1068–1070 (2001).

Lemon, B. and Tjian, R., Orchestrated response: A symphony of transcription factors for gene control, *Genes Devel.* **14**, 2551–2569 (2000).

Levine, M. and Tjian, R., Transcription regulation and animal diversity, *Nature* **424**, 147–151 (2003). [Discusses how multiple regulatory DNA sequences and transcription factors could account for differences in the complexity of organisms with similar numbers of genes.]

Nüsslein-Volhard, C., Gradients that organize embryo development, *Sci. Am.* **272**(7): 54–61 (1996).

Sumner, A.T., *Chromosomes, Organization and Function*, Blackwell Science (2003).

Turner, B.M., Cellular memory and the histone code, *Cell* **111**, 285–291 (2002).

KEY TERMS

gene expression	orphan gene	moderately repetitive sequences	negative regulator
genomics	SNP		corepressor
C value	gene cluster	selfish DNA	leader sequence
C-value paradox	centromeres	inducer	attenuation
ORF	genetic anticipation	repressor	aptamer
EST	highly repetitive sequences	operator	riboswitch
CpG island	satellite DNA	catabolite repression	heterochromatin
ncRNA	STR	positive regulator	euchromatin

Barr body
polytene chromosome
chromosome puff
chromatin-remodeling
  complex
HMG protein
histone code
HAT
bromodomain
chromodomain
epigenetic

imprinting
enhancer
silencer
enhanceosome
squelching
insulator
STAT
hormone response element
nonsense-mediated decay
antisense RNA
RNAi

siRNA
cell cycle
cyclin
tumor suppressor
apoptosis
necrosis
caspase
syncytium
blastoderm
gastrulation
imaginal disk

maternal-effect gene
segmentation gene
gap gene
pair-rule gene
segment polarity gene
homeotic selector gene
morphogen
homeodomain
Hox gene

 ## STUDY EXERCISES

1. List some of the elements responsible for the large sizes of eukaryotic genomes relative to prokaryotic genomes.

2. Explain how genes within a genome are identified.

3. Why are rRNA and tRNA genes, but not protein-coding genes, generally found in clusters?

4. Describe the regulation of the *lac* operon by *lac* repressor and CAP.

5. How does the conformation of mRNA regulate gene expression by attenuation and in riboswitches?

6. Discuss the implications of histone acetylation and DNA methylation on gene expression.

7. Explain why eukaryotic gene expression requires proteins in addition to the six general transcription factors.

8. Why is the spacing between enhancers and promoters variable?

9. Describe the steps of RNA interference.

10. Explain how p53 and pRb suppress tumor formation.

11. Summarize the events of apoptosis.

12. How do homeotic genes control embryogenesis in *Drosophila*?

PROBLEMS

1. DNA isolated from an organism can be sheared into fragments of uniform size (~1000 bp), heated to separate the strands, then cooled to allow complementary strands to reanneal. The renaturation process can be followed over time. Explain why the renaturation of *E. coli* DNA is a monophasic process whereas the renaturation of human DNA is biphasic (an initial rapid phase followed by a slower phase).

2. Explain why the organization of genes in operons facilitates the assignment of functions to previously unidentified ORFs in a bacterial genome.

3. Explain why (a) inactivation of the O_1 sequence of the *lac* operator almost completely abolishes repression of the *lac* operon; (b) inactivation of O_2 or O_3 causes only a twofold loss in repression; and (c) inactivation of both O_2 and O_3 reduces repression ~70-fold.

4. Why do *E. coli* cells with a defective *lacZ* gene fail to show galactoside permease activity after the addition of lactose in the absence of glucose?

5. Describe the probable genetic defect that abolishes the sensitivity of the *lac* operon to the absence of glucose when other metabolic operons continue to be sensitive to the absence of glucose.

6. Why can't eukaryotic transcription be regulated by attenuation?

7. Predict the effect of deleting the leader peptide sequence on regulation of the *trp* operon.

8. Red–green color blindness is caused by an X-linked recessive genetic defect. Hence females rarely exhibit the red–green colorblind phenotype but may be carriers of the defective gene. When a narrow beam of red or green light is projected onto some areas of the retina of such a female carrier, she can readily differentiate the two colors but on other areas she has difficulty in doing so. Explain.

9. Draw the molecular formulas of the covalently modified histone side chains acetyllysine, methyllysine, and methylarginine. How do these modifications alter the chemical properties of the side chains?

10. Explain why a deficiency of the vitamin folic acid could lead to undermethylation of histones and DNA.

11. Why is transcriptionally active chromatin ~10 times more susceptible to cleavage by DNase I than transcriptionally silent chromatin?

12. Is it possible for a transcription enhancer to be located within the protein-coding sequence of a gene? Explain.

13. Explain why natural selection has favored the instability of RNA.

14. Explain why RNAi would be a less efficient mechanism for regulating the expression of specific genes if Dicer hydrolyzed double-stranded RNA every 11 bp rather than every 22 bp.

15. Why is it disadvantageous for single-celled eukaryotes such as yeast to undergo apoptosis?

16. In *Drosophila*, an *esc⁻* homozygote develops normally unless its mother is also an *esc⁻* homozygote. Explain.

Many elements of modern buildings have counterparts in living systems. For example, cells contain proteins that provide structural support and mediate the movement of materials within the cell. In higher animals, proteins of the immune system guard the premises from foreign invaders. [Steve Dunwell/Stone/Getty Images.]

Protein Function Part II: Cytoskeletal and Motor Proteins and Antibodies

The preceding chapters have provided numerous examples of proteins that catalyze metabolic reactions, transduce extracellular signals, and manage the storage and expression of genetic information. In all cases, the unique three-dimensional structures of these proteins and their ability to interact specifically with other molecules allow them to carry out different physiological functions. In this chapter, we examine some additional proteins whose structures and functions are essential for cellular integrity and whose proper functioning is vital for human health. Much of this chapter is concerned with the structures and functions of the various types of cytoskeletal proteins and their associated molecular motors. The chapter ends with a discussion of the operation of the immune system and the structure, function, and generation of antibodies.

1 Actin and Microfilaments

The shape of a eukaryotic cell is determined in large part by its **cytoskeleton,** a system of protein fibers that pervades the cytoplasm. In addition to giving structural definition to the cell, the cytoskeleton undergoes changes that allow the cell to grow, change shape, and even move. Within each cell, **motor proteins** associated with cytoskeletal fibers generate mechanical force and mediate the intracellular movements of organelles.

Cytoskeletal proteins are among the most abundant proteins in eukaryotes (collagen, the most abundant vertebrate protein, forms extracellular fibers; Section 6-1C). A typical eukaryotic cell contains three types of cytoskeletal proteins whose polymeric or aggregated forms are visible by fluorescence microscopy: **microfilaments, intermediate filaments,** and **microtubules** (Fig. 28-1). These fibers differ in size, subunit structure, and dynamic properties.

1. Microfilaments, with a diameter of ~70 Å, are polymers of the protein **actin.** Globular actin molecules associate head to tail in two-stranded filaments that lengthen and shorten through the addition and removal of individual actin monomers, so that the cytoskeleton can change shape. Actin and the motor protein **myosin** are the major components of muscle.

2. Intermediate filaments, with diameters of ~100 Å, are intermediate in thickness to microfilaments and microtubules. Intermediate filaments are predominantly structural; they play no part in cell motility and have no associated motor proteins. Compared to the components of microfilaments and microtubules, intermediate filament proteins are highly heterogeneous. For example, the human genome contains over 50 intermediate filament genes. Among these are genes for the **nuclear lamins,** filaments that line the inner nuclear membrane, helping to define the shape of the nucleus and possibly serving as chromosome anchor points. In some cells, intermediate filaments are the most abundant type of cytoskeletal fiber. The intermediate filament α-keratin (Section 6-1C) is particularly abundant in the dead remnants of epidermal cells (the hard outer layers of the skin), where it accounts for up to 85% of the total protein. In general, intermediate filaments undergo fewer dynamic changes than other types of cytoskeletal fibers. This may reflect their construction from extended polypeptide chains that form coiled coils, which are less likely to undergo dissociation and reassociation than the globular subunits of microfilaments or microtubules.

(a)
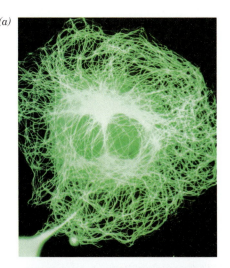

(b)
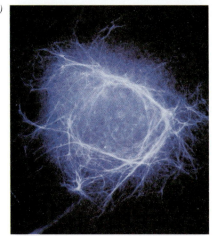

(c)
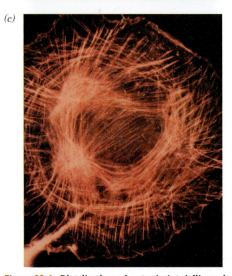

Figure 28-1 Distribution of cytoskeletal fibers in a single cell.   (a) Microfilaments. (b) Intermediate filaments. (c) Microtubules. To make these micrographs, each type of fiber was labeled with a fluorescent probe that binds specifically to one type of cytoskeletal protein. Note how the distributions of these three types of filaments differ. [Courtesy of John Victor Small, Austrian Academy of Sciences, Salzburg.]

3. Microtubules are built from the globular protein **tubulin,** which assembles to form hollow cylindrical polymers with a diameter of ~250 Å. As in microfilaments, disassembly and reassembly of microtubules can dramatically alter cytoskeletal structure. Microtubules serve as tracks for motor proteins such as **kinesin** and **dynein,** in effect forming a system of highways inside the cell.

All three types of cytoskeletal fibers interact with other proteins, and certain cross-linking proteins connect intermediate filaments to both microfilaments and microtubules. As a result, the cytoskeleton is a complex structure whose overall properties are difficult to appreciate through the traditional biochemical approach of studying isolated proteins. In the following four sections, we shall focus on actin and tubulin in order to explore how globular proteins can form fibrous structures that share some properties with the fibrous proteins keratin and collagen (Section 6-1C). We shall also discuss several types of motor proteins to illustrate the structure–function relationships involved in converting free energy to mechanical work. Although not discussed here, prokaryotic cells also contain microfilament- and microtubule-like proteins as well as specialized motor proteins.

A Actin Structure

At low ionic strengths, actin, the most abundant cytosolic protein in eukaryotes (comprising up to 15% of the protein in nonmuscle cells), occurs as an ~375-residue globular monomer called **G-actin** (G for globular) that normally binds one molecule of ATP. Under physiological conditions, however, G-actin polymerizes to form fibers known as **F-actin** (F for fibrous). Actin has a single high-affinity metal ion binding site that has a greater affinity for Ca^{2+} than for Mg^{2+}. However, because the *in vivo* concentration of Mg^{2+} is much higher than that of Ca^{2+}, this site is largely occupied by Mg^{2+} *in vivo*. Actin is one of the most highly conserved eukaryotic proteins (88% identity between yeast cytoplasmic actin and human muscle actin), with the actins expressed in the corresponding tissues of different species more closely related than the several actin isoforms expressed in different tissues within an organism. Thus, human cytoplasmic actin is more closely related to yeast cytoplasmic actin than it is to human muscle actin.

The X-ray structure of rabbit muscle actin in complex with Ca^{2+} ion and either ADP or the nonhydrolyzable ATP analog **adenosine-5′-(β,γ-imido)triphosphate (AMPPNP; also called ADPNP),**

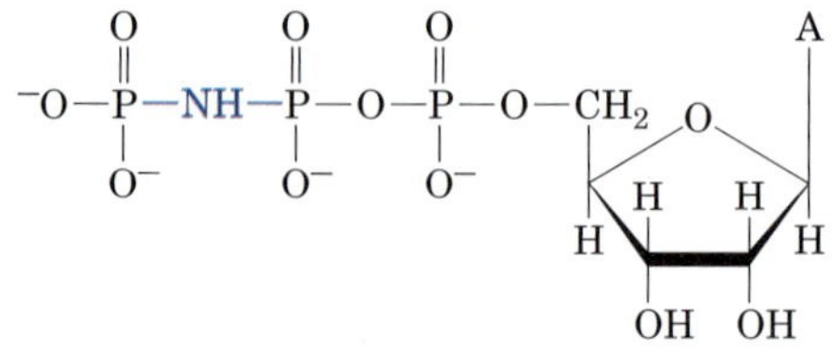

Adenosine-5′-(β,γ-imido)triphosphate
(AMPPNP)

determined by Roberto Dominguez, indicates that the protein consists of two domains, each of which is divided into two subdomains (Fig. 28-2). The ADP and AMPPNP bind at the bottom of a deep cleft between the two domains, where the Ca^{2+} ion is liganded in part by the β-phosphate of ADP or the β- and γ-phosphates of AMPPNP. Comparison of the ADP and AMPPNP structures reveals that the release of the nucleotide γ-phosphate

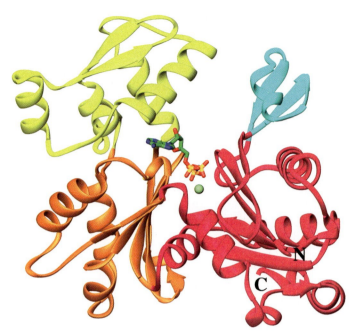

Figure 28-2 X-Ray structure of rabbit muscle actin in complex with AMPPNP and a Ca^{2+} ion. Subdomains 1 (*magenta*) and 2 (*cyan*) form the so-called small domain, whereas subdomains 3 (*orange*) and 4 (*yellow*) form the so-called large domain (which is only marginally larger than the small domain). The AMPPNP, which is drawn in stick form and colored according to atom type (C green, N blue, O red, and P gold), binds at the bottom of a deep cleft between the domains. The Ca^{2+} ion, which is represented by a light green sphere, is liganded by the β- and γ-phosphates of the AMPPNP. Note that both the N- and C-termini occur in subdomain 1. [Based on an X-ray structure by Roberto Dominguez, Boston Biomedical Research Institute, Watertown, Massachusetts. PDBid 1NWK.]

triggers a significant conformational change in subdomain 2 (cyan in Fig. 28-2), but the two structures are otherwise closely similar.

The fibrous nature of F-actin and its variable fiber lengths has thwarted its crystallization in a manner suitable for X-ray crystallographic analysis. Consequently, our current understanding of the atomic structure of microfilaments is based on electron micrographs of F-actin (Fig. 28-3a) together with low resolution models based on X-ray studies of oriented gels of F-actin into which high resolution atomic models of G-actin have been fitted (Fig. 28-3b). These models indicate that the actin polymer is a double chain of subunits in which each subunit contacts four others through interactions involving all four of its subdomains (Fig. 28-3b). Successive actin subunits within a chain are rotated ~167° relative to each other, accounting for the lumpy helical appearance of the microfilament. Each actin subunit has the same head to tail orientation (e.g., all the nucleotide-binding clefts open upward in Fig. 28-3b), so the assembled fiber has a distinct polarity. The end of the fiber toward which the nucleotide-binding sites open is known as the **(−) end** (also known as the **pointed end** from its appearance in the electron microscope when decorated with myosin), and the opposite end is the **(+) end** (also known as the **barbed end**).

B Microfilament Dynamics

In vitro, the initiation of actin polymerization is a relatively slow process because actin dimers and trimers have relatively few subunit–subunit contacts and hence little stability. However, once a stable **nucleus** of ~4 subunits has formed, additional subunits can form the full complement of interactions and hence rapidly associate with both ends of a filament. As the aggregate mass of the F-actin filaments increases, the concentration of G-actin monomers deceases until equilibrium is reached. At this stage, the F-actin subunits at the two ends of the filaments exchange with G-actin monomers in solution but there is no change in the aggregate mass of F-actin filaments, that is, the filament lengths have reached a steady state.

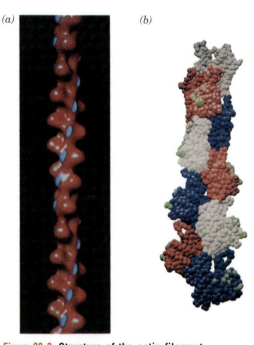

(a)   *(b)*

Figure 28-3 Structure of the actin filament. (*a*) Cryoelectron microscopy–based image. Note the bilobal appearance of each monomeric (repeating) unit. The tropomyosin binding sites (Section 28-2A) are blue. [Courtesy of Daniel Safer, University of Pennsylvania, and Ronald Milligan, The Scripps Research Institute, La Jolla, California.] (*b*) Model based on fitting the known X-ray structure of the actin monomer to the X-ray fiber diffraction pattern of F-actin. Actin monomers are shown in space-filling representation, in alternating blue, red, and white, with each amino acid residue represented by a sphere. The lowest monomer shown is oriented identically to that in Fig. 28-2. The residues that cross-linking studies indicate form the myosin binding site (Section 28-2A) are green. [Courtesy of Wolfgang Kabsch and Kenneth Holmes, Max-Planck-Institute für medizinische Forschung, Germany.]

The reaction for microfilament formation is expressed

$$\text{G-Actin} + \text{F-Actin} (n \text{ subunits}) \underset{k_{\text{off}}}{\overset{k_{\text{on}}}{\rightleftharpoons}} \text{F-Actin} (n + 1 \text{ subuits})$$

where k_{on} is the rate constant for subunit association with filaments and k_{off} is the rate constant for subunit dissociation from filaments. The G-actin concentration at the steady state is called the **critical concentration, C_c,** which is typically ~0.1 μM. Thus, at equilibrium,

$$k_{\text{on}} C_c [\text{F-actin ends}] = k_{\text{off}} [\text{F-actin ends}] \qquad [28\text{-}1]$$

so that

$$C_c = \frac{k_{\text{off}}}{k_{\text{on}}} \qquad [28\text{-}2]$$

(here the F-actin ends are reactants for both the forward and reverse reactions and hence their concentration is the same for both reactions). Above the critical concentration, G-actin polymerizes; below it, F-actin depolymerizes.

The value of k_{on} is 5- to 10-fold greater for subunit addition to the (+) end (hence its name) than for subunit addition to the (−) end. Thus, according to Equation 28-2, k_{off} for subunit dissociation from the (+) end must be larger than that from the (−) end by the same factor so that at equilibrium neither end experiences a net change in length. This makes physical sense because subunits added to either end ultimately must have identical interactions with other subunits in the filament.

Microfilaments Undergo Treadmilling. The foregoing is predicated on the assumption that the G-actin subunits that associate with F-actin are identical to those that dissociate from F-actin. However, this is not necessarily the case. Polymerization activates F-actin subunits to hydrolyze their bound ATP to ADP + P_i with the subsequent dissociation of P_i. The resulting conformational change (see above) reduces the affinity of an ADP–F-actin subunit for its neighboring subunits relative to that of ATP–F-actin; that is, the value of C_c for ADP–F-actin is greater than that for ATP–F-actin. Since F-actin–catalyzed ATP hydrolysis occurs more slowly than actin polymerization and F-actin's bound nucleotide does not exchange with those in solution, F-actin's more recently polymerized and hence predominantly ATP-containing subunits occur at its (+) end, whereas its (−) end consists mainly of less recently polymerized and hence predominantly ADP-containing subunits.

Under these conditions, the steady state (when the microfilament maintains a constant length) occurs when the net rate of addition of subunits at the (+) end matches the net rate of dissociation of subunits at the (−) end, that is, the G-actin concentration falls between the C_c values for the two ends. Then, *subunits that have added to the (+) end move toward the (−) end where they dissociate* (Fig. 28-4). This process, which is called **treadmilling,** is driven by the free energy of ATP hydrolysis and hence is not at equilibrium. Treadmilling has been directly observed *in vivo* by loading cells with a small amount of fluorescently-labeled G-actin so that the newly formed F-actin contains just a few randomly located fluorescent subunits. Each microfilament is uniquely "speckled" and can therefore be distinguished from other microfilaments as it undergoes structural changes over time.

Treadmilling Is Responsible for Cell Locomotion. In a resting cell, the rate of actin polymerization is low, and the cell maintains a steady state as described above. Following stimulation, however, the rates of actin polymer-

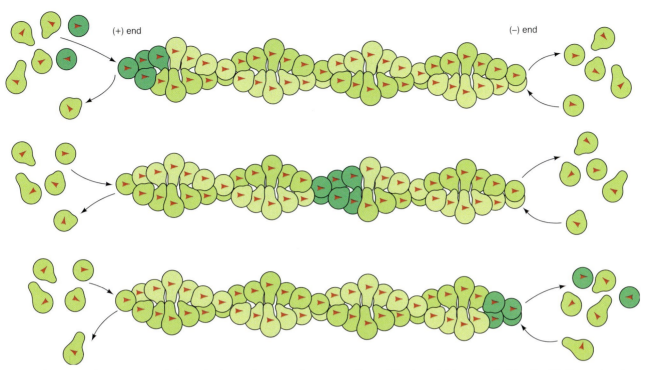

(+) end   (−) end

Figure 28-4 Microfilament treadmilling. In the steady state, actin monomers continually add to the (+) end of the filament (*left*), with eventual ATP hydrolysis, but dissociate at the same rate from the (−) end (*right*). The filament thereby maintains a constant length while its component monomers translocate from left to right.

ization and branching (see below) increase dramatically, allowing the cell to extend its cytoplasm in one direction. In this event, the (+) ends of growing microfilaments actually push the plasma membrane outward. If the cytoplasmic protrusion anchors itself to the underlying surface, then the cell can use the adhesion point for traction in order to advance further. In order to crawl, however, the trailing edge of the cell must release its contacts with the surface while newer contacts are being made at the leading edge (Fig. 28-5). In addition, as microfilament polymerization proceeds at the leading edge, depolymerization must occur elsewhere in the cell, since the pool of G-actin is limited.

Actin-mediated cell locomotion, that is, amoeboid motion, is the most primitive mechanism of cell movement. Nevertheless, virtually all eukaryotic cells undertake some version of it, even if it involves just a small patch of actin near the cell surface. More extensive microfilament rearrangements are essential for cells such as neutrophils (a type of white blood cell) that travel relatively long distances to sites of infection or inflammation.

Actin-Binding Proteins Control Microfilament Dynamics. At the pH and salt concentrations that typically occur in the cytosol, actin by itself would be almost entirely polymerized (mono- and divalent cations bind to low affinity sites on actin so as to promote its polymerization). This is not the case, however, because the rates of actin polymerization and depolymerization at each end of a microfilament are controlled by numerous other proteins. For example, **β-thymosin** (so named because it was originally isolated from mammalian thymus), a small (43 residues) and abundant protein, binds to G-actin so as to inhibit both its polymerization and its nucleotide exchange. In contrast, **profilin** binds to ADP–G-actin (i.e., actin that has just been released from a depolymerizing microfilament) so as to increase the rate

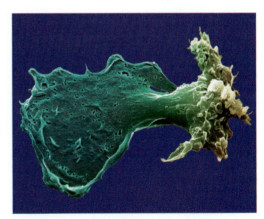

Figure 28-5 Scanning electron micrograph of a crawling cell. The leading edge of the cell (*left*) is ruffled where it has become detached from the surface and is in the process of extending. The cell's trailing edge or tail (*right*), which is still attached to the surface, is gradually pulled toward the leading edge. The rate of actin polymerization is greatest at the leading edge. [© David Phillips/Visuals Unlimited.]

of exchange of its bound ADP for ATP by over 1000-fold, thereby increasing the concentration of ATP–G-actin (the substrate for repolymerization). Moreover, profilin binds to the end of ATP–G-actin opposite its ATP-binding cleft, thereby blocking actin subunits from binding to the microfilament's (−) end but permitting them to associate with its (+) end. Since on binding to the (+) end, an actin subunit undergoes a conformational change that releases profilin, this greatly enhances the rate of microfilament elongation. The activity of profilin is regulated by several mechanisms including its phosphorylation and its binding to the membrane component phosphatidylinositol-4,5-bisphosphate (PIP$_2$), a central participant in the phosphoinositide signal transduction pathway (Section 21-3F).

ADF/cofilin (ADF for *actin depolymerizing factor*) accelerates depolymerization at the (−) end of a microfilament by preferentially binding to ADP–P$_i$–actin and promoting the dissociation of P$_i$. However, the binding of ADF/cofilin to F-actin is prevented by the phosphorylation of ADF/cofilin by a specific protein kinase, **LIM kinase,** which is itself activated by phosphorylation via signaling pathways. *In vitro,* ADF/cofilin and profilin together accelerate microfilament treadmilling by a factor of ~125, which approaches the rate observed in crawling cells.

Microfilament Branching Is Mediated by Arp2/3.

Signals that lead to actin polymerization also activate the 7-subunit, ~220-kD **Arp2/3 complex** (Arp for *actin related protein*). This protein complex initiates microfilament polymerization at random sites along a microfilament so as to form branches (Fig. 28-6). Indeed, actin filaments may be very densely branched, particularly near the periphery of that portion of a crawling cell that is being extended.

The X-ray structure of the Arp2/3 complex, determined by Thomas Pollard and Senyon Choe, reveals that **Arp2** and **Arp3,** which are both ~45% identical to actin, structurally resemble actin (Fig. 28-7). In the

(a)

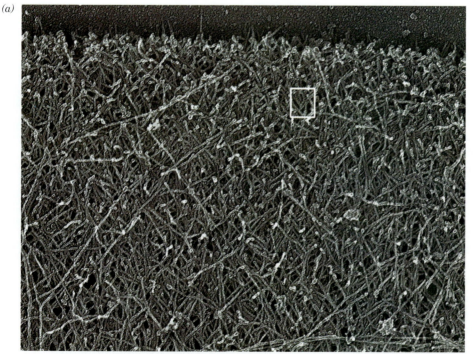

(b)

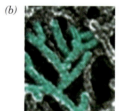

Figure 28-6 Electron micrograph of branched microfilaments.
(*a*) The leading edge of a migrating *Xenopus* keratinocyte (a type of skin cell) prepared by detergent extraction. (*b*) Magnified view of the boxed region in Part *a* in which tightly spaced multiple branches are highlighted in cyan. Note that new microfilaments branch from the parent filament at a constant angle of 70°. [Courtesy of Tatyana Svitkina, University of Pennsylvania.]

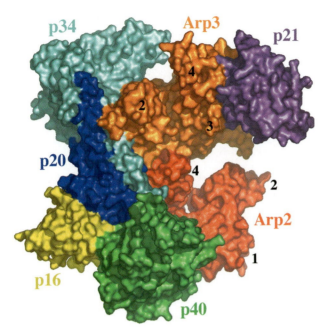

Figure 28-7 X-Ray structure of the bovine Arp2/3 complex. In this space-filling model, the subunits are drawn in different colors. Arp2 (*red*) and Arp3 (*orange*) structurally resemble actin (their subdomains are numbered; compare with Fig. 28-2) but are not aligned for polymerization. Activation of the complex causes it to bind to an existing microfilament and aligns the Arp2 and Arp3 subunits so that they can initiate the growth of a branch. [Courtesy of Thomas Pollard, Yale University. PDBid 1K8K.]

inactive Arp2/3 complex, the two Arp subunits are arranged head to tail but are not aligned as in an actin filament. The binding of the Arp2/3 complex to the side of an actin filament, as mediated by members of the **WASp/Scar** protein family (WASp for *W*iskott-*A*ldrich *s*yndrome *p*rotein; **Wiskott-Aldrich syndrome** is an X-linked immunodeficiency and bleeding disorder caused by defective WASp), induces conformational changes in the Arp2/3 complex that rotate its Arp subunits so that they more closely resemble an actin dimer. The Arp subunits thereby **nucleate** (initiate) the growth of an F-actin branch by providing a start site for the polymerization of actin subunits.

Capping and Severing Proteins Limit Microfilament Growth.   In a crawling cell, actin polymerization occurs almost exclusively at microfilament (+) ends near the cell's leading edge because the ends of microfilaments elsewhere in the cell are blocked by **capping proteins.** In effect, all the available G-actin is funneled to the uncapped ends, resulting in rapid microfilament growth. The balance between G-actin and F-actin concentrations strongly favors polymerization, so the presence of capping proteins is essential for maintaining normal microfilament dynamics.

In some cases, capping proteins also act to sever microfilaments. One such protein is **gelsolin,** which is activated by increases in intracellular Ca^{2+} concentrations. Gelsolin consists of six homologous 125- to 150-residue segments, three of which can bind to actin. The X-ray structure of an actin trimer in which each monomer is bound to a gelsolin segment 1 molecule, determined by James Spudich and Robert Fletterick, has the same general shape as F-actin but is untwisted and extended (Fig. 28-8). This form of actin has been dubbed **X-actin** (X for extended). The structure suggests that gelsolin severs a microfilament by inserting between polymerized actin subunits and altering the shape of the polymer so that intersubunit interactions are weakened. Actin subunits then dissociate from the (−) end of the newly formed microfilament ends, with gelsolin remaining associated with the (+) end as a cap. However, a local rise in the PIP_2 concentration induces gelsolin to dissociate from the filament end. ADF/cofilin may act in a similar fashion as a capping and severing protein.

Figure 28-8   Structures of X-actin and F-actin.
(*a*) The model of X-actin is based on the X-ray structure of an actin trimer bound to the 125-residue segment 1 of gelsolin (not shown). Subdomain 2 is red, subdomain 3 is purple, a hydrophobic loop (residues 263–275) is cyan, and the N-terminus is yellow. The gelsolin segment inserts between actin subunits so as to extend and untwist the actin filament, thereby promoting its breakage. (*b*) Model of F-actin colored as in Part *a* and viewed similarly to Fig. 28-3*b*. Actin substructures have the same colors in each model. The rotational symmetry and axial translation per subunit are indicated. [Courtesy of Robert Fletterick, University of California at San Francisco.]

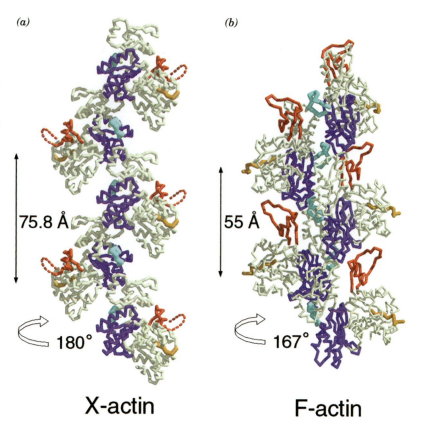

(*a*)   (*b*)

75.8 Å   55 Å

180°   167°

X-actin   F-actin

2 Muscle Contraction

Actin filaments that do not undergo dynamic changes in length are components of a contractile apparatus that includes additional proteins, notably the motor protein myosin, which binds actin to generate mechanical force. Actin–myosin systems operate during **cytokinesis** (the separation of daughter cells following mitosis), intracellular trafficking, and, most familiarly, muscle contraction. In this section, we focus primarily on the structural and chemical basis for movement in muscle.

A Structure of Striated Muscle

The voluntary muscles, which include the skeletal muscles, have a striated (striped) appearance when viewed by light microscopy (Fig. 28-9). Such muscles consist of long multinucleated cells (the muscle fibers) that contain parallel bundles of **myofibrils** (Greek: *myos,* muscle; Fig. 28-10). Electron micrographs show that muscle striations arise from the banded structure of multiple in-register myofibrils. The bands are formed by alternating regions of greater and lesser electron density called **A bands** and **I bands,** respectively (Fig. 28-11). The myofibril's repeating unit, the **sarcomere** (Greek: *sarkos,* flesh), is bounded by **Z disks** at the center of each I band. The A band is centered on the **H zone,** which in turn is centered on the **M disk.** The A band contains 150-Å-diameter **thick filaments,** and the I band contains 70-Å-diameter **thin filaments.** The two sets of filaments are linked by cross-bridges where they overlap.

A contracted muscle can be as much as one-third shorter than its fully extended length. The contraction results from a decrease in the length of the sarcomere, caused by reductions in the lengths of the I band and the

Figure 28-9   Photomicrograph of a muscle fiber.
The longitudinal axis of the fiber is horizontal (perpendicular to the striations). The alternating pattern of dark A bands and light I bands from multiple in-register myofibrils is clearly visible. [J.C. Revy, CNRI/Photo Researchers.]

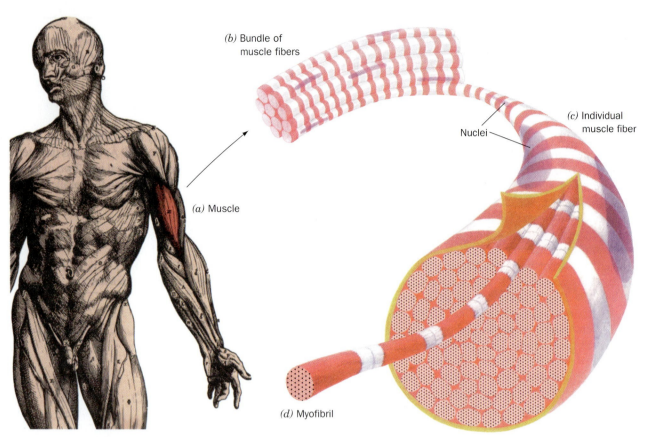

Figure 28-10 Skeletal muscle organization.   A muscle (*a*) consists of bundles of muscle fibers (*b*), each of which is a long, thin, multinucleated cell (*c*) that may run the length of the muscle. Muscle fibers contain bundles of laterally aligned myofibrils (*d*), which in turn consist of bundles of alternating thick and thin filaments.

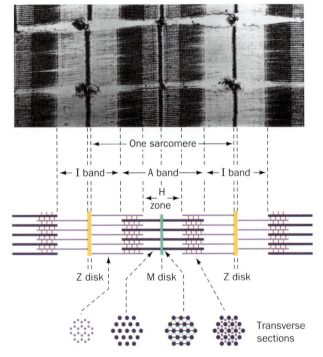

Figure 28-11 Anatomy of the myofibril.   The electron micrograph shows parts of three myofibrils, which are separated by horizontal gaps. The accompanying interpretive drawing shows the major features of the myofibril: the light I band, which contains only thin filaments; the A band, whose dark H zone contains only thick filaments and whose darker outer segments contain overlapping thick and thin filaments; the Z disk, to which the thin filaments are anchored; and the M disk, which arises from a bulge at the center of each thick filament. The myofibril's functional unit, the sarcomere, is the region between two successive Z disks. [Courtesy of Hugh Huxley, Brandeis University.]

BOX 28-1

Pathways of Discovery

Hugh Huxley and the Sliding Filament Model

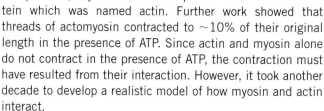

Hugh Huxley (1924–)

The mechanism of muscle action has fascinated scientists for hundreds if not thousands of years. The first close look at muscle fibers came in 1682, when Antoni van Leeuwenhoek's early microscope revealed a pattern of thin longitudinal fibers. In the modern era, research on muscle has followed one of two approaches. First, it is possible to study muscle as an energy-transducing system, in which metabolic energy is generated and consumed. This line of research received a tremendous boost in the 1930s with the discovery that ATP is the energy source for muscle contraction. The second approach involves treating muscle as a mechanical system, that is, sorting out its rods and levers. Ultimately, a molecular approach united the mechanical and energetic aspects of muscle research. The insights of Hugh Huxley made this possible.

The molecular characterization of muscle did not occur overnight. In 1859, Willi Kühne isolated a proteinaceous substance from muscle tissue that he named "myosin" (almost certainly a mixture of many proteins), but it tended to aggregate and was therefore not as popular a study subject as the more soluble proteins such as hemoglobin. A major breakthrough in muscle protein chemistry came in 1941, when the Hungarian biochemist Albert Szent-Györgyi showed that two types of protein could be extracted from ground muscle by a solution with high salt concentration (Szent-Györgyi also contributed to the elucidation of the citric acid cycle; see Box 16-1). Extraction for 20 minutes yielded a protein he named myosin A but which is now called myosin. However, extraction overnight yielded a second protein which he named myosin B but is now called actomyosin. It soon became ap-

parent that myosin B was a mixture of two proteins, myosin and a new protein which was named actin. Further work showed that threads of actomyosin contracted to ~10% of their original length in the presence of ATP. Since actin and myosin alone do not contract in the presence of ATP, the contraction must have resulted from their interaction. However, it took another decade to develop a realistic model of how myosin and actin interact.

A number of theories had been advanced to explain muscle contraction. According to one theory, the cytoplasm of muscle cells moved like that of an amoeba. Other theories proposed that muscle fibers took up and gave off water or repelled and attracted other fibers electrostatically. Linus Pauling, who had recently discovered the structures of the α helix and β sheet (see Box 6-1) ventured that myosin could change its length by shifting between the two protein conformations. Huxley formulated an elegant—and correct—explanation in his sliding filament model for muscle contraction.

In 1948, Huxley began his doctoral research at Cambridge University in the United Kingdom, in the laboratory of John Kendrew (who 10 years later determined the first X-ray structure of a protein, that of myoglobin; Section 7-1). There, through X-ray studies on frog muscle fibers, Huxley established that the X-ray diffraction pattern changes with the muscle's physiological state. Furthermore, he showed that muscle contained two sets of parallel fibers, rather than one, and that these fibers were linked together by multiple

H zone (Fig. 28-12*a*). These observations, made by Hugh Huxley in 1954 (see Box 28-1), are explained by the **sliding filament model** in which interdigitated thin and thick filaments slide past each other (Fig. 28-12*b*). Thus, during a contraction, a muscle becomes shorter, and because its total volume does not change, it also becomes thicker.

Figure 28-12 *(opposite)* **Myofibril contraction.** (*a*) Electron micrographs showing myofibrils in progressively more contracted states. The lengths of the I band and H zone decrease on contraction, whereas the lengths of the thick and thin filaments remain constant. (*b*) Interpretive drawings showing interpenetrating sets of thick and thin filaments sliding past each other. [Courtesy of Hugh Huxley, Brandeis University.]

cross-links. These observations became the germ for further research, which he carried out at MIT in 1953 and 1954. He teamed up with Jean Hanson, a Briton who was also working at MIT. Hanson made good use of her knowledge of muscle physiology and her expertise in phase-contrast microscopy, a technique that could visualize the banded patterns of muscle fibers. Huxley and Hanson observed rabbit muscle fibers under different experimental conditions, making precise measurements of the width of the A and I bands in sarcomeres (Fig. 28-11). In one experiment, they extracted myosin from the muscle fiber, noted the loss of the dark A band, and concluded that the A band consists of myosin. When they extracted both actin and myosin, all identifiable structure was lost, and they concluded that actin is present throughout the sarcomere.

When ATP was added, the muscle slowly contracted, and Huxley and Hanson were able to measure the shortening of the I band. The A band maintained a constant length but became darker. A muscle fiber under the microscope could also be stretched by pulling on the coverslip. As the muscle "relaxed," the I band increased in width and the A band became less dense. Measurements were made for muscle fibers contracted to 60% of their original length and stretched to 120% of their original length.

The key to the sliding filament model that Huxley described and subsequently refined is that the individual molecules (that is, their observable fibrous forms) do not shrink or extend but instead slide past each other. During contraction, actin filaments (thin filaments) in the I band are drawn into the A band, which consists of stationary myosin-containing fila-

ments (thick filaments). During stretching, the actin filaments withdraw from the A band. Similar conclusions were reached by the team of Andrew Huxley (no relation to Hugh) and Rolf Niedergerke, who were examining the contraction of living frog muscle fibers. Both groups published their work in back-to-back papers in *Nature* in 1954.

Hugh Huxley went on to supply additional details to his sliding filament model. For example, he showed that myosin forms cross-bridges with actin fibers. However, these bridges are asymmetric, pointing in opposite directions in the two halves of the sarcomere. This arrangement allows myosin to pull thin filaments in opposite directions toward the center of the sarcomere (Fig. 28-12).

While Huxley was describing the mechanism of muscle contraction, Watson and Crick discovered the structure of DNA, and Max Perutz made a decisive breakthrough in the use of heavy metal atoms to solve the phase problem in his X-ray studies of hemoglobin (see Box 7-2). Collectively, these discoveries indicated the tremendous potential for describing biological phenomena in molecular terms. Subsequent studies of muscle contraction have used electron microscopy, X-ray crystallography, and enzymology to probe the fine details of the sliding filament model, including the structure of myosin's lever arm, the composition of the thin filament, and the exact role of ATP in triggering conformational changes that generate mechanical force.

Huxley, H.E. and Hanson, J., Changes in the cross-striations of muscle during contraction and stretch and their structural interpretation, *Nature* **173**, 973–976 (1954).

(a)

(b)

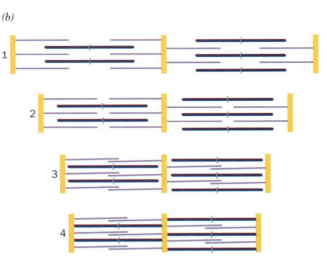

Thin Filaments Contain Actin, Tropomyosin, and Troponin. *Each thin filament consists of a double-stranded actin filament together with actin-binding proteins that help regulate the accessibility of actin subunits to myosin.* The actin filaments are anchored, via their (+) ends, to the Z-disks at each end of the sarcomere. The helical grooves of the actin filament (Fig. 28-3a) are occupied by **tropomyosin,** a homodimer of 284-residue subunits that form a coiled coil of α helices along nearly its entire length (a portion of which is shown in Fig. 6-14). Its amino acid sequence is characteristic of coiled coils such as those in α-keratin (Section 6-1C): It has a 7-residue pseudorepeat, *a-b-c-d-e-f-g,* with nonpolar residues predominantly at positions *a* and *d.* The 400-Å-long molecule associates in a head-to-tail fashion with other tropomyosin molecules to form a cable that coils along the length of the thin filament. Each tropomyosin molecule contacts seven actin monomers and binds a single heterotrimeric **troponin** molecule, whose three subunits, **TnT, TnI,** and **TnC,** are named for their *t*ropomyosin-binding, *i*nhibitory, and *c*alcium-binding properties, respectively (see below).

TnT contains an extension that interacts with tropomyosin. TnC is a Ca^{2+}-binding protein (Fig. 28-13) whose sequence and structure closely resemble those of calmodulin (Fig. 15-17). At the low ($\sim10^{-7}$ M) Ca^{2+} concentrations found in a resting muscle, TnC interacts with TnI in a way that induces TnI to bind to actin so as to block myosin from binding to actin (see below). However, at the high ($\sim10^{-5}$ M) Ca^{2+} concentrations found in contracting muscle, TnC undergoes a conformational change that releases TnI from actin and causes tropomyosin to move deeper into its binding groove, thereby exposing the actin filament's myosin-binding sites.

Thick Filaments Contain Myosin. The human genome encodes over 40 myosin proteins, comprising 12 different classes (over 18 classes are known with several occurring only in plants). Vertebrate thick filaments are composed almost entirely of **myosin II,** which consists of six polypeptide chains: two 220-kD **heavy chains** and two pairs of different **light chains** that vary in size between 15 and 22 kD, depending on their source. As is diagrammed in Fig. 28-14, the N-terminal half of each heavy chain forms an elongated globular head. Next is an ~100-Å α helix, which is stiffened by the two light chains that wrap around it. This portion of the protein functions as a lever during muscle contractions. The C-terminal half of the heavy chain forms a long fibrous α-helical tail, two of which associate to form a left-handed coiled coil. Thus, *myosin consists of a 1600-Å-long rodlike segment with two globular heads.*

Under physiological conditions, several hundred myosin molecules aggregate to form a thick filament. The rodlike tails pack end to end in a regular staggered array, leaving the globular heads projecting to the sides on both ends (Fig. 28-15). These myosin heads form the cross-bridges to thin

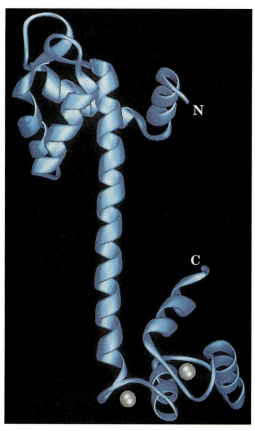

Figure 28-13 X-Ray structure of troponin C from chicken skeletal muscle. The two structurally similar domains are separated by a flexible linker, which forms an α helix in this X-ray structure. In the intact troponin trimer, the linker adopts an irregular conformation. Two Ca^{2+} ions (*silver spheres*) bind to the lower domain at low cellular $[Ca^{2+}]$. At higher $[Ca^{2+}]$, additional Ca^{2+} ions bind to the upper domain. Compare this structure to that of calmodulin (Fig. 15-17). [Based on an X-ray structure by Muttaiya Sundaralingam, University of Wisconsin. PDBid 1TOP.]

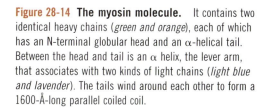

Figure 28-14 The myosin molecule. It contains two identical heavy chains (*green and orange*), each of which has an N-terminal globular head and an α-helical tail. Between the head and tail is an α helix, the lever arm, that associates with two kinds of light chains (*light blue and lavender*). The tails wind around each other to form a 1600-Å-long parallel coiled coil.

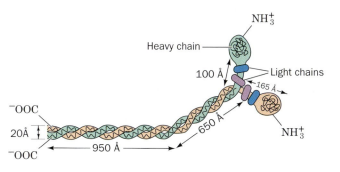

(a)

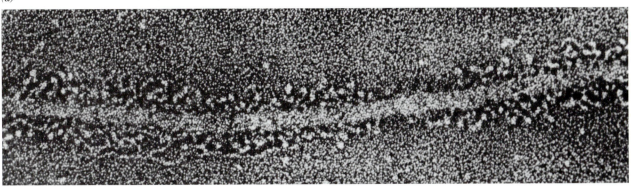

(b)

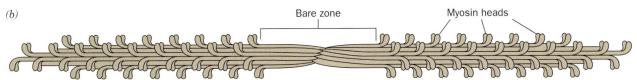

Figure 28-15 Structure of the thick filament. (*a*) Electron micrograph showing the myosin heads projecting from the thick filament. [From Trinick, J. and Elliott, A., *J. Mol. Biol.* **131**, 135 (1977).] (*b*) Drawing of a thick filament, in which several hundred myosin molecules form a staggered array with their globular heads pointing away from the filament.

filaments in intact myofibrils. The myosin head is an ATPase that has its ATP-binding site located in a 13-Å-deep V-shaped pocket. The ATP- and actin-binding sites are separated by ~35 Å, but as in hemoglobin (Section 7-3A), events at one site are communicated to the other site (Fig. 28-16). Each myosin head can bind noncovalently to an actin subunit on the thin filament, provided that access is not blocked by troponin and tropomyosin. Conformational changes resulting from ATP hydrolysis are translated into mechanical movement, as described in the following section.

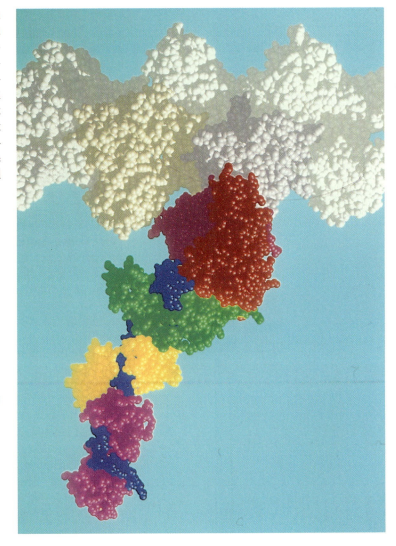

Figure 28-16 Model of the myosin–actin interaction. This space-filling model was constructed from the X-ray structures of actin and the myosin head and electron micrographs of their complex. The actin filament is at the top. The myosin head globular regions are red and green, its α-helical lever arm is blue, and the light chains are yellow and pink. The coiled-coil tail is not shown. An ATP-binding site is located in a cleft in the red domain of the myosin head. In a myofibril, every actin monomer has the potential to bind a myosin head, and the thick filament has many myosin heads projecting from it. [Courtesy of Ivan Rayment and Hazel Holden, University of Wisconsin.]

Muscle Contains Numerous Minor but Essential Proteins. Myosin and actin together account for 60 to 95% of total muscle protein. Other proteins serve to form the Z disk and the M disk and to organize the arrays of thick and thin filaments. Thus, the Z disk, to which the thin filaments are anchored, contains the proteins **α-actinin, desmin,** and **vimentin,** whereas the M disk, to which the thick filaments are attached, contains **C-protein** and **M-protein.** One of the more unusual muscle proteins, **titin,** the longest known polypeptide chain (26,962 residues), is composed of multiple repeating globular domains. Three to six titin molecules associate with each thick filament, spanning the >1-μm distance between the M and Z disks. Titin is believed to function as a molecular bungee cord to keep the thick filament centered on the sarcomere: During muscle contraction, it compresses as the sarcomere shortens, but when the muscle relaxes, titin resists sarcomere extension past the point at which the thin and thick filaments have started to slide past each other. **Nebulin,** an ~7000-residue mainly α-helical protein that is associated with the thin filament, consists almost entirely of a tandemly repeated ~35-residue motif. It is thought to set the length of the thin filament by acting as a template for actin polymerization. This length is held constant by **tropomodulin,** which caps the (−) end of the thin filament (the end not attached to the Z disk) by binding to actin, tropomyosin, and nebulin, thereby preventing further actin polymerization and depolymerization.

Duchenne muscular dystrophy (DMD) and the less severe **Becker muscular dystrophy (BMD)** are both sex-linked muscle-wasting diseases. In DMD, which has an onset age of 2 to 5 years, muscle degeneration exceeds muscle regeneration causing progressive muscle weakness and ultimately death, usually at around age 20. In BMD, the onset age is 5 to 10 years and there is an overall less progressive course of muscle degeneration and a longer (sometimes normal) lifespan than in individuals with DMD.

The ~2500-kb gene responsible for DMD/BMD, which contains 97 exons, encodes a 3685-residue protein named **dystrophin.** However, dystrophin has numerous isoforms that differ at their C-termini through alternative mRNA splicing, as well as at their N-termini through alternative transcriptional initiation sites (Section 25-3A). Dystrophin appears to be a member of the family of flexible rod-shaped proteins that includes the actin-binding cytoskeletal components spectrin (Section 9-4B) and α-actinin, each of which contains segments homologous to portions of dystrophin. Dystrophin, which has a normal abundance in muscle tissue of 0.002%, is associated with the inner surface of the muscle plasma membrane, where it functions to anchor specific transmembrane glycoproteins, much as do spectrin and ankyrin in the erythrocyte (Fig. 9-30*b*).

The dystrophin gene in most individuals with DMD/BMD contains deletions or, less frequently, duplications of one or more exons. Individuals with DMD usually have no detectable dystrophin in their muscles, whereas those with BMD mostly have dystrophins of altered sizes. Evidently, the dystrophins of individuals with DMD are rapidly degraded, whereas those of individuals with BMD are semifunctional.

B The Actin–Myosin Reaction Cycle

In accord with the sliding filament model for muscle contraction, each myosin cross-bridge to actin repeatedly detaches and reattaches itself at a new site farther along the thin filament toward the Z disk. The free energy driving muscle contraction is provided by ATP hydrolysis. The conformation of the globular myosin head changes when ATP is hydrolyzed. Small structural rearrangements in the head domain are converted to a large

swinging movement of the light-chain binding domain, which serves as a lever arm to transfer force to the thin filament. The molecular events of myosin action, as formulated by Edwin Taylor and refined by the structural studies of Ivan Rayment, Hazel Holden, and Ronald Milligan, occur as follows (Fig. 28-17):

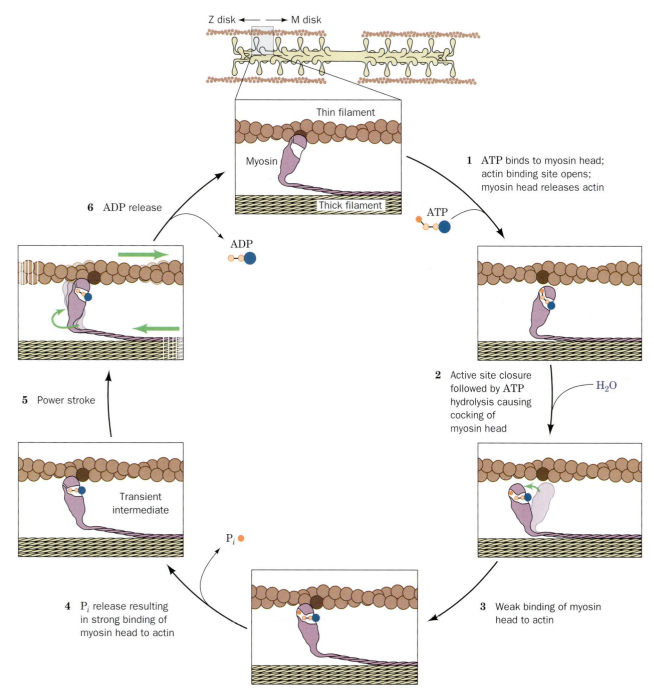

Figure 28-17 *Key to Function.* **Mechanism of force generation in muscle.** The myosin head "walks" up the actin thin filament through a unidirectional cyclic process that is driven by ATP hydrolysis to ADP and P_i. Only one myosin head is shown. The actin monomer to which the myosin head is bound at the beginning of the cycle is more darkly colored for reference. [After Rayment, I., and Holden, H., *Curr. Opin. Struct. Biol.* **3**, 949 (1993).] ✪ **See the Animated Figures.**

1. ATP binds to a myosin head in a manner that causes the actin-binding site to open up and release its bound actin.

2. The active site closes around the ATP. The resulting hydrolysis of the ATP "cocks" the myosin head so that it is approximately perpendicular to the thick filament.

3. The myosin head binds weakly to an actin monomer that is closer to the Z disk than the one to which it had been bound previously.

4. Myosin releases P_i. The departure of this group leaves a void that triggers the surrounding protein to close in. This conformational shift increases myosin's affinity for actin.

5. The resulting transient state is immediately followed by the power stroke, a conformational shift that sweeps the myosin head's C-terminal tail by an estimated 70 Å toward the Z disk relative to the actin-binding site on its head, thus translating the attached thin filament by this distance toward the M disk.

6. ADP is released, thereby completing the cycle.

This reaction mechanism converts the chemical energy of ATP to mechanical energy because the presence of a nucleotide (ATP or ADP) in the myosin binding site determines the position of the lever arm as well as the affinity of myosin for actin. The α helix of the lever arm is relatively incompressible, so it can pull the coiled-coil myosin tail along with it. The actin–myosin system has a power-to-weight ratio comparable to that of an automobile engine.

Most experiments support the view that each cycle of ATP hydrolysis moves the myosin head by 50 to 150 Å (this distance is consistent with a 70° sweep of the lever arm). Since individual actin subunits are spaced about 55 Å apart along the thin filament, the myosin head advances by at least one actin subunit per reaction cycle. Because the reaction cycle involves several steps, some of which are irreversible (e.g., ATP hydrolysis and P_i release), the entire cycle is unidirectional. The ~500 myosin heads on every thick filament asynchronously cycle through the reaction sequence about five times each per second during a strong muscular contraction. *The myosin heads thereby "walk" or "row" up adjacent thin filaments toward the Z disk with the concomitant contraction of the muscle.* Although myosin is dimeric, its two heads function independently.

Calcium Triggers Muscle Contraction. Highly purified actin and myosin can contract regardless of the Ca^{2+} concentration, but preparations containing intact thin filaments contract only in the presence of Ca^{2+}, due to the regulatory action of troponin C. Stimulation of a myofibril by a nerve impulse results in an influx of Ca^{2+} from the **sarcoplasmic reticulum** (a system of flattened vesicles derived from the endoplasmic reticulum). As a result, the intracellular $[Ca^{2+}]$ increases from ~10^{-7} to ~10^{-5} M. The higher calcium concentration triggers the conformational change in the troponin–tropomyosin complex that exposes the site on actin where the myosin head binds. When the myofibril $[Ca^{2+}]$ is low (Ca^{2+} is specifically pumped out of the myofibril by Ca^{2+}–ATPases; Section 10-3B), the troponin–tropomyosin complex assumes its resting conformation, blocking myosin binding to actin and causing the muscle to relax.

C Unconventional Myosin V

All cytoskeletal motors obey the same general mechanism in which *the enzymatic steps of ATP hydrolysis trigger conformational changes that are propagated to the rest of the structure in order for it to move relative to a*

stationary "track." In the case of myosin motors, the track is an actin filament. The various myosins differ in their number of subunits, direction of movement, and ability to travel long distances without dissociating from the actin track. It is worthwhile to consider how one of these, the variant or unconventional **myosin V,** differs in structure and function from conventional muscle myosin.

Myosin V was discovered as the product of a mutated gene in mice with reduced hair coloring. In normal mice, myosin V is involved in transporting pigment granules into the spiny protrusions of melanocytes so that the pigment can be taken up by hair-forming keratinocytes. Defects in other myosins have been linked to deafness (Box 28-2). How does myosin V act as a transport protein? Like myosin II, myosin V is a two-headed protein, but it does not aggregate to form a thick filament. Instead, it acts alone and a domain at the end of its tail binds an intracellular vesicle containing pigment molecules (cargo). The lever arm region of the protein is long enough to accommodate six light chains, giving it a longer reach than the lever arm of myosin II, which bears only two light chains. Electron micrographs of myosin V bound to an F-actin filament show that the globular heads are separated by 13 actin subunits (Fig. 28-18), corresponding to the length of the filament's helical repeat. The two heads must alternately bind to the F-actin so that the motor protein and its attached cargo can move along the track without dissociating. Biochemical experiments indicate that compared to myosin II, myosin V spends a much greater proportion of its ATP hydrolysis cycle associated with actin. The protein and its cargo move ~370 Å during each reaction cycle.

A two-headed motor protein can potentially move along a track by one of two mechanisms, known as the hand-over-hand and inchworm mechanisms. In hand-over-hand movement, the two globular heads step along the track, each one passing the other in turn, much like the feet of a person walking. Inchworm movement resembles a person walking sideways, one foot always moving in advance of the other. To distinguish between these two mechanistic possibilities in myosin V, Yale Goldman and Paul Selvin used fluorescence microscopy to follow the movement of a myosin V head domain bearing a fluorescent light chain. For each ATP hydrolyzed, the

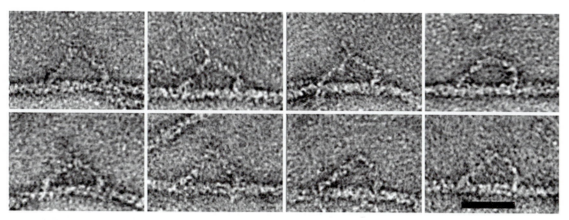

Figure 28-18 Myosin V binding to a microfilament. These electron micrographs show the two-headed myosin V, minus its cargo-binding domain, bound to F-actin. The leading head is on the right, closer to the (+) end of the track, and is separated from the trailing head by 13 actin subunits. The scale bar is 50 nm long. [Courtesy of Peter J. Knight, University of Leeds, U.K.]

BOX 28-2

Biochemistry in Health and Disease

Myosin Mutations and Deafness

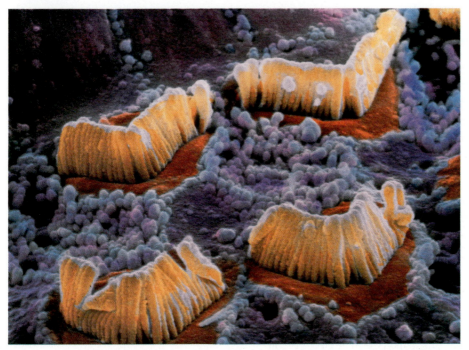

Scanning electron micrograph of stereocilia of a hair cell.

Inside the cochlea, the spiral-shaped organ of the inner ear, are thousands of hair cells, each of which is topped with a bundle of bristles known as **stereocilia.** Each stereocilium contains several hundred cross-linked actin filaments and is therefore extremely rigid, except at its base, where there are fewer actin filaments. Sound waves deflect the stereocilia at their base, mechanically triggering the opening of membrane ion channels. The resulting ionic fluxes initiate an electronic signal that is transmitted to the brain.

About half of all cases of deafness have a genetic basis, and over 100 different genes have been linked to deafness. Geneticists working on these genes are often hampered by the tendency for deaf people to marry other deaf people, so that their offspring, who usually have normal hearing, inherit not one but two different and defective genes. Among the proteins whose genes have been linked to deafness are several types of myosin. **Myosin I** appears

to cross-link actin filaments to control the tension inside each stereocilium. Indeed, the ratcheting activity of this myosin motor along the actin filaments may adjust the sensitivity of the hair cells to different sounds. Other types of myosin use their motor activity to redistribute cellular constituents along the length of the actin filaments. **Myosin VI** is notable in this regard because its lever arm rotates in the direction opposite the movement in most other myosins. Consequently, it moves its attached cargo toward the (−) end of microfilaments.

Many cases of **Usher syndrome,** the most common form of deaf-blindness in the United States, result from mutations in the gene for **myosin VIIA.** This motor protein functions in intracellular transport processes, particularly during hair cell development. Usher syndrome is characterized by profound hearing loss, **retinitis pigmentosa** (which leads to blindness), and sometimes vestibular (balance) problems. The congenital deafness of Usher syndrome reflects the failure of the hair cells to develop properly. The unresponsiveness of the stereocilia to sound waves probably also accounts for their inability to respond normally to the movement of fluid in the inner ear, which is necessary for maintaining balance. Abnormal myosin VIIA also plays a role in the blindness that often develops in individuals with Usher syndrome, usually in their second or third decade. Myosin VIIA is responsible for distributing bundles of pigment in the retina. In retinitis pigmentosa, retinal neurons gradually lose their ability to transmit signals in response to light, and in advanced stages of the disease, pigment actually becomes clumped on the retina.

[Photo by P. Motta, University La Sapienza, Rome, Italy/Science Photo Library/Photo Researchers.]

myosin head moved ~740 Å, which is consistent with the hand-over-hand model (the other head remains stationary during a single step so that the cargo moves only 370 Å per step; Fig. 28-19). These results rule out the inchworm model, which would require head displacement of no more than 370 Å.

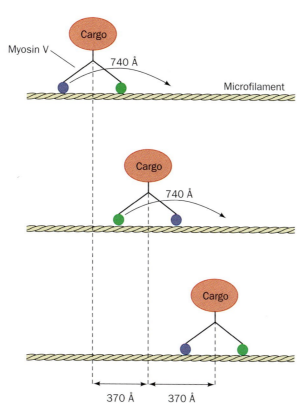

Figure 28-19 Hand-over-hand movement of myosin V. The trailing head advances by ~740 Å during each reaction cycle. Because the leading head remains in place during this step, the cargo advances in increments of ~370 Å per step. [After Yildiz, A., Forkey, J.N., McKinney, S.A., Ha, T., Goldman, Y.E., and Selvin, P.R., *Science* **300**, 2061 (2003).]

3 Tubulin and Microtubules

Like microfilaments, microtubules are cytoskeletal fibers built from small globular protein subunits. Consequently, they share with microfilaments the ability to assemble and disassemble on a time scale that allows the cell to change shape in response to external or internal stimuli. Compared to a microtubule, however, a microfilament is a thin and flexible rod. *A microtubule is about three times thicker and much more rigid because it is constructed as a hollow tube.* Bicycle frames, plant stems, and bones are built on this same principle. Cells use hollow microtubules to reinforce other elements of the cytoskeleton, to construct **cilia** (the hairlike organelles on the surfaces of many animal and lower plant cells that function to move fluid over the cell's surface or to "row" single cells through a fluid) and **flagella** (elongated single cilia that propel free cells such as sperm and certain protozoa), and to align and separate pairs of chromosomes during mitosis. The central rod among the bundle of stereocilia in the inner ear (see Box 28-2) is built from microtubules rather than actin filaments.

A The Tubulin Dimer

The basic structural unit of a microtubule is the protein tubulin. Two 40% identical, ~450-residue monomers, named **α-tubulin** and **β-tubulin,** form a dimer. A microtubule grows by the addition of tubulin dimers. The structure of the tubulin dimer, determined by Kenneth Downing and Eva Nogales using electron crystallography (Box 9-2), reveals that the core of each tubulin subunit consists of a four-stranded and a six-stranded β sheet surrounded by 12 α helices. The two subunits associate in head to tail fashion to form the dimer (Fig. 28-20). α-Tubulin and β-tubulin differ primarily in the lengths and conformations of certain loops and in the sequences at their C-termini (which are not visible in the electron crystal structure).

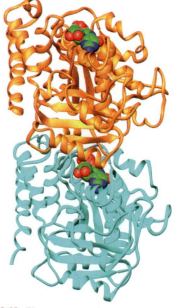

Figure 28-20 Electron crystal structure of the tubulin dimer. This 3.5-Å-resolution model of bovine tubulin reveals the structural similarities between its α (*cyan*) and β (*orange*) subunits. The GTP bound to α-tubulin and the GDP bound to β-tubulin are drawn in space-filling form with C green, N blue, O red, and P magenta. Note that the GTP is inaccessible in the dimer, whereas the GDP is more exposed to the solvent. [Based on a structure by Kenneth Downing and Eva Nogales, Lawrence Berkeley National Laboratory, Berkeley, California. PDBid 1JFF.]

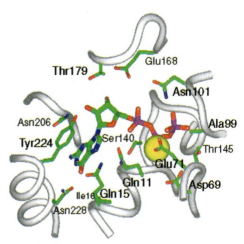

Figure 28-21 The nucleotide-binding site of α tubulin. The protein backbone is gray, and its bound GTP and its associated amino acid side chains are shown in stick form with C green, N blue, O red, and P magenta. Note how Tyr 224 stacks on the guanine base. An Mg^{2+} ion (*yellow sphere*) is coordinated by the nucleotide phosphate groups and by Asp 69 and Glu 71. [Courtesy of Kenneth Downing and Eva Nogales, Lawrence Berkeley National Laboratory, Berkeley, California.]

The contacts between α- and β-tubulin in the dimer are characterized by a large number of hydrophobic residues. Interdimer contacts in an assembled microtubule appear to involve more hydrophilic contacts, indicating that *the dimer is the most stable form of tubulin.* Indeed, the dimer dissociates only under denaturing conditions. Tubulin monomers, with their exposed hydrophobic surfaces, are therefore prone to aggregation so that their assembly into dimers requires the participation of molecular chaperones.

As it emerges from a ribosome, a newly synthesized tubulin polypeptide is bound by a hexameric chaperone named **prefoldin.** This involves two regions of the tubulin polypeptide that include surface loops. Although other molecular chaperones recognize hydrophobic patches on their target proteins (Section 6-5C), the interaction between tubulin and prefoldin involves polar groups. Intriguingly, prefoldin also binds newly synthesized actin polypeptides. The prefoldin recognition site in actin and tubulin includes a Glu–His–Gly–Ile sequence—a remarkable finding, since these proteins otherwise lack sequence similarity. After the polypeptide chain has been synthesized, prefoldin hands off the actin or tubulin to an octameric chaperone called **CCT** (*c*ytosolic *c*haperonin containing *t*ailless complex polypeptide 1). CCT facilitates the folding of cytoskeletal proteins in an ATP-dependent manner. At least five other cofactors help assemble the αβ tubulin dimer.

Tubulin Binds GTP. The N-terminal ~200 residues of both α- and β-tubulin form a dinucleotide-binding fold typical of nucleotide-binding proteins (Section 6-2C). Unlike actin and myosin, however, tubulin subunits each bind a guanine nucleotide, either GTP or GDP. Specificity for guanosine is conferred by two Asn residues that hydrogen bond with guanine's 2-amino group, which is absent in adenosine. The base fits into a hydrophobic pocket defined by nonpolar residues, including a Tyr side chain that stacks on the base (Fig. 28-21). The phosphate groups form hydrogen bonds with backbone groups and ligand a bound Mg^{2+} ion.

In the tubulin dimer, α-tubulin's bound GTP is buried at the subunit interface (Fig. 28-20) and is therefore nonexchangable. A Lys residue (Lys 254) from β-tubulin interacts with the terminal phosphate group of the GTP bound to α-tubulin. However, the GTP bound to β-tubulin is not protected in this way and remains exposed to solvent. When tubulin dimers polymerize to form a microtubule (see below), the GTP in β-tubulin is hydrolyzed to GDP + P_i. This reaction occurs only when α-tubulin from an adjacent dimer closes off the β-tubulin nucleotide-binding site. α-Tubulin has Glu rather than Lys at position 254, and this acidic residue helps catalyze GTP hydrolysis in β-tubulin. Thus, β-tubulin cannot catalyze the hydrolysis of its bound GTP until it is part of a tubulin polymer, whereupon the resulting GDP becomes nonexchangable until the subunit dissociates from the microtubule. The GTP bound to the α subunit is not hydrolyzed.

B Microtubule Dynamics

Assembly of a microtubule begins with the end-to-end association of tubulin dimers to form a short linear **protofilament** (Fig. 28-22, *left*). Protofilaments then align side to side in a curved sheet (Fig. 28-22, *middle*), which wraps around on itself to form a hollow tube that in most microtubules contains 13 protofilaments (Fig. 28-22, *right*), although this number can range from 9 to 16. The walls of the microtubule may be porous enough for small molecules to diffuse through (Fig. 28-23).

The microtubule extends as tubulin dimers add to both ends. However, microtubules are polar, as are microfilaments (Section 28-1A). *The so-called (+) end, which terminates in β-tubulin, grows about twice as fast as the (−)*

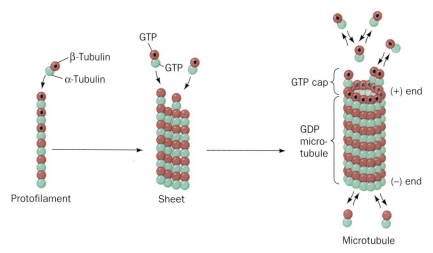

Figure 28-22   Assembly of a microtubule.
αβ Dimers of tubulin initially form linear protofilaments. These associate side by side to form a curved sheet that wraps around to form a microtubule which, in most cases, has 13 protofilaments. The microtubule then grows by addition of tubulin dimers to either end, but growth is about twice as fast at the (+) end, which terminates in the β subunits. A tubulin dimer initially has a GTP bound to both its α and β subunits but, on incorporation of the dimer into a microtubule, its β subunit (but not its α subunit) hydrolyzes its bound GTP (*black dot*) to GDP. When the rate of polymerization is greater than that of GTP hydrolysis, the (+) end will have a cap of GTP-containing β subunits, although the remaining β subunits contain GDP.

or α-tubulin end, *because tubulin dimers bind preferentially to the (+) end.* The high affinity of tubulin dimers for the (+) end permits polymerization to occur even when opposed by a compressive force. Consequently, a growing microtubule can perform work such as centering the nucleus or moving chromosomes during mitosis.

GTP Hydrolysis Destabilizes the Microtubule.   GTP hydrolysis is not required for microtubule assembly, as is demonstrated by the polymerization of tubulin subunits that have bound the nonhydrolyzable GTP analog **GMPPNP** (the guanine analog of AMPPNP). However, the resulting microtubules are essentially unable to depolymerize, which would limit their usefulness. *After each tubulin dimer adds to the (+) end of a growing microtubule, residues from an incoming α-tubulin help catalyze hydrolysis of the GTP bound to β-tubulin.* The resulting GDP destabilizes the microtubule. Nevertheless, the assembly does not fall apart as long as the (+) end is capped by a ring of β-tubulin subunits binding GTP. If this "GTP cap" dissociates, as it occasionally does, the microtubule disassembles at a rate that is ~100-fold greater than if the cap were present.

GTP hydrolysis and the departure of the phosphate group apparently trigger a conformational change in β-tubulin such that the GDP–β-tubulin bends slightly away from its α-tubulin partner in the dimer. Consequently, protofilaments containing only GTP are straight, whereas protofilaments with GDP bound to the β subunit are curved. In fact, electron micrographs of depolymerizing microtubules show protofilaments splaying out from the

Figure 28-23   Cryoelectron microscopy–based surface view of a microtubule at 8 Å resolution.
The subunits in adjacent protofilaments are slightly staggered such that they follow a shallow left-handed helix with a rise of three subunits per turn. At 8 Å resolution, the structurally similar α- and β-tubulin monomers cannot be differentiated even though many of their secondary structural elements can be discerned. Consequently, two models for the microtubule must be considered. In one model, subunits of a given type contact subunits of the same type in neighboring protofilaments to form shallowly spiraling rows of α and β subunits (as drawn in Fig. 28-22, *right*), whereas in the second model, subunits of one type contact subunits of the opposite type in neighboring protofilaments to form steeply spiraling rows of α and β subunits. Some of the electron micrographs that were used to generate the microtubule image are shown below it. [Courtesy of Kenneth Downing and Eva Nogales, Lawrence Berkeley National Laboratory, Berkeley, California.]

Figure 28-24 Electron micrograph of a depolymerizing microtubule with an accompanying interpretive drawing. Protofilaments curve and separate from the ends of the microtubule, giving it a frayed appearance. This presumably promotes the dissociation of tubulin and segments of protofilaments from the ends. [Electron micrograph courtesy of Ronald Milligan, The Scripps Research Institute, La Jolla, California.]

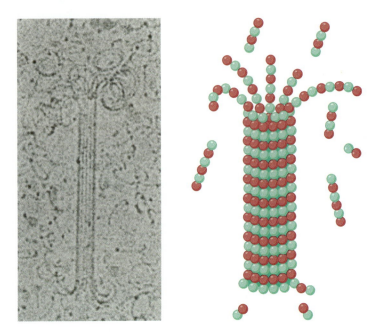

ends of the microtubule (Fig. 28-24). Depolymerization through fraying of protofilaments is probably faster than the dissociation of tubulin dimers from the microtubule end.

Microtubules Exhibit Dynamic Instability. *In vivo,* the (+) end of a microtubule undergoes random changes in length by polymerization (when it adds tubulin) and depolymerization (when it loses its GTP cap). Thus, in a phenomenon known as **dynamic instability,** some microtubules in a given population will grow while others shrink. This continual activity at the (+) end may facilitate rapid microtubule growth when needed, just as a car with an idling motor can make a faster getaway than a car with its ignition off. The balance between microtubule polymerization and depolymerization probably depends on other proteins that cap microtubule ends or bind preferentially to bent tubulin dimers to promote protofilament fraying. Such proteins are known as **microtubule-associated proteins (MAPs).**

Under the conditions that tubulin subunits add to the (+) end as fast as they dissociate from the (−) end, the microtubule undergoes treadmilling. However, *in vivo,* the (−) ends are often anchored to some sort of organizing center. Hence, most microtubule growth and regression occur at the (+) end, which is often near the cell periphery. Subunits of **γ-tubulin,** which is similar in sequence to α- and β-tubulins but does not by itself form microtubules, form rings at the (−) ends of some microtubules where they apparently function to nucleate microtubule assembly. The microtubules that form the mitotic spindle in animal cells emanate from structures called **centrioles** (Fig. 1-8), which contain the variant **δ-tubulin.** Because microtubules participate in essential cell activities, compounds that interfere with microtubule dynamics can drastically affect cell structure and function (see Box 28-3).

4 Microtubule Motors

Microtubules participate in several different forms of cell movement: The growth and rearrangement of microtubules can alter cell shape; microtubules can push or pull cellular constituents such as chromosomes dur-

ing mitosis; microtubules forced to slide past each other are responsible for the movement of cilia and flagella; and microtubules serve as tracks for vesicle- and organelle-toting transport motors. In this section we will examine two microtubule-associated motor proteins.

A Kinesins

Although kinesins were discovered only in 1985, they are nearly as well understood as myosins, with which they share important structural and functional characteristics. The kinesin superfamily comprises 16 classes, which vary as to the position of their common ATP-binding motor domain in the polypeptide chain (N-terminal, middle, or C-terminal), the nature of the remaining sequences, the number of identical and different subunits with which they associate, and their direction of transport along a microtubule [(−) to (+) or (+) to (−)]. The motor domain is less than half the size of myosin's motor domain, but structural analysis reveals some overlaps and short stretches of conserved sequence. Evidently, myosin and kinesin have a common evolutionary origin. The various classes of kinesins function to transport different types of cargos [e.g., secretory vesicles (Section 9-4E), lysosomes (20-1A), chromosomes, and mitochondria] in different portions of the cell and in different types of cells. Here we describe the first kinesin to be discovered, which is known as **conventional kinesin.**

Conventional kinesin (Fig. 28-25*a*) is an ~1000-Å-long protein that consists of two identical heavy chains (~125 kD) and two identical light chains (~65 kD) with the heavy chains forming two large globular heads and a coiled-coil tail (as do myosin heavy chains). It functions to transport vesicles and organelles in the (−) to (+) direction along microtubules (Fig. 28-25*b*). The X-ray structure of the N-terminal portion of rat conventional kinesin,

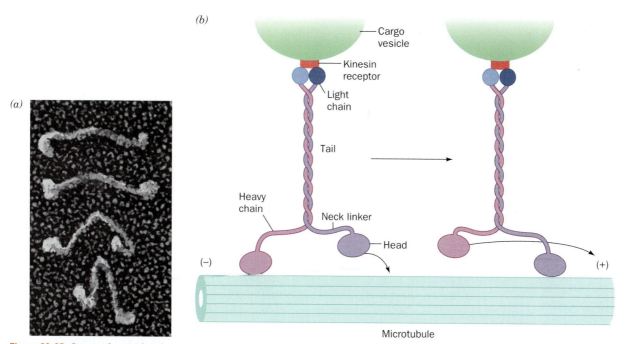

Figure 28-25 Conventional kinesin. (*a*) Electron micrograph of conventional kinesin molecules. [Courtesy of John Heuser, Washington University School of Medicine.] (*b*) Schematic diagram of vesicle transport along microtubule tracks by conventional kinesin. The as yet unidentified kinesin receptor on the vesicle surface permits kinesin to "walk" the vesicle along the microtubule from its (−) end to its (+) end.

BOX 28-3

Biochemistry in Health and Disease

Drugs That Bind Microtubules

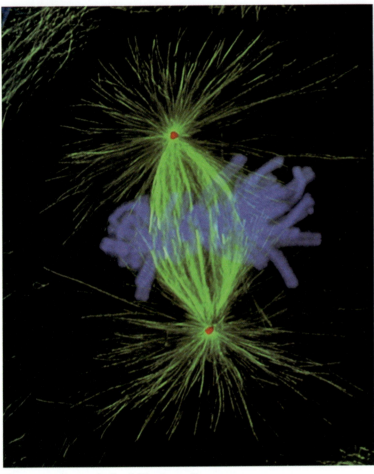

Fluorescence micrograph of a dividing lung cell.

During mitosis, paired chromosomes (blue in the micrograph at left) separate along the mitotic spindle, which consists of microtubules (green; the red filaments are keratin). **Colchicine,**

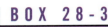

Colchicine

a product of the autumn crocus (*Colchicum autumnale*), inhibits the formation of the mitotic spindle, thereby blocking cell division. Colchicine was used over 2000 years ago to treat gout (an inflammation stemming from the precipitation of uric acid in the joints; Section 22-4B) because it inhibits the action of the white blood cells that mediate inflammation. It is still used today for this purpose as well as for treating other inflammatory conditions.

Colchicine binds to β-tubulin near the interface between the α and β subunits of the tubulin heterodimer. This increases the affinity of tubulin heterodimers for the ends of microtubules. However, the binding of one or two colchicine-bearing heterodimers to the end of

determined by Eva-Maria Mandelkow and Eckhardt Mandelkow, reveals that each 100-Å-long head consists of an eight-stranded β sheet flanked by three α helices on each side and includes a tubulin-binding site and a nucleotide-binding site (Fig. 28-26). Between kinesin's globular core and its tail is a "neck linker" containing 10 to 15 residues. This is followed by an α-helical stalk that leads into the coiled coil, which forms through the dimerization of the two heavy chains. The light chains, which bind to the end of the coiled-coil tail, bind to proteins in the membrane of a vesicle or organelle, which thereupon becomes kinesin's cargo. Although homodimeric, kinesin is asymmetric, with its two globular heads separated and rotated by ~120°. The asymmetry is due to the fact that the two heads operate in an alternating fashion so that they never have the same conformation (see below).

a microtubule inhibits both the further addition and loss of heterodimers from that end of the microtubule. Colchicine thereby disrupts the microtubule dynamics required for spindle formation and other microtubule-based processes. Consequently, in addition to its medical uses, colchicine has been an invaluable tool for investigating these processes.

Taxol

Taxol

binds to the β-tubulin subunits in microtubules, but not to free tubulin. It therefore stabilizes microtubules by preventing their depolymerization. This inhibits the dissolution of the mitotic spindle, thereby preventing cell division. The taxol–tubulin interaction appears to include close contacts between taxol's phenyl groups and hydrophobic residues such as Phe, Val, and Leu. Taxol was originally extracted from the bark of the slow-growing and endangered Pacific yew tree

(*Taxus brevifolia*), but is now obtained from other more common species of yew combined with chemical synthesis. *Taxol blocks cell division and is therefore particularly toxic to rapidly dividing cells such as tumor cells.* Indeed, it is a frequently used chemotherapeutic agent against ovarian, breast, and certain types of lung cancer. Curiously, colchicine is not selectively toxic to cancer cells.

The **vinca alkaloids vinblastine** and **vincristine,**

Vinblastine: R = CH_3
Vincristine: R = CHO

products of the Madagascan periwinkle (*Catharanthus roseus*), inhibit microtubule polymerization and are also widely used cancer chemotherapy agents.

[Photo courtesy of Alexey Khodjakov and Conly Rieder, Wadsworth Center, Albany, New York.]

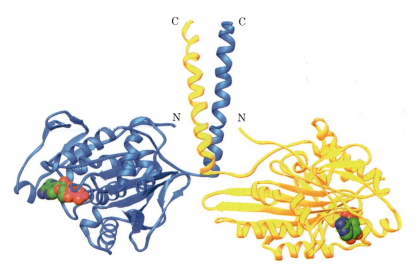

Figure 28-26 X-Ray structure of the head and neck region of a rat conventional kinesin in complex with ADP.   Each globular head terminates in a flexible neck linker. The linker connects into an α helix that winds around its counterpart in the other subunit to form a coiled coil. The ADP is drawn in spacefilling form with C green, N blue, O red, and P magenta. Two light chains at the end of the coiled-coil tail (not present in this X-ray structure) can interact with a membranous vesicle, the cargo. Note that the two chemically identical subunits do not form a symmetric structure. [Based on an X-ray structure by Eva-Maria Mandelkow and Eckhardt Mandelkow, Max-Planck Unit for Structural Molecular Biology, Hamburg, Germany. PDBid 3KIN.]

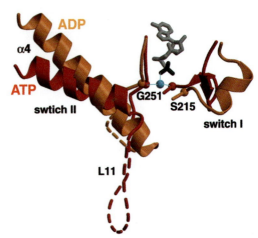

Figure 28-27 Conformational changes in switch I and switch II of the mouse kinesin KIF1A. The structures known as switch I and switch II undergo conformational changes when ATP is hydrolyzed to ADP. The ATP-bound conformation is drawn in brown and the ADP-bound conformation is drawn in tan. Disordered loops are indicated by dashed lines. ATP is shown as a gray stick model. Residues Gly 251 and Ser 215 are marked by small spheres. The blue sphere is a bridging water molecule. [Courtesy of Nobutaka Hirokawa, University of Tokyo, Japan. PDBids 1I6I and 1I5S.]

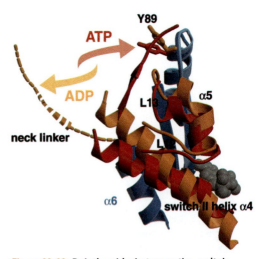

Figure 28-28 Relationship between the switch regions and the neck linker in the mouse kinesin KIF1A. The ATP-bound conformation is drawn in brown and the ADP-bound conformation is drawn in tan. Portions of the globular core are shown in blue, and ATP is represented by a gray space-filling model. The movement of the neck linker is indicated by arrows. When ATP is bound to kinesin, the neck linker docks with the head domain. Following ATP hydrolysis, the neck linker moves away. [Courtesy of Nobutaka Hirokawa, University of Tokyo, Japan. PDBids 1I6I and 1I5S.]

ATP Hydrolysis Triggers Conformational Changes in the Neck Linker. As in other motor proteins, kinesin couples the steps of ATP hydrolysis to conformational changes that allow the protein to move relative to its microtubule track. However, *kinesin cannot follow the myosin lever mechanism because its head domains are not rigidly fixed to its neck regions.* Instead of a stiff α-helical lever arm, a relatively flexible neck linker connects the ATPase domain to the coiled-coil tail. In conventional kinesin, the neck linker takes the form of two short consecutive β strands. The flexibility of kinesin's neck linker is critical for its function, because each kinesin head must swing through an arc much longer than itself.

ATP hydrolysis, followed by the dissociation of P_i, triggers a conformational change in kinesin's catalytic domain that is largely localized to two short polypeptide segments known as **switch I** and **switch II** (Fig. 28-27). Structurally analogous switch regions occur in myosin and G proteins (Section 21-3B). Movement of the switch regions triggers additional conformational changes in kinesin's neck linker.

When ATP is bound to kinesin, the neck linker docks with the catalytic core. Following ATP hydrolysis, the neck linker "unzips" so that it moves away from the globular head. The relationship between the switch regions and the neck linker is shown in Fig. 28-28. A kinesin head containing ADP (or no nucleotide at all) is free to move to a position farther along the microtubule track.

Kinesin Follows a Hand-over-Hand Mechanism. Like myosin V (Section 28-2C), kinesin moves by a stepping mechanism (Fig. 28-25b). Its two heads work in a coordinated fashion so that at least one is bound to the microtubule track at all times. According to a widely accepted model for kinesin activity (Fig. 28-29), ATP binds to the leading kinesin head, which is already bound to the microtubule. When the neck linker "zips up" against the catalytic core, the trailing kinesin head is thrown forward. The trailing head, with ADP bound, has a lower affinity for the microtubule and so is more easily detached (this helps keep kinesin moving unidirectionally). *Each kinesin head swings in turn past its partner by ~160 Å so that the net movement of the kinesin tail and its cargo is ~80 Å for each ATP hydrolyzed.* This distance corresponds to the length of a tubulin dimer along a protofilament. Most of the kinesin–tubulin contacts involve the negatively charged C-terminus of β-tubulin, which faces the outside of the microtubule.

Although kinesin and myosin respond to the events of ATP hydrolysis with conformational changes, the two proteins differ in *how* they respond. For example, ATP binding triggers the power stroke in kinesin, whereas ATP binding causes myosin to dissociate from actin and recock its lever arm (a recovery stroke). The release of P_i weakens kinesin's grip on the microtubule and loosens the neck linker, whereas P_i release from myosin causes it to bind tightly to actin and swing its lever arm forward.

Kinesin Is a Processive Motor. A free kinesin head quickly rebinds tubulin, so the heads spend most of their reaction cycle bound to the microtubule track. Kinesin can advance by hundreds of steps before dissociating from its track and is therefore a processive motor. *High processivity is essential for a transport engine such as kinesin or myosin V, because the cargo must be moved long distances to specific destinations.* A motor protein such as myosin II, which dissociates from an actin filament after a single stroke, is not processive. In a muscle cell, low processivity is permitted because the numerous myosin–actin interactions that occur more or less simultaneously keep thin and thick filaments in contact.

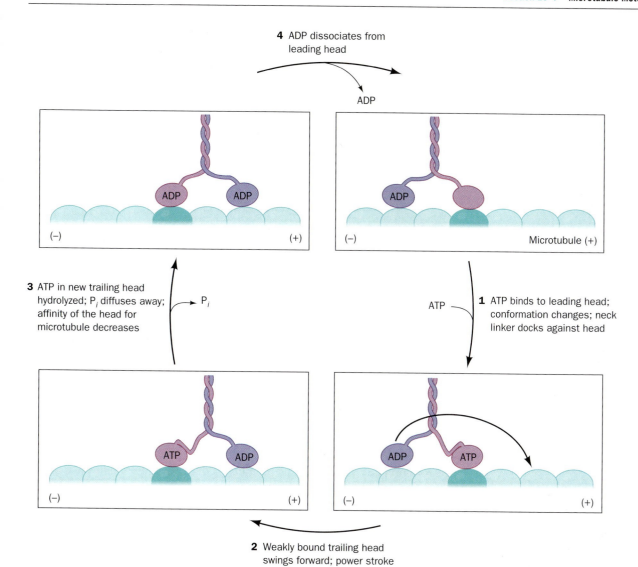

Figure 28-29 The kinesin reaction cycle. A portion of the microtubule track is shown with one tubulin subunit shaded more darkly for reference. The (+) end of the microtubule is on the right. The vesicular cargo attached to kinesin's coiled-coil tail is not shown.

Because kinesin maintains its grip on the microtubule, it cannot slide backward. Consequently, it can advance even when elastic forces in the cell oppose the movement of the bulky vesicle or organelle bound to the kinesin tail. Conventional kinesin moves toward the (+) end of a microtubule, but other kinesins, which appear to follow the same ATP-triggered mechanism, move in the opposite direction. In these proteins, the neck linker is located on the N-terminal side of the catalytic domain (it is on the C-terminal side in conventional kinesin) and so swings in the opposite direction with each cycle of ATP hydrolysis.

B Dyneins

Eukaryotic cells contain a third type of motor protein, dynein, in addition to myosin and kinesin. The 12 known mammalian dyneins can be grouped into **cytoplasmic dyneins,** which, like kinesin, participate in microtubule-

Figure 28-30 Scanning electron micrograph showing cilia lining the epithelial surface of a human bronchial tube. The rounded surfaces of a number of mucus-secreting **goblet cells** are also visible. [From Kessel, R.G. and Kardon, R.H., *Tissues and Organs: A Text-Atlas of Scanning Electron Microscopy,* p. 210, Freeman (1979). Visuals Unlimited.]

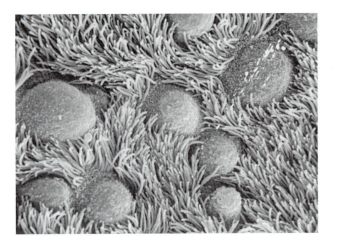

based transport, but toward the (−) end; and **axonemal dyneins,** which power the motion of cilia (Fig. 28-30) and eukaryotic flagella.

The dyneins form large (1–2 million D) proteins that consist of one or more ~520-kD heavy chains and a large and variable number of intermediate chains (50–140 kD) and light chains (8–45 kD). A dynein heavy chain has an ~350-kD globular head that contains dynein's motor unit and its ATP-hydrolyzing unit (Fig. 28-31). An ~100-Å-long stalk that protrudes from the head is predicted to form a coiled coil and ends in an ~120-residue globular domain that is implicated in binding to microtubules. The N-terminal third of the heavy chain forms a long stem that extends from the head and which sequence analysis predicts contains many short stretches of coiled coils. Two or three heavy chains are connected via their stems to a common base that contains the intermediate and light chains and which functions to bind dynein to its cargo or to the core of a cilium.

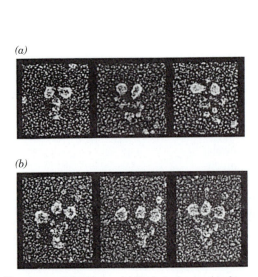

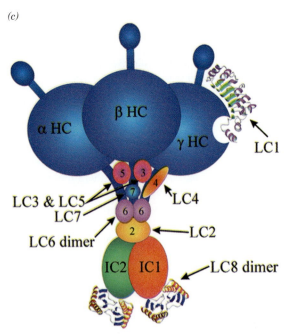

Figure 28-31 Structures of dyneins. (*a*) Electron micrographs of two-headed dynein from the outer dynein arm of sperm from the sea urchin *Strongylocentrotus purpuratus.* (*b*) Electron micrographs of three-headed dynein from the flagellar outer dynein arm of the unicellular green alga *Chlamydomonas reinhardtii.* (*c*) Model of *Chlamydomonas* dynein indicating the positions of its subunits. HC, IC, and LC are abbreviations for heavy chains, intermediate chains, and light chains. LC1 and LC8 are drawn in ribbon form because their structures are known. [Parts *a* and *b* courtesy of Ursula Goodenough, Washington University School of Medicine. Part *c* courtesy of Stephen King, University of Connecticut Health Center, Farmington, Connecticut.]

Cilia Contain Organized Sheaves of Microtubules. Cilia and flagella are flexible extensions of the plasma membrane that range in length from a few micrometers to over 2 mm. Electron micrographs (Fig. 28-32a) indicate that their core, which is known as an **axoneme,** consists of two single 13-protofilament microtubules surrounded by a ring of nine double microtubules (Fig. 28-32b), all with their (+) ends extending away from the cell. Each outer doublet consists of a ring of 13 protofilaments named subfiber A, fused to a C-shaped assembly of 10, or in some cases 11, protofilaments named subfiber B. The 11 microtubules forming an axoneme are held together by three types of connectors (Fig. 28-32b):

1. Subfibers A are joined to the central microtubules by radial **spokes,** which each terminate in a knoblike feature termed a **spoke head.**
2. Adjacent outer doublets are joined by circumferential linkers that contain the elastic protein named **nexin.**
3. The central microtubules are joined by a connecting bridge.

(a)

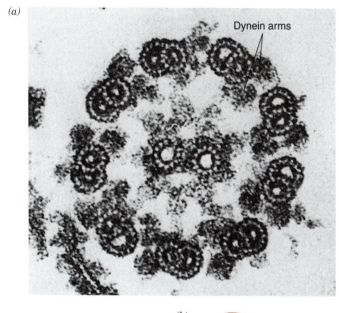

Dynein arms

(b)

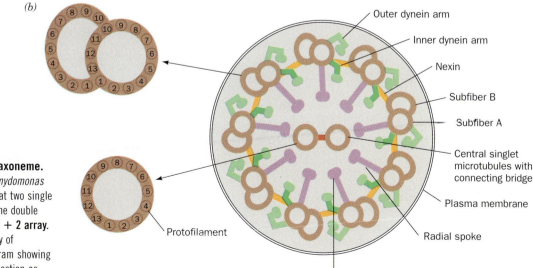

Outer dynein arm

Inner dynein arm

Nexin

Subfiber B

Subfiber A

Central singlet microtubules with connecting bridge

Plasma membrane

Radial spoke

Protofilament

Spoke head

Figure 28-32 Structure of the axoneme. (*a*) Electron micrograph of a *Chlamydomonas* flagellum in cross section. Note that two single microtubules are surrounded by nine double microtubules to form a so-called **9 + 2 array**. [Courtesy of Lewis Tilney, University of Pennsylvania.] (*b*) Schematic diagram showing the structure of a cilium in cross section as viewed from its base.

Like microfilaments in the sarcomere, axonemal microtubules have a more or less fixed length and are not subject to dynamic instability, as are cytoplasmic microtubules (Section 28-3B). Each type of connector is repeated along the length of the axoneme with its own characteristic periodicity. Eukaryotic flagella, which are longer than cilia, have the same axonemal structure (prokaryotic flagella are much smaller and have entirely different components).

Every subfiber A bears two types of arms, an **inner dynein arm** and an **outer dynein arm,** which both point clockwise when viewed from the base of the cilium (Fig. 28-32b). The inner dynein arms consist of either one- or two-headed structures and hence contain one or two heavy chains. The outer dynein arms contain either two heavy chains (e.g., in sea urchin sperm flagella; Fig. 28-31a) or three heavy chains (e.g., in *Chlamydomonas* flagella; Fig. 28-31b and c).

The dynein arms attached to one microtubule function much like myosin cross-bridges in skeletal muscle sarcomeres: They "walk" down the neighboring subfiber B toward its (−) end so that these adjacent microtubules slide past each other. However, the cross-links between neighboring microtubules prevent them from sliding past each another by more than a short distance. *These cross-links therefore convert the dynein-induced sliding motion to a bending motion of the entire axoneme.* This model is supported by electron microscopy studies showing that in straight flagella all the outer doublets have the same length and terminate at the same level but, in bent flagella, the doublets at the inside of the bend extend further than those on the outside of the bend (Fig. 28-33).

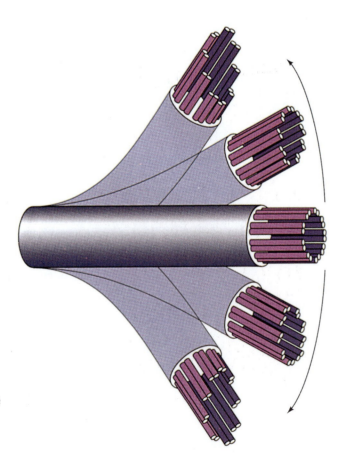

Figure 28-33 The sliding microtubule model of ciliary motion. When the cilium is straight (*center*), the ends of all the microtubules end at the same level. The cilium bends when microtubules on the inner side of the bend slide past microtubules on the outer side (*top and bottom*).

Cilia beat with a whiplike motion (resembling a swimmer's arm in the breast stroke), whereas flagella have an undulatory motion (resembling the motion of a slithering snake). In order for a cilium or flagellum to bend first in one direction and then in the other, sliding of microtubules on one side of the axoneme must alternate with sliding on the other side (Fig. 28-33). It is not clear how dynein molecules on the two sides of an axoneme are activated in the required reciprocating fashion. In addition, there must be a mechanism to coordinate the movements of the arrays of thousands of cilia in some cells that beat in synchrony, up to 40 times per second.

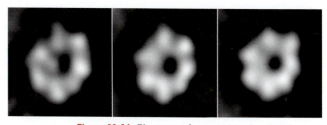

Figure 28-34 Electron micrographs of a dynein heavy chain head. The 135-Å-wide globular head is formed from seven subdomains surrounding a central cavity. [Courtesy of Michael Koonce, Wadsworth Center, Albany, New York.]

How Does Dynein Work?   Elucidating the mechanism whereby dynein transduces the chemical energy of ATP into mechanical force has been difficult because dynein's motor domain is much larger than the motor domains of myosin and kinesin. Electron micrographs indicate that the heavy chain's globular head consists of a heptameric ring of subdomains surrounding a central cavity (Fig. 28-34). The sequences of six of these subdomains resemble those of known ATPases. However, experimental evidence indicates that only one of these subdomains has ATPase activity.

The structure of the dynein heavy chain differs in the presence and absence of ATP or ADP. During the power stroke, which corresponds to the release of ADP, the angle between the stem and stalk decreases, and a linker segment between the head (ring of subdomains) and the stem is repositioned (Fig. 28-35). Thus, as in other motor proteins, conformational changes in the nucleotide-binding domain are transmitted to an attached elongated structure, here the stalk, which thereby functions as a lever arm to move the attached cargo (microtubule) relative to the rest of the motor protein.

Cilia Serve Important Functions.   A genetic defect in axonemal dynein causes **Kartagener syndrome** (also called **primary ciliary dyskinesia** and **immotile cilia syndrome**). This disorder is characterized by recurring respiratory infections resulting from the inability of ciliated cells lining the airways (Fig. 28-30) to clear bacteria-laden mucus. Moreover, males with this condition are infertile because their sperm are immotile. About half of all cases of Kartagener syndrome are accompanied by **situs inversus,** a largely benign condition in which the right–left positions of the body's organs are reversed. In normal **(situs solitus)** individuals, early embryonic cells each have a single cilium that rotates in a counterclockwise direction. Presumably, this movement affects the distribution of signaling molecules that direct organ development so that, for example, the heart develops on the left and the liver on the right. If ciliary movement is absent or impaired, the right–left axis is established at random, accounting for the 50% incidence of situs inversus.

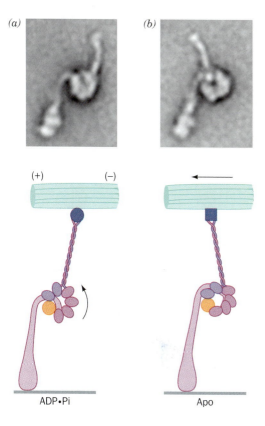

Figure 28-35 Conformational changes in dynein.   The images are based on electron micrographs of the dynein heavy chain from the flagellum of the alga *Chlamydomonas reinhardtii* and are accompanied by interpretive drawings. The heavy chains are oriented with their tails pointing down and their stalks pointing up from the central ring of subdomains. (*a*) Dynein in the presence of ADP and vanadate ion (VO_4^{3-}), which together mimic ADP + P$_i$ (VO_4^{3-} structurally resembles P$_i$). This conformational state corresponds to that following ATP hydrolysis, just before the power stroke. (*b*) Dynein in the absence of a nucleotide. This conformational state corresponds to that immediately following the power stroke. Comparison of Parts *a* and *b* indicates that the conformational change has moved the stalk to the left by ~150 Å and the tail has moved closer to the ring. [Courtesy of Stan Burgess, University of Leeds, U.K.]

5 Antibodies

All organisms are continually subject to attack by other organisms, including disease-causing microorganisms and viruses. In higher animals, these **pathogens** may penetrate the physical barrier presented by the skin and mucous membranes (a first line of defense) only to be identified as foreign invaders and destroyed by the **immune system.**

A Overview of the Immune System

Two types of immunity have been distinguished:

1. **Cellular immunity,** which guards against virally infected cells, fungi, parasites, and foreign tissue, is mediated by *T* **lymphocytes** or *T* **cells,** so called because their development occurs in the thymus.

2. **Humoral immunity** (*humor* is an archaic term for fluid), which is most effective against bacterial infections and the extracellular phases of viral infections, is mediated by an enormously diverse collection of related proteins known as **antibodies** or **immunoglobulins.** Antibodies are produced by *B* **lymphocytes** or *B* **cells,** which in mammals mature in the bone marrow.

In this section we focus on the structure, function, and generation of antibodies.

The immune response is triggered by the presence of a foreign macromolecule, often a protein or carbohydrate, known as an **antigen.** *B* cells display immunoglobulins on their surfaces. If a *B* cell encounters an antigen that binds to its particular immunoglobulin, it engulfs the antigen–antibody complex, degrades it, and displays the antigen fragments on the cell surface. *T* cells then stimulate the *B* cell to proliferate. Most of the *B* cell progeny are circulating cells that secrete large amounts of the antigen-specific antibody. These antibodies can bind to additional antigen molecules, thereby marking them for destruction by other components of the immune system. Although most *B* cells live only a few days unless stimulated by their corresponding antigen, a few long-lived **memory *B* cells** can recognize antigen several weeks or even many years later and can mount a more rapid and massive immune response (called a secondary response) than *B* cells that have not yet encountered their antigen (Fig. 28-36).

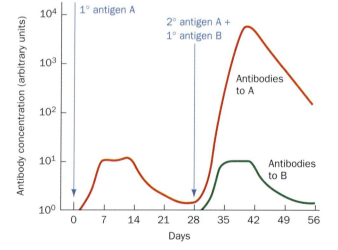

Figure 28-36 Primary and secondary immune responses. Antibodies to antigen A appear in the blood following primary immunization on day 0 and secondary immunization on day 28. Antigen B is included in the secondary immunization to demonstrate the specificity of immunological memory for antigen A. The secondary response to antigen A is both faster and greater than the primary response.

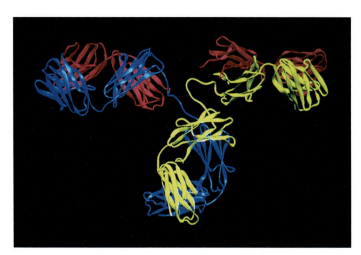

Figure 28-37 X-Ray structure of a mouse antibody. The heavy chains are yellow and blue, and the light chains are both red. The antigen-binding sites are located at the ends of the arms of the Y-shaped molecule. [Courtesy of Alexander McPherson, University of California at Irvine.]

See the Interactive Exercises.

B Antibody Structure

The immunoglobulins form a related but enormously diverse group of proteins. All immunoglobulins contain at least four subunits: two identical ~23-kD **light chains (L)** and two identical 53- to 75-kD **heavy chains (H).** These subunits associate by disulfide bonds and by noncovalent interactions to form a roughly Y-shaped symmetric molecule with the formula $(LH)_2$ (Fig. 28-37).

The five classes of immunoglobulin **(Ig)** differ in the type of heavy chain they contain and, in some cases, in their subunit structure (Table 28-1). For example, **IgM** consists of five Y-shaped molecules arranged around a central **J subunit; IgA** occurs as monomers, dimers, and trimers. The various immunoglobulin classes also have different physiological functions. IgM is most effective against microorganisms and is the first immunoglobulin to be secreted in response to an antigen. **IgG,** the most common immunoglobulin, is equally distributed between the blood and the extravascular fluid. IgA occurs predominantly in the intestinal tract and defends against pathogens by adhering to their antigenic sites so as to block their attachment to epithelial (outer) surfaces. **IgE,** which is normally present in the blood in minute concentrations, protects against parasites and has been implicated in allergic reactions. **IgD,** which is also present in small amounts, has no clearly known function. Our discussion of antibody structure will focus on IgG.

IgG can be cleaved through limited proteolysis with the enzyme **papain** into three ~50-kD fragments: two identical **Fab fragments** and one **Fc fragment.** The Fab fragments are the "arms" of the Y-shaped antibody and con-

Table 28-1 Classes of Human Immunoglobulins

Class	Heavy Chain	Light Chain	Subunit Structure	Molecular Mass (kD)
IgA	α	κ or λ	$(\alpha_2\kappa_2)_n J^a$ or $(\alpha_2\lambda_2)_n J^a$	360–720
IgD	δ	κ or λ	$\delta_2\kappa_2$ or $\delta_2\lambda_2$	160
IgE	ε	κ or λ	$\varepsilon_2\kappa_2$ or $\varepsilon_2\lambda_2$	190
IgG^b	γ	κ or λ	$\gamma_2\kappa_2$ or $\gamma_2\lambda_2$	150
IgM	μ	κ or λ	$(\mu_2\kappa_2)_5 J$ or $(\mu_2\lambda_2)_5 J$	950

[a]$n = 1, 2,$ or 3.

[b]IgG has four subclasses, IgG1, IgG2, IgG3, and IgG4, which differ in their γ chains.

Figure 28-38 *Key to Structure.* **Diagram of human immunoglobulin G (IgG).** Each light chain contains a variable (V_L) and a constant (C_L) region, and each heavy chain contains one variable (V_H) and three constant (C_H1, C_H2, and C_H3) regions. Each of the variable and constant domains contains a disulfide bond, and the four polypeptide chains are linked by disulfide bonds. The proteolytic enzyme papain cleaves IgG at the hinge region to yield two Fab fragments and one Fc fragment. CHO represents carbohydrate chains. [Illustration, Irving Geis/Geis Archives Trust. Copyright Howard Hughes Medical Institute. Reproduced with permission.]

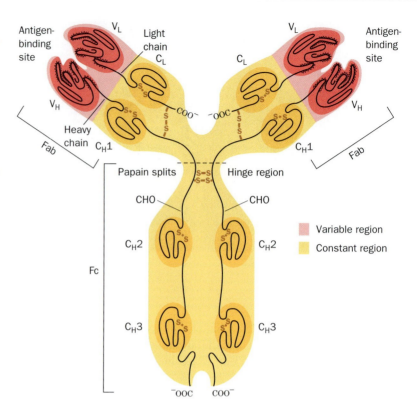

tain an entire L chain and the N-terminal half of an H chain (Fig. 28-38). These fragments contain IgG's antigen-binding sites (the "ab" in Fab stands for *a*ntigen *b*inding). The Fc portion ("c" because it *c*rystallizes easily) derives from the "stem" of the antibody and consists of the C-terminal halves of two H chains. The arms of the Y are connected to the stem by a flexible hinge region. The hinge angles may vary, so an antibody molecule may not be perfectly symmetrical (e.g., Fig. 28-37).

Although all IgG molecules have the same overall structure, IgGs that recognize different antigens have different amino acid sequences. The light chains of different antibodies differ mostly in their N-terminal halves. These polypeptides are therefore said to have a **variable region, V_L** (residues 1 to 108), and a **constant region, C_L** (residues 109 to 214). Comparisons of H chains, which have 446 residues, reveal that H chains also have a variable region, **V_H,** and a constant region, **C_H.** As indicated in Fig. 28-38, the C_H region consists of three ~110-residue segments, **C_H1, C_H2,** and **C_H3,** which are homologous to each other and to C_L. In fact, all the constant and variable regions resemble each other in sequence and in disulfide-bonding pattern. These similarities suggest that the six different homology units of an IgG evolved through the duplication of a primordial gene encoding an ~110-residue protein.

C Antigen–Antibody Binding

The immunoglobulin homology units all have the same characteristic **immunoglobulin fold:** a sandwich composed of three- and four-stranded antiparallel β sheets that are linked by a disulfide bond (Fig. 6-28*b*). Nevertheless, the basic immunoglobulin structure must accommodate an enormous variety of antigens. The ability to recognize antigens resides in three loops in the variable domain (Fig. 28-39). Most of the amino acid variation among antibodies is concentrated in these three short segments, called **hypervariable** sequences. As hypothesized by Elvin Kabat, the hypervari-

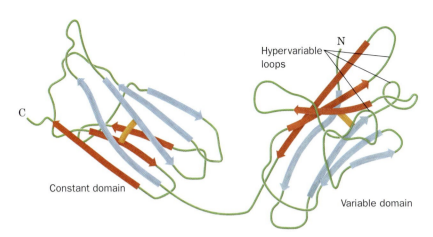

Figure 28-39 Immunoglobulin folds in a light chain.
Both the constant and variable domains consist of a
sandwich of a four-stranded antiparallel β sheet (*blue*)
and a three-stranded antiparallel β sheet (*orange*) that are
linked by a disulfide bond (*yellow*). The positions of the
three hypervariable sequences in the variable domain are
indicated. [After Schiffer, M., Girling, R.L., Ely, K.R., and
Edmundson, A.B., *Biochemistry* **12**, 4628 (1973).]

able sequences line an immunoglobulin's antigen-binding site, so that their
amino acids determine its binding specificity.

Scientists have determined the X-ray structures of Fab fragments from
monoclonal antibodies (see Box 28-4) and monospecific antibodies isolated
from patients with **multiple myeloma** (a disease in which a cancerous *B* cell
proliferates and produces massive amounts of a single immunoglobulin;
immunoglobulins purified from ordinary blood are heterogeneous and
hence cannot be used for detailed structural studies). As predicted by the
positions of the hypervariable sequences, the antigen-binding site is located
at the tip of each Fab fragment in a crevice between its V_L and V_H domains.

The association between antibodies and their antigens involves van der
Waals, hydrophobic, hydrogen bonding, and ionic interactions. Their dissociation constants range from 10^{-4} to 10^{-10} M, comparable (or even greater)
in strength to the associations between enzymes and their substrates. The
specificity and strength of an antigen–antibody complex are a function of
the exquisite structural complementarity between the antigen and the antibody (e.g., Fig. 28-40). These are also the features that make antibodies
such useful laboratory reagents (see Fig. 5-3, for example).

Most immunoglobulins are divalent molecules; that is, they can bind two
identical antigens simultaneously (IgM and IgA are multivalent). A foreign
substance or organism usually has multiple antigenic regions, and a typical
immune response generates a mixture of antibodies with different specificities. Divalent binding allows antibodies to cross-link antigens to form an
extended lattice (Fig. 28-41), which hastens the removal of the antigen and
triggers *B* cell proliferation.

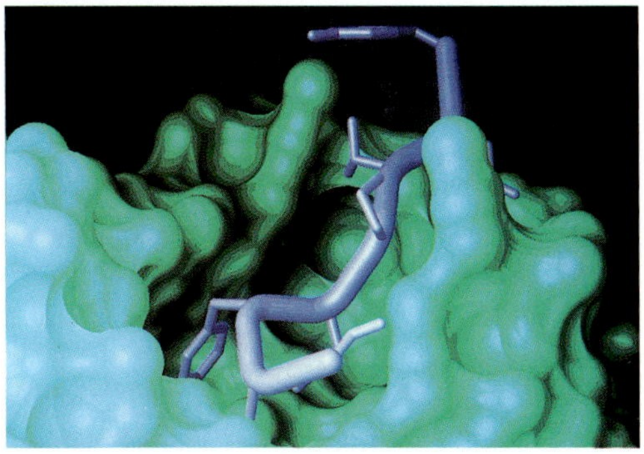

**Figure 28-40 Interaction between an antigen and
an antibody.** This X-ray structure shows a portion of the
solvent-accessible surface of a monoclonal antibody Fab
fragment (*green*) with a stick model of a bound nine-
residue fragment of its peptide antigen (*blue*). [Courtesy
of Ian Wilson, The Scripps Research Institute, La Jolla,
California. PDBid 1HMM.]

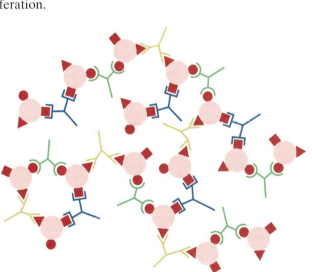

Figure 28-41 Antigen cross-linking by antibodies.
A mixture of divalent antibodies that recognizes the several
different antigenic regions of an intruding particle such as
a toxin molecule or a bacterium can form an extensive
lattice of antigen and antibody molecules.

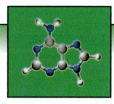

BOX 28-4

Perspectives in Biochemistry

Monoclonal Antibodies

Introducing a foreign molecule into an animal induces the synthesis of large amounts of antigen-specific but heterogeneous antibodies. One might expect that a single lymphocyte from such an animal could be cloned (allowed to reproduce) to yield a harvest of homogeneous immunoglobulin molecules. Unfortunately, lymphocytes do not grow continuously in culture. In the late 1970s, however, César Milstein and Georges Köhler developed a technique for immortalizing such cells so that they can grow continuously and secrete virtually unlimited quantities of a specific antibody. Typically, lymphocytes from a mouse that has been immunized with a particular antigen are harvested and fused with mouse myeloma cells (a type of blood system cancer), which can multiply indefinitely (see figure). The cells are then incubated in a selective medium that inhibits the synthesis of purines, which are essential for myeloma growth [the myeloma cells lack the enzyme **hypoxanthine phosphoribosyl transferase (HPRT),** which could otherwise participate in a purine nucleotide salvage pathway; Section 22-1D]. The only cells that can grow in the selective medium are fused cells, known as **hybridoma cells,** that combine the missing HPRT (it is supplied by the lymphocyte) with the immortal attributes of the myeloma cells. Clones derived from single fused cells are then screened for the presence of antibodies to the original antigen. Antibody-producing cells can be grown in large quantities in tissue culture or as semisolid tumors in mouse hosts.

Monoclonal antibodies are used to purify macromolecules (Section 5-2), to identify infectious diseases, and to test for the presence of drugs and other substances in body tissues. Because of their purity and specificity and, to some extent, their biocompatibility, monoclonal antibodies also hold considerable promise as therapeutic agents against cancer and other diseases. In fact, the monoclonal antibody known as **Herceptin** binds specifically to the growth factor receptor **HER2** that is overexpressed in about one-quarter of breast cancers. Herceptin binding to HER2 blocks its growth-signaling activity, thereby causing the tumor to stop growing or even regress.

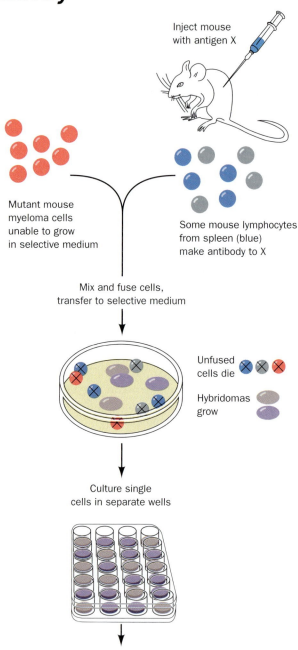

Inject mouse with antigen X

Mutant mouse myeloma cells unable to grow in selective medium

Some mouse lymphocytes from spleen (blue) make antibody to X

Mix and fuse cells, transfer to selective medium

Unfused cells die

Hybridomas grow

Culture single cells in separate wells

Test each well for antibody to X

D Generating Antibody Diversity

A novel antigen does not direct a *B* cell to begin manufacturing a new immunoglobulin to which it can bind. Rather, *an antigen merely stimulates the proliferation of a pre-existing B cell whose antibodies happen to recognize the antigen.* The immune system has the potential to produce an enormous number of different antibodies, probably $>10^{18}$. Even though this number

is so large that an individual can synthesize only a small fraction of its potential immunoglobulin repertoire during its lifetime, this fraction is still sufficient to react with almost any antigen the individual might encounter. Yet the number of immunoglobulin genes is far too small to account for the observed level of antibody diversity. The diversity in antibody sequences arises instead from genetic changes during *B* lymphocyte development.

Another remarkable property of the immune system is that its power is unleashed only against foreign substances and not against any of the tens of thousands of endogenous (self) molecules of various sorts. Virtually all macromolecules are potentially antigenic, as can be demonstrated by transplanting tissues from one individual to another, even within a species. This incompatibility presents obvious challenges for therapies ranging from routine blood transfusions to multiple organ transplants.

The mechanism whereby an individual's immune system distinguishes self from nonself is but poorly understood. It begins to operate around the time of birth and must be ongoing, since new lymphocytes arise throughout an individual's lifetime. Occasionally, the immune system loses tolerance to some of its self-antigens, resulting in an **autoimmune disease,** which may produce symptoms ranging from mild to lethal (see Box 28-5).

κ Light Chain Genes Are Assembled from Multiple Gene Segments.

Homologous recombination (Section 24-6A) in germline cells is a fundamental feature of reproduction in multicellular organisms. In cells of the immune system, **somatic recombination** (Greek: *soma,* body) is responsible for the expression of an enormous number of immunoglobulin gene products. The required antibody diversity also results from an accelerated rate of mutation during the development of antibody-producing *B* cells (see below).

One of the two types of immunoglobulin light chains, the **κ chain,** is encoded by four exons (Fig. 28-42):

1. A **leader** or L_κ **segment,** which encodes a 17- to 20-residue hydrophobic signal peptide. This polypeptide directs newly synthesized κ chains to the endoplasmic reticulum and is then excised (Section 9-4D).

2. A V_κ **segment,** which encodes the first 95 residues of the κ chain's 108-residue variable region.

3. A **joining** or J_κ **segment,** which encodes the variable region's remaining 13 residues.

4. The C_κ **segment,** which encodes the κ chain's constant region.

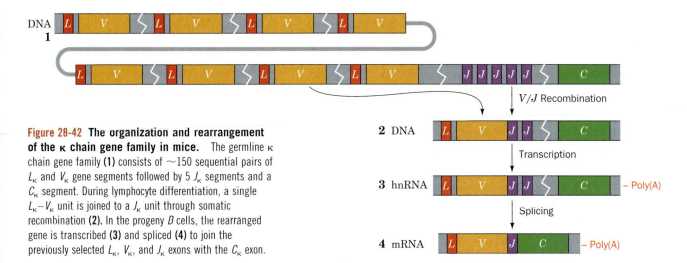

Figure 28-42 The organization and rearrangement of the κ chain gene family in mice.   The germline κ chain gene family (**1**) consists of ~150 sequential pairs of L_κ and V_κ gene segments followed by 5 J_κ segments and a C_κ segment. During lymphocyte differentiation, a single L_κ–V_κ unit is joined to a J_κ unit through somatic recombination (**2**). In the progeny *B* cells, the rearranged gene is transcribed (**3**) and spliced (**4**) to join the previously selected L_κ, V_κ, and J_κ exons with the C_κ exon.

BOX 28-5

Biochemistry in Health and Disease

Autoimmune Diseases

All the body's organ systems are theoretically susceptible to attack by an immune system that has lost its self-tolerance, but some tissues are attacked more often than others. Some of the most common autoimmune diseases are listed below. The symptoms of a particular disease reflect the type of tissue with which the autoantibodies react. In general, autoimmune diseases are chronic, often with periods of remission, and their clinical severity may differ among individuals.

The loss of tolerance to one's own antigens may result from an innate malfunctioning of the mechanism by which the immune system distinguishes self from non-self, possibly precipitated by an event, such as trauma or infection, in which tissues that are normally sequestered from the immune system are exposed to lymphocytes. For example, breaching the blood–brain barrier may allow lymphocytes access to the brain or spinal cord, and injury may allow access to the spaces at joints, which are not normally served by blood vessels. There is also evidence that some autoimmune diseases are caused by antibodies to certain viral or bacterial antigens that cross-react with endogenous substances because of chance antigenic similarities. Some diseases, such as sys-

temic lupus erythematosus, represent a more generalized breakdown of the immune system, so that antibodies to many endogenous substances (e.g., DNA and phospholipids) may be generated.

The pathological effects of autoimmune diseases reflect the complexity of the immune system, which is designed to combat a wide variety of foreign substances in different ways. The agents that mediate tissue destruction as part of the normal battle against infection, for example, can cause unregulated damage in a runaway immune response such as in an autoimmune disease. Hormones also affect the course of an immune response, which may explain why multiple sclerosis and systemic lupus erythematosus are markedly more prevalent in women than in men. The complexities of cellular and humoral immunity that make it difficult to discern the cause of autoimmune diseases also make these diseases difficult to treat. Research in this area has raised the intriguing possibility that autoimmunity may contribute to or complicate other common diseases such as atherosclerosis (Section 19-7C).

Disease	Target Tissue	Major Symptoms
Addison's disease	Adrenal cortex	Low blood glucose, muscle weakness, Na^+ loss, K^+ retention, increased susceptibility to stress
Crohn's disease	Intestinal lining	Intestinal inflammation, chronic diarrhea
Graves' disease	Thyroid gland	Oversecretion of thyroid hormone resulting in increased appetite accompanied by weight loss
Insulin-dependent diabetes mellitus	Pancreatic β cells	Loss of ability to make insulin
Multiple sclerosis	Myelin sheath of nerve fibers in brain and spinal cord	Progressive loss of motor control
Myasthenia gravis	Acetylcholine receptors at nerve–muscle synapses	Progressive muscle weakness
Psoriasis	Epidermis	Hyperproliferation of the skin
Rheumatoid arthritis	Connective tissue	Inflammation and degeneration of the joints
Systemic lupus erythematosus	DNA, phospholipids, other tissue components	Rash, joint and muscle pain, anemia, kidney damage, mental dysfunction

In embryonic tissues (which do not make antibodies), these exons occur in clusters. The κ chain gene family is made of ~150 L_κ and V_κ segments, separated by introns, with the L_κ–V_κ units separated from each other by ~7-kb spacers. This sequence of exon pairs is followed, well downstream,

by 5 J_κ segments at intervals of ~300 bp, a 2.4-kb spacer, and a single C_κ segment.

The assembly of a κ chain mRNA is a complex process involving both somatic recombination and selective mRNA splicing (Section 25-3A) over several cell generations. The first step of this process, which occurs in B cell progenitor cells, is an intrachromosomal recombination that joins an L_κ– V_κ unit to a J_κ segment and deletes the intervening sequences (Fig. 28-42). Then, in later cell generations, the entire modified gene is transcribed and selectively spliced so as to join the L_κ–V_κ–J_κ unit to the C_κ segment. The L_κ and V_κ segments are also spliced together in this step, yielding an mRNA that encodes one of each of the four elements of a κ chain gene.

Recombinational Flexibility Contributes to Antibody Diversity.

The joining of 1 of 150 V_κ segments to 1 of 5 J_κ segments can generate only $150 \times 5 = 750$ different κ chains, far less than the number observed. However, studies of many joining events involving the same V_κ and J_κ segments revealed that *the V/J recombination site is not precisely defined; these two gene segments can join at different crossover points* (Fig. 28-43). Consequently, the amino acids specified by the codons in the vicinity of the V/J recombination site depend on what part of the sequence is supplied by the germline V_κ segment and what part is supplied by the germline J_κ segment. Assuming that this recombinational flexibility increases the possible κ chain diversity 10-fold, the expected number of possible different κ chains is increased to $150 \times 5 \times 10 = 7500$.

The other type of immunoglobulin light chain, the **λ chain,** is similarly encoded by a gene family containing L_λ, V_λ, J_λ, and C_λ segments whose recombination likewise yields a large number of possible polypeptides.

Heavy Chain Genes Are Also Assembled from Sets of Gene Segments.

Heavy chain genes are assembled in much the same way as are light chain genes but with the additional inclusion of an ~13-bp **diversity** or **D segment** between their V_H and J_H segments. The human heavy chain gene family consists of clusters of ~250 different L_H–V_H units, ~10 D segments, 6 J_H segments, and 8 C_H segments (Fig. 28-44). Germline V_H, D, and J_H segments

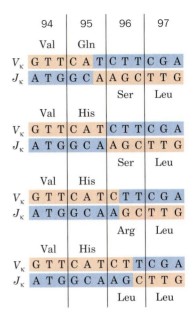

Figure 28-43 Variation at the V_κ/J_κ joint. The crossover point at which the V_κ and J_κ sequences somatically recombine varies by several nucleotides, thereby giving rise to different nucleotide sequences (*pink bands*) in the active κ gene. For example, as indicated here, amino acid 96, which occurs in the κ chain's third hypervariable region, can be Ser, Arg, or Leu.

Figure 28-44 The organization and rearrangement of the heavy chain gene family in humans. This gene family consists of ~250 sequential pairs of L_H and V_H gene segments followed by ~10 D segments, 6 J_H segments, and 8 C_H segments (one for each class or subclass of heavy chains; Table 28-1). During lymphocyte differentiation, an L_H–V_H unit is joined to a D segment and a J_H segment. In this process, the D segment becomes flanked by short stretches of random sequence called N regions. In the B cell and its progeny, transcription and splicing join the L_H–V_H–N–D–N–J_H unit to one of the 8 C_H gene segments.

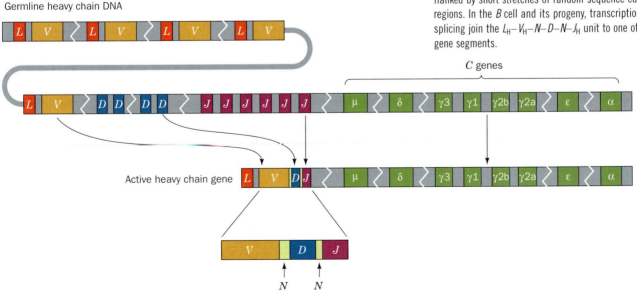

Germline heavy chain DNA

C genes

Active heavy chain gene

N   N

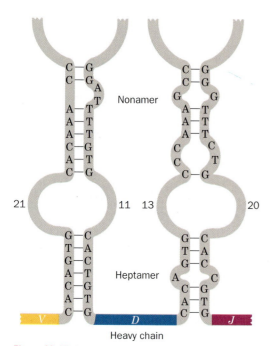

Figure 28-45 Stem–loop recombination sites in the germline heavy chain gene family. These structures mediate somatic recombination between the V_H and D segments (*left*) and between the D and J_H segments (*right*). The κ chain *V/J* recombination signal consists of a similar heptamer–nonamer stem–loop structure. The recombination system's requirement for both the 20/21- and the 11/13-bp spacers prevents it from inadvertently skipping the D segment by directly joining the V_H and J_H segments.

are joined in a particular order (D is joined to J_H before V_H is joined to DJ_H), and the joining sites are subject to the same recombination flexibility as are light chain V/J sites.

Highly conserved sequences on each side of the V, D, and J gene segments can form stem–loop secondary structures that act as recombination signals (Fig. 28-45). The structural similarities and functional interchangeabilities of these sites suggest that all **$V(D)J$ joining** reactions are catalyzed by an evolutionarily conserved **$V(D)J$ recombinatase** system. Indeed, David Baltimore discovered two proteins, **RAG1** and **RAG2** (RAG for *recombination activating genes*), that work in concert to recognize the cleavage site and participate in the required double-stranded cleavage of the DNA as well as in its subsequent joining. Under laboratory conditions, RAG1 and RAG2 can catalyze transposition, that is, the relocation of a DNA segment to another DNA molecule (Section 24-6C), consistent with the proposal that the RAG proteins originated as a transposon that appeared in vertebrates around 450 million years ago.

Assuming that recombinational flexibility contributes a factor of 100 toward heavy chain diversity, somatic recombination can generate some $250 \times 10 \times 6 \times 100 = 1.5 \times 10^6$ different heavy chains. Then, taking into account κ chain diversity (and neglecting that of λ chains), there can be as many as $7500 \times 1.5 \times 10^6 = 11$ billion different types of immunoglobulins formed by somatic recombination among ~400 different gene segments.

Somatic Mutation Is a Further Source of Antibody Diversity. Despite the enormous antibody diversity generated by somatic recombination, immunoglobulins are subject to even more variation through **somatic mutations** of two types:

1. During V_H/D and D/J_H joining, a few nucleotides may be added or removed from the recombination joints. The added nucleotides, which form so-called **N regions,** yield *NDN* units of up to 30 bp that encode enormously variable heavy chain segments of up to 10 amino acid residues (Fig. 28-44). The N regions appear to arise through the action of **terminal deoxynucleotidyl transferase,** a template-independent DNA polymerase present in the B cell progenitors that make the heavy chain joints. This enzyme is probably absent in later cell generations when the light chain joints are formed.

2. The variable regions of both heavy and light chains are more diverse than is expected on the basis of comparisons of their amino acid sequences with their corresponding germline nucleotide sequences. Indeed, these regions mutate at rates of up to 10^{-3} base changes per nucleotide per cell generation, rates that are at least a millionfold higher than the rates of spontaneous mutation in other genes. B cells and/or their progenitors apparently possess enzymes that mediate this **somatic hypermutation** of immunoglobulin gene segments. Somatic hypermutation appears to act over many cell generations after antigen stimulation, thereby tailoring antibodies to a particular antigen.

These somatic mutation processes increase the possible number of different antibodies that humans can produce by many orders of magnitude beyond the 11 billion estimated from somatic recombination alone. The diversity arising from both recombination and mutation thereby permits an individual organism to cope, in a kind of Darwinian struggle, with the rapid mutation rates of pathogenic microorganisms.

SUMMARY

1. The microfilament elements of the eukaryotic cytoskeleton are built from ATP-binding globular actin subunits that polymerize as a double chain. Polymerization is reversible, so microfilaments grow and regress as regulated by proteins that mediate capping, branching, and severing.

2. Myofibrils consist of repeating arrays of thick filaments (made of myosin) and thin filaments (made of actin, tropomyosin, and troponin).

3. The heads of myosin molecules in thick filaments form bridges to actin in thin filaments such that the detachment and reattachment of the myosin heads cause the thick and thin filaments to slide past each other during muscle contraction. The contractile force derives from conformational changes triggered by ATP hydrolysis.

4. Myosin V acts as a transport motor, advancing along a microfilament by a hand-over-hand movement.

5. GTP-binding tubulin dimers polymerize to form a hollow microtubule. β-Tubulin's hydrolysis of its bound GTP destabilizes the microtubule, contributing to dynamic instability.

6. The motor protein kinesin has two globular heads connected to flexible neck linkers, and it transports vesicular cargo along a microtubule by a processive stepping mechanism that is driven by conformational changes triggered by ATP binding and hydrolysis.

7. Dynein cross-bridges between microtubules in an axoneme drive the microtubule sliding responsible for movement in eukaryotic cilia and flagella.

8. The immune system responds to foreign macromolecules through the production of antibodies (immunoglobulins).

9. The Y-shaped IgG molecule consists of two heavy and two light chains. The two antigen-binding sites are formed by the hypervariable sequences in the variable domains at the ends of a heavy and a light chain.

10. Antibody diversity results from somatic recombination involving the selection of individual members of clustered gene sequences encoding the different segments of the immunoglobulin light and heavy chains. Diversity is augmented by imprecise V/D/J joining and by somatic hypermutation.

REFERENCES

Cytoskeletal Proteins

Cooper, J.A. and Schafer, D.A., Control of actin assembly and disassembly at filament ends, *Curr. Opin. Cell Biol.* **12,** 97–103 (2000). [Provides an overview of the principles of microfilament dynamics and some of the key protein players.]

Nogales, E., Structural insights into microtubule function, *Annu. Rev. Biophys. Biomol. Struct.* **30,** 397–420 (2001). [Discusses tubulin and microtubule structure along with other proteins and drugs that affect microtubule structure.]

Pollard, T.D., The cytoskeleton, cellular motility and the reductionist agenda, *Nature* **422,** 741–745 (2003). [Outlines some major research areas and some techniques.]

Motor Proteins

Asai, D.J. and Koonce, M.P., The dynein heavy chain: structure, mechanics and evolution, *Trends Cell Biol.* **11,** 196–202 (2001).

Endow, S.A. and Barker, D.S., Processive and nonprocessive models of kinesin movement, *Annu. Rev. Physiol.* **65,** 161–175 (2003).

Mandelkow, E. and Johnson, K.A., The structural and mechanochemical cycle of kinesin, *Trends Biochem. Sci.* **23,** 429–433 (1998). [Summarizes the motor mechanism of kinesin.]

Schliwa, M. (Ed.), *Molecular Motors,* Wiley-VCH (2003); *and* Schliwa, M. and Woehlke, G., Molecular motors, *Nature* **422,** 759–765 (2003). [Reviews of myosin, kinesin, and dynein.]

Spudich, J.A., The myosin swinging cross-bridge model, *Nature Rev. Mol. Cell Biol.* **2,** 387–391 (2001). [Summarizes the history and current state of models for myosin action.]

Vale, R.D. and Milligan, R.A., The way things move: Looking under the hood of molecular motor proteins, *Science* **288,** 88–95 (2000). [Compares and contrasts the molecular mechanisms of myosin and kinesin motors.]

Yildiz, A., Forkey, J.N., McKinney, S.A., Ha, T., Goldman, Y.E., and Selvin, P.R., Myosin V walks hand-over-hand: single fluorophore imaging with 1.5-nm localization, *Science* **300,** 2061–2065 (2003).

Antibodies

Agrawal, A., Eastman, Q.M., and Schatz, D.G., Transposition mediated by RAG1 and RAG2 and its implications for the evolution of the immune system, *Nature* **394,** 744–751 (1998).

Davies, D.R. and Cohen, G.H., Interactions of protein antigens with antibodies, *Proc. Natl. Acad. Sci.* **93,** 7–12 (1996).

Harris, L.J., Larson, S.B., Hasel, K.W., Day, J., Greenwood, A., and McPherson, A., The three-dimensional structure of an intact monoclonal antibody for canine lymphoma, *Nature* **360,** 369–372 (1992). [The first high-resolution X-ray structure of an intact IgG.]

Janeway, C.A., Jr., Travers, P., Walport, M., and Shlomchik, M.J., *Immunobiology 5,* Garland Publishing (2001).

Marrack, P., Kappler, J., and Kotzin, B.L., Autoimmune disease: why and where it occurs, *Nature Medicine* **7,** 899–905 (2001).

KEY TERMS

cytoskeleton	cytokinesis	axoneme	Fc fragment
motor protein	sarcomere	pathogen	variable region
microfilament	thin filament	immune system	constant region
intermediate filament	thick filament	cellular immunity	immunoglobulin fold
microtubule	sliding filament model	lymphocyte	hypervariability
(−) end	stereocilia	humoral immunity	monoclonal antibody
(+) end	electron crystallography	immunoglobulin	autoimmune disease
C_c	protofilament	memory B cell	somatic recombination
treadmilling	dynamic instability	Fab fragment	somatic hypermutation

STUDY EXERCISES

1. Explain how globular proteins such as actin and tubulin and fibrous proteins such as keratin and collagen can all give rise to fibrous structures.

2. What is the basis of microfilament polarity?

3. Draw a diagram of the components of a myofibril.

4. Explain the molecular basis of the sliding filament model of muscle contraction.

5. How does a microtubule differ structurally from a microfilament?

6. Compare and contrast the effect of ATP binding and P_i release on the motor activities of kinesin and myosin.

7. Compare the processivity of myosin II, myosin V, and kinesin.

8. Does dynein function more like kinesin or myosin II?

9. Identify the parts of an immunoglobulin molecule.

10. Describe how antibody diversity is generated.

PROBLEMS

1. Explain why microfilaments and microtubules are polar whereas intermediate filaments of keratin are not.

2. Explain how gelsolin could nucleate actin polymerization in a reaction mixture containing a high concentration of G-actin but no F-actin.

3. Is myosin a fibrous protein or a globular protein? Explain.

4. A myosin head can undergo five ATP hydrolysis cycles per second, each of which moves an actin monomer by ~60 Å. How is it possible for an entire sarcomere to shorten by 1000 Å in this same period?

5. **Rigor mortis,** the stiffening of muscles after death, is caused by depletion of cellular ATP. Describe the molecular basis of rigor.

6. Using Fig. 28-19 as a model, draw a diagram illustrating inchworm movement for a two-headed motor protein with a step size of 200 Å.

7. How could a microtubule-binding protein distinguish a rapidly growing microtubule from one that was growing more slowly?

8. Explain why colchicine (which promotes microtubule depolymerization) and taxol (which prevents depolymerization) both prevent cell division.

9. Early cell biologists, examining living cells under the microscope, observed that the movement of certain cell constituents was rapid, linear, and targeted (that is, directed toward a particular point). (a) Why are these qualities inconsistent with diffusion as a mechanism for redistributing cell components? (b) List the minimum requirements for an intracellular transport system that is rapid, linear, and targeted.

10. Is dynein a processive motor? Explain.

11. Give the approximate molecular masses of an immunoglobulin G molecule analyzed by (a) gel filtration chromatography, (b) SDS-PAGE, and (c) SDS-PAGE in the presence of 2-mercaptoethanol.

12. Explain why the variation in V_L and V_H domains of immunoglobulins is largely confined to the hypervariable loops.

13. Why do antibodies raised against a native protein sometimes fail to bind to the corresponding denatured protein?

14. Antibodies raised against a macromolecular antigen usually produce an antigen–antibody precipitate when mixed with that antigen. Explain why no precipitate forms when (a) Fab fragments from these antibodies are mixed with the antigen; (b) antibodies raised against a small antigen are mixed with that small antigen; and (c) the antibody is in great excess over the antigen and vice versa.

BIOINFORMATICS EXERCISES

DEVELOPED BY PAUL CRAIG
Department of Chemistry
Rochester Institute of Technology

INTRODUCTION

Over the last two decades, bioinformatics has become increasingly important in both teaching and learning biochemistry. The most obvious case is the sequencing of the human genome and those of many other species. In 1990, the determination of the sequence of a protein was often the topic of a full publication in a peer-reviewed journal such as *Science, Nature,* or *The Journal of Biological Chemistry.* Now entire genomes are the topic of individual research papers. The term "bioinformatics" is a catch-all phrase that generally refers to the use of computers and computer science approaches to the study of biological systems. This information is discussed in the text, mainly in Chapters 3 (Nucleotides, Nucleic Acids, and Genetic Information), 5 (Proteins: Primary Structure), 6 (Proteins: Three-Dimensional Structure), 12 (Enzyme Kinetics, Inhibition, and Regulation), and 13 (Introduction to Metabolism). Here we provide exercises appropriate to these chapters in order to introduce the techniques of bioinformatics that involve the use of computers, Internet-accessible databases, and the tools that have been developed to "mine" those databases.

As much as possible, the exercises are based on well established, stable web sites. If it is necessary to use less reliable sites and/or resources, attempts have been made to provide multiple sites that perform similar functions. The stable online resources that you will use most frequently include:

Genbank (http://www.ncbi.nlm.nih.gov/)

Protein Data Bank (http://www.rcsb.org)

ExPASy Proteomics Server (http://us.expasy.org/)

European Bioinformatics Institute (http://www.ebi.ac.uk/)

SCOP (http://scop.mrc-lmb.cam.ac.uk/scop/)

CATH (http://www.biochem.ucl.ac.uk/bsm/cath/)

PubMed (http://www.ncbi.nlm.nih.gov/entrez/query.fcgi)

PubMed Central (http://www.pubmedcentral.nih.gov/)

The exercises include some questions that have definite answers, but many questions may be answered in a number of ways, depending on the approach you take or the topic you select. In most cases, the answer key includes the definite answers. For some open-ended questions, a typical correct answer may be presented. Please note that because biological databases are continually being expanded by the addition of new entries, the results of some database queries—particularly quantitative results—may appear to shift slightly over time.

These exercises are available on the website (www.wiley.com/college/voet). The online format offers you easy access to resources within the web environment by simply clicking on the relevant links. You can also use your computer's cut-and-paste functions to more conveniently perform tasks involving extensive sequences of text that would otherwise need to be typed in.

CHAPTER 3 DATABASES FOR THE STORAGE AND "MINING" OF GENOME SEQUENCES

Chapter 3 is an introduction to nucleotides, nucleic acids (DNA and RNA), and the processes of transcription and translation. The exercises below are designed to introduce you to some of the relevant databases and the tools they contain for examining and comparing different bits of information (see Sections 3-4C and 3-4D). Biological databases are an important resource for the study of biochemistry at all levels. These databases contain huge amounts of information about the sequences and structures of nucleic acids (DNA and RNA) and proteins. They also contain software tools that can be used to analyze the data. Some of the software—called web applications—can be used directly from a web browser. Other software—called freestanding applications—must be downloaded and installed on your local computer.

1. Finding Databases. We'll start with finding databases.

(a) What major online databases contain DNA and protein sequences?

(b) Which databases contain entire genomes?

(c) Using your textbook and online resources (http://www.google.com), make sure you understand the meaning of the following terms: BLAST, taxonomy, gene ontology, phylogenetic trees, and multiple sequence alignment. Once you have defined these terms, find resources on the Internet that enable you to study them.

2. TIGR (The Institute for Genomic Research). Open the TIGR site (http://www.tigr.org). Find the Comprehensive Microbial Resource.

(a) What 2001 publication describes the Comprehensive Microbial Resource at TIGR?

(b) How many completed genomes from *Pseudomonas* species have been deposited at TIGR?

(c) Which *Pseudomonas* species are these?

(d) Identify the primary reference for *Pseudomonas putida* KT2440.

(e) Find the link on the Comprehensive Microbial Resource home page for restriction digests. Perform a computer-generated restriction digest on *Pseudomonas putida* KT2440 with BamH1. How many fragments form and what is the average fragment size? (See Section 3-4A for a discussion of restriction endonucleases.)

(f) In addition to microbial genomes, TIGR also contains the genomes of many higher organisms. Identify five eukaryotic genomes that are available at TIGR.

3. Analyzing a DNA Sequence. Using high-throughput methods, scientists are now able to sequence entire genomes in a very

short period of time. Sequencing a genome is quite an accomplishment in itself, but it is really only the beginning of the study of an organism. Further study can be done both at the wet lab bench and on the computer. In this problem, you will use a computer to help you identify an open reading frame, determine the protein that it will express, and find the bacterial source for that protein. Here is the DNA sequence:

```
TACGCAATGCGTATCATTCTGCTGGGCGCTCCGGGCGCAGGTAAA
GGTACTCAGGCTCAATTCATCATGGAGAAATACGGCATTCCGCAA
ATCTCTACTGGTGACATGTTGCGCGCCGCTGTAAAAGCAGGTTCT
GAGTTAGGTCTGAAAGCAAAAGAAATTATGGATGCGGGCAAGTT
GGTGACTGATGAGTTAGTTATCGCATTAGTCAAAGAACGTATCACA
CAGGAAGATTGCCGCGATGGTTTTCTGTTAGACGGGTTCCCGCGT
ACCATTCCTCAGGCAGATGCCATGAAAGAAGCCGGTATCAAAGTT
GATTATGTGCTGGAGTTTGATGTTCCAGACGAGCTGATTGTTGAG
CGCATTGTCGGCCGTCGGGTACATGCTGCTTCAGGCCGTGTTTATC
ACGTTAAATTCAACCCACCTAAAGTTGAAGATAAAGATGATGTTAC
CGGTGAAGAGCTGACTATTCGTAAAGATGATCAGGAAGCGACTGT
CCGTAAGCGTCTTATCGAATATCATCAACAAACTGCACCATTGGTT
TCTTACTATCATAAAGAAGCGGATGCAGGTAATACGCAATATTTTAA
ACTGGACGGAACCCGTAATGTAGCAGAAGTCAGTGCTGAACTGG
CGACTATTCTCGGTTAATTCTGGATGGCCTTATAGCTAAGGCGGTT
TAAGGCCGCCTTAGCTATTTCAAGTAAGAAGGGCGTAGTACCTACA
AAAGGAGATTTGGCATGATGCAAAGCAAACCCGGCGTATTAATGG
TTAATTTGGGGACACCAGATGCTCCAACGTCGAAAGCTATCAAGC
GTTATTTAGCTGAGTTTTTGAGTGACCGCCGGGTAGTTGATACTTC
CCCATTGCTATGGTGGCCATTGCTGCATGGTGTTATTTTACCGCTTC
GGTCACCACGTGTAGCAAAACTTTATCAATCCGTTTGGATGGAAG
AGGGCTCTCCTTTATTGGTTTATAGCCGCCGCCAGCAGAAAGCACT
GGCAGCAAGAATGCCTGATATTCCTGTAGAATTAGGCATGAGCTAT
GGTTCAC
```

(a) First, try to find an open reading frame in this segment of DNA. What is an open reading frame (ORF)? You can find the answer in your textbook (Section 3-4D) or online with a simple Internet search (http://www.google.com). You may also wish to try the bookshelf at PubMed (http://www.ncbi.nlm.nih.gov/entrez/query.fcgi?db=Books). In bacteria, an open reading frame on a piece of mRNA almost always begins with AUG, which corresponds to ATG in the DNA segment that codes for the mRNA. According to the standard genetic code (Table 26-1), there are three Stop codons on mRNA: UAA, UAG, and UGA, which correspond to TAA, TAG, and TGA in the parent DNA segment. Here are the rules for finding an open reading frame in this piece of bacterial DNA:

(1) It must start with ATG. In this exercise, the first ATG is the Start codon. In a real gene search, you would not have this information.

(2) It must end with TAA, TAG, or TGA.

(3) It must be at least 300 nucleotides long (coding for 100 amino acids).

(4) The ATG Start codon and the Stop codon must be in frame. This means that the total number of bases in the sequence from the Start to the Stop codon must be evenly divisible by 3 (see Section 26-1A).

Hints: Try this search by pasting the DNA sequence into a word processing program, then searching for the Start and Stop codons.

Once you have found a pair, highlight the text of the proposed ORF and use the program's Word Count function to count the number of characters between (or including) the Start and Stop codons. This number must be evenly divisible by 3. You can also use a fixed-width font such as Courier, enlarge the size of the text, and adjust the margins so that each line holds just three characters (one codon). Once you find the first ATG, delete the characters that precede it. Then search for a Stop codon that fits all on one line (is in the same reading frame as the Start codon).

(b) Admittedly, Part (a) is a tedious approach. Here is an easier one: Highlight the entire DNA sequence again and copy it. Then go to the Translate tool on the ExPASy server (http://www.expasy.org/tools/dna.html). Paste the sequence into the box entitled "Please enter a DNA or RNA sequence in the box below (numbers and blanks are ignored)." Then select "Verbose ("Met", "Stop", spaces between residues)" as the Output format and click on "Translate Sequence." The "Results of Translation" page that appears contains six different reading frames. What is a reading frame and why are there six? (Refer to Section 26-1A, the Internet, or the PubMed bookshelf for an answer.) Identify the reading frame that contains a protein (more than 100 continuous amino acids with no interruptions by a Stop codon) and note its name. Now go back to the Translate tool page, leave the DNA sequence in the sequence box, but select "Compact ("M", "-", no spaces)" as the Output format. Go to the same reading frame as before and copy the protein sequence (by one-letter abbreviations) starting with "M" for methionine and ending in "-" for the Stop codon. Save this sequence to a separate text file.

(c) Now you will identify the protein and the bacterial source. Go to the NCBI BLAST page (http://www.ncbi.nlm.nih.gov/BLAST/). What does BLAST stand for? You will do a simple BLAST search using your protein sequence, but you can do much more with BLAST. You are encouraged to work the Tutorials on the BLAST home page to learn more. On the BLAST page, select "Protein-protein BLAST." Enter your protein sequence in the "Search" box. Use the default values for the rest of the page and click on the "BLAST!" button. You will be taken to the "formatting BLAST" page. Click on the "Format!" button. You may have to wait for the results. Your protein should be the first one listed in the BLAST output. What is the protein and what is the source?

Note to instructors: You can do this exercise with any DNA sequence. You can also start from a DNA sequence directly in BLAST (use blastn) and find the genes that way. It is probably best to choose a DNA segment that encodes only one protein.

4. Sequence Homology. You will use BLAST to look at sequences that are homologous to the protein that you identified in Problem 3.

(a) First, some definitions: What do the terms "homolog," "ortholog," and "paralog" mean? Go to the NCBI BLAST page (http://www.ncbi.nlm.nih.gov/BLAST/) and choose "Protein-protein BLAST." Paste your protein sequence into the "Search" box. Before clicking on the "BLAST!" button, narrow the search by kingdom. As you look down the BLAST page, you'll see an Options section. Under "Limit by entrez query" (followed by an empty box) or "select from:" (followed by a drop-down menu), select "Eukaryota." Now click on the "BLAST!" button. Click on

the "Format!" button on the next page. Can you find a homologous sequence from yeast?

(*Hint*: Use your browser's Find tool to search for the term "Saccharomyces.") Note the Score and E value given at the right of the entry.

Can you find a homologous sequence from humans?

(*Hint*: Search for the term "Homo.") Note its Score and E value.

Most biochemists consider 25% identity the cutoff for sequence homology, meaning that if two proteins are less than 25% identical in sequence, more evidence is needed to determine whether they are homologs. Click on the Score values for the yeast and human proteins to see each sequence aligned with the *Yersinia pestis* sequence and to see the percent sequence identity. Are the yeast and human sequences homologous to the *Yersinia pestis* sequence?

(b) Use the BLAST online tutorial (http://www.ncbi.nlm.nih.gov/Education/BLASTinfo/information3.html) to discover the meaning of the Score and E value for each sequence that is reported. What is the difference between an identity and a conservative substitution? Provide an example of each from the comparison of your sequence and a homologous sequence obtained from BLAST (see Section 5-4A for a discussion of conservative substitution).

(c) BLAST uses a substitution matrix to assign values in the alignment process, based on the analysis of amino acid substitutions in a wide variety of protein sequences. Be sure you understand the meaning of the term "substitution matrix." What is the default substitution matrix on the BLAST page? What other matrices are available? What is the source of the names for these substitution matrices? Repeat the BLAST search in Problem 4(a) using a different substitution matrix. Do you find different answers?

5. Plasmids and Cloning

(a) REBASE is the Restriction Enzyme Database (http://rebase.neb.com/rebase/rebase.html), which is supported by a number of commercial restriction enzyme suppliers (restriction enzymes are described in Section 3-4A). Go to the REBASE Enzymes page (http://rebase.neb.com/rebase/rebase.enz.html) and find a restriction enzyme from *Rhodothermus marinus* (it starts with the letters Rma). What is the abbreviation for this enzyme?

Click on the enzyme's abbreviation to be taken to the page for this enzyme. Follow the links there to answer the following two questions. What is the recognition sequence for this enzyme? What are the expected and actual frequencies of restriction enzyme recognition sites for this enzyme in *Bacillus halodurans* C-125?

(b) What is a plasmid? pBR322 was one of the first plasmids to be developed for experimental work. Go to the Entrez site (http://www.ncbi.nlm.nih.gov/Entrez) and find the sequence of pBR322 by searching for the terms "pBR322, complete genome." You must select Nucleotide as your search option on the Entrez main page.

Look through the Entrez description of pBR322 and identify one gene encoded by pBR322 and name the antibiotic that it targets.

You can get Entrez to display your sequence in FASTA format by selecting this option next to the "Display" button.

(Here are two of many sites that describe the FASTA format: http://ngfnblast.gbf.de/docs/fasta.html; http://bioinformatics.ubc.ca/resources/faq/?faq_id=1). Save the pBR322 sequence in FASTA format.

(c) Go to PubMedCentral and search for a 1978 article in *Nucleic Acids Research* about restriction mapping of pBR322. Download the article in pdf format (use Adobe Acrobat to read it; you can get this program at http://www.adobe.com). What is the size of the pBR322 plasmid in number of base pairs?

How many cut sites are there for the restriction enzyme HaeIII on pBR322?

(d) Some restriction enzymes generate "blunt ends," and some generate "sticky ends." Explain the meaning of those terms and provide an example of each.

(e) Go to the RESTRICT site at the Pasteur Institute (http://bioweb.pasteur.fr/seqanal/interfaces/restrict.html). Enter your email address at the top, then input the pBR322 sequence file. Scroll down to the "Required section" and note that you have a Minimum recognition site length of four nucleotides and you have selected all the enzymes available in REBASE to digest pBR322 at the same time. Click on the "Run Restrict" button.

On the output screen, click on the "outfile.out" link. This takes you to a simple text page that lists all the cuts that were made in the pBR322 plasmid. How many pBR322 fragments did "all" the enzymes generate? (Look for the "HitCount" number on the output.out page).

What happens to the number of fragments when the minimum recognition site length is changed to six nucleotides? Why did the number change?

(f) Now change the enzyme name from "all" to "BamHI" in the enzymes box under the Required section on the RESTRICT page. How many fragments are generated? How many fragments are obtained using AvaI? What is the size of the restriction site for AvaI? How many fragments are obtained using Eco47III? What is the size of the restriction site for Eco47III?

(g) How many pBR322 fragments are produced when the three different enzymes are combined (separate the enzyme names by commas)? How large are the fragments?

(h) Use a mixture of the restriction enzymes BamHI, AvaI, and PstI to construct a restriction map of pUC18 similar to the one shown in Fig. 3-25. How does this procedure for restriction mapping differ from that used in Problem 10 at the end of Chapter 3?

(i) For the adventurous: Find an enzyme or combination of enzymes that will produce 10 fragments from pUC18. Draw a restriction map of your results.

CHAPTER 5 USING DATABASES TO COMPARE AND IDENTIFY RELATED PROTEIN SEQUENCES

1. Obtaining Sequences from BLAST. Triose phosphate isomerase is an enzyme that occurs in a central metabolic pathway called glycolysis (see Chapter 14). It is also known as an enzyme that demonstrates catalytic perfection (see Section 12-1B). For this problem, you'll start with the sequence of triose phosphate isomerase from rabbit muscle and look for related proteins in the online databases. Here is the sequence of rabbit muscle triose phosphate isomerase in FASTA format:

```
>gi|136066|sp|P00939|TPIS_RABIT Triosephosphate isomerase (TIM) (Triose-
phosphate isomerase)
```

APSRKFFVGGNWKMNGRKKNLGELITTLNAAKVPADTEVVCAPPTAYIDFARQKLDPKIAVAAQNCYKVTNGAF
TGEISPGMIKDCGATWVVLGHSERRHVFGESDELIGQKVAHALSEGLGVIACIGEKLDEREAGITEKVVFEQTK
VIADNVKDWSKVVLAYEPVWAIGTGKTATPQQAQEVHEKLRGWLKSNVSDAVAQSTRIIYGGSVTGATCKELAS
QPDVDGFLVGGASLKPEFVDIINAKQ

(a) Go to http://www.ncbi.nlm.nih.gov/BLAST and follow the link to Protein-protein BLAST (blastp) under Protein. Perform a BLAST search using the triose phosphate isomerase sequence by copying and pasting it into the Search box. Find a human homolog of rabbit muscle triose phosphate isomerase.

The first item in this record (gi|4507645|ref|NP_000356.1|) is a link to another database where this protein is described in more detail (items that begin with "gi" lead to GenBank records). The next item (triosephosphate isomerase 1 [Ho...]) is a description of the protein. Next is the score (493) followed by the E value (e-138). In bioinformatics, two proteins are called "homologs" if they arose from a common ancestor; the two proteins are called "orthologs" if they arose from a common ancestor and perform the same function in two different species. Does the NP_000356.1 entry represent a human ortholog of rabbit muscle triose phosphate isomerase? What is the percent identity between the two enzymes?

Find another human homolog to the rabbit muscle enzyme. Click on the link on the left side of the record to bring up its Genbank entry. Select "FASTA" as the display format and click on the "Display" button. Copy the FASTA text and save it to a text file (if you are using a word processor, be sure to save the file in "text only" format). Save the text file (suggested name: TIM_FASTA.txt) for later use.

(b) Instead of trying to look through the entire BLAST output to find triose phosphate isomerase homologs from plants, bacteria, and archaea, you can use some options in BLAST to narrow your search. Return to the protein-protein BLAST page and paste the rabbit muscle sequence into the "Search" box. This time, look down the BLAST page for an option to select "Archaea" and then perform the BLAST search. Select one of the resulting sequences and save it in FASTA format. Repeat this process to get FASTA-formatted sequences for triose phosphate isomerases from a bacterial and plant (Viridiplantae) source. Combine the five FASTA-formatted sequences (rabbit, human, archaea, bacterial, and plant) in a single file (suggested name: TIM_5_FASTA.txt). This must be a simple text file with individual sequences separated by a blank line.

2. Multiple Sequence Alignment. Multiple sequence alignment is a tool to identify highly conserved residues in homologous proteins. A program called CLUSTALW will perform multiple sequence alignments on protein sets that are submitted in FASTA format. CLUSTALW is available as a command line program to be executed in a UNIX environment (not very user-friendly). Fortunately, the European Bioinformatics Institute has a web interface that performs CLUSTALW alignments: http://www.ebi.ac.uk/clustalw/.

[*Note to instructors:* The EBI asks that you alert them via email if you are using this resource for your course. Respond to http://www.ebi.ac.uk/support/.]

(a) Go to the EBI site and submit your text file containing the five triose phosphate isomerase sequences in FASTA format on the input form page. There are many options for refining the alignment, but for now, use the default values. Be sure to enter your email address. The output of CLUSTALW can be accessed in many ways. The simplest version will be described here, but you are encouraged to explore other options (especially JaiView). In the simple text output, the sequences are optimally aligned and annotated: Residues that are identical in all chains are marked with an asterisk (*), those that are highly conserved are marked with a colon (:), and those that are semiconserved are marked with a period (.). From your multiple sequence alignment, how many identical residues did you find? Identify the residues, using the single-letter amino acid abbreviations found in Table 4-1. Classify these "identity" sites as polar, nonpolar, acidic, and basic amino acids. Do most of the "identities" fall into a single class of amino acids? If you plan to continue to Part (b), keep your browser open or bookmark the results page. You can learn more about CLUSTALW at a tutorial provided by EBI (http://www.ebi.ac.uk/2can/tutorials/protein/clustalw.html).

(b) Figure 5-23 of the textbook shows a phylogenetic tree, which is described as "a diagram that indicates the ancestral relationships among organisms that produce the protein." There are useful tutorials on phylogenetic trees at the Los Alamos National Laboratories web site (http://www.hiv.lanl.gov/content/hiv-db/TREE_TUTORIAL/Tree-tutorial.html), at the EBI help page (http://www.ebi.ac.uk/clustalw/tree_frame.html), and at the NCBI site (http://www.ncbi.nlm.nih.gov/About/primer/phylo.html). Complete one or all these tutorials.

Scroll down the output page from the CLUSTALW program at EBI to the tree representations of the alignments. What is the difference between a cladogram and a phylogram tree? What do these trees tell you about triose phosphate isomerase from the five different species? The tree image on the EBI site is a dynamic image, meaning that you can't just cut and paste it. If you would like to capture this image, you can use the PrintScreen button on your computer and paste the image into a simple Paint program (with Mac OSX use the program Grab for screen capture).

3. One-Dimensional Electrophoresis. Electrophoresis is a laboratory technique that is used to separate proteins and DNA molecules on the basis of size and charge. The principles of one-dimensional electrophoresis (1DE) are explained in Sections 3-4B and 5-2D. Sodium dodecyl sulfate polyacrylamide gel electrophoresis (SDS-PAGE) is the most common form of 1DE. In this technique, proteins are mixed with a reducing agent (usually dithiothreitol or 2-mercaptoethanol) and a detergent (SDS), heated for 5 minutes, then separated on an acrylamide gel. You can explore 1DE at the Electrophoresis Simulation site (http://www.rit.edu/~pac8612/electro/Electro_Sim.html), which contains a Java applet that enables you to compare the migration of an unknown protein (you can choose from seven unknowns) with a series of standards. Visit the site and report the molecular weights you determine for each of the unknowns. Also, experiment with the controls (voltage, % acrylamide, animation

speed). You can learn how to use the applet by clicking on the "How To..." button.

4. Two-Dimensional Electrophoresis. Two-dimensional electrophoresis (2DE) is also described in Section 5-2D. In the first dimension, proteins are separated by isoelectric focusing; that is, they move to a position in a pH gradient according to their isoelectric point (pI, the pH at which the net charge of the protein is 0). Then they are separated according to molecular weight by SDS-PAGE in the second dimension, as described in Problem 3. For this part of the exercise you will need to retrieve the text file containing the triose phosphate isomerase sequences from five different species (Problem 1; TIM_5_FASTA.txt).

(a) The ExPASy Proteomics Server contains many tools for analyzing data from two-dimensional electrophoresis gels, as well as a catalog of gels themselves. Go to the Primary structure analysis tools section of the ExPASy Proteomics server at http://us.expasy.org/tools/#primary. These tools will compute and predict values for a protein based only on its primary structure.

Select the Compute pI/Mw tool (http://us.expasy.org/tools/pi_tool.html). Enter the sequence for one of the five triose phosphate isomerase proteins in the data entry box. Be sure to enter only one amino acid sequence and do not include its FASTA header (e.g., >gi|17389815|gb|AAH17917.1| Triosephosphate isomerase 1 [Homo sapiens]), because the program will attempt to calculate pI and Mw values for each term entered. Record the predicted pI and molecular weight for the protein. Repeat these steps for the other four protein sequences. How similar are the pI and Mw values for the triose phosphate isomerases from the five different organisms?

(b) Now try to find triose phosphate isomerase on a published gel. Go to the SWISS-2DPAGE search page (http://us.expasy.org/cgi-bin/ch2d-search-de) and search for triose phosphate isomerase. Select the entry for the human enzyme. How do the reported values for pI and Mw compare with the theoretical values you obtained in Part (a)? If the values are different, can you suggest an explanation?

(c) The best way to identify a protein spot on a 2DE gel is to use mass spectrometry (see Section 5-3D). The ExPASy Proteomics server has tools to predict fragmentation patterns based on the primary sequence of a protein and also to identify proteins based on fragmentation patterns from actual mass spectrographs. Go to the Protein identification and characterization tool section at the ExPASy Proteomics server (http://us.expasy.org/tools/#proteome). Select the "PeptideMass" tool. Paste in one of your triose phosphate isomerase sequences, verify that "trypsin" is selected as the enzyme, and leave the other options at their default settings. Click on the "Perform" button and record the four largest fragments that you would obtain if you digested the protein with trypsin. Do the same for the four other sequences. Are any of the fragmentation patterns identical between the species?

CHAPTER 6 VISUALIZING THREE-DIMENSIONAL PROTEIN STRUCTURES

There are a number of useful free visualization tools available on the Internet. Each has strengths and weakness. For this exercise you will use a tool called Rasmol that you can download from

http://www.bernstein-plus-sons.com/software/rasmol/README.html. Rasmol is available for Macintosh, Windows, and Linux/UNIX operating systems. Install Rasmol on your computer according to the instructions on the Rasmol site (http://www.bernstein-plus-sons.com/software/rasmol/INSTALL.html). As you go through the exercises below, you are encouraged to visit any of a number of excellent Rasmol tutorials on the Internet: Gale Rhodes's tutorial at the University of Southern Maine (http://www.usm.maine.edu/~rhodes/RasTut/), Eric Martz's tutorial at the University of Massachusetts Amherst (http://www.umass.edu/microbio/rasmol/rasquick.htm), and David Hackney's tutorial (adapted to HTML by Will McClure) at Carnegie Mellon University (http://www.bio.cmu.edu/Courses/BiochemMols/RasMolTutorial/RasTut.html). If you are interested in exploring additional visualization tools, you can obtain free software via the Internet for Protein Explorer, KING (Kinemage), DeepView, CN3D, Chime, Jmol, and BioEditor.

1. Obtaining Structural Information. Review the discussion of protein secondary structure (Section 6-1). Secondary structures in proteins include alpha helices, beta sheets, and beta turns.

(a) Many programs have been written to predict secondary structures based only on the primary structure (amino acid sequence) of a protein. Here is a list of such programs that are available online:

(1) **PredictProtein** (http://www.embl-heidelberg.de/predictprotein/predictprotein.html). You can request this site to predict secondary structure from seven different web servers online. If this site is available, it will enable you to complete this problem by clicking on two or more of the optional services.

(2) **JPred** (http://www.compbio.dundee.ac.uk/~www-jpred/). If you use the JPred server, be certain to check the box under #4 to avoid comparison to known PDB structure files.

(3) **NNPredict** (http://www.cmpharm.ucsf.edu/~nomi/nnpredict.html)

For this exercise, you will use the sequence of rabbit muscle triose phosphate isomerase, given here:

```
>1R2R:A TRIOSEPHOSPHATE ISOMERASE from rabbit muscle
APSRKFFVGGNWKMNGRKKNLGELITTLNAAKVPADTEVVCAPPTA
YIDFARQKLDPKIAVAAQNCYKVTNGAFTGEISPGMIKDCGATWVVLG
HSERRHVFGESDELIGQKVAHALSEGLGVIACIGEKLDEREAGITEKVV
FEQTKVIADNVKDWSKVVLAYEPVWAIGTGKTATPQQAQEVHEKLRG
WLKSNVSDAVAQSTRIIYGGSVTGATCKELASQPDVDGFLVGGASLK
PEFVDIINAKQ
```

Submit this sequence to two of the servers listed above. You may have to wait several minutes for the results. Compare the results you receive from the different servers. Can you identify segments where the predictions are not consistent between servers?

(b) The structure of rabbit muscle triose phosphate isomerase has been determined by X-ray crystallography (Section 6-2A). Go to the Protein Data Bank web server (http://www.rcsb.org/pdb) and search for 1R2R (the PDB ID for this protein). Once you reach the Structure Explorer page for 1R2R, click on the link for Sequence Details. Scroll down the page to the section entitled "Sequence and Secondary Structure." The results shown here for the secondary structure are based on an analysis of the actual (not

predicted) three-dimensional structure, using the principles developed by Kabsh and Sander [see http://www.rcsb.org/pdb/help-results.html#sequence_details and *Biopolymers* **22,** 2577–2637 (1983)]. The secondary structure assignments are H = helix; B = residue in isolated beta bridge; E = extended beta strand; G = 3_{10} helix; I = pi helix; T = hydrogen bonded turn; S = bend. Compare your predicted secondary structure results from Part (a) with the results presented on the PDB site.

Note that the Protein Data Bank web site is undergoing revision so that some of the web addresses and specific instructions provided here may vary somewhat. For example, in the new site, you can go to the 1R2R main page and access the secondary structure information by clicking on the "Sequence Details" link on the right side of the screen under the image of the structure.

(c) Follow the link on the PDB site for "Download File." You can download the file in a number of formats, but it is best to download the file in PDB format for use with Rasmol. Save the structure file as 1R2R.pdb on your computer (suggested folder: My Documents/PDB Files). Open the Rasmol program, then use the drop-down menu File..Open to open 1R2R.pdb. You will initially see a wireframe model that simply displays all the bonds in the structure as lines. Perform the following steps to get a more informative view:

Select Display..Cartoons from the drop-down menu.

Select Colours..Structure

Now you should be able to see the alpha helix and beta sheet structures in rabbit muscle triose phosphate isomerase. How many chains are shown in the structure? What is the dominant structural feature of this protein? How does your structure compare with Fig. 6-30*c*? Take time to experiment with the other drop-down menu options on Rasmol.

In addition to drop-down menus, Rasmol also has a "command line" window that enables you to select specific atoms or parts of a structure (amino acid residues, for example) and change the way they appear. There are more details in the tutorials listed above. Eric Martz has also prepared a helpful command list (http://www.umass.edu/microbio/rasmol/distrib/rasman.htm#chcomref). Bring up the command line window and note the effects of entering the following commands:

(1) select hetero and not water (selects nonprotein parts of the structure excluding water)

(2) spacefill (a van der Waals radius representation)

(3) color cpk (standard chemistry color scheme)

What heteroatoms do you see in this structure?

Hint: Click on them to see their identities in the command line. You can also find the hetero atoms in a structure by looking at the summary information page of the pdb file. Are any substrates or inhibitors represented in this structure? Now try more commands:

(4) select protein

(5) cartoon off

(6) select sheet

(7) wireframe 30

(8) spacefill 100 (these combined commands yield a ball-and-stick structure)

Can you see the sheet structure now? If not, type the command "cartoon." What do you see?

2. Exploring the Protein Data Bank. In the first problem, you visited the Protein Data Bank (PDB). You can explore that site in more detail now (see also Section 6-6). The Protein Data Bank web site (http://www.rcsb.org/pdb) is undergoing revision so that some of the web addresses provided in this set of exercises may become outdated. The new site incorporates a "Site Search" button that will enable you to search the PDB site for teaching materials and tutorials, in addition to the standard "Search" box that can be used to find specific structures in the PDB. You may be able to find any of the materials described below using the "Site Search" button. You can also use the extensive Help files that are accessible from any page on the new PDB site.

The PDB is a repository of macromolecular structures. Perhaps the most important skill for a PDB site user is the ability to find a particular structure. There is a query tutorial at http://www.rcsb.org/pdb/query_tut.html that provides instructions on finding structures in the PDB. On the new PDB site, the query tutorial is contained in the Help files.

Each structure in the PDB is assigned a PDB ID (or PDBid), a four-character alphanumeric code that uniquely identifies that structure. So, for example, 4HHB is a hemoglobin structure and 8GCH is a chymotrypsin structure. If you know the PDB ID, then you can use that to search the PDB. You can obtain PDB IDs from research publications. Most scientists who determine macromolecular structures are highly motivated to publish their findings in journals such as *Science, Nature, Journal of Biological Chemistry, Journal of Molecular Biology,* and *Protein Science.* These journals have an agreement with the PDB that requires authors to submit their structures to the PDB before the journal will publish the results. Most of the figures in the text that contain a molecular structure include the PDB ID (PDBid) for that structure in the figure legend.

For your first PDB search, you will find a PDB ID in a journal article, then find that structure on the PDB site. Go to the *Journal of Biological Chemistry* web site (http://www.jbc.org) and search for this paper:

Parthasarathy, S., Eaazhisai, K., Balaram, H., Balaram, P., and Murthy, M.R.N., Structure of *Plasmodium falciparum* triose-phosphate isomerase-2-phosphoglycerate complex at 1.1-Å resolution. *J. Biol. Chem.* **278,** 52461–52470 (2003).

Go to the footnotes section and find the four-character PDB ID code for the *Plasmodium* protein. Next go to the Protein Data Bank main page. Type the PDB ID in the search box and click on the "Search" button to find the "Structure Explorer" page for this enzyme. You can investigate the linked resources on this page by completing the following exercises:

(a) Download the PDB (structure) file for this protein to your computer (suggested folder: My Documents/PDB Files; suggested name: 1o5x.pdb). You will need this file for Problem 3, studying the protein's structure using Rasmol.

(b) Download the protein sequence in FASTA format.

(c) Find some still images of this protein on the PDB site. You can look under the "View Structure" link on the left side of the 1o5x "Structure Explorer" page. Scroll down the "View Structure" page until you come to "Still Images." To save an image,

just right click on it (Mac users: Control click) and select the option that lets you save the file (in Internet Explorer, the command is "Download image to disk"; in Firefox, the command is "Save Image As..").

(d) Return to the "Structure Explorer" page for 1o5x. Click on the "Other Sources" link on the left side of the page. Follow the links for 1o5x to the sites at PDBSum and the IMB Jena Image Library. Collect still images from each of these sites, and be sure to keep a record of where you found each image. Suggest ways in which you can use such downloaded images.

3. Using Rasmol. In Problem 2, you saved the PDB file for 1o5x, entitled "*Plasmodium falciparum* TIM complexed to 2-phosphoglycerate." You can use Rasmol to explore this structure, focusing on identifying secondary structures and looking at the active site.

(a) Open the Rasmol program on your computer. (If you have not installed it already, please see the opening paragraph of the exercises for Chapter 6). Open the file 1o5x.pdb. You will see a wireframe image: All the bonds in the PDB structure file are shown as thin wires, colored according to Corey-Pauling-Kultun (CPK) coloring rules (oxygen is red, nitrogen is blue, hydrogen is white, and carbon is gray). There are seven drop-down menus in Rasmol: File, Edit, Display, Colours, Options, Export, and Help. Spend a few minutes trying each command in each of the menus. Perform the following operations:

(1) **File..Information.** This identifies the protein structure by name and PDB ID.

(2) **Display..Backbone.** This shows the protein backbone; the bonds actually connect alpha carbons.

(3) **Display..Cartoon.** This shows an image of the protein that clearly displays helices and sheets. Leave your image in cartoon format and move to the Colours menu.

(4) **Colours..Structure.** This shows alpha helices in magenta, beta sheets in yellow, and turns in pale blue.

(5) **Options..Labels.** This command labels all selected atoms. The view will not look good right now because all atoms are selected. If you select a single atom and then use the label command, you can attach text to that atom (call it by its name, or give it another label such as "inhibitor").

(6) **Export..GIF.** This function enables you to export a still image of the structural view you just created. Export your image as 1o5x.gif (store it somewhere accessible, such as the Desktop), then view it in a simple image viewer (Paint in Windows; Preview in OSX).

(7) **HELP..User Manual.** This is really a critical tool for using Rasmol. In order for this to work, the help file (Rasmol.hlp in Windows) must be stored in the same directory as the Rasmol program. Even then, Rasmol may ask you to find it on your computer system. The Help file is searchable. The Table of Contents

has links to major features of the program, including frequently used items such as Command Reference, Atom Expressions, and Colour Schemes. The manual is also available at http://info.bio.cmu.edu/Courses/BiochemMols/RasFrames/TOC.HTM.

(b) Open the Rasmol command line window, if it is not already visible. You will use this window to enter specific commands for viewing the structure, including highlighting the small molecules (3-hydroxypyruvic acid and 2-phosphoglyceric acid) that are bound to triose phosphate isomerase in the 1o5x file. But first you'll need to learn a little bit about viewing a PDB file.

Go to the Structure Explorer page for 1o5x at the PDB website (http://www.rcsb.org/pdb/cgi/explore.cgi?job=summary&pdbId=1O5X&page=&pid=190431099321055). Click on "Download/Display File" on the left side of the page. Under "Display the Structure File," select the "HTML" option. This shows you the complete PDB file. There is a lot of information in this file, but you'll only look at a few items. For more details, you can go to the PDB Format Description page (http://www.rcsb.org/pdb/docs/format/pdbguide2.2/guide2.2_frame.html) or click on any links you see on the HTML page for 1o5x.

Each line in a PDB file is called a "record," and the first six characters on that line tell what kind of "record" it is. In your browser, search for SEQRES. As explained in the PDB Format Description, SEQRES records contain the amino acid or nucleic acid sequence of residues in each chain of the macromolecule. Hence you can see the sequence of your protein there. For 1o5x, the first few lines of the SEQRES section are given in Table I (*below*). Each line contains 13 amino acid residues listed by their three-letter abbreviations. So residue #27 in chain A is PHE (phenylalanine). The 12th character (counting spaces) in each record is a chain identifier. If a protein contains more than one polypeptide chain, the chains are identified with a letter (in this file, there are two chains: A and B).

Anything in a PDB file that is not either protein or nucleic acid is considered a heterogen atom and is referred to with the prefix "het." So HETNAM is the label for a record that contains the name of a nonprotein, non-nucleic acid group. Search the HTML version of the 1o5x file for "HETNAM." What are the heterogen groups in this structure?

(c) Now you can display the heterogen groups in 1o5x. Go to the command line window of Rasmol and enter the command "select hetero and not water." This command selects all the heterogen atoms excluding water. The current view of 1o5x should be a cartoon diagram of the structure. To show the heterogen molecules differently, enter the command "spacefill on" and then "color cpk." This creates a space-filling representation of the two molecules and colors them according to the CPK conventions.

(d) As the last part of this exercise, you will display the active site residues, based on Figure 4 of the primary citation for this structure (see Problem 2). Figure 4a shows three residues that interact with the 2-phosphoglycerate: glutamate 165, lysine 12, and histidine 95. Select these residues by entering the command

Table I:

SEQRES	1 A 248	MET	ALA	ARG	LYS	TYR	PHE	VAL	ALA	ALA	ASN	TRP	LYS	CYS
SEQRES	2 A 248	ASN	GLY	THR	LEU	GLU	SER	ILE	LYS	SER	LEU	THR	ASN	SER
SEQRES	3 A 248	PHE	ASN	ASN	LEU	ASP	PHE	ASP	PRO	SER	LYS	LEU	ASP	VAL
SEQRES	4 A 248	VAL	VAL	PHE	PRO	VAL	SER	VAL	HIS	TYR	ASP	HIS	THR	ARG
SEQRES	5 A 248	LYS	LEU	LEU	GLN	SER	LYS	PHE	SER	THR	GLY	ILE	GLN	ASN

"select lys12,his95,glu165" in the Rasmol command line window. Then use the drop-down menu in the structure window to show the residues in a ball-and-stick format (Display..Ball & Stick). Finally, enter the command "color cpk" so that you can distinguish the atoms of the structure.

To get a better look at the intermolecular interactions, you can zoom in on the structure by pressing the "Shift" key as you move the mouse. To zoom in, drag the mouse up in the structure window. To move the image side-to-side or up-and-down, use the right-click on your mouse (Mac users: use the "Option" key) and drag the image where you want to go. Using a combination of Shift-mouse and right-click (or Option-Mouse), you can get a close-up view of the binding site for 2-phosphoglycerate.

To complete this exercise, identify and display additional residues that interact with the 2-phosphoglycerate in 1o5x. A couple of hints: Use Figures 3 and 4 from the primary citation. Also, for some reason, Rasmol won't select 2-phosphoglycerate using the command "select 2PG," but you can select it using "select 4400", which is the second way 2PG is identified in 1o5x.

4. Protein Families. The goal of this exercise is to identify a protein that shares structural homology with triose phosphate isomerase (as seen in PDB ID 1o5x) but catalyzes a different reaction. Because 1o5x is a fairly recently described structure, it is not well documented in other structural databases. Therefore you will use an earlier PDB entry on triose phosphate isomerase from *Plasmodium falciparum,* PDB ID 1ydv. You will explore two resources, CATH and SCOP (see Section 6-6).

(a) CATH. You can access the CATH homepage at http://www. biochem.ucl.ac.uk/bsm/cath/, but perhaps the easiest way to get to these resources is through the "Other Sources" link on the Protein Data Bank Structure Explorer page for 1ydv. Click on the CATH link to open a new browser window containing the CATH main page. Review the introduction to CATH before proceeding.

CATH describes proteins in a hierarchical fashion. What information is found in CATH under the following headings?

(1) Class

(2) Architecture

(3) Topology

(4) Homologous superfamily

Return to the PDB "Other Sources" page for 1ydv. Click on the "CATH" link for 1ydv (right side of the screen). Click on the "1ydvA0" link to find a list of proteins that are members of this superfamily. Explore the page to find five enzymes in this superfamily that catalyze reactions different from the reaction catalyzed by triose phosphate isomerase.

(b) SCOP. Once again, you can go directly to the SCOP homepage (http://scop.mrc-lmb.cam.ac.uk/scop/), but it may be easier to get to these resources in SCOP by going through the "Other Sources" link on the Protein Data Bank Structure Explorer page for 1ydv. Click on the "SCOP" link to open a new browser window containing the SCOP main page. Read the synopsis before continuing. Then return to the PDB page and click on the "SCOP" link for 1ydv.

SCOP provides a lineage for each protein that is classified. Follow the lineage links at the Fold level to identify five proteins that are related to triose phosphate isomerase. Are any of these proteins also in your list for Problem 4(a)?

(c) The final part of this exercise is to identify other resources that help you find proteins related to triose phosphate isomerase from *Plasmodium falciparum*. You are encouraged to follow other links from the PDB Other Sources page for 1ydv. You may also be able to find other resources by searching the Internet using the PDB ID codes. List and summarize three other resources that you find.

CHAPTER 12 ENZYME INHIBITORS AND RATIONAL DRUG DESIGN

1. Dihydrofolate Reductase. In Section 12-4, enzyme inhibitors are identified as the second largest class of drugs. The first exercise for this chapter is to find the structure of an enzyme that has a competitive inhibitor bound to its active site. First, look in the Protein Data Bank for the enzyme dihydrofolate reductase (DHFR; see Section 22-3A). How many structures do you find for DHFR? Since there are too many to analyze all at once, you can limit your search to DHFR complexed with an inhibitor. On the PDB results page, select "Refine your query" from the "Pull down to select option" menu. Enter the word "inhibitor" in the blank and click on the option for a full text search. Now go down the list to find the first human DHFR enzyme complexed with an inhibitor.

Download the PDB file and use Rasmol to visualize this structure (see the exercises for Chapter 6). Display the protein in Ribbons format, colored by structure. Then select the inhibitor (using the command "select hetero and not water") and show it as a space-filling model with CPK coloring.

2. HIV Protease. There is a description of structure-based drug design in Section 12-4A, and Box 12-4 describes two drugs that are directed at the protease from human immunodeficiency virus (HIV protease). In fact, these were some of the very first drugs to be created using structure-based drug design. The first drugs directed against HIV targeted the reverse transcriptase (Boxes 12-4 and 24-2), but the virus quickly developed resistance to these drugs. Researchers in academia and at pharmaceutical companies began studying HIV protease when initial results indicated that it might be useful as an additional drug target to delay the onset of AIDS. You are encouraged to look for recent reviews on PubMedCentral (http://www.pubmedcentral.gov) or at your local library (try journals such as *Current Opinion in Chemical Biology* and *Trends in Biochemical Sciences*) to find more information on structure-based drug design.

(a) Go to Genbank (http://www.ncbi.nlm.nih.gov/) and search for the protein sequence of HIV protease. Many mutant forms of HIV protease have been sequenced, so you may find a mutant sequence. To get the native sequence, you'll have to reverse the mutation. For example, if you have an L90M mutant sequence, you'll simply need to replace the M (methionine) at position 90 with L (leucine). Save the sequence to a file in FASTA format.

(b) Do a BLAST search with this sequence to find homologous proteins. HIV protease is a member of the aspartic protease family, which includes pepsin (Box 12-4). What other proteases also appear in your BLAST search?

(c) Now search the Protein Data Bank for HIV protease structures. How many do you find? Rather than searching through these structures to find particular inhibitors, start a new search

using the name of an inhibitor, such as saquinavir or ritonavir, both of which are mentioned in Box 12-4. Search for those terms. Does either appear in the Protein Data Bank?

(d) To take a closer look at the HIV protease complex with ritonavir, download the PDB file, 1HXW, and open it in Rasmol. Use the drop-down menu to display the protein structure as a cartoon. Then color by Structure.

The drug ritonavir is identified on the PDB Structure Explorer site for 1HXW by the abbreviation "RIT." Bring up the Rasmol command line window. Type in "select RIT" and hit return. Then type in "wireframe 100." You should see the ritonavir in wireframe format with CPK coloring. Now you can get a better look at how ritonavir affects HIV protease. Look at the picture of ritonavir in Box 12-4. Identify the feature (in red) that mimics the geometry of the transition state. Find the tetrahedral carbon atom in Rasmol. Note that when you click on an atom in the Rasmol structure window, the atom is identified in the command line window. You may want to make the protein disappear temporarily by entering "select protein; cartoon off" in the command line window (you can make it reappear with the command "cartoon on"). When you click on the correct carbon atom (in the backbone between the two phenyl groups), it will be listed in the command line window: Rasmol > Atom: C13 1865 Hetero: RIT 301.

Now identify the two aspartate residues in HIV protease that interact with ritonavir. You can use the within command to do this: Enter "select asp and within (5.0, atomno=1865)." This command selects all aspartate residues within 5.0 angstroms of atom number 1865 in the PDB structure file. Which aspartate residues are close to the carbon atom you identified above? Identify these residues by number and chain.

(e) Find the names of additional HIV protease inhibitors (using Google, for example) and see whether they occur in structures in the PDB. Explore the structures using Rasmol and identify interactions between the hydrophobic side chains on the inhibitors and the surface of HIV protease.

3. Pharmacogenomics and Single Nucleotide Polymorphisms. PubMedCentral contains an excellent article that reviews the role of pharmacogenomics in medicine and drug discovery. Go to http://www.pubmedcentral.gov and search for pharmacogenomics review. One of the articles that you should find is Chiche, J.-D., Cariou, A., and Mira, J.-P., Bench-to-bedside review: Fulfilling promises of the Human Genome Project, *Critical Care* **6,** 212–215 (2002). Oak Ridge National Laboratories (ORNL) also has an excellent site on pharmacogenomics at http://www.ornl.gov/sci/techresources/Human_Genome/medicine/pharma.shtml. Pharmacogenomics is a very broad and rapidly expanding field. This exercise is a general guide to introduce you to some relevant database and literature resources.

(a) Using the article above and the ORNL site, define the following terms: pharmacogenomics, single nucleotide polymorphism (SNP), and cytochrome P450. What is the significance of SNPs in the function of cytochrome P450 and drug metabolism (see also Section 12-4D)?

(b) Look in Entrez (http://www.ncbi.nlm.nih.gov) for cytochrome P450. You will have a number of options to explore. Explore PubMed, PubMed Central, Books, and OMIM to find documents relating to cytochrome P450 and single nucleotide polymorphisms.

(c) Return to the "Entrez" page. This time, explore the databases for information on cytochrome P450 and single nucleotide polymorphisms. Suggested sites are the Protein sequence database and SNP database. Describe the results of your exploration.

CHAPTER 13 METABOLIC ENZYMES, MICROARRAYS, AND PROTEOMICS

Here is a list of a few useful and reliable online resources about metabolism:

The Biology Project at the University of Arizona:

http://www.biology.arizona.edu/biochemistry/biochemistry.html

Metabolic Pathways of Biochemistry at George Washington University:

http://www.gwu.edu/~mpb/

Chemistry Biology Information Center at ETH Zurich:

http://www.infochembio.ethz.ch/links/en/biochem_metabolismus.html

Main Metabolic Pathways on the Internet:

http://home.wxs.nl/~pvsanten/mmp/main.htm

Kyoto Encyclopedia of Genes and Genomes (KEGG):

http://www.genome.ad.jp/kegg/metabolism.html

Enzyme Structures Database:

http://www.ebi.ac.uk/thornton-srv/databases/enzymes/

1. Metabolic Enzymes. In Chapter 12, you looked at the role of enzyme inhibitors as drugs. In this exercise, you will use some online resources to learn more about the enzymes involved and the pathways that are affected.

(a) Look in your textbook for dihydrofolate reductase (DHFR). Write out the reaction catalyzed and the pathway involved (see Section 22-3B and Box 22-1).

(b) Go to the Enzyme search page at the KEGG site (http://www.genome.jp/dbget-bin/www_bfind?enzyme) and search for the enzyme (by name). Now look for links that lead to pathways that include DHFR. Where does DHFR appear in each pathway? Is this consistent with your findings in the textbook?

(c) Go to the Enzyme Structures Database (http://www.ebi.ac.uk/thornton-srv/databases/enzymes/) and find dihydrofolate reductase. Every enzyme with a known reaction is classified in a hierarchy: EC #.#.#.# where # represents a number (see Section 11-1A). What is the enzyme classification for dihydrofolate reductase? Explore the hierarchy for dihydrofolate reductase. What does each of the numbers in the hierarchy represent?

2. Microarrays. Malcolm Campbell at Davidson University has done a remarkable job of making a high-end technology (microarrays; see Section 13-4C) available to researchers (students and faculty) at the undergraduate level.

(a) Visit Malcolm Campbell's site at Davidson and go through the following web exercise: DNA Microarray Methodology (a FLASH animation) at

http://www.bio.davidson.edu/courses/genomics/chip/chipQ.html.

(b) For more advanced background on microarrays, visit Manish Patel's microarray tutorial at

http://www.ucl.ac.uk/oncology/MicroCore/HTML_resource/tut_frameset.htm.

(c) Visit PubMed Central or PubMed and find a review article on the use of microarrays to study one of the following diseases: breast cancer, lymphoma, hypertension, atherosclerosis, or a disease that is of particular interest to you. Provide the citation for the article you found and explain how microarray technology was applied.

3. Proteomics. Proteomics (Section 13-4D) is the study of all the proteins expressed in an organism or tissue under a specific set of conditions. To gain a broader understanding of proteomics, read the following article: Graves, P.R. and Haystead, T.A., "Molecular biologist's guide to proteomics," *Microbiol. Mol. Biol. Rev.* **66,** 39–63 (2002), which is available at PubMed Central. After reading this article and reviewing other available resources, answer the following questions:

(a) What analytical techniques are used most commonly to separate proteins in proteomics?

(b) How can proteins be identified with certainty in proteomics?

(c) What is meant by the phrase "the dynamic range of protein expression"? You will need to find an additional source to answer this question; it is not addressed in the "Molecular Biologist's Guide to Proteomics." Can you find quantitative values in the literature to further define this term? Leigh Anderson has published some informative articles on the human plasma proteome; this would be a good place to look.

4. Two-Dimensional Gel Electrophoresis. A major proteomics tool is two-dimensional gel electrophoresis (2DE; see Problem 4 in the Bioinformatics exercises for Chapter 5). One of the techniques you encountered in Problem 3(a) was 2DE. One of the best bioinformatics sites on the web is the ExPASy server in Geneva, Switzerland (http://www.expasy.org), which has a database/tools site for 2DE called Swiss-2DPAGE.

(a) Open the Swiss-2DPAGE site (http://www.expasy.org/ch2d/). Follow the link to search the site by description and find out how many human proteins are catalogued there (enter "human" as the search keyword).

(b) Return to the search page and search for dihydrofolate reductase. How many listings do you find? Is a human version of DHFR catalogued here?

(c) Start over and search for *E. coli* proteins (enter "E. coli" as the search keyword). Note that if you use "E. coli," you'll get much different results than if you just use "coli." Now look for dihydrofolate reductase in *E. coli*. (Use the find function in your browser to search the *E. coli* results.) Follow the links to *E. coli* DHFR to answer the following questions:

(1) What is the theoretical molecular weight (Mw) and isoelectric point (pI) for *E. coli* DHFR?

(2) What actual values were obtained by 2DE?

(3) What peptide fragment was used to identify *E. coli* DHFR? Use the BLAST server to find out where this peptide is located in the *E. coli* DHFR sequence.

(d) Search the Protein Data Bank for structures of DHFR. Has the three-dimensional structure of this enzyme been determined? Keep searching to see if there are three-dimensional structures of *E. coli* DHFR complexed with methotrexate. What is methotrexate and how is it used in the treatment of disease? (Consult Box 22-1 and other resources for the answer.)

ANSWERS TO BIOINFORMATICS EXERCISES

CHAPTER 3

1. Finding Databases
(a) Genbank http://www.ncbi.nlm.nih.gov/entrez/
European Bioinformatics Institute http://www.ebi.ac.uk/
GenomeNet in Japan http://www.genome.jp/
(b) National Center for Genome Resources http://www.ncgr.org/pathdb/
NCBI Human Genome Resources http://www.ncbi.nlm.nih.gov/genome/guide/human/
The Institute for Genomic Research http://www.tigr.org/
(c) BLAST: http://www.ncbi.nlm.nih.gov/BLAST/
Taxonomy: http://dictionary.reference.com/search?q=taxonomy, http://mclibrary.nhmccd.edu/taxonomy/taxonomy.html,
http://www.hyperdictionary.com/dictionary/taxonomy
Gene ontology: http://www.yeastgenome.org/help/glossary.html, http://genome-www5.stanford.edu/help/glossary.shtml
Phylogenetic trees: http://en.wikipedia.org/wiki/Phylogenetic_tree, http://encyclopedia.thefreedictionary.com/phylogenetic tree
Multiple sequence alignment: http://www.ncbi.nlm.nih.gov/Education/BLASTinfo/glossary2.html, http://www.bioinfo.rpi.edu/
~zukerm/Bio-5495/multalign-html/node1.html

2. TIGR
(a) Peterson, J.D., Umayam, L.A., Dickinson, T.M., Hickey, E.K., and White, O., The comprehensive microbial resource, *Nucleic Acids Res.* **29,** 123–125 (2001). [Found on http://www.tigr.org/tigr-scripts/CMR2/CMRHomePage.spl or Nucleic Acids Research, 29:1 (2001), 123–125.]

(b) Three
(c) *Pseudomonas aeruginosa* PAO1, *Pseudomonas putida* KT2440, *Pseudomonas syringae* DC3000. [Found on http://www.tigr. org/tigr-scripts/CMR2/CMRHomePage.spl using the Genome Pages drop-down menu.]
(d) Nelson, K.T., *et al.,* Complete genome sequence and comparative analysis of the metabolically versatile *Pseudomonas putida* KT2440, *Environmental Microbiology* **12,** 799–808 (2002).
(e) 529 fragments; average size 11668 bp. Go to the Restriction Digest Tool (http://www.tigr.org/tigr-scripts/CMR2/restrict_display.pl).
(f) Select from the list of projects (some incomplete) at http://www.tigr.org/tdb/euk/.

3. Analyzing a DNA Sequence

(a) The Start codon begins at position 7. The first in-frame Stop codon that defines an ORF of at least 300 nucleotides is TAA (the ORF is 645 nucleotides in all). The Start and Stop codons are shown in boldface type:

TACGCA**ATG**CGTATCATTCTGCTGGGCGCTCCGGGCGCAGGTAAAGGTACTCAGGCTCAATTCATCATGGAGAAATACGGCATTCCGCAAATC
TCTACTGGTGACATGTTGCGCGCCGCTGTAAAAGCAGGTTCTGAGTTAGGTCTGAAAGCAAAAGAAATTATGGATGCGGGCAAGTTGGTGAC
TGATGAGTTAGTTATCGCATTAGTCAAAGAACGTATCACACAGGAAGATTGCCGCGATGGTTTTCTGTTAGACGGGTTCCCGCGTACCATTCCT
CAGGCAGATGCCATGAAAGAAGCCGGTATCAAAGTTGATTATGTGCTGGAGTTTGATGTTCCAGACGAGCTGATTGTTGAGCGCATTGTCGG
CCGTCGGGTACATGCTGCTTCAGGCCGTGTTTATCACGTTAAATTCAACCCACCTAAAGTTGAAGATAAAGATGATGTTACCGGTGAAGAGCT
GACTATTCGTAAAGATGATCAGGAAGCGACTGTCCGTAAGCGTCTTATCGAATATCATCAACAAACTGCACCATTGGTTTCTTACTATCATAAAG
AAGCGGATGCAGGTAATACGCAATATTTTAAACTGGACGGAACCCGTAATGTAGCAGAAGTCAGTGCTGAACTGGCGACTATTCTCGGT**TAA**T
TCTGGATGGCCTTATAGCTAAGGCGGTTTAAGGCCGCCTTAGCTATTTCAAGTAAGAAGGGCGTAGTACCTACAAAAGGAGATTTGGCATGAT
GCAAAGCAAACCCGGCGTATTAATGGTTAATTTGGGGACACCAGATGCTCCAACGTCGAAAGCTATCAAGCGTTATTTAGCTGAGTTTTTGAG
TGACCGCCGGGTAGTTGATACTTCCCCATTGCTATGGTGGCCATTGCTGCATGGTGTTATTTTACCGCTTCGGTCACCACGTGTAGCAAAACTT
TATCAATCCGTTTGGATGGAAGAGGGCTCTCCTTTATTGGTTTATAGCCGCCGCCAGCAGAAAGCACTGGCAGCAAGAATGCCTGATATTCCT
GTAGAATTAGGCATGAGCTATGGTTCAC

(b) The correct reading frame is listed as 5′3′ Frame 1. The protein sequence, which begins with the first Met (M) residue, is

MRIILLGAPGAGKGTQAQFIMEKYGIPQISTGDMLRAAVKAGSELGLKAKEIMDAGKLVTDELVIALVKERITQEDCRDGFLLDGFPRTIPQADAMKEA
GIKVDYVLEFDVPDELIVERIVGRRVHAASGRVYHVKFNPPKVEDKDDVTGEELTIRKDDQEATVRKRLIEYHQQTAPLVSYYHKEADAGNTQYFKLD
GTRNVAEVSAELATILG

(c) The protein is adenylate kinase. The source is *Yersinia pestis* CO92.

4. Sequence Homology

(a) The entry for a homologous yeast protein should look similar to this:

```
gi|6320432|ref|NP_010512.1|  Adk1p [Saccharomyces cerevisiae...   198   5e-50
```

The entry for a homologous human protein should look similar to this:

```
gi|33150692|gb|AAP97224.1|   GTP:AMP phosphotransferase [Homo...  160   2e-38
```

Both the yeast and human sequences are homologous to the *Yersinia pestis* sequence; their identity values are greater than 25%.
(c) The default matrix is known as BLOSUM 62 (the name BLOSUM comes from BLOcks SUbstitution Matrix). Other BLOSUM matrices and PAM (Percent Accepted Mutation) matrices are also available. Different substitution matrices yield slightly different lists of homologous sequences. Note that the same homologous yeast and human sequences should be detected, but their Scores and E values will vary with the substitution matrix used.

5. Plasmids and Cloning

(a) There are many Rma enzymes; one is Rma523I. Its recognition sequence is 5′-TTCGAA-3′. The expected and actual frequencies in *Bacillus halodurans C*-125 are about 300.
(b) The pBR322 sequence is available at http://www.ncbi.nlm.nih.gov/entrez/viewer.fcgi?db=nucleotide&val=208958. Plasmids normally contain genes that encode enzymes conferring antibiotic resistance. pBR322 encodes an enzyme that confers resistance to tetracycline. The plasmid also includes a gene for a β-lactamase that confers resistance to ampicillin.
(c) pBR322 contains 4362 base pairs; HaeIII cuts at 22 sites (see Table 2).
(d) See Section 3-4A for an explanation and examples of restriction enzymes that produce blunt and sticky ends.
(e) The hit count is 1448. The results page begins as follows:

```
#=======================================
#
# Sequence: from: 1 to: 4361
# HitCount: 1448
#
# Minimum cuts per enzyme: 1
# Maximum cuts per enzyme: 2000000000
# Minimum length of recognition site: 4
# Blunt ends allowed
# Sticky ends allowed
# DNA is linear
# Ambiguities allowed
#
#=======================================
```

Start	End	Enzyme_name	Restriction_site	5prime	3prime	5primerev	3primerev
5	8	FatI	CATG	4	8	.	.
5	8	CviAII	CATG	5	7	.	.
5	8	NlaIII	CATG	8	4	.	.
15	18	AluI	AGCT	16	16	.	.

When the Minimum recognition site length is changed to six nucleotides, the hit count is smaller (456) because the larger the recognition site, the less likely it is to occur.

(f) BamHI generates one fragment. AvaI generates one fragment; its restriction site is six nucleotides (CYCGRG, where R represents a purine and Y represents a pyrimidine). Eco47III generates four fragments; its restriction site is six nucleotides (AGCGCT).

(g) The three enzymes generate six fragments. The sizes of the fragments are 121, 141, 281, 304, 648, and 2867 nucleotides.

```
# Sequence: SYNPBR322     from: 1  to: 4361
# HitCount: 6
#
# Minimum cuts per enzyme: 1
# Maximum cuts per enzyme: 2000000000
# Minimum length of recognition site: 6
# Blunt ends allowed
# Sticky ends allowed
# DNA is linear
# Ambiguities allowed
#
#=======================================
```

Start	End	Enzyme_name	Restriction_site	5prime	3prime	5primerev	3primerev
232	237	Eco47III	AGCGCT	234	234	.	.
375	380	BamHI	GGATCC	375	379	.	.
494	499	Eco47III	AGCGCT	496	496	.	.
775	780	Eco47III	AGCGCT	777	777	.	.
1425	1430	AvaI	CYCGRG	1425	1429	.	.
1727	1732	Eco47III	AGCGCT	1729	1729	.	.

(h) Search Entrez for the pUC18 cloning vector. It should be at or near the end of the nucleotide sequences that are returned. Copy the sequence in FASTA format and repeat the steps in Problem 5(g). At the RESTRICT site, specify a six-nucleotide minimum recognition site length and enter the names of the three enzymes. Each enzyme makes one cut on the 2686-bp plasmid, but the three cuts are close together: AvaI at position 256, BamHI at position 251, and PstI at position 273.

(i) At the RESTRICT site, select "all" enzymes and set the minimum recognition site length to 8 nucleotides. This yields a set of restriction enzymes that generate a reasonable number of fragments (38). From this list you can select a few enzymes that together will generate 10 fragments, for example, BsaXI (1 fragment), BsiYI (6 fragments), BglI (2 fragments), and Sse8387I (1 fragment).

CHAPTER 5

1. Obtaining Sequences from BLAST

(a) One of the first entries on the results page should be:

```
gi|4507645|ref|NP_000356.1|   triosephosphate isomerase 1 [Ho...   493   e-138
```

NP_000356.1 is a human ortholog of rabbit muscle triose phosphate isomerase. The two proteins have 98% sequence identities. Another human homolog is:

```
gb|AAH17917.1| Triosephosphate isomerase 1 [Homo sapiens]
```

```
>gi|17389815|gb|AAH17917.1| Triosephosphate isomerase 1 [Homo sapiens]
MAPSRKFFVGGNWKMNGRKQSLGELIGTLNAAKVPADTEVVCAPPTAYIDFARQKLDPKIAVAAQNCYKVT
NGAFTGEISPGMIKDCGATWVVLGHSERRHVFGESDELIGQKVAHALAEGLGVIACIGEKLDEREAGITEK
VVFEQTKVIADNAKDWSKVVLAYEPVWAIGTGKTATPQQAQEVHEKLRGWLKSNVSDAVAQSTRIIYGGSV
TGATCKELASQPDVDGFLVGGASLKPEFVDIINAKQ
```

(b) Some possible choices for the five sequences are:

```
pdb|1R2T|B Chain B, Crystal Structure Of Rabbit Muscle Trio...

gb|AAH17917.1| Triosephosphate isomerase 1 [Homo sapiens]

ref|NP_614947.1| Triosephosphate isomerase [Methanopyrus kandleri AV19]

ref|ZP_00144330.1| Triosephosphate isomerase [Fusobacterium...

emb|CAC14917.1| triosephosphat-isomerase [Triticum aestivum]
```

2. Multiple Sequence Alignment

(a) The alignment of the five sequences selected in Problem 1(b) is shown below.

```
    CLUSTAL W (1.82) multiple sequence alignment

gi|136066|sp|P00939|TPIS_RABIT      -APSRKFFVGGNWKMN-GRKKNLGELITTLNAAKVPAD--TEVVCAPPTA 46
gi|17389815|gb|AAH17917.1|          MAPSRKFFVGGNWKMN-GRKQSLGELIGTLNAAKVPAD--TEVVCAPPTA 47
gi|11124572|emb|CAC14917.1|         --MGRKFFVGGNWKCN-GTVEQVESIVNTLNAGQIASTDVVEVVVSPPYV 47
gi|34763378|ref|ZP_00144330.1|      ---MRRLVIAGNWKMYKNNKEAVETLTQLKNLTKDIKN--VDIVIGAPFT 45
gi|20095100|ref|NP_614947.1|        --MLRVPPVIVNFKAY-SEAVGENALRLARVAAEVSEETGVEVGICPPHV 47
                                      *    :  *:*   .       :       :      .:: .*. .

gi|136066|sp|P00939|TPIS_RABIT      YIDFARQKLD-PKIAVAAQNCYKVTNGAFTGEISPGMIKDCGATWVVLGH 95
gi|17389815|gb|AAH17917.1|          YIDFARQKLD-PKIAVAAQNCYKVTNGAFTGEISPGMIKDCGATWVVLGH 96
gi|11124572|emb|CAC14917.1|         FLPTVKGKLR-PEIQVAAQNCWVKKGGAFTGEVSAEMLVNLGVPWVILGH 96
gi|34763378|ref|ZP_00144330.1|      CLSDAVKTVEGSNVKIAAENVYPKIEGAYTGEISPKMLKDIGVTYVILGH 95
gi|20095100|ref|NP_614947.1|        DLRDVVREVG-DEVTVLAQAVDAAEPGGRTGHVTPEMVVEAGADGTLLNH 96
                                      :  .   :   :: : *:      *. **.::. *: : *.  .:*.*

gi|136066|sp|P00939|TPIS_RABIT      SERRHVFGESDELIGQKVAHALSEGLGVIACIGEKLDEREAGITEKVVFE 145
gi|17389815|gb|AAH17917.1|          SERRHVFGESDELIGQKVAHALAEGLGVIACIGEKLDEREAGITEKVVFE 146
gi|11124572|emb|CAC14917.1|         SERRSLMGESSEFVGEKVAYALAQGLKVIACVGETLEQREAGSTMAVVAE 146
gi|34763378|ref|ZP_00144330.1|      SERREYFKESDEFINQKVKAVLEIGMKPILCIGEKLEERESGKTFEVLSK 145
gi|20095100|ref|NP_614947.1|        SERRMLLED----LKDVCRACINEGLLTIVCASDALAARAAG-------- 134
                                      ****  ::     ::       : *:  * * .: *  * :*

gi|136066|sp|P00939|TPIS_RABIT      QTK-VIADNVKDWS-KVVLAYEPVWAIGTGKTATPQQAQEVHEKLRGWLK 193
gi|17389815|gb|AAH17917.1|          QTK-VIADNAKDWS-KVVLAYEPVWAIGTGKTATPQQAQEVHEKLRGWLK 194
gi|11124572|emb|CAC14917.1|         QTK-AIADKIKDWT-NVVVAYEPVWAIGTGKVASPAQAQEVHANLRDWLK 194
gi|34763378|ref|ZP_00144330.1|      QIKGGLADLSKEEAGKVIIAYEPVWAIGTGKTATPEMAQETHKAVRNVLA 195
gi|20095100|ref|NP_614947.1|        ----ALSPH--------AVAVEPPELIGTGTPVSKADPEVVERSV----- 167
                                        ::            :* **  ****. .:  .: .. :

gi|136066|sp|P00939|TPIS_RABIT      SNVSDAVAQSTRIIYGGSVTGATCKELASQPDVDGFLVGGASLKP----- 238
gi|17389815|gb|AAH17917.1|          SNVSDAVAQSTRIIYGGSVTGATCKELASQPDVDGFLVGGASLKP----- 239
gi|11124572|emb|CAC14917.1|         TNVSPEVAESTRIIYGGSVTGASCKELAAQPDVDGFLVGGASLKP----- 239
gi|34763378|ref|ZP_00144330.1|      EMFGKDIADKMIIQYGGSMKPENAKDLLSQEDIDGGLVGGASLKAD---- 241
gi|20095100|ref|NP_614947.1|        -EVVKEVSEETAVLCGAGITDGSDVRAAVELGADGVLVASGVVLADDPKE 216
                                      .  :::.  :  *...:  .    : . ** **... : .

gi|136066|sp|P00939|TPIS_RABIT      EFVDIINAKQ---- 248
gi|17389815|gb|AAH17917.1|          EFVDIINAKQ---- 249
gi|11124572|emb|CAC14917.1|         EFIDIINAAAVKSA 253
gi|34763378|ref|ZP_00144330.1|      SFFEIIKAGN---- 251
gi|20095100|ref|NP_614947.1|        ALLDLISGLE---- 226
                                      :.::*..
```

Identities: R4, N11, K13, P44, A62, G71, T74, G75, M81, G87, L93, H95, S96, E97, R98, R99, G119, I123, C125, L130, R133, G136, A162, E164, P165, I169, G170, T171, G172, G209, D227, G228, L230, V231, I240

Polar (14, including Gly): N11, G71, T74, G75, G87, S96, G119, C125, G136, G170, T171, G172, G209, G228.

Nonpolar (12): P44, A62, M81, L93, I123, L130, A162, P165, I169, L230, V231, I240.

Acidic (3): E97, E164, D227.

Basic (6): R4, K13, H95, R98, R99, R133.

(b) Phylograms and cladograms are both treelike diagrams that indicate the evolutionary relationships among different species. However, a phylogram additionally indicates a sense of time or rate of evolution. The trees based on the CLUSTALW alignment of the five sequences are shown below:

Cladogram

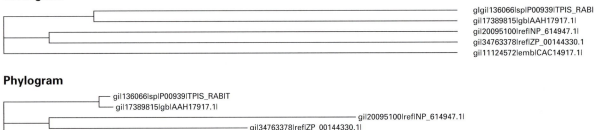

```
glgil136066lsplP00939lTPIS_RABI
gil17389815lgblAAH17917.1l
gil20095100lreflNP_614947.1l
gil34763378lreflZP_00144330.1
gil11124572lemblCAC14917.1l
```

Phylogram

```
gil136066lsplP00939lTPIS_RABIT
gil17389815lgblAAH17917.1l
gil20095100lreflNP_614947.1l
gil34763378lreflZP_00144330.1l
gil11124572lemblCAC14917.1l
```

4. Two-Dimensional Electrophoresis

(a) Answers will vary depending on the sequences entered. The pI values vary widely among the five proteins, but their Mw values are quite similar. The values for the five sequences selected in Problem 1 are shown here.

Rabbit	pI 7.09	Mw 26625.42
Human	pI 6.45	Mw 26641.43
Methanopyrus kandleri	pI 4.37	Mw 23282.43
Fusobacterium nucleatum subsp. Vincentii	pI 5.91	Mw 27564.85
Triticum aestivum	pI 5.38	Mw 26802.71

(b) Seventeen different gels have reported pI values ranging from 6.27 to 7.38 and Mw values ranging from 23,225 to 25,826. The ranges may be due to analytical error or to different processing in different tissues.

(c) Answers will vary depending on the sequences chosen for analysis. The results for the five sequences selected in Problem 1 are shown here. There are two exact identities between these specific rabbit and human enzymes.

```
rabbit muscle triose phosphate isomerase
ELASQPDVDGFLVGGASLKPEFVDIINAK 3029.5829   positions 219-247
VPADTEVVCAPPTAYIDFAR 2135.0477     positions 33-52
VAHALSEGLGVIACIGEK 1766.9469 positions 113-130
VTNGAFTGEISPGMIK 1621.8254 positions 69-84


human triose phosphate isomerase
ELASQPDVDGFLVGGASLKPEFVDIINAK 3129.5829 positions 220-248
VPADTEVVCAPPTAYIDFAR 2135.0477 positions 34-53
VAHALAEGLGVIACIGEK 1750.9519 positions 114-131
VTNGAFTGEISPGMIK 1621.8254 positions 70-85


Methanopyrus kandleri AV19 triose phosphate isomerase
TGHVTPEMVVEAGADGTLLNHSER 2520.2147 positions 76-99
AAGALSPHAVAVEPPELIGTGTPVSK 2469.3347 positions 132-157
VAAEVSEETGVEVGICPPHVDLR 2406.1969 positions 28-50
AAVELGADGVLVASGVVLADDPK 2166.1651 positions 193-215


Fusobacterium nucleatum subsp. vincentii ATCC 49256 triose phosphate isomerase
DLLSQEDIDGGLVGGASLK 1886.9705 positions 221-239
NVDIVIGAPFTCLSDAVK 1861.9728 positions 35-52
AVLEIGMKPILCIGEK 1713.9641 positions 115-130
MIIQYGGSMKPENAK 1666.8291 positions 206-220
```

```
Triticum aestivum triose phosphate isomerase
CNGTVEQVESIVNTLNAGQIASTDVVEVVVSPPYVFLPTVK 4316.2369 positions 13-53
ELAAQPDVDGFLVGGASLKPEFIDIINAAAVK 3268.7463 positions 220-251
GGAFTGEVSAEMLVNLGVPWVILGHSER 2925.4927 positions 72-99
DWTNVVVAYEPVWAIGTGK 2105.0702 positions 157-175
```

CHAPTER 6

1. Obtaining Structural Information

(a) As shown in the predictions below, the programs vary in their assignment of secondary structure (H represents helix, E represents extended beta strand, and T represents a hydrogen-bonded turn).

```
JPred

APSRKFFVGGNWKMNGRKKNLGELITTLNAAKVPADTEVVCAPPTAYIDF
-----EEEE--------HHHHHHHHHHHHH-------EEEE----HHHHH

ARQKLDPKIAVAAQNCYKVTNGAFTGEISPGMIKDCGATWVVLGHSERRH
HHHH-----EEE--------------HHHHHHHH----EEEE-------

VFGESDELIGQKVAHALSEGLGVIACIGEKLDEREAGITEKVVFEQTKVI
-----HHHHHHHHHHHHH----EEEEE---HHHH----HHHHHHHHHHHH

ADNVKDWSKVVLAYEPVWAIGTGKTATPQQAQEVHEKLRGWLKSNVSDAV
HHHHH----EEEEE-------------HHHHHHHHHHHHHHHHHH--HHH

AQSTRIIYGGSVTGATCKELASQPDVDGFLVGGASLKPEFVDIINAKQ
---EEEEE-----HHHHHHH--------E------HHHHHHHHHH--


NNPredict

APSRKFFVGGNWKMNGRKKNLGELITTLNAAKVPADTEVVCAPPTAYIDF
-----EEE--------------HHHHH----------E--------EHH

ARQKLDPKIAVAAQNCYKVTNGAFTGEISPGMIKDCGATWVVLGHSERRH
H-------HHHHHHH-HHE------------EE----HEEEE------E

VFGESDELIGQKVAHALSEGLGVIACIGEKLDEREAGITEKVVFEQTKVI
------HHHHHHHHHHHHH-H-EEHH------HHH---HEEHHHHHHHH

ADNVKDWSKVVLAYEPVWAIGTGKTATPQQAQEVHEKLRGWLKSNVSDAV
H-HHHHHHHHEE----EEEE----------HHHHHHHHHHHHE------HH

AQSTRIIYGGSVTGATCKELASQPDVDGFLVGGASLKPEFVDIINAKQ
HH--EEEE--------H-----------EEE---------EHHH----


PDB (Kabsch and Sander)

  1 APSRKFFVGG NWKMNGRKKN LGELITTLNA AKVPADTEVV CAPPTAYIDF
         EEEEE E   S    HHH HHHHHHHHHH S    TTEEEE EE   GGGHHH

 51 ARQKLDPKIA VAAQNCYKVT NGAFTGEISP GMIKDCGATW VVLGHSERRH
     HHHHS TTSE EEES   SSS S S TT    H HHHHHTT  E EEES HHHHH

101 VFGESDELIG QKVAHALSEG LGVIACIGEK LDEREAGITE KVVFEQTKVI
     TS   HHHHH HHHHHHHHTT EEEEEE     HHHHHTT HH HHHHHHHHHH

151 ADNVKDWSKV VLAYEPVWAI GTGKTATPQQ AQEVHEKLRG WLKSNVSDAV
     HHT S GGGE EEEE  GGGS SSS     HHH HHHHHHHHHH HHHHHTTHHH

201 AQSTRIIYGG SVTGATCKEL ASQPDVDGFL VGGASLKPEF VDIINAKQ
     HHH  EEE S    TTTHHHH HTTTT  EEE ESGGGGTTHH HHHHTS
```

4. Protein Families

(a) There are many possibilities. Five of them are:

PDB code 1b5t   Methylenetetrahydrofolate reductase

PDB code 1dhp   Dihydrodipicolinate synthase

PDB code 1d9e   3-deoxy-d-manno-octulosonate 8-phosphate synthase

PDB code 1ad4   Dihydropteroate synthetase

PDB code 1thf   Imidazoleglycerol phosphate synthase subunit

(b) Some related proteins are:

Ribulose-phosphate binding barrel

Thiamin phosphate synthase

Pyridoxine 5′-phosphate synthase

FMN-linked oxidoreductases

Inosine monophosphate dehydrogenase

CHAPTER 12

2. HIV Protease

(a) One possibility is:

```
>gi|55669885|pdb|1T7K|B Chain B, Crystal Structure Of Hiv Protease
Complexed With Arylsulfonamide Azacyclic Urea
PQITLWQRPLVTIKIGGQLKEALLDTGADDTVLEEMSLPGRWKPKMIGGIGGFIKVRQYDQILIEICGHKA
IGTVLVGPTPVNIIGRNLLTQIGCTLNF
```

(c) The PDB contains at least 49 HIV protease structures. At least six of them include saquinavir and three include ritonavir.

(d) Asp 25 on chains A and B.

(e) The NIAID site on HIV proteases (http://www.niaid.nih.gov/daids/dtpdb/protinh.htm) mentions the HIV protease inhibitors indinavir, nelfinavir, and amprenavir. The PDB includes at least 12 structures with indinavir, one with nelfinavir (1ohr), and one with amprenavir (1t7j).

CHAPTER 13

1. Metabolic Enzymes

(b) The KEGG Enzyme links for DHFR are:

```
EC 1.5.1.3
map00670 One carbon pool by folate
map00790 Folate biosynthesis
```

```
(c) EC 1.      Oxidoreductases
    EC 1.5.    Acting on the CH-NH group of donors
    EC 1.5.1.  With NAD(+) or NADP(+) as acceptor
    EC 1.5.1.3 Dihydrofolate reductase
```

3. Proteomics

(a) Some of the most common techniques are two-dimensional electrophoresis (2DE; see Section 5-2D) and high-performance liquid chromatography (HPLC; Section 5-2C).

(b) Proteins can be identified by mass spectrometry of proteolytic digests of the proteins (see Section 5-3D).

(c) The "dynamic range of protein expression" most frequently refers to the level of expression of different proteins: Some may be present in micromolar concentrations in cells or tissues and others may be present at femtomolar concentrations (or less).

4. Two-Dimensional Gel Electrophoresis

(a) A search for the term "human" gives about 420 hits.

(b) No. Enzymes from *E. coli* and *S. aureus* are found there.

(c) The theoretical Mw is 17,999.38, and the theoretical pI is 4.84. The actual pI values were 5.01 and 5.02, and the actual Mw values were 19,989 and 21,096. The peptide fragment used was KNIILSSQPGTDDRV, which represents residues 58–72 of *E. coli* DHFR.

(d) PDB Files of DHFR: 15

Refined query to limit to "coli": 104 structures

Refined query to limit to "methotrexate": 59 structures. Some of these actually qualify as *E. coli* DHFR complexed with methotrexate; 1DDR and 1DDS are examples.

SOLUTIONS TO PROBLEMS

CHAPTER 1

1. A   Thiol (sulfhydryl) group

   B   Carbonyl group

   C   Amide linkage

   D   Phosphoanhydride (pyrophosphoryl) linkage

   E   Phosphoryl group (P_i)

   F   Hydroxyl group

2. The cell membrane must be semipermeable so that the cell can retain essential compounds while allowing nutrients to enter and wastes to exit.

3. Concentration = (number of moles)/(volume)

   Volume = $(4/3)\pi r^3 = (4/3)\pi(5 \times 10^{-7}$ m$)^3$
   $$= 5.24 \times 10^{-19} \text{ m}^3 = 5.24 \times 10^{-16} \text{ L}$$

   Moles of protein = (2 molecules)/
   $(6.022 \times 10^{23}$ molecules $\cdot$ mol$^{-1}) = 3.32 \times 10^{-24}$ mol

   Concentration = $(3.32 \times 10^{-24}$ mol$)/(5.24 \times 10^{-16}$ L$)$
   $$= 6.3 \times 10^{-9} \text{ M} = 6.3 \text{ nM}$$

4. Number of molecules = (molar conc.)(volume)
   $$(6.022 \times 10^{23} \text{ molecules} \cdot \text{mol}^{-1})$$
   $= (10^{-3} \text{ mol} \cdot \text{L}^{-1})(5.24 \times 10^{-16} \text{ L})$
   $$(6.022 \times 10^{23} \text{ molecules} \cdot \text{mol}^{-1})$$
   $= 3.2 \times 10^5$ molecules

5. (a) Liquid water; (b) ice has less entropy at the lower temperature.

6. (a) Decreases; (b) increases; (c) increases; (d) no change.

7. (a) $T = 273 + 10 = 283$ K

   $\Delta G = \Delta H - T\Delta S$

   $\Delta G = 15$ kJ $- (283$ K$) (0.05$ kJ $\cdot$ K$^{-1})$
   $$= 15 - 14.15 \text{ kJ} = 0.85 \text{ kJ}$$

   ΔG is greater than zero, so the reaction is not spontaneous.

   (b) $T = 273 + 80 = 353$ K

   $\Delta G = \Delta H - T\Delta S$

   $\Delta G = 15$ kJ $- (353$ K$)(0.05$ kJ $\cdot$ K$^{-1})$
   $$= 15 - 17.65 \text{ kJ} = -2.65 \text{ kJ}$$

   ΔG is less than zero, so the reaction is spontaneous.

8. $K_{eq} = e^{-\Delta G^{\circ\prime}/RT} = e^{-(-20,900 \text{ J} \cdot \text{mol}^{-1})/(8.314 \text{ J} \cdot \text{K}^{-1} \cdot \text{mol}^{-1})(298 \text{ K})}$
   $= 4.6 \times 10^3$

9. $\Delta G^{\circ\prime} = -RT \ln K_{eq} = -RT \ln([C][D]/[A][B])$
   $= (8.314 \text{ J} \cdot \text{K}^{-1} \cdot \text{mol}^{-1})(298 \text{ K}) \ln[(3)(5)/(10)(15)]$
   $= 5700 \text{ J} \cdot \text{mol}^{-1} = 5.7 \text{ kJ} \cdot \text{mol}^{-1}$

   Since $\Delta G^{\circ\prime}$ is positive, the reaction is endergonic under standard conditions.

10. From Eq. 1-17, $K_{eq} = [G6P]/[G1P] = e^{-\Delta G^{\circ\prime}/RT}$

    $[G6P]/[G1P] = e^{-(-7100 \text{ J} \cdot \text{mol}^{-1})/(8.314 \text{ J} \cdot \text{mol}^{-1} \cdot \text{K}^{-1})(298 \text{ K})}$

    $[G6P]/[G1P] = 17.6$

    $[G1P]/[G6P] = 0.057$

11. $\Delta G = \Delta H - T\Delta S$

    $\Delta G = -7000 \text{ J} \cdot \text{mol}^{-1} - (298 \text{ K})(-25 \text{ J} \cdot \text{K}^{-1} \cdot \text{mol}^{-1})$

    $\Delta G = -7000 + 7450 \text{ J} \cdot \text{mol}^{-1} = 450 \text{ J} \cdot \text{mol}^{-1}$

    The reaction is not spontaneous because $\Delta G > 0$. The temperature must be decreased in order to decrease the value of the $T\Delta S$ term.

12. In order for ΔG to have a negative value (a spontaneous reaction), $T\Delta S$ must be greater than ΔH.

    $T\Delta S > \Delta H$

    $T > \Delta H/\Delta S$

    $T > 7000 \text{ J} \cdot \text{mol}^{-1}/20 \text{ J} \cdot \text{K}^{-1} \cdot \text{mol}^{-1}$

    $T > 350$ K or 77°C

13. (a) False. A spontaneous reaction only occurs in one direction. (b) False. Thermodynamics does not specify the rate of a reaction. (c) True. (d) True. A reaction is spontaneous so long as $\Delta S > \Delta H/T$.

CHAPTER 2

1. (a) Donors: NH1, NH$_2$ at C2, NH9; acceptors: N3, O at C6, N7. (b) Donors: NHl, NH$_2$ at C4; acceptors: O at C2, N3. (c) Donors: NH$_3^+$ group, OH group; acceptors: COO$^-$ group, OH group.

2. A protonated (and therefore positively charged) nitrogen would promote the separation of charge in the adjacent C—H bond so that the C would have a partial negative charge and the H would have a partial positive charge. This would make the H more likely to be donated to a hydrogen bond acceptor group.

3. From most soluble (most polar) to least soluble (least polar): c, b, e, a, d.

4. (a) Water; (b) water; (c) micelle.

5. The waxed car is a hydrophobic surface. To minimize its interaction with the hydrophobic molecules (wax), each water drop minimizes its surface area by becoming a sphere (the geometrical shape with the lowest possible ratio of surface to volume). Water does not bead on glass, because the glass presents a hydrophilic surface with which the water molecules can interact. This allows the water to spread out.

6. Water molecules move from inside the dialysis bag to the surrounding seawater by osmosis. Ions from the seawater diffuse into the dialysis bag. At equilibrium, the compositions of the solutions inside and outside the dialysis bag are identical. If the membrane were solute-impermeable, essentially all the water would leave the dialysis bag.

7. Because the bacterium and the fish have identical shapes (length/diameter = 6 for each), the ratio of their surface-to-volume ratios is the same as the inverse ratio of their lengths or diameters (e.g., 3 µm versus 30 cm). Thus, the bacterium has 100,000 times more surface area per unit volume than the fish.

8. (a)

COO⁻
|
CH
‖
HC
|
COO⁻

(b)

COO⁻
|
H—C—H
|
NH$_3^+$

(c)

COO⁻
|
H—C—H
|
NH$_2$

(d)

COO⁻
|
H—C—CH$_2$—COO⁻
|
NH$_3^+$

9. (a) pH 4, NH_4^+; pH 8, NH_4^+; pH 11, NH_3.
   (b) pH 4, $H_2PO_4^-$; pH 8, HPO_4^{2-}; pH 11, HPO_4^{2-}.

10. The increase in $[H^+]$ due to the addition of HCl is (50 mL)
    $(1\,mM)/(250\,mL) = 0.2\,mM = 2 \times 10^{-4}$ M. Because the $[H^+]$
    of pure water, 10^{-7} M, is relatively insignificant, the pH of
    the solution is equal to $-\log (2 \times 10^{-4})$ or 3.7.

11. (a) $(0.01\,L)(5\,mol \cdot L^{-1}\,NaOH)/(1\,L) =$
    $$0.05\,M\,NaOH \equiv 0.05\,M\,OH^-$$
    $[H^+] = K_w/[OH^-] = (10^{-14})/(0.05) = 2 \times 10^{-13}$ M
    $pH = -\log [H^+] = -\log (2 \times 10^{-13}) = 12.7$

    (b) $(0.02\,L)(5\,mol \cdot L^{-1}\,HCl)/(1\,L) =$
    $$0.1\,M\,HCl \equiv 0.1\,M\,H^+$$
    Since the contribution of $0.01\,L \times 100\,mM/(1L) = 1\,mM$
    glycine is insignificant in the presence of 0.1 M HCl,
    $pH = -\log [H^+] = -\log (0.1) = 1.0$

    (c) $pH = pK + \log ([acetate]/[acetic\ acid])$
    $[acetate] = (5\,g)(1\,mol/82\,g)/(1\,L) = 0.061$ M
    $[acetic\ acid] = (0.01\,L)(2\,mol \cdot L^{-1})/(1\,L) = 0.02$ M
    $pH = 4.76 + \log (0.061/0.02) = 4.76 + 0.48 = 5.24$

12. The standard free energy change can be calculated using
    Eq. 1-16 and the value of K from Table 2-4.

    $\Delta G^{\circ\prime} = -RT \ln K$
    $= -(8.314\,J \cdot K^{-1} \cdot mol^{-1})(298\,K) \ln (3.39 \times 10^{-8})$
    $= 42{,}600\,J \cdot mol^{-1} = 42.6\,kJ \cdot mol^{-1}$

13. The pK corresponding to the equilibrium between $H_2PO_4^-$
    (HA) and HPO_4^{2-} (A⁻) is 6.82 (Table 2-4). The concentra-
    tion of A⁻ is $(50\,mL)(2.0\,M)/(200\,mL) = 0.5$ M, and the con-
    centration of HA is $(25\,mL)(2.0\,M)/(200\,mL) = 0.25$ M.
    Substitute these values into the Henderson–Hasselbalch
    equation (Eq. 2-9):

    $$pH = pK + \log \frac{[A^-]}{[HA]}$$
    $$pH = 6.82 + \log \frac{0.5}{0.25}$$
    $$pH = 6.82 + \log 2$$
    $$pH = 6.82 + 0.30 = 7.12$$

14. Use the Henderson–Hasselbalch equation (Eq. 2-9) and
    solve for pK:

    $$pH = pK + \log \frac{[A^-]}{[HA]}$$
    $$pK = pH - \log \frac{[A^-]}{[HA]}$$
    $$pK = 6.5 - \log \frac{0.2}{0.1}$$
    $$pK = 6.5 - 0.3 = 6.2$$

15. Let HA = sodium succinate and A⁻ = disodium succinate.
    $[A^-] + [HA] = 0.05$ M, so $[A^-] = 0.05$ M $- [HA]$
    From Eq. 2-9 and Table 2-4,
    $\log ([A^-]/[HA]) = pH - pK = 6.0 - 5.64 = 0.36$
    $[A^-]/[HA] = $ antilog $0.36 = 2.29$
    $(0.05\,M - [HA])/[HA] = 2.29$
    $[HA] = 0.015$ M
    $[A^-] = 0.05$ M $- 0.015$ M $= 0.035$ M
    grams of sodium succinate $=$
    $(0.015\,mol \cdot L^{-1})(140\,g \cdot mol^{-1}) \times (1\,L) = 2.1$ g
    grams of disodium succinate $=$
    $(0.035\,mol \cdot L^{-1})(162\,g \cdot mol^{-1}) \times (1\,L) = 5.7$ g

16. At pH 4, essentially all the phosphoric acid is in the $H_2PO_4^-$
    form, and at pH 9, essentially all is in the HPO_4^{2-} form (Fig.
    2-18). Therefore, the concentration of OH^- required is equiv-
    alent to the concentration of the acid: $(0.1\,mol \cdot L^{-1}$ phos-
    phoric acid$)(0.1\,L) = 0.01\,mol\,NaOH\ required = (0.01$
    mol$)(1\,L/5\,mol\,NaOH) = 0.002\,L = 2\,mL$

17. (a) Succinic acid; (b) ammonia; (c) HEPES.

CHAPTER 3

1. (a) Yes; (b) no; (c) no; (d) yes.

2. Since the haploid genome contains 21% G, it must contain
   21% C (because G = C) and 58% A + T (or 29% A and
   29% T, because A = T). Each cell is diploid, containing
   90,000 kb or 9×10^7 bases. Therefore,

   $A = T = (0.29)(9 \times 10^7) = 2.61 \times 10^7$ bases
   $C = G = (0.21)(9 \times 10^7) = 1.89 \times 10^7$ bases

3. The DNA contains 40 bases in all. Since G = C, there are 7
   cytosine residues. The remainder $(40 - 14 = 26)$ must be adenine and thymine. Since A = T, there are 13 adenine residues.
   There are no uracil residues (U is a component of RNA but
   not DNA).

4. (a)

   (b)

5.

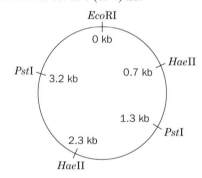

Inosine   **Adenine**

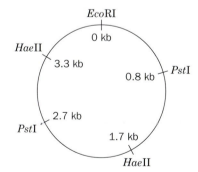

Inosine   **Cytosine**

6. The high pH eliminates hydrogen bonds between bases, making it easier to separate the strands of DNA.

7. The number of possible sequences of four different nucleotides is 4^n where n is the number of nucleotides in the sequence. Therefore, (a) $4^1 = 4$, (b) $4^2 = 16$, (c) $4^3 = 64$, and (d) $4^4 = 256$.

8. 5'–ACGT–3'   5'–CGAATC–3'
   3'–TGCAGC–5' + 3'–TTAG–5'

9. (a) *Alu*I, *Eco*RV, *Hae*III, *Pvu*II; (b) *Hpa*II and *Msp*I; (c) *Bam*HI and *Bgl*II; *Hpa*II and *Taq*I; *Sal*I and *Xho*I.

10. Let the *Eco*RI site be set at 0 (or 4) kb.

[Circular map showing:
EcoRI at 0 kb, HaeII at 0.7 kb, PstI at 1.3 kb region with 2.3 kb, HaeII, 3.2 kb PstI]

OR

[Circular map showing:
EcoRI at 0 kb, PstI at 0.8 kb, 1.7 kb, HaeII, 2.7 kb, PstI, 3.3 kb, HaeII]

11. (a) Newly synthesized chains would be terminated less frequently, so the bands representing truncated fragments on the sequencing gel would appear faint.

(b) Chain termination would occur more frequently, so longer fragments would be less abundant.

(c) The amount of DNA synthesis would decrease and the resulting gel bands would appear faint.

(d) No effect.

12. The *C. elegans* genome contains 97,000 kb, so $f = 5/97,000 = 5.2 \times 10^{-5}$. Using Eq. 3-2,
$N = \log(1 - P)/\log(1 - f)$
$N = \log(1 - 0.99)/\log(1 - 5.2 \times 10^{-5})$
$N = -2/(-2.24 \times 10^{-5}) = 8.91 \times 10^4$

13. The desired clones are colorless when grown in the presence of ampicillin and X-gal. Nontransformed bacteria cannot grow in the presence of ampicillin, because they lack the amp^R gene carried by the plasmid. Clones transformed with the plasmid only are blue, since they have an intact *lacZ* gene and produce β-galactosidase, which cleaves the chromogenic substrate X-gal. Clones that contain the plasmid with the foreign DNA insert are colorless because the insert interrupts the *lacZ* gene.

14. (a) Only single DNA strands of variable length extending from the remaining primer would be obtained. The number of these strands would increase linearly with the number of cycles rather than geometrically.

(b) PCR would yield a mixture of DNA segments whose lengths correspond to the distance between the position of the primer with a single binding site and the various sites where the multispecific primer binds.

(c) The first cycle of PCR would yield only the new strand that is complementary to the intact DNA strand, since DNA synthesis cannot proceed when the template is broken. However, since the new strand has the same sequence as the broken strand, PCR can proceed normally from the second cycle on.

(d) DNA synthesis would terminate at the breaks in the first cycle of PCR.

15. ATAGGCATAGGC and CTGACCAGCGCC.

16. (a) The genomic library contains DNA sequences corresponding to all the organism's DNA, which includes genes and nontranscribed sequences. A cDNA library represents only the DNA sequences that are transcribed into mRNA.

(b) Different cell types express different sets of genes. Therefore, the populations of mRNA molecules used to construct the cDNA libraries also differ.

CHAPTER 4

1. Ser and Thr; Val, Leu, and Ile; Asn and Gln; Asp and Glu.

2. $^+H_3N{-}CH_2{-}CH_2{-}COO^-$

3. Hydrogen bond donors: α-amino group, amide nitrogen. Hydrogen bond acceptors: α-carboxylate group, amide carbonyl.

4.

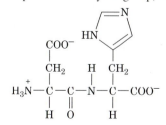

5. The first residue can be one of five residues, the second one of the remaining four, etc.
$N = 5 \times 4 \times 3 \times 2 \times 1 = 120$

6. (a) +1; (b) 0; (c) −1; (d) −2.

7. (a) $pI = (2.35 + 9.87)/2 = 6.11$
(b) $pI = (6.04 + 9.33)/2 = 7.68$
(c) $pI = (2.10 + 4.07)/2 = 3.08$

8. The polypeptide would be even less soluble than free Tyr, because most of the amino and carboxylate groups that interact with water and make Tyr at least slightly soluble are lost in forming the peptide bonds in poly(Tyr).

9.

10.

11.

Replace with CH_3 to give D-Ala

12. (2S,3S)-Isoleucine

13. (a) Glutamate; (b) aspartate

14. (a) Serine (N-acetylserine); (b) lysine (5-hydroxylysine); (c) methionine (N-formylmethionine).

15.

(a) The pK's of the ionizable side chains (Table 4-1) are 3.90 (Asp) and 10.54 (Lys); assume that the terminal Lys carboxyl group has a pK of 3.5 and the terminal Ala amino group has a pK of 8.0 (Section 4-1D). The pI is approximately midway between the pK's of the two ionizations involving the neutral species (the pK of Asp and the N-terminal pK):

$$pI \approx \tfrac{1}{2}(3.90 + 8.0) \approx 5.95$$

(b) The net charge at pH 7.0 is 0 (as drawn above).

16. At position A8, duck insulin has a Glu residue, whereas human insulin has a Thr residue. Since Glu is negatively charged at physiological pH and Thr is neutral, human insulin has a higher pI than duck insulin. (The other amino acids that differ between the proteins do not affect the pI because they are uncharged.)

CHAPTER 5

1. Peptide B, because it contains more Trp and other aromatic residues.

2. (a) Leu, His, Arg. (b) Lys, Val, Glu.

3. The protein behaves like a larger protein during gel filtration, suggesting that it has an elongated shape. The mass determined by SDS-PAGE is more accurate since the mobility of a denatured SDS-coated protein depends only on its size.

4. The protein contains two 60-kD polypeptides and two 40-kD polypeptides. Each 40-kD chain is disulfide bonded to a 60-kD chain. The 100-kD units associate noncovalently to form a protein with a molecular mass of 200 kD.

5. The protein aggregates at the higher salt concentration.

6. Because protein 1 has a greater proportion of hydrophobic residues (Ala, Ile, Pro, Val) than do proteins 2 and 3, hydrophobic interaction chromatography could be used to isolate it.

7. Dansyl chloride reacts with primary amino groups, including the ε-amino group of Lys residues.

8. (a) Gly; (b) Thr; (c) none (the N-terminal amino group is acetylated and hence unreactive with Edman's reagent)

9. Thermolysin would yield the most fragments (9) and endopeptidase V8 would yield the fewest (2).

10. Gln–Ala–Phe–Val–Lys–Gly–Tyr–Asn–Arg–Leu–Glu

11. Asp–Met–Leu–Phe–Met–Arg–Ala–Tyr–Gly–Asn

12. (a) There is one Met, so CNBr would produce two peptides.
(b) There are four possible sites for chymotrypsin to hydrolyze the peptide: following Phe, Tyr (twice), and Trp. This would yield five peptides.
(c) Four Cys residues form two disulfide bonds.
(d) Arbitrarily choosing one Cys residue, there are three ways it can make a disulfide bond with the remaining three Cys residues. After choosing one of them, there is only one way that the remaining two Cys residues can form a disulfide bond. Thus there are $3 \times 1 = 3$ possible arrangements of the disulfide bonds.

13.

14. Arg–Ile–Pro–Lys–Cys–Arg–Lys–Phe–Gln–Gln–Ala–Gln–His–Leu–Arg–Ala–Cys–Gln–Gln–Trp–Leu–His–Lys–Gln–Ala–Asn–Gln–Ser–Gly–Gly–Gly–Pro–Ser

15. Because the side chain of Gly is only an H atom, it often occurs in a protein at a position where no other residue can fit. Consequently, Gly can take the place of a larger residue more easily than a larger residue, such as Val, can take the place of Gly.

16. (a) Position 6 (Gly) and Position 9 (Val) appear to be invariant.

(b) Conservative substitutions occur at Position 1 (Asp and Lys, both charged), Position 10 (Ile and Leu, similar in structure and hydrophobicity), and Position 2 (all unchanged bulky side chains). Positions 5 and 8 appear to tolerate some substitution.

(c) The most variable positions are 3, 4, and 7, where a variety of residues appear.

CHAPTER 6

1.

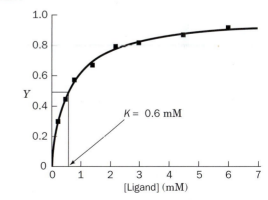

2. (a) 3.6_{13}; (b) steeper.

3. (100 residues)(1 α-helical turn/3.6 residues)

(5.1 Å/keratin turn) = 142 Å

4. (a) The first and fourth side chains of the two helices of a coiled coil form buried hydrophobic interacting surfaces, but the remaining side chains are exposed to the solvent and therefore tend to be polar or charged.

(b) Although the residues at positions 1 and 4 in both sequences are hydrophobic, Trp and Tyr are much larger than Ile and Val and would therefore not fit as well in the area of contact between the two polypeptides in a coiled coil.

5. A fibrous protein such as α keratin does not have a discrete globular core. Most of the residues in its coiled coil structure are exposed to the solvent. The exception is the strip of nonpolar side chains at the interface of the two coils.

6. The reducing conditions promote cleavage of the disulfide bonds that cross-link α keratin molecules. This helps the larvae digest the wool clothing that they eat.

7. Collagen's primary structure is its amino acid sequence, which is a repeating triplet of mostly Gly–Pro–Hyp. Its secondary structure is the left-handed helical conformation characteristic of its repeating sequence. Its tertiary structure is essentially the same as its secondary structure, since most of the protein consists of one type of secondary structure. Collagen's quaternary structure is the arrangement of its three chains in a right-handed triple helix.

8. Because collagen has such an unusual amino acid composition (almost two-thirds consists of Gly and Pro or Pro derivatives), it contains relatively fewer of the other amino acids and is therefore not as good a source of amino acids as proteins containing a greater variety of amino acids.

9. Yes, although such irregularity should not be construed as random.

10. (a) Gln; (b) Ser; (c) Ile; (d) Cys. (see Table 6-1)

11. (a) C_4 and D_2; (b) C_6 and D_3.

12. (a) Phe. Ala and Phe are both hydrophobic, but Phe is much larger and might not fit as well in Val's place.

(b) Asp. Replacing a positively charged Lys residue with an oppositely charged Asp residue would be more disruptive.

(c) Glu. The amide-containing Asn would be a better substitute for Gln than the acidic Glu.

(d) His. Pro's constrained geometry is best approximated by Gly, which lacks a side chain, rather than a residue with a bulkier side chain such as His.

13. A polypeptide synthesized in a living cell has a sequence that has been optimized by natural selection so that it folds properly (with hydrophobic residues on the inside and polar residues on the outside). The random sequence of the synthetic peptide cannot direct a coherent folding process, so hydrophobic side chains on different molecules aggregate, causing the polypeptide to precipitate from solution.

14. No

15. Hydrophobic effects, van der Waals interactions, and hydrogen bonds are destroyed during denaturation. Covalent cross-links are retained.

16. At physiological pH, the positively charged Lys side chains repel each other. Increasing the pH above the pK (>10.5) would neutralize the side chains and allow an α helix to form.

17. The molecular mass of O_2 is 32 D. Hence the ratio of the masses of hemoglobin and 4 O_2, which is equal to the ratio of their volumes, is $65,000/(4 \times 32) = 508$. The 70-kg office worker has a volume of 70 kg × 1 cm^3/g × (1000 g/kg) × $(1 \text{ m}/100 \text{ cm})^3 = 0.070 \text{ m}^3$. Hence the ratio of the volumes of the office and the office worker is $(4 \times 4 \times 3)/0.070 = 686$. These ratios are similar in magnitude, which you may not have expected.

18. Peptide c is most likely to form an α helix with its three charged residues (Lys, Glu, and Arg) aligned on one face of the helix. Peptide a has adjacent basic residues (Arg and Lys), which would destabilize a helix. Peptide b contains Gly and Pro, both of which are helix-breaking (Table 6-1). The presence of Gly and Pro would also inhibit the formation of β strands, so peptide b is least likely to form a β strand.

19. In a protein crystal, the residues at the end of a polypeptide chain may experience fewer intramolecular contacts and therefore tend to be less ordered (more mobile in the crystal). If their disorder prevents them from generating a coherent diffraction pattern, it may be impossible to map their electron density.

CHAPTER 7

1.

[Plot of Y versus [Ligand] (mM), showing a hyperbolic binding curve with data points. Dashed lines indicate $Y = 0.5$ at approximately [Ligand] = 0.6 mM. Labeled $K = 0.6$ mM. The x-axis [Ligand] (mM) ranges from 0 to 7, and the y-axis Y ranges from 0 to 1.0.]

2. b describes sigmoidal binding to an oligomeric protein and hence represents cooperative binding.

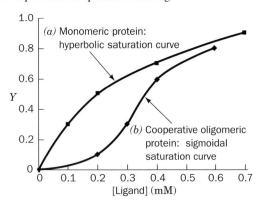

3. According to Eq. 7-6,

$$Y_{O_2} = \frac{pO_2}{K + pO_2}$$

When $pO_2 = 10$ torr,

$$Y_{O_2} = \frac{10}{2.8 + 10} = 0.78$$

When $pO_2 = 1$ torr,

$$Y_{O_2} = \frac{1}{2.8 + 1} = 0.26$$

The difference in Y_{O_2} values is $0.78 - 0.26 = 0.52$. Therefore, in active muscle cells, myoglobin can transport a significant amount of O_2 by diffusion from the cell surface to the mitochondria.

4. (a) Hyperventilation eliminates CO_2, but it does not significantly affect the O_2 concentration, since the hemoglobin in arterial blood is already essentially saturated with oxygen.

   (b) The removal of CO_2 also removes protons, according to the reaction

   $$H^+ + HCO_3^- \rightleftharpoons H_2O + CO_2$$

   The resulting increase in blood pH would increase the O_2 affinity of hemoglobin through the Bohr effect. The net result would be that less oxygen could be delivered to the tissues until the CO_2 balance was restored. Thus, hyperventilation has the opposite of the intended effect (note that since hyperventilation suppresses the urge to breathe, doing so may cause the diver to lose consciousness due to lack of O_2 and hence drown).

5. (a) Vitamin O is useless because the body's capacity to absorb oxygen is not limited by the amount of oxygen available but by the ability of hemoglobin to bind and transport O_2. Furthermore, oxygen is normally introduced into the body via the lungs, so it is unlikely that the gastrointestinal tract would have an efficient mechanism for extracting oxygen.

   (b) The fact that oxygen delivery in vertebrates requires a dedicated O_2-binding protein (hemoglobin) indicates that dissolved oxygen by itself cannot attain the high concentrations required. Moreover, a few drops of vita-

min O would make an insignificant contribution to the amount of oxygen already present in a much larger volume of blood.

6. (a) Lower; (b) higher. The Asp 99β → His mutation of hemoglobin Yakima disrupts a hydrogen bond at the α_1–β_2 interface of the T state (Fig. 7-10a), causing the $T \rightleftharpoons R$ equilibrium to shift toward R state (lower p_{50}). The Asn 102β → Thr of hemoglobin Kansas causes the opposite shift in the $T \rightleftharpoons R$ equilibrium by abolishing an R-state hydrogen bond (Fig. 7-10b).

7. The increased BPG helps the remaining erythrocytes deliver O_2 to tissues. However, BPG stabilizes the T conformation of hemoglobin, so it promotes sickling and therefore aggravates the disease.

8. (a) Because the mutation destabilizes the T conformation of hemoglobin Rainier, the R (oxy) conformation is more stable. Therefore, the oxygen affinity of hemoglobin Rainier is greater than normal.

   (b) The ion pairs that normally form in deoxyhemoglobin absorb protons. The absence of these ion pairs in hemoglobin Rainier decreases the Bohr effect (in fact, the Bohr effect in hemoglobin Rainier is about half that of normal hemoglobin).

   (c) Because the R conformation of hemoglobin Rainier is more stable than the T conformation, even when the molecule is not oxygenated, O_2-binding cooperativity is reduced. The Hill coefficient of hemoglobin Rainier is therefore less than that for normal hemoglobin.

9. As the crocodile remains under water without breathing, its metabolism generates CO_2 and hence the HCO_3^- content of its blood increases. The HCO_3^- preferentially binds to the crocodile's deoxyhemoglobin, which allosterically prompts the hemoglobin to assume the deoxy conformation and thus release its O_2. This helps the crocodile stay under water long enough to drown its prey.

10. (a)

   (b) The Hill coefficient most likely has a value between 1 (no cooperativity) and 2 (perfect cooperativity between the two subunits).

CHAPTER 8

1. (a) 4; (b) 8; (c) 16.

2. (a) and (d)

3.

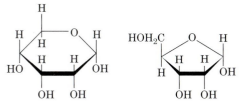

4. (a) Yes; (b) no (its symmetric halves are superimposable); (c) no.

5.

```
        O    H
         \  /
          C
          |
   HO—C—H
          |
    H—C—OH
          |
    H—C—OH
          |
   HO—C—H
          |
         CH₃
```

L-Fucose

L-Fucose is the 6-deoxy form of L-galactose.

6. α-D-glucose-(1→1)-α-D-glucose   or   α-D-glucose-(1→1)-β-D-glucose

7. 19

8.

9. One

10. Amylose (it has only one nonreducing end from which glucose can be mobilized).

11. Glucosamine is a building block of certain glycosaminoglycan components of proteoglycans (Fig. 8-12). Boosting the body's supply of glucosamine might slow the progression of the disease osteoarthritis, which is characterized by the degradation of proteoglycan-rich articular (relating to a joint) cartilage.

12. −200

13.

14. (a) There are four types of methylated glucose molecules, corresponding to (1) the residue at the reducing end of the glycogen molecule, (2) residues at the nonreducing ends, (3) residues at the α(1→6) branch points, and (4) residues from the linear α(1→4)-linked segments of glycogen. Type 4 is the most abundant type of residue.

(b)

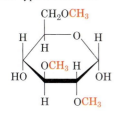

CHAPTER 9

1. *trans*-Oleic acid has a higher melting point because, in the solid state, its hydrocarbon chains pack together more tightly than those of *cis*-oleic acid.

2. Of the 4 × 4 = 16 pairs of fatty acid residues at C1 and C3, only 10 are unique because a molecule with different substituents at C1 and C3 is identical to the molecule with the reverse substitution order. However, C2 may have any of the four substituents for a total of 4 × 10 = 40 different triacylglycerols.

3. The triacylglycerol containing the stearic acid residues yields more energy since it is fully reduced.

4.

$$H_3C(CH_2)_7CH=CH(CH_2)_7-\overset{O}{\underset{\|}{C}}-O-\overset{CH_2-O-\overset{O}{\underset{\|}{C}}-(CH_2)_{14}CH_3}{\underset{CH_2-O-\overset{O}{\underset{\|}{P}}-O-CH_2CH_2NH_3^+}{\overset{|}{\underset{|}{CH}}}}$$

5. (a) Palmitic acid and 2-oleoyl-3-phosphatidylserine;

(b) oleic acid and 1-palmitoyl-3-phosphatidylserine;

(c) phosphoserine and 1-palmitoyl-2-oleoyl-glycerol;

(d) serine and 1-palmitoyl-2-oleoyl-phosphatidic acid.

6. All except choline can form hydrogen bonds.

7. No; the two acyl chains of the "head group" are buried in the bilayer interior, leaving a head group of diphosphoglycerol.

8. Both DNA and phospholipids have exposed phosphate groups that are recognized by the antibodies.

9. Steroid hormones, which are hydrophobic can diffuse through the cell membrane to reach their receptors.

10. Eicosanoids synthesized from arachidonic acid are necessary for intercellular communication. Cultured cells do not need such communication and therefore do not require linoleic acid.

11. Triacylglycerols lack polar head groups, so they do not orient themselves in a bilayer with their acyl chains inward and their glycerol moiety toward the surface.

12. The large oligosaccharide head groups of gangliosides would prevent efficient packing of the lipids in a bilayer.

13. (a) Saturated; (b) long-chain. By increasing the proportion of saturated and long-chain fatty acids, which have higher

melting points, the bacteria can maintain constant membrane fluidity at the higher temperature.

14. (a) (1 turn/5.4 Å)(30 Å) = 5.6 turns

    (b) (3.6 residues/turn)(5.6 turns) = 20 residues

    (c) The additional residues form a helix, which partially satisfies backbone hydrogen bonding requirements, where the lipid head groups do not offer hydrogen bonding partners.

15. No. Although the β strand could span the bilayer, a single strand would be unstable because its backbone could not form the hydrogen bonds it would form with water in aqueous solution.

16. (a) Inner; (b) outer. See Fig. 9-33.

17. (a) Both the intra- and extracellular portions will be labeled. (b) Only the extracellular portion will be labeled. (c) Only the intracellular portion will be labeled.

18. The mutant signal peptidase would cleave many preproteins within their signal peptides, which often contain Leu-Leu sequences. This would not affect translocation into the ER, since signal peptidase acts after the signal peptide enters the ER lumen. Proteins lacking the Leu-Leu sequence would retain their signal peptides. These proteins, and those with abnormally cleaved signal sequences, would be more likely to fold abnormally and therefore function abnormally.

19. In order for a neuron to repeatedly release neurotransmitters, the components of its exocytotic machinery must be recycled. Following the fusion of synaptic vesicles with the plasma membrane, the four-helix SNARE complex is disassembled so that the Q-SNAREs remain in the plasma membrane while portions of the membrane containing R-SNAREs can be used to re-form synaptic vesicles. This recycling process would not be possible if the R- and Q-SNAREs remained associated, and the neuron would eventually be unable to release neurotransmitters.

CHAPTER 10

1. (a) Nonmediated; (b) mediated; (c) nonmediated; (d) mediated.

2. The less polar a substance, the faster it can diffuse through the lipid bilayer. From slowest to fastest: C, A, B.

3. $\Delta G = RT \ln\dfrac{[\text{Glucose}]_{in}}{[\text{Glucose}]_{out}}$

    $\Delta G = (8.3145 \text{ J} \cdot \text{K}^{-1} \cdot \text{mol}^{-1})(298 \text{ K}) \ln\dfrac{(0.003)}{(0.005)}$

    $\Delta G = -1270 \text{ J} \cdot \text{mol}^{-1} = -1.27 \text{ kJ} \cdot \text{mol}^{-1}$

4. (a) $\Delta G = RT \ln([\text{Na}^+]_{in}/[\text{Na}^+]_{out})$

    $= (8.314 \text{ J} \cdot \text{K}^{-1} \cdot \text{mol}^{-1})(310 \text{ K})(\ln[10\text{mM}/150 \text{ mM}])$

    $= (8.314)(310)(-2.71) \text{ J} \cdot \text{mol}^{-1}$

    $= -6980 \text{ J} \cdot \text{mol}^{-1} = -7.0 \text{ kJ} \cdot \text{mol}^{-1}$

    (b) $\Delta G = RT \ln([\text{Na}^+]_{in}/[\text{Na}^+]_{out}) + Z_A \mathscr{F} \Delta \Psi$

    $-6980 + (1)(96,485 \text{ C} \cdot \text{mol}^{-1})(-0.06 \text{ J C}^{-1})$

    $-6980 \text{ J} \cdot \text{mol}^{-1} - 5790 \text{ J} \cdot \text{mol}^{-1}$

    $-12,770 \text{ J} \cdot \text{mol}^{-1} = -12.8 \text{ kJ} \cdot \text{mol}^{-1}$

5. Use Equation 10-3 and let $Z = 2$ and $T = 310$ K:

(a) $\Delta G = RT \ln\dfrac{[\text{Ca}^{2+}]_{in}}{[\text{Ca}^{2+}]_{out}} + Z\mathscr{F}\Delta\Psi$

$= (8.314 \text{ J} \cdot \text{K}^{-1} \cdot \text{mol}^{-1})(310 \text{ K}) \ln(10^{-7})/(10^{-3})$
$\quad + (2)(96,485 \text{ J} \cdot \text{V}^{-1} \cdot \text{mol}^{-1})(-0.050 \text{ V})$
$= -23,700 \text{ J} \cdot \text{mol}^{-1} - 9600 \text{ J} \cdot \text{mol}^{-1}$
$= -33,300 \text{ J} \cdot \text{mol}^{-1} = -33.3 \text{ kJ} \cdot \text{mol}^{-1}$

The negative value of ΔG indicates a thermodynamically favorable process.

(b) $\Delta G = RT \ln\dfrac{[\text{Ca}^{2+}]_{in}}{[\text{Ca}^{2+}]_{out}} + Z\mathscr{F}\Delta\Psi$

$= (8.314 \text{ J} \cdot \text{K}^{-1} \cdot \text{mol}^{-1})(310 \text{ K}) \ln(10^{-7})/(10^{-3})$
$\quad + (2)(96,485 \text{ J} \cdot \text{V}^{-1} \cdot \text{mol}^{-1})(+0.150 \text{ V})$
$= -23,700 \text{ J} \cdot \text{mol}^{-1} + 28,900 \text{ J} \cdot \text{mol}^{-1}$
$= +5,200 \text{ J} \cdot \text{mol}^{-1} = +5.2 \text{ kJ} \cdot \text{mol}^{-1}$

The positive value of ΔG indicates a thermodynamically unfavorable process.

6. (a) K^+ transport ceases because the ionophore–K^+ complex cannot diffuse through the membrane when the lipids are immobilized in a gel-like state. (b) No. An unblocked N-terminus would be a protonated amino group, which would repel another gramicidin A N-terminus rather than hydrogen bond to it, and which is less likely to be buried in the hydrophobic interior of a lipid bilayer.

7. The number of ions to be transported is

$(10 \text{ mM})(100 \text{ μm}^3)(N)$
$= (0.01 \text{ mol} \cdot \text{L}^{-1})(10^{-13} \text{ L})(6.02 \times 10^{23} \text{ ions} \cdot \text{mol}^{-1})$
$= 6.02 \times 10^8 \text{ ions}$

Since there are 100 ionophores, each must transport 6.02×10^6 ions. The time required is $(6.02 \times 10^6 \text{ ions})(1 \text{ s}/10^4 \text{ ions}) = 602 \text{ s} = 10 \text{ min}$.

8. (a) The data do not indicate the involvement of a transport protein, since the rate of transport does not approach a maximum as [X] increases.

    (b) To verify that a transport protein is involved, increase [X] to demonstrate saturation of the transporter at high [X], or add a structural analog of X to compete with X for binding to the transporter, resulting in a lower flux of X.

9. In the absence of ATP, Na^+ extrusion by the $(Na^+–K^+)$–ATPase would cease, so no glucose could enter the cell by the Na^+–glucose symport. The glucose in the cell would then exit via the passive-mediated glucose transporter, and the cellular [glucose] would decrease until it matched the extracellular [glucose] (of course, the cell would probably osmotically burst before this could occur).

10. The hyperbolic curve for glucose transport into pericytes indicates a protein-mediated sodium-dependent process. The transport protein has binding sites for sodium ions. At low [Na^+], glucose transport is directly proportional to [Na^+]. However, at high [Na^+], all Na^+ binding sites on the transport protein are occupied, and thus glucose transport reaches a maximum velocity. Glucose transport into endothelial cells is not sodium-dependent and occurs at a high rate whether or not Na^+ is present. There is not enough information in the figure to determine whether glucose transport into endothelial cells is protein-mediated.

11. (a) No; there is no glycerol backbone.

(b) Miltefosine is amphipathic and therefore cannot cross the parasite cell membrane by diffusion. Since it is not a normal cell component, it probably does not have a dedicated active transporter. It most likely enters the cell via a passive transport protein.

(c) This amphipathic molecule most likely accumulates in membranes, with its hydrophobic tail buried in the bilayer and its polar head group exposed to the solvent.

(d) The protein recognizes the phosphocholine head group, which also occurs in some sphingolipids and some glycerophospholipids. Since the protein does not bind all phospholipids or triacylglycerols, it does not recognize the hydrocarbon tail.

12. (a) A transporter similar to a porin would be inadequate since even a large β barrel would be far too small to accommodate the massive ribosome. Likewise, a transport protein with alternating conformations would not be up to the task due to its small size relative to the ribosome. In addition, neither type of protein would be suited for transporting a particle across two membranes. (In fact, ribosomes and other large particles move between the nucleus and cytoplasm via nuclear pores, which are constructed from many different proteins and form a structure that is much larger than the ribosome that spans both nuclear membranes.)

(b) Ribosomal transport might appear to be a thermodynamically favorable process, since the concentration of ribosomes is greater in the nucleus, where they are synthesized. However, free energy would ultimately be required to establish a pore (which would span two membrane thicknesses) for the ribosome to pass through. (In fact, the nucleocytoplasmic transport of all but very small substances requires the activity of GTPases that escort particles through the nuclear pore assembly and help ensure that transport proceeds in one direction.)

13. (a) Acetylcholine binding triggers the opening of the channel, an example of a ligand-gated transport protein.

(b) Na^+ ions flow into the muscle cell, where their concentration is low.

(c) The influx of positive charges causes the membrane potential to increase.

14. (a) $pH = -\log[H^+] = -\log(0.15) = 0.82$

The pH of the secreted HCl is over 6 pH units lower than the cytosolic pH, which corresponds to a $[H^+]$ of $\sim 4 \times 10^{-8}$ M.

(b) $CO_2 + H_2O \rightleftharpoons HCO_3^- + H^+$

(c)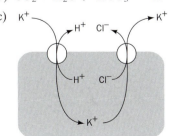

15. (a) $ATP + H_2O \rightarrow ADP + P_i$

The transporter must include a cytosolic nucleotide binding site that changes its conformation when its bound ATP is hydrolyzed to ADP. This conformational change must be communicated to the membrane-spanning portion of the protein, where the transported substrate binds. (b) Overexpression of an MDR transporter would increase the ability of the cancer cell to excrete anticancer drugs. Higher concentrations of the drugs or different drugs would then be required to kill the drug-resistant cells.

CHAPTER 11

1. b

2. (a) isomerase (alanine racemase); (b) lyase (pyruvate decarboxylase); (c) oxidoreductase (lactate dehydrogenase); (d) ligase (glutamine synthetase).

3. As shown in Table 11-1, the only relationship between the rates of catalyzed and uncatalyzed reactions is that the catalyzed reaction is faster than the uncatalyzed reaction. The absolute rate of an uncatalyzed reaction does not correlate with the degree to which it is accelerated by an enzyme.

4. There are three transition states ($X^{\ddagger}$) and two intermediates (I). The reaction is not thermodynamically favorable because the free energy of the products is greater than that of the reactants.

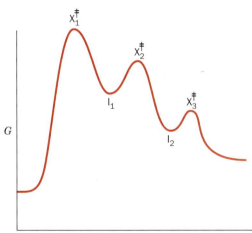

Reaction coordinate

5. The tighter S binds to the enzyme, the greater the value of $\Delta G_E^{\ddagger}$. As $\Delta G_E^{\ddagger}$ approaches $\Delta G_N^{\ddagger}$, the rate of the enzyme-catalyzed reaction approaches the rate to the nonenzymatic reaction.

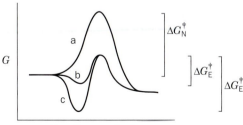

Reaction coordinate

6. At 25°C, every 10-fold increase in rate corresponds to a decrease of about 5.7 kJ·mol^{-1} in $\Delta G^{\ddagger}$. For the nuclease, with

a rate enhancement on the order of 10^{14}, $\Delta G^{\ddagger}$ is lowered about 14×5.7 kJ $\cdot$ mol^{-1}, or about 80 kJ $\cdot$ mol^{-1}. Alternatively, since the rate enhancement, k, is given by

$$k = e^{\Delta\Delta G^{\ddagger}_{cat}/RT}$$
$$\ln k = \ln(10^{14}) = \Delta\Delta G^{\ddagger}_{cat}/8.3145 \times (273 + 25)$$

Hence $\Delta\Delta G^{\ddagger}_{cat} = 80$ kJ $\cdot$ mol^{-1}

7. Glu has a pK of ~4 and, in its ionized form, acts as a base catalyst. Lys has a pK of ~10 and, in its protonated form, acts as an acid catalyst.

8. The active form of the enzyme contains the thiolate ion. The increased pK would increase the nucleophilicity of the thiolate and thereby increase the rate of the reaction catalyzed by the active form of the enzyme. However, at physiological pH, there would be less of the active form of the enzyme and therefore the overall rate would be decreased.

9. DNA lacks the 2'-OH group required for the formation of the 2',3'-cyclic reaction intermediate.

10. Two such analogs are

Furan-2-carboxylate   **Thiophene-2-carboxylate**

Both of these molecules are planar, particularly at the C atom to which the carboxylate is bonded, as is true of the transition state for the proline racemase reaction.

11. The preferential binding of the transition state to an enzyme is an important (often the most important) part of an enzyme's catalytic mechanism. Hence, the substrate binding site is the catalytic site.

12. The lysozyme active site is arranged to cleave oligosaccharides between the fourth and fifth residues. Moreover, since the lysozyme active site can bind at least six monosaccharide units, (NAG)$_6$ would be more tightly bound to the enzyme than (NAG)$_4$, and this additional binding free energy would be applied to distorting the D ring to its half-chair conformation, thereby facilitating the reaction.

13. Asp 101 and Arg 114 form hydrogen bonds with the substrate molecule (Fig. 11-19). Ala cannot form these hydrogen bonds, so the substituted enzyme is less active.

14.

Tosyl-L-alanine chloromethylketone   or

Tosyl-L-valine chloromethylketone

15. Yes. An enzyme decreases the activation energy barrier for both the forward and the reverse directions of a reaction.

16.

17. The observation that subtilisin and chymotrypsin are genetically unrelated indicates that their active site geometries arose by convergent evolution. Assuming that evolution has optimized the catalytic efficiencies of these enzymes and that there is only one optimal arrangement of catalytic groups, any similarities between the active sites of subtilisin and chymotrypsin must be of catalytic significance. Conversely, any differences are unlikely to be catalytically important.

18. (a) Little or no effect; (b) catalysis would be much slower because the mutation disrupts the function of the catalytic triad.

19. Activated factor IXa leads, via several steps, to the activation of the final coagulation protease, thrombin. The absence of factor IX therefore slows the production of thrombin, delaying clot formation, and causing the bleeding of hemophilia. Although activated factor XIa also leads to thrombin production, factor XI plays no role until it is activated by thrombin itself. By this point, coagulation is already well underway, so a deficiency of factor XI does not significantly delay coagulation.

20. As a digestive enzyme, chymotrypsin's function is to indiscriminately degrade a wide variety of ingested proteins, so that their component amino acids can be recovered. Broad substrate specificity would be dangerous for a protease that functions outside of the digestive system, since it might degrade proteins other than its intended target.

21. If the soybean trypsin inhibitor were not removed from tofu, it would inhibit the trypsin in the intestine. At best, this would reduce the nutritional value of the meal by rendering its protein indigestible. It might very well also lead to intestinal upset.

CHAPTER 12

1. (a) $v = k[A]$
   $k = v/[A]$
   $k = (5 \ \mu M \cdot min^{-1})/(20 \ mM)$
   $= (0.005 \ mM \cdot min^{-1})/(20 \ mM)$
   $= 2.5 \times 10^{-4} \ min^{-1}$

   (b) The reaction has a molecularity of 1.

2. From Eq. 13-7, $[A] = [A]_o \ e^{-kt}$. Since $t_{1/2} = 0.693/k$, $k = 0.693/14$ d $= 0.05$ d^{-1}. (a) 7 μmol; (b) 5 μmol; (c) 3.5 μmol; (d) 0.3 μmol.

3. $v = k[A]^2$
   $v = (10^{-6} \ M^{-1} \cdot s^{-1})(0.01 \ M)(0.01 \ M)$
   $v = 10^{-10} \ M \cdot s^{-1}$

4. For Reaction A, only a plot of 1/[reactant] versus t gives a straight line, so the reaction is second order. The slope, k, is 0.15 mM$^{-1} \cdot$ s^{-1}. For Reaction B, only a plot of ln[reactant] versus t gives a straight line, so the reaction is first order. The negative of the slope, k, is 0.17 s^{-1}.

Time (s)	Reaction A 1/[reactant] (mM^{-1})	Reaction B ln[reactant]
0	0.16	1.69
1	0.32	1.53
2	0.48	1.36
3	0.62	1.16
4	0.78	0.99
5	0.91	0.83

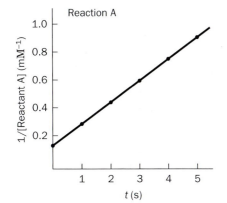

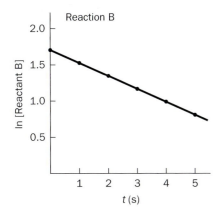

5.

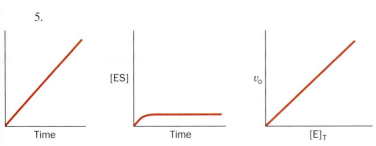

6. Enzyme activity is measured as an initial reaction velocity, the velocity before much substrate has been depleted and be-

fore much product has been generated. It is easier to measure the appearance of a small amount of product from a baseline of zero product than to measure the disappearance of a small amount of substrate against a background of a high concentration of substrate.

7. $v_o = V_{max}[S]/(K_M + [S])$
$v_o/V_{max} = [S]/(K_M + [S])$
$0.95 = [S]/(K_M + [S])$
$[S] = 0.95K_M + 0.95[S]$
$0.05[S] = 0.95\ K_M$
$[S] = (0.95/0.05)K_M = 19K_M$

8. Acetylcholinesterase, carbonic anhydrase, catalase, and fumarase.

9. Set A corresponds to $[S] > K_M$, and set B corresponds to $[S] < K_M$. Ideally, a single data set should include [S] values that are both larger and smaller than K_M.

Set A [S] (mM)	v_o (μM$\cdot$s^{-1})	Set B [S] (mM)	v_o (μM$\cdot$s^{-1})
2	0.42	0.12	0.17
1	0.38	0.10	0.15
0.67	0.34	0.08	0.13
0.50	0.32	0.07	0.11

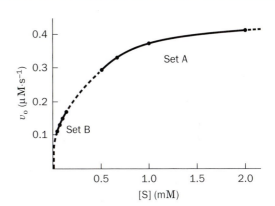

10. Construct a Lineweaver–Burk plot.

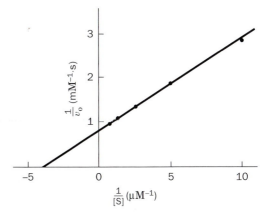

$K_M = -1/x\text{-intercept} = -1/(-4\ \mu\text{M}^{-1}) = 0.25\ \mu\text{M}$
$V_{max} = 1/y\text{-intercept} = 1/(0.8\ \text{mM}^{-1}\cdot\text{s}) = 1.25\ \text{mM}\cdot\text{s}^{-1}$

11. Comparing the two data points, since a 100-fold increase in substrate concentration only produces a 10-fold increase in reaction velocity, it appears that when $[S] = 100\ \mu M$, the velocity is close to V_{max}. Therefore, assume that $V_{max} \approx 50\ \mu M \cdot s^{-1}$ and use the other data point to estimate K_M using the Michaelis–Menten equation:

$$v_o = \frac{V_{max}[S]}{K_M + [S]}$$

$$K_M + [S] = \frac{V_{max}[S]}{v_o}$$

$$K_M = \frac{V_{max}[S]}{v_o} - [S]$$

$$K_M = \frac{(50\ \mu M \cdot s^{-1})(1\ \mu M)}{(5\ \mu M \cdot s^{-1})} - (1\ \mu M) = 9\ \mu M$$

The true V_{max} must be greater than the estimated value, so the value of K_M is an underestimate of the true K_M.

12. The experimentally determined K_M would be greater than the true K_M because the actual substrate concentration is less than expected.

13. The enzyme concentration is comparable to the lowest substrate concentration and therefore does not meet the requirement that $[E] \ll [S]$. You could fix this problem by decreasing the amount of enzyme used for each measurement.

14. Velocity measurements can be made using any convenient unit of change per unit of time. K_M is, by definition, a substrate concentration (the concentration when $v_o = V_{max}/2$), so its value does not reflect how the velocity is measured.

15. (a, b) It is not necessary to know $[E]_T$. The only variables required to determine K_M and V_{max} (for example, by constructing a Lineweaver–Burk plot) are $[S]$ and v_o. (c) The value of $[E]_T$ is required to calculate k_{cat} since $k_{cat} = V_{max}/[E]_T$.

16. (a) N-Acetyltyrosine ethyl ester, with the lower value of K_M, has greater apparent affinity for chymotrypsin. (b) The value of V_{max} is not related to the value of K_M, so no conclusion can be drawn.

17. Product P will be more abundant because enzyme A has a much lower K_M for the substrate than enzyme B. Because V_{max} is approximately the same for the two enzymes, the relative efficiency of the enzymes depends almost entirely on their K_M values.

18. (a, b) A* will appear only if the reaction follows a Ping Pong mechanism, since only a double-displacement reaction can exchange an isotope from P back to A in the absence of B. Hence, in a reaction that has a Sequential mechanism, A will not become isotopically labeled.

19. The lines of the double-reciprocal plots intersect to the left of the $1/v_o$ axis (on the $1/[S]$ axis). Hence, inhibition is mixed (with $\alpha = \alpha'$).

[S]	1/[S]	$1/v_o$	$1/v_o$ with I
1	1.00	0.7692	1.2500
2	0.50	0.5000	0.8333
4	0.25	0.3571	0.5882
8	0.125	0.2778	0.4545
12	0.083	0.2500	0.4167

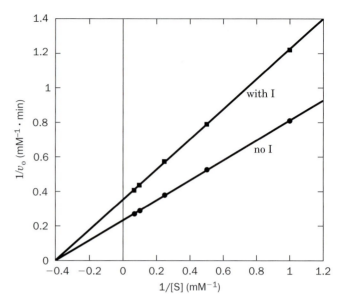

20. From Eq. 12-32, α is 3.
$$\alpha = 3 = 1 + [I]/K_I = 1 + 5\ mM/K_I$$
$$K_I = 2.5\ mM$$

21. (a) Inhibition is most likely mixed (noncompetitive) with $\alpha = \alpha'$ since it is reversible and only V_{max} is affected.

(b) Since $V_{max}^{app} = 0.8\ V_{max}$, 80% of the enzyme remains uninhibited. Therefore, 20% of the enzyme molecules have bound inhibitor.

(c) As indicated in Table 12-2 for mixed inhibition, $V_{max}^{app} = V_{max}/\alpha'$. Thus,

$$\alpha' = \frac{V_{max}}{V_{max}^{app}} = \frac{1}{0.8} = 1.25$$

From Eq. 12-35,

$$1.25 = 1 + \frac{[I]}{K_I'}$$

$$K_I' = \frac{5\ nM}{1.25 - 1} = 20\ nM$$

22. By irreversibly reacting with chymotrypsin's active site, DIPF would decrease $[E]_T$. The apparent V_{max} would decrease since $V_{max} = k_{cat}[E]_T$. K_M would not be affected since the uninhibited enzyme would bind substrate normally.

23. (a) If an irreversible inhibitor is present, the enzyme solution's activity would be exactly 100 times lower when the sample is diluted 100-fold. Dilution would not significantly change the enzyme's degree of inhibition. (b) For reversible inhibition, $K_I = [E][I]/[EI]$ so that $[E]/[EI] = K_I/[I]$. Hence, if a reversible inhibitor is present, dilution would lower the concentrations of both the enzyme and inhibitor so that the degree of dissociation of the inhibitor from the enzyme would increase. The enzyme solution's activity would therefore not be exactly 100 times less than the undiluted sample, but would be somewhat greater than this value because the proportion of uninhibited enzyme would be greater at the lower concentration.

24. Enzyme Y is more efficient at low [S]; enzyme X is more efficient at high [S].

25.

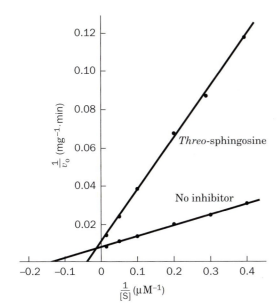

(a) K_M is determined from the x-intercept $(= -1/K_M)$. In the absence of inhibitor, $K_M = 1/0.14\ \mu M^{-1} = 7\ \mu M$. In the presence of inhibitor, $K_M^{app} = 1/0.04\ \mu M^{-1} = 25\ \mu M$. V_{max} is determined from the y-intercept $(= 1/V_{max})$. In the absence of inhibitor, $V_{max} = 1/0.008\ mg^{-1} \cdot min = 125\ mg \cdot min^{-1}$. In the presence of inhibitor, $V_{max}^{app} = 1/0.01\ mg^{-1} \cdot min = 100\ mg \cdot min^{-1}$.

(b) The lines in the double-reciprocal plots intersect very close to the $1/v_o$ axis. Hence, *threo*-sphingosine is most likely a competitive inhibitor. Competitive inhibition is likely also because of the structural similarity between the inhibitor and the substrate, which allows them to compete for binding to the enzyme active site.

CHAPTER 13

1. C, D, A, E, B

2. (a) Since $\Delta G^{\circ\prime} = -RT \ln K$,
$$K = e^{-\Delta G^{\circ\prime}/RT}$$
$$K = e^{-(7500\ J \cdot mol^{-1})/(8.3145\ J \cdot K^{-1} \cdot mol^{-1})(298\ K)}$$
$$K = 0.048$$

(b) $\Delta G = \Delta G^{\circ\prime} + RT \ln \dfrac{[B]}{[A]}$
$$\Delta G = 7500\ J \cdot mol^{-1}$$
$$+ (8.3145\ J \cdot K^{-1} \cdot mol^{-1})(310\ K)\ \ln \dfrac{(0.0001)}{(0.0005)}$$
$$\Delta G = 7500\ J \cdot mol^{-1} - 4150\ J \cdot mol^{-1}$$
$$\Delta G = 3350\ J \cdot mol^{-1} = 3.35\ kJ \cdot mol^{-1}$$

The reaction is not spontaneous since $\Delta G > 0$.

(c) The reaction can proceed in the cell if the product B is the substrate for a second reaction such that the second reaction continually draws off B, causing the first reaction to continually produce more B from A.

3. b

4. The theoretical maximum yield of ATP is equivalent to $(\Delta G^{\circ\prime}$ for fuel oxidation$)/(\Delta G^{\circ\prime}$ for ATP synthesis$)$.

(a) $(-2850\ kJ \cdot mol^{-1})/(-30.5\ kJ \cdot mol^{-1}) \approx 93$ ATP

(b) $(-9781\ kJ \cdot mol^{-1})/(-30.5\ kJ \cdot mol^{-1}) \approx 320$ ATP

5. At pH 6, the phosphate groups are more ionized than they are at pH 5, which increases their electrostatic repulsion and therefore increases the magnitude of ΔG for hydrolysis (makes it more negative).

6. The exergonic hydrolysis of PP_i by pyrophosphatase $(\Delta G^{\circ\prime} = -19.2\ kJ \cdot mol^{-1})$ drives fatty acid activation.

7. Calculating ΔG for the reaction ATP + creatine $\rightleftharpoons$ phosphocreatine + ADP, using Eq. 13-1:
$$\Delta G = \Delta G^{\circ\prime} + RT \ln \left(\frac{[\text{phosphocreatine}][\text{ADP}]}{[\text{creatine}][\text{ATP}]} \right)$$
$$= 12.6\ kJ \cdot mol^{-1} + (8.3145\ J \cdot K^{-1} \cdot mol^{-1})(298\ K)$$
$$\ln \left(\frac{(2.5\ mM)(0.15\ mM)}{(1\ mM)(4\ mM)} \right)$$
$$= 12.6\ kJ \cdot mol^{-1} - 5.9\ kJ \cdot mol^{-1} = 6.7\ kJ \cdot mol^{-1}$$

Since $\Delta G > 0$, the reaction will proceed in the opposite direction as written above, that is, in the direction of ATP synthesis.

8. Using the data in Table 13-2, we calculate $\Delta G^{\circ\prime}$ for the adenylate kinase reaction.

	$\Delta G^{\circ\prime}$
ATP + $H_2O \longrightarrow$ AMP + PP_i	$-32.2\ kJ \cdot mol^{-1}$
2 ADP + 2 $P_i \longrightarrow$ 2 ATP + 2 H_2O	2×30.5
	$= 61.0\ kJ \cdot mol^{-1}$
$PP_i + H_2O \longrightarrow$ 2 P_i	$-19.2\ kJ \cdot mol^{-1}$
2 ADP $\longrightarrow$ ATP + AMP	$9.6\ kJ \cdot mol^{-1}$

Since ΔG for a reaction at equilibrium is zero, Eq. 13-1 becomes $\Delta G^{\circ\prime} = -RT \ln K_{eq}$ so that $K_{eq} = e^{-\Delta G^{\circ\prime}/RT}$.
$$K_{eq} = \frac{[\text{ATP}][\text{AMP}]}{[\text{ADP}]^2} = e^{-\Delta G^{\circ\prime}/RT}$$
$$[\text{AMP}] = \frac{(5 \times 10^{-4}\ M)^2}{(5 \times 10^{-3}\ M)}$$
$$\times\ e^{-(9600\ J \cdot mol^{-1})/(8.3145\ J \cdot K^{-1} \cdot mol^{-1})(298\ K)}$$
$$[\text{AMP}] = 1.0 \times 10^{-6}\ M = 1.0 \times 10^{-3}\ mM$$

9. The more positive the reduction potential, the greater the oxidizing power. From Table 13-3,

Compound	$\mathscr{E}^{\circ\prime}$ (V)
Acetoacetate	-0.346
NAD^+	-0.315
Pyruvate	-0.185
Cytochrome b (Fe^{3+})	0.077
SO_4^{2-}	0.480

10. The balanced equation is
$$2\ \text{cyto}\ c\ (Fe^{3+}) + \text{ubiquinol} \rightarrow$$
$$2\ \text{cyto}\ c\ (Fe^{2+}) + \text{ubiquinone} + 2\ H^+$$
Using the data in Table 13-3,
$$\Delta \mathscr{E}^{\circ\prime} = \mathscr{E}^{\circ\prime}_{(e^- \text{ acceptor})} - \mathscr{E}^{\circ\prime}_{(e^- \text{ donor})} = 0.235\ V - 0.045\ V$$
$$= 0.190\ V$$
$$\Delta G^{\circ\prime} = -n\mathscr{F}\Delta \mathscr{E}^{\circ\prime} = -(2)(96{,}485\ J \cdot V^{-1} \cdot mol^{-1})(0.190\ V)$$
$$= -36.7\ kJ \cdot mol^{-1}$$

11. Using the data in Table 13-3:

(a) $\Delta \mathscr{E}^{\circ\prime} = \mathscr{E}^{\circ\prime}_{(e^- \text{ acceptor})} - \mathscr{E}^{\circ\prime}_{(e^- \text{ donor})} = \mathscr{E}^{\circ\prime}_{(\text{fumarate})} - \mathscr{E}^{\circ\prime}_{(\text{NAD}^+)}$
$= 0.031\ V - (-0.315\ V) = 0.346\ V$. Because $\Delta \mathscr{E}^{\circ\prime} > 0$,

$\Delta G^{\circ\prime} < 0$ and the reaction will spontaneously proceed as written.

(b) $\Delta\mathscr{E}^{\circ\prime} = \mathscr{E}^{\circ\prime}_{(e^- \text{ acceptor})} - \mathscr{E}^{\circ\prime}_{(e^- \text{ donor})} = \mathscr{E}^{\circ\prime}_{(\text{cyto } b)} - \mathscr{E}^{\circ\prime}_{(\text{cyto } a)} = 0.077$ V $- (0.290$ V$) = -0.213$ V. Because $\Delta\mathscr{E}^{\circ\prime} < 0$, $\Delta G^{\circ\prime} > 0$ and the reaction will spontaneously proceed in the opposite direction from that written.

12. Using the data in Table 13-3, for the oxidation of free $FADH_2$ ($\mathscr{E}^{\circ\prime} = -0.219$ V) by ubiquinone ($\mathscr{E}^{\circ\prime} = 0.045$ V),

$$\Delta\mathscr{E}^{\circ\prime} = \mathscr{E}^{\circ\prime}_{(\text{ubiquinone})} - \mathscr{E}^{\circ\prime}_{(\text{FADH}_2)}$$
$$= (0.045 \text{ V}) - (-0.219 \text{ V}) = 0.264 \text{ V}$$
$$\Delta G^{\circ\prime} = -n\mathscr{F}\Delta\mathscr{E}^{\circ\prime} = -(2)(96{,}485 \text{ J}\cdot\text{V}^{-1}\cdot\text{mol}^{-1})$$
$$(0.264 \text{ V}) = -50.9 \text{ kJ}\cdot\text{mol}^{-1}$$

This is more than enough free energy to drive the synthesis of ATP from ADP + P_i ($\Delta G^{\circ\prime} = +30.5$ kJ $\cdot$ mol^{-1}; Table 13-2).

13. $Z \xrightarrow{B} W \xrightarrow{C} Y \xrightarrow{A} X$

14. (a) The step catalyzed by enzyme Y is likely to be the major flux-control point, since this step operates farthest from equilibrium (it is an irreversible step). (b) Inhibition of enzyme Z would cause the concentration of D, the reaction's product, to decrease, and it would cause C, the reaction's substrate, to accumulate. The concentrations of A and B would not change because the steps catalyzed by enzymes X and Y would not be affected. The accumulated C would not be transformed back to B since the step catalyzed by enzyme Y is irreversible.

CHAPTER 14

1. (a) Reactions 1, 3, 7, and 10; (b) Reactions 2, 5, and 8; (c) Reaction 6; (d) Reaction 9; (e) Reaction 4.

2. C1 of DHAP and C1 of GAP are achiral but become chiral in FBP (as C3 and C4). There are four stereoisomeric products that differ in configuration at C3 and C4: fructose-1,6-bisphosphate, psicose-1,6-bisphosphate, tagatose-1,6-bisphosphate, and sorbose-1,6-bisphosphate (see Fig. 8-2).

3. The Zn^{2+} polarizes the carbonyl oxygen of the substrate to stabilize the enolate intermediate of the reaction.

4. (a) Glucose + 2 NAD^+ + 2 ADP + 2 P_i
$$\longrightarrow 2 \text{ pyruvate} + 2 \text{ NADH} + 2 \text{ ATP} + 2 \text{ H}_2\text{O}$$

(b) Glucose + 2 NAD^+ + 2 ADP + 2 $AsO_4^{3-} \longrightarrow$
$$2 \text{ pyruvate} + 2 \text{ NADH} + 2 \text{ ADP—AsO}_3^{2-} + 2 \text{ H}_2\text{O}$$

$$\underline{2 \text{ ADP—AsO}_3^{2-} + 2 \text{ H}_2\text{O} \longrightarrow 2 \text{ ADP} + \text{AsO}_4^{3-}}$$

Overall: Glucose + 2 NAD^+
$$\longrightarrow 2 \text{ pyruvate} + 2 \text{ NADH}$$

(c) Arsenate is a poison because it uncouples ATP generation from glycolysis. Consequently, glycolytic energy generation cannot occur.

5.   ribulose-5-phosphate   ribulose-5-phosphate
      isomerase reaction: epimerase reaction:

1,2-Enediolate   **2,3-Enediolate**
intermediate   **intermediate**

6. (a) ΔG values differ from $\Delta G^{\circ\prime}$ values because $\Delta G = \Delta G^{\circ\prime} + RT$ ln[Products]/[Reactants] and cellular reactants and products are not in their standard states.

(b) Yes. The same *in vivo* conditions that decrease the magnitude of ΔG relative to $\Delta G^{\circ\prime}$ may also decrease ΔG for ATP synthesis.

7. When [GAP] = 10^{-4} M, [DHAP] = 5.5×10^{-4} M. According to Eq. 1-17, $K = e^{-\Delta G^{\circ\prime}/RT}$

$$\frac{[\text{GAP}][\text{DHAP}]}{[\text{FBP}]} = e^{-(22{,}800 \text{ J}\cdot\text{mol}^{-1})/(8.3145 \text{ J}\cdot\text{K}^{-1}\cdot\text{mol}^{-1})(310 \text{ K})}$$

$$\frac{(10^{-4})(5.5 \times 10^{-4})}{[\text{FBP}]} = 1.4 \times 10^{-4}$$

[FBP] = 3.8×10^{-4} M
[FBP]/[GAP] = $(3.8 \times 10^{-4}$ M$)/(10^{-4}$ M$) = 3.8$

8. For the coupled reaction

$$\text{Pyruvate} + \text{NADH} + \text{H}^+ \longrightarrow \text{lactate} + \text{NAD}^+$$

$\Delta\mathscr{E}^{\circ\prime} = (-0.185 \text{ V}) - (-0.315 \text{ V}) = 0.130$ V.
According to Eq. 13-8

$$\Delta\mathscr{E} = \Delta\mathscr{E}^{\circ\prime} - \frac{RT}{n\mathscr{F}} \ln\left(\frac{[\text{lactate}][\text{NAD}^+]}{[\text{pyruvate}][\text{NADH}]}\right)$$

and $\Delta G = -n\mathscr{F}\Delta\mathscr{E}$ (Eq. 13-7). Since two electrons are transferred in the above reaction, $n = 2$.

(a) $\Delta\mathscr{E} = 0.130 \text{ V} - \dfrac{RT}{n\mathscr{F}} \ln(1) = 0.130$ V
$$\Delta G = -(2)(96{,}485 \text{ J}\cdot\text{V}^{-1}\cdot\text{mol}^{-1})(0.130 \text{ V})$$
$$= -25.1 \text{ kJ}\cdot\text{mol}^{-1}$$

(b) $RT/n\mathscr{F} = (8.3145 \text{ J}\cdot\text{K}^{-1}\cdot\text{mol}^{-1})(298 \text{ K})/$
$$(2)(96{,}485 \text{ J}\cdot\text{V}^{-1}\cdot\text{mol}^{-1}) = 0.01284 \text{ V}$$
$\Delta\mathscr{E} = 0.130 \text{ V} - 0.01284 \text{ V} \ln(160 \times 160)$
$\Delta\mathscr{E} = 0.130 \text{ V} - 0.130 \text{ V} = 0$
$\Delta G = 0$

(c) $\Delta\mathscr{E} = 0.130 \text{ V} - 0.01284 \text{ V} \ln(1000 \times 1000)$
$\Delta\mathscr{E} = 0.130 \text{ V} - 0.177 \text{ V} = -0.047 \text{ V}$
$\Delta G = -(2)(96{,}485 \text{ J}\cdot\text{V}^{-1}\cdot\text{mol}^{-1})(-0.047 \text{ V})$
$$= 9.1 \text{ kJ}\cdot\text{mol}^{-1}$$

(d) At the concentration ratios of Part a, ΔG is negative and the reaction proceeds as written. As [lactate]/[pyruvate] increases, the reaction ΔG increases even though $[\text{NAD}^+]/[\text{NADH}]$ also increases so that in Part b the reaction is at equilibrium ($\Delta G = 0$) and in Part c ΔG is

positive and the reaction proceeds spontaneously in the opposite direction.

9. (a) Pyruvate kinase regulation is important for controlling the flux of metabolites, such as fructose (in liver), which enter glycolysis after the PFK step.

(b) FBP is the product of the third reaction of glycolysis, so it acts as a feed-forward activator of the enzyme that catalyzes Step 10. This regulatory mechanism helps ensure that once metabolites pass the PFK step of glycolysis, they will continue through the pathway.

10. The three glucose molecules that proceed through glycolysis yield 6 ATP. The bypass through the pentose phosphate pathway results in a yield of 5 ATP.

11. The label will appear at C1 and C3 of F6P (see Fig. 14-30).

12. (a) Transketolase transfers 2-carbon units from a ketose to an aldose, so the products are a 3-carbon sugar and a 7-carbon sugar. (b) The products are a 4-carbon and a 7-carbon sugar. The order of binding does matter. The ketose binds first and transfers the 2-carbon unit to the TPP on the enzyme. The aldose then binds and accepts the 2-carbon unit.

13. Even when the flux of glucose through glycolysis and hence the citric acid cycle is blocked, glucose can be oxidized by the pentose phosphate pathway, with the generation of CO_2.

14. The liver enzyme is far more sensitive than the brain enzyme to the three activators. It is possible that liver PFK-1 is subject to a greater degree of regulation than brain PFK-1. Fuel must be supplied to the brain continuously and thus glycolysis is always active, but the liver has a wide variety of physiological roles and is more likely to regulate cellular pathways.

15. (a)

$$\underset{\textbf{Pyruvate}}{\begin{matrix} COO^- \\ | \\ C=O \\ | \\ CH_3 \end{matrix}} \quad + \quad \underset{\textbf{GAP}}{\begin{matrix} H \quad\quad O \\ \diagdown \;\; \diagup \\ C \\ | \\ H-C-OH \\ | \\ CH_2OPO_3^{2-} \end{matrix}}$$

(b) The pyruvate product of the aldolase reaction is not further modified. The GAP product is converted to pyruvate by the actions of the glycolytic enzymes glyceraldehyde-3-phosphate dehydrogenase, phosphoglycerate kinase, phosphoglycerate mutase, enolase, and pyruvate kinase.

(c) One ATP is consumed when glucose is converted to glucose-6-phosphate. One ATP is generated by the phosphoglycerate kinase reaction and one by the pyruvate kinase reaction (these quantities are not doubled because only one three-carbon fragment of glucose follows this route), for a net yield of one ATP per glucose. The standard glycolytic pathway generates two ATP per glucose.

CHAPTER 15

1. (a) +9 ATP, (b) +6 ATP, (c) −18 ATP.

2. (a) The equation for glycolysis is

Glucose + 2 NAD$^+$ + 2 ADP + 2 P$_i$ $\longrightarrow$
  2 pyruvate + 2 NADH + 4 H$^+$ + 2 ATP + 2 H$_2$O

The equation for gluconeogenesis is

2 Pyruvate + 2 NADH + 4 H$^+$ + 4 ATP + 2 GTP + 6 H$_2$O
  $\longrightarrow$ glucose + 2 NAD$^+$ + 4 ADP + 2 GDP + 6 P$_i$

For the two processes operating sequentially,

2 ATP + 2 GTP + 4 H$_2$O $\longrightarrow$ 2 ADP + 2 GDP + 4 P$_i$

(b) The equation for catabolism of 6 G6P by the pentose phosphate pathway is

6 G6P + 12 NADP$^+$ + 6 H$_2$O $\longrightarrow$
  6 Ru5P + 12 NADPH + 12 H$^+$ + 6 CO$_2$

Ru5P can be converted to G6P by transaldolase, transketolase, and gluconeogenesis:

6 Ru5P + H$_2$O $\longrightarrow$ 5 G6P + P$_i$

The net equation is therefore

G6P + 12 NADP$^+$ + 7 H$_2$O $\longrightarrow$
  12 NADPH + 12 H$^+$ + 6 CO$_2$ + P$_i$

3. Phosphoglucokinase activity generates G1,6P, which is necessary to "prime" phosphoglucomutase that has become dephosphorylated and thereby inactivated through the loss of its G1,6P reaction intermediate.

4. The overall free energy change for debranching is

Breaking $\alpha(1 \rightarrow 4)$ bond	$\Delta G^{\circ\prime} = -15.5 \text{ kJ} \cdot \text{mol}^{-1}$
Forming $\alpha(1 \rightarrow 4)$ bond	$+15.5 \text{ kJ} \cdot \text{mol}^{-1}$
Hydrolyzing $\alpha(1 \rightarrow 6)$ bond	$-7.1 \text{ kJ} \cdot \text{mol}^{-1}$
Total	$\Delta G^{\circ\prime} = -7.1 \text{ kJ} \cdot \text{mol}^{-1}$

The overall free energy change for branching is

Breaking $\alpha(1 \rightarrow 4)$ bond	$\Delta G^{\circ\prime} = -15.5 \text{ kJ} \cdot \text{mol}^{-1}$
Forming $\alpha(1 \rightarrow 6)$ bond	$+7.1 \text{ kJ} \cdot \text{mol}^{-1}$
Total	$\Delta G^{\circ\prime} = -8.4 \text{ kJ} \cdot \text{mol}^{-1}$

Assuming that $\Delta G^{\circ\prime}$ is close to ΔG, the sum of the two reactions of branching has $\Delta G < 0$, but debranching would be endergonic ($\Delta G > 0$) without the additional step of hydrolyzing the $\alpha(1 \rightarrow 6)$ bond to form glucose.

5. A glycogen molecule with 28 tiers would represent the most efficient arrangement for storing glucose, and its outermost tier would contain considerably more glucose residues than a glycogen molecule with only 12 tiers. However, densely packed glucose residues would be inaccessible to phosphorylase. In fact, such a dense glycogen molecule could not be synthesized because glycogen synthase and branching enzyme would have no room to operate (see Box 15-3 for a discussion of glycogen structure).

6. In the course of glucose catabolism, a detour through glycogen synthesis and glycogen breakdown begins and ends with G6P. The energy cost of this detour is 1 ATP equivalent, consumed in the UDP–glucose pyrophosphorylase step. The overall energy lost is therefore 1/38 or ~3%.

7. This mechanism allows glycogen phosphorylase activity to be regulated by the concentration of glucose so that glycogen is not broken down when glucose is already plentiful.

8. (a) Circulating [glucose] is high because cells do not respond to the insulin signal to take up glucose.

(b) Insulin is unable to activate phosphoprotein phosphatase-1 in muscle, so glycogen synthesis is not stimulated. Moreover, glycogen synthesis is much reduced by the lack of available glucose in the cell.

9. A defect in G6P transport would have the symptoms of glucose-6-phosphatase deficiency: accumulation of glycogen and hypoglycemia.

10. The conversion of circulating glucose to lactate in the muscle generates 2 ATP. If muscle glycogen could be mobilized, the energy yield would be 3 ATP, since phosphorolysis of glycogen bypasses the hexokinase-catalyzed step that consumes ATP in the first stage of glycolysis.

11. The deficiency is in branching enzyme (Type IV glycogen storage disease). The high ratio of G1P to glucose indicates abnormally long chains of $\alpha(1 \rightarrow 4)$-linked residues with few $\alpha(1 \rightarrow 6)$-linked branch points (the normal ratio is ~10).

CHAPTER 16

1. (a) Because citric acid cycle intermediates such as citrate and succinyl-CoA are precursors for the biosynthesis of other compounds, anaerobes must be able to synthesize them.

   (b) These organisms do not need a complete citric acid cycle, which would yield reduced coenzymes that must be reoxidized.

2. Citrate can be cleaved to generate an acetyl group and oxaloacetate. The oxaloacetate can then be converted to succinate to complete the cycle.

3. The decarboxylation step is most likely to be metabolically irreversible since the CO_2 product is rapidly hydrated to bicarbonate. The reverse reaction, a carboxylation, requires the input of free energy to become favorable (Section 15-4A). The other four reactions are transfer reactions or oxidation–reduction reactions (transfer of electrons) that are more easily reversed.

4. (a) The labeled carbon becomes C4 of the succinyl moiety of succinyl-CoA. Because succinate is symmetrical, the label appears at C1 and C4 of succinate. When the resulting oxaloacetate begins the second round, the labeled carbons appear as $^{14}CO_2$ in the isocitrate dehydrogenase and the α-ketoglutarate dehydrogenase reactions (see Fig. 16-2).

   (b) The labeled carbon becomes C3 of the succinyl moiety of succinyl-CoA and hence appears at C2 and C3 of succinate, fumarate, malate, and oxaloacetate. Neither C2 nor C3 of oxaloacetate is released as CO_2 in the second round of the cycle. However, the ^{14}C label appears at C1 and C2 of the succinyl moiety of succinyl-CoA in the second round and therefore appears at all four positions of the resulting oxaloacetate. Thus, in the third round, ^{14}C is released as $^{14}CO_2$.

5.

$$\underset{H_3C}{\overset{H_3C}{>}}CH - \overset{\overset{\displaystyle O}{\displaystyle \|}}{C} - S - CoA$$

6. NAD^+ ($\mathscr{E}°' = -0.315$ V) does not have a high enough reduction potential to support oxidation of succinate to fumarate ($\mathscr{E}°' = +0.031$ V); that is, the succinate dehydrogenase reaction has insufficient free energy to reduce NAD^+. Enzyme-bound FAD ($\mathscr{E}°' \approx 0$) is more suitable for oxidizing succinate.

7. Competitive inhibition can be overcome by adding more substrate, in this case succinate. Oxaloacetate overcomes malonate inhibition because it is converted to succinate by the reactions of the citric acid cycle.

8. The phosphofructokinase reaction is the major flux-control point for glycolysis. Inhibiting phosphofructokinase slows the entire pathway, so the production of acetyl-CoA by glycolysis followed by the pyruvate dehydrogenase complex can be decreased when the citric acid cycle is operating at maximum capacity and the citrate concentration is high. As citric acid cycle intermediates are consumed in synthetic pathways, the citrate concentration drops, relieving phosphofructokinase inhibition and allowing glycolysis to proceed in order to replenish the citric acid cycle intermediates.

9. To synthesize citrate, pyruvate must be converted to oxaloacetate by pyruvate carboxylase:

   Pyruvate + CO_2 + ATP + $H_2O \longrightarrow$
   $$\text{oxaloacetate + ATP + } P_i$$

   A second pyruvate is converted to acetyl-CoA by pyruvate dehydrogenase:

   Pyruvate + CoASH + $NAD^+ \longrightarrow$
   $$\text{acetyl-CoA + } CO_2 \text{ + NADH}$$

   The acetyl-CoA then combines with oxaloacetate to produce citrate:

   Oxaloacetate + acetyl-CoA + $H_2O \longrightarrow$
   $$\text{citrate + CoASH + } H^+$$

   The net reaction is

   2 Pyruvate + ATP + NAD^+ + 2 $H_2O \longrightarrow$
   $$\text{citrate + ADP + } P_i \text{ + NADH + } H^+$$

10. (a) The citric acid cycle is a multistep catalyst. Degrading an amino acid to a citric acid cycle intermediate boosts the catalytic activity of the cycle but does not alter the stoichiometry of the overall reaction (acetyl-CoA → 2 CO_2). In order to undergo oxidation, the citric acid cycle intermediate must exit the cycle and be converted to acetyl-CoA to re-enter the cycle as a substrate. (b) Pyruvate derived from the degradation of an amino acid can be converted to acetyl-CoA by the pyruvate dehydrogenase complex; these amino acid carbons can then be completely oxidized by the citric acid cycle.

11. The alternate pathway bypasses the succinyl-CoA synthetase reaction of the standard citric acid cycle, a step that is accompanied by the phosphorylation of ADP. The alternate pathway therefore generates one less ATP than the standard citric acid cycle. There is no difference in the number of reduced cofactors generated.

12. For the reaction isocitrate + $NAD^+ \rightleftharpoons \alpha$-ketoglutarate + NADH + CO_2 + H^+, we assume $[H^+]$ and $[CO_2] = 1$. According to Eq. 13-1,

$$\Delta G = \Delta G^{\circ\prime} + RT \ln\left(\frac{[\text{NADH}][\alpha\text{-ketoglutarate}]}{[\text{NAD}^+][\text{isocitrate}]}\right)$$
$$= -21 \text{ kJ} \cdot \text{mol}^{-1} + (8.3145 \text{ J} \cdot \text{K} \cdot \text{mol}^{-1})$$
$$(298 \text{ K}) \ln\left[\frac{(1)(0.1)}{(8)(0.02)}\right]$$
$$= -21 \text{ kJ} \cdot \text{mol}^{-1} - 1.17 \text{ kJ} \cdot \text{mol}^{-1} = -22.17 \text{ kJ} \cdot \text{mol}^{-1}$$

With such a large negative free energy of reaction under physiological conditions, isocitrate dehydrogenase is likely to be a metabolic control point.

13. Animals cannot carry out the net synthesis of glucose from acetyl-CoA (to which acetate is converted). However, ^{14}C-labeled acetyl-CoA enters the citric acid cycle and is converted to oxaloacetate. Some of this oxaloacetate may exchange with the cellular pool of oxaloacetate to be converted to glucose through gluconeogenesis and subsequently taken up by muscle and incorporated into glycogen.

CHAPTER 17

1. Mitochondria with more cristae have more surface area and therefore more proteins for electron transport and oxidative phosphorylation. Tissues with a high demand for ATP synthesis (such as heart) contain mitrochondria with more cristae than tissues with lower demand for oxidative phosphorylation (such as liver).

2. When NADH participates in the glycerophosphate shuttle, the electrons of NADH flow to FAD and then to CoQ, bypassing Complex I. Thus, about 2 ATP are synthesized per NADH. About three ATP are produced when NADH participates in the malate–aspartate shuttle.

3. The relevant half-reactions (Table 13-3) are

$$\text{FAD} + 2 \text{ H}^+ + 2 e^- \rightleftharpoons \text{FADH}_2 \qquad \mathscr{E}^{\circ\prime} = -0.219 \text{ V}$$
$$\tfrac{1}{2} \text{O}_2 + 2 \text{ H}^+ + 2 e^- \rightleftharpoons \text{H}_2\text{O} \qquad \mathscr{E}^{\circ\prime} = 0.815 \text{ V}$$

Since the $\text{O}_2/\text{H}_2\text{O}$ half-reaction has the more positive $\Delta\mathscr{E}^{\circ\prime}$, the FAD half-reaction is reversed and the overall reaction is

$$\tfrac{1}{2} \text{O}_2 + \text{FADH}_2 \rightleftharpoons \text{H}_2\text{O} + \text{FAD}$$
$$\Delta\mathscr{E}^{\circ\prime} = 0.815 \text{ V} - (-0.219 \text{ V}) = 1.034 \text{ V}$$

Since $\Delta G^{\circ\prime} = -n\mathscr{F}\Delta\mathscr{E}^{\circ\prime}$,

$$\Delta G^{\circ\prime} = -(2)(96{,}485 \text{ J} \cdot \text{V}^{-1} \cdot \text{mol}^{-1})(1.034 \text{ V})$$
$$= -200 \text{ kJ} \cdot \text{mol}^{-1}$$

The maximum number of ATP that could be synthesized under standard conditions is therefore $200 \text{ kJ} \cdot \text{mol}^{-1}/30.5 \text{ kJ} \cdot \text{mol}^{-1} = 6.6$ mol ATP/mol FADH_2 oxidized by O_2.

4.

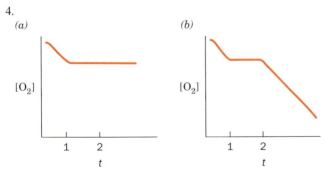

(a)   (b)

(c)   (d)

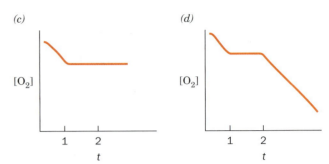

(a) O_2 consumption ceases because amytal blocks electron transport in Complex I.

(b) Electrons from succinate bypass the amytal block by entering the electron-transport chain at Complex II and thereby restore electron transport through Complexes III and IV.

(c) CN^- blocks electron transport in Complex IV, after the point of entry of succinate.

(d) Oligomycin blocks oxidative phosphorylation and hence O_2 consumption. DNP uncouples electron transport from oxidative phosphorylation and thereby permits O_2 consumption to resume.

5. $\mathscr{E}$ may differ from $\mathscr{E}^{\circ\prime}$, depending on the redox center's microenvironment and the concentrations of reactants and products. In addition, the tight coupling between successive electron transfers within a complex may "pull" electrons so that the overall process is spontaneous.

6. (a)

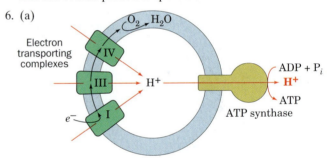

(b) An increase in external pH (decrease in $[\text{H}^+]$) increases the electrochemical potential across the mitochondrial membrane and therefore leads to an increase in ATP synthesis.

7. For the transport of a proton from outside to inside (Eq. 17-1),

$$\Delta G = 2.3 \, RT \, [\text{pH } (in) - \text{pH } (out)] + Z\mathscr{F}\Delta\Psi$$

The difference in pH is -1.4. Since an ion is transported from the positive to the negative side of the membrane, $\Delta\Psi$ is negative.

$$\Delta G = (2.3)(8.314 \text{ J} \cdot \text{K}^{-1} \cdot \text{mol}^{-1})(298 \text{ K})(-1.4) +$$
$$(1)(96{,}485 \text{ J} \cdot \text{V}^{-1} \cdot \text{mol}^{-1})(-0.06 \text{ V})$$
$$\Delta G = -7980 \text{ J} \cdot \text{mol}^{-1} - 5790 \text{ J} \cdot \text{mol}^{-1} = -13.8 \text{ kJ} \cdot \text{mol}^{-1}$$

Since $\Delta G^{\circ\prime}$ for ATP synthesis is $30.5 \text{ kJ} \cdot \text{mol}^{-1}$ and $30.5/13.8 = 2.2$, between two and three moles of protons must be transported to provide the free energy to synthesize one mole of ATP under standard biochemical conditions.

8. (a) The import of ADP (net charge -3) and the export of ATP (net charge -4) represents a loss of negative charge inside the mitochondria. This decreases the difference in electrical charge across the membrane, since the outside is positive due to the translocation of protons during electron transport. Consequently, the electrochemical gradient is diminished by the activity of the ADP–ATP translocator. The activity of the P_i–H^+ symport protein diminishes the proton gradient by allowing protons from the intermembrane space to re-enter the matrix.

   (b) Both transport systems are driven by the free energy of the electrochemical proton gradient.

9. The protonation and subsequent deprotonation of Asp 61 of the F_1F_0-ATPase's c subunits induces the rotation of the c-ring, which in turn, mechanically drives the synthesis of ATP. DCCD reacts with Asp 61 in a manner that prevents it from binding a proton and thereby prevents the synthesis of ATP.

10. In an ATP synthase with more c subunits, more proton translocation events are required to drive one complete rotation of the c ring. Consequently, more substrate oxidation (O_2 consumption) is required to synthesize three ATP (the yield of one cycle of the rotary engine), and the P/O ratio is lower.

11. DNP and related compounds dissipate the proton gradient required for ATP synthesis. The dissipation of this gradient decreases the rate of synthesis of ATP, decreasing the ATP mass action ratio. Decreasing this ratio relieves the inhibition of the electron transport chain, causing an increase in metabolic rate.

12. Hormones stimulate the release of fatty acids from stored triacylglycerols, which activates UCP1 and also provides the fuel whose oxidation yields electrons for the heat-generating electron transfer process. This cascade also amplifies the effect of the hormone.

13. The switch to aerobic metabolism allows ATP to be produced by oxidative phosphorylation. The phosphorylation of ADP increases the [ATP]/[ADP] ratio, which then increases the [NADH]/[NAD$^+$] ratio because a high ATP mass action ratio slows electron transport. The increases in [ATP] and [NADH] inhibit their target enzymes in glycolysis and the citric acid cycle (Fig. 17-29) and thereby slow these processes.

14. Glucose is shunted through the pentose phosphate pathway to provide NADPH, whose electrons are required to reduce O_2 to O_2^-.

CHAPTER 18

1. The color of the seawater indicates that the photosynthetic pigments of the algae absorb colors of visible light other than red.

2. $2\,H_2O + 2\,NADP^+ \rightarrow 2\,NADPH + 2\,H^+ + O_2$

3. The order of action is water–plastoquinone oxidoreductase (Photosystem II), plastoquinone–plastocyanin oxidoreductase (cytochrome b_6f), and plastocyanin–ferredoxin oxidoreductase (Photosystem I).

4. The label appears as $^{18}O_2$:

$$H_2^{18}O + CO_2 \xrightarrow{\text{light}} (CH_2O) + {}^{18}O_2$$

5. (a) The energy per photon is $E = hc/\lambda$, so the energy per mole of photons ($\lambda = 700$ nm) is

$$
\begin{aligned}
E &= Nhc/\lambda \\
&= (6.022 \times 10^{23}\ \text{mol}^{-1})(6.626 \times 10^{-34}\ \text{J} \cdot \text{s}) \\
&\quad (2.998 \times 10^8\ \text{m} \cdot \text{s}^{-1})/(7 \times 10^{-7}\ \text{m}) \\
&= 1.71 \times 10^5\ \text{J} \cdot \text{mol}^{-1} \\
&= 171\ \text{kJ} \cdot \text{mol}^{-1}
\end{aligned}
$$

   (b) $(171\ \text{kJ} \cdot \text{mol}^{-1})(30.5\ \text{kJ} \cdot \text{mol}^{-1}) = 5.6$

Five moles of ATP could theoretically be synthesized (at least under standard biochemical conditions).

6. (a) The relevant half-reactions are (Table 13-3):

$O_2 + 4\,e^- + 4\,H^+ \rightarrow H_2O$	$\Delta\mathscr{E}^{\circ\prime} = 0.815$ V
$NADP^+ + H^+ + 2\,e^- \rightarrow NADPH$	$\Delta\mathscr{E}^{\circ\prime} = -0.320$ V

The overall reaction is

$$2\,NADP^+ + 2\,H_2O \rightarrow 2\,NADPH + O_2 + 2\,H^+$$

$$
\begin{aligned}
\Delta\mathscr{E}^{\circ\prime} &= -0.320\ \text{V} - (0.815\ \text{V}) = -1.135\ \text{V} \\
\Delta G^{\circ\prime} &= -n\mathscr{F}\Delta\mathscr{E}^{\circ\prime} = \\
&= -(4)(96{,}485\ \text{J} \cdot \text{V}^{-1} \cdot \text{mol}^{-1})(-1.135\ \text{V}) \\
&= 438\ \text{kJ} \cdot \text{mol}^{-1}
\end{aligned}
$$

   (b) One mole of photons of red light ($\lambda = 700$ nm) has an energy of 171 kJ. Therefore, $438/171 = 2.6$ moles of photons are theoretically required to drive the oxidation of H_2O by NADP to form one mole of O_2.

   (c) The energy of a mole of photons of UV light ($\lambda = 220$ nm) is

$$
\begin{aligned}
E &= Nhc/\lambda \\
&= (6.022 \times 10^{23}\ \text{mol}^{-1})(6.626 \times 10^{-34}\ \text{J} \cdot \text{s}) \\
&\quad (2.998 \times 10^8\ \text{m} \cdot \text{s}^{-1})/(2.2 \times 10^{-7}\ \text{m}) \\
&= 544\ \text{kJ} \cdot \text{mol}^{-1}
\end{aligned}
$$

The number of moles of 220-nm photons required to produce one mole of O_2 is $438/544 = 0.8$.

7. Both systems mediate cyclic electron flows. The photooxidized bacterial reaction center passes electrons through a series of electron carriers so that electrons return to the reaction center (e.g., P960$^+$) and restore it to its original state. During cyclic electron flow in PSI, electrons from photooxidized P700 are transferred to cytochrome b_6f and, via plastoquinone and plastocyanin, back to P700$^+$. In both cases, there is no net change in the redox state of the reaction center, but the light-driven electron movements are accompanied by the transmembrane movement of protons.

8. When cyclic electron flow occurs, photoactivation of PSI drives electron transport independently of the flow of electrons derived from water. Thus, the oxidation of H_2O by PSII is not linked to the number of photons consumed by PSI.

9. (a) The buildup of the proton gradient is indicative of a high level of activity of the photosystems. A steep gradient could therefore trigger photoprotective activity to prevent further photooxidation when the proton-translocating machinery is operating at maximal capacity. (b) Photooxidation would not be a good protective mechanism since it might interfere with the normal redox balance among the electron-carrying groups in the thylakoid membrane. Releasing the energy by exciton transfer or fluorescence (emitting light of a longer wavelength) could potentially funnel light energy back to the

overactive photosystems. Dissipation of the excess energy via internal conversion to heat would be the safest mechanism, since the photosystems do not have any way to harvest thermal energy to drive chemical reactions.

10. (a) An uncoupler dissipates the transmembrane proton gradient by providing a route for proton translocation other than ATP synthase. Therefore, chloroplast ATP production would decrease. (b) The uncoupler would not affect $NADP^+$ reduction since light-driven electron transfer reactions would continue regardless of the state of the proton gradient.

11. An increase in $[O_2]$ increases the oxygenase activity of RuBP carboxylase–oxygenase and therefore lowers the efficiency of CO_2 fixation.

12. After the light is turned off, ATP and NADPH levels fall as these substances are used up in the Calvin cycle without being replaced by the light reactions. 3PG builds up because it cannot pass through the phosphoglycerate kinase reaction in the absence of ATP. The RuBP level drops because it is consumed by the RuBP carboxylase reaction (which requires neither ATP nor NADPH) and its replenishment is blocked by the lack of ATP for the phosphoribulokinase reaction.

13. The net synthesis of 2 GAP from 6 CO_2 in the initial stage of the Calvin cycle (Fig. 18-24) consumes 18 ATP and 12 NADPH (equivalent to 36 ATP). The conversion of 2 GAP to glucose-6-phosphate (G6P) by gluconeogenesis does not require energy input (Section 15-4B), nor does the isomerization of G6P to glucose-1-phosphate (G1P). The activation of G1P to its nucleotide derivative consumes 2 ATP equivalents (Section 15-5), but ADP is released when the glucose residue is incorporated into starch. These steps represent an overall energy investment of $18 + 36 + 1 = 55$ ATP.

Starch breakdown by phosphorolysis yields G1P, whose subsequent degradation by glycolysis yields 3 ATP, 2 NADH (equivalent to 6 ATP), and 2 pyruvate. Complete oxidation of 2 pyruvate to 6 CO_2 by the pyruvate dehydrogenase reaction and the citric acid cycle (Section 16-1) yields 8 NADH (equivalent to 24 ATP), 2 $FADH_2$ (equivalent to 4 ATP), and 2 GTP (equivalent to 2 ATP). The overall ATP yield is therefore $3 + 6 + 24 + 4 + 2 = 39$ ATP.

The ratio of energy spent to energy recovered is $55/39 = 1.4$.

14. These plants store CO_2 by CAM. At night, CO_2 reacts with PEP to form malate. By morning, so much malate (malic acid) has accumulated that the leaves have a sour taste. During the day, the malate is converted to pyruvate + CO_2. The leaves therefore become less acidic and hence tasteless. Late in the day, when all the malate is consumed, the leaves become slightly basic, that is, bitter.

CHAPTER 19

1. A defect in carnitine palmitoyl transferase II prevents normal transport of activated fatty acids into the mitochondria for β oxidation. Tissues such as muscle that use fatty acids as metabolic fuels therefore cannot generate ATP as needed.

The problem is more severe during a fast because other fuels, such as dietary glucose, are not readily available.

2.

L-Glycerol   **L-Glycerol-3-phosphate**

Dihydroxyacetone phosphate

3. The first three steps of β oxidation resemble the reactions that convert succinate to oxaloacetate (Sections 16-3F–16-3H).

Succinate   **Fumarate**

L-Malate   **Oxaloacetate**

4. (a) Six cycles are required.

(b) 3 Acetyl-CoA, 3 propionyl-CoA, and 1 methylpropionyl-CoA.

5. There are not as many usable nutritional calories per gram in unsaturated fatty acids as there are in saturated fatty acids. This is because oxidation of fatty acids containing double bonds yields fewer reduced coenzymes whose oxidation drives the synthesis of ATP. In the oxidation of fatty acids with a double bond at an odd-numbered carbon, the enoyl-CoA isomerase reaction bypasses the acyl-CoA dehydrogenase reaction and therefore does not generate $FADH_2$ (equivalent to 2 ATP). A double bond at an even-numbered carbon must be reduced by NADPH (equivalent to the loss of 3 ATP).

6. Oxidation of odd-chain fatty acids generates succinyl-CoA, an intermediate of the citric acid cycle. Because the citric acid cycle operates as a multistep catalyst to convert acetyl groups to CO_2, increasing the concentration of a cycle intermediate can increase the catalytic activity of the cycle.

7. Palmitate oxidation produces 129 ATP and glucose catabolism produces 38 ATP (see Section 16-4). Assuming a free energy cost of 30.5 kJ·mol^{-1} to synthesize ATP, palmitate catabolism has an efficiency of

$$129 \times 30.5/9781 \times 100 = 40\%$$

Glucose catabolism has an efficiency of

$$38 \times 30.5/2850 \times 100 = 41\%$$

Thus, the two processes have very nearly the same overall efficiency.

8. 3-Ketoacyl-CoA transferase is required to convert ketone bodies to acetyl-CoA. If the liver contained this enzyme, it would be unable to supply ketone bodies as fuels for other tissues.

9. Palmitate (C_{16}) synthesis requires 14 NADPH. The transport of 8 acetyl-CoA to the cytosol by the tricarboxylate transport system supplies 8 NADPH (Fig. 19-24), which represents $8/14 \times 100 = 57\%$ of the required NADPH.

10. The label does not appear in palmitate because $^{14}CO_2$ is released in Reaction 2b of fatty acid synthesis (Fig. 19-26).

11. The breakdown of glucose by glycolysis generates the dihydroxyacetone phosphate that becomes the glycerol backbone of triacylglycerols (Fig. 19-30).

12. This fatty acid **(linolenate)** cannot be synthesized by animals because it contains a double bond closer than 6 carbons from its noncarboxylate end.

13. The synthesis of stearate (18:0) from mitochondrial acetyl-CoA requires 9 ATP to transport 9 acetyl-CoA from the mitochondria to the cytosol. Seven rounds of fatty acid synthesis consume 7 ATP (in the acetyl-CoA carboxylase reaction) and 14 NADPH (equivalent to 42 ATP). Elongation of palmitate to stearate requires 1 NADH and 1 NADPH (equivalent to 6 ATP). The energy cost is therefore $9 + 7 + 42 + 6 = 64$ ATP.

    (a) The degradation of stearate to 9 acetyl-CoA consumes 2 ATP (in the acyl-CoA synthetase reaction) but generates, in eight rounds of β oxidation, 8 FADH$_2$ (equivalent to 16 ATP) and 8 NADH (equivalent to 24 ATP). Thus, the energy yield is $16 + 24 - 2 = 38$ ATP. This represents only about half of the energy consumed in synthesizing stearate (38 ATP versus 64 ATP).

    (b) The complete oxidation of the 9 acetyl-CoA to CO_2 by the citric acid cycle yields an additional 9 GTP (equivalent to 9 ATP), 27 NADH (equivalent to 81 ATP), and 9 FADH$_2$ (equivalent to 18 ATP) for a total of $38 + 9 + 81 + 18 = 146$ ATP. Thus, more than twice the energy investment of synthesizing stearate is recovered (146 ATP versus 64 ATP).

14. (a)

$$H_3C \overset{O}{\underset{}{\overset{\|}{\underset{14}{C}}}} - CH_2 \overset{O}{\underset{}{\overset{\|}{\underset{14}{C}}}} - O^-$$

    Acetoacetate

    See Fig. 19-21.

(b)

$$\begin{array}{c} OH \\ | \\ \overset{14}{C}H - (CH_2)_{14} - CH_3 \\ | \\ H_2N - C - H \\ | \\ CH_2OH \end{array}$$

Sphinganine

See Fig. 19-36.

15. Statins inhibit the HMG-CoA reductase reaction, which produces mevalonate, a precursor of cholesterol. Although lower cholesterol levels induce the synthesis of HMG-CoA reductase to make up for the loss in activity, some decrease in activity may still be present. Because mevalonate is also the precursor of ubiquinone (coenzyme Q), supplementary ubiquinone may be necessary.

CHAPTER 20

1. Proteasome-dependent proteolysis requires ATP to activate ubiquitin in the first step of linking ubiquitin to the target protein (Fig. 20-2) and for denaturing the protein as it enters the proteasome.

2. The urea cycle transforms excess nitrogen from protein breakdown to an excretable form, urea. In a deficiency of a urea cycle enzyme, the preceding urea cycle intermediates may build up to a toxic level. A low-protein diet minimizes the amount of nitrogen that enters the urea cycle and therefore reduces the concentrations of the toxic intermediates.

3. An individual consuming a high-protein diet uses amino acids as metabolic fuels. As the amino acid skeletons are converted to glucogenic or ketogenic compounds, the amino groups are disposed of as urea, leading to increased flux through the urea cycle. During starvation, proteins (primarily from muscle) are degraded to provide precursors for gluconeogenesis. Nitrogen from these protein-derived amino acids must be eliminated, which demands a high level of urea cycle activity.

4. Glutamate dehydrogenase, glutamine synthetase, and carbamoyl phosphate synthetase.

5. (a)

$$H_2N \overset{O}{\underset{}{\overset{\|}{C}}} - NH_2 + H_2O \rightleftharpoons 2\,NH_3 + CO_2$$

    (b) The NH$_3$ produced by the action of urease can combine with protons in gastric fluid to form NH_4^+. This could reduce the concentration of protons and therefore increase the pH.

6. Since the three reactions converting tiglyl-CoA to acetyl-CoA and propionyl-CoA are analogous to those of fatty acid oxidation (β oxidation; Fig. 19-12), the reactions are:

Tiglyl-CoA

H_2O (a hydratase)

NAD^+ (a dehydrogenase)

$NADH$

$CoASH$ (a thiolase)

Acetyl-CoA   +   **Propionyl-CoA**

7.

bond to be cleaved

8. (a) Ala, Arg, Asn, Asp, Cys, Gln, Glu, Gly, His, Met, Pro, Ser, and Val

   (b) Leu and Lys

   (c) Ile, Phe, Thr, Trp, and Tyr

9. Tryptophan can be considered a member of this group since one of its degradation products is alanine, which is converted to pyruvate by deamination.

10. In the absence of uridylyl-removing enzyme, adenylyltransferase $\cdot$ P_{II} will be fully uridylylated, since there is no mechanism for removing the uridylyl groups once they are attached. Uridylylated adenylyltransferase $\cdot$ P_{II} adenylylates glutamine synthetase, which activates it. Hence, the defective *E. coli* cells will have a hyperactive glutamine synthetase and

thus a higher than normal glutamine concentration. Reactions requiring glutamine will therefore be accelerated, thereby depleting glutamate and the citric acid cycle intermediate α-ketoglutarate. Consequently, biosynthetic reactions requiring transamination, as well as energy metabolism, will be suppressed.

11. Since only plants and microorganisms synthesize aromatic amino acids, herbicides that inhibit these pathways do not affect amino acid metabolism in animals.

12. The pigment coloring skin and hair is melanin, which is synthesized from tyrosine. When tyrosine is in short supply, as when dietary protein is not available, melanin cannot be synthesized in normal amounts, and the skin and hair become depigmented.

CHAPTER 21

1. Hyperinsulinemia would result in a decrease in blood glucose. The decrease in [glucose] for the brain would cause loss of brain function (leading to coma and death).

2. ATP generating pathways such as glycolysis and fatty acid oxidation require an initial investment of ATP (the hexokinase and phosphofructokinase steps of glycolysis and the acyl-CoA synthetase activation step that precedes β oxidation). This "priming" cannot occur when ATP has been exhausted.

3. (a) GLUT2 has a higher K_M than GLUT1 so that the rate of glucose entry into liver cells can vary directly with the concentration of glucose in the blood. A transporter with a high K_M is less likely to be saturated with its ligand and therefore would not limit the rate of transport. (b) Type I glycogen storage disease results from a deficiency of glucose-6-phosphatase so that glucose-6-phosphate produced by glycogenolysis cannot exit the cell as glucose. A defect in the glucose-transport protein GLUT2 would similarly prevent the exit of glucose (a passive transporter can operate in either direction). In both cases, the buildup of intracellular glucose-6-phosphate prevents glycogen breakdown, and glycogen accumulates.

4. (a) Because fatty acids, like glucose, are metabolic fuels, it makes sense for them to stimulate insulin release, which is a signal of abundant fuel. (b) Elevated levels of circulating fatty acids occur during an extended fast, when dietary glucose and glucose mobilized from glycogen stores are no longer available. Insulin release would be inappropriate for these conditions. A combination of abundant fatty acids and glucose, indicating the fed state, would serve as a better trigger for insulin release.

5. Insulin activates ATP-citrate lyase, which is the enzyme that converts citrate to oxaloacetate and acetyl-CoA (Section 19-4A). The activity of this enzyme is essential for making acetyl units available for fatty acid biosynthesis in the cytosol. The acetyl units, generated from pyruvate in the mitochondria, combine with oxaloacetate to form citrate, which can then be transported from the mitochondria to the cytosol for reconversion to acetyl-CoA.

6. Glutamate represents a source of fixed amino groups that can be transferred by transamination (Section 20-2A), and its carbon skeleton is the citric acid cycle intermediate α-ketoglutarate. An ability to sense the presence of glutamate in food might help ensure an adequate intake of nitrogen as well as metabolic fuel.

7. Because the GTP analog cannot be hydrolyzed, G_α remains active. Analog binding to G_s therefore increases cAMP production. Analog binding to G_i decreases cAMP production.

8. No. Although the diacylglycerol second messengers are identical, phosphatidylethanolamine does not generate an IP_3 second messenger that triggers the release of Ca^{2+}, which in turn alters protein kinase C activity.

9. In the presence of the viral protein, the cell would undergo more cycles of cell division in response to the growth factor.

10. Insulin promotes the uptake of glucose via the increase in GLUT4 receptors on the adipocyte surface. A source of glucose is necessary to supply the glycerol-3-phosphate backbone of triacylglycerols.

11. Ingesting glucose while in the resting state causes the pancreas to release insulin. This stimulates the liver, muscle, and adipose tissue to synthesize glycogen, fat, and protein from the excess nutrients while inhibiting the breakdown of these metabolic fuels. Hence, ingesting glucose before a race will gear the runner's metabolism for resting rather than for running.

12. During starvation, the synthesis of glucose from liver oxaloacetate depletes the supply of citric acid cycle intermediates and thus decreases the ability of the liver to metabolize acetyl-CoA via the citric acid cycle.

13. The leptin produced by the normal mouse will enter the circulation of the *ob/ob* mouse, resulting in decreases in its appetite and weight.

14. Adipose tissue synthesizes and releases the polypeptide hormones leptin and resistin.

15. Since PYY_{3-36} is a peptide hormone, it would be digested if taken orally. Introducing it directly into the bloodstream avoids degradation.

CHAPTER 22

1. (a) 7 ATP; (b) 8 ATP; (c) 7 ATP.

2. PRPP and FGAR accumulate because they are substrates of Reactions 2 and 5 in the IMP biosynthetic pathway (Fig. 22-1). XMP also accumulates because the GMP synthetase reaction is blocked (Fig. 22-3). Although glutamine is a substrate of carbamoyl phosphate synthetase II (the first enzyme of UMP synthesis; Fig. 22-5), the other substrates of this enzyme do not accumulate.

3. (a) The recovered deoxycytidylate would be equally labeled in its base and ribose components (i.e., the same labeling pattern as in the original cytidine). (b) The recovered deoxycytidylate would be unequally labeled in its base and ribose components because the separated ^{14}C-cytosine and ^{14}C-ribose would mix with the different-sized pools of unlabeled cellular cytosine and ribose before recombining as the deoxycytidylate that becomes incorporated into DNA. [This experiment established that deoxyribonucleotides, in fact, are synthesized from their corresponding ribonucleotides (alternative a).]

4. Hydroxyurea destroys the tyrosyl radical that is essential for the activity of ribonucleotide reductase. Tumor cells are generally fast-growing and cannot survive without this enzyme, which supplies dNTPs for nucleic acid synthesis. In contrast, most normal cells grow slowly, if at all, and hence have less need for nucleic acid synthesis.

5. Deoxyadenosine inhibits ribonucleotide reductase, thereby preventing the synthesis of the deoxynucleotides required for DNA synthesis.

6. FdUMP and methotrexate kill rapidly proliferating cells, such as cancer cells and those of hair follicles. Consequently, hair falls out.

7. The mutant cells grow because the medium contains the thymidine they are unable to make. Normal cells, however, continue to synthesize their own thymidine and thereby convert their limited supply of THF to DHF. The methotrexate inhibits dihydrofolate reductase, so THF cannot be regenerated. Without a supply of THF for the synthesis of nucleotides and amino acids, the cells die.

8. The synthesis of histidine and methionine requires THF. The cell's THF is converted to DHF by the thymidylate synthase reaction, but in the presence of methotrexate, THF cannot be regenerated.

9. The conversion of dUMP to dTMP is a reductive methylation. In the thymidylate synthase reaction shown in Fig. 22-15, THF is oxidized to DHF, so that DHFR must subsequently reduce the DHF to THF. Organisms that lack DHFR use an alternative mechanism for converting dUMP to dTMP in which the FAD cofactor of the enzyme, rather than the folate, undergoes oxidation.

10. (a) Trimethoprim binds to bacterial dihydrofolate reductase but does not permanently inactivate the enzyme. Therefore, it is not a mechanism-based inhibitor. (b) Allopurinol is oxidized by xanthine oxidase to a product that irreversibly binds to the enzyme. It is therefore a mechanism-based inhibitor of xanthine oxidase.

11. In von Gierke's disease (glucose-6-phosphatase deficiency), glucose-6-phosphate accumulates in liver cells, thereby stimulating the pentose phosphate pathway. The resulting increase in ribose-5-phosphate production boosts the concentration of PRPP, which in turn stimulates purine biosynthesis. High levels of uric acid derived from the breakdown of these excess purines causes gout.

12.

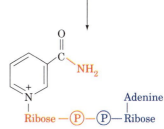

Following the scheme:

Nicotinamide

↓

-²O₃P—O—CH₂ (ribose ring with H, H, H, H and OH OH)

Nicotinamide mononucleotide (NMN)

↓

Ribose —(P)—(P)— Ribose, with Adenine

Nicotinamide adenine dinucleotide (NAD⁺)

CHAPTER 23

1. Since amino acids have an average molecular mass of ∼110 D, the 50-kD protein contains 50,000 D ÷ 110 D/residue = ∼455 residues. These residues are encoded by 455 × 3 = 1365 nucleotides. In B-DNA, the rise per base pair is 3.4 Å, so the contour length of 1365 bp is 3.4 Å/bp × 1365 bp = 4641 Å, or 0.46 μm. In A-DNA, the contour length would be 1365 bp × 2.9 Å/bp = 3959 Å, or 0.40 μm.

2. (a) Hypoxanthine pairs with cytosine in much the same way as does guanine.

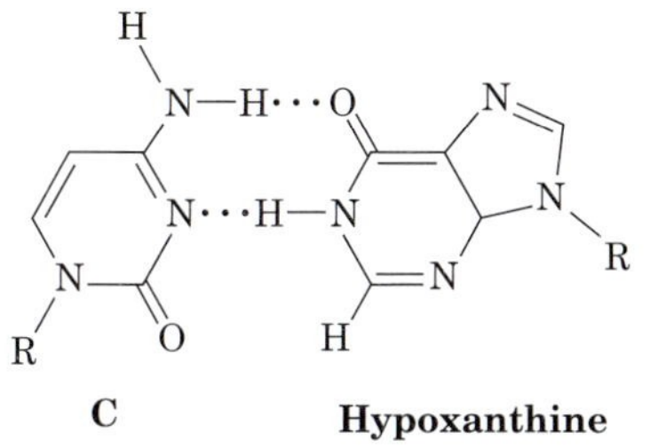

C   **Hypoxanthine**

(b)

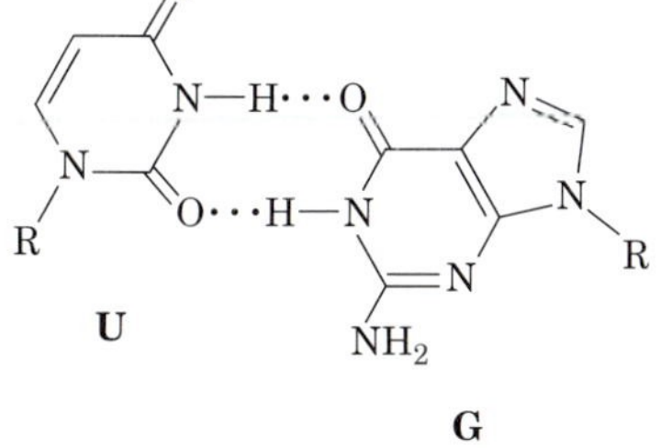

U   **G**

3. (a)

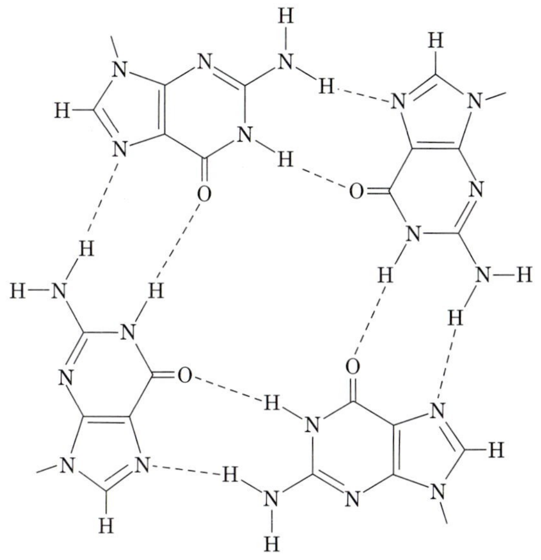

(b)

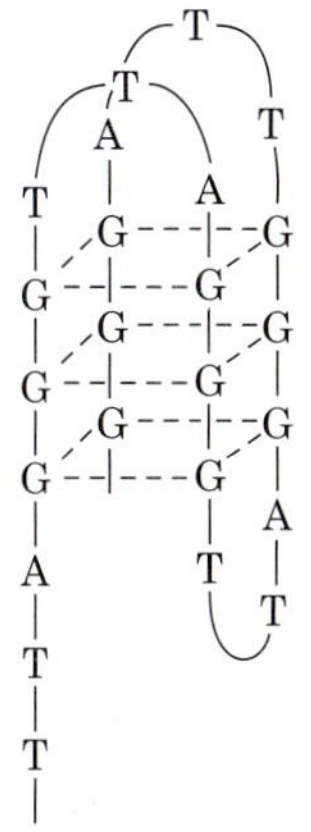

4. (a) The supercoiled molecule is more compact than the relaxed circle and therefore sediments more rapidly. (b) Both overwound and underwound DNA molecules are supercoiled (Fig. 23-11), so the ultracentrifuge cannot distinguish between them.

5. The enzyme has no effect on the supercoiling of DNA since cleaving the C2′—C3′ bond of ribose does not sever the sugar–phosphate chain of DNA.

6. In the B-DNA to Z-DNA transition, a right-handed helix with one turn per 10.4 base pairs converts to a left-handed helix with one turn per 12 base pairs. Since a right-handed duplex helix has a positive twist, the twist decreases:

$$\Delta T = -\frac{100}{10.4} + \frac{-100}{12} = -17.9 \text{ turns}$$

The linking number must remain constant ($\Delta L = 0$) since no covalent bonds are broken. Hence, the change in writhing number is $\Delta W = -\Delta T = 17.9$ turns.

7. The segment with 20% A residues (i.e., 40% A·T base pairs) contains 60% G·C base pairs and therefore melts at a higher temperature than a segment with 30% A residues (i.e., 40% G·C base pairs).

8. (a) Its T_m decreases because the charges on the phosphate groups are less shielded from each other at lower ionic strength and hence repel each other more strongly, thereby destabilizing the double helix. (b) The nonpolar solvent diminishes the hydrophobic forces that stabilize double-stranded DNA and hence lowers the T_m.

9. (a) As the temperature increases, the stacked bases melt apart so that their ultraviolet absorbance increases (the hyperchromic effect). (b) The broad shape of the poly(A) melting curve indicates noncooperative changes, as expected for a single-stranded RNA. The sharp melting curve for DNA reflects the cooperativity of strand separation.

10.

$$
\begin{array}{ccccccc}
& A & U & U & G & G & & C \\
& | & | & | & | & | & & A \\
A & A & U & A & G & C & C & \\
& & & & & & & U
\end{array}
$$

11. The largest (and therefore the heaviest) RNA forms a band closest to the bottom of the centrifuge tube; the smallest (lightest) RNA forms a band near the top of the tube.

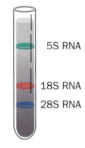

5S RNA

18S RNA

28S RNA

12. The target sequence consists of 6 symmetry-related base pairs. Since there are 4 possible base pairs ($A \cdot T, T \cdot A, G \cdot C$, and $C \cdot G$), the probability that any two base pairs are randomly related by symmetry is 1/4. Hence, the probability of finding all 6 pairs of base pairs by random chance is $(1/4)^6 = 2.4 \times 10^{-4}$.

13. (a) The contour length is 5×10^7 bp $\times 3.4$ Å/bp $= 1.7 \times 10^8$ Å $= 17$ mm.

(b) A nucleosome, which binds ~200 bp, compresses the DNA to an 80-Å-high supercoil. The length of the DNA is therefore (80 Å/200 bp) $\times$ (5×10^7 bp) $= 2 \times 10^7$ Å $= 2$ mm.

(c) The 30-nm filament contains 6 nucleosomes per turn and has a pitch of 110 Å. Therefore, the length of the DNA is (110 Å/6 nucleosomes) $\times$ (1 nucleosome/200 bp) $\times$ (5×10^7 bp) $= 4.6 \times 10^6$ Å $= 0.46$ mm.

14. Experiment 1. The restriction enzyme failed to digest the genomic DNA, leaving the DNA too large to enter the gel during electrophoresis.

Experiment 2. The hybridization conditions were too "relaxed," resulting in nonspecific hybridization of the probe to all the DNA fragments. This problem could be corrected by boiling the blot to remove the probe and repeating the hybridization at a higher temperature and/or lower salt concentration.

Experiment 3. The probe hybridized with three different mouse genes. The different intensity of each band reflects the relatedness of the sequences. The most intense band is most similar to the human *rxr-1* gene, whereas the least intense band is least similar to the *rxr-1* gene.

CHAPTER 24

1. Okazaki fragments are 1000 to 2000 nt long, and the *E. coli* chromosome contains 4.6×10^6 bp. Therefore, *E. coli* chromosomal replication requires 2300 to 4600 Okazaki fragments.

2. As indicated in Fig. *a* (*below*), nucleotides would be added to a polynucleotide strand by attack of the 3'-OH of the incoming nucleotide on the 5' triphosphate group of the growing strand with the elimination of PP_i. The hydrolytic removal of a mispaired nucleotide by the 5' → 3' exonuclease activity (Fig. *b*, *below*) would leave only an OH group or monophosphate group at the 5' end of the DNA chain. This would require an additional activation step before further chain elongation could commence.

(*a*) 3' → 5' Polymerase

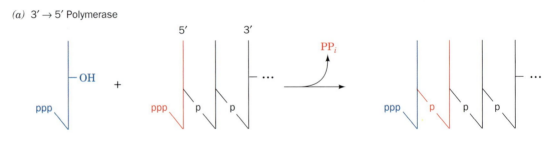

(*b*) 5' → 3' Exonuclease

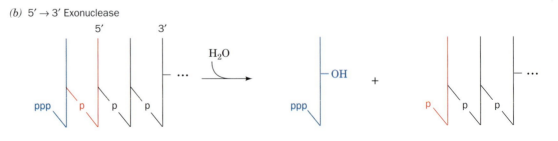

3. The drug would inhibit DNA synthesis because the polymerization reaction is accompanied by the release and hydrolysis of PP_i. Failure to hydrolyze the PP_i would remove the thermodynamic driving force for polymerization, that is, it would be reversible.

4. PP_i is the product of the polymerization reaction catalyzed by DNA polymerase. This reaction also requires a template DNA strand and a primer with a free 3′ end.

   (a) There is no primer strand, so no PP_i is produced.

   (b) There is no primer strand, so no PP_i is produced.

   (c) PP_i is produced.

   (d) No PP_i is produced because there is no 3′ end that can be extended.

   (e) PP_i is produced.

   (f) PP_i is produced.

5. The $5' \rightarrow 3'$ exonuclease activity is essential for DNA replication because it removes RNA primers and replaces them with DNA. Absence of this activity would be lethal.

6. The Klenow fragment, which lacks $5' \rightarrow 3'$ exonuclease activity (and therefore cannot catalyze nick translation), is used to ensure that all the replicated DNA chains have the same 5′ terminus, a necessity if a sequence is to be assigned according to fragment length.

7. AT-rich DNA is less stable than GC-rich DNA and therefore would more readily melt apart, a requirement for initiating replication.

8. DNA gyrase adds negative supercoils to relieve the positive supercoiling that helicase-catalyzed unwinding produces ahead of the replication fork.

9. Mismatch repair and other repair systems correct most of the errors missed by the proofreading functions of DNA polymerases.

10. The *E. coli* replication system can fully replicate only circular DNAs. Bacteria do not have a mechanism (e.g., telomerase-catalyzed extension of telomeres) for replicating the extreme 3′ ends of linear template strands.

11. (a)

5BU
(enol tautomer)   **Guanine**

   (b) When 5BU incorporated into DNA pairs with G, the result is an $A \cdot T \rightarrow G \cdot C$ transition after two more rounds of DNA replication:

$$A \cdot T \rightarrow A \cdot 5BU \rightarrow G \cdot 5BU \rightarrow G \cdot C$$

12. Base excision repair. The deaminated base can be recognized because hypoxanthine does not normally occur in DNA.

13. The cytosine derivative base pairs with adenine, generating a $C \cdot G \rightarrow T \cdot A$ transition.

Adenine

14. The triphosphatase destroys nucleotides containing the modified base before they can be incorporated into DNA during replication.

15. When 5-methylcytosine residues deaminate, they form thymine residues.

5-Methyl-C   **T**

Since thymine is a normal DNA base, the repair systems cannot determine whether such a T or its opposing G is the mutated base. Consequently, only about half of the deaminated 5-methylcytosines are correctly repaired.

16. (a) Loss of the helicase DnaB, which unwinds DNA for replication, would be lethal.

   (b) Loss of Pol I would prevent the excision of RNA primers and would therefore be lethal.

   (c) SSB prevents reannealing of separated single strands. Loss of SSB would be lethal.

   (d) RecA protein mediates the SOS response and homologous recombination. Loss of RecA would be harmful but not necessarily lethal.

17. *E. coli* contains a low concentration of dUTP, which DNA polymerase incorporates into DNA in place of dTTP. The resulting uracil bases are rapidly excised by uracil–DNA glycosylase followed by nucleotide excision repair (NER), which temporarily causes a break in the DNA chain. DNA that is isolated before DNA polymerase I and DNA ligase can complete the repair process would be fragmented. However, in the absence of a functional uracil–DNA glycosylase, the inappropriate uracil residues would remain in place, and hence leading strand DNA would be free of breaks. The lagging strand, being synthesized discontinuously, would still contain breaks, although fewer than otherwise.

18. Pol V is less processive than Pol III. When the progress of Pol III is arrested by the presence of a thymine dimer, Pol V can take over, allowing replication to continue at a high rate, although with a greater incidence of mispairings. The damage is minimal, however, since Pol V soon dissociates from the DNA, allowing the more accurate Pol III to resume replicating DNA.

CHAPTER 25

1. (a) Cordycepin is the 3′-deoxy analog of adenosine.

   (b) Because it lacks a 3′-OH group, the cordycepin incorporated into a growing RNA chain cannot support further chain elongation in the 5′ → 3′ direction.

2. The top strand in the sense strand

5′ CAACGTAACACTTTACAGCGGCGCGTCATTTGATATGATGCGCCCCGCTTCCCGATA 3′

```
        −35             −10          start
       region          region        point
```

Its TATGAT segment differs by only one base from the TATAAT consensus sequence of the promoter's −10 sequence; its TTTACA sequence differs by only one base from the TTGACA consensus sequence of the promoter's −35 sequence and is appropriately located ~25 nt to the 5′ side of the −10 sequence; and the initiating G nucleotide is the only purine that is located ~10 nt downstream of the −10 sequence.

3. The probe should have a sequence complementary to the consensus sequence of the 6-nt Pribnow box: 5′-ATTATA-3′.

4. Promoter elements for RNA polymerase II include sequences at −27 (the TATA box) and between −50 and −100. The insertion of 10 bp would separate the promoter elements by the distance of the turn of the DNA helix, thereby diminishing the binding of proteins required for transcription initiation. However, the protein-binding sites would still be on the same side of the helix. Inserting 5 bp (half of a helical turn) would move the protein-binding sites to opposite sides of the helix, making it even more difficult to initiate transcription.

5. G · C base pairs are more stable than A · T base pairs. Hence, the more G · C base pairs that the promoter contains, the more difficult it is to form the open complex during transcription initiation.

6. Transcription of an rRNA gene yields a single rRNA molecule that is incorporated into a ribosome. In contrast, transcription of a ribosomal protein gene yields an mRNA that can be translated many times to produce many copies of its corresponding protein. The greater number of rRNA genes relative to ribosomal protein genes helps ensure the balanced synthesis of rRNA and proteins necessary for ribosome assembly.

7. The cell lysates can be applied to a column containing a matrix with immobilized poly(dT). The poly(A) tails of processed mRNAs will bind to the poly(dT) while other cellular components are washed away. The mRNAs can be eluted by decreasing the salt concentration to destabilize the A · T base pairs.

8. (a) The phosphate groups of the phosphodiester backbone of the mRNA will be labeled at all sites where α-[^{32}P] ATP is used as a substrate by RNA polymerase.

   (b) ^{32}P will appear only at the 5′ end of mRNA molecules that have A as the first residue (this residue retains its α and β phosphates). In all other cases where β-[^{32}P]ATP is used as a substrate for RNA synthesis, the β and γ phosphates are released as PP$_i$ (see Fig. 25-6).

   (c) No ^{32}P will appear in the RNA chain. During polymerization, the β and γ phosphates are released as PP$_i$. The terminal (γ) phosphate of an A residue at the 5′ end of an RNA molecule is removed during the capping process.

9. The active site of poly(A) polymerase is narrower because it does not need to accommodate a template strand.

10. The mechanism of RNase hydrolysis requires a free 2′-OH group to form a 2′,3′-cyclic phosphate intermediate (Figure 11-10). Nucleotide residues lacking a 2′-OH group would therefore be resistant to RNase-catalyzed hydrolysis.

11. The mRNA splicing reaction, which requires no free energy input and results in no loss of phosphodiester bonds, is theoretically reversible *in vitro*. However, the degradation of the excised intron makes the reaction irreversible in the cell.

12. The intron must be large enough to include spliceosome binding site(s).

13. Inhibition of snRNA processing interferes with mRNA splicing. As a result, host mRNA cannot be translated, so the host ribosomes will synthesize only viral proteins.

CHAPTER 26

1. A 4-nt insertion would add one codon and shift the gene's reading frame by one nucleotide. The proper reading frame could be restored by deleting a nucleotide. Gene function, however, would not be restored if (a) the 4-nt insertion interrupted the codon for a functionally critical amino acid; (b) the 4-nt insertion created a codon for a structure breaking amino acid; (c) the 4-nt insertion introduced a Stop codon early in the gene; or (d) the 1-nt deletion occurred far from the 4-nt insertion so that even though the reading frame was restored, a long stretch of frame shifted codons separated the insertion and deletion points.

2. The possible codons are UUU, UUG, UGU, GUU, UGG, GUG, GGU, and GGG. The encoded amino acids are Phe, Leu, Cys, Val, Trp, and Gly (Table 26-1).

3. An amber mutation results from any of the point mutations XAG, UXG, or UAX to UAG. The XAG codons specify Gln, Lys, and Glu; the UXG codons specify Leu, Ser, and Trp; and the UAX codons that are not Stop codons both specify Tyr. Hence some of the codons specifying these amino acids can undergo a point mutation to UAG.

4. (a) Each ORF begins with an initiation codon (ATG) and ends with a Stop codon (TGA):

   ATGCTCAACTATATGTGA encodes *vir-2* and

   ATGCCGCATGCTCTGTTAATCACATATAGTTGA
   on the complementary strand encodes *vir-1*.

   (b) *vir-1:* MPHALLITYS; *vir-2:* MLNYM.

   (c) *vir-1:* MPHALLIPYS; *vir-2:* MLNYMGLTEHAA.

5. There are four exons (the underlined bases)

   TATAATACGCGCAATACAATCTACAGCTTC<u>GCGTA</u>
   <u>AATCGTAG</u>GTAAGTTGTAATAAATATAAGTGAGT
   ATGATAGG<u>GCTTTGGACCGATAGATGCGACCCTG</u>
   <u>GAG</u>GTAAGTATAGATTAATTAAGCACAGG<u>CATGCA</u>
   <u>GGGATATCCTCCAAATAGG</u>TAAGTAACCTTACGG
   TCAATTAATTAGG<u>CAGTAGATGAATAAACGATAT</u>
   <u>CGATCGGTTAGGT</u>AAGTCTGAT

The mature mRNA, which has a 5′ cap and a 3′ poly(A) tail; therefore has the sequence

GCGUAAAUCGUAGGCUUUGGACCGAUAG**AUG**
CGACCCUGGAGCAUGCAGGGAUAUCCUCCAAA
UAGCAGUAGA**UGA**AUAAACGAUAUCGAUCGG
UUAGGU

The initiation codon and termination codon are shown in boldface. The encoded protein has the sequence

MRPWSMQGYPPNSSR

6. Gly and Ala; Val and Leu; Ser and Thr, Asn and Gln; and Asp and Glu.

7. The assembly of functional ribosomes requires equal amounts of the rRNA molecules, so it is advantageous for the cell to synthesize the rRNAs all at once.

8. Ribosomes cannot translate double-stranded RNA, so the base pairing of a complementary antisense RNA to an mRNA prevents its translation.

9. Only newly synthesized bacterial polypeptides have fMet at their N-terminus. Consequently, the appearance of fMet in a mammalian system signifies the presence of invading bacteria. Leukocytes that recognize the fMet residue can therefore combat these bacteria through phagocytosis.

10. Prokaryotic ribosomes can select an initiation codon located anywhere on the mRNA molecule as long as it lies just downstream of a Shine–Dalgarno sequence. In contrast, eukaryotic ribosomes usually select the AUG closest to the 5′ end of the mRNA. Eukaryotic ribosomes therefore cannot recognize a translation initiation site on a circular mRNA.

11. As expected, the correctly charged tRNAs (Ala–tRNA^{Ala} and Gln–tRNA^{Gln}) bind to EF-Tu with approximately the same affinity, so they are delivered to the ribosomal A site with the same efficiency. The mischarged Ala–tRNA^{Gln} binds to EF-Tu much more loosely, indicating that it may dissociate from EF-Tu before it reaches the ribosome. The mischarged Gln-tRNA^{Ala} binds to EF-Tu much more tightly, indicating that EF-Tu may not be able to dissociate from it at the ribosome. These results suggest that either a higher or a lower binding affinity could affect the ability of EF-Tu to carry out its function, which would decrease the rate at which mischarged aminoacyl–tRNAs bind to the ribosomal A site during translation.

12. By inducing the same conformational changes that occur during correct tRNA–mRNA pairing, paromomycin can mask the presence of an incorrect codon–anticodon match. Without proofreading at the aminoacyl–tRNA binding step, the ribosome synthesizes a polypeptide with the wrong amino acids, which is likely to be nonfunctional or toxic to the cell.

13. (a) Transpeptidation involves the nucleophilic attack of the amino group of the aminoacyl–tRNA on the carbonyl carbon of the peptidyl–tRNA. As the pH increases, the amino group becomes more nucleophilic (less likely to be protonated).

(b) As the pH increases, residue A2486 would be less likely to be protonated and therefore less likely to stabilize the negatively charged oxyanion of the tetrahedral reaction intermediate. Thus, the mechanistic embellishment is inconsistent with the observed effect.

14. The enzyme hydrolyzes peptidyl–tRNA molecules that dissociate from a ribosome before normal translation termination takes place. Because peptide synthesis is prematurely halted, the resulting polypeptide, which is still linked to tRNA, is likely to be nonfunctional. Peptidyl–tRNA hydrolase is necessary for recycling the amino acids and the tRNA.

15. Aminoacylation occurs via pyrophosphate cleavage of ATP, and hence the aminoacylation of 100 tRNAs requires 200 ATP equivalents; translation initiation requires 1 GTP (1 ATP equivalent); 99 cycles of elongation require 99 GTP (99 ATP equivalents) for EF-Tu action; 99 cycles of ribosomal translocation require 99 GTP (99 ATP equivalents) for EF-G action; and translation termination requires 1 GTP (1 ATP equivalent), bringing the total energy cost to 200 + 1 + 99 + 99 + 1 = 400 ATP equivalents.

16.

	Start	Lys	Pro	Ala
5′-AGGAGCUX$_{-4}$	^A_GUG	AA^A_G	CCX	GCX-

Shine–Dalgarno sequence.
3–10 base pairs with G · U's allowed

Gly	Thr	Glu	Asn	Ser	Stop
GGX	ACX	GA^A_G	AA^U_C	UCX	UAA
				or	UAG - 3′
				AG^U_C	UGA

CHAPTER 27

1. Virtually all the DNA sequences in *E. coli* are present as single copies, so the renaturation of *E. coli* DNA is a straightforward process of each fragment reassociating with its complementary strand. In contrast, the human genome contains many repetitive DNA sequences. The many DNA fragments containing these sequences find each other to form double-stranded regions (renature) much faster than the single-copy DNA sequences that are also present, giving rise to a biphasic renaturation curve.

2. Because genes encoding proteins with related functions often occur in operons, the identification of one or several genes in an operon may suggest functions for the remaining genes in that operon.

3. (a) O_1 is the primary repressor-binding site, so *lac* repressor cannot stably bind to the operator in its absence and repression cannot occur.

(b) Both O_2 and O_3 are secondary repressor-binding sequences. If one is absent, the other can still function, resulting in only a small loss of repressor effectiveness.

(c) In the absence of both O_2 and O_3, the repressor can bind only to O_1, which partially interferes with transcription but does not repress transcription as fully as when a DNA loop forms through the cooperative binding of *lac* repressor to O_1 and either O_2 or O_3.

4. In the absence of β-galactosidase (the product of the *lacZ* gene), lactose is not converted to the inducer allolactose. Consequently, *lac* enzymes, including galactoside permease, are not synthesized.

5. Since operons other than the *lac* operon maintain their sensitivity to the absence of glucose, the defect is probably not in the gene that encodes CAP. Instead, the defect is probably located in the portion of the *lac* operon that binds CAP–cAMP.

6. In eukaryotes, transcription takes place in the nucleus and translation occurs in the cytoplasm. Hence, in eukaryotes, ribosomes are never in contact with nascent mRNAs, an essential aspect of the attenuation mechanism in prokaryotes.

7. Deletion of the leader peptide sequence from *trpL* would eliminate sequence 1 of the attenuator. Consequently, the 2 · 3 hairpin rather than the 3 · 4 terminator hairpin would form. Transcription would therefore continue into the remainder of the *trp* operon, which would then be regulated solely by *trp* repressor.

8. Red–green color blindness is conferred by a mutation in an X-linked gene so that female carriers of this condition, who do not appear to be red–green colorblind, have one wild-type gene and one mutated gene for this condition. In placental mammals such as humans, females are mosaics of clones of cells in which only one of their two X chromosomes is transcriptionally active. Hence in a female carrier of red–green color blindness, the transcriptionally active X chromosome in some of these clones will contain the wild-type gene and the others will contain the mutated gene. The former type of retinal clone is able to differentiate red and green light, whereas the latter type of retinal clone is unable to do so. Apparently, these retinal clones are small enough so that a narrow beam of light is necessary to separately interrogate them.

9.

Acetyllysine   Methyllysine   Methylarginine

In acetyllysine, the cationic side chain of Lys has been converted to a polar but uncharged side chain. In methyllysine and methylarginine, the hydrophobic methyl group partially masks the cationic character of the Lys or Arg side chain.

10. Histone and DNA methylation requires *S*-adenosylmethionine (SAM), which becomes *S*-adenosylhomocysteine after it gives up its methyl group (Fig. 20-18). *S*-Adenosylhomocysteine is converted back to methionine, the precursor of SAM, in a reaction in which the methyl group is donated by the folic acid derivative tetrahydrofolate (THF; Fig. 20-18). A shortage of this cofactor could limit cellular production of SAM, which would result in the undermethylation of histones and DNA.

11. Transcriptionally active chromatin has a more open structure due to histone modifications that help make the DNA more accessible to transcription factors and RNA polymerase as well as nucleases.

12. A sequence located downstream of the gene's promoter (i.e., within the coding region) could regulate gene expression if it were recognized by the appropriate transcription factor such that the resulting DNA–protein complex successfully recruited RNA polymerase to the promoter.

13. The susceptibility of RNA to degradation *in vivo* makes it possible to regulate gene expression by adjusting the rate of mRNA degradation. If mRNA were very stable, it might continue to direct translation even when the cell no longer needed the encoded protein.

14. A 22-bp segment of RNA, incorporating all four nucleotides, has 4^{22} or 1.8×10^{13} possible unique sequences. An RNA half this size would have only 4^{11} or 4.2×10^6 possible sequences. The shorter the siRNA, the greater is the probability that it could hybridize with more than one complementary mRNA, thereby making it less efficient in silencing a specific gene. (In the 3.2×10^9-bp human genome, a sequence of 16 bp has a high probability of randomly occurring at least once.)

15. In multicellular organisms, apoptosis of damaged cells minimizes damage to the entire organism. For a single-celled organism, survival of a genetically damaged cell is preferable in a Darwinian sense to its death.

16. The *esc* gene is apparently a maternal-effect gene. Thus, the proper distribution of the *esc* gene product in the fertilized egg, which is maternally specified, is sufficient to permit normal embryonic development regardless of the embryo's genotype.

CHAPTER 28

1. Microfilaments and microtubules consist entirely of subunits that are assembled in a head-to-tail fashion so that the polarity of the subunits (actin monomers in microfilaments and tubulin dimers in microtubules) is preserved in the fully assembled fiber. In keratin filaments, however, successive heterodimers align in an antiparallel fashion, so that in a fully assembled intermediate filament, half the molecules are oriented in one direction and half are oriented in the opposite direction (Fig. 6-15).

2. Gelsolin is a capping protein, so it must bind to one or more actin subunits at the end of a microfilament. It is likely that monomeric actin can also bind to the same site(s) on gelsolin, although with lower affinity than F-actin. In the absence of F-actin and at concentrations of G-actin high enough to favor low-affinity binding, a gelsolin–G-actin complex can form. This complex would then serve as a platform for the assembly of additional actin monomers to produce a microfilament. The law of mass action would favor polymerization since the concentration of F-actin is initially zero.

3. Myosin is both fibrous and globular. Its two heads are globular, with several layers of secondary structure. Its tail, however, consists of a lengthy, fibrous coiled coil.

4. Because many myosin heads bind along a thin filament where it overlaps a thick filament, and because the myosin molecules do not execute their power strokes simultaneously, the thick and thin filaments can move past each other by more than 60 Å in the interval between power strokes of an individual myosin molecule.

5. In the absence of ATP, each myosin head adopts a conformation that does not allow it to release its bound actin mol-

ecule. Consequently, thick and thin filaments form a rigid cross-linked array.

6.

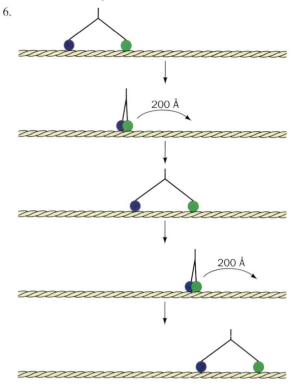

7. During rapid microtubule growth, β-tubulin subunits containing GTP accumulate near the (+) end because GTP hydrolysis occurs following subunit incorporation into the microtubule. In a slowly growing microtubule, there is time for GTP hydrolysis to occur before the microtubule grows any longer so that the β subunits at the (+) end will contain GDP. A protein that preferentially binds to β-tubulin subunits containing GTP rather than GDP could thereby distinguish fast- and slow-growing microtubules.

8. Although the drugs have opposite effects on microtubule dynamics, they both interfere with the normal formation of the mitotic spindle, which is required for cell division.

9. (a) Diffusion is a random process. It tends to be slow (especially for large substances and over long distances). Because it is random, it operates in three dimensions (not linearly) and has no directionality.

   (b) An intracellular transport system must have some sort of track (for linear movement of cargo) and an engine that moves cargo along the track by converting chemical energy to mechanical energy. The engine must operate irreversibly to promote rapid movement in one direction. Finally, some sort of addressing system is needed to direct cargo from its source to a particular destination.

10. Dynein is not processive. Like each myosin II head, each dynein arm forms a cross-bridge, binding and releasing a microtubule. The dynein heads act independently as if rowing along the microtubule.

11. (a) 150–200 kD; (b) 150–200 kD; (c) ~23 kD and 53–75 kD.

12. The loops are on the surface of the domain, so they can tolerate more amino acid substitutions. Amino acid changes in the β sheets would be more likely to destabilize the domain.

13. The antigenic site in the native protein usually consists of several peptide segments that are no longer contiguous when the tertiary structure of the protein is disrupted.

14. (a) Fab fragments are monovalent and therefore cannot cross-link antigens to produce a precipitate. (b) A small antigen has only one antigenic site and therefore cannot bind more than one antibody to produce a precipitate. (c) When antibody is in great excess, most antibodies that are bound to antigen bind only one per immunoglobulin molecule. When antigen is in excess, most immunoglobulins bind to two independent antigens.

GLOSSARY

Numbers and Greek letters are alphabetized as if they were spelled out.

A. See Absorbance.

A site. See aminoacyl site.

aa–tRNA. See aminoacyl–tRNA.

aaRS. See aminoacyl–tRNA synthetase.

ABO blood group antigens. The oligosaccharide components of glycolipids on the surfaces of erythrocytes and other cells.

Absolute configuration. The spatial arrangement of chemical groups around a chiral center.

Absorbance (A). A function of the amount of light transmitted through a solution (I) relative to the incident light (I_0) at a given wavelength: $A = \log(I_0/I)$. Also called optical density.

Acceptor stem. The base-paired region of a tRNA molecule that contains the 5′ end and the 3′ end, to which an amino acid is attached.

Accessory pigment. A molecule in the photosynthetic system that absorbs light at wavelengths other than those absorbed by chlorophyll.

Acid. A substance that can donate a proton.

Acid–base catalysis. A catalytic mechanism in which partial proton transfer from an acid or partial proton abstraction by a base lowers the free energy of a reaction's transition state. See also general acid catalysis and general base catalysis.

Acidic solution. A solution whose pH is less than 7.0 ([H^+] > 10^{-7} M).

Acidosis. A pathological condition in which the pH of the blood drops below its normal value of 7.4.

Action potential. The wave of transient depolarization and repolarization that constitutes the electrical signal generated by a nerve cell.

Active site. The region of an enzyme in which catalysis takes place.

Active transport. The transmembrane movement of a substance from low to high concentrations by a protein that couples this endergonic transport to an exergonic process such as ATP hydrolysis.

Acyl-carrier protein. A phosphopantetheine-containing protein that binds the intermediates of fatty acid synthesis as thioesters.

Acyl–enzyme intermediate. An intermediate of peptide bond hydrolysis in which the carbonyl carbon of the scissile bond is covalently bound to the enzyme nucleophile that attacked it.

Acyl group. A portion of a molecule with the formula —COR, where R is an alkyl group.

Adenylate cyclase system. A signal transduction pathway in which hormone binding to a cell-surface receptor activates a G protein that in turn stimulates adenylate cyclase to synthesize the second messenger 3′,5′-cyclic AMP (cAMP) from ATP.

Adenylylation. Addition of an adenylyl (AMP) group.

Adipocyte. Fat cell, which is specialized for the synthesis and storage of triacylglycerols from free fatty acids.

Adipose tissue. Fat cells; distributed throughout an animal's body.

ADP–ATP translocator. A membrane transport protein with an adenine nucleotide binding site that alternately allows ADP to enter and ATP to exit the mitochondrial matrix.

Adrenergic receptor. A cell surface receptor that binds and responds to adrenal hormones such as epinephrine and norepinephrine.

Aerobe. An organism that uses O_2 as an oxidizing agent for nutrient breakdown.

Affinity chromatography. A procedure in which a molecule is separated from a mixture of other molecules by its ability to bind specifically to an immobilized ligand.

Affinity labeling. A technique in which a labeled substrate analog reacts irreversibly with, and can thereby be used to identify, a group in an enzyme's active site.

Agarose. Linear carbohydrate polymers, made by red algae, that form a loose mesh.

Agarose gel electrophoresis. See gel electrophoresis.

Alcoholic fermentation. A metabolic pathway that synthesizes ethanol from pyruvate through decarboxylation and reduction.

Alditol. A sugar produced by reduction of an aldose or ketose to a polyhydroxy alcohol.

Aldol cleavage. A carbon–carbon cleavage reaction of an aldol (an aldehyde or ketone with a β hydroxyl group) that yields smaller carbonyl compounds.

Aldonic acid. A sugar produced by oxidation of an aldose aldehyde group to a carboxylic acid group.

Aldose. A sugar whose carbonyl group is an aldehyde.

Alkalosis. A pathological condition in which the pH of the blood rises above its normal value of 7.4.

Allosteric effector. A small molecule whose binding to a protein affects the function of another site on the protein.

Allosteric interaction. The binding of ligand at one site in a macromolecule that affects the binding of other ligands at other sites in the molecule. See also cooperative binding.

αα motif. A protein motif consisting of two α helices packed against each other with their axes inclined.

α-amino acid. See amino acid.

α anomer. See anomers.

α/β barrel. A β barrel in which successive parallel β strands are connected by α helices such that a barrel of α helices surrounds the β barrel.

α carbon. The carbon atom of an amino acid to which the amino and carboxylic acid groups are attached.

α cell. A pancreatic islet cell that secretes the hormone glucagon in response to low blood glucose levels.

α-glycoside. See glycoside.

α helix. A regular secondary structure of polypeptides, with 3.6 residues per right-handed turn, a pitch of 5.4 Å, and hydrogen

bonds between each backbone N—H group and the backbone C=O group that is four residues earlier.

Alternative splicing. The tissue-specific patterns of splicing of a given pre-mRNA that result in variations in the excision and retention of exons and introns.

Alzheimer's disease. A neurodegenerative disease characterized by the precipitation of β amyloid protein in the brain.

Ames test. A method for assessing the mutagenicity of a compound from its ability to cause genetically defective strains of bacteria to revert to normal growth.

Amido group. A portion of a molecule with the formula —CONH—.

Amino acid. A compound consisting of a carbon atom to which are attached a primary amino group, a carboxylic acid group, a side chain (R group), and an H atom. Also called an α-amino acid.

Amino acid composition. A tally of the number and type of amino acids in a polypeptide.

Amino group. A portion of a molecule with the formula —NH_2, —NHR, or —NR_2, where R is an alkyl group. Amino groups are usually protonated at physiological pH.

Amino sugar. A sugar in which one or more OH groups are replaced by an amino group, which is often acetylated.

Amino terminus. The end of a polypeptide that has a free amino group. Also called the N-terminus.

Aminoacyl site (A site). The ribosomal site where a tRNA with an attached aminoacyl group binds during protein synthesis.

Aminoacyl–tRNA (aa–tRNA). The covalent ester complex between an "activated" amino acid and a tRNA molecule.

Aminoacyl–tRNA synthetase (aaRS). An enzyme that catalyzes the ATP-dependent esterification of an amino acid to a tRNA with high specificity for the amino acid and the tRNA molecule.

Amphibolic. A term to describe a metabolic process that can be either catabolic or anabolic.

Amphipathic substance. See amphiphilic substance.

Amphiphilic substance. A substance that contains both polar and nonpolar regions and is therefore both hydrophilic and hydrophobic. Also called an amphipathic substance.

Amyloid. Insoluble extracellular aggregates of fibrous protein that characterize such diseases as Alzheimer's disease and the transmissible spongiform encephalopathies.

Anabolism. The reactions by which biomolecules are synthesized from simpler components.

Anaerobe. An organism that does not use O_2 as an oxidizing agent for nutrient breakdown. An obligate anaerobe cannot grow in the presence of O_2, whereas a facultative anaerobe can grow in the presence or absence of O_2.

Anammox. The anaerobic oxidation of ammonia to N_2, a process carried out by certain bacteria.

Anaplerotic reaction. A reaction that replenishes the intermediates of a metabolic pathway.

Androgen. A steroid that functions primarily as a male sex hormone.

Anemia. A condition caused by insufficient red blood cells.

Angina. Chest pain due to insufficient blood supply to the heart.

Anion exchanger. A cationic matrix used to bind anionic molecules in ion exchange chromatography.

Ankyrin. A membrane protein whose repeating structure serves as a platform for organizing other membrane skeleton components.

Anneal. To maintain conditions that allow loose base pairing between complementary single polynucleotide strands so that properly paired double-stranded segments form.

Anomeric carbon. The carbonyl carbon of a monosaccharide, which becomes a chiral center when the sugar cyclizes to a hemiacetal or hemiketal.

Anomers. Sugars that differ only in the configuration around the anomeric carbon. In the α anomer, the OH substituent of the anomeric carbon is on the opposite side of the ring from the CH_2OH group at the chiral center that designates the D or L configuration. In the β anomer, the OH substituent is on the same side.

Antenna chlorophyll. A chlorophyll group that absorbs light energy and passes it on to a photosynthetic reaction center by exciton transfer.

Anti conformation. A purine or pyrimidine nucleotide conformation in which the ribose and the base point away from each other. See also syn conformation.

Antibody. A protein produced by an animal's immune system in response to the introduction of a foreign substance (an antigen); it contains at least one pair each of identical heavy and light chains. Also called an immunoglobulin (Ig).

Anticodon. The sequence of three nucleotides in tRNA that recognizes an mRNA codon through complementary base pairing.

Anticodon arm. The conserved stem-loop structure in a tRNA molecule that includes the anticodon.

Antigen. A substance that elicits an immune response (production of antibodies) when introduced into an animal; it is specifically recognized by an antibody.

Antioxidant. A substance that destroys an oxidative free radical such as $O_2^{\cdot -}$ or OH·.

Antiparallel. Running in opposite directions.

Antiport. A transmembrane channel that simultaneously moves two molecules or ions in opposite directions. See also symport and uniport.

Antisense RNA. A single-stranded RNA molecule that forms a double-stranded structure with a complementary mRNA so as to block its translation into protein.

Antisense strand. The DNA strand that serves as a template for transcription; it is complementary to the RNA. Also called the noncoding strand.

AP site. An apurinic or apyrimidinic site; the deoxyribose residue remaining after the removal of a base from a DNA strand.

Apoenzyme. An enzyme that is inactive due to the absence of a cofactor.

Apolipoprotein. The protein component of a lipoprotein. Also called an apoprotein.

Apoprotein. A protein without the prosthetic group or metal ion that renders it fully functional. See also apoenzyme and apolipoprotein.

Apoptosis. Cell death by a regulated process in which the cell shrinks and fragments into membrane-bounded portions for phagocytosis by other cells. See also necrosis.

Aptamer. A nucleic acid whose conformation allows it to bind a particular ligand with high specificity and high affinity.

Archaea. One of the two major groups of prokaryotes (the other is eubacteria). Also known as archaebacteria.

Archaebacteria. See archaea.

Assay. A laboratory technique for detecting, and in many cases quantifying, a macromolecule or its activity.

Asymmetric center. See chiral center.

Atherosclerosis. A disease characterized by the formation of cholesterol-containing fibrous plaques in the walls of blood vessels, leading to loss of elasticity and blockage of blood flow.

ATP mass action ratio. The ratio $[ATP]/[ADP][P_i]$, which influences the rate of electron transport and oxidative phosphorylation.

ATP synthase. See F_1F_0-ATPase.

ATPase. An enzyme that catalyzes the hydrolysis of ATP to $ADP + P_i$.

Attenuation. A mechanism in prokaryotes for regulating gene expression in which the availability of an amino acid determines whether an operon (consisting of genes for the enzymes that synthesize the amino acid) is transcribed.

Attenuator. A prokaryotic control element that governs transcription of an operon according to the availability of an amino acid synthesized by the proteins encoded by the operon.

Autocatalytic reaction. A reaction in which a product molecule can act as a catalyst for the same reaction; the reactant molecule therefore appears to catalyze its own reaction.

Autoimmune disease. A disease in which the immune system has lost some of its self-tolerance and produces antibodies against certain self-antigens.

Autolysis. An autocatalytic process in which a molecule catalyzes its own degradation.

Autophosphorylation. The kinase-catalyzed phosphorylation of itself or an identical molecule.

Autoradiography. A process in which X-ray film records the positions of radioactive entities, such as proteins or nucleic acids, that have been immobilized in a matrix such as a nitrocellulose membrane or an electrophoretic gel.

Autotroph. An organism that can synthesize all its cellular components from simple molecules using the energy obtained from sunlight (photoautotroph) or from the oxidation of inorganic compounds (chemolithotroph).

Axial substituent. A group that extends perpendicularly from the plane of the ring to which it is bonded. See also equatorial substituent.

Axoneme. The core of a cilium or eukaryotic flagellum, which consists of two single microtubules surrounded by nine double microtubules.

BAC. See bacterial artificial chromosome.

Backbone. The atoms that form the repeating linkages between successive residues of a polymeric molecule, exclusive of the side chains. Also called the main chain.

Bacteria. The organisms comprising the two major groups of prokaryotes, the archaea and the eubacteria.

Bacterial artificial chromosome (BAC). A plasmid-derived DNA molecule that can replicate in a bacterial cell. BACs are commonly used as cloning vectors.

Bacteriophage. A virus specific for bacteria. Also known as a phage.

Barr body. The condensed and darkly staining inactive X chromosome in the nucleus of a mammalian cell.

Base. (1) A substance that can accept a proton. (2) A purine or pyrimidine component of a nucleoside, nucleotide, or nucleic acid.

Base excision repair (BER). The removal and replacement of a damaged nucleotide in DNA that is initiated by the removal of the base.

Base pair. The specific hydrogen-bonded association between nucleic acid bases. The Watson–Crick base pairs are A·T and G·C.

Basic helix–loop–helix (bHLH) motif. A eukaryotic protein motif that includes a basic DNA-binding region followed by two amphipathic helices connected by a loop. The second helix, which mediates protein dimerization, is often continuous with a leucine zipper motif.

Basic solution. A solution whose pH is greater than 7.0 ($[H^+] < 10^{-7}$ M).

Beer–Lambert law. The equation that describes the relationship between a solute's absorbance (A) and its concentration (c): $A = \varepsilon cl$ where ε is the solute's molar absorptivity and l is the length of the light path.

BER. See base excision repair.

Beriberi. A disease caused by a deficiency of thiamine (vitamin B_1),which is a precursor of the cofactor thiamine pyrophosphate.

βαβ motif. A protein motif consisting of an α helix connecting two parallel strands of a β sheet.

β anomer. See anomers.

β barrel. A protein motif consisting of a β sheet rolled into a cylinder.

β bend. See reverse turn.

β bulge. An irregularity in a β sheet resulting from an extra residue that is not hydrogen bonded to a neighboring chain.

β cell. A pancreatic islet cell that secretes the hormone insulin in response to high blood glucose levels.

β-glycoside. See glycoside.

β hairpin. A protein motif in which two antiparallel β stands are connected by a reverse turn.

β oxidation. A series of enzyme-catalyzed reactions in which fatty acids are progressively degraded by the removal of two-carbon units as acetyl-CoA.

β sheet. A regular secondary structure in which extended polypeptide chains form interstrand hydrogen bonds. In parallel β sheets, the polypeptide chains all run in the same direction; in antiparallel β sheets, neighboring chains run in opposite directions.

bHLH motif. See basic helix–loop–helix motif.

Bilayer. An ordered, double layer of amphiphilic molecules in which polar segments point toward the two solvent-exposed surfaces and the nonpolar segments associate in the center.

Bile acid (Bile salt). An amphiphilic cholesterol derivative that acts as a detergent to solubilize lipids for digestion and absorption.

Binding change mechanism. The mechanism whereby the subunits of the F_1F_0 ATP synthase adopt three successive conformations to convert $ADP + P_i$ to ATP as driven by the dissipation of the transmembrane proton gradient.

Bioavailability. A measure of the fraction of a drug that reaches its target tissue, which depends on the drug's dosage and pharmacokinetics.

Biochemical standard state. A set of conditions including unit activity of the species of interest, a temperature of 25°C, a pressure of 1 atm, and a pH of 7.0.

Bioinformatics. The study of biological information in the form of molecular sequences and structures; e.g., structural bioinformatics.

Biopterin. A pterin derivative that functions as a cofactor in the hydroxylation of phenylalanine to produce tyrosine.

Blastoderm. The single layer of cells surrounding a yolk-like core that forms during the early development of an insect larva.

Blunt ends. The fully base-paired ends of a DNA fragment that has been cleaved by a restriction endonuclease that cuts the DNA strands at opposing sites.

Bohr effect. The decrease in O_2 binding affinity of hemoglobin in response to a decrease in pH.

bp. Base pair, the unit of length used for DNA molecules. Thousands of base pairs (kilobase pairs) are abbreviated kb.

Bradford assay. A spectroscopic technique for determining protein concentration in solution from the absorbance of a dye bound to the protein.

Branch migration. The movement of a crossover point in a Holliday junction during DNA recombination.

Breathing. The small conformational fluctuations of a protein molecule.

Bromodomain. A protein module that binds acetylated Lys residues in histones.

Buffer. A solution of a weak acid and its conjugate base in approximately equal quantities. Such a solution resists changes in pH on the addition of acid or base.

Buffering capacity. The ability of a buffer solution to resist pH changes on addition of acid or base. Buffers are most useful when the pH is within one unit of its component acid's pK.

C-terminus. See carboxyl terminus.

C value. A measure of the quantity of an organism's unique genetic material.

C-value paradox. The occurrence of exceptions to the rule that an organism's DNA content (C value) is correlated to the complexity of its morphology and metabolism.

Cahn–Ingold–Prelog system (*RS* system). A system for unambiguously describing the configurations of molecules with one or more asymmetric centers by assigning a priority ranking to the substituent groups of each asymmetric center.

Calvin cycle. The sequence of photosynthetic dark reactions in which ribulose-5-phosphate is carboxylated, converted to three-carbon carbohydrate precursors, and regenerated. Also called the reductive pentose phosphate cycle.

Calmoduin (CaM). A small Ca^{2+}-binding protein that binds to other proteins in the presence of Ca^{2+} and thereby regulates their activities.

CaM. See calmodulin.

CAM. See crassulacean acid metabolism.

cAMP. Cyclic AMP, an intracellular second messenger.

Cap. A 7-methylguanosine residue that is posttranscriptionally appended to the 5′ end of a eukaryotic mRNA.

Capillary electrophoresis (CE). Electrophoretic procedures carried out in small diameter capillary tubes.

Carbamate. The product of a reaction between CO_2 and an amino group: $-NH-COO^-$.

Carbohydrate. A compound with the formula $(C \cdot H_2O)_n$ where $n \geq 3$. Also called a saccharide.

Carbonyl group. A portion of a molecule with the formula $\!>\!C{=}O$.

Carboxyl group. A portion of a molecule with the formula $-COOH$. Carboxyl groups are usually ionized at physiological pH.

Carboxyl terminus. The end of a polypeptide that has a free carboxylate group. Also called the C-terminus.

Carcinogen. An agent that damages DNA so as to induce a mutation that leads to uncontrolled cell proliferation (cancer).

Caspase. A heterotetrameric cysteine protease that hydrolyzes cellular proteins, including caspase zymogens, as part of the process of apoptosis.

Catabolism. The degradative metabolic reactions in which nutrients and cell constituents are broken down for energy and raw materials.

Catabolite repression. A phenomenon in which the presence of glucose prevents the expression of genes involved in the metabolism of other fuels.

Catalyst. A substance that promotes a chemical reaction without itself undergoing permanent change. A catalyst increases the rate at which a reaction approaches equilibrium but does not affect the free energy change of the reaction.

Catalytic perfection. The ability of an enzyme to catalyze a reaction as fast as diffusion allows it to bind its substrates.

Catalytic triad. The hydrogen-bonded Ser, His, and Asp residues that participate in catalysis in serine proteases.

Cataplerotic reaction. A reaction that drains the intermediates of a metabolic pathway.

Catecholamine. A hydroxylated tyrosine derivative, such as dopamine, norepinephrine, or epinephrine.

Catenate. To interlink circular DNA molecules like the links of a chain.

Cation exchanger. An anionic matrix used to bind cationic molecules in ion exchange chromatography.

C_c. See critical concentration.

CCAAT box. A eukaryotic promoter element with the consensus sequence CCAAT that is located 70 to 90 nucleotides upstream from the transcription start site.

cDNA. See complementary DNA.

CE. See capillary electrophoresis.

Cell cycle. The sequence of events between eukaryotic cell divisions; it includes mitosis and cell division (M phase), a gap stage (G_1 phase), a period of DNA synthesis (S phase), and a second gap stage (G_2 phase) before the next M phase.

Cellular immunity. Immunity mediated by *T* lymphocytes (*T* cells).

Central dogma of molecular biology. The paradigm that DNA directs its own replication as well as its transcription to RNA, which is then translated into a polypeptide. The flow of information is from DNA to RNA to protein.

Centromere. The eukaryotic chromosomal region that attaches to the mitotic spindle during cell division; it contains high concentrations of repetitive DNA.

Ceramide. A sphingosine derivative with an acyl group attached to its amino group.

Cerebroside. A ceramide with a sugar residue as a head group.

C_4 plant. A plant in which photosynthesis relies on CO_2 that has been concentrated by incorporating it into oxaloacetate (a C_4 compound).

Chain-termination procedure. A technique for determining the nucleotide sequence of a DNA using dideoxy nucleotides so as to yield a collection of daughter strands of all different lengths. Also called the dideoxy method.

Channeling. The transfer of an intermediate product from one enzyme active site to another in such a way that the intermediate remains protected by the protein.

Chaotropic agent. A substance that increases the solubility of nonpolar substances in water and thereby tends to denature proteins.

Chargaff's rules. The observation, first made by Erwin Chargaff, that DNA has equal numbers of adenine and thymine residues and equal numbers of guanine and cytosine residues.

Chemical potential. The partial molar free energy of a substance.

Chemiosmotic theory. The postulate that the free energy of electron transport is conserved in the formation of a transmembrane proton gradient. The electrochemical potential of this gradient is used to drive ATP synthesis.

Chemolithotroph. An autotrophic organism that obtains energy from the oxidation of inorganic compounds.

Chimera. See recombinant.

Chiral center. An atom whose substituents are arranged such that it is not superimposable on its mirror image. Also called an asymmetric center.

Chirality. The property of being asymmetric. A chiral molecule cannot be superimposed on its mirror image.

Chloroplasts. The plant organelles in which photosynthesis takes place.

Chromatin. The complex of DNA and protein that comprises the eukaryotic chromosomes.

Chromatin-remodeling complex. An ATP-dependent multisubunit protein in eukaryotes that transiently disrupts DNA–histone interactions so as to alter the accessibility of DNA in nucleosomes.

Chromatography. A technique for separating the components of a mixture of molecules based on their partition between a mobile solvent phase and a porous matrix (stationary phase).

Chromodomain. A protein module that binds methylated Lys residues in histones.

Chromophore. A visible light–absorbing group associated with another molecule.

Chromosome. The complex of protein and a single DNA molecule that comprises some or all of an organism's genome.

Chromosome puff. A portion of a polytene chromosome that has been decondensed to allow gene transcription.

Chylomicrons. Lipoprotein particles that transport dietary triacylglycerols and cholesterol from the intestines to the tissues.

Cis conformation. An arrangement of the peptide group in which successive C_α atoms are on the same side of the peptide bond.

Cis peptide. A conformation in which successive C_α atoms are on the same side of the peptide bond.

Cistron. An archaic term for a gene.

Citric acid cycle. A set of eight enzymatic reactions, arranged in a cycle, in which free energy in the form of ATP, NADH, and $FADH_2$ is recovered from the oxidation of the acetyl group of acetyl-CoA to CO_2. Also called the Krebs cycle and the tricarboxylic acid (TCA) cycle.

Clathrin. A three-legged protein that polymerizes to form a polyhedral structure defining the shape of membranous vesicles that travel between the plasma membrane and intracellular organelles such as the Golgi apparatus.

Clinical trials. A three-phase series of tests of a drug's safety, effectiveness, and side effects in human subjects.

Clone. A collection of identical cells derived from a single ancestor.

Cloning. The production of exact copies of a DNA segment or the organism that harbors it.

Cloning vector. A DNA molecule such as a plasmid, virus, or artificial chromosome that can accommodate a segment of foreign DNA for cloning.

Closed system. A thermodynamic system that can exchange energy but not matter with its surroundings.

Coated vesicle. A membranous intracellular transport vesicle that is encased by clathrin or another coat protein.

Coding strand. See sense strand.

Codon. The sequence of three nucleotides in DNA or RNA that specifies a single amino acid.

Coenzyme. A small organic molecule that is required for the catalytic activity of an enzyme. A coenzyme may be either a cosubstrate or a prosthetic group.

Coenzyme Q. An isoprenoid that functions in electron-transport pathways as a lipid-soluble electron carrier. Also called ubiquinone.

Cofactor. A small organic molecule (coenzyme) or metal ion that is required for the catalytic activity of an enzyme.

Coiled coil. An arrangement of polypeptide chains in which two α helices wind around each other, as in α keratin.

Cointegrate. The product of the fusion of two plasmids, which occurs as an intermediate in transposition.

Colligative property. A physical property, such as freezing point depression or osmotic pressure, that depends on the concentration of a dissolved substance rather than on its chemical nature.

Colony hybridization. A procedure in which DNA from multiple cell colonies is transferred to a membrane or filter and incubated with a DNA or RNA probe to test for the presence of a desired DNA fragment in the cell colonies. Also called *in situ* hybridization.

Combinatorial chemistry. A method for rapidly and inexpensively synthesizing large numbers of related compounds by systematically varying a portion of their structure.

Compartmentation. The division of a cell into smaller functionally discrete systems.

Competitive inhibition. A form of enzyme inhibition in which a substance competes with the substrate for binding to the enzyme active site and thereby appears to increase K_M.

Complementary DNA (cDNA). A DNA molecule, usually synthesized by the action of reverse transcriptase, that is complementary to an mRNA molecule.

Composite transposon. A genetic sequence that may include a variety of genes and is flanked by IS-like elements; such transposons apparently arose by the association of two independent IS elements.

Condensation reaction. The formation of a covalent bond between two molecules, during which the elements of water are lost; the reverse of hydrolysis.

Conjugate acid. The compound that forms when a base accepts a proton.

Conjugate base. The compound that forms when an acid donates a proton.

Conjugate redox pair. An electron donor and acceptor that form a half-reaction. Also called a redox couple.

Conservative replication. A hypothetical mode of DNA duplication in which the parental molecule remains intact and both strands of the daughter duplex are newly synthesized.

Conservative substitution. A change of an amino acid residue in a protein to one with similar properties, e.g., Leu to Ile or Asp to Glu.

Constant region. The C-terminal portion of an antibody (immunoglobulin) subunit, which does not exhibit the high sequence variability of the antigen-recognizing (variable) region of the antibody.

Constitutive enzyme. An enzyme that is synthesized at a more or less steady rate and that is required for basic cell function. Also called a housekeeping enzyme. See also inducible enzyme.

Contact inhibition. The inhibition of proliferation in cultured animal cells when the cells touch each other.

Contour length. The end-to-end length of a stretched-out polymer molecule.

Contour map. A map containing lines (contours) that trace positions of equal value of some property of the map (e.g., height above sea level, electron density).

Convergent evolution. The independent development of similar characteristics in unrelated species or proteins.

Cooperative binding. A situation in which the binding of a ligand at one site on a macromolecule affects the affinity of other sites for the same ligand. Both negative and positive cooperativity occur. See also allosteric interactions.

COPI. A type of protein that coats the membranes of intracellular transport vesicles.

COPII. A type of protein that coats the membranes of intracellular transport vesicles.

Corepressor. A substance that acts together with a protein repressor to block gene transcription.

Cori cycle. An interorgan metabolic pathway in which lactate produced by glycolysis in the muscles is transported via the bloodstream to the liver, where it is used for gluconeogenesis. The resulting glucose returns to the muscles.

Cosubstrate. A coenzyme that is only transiently associated with an enzyme so that it functions as a substrate.

Coupled enzymatic reaction. A technique in which the activity of an enzyme is measured by the ability of a second enzyme to use the product of the first enzymatic reaction to produce a detectable product.

Covalent catalysis. A catalytic mechanism in which the transient formation of a covalent bond between the catalyst and a reactant lowers the free energy of a reaction's transition state.

CpG island. A cluster of CG dinucleotides located just upstream of many vertebrate genes; such sequences occur elsewhere in the genome at only one-fifth their expected frequency.

Crassulacean acid metabolism (CAM). A variation of the C_4 photosynthetic cycle in which CO_2 is temporarily stored as malate.

Cristae. The invaginations of the inner mitochondrial membrane.

Critical concentration (C_c). The concentration of a monomeric substance, above which it assembles into a macromolecular aggregate or polymer, and below which it dissociates.

Cross talk. The interactions of different signal-transduction pathways through activation of the same signaling components, generation of a common second messenger, or similar patterns of target protein phosphorylation.

Cryoelectron microscopy (cryo-EM). A technique in electron microscopy in which a sample is rapidly frozen to very low temperatures so that it retains its native shape to a greater extent than in conventional electron microscopy.

C3′-endo. A ribose conformation in which C3′ is displaced toward the same side of the ring as C5′.

C₃ plant. A plant in which photosynthesis proceeds by the incorporation of CO_2 into three-carbon compounds.

C2′-endo. A ribose conformation in which C2′ is displaced toward the same side of the ring as C5′.

Curved arrow convention. A notation for indicating the movement of an electron pair in a chemical reaction by drawing a curved arrow emanating from the electrons and pointing to the electron-deficient center that attracts the electron pair.

Cyanosis. A bluish skin color indicating the presence of deoxyhemoglobin in the arterial blood.

Cyclic symmetry. A type of symmetry in which the asymmetric units of a symmetric object are related by a single axis of rotation.

Cyclin. A member of a family of proteins that participate in regulating the stages of the cell cycle and whose concentrations change dramatically over the course of the cell cycle.

Cytochrome. A redox-active protein that carries electrons via a prosthetic Fe-containing heme group.

Cytochrome P450. Heme-containing monooxygenases that catalyze the addition of OH groups to drugs and toxins in order to detoxify them and facilitate their excretion.

Cytokinesis. The splitting of the cell into two following mitosis.

Cytoplasm. The entire contents of a cell excluding the nucleus.

Cytoskeleton. The network of intracellular fibers that gives a cell its shape and structural rigidity.

Cytosol. The contents of a cell excluding its nucleus and other membrane-bounded organelles.

D. Dalton, a unit of molecular mass; 1/12th the mass of a ^{12}C atom.

D arm. A conserved stem–loop structure in a tRNA molecule that usually contains the modified base dihydrouridine.

Dark reactions. The portion of photosynthesis in which NADPH and ATP produced by the light reactions are used to incorporate CO_2 into carbohydrates.

Darwinian evolution. See evolution.

ddNTP. An abbreviation for any dideoxynucleoside triphosphate.

Deamination. The hydrolytic removal of an amino group.

Debranching. The enzymatic removal of side chains from a branched polymer such as glycogen.

Degenerate code. A code in which more than one "word" encodes the same entity.

Δℰ. Electromotive force. Change in reduction potential.

$\Delta G^{\ddagger}$. See free energy of activation.

ΔΨ. See membrane potential.

Denature. To disrupt the native conformation of a polymer.

Deoxy sugar. A saccharide produced by replacement of an OH group by H.

Deoxynucleotide. See deoxyribonucleotide.

Deoxyribonucleic acid. See DNA.

Deoxyribonucleotide. A nucleotide in which the pentose is 2′-deoxyribose. Also known as a deoxynucleotide.

Depolarization. The loss of membrane potential that occurs during electrical signaling in cells such as neurons.

Desaturase. An enzyme that introduces double bonds into a newly synthesized fatty acid.

Desensitization. A cell's or organism's adaptation to a long-term stimulus through a reduced response to the stimulus.

Dextrorotatory. Rotating the plane of plane-polarized light clockwise from the point of view of the observer; the opposite of levorotatory.

Diabetes mellitus. A disease in which the pancreas does not secrete sufficient insulin (also called type I, insulin-dependent, or juvenile-onset diabetes) or in which the body has insufficient response to circulating insulin (type II, non-insulin-dependent, or maturity-onset diabetes). Diabetes is characterized by elevated levels of glucose in the blood.

Dialysis. A procedure in which solvent molecules and solutes smaller than the pores in a semipermeable membrane freely exchange with the bulk medium, while larger solutes are retained, thereby changing the solution in which the larger molecules are dissolved.

Diazotroph. A bacterium that can fix nitrogen.

Dideoxy method. See chain-termination procedure.

Differential labeling. Treatment of a macromolecule with a labeling reagent in the presence and absence of a molecule to which it is thought to bind, in order to identify those portions of the macromolecule that are shielded by the bound molecule.

Diffraction pattern. The record of the destructive and constructive interferences of radiation scattered from an object. In X-ray crystallography, this takes the form of a series of discrete spots resulting from a collimated beam of X-rays scattering from a single crystal.

Diffusion. The transport of molecules through their random movement.

Diffusion-controlled limit. The theoretical maximum rate of an enzymatic reaction in solution, about 10^8 to $10^9\ M^{-1} \cdot s^{-1}$.

Dihedral angle. See torsion angle.

Dihedral symmetry. A type of symmetry in which the asymmetric units are related by a twofold rotational axis that intersects another rotation axis at a right angle.

Dimer. An assembly consisting of two monomeric units (protomers).

Dinucleotide binding fold. A protein structural motif consisting of two βαβαβ units, which binds a dinucleotide such as NAD^+. Also called a Rossmann fold.

Dipeptide. A polypeptide consisting of two amino acids.

Diphosphoryl (pyrophosphoryl) group. Two phosphoryl groups linked by a phosphoanhydride bond $(—O_3P—O—PO_3—)^{2-}$.

Diploid. Having two equivalent sets of chromosomes.

Dipolar ion. See zwitterion.

Disaccharide. A carbohydrate consisting of two monosaccharides linked by a glycosidic bond.

Dissociation constant (*K*). The ratio of the products of the concentrations of the dissociated species to those of their parent compounds at equilibrium.

Disulfide bond. A covalent —S—S— linkage.

DNA. Deoxyribonucleic acid. A polymer of deoxynucleotides whose sequence of bases encodes genetic information in all living cells.

DNA chip. See DNA microarray.

DNA glycosylase. An enzyme that initiates base excision repair of DNA by cleaving the glycosidic bond that links a nucleotide base to ribose.

DNA library. A set of cloned DNA fragments representing some or all of an organism's genome.

DNA microarray. A set of DNA segments of known sequence that are immobilized on a solid support for the purpose of hybridizing with nucleic acids in test samples. Also called a DNA chip.

dNTP. A deoxyribonucleoside triphosphate.

Dolichol. An isoprenoid that serves as a lipid-soluble carrier of an *N*-linked oligosaccharide during its synthesis in the endoplasmic reticulum.

Domain. A group of one to a few polypeptide segments comprised of around 100–200 polypeptide residues that fold into a globular unit.

Double-displacement reaction. A reaction in which a substrate binds and a product is released in the first stage, and another substrate binds and another product is released in the second stage.

Double-reciprocal plot. See Lineweaver–Burk plot.

Drug–drug interactions. Increases or decreases in the bioavailablity of a drug caused by the metabolic effects of another drug.

Dynamic instability. The random pattern of growth through monomer addition and regression through monomer removal among a population of polymeric molecules.

$\mathscr{E}$. See reduction potential.

$\mathscr{E}°'$. Reduction potential under biochemical standard conditions.

E site. See exit site.

EC classification. The Enzyme Commission's system for classifying and numbering enzymes according to the type of reaction catalyzed.

Edman degradation. A procedure for the stepwise removal and identification of the N-terminal residues of a polypeptide.

EF hand. A widespread helix–loop–helix structural motif that forms a Ca^{2+}-binding site.

Eicosanoids. C_{20} compounds derived from the C_{20} fatty acid arachidonic acid and which act as local mediators. Prostaglandins, prostacyclins, thromboxanes, leukotrienes, and lipoxins are eicosanoids.

Electrochemical cell. A device in which two half-reactions occur in separate compartments linked by a wire for transporting electrons and a salt bridge for maintaining electrical neutrality; the simultaneous activity of the half-reactions forms a complete oxidation–reduction reaction.

Electrochemical potential. The partial molar free energy of a substance (chemical potential) in the presence of an electrical potential.

Electrogenic transport. The transmembrane movement of a charged substance in a way that generates a charge difference across the membrane.

Electromotive force (emf). $\Delta\mathscr{E}$. Change in reduction potential.

Electron crystallography. A technique for determining molecular structure, in which the electron beam of an electron microscope is used to elicit diffraction from a two-dimensional crystal of the molecules of interest.

Electron density. The arrangement of electrons that gives rise to a diffraction pattern in X-ray crystallography.

Electron-transport chain. A series of membrane-associated electron carriers that pass electrons from reduced coenzymes (NADH and $FADH_2$) to molecular oxygen so as to recover free energy for the synthesis of ATP.

Electrophile. A group that contains an unfilled valence electron shell, or contains an electron deficient atom. An electrophile (electron-lover) reacts readily with a nucleophile (nucleus-lover).

Electrophoresis. See gel electrophoresis.

Electrostatic catalysis. A catalytic mechanism in which the distribution of charges about the catalytic site lowers the free energy of a reaction's transition state.

Elementary reaction. A simple one-step chemical process, several of which may occur in sequence in a chemical reaction.

ELISA. See enzyme-linked immunosorbent assay.

Elongase. An enzyme that adds acetyl units to a fatty acid previously synthesized by fatty acid synthase.

Elongation factor. A protein that interacts with tRNA and/or the ribosome during polypeptide synthesis.

Eluant. The solution used to wash a chromatographic column.

Elution. The process of dislodging a molecule that has bound to a chromatographic matrix.

emf. Electromotive force. Change in reduction potential.

Enantiomers. Molecules that are nonsuperimposable mirror images of one another. Enantiomers are a type of stereoisomer.

Endergonic process. A process that has an overall positive free energy change (a nonspontaneous process).

Endocrine gland. A tissue in higher animals that synthesizes and releases hormones into the bloodstream.

Endocytosis. The internalization of extracellular material through the formation of a vesicle that buds off from the plasma membrane; the opposite of exocytosis. See also receptor-mediated endocytosis.

Endoglycosidase. An enzyme that catalyzes the hydrolysis of the glycosidic bonds between two monosaccharide units within a polysaccharide.

Endonuclease. An enzyme that catalyzes the hydrolysis of the phosphodiester bonds between two nucleotide residues within a polynucleotide strand.

Endopeptidase. An enzyme that catalyzes the hydrolysis of a peptide bond within a polypeptide chain.

Endoplasmic reticulum (ER). A labyrinthine membranous organelle in eukaryotic cells in which membrane lipids are synthesized and some proteins undergo posttranslational modification.

Endosome. A membrane-bounded vesicle that receives materials that the cell ingests via receptor-mediated endocytosis and passes them to lysosomes for degradation.

Enediol intermediate. A reaction intermediate containing a carbon–carbon double bond and a hydroxyl group attached to each carbon.

Energy coupling. The conservation of the free energy of electron transport in a form that can be used to synthesize ATP from ADP + P_i.

"Energy-rich" compound. See "High-energy" compound.

Enhanceosome. A complex containing several transcription factors that regulates gene expression.

Enhancer. A eukaryotic DNA sequence located some distance from the transcription start site, where an activator of transcription may bind.

Enthalpy (H). A thermodynamic quantity, $H = U + PV$, that is equivalent to the heat absorbed at constant pressure (q_P).

Entropy (S). A measure of the degree of randomness or disorder of a system. It is defined as $S = k_B \ln W$, where k_B is the Boltzmann constant and W is the number of equivalent ways the system can be arranged in its particular state.

Enzyme. A biological catalyst. Most enzymes are proteins; a few are RNA.

Enzyme-linked immunosorbent assay (ELISA). A technique in which a molecule is detected, and in many cases quantified, by its ability to bind an antibody to which an enzyme with an easily detected reaction product is attached.

Enzyme saturation. A state in which the substrate concentration is so high that essentially all the enzyme molecules are in the ES form.

Epigenetics. The inheritance of patterns of gene expression that are maintained from generation to generation independent of DNA's base sequence. This occurs, for example, through the methylation of DNA.

Epimers. Sugars that differ only by the configuration at one C atom (excluding the anomeric carbon).

Equatorial substituent. A group that extends largely in the plane of the ring to which it is bonded. See also axial substituent.

Equilibrium. The point in a process at which the forward and reverse reaction rates are exactly balanced so that it undergoes no net change.

Equilibrium constant (K_{eq}). The ratio, at equilibrium, of the product of the concentrations of reaction products to that of its reactants. K_{eq} is related to $\Delta G°$ for the reaction: $\Delta G° = -RT \ln K_{eq}$. Usually abbreviated K.

Equilibrium density gradient centrifugation. A technique in which a mixture of molecules is subjected to ultracentrifugation in a concentrated solution of a dense, fast-diffusing substance such as CsCl that forms a density gradient as the centrifuge spins. Centrifugation is continued to equilibrium, thereby sepa-

rating the mixture according to the densities of its component molecules.

ER. See endoplasmic reticulum.

Erythrocyte. A red blood cell, which functions to transport O_2 to the tissues. It is essentially a membranous sack of hemoglobin.

Erythrocyte ghost. Membranous particles derived from erythrocytes, which retain their original shape but are devoid of cytoplasm.

ES complex. The enzyme–substrate complex, whose formation is a key component of the Michaelis–Menten model of enzyme action. Also called the Michaelis complex.

Essential amino acid. An amino acid that an animal cannot synthesize and must therefore obtain in its diet.

Essential fatty acid. A fatty acid that an animal cannot synthesize and must therefore obtain in its diet.

EST. See expressed sequence tag.

Ester group. A portion of a molecule with the formula —COOR, where R is an alkyl group.

Estrogen. A steroid that functions primarily as a female sex hormone.

Ether. A molecule with the formula ROR′, where R and R′ are alkyl groups.

Eubacteria. One of the two major groups of prokaryotes (the other is archaea).

Euchromatin. The transcriptionally active, relatively uncondensed chromatin in a eukaryotic cell.

Eukarya. See eukaryote.

Eukaryote. An organism consisting of a cell (or cells) whose genetic material is contained in a membrane-bounded nucleus.

Evolution. The gradual alteration of an organism or one of its components as a result of genetic changes that are passed from parent to offspring.

Exciton transfer. A mode of decay of an energetically excited molecule, in which electronic energy is transferred to a nearby unexcited molecule. Also known as resonance energy transfer.

Exergonic process. A process that has an overall negative free energy change (a spontaneous process).

Exit site (E site). The ribosomal binding site that accommodates a tRNA molecule that has just transferred its peptidyl group to an incoming aminoacyl–tRNA and is ready to dissociate from the ribosome.

Exocytosis. The release outside the cell of a vesicle's contents through the fusion of the vesicle membrane with the plasma membrane; the opposite of endocytosis.

Exoglycosidase. An enzyme that catalyzes the hydrolytic excision of a monosaccharide unit from the end of a polysaccharide.

Exon. A portion of a gene that appears in both the primary and mature mRNA transcripts. Also called an expressed sequence.

Exonuclease. An enzyme that catalyzes the hydrolytic excision of a nucleotide residue from one end of a polynucleotide strand.

Exopeptidase. An enzyme that catalyzes the hydrolytic excision of an amino acid residue from one end of a polypeptide chain.

Expressed sequence. See exon.

Expressed sequence tag (EST). A cDNA segment corresponding to a cellular mRNA, that can be used to identify genes that are transcribed.

Expression vector. A plasmid containing the transcription and translation control sequences required for the production of a foreign DNA gene product (RNA or protein) in a host cell.

Extrinsic protein. See peripheral protein.

$\mathcal{F}$. Faraday, the electrical charge of one mole of electrons.

F_1F_0-ATPase. A multisubunit protein consisting of a proton-translocating membrane-embedded component (F_0) linked to a soluble catalytic component (F_1) that catalyzes ATP synthesis in the presence of a protonmotive force. Also called ATP synthase.

Fab fragment. A proteolytic fragment of an antibody molecule that contains the antigen binding site. See also Fc fragment.

Facilitated diffusion. See passive-mediated transport.

Familial hypercholesterolemia. See hypercholesterolemia.

Fat. A mixture of triacylglycerols that is solid at room temperature.

Fatty acid. A carboxylic acid with a long-chain hydrocarbon side group.

Fc fragment. A proteolytic fragment of an antibody molecule that contains the two C-terminal domains of its two heavy chains. See also Fab fragment.

Fe–S cluster. See iron–sulfur cluster.

Feedback inhibition. The inhibition of an early step in a reaction sequence by the product of a later step.

Feedforward activation. The activation of a later step in a reaction sequence by the product of an earlier step.

Fermentation. An anaerobic catabolic process.

Fibrous protein. A protein characterized by a stiff, elongated conformation, that tends to form fibers.

First-order reaction. A reaction whose rate is proportional to the concentration of a single reactant.

Fischer convention. A system for describing the absolute configurations of chiral molecules by relating their structures to that of D- or L-glyceraldehyde.

Fischer projection. A graphical convention for specifying molecular configuration in which horizontal lines represent bonds that extend above the plane of the paper and vertical lines represent bonds that extend below the plane of the paper.

5′ end. The terminus of a polynucleotide whose C5′ is not esterified to another nucleotide residue.

Flip-flop. See transverse diffusion.

Flipase. An enzyme that catalyzes the translocation of a membrane lipid across a lipid bilayer (a flip-flop).

Fluid mosaic model. A model of biological membranes in which integral membrane proteins float and diffuse laterally in a fluid lipid bilayer.

Fluorescence. A mode of decay of an excited molecule, in which electronic energy is emitted in the form of a photon.

Fluorescence recovery after photobleaching (FRAP). A technique for assessing the diffusion of membrane components from the rate at which the fluorescently labeled component moves into an area previously bleached by a pulse of laser light.

Fluorophore. A fluorescent group.

Flux. (1) The rate of flow of metabolites through a metabolic pathway. (2) The rate of transport per unit area.

fMet. The formylated methionine that initiates ribosomal polypeptide synthesis in prokaryotes.

Footprinting. A procedure in which the DNA sequence to which a protein binds is identified by determining which bases are protected by the protein from chemical or enzymatic modification.

Fractional saturation (Y). The fraction of a protein's ligand-binding sites that are occupied by ligand. For example, Y_{O_2} is the fractional saturation of a protein's oxygen binding sites.

Fractionation procedure. A laboratory technique for separating the components of a mixture of molecules through differences in their chemical and physical properties.

Frameshift mutation. An insertion or deletion of nucleotides in DNA that alters the sequential reading (the frame) of sets of three nucleotides (codons) during translation.

FRAP. See fluorescence recovery after photobleaching.

Free energy (G). A thermodynamic quantity, $G = H - TS$, whose change at constant pressure is indicative of the spontaneity of a process. For spontaneous processes, $\Delta G < 0$, whereas for a process at equilibrium, $\Delta G = 0$. Also called Gibbs free energy.

Free energy of activation ($\Delta G^{\ddagger}$). The free energy of the transition state minus the free energies of the reactants in a chemical reaction.

Free radical. A molecule with an unpaired electron.

Functional group. A portion of a molecule that participates in interactions with other substances. Common functional groups in biochemistry are acyl, amido, amino, carbonyl, carboxyl, diphosphoryl (pyrophosphoryl), ester, ether, hydroxyl, imino, phosphoryl, and sulfhydryl groups.

Furanose. A sugar with a five-membered ring.

Futile cycle. See substrate cycle.

G. See free energy.

$G^{\ddagger}$. The free energy of the transition state. See also free energy of activation.

G protein. A guanine nucleotide–binding protein involved in signal transduction that is inactive when it binds GDP and active when it binds GTP. The GTPase activity of the G protein limits its own activity. Heterotrimeric G proteins consist of three subunits, which dissociate to form G_{α} (to which GTP binds) and $G_{\beta\gamma}$ components on activation.

G-quartet. A cyclic tetramer of hydrogen-bonded guanine groups. Stacks of G-quartets result from the antiparallel association of G-rich telomeric DNA hairpin structures.

Ganglioside. A ceramide whose head group is an oligosaccharide containing at least one sialic acid residue.

Gap genes. See segmentation genes.

Gap junction. An intercellular channel for ions and small molecules that is formed by protein complexes in the membranes of apposed cells.

Gastrulation. The stage of embryonic development in which cells migrate to form a triple-layered structure.

Gates and fences model. A model for membrane structure that includes cytoskeletal proteins that prevent or limit the free diffusion of other membrane proteins.

Gating. The opening and closing of an ion channel in response to a signal such as mechanical stimulation, ligand binding, presence of a signaling molecule, or a change in membrane voltage.

GC box. One or more copies of the sequence GGGCGG upstream of the transcription start site that act as a promoter for some eukaryotic housekeeping genes.

Gel electrophoresis. A procedure in which macromolecules are separated on the basis of charge or size by their differential migration through a gel-like matrix under the influence of an applied electric field. In polyacrylamide gel electrophoresis (PAGE), the matrix is cross-linked polyacrylamide. Agarose gels are used to separate molecules of very large masses, such as DNAs. See also SDS-PAGE and pulsed-field gel electrophoresis.

Gel filtration chromatography. A procedure in which macromolecules are separated on the basis of their size and shape. Also called size exclusion or molecular sieve chromatography.

Gene. A unique sequence of nucleotides that encodes a polypeptide or RNA; it may include nontranscribed and nontranslated sequences that have regulatory functions.

Gene cluster. A region of DNA containing multiple copies of genes, usually tRNA or rRNA genes, whose products are required in large amounts.

Gene duplication. An event, such as aberrant crossover, that gives rise to two copies of a gene on the same chromosome, each of which can then evolve independently.

Gene expression. The decoding, via transcription and translation, of the information contained in a gene to yield a functional RNA or protein product.

Gene knockout. A genetic engineering process that deletes or inactivates a specific gene in an animal.

Gene product. The RNA or protein that is encoded by a gene and that is the end point of the gene's expression through transcription and translation.

Gene therapy. The transfer of genetic material to the cells of an individual in order to produce a therapeutic effect.

General acid catalysis. A catalytic mechanism in which partial proton transfer from an acid lowers the free energy of a reaction's transition state.

General base catalysis. A catalytic mechanism in which partial proton abstraction by a base lowers the free energy of a reaction's transition state.

General transcription factor (GTF). One of a set of eukaryotic proteins that are required for the synthesis of all mRNAs.

Genetic anticipation. A pattern of inheritance of a genetic disease, in which the age of onset of symptoms decreases with each generation.

Genetic code. The correspondence between the sequence of nucleotides in a nucleic acid and the sequence of amino acids in a polypeptide; a series of three nucleotides (a codon) specifies an amino acid.

Genetic engineering. See recombinant DNA technology.

Genome. The complete set of genetic instructions in an organism.

Genomic library. A set of cloned DNA fragments representing an organism's entire genome.

Genomics. The study of the size, organization, and gene content of organisms' genomes.

Genotype. An organism's genetic characteristics.

Gibbs free energy. See free energy.

Globin. The polypeptide components of myoglobin and hemoglobin.

Globoside. A ceramide whose head group is a neutral oligosaccharide.

Globular protein. A water-soluble protein characterized by a compact, highly folded structure.

Glucocorticoid. A steroid hormone that affects a range of metabolic pathways and the inflammatory response.

Glucogenic amino acid. An amino acid whose degradation yields a gluconeogenic precursor. See also ketogenic amino acid.

Gluconeogenesis. The synthesis of glucose from noncarbohydrate precursors.

Glucose–alanine cycle. An interorgan metabolic pathway that transports nitrogen to the liver in which pyruvate produced by glycolysis in the muscles is converted to alanine and transported to the liver. There the alanine is converted back to pyruvate and its amino group is used to synthesize urea for excretion. The pyruvate is converted, via gluconeogenesis, to glucose, which is returned to the muscles.

Glycan. See polysaccharide.

Glycerophosphate shuttle. A metabolic pathway that uses the interconversion of dihydroxyacetone phosphate and 3-phosphoglycerol to transport cytosolic reducing equivalents into the mitochondria.

Glycerophospholipid. An amphiphilic lipid in which two fatty acyl groups are attached to a glycerol-3-phosphate whose phosphate group is linked to a polar group. Also called a phosphoglyceride.

Glycoconjugate. A molecule, such as a glycolipid or glycoprotein, that contains covalently linked carbohydrate.

Glycoforms. Glycoproteins that differ in the sequence, location, and number of covalently attached carbohydrates.

Glycogen. An $\alpha(1\rightarrow6)$ branched polymer of $\alpha(1\rightarrow4)$-linked glucose residues that serves as a glucose storage molecule in animals.

Glycogen storage disease. An inherited disorder of glycogen metabolism affecting the size and structure of glycogen molecules or their mobilization in the muscle and/or liver.

Glycogenolysis. The enzymatic degradation of glycogen to glucose-6-phosphate.

Glycolipid. A lipid to which carbohydrate is covalently attached.

Glycolysis. The 10-reaction pathway by which glucose is broken down to 2 pyruvate with the concomitant production of 2 ATP and the reduction of 2 NAD^+ to 2 NADH.

Glycoprotein. A protein to which carbohydrate is covalently attached.

Glycosaminoglycan. An unbranched polysaccharide consisting of alternating residues of uronic acid and hexosamine.

Glycoside. A molecule containing a saccharide and another molecule linked by a glycosidic bond to the anomeric carbon in the α configuration (α-glycoside) or β configuration (β-glycoside).

Glycosidic bond. The covalent linkage (acetal or ketal) between the anomeric carbon of a saccharide and an alcohol (*O*-glycosidic bond) or an amine (*N*-glycosidic bond). Glycosidic bonds link the monosaccharide residues of a polysaccharide.

Glycosylation. The attachment of carbohydrate chains to a protein through *N*- or *O*-glycosidic linkages.

Glyoxylate pathway. A variation of the citric acid cycle in plants that allows acetyl-CoA to be converted quantitatively to gluconeogenic precursors.

Glyoxysome. A membrane-bounded plant organelle in which the reactions of the glyoxylate cycle take place. It is a specialized type of peroxisome.

Golgi apparatus. A eukaryotic organelle consisting of a set of flattened membranous sacs in which newly synthesized proteins and lipids are modified.

Gout. A disease characterized by elevated levels of uric acid, usually the result of impaired uric acid excretion. Its most common manifestation is painful arthritic joint inflammation caused by the deposition of sodium urate.

GPCR. G protein–coupled receptor, a cell surface protein with seven transmembrane helices that interacts with an associated G protein on ligand binding.

GPI-linked protein. A protein that is anchored in a membrane via a covalently linked glycosylphosphatidylinositol (GPI) group.

Gram-negative bacterium. A bacterium that does not take up Gram stain, indicating that its cell wall is surrounded by a complex outer membrane that excludes Gram stain.

Gram-positive bacterium. A bacterium that takes up Gram stain, indicating that its outermost layer is a cell wall.

Grana (*sing.* granum). The stacked disks of the thylakoid in a chloroplast.

gRNA. See guide RNA.

Group I intron. An intron in an rRNA molecule whose self-splicing reaction requires a guanine nucleotide and generates a cyclized intron product.

Group II intron. An intron in a eukaryotic rRNA molecule whose self-splicing reaction does not require a free nucleotide and generates a lariat intron product.

Growth factor. A protein hormone that stimulates the proliferation and differentiation of its target cells.

GTF. See general transcription factor.

GTPase. An enzyme that catalyzes the hydrolysis of GTP to $GDP + P_i$.

Guide RNA (gRNA). Small RNA molecules that pair with an immature mRNA to direct its posttranscriptional editing.

H. See enthalpy.

Half-life. See half-time.

Half-reaction. The single oxidation or reduction process, involving an electron donor and its conjugate electron acceptor, that occurs in electrical cells but requires direct contact with another such reaction to form a complete oxidation–reduction reaction.

Half-time ($t_{1/2}$). The time required for half the reactant initially present to undergo reaction. Also called half-life.

Halobacteria. Bacteria that thrive in (and may require) high salinity.

Haploid. Having one set of chromosomes.

HAT. See histone acetyltransferase.

Haworth projection. A drawing of a sugar ring in which ring bonds that project in front of the plane of the paper are represented by heavy lines and ring bonds that project behind the plane of the paper are drawn as light lines.

HDL. High density lipoprotein; see lipoprotein.

Heat shock protein (Hsp). See molecular chaperone.

Helicase. An enzyme that unwinds a double-stranded nucleic acid.

Helix cap. A protein structural element in which the side chain of a residue preceding or succeeding a helix folds back to form a hydrogen bond with the backbone of one of the helix's four terminal residues.

Helix–turn–helix (HTH) motif. An ~20-residue protein motif that forms two α helices that cross at an angle of ~120°. This motif, which occurs in numerous prokaryotic DNA-binding proteins, binds in DNA's major groove to specific base sequences.

Heme. A porphyrin derivative whose central Fe(II) atom is the site of reversible oxygen binding (in myoglobin and hemoglobin) or oxidation–reduction (in cytochromes).

Hemiacetal. The product of the reaction between an alcohol and the carbonyl group of an aldehyde.

Hemiketal. The product of the reaction between an alcohol and the carbonyl group of a ketone.

Hemolytic anemia. Loss of red blood cells through their lysis (destruction) in the bloodstream.

Henderson–Hasselbalch equation. The mathematical expression of the relationship between the pH of a solution of a weak acid and its pK: $pH = pK + \log([A^-]/[HA])$.

Heptad repeat. A sequence of seven residues that is repeated in the same polymer.

Heterochromatin. Highly condensed, nonexpressed eukaryotic DNA.

Heterogeneous nuclear RNA (hnRNA). Eukaryotic mRNA primary transcripts whose introns have not yet been excised.

Heterologous DNA. A segment of DNA consisting of imperfectly complementary strands.

Heterolytic cleavage. Cleavage of a bond in which one of two chemically bonded atoms acquires both of the electrons that formed the bond.

Heteropolysaccharide. A polysaccharide consisting of more than one type of monosaccharide.

Heterotrimeric G protein. See G protein.

Heterotroph. An organism that obtains free energy from the oxidation of organic compounds produced by other organisms.

Heterozygous. Having one each of two gene variants.

Hexose monophosphate shunt. See pentose phosphate pathway.

"High-energy" intermediate. A substance whose degradation is highly exergonic (yields at least as much free energy as is required to synthesize ATP from ADP + P_i; ≥ 30.5 kJ $\cdot$ mol^{-1} under standard biochemical conditions). Also called an "energy-rich" compound.

High-pressure liquid chromatography (HPLC). An automated chromatographic procedure for fractionating molecules using precisely fabricated matrix materials and pressurized flows of precisely mixed solvents.

Highly repetitive DNA. Clusters of nearly identical sequences of up to 10 bp that are repeated thousands of times; these sequences are present at $>10^6$ copies per haploid genome. Also known as satellite DNA and short tandem repeats (STRs).

Hill coefficient. The exponent in the Hill equation. It provides a measure of the degree of cooperative binding of a ligand to a molecule.

Hill equation. A mathematical expression for the degree of saturation of ligand binding to a molecule with multiple binding sites as a function of the ligand concentration.

Histone acetyltransferase (HAT). An enzyme that catalyzes the sequence-specific acetylation of histones so as to regulate gene transcription.

Histone code. The correlation between the pattern of histone modification and the transcriptional activity of the associated DNA.

Histones. Highly conserved basic proteins that constitute the protein core to which DNA is bound to form a nucleosome.

HIV. Human immunodeficiency virus, the causative agent of acquired immunodeficiency syndrome (AIDS).

HMG protein. A member of the high mobility group (HMG) of nonhistone chromosomal proteins whose abundant charged groups give them high electrophoretic mobility.

hnRNA. See heterogeneous nuclear RNA.

Holliday junction. The four-stranded structure that forms as an intermediate in DNA recombination.

Holoenzyme. A catalytically active enzyme–cofactor complex.

Homeobox. See homeodomain.

Homeodomain. An ~60-amino acid DNA-binding motif common to many genes that specify the identities and fates of embryonic cells; such genes encode transcription factors. Also called a homeobox.

Homeostasis. The maintenance of a steady state in an organism.

Homeotic selector genes. Insect genes that specify the identities of body segments.

Homolactic fermentation. The reduction of pyruvate to lactate with the concomitant oxidation of NADH to NAD$^+$.

Homologous end-joining. A pathway in which DNA with double-strand breaks is repaired nonmutagenically through recombination with an intact homologous chromosome.

Homologous recombination. See recombination.

Homology. The evolutionary relatedness of two polypeptides.

Homolytic cleavage. Cleavage of a bond in which each participating atom acquires one of the electrons that formed the bond.

Homopolysaccharide. A polysaccharide consisting of one type of monosaccharide unit.

Homozygous. Having two identical copies of a particular gene.

Hoogsteen base pair. A form of base pairing in which thymine or uracil atom N3 hydrogen bonds to adenine atom N7 and adenine N6 hydrogen bonds to thymine or uracil O4. See also Watson–Crick base pair.

Hormone. A substance (e.g., a peptide or a steroid) that is secreted by one tissue into the bloodstream and which induces a physiological response (e.g., growth and metabolism) in other tissues.

Hormone response element (HRE). A DNA sequence to which a hormone–receptor complex binds so as to enhance or repress the transcription of an associated gene.

Hormone-sensitive lipase. An adipose tissue enzyme that releases fatty acids from triacylglycerols in response to a hormonally generated increase in cAMP. Also known as hormone-sensitive triacylglycerol lipase.

Housekeeping enzyme. See constitutive enzyme.

***Hox* gene.** A gene encoding a transcription factor that includes a homeodomain.

HPLC. See high-pressure liquid chromatography.

HRE. See hormone response element.

Hsp. Heat shock protein. See molecular chaperone.

HTH motif. See helix–turn–helix motif.

Humoral immunity. Immunity mediated by antibodies (immunoglobulins) produced by *B* lymphocytes (*B* cells).

Hybridization. The formation of double-stranded segments of complementary DNA and/or RNA sequences.

Hybridoma. The cell clones produced by the fusion of an antibody-producing lymphocyte and an immortal myeloma cell. These are the cells that produce monoclonal antibodies.

Hydration. The molecular state of being surrounded by and interacting with several layers of solvent water molecules, that is, solvated by water.

Hydrogen bond. A largely electrostatic interaction between a weakly acidic donor group such as O—H or N—H and a weakly basic acceptor atom such as O or N.

Hydrolase. An enzyme that catalyzes a hydrolytic reaction.

Hydrolysis. The cleavage of a covalent bond accomplished by adding the elements of water; the reverse of a condensation.

Hydronium ion. A proton associated with a water molecule, H_3O^+.

Hydropathy. A measure of the combined hydrophobicity and hydrophilicity of an amino acid residue; it is indicative of the likelihood of finding that residue in a protein interior.

Hydrophilic substance. A substance whose high polarity allows it to readily interact with water molecules and thereby dissolve in water.

Hydrophobic collapse. A driving force in protein folding, resulting from the tendency of hydrophobic residues to avoid contact with water and hence form the protein core.

Hydrophobic effect. The tendency of water to minimize its contacts with nonpolar substances, thereby inducing the substances to aggregate.

Hydrophobic interaction chromatography. A procedure in which molecules are selectively retained on a nonpolar matrix by virtue of their hydrophobicity.

Hydrophobic substance. A substance whose nonpolar nature reduces its ability to be solvated by water molecules. Hydrophobic substances tend to be soluble in nonpolar solvents but not in water.

Hydroxyl group. A portion of a molecule with the formula —OH.

Hyperammonemia. Elevated levels of ammonia in the blood, a toxic situation.

Hyperbolic curve. The graphical representation of the mathematical equation that describes the noncooperative binding of a ligand to a molecule.

Hypercholesterolemia. High levels of cholesterol in the blood, a risk factor for heart disease. Familial hypercholesterolemia usually results from an inherited defect in the LDL receptor.

Hyperchromic effect. The increase in DNA's ultraviolet absorbance resulting from the loss of stacking interactions as the DNA denatures.

Hyperglycemia. Elevated levels of glucose in the blood.

Hypervariable residue. An amino acid residue occupying a position in a protein that is occupied by many different residues among evolutionarily related proteins. The opposite of a hypervariable residue is an invariant residue.

Hypoxia. A condition in which the oxygen level in the blood is lower than normal.

I-cell disease. A hereditary deficiency in a lysosomal hydrolase that leads to the accumulation of glycosaminoglycan and glycolipid inclusions in the lysosomes.

IDL. Intermediate density lipoprotein; see lipoprotein.

IEF. See isoelectric focusing.

Ig. Immunoglobulin. See antibody.

Imaginal disk. A patch of apparently undifferentiated but developmentally committed cells in insect larva that ultimately gives rise to a specific external structure in the adult.

Imino group. A portion of a molecule with the formula $>$C$=$NH.

Immune system. The cells and organs that respond to microbial infection by producing antibodies and by killing pathogens and infected host cells.

Immunoaffinity chromatography. A procedure in which a molecule is separated from a mixture of other molecules by its ability to bind specifically to an immobilized antibody.

Immunoassay. A procedure for detecting, and in some cases quantifying the activity of, a macromolecule by using an antibody or mixture of antibodies that reacts specifically with that substance.

Immunoblot. A technique in which a molecule immobilized on a membrane filter can be detected through its ability to bind to an antibody directed against it. A Western blot is an immunoblot to detect an immobilized protein after electrophoresis.

Immunofluorescence microscopy. A technique in microscopy in which a fluorescence-tagged antibody is used to reveal the presence of the antigen to which it binds.

Immunoglobulin (Ig). See antibody.

Immunoglobulin fold. A disulfide-linked domain consisting of a sandwich of a three-stranded and a four-stranded antiparallel β sheet that occurs in antibody molecules.

Imprinting. The differential expression of maternal and paternal genes according to their patterns of DNA methylation.

in situ. In place.

in situ **hybridization.** See colony hybridization.

in vitro. In the laboratory (literally, in glass).

in vivo. In a living organism.

Inactivator. An inhibitor that reacts irreversibly with an enzyme so as to inactivate it.

Indirect readout. The ability of a DNA-binding protein to detect its target base sequence though the sequence-dependent conformation and/or flexibility of its DNA backbone rather than through direct interaction with its bases.

Induced fit. An interaction between a protein and its ligand, which induces a conformational change in the protein that increases the protein's affinity for the ligand.

Inducer. A substance that facilitates gene expression.

Inducible enzyme. An enzyme that is synthesized only when required by the cell. See also constitutive enzyme.

Inhibition constant (K_I). The dissociation constant for an enzyme–inhibitor complex.

Inhibitor. A substance that reduces an enzyme's activity by affecting its substrate binding or turnover number.

Initiation factor. A protein that interacts with mRNA and/or the ribosome and which is required to initiate translation.

Inorganic compound. A compound that lacks the element carbon.

Insertion sequence. A simple transposon that is flanked by short inverted repeats. Also called an IS element.

Insertion/deletion mutation. A genetic change resulting from the addition or loss of nucleotides; also called an indel.

Insulator. A segment of DNA that delimits the effective range of a transcription-regulating element.

Insulin resistance. The decreased ability of cells to respond to insulin by increasing their glucose uptake.

Integral protein. A membrane protein that is embedded in the lipid bilayer and can be separated from it only by treatment with agents that disrupt membranes. Also called an intrinsic protein.

Intercalation agent. A substance, usually a planar aromatic cation, that slips in between the stacked bases of a double-stranded polynucleotide.

Interconvertible enzyme. A protein that undergoes phosphorylation and dephosphorylation so as to modulate its activity.

Interfacial activation. The increase in activity when a lipid-specific enzyme contacts the lipid–water interface.

Intermediate filament. A 100-Å-diameter cytoskeletal element consisting of coiled coil polypeptide chains.

Intermembrane space. The compartment between the inner and outer mitochondrial membranes. Because of the porosity of the outer membrane, the intermembrane space is equivalent to the cytosol in its small-molecule composition.

Internal conversion. A mode of decay of an excited molecule, in which electronic energy is converted to heat (the kinetic energy of molecular motion).

Internal resolution site. The sequence at which site-specific recombination occurs during transposition.

Intervening sequence. See intron.

Intrinsic protein. See integral protein.

Intron. A portion of a gene that is transcribed but excised prior to translation. Also called an intervening sequence.

Invariant residue. A residue in a protein that is the same in all evolutionarily related proteins. The opposite of an invariant residue is a hypervariable residue.

Ion exchange chromatography. A fractionation procedure in which charged molecules are selectively retained by a matrix bearing oppositely charged groups.

Ion pair. An electrostatic interaction between two ionic groups of opposite charge. In proteins, it is also called a salt bridge.

Ionophore. An organic molecule, often an antibiotic, that increases the permeability of a membrane to a particular ion. A carrier ionophore diffuses with its ion through the membrane, whereas a channel-forming ionophore forms a transmembrane pore.

Iron–sulfur protein. A protein that contains a prosthetic group consisting most commonly of equal numbers of iron and sulfur ions (i.e., [2Fe–2S] and [4Fe–4S]) and that usually participates in oxidation–reduction reactions.

IS element. See insertion sequence.

Isoaccepting tRNA. A tRNA that carries the same amino acid as another tRNA.

Isoelectric focusing (IEF). Electrophoresis through a stable pH gradient such that a charged molecule migrates to a position corresponding to its isoelectric point.

Isoelectric point (pI). The pH at which a molecule has no net charge and hence does not migrate in an electric field.

Isolated system. A thermodynamic system that cannot exchange matter or energy with its surroundings.

Isomerase. An enzyme that catalyzes an isomerization reaction.

Isopeptide bond. An amide linkage between an α-carboxylate group of an amino acid and the ε-amino group of Lys, or between the α-amino group of an amino acid and the β- or γ-carboxylate group of Asp or Glu.

Isoprenoid. A lipid containing five-carbon units with the same carbon skeleton as isoprene.

Isoschizomers. Two restriction endonucleases that cleave at the same nucleotide sequence.

Isozymes. Enzymes that catalyze the same reaction but are encoded by different genes.

Jaundice. A yellowing of the skin and whites of the eyes as a result of the deposition of the heme degradation product bilirubin in these tissues. It is a symptom of liver dysfunction, bile-duct obstruction, or a high rate of red cell destruction.

Junk DNA. See selfish DNA.

K. See dissociation constant and equilibrium constant.

k. See rate constant.

k_B. Boltzmann constant (1.3807×10^{-23} J·K^{-1}); it is equivalent to R/N, where R is the gas constant and N is Avogadro's number.

kb. Kilobase pair; 1 kb = 1000 base pairs (bp).

k_{cat}. The catalytic constant for an enzymatic reaction, equivalent to the ratio of the maximal velocity (V_{max}) and the enzyme concentration ([E]$_T$). Also called the turnover number.

k_{cat}/K_M. The apparent second-order rate constant for an enzyme-catalyzed reaction; it is a measure of an enzyme's catalytic efficiency.

kD. Kilodaltons; 1000 daltons (D).

K_{eq}. See equilibrium constant.

Ketogenesis. The synthesis of ketone bodies from acetyl-CoA.

Ketogenic amino acid. An amino acid whose degradation in animals yields compounds that can be converted to fatty acids or ketone bodies. See also glucogenic amino acid.

Ketone bodies. Acetoacetate, D-β-hydroxybutyrate, and acetone; these compounds are produced from acetyl-CoA by the liver for use as metabolic fuels in peripheral tissues.

Ketose. A sugar whose carbonyl group is a ketone.

Ketosis. A potentially pathological condition in which ketone bodies are produced in excess of their utilization.

K_I. See inhibition constant.

Kinase. An enzyme that transfers a phosphoryl group between ATP and another molecule.

K_M. See Michaelis constant.

$K_M{}^{app}$. The apparent (observed) Michaelis constant for an enzyme-catalyzed reaction, which may differ from the true value due to the presence of an enzyme inhibitor.

k_{-1}. The rate constant for a reverse reaction, e.g., the breakdown of the ES complex to E + S.

Krebs cycle. See citric acid cycle.

k_2. The rate constant for the second step of a simple enzyme-catalyzed reaction that follows Michaelis–Menten kinetics, that is, the conversion of the ES complex to E + P.

K_w. The ionization constant of water; equal to 10^{-14}.

L. See linking number.

Lactose intolerance. The inability to digest the disaccharide lactose due to a deficiency of the enzyme β-galactosidase (lactase).

Lagging strand. A newly synthesized DNA strand that extends $3'\rightarrow5'$ in the direction of travel of the replication fork. This strand is synthesized as a series of discontinuous fragments that are later joined.

Lateral diffusion. The movement of a lipid within one leaflet of a bilayer.

LBHB. See low-barrier hydrogen bond.

LDL. Low density lipoprotein; see lipoprotein.

Lead compound. A drug molecule that serves as the starting point for the development of more effective drug molecules.

Leader peptide. See signal peptide.

Leader sequence. (1) A nucleotide sequence that precedes the coding region of an mRNA. (2) A signal sequence.

Leading strand. A newly synthesized DNA strand that extends $5'\rightarrow3'$ in the direction of travel of the replication fork. This strand is synthesized continuously.

Lectin. A protein that binds to a specific saccharide.

Lesch–Nyhan syndrome. A genetic disease caused by the deficiency of hypoxanthine–guanine phosphoribosyltransferase (HGPRT), an enzyme required for purine salvage reactions. Affected individuals produce excessive uric acid and exhibit neurological abnormalities.

Leucine zipper. A protein structural motif in which two α helices, each with a hydrophobic strip along one side, associate as a coiled coil. This motif, which has a Leu at nearly every seventh residue, mediates the association of many types of DNA-binding proteins.

Leukocyte. White blood cell.

Levorotatory. Rotating the plane of polarized light counterclockwise from the point of view of the observer; the opposite of dextrorotatory.

LHC. See light-harvesting complex.

Ligand. (1) A small molecule that binds to a larger molecule. (2) A molecule or ion bound to a metal ion.

Ligand-gated channel. An ion channel whose opening and closing (gating) is controlled by the binding of a specific molecule (ligand).

Ligase. An enzyme that catalyzes bond formation coupled with the hydrolysis of ATP.

Ligation. The joining together of two molecules such as two DNA segments.

Light-harvesting complex (LHC). A pigment-containing membrane protein that collects light energy and transfers it to a photosynthetic reaction center.

Light reactions. The portion of photosynthesis in which specialized pigment molecules capture light energy and are thereby oxidized. Electrons are transferred to generate NADPH and a transmembrane proton gradient that drives ATP synthesis.

Limited proteolysis. A technique in which a polypeptide is incompletely digested by proteases.

Lineweaver–Burk plot. A graph of a rearrangement of the Michaelis–Menten equation to a linear form that permits the determination of K_M and V_{max}. Also called a double-reciprocal plot.

Linker DNA. The ~55-bp segment of DNA that links nucleosome core particles in chromatin.

Linking number (L). The number of times that one strand of a covalently closed circular double-stranded DNA winds around the other; it cannot be changed without breaking covalent bonds.

Lipid. Any member of a broad class of biological molecules that are largely or wholly hydrophobic and therefore tend to be insoluble in water but soluble in organic solvents such as hexane.

Lipid bilayer. See bilayer.

Lipid-linked protein. A protein that is anchored to a biological membrane via a covalently attached lipid such as a farnesyl, geranylgeranyl, myristoyl, palmitoyl, or glycosylphosphatidyl-inositol group.

Lipid raft. A semicrystalline region of a cell membrane containing tightly packed glycosphingolipids and cholesterol.

Lipid storage disease. A defect in a lipid-degrading enzyme that causes the substrate for the enzyme to accumulate in lysosomes.

Lipoprotein. A globular particle consisting of a nonpolar lipid core surrounded by an amphiphilic coat of protein, phospholipid, and cholesterol. Lipoproteins, which transport lipids between tissues via the bloodstream, are classified by their density as high, low, intermediate, and very low density lipoproteins (HDL, LDL, IDL, and VLDL).

Liposome. A synthetic vesicle bounded by a single lipid bilayer.

Lipoyllysyl arm. An extended structure, consisting of lipoic acid linked to a lysine side chain, that delivers intermediates between active sites in multienzyme complexes such as the pyruvate dehydrogenase complex.

London dispersion forces. The weak attractive forces between electrically neutral molecules in close proximity, which arise from electrostatic interactions among their fluctuating dipoles.

Low-barrier hydrogen bond (LBHB). An unusually short and strong hydrogen bond that forms when the donor and acceptor groups have nearly equal pK values so that the hydrogen is equally shared between them.

Lung surfactant. The amphipathic protein and lipid mixture that prevents collapse of the lung alveoli (microscopic air spaces) on the expiration of air.

Lyase. An enzyme that catalyzes the elimination of a group to form a double bond.

Lymphocyte. Types of white blood cells that mediate the immune response. Mammalian B lymphocytes develop in the bone marrow, and T lymphocytes in the thymus.

Lysis. The act of breaking open (e.g., a cell) or breaking apart (e.g., bonded atoms).

Lysophospholipid. A glycerophospholipid derivative lacking a fatty acyl group at position C2 that acts as a detergent to disrupt cell membranes.

Lysosome. A membrane-bounded organelle in a eukaryotic cell that contains a battery of hydrolytic enzymes and which functions to digest ingested material and to recycle cell components.

Main chain. See backbone.

Major groove. The groove on a DNA double helix onto which the glycosidic bonds of a base pair form an angle of >180°. In B-DNA, this groove is wider than the minor groove.

Malaria. A mosquito-borne disease caused by the protozoan *Plasmodium falciparum,* which resides in red blood cells during much of its life cycle.

Malate–aspartate shuttle. A metabolic circuit that uses the malate and aspartate transporters and the interconversion of malate, oxaloacetate, and asparate to ferry reducing equivalents into the mitochondrion.

Malignant tumor. A mass of cells that proliferates uncontrollably; a cancer.

Mass spectrometry. A technique for identifying molecules by measuring the mass-to-charge ratios of molecular ions in the gas phase.

Maternal-effect genes. Insect genes whose mRNA or protein products are deposited by the mother in the ovum and which define the polarity of the embryonic body.

Matrix. The gel-like solution of enzymes, substrates, cofactors, and ions in the interior of the mitochondrion.

Mechanism-based inhibitor. A molecule that chemically inactivates an enzyme only after undergoing part or all of its normal catalytic reaction. Also called a suicide substrate.

Mechanosensitive channel. An ion channel whose opening and closing (gating) is controlled by stimuli such as touch, sound, and changes in osmotic pressure.

Mediated transport. The transmembrane movement of a substance through the action of a specific carrier protein; the opposite of nonmediated transport.

Melting curve. See melting temperature.

Melting temperature (T_m). The midpoint temperature of the melting curve for the thermal denaturation of a macromolecule.

Membrane potential ($\Delta\Psi$). The electrical potential difference across a membrane.

Membrane skeleton. The protein network that lies beneath a membrane and helps determine the cell's shape.

Memory *B* cell. A *B* cell that can recognize its corresponding antigen and rapidly proliferate to produce specific antibodies weeks to years after this antigen was first encountered.

Mercaptan. A compound containing an —SH group.

Messenger RNA (mRNA). A ribonucleic acid whose sequence is complementary to that of a protein-coding gene in DNA. In the ribosome, mRNA directs the polymerization of amino acids to form a polypeptide with the corresponding sequence.

Metabolic fuel. A molecule that can be oxidized to provide free energy for an organism.

Metabolism. The total of all degradative and biosynthetic cellular reactions.

Metabolite. A reactant, intermediate, or product of a metabolic reaction.

Metal-activated enzyme. An enzyme that loosely binds a metal ion, typically Na^+, K^+, Mg^{2+}, or Ca^{2+}.

Metal chelate affinity chromatography. A procedure in which a molecule bearing metal-chelating groups is separated from a mixture of other molecules by its ability to bind to metal ions attached to a chromatographic matrix.

Metal ion catalysis. A catalytic mechanism that requires the presence of a metal ion to lower the free energy of a reaction's transition state.

Metalloenzyme. An enzyme that contains a tightly bound metal ion cofactor, typically a transition metal ion such as Fe^{2+}, Zn^{2+}, or Mn^{2+}.

Methanogen. An organism that produces CH_4.

Micelle. A globular aggregate of amphiphilic molecules in aqueous solution that are oriented such that polar segments form the surface of the aggregate and the nonpolar segments form a core that is out of contact with the solvent.

Michaelis complex. See ES complex.

Michaelis constant (K_M). For an enzyme that follows the Michaelis–Menten model, $K_M = (k_{-1} + k_2)/k_1$; K_M is equal to the substrate concentration at which the reaction velocity is half-maximal.

Michaelis–Menten equation. A mathematical expression that describes the activity of an enzyme in terms of the substrate concentration ([S]), the enzyme's maximal velocity (V_{max}), and its Michaelis constant (K_M): $v_0 = V_{max}[S]/(K_M + [S])$.

Micro RNA (miRNA). See short interfering RNA.

Microarray. See DNA microarray.

Microfilament. A 70-Å-diameter cytoskeletal element composed of actin.

Microheterogeneity. The variability in carbohydrate composition in glycoproteins.

Microtubule. A 240-Å-diameter cytoskeletal element consisting of a hollow tube of polymerized tubulin subunits.

Mineralocorticoid. A steroid hormone that regulates the excretion of salt and water by the kidneys.

Minor groove. The groove on a DNA double helix onto which the glycosidic bonds of a base pair form an angle of <180°. In B-DNA, this groove is narrower than the major groove.

(−) end. The end of a polymeric filament where growth is slower. See also (+) end.

miRNA. Micro RNA. See short interfering RNA.

Mismatch repair (MMR). A postreplication process, in which mispaired nucleotides are excised and replaced, that distinguishes between the parental (correct) and daughter (incorrect) strands of DNA.

Mitochondria (*sing.* mitochondrion). The double-membrane-enveloped eukaryotic organelles in which aerobic metabolic reactions occur, including those of the citric acid cycle, fatty acid oxidation, and oxidative phosphorylation.

Mitochondrial matrix. See matrix.

Mixed inhibition. A form of enzyme inhibition in which an inhibitor binds to both the enzyme and the enzyme–substrate complex and thereby differently affects K_M and V_{max}. Also called noncompetitive inhibition.

MMR. See mismatch repair.

Moderately repetitive DNA. Segments of hundreds to thousands of base pairs that are present at $<10^6$ copies per haploid genome.

Modification methylase. A bacterial enzyme that methylates a specific sequence of DNA as part of a restriction–modification system.

Molar absorptivity (ε). A constant that relates the absorbance of a solution to the concentration of the solute at a given wavelength. Also called molar extinction coefficient.

Molar extinction coefficient. See molar absorptivity.

Molecular chaperone. A protein that binds to unfolded or misfolded proteins so as to promote normal folding and the formation of native quaternary structure. Also known as a heat shock protein (Hsp).

Molecular cloning. See recombinant DNA technology.

Molecular sieve chromatography. See gel filtration chromatography.

Molecular weight. See M_r.

Molecularity. The number of molecules that participate in an elementary chemical reaction.

Molten globule. A collapsed but conformationally mobile intermediate in protein folding that has much of the native protein's secondary structure but little of its tertiary structure.

Monocistronic mRNA. The RNA transcript of a single gene.

Monoclonal antibody. A single type of antibody molecule produced by a clone of hybridoma cells, which are derived by the fusion of a myeloma cell with a lymphocyte producing that antibody.

Monomer. (1) A structural unit from which a polymer is built up. (2) A single subunit or protomer of a multisubunit protein.

Monoprotic acid. An acid that can donate only one proton.

Monosaccharide. A carbohydrate consisting of a single saccharide (sugar).

Morphogen. A substance whose distribution in an embryo directs, in part, the embryo's developmental pattern.

Motif. See supersecondary structure.

Motor protein. An intracellular protein that couples the free energy of ATP hydrolysis to molecular movement relative to another protein that often acts as a track for the linear movement of the motor protein.

M_r. Relative molecular mass. A dimensionless quantity that is defined as the ratio of the mass of a particle to 1/12th the mass of a ^{12}C atom. Also known as molecular weight. It is numerically equal to the grams/mole of a compound.

mRNA. See messenger RNA.

Multienzyme complex. A group of noncovalently associated enzymes that catalyze two or more sequential steps in a metabolic pathway.

Multiple myeloma. A disease in which a cancerous B cell proliferates and produces massive quantities of a single antibody known as a myeloma protein.

Multisubunit protein. A protein consisting of more than one polypeptide chain (subunit).

Mutagen. An agent that induces a mutation in an organism.

Mutase. An enzyme that catalyzes the transfer of a functional group from one position to another on a molecule.

Mutation. A heritable alteration in an organism's genetic material.

Myocardial infarction. The death of heart tissue caused by the loss of blood supply (a heart attack).

Myofibril. The bundle of fibers that are arranged in register in striated muscle cells.

Myristoylation. The attachment of an myristoyl group to a protein to form a lipid-linked protein.

N-end rule. The correlation between the identity of a polypeptide's N-terminal residue and its half-life in the cell.

N-glycosidic bond. See glycosidic bond.

N-linked oligosaccharide. An oligosaccharide linked via a glycosidic bond to the amide group of a protein Asn residue.

N-terminus. See amino terminus.

Native structure. The fully folded conformation of a macromolecule.

Natural selection. The evolutionary process by which the continued existence of a replicating entity depends on its ability to survive and reproduce under the existing conditions.

ncRNA. See noncoding RNA.

NDP. A ribonucleoside diphosphate.

Near-equilibrium reaction. A reaction whose ΔG value is close to zero, so that it can operate in either direction depending on the substrate and product concentrations.

Necrosis. Trauma-induced cell death that results in the unregulated disintegration of the cell and the release of proinflammatory substances. See also apoptosis.

Negative cooperativity. See cooperative binding.

NER. See nucleotide excision repair.

Nernst equation. An expression of the relationship between reduction potential difference ($\Delta\mathscr{E}$) and the concentrations of the electron donors and acceptors (A, B):
$\Delta\mathscr{E} = \Delta\mathscr{E}° - RT/n\mathscr{F} \ln ([A_{red}][B_{ox}]/[A_{ox}][B_{red}])$.

Neurotransmitter. A substance released by a nerve cell that alters the activity of another nerve cell.

Neutral drift. Evolutionary changes that become fixed at random rather than through natural selection.

Neutral solution. A solution whose pH is equal to 7.0 ($[H^+] = 10^{-7}$ M).

NHEJ. See nonhomologous end-joining.

Nick translation. The progressive movement of a single-strand break (nick) in duplex DNA through the coordinated actions of a $5' \rightarrow 3'$ exonuclease function that removes residues from the $5'$ side of the break and a polymerase function that adds residues to the $3'$ side.

Nitrogen assimilation. The incorporation of fixed nitrogen (e.g., ammonia) into a biological molecule such as an amino acid.

Nitrogen fixation. The process by which atmospheric N_2 is converted to a biologically useful form such as NH_3.

NMD. See nonsense-mediated decay.

NMR. See nuclear magnetic resonance.

Noncoding RNA. An RNA molecule, such as rRNA, tRNA, or another small RNA, that is not translated.

Noncoding strand. See antisense strand.

Noncompetitive inhibition. (1) A synonym for mixed inhibition. (2) A special case of mixed inhibition in which the inhibitor binds the enzyme and enzyme–substrate complex with equal affinities ($K_I = K_I'$), thereby reducing the apparent value of V_{max} but leaving K_M unchanged.

Noncooperative binding. A situation in which binding of a ligand to a macromolecule does not affect the affinities of other binding sites on the same molecule.

Nonessential amino acid. An amino acid that an organism can synthesize from common intermediates.

Nonhomologous end-joining (NHEJ). An error-prone pathway for repairing DNA with double-strand breaks.

Nonmediated transport. The transmembrane movement of a substance through simple diffusion; the opposite of mediated transport.

Nonpolar molecule. A molecule that lacks a group with a permanent dipole.

Nonrepetitive structure. A segment of a polymer in which the backbone has an ordered arrangement that is not characterized by a repeating conformation.

Nonsense codon. See Stop codon.

Nonsense-mediated decay (NMD). The degradation of an mRNA that contains a premature Stop codon.

Nonsense mutation. A mutation that converts a codon that specifies an amino acid to a Stop codon, thereby causing the premature termination of translation.

Nonsense suppressor tRNA. A mutated tRNA that recognizes a Stop codon so that its attached aminoacyl group is appended to the polypeptide chain; it mitigates the effect of a nonsense mutation in a structural gene.

Nonshivering thermogenesis. The production of heat via fuel oxidation without the synthesis of ATP and without shivering or other muscle movement.

Northern blotting. A procedure for identifying an RNA containing a particular base sequence through its ability to hybridize with a complementary single-stranded segment of DNA or RNA. See also Southern blotting.

NSF. *N*-Ethylmaleimide-sensitive fusion protein.

nt. Nucleotide.

NTP. A ribonucleoside triphosphate.

Nuclear magnetic resonance (NMR). A spectroscopic method for characterizing atomic and molecular properties based on the signals emitted by radiofrequency-excited atomic nuclei in a magnetic field. It can be used to determine the three-dimensional molecular structure of a protein or nucleic acid.

Nuclease. An enzyme that hydrolytically degrades nucleic acids.

Nucleic acid. A polymer of nucleotide residues. The major nucleic acids are deoxyribonucleic acid (DNA) and ribonucleic acid (RNA). Also known as a polynucleotide.

Nucleolus (*pl.* nucleoli). The dark-staining region of the eukaryotic nucleus, where ribosomes are assembled.

Nucleophile. A group that contains unshared electron pairs that readily reacts with an electron-deficient group (electrophile). A nucleophile (nucleus-lover) reacts with an electrophile (electron-lover).

Nucleoside. A compound consisting of a nitrogenous base and a five-carbon sugar (ribose or deoxyribose) in *N*-glycosidic linkage.

Nucleosome. The complex of a histone octamer and ~200 bp of DNA that forms the lowest level of DNA organization in the eukaryotic chromosome.

Nucleosome core particle. The complex of histones and ~146 bp of DNA that forms a compact disk-shaped particle in which the DNA is wound in ~2 helical turns around the outside of the histone octamer.

Nucleotide. A compound consisting of a nucleoside esterified to one or more phosphate groups. Nucleotides are the monomeric units of nucleic acids.

Nucleotide excision repair (NER). A multistep process in which a portion of DNA containing a lesion is excised and replaced by normal DNA.

Nucleotide sugar. A saccharide linked to a nucleotide by a phosphate ester bond, the cleavage of which drives the formation of a glycosidic bond.

Nucleus. The membrane-enveloped organelle in which the eukaryotic cell's genetic material is located.

O-glycosidic bond. See glycosidic bond.

O-linked oligosaccharide. An oligosaccharide linked via a glycosidic bond to the hydroxyl group of a protein Ser or Thr side chain.

Oil. A mixture of triacylglycerols that is liquid at room temperature.

Okazaki fragments. The short segments of DNA formed in the discontinuous lagging-strand synthesis of DNA.

Oligomer. (1) A short polymer consisting of a few linked monomer units. (2) A protein consisting of a few protomers (subunits).

Oligopeptide. A polypeptide containing a few amino acid residues.

Oligosaccharide. A polymeric carbohydrate containing a few monosaccharide residues.

Oligosaccharide processing. The cellular pathway in which a newly glycosylated protein undergoes the enzymatic removal and addition of monosaccharide residues.

Ω loop. A loop of 6 to 16 polypeptide residues that occurs on the surface of a protein and which forms a compact globular entity.

Oncogene. A mutant version of a normal gene (a proto-oncogene), which may be acquired through viral infection; it

interferes with the mechanisms that normally control cell growth and differentiation and thereby contributes to uncontrolled proliferation (cancer).

Open complex. The separated DNA strands at the transcription start site.

Open reading frame (ORF). A portion of the genome that potentially codes for a protein. This sequence of nucleotides begins with a start codon, ends with a Stop codon, contains no internal Stop codons, is flanked by the proper control sequences, and exhibits the same codon-usage preference as other genes in the organism.

Open system. A thermodynamic system that can exchange matter and energy with its surroundings.

Operator. A DNA sequence at or near the transcription start site of a gene, to which a repressor binds so as to control transcription of the gene.

Operon. A prokaryotic genetic unit that consists of several genes with related functions that are transcribed as a single mRNA molecule.

Optical activity. The ability of a molecule to rotate the plane of polarized light.

Optical density. See absorbance.

Ordered mechanism. A Sequential reaction with a compulsory order of substrate addition to the enzyme.

ORF. See open reading frame.

Organelle. A differentiated structure within a eukaryotic cell, such as a mitochondrion, ribosome, or lysosome, that performs specific functions.

Organic compound. A compound that contains the element carbon.

Orphan gene. A gene, usually identified through genome sequencing, with no known function.

Orthophosphate cleavage. The hydrolysis of ATP that yields $ADP + P_i$.

Osmosis. The movement of solvent across a semipermeable membrane from a region of low solute concentration to a region of high solute concentration.

Osmotic pressure. The pressure that must be applied to a solution containing a high concentration of solute to prevent the net flow of solvent across a semipermeable membrane separating it from a solution with a lower concentration of solute. The osmotic pressure of a 1 M solution of any solute separated from solvent by a semipermeable membrane is ideally 22.4 atm.

Overproducer. A genetically engineered organism that produces massive quantities of a foreign DNA gene product.

Oxidation. The loss of electrons. Oxidation of a substance is accompanied by the reduction of another substance.

Oxidative phosphorylation. The process by which the free energy obtained from the oxidation of metabolic fuels is used to generate ATP from $ADP + P_i$.

Oxidizing agent. A substance that can accept electrons from other substances, thereby oxidizing them and becoming reduced.

Oxidoreductase. An enzyme that catalyzes an oxidation–reduction reaction.

Oxonium ion. A resonance-stabilized carbocation such as occurs during the lysozyme-catalyzed hydrolysis of a glycoside.

Oxyanion hole. A structure in an enzyme active site that preferentially binds and thereby stabilizes the oxyanionic tetrahedral transition state of the reaction.

Oxygen debt. The postexertion continued elevation in O_2 consumption that is required to replenish the ATP consumed by the liver during operation of the Cori cycle.

Oxygenation. The binding of molecular oxygen, e.g., to a heme group.

P site. See peptidyl site.

P/O ratio. The ratio of the number of molecules of ATP synthesized from $ADP + P_i$ to the number of atoms of oxygen reduced.

PAGE. Polyacrylamide gel electrophoresis. See gel electrophoresis.

Pair-rule genes. See segmentation genes.

Palindrome. A word or phrase or a nucleotide sequence that reads the same forward or backwards.

Palmitoylation. The attachment of a palmitoyl group to a protein to form a lipid-linked protein.

Paper chromatography. A technique for separating molecules on the basis of their rate of movement with solvent on a paper matrix.

Paralogous genes. Genes derived from a gene duplication event that evolve independently.

Partial oxygen pressure (pO_2). The concentration of gaseous O_2 in units of pressure (e.g., torr).

Passive-mediated transport. The thermodynamically spontaneous carrier-mediated transmembrane movement of a substance from high to low concentration. Also called facilitated diffusion.

Pasteur effect. The greatly increased sugar consumption of yeast grown under anaerobic conditions compared to that of yeast grown under aerobic conditions.

Pathogen. A disease-causing microorganism.

PCR. See polymerase chain reaction.

Pellagra. The human disease resulting from a deficiency of the vitamin niacin (nicotinic acid), a precursor of the nicotinamide-containing cofactors NAD^+ and $NADP^+$.

Pentose phosphate pathway. A pathway for glucose degradation that yields ribose-5-phosphate and NADPH. Also called the hexose monophosphate shunt.

Peptidase. An enzyme that hydrolyzes peptide bonds. Also called a protease.

Peptide. A polypeptide of less than about 40 residues.

Peptide bond. An amide linkage between the α-amino group of one amino acid and the α-carboxylate group of another. Peptide bonds link the amino acid residues in a polypeptide.

Peptide group. The planar —CO—NH— group that encompasses the peptide bond between amino acid residues in a polypeptide.

Peptidoglycans. The cross-linked bag-shaped macromolecules consisting of polysaccharide and polypeptide chains that form bacterial cell walls.

Peptidyl site (P site). The ribosomal site that accommodates a tRNA with an attached peptidyl group during protein synthesis.

Peptidyltransferase. The catalytic activity of the ribosome, which carries out peptide bond synthesis by promoting the nucleophilic attack of an incoming aminoacyl group on the growing peptidyl group.

Peptidyl–tRNA. The covalent complex between a tRNA molecule and a growing polypeptide chain during protein synthesis.

Peripheral protein. A protein that is weakly associated with the surface of a biological membrane. Also called an extrinsic protein.

Periplasmic compartment. The space between the cell wall and the outer membrane of gram-negative bacteria.

Peroxisome. A eukaryotic organelle with specialized oxidative functions.

Perutz mechanism. A model for the cooperative binding of oxygen to hemoglobin, in which O_2 binding causes the protein to shift conformation from the deoxy (T state) to the oxy (R state).

PFGE. See pulsed-field gel electrophoresis.

p_{50}. For a gaseous ligand, the ligand concentration, in units of pressure (e.g., torr), at which a binding protein such as hemoglobin is half-saturated with ligand.

pH. A quantity used to express the acidity of a solution, equivalent to $-\log [H^+]$.

Phage. See bacteriophage.

Pharmacokinetics. The behavior of a drug in the body over time, including its tissue distribution and rate of elimination or degradation.

Phenotype. An organism's physical characteristics.

ϕ (phi). The torsion angle that describes the rotation around the C_α—N bond in a peptide group; the dihedral angle made by the bonds connecting the $C—N—C_\alpha—C$ atoms in a peptide chain.

Phosphagen. A phosphoguanidine whose phosphoryl group-transfer potential is greater than that of ATP; these compounds can therefore phosphorylate ADP to generate ATP.

Phosphatase. An enzyme that hydrolyzes phosphoryl ester groups. See also protein phosphatase.

Phosphatidic acid. The simplest glycerophospholipid, consisting of two fatty acyl groups attached to glycerol-3-phosphate.

Phosphodiester bond. The linkage in which a phosphate group is esterified to two alcohol groups, e.g., the phosphate groups that join the adjacent nucleoside residues in a polynucleotide.

Phosphoglyceride. See glycerophospholipid.

Phosphoinositide pathway. A signal-transduction pathway in which hormone binding to a cell-surface receptor induces phospholipase C to catalyze the hydrolysis of phosphatidyl-inositol-4,5-bisphosphate (PIP_2), which yields inositol-1,4,5-trisphosphate (IP_3) and 1,2-diacylglycerol (DAG), both of which are second messengers.

Phospholipase. An enzyme that hydrolyzes one or more bonds of a glycerophospholipid.

Phosphoprotein phosphatase. See protein phosphatase.

Phosphorolysis. The cleavage of a chemical bond by the substitution of a phosphate group rather than water.

Phosphoryl group. A portion of a molecule with the formula —PO_3H_2.

Phosphoryl group-transfer potential. A measure of the tendency of a phosphorylated compound to transfer its phosphoryl group to water; the opposite of its free energy of hydrolysis.

Photoautotroph. An autotrophic organism that obtains energy from sunlight.

Photon. A packet of light energy. See also Planck's law.

Photooxidation. A mode of decay of an excited molecule, in which oxidation occurs through the transfer of an electron to an acceptor molecule.

Photophosphorylation. The synthesis of ATP from ADP + P_i coupled to the dissipation of a proton gradient that has been generated through light-driven electron transport.

Photoreactivation. The conversion of pyrimidine dimers, a form of DNA damage, to monomers using light energy.

Photorespiration. The consumption of O_2 and evolution of CO_2 by plants (a dissipation of the products of photosynthesis), resulting from the competition between O_2 and CO_2 for binding to ribulose bisphosphate carboxylase.

Photosynthesis. The reduction of CO_2 to $(CH_2O)_n$ in plants and bacteria as driven by light energy.

Photosynthetic reaction center. The pigment-containing protein complex that undergoes photooxidation during the light reactions of photosynthesis.

Phylogenetic tree. A reconstruction of the probable paths of evolution of a set of related organisms, often based on sequence variations in homologous proteins and nucleic acids; a sort of family tree.

Phylogeny. The study of the evolutionary relationships among organisms.

pI. See isoelectric point.

PIC. See preinitiation complex.

Ping Pong reaction. A group-transfer reaction in which one or more products are released before all substrates have bound to the enzyme.

Pitch. The distance a helix rises along its axis per turn; 5.4 Å for an α helix, 34 Å for B-DNA.

pK. A quantity used to express the tendency for an acid to donate a proton (dissociate); equal to $-\log K$, where K is the acid's dissociation constant. Also known as pK_a.

Planck's law. An expression for the energy (E) of a photon: $E = hc/\lambda = h\nu$, where c is the speed of light, λ is its wavelength, ν is its frequency, and h is Planck's constant (6.626×10^{-34} J·s).

Plaque. (1) A region of lysed cells on a "lawn" of cultured bacteria, which indicates the presence of infectious bacteriophage. (2) A deposit of insoluble material in an animal's tissues.

Plasmalogen. A glycerophospholipid in which the C1 substituent is attached via an ether rather than an ester linkage.

Plasmid. A small circular DNA molecule that autonomously replicates in a bacterial or yeast cell. Plasmids are often modified for use as cloning vectors.

PLP. Pyridoxal-5′-phosphate, a cofactor used mainly in transamination reactions.

(+) end. The end of a polymeric filament where growth is faster. See also (−) end.

pmf. See protonmotive force.

Point mutation. The substitution of one base for another in DNA. Point mutations may arise from mispairing during DNA replication or from chemical alterations of existing bases.

Polar molecule. A molecule with one or more groups that have permanent dipoles.

Polarimeter. A device that measures the optical rotation of a solution. It can be used to determine the optical activity of a substance.

Poly(A) tail. The sequence of adenylate residues that is post-transcriptionally appended to the 3′ end of eukaryotic mRNAs.

Polyacrylamide gel electrophoresis (PAGE). See gel electrophoresis.

Polycistronic mRNA. The RNA transcript of a bacterial operon. It encodes several polypeptides.

Polycythemia. A condition characterized by an increased number of erythrocytes.

Polyelectrolyte. A macromolecule that bears multiple charged groups.

Polymer. A molecule consisting of numerous smaller units that are linked together in an organized manner. Polymers may be linear or branched and may contain one or more kinds of structural units (monomers).

Polymerase. An enzyme that catalyzes the addition of nucleotide residues to a polynucleotide through nucleophilic attack of the chain's 3′-OH group on the α-phosphoryl group of the incoming nucleoside triphosphate. DNA- and RNA-directed polymerases require a template molecule with which the incoming nucleotide must base pair.

Polymerase chain reaction (PCR). A procedure for amplifying a segment of DNA by repeated rounds of replication centered between primers that hybridize with the two ends of the DNA segment of interest.

Polynucleotide. See nucleic acid.

Polypeptide. A polymer consisting of amino acid residues linked in linear fashion by peptide bonds.

Polyprotic acid. A substance with more than one proton that can be donated. Polyprotic acids have multiple ionization states.

Polyribosome. An mRNA transcript bearing multiple ribosomes in the process of carrying out translation. Also called a polysome.

Polysaccharide. A polymeric carbohydrate containing multiple monosaccharide residues. Also called a glycan.

Polysome. See polyribosome.

Polytene chromosome. A chromosome resulting from multiple rounds of DNA replication such that the daughter molecules remain associated and in register.

Polyunsaturated fatty acid. A fatty acid that contains more than one double bond in its hydrocarbon chain.

Porphyrias. Genetic defects in heme biosynthesis that result in the accumulation of porphyrins.

Positive cooperativity. See cooperative binding.

Posttranscriptional modification. The removal or addition of nucleotide residues or their modification following the synthesis of RNA.

Posttranslational processing. The removal or derivatization of amino acid residues following their incorporation into a polypeptide, or the cleavage of a polypeptide.

***p*O₂.** See partial oxygen pressure.

pre-mRNA. See heterogeneous nuclear RNA.

pre-rRNA. An immature rRNA transcript.

pre-tRNA. An immature tRNA transcript.

Prebiotic era. The period of time between the formation of the earth ~4.6 billion years ago and the appearance of living organisms at least 3.5 billion years ago.

Precursor. The entity that gives rise, through a process such as evolution or chemical reaction, to another entity.

Preinitiation complex (PIC). The assembly of transcription factors bound to DNA that renders the DNA available for transcription by RNA polymerase.

Prenylation. The attachment of an isoprenoid group to a protein to form a lipid-linked protein.

Preproprotein. A protein bearing both a signal peptide (preprotein) and a propeptide (proprotein).

Preprotein. A protein bearing a signal peptide that is cleaved off following the translocation of the protein through the endoplasmic reticulum membrane.

Pribnow box. The prokaryotic promoter element with the consensus sequence TATAAT that is centered at the −10 position relative to the transcription start site.

Primary active transport. Transmembrane transport that is driven by the exergonic hydrolysis of ATP.

Primary structure. The sequence of residues in a polymer.

Primary transcript. The immediate product of transcription, which may be modified before becoming fully functional.

Primase. The RNA polymerase responsible for synthesizing the RNA segment that primes DNA synthesis.

Primer. An oligonucleotide that serves as a starting point for additional polymerization reactions catalyzed by DNA polymerase to form a polynucleotide. A primer base-pairs with a segment of a template polynucleotide strand so as to form a short double-stranded segment that can then be extended through template-directed polymerization.

Primosome. The protein complex that synthesizes the RNA primers in DNA synthesis.

Prion. A protein whose misfolding causes it to aggregate and produce the neurodegenerative symptoms of transmissible spongiform encephalopathies and related diseases. Misfolded prions induce properly folded prions to misfold and thereby act as infections agents.

Probe. A labeled single-stranded DNA or RNA segment that can hybridize with a DNA or RNA of interest in a screening procedure.

Processive enzyme. An enzyme that catalyzes many rounds of a polymerization reaction without dissociating from the growing polymer.

Prochirality. A property of some nonchiral molecules such that they contain a group whose substitution by another group yields a chiral molecule.

Product inhibition. A case of enzyme inhibition in which product that accumulates during the course of the reaction competes with substrate for binding to the active site.

Proenzyme. An inactive precursor of an enzyme.

Prokaryote. A unicellular organism that lacks a membrane-bounded nucleus. All bacteria are prokaryotes.

Promoter. The DNA sequence at which RNA polymerase binds to initiate transcription.

Proofreading. An additional catalytic activity of an enzyme, which acts to correct errors made by the primary enzymatic activity.

Propeptide. A polypeptide segment of an immature protein that must be proteolytically excised to activate the protein.

Proprotein. The inactive precursor of a protein that, to become fully active, must undergo limited proteolysis to excise its propeptide.

Prostaglandin. See eicosanoids.

Prosthetic group. A cofactor that is permanently (often covalently) associated with an enzyme.

Protease. See peptidase.

Proteasome. A multiprotein complex with a hollow cylindrical core in which cellular proteins are degraded to peptides (recycled) in an ATP-dependent process.

Protein. A macromolecule that consists of one or more polypeptide chains.

Protein kinase. An enzyme that catalyzes the transfer of a phosphoryl group from ATP to the OH group of a protein Ser, Thr, or Tyr residue.

Protein module. A sequence motif of ~40–100 residues that may occur in unrelated proteins or as multiple arrays within one protein.

Protein phosphatase. An enzyme that catalyzes the hydrolytic excision of phosphoryl groups from proteins.

Proteoglycan. An extracellular aggregate of protein and glycosaminoglcyan.

Proteomics. The study of all of a cell's proteins, including their quantitation, localization, modifications, interactions, and activities.

Proto-oncogene. The normal cellular analog of an oncogene; the mutation of a proto-oncogene may yield an oncogene that contributes to uncontrolled cell proliferation (cancer).

Protofilament. A linear polymer that serves as an intermediate in the assembly of a larger fibrous structure such as a microtubule.

Protomer. One of two or more identical units of an oligomeric protein. A protomer may consist of one or more polypeptide chains.

Proton jumping. The sequential transfer of protons between hydrogen-bonded water molecules. Proton jumping is largely responsible for the rapid rate at which hydronium and hydroxyl ions appear to move through an aqueous solution.

Proton wire. A group of hydrogen-bonded protein groups and water molecules that serves as a conduit for protons to traverse a membrane.

Protonmotive force (pmf). The free energy of the electrochemical proton gradient that forms during electron transport.

Proximity effect. A catalytic mechanism in which a reaction's free energy of activation is reduced by the prior bringing together of its reacting groups.

PRPP. 5-Phosphoribosyl-α-pyrophosphate, an "activated" form of ribose that serves as a precursor in the synthesis of histidine, tyrosine, and purine and pyrimidine nucleotides.

Pseudo-first-order reaction. A bimolecular reaction whose rate appears to be proportional to the concentration of only a single reactant because the second reactant is present in large excess.

Pseudogene. An unexpressed sequence of DNA that is apparently the defective remnant of a duplicated gene.

ψ (psi). The torsion angle that describes the rotational position around the C_α—C bond in a peptide group; the dihedral angle made by the bonds connecting the N—C_α—C—N atoms in a peptide chain.

PSI. Photosystem I, the protein complex that reduces $NADP^+$ during the light reactions of photosynthesis.

PSII. Photosystem II, the protein complex that oxidizes H_2O to O_2 during the light reactions of photosynthesis.

Pulse-labeling. A technique for tracing metabolic fates, in which cells or a reacting system are exposed briefly to high levels of a labeled compound.

Pulsed-field gel electrophoresis (PFGE). An electrophoretic procedure in which electrodes arrayed around the periphery of an agarose slab gel are sequentially pulsed so that DNA molecules must continually reorient, thereby allowing very large molecules to be separated by size.

Purine nucleotide cycle. The conversion of aspartate to fumarate, which replenishes citric acid cycle intermediates, through the deamination of AMP to IMP.

Purines. Derivatives of the compound purine, a planar aromatic, heterocyclic compound. Adenine and guanine, two of the nitrogenous bases of nucleotides, are purines.

Pyranose. A sugar with a six-membered ring.

Pyrimidine dimer. The cyclobutane-containing structure resulting from UV irradiation of adjacent thymine or cytosine residues in DNA.

Pyrimidines. Derivatives of the compound pyrimidine, a planar aromatic, heterocyclic compound. Cytosine, uracil, and thymine, three of the nitrogenous bases of nucleotides, are pyrimidines.

Pyrophosphate cleavage. The hydrolysis of ATP that yields AMP + PP_i.

Pyrophosphoryl group. See diphosphoryl group.

q. The thermodynamic term for heat absorbed.

Q cycle. The cyclic flow of electrons accompanied by the transport of protons, involving a stable semiquinone intermediate of CoQ in Complex III of mitochondrial electron transport and in photosynthetic electron transport.

q_P. The thermodynamic term for heat absorbed at constant pressure.

Quantum (*pl.* quanta). A packet of energy. See also photon.

Quantum yield. The ratio of molecules reacted to photons absorbed in a light-induced reaction.

Quaternary structure. The spatial arrangement of a macromolecule's individual subunits.

R group. A symbol for a variable portion of an organic molecule, such as the side chain of an amino acid.

R state. One of two conformations of an allosteric protein; the other is the T state. The R state is usually the catalytically more active state.

Racemic mixture. A sample of a compound in which both enantiomers are present in equal amounts.

Radioimmunoassay (RIA). A technique for measuring the concentration of a molecule based on its ability to block the binding of a small amount of the radioactively labeled molecule to its corresponding antibody.

Radionuclide. A radioactive isotope.

Ramachandran diagram. A plot of ϕ and ψ values that indicates the sterically allowed conformations of a polypeptide.

Random coil. A totally disordered and rapidly fluctuating polymer conformation.

Random mechanism. A Sequential reaction without a compulsory order of substrate addition to the enzyme.

Rate constant (k). The proportionality constant between the velocity of a chemical reaction and the concentration(s) of the reactant(s).

Rate-determining step. The step with the highest transition state free energy in a multistep reaction; the slowest step.

Rate enhancement. The ratio of the rates of a catalyzed to an uncatalyzed chemical reaction.

Rate equation. A mathematical expression for the time-dependent progress of a reaction as a function of reactant concentration.

Rational drug design. See structure-based drug design.

Reaction coordinate. The path of minimum free energy for the progress of a reaction.

Reaction order. The sum of the exponents of the concentration terms that appear in a reaction's rate equation.

Reading frame. The grouping of nucleotides in sets of three whose sequence corresponds to a polypeptide sequence.

Receptor. A binding protein that is specific for its ligand and elicits a discrete biochemical effect when its ligand is bound.

Receptor-mediated endocytosis. A process in which an extracellular ligand binds to a specific cell-surface receptor and the resulting receptor–ligand complex is engulfed by the cell.

Receptor tyrosine kinase. A hormone receptor whose intracellular domain is activated, as a result of hormone binding, to phosphorylate tyrosine residues on other proteins and/or on other subunits of the same receptor.

Recombinant. A DNA molecule constructed by combining DNA from different sources. Also called a chimera.

Recombinant DNA technology. The isolation, amplification, and modification of specific DNA sequences. Also called molecular cloning or genetic engineering.

Recombination. The exchange of polynucleotide strands between separate DNA segments. Homologous recombination occurs between DNA segments with extensive homology, whereas site-specific recombination occurs between two short, specific DNA sequences.

Recombination repair. A mechanism for repairing damaged DNA, in which recombination exchanges a portion of a damaged strand for a homologous segment that can then serve as a template for the replacement of the damaged bases.

Redox center. A group that can undergo an oxidation–reduction reaction.

Redox couple. See conjugate redox pair.

Reducing agent. A substance that can donate electrons, thereby reducing another substance and becoming oxidized.

Reducing equivalent. A term used to describe the number of electrons that are transferred from one molecule to another during a redox reaction.

Reducing sugar. A saccharide bearing an anomeric carbon that has not formed a glycosidic bond and can therefore reduce mild oxidizing agents.

Reduction. The gain of electrons. Reduction of a substance is accompanied by the oxidation of another substance.

Reduction potential ($\mathscr{E}$). A measure of the tendency of a substance to gain electrons.

Reductive pentose phosphate cycle. See Calvin cycle.

Regular secondary structure. A segment of a polymer in which the backbone adopts a regularly repeating conformation.

Release factor. A protein that recognizes a Stop codon and thereby helps induce ribosomes to terminate polypeptide synthesis.

Renaturation. The refolding of a denatured macromolecule so as to regain its native conformation.

Repetitive DNA. Stretches of DNA of up to several thousand bases that occur in multiple copies in an organism's genome; they are often arranged in tandem.

Replica plating. The transfer of yeast colonies, bacterial colonies, or phage plaques from a culture plate to another culture plate, a membrane, or a filter in a manner that preserves the distribution of the cells on the original plate.

Replication. The process of making an identical copy of a DNA molecule. During DNA replication, the parental polynucleotide strands separate so that each can direct the synthesis of a complementary daughter strand, resulting in two complete DNA double helices.

Replication fork. The branch point in a replicating DNA molecule at which the two strands of the parental molecule are separated and serve as templates for the synthesis of the daughter strands.

Replicon. A unit of eukaryotic DNA that is replicated from one replication origin.

Replisome. The DNA polymerase–containing protein assembly that catalyzes the synthesis of both the leading and lagging strands of DNA at the replication fork.

Repolarization. The recovery of membrane potential that occurs during electrical signaling in cells such as neurons.

Repressor. A protein that binds at or near a gene so as to prevent its transcription.

RER. See rough endoplasmic reticulum.

Residue. A term for a monomeric unit of a polymer.

Resonance energy transfer. See exciton transfer.

Respiratory distress syndrome. Difficulty in breathing in prematurely born infants, caused by alveolar collapse resulting from insufficient synthesis of lung surfactant.

Restriction endonuclease. A bacterial enzyme that cleaves a specific DNA sequence as part of the restriction–modification system.

Restriction fragment length polymorphism (RFLP). Inherited differences in DNA sequences (polymorphisms) among members of the same species leading to variations in the sites of cleavage of the DNA by particular restriction endonucleases and hence to DNA fragments of different lengths in digests of these DNAs by these restriction endonucleases.

Restriction map. A diagram of a DNA molecule, showing the sites recognized by restriction endonucleases, constructed by analysis of the fragments generated by digestion of the DNA with those endonucleases.

Restriction–modification system. A matched pair of bacterial enzymes that recognize a specific DNA sequence: a modification methylase that methylates bases in that sequence, and a restriction endonuclease that cleaves the DNA if it has not been methylated in that sequence. It is a defensive system that eliminates foreign (e.g., viral) DNA.

Reticulocyte. An immature red blood cell, which actively synthesizes hemoglobin.

Retrotransposon. A transposon whose sequence and mechanism of transpostion suggest that it arose from a retrovirus.

Retrovirus. A virus whose genetic material is RNA that must be reverse-transcribed to double-stranded DNA during host cell infection.

Reverse transcriptase. A DNA polymerase that uses RNA as its template.

Reverse turn. A polypeptide conformation in which the chain makes an abrupt reversal in direction; usually consisting of four successive residues. Also called a β bend.

RFLP. See restriction fragment length polymorphism.

Rho factor. A prokaryotic helicase that separates DNA and RNA to promote transcription termination.

RIA. See radioimmunoassay.

Ribonucleic acid. See RNA.

Ribonucleoprotein. A complex of protein and RNA.

Ribonucleotide. A nucleotide in which the pentose is ribose.

Ribosomal RNA (rRNA). The RNA molecules that constitute the bulk of the ribosome, the site of polypeptide synthesis. rRNA provides structural scaffolding for the ribosome and catalyzes peptide bond formation.

Ribosome. The organelle that synthesizes polypeptides under the direction of mRNA. It consists of around two-thirds RNA and one-third protein.

Riboswitch. An mRNA structure that regulates gene expression through alterations in its structure triggered by the presence of the metabolite that is synthesized by the encoded protein.

Ribozyme. An RNA molecule that has catalytic activity.

Rickets. The vitamin D-deficiency disease in children that is characterized by stunted growth and deformed bones.

Rigor mortis. The stiffening of muscles after death.

RNA. Ribonucleic acid. A polymer of ribonucleotides. The major forms of RNA include messenger RNA (mRNA), transfer RNA (tRNA), and ribosomal RNA (rRNA).

RNA editing. The posttranscriptional insertion, deletion, or alteration of bases in mRNA.

RNA interference (RNAi). A form of posttranscriptional gene regulation in which a short double-stranded RNA segment triggers the degradation of the homologous mRNA molecule.

RNAi. See RNA interference.

RNAP. RNA polymerase, the enzyme that synthesizes RNA using a DNA template.

Rossmann fold. See dinucleotide binding fold.

Rotational symmetry. A type of symmetry in which the asymmetric units of a symmetric object can be brought into coincidence through rotation.

Rough endoplasmic reticulum (RER). That portion of the endoplasmic reticulum associated with ribosomes; it is the site of synthesis of membrane proteins and proteins destined for secretion or residence in certain organelles.

rRNA. See ribosomal RNA.

RS system. See Cahn–Ingold–Prelog system.

S. See entropy.

S. Svedberg, a unit for the sedimentation coefficient, equivalent to 10^{-13} s.

Saccharide. See carbohydrate.

Salt bridge. See ion pair.

Salting in. The increase in solubility of a protein (or other molecule) with increasing (low) salt concentration.

Salting out. The decrease in solubility of a protein (or other molecule) with increasing (high) salt concentration.

Salvage pathway. A metabolic pathway for converting free purines and pyrimidines to their nucleotide forms.

SAM. S-Adenosylmethionine, a nucleotide cofactor that functions mostly as a methyl group donor.

Sarcomere. The repeating unit of a myofibril, consisting of thin and thick filaments that slide past each other during muscle contraction.

Satellite DNA. DNA regions that consist of highly repetitive DNA segments; their distinctive base composition causes them to form bands known as satellites that are separate from bands of other DNA in density gradient ultracentrifugation.

Saturated fatty acid. A fatty acid that does not contain any double bonds in its hydrocarbon chain.

Saturation. The state in which all of a macromolecule's ligand-binding sites are occupied by ligand. See also enzyme saturation and saturated fatty acid.

Schiff base. An imine that forms between an amine and an aldehyde or ketone.

SCID. See severe combined immunodeficiency disease.

Scrapie. See bovine spongiform encephalopathy.

Screening. A technique for identifying clones that contain a desired gene.

Scurvy. A disease caused by vitamin C (ascorbic acid) deficiency, which results in inadequate formation of 4-hydroxyprolyl residues in collagen, thereby reducing the collagen's stability.

SDS-PAGE. Polyacrylamide gel electrophoresis (PAGE) in the presence of the detergent sodium dodecyl sulfate (SDS), which denatures and imparts a uniform charge density to polypeptides and thereby permits them to be fractionated on the basis of size rather than inherent charge.

Second messenger. An intracellular ion or molecule that acts as a signal for an extracellular event such as ligand binding to a cell-surface receptor.

Second-order reaction. A reaction whose rate is proportional to the square of the concentration of one reactant or to the product of the concentrations of two reactants.

Secondary active transport. Transmembrane transport that is driven by the energy stored in an electrochemical gradient, which itself is generated utilizing the free energy of ATP hydrolysis or electron transport.

Secondary structure. The local spatial arrangement of a polymer's backbone atoms without regard to the conformations of its substituent side chains. α helices and β sheets are common secondary structural elements of proteins.

Secretory pathway. The series of steps in which an integral membrane or secretory protein is recognized by the signal recognition particle as it emerges from the ribosome, is translocated across the endoplasmic reticulum membrane via a translocon, and in some cases is cleaved by a signal peptidase.

Segment polarity genes. See segmentation genes.

Segmentation genes. Insect genes that specify the correct number and polarity of body segments. Gap genes, pair-rule genes, and segment polarity genes are all segmentation genes.

Selectable marker. A gene whose product has an activity, such as antibiotic resistance, such that, under the appropriate conditions, cells harboring the gene can be distinguished from those that lack the gene.

Selfish DNA. Genomic DNA that has no apparent function. Also called junk DNA.

Semiconservative replication. The natural mode of DNA duplication in which each new duplex molecule contains one strand from the parent molecule and one newly synthesized strand.

Semidiscontinuous replication. The mode of DNA replication in which one strand is replicated as a continuous polynucleotide strand (the leading strand) while the other is replicated as a series of discontinuous fragments (Okazaki fragments) that are later joined (the lagging strand).

Sense strand. The DNA strand complementary to the strand that is transcribed; it has the same base sequence (except for the replacement of U with T) as the synthesized RNA. Also called the coding strand.

Sequential model of allosterism. A model for allosteric behavior in which the subunits of an oligomeric protein change conformation in a stepwise manner as the number of bound ligands increases.

Sequential reaction. A reaction in which all substrates must combine with the enzyme before a reaction can occur; it can proceed by an Ordered or Random mechanism.

Serine protease. A peptide-hydrolyzing enzyme characterized by a reactive Ser residue in its active site.

Severe combined immunodeficiency disease (SCID). An inherited disease that greatly impairs the immune system. One such defect is a deficiency of the enzyme adenosine deaminase.

Shear degradation. The fragmentation of DNA by the mechanical force of shaking or stirring.

Shine–Dalgarno sequence. A purine-rich sequence ~10 nucleotides upstream from the start codon of many prokaryotic mRNAs that is partially complementary to the 3′ end of the 16S rRNA. This sequence helps position the ribosome to initiate translation.

Short interfering RNA (siRNA). A naturally occurring RNA of 18 to 25 nucleotides that inhibits gene expression through RNA interference. Also known as micro RNA (miRNA).

Short tandem repeat (STR). See highly repetitive DNA.

Shotgun cloning. The cloning of an organism's genome in the form of a set of random fragments.

siRNA. See short interfering RNA.

Sickle-cell anemia. An inherited disease in which erythrocytes are deformed and damaged by the presence of a mutant hemoglobin (Glu 6β→Val) that in its deoxy form polymerizes into fibers.

σ factor. A component of the bacterial RNA polymerase holoenzyme that recognizes a gene's promoter and is released once chain initiation has occurred.

Sigmoidal curve. The S-shaped graphical representation of the cooperative binding of a ligand to a molecule.

Signal-gated channel. An ion channel whose opening and closing (gating) is controlled by the binding of an intracellular signaling molecule.

Signal peptide. A short (13–36 residues) N-terminal peptide sequence that targets a nascent secretory or transmembrane protein to the endoplasmic reticulum (in eukaryotes) or plasma membrane (in prokaryotes). This leader peptide is subsequently cleaved away by signal peptidase.

Signal recognition particle (SRP). A protein–RNA complex that binds to the signal peptide of a nascent transmembrane or secretory protein and escorts it to the endoplasmic reticulum (in eukaryotes) or plasma membrane (in prokaryotes) for translocation.

Signal transduction. The transmittal of an extracellular signal to the cell interior by the binding of a ligand to a cell-surface receptor so as to elicit a cellular response through the activation of a sequence of intracellular events that often include the generation of second messengers.

Silencer. A DNA sequence some distance from the transcription start site, where a repressor of transcription may bind.

Single-displacement reaction. A reaction in which a group is transferred from one molecule to another in a concerted fashion (with no intermediates).

Single nucleotide polymorphism (SNP). A single base difference in the genomes of two individuals; such differences occur every 1250 bp on average in the human genome.

Single-strand binding protein (SSB). A tetrameric protein, many molecules of which coat single-stranded DNA during replication so as to prevent formation of double-stranded DNA.

Site-directed mutagenesis. A technique in which a cloned gene is mutated in a specific manner.

Site-specific recombination. See recombination.

Size exclusion chromatography. See gel filtration chromatography.

Sliding filament model. A mechanism for muscle contraction in which interdigitated thin and thick filaments move past each other so as to shorten the overall length of a sarcomere.

Small nuclear ribonucleoprotein (snRNP). A complex of protein and small nuclear RNA that participates in mRNA splicing.

Small nuclear RNA (snRNA). Highly conserved 60- to 300-nt RNAs that participate in mRNA splicing.

Small nucleolar RNA. Eukaryotic RNA molecules of 70 to 100 nt that pair with immature rRNAs to direct their sequence-specific methylation.

SNAP. See soluble NSF attachment factor.

SNAP receptor (SNARE). A membrane-associated protein that participates in vesicle fusion; SNAREs from the two fusing membranes form a bundle of four helices that holds the membranes in close proximity.

SNARE. See SNAP receptor.

snoRNA. See small nucleolar RNA.

SNP. See single nucleotide polymorphism.

snRNA. See small nuclear RNA.

snRNP. See small nuclear ribonucleoprotein.

Soap. A salt of a long-chain fatty acid, which contains a polar head group and a long hydrophobic tail.

Soluble NSF attachment factor (SNAP). A soluble protein that partipates in membrane fusion events.

Solvation. The state of being surrounded by several layers of ordered solvent molecules. Hydration is solvation by water.

Somatic hypermutation. The greatly increased rate of mutation that occurs in the immunoglobulin genes of proliferating *B* lymphocytes and leads, over several cell generations, to antibodies with higher antigen affinity.

Somatic recombination. Genetic rearrangement that occurs in cells other than germline cells.

Sonication. Irradiation with high frequency sound waves. Such treatment is used to disrupt cells and subcellular membranous structures.

SOS response. A bacterial system that recognizes damaged DNA, halts its replication, and repairs the damage, although in an error-prone fashion.

Southern blotting. A procedure for identifying a DNA base sequence after electrophoresis, through its ability to hybridize with a complementary single-stranded segment of labeled DNA or RNA. See also Northern blotting.

Special pair. The set of two closely spaced chlorophyll groups in a photosynthetic system that undergo photooxidation.

Spectrin. A fibrous protein consisting of flexibly connected triple-helical bundles that comprises ~75% of the erythrocyte membrane skeleton.

Spherocytosis. A hereditary abnormality in the erythrocyte cytoskeleton that renders the cells rigid and spheroidal and which causes hemolytic anemia.

Sphingolipid. A derivative of the C_{18} amino alcohol sphingosine. Sphingolipids include the ceramides, cerebrosides, and gangliosides. Sphingolipids with phosphate head groups are called sphingophospholipids.

Sphingomyelin. The most common sphingolipid, consisting of a ceramide bearing a phosphocholine or phosphoethanolamine head group.

Spliceosome. A 50S to 60S particle containing proteins, snRNPs and pre-mRNA; it carries out the splicing reactions whereby a pre-mRNA is converted to a mature mRNA.

Splicing. The usually ribonucleoprotein-catalyzed process by which introns are removed and exons are joined to produce a mature transcript. Some RNAs are self-splicing.

Spontaneous process. A thermodynamic process that occurs without the input of free energy from outside the system. Spontaneity is independent of the rate of a process.

Squelching. The inhibition of the activity of a transcription factor by another transcription factor that competes with it for binding to DNA.

SR. See SRP receptor.

SRP. See signal recognition particle.

SRP receptor. The endoplasmic reticulum protein that serves as a docking point for the signal recognition particle (SRP) during the synthesis of a transmembrane or secretory protein.

SSB. See single-strand binding protein.

Stacking interactions. The stabilizing van der Waals interactions between successive (stacked) bases and base pairs in a polynucleotide.

Standard state. A set of conditions including unit activity of the species of interest, a temperature of 25°C, a pressure of 1 atm, and a pH of 0.0. See also biochemical standard state.

Starch. A mixture of linear and branched glucose polymers that serve as the principal energy reserves of plants.

STAT. A member of a family of proteins that function as signal transducers and activators of transcription (STAT) by binding to DNA in response to tyrosine phosphorylation.

State function. Quantities such as energy, enthalpy, entropy, and free energy, whose values depend only on the current state of the system, not on how they reached that state.

Steady state. A set of conditions in an open system under which the formation and degradation of individual components are balanced such that the system does not change over time.

Steady state assumption. A condition for the application of the Michaelis–Menten model to an enzymatic reaction, in which the

concentration of the ES complex remains unchanged over the course of the reaction.

Stem–loop. A secondary structural element in a single-stranded nucleic acid, in which two complementary segments form a base-paired stem whose strands are connected by a loop of unpaired bases.

Stereocilia. The cell projections, consisting largely of actin filaments, whose deflection by sound waves triggers the electronic signaling necessary for the sense of hearing.

Stereoisomers. Chiral molecules with different configurations about at least one of their asymmetric centers but which are otherwise identical.

Steroid. Any of numerous naturally occurring lipids composed of four fused rings; many are hormones that are derived from cholesterol.

Sterol. An alcohol derivative of a steroid.

Sticky end. The single-stranded extension of a DNA fragment that has been cleaved at a specific sequence (often by a restriction endonuclease) in a staggered cut such that this single-stranded extension is complementary to those of similarly cleaved DNAs.

Stop codon. A sequence of three nucleotides that does not specify an amino acid but instead causes the termination of translation. Also called a nonsense codon.

STR. Short tandem repeat. See highly repetitive DNA.

Striated muscle. The voluntary or skeletal muscles, which have a striped microscopic appearance.

Stroma. The concentrated solution of enzymes, small molecules, and ions in the interior of a chloroplast; the site of carbohydrate synthesis.

Stromal lamellae. The membranous assemblies that connect grana in a chloroplast.

Strong acid. An acid that is essentially completely ionized in aqueous solution. A strong acid has a dissociation constant much greater than unity ($pK < 0$).

Structural bioinformatics. See bioinformatics.

Structural gene. A gene that encodes a protein.

Structure-based drug design. The synthesis of more effective drug molecules as guided by knowledge of the target molecule's structure and interactions with other drug molecules. Also called rational drug design.

Substrate. A reactant in an enzymatic reaction.

Substrate cycle. Two opposing metabolic reactions that function together to hydrolyze ATP, but provide a control point for regulating metabolic flux. Also called a futile cycle.

Substrate-level phosphorylation. The direct transfer of a phosphoryl group to ADP to generate ATP.

Subunit. One of several polymer chains that make up a macromolecule.

Sugar. A simple mono- or disaccharide.

Suicide substrate. See mechanism-based inhibitor.

Sulfhydryl group. A portion of a molecule with the formula —SH.

Supercoiling. The topological state of covalently closed circular double-helical DNA that arises through the over- or underwind-ing of the double helix and which gives the DNA circle a peculiar twisted appearance. Also called superhelicity.

Superhelicity. See supercoiling.

Superoxide radical. $O_2^{\cdot-}$, a partially reduced oxygen species that can damage biomolecules through free radical reactions.

Supersecondary structure. A common grouping of secondary structural elements. Also called a motif.

Suppressor mutation. A mutation that cancels the effect of another mutation.

Surface labeling. A technique in which a lipid-insoluble protein-labeling reagent is used to identify the portion of a membrane protein that is exposed to solvent.

Surroundings. In thermodynamics, the universe other than the particular system that is of interest.

Symbiosis. A mutually dependent relationship between two organisms.

Symmetry model of allosterism. A model for allosteric behavior in which all the subunits of an oligomeric protein are constrained to change conformation in a concerted manner so as to maintain the symmetry of the oligomer.

Symport. A transmembrane channel that simultaneously trans-ports two different molecules or ions in the same direction. See also antiport and uniport.

Syn conformation. A purine nucleotide conformation in which the ribose and the base are eclipsed. See also anti conformation.

Syncytium. A single cell containing multiple nuclei that results from repeated nuclear division without the formation of new plasma membranes.

Synonymous codons. Codons that specify the same amino acid.

System. In thermodynamics, the part of the universe that is of interest; the rest of the universe is the surroundings. See also closed system, isolated system, and open system.

T. See twist.

$t_{1/2}$. See half-time.

T state. One of two conformations of an allosteric protein; the other is the R state. The T state is usually the catalytically less active state.

T→R transition. A shift in conformation of an allosteric protein induced by ligand binding.

TAFs. TBP-associated factors, which, along with TBP, constitute the general transcription factor TFIID required for the transcription of eukaryotic structural genes.

TATA box. A eukaryotic promoter element with the consensus sequence TATAAAA located 10 to 27 nucleotides upstream from the transcription start site.

Tautomers. Isomers that differ only in the positions of their hydrogen atoms and double bonds.

Taxonomy. The study of biological classification.

Tay-Sachs disease. A fatal sphingolipid storage disease caused by a deficiency of hexosaminidase A, the lysosomal enzyme that breaks down ganglioside G_{M2}.

TBP. TATA-binding protein, a DNA-binding protein that is required for transcription of all eukaryotic genes.

TCA cycle. Tricarboxylic acid cycle. See citric acid cycle.

Telomerase. An RNA-containing DNA polymerase that, using the RNA as a template, catalyzes the repeated addition of a specific G-rich sequence to the 3′ end of a eukaryotic DNA molecule to form a telomere.

Telomere. The end of a linear eukaryotic chromosome, which consists of tandem repeats of a short G-rich sequence on the 3′-ending strand and its complementary sequence on the 5′-ending strand.

Tertiary structure. The entire three-dimensional structure of a single-chain polymer, including that of its side chains.

Tetrahedral intermediate. An intermediate of peptide bond hydrolysis in which the carbonyl carbon of the scissile bond has undergone nucleophilic attack so that it has four sub-stituents.

Tetramer. An assembly consisting of four monomeric units.

Thermodynamics. The study of the relationships among various forms of energy.

Thermophile. An organism that thrives at high temperatures.

θ structure. The appearance of a circular DNA molecule undergoing replication by the progressive separation of its two strands.

THF. Tetrahydrofolate, a cofactor for reactions that transfer one-carbon units in various oxidation states.

Thick filament. The sarcomere element that is composed primarily of several hundred myosin molecules.

Thin filament. The sarcomere element that is composed primarily of actin, along with tropomyosin and troponin.

3′ end. The terminus of a polynucleotide whose C3′ is not esterified to another nucleotide residue.

Thylakoid. The innermost compartment in chloroplasts, which is formed by invaginations of the chloroplast's inner membrane. The thylakoid membrane is the site of the light reactions of photosynthesis.

Titration curve. The graphic presentation of the relationship between the pH of an acid- or base-containing solution and the degree of proton dissociation (roughly equal to the number of equivalents of strong base or strong acid that have been added to the solution).

T_m. See melting temperature.

TM protein. See transmembrane protein.

Topoisomerase. An enzyme that alters DNA supercoiling by catalyzing breaks in one or both stands, passing DNA through the break, and resealing the break.

Topology. The study of the geometric properties of an object that are not altered by deformations such as bending and stretching

Torsion angle. The dihedral angle described by the bonds between four successive atoms. The torsion angles φ and ψ indicate the backbone conformation of a peptide group in a polypeptide.

TψC arm. A conserved stem–loop structure in a tRNA molecule that usually contains the sequence TψC, where ψ is pseudouridine.

Trans conformation. An arrangement of the peptide group in which successive C_α atoms are on opposite sides of the peptide bond.

Trans peptide. A conformation in which successive C_α atoms are on opposite sides of the peptide bond.

Transamination. The transfer of an amino group from an amino acid to an α-keto acid to yield a new α-keto acid and a new amino acid.

Transcription. The process by which RNA is synthesized using a DNA template, thereby transferring genetic information from the DNA to the RNA. Transcription is catalyzed by RNA polymerase as facilitated by numerous other proteins.

Transcription factor. A protein that promotes the transcription of a gene by binding to DNA sequences at or near the gene or by interacting with other proteins that do so.

Transcriptomics. The study of all the mRNA molecules that a cell transcribes.

Transfer RNA (tRNA). The small L-shaped RNAs that deliver specific amino acids, which have been esterified to the tRNA's 3′ ends, to ribosomes according to the sequence of a bound mRNA. The proper tRNA is selected through the complementary base pairing of its three-nucleotide anticodon with the mRNA's codon, and the growing polypeptide is transferred to its aminoacyl group.

Transferase. An enzyme that catalyzes the transfer of a functional group from one molecule to another.

Transformation. (1) The permanent alteration of a bacterial cell's genetic message through the introduction of foreign DNA. (2) The genetic changes that convert a normal cell to a cancerous cell.

Transgene. A foreign gene that is stably expressed in a host organism.

Transgenic organism. An organism that stably expresses a foreign gene (transgene).

Transition. A mutation in which one purine (or pyrimidine) replaces another.

Transition state. A molecular assembly at the point of maximal free energy in the reaction coordinate diagram of a chemical reaction.

Transition state analog. A stable substance that geometrically and electronically resembles the transition state of a reaction.

Transition temperature. The temperature at which a lipid bilayer shifts from a gel-like solid to a more fluid liquid crystal form.

Translation. The process of transforming the information contained in the nucleotide sequence of an RNA to the corresponding amino acid sequence of a polypeptide as specified by the genetic code. Translation is catalyzed by ribosomes and requires the additional participation of messenger RNA, transfer RNA, and a variety of protein factors.

Translocation. (1). The movement of a polypeptide through a membrane during the synthesis of a secreted protein. (2) The movement, by one codon, of the ribosome relative to the mRNA after peptide bond synthesis.

Translocon. A multisubunit protein that forms an aqueous pore across the endoplasmic reticulum membrane for the purpose of translocating a protein as part of the secretory pathway.

Transmembrane (TM) protein. An integral protein that completely spans the membrane.

Transmissible spongiform encephalopathy (TSE). An invariably fatal neurodegenerative disease resulting from prion infection.

Transpeptidation. The ribosomal process in which a tRNA-bound nascent polypeptide is transferred to a tRNA-bound aminoacyl group so as to form a new peptide bond, thereby lengthening the polypeptide by one residue at its C-terminus.

Transposable element. See transposon.

Transposition. The movement (copying) of genetic material from one part of the genome to another or, in some cases, from one organism to another.

Transposon. A genetic unit that can move (be copied) from one position to another in a genome; some transposons carry genes. Also called a transposable element.

Transverse diffusion. The movement of a lipid from one leaflet of a bilayer to the other. Also called flip-flop.

Transversion. A mutation in which a purine is replaced by a pyrimidine or vice versa.

Treadmilling. The addition of monomeric units to one end of a linear aggregate, such as an actin filament, and their removal from the opposite end such that the length of the aggregate remains unchanged.

Triacylglycerol. A lipid in which three fatty acids are esterified to a glycerol backbone. Also called a triglyceride.

Tricarboxylic acid (TCA) cycle. See citric acid cycle.

Triglyceride. See triacylglycerol.

Trimer. An assembly consisting of three monomeric units.

Tripeptide. A polypeptide containing three amino acids.

tRNA. See transfer RNA.

TSE. See transmissible spongiform encephalopathy.

Tumor suppressor. A protein whose loss or inactivation may lead to cancer.

Turnover number. See k_{cat}.

Twist (T). The number of complete revolutions that one strand of a covalently closed circular double-helical DNA makes around the duplex axis. It is positive for right-handed superhelical coils and negative for left-handed superhelical coils.

Two-dimensional gel electrophoresis. A technique in which proteins are first subjected to isoelectric focusing, which separates them by net charge, and then to SDS-PAGE in a perpendicular direction, which separates them by size.

Tyrosine kinase. An enzyme that catalyzes the ATP-dependent phosphorylation of a Tyr side chain.

U. The thermodynamic symbol for energy.

Ubiquinone. See coenzyme Q.

Ubiquitin. A small, highly conserved protein that is covalently attached to a eukaryotic intracellular protein so as to mark it for degradation by a proteasome.

Ultracentrifugation. A procedure that subjects macromolecules to a strong centrifugal force (in an ultracentrifuge), thereby separating them by size and/or density and providing a method for determining their mass and subunit structure.

Uncompetitive inhibition. A form of enzyme inhibition in which an inhibitor binds to the enzyme–substrate complex and thereby decreases its apparent K_M and its apparent V_{max} by the same factor.

Uncoupler. A substance that allows the proton gradient across a membrane to dissipate without ATP synthesis so that electron transport proceeds without oxidative phosphorylation.

Uniport. A transmembrane channel that transports a single molecule or ion. See also antiport and symport.

Unsaturated fatty acid. A fatty acid that contains at least one double bond in its hydrocarbon chain.

Urea cycle. A catalytic cycle in which amino groups donated by ammonia and aspartate combine with a carbon atom from HCO_3^- to form urea for excretion and which provides the route for the elimination of nitrogen from protein degradation.

Uridylylation. Addition of a uridylyl (UMP) group.

Uronic acid. A sugar produced by oxidation of an aldose primary alcohol group to a carboxylic acid group.

v. Reaction velocity, typically measured as the rate of appearance of product or disappearance of reactant.

Vacuole. An intracellular vesicle for storing water or other molecules.

van der Waals distance. The distance of closest approach between two nonbonded atoms.

van der Waals forces. The noncovalent associations between molecules that arise from the electrostatic interactions among permanent and/or induced dipoles.

Variable arm. A nonconserved region of a tRNA molecule that contains 3 to 21 nucleotides and that may include a base paired stem.

Variable region. The N-terminal portions of an antibody molecule, where antigen binding occurs and which are characterized by high sequence variability.

Variant. A naturally occurring mutant form.

Vector. See cloning vector.

Vesicle. A fluid-filled sac enclosed by a membrane.

Virulence. The disease-evoking power of a microorganism.

Virus. A nonliving entity that co-opts the metabolism of a host cell to reproduce.

Vitamin. A metabolically required substance that cannot be synthesized by an animal and must therefore be obtained from the diet.

VLDL. Very low density lipoprotein; see lipoprotein.

V_{max}. Maximal velocity of an enzymatic reaction.

V_{max}^{app}. The observed maximal velocity of an enzymatic reaction, which may differ from the true value due to the presence of an inhibitor.

v_0. Initial velocity of an enzymatic reaction.

Voltage-gated channel. An ion channel whose opening and closing (gating) is controlled by a change in membrane potential.

W. (1) See writing number. (2) The number of energetically equivalent ways of arranging the components of a system.

w. The thermodynamic term for the work done by a system on its surroundings.

Water of hydration. The shell of relatively immobile water molecules that surrounds and interacts with (solvates) a dissolved molecule.

Watson–Crick base pair. A stable pairing of nucleotide bases, either adenine with thymine or guanine with cytosine, that occurs in DNA and, to a lesser extent, in RNA. See also Hoogsteen base pair.

Weak acid. An acid that is only partially ionized in aqueous solution. A weak acid has a dissociation constant less than unity ($pK > 0$).

Western blot. See immunoblot.

Wild type. The naturally occurring version of an organism or gene.

Wobble hypothesis. An explanation for the permissive tRNA–mRNA pairing at the third anticodon position that includes non-Watson–Crick base pairs. This allows many tRNAs to recognize two or three different (degenerate) codons.

Writing number (W). The number of turns that the duplex axis of a covalently closed circular double-helical DNA makes around the superhelix axis. It is a measure of the DNA's superhelicity.

X-Ray crystallography. A method for determining three-dimensional molecular structures from the diffraction pattern produced by exposing a crystal of a molecule to a beam of X-rays.

Xenobiotic. A molecule that is not normally present in an organism.

Xenotransplantation. The implantation of tissue from another species.

Y. See fractional saturation.

YAC. See yeast artificial chromosome.

Yeast artificial chromosome (YAC). A linear DNA molecule that contains the chromosomal structures required for normal replication and segregation in a yeast cell. YACs are commonly used as cloning vectors.

Ylid. A molecule with opposite charges on adjacent atoms.

Y_{O_2}. See fractional saturation.

Z-scheme. A Z-shaped diagram indicating the sequence of events and their reduction potentials in the two-center photosynthetic electron-transport system of plants and cyanobacteria.

Zinc finger. A protein structural motif, often involved in DNA binding, that consists of 25 to 60 residues that include His and/or Cys residues to which one or two Zn^{2+} ions are tetrahedrally coordinated.

Zonal ultracentrifugation. A preparative technique in which a mixture of molecules is applied to the surface of a preformed density gradient before ultracentrifugation.

Zwitterion. A compound bearing oppositely charged groups. Also called a dipolar ion.

Zymogen. The inactive precursor (proenzyme) of a proteolytic enzyme.

GUIDE TO MEDIA RESOURCES

The book website (www.wiley.com/college/voet) offers the following resources to enhance student understanding of biochemistry. These are all keyed to figures or sections in the text. They are called out in the text with a red mouse icon or margin note.

Chapter	Media Type	Title	Text Reference
2 Water	Animated Figure	Titration curves for acetic acid, phosphate, and ammonia	Fig. 2-17
	Animated Figure	Titration of a polyprotic acid	Fig. 2-18
3 Nucleotides, Nucleic Acids, and Genetic Information	Guided Exploration	1. Overview of transcription and translation	Section 3-3B
	Guided Exploration	2. DNA sequence determination by the chain-terminator method	Section 3-4C
	Guided Exploration	3. PCR and site-directed mutagenesis	Section 3-5C
	Interactive Exercise	1. Three-dimensional structure of DNA	Fig. 3-6
	Animated Figure	Construction of a recombinant DNA molecule	Fig. 3-27
	Animated Figure	Cloning with bacteriophage λ	Fig. 3-28
	Animated Figure	Site-directed mutagenesis	Fig. 3-31
	Kinemage	2-1. Structure of DNA	Fig. 3-6
	Kinemage	2-2. Watson–Crick base pairs	Fig. 3-8
4 Amino Acids	Animated Figure	Titration of glycine	Fig. 4-8
5 Proteins: Primary Structure	Guided Exploration	4. Protein sequence determination	Section 5-3
	Guided Exploration	5. Protein evolution	Section 5-4A
	Animated Figure	Enzyme-linked immunosorbent assay	Fig. 5-3
	Animated Figure	Ion exchange chromatography	Fig. 5-6
	Animated Figure	Gel filtration chromatography	Fig. 5-7
	Animated Figure	Edman degradation	Fig. 5-17
	Animated Figure	Generating overlapping peptides to determine amino acid sequence	Fig. 5-20
6 Proteins: Three-Dimensional Structure	Guided Exploration	6. Stable helices in proteins: the α helix	Section 6-1B
	Guided Exploration	7. Hydrogen bonding in β sheets	Section 6-1B
	Guided Exploration	8. Secondary structures in proteins	Section 6-2C
	Interactive Exercise	2. Glyceraldehyde-3-phosphate dehydrogenase	Fig. 6-31
	Animated Figure	The α helix	Fig. 6-7
	Animated Figure	β sheets	Fig. 6-9
	Animated Figure	Symmetry in oligomeric proteins	Fig. 6-34
	Animated Figure	Mechanism of protein disulfide isomerase	Fig. 6-42
	Kinemage	3-1. The peptide group	Fig. 6-2, 6-4
	Kinemage	3-2. The α helix	Fig. 6-7
	Kinemage	3-3. β sheets	Fig. 6-9, 6-10
	Kinemage	3-4. Reverse turns	Fig. 6-19
	Kinemage	4-1, 4-2. Coiled coils	Fig. 6-14
	Kinemage	4-3, 4-4. Collagen	Fig. 6-17
	Kinemage	5. Cytochrome *c*	Fig. 6-27, 6-32 (17-16)
7 Protein Function: Myoglobin and Hemoglobin	Animated Figure	Oxygen-binding curve of hemoglobin	Fig. 7-7
	Animated Figure	Movements of heme and F helix in hemoglobin	Fig. 7-9
	Animated Figure	The Bohr effect	Fig. 7-12
	Animated Figure	Effect of BPG and CO_2 on hemoglobin	Fig. 7-14
	Kinemage	6-1. Myoglobin structure	Fig. 7-1, 7-3

Chapter	Media Type	Title	Text Reference
	Kinemage	6-2, 6-3. Hemoglobin structure	Fig. 7-5
	Kinemage	6-3. BPG binding to hemoglobin	Fig. 7-15
	Kinemage	6-4. Conformational changes in hemoglobin	Fig. 7-9
	Kinemage	6-5. Changes at $\alpha_1-\beta_2/\alpha_2-\beta_1$ interfaces in hemoglobin	Fig. 7-10
8 Carbohydrates	Kinemage	7-1. D-Glucopyranose, α and β anomers	Fig. 8-4, 8-5
	Kinemage	7-2. Sucrose	Section 8-2A
	Kinemage	7-3. Hyaluronic acid	Fig. 8-12
	Kinemage	7-4. Structure of a complex carbohydrate	Fig. 8-17
9 Lipids and Biological Membranes	Guided Exploration	9. Membrane structure and the fluid mosaic model	Section 9-4A
	Interactive Exercise	3. Model of phospholipase A_2 and glycerophospholipid	Fig. 9-6
	Animated Figure	Secretory pathway	Fig. 9-36
	Kinemage	8-1. Bacteriorhodopsin	Fig. 9-22
	Kinemage	8-2. Photosynthetic reaction center	Fig. 9-23 (18-8, 18-9)
	Kinemage	8-3. OmpF porin	Fig. 9-24
10 Membrane Transport	Interactive Exercise	4. The K^+ channel selectivity filter	Fig. 10-8
	Animated Figure	Model for glucose transport	Fig. 10-17
11 Enzymatic Catalysis	Guided Exploration	10. The catalytic mechanism of serine proteases	Section 11-5C
	Interactive Exercise	5. Pancreatic RNase S	Fig. 11-9
	Interactive Exercise	6. Carbonic anhydrase	Fig. 11-13
	Interactive Exercise	7. Hen egg white lysozyme	Fig. 11-17
	Animated Figure	Effect of preferential transition state binding	Fig. 11-15
	Animated Figure	Chair and half-chair conformations	Fig. 11-18
	Kinemage	9. Hen egg white lysozyme—catalytic mechanism	Fig. 11-17, 11-19, 11-21
	Kinemage	10-1. Structural overview of a trypsin/inhibitor complex	Fig. 11-25, 11-31
	Kinemage	10-2. Evolutionary comparisons of proteases	Fig. 11-28
	Kinemage	10-3. A transition state analog bound to chymotrypsin	Fig. 11-30
12 Enzyme Kinetics, Inhibition, and Regulation	Guided Exploration	11. Michaelis–Menten kinetics, Lineweaver–Burk plots, and enzyme inhibition	Section 12-1
	Interactive Exercise	8. HIV protease	Box 12-4
	Animated Figure	Progress curve for an enzyme-catalyzed reaction	Fig. 12-2
	Animated Figure	Plot of initial velocity versus substrate concentration	Fig. 12-3
	Animated Figure	Double-reciprocal Lineweaver-Burk plot	Fig. 12-4
	Animated Figure	Lineweaver–Burk plot of competitive inhibition	Fig. 12-7
	Animated Figure	Lineweaver–Burk plot of uncompetitive inhibition	Fig. 12-8
	Animated Figure	Lineweaver–Burk plot of mixed inhibition	Fig. 12-9
	Animated Figure	Plot of v_0 versus [aspartate] for ATCase	Fig. 12-10
	Kinemage	11-1. Structure of ATCase	Fig. 12-12, 12-13
	Kinemage	11-2. Conformational changes in ATCase	Fig. 12-13
13 Introduction to Metabolism	Interactive Exercise	9. Conformational changes in *E. coli* adenylate kinase	Fig. 13-8
14 Glucose Catabolism	Guided Exploration	12. Glycolysis overview	Section 14-1
	Interactive Exercise	10. Conformational changes in yeast hexokinase	Fig. 14-2

Chapter	Media Type	Title	Text Reference
	Animated Figure	Interconversion of the electronic states of chlorophyll	Fig. 18-6
	Animated Figure	The Calvin cycle	Fig. 18-24
	Animated Figure	Mechanism of RuBP carboxylase	Fig. 18-26
	Kinemage	8-2. Photosynthetic reaction center	Fig. 18-8, 18-9 (9-23)
19 Lipid Metabolism	Interactive Exercise	23. Active site of medium-chain acyl-CoA dehydrogenase	Fig. 19-13
	Interactive Exercise	24. X-ray structure of methylmalonyl-CoA mutase	Fig. 19-18
	Animated Figure	Plasma triacylglycerol and cholesterol transport	Fig. 19-7
	Animated Figure	Receptor-mediated endocytosis	Fig. 19-8
	Animated Figure	The β-oxidation pathway of fatty acyl-CoA	Fig. 19-12
	Animated Figure	Comparison of fatty acid β oxidation and fatty acid biosynthesis	Fig. 19-23
	Animated Figure	Reaction sequence for biosynthesis of fatty acids	Fig. 19-26
20 Amino Acid Metabolism	Interactive Exercise	25. Ubiquitin	Fig. 20-1
	Interactive Exercise	26. The bifunctional enzyme tryptophan synthase	Fig. 20-35
	Interactive Exercise	27. *A. vinelandii* nitrogenase	Fig. 20-41
	Animated Figure	Mechanism of PLP-dependent transamination	Fig. 20-8
	Animated Figure	The urea cycle	Fig. 20-9
21 Mammalian Fuel Metabolism: Integration and Regulation	Guided Exploration	19. Hormone signaling by the adenylate cyclase system	Section 21-3C
	Guided Exploration	20. Hormone signaling by the receptor tyrosine kinase system	Section 21-3D
	Interactive Exercise	28. A heterotrimeric G protein	Fig. 21-13
	Interactive Exercise	29. Tyrosine kinase domain of insulin receptor	Fig. 21-17
	Interactive Exercise	30. Human leptin	Fig. 21-30
	Animated Figure	The Cori cycle	Fig. 21-6
	Animated Figure	The glucose–alanine cycle	Fig. 21-7
	Animated Figure	GLUT4 activity	Fig. 21-10
	Animated Figure	The Ras signaling cascade	Fig. 21-19
	Animated Figure	The phosphoinositide signaling system	Fig. 21-22
22 Nucleotide Metabolism	Interactive Exercise	31. *E. coli* ribonucleotide reductase	Fig. 22-9
	Interactive Exercise	32. Human dihydrofolate reductase	Fig. 22-17
	Interactive Exercise	33. Murine adenosine deaminase	Fig. 22-19
	Animated Figure	Metabolic pathway for de novo biosynthesis of IMP	Fig. 22-1
	Animated Figure	Control of the purine biosynthesis pathway	Fig. 22-4
	Animated Figure	The de novo synthesis of UMP	Fig. 22-5
	Animated Figure	Regulation of pyrimidine biosynthesis	Fig. 22-8
23 Nucleic Acid Structure	Guided Exploration	21. DNA structures	Section 23-1A
	Guided Exploration	22. DNA supercoiling	Section 23-1C
	Guided Exploration	23. Transcription factor–DNA interactions	Section 23-4B
	Guided Exploration	24. Nucleosome structure	Section 23-5B
	Interactive Exercise	34. An RNA–DNA helix	Fig. 23-4
	Interactive Exercise	35. Yeast topoisomerase II	Fig. 23-17
	Interactive Exercise	36. A hammerhead ribozyme	Fig. 23-27
	Interactive Exercise	37. A portion of phage 434 repressor in complex with target DNA	Fig. 23-34
	Interactive Exercise	38. *E. coli trp* repressor–operator complex	Fig. 23-35
	Interactive Exercise	39. *E. coli met* repressor–operator complex	Fig. 23-36
	Interactive Exercise	40. A three–zinc finger segment of Zif268 in complex with DNA	Fig. 23-38

Page numbers in **bold face** refer to a major discussion of the entry. F after a page number refers to a figure. T after a page number refers to a table. Positional and configurational designations in chemical names (e.g., 3-, α, *N*-, *p*-, *trans*, D-) are ignored in alphabetizing. Numbers and Greek letters are otherwise alphabetized as if they were spelled out.

A

A, *see* Adenine; Aminoacyl site
Aβ (amyloid-β protein), 170
A Antigens, 230T
AaRSs, *see* Aminoacyl-tRNA synthetases
Aa-tRNA, *see* Aminoacyl-tRNA
A bands, 1080
Abasic sites, 897
ABCA1 (ATP-cassette binding protein A1), 678
Abdominal cavity, adipose tissue in, 748
Ab initio protein design, 164
ABO blood group antigens, 230
Absolute configuration, 86
Absorbance, 99
Absorbance spectroscopy, 99
Absorption spectrum, 99
Abx (anteriobithorax) mutant, 1064
ACAT (acyl-CoA:cholesterol acyltransferase), 634F, 674
Acceptor stem (tRNA), 967F
Accessory BChl, 599
Accessory pigments, 594
Acesulfame, 216
Acetal:
    cyclic, 213
    nonenzymatic acid-catalyzed hydrolysis, 336F
Acetaldehyde, 317
    from alcoholic fermentation, 447
    geometric specificity, 316
Acetaminophen, 390, 670
Acetate, cholesterol biosynthesis from, **671–674**
Acetic acid, titration curve, 35F
Acetimidoquinone, 390
Acetoacetate:
    in amino acid degradation, 697E
    decarboxylation, 325F
    in ketogenesis, 649, 649F
    in ketone body conversion to acetyl-CoA, 650F
    from phenylalanine/tyrosine breakdown, 710–713F
    from tryptophan breakdown, 708–710
Acetoacetyl-ACP, in fatty acid synthesis, 655F

Acetoacetyl-CoA:
    in ketogenesis, 649, 649F
    in ketone body conversion to acetyl-CoA, 650F
Aceto-α-hydroxybutyrate, 722
Acetolactate, 722
Acetone, geometric specificity, 316
Acetyl-ACP, in fatty acid synthesis, 655F
Acetylcholine, hydrolysis, 342
Acetylcholinesterase, 342, 365T
Acetyl-CoA (acetyl-coenzyme A), 397, 413–414, 1036. *See also* Ketone bodies
    in amino acid degradation, 697E
    in citric acid cycle, 516F, **517–521**, 533–535
    in fatty acid synthesis, 655F
    in glyoxylate cycle, 539F
    in ketogenesis, 649, 649F, 779
    in ketone body conversion to acetyl-CoA, 650F
    mammalian metabolism, 744–745, 750
Acetyl-CoA-ACP transacylase, in fatty acid synthesis, 655F
Acetyl-coenzyme A, *see* Acetyl-CoA
N-Acetylglucosamine (NAG, GlcNAc), 220, 331, 335, 338–340
N-Acetyl-D-glucosamine, 220, 224F
N-Acetylglutamate, 697
Acetylglutamate synthase, 697
Acetyl-Lys binding sites, 1037F
ε-*N*-Acetyllysine, 89F
N-Acetylmuramic acid (NAM, MurNAc), 224F, 331, 335, 338–340
N-Acetylneuraminic acid (NANA), 213F
Acetyl phosphate, 406T, 411, 440
Acids, 33–35
    as buffering agents, 37
    conjugate bases, 33
    polyprotic, 36, 36F
    strength, 33–35
Acid–base catalysis, enzymes, **322–325**
Acid–base chemistry, **33–35**
    amino acids, **84–85**
    and proton jumping, 31–32
    standard state conventions, 17
Acidic solutions, 32
Acidosis, 37
Aconitase:
    in citric acid cycle, 516F, **526**, 535
    in glyoxylate cycle, 539F
    stereospecific enzyme, 315
Aconitate, 515
ACP (acyl-carrier protein), 653–654
Acquired immunodeficiency syndrome (AIDS), 370, 378, 379
Acridine orange, as intercalating agent, 843F, 893

Actin, 577, **1073–1080F**
    microfilament dynamics, 1075–1080
    structure, 1074–1075
Actin–myosin reaction cycle, 1086–1088
α-Actinin, 1086
Actinomycin D, 930F
Action potentials, 293–294
Action potential time course, 293F
Activation domains, 1044
Activation energy, **319–321**
Activators, regulatory proteins, 1043
Active membrane transport, 266, **286–309**
    ATP-driven, **303–307**
    endergonic process, 303
    ion-gradient-driven, **307–309**
    sodium and potassium, 305F
Active site, enzymes, 313
Active transporters, 289, **303–309**
Active transport systems, 307
Activity, 17
Activity site, 804
Acute intermittent porphyria, 728
Acute lymphoblastic leukemia, 701
Acute pancreatitis, 351
Acyl-carnitine, 637
Acyl-carrier protein (ACP), 653–654
Acyl-CoA:cholesterol acyltransferase (ACAT), 634F, 674
Acyl-CoA dehydrogenase, 638–639F
Acyl-CoA oxidase, 648
Acyl-CoA synthetases, 636, 637F
Acyl-dihydroxyacetone phosphate, in triacylglycerol biosynthesis, 659F
Acyl-dihydroxyacetone phosphate reductase, in triacylglycerol biosynthesis, 659F
Acyl-enzyme, tetrahedral intermediate, 350F
Acyl-enzyme intermediate, chymotrypsin, 347
1-Acylglycerol-3-phosphate acyltransferase, in triacylglycerol biosynthesis, 659F, 660
2-Acylglycerols, 628
Acyl phosphates, 411
N-Acylsphingosine, 663, 667F
Acyl thioester, 440
ADAR2, 951
Addison's disease, 244, 1110
Adenine (A), 41, 42T
    base pairing, 48F, 51F, 819F, 836–837
    Chargaff's rules and, 45
    as common nucleotide, 43
    modified forms in tRNA, 968F
    oxidative deamination, 892F
    stacking interactions, 837F